THE FIREFIGHTER'S HANDBOOK

ESSENTIALS OF FIREFIGHTING AND EMERGENCY RESPONSE

Second Edition

NFAAA
NATIONAL FIRE ACADEMY ALUMNI ASSOCIATION

Delmar is proud to donate a portion of the proceeds from this book to NFAAA.

THE FIREFIGHTER'S HANDBOOK

ESSENTIALS OF FIREFIGHTING AND EMERGENCY RESPONSE

Second Edition

DELMAR
THOMSON LEARNING™

Australia Canada Mexico Singapore Spain United Kingdom United States

DELMAR
THOMSON LEARNING

The Firefighter's Handbook: Essentials of Firefighting and Emergency Response, Second Edition

Thomson Delmar Learning

Vice President, Technology and Trades SBU:
Alar Elken

Editorial Director:
Sandy Clark

Acquisitions Editor:
Alison S. Weintraub

Developmental Editor:
Jennifer A. Thompson

Marketing Director:
Cyndi Eichelman

Channel Manager:
Bill Lawrensen

Marketing Coordinator:
Mark Pierro

Production Director:
Mary Ellen Black

Production Editor:
Barbara L. Diaz

Art & Design Specialist:
Rachel Baker

Technology Project Manager:
Kevin Smith

Editorial Assistants:
Jennifer Luck
Stacey Wiktorek

Printed in the United States of America

4 5 CRK 07 06 05

For more information contact Delmar Learning
Executive Woods
5 Maxwell Drive, PO Box 8007,
Clifton Park, NY 12065-8007
Or find us on the World Wide Web at
www.delmarlearning.com

Library of Congress Cataloging-in-Publication Data:

Firefighter's handbook : essentials of firefighting and emergency response.—2nd ed.
p. cm.
Includes bibliographical references and index.
ISBN 1-4018-3575-9 (alk. paper)
1. Fire extinction—Handbooks, manuals, etc.
I. Delmar Publishers.
TH9151.F458 2005
628.9′25—dc22
2003066273

ISBN: 1-4018-3575-9

NOTICE TO THE READER

CONTENTS

Dedication xvi
Foreword xvii
About Our Authors. xviii
Preface xxii
Acknowledgments xxx
NFPA 1001 Correlation Guide. xxxiii
NFPA 472 Correlation Guide xxxv
Job Performance Requirement Correlation Guide xxxviii

CHAPTER 1 OVERVIEW OF THE HISTORY, TRADITION, AND DEVELOPMENT OF THE AMERICAN FIRE SERVICE

Objectives 4
The Mission of the Fire Service. 4
Roots in the Past 4
- Ancient Beliefs 5
- Recorded History. 6
- Early History and Symbols of the Fire Service. 7
- The Middle Ages 7
- Early American History 8
- The Civil War 13
- The Industrial Revolution 14
- The Beginning of the Twentieth Century 15

Technology, Transition, and Times of Change 16
The Effects of World War II. 17
Modernization of the Fire Service 17
The Fire Service of Today 18
Lessons Learned. 19
Key Terms. 19
Review Questions. 20
Additional Resources 20

CHAPTER 2 FIRE DEPARTMENT ORGANIZATION, COMMAND, AND CONTROL

Objectives 23
Introduction 23
Fire Department Organization 23
- The Business of Fire Protection 24
- Mission Statement 24
- Organizational Structure 24

The Firefighter 26
The Company. 28
- The Engine Company 29
- The Truck Company 29
- The Rescue Company 29
- Specialty/Combination Units 31
- Emergency Medical Services 32
- The Chief Officers 32

Additional Fire Department Functions 32
- Fire Prevention and Life Safety 32
- Training 33
- Emergency Medical Services 34
- Apparatus Maintenance and Purchasing 34
- Special Operations 34

Regulations, Policies, Bylaws, and Procedures 34
- Regulations 34
- Policies 34
- Bylaws. 35
- Procedures 35

Allied Agencies and Organizations 35
Incident Management. 36
- Command and Control 36

Incident Management System (IMS) 37
- Five Major Functions of an Incident Management System 39

Lessons Learned. 42
Key Terms. 42
Review Questions. 44
Additional Resources 44

CHAPTER 3 COMMUNICATIONS AND ALARMS

Objectives 47
Introduction 47
Communications Personnel. 48
The Communications Facility 49
Computers in the Fire Service. 51
Receiving Reports of Emergencies 51
- Methods of Receiving Reports of Emergencies 53

Emergency Services Deployment 58
Traffic Control Systems. 62
Radio Systems and Procedures. 63
Arrival Reports 67
Mobile Support Vehicles. 68
Records 68
Lessons Learned. 70
Key Terms. 71
Review Questions. 72
Endnote 72
Additional Resources 73

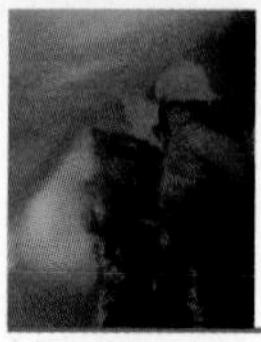

CHAPTER 4
FIRE BEHAVIOR

Objectives 76
Introduction 76
Fire Triangle, Tetrahedron, and Pyramid 77
Measurements 77
Chemistry and Physics of Fire 78
Sources of Heat 82
Chemical 83
Mechanical 83
Electrical 83
Nuclear 85
Combustion 85
Oxygen and Its Effect on Combustion 86
Vapor Pressure and Vapor Density 86
BLEVE 88
Boiling Point 89
Flammable and Explosive Limits 90
The Burning Process—Characteristics of Fire Behavior 91
Ignition Stage 91
Growth Stage 91
Fully Developed Stage 92
Decay Stage 92
Modes of Heat Transfer 94
Conduction 94
Convection 95
Radiation 96
Thermal Conductivity of Materials 97
Physical State of Fuels and Effect on Combustion . . 98
Solid 98
Liquid 98
Gas 98
Theory of Fire Extinguishment 99
Unique Fire Events 99
Thermal Layering 99
Rollover and Flashover 99
Backdraft 100
Classes of Fire 100
Lessons Learned 103
Key Terms 103
Review Questions 105
Endnotes 105
Additonal Resources 105

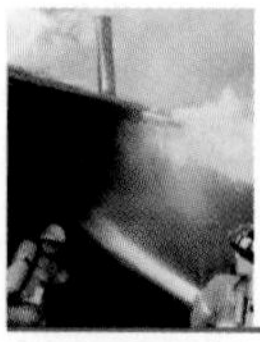

CHAPTER 5
FIREFIGHTER SAFETY

Objectives 108
Introduction 108
Safety Issues 108
Firefighter Injury and Death Causes 108
Safety Standards and Regulations 109
Accident Prevention 110
The Safety Triad 111
Procedures 111
Equipment 112
Personnel 114
Firefighter Safety Responsibilities 117
The Department 117
The Team 118
The Individual Firefighter 119
Lessons Learned 120
Key Terms 121
Review Questions 121
Endnote 122
Additional Resources 122

CHAPTER 6
PERSONAL PROTECTIVE CLOTHING AND ENSEMBLES

Objectives 125
Introduction 125
Personal Protective Equipment Factors 126
Standards and Regulations 126
Types of Personal Protective Equipment 127
Ensembles 127
Miscellaneous PPE Components 134
Care and Maintenance of Personal Protective Equipment 136
Personal Protective Equipment Effectiveness: "Street Smarts" 137
Good PPE Habits and Attitude 137
Streetsmart Suggestions 138
Lessons Learned 139
Key Terms 139
Review Questions 139
Additional Resources 140

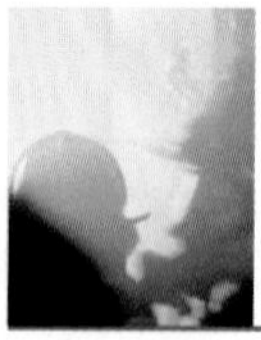

CHAPTER 7
SELF-CONTAINED BREATHING APPARATUS

Objectives 143
Introduction 143
Conditions Requiring Respiratory Protection 145
Oxygen-Deficient Environments 145
Elevated Temperatures 145
Smoke 145
Effects of Toxic Gases and Toxic Environments 146
Carbon Monoxide 146

Legal Requirements for Self-Contained Breathing Apparatus Use . . . 148
Title 29 Code of Federal Regulations, Section 1910.134 . . . 148
NFPA 1500: Standard on Fire Department Occupational Safety and Health Program . . . 148
Limitations of Self-Contained Breathing Apparatus . . . 148
SCBA Design and Size . . . 150
Limitations of the SCBA User . . . 150
Types of Self-Contained Breathing Apparatus . . . 151
Open-Circuit Self-Contained Breathing Apparatus . . . 152
Closed-Circuit Self-Contained Breathing Apparatus . . . 158
Open-Circuit Supplied Air Respirators . . . 158
Donning and Doffing Self-Contained Breathing Apparatus . . . 158
General Considerations . . . 159
Storage Case . . . 159
Seat-Mounted Apparatus . . . 162
Compartment or Side-Mounted Apparatus . . . 164
Donning the SCBA Face Piece . . . 164
Removing/Doffing the SCBA Unit . . . 166
Self-Contained Breathing Apparatus Operation and Emergency Procedures . . . 168
Safe Use of SCBA . . . 168
Operating in a Hostile Environment . . . 170
Restricted Openings . . . 170
Emergency Procedures . . . 171
Inspection and Maintenance of Self-Contained Breathing Apparatus . . . 172
Daily Maintenance . . . 172
Monthly Maintenance . . . 172
Annual and Biannual Maintenance . . . 172
Changing SCBA Cylinders . . . 172
Servicing SCBA Cylinders . . . 176
Lessons Learned . . . 180
Key Terms . . . 180
Review Questions . . . 181
Endnote . . . 182
Additional Resources . . . 182

CHAPTER 8 PORTABLE FIRE EXTINGUISHERS

Objectives . . . 185
Introduction . . . 185
Fire Classification and Risk . . . 185
Class A . . . 185
Class B . . . 185
Class C . . . 186
Class D . . . 186
Class K . . . 186
Types of Fire Extinguishers . . . 187
Types of Extinguishing Agents . . . 187
Kinds of Extinguishers . . . 188
Rating Systems for Portable Extinguishers . . . 193
Class A . . . 193
Class B . . . 194
Class C . . . 194
Limitations of Portable Extinguishers . . . 194
Portable Extinguisher Operation . . . 195
Care and Maintenance of Portable Extinguishers . . . 197
Inspection Requirements . . . 197
Lessons Learned . . . 198
Key Terms . . . 198
Review Questions . . . 199
Endnotes . . . 199
Additional Resources . . . 199

CHAPTER 9 WATER SUPPLY

Objectives . . . 203
Introduction . . . 203
Sources of Water Supply . . . 203
Groundwater . . . 203
Surface Water . . . 204
Mobile Water Supply Apparatus . . . 204
Tanks, Ponds, and Cisterns . . . 205
Water Distribution Systems . . . 205
Fire Hydrants . . . 206
Wet Barrel . . . 206
Dry Barrel . . . 207
Dry Hydrant . . . 208
Specialty Hydrants . . . 209
Valves Associated with Water Distribution Systems . . . 209
Rural Water Supply . . . 210
Portable Water Tanks . . . 210
Tender Operation . . . 211
Pressure Associated with Water Distribution Systems . . . 212
Testing Operability and Flow of Hydrants . . . 213
Determining Static, Residual, and Flow Pressures . . 214
Obstructions and Damage to Fire Hydrants and Mains . . . 215
Lessons Learned . . . 216
Key Terms . . . 217
Review Questions . . . 218
Endnotes . . . 218
Additional Resources . . . 218

CHAPTER 10
FIRE HOSE AND APPLIANCES

Objectives 221
Introduction 221
Construction of Fire Hose 221
Care and Maintenance of Fire Hose 223
Types of Hose Coupling 225
Care and Maintenance of Couplings 226
Hose Tools and Appliances 227
Coupling and Uncoupling Hose 228
Hose Rolls 230
  Straight/Storage 230
  Single Donut 231
  Twin or Double Donut 231
Hose Carries 234
  Drain and Carry 234
  Shoulder Loop Carry 234
  Single-Section Street Drag 236
Hose Loads 236
  Accordion Load 237
  Flat Load 237
  Horseshoe Load 239
  Finish Loads and Preconnected Loads 243
  Flat Load, Minuteman Load, and Triple-Layer Load 245
  Stored Hose Loads/Packs 248
  Wildland Firefighting Hose Loads 248
Advancing Hoselines—Charged/Uncharged 251
  Into Structures 253
  Up and Down Stairs 254
  Using a Standpipe System 255
  Working Hose Off Ladders 257
Establishing a Water Supply Connection 261
  From Hydrants 262
  From Static Water Supplies 264
Extending Hoselines 265
Replacing Sections of Burst Hose 268
Hose Lay Procedures 269
  Forward Lay 270
  Reverse Lay 270
  Split Lay 271
Deploying Master Stream Devices 271
Service Testing of Fire Hose 274
Lessons Learned 276
Key Terms 276
Review Questions 278
Endnotes 278
Additional Resources 278

CHAPTER 11
NOZZLES, FIRE STREAMS, AND FOAM

Objectives 281
Introduction 281
Definition of Fire Stream 281
Nozzles 281
  Solid Tip or Stream 282
  Fog 283
  Straight Stream 285
  Special Purpose 286
  Playpipes and Shutoffs 286
  Nozzle Operations 287
Operating Hoselines 288
  Small-Diameter Handlines 288
  Medium-Diameter Handlines 289
  Master Stream Devices 289
Stream Application, Hydraulics, and Adverse Conditions 289
  Direct, Indirect, and Combination Attack 289
  Basic Hydraulics, Friction Loss, and Pressure Losses in Hoselines 290
  Adverse Conditions That Affect Fire Streams 295
Types of Foam and Foam Systems 295
Foam Characteristics 295
Classification of Fuels 296
  Class A 296
  Class B 297
Application of Foam 297
  Fog Nozzle versus Foam Nozzles 300
Lessons Learned 304
Key Terms 304
Review Questions 306
Endnotes 306
Additional Resources 306

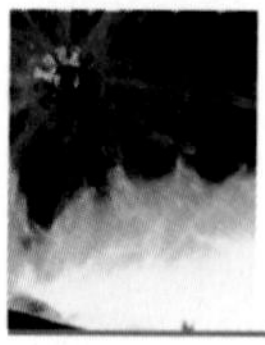

CHAPTER 12
PROTECTIVE SYSTEMS

Objectives 309
Introduction 309
Detection Systems 309
  People or Manual Systems 309
  Heat Detectors 310
  Smoke Detectors 311
  Gas Detectors 313
  Flame Detectors 313
Sprinkler Systems 314
Sprinklers and Life Safety 314

Sprinkler Head Design and Operation 315
Types of Sprinkler Systems 318
Wet Pipe Systems. 318
Dry Pipe Systems. 319
Deluge Systems 320
Preaction Systems 321
Residential Systems 321
Sprinkler System Connections and Piping 321
Control Devices for Sprinkler Systems. 324
Returning Sprinkler Systems to Service 325
Standpipe Classifications 329
Standpipe System Connections and Piping. 330
Alarms for Standpipes and Sprinklers 331
Other Protective Systems 331
Local Application and Hood Systems. 332
Total Flooding Systems 332
Fire Department Operations with Protective Systems. 333
Standpipe Operations 334
Sprinkler System Operations 335
Detector Activation Operations 335
Operations for Other Protective Systems 336
Lessons Learned 336
Key Terms. 337
Review Questions 338
Endnotes 338
Additional Resources 338

CHAPTER 13
BUILDING CONSTRUCTION

Objectives 341
Introduction 341
Building Construction Terms and Mechanics 342
Types of Loads. 342
Imposition of Loads 344
Forces 345
Structural Elements. 346
Beams 346
Columns 346
Walls 347
Connections. 347
Fire Effects on Common Building Construction Materials. 347
Wood 348
Steel. 349
Concrete 349
Masonry. 350
Composites 350
Types of Building Construction 352
Type I: Fire-Resistive. 353
Type II: Noncombustible 354
Type III: Ordinary 354
Type IV: Heavy Timber 355
Type V: Wood Frame 356
Other Construction Types. 357
Relationship of Construction Type to Occupancy Use 359
Collapse Hazards at Structure Fires 360
Trusses. 361
Void Spaces 362
Roof Structures 362
Stairs 363
Parapet Walls 363
Collapse Signs 364
Buildings under Construction. 364
Preparing for Collapse 364
Lessons Learned 366
Key Terms. 366
Review Questions 368
Additional Resources 368

CHAPTER 14
LADDERS

Objectives 371
Introduction 371
Ladder Terminology 371
Parts of a Ladder 371
Ladder Companies 373
Types of Truck-Mounted Ladders 373
Aerial Ladder. 373
Tower Ladder. 375
Articulating Boom Ladder 376
Types of Ground or Portable Ladders 376
Straight Ladder 376
Extension Ladder. 377
Roof or Hook Ladder. 378
Folding Ladder 379
A-Frame Combination Ladder 379
Pompier Ladder 381
Use and Care of Portable or Ground Ladders 381
Maintenance, Cleaning, and Inspection. 382
Cleaning Ladders. 383
Ladder Safety 383
Ladder Uses 383
Access 384
Rescue 384
Stability 384
Ventilation 384
Bridging. 384
Elevated Streams 384
Elevated Work Position 384
Ladder Selection. 384
Butt Section 385
Fly Section. 385

Special Uses 387
- Removal of Numerous Victims 387
- Chute with a Tarp 388
- Over a Fence 388
- Elevated Hose Streams 388
- Portable Pool 388
- Barrier 389
- Support 389
- Hoist Point 389
- Ventilation Fan Supports 390

Safety 390
- Overhead Obstructions 390
- Climbing Path 391
- Ground Considerations 391
- Ladder Load 391
- Working Off a Ladder 392

Miscellaneous Ladder Information 393
- Ladder Storage 393
- Apparatus Ladder Storage 393
- Ladder Apparatus Parking 393
- Ladder Painting 394
- Certification and Testing Procedures 395

Ladder Skills 395
- Carrying Ladders 395

Raising Skills 399
- Raising Ladders 399
- Rung and Beam Raises 400
- Leg Lock 408
- Carrying Tools 408
- Mounting and Dismounting 410
- Engaging the Hook on a Hook Ladder 412
- Roof Ladder Deployment 412
- Hoisting Ladders by Rope 412

Lessons Learned 415

Key Terms 415

Review Questions 416

Additional Resources 416

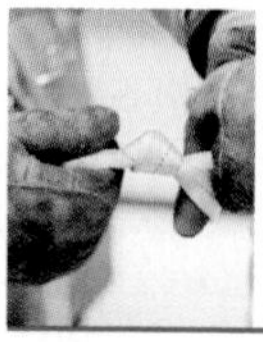

CHAPTER 15
ROPES AND KNOTS

Objectives 419

Introduction 419

Rope Materials and Their Characteristics 419
- Natural Materials 419
- Synthetic Materials 420

Construction Methods and Their Characteristics 422
- Laid (Twisted) 422
- Braided 422
- Braid-on-Braid 423
- Kernmantle 423

Primary Uses 423
- Utility 423
- Firefighting and Rescue Uses 424

Fire Service Knots 424
- Nomenclature of Rope and Knots 425
- Knots 426

Inspection 439

Maintenance 441
- Cleaning 441
- Storage 444

Rigging for Hoisting 449
- Specific Tools and Equipment 449
- Securing a Rope between Two Objects 452

Lessons Learned 457

Key Terms 457

Review Questions 458

Additional Resources 458

CHAPTER 16
RESCUE PROCEDURES

Objectives 461

Introduction 461

Hazards Associated with Rescue Operations 461

Search of Burning Structures 461
- Primary Search 465
- Secondary Search 465

Victim Removal, Drags, and Carries 467
- Carries 468
- Drags 469
- Backboard, Stretcher, and Litter Uses 478

Extrication from Motor Vehicles 482
- Tools and Equipment 482
- Scene Assessment (Size-Up) 486
- Establishment of Work Areas 486
- Vehicle Stabilization 487
- Patient Access 488
- Disentanglement 489
- Patient Removal 489
- Scene Stabilization 490

Specialized Rescue Situations and Tools 490
- Vertical Rescue 491
- Water Rescue 492
- Ice Rescue 494
- Structural Collapse Rescue 494
- Trench and Below-Grade Rescue 497
- Confined Space Rescue 498
- Rescue from Electrical Situations 498
- Industrial Entrapment Rescue 499
- Elevator and Escalator Rescue 500
- Farm Equipment Rescue 505

Lessons Learned 505

Key Terms 505

Review Questions 506

Additional Resources 507

CHAPTER 17
FORCIBLE ENTRY

Objectives 510
Introduction 510
- Knowledge 510
- Skill 511
- Experience 511

Forcible Entry Tools 511
- Striking Tools 512
- Prying and Spreading Tools 513
- Cutting Tools 515
- Pulling Tools 518
- Special Tools 518

Safety with Forcible Entry Tools 519
- Rotary and Chain Saws 520
- Carrying Tools 520
- Hand Tools 521

Maintenance of Forcible Entry Tools 521
Construction and Forcible Entry 522
- Door Construction 522
- Types of Doors 522
- Locks 526
- Additional Security Devices 530

Methods of Forcible Entry 530
- Conventional 531
- Through-the-Lock Forcible Entry 536
- Operating Lock Mechanisms 538
- Lock Variations 539

Windows 540
- Forcible Entry of Windows 540
- Glazing 542
- Types of Windows 543

Breaching Walls and Floors 544
- Techniques for Breaching Walls 544
- Techniques for Breaching Floors 545

Tool Assignments 546
Lessons Learned 546
Key Terms 546
Review Questions 547

CHAPTER 18
VENTILATION

Objectives 550
Introduction 550
Principles, Advantages, and Effects of Ventilation 550
Heat, Smoke, and Toxic Gases 551
Considerations for Proper Ventilation 551
Fire and Its By-Products 554
Flashover 556
Backdraft (Smoke Explosion) 556
Rollover 560
What Needs to Be Vented? 560
- Voids and Compartments 561
- Cocklofts 561
- Horizontal and Vertical Voids 562

Air Movement 562
Types of Ventilation 563
- Natural 563
- Mechanical 563

Mechanics of Ventilation 567
- Vertical Ventilation 567
- Horizontal Ventilation 567

Ventilation Techniques 568
- Break Glass 568
- Open Doors 569
- Effects of Glass Panes 569
- Rope and a Tool 570
- Hook or Pike Pole 570
- Iron or Halligan 570
- Ax 571
- Portable Ladder 571
- Negative Pressure Ventilation 572
- Positive Pressure Ventilation 573

Roof Ventilation 575
- Expandable Cut 575
- Center Rafter Cut (Louver) 577
- Triangular Cut 579
- Trench Cut or Strip Cut 579
- Inspection Cut 581
- Smoke Indicator Hole 581

Safety Considerations 581
- Will Ventilation Permit the Fire to Extend? 581
- Will the Escape Route Be Cut Off? 581
- Will Ventilation Endanger Others? 581
- Work in Teams 582
- Proper Supervision 582

Obstacles to Ventilation 583
- Access 583
- Security Devices 583
- Height 583
- Poor Planning 584
- Personnel Assignment 584
- Unfamiliar Building Layout 584
- Ventilation Timing 584
- Cut a Roof—Open a Roof 585

Factors Affecting Ventilation 585
- Partial Openings 585
- Partially Broken Windows 585
- Screens 586
- Roof Material 586
- Dropped or Hanging Ceilings 587
- Building Size 587
- Weather 588
- Opening Windows 588

Lessons Learned 590
Key Terms 590
Review Questions 590
Additional Resources 590

CHAPTER 19 FIRE SUPPRESSION

Objectives 593
Introduction 593
Elements of Fire Control 593
Structural Fire Components and Considerations 593
Ground Cover Fire Components and Considerations 595
Vehicular Fire Components and Considerations 599
Flammable Liquids Fire Components and Considerations . 601
Flammable Gas Fire Components and Considerations 602
Process of Fire Extinguishment 603
Proper Stream Selection 604
Tactical Considerations 606
Residential Occupancies 611
Business and Mercantile Occupancies 615
Multistory Occupancies 616
Below-Ground Structures or Basements 618
Structures Equipped with Sprinklers or Standpipes 619
Exposure Fires 620
Nonstructural Fires 620
Lessons Learned 628
Key Terms 628
Review Questions 629
Additional Resources 630

CHAPTER 20 SALVAGE, OVERHAUL, AND FIRE CAUSE DETERMINATION

Objectives 633
Introduction 633
Salvage Tools and Equipment 633
Salvage Covers 634
Floor Runner 634
Water Vacuum 635
Miscellaneous Salvage Tools 636
Maintenance of Tools and Equipment Used in Salvage 636
Salvage Cover Folds and Rolls 637
Salvage Operations 639
Safety Considerations 639
Stopping Water Flowing from Sprinkler Heads 642
Methods of Protecting Material Goods 642
Arranging of Furnishings and Salvage Cover Deployment . 642
Water Removal 644
Salvage Operations in Sprinklered Buildings 647
Post Indicator Valve and Outside Screw and Yoke Valve 647
Sprinkler Stops 648
Salvage Operations Lessons Learned 648
Overhaul Tools and Equipment 648
Common Tools 649
Carry-All 649
Overhaul Operations 649
Overhauling Roofs 650
Electronic Heat Sensors 650
Revisits of the Involved Structure 650
Debris Removal 651
Overhaul Operations Lessons Learned 651
Fire Cause Determination Concerns 652
Preservation of Evidence 653
Basics of Point of Origin Determination 653
Fire Cause Determination Lessons Learned 654
Securing the Building 654
Lessons Learned 655
Key Terms 655
Review Questions 656
Additional Resources 656

CHAPTER 21 PREVENTION, PUBLIC EDUCATION, AND PRE-INCIDENT PLANNING

Objectives 659
Introduction 659
Administration of the Fire Prevention Division 659
Fire Company Inspection Program 660
Equipment 660
Preparation for Inspections 661
Conducting the Inspection 661
Typical Violations 662
Concluding the Inspection 674
Reinspections 677
Home Inspections 677
Fire and Life Safety Education 679
Fire and Life Safety Program Presentations 680
Forms of Fire and Life Safety Programs 681
Pre-Incident Management Process 684
Deciding to Preplan 685
Site Visit 685
Diagrams 685
Seek Input from Others 685
The Finished Document 687
Lessons Learned 687
Key Terms 687
Review Questions 688
Additional Resources 688

CHAPTER 22
EMERGENCY MEDICAL SERVICES

Objectives 691
Introduction 691
Roles and Responsibilities of an Emergency Care Provider 691
- Key Responsibilities 692
- Legal Considerations for Emergency Care Providers 694
- Interacting with Emergency Medical Services Personnel 694

Safety Considerations 695
- Analyzing the Safety of the Emergency Scene 695
- Firefighter Physical and Mental Health 696
- Infection Control 696

Assessing a Patient 700
- Performing an Initial Assessment 700
- Vital Signs and the Focused History and Physical Exam 704
- Patient Findings 707

Cardiopulmonary Resuscitation/AED 707
Bleeding Control and Shock Management 709
- Internal and External Bleeding 709
- Caring for Patients with Internal Bleeding 710
- Caring for Patients with External Bleeding 710
- Types of Wounds Requiring First Aid 712
- What Is Shock? (Hypoperfusion) 713
- Recognizing the Signs and Symptoms of Shock (Hypoperfusion) 713
- Caring for Patients in Shock 713

Emergency Care for Common Emergencies 714
- Trouble Breathing 714
- Chest Pain 714
- Medical Illnesses 714
- Allergic Reactions 714
- Thermal Burns 715
- Chemical Burns 716
- Poisoning 717
- Fractures and Sprains 718

Lessons Learned 718
Key Terms 719
Review Questions 720
Additional Resources 720

CHAPTER 23
FIREFIGHTER SURVIVAL

Objectives 723
Introduction 723
Incident Readiness 723
- Personal Protective Equipment 723
- Personal Accountability 725
- Fitness for Duty 726

Safety at Incidents 728
- Team Continuity 728
- Orders/Communication 729
- Risk/Benefit 730
- Personal Size-Up 730
- Rehabilitation 730
- Rapid Intervention Teams 732

Firefighter Emergencies 733
- Rapid Escape 733
- Rapid Escape Steps 734
- Lost, Trapped, and Injured Firefighters 735
- Post-Incident Survival 736

Lessons Learned 738
Key Terms 739
Review Questions 739
Endnotes 739
Additional Resources 740

CHAPTER 24
HAZARDOUS MATERIALS: LAWS, REGULATIONS, AND STANDARDS

Objectives 743
Introduction 743
Laws, Regulations, and Standards 744
- Development Process 744

Emergency Planning 745
- State and Local Emergency Response Committees 745
- Local Emergency Response Plans 745
- Chemical Inventory Reporting 745

OSHA HAZWOPER Regulation 747
- Paragraph q 748
- Medical Monitoring 748

Standards 749
- NFPA 471 749
- NFPA 472 749
- NFPA 473 749
- Standard of Care 749

Additional Laws, Regulations, and Standards 750
- Hazard Communication 750
- Superfund Act 750
- Clean Air Act 750
- Respiratory Protection 750
- Firefighter Safety 751
- NFPA Chemical Protective Clothing 751

Lessons Learned 751
Key Terms 751
Review Questions 752
Endnotes 752
Additional Resources 752

CHAPTER 25 HAZARDOUS MATERIALS: RECOGNITION AND IDENTIFICATION

Objectives . . . 755
Introduction . . . 755
Location and Occupancy . . . 755
Placards, Labels, and Markings . . . 757
Placards . . . 757
Labels . . . 767
Other Identification Systems . . . 768
NFPA 704 System . . . 768
Hazardous Materials Information System . . . 769
Military Warning System . . . 769
Pipeline Markings . . . 770
Container Markings . . . 771
Pesticide Container Markings . . . 771
Containers . . . 771
General . . . 772
Pipelines . . . 776
Highway Transportation Containers . . . 776
Boiling Liquid Expanding Vaper Explosion (BLEVE) . . . 782
Specialized Tank Trucks . . . 784
Rail Transportation . . . 788
Bulk Storage Tanks . . . 790
Senses . . . 794
Chemical and Physical Properties . . . 795
States of Matter . . . 795
Vapor Pressure . . . 796
Vapor Density . . . 797
Specific Gravity . . . 798
Corrosivity . . . 798
Chemical Reactivity . . . 799
Flash Point . . . 800
Autoignition Temperature . . . 800
Flammable Range . . . 800
Toxic Products of Combustion . . . 800
Lessons Learned . . . 801
Key Terms . . . 802
Review Questions . . . 803
Endnotes . . . 803
Additional Resources . . . 803

CHAPTER 26 HAZARDOUS MATERIALS: INFORMATION RESOURCES

Objectives . . . 807
Introduction . . . 807
Emergency Response Guidebook . . . 807
Material Safety Data Sheets . . . 818
Using the MSDS Wisely . . . 823
Accidents and How the MSDS Relates . . . 823
MSDS in the Workplace . . . 823
Shipping Papers . . . 825
Mode of Transportation . . . 825
Facility Documents . . . 826
Computer Resources . . . 826
Chemtrec . . . 826
Reference and Information Texts . . . 827
Industrial Technical Assistance . . . 829
Lessons Learned . . . 829
Key Terms . . . 829
Review Questions . . . 830
Additional Resources . . . 830

CHAPTER 27 HAZARDOUS MATERIALS: PERSONAL PROTECTIVE EQUIPMENT

Objectives . . . 833
Introduction . . . 833
Health Hazards . . . 833
Toxicology . . . 834
Types of Exposures . . . 834
Types of Hazards . . . 834
Categories of Health Hazards . . . 834
Exposure Levels . . . 838
Types of Personal Protective Equipment . . . 840
Self-Contained Breathing Apparatus . . . 840
Chemical Protective Clothing . . . 842
Limitations of Personal Protective Equipment . . . 846
Lessons Learned . . . 849
Key Terms . . . 849
Review Questions . . . 850
Endnote . . . 850
Additional Resources . . . 850

CHAPTER 28 HAZARDOUS MATERIALS: PROTECTIVE ACTIONS

Objectives . . . 853
Introduction . . . 853
Hazardous Materials Management Processes . . . 853
Isolation and Protection . . . 853
Rescue . . . 854
Site Management . . . 857
Establishment of Zones . . . 859
Evacuations and Sheltering in Place . . . 862
Common Incidents . . . 864
Types of Releases . . . 864
Explosives . . . 866
Gases . . . 868
Flammable and Combustible Liquids . . . 871

Flammable Solids, Water Reactives, and Spontaneously Combustible Materials . . . 872
Oxidizers and Organic Peroxides . . . 874
Poisons . . . 874
Radioactive Materials . . . 875
Corrosives . . . 876
Other Incidents . . . 876
Decontamination . . . 877
Types of Decontamination . . . 877
Emergency Decontamination . . . 877
Gross Decontamination . . . 878
Formal Decontamination . . . 879
Fine Decontamination . . . 879
Mass Decontamination . . . 879
Decontamination Process . . . 881
Methods of Decontamination . . . 885
Absorption . . . 885
Adsorption . . . 885
Covering . . . 885
Dilution . . . 885
Disinfection . . . 886
Disposal . . . 886
Emulsification . . . 886
Neutralization . . . 886
Overpacking . . . 886
Removal . . . 886
Solidification . . . 886
Vacuuming . . . 886
Vapor Dispersion . . . 886
Lessons Learned . . . 887
Key Terms . . . 888
Review Questions . . . 889
Endnotes . . . 889
Additional Resources . . . 889

CHAPTER 29 PRODUCT CONTROL AND AIR MONITORING

Objectives . . . 892
Introduction . . . 892
Defensive Operations . . . 892
Absorption . . . 892
Diking and Damming . . . 893
Diverting . . . 895
Retention . . . 896
Dilution . . . 897
Vapor Dispersion . . . 897
Vapor Suppression . . . 897
Remote Shutoffs . . . 898
Air Monitoring at the First Responder Operations Level . . . 898
Regulations and Standards . . . 899
Air Monitor Configurations . . . 899
Meter Terminology . . . 900
Bump Test . . . 900
Calibration . . . 900
Reaction Time . . . 901
Recovery Time . . . 901
Relative Response . . . 901
Oxygen Monitors . . . 901
Flammable Gas Indicators . . . 903
Toxic Gas Monitors . . . 905
Other Detectors . . . 905
Carbon Monoxide Incidents . . . 906
Lessons Learned . . . 908
Key Terms . . . 909
Review Questions . . . 909
Additional Resources . . . 910

CHAPTER 30 TERRORISM AWARENESS

Objectives . . . 913
Introduction . . . 913
Types of Terrorism . . . 917
Potential Targets . . . 918
Indicators of Terrorism . . . 920
HAZMAT Crimes . . . 922
Clandestine Labs . . . 922
Incident Actions . . . 926
General Groupings of Warfare Agents . . . 927
Nerve Agents . . . 927
Incendiary Agents . . . 928
Blister (Vesicants) . . . 929
Blood and Choking Agents . . . 929
Irritants (Riot Control) . . . 929
Biological Agents and Toxins . . . 930
Radioactive Agents . . . 931
Other Terrorism Agents . . . 931
Detection of Terrorism Agents . . . 932
Federal Assistance . . . 932
Lessons Learned . . . 934
Key Terms . . . 934
Review Questions . . . 935
Endnote . . . 935
Additional Resources . . . 935

Glossary . . . 937
Acronyms . . . 957
Other Fire Science Titles from Delmar . . . 959
Index . . . 964

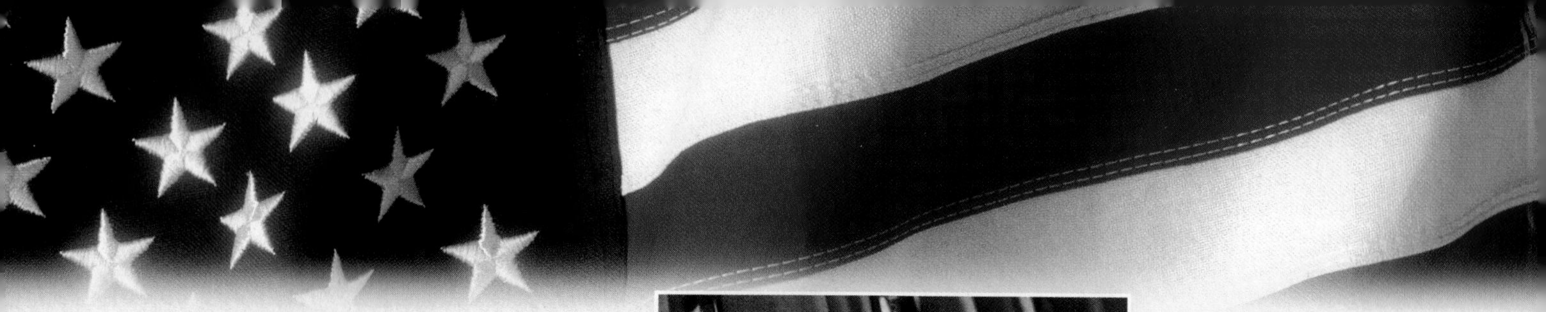

DEDICATION

Dedicated to the courageous firefighters and emergency responders who have given of themselves the greatest sacrifice, their lives. On September 11, 2001, the fire and emergency service community changed forever, and as we continue on we are left with the scar of this day and the tears of many loved ones left behind. We share in the heartache of the loss of every single firefighter and emergency responder on that day and others. Let their lives shine on in the dedication and bravery of those left to respond when the tones drop, the bells ring, and the sirens blare.

THOMSON

DELMAR LEARNING

This text is also dedicated to the driving force behind the continuation of firefighter heritage, the sharing of wisdom and experience, and the art of discovery and learning—trainers and educators. Every single classroom session, practical scenario, and review session directly affects the quality of response the fire service provides. Never underestimate the power of positive change the training and education community holds.

In honor and support of all Fire Service Educators, we are privileged to announce that Delmar Learning, a Thomson company, will donate a portion of the proceeds to the National Fire Academy Alumni Association (NFAAA) for every copy of *The Firefighter's Handbook* we sell. The NFAAA was selected as the sole recipient of this contribution because of the similarities of our missions and our belief that NFAAA makes a positive difference in the education, safety, and welfare of firefighters.

NFAAA
NATIONAL FIRE ACADEMY ALUMNI ASSOCIATION

And to every firefighter who has touched the life of someone in need and made a positive difference—you are truly the epitome of human compassion and selflessness. Don't ever stop caring.

FOREWORD

So much has changed in our country and our world since the first edition of *The Firefighter's Handbook* was published back in 2000. Of course the tragic events of September 11, 2001, have caused many changes in the lives of America's firefighters and how we do business. Sure, we used to talk about terrorism—but not with the urgency and realism that permeates today's discussions. The study of weapons of mass destruction (WMD) was an emerging specialty and today is mandatory reading for every firefighter. The changes are not limited to terrorism and WMD but have spilled over into rapid intervention team tactics, firefighter survival, and the value and hazards of live fire training.

However, some things have not changed. The value and importance of preparedness and training for every possible scenario has not changed. The principle of treating and serving the people who rely on us with respect and dignity has not changed. The idea that firefighters are held to a higher standard of service and duty has not changed. Even in the face of widespread changes in the world and in firefighting, there are some key principles that stay the same. These principles make firefighting a proud and honorable tradition and tie firefighters to the communities they serve.

This second edition of *The Firefighter's Handbook* is a reflection of the new world that today's firefighters live in. We have updated and revised many chapters and sections relating to the new threat of terrorism and the part that the fire service plays in handling this heightened threat. Other areas of study have also been improved and expanded such as self-contained breathing apparatus, firefighter safety, rescue procedures, and firefighter survival. All of the areas that we all deal with on a much more frequent basis than terrorism have received the same attention and have been noticeably improved. All of this has been done to keep you, the first line of defense, as technically prepared and tactically ready as possible. It has become apparent that every firefighter in every city and town in America must be able to respond quickly and effectively to any and every emergency that arises there. Every effort has been made to create a new and updated handbook that can be used as a reference for chief and company officers, training manual for company officers and firefighters, and textbook for candidates preparing to enter the most difficult and rewarding profession, firefighting.

John Salka, Jr.
Battalion Chief
New York City Fire Department

ABOUT OUR AUTHORS

The expertise, dedication, and passion of our contributing authors have created a text that determines a standard of excellence in the education of our nation's firefighters.

To continue in this standard of excellence for the second edition of *The Firefighter's Handbook: Essentials of Firefighting and Emergency Response,* our authors have dedicated their time to ensure that the book remains current to the 2002 edition of *NPFA Standard 1001,* as well as the changing landscape of the fire service world. Thanks to our revising authors Andrea Walter, David Dodson, Dennis Childress, Chris Hawley, and Marty Rutledge, as well as to the outstanding authors who provided the foundation of this textbook: Ron Coleman, Thomas J. Wutz, Willis T. Carter, Frank J. Miale, T. R. (Ric) Koonce III, Robert F. Hancock, Robert Morris, Geoff Miller, and Donald C. Tully.

Andrea A. Walter

Author of Chapter 22, Emergency Medical Services. Revising author of 1, 2, 3, and 4.

Andrea A. Walter is a firefighter with the Washington Metropolitan Airports Authority and a member and former officer of the Sterling Volunteer Rescue Squad. Walter has been active in the fire and emergency services community for many years, serving as the Manager of the Commission on Fire Accreditation International for the International Association of Fire Chiefs and assisting in a variety of projects with the National Volunteer Fire Council, Women in the Fire Service, and the United States Fire Administration. She has over fifteen years of experience in the fire and emergency services. In addition to being an author for this text, Walter also took on the expanded role of serving as the project's Content Editor. She is also an author and the Content Editor for Delmar Learning's *First Responder Handbook: Fire Service Edition* and the *Law Enforcement Edition.*

Ronny J. Coleman

Author of Chapter 1, Overview of the History, Tradition, and Development of the American Fire Service.

Chief Coleman is a nationally and internationally recognized member of the fire service who formerly served as the Chief Deputy Director, Department of Forestry and Fire Protection, and as California State Fire Marshal. He has served in the fire service for thirty-eight years. Previously he was Fire Chief for the Cities of Fullerton and San Clemente, California, and was the Operations Chief for the Costa Mesa Fire Department. Chief Coleman possesses a Master of Arts Degree in Vocational Education from Cal State Long Beach, a Bachelor of Science Degree in Political Science from Cal State Fullerton, and an Associate of Arts Degree in Fire Science from Rancho Santiago College. He has served in many elected positions in professional organizations, including President, International Association of Fire Chiefs; Vice President, International Committee for Prevention and Control of Fire; and President, California League of Cities, Fire Chiefs Department. He is the author of *Going For Gold*, Delmar Thomson Learning.

Thomas J. Wutz

Author of Chapter 2, Fire Department Organization, Command, and Control; Chapter 7, Self-Contained Breathing Apparatus.

Chief Wutz has been involved in the fire service for more than thirty years in both volunteer and military fire departments. He is currently Chief of Fire Services, New York State Office of Fire Prevention and Control. In this position, his duties include supervision and management of the state's outreach training program, delivered by 230 instructors assigned to fifty-seven counties. In addition, he is responsible for curriculum development and implementation of new training programs, New York State's Wireless 9-1-1 program, and state fire mobilization and mutual plan and response. On completion of a twenty-eight-year career, he recently retired as Fire Chief of the 109th Airlift Wing, New York Air National Guard. Chief Wutz is also a member of the faculty in the Fire Science Program at Schenectady County Community College, Schenectady, New York, and a firefighter with the Midway Fire Department, Town of Colonie, New York.

Willis T. Carter

Author of Chapter 3, Communications and Alarms.

Chief Willis Carter has been a member of the fire service for over thirty years. He began his career in 1972 as a firefighter with the Shreveport Fire Department, and for the past twenty-five years has served as the Chief of Communications for the department. Carter is responsible for the management and operations of the Fire Communications Center, which serves as the Primary Public Safety Answering Point (PSAP) for the Caddo Parish, Louisiana, 9-1-1 system.

In addition to his work in the fire service, he is active at the national level. He has served as past president of the Louisiana Chapter of the Association of Public Safety Communications Officials (APCO) and currently serves as Executive Council representative for the state of Louisiana. His other work with APCO includes serving as Chairperson for the Membership Task Force, and as a member of APCO Project 37 Team (Telecommunicator

Certification Program). He also serves as a member of APCO Bulletin Editorial Advisory Board. In addition to his work with APCO, he is a member of the International Fire Chief's Association and the National Emergency Number Association. He is also an Assessment Team Leader for the Commission on Law Enforcement Accreditation (CALEA). Carter led the effort by the Shreveport Fire Department Communications Center to become the first Public Safety Communications Center in the state of Louisiana to achieve accredited status through CALEA.

Frank J. Miale

Author of Chapter 4, Fire Behavior; Chapter 14, Ladders; and Chapter 18, Ventilation.

Miale, a Battalion Chief with over thirty years in the FDNY, recently retired. A twenty-five-year active member in his local Volunteer Lake Carmel Fire Department, he maintains a busy role as treasurer and training instructor. A former high school teacher, he holds two Bachelor of Science degrees with several concentrations in Education, Biology, and Fire Administration. During his career in the FDNY, he taught at the NYC fire academy, participated in the introduction of a communication system using apparatus-mounted computers, and headed a special Emergency Command Unit while an active line officer. Formerly the Training Officer for the 27th Battalion in the FDNY, he taught many ladder company and ventilation courses throughout the country. His career was spent primarily in busy ladder companies in Brooklyn, Harlem, and the South Bronx sections of New York City prior to promotion to Chief Officer. He is the recipient of nine awards for courage and valor, including two department medals from the FDNY, and has been published many times in *WNYF, Fire Command,* and *Fire Service Today.*

David W. Dodson

Author of Chapter 5, Firefighter Safety; Chapter 6, Personal Protective Clothing and Ensembles; and Chapter 23, Firefighter Survival. Revising author of Chapters 5, 6, 9, 10, 11, 12, 13 ,16, and 23.

Dodson is a twenty-four-year fire service veteran. He started his fire service career with the U.S. Air Force. He served at Elmendorf AFB in Alaska and spent two years teaching at the USAF Fire School. After the USAF, Dodson spent almost seven years as a Fire Officer and Training/Safety Officer for the Parker Fire District in Parker, Colorado. He became the first Career Training Officer for Loveland Fire and Rescue in Colorado and rose through the ranks, including time as a HAZMAT Technician, Duty Safety Officer, and Emergency Manager for the city. He accepted a Shift Battalion Chief position for the Eagle River Fire District in Colorado before starting his current company, *Response Solutions,* which is dedicated to teaching firefighter safety and practical incident handling. Chief Dodson has served on numerous national boards including the NFPA Firefighter Occupational Safety Technician Committee and the International Society of Fire Service Instructors (ISFSI). He also served as president of the Fire Department Safety Officers' Association. In 1997, Dodson was awarded the ISFSI "George D. Post Fire Instructor of the Year." He is also the author of *Fire Department Incident Safety Officer,* published by Delmar Learning, a Thomson Company.

T. R. (Ric) Koonce, III

Author of Chapter 8, Portable Fire Extinguishers; Chapter 9, Water Supply; Chapter 10, Fire Hose and Appliances; Chapter 11, Nozzles, Fire Streams, and Foam; and Chapter 12, Protective Systems.

Koonce is an Assistant Professor and Program Head of Fire Science Technology at J. Sargeant Reynolds Community College in Richmond, Virginia. He is a retired Battalion Chief with the Prince George's County (Maryland) Fire Department and has over thirty years of fire service experience. He is an adjunct instructor for the Virginia Department of Fire Programs. He holds two associate degrees, a Bachelor of Science degree in Fire Service Management from University College of the University of Maryland, and a Certificate of Public Management from Virginia Commonwealth University.

Robert F. Hancock

Author of Chapter 15, Ropes and Knots; and Chapter 16, Rescue Procedures.

Hancock is Assistant Chief/Administration with Hillsborough County Fire Rescue in Tampa, Florida, a department that services an area of 931 square miles and a population of over 600,000 with 615 career personnel and 205 volunteers and a $42.7 million budget. He was hired in November 1974 as a firefighter and was promoted through the ranks to his present position in October 1993. He was awarded an Associate of Science Degree in Fire Science, with honors, from Hillsborough Community College. He graduated from the Executive Fire Officers Program at the National Fire Academy and has been certified as an instructor with the State of Florida since 1983. Hancock is chairman of the Florida Fire Chiefs' Disaster Response Communications Sub-Committee, charged with identifying short- and long-term solutions to the disaster response communication issue statewide. He is Rescue Series Editor, a contributing author for Delmar Thomson Learning, and a member of Florida EDACS PS User's Group, serving as President for 1999.

Robert Morris

Author of Chapter 17, Forcible Entry.

Morris is a thirty-year veteran of the New York City Fire Department and has been assigned to some of the busiest fire companies in New York City, including Ladder Company 42, Engine 60 in the Bronx, and Rescue Company 3 in Manhattan. After serving in the Bronx and Harlem, he served as Company Commander of Ladder Company 28. Captain Morris is currently Company Commander of Rescue Company 1 in Manhattan. He is an Instructor with the Connecticut State Fire Academy, the New York City Fire Department Institute, a national lecturer, and a Contributing Editor for *Firehouse Magazine*. Captain Morris is the recipient of seventeen meritorious awards, including three department medals.

Dennis Childress

Author of Chapter 19, Fire Suppression.

Childress is with the Orange County Fire Authority in Southern California and has been in the fire service for just over thirty-five years. He is a Certified Chief Officer with the state of California, and he holds an Associate of Arts degree in Fire Science and a Bachelor of Science degree in Fire Protection Administration. He holds a seat on the Board of Directors for the Southern California Fire Training Officers Association, chairs the California State Firefighters Association Health and Safety Committee, and sits on the Statewide Training and Education Advisory Committee for the State Fire Marshal's office. He is a principal member of the NFPA 1500 Fire Service Occupational Safety and Health Committee and the NFPA 1561 Standard for Emergency Services Incident Management System Committee. He has authored a number of articles in fire service publications over the years, and he has also been an instructor in Fire Command and Management in the California State Fire Training System for over twenty years as well as an instructor for the National Fire Academy. He is the original author of the *Workbook to Accompany Firefighter's Handbook: Essentials of Firefighting and Emergency Response.*

Geoff Miller

Author of Chapter 20, Salvage, Overhaul, and Fire Cause Determination.

Miller is a twenty-nine year veteran of the fire service and is currently a line Battalion Chief with the Sacramento Metropolitan Fire District in California. Previous assignments have included four years as the district's Training Officer, ten years as a line Captain, and two years as an Inspector. He has been involved in several California Fire Fighter I and II curriculum development workshops as well as participating on the rewrite of Fire Command 1A and 1B. He was also on the IFSTA Material Review Committee for Fire Department Company Officer, third edition. He is happily married, has two daughters, and lives in El Dorado Hills, California.

Donald C. Tully

Author of Chapter 21, Prevention, Public Education, and Pre-Incident Planning.

Tully is a member of the Orange County, California, Fire Authority. With thirty years in the fire service, he has also been a Division Chief/Fire Marshal in Buena Park and Westminster, California, for ten years, and a Fire Technology Instructor at Santa Ana College, California. He is Past President of the Orange County Fire Prevention Officers' Association and was a member of IFSTA's Fire Investigation Committee. He also served as a member of NFPA Committees 1221 (CAD Dispatch and Public Communications) and 72 (Fire Alarms), and as a member of the California State Fire Marshal Committees on Fire Sprinklers and Residential Care Facilities (ad hoc committees). He is a California State Certified Chief Officer, Fire Officer, Fire Investigator, Fire Prevention Officer, and Fire Service Instructor and Technical Rescue Specialist.

Chris Hawley

Author and revising author of Chapter 24, Hazardous Material: Laws, Regulations, and Standards; Chapter 25, Hazardous Materials: Recognition and Identification; Chapter 26, Hazardous Materials Information Resources; Chapter 27, Hazardous Materials: Personal Protective Equipment; Chapter 28, Hazardous Materials: Protective Actions; Chapter 29, Product Control and Air Monitoring; and Chapter 30, Terrorism Awareness.

Hawley is a retired Fire Specialist with the Baltimore County Fire Department. Prior to this post, he was assigned as the Special Operations Coordinator and was responsible for the coordination of the Hazardous Materials Response Team and the Advanced Technical Rescue Team along with the team leaders. He has served on development teams for local, state, and federal projects, including the National Fire Academy, and provides HAZMAT and terrorism response training nationwide. Hawley is also the author of four Delmar Learning textbooks, including *Hazardous Materials Response and Operations, Hazardous Materials Incidents, Haz-Mat Air Monitoring and Detection Devices,* and *Fire Department Response to Sick Buildings.* He is currently the owner of FBN Training, which provides a wide variety of emergency response training, including hazardous materials, confined space, technical rescue, and emergency medical services, as well as consulting services to emergency services and private industry.

Marty L. Rutledge

Revising author of Chapters 7, 8, 14, 15, 20, and 21.

Rutledge is a Firefighter/Engineer, ARFF Specialist, and EMS Program Manager for Loveland Fire and Rescue in Loveland, Colorado. He is a member of the Fire Certification and Advisory Board to the Colorado Division of Fire Safety, as well as serving as the State First Responder program coordinator. He is also a member of the Colorado State Fire Fighter's Association and has over thirteen years of fire and emergency services experience in both volunteer and career ranks. Rutledge has authored and served as technical expert for a supplementary firefighter training package for Delmar Learning's *The Firefighter's Handbook* and as co-author for Delmar Learning's *First Responder Handbook: Fire Service Edition* and *First Responder Handbook: Law Enforcement Edition.*

PREFACE

Welcome new recruits, volunteers, and experienced firefighters to the second edition of *The Firefighter's Handbook: Essentials of Firefighting and Emergency Response.* Within the pages of this textbook you will find all the information you need to successfully complete the Firefighter I and II certification courses, as well as a guide to refresh your knowledge and skills as you continue on as a firefighter.

The Firefighter's Handbook is a comprehensive guide to the basic principles and fundamental concepts involved in firefighting, emergency medical services, and hazardous materials operations, as defined by the 2002 edition of the *National Fire Protection Association 1001 Standard for Firefighter Professional Qualifications.* The text can be used by both new and experienced firefighters and hazardous materials technicians to study the basic skills required to perform a wide variety of firefighting and emergency service activities.

As a firefighter you will make a difference in the lives of many. Use your knowledge, practice your skills, and above all—be safe.

Development of This Text

The Firefighter's Handbook was created to fill a void in the firefighting and emergency service education system. Through the dedication of our authors, content and technical reviewers, as well as our Fire Advisory Board members, the second edition of *The Firefighter's Handbook* continues to remain up to date with the changing landscape of the fire service world. This text is designed to meet and exceed the requirements set forth in the 2002 edition of *NFPA Standard 1001,* as well as the Hazardous Materials Awareness and Operations Level of the 2003 Edition of *NFPA Standard 472,* and presents the information in a realistic and challenging way. The content is written in a clear and concise manner, and step-by-step photo sequences illustrate the need-to-know Job Performance Requirements that are so critical to hands-on training. A special emphasis on safety and the development of critical thinking skills through featured text boxes ensures that both aspiring and experienced firefighters have the knowledge they need to effectively respond to fires and other emergencies.

Organization of This Text

The Firefighter's Handbook: Essentials of Firefighting and Emergency Response, Second Edition, consists of thirty chapters, including coverage of hazardous materials and terrorism. All the essential information—from the history of the fire service to the governing laws and regulations, from the use of apparatus and equipment to the practice of procedures, from understanding fire behavior and building construction to effective planning and prevention measures—is covered in this text. The chapters are set up to deliver a straightforward, systematic approach to training, and each includes an outline, objectives, introduction, lessons learned, key terms, review questions, and a list of additional resources. Also included at the front of the book is an NFPA 1001/472 Correlation Guide and Job Performance Requirement Guide that correlates the requirements outlined in the Standard to the content of *The Firefighter's Handbook* by chapter and page references. These resources can be used as a quick reference and study guide.

The Firefighter's Handbook: Basic Essentials of Firefighting

New for the release of the 2002 edition of the *NFPA Standard 1001* is the addition of *The Firefighter's Handbook: Basic Essentials of Firefighting* to the fire science list from Delmar Learning. This book retains all the features of the original text, but excludes the coverage of hazardous materials. This is an excellent choice for fire departments, academies, and schools in which hazardous materials are taught in a separate course with separate learning materials.

The *Basic Essentials of Firefighting* textbook meets the requirements of the *NFPA Standard 1001* when taught in conjunction with a hazardous materials course that meets the requirements of *NFPA Standard 472,* Awareness and Operations levels.

Order #: 1-4018-3582-1

New to This Edition

The Firefighter's Handbook: Essentials of Firefighting and Emergency Response, Second Edition, contains many new updates and additional information to address the needs of the fire service today:

- *Safety:* Safety—when responding to incidents in apparatus, and while performing scene assessment on vehicle accidents—is thoroughly covered. New sections on unique fire events and "reading smoke" encourage firefighters to apply an understanding of basic fire behavior when responding to structural fires. In addition, expanded content on the two in/two out rule educates firefighters on how to rescue their own in emergencies.
- *Current Technology:* This text offers information on the latest technology, including information on up-to-date communication systems and a new section on thermal imaging cameras.
- *Building Construction:* Chapter 13 was completely revised to address new building structures and additional considerations in structural collapse. Also included is a special section dedicated to the expert in the field, Francis L. Brannigan.
- *Ladders:* Chapter 14 was thoroughly revised to reflect a variety of procedures utilizing ladders in rescue situations.
- *Hazardous Materials and Terrorism:* In light of the events of September 11, 2001, and the worldwide terrorist attacks that followed, Chapters 24–30 have been thoroughly revised to reflect the latest threats to our nation. Expanded coverage of BLEVE and decontamination provides necessary information to firefighters responding to HAZMAT incidents. Valuable information related to terrorist activities, such as how to determine if a threat is credible, new coverage on hazardous material crimes, including a discussion on drug labs, as well as current statistics and events bring into focus the changing world of the fire service.
- *New Photos:* The inclusion of new photos brings the text up to date with the latest in apparatus, tools, equipment, and procedures in the fire service.
- *New Feature Text Boxes:* Additional text boxes and featured articles provide helpful tips, advice on safety, and important information for firefighters.

FEATURES

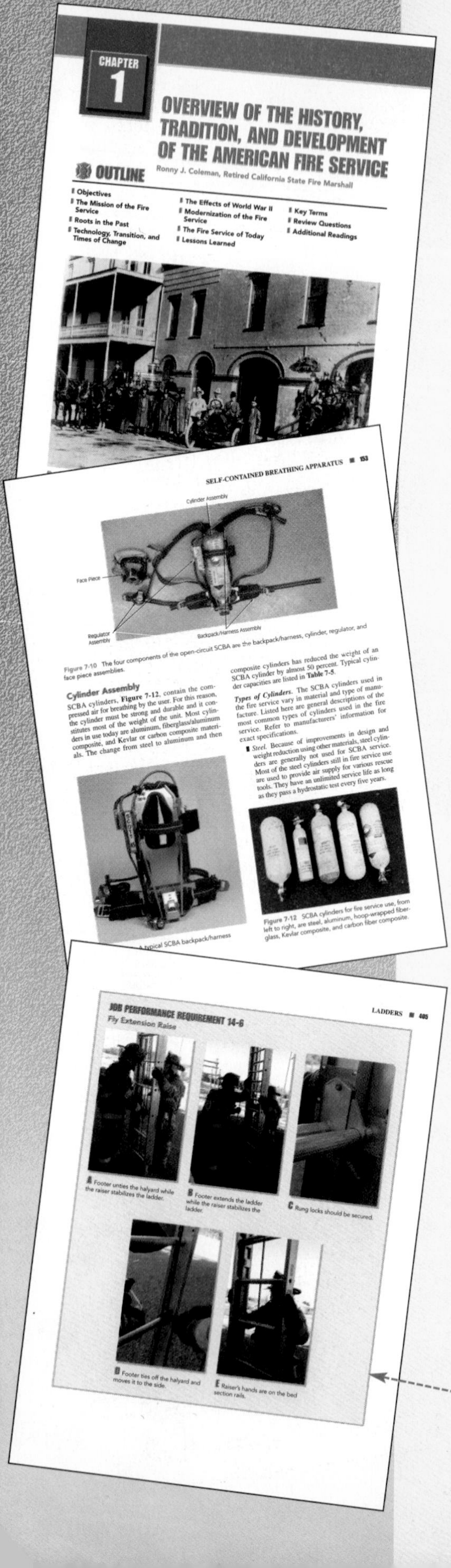

The Firefighter's Handbook contains a number of features that set it apart from other basic fire service texts. What is unique about this text is that it offers a realistic approach to the world of firefighting and emergency response. It is comprehensive in coverage, including all the need-to-know information, but presents the content in a clear and concise manner, so it is easy to read, follow, and understand. What is essential is not only acquiring the knowledge, but more importantly, putting it into practice. *The Firefighter's Handbook* emphasizes this fact through the use of step-by-step photo sequences and text descriptions to illustrate critical Job Performance Requirements. Firefighters throughout the nation also continue to contribute their stories to the book, bringing insight, advice, and experience to the text.

The Firefighter's Handbook continues to recognize that today's firefighters must respond to more than just fire emergencies. Firefighters are expected to effectively respond to medical emergencies as well as hazardous materials and terrorism incidents. In response, the textbook includes a chapter on emergency medical services, as well as seven chapters on all the latest information on hazardous materials and terrorist activities. Tips for safety that are vital to survival and critical to the success of the operation on the scene of an incident are integrated throughout the chapters.

In addition, *The Firefighter's Handbook* provides instructors with handy reference tools and supplemental materials that alleviate those heavy preparation commitments. A new emphasis on the modularized lesson plans in the Instructor's Curriculum Kit and Instructor's Curriculum CD-ROM for departments provides instructors with flexible lecture outlines that will meet individual department or academy schedules.

How to Use This Text

The following suggests how you can best utilize the features of this text to gain competence and confidence in learning firefighting essentials.

NFPA Standard 1001/472 Correlation Guide

This grid provides a correlation between *The Firefighter's Handbook* and the requirements for Firefighter I and II, as stipulated by the *NFPA Standard for Firefighter Professional Qualifications,* 2002 Edition, Chapters 5 and 6. These sections from the Standard are correlated to the textbook chapters and are referenced by page numbers.

Job Performance Requirement Correlation Guide

This grid provides an outline of the Job Performance Requirements stipulated by the *NFPA Standard for Firefighter Professional Qualifications,* 2002 Edition, Chapters 5 and 6, and the corresponding step-by-step photo sequences illustrated in *The Firefighter's Handbook.* Additional supplemental skills are also included to encourage practice of essential skills for on-the-job training. A quick reference for studying and reviewing important hands-on skills, this grid correlates the JPR to the textbook chapters and references the page numbers.

Job Performance Requirements

Step-by-step photo sequences illustrating important procedures are integrated throughout the chapters. These are intended to be used as a guide in mastering the job performance requirement skills and to serve as a review guide reference.

Street Stories

Each chapter opens with a personal experience written by noted contributors from across the nation. These personal accounts draw you into the minds of those who wrote them and help to reinforce to you why the subsequent chapter can make all the difference in the world.

Streetsmart Tips

As is true of any profession, sometimes experience can be the best teacher. These tips are power-packed with a wide variety of hints and strategies that will help you to become a streetsmart firefighter.

Firefighter Facts

These boxes offer a detailed snapshot of facts based on firefighting history, experience, and recorded data to provide thought-provoking information.

Notes

The Note feature highlights and outlines important points for you to learn and understand. Based on key concepts, this "must know" material is an excellent study resource.

Safeties and Cautions

As a firefighting professional, you will face situations in which you will need to react immediately in order to ensure your safety and the safety of those with you. These tips emphasize when and how to react safely.

Lessons Learned

The Lessons Learned feature summarizes the main points presented in the chapter and is ideal for review purposes.

Review Questions

At the end of every chapter, the review questions assist you with the learning process and help you develop the necessary critical thinking skills.

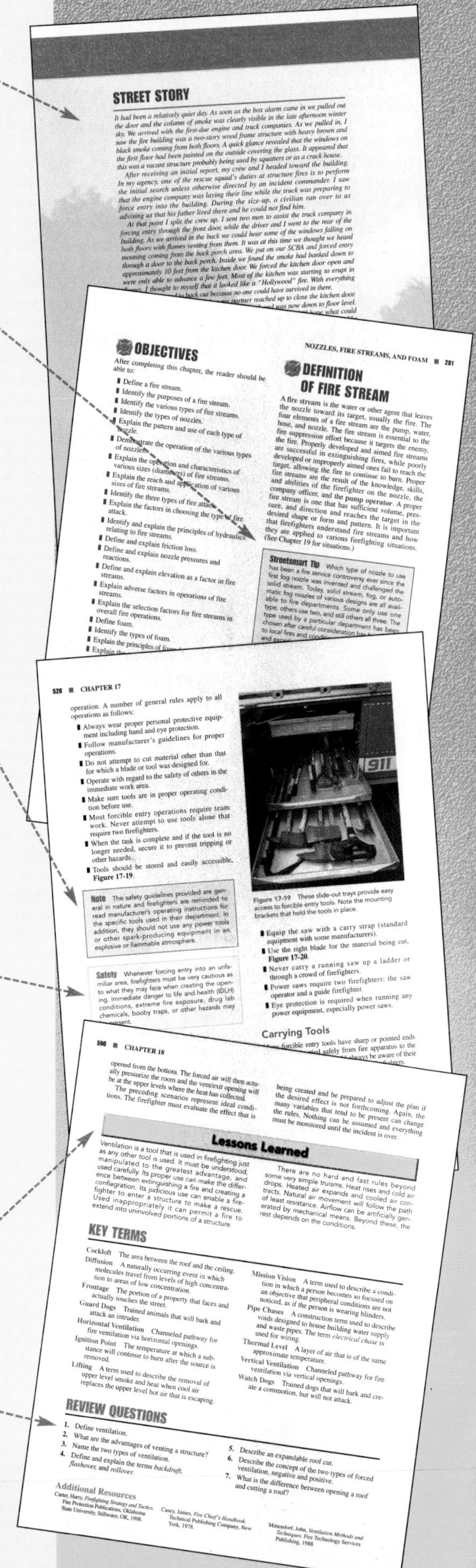

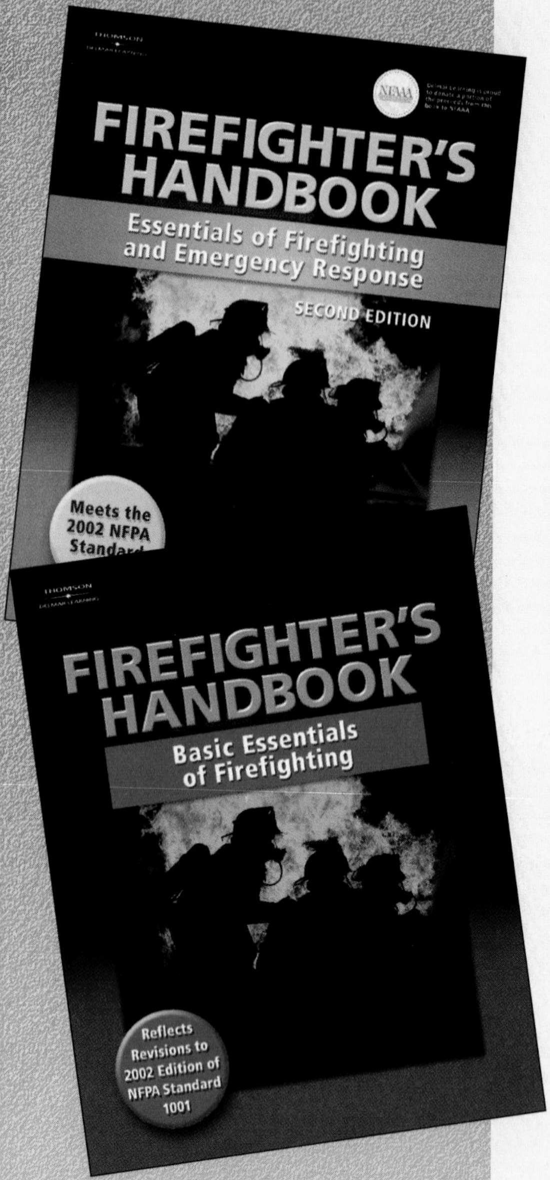

Supplemental Curriculum Package

The Firefighter's Handbook

This text was created not only as a stand-alone manual for firefighters, but as a special package of materials for the full instructional experience. The supplement package provides a variety of tools for students and instructors to enhance the learning experience.

Instructor's Curriculum Kit

The *Instructor's Curriculum Kit* is designed to allow instructors to run programs according to the standards set by the authority having jurisdiction where the course is conducted. It contains the information necessary to conduct Firefighter I, Firefighter II, hazardous materials awareness, and hazardous materials operations courses. It is divided into sections to facilitate its use for training:

Administration: Provides the instructor with an overview of the various courses, student and instructor materials, and practical advice on how to set up courses and run skill sessions.

Modularized Lesson Plans: Are ideal for instructors, whether they are teaching at fire departments, academies, or longer-format courses. Each section presents learning objectives, recommended time allotment, equipment and reading assignments for each lesson and outlines key concepts presented in each chapter of the text with coordinating PowerPoint slides, textbook readings, and Job Performance Requirement and Supplement Skill sheets.

Equipment Checklist: Offers a quick guide for ensuring the necessary equipment is available for hands-on training.

Job Performance Requirement and Supplemental Skills Sheets: Outline important skills that each candidate must master to meet requirements for certification and provide the instructor with a handy checklist. Also includes a *Job Performance Requirement Correlation Guide,* which cross-references the Standard with the *Job Performance Requirements* outlined in *The Firefighter's Handbook.* These guides can be used for quick reference when reviewing important skills and for studying for the Firefighter I and II certification exam.

Progress Log Sheets: Provide a system to track the progress of individual candidates as they complete the required skills.

Quick Reference Guides: Contain valuable information for instructors. Included are three grids: *NFPA Standard 1001/472 Correlation Guide* used to cross-reference *The Firefighter's Handbook* with standards 1001/472, *New Edition Correlation Guide* used to cross-reference the revisions between the first and second editions of *The Firefighter's Handbook,* and a *Comparison Guide* that correlates *The Firefighter's Handbook, Second Edition,* to the *IFSTA/Essentials, Fourth Edition,* text.

Answers to Review Questions: Included for each chapter in the text.

Additional Resources: Offer supplemental resources for important information on various topics presented in the textbook.

Also includes: Instructor's Curriculum CD-ROM

Order #: 1-4018-3576-7

Instructor's Curriculum CD-ROM

Available in the *Instructor's Curriculum Kit* and as a separate item, the *Instructor's Curriculum CD-ROM* ensures a complete, electronic teaching solution for *The Firefighter's Handbook: Essentials of Firefighting and Emergency Response, Second Edition.* Designed as an integrated package, it includes the following:

- *PowerPoint Presentations* outline key concepts from each chapter, and contain graphics and photos from the text, as well as video clips, to bring the content to life.
- A *Testbank* containing hundreds of questions helps instructors prepare candidates to take the written portion of the certification exam for Firefighter I and II.
- A searchable *Image Library* containing hundreds of graphics and photos from the text offers an additional resource for instructors to enhance their own classroom presentations or to modify the PowerPoint provided on the CD-ROM.
- *Complete Curriculum* available in Word to allow instructors to add their own notes or revise to meet the requirements of the Authority Having Jurisdiction.

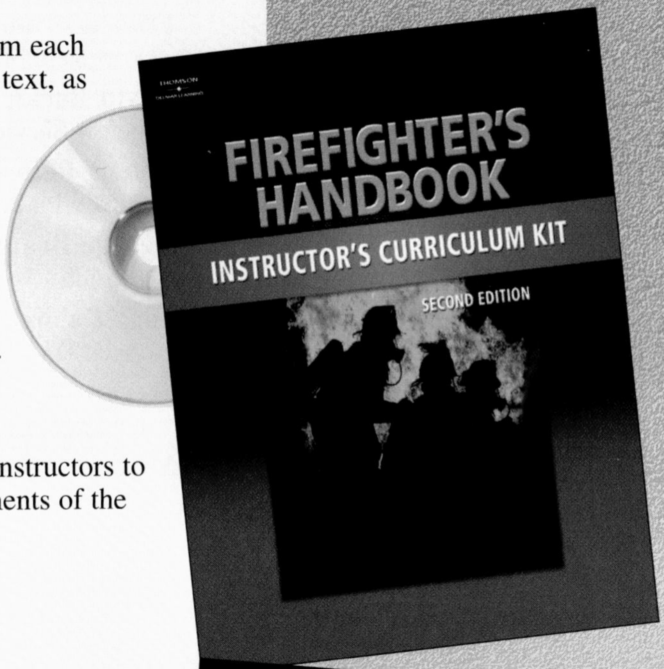

Order #: 1-4018-7175-5

Student Workbook

This is helpful in the classroom setting as a guide for study and a tool for assessing progress. The workbook consists of questions in multiple formats, including new and revised questions to support the second edition of *The Firefighter's Handbook.*

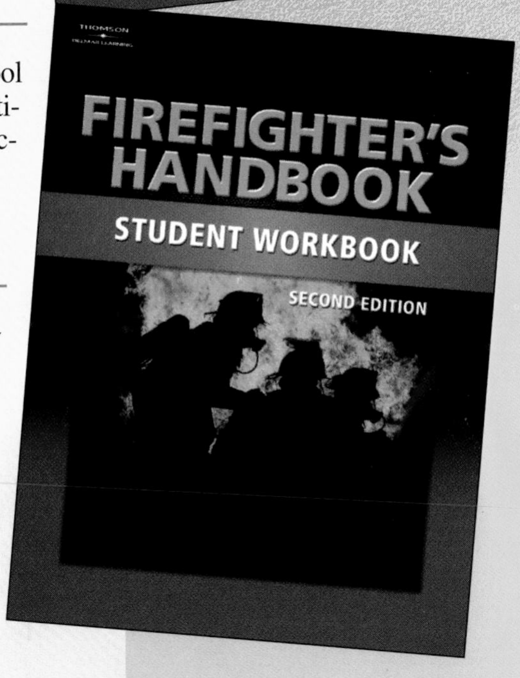

Multimedia

The Basic Firefighting Series Video Set shows introductory firefighting concepts to enhance learning with a visual presentation. Instructor's Guide with pre and post-tests included.

Order #: 0-7668-4099-9

The Basic Firefighting Series DVD contains the same great information and instructor resources in DVD format.

Order #: 1-4018-8298-6

ACKNOWLEDGMENTS

The Firefighter's Handbook: Essentials of Firefighting and Emergency Response, Second Edition, remains true to the tradition of Delmar to remain dedicated to the individuals we serve—among them, both aspiring and experienced firefighters. However, we would not be able to accomplish this without the contributions of many professionals whose passion, commitment, and hard work have helped shape a book of which we all can be proud.

Delmar and the contributing authors would like to extend a special thanks to those who served on the Editorial Review Board, as well as those who are members of our Fire Advisory Board. Their expertise and objectivity has provided invaluable insight:

Mike Arnhart, Deputy Chief, High Ridge Fire District, High Ridge, Missouri

Francis Brannigan, (Building Construction chapter) SFPE & Fellow, Port Republic, Maryland

Kevin Barkley, Fire/HazMat Coordinator, Saratoga County Emergency Services, Galway, New York

Robert Bettenhausen, Chief Fire Marshal, Village of Tinley Park, Tinley Park, Illinois

George Braun, Lieutenant, Gainesville Fire Rescue, Gainesville, Florida

Tony Calorel, Senior Instructor, Burlington County Emergency Services Training Center, West Hampton, New Jersey

Steve Chickerotis, Chief of Training, Chicago Fire Department, Chicago, Illinois

Michael J. Connors, Assistant Fire Chief, Naperville Fire Department, Naperville, Illinois

Lee Cooper, Fire Service Specialist, Wisconsin Indianhead Technical College, New Richmond, Wisconsin, and President, Fire Instructors Association of Minnesota, Bloomington, Minnesota

Richard W. Davis, Lieutenant, Loveland Fire and Rescue, Loveland, Colorado

Peter Evers, Captain, Auburn City Fire Department, Auburn, California

Tom Feierabend, Director of Fire Technology, Mount San Antonio College, Walnut, California

Herald Good, Associate Instructor, Virginia Department of Fire Programs, Newport News, Virginia

Craig Hanna, Training Chief, Sioux Falls Fire Rescue, Sioux Falls, South Dakota

Attila Hertelendy, Instructor, University of Nevada Fire Science Academy, Carlin, Nevada

Al Ionnone, Director of Fire Technology, American River College, Sacramento, California

Kim Johnson, MBA, Fitness Trainer, 24-Hour Fitness

John Kingyens, Training Safety Officer, Sarnia Fire Rescue, Canada

Kent D. Neiswender, Supervisor, Office of Training and Certification, New Jersey Division of Fire Safety, Trenton, New Jersey

Bob Leigh, Battalion Chief, Aurora Fire Department, Aurora, Colorado

Ron Marley, Fire Technology Director, Shasta College, Redding, California

Pat McAulliffe, Director of Fire Science, Collin Community College Fire Academy, McKinny, Texas

Peter McBride, Safety Battalion Chief, Ottawa Fire, Canada

David P. Pritchett, Director Georgia Fire Academy, Forsyth, Georgia, and Training Captain, City of Jackson, Jackson, Georgia

Taylor Robertson, District/Training Chief, City of Eugene Fire and EMS Department, Eugene, Oregon

Chris Reynolds, Battalion Chief, Hillsborough County Fire Rescue, Tampa, Florida

John J. Salka, Jr., Battalion Chief, New York City Fire Department

Bob Sanborn, Captain, Bowling Green Fire Department, Bowling Green, Kentucky

Randy Scheerer, Battalion Chief, Newport Beach Fire Department, Newport Beach, California

R. Peter Sells, Chief Training Officer, Toronto Fire Services, Canada

Billy Shelton, Curriculum and ARFF Manager, Virginia Department of Fire Programs, Richmond, Virginia

William Shouldis, Deputy Chief, Philadelphia Fire Department, Philadelphia, Pennsylvania

Fred C. Windisch, Fire Chief, Ponderosa VFD, Houston, Texas, and Chairman Volunteer Chief Officers Section, International Association of Fire Chiefs

We would also like to recognize those individuals who contributed content:

Mike West (Thermal Imaging Article/Photos), Lieutenant, South Metro Fire and Rescue, Instructor, SAFE-IR, Colorado

Tom Wutz (Original Author/Building Construction Chapter), Deputy Chief of Fire Services, New York State Office of Fire Prevention and Control

The majority of photographs shown in this book are the result of numerous days of photo shoots at various locations. A special acknowledgment is owed to four very patient firefighters who facilitated numerous days of shooting. Our appreciation is extended to Kevin P. Terry, Assistant Chief, Fuller Road Fire Department, Patrolman, Town of Colonie Police Department, and New York Regional Team-1 (NYRRT-1) Logistics Liaison; Steven M. Leonardo, Past Chief, Shaker Road, Loudonville Fire Department, Patrolman, Town of Colonie Police Department, and New York Regional Team-1 (NYRRT-1) Rescue Specialist; Warren E. Carr, Jr., Past Chief, S.W. Pitts Hose Company, Inc., Latham, New York, and New York Regional Response Team-1 (NYRRT-1) Team Leader; and Mike Kelleher, Troy Fire Department, Troy, New York, and New York Regional Response Team-1 (NYRRT-1) Team Manager. Your knowledge and willingness were invaluable.

Credit is also owed to the many departments, models, and photographers who shared their time, expertise, equipment, and photographs with us: Fuller Road Fire Department, Albany, New York; S.W. Pitts Hose Company, Inc., Latham, New York; Troy Fire Department, Troy, New York; Albany Fire Department, Albany, New York; Metropolitan Washington Airports Authority, Washington, DC; Loveland Fire and Rescue, Loveland, Colorado; Poudre Fire Authority, Fort Collins, Colorado; South Metro Fire and Rescue, Greenwood Village, Colorado; Hillsborough County Fire Rescue, Tampa, Florida; Baltimore County Fire Department, Baltimore, Maryland; Fairfax County Department of Fire and Rescue, Fairfax, Virginia; Sterling Park Rescue Squad and Sterling Volunteer Fire Department, Sterling, Virginia; Ashburn Volunteer Fire and Rescue Department, Ashburn, Virginia, and Loudoun County Department of Fire Rescue Services, Leesburg, Virginia. Our photographers included Michael Dzaman, Dzaman Photography, Latham, New York; Rick Fulford, Cockeysville, Maryland; Rick Michalo, Brentt Sporn, California Fire Photos, Anaheim, California; and Captain Pete Evers, City of Auburn Fire Department, California.

We also would like to recognize those who participated in our focus groups, reviewed various material, or were just there to answer questions: Randy Napoli, Chief, Bureau of Fire Standards and Training, Division of State Fire Marshal, Ocala, Florida; Dave Edmunds, Sarasota County Technical Institute, Sarasota, Florida; Mike Brackin, State Fire Academy, Jackson, Mississippi; David Pritchett, Georgia Fire Academy, Forsyth, Georgia; David Herndon, Georgia Fire Academy, Forsyth, Georgia; Claude Shew, North Carolina Fire and Rescue Commission, North Carolina; David Fultz, Louisiana State Fire Academy, Baton Rouge, Louisiana; Larry McCall, Florida State Fire Academy, Ocala, Florida; Timothy Dunkle, Pennsylvania State Fire Academy, Pennsylvania; Gregory Kirt, Michigan Firefighters Training Council, Lansing, Michigan; Ron Coleman, Elk Grove, California; Michael Richwine, Department of Forestry and Fire Protection, Ione, California; Michael Ridley, Elk Grove Community Services District Fire Department, Elk Grove, California; Mark Lewandowski, Connecticut Fire Academy, Windsor

Locks, Connecticut; John Pangborn, Jersey City Fire Department, Jersey City, New Jersey; Doug Hall, Red Rocks Community College, Red Rocks, Colorado; Michael Forgy, Fairfax Volunteer Fire Department, Fairfax City, Fairfax, Virginia.

We also give our sincere appreciation to those who shared with us their stories that you see at the beginning of each chapter: Andrea Walter, Lead Author; Mike Smith, District of Columbia Fire and EMS Department, Washington, DC; Mike Kelleher, Troy Fire Department, Troy, New York; Fred Windisch, Ponderosa Volunteer Fire Department, Houston, Texas; Randy Sheerer, Newport Beach Fire Department, Newport Beach, California; Gordon Sachs, Fairfield Fire and EMS, Fairfield, Virginia; Mike Gala, FDNY, New York, New York; Chief Bernard Lach (ret.) Torrington Fire Department, Torrington, Connecticut; Michael Arnhart, High Ridge Fire Protection District, High Ridge, Missouri; Battalion Chief Billy Goldfelder, Loveland Fire Department, Loveland, Ohio; Peter F. Kertzie, Buffalo Fire Department, Buffalo, New York; Paul LePore, Long Beach Fire Department, Long Beach, California; Lieutenant Michelle Steele, Miami-Dade Fire Rescue, Miami, Florida; Richard Arwood (ret.), Memphis Fire Department, Memphis, Tennessee; Battalion Chief Frank Montagna, FDNY, New York, New York; William Shouldis, Philadelphia Fire Department, Philadelphia, Pennsylvania; James P. Smith, Philadelphia Fire Department, Philadelphia, Pennsylvania; Michael Ramsey, Oran Fire Protection District, Oran, Iowa; James Angle, Palm Harbor Fire Department, Palm Harbor, Florida; Mary K. Marchone, Montgomery County Fire and Rescue Services; John J. Salka, FDNY, New York, New York; Rob Schnepp, Alameda County Fire Department, San Leandro, California; Mike Callan, Callan & Company, Ltd., Middlefield, Connecticut; Jan R. Dunbar, Sacramento Fire Department, Sacramento, California; Lieutenant Julius Stanley, Chicago Fire Department, Chicago, Illinois; Frank Docimo, Turn of River Fire Department, Turn of River, Connecticut; Joseph DeFrancesco, Madison County Fire and Rescue, Madison, New York; David Mitchell, Fayettville Fire Department, Fayettville, Arkansas; Greg Noll, Hildebrand & Noll Associates, Lancaster, Pennsylvania; Greg Socks, Washington County Special Operations, Hagerstown, Maryland; Tom Creamer, Worcester Fire Department, Worcester, Massachusetts; and Rick Townsend, Sierra Vista Fire Department, Sierra Vista, Arizona.

And to those we rarely take the time to recognize because this is their job, a special thanks. The Delmar Learning team developed, produced, and marketed *The Firefighter's Handbook* setting an example for not only getting the job done, but having the creativity and fortitude to go above and beyond. Our appreciation to Alison Weintraub, Mary Ellen Black, Jennifer Thompson, Rachel Baker, Toni Hansen, Barbara Diaz, Jennifer Luck, Erin Coffin, Mark Pierro, Fair Huntoon, Cindy Eichelman, Sandy Clark, and Alar Elken.

How to Contact Us

At Delmar Learning, listening to what our customers have to say is the heart of our business. If you have any comments or feedback on *The Firefighter's Handbook,* you can e-mail us at firescience@delmar.com or fax us at 518-881-1262, Attention: Fire Rescue Editorial.

For additional information on other titles that may be of interest to you, or to request a catalog, see our listing on pages 959–963, or visit www.firescience.com.

NFPA 1001 CORRELATION GUIDE

Entrance Requirements			
NFPA 1001 Section	**NFPA 1001 Description**	***Firefighter's Handbook* Chapter Reference**	***Firefighter's Handbook* Page Reference**
4.3	Emergency Medical Care	22	696, 707, 709

Firefighter I			
NFPA 1001 Section	**NFPA 1001 Description**	***Firefighter's Handbook* Chapter Reference**	***Firefighter's Handbook* Page Reference**
5.1	General	Not Applicable	
5.1.1	General	24–30	See hazmat grid.
5.1.1.1	General	2, 5, 15	23, 26, 24, 34, 35, 109, 419, 423, 424, 439, 449
5.1.1.2	General Skill Requirements	2, 6, 13, 15, 20	137, 449, 432, 428, 433, 427, 430, 427
5.2	Fire Department Communications	3	51, 58, 63
5.2.1	Fire Department Communications	3	51, 63
5.2.2	Fire Department Communications	3	51
5.2.3	Fire Department Communications	3	63
5.3	Fireground Operations	4, 7-20, 23	74, 141, 721
5.3.1	Fireground Operations	7	158, 168, 145, 150, 172, 170
5.3.2	Fireground Operations	5, 23	120, 723
5.3.3	Fireground Operations	16, 19	624, 486
5.3.4	Fireground Operations	13, 17, 18	339, 522, 511, 530
5.3.5	Fireground Operations	7, 16, 19, 23	725, 733, 168, 591, 459
5.3.6	Fireground Operations	14	369
5.3.7	Fireground Operations	11, 19	279, 622
5.3.8	Fireground Operations	10, 11, 19	219, 279, 603, 627
5.3.9	Fireground Operations	7, 14, 16, 17	170, 383, 459, 511
5.3.10	Fireground Operations	4, 10, 11, 19	74, 219, 279, 591
5.3.11	Fireground Operations	4, 14, 18, 19	79, 369, 548, 591
5.3.12	Fireground Operations	4, 13, 14, 15, 18	79, 339, 369, 449, 591
5.3.13	Fireground Operations	11, 20	279, 631
5.3.14	Fireground Operations	12, 20	307, 631
5.3.15	Fireground Operations	9, 10	201, 219
5.3.16	Fireground Operations	4, 8	100, 183
5.3.17	Fireground Operations	Instructor's Guide	Instructor's Guide

NFPA 1001 Section	NFPA 1001 Description	*Firefighter's Handbook* Chapter Reference	*Firefighter's Handbook* Page Reference
5.3.18	Fireground Operations	19	591
5.3.19	Fireground Operations	19	595
5.4	Rescue Operations	Not Applicable in FF I	—
5.5	Prevention, Preparedness, Maintenance	21	657
5.5.1	Prevention, Preparedness, Maintenance	21	657
5.5.2	Prevention, Preparedness, Maintenance	21	657
5.5.3	Prevention, Preparedness, Maintenance	7, 14, 15, 17, 20	172, 382, 439, 441, 511, 636
5.5.4	Prevention, Preparedness, Maintenance	10	223, 226

Firefighter II

NFPA 1001 Section	NFPA 1001 Description	*Firefighter's Handbook* Chapter Reference	*Firefighter's Handbook* Page Reference
6.1	General	Not Applicable	—
6.1.1	General	Not Applicable	—
6.1.1.1	General Knowledge Requirements	2, 5	37, 106
6.1.1.2	General Skill Requirements	2	37
6.2	Fire Department Communications	3	45
6.2.1	Fire Department Communications	3	68
6.2.2	Fire Department Communications	3	45
6.3	Fireground Operations	4, 9-11, 13, 14, 16-21	74, 201, 339, 369, 459
6.3.1	Fireground Operations	11, 19	279, 625
6.3.2	Fireground Operations	10, 11, 13, 16, 17, 18, 19	219, 279, 339, 459, 508, 548, 591
6.3.3	Fireground Operations	4, 19	86, 88, 625
6.3.4	Fireground Operations	20	652
6.4	Rescue Operations	16	459
6.4.1	Rescue Operations	16	482
6.4.2	Rescue Operations	16	490
6.5	Prevention, Preparedness, Maintenance	9, 10, 12, 17, 21	201, 274, 307, 511, 657
6.5.1	Prevention, Preparedness, Maintenance	9, 12, 21	201, 307, 657
6.5.2	Prevention, Preparedness, Maintenance	17	511
6.5.3	Prevention, Preparedness, Maintenance	10	274
6.5.4	Prevention, Preparedness, Maintenance	9	213

NFPA 472 CORRELATION GUIDE

Competencies at the Awareness Level			
NFPA 472 Section	**NFPA Description**	***Firefighter's Handbook* Chapter References**	***Firefighter's Handbook* Page Reference**
4.1	General	Not Applicable	
4.1.1	Introduction	Not Applicable	
4.1.1.1	Introduction	24–30	741–935
4.1.1.2	Introduction	24–30	741–935
4.1.2	Goal	24	747–752
4.1.2.2	Goal	24	747–752
4.2	Analyzing the Incident	24, 25, 26	744–745, 753–804, 805–830
4.2.1	Detecting the Presence of Hazardous Materials	25	753–804
4.2.3	Collecting Hazard Information	24, 25, 26, 29, 30	744–745, 753–774,. 805–830, 890–897, 905–910
4.3	Planning the Response	Not Applicable	
4.4	Implementing the Planned Response	24, 28	744–752, 851–889
4.4.1	Initiating Protective Actions	24, 28	744–752, 851–889
4.4.2	Initiating the Notification Process	24, 28	744–752, 851–889
4.5	Evaluating Progress	Not Applicable	
4.6	Terminating the Incident	Not Applicable	

Competencies at the Operational Level			
NFPA 472 Section	**NFPA Description**	***Firefighter's Handbook* Chapter References**	***Firefighter's Handbook* Page Reference**
5.1	General	Not Applicable	
5.1.1	Introduction	Not Applicable	
5.1.1.1	Introduction	24–30	741–935
5.1.1.2	Introduction	24–30	741–935
5.1.2	Goal	24–30	741–935
5.1.2.1	Goal	24–30	741–935
5.2	Analyzing the Incident	24, 25, 26	745–746, 753–804, 805–830
5.2.1	Surveying the Hazardous Materials Incident	24, 25	745–752, 753–804
5.2.2	Collecting Hazard and Response Information	24, 25, 26, 29	745–752, 805–830, 890–910

NFPA 472 Section		NFPA Description	*Firefighter's Handbook* Chapter References	*Firefighter's Handbook* Page Reference
5.2.3		Predicting the Behavior of a Material and Its Container	24, 28, 29	768–804, 864–889, 890–898
5.2.4		Estimating the Potential Harm	25, 27, 28	770–801, 864–877, 890–898
	5.3	Planning the Response	24, 27, 28, 29, 30	744–745, 750–751, 840–848, 851–889, 890–898, 911–932
5.3.1		Describing Response Objectives for Hazardous Materials Incidents	28, 29, 30	851–864, 890–898, 911–926, 933
	5.4	Implementing the Planned Response	28, 29, 30	851–864, 890–898, 911–926, 933
5.4.1		Establishing and Enforcing Scene Control Procedures	28, 30	851–864, 933
5.4.2		Initiating the Incident Management System	28, 30	851–864, 933
5.4.3		Using Personal Protective Equipment	27	840–848
5.4.4		Performing Defensive Control Actions	29	890–898
	5.5	Evaluating Progress	28	851–864
5.5.1		Evaluating the Status of Defensive Actions	28	851–864
5.5.2		Communicating the Status of the Planned Response	24, 28	750, 851–864
	5.6	Terminating the Incident	Not Applicable	

Competencies at the Technician Level

NFPA 472 Section		NFPA Description	*Firefighter's Handbook* Chapter References	*Firefighter's Handbook* Page Reference
	6.1	General	Not Applicable	
6.1.1		Introduction	Not Applicable	
6.1.1.1		Introduction	Not Applicable	
6.1.1.2		Introduction	Not Applicable	
6.1.2		Goal	24–30	740–935
	6.2	Analyzing the Incident	24, 25, 26	744–745, 753–804, 805–830
6.2.1		Surveying the Hazardous Materials Incident	24, 25	744–745, 750–751, 753–804
6.2.2		Collecting and Interpreting Hazard and Response Information	24, 25, 26, 29, 30	744–745, 753–771, 805–830, 898–906, 932
6.2.3		Describing the Condition of the Container Involved in the Incident		

NFPA 472 Section		NFPA Description	*Firefighter's Handbook* Chapter References	*Firefighter's Handbook* Page Reference
6.2.4		Predicting Likely Behavior of Materials and Their Containers When Multiple Materials		
6.2.5		Estimating the Likely Size of an Endangered Area	29	898–906
	6.3	Planning the Response		
6.3.1		Identifying Response Objectives		
6.3.2		Identifying the Potential Action Options		
6.3.3		Selecting Personal Protective Equipment	27	838–848
6.3.4		Selecting Decontamination Procedures	28	877–887
	6.4	Implementing the Planned Response		
6.4.1		Performing Incident Management Duties		
6.4.2		Using Protective Clothing and Respiratory Protection	27	838–840
6.4.3		Performing Control Functions Identified in Plan of Action		
	6.5	Evaluating Progress		
6.5.1		Evaluating the Effectiveness of the Control Functions		
	6.6	Terminating the Incident		
6.6.1		Assisting in the Debriefing		
6.6.2		Assisting in the Incident Critique		
6.6.3		Providing the Reports and Documentation		

JOB PERFORMANCE REQUIREMENT CORRELATION GUIDE

Designation	No.	Description	NFPA 1001 References	Firefighter's Handbook Chap. Ref.	Firefighter's Handbook Page Ref.
Supplemental Skill	6-1	Don Protective Clothing	5.1.1.2, 5.3, 6.3, 6.4	6	N/A
JPR	7-1	Donning Self-Contained Breathing Apparatus, Over the Head Method	5.3.1, 5.3, 6.3	7	161
JPR	7-2	Donning Self-Contained Breathing Apparatus, Coat Method	5.3.1, 5.3, 6.3	7	163
JPR	7-3	Donning Self-Contained Breathing Apparatus, Seat Method	5.3.1, 5.3, 6.3	7	165
JPR	7-4	Donning Self-Contained Breathing Apparatus Face Piece	5.3.1, 5.3, 6.3	7	167
JPR	7-5	Self-Contained Breathing Apparatus, Daily Inspection	5.3.1, 5.3, 6.3	7	173
JPR	7-6	SCBA Cylinder Replacement Procedure	5.3.1, 5.3, 6.3	7	175
JPR	7-7	SCBA Cylinder Replacement Procedure, Firefighter Wearing SCBA	5.3.1, 5.3, 6.3	7	176
JPR	7-8	Servicing an SCBA Cylinder Using a Cascade System	5.3.1, 5.3, 6.3	7	178
JPR	7-9	Servicing an SCBA Cylinder Using a Compressor System	5.3.1, 5.3, 6.3	7	179
JPR	8-1	Operation of Portable Fire Extinguisher	5.3.16	8	196
JPR	10-1	Coupling Hose—One-Person Foot-Tilt Method	5.3, 6.3	10	229
JPR	10-2	Coupling Hose—One-Person Over-the-Hip Method	5.3, 6.3	10	229
JPR	10-3	Coupling Hose—Two-Person Coupling Method	5.3, 6.3	10	230
JPR	10-4	Uncoupling Hose with Spanners	5.3, 6.3	10	230
JPR	10-5	Uncoupling Hose—One-Person Knee Press	5.3, 6.3	10	231
JPR	10-6	Storage Hose Roll	5.5.4	10	232
JPR	10-7	Single-Donut Hose Roll (Option 1)	5.5.4	10	232
JPR	10-8	Single-Donut Hose Roll (Option 2)	5.5.4	10	233

Designation	No.	Description	NFPA 1001 References	Firefighter's Handbook Chap. Ref.	Page Ref.
JPR	10-9	Twin-Donut Hose Roll	5.5.4	10	233
JPR	10-10	Hose Drain and Carry	5.5.4	10	234
JPR	10-11	Hose Carry—Shoulder Loop	5.5.4	10	235
JPR	10-12	Hose Drag	5.5.4	10	236
JPR	10-13	Accordion Hose Load	5.5.4	10	238
JPR	10-14	Advancing an Accordion Load	5.3, 6.3	10	239
JPR	10-15	Demonstrate a Flat Hose Load	5.5.4	10	240
JPR	10-16	Advancing a Flat Hose Load from a Supply Bed	5.3, 6.3	10	241
JPR	10-17	Demonstrate a Horseshoe Hose Load	5.5.4	10	242
JPR	10-18	Advancing a Horseshoe Hose Load	5.3, 6.3	10	243
JPR	10-19	Demonstrate a Reverse Horseshoe Load for an Attack Line	5.5.4	10	244
JPR	10-20	Advancing a Flat Load from a Preconnected Bed	5.3, 6.3	10	246
JPR	10-21	Loading a Minuteman or Slot Load	5.5.4	10	247
JPR	10-22	Advancing the Minuteman Load	5.3, 6.3	10	248
JPR	10-23	Perform the Triple-Layer Load	5.5.4	10	249
JPR	10-24	Advancing the Triple-Layer Load	5.3, 6.3	10	250
JPR	10-25	Modified Gasner Bar Pack	5.3.19, 5.5.4	10	252
JPR	10-26	Advancing a Charged Hoseline Up a Stairwell	5.3.10	10	255
JPR	10-27	Advancing an Uncharged Hoseline Over a Ladder	5.3.10	10	259
JPR	10-28	Advancing an Uncharged Hoseline Over a Ladder at Entry Point of Building	5.3.10	10	260
JPR	10-29	Soft Sleeve Hydrant Connection	5.3.10	10	264
JPR	10-30	Hard Sleeve Hydrant Connection	5.3.15	10	265
JPR	10-31	Assemble and Connect Equipment for Drafting	5.3.15	10	266
JPR	10-32	Extending a Hoseline with a Break-Apart Nozzle	5.3.10, 5.3, 6.3	10	267

Designation	No.	Description	NFPA 1001 References	Firefighter's Handbook Chap. Ref.	Firefighter's Handbook Page Ref.
JPR	10-33	Extending a Hoseline Using a Hose Clamp	5.3.10, 5.3, 6.3	10	268
JPR	10-34	Wildland Hose Advancing and Extension	5.3.19	10	269
JPR	12-1	Using "Stops" to Stem the Flow of Water from a Sprinkler Head	5.3.14	12	327–328
JPR	14-1	Ladder Suitcase Carry	5.3.6, 5.3.12, 5.3.13	14	396
JPR	14-2	Shoulder Carry	5.3.6, 5.3.12, 5.3.13	14	397
JPR	14-3	Flat Ladder Carry	5.3.6, 5.3.12, 5.3.13	14	398
JPR	14-4	Two-Person Rung Raise	5.3.6, 5.3.12, 5.3.13	14	401
JPR	14-5	Two-Person Beam Raise	5.3.6, 5.3.12, 5.3.13	14	403–404
JPR	14-6	Fly Extension Raise	5.3.6, 5.3.12, 5.3.13	14	405
JPR	14-7	Lowering a Ladder into a Building	5.3.6, 5.3.12, 5.3.13	14	406
JPR	14-8	One-Person Ladder Raise	5.3.6, 5.3.12, 5.3.13	14	407
JPR	14-9	Use Ladder Leg Lock	5.3.6, 5.3.12, 5.3.13	14	408
JPR	14-10	Carry Tools Up and Down Ladder	5.3.6, 5.3.12, 5.3.13	14	409
JPR	14-11	Securing/Heeling a Ladder	5.3.6, 5.3.12, 5.3.13	14	411
JPR	14-12	Engaging the Hooks on a Roof Ladder	5.3.6, 5.3.12, 5.3.13	14	413
JPR	14-13	Hoisting Ladders	5.1.1.1	14	414
JPR	15-1	Half Hitch Around an Object	5.1.1.2	15	427
JPR	15-2	Tie Overhand Safety	5.1.1.2	15	427
JPR	15-3	Tie a Clove Hitch in the Open	5.1.1.2	15	428
JPR	15-4	Tie a Clove Hitch Around an Object	5.1.1.2	15	429
JPR	15-5	Tie a Becket Bend Knot	5.1.1.2	15	430
JPR	15-6	Tie a Double Becket Bend	5.1.1.2	15	431
JPR	15-7	Tie a Bowline Knot	5.1.1.2	15	432
JPR	15-8	Tie a Figure Eight Knot	5.1.1.2	15	434
JPR	15-9	Tie a Follow-Through Figure Eight	5.1.1.2	15	435
JPR	15-10	Tie a Figure Eight Knot on a Bight	5.1.1.2	15	436
JPR	15-11	Tie a Rescue Knot	5.1.1.2	15	437–438
JPR	15-12	Tie a Water Knot	5.1.1.2	15	439
JPR	15-13	Coiling a Rope	5.5.3	15	446
JPR	15-14	Rope Storage Bag	5.5.3	15	448

Designation	No.	Description	NFPA 1001 References	Firefighter's Handbook Chap. Ref.	Page Ref.
JPR	15-15	Hoist an Ax	5.3.12	15	450
JPR	15-16	Pike Pole Hoist	5.3.12	15	451
JPR	15-17	Hoist a Charged Hoseline	5.3.12	15	452–453
JPR	15-18	Hoist an Uncharged Hoseline	5.3.12	15	453–454
JPR	15-19	Hoisting Small Equipment	5.3.12	15	454
JPR	15-20	Hoisting a Ladder	5.3.12	15	455
JPR	15-21	Rope Between Two Objects	5.1.1.2	15	456–457
JPR	16-1	Firefighter's Carry	5.3.9	16	468
JPR	16-2	Extremity Carry	5.3.9	16	470
JPR	16-3	Seat Carry	5.3.9	16	471
JPR	16-4	Blanket Drag	5.3.9	16	472
JPR	16-5	Clothing Drag	5.3.9	16	473
JPR	16-6	Webbing Sling Drag	5.3.9	16	474
JPR	16-7	Sit and Drag Method	5.3.9	16	475
JPR	16-8	Firefighter's Drag	5.3.9	16	476
JPR	16-9	Rescue of a Firefighter Wearing SCBA	5.3.9	16	477
JPR	16-10	Placing a Patient on a Blackboard	5.3.9	16	478–479
JPR	16-11	Placing a Patient on an Ambulance Stretcher	5.3.9	16	481
JPR	17-1	Conventional Door Opening Away from Team	5.3.4, 6.3.2	17	534
JPR	17-2	Conventional Door Opening Toward Team	5.3.4, 6.3.2	17	535–536
JPR	17-3	"Through the Lock" Wrenching Lock	5.3.4, 6.3.2	17	537
JPR	17-4	"Through the Lock" K Tool	5.3.4, 6.3.2	17	538
JPR	18-1	Horizontal Ventilation from Above	5.3.11	18	570
JPR	20-1	Salvage Cover Roll	5.3.14	20	638
JPR	20-2	Preparing a Folded Salvage Cover for a One-Firefighter Spread	5.3.14	20	640
JPR	20-3	Preparing a Folded Salvage Cover for a Two-Firefighter Spread	5.3.14	20	641
JPR	20-4	Salvage Cover Shoulder Toss	5.3.14	20	644
JPR	20-5	Salvage Cover Balloon Toss	5.3.14	20	645
JPR	22-1	Removing Gloves	4.3	22	699
JPR	28-1	Steps in the Decontamination Process	5.1.1, 6.1.1	28	883–884

CHAPTER 1

OVERVIEW OF THE HISTORY, TRADITION, AND DEVELOPMENT OF THE AMERICAN FIRE SERVICE

Ronny J. Coleman, Retired California State Fire Marshall

OUTLINE

- Objectives
- The Mission of the Fire Service
- Roots in the Past
- Technology, Transition, and Times of Change
- The Effects of World War II
- Modernization of the Fire Service
- The Fire Service of Today
- Lessons Learned
- Key Terms
- Review Questions
- Additional Resources

Photo courtesy of Marysville Fire Department (California)

STREET STORY

When I began my work in the fire and emergency services, I was confused by the many traditions, symbols, and practices that have come straight out of history. In these high-tech times, when industry technology seems to change almost weekly, it seemed odd to find these traditions still alive and well in the fire service.

As I learned more about the history of the fire service, I realized how valuable these pieces of history are to what we do every day. It is from this history that the fire service derives its pride, honor, integrity, and courage. The symbolism ties our modern-day practices to those of the early firefighters and emergency responders. It provides firefighters and emergency responders with a unique sense of belonging, perspective, unity, and promise for the future.

The history of firefighting and emergency response also gives modern-day firefighters the safety and protection that we need to survive and thrive in the industry. This history, including knowledge of the incidents in which firefighters have fallen in the line of duty, provides us with the information and expertise we need to ensure that firefighters can perform their duties safely with safer personal protective equipment, technological advances, and improved operational practices.

The fire service has changed dramatically over the centuries, but some key elements still remain strong. Would Ben Franklin ever have expected that fire departments in America would be responding to terrorism and hazardous materials incidents and training firefighters on how to protect themselves from infectious diseases? No, of course not. But he would have expected that whatever firefighters do, it would be the best that they could.

The types of emergency calls may have changed with society, but our dedication to protecting our communities and our brother and sister firefighters has not. I hope it never does.

The history of the fire service helps modern firefighters learn from the successes and failures of the past and unites us in a common tradition of serving those in need.

—Street Story by Andrea A. Walter, Lead Author, *Firefighter's Handbook*

OBJECTIVES

After completing this chapter, the reader should be able to:

- Describe the role of the firefighter in the fire service.
- Define the importance of the mission of the fire service and the purpose of a mission statement.
- Identify the major events that have altered the history of the fire service.

THE MISSION OF THE FIRE SERVICE

In general, the tasks of firefighters are the same the world over: to save lives and property from fire. However, not all fire departments approach that goal in the same manner. Many variables are considered by the agencies that provide fire protection. Some agencies only protect airports, whereas others do structural firefighting and provide emergency medical services (EMS). Fire departments that provide EMS may only perform basic life support (BLS), whereas others may provide advanced life support (ALS).

The term *mission,* used in the context of modern fire department management, is usually synonymous with an agency's purpose for existence, or even its legal authority to act in a certain manner. A **mission statement** is a written declaration by a fire agency describing the things that it intends to do to protect its citizenry or customers. When we look at the variety of situations under which the fire service functions, we begin to understand why one mission statement will not cover all agencies. The size and organizational structure of fire departments range from small volunteer companies that protect only a few hundred souls to metropolitan departments that protect millions.

Many fire agencies have a written mission statement. Firefighters should understand their departments' mission statements and how they contribute to their departments' successes. Mission statements vary from very general to very specific. Once written and posted, it is the responsibility of every individual in that department to attempt to achieve its goals. Each and every firefighter, from the beginning of civilization to the modern fire service, has contributed something to the accomplishment of an agency's mission. That is one of the legacies of the fire service.

ROOTS IN THE PAST

Somewhere, lost in the ancient past, is the name of the very first firefighter. That person was probably thrust into the job of fighting fire out of sheer desperation, to protect himself, his family, or his possessions. It was a courageous act then, and it has remained a courageous act as the fire problem has passed from those obscure ancient events to the streets of small towns and villages, suburban cities, metropolitan areas, airports, harbors, marinas, and wildlands, all of which are protected by contemporary firefighters.

> **Note** What is important to recognize is that the task of being a firefighter is an old, yet constantly evolving occupation.

The fire service contains a great deal of tradition. These traditions are often in conflict with the constant changes required in the fire service to keep up with the task of combating fire in a modern society. Being a firefighter today is much different from the past, but practically everything firefighters and fire service organizations do to save lives and protect property is rooted in events from the past. That is why it is important to look at the legacy and heritage of the fire service for the fundamental reasons firefighters do many of the things they do.

> **Firefighter Fact** The fire service of today is a direct result of an evolution in the methodology, technology, and responsibility of a service that has been vital to communities since the beginning of civilization.

Present-day firefighters need to recognize that what they do today is going to be the basis for the processes future firefighters will use. Understanding the history of the fire service is a lot like climbing a ladder, as shown in **Figure 1-1**. The ladder has to be well grounded in order to be stable, and there must be strong beams to support the ladder. There are rungs to stand on as the ladder is climbed. The two beams of the fire service's history are courage and commitment. The bottom rungs represent the experience and knowledge gained from past years. The rung we are standing on is the present. The ones that are beyond

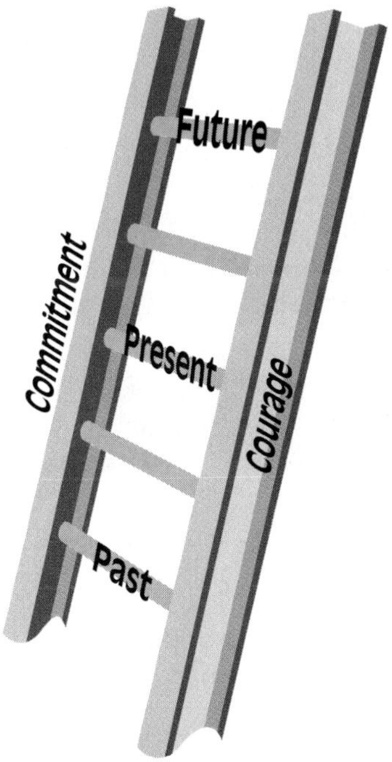

Figure 1-1 Understanding the history of the fire service is like climbing a ladder.

our reach are the future. The ladder is being raised to achieve the goal of saving lives and protecting property. We anticipate that we will always be climbing that ladder in an unending quest to achieve that goal.

The history, traditions, and development of the fire service is a pageant that covers a great deal of time. Significant events have shaped the development of fire service technology and methods. Individual people have actually shaped fire protection philosophy through their actions. Fire service history is a colorful and exciting cavalcade of personalities, tools, equipment, and apparatus that have been used to combat both the ordinary fire and the **conflagration**. The legacy and heritage of the fire service is rich with experience.

Someone once made a joke that the fire service is "two hundred years of tradition, unhampered by progress." Nothing could be more inaccurate. For the last 350 years, the fire service in this country has advanced on all fronts. What has not changed a great deal is the nature of individuals who have chosen to follow a fire and emergency service career, either as a paid, career firefighter or as a volunteer. Granted, in the last few decades there have been more changes in the firefighters' working environment than at any other time in the history of firefighting. But a brief review of history reveals that the process of change has been there for a long time. We just have to search for the stepping stones and recognize that they are incremental steps of change.

> **Note** The fire service has evolved into what it is today because it is a service for people. What communities need and want from their fire service determine what firefighters should or could be providing them. That is what makes the role of the firefighter so exciting and fulfilling. It is a job that not everyone can do. It is not a job that remains static over time. It is a constantly challenging and stimulating environment where the past collides with the future in the actions of those who serve today.

Although many fire agencies have their own mission statement, that does not mean they are all different in all ways. Some factors are common to all agencies. For example, fire agencies all have an organizational structure, an inventory of facilities, apparatus, equipment, and methods for responding to emergencies. They almost all conduct a variety of programs to achieve their goal of protecting life and property. These programs, in general, fall into the categories of fire and emergency operations, fire prevention, arson investigation, training, emergency medical services, communications, and maintenance to name a few. These elements of the fire service are all founded on events and decisions from the past.

Ancient Beliefs

The invention of fire was a turning point in the human race's quest for safety and security. In actuality, humans did not invent fire at all. Fire is a natural phenomenon that results from a number of natural occurrences like lightning or volcanic activity. Even spontaneous combustion requires no human help. What the human race did was to recognize the value of fire for heating their food and warming their homes or living area, and then eventually using it to forge metals and create engines of commerce and weapons of war. Although the human race did not invent fire, the greatest initial challenge was to learn to manage fire and to prevent it from destroying its user.

Records from ancient history clearly indicate that fire was considered a force of nature that was to be feared. The Greeks believed that fire was given to common humans as a gift by one of their gods, Prometheus. A number of religions regarded fire and flame as a deity. The ancient myth of the Phoenix bird was based on the consequences of fire. The Phoenix, as shown in **Figure 1-2**, reportedly lived for 1,000 years. When it reached the end of its life cycle, the bird reportedly built a nest of wooden sticks and

Figure 1-2 The Phoenix rising from the ashes.

set fire to it. When the ashes cooled, a small worm emerged from the ruins of the fire and turned, once again, into a beautiful, peacock-like bird that lived again for another 1,000 years. This legend was based on the idea that fire was a destroyer of monumental power, but also a giver of life. The concept of a resurrection from the flames created the idea that out of death and destruction comes hope.

As human philosophy evolved, the human race continued to experiment and improve on the use of fire as a tool instead of worshipping it as a sign from the gods. Archaeologists have proven not only that ancient cultures used fire, but also that fire destroyed their homes, consumed their property, and in general did a great deal of harm to early villages and towns.

Recorded History

The Romans worshipped a goddess, Vesta, who was the protector of the hearth fire. In the cities of ancient Rome, Vesta was the only goddess with no statue. Instead, a flame was kept burning on her altar constantly. The Romans worshipped fire and respected its impact, for they suffered a lot of fires in their city. Hero of Alexandria, a Roman leader, was credited with creating the first fire pump, which was based on a syringe-like design that squirted water when the plunger was pushed, **Figure 1-3**.

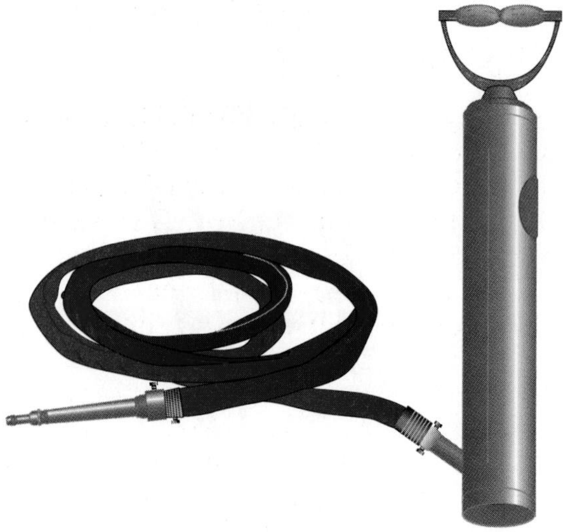

Figure 1-3 An early European hand-operated pump based on the ancient Roman design.

The first record of a truly organized fire department began with actions taken by the Roman Empire to protect their capital. In 22 B.C. a group of individuals, called magistrates, was responsible for maintaining a watch over slaves during the night. Naturally, they were in a good position to report the outbreak of fire. Wealthy Roman citizens expanded that group to become what was called the "familia publica," which were organized along military lines, but they were not very efficient. After a major fire occurred in Rome in A.D. 6 a permanent fire brigade, called the "Cohortes Vigilum," or Vigiles, was established. They were housed in large barracks and toured the streets at night looking for fires.

The Emperor Augustus created seven cohorts, which were seven military units, each under the command of a tribune, and the groups were under a prefect, or an officer of equestrian rank. The prefect was an important person in the Roman hierarchy. Each unit initially had 500 men. Later this was changed to a group of 1,000 men. The Vigiles were further broken down into a variety of specialists. There were men who worked the pumps and men with grappling hooks who tore off roofs and tore down walls to get to the fires. There were also individuals who directed the fire stream provided by the pump-like device. In spite of the presence of this brigade, Rome burned many more times during the next several centuries. Many Roman leaders attempted to use fire regulations to control the buildings that were breeding these destructive fires. Despite these losses nothing slowed down humans' use of fire. Its use was expanded in areas of commerce and industry, such as for the mining and smelting of ore.

With the collapse of the Roman Empire, the Vigiles passed out of use. Europe was not to see any dedicated fire forces for another 1,000 years. This did not mean that no one was fighting fires, but it did mean that there were no organized or collective efforts. The legacy left by the Romans was the idea that firefighting required organizational effort, that there were specialized tasks to be performed, and that a command and control structure, similar to the military, was a good model.

Early History and Symbols of the Fire Service

The next organized groups of individuals devoted to the saving of lives and protection of property arose during the Crusades. Bands of knights going to the Holy Land to search for the Holy Grail were organized into "orders." One such order was entitled the Order of Saint John of Hospitaliers. This band of knights dedicated themselves to the treatment of serious wounds caused on the battleground.

During this same time period there was an order called the "Knights of Malta." These brave knights did not go into combat to kill, but to save. They developed a reputation for saving lives by serving as stretcher-bearers for the victims of battles. After a battle was over they carried the victims back to the crude battlefield hospitals, often using shields as stretchers. In the age of the Crusades, one of the only ways to tell friend from foe was by a symbol on the uniform and another on the shield carried by soldiers. These symbols were used to denote that a person had the training and the responsibility to act on behalf of the order they represented. The symbol for the Order of Saint John was a red cross. The symbol for the Knights of Malta was the Maltese cross. Both emblems became symbolic for saving lives and property. Today, these symbols still exist in the fire and emergency services, **Figure 1-4**.

The Middle Ages

In the Middle Ages, fire was a serious threat to every city. History books are full of stories of the devastation of major European cities in the years between the end of the Roman Empire and the beginning of the Middle Ages. During the next ten centuries, technological advances were created all around the globe, but they were not widely available to European cultures. For example, in China, fire pumps were created that were based on the syringe-like concept mentioned earlier. The Chinese also created fire brigades to protect their largest cities. Because of limited contact between Europe and the Orient, these advances were not copied.

The English, from about A.D. 1100 to 1600, devised a whole set of regulations about fire. For example, in 1189, a law required that homeowners had to have ladders prepared at their homes to help their neighbors. Other laws required buckets and barrels of water to be kept handy. **Arson** remained a serious crime during this period, even though the punishment for arson was to be burned alive.

Fires continued to devastate, even after the creation of the concept of fire prevention. London burned in 1666. Originally labeled the Pudding Lane fire for the street on which it began, it has since been called the Great Fire of London. Unfortunately, even with the proliferation of business and industry in London, the concept of fire protection had been allowed to lag.

The community leaders who were creating the commerce, business, and industrial growth were also placing a new emphasis on the need to eliminate conflagration. The next 200 years would see the creation of almost all of the basic institutions that provide fire protection in the modern age. These components would consist of the organization of fire departments, the creation of a fire insurance industry, and the rise of technology to both prevent and combat fires.

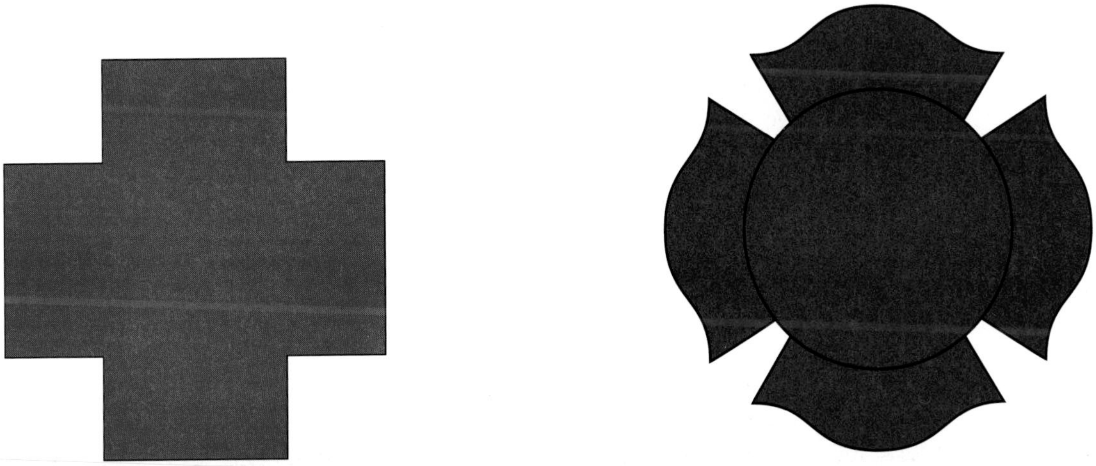

(A) (B)

Figure 1-4 Symbols taken from history: (A) red cross and (B) Maltese cross.

Early American History

Most of the settlers of this country's colonies were of English, Irish, German, French, or other European background. Therefore, the first steps to providing fire protection were based on the practices from these same countries. Among the very first laws put on the books in the New World were ones dealing with fire prevention. Among those regulations were prohibitions on the construction of wooden chimneys and limitations on the candle-making process. Most of the early settlements were constructed of wooden buildings and were extremely vulnerable to fire spread. Therefore, it became important for the colonists to prevent fires from happening rather than try to combat them.

History books contain many stories of the early settlements in New England that burned to the ground, leaving their inhabitants exposed to the elements. Although fire protection in these early years consisted mostly of active fire prevention, it was not always effective. Fire suppression was essentially done by bucket brigades, in which buckets of water were passed down a line of people from the source of the water to the fire.

The King of England authorized the firm of Ryley and Mabb to create a new idea called "insurance" in 1637. This unique idea was far from successful at the time. Little or no attention was paid to the idea until the Great Fire in 1666. Nonetheless, the insurance companies went into business and marked the occupancies they protected with signs on sheets of metal telling the firefighters which company held the insurance policy on the building. These were called **firemarks**, as shown in **Figure 1-5**, and were used in the colonies by the early settlers.

The development of fire suppression capability became a pressing need. As early as 1647 the city of Boston sent an order to London to purchase a fire pump. Unfortunately, the pump did not arrive in time for the fire of 1653 that almost destroyed the town. That fire was fought by the bucket brigades. During the colonization period, most cities realized that they were entirely on their own against fire, and they began to prepare for it by developing some specific strategies to cope in their individual communities. Fire protection was initially viewed as a local problem. Because of the close relationship with Europe, the exchange of ideas was soon flowing. Boston, in the early 1700s, designated specific people to serve as **fire wardens**. These individuals were not unlike the Romans who walked the streets at night looking for fire. They were given big wooden rattles, as shown in **Figure 1-6**, and told to make as much noise as they could if they found a fire. They were soon labeled the "Rattlewatch." By 1718, these concepts had jelled into the idea of creating **fire societies**. Fire societies consisted of groups of people who voluntarily banded together to deal with a community's fire problems. It was to

Figure 1-5 Firemarks were used to show which insurance company protected a home or building. *(Courtesy CIGNA Museum and Art Collection)*

were hired to protect the insured homes and buildings of certain insurance companies. The fire marks were important indicators because they allowed the fire companies to know whether or not they should extinguish the fire in a home or not! These metal signs were made from lead castings and had the company's logo on the surface. They were attached to the building that had the policy. In Philadelphia the Contributionship's directors saw the value of fire marks and continued the practice even though volunteer fire companies extinguished fires regardless of insurance. The fire mark of The Philadelphia Contributionship, four clasped hands on a wood shield, is known as the Hand in Hand, **Figure 1-8**. Interestingly, most of the early insurance companies are still active, but many of their fire marks were removed from houses during the Revolutionary War in order to make bullets out of the lead.

Fire was used as a weapon of war in the colonies during the two wars fought over independence. Major fires occurred in most of the cities when they were occupied by British military forces, but because many volunteers had joined the rebel military forces themselves, there were few firefighters to cope with local fires. When the war ended, almost all of the fire equipment that existed prior to the war had been either destroyed or badly damaged. Early America, which needed to build a federal government, also had to spend some time in rebuilding the fire service. New companies were formed, volunteers were recruited, and the remains of the cities destroyed by fire were replaced by new homes, new businesses, and an increasing population.

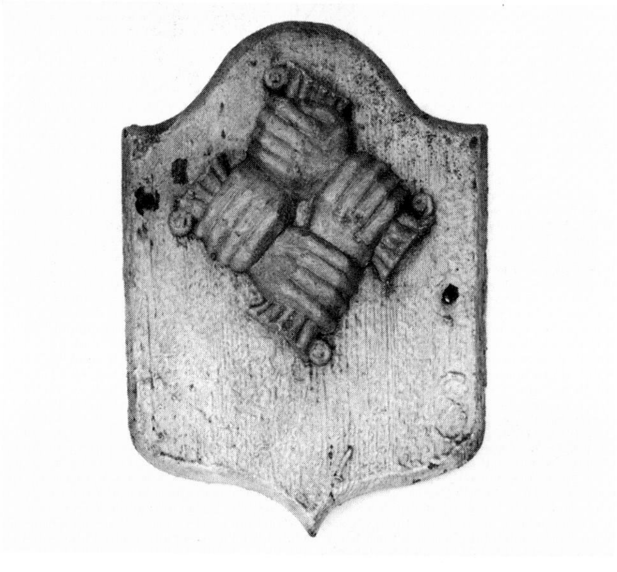

Figure 1-8 Fire mark for the Philadelphia Contributionship for the Insurance of Houses from Loss by Fire, organized 1752. *(Courtesy of The Philadelphia Contributionship for the Insurance of Houses from Loss by Fire)*

Note With the close of the war for independence, the American fire service entered a new age. With the end of one century, fire protection had become a public need. With the beginning of the new century, fire was becoming an important economic consideration. Fire protection went from being the concern of only the larger cities to being a concern for smaller communities as well. By the beginning of the 1800s, there were very few settlements that did not have a volunteer fire company, and most even had some form of pumping apparatus powered by volunteers.

The American fire service began to evolve away from European practices, customs, and styles. The idea of a distinctive American fire service began to emerge. Changes began to occur, but not without traces of the legacies of the past. For example, many of the volunteer fire companies were organized around the ethnicity of the community they served. The Hibernia Fire Company consisted mostly of those of Irish descent. The hat for the Hibernia Fire Company is shown in **Figure 1-9.** There were also African-American fire companies, including one from 1815, which had to quickly disband. Some required a person to pay extensive dues and were therefore only open to the wealthy. There were volunteer fire companies for just about every ethnic, social, and political orientation in society.

The volunteer fire companies were often made up of soldiers who had fought in the war. When the smoke of battle faded, many young men converted their sense of community service from combating Redcoats to combating fire. Among the traits that attracted the volunteer were the

Figure 1-9 Firehat illustrating the Hibernia Fire mark. *(Courtesy of The Philadelphia Contributionship for the Insurance of Houses from Loss by Fire)*

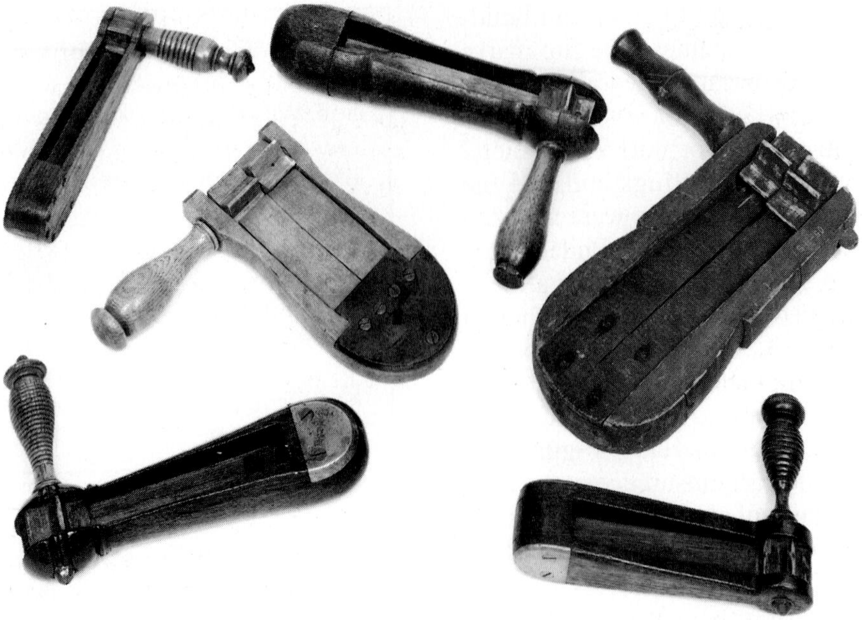

Figure 1-6 Examples of rattles used by the Rattlewatch. *(Courtesy CIGNA Museum and Art Collection)*

be the beginning of one of the American fire service's most colorful ages, the era of the volunteer firefighter. Many of these volunteer firefighters were also destined to play an increasingly important role in history as founding fathers of the American political system.

> **Firefighter Fact** Benjamin Franklin, then a businessman from Philadelphia, began his influence over the early American fire service by being a cofounder of the Union Fire Company in 1736. Supposedly, he had witnessed the Boston Towne House fire, one of the largest conflagrations in the 1700s. He modeled the Union Fire Company after Boston's Fire Society. Franklin, known for his adventures in politics, science, publishing, and other fields, is considered by many to be America's first fire chief, **Figure 1-7**.

As the need to develop a fire service grew in this country, Franklin was joined by many other key historical figures such as George Washington, Thomas Jefferson, John Hancock, and Samuel Adams. These individuals were involved in the creation of fire companies or the acquisition of fire equipment from Europe.

The colonies in New England suffered terrible losses during the next few decades. A review of history books reveals that almost every major city was struck by serious fires, many reaching conflagration size. Boston, New York, and Philadelphia each experienced severe losses of life and property several times.

Around 1752, Benjamin Franklin, once again following European practice, started up his own insurance company, the Philadelphia Contributionship, in concert with other Philadelphians. One of the ideas he borrowed from Europe was the concept of placing firemarks on the outside of the structure to let fire suppression crews know that the contents were insured. In England, individual engine companies

Figure 1-7 Benjamin Franklin, commonly referred to as "America's first fire chief." *(Courtesy CIGNA Museum and Art Collection)*

excitement of firefighting and the camaraderie of the organization.

From the beginning of the 1800s to about 1850 many advances were made in fire protection, but almost all of the emphasis was on fire suppression. What was most significant in this era was the development of larger, more powerful hand pumpers, as shown in **Figure 1-10**, and the development of the American fire helmet. Although the hand pumper owed its origins to England's early models, American versions were more elaborate and, for lack of a better term, more flamboyant. Volunteer fire companies vied with each other to see who could design and build the most powerful and yet beautiful machine.

The American fire helmet was developed by a leather hat maker named Andrew Gratacap. The original hats were shaped like top hats and had the name of the fire company painted on them. Later, the idea of a short front brim and an extended back brim was incorporated into the design to assist in water runoff. Firefighters also copied the large front piece off the Hessian soldier military helmets to create the distinguished American fire helmet profile. Europeans tended to use military-style helmets, so the American fire helmet began to be a symbol of the departure from the past, **Figure 1-11**.

Figure 1-11 An early American fire helmet. *(Courtesy CIGNA Museum and Art Collection)*

The time period from 1800 to 1850 was one of great expansion and development in American society. The thirteen tiny colonies that created the United States were soon dwarfed by the states that were carved out of the rest of the country. City after city was established. The country grew quickly as western expansion resulted in the entire continent being settled in about 100 years. Cities like Chicago and Detroit became major trading centers. The growth of the fire service during this period was more along the lines of expansion of numbers of volunteer personnel and new fire departments and less along the lines of technological advance. Volunteer fire companies of the era became heavily involved in local politics, and individual fire companies became known as social clubs for younger, often aggressive and boisterous men.

This led to a time in the fire service that has been called the "Rowdies and Rum" era. During this era,

Figure 1-10 This early hand pumper, called the "Red Rover," was used by the Howard Engine Company in New York. *(Courtesy CIGNA Museum and Art Collection)*

manually operated hand pumpers required a tremendous number of people to keep a fire stream in operation for any lengthy period of time. Some larger pumpers required the service of more than 100 individuals for the period it took to control a large building fire. Ladders at this time were limited in height to about four stories and were hauled to the scene of fires by a team of men pulling on a long tongue of rope.

Firefighter Fact With an increase in the sociopolitical nature of volunteer fire companies, and in some situations, outright physical competition between fire companies, there was often more violence between fire companies than confrontation with the fire. Incidents often occurred that involved exchanges of blows between firefighters over who had the right to be on a specific fire. There were even cases of volunteer companies lying in wait to ambush another company on the way to a fire.

Figure 1-12 A reproduction of a Currier and Ives lithograph entitled "Rushing to the Conflict," published in 1858. *(Courtesy CIGNA Museum and Art Collection)*

On the other hand, this era was also responsible for the creation of the image of the firefighter as being an important part of society. Currier and Ives produced a series of lithographs during this time that are still popular today. This series, which focused on firefighters and firefighting activities, included scenes at the firehouse, scenes of responding fire apparatus, and large fire scenes. These pictures also chronicled the changes that occurred during that period, showing both hand-drawn and horse-drawn firefighting equipment. One of the prints that demonstrates an image that resulted in a symbol was a print showing a **foreman**, which was a leader of a fire company, holding a **speaking trumpet** aloft, **Figure 1-12**. The foreman's trumpet was originally used to allow the crew boss to shout orders over the noise of the firefighting activity. The trumpet became the icon of authority within the fire service.

The mid-1800s was a period of both social and economic change. Invention after invention was brought forth to both simplify and complicate life. The railroad was created, printing presses became widely available, petroleum was discovered, and America's fire problems continued to become more complicated. Not all fire companies evolved in parallel to the need.

Arson was a common crime in this era. It was even called the "working man's vengeance" because arson was used as a retribution against unfair employers. Authorities searched for a measure that would cut down on the crime. In 1856, Philadelphia Mayor Richard Vaux, a former foreman of a fire company, established a special arson investigation unit. Although rewards for arsonists had been offered as early as 1723 in Boston, specially trained individuals who could track down arsonists did not come into being until it became clear that arson was a crime against society instead of against an individual.

One event that significantly affected the fire service was a riot by firefighters at a fire in Cincinnati, Ohio, in the 1850s. The city leaders were so distressed at the behavior of their volunteers that they instituted the first full-time, paid firefighting force in 1853. To cope with the need to provide the necessary energy to pump the water previously pumped manually by the volunteers, they provided a new technology—the steam-powered fire engine. The volunteers were outraged.

Nonetheless, during the next decade, city after city began to realize that their fire protection needs now required a fire department with paid firefighters. The conflict over creating a new way of fighting fire was a double-edged sword. On the one hand, moving to full-time paid firefighters on fire apparatus removed a lot of the political turmoil from the ranks, but caused a great deal of conflict for local officials. Adopting steamer technology was a boost to the development of fire streams to match the growing size of buildings. This technology provided the power to reach higher and larger buildings, but required a lot more commitment of community resources, especially in the way of funding.

Firefighter Fact It should be noted that this transition to paid forces was certainly not universal. In fact, many communities chose to retain their volunteer fire forces. Many volunteer fire departments began to embrace the technology being used by the paid fire service agencies. Today, the volunteer fire force still constitutes more than half of the fire service in the United States.

The Civil War

The Civil War started at a crucial time in the history of the American fire service. While the fire service had a background in paramilitary organizational structures, the Civil War played a key part in reestablishing the significance of command and control. It also created the environment for the establishment of many of the organizational practices for a paramilitary organization.

Earlier in this chapter it was noted that the Revolutionary soldier returned to become a volunteer; this was not true of the Civil War. The fire service had provided a huge reserve of candidates to become soldiers in the Civil War, including entire units, for example, the 11th New York Fire Zouaves, the 73rd New York Infantry, and the 72nd Philadelphia Fire Zouaves. A monument dedicated to the 73rd New York Infantry, shown in **Figure 1-13**, stands on the battlefield in Gettysburg, Pennsylvania, the site of a crucial Civil War battle.

For the most part, the Civil War did not provide any specific advances in firefighting technology itself. Because thousands of volunteer firefighters went off to that war, on both sides, and were either killed or so severely injured that they could no longer serve their communities, when the war was over most of the country's experienced firefighters were gone.

Shortly after the Civil War was over, hundreds of veterans returned to communities looking for jobs and some opportunities for community service. Those cities that were converting their volunteer forces over to paid personnel saw the military veterans as excellent choices. They had demonstrated their courage in combat, were familiar with the technology of the emerging industrial age, and were well disciplined.

Figure 1-13 Located at the battlefield in Gettysburg, Pennsylvania, this monument is dedicated to the 73rd New York Infantry. *(Photo by Michael O. Forgy)*

Another factor that played into this time period was that a new skill was needed in using steam fire apparatus. Most of the steamers were so heavy they could not be pulled by firefighters alone. Horses had been introduced to provide the muscle to move the steamers. Many soldiers, experienced in use of horses for pulling artillery batteries and operating logistical support wagon trains, went right into service driving teams of horses to haul steamers.

One of the most recognizable symbols of the fire service today was introduced at that time: the firehouse dog—specifically the Dalmatian, as shown in **Figure 1-14**. Although firefighters had had mascots in firehouses for years, the Dalmatian was not there as a pet. The breed was originally developed as a companion for horses, to calm them and chase away any small animals in their path. Dalmatians trotting alongside a team of horses, was already a common sight in other commercial applications. It is very unlikely that the firefighters who brought these dogs into the stations would have predicted that fifty years later, the horses would be gone and the dogs would still be there!

The Civil War veteran brought many new ideas into the firehouse. Among these were the concept

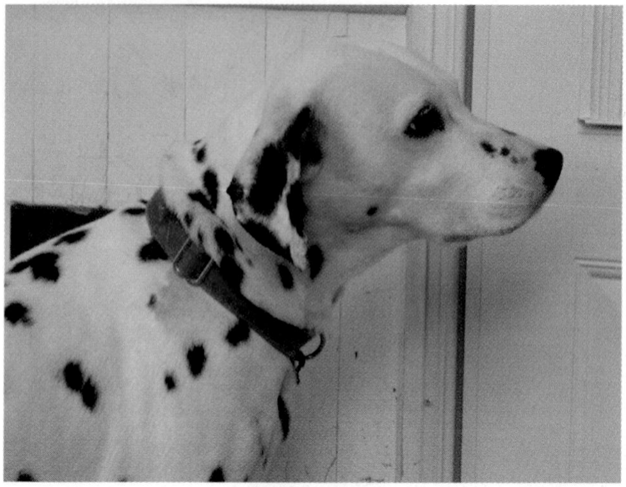

Figure 1-14 The Dalmatian has become a symbol of the fire service in America.

of military rank structure, command and control similar to infantry tactics of the war, and even the coloration and design of firefighters' uniforms. Prior to the war, uniforms looked more like costumes, with color and flair being the lead criteria. Red, green, and yellow were commonly used in volunteer uniforms. After the war, uniforms tended to be more military-like, almost exclusively blue in color. The use of gold for top-level insignias and silver for lower level ranks came from the military system. The speaking trumpet went out of use as a tool and became a symbol of authority on collar brass. Prior to the war a person in charge of a group of firefighters was commonly called a foreman. After the war, terms such as lieutenant, captain, sergeant, and battalion were borrowed from the military and became common in the fire service.

The end of the Civil War also brought a new wave of professionalism and organizational structure to the fire service. In a few short years, the fire chiefs, who were often called fire engineers, had banned together to create the National Association of Fire Engineers (which was to become the International Association of Fire Chiefs) in 1873. A host of state and local fire chiefs' organizations were started at about the same time.

None of the advances in either methodology or technology would prevent the continuous exposure to large losses from fires. Among the most famous fires of this era were the Great Chicago Fire and the Peshtigo (Wisconsin) forest fire (1871). Both fires were major catastrophes and neither the presence of courageous firefighters nor powerful steamers prevented the loss of lives and property. Fire was proving to be a greater adversary than anticipated.

The Industrial Revolution

The 1870s through the turn of the century has often been called the "Industrial Revolution." America moved from a farming society of a large number of small towns and villages into a world leader in industry and commerce. Part of that revolution was the requirement that advances were needed to deal with fire.

Firefighter Fact One of the largest industries that took hold in the United States was the textile industry. Cotton from the South was sent to the North to be made into cloth to sell to the rest of the world. In New England, these textile factories were burning to the ground with regularity when a young man named Henry Parmalee invented a device called an **automatic sprinkler system**. Sprinkler systems were installed in factory after factory and the losses began to decrease.

A new organization was formed consisting of individuals that were involved in designing and installing automatic sprinklers. They called themselves the National Fire Protection Association (NFPA). They took the task of forming this new concept into a "standard," which was given the number 1. It was the first, but certainly not the last, national fire code of the United States.

As the technology of industry increased, the technology of firefighting tended to increase too. Almost all of the basic research in **fire engineering** today was started during this period. The first **fire alarm** systems were created in the 1870s. Much of the basic information on **fire hydraulics** was done during this time. The first **aerial apparatus**, a spring-loaded device, was designed by San Francisco firefighter Daniel Hayes. The fire service of this era was quick to adopt the new ideas coming from the inventors and innovators of the industrial world. The steam-powered fire pump was one of the best examples, as shown in **Figure 1-15**. Steam power was not invented for the fire service. It was adapted for use in the fire service when it was needed. Its primary purpose was to be an inexpensive source of power to operate machinery. Steam engines ran trains and boats and operated machinery in factories. The trend that was created during the industrial age was to link the technological advances of the firefighter with the advances of society in general. At the outset of this phenomenon, the time frame for adoption was just slightly behind the creation of the concept in general society. It has never been that closely linked since.

Figure 1-15 An example of an early steam-powered fire engine. *(Courtesy CIGNA Museum and Art Collection)*

The Beginning of the Twentieth Century

The nation's fire problem did not go away as a result of technological innovations. To the contrary, it got more complicated. Career firefighters recognized many of these changes in working conditions. They banned together to form a labor union, the International Association of Fire Fighters (IAFF), near the turn of the century.

Although the capacity to fight fires was increased by technology, the very equipment, materials, and processes used to modernize the country created new fire problems. Fire continued to devastate cities and towns. In the late 1890s and into the turn of the century, fire took a terrible toll. There were so many large fires that the insurance industry almost went bankrupt. As a result they formed a group called the National Board of Fire Underwriters (NBFU) and started to evaluate the level of fire defense in different cities to see which ones were ready to deal with conflagration and which were not. This concept would have a lasting effect on the fire service.

As a result, fire departments and insurance organizations began to place more emphasis on the subject of fire prevention. Fire and building codes began to place emphasis on such things as fire-resistive construction techniques and fire control processes. A series of large loss of life fires during this era brought the subject of fire prevention into the forefront of fire science. Many fires, such as the Iroquois Theater, the Triangle Shirtwaist, and the *General Slocum* ferry fires, took the headlines away from courageous firefighting efforts and placed an emphasis on eliminating the reason for the fires occurring in the first place. This has sometimes been called the "theory of catastrophic reform."

> **Note** Almost all of the advances in fire prevention that have created the fire and building codes of today were created during this time frame. Simple concepts such as adequate exits, sprinkler protection for high-risk occupancies, exit requirements for assembly occupancies, and limitations on processes that used flammable or hazardous materials were all created in a short time frame by the leadership of the fire service of the time.

During the 1920s, Fire Chief Ralph Scott of the Los Angeles City Fire Department expressed a concern about the quality and quantity of fire service training. It had been almost forty years since the concept of having a training system for firefighters had been created and he suggested that it needed an overhaul. Chief Scott, who was the president of the International Association of Fire Chiefs, created an Education Committee and made himself the chair. While participating in a regional conference on trade and education he had learned of federal legislation that funded vocational training. Scott requested and received money to write a job analysis for the occupation of a firefighter.

Most of the metropolitan fire departments had training programs in place at the time. What was lacking, however, was an organized, systematic listing of tasks, knowledge, and skills based on current practices and needs in the fire service. The document that was produced by this project was called Firefighting Bulletin Number 155, series 44, Federal Board of Vocational Education, 1931. This was the landmark document on which almost all subsequent fire service training and education has been based. The report was validated by representatives from fire agencies all over the United States. It created the minimum standards for the training of entry-level fire personnel and created a body of knowledge that resulted in the industrial education system recognizing that firefighting was a separate and distinct occupation. Since then, other studies have been conducted and vast improvements have been made in the standards, but they all started with this bulletin.

TECHNOLOGY, TRANSITION, AND TIMES OF CHANGE

What was really remarkable about the first quarter of the twentieth century was the speed with which change impacted the fire service. In fact, the speed of change increased rapidly. The result was that during this era there were often many different levels of technology in the fire service at the same time. For example, the steamer era did not end with the creation of the internal combustion engine. Steamers were in service on a parallel track for many years after gasoline engines became available. Steamers had to be hauled with horses, and when the internal combustion engine first came along it was not used to pump water, but rather to replace the horses. This time period saw fire stations that had both horse-drawn and motorized fire apparatus in the stations next to one another.

Prior to World War I, the internal combustion engine was a novelty. After the war it became the power source of choice. Initially horses were replaced on steamers, not aerial apparatus. Aerials were still cranked up by hand or required spring power to elevate. One of the more unique motorized pieces of equipment was the early adaptation of soda-acid extinguisher capacity to small vehicles. They were often sent ahead to try to perform an initial attack on small fires.

A whole generation of fire apparatus arose that consisted of hybrids of steam power and internal combustion. During that era there were hose wagons to haul hose, and they worked very well with horses. Once the internal combustion engine was adapted to not only haul the apparatus, but to operate the pump too, hose wagons were then incorporated into engine companies. Once the horsepower of combustion engines improved, water tanks were added to the chassis. Prior to this time, several pieces of equipment had to be assembled to fight a fire.

The internal combustion engine made it possible to combine three fire apparatus into just one. The same vehicle could carry water, have a pump, and also carry the amount of hose and equipment required to apply the water. Basically this is the design of even the most modern of pumpers, the **triple combination engine company**, as shown in **Figure 1-16**.

Figure 1-16 Triple combination engine companies can carry water and hoses and other equipment as well as pump the water. *(Owned and photographed by William Killen)*

As tools were added to the apparatus, they became larger and more complicated. Aerial ladders were converted from horse drawn with the advent of hydraulic systems. What is interesting to note about this transition is that the fire service improved quickly on its ability to get to the fire, but did not make many advances in its capacity to fight the fire once on scene. The control panel of a steamer in the 1880s describes the pump as having a capacity of 1,250 gallons per minute (gpm) at 150 pounds per square inch (psi). That is about the average capacity for fire engines today.

A review of the apparatus designs from the beginning of the twentieth century to current times demonstrates that the fire service tends to adopt any technological improvement as soon as it has proven its dependability. Fads and trends are not as important to the firefighter as reliability and predictability. New ideas must go through a test of both time and experience to demonstrate their survivability. As a result, even with hundreds of millions of dollars being devoted to designing and building fire apparatus, the contemporary fire service is usually on the trailing edge of technology instead of the leading edge.

THE EFFECTS OF WORLD WAR II

World War II also had its effects on the American fire service. Once again, large numbers of firefighters were called to serve their country. Another impact was not as obvious, but was far reaching. The war accelerated the need to deal with fire. Fire was still a weapon of war, and the United States devoted a fairly large amount of resources to prepare for it. In Europe, the fire service had already suffered terribly in the war, on both sides of the conflict. In the United States, several military projects resulted in findings that would influence the postwar fire service. One was the research that resulted in the development of the indirect attack method. This work, performed by Fire Chief Lloyd Layman, was distributed to the fire service after the war in the form of two books, *Firefighting Tactics and Strategy* and *Attacking and Extinguishing Interior Fires.* Research into fighting flammable liquid fires resulted in improved foams for use by the fire service. Personal protective clothing used for firefighting was also improved. Fire nozzle technology was improved drastically with the development of advanced versions of fog nozzles.

No records were kept of the number of firefighters that lost their lives in World War II, but it can be anticipated that the fire service suffered a loss of both experience and knowledge as a result of the conflict. Firefighters who did return home after World War II to reenter the department and the veterans that returned home trying to get jobs in the fire service both had a frame of reference that would affect the fire service for generations. This group saw the fire service as being even more paramilitary than ever in the past.

Innovation after innovation was created from the war effort. The military buildup also resulted in an increased need for domestic fire service agencies to protect new factories and increased residential developments. Many innovations for the fire service emerged: the availability of radios for fire engines to improve communications and of diesel engines to improve road performance. These innovations and the growth of the fire service placed new demands on the fire service.

From the 1950s to the end of the twentieth century, the fire service saw more changes than in the previous 200 years. Most of the hardware and basic components of the service have evolved fairly slowly. The process of accelerating change has more to do with the actual duties of the firefighter and fire agency staffing. For example, the role of EMS has grown rapidly in the fire service; many departments are much more involved in **hazardous materials** response, **Figure 1-17**, than ever before; the role of **search and rescue** has grown in view of major earthquakes and structural collapses; and fire service agencies are becoming more and more involved as first responders in **terrorism** incidents. More importantly, the fire service has also evolved into a profession that embraces gender and ethnic diversity. Members of the fire service of today tend to think of themselves as being the "modern fire service"—but then so has every other generation.

MODERNIZATION OF THE FIRE SERVICE

As used in this text, the term *modernization* is a process, not a place in time. The process of change has been constant in the fire service. What has not changed has been the basic mission of firefighters. When it comes to describing modernization in this text, we are talking only about the present. What is being done today is necessary to keep the fire service modern. This gives rise to two concepts that all firefighters should be aware of as they enter the service: information half-life and technological obsolescence. Information half-life is how long it takes for 50 percent of the information a firefighter has to become obsolete, inaccurate, or ineffective. Information half-life is not the same for all generations of firefighters.

Figure 1-17 The fire service has expanded into many areas, including hazardous materials response.

A member of the fire service in Ben Franklin's day would have used the same tools that his grandfather had used. Moreover, his grandchildren would not have used much more effective equipment. The information half-life in the 1700s for the fire service was about 100 years. The information half-life in the last 100 years has decreased almost every generation. While there is no scientific study to determine what the half-life is today, it is a fact that the body of knowledge regarding fire protection is changing rapidly. It can be anticipated that fire science information will change a great deal more in the future than it has in past generations.

The second phenomenon is called *technological obsolescence.* This term means that any given technology will only be useful for a certain period of time before it is replaced by another. Most technology goes through an initial stage of being controversial. Then it becomes superior to the one that it replaces. After it serves a period of time as "state of the art," it too has the possibility of becoming less than adequate. Eventually it can become old-fashioned and is replaced. Steamers were once superior to hand pumpers, internal engines replaced horses, diesel replaced gasoline, and so on. All obsolete technology becomes a source for museums and collections of memorabilia. All status quo technology is just one invention away from obsolescence.

The lesson to be learned here is that the fire service needs to be alert to any indications that its current technology is not doing the job and be prepared to adopt any improvements. It is equally important that firefighters do everything they can to be competent and effective in the knowledge and skills needed to perform the job of firefighter safely.

THE FIRE SERVICE OF TODAY

Although fire protection remains essentially a function of local government, the second half of the twentieth century saw efforts to create a national focus on the fire problem in the United States and bring together training and information sources at state and national levels to improve the business of fire protection. In the 1970s the federal government recognized that the country's fire problem was changing and commissioned a panel of exports to study it. This group, called the National Commission on Fire Protection and Control, published a document called *America Burning.* As a result of that document a lot of positive change occurred, including the creation of the United States Fire Administration and the development of the National Fire Academy.

In the twenty-first century, the fire service continues to grow and develop. There are an estimated 30,020 fire departments in the United States with an estimated 1,078,300 firefighters. The actual numbers vary because new communities are created that require fire protection, and sometimes several fire organizations band together as a result of mergers or consolidations. Many fire departments are still totally volunteer, while others are growing larger due to population increases and increased density in some urban areas. In some areas, fire and rescue services may be provided by private, for-profit companies. Fire departments are increasingly complex in this day and age, with a diversification in the services they provide. In addition, fire departments are also more gender and ethnically diverse, offering new insights into the way they provide services to their communities.

According to the United States Fire Administration, a fire department in this country responds to a fire approximately every eighteen seconds. Even with the incredible advancements in fire service tactics, training, and technology, the United States still has the fourth highest fire death rate of all industrialized countries, with someone dying in a fire an average of every two hours and someone injured by fire approximately every twenty-three minutes. Fire kills more Americans than all natural disasters combined. America is still burning, and the fire service continues to race to meet these ever-evolving challenges.

The late twentieth century and the early twenty-first century also laid a new obstacle before the fire service, the threat of terrorism in this country. The attacks on the World Trade Center and the Pentagon on September 11, 2001, shed new light on the fire service by demonstrating that firefighters are not immune from the horrible acts of terrorists. Although this event will remain fresh in every American's mind, it is very possible that the events of that day will shape the fire service in new and different ways well into the future.

The fire service of today has a very diverse, complicated system of delivery, as a result of the contributions of all past generations. The basic mission of saving lives and protecting property has not changed. Only the manner and means to cope with the problem have changed. Nonetheless, the fire service of today still contains a great deal of the past in its inventory of tools, equipment, and methods. If we go back to the beginning of this chapter and revisit the term mission, it may be possible to reconcile the past with the present. The fire service is today what it has become from past experiences.

Lessons Learned

What does the future hold for firefighters as they begin their journeys? In one sense a firefighter's career is like climbing that ladder we described earlier in the chapter. There are a lot more rungs to be climbed as firefighters pursue their careers. And, while we stand on only one rung at a time, we must be able to place our hands on those rungs that are higher.

No one should attempt to predict the future without giving due credit to the past. That is what this chapter has been all about. The future will contain a lot of challenges and opportunities. The future will contain some changes that will be difficult to deal with, just as they were for previous generations. The future will contain a requirement that firefighters continue to learn and develop skills that did not even exist the day they entered the fire service, not unlike those from the past.

> **Ethics** What is predictable about the future is that firefighters will likely be called on to deal with whatever dilemma or disaster strikes the community they serve. They must be prepared.

The motto of the Roman fire brigade was "Semper Vigilans" or "Always Vigilant." It was good enough for then. It is good enough for today.

KEY TERMS

Aerial Apparatus Fire apparatus using mounted ladders and other devices for reaching areas beyond the length of ground ladders.

Arson A malicious fire or fires set intentionally by humans for vengeance or profit.

Automatic Sprinkler System A system of devices that will activate when exposed to fire, connected to a piping system that will supply water to control the fire. Typically, an automatic sprinkler system is also supported by firefighters when they arrive on the scene.

Conflagration A large and destructive fire.

Fire Alarm Notification to the fire department that a fire or other related emergency is in progress, which results in a response.

Fire Engineering The study of fire, fire behavior, fire extinguishment, and suppression.

Fire Hydraulics The principles associated with the storage and transfer of water in firefighting activities.

Fire Societies Groups of people who voluntarily banded together to deal with a community's fire problems.

Fire Wardens Designated community individuals who walked the streets at night looking for fire and carrying large wooden rattles with which to signify a found fire.

Firemark Signs on sheets of metal telling firefighters which company held the insurance policy on a home or building.

Foreman Individual designated as the leader of an early fire company; a predecessor to the modern title of fire chief.

Hazardous Materials Chemicals that are flammable, explosive, or otherwise capable of causing death or destruction when improperly handled or released.

Mission Statement A written declaration by a fire agency describing the things that it intends to do to protect its citizenry or customers.

Search and Rescue Attempts by fire and emergency service personnel to coordinate and implement a search for a missing person and then effect a rescue.

Speaking Trumpet Trumpet used by a foreman or crew boss to shout orders above the noise of firefighting activities.

Terrorism Acts of violence that are arbitrarily committed against lives or property and intended to create fear and anxiety.

Triple Combination Engine Company Fire apparatus that can carry water, pump water, and carry hose and equipment.

REVIEW QUESTIONS

1. What is a mission statement? Why do fire and emergency services departments have mission statements? What is the mission statement for your department?
2. What is a firemark? What was the purpose of the firemark in early American history?
3. What was the purpose of the "Rattlewatch"?
4. What contributions did Benjamin Franklin make to the American fire service? What other American historical figures played a role in the development of the American fire service?
5. List some symbols present in today's fire service and describe their historical significance. For example, why is the Dalmatian considered a "firehouse dog"?
6. What do you think the future of the American fire service will look like in 5 years? In 10 years? In 100 years?
7. Discuss what effects you think terrorism has had on the fire service.

Additional Resources

Bureau of Alcohol, Tobacco, and Firearms, Arson and Explosives National Repository http://www.atf.treas.gov/aexis2/index.htm

Goudsblom, Johan, *Fire and Civilization.* Penguin Press, New York, 1992.

Great Fires of America. Country Beautiful Corporation, Waukesha, WI, 1973.

Gurka, Andrew G., *Hot Stuff Firefighting Collectibles.* L-W Book Sales, Gas City, IN, 1994.

National Center for Injury Prevention and Control http://www.cdc.gov/ncipc

National Institute for Occupational Safety and Health, Fire Fighter Fatality Investigation and Prevention Program http://www.cdc.gov/niosh/firehome.html

National Fire Information Council http://www.nifc.org

National Interagency Fire Center http://www.nifc.gov

Smith, Dennis, *History of Firefighting in America.* Dial Press, New York, 1978.

United States Department of Transportation, Office of Hazardous Materials Safety http://hazmat.dot.gov

United States Fire Administration http://www.usfa.fema.gov

CHAPTER 2

FIRE DEPARTMENT ORGANIZATION, COMMAND, AND CONTROL

Thomas J. Wutz, Midway Fire Department and New York State Office of Fire Prevention and Control

OUTLINE

- Objectives
- Introduction
- Fire Department Organization
- The Firefighter
- The Company
- Additional Fire Department Functions
- Regulations, Policies, Bylaws, and Procedures
- Allied Agencies and Organizations
- Incident Management
- Incident Management System (IMS)
- Lessons Learned
- Key Terms
- Review Questions
- Suggested Readings
- Additional Resources

STREET STORY

Winters in Washington, D.C., can be fickle. One year the temperature could be 70 degrees during January, or it could easily be 17 degrees with three feet of snow on the ground. On January 13, 1982, it was the latter. When people woke up that morning to blowing snow and sub-teen temperatures, they may have expected rough weather but little did they realize that by afternoon rush hour the city would experience a major subway crash in the tunnel with seven fatalities and that Air Florida flight 90 would collide, with seventy-nine souls on board, into the 14th Street Bridge that links Washington with Virginia across the Potomac River.

Concepts such as command *and* control *and tools such as the* Incident Management System (IMS) *were not readily employed in 1982. The incident was under the command of a senior fire department officer. The outcome would depend heavily on the abilities of that officer because he was responsible for all aspects of command. I was a young officer in command of a ladder company that day. Before the day was over I would have the fortune of responding to both the subway and the airplane crashes. My clearest recollection of the incidents that day is that there was a great amount of confusion. Many times orders were issued only to be countermanded or withdrawn. A jurisdiction dispute existed between many of the responding entities. The number of casualties was compounded by the fact that many of the dead and injured had to be removed from the wreckage in the subway and at the 14th Street Bridge. The aircraft itself went into the water with no flames but it took most of the victims with it. The weather was the final trump card played against us that day, making traveling and responding extremely difficult. This was a very taxing and difficult operational day for anyone to handle and should not reflect on the commanders that fateful day, but recently a panel of chief officers who had experience in the implementation of IMS and the concepts of command and control revisited that day in the form of an exercise. No facts were given except for the pertinent data of resources available, time of day, weather, and the scene as reported by the first arriving units. Needless to say, the outcome from a command point of view was much more successful this second time. The IMS can be considered one of the high points in tool development for the fire service of the twentieth century.*

The fire service continues to develop tools, and concepts continue to evolve as the need dictates. Some of these tools, such as the IMS, will be used well into the next millennium; that is, if we embrace and use them wisely. The leaders of the fire service must ensure that those who follow us maintain an open mind and continue to educate themselves in order to push the concept of command and control to levels where the safety of our people will be ensured and our operations will be successful in protecting civilians and property.

—Street Story by Mike Smith, Battalion Fire Chief, District of Columbia Fire and EMS Department

OBJECTIVES

After completing this chapter, the reader should be able to:

- Describe a typical fire department organization and mission statement.
- Define the functions of a firefighter and list the common tasks a firefighter must be able to perform.
- Explain what a standard operating procedure is and list five general areas covered by SOPs.
- List five rules and regulations of an organization and describe how they apply to the firefighter.
- List and define the five major components of an incident management system.
- Describe duties and responsibilities in assuming and transferring command within the incident management system.
- List five allied agencies that assist with fire department operations and describe their functions.

INTRODUCTION

The American fire service has a history and tradition that is more than 250 years old, dating back to Ben Franklin who organized the Union Fire Company in 1736. That basic organizational structure is still in use today. Depending on the size and needs of a community, a fire department will consist of a number of companies in one or a number of stations throughout a community. The number of stations depends on the geography of the response area, travel time, and distance.

Companies are usually divided into functions such as engine, ladder, or truck companies and specialized units such as rescue or hazardous materials companies. In addition, many fire departments provide either basic or advanced life support emergency medical services. The organization is designed to establish a clear division of work assignments:

- To assign responsibility for a specific response area
- To assign responsibility for a specific activity
- To eliminate duplication of work and confusion
- To establish an adequate level of equipment and personnel response to control the emergency scene or incident.

This chapter describes the many different roles a firefighter may have in the organization of a fire department. The firefighter is the basic unit of the organization, and a well-trained firefighter is essential for the department to accomplish its mission.

FIRE DEPARTMENT ORGANIZATION

Fire departments, as shown in **Figure 2-1**, are like any other organization in that they have a reason for existing and a structure for operations. The mission

Figure 2-1 A typical fire department with apparatus and facilities. It is necessary to look into the fire department's organization to understand how it functions.

statement should communicate the reason for being, and the organizational structure defines the chain of command and authority.

The Business of Fire Protection

Providing fire protection and life safety services is a business. Like businesses of all sizes and types, the fire service strives to provide the highest quality services as effectively and efficiently as possible while working to meet customer or community expectations. The fire service also works to meet these goals while protecting the safety of firefighters, emergency responders, and the public. Also similar to businesses, the fire and emergency services have adopted a quality mind-set and a focus on customer service in the provision of their emergency and nonemergency services.

Mission Statement

Each fire department should have a mission statement. This will provide the members of the organization with a clear and defined purpose of the type and level of service the department will provide. Also, the mission statement must be specific so the public—the people depending on the entire department for help—know what to expect from the organization and understand what duties firefighters can and cannot perform.

> **Streetsmart Tip** The fire department must communicate its mission statement to the public. A few ways this can be accomplished are during fire prevention inspections and fire safety information presentations to community groups.

The mission statement provides a clear and concise statement of what the organization intends to do or accomplish. The statement should contain the type of services or product to be delivered and who will receive this service or product.

Sample Mission Statement 1

The Midway Fire Department is organized to deliver fire prevention, fire suppression (extinguishment), and rescue services to the citizens of its protection area. This will include response to conduct vehicle extrication, hazardous materials mitigation, and basic life support emergency medical services.

The sample mission statement describes what the Midway Fire Department does. The mission statement does not limit what an organization will do; it provides the focus for the service it will provide to the people in its community. This is especially true for an emergency response organization when a person calls for help.

If the Midway Fire Department decides to expand its services to include additional activities, its mission statement would read as follows:

Sample Mission Statement 2

The Midway Fire Department is organized to deliver fire prevention, life safety, fire suppression (extinguishment), and rescue services to the citizens of its protection area. This will include response to conduct vehicle extrication, hazardous materials mitigation, confined space rescue, advanced life support emergency medical services, disaster response, and fire life safety code enforcement.

Again the services provided by this organization are very specific, and the firefighters and public know what to expect from the fire department. The second mission statement shows that the Midway Fire Department has decided to provide advanced life support, confined space rescue, disaster response, and code enforcement services.

Organizational Structure

To accomplish its tasks, a fire department must have some type of organizational structure. This structure may be as complex as that shown in **Figure 2-2** or as simple as that of **Figure 2-3**. Both figures show a structure for internal organization and how functions and responsibilities are delegated in the organization. Neither of these shows the receiver of the fire department's services, the citizens they protect, or how the fire department is connected to the community. Another way of defining the fire department organization could look like **Figure 2-4**. This structure shows circles of interdependence the organization has with its community. The fire department depends on the community and governing body for support. In turn, the community depends on the fire department for services and the governing body depends on the department to provide those services. The firefighter is the key position in the organization and is essential for the department to fulfill its mission.

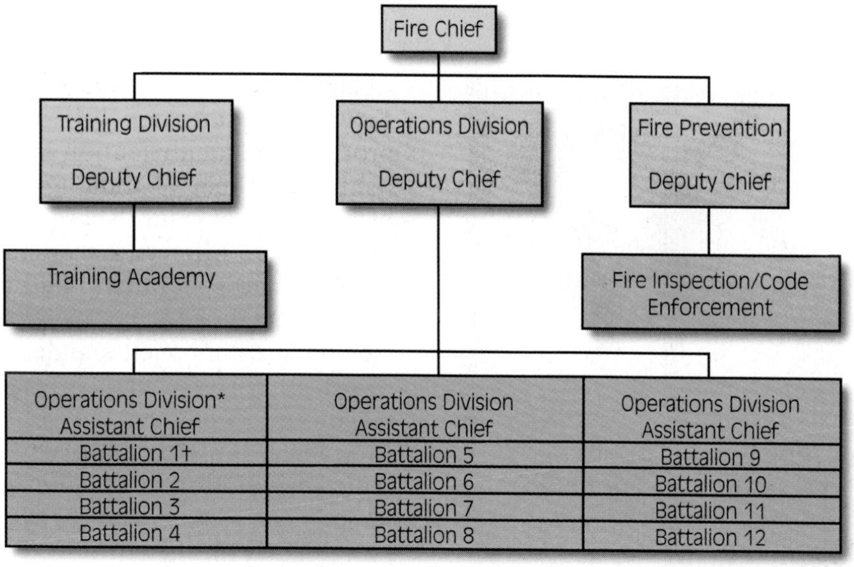

*Each division has one heavy rescue company.
† Battalion consists of four engine companies and one ladder company.

Figure 2-2 The organizational structure for a medium to large fire department shows the division of work assignments and chain of command. Most large departments will have a structure like this with separate divisions for suppression, training, and fire prevention.

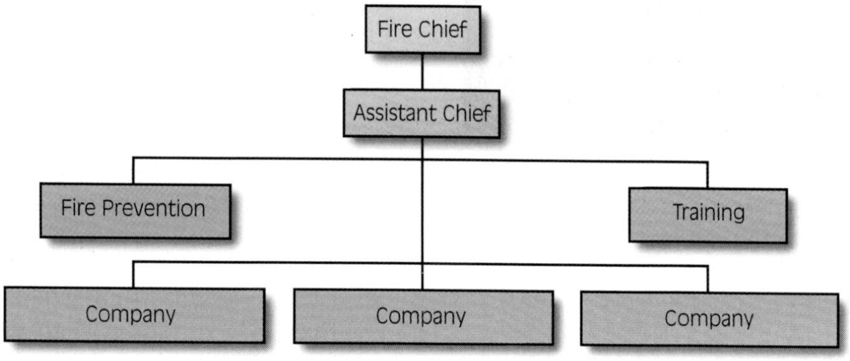

Figure 2-3 The organizational structure for a small fire department will have a more direct chain of command and structure. In this size of organization, one individual may be responsible for a number of functions.

Figure 2-4 Different from most organizational charts, this organizational structure shows the interdependence of the community, fire department, governing body, and the firefighters.

Figure 2-5 The firefighter is the individual who makes a department operate.

THE FIREFIGHTER

A firefighter is an individual trained to perform the function of fire prevention and suppression. Firefighters must possess both the knowledge and skills to be able to perform safely and effectively at an emergency incident, **Figure 2-5.**

Depending on the mission of the fire department or organization, there may be other areas that firefighters must be knowledgeable about in addition to firefighting. A firefighter may be required to be an **emergency medical technician** (EMT) or **paramedic** (EMT-P), a **hazardous materials technician**, a **rescue specialist**, **Figure 2-6,** a fire investigator, or a fire prevention officer. Many other positions are available in the firefighting community, and as firefighters begin their training they can start establishing personal and career goals. The only limitation to advancement and personal growth is a failure to take advantage of training and learning opportunities.

There are many organizations in the fire and emergency service community that assist in the determination of the necessary qualifications firefighters need to perform their jobs. The **National Fire Protection Association (NFPA)**, through its professional qualification committee, has established a number of training standards for various department positions. These standards address firefighters, fire officers, hazardous materials responders, rescue specialists, and driver/operators to name a few. In addition, many state and local agencies have established standards and qualification requirements for firefighting and emergency response occupations and volunteer positions. All firefighters must be familiar with the standards in their jurisdictions and determine what types of training and certification are necessary to maintain a position as a firefighter.

> **Firefighter Fact** Two national organizations provide certification of state and local firefighter training programs in accordance with NFPA standards. These are the National Board on Fire Service Professional Qualifications and the International Fire Service Accreditation Congress (IFSAC).

Through a consensus development process, the NFPA provides the *Standard for Fire Fighter Professional Qualifications,* also known as **NFPA 1001**, **Figure 2-7.** This standard outlines the minimum knowledge, skills, and abilities firefighters must have at two different levels, Firefighter I and Firefighter II. In addition to knowledge and skills gained through training and experience, firefighters must also be in good physical condition to perform the demanding and strenuous tasks involved in firefighting and emergency response.

> **Firefighter Fact** According to the United States Fire Administration, 102 firefighters died in the line of duty in 2002. Of these line-of-duty deaths, 31 (nearly one-third of the total) were from heart attacks. Regular exercise, physical conditioning, and a healthy lifestyle can help save firefighters' lives and reduce the number of line-of-duty deaths from heart attacks.

Listed here are the typical requirements an individual trained in structural fire suppression would be expected to meet:

- Know the department's organizational structure and operating procedures.
- Perform all duties safely.
- Know the department's response area or district, including streets and hazards.

(A)

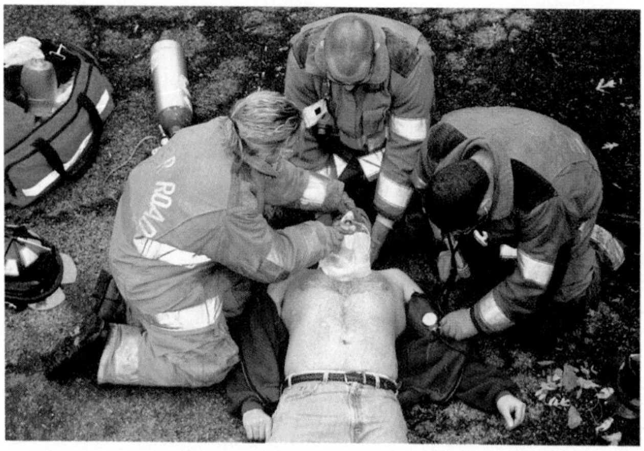

(B)

(C)

Figure 2-6 Some positions available to firefighters are (A) rescue specialist, (B) paramedic, and (C) hazardous materials technician.

Figure 2-7 National Fire Protection Association 1001, *Standard for Fire Fighter Professional Qualifications*, provides the basis for firefighter training.

- Maintain firefighting equipment, especially personal protective equipment.
- Respond to alarms of fire as a member of a trained unit/company.
- Use self-contained breathing apparatus and personal protective equipment.
- Rescue people endangered by the fire.
- Use fire department tools to conduct forcible entry, ventilation, and fire extinguishment at structure, vehicle, and ground cover fires.
- Conduct overhaul operations at a fire scene.
- Conduct fire prevention inspections.
- Present fire safety information to the community.
- Provide basic life support activities such as CPR or control of bleeding.

Depending on the organization's mission statement and responsibilities, the firefighter may also be required to perform additional duties or functions. A sampling of these and the corresponding training standard are shown in **Table 2-1** and **Figure 2-8**.

Additional Duty Training Standards

POSITION DESCRIPTION	TRAINING STANDARD
Fire apparatus driver/operator	NFPA 1002
Airport firefighter	NFPA 1003
Public safety communications telecommunicator	NFPA 1061
Paramedic/ emergency medical technician	State or local standards

TABLE 2-1

THE COMPANY

Historically the basic unit of a fire department is the **company**, a team of firefighters with apparatus assigned to perform a specific function in a designated response area. A company is typically designated as an engine, ladder, truck, squad, or rescue depending on the type of apparatus, equipment, and firefighters on board. The company officer, often a captain, is responsible for all activities of the company. The company officer may have lieutenants or other officers assigned to assist with company tasks and objectives.

> **Streetsmart Tip** Depending on history, traditions, and/or geography, the titles used to designate a position in the fire department may vary from those presented in this text. Not all organizations have the position of lieutenant and some use a military designation such as sergeant or private. Firefighters should be familiar with the rank structure and titles used in their own organization.

(A)

(B)

(C)

Figure 2-8 Firefighters may be required to perform additional duties as (A) dispatcher, (B) airport firefighter, or (C) apparatus driver/operator.

Company officers are supervisory-level positions and are responsible for both firefighters and administrative duties, **Figure 2-9.** Requirements for the position of company officer are addressed in a separate training standard, NFPA 1021, *Standard for Fire Officer Professional Qualifications.*

Just as firefighters and company officers have standards that help in determining the knowledge, skills, and abilities to complete their jobs, so does the apparatus they ride to emergency incidents. NFPA 1901, *Standard for Automotive Fire Apparatus,* outlines minimum equipment, specifications,

(A)

(B)

Figure 2-9 Company officers perform both (A) firefighting and (B) administrative duties.

and capabilities for various types of firefighting apparatus. **Table 2-2** outlines some of the basic requirements for common types of fire apparatus.

The Engine Company

The **engine company**, as shown in **Figure 2-10,** is organized to provide firefighters who deliver water at the fire scene, deploy hoselines, and attack and extinguish fires in vehicles and structures.

An engine company often uses an apparatus referred to as an engine or pumper. NFPA 1901 states that a pumper should have a permanently mounted fire pump with a capacity of at least 750 gallons per minute (gpm). NFPA 1901 also states that a pumper should carry no less than 300 gallons of water. Other specific hose and ladder requirements are listed in **Table 2-2.**

In some organizations, the engine company may also perform rescue or squad functions, as detailed later in this section. This increases the amount of equipment necessary on the apparatus and may result in a different company designation, such as a rescue engine.

The Truck Company

The primary functions of the **truck company**, as shown in **Figure 2-11** (also known as a ladder company), are to carry firefighters for forcible entry, search and rescue, ventilation, provision of ladders, securing of utilities, and overhaul functions at a fire scene.

NFPA 1901 states that the apparatus used by truck companies, an aerial fire apparatus, should have a minimum of 115 feet of ground ladders, including at least one attic ladder, two straight ladders with folding roof hooks, and two extension ladders. Aerial fire apparatus may or may not be equipped with a fire pump. Aerial fire apparatus are also required to carry an extensive list of hand and power tools to accomplish the many tasks that are associated with truck company operations.

In addition to traditional truck company operations, in some jurisdictions, truck companies may also perform the functions of a rescue company, as detailed in the next subsection. Regardless of apparatus, all fire departments must be able to perform the functions associated with the truck company.

The Rescue Company

The **rescue company** (also known as a squad or squad company) is shown in **Figure 2-12.** It is organized to provide specially trained firefighters and specialized rescue equipment at an incident scene. The apparatus designated as a rescue company carries tools to conduct forcible entry and search and rescue at fire operations. Most rescue companies also carry specialized tools to conduct vehicle extrications, confined space rescue, rope

NFPA 1901: Standard for Automotive Fire Apparatus

OVERVIEW OF BASIC APPARATUS REQUIREMENTS
PUMPER
Fire pump with minimum rated capacity of 750 gpm.
Water tank with a minimum of 300 gallons.
Ground ladders, to include at least one attic ladder, one extension ladder, and one straight ladder with roof hooks.
A minimum of 15 feet of soft suction hose or 20 feet of hard suction hose.
A minimum of 800 feet of 2 1/2-inch or larger diameter fire hose.
A minimum of 400 feet of 1 1/2-inch, 1 3/4-inch, or 2-inch fire hose.
AERIAL FIRE APPARATUS
A minimum of 115 feet of ground ladders, to include at least one attic ladder, two straight ladders with folding roof hooks, and two extension ladders.
Apparatus may or may not have a fire pump and water tank.
MOBILE WATER SUPPLY APPARATUS
A minimum of 1,000-gallon water tank.
Apparatus may or may not have a fire pump. If the apparatus does have a pump, there are minimum requirements for the fire hose that must be carried.
QUINT
A permanent fire pump with a minimum rated capacity of 1,000 gpm.
Water tank with a minimum of 300 gallons.
A minimum of 85 feet of ground ladders, to include at least one attic ladder, one extension ladder, and one straight ladder with roof hooks.
A minimum of 15 feet of soft suction hose or 20 feet of hard suction hose.
A minimum of 800 feet of 2 1/2-inch or larger diameter fire hose.
A minimum of 400 feet of 1 1/2-inch, 1 3/4-inch, or 2-inch fire hose.

TABLE 2-2

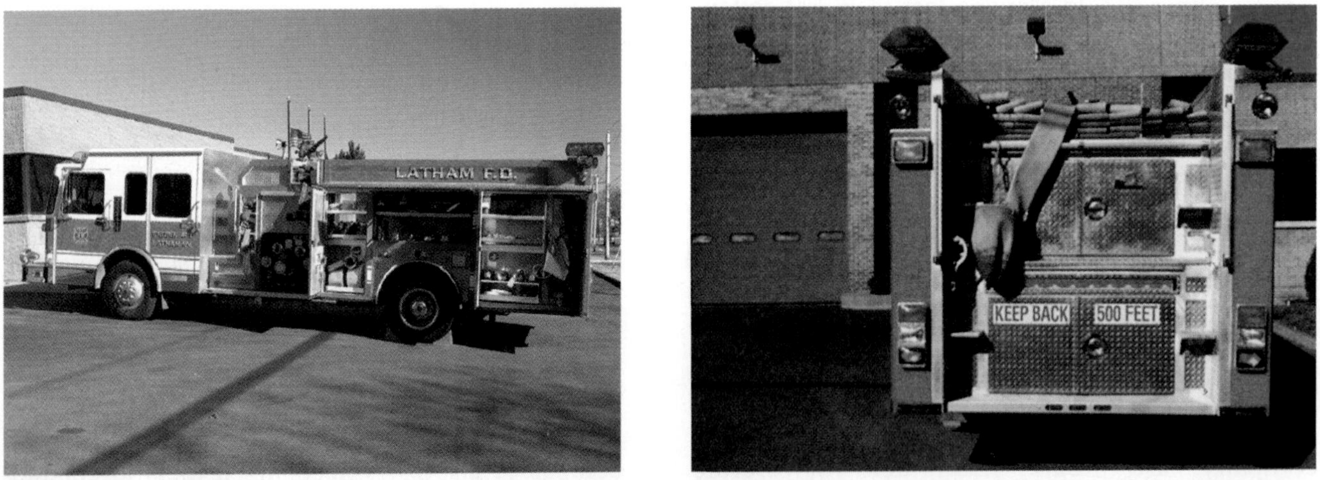

Figure 2-10 The engine company provides the firefighters with the tools, equipment, and hose necessary for fire suppression operations.

Figure 2-11 The ladder or truck company is designed to carry firefighters and for forcible entry, search and rescue of building occupants, ventilation, ladder provision, and overhaul functions at a fire scene.

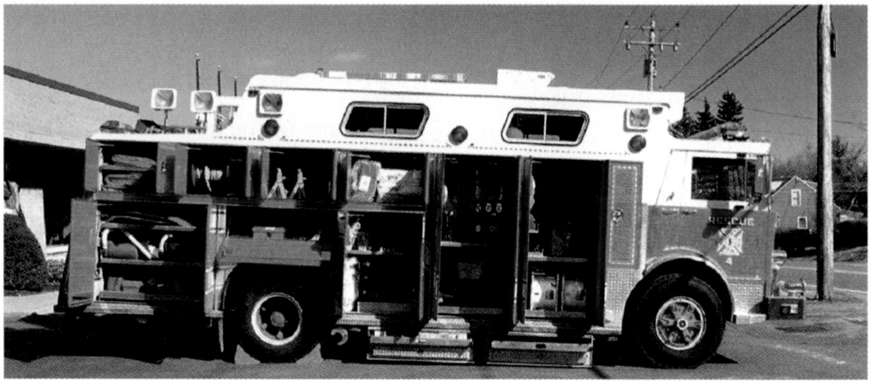

Figure 2-12 The rescue company is designed to carry specialized equipment for rescue operations and may provide support equipment such as light towers.

rescue, and other forms of technical rescue operations. These specialized functions depend on local requirements and the mission of the organization.

Specialty/Combination Units

Because the fire service has a great amount of diversity in the types of services provided to communities, and an equally great diversity in the way these services are delivered, there are a wide variety of unique fire and emergency service apparatus that are used in departments worldwide.

Combination units are typically a blend of two major company functions into one piece of apparatus. For example, many departments use a **quint** to deliver fire suppression services. A quint is a combination of an engine and ladder company and can perform either function as necessary on a fire scene. According to NFPA 1901, a quint must have a permanently mounted fire pump and water tank, an area to store hose, an aerial device (ladder or platform) with a permanently mounted waterway, and at least 85 feet of ground ladders.

Another type of specialty unit, a mobile water supply apparatus, is used in areas where a water supply is not present, such as hydrants or natural water sources. Mobile water supply apparatus, also known as tankers or tenders, carry a minimum of 1,000 gallons of water per NFPA 1901, and they may or may not have a fire pump and fire hose.

Specialty units provide unique types of services at emergency scenes. Examples of specialty units include wildland fire or brush fire response units, **aircraft rescue and firefighting (ARFF)** apparatus, light and air units, hazardous materials units, mass casualty response units, and water rescue and firefighting apparatus. These units are specially equipped to handle unique emergency situations, and the personnel who operate on these units are specially trained.

Emergency Medical Services

Many fire departments provide either basic life support (BLS) or advanced life support (ALS) **emergency medical services**. This may be an additional duty assigned to an existing engine, truck, or rescue company or a specific unit designated as a BLS/ALS response unit. Also, fire departments may operate ambulances to provide transport services.

The Chief Officers

The chief of the department, as shown in **Figure 2-13**, is the person ultimately responsible for the operations and administration of the fire department. To carry out these responsibilities, the chief may have a number of deputy, division, assistant, or battalion chiefs.

The rank structure and position designation depend on the size, needs, and in some cases the history of an individual fire department. Not all fire departments will have all of the positions listed next because some designations serve the same function:

- *Chief of department:* Responsible for all department functions.
- *Deputy chief:* A staff position designated by the chief of department for a specific function, such as personnel, training, or administration. Deputies usually have the authority to act in the absence of the chief.
- *Assistant chief:* Similar to a deputy chief; the assistant may be a staff position responsible for a functional area. In some departments, an assistant chief may be a higher rank than a deputy chief.
- *Division chief:* Usually an operational position responsible for a large geographic area or a number of battalions.
- *Battalion/district chief:* An operational position responsible for a specific number of companies.

Figure 2-13 Chief officers are responsible for many functions, including command of emergency operations.

The number of officers in any department depends on the size of the organization and the necessity to maintain unity of command, span of control, and division of task responsibilities. Chief officers are management-level positions and requirements for these positions are addressed in a separate training standard, NFPA 1021, *Standard for Fire Officer Professional Qualifications.*

ADDITIONAL FIRE DEPARTMENT FUNCTIONS

Because of the complex issues and tasks facing a fire department, many additional functions are assigned to its operations. Some of these, such as training or fire prevention, are necessary for day-to-day operations. Additional sections may be established to manage complex issues such as hazardous materials, urban/technical search and rescue, water rescue, or the delivery of emergency medical services.

Fire Prevention and Life Safety

One of the prime missions of all fire departments is the prevention of fires. Preventing fire not only reduces the risk to a community both in lives and property but also reduces the exposure of firefighters to a very dangerous and hazardous duty. The organization of a fire prevention office is usually divided into two functions: code enforcement/inspection services and fire/life safety education. A chief-level officer usually heads the fire prevention office.

Fire department personnel, as shown in **Figure 2-14**, should conduct fire prevention inspections. This provides the organization with increased knowledge of the hazards present in a community and provides the opportunity to develop prefire plans.

Public fire/life safety education activities are usually assigned to the fire prevention office and include fire prevention education and fire survival programs. In addition to code enforcement, fire prevention programs are designed to prevent the start of

Figure 2-14 The most important function of the fire prevention staff is to conduct inspections to prevent fires or ensure early control by fire suppression systems.

a fire. If an area were to experience a high number of fires starting because of careless disposal of smoking materials, an education program on proper disposal would be delivered to the public. Fire survival programs educate the public on what to do after a fire has started. An example of these activities includes education on installation and maintenance of smoke detectors or **exit drills in the home (EDITH)**.

Staff assigned to fire prevention/fire inspection duties should meet the requirements as addressed in NFPA 1031, *Standard for Professional Qualifications for Fire Inspector and Plan Examiner.*

Training

> **Street Smart Tip** Training must be a continuing function in all fire departments.

Training begins with basic firefighter or probationary training, as shown in **Figure 2-15**, and continues with proficiency training as new tools, equipment, or techniques become available. Depending on the size, organizational structure, and other responsibilities, a chief-level officer usually heads the training division or group. Regardless of size, all fire departments must have a designated training officer or division.

The training division administers and documents all training activities for the department and will usually have a number of instructors assigned to present specific subjects. This division may also be responsible for conducting mandatory employee

Figure 2-15 Training must be a continuing function in all fire departments regardless of size or area served.

safety training required by the Occupational Safety and Health Administration (OSHA). This program requires firefighters to be trained to perform assigned tasks in a safe manner. Some of the required topics for firefighting are respiratory protection, response safety, and workplace safety.

Training officers and instructors should meet the requirements for these positions as addressed in NFPA 1041, *Standard for Fire Service Instructor Professional Qualifications.*

Emergency Medical Services

As mentioned previously in this chapter, and depending on the size of the organization or its jurisdiction, the EMS function may be a separate division within the fire department with a chief-level officer responsible for its activities.

Apparatus Maintenance and Purchasing

Large departments may have a fire apparatus maintenance or repair shop that is usually responsible for all vehicle repairs, maintenance, and purchasing. Depending on local policy this organization may be headed by a fire department officer or a nonuniform staff member knowledgeable in the areas of vehicles, heavy equipment, and fire apparatus.

Special Operations

Depending on the size of its community or potential hazards present, a fire department may establish a special operations section, **Figure 2-16**, which will deliver or support services such as hazardous materials mitigation, high-rise operations, air operations, confined space rescue, and swift water or ice rescue. A chief-level officer usually heads these specialized functions.

Individuals assigned to these areas must have specific knowledge and skills for their specialty and this usually requires many additional hours of training.

Figure 2-16 Trench rescue is one of many specialized operations requiring additional equipment and training.

REGULATIONS, POLICIES, BYLAWS, AND PROCEDURES

All organizations must have regulations, policies, bylaws, and procedures to operate in an effective manner. This is especially true in a fire department to ensure an adequate and effective emergency response. These regulations are implemented and used to establish the daily and emergency operations of the organization.

Regulations

Regulations are the rules that determine how an organization operates and are usually established by top-level management or a governing body. Regulations address topics such as time and attendance, uniform, and use of department equipment and vehicles. These are usually very specific in nature and may carry some type of disciplinary action for violation.

In addition, state or federal government agencies, such as OSHA, may establish regulations governing fire department activities. These usually deal with such issues as health and safety, antidiscrimination laws, and work hours.

> **Note** Regulations established by the governing body of a fire department are internal in nature and design, whereas regulations established by state or federal agencies have the force of law, with some penalty or fine charged for violations.

Policies

Policies are formal statements or directives established by fire department managers to provide guidance for decision making. Policies will usually be general in nature and provide a framework for administering the day-to-day department activities. Topics addressed by policies include types of response levels, staffing of companies, and the release of information at an incident.

Bylaws

Volunteer fire departments provide a large amount of fire protection, both in the United States and Canada. These departments may be organized as previously described in this chapter, as independent corporations, or as some combination of both types of organizational structure. A fire corporation will usually be organized as a not-for-profit organization with a governing body of a president and vice president, or a board of directors. The board of directors or membership will establish bylaws in place of regulations or policies on how the department business structure is organized. Generally, fire and emergency operations are not covered as part of the bylaws.

Procedures

Standard operating procedures (SOPs) provide specific information and instructions on how a task or assignment is to be accomplished.

SOPs are established so that all members of a department will perform the same function with the same level of uniformity. Also, these procedures are generally tactical in nature because, in most instances, they address emergency operations. In addition, all SOPs must be distributed to all department members. Some topics addressed by SOPs include cleaning and use of protective equipment, apparatus response policies, use of specific tactics or tools, and how to establish a water supply at a fire scene.

A note on terminology: This text uses the term *standard operating procedure* to describe documentation outlining actions or procedures to be followed. Many departments use the term *standard operating guideline (SOG)* to describe this documentation. SOGs are often thought of as providing guidance to company officers and firefighters when operating on emergency scenes as opposed to outlining strict parameters for actions. Each department is different in its use of SOP and SOG terminology and even different in its interpretation of the flexibility of the documents. It is up to the individual firefighter to understand what terms are used by his or her department.

There are a variety of ways in which SOPs are developed and documented. In general, the following should be true:

- SOPs address the who, what, when, where, and how of a topic.
- Firefighter safety is the first consideration in all procedures.
- SOPs are brief, clear, and concise.
- Lengthy SOPs are broken down into smaller sections.
- SOPs cover dynamic topics and must be reviewed often, at least once every three years.
- A method for changing SOPs exists if they should become obsolete, inaccurate, or ineffective.

Figure 2-17 shows a sample of an SOP that might be found in a typical fire service organization.

Sample Standard Operating Procedure

Subject: Emergency Vehicle Response Section: 303
Date: 01/01/2001
Purpose: To provide requirements for emergency response of Delmar Fire Department apparatus and vehicles.
Scope: This procedure applies to all drivers/operators of Delmar Fire Department vehicles.

1. Apparatus will not exceed the posted speed limit at any time.
2. Seat belts must be used by all personnel riding on apparatus.
3. No riding is allowed on the rear step area nor standing in the crew compartment while the vehicle is in motion.
4. Apparatus must come to a complete stop before attempting to proceed in any of the following conditions:
   - Intersection against a red traffic light
   - All railroad grade crossings
   - Blind intersections
   - Stop signs
   - Any condition when the operator cannot account for all traffic lanes
5. A spotter shall be located at the rear of the apparatus at any time the vehicle moves in reverse.
6. Operators are responsible for any traffic citation issued to them.

Figure 2-17 A sample standard operating procedure.

ALLIED AGENCIES AND ORGANIZATIONS

During the course of any operation, the fire department and firefighter interact with many different people and organizations. Depending on the local jurisdiction, a few of these agencies could be police/law enforcement, public works or highway department, utility companies, environmental conservation/protection agencies, and private business. Depending on the nature of the incident, these organizations may be assisting with the operation or they may be receiving fire department services.

- *Police:* Provide assistance with traffic management and scene security at all incidents. May be the assisting or lead agency in fire investigations, bomb threats, or terrorism incidents.
- *Public works:* Provide assistance with traffic management and maintain public water system for fire protection. May maintain fire station facilities and provide assistance in securing materials (earth, sand) for the control of hazardous material incidents.
- *Utility companies:* Secure building and street utilities during or after an incident.
- *Environmental protection:* Depending on the incident they may be a lead or assisting agency at a hazardous materials spill.
- *Private business:* May provide supplies for emergency operations or may be the receiver of fire department services.

INCIDENT MANAGEMENT

Fire departments respond to millions of emergency incidents every year. These incidents are the operations for which a firefighter trains. Some incidents require a single company; some require multiple companies. Extremely large incidents such as wildfires or natural disasters may require mutual aid assistance from outside the authority having jurisdiction and involve fire agencies on a national basis.

> **Note** Mutual aid or assistance agreements are prearranged written agreements of the type and amount of assistance one jurisdiction will provide to another in the event of a large-scale fire or disaster. Mutual aid plans may include the fire department resources from a town, county, or entire state. The key to understanding mutual aid is that it is a reciprocal agreement. Many fire departments use automatic mutual aid to provide for immediate dispatch to an incident. This may be to supplement staffing or equipment required for high hazard facilities such as a hospital.

Command and Control

To manage incidents, firefighters must understand the concept of command and control and how it is applied at the emergency incident. The basic tactical unit at any incident is the company or single resource. As the size or complexity of the incident grows, additional companies and resources respond. To maintain command and control, the command officer must have the ability to manage effectively the increasing number of companies operating at an incident. This is accomplished by using **unity of command** and span of control.

Unity of command means having one designated leader or officer in charge of an operation, company, or single resource. Every operating resource has a single designated supervisor. On the initial response this could be a company officer or a firefighter. All units responding to the incident report to command and receive instructions on operations. As the incident becomes larger, command may transfer to a high ranking officer. This is accomplished by briefing the new command on the incident, actions taken, and units operating.

Span of control refers to the number of resources (people, companies, etc.) that any one person supervises. For emergency operations the span of control is usually three to seven, with the optimum number being five. This allows for effective management and safety of all operations.

Assuming Command

The first unit arriving at an incident will usually assume command of the incident and give a status report. This is accomplished with a radio announcement, for example: "Engine 438 arriving, establishing Maple Avenue Command. We have a two-story wood frame residential structure with smoke showing from the second floor. Two occupants are reported in the structure." This advises all other responding units of a fire condition in an occupied residential building. Generally, SOPs will dictate the tasks assigned to other responding units.

> **Streetsmart Tip** Local and regional policies may dictate the format of the radio status report, and firefighters should be familiar with this format and structure. In addition, the locality or region may have special terminology for radio status reports, such as location designations. In some areas, the address side of a structure may be considered side 1, whereas in other areas this may be considered side A. Firefighters should be familiar with these specific requirements and local terminology.

Transfer of Command

Transfer of command is the process of briefing an individual of equal or higher experience or authority so that person can assume command of an incident. As noted earlier, the first arriving unit will usually assume command of an incident and as the incident becomes more complex, the command structure will expand and the incident commander (IC) will usually change. The information necessary for a complete transfer of command is known as a status or situation status report and must include the following information:

- What has happened or the type of incident
- Incident action plan (strategy and tactics)
- Current units/resources operating, status, and location
- Civilians injured or trapped
- Firefighters injured or trapped
- Actions that have been accomplished
- Ability to control the incident with available resources

Transfer of command should occur during a face-to-face meeting, but under extreme conditions transfer may be accomplished by radio or telephone. Transfer can only be accomplished when the person assuming command is at the incident scene. Also, the individual assuming command must acknowledge receipt of the information, confirm its accuracy, and announce the transfer of command to avoid any confusion with the previous IC. The more complex an incident is, the more complex this briefing will be, possibly involving additional staff or technical experts.

INCIDENT MANAGEMENT SYSTEM (IMS)

The **incident management system** is exactly what the name implies: a systematic approach for the command, control, and management of an emergency incident. It provides a command structure and designated responsibilities for the functions that must be addressed to stabilize any incident, from a room and contents fire to a large-scale natural disaster.

The concept of incident command, as shown in **Figure 2-18**, has evolved over the years and is based on the FireScope Project from California. As a result of numerous and large wildland fires, the California Department of Forestry developed the system to provide a unified system for use by all agencies responding to an incident. As it has evolved, the application of the IMS has taken on many names such as Fire Ground Command, Incident Command System, National Fire Academy (U.S. Fire Administration), or the Incident Command System, National Wildfire Coordinating Group.

Regardless of the name of the system used in a particular jurisdiction, to function properly, it must contain the following components:

- **Common terminology**: The designation of a term that is the same throughout an IMS.

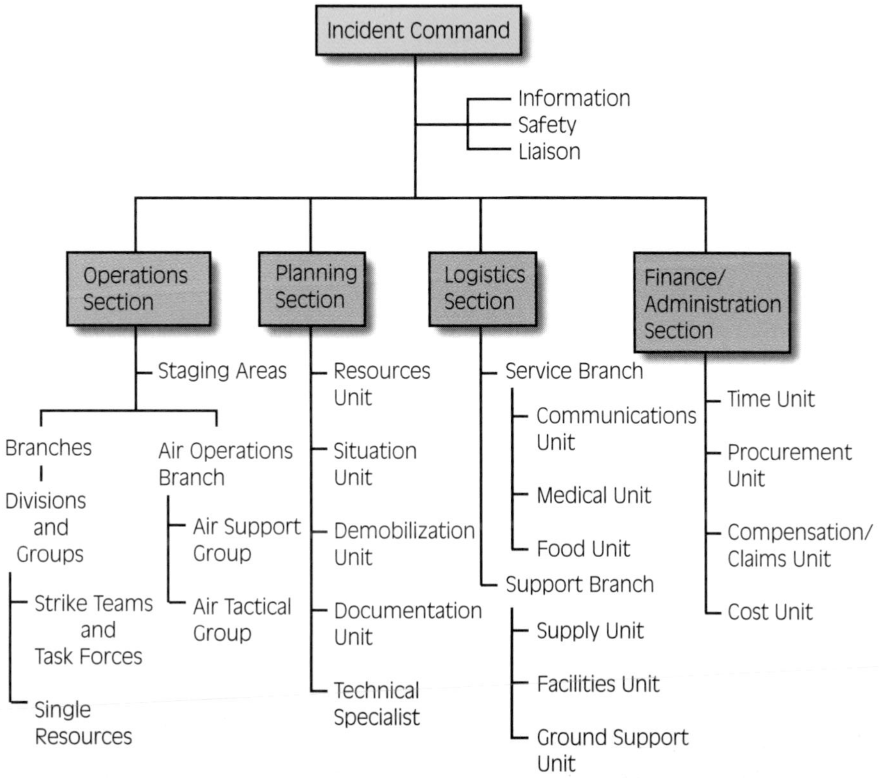

Figure 2-18 This chart of a typical incident management system shows the modular organization necessary to manage an incident.

- **Modular organization**: The ability to start small and expand if an incident becomes more complex.
- **Integrated communications**: The ability of all units or agencies to communicate at an incident. This may be as simple as assigning a tactical radio channel to an incident or developing a communications plan for use by many agencies at a large-scale incident.
- **Consolidated incident action plans**: The strategic goals to eliminate the hazard or control the incident. All companies or agencies working an incident must have the same objective.
- **Span of control**: The ability of one individual to supervise a number of other people or units. At an emergency incident the range for safe supervision is usually three to seven people or units. The ideal number is five.
- **Designated incident facilities**: These may be as simple as a command post established at a chief's vehicle **Figure 2-19A**, or complex enough to include a staging area, rehabilitation area for firefighters, feeding facilities, and office space for other agencies. A specially designed mobile command post is shown in **Figure 2-19B**. In some instances it is possible to designate these facilities prior to an incident.
- **Resource management**: Common designators are used for all resources assigned to an incident. Resources are assigned and managed in one of the following ways:
    - *Single resource:* Personnel or vehicle and required equipment.
    - *Task force:* A task force is any combination of single resources within the span of control guidelines, assembled for a common task or assignment. A structural assignment task force may consist of three engines, two trucks, and one rescue company.
    - *Strike team:* A combination of a specific number of units of the same kind and type. A water supply strike team would be five Type 1 engine companies.

Note Two key concepts that firefighters and officers must remember for all incidents are that the incident commander maintains responsibility for safe control of an incident and that the IC may delegate authority for a function or task but never responsibility. Also, any function or task that the IC does not delegate remains with the IC to accomplish.

(A)

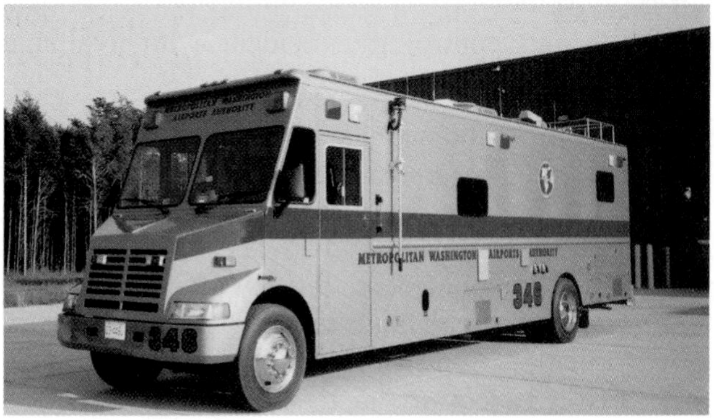

(B)

Figure 2-19 Depending on an incident's complexity, a command post may be as simple as (A) the command officer's vehicle or (B) a specially designed mobile command post.

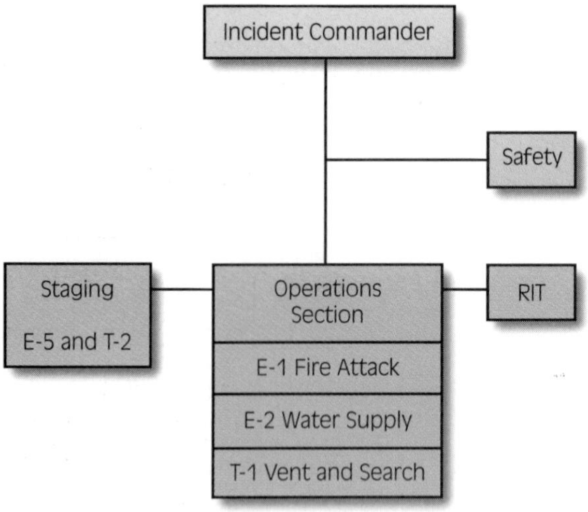

Figure 2-20 Command organization for a small structure fire incident.

As shown in **Figure 2-20**, the command structure for a small structure fire (such as a single-family residential fire) involves an IC who has designated a safety officer and retained responsibility for operations. The IC is always responsible for all functions and unassigned tasks. The IC, as operations chief, has established a staging area and designated a fire attack group, a water supply group, a ventilation/search group, and a **rapid intervention crew (RIC)**. These firefighter rescue teams may also be known as a **firefighter assist and search team (FAST)** or a **rapid intervention team (RIT)**.

Five Major Functions of an Incident Management System

An incident command or management system has five major functional areas, four of which may or may not be established depending on incident complexity. All incidents will always have an incident commander, who may establish the operations, planning, logistics, and finance/administration sections. When these positions are established, an individual is designated as the chief and reports directly to the IC. In addition, the IC may establish three command staff positions—safety officer, liaison officer, and public information officer—to assist with the incident management. These positions are not considered under the span of control guidelines. During incidents involving long **operational periods** or other agencies these functions should be established. The managerial level of the organization structure used should follow a format similar to that shown in **Figure 2-21**.

Command

The incident commander is responsible for developing the **strategic goals** for control of an incident. This may be a chief-level officer, captain, or firefighter, depending on the staffing of the fire department that responds and who arrives first. The IC has the authority to request and assign resources to an incident, the power to establish functional areas to control the incident, and the responsibility for the safety of all responders. The IC manages the incident, not the **tactics**.

Operations

The operations section chief is responsible for implementing the tactical assignments to meet the strategic goals established by the IC. The operations chief reports directly to the IC. Depending on the size and complexity of the incident, the operations section may be a single unit or a number of **branches** or **divisions**.

Staging

Staging is part of the operations section under the direct control of the operations chief with the assistance of the staging area manager. All apparatus and personnel assigned to staging are committed to the incident and must be available for deployment within three minutes.

Planning

The planning section chief is responsible for the development of the incident action plan. This plan is based on evaluation of the information and particulars of an incident. The planning section is also responsible for tracking the status of all resources used at an incident.

The planning section provides the IC with situation status report updates. These reports detail what

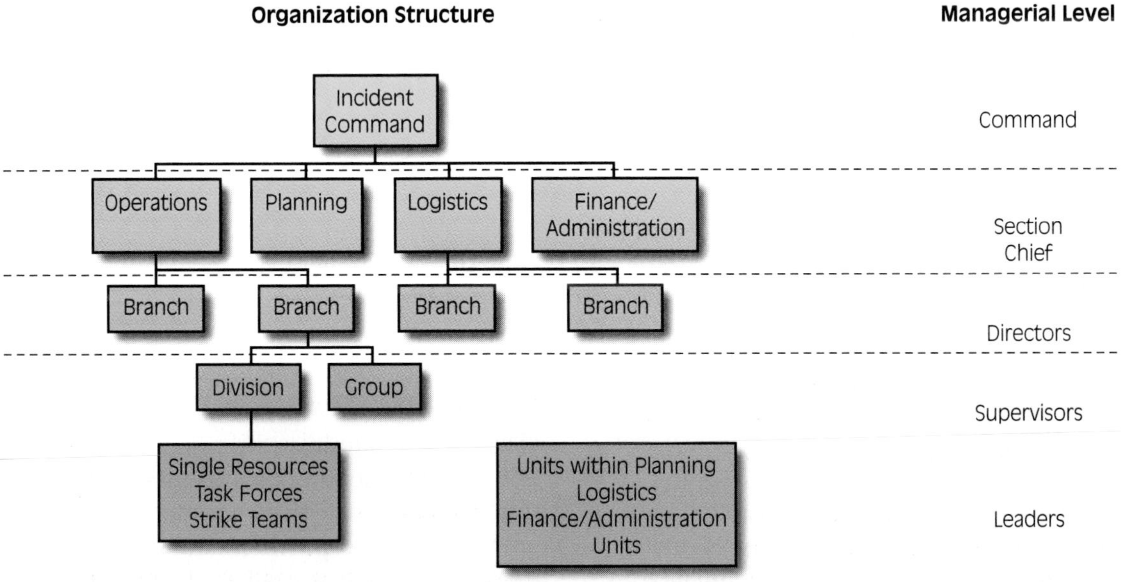

Figure 2-21 Organizational structure showing the managerial level for an incident.

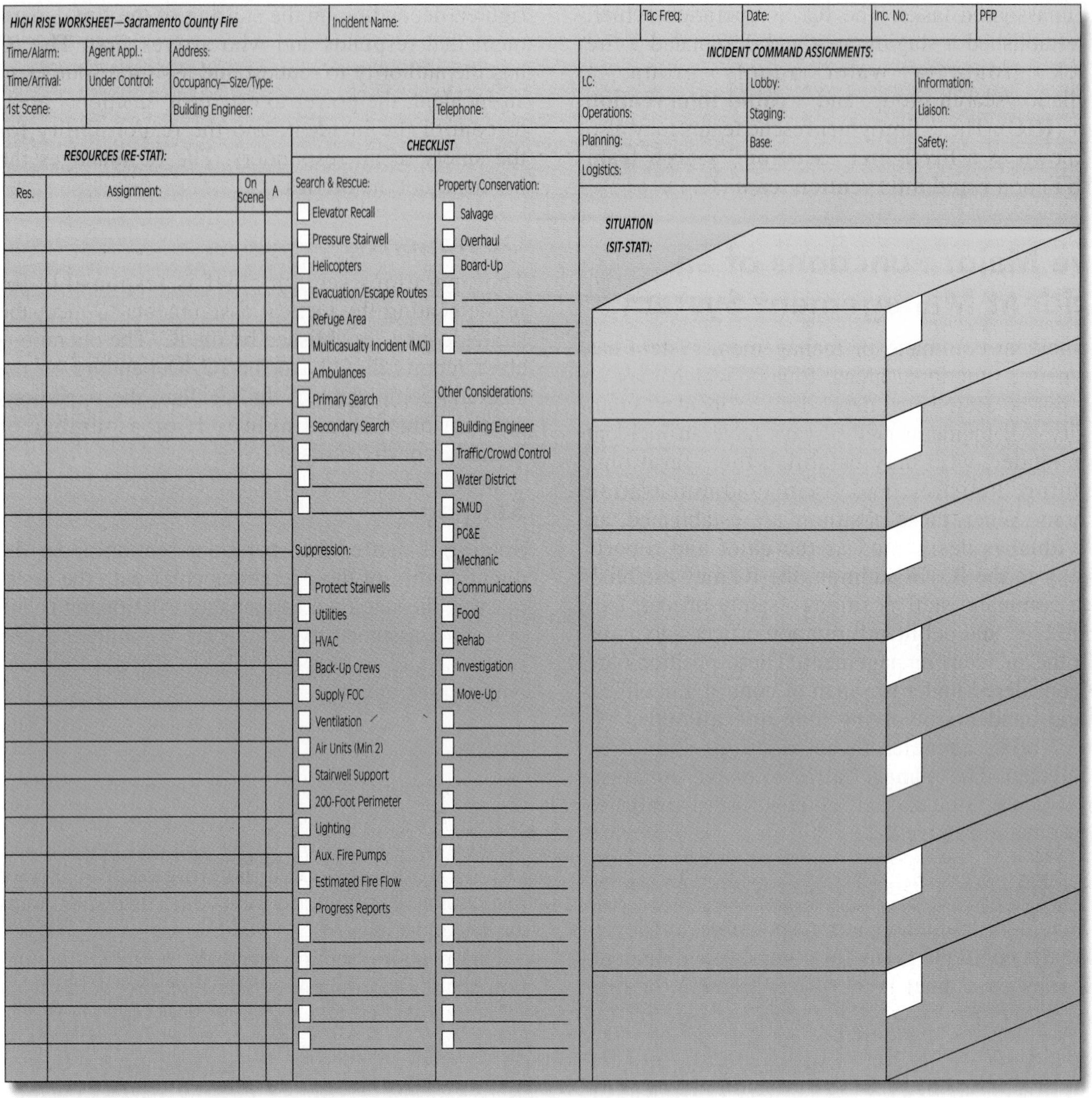
HIGH RISE WORKSHEET—Sacramento County Fire | Incident Name: | Tac Freq: | Date: | Inc. No. | PFP

Time/Alarm: | Agent Appl.: | Address:

Time/Arrival: | Under Control: | Occupancy—Size/Type:

1st Scene: | Building Engineer: | Telephone:

INCIDENT COMMAND ASSIGNMENTS:

IC: | Lobby: | Information:

Operations: | Staging: | Liaison:

Planning: | Base: | Safety:

Logistics:

RESOURCES (RE-STAT):

Res. | Assignment: | On Scene | A

CHECKLIST

Search & Rescue: Elevator Recall; Pressure Stairwell; Helicopters; Evacuation/Escape Routes; Refuge Area; Multicasualty Incident (MCI); Ambulances; Primary Search; Secondary Search

Suppression: Protect Stairwells; Utilities; HVAC; Back-Up Crews; Supply FDC; Ventilation; Air Units (Min 2); Stairwell Support; 200-Foot Perimeter; Lighting; Aux. Fire Pumps; Estimated Fire Flow; Progress Reports

Property Conservation: Salvage; Overhaul; Board-Up

Other Considerations: Building Engineer; Traffic/Crowd Control; Water District; SMUD; PG&E; Mechanic; Communications; Food; Rehab; Investigation; Move-Up

SITUATION (SIT-STAT):

Figure 2-22 Tactical worksheets provide the incident commander with a guide for managing an incident.

has happened at the incident, injuries, how the incident is being controlled, resources in use, and suggested actions for stabilization of the incident. For large-scale incidents or incidents of long duration, the planning section will develop an incident action plan for approval and use by the IC. These plans are usually for operational periods of eight to twelve hours in length.

Logistics

The logistics section chief is responsible for securing the facilities, services, equipment, and materials for an incident. Usually this is accomplished by using a support and service section. The support section is responsible for medical support for incident responders, communication, and food services. The services section is responsible for supplies (food, medical, and specialized extinguishing agent), facilities, and the resources to deliver these items.

Finance/Administration

The finance/administration section chief is responsible for documenting cost of materials and personnel for the incident. This is an important function, especially when the potential for reimbursement exists from areas such as state or federal disaster declarations and insurance companies. Generally this func-

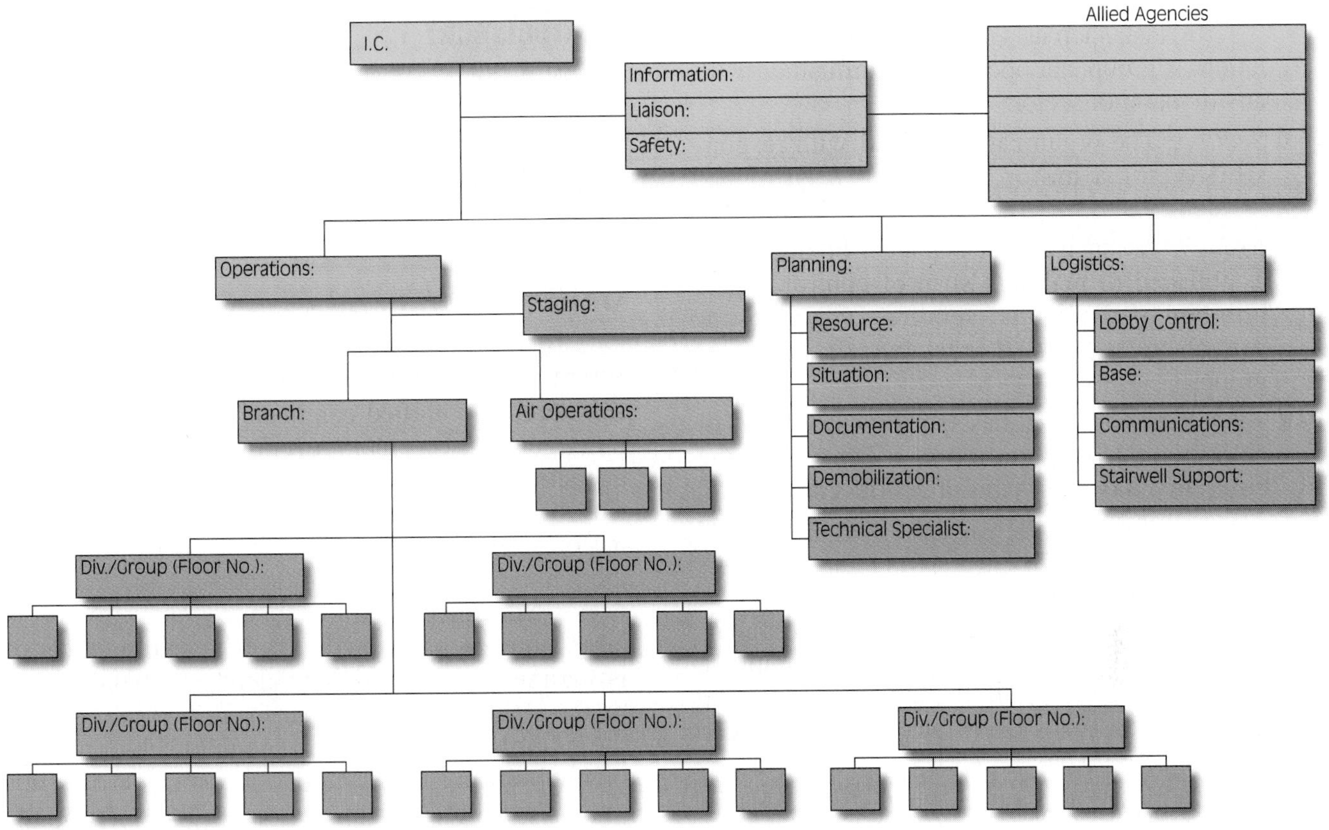

Figure 2-22 Continued

tion will be established for large-scale or long-duration incidents such as natural disasters.

Command Staff Positions

An incident commander has the ability to create staff positions to assist with an incident. These individuals report directly to the IC.

Safety Officer

The safety officer is responsible for the safety of all responders at an incident. The incident safety officer reports directly to the IC and has the authority to stop any activities that pose an immediate danger to incident responders. All other safety issues are channeled through the IC. For large-scale incidents the safety officer will develop an incident safety plan to be incorporated as part of the incident action plan.

Liaison Officer

The liaison officer is responsible for communication and contact between other agencies that respond. These may include environmental protection, police, highway or transportation, and state and federal government agencies. This position is especially important for long-duration, hazardous materials, or mass casualty incidents.

Streetsmart Tip The PIO position is very important for incidents involving evacuation of people or an unusual event. The news media is a valuable resource for conducting evacuation operations. Another benefit from this is the development of a relationship with the media for publicizing other events such as fire safety education or recruitment campaigns.

Firefighter Fact To implement an incident management system many fire departments use tactical worksheets like that shown in **Figure 2-22**.

Public Information Officer

The public information officer (PIO) is responsible for providing factual and accurate information concerning an incident to the news media. Only one information officer is appointed for each incident.

Incident Command Designations

- *Division:* The division is responsible for all operations within an assigned geographic area. At structural incidents a division may be designated for each floor or level of a building.

- *Group:* A functional designation to conduct a specific task such as search and rescue or ventilation. A group can operate in coordination with any division or sector.
- *Section:* The organizational level with responsibility for a major functional area of the incident.
- *Branch:* As an incident expands, branches are established to maintain span of control over a number of divisions, sectors, or groups. A branch must have at least two divisions or groups.
- *Type:* A resource that has specific capabilities or equipment. As an example, a Type 1 engine company has a minimum 1,000-gpm pump, 400-gallon water tank, 1,200 feet of 2½-inch hose, and a 500-gpm master stream appliance with four firefighters.
- *Task force:* Any combination of single resources assembled for an assignment. As an example, a task force for a structure fire assignment may be three engines, two ladder trucks, and one rescue company.
- *Strike teams:* Designation for a set number of resources of the same type and kind. A water supply strike team would be five Type 1 engine companies.
- *Crew:* A specific number of firefighters with an assigned task, but usually without apparatus. A ventilation crew may consist of six firefighters with tools but not assigned to a truck company.

Streetsmart Tip Incident command terms and designations may vary by jurisdiction. For example, "sector" and "division" are interchangeable terms used for tactical assignments. Firefighters must know the terms used in their organization!

Unified Command

One of the benefits of any IMS is the ability to establish what is known as a **unified command** structure. The unified command structure is used to manage an incident involving multiple response agencies or when multiple jurisdictions have responsibility for control of an incident. Unified command has only one IC, but it allows for agencies with jurisdiction to be part of the command structure or team.

For example, a natural disaster may involve a large number of agencies from multiple levels of government. The initial incident commander may be the chief of the fire department; however, due to the complex nature of the incident, law enforcement, building code, and social service officials may be needed for legal reasons or for their expertise. Individuals from these agencies may be designated deputy ICs or may assume command as incident problems and priorities involve their specific discipline. Regardless of how complex or large an incident is, the incident will have one and only one IC. The use of unified command structure requires knowledge and training before the incident!

Lessons Learned

In the past few years, five notable incidents have resulted in the line-of-duty death of sixteen firefighters. The investigation report from each incident cited the lack of an organized and effective fire department command structure as one of the factors resulting in these deaths. To survive on the fire scene, firefighters must know the roles and responsibilities of the personnel, how their fire department command structure works, and be part of that command structure.

KEY TERMS

Aircraft Rescue and Firefighting (ARFF) Of or pertaining to firefighting operations involving fixed or rotary wing aircraft.

Branch The command designation established to maintain span of control over a number of divisions, sectors, or groups.

Common Terminology The designation of a term that is the same throughout an IMS.

Company A team of firefighters with apparatus assigned to perform a specific function in a designated response area.

Consolidated Incident Action Plan The strategic goals to eliminate the hazard or control the incident.

Division Command designation responsible for operations within an assigned geographic area.

Emergency Medical Services The delivery of prehospital medical treatment.

Emergency Medical Technician (EMT) An individual trained and certified to provide basic life support emergency medical care.

Engine Company The unit designation of a group of firefighters assigned to a piece of apparatus designed to deliver water to the fire scene.

Exit Drills in the Home (EDITH) A fire survival program to encourage people to practice fire drills from their home or residence.

Firefighter Assist and Search Team (FAST) A company designated to search for and rescue trapped or lost firefighters. May also be called a rapid intervention team (RIT).

Hazardous Materials Technician An individual trained to meet the requirements of CFR OSHA 1910.120, *Technician Level for Hazardous Materials Response.*

Incident Management System (IMS) An organized, systematic method for the command, control, and management of an emergency incident.

Integrated Communications The ability of all units or agencies to communicate at an incident.

Modular Organization The ability to start small and expand if an incident becomes more complex.

Mutual Aid or Assistance Agreements Prearranged written agreements of the type and amount of assistance one jurisdiction will provide to another in the event of a large-scale fire or disaster. The key to understanding mutual aid is that it is a reciprocal agreement.

National Fire Protection Association (NFPA) A not-for-profit membership organization that uses a consensus process to develop model fire prevention codes and firefighting training standards.

NFPA 1001 *Standard for Fire Fighter Professional Qualifications,* a national consensus training standard establishing the job performance requirements of tasks to be performed by firefighters.

Operational Period The time frames for operations at an incident. At large-scale or complex incidents these will usually be eight- to twelve-hour time frames.

Paramedic (EMT-P) An individual trained and certified to provide advanced life support emergency medical care, including drug therapy.

Quint A combination fire service apparatus with components of both an engine company and a truck company.

Rapid Intervention Crew (RIC) See *Rapid Intervention Team.*

Rapid Intervention Team (RIT) A company designated to search for and rescue trapped or lost firefighters. Depending on location, may also be called a FAST.

Rescue Company The unit designation of a group of firefighters assigned to perform specialized rescue work and/or tactics and functions such as forcible entry, search and rescue, ventilation, and so on.

Rescue Specialist A firefighter with specialized training and experience in areas such as high angle rope rescue, confined space, trench, or structural collapse rescue.

Span of Control The ability of one individual to supervise a number of other people or units. The normal range is three to seven units or individuals, with the ideal being five.

Staging Part of the operations section where apparatus and personnel assigned to the incident are available for deployment within three minutes.

Standard Operating Procedure (SOP) Specific information and instruction on how a task or assignment is to be accomplished.

Strategic Goals The overall plan developed and used to control an incident.

Tactics The specific operations performed to satisfy the strategic goals for an incident.

Truck Company The unit designation of a group of firefighters assigned to perform tactics and functions such as forcible entry, search and rescue, ventilation, and so on.

Unified Command The structure used to manage an incident involving multiple response agencies or when multiple jurisdictions have responsibility for control of an incident.

Unity of Command One designated leader or officer to command an incident.

REVIEW QUESTIONS

1. Write a mission statement for a fire department including the level of services provided and to whom the services will be delivered.
2. Draw an organizational diagram for a fire department including the chain of command and who is responsible for training, fire prevention, apparatus maintenance, and fire scene operations.
3. List five functions or tasks performed by firefighters in a fire department.
4. Explain the reason for standard operating procedures.
5. List five areas addressed by standard operating procedures.
6. Explain where to find the rules and regulations for a particular firefighting organization and identify who is responsible for developing and enforcing them.
7. List five rules and regulations of a firefighting organization and describe how they apply to the firefighter.
8. Research a jurisdiction's mutual aid agreement and list the apparatus, specialized equipment, and personnel available for response.
9. List and define the five major components of an incident management system.
10. Describe duties and responsibilities of assuming and transferring command for the IMS used by a particular organization.
11. List and describe the functions of five allied agencies that assist a fire department.

Additional Resources

Diamantes, David, *Fire Prevention: Inspection & Code Enforcement.* Delmar Learning, a part of the Thomson Corporation, Clifton Park, NY, 1998.

Incident Command System National Fire Academy (available through state training agencies).

Incident Command System National Training Curriculum I-100 and I-200 courses (available through state training agencies).

International Fire Service Accreditation Congress, http://www.ifsac.org

Klinoff, Robert, *Introduction to Fire Protection.* Delmar Learning, a part of the Thomson Corporation, Clifton Park, NY, 1997.

National Board on Fire Service Professional Qualifications, http://www.npqs.win.net

Smoke, Clinton H., *Company Officer.* Delmar Learning, a part of the Thomson Corporation, Clifton Park, NY, 1999.

United States Fire Administration, http://www.usfa.fema.gov

CHAPTER 3

COMMUNICATIONS AND ALARMS

Willis Carter, Chief, Shreveport Fire Department

OUTLINE

- Objectives
- Introduction
- Communications Personnel
- The Communications Facility
- Computers in the Fire Service
- Receiving Reports of Emergencies
- Emergency Services Deployment
- Traffic Control Systems
- Radio Systems and Procedures
- Arrival Reports
- Mobile Support Vehicles
- Records
- Lessons Learned
- Key Terms
- Review Questions
- Endnote
- Additional Resources

Photo courtesy of Scot Smith, Smith Photographic, Shreveport, Louisiana

STREET STORY

Several years ago I had the unique opportunity to spend time working in the 9-1-1 communications center. It was a very eye-opening experience for me since I had always been a firefighter in the field. Spending time in the center allowed me to realize the role that communications personnel play in the overall fire and emergency services system.

As a firefighter, I would get frustrated when calling dispatch on the phone for alarm numbers and times and the dispatcher would say, "We are busy—can you call back?" I would think to myself, "How long can it take to give me times and a number? I have to get these reports done, and the dispatchers are holding me up." When responding to emergency calls, if the dispatch information was not completely accurate, the field personnel would often blame the dispatchers and communications personnel. We often did not consider that maybe the caller did not provide accurate information.

After spending time in the communications center, I realized a new appreciation for what communications personnel do on a daily basis. While it may be frustrating to be told to call back for information, I realized that dispatchers were dealing with units calling in from twenty-four other stations. While they were dealing with these phone calls, they also had to answer and dispatch 9-1-1 emergency calls, answer inquiries from the media calling about incidents, and of course talk to people calling 9-1-1 for nonemergency information. I also came to understand that the communications personnel often have difficulty in obtaining accurate information from callers. Many times this is because the callers are upset, frantic, or just do not have good information to pass on to the communications center. The information given to the responding units is only as good as the information available to the dispatchers.

My experience in the communications center taught me a great deal about the role of the communications personnel. They are truly the critical link between the public and the fire rescue system. They are a dedicated group who try to do the best job possible, even under difficult circumstances and situations. They take the 9-1-1 calls and dispatch the units and then try to anticipate the needs of the units throughout the incident.

Do not make the same mistake that I did. Visit your communications center and experience what they do for yourself. And when they ask you to call back because they are busy, just do it with a smile.

—Street Story by James Angle, Chief, Palm Harbor Fire Department, Palm Harbor, Florida

OBJECTIVES

After completing this chapter, the reader should be able to:

- Demonstrate the proper method of answering a nonemergency administrative call.
- Demonstrate the proper method of answering an emergency call and effectively obtaining full and complete information, and promptly relaying that information to the communications center.
- Demonstrate the proper method of operating a mobile radio.
- Demonstrate the proper method of operating a portable radio.
- Complete a basic incident report.

INTRODUCTION

"Nine-one-one. What is your emergency?"

"Oh, please send help! There has been a terrible accident!"

"What is the location of the accident?"

"It's at the intersection of Main and. . . ."

Terrifying emotionally wrenching scenarios such as this unfold thousands of times each day in **emergency communications centers** all across America. To get the information emergency responders need to do their jobs effectively, **telecommunicators** must be prepared to communicate effectively with citizens and relay accurate information to first responders.

In this chapter we discuss many aspects of fire service communications. It is safe to say that throughout history, effective communications have been essential to all successful endeavors. Public safety communications is certainly no exception to this. What is **communications**? Although communications takes on many forms, conveyance of information through the medium of speech is considered one of the most universal functions. It is estimated that interactions involving speech account for 74 percent of our communications time.[1] Yet effective communication is still a challenge for most of us in today's busy and complex workplace.

> **Note** From the standpoint of fire service emergency communications center operations, the communications process must include four basic elements: information from the caller must be *received, understood, recorded* accurately, and *communicated* to emergency responders.

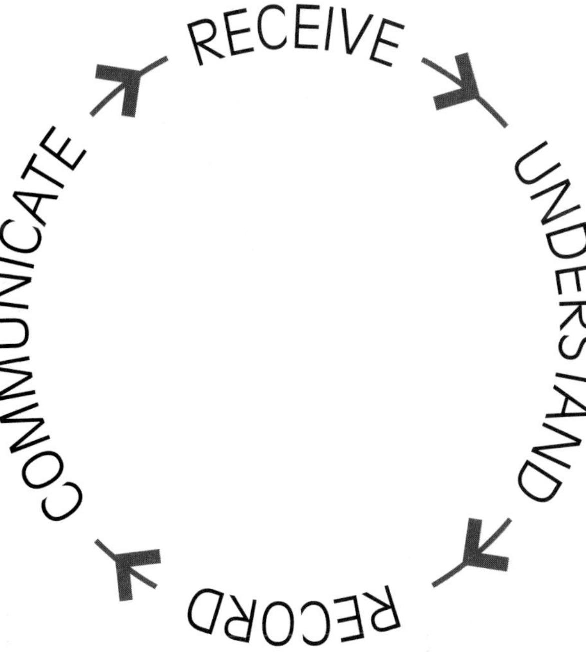

Figure 3-1 The communications process must be complete and clearly understood in order to be effective.

The communications process is illustrated in **Figure 3-1**. Reports from citizens can usually be categorized as true emergencies, perceived emergencies, nonurgent reports, or requests for information.

Despite our knowledge of the importance of complete communications, in many cases complete communication does not occur between individuals. Ironically, it has not been until recently that many fire service leaders have come to realize the importance of good, reliable communications and the impact that successful communication can have on the delivery of the services provided by their departments. As a result of a greater awareness of the need to maintain effective communications, many fire service leaders have started taking proactive measures to ensure that the quality of the communications process is maintained. These measures include:

1. Teaching communications skills to employees.
2. Upgrading communications systems.
3. Incorporating modern technology into the daily operational routines of their departments.

While it takes many special skills and various personnel to provide total service, the communications center is the heart of the operations. As a result, fire service leaders are putting more emphasis on having qualified staff managing and operating emergency dispatch systems. Due to this increased awareness, greater emphasis is being placed on the selection and training of the telecommunicators who staff communications centers in larger departments, and also for those smaller volunteer departments who rely on volunteers to perform these duties.

> **Firefighter Fact** A communications center may also be referred to in some instances as a PSAP, public safety answering point.

COMMUNICATIONS PERSONNEL

Although the names may have changed and the roles and responsibilities expanded, the basic role of the telecommunicator has remained remarkably the same. The primary role of the telecommunicator is to receive emergency requests from citizens, evaluate the need for a response, and ultimately sound the alarm that starts first responders on their way to the scene of an emergency.

Receiving calls is only the first challenge facing telecommunicators. This is the beginning of what can be a very complex process of gathering information and deploying emergency apparatus, personnel, and equipment. A myriad of work goes on behind the scenes of communications centers. Telecommunicators provide a variety of ancillary services to support field operations. In some cases a telecommunicator serves as the public relations officer for the department. After all, a telecommunicator is the first person to speak to the caller in need of emergency services. A telecommunicator is often the first person contacted by the news media when updates are required. Some departments issue departmental news and updates from their communications center, whereas others require that this type of information be distributed by the chief or the person acting as the public information officer. In either case, the news media can be the department's best friend or worst enemy; they should be handled with care.

In some instances local protocols may dictate that after initial contact by the citizen, the telecommunicator will maintain contact with the caller, enabling the telecommunicator to provide valuable **prearrival instructions**. As a result of the expanding role of the telecommunicator, fire service leaders are reconsidering the complexity of emerging technology and the importance of the *individuals* hired to operate these systems.

> **Firefighter Fact** A telecommunicator's job has been accurately defined by some as hours of boredom punctuated by moments of terror.

The NFPA 1061 standard, *Standard for Professional Qualifications for Public Safety Telecommunicator,* clearly outlines behavioral characteristics or traits that a person should possess to be a viable candidate for a position in the emergency communications center. Among these traits are the ability to perform multiple tasks, the ability to make decisions based on common sense and standard values, the ability to maintain composure in high stress conditions, and the ability to remember details and recall information easily. Candidates must also be able to exercise voice control, including the ability to maintain balanced tone, modulation, volume, and inflection while communicating. Only a tried, trusted, and complete training program will ensure that telecommunicators are armed with the knowledge and skills necessary to do the job.

A quality training program must be followed by a comprehensive work performance evaluation program to ensure fire service leaders that their communications centers are staffed by well-trained personnel who are able to apply the training they receive and develop the skills necessary to perform the required tasks. One organization that has been a leader in the field of communications training for many years is the **Association of Public Safety Communications Officials–International, Inc. (APCO)**. APCO has established minimum training standards that are being adopted by many states throughout the country. Initial certification, through a tested program coupled with a strong continuing education program, is paramount to the efficient and effective operations of a modern communications center.

An adequate staffing level at communications centers is the next most significant concern of fire service leaders. Most communications center managers will readily attest that there are never enough personnel when a **mass casualty** or **multiple-alarm incident** occurs. Communications managers rely on historical data to produce staffing models that closely resemble actual staffing requirements. By using some widely accepted formulas, managers are able to closely replicate staffing needs.

Note The NFPA states that "Ninety-five percent of alarms shall be answered within 30 seconds, and in no case shall the initial operator's response to an alarm exceed 60 seconds." It additionally recommends that "the dispatch of the appropriate fire services shall be made within 60 seconds after the completed receipt of an emergency alarm."

To meet this requirement, some departments with limited resources and low call volumes need only ensure that at least one person is on duty at all times to answer emergency phones. Regardless of size, fire departments must provide well-trained personnel to serve in the telecommunicator role because those individuals who answer calls from citizens have a direct impact on the overall response time of the agency.

THE COMMUNICATIONS FACILITY

The importance of staffing levels and having properly trained personnel in a communications center in order to process emergency calls within the time prescribed by the NFPA has been discussed. Now we turn our attention to the places where these telecommunicators work. If the facility is intended to house emergency communications operations, there are several important considerations.

Firefighter Fact The first national emergency telephone number, 999, was introduced in Great Britain in 1930.

Communications centers have many different configurations. Some are served by **9-1-1** systems, and some rely on a regular seven-digit telephone number to receive calls. Many serve only a single agency, whereas others have joint authority over police, fire, emergency medical services, and even utilities. Many communications centers are nothing more than a small alcove in a larger administrative office, with only one person responsible for communications duties for a single agency such as police. That same person might also book prisoners, answer phones, serve walk-up customers, and perform other combinations of job functions as assigned. Another example is a small volunteer fire department that has only one person on duty to answer calls and dispatch personnel. This is common in jurisdictions with low call volume activity. As shown in **Figure 3-2**, other facilities can be larger and constructed expressly for the purpose of serving as an emergency communications center. These are generally found in larger metropolitan areas with high volumes of emergency responses.

Regardless of the size or type of facility, all serve the one common goal of receiving and disseminating information that can be both emergency and nonemergency in nature.

The NFPA also has standards (NFPA 1221) for location and construction of emergency communi-

Figure 3-2 Exterior view of a 9-1-1 emergency communications center that houses fire/EMS, police, and sheriff communications operations. *(Courtesy of Caddo Parish 9-1-1)*

cations centers. Modern emergency communications facilities are typically constructed in areas where there is little risk of damage either by natural or man-made hazards. Communications centers should not be constructed in low-lying areas prone to flooding or along coastlines prone to hurricanes or high winds. In like terms, most agree that they should be located in areas of limited traffic and limited exposure to man-made hazards such as terrorism, arson, and so on. The NFPA suggests that communications centers that are built closer than 150 feet to existing structures should receive special attention regarding protection from collapse. Therefore, parks and other open areas are well-suited locations for these types of facilities.

Special attention should also be given to protecting the facility from vandalism and the effects of civil disturbance. Only authorized personnel should be permitted access to the communications center. Communications centers should have a limited number of outside windows, and all outside entrances should be monitored by security systems. The facility should also be monitored by a fire alarm system that complies with **NFPA 72,** National Fire Alarm Code.

Caution Every communications facility should be supported by a backup location in the event that the primary facility encounters problems that render it inoperable and result in evacuation.

Emergency communications centers should be equipped with backup power supplies that also meet the requirements set forth by the NFPA. An example of a backup power generator is shown in **Figure 3-3**. In some cases, backup power generators must be able to maintain the operations of a large communications center for prolonged periods of time. Some emergency generators are fueled by diesel fuel, which is stored on site as shown in **Figure 3-4**. It is also important to note that liquid petroleum or natural gas may fuel some power-generating equipment. Most backup power supplies are designed with electrical load distribution, which provides separation of power between emergency and nonemergency circuits. In this type of arrangement, where multiple generators are in use, they are typically installed with a switching device that automatically transfers the *emergency* loading to the first generator to achieve operating speed.

Uninterruptible power supplies (UPS) are also used in some of the more modern communications

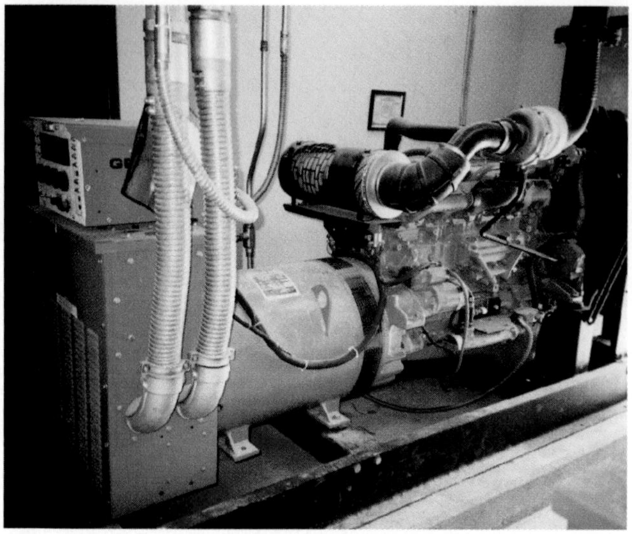

Figure 3-3 The emergency communications center is protected from power outages by use of backup power generators. Generators of this type can provide continuous power to operate the communications center during periods of power outages. *(Photo courtesy of Caddo Parish 9-1-1)*

centers. These units use battery power to provide uninterrupted power to the critical communications and computer equipment used in the communications center. Properly designed and regularly tested power backup systems that employ automatic switching devices that do not interrupt the opera-

Figure 3-4 A 500-gallon supply of fuel is maintained to ensure that uninterrupted power can be provided for a minimum of twenty-four hours, and for longer periods if necessary. Liquefied petroleum gas and natural gas (not shown) may also be used to power backup generators. *(Photo courtesy of Caddo Parish 9-1-1)*

Figure 3-5 UPS systems such as this help maintain uninterrupted service should a power outage occur.

Figure 3-6 Computer-aided dispatch workstations such as this one are found in many communications centers that have high activity levels.

tions are extremely important, especially when modern computerized systems are in use. A large UPS system is shown in **Figure 3-5**.

COMPUTERS IN THE FIRE SERVICE

The proliferation of new and more technically advanced computer hardware, along with the development of highly specialized computer software designed especially for use by public safety agencies, has made it possible for many departments to incorporate the use of computers into their communications systems. Many fire departments have found that the use of **computer-aided dispatch (CAD)** systems, **Figure 3-6**, has made it easier to handle increased call volumes.

The computer can keep track of the location of active incidents and which units have been assigned to respond and therefore help telecommunicators to better manage their resources. At the same time, computers can create and store valuable records on each incident and other departmental activities. Computers have greatly improved the way in which statistical analyses of departmental activity are performed. Additionally, computers can provide remote locations with access to a variety of information that is stored in a main **database** that can be accessed when needed. This information may include maps, hazardous materials information, prefire plans, policies and procedures, mutual aid instructions, or any other information that may be useful to firefighters. This type of information is often stored on **mobile data computers** in all types of fire apparatus and **command vehicles**. Computers also allow fire personnel to access off-site databases that may be beneficial either for training or incident mitigation.

RECEIVING REPORTS OF EMERGENCIES

The call-taking process consists of receiving a report, interviewing, and referral or dispatch composition. **Figure 3-7** provides a visual representation of the various stages of call processing. Whether a highly trained telecommunicator or a firefighter is performing this duty, speed is very important during the interview portion of the call-taking process. Telecommunicators must be able to prioritize incoming calls in order to ensure that the most important call gets the fastest attention. Incoming telephone calls should be answered in the following priority: (1) 9-1-1 and other emergency lines, (2) direct lines, and (3) business or administrative lines. Telecommunicators should answer incoming lines promptly and determine if the caller has an emergency. If not, and another emergency line is ringing, the telecommunicator should put the caller on hold and answer the other incoming call. However, speed is of little consequence if accurate and complete information is not obtained from the caller and relayed to emergency response personnel.

When receiving reports of emergencies by telephone, telecommunicators should always speak clearly and slowly with good volume. If questions to the caller have to be repeated, valuable time is

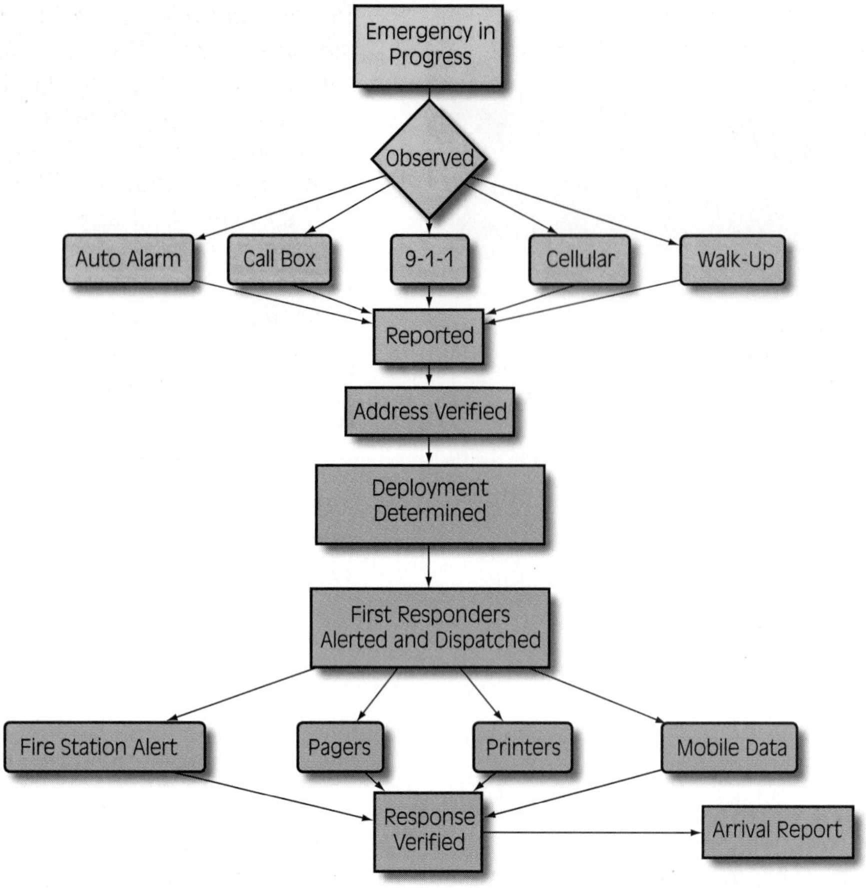

Figure 3-7 This figure illustrates the work flow of call processing by a public safety telecommunicator.

wasted. When speaking with a citizen who is reporting an emergency, the telecommunicator's voice should project authority and knowledge. The telecommunicator should use plain, everyday language at all times when speaking to the public and maintain a polite and friendly tone. It is never appropriate to argue with a caller. The citizen is the "customer."

Time is of the essence when answering and processing requests for service. The 9-1-1 and other emergency lines should always be the top priority. When answering incoming lines, the telecommunicator should always identify the department and ask, "Do you have an emergency?" or "What is your emergency?" It is critically important for the telecommunicator to control the conversation in order to obtain the necessary information in the least amount of time. In many cases the caller knows what has happened but finds it difficult to relay key elements of the situation to emergency service providers. The key information must be extracted from the caller, even in cases where a caller is hysterical or in severe emotional distress. A calm and authoritative tone of voice is helpful. In most cases, asking short, specific questions (name, address, telephone number, location) is the most effective tactic.

In the case of nonemergency calls, the telecommunicator should make every effort to accommodate the requests of the callers, which includes transferring them to another agency that may be better suited to provide assistance. It is important from a public relations standpoint to always offer to transfer the caller if the telephone system is capable of this function. Prior to transferring the call, the telecommunicator should provide the number to the caller in case the transfer process is interrupted or fails.

Streetsmart Tip A professional telecommunicator who speaks with confidence and authority will be able in most cases to gain the trust of the caller, who will then be more apt to follow the instructions issued by the call taker.

From time to time a caller may hang up, be disconnected, or simply drop the telephone before providing all of the needed information. This could be the result of a medical problem, an equipment failure, or simply the fact the caller realized he or she dialed a wrong number.

Caution Telecommunicators should diligently follow up on all incomplete calls.

How this is done varies greatly from one department to another depending on the type of telephone system available, but the basic fundamentals remain the same.

When processing an emergency call, telecommunicators should always attempt to obtain the following minimum information:

- Location of the emergency
- Nature of emergency
- Callback number
- Caller's location and situation

Without a doubt the most important piece of information is the *location* of the incident. *What* is happening is of little importance if the call taker doesn't know *where* to send help. Once the caller has provided the location, the telecommunicator should attempt to secure any additional information that the caller can provide with respect to landmarks such as cross-streets or easily recognizable buildings, if this can be done safely.

Life safety is of primary importance. Therefore, the call taker should always try to determine if the caller is in danger. If so, the call taker should provide prearrival instructions in accordance with local protocols.

Caution Getting and verifying callback numbers of the caller reporting the emergency is extremely important in the event it is necessary to recontact the caller in order to obtain additional information concerning the incident.

This information can be relayed to field units via radio as it is received and updated. Callers should be asked to provide all the information they can about an incident as long as they can do so *safely.*

In addition to a callback number, the caller's proximity to the location of the incident should be noted. This information is sometimes beneficial when attempting to locate incidents by following the directions of the caller. It is helpful if the first telecommunicator and the first responders are able to visualize the location of the caller in proximity to the locations of the incident. In most dispatch systems, once sufficient address and incident type information is verified, deployment of emergency apparatus and personnel can be initiated. As a rule, the average citizen will only find a need to report an emergency once in a lifetime.

Therefore, in emergency situations citizens may be very excited and unable to formulate and convey clear and concise information. A telecommunicator providing clear, calm instructions can reduce the normal panic associated with this activity and thereby effect faster and more accurate responses.

Streetsmart Tip The caller is the telecommunicator's set of eyes and ears at the emergency scene.

The information callers provide can be extremely valuable but only if the call taker is able to ask the right questions that will generate meaningful responses. After deployment of the appropriate responders, communications center personnel should retrieve any pre-incident plans, SOPs, hazardous conditions, or any other information that may be available on site at the communications center and relay all pertinent information to the emergency responders.

In the case of emergency medical calls, much more information may be requested from the caller in accordance with **emergency medical dispatch** protocols. Emergency medical dispatch systems are designed to meet the needs of specific jurisdictions by providing the telecommunicators with a set of standard questions and predetermined actions used to evaluate the situation. A response to an emergency medical incident must be based on a telecommunicator's ability to recognize and react to the "symptoms" being displayed by the victim, rather than an attempt on the part of the telecommunicator to make a "diagnosis" of the injury or illness.

Methods of Receiving Reports of Emergencies

Reports of emergencies can be received in a variety of ways. We briefly discuss some of the most common means in this section:

- Conventional telephones
- Wireless or cellular telephones
- Emergency call boxes
- Automatic alarms

- TDD equipment for the hearing impaired
- Still alarms or walk-ups

Receiving Reports by Telephone

Throughout the country, conventional residential and business telephones are the most commonly used method of reporting emergencies. However, the use of cellular telephones is becoming much more popular as they become available to larger cross sections of citizens, as discussed in the next subsection.

According to the **National Emergency Number Association** (NENA), nearly 93 percent of the population of the United States is covered by some type of 9-1-1, and 95 percent of 9-1-1 coverage is enhanced 9-1-1. Approximately 96 percent of the geographic United States is covered by some type of 9-1-1.

> **Caution** To reduce confusion, 9-1-1 should always be pronounced "nine-one-one," not "nine-eleven."

Incidents have been reported in which a panicked citizen has cited problems locating the "eleven" on the telephone touch pad. Regardless of whether **basic 9-1-1** or **enhanced 9-1-1** is used, these systems provide access to emergency services via the use of a simple, easy-to-remember three-digit number.

In addition to voice communication, enhanced 9-1-1 service provides emergency communications centers with the telephone number and address from which the call is originating. This decreases the time necessary to determine the caller's location when callers are either unwilling or unable to provide this information. This feature eliminates the need to trace or research telephone company records to associate an address with a telephone number. Again, features like this provide valuable assistance quickly and effectively. They are extremely important to emergency call takers, especially in cases where the caller is very excited, the caller is not familiar with an area, or for other reasons the caller is unable to communicate clearly the location of an emergency.

> **Caution** When considering the value of 9-1-1, also be reminded that the caller may not be calling from the address of the emergency, so address information must be verified.

Both basic and enhanced 9-1-1 service are available through conventional business and residential telephone service. In addition to conventional or wired telephones, wireless communications devices are growing more popular as a means of communicating emergencies.

Receiving Reports via Cellular Telephones

Wireless systems, such as cellular telephones, are proving very beneficial to the process of reporting emergencies. They are widely available to a large cross section of citizens and are very mobile. Any 9-1-1 calls initiated by a citizen using a cellular telephone are routed to a predetermined answering point for processing. In some areas the state highway patrol or state police is designated to receive all 9-1-1 cellular calls for a specific area or jurisdiction. In other areas of the country, 9-1-1 calls that originate from cellular telephones are routed to a communications center that may serve either a single agency or multiple agencies and even multiple jurisdictions.

Although the use of cellular telephones is proving very beneficial, it also has some negative points. The use of cellular telephones has caused significant increases in communications center call volumes. For example, on today's busy highways, one accident usually generates multiple reports by motorists using cellular telephones to access 9-1-1. Cellular callers are also more likely not to know their exact location, because they may be traveling along busy highways in an unfamiliar area. In these types of situations, landmarks and direction of travel are very important to the telecommunicator. In the case of small departments, the increase in the number of calls generated by cellular phones can have a budgetary impact. Also, in some cases, the telecommunicator will be criticized unfairly for not getting "accurate" information when in fact some of the things discussed earlier in this chapter are the cause for inaccurate information. Government-mandated upgrades in technology require cellular manufacturers to provide a means by which the location of wireless callers can be determined and provided to the emergency call taker. This greatly reduces the risk of location errors during the processing of emergency reports received via cellular phones.

New technologies created for automobiles provide citizens with specialized communication systems built into vehicles that will contact emergency services with the press of a button or when certain safety systems, such as air bags, are activated within the vehicle. This new wireless emergency reporting technology does not provide much information about the nature of the emergency; however, using satellite technology it can provide the communications center with an exact position of the vehicle reporting the emergency. As this system is used and tested, its popularity in new automobiles may grow in the future.

Receiving Reports via Municipal Fire Alarm Systems

Municipal fire alarm systems are those systems that allow a coded or voice message to be generated from an alarm box typically located in highly visible, easily accessible areas that are open to the general public. Systems of this type came into use during the late 1800s and many are still in use today with few upgrades and modifications.

> **Firefighter Fact** **Emergency call boxes** were first installed in the United States in 1852 and are still used in many parts of the country as a means of reporting emergencies.

According to the Boston Fire Department, the first emergency call box was placed in operation in Boston, Massachusetts, on April 28, 1852. This system is still in operation today and has seen only one upgrade since its installation. While Boston finds it beneficial to use a system of this type, some cities have discontinued them largely due to high rates of false alarms. As seen in **Figure 3-8**, call boxes can be operated via "hardwired" systems or they can be of the wireless solar-powered variety that can be installed in remote locations not serviced by electrical power or conventional telephone lines. An example of this type of technology is shown in **Figure 3-9**.

Some call boxes simply transmit a preset identification number to the communications center without providing a means for the reporting party to communicate verbally with the communications center. Others are equipped with signal switches that allow the caller to select the type of emergency being reported by pressing the appropriately labeled button, **Figure 3-10**.

Figure 3-8 "Fire alarm" boxes of this type came into being during the late 1800s and some are still in use today. *(Photo courtesy of Shreveport Fire Department)*

Figure 3-9 More modern technology is being used as solar-powered cellular call boxes go up in areas with limited access to power and communications networks. This type of call box supports voice communications with emergency communications centers. *(Photo courtesy of Caddo Parish 9-1-1)*

Figure 3-10 Some call boxes are equipped with signal switches that allow the caller to select the type of emergency being reported.

Receiving Reports via Automatic Alarm Systems

There are two types of public alarm systems as defined by the NFPA. A **Type A reporting system, Figure 3-11,** is defined by NFPA as "a system in which an alarm from a fire alarm box is received and is retransmitted to fire stations either manually or automatically." A **Type B reporting system, Figure 3-12,** is defined by NFPA as "a system in which an alarm from a fire alarm box is automatically transmitted to fire stations and, if used, to outside alerting devices." Properly designed and installed automatic fire alarm systems can be the key element to any building's overall safety. Automatic alarm monitoring systems are typically comprised of five common types:

1. *Local protective signaling system:* An alarm system operating in the protected premises.
2. *Auxiliary protective signaling system:* An alarm system that utilizes a municipal fire alarm box to transmit a fire alarm from a protected property to a fire communications center.
3. *Remote station protective signaling system:* An alarm system that connects a protected premise over leased telephone lines to a remote station such as a fire communications center.
4. *Central station protective signaling system:* An alarm system that connects a protected premise to a privately owned monitoring site that monitors the lines constantly for any indication of fire or other trouble signals.
5. *Proprietary protective signaling system:* An alarm system that protects contiguous or noncontiguous properties with common ownership from a location on the protected property.

Alarm systems typically consist of a system of sensors designed to detect smoke, heat, or a combination of both and also manual stations that can be activated by occupants in the event a fire is detected.

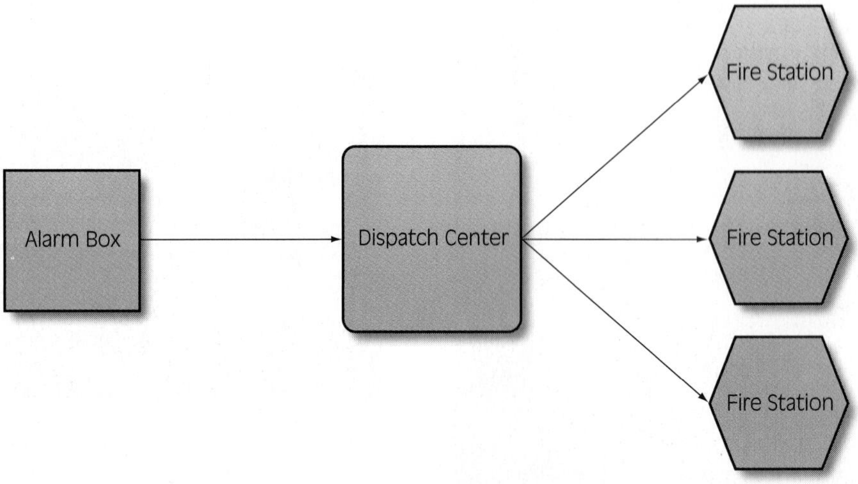

Figure 3-11 Type A municipal alarm systems typically transmit an alarm from a call box to a communications center where the alarm is retransmitted to emergency responders.

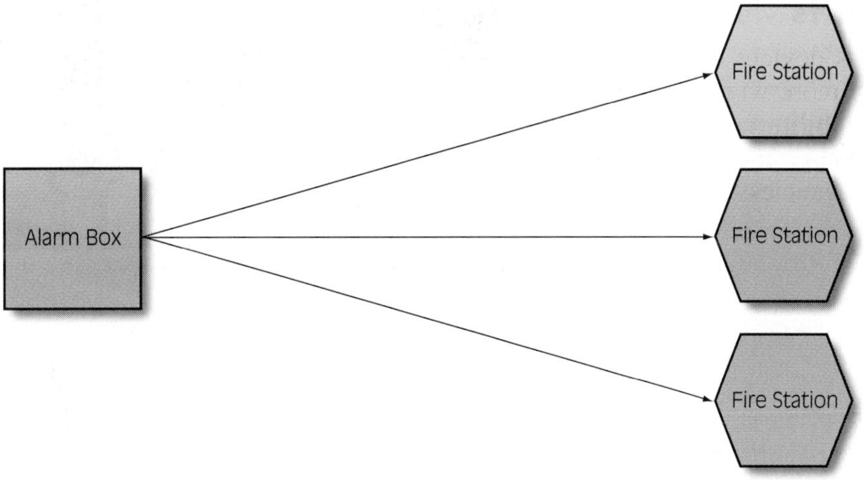

Figure 3-12 Type B municipal alarm systems typically transmit alarms directly from a call box to first responders.

Detecting devices protect specific and limited areas. Initiating the signal for any fire alarm system is accomplished by many means: manual pull boxes, heat detectors, and smoke detectors. **Figure 3-13** shows a manual pull box, and **Figure 3-14** shows both a heat and a smoke detector. Fire alarms are covered in greater detail later in Chapter 12.

Figure 3-13 Manual pull box stations are used to activate local alarms.

Figure 3-14 Heat detectors are designed to activate fire alarms and automatic extinguishing systems when ambient room temperatures reach certain levels. Smoke detectors monitor air quality and can also activate automatic alarm systems. This photo shows both in a typical installation.

Receiving Reports via TDD

Telecommunications devices for the deaf (**TDDs**) are becoming much more widely used by individuals with hearing disabilities. The **Americans with Disabilities Act** (ADA) entitles citizens to equal service from public agencies.

> **Firefighter Fact** ADA regulation 35.162 states: "The legislative history of Title II specifically reflects congressional intent that public entities must ensure that telephone emergency services, including 9-1-1 services, be accessible to persons with impaired hearing and speech through telecommunication technology."

Therefore, communications centers are required to remain ready to receive calls via specialized equipment designed to allow citizens to communicate with telecommunicators through the use of a keyboard using text messages instead of voice communications, **Figure 3-15**.

Figure 3-15 Devices such as this assist public safety agencies in communicating with citizens who have hearing impairments.

Receiving Reports via Still Alarm or Walk-Ups

From time to time citizens may report an emergency directly to the personnel at the fire station. Receiving complete and accurate information is just as important in these cases as those that are received at the communications center by telecommunicators. The section titled Receiving Reports of Emergencies (earlier in this chapter) provides information that can be applied to deal with a citizen who is reporting an emergency in person. **Figure 3-16** shows a firefighter relaying information to the communications center that was received from a citizen who actually stopped at the fire station and reported an emergency. Although the protocols of different departments may vary with respect to how these types of reports are handled, it is important to always obtain as much information as possible from the person reporting the emergency. Once this is done, and following departmental protocols, the person taking the report should immediately activate the emergency notification system and relay all of the information to the communications center. How specific emergency notification systems are activated is covered in departmental protocols. The same is true in the case of notifying the communications center that a report of an emergency has been received. Some departments may use **ringdown circuits**, **base radio**, or **mobile radio** to communicate with their communications center. It is very important for firefighters always to notify the communications center of location and assignment when reports are received directly from citizens. Telecommunicators can then ensure that any necessary support and assistance are provided.

Figure 3-16 A firefighter relays information from the fire station to the communications center via direct telephone circuit.

EMERGENCY SERVICES DEPLOYMENT

Once an emergency is recognized and subsequently reported or relayed to the communications center, the next step is to determine what action must be taken. A variety of methods is used to accomplish this throughout the country; however, some elements of the process are essential regardless of jurisdiction or geographic location. Identification of the situation, address verification, and unit selection must occur in order for telecommunicators to deploy the appropriate types and numbers of emergency responders.

As stated earlier, the most important information to obtain from the caller is the address. However, to deploy the most effective emergency response, the telecommunicator must also know the nature of the emergency. Emergency response organizations typically identify the most common types of situations and preassign a standard response to each such situation. An example would be a fire department that predetermines the routine response to all single-family dwelling fires will be three engines, one ladder, and one district chief. Based on this plan, the telecommunicators have baseline criteria from which to develop a **deployment plan**. In the modern fire service, deployment plans are based not only on apparatus types, but also take into consideration what equipment is carried on the apparatus, the number of personnel, and their skill levels.

The process of deploying apparatus and personnel varies greatly from one department to another. For instance, in smaller communities with a low volume of emergency response activity, a manual **run card system** may be sufficient to manage the emergency response deployment process. Manual run card systems, similar to the one shown in **Figure 3-17**, are typically comprised of a card file containing street and location information relating to a jurisdiction and predetermined unit assignments for each location.

As discussed earlier, jurisdictions with high volumes of activity may require automated systems to assist in the deployment process. CAD systems are widely used today and provide a very sophisticated method of assessing resources and making recommendations for deployment of equipment and personnel.

Regardless of which method of deployment is used, whether it is accomplished manually or through the use of a CAD system, a predetermined deployment of apparatus must exist. An example of a deployment table is shown in **Figure 3-18**.

The basic elements of the deployment process remain the same in either the manual or automated systems: Verify the location and nature of the emergency and determine what resources are available.

> **Firefighter Fact** Enhanced 9-1-1 systems provide number identification and location information even in those cases where computer-aided dispatch is not being used.

STREET NAME: Delmar Ave. BOX NUMBER: 1128

BLOCK RANGE FROM: 1000 TO: 1500

BLOCK	ENGINES	TRUCKS	RESCUE	D/CHIEF	A/CHIEF	MEDIC	OTHER
1000	1, 5, 4, 7, 10	1, 7, 10	1, 9	1, 2, 3	1	1, 4, 5	
1100	1, 5, 4, 7, 10	1, 7, 10	1, 9	1, 2, 3	1	1, 4, 5	
1200	1, 5, 4, 7, 10	1, 7, 10	1, 9	1, 2, 3	1	1, 4, 5	
1300	1, 4, 5, 7, 10	1, 7, 10	1, 9	1, 2, 3	1	1, 4, 5	
1400	1, 5, 4, 7, 10	1, 7, 10	1, 9	2, 1, 3	1	1, 5, 4	
1500	1, 5, 4, 7, 10	1, 7, 10	1, 9	2, 1, 3	1	1, 5, 4	

STREET NAME: Delmar Ave BLOCK RANGE FROM: 1000 TO: 1500

Figure 3-17 Run cards show the response assignment of a variety of apparatus to specific street addresses.

TYPE	DISP. CODE	DESCRIPTION	RESPONSE
Fire	01	Single Company Response	1E
	11	Structure Fire/Alarm (Residential)	2E-1D-1R-2T
	21	Structure Fire/Alarm (Commercial)	3E-1D-1R-1T
	51	Hospitals	4E-2D-2R-2T
EMS	08	Medical (Noncritical)	1E-(COLD)
	18	MVA	1M-1E

Legend: E — Engine or Pumper
D — Chief Officer
R — Rescue Unit
T — Truck or Aerial
M — Medic Unit
(Cold) — Nonemergency

Figure 3-18 Deployment tables such as the example shown are used to identify the appropriate apparatus response for the type of dispatch.

Once telecommunicators receive a report of an emergency situation, verify the location, and determine the appropriate deployment scheme, the next step is to notify the emergency responders. As is the case with deployment plans, this process also varies greatly from agency to agency. Volunteer departments may rely on personal pagers or use an automatic telephone system that "rings" multiple telephones on a common circuit and, in some cases, a system of sirens to alert them of an emergency. **Figure 3-19** shows a volunteer wearing a personal pager.

A variety of **home alerting devices** are also available. These are used by some agencies that operate either an all-volunteer or combination volunteer and paid department. A common example of this type of alerting device is shown in **Figure 3-20**.

In departments where fire stations are staffed twenty-four hours a day, some type of **fire station alerting system** is usually employed. Fire station alerting can be accomplished in a variety of ways, but should always comply with NFPA standards. In some systems a voice message is transmitted from the communications center to a fire station via the use of a vocal alarm system. These systems typically operate either via some type of control unit connected to leased telephone circuits or a radio transmitter. **Figure 3-21** shows a receiver that would be located at the fire station. The telecommunicator, using the control device, which may be similar to the **encoder** shown in **Figure 3-22**, decides the appropriate fire stations to notify and activates the system. Normally some type of distinctive tone is transmitted via a public address system within the fire station to alert personnel of an incoming message.

This alert tone is followed by voice instructions over the PA from the telecommunicator. Some fire station alerting systems perform additional functions such as turning on selected lights in the fire station, opening apparatus bays, turning off appliances, and controlling traffic signals. Some systems are capable of a *zoned* alert that can notify only specified areas of a station such as those stations that house both fire and EMS units and personnel.

Departments that utilize CAD systems can enhance this method by installing "tear and run" printers in each fire station. These printers provide the responders with a hardcopy printout showing details of the incident and location. A printer of this type is shown in **Figure 3-23**.

Some departments also employ mobile data terminals and mobile data computers in their dispatch and deployment process. These units allow dispatch information to be transmitted directly to the apparatus on a display screen or, in some cases, directly to mobile printers. Modern communications equipment such as this can provide two-way confidential information flows between the communications center and emergency responders. **Figure 3-24** shows a **mobile**

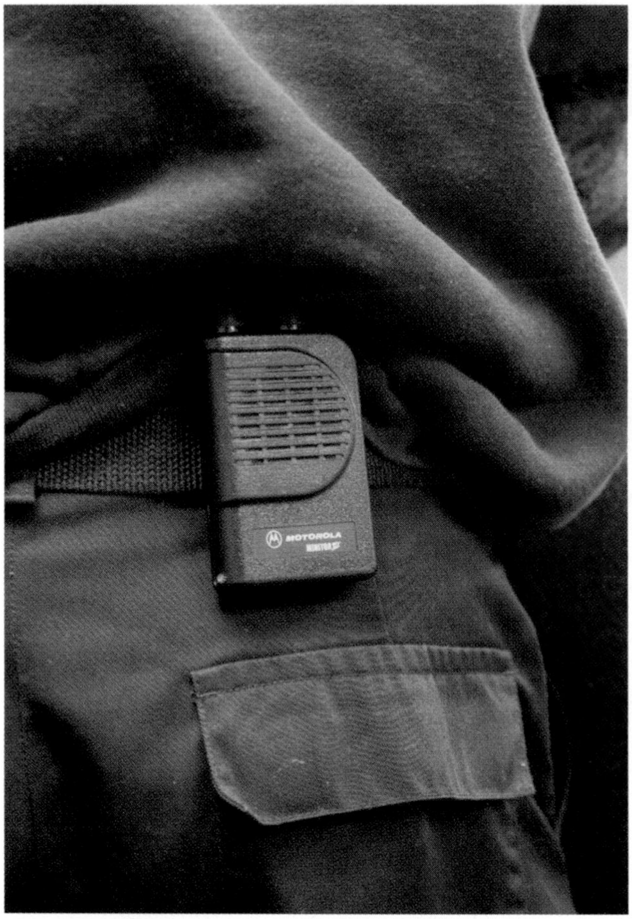

Figure 3-19 Some departments use personal pagers to alert personnel of the need to respond to emergencies.

Figure 3-21 This is yet another device that is used to alert emergency responders. This unit is equipped with relays that activate station lights and open station doors in preparation for a response. *(Photo courtesy of Shreveport Fire Department)*

Figure 3-20 This device is typically found in fire stations and also in some private homes and is used to call out emergency responders.

Figure 3-22 Encoders such as this are used to control many paging systems.

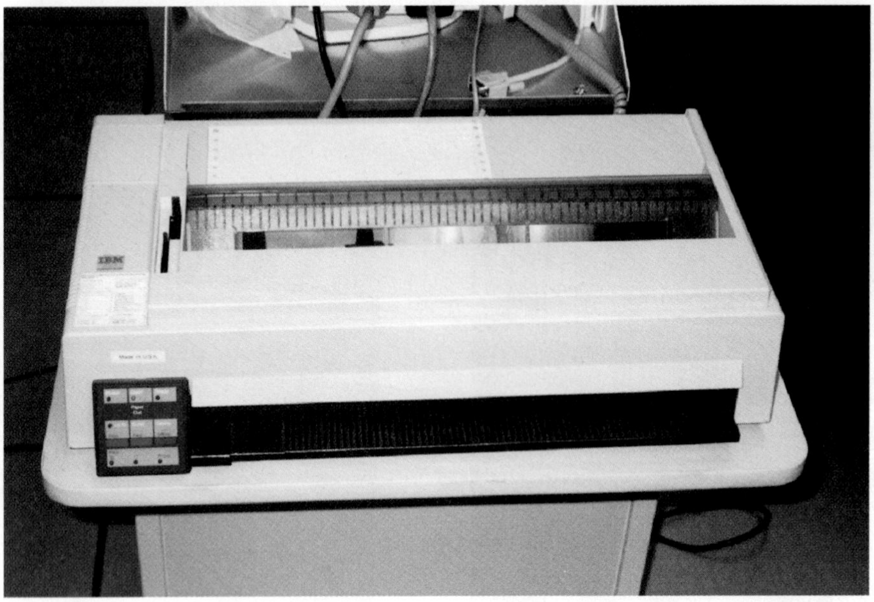

Figure 3-23 Printers such as these are used to relay incident information to first responders at the time of dispatch. They are typically referred to as "tear and run" printers. *(Photo courtesy of Caddo Parish 9-1-1)*

Figure 3-24 Mobile data terminals such as this one are used by some departments and provide information to emergency responders while en route to incidents. Unit status can also be managed with systems of this type. *(Photo courtesy of Shreveport Fire Department)*

data terminal widely used in the late 1980s and mid- to late 1990s, and **Figure 3-25** shows a more modern mobile data computer, which is replacing the older model mobile data terminals in some departments.

Regardless of the type of fire station alerting system in use, it is critically important that accurate information pertaining to an emergency situation be transmitted from the communications center to first responders in a clear, concise, and expedient manner. Again, speed and accuracy are of the utmost importance.

Streetsmart Tip Remember that the response times of first responders are directly impacted by the amount of time required for telecommunicators to receive information, verify location, determine deployment, compose a dispatch, and transmit this information to first responders.

TRAFFIC CONTROL SYSTEMS

Caution Thousands of collisions resulting in tragic accidents involving emergency response vehicles are evidence of the dangerous nature of emergency responses.

Figure 3-25 Mobile data computers such as this one are replacing the less powerful mobile data terminal as shown in Figure 3-24. Units such as this are capable of storing information on board the apparatus for easy retrieval. *(Photo courtesy of Shreveport Fire Department)*

Careful planning and use of the technology described in this section can create a safer emergency response path without undue or prolonged disruption of normal traffic flows. However, there is no better way for emergency responders to reduce traffic accidents than by exercising prudent judgment and applying safe driving practices at all times.

To speed the response of emergency responders as well as reduce stress and increase the safety of both emergency responders and the general public, some jurisdictions utilize various types of emergency preemption systems to control traffic signals and provide a safe transition to a priority right-of-way for emergency vehicles. Systems of this type are designed to recognize an emergency response vehicle and actually allow it to change the traffic control signals on its route to allow clear passage for the emergency responders. A variety of these systems are in operation today and each uses slightly different technology. Some operate using radio-frequency while others communicate with remote signal detection devices through the use of microwave or laser technology mounted on the vehicle.

RADIO SYSTEMS AND PROCEDURES

Once apparatus and personnel are deployed to emergency situations, the function of fire communications personnel then becomes that of providing support for the field units deployed. The primary link between the communications center and field units is the radio system. The use of radios in the fire service serves to carry both verbal and digital messages. Radio systems have various components; however, every system must have at minimum a base station and antenna capable of transmitting at a power or signal strength necessary to provide coverage to all parts of a jurisdiction.

The radio-frequencies that have commonly been used by the fire services are VHF low band frequencies, 33 to 46 MHz; the VHF high band frequencies 150 to 174 MHz; and the UHF frequencies, 450 to 460 MHz. These frequency ranges have provided reliable fire service communications for many years. However, as the result of growth, the fire service has experienced severe difficulties when attempting to add frequencies to their radio systems. The **Federal Communications Commission (FCC)** closely monitors frequency allocations, and in some cases additional frequencies in the ranges mentioned are simply not available. As the result of a need for additional frequency spectrum, other frequencies have been approved for use by the fire service. The 800-MHz frequency range is being used successfully by some departments for voice communications. Voice communications via the 800-MHz frequency range are limited by the FCC to systems using trunking technology. This type of radio system is discussed next.

One common radio system used by fire departments is a simplex system that uses only one frequency to transmit outgoing messages and to receive incoming messages, **Figure 3-26**. The advantage is simplistic design, resulting in decreased system cost. The primary disadvantage of systems of this type is the limited range and interference between multiple

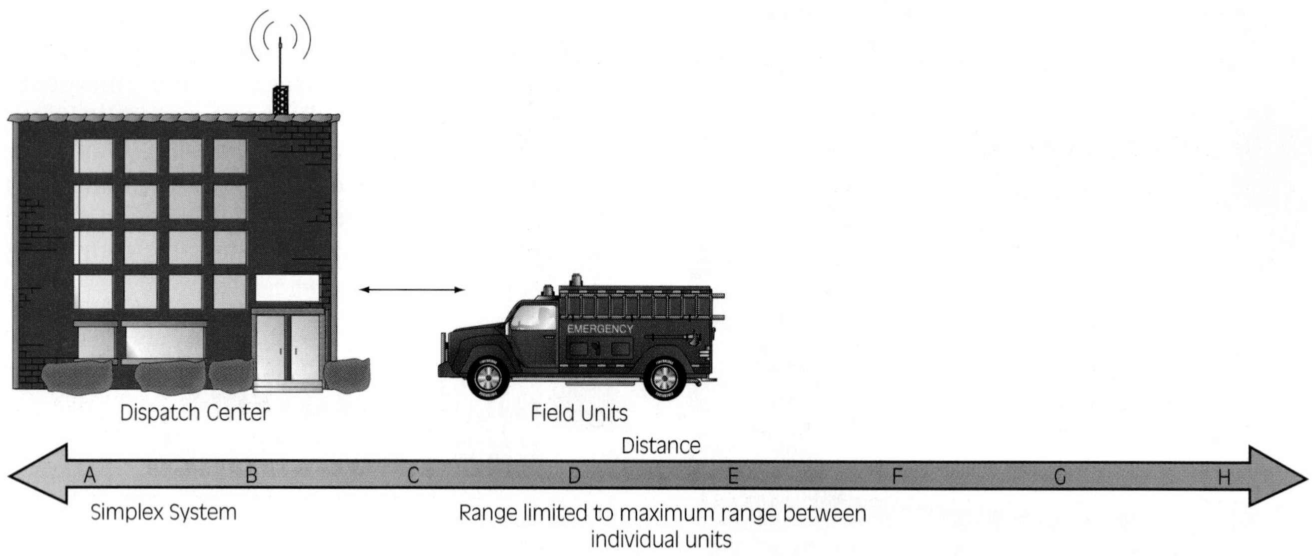

Figure 3-26 Simplex radio system designs such as this one are reliable and relatively inexpensive to install. However, they are limited with respect to range of operations.

units in the same system attempting to access the base station simultaneously.

A more complex system, shown in **Figure 3-27**, is a duplex system that uses two frequencies per channel, transmitting outgoing messages on one and receiving incoming messages on the other. This system uses base repeater stations whereby a fixed control station transmits an outgoing message that is received at a repeater site and retransmitted to mobile and portable units in the field. The benefits of systems of this type are more range and the elimination of the self-interference found in simplex systems. The disadvantages are more complex system design, the need for multiple frequencies, system cost, and the ongoing maintenance costs associated with the system.

Multisite trunking systems as depicted in **Figure 3-28** use computer processors that make the most efficient use of radio spectrum. Multiple transmitters operating on different channels are controlled by microprocessors that sense available channels and reallocate their use as needed. In duplex systems of this type, user transmissions are transmitted on one frequency, received by the field unit on another, and retransmitted back to the communications center on the other. The user does not notice the fact that the system is "changing" frequencies with each transmission. However, this more efficient allocation of radio-frequency resources allows the use of fewer frequencies by individual agencies. Several agencies can operate

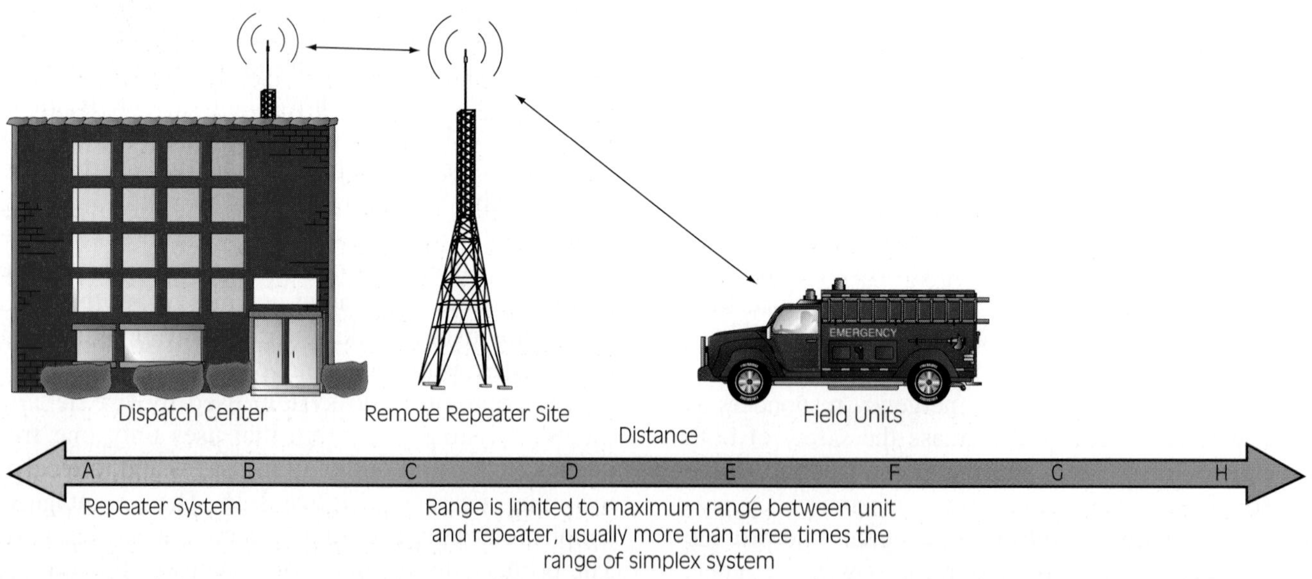

Figure 3-27 A slightly more advanced duplex design using multiple transmitters extends the operating range of the simplex system shown in Figure 3-26.

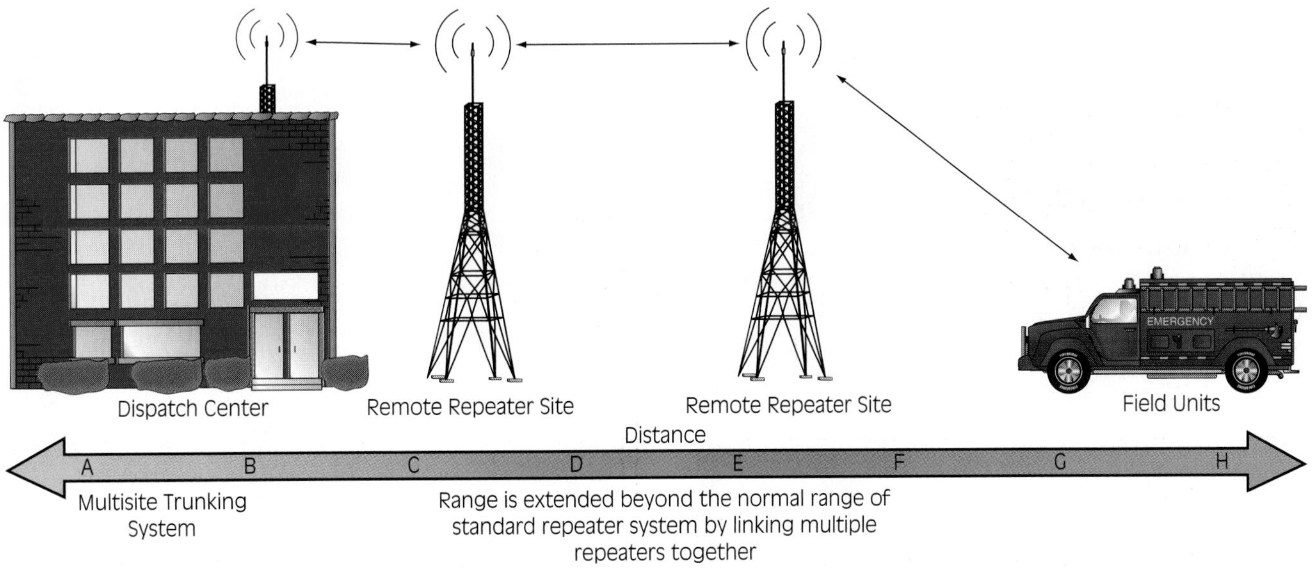

Figure 3-28 Multisite trunked radio systems provide perhaps the best coverage and also offer direct benefits associated with the most efficient use of radio resources.

simultaneously on the same trunked radio system and not interfere with each other. The benefits of systems of this type are expanded range, more efficient frequency use, and the ability of multiple agencies to operate on one system. Disadvantages are more complex system design and overall higher system cost.

> **Streetsmart Tip** The proper operation of the radio system, regardless of type, is of primary importance if information to and from the scene of the emergency is to be relayed in a timely and accurate manner.

Proper radio discipline is very important during active incidents.

> **Caution** On-scene personnel must listen before they talk. Routine or nonemergency traffic should not interfere with emergency communications. Transmissions should occur when airwaves are clear.

When using any nontrunked two-way radio, it is important to depress and hold the "push to talk" button at least two seconds before talking to avoid **clipping** the first part of the message. The same is true at the end of the message. The user should always pause briefly before releasing the "push to talk" button to avoid clipping the end of a message. Before keying any radio, users should know what they are going to say. The golden rule is to be brief but be concise. Firefighters should avoid touching any radio antenna during transmission to avoid burns that can result from radio-frequency energy.

> **Firefighter Fact** Some radios are equipped with a time-out feature that forces brevity by interrupting transmissions that are longer than the preset duration allowed. This feature also prevents long-term open microphones from accidental keying.

The user should never attempt to operate a radio while eating or chewing anything and should never use slang or jargon when transmitting a message over the radio. The message must be clearly understood. During large-scale incidents, radio economy is important. When using mobile radios, the user should speak clearly (not shout) across the microphone, as shown in **Figure 3-29**, as opposed to speaking directly into the microphone as pictured in **Figure 3-30.**

When using portable radio equipment, the unit should be held perpendicular to the ground with the antenna facing skyward to allow for better radio wave distribution. The user should speak across the microphone as shown in **Figure 3-31**. The use of a portable radio that is located in a "radio pocket" or belt clip impairs the performance of the unit. **Figure 3-32** shows the improper position of the portable radio when transmitting. In transmitting on both mobile and portable radios holding the microphone one to two inches from the mouth and at a 45° angle is important for a clear transmission.

Figure 3-29 This figure shows the proper use of a mobile radio microphone.

Figure 3-31 In this figure, the user has positioned the portable radio properly and is speaking across the microphone.

Figure 3-30 Improper use of a mobile microphone. The radio microphone should not be held directly in front of the operator's mouth when transmitting.

Figure 3-32 This figure shows the improper positioning of a portable radio.

Microphones should never be left on the seat or in other locations where they might accidentally be keyed up. Microphones and portable radios should be placed in their appropriate storage locations or in a protected location. Local agency protocols and policy will dictate how radios are to be operated in a particular jurisdiction. However, the following is a simple illustration of the "call up" method that is used by some agencies:

Telecommunicator:	"Engine One, this is dispatch."
Engine One:	"Dispatch, this is Engine One."
Telecommunicator:	"Engine One, respond to 1234 Main Street for a trash fire."
Engine One:	"Dispatch, Engine One is responding for a trash fire at 1234 Main Street."
Engine One:	"Dispatch, Engine One is on the scene and establishing command. We have a small trash fire in the rear of the building, no exposures, no assistance needed."
Telecommunicator:	"Engine One, dispatch copies."

Ten codes are a set of radio signals preceded by the number "10" that make up a predetermined message. Considerable debate exists as to whether radio codes or clear speech should be used, but in most cases that decision is made based on the needs of each specific agency. Use of radio signal codes provides a somewhat more confidential and cryptic means of communicating. However, such codes must be learned and remembered and are not usually standardized from department to department, making communications during multijurisdictional responses somewhat problematic.

Clear speech, on the other hand, is exactly what the name implies. Clear speech is used to convey information and issue instructions. The use of clear speech eliminates much of the confusion associated with the use of radio codes and, although somewhat lengthier, in most cases is easily understood by all. However, even in the "clear speech" environment, the same phrase may carry a different meaning from agency to agency. For example, "in-service" for some agencies means that a crew is *going to work at an incident*. On the other hand, in some agencies "in-service" means that the crew is *available* for an assignment. Thus, it is important not to assume anything during radio communications.

Another important function associated with the radio systems is the issuance of emergency evacuation signals to on-scene personnel who may be subject to imminent danger. Some departments use the radio system to transmit electronic *tones* that are intended to attract the attention of firefighters and alert them of the need to evacuate to a safe area. This works relatively well in most cases. However, a typical emergency scene is very noisy and not everyone on the scene has a portable radio. Another system used by some departments employs the use of apparatus air horns. Three bursts on the air horn indicate the need to evacuate. Regardless of the type of system used, it is critically important that firefighters learn it and be familiar with how it is used. Firefighter safety is paramount in all operations.

The use of two-way radios has grown greatly over the years. Routine administrative traffic, which is necessary in the daily operations of a department, should never be allowed to interfere with emergency operations. In departments that have access to multiple radio channels, emergency operations should be assigned to a separate channel dedicated for use on that scene only. This greatly improves the ability for incident commanders to communicate with on-scene personnel and also minimizes the threat of interference from some other source.

Radios and radio systems are evolving as the advances in technology grow. Newer radio systems can identify radios assigned to a particular apparatus to help in tracking transmissions. As a result of the technology, some radios incorporate an emergency feature that alerts the emergency communications center when a crew is in trouble or encountering an emergency. Firefighters should spend time familiarizing themselves with the radio system in use in their department and the many features associated with the equipment and the system.

Note In summary, firefighters must consider several important factors when transmitting messages across mobile or portable radios. First, the information provided to an emergency communications center or other emergency service units must be accurate, clear, and complete. The radio transmissions must also be within the time parameters established by the department or local jurisdiction if such policies and procedures exist. Firefighters should consult their department or jurisdiction's policies on time parameters for radio communications, specific terminology, and proper designations for units operating on a radio system.

ARRIVAL REPORTS

Until field units arrive on the scene, for all practical purposes, the communications officer is the incident commander. The first unit arriving on the scene should establish command and give a brief arrival report.

Note The initial incident commander remains in command of the incident until command is either assumed by a higher ranking officer or transferred to another officer.

This arrival report contains pertinent, but brief, information about the on-scene conditions. The arrival report should be given using clear, precise, language. The report should contain, at minimum, the following information: (1) the correct address, (2) a situation evaluation, (3) where the emergency is located in the building, (4) some information about the building as well as its potential occupants, (5) a request for any other agency support such as law enforcement, (6) the location of the on-scene command post, and (7) the identity of the incident commander. Arrival reports may also contain a very brief action plan for the incident.

In the case of most routine emergency medical incidents, this much detail is not necessary. However, in the event of a major accident or other incidents involving multiple patients the same type of detailed information should be provided along with any additional information such as a quick assessment of the number of patients and general types of injuries involved.

Status reports are very important to the overall success of any major scene operations. Most fire command officers agree that the first status report from the field should be made no longer than ten minutes into the incident, and every ten to fifteen minutes thereafter until the situation is brought under control. Communications centers may implement SOPs that call for "time marking" incidents at important thresholds for improved documentation and reporting purposes. Some CAD mobile data computer systems perform this function automatically.

MOBILE SUPPORT VEHICLES

Major events involving fire and EMS sometimes call for the use of **mobile support vehicles** (MSVs) or mobile communications units. These vehicles greatly enhance the overall effectiveness of the communications system in use at the scene of major incidents. Coordination of the communications process is absolutely necessary in order to manage large-scale operations effectively. MSVs provide an on-scene command post from which operations can be directed. The need to deploy vehicles of this type is usually determined by the size of the incident and the projected duration of activities. MSVs are normally highly specialized vehicles designed exclusively for use as on-scene command posts. MSV size depends greatly on the jurisdiction that it serves. The vehicle should have radios that allow communications on all of the frequencies that may be in use by the command agency and any additional jurisdictions that may be called in for mutual aid. Telephone service to vehicles such as these can be provided by the local telephone company via temporary connections or through the use of cellular technology. The MSV should be equipped to operate on both battery power as well as standard 120-V current. **Figure 3-33** shows an example of an MSV that is the result of modifications to a public transit bus. Conversions such as these, although often time consuming, are for the most part a much less expensive way to incorporate the use of this type of vehicle into the operations of a department. Custom-built units from the factory are also available, as shown in **Figure 3-34.** This particular unit was custom built to serve both fire and police operations through a cooperative agreement between the services.

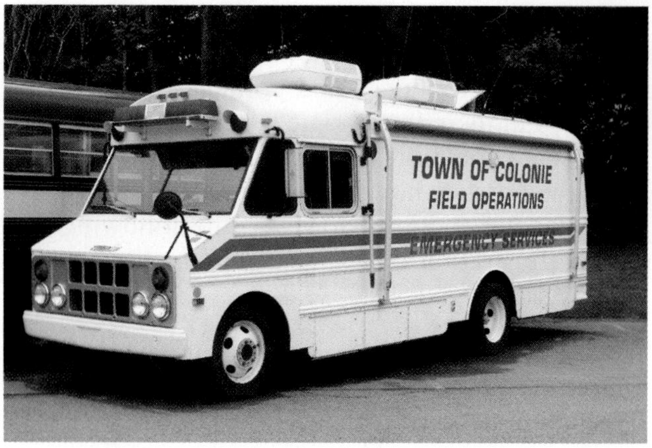

Figure 3-33 The vehicle pictured here is the result of a conversion project performed on a city transit system bus. The resulting mobile support vehicle cost much less than a custom-built factory unit.

RECORDS

Complete and accurate communications center records should be maintained on all responses. It is considered routine practice in most communications centers to record all emergency telephone and radio traffic either in some type of manual log book or on magnetic or digital recording devices for future reference. Examples of both a tape logging device and a digital recording device are shown in **Figure 3-35** and **Figure 3-36**, respectively.

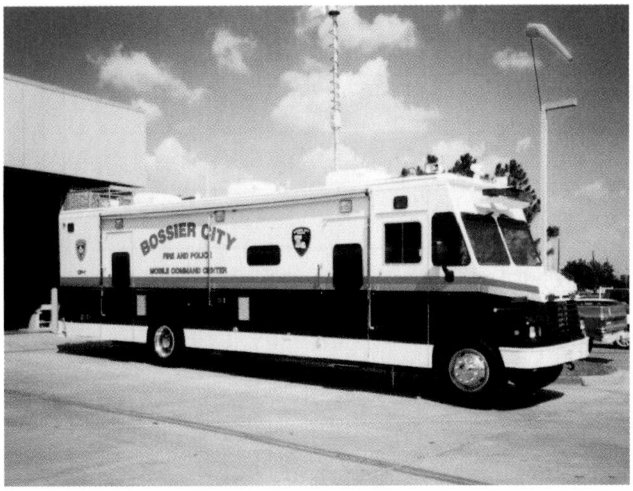

Figure 3-34 Unlike the vehicle shown in Figure 3-33, the vehicle shown here is a custom-built mobile support vehicle. *(Photo courtesy of Bossier City Fire Department)*

Figure 3-35 Magnetic tape logging recorders are used by some communications centers to record telephone and radio activity. *(Photo courtesy of Shreveport Fire Department)*

Figure 3-36 Digital logging recorders such as this one are taking the place of magnetic tape loggers in some modern communications centers.

In most states fire reports are considered public record and as such are available to newspaper reporters, insurance adjusters, lawyers, and others. Records of this type speak for the department. A well-written report will be very valuable when years later a firefighter is called to testify at a deposition regarding legal action resulting from an incident.

The following is the minimum information that should be recorded and maintained as the legal record of an incident. This information can be used to complete a more detailed National Fire Incident Report or other appropriate reports approved for use by a particular state or jurisdiction.

- *Time call received:* The time the telecommunicator received the call from the caller.
- *Units dispatched:* The unit numbers of the units dispatched to the incident.
- *Dispatch times:* The time of the initial dispatch and the times of any additional units subsequently dispatched.

Inc. Number:	98Ø7670	Under Control Time:	10:20
Address:	1234 Delmar Ave		
Location:	CITY HALL	Dispatch Code:	
Complaint:	FIRE ALARM (21)		
Reported By:	JOE SMITH	Call Back Number:	555-5555

UNIT	I/ROUTE	ARRIVE	TO HOSP	AT HOSP	COMP	I/SERVE
EØ1	1005	1007			1025	—
EØ4	1005	1008			1020	—
EØ5	1005	1006			1020	—
DØ1	1005	1015			1020	—
TØ1	1005	1007			1020	—
RØ1	1005	1007			1020	—

Figure 3-37 A manual incident card that is used to maintain a record of event history.

- *Arrival times:* The arrival times of all units at the scene and the unit initiating the incident management system.
- *Command post information:* Commander, location, and so on.
- *Requests:* For example, for additional fire department units or for other agencies or services.
- *All clear time.*
- *Under control time.*
- *Back in service times for all units.*

Figure 3-37 shows an example of a "manual" dispatch card that is used to record incident information. This type of logging information is typically found in smaller departments and also as backup systems for departments that rely on CAD and other automatic systems.

Lessons Learned

Whether a telecommunicator in a large emergency communications center or a firefighter answering an emergency call at a fire station, the manner in which calls are answered and information processed has a direct impact on citizens' impressions of the department. In the case of emergency incidents, information that is accurately collected and rapidly transmitted to first responders is paramount to the successful resolution of the incident. The ability to answer incoming calls quickly, gain control of the call, and, in some instances, calm the caller is a very important aspect of this process. Remember that the telecommunicator is the first person "on the scene." Utilize the caller's eyes and ears. A telecommunicator's knowledge and skills, combined with the ability to make wise use of all available resources, will play a vital role in the successful outcome of an emergency incident.

KEY TERMS

9-1-1 Emergency telephone number that provides access to the public safety services in the community, region, and, ultimately, nation.

Americans with Disabilities Act Public law that bars discrimination on the basis of disability in state and local services. Enacted in 1990.

Association of Public Safety Communications Officials–Int., Inc. (APCO) International not-for-profit organization dedicated to the advancement of public safety communications. Membership is made up of public safety professionals from around the world.

Base Radio Radio station that contains all of the antennas, receivers, and transmitters necessary to transmit and receive messages.

Basic 9-1-1 Telephone system that automatically connects a person dialing the digits "9-1-1" to a predetermined answering point through normal telephone service facilities. Number and location information is not normally provided in *basic* systems.

Clipping Term associated with the use of two-way radios that is used to describe instances when either the first part of a message or the last part of a message is cut off as the result of either speaking before pressing the transmit key or releasing the transmit key prior to the end of a transmission.

Command Vehicle Typically used by operations chief officers in the fire service.

Communications Sending, giving, or exchanging of information.

Computer-Aided Dispatch Computer-based automated system that assists the telecommunicator in assessing dispatch information and recommends responses.

Database Organized collection of similar facts.

Deployment Plan Predetermined response plan of apparatus and personnel for specific types of incidents and specific locations.

Emergency Call Box System of telephones connected by private line telephone, radio-frequency, or cellular technology usually located in remote areas and used to report emergency situations.

Emergency Communications Center Facility either wholly or partially dedicated to being able to receive emergency and, in some instances, nonemergency reports from citizens. Centers such as these are sometimes referred to as fire alarm, headquarters, dispatch, or a public safety answering point (PSAP).

Emergency Medical Dispatch System designed for use by telecommunicators to assist them in evaluating patient symptoms using predetermined criteria and responses.

Encoder Device that converts an "entered" code into paging codes, which in turn activate a variety of paging devices.

Enhanced 9-1-1 Similar in nature to basic 9-1-1 but with the capability to provide the caller's telephone number and address.

Federal Communications Commission Government agency charged with administering the provisions of the Communications Act of 1934 and the revised Telecommunications Act of 1996 and is responsible for nonfederal radio-frequency users.

Fire Station Alerting System System used to transmit emergency response information to fire station personnel via voice and/or digital transmissions.

Home Alerting Devices Emergency alerting devices primarily used by volunteer department personnel to receive reports of emergency incidents.

Mass Casualty EMS incidents that involve more than five patients.

Mobile Data Computer Communications device that, unlike the mobile data terminal, does have information processing capabilities.

Mobile Data Terminal Communications device that in most cases has no information processing capabilities.

Mobile Radio Complete receiver/transmitter unit that is designed for use in a vehicle.

Mobile Support Vehicle Vehicle designed exclusively for use as an on-scene communication center and command post.

Multiple-Alarm Incident Involves the response of additional personnel.

National Emergency Number Association Not-for-profit organization founded in 1982 and made up of more than 6,000 members. The association fosters technical advancement, availability, and implementation of a universal emergency telephone number system.

NFPA 72 National Fire Alarm Code.

Prearrival Instructions Self-help instructions intended to enhance the overall safety of the citizen until first responders arrive on the scene.

Ringdown Circuits Telephone connection between two points. Going "off-hook" on one end of the circuit causes the telephone on the other end of the circuit to "ring" without having to dial a number.

Run Card System System of cards or other form of documentation that provides specific information on what apparatus and personnel respond to specific areas of a jurisdiction.

TDD Device that allows citizens to communicate with the telecommunicator through the use of a keyboard over telephone circuits instead of voice communications.

Telecommunicator Individual whose primary responsibility is to receive emergency requests from citizens, evaluate the need for a response, and ultimately sound the alarm that sends first responders to the scene of an emergency.

Type A Reporting System System in which an alarm from a fire alarm box is received and retransmitted to fire stations either manually or automatically.

Type B Reporting System System in which an alarm from a fire alarm box is automatically transmitted to fire stations and, if used, to outside alerting devices.

REVIEW QUESTIONS

1. Throughout history, what element has been essential to all of successful endeavors?
2. List four basic elements of communications.
3. What is the primary role of the telecommunicator in relation to fire and EMS apparatus?
4. What is the most important part of call processing for a telecommunicator or firefighter?
5. List the order in which ringing lines should receive priority.
6. When speaking with citizens who are reporting emergencies, what should the call taker's voice project?
7. When processing a nonemergency call, and an emergency line rings, what steps should the telecommunicator take?
8. Telecommunicators or others who answer emergency telephone lines must extract key information from individuals who are, in some instances, hysterical and in severe emotional distress. How do telecommunicators extract information in such a case?
9. What is the most important information to learn from a caller reporting an emergency?
10. What other information is important in the event it becomes necessary to recontact the caller?
11. What are emergency medical dispatch protocols designed to accomplish?
12. Name the five most common technologies used to report emergencies.
13. What is the primary benefit of basic or enhanced 9-1-1 service?
14. What sector of our society would have the most need for TDD equipment?
15. What is a preassigned standard response to a specific situation called?
16. When operating radio equipment, emergency personnel should always do what before talking?
17. How should a user speak when using a mobile radio?
18. True or false: Routine administrative radio traffic always gets priority attention.
19. In most states, fire reports are considered _____ .
20. True or false: The actions of the person answering an emergency telephone call are vital to the successful outcome of the emergency incident.

Endnote

1. *NFPA 1061*. National Fire Protection Association, Quincy, MA.

Additional Resources

Basic Telecommunicator Course. Association of Public Safety Communications Officials, South Daytona, FL. http://www.apcointl.org

Bukowski, Richard W., and Robert J. O'Laughlin, *Fire Alarm Signaling Systems.* National Fire Protection Association, Quincy, MA, and Society of Fire Protection Engineers, Boston, MA, 1997.

Bunker, Merton W., Jr., *National Fire Alarm Code Handbook, 1996 Edition.* National Fire Protection Association, Quincy, MA.

Fire Department Communications Manual, A Basic Guide to System Concepts and Equipment. U.S. Fire Administration, Federal Emergency Management Agency, Washington, DC, 1995.

National Emergency Number Association http://www.nena9-l-l.org

Nathonal Academies of Emergency Dispatch http://www.emergencydispatch.org

NFPA 1061. National Fire Protection Association, Quincy, MA.

NFPA 1221. National Fire Protection Association, Quincy, MA.

Pivetta, Sue, *9-1-1 Emergency Communications Manual,* 3rd ed. Kendall/Hunt Publishing Company, Dubuque, IA, 1997.

Project 33, Telecommunicator Training Standards. Association of Public Safety Communications Officials, South Daytona, FL, August 1996.

CHAPTER 4

FIRE BEHAVIOR

Frank J. Miale, Retired Battalion Chief, New York City Fire Department and Lake Carmel Volunteer Fire Department

OUTLINE

- Objectives
- Introduction
- Fire Triangle, Tetrahedron, and Pyramid
- Measurements
- Chemistry and Physics of Fire
- Sources of Heat
- Combustion
- Oxygen and Its Effect on Combustion
- Vapor Pressure and Vapor Density
- Boiling Point
- Flammable and Explosive Limits
- The Burning Process—Characteristics of Fire Behavior
- Modes of Heat Transfer
- Thermal Conductivity of Materials
- Physical State of Fuels and Effect on Combustion
- Theory of Fire Extinguishment
- Unique Fire Events
- Classes of Fire
- Lessons Learned
- Key Terms
- Review Questions
- Endnotes
- Additional Resources

STREET STORY

It had been a relatively quiet day. As soon as the box alarm came in we pulled out the door and the column of smoke was clearly visible in the late afternoon winter sky. We arrived with the first-due engine and truck companies. As we pulled in, I saw the fire building was a two-story wood frame structure with heavy brown and black smoke coming from both floors. A quick glance revealed that the windows on the first floor had been painted on the outside covering the glass. It appeared that this was a vacant structure probably being used by squatters or as a crack house.

After receiving an initial report, my crew and I headed toward the building. In my agency, one of the rescue squad's duties at structure fires is to perform the initial search unless otherwise directed by an incident commander. I saw that the engine company was laying their line while the truck was preparing to force entry into the building. During the size-up, a civilian ran over to us advising us that his father lived there and he could not find him.

At that point I split the crew up. I sent two men to assist the truck company in forcing entry through the front door, while the driver and I went to the rear of the building. As we arrived in the back we could hear some of the windows falling on both floors with flames venting from them. It was at this time we thought we heard moaning coming from the back porch area. We put on our SCBA and forced entry through a door to the back porch. Inside we found the smoke had banked down to approximately 10 feet from the kitchen door. We forced the kitchen door open and were only able to advance a few feet. Most of the kitchen was starting to erupt in flames. I thought to myself that it looked like a "Hollywood" fire. With everything burning, we started to back out because no one could have survived in there.

As we backed out of the kitchen my partner reached up to close the kitchen door. As he was doing this, the smoke became jet black and was now down to floor level. The temperature became unbelievably hot. I thought to myself "I hope what could happen doesn't," but it did. The porch flashed over. It was incredible. It was like having a giant orange flashbulb go off in your face. The entire porch was in flames. As we turned and headed back toward the door, we kept the wall to the left because, on the way in, the wall had been on our right. We came to the back wall and turned right to where the door should have been. What we didn't realize was that the inside panel of the door was covered by plywood with no doorknobs, and it was secured with latches on the inside because the resident was concerned about break-ins. When we could not find the door, I actually became afraid that we would not get out alive. My partner made a Mayday call giving our location while I searched for the door. The pain from the heat was like being stung by a thousand bees. At one point I just wanted to stand up and try to run where I thought the door was but I told myself "Stand up and you've really had it. Stay calm, stay together, and they'll get us out." I thought of my son who was only six weeks old at the time, my partner's kids, and our wives, and I said to myself, "I'm not dying in this place."

We stayed together in the area by the rear wall, when suddenly I felt someone grab me by the air-pack harness and throw me out the door. As I landed on the ground I turned and saw my partner come flying out the door. Thankfully, one of the captain/paramedics had seen us enter the building and heard our Mayday call.

Because we were wearing our full protective ensemble—boots, bunker pants, hoods, coats, gloves, helmets, and SCBA—we were fortunate enough to have sustained only minor burns. Our training also saved us by prompting us to create a search pattern that led us back to the door, make a Mayday call, and stay together.

—Street Story by Mike Kelleher, Captain, Troy Fire Department, Troy, New York

OBJECTIVES

After completing this chapter, the reader should be able to:

- Describe the chemistry and physics of fire.
- Identify the sources of heat.
- Describe the characteristics of fire.
- Describe the effect of oxygen on fire.
- Define combustion.
- Describe vapor pressure and vapor density.
- Describe the meaning of flammable and explosive limits.
- Describe the three types of heat transfer.
- Describe the significance of the thermal conductivity of materials.
- Describe fuel types and their effect on combustion.
- Describe the basis for the theory of fire extinguishment.
- Identify the classes of fire and methods of extinguishment.
- Explain thermal layering, flashover, and backdraft.

INTRODUCTION

Humans have been familiar with fire all of their lives. From their earliest memories, people have been warned against its dangers, entertained by its explosions and sound, comforted by its warmth, frightened by its power, mystified by its characteristics, and assisted by its light. In the history of mankind, fire has played a major role as a tool in the development of society. Sometimes an ally, sometimes an enemy, much has been learned about it, especially in the last thirty years, **Figure 4-1**.

So what is **fire**? Burning, also called **combustion**, is a simple chemical reaction. It is described as "a rapid, persistent chemical change that releases heat and light and is accompanied by flame, especially the exothermic oxidation of a combustible substance."[1] It is also referred to as a process of "rapid oxidation with the development of heat and light,"[2] and as "a reaction that is a continuous combination of a fuel (reducing agent) with certain elements, prominent among which is oxygen in either a free or combined form (oxidizing agent)."[3]

As more research into the mystery of combustion is accomplished, more is understood.

> **Firefighter Fact** Once thought to be simply a gift from the gods, combustion is now understood to be a complex chemical reaction.

When fighting a foe, one of the best weapons one can have is knowledge and understanding of the enemy. Military training and warfare are predicated on such information. Entire units in a military operation are devoted to the mission of uncovering, analyzing, and developing intelligence to explain the needs, capabilities, and probable actions of the enemy. The information learned dictates the battle plan. Firefighters need to know

Figure 4-1 Since its discovery, fire has been considered both an ally and an enemy.

about the behavior of fire as one of the elements in understanding the enemy. This chapter examines and discusses the process of burning. Once we know what causes fire to begin, grow, and spread, the means employed to extinguish it will become more understandable.

FIRE TRIANGLE, TETRAHEDRON, AND PYRAMID

The combustion process was once depicted as a triangle with three sides. Each side represented an essential ingredient for fire. Heat, fuel, and oxygen were thought to be the essential elements. As the scientific study of fire progressed, it became evident that a fourth ingredient was necessary. That fourth element was the actual chemical reaction that permitted flame propagation. A new four-sided figure was created to represent the essential ingredients for fire, the **fire tetrahedron**. The fire tetrahedron, **Figure 4-2**, is a pyramid shape describing the heat, fuel, oxygen, and chemical reaction necessary for combustion.

The four elements of the fire tetrahedron must be present in order to support the combustion process. Therefore, removing one or more of the elements in the fire tetrahedron will result in an end to the combustion process. Basic firefighting strategies are based on this principle of removing an element of the fire tetrahedron to halt the combustion process and put out the fire. An example of this is a simple fire in a pan on the stove. Placing a lid on the pan removes the oxygen source, thereby stopping the combustion process and putting out the fire.

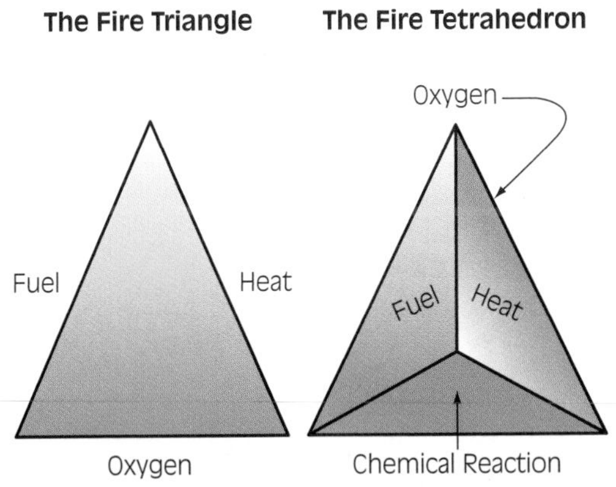

Figure 4-2 The old and new ways of visualizing the combustion process, the fire triangle and the fire tetrahedron.

MEASUREMENTS

In the field of firefighting, knowing units of measurements is essential. For example, firefighters need to know that a fire stream putting out 175 gallons per minute of water is releasing 175 times the amount of water that a stream putting out 1 gallon per minute is releasing. Furthermore, it is essential for a firefighter to know that a fire that is generating many thousands of **British thermal units (Btus)** will need a fire flow output of hundreds of gallons per minute to extinguish the fire. Measurement is an important part of firefighting.

Some types of measurement are part of our everyday language such as minutes, gallons, liters, watts, feet, centimeters, pounds per square inch, horsepower, degrees, and many others, **Table 4-1** and **Figure 4-3**.

Firefighters use terms for heat, electricity, volume, length, energy output, concentrations, and weight, an example of which is shown in **Figure 4-4**. To a firefighter, it is important to know the forms of measurement that describe these elements and understand the limits of each as it relates to safety of the firefighter and the team. For example, a 1,000-gallon pump on an engine is capable of putting out about 1,000 gallons per minute. If a particular hoseline requires a water flow of 250 gallons per minute to be

Measurements

ENGLISH SYSTEM	METRIC SYSTEM
Length	
Inches	Millimeters
Feet	Centimeters
Yards	Meters
Miles	Kilometers
Volume	
Ounces	Milliliters
Pints	Liters
Quarts	
Gallons	
Weight	
Ounces	Grams
Pounds	Kilograms
Tons	

TABLE 4-1

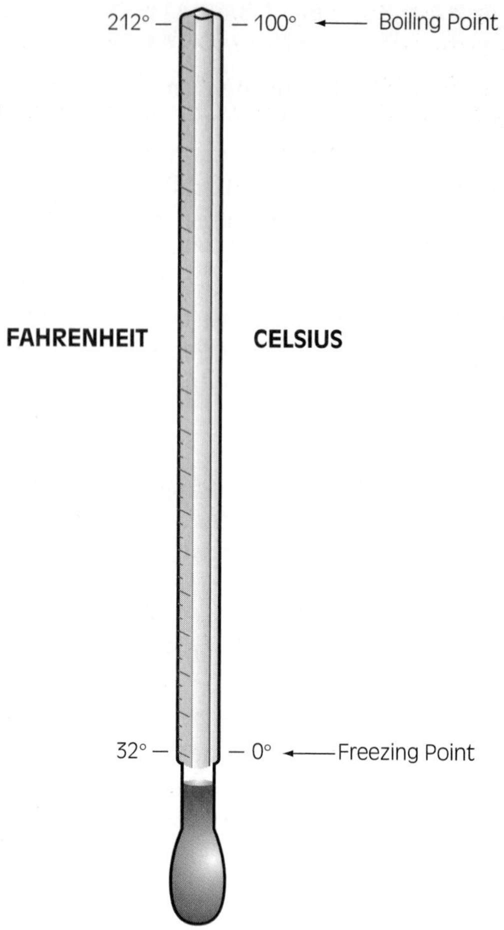

Figure 4-3 Several types of scales are used to measure heat. The two most common are the Fahrenheit and Celsius scales.

Note Different types of measurements are used in this text. The English measurement system and the metric system **(Table 4-1)** can be converted from one to the other. The key to metric measurement is in the terms that precede the measurement. *Milli-*, *centi-*, and *kilo-* are terms that mean thousandth, hundredth, and thousand, respectively. A meter is a measurement of length. (For example, a millimeter is one thousandth of a meter, a centimeter is one hundredth of a meter, and a kilometer is 1,000 meters.) The English version is a little more complicated, and memorization of the terms and multiples is required. (For example, there are 12 inches in a foot and 3 feet in a yard.)

effective, then the 1,000-gallon pumper could feed water to four of these hoses. A fifth hoseline of the same size would be beyond the capability of this pumper. A separate water supply source would be required to deliver the proper streams.

CHEMISTRY AND PHYSICS OF FIRE

The universe is made of a substance that is referred to as **matter**. Matter is defined as "something that occupies space and can be perceived by one or more senses." Some forms of matter are so small that they cannot actually be touched by a human hand, felt by the human sense of touch, or even seen with the human eye. But matter is the basis for the existence of the universe.

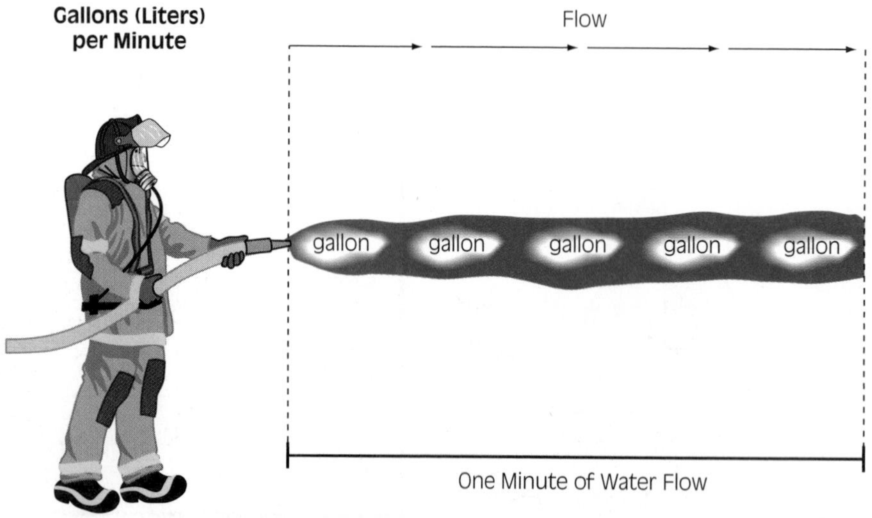

Figure 4-4 A way to measure water flow is by gallons (liters) per minute.

As matter interacts, substances are formed, changed, and destroyed. Matter doesn't disappear, it merely changes form. For example, a piece of iron that is left outside in the weather for years will eventually deteriorate into a form of dust and then disappear. Although it may seem as if the substance that makes up the iron has been destroyed, it has actually just changed form.

A house can be disassembled piece by piece and, at some point, simply become a pile of wood, shingles, nails, masonry, glass, roofing, metal, wire, pipes, and ducts. These components make up the house, and each component can be broken down further.

The wood contained in the house is made up of a material called *cellulose.* Because the wood was once a living organism, it is made up of millions of cells. The cells, in turn, are made up of compounds. A **compound** is composed of **molecules**, which are in turn composed of atoms of two or more elements in chemical combination, **Figure 4-5**. These compounds are **organic**, which refers to substances that are or were once living organisms. Organic substances generally contain the elements of carbon, hydrogen, and oxygen in their makeup. Examples of organic substances are hair, gasoline, wood, and plastics. It should be understood that something can be organic although it has never been alive itself, but is made up of chemicals that were once alive. For example, a plastic substance, although never actually alive, is made from oil more commonly referred to as **hydrocarbons**. The oil is left over from the decay, stabilization, and collection of millions of years of prehistoric life-forms that pooled into oil fields and has been tapped and drawn from wells. Once alive, the material is now made into nonliving materials by natural chemical actions.

Organic compounds are made up of many different forms of substances. These substances are various forms of chemicals that exist in the universe. Through combinations of these chemicals, substances are formed. Chemicals are made up of combinations of molecules. Molecules are joined and separated by bonding actions that use a form of electricity as energy.

Molecules are made up of atoms joined in various combinations. Atoms are bound together in much the same way that molecules join. In fact, it is the molecules with atoms that have loose ends that combine with other molecules with atoms with loose ends when conditions are right. The combination and separation of these molecules and atoms provide the basis for what causes oxidation or combustion.

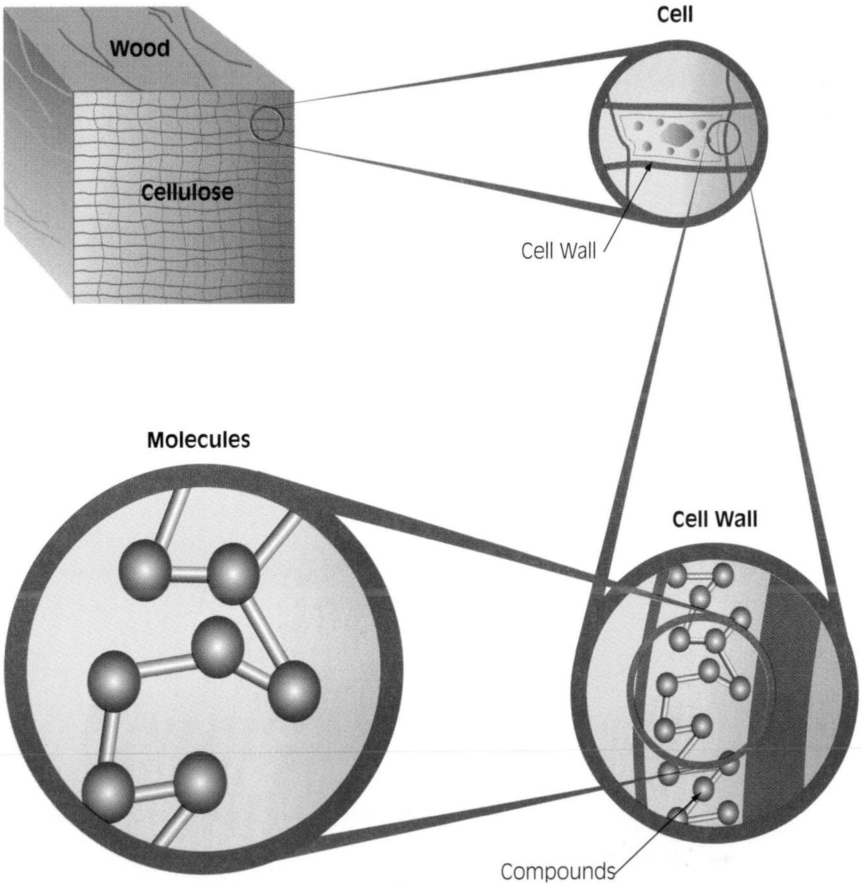

Figure 4-5 A piece of wood broken down to the molecular level.

Atoms are made up of still smaller particles called protons, neutrons, and electrons. The nucleus, or core, of the atom is made up of the protons and neutrons. The electrons encircle or orbit the nucleus, creating an electron cloud around the protons and neutrons, **Figure 4-6**.

> **Note** The numbers of protons and neutrons and the electrons circling them in various configurations are the determining factors in what properties that element will exhibit.

Atoms lacking electrons will tend to be quick to link up and form molecules. Atoms that are stable will tend to be those that have a satisfactory balance between the electron/proton ratio.

The term *organic* was described as a substance that is or was once living. In the universe, there are also substances that were never alive and the term used to describe that type of substance is **inorganic**. For example, iron, sulfur, granite, quartz, and silicon are all examples of inorganic substances. They are generally termed *minerals* and for the most part do not contribute to the combustion process.

The differences between organic and inorganic substances become important when discussing the chemistry of fire. As a rule of thumb, only organic materials will burn. There are some exceptions, which are discussed in a later section.

The term **bond** describes the "atomic glue" that holds molecules together. It is in this bond that fire (or combustion) has its origins. When molecules and atoms are joined, a certain amount of heat is absorbed into the bond as part of the mechanism that keeps the elements together. This is a process known as an **endothermic reaction**, **Figure 4-7**. This means that heat is absorbed when the bond is created. When a bond is broken, an **exothermic reaction** occurs, **Figure 4-8**. This type of reaction causes the release of heat. The energy released comes in the form of heat and light. If the release is rapid enough to sustain a continuous reaction, we see it as fire and feel the release of the heat. When this occurs, combustion is taking place. Fighting a fire is actually a process of breaking up a chain reaction. A look at the chemical reaction of combustion will help lead to a better understanding of what is achieved when fires are fought and extinguished.

A hydrocarbon is a compound that is made up of at least two elements, hydrogen and carbon. Chemical compounds can often be deciphered by their names. Looking closely at the hydrocarbon itself, one can see the two elements described in the name: *hydro* (hydrogen) and *carbon* (carbon). The ability

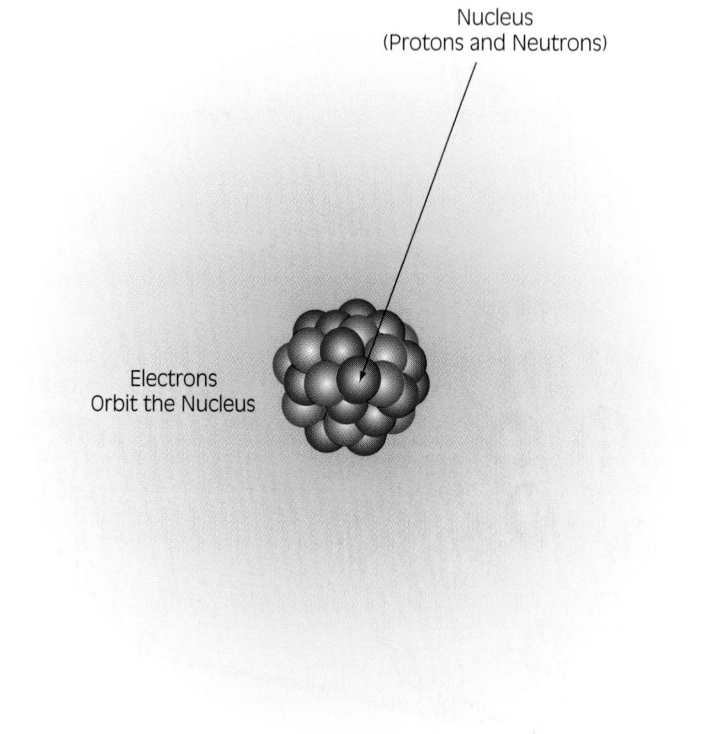

Figure 4-6 An atom is made up of protons, neutrons, and electrons.

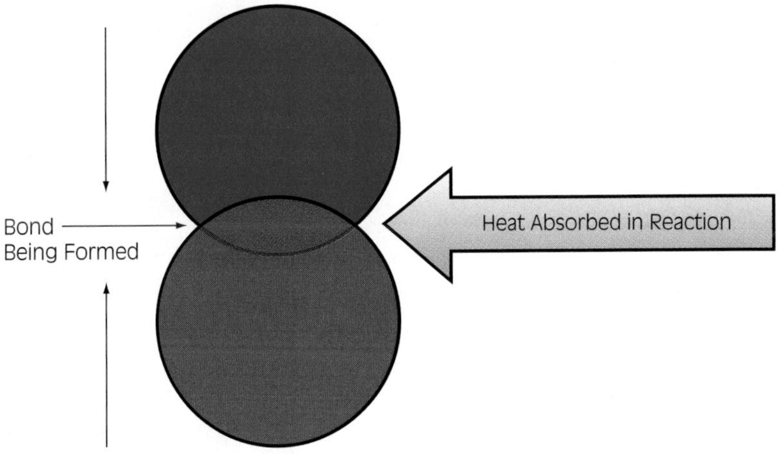

Figure 4-7 An endothermic reaction.

to recognize a chemical compound can tip off a firefighter as to what extinguishment agent to employ.

Organic compounds all contain some form of hydrocarbons. The number of carbon atoms in combination with the number of hydrogen atoms will determine the properties of the substance and, more importantly, how it will react under varying conditions.

To understand the process of combustion, a basic knowledge of chemistry is helpful. Imagine a jigsaw puzzle that has common joints fashioned in such a way that only certain pieces can be joined with other pieces. In **Figure 4-9**, the hydrocarbon elements are joined to form the compound methane. In its bonds between the atoms, the substance methane holds forces that, when split apart, will emit heat and light.

Most chemicals that are stable will maintain their form unless something presents itself to change it. An **oxidizer** acts as a catalyst in the breakdown of otherwise stable molecules. An oxidizer possesses a chemical property that can pull apart a molecule and break apart the bond that previously existed. The emission of light and, more importantly, heat then causes the chemical to break apart other compounds, letting loose more light and heat, which causes other bonds to break apart, and so on. If the process is able to continue a self-sustaining chemical reaction, combustion results. One example of an oxidizer is oxygen. The unique characteristic of these elements is that they all have the same chemical properties in attraction and combination with other chemicals. It is this property that makes oxidizers react in a similar fashion.

Streetsmart Tip The key to the chemicals that promote and support combustion is found in the suffix of the word (**Table 4-2**). It is important to watch for chemicals that end in *-ines*, *-ates*, and *-ides*. In chemistry, these endings identify oxidizers. Firefighters must be aware of these chemicals and the potential effect on combustion when discovered at a fire scene or during an inspection.

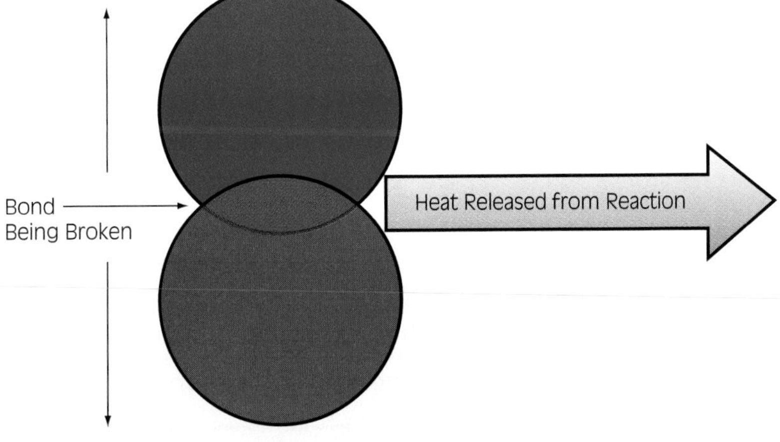

Figure 4-8 An exothermic reaction.

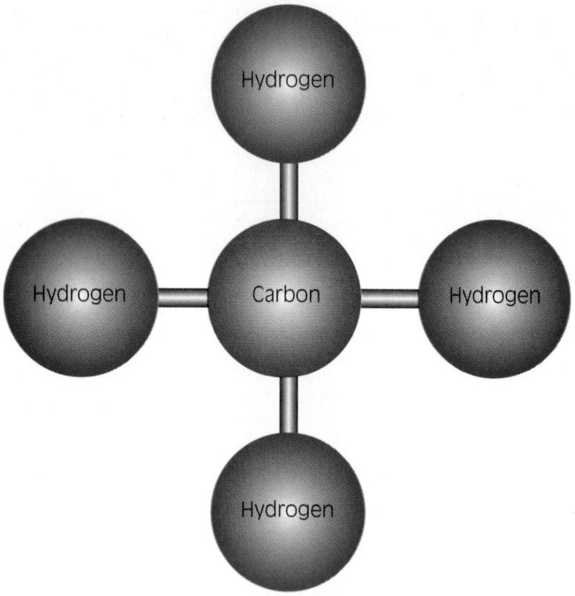

Figure 4-9 A methane (CH_4) molecule.

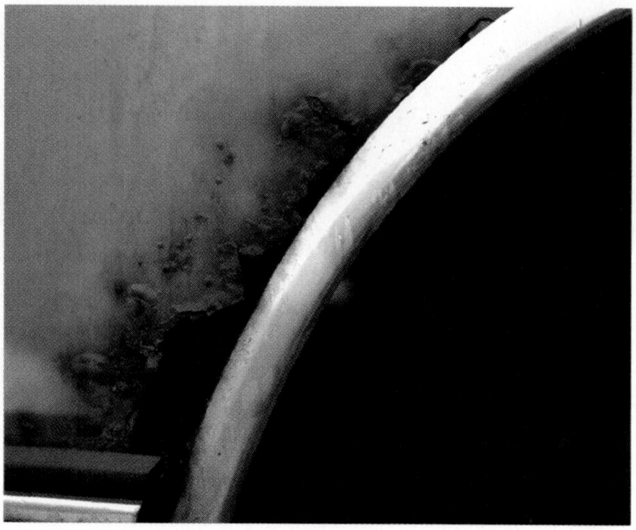

Figure 4-10 Rusting metal is a common example of the oxidation process.

If the reaction is slow, a gradual deterioration of the material will occur over time almost without notice. An example of an oxidizing process is iron that is undergoing a rusting process, **Figure 4-10**. The oxygen of the atmosphere combines with the properties of the iron and gradually pulls out the bonds that are keeping the iron atoms together. Although not visible, heat and light are emitted during this process. If sensitive monitoring instruments were affixed to the piece of iron, it would be possible to measure the heat and light.

The oxidation process is much more evident during combustion. The presence of oxygen in combination with heat from a previous bond break causes a rapid chain reaction to take place. Because of the speed at which the reaction takes place, the light is visible and the heat emitted can be perceived by the senses. This is the process of combustion.

Firefighter Fact In oxidizing reactions, such as combustion, electrons are transferred from one atom to another. In these reactions, there are reducing agents and oxidizing agents. Reducing agents lose electrons in the chemical reaction. In the fire tetrahedron, fuel is considered the reducing agent in the chemical reaction of fire.

SOURCES OF HEAT

Heat is an energy form that, in essence, powers the universe. It is heat that forges atomic-level synthesis of compounds, and it is heat that is released when those bonds are separated. Heat, because it is an energy force, can be neither created nor destroyed. It is merely the recognizable physical manifestation of energy as it changes from one form to another.

Chemicals and Their Names

SUBSTANCE	CHEMICALS	DESCRIPTION	SUFFIX
Chlor*ine*	Cl	Chlorine by itself (Cl)	*-ine*
Sodium Chlor*ide*	Na Cl	Chlorine with sodium (Na)	*-ide*
Sodium Chlor*ite*	Na Cl O_2	Chlorine with sodium and two oxygens (O)	*-ite*
Sodium Chlor*ate*	Na Cl O_3	Chlorine with sodium and three oxygens	*-ate*

TABLE 4-2

The phenomenon of heat comes from four basic sources, all of which are important to understand because they can all be initiators of fire. Although some sources of heat are very common and are experienced every day, other sources are not well understood. The four sources of heat are chemical, electrical, mechanical, and nuclear.

Chemical

The most common of the four sources of heat that firefighters deal with on a regular basis is the chemical reaction that releases heat as a by-product. Anything that burns does so through a chemical reaction in which heat is released as the bonds of the molecules break down. Anything from organic cellulose to petrochemicals will burn through a process called **pyrolysis**, which is decomposition or transformation of a compound caused by heat. The decomposition is the breaking down of the compound.

Caution The transformation of a compound is the reconstruction of the molecules to form another compound. It is important to understand that in the process of transformation, chemicals can be created that are extremely hazardous to health. It is also essential to possess a respect for the invisible by-products of burning. In addition to heat and light, products of combustion include water vapor, carbon particles, carbon monoxide, sulfur dioxide, and hydrogen cyanide. These products of combustion can cause damage to and diseases of the lung tissue from the heat, toxic chemicals, and particulate matter. The use of protective breathing apparatus in fires is required because of this danger, **Figure 4-11**.

Figure 4-11 The breakdown or transformation of substances can release hazardous chemicals. This is one reason why firefighters wear self-contained breathing apparatus.

Mechanical

Another form of heat source is mechanical. Friction causes heat that can reach levels hot enough to ignite surrounding combustible material. The buildup of heat from friction is often the cause of fire in machinery. Because heat from friction can be produced whenever any rubbing or compression occurs, **Figure 4-12**, industrial processes that employ this type of action have means by which to carry the heat away from the source. These heat transfer mediums can be simply water or a chemical solution designed in various forms to be a coolant.

The breakdown of this cooling mechanism will cause the two materials rubbing against each other to heat up to a point where the surrounding materials can ignite. Understanding how this can result in a fire becomes important when inspections for fire hazards are taking place. When extinguishing fire, ensuring that the source of the heat has been stopped becomes significant. Although a fire can be extinguished even if the heat source is still pumping in heat, the likelihood of keeping the fire extinguished is reduced.

Electrical

As a source of heat, electricity is probably the most recognized. Electrical power is used to heat homes and cook food. Nowhere is electricity as a heat

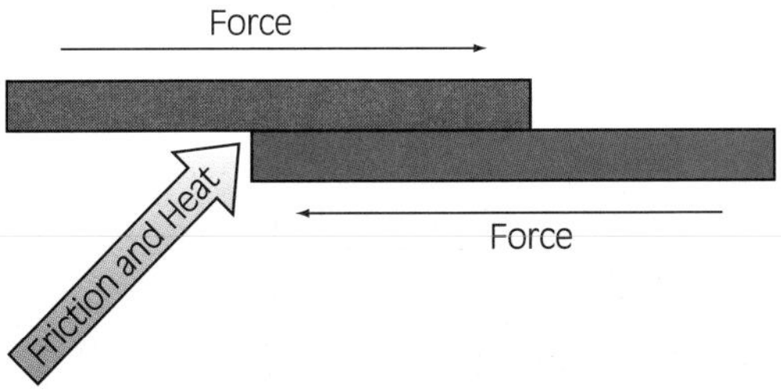

Figure 4-12 Heat from friction can be produced whenever any rubbing or compression occurs.

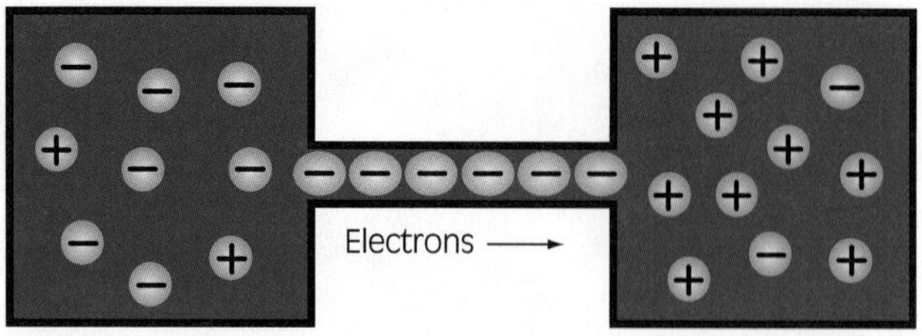

Figure 4-13 Electricity is simply a flow of electrons from a place where there are many to a place where they are lacking.

source more evident than in the home where space heaters with glowing orange-red metallic strips radiate heat into a room.

An electrical current can generate heat. Electricity is simply a flow of electrons from a place where there are electrons to a place where they are lacking, **Figure 4-13**. The place where electrons are in abundance is referred to as being *negatively charged*. Conversely, the place where electrons are lacking is referred to as being *positively charged*. It is the nature of electricity to attempt to equalize charges by sending electrons from where they are in abundance to where they are lacking. This is most commonly done by way of an **electrical conductor**. An electrical conductor permits the flow of electrons from one place to the other. When this electrical flow occurs, one electron does not exactly jump from one atom to the other. What occurs is more like a relay of electrons where one jumps onto an atom, which releases another to move to the next atom. When electrons are traveling along a conductor, the atoms that make up that conductor trade off their electrons from one atom to the next. During this activity, billions of electrons are jumping and, inevitably, collisions occur. When the collisions send the electrons astray, they collide with molecules and cause them to break apart. With the breakdown of the molecule, the energy that held the molecule together is broken and the bond is released as heat. Depending on the density of the conductor and the insulation surrounding it, the heat generated differs greatly. If a substance heats up easily, it is generally not a good conductor because the greater the amount of heat, the more inefficient is the action of the electrons milling about, causing much confusion at the atomic level. An efficient conductor will have less heat because more electrons are busy being transferred and there are fewer collisions.

Firefighter Fact A general rule of thumb is that the heavier the material, the more efficient it is as a conductor. For example, metal is a good conductor, but cotton is not.

It is not simply the weight of the substance that determines this "rule of thumb" of conduction capability; it is also determined by **density**. Usually, the more dense a substance is, the better its conductivity. Conduction is discussed in greater detail later in this chapter.

Electrical energy is a heat source. As firefighters, it is important to recognize forms of electrical energy. Obviously, lightning, arcing, and wiring outlets in the home are sources but also included in this classification are static electricity and induction heating. Static electricity occurs when dissimilar materials are rubbed, scraped, or suddenly joined or separated. This action creates heat in the form of mechanical heat, but it also creates a different potential in electrical charges due to the tearing apart of the surface at the molecular level. Electrons from one substance are taken by another and when enough of them collect, they will attempt to equalize by jumping the gap in the form of a static charge.

Firefighter Fact Static electricity is a small but powerful electrical charge that can emit temperatures of more than 2,000°F. Although this heat level is high as far as a temperature reading goes, it dissipates very quickly and offers no particular hazard to ordinary combustibles. If in the presence of a flammable gas, however, the small charge can ignite the gas with explosive results.

With one molecule breaking apart from the heat, more heat is released and that causes other molecules to break apart in like fashion. The resulting chain reaction will continue until all of the fuel is spent. If there is enough sustained heat, ordinary surrounding combustibles will have been exposed to the heat long enough to break down and generate their own heat for combustion.

The other type of induction heat can be found in a microwave oven. In this case, waves of alternating electrical energy subject the food to exposure. The current waves, called *microwaves,* alternate direction at high speeds resulting in molecular bombardment of the substances' internal electrons attempting to join the flow. The resultant collisions of the electrons release heat energy as the molecular bonds break apart and heat the surrounding material.

Streetsmart Tip As firefighters, knowing the process of combustion will provide a direction to the solution. A fire in a microwave oven requires the removal of outside power before any further internal action can be taken.

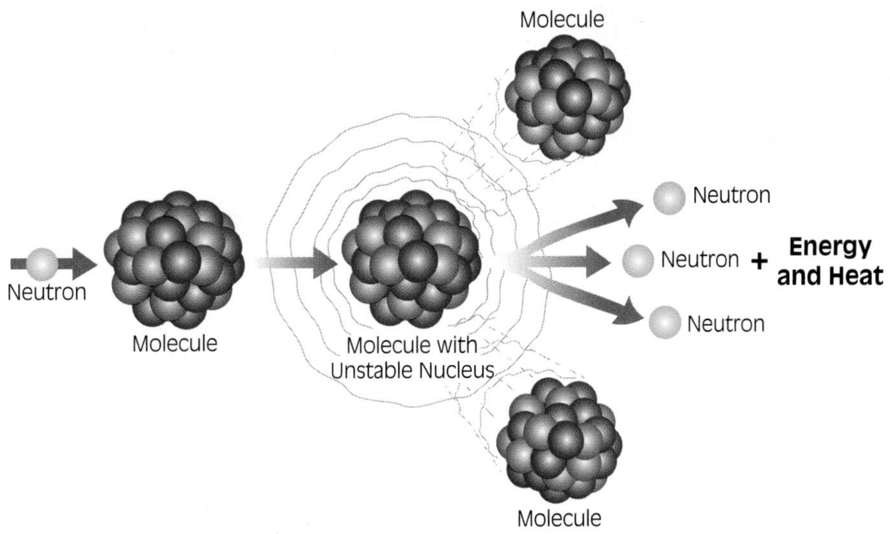

Figure 4-14 The process of creating nuclear energy and heat.

Nuclear

The last heat source is nuclear, **Figure 4-14**. Used in the atomic bomb, nuclear power has been hotly debated for years as to whether it is a safe source of energy. Essentially, all forms of atomic energy that we use are first converted to heat. The means by which nuclear energy generates heat is essentially the same as for the other forms already described. Radioactive materials are very unstable and are constantly breaking down as they seek to form a stable molecular composition. In that process, atomic particles are randomly flying in all directions. When encased in a heavy shield called a *core* made of very dense materials such as lead, the particles remain confined within the radioactive material. When activity is desired, the radioactive material, usually in the form of rods, is pulled out of the protective surrounding core, and a controlled amount of energy is permitted to be transferred into the surrounding water. This heat release into the water then turns to steam. The steam then turns turbines that generate electricity. As the need for steam is increased or reduced, the rods are extracted from or reinserted into the protective core. When fully inserted, the nuclear bombardment is confined.

As firefighters, little can be done to safely fight a fire that has been caused by a nuclear heat source. In the nuclear accident that occurred at Chernobyl (Ukraine) in 1986, a meltdown of the core occurred as a result of the water level being too low to accept the heat from the rods. The heat then melted the core and exposed the rods. Once exposed, the rods have no controlling mechanism and runaway reaction results, appropriately called a *meltdown.*

If a nuclear heat source causes a fire, there is little to do but protect uninvolved areas, evacuate, and let more skilled technicians handle the emergency. On-site personnel are trained to act in such an emergency. The real danger inherent in a nuclear fuel fire results from exposure to radiation. Without proper protection, a firefighter can sustain a serious long-term illness or even a fatal injury from radiation exposure.

In discussing the source of heat energy, one common thread winds through the subject. In all cases, the heat source is generally the initiator of the fire that ignites surrounding combustibles, be they solid, liquid, or gas. Once removed as a heat source, the extinguishment of the fire is simply a matter of removal of the oxygen or fuel or a breakup of the flame's chain reaction.

COMBUSTION

Combustion is often confused with the term *fire.* It is not important to make the distinction a critical element at this level of learning, but it is important to know that there is a difference on a chemical level. Fire is a chemical reaction that is a self-sustaining process that emits light and heat as a by-product of that reaction.

In combustion, the released heat energy is reinvested in the process, causing the continued reaction to occur repeatedly. Unchecked, with involvement of greater access to fuel, oxygen, and heat, the growth will accelerate. The products of combustion, **Figure 4-15**, can be hazardous and deadly to firefighters.

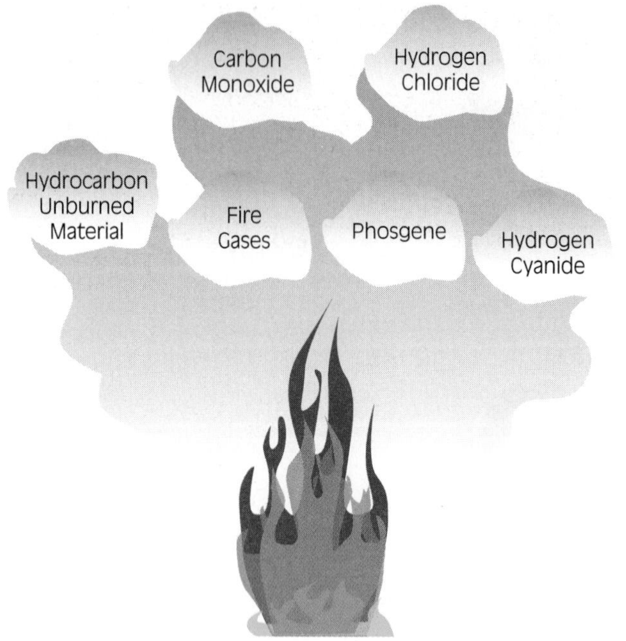

Figure 4-15 The products of combustion can be deadly.

Figure 4-16 The speed of combustion is affected by the amount of oxygen present.

OXYGEN AND ITS EFFECT ON COMBUSTION

Oxygen is an important element in the sicence of fire. It acts as a catalyst to the combustion process. The chemical reaction that occurs during combustion, called oxidation, is the process of oxygen bonding to other elements. The amount of oxygen present affects the process of oxidation, thus affecting the process of combustion.

> **Firefighter Fact** The concentration of oxygen in the atmosphere is important to the combustion process. Oxygen occurs normally in the Earth's atmosphere in a concentration of approximately 21 percent. Combustion can occur in oxygen concentrations of 14 percent or greater. In concentrations less than 14 percent, there is not enough oxygen present to support combustion.

The amount of oxygen present can also affect the rate at which oxidation occurs. It can be very slow, as is the case with rusting, or it can occur very quickly, as is the case in combustion. With a diminished amount of oxygen, the combustion process is slowed. With an abundance of oxygen, the chemical reaction is accelerated, **Figure 4-16.** In addition, the presence of oxygen can affect a material's combustibility. Higher concentrations of oxygen can cause some materials to ignite spontaneously or permit materials to burn that would not burn under otherwise normal oxygen concentrations.

> **Streetsmart Tip** Oil in the presence of high levels of oxygen will ignite spontaneously. For this reason, any pipe, gauge, or fitting that carries oxygen posts the warning to "use no oil." Nomex, a material that is used as a fire protection component in fire protection equipment, will ignite and burn in high levels of oxygen.

VAPOR PRESSURE AND VAPOR DENSITY

Pressure is defined as the application of continuous force by one body on another body that it is touching. In terms of liquid or gas, it is the amount of force applied to the surrounding container or in the open atmosphere at a given altitude. *Vapor pressure* is the measurable amount of pressure being exerted by a liquid substance as it converts to a gas and exerts pressure against a confined container, **Figure 4-17**.

A liquid is a collection of molecules that occupy the same space in a fluid state. In normal molecular activity, these molecules are constantly circulating and colliding with either one another or the sides of the vessel in which they are contained. However, when the molecules circulate beyond the upper boundary of the sur-

Figure 4-17 Vapor pressure. Container on left has less vapor pressure than container on right.

face, they escape into the space above the liquid. Some return to the liquid in their random paths. The molecules that escape and do not return tend to fill the surrounding area with the molecules of that liquid in a gaseous state. This is called **diffusion**.

If the weight of the gas is lighter than the surrounding atmosphere, the molecules will continue to escape and return with more molecules rising and escaping than returning. If this were to continue, eventually total **evaporation** of the liquid would occur. If the gas were heavier than air, the molecules would remain in the area between the upper surface of the liquid and the top of the open vessel. At the point where there are as many molecules being liberated from the liquid as being reabsorbed, it is referred to as being in a state of **equilibrium**. Of course, if the gas were lighter than air, this state would never be achieved because the molecules would be constantly escaping to higher levels. The rate of molecular escape or evaporation is highly dependent on many factors, most notably temperature and pressure. The liberation of molecules can be curtailed if the pressure of the surrounding atmosphere is higher than normal because it might exceed the liquid's ability to diffuse. Or, if the temperature is lower and the molecular activity is reduced, a reduction in the number of molecules being liberated from the liquid could result.

Air pressure changes at different altitudes and must be considered when examining this area of study. At sea level, air pressure is 14.7 psi (pounds per square inch). If a vapor pressure exceeds 14.7 psi, it will escape if it is at sea level.

Vapor density describes the weight of a gas as compared to normal air and is identified as a number. A normal concentration of air at sea level (14.7 psi) is designated as "1" when measured at 0°C (32°F). Gases that weigh less than the same volume of air will be lighter and tend to rise and their vapor

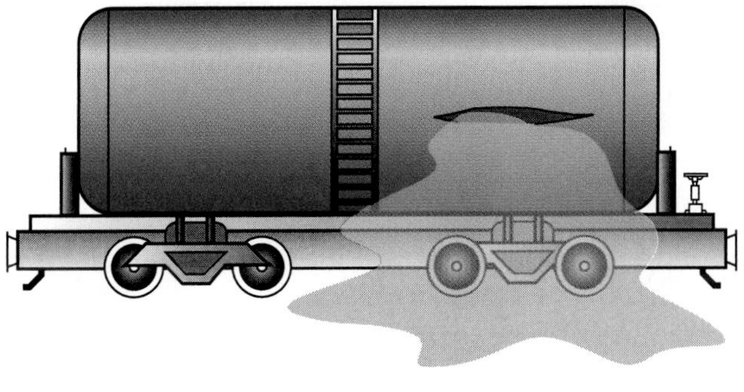

Vapor Density Greater than 1

Vapor Density Less than 1

Figure 4-18 Vapor density. Vapors lighter than air rise and those heavier than air sink.

density is described in terms of a number less than 1. If the gas is heavier than air, the number will be greater than 1, **Figure 4-18**. The importance of this to the firefighter is knowing that a particular gas that is heavier than air (a vapor density of more than 1) will tend to collect in low areas such as depressions, basements, cellars, sewers, ravines, or the like, **Table 4-3**. Normal air pressure can act as a cap to an open vessel. This is where the terms *vapor pressure* and *vapor densities* are often confused.

> **Firefighter Fact** Vapor density can be a valuable number to a firefighter, **Table 4-3**. The vapor density of a substance can display whether it is heavier or lighter than air. Propane, with a vapor density of 1.6, will tend to collect in low areas, depressions, or cellars. Methane, on the other hand, with a vapor density of 0.6, is lighter than air and will rise and tend to mix and dilute with atmospheric air.

Vapor pressure is the force exerted on the sides of a closed container. Vapor density is a function of the weight of the gas. Although different, they are similar

Vapor Density of Normal Atmospheric Air at Sea Level Is Designated as 1		
	VAPOR DENSITY	**EFFECT**
Propane	1.6	Will sink and collect at low points
Methane	0.6	Will rise

TABLE 4-3

in some respects. The pressure that exists between the top of the liquid and the sides of the container vessel occurs as a result of the liquid's conversion to a gas and is the vapor pressure. Conventional gauges can measure this pressure, and determinations can be made from those data if the vapor pressure poses a threat. In a sense, a container that is open at the top is capped at the top by the weight of air acting as a lid.

If the liquid vaporizes and creates its respective vapor pressure, which turns out to be a pressure that is less than the ambient air (14.7 psi), the weight of the air will keep the substance "capped" in the open vessel. In this case, the vapor density of the substance would be less than air and the vapor pressure not enough to overcome the weight of the air holding it in the open vessel.

Conversely, if the vapor pressure is greater than the ambient air, the vapor will escape and not be measurable at all because no pressure is being created. The gas escapes as it evaporates through the "cap" of the weight of the air. In this case the vapor pressure is greater than the 14.7 psi of air. However, higher altitudes will exhibit different air pressures, so although numbers will be different, the principles will remain the same. Therefore, at higher altitudes, vapor escape will occur at lower vapor pressure levels. Evaporation of a liquid will occur more quickly and diffusion will be accelerated.

Note Density—the weight of a material—is dependent on the number of molecules and atoms that occupy a given volume. The more molecules crushed into a given volume, the denser it is. As an added factor, the heavier each individual molecule is, the heavier the substance will be. So a heavily loaded volume with a bunch of heavy molecules will result in a weighty material. The more molecules there are, the easier it is for them to collide with one another and pass on the heat energy. The more dense the material, the greater its ability to conduct heat.

BLEVE

A **boiling liquid expanding vapor explosion (BLEVE)** occurs when the vessel holding liquid ruptures as a result of pressure being exerted on its sides when the liquid it holds boils and the resulting pressure exceeds the container's ability to hold it, **Figure 4-19**. This usually occurs when heat is

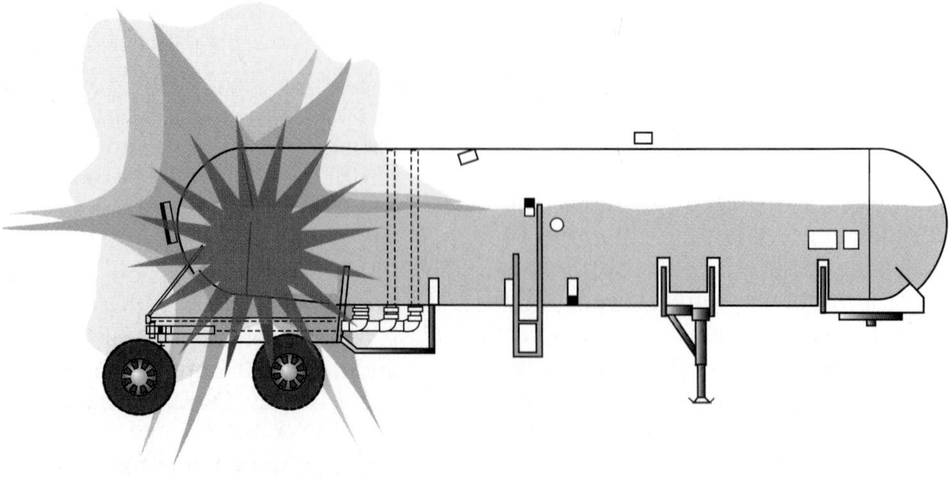

Figure 4-19 A BLEVE.

applied to the closed container, causing the liquid inside to boil. The vapor pressure in the closed container is increased until it reaches a point where the vessel can no longer withstand the pressure exerted against it. When the vessel fails, the ensuing release of vapor and liquid can be very violent, causing the explosive release of tremendous forces.

> **Firefighter Fact** The most common example of a BLEVE is popcorn. The liquid inside the hard kernel shell heats, boils, and exerts pressure against its container until the shell fails. The result is a cooked kernel that has escaped its container while the pressure inside has escaped and equalized.

Even if the liquid in a vessel is not flammable, the container rupture could still be violent, resulting in a force that can send container fragments great distances accompanied by a shock wave. If the liquid is flammable, the container failure will exhibit a fireball that adds in a fire extension element to the problem. If the liquid is a hazardous material, still another set of variables must be addressed.

BOILING POINT

All materials obey the laws of nature and exist in one of three forms or states: solid, liquid, or gaseous. Pressure and temperature will affect the state of matter, **Figure 4-20**. As a point of reference, the words *normal* and *ambient* are often used to describe the conditions under which a substance exists. To standardize the understanding of scientific description, 70°F was established as the reference point for temperature and 14.7 psi for pressure. This set of parameters is termed *normal.* When a substance being described is referred to as being in its normal state, it is understood that it is at that temperature and pressure.

If the pressure remains the same and the temperature changes, the form of the substance can change. Similarly, a change in pressure while the temperature remains the same can cause a change in the state of the substance. Gases compressed to liquids under great pressure generate heat, and liquids that boil absorb heat. When a material absorbs heat, it usually does so at its boiling point.

The state of conversion from a liquid to a gas under normal atmospheric conditions is called *evaporation.* Evaporation is simply the state where a liquid's molecules, which are constantly sending off molecules into the surrounding area and accepting them as they randomly return, wind up with a greater loss rate than recovery rate. Eventually, the entire supply of the liquid's molecules will escape into the surrounding atmosphere. This phenomenon occurs as long as the surrounding atmospheric pressure is less than that of the gas above the liquid. If the liquid is heated, the molecules move about more quickly, crash into the sides of the container vessel, and increase pressure on the vessel. When the molecules collide into the sides, they rebound into the liquid, eventually escaping over the surface of the liquid. The increased activity of the molecules over the surface increases the pressure of that vapor. When the pressure exerted overcomes the surrounding atmospheric pressure, then the boiling point of that liquid has been reached. Water is the liquid state and steam is the gas state, **Figure 4-21**.

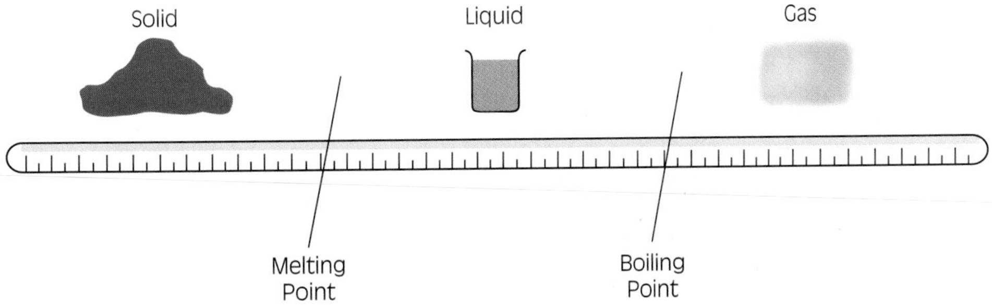

Figure 4-20 States of matter, in many cases, are temperature dependent.

Figure 4-21 Water turns to steam at the boiling point.

FLAMMABLE AND EXPLOSIVE LIMITS

A combustible material must meet certain requirements before it can oxidize. Depending on the molecular makeup of the compound, the presence of oxygen is required at different concentrations. This relative concentration is described in terms of percentages.

If there are 100 gallons of a liquid made up of 50 gallons of water, 35 gallons of detergent, and 15 gallons of a degreasing solution, we could say that the percentages of each substance in the total mixture are represented as 50 percent water, 35 percent detergent, and 15 percent degreaser. The number 100 was used for simplification. Whatever the total volume happens to be, the percentage of the components of that mixture can be described as a percentage of the total. A percentage is determined by dividing the amount of the substance to be measured by the total amount in the mixture, **Figure 4-22**.

The same is true for gases. Because combustion can only take place when a substance is a gas, the levels of the gases are described in terms of percentages. Through chemical analysis and scientific testing, it has been determined that gases can only ignite when certain concentrations of that substance are present in air. If enough combustible gas is not present, it is said that the mixture is too lean to burn. If there is too much gas, it is said to be too rich to burn. When a concentration of a gas falls into the range where it can ignite, it is said to be within its **flammable** or **explosive limits**.

Flammable limits can change depending on the temperature. The limits can contract or expand depending on the surrounding conditions. Comparing the flammable or explosive limits of one substance with another will assist the firefighter in understanding the relative volatility of that substance. Referring to **Table 4-4**, compare the flammable limits of gasoline, natural gas, and carbon monoxide. For example,

Determining Percentage of Substance in Normal Air Concentrations

$$100\text{ gal}\,\overline{)\,50.00}\text{ gal} = 50\% \quad (\text{quotient } .50)$$

$$100\text{ gal}\,\overline{)\,35.00}\text{ gal} = 35\% \quad (\text{quotient } .35)$$

$$100\text{ gal}\,\overline{)\,15.00}\text{ gal} = 15\% \quad (\text{quotient } .15)$$

To determine the percentage of a substance in a whole, divide the total amount of the sample into the portion being examined. Take the division to the second decimal level. Then merely convert it to a percentage.

In the sample illustrated above, one hundred (100) gallons of liquid has fifty (50) gallons of water in it. Take the one hundred gallon number and divide it into the amount of the substance, in this case 50 gallons of water. The result is .50, or 50%. Therefore, of the 100 gallons of liquid, 50% is water.

Figure 4-22 Determining the percentage of a substance in normal air concentrations.

Flammable Limits of Some Materials

	LOWER	UPPER
Acetone	2.6	12.8
Butane	1.9	8.5
Kerosene	0.7	5
Natural gas	6.5	17
Gasoline (92 octane)	1.5	7.6
Carbon monoxide	12.4	74

TABLE 4-4

a concentration of carbon monoxide at 37 percent is very significant if the reading was taken in a 50,000 square foot undivided building space. On the other hand, a reading of 37 percent in a 1 cubic foot container is much less dangerous and might only emit a slight "pop" if ignited, whereas in the 50,000 square foot building space, an explosion could lift the roof off its supports and blow out concrete walls.

Firefighter Fact Flammable limits of the materials in **Table 4-4** demonstrate how different the flammable limits of different substances can be. Especially notable are the upper and lower limits of carbon monoxide, a substance found at every fire. It will ignite at a much greater range than natural gas.

Specialized instruments have been developed to display the measurements of flammable and explosive limits. Through the use of these instruments, firefighters can better evaluate their surroundings, have a better understanding of the enemy, and be better prepared to engage in firefighting activities safely.

THE BURNING PROCESS—CHARACTERISTICS OF FIRE BEHAVIOR

The process of burning occurs in clearly defined stages. By recognizing the different stages, a firefighter can better understand the process of burning and fight the fire at different levels and with different tactics and tools. As is true in any combat situation, knowledge of the enemy's needs and practices leads to the use of different tactics and practices to attack that enemy successfully.

Note Burning materials follow a specific sequence of events from start to finish. Interrupt any of the sequential steps and fire can be minimized or extinguished entirely. Older texts refer to these stages as the incipient stage, the smoldering stage, and the free-burning stage. Today's firefighting science has redefined the stages of fire burn. They are the ignition, growth, fully developed, and decay stages, **Figure 4-23**.

Ignition Stage

When a substance begins to heat up, it liberates gases that can burn. The preliminary heating usually comes from an outside source such as a match or spark from a fire already burning. Heating can also occur through conduction, convection, or radiation from another fire or heat source. At some point, the amount of heat being created exceeds the amount of heat that is dispersed and the combustion feeds itself the necessary heat to sustain a burn.

When the necessary ingredients of a self-sustaining chemical reaction are present, ignition occurs. **Ignition** is that point where the need for outside heat application ceases and the ability for the material to sustain combustion comes from the heat generation of the material itself.

The ignition stage of the fire is, very simply, the point at which the four elements in the fire tetrahedron come together, materials reach their ignition temperatures, and a fire is started. At the ignition stage, the fire is typically very small and limited in area.

Growth Stage

From the point of ignition, fire begins to grow. Starting out as a spark or a small flame, other combustibles heat up, liberate flammable gases, and ignite, spreading the chain reaction to other flammables and resulting in an increase in size. (The growth stage was formerly considered the incipient and smoldering stages.)

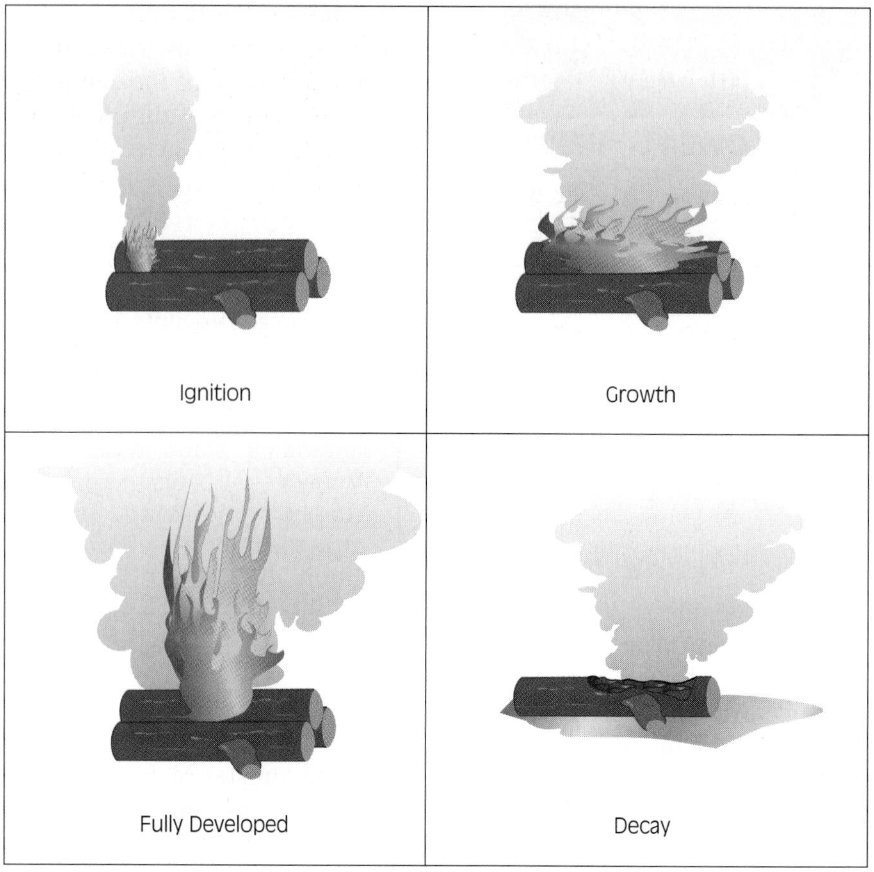

Figure 4-23 Four stages of fire.

The speed of the growth and ultimately the size of the fire are dependent on several factors:

- *Oxygen supply:* The amount of oxygen will have a direct effect on the speed of growth and the size of the fire. Any limitation of the oxygen supply will curtail the growth and can even result in extinguishment.
- *Fuel:* The size of the fire will naturally depend on the amount of fuel available to burn.
- *Container size:* In a structure, the container would be the surrounding walls and obstructions. A large container would permit dissipation of heat and slow the growth of the fire. In an open and unconfined area it even serves to inhibit fire ignition. With a container, heat can radiate back down and back into a space, further heating up uninvolved fuel sources.
- *Insulation:* Heat that is radiated back into unburned areas will accelerate growth. If the container wall and ceiling are insulated, the amount of heat kept trapped in the container and not permitted to escape serves to further "reinvest" the heat produced back into the fire formula.

Fully Developed Stage

This stage is recognized as the point at which all contents within the perimeter of the fire's boundaries are burning. In a structure this would mean the entire contents of a room. In an outside fire, it would mean all combustible material within the fire's furthest reach. The fully developed stage is regulated by one of two features. In a structure, the speed and extent of a fully developed stage is controlled or regulated by the amount of air that can be introduced or supplied to the fire area, making it an air-dependent fire. In an outside fire, because the amount of air is unlimited, the amount of fuel will dictate the size of the fire, making it a fuel-dependent fire. The fully developed phase is where a phenomenon called flashover can occur, **Figure 4-24.** Flashover is detailed later in this chapter.

Decay Stage

When the point at which all fuel has been consumed is reached, the fire will begin to diminish in size. Ultimately, the fire will extinguish itself when the fuel or oxygen supply is exhausted. Obviously,

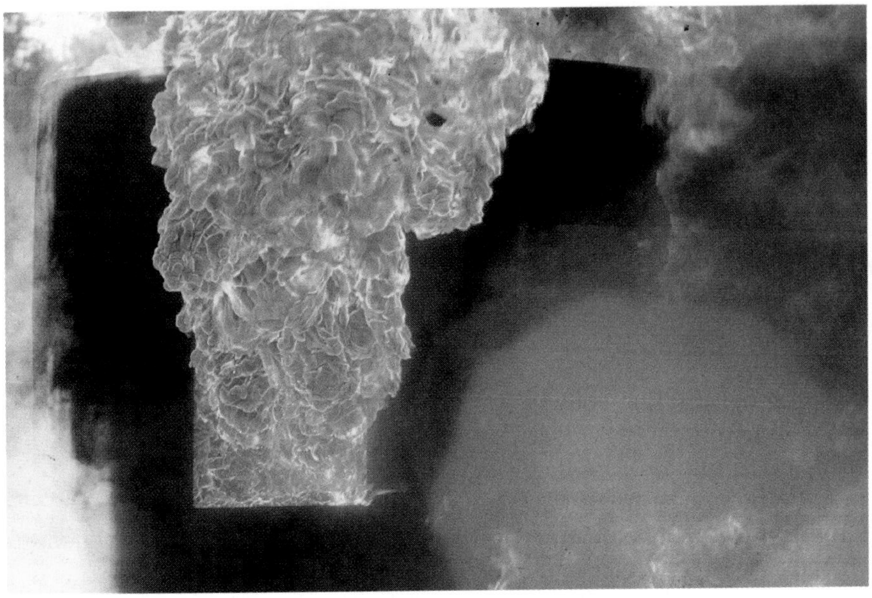

Figure 4-24 A flashover occurs suddenly when all or most of the combustible materials reach their ignition temperature nearly simultaneously.

this can take a considerable period of time. The bright array of flames will diminish, becoming a series of separate flame fronts, **Figure 4-25**. Then the flames will disappear until only glowing embers are visible. Eventually, those too will disappear. What is witnessed in a fireplace from ignition to final extinguishment is the same series of events that occurs on a much larger scale in a structure fire or an open wildfire. As the decay stage progresses, even a fire that was air controlled will become a fuel-dependent fire as the amount of fire decreases. At some point the amount of fire will be small enough to be fully supplied by the air supply. At that point the fire's future will depend on the amount of fuel left.

These factors will dictate the tactics that will be employed when fighting a fire. In some cases, such as a fire in the hold of a ship, cutting off the oxygen is usually the tactic of choice. If a ship's watertight spaces can be sealed, the supply of oxygen can be cut off. Ultimately, this will cause the fire to self-extinguish. In fighting a wildfire, use of a fire break to cut off access to additional fuel might be the choice of attack.

Figure 4-25 The decay stage.

Note The type of fire, the characteristics of the burn, and the manner in which fuel is being supplied are the determining factors in deciding the best attack.

MODES OF HEAT TRANSFER

Heat is a by-product of combustion that is of significant importance to the firefighter. It is heat that causes fire to sustain its combustion and, more importantly, to extend. When heat given off as a product of combustion is exposed to an unheated substance, certain changes occur that can make the new substance a contributing factor in extending a fire. Therefore, knowing how heat is transmitted from one place to another is the first step in knowing how to control the extension of fire, one of the first steps in extinguishing it.

Caution Heat, as a by-product of the combustion process, can be dangerous to firefighters, causing dehydration and heat exhaustion as well as burns. Each of these conditions can be mild to severe and ultimately life threatening. The amount of energy released as heat over a certain period of time is called a *heat release rate.* Heat release rate is influenced by the quantity and type of burning materials. The greater the heat release rate for a particular fire, the greater the danger to firefighters for heat-related injuries.

The three modes by which heat transfers its energy from one substance to another are through conduction, convection, and radiation, **Figure 4-26**. The type of substance being heated, the distance covered by the material being heated, and the ability of the substance to retain the heat will be factors in the spread of fire.

Conduction

When a hot object transfers its heat, conduction has taken place. The transfer could be to another object or to another portion of the same object. Combustion occurs on the molecular level. When an object heats up, the atoms become agitated and begin to collide with one another. A chain reaction of molecules and atoms, like a wave of energy, occurs and causes the agitated molecules to pass the heat energy to areas of nonheat. As the heat increases, so do the waves of energy passing to the cooler areas, **Figure 4-27**. As more agitation occurs, the substance heats up and unless the heat energy can be dissipated in a **heat sink**, the overall and internal temperature of the substance increases until it reaches its boiling point if it is a liquid, or its ignition temperature if it is a solid.

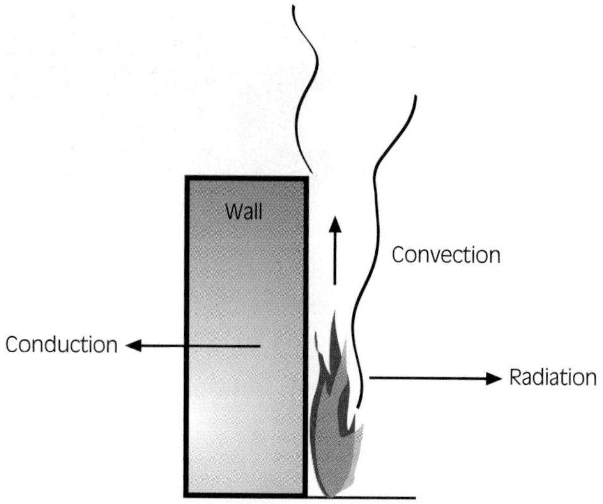

Figure 4-26 Examples of heat transfer in fire.

When examining conduction as a heat transfer vehicle, it is important to understand that heat is conducted through materials at different rates. The rate of heat transfer will depend on several factors, the most significant being density. If a given volume of any substance is weighed, we can determine its relative density.

The more dense a material, the better a conductor it will be. Because density is a function of weight, the heavy substances are generally better conductors. Metals, being among the most dense in the universe, are therefore generally better conductors of heat, and the heavier the metal, the better the ability to transfer heat. Concrete, a dense material made up of rock, sand, and cement, will also conduct heat through it with enough transfer ability to ignite combustibles. Essentially, the ability of a material to transfer heat is dependent on that material's ability to keep the heat energy accumulating faster than it can be dissipated.

If the denser materials are better conductors, then conversely, the less dense materials are better insulators. Lighter materials whose ignition temperature is very high become the best insulators and, therefore, the poorest conductors. Mineral wool, essentially a form of rock spun into a web-like material with a lot of air space between fibers, is one of the best insulators. Because heat energy uses the collision chain

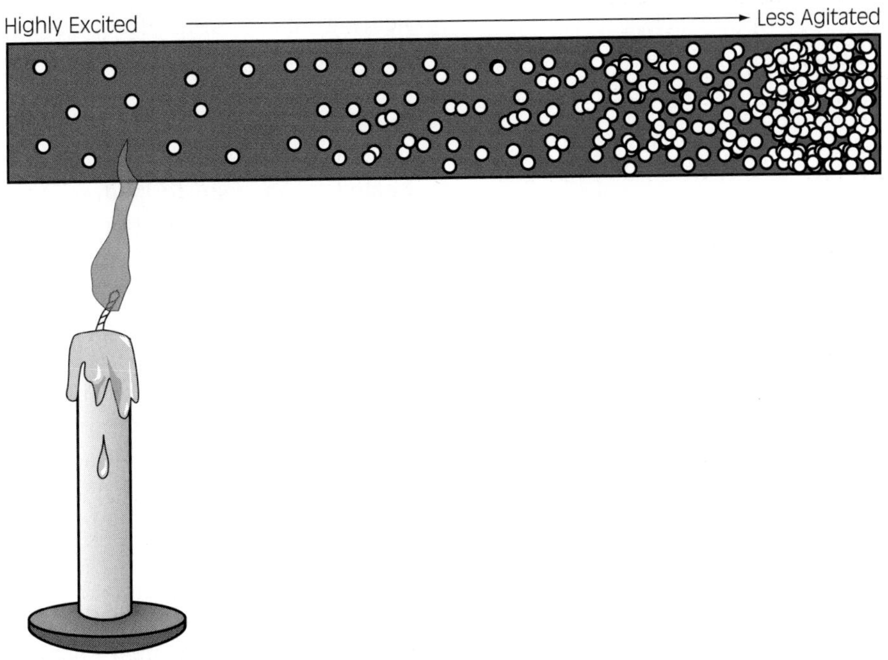

Figure 4-27 Conduction. A metal bar is heated at one end, causing the molecular activity that transfers the heat.

reaction to pass heat along, it stands to reason that energy will dissipate in the absence of collisions. There is no conduction whatsoever in a vacuum because there are no molecules or atoms to contribute to the collision process. The fewer such molecules, the poorer a conductor the material will be. Therefore, besides the noncooperation of the material in the mineral wool, the air spaces between the strands of mineral wool serve to limit conduction further, making it an excellent insulator and a poor conductor.

The relative ability of a substance to be a good or poor conductor is a function of time and application of heat. If the application of heat to a substance over a given period of time exceeds its ability to dissipate or shed that heat, the substance will begin to heat up. If it is a poor conductor because of its light density, it will be able to shed the heat without having its internal temperature rise. If it is a good conductor, it will absorb the heat and transfer that heat to distant locations internally. A piece of heated steel will take a long time to cool because the heat energy must come to the surface and dissipate in the less dense air. Mineral wool, conversely, will cool almost as quickly as the heat source is removed because of its excellent ability to dissipate heat, owing to its light density.

Convection

Air that is hotter than its surroundings rises. Air that is cooler than its surroundings sinks. Air is made up of many molecules floating about freely. Even so, it still has weight. Some molecules are made up of the same element. For example, oxygen in its natural state will combine with another oxygen atom to form a stable oxygen molecule. In a given volume, air at a given temperature will have the same density. When heated, as in conduction theory, the molecules become agitated and begin to collide with one another. In the process, the molecules are demanding more space to accommodate the vibrations and they push into one another as they seek that space, **Figure 4-28.** When that happens, the density of a given volume is reduced and it weighs less. Because it weighs less, it rises until it reaches equilibrium—the level at which the weight is the same as the surrounding atmosphere. On the way up, it mixes with cooler air around the perimeter of the column. Once there, the agitation reaches its maximum. As the molecules calm down, they start to return to their original density and become less buoyant. As the buoyancy decreases, that particular volume of air drops. If it is heated again by the rising heat energy, its agitation increases again. If it falls outside the **thermal plume**, it drops down to the bottom of the plume.

In theory, an unobstructed plume that is unaffected by any other influences such as horizontal air current would behave in the mushroom fashion. In a structure, the flow and rate of convection would be accelerated when the heated air is confined and must seek a "cooler" space after meeting a vertical obstruction such as a ceiling. The "mushroom"

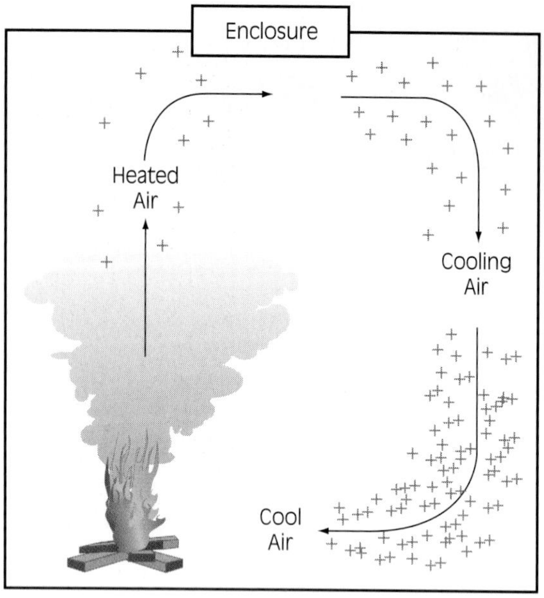

Figure 4-28 Convection. As air is heated, its molecules become excited and seek more space.

heating and cooling effect contributes to horizontal spread of fire at upper levels, **Figure 4-29**.

Convection currents created by heat that occur in the air also occur in liquids in the same fashion. The process of evaporation is accelerated in a liquid heated in an open-topped container. The molecules that are heated tend to bounce off the sides and reintroduce the energy back into the liquid. The reintroduced energy then causes more molecules to collide until the energy forces them over the top of the container where they escape. This is also the same phenomenon that causes pressure in a closed container to rise. In the case of the closed container, there is no escape and the reintroduced energy is kept in the contained vessel.

Radiation

The last form of heat transfer occurs by radiation. As discussed earlier, heat energy can be transmitted directly when molecules collide with one another and cause the waves of heat energy to travel. It stands to reason that in the absence of molecules, such as would be the case in a vacuum, heat energy would not be able to travel. However, this is not the

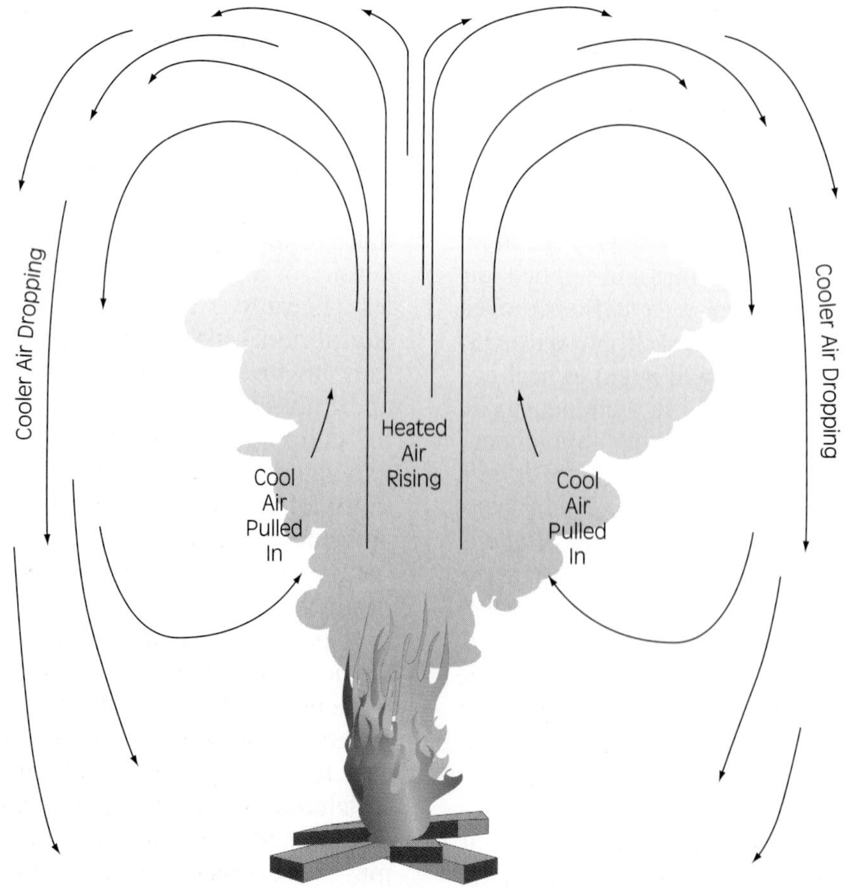

Figure 4-29 Fire plume.

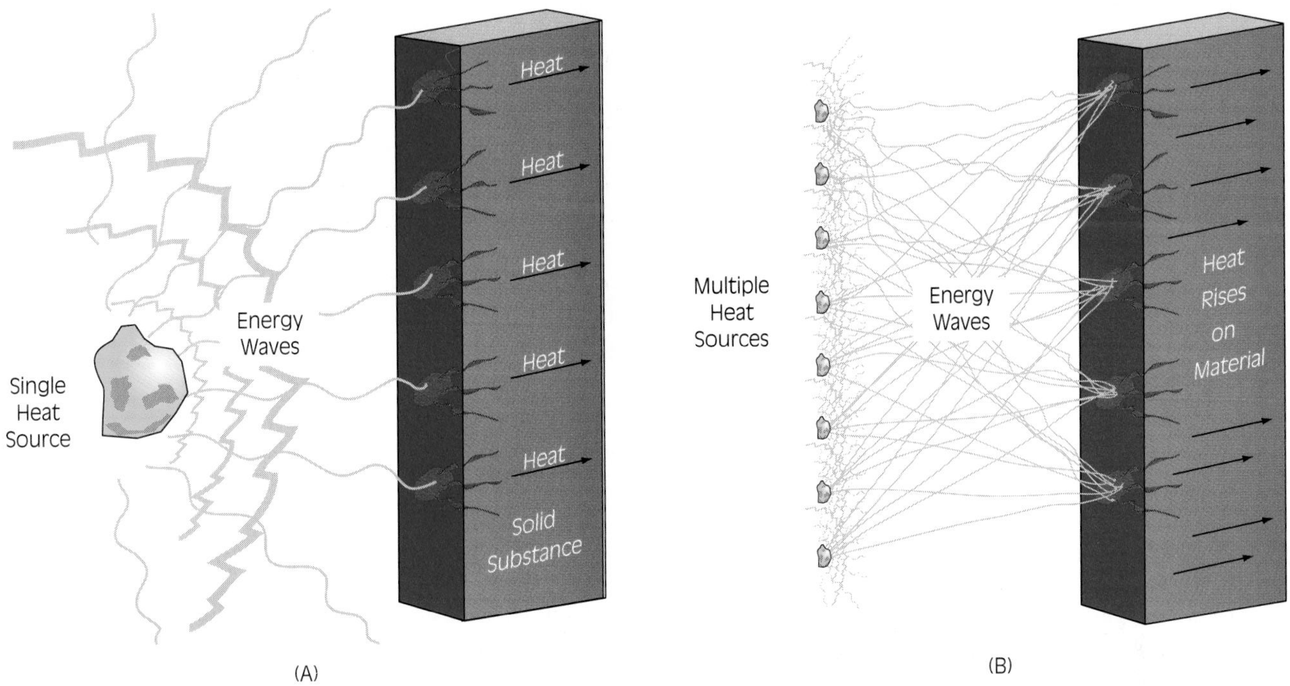

Figure 4-30 (A) Radiation, single heat source. (B) Radiation, multiple heat sources.

case. When combustion occurs, light is produced. Light travels by way of lightwaves. These lightwaves range from ultraviolet to infrared. Contained in the lightwaves are electromagnetic forces that can actually travel across vacuums and deposit themselves on remote objects. An example of this is the sun and its ability to heat the Earth.

Fire produces the same infrared lightwaves and, with enough concentration, can permit fire to jump from the source to a distant object, heat it up, and, if intense enough, cause it to ignite, **Figure 4-30**. Several factors need to be in place before this can occur: the source must be strong enough to sustain the bombardment of the lightwaves; the recipient must be able to absorb the heat energy; and the object must be able to build up the heat without it dissipating away through its own heat sink.

Radiation is a major contributor to flashover, in which heat buildup at upper levels in a compartment radiates heat down into the room. Eventually the objects in the room at lower levels approach and reach their respective ignition temperatures through this mode.

Firefighters must cope with this phenomenon because fire can extend through radiation as much as through conduction or convection. Water is used to absorb the heat and carry it away so it cannot serve to heat surrounding materials. Water absorbs the accumulating heat from the exposed objects, thereby depriving them from reaching ignition temperature. By controlling this element of the fire tetrahedron, the fire is prevented from extending. To a limited extent, water can also be employed to combat conduction extension. In that case, the water carries away the heat, keeping the unburned or unheated portion of the material from reaching the ignition temperature or permitting it to transfer the heat to another substance.

THERMAL CONDUCTIVITY OF MATERIALS

All matter will conduct heat. The ability of a material to conduct thermal energy depends on its density. Because heat transmission is actually a transfer of agitation from one molecule to another, for heat to be conveyed from one place to another, molecules must be present. The less dense a material, the more difficult it is for heat to be transferred through it. In the absence of all molecules—a vacuum—there is no transmission of heat. The more dense a material is, the greater the thermal conductivity potential. If a substance has a great deal of open space between its molecules, the molecules can become agitated without transferring that agitation to other molecules. The same holds true for material in a larger scope beyond the molecular level. Materials such as cellulose or mineral wool, which have voids in their makeup, contain a great deal of space, and these spaces serve to insulate the heat from being transferred from the heated side of the material to the unheated side.

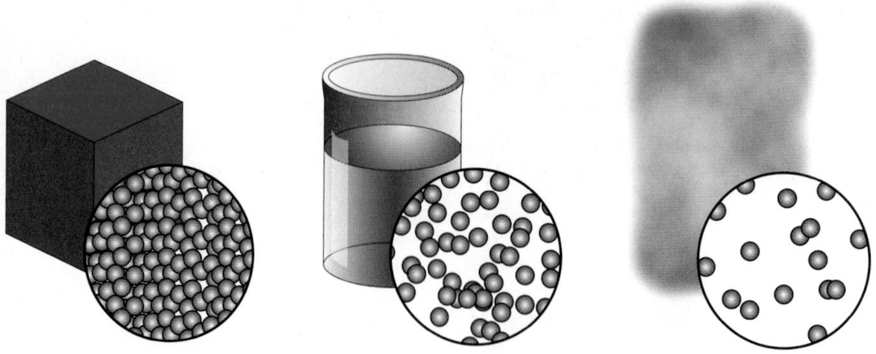

Figure 4-31 States of matter.

PHYSICAL STATE OF FUELS AND EFFECT ON COMBUSTION

Matter can be found in one of three states: solid, liquid, or gas. To take part in the combustion process, most fuels must be in their gaseous state. Although this is not the case in every single situation, in most cases, a molecular free-floating state must exist before oxidation reactions can take place. In many cases, wood for example, the solid matter is turned to gas and it combusts almost simultaneously. Knowledge of a material in its physical state and what state it is required to be in before combustion takes place is of paramount interest to the firefighter. The physical state can affect combustion, **Figure 4-31**.

Solid

The molecules in a solid material are packed closely together. That criterion gives the material its density and, to a great extent, its physical state. The molecules are bound together by bonds that maintain the form and shape. When heat is applied, the molecules become agitated and begin to collide with one another. In some cases, this causes the molecules to break apart and free up the components of the material. Among some of the molecules being broken apart are those that will readily combine with oxygen and become oxidized. The others are left to free float for the time being. As the bonds with oxygen are formed, a heat-producing exothermic reaction occurs, and the released heat causes the free-floating molecules to develop a greater affinity for the oxygen. These newly formed by-products of oxidation then release even more heat, and the chain reaction called combustion is under way.

Heat that is produced causes additional molecules to break away from the solid physical state, become a gas, and combust. The additional heat causes more reactions to occur, and a free-burning self-sustained chain reaction results. In the case of the solid material, the applied heat is absorbed into the mass and dissipates to some degree. When the amount of heat applied exceeds the ability of the mass to dissipate that heat, chemical breakdowns occur and the self-sustaining burning reaction results and continues until it either is interrupted or runs out of fuel.

Liquid

In a burning liquid, the ability to burn is dependent on the substance's ability to place its molecules into suspension. A liquid cannot burn unless it is in suspension, also referred to as **atomization**. That liquid will then become engaged in the self-sustaining combustion process if it has the affinity for oxygen and will permit oxidation. Not all liquids possess that property. Some will boil, evaporate, and never engage in combustion. Water is a prime example.

Similar to a solid, a liquid will act as a heat sink and dissipate the introduced heat into the cooler areas of the liquid. When the entire pool of liquid is heated or the ability to dissipate is overcome by the application of heat, a rise in temperature of the liquid will occur, resulting in boiling or **flash point** ignition. In most cases, there is a physical state transformation from solid to liquid to gas before combustion can take place.

Gas

A gas apart from the other two states is essentially primed for combustion. Its physical property is in a ready-made state that will permit a chemical reaction.

Awareness of these properties and their place in the combustion process is another weapon in the arsenal for fighting fire. Knowing that keeping a liquid cool enough to prevent **vaporization** or that by excluding oxygen ignition can be minimized or even prevented becomes a valuable piece of information when dealing with an incident such as a tank truck fire or a liquid fuel spill.

> **Note** Remember that as a substance changes from one state to another, heat is either given off (exothermic reaction) or absorbed (endothermic reaction).

We see heat absorbed when water turns to steam, and a carbon dioxide extinguisher will actually create snow and collect frost on the cylinder wall when the liquid is released as a gas. Essentially, when pressure is applied heat is released. If enough pressure is applied to a gas, it turns to a liquid (e.g., carbon dioxide gas to liquid carbon dioxide). Liquid CO_2 is used in carbon dioxide extinguishers. When additional pressure is applied, the liquid turns to a solid with the release of additional heat (liquid carbon dioxide to dry ice). Conversely, when the solid becomes a liquid, it absorbs great quantities of heat (ice to water) and again, when the liquid becomes a gas (water to steam).

The heat-absorbing capabilities of the water-steam conversion are employed when extinguishing fire in ordinary combustibles. Water robs the fire of its heat and reduces the heat level so that self-sustaining combustion is no longer possible.

THEORY OF FIRE EXTINGUISHMENT

The way to stop a fire is to remove one of its essential ingredients. Heat, fuel, oxygen, and a continuing chemical reaction are the four elements in the fire tetrahedron. Remove one of them and the fire will collapse.

UNIQUE FIRE EVENTS

There are several unique events that can occur in a fire within a compartment or structure that firefighters must be familiar with, including thermal layering, rollover, flashover, and backdraft.

Thermal Layering

Hot air and gases from fire rise and will continue to rise until they reach equilibrium, or balance, with the surrounding atmosphere. **Thermal layering** occurs when the gases produced by fire stratify into layers based on their temperatures. The hottest gases rise to the highest point in the compartment or room, and the gases with the lowest temperatures stay closer to the bottom of the container, or nearer to the floor.

When confined to a structure, the hottest of the gases and air will accumulate near the ceiling of the compartment or room, and as the amount of by-products from fire increases, the gases, heated air, and smoke will **bank down** until they can find an escape route.

Thermal layering is an important concept for firefighters because it affects how to enter and function in a room or area that is on fire or where the fire has just been extinguished. Firefighters must stay low to the floor in attacking structural fires because the hottest gases rise and stay at the highest layers in the compartment.

Rollover and Flashover

As the contents of a closed compartment or room burn, the superheated gases stratify and thermal layering occurs. These superheated gases can ignite in the uppermost thermal layers and cause the fire to travel across the top of the compartment. This is known as **rollover**, or flameover. This situation can be dangerous for firefighters because it can occur quickly and cause the fire to travel over the top of the firefighting team and sometimes get behind the team, affecting the firefighers' egress point from the compartment. Also, a rollover can cause the fire to spread to unaffected areas of the compartment, expanding the amount of area involved.

As a fire continues to burn in a compartment, more and more heated gases are generated, and the hottest layer of gases becomes larger, expanding down to the container. These superheated gases raise the temperature of all the unburned contents of the compartment. When the gases and fire have brought the entire contents of a container to their ignition temperature, a flashover can occur. In a **flashover**, the entire contents of the compartment ignite almost simultaneously, generating intense heat and flames. This rapid change in the compartment occurs in seconds and is a very dangerous and often deadly occurrence for firefighters. The tactics and skills covered in this text will assist firefighters in strategies to avoid flashover situations through the extinguishment of fire and ventilation of the compartment.

Backdraft

The fire tetrahedron shows that there are four elements that must be present to support combustion: fuel, oxygen, heat, and a continuing chemical reaction. A fire burning in a closed container, such as a room in a structure, can use all the available oxygen in the compartment, slowing the combustion process and the fire. The introduction of a source of oxygen to the compartment, such as opening a door or window to a room, will bring together the four elements of the fire tetrahedron again, causing the room to violently burst into flames. A **backdraft** is a dangerous situation. Firefighters should look for the signs of a potential backdraft situation when they are on the scene of fires in closed areas or structures. Signs of backdraft are covered in Chapter 18 of this text.

CLASSES OF FIRE

To better assist both the firefighter and the non-firefighter alike, fires have been classified into different types based on the substance burning. This is primarily done so that the correct extinguishing agent can be applied to the fire by the firefighter.

- *Class A:* Class A type fires are made up of ordinary combustibles such as cellulose, rubber, or plastic. Combustibles such as paper, wood, cloth, rubber, and other organic solids including petrochemical solids (plastics) make up this class.
- *Class B:* Class B type fires are fueled by liquids, gases, or grease-type fuels. Oil, gasoline, alcohol, and other liquids are the more common types found in this class of fuel.
- *Class C:* Class C type fires are basically fueled by electricity. In this case, the electricity is actually the heat source that propagates the fire and often communicates to other fuels of the Class A or B type to sustain the burning process. If no other fuels are involved, merely shutting off the flow of electricity is enough to extinguish the fire.
- *Class D:* Class D, a less common fire type, is fueled by metals. A particular class of heavy metals, which can be identified on the periodic table of the elements and found mostly in the alkali metal group, will burn. Most common metals in the group are magnesium, titanium, zirconium, sodium, and potassium.

Firefighter Fact In 1998 a new classification of fire, Class K, was created to cover fires in combustible cooking fuels such as vegetable or animal oils and fats. Class K fires are similar to Class B fires, although the oils and fats have some special characteristics. Class K fires and extinguishing agents will be covered in detail in Chapter 8.

Because each type of fuel has different burning characteristics, the method of extinguishing them differs, **Figure 4-32**.

The Class A fire, made up of what are considered normal combustibles, is extinguished by cooling the fire. The application of water cools the fire by absorbing heat as water is converted to steam. When enough of the heat is removed, the temperature of the fire is lowered below the ignition temperature of the substance and extinguishes the fire.

In a Class B fire, the fuel is a liquid, grease, or gas. As the temperature of ignition is approached, the liquid fuel vaporizes into gas and ignites. The gaseous materials are already in that state. The application of a smothering agent is used to prevent oxygen from getting to the fuel and propagating the chain reaction of fire by removing the oxygen element of the fire tetrahedron. In this case, the fire collapses due to a lack of oxygen.

A Class C fire, fueled by electricity, is overcome by the removal of the flow of electricity—the source and sustainer of heat. In this case the removal of the fuel, electricity, is the action taken to break down the fire tetrahedron and put the fire out. Keep in mind that in most cases an electrical fire is only the initiator. The fire then communicates to surrounding combustibles and can become a combination Class A or B in addition to Class C fire. Once the electrical power flow is removed, it becomes a straight Class A or B fire and is extinguished accordingly. In other cases, removal of the electrical current will extinguish the fire without any further effort. So here removal of what essentially is the fuel, one element of the tetrahedron, serves as the mechanism by which Class C fires are extinguished.

The Class D heavy metal fire is a chemical reaction fire. Almost all metals will burn if conditions are correct. Although some merely oxidize very slowly as in metal rusting, others will burn with great violence producing high heat and brilliant light. Among the combustible metals of significance are thorium, titanium, plutonium, hafnium, lithium, magnesium, zirconium, zinc, uranium, sodium, potassium, and calcium.

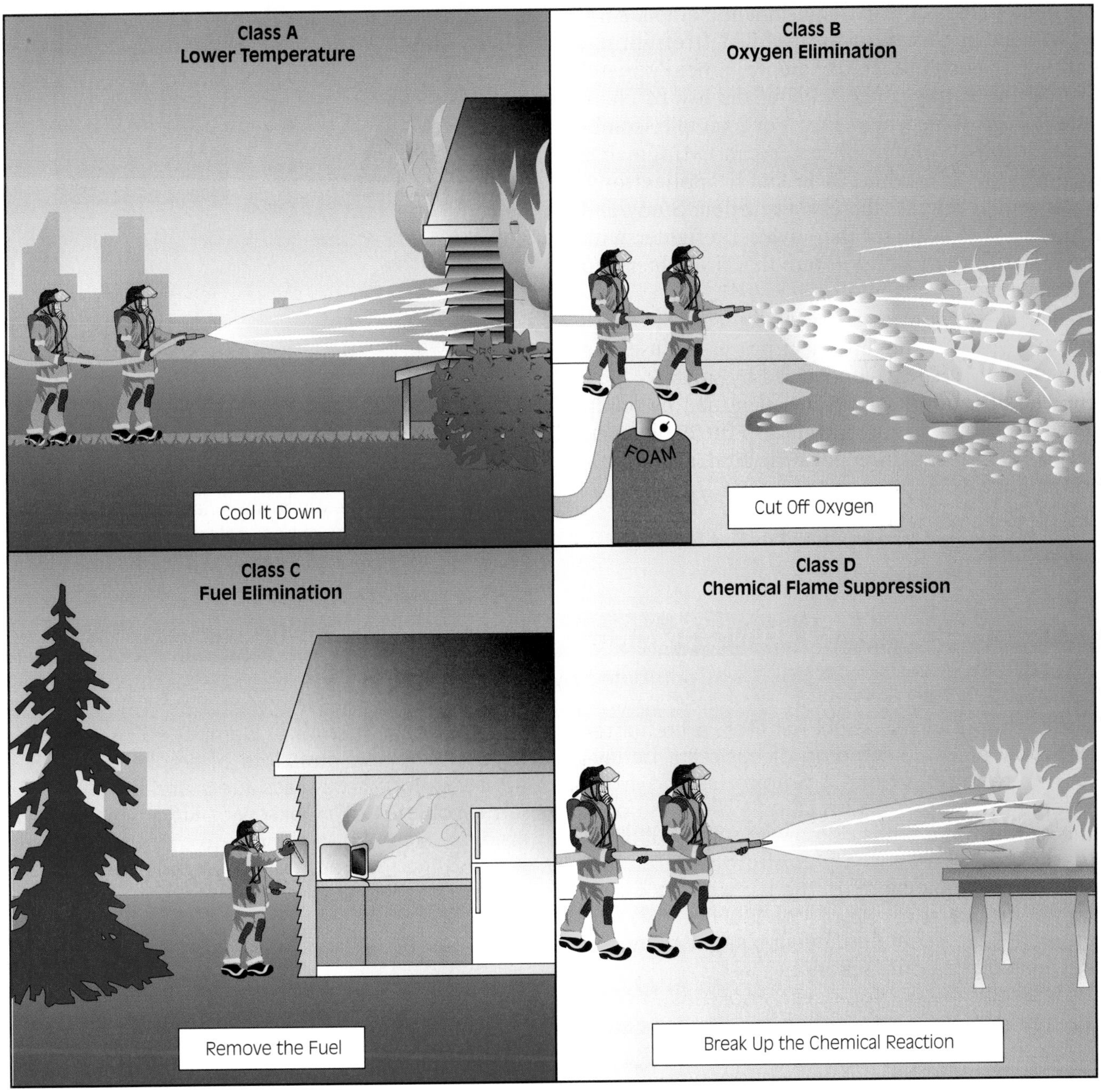

Figure 4-32 Four methods of extinguishing fires that closely follow the classification assigned to the material involved.

The combustible heavy metals differ somewhat in their reactions under fire. In some cases, the mere presence of water will cause a violent reaction, releasing heat and brilliant light. In other cases, the mere presence of air will cause the reaction. Some metals will burn so hot that the water molecule will actually be broken down into its component hydrogen and oxygen atoms. Then the hydrogen burns away in the presence of oxygen creating a brilliant and hot fire. Each metal's characteristics must be evaluated. Fortunately, these metals are not found in great abundance in normal occupancies. They are usually found in industrial processes and their presence should be known in advance to responding firefighters.

Other metals are also subject to burning but usually only when the material is in fine shaving form. For example, steel will not ignite easily, but a piece of steel wool will ignite rapidly when flame is applied.

While the classes of fires are essentially predicated on the fuel type, the classification follows the same track as the extinguishment method. And, as with most things in the field of firefighting, nothing is absolute. For example, a heavy metal fire can be extinguished without the use of chain reaction–breaking chemicals. For example, sometimes a huge volume of water can extinguish a magnesium fire, whereas in small applications, water will accelerate the chain reaction. Study and exposure will acquaint the novice firefighter with the basics. Experience will turn that firefighter into a knowledgeable veteran.

Classed primarily as a means by which to identify the type of extinguishment process to use, the extinguishment agents are labeled by several codes. The classification symbols are identified by letter, shape, and color and are attached to fire extinguishers for better recognition, identification, and utilization, **Figure 4-33**.

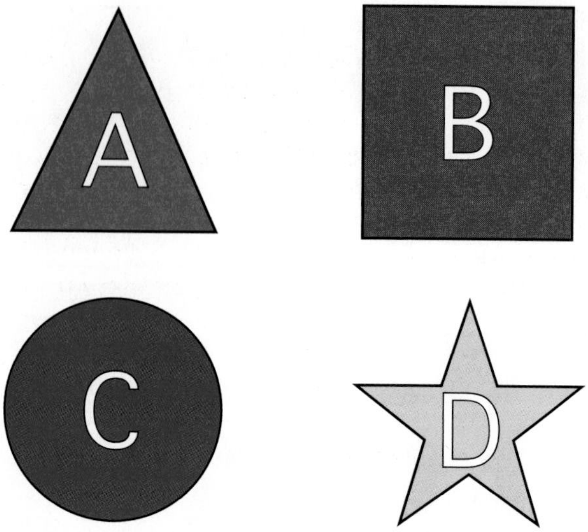

Figure 4-33 Fire extinguisher classification symbols may be displayed by shape, color, and/or letter.

Reading Smoke at Structural Fires

Contributed by David Dodson

When responding to structural fires, a firefighter can apply an understanding of basic fire behavior by "reading smoke." Smoke issuing from multiple openings of a structure is often the only clue as to what a fire is doing within the building. Reading smoke can help firefighters discover clues about the location of the fire within a building as well as the severity of the fire and the potential for a hostile fire event such as flashover or smoke explosion/backdraft.

Typically, firefighters view smoke as "light" or "heavy." While this is fine for a rapid radio report, it is not descriptive enough when trying to understand what is actually happening with a fire. Smoke from a structural fire has four attributes that must be analyzed. These are *volume*, *velocity*, *density*, and *color*. All four of these factors of smoke are comparative—each opening that issues smoke is compared to the others—thus painting a story about the fire behavior and the location of fire within the building.

Volume

Smoke volume is an indicator of the amount of fuels that are "off-gassing" within a given space. In itself, smoke volume tells very little about the fire. A hot, clean-burning fire will emit very little visible smoke—yet a hot, fast-moving fire in a poorly ventilated building will show a tremendous volume of smoke. Dampened material will burn slowly and emit lots of smoke (typically a lighter color). Low-mass fuels and plastics emit large volumes of smoke with little heat. A large volume of smoke may indicate that the container holding the fire (room, area, or other portion of a structure) is full and needs ventilation.

Velocity

The "speed" that smoke leaves a building is referred to as velocity. In actuality, smoke velocity is an indicator of pressure that has built up within the building. Only two things can create smoke pressure: heat and restricting the volume of smoke within a container (a room or building). Velocity caused by heat will typically rise and slow gradually after it leaves the building. Velocity caused by restricted volume will immediately slow and balance with the outside atmosphere. If the velocity of the smoke leaving an opening is agitated or turbulent (some call this boiling or rolling smoke), a flashover is likely to occur. This flow is caused by the rapid expansion of the smoke due to heat. In other words, the box cannot hold the rapid expansion and therefore a turbulent flow of smoke is created. In these cases, the structure must be ventilated and cooled. Persons in these conditions have little chance of surviving due to smoke toxicology and thermal exposure.

Density

Incomplete burning causes smoke density. Some refer to this as smoke thickness. Smoke density is indicative of the amount of fuel that is laden within the smoke (particulates, gases, and aerosols). The greater the smoke density, the more likely a hostile fire event, such as flashover or rapid fire spread, can occur. In essence, the thicker the smoke, the more spectacular the flashover or fire spread. Thick, black smoke within a compartment reduces the chance of life sustainability due to the toxicology of the smoke. A few breaths of thick, black smoke will render a victim unconscious and cause death within minutes.

Color

For single-fuel fires, smoke color may indicate the type of material burning. In typical residential and commercial fires, it is rare that a single fuel source is emitting smoke. Smoke color can, however, tell the firefighter what stage of burning is taking place or where the fire is within a building. Virtually all solid materials will emit a white "smoke" when first heated. This white smoke is actually moisture. As a material dries out and breaks down, the color of the smoke will change. Natural materials will change to tan or brown, while plastics and painted surfaces will turn to gray. Gray is a result of moisture and hydrocarbons (black) mixing. All materials will eventually off-gas a black smoke—flame contact will cause materials to give off black smoke right away. As smoke leaves a fuel that is ignited, it heats up other materials and the moisture from those objects can cause black smoke to turn gray. As smoke travels, carbon content from the smoke will deposit along surfaces and objects. This also lightens the smoke color.

By combining these smoke attributes, some basic observations about the fire can be made before firefighters even enter a structure. Smoke velocity and color differences from opening to opening help firefighers find the location of the fire. Faster/darker smoke is closer to the fire seat, whereas slower/lighter smoke is farther away. If smoke from multiple openings is a constant color and velocity, firefighters should start thinking that the fire is deep-seated within the building. In these cases, the smoke has traveled some distance or has been pressure-forced through close doors or seams (walls/concealed spaces) prior to leaving the building. "Black fire" is a good phrase to describe smoke that is high volume, turbulent velocity, ultradense, and black. Black fire is a sure sign of impending autoignition and flashover.

Other factors influence smoke and may cause a false read. Wind, thermal balance, fire streams, ventilation openings, and sprinkler systems change the appearance of smoke. These four simple factors—volume, velocity, density, and color—can help firefighters refine their ability to read smoke and ultimately help protect their own safety and predict what fires will do next within a building.

Lessons Learned

An understanding of fire behavior is a fundamental requirement for anyone who has aspirations in the field of firefighting. Like the chemist who must know and understand measurements, or the tennis player who must know the rules of the game and employ them in a strategy, so must the firefighter know and understand what to expect fire will do based on information obtained on arrival at a fire scene. Because so many factors will be unknown in the initial stages of a fire, such as structural components, contents in the structure, effect of wind or weather, and other elements, it is incumbent on the firefighter to be armed with as much knowledge as possible about fire behavior so trends of fire spread in a given situation can provide the basis for strategy and tactics to be employed while more information is obtained.

KEY TERMS

Atomization The separation of atoms and molecules into an unconnected state where they are in suspension rather than in liquid form.

Backdraft A sudden, violent reignition of the contents of a closed container fire that has consumed the oxygen within the space when a new source of oxygen is introduced.

Bank Down A condition in which the heat, smoke, and fire gases have reached the uppermost level in a compartment and, instead of continuing up, begin to push down from the ceiling toward the floor.

Boiling Liquid Expanding Vapor Explosion (BLEVE) Describes the rupture of a container when a confined liquid boils and creates a vapor pressure that exceeds the container's ability to hold it.

Bond A substance or an agent that causes two or more objects or parts to bind.

British Thermal Unit (Btu) A measurement of heat that describes the amount of heat required to raise 1 pound of water 1°F.

Combustion The chemical action in which heat and light are produced and the heat is used to maintain the chemical chain reaction to continue the process.

Compound A combination of substances joined in a chemical bond that exists in a proportional amount and cannot be separated without chemical interaction.

Density The mass per unit volume of a substance under specified conditions of pressure and temperature.

Diffusion A process in which liquids or gases mix with one another by natural processes stemming from molecular movement.

Electrical Conductor Any material that will permit electricity to flow through it.

Endothermic Reaction A chemical reaction that absorbs heat.

Equilibrium When referring to gas or liquids, a state where a balance has occurred in mixture or weight.

Evaporation A process in which the molecules of a liquid are liberated into the atmosphere at a rate greater than the rate at which the molecules return to the liquid. Ultimately the liquid becomes fully airborne in a gaseous state.

Exothermic Reaction A chemical reaction that liberates heat.

Explosive Limits A concentration of a gas or liquid that is not too rich or too lean to ignite with force.

Fire Also called combustion, it is the chemical action in which heat and light are produced and the heat is used to maintain the chemical chain reaction to continue the process.

Fire Tetrahedron Four-sided pyramid-like figure describing the heat, fuel, oxygen, and chemical reaction necessary for combustion.

Flammable Limits The concentration level of a substance at which it will burn.

Flash Point The temperature at which a liquid will liberate a flammable gas.

Flashover A sudden event that occurs when all the contents of a container reach their ignition temperature nearly simultaneously.

Heat Sink The term used to denote a place where heat is drained away from a source.

Hydrocarbon Any of numerous organic compounds, such as benzene and methane, that contain only carbon and hydrogen.

Ignition The point at which the need for outside heat application ceases and a material sustains combustion based on its own generation of heat.

Inorganic A substance that is not of any living organism.

Matter Something that occupies space and can be perceived by one or more senses; a physical body, a physical substance, or the universe as a whole. Something that has mass and exists as a solid, liquid, or gas.

Molecule The smallest particle into which an element or a compound can be divided without changing its chemical and physical properties; a group of like or different atoms held together by chemical forces.

Organic A substance derived from living organisms.

Oxidizer A catalyst in the breakdown of molecules.

Pyrolysis Decomposition or transformation of a compound caused by heat.

Rollover A phenomenon where the burning of superheated gases from fire extends into the top areas of the compartment in the upper thermal layers.

Thermal Layering The stratification of gases produced by fire into layers based on their temperature.

Thermal Plume A column of heat rising from a heat source. A fully formed plume will resemble a mushroom as the upper level of the heat plume cools, stratifies, and begins to drop outside the rising column.

Vaporization The process in which liquids are converted to a gas or vapor.

REVIEW QUESTIONS

1. Why is knowing the sources of heat important to the firefighter?
2. What are the four modes of heat sources and why are they important?
3. Does an increasing span for a flammable limit denote a more hazardous material?
4. What are the three modes of heat transfer and why are they important?
5. What are the three states in which fuel may be found?
6. What is the basis for classification of fire into classes?
7. What is the significance of thermal balance and imbalance in a fire?

Endnotes

1. The American Heritage® Dictionary (CD-ROM), based on *The American Heritage® Dictionary of the English Language,* Third Edition © 1992 by Houghton Mifflin Company. Licensed from and portions © 1994 INSO Corporation.
2. *Fire Protection Handbook,* 18th ed., pp. 4–9. National Fire Protection Association, Quincy, MA, 1997.
3. *Fire Protection Handbook,* 18th ed., pp. 2–21. National Fire Protection Association, Quincy, MA, 1997.

Additional Resources

Bettelheim, Frederick, William H. Brown, and Jerry March, *Introduction to General, Organic, and Biochemistry,* 7th ed., Brooks/Cole, a part of the Thomson Corporation, Belmont, CA, 2004.

Bevelacqua, Armando S., *Hazardous Materials Chemistry,* Delmar Learning, a part of the Thomson Corporation, Clifton Park, NY, 2001.

Casey, James F., *The Fire Chief's Handbook,* Reuben H. Donnelley Corp., New York, 1978.

Cracolice, Mark S. and Edward I. Peters, *Introductory Chemistry: An Active Learning Approach,* 2nd ed., Brooks/Cole, a part of the Thomson Corporation, Belmont, CA, 2004.

Fire Protection Handbook, 18th ed., National Fire Protection Association, Quincy, MA, 1997.

Joesten, Melvin, D., James L. Wood, and Mary E. Castellion, *World of Chemistry: Essentials,* 3rd ed. Brooks/Cole, a part of the Thomson Corporation, Belmont, CA, 2004.

Klinoff, Robert, *Introduction to Fire Protection,* Delmar Learning, a part of the Thomson Corporation, Clifton Park, NY, 1997.

Lowe, Joseph D., *Wildland Firefighting Practices,* Delmar Learning, a part of the Thomson Corporation, Clifton Park, NY, 2001.

Masterton, William L. and Cecil N. Hurley, *Chemistry: Principles and Reactions,* 5th ed., Brooks/Cole, a part of the Thomson Corporation, Belmont, CA, 2004.

McMurry, John E., *Organic Chemistry,* 6th ed., Brooks/Cole, a part of the Thomson Corporation, Belmont, CA, 2004.

Quintiere, James G., *Principles of Fire Behavior,* Delmar Learning, a part of the Thomson Corporation, Clifton Park, NY, 1998.

Skoog, Douglas A., Donald M. West, James F. Holler, and Stanley R. Crouch, *Fundamentals of Analytical Chemistry,* 8th ed., Brooks/Cole, a part of the Thomson Corporation, Belmont, CA, 2004.

Whitten, Kenneth W., Raymond E. Davis, Larry M. Peck, and George Stanley, *General Chemistry,* 7th ed., Brooks/Cole, a part of the Thomson Corporation, Belmont, CA, 2004.

CHAPTER 5

FIREFIGHTER SAFETY

David Dodson, Lead Instructor, Response Solutions, Colorado

OUTLINE

- Objectives
- Introduction
- Safety Issues
- The Safety Triad
- Firefighter Safety Responsibilities
- Lessons Learned
- Key Terms
- Review Questions
- Endnote
- Additional Resources

STREET STORY

I was at work one day when the phone rang. The caller stated that my son Scott, who happens to be both a career and volunteer firefighter, was injured at a fire on his career job. It is the phone call a parent always dreads. I immediately had a rush of adrenaline, an increase in respiration, and became flushed. I asked, "How bad?" The caller informed me that luckily the injuries appeared to be minor.

Of course I made a phone call or two to verify the information and to explain the situation to my wife. I was informed that Scott had some burns on his shoulders and hand, where there would be some scarring, and that he would miss a shift or two, but he would be O.K. "O.K.?" I guess that term takes on a new definition to a father who is 45 miles away and worried!

I left work early to coincide with Scott's arrival at home. He had large bandages around his shoulders and his hand, and he was beginning to feel some pain. We talked about the fire, and he told me it was a typical garden apartment-type fire and that his company had been assigned on automatic mutual aid. He and his lieutenant were in search mode because residents reported an eleven-year-old was trapped. Their backup line was right behind them, but it was not quite ready and they were; seconds became hours!

During their search, crawling on hands and knees to stay below the heat and smoke as they had learned, Scott felt something very hot on the back of his right hand, a sharp pain. He turned to his lieutenant and both of them shouted at the same time, "Let's get out of here now. It is too hot!" They exited quickly.

They were "greeted" at the door, and their coats were smoking and discolored around the shoulders. The firefighter at the door began patting the smoking gear, stating it was "on fire." The air barrier was now gone, and both the lieutenant and Scott felt the heat, big time! They both ended up with second-degree burns on both shoulders, and the hand burn resulted from the inside Nylon label on the gloves melting, even though they were compliant gloves!

The good news is that the properly designed and worn personal protective clothing prevented serious injuries, maybe even third-degree, debilitating burns. The coats, pants, and gloves could be replaced, but the scar tissue is still there. Scott received a very nice leather jacket from the PPE outer shell material manufacturer, and the damaged PPE pictures can be seen at major fire service conferences across the country.

As Scott's parents, it was a difficult day for both of us, and emotions were running high. I think sometimes we forget how our choice to be a member of the fire service can have such an impact on our family members. His mother was very concerned about the inevitable scarring and strongly suggested that he really needed to quit doing that! Scott looked her in the eye and with deep conviction in his voice stated, "Mom, it's my job." Kind of made me shiver. But I know one thing. Scott and his lieutenant learned a lot that day—about safety, about PPE, and that on the fireground you never know what might come next. As he shares what he learned with other firefighters, foremost on his mind is what he might teach to his son, Cody, the next generation of Windisch firefighters.

—Street Story by F. C. (Fred) Windisch, Fire Chief, Ponderosa VFD, Houston, Texas

OBJECTIVES

After completing this chapter, the reader should be able to:

- Define risk management.
- List the leading causes of death and injury in the fire service.
- List the NFPA standards that affect and pertain to firefighter occupational safety.
- List the five components that make up the accident chain.
- List the three key components of the safety triad.
- Discuss the difference between formal and informal procedures.
- Name the three factors that influence the equipment portion of the safety triad.
- Name the three factors that influence the personnel portion of the safety triad.
- Name the three partners that work together to achieve firefighter safety.

INTRODUCTION

Simply put, the firefighting profession is one of significant **risk**—that is, one filled with the potential to be seriously injured or killed. Typically, this potential for death has created the aura that firefighters are hero types. Unfortunately, the "hero badge" has led to unnecessary injuries and deaths. Today's firefighter understands that certain risks have no tangible benefit. In simple terms, this understanding is called **risk management**. Firefighters practice risk management when they ask questions of risk/benefit. Why risk a team of firefighters for a building that is basically lost? Why cross over a collapse-zone barrier tape just to get a better angle for a fire stream?

To better understand risk management and to improve firefighter safety, the firefighter needs to look at the common causes of injuries and deaths associated with firefighting. Additionally, an understanding of the forces that combine to make firefighting safer is imperative. With this knowledge, the firefighter can fill the "hero" role with intelligence and wisdom rather than recklessness.

SAFETY ISSUES

One way to keep firefighter injury and death events to a minimum is to understand what events and circumstances typically lead to injury.

> **Firefighter Fact** Firefighter injury and death information is invaluable and has, in many notable cases, prevented additional injuries. For example, the Hackensack (NJ) Fire Department lost five firefighters after a roof collapsed on interior firefighters battling an attic fire in an automobile service shop. This shop had a bowstring-trussed roof. Following the incident and subsequent investigations, recommendations were made by various fire service organizations to help prevent a similar incident.
>
> One Oregon chief publicly thanked the fire service organizations for their recommendations—he experienced a similar situation and withdrew his firefighters ten minutes before the roof collapsed.

A study of injury causes has inspired fire and safety professionals to create standards and regulations to help prevent injuries and deaths. These standards and regulations directly affect some of the training and tactics the fire service employs today. The firefighter who understands simple accident prevention steps is actually helping the fire service address safety issues.

Firefighter Injury and Death Causes

A study of firefighter injury and death statistics reveals that approximately one-half of all duty deaths and injuries occur at the incident scene. The other half is split between training, response to/returning from an incident, and "other" duties. This helps explain *where* firefighters get hurt and killed. The next question is what *caused* the injuries and deaths. **Figure 5-1** shows the leading causes of death, and **Figure 5-2** shows the leading causes of injuries.

Firefighter death statistics also track the nature of the injury that caused the death. These statistics have shown that heart attacks (as a result of stress) are the leading type of death-producing injury. Internal trauma, crushing injuries, and asphyxiation follow heart attacks. Data suggest that firefighter duty deaths are not noticeably decreasing—even with advances in equipment, training, and uniform procedures over the past decade. According to the U.S. Fire Administration, firefighter fatalities as a result of fire-related causes (burns, asphyxiation, and structural collapse) have actually increased.[1] This trend suggests that even more emphasis needs to be placed on firefighter safety during actual fires.

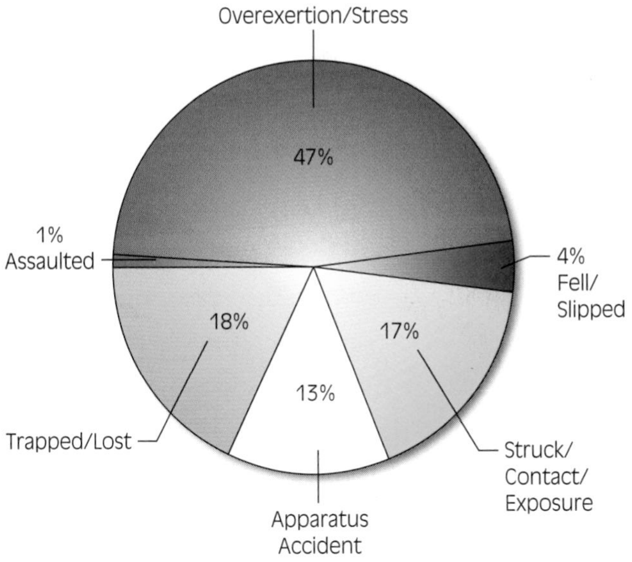

Figure 5-1 Firefighter deaths by immediate cause, 1990–2000. *(Source: U.S. Fire Administration, April 2002, Publication FA-220)*

> **Note** Unlike firefighter fatalities, firefighter injuries have not noticeably decreased—meaning, the fire service needs to increase accident prevention and risk management practices.

Safety Standards and Regulations

In 1970, the William Stieger Act was passed by congress and signed into law by President Nixon. This act created the **Occupational Safety and Health Administration (OSHA)**, which is part of the Department of Labor. OSHA is responsible for the enforcement of

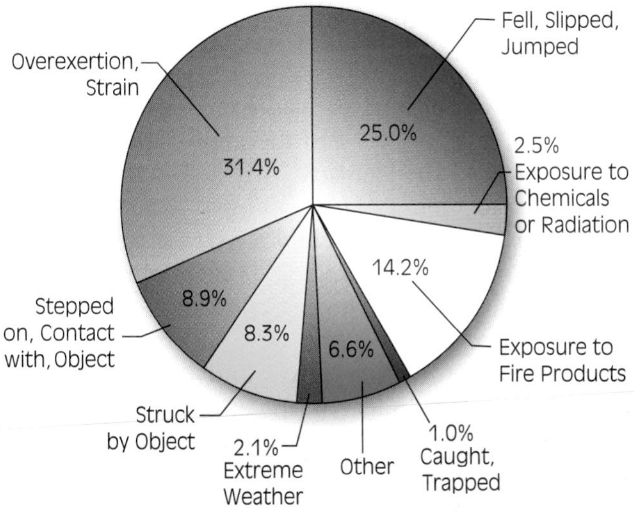

Figure 5-2 Firefighter injuries by cause—2000 sample. *(Source: NFPA, NFPA Journal, Nov/Dec 2001)*

NFPA 1500 Topics

Administration
Training and Education
Vehicles, Equipment, and Drivers
Protective Clothing and Equipment
Emergency Operations
Facility Safety
Medical and Physical
Member Assistance and Wellness
Critical Incident Stress Program

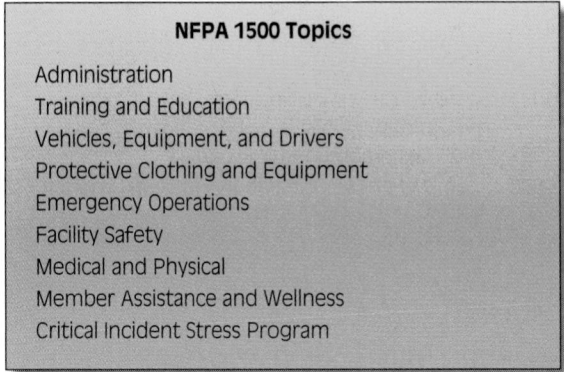

Figure 5-3 NFPA 1500, *Standard on Fire Department Occupational Safety and Health Program*, covers a multitude of topics.

safety-related regulations in the workplace. These regulations are part of the **Code of Federal Regulations (CFR)**. OSHA CFRs outline specific processes and procedures to be followed to help reduce the chance of injury. Interesting enough, most fire departments and public agencies were exempted from complying with OSHA CFRs when they were first developed. This created a double standard for occupational safety. In the 1980s, states began to address this issue by creating state OSHA plans that were compulsory for public agencies. For the first time, fire departments were obligated to follow OSHA CFRs adopted by an individual state. Unfortunately, the spontaneous nature of firefighting did not fit well into defined OSHA procedures and processes. Further, the firefighter injury and death trends of the 1970s and 1980s suggested that firefighting was the *most dangerous profession* in America. With this in mind, fire service representatives came together and helped write a comprehensive occupational health and safety standard for the fire service.

NFPA 1500, *Standard on Fire Department Occupational Safety and Health Program,* outlines consensus requirements and procedures for the safety of fire service personnel—much like OSHA CFRs do for industry. **Figure 5-3** shows the topical areas addressed by NFPA 1500. This standard was written to help fire departments address a whole host of safety issues.

> **Note** Currently, OSHA looks at NFPA standards as a guideline to address issues not directly covered by CFRs.

This OSHA/NFPA alliance is furthering the importance and accountability placed on firefighter safety. A classic example of this alliance is the requirement

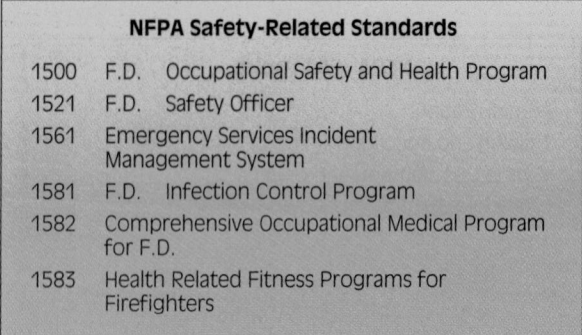

NFPA Safety-Related Standards

1500	F.D. Occupational Safety and Health Program
1521	F.D. Safety Officer
1561	Emergency Services Incident Management System
1581	F.D. Infection Control Program
1582	Comprehensive Occupational Medical Program for F.D.
1583	Health Related Fitness Programs for Firefighters

Figure 5-4 The NFPA 1500 series specifically addresses firefighter safety and wellness issues.

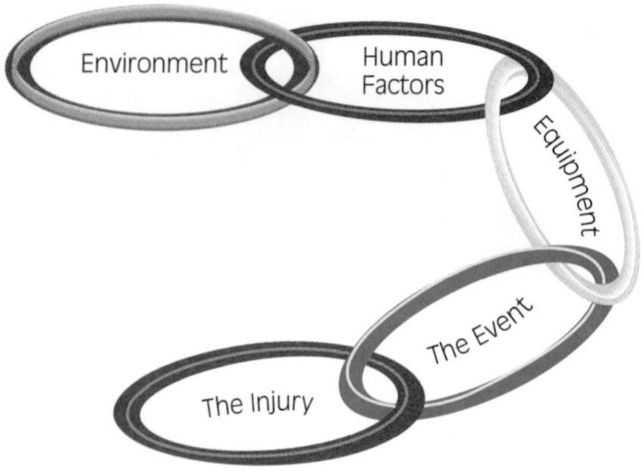

Figure 5-5 Accident prevention is simply the efforts expended to break the accident chain.

and clarification given to safe initial fire attack operations. The two-in/two-out rule is a concept underscored in OSHA 1910.134 and NFPA 1500. In essence, the rule states that before firefighters engage in interior fire attack operations, they must have two equipped and ready firefighters who form a team to make that attack. Additionally, two equipped and ready firefighters must be outside to monitor the progress of the interior crew and serve as an initial rapid intervention team should the interior firefighters experience an emergency. It is also important to note the outside team will likely need to perform some support activities (incident command or pump operations). The persons performing these tasks must be immediately available to rescue the interior team. These two are not considered a dedicated rapid intervention team—as more resources arrive, a dedicated rapid intervention team is required. This concept is covered more specifically in Chapters 19 and 23.

NFPA has also created other standards that address safety. It can be argued that all NFPA standards address safety, but specifically, NFPA has created a 1500 series that focuses on safety issues, **Figure 5-4**. The **National Institute of Occupational Safety and Health (NIOSH)** has also entered the firefighter safety equation. NIOSH does not set regulations or standards—they write recommendations based on the investigation of firefighter fatalities. In the 1990s, President Clinton signed legislation that mandated the investigation of incident-related firefighter fatalities. NIOSH is tasked with preparing recommendations so that other fire departments can take preventive steps to reduce the chance of an injury or fatality.

Accident Prevention

An **accident** can be defined as the result of a series of events and conditions that lead to an unsafe situation resulting in injury and/or property damage. This series of events and conditions is called the **accident chain**, **Figure 5-5**. The accident chain is made up of five components:

1. *The environment:* Included here are physical surroundings such as weather, surface conditions, access, lighting, and physical barriers.
2. *Human factors:* This includes human and social behaviors, training (or lack of training), fatigue, fitness, and attitudes.
3. *Equipment:* Included here are apparatus, personal protective equipment (PPE), maintenance and serviceability, proper application, and equipment limitations.
4. *The event:* The event is the intersection of the foregoing components. Something or someone had to bring those components together in such a way to create the unsafe situation.
5. *The injury:* This last part of the chain deals with the actual injury (or property damage) associated with the accident. A "near miss" or "close call" is an accident without injury or physical damage.

The key to accident prevention is awareness of these five components and the ability to "break the chain" prior to the last link. Any action designed to break the accident chain is known as an **intervention**. Intervention is typically a *reactive* action. Any strategy designed to reduce the potential of creating an accident chain is known as **mitigation**. Mitigation is typically a *proactive* action. Knowing the difference between mitigation and intervention is important because these terms are being used more and more frequently in all aspects of the fire service, **Figure 5-6**.

Most often, firefighters and fire officers are instrumental in preventing accidents through intervention.

Intervention (Reactive)	vs.	Mitigation (Proactive)
• Incident Safety Officers • Collapse Zones • Rehab		• Protective Equipment • Training • Physical Fitness

Figure 5-6 Accident prevention relies on *proactive* and *reactive* efforts.

> **Safety** A firefighter's awareness of surroundings, proficiency in tasks, appropriate use of equipment, and, most importantly, positive attitude about safety will ultimately prevent injuries.

THE SAFETY TRIAD

Most fire service operational environments are made up of three key components: procedures, equipment, and personnel. To mitigate injuries, each of these components needs to be addressed. This mitigation effort is called the safety triad, **Figure 5-7**.

(A)

(B)

(C)

Figure 5-7 The safety triad includes (A) procedures, (B) equipment *(Courtesy of Richard W. Davis)*, and (C) personnel.

Procedures

As stated, the three basic components or elements of the safety triad come together to form an operational environment like the one found at working structure fires or multicar extrications. If the operation is to be "safe," each component needs to be addressed.

Procedures or processes are the structures from which all activity at an incident begins. The first-arriving engine at a fire alarm activation will most likely follow a set of procedures or start a series of processes to investigate the alarm. Procedures can be classified as *formal* or *informal*. Formal procedures are those that are in writing as standard operating procedures (SOPs). In some departments, formal procedures are derived from standard evolutions or lesson plans. These evolutions and lessons can be drilled periodically, on a rotating basis, to ensure that a crew's response to a given situation is appropriate. Many of these procedures and evolutions have been written down to help ensure consistency and, therefore, a greater margin of safety.

Informal procedures are those processes and operations that are obviously part of the routine of a given department, but may not be written. Informal procedures are passed on through new member training as well as day-to-day operations. One example of an informal procedure is the practice of placing a full-face hood across bunker boots so that the firefighter must don the hood before putting on the boots. Another example is the company officer who talks through assignments at the start of each shift or when new members are assigned to the crew.

1. 1.1 Incident Command System
2. Emergency Ground Operations
   - 2.1 Rapid action group
   - 2.2 Gas/odor investigation
   - 2.3 Auto alarms
   - 2.4 Train fires
   - 2.5 Vehicle fire
   - 2.6 Fires at postal facilities
   - 2.7 Emergent driving procedure
   - 2.8 Kaneb pipeline response
   - 2.9 Volunteer and fire apparatus placement for motor vehicle accidents (MVAs)
   - 2.10 Operations involving Thompson Valley Ambulance
   - 2.11 Minimal staffing for interior firefighting
   - 2.12 Foreground formation and activation of companies
   - 2.13 Standard fire attack procedures/dwelling fires
3. Alarm Levels/Dispatching
   - 3.1 City alarm level assignments
   - 3.2 Rural alarm level assignments
   - 3.3 Fire resource officer
   - 3.4 Fire alarm panel operation and response policy
   - 3.5 Mutual/automatic aid agreement
   - 3.6 Staffing considerations during adverse weather conditions
   - 3.7 Cancellation procedures for emergency medical service (EMS) and MVA incidents
4. Hazardous Materials
   - 4.1 Hazardous materials operations
5. Emergency Medical Services
   - 5.1 Duties for non-EMS-certified personnel
6. Aircraft Rescue and Firefighting (ARFF)
   - 6.1 ARFF standby policy
7. Technical Rescue and Special Operations
   - 7.1 Vehicle extrication
   - 7.2 Rope
   - 7.3 Trench
   - 7.4 Collapse rescue
   - 7.5 Confined space
   - 7.6 Farm equipment and industrial rescue
   - 7.7 Loveland dive rescue standard operating procedures
   - 7.8 Use of Civil Air Patrol

Figure 5-8 Sample SOP index.

> **Safety** Both formal and informal procedures can be responsible for the overall safety of a department.

Typically, SOPs are arranged in a topical format, **Figure 5-8**. Individual SOPs are formatted for easy reading and reference, **Figure 5-9**.

Equipment

In the past few years, the fire service has seen the release of vast amounts of new equipment designed uniquely for improved "safety." Equipment helps makes an operation more safe but is arguably the least important factor in the safety triad of procedures, equipment, and personnel.

> **Safety** Making equipment safe is addressed in three ways: equipment selection, equipment inspection and maintenance, and equipment application.

Each of these is important as it relates to safety.

Equipment Selection

Most critical equipment used in firefighting (PPE, ladders, hose, etc.) is designed and built to meet NFPA standards. NFPA standards are written with safety paramount to the topic. The fire department that purchases equipment that meets NFPA standards is buying equipment that has a certain safety element built into it. **Figure 5-10** lists common equipment that is designed to improve safety.

Purpose:
To establish policy and direction to all department members regarding minimal staffing and resource allocation for safe and aggressive interior structural firefighting.

Responsibility:
It is the responsibility of all officers and firefighters engaged in firefighting operations to adhere to this policy. The Incident Commander is accountable for procedure included within this policy.

Procedure:

1. This policy is applicable to situations where the Incident Commander (IC) has made a tactical decision to initiate an ***offensive fire attack***, by firefighters, inside the structure. Additionally, tactical firefighting assignments that expose firefighters to an atmosphere that is immediately dangerous to life and health (IDLH) dictate the application of this policy.[1]

2. Prior to initiating interior fire attack or exposure of firefighters to an IDLH atmosphere, a ***minimum of four (4) firefighters shall assemble on scene***.[2] These four members shall utilize a "two-in, two-out" concept.

3. The "***two-in***" firefighters that enter the IDLH atmosphere shall remain as partners in close proximity to each other, generally fulfilling the operational role as the ***FIRE ATTACK GROUP***. As a minimum, the "***two-in***" firefighters entering the IDLH atmosphere shall have full PPE, with SCBA and PASS devices engaged, and have between them a two-way portable radio, forcible entry tool, and flashlight or lantern.

[1] An ***IDLH atmosphere*** can be defined as an atmosphere that would cause immediate health risks to a person who did not have ***Personal Protective Equipment (PPE)*** and/or ***Self-Contained Breathing Apparatus (SCBA)***. This includes smoke, fire gases, oxygen deficient atmospheres, or hazardous materials environments. For Loveland Fire and Rescue application, an IDLH atmosphere can be further defined as an environment that is ***suspected*** to be IDLH, has been ***confirmed*** to be IDLH, or ***may rapidly become*** IDLH. The use of full protective equipment including an activated SCBA and an armed PASS device is mandatory for anyone working in or near an IDLH atmosphere.

[2] The firefighters must be SCBA qualified and capable of operating inside fire buildings without immediate supervision.

Figure 5-9 Sample SOP format.

Firefighter Safety Equipment

PPE	Apparatus
Accountability Tags	Enclosed Cabs
Nomex/PBI Clothing	Headsets
EMS Gloves/Masks	Automatic Vehicle Locator
SCBA	Reflective Trim
Goggles	Swing-Down Tool Trays

Tools

Multi-Gas Detectors
Two-Way Radios
Infrared Cameras
Disposable EMS Adjuncts
Rehab Kits

Figure 5-10 Fire departments spend thousands of dollars on equipment designed to enhance firefighter safety.

Equipment Inspection and Maintenance

For equipment to be safe, it must be inspected and maintained. Firefighters spend many hours each week ensuring that their equipment is clean, properly functional, and ready for rough service during critical situations, **Figure 5-11**.

Because many different firefighters may use and maintain a given piece of equipment, complete documentation of repairs and maintenance is essential. Further, a complete set of guidelines is often developed or adopted for essential equipment. These guidelines include:

- Selection
- Use

Figure 5-11 Ensuring that equipment is clean, functional, and well maintained is essential to firefighter safety.

- Cleaning and decontamination
- Storage
- Inspection
- Repairs
- Criteria for retirement

Equipment Application

Choosing the right tool for a given job is paramount for safety. It is imperative to use equipment in the manner for which it was designed. Using an ax as a prying tool can cause the handle to shatter. At times, the ingenuity and inventiveness of firefighters facing an unusual situation places equipment in a position for which it may not be intended. Extra care and insight are required here.

> **Safety** Firefighters must always prepare for the worst—including equipment failure.

Personnel

Human factors are often cited as the cause of injuries and deaths. To mitigate these injuries and deaths, the safety triad must address personnel issues. A firefighter's training, fitness/health, and attitude all factor into the safety equation.

Training

> **Note** Without a doubt, the single most important step that can be taken to reduce firefighter injuries is that of regular training. It is widely accepted that once a person enters the fire service, the training *never* ends.

Unfortunately, many firefighters become comfortable in their knowledge or position and let the basics slip away unknown. Each and every skill, knowledge item, or behavior that is learned in basic academy should be drilled, tested, or confirmed every year. In a perfect example, the New York City Fire Department (FDNY) launched a huge "back-to-the basics" program after a series of firefighter duty deaths within a year. The FDNY line leadership felt that many of the deaths could have been avoided had firefighters and fire officers practiced the "basics." Many departments utilize company drills on a twice-monthly basis in order to ensure that the "basics" are practiced, **Figure 5-12**.

Because regular training is the single most important step in firefighter safety, the firefighter must strive to retain the information and skills that are presented in training sessions. Here are some strategies for firefighters to help retain essential skills and information and thereby improve safety:

- *Always take notes and keep handouts.* Establish a library and/or filing system for the volumes of notes and information collected during fire service tenure. These notes will provide a handy reference for future recall and assist the firefighter in imprinting information.
- *Envision the application of all training.* After a training session, firefighters should visualize themselves using the skill or information on an actual emergency or scenario. This can be achieved alone or as part of a group. Firefighters should ask themselves what building, situation, or series of events will place them in a position to use the given training.
- *Acknowledge that skills will diminish over time if they are not used.* An old fire service adage says firefighters need to spend 95 percent of their training time practicing for what they do

Figure 5-12 Basic skills need to be practiced on a regular basis.

5 percent of the time. Arguments can be made on which training subjects or behaviors are most important to safe operations; however, a common list can be created based on firefighter injury and death statistics. **Figure 5-13** lists training subjects that directly affect incident safety—if these subjects are practiced and appropriately applied, incident operations will become safer. An expectation of the depth of understanding and the methods needed to achieve this are also suggested.

Firefighter Fitness and Health

The safety and well-being of any firefighter increases with the health of the individual firefighter. Much has been written on the benefits of healthy firefighters—most of which centers on *physical* health.

> **Caution** "Stress" continues to lead the list of causes of firefighter duty deaths and is a significant contributor in injuries.

Essential Training Subjects for Increased Incident Safety

Subject	Degree of Understanding
• Personal Protective Equipment	Mastery
• Accountability Systems	Mastery
• Company Formation and Team Continuity	Mastery
• Fire Behavior and Phenomena	Proficient
• Incident Management Systems	Proficient
• Apparatus Driving	Proficient Under Stress
• Fitness and Rehabilitation	Practitioner

Figure 5-13 Firefighter safety is directly related to the training of essential topics.

To handle the inherent stress of firefighting, each firefighter's body must be accustomed to and capable of handling stress. Additionally, firefighters need to protect themselves from, and prevent the spread of, communicable diseases and infections. Keys to improving physical health, and therefore firefighter safety, include these:

- *Annual health screening for all firefighters and line officers.* These physicals should also include vaccination and immunization offerings, blood screening, and stress testing.
- ***Work hardening*** *and mandatory ongoing fitness programs.* A personal devotion to physical fitness is essential to firefighter safety. Fitness includes improving flexibility, strength, and cardiovascular endurance. Work hardening is the effort and physical training put forth to better perform physical tasks without overstressing or injuring an individual. The firefighter who addresses these fitness areas will breeze through any "fit-for-duty" agility tests that a fire department may require as a condition of employment or association. Improper lifting techniques and slip and fall accidents are two of the most common causes of injury on duty. *Work hardening helps prevent these injuries through strength, balance, and coordination improvement,* **Figure 5-14.**
- *Firefighter fueling (nutrition) education.* Nutrition is important not only to lifestyle changes (like losing excessive fat) but also to incident readiness.

> **Caution** Diet fads and meal-replacement aids may help an individual lose weight but often interfere with maintaining energy during working incidents.

Figure 5-14 Work hardening must become a habit—leading to increased safety and professional tenure.

This is especially true with the sporadic nature of the firefighter's energy demands. Chapter 23, Firefighter Survival, more clearly defines appropriate fueling of the firefighter.

Attention to *physical health* is indeed important, but *mental health* is also important to firefighter safety. Keys to addressing mental health include the following:

- *Training and understanding of* ***critical incident stress management (CISM)***. Knowing the signs and symptoms of critical incident stress is important, **Figure 5-15**. Most departments have developed a process to have a **critical incident stress debriefing (CISD)** team activated or available for unusual events or at the request of one or more responders.
- ***Making an employee assistance program (EAP)*** *available*. An EAP offers individuals a confidential approach to dealing with situations and problems that can affect job performance. Drug and alcohol dependence, depression, worker relationships, and job stress are a few of the issues that can be addressed through an EAP.

Signs and Symptoms of Critical Incident Stress

Numbing and Withdrawal
Reexperiencing the Event
Flashbacks
Depression
Sleep Difficulties
Substance Abuse
Guilt
Family Problems
Low Job Efficiency
Fear of Next Incident
Anger and Hostility

Figure 5-15 Critical incident stress is inevitable and unpreventable. CISM can be addressed through peer support and formal debriefings.

Attitude

Of all the "people" factors affecting safety, attitude is the hardest to address. Many factors affect the attitude of a given individual, not to mention the fact that attitudes are dynamic. Here are some factors that affect safety attitudes:

- *The fire department's safety "culture."* The culture of an organization is reflected in the ideas, skills, and customs that are passed through generations. If safety has not been given a high priority by administrators and members, a poor safety attitude will be reflected in daily activities. The inverse is also true.
- *The fire department's history.* A department that has experienced a firefighter duty-related death or serious injury is more likely than not to have increased safety awareness. Likewise, some departments have taken from others' experiences—this is known as having a **vicarious experience**. Realizing that a potential for injury or death exists is an important first step in developing a proactive safety attitude.
- *The example set by others.* Firefighters and line officers display their safety attitudes in what they do rather than what they say. These examples will shape the attitudes of others, **Figure 5-16**.

Ethics It is best to shape a positive safety attitude early in a career rather than later. Likewise, the exciting period just following a promotion is ideal for setting in good habits.

Figure 5-16 Developing a positive safety attitude and practicing safe habits will demonstrate safe examples to others.

The following are steps a firefighter can take to create a positive safety attitude:

- *Practice good habits.* Each and every incident response and training activity results in opportunities to practice and reinforce good habits. PPE use, readiness, team formation, and assignment follow-through are all items that lead to a more positive safety attitude. These habits will serve the firefighter well when the incident becomes more complex or dangerous.
- *Learn from others.* Each month, trade magazines and investigative reports outline circumstances that have led to a firefighter injury or death. These serve as a constant reminder that an injury can happen. Firefighters can learn these lessons vicariously. Significant fatality and injury reports can be found at the NIOSH Web site, http://www.cdc.gov/niosh.
- *Be vigilant.* Inactivity, complacency, and overconfidence have been cited as primary factors in many firefighter fatalities. Often, merely visualizing an injury will help firefighters regain focus and obtain vigilance in checking their safety attitude.

FIREFIGHTER SAFETY RESPONSIBILITIES

Firefighter safety is dependent on the efforts of everyone. Specifically, responsibility for firefighter safety rests in one of three areas: the department itself (administration), the working team, and the individual firefighter, **Figure 5-17**. Knowing these responsibilities—especially where individuals fit—will help achieve firefighter safety. Firefighter safety is achieved if, and only if, all "partners" hold up their end of the safety partnership.

The Department

The responsibility for firefighter safety ultimately rests with the department's leadership. This is an incredible weight to carry. Fire chiefs must impose rules, regulations, and expectations in order to address the issue of firefighter safety. Additionally, a department's leadership must obtain budget appropriations to purchase equipment and apparatus that will ultimately help make firefighting safer. Often, a fire chief will appoint a health and safety officer to specifically address firefighter safety issues. In a nutshell, NFPA 1500 outlines what firefighter safety components are needed as part of an occupational safety and health program.

To hold up the department's end of the safety partnership, the fire chief, health and safety officer, or other administrative officer should create a health and safety committee, develop standard procedures and guidelines, and implement a risk management plan. Additionally, the department should research and purchase safety equipment, as well as develop and deliver hazard awareness training. The following paragraphs address each of these department responsibility areas.

Creation of a Health and Safety Committee

This group is usually made up of personnel from all levels. The focus here is to identify and create solutions for safety issues through a committee process.

Development of SOPs

Defining a proper and expected level of procedure or behavior is an important step in addressing operations that can cause injury. It is common to have safety SOPs that address these areas:

- PPE
- Firefighter injuries
- Training safely

(A) (B) (C)

Figure 5-17 Firefighter safety is dependent on all partners holding up their responsibilities: (A) administration, (B) teams, and (C) individual firefighters.

- Zoning
- Evacuation
- Emergent driving
- Slip and fall hazards
- Self-contained breathing apparatus (SCBA) use and care
- Accountability systems
- Incident command
- Riding apparatus
- Lifting heavy objects
- Power tool use
- Infection control
- Hazard reporting.

Implementation of a Risk Management Plan

NFPA 1500 requires that the department develop, implement, and train all firefighters on a risk management plan. This plan is developed based on local needs and includes expectations from the community. Typically, though, the risk management plan acknowledges that certain risks are inherent in firefighting. These risks can be categorized, however. An example of this risk/benefit philosophy follows:

- Significant risk to the life of a firefighter shall be limited to those situations where the firefighter can potentially save endangered lives.
- Situations endangering valued property shall cause firefighters to take a calculated and weighted risk.
- Where no life or valued property can be saved, no risk shall be taken by firefighters.

Research and Purchase of Appropriate Apparatus and Equipment

Firefighting requires rugged, specially designed equipment that ensures a certain level of reliability and safety for the user. NFPA standards outline some of the requirements that make apparatus and equipment safer. This equipment is often very expensive and requires a significant effort to obtain funds.

Development and Delivery of Hazard Awareness Training

Virtually all training is designed to help firefighters operate in a safe manner. The training effort must also inform firefighters of the hazards they may face on any particular type of incident.

The Team

The department as a whole cannot be effective in its firefighter safety effort without the support of a team approach to handling incidents. Therefore, the team that handles an incident must hold up its part of the safety partnership. This team obviously includes the individual firefighter. To ensure safety, the team should follow these procedures:

1. Utilize an **incident management system (IMS)**. The IMS should detail lines of authority and communication, be reasonable in its span of control, and include an action plan. Company officers, sector officers, or group supervisors need to communicate progress and status. Orders need to be clear and accomplished within an acceptable risk environment.

2. Work together and remain intact. The "buddy" concept is imperative for firefighter safety.

> **Caution** The separation of members within a team is a contributing factor to firefighter fatalities.

3. Look after each other. *Team members who continuously watch each other will find and address conditions that will lead to injury.* Aggressive rehabilitation will provide hydration, rest, and nourishment—thereby increasing the time the team can perform safely. At the team level, issues of PPE and SCBA use or misuse should be addressed. Likewise, the team must watch each other for signs of fatigue, overaggressiveness that is dangerous, and freelancing.

The Individual Firefighter

The individual firefighter holds the final key to making the safety partnership work. The following paragraphs outline *how* a firefighter can contribute to individual safety and the safety of the team.

Be Ready

> **Streetsmart Tip** A firefighter's readiness includes not only wearing PPE and SCBA as appropriate, but also being ready mentally and physically.

Reporting to duty or an incident with a physical limitation will increase everyone's danger. Firefighters working while sick or under the influence of certain over-the-counter and prescribed drugs can experience less than clear thinking as well as injury. Likewise, mental barriers and "baggage" will likely impair a firefighter's ability to think and act.

Understand and Act within the Chain of Command

For an incident management system to work, each individual must fill a role and not operate outside of it. Each person is responsible to someone else. Crossing these lines of responsibility causes confusion—a deadly proposal. In addition to understanding the chain of command, firefighters must clearly understand their orders (tasks). It is good practice to repeat the orders that have been given and to ask for clarification if those orders are not clear.

Perform as Trained

Incident task needs may place a firefighter in a position to perform a skill or task that the individual may or may not have been trained for. This should not be allowed to happen.

> **Safety** If a firefighter has not been trained for a task that is required, the firefighter should tell the direct on-scene supervisor.

The incident scene is rarely the place for on-the-job training of critical skills. Firefighters should practice individual skills—knot tying, ladder raising, and tool use can deteriorate over time. Additionally, the department has outlined SOPs to be followed—firefighters should do so.

Do Not Freelance

Fire and rescue work is a team effort. A leader guides a team and the leader must follow the incident commander's action plan. Working alone or outside the action plan endangers individuals and the team, **Figure 5-18**.

Figure 5-18 Freelancing endangers individuals and the team. *Never work alone!*

Use an Incident Engagement Checklist

At the individual level, firefighter safety can be achieved if firefighters practice a standard approach to incidents—that is, doing things in a standard way all the time. Using an engagement checklist, **Figure 5-19,** helps achieve a standard approach to incidents.

Safety Riding the fire apparatus is one of the most common dangers the firefighter experiences. Routinely, firefighters are injured or killed while riding to and from incidents. This may seem incongruent to a profession dedicated to preserving life. A firefighter can minimize the chance of injury by following a few simple steps. The first is to always wear a seat belt! While most departments have purchased apparatus with fully enclosed cabs, some firefighters still "jump on" the tailboard when the fire apparatus is shuttling people during training drills and hose-laying activities. The firefighter must resist the temptation to ride the tailboard; this is considered a major safety violation for most fire departments. Finally, if a firefighter rides in fire apparatus that is not equipped with a fully enclosed crew compartment (open jump seat), it is important to use eye and ear protection as well as any installed safety bar or restraint device. In all cases, it is common practice for the firefighter to remain seated and belted until the officer or driver gives instructions to dismount.

Incident Engagement Checklist

- Don personal protective equipment appropriate for the response.
- Mount apparatus using handholds.
- Buckle seat belt, don headset—report you are ready.
- Listen to radio reports and listen for details.
- Mentally run through tasks that might be required for the type of incident.
- Don SCBA or collect tools, gloves, etc.., if appropriate and if this can be done with seat belt on.
- Upon arrival, listen for orders.
- Unbuckle only when orders have been given.
- Prior to leaving the apparatus, look out the window and make sure the path is clear. Watch for traffic when dismounting.
- Close all doors and compartments after retrieving tools and/or equipment.
- Scan the environment for overhead wires, trip/fall hazards, and unusual circumstances.
- Know who you are reporting to—make sure your accountability tag/passport is processed.
- Proceed with orders.
- Make sure you can be seen and heard by other firefighters at all times during an incident.

Figure 5-19 Firefighters should perform a mental "incident engagement checklist" for every response.

Lessons Learned

The issue of firefighter safety is dependent on many factors. The fact that roughly half of injuries and deaths occur at incident scenes demonstrates that firefighting is a dangerous profession. The majority of these deaths and injuries are a result of stress. To help prevent injuries and death, the fire service is required to follow federal regulations called OSHA CFRs. In addition, NFPA standards, like NFPA 1500, have been written and adopted to further the firefighter safety effort.

Accident prevention is actually the process of mitigating hazards or intervening with the accident chain. Within any operational environment, the safety triad is used to address issues. To make sure the safety triad is effective, a partnership between the department administration, the working teams, and the individual firefighter is formed. Each of these partnerships carries equal weight in creating a safe atmosphere. It is then the individual's challenge to develop safe habits and be ever mindful of the events and conditions that cause injury.

KEY TERMS

Accident The result of a series of events and conditions that lead to an unsafe situation resulting in injury and/or property damage.

Accident Chain A series of events and conditions that can lead to or have led to an accident. These events and conditions are typically classified into five areas: environment, human factors, equipment, events, and injury.

Code of Federal Regulations (CFR) The documents that include federally promulgated regulations for all federal agencies.

Critical Incident Stress Debriefing (CISD) A formal gathering of incident responders to help defuse and address stress from a given incident.

Critical Incident Stress Management (CISM) A process for managing the short- and long-term effects of critical incident stress reactions.

Employee Assistance Program (EAP) A defined program that offers professional mental health and other health services to employees.

Incident Management System (IMS) An expandable management system used to deal with a myriad of incidents. Helps achieve the highest level of accountability and effectiveness for incident handling. Limits span of control and provides a framework to help make tasks manageable.

Intervention The act of intervening; to come between as an influencing force. Typically a reactive action.

Mitigation Actions taken to eliminate a hazard or make a hazard less severe or less likely to cause harm. Typically a proactive action.

National Institute for Occupational Safety and Health (NIOSH) A federal institute tasked with investigating firefighter fatalities and making recommendations to prevent reoccurrence.

Occupational Safety and Health Administration (OSHA) The federal agency, under the Department of Labor, that is responsible for employee occupational safety.

Risk The chance of injury, damage, or loss; hazard.

Risk Management The process of minimizing the chance, degree, or probability of damage, loss, or injury.

Vicarious Experience A shared experience by imagined participation in another's experience.

Work Hardening A phrase given to the effort and physical training designed to prepare an individual to better perform the physical tasks that are expected of the individual. Work hardening is key in preventing injuries resulting from typical firefighting tasks.

REVIEW QUESTIONS

1. Define risk management and the concept of risk/benefit.
2. List the leading causes of death and injury in the fire service.
3. List the NFPA standards that affect and pertain to firefighter occupational safety.
4. List the five components that make up the accident chain and discuss the difference between mitigation and intervention.
5. List and briefly describe the three key components of the safety triad.
6. Explain the difference between formal and informal procedures.
7. Name the three factors that influence the equipment portion of the safety triad.
8. Name the three factors that influence the personnel portion of the safety triad.
9. Name the three partners that work together to achieve firefighter safety and give examples of how the individual firefighter can contribute to firefighter safety.

Endnote

1. U.S. Fire Administration, Firefighter Fatality Retrospective Study, FA-220, USFA, Washington, DC, April 2002.

Additional Resources

Angle, James S., *Safety in the Emergency Services.* Delmar Learning, a part of the Thomson Coproration, Clifton Park, New York, 1999.

Dodson, David W., *Fire Department Incident Safety Officer.* Delmar Learning, a part of the Thomson Coproration, Clifton Park, New York, 1999.

National Institute for Occupational Safety and Health: http://www.cdc.gov/niosh– to obtain firefighter fatality reports and recommendations.

NFPA 1500: Standard on Fire Department Occupational Safety and Health Program, National Fire Protection Association, Quincy, MA, 2002.

Risk Management Practices for the Fire Service, FA-166. United States Fire Administration, Emmitsburg, MD, 1996.

CHAPTER 6

PERSONAL PROTECTIVE CLOTHING AND ENSEMBLES

David Dodson, Lead Instructor, Response Solutions, Colorado

OUTLINE

- Objectives
- Introduction
- Personal Protective Equipment Factors
- Types of Personal Protective Equipment
- Care and Maintenance of Personal Protective Equipment
- Personal Protective Equipment Effectiveness: "Street Smarts"
- Lessons Learned
- Key Terms
- Review Questions
- Additional Resources

STREET STORY

As the department's training officer, I served as the safety officer on all major incidents. One afternoon while working in my office at our suppression headquarters station, I heard Engine 4 dispatched to a vegetation fire. Engine 4 arrived on scene to report approximately one acre involved, with the fire rapidly spreading. They requested two additional engines. The paramedics I was working with decided to drive down the street to observe, as we were on the opposite side of the involved area. I decided to ride with them. I did not plan on this being a major incident and did not take any protective clothing with me.

As we approached the area across from the fire, we noticed that the fire had spread across the ravine and was now threatening homes and utility services. An engine had laid lines, and one firefighter was attempting to put a 2½-inch line into service to protect the exposures. The incident commander was calling for two strike teams. The paramedics were given an assignment to help extend lines to the right flank. I jumped in to assist with the exposure line. My protective equipment consisted of slacks, a dress shirt, a tie, and shoes. Within minutes we were exposed to heat and smoke thick enough to cause difficulty breathing. Through the smoke an engineer brought me his turnout coat and helmet. He later gave me his turnout boots and stood at the pump panel in stockinged feet.

Not only was I completely ineffective as a safety officer, I was a safety hazard myself! By not having my protective gear, I created a safety hazard for myself and took away safety gear from another firefighter. The few minutes it would have taken for me to grab my gear would have been more than worthwhile. Our protective clothing is as essential to us as a scalpel is to a surgeon. The lesson I learned was never to respond to an incident without my safety gear and never to enter the hazard environment without it. Fortunately, no one was injured and the fire was extinguished with no loss of structures.

—Street Story by Randy Scheerer, Battalion Chief, Newport Beach Fire Department, Newport Beach, California

OBJECTIVES

After completing this chapter, the reader should be able to:

- Describe the role of personal protective equipment for firefighters.
- Define the relationship between PPE and national standards and regulations.
- List the components and unique elements of structural, proximity, and wildland PPE ensembles.
- Describe a serviceability inspection of structural PPE.
- Describe the conditions and damage that render structural PPE unserviceable.
- Given a structural PPE ensemble, appropriately don the ensemble within one minute.
- Demonstrate a team check following PPE donning.

INTRODUCTION

Firefighters and emergency medical providers respond to incidents that are often immediately dangerous to life and health. The vernacular for this is IDLH, **Figure 6-1**. Often, the difference between injury and safety is determined by the responder's personal protective equipment (PPE). It is important to note, however, that PPE provides a minimum level of protection and should be considered the *last resort* of protection for firefighters and emergency responders operating at an incident. Proper fire streams, **zoning**, and sound tactics and procedures should provide a greater measure of safety for teams, especially in IDLH atmospheres. Simply stated, PPE is the *first* thing a firefighter puts on to deal with an incident and the *last* thing taken off when the incident is over.

Personal protective equipment for the firefighter can take on many forms. Clothing, helmets, hoods, gloves, harnesses, **personal alert safety systems (PASS)**, and self-contained breathing apparatus (SCBA) are just some of examples of the PPE used by firefighters. Further, defined collections of PPE make up ensembles that are designed for specific firefighter hazards. These ensembles include structural, proximity, and wildland PPE. Other ensembles may include technical rescue, ice rescue, and swift-water rescue gear. Each of these ensembles is discussed later in this chapter. Ensembles such as EMS infection control and hazardous materials PPE are used by many fire departments. Additional information on these can be obtained locally or in Chapter 27 of this book. It is important to note that each ensemble or piece of protective equipment is designed to meet a minimum level of safety and each has specific limitations that govern performance of the equipment.

Caution Failure to operate within a PPE component's designed limitations can lead to an injury, illness, or perhaps death of the user.

Pushing the limitations of a PPE component is not the only way PPE-related injuries occur. Injuries and illnesses have been suffered by the firefighter who fails to properly don and secure PPE—usually because the wearer was trying to create a "macho" image or skipped complete donning in the haste to perform a task.

This chapter starts with a discussion of the factors that influence PPE design and use, including national standards and regulations. Next, the various

Figure 6-1 A hostile fire within a structure creates an IDLH environment. Personal protective equipment can help the firefighter work in an IDLH environment.

types of firefighter PPE are outlined followed by care and maintenance of PPE. This chapter then discusses how firefighters can effectively use PPE and keep their first and last defense intact. SCBA—although an important component of PPE—is discussed in Chapter 7. Other PPE items such as those for emergency medical operations and hazardous materials operations are discussed in later chapters.

PERSONAL PROTECTIVE EQUIPMENT FACTORS

Firefighter personal protective equipment has evolved significantly during the past two decades. The reasons for these recent improvements are varied. A study of the history of injuries, evolution of materials, and sound risk management practices have led to vast improvements in protective equipment. Modern PPE has been developed as a result of the direct efforts of labor groups (most notably, the International Association of Firefighters, IAFF), other fire service membership associations, equipment manufacturers, and government entities such as NASA. These efforts usually come together as a result of a consensus process that establishes minimum standards. The National Fire Protection Association (NFPA) provides the forum for this consensus building. Although NFPA standards certainly influence PPE design and use, the federal government has also become involved in the PPE equation through the development of regulations and guidelines.

Standards and Regulations

The NFPA has developed numerous standards for firefighter protective equipment and ensembles, **Figure 6-2**. Typically, these standards cover component parts, manufacturers' quality assurance and labeling requirements, design requirements, user information, performance requirements, and test methods. The specific NFPA standards listed in **Figure 6-2** primarily address the design, performance, and manufacturing of PPE. Additionally, NFPA has developed an important PPE "use" standard. NFPA 1500, *Standard on Fire Department Occupational Safety and Health Program,* dedicates a whole chapter to the use, care, and maintenance of many forms of protective equipment.

NFPA Standards That Address PPE and Ensembles	
1500	Fire Department Occupational Safety and Health Program
1971	Protective Ensemble for Structural Firefighting
1975	Station/Work Uniforms for Firefighters
1976	Protective Clothing for Proximity Firefighting
1977	Protective Clothing and Equipment for Wildland Firefighting
1981	Open-Circuit Self-Contained Breathing Apparatus for the Fire Service
1982	Personal Alert Safety Systems
1983	Life Safety Rope and System Components
1991	Vapor-Protective Hazardous Ensembles/Materials Emergencies
1999	Protective Clothing for Medical Emergency Operations

Figure 6-2 NFPA develops consensus standards that address PPE.

All equipment worn by a firefighter should meet current applicable standards. Firefighters should check PPE components for a conspicuous and permanently attached label that verifies that a component meets an applicable NFPA standard, **Figure 6-3**.

The federal government is also involved in PPE use. Of primary importance is the involvement of the Occupational Safety and Health Administration (OSHA). OSHA is responsible for the development and enforcement of regulations—namely, the Code of Federal Regulations (CFRs)—that govern safe work practices. At the firefighter level, this involvement essentially means that failure to use appropriate PPE in certain applications (e.g., IDLH atmospheres) means that a federal regulation has been violated, leading to a potentially expensive fine and disciplinary actions against the firefighter and the department. Other government and ancillary agencies that are involved in protective equipment issues include the Environmental Protection Agency, Centers for Disease Control and Prevention, American National Standards Institute, American Society for Testing and Materials, and the National Institute for Occupational Safety and Health.

TYPES OF PERSONAL PROTECTIVE EQUIPMENT

Ensembles

Within the fire service, many personal protective ensembles have been developed. The NFPA has developed minimum ensemble standards for structural firefighting PPE, proximity PPE (commonly used for aircraft rescue and firefighting), and wildland PPE. NFPA also addresses specialized ensembles that exist for technical rescue such as Urban Search and Rescue (USAR). The important point to underscore is the word *ensemble*. Frequently, firefighters short-circuit the concept of an "ensemble" in the use of their PPE. An action as simple as forgetting to secure the top fastener of a protective coat violates the concept of an ensemble and can lead to an injury.

> **Safety** To be effective, all components of a PPE ensemble must be utilized as recommended by the manufacturer. Failure to complete an ensemble can cause an injury.

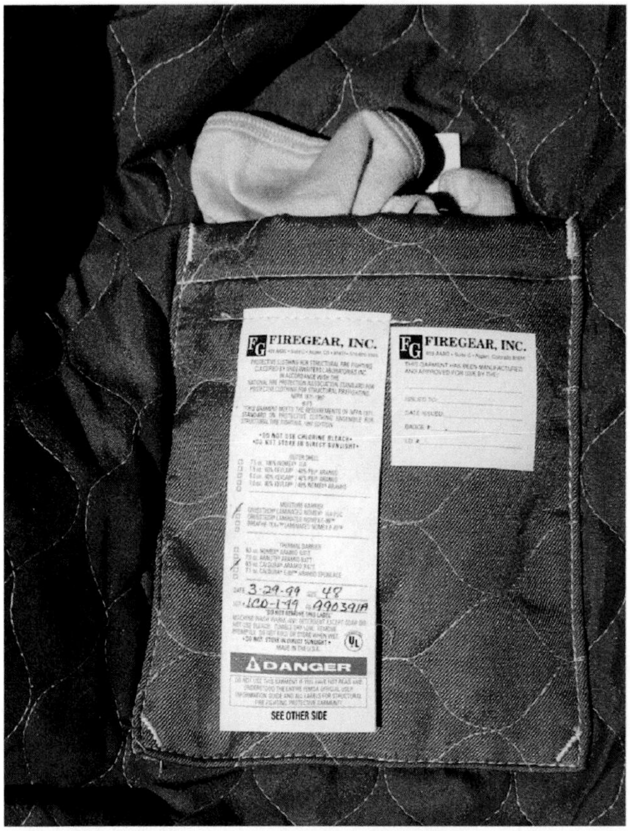

Figure 6-3 NFPA-compliant PPE components should have a permanently affixed label.

Structural Ensemble

The protective ensemble for structural firefighting includes all of the components listed and shown in **Figure 6-4**. For the sake of clarification, NFPA defines structural firefighting as "the activities of rescue, fire suppression, and property conservation in buildings, enclosed structures, aircraft interiors, vehicles, vessels, or like properties that are involved in a fire or emergency situation."

Structural PPE is commonly referred to as **bunkers**. The term *bunkers* was originally associated with short boots and protective pants that

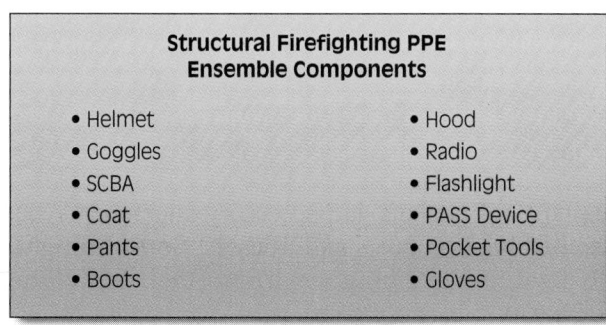

Structural Firefighting PPE Ensemble Components

- Helmet
- Goggles
- SCBA
- Coat
- Pants
- Boots
- Hood
- Radio
- Flashlight
- PASS Device
- Pocket Tools
- Gloves

Figure 6-4 A full structural firefighting ensemble includes more than the NFPA minimum required components.

duty firefighters would only wear at night—donned in the bunk room of a fire station. During the day, these same firefighters wore long coats and three-quarter pull-up boots—a practice no longer allowed by NFPA 1500 due to the inadequate protection offered to the wearer. (Many serious burns have been experienced by firefighters using this antiquated type of clothing.) Today, many terms are used to refer to the structural firefighting ensemble. These terms include the previously mentioned *bunkers* as well as *turnouts* and *structural gear.* Some highlights of the structural ensemble are discussed next.

Coats and Trousers. The heart of structural PPE is the coat and pant combination. Both components rely on a layered protection system that includes a fire-resistive outer shell, vapor barrier, and thermal barrier.

Safety All three layers of the pant and trouser assembly must be intact—failure to wear the entire layer system may cause injury or death during fire suppression operations.

The three layers help the coat and pant meet thermal protective criteria—insulation that minimizes the chance that the wearer will be burned. This criterion is called **thermal protective performance (TPP)**. The minimal TPP for structural coats and pants is 35. In simple terms, the wearer will sustain life threatening injuries after 35 seconds of flame contact exposure. Pratically speaking, the wearer has half of that—17 seconds—before burns will likely to occur. Dynamic influences such as dirty gear, moisture (including perspiration), and fabric compression (from an SCBA) can reduce the TPP of the clothing.

The coat and trouser combination has reflective trim to increase the visibility of the wearer to others, **Figure 6-5**. Flaps, enclosures, wristlets, and fastening devices are all designed to seal the ensemble and to provide a protective interface with gloves, hoods, boots, and helmets. Failure to "seal" the ensemble invites injury in hazardous environments. Trousers should have heavy-duty suspenders to help keep the pants from sagging when they become wet with perspiration or water.

Shoulder padding, knee reinforcements, and various types of pockets and accessory attachments can also be found on coats and trousers. Some firefighters have had webbing sewn into their structural trousers to create a "sit harness" for certain rescue situations. This customizing should only be performed by a certified manufacturer that understands and complies with NFPA standards for both structural PPE and life safety system components (NFPA 1983, *Standard on Fire Service Life Safety Rope and System Components*).

Figure 6-5 Reflective trim increases firefighter visibility to others during low-light activities.

Helmets. The classic firefighter's helmet was designed to help shed water and to prevent hot embers from falling down on the firefighter's neck, back, and ears. This classic design remains in most styles of helmets. Newer helmet safety features exceed those of yesteryear in many ways and include impact resistance; thermal insulation; earflaps for a layered interface with hoods, coats, and SCBA face pieces; chin straps; and clear or tinted face shields and/or eye protection accessories. The chin strap is an important part of the helmet—it must be utilized to take full advantage of the helmet's safety feature. Failure to utilize the chin strap is an invitation to injury or loss of the helmet during critical operations.

Gloves. Hand protection is essential in the structural ensemble. Gloves meeting NFPA standards for structural firefighting must provide thermal protection as well as protection from cuts, punctures, and scrapes. Unfortunately, dexterity is almost always reduced when wearing structural gloves. This is a common complaint of firefighters.

Streetsmart Tip Two things can improve dexterity with gloves: good fit and practice. It is important to note that dexterity when wearing gloves increases with practice—thanks to muscle memory.

The simple act of practicing hands-on tool use with gloves on can increase muscle memory and reduce the frustration of lost dexterity.

Footwear. The firefighter's choices for approved and effective footwear are growing. Now firefighters can choose from traditional rubber-like boots or leather-type boots. Each has advantages and disadvantages, **Figure 6-6**. Both must meet NFPA standards for foot protection.

Protective Hoods. An important interface that creates an encapsulating link to the firefighter's helmet, coat, and SCBA face piece is the protective hood. Hoods are made of fire-resistive, form-fitting cloth that protects the face, ears, hair, and neck in areas not covered by the helmet, earflaps, and coat collar, **Figure 6-7**. It is important to note that structural protective hoods have a TPP less than that of a structural coat (20 for a hood versus 35 for a coat). Additionally, hoods are designed to be worn *over* the straps of an SCBA face piece. This practice helps protect the SCBA face piece straps and helps keep the hood from binding.

Miscellaneous Structural PPE. While not necessarily covered by the NFPA standards, many seasoned firefighters feel that certain accessories are as much a part of the full PPE ensemble as the coat, pants, and helmet. These items include primary eye protection (goggles or safety glasses), hearing protection, PASS devices, and pocket tools, **Figure 6-8**.

Proximity Ensemble

The proximity firefighting PPE ensemble is most often associated with aircraft rescue and firefighting (ARFF). Proximity gear utilizes an aluminized coating to help reflect radiant heat, **Figure 6-9**.

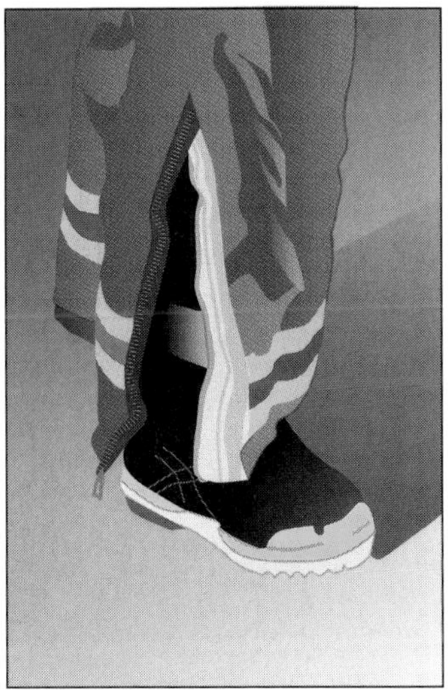

Common Rubber Boot

- Easy to Don
- Excellent Water Repellency
- Easy Decontamination
- Inexpensive
- Sloppy Fit

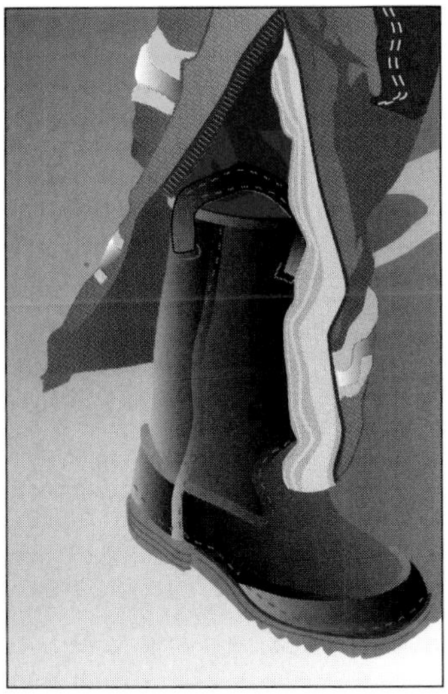

Leather Pull-Up Boot

- Lightweight
- Durable
- Comfortable
- Minimal Ankle Support

Leather Lace-Up Boot

- Tight Fit
- Ankle Support
- Durable
- Expensive

Figure 6-6 Firefighters have choices for structural firefighting footwear—each with its own advantages and disadvantages.

Figure 6-7 A protective hood provides an interface layer that links the coat collar, helmet flaps, and opening for an SCBA mask.

Although similar in many ways to structural PPE, the proximity gear must meet more stringent heat reflection and wearer insulation standards.

> **Note** It is important to note that proximity PPE is *not* designed for fire entry (to be totally enveloped by fire).

The aluminized fabric allows the wearer to get closer to fires that emit extreme radiant heat such as that found with petroleum-based fires. Bulk fuel facilities, airports, and certain chemical plants are complexes where the assigned fire attack teams may be required to wear proximity ensembles.

In addition to the aluminized fabric, proximity PPE features full face shields that are coated with an anodized gold material to help create a "mirrored" reflective surface. Without this special coating, the wearer could receive radiant facial burns, and the face shield could quite possibly melt.

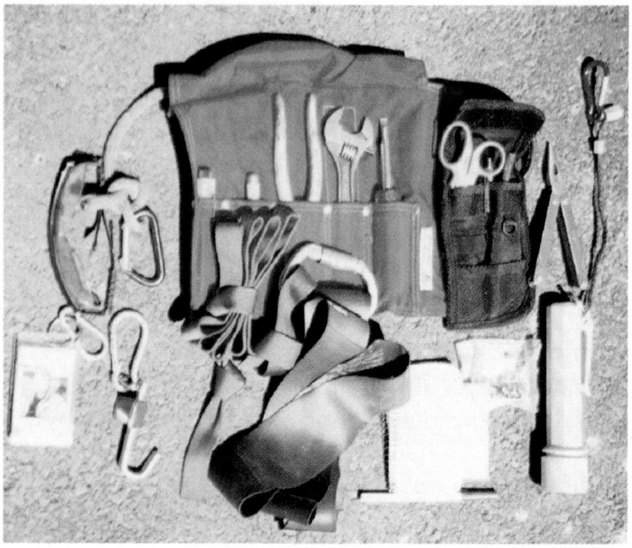

Figure 6-8 Personal protective equipment can also include many tools that firefighters pocket.

Wildland Ensemble

Wildland firefighting conditions are unique in that firefighting operations are often outdoors, often require prolonged physical effort, and are typically accomplished when ambient temperatures are high. Fighting wildland fires with a structural ensemble can invite strained necks, heat stroke, and sprained ankles. Wildland PPE (called **brush gear** by many) addresses the specific needs of the wildland firefighter: It is lightweight and provides breathability, firm ankle support, and hot ember protection, **Figure 6-10**.

The wildland PPE ensemble is designed to be worn over undergarments. These undergarments (long-sleeve T-shirt, pants, and socks) should be 100 percent cotton or of a fire-resistive material, **Figure 6-11**.

> **Caution** Synthetic material should never be worn under wildland PPE—most synthetic materials can melt and cause severe burns to the wearer.

Other highlights of the wildland PPE ensemble are discussed next.

Lightweight Jacket/Shirt and Trousers. These garments are typically made of a fire-resistive material or a treated cotton. Wool has also been used for wildland PPE garments. Once again, these protective elements need to be worn over undergarments in order to increase thermal insulation.

(A) (B)

Figure 6-9 Proximity firefighting ensembles can utilize (A) a special helmet or (B) a full hood. Either can interface with an SCBA. Note the gold-anodized visors.

Figure 6-10 Wildland PPE is lightweight, but still provides protection from hot, flying brands.

Figure 6-11 Wildland PPE includes cotton or fire-resistive undergarments. Long-sleeve T-shirts are a *must!*

Footwear. Lace-up leather boots that rise well above the ankle (8 to 10 inches) help protect the wearer from cuts, snakebites, and burns. Additionally, a good fitting, tightly laced boot can help prevent ankle sprains and reduce foot fatigue.

Fire Shelter. A **fire shelter** is another unique component in the wildland PPE ensemble, **Figure 6-12**. The fire shelter must be carried in a case that protects the aluminized fabric from being crushed, yet still allows the shelter to be deployed quickly. A fire shelter is a last-resort protective device for firefighters caught or trapped in an environment where a firestorm or blow-up is imminent.

Web Gear. Although not listed in the NFPA standard for wildland PPE, **web gear** is essential for the wildland firefighter. Web gear consists of a belt (with or without shoulder support straps) that can carry a fire shelter, water bottles, flares (fusees), a radio, and other assorted gear to help the wildland firefighter, **Figure 6-13**. Often, a web gear setup includes a detachable day sack that can carry meals and maps, or even overnight sleeping gear and personal items.

Ice Rescue Ensemble

In areas where frozen lakes and recreation ponds exist, the fire department may offer an ice rescue service. The ice rescue ensemble includes a buoyant, insulated suit that protects the wearer from submersion and the freezing cold water should the rescuer break through the ice. The suit has a form-fitting face seal and hood to minimize the amount of water that can enter the suit. Typically one-piece, the suit has sealed gloves and boots built in. To complete the ensemble, the suit should be worn with a simple chest harness and lightweight helmet with face screen (similar to a football face mask), **Figure 6-14A.**

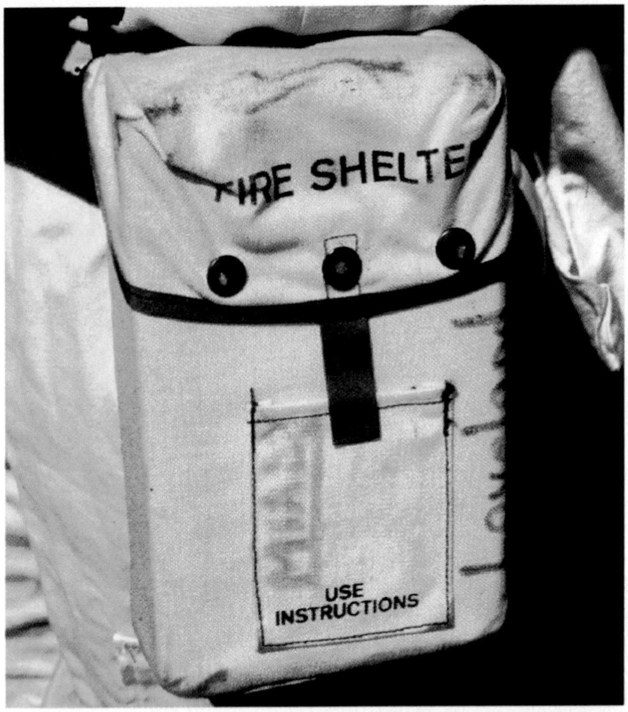

(A)

(B)

Figure 6-12 (A) A wildland fire shelter is an essential PPE component. (B) Training for rapid deployment can help save a firefighter's life.

Figure 6-13 Wildland web gear is designed to keep personal items, water, flares, and the fire shelter within easy reach.

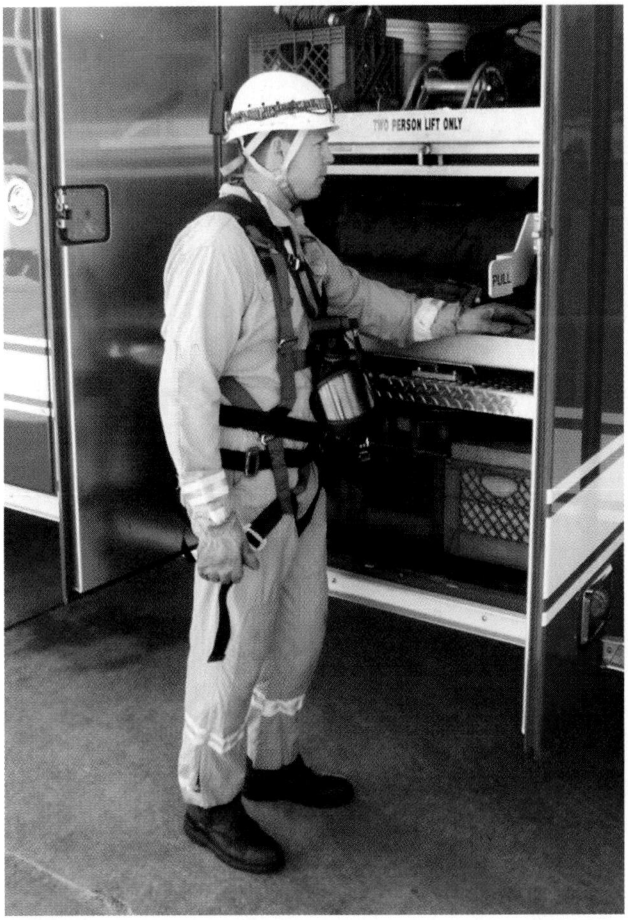

(A)

(B)

Figure 6-14 Specialized ensembles include (A) the technical rescue ensemble, (B) the ice rescue ensemble, and (C) the swift-water ensemble. *(Courtesy of Ron Marley)*

(C)

Technical Rescue Ensemble

Rescue operations such as confined space, collapse, rope, and trench do not necessarily require full structural PPE. In these cases, a lightweight yet durable ensemble is often used to protect rescuers. Typically, this ensemble consists of a durable pant and overshirt (or coverall) coupled with lace-up leather boots, tight-fitting durable gloves, lightweight helmet and eye protection, and a harness, **Figure 6-14B.**

Swift-Water Ensemble

The use of structural PPE near swift water can introduce *more* danger to the firefighter if swept into the stream. While specific intense training can teach firefighters how to "float" in structural PPE, it is not a preferred choice for swift-water environments. The swift-water ensemble includes a typical work uniform overlaid with a personal flotation device (PFD), harness with throw-rope bag, lightweight helmet with face cage, and no-slip gloves **Figure 6-14C.**

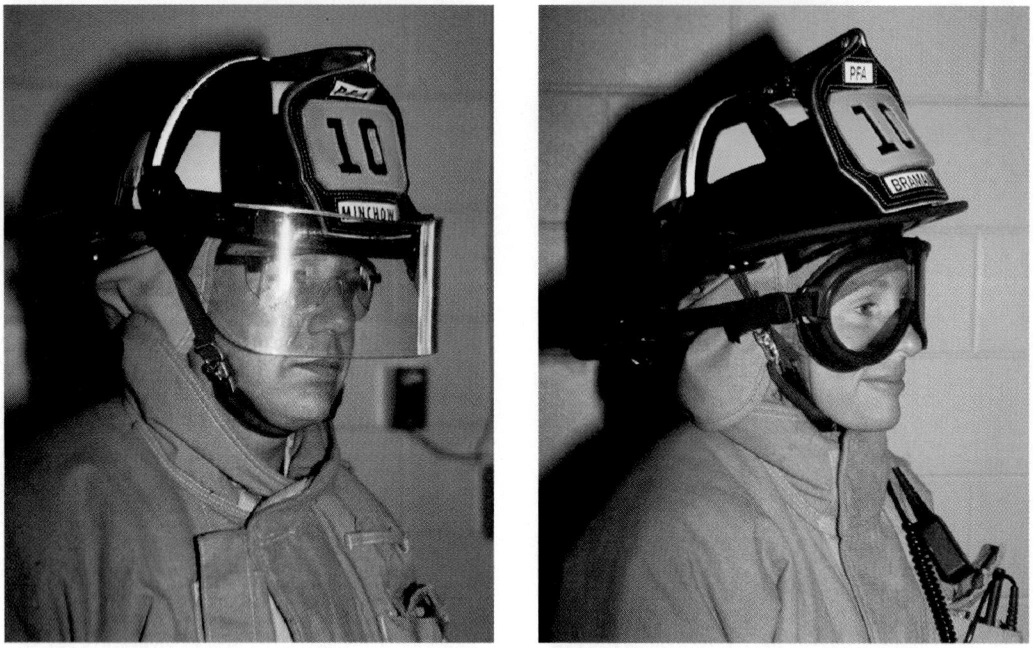

(A) (B)

Figure 6-15 Different types of eye protection are available for firefighters. (A) Wearing safety glasses under a face shield improves eye protection. (B) A face shield alone may not protect the eyes.

Miscellaneous PPE Components

Firefighters use many different types of PPE components to increase their personal safety in various hazardous environments. The following is a list of some of these items.

Eye Protection

Firefighters use many different forms of eye protection. Goggles, safety glasses, and wrap-around shields are examples. An SCBA face shield can provide primary eye protection although a structural helmet face shield may not. The helmet face shield—because of its wide facial opening—may not prevent windblown particulate or liquid splashes from getting into the eyes, **Figure 6-15**.

Hearing Protection

Many typical rescue and firefighting operations can expose firefighters to noise levels above 100 dB, which can cause hearing damage (including permanent damage) after a short-duration exposure. For this reason, firefighters should always have rapid access to hearing protection. Hearing protection takes on many forms—foam plugs, rigid earmuffs, and headsets are common examples. Many fire apparatus are equipped with a technology that combines hearing protection/intercom/radio microphone into a single headset. These combination systems make communication within the cab of a responding apparatus efficient and effective, **Figure 6-16**.

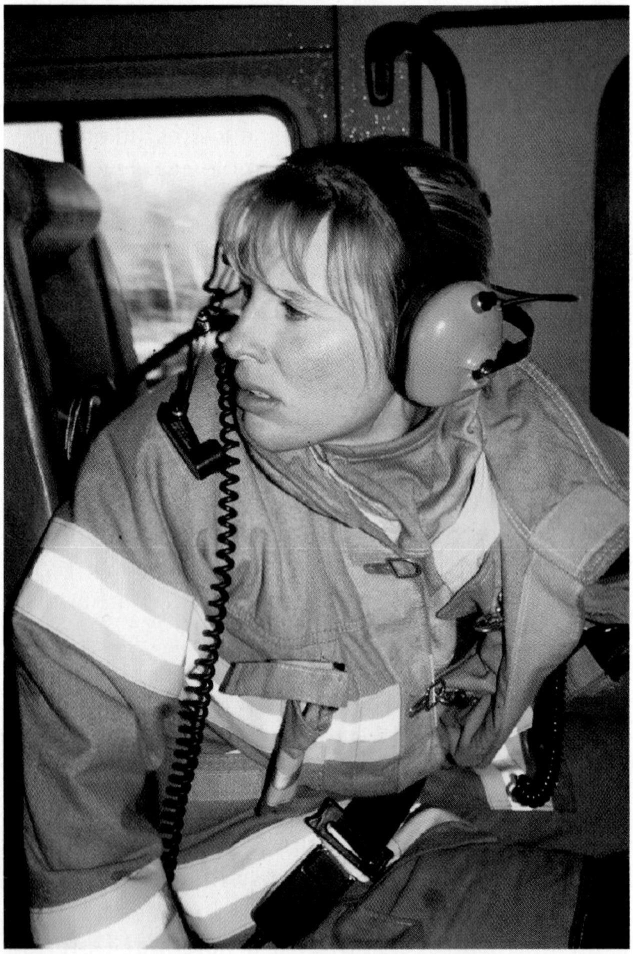

Figure 6-16 Combination headsets protect the wearer's hearing and also increase communication effectiveness between crew members and other responders.

Personal Alert Safety System (PASS)

A PASS device is a small, motion-sensitive unit that is typically battery powered and includes a loud audible warning signal. Some devices include a small flashing beacon or strobe. The PASS device, when worn on a firefighter, senses the firefighter's motion. As long as the device senses motion, the alarm will not sound. Inactivity for thirty seconds causes the device to send a "chirp" or other reminder to the wearer. If the wearer fails to move, the device will go into alarm mode and emit a loud noise to signal or warn other firefighters that the wearer may be in trouble. Most new PASS devices are an integral part of the SCBA unit, although some fire departments issue PASS devices to individual firefighters, **Figure 6-17**. The biggest problem with PASS devices results when wearers simply forget to turn their units on—this simple mental lapse has contributed to numerous firefighter fatalities.The National Fire Protection

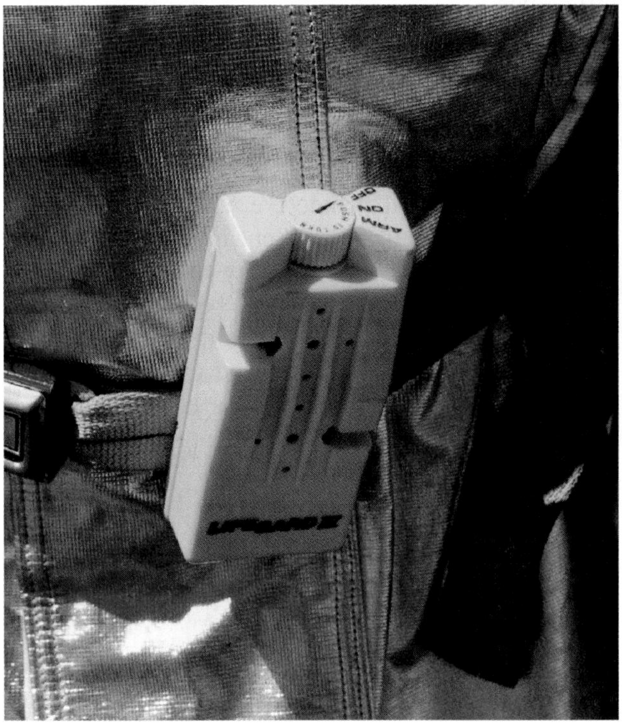

(A)

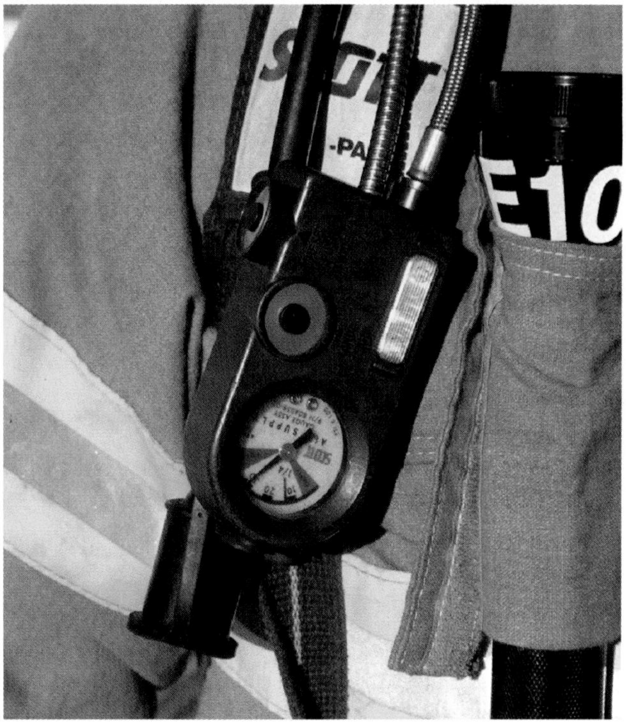

(B)

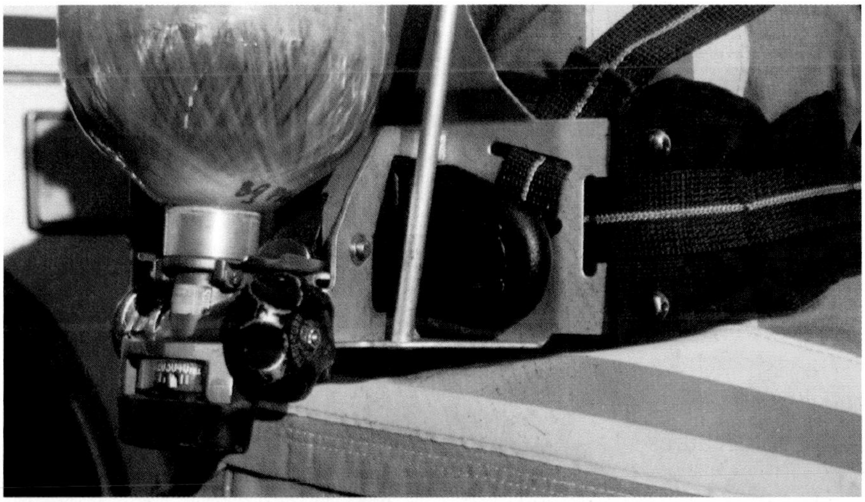

(C)

Figure 6-17 PASS devices can help save a firefighter's life—but they *must* be activated. (A) A manual PASS device *must* be armed by the wearer. (B and C) An integrated PASS device on an SCBA, which is activated when the wearer opens the SCBA bottle.

Association requires that PASS devices integrated with SCBA activate automatically when the SCBA air supply is turned on.

Safety Arming a manual PASS device can save a firefighter's life!

Work Uniform

Firefighters assigned to emergency response duty are often required to wear a uniform—a uniform that meets NFPA standards. NFPA 1975, *Standard on Station/Work Uniforms for Fire Fighters,* outlines manufacturers' guidelines and requirements for work uniforms. The underlying message is that a work uniform can add a protective measure to firefighters engaged in support activities or station duties. A work uniform that meets NFPA standards is *not* designed to protect the wearer from IDLH atmospheres, but can add another layer of reasonable protection under wildland, proximity, and structural ensembles.

CARE AND MAINTENANCE OF PERSONAL PROTECTIVE EQUIPMENT

The key to maintaining personal protective equipment in a high state of readiness is simple: Follow the specific instructions given by the manufacturer. NFPA requires manufacturers to clearly label care instructions for cleaning each piece of equipment. In addition, manufacturers should provide the user with specific instructions and information that address the following:

- Safety considerations
- Limitations of use
- Marking recommendations and restrictions
- Warranty information
- Sizing/adjustment procedures
- Recommended storage practices
- Inspection frequency and details
- Donning and doffing procedures
- Interface issues

Specific to cleaning and maintenance, the manufacturer will provide cleaning instructions and precautions, including a warning that the user should not wear equipment that is not thoroughly clean and dry.

Caution Wearing clothing that is not dry can lead to reduced thermal resistance and to burns.

Equipment exposed to biological and chemical contaminants should be decontaminated in accordance with manufacturers' instructions, **Figure 6-18**. NFPA 1581, *Standard on Fire Department Infection Control Program,* requires that clothing be cleaned every six months as a minimum. Obviously, PPE that is dirty or contaminated should be cleaned immediately. Fire departments should have specific cleaning equipment or a cleaning service that can clean equipment in accordance with manufacturers' guidelines. Washing structural or wildland gear along with linens or other household items should be forbidden because cross-contamination can result.

Finally, personal protective equipment must be routinely inspected and, when appropriate, retired and disposed of as suggested by the manufacturer. This last maintenance check is best accomplished through a team process that involves the individual firefighter, his or her officer, and the organization. This partnership process creates checks and balances that help ensure that PPE is serviceable for the user.

Figure 6-18 Structural and wildland clothing should be washed in accordance with manufacturers' recommendations. Dedicated washers that are used for PPE *only* can help lower cross-contamination of other washables.

PERSONAL PROTECTIVE EQUIPMENT EFFECTIVENESS: "STREET SMARTS"

To maximize the effectiveness of all PPE, firefighters must develop "automatic behaviors." Simple steps can help achieve these behaviors. This final—and perhaps most important—section of the chapter contains suggestions to help firefighters develop good habits that will reduce their chance of injury.

Good PPE Habits and Attitude

Fire departments spend literally thousands of dollars equipping firefighters with PPE for the hazards they may face. Unfortunately, many firefighters are still injured—all too often because the individual firefighter has failed to properly utilize the PPE. This unfortunate result can be eliminated simply by good PPE habits and a positive attitude toward safety. The proper PPE attitude starts with a simple concept:

> **Safety** Firefighters should don all PPE necessary for the potential worst case scenario.

Granted, this approach may lead to "overdressing" for an incident. In these cases, the firefighters' company officer, incident commander, or incident safety officer may allow firefighters to "dress-down." The inverse, however, is unacceptable. Firefighters who take a "wait and see" attitude to decide what level of PPE they are going to need have set themselves up to shortcut their PPE ensemble if, on arrival at the scene, the situation requires immediate lifesaving actions. It is better to be prepared first—then act.

Another key to proper and effective use of PPE is the development of good habits that include fast, proper, and complete donning of the appropriate PPE ensemble, **Figure 6-19**. Unfortunately, many firefighters begin shortcutting their use and completeness of PPE donning as days, weeks, months, and years of firefighting experience are accrued. This is one of the travesties that has kept injury rates high for firefighters.

Establishing good habits takes nothing more than self-discipline and practice. The benefits of self-discipline applied to PPE completeness pay a dividend in the form of **acclimation**. PPE acclimation simply refers to "getting used to" the wearing of PPE. The restrictive, encompassing, and stifling heat initially associated with most firefighter PPE simply becomes less of an issue and physically less taxing with acclimation. No firefighter particularly enjoys donning full structural PPE in 100°F heat. The firefighter who is acclimated physically and mentally to the PPE will, however, accept the merits of the PPE and have less stress—and certainly less risk—than the firefighter who complains that the gear is "too hot" for the weather. The firefighter with good PPE habits and a positive PPE attitude knows that acclimation, rest, and on-scene **rehab** will beat the discomfort that some firefighters complain about.

> **Streetsmart Tip** Hydration prior to wearing PPE ensembles can help prevent stress-related injuries. Firefighters should have a bottle of water available near their gear or at their seat on the apparatus, and take a few drinks as they respond to the emergency.

When donning PPE, the firefighter should follow the manufacturer's instructions and ensure that shortcuts have been eliminated. Firefighters who have experienced a rash of false alarms (i.e., unintentional

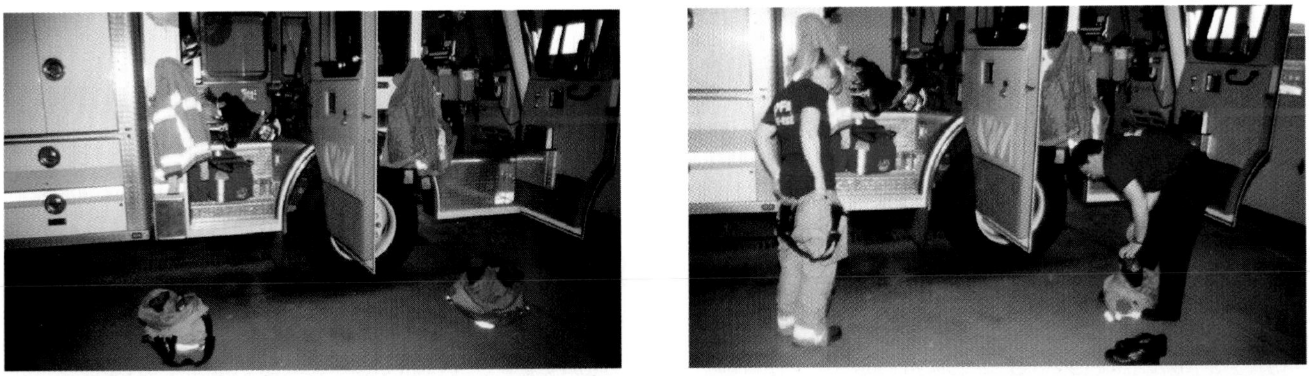

Figure 6-19 Duty personnel should set up their gear for rapid—and complete—donning. Establish good habits to help eliminate shortcuts.

fire alarm activations in a building) seem to be particularly guilty when it comes to shortcutting their PPE. The firefighter with an appropriate PPE attitude will view each one of these incidents as an opportunity to fine-tune donning procedures and achieve **mastery** in PPE donning. Further, it makes sense that a firefighter who has appropriately donned the PPE is in a strong readiness position to be immediately assigned duties in IDLH atmospheres. Firefighters must resist the temptation to get caught in the "cry wolf" syndrome. They must always be ready!

Streetsmart Suggestions

Daily, firefighters utilize PPE as their first and last defense against injury and illness. From these experiences come many "streetsmart" suggestions to help new firefighters:

- Keep PPE clean, inspected, and serviceable—an ownership attitude!
- Practice "team checks." Firefighters should check each others' PPE for readiness, **Figure 6-20**.
- When on duty in a station, position PPE for rapid, complete donning. A practical example is to place a structural hood across bunker pants and boots as a reminder to don the hood first so it will interface appropriately under the bunker coat.
- When doffing PPE, take the time to prepare it for the next response. Set the clothing up for rapid donning—preparation time spent now speeds the next donning.
- Do not let the urgency of the situation override prudent judgment.

Figure 6-20 Firefighting teams should check each others' PPE for complete donning. This "team check" can help prevent a burn or other injury.

- PPE also includes a flashlight, personal tool, radio, earplugs, safety glasses/goggles, accountability tag (see Chapter 23), and a partner, **Figure 6-21**.
- Practice doesn't make perfect—*perfect practice makes perfect!* Firefighters need to use PPE the right way every time to cement good habits. Can firefighters don their structural PPE ensemble in less than a minute (not including SCBA)? Can they add the SCBA in less than a minute? The most important question: Can they do both *correctly* in less than two minutes?
- Be the PPE success example—not an injury statistic.
- When wearing any PPE ensemble, the firefighter must increase water consumption to help stay hydrated.

Mastery of PPE skills allows the firefighter to concentrate on the important task at hand.

Figure 6-21 A full structural protective system includes NFPA-compliant gear, as well as a flashlight, forcible-entry tool, radio, SCBA, and a *partner!*

Lessons Learned

Firefighter personal protective equipment and ensembles are the first and last defense against injuries and illness in a profession that is extremely hazardous. The firefighter who develops good PPE habits and a positive PPE attitude is likely to be quicker, more prepared, and better acclimated to the real-world confines of the PPE. Shortcutting the use of PPE is a dangerous practice that can contribute to injuries.

Through national consensus standards like the NFPA process, firefighter protective equipment has advanced significantly in the past two decades. NFPA standards address the performance requirements and manufacturing criteria for PPE as well as "use" standards for firefighters and fire departments. Federal government entities also influence PPE use and guidelines.

Different PPE ensembles exist for firefighters, including structural, proximity, and wildland ensembles. The key with any ensemble is the simple concept that an entire ensemble must be donned completely in order for the ensemble to protect the wearer. A shortcut can open the door to an injury. SCBA, harnesses, hearing protection, eye protection, and other PPE components must also be considered to protect the firefighter from hazards.

Care and maintenance of PPE are essential to its readiness. The firefighter must understand manufacturers' instructions and information regarding the fit, use, cleaning, and serviceability of assigned PPE. A sense of ownership is important here.

Finally, PPE effectiveness is dependent on the user's (individual firefighter's) good PPE habits and positive PPE attitude. Donning gear for the worst-case scenario prior to incident engagement is preferable to a "wait and see" approach. The firefighter must *be ready!*

KEY TERMS

Acclimation The act of becoming accustomed or used to something. Typically achieved through repeated practice within a given set of conditions.

Brush Gear Another term for a wildland personal protective ensemble.

Bunkers A slang term that is used mostly to describe the components of a structural firefighting ensemble. The original use of the term *bunkers* referred only to the pant/boot combination that firefighters wore at night and placed next to their "bunks" for rapid donning.

Fire Shelter A last-resort protective device for wildland firefighters caught or trapped in an environment where a firestorm or blowup is imminent.

Mastery The concept that an individual can achieve 90 percent of an objective 90 percent of the time.

Personal Alert Safety System (PASS) A device that emits a loud alert or warning that the wearer is motionless.

Rehab A shortened word meaning *rehabilitation.* Rehab typically consists of rest, medical evaluation, hydration, and nourishment.

Thermal Protective Performance (TPP) A rating level, expressed in seconds, used to characterize the protective qualities of a PPE component before serious injury is experienced by the wearer.

Web Gear The term given to a whole host of personal items carried on a belt/harness arrangement worn by wildland firefighters. Items include water bottles, a fire shelter, radio, and day sack.

Zoning A term given to the establishment of specific hazard zones; that is, hot zone, warm zone, cold zone. Also collapse zones.

REVIEW QUESTIONS

1. Explain what is meant by the phrase "PPE is the first and last resort."
2. List four government and ancillary agencies or associations that influence PPE design or use.
3. Name the agency responsible for developing consensus firefighter PPE standards. Name the agency that writes PPE use regulations.

4. List the components of a structural PPE ensemble.
5. Describe the importance of TPP and "sealing" a structural ensemble.
6. Describe how a proximity ensemble differs from other ensembles.
7. List the components of a wildland PPE ensemble.
8. Describe the operational concept of a PASS device.
9. Explain the relationship of a work uniform and undergarments to PPE ensembles.
10. List three important guidelines relative to the washing of PPE garments.
11. Define the concept of acclimation in regards to PPE.
12. List five "street smart" suggestions for improved effectiveness with PPE.

Additional Resources

Angle, James S., *Safety in the Emergency Services.* Delmar Learning, a part of the Thomson Corporation, Clifton Park, NY, 1999.

Dodson, David W., *Fire Department Incident Safety Officer.* Delmar Learning, a part of the Thomson Corporation, Clifton Park, NY, 1999.

Emergency Incident Rehabilitation, FA-114, United States Fire Administration, Emmitsburg, MD, 1992.

NFPA 1500: Standard on Fire Department Occupational Safety and Health Programs. National Fire Protection Association, Quincy, MA, 2002.

NFPA 1971: Standard on Protective Ensemble for Structural Firefighting. National Fire Protection Association, Quincy, MA, 2002.

NFPA 1976: Standard on Protective Clothing for Proximity Firefighting. National Fire Protection Association, Quincy, MA, 2000.

NFPA 1977: Standard on Protective Clothing and Equipment for Wildland Firefighting. National Fire Protection Association, Quincy, MA, 1998.

NFPA 1851: Standard on Selection, Care, and Maintenance of Structural Fire Fighting Protective Ensembles, National Fire Protection Association, Quincy, MA, 2001.

Risk Management Practices for the Fire Service, FA-166, United States Fire Administration, Emmitsburg, MD, 1996.

CHAPTER

7

SELF-CONTAINED BREATHING APPARATUS

Thomas J. Wutz, Midway Fire Department and New York State Office of Fire Prevention and Control

OUTLINE

- Objectives
- Introduction
- Conditions Requiring Respiratory Protection
- Effects of Toxic Gases and Toxic Environments
- Legal Requirements for Self-Contained Breathing Apparatus Use
- Limitations of Self-Contained Breathing Apparatus
- Types of Self-Contained Breathing Apparatus
- Donning and Doffing Self-Contained Breathing Apparatus
- Self-Contained Breathing Apparatus Operation and Emergency Procedures
- Inspection and Maintenance of Self-Contained Breathing Apparatus
- Lessons Learned
- Key Terms
- Review Questions
- Endnote
- Additional Resources

STREET STORY

It was early one morning, close to 8 a.m., when we got a call that a house in our area had a leak in the gas main outside. Apparently the leak was very close to the foundation of the house and had filtered through the cellar. Before we arrived, the house blew apart—it was totally lifted off its foundation and then settled. By the time we got there, you could see little flames all around the front lawn, which meant the ground was saturated with natural gas. So we put our SCBA masks on immediately. Then we went into the building to extinguish a fire in the cellar.

It was precarious because the destruction left the ceilings only 5 feet instead of 8 feet, and it was completely dark. It wasn't that big of a fire, but we were exerting a lot of energy because we were crawling and pulling the hose with us. Under normal conditions a tank of air will last you maybe half an hour, but the harder you work—if you're huffing and puffing—the more air you end up using. So in this case, we'd been down there for maybe twenty minutes, when the bell went off on my air supply, meaning I had about five minutes to get out of there. Sometimes it's hard to tell if it's your bell or someone else's, so I reached back and felt the bell on my tank—it was vibrating, which indicated it was mine.

Our hand lights stopped operating properly. And my partner and I didn't know how we'd get out since we couldn't see anything—then we had an idea. We grabbed onto the hoseline and followed it back, crawling out. If I would have run out of air, I would have had to put my face as close to the floor as possible so I could get just a little oxygen (gas rises). It would have been dangerous, because there would be the possibility I still could have been overcome if the air had been toxic with gas fumes. That's why it's important to stick with your partner and wear your SCBA in any situation—you never know whether what's burning is toxic.

—Street Story by Bernard Lach, Retired Chief, Torrington Fire Department, Torrington, Connecticut

OBJECTIVES

After completing this chapter, the reader should be able to:

- List two conditions requiring respiratory protection.
- List and explain the effects of oxygen deficiency and toxic gases on the human body.
- List one legal requirement for use of self-contained breathing apparatus (SCBA).
- List two types of SCBA.
- List four components of the SCBA used by the authority having jurisdiction.
- Demonstrate two different SCBA donning procedures at 100 percent accuracy in the time limit established by the authority having jurisdiction.
- Demonstrate routine inspection procedures of SCBA in accordance with manufacturers' instructions.
- Demonstrate after-use maintenance and servicing of SCBA in accordance with manufacturers' instructions.
- Demonstrate the servicing of an SCBA cylinder with the air-filling system used by the authority having jurisdiction.

INTRODUCTION

Self-contained breathing apparatus (SCBA) is one of the most important items of personal protective equipment used by firefighters and rescue personnel. SCBA allows firefighters to enter hazardous atmospheres to perform critical interior operations including offensive fire attack; search, rescue, and removal of victims; ventilation; and overhaul. In addition, SCBA is used at non-fire incidents, such as hazardous material or confined space rescue situations involving toxic fumes or oxygen-deficient atmospheres.

The **respiratory system** of the human body is most vulnerable to injury, especially from toxic conditions and gases encountered during firefighting operations. To protect firefighters in this environment, fire departments must have a respiratory protection policy or "mask rule." This policy should require all personnel to not only wear, but use their SCBA during operations where an **immediate danger to life and health (IDLH)** atmosphere may be encountered. Because it is impossible to predetermine all IDLH atmospheres, this policy must include operations during interior or exterior fire attack, such as structure, vehicle, and dumpster fires, below-grade or confined space rescue, and hazardous materials incidents. Respiratory protection provided by SCBA is necessary, even during exterior defensive operations as shown in **Figure 7-1**.

Figure 7-1 Large volumes of smoke require the use of SCBA, even for exterior operations as shown here at a tire storage facility.

> **Safety** Hydrogen cyanide is a deadly gas and has a distinct odor of almonds. Any inhaled toxic gas can directly cause disease of the lung tissue. SCBA, keen senses, incident command, and a thorough safety program are vital to keep firefighters safe from these deadly toxins and other safety hazards.

Failure to understand and know how to use SCBA properly could result in injury or death to a firefighter, failed rescue attempts of victims, or deterioration of the emergency incident. In addition to the short-term effects experienced during the emergency incident, firefighters may suffer long-term health problems due to repeated exposure to toxic environments. The tasks of firefighters can only be accomplished effectively through the proper use of personal protective equipment (PPE), **Figure 7-2.**

Syracuse, New York, and Lubbock, Texas, are two cities separated by 1,700 miles. Although they are separated geographically, they have a common connection: line-of-duty deaths of firefighters. Within a year's time from April 1978 to March 1979, these two cities lost a total of seven firefighters in structure fire incidents.

At 0046 hours on April 9, 1978, the Syracuse Fire Department responded to a structure fire. The

Figure 7-2 These firefighters in full protective equipment including SCBA are ready to begin interior firefighting operations.

fire involved a three-story, wood frame, balloon construction apartment building that had been converted from a single-family residence prior to 1957. On arrival, the first engine company reported light smoke showing from the second and third floors. As firefighters searched those floors, fire in concealed spaces progressed. Light smoke conditions deteriorated. An extensive fire developed on the third floor and in the attic, and the atmosphere above the second floor became untenable. At approximately 0059 hours, firefighters were ordered to evacuate the third floor. For unknown reasons four firefighters did not leave. All four became trapped and died.

Less than a year later, on March 25, 1979, the Lubbock Fire Department responded to a structure fire at 0432 hours. This incident was in a one-story building of ordinary construction, occupied by a restaurant undergoing renovations. Firefighters attacked a fire in the kitchen area, bringing it under control in about twenty minutes. During the overhaul phase, three firefighters entered the building, became disoriented, and were overcome by carbon monoxide gas and died.

During the investigation of these incidents, questions concerning the design, duration, and use of the SCBA were reviewed, along with training the firefighters had received for its use. Since these deaths, many changes have occurred in SCBA including design and use. The most notable of these changes follow:

- *SCBA weight.* The weight of a standard thirty-minute SCBA unit used in the fire service has been reduced by 15 to 17 pounds, approximately 40 percent since 1980. This reduces the physical stress on the firefighter.
- ***Positive pressure.*** A constant supply of air is delivered, pressurizing the face piece, keeping toxic gases from entering. This pressure (1½ to 2 psi, depending on the manufacturer), which is slightly above atmospheric pressure, also helps maintain face piece seals.
- *Improved design.* Breathing tube, regulator, and harness designs have been improved to limit catastrophic failure during firefighting operations.
- *Improved regulator design.* New regulator designs provide increased airflow and redundancy, providing a backup in the event of a regulator/pressure-reducing failure.
- *SCBA maintenance.* Field- and factory-level maintenance programs have been implemented to ensure proper operation of SCBA units.
- *PASS devices.* Personal Alert Safety Systems (PASS) are audible warning devices that incorporate a motion detector to sense movement. The PASS will automatically sound an alarm signal (and usually a small flashing strobe) if movement is undetected for thirty seconds. This alarm alerts other firefighters to an unconscious or incapacitated firefighter. The device can also be activated manually to signal others that a firefighter is in trouble and needs help. Newer SCBA units have the PASS device incorporated in the SCBA design, while older removable PASS devices are attached to the SCBA unit, usually somewhere on the harness. These devices are in common use and are required by NFPA 1500.
- *Training.* **Mask confidence or "smoke divers" training** programs develop the firefighter's knowledge of and confidence in using an SCBA unit.
- *Increased regulation.* Implementation of **respiratory protection programs** is required by OSHA 29 CFR 1910.134 or NFPA 1500 and NFPA 1404.

These improvements are only as effective as the training firefighters receive and the proficiency they develop using SCBA. Technology, regulations, and mandates are only as good as an individual firefighter's commitment to use SCBA.

Note All SCBA used or manufactured for fire service use must be positive pressure.

CONDITIONS REQUIRING RESPIRATORY PROTECTION

Four conditions that present respiratory hazards are commonly found at fire or other emergency incidents:

- Oxygen deficiency
- High temperatures
- Smoke or other by-products of combustion
- Toxic environments.

These conditions can be found separately, such as an oxygen-deficient atmosphere in a confined space situation, or combined in a fire incident. Regardless of the incident, any time the potential for these environments exists or develops, firefighters must use their SCBA.

Oxygen-Deficient Environments

The human body and fire are similar as both require oxygen to survive. Fire (combustion) consumes oxygen and produces toxic gases that may displace or dilute the available oxygen. Atmospheres with oxygen concentrations below 19.5 percent are classified as **oxygen-deficient atmospheres.** As shown in **Table 7-1**, decreased oxygen affects the human body by causing muscular impairment, mental confusion, and eventually death. This in combination with toxic gases is the cause for most fire deaths for individuals lacking the protection of SCBA.

Streetsmart Tip Installed fire-extinguishing systems such as total flooding carbon dioxide or halon systems create an oxygen-deficient atmosphere. When entering a structure or area where this type of system has activated, even without a fire, the firefighter must have the protection of SCBA.

Elevated Temperatures

The respiratory system of the human body is extremely delicate and sensitive to elevated temperatures. Even air temperatures related to a recreational activity such as a sauna, commonly 120° to 130°F, taken into the lungs may cause a serious decrease in blood pressure and circulatory system problems. Inhaling of heated gases will cause an accumulation of fluid in the lungs, **pulmonary edema**, which can cause death from **asphyxiation**. In addition, the damage caused by inhaling heated air or gases is long term and not reversible by treatment of fresh, cool air.

Firefighter Fact Temperatures in a structure fire reach 1,000° to 2,400°F. One unprotected breath at this temperature level will cause death or severe damage to a firefighter's respiratory system.

Smoke

Smoke is the combination of unburned products of combustion, particles of carbon, tar, and associated gases such as carbon monoxide, carbon dioxide, sulfur dioxide, and hydrogen cyanide. This

Effects of Hypoxia (Reduced Oxygen)

OXYGEN PRESENT/AVAILABLE (%)	SYMPTOMS
21	Normal conditions, no effect
19.5	OSHA definition as oxygen deficient
17	Some muscular impairment, increased respiratory rate
12	Dizziness, headache, rapid fatigue
9	Unconsciousness
7 to 6	Death within a few minutes

Source: Browne & Crist, *Confined Space Rescue*, page 8, Delmar Learning, a part of the Thomson Corporation, Clifton Park, NY, 1999.

TABLE 7-1

combination of materials is an irritant to the respiratory system and, in many cases, small inhaled quantities may be fatal. In addition, the temperature of smoke may cause burns to the respiratory system.

EFFECTS OF TOXIC GASES AND TOXIC ENVIRONMENTS

The combustion process produces toxic gases and irritants that affect both the short- and long-term health of firefighters operating in hazardous environments. In addition, when products of combustion combine, they may form toxins that are lethal. A swimming pool at a residential occupancy, **Figure 7-3,** indicates storage of chemicals such as chlorine, which will produce a poisonous gas. Firefighters must understand that there is no "routine" fire. Even a light smoke condition at this type of structure could be deadly.

Some of the common gases produced in a fire affect not only the respiratory system, but the circulatory system as well, **Table 7-2.** In addition to the toxic products of combustion pesticides and herbicides present at an agricultural occupancy will produce many additional toxins, thus requiring a higher level of protection for firefighters, **Figure 7-4.**

Figure 7-3 This swimming pool is a warning to firefighters of the possible presence of chemicals in storage in this residence.

Carbon Monoxide

Carbon monoxide (CO) is produced in great quantities during the combustion process and is one of the most lethal gases found in a fire. This colorless and odorless gas is always present, especially during incomplete combustion or in areas of poor ventilation. Its presence is of great concern, especially during the overhaul stage when respiratory protection habits may be lax.

Toxic Gases Formed as Products of Combustion

GAS	TOXICOLOGICAL EFFECT	PRODUCED BY	IDLH (PPM)*
Carbon dioxide	Displaces oxygen	Free burning	40,000
Carbon monoxide	Displaces oxygen	Incomplete combustion	1,200 to 1,500
Hydrogen cyanide	Chemical asphyxiant	Wool, silk, nylon, polyurethane	50
Hydrogen chloride	Respiratory irritant	Polyvinyl chloride (PVC), building materials, furnishings	50
Nitrogen dioxide	Pulmonary irritant	Small quantities from fabrics, large quantities from cellulose nitrate	20
Phosgene	Poison	Burning refrigerants; produces hydrochloric acid when inhaled	25

*ppm = parts per million, ratio of volume of gas compared to volume of air.

TABLE 7-2

Figure 7-4 Agricultural occupancies will usually have a large supply of chemicals that are extremely hazardous during firefighting operations.

> **Caution** Firefighters should always use their SCBA, even when the fire is thought to be extinguished. Lethal amounts of carbon monoxide may be present.

Ordinarily, 98 percent of the oxygen carried by the blood is bound to the hemoglobin, which is contained in the red blood cells. CO attaches itself to the red blood cells and prevents oxygen from bonding with the hemoglobin. In fact, CO combines with blood almost 218 times easier than oxygen. It does not act directly on the body, but prevents oxygen from being distributed to the body, causing **hypoxia**, followed by death. **Table 7-3** shows the stages and symptoms of this process.

Intense physical activities during firefighting operations, **Figure 7-5**, require the respiratory system to increase the respiratory rate to deliver

Figure 7-5 The light smoke condition present during overhaul will contain large amounts of carbon monoxide requiring SCBA protection.

Symptoms of Carbon Monoxide Poisoning

SYMPTOMS	CARBON MONOXIDE CONCENTRATION (PPM)*
Mild headache	1,000
Headache, nausea	1,300
Unconsciousness after 1 hour†	1,500
10-minute exposure, dizziness, nausea	3,200
Fatal, less than 1-hour exposure	4,000
Danger of death in 1 to 3 minutes	10,000

*ppm = parts per million, ratio of volume of gas compared to volume of air.
†IDLH level.

TABLE 7-3

increased oxygen to the muscles and organs of the body. If a firefighter is working unprotected (without the protection of SCBA) in an atmosphere of CO, the increased respiration augments the ability of CO to incapacitate the firefighter. CO also does not metabolize out of the blood very quickly. During repeated exposures to low levels of CO during a shift, or during a number of fires, the blood level of CO is compounded. CO symptoms range from a mild headache to death, **Table 7-3.**

LEGAL REQUIREMENTS FOR SELF-CONTAINED BREATHING APPARATUS USE

Toxins are always present in the by-products of combustion. Common sense dictates that firefighters should use SCBA on every fire scene—from start to finish. For the firefighter's safety and health, a number of regulations have been developed for SCBA use. A number of organizations have also established regulations and standards concerning the design and use of SCBA, **Table 7-4.**

> **Firefighter Fact** Depending on the laws, rules, and regulations of a particular state or local government, OSHA standards may or may not be legal requirements. In general, the U.S. Department of Labor OSHA regulations do not apply to municipal employees, including firefighters. However, a number of states, **Figure 7-6**, have adopted their own plans, which must be as stringent as the federal requirements. In these states, the requirements do apply to municipal employees, including firefighters. Generally, volunteer firefighters are considered employees for the purposes of health and safety regulation. Simply said, firefighters must know the rules and regulations that apply to their organization.

Title 29 Code of Federal Regulations, Section 1910.134

OSHA 29 CFR 1910.134 establishes the standard for all entries into IDLH atmospheres. The April 1998 revision contains requirements related to interior structural firefighting and defines interior structural firefighting as an IDLH atmosphere. In addition to requiring the use of SCBA, the standard also establishes requirements for a complete respiratory protection program and regular medical evaluation of employees designated to wear and use SCBA.

Although employers (fire departments) are responsible for providing a safe and healthy work environment, all firefighters have a duty to understand and follow these regulations.

NFPA 1500: Standard on Fire Department Occupational Safety and Health Program

The National Fire Protection Association has established **NFPA 1500**, *Standard on Fire Department Occupational Safety and Health Program.* This standard covers many safety and health-related issues for firefighters, including respiratory protection. The difference between this standard and the OSHA regulations is that a government authority (city, town, county, or state), called the **authority having jurisdiction (AHJ)**, must adopt the standard as policy for the fire department.

The NFPA has two additional standards dealing with SCBA: **NFPA 1404**, *Standard for a Fire Department Self-Contained Breathing Apparatus Program,* and **NFPA 1981**, *Standard on Open-Circuit Self-Contained Breathing Apparatus for Fire Service.* The NFPA 1404 standard establishes minimum requirements for fire department respiratory protection programs including the use of, training, safety, emergency procedures, maintenance, and breathing air for SCBA used in the fire service. The NFPA 1981 standard establishes design and performance criteria, test methods, and certification for open-circuit SCBA intended for fire service use.

Both 29 CFR 1910.134 and NFPA 1500 contain similar requirements for respiratory protection. These are nationally recognized standards, and fire departments in jurisdictions without mandatory respiratory protection requirements should adopt or reference these for operations requiring the use of SCBA.

LIMITATIONS OF SELF-CONTAINED BREATHING APPARATUS

As with any other tool used by the fire service, SCBA has a number of limitations that firefighters must understand if they are to use the unit effectively and safely. These limitations are the SCBA

Organizations Concerned with SCBA Design and Use

ORGANIZATION	STANDARD	APPLICATION
National Institute for Occupational Safety & Health (NIOSH)	42 CFR Part 84	Requirements for design, testing, and certifying SCBA
Occupational Safety & Health Administration	29 CFR 1910.134	Respiratory protection programs for SCBA use
Occupational Safety & Health Administration	29 CFR 1910.156	Fire Brigade Standard, references 1910.134
National Fire Protection Association	NFPA 1404	Standard for a Fire Department SCBA Program
National Fire Protection Association	NFPA 1500	Standard on Fire Department Occupational Safety and Health Program
National Fire Protection Association	NFPA 1981	Standard on Open-Circuit SCBA for the Fire Service

TABLE 7-4

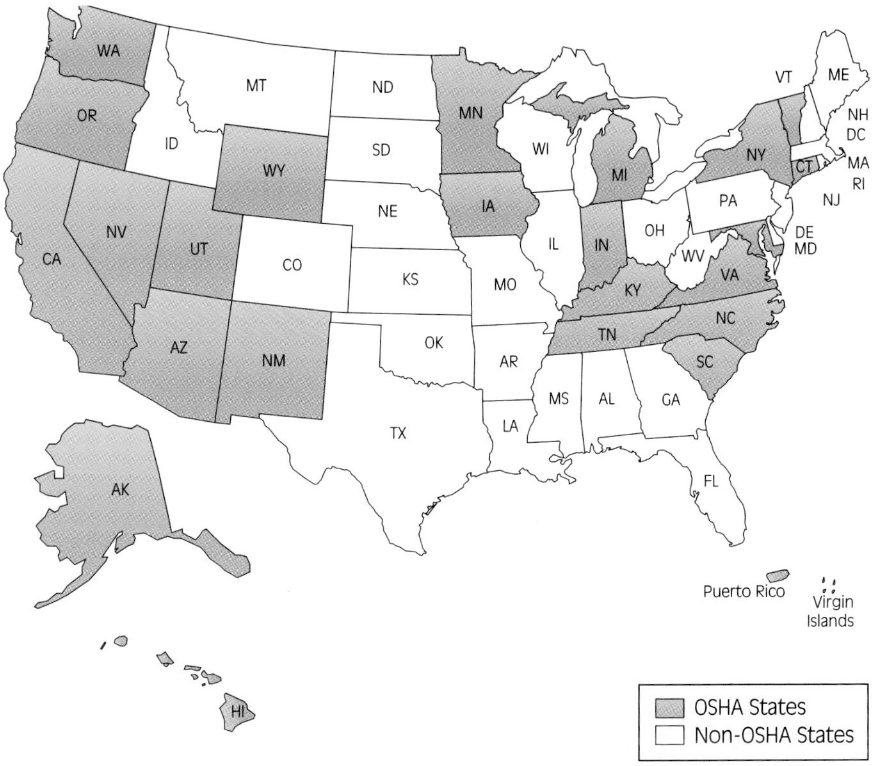

Figure 7-6 States requiring respiratory protection plans in accordance with OSHA's 29 CFR 1910.134 regulations.

unit itself (i.e., its size, weight, and limited air supply) and the physical and physiological condition of the user. Firefighters must be well trained to successfully complete all tasks and assignments requiring the use of SCBA, **Figure 7-7**.

SCBA Design and Size

The design and size of SCBA units vary greatly by manufacturer and by the age of the unit. Older units placed in service as little as ten years ago may be heavier and bulkier than newer ones. Firefighters must be conditioned to the added weight and bulk of the SCBA and the personal protective equipment they must wear. Depending on the activity level required and the physical condition of the user, *an SCBA cylinder with a rating of thirty minutes of air supply will usually be consumed more rapidly.* This factor limits the distance a firefighter may advance into a building or fire and requires frequent crew rotations at large incidents. Other concerns include:

- Visibility is restricted, peripheral vision is reduced, and fogging of the face piece can occur.
- Depending on the manufacturer's style, the age of the unit and the type of cylinder used, SCBA units will add 23 to 35 pounds of weight and 9 to 15 inches to the profile of the firefighter. This will limit the mobility of the firefighter.
- Unless the SCBA unit is equipped with a voice amplification accessory, the firefighter's voice will be muffled and difficult for others to understand, especially if the firefighter is trying to communicate by radio.
- The quantity of air is limited. It is vital that a firefighter know the status of the cylinder air supply prior to its use.

Streetsmart Tip The Philadelphia Fire Department conducted extensive testing in a firefighting skills proficiency course with 750 firefighters using SCBA. The average air consumption for SCBA rated for thirty minutes was less than fifteen minutes—from full cylinder to low air warning. Forcible entry and ventilation are two tasks requiring increased physical activity, **Figure 7-8**.

Limitations of the SCBA User

Firefighters themselves can be limited in their use of SCBA. A firefighter's physical, mental, and emotional condition can cause usage problems. When these limitations are coupled with the confines of the SCBA unit there can be serious problems.

Physical Limitations

- The total protective envelope including SCBA and PPE adds approximately 40 to 50 pounds of weight to the firefighter.
- The additional weight and bulk affect agility and mobility, requiring a high level of physical strength and endurance.

Figure 7-7 Continuous training with SCBA is one of the keys to effective firefighting operations.

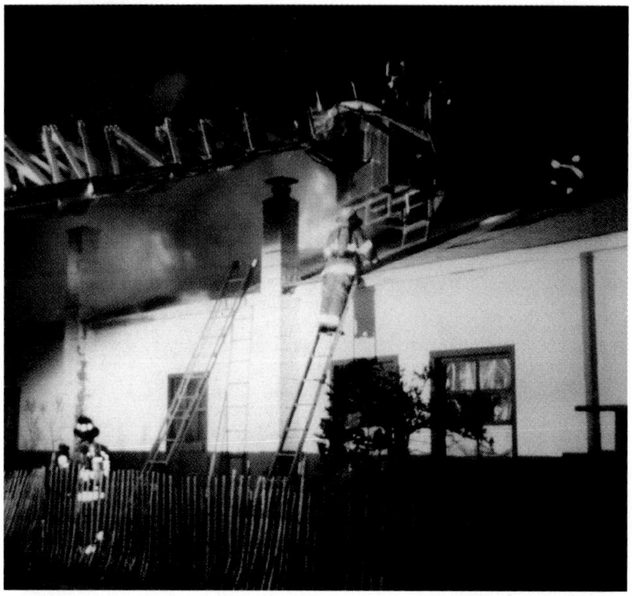

(A)

(B)

Figure 7-8 (A) Ventilation and (B) forcible entry are physically demanding. They produce increased respiration rates and air consumption.

- Even though face pieces have been fitted and tested, weight loss or a twenty-four-hour growth of facial hair may affect the ability to obtain a good seal.
- In addition to strength and endurance, the stress of firefighting in elevated temperatures requires additional cardiovascular and respiratory conditioning. Lack of conditioning will increase stress on the body.

> **Firefighter Fact** According to the U.S. Fire Administration, in the year 2000, seventy-two deaths occurred either at an emergency scene, or responding to or from the scene, with fourteen of the deaths occurring on wildland fires. There were also thirty nonemergency scene deaths involving training, administrative, or other duties. Forty-five (62.5 percent) of these fatalities were heart attack or stroke related. It is important for all fire departments to institute a program of physical fitness and nutrition training and regular medical examinations to combat firefighter stress and overexertion. Firefighters should be taught how to recognize their own physical and mental limitations and how to react when other company members reach their limitations.

Physiologic Limitations

- Lack of confidence in the SCBA unit and its ability to protect the firefighter may cause anxiety, increasing breathing rate.
- The degree of training or experience users have with SCBA affects their self-confidence and ability to function.
- Increased physical stress may cause anxiety.
- Emotional conditions, such as fear of being confined, excitement, or claustrophobia, may increase the user's breathing rate and air consumption.

These problems are addressed with proper training and education. Training in SCBA use is not a one-time deal; it must be continuous so firefighters maintain their proficiency and confidence with SCBA.

TYPES OF SELF-CONTAINED BREATHING APPARATUS

Two different types of SCBA are in general use in today's fire service: **open-circuit** and **closed-circuit** systems. In an open-circuit SCBA, exhaled air is vented to the outside atmosphere, whereas in a closed-circuit SCBA, the exhaled air stays in the system for filtering, cleaning, and circulation. The open-circuit SCBA, **Figure 7-9**, is the most commonly used for firefighting operations. The closed-circuit type is sometimes used for specialized rescue incidents.

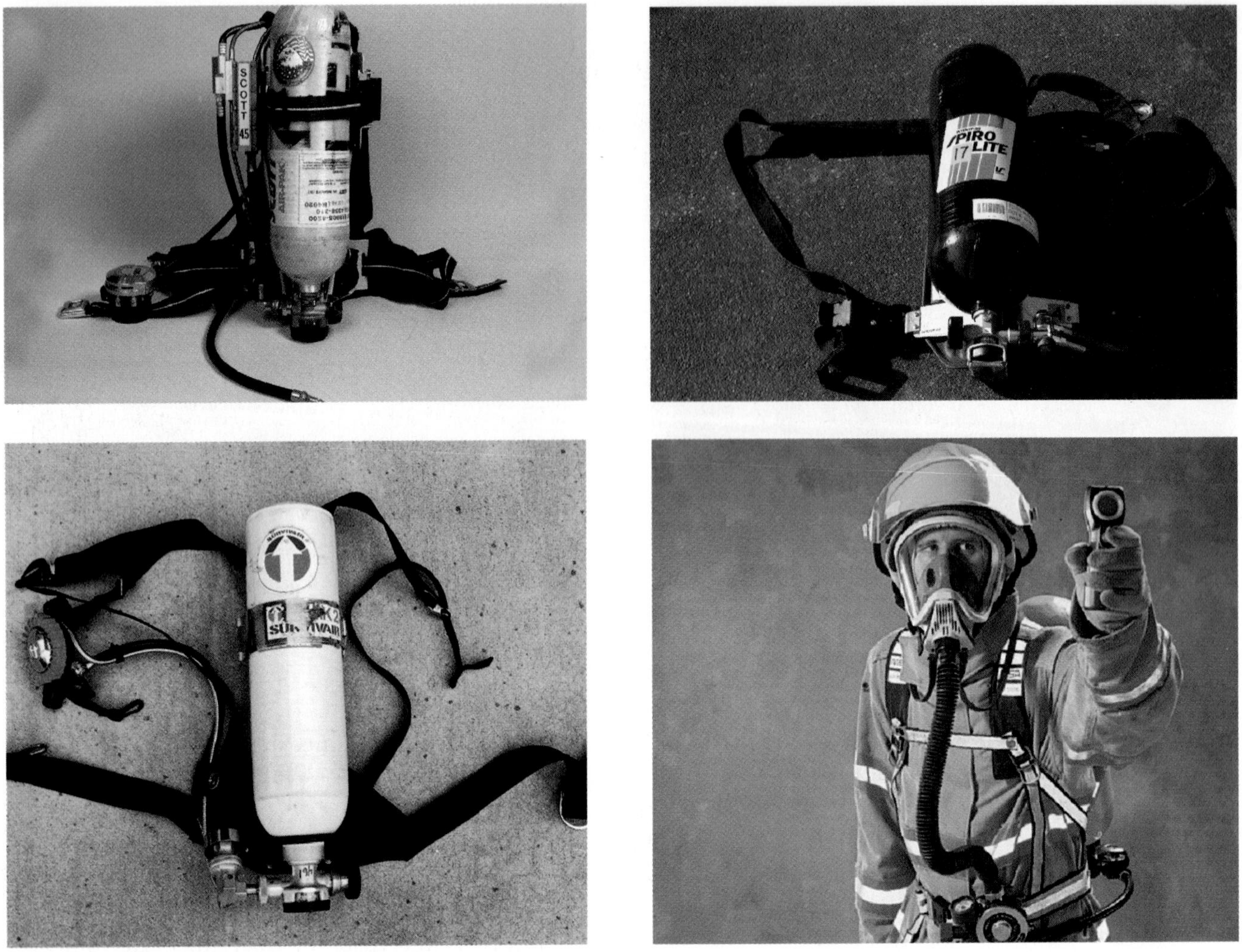

Figure 7-9 A number of manufacturers produce open-circuit SCBA for fire service use. *(Bottom right photo courtesy of Mine Safety Appliance Company)*

Open-Circuit Self-Contained Breathing Apparatus

A number of manufacturers supply SCBA for fire service use. Regardless of the manufacturer, SCBAs for fire service use are designed and built in accordance with **NIOSH** and NFPA standards. Certain parts of an SCBA unit may be interchangeable with parts from another manufacturer; however, this practice will void the NIOSH and manufacturers' certifications.

Generally, as shown in **Figure 7-10**, SCBA has four basic assembly components:

- *Backpack and harness:* The backpack holds the air cylinder and the harness allows the unit to be worn.
- *Cylinder:* Includes cylinder, valve, and cylinder pressure gauge.
- *Regulator:* Includes the high-pressure hose from the cylinder, the regulator, and end of service time indicator (EOSTI).
- *Face piece assembly:* Includes face piece, exhalation valve, and head harness.

Backpack and Harness Assembly

The backpack assembly, **Figure 7-11**, is designed to hold the air cylinder and provide the straps for securing the SCBA unit to the firefighter. It will consist of a metal or a high-temperature-resistant plastic frame, a mechanism for securing the air cylinder, and shoulder and waist straps. Depending on the manufacturer's design and style, the regulator may be attached to the waist straps. The straps are adjustable to the size of the user and designed to distribute the weight of the unit. Most models in use today are designed to have the greatest portion of the unit's weight carried on the hips by the waist straps.

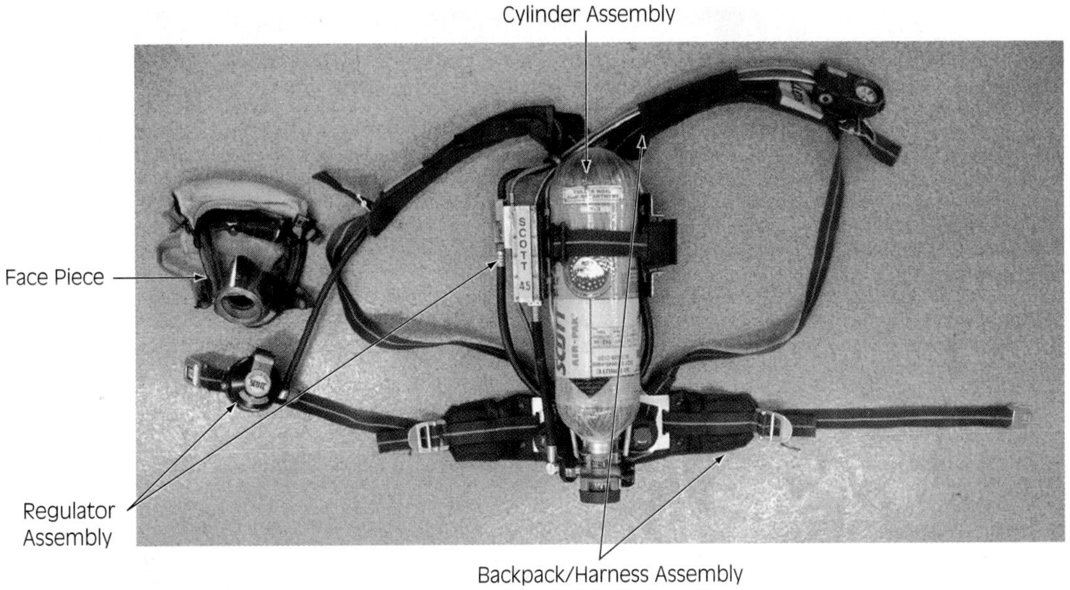

Figure 7-10 The four components of the open-circuit SCBA are the backpack/harness, cylinder, regulator, and face piece assemblies.

Cylinder Assembly

SCBA cylinders, **Figure 7-12**, contain the compressed air for breathing by the user. For this reason, the cylinder must be strong and durable and it constitutes most of the weight of the unit. Most cylinders in use today are aluminum, fiberglass/aluminum composite, and Kevlar or carbon composite materials. The change from steel to aluminum and then composite cylinders has reduced the weight of an SCBA cylinder by almost 50 percent. Typical cylinder capacities are listed in **Table 7-5**.

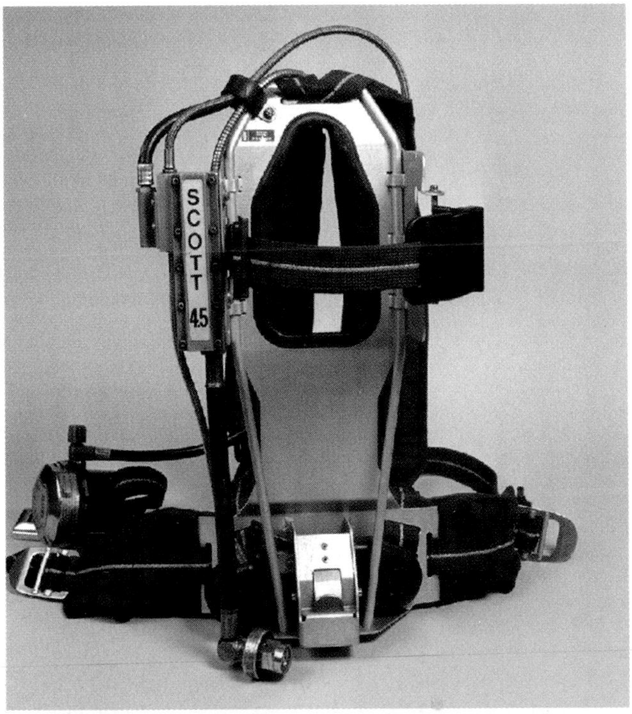

Figure 7-11 A typical SCBA backpack/harness assembly.

Types of Cylinders. The SCBA cylinders used in the fire service vary in material and type of manufacture. Listed here are general descriptions of the most common types of cylinders used in the fire service. Refer to manufacturers' information for exact specifications.

- *Steel.* Because of improvements in design and weight reduction using other materials, steel cylinders are generally not used for SCBA service. Most of the steel cylinders still in fire service use are used to provide air supply for various rescue tools. They have an unlimited service life as long as they pass a hydrostatic test every five years.

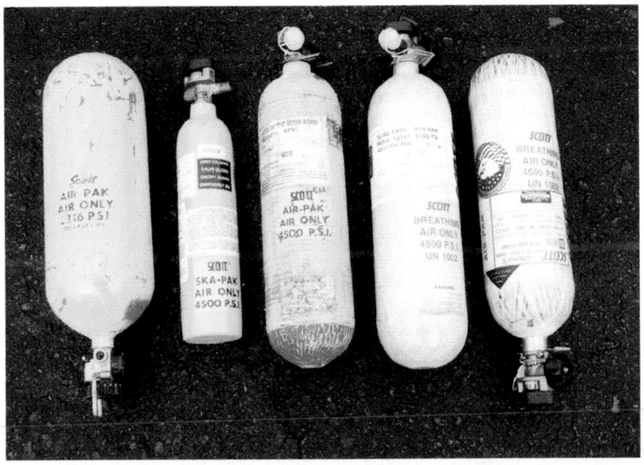

Figure 7-12 SCBA cylinders for fire service use, from left to right, are steel, aluminum, hoop-wrapped fiberglass, Kevlar composite, and carbon fiber composite.

SCBA Air Cylinder Capacities

RATED DURATION (MINUTES)	MATERIAL	CYLINDER PRESSURE (PSI)	CAPACITY-COMPRESSED AIR (FT^3)
30	Aluminum or composite	2,216	44
30	Aluminum or composite	4,500	44
45	Carbon fiber	3,000	65
45	Aluminum or composite	2,216	65
45	Aluminum or composite	4,500	65
60	Aluminum or composite	4,500	88

TABLE 7-5

- *Aluminum.* Aluminum cylinders are used for thirty-minute (2,216-psi) rated SCBA units. This type of cylinder contains 44 cubic feet of compressed air and weighs about 22 pounds when charged. These cylinders have an unlimited service life as long as they pass a hydrostatic test every five years.
- *Fiberglass (hoop-wrapped).* Fiberglass cylinders are used for thirty-minute (2,216-psi) rated SCBA units and are constructed with an aluminum inner shell. The cylinder sides are wrapped with fiberglass for strength. This type of cylinder contains 44 cubic feet of compressed air and weighs about 16 pounds when charged. These cylinders have a service life limited to fifteen years and must have a hydrostatic test every three years. In accordance with current U.S. Department of Transportation (USDOT) regulations, these cylinders are destroyed at the end of their service life.
- *Fiberglass (full-wrapped).* Full-wrapped fiberglass cylinders are used for thirty-, forty-five-, and sixty-minute (4,500-psi) rated SCBA units. They are constructed with an aluminum inner shell and fully wrapped with fiberglass for strength. These cylinders contain 44 to 88 cubic feet of compressed air and weigh 16 (44 ft^3) to 26 (88 ft^3) pounds when charged. These cylinders have a service life limited to fifteen years and must have a hydrostatic test every three years. In accordance with current USDOT regulations, these cylinders are destroyed at the end of their service life.
- *Kevlar/carbon composites.* Composite cylinders are used for thirty-, forty-five-, and sixty-minute (4,500-psi) rated SCBA units and are constructed with an aluminum inner shell and fully wrapped with Kevlar or carbon fibers for strength. These cylinders contain 44 to 88 cubic feet of compressed air and weigh 12 (44 ft^3) to 22 (88 ft^3) pounds when charged. These cylinders have a service life limited to fifteen years and must have a hydrostatic test every three to five years—depending on the manufacturer of the cylinder. In accordance with current USDOT regulations, these cylinders are destroyed at the end of their service life.

Caution In the past few years the failure of composite-type cylinders has occurred with increasing frequency. In three instances, cylinders failed while stored on fire apparatus. Investigations concluded that the failures were caused by stress-corrosion cracking of the fiberglass wraps resulting from exposure to a strong corrosive agent. Fiberglass composite cylinders are particularly at risk for stress-corrosion cracking because the fibers are under constant tension due to the internal pressure. When the structural integrity of the overwrap is weakened, a catastrophic failure of a cylinder can occur that may result in serious injury or death. Persons responsible for maintenance of composite cylinders should take measures to ensure that they do not come in contact with strong corrosive agents. In addition, cylinders should be washed only with a mild soap and water solution, and all recommendations of the cylinder manufacturer or distributor with regard to maintenance, requalification, and use must be carefully followed. In addition, during the 1980s a number of aluminum cylinders developed catastrophic failures at the neck area, near the connection of the cylinder valve. Even with regulations and testing, failures do occur and firefighters must be aware of the correct maintenance procedures for SCBA cylinders.

SCBA Cylinder Failures. The rated duration of all SCBA cylinders is based on laboratory testing; actual duration will vary significantly with each individual user. One of the most important factors affecting air supply duration is the physical and physiologic condition of the user. Firefighting is physically demanding work, in a very harsh environment, and when combined with the stress of searching for victims usually leads to increased breathing and consumption of air. For this reason even firefighters in top physical condition may only be able to operate for fifteen to twenty minutes with a unit rated for thirty minutes.

Cylinder Testing. The USDOT regulates compressed gas cylinders, including those used for SCBA, and requires hydrostatic testing. This test is done to ensure that the cylinder is capable of withstanding its rated pressure and capacity and the stress created when the cylinder is being filled. Cylinder testing is usually accomplished by an outside vendor; however, some large fire departments may have their own service unit that conducts these tests.

Once the test is complete, each cylinder is labeled or stamped as shown in **Figure 7-13**. This label must show the licenses of the testing organization and the date of the most recent test. One should never attempt to fill a cylinder with an out-of-date test label.

Caution In 1993 a firefighter filled a composite-type SCBA cylinder that was beyond its useful service life of fifteen years. After filling the cylinder, he placed it back on the apparatus and then a catastrophic failure occurred. The cylinder valve exploded out of the cylinder and killed the firefighter instantly. One should never service an SCBA cylinder that is out of date for hydrostatic testing or a composite cylinder that has passed its fifteen-year service life.

Air Quality for SCBA Use

The quality of the compressed breathing gas used in open-circuit SCBA has a direct effect on the performance of this equipment. Many standards address the quality or purity of compressed air used for breathing purposes. The most generally accepted are those established by the Compressed Gas Association in its pamphlet G-7.1-1989, which are incorporated by reference in OSHA 29 CFR 1910.134 and NFPA 1500. Briefly, the minimum quality of air used for SCBA is Grade D, as established in the pamphlet. This classification establishes maximum allowable quantities of impurities allowed in breathing air. In accordance with this specification the supply source (purifier/compressor) must be tested at least every three months and certified as shown on the certificate in **Figure 7-14**. NFPA 1404 has one exception to the Grade D breathing air requirement and that is a lower maximum moisture content of 24 ppm or drier.

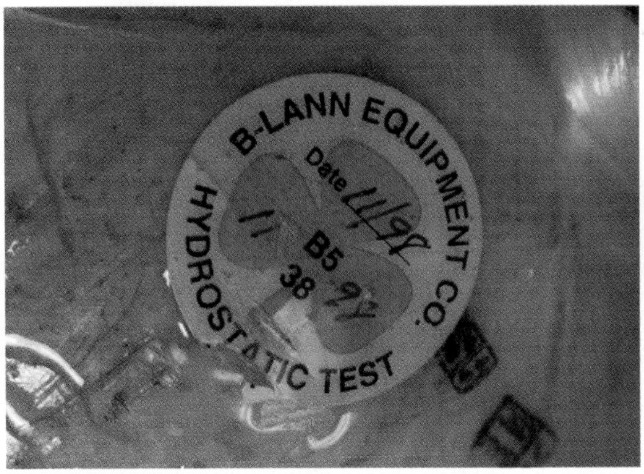

Figure 7-13 Depending on the type of material, SCBA cylinders are required to be hydrostatically tested every three to five years.

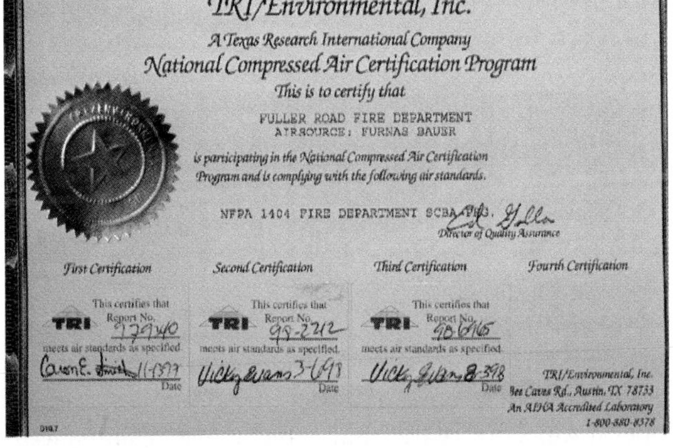

Figure 7-14 All air systems for SCBA use must be tested every three months and certified as shown by this air test quality certificate.

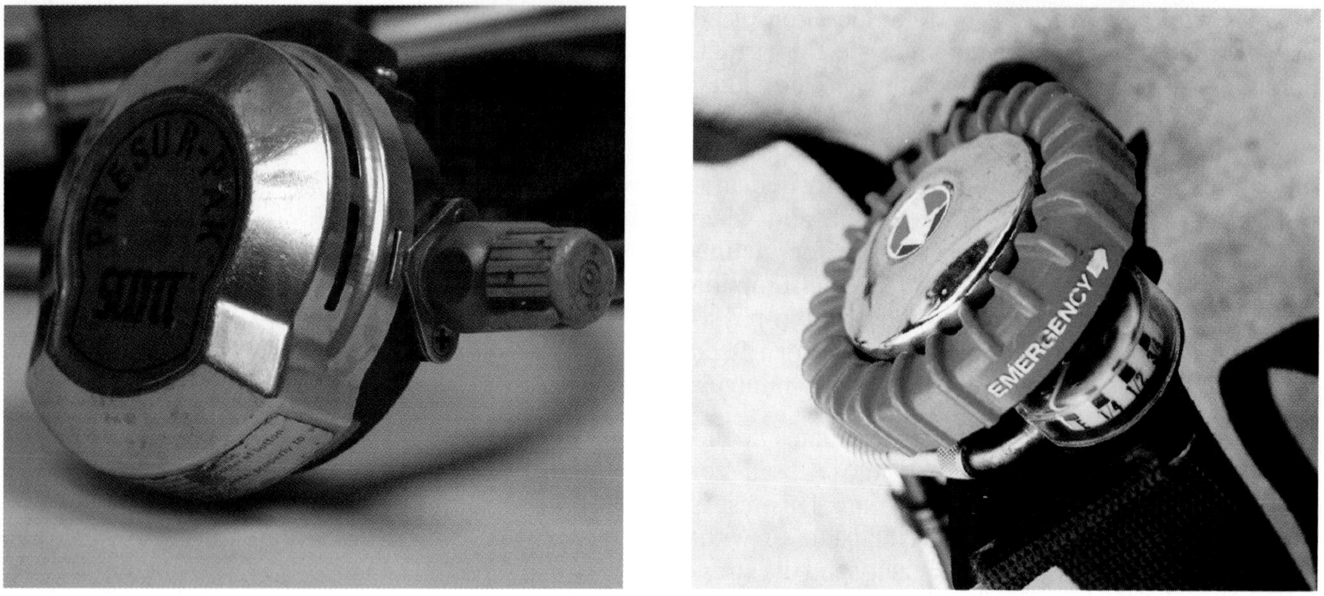

(A) (B)

Figure 7-15 SCBA designed for the fire service will use either (A) a face piece–mounted regulator or (B) a waist-mounted regulator.

This air is obtained from an air cylinder refill system owned by a fire department or from a commercial vendor. These systems, commonly called *cascade systems,* consist of a compressor, purifier, air quality monitor, storage system, booster compressor, control panel, and cylinder fill containers.

Regulator Assembly

Depending on the manufacture and style of the SCBA unit, the regulator assembly, **Figures 7-15A** and **B**, is attached to the face piece or waist straps. The regulator reduces the high-pressure air from the cylinder to a low, slightly above atmospheric pressure and controls the flow of air to the face piece. The regulator contains a diaphragm that is activated by the breathing action of the user. This action creates a pressure differential, which opens the diaphragm and allows air to flow to the face piece. In addition, all SCBA units designed for fire service use also maintain a constant positive pressure airflow to the face piece. The exhalation action moves the diaphragm to the closed position and opens exhalation valves, and the exhaled air is vented outside the face piece.

Again depending on the manufacture and style of the SCBA unit, the regulator will have color-coded valves to control both normal and emergency operation. One is the main line valve, usually colored yellow for normal operation; the second is the bypass valve, usually colored red for emergency operation.

During normal operation the main line valve is fully open, allowing maximum air to flow, and the bypass valve is in the fully closed position. The valves should remain in these positions for proper operation of the unit. In the event of a malfunction of the regulator or high-pressure reducer, the bypass valve must be manually operated.

The regulator may have a pressure gauge that is part of the regulator body or attached to the shoulder straps. This gauge should provide the same readings as those shown on the cylinder valve. There may be some difference, but the two gauges must read within 100 psi of each other or the unit should be taken out of service. Depending on the manufacturer, the increments shown on the gauge may be in percentages of a full cylinder or will display the actual psi remaining in the cylinder.

SCBA units designed for fire service use have a low air supply warning alarm, which is usually part of the regulator assembly. This alarm sounds when the cylinder pressure decreases to approximately one-fourth of the rated capacity of the cylinder. Depending on the manufacturer, some alarms on face piece–mounted regulators will also vibrate the face piece assembly to warn the firefighter of the low air supply. The 1997 edition of NFPA 1981 requires SCBA units to be equipped with two different types of low air alarms. The two alarms must function independently, failure of one alarm must not affect the operation of the second, and each must alert different senses, for instance, an audible alarm accompanied by a visual or tactile (vibration) alarm.

(A) (B)

Figure 7-16 Two types of SCBA face pieces. The regulator connects to (A) the face piece, and (B) the hose connects to the regulator.

> **Streetsmart Tip** When using SCBA, firefighters must work in teams of a minimum of two, and *all* team members must exit the hazardous environment when a low air supply alarm sounds. A firefighter must never leave a partner or allow a partner to leave alone.

Face Piece Assembly

The face piece assembly, **Figures 7-16A** and **B**, provides fresh air to the firefighter wearing an SCBA unit. In addition, protection from the hazardous environment is provided to the face and eyes. The assembly consists of a flexible rubber or silicon mask with a lens, exhalation valves, and a harness with adjustable straps. Most manufacturers provide a number of different sizes for proper fit without leakage. As mentioned later in this chapter, 29 CFR 1910.134 and NFPA 1500 require an annual fit test to ensure proper sizing of the face piece. Depending on manufacturer and style, the face piece will have a low-pressure air hose to connect to the regulator or the regulator will connect directly to the face piece.

The exhalation valve is the outlet for exhaled breath and prevents toxic gases from entering the face piece. These valves must be inspected regularly because dirt or moisture in a cold climate may keep it partially open, allowing toxic gases to enter the face piece. The following section on donning procedures explains how to check the operation of the exhalation valves.

The face piece assembly may have a number of options depending on manufacturer and style. These include a "nose cup" to prevent fogging and "voice amplifiers" to facilitate communication. The NIOSH regulation requires the nose-cup option in cold climates.

The last part of the face piece assembly is the harness and straps. This provides a tight fit to the head and prevents the face piece from loosening during firefighting operations. Again there are variations with different manufacturers, but the types in use today are a web style or headnet style. When the face piece is stored, the harness straps must be in the full out position to reduce wear on the straps and facilitate quick donning. Also, regardless of manufacturer, when donning the face piece, the straps must be pulled straight back from the head to ensure proper fit.

Each SCBA manufacturer has specific instructions for cleaning individual face pieces. Failure to follow these instructions may result in damage to the face piece and potential failure during use. An SCBA face piece must be cleaned after each use or regularly to remove dust and particles and to prevent the spread of communicable diseases. To minimize this problem, many fire departments issue each individual an SCBA face piece.

Closed-Circuit Self-Contained Breathing Apparatus

Generally, closed-circuit SCBA is not used for firefighting operations. For services that do utilize this system, the most common use is for hazardous materials incidents. Below-grade rescue use requires an extended air supply. These units are available with air supplies that range from thirty minutes to approximately four hours.

Closed-circuit SCBAs contain a cylinder of oxygen, a filter system, a regulator, and valves. They work on the principle of cleaning and filtering exhaled breath and adding pure oxygen to continue operation. The duration of the air supply is based on the filtering/cleaning and oxygen capacity of the unit.

Open-Circuit Supplied Air Respirators

Open-circuit **supplied air respirators** (**SARs**), also called airline respirators, **Figure 7-17**, are similar to SCBA units, except that the air supply cylinder is remote from the user. Air is supplied in the same manner as for a regular SCBA, but the hose connecting the cylinder and the SCBA unit may be 100 to 200 feet long, **Figure 17-7A.** These types of units are not used for firefighting operations, but are commonly used for hazardous materials incidents or confined space rescues. They provide the user with a long-duration air supply with mobility and agility. This type of unit must be equipped with an SCBA escape unit with duration of approximately five to ten minutes, **Figure 7-17B**.

DONNING AND DOFFING SELF-CONTAINED BREATHING APPARATUS

Depending on fire department apparatus and procedures, SCBA may be placed in service in a number of ways. The most common donning procedures are from a seat-mounted position in the apparatus, from a side compartment on the apparatus, or from a storage case. Regardless of how the SCBA unit is stored or mounted, firefighters should always refer to the manufacturers' instructions for specific procedures.

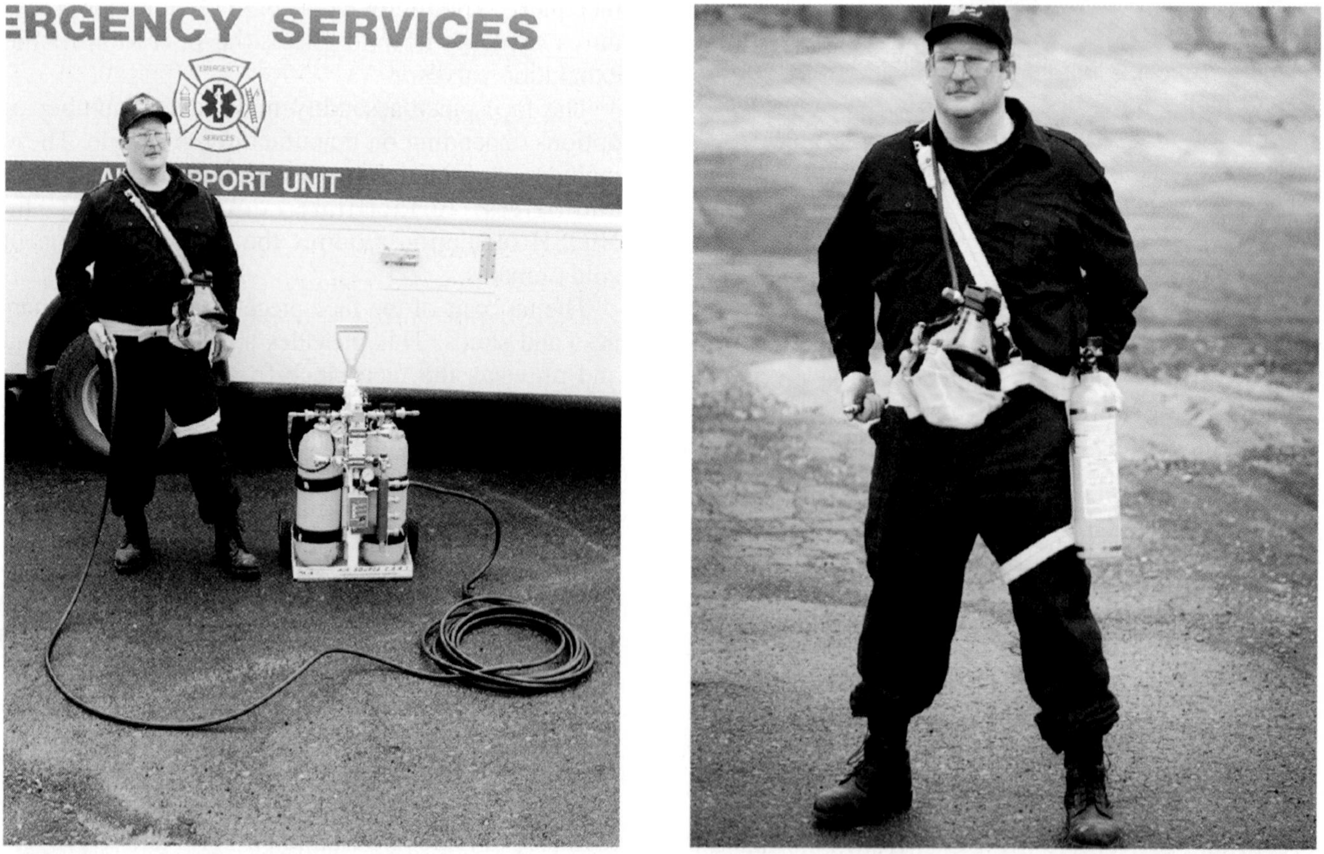

(A) (B)

Figure 7-17 (A) A supplied air respirator with face piece, hose, and air supply. (B) Note the emergency escape cylinder.

Standard Daily Checks for SCBA

SCBA COMPONENT	CHECK FOR OPERATION
Cylinder gauge	Cylinder is 100% full; usually a green shaded area on gauge.
Cylinder valve (slowly open two to three full turns)	Listen for audible low alarm to activate. If it does not activate or continues to sound, place unit out of service.
Regulator or remote gauge and cylinder gauge	Compare gauges. They should read within 100 psi of each other.
PASS device	If unit has an integral PASS device, check for operation. If PASS device is a separate unit, check operation.
Valves	Check all valves for operation and proper position.
Valves and regulator	Close cylinder valve; bleed down system.
Face piece	Check that face piece is clean, free from cracks or deterioration; check condition of hose.
Backpack and harness	Check harness assembly for deterioration and that straps are fully extended. Check condition of straps and buckles. Make sure cylinder is secured.

TABLE 7-6

General Considerations

Before using any SCBA unit, regardless of manufacturer, model, or donning method, operational safety checks must be performed. As shown in **Table 7-6** and **Figure 7-18**, these checks must be conducted on a daily or regular basis, or immediately prior to using the SCBA unit.

Once these checks are completed and the unit operates properly, the SCBA is ready for use. If any of the components listed in **Table 7-6** does not operate properly, is broken, or is damaged, the unit is taken out of service immediately.

Storage Case

Generally, two methods are used to don SCBA units that are stored in their cases. These are the "over the head" and the "coat" methods, and the method used is usually a matter of personnel preference and training.

Firefighters should be proficient in donning SCBA in a number of different ways. Some departments have SCBA units mounted in the seat backs of the apparatus. Other services mount their SCBA units in apparatus compartments. Still other services store the SCBA units in hard-sided cases that are stored in compartments. Firefighters must be familiar and proficient with the proper donning method of their service. Firefighters must also be familiar with the "over the head" of "coat" donning method for quick response to an emergency or for donning after rest and rehabilitation during an emergency incident. For safety, the SCBA unit should only be donned once the apparatus has arrived on scene.

Donning Self-Contained Breathing Apparatus, Over the Head Method

These instructions and figures show the use of a face piece–mounted regulator. The procedure for a waist strap–mounted regulator is generally the

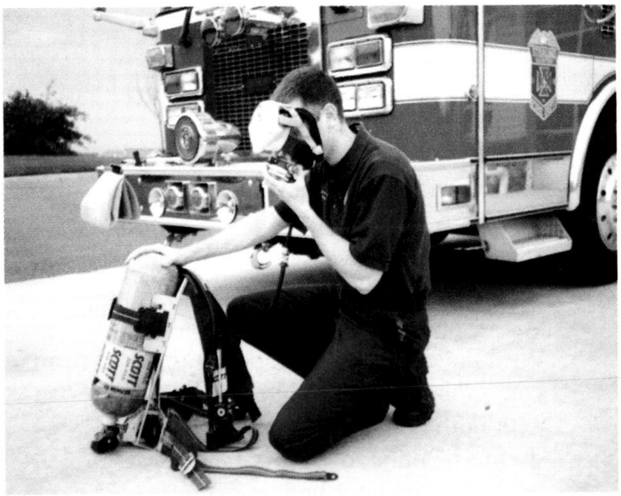

Figure 7-18 Firefighters must perform regular checks of SCBA to ensure the unit's ability to operate.

same, with the exception of donning the face piece. Refer to the face piece instructions for specific details. In addition, the hand position referred to may vary for left-handed users.

1. Check the air supply of the SCBA unit. The cylinder gauge must indicate the cylinder is full (100 percent), **JPR 7-1A**. Full reading will depend on the size of the unit, but on all units, the indicator needle will be in the green shaded area.
2. Slowly open the cylinder valve two to three full turns and listen for the low air alarm to activate as the regulator pressurizes. After the alarm sounds, *open the cylinder valve fully.* If it does not sound, or does not stop sounding, place the unit out of service (follow local SOP) and use another unit, again completing the first two steps.

> **Caution** On some units, if the high-pressure line was not bled down after the last use, the low air alarm may not activate. If the low air alarm does not activate, the firefighter should close the cylinder valve, bleed all air from the system, close all valves, again open the cylinder valve two to three turns, listen for the low air alarm to activate, and follow the rest of step 2 as described.

3. Compare the cylinder gauge to the regulator gauge, **JPR 7-1B**. These gauges must be within 100 psi of each other.
4. Spread and fully extend the harness assembly straps (shoulder and waist straps).

> **Streetsmart Tip** It is good practice to form the habit of readying the SCBA unit for use as it is put into storage on the apparatus. Fully extended and dressing all straps on the unit can save time on an emergency scene. Once the firefighter is tangled in an SCBA it might take precious time and possibly the assistance of another firefighter to untangle the mess.

5. With the cylinder valve pointing away from you, grasp the back plate and/or cylinder with both hands, **JPR 7-1C**. All straps should be outside of your hands.
6. With the regulator hanging freely, lower your head and lift the back plate/cylinder overhead, **JPR 7-1D**. Using your elbows to extend through the loops formed by the shoulder straps, pull the shoulder straps into the body and grasp them with the hands. Let the shoulder straps slide through the hands as the unit slides into place.
7. Balance the unit on your back by leaning forward, and tighten the shoulder straps by pulling out and down, **JPR 7-1E**. Grasp the waist strap and buckle the belt, adjusting the belt for a firm fit on the hips.

> **Caution** Firefighters should always refer to the specific donning instructions for the particular SCBA unit used by a fire department. Some units have a "chest" or "cross" strap, which should be fastened before the shoulder straps are tightened. In addition, some manufacturers recommend loosening the shoulder straps slightly so the weight of the SCBA is carried on the hips.

Once the SCBA unit is in place, refer to the section on the SCBA face piece for detailed donning instructions.

Donning Self-Contained Breathing Apparatus, Coat Method

These instructions and figures show the use of a face piece–mounted regulator. The procedure for a waist strap–mounted regulator is generally the same, with the exception of donning the face piece. Refer to the face piece instructions for specific details. In addition, the hand position referred to may vary for left-handed users.

1. Check the air supply of the SCBA unit. The cylinder gauge must indicate the cylinder is full (100 percent). Full reading will depend on the size of the unit, but on all units, the indicator needle will be in the green shaded area.
2. Slowly open the cylinder valve two to three full turns and listen for the low air alarm to activate as the regulator pressurizes. After the alarm sounds, *open the cylinder valve fully.* If it does not sound, or does not stop sounding, place the unit out of service (follow local SOP) and use another unit, again completing the first two steps.

On some units, if the high-pressure line was not bled down after the last use, the low air alarm may not activate. If the low air alarm does not activate, close the cylinder valve, bleed all air from the system, close all valves, again open the cylinder valve two to

JOB PERFORMANCE REQUIREMENT 7-1

Over the Head Method

A Checking the air supply is the first step before using SCBA.

B Always compare the cylinder and regulator gauges. They must be within 100 psi of each other.

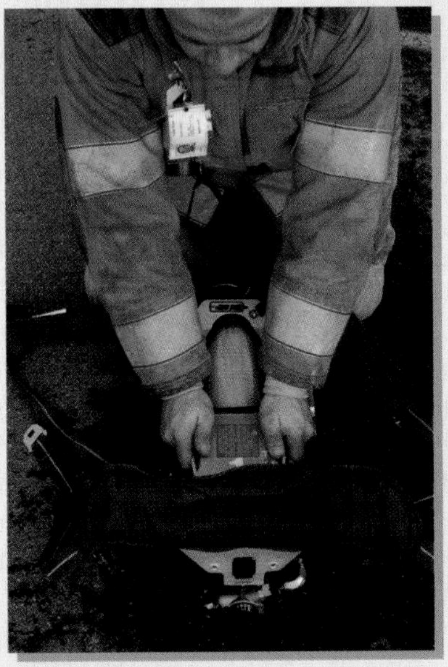

C Grasp the back plate with both hands. Harness straps should be outside of your hands.

D Lift the back plate/cylinder overhead, using your elbows to extend through the loops formed by the shoulder straps.

E Balance the unit on your back by leaning forward, and tighten the shoulder straps by pulling down and out.

three turns, listen for the low air alarm to activate, and follow the rest of step 2 as described.

3. Compare the cylinder gauge to the regulator gauge. These gauges must be within 100 psi of each other.
4. Spread and fully extend the harness assembly straps (shoulder and waist straps).
5. Position the SCBA unit with the cylinder valve toward you. Using your left hand, grasp the left shoulder strap at the top of the back plate/harness assembly and grasp the lower portion of the same strap with your right hand, **JPR 7-2A**. When the unit is positioned correctly, the left shoulder strap will be on your right as you face the SCBA unit.
6. Lift the SCBA unit, swing it around the left shoulder and onto your back, **JPR 7-2B**. Both hands are still grasping the shoulder strap.
7. Holding the left shoulder strap with your left hand, release your right hand and put your right arm through the right shoulder strap.
8. Balance the unit on your back by leaning forward. Tighten the shoulder straps by pulling out and down on the straps, **JPR 7-2C**. Grasp the waist strap and buckle the belt, adjusting the belt for a firm fit on the hips.

Once the SCBA unit is in place, refer to the section on the SCBA face piece for detailed donning instructions.

Remember, always refer to the specific donning instructions for the particular SCBA unit used by a fire department.

Seat-Mounted Apparatus

With the use of enclosed firefighter riding positions becoming more common, many fire departments are mounting SCBA units at the seat position, as shown in **Figure 7-19**. This method allows for quick donning and the unit is readily available for regular inspection. Various types of mounting brackets and hardware are used for this type of mount, and firefighters must be familiar with the style used in their department. The specific steps for donning the SCBA unit from this position are detailed in the skill given later.

Three important safety requirements must be observed if SCBA units are mounted at the seat position. These are storing of the face piece, donning the unit while the vehicle is moving, and checking the cylinder gauge. The SCBA face piece should never be left connected to the regulator for storage. Ideally, the face piece should be stored in a bag to maintain cleanliness and keep the unit free of dust particles, which could affect operation or injure the user's eyes.

Figure 7-19 Many fire departments are mounting SCBA units at the seat position, allowing for easy access for inspection and quick donning.

In addition, a firefighter should never attempt to stand or don the SCBA unit while the apparatus is in motion. Donning gear while the vehicle is in motion may require loosening of the seat belt, and the firefighter may be thrown around the enclosed cab in the event of a sudden stop or turn. *Firefighters must always remain seated with seat belt fastened when the vehicle is moving.*

From the seated position, it is extremely difficult, if not impossible, to check the cylinder gauge and compare it to the regulator gauge. If this is not accomplished prior to response, then after dismounting the apparatus firefighters should use the buddy system to check each other's gauges, **Figure 7-20**.

Streetsmart Tip It is a good rule of thumb for all firefighters to remain seated with seat belts fastened until the driver/operator of the apparatus has stopped the vehicle and set the parking brake.

JOB PERFORMANCE REQUIREMENT 7-2

Coat Method

A With your left hand, grasp the left shoulder strap at the top of the back plate/harness assembly and grasp the lower portion of the same strap with your right hand. When the unit is positioned correctly, the left shoulder strap will be on your right as you face the SCBA unit.

B Lift the SCBA unit, then swing it around the left shoulder and onto your back. Both hands are still grasping the shoulder strap.

C Balance the unit on your back by leaning forward. Tighten the shoulder straps by pulling out and down. Grasp the waist strap and buckle the belt, adjusting the belt for a firm fit on the hips.

Figure 7-20 After dismounting the apparatus, firefighters use the buddy system to check each other's cylinder gauges.

Donning Self-Contained Breathing Apparatus, Seat-Mounted Apparatus

These instructions and figures show the use of a face piece–mounted regulator. The procedure for a waist strap–mounted regulator is generally the same, with the exception of donning the face piece. Refer to the face piece instructions for specific details. In addition, the hand position referred to may vary for left-handed users.

1. Check the cylinder valve prior to response, if possible. *Do not attempt to check the unit when the vehicle is in motion.* Using the buddy system, turn the unit on and check the cylinder gauge when you dismount the apparatus.
2. With straps fully extended, place the right arm between the right shoulder strap and back plate assembly, **JPR 7-3A**. Repeat this action for the left arm.
3. Tighten the shoulder straps by pulling out and down on the straps, **JPR 7-3B**. Grasp the waist strap and buckle the belt, adjusting the belt for a firm fit on the hips. *Do not entangle the waist belt with the vehicle seat belt.*
4. Dismount the apparatus, **JPR 7-3C**, and recheck shoulder and waist straps.

Always refer to the specific donning instructions for the particular SCBA unit used by a fire department. Some units have a "chest" or "cross" strap, which should be fastened before the shoulder straps are tightened. In addition, some manufacturers recommend loosening the shoulder straps slightly so the weight of the SCBA is carried on the hips.

5. With a partner, **JPR 7-3D** and **E**, slowly open the cylinder valve two to three full turns and listen for the low air alarm to activate as the regulator pressurizes. After the alarm sounds, *open the cylinder valve fully.* If the alarm does not sound, or does not stop sounding, place the unit out of service (follow local SOP) and use another unit, again completing the first two steps. Compare the cylinder gauge to the regulator gauge. These gauges must read within 100 psi of each other.

On some units, if the high-pressure line was not bled down after the last use, the low air alarm may not activate. If the low air alarm does not activate, close the cylinder valve, bleed all air from the system, close all valves, again open the cylinder valve two to three turns, listen for the low air alarm to activate, and follow the rest of step 2 as described.

Once the SCBA unit is in place, refer to the section on the SCBA face piece for detailed donning instructions.

Compartment or Side-Mounted Apparatus

Some fire departments carry their SCBA units mounted on the side of apparatus or in compartments. Depending on the height of the apparatus compartment and the bracket used to mount the SCBA, the donning method may be similar to that for the seat-mounted position except that the firefighter is standing rather than sitting. If the cabinet height or mounting bracket position of the SCBA does not allow for ease of donning while standing, the firefighter should remove the SCBA unit and use the "coat" or "over the head" method of donning.

To don the SCBA unit from a compartment or side-mounted storage position, it is recommended to follow the donning instructions for the method best suited for the particular mounting style.

Donning the SCBA Face Piece

Most SCBA face pieces are donned in a similar manner, with the difference being in the style of head straps and the location of the regulator. Proper donning of the SCBA face piece is essential to

JOB PERFORMANCE REQUIREMENT 7-3

Seat-Mounted Apparatus

A With straps fully extended, place the right arm between the right shoulder strap and back plate assembly. Repeat this action for the left arm.

B Tighten the shoulder straps by pulling out and down on the straps. Grasp the waist strap and buckle the belt, adjusting the belt for a firm fit on the hips.

C Dismount the apparatus and recheck shoulder and waist straps.

D With a partner, slowly open the cylinder valve two to three full turns and listen for the low air alarm to activate as the regulator pressurizes.

E *Open the cylinder valve fully* and compare the cylinder gauge to the regulator gauge.

protect the firefighter from the effects of toxic gases and hazardous atmospheres. Each firefighter must be fitted for the face piece to be used with a particular manufacturer's SCBA as required by OSHA 29 CFR 1910.134:

> 29 CFR 1910.134 (f)(1)
> The employer shall ensure that employees using a tight-fitting facepiece respirator pass an appropriate qualitative fit test (QLFT) or quantitative fit test (QNFT) as stated in this paragraph. 1910.134(f)(2)
> The employer shall ensure that an employee using a tight-fitting facepiece respirator is fit tested prior to initial use of the respirator, whenever a different respirator facepiece (size, style, model or make) is used, and at least annually thereafter.[1]

In addition, 29 CFR 1910.134, NFPA 1404, and SCBA manufacturers' recommendations prohibit any facial hair, which may interfere with proper fit and seal of the face piece. Again, all firefighters must be familiar with the specific equipment used by their fire department.

Caution No glasses or facial hair are allowed between the skin and the sealing surface of the SCBA.

Donning Self-Contained Breathing Apparatus Face Piece

1. With the head straps/harness in the full out position, hold the head harness with one hand while placing the face piece on the face, with the chin properly located in the chin pocket, **JPR 7-4A**.
2. Fit the face piece to the face, pull the head harness over the head, and make sure straps are lying flat against the head with no twists.
3. Tighten the neck or lower straps by simultaneously pulling both straps to the rear, **JPR 7-4B**.
4. Check the fit of the head harness by stroking the harness down the back of the head using one or both hands. Retighten the neck straps at this time, **Figure 7-4C**.
5. Adjust the temple straps by simultaneously pulling both straps to the rear.
6. Check for proper seal by attaching the regulator and inhaling a breath, activating the regulator, **JPR 7-4D.** Hold that breath for about five seconds and listen and feel for any air leaks. If leaks occur, repeat steps 2 through 5. In addition, once the regulator or breathing tube is connected, listen for air leaks from the positive pressure mode. If leaks occur, repeat steps 2 through 5.

Streetsmart Tip It is common to see firefighters testing the seal of a face piece by placing a hand over the opening and inhaling. The vacuum will draw the mask to the face. This method is not accurate, as pressure can be inadvertently applied to the face piece, pushing it toward the face.

7. Pull the protective hood on over the entire head to cover exposed skin, **JPR 7-4E.** Ensure all long hair is inside the protective hood. Also ensure that the protective hood is tucked completely down inside the bunker coat. Pull up and fasten the bunker coat collar. Place the helmet on, and adjust it for the increased size of the head, due to the face piece and protective hood. Fasten the helmet strap. Make sure the helmet shroud is pulled over the neck and coat collar. Place protective gloves on.
8. As you approach the hazardous atmosphere, install regulator on face piece or connect the hose to the regulator for a waist strap–mounted regulator. Airflow to the face piece will begin with the breathing action. Again, listen for air leaks from the positive pressure mode. If leaks occur repeat steps 2 to 5.

Streetsmart Tip Firefighters should always work in pairs, and each firefighter should check the other for any situations that are unsafe. One firefighter should ensure all protective equipment is properly donned and that all areas of the skin are covered. All layers of PPE should interface with the additional layers. The roles are then reversed and the second firefighter is checked for safety.

Caution It is important to tighten either neck or lower straps first, pulling straight back. Pulling straps outward may damage the straps and prevent proper seal. The temple straps are adjusted by pulling the strap ends toward the rear of the head. Regardless of the type of regulator used, one should always check for proper face piece seal. A firefighter who does not get a leak-free seal should not enter the hazardous environment.

Removing/Doffing the SCBA Unit

Upon exiting the hazardous atmosphere, firefighters should remove the SCBA unit and rest. Generally, to remove the unit, the donning process is reversed

JOB PERFORMANCE REQUIREMENT 7-4

Face Piece

A With the head straps/harness in the full out position, hold the head harness with one hand while placing the face piece on the face, with the chin properly located in the chin pocket.

B Tighten the neck or lower straps by simultaneously pulling both straps to the rear.

C Check the fit of the head harness by stroking the harness down the back of the head. Retighten the neck straps as necessary.

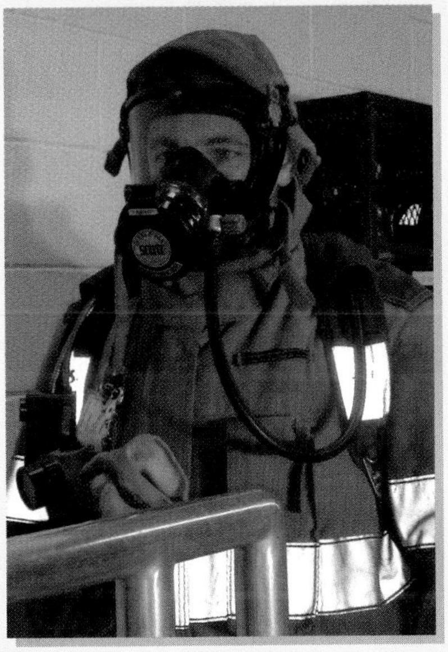

D Check for proper seal by attaching the regulator and inhaling a breath, activating the regulator. Hold that breath for about five seconds and listen and feel for any air leaks.

E Pull on protective hood to cover exposed skin and head. Place helmet on head with earflaps down. Pull up coat collar and buckle. Snug or buckle helmet straps. Place gloves on hands.

following manufacturers' instructions for the type or model of unit used. If awaiting another assignment, the face piece should be removed to allow normal breathing and to conserve air. The regulator or face piece must not be contaminated by laying it on the ground or other dirty environment. This may cause injury if firefighters need to use the unit again.

Depending on local SOP, once an assignment with SCBA is complete, firefighters should report to rehabilitation for rest, fluids, and monitoring of vital statistics. Many respiratory protection programs limit firefighters to consumption of two SCBA cylinders, after which mandatory rehabilitation is required. Again, this rest period for fluids and monitoring of vital statistics reduces the physical and mental stress encountered during firefighting operations.

SELF-CONTAINED BREATHING APPARATUS OPERATION AND EMERGENCY PROCEDURES

In accordance with 29 CFR 1910.134, NFPA 1404, and NFPA 1500, fire departments must establish respiratory protection programs for firefighters using SCBA. In addition, the firefighters must be proficient in the safe use of SCBA, donning and doffing procedures, individual limitations, and limitations of the SCBA unit.

Safe Use of SCBA

Safe use of SCBA is essential to firefighter survival during operations requiring its use. Firefighting tasks are both mentally and physically demanding and firefighters must be in excellent physical condition. The SCBA unit and protective equipment add weight and bulk to the firefighter, causing increased exertion with loss of body fluids through perspiration. These actions increase during firefighting operations and firefighters must be aware of them and of the symptoms of heat stress and their own limitations and abilities.

The following items are essential for maximum safety while using SCBA:

- All firefighters using SCBA must be certified physically fit for respirator use in accordance with local policy or 29 CFR 1910.134.
- Fire departments must establish an accountability system, **Figure 7-21**, to track personnel entering an IDLH atmosphere.
- Firefighters must work in teams of two as a minimum. If one member of the team must leave for any reason, all members of the team must exit the IDLH.
- In accordance with OSHA CFR 29 1910.134, the "two in/two out" rule states that a rescue team of two firefighters must be available and ready to rescue or assist the firefighters operating in the IDLH.
- PASS devices, **Figures 7-22A** and **B**, must be activated in order to function. If the PASS is an integral part of the SCBA unit, it must be activated before entering a hazardous environment.
- Fire departments should establish policies for firefighter rehabilitation during operations

Firefighter Fact New SCBA designs incorporate a variety of innovative technologies to provide maximum safety for the user. Some SCBAs incorporate the PASS device or distress alert as an integral component of the unit. These systems offer redundant visual and audible warning alarms, with the audible alarm located on the harness assembly and the visual warning, a flashing LED, located on the regulator or remote gauge.

Other new designs include a technology that projects a display of the level of air in the cylinder to the face piece, straight ahead in the field of vision of the firefighter. This "Head's Up Display (HUD)" allows a firefighter to keep the head up, look straight, and continue to perform firefighting or other duties, while still being able to see the level of air remaining in the SCBA. The HUD uses an LED indication for air levels of full three-quarter, one-half, and one-quarter. Further, the HUD will flash visual notification of one-half and one-quarter remaining air supply. The HUD may also include a visual indicator of temperature. The HUD may be retrofitted into older SCBA units.

Another new design feature is the Rapid Intervention Crew Universal Air Connection (RIC/UAC.) The RIC/UAC is a common style connector for use in emergency situations, such as rescuing a trapped firefighter. Rapid intervention teams (RIT) can carry an extra SCBA cylinder and recharge the trapped firefighter's air, as well as their own, during the rescue mission. The common connection is mounted on all compliant SCBA within 4 inches of the cylinder valve outlet and has a relief valve to prevent overfilling of lower pressure cylinders, such as 2,216-psi cylinder if a 4,500-psi cylinder is used as a filling source.

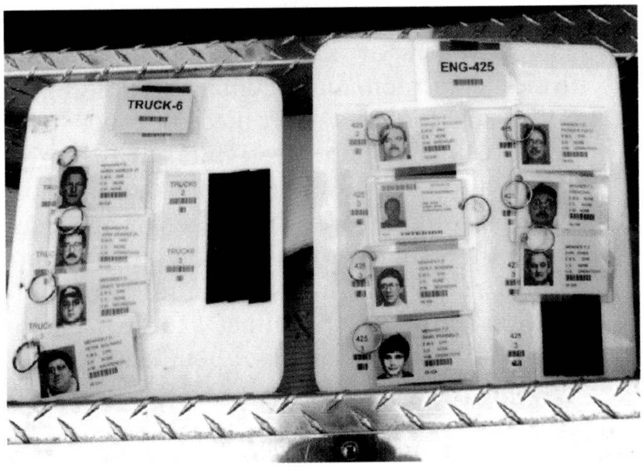

Figure 7-21 A typical accountability system to track personnel entering an IDLH atmosphere.

requiring SCBA use. This policy should include the maximum number of cylinders a firefighter can use before mandatory rehabilitation.

- During rehabilitation, EMS personnel should monitor vital signs, **Figure 7-23**, and firefighters must hydrate to replace body fluids.
- Air consumption will vary with each individual's physical condition, the level of training, the task performed, and environment.
- The SCBA face piece should never be removed in a contaminated environment.
- Depending on an individual's air consumption and the amount of time required to exit a hostile environment, the low air alarm may not provide adequate time to exit. The individual should not panic, but move deliberately to a safe environment. All team members must exit together, even if only one low air alarm is sounding.

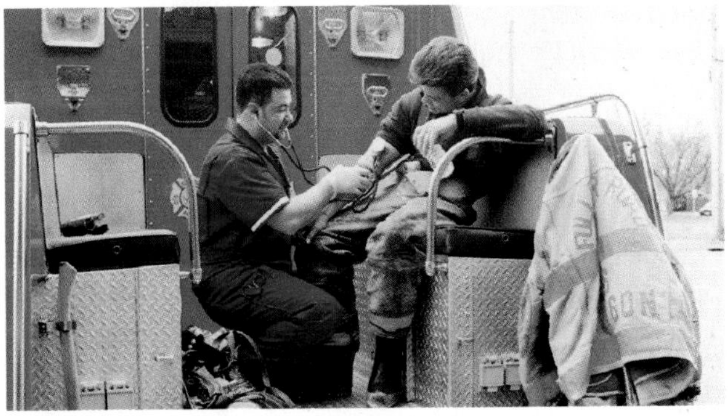

Figure 7-23 EMS personnel should monitor vital signs, and firefighters must hydrate to replace body fluids during firefighting operations.

Streetsmart Tip Air consumption: All firefighters must be aware of the amount of air an individual consumes using SCBA. This is accomplished by performing an air consumption activity. Firefighters in full protective equipment and SCBA perform simulated firefighting activities, such as advancing hoselines, search and rescue, and ventilation. The time is noted when the firefighter begins the activity, when the low air alarm sounds, and when the air supply is depleted. The air consumed divided by the time of the activity will provide an *approximate rate* (this is a controlled environment and actual fireground operations will result in higher air consumption rates) of air consumption and the amount of time/air supply available to the firefighter to exit after the low air supply alarm sounds. Each firefighter should perform the test a minimum of two times to provide an average rate. In addition, the test should be conducted on an annual basis to observe changes that may affect the air consumption rate.

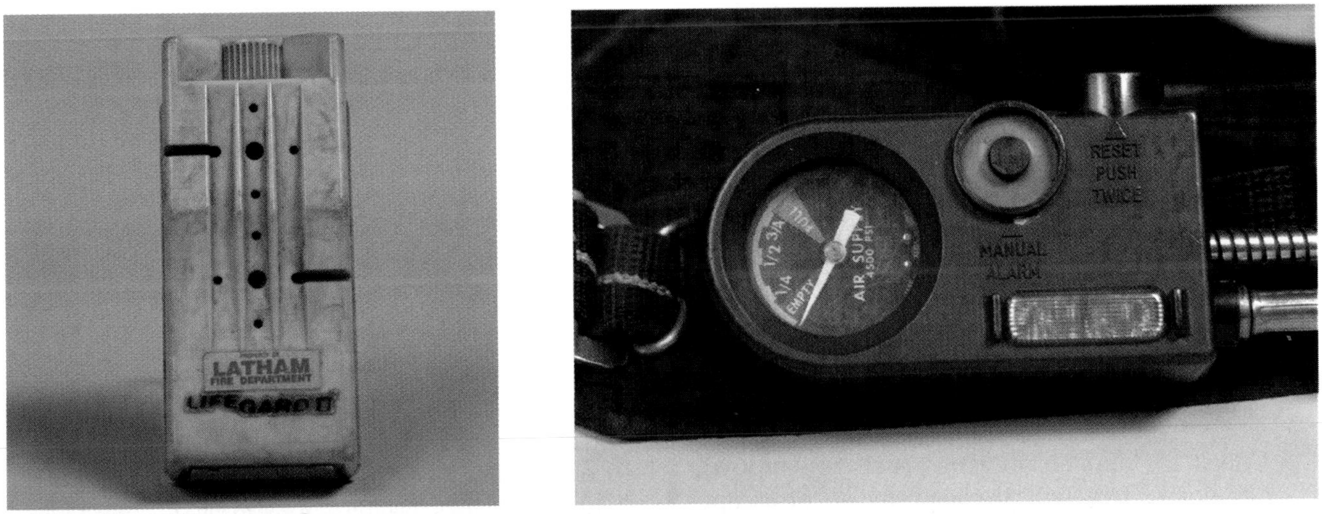

(A) (B)

Figure 7-22 PASS devices used with SCBA include (A) the independent self-contained unit that attaches to the harness and (B) the integral type that is part of the SCBA unit.

Figure 7-24 Firefighters should always remain low. Heat and smoke from the fire will rise; floor level provides the best visibility.

Operating in a Hostile Environment

Firefighters will use SCBA in hostile environments and toxic atmospheres that will have limited visibility. During firefighting operations requiring the use of SCBA, teams conducting search or fire attack must be able to function effectively. Firefighters should follow these general rules:

- Always check in with the accountability/company officer when entering or exiting a building or when an assignment is complete.
- Always remain low, **Figure 7-24**. Heat and smoke from the fire will rise, so floor level provides the best visibility. This will also reduce the risk of injury to a firefighter from walking into a hole in the floor or falling into a pit or shaft.
- Check the environment; extend a gloved hand above the head to determine the heat conditions above. This can also be done with a tool, extending it above the head and checking the end for heat.
- Never remove the face piece.
- Maintain an awareness of location, floor level, front or rear of building.
- Ventilate as you advance, as long as it will not spread the fire. Ventilation allows the products of combustion to escape and provides a better environment.
- Check for outside openings such as windows and doors. This will provide a means of escape in an emergency and provide the firefighters' location to outside personnel.
- *Always* maintain direct contact with other team members. It is also acceptable for both firefighters to remain in contact by holding on to the same tool or a short piece of rope. There are also times where firefighters will be following a rope or hoseline into or out of a hostile environment. In this case, or in cases where direct contact is broken, firefighters should remain in constant voice communication until direct contact can be reestablished or until they are safely out of the IDLH atmosphere.
- Never enter a hostile environment alone.

Restricted Openings

As a general rule, areas with restricted openings that a firefighter cannot fit into should be probed with a tool. SCBA removal, or the loosening of straps and repositioning of the SCBA unit, can reduce the profile of the firefighter. These are last-resort measures and only used for escaping through a wall or a small opening under emergency conditions. Firefighters should remove or loosen only those parts of the SCBA unit necessary to negotiate the restricted opening. The exact procedures for accomplishing a reduced profile depend on the type of SCBA unit used. The following rules must be observed:

- Never remove the face piece to reduce the profile.
- Loosen straps and rotate the SCBA unit under the arm along the rib cage as a first step, **Figure 7-25.**
- As a last resort, perform a "full escape" by removing the harness assembly and holding the SCBA unit in front of you.
- Always maintain contact with the SCBA unit, **Figure 7-26**. Hold onto the shoulder straps and regulator assembly or regulator on a waist-mounted unit.
- Maintain control of the unit. Do not move it away or allow the face piece to be pulled away.
- Practice this procedure with an obscured face piece to accomplish it by feel.

Caution The full escape procedure should only be used as a last resort for an emergency escape from a hostile environment. It may void the NIOSH certification for the SCBA; however, it is a tactic that can save lives.

Figure 7-25 To reduce profile with an SCBA unit, the firefighter loosens straps and rotates the SCBA unit under the arm along the rib cage.

Figure 7-26 Firefighters should only remove the SCBA unit as a last resort to escape from a life-threatening situation.

Emergency Procedures

SCBA units can be damaged and malfunction or the firefighter may become tangled in debris while operating. If this happens, a number of emergency procedures exist that will assist in the safe escape from the hazardous environment. Above all, firefighters must remain calm and rely on their training and knowledge.

Firefighters must be familiar with and practice the emergency procedure used by their fire department for failure of the unit. In the event of an SCBA unit malfunction, firefighters should follow these guidelines:

- Remain calm; rely on your training.
- Immediately exit the hazardous environment. All team members must exit—never leave a firefighter alone.
- Manually activate the PASS device.
- If equipped with a portable radio, announce a "Mayday" message, giving current location if possible.
- In case of a regulator failure, breathe using the bypass valve. Close the main line valve, open the bypass valve, breathe, and close the bypass valve to conserve air. This procedure may vary by manufacturer.
- For a face piece failure, use the bypass valve to increase airflow to keep the face piece free from toxic gases.
- If necessary, use the protective hood as a filter. *Use the "buddy breathing" attachment on an SCBA as a last resort because this will deplete the air supply at least twice as fast.* Many SCBA units are equipped with an emergency escape breathing support system, or a "buddy breathing" attachment. Current NIOSH and NFPA standards do not allow the use of this option. The use of these devices voids the NIOSH certification and use cannot be recommended at this time. Fire departments should review SOPs concerning "buddy breathing" and determine the best option for their needs.
- Practice these procedures during training. During an emergency is not the time to try something new.

If the firefighter becomes entangled in debris, other team members will assist in freeing the trapped firefighter. This is accomplished by cutting away the debris or removing the SCBA unit as a last resort.

> **Streetsmart Tip** Many firefighters carry a small tool or wire cutters to free themselves from mattress springs or suspended ceiling tie wires, some of the most common debris causing entanglement.

INSPECTION AND MAINTENANCE OF SELF-CONTAINED BREATHING APPARATUS

As with any piece of equipment used in the fire service, SCBA units must be ready to go on a moment's notice. Considering the importance of SCBA to the firefighter, inspection must be completed on a daily or regular basis.

There are a number of variations of maintenance procedures for inspection and servicing of SCBAs. Always follow the manufacturers' instructions and recommendations provided with the unit. The step-by-step instructions given in this chapter are intended for training purposes and may differ from the procedures recommended for the SCBA used by a particular organization.

Daily Maintenance

SCBA units should be checked daily to ensure they are secured and ready for operation. When an SCBA is used during an emergency scene or a training exercise, the unit should be serviced and checked in the same manner.

Daily Inspection of SCBA

1. Check to ensure the cylinder is full, **JPR 7-5A.**
2. Open the cylinder valve slowly, two to three full turns, until the low air alarm is activated. When the alarm stops sounding, open the cylinder valve completely. If the alarm does not activate or continues to sound, place the unit out of service following department procedures.

Depending on the manufacturer, some units—if the high-pressure line was not bled down after use—will not have activation of the low air alarm. In this case, close the cylinder valve, bleed all air from the system, close all valves and open the cylinder valve as described.

3. Compare the cylinder and regulator gauges, **JPR 7-5B.** Gauge readings should be within 100 psi.
4. Check to see that all hose connections are tight and free from leaks.
5. Check the bypass and the main line valves for operations.
6. Close the cylinder valve and (depending on manufacturer's instructions) drain all air from the system. Note that the low air warning alarm activates when the regulator gauge reads about 20 to 25 percent.
7. Ensure the main line valve (if the unit has one) is open and that the bypass valve is fully closed, **JPR 7-5C** and **D.**
8. Check the condition of the face piece, if one is assigned to the unit. Ensure cleanliness and check for cracks, for missing or broken straps, and that the diaphragm is in place. Clean face piece per local department policy, normally using a 1 percent bleach water solution.
9. Check the harness assembly, condition of belts, and cylinder fastening device.
10. Check the operation of the PASS device.

Monthly Maintenance

The monthly SCBA check contains all elements of the daily check but adds several checks of the mechanics of the system. Firefighters should be trained by their department to conduct the monthly maintenance. An example of a monthly check sheet is shown in **Figure 7-27.** Any irregularities should be noted and repaired, or the SCBA should be pulled from service until a department technician can repair the unit.

Annual and Biannual Maintenance

NIOSH and SCBA manufacturers require a number of different functional tests of SCBA units. Only a manufacturer's authorized or trained service personnel shall conduct these tests. Firefighters should refer to the instructions for the SCBA units used.

Changing SCBA Cylinders

SCBA cylinders must be changed after use, following local SOPs. Depending on the size and air consumption rate of a firefighter, allowing a cylinder at 90 percent full to remain in service could mean a loss of two to five minutes of air supply. This time may be the difference in being able to successfully exit or escape from a hazardous situation.

The firefighter should follow these procedures for cylinder replacement:

Cylinder Replacement Procedure

1. Have full air cylinder ready for use.
2. Place the SCBA unit on the ground with the cylinder valve toward you, **JPR 7-6A.**

JOB PERFORMANCE REQUIREMENT 7-5

Daily Inspection of SCBA

A Check to ensure that the cylinder is full.

B Compare cylinder gauge and regulator gauge. Gauge readings should be within 100 psi of each other.

C and **D** After checking for proper operation, ensure air is purged from the system, the main line valve (if the unit has one) is open, and the bypass valve is fully closed.

3. Push in the knob on the cylinder valve and close the cylinder valve.
4. Bleed residual air in the high-pressure line by slowly opening the bypass valve. When the flow of air from the face piece stops, close the valve. On some units, the pressure must be released by "breathing down" the regulator or opening the main line valve. Refer to the manufacturer's instructions for the unit in use.
5. Disconnect the high-pressure coupling from the cylinder, **JPR 7-6B**. If the knob is difficult to turn, the high-pressure line may still be pressurized; repeat step 5 and attempt to disconnect the coupling again. Lay the coupling on the back plate to prevent dirt or grit from contaminating the threads or seat.
6. Release the cylinder clamp or locking mechanism used on the unit and slide the empty cylinder down out of the harness back plate assembly, **JPR 7-6C**.
7. Inspect the O-ring on the seat of the high-pressure hose for any damage, scratches, or foreign matter, **JPR 7-6D**. If the O-ring is damaged, replace it in accordance with the manufacturer's instructions.

SCBA FIELD MAINTENANCE SHEET

DATE: ______________ REDUCER #: ____________________ LOCATION: ____________________

Air-Pak shows no use; visual and function checked only _____

2.2 / 4.5 30 DAY / REPAIR

(REGULATOR ASSY.)

REG. COVER:	OK	Y / N	RETAINING RING:	OK	Y / N
DIAPHRAGM:	OK	Y / N	REG. GASKET:	OK	Y / N
PURGE VALVE:	OK	Y / N	DONNING SWITCH:	OK	Y / N
THUMB LATCH:	OK	Y / N	REG. HOSE:	OK	Y / N

(BACKFRAME ASSY.)

SHOULDER HARNESS:	OK	Y / N	WAIST BELTS:	OK	Y / N
REMOTE GAUGE:	OK	Y / N	ALL HOSES:	OK	Y / N
EBSS EQUIPMENT:	OK	Y / N	PACK ALERT:	OK	Y / N

(FUNCTION TEST & LEAK TEST)

BREATHING:	OK	Y / N	PURGE VALVE:	OK	Y / N
LOW AIR ALARM:	OK	Y / N	REMOTE GAUGE:	OK	Y / N
DONNING SWITCH:	OK	Y / N	ANY LEAKAGE:	OK	Y / N

(BOTTLE CONDITION)

Note any excessive wear or damage to bottle and/or valve. Bottle # ______________

*LIST ANY PROBLEMS NEEDING REPAIRS: ____________________

____

____

*LIST ANY PARTS REPLACED: ____________________

____

____

Rev: 07-04-02

EQUIP. # ______________

Figure 7-27 SCBA Field Maintenance Sheet.

8. Replace with a fully charged cylinder. Reverse the process in step 6 by sliding the cylinder into the back plate assembly. Do not secure the cylinder clamp until the high-pressure line is aligned and connected.
9. Align the high-pressure hose with the cylinder valve assembly and connect the hose. This connection should be hand tightened.
10. Lock the cylinder into place on the back plate assembly.

JOB PERFORMANCE REQUIREMENT 7-6

Cylinder Replacement Procedure

A Place the SCBA unit on the ground with the cylinder valve toward you.

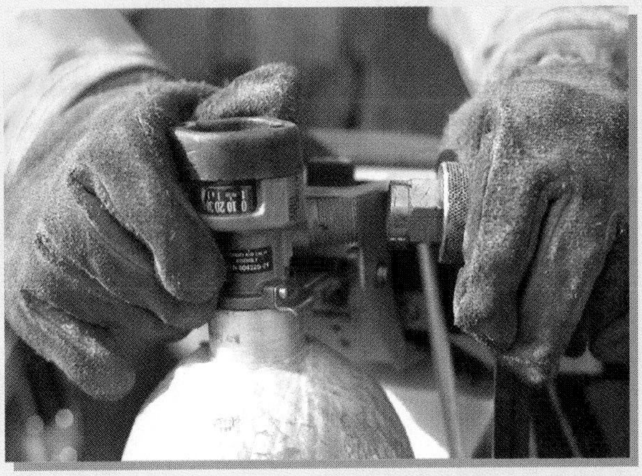

B Disconnect the high-pressure coupling from the cylinder. If the knob is difficult to turn, the high-pressure line may still be pressurized.

C Release the cylinder clamp or locking mechanism used on the unit and slide the empty cylinder down out of the harness back plate assembly.

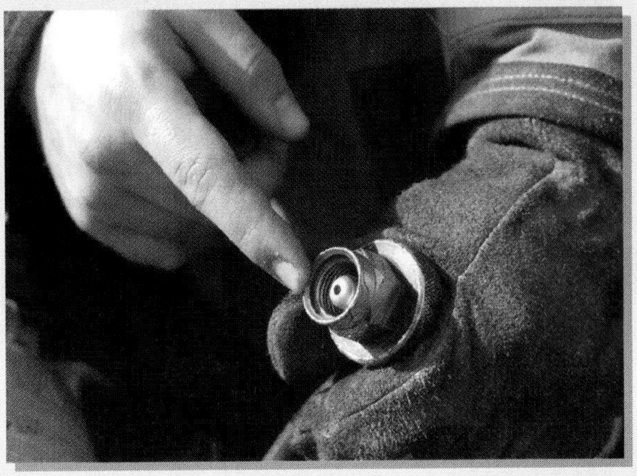

D Inspect the O-ring on the seat of the high-pressure hose for any damage, scratches, or foreign matter.

11. Open the cylinder valve, check for air leaks, and compare the cylinder valve reading with the regulator reading. These should be within 100 psi of each other.
12. Close the cylinder valve, bleed all air from the system, ensure that the main line and by-pass valves are in the proper position, and place the unit in its mounting bracket or storage case.

Other units may have an integrated PASS device; once the pressure is off the system, the PASS device should be shut off.

Additional Steps for Two-person SCBA Cylinder Replacement

There are many times on a fire scene where firefighters will require other firefighters to change out a cylinder while they are wearing the SCBA. This can be done safely using the same sequence of events, with the following exceptions:

1. While wearing the SCBA, stop and lean forward, **JPR 7-7A** or kneel on the ground, **JPR 7-7B.** Ensure that your head is pointed down and out of the way of the cylinder as it is changed.

JOB PERFORMANCE REQUIREMENT 7-7

Additional Steps for Two-person SCBA Cylinder Replacement

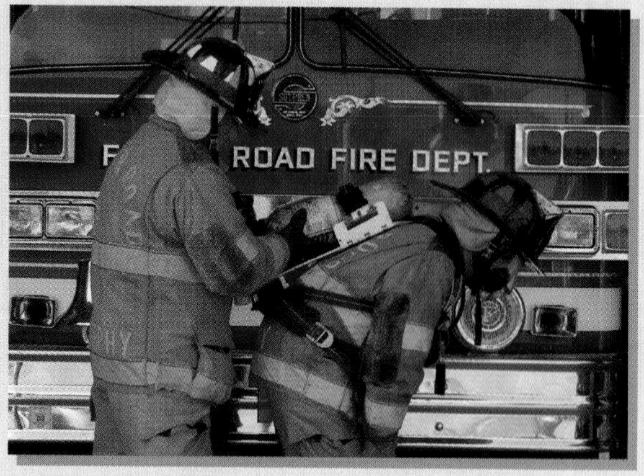

A Stop and lean forward in a stable position.

B In some cases, it will be easier to kneel on the ground.

C Always know the amount of air in your cylinder.

2. The firefighter changing out the cylinder should follow the steps for removing the cylinder as noted in steps 1 through 7 above, taking care not to hit the head of the firefighter wearing the pack while removing the cylinder.
3. Before replacing a full cylinder, the firefighter changing the cylinder should ensure that it is full. It is good practice to quickly show the bottle gauge to the firefighter wearing the SCBA unit; this way both firefighters have seen the gauge and know the cylinder is full, **JPR 7-7C.**
4. The firefighter changing the cylinder should replace the cylinder as noted in steps 8 through 11 of JPR 7-6.

Servicing SCBA Cylinders

When the cylinder capacity is below full, the cylinder must be serviced. As noted earlier in this chapter, filling SCBA cylinders is completed using a

Figure 7-28 A cascade system is one of the systems available to service SCBA cylinders. These may be fixed or mobile units.

cascade system, **Figure 7-28**, or compressor/purifier system, **Figure 7-29**. Regardless of how this is accomplished, the following safety precautions must be followed:

- The air source being used must be tested and certified meeting the requirements of Compressed Gas Association Pamphlet G-7.1-1989, OSHA 29 CFR 1910.134, and NFPA 1500.
- All cylinders must have a current hydrostatic test date: no older than five years for aluminum/ steel cylinders, or three years and less than fifteen years from manufacture date for composite cylinders.
- All fill stations must have fragmentation containment devices in case of cylinder failure.
- All manufacturers' recommendations should be followed, especially for recalls or safety notices concerning cylinder capacity.

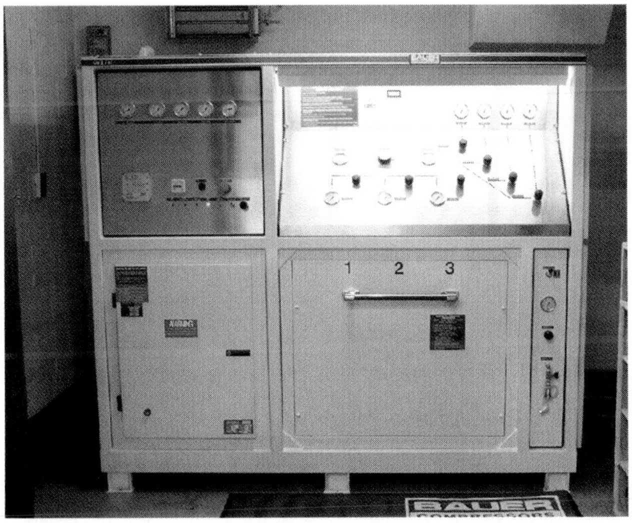

Figure 7-29 A compressor/purifier system is the second type of system used to service SCBA cylinders. These are usually located at stations or other fixed facilities; however, larger departments may have mobile units.

- Fill rate may vary; 300 to 600 psi per minute is considered an acceptable range.

Servicing an SCBA Cylinder Using a Cascade System

Many organizations require certification before using a cascade system or air compressor/purifier system. Do not attempt this skill without training on the specific system used by a particular fire department and follow the procedures recommended by the system manufacturer. Never attempt to fill a cylinder above its rated capacity and never attempt to fill bottles with different rated capacities at the same time.

1. Check the hydrostatic test date of the cylinder, **JPR 7-8A**. Remove from service if test is out of date or usable service life for composite cylinders has passed.
2. Inspect the cylinder for any signs of physical damage, discoloration from heat, gouges, or nicks in the cylinder surface. *Never fill an SCBA cylinder that has visible signs of damage or an out-of-date hydrostatic test.*
3. Place the cylinder in the fragmentation containment device, **JPR 7-8B**.
4. Connect the fill hose to the cylinder, and if it is equipped with a bleed valve, close it.
5. Open the SCBA cylinder valve.

Caution A firefighter should never attempt to fill an SCBA cylinder without fragmentation protection. It is important never to fill a composite SCBA cylinder in a water-filled fragmentation protection device. Water may infiltrate between the composite layers, causing a weakness and failure.

JOB PERFORMANCE REQUIREMENT 7-8
Servicing an SCBA Cylinder Using a Cascade System

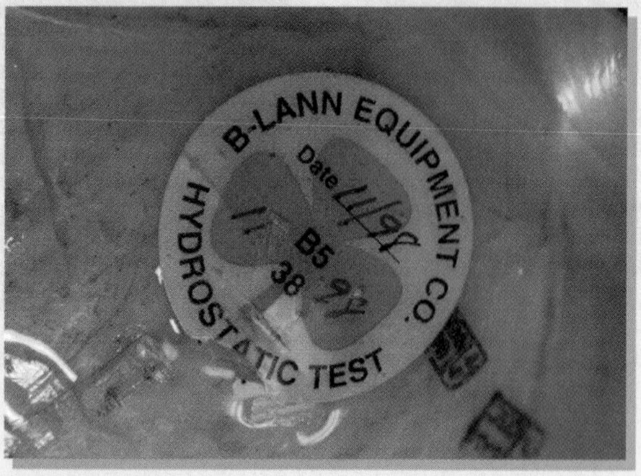

A Check the hydrostatic test date of the cylinder and inspect the cylinder for physical damage, discoloration from heat, gouges, or nicks in the cylinder surface.

B Place the cylinder in the fragmentation containment device. Connect the fill hose to the cylinder, and if it is equipped with a bleed valve, close it. Open the SCBA cylinder valve.

C Observe the cylinder gauge and watch that the cylinder fills at approximately 300 to 600 psi per minute. Close the cascade valve when pressure equalizes. If cylinder is not full—open cascade valve on cylinder with next highest pressure.

6. Open the valve at the cascade system manifold or the fill hose valve, depending on how the system is designed. Open both valves if the system has both.
7. Open the valve of the cascade cylinder with the lowest pressure. This pressure must be higher than the pressure in the cylinder being filled. *Control the fill rate of air to avoid excessive heating or chatter in the cylinder. If the cylinder heats or chatters, reduce the fill rate.*
8. Observe the cylinder gauge and watch that the cylinder fills at approximately 300 to 600 psi per minute, **JPR 7-8C**.
9. Close the cascade cylinder valve when the pressure equalizes with the pressure in the SCBA cylinder. If the SCBA cylinder is not full, open the valve on the cascade system

JOB PERFORMANCE REQUIREMENT 7-9

Servicing an SCBA Cylinder Using a Compressor/Purifier System

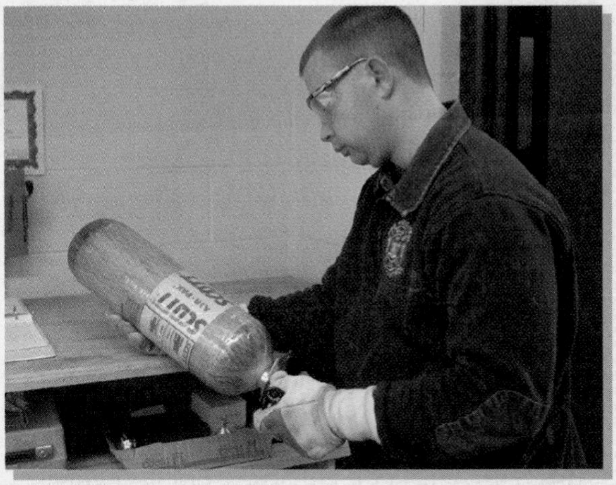

A Check the hydrostatic test date of the cylinder. Remove from service if the cylinder test date is not current or, if a composite cylinder, it is beyond the fifteen-year service life.

B Place the cylinder in the fragmentation containment device.

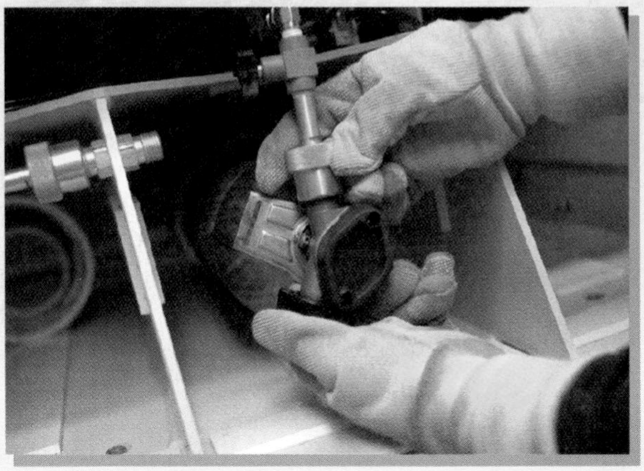

C Connect the fill hose to the cylinder and close the bleed valve.

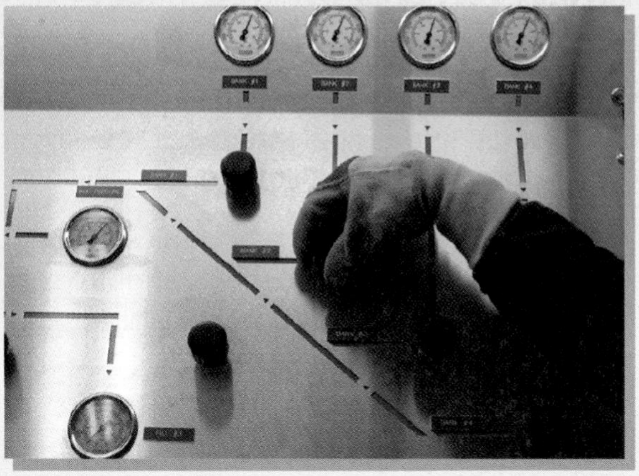

D Operate the compressor/purifier system in accordance with the manufacturer's instructions.

cylinder with next highest pressure. Repeat this step until the SCBA cylinder is full.

10. Close all valves on the cascade system, then close the SCBA cylinder valve.
11. Open the cascade system fill hose bleeder valve and bleed off excess line pressure. *Failure to follow this step could result in injury from the fill line as it discharges high-pressure air or blows off the O-ring seal on the end of the fill hose.*
12. Disconnect the cascade system fill hose from the cylinder valve, remove the cylinder from the fragmentation container, and place in storage.

Servicing an SCBA Cylinder Using a Compressor/Purifier System

Many organizations require certification before using either a cascade system or air compressor/purifier system. Do not attempt this skill without training on the specific system used by a particular fire department and follow the procedures recommended by the system manufacturer. Never attempt to fill a cylinder above its rated capacity.

1. Check the hydrostatic test date of the cylinder, **JPR 7-9A**. Remove from service if the cylinder test date is not current

or, if a composite cylinder, it is beyond the fifteen-year service life.
2. Inspect the cylinder for any signs of physical damage, discoloration from heat, gouges, or nicks in the cylinder surface. *Never fill an SCBA cylinder that has visible signs of damage or an out-of-date hydrostatic test.*
3. Place the cylinder in the fragmentation containment device, **JPR 7-9B**.
4. Connect the fill hose to the cylinder, and if equipped with a bleed valve, close valve, **JPR 7-9C**.
5. Open the SCBA cylinder valve.
6. Operate the compressor/purifier system in accordance with the manufacturer's instructions and open the outlet valve, **JPR 7-9D**.
7. Set the cylinder pressure adjustment on the compressor/purifier system to the correct pressure to fill the cylinder. Never attempt to fill a cylinder above its rated capacity.
8. Open the manifold valve and again check the fill pressure.
9. Open the fill station valve to begin filling the cylinder.
10. Observe the cylinder gauge and watch that the cylinder fills at approximately 300 to 600 psi per minute. Control the fill rate of air to avoid excessive heating or chatter in the cylinder. If the cylinder heats or chatters, reduce the fill rate.
11. Close the fill station valve when the cylinder is full.
12. Close the SCBA cylinder valve.
13. Open the fill hose bleeder valve and bleed off excess line pressure. *Failure to follow this step could result in injury from the fill line as it discharges excess air pressure or damages the O-ring seal.*
14. Disconnect the fill hose from the cylinder valve, remove the cylinder from the fragmentation container, and place in storage.

Lessons Learned

The SCBA unit is to a firefighter what a weapon is to a soldier. There are many events in military history where soldiers have lost their lives and battles because of lack of knowledge of their weapons. The same holds true for firefighting.

> **Streetsmart Tip** Firefighters should be encouraged to perform a "buddy check" with their partner or team members before entering a hazardous environment. Firefighting is teamwork and this teamwork can be a lifesaver.

As with any firefighter skill, there is no substitute for proper training with SCBA. In addition to initial training and certification in SCBA use, continued practice and advanced training are necessary to ensure that firefighters maintain proficiency. The same is true for maintenance because some firefighters have a tendency to treat SCBA as a tool, such as a ladder or ax, rather than as the vital piece of protective equipment that it is. To prevent failures from occurring and endangering lives, it is imperative that firefighters thoroughly inspect and test the function of SCBA as often as possible.

Each year a number of line-of-duty deaths are documented where firefighters for various reasons have died because they did not rely on their training with self-contained breathing apparatus. Regardless of the nature of the alarm, firefighters must always be prepared to go in harm's way; to do so safely, they must be knowledgeable and proficient in the use of SCBA.

KEY TERMS

Asphyxiation Condition that causes death due to lack of oxygen and an excessive amount of carbon monoxide or other gases in the blood.

Authority Having Jurisdiction (AHJ) The responsible governing organization or body having legal jurisdiction.

Carbon Monoxide (CO) Colorless, odorless, poisonous gas that when inhaled combines with the red blood cells excluding oxygen.

Closed-Circuit SCBA A type of SCBA unit in which the exhaled air remains in the system to be filtered and mixed with oxygen for reuse.

Hypoxia A deficiency of oxygen.

Immediately Dangerous to Life and Health (IDLH) The maximum level of danger one could be exposed to and still escape without experiencing any effects that may impair escape or cause irreversible health effects.

Mask Confidence or "Smoke Divers" Training Training courses designed to develop a firefighter's skills and confidence for using SCBA.

NFPA 1404 National Fire Protection Association standard created by the Fire Service Training Committee detailing the requirements for fire service SCBA programs, including training and maintenance procedures.

NFPA 1500 National Fire Protection Association standard created by the Technical Committee on Fire Service Occupational Safety and Health that addresses a number of issues concerning protective equipment.

NFPA 1981 National Fire Protection Association standard specific to open-circuit SCBA for fire service use that contains additional requirements above the NIOSH certification.

NIOSH National Institute for Occupational Safety and Health, 42 CFR Part 84, sole responsibility for testing and certification of respiratory protection including fire service SCBA.

Open-Circuit SCBA A type of SCBA unit in which the exhaled air is vented to the outside atmosphere.

OSHA 29 CFR 1910.134 Standard establishing minimum medical, training, and equipment levels for respiratory protection programs.

Oxygen-Deficient Atmosphere An atmosphere with an oxygen content below 19.5 percent by volume.

Positive Pressure A feature of SCBA providing a continuous supply of air, delivered by the regulator to the face piece, keeping toxic gases from entering. This pressure (1½ to 2 psi, depending on the manufacturer) is slightly above atmospheric pressure.

Pulmonary Edema Fluid filling the lungs causing death by drowning.

Respiratory Protection Programs Management programs designed to ensure employee respiratory protection as required by OSHA 29 CFR 1910.134 and NFPA 1500.

Respiratory System The system of the human body that exchanges oxygen and waste gases to and from the circulatory system.

Self-Contained Breathing Apparatus (SCBA) A type of respiratory protection in which a self-contained air supply and related equipment are worn or attached to the user. Fire service SCBA is required to be of the positive pressure type.

Supplied Air Respirator (SAR) A type of SCBA in which the self-contained air supply is remote from the user, and the air is supplied by means of air hoses.

REVIEW QUESTIONS

1. List two conditions requiring respiratory protection and explain the effects of these conditions on the human body.
2. List and explain the effects of oxygen deficiency on the human body.
3. List and explain one legal requirement for use of SCBA.
4. List two different types of operation of self-contained breathing apparatus and describe each.
5. List the four basic assemblies of an SCBA unit.
6. Explain and demonstrate two different SCBA donning procedures at 100 percent accuracy in the time limit established by the authority having jurisdiction.
7. Explain and demonstrate routine inspection procedures for SCBA used by the authority having jurisdiction in accordance with the manufacturer's instructions.

8. Explain and demonstrate maintenance and servicing procedures for the SCBA used by the authority having jurisdiction in accordance with the manufacturer's instructions.

9. Explain and demonstrate the SCBA cylinder servicing equipment and procedures used by the authority having jurisdiction in accordance with the manufacturer's instructions.

Endnote

1. U.S. Department of Labor, Washington, DC. 29 CFR 1910.134 (f)(1).

Additional Resources

Because of the special nature of SCBA, readers should see specific manufacturers' instructions.

CHAPTER

8

PORTABLE FIRE EXTINGUISHERS

Ric Koonce, J. Sargeant Reynolds Community College

OUTLINE

- Objectives
- Introduction
- Fire Classification and Risk
- Types of Fire Extinguishers
- Rating Systems for Portable Extinguishers
- Limitations of Portable Extinguishers
- Portable Extinguisher Operation
- Care and Maintenance of Portable Extinguishers
- Inspection Requirements
- Lessons Learned
- Key Terms
- Review Questions
- Endnotes
- Additional Resources

STREET STORY

I remember going to a fire several years ago as a firefighter assigned to a ladder company. I was the can man for the tour, which meant my tool assignment was a 6-foot wooden hook and a 2½-gallon water extinguisher. We were out on building inspection duty, and an alarm came in that we were not assigned to; however, because we were only six blocks away the dispatcher told us to respond. As we were responding we were informed of children possibly trapped in a rear bedroom.

We were the first company to arrive, and heavy smoke was pushing from the second-floor front window of a four-story apartment house. We proceeded to the fire apartment, and as we forced entry, heavy smoke was pushing out of the apartment. Without the luxury of a hoseline, we proceeded to crawl down a narrow smoke- and heat-filled hallway. The heat was tremendous, and the lieutenant kept pushing me down the hall, telling me we were getting closer to the fire.

At this time, we were able to see fire lapping out of a bedroom door. I was instructed to use my water extinguisher to hit the fire both around the top of the door and the ceiling, driving the fire back into the bedroom. The lieutenant then took my 6-foot hook, grabbed the bottom of the bedroom door with it, and pulled it closed. Fire was now lapping from around the door, and I was able to give quick short blasts of the extinguisher and knock it down. The forcible entry man climbed over our backs and with the lieutenant was able to search the rear bedrooms, finding no one. My job now was to stay at the door to the fire and protect lives until a hoseline was stretched and operating. It only took about two minutes to get the hoseline operating and for me to join my company in the search effort, but it felt like a lifetime. I have been to many fires since that day, both as a firefighter and an officer, and I have seen much accomplished with a little 2½-gallon water extinguisher. It is an underrated piece of equipment that when properly used is worth its weight in gold.

—Street Story by Mike Gala, Jr., Lieutenant, Ladder 148, FDNY

OBJECTIVES

After completing this chapter, the reader should be able to:

- Explain the five classes of fire and the risks associated with each class.
- Identify the kinds of fire extinguishers used for each class.
- Explain the rating systems of portable extinguishers for Classes A, B, and C.
- Identify the limitations of portable extinguishers.
- Demonstrate the operation of portable fire extinguishers.
- Explain the care and maintenance of portable fire extinguishers.
- Discuss the inspection requirements of portable fire extinguishers.
- Given a fire scenario or an actual fire, choose the fire extinguisher of correct size, agent, and rating to extinguish the fire.

INTRODUCTION

Portable fire extinguishers are designed to fight small fires or unusual fires that are not easily extinguished with water. There are also some small fires that firefighters cannot reach quickly with hoselines. The use of a portable fire extinguisher can knock down a small fire or control a larger one until firefighters can stretch, or advance, a hoseline to the fire area. Citizens have also used fire extinguishers to control a small fire until the fire department arrives. When used by untrained persons, fire extinguishers have created deadly delays in alerting the fire department. Therefore, it is very important that firefighters be proficient in using fire extinguishers and in teaching proper extinguisher use to any interested citizens.

Firefighters usually use hose streams to fight fires; extinguishers are used only occasionally. However, this should not prevent firefighters from learning how to effectively use them. Firefighters must practice extinguisher usage to retain the knowledge and skill required to use the unit effectively. There are four basic steps in extinguisher use: Pull the pin, aim the nozzle, squeeze the handle, and sweep the base of the fire with the extinguishing agent. These steps can be practiced using a pressurized water extinguisher, which is easily refilled and very cost effective to use. These steps carry over to other types of extinguishers. It is important for firefighters to be knowledgeable about fire extinguisher types, extinguishing agents, and the five classes of fire. Throughout the fire service, the phrase "Use the right tool for the right job" is prevalent.

Fire extinguishers come in a variety of types and sizes. Firefighters should know the extinguishers carried on each apparatus, as well as those inside the various buildings of the response district. During building inspections, firefighters should preplan special hazards and locations where an extinguisher would be a valuable tool to control a fire. Fire prevention will dictate the correct extinguishers for the occupancy. Firefighting skill will dictate the use of the right extinguisher for the conditions present during a fire. The general public will often ask firefighters questions about firefighting and fire extinguisher use. Firefighters must be prepared to answer questions and train citizens to use portable fire extinguishers, when requested.

FIRE CLASSIFICATION AND RISK

The type or nature of the material burning, that is, its fuel, defines the fire. Fuel is the key ingredient because as it varies so does the fire. The different types of fuels classify types of fires, and these classes of fire are used to identify the extinguishers and extinguishing agents used to put them out. An understanding of the fire classes leads to selection of the proper unit and agent. There are four traditional fire classes and one new one; however, the first three are the most common and the ones primarily covered by this chapter. Firefighters should be aware of places in their community with these types of fuels. This is the beginning of the process of planning for and dealing with potential emergencies and creating a means to deal with them. (For more information on fire classes, see Chapter 4.)

Class A

Class A fires involve ordinary combustibles such as wood, paper, cloth, plastics, and rubber. These fuels can be extinguished with water, water-based agents or foam, and multipurpose **dry chemicals**. Because of its availability, water is usually used by the fire department.

Class B

Class B fires involve flammable and combustible liquids, gases, and greases. Common products are gasoline, oils, alcohol, propane, and cooking oils. Pressurized flammable liquids and gases are special fire hazards that should not be extinguished unless the fuel can be immediately shut off. Flammable liquids that are flowing horizontally plus dripping or overflowing their container and spilling vertically (creating a three-dimensional flow), such as an overflowing tank, are also considered special hazards. *Special hazards*[1] refers to situations for which fire extinguishers have not been tested and therefore may be inadequate; it is important to carefully evaluate the situation prior to attacking these types of fires. Some solids under fire conditions may melt and act like flammable liquids. Common extinguishing agents for Class B fires are carbon dioxide (CO_2), regular and multipurpose dry chemical, and foam.

Safety Pressurized flammable liquids and gases should be extinguished by stopping the flow of the fuel. In some cases, firefighters need to extinguish the fire to shut off control valves. These situations are extremely dangerous because firefighters have to work in a flammable atmosphere and any source of ignition may reignite that atmosphere with catastrophic results. This entry into a flammable atmosphere should be the last available option.

Class C

Class C fires involve energized electrical equipment, which eliminates the use of water-based agents to extinguish them. The recommended method of fighting these fires is to turn off or disconnect the electrical power and then use an appropriate extinguisher, depending on the remaining fuel source. Class C extinguishers have extinguishing agents and hoses with nozzles that will not conduct electricity. Class C only extinguishers are not made. Class C agents include carbon dioxide (CO_2) and regular and multipurpose dry chemicals.

Class D

Class D fires involve combustible metals and alloys such as magnesium, sodium, lithium, and potassium. These metals can be found in some lightweight motor vehicle engine components or lawn mower bodies. Great care must be used when attempting to extinguish a fire in these types of fuels. Water and other extinguishing agents can react violently when applied to burning combustible metals and can endanger nearby firefighters. Also, because of the differences in the metal and alloy fuels, there is no universal Class D extinguishing agent that works on all Class D materials; what works well on one metal may be relatively ineffective on a different alloy. Personnel must use the correct and uncontaminated (clean, dry, and without any other foreign materials in it—basically as it comes from the factory) extinguishing agent for each different Class D material. Facilities that use or store these materials are required to maintain adequate amounts of the proper extinguishing agents to combat any potential fire situation. Class D agents are called **dry powders** and should not be confused with *dry chemicals,* which, although dry and powdery, are not the same. (Some of these agents are dry sand, phosphate salts, or silica.) Other special agents such as Lith-X and Met-Ex are not commonly used and are only mentioned here. Firefighters using these special agents locally should seek additional information and training on their use.

Streetsmart Tip Many facilities that use or process Class D fuels have frequent fires involving these materials. The local fire department may or may not be called to deal with these routine fires, especially if a fire brigade is on-site. Firefighters who do respond to these types of fires on a regular basis must ensure that they follow proper safety precautions with *each* fire incident and ensure they do not become complacent or relaxed about dealing with this dangerous class of fire.

One fire department that regularly responds to a facility with a Class D material had successfully extinguished several fires over a period of a couple of weeks. When returning to the facility for another fire, they began their operations as they had previously done. While applying the extinguishing agent to the surface of the fuel, a violent explosion occurred that expelled some of the molten material onto several firefighters, causing injuries. The resulting investigation revealed that water had been used near the tank and had contaminated both the extinguishing agent and the shovels used to apply it.

Class D materials are always considered hazardous materials. Proper hazardous materials procedures must be followed at *each* incident. Familiarity is the breeding ground of complacency and can result in injuries or worse.

Class K

Class K is a new classification of fire as of 1998 and involves fires in combustible cooking fuels such as vegetable or animal oils and fats. Its fuels are similar to Class B fuels but involve high-temperature cooking oils and therefore have special

Figure 8-1 Class K equipment.

Figure 8-2 Various types of fire extinguishers.

characteristics. Typically, firefighters have used Class B extinguishers on these types of fires, but they have been less effective on deep layers of cooking oils. Class K agents are usually **wet chemicals**, water-based solutions of potassium carbonate–based chemical, potassium acetate–based chemical, potassium citrate–based chemical, or a combination.[2] These agents are usually used in fixed systems, **Figure 8-1,** but some extinguishers are available.

TYPES OF FIRE EXTINGUISHERS

Many types of fire extinguishers are available for purchase and use, **Figure 8-2.** The best type of extinguisher depends on many factors and each should be considered prior to placing an extinguisher in use. *The wrong extinguisher can be worse than no extinguisher.*

Factors for selecting an extinguisher are the type of fuel, the person using the extinguisher, and the building or place where it will be used. The first factor to consider when choosing an extinguisher is the type and amount of fuel present. This will provide clues as to the type of fire to anticipate. The amount of fuel determines fire size, and the wrong size extinguisher will not completely extinguish the fire. The user of the extinguisher and the occupancy in which it will be used represent another factor. Are potential users trained to combat a fire effectively or will they use improper technique and exhaust the extinguisher before the fire is knocked down? What will the people in the building do with or to the extinguisher while it is stored? Will the extinguisher be tampered with, stolen, or otherwise ineffective when needed?

Another factor is the type of building construction and occupancy. What hazards or other conditions exist? What building areas need to be protected to avoid further problems? These potential hazards set the fire code requirements for selecting and placing the extinguishers. Environmental conditions must be examined for effects on the fire and the extinguishing agent. Temperature may eliminate using water-based agents that may freeze; corrosive atmospheres may require special protection for the user; the wind may blow away the extinguishing agent before it reaches the fire; and a confined space may create an unsafe or nonsurvivable atmosphere for the user.

A final factor would be the type of equipment protected. Often this last factor is given too much consideration. Delicate equipment and high-value items may require special considerations as they may be irreparably damaged in an extinguishing operation. It should be stressed, however, that the main objective is to extinguish the fire completely and effectively. Most extinguishing agents can be cleaned from equipment, whereas damage from a fire can render equipment permanently destroyed. It is important to remember that certain extinguishing agents are corrosive to certain substances. This should be a factor when choosing an extinguisher for the occupancy.

Types of Extinguishing Agents

Water is the basic fire-extinguishing agent for Class A materials. Its ability to absorb heat is one of the prime reasons for its effectiveness. The disadvantages are that it is subject to freezing and is ineffective—and sometimes dangerous—on other classes of fuels. **Loaded stream** combats the freezing problem by adding an alkali salt as an antifreezing agent. Water-based foam extinguishers for use on Class B fires have either **Aqueous Film-Forming Foam (AFFF)** or **Film-Forming Fluoroprotein Foam (FFFP)** of both the regular and **polar solvent type**. These agents are effective in cooling and smothering the fire and creating a vapor barrier. (See Chapter 11 for more foam information.)

Carbon dioxide (CO_2) is an inert gas stored under pressure as a liquid capable of being self-expelled. The colorless and odorless gas is effective in smothering a Class B or C fire. CO_2 works best in enclosed or semienclosed areas as the agent can be easily blown away by the wind. Personnel should take caution to avoid oxygen deprivation when CO_2 is used in small enclosed areas. This clean agent is still popular with those concerned about property damage, although in well-ventilated areas it can be ineffective. Some sensitive electrical equipment can be thermoshocked by CO_2.

Dry chemical extinguishing agents are particles propelled by a gaseous medium for distribution. There are three general categories: sodium bicarbonate–based, potassium–based, and multipurpose dry chemicals. The first two categories work on Class B and C fires, and are called regular dry chemical agents. Sodium bicarbonate, similar to baking soda, is the agent in the first category and is highly effective for cooking grease fires. The second category includes potassium bicarbonate, potassium chloride, and urea-based potassium bicarbonate—all of which are usually more effective than sodium bicarbonate. The third category is known as a multipurpose dry chemical agent—effective on Class A, B, and C fires. These multipurpose dry chemicals are monoammonium phosphate.[3] Dry chemicals are very effective due to their coating action, which reduces the chances of reignition. The coating action is a drawback, however, when protecting sensitive items such as computer mainframes.

Wet chemical agents are water-based solutions of potassium carbonate–based chemical, potassium acetate–based chemical, potassium citrate–based chemical, or a combination of those chemicals.[4] They are used for special applications, particularly Class K fires.

Clean agents are the agents used to replace halon or halogenated hydrocarbon extinguishers. These agents are thought to cause damage to the Earth's ozone layer and have been banned by an international treaty. These agents may still be in limited use, but must be phased out and replaced with clean agent systems. Clean agents are gases that do not conduct electricity or leave a residue, and are nonvolatile. They are divided into two classes—halocarbon agents and inerting gases—and are currently rated as somewhat ineffective for local application. Research on their application continues.[5]

Kinds of Extinguishers

Many types of portable fire extinguishers are in use today. Some are small and handheld, while others are so large that they require a wheeled cart to move them. Most extinguishers operate on the same basic principle of storing and expelling an extinguishing agent. Fire extinguishers are labeled to make their firefighting rating quick and easy to identify. The older versions of fire extinguishers are labeled with colored geometrical shapes with letter designations, **Figure 8-3A.** Newer fire extinguishers are labeled with a picture label system, **Figure 8-3B.** Many fire extinguishers are multiuse, that is, a single extinguisher may be used to fight more than one class of fire, **Figure 8-3C.** Class A and B fire extinguishers have a numerical rating discussed later in this chapter. Listed here are current kinds of fire extinguishers for each type of agent:[6]

1. *Water type* for Class A fires:
   a. Pump type
   b. Pressurized water
   c. Pressurized loaded stream
2. *Foam extinguishers* for Class A and B fires; stored pressure
3. *Carbon dioxide* for Class B and C fires; self-expelling gas, stored pressure
4. *Halon and clean agents* for Class B and C, or Class A, B, and C
5. *Dry chemical* for Class B and C or Class A, B, and C fires (also some dry powder, Class D extinguishers)
   a. Stored pressure
   b. Cartridge-operated type
6. *Wet chemical* for Class K fires; stored pressure

Pump-type extinguishers are hand-pumped devices of two designs, depending on whether the pump is internal or external to the tank. The pump tank extinguisher has the pump housed inside the water tank area and each stroke pumps water out of the tank. This type of extinguisher is simple to operate and maintain, **Figure 8-3D**. The second type is designed for wildfires and is a backpack pump that has the tank carried on the back with the hose and pump in front of the operator, **Figure 8-4** and **Figure 8-5**. Both usually come in a 2½-gallon size.

Pressurized water, pressurized loaded stream, and stored pressure extinguishers operate by means of

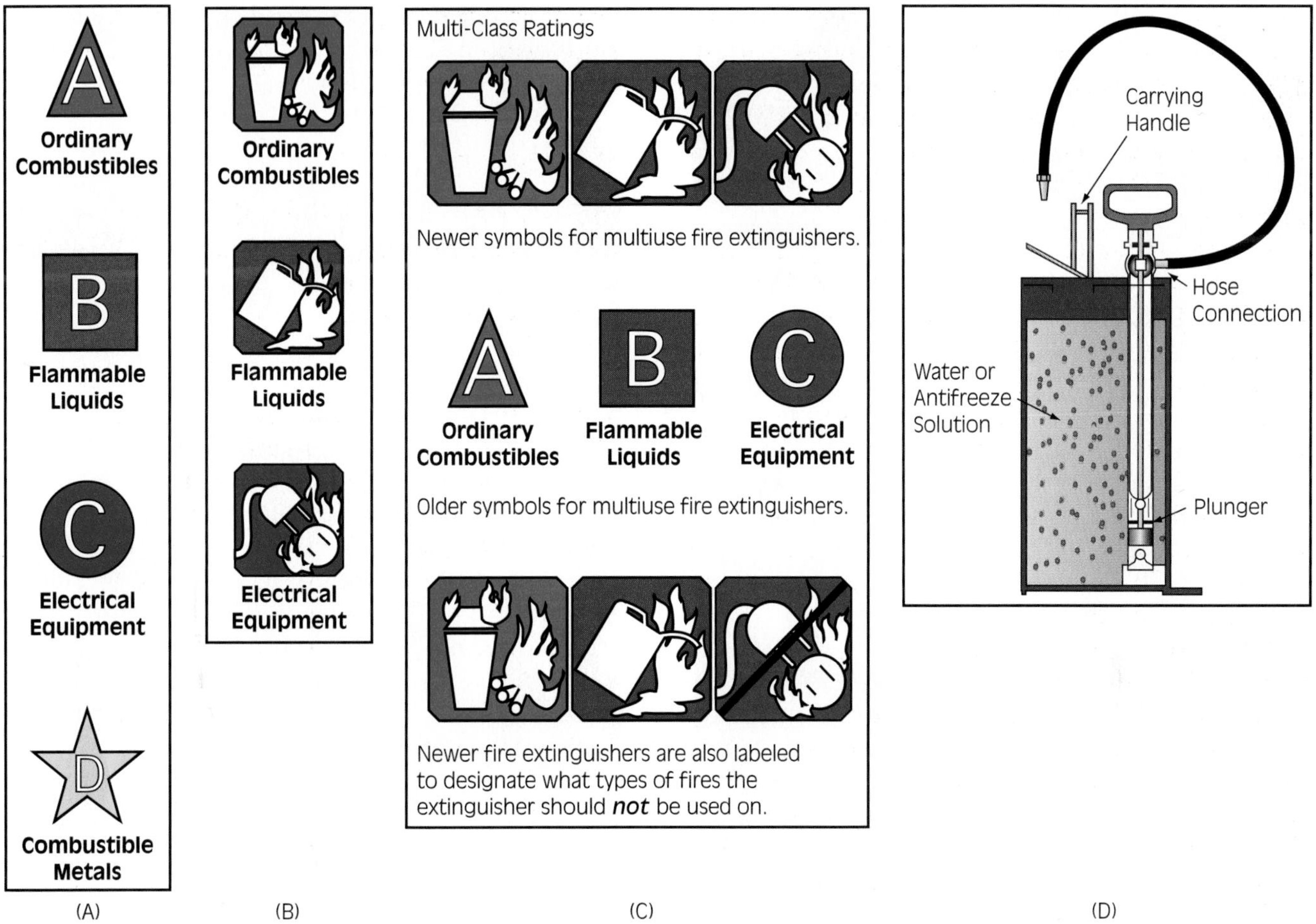

Figure 8-3 (A) Older versions of fire extinguishers are labeled with colored geometrical shapes with letter designations. (B) Newer fire extinguishers are labeled with a picture label system. (C) Many fire extinguishers can be used to fight more than one type of fire. (D) Pump tank extinguisher.

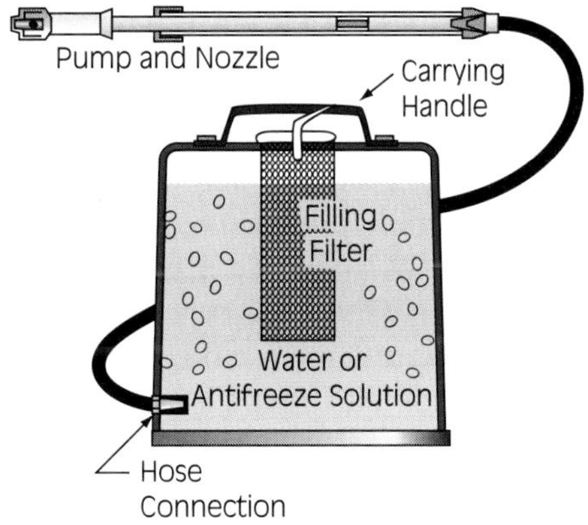

Figure 8-4 Backpack pump tank fire extinguisher.

an expelling gas that propels the agent out of the container. The differences are seen in the type of container and hose and nozzle assemblies. The container differences result from the various pressures required to store the agent. The hose and nozzle differences relate to the agent and its special need for application. Some agents may be dispensed over a wide area, others over a smaller area; some foam needs an air-aspirating nozzle or may require the nozzle to be nonconductive.

The basic principle is that the agent is inside the container with a pressurizing gas above it under constant pressure. Some use air or nitrogen as an expelling gas that is added to the container with the agent. Other agents, like carbon dioxide, are their own expelling gases. Most extinguishers have a gauge to measure the pressure of the gas, but CO_2 extinguishers do not. **Figures 8-6** through **8-15** show the various extinguishers and how they store their product.

Cartridge-operated extinguishers are used for some regular and multipurpose dry chemical and most dry powder Class D extinguishers. They are similar to stored pressure extinguishers except that instead of being under constant pressure, the expelling gas is stored in a cartridge on the side of the container. When the puncturing lever is

Figure 8-5 A backpack pump tank fire extinguisher. *(Courtesy of Fred Schall)*

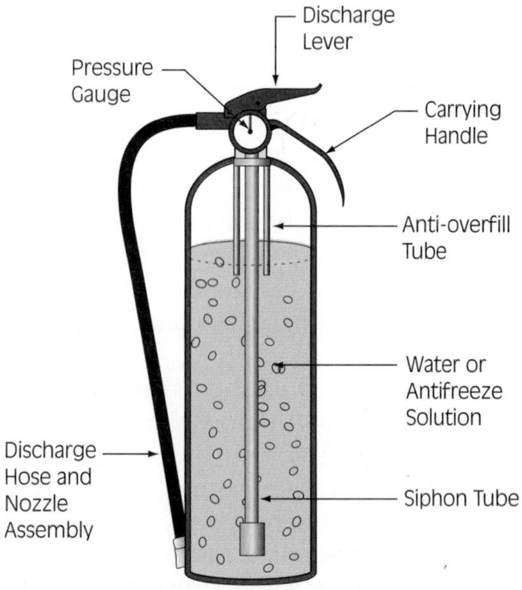

Figure 8-6 Stored pressure water extinguisher.

depressed, it ruptures a disk and the gas is expelled into the tank, which pressurizes it. The operator controls the flow of agent by a nozzle mounted at the end of the hose. Cartridge-type extinguishers are used when the agent may cake excessively and needs to be stirred. The agent is accessible without charging the pressure or dumping the agent. This type is also easy to refill by adding the agent into the tank and a new cylinder to the charging mechanism, **Figure 8-16** and **Figure 8-17**.

Note Inverting extinguishers were popular many years ago. These extinguishers are no longer manufactured and are seldom seen today.

Figure 8-7 Stored pressure water extinguisher.

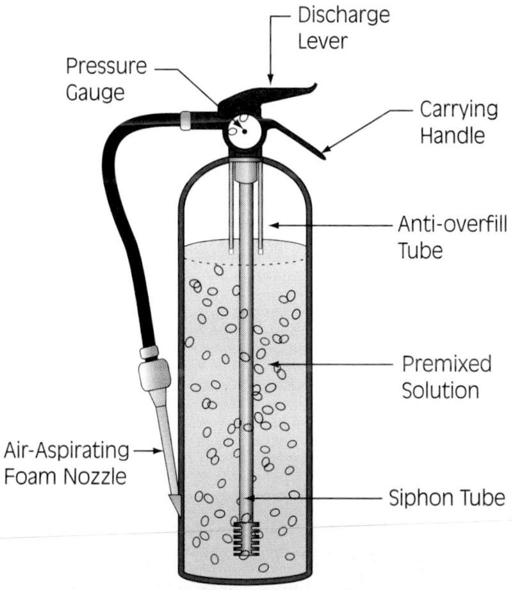

Figure 8-8 Stored pressure AFFF or FFFP extinguisher with air-aspirating nozzle.

Figure 8-9 Stored pressure foam extinguisher.

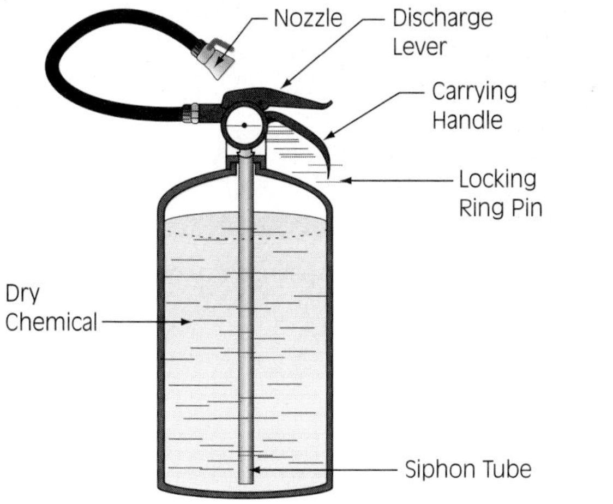

Figure 8-10 Stored pressure dry chemical extinguisher.

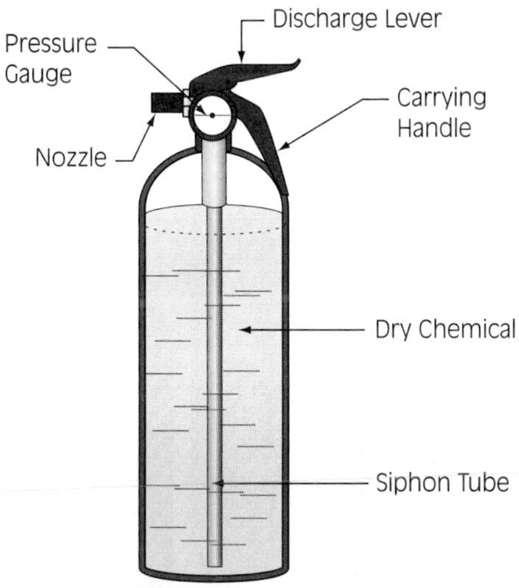

Figure 8-11 Stored pressure dry chemical extinguisher with fixed nozzle.

Figure 8-12 Stored pressure dry chemical extinguisher.

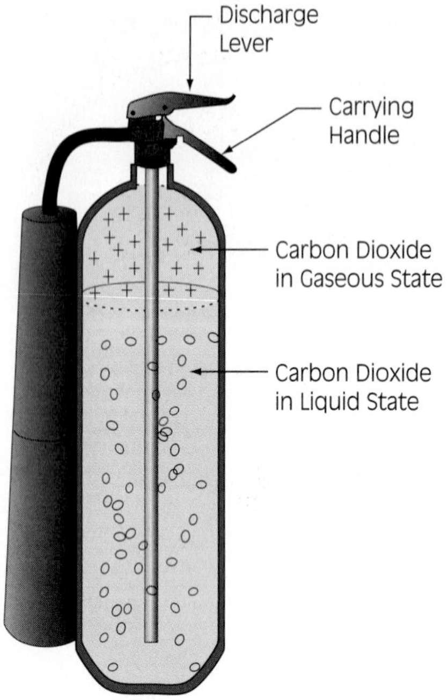

Figure 8-14 Carbon dioxide extinguisher.

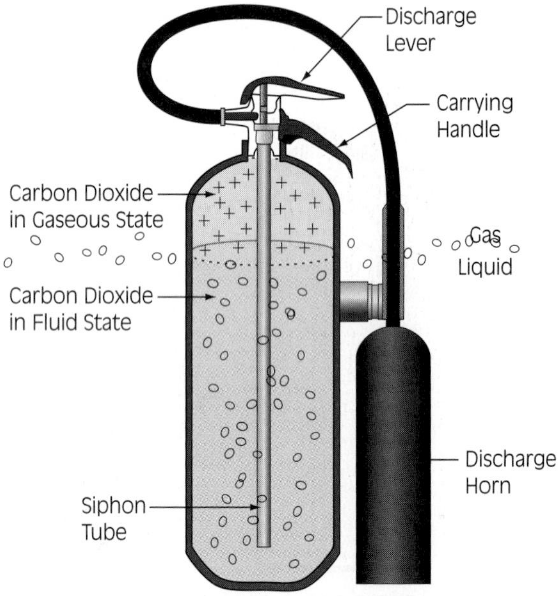

Figure 8-13 Carbon dioxide extinguisher.

Figure 8-15 Carbon dioxide extinguishers.

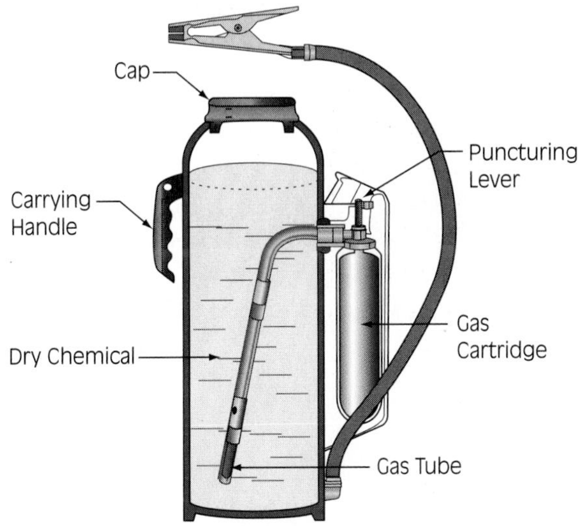

Figure 8-16 Cartridge-operated dry chemical extinguisher.

Figure 8-17 Cartridge-operated dry chemical extinguisher.

RATING SYSTEMS FOR PORTABLE EXTINGUISHERS

Each class of fuel is subjected to a separate type of extinguisher test for its class. The testing of extinguishers is usually conducted by an independent testing agency such as Underwriters Laboratories or Factory Mutual Research Corporation, or a government agency such as the Coast Guard. This section covers the tests for Classes A, B, and C. Classes D and K have special tests that are more specific to their fuels or application. The ratings are noted on the label of the extinguisher as is a symbol to show the class(es) of fire on which it works. (See Chapter 4 for these symbols.)

Class A

The testing of Class A extinguishers and agents uses a wood cribbing test. The cribbing is set on fire, allowed a pre-burn period, and then attacked with the extinguisher. For a 1-A rating, the extinguisher should extinguish a wood crib of about one cubic foot (0.03 cubic meters), **Figure 8-18**. The

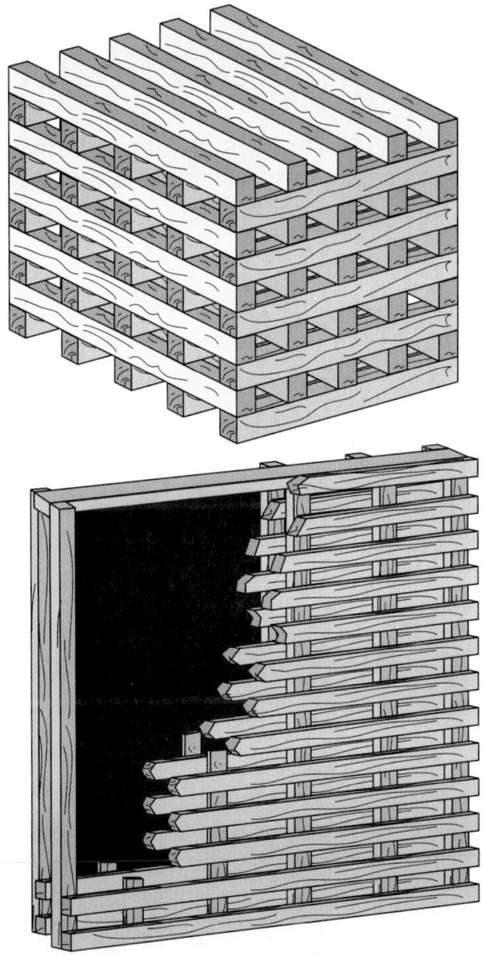

Figure 8-18 Wood cribbing for Class A extinguisher test.

ratings increase as the amount of fire suppressed increases; for instance, a 2-A extinguisher will put out twice the fire of a 1-A. In a wood paneling and excelsior test, the excelsior is ignited and allowed to spread to wood paneling before it is extinguished. Larger scale tests use only the wood cribbing and are conducted outside, while smaller tests are done indoors.

Class B

The test for Class B extinguishers involves igniting a pan of a flammable liquid (heptane), allowing a pre-burn period, and attacking the fire. The size of the pan determines the rating, **Figure 8-19**. For example, a 4-square-foot (0.37-square-meter) pan that is extinguished should yield a rating of 4-B. The rating assumes that an inexperienced operator using a 20-B rate extinguisher on a 20-square-foot fire should find the results adequate. The number rating compares approximately to the square footage to be extinguished, although larger units will not offer a direct relationship, as larger fires require more agent per area than the smaller ones.

Class C

The testing of Class C extinguishers and agents tests only the conductivity of the agent and the nozzle or hose and nozzle combination, **Figure 8-20**. There is no actual fire test for Class C agents or extinguishers. No numbers are assigned with the Class C rating.

Figure 8-19 Flammable pan for Class B extinguisher test.

LIMITATIONS OF PORTABLE EXTINGUISHERS

Fire extinguishers have limited capabilities, and attempting to exceed those capabilities can increase damage and cause injuries. They are designed for

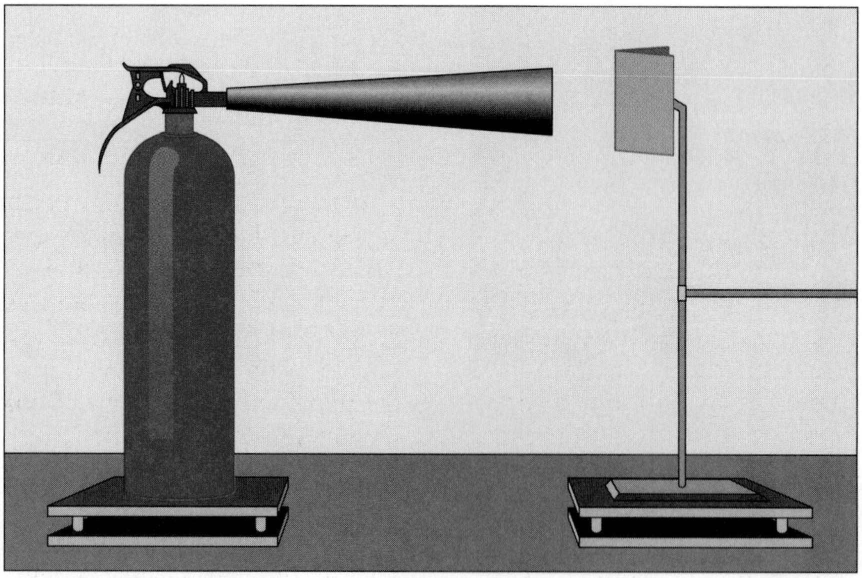

Figure 8-20 Class C test for conductivity.

specific purposes and are usually a first-aid method for fire extinguishment. Fire extinguishers are designed and rated with certain types and sizes of fires in mind; using the wrong class or the wrong size extinguisher may cause problems. When thinking of size, it is usually best to pick the larger size. However, picking the largest fire extinguisher available to put out a small fire can make it an expensive fire. The wrong class extinguisher will not do the job, will waste the agent, and can cause a reaction or electrical shock.

Note Using the wrong class extinguisher will not do the job, wastes the agent, and can cause a reaction or electrical shock. *It is important to pick the right extinguisher for the job.*

PORTABLE EXTINGUISHER OPERATION

Suppose a wastebasket is on fire and the firefighter has chosen the correct extinguisher for putting the fire out. What is the right way to use it? The operation of most fire extinguishers is covered by the acronym *PASS* for the four simple steps in using an extinguisher: *P* is for pull the pin, *A* is for aim the nozzle, *S* is for squeeze the operating handle, and the second *S* stands for sweep the nozzle across the base of the fire.

Note PASS, the four steps for using a fire extinguisher:

***P**ull* the pin.
***A**im* the nozzle.
***S**queeze* the handle.
***S**weep* the base of the fire.

Step one is *pull the pin* that prevents the handle from operating. Most extinguishers have the pin mounted at the operating handle to prevent it from being accidentally squeezed, **JPR 8-1A**. Cartridge-operated extinguishers have a pin at the activation lever for the cartridge, not on the operating handle at the nozzle. The pin is held in place by a wire or plastic tie; a simple tug removes this wire and the pin.

The second step is to *aim the extinguisher* toward the base of the fire closest to the firefighter. Firefighters must remember to have a path of escape behind them and *not allow the fire to get between them and the exit,* **JPR 8-1B**. The object of aiming at the base of the fire is to sweep the fire away from the firefighter, confining it and driving it back to its origin. Aiming at other points may cause the fire to spread beyond the capability of the extinguisher. The extinguisher must be aimed from the proper distance, which often is set by the room size and fire size. Various extinguishers can reach a variety of distances up to 50 feet (15 meters), but reaching and being effective are two different subjects. The practical effective range of a pressurized water or foam extinguisher is about 20 feet (6 meters), dry chemical units from 15 to 20 feet (4.5 to 6 meters), and a CO_2 type from 10 to 15 feet (3 to 4.5 meters). A final aiming caution when using a CO_2 extinguisher is to make sure that the aiming hand is on the handle, not on the horn where the cold can freeze the skin.

The third step is to *squeeze the handle* together to apply the agent, **JPR 8-1C**. The firefighter should continue to squeeze the handle for enough time to cover the whole fire area. Some short blasts are okay but too many will empty the extinguisher without extinguishing the fire.

Last, *the firefighter sweeps the area of the fire* by carefully keeping the nozzle aimed at the base of the fire, continuing to push the fire back, **JPR 8-1D**. The sweeping motions help extinguish the entire fire instead of creating a pocket that may allow the fire to circle around the firefighter. Foam extinguishers can also use the methods of regular foam application that are covered in Chapter 11. Bending over or sweeping the extinguishing agent at a low angle allows the agent to spread faster across the fire.

Safety Firefighers must remember to have a path of escape behind them; *they cannot allow the fire to get between them and the exit.*

If the extinguisher does not put out the fire, the firefighter should not deploy a second extinguisher. Multiple extinguishers used at the same time, however, are more effective than using one at a time. Many buildings have been lost due to improper "one at a time" use of multiple extinguishers, and the untrained individuals are quickly forced to exit the building due to the fire's increasing intensity. In this situation, the fire department is not notified until the fire has increased greatly in size. Extinguishers are meant to combat small fires in the growth stage. Large fires are beyond the capabilities of portable extinguishers.

JOB PERFORMANCE REQUIREMENT 8-1

Portable Extinguisher Operation

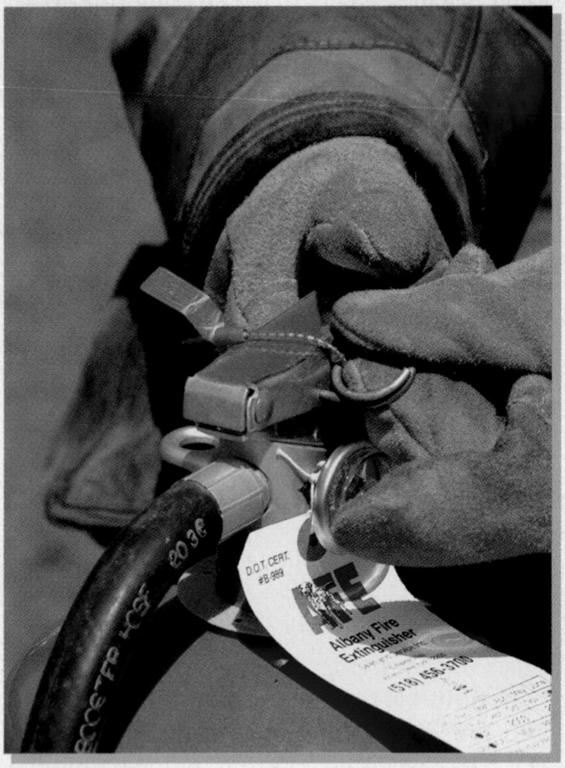

A Pull the pin.

B Aim the nozzle.

C Squeeze the handle.

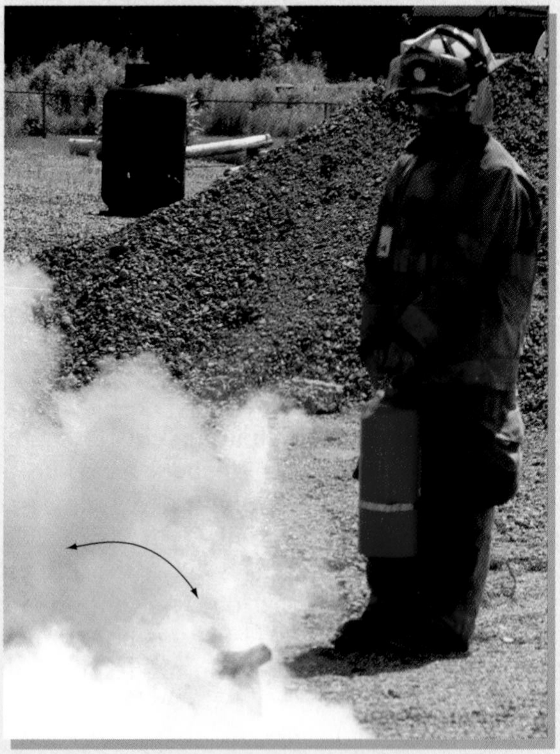

D Sweep the base of the fire.

Note Stored-pressure water extinguishers are designed to be carried in an upright position when approaching a fire. The firefighter is cautioned to keep the hose in hand at all times.

Streetsmart Tip Civilians should be taught to call the fire department before attempting to use an extinguisher.[7] Often they plan to use just one extinguisher and then call the fire department. They almost get the fire out and run for another extinguisher, which again fails to put out the fire. One group of maintenance personnel used more than twenty extinguishers on an apartment fire, but when the first arriving officer, on a 1,000-gpm pumper, found fire coming through the roof, he called for a third-alarm assignment. The civilians got caught up in fighting the fire, rather than calling for help.

Firefighters must be adept at using all types of firefighting equipment. This includes portable fire extinguishers. However, in many services this tool is not used as often as a hoseline. It is, therefore, extremely important that firefighters train using portable fire extinguishers to stay proficient with the use of this tool and to effectively teach firefighting techniques to the general public.

Streetsmart Tip In an emergency, a firefighter may use a CO_2 extinguisher on scenes where dangerous dogs or other animals are present. Normally the noise of the CO_2 extinguisher discharging will scare the animal. The cold cloud of agent will usually further frighten the animal into retreat. *Note:* Discharging the CO_2 extinguisher directly onto the animal can render permanent harm to the animal. The extinguisher should be thought of as a reasonable force tool. Only the amount of force (agent) to get the animal to retreat should be used and no more. This choice of protection should be a last resort for firefighters or other persons in danger.

CARE AND MAINTENANCE OF PORTABLE EXTINGUISHERS

Care and maintenance of fire extinguishers is fairly straightforward. Simple inspections (see next section) and careful storage prevent most problems. The extinguisher should be inspected carefully when first placed in service and properly stored in a cabinet or on a bracket. Keeping apparatus-mounted extinguishers in their brackets will prevent damage from falling or from striking other equipment. New apparatus standards require this and it is a good practice on older vehicles. Routine apparatus equipment checks should require the extinguisher to be visually examined for proper pressure and visible damage and to ensure the pin seal is in the proper place. If any problems are noted, the unit should be removed and sent for repairs. Vehicle operators should periodically remove the unit from its bracket to hand test its weight and to do a better visual check. Dry chemical or dry powder extinguishers carried on apparatus should occasionally be rotated upside down and shaken to keep powders from packing at the bottom as a result of vehicle vibration. Cleaning of any dirt or grit will help keep the unit in good working order. Finally, any extinguisher that has been discharged, that has low pressure on the gauge, or that has a broken seal should be removed from the vehicle until thoroughly examined and/or recharged.

INSPECTION REQUIREMENTS

Some very popular fire extinguishers of the past have now been declared obsolete and should be removed from service. These obsolete extinguishers include soda acid, chemical foam (except film-forming foam), vaporizing liquids such as carbon tetrachloride, cartridge-operated water or loaded stream, and any copper or brass extinguisher with solder or rivets (except pump tanks). Some of these have been removed because of the hazards of the extinguishing agents. Most of these extinguishers have the potential to explode when pressure is applied during use. Firefighters should never attempt to operate an obsolete extinguisher and should instead have them decommissioned by trained technicians.

The inspection of fire extinguishers by firefighters is usually a visual inspection, **Figure 8-21**. If something does not look right, the unit should be removed and replaced. Extinguishers in buildings should be checked every thirty days, and extinguishers on apparatus should be inspected each time the vehicle is inspected. Extinguishers that are new or old ones being returned to service should be examined prior to placing back on apparatus. The visual inspection should first check that the unit has a tag indicating its last service and/or inspection. Annual inspection and maintenance by a service technician is recommended. The pin seal should be firmly in place. If the unit has a gauge, it should register the proper pressure. If a gauge is not present, lifting the unit or weighing it will determine if it is properly filled. Some extinguishers such as carbon dioxide extinguishers can only be truly checked by weighing; the filled weight is stamped on the cylinder. A careful look at the unit should detect any damage, dents, or corrosion to the shell, hose, nozzle, and other parts. If any defects are present, the unit should be removed and replaced. A last check should be to see if the unit is within the time frame for its hydrostatic testing, which is a pressure test to ensure that the unit will not explode when operated. The test period ranges from five to twelve years; firefighters should consult their service center or the NFPA standard to verify extinguishers are compliant.[8]

Figure 8-21 Conducting visual inspections. *(Courtesy of Fred Schall)*

Lessons Learned

Fire extinguishers can be used as initial response tools or to fight fires in special situations. Firefighters must know how to use fire extinguishers and to teach the public how to use them. Part of that knowledge includes classifying fires by their fuels, especially ordinary combustibles, flammable liquids and gases, and energized electrical equipment. Knowing the classes of fire is the first step toward choosing the right extinguisher and the correct size. The four-step process for using an extinguisher is *PASS:* pull, aim, squeeze, and sweep.

KEY TERMS

Aqueous Film-Forming Foam (AFFF) A synthetic foam that as it breaks down forms an aqueous layer or film over a flammable liquid.

Carbon Dioxide (CO_2) An inert colorless and odorless gas that is stored under pressure as a liquid that is capable of being self-expelled and is effective in smothering Class B and C fires.

Class A Classification of fire involving ordinary combustibles such as wood, paper, cloth, plastics, and rubber.

Class B Classification of fire involving flammable and combustible liquids, gases, and greases. Common products are gasoline, oils, alcohol, propane, and cooking oils.

Class C Classification of fire involving energized electrical equipment, which eliminates using water-based agents.

Class D Classification of fire involving combustible metals and alloys such as magnesium, sodium, lithium, and potassium.

Class K A new classification of fire as of 1998 that involves fires in combustible cooking fuels such as vegetable or animal oils and fats.

Dry Chemicals Dry extinguishing agents divided into two categories. Regular dry chemicals work on Class B and C fires; multipurpose dry chemicals work on Class A, B, and C fires.

Dry Powders Extinguishing agents for Class D fires.

Film-Forming Fluoroprotein Foam (FFFP) Foam that incorporates the features of AFFF and fluoroprotein foams with good resistance and the film-forming barrier.

Loaded Stream Combats the water freezing problem by adding an alkali salt as an antifreezing agent.

Polar Solvent Type of Foam or Alcohol-Resistant Foam Foam that is compatible with alcohol and/or polar solvents by creating a polymeric barrier between the water in the foam and the polar solvent.

Wet Chemicals Extinguishing agents that are water-based solutions of potassium carbonate–based chemicals, potassium acetate–based chemicals, potassium citrate–based chemicals, or a combination.

REVIEW QUESTIONS

1. Name the four traditional classes of fire, give an example of each and an extinguishing agent for each class.
2. Name the new class of fire, describe its materials, and name an extinguishing agent.
3. What are the three categories of agents of dry chemicals?
4. What is a wet chemical?
5. How does carbon dioxide extinguish a fire?
6. Why are halons no longer used for fire extinguishers?
7. How does a stored pressure extinguisher work?
8. What are the two types of pump-type extinguishers?
9. How are Class A extinguishers rated?
10. What does a 10-B rating mean?
11. What is a Class C extinguisher tested for?
12. Describe the four steps for operating an extinguisher.
13. Describe a visual inspection process on a fire extinguisher.

Endnotes

1. "Special hazard" is a term that has several meanings. In this case, it denotes special fire situations that are so unique that it would be impossible to test and rate the effectiveness of a fire extinguisher for each special case.
2. *NFPA 17A: Standard for Wet Chemical Extinguishing Systems, 1998 Edition,* p. 17A-5. National Fire Protection Association, Quincy, MA.
3. Gagnon, Robert, *Design of Special Hazard and Fire Alarm Systems,* pp. 149–150. Delmar Learning, a part of the Thomson Corporation, Clifton Park, NY, 1998.
4. *NFPA 17A: Standard for Wet Chemical Extinguishing Systems, 1998 Edition,* p. 17A-5. National Fire Protection Association, Quincy, MA.
5. Gagnon, Robert, *Design of Special Hazard and Fire Alarm Systems,* pp. 101–102. Delmar Learning, a part of the Thomson Corporation, Clifton Park, NY, 1998.
6. Gagnon, Robert, *Design of Special Hazard and Fire Alarm Systems,* p. 30. Delmar Learning, a part of the Thomson Corporation, Clifton Park, NY, 1998.
7. OSHA regulations limit untrained or limited trained persons to fight only incipient fires.
8. *NFPA 10: Standard for Portable Fire Extinguishers, 1998 Edition,* Chaps. 4 and 5. National Fire Protection Association, Quincy, MA.

Additional Resources

From Delmar Learning, a part of the Thomson Corporation:

Diamantes, David, *Fire Prevention, Inspection and Code Enforcement.* Delmar Learning, a part of the Thomson Corporation, Clifton Park, NY, 1997.

Gagnon, Robert, *Design of Water-Based Fire Protection Systems.* Delmar Learning, a part of the Thomson Corporation, Clifton Park, NY, 1996.

Gagnon, Robert, *Design of Special Hazard and Fire Alarm Systems.* Delmar Learning, a part of the Thomson Corporation, Clifton Park, NY, 1998.

Klinoff, Robert, *Introduction to Fire Protection.* Delmar Learning, a part of the Thomson Corporation, Clifton Park, NY, 1997.

Sturtevant, Thomas, *Introduction to Fire Pump Operations.* Delmar Learning, a part of the Thomson Corporation, Clifton Park, NY, 1997.

From NFPA:

NFPA 10: Standard for Portable Fire Extinguishers, 1998 Edition. National Fire Protection Association, Quincy, MA.

NFPA 17: Standard for Dry Chemical Extinguishing Systems, 1998 Edition. National Fire Protection Association, Quincy, MA.

NFPA 17A: Standard for Wet Chemical Extinguishing Systems, 1998 Edition. National Fire Protection Association, Quincy, MA.

CHAPTER

9

WATER SUPPLY

Ric Koonce, J. Sargeant Reynolds Community College

OUTLINE

- Objectives
- Introduction
- Sources of Water Supply
- Water Distribution Systems
- Fire Hydrants
- Valves Associated with Water Distribution Systems
- Rural Water Supply
- Pressure Associated with Water Distribution Systems
- Testing Operability and Flow of Hydrants
- Determining Static, Residual, and Flow Pressures
- Obstructions and Damage to Fire Hydrants and Mains
- Lessons Learned
- Key Terms
- Review Questions
- Endnotes
- Additional Resources

Courtesy of Fred Schall

STREET STORY

Several years ago, I was a brand new assistant fire chief in a rural area of Florida. Having come from the New York metro area, "rural water supply" was new to me. In New York, if you needed water, you just used a hydrant! But this was not the case in the new department.

We were dispatched to a reported building fire. As I approached the scene, I found a very large single-family dwelling with heavy fire involvement in approximately 30 percent of the structure. I knew that there were no hydrants in that part of the district, so I started to radio the Emergency Communications Center for multiple mobile water supply apparatus. I wanted every mobile water supply apparatus for miles around. And I got them. Eventually.

As the incident progressed, I was interrupted two or three times by some of our firefighters who were mumbling "something" about water supply. I was busy and made it clear that I had it covered. As one might expect, the fire progressed faster than the arrival of the mobile water supply apparatus, and for all intents and purposes, the building burned to the ground. Not a good night.

In the mop-up stages of the fire, I went up to some of the firefighters who were speaking to me earlier and I asked them what they were trying to tell me. What they'd been trying *to tell me was that a major river was less than 100 feet away from the building—and they were going to suggest we draft out of that river for our water supply. Hmmm. Great idea.*

I learned a few lessons from that fire. The first was that if you're a Yankee, and you burn down a 150-year-old "pride of the South" building, you will hear about it for 20+ years! But more importantly, I learned there's no excuse not to be prepared when you know your district has areas without municipal water supply. You know what areas you protect right now—and certainly know where you have and don't have municipal water supply. Plan ahead for rural water supply before the fire comes in. Know what mobile water supply apparatus will be due to respond and have them respond on the first alarm. And drill with them before a fire to make sure that your radios and hose connections are all compatible. At the fire scene is too late to find out. Also, a few seconds of listening to the firefighters may have made a big difference in our tactical operation. This fire incident now prompts me to do a walk-around on almost any incident I'm involved with. The walk-around gives me a view of the big picture and allows me to determine our strategy—had I done it that night, it would have been clear to me that water, and lots of it, was flowing behind the building.

—Street Story by Battalion Chief Billy Goldfeder, E.F.O., Loveland-Symmes Fire Department, Loveland, Ohio

OBJECTIVES

After completing this chapter, the reader should be able to:

- Explain the value of proper water supply to other goals of firefighting.
- Identify sources of water supply for drinking and firefighting.
- Explain the difference between groundwater and surface water.
- Explain the purpose of mobile water supply apparatus.
- Explain the features of water distribution systems.
- Identify types of fire hydrants and their uses.
- Identify valves associated with water distribution systems.
- Explain how to operate a rural water supply.
- Explain a portable water tank operation.
- Explain tender operations.
- Identify the proper pressures associated with water distribution systems.
- Conduct a test of the operability and flow of fire hydrants.
- Determine the static, residual, and flow pressures of water sources.
- Identify cause of obstructions and damage to fire hydrants and mains.

INTRODUCTION

Water supply is one of the most critical elements of firefighting *because water is the most common extinguishing agent.* In considering the expression "the first five minutes of fighting a fire sets the stage for the entire battle," remember that without water it is not even a fight. Water supply is important in areas with a water distribution system and even more important in areas without a system because one must be created. Firefighters should understand the basics of water supply and how to establish one because without a water supply, hoses, appliances, and fire streams are useless.

Where the water comes from, how much is available, and how it is delivered are key questions. Water supply dictates the **fire flow capacity** or amount of water that can be flowed. The **fire flow requirement** is the amount of water required for putting out the fire.[1] Knowing the amount of water available and how much is needed lets firefighters select the right strategy, tactics, hose appliances, and fire streams. All are interrelated, and facing a large fire without sufficient amounts of water leads to a situation in which a fire will continue to burn.

Water is the most common extinguishing agent. It is economical, readily available, and abundant. It has the ability to absorb large quantities of heat. It is inexpensive when getting it directly from a river or pond. From a municipal water supply, the additional needs for fire protection make it more expensive but not as much as other extinguishing agents. Its availability and abundance often depend on the fire's location. Desert or rural areas have limited or no water. A city can have areas where the availability of water is limited. In such situations, firefighters must haul in the necessary water. Finally, water's ability to absorb large quantities of heat makes it extremely effective in fighting fire. Water absorbs such large amounts of heat from the fire that it cools the fuel below its ignition temperature. (More information on this can be found in Chapter 4.)

SOURCES OF WATER SUPPLY

An understanding of the water supply situation for firefighting requires first knowing where the water comes from and how it gets from that point to the fire scene. When using a lake as the water source, it is pretty obvious, but other cases are not as clear. How does the water get to a fire hydrant and what factors are involved in making sure that enough water is available for the needs? A wide range of natural and man-made factors affects water sources. The weather is probably the greatest of these factors. The Earth's water is in a constant cycle of change. The sun evaporates water into the atmosphere. It condenses into clouds that carry it to other places where it eventually falls as rain. Some areas get so much rain they are called rain forests, whereas other places get none and are deserts. Some areas alternate between rainy and dry seasons; others have plenty of water but it is frozen some or all of the time. These are just some problems encountered with water supply.

Groundwater

Most of the Earth's freshwater supply is groundwater. Groundwater is water that seeps into the ground from rain and other surface sources. It collects in pockets of the Earth or permeates into layers of water-bearing soil. These pockets are called **aquifers** and some are large underground rivers. The level of the water under the Earth's surface is the **water table**, which can rise and fall.

Springs are groundwater sources that naturally flow to the surface; others are under great pressure and when drilled may force their way to the surface. This pressure may decrease over the years such that pumps will be needed to draw the water to the surface, as is the case with most wells, **Figure 9-1**. Shallow wells are closer to the Earth's surface and are cheaper to drill but are more prone to changes in the water table and are subject to contamination and saltwater infiltration. Deep wells may penetrate through several layers of water before finding an aquifer. They are a more predictable water source with less chance of contamination. Wells used for municipal water supply and fire protection must be of sufficient volume; often multiple wells are employed. Domestic and farm wells are insufficient in volume and pressure for firefighting.

Surface Water

Surface water is the world's most common source of water—almost 75 percent of the Earth is covered by water. Unfortunately, most is found in the oceans and seas with limited availability. Other natural sources of surface water are rivers, lakes, and ponds, **Figure 9-2**. Man-made surface sources include lakes, ponds, reservoirs, swimming pools, and water tanks. Surface water may be fresh, salt, or brackish water, and many surface water sources are influenced by tidal changes. **Tidal changes** are the rising and falling of the surface water levels due to the gravitational effects between the Earth and the moon. In some areas, these changes are insignificant but in others a more than 40-foot difference exists between high and low tide. When using a tidal water source, it is important to understand the effects of the tide and know when the changes will occur or the water source may not be available when it is needed.

Figure 9-2 A surface water source. *(Photo courtesy of Donald Fischer)*

Mobile Water Supply Apparatus

For many small fires and in those areas without a water distribution system, the water tank on fire apparatus supplies the water, having gotten its water from another source. The tank sizes of fire apparatus vary, but most engines today have 500-gallon or larger tanks. If the vehicle has a tank of over 1,000 gallons and is primarily used to transport water, it is considered a mobile water supply apparatus, **Figure 9-3**. The term **water tender** is used to describe mobile water supply apparatus, although some jurisdictions use the term **tanker**. The fire service is still divided on this language to some degree. The National Fire Service Incident Management System Consortium uses the term *water tender* to describe land-based water supply apparatus in its *Model Procedures Guide for Structural Firefighting*. This guide has been endorsed by the International Association of Firefighters (IAF), International Association of Fire Chiefs (IAFC), National Wildfire

Figure 9-1 Well pump with storage tanks.

Figure 9-3 Mobile water supply apparatus.

Coordinating Group (NWCG), and Fire Resources of Southern California Organized for Potential Emergencies (FIRESCOPE) groups. The guide refers to a tanker as an aircraft capable of carrying and dropping water for firefighting operations. Although regional, some fire departments use *tanker* to describe the land-based water supply apparatus and **tender** to describe hose-carrying apparatus. This chapter will use the term *tender,* because it is becoming the preferred standard term for ground vehicle units, while tankers are understood to be air units, such as helicopters and airplanes. Tenders might or might not have pumps and hose. Mobile water supply apparatus range from 1,000 to over 8,000 gallons with some units being tractor-trailer units. Tenders combined with **portable water tanks** can efficiently provide large volumes of water to a fireground operation. Tenders may have a small booster or attack pump of at least 250 gpm, a fire pump of at least 750 gpm, or a transfer pump of at least 250 gpm. Transfer pumps use a vacuum-operated pump to draw and expel their load.

Safety Mobile water supply apparatus have been involved in a large percentage of fire department vehicle accidents and firefighter injuries and fatalities. The rural environment that requires these large, heavy vehicles has older, narrower, crowned roads that are often driven by persons with less experience, training, and supervision. Some fire departments cannot afford a new vehicle so they modify an old delivery tank truck into a water supply unit without the necessary safety modifications such as extra baffles and heavier brakes. Another problem is that a gasoline tank truck with 7,600 gallons of water has a 20 percent heavier load than if carrying lighter weight gasoline. Fire departments using this type of apparatus need to examine all aspects of this problem and work to provide safe and effective fire protection.[2]

Tanks, Ponds, and Cisterns

An additional source of water can come from a developed source of water such as water tanks, ponds, and cisterns. Water tanks may be underground, ground level, or elevated and may be connected to a piping system, have a **dry hydrant** or other connection, or have just a drafting point. Ponds may be developed for fire protection reasons and may be lined or unlined with or without dry hydrants. A **cistern** is an underground water tank made from natural rock or concrete. Cisterns can store large quantities of water—30,000 gallons or more—in areas without other water supplies or as backup. Connections are dry hydrants, drafting pit, or other type. Firefighters should also be aware of other sources of water available to them such as swimming pools and any other body of water that can be accessed for firefighting.

WATER DISTRIBUTION SYSTEMS

How does the water get to the fire? This is a good question in areas with fire hydrants because the source may be many miles away. Water distribution systems have several components, including a method of getting the water, filtration or treatment processes, and a method of storage, supply, and distribution. The important parts for firefighters are the supply and distribution system, including storage facilities. The water can be obtained from a surface or groundwater source. Water supply and distribution systems should be designed to meet the community's domestic, industrial, and fire protection needs at peak periods.

Small groundwater systems need a well with a pumping station that can also treat and store the water and that connects to a local supply system. In larger well or surface systems, multiple supply, processing, and storage units with massive feeder and distribution lines are often used.

Water is supplied in three ways. The first is *gravity fed,* **Figure 9-4A,** in which the water source is at a higher elevation than where the water is used, such as in a mountain reservoir feeding a valley city. The next is a *pumped* system, **Figure 9-4B**, in which pumps are used to draw water from a well, river, or reservoir and pump it through the system. The third type is a *combination gravity-pumped system,* **Figure 9-4C**, where part of the system relies on pumps and the remainder on gravity. This is the most common type and even good gravity systems find areas where distance or elevation requires additional pumping and storage for maintaining pressure. It is considered to be a combined system because elevated water tanks that are filled by pumps in turn supply the piping system by gravity.

After treatment for drinking, water goes into the distribution system or water mains. The mains are divided into feeders and distribution lines. Primary feeders supply secondary feeders and then distribution lines. Primary and secondary feeder lines in large cities can be measured in feet; some of New York City's aqueducts are 20 feet in diameter. Some older and larger cities have separate water systems

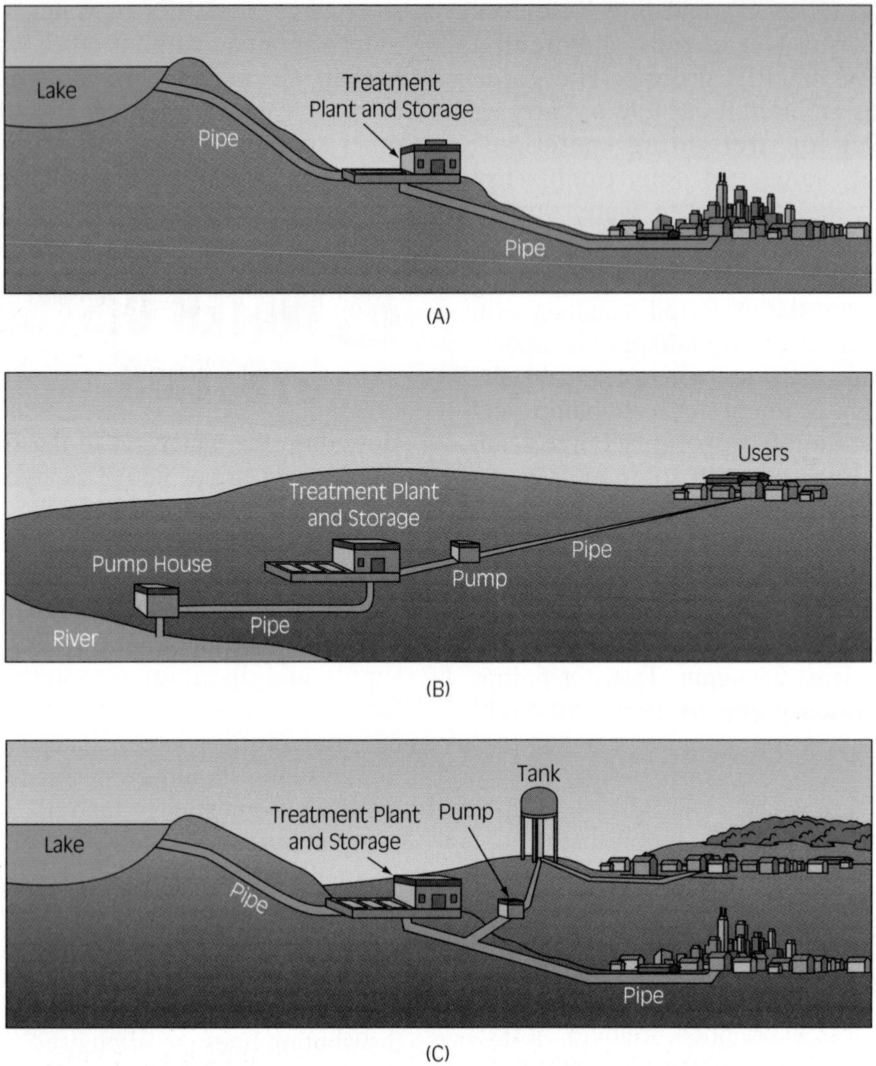

Figure 9-4 (A) Gravity-fed water distribution system. (B) A direct pump water distribution system. (C) A combination gravity-pumped water distribution system.

just for fire protection. Distributor lines are the water mains with the building connections and fire hydrants, and for fire protection; they range in size from 6 to 16 inches. Good water distribution systems are interconnected into *loops* and *grids* that allow multidirectional supply and reduce dead-end mains that cannot provide adequate pressure or volume, **Figure 9-5**.

> **Streetsmart Tip** Firefighters should know the location of dead-end water mains and, where possible, avoid using them. Connecting to a dead-end water main may not provide adequate water, or if another unit connects to another hydrant on the same dead-end main, it may rob the original unit of what was a good water supply.

FIRE HYDRANTS

Fire hydrants allow the fire department to access water supply systems. A fire hydrant is also known as the "plug." The term *plug* is a throwback to days when firefighters drilled a hole in wooden water mains, then capped the hole with a plug when they were finished. The two major hydrant types are wet and dry barrel hydrants. Another hydrant type is a dry hydrant that is neither dry nor a true hydrant but a pipe system for drafting from a static water source. Some specialty hydrants are also used.

Wet Barrel

Wet barrel hydrants, **Figure 9-6** and **Figure 9-7**, have water in the barrel up to the valves of each outlet. They are used in areas that are not subject to

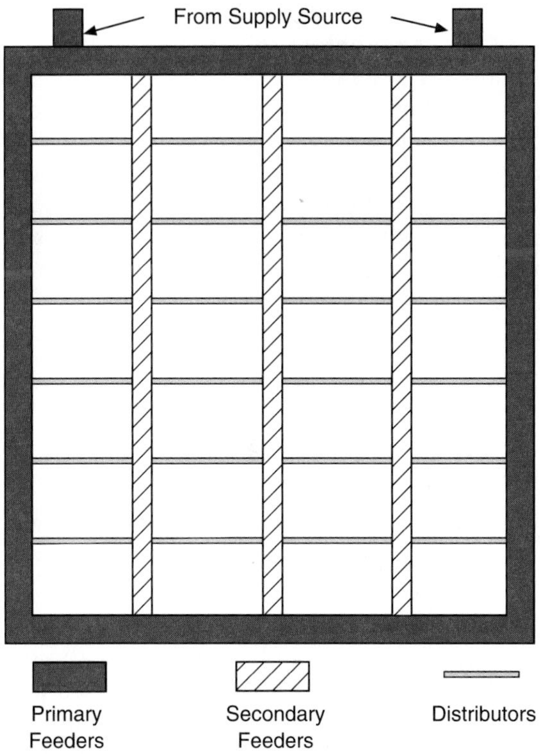

Figure 9-5 Grid or looped system.

Figure 9-6 Multivalved wet barrel hydrant.

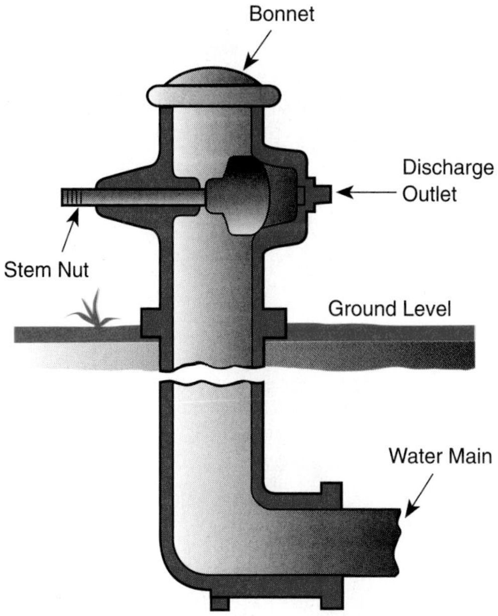

Figure 9-7 Typical schematic of wet barrel hydrant.

freezing temperatures, primarily California and Florida. The wet barrel hydrant allows each outlet to be controlled by a separate valve and can have additional lines taken off or supplied if an outlet is available. This additional connection does not require the flow through other outlets to be stopped. Some also have a main control valve that can control the flow to all outlets.

Dry Barrel

Dry barrel hydrants, **Figure 9-8** and **Figure 9-9**, are used in areas where freezing temperatures could damage the hydrant. It uses a valve at the base of the hydrant to control water flow to all outlets. The base and this valve are below ground at the level of the water main. Connecting additional lines to a flowing dry barrel hydrant requires the entire hydrant to be shut down until the new connect is made and then reopening the valve or adding a gate on one of the outlets prior to charging the hydrant. When the valve is fully closed or partially opened, a drain allows the water to drain from the hydrant preventing damage from freezing. Firefighters should ensure that the hydrant is operated and shut down with the valve in either the fully opened or closed position. When this has not been done, it has caused undermining of the hydrant and ground with damage to nearby roads, buildings, and fire apparatus.

Figure 9-8 Typical dry barrel hydrant.

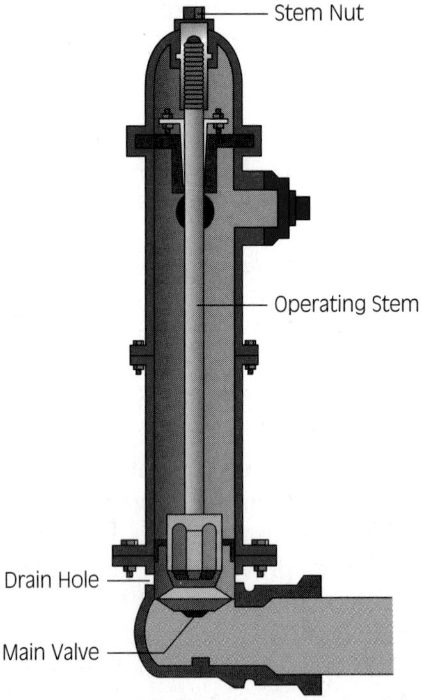

Figure 9-9 Typical schematic of dry barrel hydrant.

Caution Firefighters should ensure that fire hydrants are operated and shut down with the valve in either the fully opened or closed position. When the valve is not fully opened or closed, undermining of the hydrant and ground with damage to nearby roads, buildings, and fire apparatus has occurred, plus it may restrict volume. Also, anytime a valve is operated, it should be done slowly to prevent a water hammer.

Dry Hydrant

A dry hydrant is not really a fire hydrant but a connection point for **drafting** from a static water source such as a pond or stream, not a pressurized one. A dry hydrant is a pipe system with a pumper suction connection at one end and a strainer at the other, **Figure 9-10** and **Figure 9-11**. They are used primarily in rural areas with no water system but may be found in urban and suburban areas as a backup or where certain buildings may be of great distance from other water sources but a pond or stream is located nearby. Placing dry hydrants allows fire departments to have better access to the water source; some even have graveled areas for parking and a turnaround point for the apparatus. The landowner benefits by having a ready source of water in case of fire and may even get a discount on insurance.

Figure 9-10 Dry hydrant.

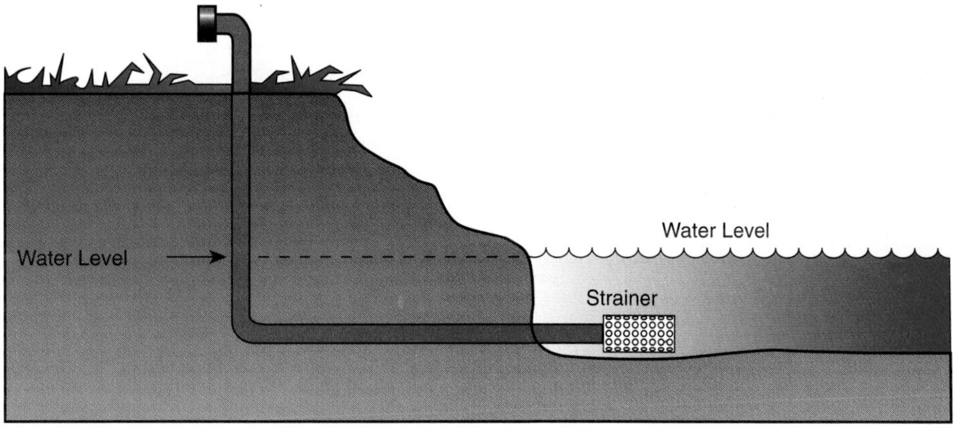

Figure 9-11 Schematic of dry hydrant installation.

Specialty Hydrants

Wall hydrants are hydrants mounted on the wall of a building after the water line has been run into the building, **Figure 9-12**. Firefighters should make sure that they are connecting to a wall hydrant and not the fire pump test connection outlets. A wall hydrant has a direct connection to the water supply system, whereas a fire pump test connection is an inspector's way to test the flow of the fire pump. The test connection will not supply any additional water; it will just rob the water from the fire pump.

Figure 9-12 Wall fire hydrant.

A flush-type hydrant is mounted below grade level and is usually found in a pit, vaults, or valve box. The purpose of mounting the hydrant below grade allows access to a water source in areas where an above-grade hydrant would interfere with operation of the facility. Airports, shipping terminals, and many European cities utilize flush-mounted hydrants.

High-pressure hydrants are hydrants that are connected to a separate high-pressure water system used only for fire protection purposes. Some larger cities, San Francisco is one, have these types of systems installed in the downtown commercial and industrial areas to provide high volume and pressure for large fires. Some of these systems draw their water directly from the city harbor, and the pumps are often maintained by the fire department. These hydrants are either of the wet or dry barrel design.

VALVES ASSOCIATED WITH WATER DISTRIBUTION SYSTEMS

The valves in public water systems are usually nonindicating type **gate valves** and **check valves**. The gate valves are butterfly valves that are opened and closed to control water flow, **Figure 9-13**. Nonindicating gate valves are installed at interconnections of water mains, at intermediate points on long sections of water mains, and before each hydrant and major building connection. Check valves are installed to control water flow in one direction, typically when different systems are interconnected. **Backflow preventers** are a check valve or pair of

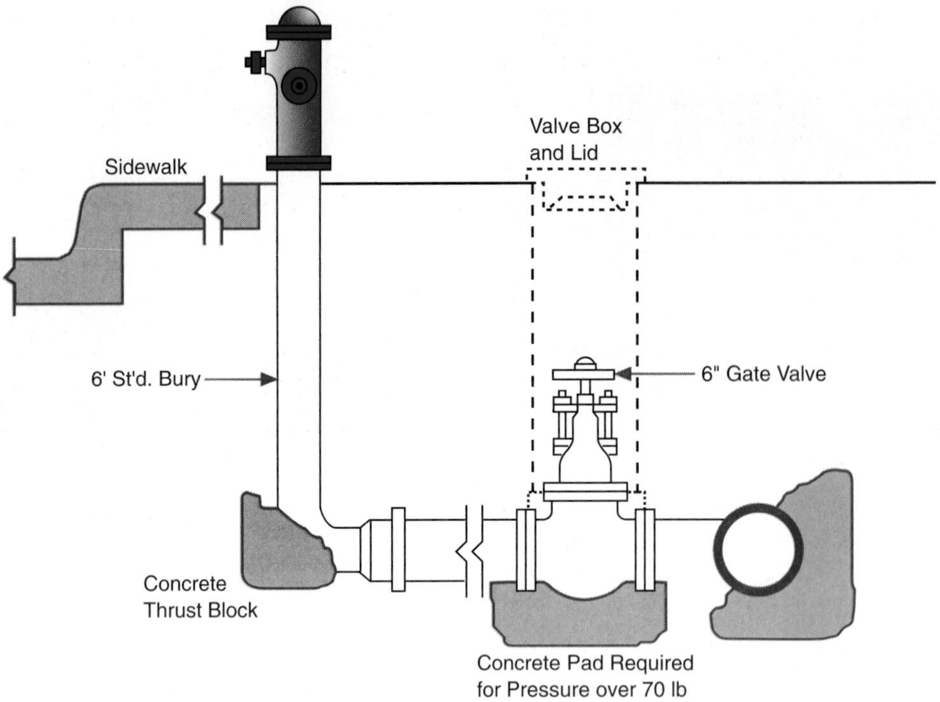

Figure 9-13 Hydrant with plumbing. Note the location of a gate valve between the water main and hydrant.

check valves that prevent a backflow of water from one system into another and are mostly required where a building water or fire protection system connects with the public water system. Backflow preventers are being required for environmental and health reasons. The backflow preventer is installed on the building's private water piping after the building's main water shutoff. Private water valves are of the indicating type such as a post indicator valve (PIV), wall indicator valve (WIV), or an outside stem and yoke valve (OS&Y). These valves are commonly associated with standpipe and sprinkler systems.

RURAL WATER SUPPLY

Rural water supply or tender-based water supply operations can occur anywhere, not just in rural areas. Urban and suburban areas often have undeveloped areas or other places where the hydrants are just too far away to use effectively. Limited access highways and large bridges can be found to have water supply problems. Even redundant water supply systems can suffer a catastrophic event that knocks out or limits the water supply. The basics of rural water supply should be understood by all firefighters. Rural firefighters should find it easier to adjust to water supply operations in an urban area than urban firefighters could adjust to nonhydrant areas.

Rural water supply operations require careful coordination and control. A water supply officer should be part of the incident command system with full authority over tender operations. Firefighters assigned to water supply operations may find the work less glamorous than actual firefighting, but the nozzle person cannot put out the fire without this vital support.

Portable Water Tanks

Since tenders are designed to transport water, they need to be able to quickly drop off their load of water and return to the **fill site** as soon as possible. To speed the operation, each tender should have a portable water tank with a capacity equal to or greater than its tank size, **Figure 9-14**. The collapsible or inflatable tanks are filled by unloading or dumping the tender, allowing it to return to the fill site and bringing back an additional load of water. The portable tank is either set up next to the attack engine or, if space is not available there, it is assigned an engine that drafts from the tank and supplies the attack engines. Depending on the size of the fire and availability of tanks and tenders, several tanks may be used to increase the fire flow. Many large modern tenders use a **jet dump** or **jet siphon**

Figure 9-14 Portable water tanks are an essential piece of equipment for shuttle operations.

Figure 9-16 Tender at dump site dropping water directly into portable tank. *(Photo courtesy of Donald Fischer)*

to speed the unloading of tanks, **Figure 9-15**. The jet dump or siphon draws the water out as opposed to just using gravity in a nonassisted system.

The portable tank is erected or inflated at the **dump site** in a location where the loaded tenders can offload their water and an engine can place its suction hose into the tank to draft and supply the water to attack engines. Placing a heavy tarp on the ground prior to opening the portable tank can help protect the portable tank liner before water is dumped. In larger operations, multiple tanks may be set up together. Departments that use portable tanks and ones that may support those companies should regularly practice tender shuttle operations.

Tender Operation

A tender operation is a **shuttle operation** that involves tenders moving large quantities of water between a dump site and a fill site. The dump site is where the water is delivered for quick unloading, and because the tenders arrive already full, it is set up first. Using portable tanks and a supply engine to draft from the tanks speeds the process and allows the tender to return for another load. Personnel at the dump site connect hoselines or operate dump valves to fill the portable tanks and help set up the drafting operation for supplying the attack pumpers, **Figure 9-16**. Dump sites should be selected for availability to unload multiple tenders, turnaround area for the tenders, operational area, continued access to the fireground, and safety of personnel. The fill site is the location of the water source where the tenders are filled, **Figure 9-17**. The fill site should be staffed with enough personnel to quickly connect hoselines to fill the tender and may require a pumper to draft from a static source, **Figure 9-18**. The fill site should be picked for an adequate water source including filling multiple units, the ability to either turn around the vehicles or, better yet, have a drive-through site, and the safe operation of the site.

Figure 9-15 Jet siphon valve. *(Photo courtesy of Donald Fischer)*

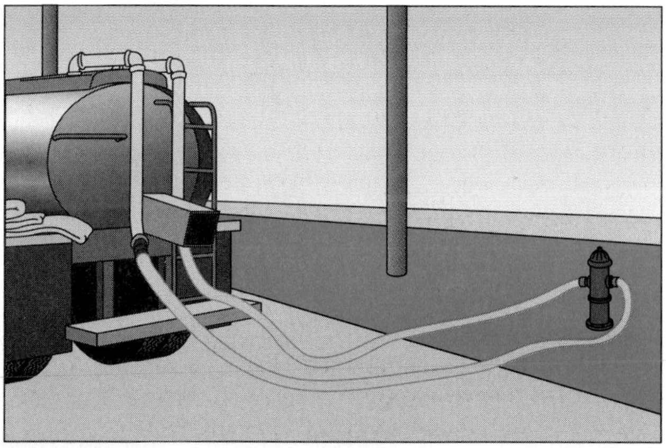

Figure 9-17 Tender at fill site with pressurized water source.

Figure 9-18 A pumper at fill site drafting water. *(Photo courtesy of Donald Fischer)*

Shuttle operations control the fire flow capacity of the incident, **Figure 9-19**. Fire flow rates can be calculated by adding the time to fill a tender, turn the unit around and return to the dump site, drop the water, and turn around and return to the fill site divided by the quantity of water carried. If a 1,000-gallon (3,785-L) tender[3] has a 10-minute shuttle time, the flow rate cannot exceed 100 gpm (378 L/min) [1,000 gallons (3,785 L) ÷ 10 minutes]. Reducing the time or adding more tenders or larger tanks would increase the flow rate. With five 2,000-gallon (7,570-L) tenders operating on 10-minute cycles, the flow rate would be 1,000 gpm (3,785 L/min) [10,000 gallons (37,850 L) ÷ 10 minutes]. Tender operations cannot be sped up by increasing vehicle speed on the highway. Time should only be gained at dump and fill sites by increased efficiency of personnel and equipment.

Safety It is important to note that attempts to lower shuttle time cannot and should not be done by increasing vehicle speed. Time should only be gained at dump and fill sites by increased efficiency of personnel and equipment.

PRESSURE ASSOCIATED WITH WATER DISTRIBUTION SYSTEMS

All of the Earth's water is under **pressure**. (See Chapter 11 for more discussion on pressure.) Even seawater is under **atmospheric pressure**, which is 14.7 pounds per square inch (psi) (101 kPa) at sea level, **Figure 9-20**. Drafting water from a lake or the sea is accomplished by taking advantage of this atmospheric pressure. Creating a partial vacuum or low atmospheric pressure area inside a pump causes the atmospheric pressure on the water's surface to force the water up the suction hose and into the pump, which adds pressure and pumps it out. Pressure is the force, or weight, of water measured over an area, **Figure 9-21**. A 1-foot (0.305-meter) column of water 1 inch square weighs 0.433 pounds (0.19 kg), creating a pressure at the bottom of 0.433 psi (3 kPa). Pressure can also be expressed as feet of head with the 1-foot column having a head of 1 foot (0.305 meters) and 2.31 feet (0.7 meter) of head equaling 1 psi (6.895 kPa). An atmospheric pressure of 14.7 psi (101 kPa) would create a head of almost 34 feet (10.4 meters) [14.7 psi × 2.31 ft/lbs = 33.9 feet] if a perfect vacuum was created or the ability to lift a column of water 33 feet (10.1

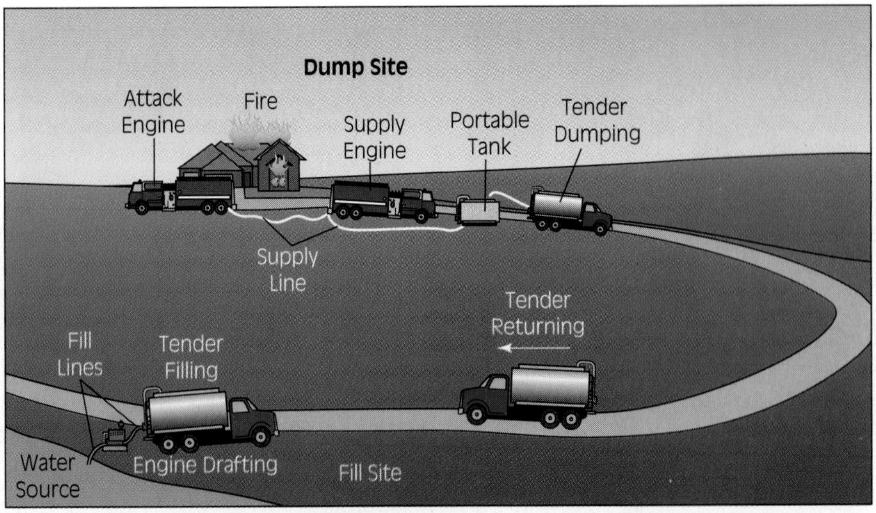

Figure 9-19 Tender shuttle operation.

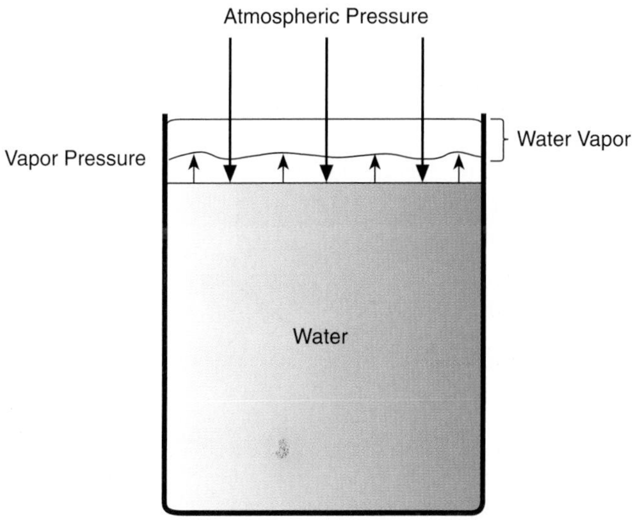

Figure 9-20 Atmospheric pressure being exerted on container of water.

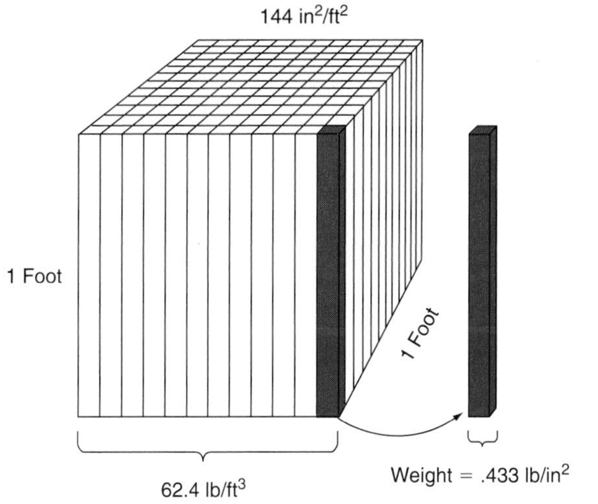

Figure 9-21 Pressure is expressed in pounds per unit area (psi).

meters). Practicality limits this ability to about 30 feet (9 meters). A tank in a distribution system that stores water 100 feet (30 meters) high would create 100 feet (30 meters) of head or 43 psi (296 kPa) [100 feet ÷ 2.31 ft/lbs = 43.3 psi or 100 feet × 0.433 lbs/ft = 43.3 psi]. Pressure in a nonflowing closed system is equal at all points, but while flowing, it is reduced by friction created by its movement and the loss of pressure at the opening. Pressure in an open system is subject to atmospheric pressure.[4]

Water distribution systems are supplied under pressure to meet domestic needs and fire protection. This dual use means the pressure is never as high as firefighters would prefer because it could cause leaks in private plumbing. The average pressures in the United States are between 65 and 80 psi (448 to 551 kPa) with a high of 150 psi (1,034 kPa) before damaging plumbing and a usual minimum of 20 psi (137 kPa).[5] The 20-psi (137-kPa) figure is the recommended low **residual pressure** when pumping from a hydrant; pressures below this may create a vacuum in part of the supply system that could cause damage to the system. Fire departments should know the normal and flowing pressures and capacity of the water distribution system including areas that are high or low in either category. High-flow and/or high-pressure areas are an advantage, whereas low-pressure and/or low-flow areas should be avoided if possible.

TESTING OPERABILITY AND FLOW OF HYDRANTS

Testing should be conducted on fire hydrants periodically to ensure that they are operable and to determine the flow rate of the hydrants, **Figure 9-22**. Testing prevents the fire department from finding out that the hydrants do not work or are insufficient while operating at a fire. Testing of the hydrant may be the responsibility of the water department, but the fire

Figure 9-22 Firefighters inspecting and servicing a hydrant.

Figure 9-23 Typical pump panel gauges are Bourdon gauges.

department should at least have access to the results of the testing. Some fire departments do the testing and maintenance of the hydrants, whereas others have joint programs with the water department. All testing should be coordinated between the two agencies.

The first set of tests involves the operability of the hydrants. On wet and dry barrel hydrants, the test starts with a visual inspection for damage. Then after ensuring the hydrant is off, inspectors remove all of the caps and check the threads and gaskets for damage. The next task would be to check that all valves on the hydrant allow water to flow. A dry barrel hydrant should be checked to ensure the drain valve is working properly by determining that the water level is dropping inside the barrel after the hydrant has been shut off. For a dry hydrant, a visual inspection of the piping, caps, and gaskets should be conducted. A flow test would be the next step to ensure water is available, followed by backflushing of the hydrant to clear any debris from the strainer. Backflushing involves pumping into the outlet of the hydrant so that water flows out of its intake. *This should not be done on other types of hydrants, because it will contaminate the drinking water supply.* Replacing any caps, oiling any moving parts, and painting may complete this inspection process.

The next tests are flow tests that should be conducted during the normal operations of the supply system. There are two ways to do a flow test in a pressurized system; one is a fireground method and the other is more extensive and is done for fire department planning purposes and insurance ratings. The fireground method uses the gauges on the pumper using the hydrant while the other requires a 2½ inch (63-mm) cap with a **Bourdon gauge**, **Figure 9-23**, hydrant wrenches, **Pitot gauges**, **Figure 9-24**, a ruler, and paper to record the data.[6] The tests are conducted in the next section.

Figure 9-24 Hydrant testing with a Pitot gauge.

DETERMINING STATIC, RESIDUAL, AND FLOW PRESSURES

The fireground method of flow testing involves connecting a pumper to a hydrant and turning it on. Prior to charging any lines, the **static pressure** is read on the main intake compound gauge. Static pressure is the pressure in the system with no hydrants or water flowing. The pump operator then charges the first line with the desired volume, noting

Percentage Drop Measurements		
PERCENTAGE DROP 1	**PERCENTAGE DROP 2**	**AMOUNT OF ADDITIONAL WATER**
0–10% drop	0–5% drop	Three times the amount
11–15% drop	6–10% drop	Two times the amount
16–25% drop	11–20% drop	One times the amount

TABLE 9-1

the pressure first. With this flow going, the operator again reads the intake gauge and gets the residual pressure or the remaining pressure left in the system after the flow and friction loss from the flow. The pump operator then compares the percentage of pressure drop from static to residual and determines the amount of additional volumes that may be pumped from that hydrant. Two comparisons are used for percent drop as shown in **Table 9-1**.

Consider this example using the first column of drop: a static pressure of 50 psi (345 kPa) with a flow of 500 gpm (1,893 L/min) at a residual of 45 psi (310 kPa). Percent drop is $50 - 45 = 5 \div 50 =$ 10 percent. With 10 percent drop, the first column would allow an additional three times or 1,500 gpm (56,775 L/min) to come from the hydrant. The second column is more conservative and would allow only an additional 1,000 gpm (3,785 L/min) to be taken. Department SOPs should define which chart is to be used. In any case, the residual pressure should not drop below 20 psi (137 kPa).

The second test involves testing multiple hydrants and is not conducted during fire operations. The first hydrant selected is used with the cap gauge to take a static pressure with no hydrants flowing, and then use continues while measuring the system's residual pressure. The second hydrant is opened and the flow measured, preferably at a 2½-inch (63-mm) outlet, with a Pitot gauge while the residual reading at the first hydrant is taken. The Pitot gauge is inserted into the stream's midpoint and out about half the distance of the diameter, **Figure 9-25** and **Figure 9-26**. This process would continue if additional hydrants are to be tested, gathering each flowing hydrant pressure on the Pitot gauge and at the same time gathering the residual on the first hydrant, **Figure 9-27**. The next step involves the calculations of discharge for the second hydrant and beyond. Using a chart is easiest or the following formula can be used:

$$Q = 29.83\, c\, d^2 \sqrt{p}$$

where Q is quantity in gallons, c is the coefficient of the outlet, d^2 is the diameter of the outlet in inches squared, and $\sqrt{p}$ is the square root of the pressure. Each type of outlet or nozzle has a coefficient with most hydrants having a number of 0.90, 0.80, or 0.70.[7] The way to determine the correct coefficient for hydrants is simply to feel the inside edge of the discharge orifice. A smooth or rounded edge uses the .9 coefficient. A square edge requires a .8, and a rough lip or edge uses .7. For example, for a hydrant with a 2½-inch outlet with a coefficient of 0.90 and a Pitot reading of 49 psi, we would get:

$$\begin{aligned} Q &= 29.83\, c\, d^2 \sqrt{p} \\ &= 29.83 \times (0.9) \times (2.5^2) \times (\sqrt{49}) \\ &= 29.83 \times (0.9) \times (6.25) \times (7) \\ &= 1{,}175 \text{ gpm} \end{aligned}$$

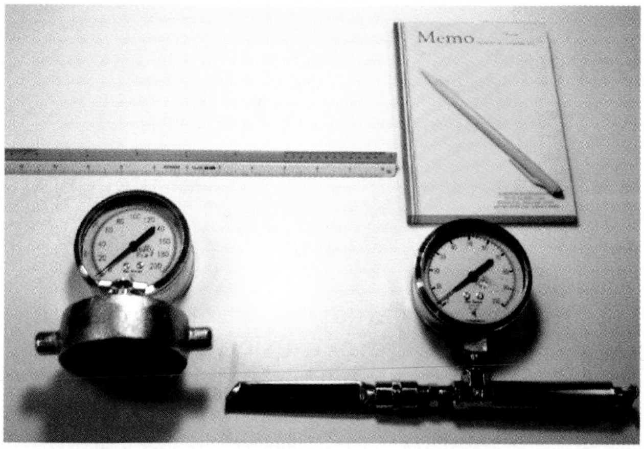

Figure 9-25 Pitot gauge and cap gauge with Bourdon gauge.

OBSTRUCTIONS AND DAMAGE TO FIRE HYDRANTS AND MAINS

Obstructions and damage can occur to fire hydrants and water mains as they can to any other type of system. Nature, vandals, accidents, and

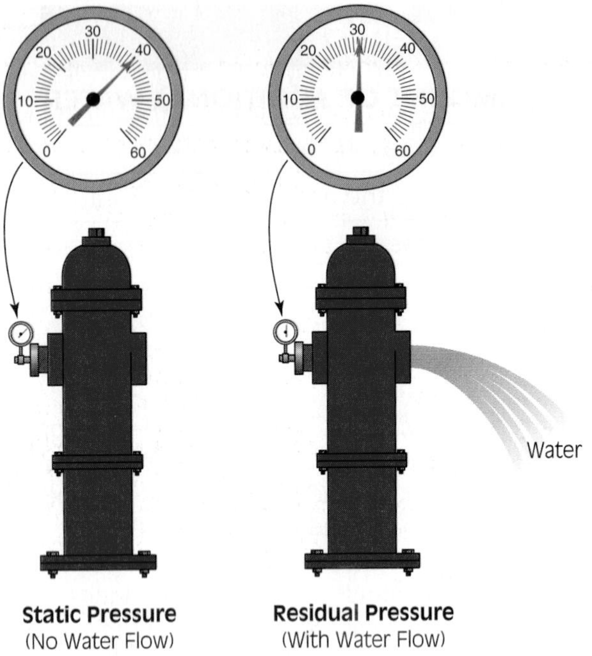

Figure 9-26 Static and residual pressures.

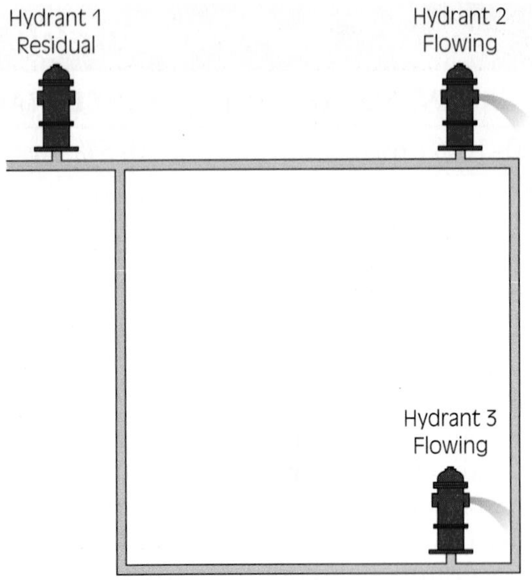

Figure 9-27 Flow testing diagram showing flowing of two hydrants with another used for residual pressures.

improper actions by members of the fire department can cause problems. The natural causes can be sudden events or an event occurring over a long term. Sudden events such as an earthquake or flood can undermine or break mains or hydrants. Other natural events are long term such as tree roots working into the piping to block it or crack the piping. Encrustation and blockage of the piping by minerals or organisms that survive the treatment process can slowly build up and restrict or completely block the flow. Vandals can do damage by opening or closing valves that can shut off the flow of water or undermine a hydrant. Opening a hydrant in freezing weather without allowing water to flow can cause the hydrant to freeze and crack. Vandals can also put debris into the system damaging the hydrant or the pump; the firefighter should check the outlet prior to connecting the hose and flush it if necessary. Accidents can crack pipes or break off a hydrant. For example, a bank building began to lose its front wall after a broken hydrant and main undermined the street, sidewalk, and then the foundation of the bank. All this damage could have been prevented if the engine company had closed off the gate valve on the water main at the hydrant instead of calling for the water company. Damage from improper actions by firefighters includes opening and closing valves or hydrants too quickly, thus creating a **water hammer**, failing to open or close a hydrant or other valve fully, or cross-threading the threads on the connections.

Lessons Learned

Water is the most common fire extinguishing agent used by firefighters and it must be supplied in sufficient quantity to accomplish extinguishment. Firefighters must understand the relationship between the amount of water that can be supplied and the amount needed. If they do not have enough water, their efforts will not be effective.

Supplying the water needed requires an understanding of where water comes from, how it gets to the firefighter, or where they need to go to get it. This leads to learning about water sources and distribution systems. Distribution systems have valves and hydrants that must be operated. The more experienced firefighters must know about pressure in the system and determining flow. They must also know about problems with the system.

Firefighters in areas without a distribution system must create one using mobile supply apparatus, portable water tanks, and shuttle operations. They quickly learn the value of water because if they run out, they must wait until the tender gets back before they can continue to fight the fire.

KEY TERMS

Aquifer A formation of permeable rock, gravel, or sand holding water or allowing water to flow through it.

Atmospheric Pressure The pressure exerted by the atmosphere, which for Earth is 14.7 pounds per square inch at sea level.

Backflow Preventers A check valve or set of valves used to prevent a backflow of water from one system into another. Required where a building water or fire protection system connects with the public water system. Backflow preventers are being required for environmental and health reasons.

Bourdon Gauge The type of gauge found on most fire apparatus that operates by pressure in a curved tube moving an indicating needle.

Check Valves Valves installed to control water flow in one direction, typically when different systems are interconnected.

Cistern An underground water tank made from natural rock or concrete. Cisterns store large quantities of water—30,000 gallons or more—in areas without other water supplies or as a backup supply.

Drafting The pumping of water from a static source by taking advantage of atmospheric pressure to force water from the source into the pump.

Dry Hydrant A piping system for drafting from a static water source with a fire department connection at one end and a strainer at the water end.

Dump Site The area where tenders are unloaded or their load dumped.

Fill Site The area where tenders are filled or get their water.

Fire Flow Capacity The amount of water available or amount that the water distribution system is capable of flowing.

Fire Flow Requirement A measure comparing the amount of heat the fire is capable of generating versus the amount of water required for cooling the fuels below their ignition temperature.

Gate Valves Indicating and nonindicating valves that are opened and closed to control water flow.

Jet Dump A device that speeds the process of dumping a load of water from a tanker/tender.

Jet Siphon A device that speeds the process of transferring water from one tank to another.

Pitot Gauge A device with an opening in its blade-shaped section that allows water to flow to a Bourdon gauge and registers the flowing discharge pressure of an orifice.

Portable Water Tanks Collapsible or inflatable temporary tanks for the storage of water that is dumped from tankers or tenders. Usually carried by the tender to set up a dump site.

Pressure The force, or weight, of a substance, usually water, measured over an area.

Residual Pressure The pressure in a system after water has begun flowing.

Shuttle Operation The cycle in which mobile water supply apparatus is dumped, moves to a fill site for refilling, and is returned to the dump site.

Static Pressure The pressure in the system with no hydrants or water flowing.

Tanker The term given to aircraft capable of carrying and dropping water or fire retardant. Some departments still use the term to describe land-based water apparatus.

Tender The abbreviated term for *water tender.* A water tender is defined as a land-based mobile water supply apparatus. Some departments still use the term *tender* to describe a hose-carrying support apparatus.

Tidal Changes The rising and falling of the surface water levels due to the gravitational effects between the Earth and the moon. In some areas, these changes are insignificant, but in others there is more than 40 feet of difference between high and low tide.

Water Hammer A sudden surge of pressure created by the quick opening or closing of valves in a water system. The surge is capable of damaging piping and valves.

Water Table The level of groundwater under the surface.

Water Tender The term given to land-based water supply apparatus.

REVIEW QUESTIONS

1. How does the fire flow capacity relate to the fire flow requirements?
2. What are the four reasons for using water for firefighting?
3. Where is most freshwater found?
4. Name three surface water sources.
5. What are the two names for mobile water supply apparatus?
6. Water is distributed or supplied through a supply system in what three ways?
7. What is a dead-end main and why should it be avoided?
8. Explain the two main types of fire hydrants and where they are found.
9. Why is a dry hydrant different from other hydrants?
10. What purpose does a backflow preventer serve?
11. What does a gate valve do?
12. Describe a portable water tank.
13. Explain a tender shuttle.
14. What is pressure?
15. What is the difference between static and residual pressure?
16. If a hydrant's water pressure drops when flowing a line by 9 percent, how many more lines can be taken from that hydrant?

Endnotes

1. See Sturtevant, Thomas, *Introduction to Fire Pump Operations,* Chap. 11. Delmar Learning, a part of the Thomson Corporation, Clifton Park, NY, 1997; *NFPA 1231: Standard on Suburban and Rural Water Supply;* or the NFPA's *Fire Protection Handbook,* 18th ed., Chap. 6-5 for additional information.
2. See many of the fire service publications about this safety issue. To examine closely the accident, injury, and fatality statistics, see the *NFPA Fire Journal* issues containing annual firefighter injuries and fatalities surveys.
3. Tank size is the total capacity of the tank, but often the tender cannot actually offload its capacity or the time to do so would not be productive. Departments operating tenders should conduct tests in which the tender is loaded and offloaded as if under fireground conditions and weigh the vehicle and divide the offloaded amount by 8.33 lb/gallon (or 1 kg/L) to determine its operating capacity.
4. See Sturtevant, Thomas, *Introduction to Fire Pump Operations,* Chap. 11. Delmar Learning, a part of the Thomson Corporation, Clifton Park, NY, 1997, for more information.
5. *Fire Protection Handbook,* 18th ed., pp. 6–72.
6. *NFPA 291: Recommended Practices for Fire Flow Testing and Marking of Fire Hydrants, 2002 Edition.*
7. *NFPA 291: Recommended Practices for Fire Flow Testing and Marking of Fire Hydrants, 2002 Edition.*

Additional Resources

Crapo, William F., *Hydraulics for Firefighting.* Delmar Learning, a part of the Thomson Corporation, Clifton Park, NY, 2002.

From Delmar Learning, a part of the Thomson Corporation:

Gagnon, Robert, *Design of Special Hazard and Fire Alarm Systems.* Delmar Learning, a part of the Thomson Corporation, Clifton Park, NY, 1998.

Klinoff, Robert, *Introduction to Fire Protection,* 2nd ed. Delmar Learning, a part of the Thomson Corporation, Clifton Park, NY, 1997.

Sturtevant, Thomas, *Introduction to Fire Pump Operations.* Delmar Learning, a part of the Thomson Corporation, Clifton Park, NY, 1997.

Other Publishers:

Cote, Arthur, and John Linville, *Fire Protection Handbook,* 18th ed. National Fire Protection Association, Quincy, MA, 1997.

NFPA 22: Standard for Water Tanks for Private Fire Protection, 1998 Edition. National Fire Protection Association, Quincy, MA.

NFPA 24: Standard for the Installation of Private Fire Service Mains and Their Appurtenances, 2002 Edition. National Fire Protection Association, Quincy, MA.

NFPA 1410: Standard on Training for Initial Emergency Scene Operations, 2000 Edition. National Fire Protection Association, Quincy, MA.

NFPA 291: Recommended Practices for Fire Flow Testing and Marking of Fire Hydrants, 2002 Edition. National Fire Protection Association, Quincy, MA.

NFPA 1142: Standard on Water Supplies for Suburban and Rural Firefighting, 2001 Edition. National Fire Protection Association, Quincy, MA.

NFPA 1901: Standard on Automotive Fire Apparatus, 1999 Edition. National Fire Protection Association, Quincy, MA.

CHAPTER 10

FIRE HOSE AND APPLIANCES

Ric Koonce, J. Sargeant Reynolds Community College

OUTLINE

- Objectives
- Introduction
- Construction of Fire Hose
- Care and Maintenance of Fire Hose
- Types of Hose Coupling
- Care and Maintenance of Couplings
- Hose Tools and Appliances
- Coupling and Uncoupling Hose
- Hose Rolls
- Hose Carries
- Hose Loads
- Advancing Hoselines—Charged/Uncharged
- Establishing a Water Supply Connection
- Extending Hoselines
- Replacing Sections of Burst Hose
- Hose Lay Procedures
- Deploying Master Stream Devices
- Service Testing of Fire Hose
- Lessons Learned
- Key Terms
- Review Questions
- Endnotes
- Additional Resources

STREET STORY

My station covers an area that has lots of old abandoned homes with wood construction. One evening a call came in as a house fire, and as we approached from several blocks away we could see the smoke. When we arrived, the house was fully involved. We were told that a homeless person had taken residency in this shack—it had no power. As we attempted to go in, we found out that he was a major pack rat. There were several appliances, including washing machines and refrigerators, and parts throughout the front of the house, and that made it nearly impossible to gain access. We had to squeeze through the door sideways and walk over things. Inside there were more piles: candles, cans, papers, etc.

It was an incredible obstacle just for the the rescue crew to get to the victim. (Eventually they found him behind a locked door. He wasn't breathing but he did have a pulse. He also had a gun in one hand and a knife in the other!) In terms of advancing the hose, we decided to do it as a team. We needed an action plan. Normally it's a three-person job, but in this case it took two entire crews. Instead of one person at the front, we had somebody every 4 or 5 feet, because otherwise as soon as you pulled you'd get stuck on something.

That day was really long and by the end of the incident more than half of the house was destroyed. But we really learned to adapt to a difficult situation. Most important is remembering and applying the basic skills you know to that specific incident. I remember learning hoseline rolling in training, but it's not something you use much in application. In this case, though, it was the best way to get the job accomplished. I also learned the importance of teamwork. You may come in on a fire assigned to one job, but when there's an issue like this, you all have to work together on one part.

—Street Story by Michelle Steele, Lieutenant, Miami-Dade Fire and Rescue, Miami, Florida

OBJECTIVES

After completing this chapter, the reader should be able to:

- Identify and explain the construction of fire hose.
- Demonstrate the care and maintenance of fire hose.
- Identify the types of hose couplings and threads.
- Demonstrate the care and maintenance of hose couplings.
- Identify and explain the use of hose tools and appliances.
- Demonstrate the coupling and uncoupling of fire hose.
- Demonstrate the rolling, carrying, and loading of fire hose.
- Demonstrate the advancing of fire hoselines, both charged and uncharged.
- Demonstrate the establishment of a water supply connection.
- Demonstrate the extending of hoselines.
- Demonstrate the replacement of burst hose sections.
- Demonstrate the procedures for laying hoselines for water supply.
- Demonstrate the deployment of master stream devices.
- Demonstrate the service testing of hose.

INTRODUCTION

This chapter introduces the firefighter to one of the most important tools for fighting fires: hose. Hose is the tool used to move water from one place to the fire. The firefighter should know about the hose, how it is made, and how to care for it. More importantly, the firefighter must know how to properly store it on the apparatus and how to quickly move it from that apparatus to the location where the firefighting will take place. The techniques learned in this chapter and the next one on fire streams and nozzles are the basic techniques of fire control. These chapters, along with many others in this book, are the foundation of firefighting in a safe, efficient, and effective manner. Firefighters should learn them well and practice them often. Many experienced fire officers often comment when things do not go well on the fireground of the need to "get back to the basics." These are those basics.

Fire hose is a flexible conduit used to convey water or other agent from a source to the fire. Early firefighters used bucket brigades to supply the water to the engines. Top-mounted nozzles sprayed the water on the fire. Leather hose was invented to first allow firefighters to advance nozzles from the engine to the fire and then to supply the engine. Today, many different materials have replaced leather hose but the basic tasks performed for supply and attack remain the same. Firefighters, fire brigade members, wildland firefighters, and some building occupants use hose today.

Couplings, adapters, and appliances are used in modern firefighting to connect hose and adapt it to different sizes or ways of operating. Hose couplings were a problem at several major fires at the beginning of the twentieth century. The different fire departments fighting these fires discovered that their hose couplings did not match and the hoses could not be connected. Today, departments that do not use National Standard Hose Threads use adapters to make connections between their threads and the standard ones. Adapters and appliances have been created or made lighter weight to make the job of firefighting easier or better. This chapter covers many of these, but others may be used locally to solve special problems.

Firefighter Fact Large conflagrations in Boston in 1872 and Baltimore in 1904 resulted in many large cities sending fire units to help fight the fire only to discover that their hose couplings could not be connected.

CONSTRUCTION OF FIRE HOSE

Fire hose has two components, the hose itself and the couplings that connect the hose sections or appliances used with them. Fire hose is made in three types of construction: wrapped, braided, and woven, **Figure 10-1**. In wrapped construction, a fabric or mesh material is then impregnated with rubber or a plastic and then wrapped around a rubber **liner** and covered with rubber or plastic. Braided construction uses a yarn braided over the rubber liner, and a rubber or plastic layer separates it from the next layer. A rubber or plastic cover encases the entire hose. Woven hose has a rubber liner and one or more outer layers called a **jacket** that is woven of cotton, synthetic, or blended materials. Some hoses have jackets of rubber or plastic-impregnated woven construction. The jacket is designed to protect the

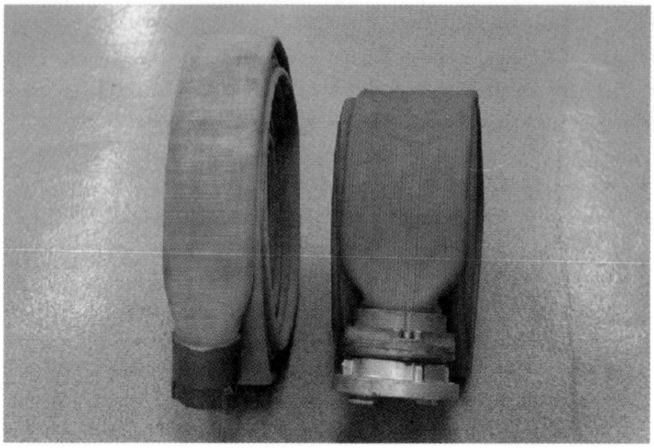

Figure 10-1 Woven and rubber-coated fire hose.

Figure 10-2 Booster hose on reel.

hoselining from heat and mechanical damage. In addition, **hard suction hose** has a plastic or wire **helix** to provide addition support and prevent collapse under a vacuum when drafting. Some newer synthetic types of hose combine the jacket and liner. Fire hose is divided into attack, supply, occupant use, forestry, and hard suction hose. Each type of hose has certain requirements for construction, size, and pressure. Most hose comes in 50- or 100-foot (15- or 30-m) lengths except as noted.

> **Firefighter Fact** Cotton and other natural fibers replaced leather hose. Rubber liners were added in the late 1800s. Prior to the rubber liner, firefighters had to wait for the fibers to absorb enough water to cause them to swell and slow the leakage of water. Some fire departments and building occupant standpipe hoses used unlined single-jacket hose. Current standards now recommend lined hose.

Booster hose is smaller diameter, rubber-coated hose of ¾- or 1-inch (19- to 25-mm) size usually mounted on reel that can be used for outside fires or overhaul operations after the fire is out, **Figure 10-2**. It is not to be used for structural firefighting. Booster lines have a limited flow rate of up to 30 gpm (114 L/min) and, although easy to maneuver and control, they are unsafe for structural firefighting. Many departments have stopped purchasing this type of hose.

Attack hose is used by trained personnel to fight fires. This hose is connected to nozzles and distributors, master stream appliances, standpipes and sprinkler systems, hydrants, and fire pumps. Attack hose is a minimum of 1½ inches (38 mm) and can include large-diameter hose when supplying master stream devices. NFPA requires pumpers to carry 400 feet of attack hose. Attack hose is usually service tested annually at 250 psi (1720 kPa), and its use above this pressure may result in hose damage or personnel injury. Attack hose can be either single- or double-jacketed hose with the most common being double-jacketed for additional protection of the hose, **Figure 10-3**. Standpipe packs often use single-jacket hose to reduce the weight of the pack. **Medium-diameter hose (MDH)** is either 2½- or 3-inch (63- or 75-mm) hose, and sizes smaller than that are called **small lines** or **small-diameter hose**.

Supply hose or **large-diameter hose (LDH)** is larger hose [3½ inches (90 mm) or bigger] used to

Figure 10-3 Various hoselines in a hose bed.

move water from one engine to another or from the water source to a portable hydrant for distribution. Common sizes are 4 and 5 inches (100 to 125 mm) with some 6 inch (150 mm). Supply hose uses its larger diameter to move water at lower pressure. NFPA requires pumpers to carry at least 800 feet of 2½-inch or larger supply line.

Hard suction hose is rubber or plastic-coated hose with helix bands that resist collapse from the vacuum created by drafting. It comes in sizes from 2½ to 6 inches (65 to 150 mm) in diameter and is usually matched to the size of the main intake of the pump for which it will be used. Because of its more rigid construction, it is less flexible and movable than other hose. Hard suction hose is tested on its ability to withstand a vacuum. Hard suction hose is standard in 10-foot (3-m) lengths.

Soft suction hose is large-diameter woven hose used to connect a pumper to a hydrant and is also known as a soft sleeve, **Figure 10-4**. It is also sized according to the size of the intake of the pumper and usually comes in sizes from 4½ to 6 inches (112 to 150 mm). The lengths of a section vary: 20, 25, and 30 feet (6, 8, and 9 m). NFPA requires pumpers to carry at least 15 feet of soft sleeve intake hose.

Occupant use hose is hose used in standpipe systems for building occupants to fight incipient fires. It is usually 1½-inch (38-mm)-diameter single-jacket hose similar to attack hose. Older types of this hose may be unlined. Most fire departments do not use the occupant use hose, and if they use the standpipe system, they replace it with their own hoselines.

Forestry hose is specially designed for use in forestry and wildland firefighting, **Figure 10-5**. It comes in 1- and 1½-inch (25- and 38-mm) sizes and should meet U.S. Forestry Service specifications. Forestry hose is woven hose and comes in 50-, 75-, and 100-foot (15-, 23-, and 30-m) lengths.

Figure 10-4 A soft suction hose.

CARE AND MAINTENANCE OF FIRE HOSE

The care and maintenance of fire hose includes some basic measures. The care of hose begins with the careful placement and folding of the dry hose on the apparatus to prevent sharp turns or bends that can damage the liner (see Hose Loads section later). Previously used hose should be folded at different places to prevent overstressing one area of the hose. Hose standards still recommend that the hose be changed every thirty days with a visual inspection at the time of change but many departments no longer do so. The **hose bed** should be designed to allow the circulation of air to allow drying. This prevents mildew of natural fiber hose and vehicle rusting. The use of a hose bed cover will prevent rain and other road contaminants from getting into hose. The cover also blocks sunlight, which can affect rubber and plastic-coated hose. Wet hose should be thoroughly dried prior to returning it to apparatus but the newer synthetic hoses are often loaded wet.

When used on the fireground or for training, several steps can be taken to reduce damage to hose. Avoid laying hose over sharp or rough corners; use a hose roller or a salvage cover to prevent damage. Remove any kinks in hose prior to charging or soon afterward. Do not allow traffic to run over the hose, which could damage it, especially the couplings, which could become misshapen. Vehicles cause more damage to uncharged hose than charged hose because the inner liner is not protected when uncharged. Clean dirt and grit from hose with a brush or plain water to prevent these materials from creating abrasions on the hose. Firefighting exposes hose to heat and burning embers plus broken glass, nails, and other sharp objects that can quickly damage it. Avoid heat, burning embers, and chemicals, especially gasoline and oil. Mild chemicals can be removed using detergent, but large amounts or strong chemicals should be removed using hazardous materials decontamination procedures. Care should be taken to prevent hose from freezing. If it does freeze, it should be carefully moved to a place to thaw prior to any folding or bending. Any hose that has been subjected to any damage should be service tested prior to returning to service.

Note The most common damage to hose is wear and tear due to dragging hose—this is especially damaging to couplings. Excessive pressure and water hammers should be avoided, as this can cause sudden rupture of the hose or separation of the hose from the coupling.

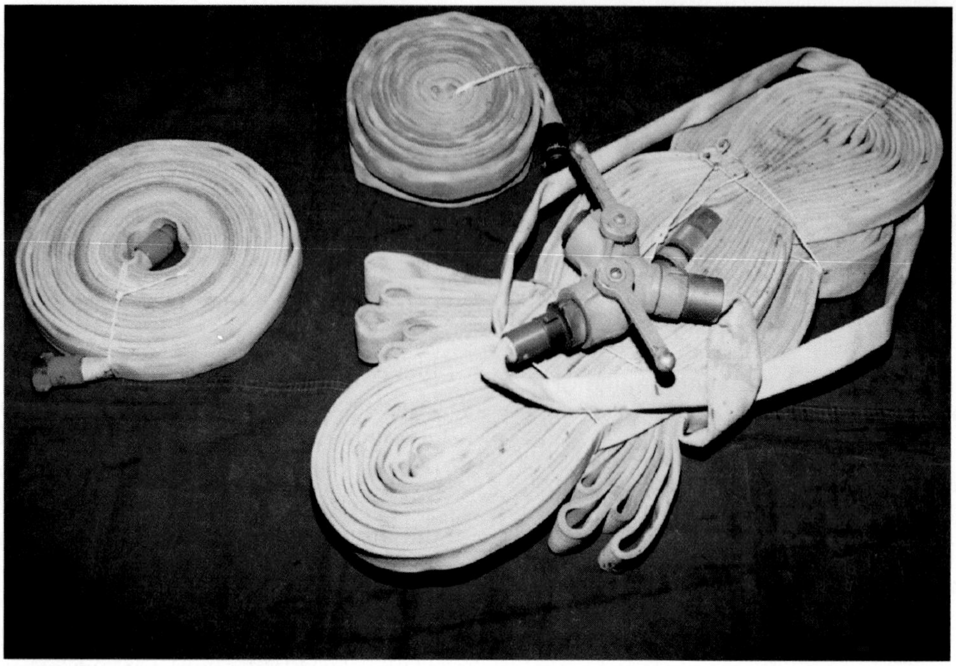

Figure 10-5 Hose packs and rolls for wildland firefighting.

Using a long, small-diameter hose and overpumping a hoseline at the incorrect pressure are common causes of excessive pressure. A water hammer is the pressure surge that occurs when water valves are rapidly opened or closed. Water that is not flowing has its pressure equalized throughout the whole system. Slamming a valve shut or opening it quickly allows this pressure to rapidly act on the parts of the system and damage hose, valves, pumps, and the water supply system including water mains, hydrants, and valves. Water hammers result from mistakes made by people operating valves and nozzles too quickly. The correct action is to operate all valves slowly and let the pressure either bleed off or build up slowly without the surge.

Dirty or contaminated hose can be rolled up and returned to the station for cleaning, or a portable washer can be used on scene or in a nearby parking lot. Cleaning may be done by placing the hose on a flat surface, rinsing with water, and scrubbing with mild detergent. A thorough rinsing and hanging in a hose tower or storage rack until dried will get the hose ready for reuse. Commercial hose washers are also available, but are not capable of adequately cleaning the threads of a coupling. Some work by connecting the hose washer to a hydrant and as the hose is thread through it is rinsed, **Figure 10-6**. This type still requires hand scrubbing by personnel. Another type has the ability to scrub, wash, rinse, and reroll the hose, **Figure 10-7**. Hose drying cabinets are also available that have heaters to dry the hose. Care must be taken not to overheat and damage the hose. Stored hose also requires maintenance. Occasionally water should be flowed through stored sections to prevent the liner from drying and cracking, and then it should be redried and stored.

The final part of the care of hose involves a regular visual inspection by firefighters after each use. Firefighters should carefully inspect the outer cover and couplings for any damage. Coupling care is discussed later but the outer cover should be checked as hose is reloaded for discoloration and abrasions. Further care is conducted during the annual service test.

Figure 10-6 Hose washer on hydrant.

Figure 10-7 An automatic hose washer.

TYPES OF HOSE COUPLING

Couplings allow hose and appliances to be joined or connected using a set of connection devices. Couplings are divided into two types: threaded and nonthreaded. Threaded couplings use a screw thread that secures the two sections of hose together, while nonthreaded couplings use locks or cams. Couplings are made of brass, aluminum, or an alloy called pyrolite, which is lighter than brass but more resistant to bending.

Threaded couplings are further separated into two different types: one with external threads, the male, and the other with internal threads, the female, with a matching thread type. A threaded hose coupling set is typically of three-piece construction with a hose bowl on each piece and a swivel attached to the internal coupling. The bowls are attached to the hose using an expansion ring with a tail gasket to prevent leaks. The internal swivel also has a swivel or thread gasket to stop leakage where the threads connect, **Figure 10-8**.

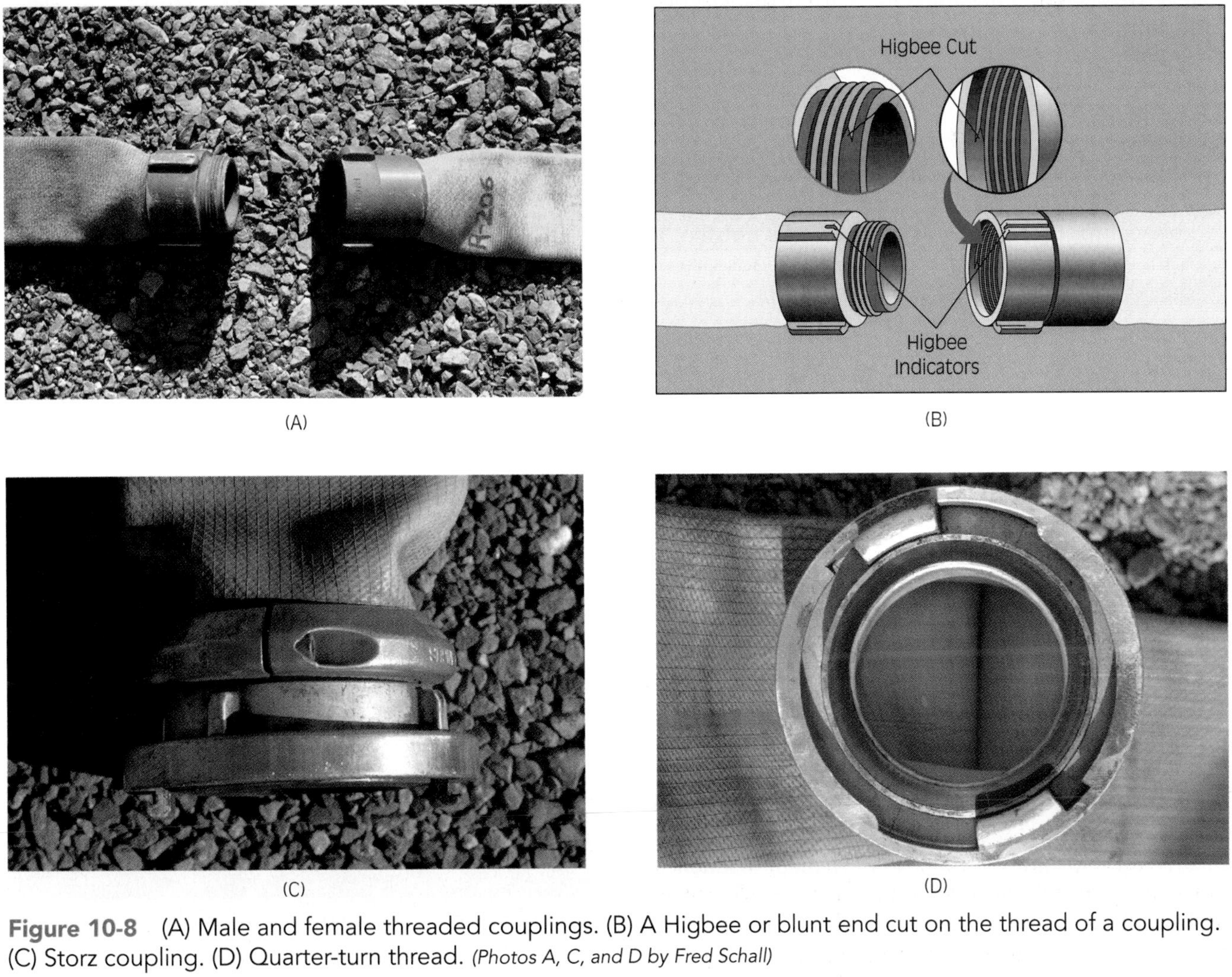

(A) (B) (C) (D)

Figure 10-8 (A) Male and female threaded couplings. (B) A Higbee or blunt end cut on the thread of a coupling. (C) Storz coupling. (D) Quarter-turn thread. *(Photos A, C, and D by Fred Schall)*

Threaded couplings are of common design but with many nonstandard types of threads. Today, although a National Standard Thread (NST) (also called National Hose Thread, NHT) exists, there are other "standard" hose threads. Some include National Pipe Thread, Eastern Hose Thread, Pacific Coast Thread, Quebec Standard Thread, and Canadian Standards Association. New York and Chicago each have both a fire department thread and a city hose thread. Personnel checking new hose and connection equipment should ensure the correct threads are used on all hose, appliances, connections, and so on. Finding out the wrong threads are present while trying to make a connection during a fire is not the right time. Hose threads allow the couplings to screw or fasten together. Regular pipe threads normally screw grooves that go to the end of the fitting. National Standard and other fire hose type threads have blunt end threads. The blunt end threads can be started at only one point, thus reducing the possibility of cross-threading the couplings. These blunt ends are called a **Higbee cut**. A groove is notched into coupling lugs to help firefighters align the Higbee cuts. This notch is called the Higbee indicator.

> **Caution** Personnel checking new hose connection equipment should ensure that the correct threads are used on all hose, appliances, connections, and so on. During a fire is not the right time to find out the wrong threads are present while trying to make a connection.

Nonthreaded couplings use locks or cams to secure the connection. Both couplings are identical and either end can be used to connect to another coupling. The couplings are aligned and twisted one-quarter turn, locking them together. **Storz couplings** are the most popular of these types but there are other kinds. Some nonthreaded couplings use expansion rings and others are attached by wire or straps on the exterior of the hose that bind the hose to the couplings.

To assist firefighters in making, tightening, or breaking the connection of a coupling set, lugs or handles are placed on the couplings. The most common type of lug is the rocker lug, but pins and recessed pins are also used. Handles are used mostly on hard and soft sleeves. Spanner wrenches are used with the lugs to assist in tightening and loosening the couplings.

CARE AND MAINTENANCE OF COUPLINGS

Care and maintenance of hose couplings requires keeping them clean and preventing mechanical damage. Proper storage, rolling, and simply paying attention to where the coupling is placed can solve most problems. Mechanical damage prevention starts with caution being used to prevent dropping the couplings, which can damage the threads or the coupling itself. Couplings should not be dragged. Whenever possible, personnel should carry hose near the couplings to avoid damaging them. They should avoid the placement of hose couplings in vehicle traffic areas where the couplings may be struck or rolled over. Vehicles can easily crush couplings or misshape them, preventing the free operation of the couplings or breaking them entirely. Good storage and rolling methods can prevent damage to exposed threads, **Figure 10-9**.

The care of couplings should include a visual inspection each time the hose is reloaded. The cou-

Figure 10-9 Hose storage rack.

plings should be checked for any damage to the coupling and threads, the proper operation of the swivel, and the presence and condition of the gasket. The gasket should be pulled out and checked for cracks and pliability. One last check would be for any movement or slippage of the hose from the coupling. If dirt or grit is detected in the coupling, rinse it and take a brush to the coupling and threads to clean it. Remember, dirt and grit may prevent tightening or loosening of a connection.

HOSE TOOLS AND APPLIANCES

Hose tools are accessories that help firefighters move or operate hoselines. Appliances are devices that water flows through, including adapters and connectors. There are many types and varieties of such items; this chapter covers the more common ones. Many of these devices must be the correct size to work with the size(s) of hose being used.

Tools include rope hose tools, wrenches, rollers, clamps, and other items, **Figure 10-10**. A **rope hose tool** is about 6 feet (2 m) of ½-inch (13-mm) rope spliced into a loop with a large metal hook at one end and a 2-inch (50-mm) ring at the other. It is used to tie in hose and ladders, carrying hose, and many other tasks requiring a short piece of rope. Some departments have an engine or spanner belt to accomplish the same tasks. A **hose strap** is a variation of the rope hose tool, although shorter in length. The hose strap is perhaps the most useful tool in handling a charged hoseline due to its effective design of a strap, cinch clip, and forged handle. **Spanner wrenches** come in several sizes and are used to tighten or loosen couplings. They may also be useful as a pry bar, door chock, gas valve control, and for many other uses. A **hydrant wrench** is the tool used to operate the valves on a hydrant and may also be used as a spanner wrench. Some are plain wrenches and others have a ratchet feature to speed the operation of the valve.

A **hose roller** or **hoist** has a metal frame, with a securing rope, shaped to fit over a windowsill or edge of a roof with two rollers to allow the hose to roll over the edge, preventing chafe. A **hose clamp** is a device used to control the flow of water by squeezing or clamping the hose shut. Some work by pushing a lever, which closes the jaws of the device, and others have a screw mechanism or hydraulic pump that closes the jaws. Hose clamps create a pressure buildup in the hose. It is important to place the clamp at least 5 feet from any coupling (toward the supply side) and at least 20 feet from an apparatus. Another device used for stopping leaks without shutting down the line is a **hose jacket**. Hose jackets may be a metal or leather device that is fitted over the leaking area and either clamped or strapped together to control the leak. A **hose bridge** is a device that allows vehicles to pass over a section of hose without damaging it. A **hose cart** is a handcart or flat cart modified to be able to carry hose and other equipment around large buildings. Some departments use them for high-rise situations.

Figure 10-10 Various hose tools.

Appliances allow firefighters to connect various hoses and nozzles, **Figure 10-11**. Nozzles and foam equipment are covered in Chapter 11, and master stream devices are covered later in this chapter. A **double female** is a device that allows two male ends of hose to be connected, and a **double male** does the same for two female ends. A double male is used to connect the two female thread couplings when doing a split lay. An **increaser** is used to connect a smaller hose to a larger one, and a **reducer** connects a larger hose to a smaller one. An **adapter** is a device that adapts or changes one type of hose thread to another, allowing connection of two different lines. Adapters have a male end on one side and a female on the other with each side being a different thread type, for example, an iron pipe to national standard adapter.

Gate valves have an inlet and outlet of the same size with a gate to control water flow; they can be on apparatus or in hoselines. Large-diameter hose requires an **intake relief valve** at the receiving engine. These valves may function as a combined overpressurization relief valve, a gate valve, and an air bleed-off. A **wye** is a device that divides one hoseline into two or more. The wye lines may be the same size or smaller size, and the wye may or may not have gate control valves to control the water flow. A variation of the wye is the **water thief**, which has one inlet and one outlet of the same size plus two smaller outlets with all of the outlets being gated. The standard water thief usually has a 2½-inch (65-mm) inlet with one 2½-inch (65-mm) and two 1½-inch (38-mm) outlets. A **portable hydrant** or **manifold** is like a large water thief and may have one or more intakes and numerous outlets to allow multiple hoselines to be utilized with or without a pumper being directly at the fire location. A **siamese** is a device that connects two or more hoselines into one line. Siamese valves either have clapper valves or gate valves to prevent loss of water if only one line is connected. Clapper valves are spring-loaded and operate automatically, whereas the gate valve requires that a firefighter operate the valve to the open position. **Hydrant valves** or **switch valves** are used on a hydrant to allow an engine to connect and charge its supply line immediately but also allow an additional engine to connect to the same hydrant without shutting down the hydrant. A **strainer** is placed over the end of a suction hose to prevent debris from being sucked into the pump. Some strainers have a float attached to keep them at or near the water's surface. A strainer designed to draw water from a shallow, flat surface is typically called a "duck's foot." The duck's foot strainer is most often used for drafting from portable water tanks. A different style of strainer or screen is located on each intake of a pump. **Distributor pipe** or **extension pipe** allows a nozzle or other device to be directed into holes to reach basements, attic, and floors when access by personnel is not possible. The distributor pipe has self-supporting brackets that help hold it in place when in use. A **hose cap** does not really let water flow through it, but rather caps the end of a hoseline or appliance to prevent water flow.

Figure 10-11 Various hose appliances.

COUPLING AND UNCOUPLING HOSE

Connecting hose couplings can be accomplished in several ways, depending on the number of personnel available. The use of spanner wrenches can assist in tightening and loosening hose couplings. Most threaded couplings should be tightened until they are "hand tight." Arbitrarily tightening threaded couplings with a spanner wrench causes stress and premature wear on gaskets. Spanners should be used to tighten leaking couplings or to assist the uncoupling of hose.

The first coupling technique is a foot-tilt method for one person. With the male coupling lying flat on the ground, place the left foot directly behind the coupling, stabilizing it and raising it upward, **JPR 10-1A.** With the female coupling in the hands, bend down and bring the two couplings together. When the couplings are aligned, turn the swivels first left to align the Higbee cut and then to the right to tighten, **JPR 10-1B.**

JOB PERFORMANCE REQUIREMENT 10-1

One-Person Foot-Tilt Method

A With the male coupling lying flat on the ground, place the left foot directly behind the coupling, stabilizing it and raising it upward.

B With the female coupling in the right hand, bend down and bring the two couplings together. When aligned, turn the swivels to the right.

Note Remember when making a connection, "righty tighty-lefty loosey."

A second one-person method is the over-the-hip method. Hold the hose with the female coupling slightly below the waist, **JPR 10-2A.** Pick up the male coupling, align the two couplings, and swivel the female coupling to make the connection, **JPR 10-2B**.

In the two-person method, the two firefighters face each other with each holding a coupling. The couplings are brought together, aligned, and swiveled, **JPR 10-3**. The firefighter with the female coupling watches and adjusts the alignment of the couplings.

Uncoupling hose is usually done in the opposite manner in which it was connected, but sometimes a connection is too tight to easily break the connection. If the hose has been charged, it should be

JOB PERFORMANCE REQUIREMENT 10-2

One-Person Over-the-Hip Method

A Hold the hose with the female coupling slightly below the waist.

B Pick up the male coupling, align the two couplings, and swivel the female coupling to make the connection.

JOB PERFORMANCE REQUIREMENT 10-3

Two-Person Coupling Method

Two firefighters face each other, each holding a coupling. The couplings are brought together, aligned, and swiveled.

uncharged and the pressure bled off. The first recommendation is to get a pair of spanner wrenches. With the hose lying flat on the ground, place the spanners on the lugs going in opposite directions, **JPR 10-4A.** Push the spanners downward to break the connection, **JPR 10-4B.** If spanners are unavailable, one of the following methods can be used.

The one-person uncoupling method is known as the knee-press method. Fold the hose coupling over on itself on the ground and press your knee down on the coupling, **JPR 10-5A.** Try to twist the coupling to the left, breaking the connection, **JPR 10-5B**.

The two-person method is called the stiff-arm method and is similar to the two-person coupling, but the firefighters brace themselves and twist the coupling while holding their arms stiff with elbows slightly bent, **Figure 10-12**.

HOSE ROLLS

Hose is rolled either to store it or to have it ready for use with hoselines or appliances.

Straight/Storage

The straight or storage hose roll is the easiest to work with. Start with the hose flat on the ground. From the male end, to protect the threads, roll it straight to the opposite end, **JPR 10-6**. This is called a storage roll because of its common use, and it is often used when picking up after a fire. An exception used by some departments is to store damaged hose with the male end out and/or tying a knot at the end.

JOB PERFORMANCE REQUIREMENT 10-4

Uncoupling Hose with Spanners

A With the hose lying flat on the ground, place the spanners on the lugs going in opposite directions.

B Turn the spanners downward to break the connection.

JOB PERFORMANCE REQUIREMENT 10-5

Knee-Press One-Person Uncoupling Method

A Fold the hose coupling over on itself on the ground and press your knee down on the coupling.

B Twist the coupling to the left, breaking the connection.

Single Donut

The single-donut hose roll is used when access to either or both couplings may be needed. There are several ways to do a donut roll. The first method has the hose lying flat. Fold the hose on top of itself with the male coupling on top about 3 feet (1 m) short of the female coupling, **JPR 10-7A.** Starting at the fold, roll the hose toward the couplings with the extra hose on the female end protecting the male coupling, **JPR 10-7B**. Leave a small space at the center of the roll to provide a handhold, **JPR 10-7C.** A second firefighter can assist in guiding the hose and adjusting the slack as it is rolled.

Figure 10-12 Two-person stiff-arm method. The firefighters brace themselves and twist the coupling while holding their arms stiff with elbows slightly bent.

The second method used has the hose laid out flat. Starting at a point off-center about 6 feet (2 m) toward the male coupling, roll the hose toward the female coupling, **JPR 10-8A.** Once again the extra hose will protect the male coupling and the female coupling will be exposed. The handhold space and the extra firefighter are also useful with this method, **JPR 10-8B.**

Twin or Double Donut

The twin or double-donut roll is used for special applications and works best with 1½- to 2-inch (38- to 50-mm) hose. First the hose is laid flat with both couplings at one end and each half lying parallel to the other, **JPR 10-9A.** At the center, the loop is folded over the top of both halves. The roll is started toward the couplings at the same time, **JPR 10-9B** and **C.** At the end, the roll may be tied together for carrying, **JPR 10-9D**. The twin donut can be secured by using the hose itself. This is called a self-locking roll. To accomplish this, extend the amount of hose that is used for the starting fold and loop. Allow this excessive hose to "flop" as the twin donuts are rolled. When finished, use the extra hose at the center to form a bight around the two end couplings.

JOB PERFORMANCE REQUIREMENT 10-6

Straight or Storage Hose Roll

A Lay the hose flat on the ground. Starting at one end, usually the male end to protect the threads, roll it straight to the opposite end.

B The finished roll.

JOB PERFORMANCE REQUIREMENT 10-7

Single-Donut Hose Roll

A The first method has the hose lying flat. Fold the hose on top of itself with the male coupling on top and about 3 feet (1 m) short of the female coupling.

B Starting at the fold, roll the hose toward the couplings with the extra hose on the female end protecting the male coupling. A second firefighter can assist in guiding the hose and adjusting the slack as it is rolled.

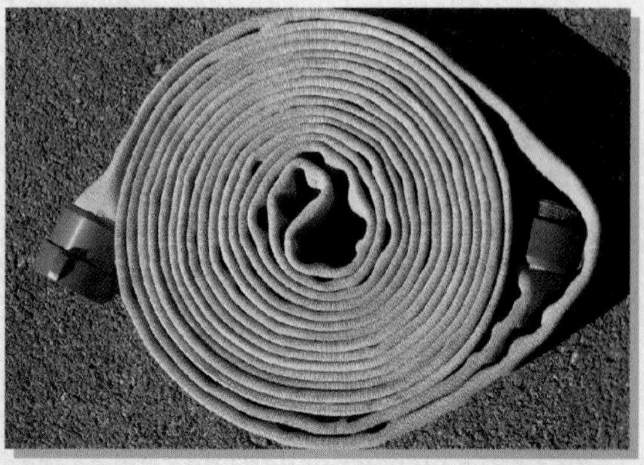

C Leaving a small space at the center of the roll allows a handhold.

JOB PERFORMANCE REQUIREMENT 10-8

Single-Donut Hose Roll

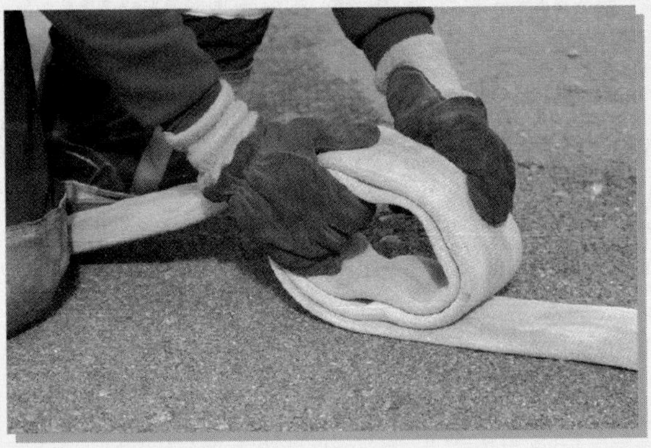

A Lay the hose out flat. Starting at an off-centered point about 6 feet (2 m) toward the male coupling, roll the hose toward the female coupling.

B The handhold space and the extra firefighter are also useful with this method.

JOB PERFORMANCE REQUIREMENT 10-9

Twin Donut Roll

A First lay the hose flat with both couplings at one end and each half laid parallel to the other.

B At the center, fold the loop over the top of both halves.

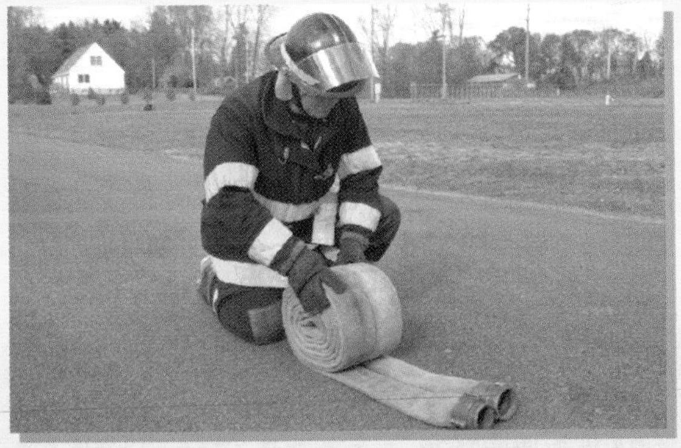

C Start the roll toward the couplings at the same time.

D At the end, you may tie the roll together for carrying.

HOSE CARRIES

Drain and Carry

The drain and carry method is used to combine the two steps of draining and carrying a section of hose into one operation. This is usually done with one section of hose. The firefighter starts at one end of the hose and with the coupling held waist height feeds the hose over the shoulder and back down to the waist, **JPR 10-10A.** A fold is created and the hose is laid on itself back to the front, **JPR 10-10B.** The firefighter continues to walk forward folding and refolding the hose at the waist until finished. The hose can then be carried to the new location, **JPR 10-10C**.

Shoulder Loop Carry

In the shoulder loop carry, the firefighter places the nozzle or end of hose over the shoulder resting against the back **JPR 10-11A.** The firefighter walks forward about 3 feet (1 m), picks up the hose, and forms a bight to bring the hose back up

JOB PERFORMANCE REQUIREMENT 10-10

Drain and Carry

A Start at one end of the hose. With the coupling held waist height, feed the hose over the shoulder and back down to the waist.

B Create a fold and lay the hose on itself back to the front.

C Continue to walk forward folding the hose at the waist until finished so it can be carried to the new location.

and over the shoulder, creating a loop, **JPR 10-11B.** (This is similar to rolling an electrical cord around one's arm, but with bigger loops.) The action is continued as each section is picked up and carried forward, **JPR 10-11C.** If the firefighter needs to move in the opposite direction, the loops are collected and raised by the hands and the firefighter rotates to the opposite direction and returns the hose to the opposite shoulder moving forward in the new direction, **JPR 10-11D** and **E**.

JOB PERFORMANCE REQUIREMENT 10-11

Shoulder Loop Carry

A Place the nozzle or end of the hose over the shoulder, where it rests against the back.

B Walk forward about 3 feet (1 m), pick the hose up, and form a bight to bring the hose back up and over the shoulder, creating a loop.

C Continue as each section is picked up and carried forward.

D If you need to move in the opposite direction, collect and raise the loops with your hands and rotate to the opposite direction.

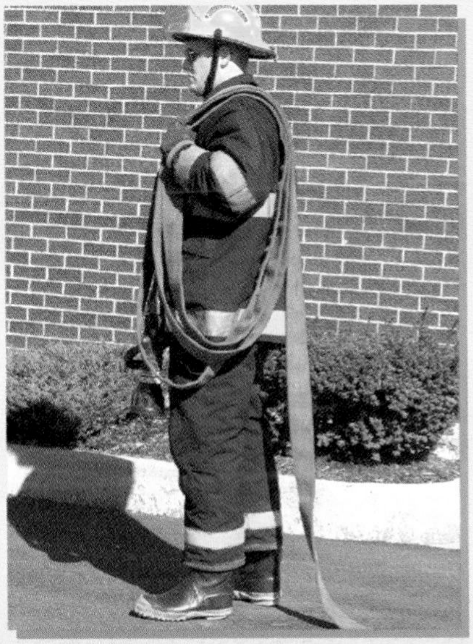

E Return the hose to the opposite shoulder as you move forward in the new direction.

Single-Section Street Drag

The simple hose drag technique can move one or two hoselines. The firefighter puts a section of hose over a shoulder with the coupling in front at waist height. The firefighter then walks away dragging the line behind, **JPR 10-12A.** Placing a line over each shoulder allows the firefighter to pull two lines, **JPR 10-12B.** If additional sections are needed, additional firefighters can do the same with the following sections until the desired amount of hose is stretched, **JPR 10-12C**. Note that this technique can cause damage to the coupling being dragged.

HOSE LOADS

Apparatus hose loads are designed so that the engine can carry out any of the required fire attack or water supply evolutions. These include water supply, small and medium attack lines, and supply for protective systems and master stream devices. Some of these lines are made up in advance and are preconnected to the pump, while others are made up depending on the incident. A well-trained engine crew should be able to perform any of these required tasks quickly and efficiently. Part of this efficiency is how the crew trains and performs while the design and layout of the hose bed, the type of hose, and its couplings determine other parts. Hose loads should consider the need to have the right hose coupling or adapter at the end that will be used farthest from the engine.

Note Time and care taken during the loading of hose will be repaid many times over when pulling the hose loads off during an incident.

JOB PERFORMANCE REQUIREMENT 10-12

Hose Drag

A Put the end of a section of hose over your shoulder with the coupling in front at waist height and walk away, dragging the line.

B Place a line over each shoulder to pull two lines.

C If additional sections are needed, additional firefighters can do the same to the following sections until the desired amount of hose is stretched.

Figure 10-13 A dutchman is a short fold of hose or a reverse fold that is used when loading hose and a coupling comes at a point where a fold should take place or when two couplings end up on top of each other. The dutchman moves the coupling to another point in the load.

A **dutchman** is a short fold of hose or a reverse fold that is used when loading hose and a coupling comes at a point where a fold should take place or when two sets of couplings end up on top of or next to each other. The dutchman moves the coupling to another point in the load, **Figure 10-13**.

Accordion Load

Accordion hose loads can be used for preconnected, made-up hoselines or for providing additional supply line. The accordion load is ideal for making up **shoulder loads**. Care should be used with the accordion load because overpacking the hose bed can stress and crack the liner of the hose. For use with preconnected lines, the accordion load should be started with the female coupling at the pump outlet or the front of the bed, **JPR 10-13A.** The hose is placed in the bed on its side rather than lying flat. The hose is brought forward in the bed until it reaches the rear, where it is folded and returned to the front where the process is repeated, **JPR 10-13B.** When the layer of the hose fills the bed, the hose is brought up at the rear and folded back on itself starting the next layer, again on the left side, **JPR 10-13C.** Watch the placement of hose couplings near the wall of the hose bed; they should be on their way out or have enough room to flip around when pulled from the bed. Continue until the bed is filled or the desired amount is loaded. Add a nozzle if it is a preconnected line. If a supply line, it may be connected to another bed or left with the male end exposed. The male end can be loaded into the bed first if the hose is being used for a forward lay. A blind cap, double male, or double female can be added to protect threads or to allow connection to a hydrant, **JPR 10-13D**.

Making up a hoseline from an accordion hose load is an easy task. A nozzle or other appliance is attached to the end of the hose and the desired amount is selected, **JPR 10-14A.** The firefighter pulls the amount of hose about halfway out of the bed and places it on one shoulder with the nozzle at the bottom. Use care not to have too much of the hose in front, which could cause tripping; it is better to drag the excess behind, **JPR 10-14B.** Stepping away from the bed removes the rest of the load. Pulling another number of folds and dropping them to the ground will allow some hose to be dragged prior to using the shoulder load, **JPR 10-14C.** When the point is reached where the shoulder load is to be deployed, the firefighter allows the hose to flake off one fold at a time while moving forward, **JPR 10-14D.** If additional hose is needed, another firefighter may repeat this process with the attached hose instead of the nozzle.

Flat Load

The flat load is used for supply lines and some attack lines. It involves simply laying the hose flat by starting at either end of the hose depending on use. An example of a flat load for water supply would be to start with the male end of the hose on the right side of the bed at the rear, **JPR 10-15A.** This will allow this hose bed to be connected to another one for long hose lays. Lay the hose flat until the front is reached where the hose is folded over and brought to the rear. When back at the rear, turn the hose off the first layer and lay it flat next to the first row, **JPR 10-15B** and **C.** When the edge of

JOB PERFORMANCE REQUIREMENT 10-13

Accordion Load

A For use with preconnected lines or for making up lines, start the accordion load with the female coupling at the pump outlet or the front of the bed.

B Place the hose in the bed on its side rather than lying flat. Bring the hose forward in the bed until it reaches the rear, where it is folded and returned to the front. Repeat the process.

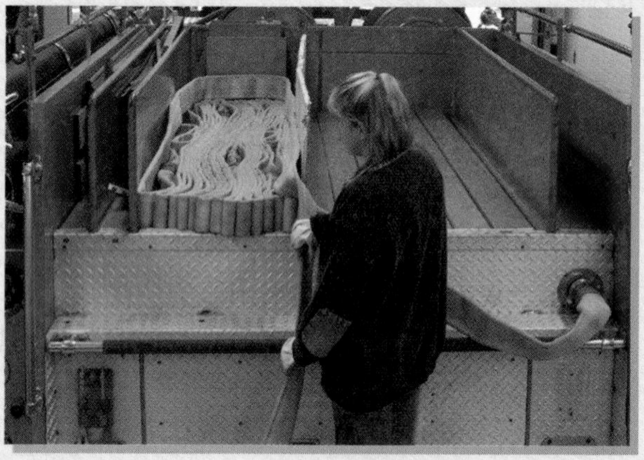

C When the layer of hose fills the bed, bring the hose up at the rear and fold it back on itself to start the next layer.

D Continue until the bed is filled or the desired amount is loaded. Add a nozzle if it is a preconnected line. If a supply line, add a blind cap and adapters.

the hose bed is reached, a second layer is started by doing a double layer and then moving toward the right again with the rows over the top of the first layer's rows, **JPR 10-15D** and **E**.

Making a shoulder load from a flat load is slightly more difficult than the accordion load because an accordion load must be created. The nozzle or appliance is again attached. Next the firefighter gathers the amount of hose to be placed on the shoulder, **JPR 10-16A.** Sometimes this can be done by simply twisting the folds of the hose to the side, creating a layer of accordion-style hose and following the accordion load directions, **JPR 10-16B.** When this cannot be done, the firefighter must begin loading the folds one at a time on the shoulder. Place the fold with the nozzle side down over the shoulder and pull the hose until it is properly positioned. Grab the next fold of the hose and place it on top of the first until the desired amount is reached. Step away from the bed and again pull some extra folds. The hose should now deploy like an accordion load, **JPR 10-16C** and **D**.

JOB PERFORMANCE REQUIREMENT 10-14

Advancing an Accordion Load

A Attach a nozzle or other appliance to the end of the hose and select the desired amount.

B Pull the amount of hose about halfway out of the bed and place it on the shoulder with the nozzle at the bottom. Use care not to have too much of the hose in front, which could cause tripping; it is better to drag the excess behind.

C Pull another number of folds and drop them to the ground to allow some hose to be dragged prior to using the shoulder load.

D When you reach the point where the shoulder load is to be deployed, allow the hose to flake off one fold at a time while moving forward.

Horseshoe Load

The horseshoe load also has several applications. The load starts at a rear corner of the hose bed with the hose on its side, **JPR 10-17A.** The hose is laid toward the front wall of the hose bed and then along the wall to the other side and back toward the rear. At the rear, the hose is folded alongside itself and heads back to where it started. It is folded again and another "U" is formed, **JPR 10-17B** and **C.** When the bed is filled a new layer may be started by laying the hose flat to the front of the bed where it is turned again on its side and moved to the rear on the same side as originally started, **JPR 10-17D** and **E.**

Making a shoulder load from the horseshoe bed is somewhat similar to the accordion load. Care should be taken when reaching the outer edges of the hose bed as the increasing length of hose between folds may cause problems advancing the line. Again the nozzle is placed on the hose and the desired amount selected, **JPR 10-18A.** The firefighter pulls the hose and twists it into place on the shoulder and stepping away pulls the hose out of the bed. The remaining steps are as given earlier, **JPR 10-18B** and **C.**

JOB PERFORMANCE REQUIREMENT 10-15

Flat Load

A To use the flat load for water supply, start with the male end of the hose on the right side of the bed at the rear.

B Lay the hose flat until the front is reached where the hose is folded over and brought back to the rear.

C When back at the rear, turn the hose off the first layer and lay it flat next to the first row.

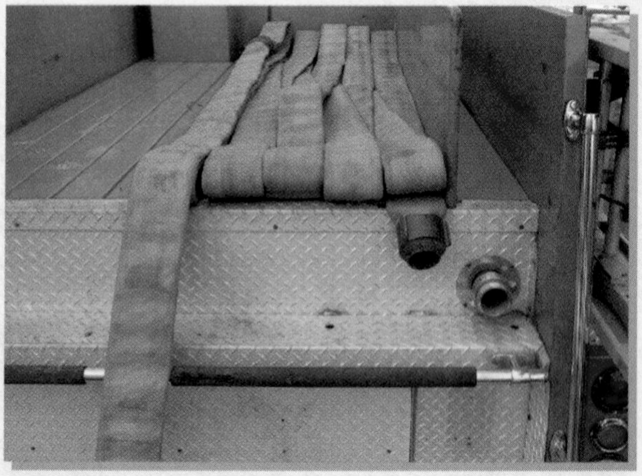

D When the edge of the hose bed is reached, do a double layer and then move toward the right again with the rows over the top of the first layer's rows to start a second layer.

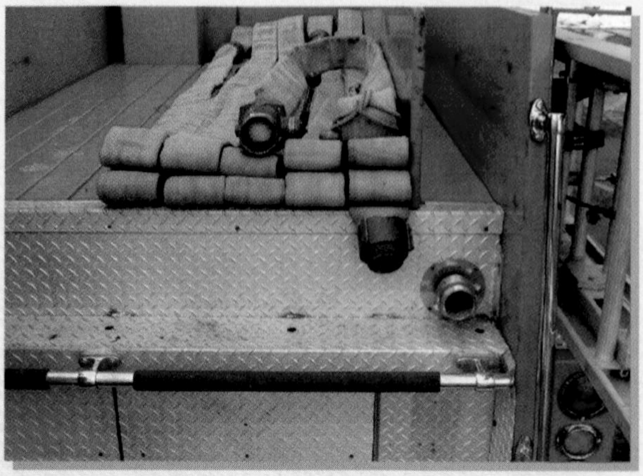

E The finished load.

JOB PERFORMANCE REQUIREMENT 10-16

Advancing a Flat Load from a Supply Bed

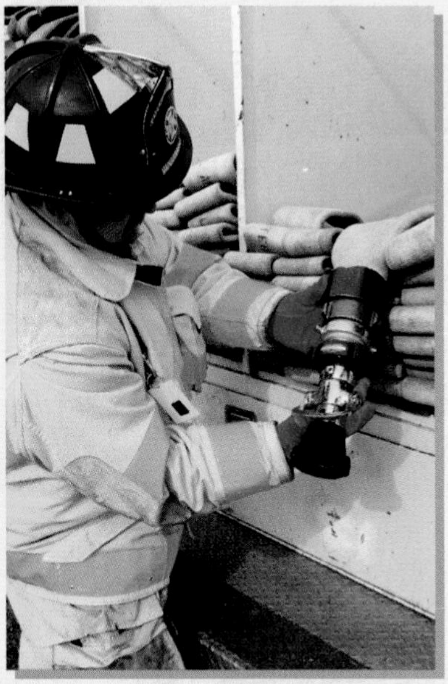

A Attach the nozzle or appliance and begin to gather the amount of hose needed to be placed on the shoulder.

B Sometimes this can be done by simply twisting the folds of the hose to their sides, creating a layer of accordion-style hose and following the accordion load directions.

C When this cannot be done, you must begin loading the folds one at a time on the shoulder. Place the fold with the nozzle side down over the shoulder and pull the hose until it is properly positioned. Grab the next fold of the hose and place it on top of the first until the desired amount is reached.

D Step away from the bed and again pull some extra folds. The hose should now deploy like an accordion load.

JOB PERFORMANCE REQUIREMENT 10-17

Horseshoe Load

A Start the load at a rear corner of the hose bed with the hose on its side.

B Lay the hose toward the front wall of the hose bed and then along the wall to the other side and back toward the rear.

C At the rear, fold the hose alongside itself and head back to where it started. Fold it again and another "U" is formed.

D When the bed is filled, start a new layer by laying the hose flat to the front of the bed where it is turned again on its side and moved to the rear on the same side as originally started.

E The finished bed.

JOB PERFORMANCE REQUIREMENT 10-18

Advancing a Horseshoe Load

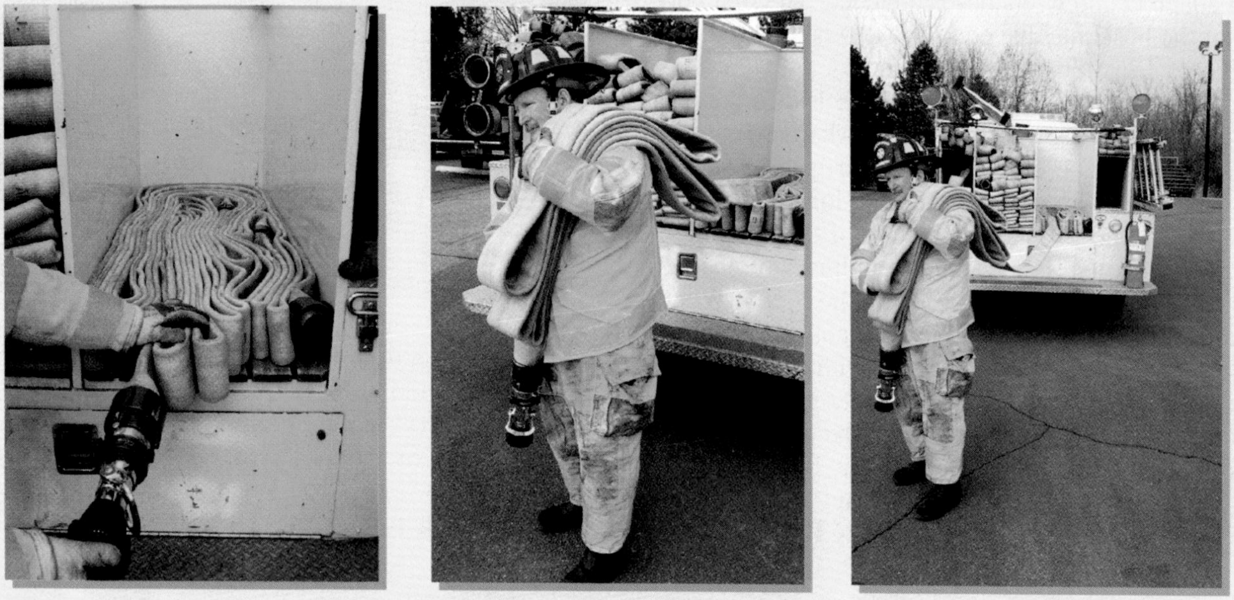

A Place the nozzle on the hose and select the desired amount.

B Pull the hose and twist it into place on the shoulder.

C Step away to pull the hose out of the bed.

Finish Loads and Preconnected Loads

Finish loads and preconnected loads can utilize the three methods of loading hose just discussed. Some use combinations of these loads in different layers to assist firefighters in advancing hose quickly from an engine. Finish loads can be used for assisting in laying supply line or attack lines. Preconnected loads are usually used for attack lines. Preconnected loads are typically preattached to a pump discharge. Therefore, the hose must be loaded in such a way that the entire hose load can be pulled off and deployed easily. Failure to clear all the hose from the bed will cause kinking of the hose and jamming of the hose in the bed when the pump operator charges the line. Apparatus damage may also occur if hose loaded in a bed is charged prematurely.

A straight finish load is used usually with a straight hose lay and simply involves taking the final length or two of a load and laying it flat across the top of the load. A rope with adapters, a spanner wrench, and a hydrant wrench attached allows the layout person to quickly have all the necessary tools and enough hose to make the hydrant connection, **Figure 10-14**. A reverse horseshoe load can be used

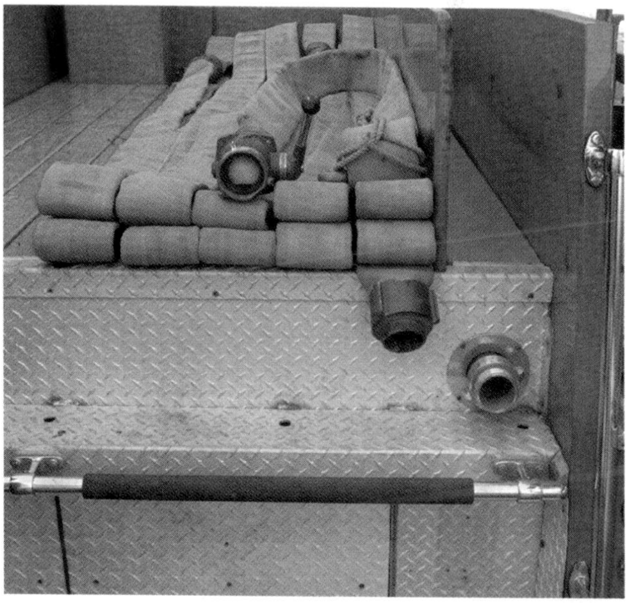

Figure 10-14 A straight finish load simply involves taking the final length or two of a load and laying it flat across the top of the load. A rope with adapters, a spanner wrench, and a hydrant wrench attached allows the layout person to quickly have all the necessary tools and enough hose to make the hydrant connection.

for laying out or making an attack line. For laying out, a horseshoe load is made on top of the hose load but in the reverse direction (front to back), and at the center point of the "U" of the horseshoe, the rope with adapters and wrenches is attached. The first portion of the hose may have to have a twist in it to get it to change direction, **Figure 10-15**.

Similarly an attack line can be attached to the end of a hose load. This works when a **backstretch** or **flying stretch** will be utilized. A wye or gate valve can be installed at the end of the hose load before the horseshoe. (One will have to be used if changing hose size.) The reverse horseshoe, including any needed twist, is created again on top of the hose bed. When the amount of hose is loaded, a nozzle is attached and laid in the center of the horseshoe with the nozzle at the rear of the hose bed. The firefighter grabs the nozzle and a few lengths of hose over the shoulder. The horseshoe load is then grabbed at the bottom of the "U," pulling it out of the hose bed and allowing it to flake out while moving to the fire, **JPR 10-19A** and **B**.

Figure 10-15 A reverse horseshoe load for laying out is made on top of the hose load but in the reverse direction (front to back), and at the center point of the "U" of the horseshoe the rope with adapters and wrenches is attached. The first portion of the hose may have to have a twist in it to get it to change direction.

JOB PERFORMANCE REQUIREMENT 10-19

Reverse Horseshoe Load for an Attack Line

A For an attack line, create the reverse horseshoe again on top of the hose bed. When the amount of hose is loaded, attach a nozzle and lay it in the center of the horseshoe with the nozzle at the rear of the hose bed.

B Grab the nozzle and place a few of the hose folds over the shoulder. Grab the horseshoe load at the bottom of the "U," pulling it out of the hose bed and allowing it to flake out while moving to the fire.

Preconnected lines can be made up using the accordion or horseshoe loads, and some use a combination. Since they are preconnected to a discharge they will always start with the female end where the discharge is located and may require some twists to start. It is possible to use preconnected loads with all accordion or horseshoe layers. An example of a combination load would be a horseshoe bottom layer, with the top layer accordioned. Other combinations include horseshoe, accordion, accordion layers, or alternating horseshoe and accordion layers, **Figure 10-16A** and **B**. The combinations are often designed based on a specific need at a certain property within the response district. When using multiple layers or various hose loads, it is common to place **ears** on the hose to assist in pulling the layer, **Figure 10-17**.

Flat Load, Minuteman Load, and Triple-Layer Load

Preconnected loads must rapidly remove the hose from a slot or bed. Many preconnected loads exist, and there are many innovative ways to accomplish the rapid deployment of fire attack lines. The following loads are among the more popular. The flat load, as a preconnect, is also based on the flat load described earlier and starts at the discharge. The hose is laid flat as described earlier. At a point from one-third to one-half the length of the line an ear or row of ears should be added to assist in pulling the line, **JPR 10-20A.** The line is advanced by grabbing the nozzle and placing it over the shoulder with the other hand reaching around and pulling the ear(s). The firefighter then walks away pulling the line behind, **JPR 10-20B** and **C**.

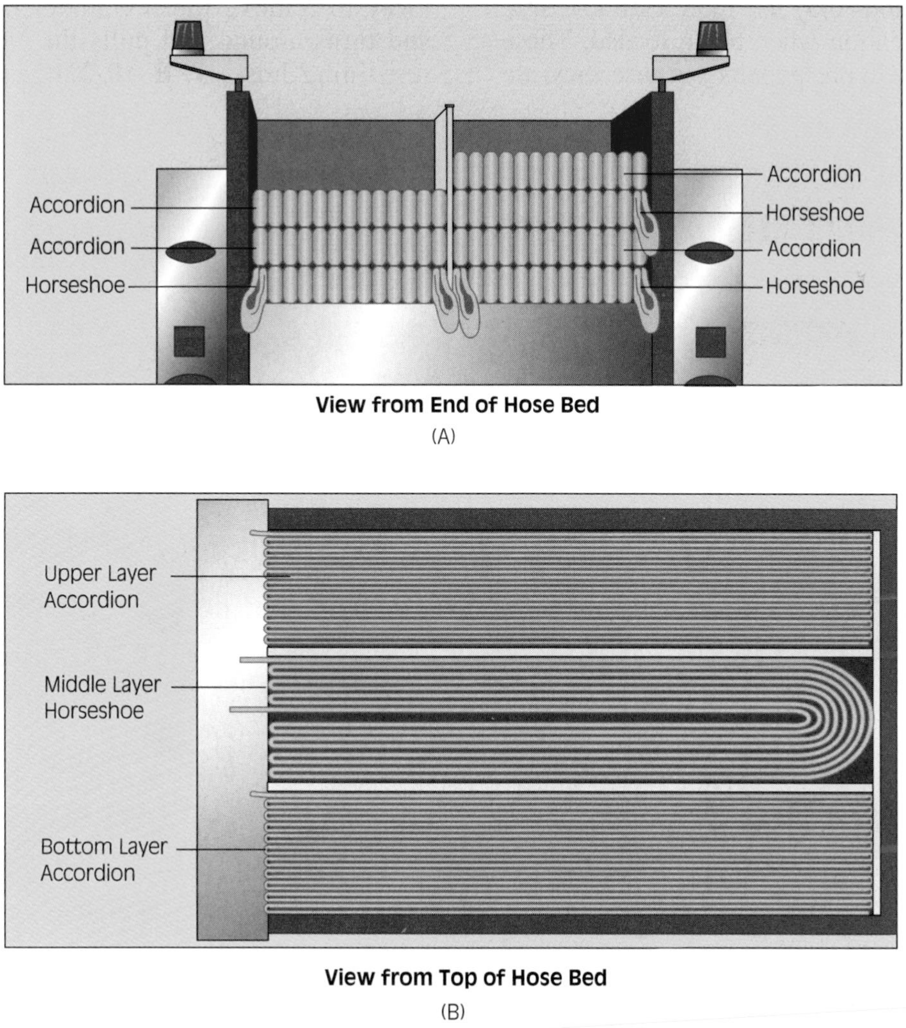

Figure 10-16 Preconnected combination loads include horseshoe, accordion, accordion layers or alternating horseshoe and accordion layers. (A) Horseshoe, accordion, accordion layers. (B) Alternating horseshoe and accordion layers.

Figure 10-17 Ears on hose load.

The minuteman load is a preconnected load using a narrower section of the hose bed. This narrower section is called a **slot load** and can allow more hoselines to be placed in the same space. Some longer length versions may use more than one slot, and all require attention when being loaded. These loads have become more popular because they are quicker and more efficient than some of the previous loads described, **Figure 10-18**.

The minuteman load has the female end connected to the discharge, and several sections of hose are flat loaded into the slot with the last few feet (about a meter) placed aside out of the bed. An ear may be placed one or two folds from the bottom to assist in pulling, **JPR 10-21A.** Next a section of hose with the nozzle is started and the hose flat loaded above that. The nozzle will therefore be in the middle of the slot at the point where it will be pulled, **JPR 10-21B.** When the designated number of sections is finished, the set-aside piece is connected to the pieces with the nozzle attached and the connected sections are on the top layer, **JPR 10-21C**.

To advance the minuteman load, the firefighter lifts the nozzle and several layers of hose while pulling them out and placing them midway on the shoulder, **JPR 10-22A.** The firefighter then steps away to remove the remainder of the top layers and turns around and pulls the ear to remove the remaining hose, **JPR 10-22B** and **C.** When the

JOB PERFORMANCE REQUIREMENT 10-20

Advancing the Flat Load from a Preconnected Bed

A Start the flat load at the discharge with the hose laid. At a point from one-third to one-half the length of the line, an ear or row of ears should be added to assist in pulling the line.

B To advance the line, grab the nozzle and place it over the shoulder with the other hand reaching around and pulling the ear(s).

C Walk away, pulling the line behind.

Figure 10-18 The minuteman load or slot load is a preconnected load using a narrower section of the hose bed, which allow more hoselines to be placed in the same space. Some longer length versions may use more than one slot.

bottom sections are fully stretched out, the shoulder load is allowed to flake out toward the fire, **JPR 10-22D**.

The triple-layer load is another modified flat load where the hose is folded over on itself three times. All the hose is connected to the discharge, and the nozzle is attached at the opposite end with the hose stretched out its full length, **JPR 10-23A.** At a point two-thirds of the way from the hose bed, a fold is made and the hose is doubled on itself back toward the engine. This fold creates three layers of hose. One is the original bottom layer, one the folded back middle layer, and the top third layer has the hose with the nozzle attached. These three layers are now treated as one for loading on the engine, **JPR 10-23B.** The triple-layered hose is loaded into the slot as a single layer. A fold is made at the edge of the hose bed and the hose doubled over creating two other layers. This continues until all of the hose is loaded. The nozzle's layer should be at the edge of the bed for pulling, **JPR 10-23C.**

JOB PERFORMANCE REQUIREMENT 10-21

The Minuteman or Slot Load Being Loaded

A This is a narrower version of the flat load with a slight modification. Connect the female end to the discharge and flat load several sections of hose into the slot with the last few feet (about a meter) placed aside out of the bed. An ear may be placed one or two folds from the bottom to assist in pulling.

B Next, start a section of hose with the nozzle, and load the hose flat above that. The nozzle will therefore be in the middle of the slot at the point where it will be pulled.

C When the designated number of sections is finished, connect the set-aside piece to the pieces with the nozzle attached. The connected sections are on the top layer.

JOB PERFORMANCE REQUIREMENT 10-22

Advancing the Minuteman Load

A Lift up the nozzle and layers above it while pulling them out and placing them midway on the shoulder.

B Step away to remove the remainder of the top layers.

C Turn around and pull the ear to remove the remaining hose.

D When the bottom sections are fully stretched out, allow the shoulder load to flake out toward the fire.

Advancing the line has the firefighter grab the layer with the nozzle and place it on the shoulder, **JPR 10-24A.** Either another firefighter grabs the next layer or the nozzleperson pulls them out of the slot and the hose is stretched to the fire, **JPR 10-24B** and **C**.

Stored Hose Loads/Packs

Apparatus typically carry stored hose rolls and special application hose packs. Hose rolls are just that—extra sections of rolled hose for replacing damaged hose or short sections to help assemble an evolution. These hose rolls can be stored as a straight roll, donut roll, or double donut. Hose packs can be numerous in design and makeup. High-rise (or standpipe) packs are hose loads that are preassembled and bundled to be easily carried into a building, **Figure 10-19.** High-rise packs may also include a tool and appliance bag containing adapters, spanners, and hose straps. Wildland packs are covered in the next section.

Wildland Firefighting Hose Loads

Wildland firefighting often requires firefighters to stretch hoselines a great distance from the engine while fighting the fire. To accomplish this task, the hose is rolled and bundled together to allow it to be

JOB PERFORMANCE REQUIREMENT 10-23

The Triple-Layer Load

Another modified flat load where the hose is folded over on itself three times.

A Connect all the hose to the discharge and attach the nozzle at the other end with the hose stretched out its full length.

B At a point two-thirds of the way from the hose bed, make a fold and double the hose on itself back toward the engine. Make another fold, the last part with the nozzle, to create the third layer. These three layers are now treated as one for loading on the engine.

C Load the hose in the slot in a layer and make a fold, creating another layer, until all hose is loaded. The nozzle's layer should be at the edge of the bed for pulling.

JOB PERFORMANCE REQUIREMENT 10-24

Advancing the Triple-Layer Load

A Grab the layer with the nozzle and place it on the shoulder.

B Pull the layers out of the slot, or another firefighter can grab the next layer.

C Stretch the hose to the fire.

carried on the firefighters' backs while carrying tools or stretching and advancing the line. Placing two bundles together allows each firefighter to carry 200 feet (60 m) of 1-inch (25-mm) or 1½-inch (38-mm) hose.

A standard wildland hose load is the modified Gasner bar pack, which provides ease of rolling and stretching the line, convenience of carrying with hands free, and protection for the couplings. The first step starts at the male end of the hose [100 feet (30 m)] and the firefighter forms a roll 30 inches (0.8 m) in diameter on flat ground, **JPR 10-25A.** Continue to roll the hose in the direction of the rows and make the rolls as tight as possible, **JPR 10-25B.** The completed field roll, with the female coupling on the outside, is ready for packing. Grab the outside of the hose roll near the female coupling and pull toward the center of the roll, **JPR 10-25C.** The fold should now protect the female end. The top layer of hose can now be tied to the bottom layer using a piece of rope 32 inches long (0.9 m). Two rolls of hose may be placed together and tied off with rope on the outside of the rolls with a shoulder strap to allow carrying, **JPR 10-25D.**

A modification would be to add a gated wye or tee at the end of each roll of hose. To do this merely requires that when the roll is closed together, the female end is left to one side, on the outside, and the gated wye or tee is attached after tying the roll off. When placing the two rolls together, they should be placed so that the gated wyes are on opposite sides of the pack, **Figure 10-20.**

Figure 10-19 High-rise (standpipe) hose packs are preassembled in a bundle for deployment.

ADVANCING HOSELINES—CHARGED/UNCHARGED

The engine company's purpose is to advance hoselines to the fire's seat to apply water for extinguishment. These tasks should be accomplished in the most efficient manner, whether around, onto, or into buildings. The following evolutions are based on a four-person crew with one person being at the layout position and connecting to the water source. As crew situations change, the crew positions may also change. Remember that two in/two out firefighter safety rules apply, and conducting any fireground operation without sufficient personnel for a firefighter rescue team can be highly dangerous and unlawful. (See Chapters 2 and 5 for more information.)

When advancing hoselines in any of the evolutions described, the nozzleperson will advance the first shoulder load with the nozzle. The officer takes the second position on the line if it is a two- or three-person line. The engine driver will take the third position on a three-person line. If additional personnel are available, they will be substituted for the officer and driver. The driver is responsible for clearing the hose bed prior to charging any of the lines.

The officer determines the length of the hoseline by figuring the distance from the engine to the location of the fire. Upon reaching the fire area, there should be about 50 feet (15 m) of hose left for movement around the fire area. When taking a hoseline up the outside of a building a 50-foot (15-m)

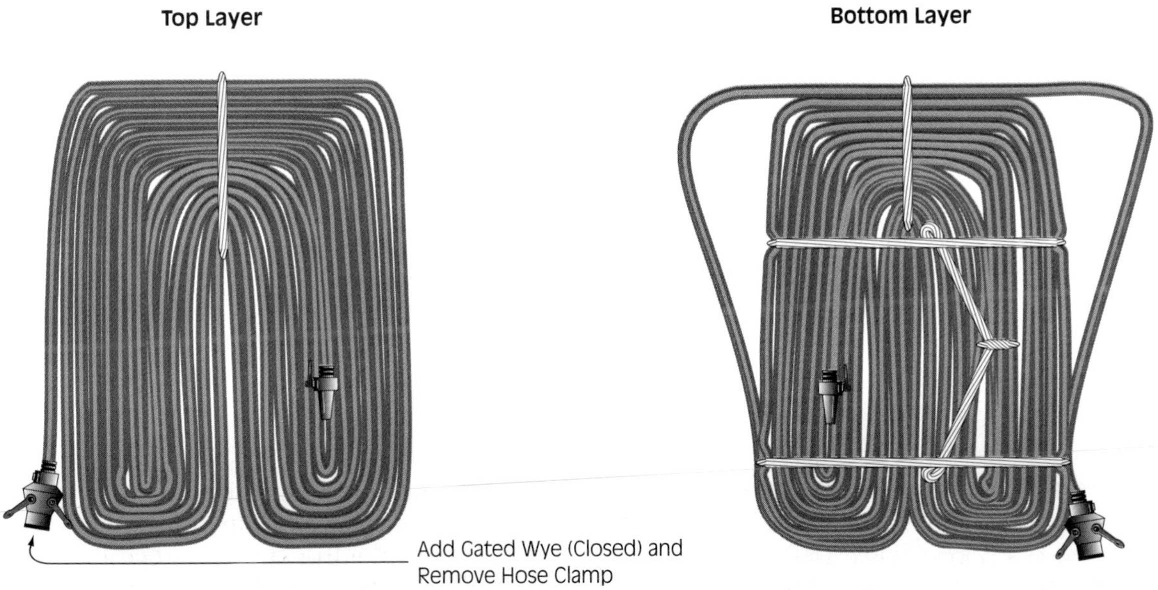

Figure 10-20 Modified Gasner bar pack with added gated wyes.

JOB PERFORMANCE REQUIREMENT 10-25

Modified Gasner Bar Pack

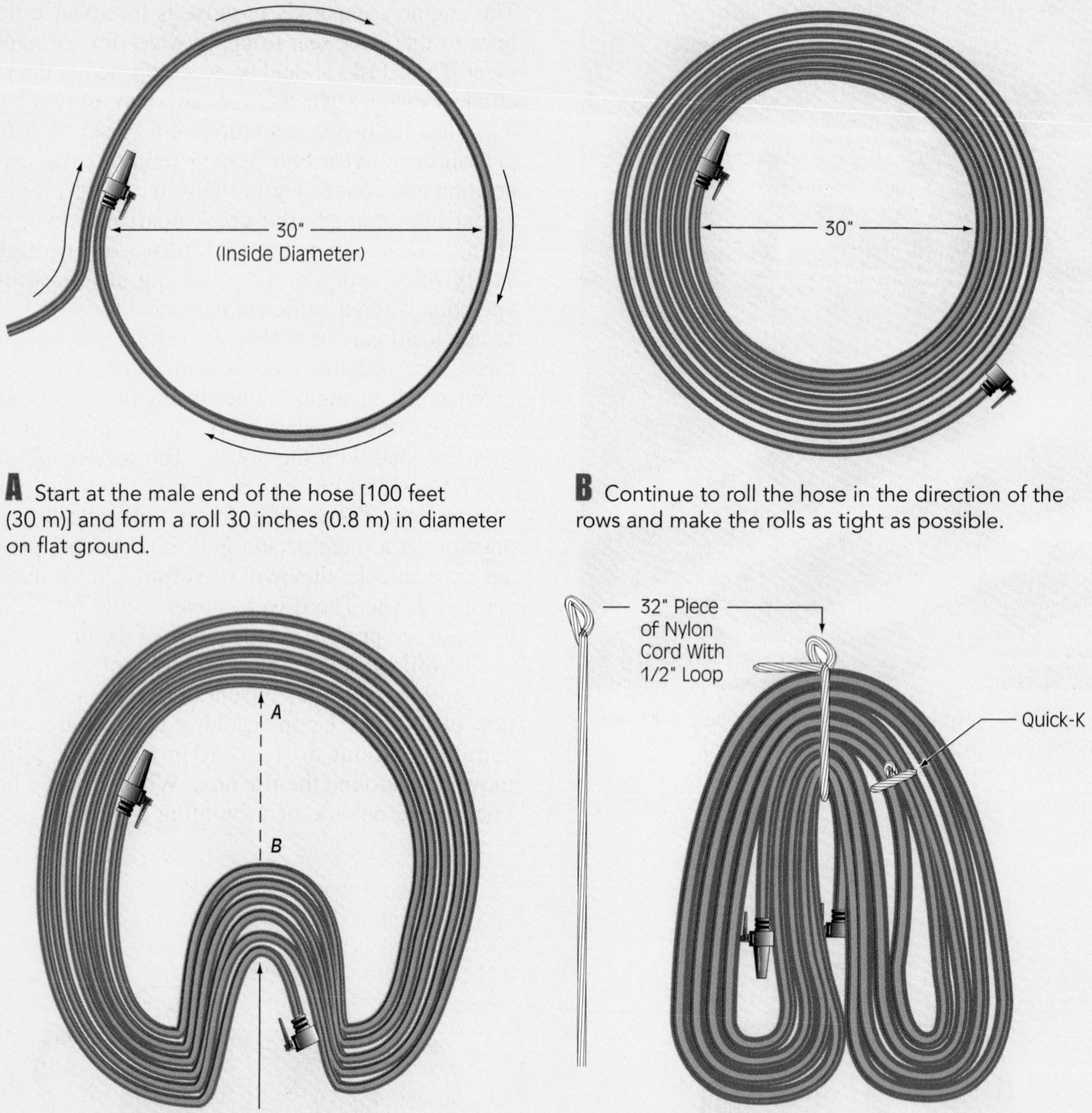

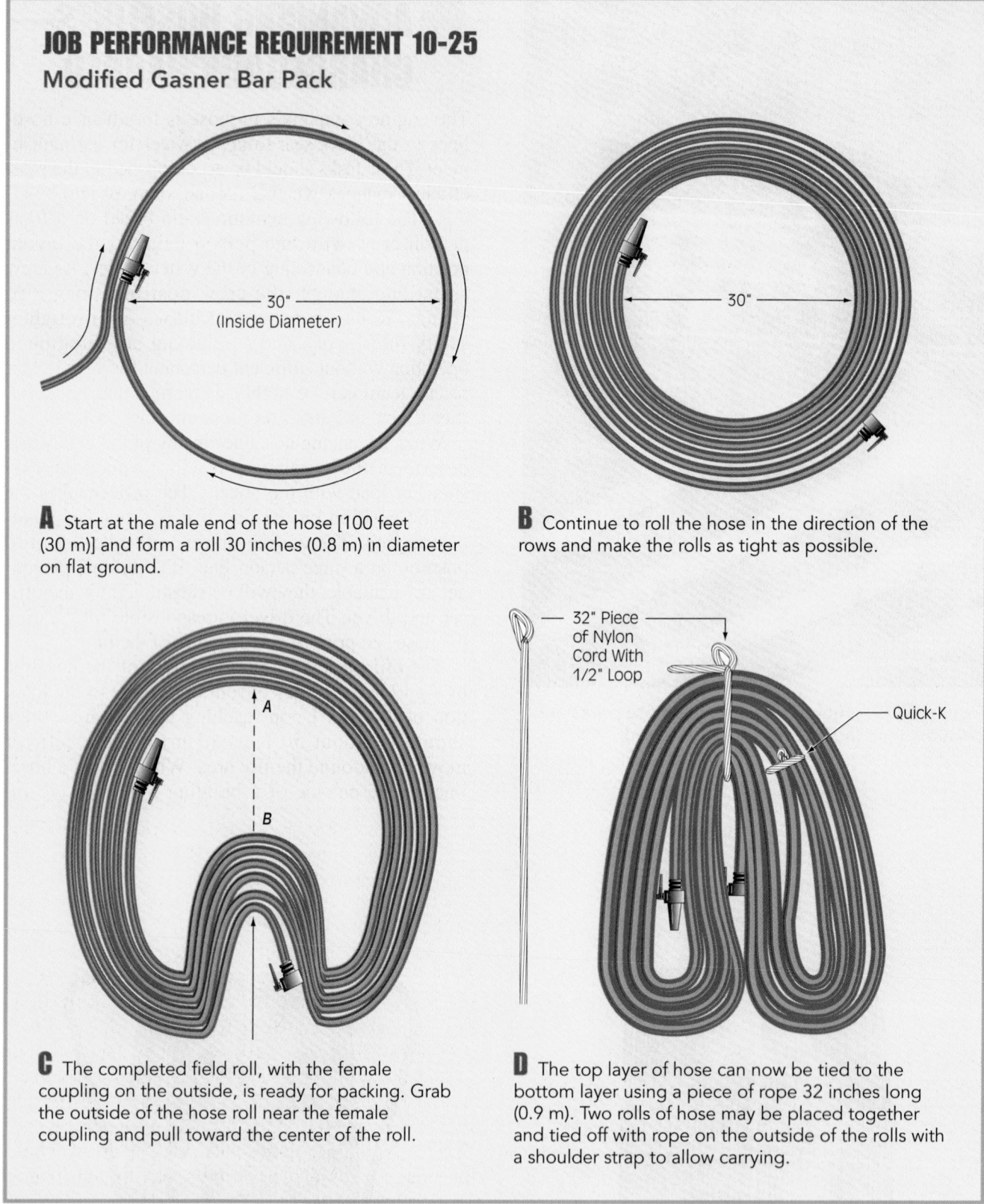

A Start at the male end of the hose [100 feet (30 m)] and form a roll 30 inches (0.8 m) in diameter on flat ground.

B Continue to roll the hose in the direction of the rows and make the rolls as tight as possible.

C The completed field roll, with the female coupling on the outside, is ready for packing. Grab the outside of the hose roll near the female coupling and pull toward the center of the roll.

D The top layer of hose can now be tied to the bottom layer using a piece of rope 32 inches long (0.9 m). Two rolls of hose may be placed together and tied off with rope on the outside of the rolls with a shoulder strap to allow carrying.

section will reach four floors, while taking the same section inside the building via a stairwell will reach only one individual floor, **Figure 10-21**.

When operating a charged hoseline in one location, if the line is then ordered moved to another position, such as to another floor, it is best for time and efficiency to shut down, drain, and advance the line as a dry line if fire conditions allow. When the new location is reached, the line is then recharged.

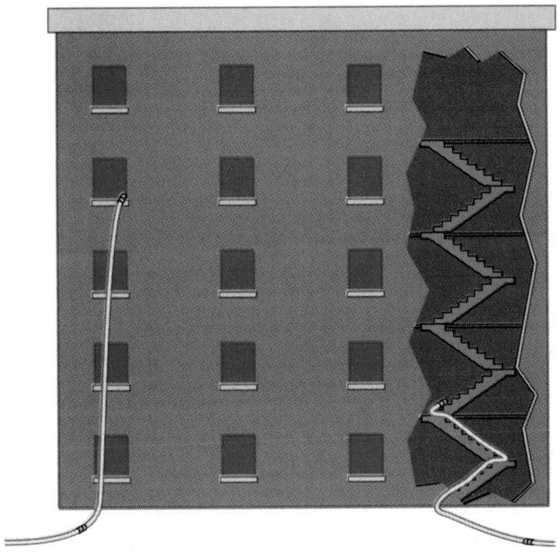

Figure 10-21 When taking a hoseline up the outside of a building a 50-foot (15-m) section will reach four floors, while taking the same section inside the building via a stairwell will reach only one individual floor.

Into Structures

Advancing a hoseline into a structure requires careful placement of the pumper and hoseline, proper selection of the correct size and length hoseline, and skillful execution by the hose crew. Park the engine at a safe location that allows good access for hoselines without hindering any other apparatus placement, especially a ladder company, and allows for any possible changes in fire conditions and equipment usage, **Figure 10-22**. The hoseline must be placed to attack the fire while also being at a position to prevent any fire spread. The hoseline should be large enough to suppress the fire, a minimum of 1½ inches (38 mm). The length has already been discussed. Finally, it is the skillful crew that carefully and safely advances the line, ensuring that they are ready to combat the fire.

To advance the line into a structure, the crew selects a hoseline and properly removes it from the engine, deploying it toward the entrance. A careful evaluation of the conditions is required prior to entering any structure or room. Structural firefighting involves the possibility of flashover, backdrafts, and building collapse. Entering a burning building without having a safe plan of attack and emergency escape can be fatal.

Caution A careful evaluation of the conditions is required prior to entering any structure or room. Structural firefighting involves the possibility of flashover, backdrafts, and building collapse. Entering a burning building without having a safe plan of attack and emergency escape can be fatal.

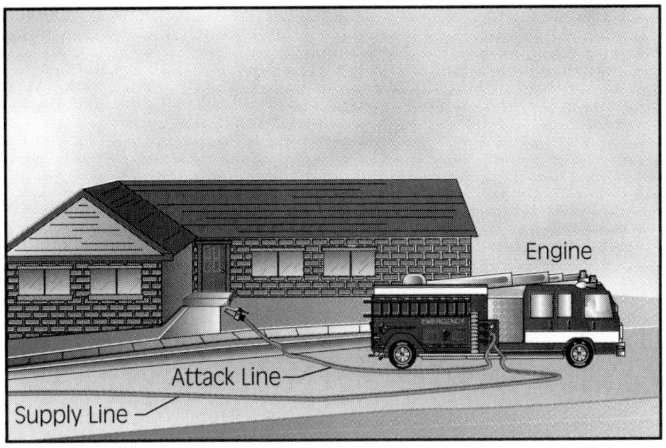

Figure 10-22 Park the engine to allow good access for hoselines without hindering any other apparatus placement.

At the entrance doorway, there should still be plenty of hose available. If the fire is not near this door, the line may be extended into the building as a shoulder load. However, when reaching the entrance or doorway to the area of the fire, the excess hose is removed from the shoulder and carefully laid in a manner to allow it to be easily pulled into the building or room. This would include straightening the hose and, if necessary, looping it back and forth so that it will pull without kinking. The nozzleperson and other firefighters should all be on the same side of the hoseline and properly spaced on the line for easier movement of the hose. The nozzleperson is at the nozzle, the officer or second firefighter a few feet (a meter or so) back, and the next firefighter, if extra ones are available, between the midpoint of the first section and the coupling, **Figure 10-23**. Prior to entering the structure, the door is checked for heat with the back of the hand with the glove off. With the line charged, the firefighter opens the nozzle to bleed off air from the hose and selects the proper pattern. (See nozzle operation in Chapter 11.) Crouching in a low or kneeling position while positioned to the side of the door, the firefighter opens the door and moves the line forward. When advancing the line past doorways, around corners, and furniture, remember that the extra time spent ensuring the line can smoothly move past will be saved when the line runs short and the crew has to wait while someone crawls back to free it.

Hoselines are usually advanced into buildings but to attack some fires the hose may have to be brought through an adjoining property. In these cases with the fire near the building exit, the doorway going outside should be treated as if it were the doorway going into the building and the steps just cited followed.

Figure 10-23 The nozzleperson is at the nozzle, the officer or second firefighter a few feet (a meter or so) back, and the next firefighter, if there is one, between the midpoint of the first section and the coupling.

Up and Down Stairs

When the fire is on a level other than the ground floor, firefighters will need to advance the line up or down stairs. The method used to advance the line to another floor depends on whether the fire is in an open stairwell or has taken control of the landing on the other floor. The first method discussed is for situations where the fire is not involving the stairs or landing, and the second is for situations where there is fire in the stairs.

If the fire does not involve the stairs, the best method is to advance an uncharged line to the fire floor. Make sure that the proper length and size of hoseline is chosen considering the length and size of the structure. The hoseline is advanced from the engine as described and the crew carries the shoulder loads into the building and up the stairs, allowing it to play out as needed, **Figure 10-24**. Use caution so that the hose does not slip back down the stairs. Do not allow it to get caught on stair handrails where it can become wedged between them. To help avoid this and the resulting kinking of the hose, lay the hose against the outside edge or wall of the stairs. If it is necessary to run the hoseline up between the handrails, it should be done in a vertical manner and tied off with a rope hose tool or strap at the upper level. The hose tool is passed around the hose and pulled snugly, near a coupling if possible. The rope or strap is wrapped around the railing several times and secured back on itself. When the desired floor is reached, the shoulder load is carefully deployed to avoid kinks. Do not just drop it into a pile; after it is charged the pressure will be reduced by the kinks and the "pile of spaghetti" will make movement on the landing nearly impossible. One ideal solution for the extra hose, if possible, is to run the line up the stairs toward the next level and back down to the fire floor. This creates a large loop that will be easy to pull, especially since gravity will help it move down, **Figure 10-25**. With the hoseline

Figure 10-24 A crew advancing an uncharged hoseline up stairs.

Figure 10-25 Extra hose on stairs creates a large loop that will be easy to pull.

in place, the nozzleperson has the line charged, bleeds the nozzle, and is ready to advance onto the floor.

If the stairwell or the landing on the desired floor is involved with fire, the crew will need to advance a charged line. The line is brought to the entrance or landing as described earlier, charged, and advanced at the point where the attack is to be made, **JPR 10-26A.** Care is taken to ensure that sufficient hose and personnel are available to move to the next floor without having to stop on the stairs, **JPR 10-26B.** The nozzle is operated to darken down the fire at the next landing, then shut down, and quickly advanced to that level, where it is opened again to secure the landing and moved forward to maintain the floor, **JPR 10-26C.** While advancing a hoseline on a stairwell is always a challenge, it is more so when going down the stairs. Firefighters going down the stairs will encounter the chimney effect of rising heat, hot gases, and possibly fire. "Making the floor" when the fire is on a lower level is a true test of a fire attack crew.

Using a Standpipe System

Advancing hoselines using a standpipe system involves two different hoseline evolutions. The first is the engine driver connecting to the fire department connection on the structure, and the second is

JOB PERFORMANCE REQUIREMENT 10-26

Advancing a Charged Hoseline Up a Stairwell

A Apply water from bottom of stairs to darken down fire.

B Shut down nozzle and advance quickly up stairs.

C At top of stairs, open nozzle and continue fire attack.

the hose crew connecting to the standpipe outlet and advancing the hoseline to attack the fire.

The pumper first establishes a water supply from the nearest hydrant whenever possible. Pre-emergency planning should have determined the water supply for each standpipe system, including water supply and any necessary hose evolutions if the fire department connection is not near the building entrance. The engine driver advances a supply line of medium- or large-diameter hose to the fire department connection. Local standard operating procedures (SOPs) should outline what size hose should be connected. Depending on the length of the lay and the style of hose load, the driver can drag the hose or make up a shoulder load. The hose load and type of coupling will determine the need for any adapters; the male end is needed at the siamese. Reaching the siamese, the driver removes all caps, usually two, and checks the operation of the clapper valve, **Figure 10-26**. If the clapper is inoperative but open, then advance a second line to the connection prior to charging the system. The driver connects the hose first to the left outlet, as the rotation of the coupling will be easier to do when a second line is connected, **Figure 10-27**. Returning to the engine, straightening any kinks, the driver disconnects the hose from the bed and connects to the pump's discharge outlet. The line(s) are charged to the proper pressure. If a second line has not already been run, it should now be done.

On occasion the fire department connection will be damaged or inoperable. Several steps can be taken to still use the system. Pick the one that gets the system into service the quickest. If the siamese swivel does not turn, place a double male into the siamese and add a double female to connect the hose, **Figure 10-28**. If the siamese cannot be used at all, the supply line must be extended into the building to a first floor outlet. To connect the hose, a double female will be required, **Figure 10-29**. After connecting the hose but before opening the outlet, the driver must return to the engine and break the supply line and connect to the discharge outlet and charge the line to the proper pressure. The driver then returns to the outlet and opens the valve, **Figure 10-30**. Opening the outlet valve prior to breaking, connecting, and charging the line will result in the standpipe draining into the hoseline and charging the hose bed. A second line should then be run to the next most convenient outlet following the steps outlined earlier.

The advancing of the interior hoseline to attack the fire should use a standpipe pack that includes the attack hose and nozzle, a gated wye, a short section of supply hose, and necessary wrenches and adapters. The crew proceeds first to the floor below the fire with the required equipment and then to the

Figure 10-26 Driver checking standpipe connection and clapper valve.

Figure 10-27 Driver attaching second supply line to FD connection. Note the first line is on left inlet.

Figure 10-28 If siamese swivel is inoperable, place a double male into the siamese and add a double female to connect the hose.

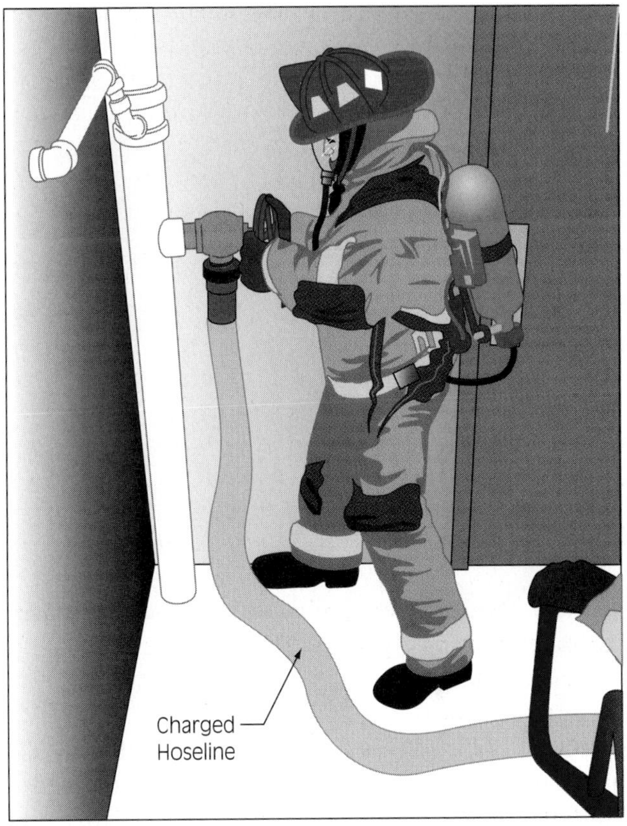

Figure 10-30 After connecting the hose at the outlet and the discharge outlet of the pump, the line is charged to the proper pressure. The driver then returns to the outlet and opens the valve.

Figure 10-29 If the siamese cannot be used at all, the supply line must be extended into the building to a first floor outlet. To connect the hose a double female will be required.

fire floor. Depending on fire conditions and location of the standpipe outlets in relation to the fire, an outlet on a lower floor can be chosen. The crew disassembles the hose pack and connects to the outlet, **Figure 10-31** and **Figure 10-32**. The crew stretches the hose either in the stairwell or hallway depending on fire location and conditions. The crew advances the line similar to a line from a stairs or door.

Working Hose Off Ladders

Advancing a hoseline up a ladder can be done with the line charged or uncharged. The best and safest manner is to advance an uncharged hoseline up the ladder and into the building or onto a fire escape before it is charged. The other method advances a charged hoseline up a ladder and either into the building or for operation from a ladder. This is the least preferred option and should be done using great caution when conditions require this method.

When advancing an uncharged hoseline over a ladder, start with the nozzleperson and or firefighters removing and advancing the shoulder loads as described earlier to the base of the ladder. The shoulder load is laid on the left side of the ladder,

Figure 10-31 Engine crew with standpipe equipment.

JPR 10-27A. The nozzleperson takes the nozzle and pulls it under the left armpit, across the chest, and over the right shoulder, allowing the nozzle to rest in the small of the back, **JPR 10-27B.** The nozzleperson, with both hands free for climbing, begins climbing the ladder to a point about 20 feet (6 m) up and stops. The next firefighter places the hoseline over the left shoulder at the next coupling and begins to climb the ladder. The two firefighters coordinate and maintain their distance on the ladder and hose, **JPR 10-27C** and **D**. Additional lengths and firefighters are added as needed and space allows.

Upon reaching the entry point of the building—a window, balcony, or the roof—the firefighters stop. The nozzleperson removes the nozzle and places it over the tip of the ladder into the building, **JPR 10-28A.** The nozzleperson climbs into the building, using the top rung of the ladder as a hose roller, and pulls the hoseline up as the next firefighter climbs, **JPR 10-28B.** The nozzleperson must complete pulling up the section as the next firefighter arrives at the tip, **JPR 10-28C**. The second firefighter pulls the remainder of the hose into the building while the nozzleperson advances the line. To remove the line from the building after use, the line is drained and the process of advancing the hose is reversed.

Another method to advance the line over a ladder involves using rope hose tools or straps wrapped around the line just below the nozzle and at the next coupling and the hose tool looped over the shoulder of the firefighters carrying the line up the ladder, **Figure 10-33**. Distances on the hose and ladder are the same as for the previous method.

Figure 10-32 Engine crew deployed with hose connected to standpipe.

Safety The best and safest manner is to advance an uncharged hoseline up the ladder and into the building or onto a fire escape before it is charged.

Advancing a charged hoseline over a ladder requires more firefighters to advance the hoseline, because the line is heavier and climbing with the line is very dangerous. It is usually done when the line is to be operated from the ladder rather than advanced into the building. The hoseline is actually passed by the firefighters rather than advanced by one or more firefighters. The hoseline is brought to the base of the ladder and charged and bled. Firefighters then climb the ladder spacing themselves at the opening and then at the feet of the next firefighter. When they are in place, the hoseline is passed up the ladder from one firefighter to the next until it reaches the opening, **Figure 10-34**.

JOB PERFORMANCE REQUIREMENT 10-27

Advancing an Uncharged Hoseline Over a Ladder

A The shoulder load is laid on the left side of the ladder.

B The nozzleperson takes the nozzle and pulls it under the left armpit, across the chest, and over the right shoulder, allowing the nozzle to rest in the small of the back.

C The next firefighter places the hoseline over the left shoulder at the next coupling and begins to climb the ladder.

D The two firefighters coordinate and maintain their distance on the ladder and hose.

JOB PERFORMANCE REQUIREMENT 10-28

Advancing an Uncharged Hoseline at Entry Point of Building

A The nozzleperson removes the nozzle and places it over the tip of the ladder into the building.

B The nozzleperson in the building, using the top rung of the ladder as a hose roller, pulls the hoseline up as the next firefighter climbs.

C The nozzleperson must complete pulling up the section as the next firefighter arrives at the tip.

Figure 10-33 Another method to advance the line over a ladder involves using rope hose tools or straps wrapped around the line just below the nozzle and at the next coupling. The hose tool or strap is looped over the shoulder of the firefighters to carry the line up the ladder.

> **Caution** Advancing a charged hoseline over a ladder requires more firefighters to advance the hoseline, because the line is heavier and climbing with the line is very dangerous.

If the line is to be used from the ladder, the nozzleperson puts the nozzle through the top two rungs of the ladder allowing 1 foot (0.3 m) to be extended beyond the rung for movement and the line secured with a rope hose tool. The rope hose tool is wrapped around the hoseline snugly and then two or three wraps are taken around the second rung and the hose tool secured on itself. The hoseline is also secured by rope hose tools every 20 feet (6 m) and at the base of the ladder, **Figure 10-35**. Operating a hoseline from a ground ladder requires the ladder to be securely tied in and heeled. (See Chapter 14.)

Figure 10-34 Firefighters passing a charged hoseline up a ladder from one firefighter to the next until it reaches the opening.

> **Safety** Operating a hoseline from a ground ladder requires the ladder to be securely tied in and heeled.

An uncharged hoseline can also be advanced to the tip of a ladder, secured in the same manner as the charged line, and then operated. The same safety precautions apply plus the line should be charged in a very careful and slow manner.

ESTABLISHING A WATER SUPPLY CONNECTION

Several different methods exist for establishing a water supply depending on the type of water source (static or hydrant), style of hydrant, the hose lays

Figure 10-35 A charged hoseline on a ladder. The hoseline is also secured by rope hose tools every 20 feet (6 m) and at the base of the ladder.

used (discussed later), and whether a pumper will be used at the water source. Static water sources obviously require a pumper at the source, but a hydrant source may not, depending on the volume and pressure of the hydrant and its distance from the fire. A supported fire hydrant is one that has a pumper at the fire and another pumper at the hydrant.

Firefighter Fact Early fire department pumpers needed one unit to do the pumping at the water source and another to carry the hose to the fire. The pumper was called the pumper and the hose cart was called a wagon. Even after modern engines were capable of supplying adequate pressure and volume, some fire departments continued to operate a two-piece engine company using a wagon–pumper combination until staffing requirements limited the concept.

Firefighters need to be able to connect directly to a fire hydrant or assist the engine driver in making the connections to hydrant and static sources.

From Hydrants

Using an unsupported hydrant requires a hoseline to be connected to a fire hydrant without an engine at the hydrant. This can be done using a single hoseline, multiple lines, or with adapters or valves to allow additional lines or a pumper to be connected later. Most departments use some type of finished hose load for a layout load. This load has a roll or several folds of hose to make any connections, a hydrant wrench and necessary adapters, and a rope to tie it all together and to allow the hydrant to be wrapped. "Wrapping the hydrant" allows the hose to be pulled far enough to reach the hydrant and helps prevent the hose from being dragged by the engine as it pulls away. The engine stops at the hydrant or layout point and the officer gives instruction for the hose lay. The layout person dismounts the vehicle and removes the layout load from the hose bed. Any additional adapters should also be removed from the engine. The layout person pulls sufficient hose and advances it to the hydrant and wraps either the hose or rope around the hydrant to secure it. The firefighter should be positioned to prevent being pinned between the hose and the hydrant. At this point the engine can move forward completing the hose lay, **Figure 10-36**.

Safety Caution should be used to ensure that the layout person is safely away from the vehicle prior to moving it. Firefighters have been struck by their own engine or other vehicles while laying out. Many departments now require the officer to dismount the vehicle to supervise this operation, and the engine does not move until the officer has remounted the vehicle.

After unwrapping the hydrant and removing the rope and tools, the hose is stretched out to remove any kinks. The layout person picks the proper hydrant outlet and removes the cap. This outlet depends on hydrant style and size and SOPs. On a wet barrel hydrant, some departments utilize a gate valve or a switch valve for additional lines or a pumper to be connected without shutting down the hydrant, **Figure 10-37**. To remove the cap, the hydrant wrench may be needed. The hydrant itself should then be briefly flushed prior to connecting the hose. The hose or adapter is leveled at the outlet and the coupling is screwed on, carefully ensuring that the threads are not crossed. The coupling should be hand tightened, although wrench tightening may be needed after turning the hydrant on, **Figure 10-38**. The hydrant valve is opened when water is called for and leaks and kinks are removed as the layout person returns to the engine.

Figure 10-36 The layout person pulls the layout section and enough hose to reach and wrap the hydrant.

Figure 10-37 A supply line with a switch valve connected to hydrant.

> **Streetsmart Tip** Care should be taken in removing kinks in large-diameter hose (LDH). For safety reasons, the firefighter should always try to "kick" the kink out of LDH. If it is necessary to lift the LDH to remove a kink, it should be done on the charged and pressurized side of the hose (toward the hydrant). It is important to use a strong lifting position (using the legs and a tight "core" posture).

An engine can be connected directly to a fire hydrant or to a switch valve already in service to supply attack lines or to support another engine. A soft sleeve hose is typically connected to the large hydrant outlet or the switch valve. The hard sleeve may also be used and other supply hose as well. Remember that the larger the hose, the more water that can be moved through it. The engine should be set up with gate valves at the inlet to speed up the process; in fact, some fire departments carry the soft sleeve preconnected. Care should be taken to make sure that the hose is connected to an inlet rather than a discharge valve of the pumper.

Connecting a soft sleeve to a hydrant starts with proper parking of the engine. The charged line should

Figure 10-38 The layout person making the hose connection to a hydrant.

be without kinks and undue stress on the hose or couplings. If the hose rests on the ground, it should have a chafing block between it and the ground. The hose is removed from its compartment or bed and stretched from the engine to the hydrant, **JPR 10-29A.** The outlet cap is removed, the hydrant flushed, and the coupling connected, **JPR 10-29B.** If the line is not preconnected, the hose must be attached to the engine's inlet. If a gate valve is on the inlet, the hose is connected directly to it. If not preconnected, the inlet cap must be removed. This will cause water in the pump to flow out; the tank-to-pump valve should be closed. The sleeve is then connected to the inlet, **JPR 10-29C**. The hydrant valve can now be opened slowly to avoid a water hammer. The gate valve on the engine or switch valve on the hydrant must also be opened if used.

Connecting a hard sleeve to a hydrant can also be done, but it is not recommended due to the chance of drawing a vacuum on the hydrant or piping and causing damage. The positioning of the engine is even more critical with the hard sleeve. The best way to do this is to stop the engine just prior to the hydrant. If the side intake is used, the engine should be at a slight angle. The hard sleeve is removed from its bed and attached to the inlet of the engine, **JPR 10-30A.** The engine is then moved forward toward the hydrant with several firefighters carrying and positioning the sleeve, **JPR 10-30B.** When the hard sleeve is aligned with the outlet, the engine is stopped and the sleeve connected, **JPR 10-30C**.

Medium- or large-diameter hose is used to connect an engine to a hydrant as a standard operation or used when the soft sleeve cannot reach the hydrant. The connection would be similar to connecting a layout line to the hydrant with a shorter distance.

From Static Water Supplies

The supply of water from a static source requires use of an engine and its hard sleeves to draft the water. The engine has to be able to position itself close enough to place the hard sleeve and strainer into the water or else reach the dry hydrant connection. Connecting a hard sleeve to a dry hydrant, not a dry barrel hydrant, is the same procedure as connecting to a regular hydrant without the normal concerns surrounding creation of a vacuum. A vacuum has to be created and it is important to make a tight connection to prevent leakage and loss of the vacuum.

The placement of a hard sleeve and strainer directly into a water source is somewhat similar to using it on a hydrant. The engine is placed near the water source but slightly back and at an angle, **JPR 10-31A.** The hard sleeve and strainer are attached to the pumper's inlet and the engine moves forward as

JOB PERFORMANCE REQUIREMENT 10-29

Soft Sleeve Hydrant Connection

A Stretch a non-preconnected soft sleeve for making a connection.

B Connect to the steamer connection on the hydrant.

C Connect to the steamer connection on the pumper.

JOB PERFORMANCE REQUIREMENT 10-30

Hard Sleeve Hydrant Connection

A Connect hard sleeve to engine inlet.

B Move engine and sleeve toward hydrant.

C Attach sleeve to hydrant outlet.

the crew positions the sleeve into the water. A rope is attached to the strainer to assist in its placement, **JPR 10-31B** and **C.** If a nonfloating strainer is used, the rope also keeps the strainer from resting on the bottom. Nonfloating strainers should be well under the water's surface to prevent drawing in of air. When in place the engine is stopped, the line is secured, and drafting operations begin, **JPR 10-31D** and **E**.

EXTENDING HOSELINES

Despite the best efforts of judging the needed length of a hoseline, there will be occasions when the line comes up short and will need to be extended. Wildland firefighters encounter this often, not because of misjudgment, but because they have successfully driven the fire back beyond the length of their line. Whatever the case, all firefighters should be familiar with techniques used to extend their hoselines. While the pump operator can shut down the line when the additional hose is needed, it is easier to extend the line using either a break-apart nozzle or a hose clamp.

The preferred method of extending a hoseline is to use a break-apart nozzle. The additional hose is brought to the nozzle end of the hoseline and the line shut down with the control handle. The nozzle tip is removed and the additional hose added if it has 1½-inch (38-mm) couplings, **JPR 10-32A** and **B.** If extending a line with hose couplings larger than 1½ inch (38 mm), then a 1½-inch (38-mm) to 2½-inch (65-mm) increaser will be required at the control handle, **JPR 10-32C.** The additional line with the nozzle tip attached is advanced and charged from the control valve, **JPR 10-32D**.

JOB PERFORMANCE REQUIREMENT 10-31

Setting up for Drafting Operation

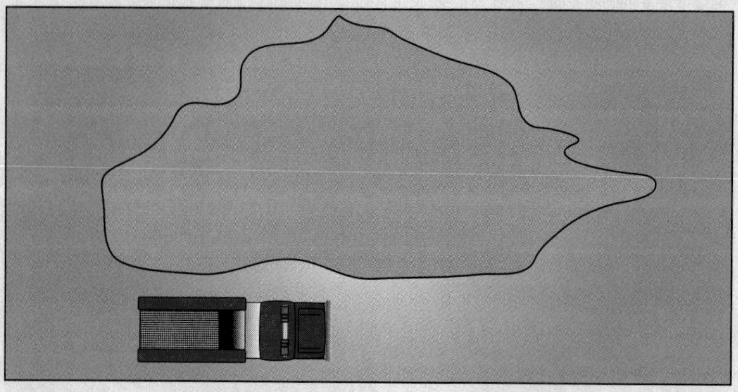

A Pre-position the engine at an angle to the water source.

B Connect the hard sleeve to the engine inlet.

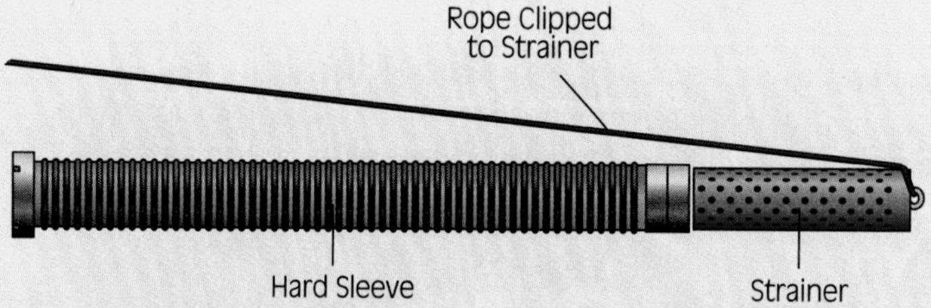

C Attach a strainer and guide rope to the other end of the sleeve.

D Move the engine and sleeve toward the water source.

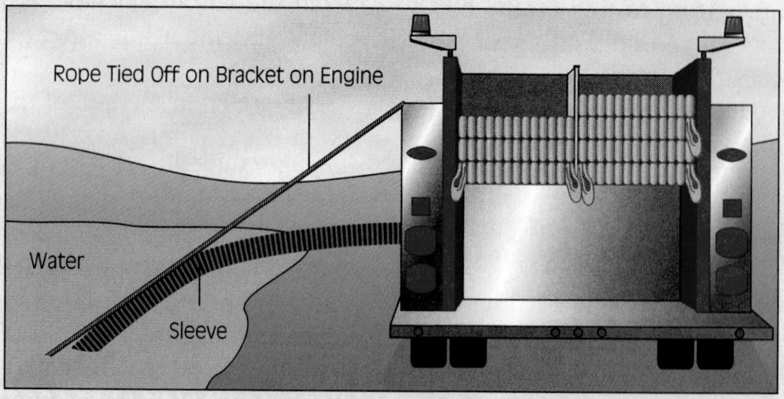

E Ensure that sleeve is in position and strainer is tied off for operations.

JOB PERFORMANCE REQUIREMENT 10-32

Extending a Hoseline with a Break-Apart Nozzle

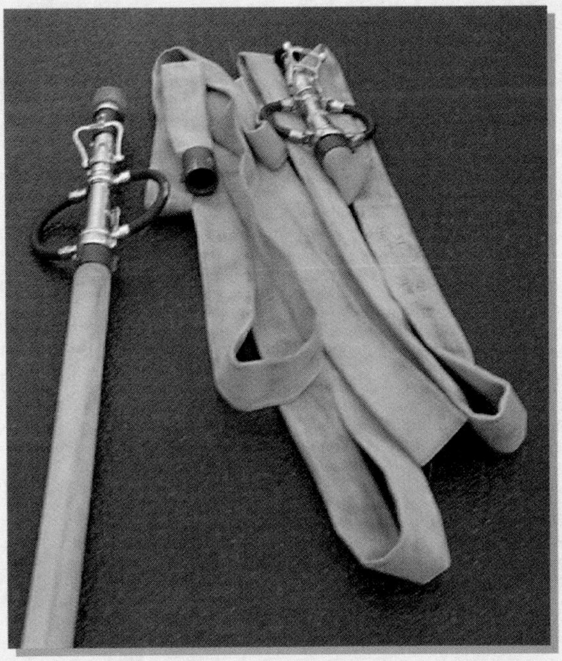

A Bring additional hose to the end of the hoseline and shut the line down. Remove the nozzle tip and add the additional hose if it has 1½-inch (38-mm) couplings.

B Remove the nozzle from the break-apart handle and connect hose.

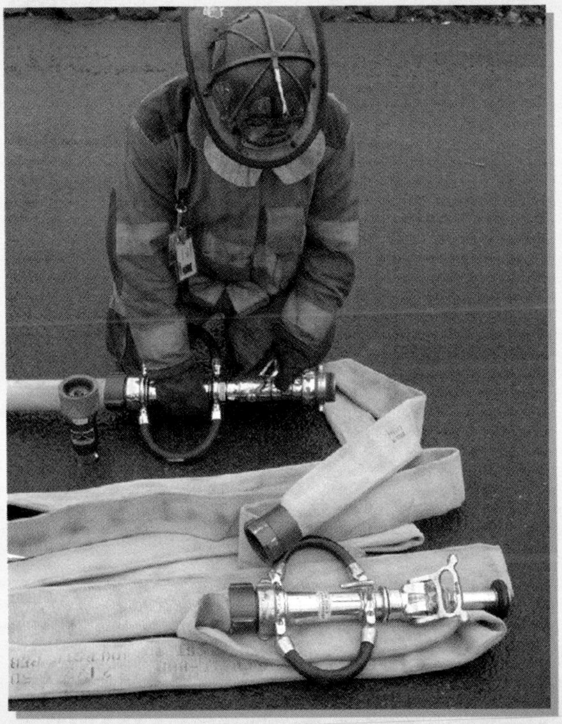

C If extending a line with hose couplings larger than 1½ inches (38 mm), then a 1½-inch (38-mm) to 2½-inch (65-mm) increaser will be required at the control handle.

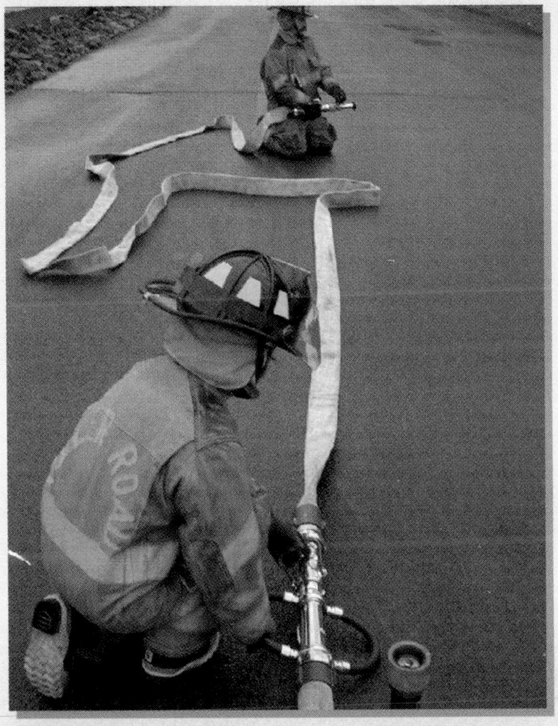

D The additional line with the nozzle tip attached is advanced and charged from the control valve.

The other method is to use a hose clamp. The clamp and additional hose are brought up to the nozzle end of the hose. Clamping a hose can be dangerous, so firefighters need to ensure that the clamp is safely operated and is fully locked before releasing the clamp. The firefighter is positioned to the side of the clamp to prevent injury if it slips. The hoseline is clamped just before the nozzle and the nozzle removed. The additional line is attached as is the nozzle and the line is then advanced. When water is called for, a firefighter removes the hose clamp, **JPR 10-33A** and **B**.

Wildland firefighters often extend their hoselines as they advance on a fire. Some of their hose loads plan for this with the addition of a gated wye at each layer of hose. To extend these lines, the firefighters can simply go back to the last connection and close the gate until the additional hose is added, the nozzle reattached, and the gate reopened. They also use the hose clamp method, **JPR 10-34A–D**.

REPLACING SECTIONS OF BURST HOSE

The bursting of a section of hose is a dangerous situation that stops the flow of water to the fire and can injure firefighters and bystanders while causing property or water damage. It is a situation that requires immediate attention, especially when firefighters are operating under fire conditions. The hoseline must immediately be shut down.

The other firefighter should then remove the burst section and replace the section or if enough hose is available, reconnect the two adjacent sections. If enough hose is not available, a new section must quickly be brought and connected. The line may then be recharged.

Typically, the pump operator will shut down the line. If this is not possible, a hose clamp can be used

JOB PERFORMANCE REQUIREMENT 10-33

Extending a Hoseline Using a Hose Clamp

A Place and clamp hose clamp on hose behind nozzle.

B Remove nozzle, add additional hose, and release hose clamp.

JOB PERFORMANCE REQUIREMENT 10-34

Wildland Hose Advancing and Extension

A A preconnected line is pulled to begin the hose lay. As the line is moved forward, it is operated for suppression and crew protection.

B When the line is almost fully extended, a second line is rolled back in preparation for the next length. A working area is created with the water before the line is clamped off.

C With the line clamped, the nozzle is removed and placed on the new length, and the hoses are connected.

D The clamp is released and the line advanced. When the line is again fully extended, the team repeats the clamping and extending process.

Photos by Brentt Sporn, California Fire Photo

to stop the water flow. If no clamp is available, a firefighter can fold the hose twice over itself and kneel down to hold pressure buildup in the kinks. This will stop the flow sufficiently to have another firefighter replace the burst section.

HOSE LAY PROCEDURES

Hose lay procedures bring the water to the fire location by placing supply hose between the water source and the attack engine. SOPs should cover preferred hose lays and water supply operations, and apparatus is then designed for these preferences. The direction of the engine's travel to or from the water source gives each procedure its name. Supply lines and the hose beds on apparatus are designed to use at least one of the following three hose lays techniques; many are designed to allow all three. Any needed adapters are either attached to the supply line or readily accessible to the layout person. Each of these may work with either a hydrant or static source of water with the static source needing a supply engine at the source.

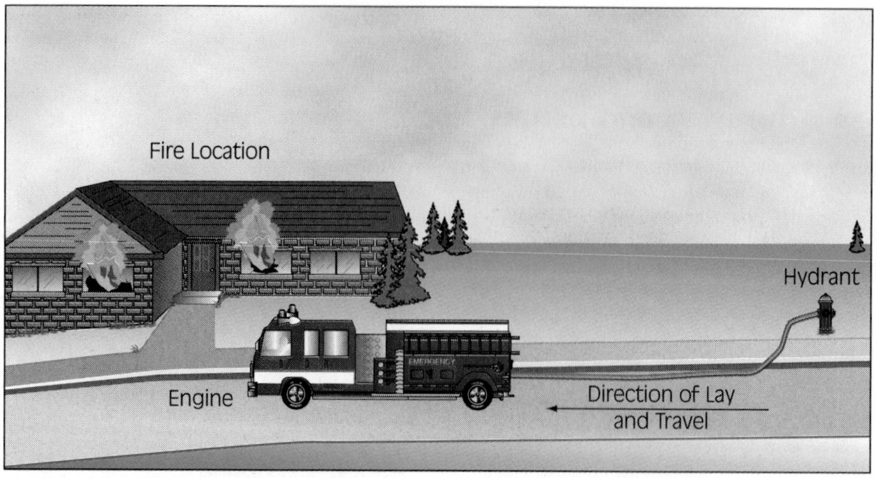

Figure 10-39 The forward or straight hose lay.

Forward Lay

Forward lay refers to the engine stopping first at the water source to drop off a supply line(s) and then advancing to the location of the fire, **Figure 10-39**. The forward lay is a preferred lay because a water supply can be established by connecting and charging the hydrant and may need no additional engines for the water supply. The attack engine is closer to the fire with access to additional hoselines and equipment. The forward lay works best when a water source is located in the approach path of the engine. An additional engine may still be required at the water source if needed to support the pressure or volume and will be required if the source is a static water source. The forward lay can also be used with a portable water tank operation where the fire is off the main road and the tank dump site is on the main road.

Reverse Lay

A reverse lay is the opposite of the forward lay with the supply line being dropped off at the fire location and the engine laying the hose toward the water source, **Figure 10-40**. Typically, the first arriving engine is used as the attack engine with the second engine doing the reverse lay and being the support or supply engine. Reverse lays can be used where a manifold, attack lines, and equipment can be dropped off at the fire with the engine going to the water source. This is less preferred but may be necessary in areas with few responding units and poor or static water sources.

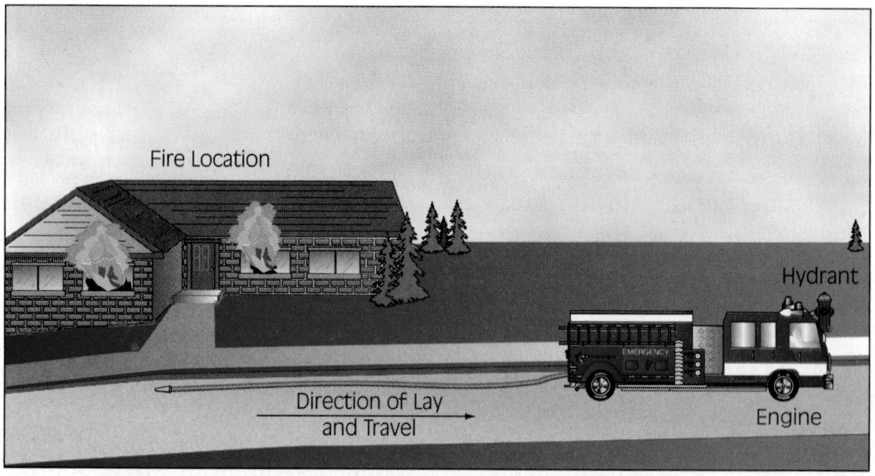

Figure 10-40 The reverse hose lay.

Figure 10-41 The split hose lay.

Split Lay

The split lay is used where the fire and the water source are in two different directions, such as on two different streets, thus needing to split the lay between two engines. The split lay has the first engine laying its line from a point or intersection to the fire location and the second engine laying its line from the point to the water source, **Figure 10-41**.

DEPLOYING MASTER STREAM DEVICES

Master streams or **heavy appliances** are non-handheld water applicators capable of flowing over 350 gallons of water per minute (1325 L/min). Four basic types of master stream devices are used, although the names are sometimes interchanged. The wagon pipe is a permanently mounted master stream device on an engine that has either prepiped water connection or needs a short section of hose to connect it to the pump, **Figure 10-42**. A similar looking device called a deluge set is not permanently mounted so it can also be removed and operated from the ground. Different models of deluge sets have one base that can be used both mounted on the apparatus or on the ground while others have separate bases for each mounting. A monitor pipe is a permanently attached master stream device with a prepiped waterway on an aerial device such as an aerial ladder or platform. A ladder pipe is a non-permanently mounted device that needs a hoseline for its waterway on an aerial ladder or platform. Some ladder trucks have both a monitor pipe and a ladder pipe, **Figure 10-43**.

Safety Master stream appliances are often called the "artillery" of the fire service. Because of the power of these streams, great care should be used in the setup and operation of these appliances. One long-standing rule states that no master stream shall be introduced into a structure where interior firefighters are operating. Individual companies should be well versed in setting up a master stream to protect exterior exposures or to make a quick knock on a fire before firefighters enter a structure. Orders to charge a master stream after fire attack operations have begun should come only from the incident commander or operations section chief as part of a well-communicated and coordinated incident action plan (IAP).

When operating master streams with solid stream tips, the following rules apply for figuring reach, vertically and horizontally. For every foot (0.3 m) of vertical reach needed, move the device 1 foot (0.3 m) away. For horizontal reach, each pound of pressure (6.89 kPa) equals 1 foot (0.3 m) of reach. An angle of 45 degrees is about the maximum angle that can produce an efficient stream. The reach is generally limited to three floors above the nozzle's height.

The wagon pipe is a permanently mounted device that usually has a prepipe water connection needing no hoselines to be attached to it. Some departments refer to wagon pipes as a "deck gun." If it is without permanent piping, then either short or regular sections of supply hose can be run from the discharge to the inlets of the device. It is about 9 feet (2.8 m) in the air and can reach about four stories with a solid stream nozzle. A wagon pipe can also be directed at lower angles than a ground-mounted deluge set.

(A)

(B)

Figure 10-42 (A) A wagon pipe. (B) A deluge set and a wagon pipe operating.

The monitor pipe has a prepiped waterway with either a direct discharge valve if the unit also has a pump or needs a supply hoseline(s) if not equipped with a pump. If a hoseline from an engine is needed, the monitor pipe has a siamese or manifold that can be supplied similarly to a standpipe system, as described earlier.

Deluge sets when operated from the top of an engine may have prepiped water connections to a

Figure 10-43 A ladder truck with both a monitor pipe and ladder pipe.

pump and need no supply lines attached. Some newer ones come with an extension device or are able to be adjusted in several positions that extend their reach several feet. Supply lines can be run up from the discharge to the inlet if no prepipe water connection is available. When securely attached to the engine, they can also be operated at low angles or can reach four stories. Portable deluge sets operated from the ground may not be operated at an angle lower than 25 degrees. Most deluge sets have a pin or other device that prevents them from going below this angle, and this pin should only be removed when the set is bolted down to the apparatus. Going below this angle can cause the set to become unstable.

Safety Portable deluge sets operated from the ground may not be operated at an angle lower than 25 degrees. Most deluge sets have a pin that prevents them from going below this angle, and this pin should only be removed when the set is bolted down to the apparatus. Going below this angle can cause the set to become very unstable.

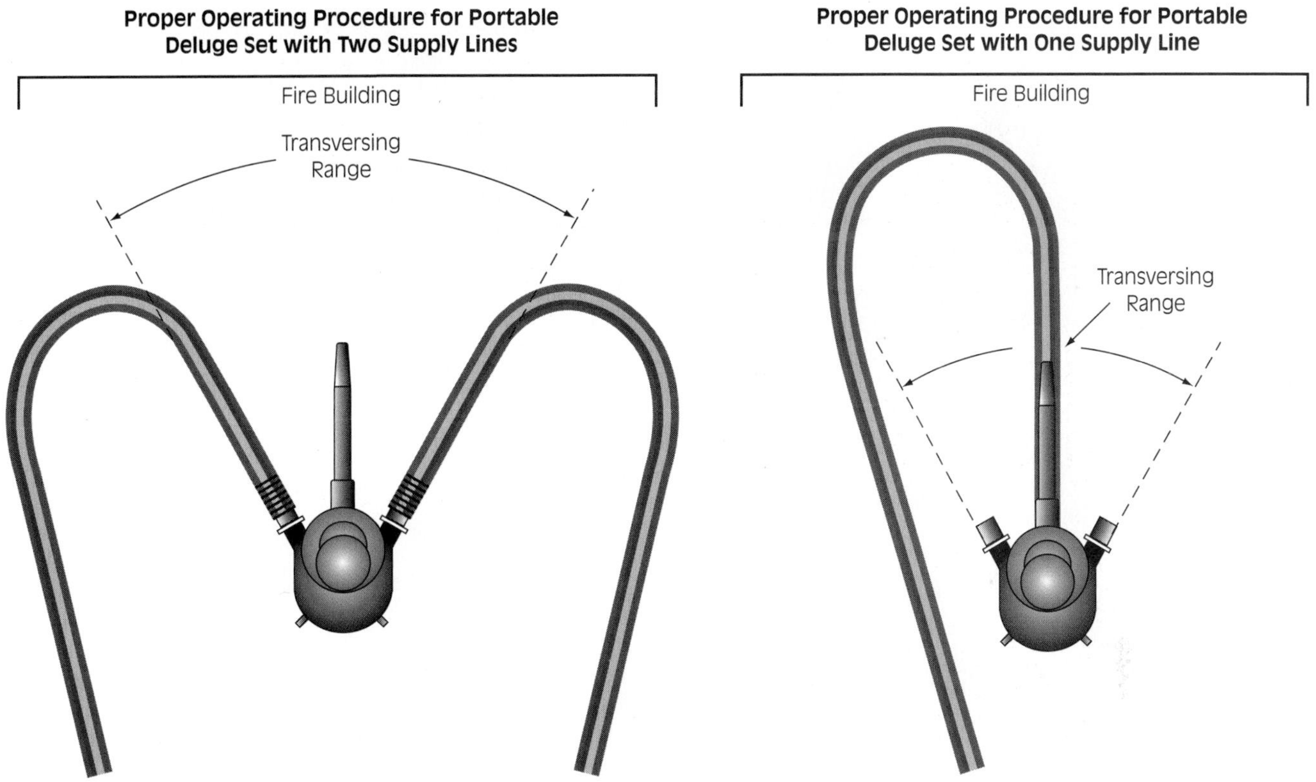

Figure 10-44 Proper operations of portable deluge sets.

When operating a portable deluge set on the ground, the intakes normally should be facing the fire building to counteract the nozzle reaction. Some newer balanced flow models do not need to have their intakes facing the building. Hoselines connected to the intakes should be balanced among the intakes and if only one hose is run, the nozzle shall be operated directly over that intake. When operating the set at the minimum 25-degree angle, the nozzle should not pass beyond the two outboard intakes, **Figure 10-44**. The advancing of medium- or large-diameter hoselines to the deluge set is also similar to the standpipe siamese described earlier, **Figure 10-45**.

The ladder pipe is not permanently mounted and therefore needs a hoseline for rigging it up the ladder and another for supply. The ladder pipe is often assembled, completely or partially, with its siamese, hose for the ladder, guide ropes, and the pipe itself. The ladder pipe is removed from its bracket and mounted on the aerial ladder in the center. The nozzle and **stream straighter** are adjusted to the operating position and guide ropes attached. If the hose is also attached, it is deployed down the aerial to the turntable and the ground. If the hose is not attached, it is often run up the ladder from the turntable and attached to the pipe. In either case the hose is run in the center of the ladder and secured at the top with a rope hose tool. The bottom of the hose and the

Figure 10-45 A portable deluge set in operation.

(A) (B)

Figure 10-46 Ladder pipe operations. (A) Ladder pipe on ladder. (B) Ladder pipe hose with rope hose tools secured to the hose. Notice the hose is run in the center of the ladder.

siamese are placed on the ground and the aerial is elevated to its operating position. Another rope hose tool is attached at the base of the aerial, **Figure 10-46**. A supply line(s) is run from an engine to the siamese and attached. When water is called for, the lines are charged slowly and carefully to ensure everything is properly connected.

SERVICE TESTING OF FIRE HOSE

Fire hose needs to be tested prior to being placed in use and then retested annually during its lifetime. Hose also should be tested after being damaged and after repairs have been made. To ensure accuracy of the testing program and for personnel safety, a record-keeping system must be utilized. The record system should have identification numbers, dates of testing, repairs, and other important notations about each section.[1]

The testing of hose begins with a visual inspection of the hose coupling. This inspection should be done with the annual test and during routine reloading and reconnection of hose sections. The inspection should check thread damage. The coupling should not be misshapened and it should swivel freely. Missing, worn, or damaged gaskets should be replaced and the inspector should note any missing lugs, slippage of the hose, or loose collars.[2] As the hose is loaded, it should also be visually inspected for any type of damage. Any damaged or suspect sections should be removed from service until tested or repaired.

The service test consists of testing the hose under pressure or, in the case of a hard sleeve, under a vacuum. These tests can be done with a hose testing machine, a fire apparatus pump, or a stationary pump. Pressure testing is designed to check for hose failure, and proper safety precautions should always be taken. Firefighters should follow manufacturers' instructions or NFPA standards. The hose is laid out in a straight line and a visual inspection conducted. The hose is marked, using a new color each year, at the point where the couplings are connected to check for any slippage of the coupling under pressure, **Figure 10-47**. A hose test valve, a gate valve with a ¼-inch (6.4-mm) opening drilled into it, is placed on the discharge of the pump. This hose test valve is designed to limit the flow rate of the water into the hose to reduce chances of injury if the hose fails during the test.

The hoselines are attached to the pumping device with no more than 300 feet (91 m) of length per line and a nozzle or test cap with a bleeder valve at the other end. The hoseline is charged to about 45 psi (310 kPa) with the nozzle open to bleed off any air

Figure 10-47 The hose is marked at the point where the couplings are connected to check for any slippage of the coupling under pressure.

and then closed. The hoseline and couplings are checked for leaks with couplings tightened with a spanner if necessary. The hose is now ready for the pressure testing.

Each new length of hose made after 1987 has its service test pressure rating stamped on it, **Figure 10-48**. In most cases, this is about 110 percent above its maximum operating pressure and maintained for at least three minutes.[3] The highest pressure is usually about 250 psi (1720 kPa) as the fire apparatus pumps are usually tested with this as their maximum pressure. SOPs should define maximum operating pressures for hoselines and pumps for each department. After the test is completed the hose has the current date written on it, and it is cleaned, dried, and returned to service or storage.

> **Safety** Hose testing is a destructive process that identifies weak hose by causing it to burst. Firefighters should secure the nozzle end of the hose and operate hose valves from a distance to prevent personal injury if the hoseline breaks and begins to whip about.

Hard sleeves are tested by being connected to a suction source and capped at the opposite end with a transparent cap. A vacuum of 22 inches of mercury (74.5 kPa) must be reached and maintained for 10 minutes while the inner lining is viewed for any collapse. If the hard sleeve is used under positive pressure, it must pass a service test of 165 psi (1138 kPa).[4]

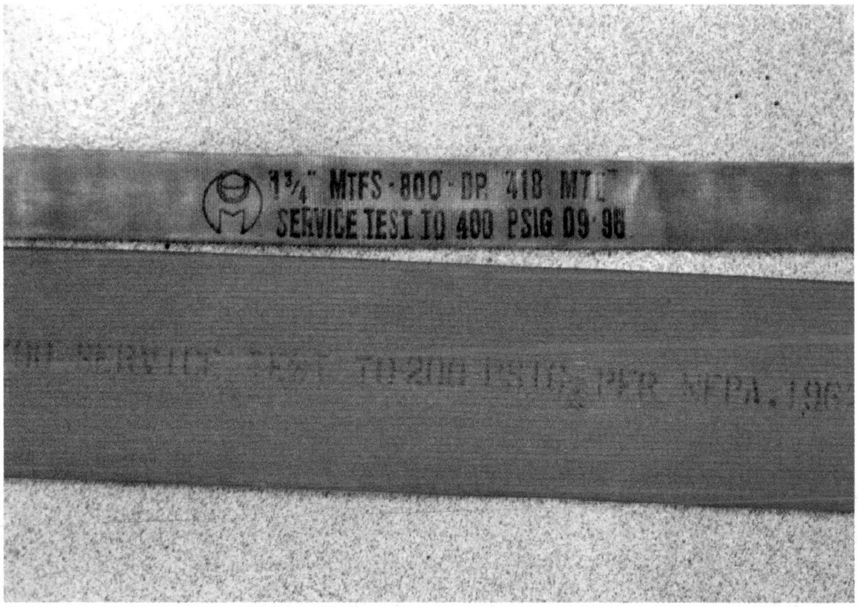

Figure 10-48 Each new length of hose made after 1987 has its pressure testing rating stamped on it.

Lessons Learned

Fire hose, adapters, and appliances allow firefighters to move water at a distance from its source and from the apparatus's pumps. Without these valuable tools, firefighters would be extremely limited in the ability to move water and to conduct most fire suppression operations, especially interior fire attack. Firefighters must understand the proper use and care of these tools and how to connect, advance, and operate these tools. These are the basics of firefighting, and the best method of learning these basics is practical application.

KEY TERMS

Adapter Device that adapts or changes one type of hose thread to another, allowing connection of two different lines. Adapters have a male end on one side and a female on the other with each side being a different thread type, for example, an iron pipe to national standard adapter.

Attack Hose Small- to large-diameter hose used to supply nozzles and other applicators or protective system for fire attack. Attack hose commonly means handheld hoselines from 1½ to 2½ inches (38 or 63 mm) in diameter.

Backstretch or **Flying Stretch** An attack line lay where the engine is at the hydrant and the line is stretched back from the engine to the fire. The flying stretch is a version of the backstretch where the engine stops in front of the fire, the attack portion is removed, and the engine proceeds to the hydrant.

Booster Hose Smaller diameter, flexible hard-rubber-coated hose of ¾- or 1-inch (19- to 25-mm) size usually mounted on reel that can be used for small trash and grass fires or overhaul operations after the fire is out.

Distributor Pipe or **Extension Pipe** Devices that allow a nozzle or other device to be directed into holes to reach basements, attic, and floors that cannot be accessed by personnel. The distributor pipe has self-supporting brackets that help hold it into place when in use.

Double Female Allows the two male ends of hose to be connected.

Double Male Used to connect two female thread couplings.

Dutchman A short fold of hose or a reverse fold that is used when loading hose and a coupling comes at a point where a fold should take place or when two sets of couplings end up on top of or next to each other. The dutchman moves the coupling to another point in the load.

Ears Elongated folds or flaps at the ends of a layer of hose to assist in pulling that layer.

Fire Hose A flexible conduit used to convey water or other agent from a water source to the fire.

Forestry Hose Specially designed hose for use in forestry and wildland firefighting. It comes in 1- and 1½-inch (25- and 38-mm) sizes and should meet U.S. Forestry Service specifications.

Hard Suction Hose A special type of hose that does not collapse when used for drafting.

Helix The metal or plastic bands or rings used in hard suction hose to prevent its collapse under drafting conditions.

Higbee Cut The blunt ending of the threads of fire hose couplings that allows the threads to be properly matched, avoiding cross-threading.

Hose Bed The portion or compartment of fire apparatus that carries the hose.

Hose Bridges Devices that allow vehicles to pass over a section of hose without damaging it.

Hose Cap Does not allow water to flow through it. Instead, it caps the end of a hoseline or appliance to prevent water flow.

Hose Cart A handcart or flat cart modified to be able to carry hose and other equipment around large buildings. Some departments use them for high-rise situations.

Hose Clamp A device to control the flow of water by squeezing or clamping the hose shut. Some work by pushing a lever that closes the jaws of the device, and others have a screw mechanism or hydraulic pump that closes the jaws.

Hose Jackets Metal or leather devices used for stopping leaks without shutting down the line that is fitted over the leaking area and either clamped or strapped together to control the leak.

Hose Roller or **Hoist** A metal frame, with a securing rope, shaped to fit over a windowsill or edge of a roof with two rollers to allow the hose to roll over the edge, preventing chafe.

Hose Strap A short strap with a forged handle and cinch clip attached. Used to help maneuver hose and attach hose to ladders and stair rails.

Hydrant Valves or **Switch Valves** Valve used on a hydrant that allows an engine to connect and charge its supply line immediately but also allows an additional engine to connect to the same hydrant without shutting down the hydrant, and increases the flow of the hydrant.

Hydrant Wrenches Tools used to operate the valves on a hydrant. May also be used as a spanner wrench. Some are plain wrenches and others have a ratchet feature to speed the operation of the valve.

Increaser Used to connect a smaller hose to a larger one.

Intake Relief Valve Required on large-diameter hose at the receiving engine that functions as a combined overpressurization relief valve, a gate valve, and an air bleed-off.

Jacket The outer part of the hose, often a woven cloth or rubberized material, which protects the hose from mechanical and other damage.

Liner The inner layer of fire hose, usually made of rubber or a plastic material, that keeps the water in the tubing of the hose.

Master Stream or **Heavy Appliances** Non-handheld water applicator capable of flowing over 350 gallons of water per minute (1325 L/min).

Medium-Diameter Hose (MDH) Either 2½- or 3-inch (63- or 75-mm) hose.

Occupant Use Hose Hose that is used in standpipe systems for building occupants to fight incipient fires. It is usually 1½-inch (38-mm) single-jacket hose similar to attack hose.

Portable Hydrant or **Manifold** Like a large water thief and may have one or more intakes and numerous outlets to allow multiple hoselines to be utilized with or without a pumper at the fire location.

Reducers Used to connect a larger hose to a smaller one.

Rope Hose Tool About 6 feet (2 m) of ½-inch (13-mm) rope spliced into a loop with a large metal hook at one end and a 2-inch (50-mm) ring at the other. Used to tie in hose and ladders, carry hose, and perform many other tasks requiring a short piece of rope.

Shoulder Load Hose load designed to be carried on the shoulders of firefighters.

Siamese A device that connects two or more hoselines into one line with either a clapper valve or gate valve to prevent loss of water if only one line is connected.

Slot Loads Narrow section of a hose bed where hose is flat loaded in the slot.

Small Lines or **Small-Diameter Hose** Hose less than 2½ inches (63 mm) in diameter.

Soft Suction Hose Large-diameter woven hose used to connect a pumper to a hydrant. Also known as a soft sleeve.

Spanner Wrenches Used to tighten or loosen couplings. They may also be useful as a pry bar, door chock, gas valve control, and so on.

Storz Couplings The most popular of the non-threaded hose couplings.

Strainers Placed over the end of a suction hose to prevent debris from being sucked into the pump. Some strainers have a float attached to keep them at or near the water's surface. A different style of strainer or screen is located on each intake of a pump.

Stream Straighter A metal tube, commonly with metal vanes inside it, between a master stream appliance and its solid nozzle tip. The purpose is to reduce any turbulence in the stream, allowing it to flow straighter.

Supply Hose or **Large-Diameter Hose (LDH)** Larger hose [3½ inches (90 mm) or bigger] used to move water from the water source to attack units. Common sizes are 4 and 5 inches (100 to 125 mm).

Water Thief A variation of the wye that has one inlet and one outlet of the same size plus two smaller outlets with all of the outlets being gated. The standard water thief usually has a 2½-inch (65-mm) inlet with one 2½-inch (65-mm) and two 1½-inch (38-mm) outlets.

Wye A device that divides one hoseline into two or more. The wye lines may be the same size or smaller size, and the wye may or may not have gate control valves to control the water flow.

REVIEW QUESTIONS

1. Describe the various types of hose construction and how each type is used by firefighters.
2. What is the difference between single- and double-jacketed hose?
3. What are the different types of hose couplings important to firefighters?
4. What is a hose clamp?
5. What types of wrenches are used with hose?
6. What are double males and double females used for?
7. Describe the foot-tilt method of coupling hose.
8. What is a dutchman?
9. Describe flat, accordion, and horseshoe loads.
10. How is a hoseline advanced up a stairwell?
11. Describe connecting a supply line to a hydrant.
12. How can a hoseline be extended without shutting down the line at the pump?
13. What is a forward lay?
14. What is a deluge set?
15. What is a triple-layer load?
16. Describe the testing of fire hose.

Endnotes

1. *NFPA 1962: Standard for Care, Use and Testing of Fire Hose Including Couplings and Nozzles,* pp. 1962-7. National Fire Protection Association, Quincy, MA, 1998.
2. *NFPA 1962: Standard for Care, Use and Testing of Fire Hose Including Couplings and Nozzles,* pp. 1962-7. National Fire Protection Association, Quincy, MA, 1998.
3. *NFPA 1962: Standard for Care, Use and Testing of Fire Hose Including Couplings and Nozzles,* pp. 1962-9. National Fire Protection Association, Quincy, MA, 1998.
4. *NFPA 1962: Standard for Care, Use and Testing of Fire Hose Including Couplings and Nozzles,* pp. 1962-9. National Fire Protection Association, Quincy, MA, 1998.

Additional Resources

Richman, Harold, *Engine Company Fireground Operations,* 2nd ed. National Fire Protection Association, Quincy, MA, 1986.

Smoke, Clinton H., *Company Officer,* Appendix B: Suggested Training Evolutions. Delmar Learning, a part of the Thomson Corporation, Clifton Park, NY, 1999.

Sturtevant, Thomas, *Introduction to Fire Pump Operations,* Chapter 4: Hose, Nozzles, and Appliances. Delmar Learning, a part of the Thomson Corporation, Clifton Park, NY, 1997.

CHAPTER 11

NOZZLES, FIRE STREAMS, AND FOAM

Ric Koonce, J. Sargeant Reynolds Community College

OUTLINE

- Objectives
- Introduction
- Definition of Fire Stream
- Nozzles
- Operating Hoselines
- Stream Application, Hydraulics, and Adverse Conditions
- Types of Foam and Foam Systems
- Foam Characteristics
- Classification of Fuels
- Application of Foam
- Lessons Learned
- Key Terms
- Review Questions
- Endnotes
- Additional Resources

Photo by Fred Schall

STREET STORY

A few of the senior firefighters in my engine company had told me that when I was the nozzle man to always keep the nozzle pointed up and in the ready position when waiting to move into a burning room. In this way it will be ready to quickly knock down any fire that might roll out the top of the door when the door was forced open by the forcible entry team. If the nozzle was pointed downward, toward the floor, I might not get it pointing up and open it in time to protect the firefighters in the hall and, if the hallway was crowded with firefighters, it might not be possible to raise the nozzle quickly if it was blocked by people. Also, anytime I was in nozzle position, I was told never to put the nozzle down, even when things seemed under control, because you never know. As nozzle man, I was expected to be ready to open the line immediately, should the need arise

This lesson served me well in a cellar fire in an apartment building that we were called to in the early 1970s. The fire involved the gas heating unit and had spread to most of the cellar. I had extinguished the fire, and the truck company reported that the gas to the heating unit was turned off. There were three firefighters overhauling in the room containing the heating unit. I was kneeling at the door to the room with the nozzle in my hands pointed up and forward toward the ceiling when suddenly the entire room lit up—leaking gas had ignited, filling the room with flame. The sudden burst of flame knocked me on my back. Instinctively, I pulled back on the nozzle's handle, opening the nozzle and directing the water from my hose stream into the room and onto the ceiling of the room, quickly extinguishing the flames and allowing the firefighters in the room to scramble to safety. Luckily, I had the nozzle in ready position even though the fire was out and everything had seemed to be under control. While extinguishing a gas flame is not the proper way to handle a gas fire, in this case it was necessary to allow firefighters to escape the flaming room. Proper training and attention to seemingly minor details can save lives. Never let your guard down. If you are the nozzle man, stay alert, hold onto the nozzle, and always be ready to open it.

—Street Story by Frank Montagna, Batallion Chief, FDNY, New York City

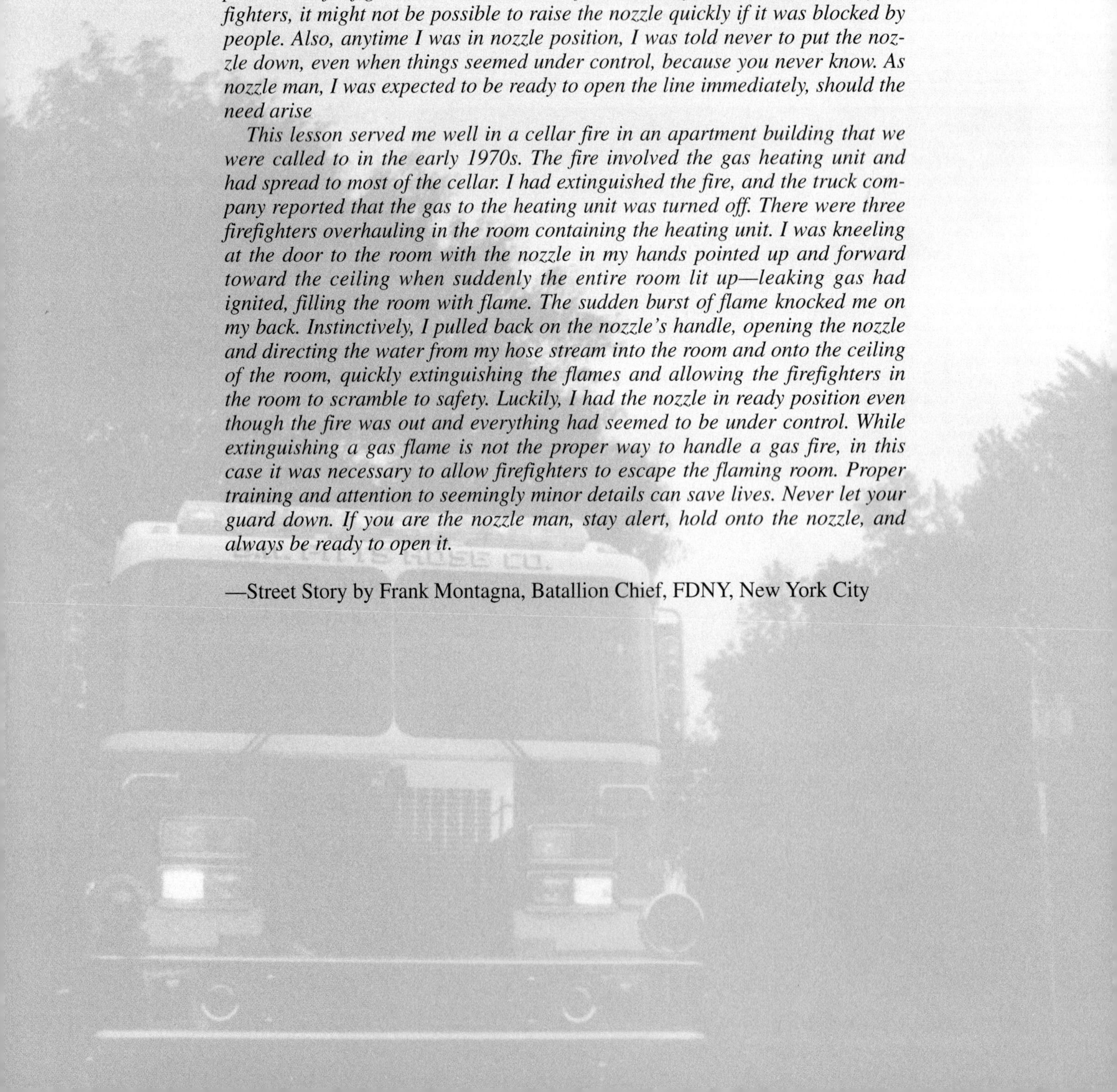

OBJECTIVES

After completing this chapter, the reader should be able to:

- Define a fire stream.
- Identify the purposes of a fire stream.
- Identify the various types of fire streams.
- Identify the types of nozzles.
- Explain the pattern and use of each type of nozzle.
- Demonstrate the operation of the various types of nozzles.
- Explain the operation and characteristics of various sizes (diameters) of fire streams.
- Explain the reach and application of various sizes of fire streams.
- Identify the three types of fire attack.
- Explain the factors in choosing the type of fire attack.
- Identify and explain the principles of hydraulics relating to fire streams.
- Define and explain friction loss.
- Define and explain nozzle pressures and reactions.
- Define and explain elevation as a factor in fire streams.
- Explain adverse factors in operations of fire streams.
- Explain the selection factors for fire streams in overall fire operations.
- Define foam.
- Identify the types of foam.
- Explain the principles of foam for fire suppression.
- Explain the operation of foam-making equipment.

INTRODUCTION

Fires are usually extinguished using water to cool the heat produced. Foam is added to improve water's extinguishment ability or on fuels where plain water is ineffective. Water and foam are delivered using nozzles and fire streams to reach the seat of the fire at the proper quantity. This chapter examines fire streams of water and foam, various nozzles and appliances, attack applications, and basic hydraulic principles. Proper selection of the right nozzle gives the attack crew the tool needed to fight the fire successfully, and no one nozzle or extinguishing agent is perfect for every fire situation.

DEFINITION OF FIRE STREAM

A **fire stream** is the water or other agent that leaves the nozzle toward its target, usually the fire. The four elements of a fire stream are the pump, water, hose, and nozzle. The fire stream is essential to the fire suppression effort because it targets the enemy, the fire. Properly developed and aimed fire streams are successful in extinguishing fires, while poorly developed or improperly aimed ones fail to reach the target, allowing the fire to continue to burn. Proper fire streams are the result of the knowledge, skills, and abilities of the firefighter on the nozzle, the company officer, and the **pump operator**. A proper fire stream is one that has sufficient volume, pressure, and direction and reaches the target in the desired shape or form and pattern. It is important that firefighters understand fire streams and how they are applied to various firefighting situations. (See Chapter 19 for situations.)

> **Streetsmart Tip** Which type of nozzle to use has been a fire service controversy ever since the first fog nozzle was invented and challenged the solid stream. Today, solid stream, fog, or automatic fog nozzles of various designs are all available to fire departments. Some only use one type, others use two, and still others all three. The type used by a particular department has been chosen after careful consideration has been given to local fires and conditions and based on testing and experimenting with the various kinds of nozzles available, and the training and experience of firefighters using the nozzles. Equipment changes and improves regularly, so fire departments should constantly reevaluate their equipment and tactics to ensure that they are using the best equipment for the conditions encountered. Nozzles are seldom used at their maximum performance capabilities, but are often used to maximize the performance of their operators. Regardless of which type of nozzle is used, firefighters should learn how to use it in the best manner possible.

NOZZLES

Nozzles are the appliances that apply the water or extinguishing agent. *Webster's Eleventh Collegiate Dictionary* states a nozzle "a short tube with a taper or constriction used (as on a hose) to speed up or direct a flow of fluid." The two basic types of nozzles are **solid stream** (also called a smooth bore, straight bore, or solid tip) and **fog nozzles** with different styles available for each kind, especially fog nozzles. **Combination nozzles** are capable of providing straight stream and spray patterns, which can be varied or adjusted by the operator. Some fire departments primarily use either solid stream or combination fog nozzles; however, firefighters should note that each type of nozzle has its advantages and disadvantages.

Safety Selecting the proper type of nozzle to match the fire situation improves fire operations and personnel safety.

Figure 11-1 Nozzles showing the stream shape for straight, solid, and wide pattern streams.

Important factors in nozzle selection are the nozzle pressure, nozzle flow, nozzle reach, stream shape, and nozzle reaction. **Nozzle pressure** is the pressure required for effective nozzle operation and relates to flow and reach. Nozzles are designed to operate at a certain pressure, usually 50, 75, 80, or 100 psi (345, 517, 552, or 690 kPa). Pressure is measured in pounds per square inch (psi) or kilopascals (kPa). **Nozzle flow** is the amount or volume of water that a nozzle will provide at a given pressure. Flow is critical because the amount of water provided determines the amount of heat absorbed or cooled. Some nozzles only flow a set volume at a set pressure, while others can be adjusted manually or automatically adjust their flow. Flow is measured in gallons per minute (gpm) or liters per minute (L/min).

Nozzle reach is the distance the water will travel after leaving the nozzle. Greater reach is important in large rooms or during exterior fire operations. Reach is a function of the pressure, which is converted to velocity or speed, of the water leaving the nozzle. Reach is measured in feet or meters. Reach is affected by the other factors, especially shape, pressure, wind direction, gravity, and friction of the air. The angle of the nozzle can affect the reach: Maximum horizontal reach is optimum at 32 degrees, while maximum vertical reach is obtained at 65 to 70 degrees. The objective is to reach the fire with maximum effect, not to be at the maximum distance from the fire.

Stream shape, also called stream pattern, is the arrangement or configuration of the droplets of water or foam as they leave the nozzle, **Figure 11-1**. The shape of the pattern helps determine the reach of the fire stream. **Nozzle reaction** is the force of nature that makes the nozzle move in the opposite direction of the water flow, **Figure 11-2**. The nozzle operator must counteract or fight the backward thrust exerted by the nozzle to maintain control of the nozzle and to direct it to the correct location. The nozzle pressure and stream shape affect nozzle reaction.

Solid Tip or Stream

Solid tip, solid stream, or smooth bore nozzles deliver an unbroken or solid stream of water at the tip and toward the fire, **Figure 11-3** and **Figure 11-4**. The solid stream nozzle can deliver its water as a solid mass or cone of water or, when bounced off a ceiling, wall, or other object, as large water droplets. This solid mass breaks or shears apart the farther the

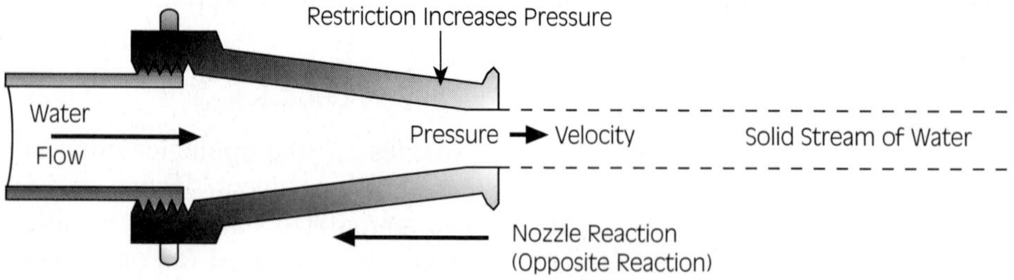

Figure 11-2 Nozzle reaction is the force of the water as it escapes and pushes back toward the nozzle and nozzleperson.

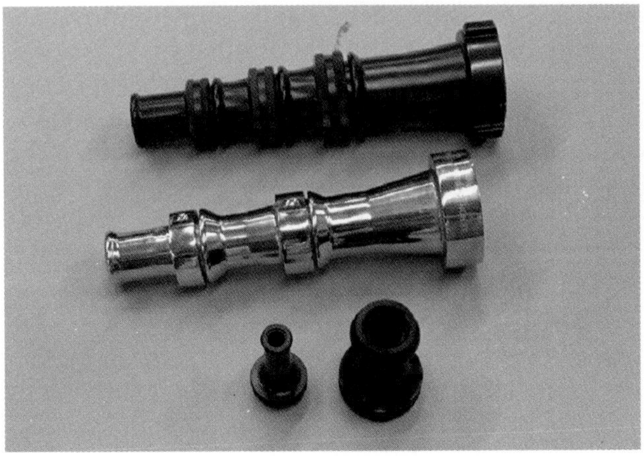

Figure 11-3 Various solid tips (stacked).

water travels. The solid stream nozzle's flow is a factor of the tip size at a certain nozzle pressure. Excessive or reduced nozzle pressures have adverse effects on stream performance. Handlines use tips from ¾ to 1¼ inches (19 to 32 mm) at 50 psi (345 kPa), and master streams use tip sizes of 1¼ inches (32 mm) and larger at 80 psi (552 kPa).

Solid stream handlines can reach over 70 feet (21 m) and master streams about 100 feet (33 m). They have the ability to penetrate through the fire's heat without absorbing that heat before reaching the target. A solid stream has less effect on a room's thermal balance, that is, the layers of heated gases in a burning room (see Chapter 4 on fire behavior), and produces minimal amounts of steam compared to narrow and wide fog patterns. Disruption of the thermal balance of a room can cause steam burns to firefighters as heavier water vapor descends onto firefighters after discharging the water. Fog streams are more apt to do this. Many fire departments have tried to minimize this chance of burns through the utilization of solid stream nozzles for aggressive, interior fire attack. It has good penetration into piles of materials to quench the fire. Smooth bore nozzles are more durable and easier to maintain than fog nozzles because they have fewer parts. The disadvantages of a solid stream are that there is no volume control other than changing tip sizes or adjusting the shutoff, the lack of fog protection when working close to the fire, and a higher nozzle reaction at the same pressure than a fog nozzle.

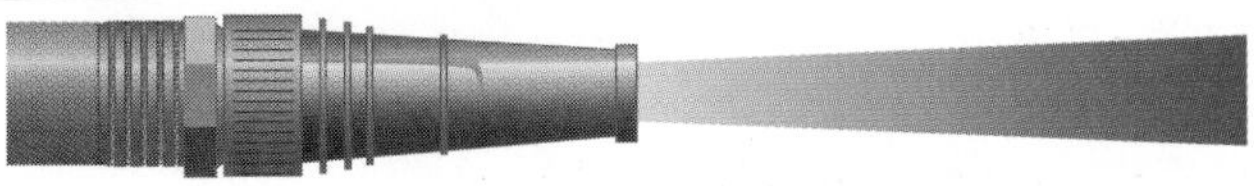

Figure 11-4 The flow pattern of a solid tip nozzle.

Fog

Fog nozzles deliver either a fixed spray pattern or a variable combination pattern with both **straight stream** and spray patterns that can be adjusted by the nozzleperson. Fixed spray pattern nozzles are of the impinging design in which the nozzle has a series of holes at the end that creates a water spray, **Figure 11-5**. The variable fog patterns vary from the straight stream pattern, which is similar to a solid stream pattern, to a wide-angle fog pattern of at least 100 degrees, **Figure 11-6** and **Figure 11-7**. The different types of combination fog nozzles depend on the variations allowed and include constant volume or set gallonage nozzles; variable, adjustable, or selectable gallonage nozzles; and automatic or constant pressure nozzles, **Figure 11-8**. The **constant** or **set volume nozzle** has one set volume at a set pressure, for example, 60 gpm at 100 psi (227 L/min at 690 kPa), and the only adjustment is the pattern. **Variable, adjustable,** or **selectable gallonage nozzles** allow the nozzleperson to select from two or three flow choices and the pattern. This allows the flow choice to be made at the nozzle closest to the fire. The nozzleperson or officer should ensure that the pump operator is aware of this flow choice to maintain the correct pressure.

The **automatic** or **constant pressure nozzle** has a flow that can be adjusted by the pump operator, who increases the pressure, which in turn increases the gallons flowing. The pattern can be adjusted by the nozzleperson. Some newer automatic nozzles have a dual-pressure option for normal and low-pressure situations that is controlled by the nozzleperson. The automatic feature has a spring mechanism built around the baffle that reacts to

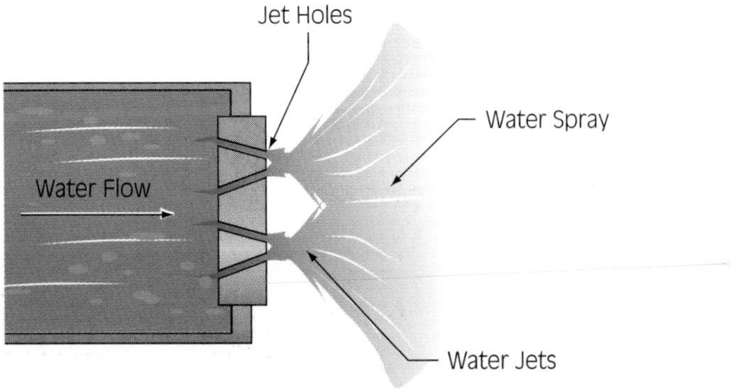

Figure 11-5 A fixed fog nozzle pattern of impinging design.

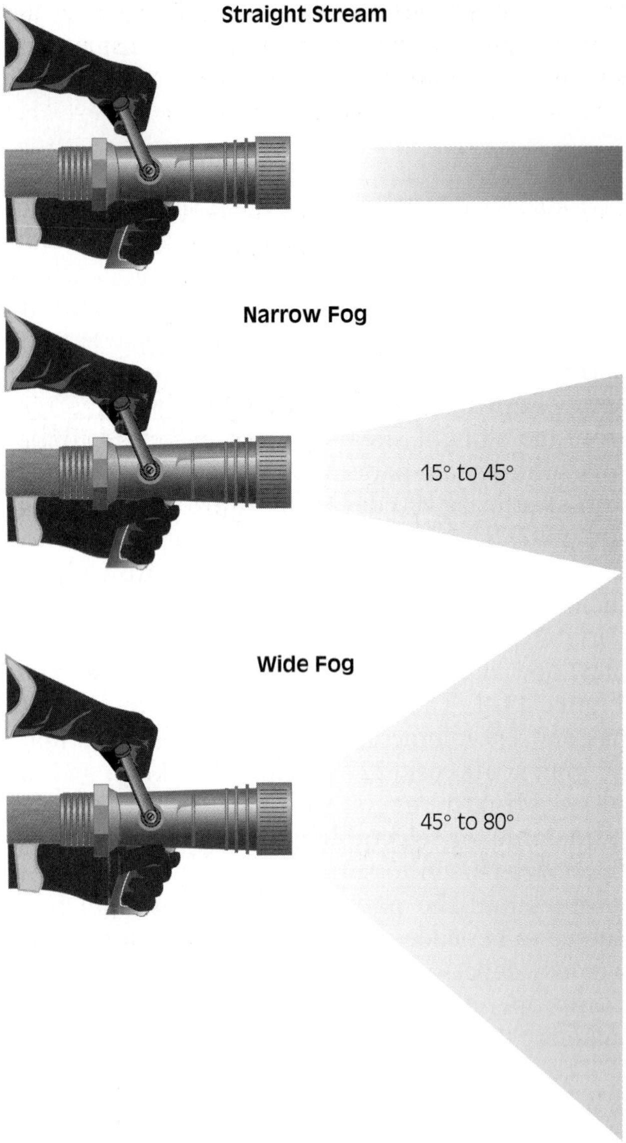

Figure 11-6 Variable combination fog nozzle patterns. From top to bottom: straight stream, narrow fog, and wide fog.

increased pressure, hence adjusting the flow and resultant reach of the nozzle, **Figure 11-9**. The disadvantage is that the pump operator, who is farther away from the fire, controls the maximum flow, while gating down at the nozzle will reduce the flow. The officer or nozzleperson should be in contact with the pump operator to control this flow. Fog nozzles have traditionally operated at 100 psi (690 kPa) but new low-pressure nozzles operating at 50 or 75 psi (345 or 517 kPa) have been approved. The lower nozzle pressure gives the same volume, but nozzle reaction is reduced and greater lengths of hose can be used when at the maximum pump pressure. On the low-pressure settings, additional flow is available at higher pressures.

Fog nozzles provide good reach that varies with the pattern from 25 to over 100 feet (7.5 to +33 m). They also provide good penetration. Fog nozzles that are adjustable provide personnel protection because of the screening effect between them and the fire (convected and radiated heat). Fog provides better heat absorption, but this can cause the water to change to steam before reaching the seat of the fire. Fog streams can produce more steam, which can extinguish hidden fire and is good for indirect attack. Fog streams can be used as a fan due to their ability to move large volumes of air. This advantage is seen when used at a window for horizontal ventilation, but can create problems for firefighters by drawing the fire's heat back toward them, **Figures 11-10 A** and **B**. The fog nozzle can be used to assist horizontal ventilation. Air movement is created to swirl and mix conditions including affecting the thermal balance, which can pull heat down to the firefight-

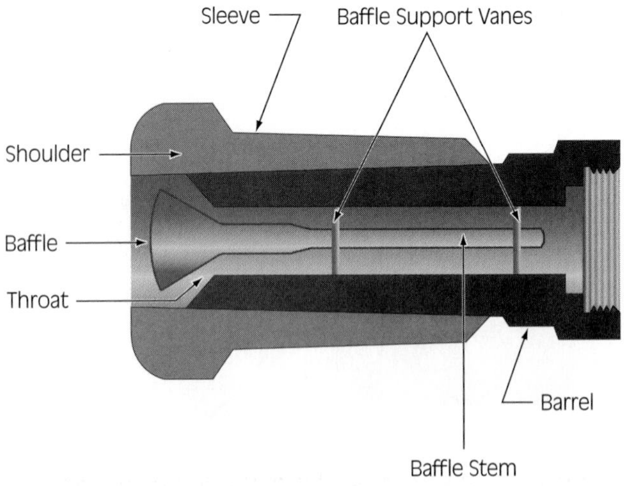

Figure 11-7 Parts of a fog nozzle.

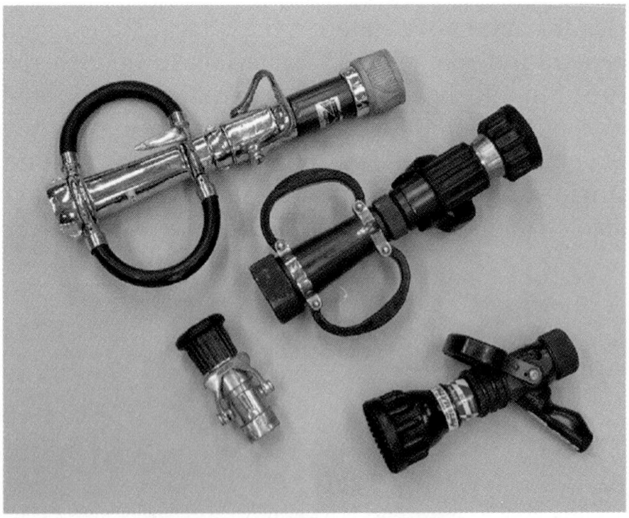

Figure 11-8 Various styles of fog nozzles.

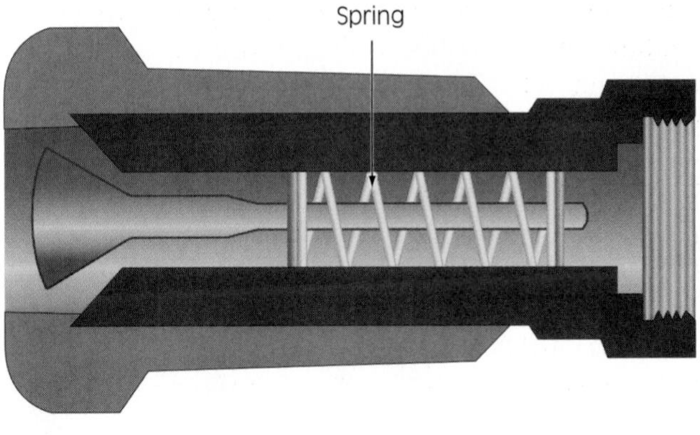

Figure 11-9 An automatic or constant pressure fog nozzle.

ers at the floor level. For this reason, firefighters should use straight streams when attacking fires in high-heat environments or when trying to cool interior compartments while advancing toward the fire. A 1½-inch (38-mm) fog nozzle moves more air than a 14-inch (0.36-m) smoke ejector, and a 2½-inch (63-mm) fog nozzle moves more air than a 24-inch (0.6-m) smoke ejector.

Straight Stream

The straight stream nozzle pattern creates a hollow type stream that is similar in shape to the solid stream pattern, but the straight stream pattern must pass around the baffle of the nozzle. This creates an opening in the pattern, which may allow air into the stream and reduce its reach, **Figure 11-11**. Newer fog nozzle designs, especially the automatic nozzles, only have this hollow effect from the tip, and it is a short distance to refocus the stream to create a solid stream with good reach and penetration abilities. In fact, some are better than solid stream nozzles.

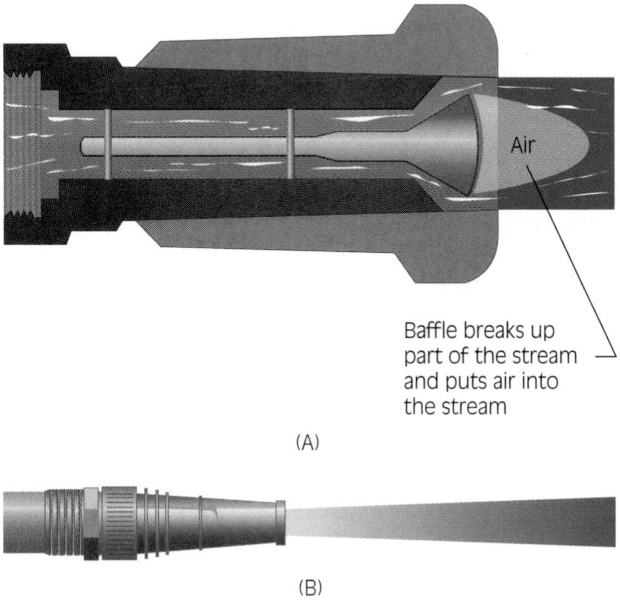

Figure 11-11 Comparison of (A) straight and (B) solid streams at tip.

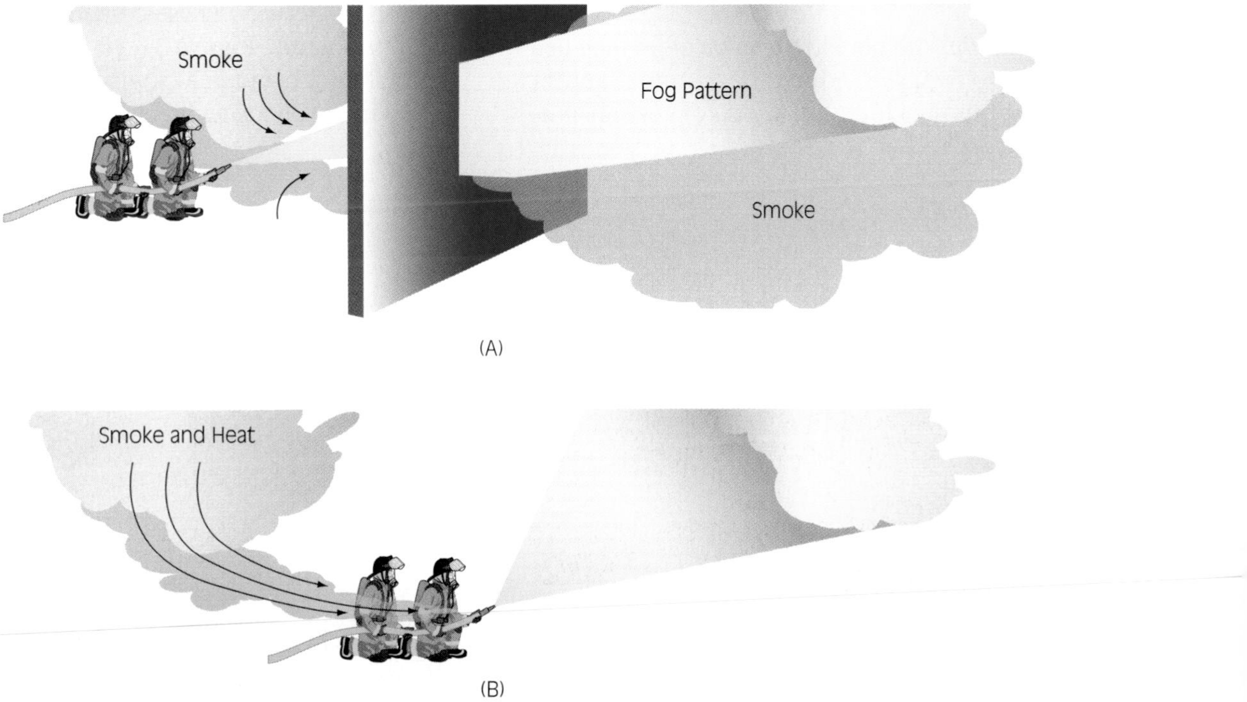

Figure 11-10 (A) One advantage of a fog nozzle is that, with ventilation at a window, it draws out the heat and smoke. (B) The disadvantage is that heat and gases are drawn onto the firefighters at the nozzle.

Special Purpose

Special-purpose nozzles were developed for use in limited types of situations by some fire departments. Special-purpose nozzles are not often used but firefighters should know when and how to use them. **Cellar nozzles** and **Bresnan distributors** can be used to fight localized fires in basements or cellars when firefighters cannot make a direct attack on the fire, **Figures 11-12A** and **B**. The cellar nozzle has four spray nozzles, and the Bresnan distributor has six or nine solid tips or broken stream openings that are designed to rotate in a circular spray pattern.

Piercing nozzles were originally designed to penetrate the skin of aircraft and now have been modified to pierce through building walls and floors, **Figure 11-13**. Some have striking points that allow them to be driven through the material. Other nozzles include high-pressure fog that can operate at 1,000 psi (6,895 kPa) and industrial and forestry nozzles that are combination fog nozzles with a shutoff built into the design.

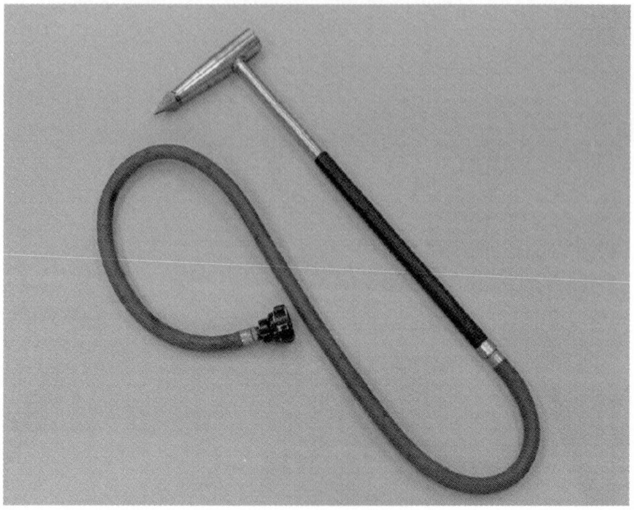

Figure 11-13 Piercing nozzle.

Another special nozzle is a **water curtain nozzle**, which is designed to spray water to protect against exposure to heat, **Figure 11-14**. When using a water curtain, its spray should be directed on the object being protected to absorb the heat, not just up into the air. This is because radiant heat passes through clear materials, such as water, without being absorbed.

Playpipes and Shutoffs

Originally nozzles did not have a valve other than that at the pump. Today, Underwriters' and Factory Mutual playpipes, which are used for testing purposes, not firefighting, are still made without a shutoff mechanism, **Figure 11-15**. Some nozzles or just nozzle tips are placed on hoselines for occupants' use, not for firefighters or fire brigades. The shutoff at the nozzle places the water flow control with the nozzleperson. The most commonly used is the lever type, bale, or handle, which operates in a line with the waterway, usually by moving a ball valve. The shutoff is opened by pulling back on the lever and closed by pushing the handle toward the nozzle. It can come built into the nozzle or as a break-apart type as a shutoff, pistol grip, or playpipe. The rotating type operates by either rotating a gate a quarter turn or rotating from a seat to open the waterway. These come either built in or as a separate shutoff, **Figure 11-16**. The nozzle is opened by rotating it counterclockwise or to the left, and it is closed in the opposite direction.

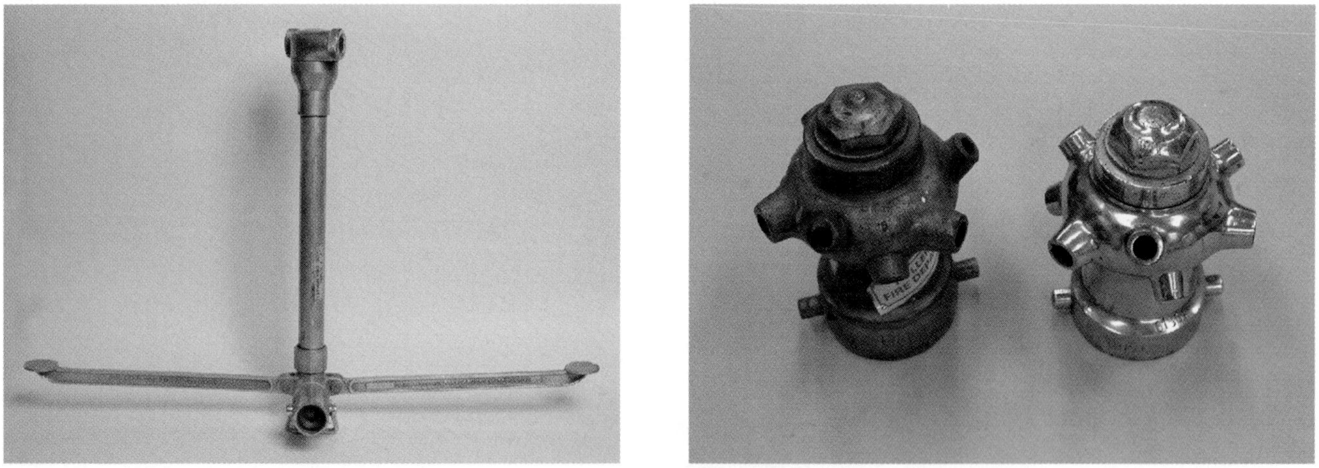

(A) (B)

Figure 11-12 (A) Cellar nozzle and (B) Bresnan distributor.

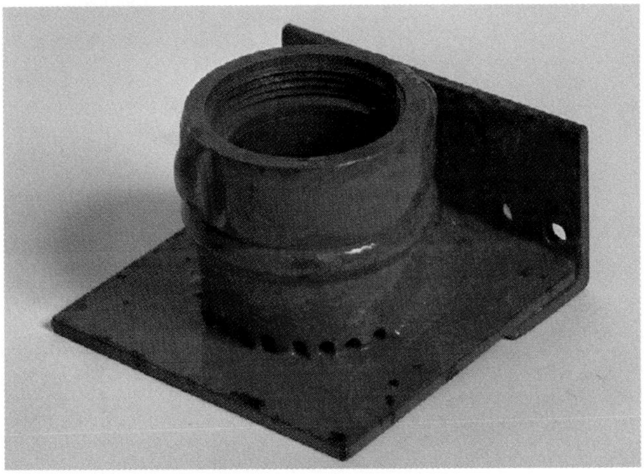

Figure 11-14 Water curtain nozzle.

Nozzle Operations

> **Note** Handline nozzles usually have shutoffs, but many other appliances do not provide shutoffs at the nozzle and the only control is the ability to aim it.

Some older appliance nozzles without shutoffs have adjustable fog patterns, and they are operated similarly to nozzles described in the following paragraphs. Appliances with shutoffs are also operated as described next.

Solid stream nozzles are simple to operate with the nozzleperson controlling the shutoff, tip size, and aim. The typical shutoff for a solid tip nozzle is a lever and can be fully opened or closed or sometimes partially opened. The other control is selecting the tip or nozzle size to match the desired flow, coordinated with the pump operator for proper pressure. Screwing them on or off is one way to change tips; many are designed to stack onto each other. Even with this stacking effect, the firefighter needs to be able to carry the nozzle tips when switching to a different size.

Fog nozzles have either the lever-type open/close shutoff, which is the most common, or the rotating

Figure 11-15 Underwriters' playpipe.

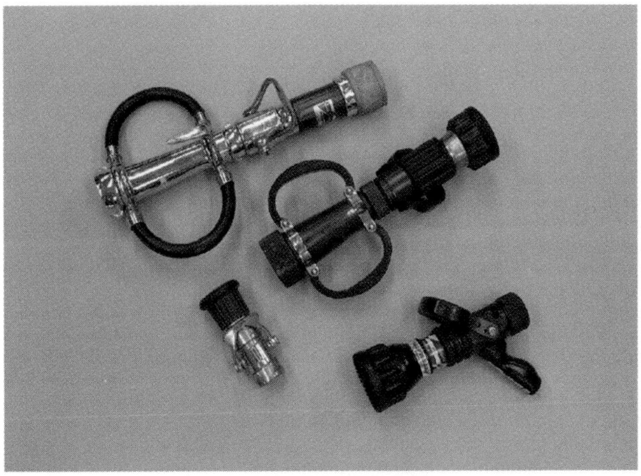

Figure 11-16 Various shutoff devices for nozzles. From bottom right to top left: built-in lever with pistol grip, built-in lever, rotating with playpipe, and break-apart playpipe.

type. In addition, the fog pattern can be adjusted by rotating the nozzle barrel counterclockwise to move from straight stream to narrow fog to wide-angle fog, **Figure 11-17**. Rotating clockwise adjusts the pattern the opposite way. Fog nozzles with variable gallonage have an addition ring on the collar that rotates

Figure 11-17 Close-up of rotating type of nozzle shutoff.

Figure 11-18 Close-up of combination fog nozzle showing the adjustments for selecting patterns and for selecting gallonage.

from one gallonage to the next, **Figure 11-18**. Both gallonage and pattern adjustments can be detected in the dark because the nozzle clicks at each position. Fog nozzles have more applications and many firefighters consider them more effective than smooth bore nozzles.

OPERATING HOSELINES

Advancing hoselines was covered in Chapter 10 including some of the initial operation of the nozzle. Included was straightening the hose, having firefighters properly spaced on the same side of the line, and with the line charged, bleeding off air from the hose and nozzle, and selecting the proper pattern. Most hoselines are operated from a crouching or kneeling position, but lying, standing, or sitting positions are also used, **Figure 11-19**, **Figure 11-20**, and **Figure 11-21**. Chapter 19, on fire suppression, has a table that highlights some

Figure 11-19 Firefighter operating nozzle in the crouching or kneeling position.

of the characteristics of various size hose streams and covers hoseline operations in fire situations.

Small-Diameter Handlines

Small-diameter handlines are typically 1½, 1¾, or 2 inches (38, 44, or 50 mm) in diameter and flow from 100 to over 250 gpm (378 to over 946 L/min).[1] When flowing at the lower volumes, these lines can be operated by one person; larger volumes require two persons.[2] Both fog and solid tip nozzles can be used for small lines. Small lines are popular because of their ease of mobility, the number of personnel required to operate them, and their ability to extinguish one to three typical rooms of fire with their flow.

Figure 11-20 Firefighter operating nozzle in the lying position.

Figure 11-21 Firefighter operating nozzle in the sitting position. Notice that the hoseline is placed in a loop to counter nozzle reaction.

Medium-Diameter Handlines

Medium-diameter hose for handlines is 2½-inch (75-mm) hose and can be used with solid tip and fog nozzles, flowing from 165 to 325 gpm (625 to 1,230 L/min). This is the standard size hoseline for the fire service, used for calculating friction loss, pump capacity, and appliance discharge values. Many fire departments abandoned the 2½-inch hose as a fire attack tool, advocating 1¾-inch and 2-inch lines because of increased maneuverability. However, increased Btu development of fires due to plastics and low-mass fuels has caused many to reintroduce the 2½-inch hoseline to increase gpm flow to quench the higher heat. This is especially important in large, commercial structures or buildings with high fire loading (like an auto body repair shop or high rack storage).

Medium-size hoselines usually require two or more personnel to operate them because they are less mobile than small lines, which has been a major drawback. Some departments have gone to 3-inch (75-mm) hoses with 2½-inch (65-mm) couplings, while other departments do not use any handline over 2½ inch (65 mm).

Master Stream Devices

Master stream devices (see Chapter 10) are capable of flowing over 350 gpm (1,325 L/min), and some have capabilities of many thousands of gallons per minute. This is the artillery of the fire service and is used when large volumes are required. These devices must be mounted or secured properly and safety should be a major concern during their operation. Specific operating instructions are necessary for each type of device. They require only one person to operate them, but have either poor or no mobility.

STREAM APPLICATION, HYDRAULICS, AND ADVERSE CONDITIONS

Application of fire streams depends on the method of fire attack and conditions encountered, including environmental factors and water supply. The fire stream must have the proper pressure and flow, and an understanding of hydraulics is needed to provide those factors. Improper hydraulic calculations are the leading causes of poor fire streams.

Direct, Indirect, and Combination Attack

Note The method of stream application or fire attack depends on the fire's fuel, its location, and the equipment of the fire department, especially the size and type of hoseline and nozzle.

Three methods of fire attack exist: direct, indirect, and combination. **Direct fire attack** is used to attack the fire by aiming the flow of water directly at the seat of the fire, **Figure 11-22**. Direct fire attack is used on most deep-seated fires that require penetration by the hose stream.

Indirect fire attack is used to attack interior fires by applying a fog stream into a closed room or compartment, converting the water into steam to extinguish the fire, **Figure 11-23**. Firefighters apply the water at the doorway and then close the door, allowing the steam to put the fire out. The closed compartment is needed to contain the steam, thereby smothering the fire. The estimated quantity of water applied is the amount needed for total conversion to

Figure 11-22 Firefighter directly attacking a fire.

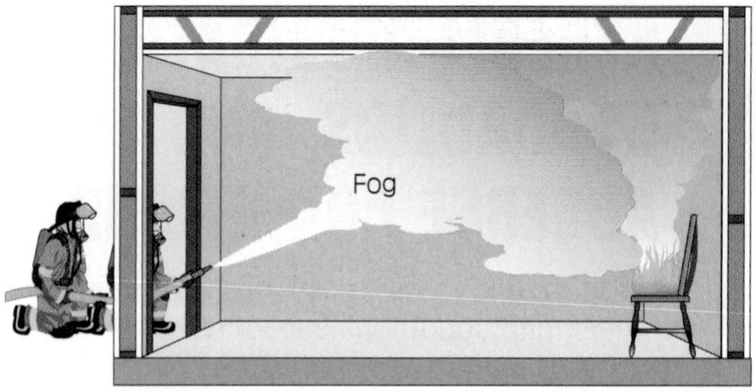

Figure 11-23 Firefighter using indirect attack by applying water into room and then closing the door.

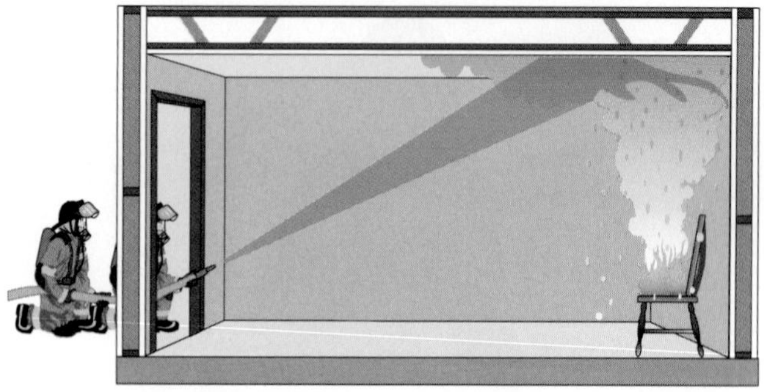

Figure 11-25 Firefighter using combination fire attack.

steam to fill the room. As water is converted to steam at 212°F (100°C), it expands 1,600 to 1,700 times its volume; 1 ft^3 of water (0.028 m^3 or 7.48 gallons or 28 L) would fill a 1,700-ft^3 (48.1-m^3) room, **Figure 11-24**. At 1,000°F (500°C), this expansion is over 4,000 times. Because the entire space is filled with steam, the indirect attack should not be used when people are in the space.

A **combination fire attack** uses a blend of the direct and indirect fire attacks, with firefighters applying water to both the fuel and the atmosphere of the room, **Figure 11-25**. To achieve this, the nozzleperson opens the nozzle and directs the stream toward the ceiling and then the fire with a circular, "Z," or "T" motion. This puts water on the seat of the fire and the atmosphere, while creating limited amounts of steam to extinguish the fire in any hidden areas. Ventilation with this combination attack controls the flow of fire gases and steam, improving the survival chances of victims, and makes this attack the type typically used by firefighters in structural firefighting.

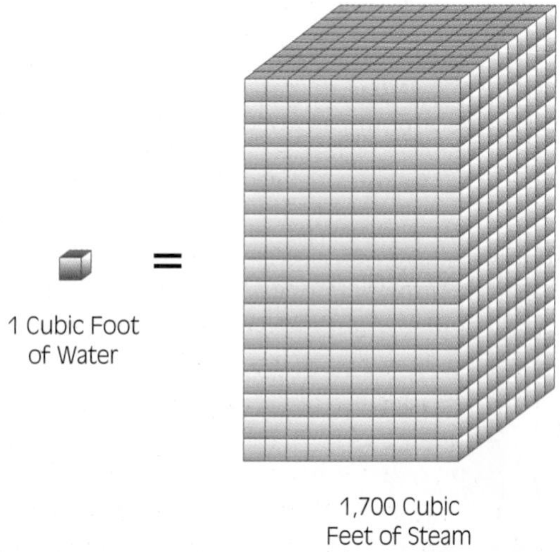

Figure 11-24 One cubic foot of water in liquid form expands 1,700 times when converted to steam at 212°F.

Basic Hydraulics, Friction Loss, and Pressure Losses in Hoselines

Hydraulics is the study of fluids at rest and in motion, which describes the flow pattern of water supply and fire streams. Water moving through a hoseline and out a nozzle is an example of water in motion, while use of the nozzle shutoff leads to water at rest. An effective fire stream must have sufficient volume and pressure and be delivered in the correct form. This relationship begins with the water supply into the fire pump and then out the lines to the nozzle tip and to the fire (see Chapter 9 for more information). Hydraulic principles are a whole field of study; this section examines only the basics.

Pressure is force divided over an area, usually expressed in pounds per square inch (psi) or kilopascals (kPa) where 1 psi = 6.895 kPa. Pressure is required to lift, push, or move water. Force is a measurement of weight and is measured in pounds (kilograms). The pull of gravity on a mass of water creates a force. Other external pushing and pulling, like the actions of a fire pump, applied to a substance (water) can also create motion. Water weighs 62.4 pounds per cubic foot (1,000 kg/m^3), which creates a force of 62.4 pounds (28.3 kg). This would also create a pressure of 62.4 pounds per square foot (or 1,000 kg/m^2) or 0.434 pounds per square inch or psi (3 kPa), **Figure 11-26**. Pressure can also be measured in feet (meters) of head or the height of a column of water. Thus, a column of water 100 feet (30 m) would create a head of 100 feet (30 m) or a pressure of 43.4 psi (300 kPa).

Firefighters should be familiar with the several types of pressure: Atmospheric pressure is the pressure exerted by the atmosphere or the weight of the air at the Earth's surface. At sea level this pressure is 14.7 psi (101 kPa) and gauges reading this pressure show **absolute pressure**, which is indicated as pounds per square inch absolute (psia). **Gauge pressure** normally measures pres-

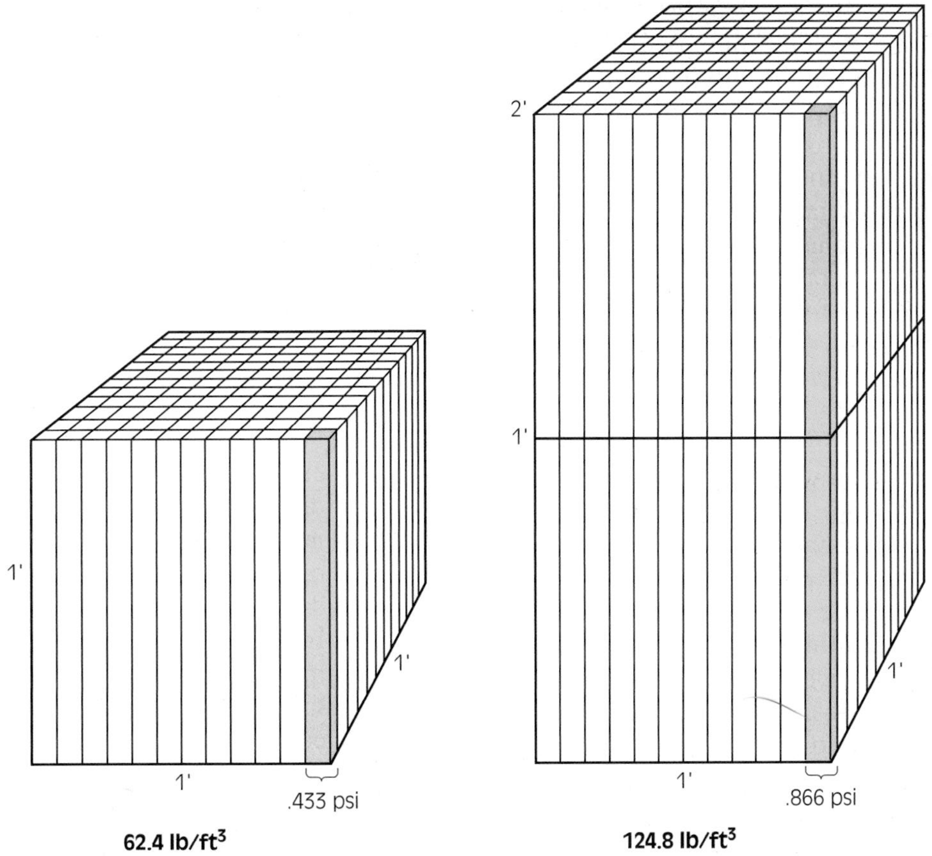

Figure 11-26 Water in a container (like a hose) exerts pressure at the lowest point.

sure minus atmospheric pressure and is measured in psi or psig (pounds per square inch gauge). Most fire department readings are from gauge pressure and begin at zero, **Figure 11-27**. **Vacuum (negative) pressure** is the measurement of pressure less than atmospheric pressure, which is usually read in inches of mercury (in. Hg or mm Hg). Fire apparatus capable of drafting have at least one gauge that measures vacuum pressure, called a compound gauge. Today, it is common for all gauges on apparatus to be of the compound type. **Head pressure** measures the pressure at the bottom of a column of water in feet (meters) (see example given earlier) and head pressure can be gained or lost when water is being pumped above or below the level of the pump. A head of 2.31 feet (0.7 m) would equal 1 psi (6.895 kPa). This is also called elevation pressure and is usually rounded to 5 psi (35 kPa) per floor when pumping water up or down in a building. **Velocity pressure** is the pressure in a hose being converted to velocity or speed as it leaves an opening. When leaving a nozzle, this becomes the nozzle pressure. Chapter 9 discussed static pressure, the pressure with no water flowing, and residual pressure, the pressure in a system after water has been flowing.

Figure 11-27 Typical pump pressure gauges.

Flow is the rate and quantity of water delivered and is usually measured in gallons per minute or liters per minute (1 gpm = 3.785 L/min). The **needed** or **required flow** is the amount of water required to put out the fire and is determined by what and how much is burning. The **available flow** is the amount of water that can be moved to extinguish the fire. The water supply, pump(s) and their capability, the size and length of hose, and the type(s) of nozzles determine available flow. Next to be determined is the **discharge flow**, the amount of water flowing from the discharge side of the pump. The discharge flow is the flow of each hoseline and of all hoselines and outlets from the pump or the total gallons per minute. Water flow can be determined using a flowmeter or pressure gauge. The flowmeter measures the flow of water and is simpler to use than the pressure gauge, **Figure 11-28**. Flowmeters have only recently become effective for fire apparatus use. Pressure gauges have been primarily used for fire apparatus for determining pressure requirements to generate a certain flow. (Actual flow measurement calculations were discussed in Chapter 9; see Eq. 9-1.) The pressure gauge method requires more complex calculations using flow rate and pressures of the nozzle and any **friction** in the system, that is, the **friction loss**. Note that friction loss in hose and appliances, including nozzle pressures, and elevation losses or gains only occur when water is in motion or flowing. When all water flow stops, the system equalizes at the highest pressure in the system or hoseline. A sudden stop of water flow creates a water hammer or pressure surge that could damage equipment, piping, and hose. (See Chapter 10 for more on water hammer.)

After determining what flow is needed, the pump operator adds the pressures of the various parts of the pump to nozzle system and increases engine pressure to generate that flow. Errors made in this formula can create too much or too little pressure, which affects nozzle performance. Hose and appliances are constantly being improved. When new equipment is purchased, water flow and pressure evaluations should be conducted to revalidate these formulas. To calculate the discharge pressure of a pump, the following formula is used:

$$EP = NP + FL \pm E + SA \qquad \text{(Eq. 11-1)}$$

where:

EP = engine pressure, also called pump discharge pressure.

NP = nozzle pressure, the pressure required to operate the nozzle and the pressure as it discharges from the nozzle.

FL = friction loss in the hose(s). (There can be multiple cases of friction losses, such as the hose between the pump and a standpipe connection and also the attack line hose.)

E = elevation gain or loss from moving the weight of the water up or down in relation to the pump.

SA = friction loss in any special appliance, not hose or elevation. It is also called appliance friction loss.

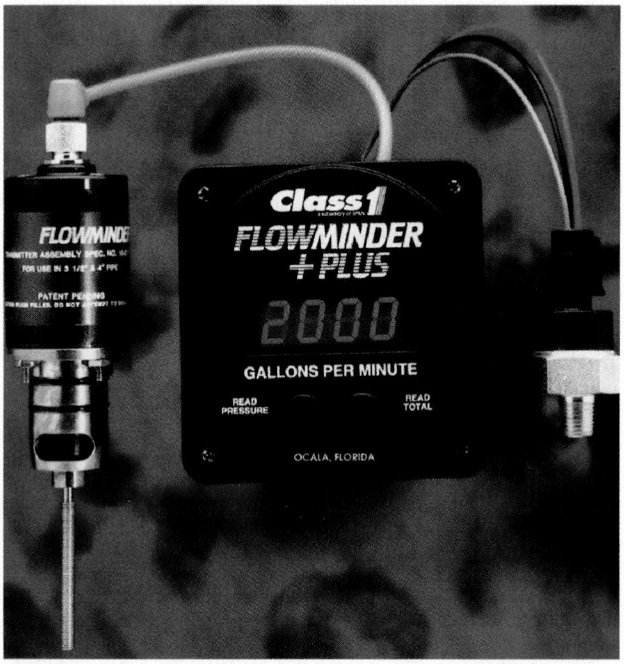

Figure 11-28 Flowmeters eliminate the need for friction loss calculations during pump operations. *(Photo courtesy of Class One)*

Each of the formula's components is either a given number or must be calculated. Engine pressure is the addition of the other components and is typically the highest value for any of the hoselines. If more than one hoseline is pumped, the lower pressured line is gated down. In reality, the engine pressure is figured for each line, the highest one being the main pump pressure. Nozzle pressure is usually a given value for each type of nozzle. **Table 11-1** gives the typical values.

Nozzle pressure pushes the water out of the nozzle, but it also pushes the nozzle in the opposite direction against the firefighter or whatever is holding it. A higher nozzle reaction means more effort is required by the nozzleperson and hose crew to control the nozzle and hoseline to keep them from getting away or when advancing an opened line. Now it is easier to understand why operating a nozzle in the sitting or lying position is less stressful than the standing position—the ground helps absorb this reaction. Nozzle reaction can make an unsecured hoseline unsafe and dance like a wild snake. It is important to keep a firm grip. Nozzle reaction is a relationship between the nozzle pressure and the type of nozzle. Nozzle reaction for a smooth bore nozzle is:

Nozzle Types and Pressure	
TYPE OF NOZZLE	**NOZZLE PRESSURE, psi (kPa)**
Smooth bore handline	50 (345)
Fog handline, normal	100 (690)
Fog handline, mid pressure	75 (517)
Fog handline, low pressure	50 (345)
Smooth bore master stream	80 (552)
Fog master stream	100 (690)

TABLE 11-1

$$NR = 1.57 \times d^2 \times NP \quad \text{(Eq. 11-2)}$$

where

NR = nozzle reaction
1.57 = a constant (some round to 1.5)
d = diameter of the nozzle tip in inches
NP = nozzle pressure

For example, a 1-inch nozzle with 50 psi of nozzle pressure flowing at 210 gpm would have a nozzle reaction of 78.5 lb.

$$\begin{aligned} NR &= 1.57 \times d^2 \times NP \\ &= 1.57 \times (1\text{ in.})^2 \times 50\text{ psi} \\ &= 1.57 \times 1\text{ in.}^2 \times 50\text{ psi} \text{ or } 1.57 \times 50\text{ psi} \\ &= 78.5\text{ lb} \end{aligned}$$

The nozzle reaction for a fog nozzle is figured using the following formula:

$$NR = \text{gpm} \times \sqrt{NP} \times 0.0505 \quad \text{(Eq. 11-3)}$$

where NR = nozzle reaction
gpm = gallons per minute
$\sqrt{NP}$ = square root of the nozzle pressure
0.0505 = constant

A fog nozzle flowing 200 gpm at 50 psi of nozzle pressure would have a nozzle reaction as follows:

$$\begin{aligned} NR &= \text{gpm} \times \sqrt{NP} \times 0.0505 \\ &= 200\text{ gpm} \times \sqrt{50}\text{ psi} \times 0.0505 \\ &= 200\text{ gpm} \times 7.07 \times 0.0505 \\ &= 71.4\text{ lb} \end{aligned}$$

A fireground formula for fog nozzles operating with a nozzle pressure of 100 psi is

$$NR = \text{gpm} \times 0.5 \quad \text{(Eq. 11-4)}$$

Note that, as previously stated, at the same pressure and gpm, the smooth bore nozzle has a higher nozzle reaction than the fog nozzle.

Friction loss is the loss of energy from the turbulence or rubbing of the moving water through the hose. The friction loss is compensated for by the pump operator increasing the pump pressure and providing the correct pressure to the nozzle. Friction loss is based on four principles. The first is that the friction loss varies directly with the length of the hose if all other variables are held constant. This means that if the length of the hose doubles, so does the friction loss. Second, with all other variables held constant, friction loss varies approximately with the square of the flow. Therefore, if the flow rate doubles, then the friction loss will increase four times or be squared ($2^2 = 2 \times 2 = 4$). Third, when the flow remains constant, friction loss varies inversely with the hose diameter. This means that at the same flow, the friction loss will decrease as the size of the hose diameter gets larger. The fourth and final principle states that for any given velocity, the friction loss will be about the same regardless of the water pressure. This says that the speed of the water flow rather than the pressure is what determines friction loss.[3] Friction loss can be calculated using Eq. 11-5, which factors these principles into the formula:

$$FL = Q^2 \times c \times L \quad \text{(Eq. 11-5)}$$

where
FL = friction loss of a hoseline
Q = quantity of water in hundreds of gallons per minute, that is, the flow:
Q = gpm/100 (Eq. 11-6)
c = friction loss coefficient for the diameter of the hose with 2½-inch (65-mm) hose being used as the standard (see **Table 11-2**)
L = length of hose in hundreds of feet:
L = hose length/100 (Eq. 11-7)

Friction Loss Coefficients

HOSE DIAMETER, INCHES (MM)	COEFFICIENT VALUE (c)
1½ (38)	24
1¾ (44)	15.5
2 (50)	8
2½ (65)	2
3 (76) with 2½-inch (65-mm) couplings	0.8
4 (101)	0.2
5 (127)	0.08
6 (152)	0.05

Note: Use caution because the hose diameter required may not be the same as the labeled hose diameter.

TABLE 11-2

Streetsmart Tip When discussing friction loss formulas and the values of the coefficient *c*, many firefighters will use different formulas. The formula FL = $Q^2 \times c \times L$ was once given as FL = $2Q^2 \times c \times L$ but with different values for *c*. The different values depended on whether the coefficient had already been multiplied by 2. Many older firefighters remember the formula as FL = $(2Q^2 + Q) \times c \times L$, which has changed over time as hose construction has reduced the friction created in the line.

An example calculating the friction loss formula is given next, using a 2-inch hoseline that is 150 feet long with a flow of 200 gpm, **Figure 11-29**.

$$\begin{aligned} \text{FL} &= Q^2 \times c \times L \\ &= (200/100)^2 \times 8 \times (150/100) \\ &= (2)^2 \times 8 \times 1.5 \\ &= (4) \times 8 \times 1.5 \\ &= 48 \text{ psi} \end{aligned}$$

Elevation can be a positive number if the nozzle is above the level of the pump or a negative number if below the pump. If even, elevation is not a factor. Elevation is equal to head pressure of 0.434 psi per foot (9.81 kPa/m) of height. [Note: This is usually rounded to 0.5 psi per foot or 5 psi per floor (10 kPa/m) for practical fireground purposes.]

Special appliances or appliance friction loss is the friction loss created by movement of water through the valves and turns of the appliances. Special appliance friction losses vary with each device. Examples include wagon pipe or deluge set, 20 psi; ladder pipe, 50 psi; and 2½ to 2½ siamese or wye, 5 psi. Manufacturer's specifications or department SOPs should be consulted for guidance in determining these friction loss factors.

The next example uses the preceding example's 2-inch hoseline that is 150 feet long flowing at 200 gpm, with a fog nozzle and operating on the second floor. Using that friction loss equation, engine pressure (EP) would be:

Figure 11-29 Example for friction loss and engine pressure calculations.

$$
\begin{aligned}
EP &= NP + FL \pm E + SA \\
&= 100 + 48 + 5 + 0 \\
&\text{or shorter } EP = 100 + 48 + 5 \\
&= 153 \text{ psi}
\end{aligned}
$$

Adverse Conditions That Affect Fire Streams

Adverse conditions that affect fire streams are of two types: natural and man-made. The major natural factor affecting a fire stream is the wind and wind direction. Anyone who has tried to wash a car on a windy day knows the wind will break up a stream and deflect it from its target unless the flow is being aimed downwind. This also occurs with fire streams. Rain, snow, hail, and objects such as tree branches and wires are solid or semisolid objects that deflect and break up hose streams. Gravity and air friction are also natural factors, especially the farther the travel distance of the stream. These natural causes cannot be removed but getting the stream closer to its target or in a better position allow these effects to be reduced.

> **Streetsmart Tip** Man-made causes are typically from improper operation of the pump or nozzle. Except on automatic nozzles, too much and too little pressure are both causes of poor fire streams as is failing to operate the nozzle properly, including fully opening the shutoff valve. Automatic nozzles use the shutoff to control the flow while still providing a proper stream.

These problems can be remedied by following the correct procedures for the nozzle being used.

TYPES OF FOAM AND FOAM SYSTEMS

Foam is an aggregate of gas-filled bubbles formed from aqueous solutions of specially formulated concentrated liquid foaming agents.[4] The bubbles are filled with a gas, usually air, creating a blanket over the surface of the fuel to cool and smother the fire, while sealing the escaping vapors. The mechanical action of mixing the foam concentrate in the water makes a foam solution to which air is added. Combining these three components makes the foam lighter than the flammable liquids and gives it the ability to float over their surface, **Figure 11-30**.

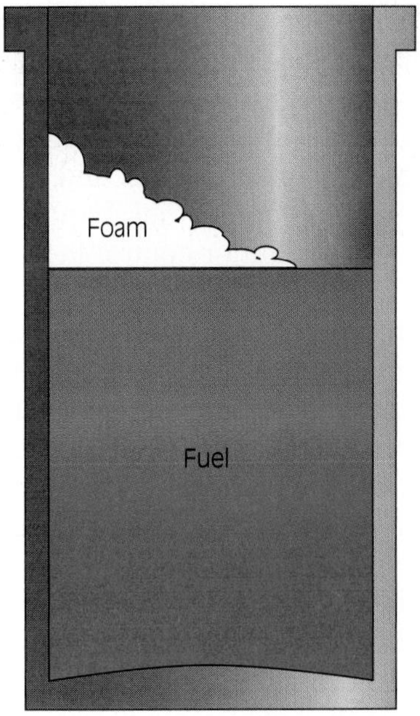

Figure 11-30 Foam applied on Class B fuel.

Class A foam is an aggregate of gas-filled bubbles formed from aqueous solutions of detergent or soap-based surfactants used to penetrate ordinary combustible materials and to keep the fuel wet that is unable to burn.

FOAM CHARACTERISTICS

Foam's ability to extinguish fires is based on several characteristics. The various foams available differ in their abilities to supply these characteristics. **Application rate** is the amount of foam or foam solution needed to extinguish a fire and is usually expressed in gallons per minute per square foot (gpm/ft^2) or liters per minute per square meter (L/min/m^2). Application rates vary depending on the fuel type and severity and fuel depth. See NFPA Standard 11 for recommended application rates.

Application rates must be maintained for a minimum amount of time, and additional foam will need to be reapplied to any remaining fuel to prevent reignition. The use of application rates allows firefighters to preplan for probable incidents or assess needs for foam prior to beginning a foam attack. For example, a shallow spill fire of 1,000 ft^2 (90 m^2) of regular gasoline would require an application rate of 0.1 gpm/ft^2 using a 3 percent foam concentrate with a minimum run time of 15 minutes. This would equal a need for 100 gpm of foam solution per minute or 3 gallons of foam concentrate for 15 minutes for a total of 45 gallons of foam concentrate required.

Heat resistance is the ability of the foam to stand up to the heat of the fire or to hot surfaces near the fire. **Knockdown speed** is how fast the foam spreads across the surface of a fuel. **Fuel resistance** is the ability to tolerate the fuel and to avoid being saturated or picking up the fuel. **Vapor suppression** is the ability to contain or control the production of fuel vapors.

Protein foam is made from chemically broken down natural protein materials, such as animal blood, that have metallic salts added for foaming. The foam created is stiff, elastic, and has a high water retention rate and heat resistance. It is highly effective in vapor suppression but can be blown away by the wind, does not spread easily, and has a limited shelf life. It cannot be saturated or dipped into the fuel (poor fuel resistance), has slow knockdown time, and breaks down when used with dry chemical extinguishing agents. It is applied in 3 and 6 percent concentrations.

Fluoroprotein foam was designed as improved protein foam and has a fluorinated surfactant added. This surfactant allows the foam to be dipped into the fuel. In fact, fluoroprotein can be subsurface injected into a fuel tank to rise through the fuel and to the surface. It also has the ability to work with dry chemical extinguishing agents. It has high heat resistance and vapor suppression capabilities and a better knockdown time than protein foam. Fluoroprotein foam is popular at fuel processing and storage facilities and is applied in 1, 3, and 6 percent concentrations.

Aqueous film forming foam (AFFF) is made from fluorochemical surfactants and synthetic foaming agents that have a quick drain-down time. This feature creates a liquid that forms a film or layer of water that spreads quickly over the surface of the flammable liquid, **Figure 11-31**. This film provides a faster knockdown time but gives up some heat resistance, fuel tolerance, and vapor suppression compared to fluoroprotein foam. AFFF can be applied with regular fog nozzles and comes in alcohol-type concentrates (ATC) (see later discussion). Because of its effectiveness, ease of use,

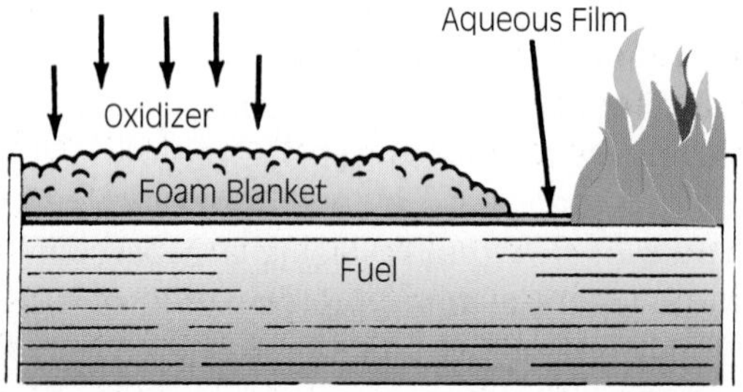

Figure 11-31 AFFF applied on Class B fuel. Note the film barrier on the surface.

and reduced need for special applicators, it is carried by most fire departments. AFFF can also be used with dry chemical extinguishing agents. The quick drain-down does require continued application with a poorer resistance to burnback than fluoroprotein foam. It is also applied in 1, 3, and 6 percent concentrations.

Fluoroprotein film-forming foam (FFFP) combines protein with the film-forming fluorosurfactants of AFFF to improve on the qualities of both types of foam. It has the fast film-forming capabilities of AFFF and the burnback and heat resistance of protein foams. It has some alcohol resistance.

Detergent-type foams use synthetic surfactants to break down the surface tension of water and create a foaming blanket. These foams originally were good at penetrating into Class A materials but lacked heat resistance and stability. Newer types have overcome these drawbacks, and these types are used both as an active firefighting agent and as a protective barrier against fire spread. A special type of detergent foam is used for high expansion foam, which is used to fill up entire areas such as mine shafts or buildings.

CLASSIFICATION OF FUELS

Foams are used for Class A and B fires and some specific considerations affect their use.

Class A

Class A fuels are ordinary combustibles that have traditionally been extinguished with water. Sometimes piles of Class A materials are extinguished using a wetting agent, often a detergent-like substance, to help soak through the piles. Today, a new class of foams is available for Class A materials that uses a detergent base that extinguishes by reducing the surface tension of the water, allowing greater penetration into the materials. The foamy water solution has the ability to cling to the sides of objects, thus enhancing protection. This clinging ability is now being used to protect homes in urban interface areas during wildland fires, **Figure 11-32**. Urban fire departments are using it for interior firefighting and have reported faster fire extinguishment with less water. Class A foams use application ranges of 0.03 to 1.0 percent, which require a separate and more accurate proportioning system than Class B foams. Some Class B foams like AFFF have a detergent action that can be used on Class A materials.

Disadvantages of Class A foams include cost of equipment and agent, additional possibilities of

Figure 11-32 Firefighters applying Class A foam in wildland–urban interface. (Photo courtesy of Ansul, Inc.)

equipment failure, possible effects on the environment and fire investigation laboratory tests.[5] Other potential problems are more difficult salvage operations and a customer service problem of explaining the milky residue on furniture and possessions to the homeowner.

Class B

Class B fuels include flammable liquids in two categories: hydrocarbons and polar solvents. Because gases should be extinguished by shutting off the flow of the fuel, firefighters do not use foam on them.

Hydrocarbons and Polar Solvents

Hydrocarbons cover a wide range of substances in forms from gaseous to liquid to semisolid and solid. Common examples are heating oil, diesel fuel, gasoline, kerosene, petroleum jelly, paraffin, and asphalt. These fuels do not mix with water; they are not **miscible** or water soluble and the best method to extinguish large quantities is foam. These foams work by cooling and smothering to extinguish the fire and provide a vapor barrier.

Polar solvents mix with water and this ability to mix with water causes either a breakdown of the foam or mixing of flammable vapors into the bubbles of ordinary foams. These liquids include alcohols, both methyl and ethyl alcohol, lacquer thinners, acetone, ketones, acrylonitrile, and acetates. Perhaps the most commonly used is the ethyl alcohol, which when added to water with some flavoring makes alcoholic beverages such as whiskey. Many gasoline blends today have polar solvents in solution and require alcohol-resistant foams for effective extinguishment. To prevent the breakdown of ordinary foams, special foams called polar solvent-type, alcohol-resistant concentrate, or alcohol-type foams have been developed. These polar solvent-type foams create a **polymeric barrier** that separates the polar solvent from the liquid of the foam, **Figure 11-33**.

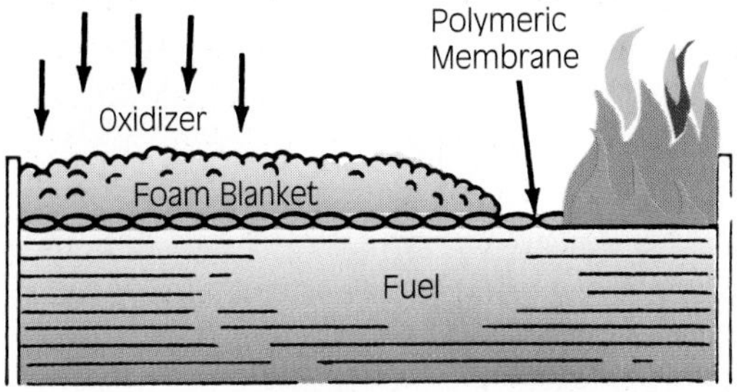

Figure 11-33 Polar solvent or alcohol-type foam applied on Class B polar solvent fuel. Note the polymeric film barrier on the surface.

APPLICATION OF FOAM

Foam is a mixture that requires a device to proportion, meter, or mix the foam concentrate into the water. Air must then be added to the foam solution to complete the process. There are several ways of adding air to the foam concentrate. Concentrations are usually expressed as a percentage of foam concentration to water in the solution. For example, a 3 percent concentration would have three parts of foam concentrate to every ninety-seven parts of water. In premixed systems, the foam concentrate is added directly to the tank, and the solution is pumped as the tank is drained. The problems are the resulting limited supply, the size of the tank, and the possible damage caused to the pump and valves by the foam solution.

One common proportioner is an **eductor**, **Figure 11-34,** which works on the **venturi principle**. In a venturi eductor, water enters its inlet which is then reduced down, causing the water to move faster through the smaller opening. As the speed increases, the pressure drops and creates suction. At the point of suction, the foam concentrate pick-up tube is attached and the concentrate is drawn up

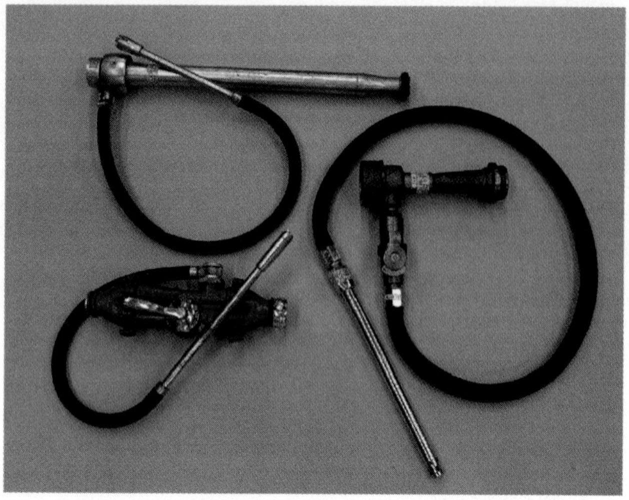

Figure 11-34 From right to top left clockwise: an in-line eductor, bypass eductor, and a foam eductor nozzle.

the tube to mix with the water, **Figure 11-35**. **Figure 11-36** shows how to connect and operate an in-line eductor.

Several types of eductors can be permanently piped into or added to a hoseline. One that is always piped through the venturi is an **in-line eductor,** and one that has a separate waterway and valve to allow plain water to pass by the venturi is called a **bypass eductor**, **Figure 11-34** and **Figure 11-37**.

Eductors must have the proper gpm flow, have the correct pressure, be kept clean, and not have any back-pressure situations such as hose kinks, too much hose, a partially opened nozzle, or too much elevation to work properly. Firefighters must ensure that the proper setup is used to match the eductor, nozzle, and hoseline to give this correct flow. The eductor and nozzle must be clean of dried foam concentrate and other debris. Things as simple as a few pebbles in the nozzle baffle or kinks in the hose can reduce the flow and prevent foam pickup. The eductor must be set at the correct percentage for the concentrate.

Practice and maintenance are often keys to good foam operations. An around-the-pump proportioner uses an eductor between the intake and discharge side of the pump. The discharged water pressure is pumped through the eductor to siphon the concentrate into the intake side of the pump and through the pump. This system allows the pump to operate at full flow, but may cause damage to the pump and valves without proper backflushing of the entire pump. Balanced pressure systems use one of two

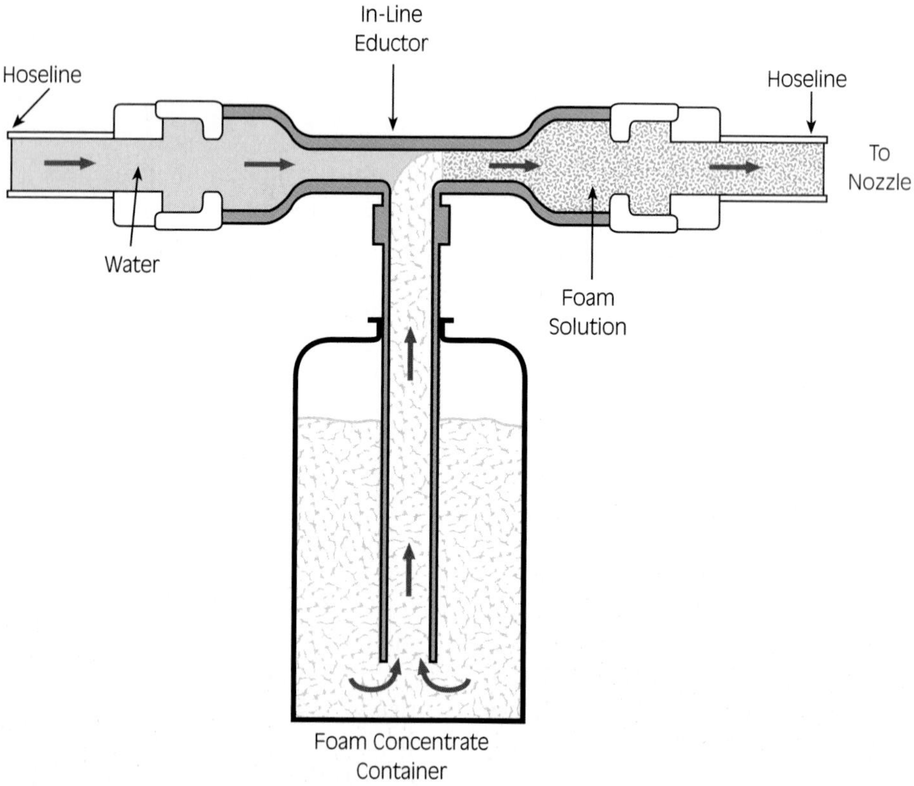

Figure 11-35 Foam eductor using the venturi principle.

Figure 11-36 The in-line eductor is connected either directly to the pump discharge or to a supply hoseline. Eductors require a specific amount of hose between the eductor and the nozzle, plus the gallonage of the nozzle must match the eductor. The eductor tube is placed into the can of concentrate, and any necessary metering adjustments are made. The eductor only operates when water is flowing.

Figure 11-37 Close-up of bypass eductor bypass valve.

foam pumping methods, in which the foam is pumped under pressure into a metering chamber that balances the pressure of the concentrate and water controlling the flow of the foam solution, **Figure 11-38**.

In **compressed air foam systems (CAFS)** or dual-injection systems, the concentrate is in a separate foam tank and a foam pump pumps the concentrate directly into the hoseline, which is metered by a flow-metered microprocessor, **Figure 11-39**. After the foam solution is created, an air compressor line injects air into the hoseline to create a light and

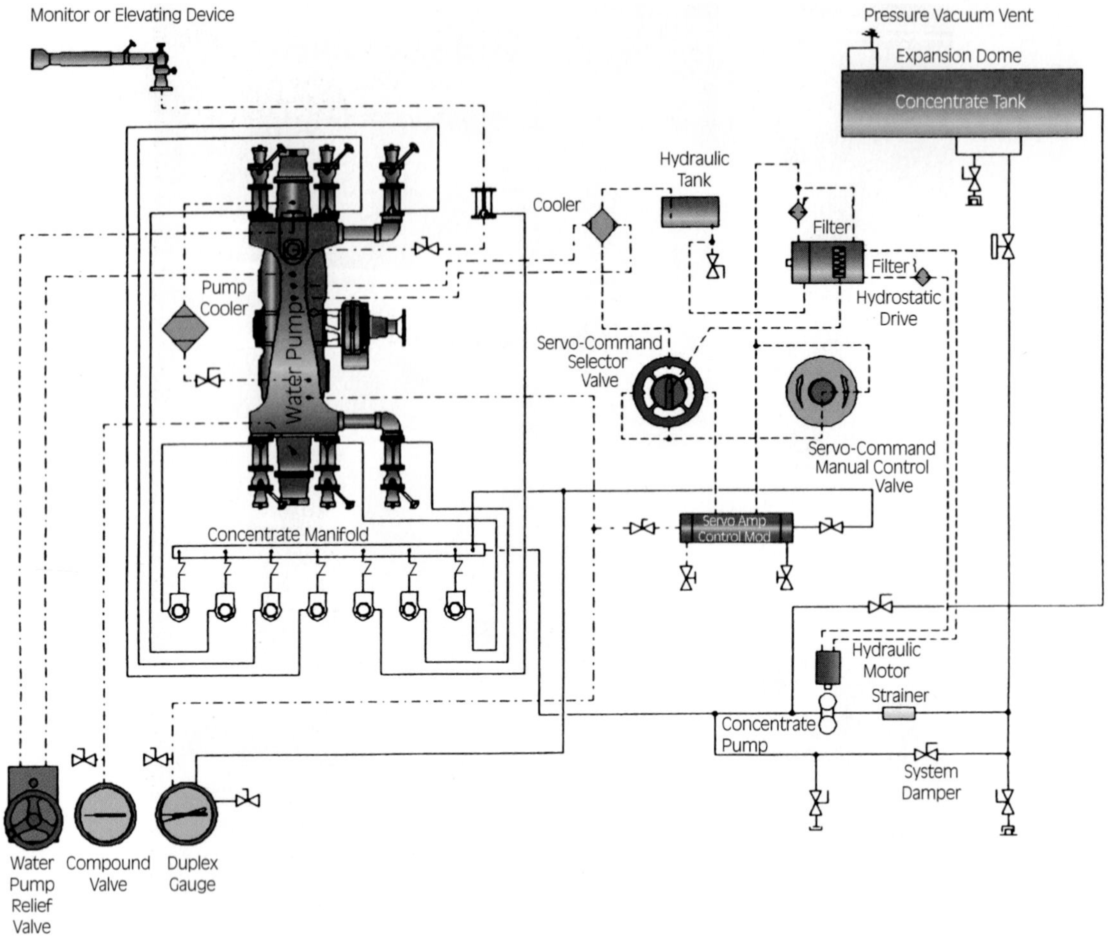

Figure 11-38 Balanced pressure demand–type foam proportioner system.

fluffy foam, **Figure 11-40**. These systems are used for Class A foams.

To finish making the foam, air must be added to the foam solution, normally at the nozzle. The various types of foam nozzles, regular fog nozzles, and foam generators have the ability to aspirate various quantities of air into the foam solution. The goal is to get the correct expansion ratio, that is, amount of air added to the solution. The expansion ratio is expressed as the volume of air added to the volume of solution. For example, 10:1 refers to 10 volumes of air added to 1 volume of solution. Foam's expansion ratios are in three ranges: low, medium, and high. Low expansion foam has an air to solution ratio of up to 20 to 1, or as much as 20 ft^3 (0.56 m^3) of finished foam for each cubic foot (0.028 m^3) of foam solution. Medium expansion foam has ratios ranging from 20:1 to 200:1, and high expansion foam is rated from 200:1 and higher (the upper limit is about 1,000:1). The creation of medium and high expansion foams requires a special foam generator. This special unit has a nozzle that sprays the foam solution onto a screen or mesh and a fan that pushes large volumes of air to create millions of bubbles. These foam generators may be gas, electric, hydraulic, or water powered, **Figure 11-41** and **Figure 11-42**.

Fog Nozzles versus Foam Nozzles

Originally, foam making required a special foam nozzle to properly aspirate the air into the foam solution. Some of these foam nozzles incorporated the eductor into the nozzle, combining all of the foam-making steps. Some of these combined eductor nozzles are still in use but because they require having the foam concentrate containers and the personnel to move the containers at the nozzle, these devices are not very popular. Foam nozzles are designed today to aspirate the proper amounts of air and apply the foam to the fuel. They do this by having air vents or ports built into the nozzle, and the movement of the foam solution past these ports draws the air into the solution, and it mixes both

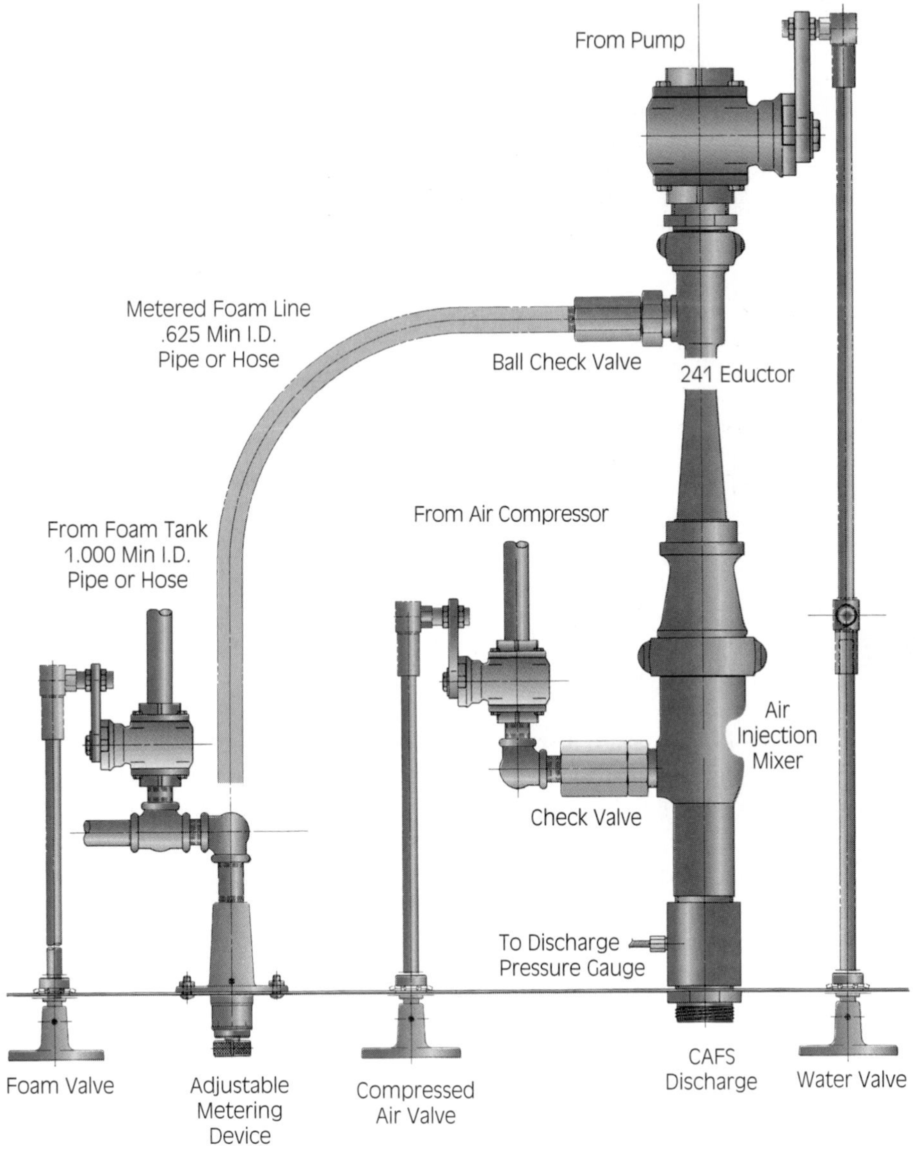

Figure 11-39 Typical compressed air foam system (CAFS).

Figure 11-40 Foam from CAFS being applied. *(Photo courtesy of Ansul, Inc.)*

Figure 11-41 Handline-operated medium expansion foam generator. *(Photo courtesy of Task Force Tips)*

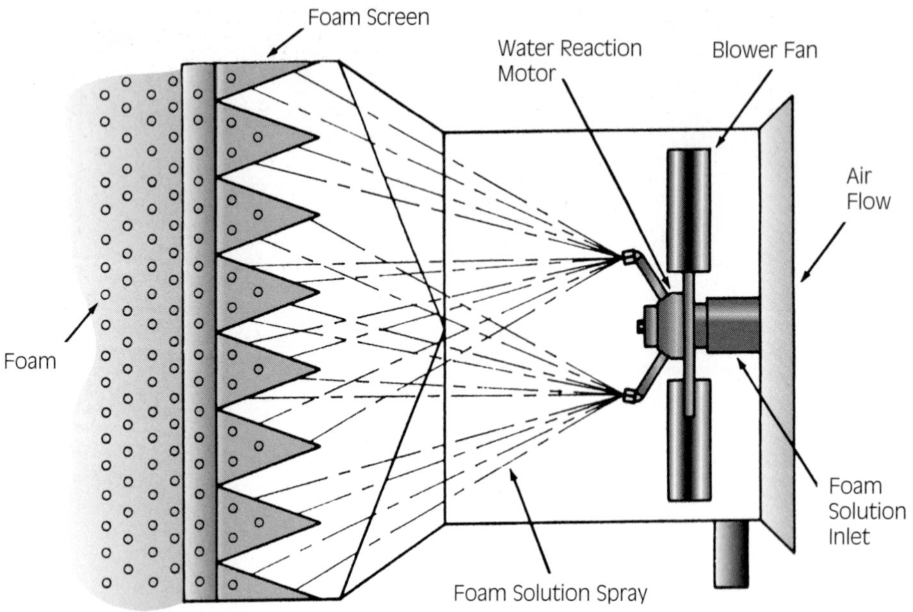

Figure 11-42 High expansion foam generator.

before and after the tip, **Figure 11-43**. Foam nozzles are designed for low and medium expansion foams usually having expansion ratios of 8:1 to 20:1 in the low range and up to 50:1 in the medium range. Foam nozzles are recommended with protein and fluoroprotein foams.

When AFFF was introduced, one of its additional advantages was that it did not require a special foam nozzle for application. Fog nozzles typically used by the fire department could be used. This helped reduce the costs of purchasing an effective foam system, and many departments began carrying foam concentrate and foam eductors on all pumpers. While the cost is lower, so is the expansion ratio, with 8:1 being at the high end. To increase this expansion efficiency while keeping costs lower, some manufacturers now have clip-on or snap-on foam nozzle adapters that attach to the fog nozzle and make it a foam nozzle, **Figure 11-44**. These are effective in increasing the expansion ratio and are almost as good as a specially designed foam nozzle. CAFSs, by preinduction of air, are effective with a fog nozzle and even with solid tip nozzles.

Foam from nozzles is applied using one of three techniques. The first is the *bank-in technique,* in which the foam strikes the ground before the fire and rolls into the fire, **Figure 11-45**. The second is the *bank-back* or *bounce-off technique*, in which the foam is banked off a wall or other object and rolls back into the fire, **Figure 11-46**. The third technique is the *raindown* or *snowflake technique,* in which the foam is sprayed high into the air over the fire and it floats down onto it, **Figure 11-47**.

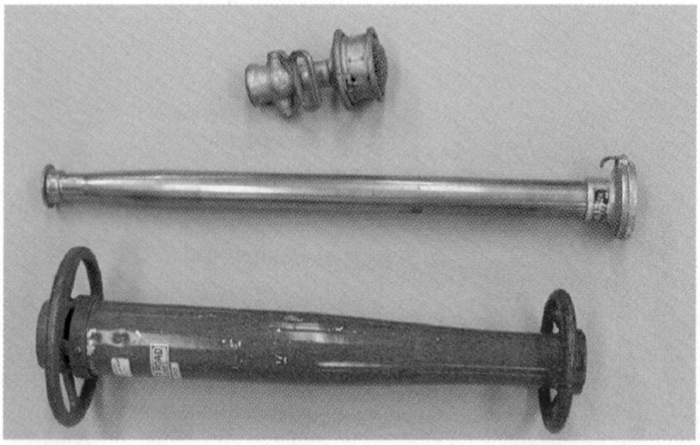

Figure 11-43 Typical foam nozzles.

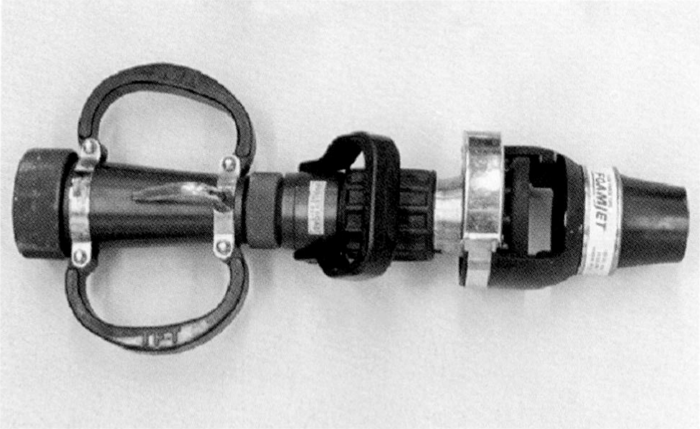

Figure 11-44 Clip-on foam attachment for fog nozzle.

Figure 11-45 The bank-in technique of foam application.

Figure 11-46 The bank-back technique of foam application.

Figure 11-47 The raindown technique of foam application.

Lessons Learned

Fire streams are made of water that leaves a nozzle and heads toward the target. The nozzle defines the characteristics of the fire stream. The two main types of nozzles are solid tip and fog, and they need to match the fire conditions and fire department resources. Each type of nozzle has its advantages and disadvantages; firefighters should use the one that will do the best job of suppressing the fire. Proper use of nozzles requires shutoffs and even some special types of nozzles. Fire streams and fire conditions determine how firefighters will attack a fire.

An understanding of fire streams is not possible without understanding the basic hydraulics of moving the water from a source to the fire. Correct hydraulic calculations result in a good fire stream being delivered in the proper amount, with a good shape. To do this properly requires an understanding of the pressures and the amount of friction loss in the system. Calculations of the flow, friction loss, and pressure are combined as needed by the hoseline(s) and nozzle(s) being used.

When the fuels involved in a fire are not compatible with water, other agents must be used to fight the fire and, commonly, this other agent is foam. Firefighters should know the different kinds of fuels and types of foams available. Foam requires special equipment to create and apply it, and some special application techniques are used.

KEY TERMS

Absolute Pressure The measurement of pressure, including atmospheric pressure. Measured in pounds per square inch absolute.

Application Rate Amount of foam or foam solution needed to extinguish a fire. Usually expressed in gallons per minute per square foot or liters per minute per square meter.

Automatic or **Constant Pressure Nozzle** Nozzle with a spring mechanism built in that reacts to pressure changes and adjusts the flow and resultant reach of the nozzle.

Available Flow Amount of water that can be moved to extinguish the fire. Depends on the water supply, pump(s) and their capabilities, and the size and length of hose.

Bresnan Distributors Has six or nine solid tips or broken stream openings designed to rotate in a circular spray pattern. Used to fight fire in basements or cellars when firefighters cannot make a direct attack on the fire.

Bypass Eductor Eductor with two waterways and a valve that allows plain water to pass by the venturi or through the venturi to create foam solution.

Cellar Nozzles Has four spray nozzles designed to rotate in a circular spray pattern for fighting fires in basements or cellars when firefighters cannot make a direct attack on the fire.

Combination Fire Attack A blend of the direct and indirect fire attack methods, with firefighters applying water to both the fuel and the atmosphere of the room.

Combination Nozzle A spray nozzle that is capable of providing straight stream and spray patterns, which are adjustable or variable by the operator. Most fog nozzles used today are combination nozzles.

Compressed Air Foam Systems (CAFS) A foam system where compressed air is injected into the foam solution prior to entering any hoselines. The fluffy foam created needs no further aspiration of air by the nozzle.

Constant or **Set Volume Nozzle** Nozzle with one set volume at a set pressure. For example, 60 gpm at 100 psi (227 L/min 690 kPa). The only adjustment is the pattern.

Detergent-Type Foams Use synthetic surfactants to break down the surface tension of water to create a foaming blanket.

Direct Fire Attack An attack on the fire made by aiming the flow of water directly at the material on fire.

Discharge Flow Total amount of water flowing from the discharge side of the pump.

Eductor Device that siphons a liquid from a container into a moving stream.

Fire Stream The water or other agent as it leaves the hose and nozzle toward its objective, usually the fire.

Flow The rate or quantity of water delivered, usually measured in gallons per minute or liters per minute (1 gpm = 3.785 L/min).

Fluoroprotein Film-Forming Foam (FFFP) Combines protein with the film-forming fluorinated surfactants of AFFF to improve on the qualities of both types of foam.

Fluoroprotein Foam Designed as an improved protein foam with a fluorinated surfactant added.

Foam An aggregate of gas-filled bubbles formed from aqueous solutions of specially formulated concentrated liquid foaming agents.

Fog Nozzle Delivers either a fixed spray pattern or variable combination of straight stream and spray patterns.

Friction Caused by the rubbing of materials against each other while in movement and converts or robs some of the movement energy into heat energy.

Friction Loss Measurement of friction in a system such as a hoseline.

Fuel Resistance Ability to tolerate the fuel and to avoid being saturated by or picking up the fuel.

Gauge Pressure Measures pressure without atmospheric pressure. Normally fire department gauges do not measure atmospheric pressure. Gauge pressure is measured in psi or psig.

Head Pressure Measures the pressure of a column of water in feet (meters). Head pressure gain or loss results when water is being pumped above or below the level of the pump. A head of 2.31 feet (0.7 m) would equal 1 psi (6.895 kPa).

Heat Resistance The ability of foam to stand up to the heat of the fire or to hot surfaces near the fire.

Hydraulics The study of fluids at rest and in motion.

Indirect Fire Attack An attack made on interior fires by applying a fog stream into a closed room or compartment, thus converting the water into steam to extinguish the fire.

In-Line Eductor Eductor in which the waterway is always piped through a venturi.

Knockdown Speed Speed with which foam spreads across the surface of a fuel.

Miscible Having the ability to mix with water.

Needed or **Required Flow** Estimate of the amount of water required to extinguish a fire in a certain time period. Based on the type and amount of fuel burning.

Nozzle A tapered or constricted tube used to increase the speed or change the direction of water or other fluids.

Nozzle Flow The amount or volume of water that a nozzle will provide. Flow is measured in gallons per minute or liters per minute.

Nozzle Pressure The pressure required to effectively operate a nozzle. Pressure is measured in pounds per square inch or kilopascals.

Nozzle Reach The distance the water will travel after leaving the nozzle. Reach is a function of the pressure, which is converted to velocity or speed of the water leaving the nozzle.

Nozzle Reaction The force of nature that makes the nozzle move in the opposite direction of the water flow. The nozzle operator must counteract the thrust exerted by the nozzle to maintain control.

Piercing Nozzles Originally designed to penetrate the skin of aircraft and now have been modified to pierce through building walls and floors.

Polymeric Barrier A separation barrier made up of polymer or a chain of molecules linked in a series of long strands. This separates a polar solvent from an ATC foam blanket.

Protein Foam Made from chemically broken down natural protein materials, such as animal blood, that have metallic salts added for foaming.

Pump Operator A generic term to describe the person responsible for operating a fire apparatus pump. Other commonly used titles include motor pump operator, engineer, technician, chauffeur, and driver/operator.

Solid Stream Nozzles Type of nozzle that delivers an unbroken or solid stream of water to the fire. Also called solid tip, straight bore, or smooth bore.

Straight Stream A nozzle pattern that creates a hollow stream, similar in shape to the solid stream pattern, but the straight stream pattern must pass around the baffle of the nozzle. Newer fog nozzle designs, especially the automatic nozzles, only have this hollow effect from the tip on and, hence, create a solid stream with good reach and penetration abilities, some better than solid stream nozzles.

Stream Shape The arrangement or configuration of the water or other agent droplets as they leave the nozzle.

Vacuum (Negative) Pressure The measurement of the pressure less than atmospheric pressure, which is usually read in inches of mercury (in. Hg or mm Hg) on a compound gauge.

Vapor Suppression Ability to contain or control the production of fuel vapors.

Variable, Adjustable, or **Selectable Gallonage Nozzle** Nozzle that allows the nozzleperson to select the flow, with usually two or three choices, and the pattern.

Velocity Pressure The forward pressure of water as it leaves an opening.

Venturi Principle A process that creates a low-pressure area in the induction chamber of the eductor and allows the foam concentrate to be drawn into and mixed with the water stream.

Water Curtain Nozzle Designed to spray water to protect exposures against heat by wetting the exposure's surface.

REVIEW QUESTIONS

1. Define a fire stream.
2. Explain what a nozzle is and how it works.
3. What is the amount or volume of water leaving the nozzle called?
4. Explain the term *nozzle reaction.*
5. Explain the difference between a solid tip nozzle pattern and a straight stream nozzle pattern.
6. A selectable or variable nozzle allows the operator to adjust what feature?
7. What is the purpose of a Bresnan distributor?
8. Why should the pattern of a water curtain be applied onto a solid object?
9. What is the most common type of shutoff device for a nozzle and how does it work?
10. Explain the difference between a direct and indirect attack.
11. What is the difference between atmospheric and gauge pressure?
12. Identify common nozzle pressures for solid tip and fog nozzles.
13. To reduce the friction loss in hose, what two things can be done?
14. Define foam and name the three major components.
15. How does AFFF work on a flammable liquid fire?
16. If a foam concentration was 6 percent, how many gallons of concentrate would be added to how many gallons of water to equal 100 gallons?

Endnotes

1. A 1½-inch (38-mm) hoseline is the minimum size hoseline for structural firefighting.
2. Although the hoseline can be operated by one person, safety regulations require operation with a second team member.
3. Sturtevant, Thomas, *Introduction to Fire Pump Operations,* pp. 228–232. Delmar Learning, a part of the Thomson Corporation, Clifton Park, NY, 1997.
4. Cote, Arthur and John Linville, *Fire Protection Handbook,* 18th ed., pp. 6-349. National Fire Protection Association, Quincy, MA, 1997.
5. Stern, Jeff and J. Gordon Routley, *Class A Foam for Structural Firefighting.* U.S. Fire Administration, Emmittsburg, MD, December 1996.

Additional Resources

Gagnon, Robert M., *Design of Special Hazard and Fire Alarm Systems.* Delmar Learning, a part of the Thomson Corporation, Clifton Park, NY, 1998.

Klinoff, Robert, *Introduction to Fire Protection.* Delmar Learning, a part of the Thomson Corporation, Clifton Park, NY, 1997.

Sturtevant, Thomas, *Introduction to Fire Pump Operations.* Delmar Learning, a part of the Thomson Corporation, Clifton Park, NY, 1997.

Crapo, William F., *Hydraulics for Firefighting,* Delmar Learning, a part of the Thomson Corporation, Clifton Park, NY, 2002.

NFPA 11: Standard for Low, Medium, and High Expansion Foam Systems. National Fire Protection Association, Quincy, MA, 2002.

NFPA 1961: Standard on Fire Hose. National Fire Protection Association, Quincy, MA, 2002.

NFPA 1962: Standard for Care, Use and Testing of Fire Hose Including Couplings and Nozzles. National Fire Protection Association, Quincy, MA, 1998.

NFPA 1964: Standard for Spray Nozzles (Shutoff and Tip). National Fire Protection Association, Quincy, MA, 1998.

Stern, Jeff and J. Gordon Routley, *Class A Foam for Structural Firefighting.* U.S. Fire Administration, Emmittsburg, MD, December 1996.

CHAPTER 12

PROTECTIVE SYSTEMS

Ric Koonce, J. Sargeant Reynolds Community College

OUTLINE

- Objectives
- Introduction
- Detection Systems
- Sprinkler Systems
- Sprinklers and Life Safety
- Sprinkler Head Design and Operation
- Types of Sprinkler Systems
- Sprinkler System Connections and Piping
- Control Devices for Sprinkler Systems
- Returning Sprinkler Systems to Service
- Standpipe Classifications
- Standpipe System Connections and Piping
- Alarms for Standpipes and Sprinklers
- Other Protective Systems
- Fire Department Operations with Protective Systems
- Lessons Learned
- Key Terms
- Review Questions
- Endnotes
- Additional Resources

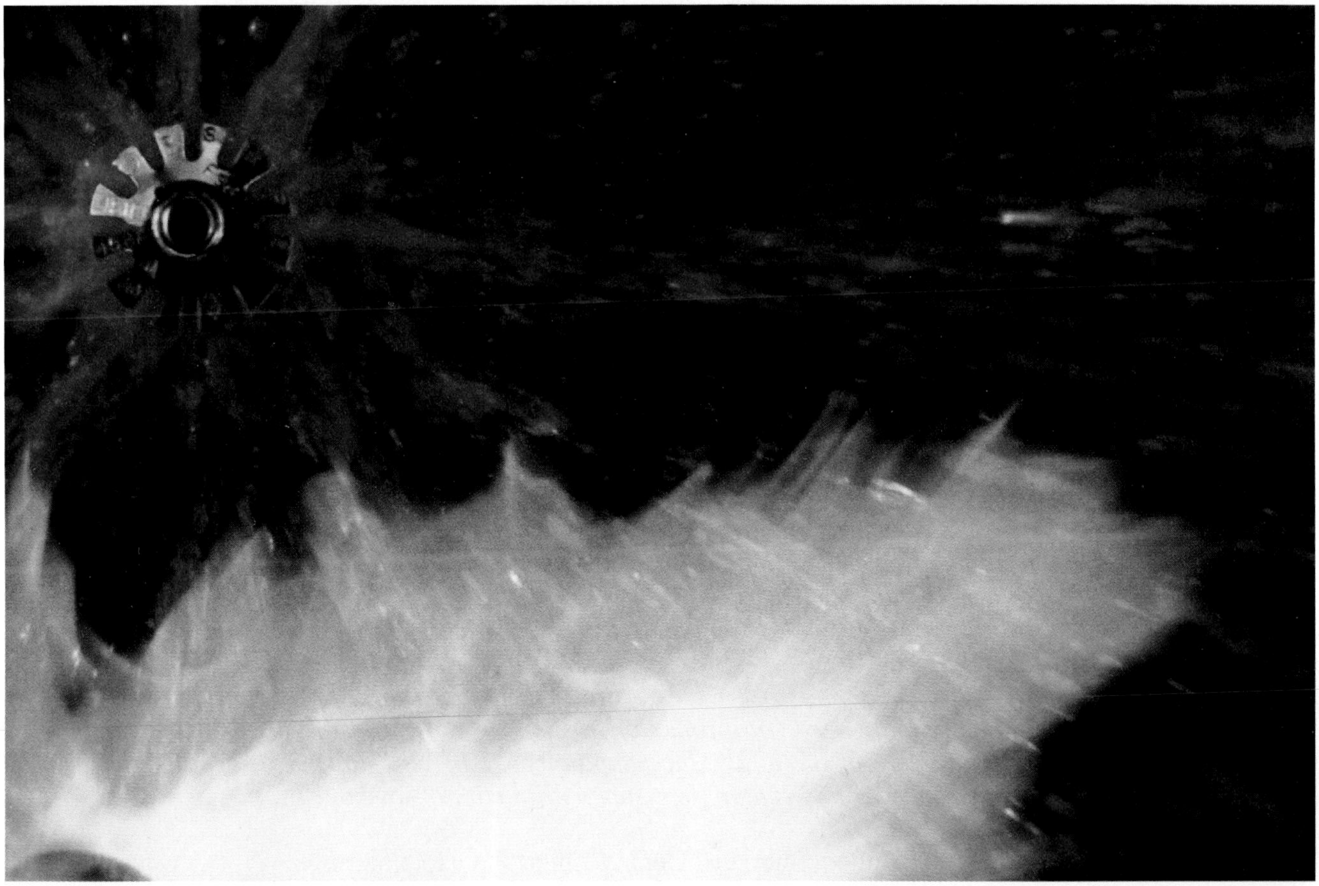

STREET STORY

One winter night in the heart of downtown Philadelphia an entire fire protection system failed; a trio of firefighters was lost, a thirty-eight-story skyscraper was destroyed, and the confidence of those dependent on fixed fire protection systems was shaken. It was the devastating No. 1 Meridian Plaza fire that changed my view on active and passive protective devices.

Upon arrival of the first-alarm companies, fire was venting from the twenty-second floor. At the onset it was a controllable fire. Then the fire protection systems began to quickly fail. Initially, it was the electrical systems. The primary and secondary wires "shorted" and placed the entire high-rise office building into total darkness. With the elevator stalled, access to the upper floors was limited to climbing the three fire towers. After firefighters hauled their equipment to the twentieth floor they discovered the pressure-reducing valves on the standpipe system were improperly set. Generating an effective fire streams system was impossible. Later, without an operable HVAC the core temperature in the building began to escalate. Pipe chases for the bathroom plumbing were never sealed. These "poke-holes" permitted the toxic smoke to spread in all directions. Smoke began to bank down, and the ventilation group became disoriented and lost and perished on a smoke-charged upper floor. With conditions rapidly deteriorating, a partially sprinklered structure, crews working with only hand lights, and practically no way to ventilate the building, the odds did not favor the men and women of the Philadelphia Fire Department.

After twelve alarms were struck, bringing more than 300 firefighters to the scene, and nineteen long hours, the blaze was contained to eight floors. In the end, ten sprinkler heads stopped the upward progress of the flames on the thirtieth floor. The lessons from No. 1 Meridian Plaza are plentiful, but one undisputed fact is that fire protection systems must be properly designed, installed, and maintained.

Recently, at the Canarsia Senior Citizen Center on Vandalla Avenue in New York City, a self-closing steel fire door remained ajar and a sprinkler system did not operate in a tenth-floor common hallway. When heavy winds pushed the flames like a blowtorch from the upper floor apartment into the hallway, a search and rescue group was engulfed in the fireball. Tragically, again three firefighters were killed because essential components of the fire protection system did not properly function.

Daily across this nation, emergency responders are witness to incidents that reinforce the need for knowledge of these basic systems. Somewhere in the nation this week a rate-of-rise alarm in a commercial building warned of an incipient fire, a sprinkler head activated in a factory and extinguished a minor fire before it spread to a manufacturing process, a closed fire-rated door in a hospital contained the smoke to a particular room before a massive evacuation was needed, a damper over a kitchen grill closed to prevent the involvement of a metal duct that crisscrossed vital structural supports in a restaurant, and the annunciator panel in an enclosed shopping mall pinpointed the exact location of a potential problem. Unfortunately, when these fire protection systems malfunction, the lives and safety of the occupants and emergency responders are at a much higher risk.

What is the price of success? In the fire rescue service, personnel safety and operational efficiency are surely two of the keys; often both hinge on fire protection systems. Firefighters often see firsthand the economic value of a properly maintained and tested fire protection system. I know both the economic and emotional value of these auxiliary appliances. I believe that an in-depth knowledge of sprinkler and standpipe systems will help to "save our own."

—Street Story by William Shouldis, Deputy Chief, Philadelphia Fire Department, Philadelphia, Pennsylvania

OBJECTIVES

After completing this chapter, the reader should be able to:

- Identify the value of protective systems in protecting life and property.
- Identify and explain the operation of the various types of detection devices.
- Explain and recognize the types of sprinkler heads and how they operate.
- Identify the various types of sprinkler systems and the components of each type.
- Identify the piping arrangements of sprinkler systems and connections.
- Demonstrate how to connect to a fire department connection.
- Identify control valves for systems and explain their operation.
- Explain the methods used to return a sprinkler system to service.
- Demonstrate techniques for stopping a flowing sprinkler head.
- Identify standpipe classes and types of systems.
- Identify piping and connections for standpipe systems.
- Demonstrate how to connect supply and attack lines to standpipe connections.
- Identify alarm systems for protective systems.
- Explain fire department procedures at protective properties.
- Identify other protective systems, their components, and their benefits and hazards.

INTRODUCTION

Protective systems help protect lives and property by detecting and/or suppressing fires. Detection systems detect the presence of a fire and alert building occupants and/or the fire department and may also activate a suppression system. Suppression systems are devices that help firefighters in controlling fires. Some systems are designed to automatically detect and suppress a fire, while others assist firefighters and occupants in putting out the fire. They are also called **auxiliary appliances**. Regardless of the term used, firefighters must know about these systems and how to use them. Great efforts have been expended to require these systems in building and fire prevention codes. To builders, owners, and occupants, they are expensive to purchase and maintain, and the expectation is that the fire department will use them correctly.

> **Note** Remember that these devices can make the job of firefighting safer and easier.

Detection systems are varied. Some require people to detect the fire and manually activate an alarm. Others are highly complex systems that can detect a fire almost at its ignition and sound an alarm. They may also activate suppression systems.

The two main suppression systems are sprinklers and standpipes. These are devices that allow firefighting's favorite extinguishing agent, water, to be used. When the situation dictates that another agent may be more effective, other types of protective systems are required. Some use the agents found in portable fire extinguishers to cover room or building size areas. These systems include dry and wet chemicals, inerting gases, and halogenated agents.

Detection systems, suppression systems, and alarm systems (covered briefly in Chapter 3) can all be tied together to form a protective ensemble for building occupants. This chapter describes individual detection and suppression components as well as the interface of each. It also covers fire department operations using protective systems.

DETECTION SYSTEMS

Detection systems are designed to notify people of a fire or other problem. Simple detection systems can actually be viewed as a warning system requiring a person to recognize a danger. More complex systems use a series of devices to automatically detect a fire or event and initiate a warning alarm. Fire detection and warning can also come from the initiation of a suppression system. For example, a sprinkler activates, tripping a water flow alarm that can notify occupants. Regardless of how complex or simple, detection systems are designed to notify *people* that a potentially life-endangering event is happening. By understanding the basic principles of operation of various detection systems, firefighters more effectively utilize these systems to perform their job. Communities *expect* firefighters to understand these systems.

People or Manual Systems

People can alert other building occupants and can call the fire department after discovering a fire. Some buildings are equipped with a manual fire alarm system that requires a person to discover the fire and then pull a lever or push a button. A restaurant cooking area or a laboratory hood with a local application system that sounds an alarm and activates the suppression system can have a manual switch for people to operate. Street box systems are also manual fire alarm systems.

Manual systems suffer from two typical problems. The first is that if they are to work, a person must be present, awake, and alert to discover the fire and then activate the system. Unfortunately, sleeping people and pets make poor fire detectors, because all of their detection senses are also asleep. The second problem is that many manual fire detection systems are local only, meaning that they will alert any building occupants but will not summon any additional help. A public education program and proper signs at the pull station can help teach people about the need to call the fire department.

Heat Detectors

Heat detectors operate by detecting the heat of a fire at a fixed temperature or as the rising temperature builds at a rapid rate. These can be used as part of a suppression system such as a sprinkler system, can operate a fire protection device such as closing a door, or are merely detection devices that operate an alarm. Heat detectors are slow to detect fire and hence should not be used for life safety. They are, however, inexpensive and have a low rate of false alarms. Heat detectors can be of the spot type, which covers just one area, or of the line type, which covers a larger area.

The *rate-of-rise heat detector* measures temperature increases above a predetermined rate. The pneumatic or air-sensitive device uses a diaphragm with a small hole or orifice that acts as a relief valve for slow temperature increases. When a fast temperature increase occurs, the diaphragm is pushed toward a contact point and the alarm is activated, **Figure 12-1**.

The *fixed temperature heat detector* is usually electrically operated with a bimetallic strip of two metals that expand at different rates, eventually bending to touch a contact point and complete the alarm circuit, **Figure 12-2**. Another fixed temperature type uses a metal element similar to solder that melts at a fixed temperature and breaks a circuit or

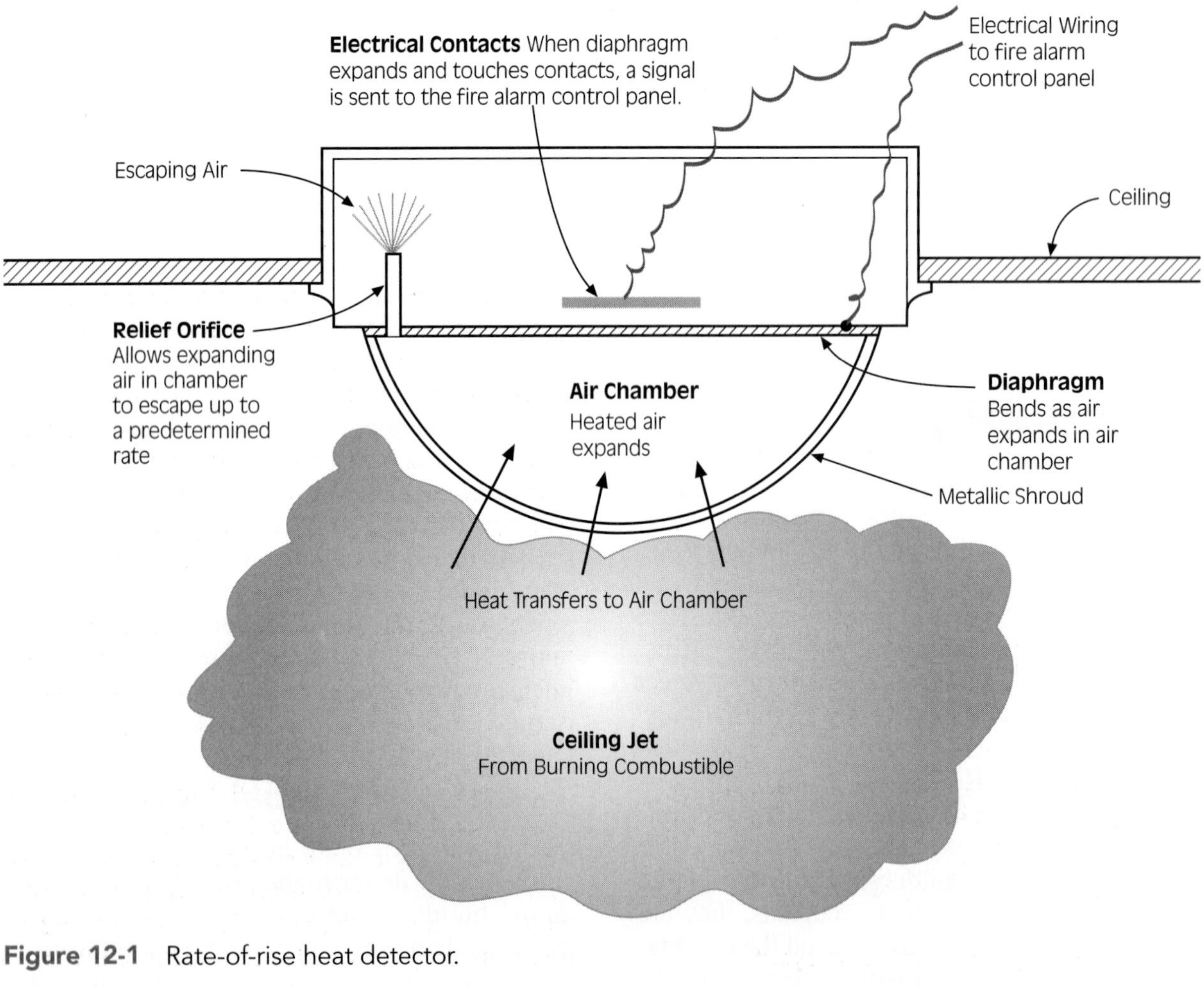
Electrical Contacts When diaphragm expands and touches contacts, a signal is sent to the fire alarm control panel.
Electrical Wiring to fire alarm control panel
Escaping Air
Ceiling
Relief Orifice Allows expanding air in chamber to escape up to a predetermined rate
Air Chamber Heated air expands
Diaphragm Bends as air expands in air chamber
Metallic Shroud
Heat Transfers to Air Chamber
Ceiling Jet From Burning Combustible

Figure 12-1 Rate-of-rise heat detector.

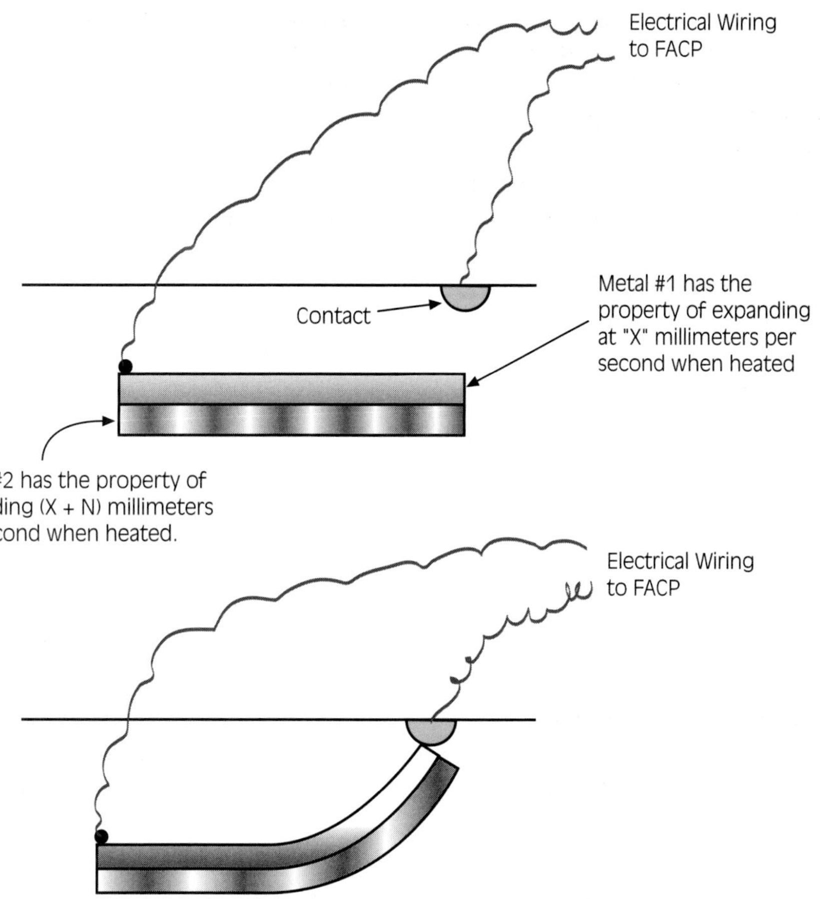
Electrical Wiring
to FACP
Metal #1 has the
property of expanding
at "X" millimeters per
second when heated
Contact
Metal #2 has the property of
expanding (X + N) millimeters
per second when heated.
Electrical Wiring
to FACP
Since metal #2 expands at a rate faster than metal #1, the
bimetallic strip will deform toward the contact when heated
until the circuit is closed, and a signal is sent to the FACP.

Figure 12-2 Fixed temperature heat detector.

opens a cap. Line-type detectors can be either pneumatic or electric, **Figure 12-3**.

Smoke Detectors

Smoke and toxic gases are the leading killers of people in the home. Smoke detectors have greatly improved residential life safety by quickly detecting

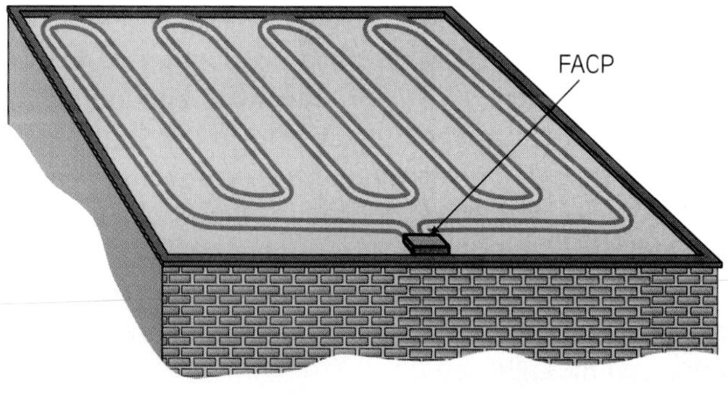
FACP

Figure 12-3 Line-type heat detector layout.

and alerting building occupants to the harmful effects of fire. Smoke detectors are the most prevalent automatic detection system. Even with the abundance of smoke detectors installed in homes across North America, many occupants do not understand how they work—or how to fix problems. It is important that firefighters understand the operation of smoke detectors. It is *essential* that they can describe problems and suggest solutions to the building owner or occupant.

Smoke detectors are hard or permanently wired, battery operated, or a combination of hard wiring with a battery backup. Smoke detectors work primarily on two principles: ionization and photoelectric.

Ionization detectors are the most common type of smoke detector—the kind installed in most homes and apartments since the early 1980s. These detectors use a radioactive element that emits ions into a chamber. The positive and negative ions are measured on an electrically charged

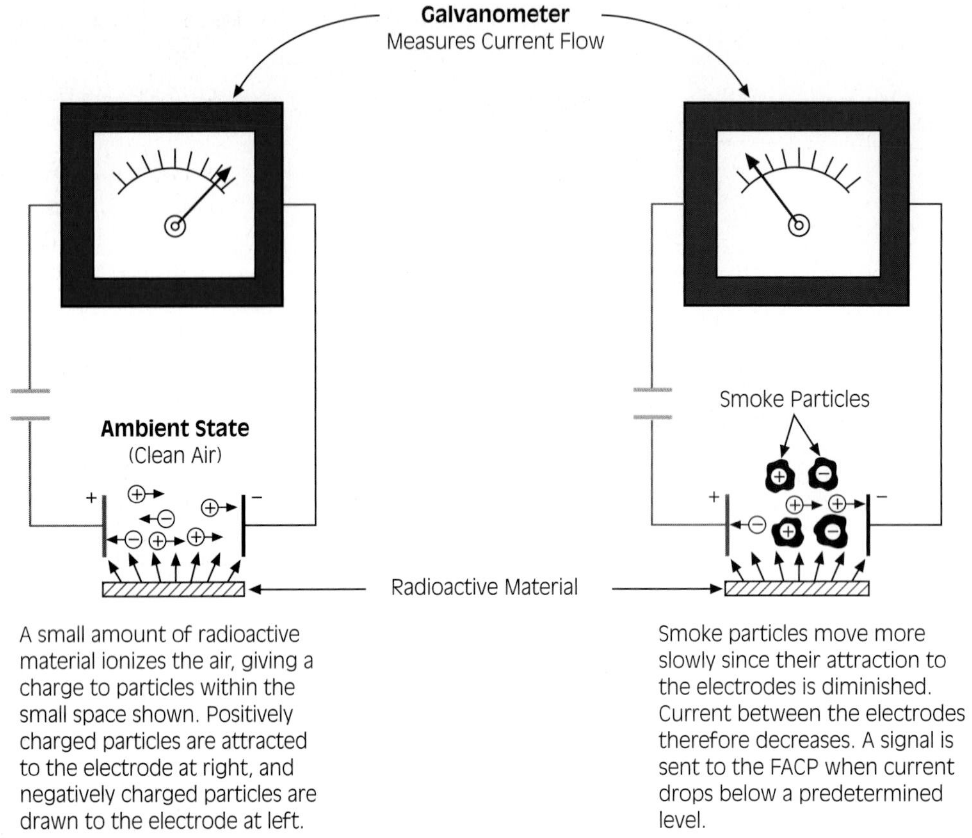

Galvanometer
Measures Current Flow
Ambient State
(Clean Air)
Smoke Particles
Radioactive Material
A small amount of radioactive material ionizes the air, giving a charge to particles within the small space shown. Positively charged particles are attracted to the electrode at right, and negatively charged particles are drawn to the electrode at left.
Smoke particles move more slowly since their attraction to the electrodes is diminished. Current between the electrodes therefore decreases. A signal is sent to the FACP when current drops below a predetermined level.

Figure 12-4 Ionization-type smoke detector.

electrode. When smoke enters the chamber, the flow of ions to the electrode changes and an alarm is activated, **Figure 12-4**. Ionization detectors are spot-type devices that are inexpensive and are very quick to react, especially to small smoke particles from flaming fires. Ionization smoke detectors can detect invisible smoke—certain fire gases that are colorless can disrupt ion flow within the chamber. Ionization smoke detectors are quite reliable but can be activated by cooking, steam (even from a hot shower), electrostatic discharge (from a thunderstorm), and dust buildup. Some people may classify these activations as false alarms. In actuality, these activations indicate that the detector is working as designed. Most ionization detectors will reset themselves once the smoke has cleared. Sometimes it is necessary to "blow out" the detector to clear the internal chamber of smoke products and dust. Ionization detectors that are activated by an electrostatic discharge may not reset; the detector will likely need to be replaced. Smoke detectors that "chirp" typically indicate that the battery is low.

Photoelectric detectors use two different methods to detect the smoke of a fire and can be spot or line detectors. The first is the *light obscuration detector,* which has a light beam aimed at a light sensor. When the smoke particles obscure or block the light beam, the light sensor notes the loss of light and activates the alarm, **Figure 12-5**. The *light-scattering detector* operates by having a light beam aimed at the end of the chamber with a light sensor in an angled-off chamber, **Figure 12-6**. When the smoke enters the chamber, the smoke particles scatter the light and the scattered light strikes the light sensor, activating the alarm. Photoelectric detectors are activated by visible smoke. They make a better smoke detector for areas where some smoke may be developed but is normally dissipated visually, such as in cooking areas and where fuel-powered equipment is working. Like ionization detectors, dust and steam can cause activation of the photoelectric detector. If the smoke causing activation of the photoelectric detector is especially oily (like from a grease fire), the detector may not reset once the smoke clears. A film has likely obscured the light sensor.

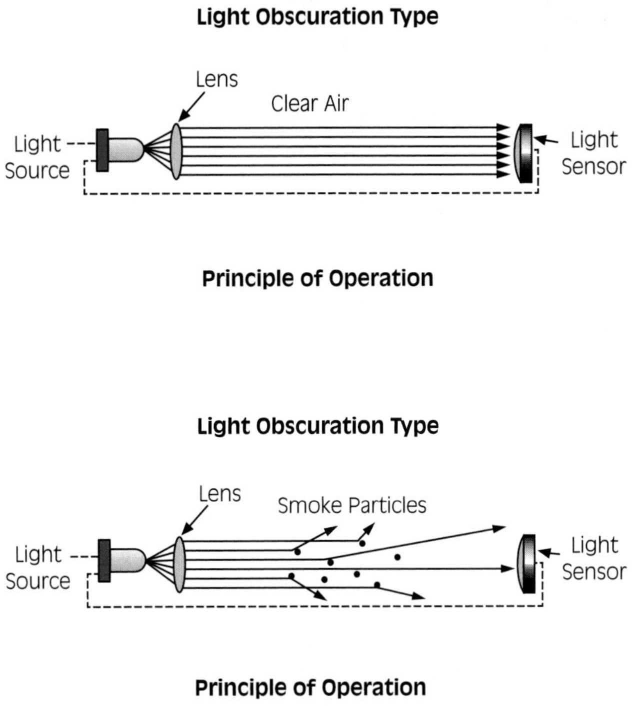

Figure 12-5 Light obscuration photoelectric smoke detector.

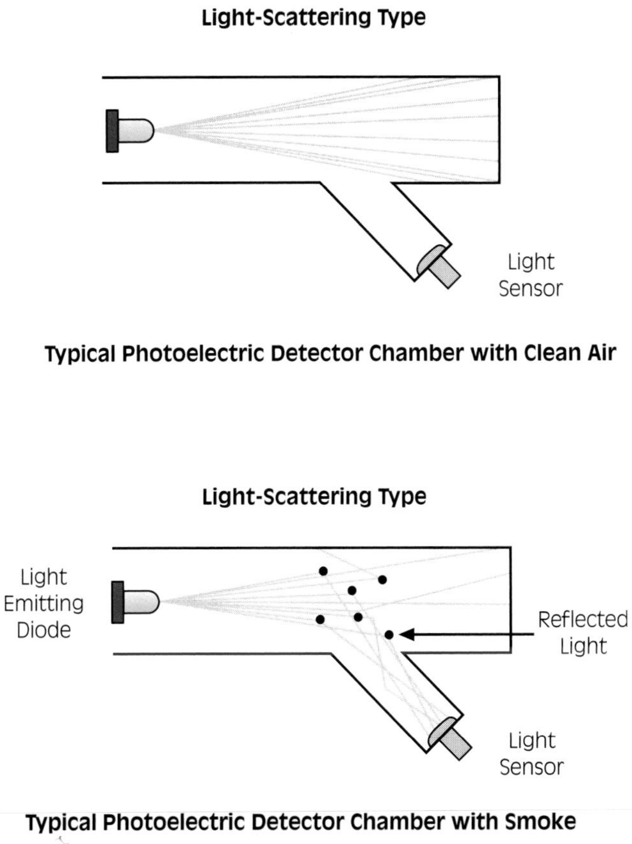

Figure 12-6 Light-scattering photoelectric smoke detector.

Gas Detectors

Gas-sensing detectors are designed to detect the presence of a certain gas or gases prior to the gases reaching a concentration that can cause danger. A flammable gas detector would detect the presence of the gas, such as a petroleum product, before it reached its ignitable concentration. Other gases would be detected prior to reaching a concentration that could cause death or injury, such as carbon monoxide. Gas detectors can be permanently mounted devices or portable units.

Carbon monoxide (CO) detectors for the home are becoming popular and are even required in some communities. This popularity has placed an additional service load on fire departments. Like smoke detectors, CO detectors have saved many lives. CO detectors function using several different methods. Consulting manufacturers' literature is perhaps the best approach to understand how each operates and how each needs to be reset. Most CO detectors, however, provide an early warning alarm (like an intermittent beep) when low-level CO is detected. Some CO detectors have an actual LED display that indicates the level (parts per million) of CO being detected. Often, persons reporting *smoke* detector activation may actually be hearing a CO detector activation—and vice versa. CO is odorless and colorless. Even though smoke is not visible, dangerous levels of CO may be present. A firefighter must be careful not to get caught investigating a smoke alarm activation only to find a CO detector has alarmed to dangerous CO levels. All detector activations should be treated as a worst-case scenario. Many fire departments carry gas detectors to check for flammable gases, carbon monoxide and other hazardous materials, and the oxygen content of the air.

Flame Detectors

Flame detection devices that detect the flames or lightwaves of the fire are mainly of three types: ultraviolet (UV), infrared (IR), or a combined ultraviolet-infrared (UV-IR) detector. These detectors are extremely rapid and are to protect petroleum and chemical facilities where fast and high temperature fire can occur.

Ultraviolet flame detectors detect the radiation of high-intensity flaming fires by detecting the lightwaves emitted in the UV spectrum (below 4,000 angstroms), which is below the visual light wavelengths. The UV detectors must have proper shielding from light sources, especially sunlight and welding. Dust or moisture can also fog the lens of the UV detector and cause detection problems.

Infrared flame detectors detect the infrared radiation (8,500 to 12,000 angstroms) of a fire using a photocell, which is above the visual light wavelengths. IR detectors are effective in rapidly spreading fires, but must be screened from sunlight and have a limited distance range. Beyond that range, unwanted signals will occur.

Combined *ultraviolet-infrared flame detectors* are used to rapidly detect the flames of a fire but without the false alarm problems of either a UV or IR detector. This is because the detector requires that both the UV and IR components be activated before an alarm is sent. Because the weakness of one does not affect the other, the double positive usually indicates a fire.

SPRINKLER SYSTEMS

Sprinkler systems are designed to automatically distribute water through sprinklers that are placed at set intervals on a system of piping, usually in the ceiling area, to extinguish or control the spread of fires. Most sprinkler heads detect the heat of a fire and begin to apply water directly over the source of the heat. Sprinkler heads, unless deluge-type heads, are heat-sensitive devices that react to a fixed temperature and disperse water in a specific pattern and quantity over a set area. Sprinkler systems are highly effective. In fact, some people describe the benefit of sprinklers as similar to having a firefighter constantly on duty in the protected building. The NFPA publishes standards for installation of sprinkler systems and their inspection and maintenance.[1]

SPRINKLERS AND LIFE SAFETY

Sprinkler systems were originally designed in the late 1800s to protect property, especially businesses and factories, from total loss from fires. They are almost 100 percent effective. The times when they do not work properly usually involve human action such as improper maintenance or turning off of the water supply. In the early 1900s, the idea that sprinklers might be able to save lives was beginning to take shape.

> **Firefighter Fact** New York City's tragic Triangle Shirtwaist Factory fire in 1911 killed 146, mostly women, many of whom jumped to their deaths trying to avoid being burned. This fire was the beginning of the concept of adding sprinklers to buildings for **life safety** and providing occupational safety for workers.

America's fire loss statistics show that most structure fires, most fire damage, and, of greatest importance, most fire injuries and fatalities occur in residential properties. Yet the American home is one of the least protected properties. It is usually built of wood and other combustible materials. Until the mid-1970s most American homes did not have smoke detectors, and only a few non-high-rise residential units had sprinklers. Recognizing the potential benefit of these types of systems, the U.S. Fire Administration, National Fire Protection Association, International Association of Fire Chiefs, and Factory Mutual, among many others, led the research efforts to develop faster responding and less expensive residential sprinkler systems. The first standard on residential sprinkler systems was developed in 1975.[2] Some cities and counties now have mandatory sprinkler requirements in all newly constructed residential structures, and many states require them in multifamily dwellings. There is still, however, a long way to go to place them in all new residential buildings. The American heritage of protecting personal property rights often allows citizens the right to die in their own homes, the least protected and inspected of any occupancy classes.

> **Note** Firefighters must realize that protective systems are designed for specific purposes and will not completely defend a property unless they are designed with that goal in mind.

Residential sprinklers are designed for life safety and not necessarily to protect property. While a residential sprinkler system may completely extinguish a fire in a protected area, unprotected areas are allowed in the design of such systems. Fires that may occur in these unprotected areas can gain great headway or move to other unprotected areas and the system may not be able to prevent a total loss. Fires have occurred in apartment buildings, motels, and hotels protected by a life safety sprinkler system, yet resulting in a total loss of the property. When looking at these fires and rating the effectiveness of the protection, the question is, Were there any occupant fatalities or injuries? If the answer is no, the system did what it was designed to do.

Sprinkler systems are also not totally effective in life safety in some other types of fires. These include slow burning and smothering fires that produce fatal quantities of smoke without generating sufficient heat to activate the system, flash fires that rapidly engulf a person, or a fire in the immediate vicinity of the victim where the fire inflicts the injuries prior to system activation. Another situation is one in which the victims are disabled or unable to remove themselves from the fire area. Protecting against all situations is impossible, but residential sprinkler systems combined with an effective smoke detection system would allow maximum life safety in the home.

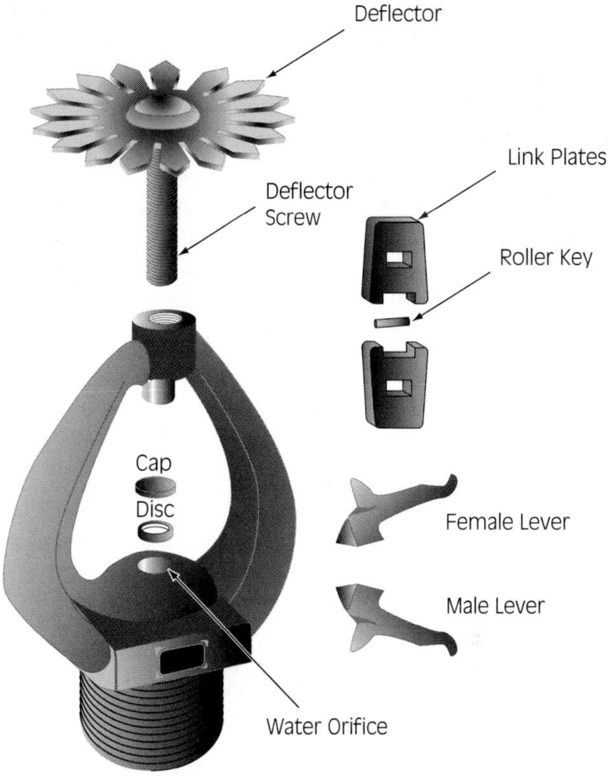

Figure 12-7 Sprinkler head parts.

SPRINKLER HEAD DESIGN AND OPERATION

Sprinklers or sprinkler heads are the key components of the system, **Figure 12-7**. Most important are the heat-sensitive parts that usually detect heat and apply water to the fire. (Note the word *usually*. Some types of sprinkler systems such as a deluge-type system have separate detection devices but the head is still the applicator. See deluge system discussion later). Sprinkler heads must be appropriate in design and performance, orientation, application or environment, and temperature for the type of property to be protected.

Sprinkler heads come in many designs and this affects their performance. One main difference in sprinkler heads is the new and old-style sprinkler head, **Figure 12-8**. Sprinklers designed and manufactured up to the early 1950s are called *old-style sprinklers*. They deflected the majority of the water toward the ceiling to further break up its pattern. The effect was that the water was concentrated in a smaller area.

The new style or standard sprinklers have a much more even flow across the coverage area and do not bounce the water off the ceiling, **Figure 12-8**. Standard sprinklers are marked with *SSU* (Standard Sprinkler Upright) or *SSP* (Standard Sprinkler Pendent) on the deflector. Orientation means up, down, or sideways. Upright sprinklers have the head vertically above the piping with the deflector at the top, while pendents have the head suspended below the piping with the deflector at the lowest point, **Figure 12-9**. The design of sidewall sprinklers allows them to be placed near the wall and, although the deflectors look bent, they provide the correct pattern. Sidewall heads may be pendent, upright, or horizontal. Each must be properly positioned.

Application and environment may require corrosion-resistant heads, special dry head with extension piping, rack storage head, or decorative head. The orifice or the size of the water opening varies from ¼ to ¾ inch (6.4 to 19 mm) depending on the occupancy protected. Sizes other than ½ and 17/32 inch (12.7 and 13.5 mm) are noted on the sprinkler frame.[3] Temperature ratings range from ordinary at 135°F (57°C) to ultra high at 625°F (343°C) with the most common temperature being 165°F (74°C),[4] **Table 12-1**.

The operation of a sprinkler head begins with the fire's heat raising the temperature of the sprinkler

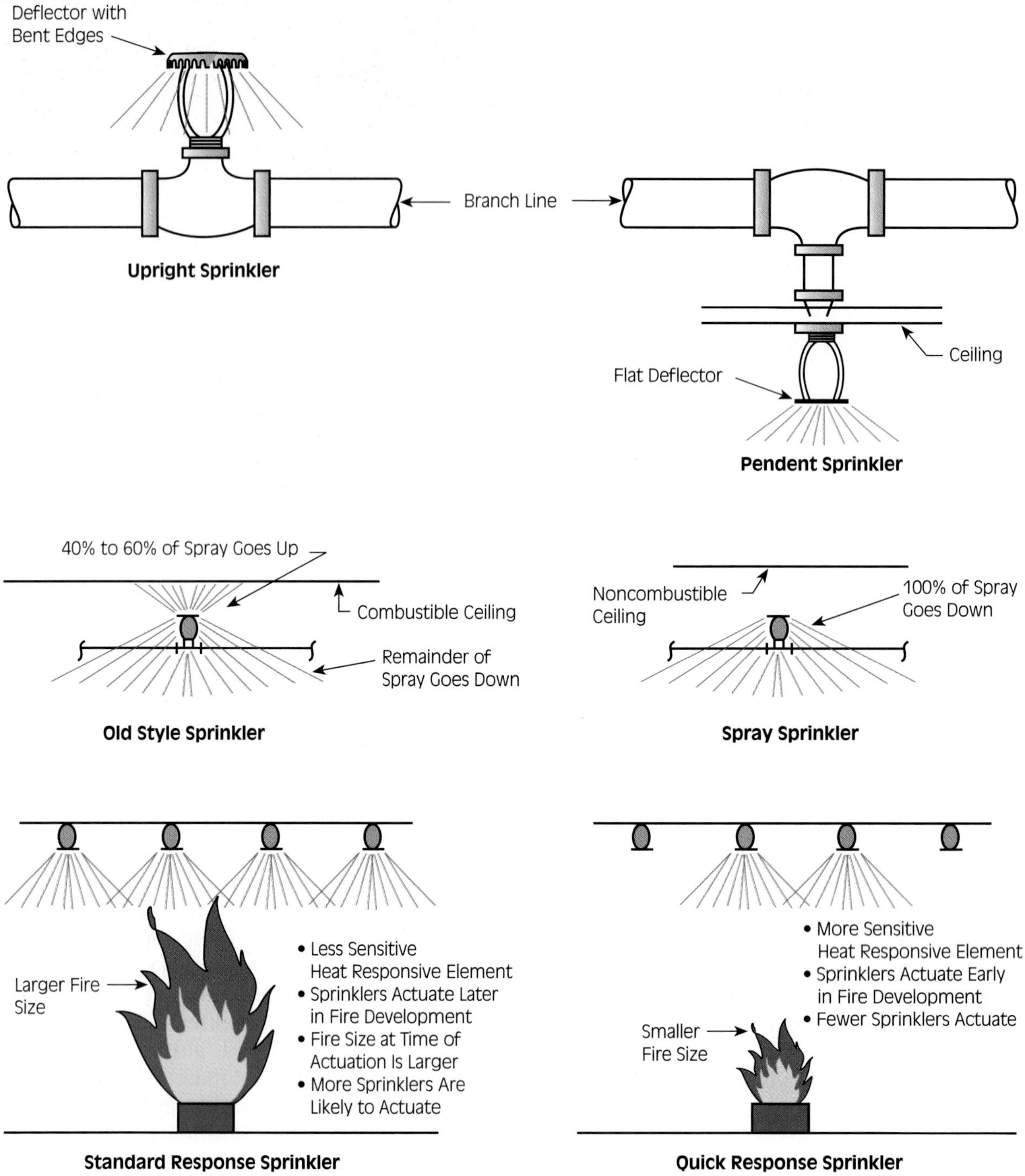

Figure 12-8 Sprinkler differences.

head and its fusible element to its fusing point. The fusible elements can be of three types. The first is a fusible link that has a metal or solder that melts at a fixed temperature. The second is a bulb filled with a liquid, leaving only a small air bubble, which expands and bursts at a fixed temperature. The third type uses a chemical pellet that melts at a fixed temperature. The fusible element melts or bursts and the levers holding the cap then fall out. The cap pops out and the water flows, striking the deflector and spraying into the designed pattern, **Figure 12-10**. Another newer innovation is an on/off head that operates at a fixed temperature, but when the fire's temperature drops, it operates a spring to close the waterway and can reopen if the temperature rises again.

(A)

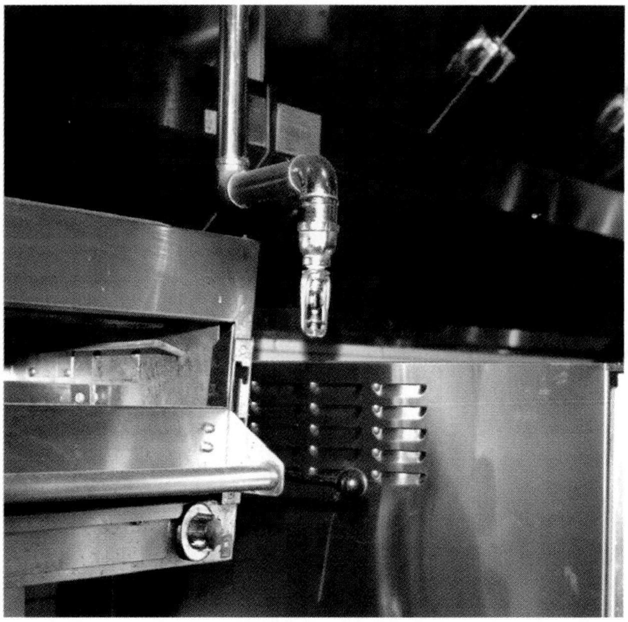

(B)

(C)

Figure 12-9 Types of sprinklers: (A) upright, (B) pendent (without guard), and (C) pendent sidewall (with guard). *(Photos by Fred Schall)*

Temperature Ratings, Classifications, and Color Coding

TEMPERATURE RATING °F	°C	TEMPERATURE CLASSIFICATION	COLOR CODING	GLASS BULB COLORS
135–170	57–77	Ordinary	Uncolored or black	Orange or red
175–225	79–107	Intermediate	White	Yellow or green
250–300	121–149	High	Blue	Blue
325–375	163–191	Extra high	Red	Purple
400–475	204–246	Very extra high	Green	Black
500–575	260–302	Ultra high	Orange	Black
625	343	Ultra high	Orange	Black

Note: A portion of Table 6-10c is reprinted with permission from *Fire Protection Handbook,* 18th edition. Copyright © 1997, National Fire Protection Association, Quincy, MA 02269. This reprinted material is not the complete and official position of the National Fire Protection Association on the referenced subject, which is represented only by the complete table in its entirety.

TABLE 12-1

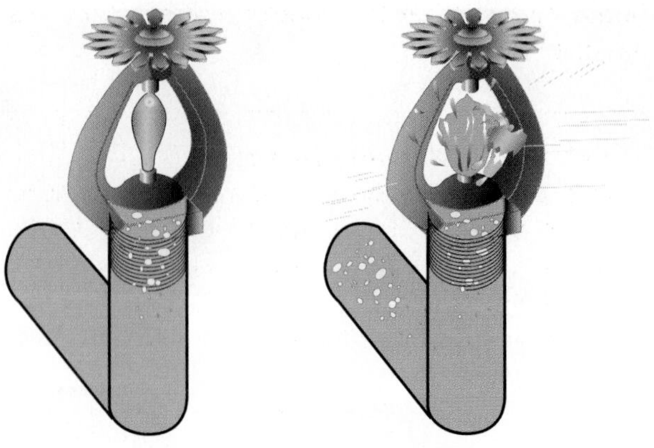

Figure 12-10 Sprinkler head closed and opened or operating.

TYPES OF SPRINKLER SYSTEMS

The four major types of sprinkler systems and the residential system are discussed in the following subsections. Specialty sprinkler systems include some combination-type sprinkler systems and systems that cannot meet the standards for some reasons. They may have an inadequate water source or supply or may be a partial or outside system. Even if a system does not completely meet a standard, it provides a higher measure of protection than if no protection were available. Fire departments using special systems should become familiar with their limitations.

Wet Pipe Systems

A **wet pipe sprinkler system** has automatic sprinklers attached to pipes with water under pressure all the time, **Figure 12-11**. This allows the quickest response when the head is opened. The wet pipe system is the simplest sprinkler system in design and operation. The main or alarm valve is a one-way check or clapper valve that prevents water from reentering the water supply and, when closed, shuts off the water flow to the alarm line. Both sides of the alarm valve have pressure gauges that register the water pressure of the supply and the system. The system side gauge should read a slightly higher pressure, because the reclosing of the clapper valve would trap any pressure surges. The alarm line piping usually has a **retard chamber** that acts to prevent false alarms from a sudden pressure surge in the water supply, **Figure 12-12**. The chamber collects a small volume of water before allowing a continued flow to the alarm device. The water from a surge is drained from a small hole in the bottom of the collection chamber. A waterflow indicator, a vane or paddle in the waterway, detects the water flow and activates an alarm signaling system.

A wet pipe system has three more valves. The first is the control valve, which is used to shut off the supply of water to the system and is usually an **outside stem and yoke (OS&Y) valve**, **Figure 12-13**. The second valve is the main drain, which allows the system to be drained for maintenance or to be restored from a fire, and the last valve is an inspector test valve. The test valve is at the farthest end of the system and is used to simulate the flow of a single head and to measure the response time of the system.

The operation of a wet pipe system starts with the fusing or bursting of a sprinkler head, which causes it to begin applying water to the fire. As the water pressure in the system begins to fall, the main check valve opens and water flows into the system and the alarm line, filling the retard chamber, and then activating the automatic alarm and water motor gong. The alarm signal may be used to notify the fire department or an alarm company. *After ensuring the fire is out or completely under control,* the control valve is closed. Sprinkler maintenance personnel replace the head and restore the system. When the system is shut down, a firefighter with a radio

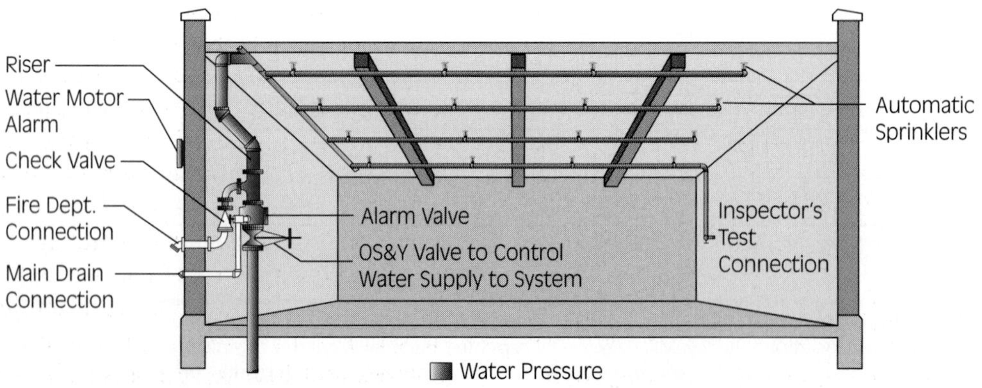

Figure 12-11 Wet pipe sprinkler system.

should be posted at the control valve and be ready to reopen the valve.

> **Note** Some wet pipe systems are located in or have parts of the system located in areas subject to freezing temperatures. To protect these systems, an antifreeze solution can be added to the water in the system. These systems require special attention to restore and maintain.

> **Streetsmart Tip** One of the reasons that sprinklers have not been 100 percent effective is that firefighters or building personnel shut off the control valve prior to the complete extinguishment of the fire. The fire returns and, before the system is turned back on, it gains enough headway to cause additional loss. Building personnel are often overly afraid of water damage to property. Most water-damaged materials can be salvaged but fire-damaged materials are usually destroyed. *Firefighters should not close down any sprinkler system until making sure the fire is out or completely under control. They should keep a hoseline in place to make sure it stays that way.*

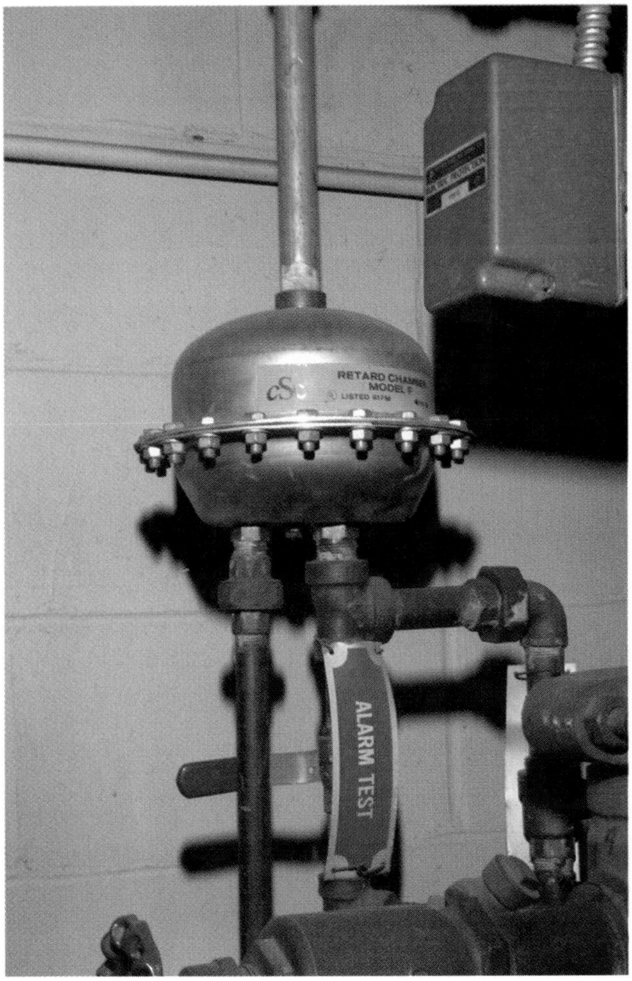

Figure 12-12 Retard chamber.

Dry Pipe Systems

In **dry pipe systems**, air under pressure replaces the water in the system to protect against freezing temperatures. The system uses a dry pipe valve to keep pressurized air maintained above with the supply water under pressure below the valve, **Figure 12-14** and **Figure 12-15**. A small amount of water at the seat of the valve, called the priming water, maintains the seal at the valve and is filled to the priming level. The clapper valve has a locking mechanism that keeps the clapper open until it is manually reset to prevent **water columning**.

Dry pipe valves use an air differential system having a smaller air pressure maintained over the larger head surface of the clapper valve, which keeps back the higher water pressure exerted on the smaller water side of the clapper valve. If water were allowed to fill the riser above the clapper, the water column's weight would never allow the clapper to be forced open and make the system inoperative. When a sprinkler head is fused by heat, air is first discharged. As the air pressure drops below the pressure of the supply water, the clapper valve is opened and locked. Because air is in the system rather than water, dry pipe systems are slightly

Figure 12-13 OS&Y valves chained in the opened position.

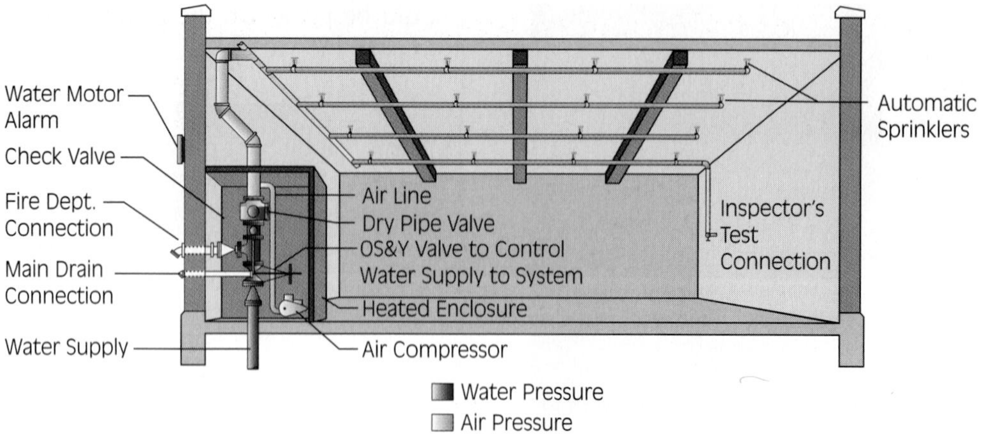

Figure 12-14 Dry pipe system schematic.

slower to activate than wet pipe systems. Most systems have either an **exhauster** or an **accelerator** to speed up the operation of the dry pipe valve. The exhauster detects the decrease in air pressure and helps bleed off air. The accelerator detects the decrease in air pressure and pipes air pressure below the clapper valve, speeding its opening. Drain and alarm valves are similar to wet pipe systems. Dry pipe systems are used in unheated buildings, in buildings that refrigerate or freeze materials, and in outdoor applications where freezing temperatures occur, but the valve room must be heated.

The dry pipe system is more complex in design than a wet pipe system and also harder to restore because it requires the dry pipe valve cover to be opened after draining the system and resetting the lock on the valve. The valve is primed, air pressure charged in the lines, and the control valve opened carefully to prevent retripping and creation of a water column.

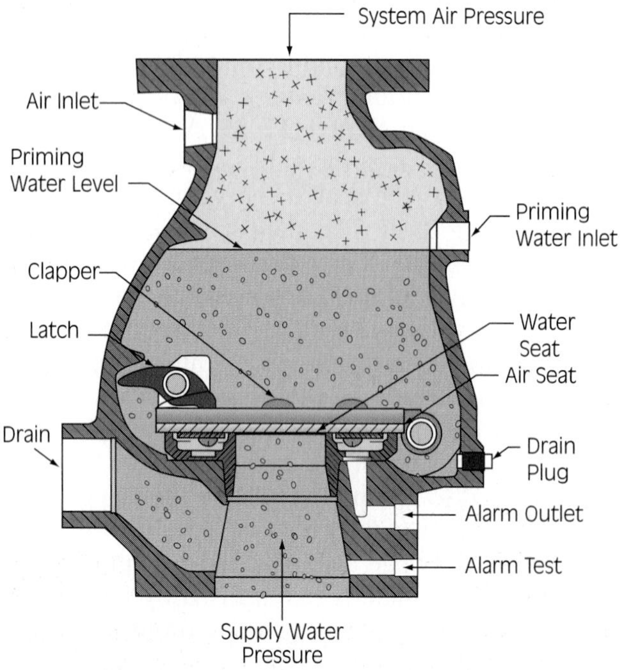

Figure 12-15 Dry pipe valve holds back the water with air pressure.

Deluge Systems

Deluge systems are designed to protect areas that may have a fast-spreading fire that could engulf an entire area. Petroleum-handling facilities, aircraft hangars, some manufacturing facilities, and hazardous materials storage areas are all examples of occupancies that may have a deluge system. The essential difference between a standard wet system and a deluge system is the individual sprinkler head. Deluge systems utilize open heads, without any fusible elements. When a fire is detected, the deluge system delivers water (or foam solution) to all heads in the area—allowing total coverage of the area.

Operationally, the deluge system must interface with a detection system. Once the detection system activates, it sends a signal to a "deluge valve" which opens, sending water or foam solution to all the open heads, **Figure 12-16.** Most municipal water systems lack the pressure to effectively supply the numerous open heads in the deluge system. For this reason, deluge systems also incorporate a fire pump to boost pressure in the system. Likewise, deluge systems that deliver foam solution require a foam supply tank, foam pump, and foam mixing device.

As can be imagined, deluge system activation will cause tremendous quantities of water to flow. While this is essential to fire control of high hazards, activation of the system in absence of a fire will likely cause much water damage and a significant cleanup effort. For this reason, deluge system activation usually requires activation of several detectors. For example, several flame detectors must activate before a signal is sent to the deluge valve. Some sys-

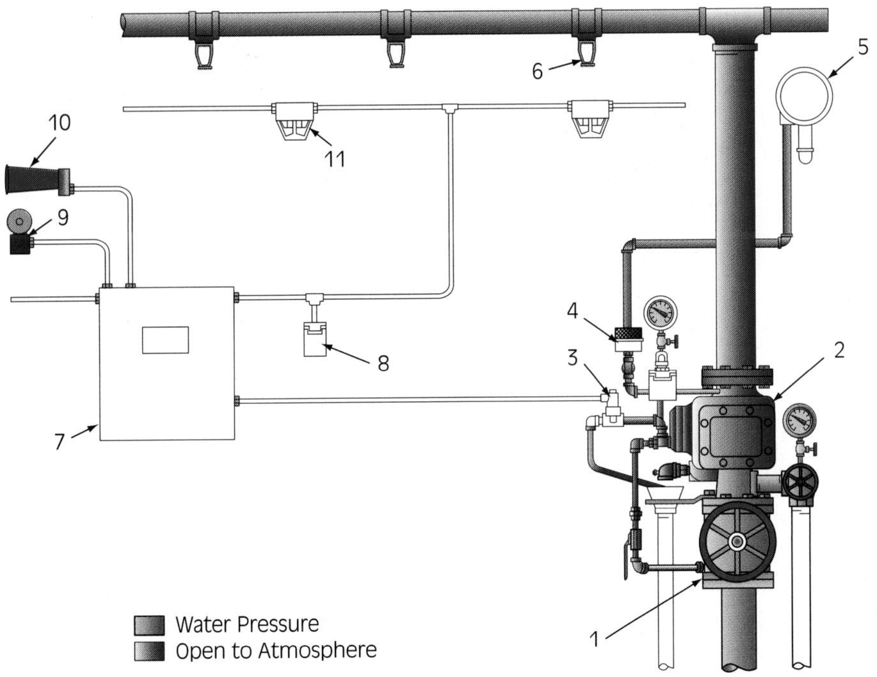

Figure 12-16 Deluge system schematic. 1, OS&Y valve; 2, deluge valve with basic trim; 3, solenoid valve and electric actuation trim; 4, pressure alarm switch; 5, water motor alarm; 6, spray nozzles or open sprinklers; 7, deluge releasing panel; 8, electric manual control stations; 9, fire alarm bell; 10, trouble horn; 11, heat detectors. *(© Copyright Simplex Grinnell. All rights reserved)*

tems require detection from different kinds of devices (i.e., activation of a flame detector and heat detector). To prevent unneeded damage from activation, the deluge system may utilize a manual override alarm and "deadman" switch. In these systems, an alarm is sent telling occupants that the deluge system is ready to fire. If someone does not push a button in a prescribed time, the system will deluge the area. A person who does push the hold button cannot release the button until other persons reset the detectors that have activated. If the person does release the button prior to system reset, the deluge valve will open—hence the deadman label.

Preaction Systems

Preaction systems are similar to the dry pipe and deluge system. The system has closed piping and heads with air under no or little pressure, but the water does not flow until signaled open from a separate fire detection system, **Figure 12-17**. The preaction valve then opens and allows water to flow through the system to the closed heads. When an individual head is heat activated, it opens and water attacks the fire. Preaction systems are used in areas where the materials protected are of high value and water damage would be expensive, such as computer rooms and historical items.

Residential Systems

Residential sprinkler systems are smaller and more affordable versions of wet or dry pipe sprinkler systems, **Figure 12-18**. They are designed to control the level of fire involvement such that residents can escape. The water supply is combined with the domestic water supply, and flow rates are designed for one or a few heads in operation. Residential sprinkler heads have a faster response time and use a lighter and smaller pipe than wet/dry pipe systems, and were the first to use plastic piping. Residential systems use a check valve, waterflow alarms, and drains similar to the bigger systems and are not required to have a fire department connection, although some do have one. Some residential systems use antifreeze to protect all or part of the system.

SPRINKLER SYSTEM CONNECTIONS AND PIPING

The connections and piping for a sprinkler system provide the water from its source to the heads, and they comprise most of the system. The main water supply can come from a public or private water

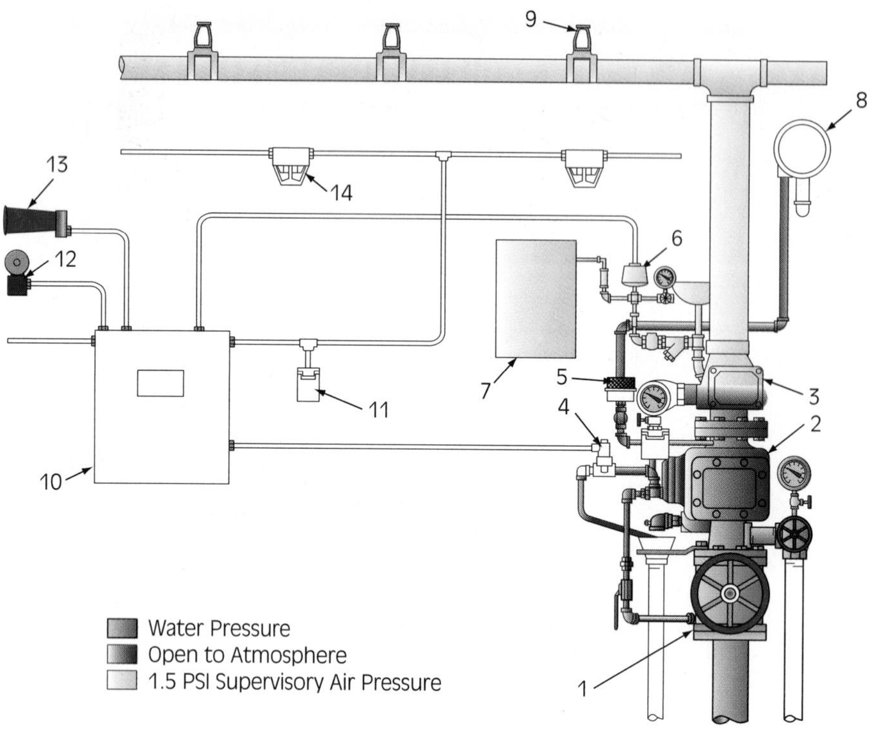

Figure 12-17 Preaction system schematic. 1, OS&Y valve; 2, deluge valve with basic trim; 3, check valve; 4, solenoid valve and electric actuation trim; 5, water pressure alarm switch; 6, 1.5-psi low air pressure alarm switch; 7, 1.5-psi supervisory air pressure control; 8, water motor alarm; 9, automatic sprinklers; 10, deluge releasing panel; 11, electric manual control stations; 12, fire alarm bell; 13, trouble horn; 14, heat detectors. *(© Copyright Simplex Grinnell. All rights reserved)*

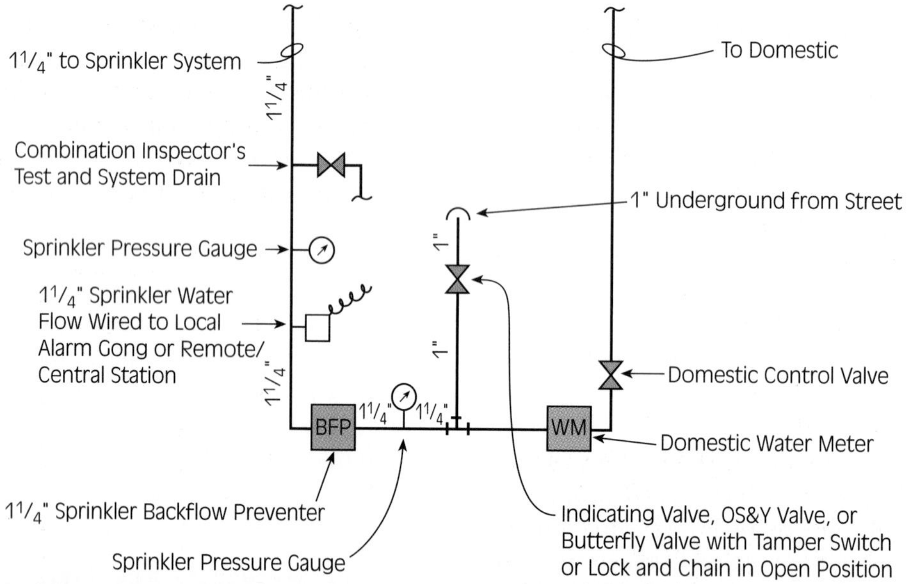

Figure 12-18 Schematic of residential sprinkler system.

company, the protected property's own supply system, a variety of tanks such as gravity or suction, or a cistern, pond, or stream. Depending on the requirements of the system and the water source, a fire pump may be included. A secondary water source is a fire department siamese connection, which allows pumpers to supplement the water supply, **Figure 12-19**.

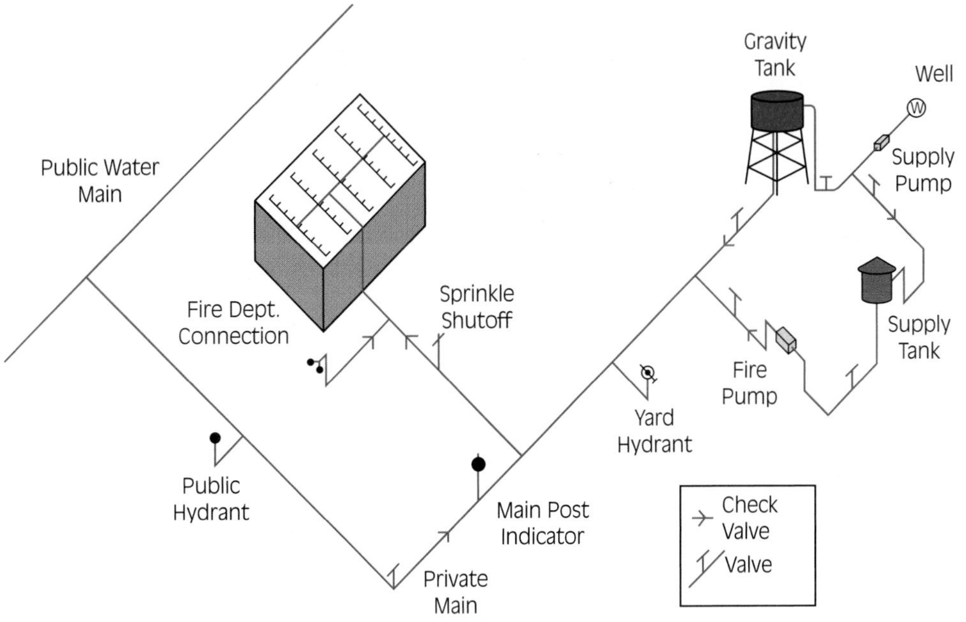

Figure 12-19 Various water supply systems to a sprinkler system.

> **Note** A serious consideration is where the fire department gets its water supply to determine if water is being robbed from the sprinkler system.

It is best to use a different water main or supply source than the one being used by the sprinkler system. The fire department connection is usually piped in past the main control valve and can supply the system even if the main control valve is closed.

Another required valve in most systems is a backflow preventer, a set of one-way check valves designed to keep water in a sprinkler or other system from reentering the public water supply. These are an environmental requirement, and most codes require and mandate them even on old systems. The system may have other control valves and check valves and will have a water control valve if connected to a public water supply.

Next in line is the main control valve, which is of the wet, dry, deluge, or preaction type as described earlier. Above that is the riser. The riser pipe feeds the mains that feed the cross mains to the branch lines, **Figure 12-20**. The branch lines have the sprinkler heads attached to them. Some larger systems may have sectional valves that divide the system into floors or other subareas of the system. The location of these valves can be important when shutting down the system or doing maintenance. They should be kept locked and supervised for tampering. Tamper alarms alert the alarm company whenever someone operates the valve, and unauthorized valve operations are quickly checked to prevent an arsonist from disabling the system prior to setting a fire.

Firefighters should be able to properly connect a supply line to a fire department connection for either a sprinkler or standpipe. To do this requires the correct amount of hose to reach from the pumper's discharge gate to the connection, a spanner wrench to tighten the couplings, and any adapters needed to complete the connection. The firefighter should check the connection for damage, remove the cap cover, and check inside the siamese for debris, damage, and operation of a clapper valve and an O-ring gasket. If no clapper valve is provided, all necessary connections must be made prior

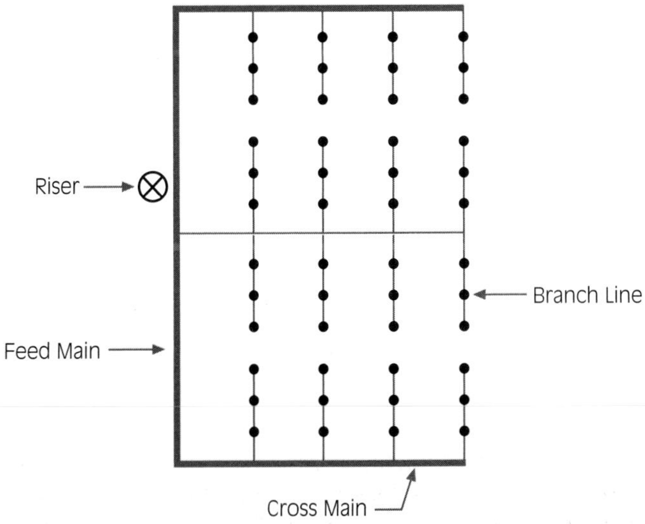

Figure 12-20 Sprinkler piping diagram.

to charging the system. When connecting to a siamese, the outlet on the far left should be chosen first, because this will allow better access for using the spanner wrench to tighten the coupling. After completing the connection, the pump operator should be advised that the line is ready for charging.

> **Streetsmart Tip** When connecting to a siamese, the firefighter should choose the outlet on the far left first, because this will allow better access for using the spanner wrench to tighten the coupling. Some connections only have clappers installed on the right side and must be connected on the left side first.

CONTROL DEVICES FOR SPRINKLER SYSTEMS

The three main control devices for sprinkler and standpipe systems are the outside stem and yoke valve (OS&Y), the **post indicator valve (PIV)**, and the **wall indicator valve (WIV)**, **Figure 12-21**, **Figure 12-22**, and **Figure 12-23**. The valves have either a butterfly or gate valve type. The names come from the appearance of the valves. The OS&Y valve has a wheel on a stem housed in a yoke or housing. When the stem is exposed or outside, the valve is open. These valves must have a chain lock on them to prevent tampering. If the system is

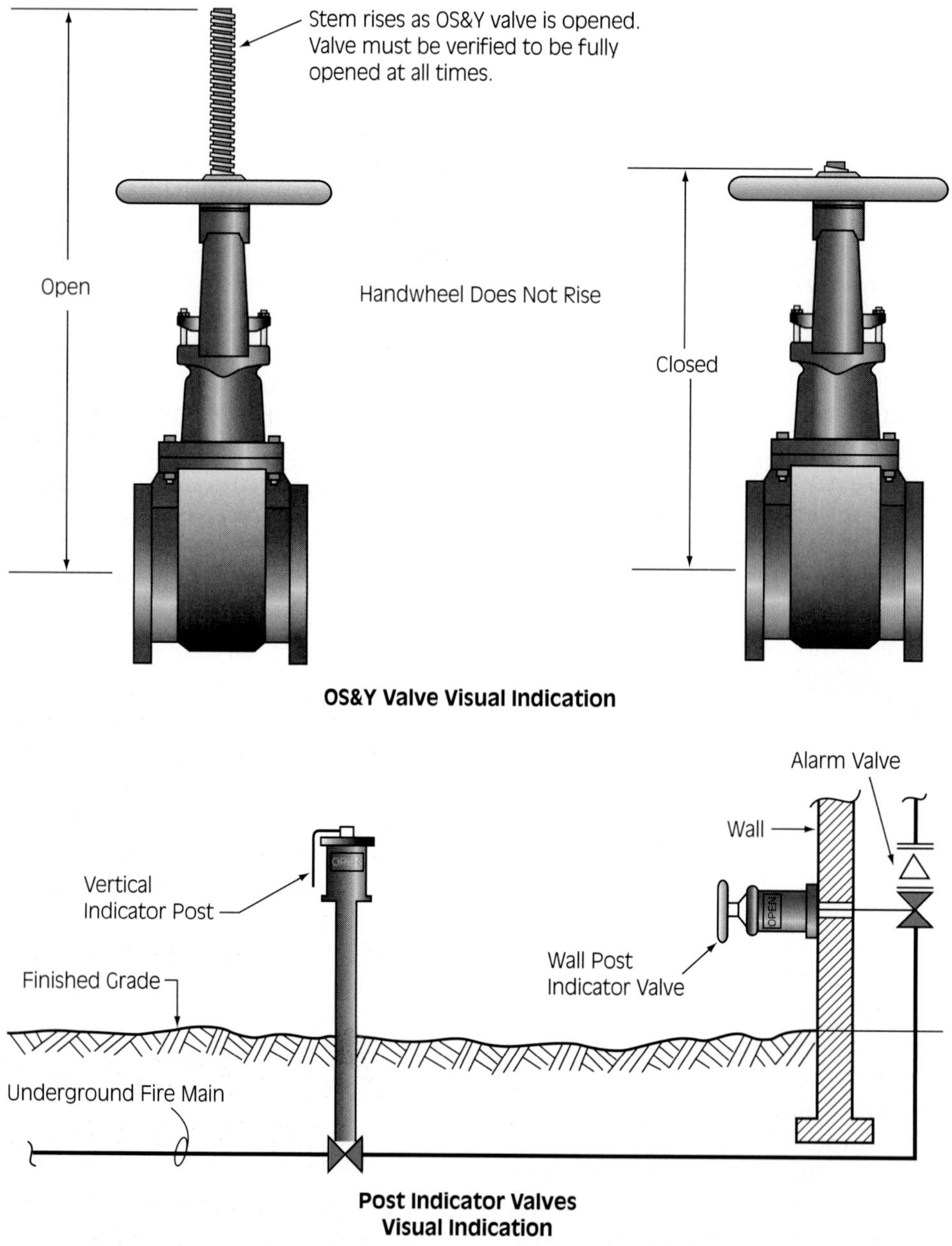

Figure 12-21 OS&Y valves and post indicator valves.

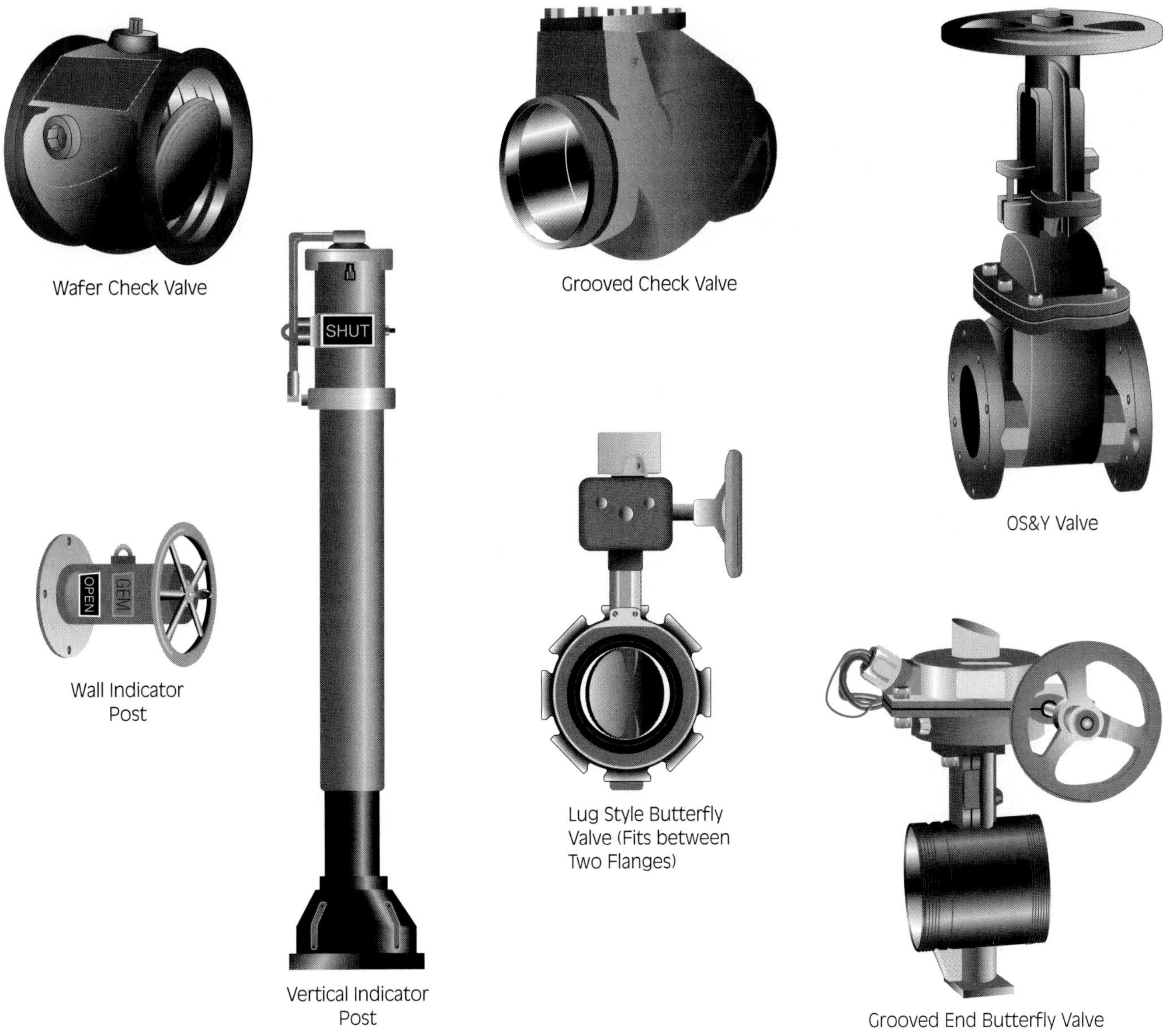

Figure 12-22 Fire protection valves.

Figure 12-23 Fire department connection and post indicator valve. The box on the side of the PIV is a tamper alarm switch box.

supervised, a tamper switch will also be found on the stem. PIVs and WIVs are very similar, but one is mounted on a post in the ground, and the other is mounted on a wall. Both valves are housed in a metal case with a small window, reading either "OPEN" or "SHUT." A wrench or a wheel controls these valves; a padlock and chain are used to lock them opened. Some water control valves may also be of the nonindicating type; public water system valves usually are of the nonindicating type.

RETURNING SPRINKLER SYSTEMS TO SERVICE

Fire departments used to regularly restore sprinkler and other protective systems to service for building owners. Many also regularly tested these systems.

Many engine and ladder companies carried a sprinkler kit with wrenches, extra heads, stops, and other equipment to do this. Today many departments no longer provide this service; instead they either stop the flowing head or shut down the system, letting the owner contact a fire protection service company. This has become necessary for liability issues. Fire departments should have local SOPs that address the response, operations, and return-to-service guidelines for protective systems. Firefighters must understand how to properly shut down either individual heads or the entire system to stop excessive water damage until the heads can be completely restored.

When the fire is completely out, the first step to restore the system is to shut down any pumper supplying the system. Firefighters should not disconnect the hoselines at this point; the fire may reignite so hoselines should stay connected until overhaul is complete. If the system is to be restored immediately or a large number of heads have opened, either the main sprinkler or a sectional valve should be shut down and drained. The sprinkler head is replaced with an identical one from the spare heads that are required to be kept in a cabinet in the sprinkler control room. The sprinkler valves must be reset or reopened.

The simplest—and often quickest—way to stop water flow from an individual sprinkler head is to insert a stop. Sprinkler stops may include manufactured sprinkler tongs or improvised wedges, dowels, or clamps, **Figure 12-24.** While it may seem easy, stopping water flow from a sprinkler head requires practice to effectively stop the flow and establish a leak-free seal, **Figure 12-25.** The firefighter assigned to stop sprinkler flow at the head will get seriously wet. It is important to use caution when climbing a wet ladder and to make sure eye protection is used to prevent injury. Following a step-by-step plan will minimize the exposure and speed the water stoppage, **JPR 12-1.**

Figure 12-24 Sprinkler tongs and wedges.

Figure 12-25 Sprinkler tongs and wood wedges stopping sprinkler flow.

JOB PERFORMANCE REQUIREMENT 12-1

Using "Stops" to Stem the Flow of a Sprinkler Head

A Empty your PPE pockets of radios or other items that can be damaged from getting soaked.

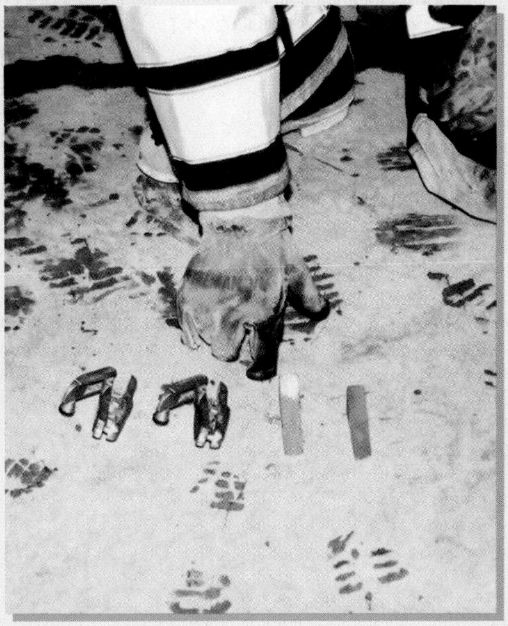

B Select the right stop for the head discharging water. Remember, simple wedges may not fit into recessed sprinkler heads found in some finished ceilings.

C Choose a stable ladder, desk, or other platform to reach the head.

D Put your face shield down or don eye protection to help you protect your eyes from the deluge when you climb the platform and secure the wedges into place.

Photos courtesy Captain Pete Evers, City of Auburn Fire Department, CA

(continued)

JOB PERFORMANCE REQUIREMENT 12-1

Using "Stops" to Stem the Flow of a Sprinkler Head (Continued)

E Most times you will be unable to "see" through the rain of water and be required to stop the water flow blind. Feel your way, and do not rely on sight.

F Carefully reach up to the head with the stop.

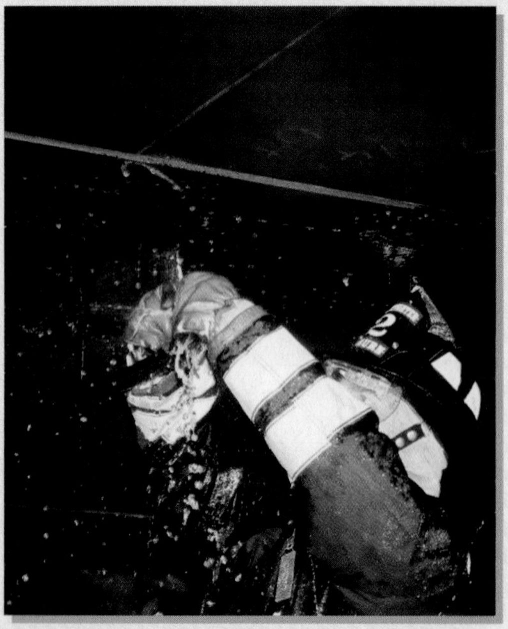

G Apply the stop and tighten it until the flow stops.

Photos courtesy Captain Pete Evers, City of Auburn Fire Department, CA

H Sometimes, using two opposing wedges is easier than trying to force in a single wedge. The sliding motion from two directions overcomes water pressure quicker and provides a better seal than a single wedge.

Streetsmart Tip Firefighters will find that carrying a small assortment of tools and equipment in their turnout gear will be invaluable. A flashlight is essential. Tools should include straight and Phillips screwdrivers, a pair of pliers, an adjustable wrench, and a pocketknife. Other items would include several wooden wedges that can be used for sprinkler stops, door chocks, and other tools that come to mind. Fire departments should also provide search tags, utility shutoff tags, and for firefighters acting as officers, pocket-size versions of a command chart.

The other way to shut off the water flow from opened sprinklers is to shut down either the main sprinkler valve or a floor or sectional valve. This task will require a key or set of bolt cutters because main valves are locked as discussed earlier. If the main valve is an OS&Y valve, it is turned until the stem is no longer exposed. If it is a PIV or WIV valve, it is turned until it reads "SHUT." Personnel with a portable radio should stand by the valves until the system is recharged or until the fire is determined to be out. If the entire system is shut down or if there is a sectional drain, that valve can be opened. After draining the system, this valve should be closed. Once the system is shut down and drained, the heads may be replaced. Replacement of the heads is recommended as soon as possible because in the event of an additional fire, even with the system shut down, the fire department can pump through the fire department connection or turn the system back on and have it operational for firefighting. Refilling a system should be done slowly to prevent damaging the system.

To avoid liability of restoring sprinkler systems, some fire departments *do not* replace heads and restore the system. In these cases, the fire department should require that a **fire watch** be established until the system has been restored by a qualified technician. Some occupancies such as hotels, public assemblies, and high-rises *require* a sprinkler system for life safety. In these cases, the fire department should *not* allow the building to be occupied without an approved fire watch system or sprinkler restoration.

STANDPIPE CLASSIFICATIONS

Standpipe systems are designed to allow firefighters to fight fires in larger buildings by prepiping water lines for fire streams through the building. Some systems allow both occupants and firefighters to use the system, and the systems for building occupants have hoselines and nozzles attached. Standpipe systems are not an automatic; they must be manned by people, who use them to attack the fire. They are used in high-rise buildings; large commercial, retail, and industrial buildings; places of public assembly; and other areas where advancing hoselines from hydrants or fire apparatus would be difficult due to the size of the building. Tunnel systems, such as subway systems, and shopping malls have horizontal standpipe systems. Large does not always mean tall—providing access is the key goal.

Standpipe systems are classified according to the type of intended user and each class sets requirements for volume and size of outlets. *Class I systems* are designed for use by the fire department or trained personnel such as a fire brigade, **Figure 12-26**. The systems use 2½-inch (65-mm) outlets. A single standpipe should flow a minimum of 500 gpm (1,890 L/min) plus an additional 250 gpm (945 L/min) for each additional riser. No hose is provided for this class. Outlets are found in stairwells on each floor, near the exits, or in the hallways on each floor near the stairs.

A *Class II system* is designed for use by untrained building occupants and has a 1½-inch (38-mm) outlet with a minimum flow of 100 gpm (378 L/min), **Figure 12-27**. Hose is provided and is either 1 or 1½ inch (25 or 38 mm). Firefighters wishing to use a Class II system should not use the hose provided, because it is not tested and may be single-jacketed unlined hose with a nozzle that has no shutoff. Firefighters using this system should replace the hoseline with one of their own.

A *Class III system* is used by either users of the other two classes but meets the Class I requirements for flow. It has both 1½-inch (38-mm) and 2½-inch (65-mm) outlets or may have a 2½-inch (65-mm) outlet with a 2½- to 1½-inch (65- to 38-mm) reducer. All of the flows given are at the top outlet with a residual pressure of 65 psi (448 kPa).[5] Any of these systems may have a **pressure-regulating device**, which is designed to maintain or reduce the pressure to a set amount regardless of the height of a building. Standpipes have height limits on each section [about 280 feet (88 m)], but the pressure difference in an unregulated system could vary by almost 140 psi (966 kPa) from top to bottom.

Caution Firefighters wishing to use a Class II or III standpipe system should not use the hose provided, because it is not tested and may be single-jacketed unlined hose with a nozzle that has no shutoff. Firefighters using this system should replace the hoseline with one of their own.

Figure 12-26 Class I standpipe system.

Caution Pressure-regulating valves on standpipe systems may cause firefighters problems if improperly installed or reset. Some are preset at the factory and must be installed on the correct floor to properly regulate the pressure. Others are adjustable and can be reset after installation. Improperly installed pressure-regulating devices have contributed to some firefighter fatalities in high-rise firefighting. Careful inspection and testing of these devices is highly recommended on a regular basis due to maintenance problems. (See Street Story in this chapter.)

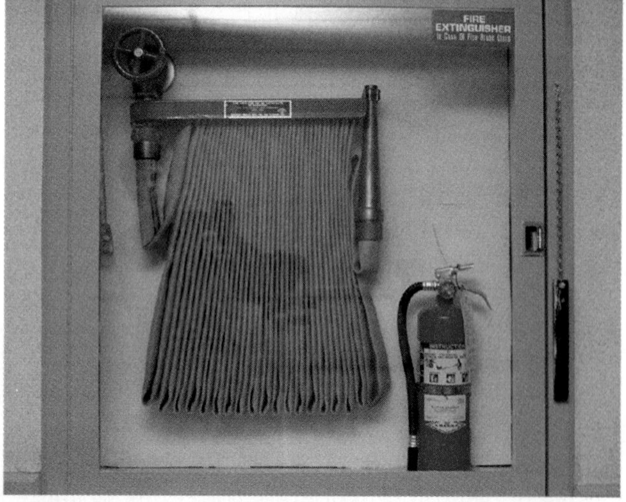

Figure 12-27 Class II standpipe system.

The types of standpipe systems are differentiated based on their water supply. An automatic wet pipe system has a water supply that is ready to conduct firefighting, whereas automatic dry and semiautomatic dry pipe systems use dry pipe until a hose station (an outlet) is opened to allow water to flow into the system. The semiautomatic system requires an activation valve to be opened manually. The manual dry pipe is completely dry and relies on the fire department connection for water supply, whereas a manual wet pipe system has the piping filled with water but needs the fire department connection for firefighting supply.[6]

STANDPIPE SYSTEM CONNECTIONS AND PIPING

Standpipe systems can range from very simple to highly complex ones with multiple connections, risers, sources of water supply, and many outlets on each floor or area. A manual dry system can be a single riser having a fire department connection and a drain at the bottom and an outlet on each floor that is capped at the top. The wet manual type may just add a small water line to maintain water in the piping. The other systems become more complex as a water supply is added and are found in larger buildings with more requirements.

Some standpipe systems are combined with the building sprinkler system with outlets coming off a branch line.

The components of a standpipe system are the piping, outlets with hose and other attachments, the valves, the fire department connection, and any monitoring devices. The piping includes the riser or risers and any attachments needed to plumb between the water sources and the uppermost or furthest outlet. The piping must be capable of discharging the required volume and pressure with minimal friction loss. The outlets are usually placed in the building stairwell or may be a wall-mounted cabinet. Attached to the outlet would be any pressure-regulating device required. If it were a Class II or III system, a hoseline and nozzle would be attached. The hose connection usually has a wheel-type handle to operate the valve and a cap covering the threads. The nozzles may be a combination fog-straight stream or a solid tip nozzle and quite often have no shutoff, which could confuse the untrained user.

Standpipe valves are similar to the types used on sprinkler systems, including gate valves of various types and check valves, with the addition of hose valves at the outlets. If a water supply is provided, an indicating gate valve is required and a check valve or backflow preventers. A check valve is also added with a drain on the fire department connection line to allow drainage and prevent freezing. Gate valves should also be installed on each section or separate riser to allow maintenance work to be performed without shutting down the entire system. Drain valves are installed on each section. The fire department connection(s) is an inlet or siamese device with protective caps and a raised letter sign marked "Standpipe." If there are multiple connections that are not interconnected, the sign should state the zone covered. Monitoring systems are covered in the next section.

ALARMS FOR STANDPIPES AND SPRINKLERS

Alarm and monitoring systems are found in most sprinkler systems and many standpipe systems, especially Class II and III systems. Sprinkler systems are both fire detection and suppression devices and are designed to notify at least the building occupants. Occupant-used standpipe systems should be monitored in case the occupants forget to notify the fire department prior to fighting the fire. Most of these systems require monitoring to prevent tampering or deliberate attempts to disable the system. These may be electronically or mechanically activated by the movement of a gate valve. Waterflow alarms are also electrical or mechanical. The most familiar mechanical one is the water motor gong, which operates like a waterwheel with the flow of the water causing it to turn and strike a gong, **Figure 12-28**. The water motor gong, unless connected to another alarm, is a local alarm only. Other waterflow devices are hooked to an electronic sensor that is connected to a paddle or vane inside the riser. A monitoring alarm company notifies the fire department about fires, but its own personnel respond to tamper alarms.

Figure 12-28 Water motor gong.

OTHER PROTECTIVE SYSTEMS

Many other types of protective systems are used today. Some are rather simple, for instance, those designed to protect the grill, fryer, and ductwork of a local restaurant, whereas others are extremely complex and are designed to prevent or suppress an explosion in highly hazardous locations.

Figure 12-29 Local application system (notice nozzles) protecting cooking area and ductwork.

Figure 12-30 Dry chemical nozzles are located above the island and at ground level.

Firefighters need to understand the most common types of protective systems. Fire departments responding to the more complex systems should ensure that all firefighters are familiar with them. Firefighters should be aware of how these systems operate, what the hazards of the fires and extinguishing agents are to themselves and any occupants, and how to safely operate these systems to ensure that the fire is extinguished and the building is safe.

Local Application and Hood Systems

One of the most common types of protective systems is a **local application system**, especially the systems protecting the cooking areas of restaurants, **Figure 12-29**. Local application means that the system is designed to protect only a certain or local portion of the building, usually directly where the hazard will occur or spread. Local applications are also used in laboratory hoods, paint booths, and other small hazardous locations, **Figure 12-30**. Most of the local application systems use dry or wet chemical agents with the most popular being an ABC dry chemical. Higher temperature cooking units and oils have created the need for a new Class K classification, and these systems also use dry or wet chemicals. Hood systems use a heat-sensitive device or a manual switch for activation.

Total Flooding Systems

Total flooding systems are used to protect an entire area, room, or building. The total flooding system discharges an extinguishing agent that completely fills or floods the area with the extinguishing agent to smother or cool the fire or break the chain reaction. Total flooding systems can use carbon dioxide or other inert gases, halogenated or clean agents, dry chemicals, or foam as extinguishing agents (see Chapter 8 for a discussion of these agents). They are effective as long as the proper amount discharges, the area is contained to prevent loss of agent, or the fire goes out prior to ventilation of the agent. There are hazards associated with each type of application with which firefighters and building personnel should be familiar.

Caution Carbon dioxide (CO_2) and other inert gases such as nitrogen can be used to smother a fire. CO_2 is most often used for firefighting, while nitrogen is used to inert an atmosphere to allow welding or flammable activities to take place near combustible materials. CO_2 systems can be either high- or low-pressure storage systems and are designed to fill the area with a concentration of the gas that extinguishes the fire, **Figure 12-31**. Unfortunately, this smothering effect is also fatal to any living beings in the area, and an evacuation period is built into the system activation. Firefighters, using SCBA, may enter the area to ensure the fire is extinguished and all persons have been safely removed. This agent is very clean and is often used in computer and electronics areas. If the fire is out, ventilation of the area and restoration of the system are the only cleanup activities required.

Caution Halon and the newer clean agents are also used for total flooding applications. Depending on the toxicity and concentration of agent needed, some of these systems may even be used when people are still occupying the area. Firefighters should still use SCBA when entering areas where even safe agents have been discharged to ensure that the fire is out. As with CO_2, electronics and other high-cost or sensitive valuables are protected by these systems. The halons are being phased out of use, but many buildings will continue to have them for the foreseeable future. The replacement clean agents are in use now and will continue to take over these types of properties.

Caution Dry chemicals and foam flooding systems are found in areas with highly hazardous processes and fast-moving fires. Dry chemical total flooding systems discharge enough agent to cover all surface areas and sufficiently extinguish the fire. A caution is in order when dealing with a deep-seated Class A fire, because the lasting effects of these agents may not completely extinguish these types of fires. Foam systems use high-expansion foam generators to completely fill the area with foam. These are effective in fighting flash-type fires and will also work in Class A materials. High-expansion foam was designed for fighting coal mine fires, and some fire departments have used it to fight large commercial basement fires when access for personnel was inadequate. If the system works properly, firefighters using SCBA and hoselines merely need to mop up any remaining hot spots.

FIRE DEPARTMENT OPERATIONS WITH PROTECTIVE SYSTEMS

Standpipe and sprinkler systems plus any other protective system should be part of the fire department overall strategic plan to provide fire protection for its community. This strategic plan recognizes the community's hazards and tries to keep those hazards within certain limits. When properties or processes create hazards beyond those limits, protective systems are required. The fire department

Figure 12-31 Low-pressure carbon dioxide extinguishing system.

must survey its hazards and also its resources, both public and private. A program of maintenance, inspections, and pre-emergency planning keeps the department in compliance with the strategic plan. The property owner does maintenance, the fire prevention bureau does the inspections, and fire companies the preplanning. Preplanning identifies hazards and resources, and some key resources are protective systems. Fire companies should know where every protective system is and how it operates. For standpipe and sprinkler systems, fire company personnel should know the type of system, the location of all fire department connections, two closest water supply points, and the location of key valves. This information should be noted on area maps, pre-emergency plans, and dispatch information. Knowing where the systems are and how to use them is the first step in coordinating a proper action plan. The next step is having SOPs that give a recommended course of action in buildings with these systems.

From an operational point of view, protective systems can be separate components requiring the fire department to look in different areas to understand what has been activated. In newer systems, the protective systems are integrated into a single "smart" system that includes an annunciator panel, fire alarm control panel, and system override controls. Typically, annunciator panels are near the primary entrance to the building and simply "announce" what has been activated and where. The fire alarm control panel and other protective system controls are in a locked room and allow the fire department to reset, silence, or otherwise control the system. Large occupancies such as a high-rise may have a complete fire command center where the fire department can monitor and/or control all protective systems as well as HVAC systems and building intercom/communications systems. Some of the operational procedures for protective systems are addressed in this chapter.

Standpipe Operations

Standpipe operations start with establishing a water supply to the fire department connection with a minimum of one hoseline. Additional hoselines are added as needed. The pump operator should immediately charge the standpipe system for the pressure required at the reported fire level. The pressure should account for the required nozzle pressure, friction loss in the supply and attack lines and the standpipe piping, plus any elevation. The first arriving unit should have its personnel check the annunicator panel and then go to the reported location of the fire. Methods of going to this location will vary depending on the situation but may involve use of stairs or elevators and should be covered in SOPs. Firefighters using elevators should do so only if they are equipped with firefighter service and use this control system. They should *not* take the elevator directly to the reported fire floor because this can be hazardous. They should stop at least two floors below.

Personnel should have full personal protective equipment, standpipe pack, and forcible entry equipment. The suggested standpipe pack equipment is a minimum of 150 feet (46 m) of 1½-inch (38-mm) hose or larger [many departments are now using 1¾ or 2 inch (45 and 50 mm)], a 2½- to 1½-inch (65- to 38-mm) gated wye with 1½- to 2½-inch (38- to 65-mm) increaser, 5 to 10 feet (1½ to 3 m) of 2½-inch (65-mm) hose, spanner wrench, adjustable wrench, and valve wheel. Additional sections of hose are recommended if personnel allow, **Figure 12-32**. Personnel should connect their standpipe pack to the nearest and safest outlet in relation to the fire location. This may require using the floor below and advancing the hoseline toward the fire location. Depending on the fire's location, the line may be charged prior to entering the hallway or delayed until reaching the door of the room or unit on fire. Additional lines may be run off the gated wye, another standpipe outlet on that floor, or another floor.

Caution Firefighters using elevators should do so only if they are equipped with firefighter service and use this control system. They should *not* take the elevator directly to the reported fire floor because this can be hazardous. They should stop at least two floors below.

Figure 12-32 Hose pack for standpipe use.

> **Streetsmart Tip** When advancing a hoseline from a standpipe system in a stairwell, or any hose in a stairwell, extra hose should be looped up the stairs and then down so that when pulled, gravity will assist in pulling the hose down instead of pulling it up the stairs.

Sprinkler System Operations

Sprinkler system operations start with an investigation of the building to determine if water is flowing or a fire is present. Typically, the first arriving officer checks the annunciator panel and then directs the crew to pull fire attack lines (or begin standpipe operations). Typically, SOPs require the first arriving crew to start fire attack while the pump operator sets up to support the sprinkler system. Establish a water supply to the fire department connection with a minimum of one medium- or large-diameter hoseline. Additional hoselines are added as needed; the second line can be a backup for the first. The pump operator should charge the sprinkler system when smoke or fire is showing, the water motor gong is operating, or when ordered by the officer-in-charge. The system should be charged immediately if it is combined with a standpipe system. Sprinkler systems are normally charged and maintained at 150 psi (1035 kPa). Personnel should advance hoselines into the fire area and conduct extinguishment and overhaul operations. If no fire is found, the officer may direct firefighters to stop sprinkler flow as discussed earlier in this chapter. It is important to note that all sprinkler system alarms should be treated as an actual fire until confirmed otherwise. Just because a building is fully sprinkled does not mean the fire department should prepare to fight fire differently. Being fully prepared with PPE, tools, radios, and so forth, should take precedence.

Detector Activation Operations

Operationally, the generic "fire alarm activation" response can range from investigating an alarm malfunction to a full-blown fire operation. Histories of false alarms have caused some firefighters to take a less than ready approach to their operations. Not only is this dangerous, it can be fatal. Most experienced firefighters can tell a story of arriving at a fire alarm activation, only to be surprised that an actual fire existed. These firefighters now respond to all alarms just as if an actual fire were awaiting them. It cannot be overemphasized: Firefighters must treat all fire alarm activations as an actual fire and prepare for the worst.

First arriving firefighters at an alarm activation should dismount the apparatus wearing full PPE and carrying tools and radio. Typically, the first step at an alarm activation is to investigate. At residential structures, firefighters should meet with the building occupant if no fire signs or conditions exist. In cases where nobody is awaiting fire department arrival, the firefighters should systematically walk around the structure and check for smoke or other indicators of an internal fire. Local SOPs should address further steps if no visible signs exist. These SOPs can range from notifying neighbors, to keeping an eye on the structure, to forcible entry and interior search. If access can be made to the structure, caution should be employed. Smoke detector and CO detector activation may sound similar, but the CO can incapacitate unprepared firefighters with little warning.

At commercial structures, the first arriving crew should first check the annunciator panel to ascertain which device has activated. If no annunciator exists, firefighters may have to access the fire alarm control panel if no obvious signs are found. After business hours, firefighters should perform a walk-around to look for signs. If the building is so equipped, the officer will then access the lockbox and enter the structure. Anytime a lockbox is utilized, the officer should radio dispatch that locked entry is being made. After checking the annunciator panel, the firefighter should proceed to the indicated zone and investigate. If any signs of smoke or fire are visible, the operation resorts to a fire attack mode.

Buildings equipped with strobes and loud audible warning devices can be annoying to occupants if they see no signs of fire. Firefighters must not get trapped in a "false alarm mentality" and arbitrarily silence these alarms to investigate the activated zone. Premature silencing may send a message to occupants that they can return to their business. Using simple foam earplugs (which should be in the pocket of firefighters' PPE) can reduce the potential of temporary hearing loss while investigating the alarm. Once the activated detector is found, the immediate area can be checked for potential causes. If no fire signs are found, the firefighters can then take steps to clear the detector (discussed previously) and try to reset the alarm. Systems that will not reset become the property owners' responsibility to correct. Often, the fire department needs to advise the building representatives to contact their alarm company or a repair technician. All fire department actions must be documented on the incident report.

Operations for Other Protective Systems

Local SOPs should address the operations for total flooding, foam, dry chemical, and other unique systems. Some general procedures, however, can be addressed here. Total flooding systems that have discharged present a hazard to occupants and firefighters. Firefighters must never enter a total flooding environment without first engaging their SCBA. It is essential to be prepared for the worst. The introduction of air into total flooding environments can reignite the fire. Hood systems that have discharged will require an overhaul effort to ensure that the fire is totally out. Firefighters should check ceiling spaces and the rooftop ventilator to make sure the fire has not extended out of the hood and ducts. Dry chemical systems are quite effective for fire control but require a large cleanup effort. This cleanup is not the responsibility of firefighters although it is important to notify occupants that breathing dry chemical can cause lung irritation. Activation of suppression systems may cause secondary damage to electrical circuits and other building infrastructure. It is important to be aware of this potential and protect firefighters and occupants accordingly.

Lessons Learned

Protective systems are devices designed to either automatically detect or suppress a fire or to assist people in extinguishing the fire. They can apply water or other extinguishing agents on a fire. Protective systems have been credited with saving both lives and property and are essential fire protection tools. Firefighters need to understand the value and operation of these systems to protect their communities.

Sprinkler systems are used for detection and suppression and can apply water or foam to extinguish the fire. Most sprinkler systems have a closed sprinkler head that thermally detects the fire, causing it to open and begin spraying water on the fire while sounding an alarm. The heads are designed to react to set temperatures and to spray water over a certain pattern. The four main types of systems are wet pipe with water in the pipes for immediate action, dry pipe for areas with freezing temperatures, deluge systems for fast-spreading fires, and preaction systems to prevent accidental discharge of water onto valuable property. Each type of system uses a different type of alarm valve but many of the other features are similar. Gate and check valves such as OS&Y or PIV valves are commonly used. Firefighters are still needed in sprinkler-protected property to finish putting out any fire that the system could not.

Standpipe systems facilitate manual fire suppression in which people do the firefighting. Standpipes supply water in large buildings where it would be hard to drag hoselines to and from fire apparatus, such as in high-rise and other large buildings. Building occupants can also use some standpipe systems, which are a system of pipes with a water supply connection at one end and a series of outlets, some with hose and nozzles, at the other. Standpipe operations require careful coordination of firefighters due to the size of the buildings involved.

The last group of protective systems detects fire and applies other extinguishing agents to fires in proximity to the hazard or fills an entire area. Extinguishing agents range from carbon dioxide to wet and dry chemicals to halogenated agents. Restaurants and kitchens usually have a hood system protecting the cooking areas, while other systems are used to protect high hazards or high value items. These protective systems are highly specialized but firefighters must understand their operation and any hazards of the agents.

Fire department operations at buildings with protective systems are typically outlined in local SOPs. Firefighters should treat all fire alarm activations as the worst-case scenario. This means investigating with full PPE, tools, and radios. Operations involving standpipe and sprinkler systems require the fire department to support the system with a water supply and fire attack hoselines. Stopping water flow from sprinkler systems should happen only after the fire has been extinguished. Hoselines should stay in place while the sprinkler is shut down. Firefighters need to practice using stops to stem the flow of activated sprinkler heads.

KEY TERMS

Accelerator A device to speed the operation of the dry pipe valve by detecting the decrease in air pressure. It pipes air pressure below the clapper valve, speeding its opening.

Auxiliary Appliances Another term for protective devices, particularly sprinkler and standpipe systems.

Deluge Systems Designed to protect areas that may have a fast-spreading fire engulfing the entire area. All of its sprinkler heads are already open, and the piping contains atmospheric air. When the system operates, water flows to all heads, allowing total coverage. The system uses a deluge valve that opens when a separate fire detection system senses the fire and signals to trip the valve open.

Dry Pipe System Air under pressure replaces the water in the system to protect against freezing temperatures. The sprinkler control valve uses a dry pipe valve to keep pressurized air maintained above with the supply water under pressure below the valve.

Exhauster A device to speed the operation of the dry pipe valve by detecting the decrease in air pressure. It helps bleed off air.

Fire Watch An organized patrol of a protected property when the sprinkler or other protection system is down for maintenance. Personnel from the property regularly check to make sure a fire has not started and assist in evacuation and prompt notification of the fire department.

Life Safety Term applied to the fire protection concept in which buildings are designed to allow for the escape of building occupants without injuries. Life safety usually makes the building more fire resistant, but this is not the main goal.

Local Application System Designed to protect only a certain or local portion of the building, usually directly where the hazard will occur or spread.

Outside Stem and Yoke (OS&Y) Valve Has a wheel on a stem housed in a yoke or housing. When the stem is exposed or outside, the valve is open. Also called an outside screw and yoke valve.

Post Indicator Valve (PIV) A control valve that is mounted on a post case with a small window, reading either "OPEN" or "SHUT."

Preaction System Similar to the dry pipe and deluge systems. The system has closed piping and heads with air under no or little pressure, but the water does not flow until signaled open from a separate fire detection system. The preaction valve then opens and allows water to flow through the system to the closed heads. When an individual head is heat activated, it opens and water attacks the fire. Usually used when water can cause a large dollar loss.

Pressure-Regulating Device Designed to control the head pressure at the outlet of a standpipe system to prevent excessive nozzle pressures in hoselines.

Residential Sprinkler System Smaller and more affordable version of a wet or dry pipe sprinkler system designed to control the level of fire involvement such that residents can escape.

Retard Chamber Acts to prevent false alarms from a sudden pressure surge in the water supply by collecting a small volume of water before allowing a continued flow to the alarm device. The water from a surge is drained from a small hole in the bottom of the collection chamber.

Sprinkler Systems Designed to automatically distribute water through sprinklers placed at set intervals on a system of piping, usually in the ceiling area, to extinguish or control the spread of fires.

Standpipe Systems Piping systems that allow for the manual application of water in large buildings.

Total Flooding System Used to protect an entire area, room, or building by discharging an extinguishing agent that completely fills or floods the area with the extinguishing agent to smother or cool the fire or break the chain reaction.

Wall Indicator Valve (WIV) A control valve that is mounted on a wall in a metal case with a small window, reading either "OPEN" or "SHUT."

Water Columning A condition in a dry pipe sprinkler system in which the weight of the water column in the riser prevents the operation of the dry pipe valve.

Wet Pipe Sprinkler System Has automatic sprinklers attached to pipes with water under pressure all the time.

REVIEW QUESTIONS

1. What are the two most common styles of smoke detectors?
2. How does an ionization detector operate?
3. Why should firefighters treat a smoke detector activation like a CO detector activation?
4. What is an upright and pendent sprinkler?
5. What happens when a fire occurs in a sprinkled room?
6. What is an OS&Y valve and how does it work?
7. What is a fire department connection and how is hose connected to it?
8. What are the steps to restoring a wet pipe sprinkler system?
9. Explain how to stop a sprinkler head from flowing.
10. Name the three classes of standpipes and who can use them.
11. What is a waterflow alarm?
12. What equipment should be carried by firefighters in a high-rise fire?
13. What is the hazard with a carbon dioxide total flooding extinguishing system?
14. What is the difference between an annunciator panel and a fire alarm control panel?
15. What precautions should firefighters take when using an elevator to access upper floors?
16. When should firefighters silence the audible fire alarm devices in a building?

Endnotes

1. See NFPA Standards 13, 13D, 13R, and 25 for more information.
2. *NFPA 13D: Standard for the Installation of Sprinkler Systems in One- and Two-Family Dwellings and Manufactured Homes, 2002 Edition*. National Fire Protection Association, Quincy, MA.
3. Cote, Arthur and John Linville, *Fire Protection Handbook,* 18th ed., pp. 6-145. National Fire Protection Association, Quincy, MA, 1997.
4. *NFPA 13: Standard for the Installation of Sprinkler Systems, 2002 Edition,* 13-11. National Fire Protection Association, Quincy, MA.
5. Cote et al., *Fire Protection Handbook,* pp. 6-249–6-250.
6. Ibid., pp. 6-250.

Additional Resources

Diamantes, David, *Fire Prevention, Inspection and Code Enforcement.* Delmar Learning, a part of the Thomson Corporation, Clifton Park, NY, 1997.

Gagnon, Robert, *Design of Special Hazard and Fire Alarm Systems.* Delmar Learning, a part of the Thomson Corporation, Clifton Park, NY, 1998.

Gagnon, Robert, *Design of Water Bases Fire Protection Systems.* Delmar Learning, a part of the Thomson Corporation, Clifton Park, NY, 1998.

Klinoff, Robert, *Introduction to Fire Protection.* Delmar Learning, a part of the Thomson Corporation, Clifton Park, NY, 1997.

NFPA 13: Standard for the Installation of Sprinkler Systems, 2002 Edition. National Fire Protection Association, Quincy, MA.

NFPA 13D: Standard for the Installation of Sprinkler Systems in One- and Two-Family Dwellings and Manufactured Homes, 2002 Edition. National Fire Protection Association, Quincy, MA.

NFPA 13E: Guide for Fire Department Operations in Properties Protected by Sprinkler and Standpipe Systems, 2000 Edition. National Fire Protection Association, Quincy, MA.

NFPA 13R: Standard for the Installation of Sprinkler Systems in Residential Occupancies up to and Including Four Stories in Height, 2002 Edition. National Fire Protection Association, Quincy, MA.

NFPA 14: Standard for the Installation of Standpipe, Private Hydrants, and Hose Systems, 2003 Edition. National Fire Protection Association, Quincy, MA.

NFPA 25: Standard for the Inspection, Testing, and Maintenance of Water-Based Fire Protection Systems, 1998 Edition. National Fire Protection Association, Quincy, MA.

Sturtevant, Thomas, *Introduction to Fire Pump Operations.* Delmar Learning, a part of the Thomson Corporation, Clifton Park, NY, 1997.

CHAPTER 13

BUILDING CONSTRUCTION

David Dodson, Lead Instructor, Response Solutions, Colorado

OUTLINE

- Objectives
- Introduction
- Building Construction Terms and Mechanics
- Structural Elements
- Fire Effects on Common Building Construction Materials
- Types of Building Construction
- Collapse Hazards At Structure Fires
- Lessons Learned
- Key Terms
- Review Questions
- Additional Resources

STREET STORY

A firefighter's worst fear is being trapped in a collapsing building and burning to death. Real-life encounters become implanted in our memory. It is one thing to read or hear about certain occurrences, it is another thing to have lived through one.

As a chief officer arriving at a fire scene, I encountered a large body of fire on the top floor of a three-story building. The first-due companies were initiating an offensive attack. My initial reaction was that the tactics being employed were correct. I surveyed the scene by doing a 360-degree walk-around of the building. I realized that conditions were not improving. I contacted the interior sector officer, and he informed me that interior conditions were deteriorating. I had noticed the following collapse indicators in my size-up: a large volume of unabated fire despite the aggressive interior attack; heavy smoke conditions indicating that the fire was probably attacking the building's structural supports; the ordinary constructed building contained a corbelled brick cornice, which added an eccentric load to the bearing wall supporting it; and lack of progress in the interior attack.

I had witnessed similar indicators in previous collapses. I ordered the units to withdraw from the building, set up a collapse zone, and initiated a defensive attack. The building's masonry wall collapsed shortly after the units' removal. The ensuing investigation found that the collapse occurred due to the combination of the collapse indicators.

Knowledge of building construction is vital before firefighters can attempt to fight a fire within a structure. Each type of construction contains inherent problems or positive features that can impede or assist in our ability to control a building fire. Examine the building. If it contains collapse indicators, decide their potential impact on the building's stability. If in doubt, it is better to err on the side of safety. Remember that once all civilians are removed from a structure, the life safety of firefighters is our utmost consideration.

—Street Story by James P. Smith, Deputy Chief, Philadelphia Fire Department, Philadelphia, Pennsylvania

OBJECTIVES

After completing this chapter, the reader should be able to:

- Describe the relationship between loads, imposition of loads, and forces.
- List and define four structural elements.
- Identify the effects of fire on five common building materials.
- List and define the five general types of building construction.
- List and define hazards associated with alternative building construction types.
- List five building collapse hazards associated with fire suppression operations.
- List five indicators of collapse or structural failure that might be found during fire suppression operations.

INTRODUCTION

Many fire departments pride themselves in their ability to launch aggressive interior structural fire attacks. Unfortunately, many firefighters are injured and killed when that same structure collapses, **Figure 13-1.** Often, buildings collapse without a "visual" warning such as sagging floors and roofs, leaning walls, and cracks. To keep from getting trapped in a collapse, firefighters must understand the types of structures they enter from the perspective of how the buildings are assembled, what materials are used, and how buildings react to fire. Additionally, firefighters must understand how fire travels through a building and choose appropriate tactics to stop the fire before key structural elements are attacked by the fire. Many firefighter fatality investigations conclude that fire departments need more training and education on building construction and the effects of fire on buildings. This chapter introduces several key topics regarding building construction and how fire affects buildings. It is important to note, however, that this chapter is merely an introduction. Firefighters must bridge the information in this chapter with a long-term commitment to study and research building construction and, more importantly, to explore the buildings within their jurisdiction.

Figure 13-1 This collapse happened seconds after firefighters were repositioned.

Streetsmart Tip Firefighters must realize that most of the buildings they will work in are already in place, and it is very difficult to determine the type of construction and fire-resistive rating by driving by or standing in front of a structure. Conducting in-service inspections, and securing the building owner's permission to walk through a building and get an "inside view" are ways of learning more about the structures in a particular jurisdiction.

If new construction is being conducted in a firefighter's response area or jurisdiction, contact the building department and secure a set of the plans. The fire department should be involved in the plan review process; however, this is not always the case. Also, take photographs of how the building is built and what materials are used for future training references. Many new materials and construction methods are introduced regularly. Firefighters should find out what materials are in the buildings in their jurisdictions.

This chapter begins by exploring some basic terms and mechanics of building construction, and then examines structural hierarchy and fire effects on materials. That information is then applied to classic and new construction types. The chapter concludes with a look at collapse hazards associated with structural fires.

BUILDING CONSTRUCTION TERMS AND MECHANICS

Firefighters need a basic understanding of certain terms and concepts associated with building construction. Obviously, buildings are constructed to provide a protected space to shield occupants from elements. The building must be built to resist wind, snow, rain, and still resist the force of gravity. Additionally, the intended use of the building can add a tremendous amount of weight, placing more stress on the building's ability to resist gravity. In building terms, these elements create building **loading**. Loads are then *imposed* on building materials. This imposition causes stress on the materials, called *force.* Forces must be delivered to the earth in order for the building to be structurally sound. With this basic understanding, we can start to define terms and mechanics.

Types of Loads

Loads can be divided into two broad categories as it relates to building construction: **dead loads** and **live loads**. Dead loads include the weight of all materials and equipment that are permanently attached to the building. Live loads include equipment, people, movement, and materials not attached to the structure. Dead loads and live loads can be more specifically described using the following terms:

Concentrated load: A concentrated load is a load that is applied to a small area, **Figure 13-2.** An example of a concentrated load is a heating, ventilation, and air-conditioning (**HVAC**) unit on a roof.

Distributed load: A distributed load is a load applied equally over a broad area, **Figure 13-3.** Examples of this include snow on a roof or a hoist attached to numerous roof supports.

Impact load: Impact load is a load that is in motion when applied, **Figure 13-4.** Crowds of people, fire streams, and wind gusts are examples of impact loads.

Figure 13-2 The steel stairs and air-conditioning unit apply a concentrated load on this roof structure. Also note the potential instability of the air-conditioning unit placed on cement blocks.

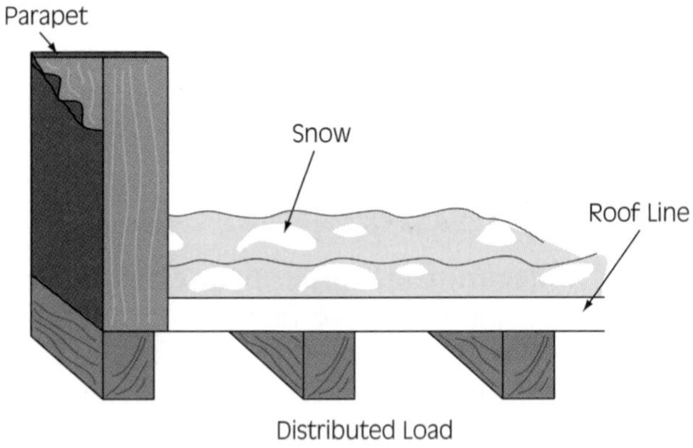

Figure 13-3 The weight of snow is a distributed load on this roof structure.

Figure 13-4 This ladder pipe operation is applying an impact load to the wall of this structure. As the wall weakens, it will eventually collapse. *(Photo courtesy of William H. Schmitt, Jr.)*

Historically Significant Building Collapses

Contributed by Dave Dodson

Many firefighters have been killed as a result of building collapse during firefighting operations. With each of these tragic losses, lessons can be learned. The following is a brief look at some of the more significant collapses. Each of these events should be researched to find all the contributing factors that led to the event. The italicized lessons are perceptions that are shared in the spirit of preventing firefighter injuries and death.

New York City, 1966

Firefighters responded to a commercial structure fire only to discover that the building shared a basement with another building. The concealed fire advanced rapidly and undetected. The ensuing firefight trapped and killed twelve firefighters. *It is vitally important to preplan buildings prior to a fire event. Older buildings may have access ways sealed from adjoining buildings. Shared utilities and other hidden voids can facilitate fire spread.*

The Boston Vendome Hotel Fire, 1972

Nine firefighters died during overhaul of a fire in an old, remodeled hotel. The investigation revealed that a masonry wall had been breached to make way for an air duct. Just above the breach, a column carried the load from floors above. A corner of the five-story building collapsed, trapping the firefighters. There were no obvious signs of impending collapse. *The Vendome building was brought back from disrepair in 1971, and many alterations were made that were unknown to the fire department. Firefighters should take an interest in the construction activities in and around buildings. Remodeling and restoration can compromise structural elements.*

Detroit, Abandoned Building, 1980

A fire was reported in a large, abandoned building that was scheduled for demolition. Responding firefighters found "light smoke" showing. The fire escalated rapidly due to the poor interior conditions and wide-open spaces. One firefighter died while trying to escape. Two other firefighters died when a firewall collapsed. *Abandoned buildings are much like a building under construction. Firefighters cannot take anything for granted. Rapid fire spread and suspect integrity should be the order of the day. Defensive operations must respect collapse zones.*

Hackensack, New Jersey, Ford Dealer, 1988

A fire was discovered in the attic space above an automobile repair garage. Responding firefighters launched an aggressive attack. The bowstring truss roof space was being used as a parts storage area, placing additional load on the structure. During firefighting operations, the roof collapsed, trapping and killing five firefighters. *Fighting fires in truss spaces is like playing Russian roulette. Trusses help form a wide, open space beneath. Clear spans are a warning sign of quick collapse should the truss space be involved in fire. Where there are no occupants to rescue, firefighters should reduce their risk and fight fire from safe attack points.*

Orange County, Florida, 1989

Firefighters responded to a fire in a single-story commercial structure. Interior conditions were described as light smoke and no heat. The fire had gained headway in a truss space above the ceiling. The tile-covered roof collapsed twelve minutes after firefighters arrived and killed two firefighters. *A fire can be roaring over firefighters' heads without them being aware. Firefighters should routinely inspect the ceiling space above their heads for fire. Once fire or heavy, dark smoke conditions are found in truss spaces, tactics should change. Tile roof coverings are quite attractive but show very little signs that the roof supporting them is about to collapse.*

Brackenridge, Pennsylvania, 1991

Firefighters were attempting to attack a fire in the basement of a large commercial building with concrete floors supported by steel columns. During the attack, the floor collapsed, trapping and killing four firefighters. *Basement fires present many difficult challenges to firefighters. Limited access, trapped heat and smoke, and the storage nature of basements must be factored in fire attack. Unfinished basements allow the fire to attack the floor above and floor supports rather quickly. Unprotected steel exposed to fire will soften quickly, leading to collapse.*

(continued)

Historically Significant Building Collapses (continued)

One Meridian Plaza, Philadelphia, Pennsylvania, 1991

Three firefighters died when they became disoriented and ran out of air while fighting fire in the high-rise building. The fire started on the twentieth floor and ran up to the thirtieth floor where a sprinkler system extinguished the fire. Although the firefighters did not die from a collapse, this event is significant in that fire officers feared a catastrophic collapse due to stress cracks found in the concrete stair towers and withdrew their firefighters. *There are many lessons learned from this event. The fatality and fire investigation details many building construction issues associated with high-rise firefighting. It is available from the U.S. Fire Administration (http://www.fema.gov/usfa).*

Mary Pang Fire, Seattle, Washington, 1995

An arson fire in a multiuse commercial building caused the deaths of four firefighters when a portion of the floor collapsed. The building had been altered several times, and a lightweight "pony wall" had been used to replace a portion of a load-bearing wall. The building had a confusing layout including entry points at two different elevations (like a walkout basement). The advanced fire was not apparent from the "front" side of the building. An aggressive interior attack was under way when the floor collapsed. *Buildings that have gone through several owners and occupancy changes should always be suspect. Prefire planning helps uncover hazards that could change firefighting tactics. Firefighters should make a habit of reporting fire and building conditions and observations.*

Stockton, California, 1997

Two firefighters died when a home addition collapsed during interior firefighting operations. The homeowners had built a large, two-story clear-span addition on the back of the home. Firefighters entered the front of the building and found a heavy fire and dense smoke conditions. *Homeowners do not necessarily follow established building practices and codes when making additions. A "360" prior to fire attack can uncover significant hazards when "reading" the building. Firefighters cannot preplan every home, so they must rely on their ability to read buildings and read smoke conditions.*

World Trade Center, New York City, 2001

Terrorists hijacked two large airliners and hit the twin towers of the World Trade Center. The Fire Department of New York (FDNY) responded and began the biggest rescue effort in fire service history. The high-rise towers collapsed, killing 343 of FDNY's bravest. Thousands of civilians also died in the collapse. *Steel high-rise construction relies on fire-resistive coatings to protect the steel. The combination of burning jet fuel and the trauma of the aircraft strikes rendered the steel unprotected. Failure came much quicker than the four-hour time limit prescribed for Type I fire-resistive construction found in high-rises. Firefighters should never rely on fire protection time ratings when making decisions for fire attack. History will always remember that the FDNY firefighters died trying to rescue trapped civilians. They have the undying respect of people throughout the world.*

Design load: Design loads are loads that an engineer has planned for or anticipated in the structural design.

Undesigned load: Undesigned load is a load that was not planned for or anticipated. Buildings that are altered or are being used for occupancy other than original intent create an undesigned load. One common example is a residential structure that is converted to a print shop or legal office. These buildings were not designed to hold the additional live loads caused by the change in occupancy.

Fire load: A fire load is the number of British thermal units (Btus) generated when the building and its contents burn. It is important to note that the construction industry does not recognize *fire load* in its vocabulary—it is a fire engineering term.

Imposition of Loads

Loads must be transmitted to structural elements. This is called *imposition of loads.* Terms associated with imposition include axial load, eccentric load, and torsion load, **Figure 13-5.**

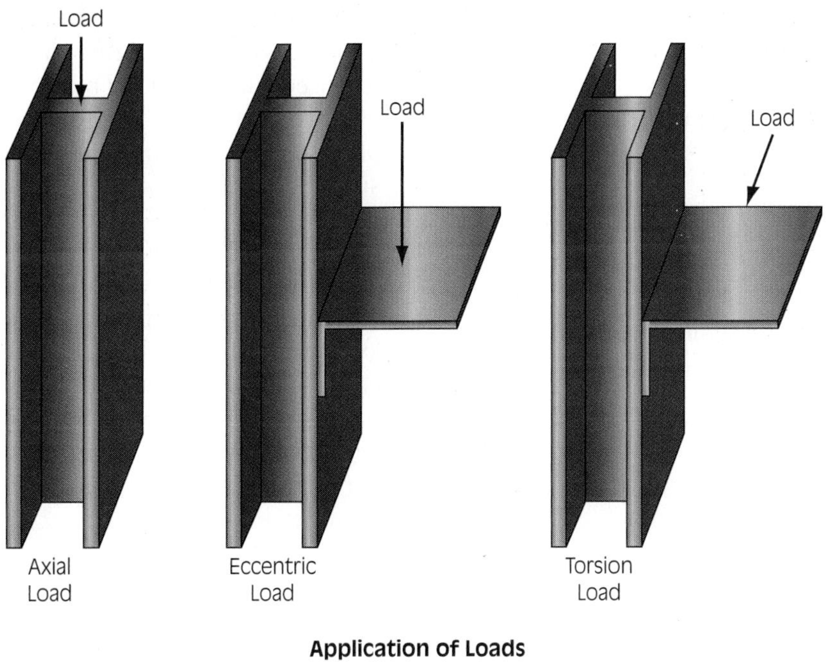

Figure 13-5 There are three types of loads that can be transmitted through a structural member: axial, eccentric, and torsion.

An **axial load** is a load that is transmitted through the center of an element and runs perpendicular to the element. **Eccentric load** is applied perpendicular to an element and, subsequently, does not pass through the center of the element. **Torsion load** is a load that is applied offset to an element, causing a twisting stress to the material.

Streetsmart Tip As a building burns, the structural elements decompose and lose their strength. This causes a change in the forces and the way the design loads are applied, leading to structural failure and collapse.

Forces

Loads imposed on materials create stress and strain on the materials used to make the element. Stress and strain are defined as forces applied to materials. These forces are defined as compression, tension, and shear, **Figure 13-6.** In **compression**, forces tend to push materials together. **Tension** occurs when forces tend to pull a material apart. **Shear** occurs when a force tends to "tear" a material apart—the molecules of the material are sliding past each other.

All loads—and the forces they create—must eventually pass through the structure and be delivered to the earth through the foundation of the building. Under normal conditions, structures will resist failure. Under fire conditions, the materials used to resist forces start breaking down. Eventually, gravity takes over and pushes the building to the earth.

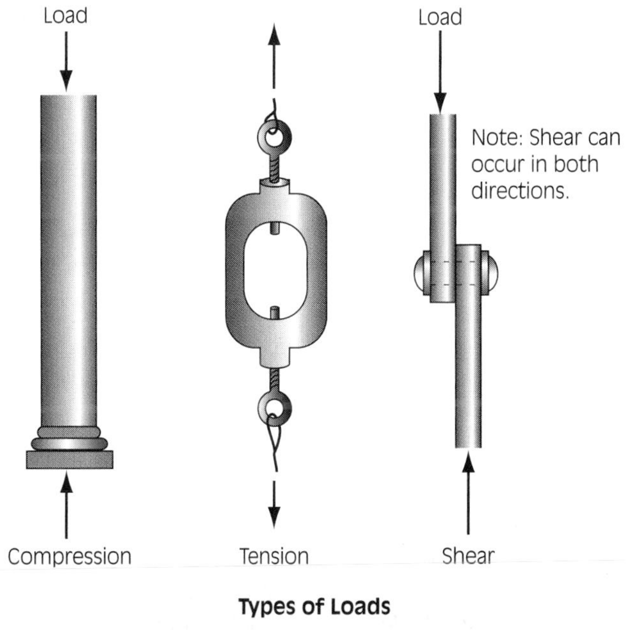

Figure 13-6 Loads are applied to a structural member as compression, tension, and shear forces.

The time it takes for gravity to overcome the structure during a fire is not predictable. A number of variables determine the amount of time a material can resist gravity and fire degradation. These include:

- Material mass
- Surface-to-mass ratio
- Overall load being imposed
- Btu development (fire load)
- Type of construction (assembly method)
- Alterations (undesigned loading)
- Age deterioration/care and maintenance of the structure
- Firefighting impact loads
- Condition of fire-resistive barriers

Streetsmart Tip Surface-to-Mass Ratio: **Surface-to-mass ratio** is defined as the exposed exterior surface area of a material divided by its weight. In simple terms, smaller, lighter structural members will have a large surface with small mass when compared to larger structural members capable of carrying an equal load. A 3- × 14-inch solid wood beam may carry the same design load as six 2- × 4-inch parallel chord trusses. The trusses have much more wood surface exposed and are more likely to ignite and burn rapidly.

The larger the surface area, the smaller the mass, the quicker it will burn or fail. Also, in combustible construction, the large surface area provides more fuel for the fire. Surface-to-mass ratio may also be applied to lightweight steel or pre-engineered buildings. These lightweight structural steel members will absorb heat quickly, and the steel elements will lose their strength and fail, causing collapse.

STRUCTURAL ELEMENTS

Buildings are an assembly of structural elements designed to transfer loads to the earth. Structural elements can be defined simply as beams, columns, and walls. Each of these elements must be connected in some fashion in order to effectively make the load transfers to the building foundation, which delivers the building live and dead loads to earth.

Beams

A **beam** is a structural element that delivers loads perpendicular to its length. Obviously, something must support the beam—usually a wall or column. It stands to reason that beams are used to create a covered space. In doing so, the beam is subjected to an eccentric load. This load causes the beam to deflect. The top of the beam is subjected to a compressive force whereas the bottom of the beam is subjected to tension, **Figure 13-7.** The distance between the top of the beam and the bottom of the beam dictates the amount of load the beam can carry. I beams are very typical and usually refer to the use of steel to form a beam. The top of the I is known as the top **chord;** the bottom of the I is called the bottom chord. The material in between is known as the **web.** There are numerous types of beams although the principal method of load transfer remains the same. A few types of beams include:

- **Simple beam:** A beam supported at the two points near its ends.
- **Continuous beam:** A beam that is supported in three or more places.
- **Cantilever beam:** A beam that is supported at only one end—or a beam that extends over a support in such a way that the unsupported overhang places the top of the beam in tension and the bottom in compression.
- **Lintel:** A beam that spans an opening in a load-bearing masonry wall—such as over a garage door opening. In wood construction, the same beam is often called a *header.*
- **Girder:** A beam that supports other beams.
- **Joist:** A wood framing member used to support floors or roof decking. A **rafter** is a joist that is attached to a ridge board to help form a peak.
- **Truss:** A series of triangles used to form a structural element that in many ways is really a "fake" beam. That is, a truss uses geometric shapes, lightweight materials, and connections to transfer loads just like a beam. Trusses will be covered in detail later in this chapter.
- **Purlin:** A series of wood beams placed perpendicular to steel trusses to help support roof decking.

Columns

A **column** is any structural component that transmits a compressive force parallel through its center. Columns typically support beams and other columns, **Figure 13-8.** Columns are typically viewed as the vertical supports of a building; however, columns can be diagonal or even horizontal. The guiding principle is that a column is totally in compression.

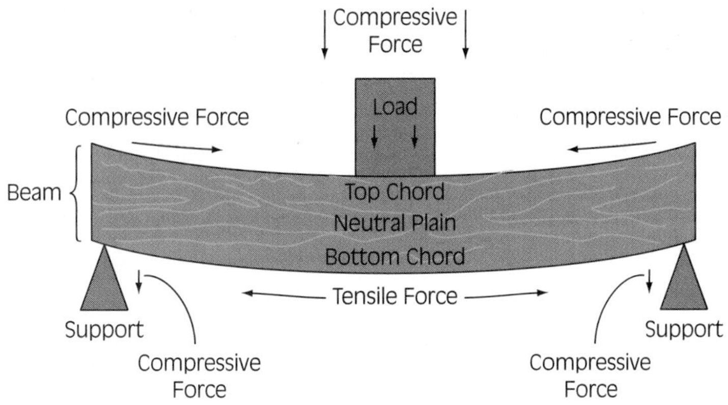

Figure 13-7 A beam transfers a load perpendicular to the load—creating compressive and tensile forces within the beam.

Walls

A wall is also a component that transmits compressive force through its center. Simply put, a wall is a really long, but slender, column. Walls are subdivided into two categories: **load-bearing** and non-load-bearing. A load-bearing wall carries the weight of beams, other walls, floors, roofs, or other structural elements as well as the weight of the wall itself. A non-load-bearing wall need only support its own weight. A partition wall is an example of a non-load-bearing wall.

Connections

As mentioned previously, beams, columns, and walls must be connected in some fashion in order to effectively transfer loads. Often, the connection is the weak link as it relates to structural failure during fires. The connection point is often a small, low-mass material that lacks the capacity to absorb much heat, thereby failing quicker than an element that has more mass such as a column or wall. Connections fall into three categories: pinned, rigid, and gravity. Pinned connections use bolts, screws, nails, rivets, and similar devices to transfer load. Rigid connections refer to a system where the elements are bonded together such that all the columns (or load-bearing walls) are bonded to all the beams. Typically, failure of one element will cause the loads to be transferred to other elements. Gravity connections are just that—the load from an element is held in place by gravity alone.

Figure 13-8 This column is supporting a beam, flooring, and another column. Columns are subjected to compressive forces.

Together, structural elements defy gravity and make a building sound. A series of columns and beams used to hold up a building are often referred to as the *skeletal frame. Post and beam* describes the same concept. Beams resting on walls are simply called wall-bearing buildings. One factor that has not been discussed is the suitability of the materials used to form structural elements. The next section covers these materials and how they act during fires.

FIRE EFFECTS ON COMMON BUILDING CONSTRUCTION MATERIALS

Many factors determine which material is used to form structural elements. Quality, cost, application, engineering capabilities, and adaptability all play in the suitability of a material. In some cases, the material chosen for a structural application

Tribute to the Ol' Professor

Contributed by Dave Dodson

Francis L. Brannigan is a true friend to the fire service. For over thirty-five years Mr. Brannigan has shared his knowledge of fire and, more specifically, effects of fire on buildings. His book *Building Construction for the Fire Service* (see the Additional Resources section at the end of this chapter) is a must-read for any firefighter and critical reading for anyone wanting to promote into fireground decision-making positions.

The fire service has affectionately called Mr. Brannigan the "Ol' Professor." His teachings have saved untold numbers of firefighters. I will never forget my first exposure to Mr. Brannigan. It was at a national conference in Cincinnati. The Ol' Professor was teaching a daylong class on steel buildings. As a young and inexperienced firefighter, I was all ears. The class taught me two things. First, I had much to learn about fire effects on building construction. I was way behind in my knowledge, and the Ol' Professor motivated me to make a never-ending knowledge quest to understand buildings. Second, I realized that reading buildings and reading smoke were the keys to rescuing people and putting the "wet stuff on the red stuff."

Over the years, Mr. Brannigan has coined many powerful—and lifesaving—phrases. These bits of advice have remained part of the teachings of many fire instructors. Among my favorites:

Trusses

"BEWARE the TRUSS!"

"A truss is a truss, is a truss . . ."

—in reply to the notion that a bowstring truss is more dangerous than other trusses.

"The bottom chord of a truss is under tension—It's like you hanging on a rope. If the rope gets cut, you will fall. So it is with a truss."

"Failure of one element of a truss may cause the entire truss to fail—failure of one truss can cause other trusses to fail."

Columns

"The failure of a column is likely to be more sudden than failure of a beam."

"The slightest indication of column failure should cause the building to be cleared immediately."

On Collapses

"There is a tendency among those concerned about building stability to make light of partial collapse . . . a partial collapse is very important to at least two groups—those under it and those on top of it!"

"From an engineering point of view, (lightweight, trussed) buildings are made to be disposable . . . we don't make disposable firefighters!"

In response to those who claim a building collapsed without warning during a fire:

"The warning is the brain—in your ability to understand buildings and anticipate how they will react to a fire."

The fire service is indebted to the Ol' Professor. Thanks for all you have done!

---

needs to meet fire resilience criteria. Regardless, the firefighter needs to understand how these materials react to fire. In the past, the fire service looked at the characteristics of four basic material types: wood, steel, concrete, and masonry. Each of these materials can be found together or separately. Each material reacts to fire in a different way, **Table 13-1.** Now, advanced material technology has found its way into structural elements. Buildings are being assembled using plastics, graphites, wood derivatives, and other composites. This section covers the four basic building materials as well as some of the new composites.

Wood

Wood is perhaps the most common building material. It is used in millions of residential and commercial buildings. Wood is relatively inexpensive, easy to manipulate, and a replenishable natural resource (although that can be argued). Wood has marginal resistance to forces compared to its weight, but it does the job for most residential and small commercial buildings. Wood also burns—and in doing so gives away its mass. The more mass a section of wood has, the more material it must burn away before strength is lost. This is true of native wood—that is, wood that

Performance of Common Building Materials under Stress and Fire

MATERIAL	COMPRESSION	TENSION	SHEAR	FIRE EXPOSURE
Brick	Good	Poor	Poor	Fractures, spalls, crumbles
Masonry block	Good	Poor	Poor	Fractures, spalls
Concrete	Good	Poor	Poor	Spalls
Reinforced concrete	Good	Fair	Fair	Spalls
Stone	Good	Poor	Fair	Fractures, spalls
Wood	Good w/grain; poor across grain	Marginal	Poor	Burns, loss of material
Structural steel	Good	Good	Good	Softens, bends, loses strength
Cast iron*	Good	Poor	Poor	Fractures

*Some cast iron may be ornamental in nature and not part of the structure or load bearing.

TABLE 13-1

has been cut from a tree. Engineered wood can react differently when exposed to heat from a fire. Engineered wood includes a host of products that take many pieces of native wood and glue them together to make a sheet, longer beam (trees only grow so tall!), or stronger column. Plywood delaminates when exposed to fire. Some newer wood products such as composites, which are discussed later in this chapter, present safety concerns for all firefighters.

Steel

Steel is a mixture of carbon and iron ore heated and rolled into structural shapes to form elements for a building. Steel has excellent tensile, shear, and compressive strength. For this reason, steel is a popular choice for girders, lintels, cantilevered beams, and columns. Additionally, steel has high factory control. It is easy to change its shape, increase its strength, and otherwise manipulate it during production.

As it relates to fires, steel loses strength as temperatures increase. The specific range of temperatures depends on how the steel was manufactured. Cold drawn steel, like cables, bolts, rebar, and lightweight fasteners, loses 55 percent of its strength at 800°F. Extruded structural steel used for beams and columns loses 50 percent of its strength at 1,100°F. Structural steel will also elongate or expand as temperatures rise. At 1,000°F, a 100-foot-long beam will elongate 10 inches. Imagine what that could do to a building. If a beam is fixed at two ends, it will try to expand—and likely deform, buckle, and collapse. If the beam sits in a pocket of a masonry wall, it will stretch outward and place a shear force on the wall—which was designed only for a compressive force. *This could knock down the whole wall!*

Because steel is an excellent conductor of heat, it will carry heat of a fire to other combustibles. This can cause additional fire spread, sometimes a considerable distance from the original fire.

Caution Steel softens, elongates, and sags when heated, leading to collapse. Cooling structural steel with fire streams is just as important as attacking the fire.

Concrete

Concrete is a mixture of portland cement, sand, gravel, and water. It has excellent compressive strength when cured. The curing process creates a chemical reaction that bonds the mixture to achieve strength. The final strength of concrete depends on the ratio of these materials, especially the ratio of water to portland cement. Because concrete has poor tensile and shear strength, steel is added as reinforcement. Steel can be added to concrete in many ways. Concrete can be poured over steel rebar and become part of the concrete mass when cured. Cables can be placed through the plane of concrete and be tensioned, compressing the concrete to give it required strength. Cables can be pretensioned (at a factory) or posttensioned (at the job site). *Precast concrete* refers to slabs of concrete that are poured at a factory and then shipped to a job site. Precast slabs are "tilted up" to form load-bearing walls—thus the term *tilt-up construction.*

All concrete contains some moisture and continues to absorb moisture as it ages. When heated, this moisture content will expand, causing the concrete to crack or spall. **Spalling** refers to a large pocket of concrete that has basically crumbled into fine particles, taking away the mass of the concrete. Reinforcing steel that becomes exposed to a fire can transmit heat within the concrete, causing catastrophic spalling and failure of the structure. Unlike steel, concrete is a heat sink and tends to absorb and retain heat rather than conduct it. This heat is not easily reduced. Concrete can stay hot long after the fire is out, causing additional thermal stress to firefighters performing overhaul.

Masonry

Masonry is a common term that refers to brick, concrete block, and stone. Masonry is used to form load-bearing walls because of its compressive strength. Masonry can also be used to build a **veneer** wall. A veneer wall supports only its own weight and is most commonly used as a decorative finish. Masonry units (blocks, bricks, and stone) are held together using **mortar**. Mortar mixes are varied but usually contain a mixture of lime, portland cement, water, and sand. These mixes have little to no tensile or shear strength. They rely on compressive forces to give a masonry wall strength. A lateral force that exceeds the compressive forces within a masonry wall will cause quick collapse of the wall.

Streetsmart Tip **Masonry Walls Collapse:** Masonry has very little lateral stability, and in many cases the roof or floor structure of a building holds the walls in place. Steel beams or joists will expand during a fire, creating lateral loads that the walls were not designed to withstand. In addition, wood roof structures will burn away or collapse during a fire, leaving little lateral support. The effects of the fire, pulling forces of the collapsing wood structure, or the force of a hose stream may cause the wall to collapse. Remember, the designer and contractor did not plan for the building to burn down.

Figure 13-9 Prior to a fire, the effects of age will take their toll on masonry walls. How stable is this wall with joint deterioration and lack of full mortar bond?

Brick, concrete block, and stone have excellent fire-resistive qualities when taken individually. Many masonry walls are typically still standing after a fire has ravaged the interior of the building. Unfortunately, the mortar used to bond the masonry is subject to spalling, age deterioration, and washout. Whether from age, water, or fire, the loss of bond will cause a masonry wall to be very unstable, **Figure 13-9.**

Composites

New material technologies have introduced some interesting challenges for the firefighting community. Composites are a combination of the four basic materials listed above as well as various plastics, glues, and assembly techniques. Of particular interest are the many wood products that are widely used for structural elements.

Lightweight wooden I beams (joists) are nothing more than wood chips that are press-glued together into the shape of an I beam, **Figure 13-10.** While structurally strong (stronger than a comparable solid wood joist), the wooden I beam fails quickly when heated. Actually, no fire contact is required. Ambient heating causes the binding glue to fail, leading to a quick collapse. The bottom of a beam is under tensile forces. If the bottom of the beam falls off, due to glue failure, the beam will immediately snap and collapse.

New products, known as FiRP (fiber-reinforced products) are becoming common in the construction industry. FiRP can be plastic fibers mixed with wood to give the wood increased tensile strength. As with most plastics, fire exposure can cause quick failure as the plastic melts.

The mixture of steel and wood as a structural element can cause rapid collapse because steel expands

Figure 13-10 To save on materials and cost, the use of composites or engineered wood structural members is becoming popular. Shown here is a typical wood I beam with 2- × 3-in. flanges and a ⅜-in. structure board web. In addition to providing a large surface-to-mass ratio, the flanges and web are fastened with glue, which may deteriorate quickly under fire conditions.

Figure 13-11 A composite truss. Rapid heating will cause the steel to separate from the wood chords.

(A)

(B)

Figure 13-12 These are wall panels that are load-bearing (A). Expanded polystyrene is sandwiched between OSB sheets. Fire can easily enter the wall space. (B) Failure of the wall panel will cause instability in the roof structure.

faster than wood. This causes stress at the intersection of the two materials, **Figure 13-11.**

Structural insulated panels (SIP) are another interesting composite. This technique is characterized by large wall panels made of expanded polystyrene sandwiched between two sheets of oriented strand board (OSB). OSB is a sheet of wood chips bonded by glue, **Figure 13-12A** and **B.** The OSB is covered with a typical wall finish. It is anticipated that a fire will cause rapid deterioration of the load-bearing panel.

To review, this chapter has so far explored the basic terminology, mechanics, elements, and materials used in the construction of buildings. It has also discussed a bit about how fire attacks materials and causes failure. The next section explores the various methods used to assemble buildings.

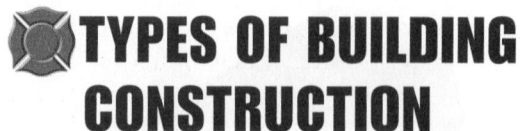

TYPES OF BUILDING CONSTRUCTION

Over time, five broad categories of building construction types have been developed to help classify structures. These categories give firefighters a basic

Types of Construction from NFPA 220

	TYPE I		TYPE II			TYPE III		TYPE IV	TYPE V	
	443	332	222	111	000	211	200	2HH	111	000
Exterior Bearing Walls—										
Supporting more than one floor,										
columns, or other bearing walls	4	3	2	1	0	2	2	2	1	0
Supporting one floor only	4	3	2	1	0	2	2	2	1	0
Supporting a roof only	4	3	1	1	0	2	2	2	1	0
Interior Bearing Walls—										
Supporting more than one floor,										
columns, or other bearing walls	4	3	2	1	0	1	0	2	1	0
Supporting one floor only	3	2	2	1	0	1	0	1	1	0
Supporting roofs only	3	2	1	1	0	1	0	1	1	0
Columns—										
Supporting more than one floor,										
columns, or other bearing walls	4	3	2	1	0	1	0	H[1]	1	0
Supporting one floor only	3	2	2	1	0	1	0	H[1]	1	0
Supporting roofs only	3	2	1	1	0	1	0	H[1]	1	0
Beams, Girders, Trusses										
& Arches—										
Supporting more than one floor,										
columns, or other bearing walls	4	3	2	1	0	1	0	H[1]	1	0
Supporting one floor only	3	2	2	1	0	1	0	H[1]	1	0
Supporting roofs only	3	2	1	1	0	1	0	H[1]	1	0
Floor Construction	3	2	2	1	0	1	0	H[1]	1	0
Roof Construction	2	1½	1	1	0	1	0	H[1]	1	0
Exterior Nonbearing Walls	0	0	0	0	0	0	0	0	0	0

[shaded box] Those members that shall be permitted to be of approved combustible material.

[1]"H" indicates heavy timber members; see text for requirements.

Source: Reprinted with permission from NFPA 220, *Types of Building Construction*, copyright © 1995, National Fire Protection Association, Quincy, MA 02269. This reprinted material is not the complete and official position of the National Fire Protection Association on the referenced subject which is represented only by the standard in its entirety.

TABLE 13-2

understanding of the arrangement of structural elements and the materials used to construct the building. Unfortunately, these broad classifications are dangerously incomplete for firefighters and may lead to deadly assumptions about the makeup of a building. As stated before, firefighters need to explore the buildings within their jurisdiction to determine how buildings are assembled.

It is important to note that buildings are built to meet certain codes. These codes are designed to give occupants time to escape during a fire. Concrete is **fire resistive**—meaning it has some capacity to withstand the effects of fire. Other materials, like steel and wood, need fire-resistive assistance to give occupants a chance to escape. Building codes outline **fire-resistive ratings**, occupancy classifications, and means of egress based on five general types of buildings. The features of each type of construction will be discussed shortly, but first it is important to understand fire resistance for structural elements. **Table 13-2** outlines the number of hours that a structural element needs to be protected for the five types of construction. Simply put, firefighting time is not part of the fire-resistive and building construction equation. Fire-resistive ratings are established in a laboratory. In the real world, fire resistance ratings could "underperform" due to many factors. For example, a structural element with a two-hour fire rating may fail in thirty minutes if it was not assembled correctly or if improperly inspected. Fire resistance for structural members can be achieved by various methods including drywall (gypsum wallboard), spray-on coatings, and concrete, **Figure 13-13.** Aging, alterations, and wear can damage fire-resistive methods to the point that structural elements have no fire resistance protection.

Figure 13-13 This parking garage is of Type II construction. The protective coating applied to the structural steel may increase the fire-resistive rating, but note that the unprotected corrugated metal flooring and interior steel structure are not protected. These unprotected structural members may fail early in a fire. *(Photo courtesy of William H. Schmitt, Jr.)*

The following paragraphs outline the basic definition of each building type, its general configuration, and some historical fire spread problems associated with each. Also included are some construction methods that do not fit into the five common types.

Type I: Fire-Resistive

Type I fire-resistive construction is a type in which structural elements are of an approved noncombustible or limited combustible material with sufficient fire-resistive rating to withstand the effects of fire and prevent its spread from story to story. Concrete-encased steel, **Figure 13-14,** monolithic-poured cement, and steel with spray-on fire protection coatings are typical of Type I, **Figure 13-15.** Generally, the fire-resistive rating must be three to four hours depending on the specific structural element. Fire-resistive construction is used for high-rises, large sporting arenas, and other buildings where a high volume of people are expected to occupy the building.

Most Type I buildings are typically large, multistoried structures with multiple exit points. Fires are difficult to fight due to the large size of the building and the subsequent high fire load. Type I buildings rely on protective systems to rapidly detect and extinguish fires. If these systems do not contain the fire, a difficult firefight will be required. Fire can spread from floor to floor on high-rises as windows break and the next floor windows fail, allowing the fire to jump. Fire can also make vertical runs through utility and elevator shafts. Regardless, firefighters are relying on the fire-resistive methods to protect the structure from collapsing. The collapse

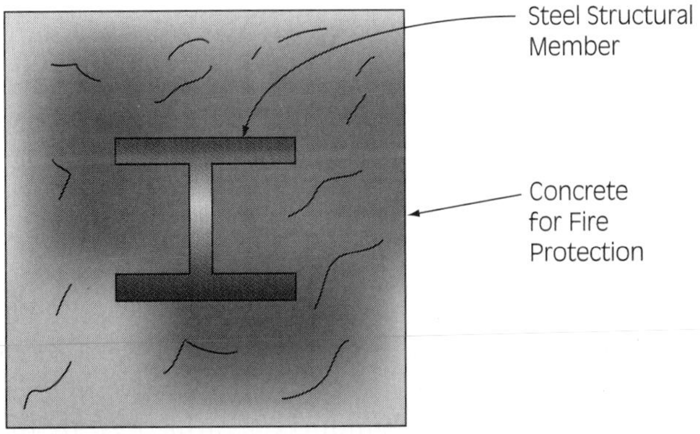

Figure 13-14 To achieve a Type I fire-resistive rating, structural steel members are encased with concrete to prevent failure from the effects of a fire.

Figure 13-15 A typical Type I building, with structural members designed to resist the effects of fire for three to four hours. This building is of reinforced concrete construction.

Figure 13-16 Buildings of Type II construction will have structural elements with little or no protection from the effects of fire. Remember, in the event of a fire, these unprotected steel structural members may fail and collapse with little warning.

of fire-resistive structures can be massive, as we are reminded from the World Trade Center collapse in New York City.

Type II: Noncombustible

Type II noncombustible construction is a type in which structural elements do not qualify for Type I construction and are of an approved noncombustible or limited combustible material with sufficient fire-resistive rating to withstand the effects of fire and prevent its spread from story to story. More often than not, Type II buildings are steel, **Figure 13-16.** Modern warehouses, small arenas, and newer churches and schools are built as noncombustible. Because the steel is not required to have significant fire-resistive coatings, Type II buildings are susceptible to steel deformation and resulting collapse. Fire spread in Type II buildings is influenced by the contents. While the structure itself will not burn, rapid collapse is possible from the content Btu release stressing the steel.

Suburban strip malls with concrete block load-bearing walls and steel roof structures can be classified as Type II. Fires can spread from store to store through wall openings and shared ceiling and roof support spaces. The roof structure is often of lightweight steel that fails rapidly. More often than not, the fire-resistive device used to protect the roof structure is a dropped-in ceiling. Missing ceiling tiles, damaged drywall, and utility penetrations can render the steel unprotected. These buildings may have combustible attachments such as facades and signs as well as significant content fire loading.

Type III: Ordinary

The term **Type III ordinary construction** is often misapplied to wood frame buildings. By definition, ordinary construction includes buildings where the load-bearing walls are noncombustible, and the roof and floor assemblies are wood. Most commonly, this is load-bearing brick or concrete block with wood roofs and floors. Ordinary con-

Figure 13-17 Buildings of Type III, ordinary construction, are common throughout North America. These typical "Downtown USA" buildings provide many challenges to firefighters, such as void spaces and common walls allowing rapid fire extension and little structural protection, with early collapse during firefighting operations.

struction is prevalent in most downtown or "main street" areas of established towns and villages, **Figure 13-17.** Firefighters have long called ordinary construction "taxpayers." This slang is derived from landlords who built buildings with shops and/or restaurants on the first floor with apartments above in order to maximize income to help pay property taxes. Newer Type III buildings include strip malls with block walls and wood truss roofs, **Figure 13-18.**

Ordinary construction presents many challenges to firefighters. In older buildings, numerous remodels, restorations, and repairs have created suspect wall stability and hidden dangers.

Firefighter Fact Sagging or bowing load-bearing walls are often pulled back in alignment by tightening a steel rod that runs through the building from wall to wall. A small interior fire can elongate this steel and cause catastrophic wall failure. These buildings can be spotted by decorative stars or ornaments (called *spreaders*) on the outside brick wall.

Ordinary construction has many void spaces where fire can spread undetected. Common hallways, utilities, and attic spaces can communicate fire rapidly. Masonry walls hold heat inside, making for difficult firefighting. Wood floors and roof beams are often gravity fit within the masonry walls. These can release quickly and cause a general collapse, leaving an unsupported masonry wall. Older Type III buildings have structural mass; therefore, they burn for a long time.

Type IV: Heavy Timber

Type IV heavy timber construction can be defined as those buildings that have block or brick exterior load-bearing walls and interior structural members, roofs, floors, and arches of solid or laminated wood without concealed spaces. The minimum dimensions for structural wood must meet the criteria in **Table 13-3.** Heavy timber buildings, as the name suggests, are quite stout and are used for warehouses, manufacturing buildings, and some older churches, **Figure 13-19.** In many ways, a Type IV building is like a Type III—just larger dimension lumber instead of common wood beams and trusses.

Figure 13-18 One of the most common uses of Type III, ordinary construction, is the "strip mall" with masonry walls and lightweight steel or wood trusses. Common problems associated with this type of construction are void spaces allowing for rapid fire extension and collapse of lightweight structural elements.

Heavy Timber Dimensions

TYPE OF ELEMENT	USE	SIZE
Column	Supporting floor load	8- × 8-in. minimum any dimension
Column	Supporting roof load	6-in. smallest dimension, 8-in. depth minimum
Beams and girders	Supporting floor load	6-in. width and 10-in. depth minimum
Beams, girders, and roof framing	Supporting roof loads only	4-in. width minimum, 6-in. depth minimum
Framed or laminated arches	As designed	8-in. minimum dimension
Tongued and grooved planks	Floor systems	3-in. minimum thickness with additional 1-in. boards at right angles
Tongued and grooved planks	Roof decking	2-in. minimum thickness

TABLE 13-3

Figure 13-19 Type IV buildings, heavy timber construction, have large wood structural elements with great mass. The mass of these structural members requires a long burn time for failure. The connections, usually steel, are the weak points in this type of construction.

Some firefighters mistakenly call Type IV buildings "mill construction." Mill construction is a much more stout, collapse-resistive building that may or may not have block walls. A new Type IV building is hard to find. The cost of large-dimension lumber and laminated wood beams makes this type of construction rare.

Fire spread in a heavy timber building can be fast due to wide-open areas and content exposure. The exposed timbers contribute Btus to the fire. Because of the mass and large quantity of exposed structural wood, fires burn a long time. If the building housed machinery at one time, oil-soaked floors will add more heat to the fire and accelerate collapse. Once floors and roofs start to sag, heavy timber beams may release from the walls. This is accomplished by making a fire-cut on the beam, and the beam is gravity fit into a pocket within the exterior load-bearing masonry wall, **Figure 13-20.** As the floor sags, it loses its contact point with the wall and simply slides out of its pocket without damage to the wall. It is important to recall that a free-standing masonry wall has little lateral support and requires compressive weight from floors and roofs to make it sound.

Type V: Wood Frame

Type V wood frame construction is perhaps the most common construction type. Homes, newer small businesses, and even chain hotels are built primarily with wood, **Figure 13-21.** Older wood frame buildings were built as **balloon frame—**

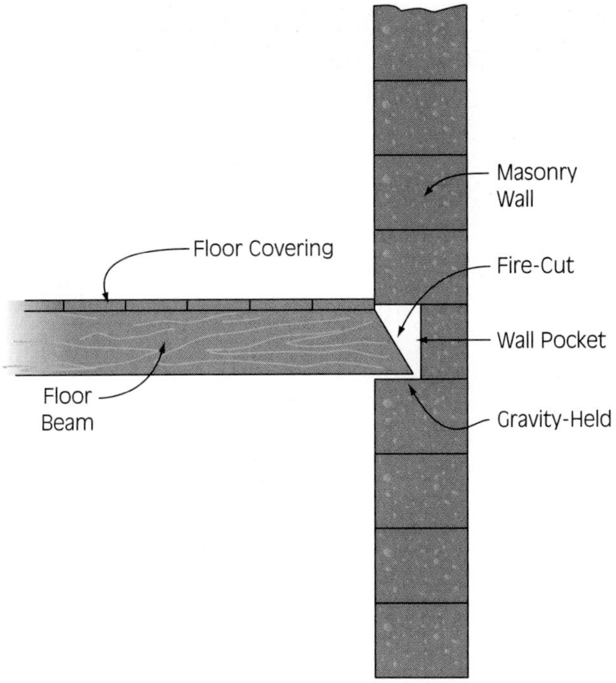

Figure 13-20 Wood and heavy timber beams were often "fire-cut" so that a fire-damaged, sagging floor would simply slide out of the wall pocket in order to preserve the wall.

Figure 13-21 The wood frame structure, Type V construction, is the most common type of construction in North America.

that is, wood studs ran from the foundation to the roof and floors were "hung" on the studs. As can be envisioned, fire could enter the wall space and run straight to the attic. In the early 1950s, builders started using a **platform framing** arrangement where one floor was built as a platform for the next floor. This created **fire stopping** to help minimize fire spread. Newer wood frame buildings utilize lightweight wood trusses for roofs and floors. This is akin to a "horizontal balloon frame" that can allow quicker lateral fire spread. Coupled with high surface-to-mass wood exposure, collapse becomes a real possibility, **Figure 13-22.** Some codes require truss spaces to have fire stopping every 500 square feet. Even with this fire stopping, it remains dangerous to step onto the 500 square feet where the fire is. Wood frame structures may appear more like a Type III ordinary building because of a brick-wall appearance. Remember, brickwork may be a simple veneer to add aesthetics.

To protect structural members from a fire, wood frame construction typically uses gypsum board (drywall or the brand name Sheetrock). Once finished, wood frame buildings typically have many rooms that can help compartmentalize content fires. Fire that penetrates wall, floor, or attic spaces becomes a significant collapse threat, especially in newer buildings. Often, the only warning that fire has penetrated these spaces is the issuance of smoke from crawl space vents, gable end vents, and eaves.

Other Construction Types

As mentioned earlier, the five broad building types can actually lead to dangerous assumptions. Newer construction and alternative building methods may not fit cleanly into one of the above types. Some buildings are actually two types of construction. For example, a particular restaurant located in Colorado is built as a Type II noncombustible yet is topped with a large wood frame structure to hide rooftop HVACs and cooking vent hoods, **Figure 13-23A** and **B.** The square feet space of the false dormers and wood frame structure exceeds that of most homes.

New lightweight steel homes resemble wood frame homes. These buildings are actually a "post and beam" steel building with lightweight steel studs to help partition the home. OSB is added to the studs to help make the house more "stiff" and increase wind-load strength, **Figure 13-24.** Another interesting construction type uses foam blocks to make a form for a lightweight concrete mud mixture. The concrete is not contiguous—there are many voids, utility runs, and foam block spacers (made of plastic or galvanized steel), **Figure 13-25.** These structures are called "ICF" or insulated concrete formed." However, it is important not to be fooled by any claims that these buildings are concrete or are less combustible. In reality, these composite buildings are assembled with plastics, polystyrene, lightweight steel, and lightweight concrete. When finished, these buildings may resemble wood frame or even ordinary construction.

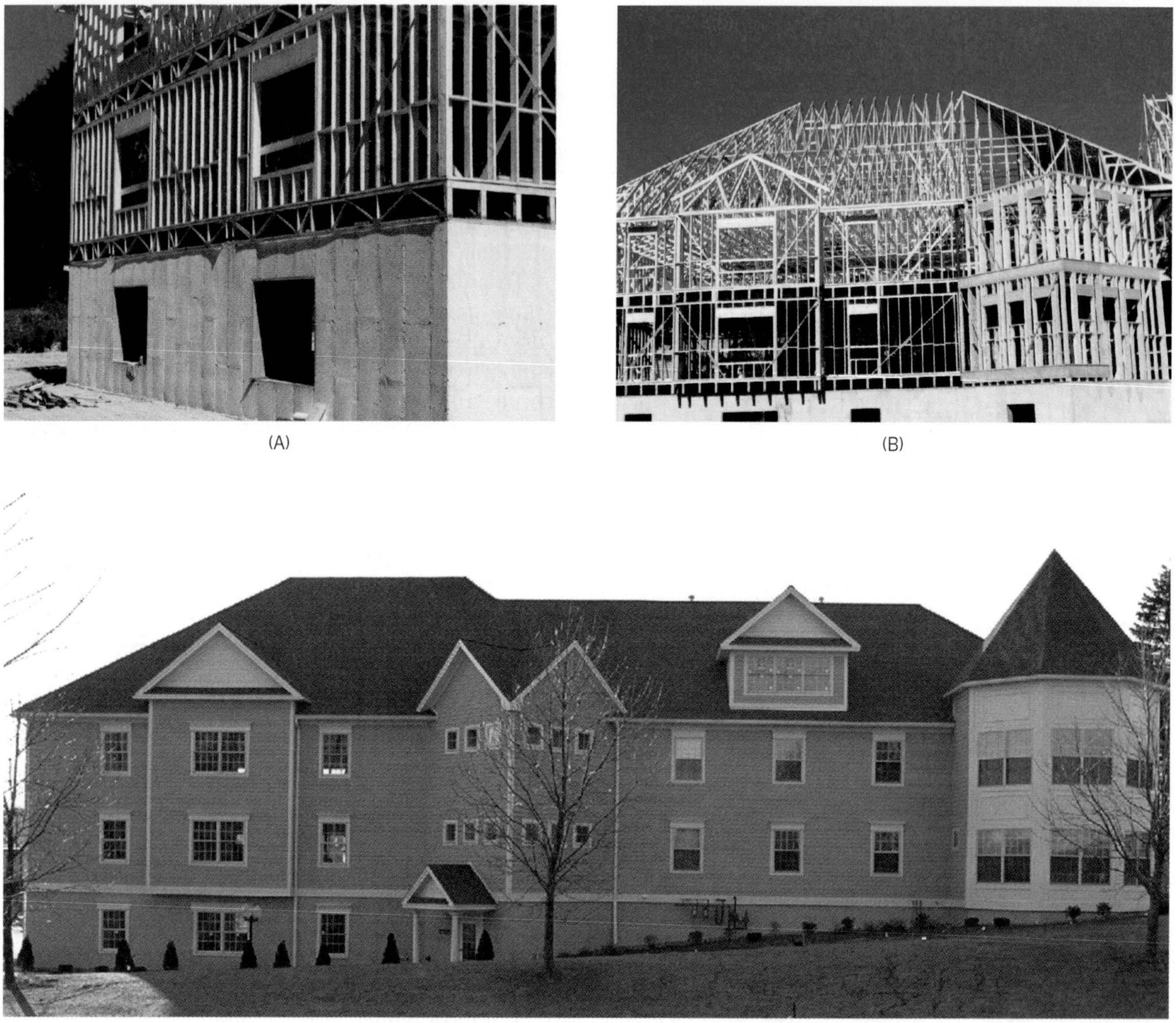

(A) (B) (C)

Figure 13-22 (A, B) Note the void spaces at the first and second floor levels and in the attic area that are created by the use of truss systems. (C) The building after it was completed. Firefighters should survey the buildings in their area before they are completed.

Extended window and door jambs are clues that indicate the wall is thicker than that of typical wood or masonry built buildings.

The fire service has very little research information on the stability of these new types of buildings during fires. One thing is certain: Firefighters should expect rapid collapse due to the low-mass, high surface-to-mass exposure of structural elements.

Manufactured buildings can be defined as those structures that are built at a factory and then trucked to a job site. These building are quite light with little mass. Where a stick-built home uses 2 × 4 or 2 × 6 lumber, the manufactured home uses 1 × 2 and 2 × 2 lumber. These buildings use galvanized strapping to give required strength. In any case, these buildings burn quickly and collapse equally fast.

(A)

(B)

Figure 13-23 (A) The decorative roof assembly is a Type V wood frame structure while the occupancy space is Type II noncombustible. (B) This building uses two types of construction.

Figure 13-24 This lightweight steel home is built similar to a Type V. OSB sheeting gives the steel rigidity to torsional loads like wind.

Relationship of Construction Type to Occupancy Use

Before considering the basic types of construction, many officials and builders first look at the anticipated use of the building—its occupancy type. **Occupancy classifications** are called many different names around the country, but they are usually broken down into five basic arenas: residential, commercial, business, industrial, and educational. Each of these general occupancies has a number of hazards that firefighters must understand, **Table 13-4.** Remember, a building may have been built for one type of occupancy only to be sold and converted to another occupancy type for which it may not have been designed. Firefighters should go out and explore the buildings in their community.

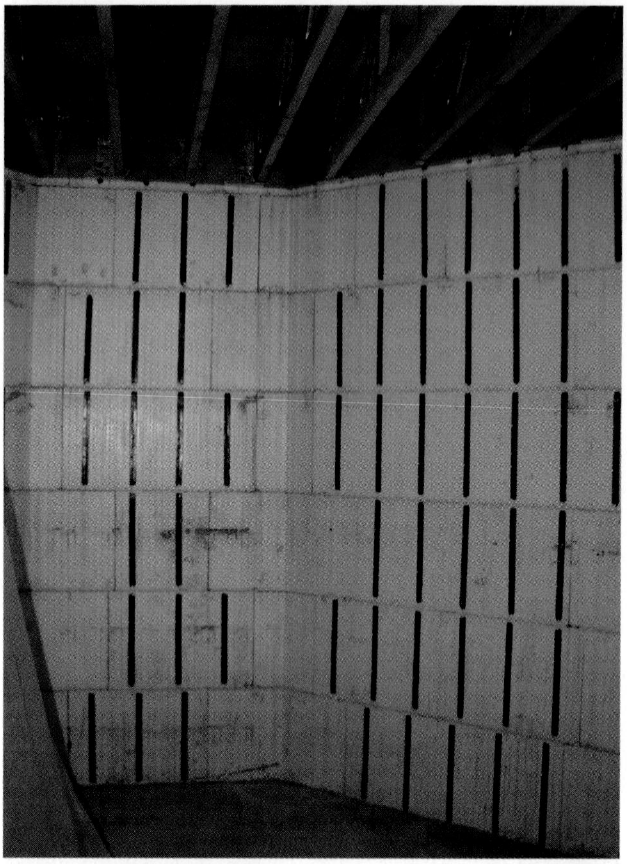

(A)

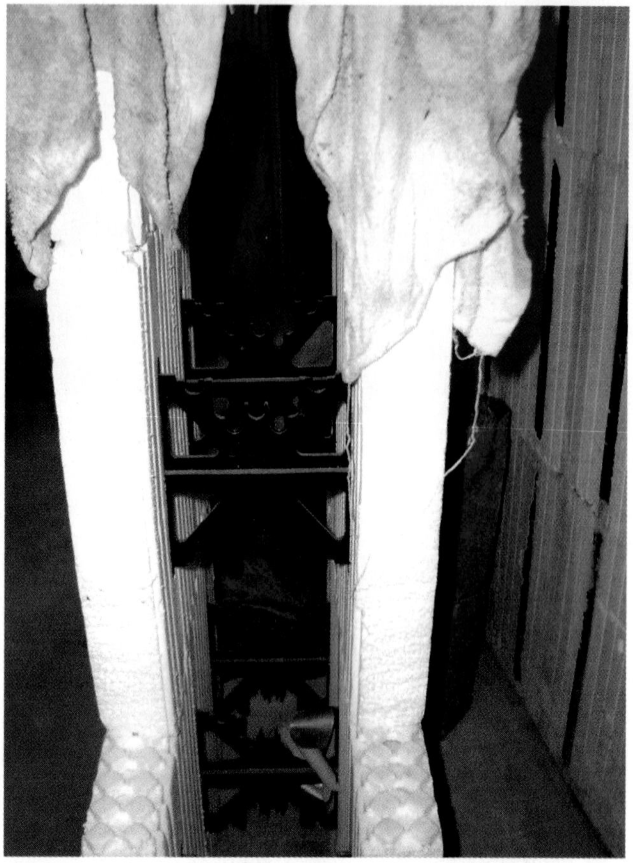

(B)

Figure 13-25 (A) This wall is a load-bearing foam block unit filled with a lightweight concrete mud mix. (B) Note the black plastic spacers that will fail early in a fire.

Typical Hazards Associated with Occupancies

OCCUPANCY	TYPE OF CONSTRUCTION	HAZARDS
Residential	Type V, most common	Fire loading, truss construction, owner alterations, rapid fire extension in void spaces
Commercial	Type III, most common	Fire loading, truss construction, rapid fire extension in void spaces, unknown occupancy change
Educational	Type II, most common	Unprotected structural steel, collapse, high fire load in some areas
Business	Types II and III, most common	Unknown change in occupancy, high fire load, difficult to ventilate
Industrial	Types I and II, most common	Hazardous materials, difficult to ventilate

TABLE 13-4

COLLAPSE HAZARDS AT STRUCTURE FIRES

It cannot be overstated that firefighters have to understand the buildings in their jurisdiction. Constant reading, study, and site visits will help them "read" buildings. Reading buildings is essential to anticipating collapse proactively. This section addresses some specific collapse threats that the fire service has experienced throughout history and the importance of understanding buildings and how they react at structure fires.

Figure 13-26 Wood trusses provide a large surface-to-mass ratio, fuel load, and void spaces—three of the worst conditions a firefighter will encounter during structural firefighting operations.

Figure 13-27 A typical parallel chord truss. The gusset plates on this truss are pressed into the wood. In addition to the decomposing of the wood element, the light-gauge steel plates will deform and pull away from the wood under fire conditions.

Trusses

Truss roof collapses have killed many firefighters. As stated previously, a truss is actually a fake beam. A truss uses geometric shapes (the triangle) to create a structural element similar to a beam. A wood truss can actually be stronger than a like-sized solid wood beam, and does so with less material, **Figure 13-26.** It is this loss of material and subsequent increase in exposed surface area that make them so vulnerable during fires. Trusses rely on each and every part of the truss to carry a portion of the imposed load. Like a beam, the top of the truss (called the top chord) is typically under a compressive force. The bottom chord is under tension. In between the two chords, connecting members (the web) transfer the two forces creating stress and strain. Failure of one part of the truss will likely cause the whole truss to fail. This distributes the weight of the failed truss to other trusses—which may not have the capacity to take that weight—thereby starting a domino effect collapse. Trusses come in many styles and shapes. Bowstring truss, parallel chord truss, and open web joists are some of the more common names. Trusses are also classified by the type of material used in assembly.

Wood Trusses

Wood trusses are an assembly of many pieces of wood. Some may even be press-glued particles. These pieces are connected using **gusset plates**. A gusset plate is a simple galvanized steel plate (very thin) with perforations punched into the plate. The perforations are used to pierce into wood fibers to hold pieces together. These perforations only penetrate the wood a fraction of an inch (3/8 inch is typical), **Figure 13-27.** During fires, the steel gusset heats up and transfers heat into the very wood fibers that are being held. If the heating is slow (like a smoke-filled attic), the wood decomposes, allowing the gusset plate to fall out. If the heating is fast, like sudden exposure to flame, the steel expands too quickly for the wood and the gusset simply pops out. Either way, truss failure is imminent. Sometimes the truss gingerly stays together because of the weight of roofing or flooring materials, yet a sudden force, like a walking firefighter, will cause the truss to disassemble and suddenly collapse.

Wood trusses are mass-produced at a factory where quality control may not be adequate. Further, the truss gusset plates may vibrate or be damaged while being delivered to a job site. Once on the job site, contractors may use shortcuts to lift the truss into position, furthering the damage to gusset plates.

Steel Trusses

Steel trusses are no less susceptible to collapse than wood trusses. Like wood, steel trusses are an assembly of pieces—typically angle iron for the chords and cold-drawn round stock for the web. The pieces are tack-welded together to form the truss unit. While not a true joist by definition, many call the common steel truss an open web steel joist, **Figure 13-28.** The term *bar joist* is also used to describe an open web steel joist. These trusses expose a large surface area to heat during fires. Given the lack of mass, the truss heats quickly and will soften and expand. The expansion can cause wall movement. (Remember, masonry walls must be loaded axial with compressive force.) Lateral movement can cause wall collapse. If the wall does not move, the steel truss will twist and buckle to allow expansion. It is very important to keep steel trusses cool.

Figure 13-28 Unprotected open web steel joists present a large surface area to absorb the heat of a fire, expand, and collapse. Structural steel will lose 50 percent of its strength at temperatures of 1,000°F.

Figure 13-29 Lightweight floor truss systems have many void spaces. This could be called "horizontal balloon frame."

Void Spaces

Trusses create large void spaces. The area between the chords of trusses will allow fires to spread horizontally, **Figure 13-29.** Some codes require fire stopping in floor truss spaces but may allow wide-open attic spaces. Fires can start in void spaces due to electrical and other utility problems. In Type III ordinary construction, voids are numerous. Some voids may pass through masonry walls, causing fire spread from one store to the next in a row of buildings. The obvious collapse danger with void spaces is that the fire may be undetected with simultaneous destruction of structural elements.

Roof Structures

The roof of a building can be flat, pitched, or inverted. Many factors help determine how and why the roof is built the way it is. Sometimes the roof is designed just to hide rooftop HVACs. Other times the roof shape is designed to shed snow, accommodate a vaulted ceiling, or merely give the building character, **Figure 13-30.** As it relates to

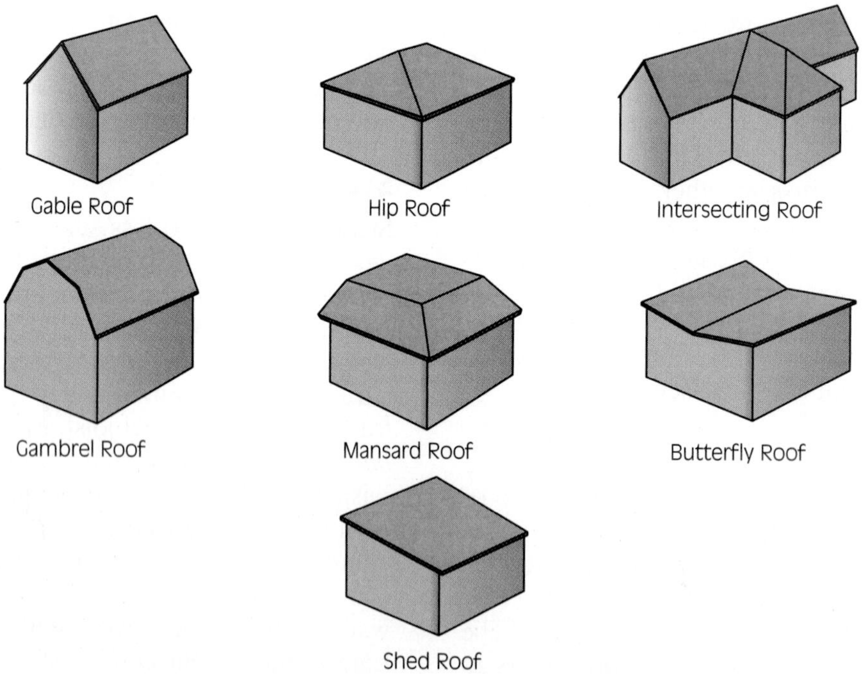

Figure 13-30 Some common roof framing styles used in wood frame or ordinary construction.

structural collapse, the roof style may allow a large volume of fire to develop. Other roofs, like the mansard, have many concealed spaces. Dormers are protrusions from a roof structure. Dormers can be used to introduce daylight into a roof space that is converted into a living space, **Figure 13-31.** Other dormers are actually aesthetic (false) and can fool ventilation crews attempting to relieve heat from a roof space.

Stairs

First arriving firefighting crews rely on internal stairways to help gain access for rescue and fire attack. For years, firefighters have found stairways to be durable and a bit stronger than other interior components. This is a dangerous assumption in newer wood frame buildings. Stairs are now being built offsite and simply hung in place using light metal strapping, **Figure 13-32.** Additionally, stairs are being made using lightweight engineered wood products that fail quickly when heated. Remember, press-glued wood chip products can fail from the heat of smoke—no flame is required.

Figure 13-31 An internal view of dormers in a mansard roof. Fire can run through the many voids between the exterior and the roof.

Parapet Walls

A **parapet** wall is the extension of a wall past the top of the roof. Parapets are used to help hide unsightly roof equipment and HVACs and give a building a finished look. Typically masonry, these walls are free-standing with little stability. Collapse may be caused by the failure of the roof structure, **Figure 13-33.** Business owners hang signs, utility connections, and other loads on the parapet. During a fire, the steel cables and bolts holding these will weaken and subsequently pull down the parapet, **Figure 13-34.**

Figure 13-32 This prefab stair assembly is hung in place by thin metal strapping. Note the staples and plastic shims that can quickly fail under fire conditions.

Figure 13-33 This is the scene of a typical "parapet" wall failure common in Type III ordinary construction.

Figure 13-34 This electrical service entrance and weather head may be the eccentric load causing an early failure of this parapet wall.

Collapse Signs

Firefighters must rely on building material knowledge, building construction principles, and an understanding of fire effects on buildings in order to predict or anticipate collapse. Waiting for a visual sign that a building will collapse is dangerous, especially in newer buildings. There are, however, some factors and observations that can be used to help anticipate collapse. These include:

- Overall age and condition of the building
- Deterioration of mortar joints and masonry
- Cracks, in anything
- Signs of building repair including reinforcing cables and tie-rods
- Large open spans
- Bulges and bowing of walls
- Sagging floors
- Abandoned buildings
- Large volume of fire
- Long firefighting operations—remember gravity?
- Smoke coming from cracks in walls
- Dark smoke coming from truss roof or floor spaces (Brown smoke indicates that wood is being heated significantly; black smoke means combustibles have ignited or are near ignition.)
- Multiple fires in the same building or damage from previous fires

Buildings under Construction

Buildings are especially unsafe during construction, remodeling, and restoration. The word *unsafe* applies not only to fire operations but also to rescues, odor investigations, and on-site inspections. Buildings need only meet fire and life safety codes when they are completed. During construction, many of the protective features and fire-resistive components are incomplete. Additionally, stacked construction material may overload other structural components. This is not to say contractors are using unsafe practices, but to underscore that exposed structural elements, incomplete assemblies, and material stacks will contribute to a rapid collapse if a fire were to develop.

> **Streetsmart Tip** **Beware of Buildings under Construction:** A building as a complete unit has a number of interdependent parts. During construction, these parts may not be fully connected (steel), may not be at their full design strength (concrete), and may lack any type of fire protection (gypsum board or concrete). Also, many structural elements may be held in place by scaffoldings, false work, or forms. Because of this, a building under construction is exposed to early collapse due to the effects of fire or other elements such as high winds.

Historical building restoration and general remodel projects in buildings are similar to buildings under construction. Firefighters may find temporary shoring up of walls, floors, and roofs while other structural components are being updated, replaced, or strengthened. Contractors may use simple 2 × 4s to temporarily shore up heavy timber, leading to disastrous results during fire conditions. The best approach for firefighters to take when responding to fires in buildings under construction is to be defensive. They should make sure everyone is out and accounted for and then attack the fire from a safe location. A building under construction can be replaced—a firefighter's life cannot.

Preparing for Collapse

There are no time limits for firefighting operations within a building. Tests have shown that the age-old "twenty-minute rule" used by previous fire officers is no longer accurate. Roofs and walls can collapse within minutes of fire involvement given certain conditions. An overloaded (due to improper storage or other factors) truss can collapse immediately when heated.

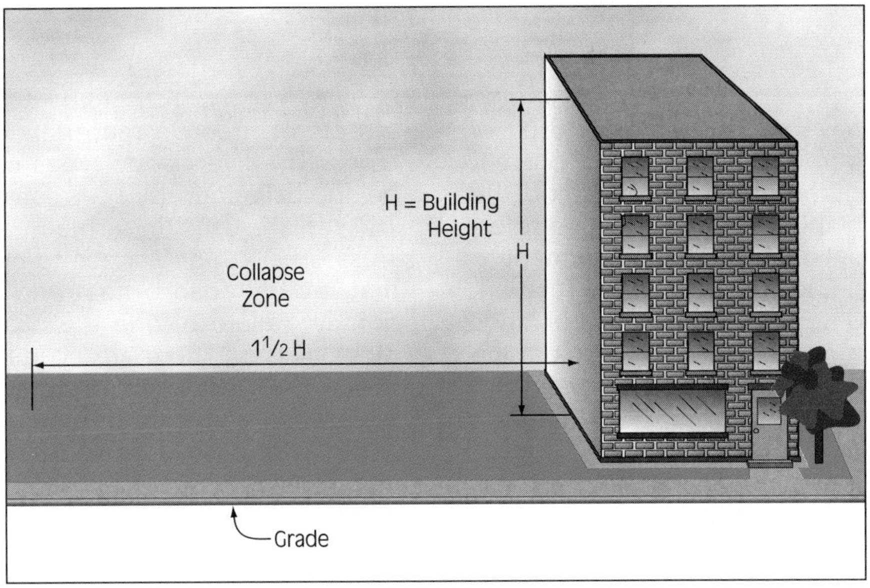

Figure 13-35 A minimum collapse zone should be 1½ times the height of the building.

Streetsmart Tip **Collapse:** Every firefighter must understand two rules about structural collapse during fire operations. The first is that the potential for structural failure during a fire always exists. Do not set artificial time limits based on experience. The second rule is to establish a collapse zone, as shown in **Figure 13-35,** which is an area around and away from a building where debris will land if the building fails. As an absolute minimum this distance must be at least 1½ times the height of the building. The walls may crumble into a pile or they may tip out the full height of the building. Also you need to provide extra room for cascading debris.

Once a building has been searched for occupants, the risks firefighters take to control the fire should be reduced—after all, it is now a property issue. Many firefighters have been killed fighting interior fires, only to have the building torn down after the investigation. Outside (defensive) firefighting operations can be equally dangerous if firefighters wander into the **collapse zone**, **Figure 13-36.**

Figure 13-36 These photos show the effects of fire on a masonry wall. Note the debris and distance the bricks fell away from the building. Firefighters should always establish a collapse zone, as shown in Figure 13-35.

Lessons Learned

Many firefighters have been killed as a result of building collapse from structural fires. To prevent future deaths, firefighters must understand the buildings in which they fight fires. This understanding comes from a long-term commitment to read and study building construction information. Additionally, firefighters must get into buildings within their jurisdictions to survey and explore the way buildings are assembled, remodeled, and used in the real world. Knowledge of building construction starts with an understanding of loads, forces, and materials found in the structural makeup of buildings. Firefighters also study the effects of fires on materials and construction types. The five classic types of construction are being challenged by new construction methods. Trusses are used in virtually all new buildings. Trusses have high surface-to-mass characteristics that rapidly absorb heat and subsequently fail quickly. Failure of one truss can cause failure of other trusses. There are no rules for how long a building will last while on fire. Many factors determine when materials and construction design fail and gravity pushes down the building. Buildings under construction are losers from a firefighting point of view—they collapse quickly.

KEY TERMS

Axial Load A load passing through the center of the mass of the supporting element, perpendicular to its cross section.

Balloon Frame A style of wood frame construction in which studs are continuous the full height of a building.

Beam A structural member subjected to loads perpendicular to its length.

Cantilever Beam A beam that is supported at only one end.

Chord The top and bottom components of a beam or truss. The top chord is subjected to compressive force; the bottom chord is subjected to tensile force.

Collapse Zone The area around a building where debris will land when it falls. As an absolute minimum this distance must be at least 1½ times the height of the building.

Column A structural element that is subjected to compressive forces—typically a vertical member.

Compression A force that tends to push materials together.

Concentrated Load A load applied to a small area.

Continuous Beam A beam that is supported in three or more places.

Dead Load The weight of the building materials and any part of the building permanently attached or built in.

Design Load A load the engineer planned for or anticipated in the structural design.

Distributed Load A load applied equally over a broad area.

Eccentric Load A load perpendicular to the cross section of the supporting element that does not pass through the center of mass.

Fire Load The amount of heat energy released when combustibles burn in a given area or building—expressed in British thermal units (Btus).

Fire Resistive The capacity of a material to withstand the effects of fire.

Fire-Resistive Rating The time in hours that a material or assembly can withstand fire exposure. Fire-resistive ratings are usually provided for testing organizations. The ratings are expressed in a time frame, usually hours or portions thereof.

Fire Stopping Pieces of material, usually wood or masonry, placed in stud or joist channels to slow the extension of fire.

Girder A large structural member used to support beams or joists—that is, a beam that supports beams.

Gusset Plate A connecting plate used in truss construction. In steel trusses, these plates are flat steel stock. In wood trusses, the plates are either light-gauge metal or plywood.

HVAC Acronym for heating, ventilation, and air-conditioning unit. HVACs are typically a rooftop unit on commercial buildings. Buildings may have one or dozens of these units.

Impact Load A load that is in motion when it is applied.

Joist A wood framing member that supports floor or roof decking.

Lintel A beam that spans an opening in a load-bearing masonry wall.

Live Load The weight of all materials and people associated with but not part of a structure.

Load-Bearing Wall Any wall that supports other walls, floors, or roofs.

Loading The weight of building materials or objects in a building.

Mortar Mixture of sand, lime, and portland cement used as a bonding material in masonry construction.

Occupancy Classifications The use for which a building or structure is designed.

Parapet The projection of a wall above the roofline of a building.

Platform Framing A style of wood frame construction in which each story is built on a platform, providing fire stopping at each level.

Purlins A series of wood beams placed perpendicular to steel trusses to help support roof decking.

Rafter A wood joist that is attached to a ridge board to help form a peak.

Shear A force that tends to tear a material by causing its molecules to slide past each other.

Simple Beam A beam supported at the two points near its end.

Spalling Deterioration of concrete by the loss of surface material due to the expansion of moisture when exposed to heat.

Surface-to-Mass Ratio Exposed exterior surface area of a material divided by its weight.

Tension A force that pulls materials apart.

Torsion Load A load parallel to the cross section of the supporting member that does not pass through the long axis. A torsion load tries to "twist" a structural element.

Truss A rigid framework using the triangle as its basic shape to emulate a beam.

Type I, Fire-Resistive Construction Type in which the structural members, including walls, columns, beams, girders, trusses, arches, floors, and roofs, are of approved noncombustible or limited combustible materials with sufficient fire-resistive rating to withstand the effects of fire and prevent its spread from story to story.

Type II, Noncombustible Construction Type not qualifying as Type I construction, in which the structural members, including walls, columns, beams, girders, trusses, arches, floors, and roofs, are of approved noncombustible or limited combustible materials with sufficient fire-resistive rating to withstand the effects of fire and prevent its spread from story to story.

Type III, Ordinary Construction Type in which the exterior walls and structural members that are portions of exterior walls are of approved noncombustible or limited combustible materials, and interior structural members, including wall, columns, beams, girders, trusses, arches, floors, and roofs, are entirely or partially of wood of smaller dimension than required for Type IV construction or of approved noncombustible or limited combustible materials.

Type IV, Heavy Timber Construction Type in which exterior and interior walls and structural members that are portions of such walls are of approved noncombustible or limited combustible materials. Other interior structural members, including columns, beams, girders, trusses, arches, floors, and roofs, shall be of solid or laminated wood without concealed spaces.

Type V, Wood Frame Construction Type in which the exterior walls, bearing walls, columns, beams, girders, trusses, arches, floors, and roofs are entirely or partially of wood or other approved combustible material smaller than the material required for Type IV construction.

Undesigned Load A load not planned for or anticipated.

Veneer A covering or facing, not a load-bearing wall, usually with brick or stone.

Web The portion of a truss or I beam that connects the top chord with the bottom chord.

REVIEW QUESTIONS

1. What are three ways loads are imposed on materials?
2. List the three types of forces created when loads are imposed on materials.
3. Name three kinds of beams.
4. Explain the effects of fire on steel structural elements.
5. How does a masonry wall achieve strength?
6. List and define the five common types of building construction. Give an example of each type that is located in your district or response area.
7. List the three parts of a truss and explain what forces are being applied to each.
8. List three buildings in your district or response area that have truss construction.
9. List how fire affects the four more common building materials in use today.
10. Diagram and label four different roof shapes.
11. List and describe eight conditions or observations that might indicate potential structural collapse.

Additional Resources

Brannigan, Francis, *Building Construction for the Fire Service,* 3rd ed. National Fire Protection Association, Quincy, MA, 1992.

"High Rise Office Building Fire, One Meridian Plaza, Philadelphia, PA," Technical Report Series, Report #049, U.S. Fire Administration, Washington, DC.

"Without Warning, A Report on the Hotel Vendome Fire," Boston Sparks Association.

Building Construction for Fire Suppression Forces—Wood & Ordinary. National Fire Academy, National Emergency Training Center, Emmittsburg, MD, 1986.

Angle, James; Harlow, David; Gala, Michael; Maciuba, Craig; and Lombardo, Williams; *Firefighting Strategies and Tactics.* Delmar Learning, a part of the Thomson Corporation, Clifton Park, NY, 2001.

Diamantes, David, *Fire Prevention: Inspection & Code Enforcement,* 2nd ed. Delmar Learning, a part of the Thomson Corporation, Clifton Park, NY, 2003.

CHAPTER 14

LADDERS

Frank J. Miale, Retired Battalion Chief, New York City Fire Department and Lake Carmel Volunteer Fire Department

OUTLINE

- Objectives
- Introduction
- Ladder Terminology
- Ladder Companies
- Types of Truck-Mounted Ladders
- Types of Ground or Portable Ladders
- Use and Care of Portable or Ground Ladders
- Maintenance, Cleaning, and Inspection
- Ladder Safety
- Ladder Uses
- Ladder Selection
- Special Uses
- Safety
- Miscellaneous Ladder Information
- Ladder Skills
- Raising Skills
- Lessons Learned
- Key Terms
- Review Questions
- Additional Resources

STREET STORY

With newer and more technical equipment being introduced to the fire service almost on a daily basis, it is hard to stay focused on basic firefighting. Ground ladders are as basic as you can get; however, they are still as important today as they ever were. When you place a ladder to a window, you have an instant exit for both occupants and firefighters. Place a roof ladder on a roof and you have a safe place to start ventilation. Aside from the obvious gravity-defying uses for ladders, others are breaking out windows, and ice rescue. Of course, do not forget about the occasional cat stuck in a tree. As long as rescuing cats does not interfere with more serious firefighting duties, this type of rescue provides training and good public relations. The more you use ladders, the more uses you will find for them.

New firefighters often wonder why they must practice raising ground ladders over and over, almost to the point where they no longer have to consciously think about what they are doing. That is exactly the point. Anytime, day or night, snow or rain, heat or cold, firefighters must use ground ladders with practiced skill and also must understand the features and limitations of them.

This concept became apparent to me several years ago at the scene of a 2½-story wood frame occupied structure fire. As my crew approached the fireground, we noticed a man visible through the lower sash of a double-hung window on the second floor. His outline was obscured by thick black smoke with an orange tint to it, which meant that his seconds were numbered. From a different vantage point, another crew observed the same scene. In an instant, both groups bolted for their respective ladder trucks to retrieve the appropriate ground ladders needed to effect the rescue. Both crews returned with exactly the same sizes and types of ground ladders for the job. This was no coincidence. The members of both crews had received their training at different times and by different instructors, but the proper use of ground ladders had been embedded in the minds of both crews, so they responded to the task in the same way.

Keep in mind that the rescue was successful because of ground ladders, not a thermal imaging camera, not an articulated telescoping tower ladder, not any other specialized expensive equipment—just ground ladders and firefighters who knew what they were doing. Of course, there are situations that call for special equipment, but the proper knowledge of and use of ground ladders will save lives and reduce property damage at almost every fire.

—Street Story by Peter F. Kertzie, Captain, Buffalo Fire Department, Buffalo, New York

OBJECTIVES

After completing this chapter, the reader should be able to:

- Name the parts of a ladder.
- Describe the many functions for which a ladder can be used.
- Name the different types of mounted ladder apparatus.
- Describe the function of the different types of ground ladders.
- Describe the care of ladders.
- Cite maintenance, cleaning, and inspection functions for ladders.
- Exhibit ladder operation safety.
- Name different types of ladder uses.
- Describe the ladder selection process.
- Describe the concepts behind different ladder-raising techniques.
- Cite safety concerns of ladders and their use.
- Demonstrate skills associated with ladders, such as raising, leg locks, rope handling, mounting and dismounting of ladders, and use of roof ladders.
- Describe fundamentals of ladder placement.
- Determine how far away from a building a ladder should be placed.

INTRODUCTION

Ladders are used in the fire service for many purposes besides providing access to elevated locations. Used with ropes, they can assist in shoring a wall as an emergency support. When bound and tied together and placed on their side with tarps draped over them, they can be used to build emergency walls for liquid pools. Covered with tarpaulins, they can be used to channel water to drains. Placed over openings, they can act as barriers to prevent accidental falls into holes. When used with handlines, they can serve to provide elevated stream penetration. Some types of doors can even be forced open with ladders, a technique that was used as a standard form of forcible entry in the past. When lashed together in an A-frame setup, they can be used as a hoist point for pulley placement. These applications are described in this chapter.

Ladders were originally constructed of wood. As the need for greater height became evident, the supporting solid wood beams were replaced with newer technology truss-type beams. This truss construction is a design that removes the nonsupport portions of a solid structural member without reducing strength but significantly reducing weight. While this design permitted longer ladders to be constructed, there were still limits. Introduction of lightweight high-strength aluminum replaced the wood, first in solid beam construction, then in truss construction. Later, higher strength aluminum alloys were developed and later still, fiberglass. Along with the change in the ladder's material makeup, new design technology continued to meet the needs. For instance, several smaller ladders were arranged to be adjustable by sliding them against one another through connecting channels. These extension ladders were designed to reach a range of heights. As ladders got longer and longer, additional features such as tormentor poles were designed into the ladder to assist in raising.

Ladders can be used for many purposes besides climbing. This chapter describes many innovations that have evolved from hands-on practical use.

LADDER TERMINOLOGY

A ladder is defined as "a structure consisting of two long sides crossed by parallel rungs, used to climb up and down" and as "a means of ascent and descent." Today, fire departments boast many different ladder types and lengths.

There are many parts to a ladder. It is important for every firefighter to know each part by its name, **Figure 14-1**.

In different departments, the same ladder part might have a different name.

> **Safety** Common terminology usage will reduce miscommunication when passing along information or requesting assistance.

Parts of a Ladder

- *Beam:* The side of the ladder. It is the rail that runs the full length of the ladder from top to bottom from which rungs span, **Figure 14-1A.**
- *Bed section:* The part of the ladder that is the foundation, usually the part of the ladder that is in touch with the ground or attached to the body of an aerial ladder truck. It is the section from which all other sections (see Fly section entry) are raised in extension ladders. This is also called the *bed ladder.*

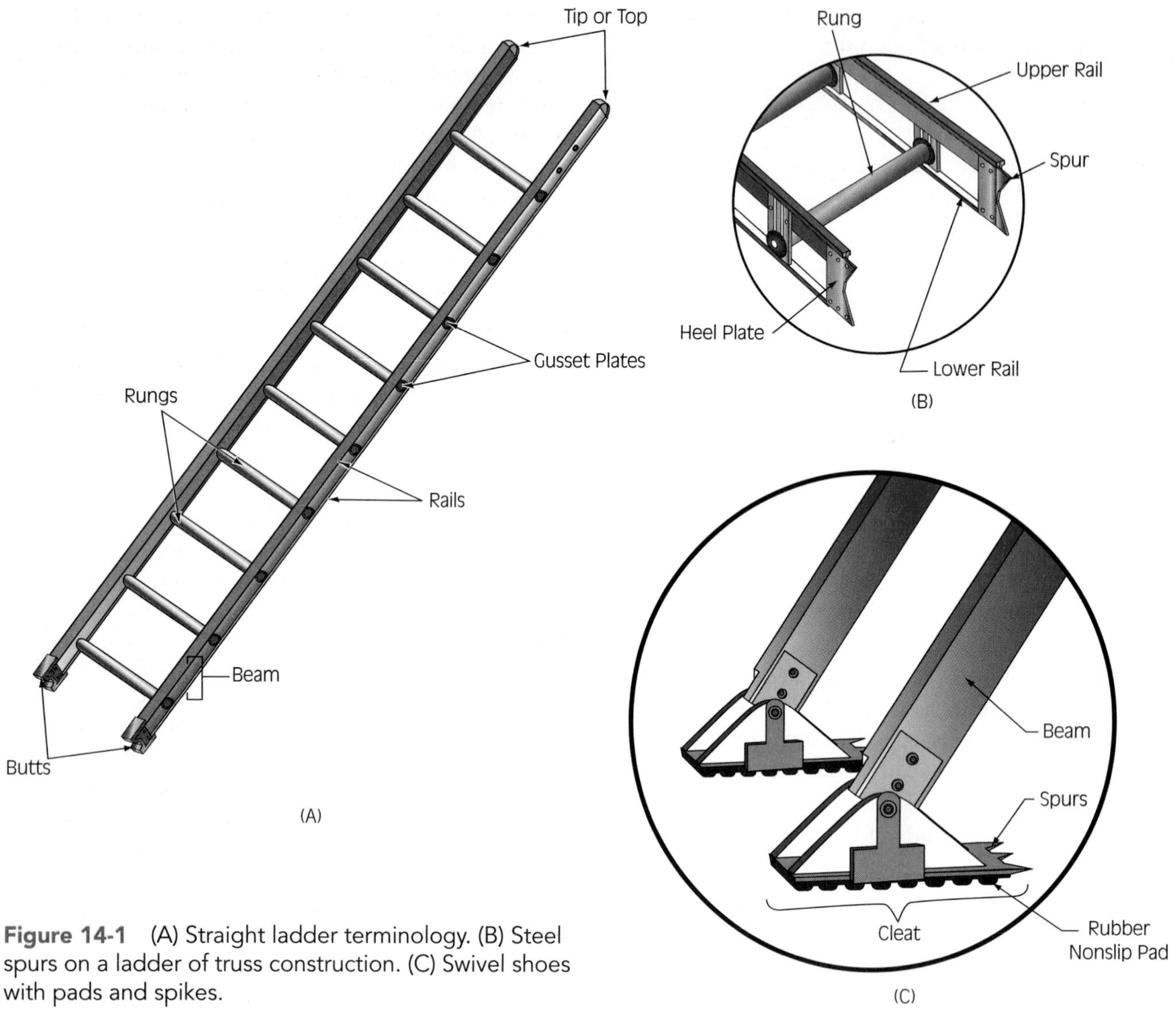

Figure 14-1 (A) Straight ladder terminology. (B) Steel spurs on a ladder of truss construction. (C) Swivel shoes with pads and spikes.

- *Heel (also called foot, base, or butt):* The bottommost part of a ladder. It is usually a reinforced section with points, spurs, or rubber pads to reduce slipping on various surfaces.
- *Spurs, spikes, cleats, shoes, and butt plates:* The pointed shoes that are attached to the base of a ladder to dig into the surface and prevent slippage during use, **Figure 14-1B**.
- *Dogs, pawls, rung locks, or ladder locks:* The mechanisms in an extension ladder that ride up along with a fly section and engage a rung of the section from which they extend to prevent retraction through the use of a spring-loaded lock.
- *Fly section:* The section (or sections) of an extension ladder that is (are) raised. Also called the *fly ladder.*
- *Pads:* Nonslip pieces of rubber or plastic that attach to the bottom of a ladder, usually in a swivel-type foot designed to lie flat against the ground or floor to prevent slippage on smooth surfaces, **Figure 14-1C**.
- *Guides/channels:* Channels of the bed ladder that permit the fly sections to ride up and maintain stability.
- *Halyard:* The cable or rope, made of nylon, hemp, or steel, that is used to raise or lower the fly section out of or into the bed section through the use of pulleys.
- *Sensor label:* A heat-sensitive label affixed to the ladder to alert firefighters that the ladder has been exposed to a potentially damaging heat level and that testing should be performed before it is used again.
- *Hooks:* Retractable hooks that permit certain types of ladders to be placed on a slanted roof surface and used for footing stability.
- *Rung locks:* See Dogs.
- *Pawls:* See Dogs.

- *Protection plates:* Reinforced metal that is built up at chafing points to avoid weakening created by rubbing and friction wear.
- *Pulley:* A wheel with a groove through which the halyard passes. It is used for raising or lowering the fly sections of extension ladders.
- *Rails:* Running lengthwise, the upper and lower surfaces of the beams.
- *Rungs:* The "steps" of a ladder that connect one beam to the other. Generally round, they can also be flattened on the upper side for more secure footing.
- *Stops:* Limiters built into the bed section to prevent the extension fly sections from being overextended. They are found in the shape of solid blocks or angled metal.
- *Tie-rods:* Found on wooden ladders, these metal rods secure the two beams and prevent them from spreading apart when the rungs are doweled into the beams. These rods prevent the rungs from pulling out, which would result in a ladder collapse.
- *Tip:* The top of the ladder. In an extension ladder, it would be the top of the fly section that attains the greatest height.

LADDER COMPANIES

In the past, most fire departments employed a single apparatus and it was the "fire department." Eventually as the number of fire companies grew, they became specialized to perform specific duties. The tasks required were logically associated with the apparatus. Ladder companies are the apparatus that carry ladders and other devices, tools, and personnel to upper levels. **Tower ladders** and **articulating boom ladders** are included in this category, although technically, they are not ladders. These units, however, carry ladders that are more traditional in construction and use.

As more and more fire departments created special units to serve particular functions, the personnel assigned to those apparatus were trained to operate the equipment on that vehicle. With specialized tools designed to complete a specific function, a certain "division of labor" developed in which firefighters on certain types of apparatus performed designated functions. Engine companies carried hose, nozzles, and appliances that served the water delivery function. Ladder companies, originally developed to gain access to upper floors, became responsible for tasks associated with entry. That included forcible entry of the fire area, access to the roof for ventilation operations, and access to upper floors for entry, search, and rescue. Eventually, the need for access to higher levels of the building became evident. Higher reaching ladders, better designed tools, and a wider measure of knowledge became the trademark of the ladder company and its personnel.

TYPES OF TRUCK-MOUNTED LADDERS

Many types of ladder trucks are used by the fire service today. Each was invented and designed to serve a particular function and has been named by its function.

Aerial Ladder

An aerial ladder is an apparatus-mounted ladder capable of reaching heights of generally up to 100 feet with some that even go beyond 110 feet, **Figure 14-2**. An aerial ladder with a reach capability of 144 feet was once tested in field service. It was taken out of service after a short time because of limited use, safety concerns, and maintenance.

Because of the heavy-duty use and reach, most of these ladders are very heavy and are constructed of some combination of steel, aluminum, and other metallic alloys designed for strength using a truss-style construction for maximum strength.

Aerial ladders are designed so that various sections slide out from one another to produce greater reach, **Figure 14-3**. Each **fly ladder** section is designed to overlap the section below sufficiently to maintain stability. It is usually made of a steel/aluminum alloy for lighter weight. The lowest section of the ladder is called the **bed ladder**. It is usually made of steel and is attached to the main body of the apparatus by a combination of pins and pistons that allow it to be raised

Figure 14-2 Rearmount aerial ladder.

Figure 14-3 Extended aerial ladder. Note the sliding sections.

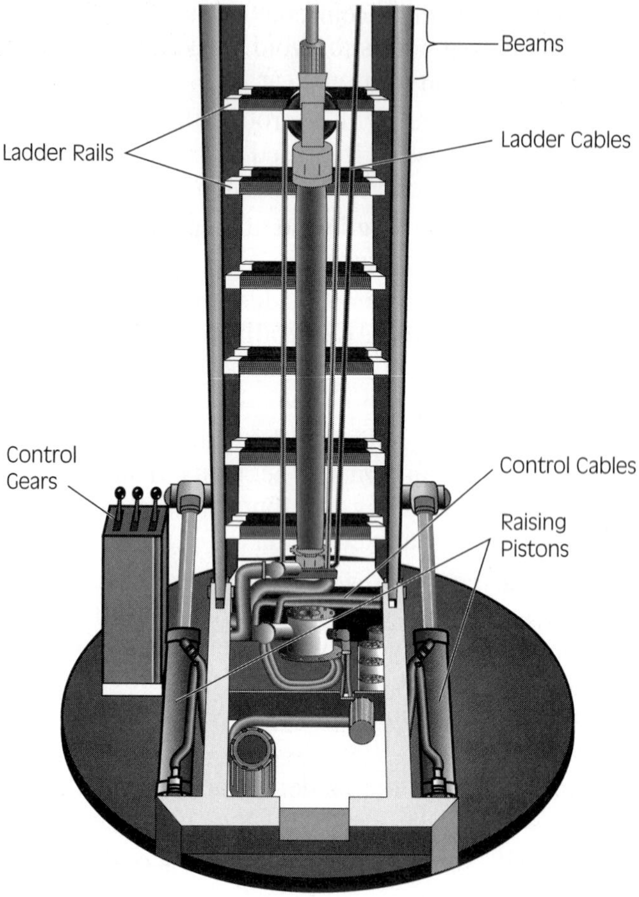

Figure 14-4 Aerial ladder raising mechanisms as seen from under a raised bed ladder.

from the horizontal to the vertical positions. The bed ladder is attached to a **turntable**, a 360-degree rotatable platform that is attached to the framework of the apparatus. The fly sections of an aerial ladder are the sections that extend out from the bed ladder and one another to reach whatever height is desired.

The bed ladder, including the **nested** fly sections, is raised out of the bed of the apparatus through the use of **hydraulic pistons**, **Figure 14-4**. The turntable rotates the ladder to align with the target, and the fly sections are extended to reach the target.

Proper terminology for use of these ladders is very important, **Figure 14-5**. The same word must mean the same thing to all people. **Table 14-1** describes the terms that should be used to identify the operation desired.

> **Safety** To avoid confusion firefighters should use standard terminology when working with ladders.

The word *raise* should refer only to the raising of the bed ladder out of the bed and not to making the ladder longer. The order to make the ladder longer would be to *extend* the ladder.

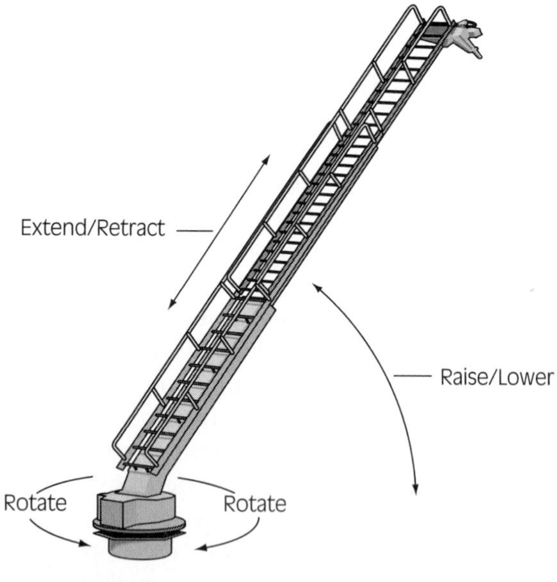

Figure 14-5 Ladder positioning terminology.

Terminology for Ladder Raising	
Raise:	Lifts the ladder from horizontal to vertical angles.
Lower:	Lowers the ladder from vertical angles to horizontal.
Rotate:	Turns ladder right and left.
Extend:	Increases the length and reach by extending fly sections.
Retract:	Decreases length and reach by nesting the fly sections.

TABLE 14-1

Note In many cases, the operator of the ladder might not be in visual contact with the target and is being directed by a forward observer. In cases such as this, common terminology becomes essential.

The ladder's reach is achieved through the use of cables and pulleys. Using the physics of mechanical advantage associated with block and tackle, a continuous cable is spun off a drum and woven through the various fly sections of the ladder, **Figure 14-6**. As the first fly section moves out from the bed ladder, the cable and pulley configuration pulls out the second and third sections at the same time. When in operation, each fly section moves faster than its parent section. This permits quick placement of the ladder at the target. Depending on the pulley and cable arrangement, the upper fly section can be moving three times faster than the lowest fly section.

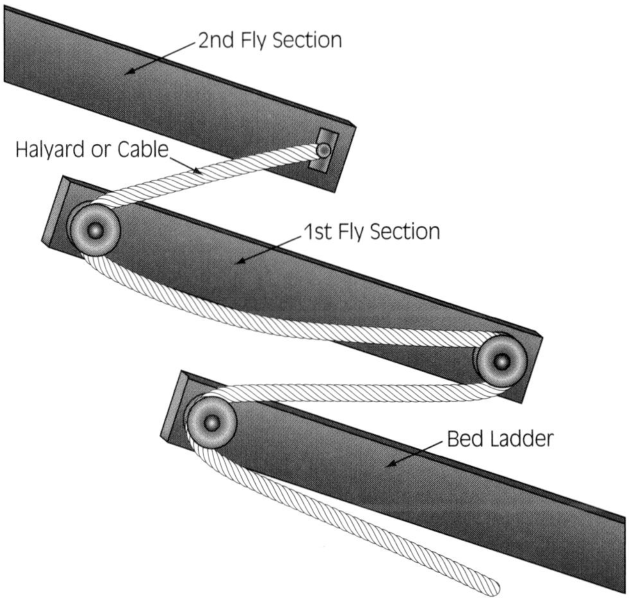

Figure 14-6 Typical multifly halyard hoisting pulley arrangement. As the halyard is pulled, the fly sections are hoisted in unison with all the rungs aligning at each locking position.

Once the mainstay of a fire department's high elevation needs, the aerial ladder has become only one of the fire service's high elevation tools. Today's fire service has other apparatus such as tower ladders and articulating boom ladders that are equally capable, and, in some cases, better designed for increased versatility. Newer designs incorporate features of both the tower and the articulating boom.

Tower Ladder

The tower ladder is often found as a standard piece of equipment in any moderate to large fire department. It is still referred to as a ladder company because it carries a complement of ground and portable ladders. Performing many of the same functions as the aerial ladder, in many departments it has replaced older aerial ladders as they were retired from service.

The tower ladder, **Figure 14-7,** has a telescopic boom with a mounted basket capable of holding from 750 to more than 1,000 pounds. Like the aerial ladder, it is capable of rotation, extension, and retraction. The fly sections of the boom telescope out of the bed boom section. It has an advantage over the aerial ladder in that it can hold many people at the same time and is capable of many uses including use as a work platform, an observation vantage point, and an elevated water stream appliance. The tower ladder can remove many trapped victims at the same time with limited commitment of resources. One firefighter at the turntable as a safety person and one in the basket can safely remove many victims, including children and small animals from an elevated location such as a window or a

Figure 14-7 Tower ladder.

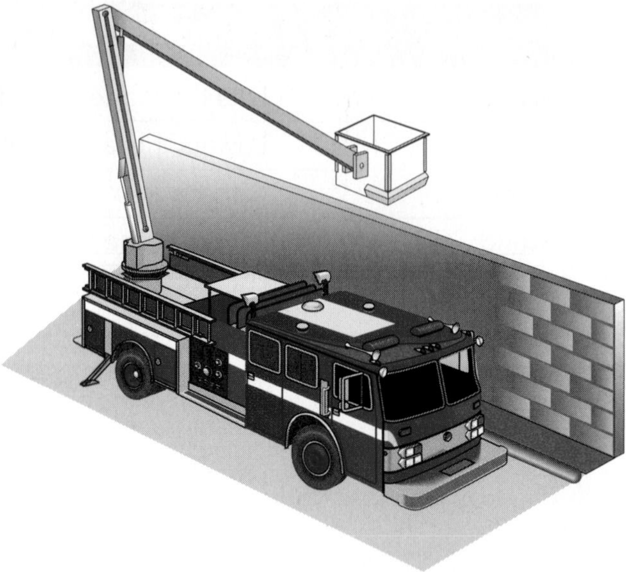

Figure 14-8 A snorkel basket can reach places not possible with other types of apparatus.

rooftop. A ground ladder would generally require the use of a separate firefighter for each victim removed to assist in climbing down the ladder.

In addition, the tower ladder affords people who suffer from a fear of height, psychologically paralyzed victims, and persons with physical and mental disabilities a safe escape.

The tower ladder is a very versatile piece of equipment that achieves many fireground elevation needs. It has a tendency, however, to take slightly longer to place into operation and position than the aerial ladder.

Articulating Boom Ladder

The articulating boom (sometimes known as a snorkel ladder) truck was among the first designs for elevated platform use in the fire service. First appearing in the early 1960s, it broke new ground that eventually led to changes in fire service tactics and equipment design. Some of the original designs had a ladder attached to the booms for escape, but its use proved very impractical because of the angles and accessibility shortcomings. Although not a true ladder, it has been used to replace aging aerial ladders. Newer designs have incorporated elements of the telescopic and articulating boom designs, **Figure 14-8**.

Through the use of several articulating booms, a snorkel ladder uses balance and individual extension and retraction capability to place the bucket into places that are not reachable by tower ladders due to obstructions. By positioning the booms at different angles, the basket can be lowered behind an obstruction. Tower and aerial ladders would be unable to reach the same objective.

An articulating boom ladder can also be used as an elevated water application platform or observation point and can simultaneously remove several victims. The disadvantage of a snorkel-type ladder is that it requires observation of many possible points of contact. Each articulation joint is capable of striking an object, building, electric wires, or other obstacle.

TYPES OF GROUND OR PORTABLE LADDERS

Carried on all ladder truck apparatus is a complement of ground ladders, sometimes referred to as portable ladders. The most common types are straight ladders, extension ladders, and various types of specialized ladders. Although standard ladder lengths are often found on an apparatus, each truck might contain different lengths depending on the department and its needs. Ladders are also carried on engine company apparatus for limited applications.

Straight Ladder

Also referred to as a wall ladder, the straight ladder is a fixed length ladder. Usually found in lengths between 12 and 20 feet, they are generally long enough to gain access into first-floor and second-floor windows, **Figure 14-9**. The surrounding topography of the ground will affect this capability. Beyond the height of the fixed ladders, **extension ladders** may be employed. The straight ladder is generally light, can be carried by one person, and can often be raised by one person using special techniques designed to perform that function. Straight ladders can be used for access, ventilation of upper floor windows, and escape.

Figure 14-9 Straight wall ladder.

In departments without ladder companies, several ladders chosen to serve the needs of the response area are carried on the pumping engines or tankers. Several lengths of ladders can be carried nested into one another and secured to the apparatus by restraining devices. This arrangement maximizes carry capability with minimal storage space use, **Figure 14-10**.

Figure 14-10 Ground ladder nested on the side of a ladder truck.

Extension Ladder

An extension ladder consists of two or more ladders that operate as a unit. The bed ladder acts as the nest for the movable fly ladder(s). In the two-piece extension ladder, the fly ladder slides in channels built into the bed ladder, **Figure 14-11**. Some ladders that reach beyond 25 feet can have two or more fly ladders with each ladder running through channels of the ladder beneath it. Extended by use of a rope **halyard**, these ladders are designed to maximize mechanical advantage through the use of pulleys. These pulleys are arranged so that as the fly ladders extend, the rungs of the various sections come to rest in alignment at each stop. Each level that the ladder can reach is locked into place on the bed ladder by the use of rung locks, also called dogs or pawls. These locks secure the unit into place. The value of an extension ladder is that one ladder can be used to reach various heights by moving the fly ladders up and out from the bed ladder. One adjustable ladder can serve the function of several fixed length ladders. In addition, if the need for different heights is required at a remote location from the apparatus, the same ladder can be used. Access to a second-story window and then to a third or even a roof can be accomplished without the need for a second ladder. Primarily carried on the ladder truck, they are also found on pumpers for use when a ladder company is not on the scene.

When an extension ladder exceeds 40 feet, it is required to be equipped with **staypoles,** also called tormentor poles. Called Bangor ladders or pole ladders, these ground ladders generally do not exceed 50 feet. The staypoles are used only for raising, and, once the ladder is in a raised position, they are not

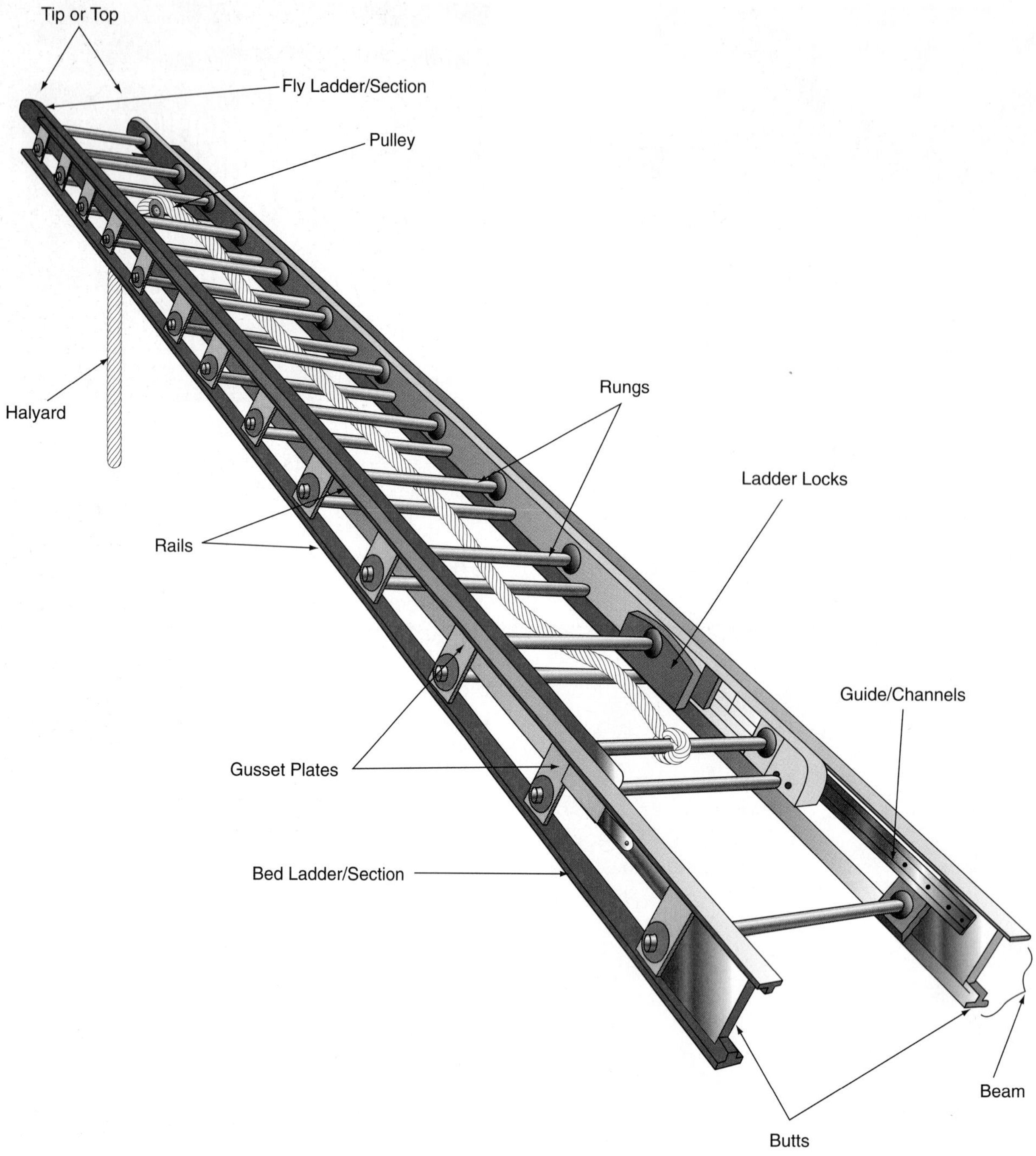

Figure 14-11 Extension ladder terminology.

used to support any weight on the ladder. Because of the weight and size of these ladders, the staypoles are needed to push up the tip of the ladder in the initial stages of raising. Used primarily for elevated access in remote locations where aerial apparatus cannot be placed, they require up to six firefighters and are generally used when no other means of access is available, **Figure 14-12**.

Roof or Hook Ladder

The roof or hook ladder is basically a straight wall ladder that possesses a set of retractable hooks at the tip end, **Figure 14-13**. Used when operating on a sloped roof, it enables a firefighter to work with more secure footing. When extended, the hooks are placed over the ridge of a peaked roof

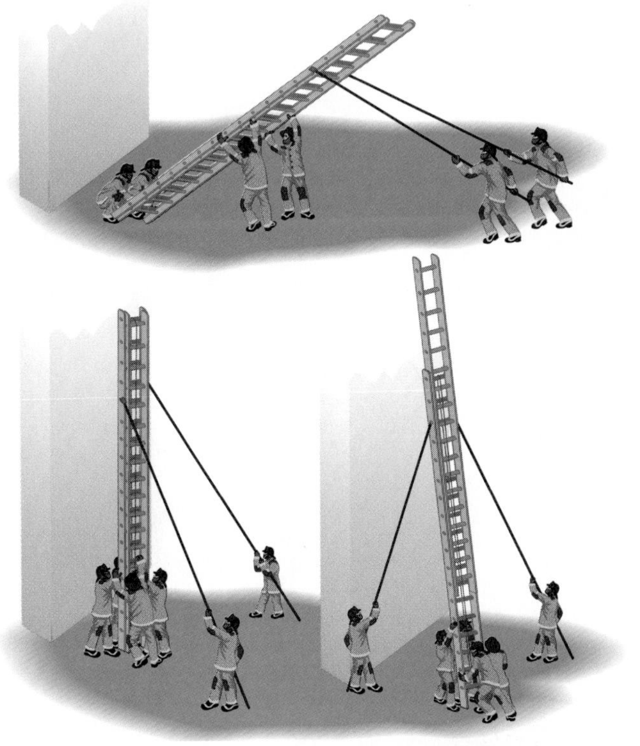

Figure 14-12 Bangor ladder raise (also called a pole ladder). The staypoles are used only for raising.

Figure 14-13 Hook ladder with retracted hooks being used as a straight ladder.

while the ladder rests on the sloped roof. The rungs are then used by the firefighter to gain a foothold where it would otherwise be impossible. In the retracted position, the hooks remain out of the way, enabling this ladder to be used as a standard straight wall ladder. A roof or hook ladder is usually found in lengths of from 12 to 24 feet. A hook ladder is not designed to be used as a hanging ladder.

Folding Ladder

The folding ladder is known by many names. Also called a suitcase ladder, an attic ladder, or a closet ladder, its function is to enable transport into narrow and confined places where a ladder is required. Although usually carried in a 10-foot model, it is available in lengths from 8 to 16 feet. It is a straight ladder that can provide access to attic spaces through hatches from the top floor of a structure. Its collapsible feature is very useful when attic access hatches are found in closets. When bedded (or folded), it is very portable through tight corners and into narrow spaces, **Figure 14-14A**. It can be used to remove trapped occupants through an elevator car's roof hatch. It is an ideal ladder to use to reach a sprinkler shutoff located high off the floor in a large open area of a warehouse. Footpads are provided as a safety feature to prevent slipping because this ladder is very often used indoors on structure floors, **Figure 14-14B**.

A-Frame Combination Ladder

An A-frame ladder is a combination ladder that can be used in various configurations. When nested, it is easily stored and acts as a mini-extension ladder, **Figure 14-15A**. Without the use of halyards, the fly ladder can be manually raised to the desired level and locked into place, **Figure 14-15B**. When fully articulated and extended, it can be locked into place as a full fixed straight ladder. When articulated into an A shape, it becomes a stepladder, **Figure 14-15C**. Each position is provided with locking mechanisms to secure the ladder into that configuration. Like the folding ladder, its greatest asset is its mobility in tight places because it can

(A) (B)

Figure 14-14 Folding ladder (also called a suitcase ladder and attic ladder). (A) Folded. (B) Opened.

(A) (B) (C)

Figure 15-15 Combination A-frame ladder. (A) Used as a short extension ladder. (B) Being converted from an extension ladder to an A-frame stepladder. (C) In the A-frame stepladder mode.

reach locations that would otherwise be unsafe or difficult with conventional length ladders.

Pompier Ladder

The Pompier ladder, also called the scaling ladder, is no longer an approved ladder according to **NFPA Standard 1931,** *Design and Verification Test for Fire Department Ground Ladders.* Because it is still used for training in some places, it is described here, **Figure 14-16**. The ladder is a single-beam configuration with rungs emanating from the central beam axis. On the top is a large hook that is designed to actually break through the window of an upper floor and hook onto the windowsill or roof parapet. The firefighter then uses this hanging ladder to climb up to the floor where the hook is attached. Then, while straddling the windowsill, the firefighter lifts up the ladder and, in a hand-over-hand motion, raises it up to the next window level, hooks it in, and repeats the climbing process. Using this procedure, a firefighter can literally climb up an entire building one floor at a time. In years past this ladder was used successfully in many rescues where the victim was out of reach of standard ladders.

Safety Because of the danger in removing a victim with a Pompier ladder and the provision of other safer methods of reaching victims through the interior with breathing apparatus and from the exterior with ropes, this ladder was removed from front-line service.

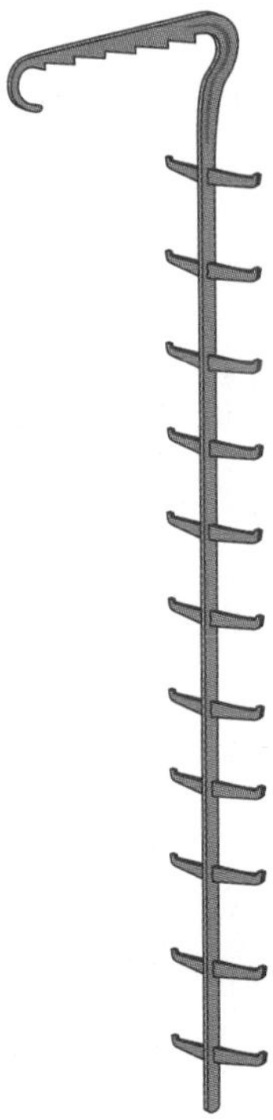

Figure 14-16 Pompier (or scaling) ladder.

USE AND CARE OF PORTABLE OR GROUND LADDERS

Because of the harsh environment in which ladders will be used on a regular basis, as much care as possible must be employed to prevent the avoidable causes of ladder damage. See **Table 14-2** for a list of some tips that should be second nature to all firefighters.

Ground Ladder Tips
• Store ladders in clean, dry places.
• Store ladders with multiple support points to prevent bowing or sagging.
• Avoid moisture and keep ladders as dry as possible.
• Do not place ladders in a location where they are subjected to heat or exhaust.
• Do not lean ladders on movable objects such as apparatus.
• Do not place ladders in out-of-sight areas where they can become tripping hazards.
• When left unattended, secure ladders at their tips to prevent them from toppling.
• Paint will tend to hide defects. Paint only 14 inches of the tips or heels for identification.

TABLE 14-2

NFPA Standard 1932, *Use, Maintenance and Service Testing of Fire Department Ground Ladders,* covers the use, maintenance, and service testing of ground ladders.

MAINTENANCE, CLEANING, AND INSPECTION

Because of the nature of firefighting, ladders will be subjected to a great deal of misuse, for instance, in emergency situations where on-site adaptations are employed when time is short. As a result, extra vigilance must be exercised when inspecting ladders, **Table 14-3**. With high levels of heat, overloading, and general rough handling, fire service ladders experience abuse that warrants special inspection attention. Ladders should be inspected at regular intervals and after each use—before they are stowed on the apparatus. NFPA Standard 1931 sets out the standards to which fire service ladders must conform. Also required is a certification label affixed to the ladder that provides evidence that the ladder is in conformance with the NFPA standard. Any ladder that needs repair should be immediately taken out of service.

Defects are uncovered through regular inspection and maintenance, **Table 14-4**. When a ladder is put through regular maintenance routines, it is generally

General Inspection Guidelines
• Check halyard ropes for undue wear. Replace if rope is kinked or frayed.
• Check rivets for tightness.
• Make sure that rub plates are secure, without burrs, and not worn out.
• Check support plates for tightness.
• Check the heat sensor label for presence and condition.
• Check for splitting or cracks, which are cause for repair or replacement.
• Make sure that bolts are secure and not overtightened.
• Make sure that rungs have no play or movement.
• Examine any discoloration closely for possibility of heat damage.
• Check for cracks in welds, which are cause for concern.
• Check for any wavy or deformed areas on the surface of the ladder that might indicate damage.

TABLE 14-3

Extension Ladder Inspection Guidelines
• Ensure that dogs are freely operating and springs are operational and in place.
• Make sure rope halyard is not knotted, kinked, or worn.
• Make sure there is no excessive play in the halyard.
• Ensure that pulley is not out of round and operates freely.
• Ensure that fly ladders operate and slide freely in channels.
• Ensure that any operating latches on Bangor ladders do not bind.
• Ensure that the halyard cable of multiple-fly extension ladders are snug when in bedded position. Inspect and replace cable if it is kinked or frayed.
• In the bedded and extended positions, ensure that all rungs line up when pawls engage.
• With any of the special folding type or articulating ladders, check that the hinges are secure.

TABLE 14-4

verified that it is in a state of readiness. Any work beyond general maintenance should be performed by trained ladder repair technicians. In addition, detailed maintenance records need to be kept to confirm that required inspections are being performed regularly.

Cleaning Ladders

The act of cleaning a ladder is not just a function of improving its appearance. Dirt can cake on moving parts and prevent proper operation. Dirt and caustic substances can also act as an abrasive, causing accelerated wear of moving parts. Generally, warm, soapy detergent and a scrub brush will remove most dirt. During the cleaning process, closer inspection of the ladder will occur as opposed to a general overview inspection. The nooks and crannies of the ladder will be observed, and defects are more likely to be uncovered. If a buildup of tar or grease is discovered, proper solvents and thorough cleaning are indicated with renewed lubrication of moving parts if warranted. Manufacturer's recommendations should be consulted for the proper choice of a cleaning agent for use with ladder material.

LADDER SAFETY

In most cases, ladder safety is equated with common sense.

> **Safety** Something that would not be done on the ground should not be done on a ladder.

Overreaching, overbalancing, or overloading ladder limits are the most common causes of ladder-related injuries in general use. At fires, a host of new dangers is added to the list. Ladder placement is a critical element. The obvious is generally recognized, but placement of a ladder where it might be a hazard later in the operation is one that is easily and often overlooked. Placing a ladder in front of a lower floor window or door is not a good practice at any time. It becomes a critical danger if and when fire erupts out that lower window or door, exposing the ladder and cutting off the escape route of firefighters using the ladder.

The use of gloves reduces the possibility of pinching fingers or skin, which can occur with articulating, folding, or extension ladders. When climbing, placing, or positioning a ladder with debris falling, the firefighter must be looking toward the target. Eye protection is essential.

The use of the correct ladder is important. With an extension ladder, the length is adjustable. With a fixed straight ladder, a target that is beyond the ladder length should not be reached by decreasing the angle to the building or by standing on a set of rungs closer to the tip.

All ladders are electrical conductors despite their construction material. Moisture on the surface negates any composite insulation. Overhead wires must always be suspected of being live and improperly insulated. The quality of the insulation should never be trusted. Age, weather, or heat from the fire can reduce the insulation value to nothing.

When moving on a ladder, firefighters should always keep at least three limbs in contact with the ladder. They should not move to the next rung (hand or foot) until the previous limb is in place. Some firefighters choose to slide one hand along the underside of the beam while using the other hand to move from rung to rung. This movement makes the three-limb contact rule easier for them. Others choose to move their hands alternately from rung to rung. No matter which method the firefighters use, movement up or down a ladder should be rhythmic, safe, and smooth. Firefighters carrying a tool up or down a ladder should slide their free hand along the underside of the beam while ascending or descending, as described above. A ladder should always be butted (also referred to as *footed, heeled,* or *tended*) at the bottom by another firefighter. In the event a ladder must be left unattended, it should be secured by a short rope or cord at the tip to prevent toppling. Finally, whenever the job requires reaching and working off a ladder, the firefighter should be secured to the ladder with a ladder belt or the firefighter should utilize a leg lock. (The leg lock is described later in this chapter in the Ladder Skills section.)

> **Safety** When ascending ladders, firefighters should keep their eyes focused level and ahead, occasionally looking up the climbing path for hazards and to view the objective.

LADDER USES

Ladders are used for many purposes in the fire service. Used primarily for climbing, they can also be used as a shoring tool, a fence, a means to hold back loose debris, or a chute to channel water with a tarp—and these are just some of the other available applications. Some of the more exotic uses are covered later in the Special Uses section.

Access

The most obvious use of a ladder is for access. A ladder can provide a path to an otherwise inaccessible opening or height. The first image often pictured is an elevated climb, but a ladder can be used to descend into an opening or as a bridge between two points at the same level.

Rescue

The use of a ladder for rescue is probably the most recognizable use at a fire scene. The drama in extracting a victim from the "jaws of death" is played out in most daily newspapers and television news programs. Not to diminish the importance of this role, it is probably the use of ladders that is employed *least often!*

Stability

The use of a special ladder called a roof or hook ladder will provide footing stability to a firefighter working on a sloped roof.

Ventilation

Using a ladder for ventilation can take place in one of two ways. A firefighter can use the ladder to remove glass with a tool from an elevated position. Also, the ladder itself can be used as the tool that takes out the glass. As always, safety must be paramount in either use.

Bridging

A ladder can be an effective bridge between two points. It can be used to support weight over a weakened floor or afford stable footing over a collapsed stairway. It has been successfully used to reach victims through windows from across a shaft, such as an alley. When bridging, a bedded extension ladder is the safest to use. By using it in this fashion, the firefighter is afforded the benefit of two ladders that support one another. When bridging, the portion of the ladder that rests on the support points should be at least 5 feet on each end of the ladder for each 10 feet spanned. In other words, the portion of the ladder that rests should be one-half of the distance spanned.

Elevated Streams

The use of a ground ladder for an elevated stream has lost its place in the fire service over the years with the introduction of the tower ladder and **ladder pipe**. It is still, however, an option if needed. The application of water from an exterior location off a ground ladder may be employed when no other approach to the fire is available. This certainly does not advocate the practice of exterior firefighting over the interior approach, but in some cases there might be an advantage to such a practice. For example, in a structure that has had its interior stairs burned away with no other access, firefighters might find the elevated handline stream an asset. Apparatus-mounted appliances in the form of tower ladders, ladder pipes, or deck guns might be unable to be positioned for remote position water application.

Elevated Work Position

A ladder can serve as an exterior work platform. Just as a painter uses a ladder for reach, so can a firefighter. There may be a need to remove something or check for heat during overhauling.

LADDER SELECTION

> **Note** One of the first considerations that will have to be made is the location where the ladder will be placed. The second will be the length of the ladder needed to reach the objective. A general rule of thumb for ladder selection involves the normal heights of the floors of residential and commercial structures. Residences measure approximately 8 to 10 feet from floor to floor. The average commercial story will average approximately 10 to 12 feet from floor to floor. This information is essential for choosing the correct length ladder to use. When in doubt, firefighters should use a longer ladder.

Once a general tactical attack plan is formulated, ladder selection can take place. Among the many ladder choices will be straight, extension, aerial, and tower. Once the target is identified, a secondary set of considerations should be entertained. What length ladder is necessary, and what will be done with the ladder after the initial use? When being used as an access path to a fire or search area it must be left in place. Or will the firefighter be using it at several locations, as would be the case for ventilating a series of upper level locations? Whereas a straight ladder would be ideal in the first application, an extension ladder might be more appropriate if different heights will be necessary after the first goal is attained. The ease of portability and use of the straight ladder might have to be sacrificed in order to meet the needs of the second and third goals. Some of the considerations that generally need to be entertained are as follows:

- *Ground condition.* Soft, hard, muddy, gravel, concrete, slippery.
- *Height needed.* Single height, several heights, same level but different ground slope.
- *Purpose.* Access will require stationary placement. Rescue might entail a number of people and also ventilation.
- *Slope of ground.* Might prevent setting up at some locations. Different heights from front to rear.
- *Accessibility of location.* Ladder portability through passageways to rear or sides, over the roof, and down the rear of the structure, or over fences.
- *Available personnel.* The number of personnel available for assistance might affect the choice of ladders.
- *Overhead considerations.* Electrical wires, porch overhangs, tree limbs, clotheslines, signs.
- *Raising space considerations.* **Tip arc** clearance will be necessary for raising, or the ladder might need to be raised first and then carried to the location in vertical mode.
- *Stability.* Stability of the structure where the ladder's tip will rest.

Because ladder work usually requires two people and not always the same two people, specific guidelines must be established in the raising of a department's ladders. There are many ways to raise a ladder, and the individual past history of each firefighter who may have worked for a utility company, construction, or in general household duties will present an array of different raising techniques. One standard technique should be established so that each firefighter will be able to perform the same function, same foot location, same hand placement, and so on. At any time, the firefighter can be interchangeable but the routine will remain the same. This eliminates wasted effort and unnecessary discussions on how to perform the skill at the scene.

The skill should be practiced and become second nature to all firefighters. This in no way implies that discussion and communication should be curtailed, but, on operations such as this, discussion should not be necessary beyond the identification of the target and a general "thinking out loud" so each firefighter knows what the other is going to do.

Butt Section

If the butt or heel of the ladder will be placed on flat ground, there generally will not be a problem deploying a ladder. However, if the ground slopes, raising a ladder might be impossible. The practice of chocking one beam with fillers such as wood or chocks is dangerous and should not be employed under any circumstances.

> **Caution** Ladder checking is dangerous! One shift of the ladder as weight is being distributed can cause the entire ladder to topple. The operation is too dangerous and should not be permitted.

The point where the butt is positioned should be directly under the target with an appropriate distance from the wall. It is easier to adjust the ladder's angle to the building than to move the raised ladder laterally. When transporting a ladder, the heel or butt should be carried in the direction of the target. Once the target is reached, the butt can be planted and the ladder raised almost without hesitation. If the butt were carried in reverse, the lead firefighter (carrying the tip end) would have to pass the target before the butt can be planted. This practice wastes valuable time and requires additional effort.

Fly Section

Just as the butt of the ladder dictates the ladder placement, the tip of the fly dictates how the ladder will be used. There are several specific locations where the placement of the tip will be important and contribute to the success of the operation and achievement of the goals.

Windows

When the tip is placed at a window, two general guidelines should be followed for placement:

1. If the ladder will be used for access or escape, the tip should not extend into the window frame, **Figure 14-17A.** For every inch the ladder penetrates the opening, the size of the opening is reduced. Furthermore, the higher the extended ladder is into the opening, the more the victims or rescuers will be exposed to the heat escaping the window at the upper level as they attempt to mount the ladder. The ideal location for access is for the ladder tip to be level or slightly below (no more than a few inches) the windowsill. This permits the firefighter to climb in face first over the windowsill below any heat. In

(A) (B)

Figure 14-17 (A) Ladder placed with the tip below the windowsill. (B) Ladder placed with the tip at the top of the windowsill to either side.

addition, this ladder placement will provide some measure of safety to a victim or firefighter if immediate escape is required. The ladder location will at least provide a place on which to land if immediate escape is absolutely required.

2. Placement of the ladder with the tip at the top of the window frame to either side of the window, **Figure 14-17B,** has several positive and negative aspects. On the positive side, it affords the firefighter a relatively safe location from which to perform ventilation. The firefighter can remove windows with a tool while remaining below the venting heat. However, to make access once the window is opened, the firefighter must negotiate the distance from the window to the ladder that is created by the angle of the window to the wall. As the distance from the tip increases, the distance of the beam from the building increases. At the sill level, depending on the height of the ladder, the distance could be from 18 to 24 inches. This gap can create difficulty in victim removal or access to a window that is spewing heat and smoke.

Roof Level

When placed for access to a roof, the tip should extend above the roof level approximately five rungs. That will provide about 5 feet of visible ladder over the roof. There are several reasons for this overextension. For one, the firefighter should climb onto the roof without overreaching and causing imbalance. The extended tip affords a handhold that can be used while dismounting the ladder. Second, the tip will be visible to any roof occupants that need to find it, especially if it is an escape route.

> **Note** Some ladder companies have installed mini-strobe lights on the tips of aerial ladders that only illuminate in the raised position for visibility at night or in smoke.

Third, the upper rungs of the ladder might be needed as a tie-off point for a rope where one is needed and the roof structures offer no alternatives. This practice should be employed only when absolutely necessary, such as when a lifesaving effort is involved. It should not be used merely as a tie-off point for rope in any other application because it would immobilize the ladder, which might be needed for a rescue effort elsewhere.

Fire Escapes

When raised to a fire escape, the position of the ladder will depend on the intended use for the given situation:

1. If the ladder tip is on the fire escape itself, the tip should be level with or below the upper rail so it can be mounted by straddling the rail and stepping onto the ladder.

2. If the ladder tip can be placed against a wall adjoining the fire escape, it should be several rungs above the fire escape rail so a person can swing a leg over the rail and step onto a rung while holding on to an upper rung. This is the preferred method. It will provide a victim with a stronger sense of stability.

When placing a ladder on a fire escape for victim removal, the priority and sequence should be that the first ladder is placed opposite the drop ladder. The second ladder is placed on the same side as the drop ladder on the floor above, **Figure 14-18**. In this manner, the evacuees can mount ladders at three points without having to wait for the person in front to clear the area. Additionally, if all the escape ladders can only be served by a single file access, removal will only occur as fast as one person can climb on a ladder. That speed is usually slow.

Figure 14-18 Ladder placement at fire escapes: (1) Drop ladder or stairway. (2) Opposite drop ladder or stairway. (3) Balcony above drop ladder on the same side of balcony. This provides for maximum escape flow.

SPECIAL USES

Ladders can be used as tools or as a portable stairs as long as their integrity is not compromised. Whether for venting upper floor windows with the tip or as a fence, the firefighter should be alert to visualizing adaptive uses for a ladder to achieve the firefighting goal.

Removal of Numerous Victims

The usual method of removing a victim is to raise the ladder, ascend and secure the victim onto the ladder, and then descend escorting the victim. When multiple victims need removal, taking one person at a time can be a very time-consuming process, even if several ladders and firefighters are available. A method to "keep the flow going" is to place two or more ladders at the escape point. This will usually be a large area where victims might be collecting such as a roof of a lower building, an area of escape refuge, or perhaps a school classroom where children have been brought. One ladder is used by the firefighters strictly to ascend, and the others are used to descend escorting a victim. The one "supply" ladder can service many escape ladders, **Figure 14-19**.

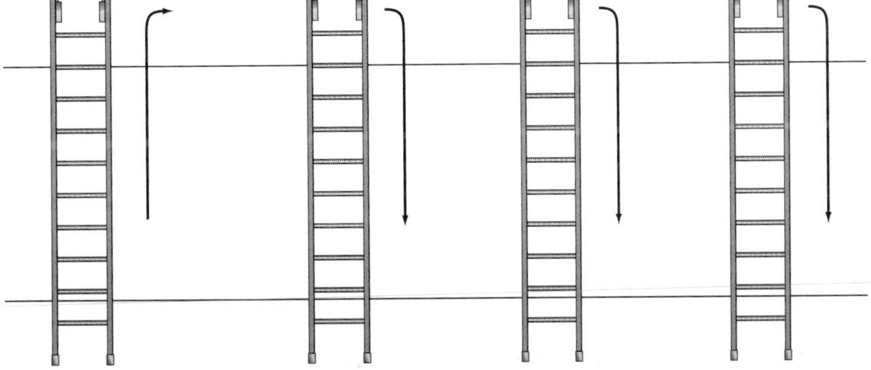

Figure 14-19 Multiple ladder removal technique.

Figure 14-20 Ladder with a salvage cover, plastic sheet, or tarpaulin used as a chute to divert and discharge water.

Figure 14-21 Ladders can be used to climb over a high fence.

Chute with a Tarp

If water from a leak or as part of an upper floor firefighting operation is threatening lower floors with damage, a ladder with a tarp draped over it can be used as a makeshift water chute to direct water coming through the ceiling out a window, **Figure 14-20**.

Over a Fence

Two short ladders tied together at the tip in an A-frame format can be used to climb over a fence that might impede operations and for some reason cannot be cut away, **Figure 14-21**.

Elevated Hose Streams

In a location that could not be approached by conventional apparatus with elevated stream capability, a handline off a portable ground ladder was a standard firefighting technique in the days before mechanized master streams. It can still be effective when no other approach is available, **Figure 14-22**. When using this technique, basic safety practices must be employed. The firefighter and hose must be secured to the ladder, and the ladder must be stabilized at the base or at the tip. Each department must develop this operation in a form that uses its individual inventory of tools without compromising safety.

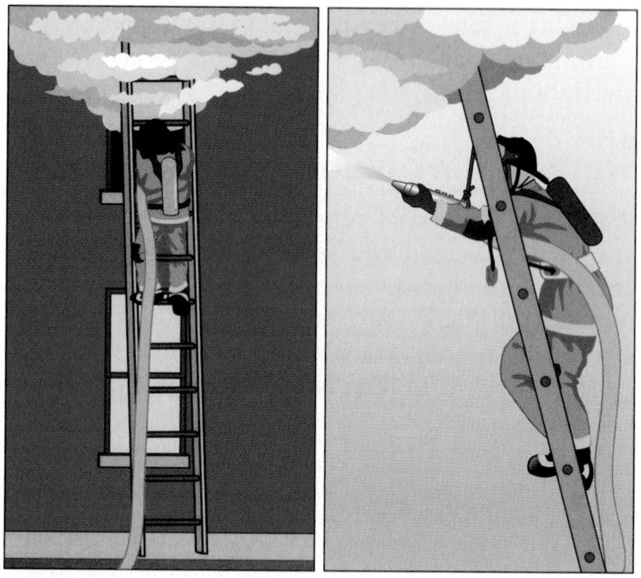

Figure 14-22 A handline can be used off a ground ladder for difficult-to-reach areas. Note the use of a ladder belt.

Portable Pool

If an emergency pool is needed, three or four ladders tied together to form a crib can be lined with a tarp and filled with water, either for firefighting use or to capture runoff, **Figure 14-23**.

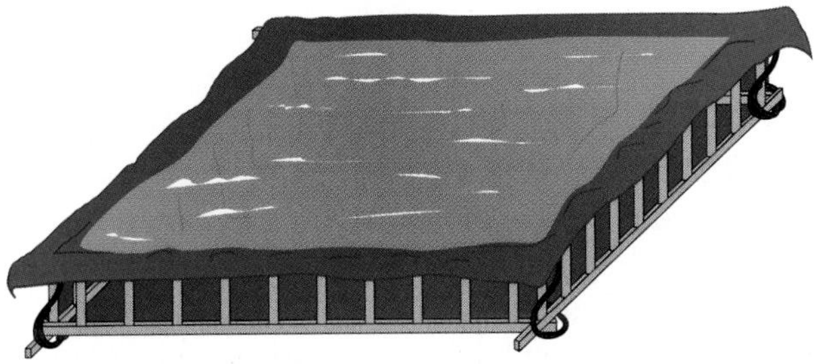

Figure 14-23 Ladders can be turned into an emergency water pool or collection area.

Figure 14-24 A ladder can be used as a barrier.

Barrier

A ladder mounted and tied off to secure objects can be used as a makeshift fence in cases where a barrier tape will not suffice. This would be especially useful where, for example, a large hole needs to be barricaded and an individual not paying close attention might accidentally compromise the presence of flimsy tape. The ladder barrier will offer a positive visual as well as physical mechanism to prevent a pedestrian from passing, **Figure 14-24**.

Support

A dangling sign or cornice that needs to be supported across a span and is beyond the capability of a mere rope can be supported by a ladder that spans the unsafe structural component. Using ropes, each end can be tied off to secure objects only as an emergency structural stabilizer. This temporary emergency measure should be replaced as soon as possible with tools designed for the situation. The ladder must be thoroughly inspected before being placed back into service, **Figure 14-25**.

Hoist Point

A set of ladders tied off at the tip and at the base into an A-frame configuration can be used as an emergency hoist point if a pulley and rope are attached to the apex of the ladder triangle. Ropes are lashed at the beam intersections and where the rungs run parallel. In addition, the support for the pulley is lashed to the rungs, creating a stable base, **Figure 14-26**. Special care should be used to make sure the ladder weight limits are not exceeded.

Figure 14-25 Ladder used as a shoring tool. A ladder secured to substantial objects by ropes can assist in stabilizing a structural defect as an emergency measure.

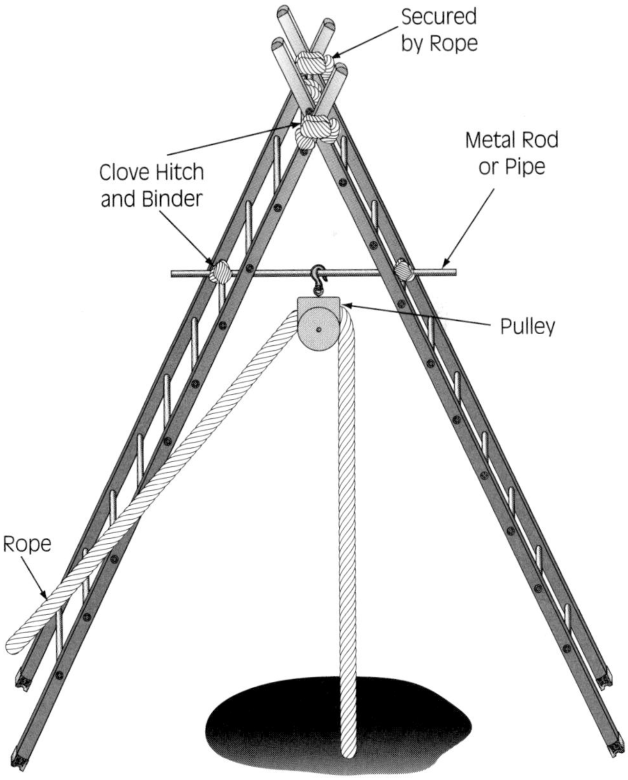

Figure 14-26 A-frame hoist.

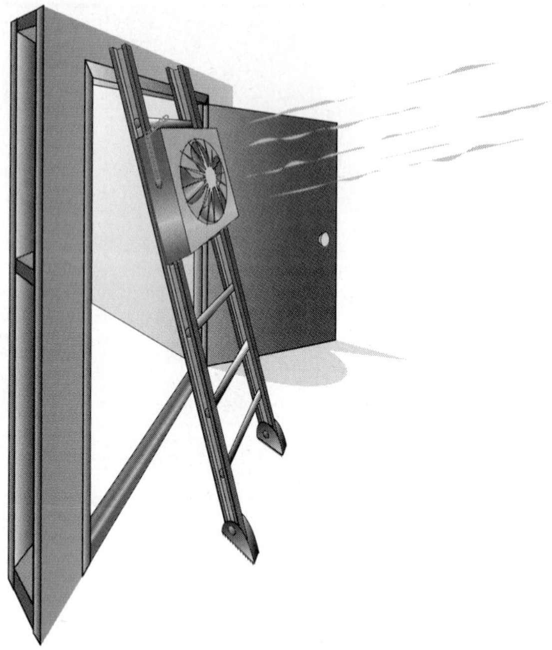

Figure 14-27 A ladder can be used to support a fan in a doorway.

Ventilation Fan Supports

A short ladder spanning an opening—such as a cellar opening or a hole made in a floor—can support a ventilation fan or blower. It can also be used to hang a fan at an upper level opening if such fan placement is desired, **Figure 14-27**. The fan can be placed on the ladder, in front of the opening, and operated from an exterior location.

These are just some of the examples where a ladder can be used as other than a climbing tool. As is always the case, innovation must be tempered with safety, and the weight limits of a ladder should never be exceeded. *In addition, whenever a ladder is used in a fashion other than that for which it was designed, a thorough inspection is essential before the ladder can be placed back into service.*

SAFETY

Even though ladder use is common in the fire services, firefighters must always be conscious of safety concerns and potentially dangerous situations. Some common safety concerns are listed in this section. Dangerous situations change from one fire scene to another. Firefighters must learn to quickly scan the area in which a ladder will be deployed. Anything considered a hazard should be removed. If removal is not possible, the ladder should be deployed elsewhere.

Overhead Obstructions

Tree limbs, structural overhangs, television and telephone wires, and, of course, overhead electrical lines are all potential injury producers. The path in which the ladder will be raised must be visualized before setting the ladder base and before the ladder is lowered.

Safety All overhead obstructions must be considered when placing a ladder.

Overhead electrical wires are always a hazard. Although insulated when installed, weathering can create cracks, which in turn can provide total exposure to a bare current-carrying cable underneath. There are many other dangers to be alert for when raising a ladder. If a fire has caused the insulation of a wire to be burned away and the wire is in contact with electrically conductive material (aluminum siding, a fire escape railing, television cable wires, etc.), this material could be electrically charged, **Figure 14-28.** If a ladder is placed against an energized material, the ladder could become an immediate path to ground. If the firefighter is in contact with the ladder, the ground path may course electricity through the firefighter's body.

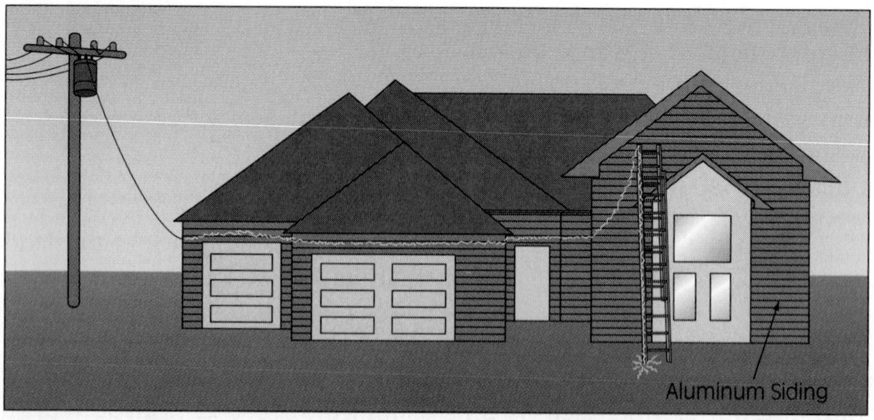

Figure 14-28 A ladder can make an electrical connection to ground.

Safety Overhead wires have always been a safety consideration when raising ladders. There are situations where a ladder must be placed in a certain region in the presence of overhead wires. This presence will require an alternate raising tactic or site selection. A safety officer should always be on scene when ladders are raised in the vicinity of overhead wire hazards.

Climbing Path

The climbing path is the imaginary passageway a firefighter climbs through while ascending a ladder. The path is not only the straight line created by the ladder butt to the ladder tip, but also the pass-through "tunnel" area that is necessary for a person to climb, **Figure 14-29.** If a firefighter is required to alter the normal climbing angle or squeeze through a tight space created by wires, tree limbs, or structural components, the climbing path is obstructed. The space a breathing apparatus cylinder will occupy must be accounted for when estimating the space needed for the climbing path.

Figure 14-29 "Climbing path" pass-through area.

Ground Considerations

An assurance that the ladder will be stable must be confirmed before raising. The ground under the ladder must be level and not create a dangerous lateral lean. The longer the ladder, the more a lateral lean will be exaggerated, **Figure 14-30**. What might be an acceptable lateral lean at a very low level might be dangerously beyond a safe position at an extended height.

Ladder Load

The number of people permitted on a ladder at one time will vary. The load capacity is based on weight, not the number of people. The load limits

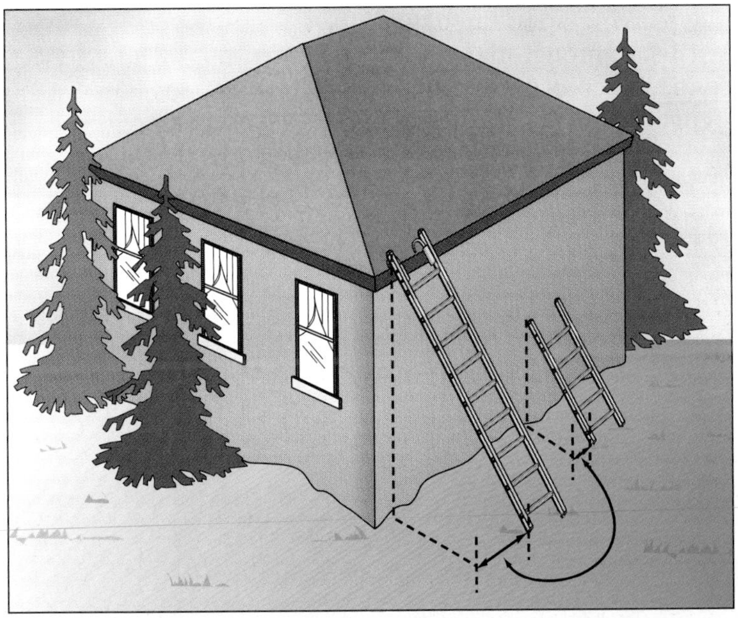

Figure 14-30 Uneven ground effect is magnified as the ladder increases in height.

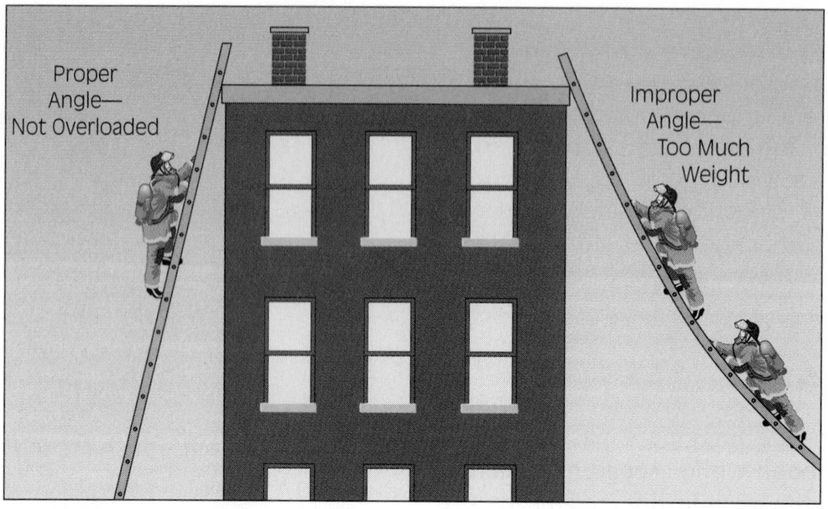

Figure 14-31 Ladders must be positioned properly and not overloaded.

of a ladder should always be understood, known, and not exceeded, **Figure 14-31**. Because each type of ladder can have different construction and weight-carrying capacity or be built differently by various manufacturers, firefighters should be aware of their department's equipment or at least know where to look for the capacity statement. The recommended maximum load will be found on the manufacturer's label affixed to the ladder, **Figure 14-32.**

Figure 14-32 Ladder load limit is found on the label attached to the ladder.

Working Off a Ladder

When working off a ladder, it is imperative that the firefighter be secured to the ladder. Ladder belts or safety harnesses have the ability to hook into a rung to secure the firefighter should balance be lost. Another method of locking into a ladder is by use of a leg lock. The firefighter inserts a leg through the rungs and articulates it back through the next lower rung opening. Then the instep of the foot is locked onto the beam, **Figure 14-33**. This maneu-

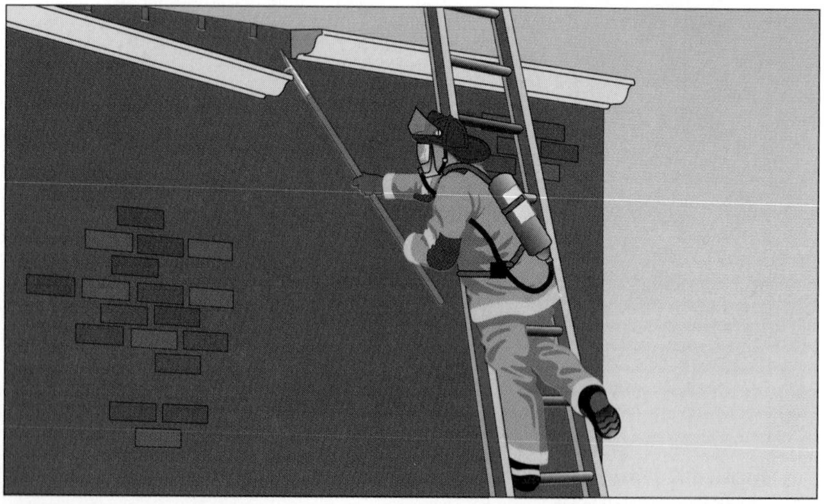

Figure 14-33 Leg lock.

ver enables the firefighter to operate with both hands free of the ladder.

MISCELLANEOUS LADDER INFORMATION

Ladder Storage

When ladders are stored for extended periods it is important that they be supported by more than just two support points. Ideally, ladders should be stored on a flat surface with the ladder on its beam or flat on both beams. By avoiding specific contact points, the tendency for a ladder to sag is reduced. For ladders that are often removed and replaced, a storage technique not using the same contact points will reduce any wear and point weakening. A ladder that is continually hung on its upper beam and constantly slid on and off along a hanger could develop wear spots and weak points at those locations on the beam.

Apparatus Ladder Storage

Under ideal conditions, ladders should be stored under cover in compartments, **Figure 14-34**. With protection, ladders will not be subjected to weather that could accelerate deterioration of halyards or permit caking of grime or debris on moving parts. It will also prevent accumulation of ice, snow, or water on the ladders during inclement weather. This is especially relevant during below-freezing conditions when ice can accumulate on ladder rungs or extension ladder operating parts.

Ladder Apparatus Parking

Most new ladder apparatus have ladder storage compartments that unload at the rear of the apparatus. Very often, units responding to the emergency

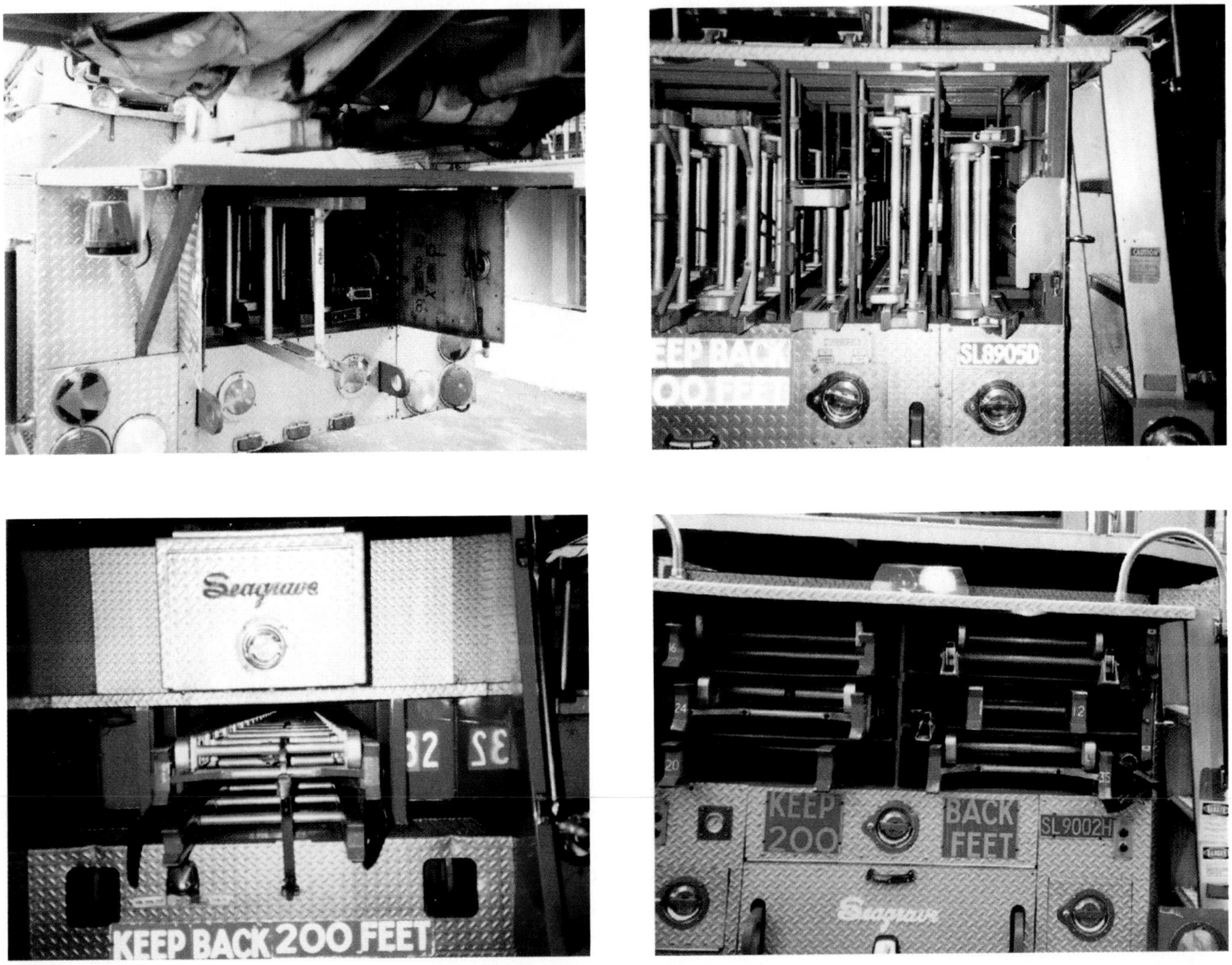

Figure 14-34 Various ground ladder storage practices.

place their apparatus directly behind a parked ladder apparatus, impeding ground ladder access, and, in some cases, preventing use of ladders all together, **Figure 14-35A**. The best way to prevent the actions of another from affecting the operation is to be proactive. When parking the ladder apparatus, the ladder apparatus driver should place the apparatus at an angle to the fire building. The ground ladders will still be able to be removed, and the turntable ladder position will remain unaffected, **Figure 14-35B**. In addition, all department apparatus drivers should be made aware of this so that, in their haste to place their own company into operation, they do not block the activities of another.

Ladder Painting

Ladders should never be painted as a means of maintenance, especially wooden ladders. With good intent, ladders are painted in their entirety with a specific color for unit identification or for general appearance, but paint obscures imperfections and hides defects that would otherwise be discovered during inspections. However, small areas of ladders should be painted for identification, visibility, and quick reference.

- *Identification.* Some departments use a color code for individual companies to quickly identify equipment during the postoperation fireground take-up (breakdown) period. It is acceptable to paint the base of a ladder up to the first rung for this purpose, but no part of any structural connecting plates, operating parts, or rivets should be painted.
- *Visibility.* The ladder may be painted from the tip down to the top rung for visibility. Yellow, white, and/or reflective paint helps make the tip visible in dark conditions and can be a genuine aid in raising. It is also a valuable feature for the

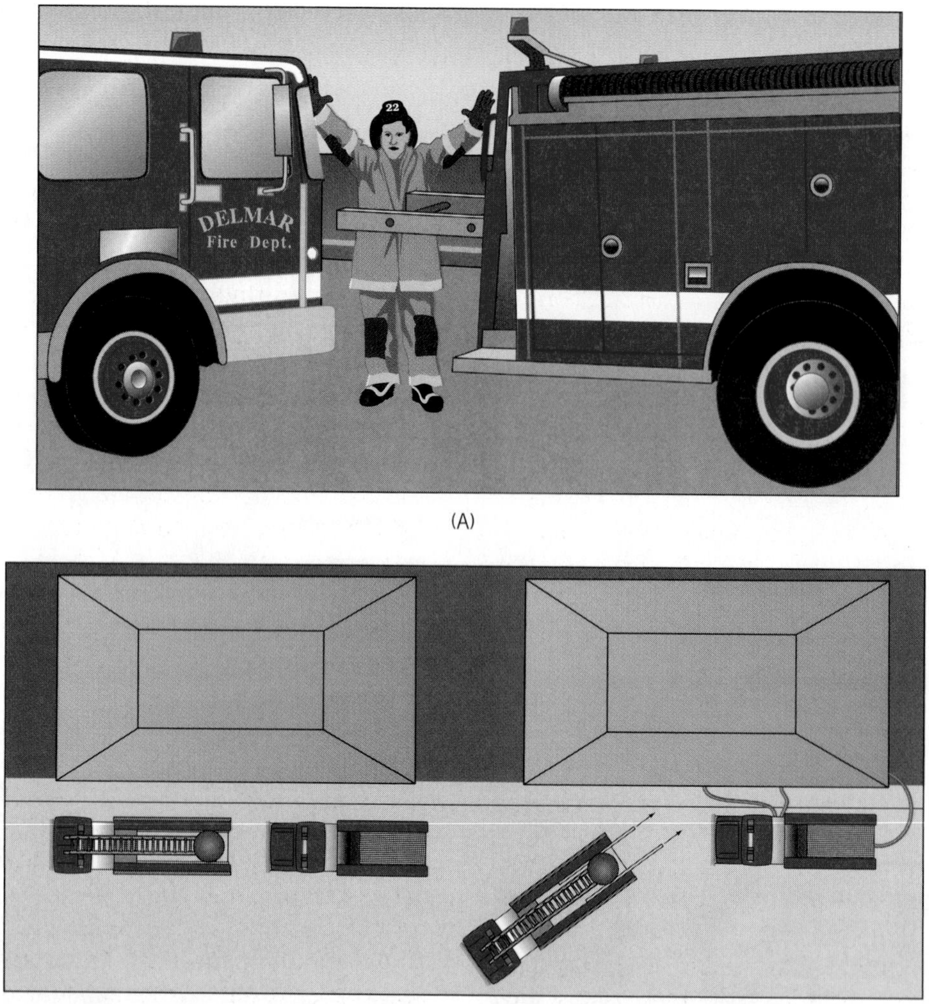

(A)

(B)

Figure 14-35 (A) It is important to leave room to remove portable ground ladders. (B) Parking apparatus can impede ladder access, but parking on an angle can be a simple solution.

firefighters using a raised ladder as an escape path from a roof or upper level location. Aerial ladder tips can also be painted to serve this safety function.

- *Quick reference.* The ladder size should be affixed on places that would be instantly visible. In addition to enabling the size to be ascertained immediately when obtaining a ladder, it would also be visible from the ground when the ladder is in a raised position. The use of ladder length identification numbers on the butt (or base) end of the ladder will assist in choosing the correct ladder from the ladder storage bed.
- *Hoist points.* One type of rope hoist method employs using a bowline loop inserted through a specific set of rungs. Small painted identifier strips may encircle the rungs involved for quick recognition (see Ladder Skills section).

Certification and Testing Procedures

NFPA Standard 1931, *Design and Verification Tests for Fire Department Ground Ladders,* outlines specific procedures for ladder testing and certification. This standard requires that a ladder manufacturer attach a certification label to each ladder verifying that the ladder was manufactured according to the NFPA standard and to OSHA guidelines.

When a ladder model is designed, it is subjected to rigid testing to ensure it will withstand the rigorous tasks required of it on the fireground. Ladder weight loads are tested while the ladder is in a horizontal position and resting on one beam, supported only at the tip and the butt, and again while the ladder is in the climbing position of 75 degrees. The ladder is tested to ensure that both butt spurs remain on the ground under simulated climbing conditions and that the rungs do not fail under simulated fireground operations. The ladder is tested to ensure that swaying and twisting will not result and that the butt spurs will not move across a surface. Special-use ladders (roof, pole combination ladders, etc.) are tested to ensure they are capable of accomplishing their specialized functions. It is important to note that a *design model ladder* is tested and certified, not each individual ladder. After the testing process, the ladder is discarded. The ladder label will attest that the model of ladder is in compliance with the NFPA standard and OSHA guidelines. For more detailed information on ladder certification and testing, please consult NFPA Standard 1931.

LADDER SKILLS

Moving a ladder from point A to point B and setting it in position may seem to be a simple task. However, without common terminology and technique, the moving and positioning of a ladder can be chaotic. It is important that firefighters understand commands given to carry, position, and raise a ladder.

Carrying Ladders

Several techniques are used for carrying ladders. Each technique has its own advantages and disadvantages. For short distances, the suitcase carry is sufficient. If longer distances are involved, the shoulder carry is more suitable to reduce fatigue. The butt (or base) firefighter should be the lead and the tip firefighter should be in the rear. Once the raising point is reached, the firefighters plant the base and raise the ladder in almost one motion.

Carrying Ladders on the Ground: The Suitcase Carry

The suitcase carry is used primarily as a short carry maneuver. Although it is very effective and a quick method of moving a ladder, carrying a ladder in this manner for long distances can be fatiguing. With the butt of the ladder facing the direction of travel the firefighters position themselves on the same side of the ladder facing in the direction of travel; one firefighter is at the tip and one is at the butt, **JPR 14-1A.** The tip firefighter controls the operation and gives the commands. On command, both firefighters raise the ladder to stand on one beam, **JPR 14-1B.** Grasping the upper ladder beam with the hand closest to the ladder, in unison, the firefighters lift and carry the ladder to the objective as if it were a suitcase, **JPR 14-1C.**

The firefighter at the butt is leading and can use the other hand to push away minor obstructions, open gates, and so on. The firefighter at the tip is the guide. It is up to this guide firefighter to ensure that turns are not too sharp or that angles are not created that cannot be negotiated by the tip firefighter. Communication is key, and the guide should dictate an alternate route if the travel path is unacceptable or if another carry method is necessary. If the ladder is heavy or the distance traveled is lengthy, a shoulder carry should be used. However, if the suitcase carry is selected a third firefighter can assist, **JPR 14-1D.**

The Shoulder Carry

The shoulder carry is useful for operations where firefighters will have to carry a ladder a good distance. The tip firefighter controls the operation and gives the commands. With the ladder flat on the ground, the two firefighters position themselves, facing opposite the direction of travel (facing the tip) on one side of the ladder beam. One is at the tip and one is at the butt, **JPR 14-2A.** On command, both firefighters squat down and grasp a rung and raise the ladder to stand on one beam, **JPR14-2B.** In unison, the firefighters lift the ladder using their legs for the power maneuver, **JPR 14-2C.** Both firefighters pivot in unison toward the direction of travel (facing the butt), place their free arm through the space between two rungs, and place the underside of the upper beam to rest on their shoulders, **JPR 14-2D.** The hand of the arm that was inserted through the rungs reaches to the next rung forward and grasps it, **JPR 14-2E.** The ladder is then carried to the objective. The tip firefighter is the guide. The free hand of either firefighter may be used to push away obstructions. In the case of large ladders, three or more firefighters might be needed to perform the carry. The firefighters must use on-the-spot adjustments to account for different firefighter body heights when positioning along the ladder.

Flat Carry

The three-person flat carry is a particularly useful technique when great disparity exists in the height of the firefighters. It also helps to evenly distribute the weight of the ladder. With the ladder flat on the ground, the two firefighters who are about the same height will position themselves, facing the direction of travel, on one side of the ladder beam. One is at the tip and one is at the butt. The firefighter who is taller or shorter locates midway on the other beam,

JOB PERFORMANCE REQUIREMENT 14-1

The Suitcase Carry

A Position of members before lifting.

B Raise the ladder to stand on one beam.

C Suitcase carry.

D Three-person suitcase carry.

JOB PERFORMANCE REQUIREMENT 14-2

The Shoulder Carry

A Two firefighters position on the same side of the ladder. One firefighter is at the tip of the ladder, and one firefighter is at the butt. Both firefighters are facing opposite the direction of travel, and both face the tip.

B On command, both firefightrs squat down and grasp a rung and raise the ladder to stand on one beam.

C In unison, the firefighters lift the ladder using their legs for the power maneuver.

D Both firefighters pivot in unison toward the direction of travel, rest the beam on their shoulders.

E The firefighters place an arm through the space between two rungs and grasp the next rung forward.

JOB PERFORMANCE REQUIREMENT 14-3
The Flat Carry

A Three-person flat carry.

B Four-person flat carry.

JPR 14-3A. The tip firefighter controls the operation and gives the commands. On command, all three firefighters squat down and grasp a beam with the hand nearest the ladder. Using their legs for the power maneuver the firefighters stand with the ladder. The ladder will slant up or down toward the odd-sized person, but the weight will be evenly distributed. As with the other carries, the tip firefighter is the guide. The butt firefighter leads the way to the objective.

The sequence is basically the same for the four-person flat carry, except that two firefighters are at the tip and two firefighters are at the butt. Either the beam or the rung can be grasped during the carry, depending on personal choice, **JPR 14-3B.**

Single Carry

A single firefighter can carry a small ladder individually by hoisting it up onto the shoulder and inserting an arm through the ladder rungs and grasping the next forward rung set or over the upper beam for weight balance, **Figure 14-36**. When initially placed on the shoulder, the tip should be slightly elevated with the greater weight on the butt (or base) of the ladder. The ladder should be balanced so that it can be adjusted by moving the hand on the rung or the

Figure 14-36 Single-person ladder carry.

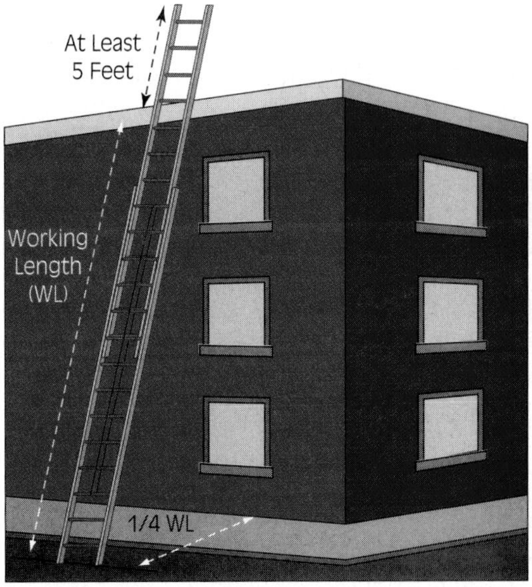

Figure 14-37 Working length is the distance from the ground to the point of contact with the building. Ladder is placed at a point approximately one-quarter of the working length.

beam to make a final weight equalization adjustment. When being carried, the butt (or base) of the ladder should be slightly dipped for maximum visibility and control.

Carrying Victims

Many types of carry techniques are used when dealing with a victim on a ladder. The level of consciousness, fear on the part of the victim, and trust in the firefighter are factors that will aid in removing a victim down a ladder. This is covered in greater detail in Chapter 16, Rescue Procedures.

RAISING SKILLS

Several considerations must be weighed when raising a ladder. The heel or foot of the ladder must be a calculated distance from the building for stability. The distance will depend on the target height. With the correct distance from the building for the given height, the ladder will be at a safe climbing angle, ideally about 75 degrees. For most people, this angle will be naturally comfortable and not require excessive bending or reaching to accomplish tasks performed while on the ladder. In addition, the ladder is designed to carry its rated weight at this angle. At lower angles, the ladder will actually bend due to the weight it is carrying. At higher angles, climbing and attempting any tasks while on the ladder may be unsafe. The distance between the ground and the point of contact with the structure is called the **working length, Figure 14-37.** If the ladder tip goes beyond the building when it is placed against the roof, the working length is from the ground to the point where the ladder contacts the roof level.

Raising Ladders

The number of firefighters needed to raise a ladder will vary. Depending on its size, as many as six firefighters may be needed to raise one ladder. Usually one, two, or three firefighters will be able to raise almost any of the ladders carried by a specific department. When conducting ladder operations, common terminology and practice is the key. There is no specific set of commands or one correct way to raise a ladder. Ladder operations depend on factors such as the specific goal to be attained, the size of the ladder, the manpower available, the topography of the area, and so on. However, it is very important that every firefighter from the same department use like methods and common ladder terminology. All firefighters must attempt to accomplish the same objective in the same way.

Most firefighting situations call for two firefighters to raise a ladder. The average ladder used at a fire is in the 18- to 35-foot range. The two people required are referred to by the functions they perform. The pivot firefighter is referred to as the *heel, butt,* or *footer,* while the firefighter raising the ladder is called the *raiser, tip,* or *fly.*

Note There are many different ladder commands used. The actual verbiage is not important. What is paramount is that common terminology be used. Ladder commands must be consistent for each agency, and interagency operations should be considered when deciding what actual commands the department will choose. When a ladder command is given, usually by the tip firefighter, every member of the operation must be thinking of the same potential actions required to accomplish the goal.

Streetsmart Tip The standard formula for determining the proper distance the foot or butt should stand out from the building is *one-quarter of the working distance* of the ladder from the base of the wall. A good way to estimate this is to stand with the toes against the foot of the beams and then, with outstretched arms, reach for a rung at about shoulder height. If the firefighter can grasp the rung, the angle is within a range that is acceptable. Remember that this is a rule of thumb for the *average* firefighter. An exceptionally tall or short firefighter might have to make some adjustments. The time for this adjustment calculation is during drills, not during the fireground operations.

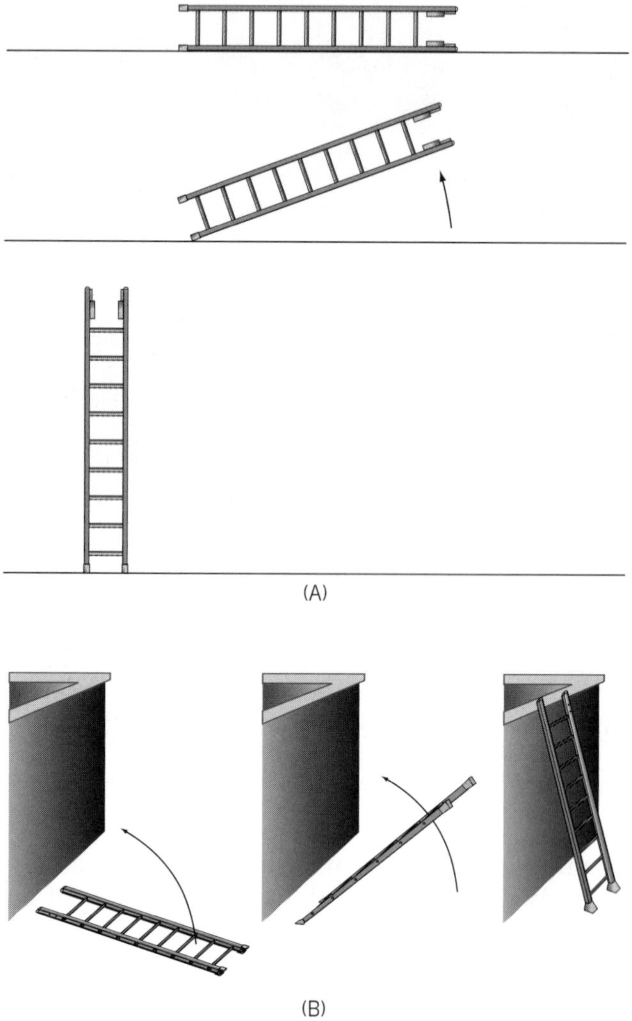

Figure 14-38 (A) Beam raise. (B) Rung raise.

Rung and Beam Raises

There are two methods of raising a ladder to the vertical position, **Figures 14-38A** and **B.** In one method, called the beam raise, the raiser brings the ladder up using the beam in a hand-over-hand motion. In the rung raise, the raiser uses the same approach but turns the ladder 90 degrees and utilizes the rungs for raising in a hand over hand motion. The beam raise is a good technique when the approach to the objective is parallel to the wall to be scaled. The beam raise is especially helpful when buildings are in close proximity with limited access. This would also be true where the presence of overhead obstructions would impede the operation. The rung raise is a good technique to use when the approach to the objective is perpendicular to the wall to be scaled.

Two-Person Rung Raise

With the rung raise, the ladder is brought to the raising point and positioned perpendicular and out from the wall approximately one-quarter of the working length from the wall or objective. The footer positions the ladder on the ground and places a foot at each beam to plant the pivot point of the raise. The toes of the boots are snubbed up against the heel of the ladder for stability. The raiser is positioned approximately two-thirds of the length of the ladder away from the base and toward the tip, **JPR 14-4A.** This distance can be adjusted depending on the length of the ladder and the comfort level of the firefighter.

When ready, the raiser grasps the beams or the rungs and raises the ladder to hip level using the strength of the legs as lifting mechanisms, **JPR 14-4B.** Using the upper body with the back straight and legs slightly bent, the raiser lifts the ladder overhead while stepping under the ladder and pivoting to face the footer in one fluid motion, **JPR 14-4C.** Using a hand-over-hand motion, the raiser with outstretched arms, moves from rung to rung while walking toward the footer until the ladder is in the vertical position, **JPR 14-4D.** The footer constantly surveys the area around the

JOB PERFORMANCE REQUIREMENT 14-4

Two-Person Rung Raise

A Rung raise—ready.

B Tip lifts to hip level.

C Swings underneath and lifts to full arm length.

D Tip walks to butt until ladder is in vertical position.

ladder for hazards that could impede the operation. The ladder is then lowered onto the building.

Safety The hands may slide along the beam rather than from rung to rung to prevent a missed grip on a rung that would permit the ladder to fall onto the firefighter underneath. The sliding hands move and maintain contact with the ladder at all times, thus minimizing the possibility of losing control of the ladder.

Two-Person Beam Raise

The ladder is placed down on the beam parallel to the building approximately one-quarter the working distance of the ladder. The footer is positioned in such a way that the firefighter's back is to the structure. The footer places a foot on the beam of the ladder to plant the pivot point of the raise. The toe of the boot is snubbed up against the heel of the ladder for stability and to secure the pivot point. The raiser is positioned approximately two-thirds of the length of the ladder away from the base and toward the tip, **JPR 14-5A,** adjusting this position as necessary. When ready, the raiser grasps the beams or the rungs and raises the ladder to hip level using the strength of the legs as lifting mechanisms, **JPR 14-5B.** Using the upper body with the back straight and legs slightly bent, the raiser then lifts the ladder overhead while stepping under the ladder and pivoting to face the footer in one fluid motion, **JPR 14-5C.** Using a hand-over-hand motion, the raiser moves along the beam with outstretched arms while walking toward the footer, **JPR 14-5D,** until the ladder is in the vertical position. Simultaneously, the footer uses both hands to steady the ladder by moving up the beam in a hand-over-hand method as the ladder ascends to the vertical position, **JPR 14-5E.** The ladder is then lowered into the building.

Fly Extension Raise

If the ladder is an extension ladder, there are additional factors and steps to consider before lowering the ladder into the building. Any extension ladder over 25 feet will usually require three persons to carry it to the objective because of the sheer weight of the ladder and its contribution to instability. The third firefighter can assume a raiser position if the ladder is too heavy for one firefighter to raise. When positioning an extension ladder, the fly section is positioned away from the building, **Figure 14-39.** Once the ladder is positioned and vertical, the footer unties the halyard while the raiser accepts the weight and stabilizes the ladder, **JPR 14-6A.** The footer extends the ladder to the desired height by pulling down on the rope, **JPR 14-6B.** The footer must watch to ensure the fly section is locked and secured in place once the desired height is reached, **JPR 14-6C.** The footer then secures the rope around the rungs and ties a bowline knot in the halyard as a safety precaution and slides the rope to the side against the beam so as not to interfere with the climbing, **JPR 14-6D.** For personal safety, it is important that the raiser hold the weight of the ladder by grasping the bed section of the ladder on the beams where the sliding fly section can operate free of any hand contact, **JPR 14-6E.**

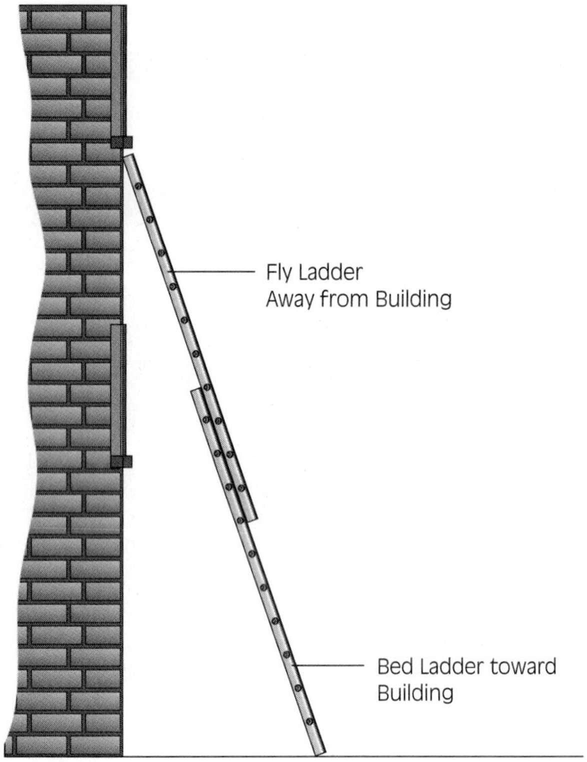

Figure 14-39 Extension ladder. The fly ladder is positioned away from the building, the bed ladder toward the building.

Safety When raising the fly ladder with the halyard, the firefighter should grasp the rope so that the heel of the hand is the uppermost part of the hand on the rope. The hand is twisted so that it creates a 90-degree angle with the rope with the balance of the rope passing through the fist and out between the firefighter and the thumb, **Figure 14-40.**

JOB PERFORMANCE REQUIREMENT 14-5

Two-Person Beam Raise

A Firefighters position the ladder on the beam, parallel to the building and approximately one-quarter the working distance of the ladder. The footer's back is to the structure. The footer snubs the heel of the ladder for stability and to secure the pivot point. The raiser positions approximately two-thirds of the length of the ladder away from the base and toward the tip.

B The raiser grasps the beams or rungs and rises the ladder to hip level.

C The raiser lifts the ladder overhead while stepping under the ladder and pivoting to face the footer in one fluid motion.

(continued)

JOB PERFORMANCE REQUIREMENT 14-5

Two-Person Beam Raise (Continued)

D Using a hand-over-hand motion, the raiser, with outstretched arms, moves along the beam while walking toward the footer until the ladder is in the vertical position.

E The ladder is then lowered into the building.

Figure 14-40 Hand grasps rope with heel of hand at 90 degrees to rope for better grip.

Lower Ladder into Building

From the raising procedure, the footer is left standing between the ladder and the structure, while the raiser is standing on the outside of the ladder facing the building. The raiser grasps a set of rungs at shoulder height and places one foot on the lowest rung. The footer grasps both beams at shoulder height, **JPR 14-7A.** As the raiser lowers the ladder into the structure, the footer accepts the weight of the ladder in a controlled maneuver until the ladder is resting against the building, **JPR 14-7B.** If any angle adjustment is necessary, the firefighters can lift the ladder off the ground by each grasping a rung while positioned on either side of the ladder beams. The ladder can then be safely adjusted by lifting and pulling it away from the building until

JOB PERFORMANCE REQUIREMENT 14-6

Fly Extension Raise

A Footer unties the halyard while the raiser stabilizes the ladder.

B Footer extends the ladder while the raiser stabilizes the ladder.

C Rung locks should be secured.

D Footer ties off the halyard and moves it to the side.

E Raiser's hands are on the bed section rails.

JOB PERFORMANCE REQUIREMENT 14-7

Lower Ladder onto Building

A The raiser grasps a set of rungs at shoulder height and places one foot on the lowest rung. The footer grasps both beams at shoulder height.

B As the raiser lowers the ladder into the structure, the footer accepts the weight of the ladder in a controlled maneuver until the ladder is resting against the building.

the desired angle is attained. Fine adjustments can be made by moving the ladder out and letting the ladder tip ride down the wall. In like manner, the tip can be pushed up the wall if the angle needs to be decreased.

One-Person Raise

In some cases the situation might call for only one person to raise the ladder. The condition in which a firefighter would be working alone is rare and generally prohibited by NFPA standards. However, it is possible for a ladder to be raised by one person provided the ladder is of a length and weight that is not beyond the capability of that person. Under normal two-person raising conditions, one firefighter will secure the pivot of the ladder foot or butt while the other raises the tip section. With only one person, the base pivot firefighter is absent. A substitute can be employed through the use of a wall, fence, bush, or anything that is substantial enough to act as a foot for the ladder. The ladder is placed on the ground in a rung raise position with the ladder heel against the substitute foot, **JPR 14-8A.** The firefighter stands at the butt of the ladder facing the building and raises the butt of the ladder with a beam in each hand to the hip level, **JPR 14-8B.** With an upward swing the firefighter raises the ladder to an overhead position and walks the ladder into the building using a sliding hand motion along the beams to support the weight of the ladder, **JPR 14-8C,** as the ladder ascends into a vertical position. When the ladder is in a vertical position it will be against the building from base to tip, **JPR 14-8D.** The firefighter must continue to support the weight of the ladder. The firefighter then sets the ladder angle by lifting the ladder off the ground and backing away while the tip rests against the building until the desired angle is attained. The

JOB PERFORMANCE REQUIREMENT 14-8

One-Person Raise

A Single-person raise—ready.

B Raise ladder to hip level.

C Raise ladder overhead and walk toward wall using a sliding hand motion on beams.

D Bring ladder to vertical position.

E Rule of thumb for ladder angle: toes at base, hands on beam at shoulder level.

ladder butt is then placed on the ground. The ladder will be at approximately the correct angle when the firefighter's toes are at the base and the outstretched arms are able to reach the beams at shoulder length, **JPR 14-8E.** As normal fireground take-down operations are not urgent, there should never be a reason for a firefighter to bring down a ladder alone. Take-down should be accomplished as described in the previous ladder raise sections.

Leg Lock

The leg lock is used to secure the firefighter to the ladder when both hands must be used to perform a task and a ladder belt is not available. The firefighter first chooses the desired height, **JPR 14-9A**, and climbs up one additional rung. Then the locking leg is inserted into the space between the rungs, **JPR 14-9B**, and over the rung that was at knee height.

Using the supporting leg as an assist, the firefighter lowers the entire body until the leg inserted between the rungs can be inserted through the next lower set of rungs such that the foot projects beyond the ladder beam, **JPR 14-9C**. Twisting the ankle slightly, the toes are projected beyond the ladder beam. Then the instep of the foot becomes the locking mechanism to the ladder or, alternatively, just hooking a lower rung of the ladder. Then, the desired operation may be performed with the stability of the firefighter in place. To remove the leg lock, perform the leg lock steps in reverse.

Carrying Tools

When carrying tools on a ladder, a certain amount of security is sacrificed. The positive grip that would be afforded by the hand holding the tool is negated. When possible, tools should be passed up to another firefighter first. Another technique is to hang the tool on an upper rung, **JPR 14-10A**, and then climb up to it, **JPR 14-10B**, reposition the tool to another upper rung and again climb up to it, **JPR 14-10C**. One last—but not the best—technique is to ride the tool up the rails while climbing, **JPR 14-10D**. This last technique is suited for climbing an aerial ladder where high side rails are available. A ground ladder is not a good candidate for this method, and special care must be exercised.

JOB PERFORMANCE REQUIREMENT 14-9
The Leg Lock

A Leg lock—ready.

B Step up one rung higher and insert other leg into rung space.

C Weave leg back through next rung opening and hook instep on beam or lower rung.

JOB PERFORMANCE REQUIREMENT 14-10

Carrying Tools

A Hook tool as high as comfortable.

B Climb up to top of tool so tip of tool is at hip level.

C Lift tool and push it to a higher rung level.

D On ladder with rails, slide tool up to top of rail to help maintain balance.

Mounting and Dismounting

Getting on and off a ladder is the most difficult action for the uninitiated. It requires releasing a grip on the very thing that is supporting the climber off the ground. As height increases, so does the fear of falling. A person who has a fear of heights dreads the very act of moving from one rung to another.

Before a firefighter ascends a ladder it should be secured. If a ladder is not secured with a rope, it should be heeled by another firefighter. There are two methods of heeling a ladder. Each method uses the objective to assist in steadying the ladder by applying pressure on the ladder toward the objective. In these operations, the heeling firefighter should be wearing full protective gear including gloves and a helmet with eye protection. A firefighter may stand under (or inside) the ladder, grasping a beam with each hand at about eye level, **JPR 14-11A.** The firefighter then leans away from the ladder, pulling it toward the objective. This firefighter must be disciplined to look forward and *never* look up as falling objects could strike the firefighter, and neck, facial, or other serious injuries could result.

A firefighter may also heel a ladder from the outside of the ladder, facing the building or objective. After another firefighter begins to ascend the ladder, the heeler moves into position, grasps a beam with each hand at about eye level, and chocks the ladder with a foot by snubbing the toe of the boot against the butt, **JPR 14-11B,** or by placing a foot on the lower rung, **JPR 14-11C.** The heeler pushes the ladder toward the objective. This firefighter must be prepared to move out of the way as a descending firefighter nears the bottom of the ladder.

Again, to be as safe as possible, at least three limbs should be in contact with either the ladder or the objective at all times if practical. Only one limb should move at a time, and the next limb to move should not do so until the previous one is secure. Whether mounting a ladder from an elevated location or dismounting the ladder to a structural item such as a fire escape or a window, a few simple commonsense actions must be followed. First, the firefighter should make sure to step onto a secure and stable objective. Using the hand-move first will permit the climber to get an immediate "feel" for whether the target is unstable, shaky, or questionable. This action will occur while the full weight of the firefighter is still on the ladder. The next foot movement permits the firefighter to gradually apply some weight onto the target while one hand and foot are still on the ladder. If at any time the objective becomes unstable as more weight is shifted, the firefighter can still pull back and retreat to the safety of the stable ladder. Once the second foot is on the objective and secure, the second hand can release its grip from the ladder.

> **Caution** Carelessness on a ladder can result in severe injury or death. Simple rules that should become the standard must be followed and be second nature to the ladder climber.

If the dismount will be made onto a balcony, a parapet, or a roof where there is no structural element to grasp, the firefighter should first check the structural stability with a tool. If stability seems to be intact, the firefighter can move one foot onto the objective and, while still grasping the ladder with both hands, shift weight to that foot. If any unstableness develops, immediate retreat is possible without a loss of balance. If stable, the weight shift can continue until the majority of the firefighter's weight is on the objective. When ensured that stability is present, the firefighter can move the other foot to the objective and let go of the ladder.

Window Dismount

When climbing into a window from a ladder, two methods are used. The method chosen should be used based on the conditions. When the ladder is placed on the side of the window, a step-in method is satisfactory provided the ladder is close enough to the building. It is always better to step down onto the windowsill so that if there is structural instability, weight can be withdrawn. If the step-up method is used, the firefighter's weight is shifting onto the structure, and, if failure occurs, the firefighter might already be committed and fall from the ladder. Mounting the ladder is essentially the same process in reverse. It is best to step up to and on the ladder rung rather than step down to the rung. This permits a more positive grip and less chance of loss of balance.

If entry into the window must be made under heat and smoke conditions, the ladder's tip is placed under the windowsill and the firefighter actually climbs right over the tip of the ladder and into the window. This method keeps the firefighter low. Exit from the window would require a

JOB PERFORMANCE REQUIREMENT 14-11

Heeling A Ladder

A Heeler assuming the inside position to secure the ladder.

B Heeler assuming the outside position to secure the ladder.

C Alternate foot position for outside heeling.

somewhat different approach. The firefighter would sit on the sill and roll over onto the ladder and descend facing the ladder.

Ornamental Works

When mounting or dismounting a ladder from an ornamental works such as a fire escape or a raised parapet wall at the front of the building, special care should always be observed to account for structural weakening from fire, weather, or just plain deterioration from lack of maintenance. The weight shift from the ladder should be slow and calculated to permit quick withdrawal should structural stability become questionable.

Engaging the Hook on a Hook Ladder

The hook (or roof) ladder is held hip high, **JPR 14-12A**. Using the palm of the hand, **JPR 14-12B**, the hook is depressed against its spring-loaded resistance. Once at the limit of the hook's travel, it is rotated 90 degrees to the perpendicular position from the ladder, **JPR 14-12C**. The hook is able to rotate in either direction.

When fully rotated, the pressure on the hook is gradually released, and it is permitted to return to its limit in the up/open position, **JPR 14-12D**. The same procedure is used for the other hook.

Roof Ladder Deployment

A ladder of any size or type is first raised to the eave of the roof directly under the desired access point. The hook ladder (with hooks retracted) is raised alongside the prepositioned ladder using the beam-raise method. One firefighter climbs the non-hook ladder. While secured into the ladder by use of a life belt or a leg lock, the firefighter at the tip of the ladder engages the hooks while the heel of the hook ladder is still in contact with the ground. The footer maintains the stability of the climbing ladder and assists with the hoisting of the hook ladder if a third firefighter is not available or if the climbing ladder is secured. The hook ladder is slid up the climbing ladder on its beam until it reaches its balance point on the edge of the roof. At that point it can be gently tilted until it rests firmly on the roof on one beam. The ladder is pushed up the slope of the roof on its beam until the hooks clear the peak. The hook ladder is then rotated so the hooks are down. Once over the peak, a firm tug is employed to secure the hook ladder into the roof.

Hoisting Ladders by Rope

On occasion, the need to use a ladder from an elevated location might arise. For example, there might be a window that can be reached from the flat roof of a one-story building, or the only way to get to the roof of a four-story building is to climb up a ladder from the roof of a three-story building. In any case, the need to get a ladder to an elevated location might be present. All ladders are treated the same, and the same technique can be used for any ladder that has two beams and rungs. When lifting an extension ladder, the fly ladder should be nested or bedded before raising. A utility rope of sufficient strength is used for this operation. A bowline knot is tied at one end at a point that will create a loop that is 9 feet in circumference or at least sufficient for the bowline knot to rest above the next set of rungs, **JPR 14-13A.** The middle of the ladder is ascertained by counting rungs. Once the middle is established, the firefighter counts up two rungs and passes the rope through the hole that the rungs create. Strips of electrical tape or painted bands can be used to semipermanently identify these rungs.

With the ladder on the ground, the butt (or base) of the ladder is gently lifted and the rope slid up to the marked rungs while the slack rope is taken up, **JPR 14-13B**. The knot should be approximately in the center of the ladder between the beams. The ladder is then rolled over so that the rope runs under the ladder rungs, **JPR 14-13C**. The rope is arranged to be between the ladder and the building. Then the command to hoist may be given.

The tip will tend to lean away from the building and prevent the ladder tip from snagging on any projections caused by windowsills, protruding brickwork, or any other structural element that might cause difficulty, **JPR 14-13D**. Once the rope knot is at the point of raising, the ladder is pivoted on the edge of the structure and gently pulled down on the roof. Then the remaining portion of the ladder is pulled onto the roof. When lowering the ladder, the procedure is reversed. However, during lowering, the ladder is placed between the rope and the building to avoid any snagging on building protrusions. The tip will drag along the building, and the base or heel of the ladder will tend to stay out from the wall a few inches and ride over any obstructions.

JOB PERFORMANCE REQUIREMENT 14-12

Engaging the Hook on a Hook Ladder

A Raise hook ladder to hip level on ground.

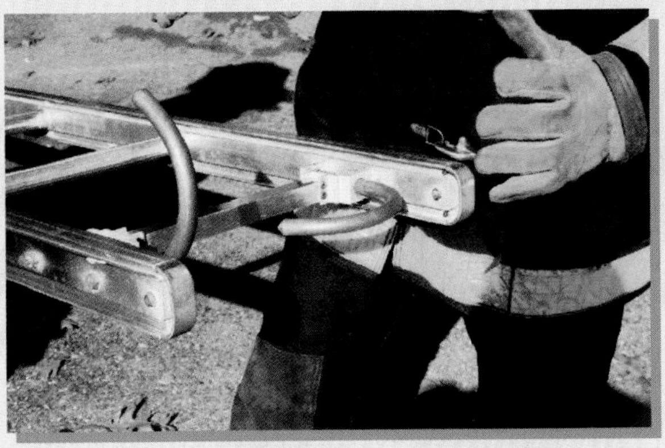

B Depress hook with the palm of the hand.

C Push hook in and turn 90 degrees to right or left.

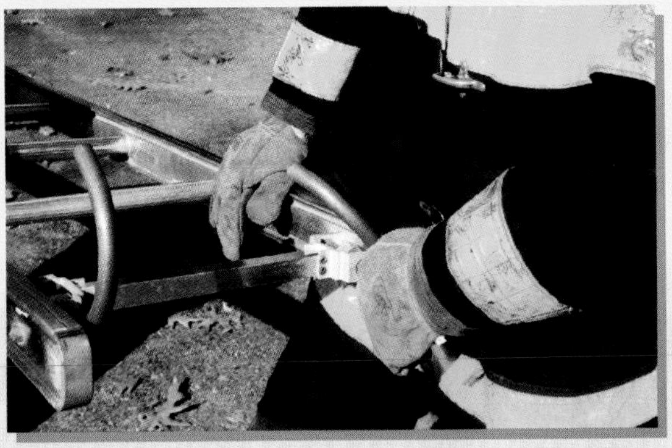

D When fully rotated, release hook into extended position.

JOB PERFORMANCE REQUIREMENT 14-13

Hoisting Ladders by Rope

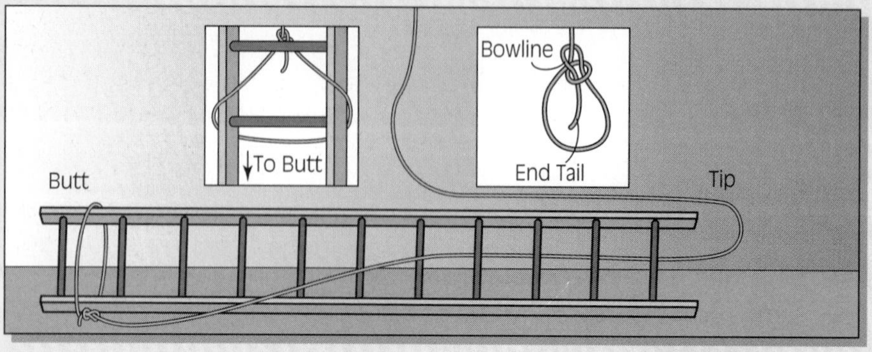

A A bowline knot is tied at one end at a point that will create a loop. The rope is passed between the rungs at a point a little more than half the distance of the ladder.

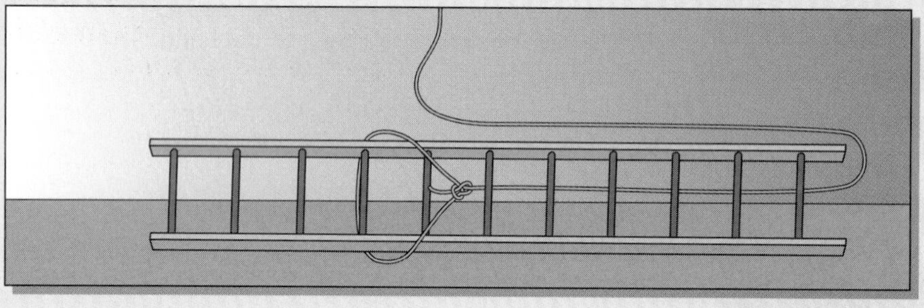

B With the ladder on the ground, the butt (or base) of the ladder is lifted and the rope slid up the beams while the slack rope is taken up.

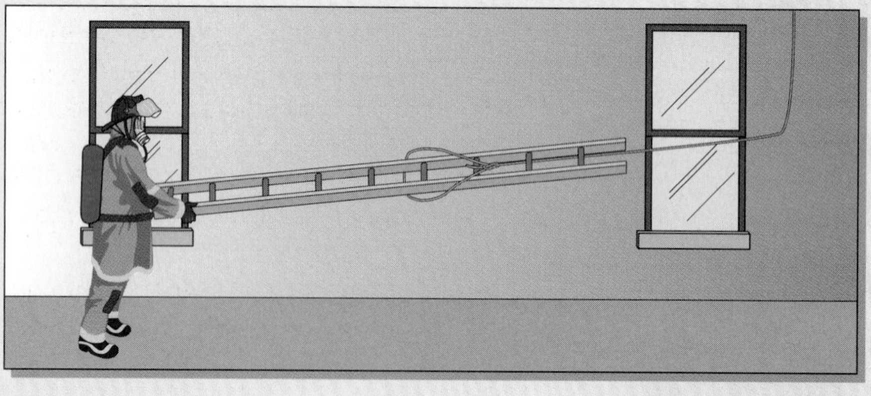

C The ladder is rolled over so that the rope runs under the ladder rungs.

D The rope is arranged to be between the ladder and the building and the ladder may be hoisted.

Lessons Learned

Ladders have many applications for use in the fire service beyond just climbing. With safety always in the forefront, innovative techniques can provide greater uses for ladders. However, it is important to remember what use a ladder was designed for and where the strengths and weaknesses lie. Any use other than such requires a thorough inspection of the ladder before returning it to service on the apparatus.

The different types of ladders, from truck-mounted aerial and tower ladders to portable ground ladders, are designed for a specific use. Roof ladders are used to stabilize footing on sloped roofs, and folding ladders are used for tight space application. New designs are always being developed. Some new designs in truck-mounted ladders include combination aerial ladder/bucket combinations and telescopic boom/articulating boom combinations. Ground ladders are also undergoing constant redesign. Collapsible telescopic ground ladders that save storage space and can be hand-carried into tight places and then extended are currently on the drawing board and might be introduced in the new generation of ground ladders. From climbing to bridging and from hoisting to shoring, ladders have many uses that can provide a superior application when used with judicious common sense.

Ladder use is packed with additional dangers that each firefighter must respect. Maintenance duties, inspection, and recording results all have a place where safety is the underlying motive and the overlying concern.

KEY TERMS

Articulating Boom Ladder An apparatus with a series of booms and a platform on the end. It is maneuvered into position by adjusting the various boom sections into place to position the platform at the desired location.

Bed Ladder The nonextending part of an extension ladder.

Extension Ladder A ladder consisting of two or more sections that has the ability to be extended to a desired height through the use of a halyard.

Fly Ladder That portion of a ladder that extends out from the bed ladder. Also called *fly section.*

Halyard A rope or cable that is used to raise the fly ladders of an extension ladder.

Hydraulic Pistons Mechanical rams that operate by pressure exerted through the use of a liquid, usually some form of oil.

Ladder Pipe An appliance that is attached to the underside of an aerial ladder for an elevated water application.

Nested The state when all the ladders of an extension ladder are unextended.

NFPA Standard 1931 The standard issued by the National Fire Protection Association that governs fire service ladder testing and certification.

Staypoles The stabilizer poles attached to the sides of Bangor ladders that are used to assist in the raising of this type of ladder. Once raised, they are not used to support the extended ladder.

Tip Arc The path that a ladder's tip will take while being raised.

Tower Ladder An apparatus with a telescopic boom that has a platform on the end of the boom or ladder. It can be extended or retracted and rotated like an aerial ladder.

Turntable The rotating platform of a ladder that affords an elevating ladder device the ability to turn to any target from a fixed position.

Working Length The length of the ladder that spans the distance from the ground to the point of contact with the structure. This does not include any distance the ladder might go beyond the point of contact as would be the case when the tip extends beyond the roof.

REVIEW QUESTIONS

1. What is the main difference between a wall ladder and an extension ladder?
2. What is the function of the hooks on a roof ladder?
3. What is the main function of cleaning and maintenance duties?
4. What is the main use of ladders at fire operations?
5. Under what conditions is a ground ladder used as an elevated water stream application device?
6. Describe the rules for mounting and dismounting a ladder.
7. When using a ladder for access, why should it be left in position?
8. What is the "climbing path"?
9. How far away from the building should the ladder be placed?
10. Describe the two techniques a firefighter can use to heel or foot a ladder.
11. When stabilizing an extension ladder, the raiser who is stabilizing the ladder must take care to ensure proper hand placement. Why?
12. What are the hazards a firefighter should look for when raising a ladder? What additional resource is advisable when a ladder must be raised in the area of charged power lines or other overhead hazards?

Additional Resources

Casey, James F., *Fire Chief's Handbook.* Dun-Donnelley Publishing Corporation, New York, 1978.

NFPA 1931: Standard on Design and Verification Tests for Fire Department Ground Ladders, National Fire Protection Association, Quincy, MA.

Training Bulletins of the Fire Department of the City of New York, *Firefighting Procedures* and *Ladder Company Tactics* (vols. 1–6), New York, 1997.

Walsh, Charles V., *Firefighting Strategy and Leadership.* McGraw-Hill Book Company, New York, 1963.

CHAPTER 15

ROPES AND KNOTS

Robert F. Hancock, Hillsborough County Fire Rescue

OUTLINE

- Objectives
- Introduction
- Rope Materials and Their Characteristics
- Construction Methods and Their Characteristics
- Primary Uses
- Fire Service Knots
- Inspection
- Maintenance
- Rigging for Hoisting
- Lessons Learned
- Key Terms
- Review Questions
- Additional Resources

STREET STORY

For many rookie firefighters, the prospect of memorizing the array of knots taught in recruit school seems daunting and relatively unimportant. Advancing hoselines into burning buildings and performing other fireground operations are much more exciting and seemingly more important skills to master.

However, a number of experiences in my twenty-five-year career as a firefighter have convinced me of the necessity of not just learning the knots but practicing them over and over until their tying becomes almost instinctual.

As a young chief I once had the experience of ordering a vent saw to the roof of a three-story building. I watched as a young firefighter started and briefly warmed up a K12 saw according to department protocols and then shut off the saw and tied it into the center of the rope hanging down from the roof so that the tail of the rope could be used as a tag line. Confident that the saw would soon arrive at the roof to cut the vent hole that would allow the hose team to advance to the top floor, I turned to answer a question. The startled expression etched on this captain's face was matched only by the obvious anguish of the firefighter whose knot had slipped loose, sending the saw crashing to the ground from the eaves of the roof. Luckily no one was in the saw's path, and that rookie survived his lesson to become a department chief and a master of fire service knots!

My first high-angle rescue also taught a similar lesson. The call was for a teenager who had fallen to the bottom of a 110-foot gorge. After reaching the unconscious but still breathing patient and stabilizing his multiple fractures and bleeding, a decision was made to secure him to a backboard and haul him back up the cliff face in a Stokes basket.

The necessary ropes and webbing slings were lowered to our position, and I proceeded to diamond lash the young man into the litter and attach haul lines to the litter bridle in the way that I had been trained. As the litter was being hauled up the cliff face, my partner and I used tag lines affixed to each end of the litter to pull the basket away from the ledges and snags. Everything was going according to plan until the litter was approximately halfway up the cliff. The patient suddenly regained consciousness. The combination of his being disoriented by his surroundings coupled with multiple facial injuries that further compromised his airway caused him to begin thrashing around quite violently against his restraints. Visions of this panicked young man clawing his way out of the litter and plunging to his death on rocks at the bottom of the falls made the few minutes that it took to complete the haul to the top of the cliff an eternity. I replayed the tying of the knots that was used to secure the patient over and over again in my mind as the litter inched its way up the cliff face.

Fortunately, the knots had all been tied correctly and safetied in just the way I had been taught, and the patient survived both the fall and the rescue, eventually making a full recovery. Since that day, a rope short has hung from the top drawer of the desk in my office, and I will often throw knots when some phone caller has me on hold.

Mastery of the unique basic skills associated with an occupation is one of the defining characteristics of any profession. Like suturing for physicians or writing legal briefs for lawyers, knot tying is one of the most basic and perhaps most critical of the core skills associated with our profession. Take pride in your ability to tie knots correctly, safely, and efficiently, for you might be asked to on your next call.

—Street Story by Joseph De Francesco, Coordinator, Madison County Fire and Rescue Services

OBJECTIVES

After completing this chapter, the reader should be able to:

- Identify the different materials that fire service rope is constructed from and their characteristics.
- Describe the differences between life safety and utility ropes.
- Define the basic terminology used when discussing ropes and knots.
- Identify the basic knots used by the fire service, how to tie each of them, and their uses.
- Describe the proper methods of inspection, maintenance, and storage of ropes.
- Describe the method of rigging basic firefighting equipment to be hoisted.
- Explain reasons for placing rope out of service.

INTRODUCTION

Rope is one of the most important and routinely used tools in the fire service. In this chapter the firefighter will learn how to select the proper rope for a given application based on the material that it is constructed of and the type of construction method. Also the firefighter will learn the primary uses of rope in the fire service, how to tie the knots utilized in the fire service, and how to properly inspect, maintain, and store rope.

ROPE MATERIALS AND THEIR CHARACTERISTICS

Rope is constructed of a wide variety of natural and synthetic materials. Each material has a different set of characteristics that impact its appropriateness or inappropriateness for utilization in the fire/rescue service. Some materials present characteristics that make them perfect for specific applications while rendering them useless for other applications.

The earliest ropes made by man and used in the fire service were made of natural materials. The use of natural material ropes by the vast majority of departments continued until the 1980s. As happens historically in the fire service (unfortunately), a series of incidents occurred that resulted in firefighters being killed or seriously injured while using natural fiber ropes. These incidents forced the whole fire service to reexamine the type of rope materials being utilized as **life safety lines**. This review ultimately resulted in the drafting of NFPA 1983, **Figure 15-1**, which deals with life safety ropes, harnesses, and hardware. NFPA 1983 established, for the first time (effective date June 6, 1985), minimum standards for this type of equipment if it is to be used by firefighters during the performance of their duties.

The following section presents the types of materials and characteristics of the different ropes used in the fire service.

Natural Materials

The materials that fall into the natural materials category include manila, sisal, and cotton. Because they are all natural materials, they share some of the same poor characteristics with regard to rot, mildew, abrasion resistance, natural deterioration/degradation due to age, a very low strength-to-weight ratio (when compared to synthetic materials), and a low **shock load** absorption capability.

> **Safety** Any ropes manufactured from natural materials should be used only as utility lines (ladder halyards, hoisting lines, etc.) and never in a life safety situation.

Figure 15-1 2001 Edition of NFPA 1983 Standard.

Manila

Manila rope, **Figure 15-2**, is made from the fibers that grow in the leafstalk of the abaca plant, which grows predominantly in the Philippines (Manila), hence the name. These are relatively short fibers that are twisted tightly together during the manufacturing process. This twisting creates the bond, through friction, that gives the rope its strength. Although the rope appears to be one continuous length, it is actually innumerable short fibers twisted together.

Manila rope is available in different types, with Type 1 being the higher quality and the one that was used most often by fire departments. Type 1 manila rope is manufactured from the higher quality inner fiber of the abaca plant leafstalk and can be identified by the colored string or ribbon that is twisted into the strands of the rope.

For years it was a common practice to soak new manila rope in water to make it limber. In actuality, this soaking reduced its strength by approximately 50 percent (which it never regained). This same reduction in strength occurs when a manila rope is stored in a compartment or area that has a high degree of humidity.

Sisal

Sisal is another fiber obtained from plant leaves, in this case the agave plant, native to southern Mexico. Ropes made of sisal fiber have approximately 25 percent less tensile strength than manila fiber ropes of similar diameter.

Cotton

Cotton fiber rope is made from the seed hairs obtained from the cotton boll. As can be expected, this rope tends to be very soft and pliable. It is also much lower in tensile strength (approximately 50 percent) when compared to manila, due in part to the shortness of the fibers relative to manila or sisal. It is very susceptible to damage from abrasion.

Figure 15-2 Type 1 manila rope.

Synthetic Materials

The primary synthetic materials utilized in the manufacture of ropes are nylon, polypropylene, polyethylene, and polyester. In the fire service today ropes made of these materials are the rule rather than the exception. These materials have excellent properties resisting rot, mildew, and natural degradation due to age. They are also much more resistant to physical damage and damage from abrasion. One of the major advantages/differences between these materials and the natural materials is that the fibers making up the rope are continuous from end to end, as opposed to the short fibers inherent to natural material ropes. This difference is readily identifiable in **Figure 15-3**.

Some of the other significant advantages that ropes manufactured from synthetic materials have over those manufactured from natural materials are a high strength-to-weight ratio, much greater shock load absorption, higher resistance to acids (polypropylene and polyethylene—not nylon), and no permanent loss of significant strength when they become wet. The temperature at which the various rope materials begin to lose strength and either melt or char is shown in **Table 15-1**.

Nylon

The material nylon was introduced in 1938 by E. I. Du Pont de Nemours and Company. It began to be heavily used to make ropes during World War II due to the shortage of natural fibers. The simplest example of nylon line is monofilament (single filament) fishing line, **Figure 15-4**. Nylon ropes are constructed of this same type of filament configured in multifilament bundles.

Figure 15-3 The manila fibers can be seen sticking out, while the nylon kernmantle rope is smooth and unbroken.

Rope Strengths

	TEMPERATURE AT WHICH THERE IS A LOSS OF STRENGTH (°F)	CHAR OR MELTING TEMPERATURE (°F)
Manila	180	375
Nylon	300	400–500
Polypropylene	200	275–300
Polyethylene	230	285
Polyester	300	450–650

TABLE 15-1

> **Note** Nylon has the following positive properties: high melting point (400–500°F), excellent abrasion resistance, can be bent sharply, high tensile strength (3 to 3½ times that of equal size manila), and a resistance to most chemicals (not acids).

Negative properties of nylon include susceptibility to damage by acids (*particularly battery acid*), loss of up to 25 percent strength when wet and/or frozen, stretches (elongates) when under load (use caution when utilizing to stabilize objects), and will not float on water.

Polyester

The most significant differences between nylon and polyester materials are that polyester has a good resistance to both acids and alkalis (however, an 80%+ acid and 10%+ base will affect it), it has low elongation under load, and its strength is not negatively affected by being wet or frozen. However, polyester does not handle shock loading very well. A correlation can be drawn between its low elongation characteristic and its ability to handle shock loading.

Polypropylene

Ropes constructed from polypropylene, **Figure 15-5**, are primarily used for water rescue operations by the fire service. This is due to the fact that water has no effect on their strength, and even more beneficial is the fact that they will float. Ropes of this material may be utilized in industrial operations where there is a high likelihood of exposure to chemicals and/or acids due to their high resistance to these materials. Polypropylene also has good resistance to rot and mildew.

Polypropylene ropes have a low melting point, have low resistance to abrasion, are susceptible to damage from sunlight, and have a relatively low breaking strength.

Polyethylene

Polyethylene ropes, **Figure 15-6**, have properties that are very similar to those of polypropylene.

Figure 15-4 Fishing line is the simplest example of nylon line.

Figure 15-5 Polypropylene rope floating on water.

Figure 15-6 Brightly colored polyethylene rope floating on water.

However, they will float indefinitely and can be purchased in bright, highly visible colors if desired.

CONSTRUCTION METHODS AND THEIR CHARACTERISTICS

When natural fiber ropes dominated the fire service, they were basically constructed utilizing the laid (twisted) method, **Figure 15-7**. When synthetic materials were introduced, a variety of other construction methods were also introduced.

Modern ropes utilize a number of different construction techniques (braided, braid-on-braid, and kernmantle), each with its own particular set of characteristics, which in conjunction with the type of fiber used combines to determine a given rope's properties. The two broad categories that rope falls into are **static** or **dynamic**. Static ropes have very little (less than 2 percent) elongation at normal safe working loads. This characteristic is very beneficial in the rescue field and the primary reason why most departments use static rope for rescue operations. Dynamic rope on the other hand has a much higher degree of elongation (10 to 15 percent) at normal safe working loads. This is the primary reason why this is the type of rope most preferred by mountaineers and others where fall protection is the primary purpose of a rope.

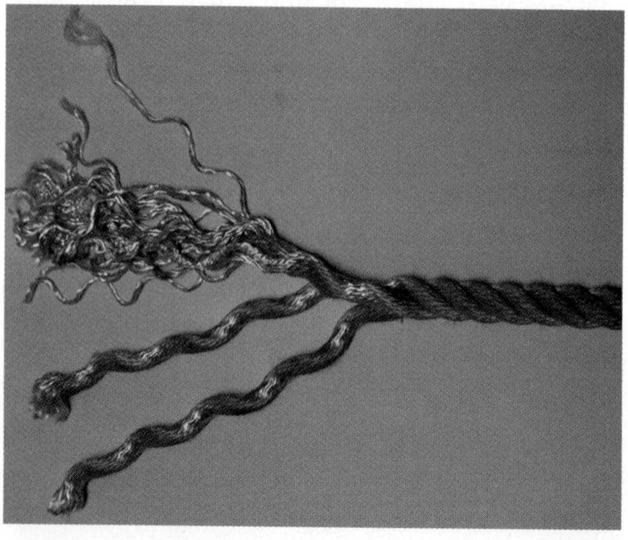

Figure 15-7 Example of laid construction method.

Laid (Twisted)

As previously mentioned, the laid method is the most common type of construction for natural fiber ropes. It is also utilized with synthetic fibers, but is by no means the construction method of choice for rescue rope. Laid ropes are formed by twisting individual fibers together to form strands or bundles. These strands (three or more for larger ropes) are then laid (twisted) together to form the finished rope. One of the major drawbacks to this type of construction method is that with all of the twisting to form the strands every fiber is exposed on the outside of the rope. This means that every fiber has potential to be damaged by abrasion, sunlight, chemicals, and so on.

Another type of laid rope is used primarily for mountaineering and called *hard laid.* It is generally much stiffer than other types and tends to get stiffer the more it is used. It is also difficult to form knots in, and one has to be very careful to use safety knots, especially when using a bowline knot. This stiffness also makes it difficult to handle and store.

A characteristic that is common to both types of laid ropes is that it tends to impart and/or accentuate spinning and subsequent twisting and possibly knotting at the bottom end of the rope when used for a rappel rope.

The one real advantage of laid rope is that, because all fibers are exposed, it is easy to inspect. Again, it is important to remember that this also makes all of the exposed fibers susceptible to damage.

Braided

The braided method of rope construction, **Figure 15-8**, is utilized predominantly with synthetic fibers, although there are some natural fiber braided ropes. Braided ropes are formed by weaving small bundles (not twisted) of fibers together, much the same as braiding hair, uniformly and systematically. These ropes are generally smooth to the touch and have a good degree of flexibility.

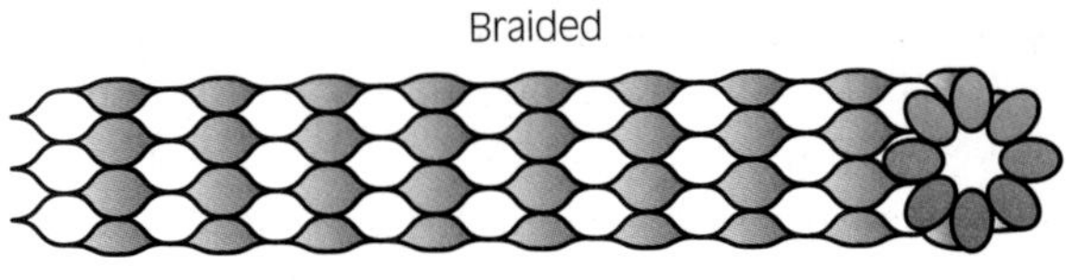

Figure 15-8 Example of braided construction method.

While braided rope does not induce or accentuate spinning during rappelling, the braiding process exposes every fiber to abrasion, sunlight, and other forms of physical damage.

Braid-on-Braid

As can be assumed from the name, this type of rope is formed by braiding a sheath over a smaller braided core, **Figure 15-9**. This type of construction often results in an approximately 50/50 split between the core and the sheath in the total strength of the rope, although there is at least one manufacturer that has designed braid-on-braid rope so that the core maintains 80 percent of the strength. Ropes of this construction method tend to be quite dynamic (stretchy) and most have a soft sheath that is more susceptible to damage from abrasion.

Kernmantle

The term *kernmantle,* when separated, very accurately describes this construction method, **Figure 15-10**. The first part, **kern**, is a derivative of the term *kernel,* which is defined as "the central, most important part of something; core; essence." The second part of the word, **mantle**, is defined as "anything that cloaks, envelops, covers, or conceals." In a kernmantle rope, the kern generally carries the vast majority of the load, accounting for approximately 75 percent of the strength of the rope, with the mantle making up the remaining 25 percent. The mantle's main purpose is to protect the kern from physical damage.

Kernmantle ropes can be either dynamic or static depending on the configuration of the fibers of the kern. When the kern fibers are twisted together and the mantle woven over them, the result is a dynamic kernmantle rope. When the kern fibers run parallel continuously and the mantle is woven over them, the result is a static kernmantle rope.

Kernmantle ropes tend to be quite resistant to abrasion and other forms of physical damage since the main strength of the rope, the kern, is protected by the mantle. However, since the vast majority of the kernmantle ropes used in the fire service are manufactured from nylon, they are subject to the possible sources of damage listed previously in the synthetic materials section.

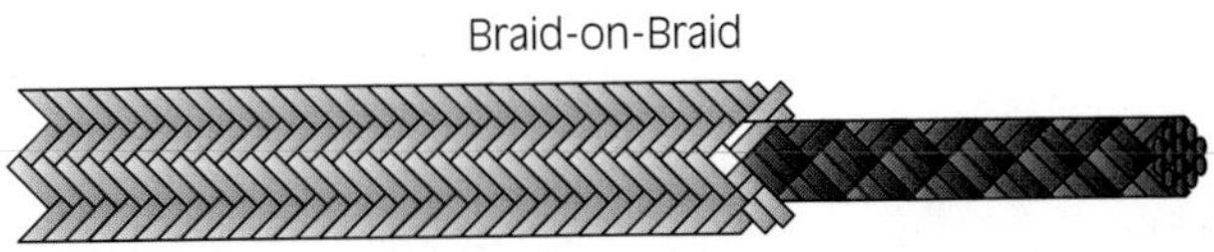

Figure 15-9 Example of braid-on-braid construction method.

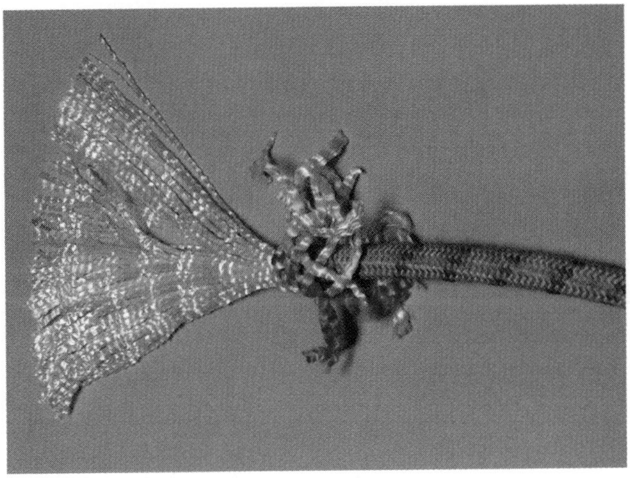

Figure 15-10 Example of kernmantle construction method with the outer kern cut and pulled back at the end showing the inner mantle section.

> **Caution** Care must be taken that lifelines are not stored in an area that has the potential of exposing them to acids, such as near the batteries or battery box vent of an apparatus.

PRIMARY USES

Although ropes have many uses, the fire service tends to utilize them on a regular basis for a few tasks and operations that can be divided into two classifications: utility and life safety. These classifications are discussed in detail in the following section. If a department/area has a particular evolution not shown here as a part of their standard operating procedures, this is by no means intended to intimate that they are wrong or improper.

A firefighter should be familiar with department standard rope use, as well as a broad range of *possible* uses. Firefighters are faced with new and different challenges and unusual situations every day. Experience and a broad knowledge base are vital assets of firefighting.

Utility

Rope used only for utility purposes has no standard governing materials, required strength, safety factors, number of uses, associated hardware, and so on, so it is the responsibility of the department to determine what type of material and construction method is desired for its utility ropes. The department should use good judgment and look at the tasks for which the ropes will be used.

Safety It is not only inappropriate, it is dangerous and contrary to NFPA 1983, *Standard on Fire Service Life Safety Rope and System Components,* to use the same rope for both utility and life safety operations.

Some of the tasks for which **utility ropes** are used are shown in **Figure 15-11**. There are many uses too numerous to list, but firefighters should become intimately familiar with the common uses within their own department.

Firefighting and Rescue Uses

Ropes used for structural search and rescue guide ropes and other non-life-supporting operations do not fall into the category of life safety ropes according to NFPA 1983. However, it would be prudent to utilize either a life safety rope or a rope used specifically for these purposes, as opposed to utilizing everyday utility rope.

Ropes, harnesses, and hardware utilized anywhere there will be a life supported, as shown in **Figure 15-12**, must comply with NFPA 1983, which sets minimum standards for equipment used in these types of situations. NFPA 1983 categorizes life safety ropes as **one- or two-person ropes** and sets minimum **tensile strength** requirements for each.

Safety A one-person rope requires a minimum tensile strength of 4,500 pounds, and a two-person rope requires a minimum tensile strength of 9,000 pounds.

While this chapter does not go into depth on life safety rope operations, firefighters need to know the basics in order not to endanger themselves, other firefighters, or the persons they are attempting to rescue.

FIRE SERVICE KNOTS

While some of the knots covered here have been around for as long as people have been tying vines together, some were introduced along with the switch to synthetic fibers and modern construction techniques; some of the knots that have been used for years in the fire service do not hold well in these

(A) (B) (C)

Figure 15-11 (A) Utility rope hoisting a pike pole. (B) Hoisting a ladder. (C) Hoisting a charged hoseline.

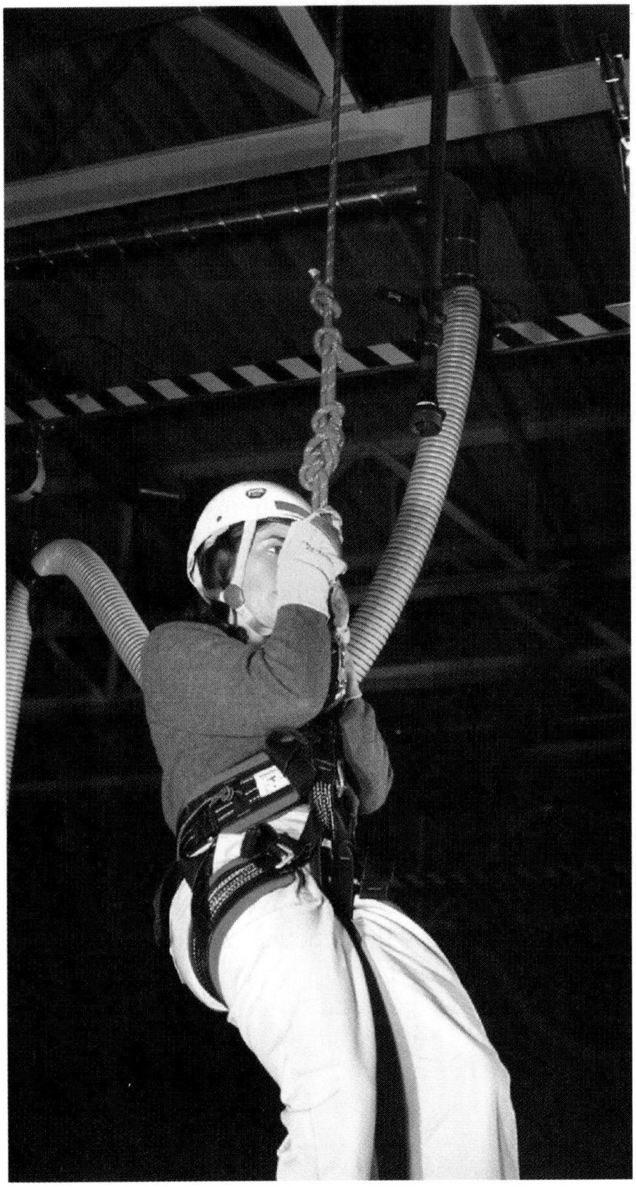

Figure 15-12 Life safety rope.

new ropes. The bowline is an excellent example. For years, the bowline was the standard knot for fire service application. However, this knot was found not to work as well with modern synthetic rope. Cavers and mountaineers used synthetic ropes for many years. Because of synthetic rope characteristics and life safety applications, these mountaineers began to utilize and perfect their technique. With the fire service switch to synthetic ropes, these knot-tying techniques were incorporated into department applications.

Nomenclature of Rope and Knots

To understand knot tying instructions, firefighters need to know the basic terms used to describe the parts of a rope and knot.

Parts of a Rope

A rope can be divided into three separate and distinct parts, **Figure 15-13.**

> **Note** It helps to think of the rope as having two ends and a part or portion between the ends.
>
> 1. The **working end** is the end of the rope utilized to tie the knot.
> 2. The **standing part** is between the working end and the running end.
> 3. The **running end** is used for work such as hoisting a tool.

Figure 15-13 The three parts of a rope: working end, standing part, and running end.

Figure 15-14 Left to right: a round turn, a bight, and a loop.

Elements of a Knot

Just as firefighters need to know the basic terms used to describe the parts of a rope, they also need to know the terms used to describe the elements that are combined to form a knot, **Figure 15-14**. A **round turn** is formed by continuing the loop on around until the sections of the standing part on either side of the round turn are parallel to one another. A **bight** is a doubled section of rope, usually made along the standing part. A bight forms a U-turn in the rope that does not cross itself. A **loop** is a turn in the standing part that crosses itself and results in the standing part continuing on in the original direction of travel.

Streetsmart Tip Studying the photos in the Job Performance Requirements (JPRs) and reading the associated directions will assist firefighters with the skills necessary to master knot-tying techniques. It is helpful to have another firefighter read the instructions step by step and cross-check the steps with the photographs in the JPRs.

Knots

The following pages provide step-by-step instructions for tying the knots that are utilized most often by the fire service. They are tried and true and straightforward. Fellow firefighters should not have any problem recognizing the knot when it needs to be untied. Later, this chapter shows how each of these knots can be used for various tasks that may be required at any given incident.

It is important to point out that the step-by-step instructions are intended to teach firefighters the steps necessary to tie the various knots. Very seldom are firefighters called on to tie any of these knots while standing in a brightly lighted, air-conditioned classroom or apparatus bay.

Streetsmart Tip For the purposes of consistency and safety, it is the firefighter's responsibility to learn and practice tying knots in all of the different types of situations: in the open, around an object (both vertical and horizontal), in low visibility, when wet, with gloves on, and so forth.

All knots should be neat and tight. The terms used to describe this practice are to *dress* and *set* a knot. **Dressing** a knot is the practice of making sure that all parts of the knot are lying in the proper orientation to the other parts and look exactly as the pictures indicated. **Setting** a knot is the finishing step of making sure the knot is snug in all directions of pull. This is an important step because a perfect knot, if not set, will slip when a load is applied. This is due to the lack of appropriate friction and cinching effects that cause knots to hold. **Figure 15-15** shows knots incorrectly and correctly dressed and set.

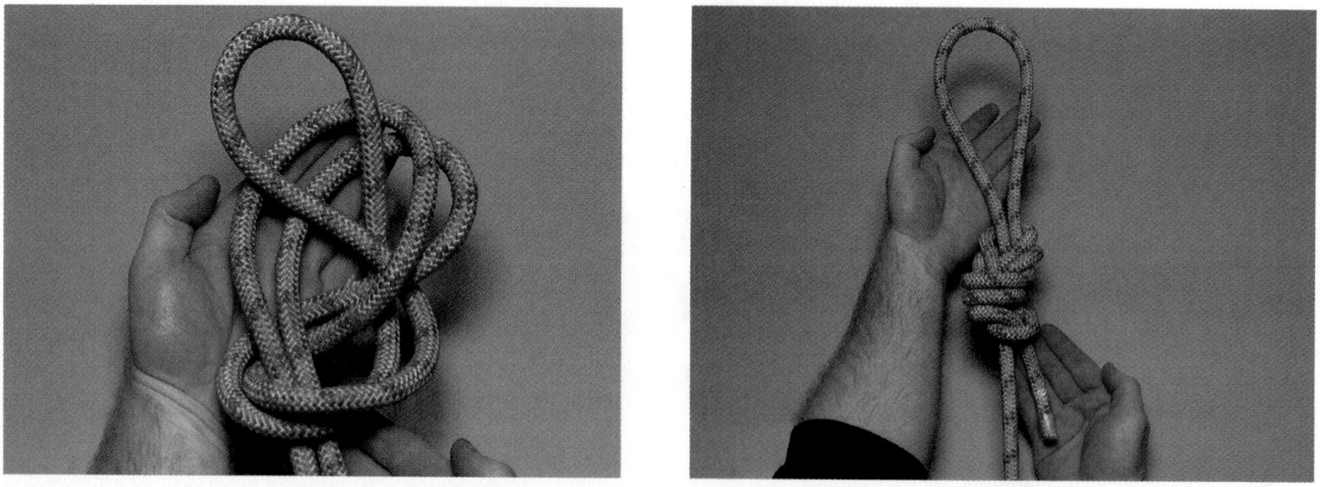

(A) (B)

Figure 15-15 (A) A loose and sloppy knot. (B) A knot properly dressed and set.

JOB PERFORMANCE REQUIREMENT 15-1
Making a Half Hitch Around an Object

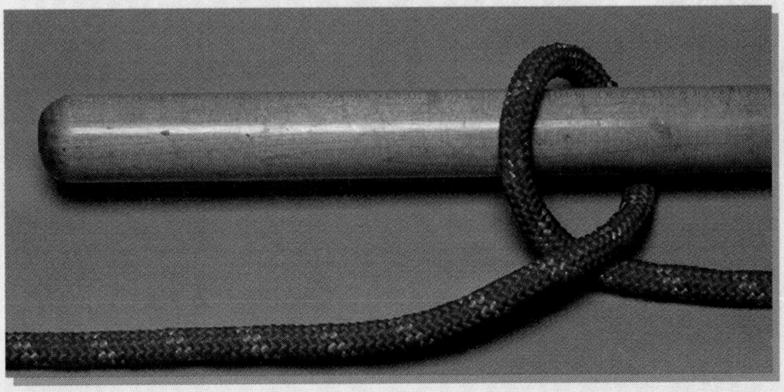

Make a round turn in the standing portion of the rope, and slide the round turn down over the object being hoisted.

Half Hitch

A half hitch, **JPR 15-1**, is almost always utilized in conjunction with some other knot and is used to maintain the proper orientation of the object being hoisted.

1. Make a round turn in the standing portion of the rope.
2. Slide the round turn down over the object being hoisted, making sure that the running end passes under the working end. If necessary, on particularly long objects such as pike poles more than one half hitch can be used.

Overhand (Safety) Knot

An overhand knot, **JPR 15-2**, is generally used to secure the loose end of the working end after tying a knot. All of the knots described in this chapter should be secure in and of themselves, but the addition of this simple knot provides an extra measure of security.

> **Note** A knot should not be considered complete until a safety knot is tied to back it up.

1. Take the loose end of the working end after tying your primary knot and secure it by making a round turn around the standing part and bringing the loose end through between this round turn and the primary knot.

Clove Hitch

The clove hitch is used to attach a rope to an object, such as a pole, tree, or fence post, or to a tool, such as a pike pole or hoseline. A clove hitch can be tied anywhere along the rope if needed. It will hold

JOB PERFORMANCE REQUIREMENT 15-2
Tying an Overhand Safety Knot Around an Object

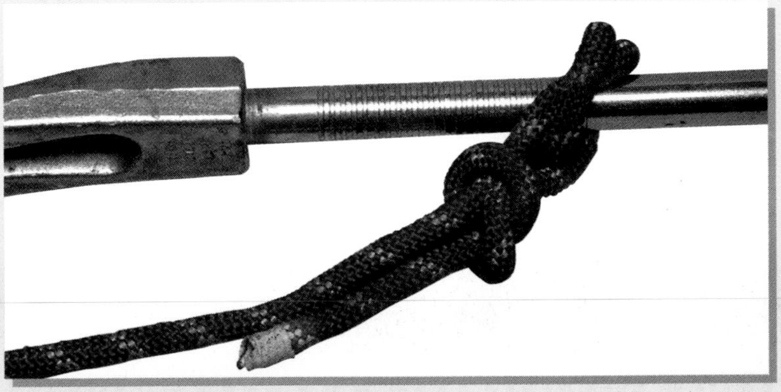

Take the loose end of the working end after tying the primary knot, and secure it by making a round turn around the standing knot and bringing the loose end through.

equally well when tension is applied from either direction—if tied correctly.

There are basically two ways of tying a clove hitch. Step-by-step instructions and figures are provided for each method. The first way, which is fastest, is to tie a clove hitch in hand (sometimes referred to as in the open) and then slip it over the object. In many cases, however, this is not possible, so it is important that firefighters also know how to tie the clove hitch around an object.

Clove Hitch Tied in the Open (JPR 15-3)

1. Hold the rope in your hands with the working end to your right, **JPR 15-3A.**
2. Form a loop with the working end passing in front of the standing part (between you and the rope).
3. Move the rope down to the right and form another loop exactly like the one formed in step 2, **JPR 15-3B.**

JOB PERFORMANCE REQUIREMENT 15-3

Clove Hitch Tied in the Open

A Form a loop with the working end passing in front of the standing part (between you and the rope).

B Move the rope down to the right and form another loop exactly like the one formed in step A.

C Place the loop formed in step B behind the loop formed in step A. You now have a clove hitch.

D Keeping the clove hitch loops in the proper orientation to one another, slide them over the object to be secured.

E Finish with a safety knot.

4. Place the loop formed in step 3 behind the loop formed in step 2. You now have a clove hitch, **JPR 15-3C.**
5. Keep the clove hitch loops in the proper orientation to one another and slide them over the object to be secured, **JPR 15-3D.**
6. Finish with a safety knot, **JPR 15-3E.**

Clove Hitch Tied around an Object (JPR 15-4)

1. With the working end, make a round turn around the object, **JPR 15-4A.**
2. Cross the working end over the standing part and make another round turn around the object.
3. Bring the working end under at the point you crossed with the first round turn. At this point your round turns should be side by side with the standing part and the loose end of the working end opposing one another, **JPR 15-4B.**
4. Finish with a safety knot, **JPR 15-4C.**

Becket Bend and Double Becket Bend

Also known as the sheet bend and double sheet bend, the becket bend and the double becket bend are very useful knots for tying ropes together. The becket bend is utilized to tie ropes of equal diameter together, while the double becket bend is used most often when tying ropes of unequal diameter.

Tying a Becket Bend (JPR 15-5)

1. Form a bight in the running end of the line that needs to be extended, **JPR 15-5A.**
2. Utilizing the working end of the rope being added, bring it up through the bight, **JPR 15-B.**

JOB PERFORMANCE REQUIREMENT 15-4

Clove Hitch Tied Around an Object

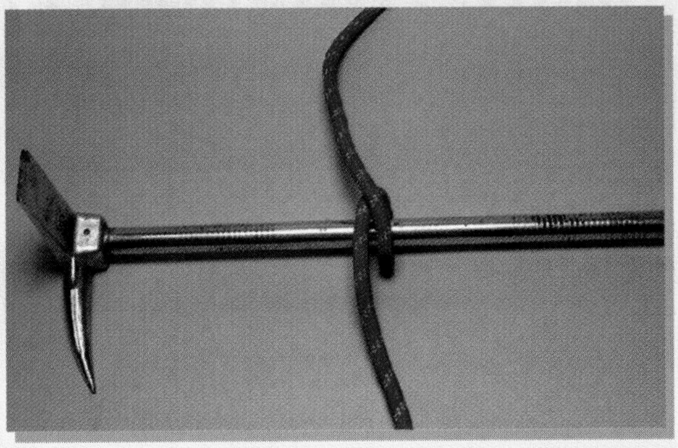

A With the working end, make a round turn around the object. Cross the working end over the standing part and make another round turn around the object.

B Bring the working end under at the point you crossed with the first round turn. At this point your round turns should be side by side with the standing part and the loose end of the working end opposing one another.

C Finish with a safety knot.

3. Take this same end around both sides of the bight, forming a loop around the bight, **JPR 15-5C.**
4. Still utilizing this same end, pass it between itself and the rope in which you formed the bight.
5. When dressed and set, the end you were working with should be at approximately 90 degrees to the rope, **JPR 15-D.**
6. Finish with a safety knot, **JPR 15-5E.**

Tying a Double Becket Bend (JPR 15-6)

1. Form a bight in the running end of the larger line, **(JPR 15-6A).**
2. Utilizing the working end of the smaller rope, bring it up through the bight, **JPR 15-6B.**
3. Take this same end all the way around both sides of the bight forming a round turn around the bight, **JPR 15-6C.**
4. Continue around the bight a second time, **JPR 15-6D.**

JOB PERFORMANCE REQUIREMENT 15-5

Tying a Becket Bend

A Form a bight in the running end of the line needing to be extended.

B Utilizing the working end of the rope being added, bring it up through the bight.

C Take this same end around both sides of the bight, forming a loop around the bight. Still utilizing this same end, pass it between itself and the rope in which you formed the bight.

D When dressed and set, the end you were working with should be at approximately 90 degrees to the rope.

E Finish with a safety knot.

5. Still utilizing this same end, pass it between itself and the rope in which you formed the bight. At this point the rope you have been working with should have formed what looks like the number eight (8), **JPR 15-6E.**
6. When dressed and set, the end you were working with should be at approximately 90 degrees to the rope, **JPR 15-6F.**
7. Finish with a safety knot.

Bowline Knot

Although it was the mainstay of fire service knots for years, the advent of synthetic fiber ropes has greatly reduced the utilization of the bowline knot. This is due in large part to the fact that the bowline is an inherently loose knot, which is beneficial when it needs to be untied but has a dangerous tendency to slip when used on synthetic fiber ropes.

JOB PERFORMANCE REQUIREMENT 15-6
Tying a Double Becket Bend

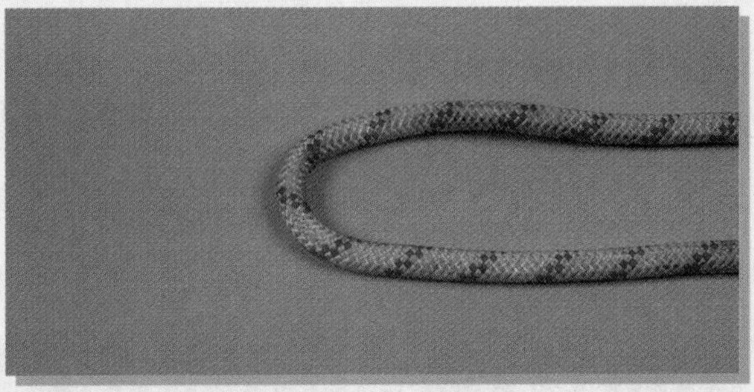

A Form a bight in the running end of the larger line.

B Utilizing the working end of the smaller rope, bring it up through the bight.

C Take this same end all the way around both sides of the bight, forming a round turn around the bight.

D Continue around the bight a second time.

E Still utilizing this same end, pass it between itself and the rope in which you formed the bight.

F When dressed and set, the end you were working with should be at approximately 90 degrees to the rope. Finish with a safety knot.

Caution If a bowline is not properly dressed, set, and secured with a safety knot there is a possibility of it inverting and becoming a slip knot.

Tying a Bowline (JPR 15-7)

1. Holding the standing part in your hand with the working end coming toward you, form a loop in the rope with the working end on your side when it passes the standing part, **JPR 15-7A.**
2. Still utilizing the working end, bring it up through the loop formed in step 1, **JPR 15-7B.**
3. Pass this same end around behind the standing part and then bring it back down through the loop you brought it up through.
4. Be sure that the working end is on the inside of the loop that you have just formed with the completed knot. If left on the outside there is a much greater chance of it getting caught and the knot inverting to a slip knot, **JPR 15-7C.**
5. Finish with a safety knot, **JPR 15-7D.**

Figure Eight Knots

The figure eight knots came into use in the fire service during the same time line as synthetic fiber rope. This was a natural transition because cavers and

JOB PERFORMANCE REQUIREMENT 15-7

Tying a Bowline

A Holding the standing part in your hand with the working end coming toward you, form a loop in the rope with the working end on your side when it passes the standing part.

B Utilizing the working end, bring it up through the loop formed in step 1.

C Pass this same end around behind the standing part and then bring it back down through the loop you brought it up through. Be sure that the working end is on the inside of the loop that you have just formed with the completed knot. If left on the outside there is a much greater chance of it getting caught and the knot inverting to a slip knot.

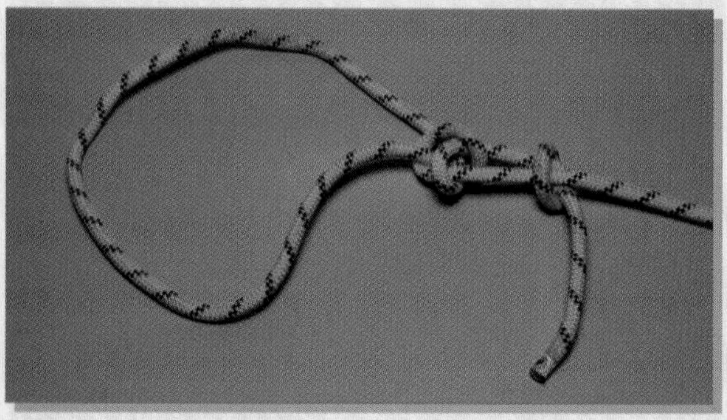

D Finish with a safety knot.

mountaineers, from whom the fire service adopted synthetic fiber ropes, had been utilizing figure eight knots for years. Figure eight knots have several good features. They are very simple to tie; they form a knot that will not slip or pull from the rope (as a bowline knot will); and they place less stress on the rope, as the turns in the figure eight are not as sharp as in the other knots. Figure eight knots are the preferred knots when working with synthetic rope.

The basic name *figure eight knot* is used to describe many different knots with a modifier added to the basic name to delineate one from the other. This section covers the basic figure eight, the follow-through figure eight, and the figure eight on a bight.

Basic Figure Eight Knot

The basic figure eight knot shown in **JPR 15-8** is useful when an "end of the line" knot is needed, such as when a rappel rope is not long enough to reach the ground or solid landing. This knot is also useful when joining two ropes of equal diameter together.

Tying a Figure Eight (JPR 15-8)

1. Form a bight in the working end of the rope, holding the bight out in front of you by the apex so that both sides of the bight are hanging parallel, **JPR 15-8A.**
2. Take the short end (working end) of the rope and wrap it around the standing part from front to rear, **JPR 15-8B.**
3. After the working end comes around the standing part, you will have formed an eye. Now take the working end through the eye from front to rear, **JPR 15-8C.**
4. If you are joining two ropes, at this point you will take the working end of the second rope and, coming from the opposite direction, trace the path of the original rope as it flows through the knot. You should end up with both working ends lying parallel to the opposite rope's standing part, **JPR 15-8D.**
5. Finish with a safety knot with each working end around the opposite rope's standing part, **JPR 15-8E.**

Follow-Through Figure Eight Knot

The follow-through figure eight knot shown in **JPR 15-9** is very useful when attaching a utility or life safety line rope to an object that does not have a free end available.

Tying a Follow-Through Figure Eight (JPR 15-9)

1. Follow the directions for tying a basic figure eight; however, you need to make sure that you have an adequate amount of rope left on the working end side of the basic knot to go around the object you are securing the rope to, **JPR 15-9A.**
2. If you are securing the rope to an object, wrap the working end around the object at this point, **JPR 15-9B.** (From this point on the procedure is very similar to the procedure for joining two ropes together.)
3. As the name implies, you now follow the other rope through the knot, coming from the opposite direction. In effect, you have formed a figure eight on a bight (the next knot described) around an object, **JPR 15-9C.**
4. You should conclude with the end of the working end parallel to the standing part, **JPR 15-9D.**
5. Finish with a safety knot, **JPR 15-9E.**

Figure Eight on a Bight

The figure eight on a bight shown in **JPR 15-10** is utilized when the object being attached to has a free end available to place the rope over and is very useful in forming a loop in the line that will not slip or cinch up. This knot can also be used when a loop is needed in the middle of a rope.

Tying a Figure Eight on a Bight (JPR 15-10)

1. Start by forming a bight. The size of the bight needs to be relative to the object the finished loop is going to be placed over or around, **JPR 15-10A.**
2. Grab both ropes so that the bight is hanging down from one side of your hand while the standing part and the end of the working part are hanging out of the opposite side of your hand. Basically you are forming a second bight of doubled ropes, **JPR 15-10B.**
3. Hold this bight out in front of you by the apex so that both sides (four parts of the same rope) of the bight are hanging parallel, **JPR 15-10C.**
4. Take the first bight and wrap it around the standing part from front to rear, **JPR 15-10D.**
5. After the bight comes around the standing part, you will have formed an eye. Now take the bight through the eye from rear to front, **JPR 15-10E.**
6. You have now formed an easy, reliable loop that will not slip. Do not forget to dress and set your knot and apply a safety with the end of the working end, **JPR 15-10F.**

JOB PERFORMANCE REQUIREMENT 15-8

Tying a Figure Eight

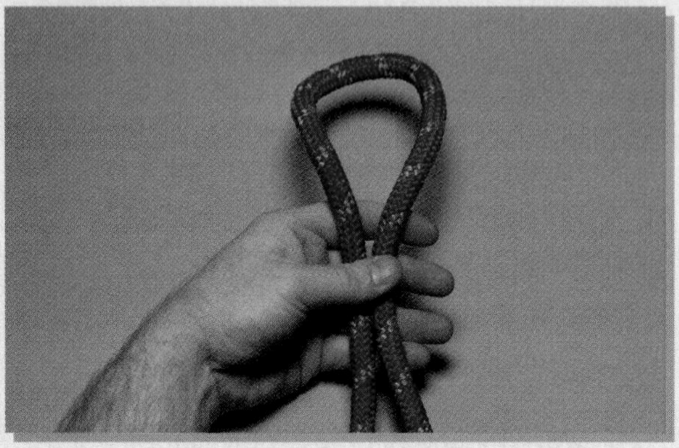

A Form a bight in the working end of the rope, holding the bight out in front of you by the apex so that both sides of the bight are hanging parallel.

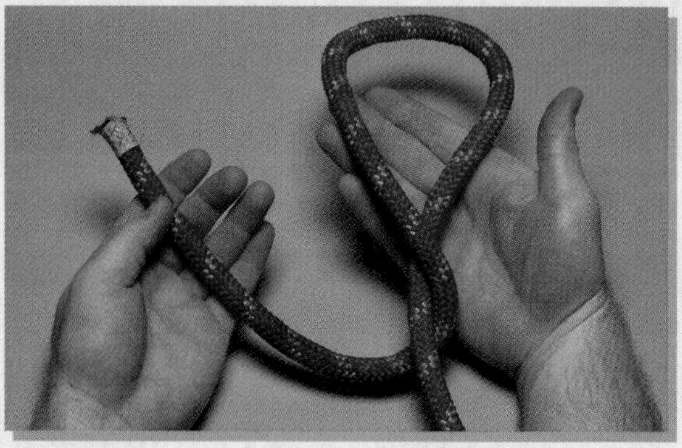

B Take the short end (working end) of the rope and wrap it around the standing part from front to rear.

C After the working end comes around the standing part you will have formed an eye. Now take the working end through the eye from front to rear.

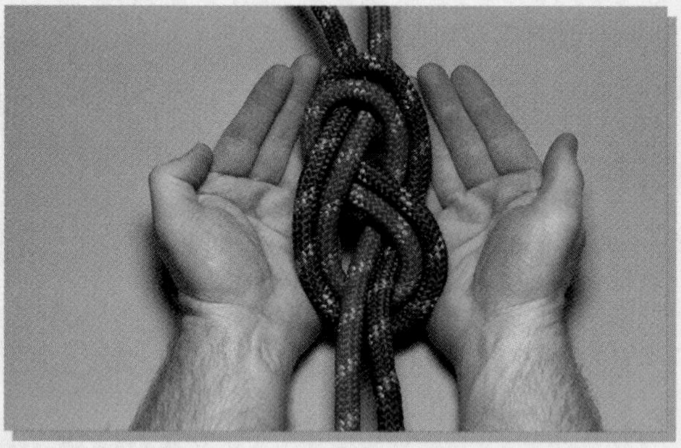

D If you are joining two ropes, at this point you will take the working end of the second rope and, coming from the opposite direction, trace the path of the original rope as it flows through the knot. You should end up with both working ends lying parallel to the opposite rope's standing part.

E Finish with a safety knot with each working end around the opposite rope's standing part.

JOB PERFORMANCE REQUIREMENT 15-9

Tying a Follow-Through Figure Eight

A Follow the directions for tying a basic figure eight; however, you need to make sure that you have an adequate amount of rope left on the working end side of the basic knot to go around the object you are securing the rope to.

B If you are securing the rope to an object, wrap the working end around the object at this point.

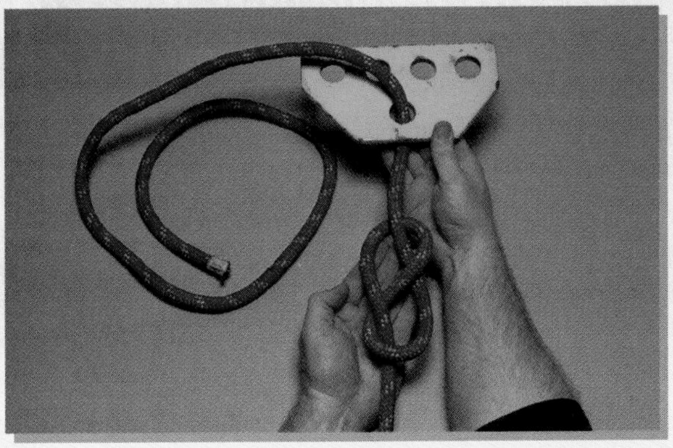

C As the name implies, you now follow the other rope through the knot, coming from the opposite direction. In effect, you have formed a figure eight on a bight around an object.

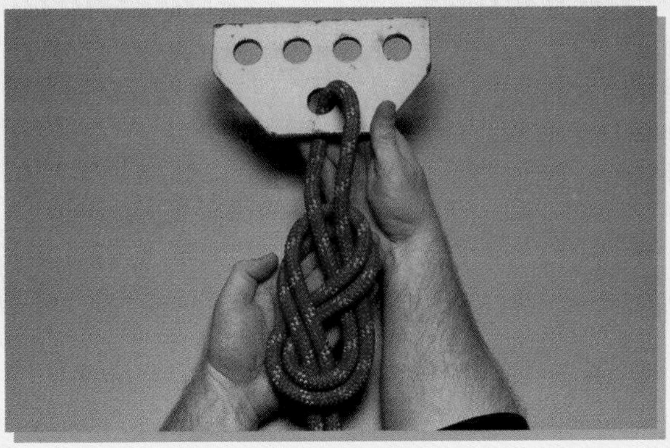

D You should conclude with the end of the working end parallel to the standing part.

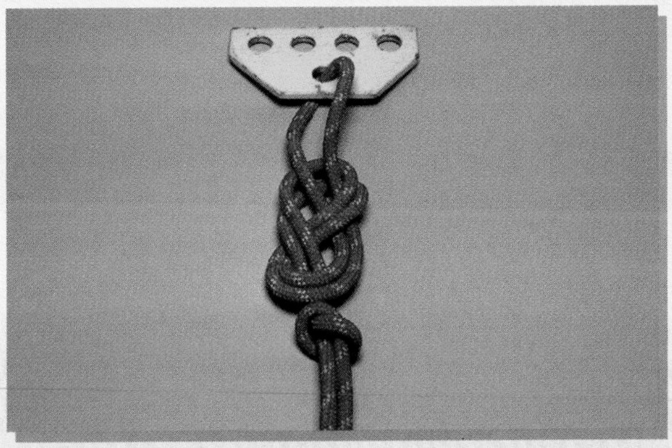

E Finish with a safety knot.

JOB PERFORMANCE REQUIREMENT 15-10

Tying a Figure Eight on a Bight

A Start by forming a bight. The size of the bight needs to be relative to the object the finished loop is going to be placed over or around.

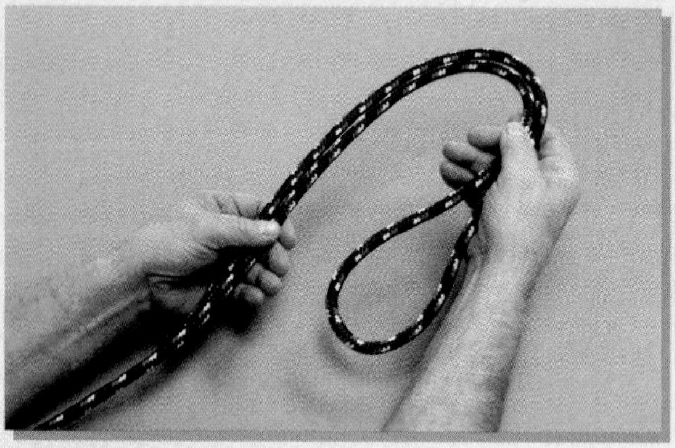

B Grab both ropes so that the bight is hanging down from one side of your hand while the standing part and the end of the working part are hanging out of the opposite side of your hand. Basically, you are forming a second bight of doubled ropes.

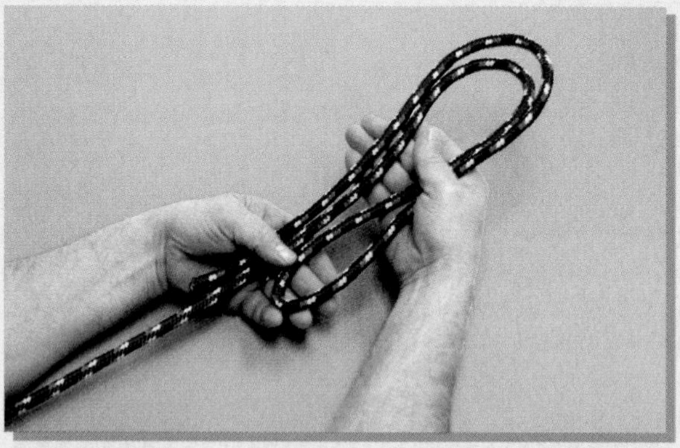

C Hold this bight out in front of you by the apex so that both sides (four parts of the same rope) of the bight are hanging parallel.

D Take the first bight and wrap it around the standing part from front to rear.

E After the bight comes around the standing part, you will have formed an eye. Now take the bight through the eye from front to rear.

F You have now formed an easy, reliable loop that will not slip. Do not forget to dress and set your knot and apply a safety with the end of the working end.

Rescue Knot

Almost every department and/or recruit school has a knot or combination of knots that they recognize as the best knot to be used as a rescue knot. The rescue knot described in this section is by no means intended to imply that their choice is incorrect or to disparage their choice in any way. The knot presented here and shown in **JPR 15-11** is one of the easiest, fastest, and surest among the many that have been tried over the years. It should come as no surprise that the primary knot utilized is a member of the figure eight family, a figure eight on a bight.

This rescue knot can be used on oneself, a conscious patient, or an unconscious patient. The basic method of applying this rescue knot does not change with the patient; however, the difficulty factor does go up if the patient is unconscious. The following instructions are based on applying the rescue knot to oneself. However, it is also important to practice the rescue knot on other people to simulate victims needing rescue. An emergency scene is not the proper venue for learning skills.

Tying a Rescue Knot (JPR 15-11)

1. Tie a figure eight on a bight. Your initial bight should be approximately 6 feet long. This will allow you to tie the figure eight on a bight, including a safety at the working end, and still have a finished loop approximately 4 to 5 feet long. This loop is going to need to be long enough to go around the victim's waist and back up between the legs. The loop does not need a precise measurement, but it is important to remember that the length cannot be adjusted once the knot is tied. Time should be spent sizing and applying this knot on different size victims in practice scenarios. Firefighters should rehearse this skill until they feel comfortable and can be consistently successful with this maneuver on the first try, **JPR 15-11A.**
2. Place the loop on the floor or step into it with the actual figure eight knot behind you. Raise the loop to waist level, being careful not to get above the top of the hip bones, **JPR 15-11B.**

JOB PERFORMANCE REQUIREMENT 15-11

Tying a Rescue Knot

A Tie a figure eight on a bight. Your initial bight should be approximately 6 feet long. This will allow you to tie the figure eight on a bight, including a safety at the working end, and still have a finished loop approximately 4 to 5 feet long.

B Place the loop on the floor or step into it with the actual figure eight knot behind you. Raise the loop to waist level being careful not to get above the top of the hip bones.

C Reach down between your legs pulling the standing part of the rope up in front of you. The standing part should not be between your body and the loop but on the outside of the loop.

(continued)

JOB PERFORMANCE REQUIREMENT 15-11

Tying a Rescue Knot (Continued)

D Continue to hold the rope in front of you and raise your hand, holding the rope above your head. Reach out with the other hand and grasp both sections of rope.

E Make three full twists with the hand over your head and form a loop. Place the loop over your head and shoulders so that it comes to rest under your arms.

F Keep the standing part pointed down and adjust the slack out of the rope between the figure eight knot coming from your legs and the twists forming the loop around your chest. Raise the standing part up over your head and allow the twists to wrap around the rope. Cinch down on the rope, forming the loop around your chest. Arrange the ropes between your legs as necessary.

3. Reach down between your legs, pulling the standing part of the rope up in front of you. The standing part should not be between your body and the loop but on the outside of the loop, **JPR 15-11C.**
4. Continuing to hold the rope in front of you, raise your hand holding the rope above your head. Reach out with the other hand and grasp the ropes, both the one coming from between your legs and the one hanging down from the hand over your head, **JPR 15-11D.**
5. With the hand over your head make three full twists, forming a loop. Place the loop over your head and shoulders so that it comes to rest in your underarms, **JPR 15-11E.**
6. Keeping the standing part pointed generally down, adjust what slack you can out of the rope between the figure eight knot coming from between your legs and the twists forming the loop that is now around your chest, **JPR 15-11F.**
7. Raise the standing part up over your head, allowing the twists to wrap around themselves and cinching down on the rope forming the loop around your chest. This is important to keep the loop from cinching down on your chest.
8. Arranging the ropes that come up between your legs as your weight is lifted by the rope will help ease the discomfort somewhat. Remember this tip when lifting a conscious or unconscious patient.

Water Knot

Webbing has become very popular and is carried by many firefighters in their personal protective clothing for use as an emergency harness or hose holding strap.

> **Safety** The water knot, **JPR 15-12**, is the only knot that is recommended for use when tying webbing.

This is one knot that absolutely must be kept neat and be tightly set.

Tying a Water Knot (JPR 15-12)

1. Tie a simple overhand knot in one end of the webbing, making sure the webbing lies flat as it crosses back over itself. Leave enough tail beyond the overhand knot to be able to place a safety knot on that side of the water knot when finished, **JPR 15-12A.**
2. Take the other end of the webbing and thread it through in the opposite direction, again making sure that the webbing lies flat as it crosses itself. Make sure you feed enough webbing through so that you are able to place a safety knot on this side of the water knot, **JPR 15-12B.**

INSPECTION

As with any emergency service tool, all ropes must be inspected and properly maintained to ensure they are in good shape for use during an emergency incident. These inspections should be a matter of department policy and done on a regular basis; many departments conduct monthly inspections. It is also recommended that all inspections of life safety ropes be logged on a form specifically designed for this purpose. These forms can be purchased from most rescue supply houses or the department can develop its own. An example is shown in **Figure 15-16**. In order for an inspection log to be useful, the ropes themselves need to be individually identified.

If a life safety rope is found to be damaged or suspect it should be immediately removed from service. According to NFPA 1983 a life safety rope must also be removed from emergency operation service once it has been used during an actual

DELMAR ROPE USAGE & HISTORY

SERIAL NUMBER:	I.D. MARKING:	LENGTH: 155′	DIAMETER: 1/2″
DATE OF MFG.: 6-6-94	ISSUE DATE: 3-11-95	DATE IN SERVICE: 3-11-95	
FIBER: Kernmantle	COLOR: Red	CONSTRUCTION: E-2-Bend	MFG'S LOT NUMBER: 06F94421KB

INSPECT ROPE FOR DAMAGE OR EXCESSIVE WEAR EACH TIME IT IS DEPLOYED AND AGAIN AFTER EACH USE. IMMEDIATELY RETIRE ALL SUSPECT ROPES.

DATE USED	INCIDENT LOCATION	TYPE OF USE	ROPE EXPOSURE	DATE INSPECTED	INSPECTOR'S INITIALS	ROPE CONDITION & COMMENTS
9/16/95	Shaker High	Training	Ground, Grass + Dirt Some Sunlight	9/16/95	V.G. F. Carr, Jr.	Good No Damage
11/18/95	Vecky	Training	Ground, Dirt, Grass Rocks, MLN	11/18/95	F. Carr, Jr.	Good No Damage
7/15/96	Boght	Training	Dirt, Grass, Rocks	7/15/96	F. Carr, Jr.	Good No Damage

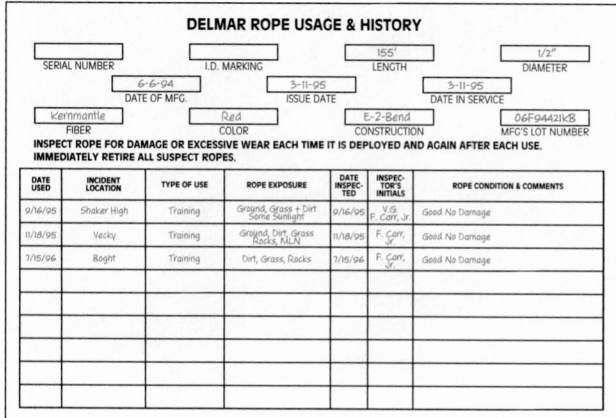

Figure 15-16 A standard life safety rope inspection form.

JOB PERFORMANCE REQUIREMENT 15-12

Tying a Water Knot

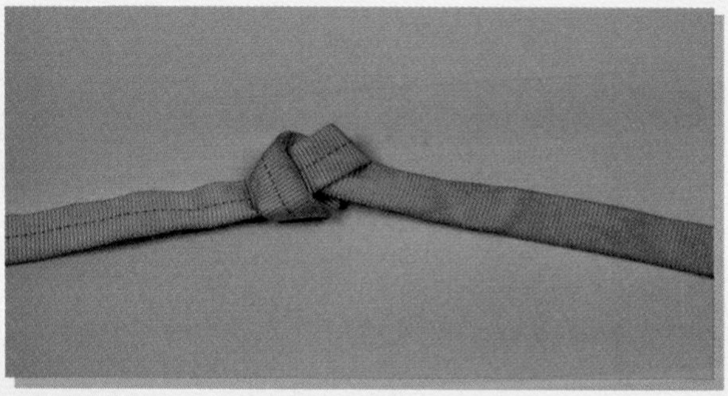

A Tie a simple overhand knot in one end of the webbing, making sure the webbing lies flat as it crosses back over itself. Leave enough tail beyond the overhand knot to be able to place a safety knot on that side of the water knot when finished.

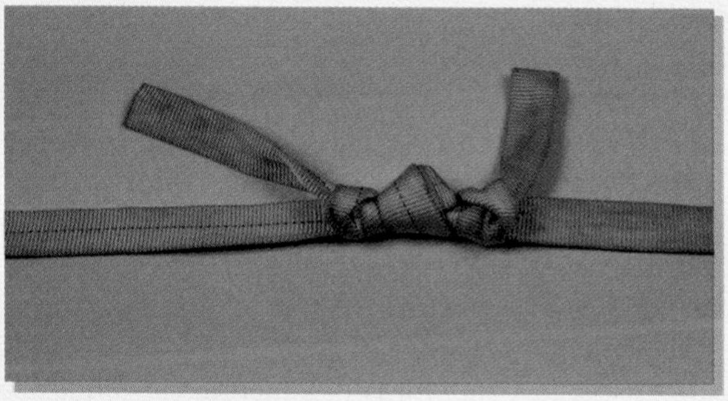

B Take the other end of the webbing and thread it through in the opposite direction, again making sure that the webbing lies flat as it crosses itself. Make sure you feed enough webbing through so that you are able to place a safety knot on this side of the water knot.

Figure 15-17 It is very important that rope is inspected as it is being put back into a rope bag.

emergency operation. This is mainly due to the unavoidable rigors that can be placed on a life safety rope during emergency operations, such as shock loading, overloading, chafing, abrasion, and exposure to chemicals/acids. These ropes, although removed from actual emergency operations, can be used for training as long as they are inspected and determined to be undamaged. *Training ropes should be inspected after every use.*

Ropes should be inspected along their entire length. This can be done when they are being placed back in storage, coiled, or bagged, **Figure 15-17**. It is best to inspect ropes after they have been cleaned. Generally, all ropes should be visually inspected for abrasion, laceration, chemical exposure, melting, and excessive fuzzing, many of which can be seen in **Figure 15-18**. They should also be inspected **tactilely** by running the rope through the hands. For this reason, firefighters should not wear gloves when inspecting and storing a rope, because they are feeling for foreign material

Figure 15-18 Damaged rope.

embedded in the rope, slippery spots (possible chemical damage), voids in the center fibers (kernmantle and braid-on-braid), stiff or hard spots, and soft spots.

Laid (Twisted)

When inspecting laid ropes, firefighters should look and feel for all of the items just mentioned. In addition, laid ropes should be untwisted at random intervals to inspect the inside between the strands, **Figure 15-19**. When untwisting the strands of a natural fiber rope, it is important to watch for small particles such as sand or gravel that may damage the rope fibers. If a rope is stored wet or stored in a very humid location there may be signs of mold, mildew, or rot. The presence of mildew is not necessarily a terminal problem if found early enough. If these signs are present, the rope should be

Figure 15-19 It is important to twist apart a laid rope to inspect between strands.

removed from service and the problem should be addressed as soon as possible.

Braided

Braided rope should be inspected visually and tactilely as described previously. It cannot be twisted apart, so inspecting the inside is not possible. With braided rope, all strands will appear on the surface somewhere along the rope.

Braid-on-Braid

It is very important to remember, when inspecting braid-on-braid rope, that there is no way to see the inside braided rope. The person doing the inspection needs to pay particular attention to the tactile part of the inspection, feeling for possible types of damage previously listed. Another problem with this type of rope is that the outside braid will sometimes slip over the inner braid, causing the rope to invert on itself. A rope with this problem should be immediately removed from service. The outside braid typically represents 50 percent of the strength, so any damage to the outside braid will have a significant impact on the remaining strength.

Kernmantle

As with braid-on-braid rope, there is no way to see the kern portion, which typically represents 75 percent of the total strength of a kernmantle rope. The tactile inspection is the best and only way to discover damage to the kern. It is possible (but unusual) for the kern to be damaged without damaging the mantle. One should be alert for any change in the regular uniform feel of the rope as it is run through ungloved hands. Irregularities that can be discovered in a thorough tactile inspection include flat spots, voids, bunches, stiffness, and limpness. The key is to search for a different feel. If a minor imperfection seems to have been located, it may not necessitate the removal of the rope, but would certainly indicate the need for another thorough inspection.

Streetsmart Tip Another way to inspect the rope, especially if a minor imperfection seems to have been located, is to tie it off to an object and do a thorough tactile inspection while placing slight tension on it. This may cause the minor imperfections to be more pronounced and therefore easier to feel.

MAINTENANCE

As with all firefighting and life safety tools and equipment, proper maintenance of utility and life safety ropes is required in order to ensure that they are available, safe, and ready for immediate use at an emergency scene. The maintenance of ropes is not difficult or complicated; it is generally limited to cleaning, inspecting, and storing.

Occasionally the firefighter may be called on to assist in placing new rope in service. *If the rope was purchased in bulk, it is important that the manufacturer's instructions with regard to cutting, sealing, whipping, and so on, be carefully adhered to.* The same goes for cleaning and storage of ropes. Most new ropes (especially life safety ropes) purchased from a rescue equipment dealer come with instructions and warnings, either attached to or packaged with the rope. An example is shown in **Figure 15-20**.

Cleaning

As mentioned in the previous section, the best policy for cleaning is to follow the manufacturer's instructions; however, since not all rope comes with instructions, the following are some general guidelines for the cleaning of rope.

Natural Materials

The cleaning of natural fiber rope is very difficult because water cannot be utilized. Remember from the earlier discussion that natural fiber ropes lose approximately 50 percent of their strength when wet and do not regain the lost strength when they dry. Firefighters are limited to brushing off the loose dirt and foreign materials with a stiff broom or brush,

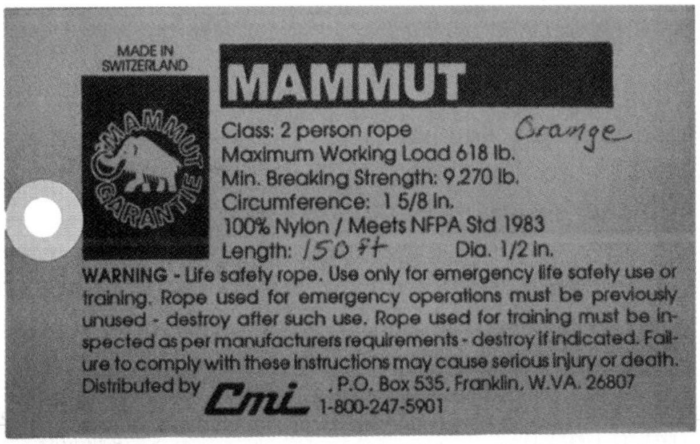

Figure 15-20 A standard manufacturer's warning/instruction label.

Figure 15-21 Manila rope is best cleaned by laying it out and brushing it off.

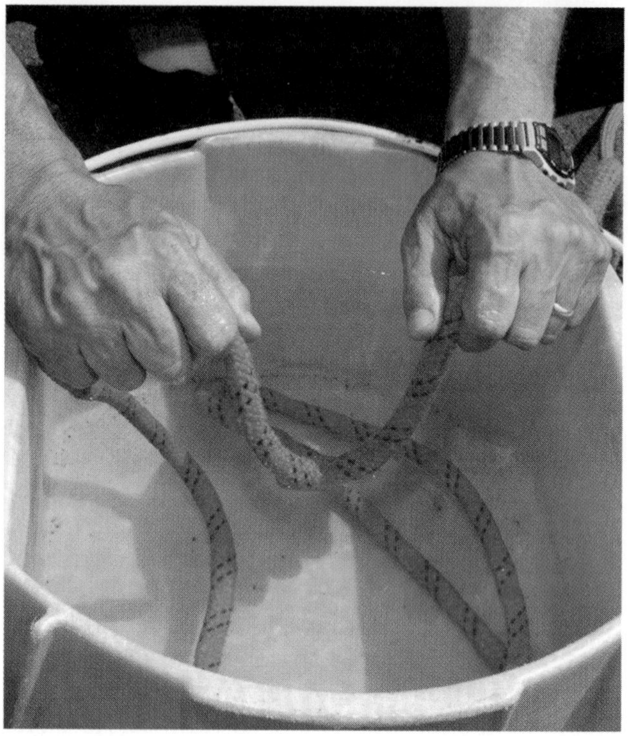

Figure 15-22 A kernmantle (nylon) rope being washed in a bucket.

Figure 15-21. This is another reason why the fire service is moving away from the use of natural fiber rope, even as utility rope.

Synthetic Materials

Ropes manufactured from synthetic materials can be cleaned in a number of ways. Once again it is always the best policy to follow the manufacturer's instructions. In the absence of these, however, the following procedures can be utilized:

1. Use tap/cold water (high-temperature hot water may damage rope).
2. If detergent use is necessary, use only mild detergent that has been well diluted. (Use absolutely no bleach or any cleaning product containing bleach.)

Hand Washing

The most basic method of washing rope is to wash it by hand, using a large utility sink, bucket, or hose and scrub brush. The first two methods (sink or bucket) require immersing the rope to be washed in the sink or bucket, letting it soak for a few minutes, and then agitating and scrubbing it with the hands or small brushes, **Figure 15-22**. The third method (hose and scrub brush) described is to lay the rope out on the apparatus bay floor or apron, wet it with a garden hose, and scrub it with a brush or broom. If detergent is needed, it can be mixed in a bucket and the brush or broom can be dipped into it.

If detergent is used with any of these methods, all residue must be removed. If the sink or bucket method is used, this can be accomplished by changing the water or using other sinks or buckets full of fresh water to thoroughly rinse the rope. With the hose method, the rope must be rinsed thoroughly by occasionally moving it once or twice so that the soap trapped underneath it can be rinsed away.

Rope Washer

A number of different rope washers are available for purchase from rescue equipment suppliers. These are generally small PVC devices that fasten directly to a hose bib or to the male end of a garden hose, **Figure 15-23**. They are designed so that the water is directed at the rope from all different directions at the same time. The rope is inserted in one end (the direction of travel is usually marked) of the washer and slowly pulled out the other end. Without some method of injecting detergent into the hose stream, detergent cannot be used with rope washers.

Washing Machine

If using a clothes washing machine to wash rope, it is very important to make sure that the machine itself cannot damage the rope. Only a front-loading machine (with a glass window) should be used. These machines do not have an agitator for the rope to get caught on or wrapped around. Placing the rope in a large mesh bag before placing it in a machine, **Figure 15-24**, is the best way to keep it from getting fouled in the machinery and terribly knotted. If a large mesh bag is not available, the rope can be "chained," **Figure 15-25**, to try to keep it from getting fouled and/or knotted. Once again the previous directions regarding use of detergents should be followed and all residue rinsed out of the rope.

Drying Ropes

No matter what cleaning method is utilized, the rope must be completely dry prior to being stored. A few different methods of drying a rope are discussed next.

Laying Flat to Dry. The rope can be laid out on the apparatus bay floor or any other area that is clean, dry, and out of direct sunlight. Rope should be turned as it dries so it dries uniformly throughout.

Hanging to Dry. This is one of the easier methods of drying ropes. They can be hung in a hose tower, from the bar joists/trusses of the apparatus bay, or from any other place that is clean, dry, and out of direct sunlight. The rope must not come in contact with roofing tar, paint, bug spray, or other contaminants that may have been sprayed on bar joists or trusses. It is important to use care when removing

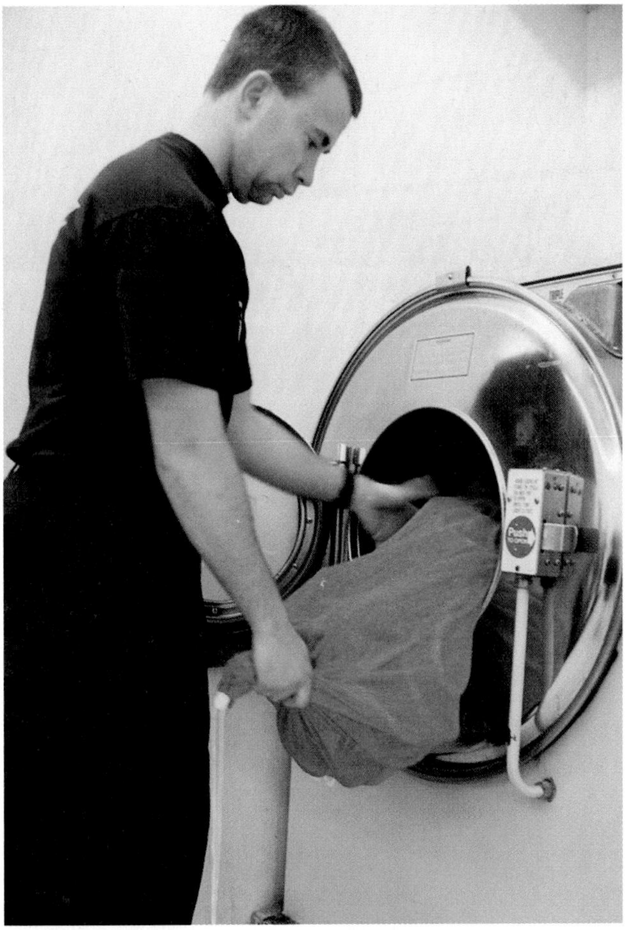

Figure 15-24 A kernmantle rope can be placed in a mesh bag and washed in a front-loading washing machine.

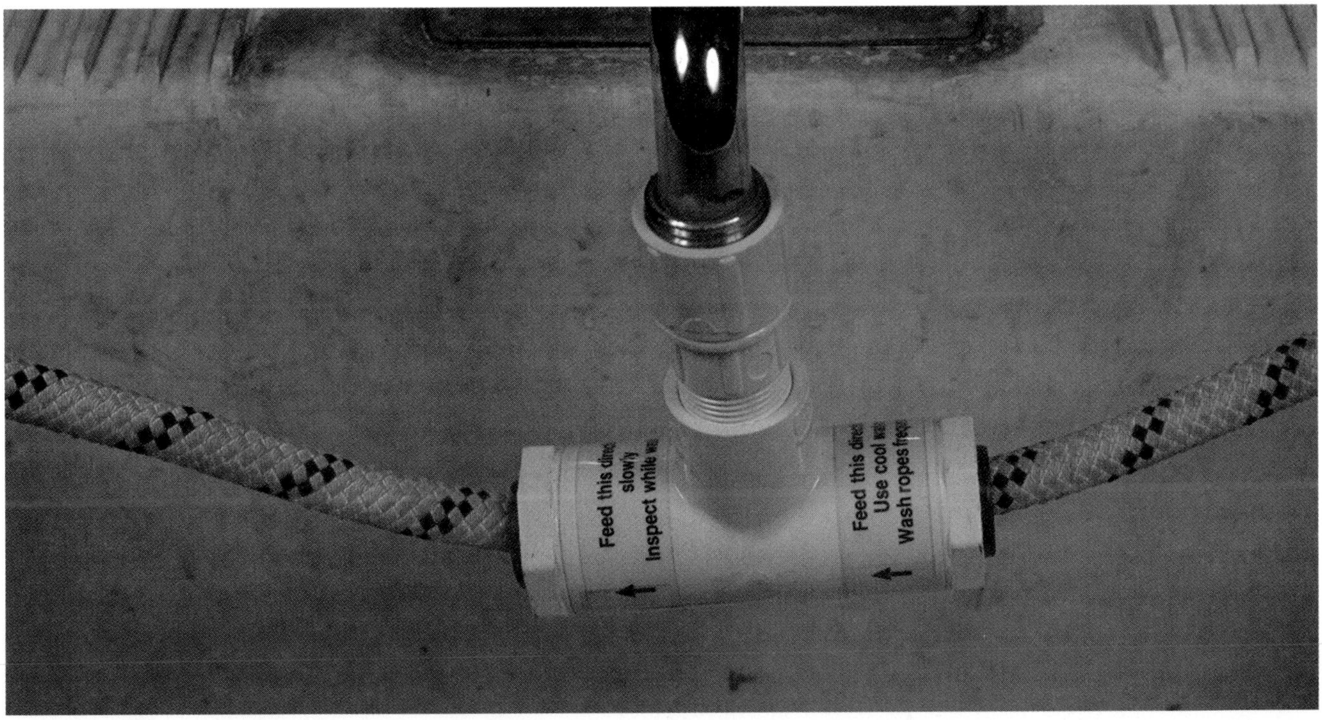

Figure 15-23 A commercial (PVC) rope washer being used to wash a kernmantle rope.

Figure 15-25 A kernmantle rope can also be "chained" and washed in a front-loading washing machine.

Figure 15-26 Properly identified, bagged, and stored life safety ropes.

the rope so that it does not get snagged or otherwise damaged by the truss plates on wood trusses or slag from welds on steel bar joists.

Machine Drying. Some departments utilize and some texts list clothes dryers as a viable alternative for drying rope.

> **Caution** Extreme caution is necessary if a clothes dryer is to be used.

Not only is there the same issue as with the washer (i.e., the possibility of the rope becoming entangled in the machinery), but with a dryer (gas or electric) it is very difficult to tell what the interior temperature is, even if there is a "low-temperature" setting. The temperature could surpass the manufacturer's recommendations.

Storage

> **Note** It is very important that rope be stored in a manner that allows for quick identification, access, and deployment at emergency operations, **Figure 15-26**.

Quick identification is very important so that a firefighter can rapidly select the proper type (utility or life safety) and length required for a given application. There are many ways to accomplish quick identification, **Figure 15-27**. Many departments use different material types, sizes, or colors of bags or tags to differentiate between utility ropes, life safety ropes, and ropes of different lengths. While there is no "right" way, to comply with NFPA 1983, each department needs to establish a policy.

Figure 15-27 Different types, sizes, and colors of bags are useful for quickly identifying different sizes and types of ropes.

Quick access is usually not a problem on fire or rescue apparatus, especially if there is a routine inspection policy in place that keeps the ropes from being buried behind other equipment. Remember that ropes must be stored away from areas where they might be exposed to battery acid, fuel (including fumes from either), or sunlight.

The issue of quick deployment has been addressed for the most part by the storage of ropes in specially designed bags, although some departments still utilize the coiling method to store their ropes. While either method, **Figure 15-28**, will work to keep the rope neatly stored, the bags have some significant advantages. Depending on its construction, a bag offers protection from dirt, liquid contaminants, sunlight, abrasion, and so on. If a rope is properly bagged, it is easier to deploy than a rope that is coiled.

Figure 15-28 The two most common methods of storing life safety ropes (bagged or coiled).

Coiling

The standard coil used by fire departments for years will work with either natural or synthetic ropes equally well. If the coiled method of storage is selected by a department, it is imperative that all members of the department be familiar with the process of both coiling and deploying from the coil. The proper process for coiling a rope is shown in **JPR 15-13.**

Some departments have built special apparatus to assist with the coiling of their ropes, but this is not really necessary. It is possible to use improvised standards and common items to coil a rope. The heel or tip of a ground ladder (do not use the tip of a roof ladder), a four-legged kitchen chair, or any item that has parallel posts (legs) approximately 18 to 24 inches apart will work for rope coiling. The distance between the standards and the number of wraps per layer are dependent on the size and length of the rope being coiled.

Coiling a Rope (JPR 15-13)

1. Measure off an amount of rope equal to approximately three times the distance between the standards to be used in tying the finished coil, **JPR 15-13A.**
2. Starting with this point measured at one of the standards, wrap the rope around the standards, forming the loops and moving toward you until you have nine to twelve loops on the standards. Then form a second layer on top of the first, moving away from you. Continue in this manner until approximately twice the amount measured off in step 1 remains, **JPR 15-13B.**

JOB PERFORMANCE REQUIREMENT 15-13

Coiling a Rope

A Measure off an amount of rope equal to approximately three times the distance between the standards to be used in tying the finished coil.

B Starting with this point at one of the standards, wrap the rope around the standards until you have nine to twelve loops on the standards. Then form a second layer on top of the first, going back the other way. Continue in this manner until approximately twice the amount measured off in step 1 remains.

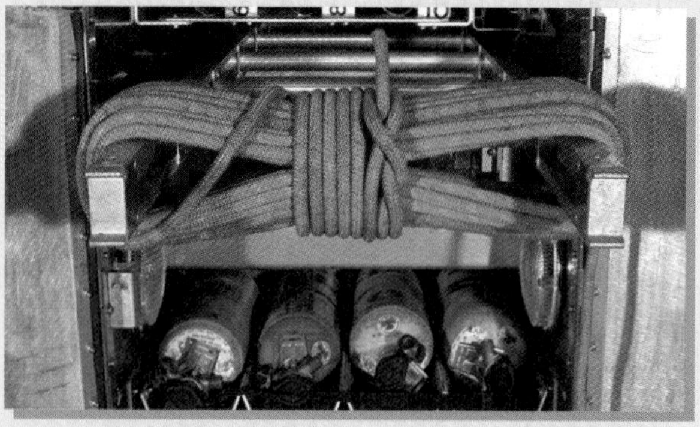

C Begin wrapping this remaining rope around the coils that you formed in step 2. Finish these wraps off with a clove hitch.

D Utilizing the rope measured off in step 1, form a bight, leaving the loose end of the rope coming out toward you. Feed the bight from the back side to the front side on the wraps. You should now have the loose end and the bight sticking out toward you.

E Place the loose end through the bight and adjust the bight until it is snug where it passes through the bight.

3. Begin wrapping this remaining rope around the coils that you formed in step 2. Finish these wraps off with a clove hitch, **JPR 15-13C.**
4. Utilizing the rope measured off in step 1, form a bight leaving the loose end of the rope coming out toward you. Feed the bight from the back side to the front side on the wraps. You should now have the loose end and the bight sticking out toward you, **JPR 15-13D.**
5. Place the loose end through the bight and adjust the bight until it is snug where it passes through the bight, **JPR 15-13E.**

To deploy a rope that has been coiled in this manner, the firefighter pulls the loose end out of the bight and pulls the bight out from between the loops. Next, the firefighter pulls a couple of loops out of the center and drops the coil if in an elevated position or feeds the rope out, pulling it from the center, as needed. A firefighter must be aware that if the coil is dropped, it may hang up before it reaches the ground, either on itself or on some structural member. If extending the rope horizontally, **Figure 15-29**, it is usually easier if one person holds the coil on the ground while another person pulls the rope out of the coil.

Figure 15-29 A coiled rope in the process of being deployed.

Bagging

The utilization of special bags for the storage of rope is another of the things that came about with the switch to synthetic fiber ropes. Ideally rope bags should have attached watertight storage for the rope log. If they do not, the rope and bag need to be clearly identified and associated with a rope log on file in the apparatus or at the station.

The process for placing a rope in a rope bag, **JPR 15-14**, is very simple (especially when compared to coiling a rope) and can be easily accomplished by one or two persons. Many bags have holes in the bottom for drainage and/or to feed the rope out of. Department policy should determine whether the rope is fed out of the bottom hole.

Bagging a Rope (JPR 15-14)

1. If policy is for the rope to be fed out the bottom hole, start by feeding enough rope out the bottom to tie a basic figure eight knot. If not, tie a basic figure eight knot in the end of the rope and place it inside at the bottom of the bag, **JPR 15-14A.**
2. Begin placing the rope in the bag by sliding your hand approximately 12 to 18 inches up the rope at a time and "stuffing" it in the bag. *Do not coil the rope in the bag; it will hang up almost every time if you do,* **JPR 15-14B.**
3. Continue this until you are at the end of the rope. Tie a basic figure eight knot in the end and place it on top of the rope in the bag. Secure the bag closed and you are finished, **JPR 15-14C.**

The process for deploying a rope stored in a bag is very straightforward. If the policy is to deploy through the bottom hole, the firefighter grasps the figure eight tied just outside the hole and either drops the bag or feeds the rope out of the bag as needed. As with a coil, if the bag is dropped, the rope may hang up on either itself or the structure. Feeding the rope out allows the firefighter to keep the bag in case a problem develops, **Figure 15-30**. If deploying the rope horizontally, the firefighter can either have someone take the end of the rope or the bag and walk in the desired direction of deployment. The rope will feed out as the person walks.

JOB PERFORMANCE REQUIREMENT 15-14

Bagging a Rope

A If policy is for the rope to be fed out the bottom hole, start by feeding enough rope out the bottom to tie a basic figure eight knot. If not, tie a basic figure eight knot in the end of the rope and place it in the bottom of the bag.

B Begin placing the rope in the bag by sliding your hand approximately 12 to 18 inches up the rope at a time and "stuffing" it in the bag. *Do not coil the rope in the bag; it will hang up almost every time if you do.*

C Continue doing this until you are at the end of the rope. Tie a basic figure eight knot in the end and place it on top of the rope in the bag. Close the bag and you are finished.

Figure 15-30 A bagged lifeline in the process of being deployed.

Figure 15-31 A firefighter deploying a rope from an elevated position in order to hoist a tool.

RIGGING FOR HOISTING

One of the primary uses of rope on an emergency scene is for hoisting of tools and equipment to the needed elevation/location. This is not the type of job that requires a life safety rope. Instead, a much smaller diameter rope can be utilized, which translates to a lighter, easier-to-carry rope. Ropes used for hoisting can be stored either coiled or bagged. It is usually much easier if the rope is fed out by the person who will be hoisting the item, **Figure 15-31**, as opposed to dropping the whole bag or trying to pass the end of the rope up to the person at the higher elevation.

This section presents examples of how to hoist a few specific tools and equipment and the proper knots to utilize when doing so.

> **Note** If a firefighter has good knowledge and is skilled in tying the knots presented in this chapter, there should be no problem with hoisting almost any type of tool or equipment.

In general, anything that has a closed handle can be hoisted with a figure eight or bowline, while longer cylindrical tools (i.e., ax and pike pole) can be hoisted using a clove hitch and half hitches.

Some departments have policies requiring the use of **tag/guide lines**, which are guide ropes held and controlled by firefighters on the ground, **Figure 15-32**. There are some circumstances in which tag lines should be used whether required or not. Examples of these circumstances are when an overhang(s) exists that the item is likely to get caught on, when the item rubbing against the side of the structure may be damaged, when there is a strong wind that may cause the item to blow out of control, or any time firefighters think a tag/guide line is necessary. All knots must be dressed, be set, and have safeties.

Specific Tools and Equipment

Ax

A small figure eight on a bight with a half hitch up the handle is the easiest and quickest way to hoist an ax.

Figure 15-32 Use of a tag/guide line.

Hoisting an Ax (JPR 15-15)

1. Tie a figure eight on a bight forming a small loop. Drop the loop over the ax handle, **JPR 15-15A.**
2. Take the loop around the head of the ax, bringing it back up, paralleling the handle. Place a half hitch approximately 6 to 8 inches below the handle, **JPR 15-15B** and **C.**

Pike Pole

Pike poles should be hoisted point up.

Hoisting a Pike Pole (JPR 15-16)

1. To hoist a pike pole point up, place a clove hitch near the end of the handle, **JPR 15-16A.**
2. Place two half hitches around the handle between the clove and the point, with the last one being located immediately below the head, **JPR 15-16B.**
3. Raise the pole with the head at the highest point, **JPR 15-16C.**

JOB PERFORMANCE REQUIREMENT 15-15

Hoisting an Ax

A Tie a figure eight knot and slip the knot over the ax handle.

B Wrap the rope over the head and back toward the handle.

C Place a half hitch near the end of the handle.

JOB PERFORMANCE REQUIREMENT 15-16

Hoisting a Pike Pole

A Tie a clove hitch near the end of the handle.

B Place half hitches on the pole. Put one immediately below the pike pole head.

C Hoist the pole up the building.

Hoselines

Hoselines can be hoisted either charged or uncharged. A charged hoseline is going to be dramatically heavier than an uncharged one.

Hoisting a Charged Hoseline (JPR 15-17)

1. With the nozzle bale in the closed (forward) position, tie a clove hitch around the hoseline 18 to 24 inches behind the nozzle, **JPR 15-17A.**
2. Form a bight in the rope and feed it through the bale from the coupling side, flipping it over before slipping it over the nozzle tip. You will have formed a half hitch by doing so, **JPR 15-17B** and **C.** Hoisted by this method, the rope will actually hold the nozzle closed should the bale be caught on something while being hoisted, **JPR 15-17D** and **E.**

Hoisting an Uncharged Hoseline (JPR 15-18)

1. Fold the nozzle back on the hose approximately 3 to 4 feet. Tie a clove hitch around the hose and nozzle to hold them together, **JPR 15-18A.**
2. Place a half hitch around the end of the hose approximately 6 inches from where it is folded back, **JPR 15-18B** and **C.**

Smoke Ejector, Chain Saw, Rotary Saw

All of these items and a host of others used on emergency scenes have closed handles or support pieces that can have rope tied around them for hoisting.

Hoisting Small Equipment (JPR 15-19)

1. Tie a follow-through figure eight through the closed handle.
2. The use of a tag line is highly recommended for items of this type, which tend to be heavy and hard to control with only the hoisting line from above.

Ladders

Both ground ladders and roof ladders are hoisted on a regular basis at emergency scenes. The hoisting procedure is the same for both types. Once

JOB PERFORMANCE REQUIREMENT 15-17

Hoisting a Charged Hoseline

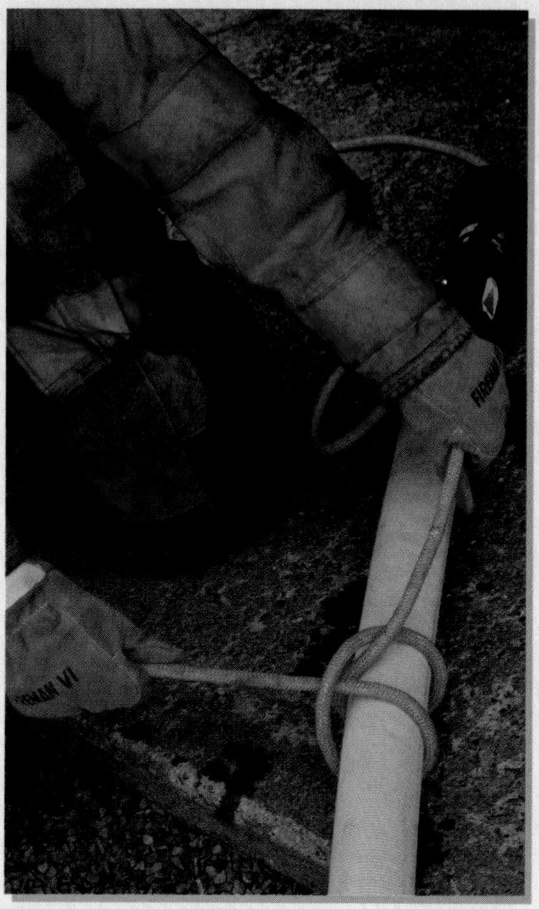

A Tie a clove hitch around the charged hoseline 18 to 24 inches behind the nozzle.

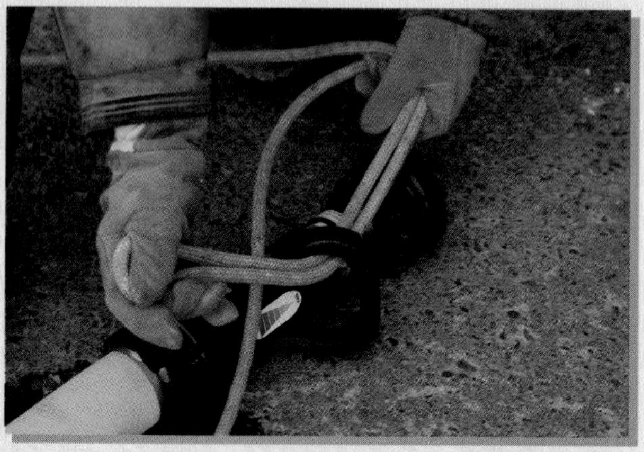

B Form a bight in the rope and feed it through the bale.

C Form a half hitch.

again the use of a tag line is recommended, especially in those cases where the firefighter on the ground will not be able to guide the bottom of the ladder as it is hoisted.

Hoisting a Ladder (JPR 15-20)

1. Tie a large figure eight on a bight forming a loop approximately 3 to 4 feet long, **JPR 15-20A.**
2. Go approximately one-third the length down the ladder and put the loop through the rungs, **JPR 15-20B.**
3. Pull the loop up and slip it around the top of the ladder allowing it to slide back down, securing the ladder, **JPR 15-20C** and **D.**

Securing a Rope between Two Objects

While the use of rope(s) to cordon off an area is not a common practice in many departments today, the need to secure a rope between two objects may arise at any emergency scene. If this need arises, a rope may be used as a barrier using one of two

JOB PERFORMANCE REQUIREMENT 15-17

Hoisting a Charged Hoseline (Continued)

D Slip the half hitch over the nozzle.

E The charged hoseline is rigged for hoisting.

JOB PERFORMANCE REQUIREMENT 15-18

Hoisting an Uncharged Hoseline

A Fold the nozzle back on the hose approximately 3 to 4 feet. Tie a clove hitch around the hose and nozzle together.

B Place a half hitch around the end of the hose, approximately 6 inches below the bend.

(continued)

JOB PERFORMANCE REQUIREMENT 15-18

Hoisting an Uncharged Hoseline (Continued)

C The uncharged hoseline is rigged for hoisting.

JOB PERFORMANCE REQUIREMENT 15-19

Hoisting Small Equipment

A rotary saw (closed handle) being hoisted with a tag line.

methods. A figure eight on a bight may be used to secure an anchor point.

Tying a Rope between Two Objects

1. Starting at one end, secure the rope to a solid object (anchor) by forming a figure eight on a bight and sliding the rope over the anchor, **JPR 15-21A.** If the rope cannot be placed over the anchor, utilize a follow-through figure eight to secure the anchor point, **JPR 15-21B.**
2. Lay the rope out, keeping it as straight as possible to minimize slack, until reaching the other objective to be tied, **JPR 15-21C.** Measure approximately one-third of the standing part of the rope toward the anchor and tie a figure eight on a bight resulting in a loop 6 to 12 inches long, **JPR 15-21D.**
3. Wrap the running end of the rope around the objective and bring it to the figure eight on a bight tied in step 2, **JPR 15-21E.**
4. Thread the running end through the loop and pull it back toward the objective, tightening the rope as necessary, **JPR 15-21F.**
5. Using the running end, tie three or four consecutive half hitches around both sections of rope, **JPR 15-21G.**

JOB PERFORMANCE TASK 15-20

Hoisting a Ladder

A Tie a large figure eight on a bight forming a loop approximately 3 to 4 feet long.

B Go approximately one-third the length down the ladder and pull the loop through the rungs.

C Pull the loop up and slip it around the top of the ladder.

D Allow the loop to slip down the ladder. Attach a tag line. Ladder is ready to be hoisted.

JOB PERFORMANCE REQUIREMENT 15-21

Tying a Rope Between Two Objects

A Secure the rope to an anchor using a figure eight on a bight.

B Secure the rope to an anchor using a follow-through figure eight.

C Lay the rope out to the objective.

D Measure approximately one-third of the standing part and tie a figure eight on a bight resulting in a loop 6 to 12 inches long.

E Wrap the running end of the rope around the objective and bring it to the figure eight on a bight tied in step 2.

F Thread the running end through the loop and pull it back toward the objective, tightening the rope as necessary.

JOB PERFORMANCE REQUIREMENT 15-21

Tying a Rope Between Two Objects (Continued)

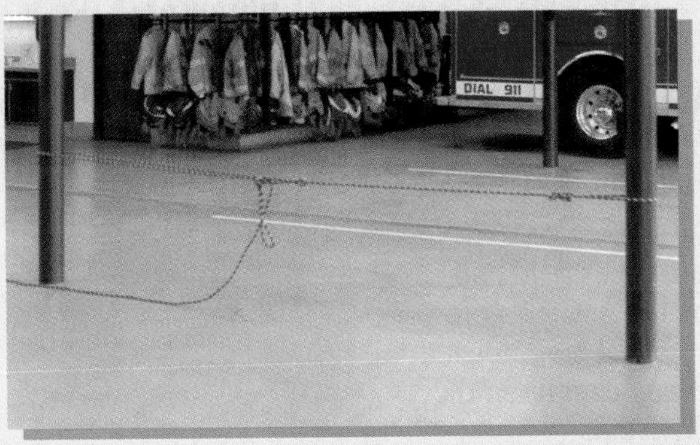

G Using the running end, tie a minimum of one half hitch around both sections of rope. Multiple half hitches will secure the rope and ensure that the knot is not accidentally untied.

Lessons Learned

This chapter presented basic information and introduced skills to accomplish basic tasks required at emergency scenes. It has not attempted to present all of the information about skills that firefighters may need as they progress in their careers and become involved in more technical and complex emergency operations. Firefighters are encouraged to regularly practice the basics learned in this chapter and pursue further information and training to expand their knowledge, skills, and abilities in this very interesting and exciting subject.

KEY TERMS

Bight A doubled section of rope, usually made along the standing part, that forms a U-turn in the rope that does not cross itself.

Dressing The practice of making sure that all parts of a knot are lying in the proper orientation to the other parts and look exactly as the pictures herein indicate.

Dynamic A rope having a high degree of elongation (10 to 15 percent) at normal safe working loads.

Kern A derivative of the term *kernel,* which is defined as "the central, most important part of something; core; essence."

Life Safety Line According to NFPA 1983, rope dedicated solely to the purpose of constructing lines for supporting people during rescue, firefighting, or other emergency operations, or during training evolutions.

Loop A turn in the standing part that crosses itself and results in the standing part continuing on in the original direction of travel.

Mantle Anything that cloaks, envelops, covers, or conceals.

One- or Two-Person Rope According to NFPA 1983, a one-person rope requires a minimum tensile strength of 4,500 pounds, and a two-person rope requires a minimum tensile strength of 9,000 pounds.

Round Turn Formed by continuing the loop on around until the sections of the standing part on either side of the round turn are parallel to one another.

Running End End of the rope that is not rigged or tied off.

Setting The finishing step, making sure that the knot is snug in all directions of pull.

Shock Load A load or impact being transferred to a rope suddenly and all at one time.

Standing Part The part of a rope that is not used to tie off.

Static A rope having very little (less than 2 percent) elongation at normal safe working loads.

Tactilely Using the sense of touch to feel for any differences or abnormality.

Tag/Guide Lines Tag lines are ropes held and controlled by firefighters on the ground or lower elevations in order to keep items being hoisted from banging against or getting caught on the structure as they are being hoisted.

Tensile Strength Breaking strength of a rope when a load is applied along the direction of the length, generally measured in pounds per square inch.

Utility Rope Rope used for utility purposes only. Some of the tasks utility ropes are used for in most every fire department are hoisting tools and equipment, cordoning off areas, and stabilizing objects. Also used as ladder halyards.

Webbing Nylon strapping, available in tubular and flat construction methods.

Working End The end of the rope that is utilized to secure/tie off the rope.

REVIEW QUESTIONS

1. What NFPA standard addresses life safety ropes and system components?
2. Polypropylene and polyethylene are two of the four synthetic materials discussed in this chapter. What is the major distinguishing feature between these two materials and nylon and polyester, the other two discussed?
3. Modern ropes manufactured from synthetic materials fall into two broad categories, static and dynamic. What is the major difference between the categories?
4. In a kernmantle rope the kern carries the vast majority of the load. What is the role of the mantle?
5. According to NFPA 1983, there are two classifications of life safety ropes. What are they and what are the minimum requirements for each?
6. Three terms are commonly utilized to describe parts of a rope. What are they and what does each mean?
7. To "dress" and to "set" a knot means to do what to it?
8. What is the best feature of figure eight knots?
9. What is the only knot recognized for tying webbing?
10. According to NFPA 1983, how many times can a life safety rope be utilized for emergency service after having been used during an emergency?
11. When storing ropes, it is important that the rope be stored in a manner that allows for what three things at emergency scenes?
12. If a rope, coiled or bagged, is dropped from an elevated position, what are some of the negative things that could occur?

Additional Resources

Frank, James and Jerrold Smith, *Rope Rescue Manual.* California Mountain Company Limited, Santa Barbara, CA, 1987.

Padgett, Alan and Bruce Smith, *On Rope.* National Speleological Society, Huntsville, AL, 1987.

Setnicka, Tim, *Wilderness Search and Rescue.* Appalachian Mountain Club, Boston, MA, 1980.

CHAPTER 16

RESCUE PROCEDURES

Robert F. Hancock, Hillsborough County Fire Rescue

OUTLINE

- Objectives
- Introduction
- Hazards Associated with Rescue Operations
- Search of Burning Structures
- Victim Removal, Drags, and Carries
- Extrication from Motor Vehicles
- Specialized Rescue Situations and Tools
- Lessons Learned
- Key Terms
- Review Questions
- Additional Resources

STREET STORY

We had just sat down for a steak dinner at 6 p.m. when the alarm came in. Over the enhanced 9-1-1, we got a dispatch that there was a person trapped in an elevator. We do two or three elevator rescues per week, but we knew this one was unusual because we heard the communications center dispatch a medic unit on the EMS channel. When we arrived at the scene—a historical home downtown that was being renovated—one person met us and said that the elevator was stuck between floors and someone's head was caught. Apparently, four kids, one of them a fourteen-year-old boy, had been starting the elevator, holding the door open, and trying to jump in and catch the car as it was moving down between flights. The final time, this boy got his legs and torso in but he'd gotten his head pinned between the floor landing and the top of the elevator car.

There was a stained-glass transom window over the elevator doorway on the first floor so we could see into the shaft that there were three hysterical kids in there, and the boy's legs were hanging into the car. The kids were holding his body up.

We had a three-person engine company, so we split up to do the size-up from a few different angles. We shut the power off first. In a rescue, the first thing to consider is not only the safety of the victim, but the safety of all the firefighters involved, and the big threat was that this elevator could move and we could get hurt. At first, it seemed the elevator was inaccessible, but then I realized I could make it in by breaking the transom window and climbing in the car. (If you had ever met me, you would never believe that at 6 feet 4 inches and 300 pounds, I made it through a 2-foot by 2-foot window!) The other two firefighters proceeded upstairs to evaluate the situation from there. They could only see the top of the kid's head; the area in which his head was compressed was only about 2 inches wide. I could see his neck and chin; it was obvious he was unconscious and not breathing. I concentrated on calming the hysterical kids, telling them everything would be all right. This was traumatic for them.

The team above determined that they would get something under the roof of the elevator car to pry it up. The other firefighters grabbed a hydraulic rescue tool, called a ladder and rescue for backup, and proceeded to pry enough to create an opening. Then they used some cribbing as a wedge to maintain the space, so it wouldn't spring back. I was able to pull the boy down into the elevator with me. I gave him a couple of rescue breaths before I passed him out the same window I had come in to the EMS crew that was waiting outside. Obviously you want to immobilize people as much as possible to prevent spinal injuries, but in this situation we couldn't get a backboard in that space, and the priority was to get him breathing, so we wanted to get him out right away. So I supported him as much as possible, passing him out feet first. As the EMS crew grabbed his legs, I held his shoulders, back, and head. Once they got him out, he was fully immobilized, intubated, and rushed to the hospital.

He survived the event—within ten days he walked into our fire station.

—Street Story by Mike Wisko, Acting Captain, Galveston Fire Department, Engine 1, B Shift, Galveston, Texas

OBJECTIVES

After completing this chapter, the reader should be able to:

- Recognize the hazards associated with various rescue operations.
- Describe the differences between primary and secondary searches.
- Demonstrate the proper procedures for victim drags and carries.
- Define the proper terminology utilized during motor vehicle extrication operations.
- Demonstrate proper and safe operation of vehicle extrication tools and equipment.
- Explain the various types of specialized rescue situations presented and the specific hazards associated with each of them.

INTRODUCTION

The term **rescue** in the emergency services has many meanings. In this chapter, *rescue* describes those actions that trained firefighters perform at emergency scenes to remove someone from imminent danger or to **extricate** them if they are already entrapped. Rescue is a very broad subject, and this chapter is going to touch on various rescue situations firefighters may find themselves confronted with. It is not the intent to make the reader an "expert" in all rescue situations discussed, but rather to bring the reader to an awareness level in order to recognize a situation for what it is, be aware of the dangers associated with it, and apply safe procedures to any potential rescue situation that may be encountered. The chapter does, however, cover building search, victim removal, and vehicle extrication in more detail since these are the areas that are most common to daily firefighting operations.

Streetsmart Tip Teamwork and safety are the key points to remember in any rescue operation.

HAZARDS ASSOCIATED WITH RESCUE OPERATIONS

Hazards are associated with every type of rescue operation. The most common hazards are presented along with discussions of the various rescue topics later in this chapter.

When firefighters are involved in a rescue operation, one of the biggest dangers that they must be aware of is the focusing of attention on a particular problem without proper regard for possible consequences or alternative approaches. This is generally referred to as **tunnel vision**. It is very easy to develop tunnel vision when a rescuer is involved in an unusually complex and/or lengthy rescue. Tunnel vision can keep, and in many cases has kept, the rescuer from seeing an obvious solution or more often an impending danger.

Note An injured rescuer does more than just add another patient to be cared for. As can be seen in **Figure 16-1,** it is not uncommon for the remaining rescue personnel to lose focus on the person originally injured and focus on the injured rescuer.

SEARCH OF BURNING STRUCTURES

Searching burning structures is one of the most dangerous rescue situations regularly faced by the majority of firefighters. As with other situations that firefighters respond to which are discussed in this chapter and elsewhere in this text, the best way to reduce the danger while searching involved structures is through training, practicing, and planning.

Safety Any time that firefighters enter a structure that is burning or in danger of becoming so, they must always wear appropriate personal protective equipment (PPE), including self-contained breathing apparatus (SCBA) and a personal alert safety system (PASS) device, for the known or potential threat/danger.

Figure 16-1 When a member of the rescue crew is injured, it is hard to maintain focus on the person originally injured. *(Courtesy of William Schmitt, Jr.)*

Firefighters must always work in teams of two or more when entering an involved structure for any reason (i.e., interior firefighting, search and rescue, ventilation). In addition to this search team, a minimum of two firefighters must be standing by immediately outside in full protective clothing and SCBA with a charged hoseline (although the charged hoseline is not required, it is a good logical practice) ready to come in and assist the search team should a problem develop, **Figure 16-2**. This is commonly referred to as the **two in/two out** rule. Even if the search team also has a hoseline with them, this recommendation remains in effect.

As firefighters approach a structure that is going to be searched, they should perform a size-up to determine a "rescue profile." The rescue profile helps the firefighters prioritize the probability, location, and status of potential victims as well as the avenues of access and egress. While life, safety, and rescue always remain a priority, there are structural fire environments that are a "recovery" environment. Simply stated, a recovery environment means there is little to no chance of saving a victim due to fire and smoke conditions and/or building collapse potential. Factors that should be evaluated to determine the rescue profile include:

Occupancy type/time of day: Residential structures can be occupied at any time, although typical sleeping hours present the greatest indication that victims may need rescue. Commercial occupancies seldom have sleeping occupants and, therefore, a greater chance exists that the occupants have self-evacuated. In some high-cost resort areas, employees may sleep on the premises due to the high cost of domicile, therefore increasing the rescue profile. Typically, victims in a commercial structure are found in the path of exits, whereas residential victims can be anywhere. Children are likely to hide from the smoke and fire—and may be in closets, in cabinets, under beds, and even in their parents' bed.

Fire/smoke conditions: Turbulent smoke (pressurized with high heat) is an unsurvivable environment for occupants due to pain threshold and toxicity—a recovery environment versus a rescue environment. Other areas with less dense and lighter colored smoke indicate a higher rescue profile. Post-flashover compartments (rooms) are a recovery environment. Rooms adjacent and above an involved room have a higher rescue priority, assuming the smoke in these areas is not turbulent. Remember, flashover of one room is likely to cause rapid and rolling ignition of dense smoke.

Activity clues: Cars in the driveway or garage, toys strewn about, shoveled snow and footprints, and open windows provide clues that a home is occupied, **Figure 16-3.** If home

Figure 16-2 A rapid intervention team should stand by for immediate action should an interior team need rescue.

Figure 16-3 A street side view of a typical residential occupancy. Note the clues that can be spotted in this photo: cars in driveway, toys, bicycles, and so on.

Figure 16-4 Well-equipped interior structural firefighting/search and rescue crews need a minimum of full PPE, SCBA, PASS, forcible entry tool, flashlight, portable radio, and thermal imaging camera.

occupants are not out waiting for the fire department, the rescue profile is high. Conversely, no cars, boarded-up windows, signs of neglect, and vacancy signs should indicate a lower rescue profile.

In addition to the protective clothing and equipment listed earlier, firefighters should carry with them a forcible entry tool (ax, pry bar, Halligan tool, etc.), flashlight, portable radio, and thermal imaging camera, **Figure 16-4**. The forcible entry tool can be useful in gaining access to locked or blocked rooms within the structure and is also useful in extending the searching firefighters' reach under and behind objects such as beds, dressers, and tables and can assist the firefighter in creating an emergency egress if necessary. The flashlight is useful in searching if the smoke is not too thick and can be useful to signal a rescue/backup crew should a firefighter get into trouble. The portable radio is helpful for keeping the incident commander informed of progress, fire/smoke conditions at a location, and the results of a search. It is very useful in communicating with a rescue/backup crew if a firefighter becomes disoriented or lost. Thermal imaging cameras and devices also allow the search team to see through the smoke, thereby increasing the speed with which searches can be accomplished. These devices are discussed later in this chapter.

In many single-family residential structures, it may be possible to conduct a search thoroughly and safely without a **guideline/lifeline** or hoseline by using the wall as a reference, **Figure 16-5**. Utilizing a wall in this manner is referred to as conducting a "right-hand" or a "left-hand" search, meaning that the firefighter maintains constant contact between the wall and that side of the body. By doing this, the firefighter will return to the point of entry while searching all the way around the room. However, this is not the case in larger mercantile, commercial, or industrial occupancies. In these types of occupancies, it is mandatory that a guideline/lifeline be

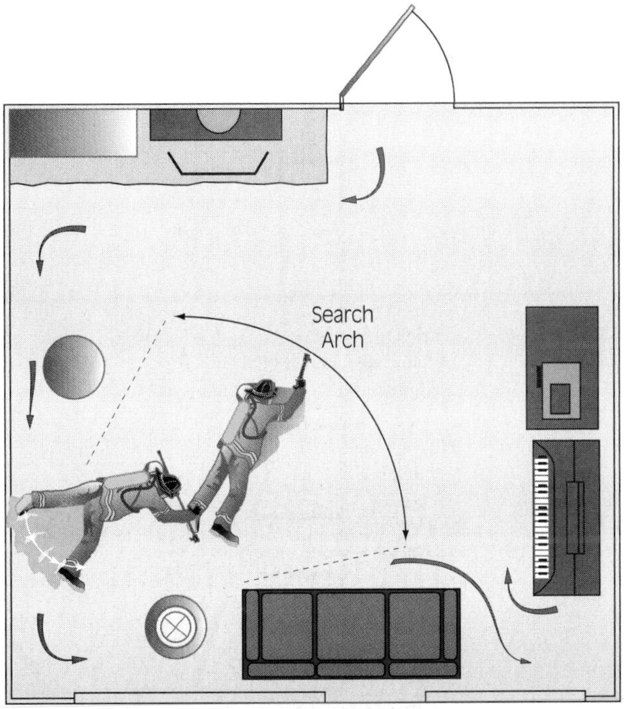

Figure 16-5 During an interior search, firefighters should stay in contact with a wall. If visibility is hampered, firefighters can reach into the center of the room using a tool and a "human chain" technique.

utilized. It is impossible to conduct a thorough search with an acceptable level of safety without one. It is very easy to become disoriented and completely turned around when in a large open area such as a department store, warehouse, or industrial plant. As shown in **Figure 16-6**, these types of occupancies generally not only have large open spaces, but often have obstructions (shelf units, machines, displays, etc.) at various and random locations within the open spaces.

Ropes are now available that are specifically designed to be utilized as guidelines. They have reflective material woven into the outer jacket of the rope to make them more visible. Ropes that are used as guidelines should not be used as lifelines since they are subjected to damage, both chemical and mechanical, while being dragged through structures.

> **Streetsmart Tip** Another way for firefighters to find their way back to where they entered a structure is to follow the hoseline back out. To be able to do this in firefighting conditions, it must be practiced during training until firefighters can tell which way the hose is going by feeling the couplings. The male coupling will be pointing toward the nozzle, away from the entry/exit.

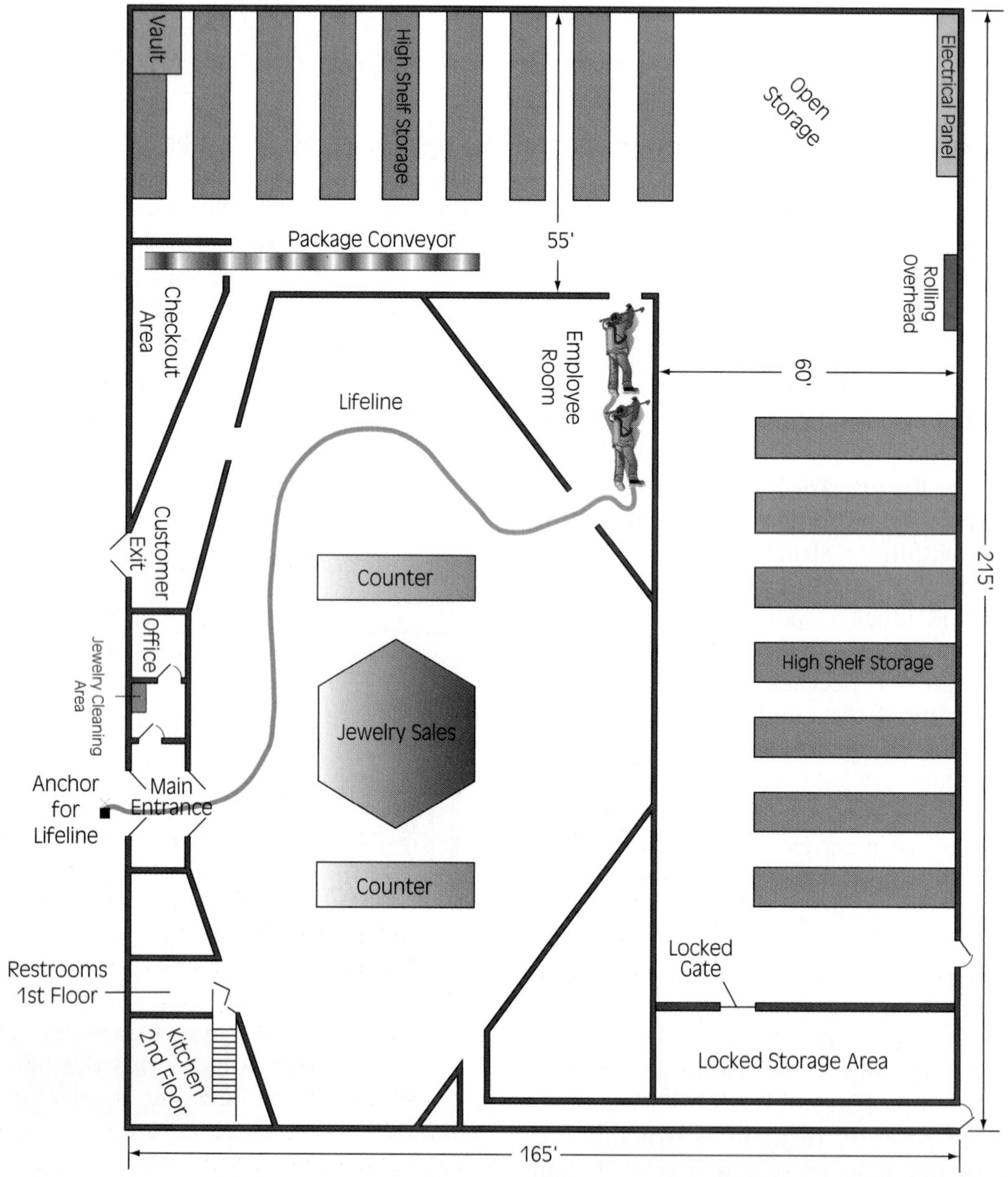

Figure 16-6 Commercial/industrial occupancies have large areas and various obstructions, machines, and storage. Firefighters must use a guideline/lifeline to help find their way out. The lifeline will also help an RIT find interior crews experiencing an emergency.

The biggest asset to conducting a safe and successful search is to have practiced and trained together prior to the actual emergency and to have a plan prior to entering the structure. "Having a plan" means that the search team members know who is in charge, in what direction they are going, on what side (left or right) to keep the wall, and any other pertinent information. Team members must stay together. If searching a small room or area with decent visibility, one team member may remain at the entrance but must be able to stay in constant voice communication with the other member so that a reference point can be maintained. Another way of covering more area is for one member to remain in contact with the wall while the other member holds onto the first member's arm, leg, or forcible entry tool, effectively doubling the distance that can be reached without leaving the wall. If a room is too hot to be entered, a firefighter can use a forcible entry tool to probe through the doorway or window.

Searching a building is completed in two different operations: the primary search and the secondary search. These operations are also two of the tactical benchmarks utilized by the incident commander.

Primary Search

The primary search is the first and most dangerous of the two. During the primary search, the team is often ahead of the attack lines and may be above the fire (the most dangerous place). They search the areas that are most likely to have victims in a rapid, but thorough, manner. In residential occupancies at night, these would be bedrooms, closets, near doorways, and bathrooms. In commercial occupancies during operating hours, victims would be expected near the exits or in offices or restrooms.

At times, the primary search is conducted from outside the structure, especially in residential structures. Called a *window search,* this primary search takes advantage of speed by opening windows of rooms uninvolved and doing a quick look into the room. This can be especially advantageous if the main path into the building is heavily charged with fire and/or turbulent, dense smoke. It is important to remember that "opening" the structure can increase fire intensity—although the accompanying ventilation can help relieve pressure and smoke conditions that are the main threat to those needing rescue. This approach is compatible with a *vent for life* strategy. While inside a structure, firefighters may "vent as they go" to help relieve smoke pressure, but only if such will not cause significant fire spread.

Often, it is quicker and safer for firefighters to ladder and break out an upstairs (bedroom) window and enter for a primary search versus a blind crawl through thick smoke and heat to find a stairway. A fire on the main floor will convect heat up open stairways and make it untenable for victims and quite dangerous for firefighters. Flashover of a room can lead to smoke cloud ignition that will literally blowtorch up the stairway. If firefighters are using stairways, they must stay low. When descending stairs, firefighters should stay low by crawling on their hands and knees and proceed feet first.

When conducting the primary search, visibility is often obscured by smoke conditions and darkness. The best visibility and portable light penetration will be closer to the floor. Additionally, the noise created by SCBA, radios, crawling, running into furniture and obstacles, as well as general fireground activities can mask the simple whimper of a child or the tapping of a trapped occupant who has little or no strength left. Firefighters should occasionally pause and listen for cries or signals for help.

Once the primary search is completed, crews should notify the incident commander that the assigned search is complete. Once all search crews report a complete search, the incident commander will broadcast an all clear. This is an incident tactical benchmark. Following an all clear, firefighters should reduce the degree of risk they take to help keep risk/benefit in balance.

Streetsmart Tip Many people act in unpredictable ways when a fire occurs. Because they are scared, they will try to hide from the fire, which is why firefighters must check under beds and in closets and restrooms thoroughly.

Secondary Search

The secondary search is usually conducted once the fire is out or at least well under control. This search can be much more thorough since there is no immediate fire danger, and the visibility is much better once the smoke is markedly reduced. During the secondary search, the team can search through debris that has fallen or been knocked over. It is also possible during this search to locate areas that may still need to be extinguished. If the fire was extinguished before consuming the room contents, when firefighters return to do the secondary search, they may be able to see where the primary searchers missed areas. This an excellent learning opportunity.

Thermal Imaging Cameras

Mike West, Lieutenant, South Metro Fire and Rescue, Colorado, and Instructor for SAFE-IR

Introduction to Thermal Imaging Cameras

Firefighters have long fought the lack of visibility at structural fires. Many operations are conducted blindly due to dense smoke and darkness. Firefighters are beginning to overcome this issue with the use of technology. Specifically, thermal imaging cameras (TICs) help firefighters "see" in low-visibility environments. The TIC uses an electronic detector to receive heat energy and then turn that energy into a picture that is viewed by the firefighter on a display screen. TICs do not require visible light to operate and work well in most low-visibility environments. The military developed much of the thermal imaging technology used today, and some fire departments have used TICs for nearly twenty years. Its popularity with most departments, however, has increased only in the last few years.

Fundamental Operational Concept

All objects emit heat energy. Some things, like humans, animals, or fire, create their own energy. Humans, for instance, emit energy at about 98.6°F. Other objects simply absorb and reflect the heat around them. For example, furniture, walls, flooring, and just about anything that is not "living" inside a building will absorb and reflect heat. All TICs have detectors used to gather heat energy. Currently, there are several types of image detectors available, but all process the detected heat energy into a picture or image representing temperature differences. Images are displayed as black, white, and shades of gray. In most TICs, white represents the warmest object in the scene and black represents the coolest. Some TICs also display colors that are associated with specific temperatures. TICs are typically battery powered. The size, type, and display of the TIC will impact how long the battery can last before needing recharging. It is important that firefighters using the imager understand the specific camera and detector used in their department—all TICs have strengths and limitations and can vary in how they depict different scenes. Only through instruction and training will firefighters learn how to interpret the pictures of the specific camera their department uses. Thermal imagers have handheld or helmet-mounted housings. Some fire departments have fixed-mount thermal imagers on their apparatus. For instance, an airport rescue and firefighting apparatus may use fixed-mounted TICs to locate downed planes in dense fog. Technology already exists that allows some TICs to transmit the image to remote displays so outside fire officers can see what the interior crew is seeing on their TIC.

(Photo courtesy of SAFE-IR)

(Photo courtesy of SAFE-IR)

Thermal Imaging Cameras (Continued)

Uses

TICs allow the firefighter to see through smoke. Because of this, thermal imagers are excellent search and rescue tools. It is important to note that TICs *do not* take the place of standard search techniques like following a wall, rope, or hoseline. The camera, like any tool, can fail, leaving the firefighting crew in a position to be lost. Maintaining good orientation is essential with or without the use of TICs. Thermal imaging cameras can be used to direct hose streams toward the seat of the fire and are helpful in overhaul by locating areas that may have hidden fire. A TIC used on the roof can point crews toward the hottest point, making vertical ventilation more effective. At odor investigations, they can help locate overheated light ballasts and electrical motors. TICs are used at nonfire emergencies as well. At HAZMAT calls, imagers can monitor fluid levels inside single-wall tanks, and they can monitor foam blankets on hydrocarbon spills. TICs have also been used to search vegetation for victims that may have been ejected from a vehicle accident. In all operations, it is important to use standard safety rules and local departmental SOPs in the performance of the task.

Training

Training is vital to use the TIC effectively and safely. The training should be specific to the camera used by the fire department and should include classroom orientation and hands-on usage. Only through training and repeated use can the firefighter become proficient. Each fire department should use the manufacturer's literature and guidelines as a starting point for their training.

VICTIM REMOVAL, DRAGS, AND CARRIES

Victims must be removed as carefully and expeditiously as possible. The goal should be not to cause any further injury or aggravation of existing injuries during the rescue process. Many times it is not possible to utilize all of the patient handling and immobilization skills that firefighters have learned due to the imminent danger presented by the heat, smoke, and gases in a structural fire. Other types of rescue situations also sometimes prevent the rescuer from using all the care that the person would like to due to a continuing hazard, or being in a confining area or other hostile environment.

All carries and drags place additional stress on the rescuer's musculoskeletal system. Training and work hardening prepare the firefighter to perform these victim removal techniques. Of particular importance with all drags and carries is the need to keep a tight core. Simply put, firefighters performing drags and carries need to do so by tightening the core muscles around the hips, back, and torso. This creates a "power center" that aids in balance, strength, and injury prevention. Ideally, the firefighter should keep the spine (back) in a neutral position (straight and tight) and use the legs and buttocks for leverage and lifting power, **Figure 16-7.**

Figure 16-7 To avoid injury, firefighters should lift heavy objects using a "tight core" and leverage from the legs and buttocks.

Carries

Firefighter's Carry

The firefighter's carry can be utilized on both conscious and unconscious patients. It is a one-rescuer operation. While this is not the preferred method, it is very effective when one rescuer must carry an unconscious patient. As a rule, the firefighter's carry can be used to carry someone who weighs *less* than the rescuer. Obviously, very strong (muscular) firefighters may be able to carry someone who is heavier than they are, although safety should be paramount. By following the steps given next, it is possible for a single rescuer to pick up and carry the patient without any assistance.

Firefighter's Carry (JPR 16-1)

1. Lay the patient on the back, with arms laid alongside the torso, knees bent, and

JOB PERFORMANCE REQUIREMENT 16-1

Firefighter's Carry

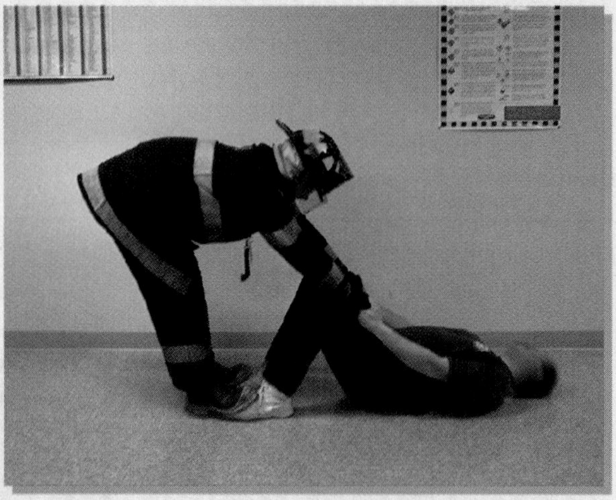

A Lay the patient on the back, with arms laid alongside torso, knees bent, and feet pushed back close to buttocks. Stand in front of the patient with your feet holding the patient's feet in place.

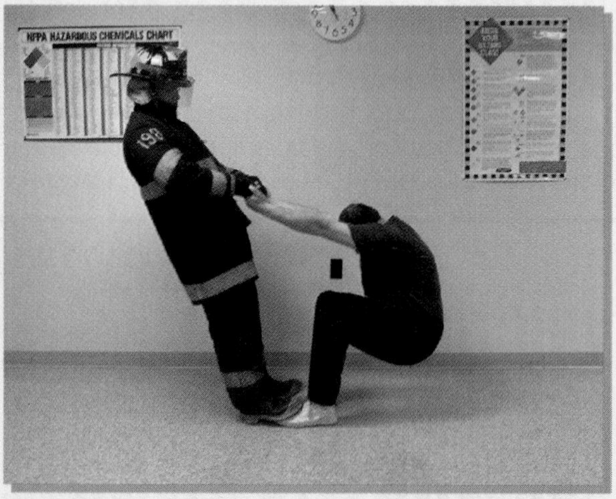

B Reach down and grasp the patient's hands and lean back as you stand, using your weight to help lift the patient. (It may be necessary to rock back and forth a couple of times in order to build sufficient momentum.)

C When you are ready, pull hard on the patient. As the patient comes up, duck your shoulder into the midsection and stand up. As you stand, wrap one arm around the patient's legs and maintain your grasp on the arm that will now be across your back/shoulder.

feet pushed back close to the buttocks, **JPR 16-1A.**
2. Stand in front of the patient with your feet holding the patient's feet in place.
3. Reach down and grasp the patient's hands and lean back as you stand using your weight to help lift the patient, **JPR 16-1B.** (It may be necessary to rock back and forth a couple of times in order to build sufficient momentum.)
4. When you are ready, pull hard on the patient. As the patient comes up, duck your shoulder into the midsection and stand up. Remember to keep a tight core and use the legs and buttocks for power. As you stand, wrap one arm around the patient's legs and maintain your grasp on the arm that will now be across your back/shoulder, **JPR 16-1C.**

Extremity Carry

The extremity carry (sometimes referred to as the cross-arm carry) can be utilized on both conscious and unconscious patients and requires two rescuers. However, it is much easier on the rescuers and patient than the firefighter's carry.

Extremity Carry (JPR 16-2)

1. Lay the patient on the back, with arms laid across the torso, knees bent, and feet pushed back approximately halfway to the buttocks, **JPR 16-2A.**
2. For an unconscious patient, it may be easiest to have the rescuer at the patient's feet reach down, grasp the patient's hands, and pull them into a sitting position using the rescuer's weight. Have a conscious patient assume a sitting posture with feet pulled back, lifting the knees above the floor.
3. The rescuer at the patient's head squats behind the patient, sliding arms under the patient's armpits and grasping the wrist of the patient's opposite arm (if possible). At the same time the rescuer at the patient's feet squats between the patient's feet and grasps the patient's legs under the knees (if possible) or as close as possible below the knees, **JPR 16-2B.**
4. When ready to lift, the rescuers need to communicate clearly with each other so that they can stand as one and carry the patient to safety, **JPR 16-2C.**

Seat Carry

The seat carry can be utilized on conscious patients only and requires two rescuers.

Seat Carry (JPR 16-3)

1. The rescuers face each other. Each rescuer grasps his own right forearm just above the wrist. The rescuers then grasp each other's left forearm just above the wrist, forming a square "seat," **JPR 16-3A.**
2. The rescuers lower the "seat" that they have just formed, allowing the patient to sit on the seat with arms across the shoulders of each rescuer, **JPR 16-3B.**
3. When ready to lift, the rescuers need to communicate clearly with each other so that they can stand as one and carry the patient to safety.

Drags

Rescuers can move a patient by placing him or her on a blanket, bunker coat, salvage cover, and so on, or by using the patient's own clothing as a handhold or by the utilization of a **webbing sling**. Each of these various types of drags is described below. All of the drags shown here can be carried out by a single rescuer.

Blanket Drag

As mentioned, the blanket drag can also be accomplished utilizing a bunker coat, salvage cover, or any other type of material of adequate size. The steps and procedures are the same regardless of the material being used.

> **Streetsmart Tip** A rescuer should always drag a patient head first.

Blanket Drag (JPR 16-4)

1. With the patient lying face up, lay the material the patient is to be placed on along one side of the patient's body with just over one-half of the material gathered close to the patient's side, **JPR 16-4A.**
2. Kneel on the opposite side (from the blanket) of the patient, extend the patient's arm (on your side) above the head if injuries permit, reach across the patient with one hand just above the waist and one just below the hips, and roll the patient toward you, **JPR 16-4B.**
3. While supporting the patient, now on the side, tuck the gathered blanket material close to the body.
4. Roll the patient back onto the blanket material, return the arm to the side, and

JOB PERFORMANCE REQUIREMENT 16-2

The Extremity Carry

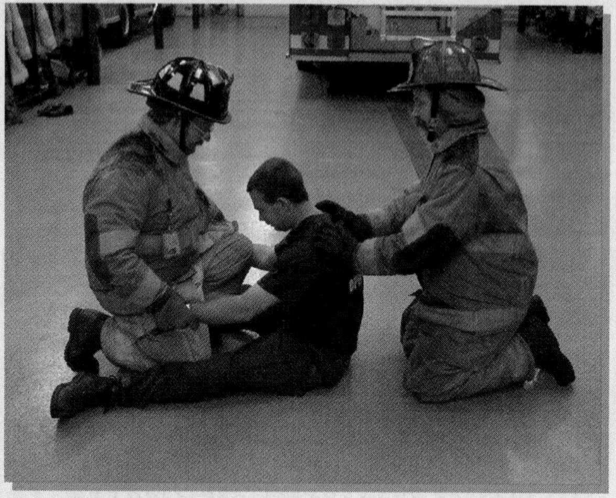

A Lay the patient on his back, with arms laid across torso, knees bent, and feet pushed back approximately halfway to buttocks. For an unconscious patient, it may be easiest to have the rescuer at the patient's feet reach down, grasp the patient's hands, and pull him into a sitting position using the rescuer's weight. Have a conscious patient assume a sitting posture with feet pulled back, lifting the knees above the floor.

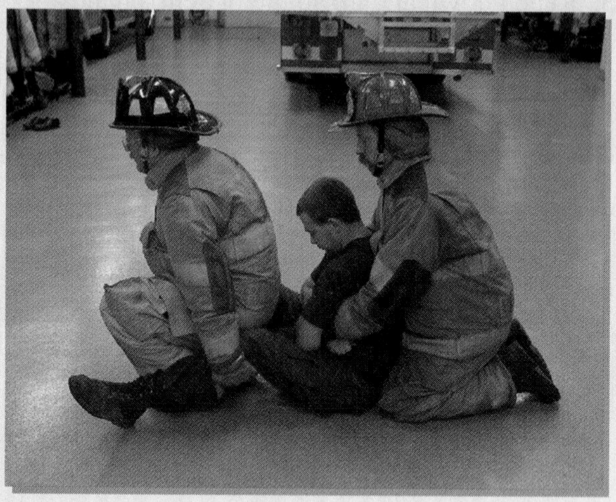

B The rescuer at the patient's head squats behind the patient, sliding arms under the patient's armpits and grasping the wrist of the patient's opposite arm (if possible). At the same time, the rescuer at the patient's feet squats between the patient's feet and grasps the patient's legs under the knees (if possible) or as close as possible below the knees.

C When ready to lift, the rescuers need to communicate clearly with each other so that they can stand as one and carry the patient to safety.

JOB PERFORMANCE REQUIREMENT 16-3

The Seat Carry

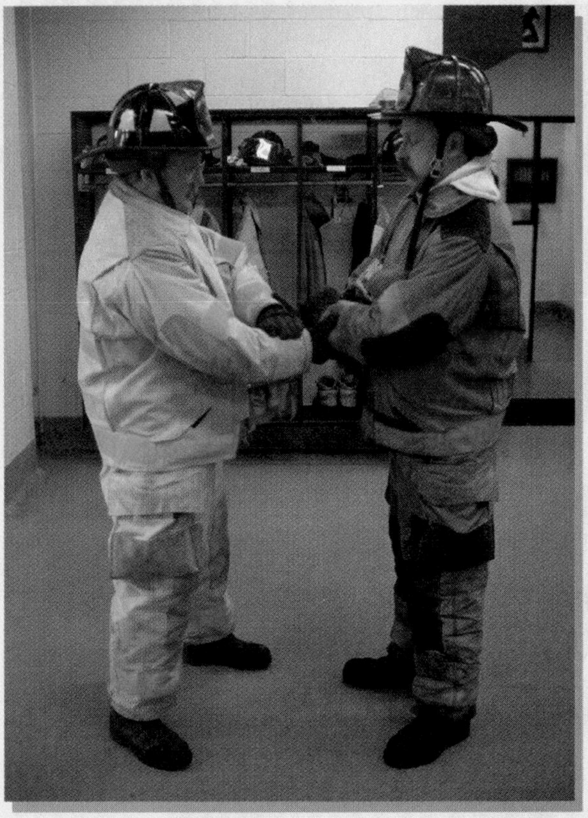

A The rescuers face each other. Each rescuer grasps his own right forearm just above the wrist. The rescuers then grasp each other's left forearm just above the wrist, forming a square "seat."

B The rescuers lower the "seat" that they have just formed, allowing the patient to sit on the seat with arms across the shoulders of each rescuer. When ready to lift, the rescuers need to communicate clearly with each other so that they can stand as one and carry the patient to safety.

wrap the blanket snugly around the patient, supporting the head and neck as much as possible, **JPR 16-4C.**

5. The patient is now ready to be dragged head first with the head and shoulders raised slightly off the floor, **JPR 16-4D.**

Clothing Drag

If a patient is wearing substantial clothing such as protective clothing or a heavy jacket, it may be possible to utilize this clothing to drag a patient to safety.

Clothing Drag (JPR 16-5)

1. Place the patient on the back arranging clothing to provide support to the head and neck. Be careful that the patient's ability to breathe is not compromised, **JPR 16-5A.**
2. Grasp the top of the patient's clothing on each side of the patient's head, supporting the head on your forearms. When using this method, it is important to keep the patient's head close to the floor to avoid causing the head to be pushed downward toward the chest, possibly causing difficulty breathing, **JPR 16-5B.**

JOB PERFORMANCE REQUIREMENT 16-4

The Blanket Drag

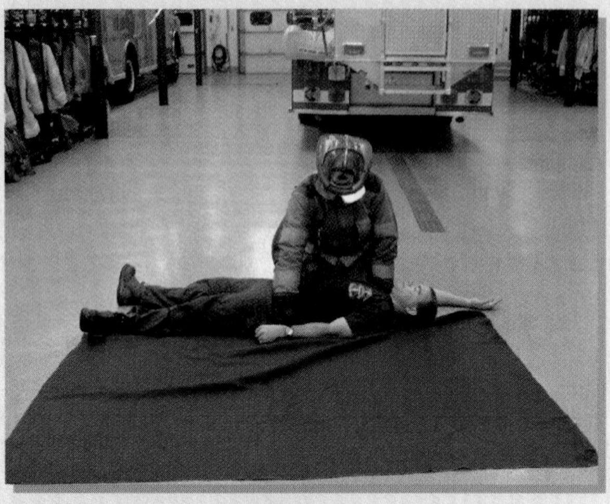

A With the patient lying face up, lay the material the patient is to be placed on along one side of the patient's body with just over one-half of the material gathered close to the patient's side.

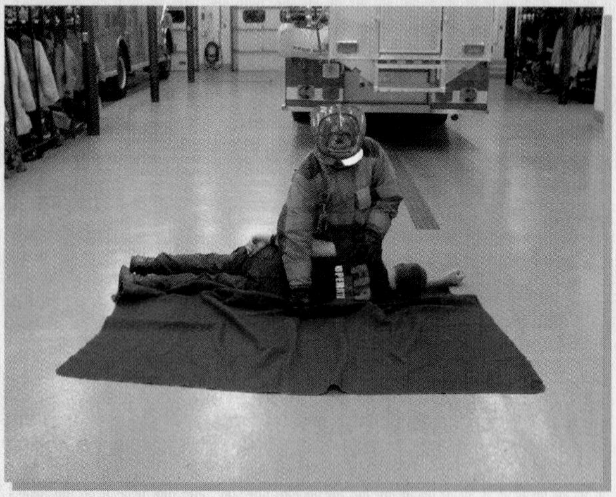

B Kneel on the opposite side (from the blanket) of the patient, extend the patient's arm (on your side) above the head if injuries permit, reach across the patient with one hand just above the waist and one just below the hips, and roll the patient toward you. While supporting the patient, now on the side, tuck the gathered blanket material close to the body.

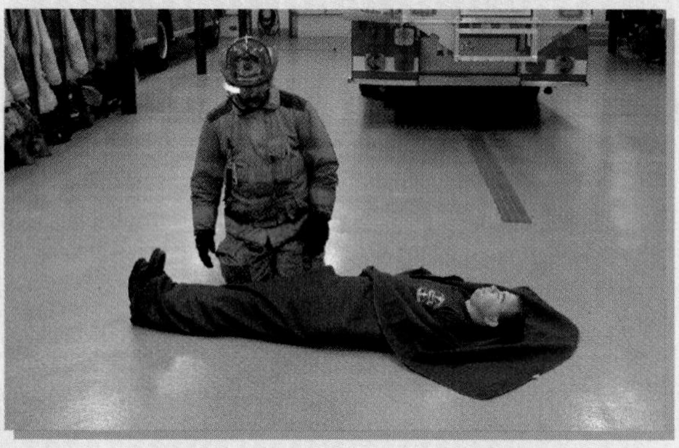

C Roll the patient back onto the blanket material, return the arm to the side, and wrap the blanket snugly around the patient, supporting the head and neck as much as possible.

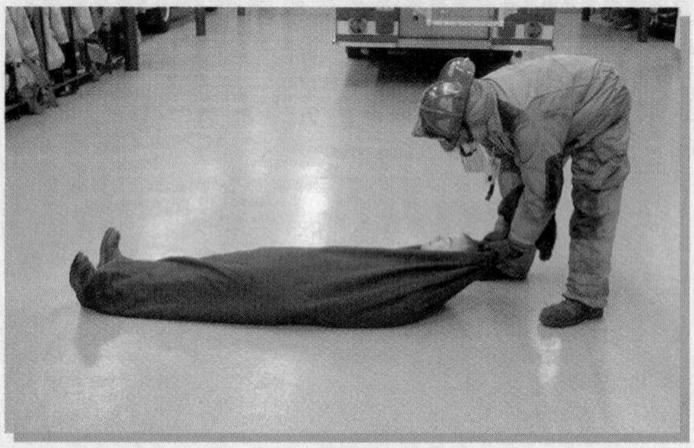

D The patient is now ready to be dragged head first with the head and shoulders raised slightly off the floor.

Webbing Sling Drag

For this drag the rescuer is going to need to have a piece of webbing or rope (although rope does not work nor carry as easily) approximately 12 to 15 feet long. The webbing drag enables a rescuer who is significantly smaller than the victim to perform a rescue.

Webbing Sling Drag (JPR 16-6)

1. Tie the webbing (using a water knot with safety) end to end forming a continuous loop, **JPR 16-6A.**
2. With the patient lying on the back, place the loop under each arm, coming up under the armpits.

JOB PERFORMANCE REQUIREMENT 16-5
The Clothing Drag

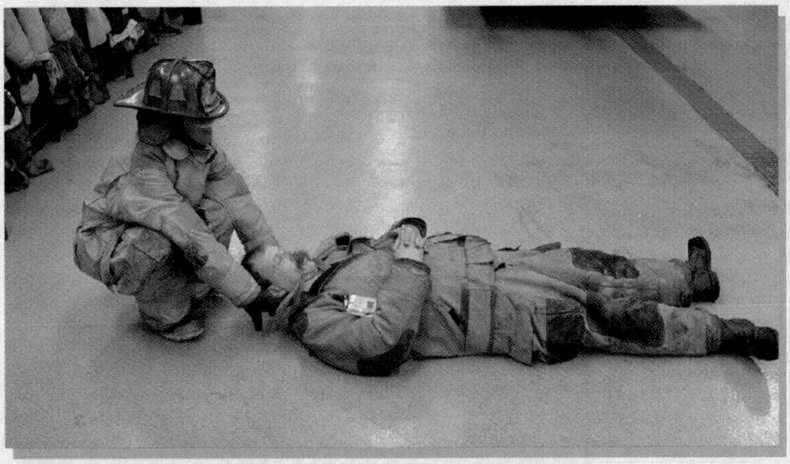

A Place the patient on the back, arranging the clothing to provide support to the head and neck. Be careful that the patient's ability to breathe is not compromised.

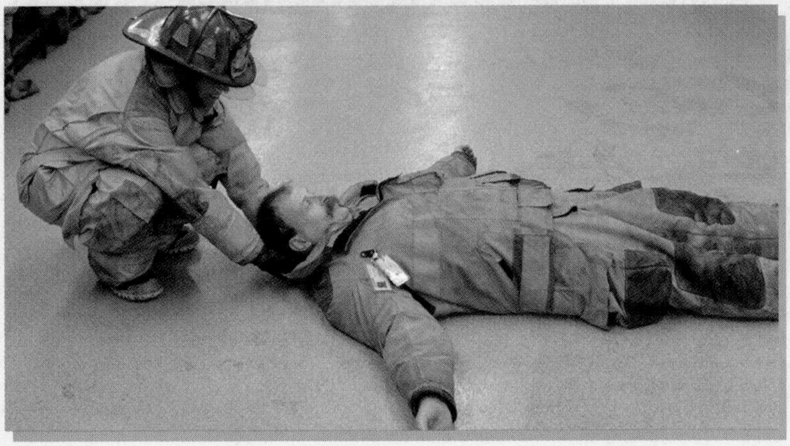

B Grasp the top of the patient's clothing on each side of the patient's head, supporting the head on your forearms. When using this method, it is important to keep the patient's head close to the floor to avoid causing the head to be pushed downward toward the chest, possibly causing difficulty breathing.

3. You will then have two loops stretched out above the patient's head. Adjust the loops so that the loop that comes out the underside of the armpits is snug against the patient's back. Feed the longer loop through between the body and the other webbing loop, **JPR 16-6B.**
4. When the patient is pulled by the long loop, the webbing should snug up under the armpits and provide some support to the patient's head, **JPR 16-6C.**

Sit and Drag Method

This method can be useful when the firefighter experiences trouble moving a patient due to size difference, sheer weight, or any other problem that may be encountered.

Sit and Drag Method (JPR 16-7)

1. Place the patient in a face-up position, **JPR 16-7A.**
2. Assume a sitting position at the head of the patient with legs to each side and hands on either side of the patient.
3. Grasp the patient under the arms. Pull the patient close to you and move under the patient so that the patient's head and back rest against your chest, and your thighs are under the patient's armpits with the patient's arms on the outside of your legs, **JPR 16-7B.**
4. Remove the patient from the area by sliding backward in the sitting position and using your legs to drag the patient along, **JPR 16-7C.**

JOB PERFORMANCE REQUIREMENT 16-6

Webbing Sling Drag

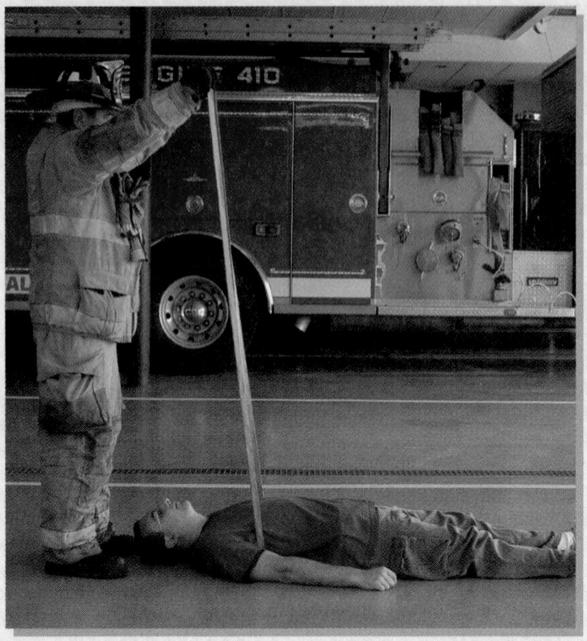

A Tie the webbing (using a water knot with a safety) end to end, forming a continuous loop. With the patient lying face up, place the loop under each arm, coming up under the armpits.

B You will then have two loops stretched out above the patient's head. Adjust the loops so that the loop that comes out the underside of the armpits is snug against the patient's back. Feed the longer loop through between the body and the other webbing loop.

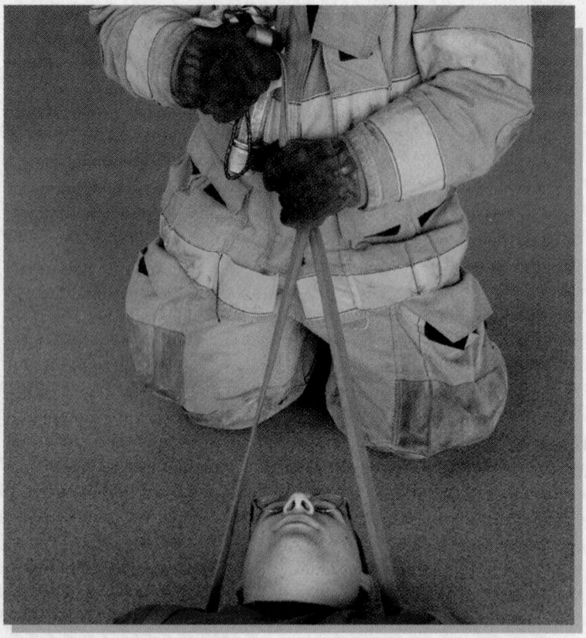

C When the patient is pulled by the long loop, the webbing should snug up under the armpits and provide some support to the patient's head.

JOB PERFORMANCE REQUIREMENT 16-7

Sit and Drag Method

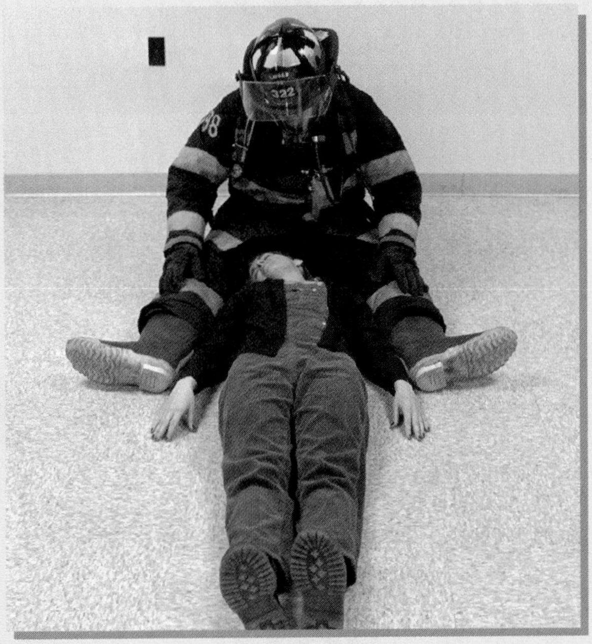

A Place the patient in a face-up position. Assume a sitting position at the head of the patient with legs to each side and hands on either side of the patient.

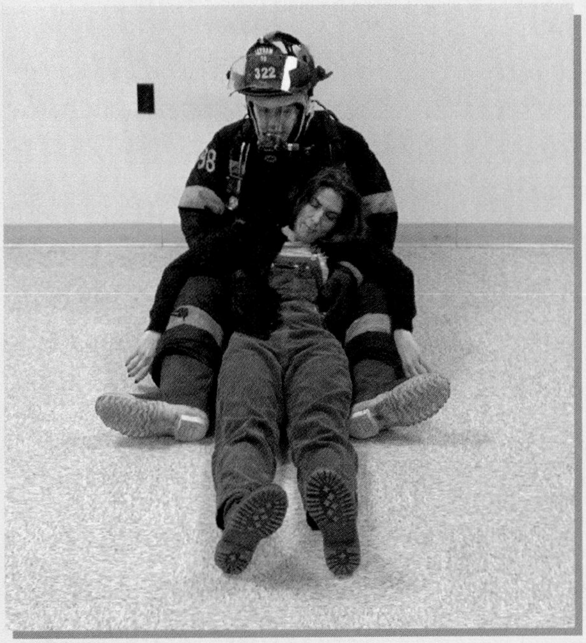

B Grasp the patient under the arms. Pull the patient close to you and move under the patient so that the patient's head and back rest against your chest, and your thighs are under the patient's armpits with the patient's arms on the outside of your legs.

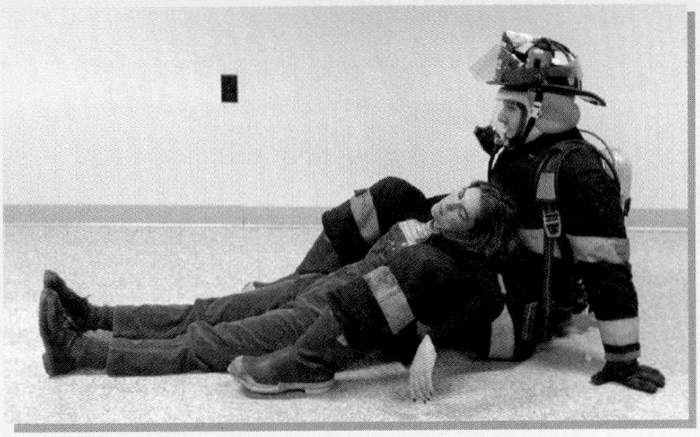

C Remove the patient from the area by sliding backward in the sitting position and using the legs to drag your patient along.

JOB PERFORMANCE REQUIREMENT 16-8

Firefighter's Drag

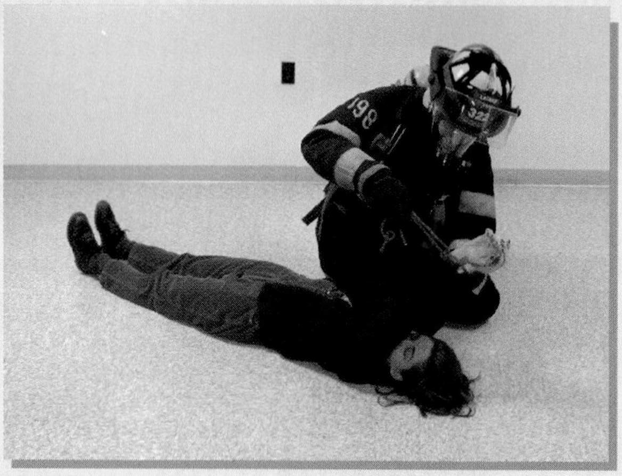

A Place the patient in a face-up position. Use a piece of rope, webbing, belt, handkerchief, or other available material to tie the patient's wrists together.

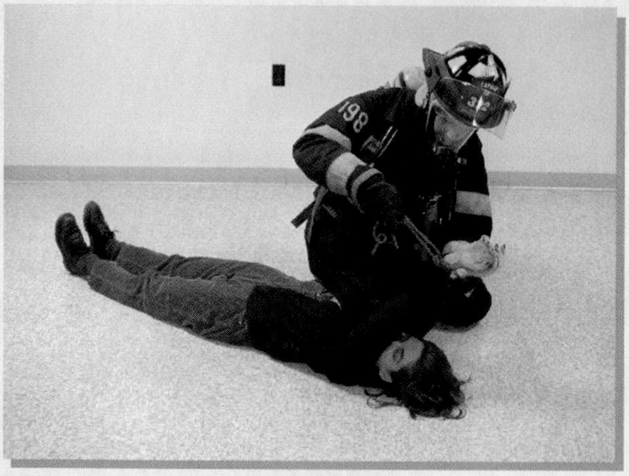

B Straddle the patient, facing the patient, and place the patient's bound wrists over your head and behind the neck.

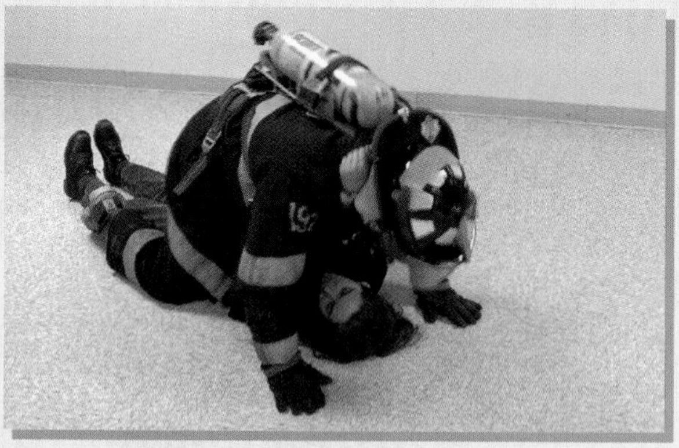

C Crawl on hands and knees while dragging the patient out of the area.

Firefighter's Drag

This is another useful method of dragging a patient from a dangerous/hazardous area.

Firefighter's Drag (JPR 16-8)

1. Place the patient in a face-up position, **JPR 16-8A.**
2. Use a piece of rope, webbing, belt, handkerchief, or other available material to tie the patient's wrists together.
3. Straddle the patient, facing the patient, and place the patient's bound wrists over your head and behind the neck, **JPR 16-8B.**
4. Crawl on hands and knees while dragging the patient out of the area, **JPR 16-8C.**

Note For pregnant or large patients, the rescuer will need to drag the patient while crawling alongside rather than straddling the patient.

Rescue of a Firefighter Wearing a SCBA

This method works very well for removing an unconscious or incapacitated firefighter who is wearing a SCBA whether it is functioning or not.

JOB PERFORMANCE REQUIREMENT 16-9

Rescue of a Firefighter Wearing a SCBA

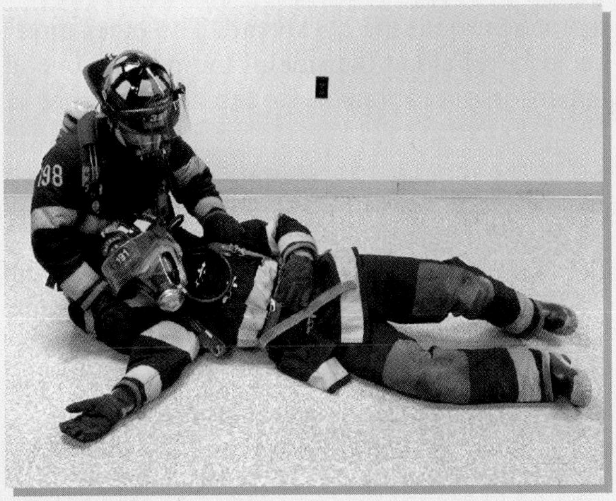

A Determine if the SCBA is functioning. If the SCBA is *not* functioning, disconnect the low-pressure tube and place inside the firefighter's coat or remove the regulator from the face piece, leaving the face piece in place.

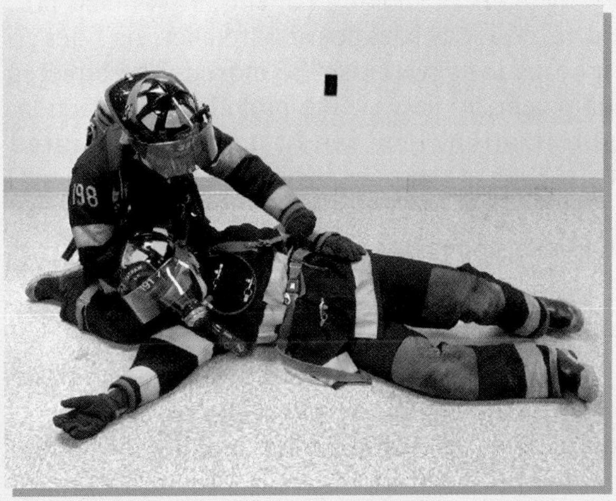

B Roll the firefighter onto the side, ensuring that the air supply is not compromised.

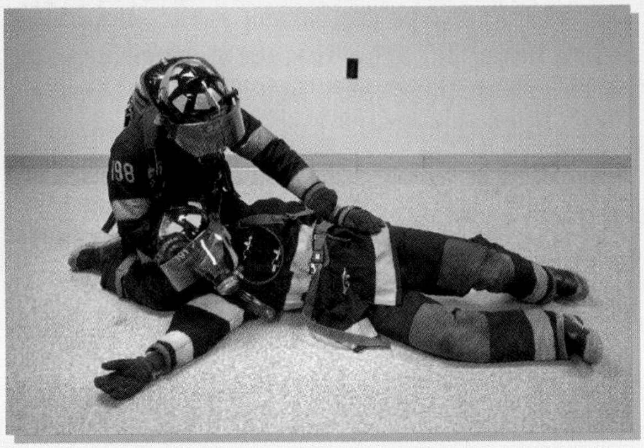

C Verify that the SCBA is securely fastened on the firefighter.

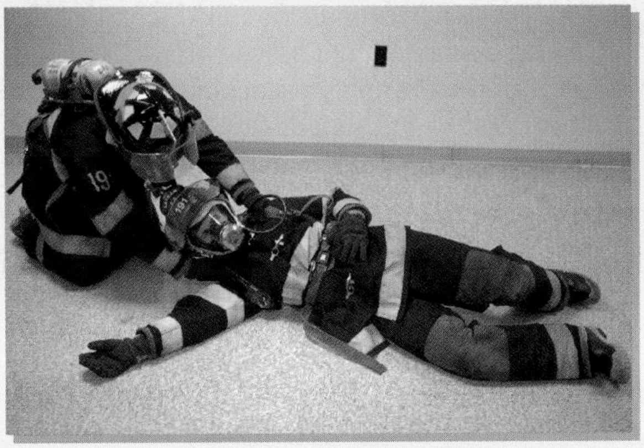

D Grasp the shoulder straps of the firefighter's SCBA and drag the firefighter from the area.

Rescue of a Firefighter Wearing a SCBA (JPR 16-9)

1. Determine if the SCBA is functioning. If SCBA is *not* functioning, disconnect the low-pressure tube and place inside the firefighter's coat or remove the regulator from the face piece, leaving the face piece in place, **JPR 16-9A.**
2. Roll the firefighter onto the side, ensuring that the air supply is not compromised, **JPR 16-9B.**
3. Verify that the SCBA is securely fastened on the firefighter, **JPR 16-9C.**
4. Grasp the shoulder straps of the firefighter's SCBA and drag the firefighter from the area, **JPR 16-9D.**

Streetsmart Tip The rescuer must remember that none of these drags provides **spinal immobilization** and are intended to be utilized only in situations where greater harm will come to the patient if not immediately moved. If the patient is wearing a functioning SCBA, the rescuer needs to be careful not to break the face piece seal.

Backboard, Stretcher, and Litter Uses

While the carries and drags discussed are emergency methods of moving patients, it is much preferred to utilize a backboard, stretcher, or litter to transport patients who must be moved. These pieces of equipment are designed to provide the protection, immobilization, and safety needed by injured patients. Each of these items has specific characteristics that make a particular piece of equipment more appropriate for a given situation than the other.

Backboards

Backboards (also known as long spine boards) are designed to provide the maximum in spinal immobilization. They are manufactured (commercially and homemade) from many different types of materials, the most common of which is plywood. However, the more modern backboards are manufactured using materials that allow the patient to remain on the backboard while being x-rayed. This minimizes the amount of movement the patient is subjected to and therefore reduces the possibility of aggravation of injuries. These newer materials also simplify the cleanup and decontamination process.

Placing a patient who is suspected of having a spinal injury onto a backboard takes a coordinated team effort. It is best to have four personnel available for this task. The rescuer at the patient's head is responsible for maintaining traction (in line with spine) on the patient's cervical spine and is also the person in charge of directing the process of placing the patient onto the backboard. The other three rescuers kneel along the side of the patient, one at the upper torso area, one at the hip area, and one at the knee/shin area.

Placing a Patient on a Backboard (JPR 16-10)

1. While the rescuer at the patient's head is maintaining traction, the rescuer at the upper torso places a cervical collar (or some other immobilization device) on the patient. *Even with an immobilization device in place, traction must be maintained at all times during this process,* **JPR 16-10A.**
2. The backboard is laid alongside the patient on the opposite side from the three kneeling rescuers. When ready the rescuer at the patient's head directs the others to "prepare to roll." The three rescuers reach across and grasp the patient at the appropriate locations: The rescuer at the torso grasps the patient at the shoulder and upper arm area; the rescuer at the hips grasps the patient just above and below the hip area; the rescuer at the knee/shin

JOB PERFORMANCE REQUIREMENT 16-10

Placing a Patient on a Backboard

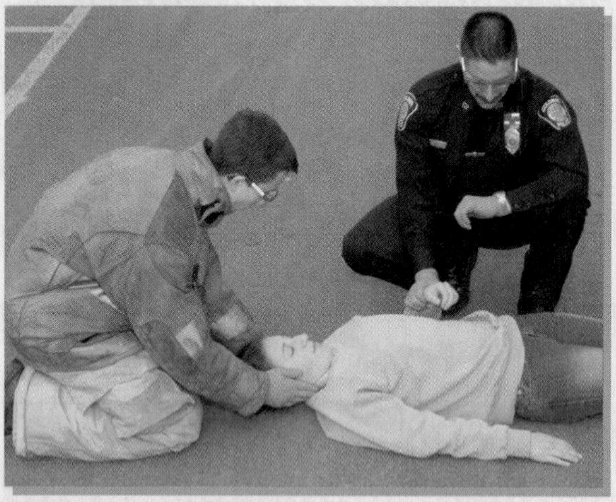

A A firefighter maintains manual stabilization while another first responder checks pulse, movement, and sensation.

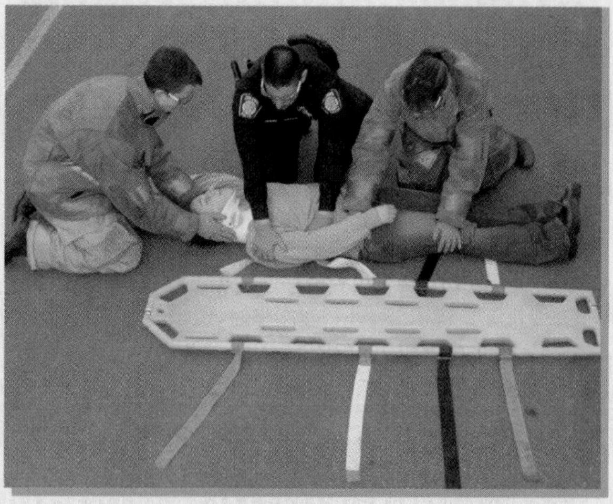

B A cervical collar is applied. While a firefighter maintains manual stabilization, two other firefighters take positions at the patient's shoulders and pelvis, reaching across the patient and grasping the patient's shoulders and pelvis, respectively.

JOB PERFORMANCE REQUIREMENT 16-10

Placing a Patient on a Backboard (Continued)

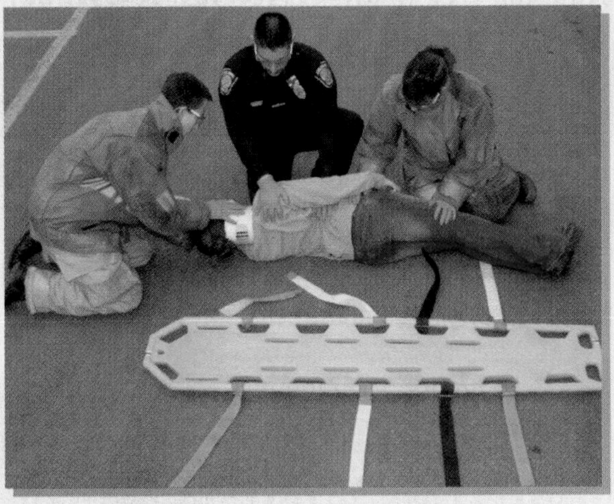

C On the command of the firefighter maintaining manual stabilization, the team rolls the patient onto the patient's side.

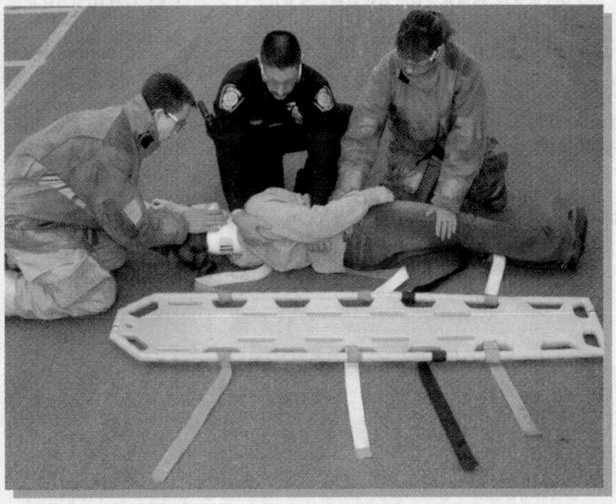

D One firefighter places the backboard under the patient with the bottom of the backboard at the patient's knees.

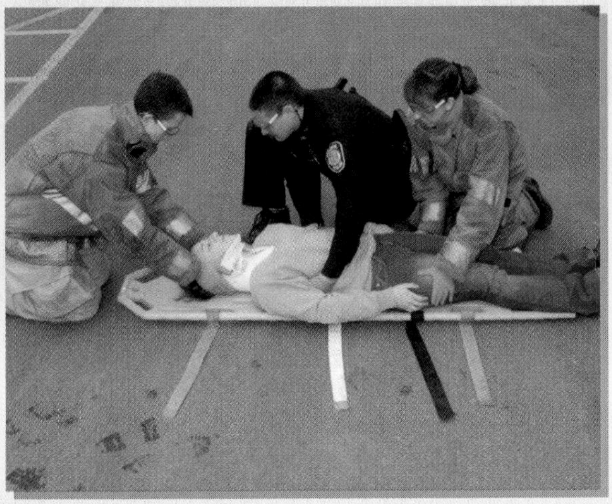

E On command, the team rolls the patient back onto the backboard, and the patient is pulled up to the center of the board using a long axis drag.

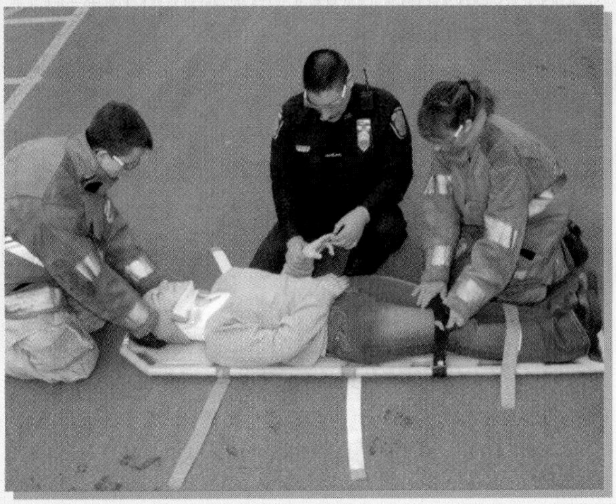

F Once the patient is centered on the backboard, the firefighter secures the patient to the backboard and reassesses distal pulses, movement, and sensation.

area grasps the patient just above and below the knee, **JPR 16-10B.**

3. When ready the rescuer at the head directs the others to "roll patient." At this time the three others roll the patient toward them on to the side. During all of these maneuvers it is essential for the rescuer at the head to rotate the patient's head along with the body, maintaining traction, **JPR 16-10C.**
4. With the patient on the side, one of the rescuers (usually the hip area rescuer) reaches over the patient and slides the backboard up tight against the patient's body. When this is completed, the rescuer at the head directs the others to "prepare to lower" and then "lower patient." At this command the three rescuers allow the patient to roll back down on top of the backboard, **JPR 16-10D.**
5. At this point the patient should be on the backboard; however, it is usually necessary to move the patient in order to

center the patient on the board. This needs to be done very carefully so as not to aggravate any injuries. The patient must be moved as a unit, meaning the patient's body is moved all at the same time. The safest way to accomplish moving the patient to the center of the backboard is to move the patient along the long axis of the body rather than try to move sideways, **JPR 16-10E.**

The rescuer at the head remains in charge. When ready to move the patient, the rescuer at the head directs the others to "prepare to slide." At this point the rescuers reach across or straddle the patient, grasping the upper arm/chest, waist, and lower legs. When ready to slide, the rescuer at the head directs the others to "slide." All of the rescuers then slide the patient down toward the foot of the board while moving slightly toward the center. When the patient has been moved enough, the rescuer at the head directs them to "stop." The patient now needs to be moved back up on the board to the proper location both top-to-bottom and side-to-side. This is accomplished by reversing the just completed movement, **JPR 16-10F.** The patient's distal pulses, movement, and sensation are then reassessed.

6. With the patient now in the proper location, the head is supported on both sides by sandbags, towels, commercial head blocks, or some other acceptable manner. The actual method of strapping the patient to the board varies widely from department to department, but whatever the particular method, the patient must be securely fastened to the board when complete. The strapping method shown in **JPR 16-10** is one that has been found to work very well and can be accomplished in an expeditious manner when practiced.

Stretchers

Stretcher is a universal term that can be applied to the patient-carrying device for ambulances as well as the common army litter. Other terms and slang words are used for the ambulance-type stretcher. Perhaps most used is the term cot, although pram and wheels are still prevalent. The venerable army litter is still used in some situations.

A variety of different types of stretchers are manufactured by various companies and are in use today. It would be very difficult and unnecessary to describe in detail the operation of all of the different units available. However, it is imperative that firefighters know the correct method of operating the type of stretcher(s) that is used in their community.

The most common methods of placing patients onto these types of stretchers are the extremity carry, by utilizing a backboard, or by having the patient lie directly onto the stretcher. Regardless of which of these methods is utilized, the rescuers need to assist and support the patient onto the stretcher. Once on the stretcher, the patient must be secured as soon as possible.

Streetsmart Tip Here are some safety rules that apply to the operation of all types of ambulance stretchers, **JPR 16-11**:

1. Always make sure the patient is strapped securely prior to lifting, lowering, or moving, **JPR 16-11A.**
2. Make sure that your partner is ready and understands what movement is desired.
3. Any time that a stretcher is being moved a minimum of two rescuers should work together, **JPR 16-11B.**
4. Do not attempt to roll a stretcher across uneven or rough terrain. It should be carried by an adequate number of personnel to make it safe. If at all possible, no rescuers should be backing up as they transport it.
5. The stretcher should be placed in the transport unit by the personnel responsible for operating the unit. Many times when firefighters or others attempt to assist, the stretcher gets unbalanced or in worst cases turns over, **JPR 16-11C.**

The army stretcher is most commonly utilized in mass casualty incidents or at events where transport units are either not available or their use is not practical (large crowds, stadiums, evacuations, etc.). A patient may be placed on an army stretcher using the extremity carry, by having the patient lie directly on the stretcher, or by using the method just given for placing a patient on a backboard. Whichever method is utilized, the patient must be securely strapped to the stretcher as soon as possible. Rescuers should not back up when carrying a patient on an army stretcher. The handles allow room for both rescuers to face and walk in the same direction.

JOB PERFORMANCE REQUIREMENT 16-11

Placing a Patient on an Ambulance Stretcher

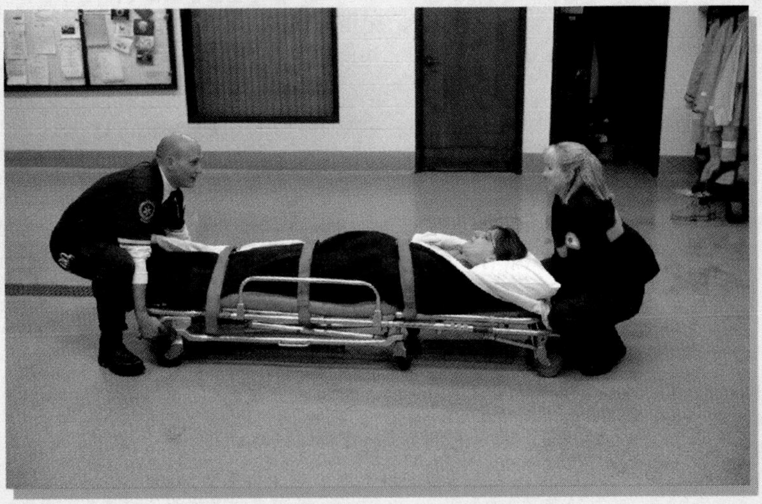

A Always make sure the patient is strapped securely prior to lifting, lowering, or moving. Make sure that your partner is ready and understands what movement is desired.

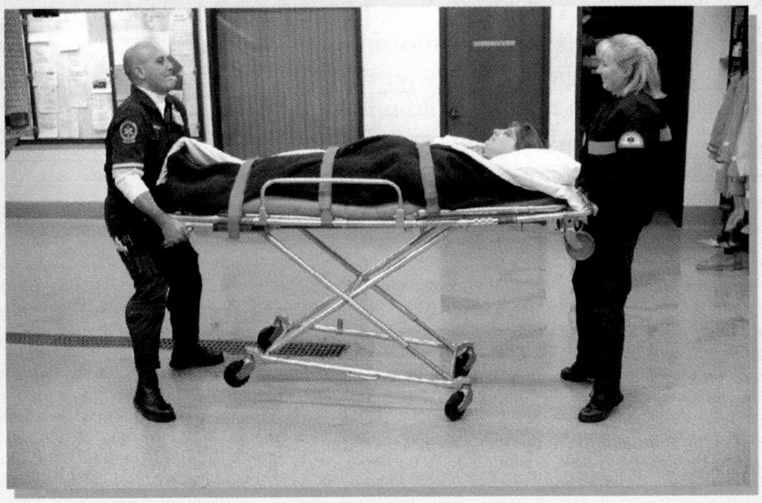

B Any time that a stretcher is being moved, a minimum of two rescuers should work together.

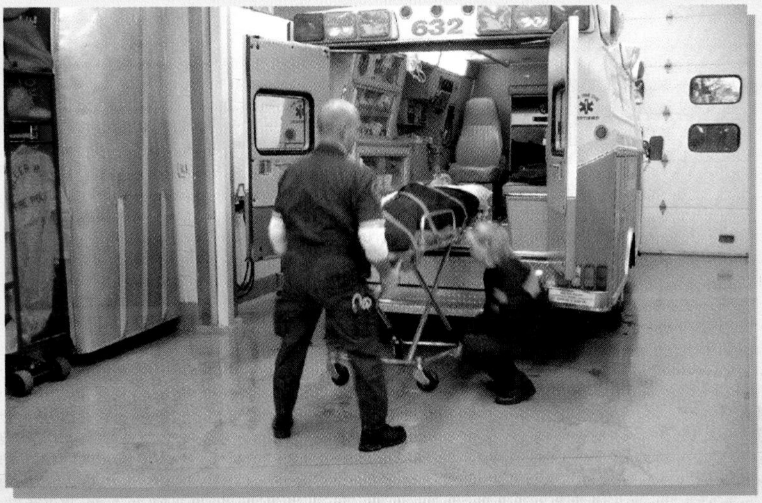

C The stretcher should be placed in the transport unit by the personnel responsible for operating the unit. Many times when firefighters or others attempt to assist, the stretcher gets unbalanced or in worst cases, turns over.

EXTRICATION FROM MOTOR VEHICLES

As was discussed in the introduction to this chapter, motor vehicle crashes are probably the most common rescue situation that today's firefighters respond to. The firefighters' most valuable tool at vehicle crash incidents is their knowledge, experience, and skill—not, as is often believed, more powerful tools. For the purpose of this section, a good working definition of extrication is "to set free, release or disentangle a patient from an entrapment."

Extrications can be as simple as removing a window, unlocking a door, or sliding a seat back, or as complicated as forcing and/or removing vehicle doors, roofs, dashes, seats, and so on. In general, as vehicles become larger and heavier (i.e., buses, tractor trailers, trains) extrication becomes more difficult due to the heavier structural components and severity of the crashes.

Operations at an extrication incident should follow a predetermined sequence of events. The following order of procedures works very well and ensures that none of the essential procedures or operations are overlooked:

1. Scene assessment (size-up)
2. Establishment of work areas
3. Vehicle stabilization
4. Patient access
5. Disentanglement
6. Patient removal
7. Scene stabilization

These procedures/operations are discussed in further detail later in this section.

Tools and Equipment

Tools and equipment utilized at vehicle crash incidents range from the most basic of firefighting tools, such as axes and pry bars, to the much more complex and specialized power hydraulic tools, air bags (low and high pressure), and battery-powered saws. A statement made earlier bears repeating here: The firefighters' most valuable tool at vehicle crash incidents is their knowledge, experience, and skill—not, as is often believed, more powerful tools. While power hydraulic tools, air bags, and the other advanced tools and equipment available today have certainly increased a firefighter's capabilities, they cannot replace the knowledge, skill, and ability developed through training and experience.

Power Hydraulic Tools

Many different companies manufacture power hydraulic tools today, but they all operate on basically the same principle. A hydraulic pump is powered by a gasoline engine, an electric motor, an air-driven motor, or the apparatus engine itself through a power take-off. Some manufacturers offer a manual hydraulic pump as a backup should the primary power hydraulic pump fail. The hydraulic pump, **Figure 16-8**, provides the required fluid and pressure to operate the variety of spreaders, cutters, and rams available. Because different companies' tools operate at different pressures (5,000 and 10,000 psi predominantly) and use different types of fluid, most of the manufacturers use hoses that will not connect to anything other than a compatible tool.

If a department has more than one type of power tool, personnel adding hydraulic fluid must be sure they have the right hydraulic fluid for the specific tool. The wrong type of fluid can cause serious damage.

Figure 16-8 Gasoline engine-powered hydraulic pumps for extrication equipment (left to right): Hurst, Genesis, Amkus. *(Photo courtesy of Rick Michalo)*

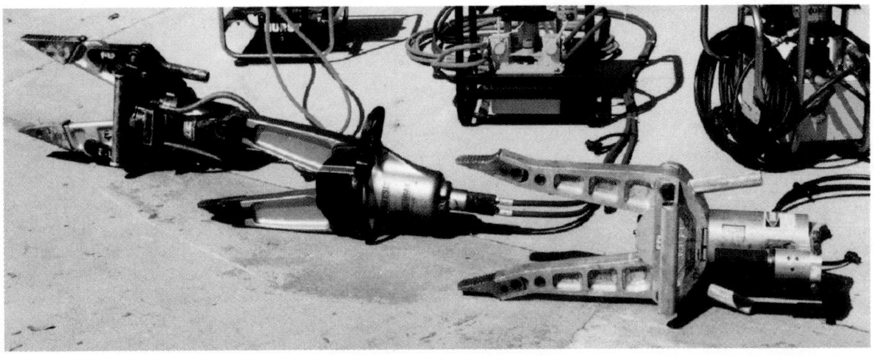

Figure 16-9 Power hydraulic spreaders (left to right): Hurst, Genesis, Amkus. *(Photo courtesy of Rick Michalo)*

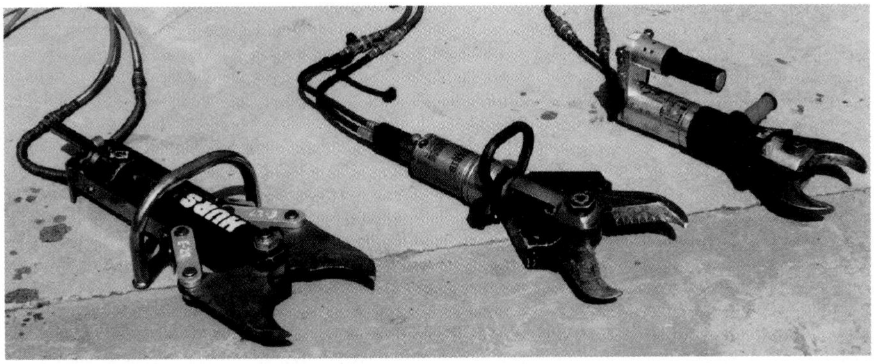

Figure 16-10 Power hydraulic cutters (left to right): Hurst, Genesis, Amkus. *(Photo courtesy of Rick Michalo)*

Spreaders were the first power hydraulic tool that became available to firefighters, and they are still widely used today. Spreaders can be used to both push and pull (if they are equipped with a chain attachment); they can be used to spread things apart or squeeze them together. Spreaders, **Figure 16-9**, come in various sizes of spread up to 32 inches with spreading forces up to 60,000 psi. Generally the pulling forces are less than the pushing forces for the same tool.

Cutters (also known as shears) were first introduced as a separate attachment to go on the spreader arms, either at the ends sticking out or coming back between the arms. *All cutters of this type should be removed from service since there is a possibility of the blades becoming crossed and shattering.* Modern cutters, **Figure 16-10**, are completely separate tools and come in a variety of sizes also. The smaller cutters are designed for working in close places to cut pedal supports or steel rods. Larger cutters have a tip opening in excess of 7 inches and a cutting force of greater than 80,000 psi.

Rams were the next hydraulic tool to appear with limited available sizes, but that has changed such that rams that measure as small as 10 inches extended up to more than 60 inches extended are available, **Figure 16-11**. Some rams are designed to both push and pull; however, the pulling strength is generally one-half that of the pushing strength.

Combination tools combine the functions of the spreader and the cutter. This type of tool incorporates the cutting blades on the inside of the spreader arms, **Figure 16-12**. These tools are generally less powerful than the individual stand-alone spreader or cutter.

Figure 16-11 Power hydraulic rams of different sizes (left to right): Hurst, Genesis, Amkus. *(Photo courtesy of Rick Michalo)*

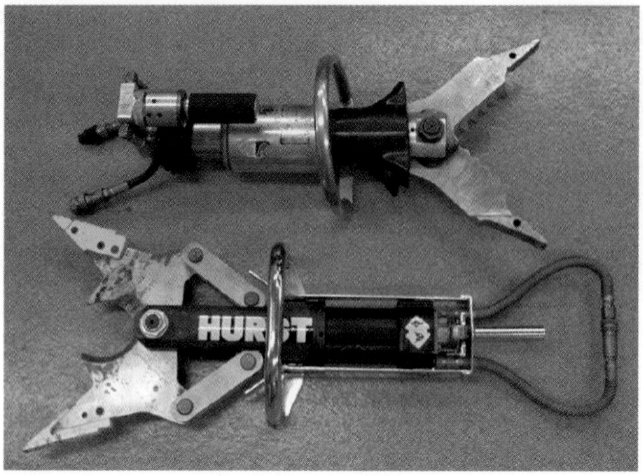

Figure 16-12 Power hydraulic combination tools: Amkus (top); Hurst (bottom).

Safety Firefighters should remember these important safety precautions when operating power hydraulic tools, **Figure 16-13**:

- Wear full protective clothing, including eye protection.
- Operating a tool is a one-person task.
- Do not put hands or arms inside the arms or blades of an operating tool.
- Make sure you are balanced.
- Watch the movement of the operating handle, ensuring that hands or the hydraulic lines do not get pinched.

Figure 16-13 It is important to be properly dressed and balanced when operating a power hydraulic tool at a vehicle extrication scene. *(Photo courtesy of Rick Michalo)*

Air Bags

Air bags used in rescue operations come in high-pressure and low/medium-pressure styles, **Figures 16-14A** and **B**. Each style has operational properties that are peculiar to that style, and therefore, each style serves a specific purpose in the rescue field.

High-pressure bags operate at a maximum inflation pressure of approximately 130 psi. They come in sizes ranging from 6 by 6 inches up to the largest ones measuring upwards of 36 by 36 inches. The larger bags are capable of lifting up to approximately 80 tons. One of the major advantages of these bags is that when deflated they are only 1 inch thick, allowing them to be inserted into very small spaces when necessary, **Figure 16-15**. The biggest drawback is that even the largest bag will only lift a maximum of approximately 20 inches.

Caution Although it is possible to stack high-pressure bags, extreme caution must be exercised to keep them from shifting under a load and being expelled at high velocity.

Low/medium-pressure bags operate at 7–10 psi/12–15 psi inflation pressure, respectively, depending on specific type and brand. These bags are bulky when deflated, not getting much less than 3 to 6 inches thick, but can lift heavy loads a considerable distance. Some of the larger bags can lift loads in excess of 6 feet, **Figure 16-16**.

Safety Firefighters should remember these important safety precautions when operating air bags, **Figure 16-16**.

- Bags must be on or against a solid base.
- Never inflate bags against sharp or hot objects.
- Make sure that air supply is adequate. (Do not use oxygen.)
- You must crib as you lift.
- Do not stack more than two bags, and if they are different sizes, the smaller bag goes on top.
- Operate controls carefully, filling bags in a controlled manner.

(A)

(B)

Figure 16-14 (A) A typical high-pressure air bag set. (B) A typical low-pressure air bag set. *(Photo courtesy of Rick Michalo)*

Figure 16-15 A high-pressure bag can be placed in a very small space at a vehicle extrication scene. *(Photo courtesy of Rick Michalo)*

Figure 16-16 A low/medium-pressure bag can lift a load 3 to 4 feet. Load must be properly cribbed as it is lifted. *(Photo courtesy of Rick Michalo)*

Air Chisels and Reciprocating Electric Saws

Standard air chisels have been used in the rescue field for quite some time. These chisels are designed to be operated at between 100 and 150 psi. However, newer versions of the air chisel, **Figure 16-17**, have been designed specifically for rescue operations and operate at up to 300 psi. The performance of these new air chisels is markedly improved.

Reciprocating electric saws are one of the newer tools making their way into the extrication field. One of the drawbacks with the earlier versions was that it was necessary to have 120 volts to operate them. However, with the evolution of the high-capacity battery-powered saws on the market today, they are finding widespread acceptance and usage in the rescue field, **Figure 16-18**.

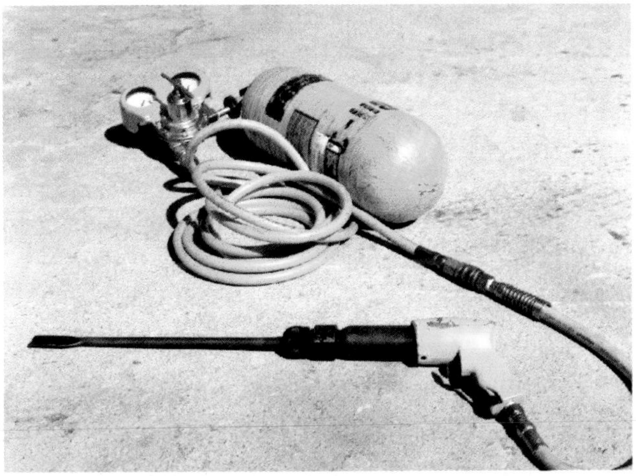

Figure 16-17 A high-pressure air chisel kit. *(Photo courtesy of Rick Michalo)*

Figure 16-18 A battery-powered reciprocating saw.

Safety Firefighters should remember these important safety precautions when operating air chisels and reciprocating electric saws:

- Do not use in hazardous or flammable atmospheres because they can cause sparks.
- Make sure air supply is adequate. (Do not use oxygen.)
- Make sure that no victims or rescuers are in the way when cutting.
- Use caution cutting hardened steel because blades may chip and/or break.

Scene Assessment (Size-Up)

Scene assessment should be a predetermined sequence of steps or actions that is used in evaluating a crash scene. This evaluation is usually carried out by the officer; however, a good firefighter who conducts an assessment will be able to foresee many of the needs and requests of the officer.

Note While it is important that this assessment be done quickly, it is more important that it be done accurately and completely.

Scene Safety

Good scene assessment considers many facts and probabilities. Specifically, the following items need to be taken into consideration:

- Traffic (see safety box)
- Number and type of vehicles involved
- Potential number and apparent extent of patient injuries
- Hazardous conditions (fire, HazMat, electrical, water, structural integrity, weather, crowds)
- Degree of entanglement

Safety Many firefighters have been injured and killed by traffic and secondary crashes at vehicle incidents. For this reason, fire and rescue responders need to size up traffic conditions as a *priority*. Blind spots, poor visibility (weather/darkness), road surfaces, bridges, barriers, congestion, and passing traffic speeds are all factors that need to be assessed.

The results of this assessment will help determine the need for additional resources or assistance that might include more ambulances, law enforcement, specialized equipment, power company response, or other ancillary needs. It is important to mention that the scene assessment is an ongoing process.

Establishment of Work Areas

Ideally, the fire department would like to shut down all traffic in and around the area of the vehicle incident. Unfortunately, the congestion this may cause can create secondary hazards and reduce the ability of additional responders to reach the scene. Many state and local law-enforcement agencies will work at all costs to keep traffic flowing. Because of this, it is essential that the fire department create protected work areas for rescuers and responders. Work areas are established using a combination of traffic barriers, traffic-calming strategies, and hazard zoning. Based on the scene assessment, the first arriving apparatus should be positioned to create a traffic barrier to help shield the greatest number of rescuers. Additional apparatus can increase the size of the barrier or be positioned *past* the scene so that they are also screened from traffic by the first arriving apparatus. While various positioning strategies exist, firefighters need to understand the logic and process of vehicle positioning so that they can work within the protection and limitations afforded by the barrier, **Figure 16-19.**

Traffic Barrier

A large fire apparatus such as an engine or heavy rescue can form the initial barrier by stopping in a position that shields the area where firefighters need to work. The apparatus should park with a slight diagonal angle to help increase the work area and make the apparatus appear larger to approaching traffic. The diagonal park also helps approaching traffic recognize that the fire apparatus is *not* moving. Even with a barrier in place, firefighters should mentally preplan an escape route should the barrier apparatus get hit by traffic.

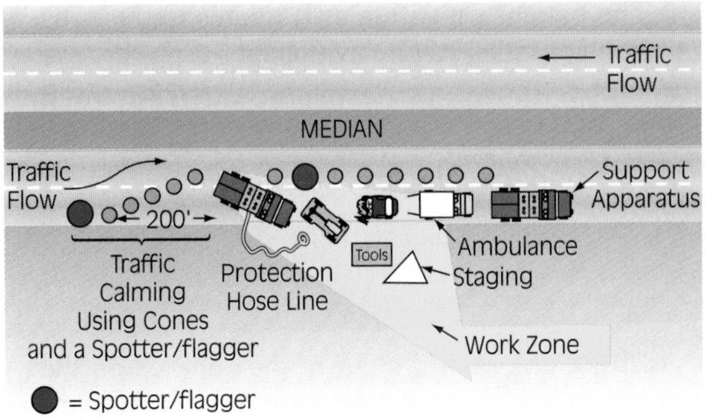

Figure 16-19 The first arriving large fire apparatus should position to create a traffic barrier and work zone. Cones and a spotter/flagger can help with traffic calming.

Traffic Calming

Efforts expended to warn approaching traffic of an upcoming hazard are known as traffic-calming strategies. Typically, law-enforcement agencies and the Department of Transportation will handle this task. However, in the early stages of the incident, the fire department can begin some simple traffic-calming strategies to slow traffic. Placing traffic cones on the street to help guide traffic is a popular strategy.

> **Note** Traffic cones should never be used to create a barrier. Cones should be placed well behind the traffic barrier (apparatus). Some departments still use flares (fusees) to help with traffic calming. These devices are incendiary. It is important to make sure they cannot roll and will not ignite flammable gases, liquids, or vegetation.

Other traffic-calming strategies include:

- Deploying portable Accident Ahead signs.
- Switching off white lights and strobe warning lights that cause night blindness to approaching traffic.
- Utilizing arrow sticks (flashing amber lights that signal in which direction traffic needs to move).
- Positioning a firefighter to wave down traffic speeds. This is done using a wand light or other highly visible device. The firefighter waves the wand up and down in a full arm length waving motion. This firefighter is exposed to traffic, but can also serve as a spotter for other rescuers. It is important to make sure an escape route is planned.

Hazards Zoning

Hazards that are identified in the scene assessment must be addressed. Initially, the fastest way to address the hazard is to create an exclusion zone (verbally or using flagging and/or traffic cones). Electrical wires will arc and move as automated systems try to reenergize the line. Pad-mounted transformers and electrical boxes can create a ground gradient of energy when they are damaged. Firefighters should treat them all as if they are energized and keep a 10-foot ring clear around the box. Damaged electrical poles can suddenly fall, dropping energized lines. Placing a lookout and preplanning an escape route can help zone these hazards. Hazardous materials should be treated as worst-case and zoned accordingly.

Most fire departments require that a fire attack line be pulled and charged as a standby measure whenever a vehicle extrication operation is under way. The firefighter assigned to this protective task can also serve as a lookout and remind other firefighters when they wander into an exclusion zone.

The vehicle(s) involved in the incident will likely present additional hazards. Fuel, electrical hazards, cargo, fire risk, and vehicle systems can be compromised in a motor vehicle incident. Understanding all the potential hazards a vehicle incident can present is well past the scope of this book; however, some hazards deserve a mention. Fuel systems on most vehicles are pressurized. If part of the fuel system is compromised, the fuel will escape under pressure initially.

Newer hybrid and alternative fuel systems present unique hazards ranging from asphyxiation to electrocution to toxicity. Alternative fuels can include compressed natural gas and propane. Hybrid systems include hydrogen and high-voltage battery systems combined with traditional fuels to increase engine efficiency. Each of these presents a unique set of hazards to responders. Firefighters should be encouraged to pursue understanding of new technologies through constant study of trade journals and extrication-specific training.

Vehicle Stabilization

Vehicle stabilization can be as simple as putting the transmission in gear or extremely complex, requiring cribbing, air bags, ropes, or other tools and equipment. The rescuer needs to use the information gathered during the assessment process to determine the amount of stabilization required. In any case, if a vehicle has an injured person inside, it must be stabilized by taking the weight off the vehicle's **suspension system**. (*Note:* Deflating the tires will not accomplish this.) This can be accomplished by using

cribbing, **Figure 16-20**. In more complex situations, stabilization may require the use of a variety of tools and equipment, **Figure 16-21**. Stabilization struts, cables, "step chocks," winches, and booms are just a few examples of the many tools that can be used to help stabilize vehicles. Tow-vehicle and vehicle-recovery businesses may have equipment that can help with vehicle stabilization. Utilizing the expertise of vehicle-recovery technicians can be helpful in unusual stabilization situations. It is important to have established previous communication and training with these technicians so that safety and accountability issues are clear.

Patient Access

Accessing the patient refers to providing a pathway for the rescuer assigned to evaluate and care for the patient, **Figure 16-22**. Many times this can be through a doorway or window that can be quickly broken or removed. Once access is gained:

- The patient can be evaluated.
- Life support activities can be initiated.
- The patient's position as it relates to the extrication activities can be evaluated.
- The patient can be protected from further injuries.
- Patient **packaging** can be initiated.

Figure 16-20 A properly cribbed vehicle using step and box cribbing styles. *(Photo courtesy of Rick Michalo)*

Figure 16-21 A properly cribbed vehicle using a combination of methods. *(Photo courtesy of Rick Michalo)*

Safety More and more of the vehicles on the road today have supplemental restraint systems (SRS) or, as most of us know them, air bags. These systems were initially located in the steering wheel or dashboard on the passenger side, but they can now be found in many different areas of a vehicle. Either currently or soon to be in production are bags that deploy from the seat cushions, the headrest, the head liner above the window, the dashboard under the steering wheel, and other locations. Most of these systems are deployed by electrical sensors that react to sudden directional impacts of sufficient force. SRS devices that have *not* deployed should be considered "live." Firefighters can be severely injured should the device activate during patient disentanglement or removal. Although disconnecting the vehicle battery will in some cases render the system inoperative, manufacturers have various backup or stored energy systems in case the battery was destroyed during a crash. Some of the best ways to learn the different systems and how they operate is to read fire rescue periodicals, visit various dealerships as training exercises, and attend specialized training.

Figure 16-22 Many times access to the patient can be made by removing the rear window. Note that the vehicle is properly cribbed, and the glass edges the patient attendant has to crawl over are covered. *(Photo courtesy of Rick Michalo)*

Disentanglement

The process of disentangling a patient begins with the rescuer who is tending the patient advising the incident commander as to the extent of injuries and mechanism of entrapment. The best pathway for patient removal is determined, keeping in mind that the pathway must provide ample working space and adequate space for removal of the packaged patient.

Once the pathway is selected the method or combination of methods for disentanglement must be selected. The options are:

Disassembly: The actual taking apart of the vehicle components.
Distortion: The bending of sheet metal or components, **Figure 16-23**.
Displacement: The relocating of major parts (i.e., doors, roof, dash, steering column), **Figure 16-24**.
Severance: The cutting off of components (i.e., brake pedal, steering wheel), **Figure 16-25**.

The method of disentanglement selected from those given here will provide information to the rescuer in selection of the proper tools to accomplish the desired goal. The safety of the rescuer(s) and the patient must always be at the forefront.

Entire texts have been written on effective disentanglement. Tried and tested extrication methods can be rendered ineffective with newer vehicle and material technologies. Training, tool capability understanding, practice, and inventiveness are the keys to effective disentanglement.

Patient Removal

Once the pathway has been created and made as safe as possible by covering sharp edges left by cut or torn metal, covering any broken glass that the patient could come in contact with, and removing all tools that may be in the way, the properly packaged patient should be carefully removed, **Figure 16-26**, again remembering that the goal is to minimize any aggravation of existing injuries.

Figure 16-24 The vehicle roof being displaced to allow better access to the patient. Note that the roof must be secured once it has been folded back to prevent it blowing back over on the patient and/or rescuers. *(Photo courtesy of Rick Michalo)*

Figure 16-23 A hydraulic ram is being used to distort (bend) the dashboard up and away from the patient. Note the long spine board being used for patient protection, victim cover in place, and attendant maintaining cervical support. *(Photo courtesy of Rick Michalo)*

Figure 16-25 A steering column being severed to allow for easier removal of patient. Note that protection must be provided between steering wheel and patient when performing this procedure. *(Photo courtesy of Rick Michalo)*

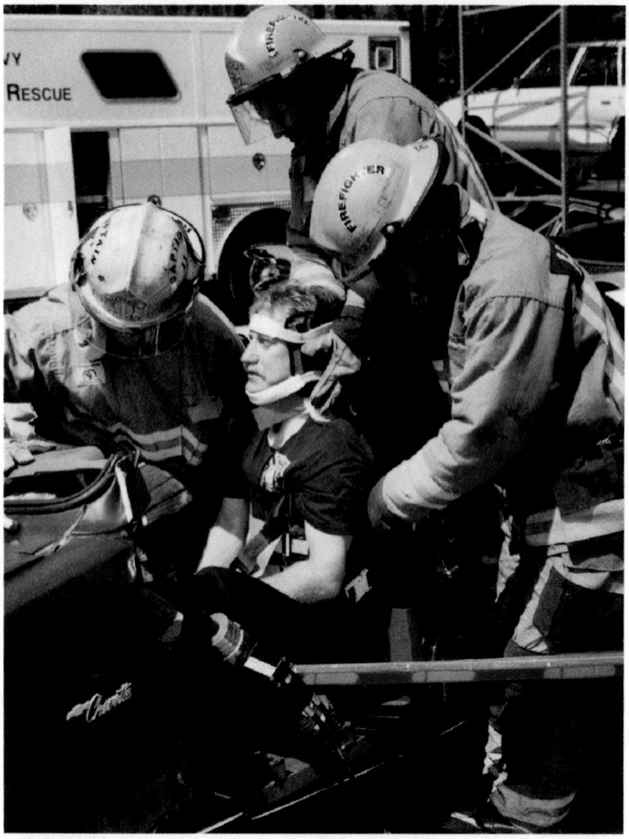

Figure 16-26 A properly packaged patient being carefully removed from a two-vehicle accident and placed on to a long spine board. *(Photo courtesy of Rick Michalo)*

Patient removal should be made only with the direct supervision of a certified EMS responder—typically a paramedic or experienced EMT. Appropriately, the supervising EMS provider should be focused on patient care. Often, the removal of the patient and subsequent movement toward an ambulance takes the rescuers and patient out of the work area protected by traffic barriers. Spotters should watch for traffic during patient movement, and escape routes must be planned.

Scene Stabilization

After patient removal, firefighters must refocus their energy toward securing the incident scene. Although this is not as exciting as the actual rescue, firefighters must continue to exercise safe practices. Extreme attention is required to release tools that are holding the vehicle—many components may "spring back" when the force of the rescue tool is relieved. Cribbing used to shore and support the vehicle might need to be retrieved, creating a potentially unstable vehicle again. At times, it may be best to allow tow vehicles to connect to the vehicle prior to cribbing removal. As tools are being removed from the wreckage, firefighters should inspect the cutting/contact surfaces, hydraulic lines, and safety features for damage and report any problems or observations to their officer. Additionally, the scene stabilization effort may include the following:

Vehicle Recovery

Fire resources are often asked to stand by for the tow vehicle to overturn or "hook" the damaged vehicle. The tow-vehicle operator must be advised of vehicle hazards such as fuel leaks and supplement restraint systems (SRS air bags) that have *not* deployed. Local SOPs may require that the battery of a damaged vehicle be disconnected. Only trained personnel should perform this disconnect. Tensioned tow cables introduce another hazard. It is a good practice to "envision" the tow cable breaking and stay clear of the path it will recoil. Likewise, it is important to stay clear of the vehicle as it is being moved and to predict its path should the tow cable break.

Fluid/Parts Cleanup

The use of a dry absorbent is typical for oils, engine coolant, and diesel fuel. Gasoline is best absorbed with specially designated "fuel pads" and disposed of in accordance with local SOPs. When handling fuel pads, it is best to use gloves that can be decontaminated. Battery acids are extremely caustic and should be handled with diligence. Vehicle parts that have been strewn about can be sharp. Law-enforcement officials may want strewn parts to be left in place to assist with accident reconstruction investigations. If not, carefully place all loose parts back into the wrecked vehicle.

Once the scene has been stabilized, the responding crews must secure all deployed equipment and pick up traffic-calming devices. This exposes rescuers to traffic hazards again. As a rule, they should not put their back to traffic—they should pick up traffic cones while facing traffic. Often, spotters are needed to help apparatus back out of position and into traffic. These spotters must also have a traffic lookout watching their back.

SPECIALIZED RESCUE SITUATIONS AND TOOLS

The previously discussed situations are the ones that firefighters are most commonly faced with. There are a variety of other rescue situations, however, that a firefighter may be called on to respond to. The following sections discuss a number of these, focusing on enabling firefighters to recognize the situation, call for the necessary assistance, not cause the situation to worsen for the victim, and not become victims themselves. Many of the techniques discussed earlier can and will be very beneficial to a firefighter responding to these special situations.

Vertical Rescue

Vertical rescue deals with a victim who is either above or below normal ground level and beyond the reach of available ladders or other practical means. Chapter 15 on rope discussed life safety rope; the requirements as far as size, usage, and care; and the appropriate knots and their uses. That chapter also referenced NFPA 1983, the standard that addresses life safety rope. In addition to rope, the standard also addresses **hardware** and **harnesses** utilized in rescue situations. Harnesses are classified into three categories:

Class 1: A single belt with an adjustable buckle that goes around the wearer's waist and some type of hook, **Figure 16-27**. This harness is commonly known as a ladder belt, and the hook is used to secure a firefighter to a ladder. It is intended to hold a firefighter in place, not provide fall protection. If a fall of more than 2 feet were to occur, serious injury can be expected to the abdominal and spine area. It is not acceptable as a life safety harness.

Class 2: A Class 2 harness consists of a waist strap and straps that go around each leg, creating a seat-type arrangement, **Figure 16-28**. Class 2 harnesses are required to be capable of supporting a rescuer and patient at the same time.

Class 3: A Class 3 harness consists of the same type of seat arrangement as the Class 2 harness, but has additional pieces that go over the shoulders and have additional attachment points, usually at the top of each shoulder and in the center of the upper back, **Figure 16-29**. They are designed to keep wearers in an upright position or to keep them from slipping out of the harness if they are using the waist attachment point and get in a head-down position. Class 3 harnesses are also required to be capable of supporting a rescuer and patient at the same time.

As with all rescue equipment, proper maintenance, cleaning, and inspection are required for life

Figure 16-27 A Class 1 harness.

Figure 16-28 A Class 2 harness.

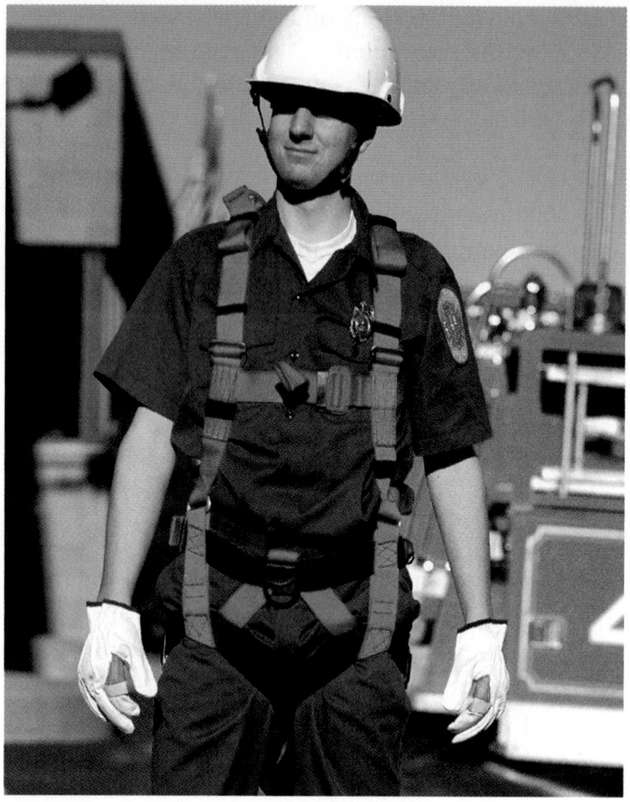

Figure 16-29 A Class 3 harness.

safety harnesses. They should be inspected and cleaned following the same general criteria used to inspect life safety ropes with particular attention paid to the stitching.

If a patient is going to be suspended in a harness for any reason, it is highly recommended that a Class 3 harness be used. In emergency situations where waiting to get the proper harness could cause greater harm, a patient may be lowered or raised utilizing the rescue knot described in the previous chapter or the department-approved rescue knot.

If a patient is to be lowered using a life safety rope it is imperative that the personnel on scene have the training, skill, and equipment to do so safely. Many fire/rescue departments have developed special teams to respond to these types of incidents. The teams have the proper equipment, and the personnel receive specialized training and are required to maintain proficiency. However, firefighters should be familiar with the basic equipment and techniques utilized by their department at this type of incident.

Water Rescue

Water rescue operations can be very dangerous for the rescuers. In many cases, the victim(s) can be seen, which results in the emotions of the bystanders being very high. They may expect the firefighters to jump in the water immediately to rescue the person in trouble. That is the *last* thing a rescuer should do, and it should never be done without a personal flotation device (PFD).

Figure 16-30 A firefighter using the reach method to rescue a victim in the water. Note the PFD and the use of a pole to extend the firefighter's reach.

Safety Every firefighter at the scene of a water rescue incident should have a PFD on if at all possible. If not, those without PFDs should be kept well back from the water's edge. If they are allowed to be near the water's edge and something happens in front of them, they may go into the water without the proper equipment.

Following in priority order are the methods and procedures that firefighters should utilize to rescue a victim from static (lake) water and slow-moving rivers and streams:

1. *Reach.* If the victim can be reached without the rescuer entering the water, a pike pole, pool rescue pole, or anything that can be used to extend the rescuer's reach can and should be utilized, **Figure 16-30**.
2. *Throw.* If the victim cannot be reached directly but is within throwing range of a rescue throw bag or flotation device attached to a rope, then one should be thrown and the rope utilized to pull the victim to safety, **Figure 16-31**.
3. *Row.* If the victim cannot be reached by either of the preceding methods and there is a suitable boat available, the rescuer(s) should utilize it. Extreme caution should be exercised on approaching the victim. In a

Figure 16-31 A firefighter using the throw method to rescue a victim in the water. Note the PFD and the underhand throwing technique.

Figure 16-32 A firefighter using the row method in a small boat to rescue a victim in the water. Note the extra PFD in the boat for the victim. Do not attempt to lift the victim into a small boat; instead have the victim hang on to the side.

panic, the victim could capsize the boat, resulting in the rescuers now becoming potential victims. The best course of action is *not* to try lifting the victim into the boat at all (unless it is designed for rescue operations), but to hold onto the person and return to shore, **Figure 16-32**.

4. *Go*. If none of the preceding options is possible, the absolute last method is for the rescuer to enter the water, **Figure 16-33**. Extreme caution must be exercised when approaching the victim. In a panic, the victim could injure the rescuer or make it difficult for the rescuer to assist him or her. If possible the rescuer should carry an additional PFD to the victim. In still or extremely slow-moving water it is a good idea for the rescuer to swim out holding a **tether line**. This will allow the rescuers on shore to pull the rescuer and victim back to shore.

Safety A rescuer should not use an attached tether line in moderate to fast moving water. It can pull the rescuer under.

Swift-water rescue is a specialized field requiring additional training, practice, and protective equipment. The force of a fast moving stream is considerable, even in shallow currents. As mentioned before, tether lines increase the danger to rescuers. Special rigging and procedures are required for a successful rescue. Local SOPs and training are required for swift-water rescues; however, some basic rescuer safety procedures can be discussed here. First, all rescuers must wear PFDs. They should try to encourage the victim to keep afloat until a swift-water team can arrive and set up. If an immediate rescue attempt must take place, rescuers can place themselves at the shore and attempt the throw method, making sure a float is attached to the end of the rope. A PFD is a great float and will help calm the victim. Rescuers should throw the rope well upstream of the victim and allow the current to drift the rope toward the victim. Once the victim has the rope, rescuers can encourage the victim

Figure 16-33 A firefighter using the go method to rescue a victim in the water. Note that the firefighter is carrying an extra PFD for the victim.

to don the PFD. The rope is *not* used as a tether; it acts merely as a guide rope. The victim holds the rope, and the rescuers allow plenty of slack in the line. As the victim drifts into the stream, the rescuers release just enough slack to let the victim swing toward the shore. Once again, this is a last-ditch method. Properly trained swift-water technicians are the key to a successful rescue.

Ice Rescue

Many of the safety precautions and rescue procedures just presented with regard to water rescue apply to ice rescue as well. The procedures and hazards are very similar with the added element of extreme cold. In place of or in addition to PFDs, rescue personnel working at an ice rescue scene should have thermal rescue suits, which provide the required personal flotation as well as protection from the extreme cold. The rescuers in **Figure 16-34** are wearing thermal suits.

If rescuers must go out onto the ice themselves, they need to use whatever is available to distribute their weight over as large an area as possible. Some items that work well for this are ladders, sheets of plywood, planks of wood, and backboards—even folded salvage covers would be better than nothing.

> **Caution** A rescuer should not approach too close to the hole itself. The ice is obviously not stable in that area.

The rescuer should use the reach (with extension) or throw method to keep from having to be right up next to the hole. Ice rescue victims may not have the strength or determination to grab a rope or pole. The effects of the cold water quickly rob the human body of energy. In these cases, firefighters with appropriate thermal flotation PPE should try to approach the victim in such a way as *not* to disturb the ice the victim is clinging to. The best approach is to come in behind the victim and use a rescue collar or other rope device to secure the victim. It is essential to coach the victim to hold tight to the rope but not to the rescuer. Again, these rescue skills are past the scope of this book. All firefighters should train and practice under the supervision of a competent instructor.

Structural Collapse Rescue

Structural collapse, while not a common incident, may occur for any number of reasons: weakening due to age or fire, environmental causes (earthquake, tornado, hurricane, flooding, rain, or snow buildup on roofs), or an explosion (accidental or intentional). Whatever the cause, the result can be any number of victims trapped, injured, or killed. One of the firefighter's first priorities at this type of incident is to determine the number of possible victims. This information along with the construction type, size, and occupancy of the collapsed structure will have a direct impact on the assistance required.

(A) (B)

Figure 16-34 (A) A ladder, a sheet of plywood, or a specialized ice rescue boat can be used to spread a rescuer's weight over a larger area. (B) Using a throw bag or pike pole to reach out to a victim who has fallen through the ice prevents the rescuer from having to approach the weakened area too closely. *(Photo courtesy of Halfmoon-Waterford Fire District 1)*

Safety Safety of the rescuers needs to be the number one consideration as in all rescue operations. Some of the particular hazards to be aware of at collapse scenes are secondary collapse, live electrical wires, and gas leaks.

Rescue of victims who are not trapped or are lightly trapped should be done as soon as safely possible. It is very common for this process to already be under way by civilians from the immediate area when a rescue unit arrives. It is important to remember that as civilians and rescuers are climbing on and over debris there is a possibility that other victims may be trapped under that debris, **Figure 16-35**.

Structural collapses are generally classified into three different types, which are described next. However, it is not uncommon to encounter more than one type at any given collapse incident.

Pancake Collapse

A pancake collapse occurs when both sides of the supporting walls or the floor anchoring system fails. It is characterized by the roof or upper floor(s) falling parallel onto the one below, often causing a domino effect, which may result in several or all of the floors collapsing one upon the other until the lowest level is reached. This type of collapse often results in many small **voids** being created by debris between floors supporting the upper floor, **Figure 16-36**.

Lean-To Collapse

A lean-to collapse occurs when only one side of the supporting walls or floor anchoring system fails. It is characterized by one side of the collapsed roof or floor remaining attached or supported by the wall or anchoring system that did not fail. This type of collapse usually results in a significant void being created near the remaining wall, **Figure 16-37**.

V-Type Collapse

A V-type collapse occurs when the center of the floor or roof support system is overloaded (improper placement of stock, buildup of snow or rain, etc.) or becomes weakened for some reason (fire, rot, termites, improper removal of support beams, etc.). It is characterized by both sides remaining attached or supported by the walls or anchoring system but collapsing in the center. This type of collapse will

Figure 16-35 Firefighters need to remember that victims may be buried under the debris that they are climbing on and over at a collapse scene.

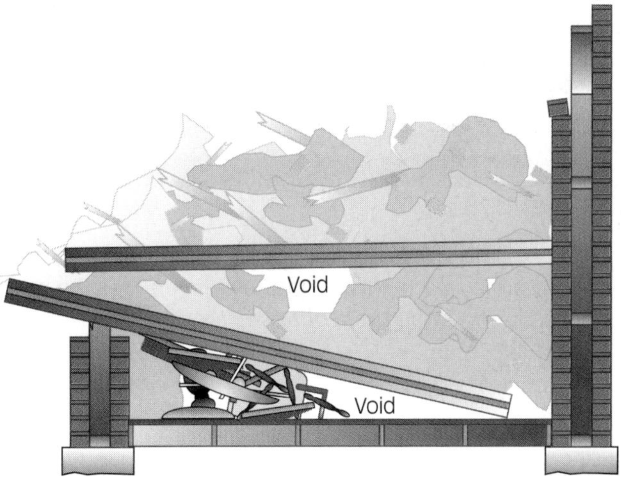

Figure 16-36 A pancake collapse. Note voids where survivors may be located that have been created by debris during structural collapse.

Figure 16-37 A lean-to collapse. Note voids where survivors may be located that have been created by debris during structural collapse.

usually result in voids being left on each side near the supporting walls, **Figure 16-38**.

It is important for the firefighter to know where these voids can be anticipated in each of the different types of collapses since this is where the greatest likelihood of finding survivors exists. Knowing where the survivors may be is just part of the problem. The firefighter must also know how to make the area as safe as possible prior to entering. Having dealt with the issues previously discussed (live wires, gas leaks, etc.) the firefighter needs to have a basic knowledge of cribbing, **shoring,** and **tunneling**: examples of each can be seen in **Figure 16-39**.

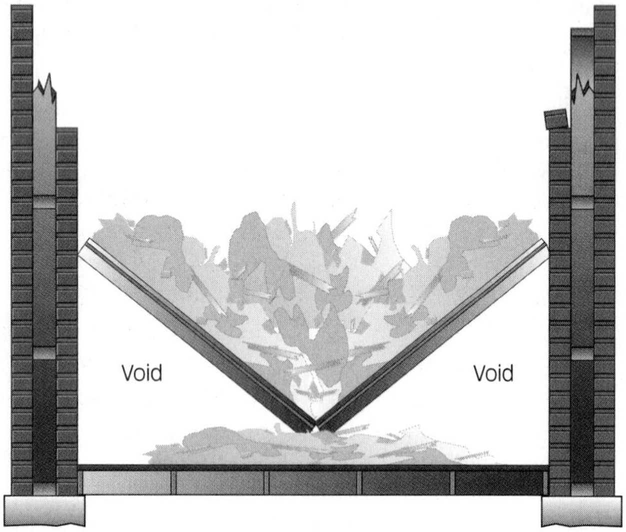

Figure 16-38 A V-type collapse. Note voids where survivors may be located that have been created by debris during structural collapse.

Safety It is important to remember in all of these operations that the firefighter is not lifting the fallen members, but merely supporting them where they are. To lift them may cause further collapse.

- *Cribbing.* As previously defined, cribbing is the use of various dimensions (2×4, 2×6, 4×4, 6×6, etc.) of lumber arranged in systematic stacks (pyramid, box, step, etc.) to support an unstable load. The same principle applies to the use of cribbing in collapse operations; however, it is common to use timbers of larger dimensions.
- *Shoring.* Shoring is the use of timbers to support and/or strengthen weakened structural members (roofs, floors, walls, etc.) in order to avoid a secondary collapse during the rescue operation. These can be very complex operations and

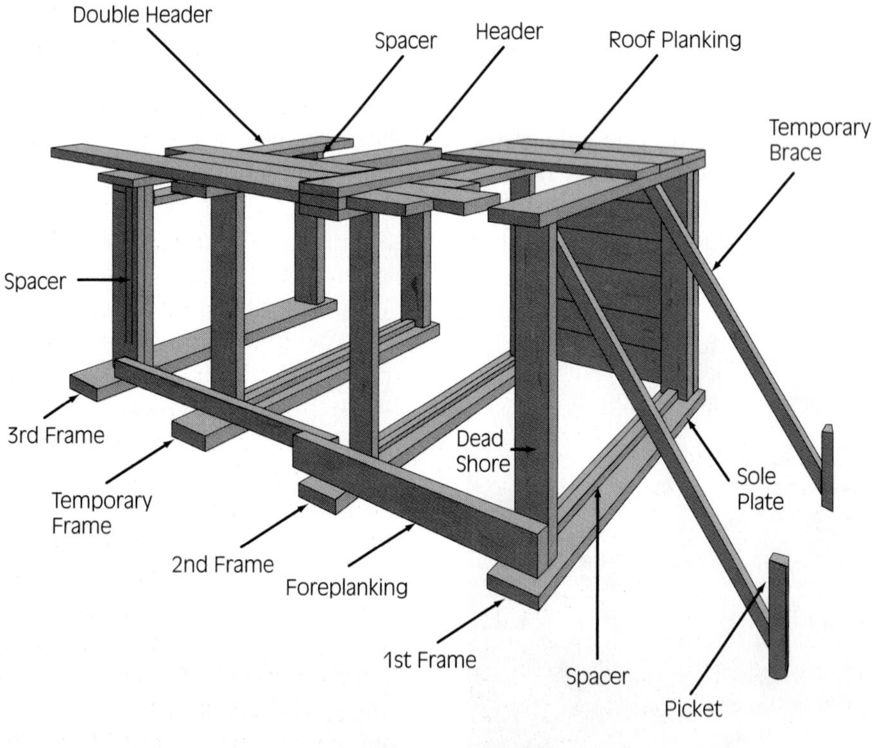

Figure 16-39 The proper usage and placement of cribbing to support fallen members, shoring to hold back the sides and roof debris, and tunneling supports are required prior to entering trench or collapse areas.

should not be undertaken without the input, advice, and guidance of experts in the engineering field, preferably someone with experience with collapsed structures.

- *Tunneling*. Tunneling may be required to reach survivors located in the voids created during the collapse. This is a very dangerous and time-consuming process and should not be undertaken unless all other avenues have been tried. Tunneling should not be utilized as a general search process; a specific destination is necessary along with a good indication that survivor(s) will be located when the destination is reached.

Trench and Below-Grade Rescue

Most trench or below-grade incidents occur at construction sites, utilities (gas, water, sewer, etc.), maintenance sites, or well digging sites, an example of which can be seen in **Figure 16-40**. However, every year a number of incidents are responded to that involve children digging a large hole for whatever reason they can think of. There are also a number of incidents that do not involve trenches or below-grade operations at all but are very similar in the challenges they present to rescuers and the life-threatening entrapment of the victim. These incidents occur in grain silos, cement hoppers, sawdust collectors, fertilizer hoppers, etc.—basically anywhere that a victim can be **engulfed** by a loose granular-type product. What makes the entrapment by dirt, sand, or any of the listed types of products so life threatening is not only that the engulfing material could cover the victim's mouth and nose, thereby compromising the airway, but second and possibly even more life threatening is that the engulfing material gets packed in around the victim's chest, preventing the ability to inhale. This is why a victim succumbs to **asphyxiation** even though the head is not buried.

Figure 16-40 An actual trench rescue operation. Note the amount and variety of equipment that can be seen.

The most immediate danger to the victim is usually a compromising of the ability to breathe. It follows then that the first priority on reaching a victim is to uncover the victim's head and as much of the chest as possible. Supplementing respiration with oxygen should also occur as soon as safely possible.

Even with the need for haste as described in the previous paragraph, rescuers should assume that, because the fire department has been called to the scene for a rescue, proper safety procedures were not being utilized to begin with. With this in mind, the rescue crew must approach the scene very carefully keeping their safety in the forefront. The rescuers should use **ground pads** or something else to distribute their weight over a larger area (the same as with ice rescue discussed earlier) when approaching the caved-in area, to prevent causing more material to fall in on the victim or others.

Safety Even if coworkers or others are already attempting to assist the victim(s), the fire department must resist the urge to jump in and help until the area is made safe.

For a trench or below-grade incident to be resolved as safely as possible, the previously mentioned ground pads should be in place, all nonessential personnel and equipment should be kept away from the site, the cave-in area must be shored up, the air quality needs to be monitored, fresh air needs to be introduced into the area, and adequate egress needs to be provided for the rescuers, many of which can be seen in **Figure 16-41**.

Upon entering the caved-in area, the rescuer needs to have appropriate personal safety clothing and equipment and the correct tools to rescue the victim. In the case of a cave-in, power equipment and tools should not be utilized until the exact number, location, and position of the victim(s) are determined. Even then they must be used very cautiously, and the effect they are having on the possibility of further cave-in closely monitored. In most cases the safest (for victim and rescuers) and most effective method is to dig with small hand tools or the hands only.

While one rescue team is in the process of excavating the victim, there should be another team topside addressing the need to remove the victim once freed. It can be anticipated that the victim is going to have injuries such as trauma to the head, spine, torso, and extremities. The removal team needs to be factoring these possible injuries into the removal

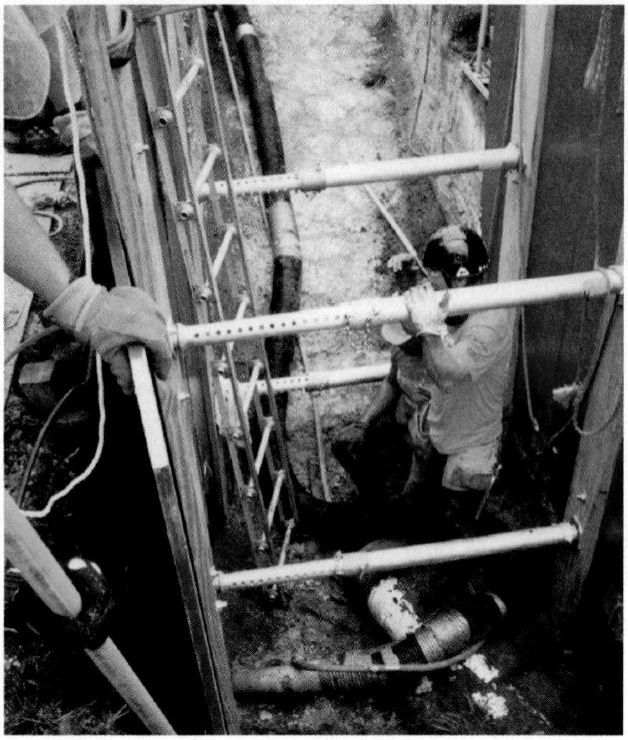

Figure 16-41 A trench rescue operation requires that all necessary safety equipment/precautions be in place prior to rescuers entering the trench.

Figure 16-42 The topside team needs to be prepared with any immobilization and hauling equipment that may be required.

plan and be prepared to provide adequate immobilization and hauling systems, **Figure 16-42**.

Confined Space Rescue

A **confined space** is defined as a space that is large enough to be entered but is not designed for continuous occupancy. Confined spaces come in many different forms; they can be found at grade level, below grade, above grade, and on board ships. Typical confined spaces are utility vaults, storm water or sewer culverts, septic tanks, industrial boilers, storage tanks, grain silos, cement hoppers, sawdust collectors, fertilizer hoppers, and areas aboard ships. As can be seen from this list, confined spaces can and do exist in every department's jurisdiction.

> **Note** It is estimated that approximately one-half of the people who die each year in confined space incidents do so while attempting to "rescue" a victim. Although the majority of these "rescuers" are civilians or coworkers, unfortunately some are firefighters.

The most common hazard encountered at a confined space incident is an oxygen-deficient atmosphere, followed closely by toxic and/or hazardous vapors. Entry into a confined space should never be attempted without having first sampled the atmosphere and confirmed it as tenable or identified the hazards. If the atmosphere is found to be oxygen deficient, this situation can be improved by the use of an air blower and flexible duct.

Whenever members of a rescue team enter a confined space, they must utilize appropriate personal protective clothing and equipment (for the known and possible hazards), including lifelines for each member and a radio or other communication system between the rescue team and the operations chief. The space must be constantly monitored, preferably by individual monitors on each entry team member, but at a minimum by one member of the entry team assigned this task. A backup crew must be fully equipped and dressed out to act as a rescue team for the entry team if a problem develops or the entry team needs assistance for any reason.

Rescue from Electrical Situations

Rescuers responding or arriving at an incident scene involving electrical wires, transformers, transfer stations, and so on, must assume these items are energized until advised otherwise by the utility company representative at the scene. Very often wires will appear to be de-energized because they are lying still on the ground and no arcing is visible. This is not necessarily the case; the wire may be well grounded at that time or there may be a time delay reset breaker on the line that will reset itself after a set amount of time and may do so a number of times before finally staying in the off position. Any victim who is in contact with an electrical wire must also be considered energized and cannot be touched. Firefighters need to remember that none of their protective clothing and/or equipment is designed to protect them from electrical current.

Vehicle accidents involving electrical equipment like pad transformers and downed wires are especially dangerous. It is best to leave victims in the vehicle and not approach. Often, the vehicle may be energized—yet the passengers are safe as long as they do not exit. A victim in an energized car is like a bird on an overhead power wire—that is, as long as the ground is not touched, no electrocution will take place. Anything or anyone who touches the vehicle and ground at the same time will cause an easier path for electricity to seek ground. Firefighters attempting to approach involved electrical equipment may feel a tingling sensation in their boots. This is a dangerous sign known as ground gradient. When the tingling is felt, firefighters must back away using a shuffle foot motion to keep their feet in contact with the ground. In all cases, the fire department shall verify that power company technicians have been dispatched.

If the electrical contact is within a building, it should be possible to shut the power off at the main disconnect. It is recommended that the main disconnect be utilized as opposed to individual breakers because sometimes circuits or equipment may be receiving electricity from more than one breaker. If this (shutting off main disconnect) method is used, a firefighter should be stationed at the disconnect, **Figure 16-43**, to ensure that no one turns the power back on either accidentally or because the person did not know a rescue was in progress.

Firefighters should never enter an enclosed electrical transfer station, substation, or vault without a company representative on scene to ensure that all electrical power is shut off. Every wire, bracket, frame, transformer, and so on, within the enclosed area should be assumed to be energized prior to the on-scene company representative confirming otherwise.

Some departments carry specialized equipment such as hot sticks, insulated cutters, and insulated lineman's gloves on their apparatus. If these items are to be utilized they must be stored, tested, and certified in accordance with the very strict standards that apply to such equipment. The personnel assigned to utilize this equipment must also receive specialized training in the use and care of this equipment.

Most fire departments have recognized that this is a highly specialized and dangerous operation and best left to the professionals with the utility company.

Any time large electrical wires are going to be cut (energized or not) one of the dangers to be aware of is what is known as **reel coil**. This is memory that the wire develops from being placed on the wooden spool as it is being manufactured. This reel coil will cause the wire to recoil (spring away) from the point at which it was cut, attempting to resume the coiled

Figure 16-43 Rescuers must take control of the electrical disconnect and remain in contact with the rest of the team via portable radio.

form again. This recoiling can be very quick and powerful in heavier wire and can cause severe injury from the impact even if it is de-energized.

Industrial Entrapment Rescue

Extricating a victim from an industrial entrapment is usually a very complex process. In most cases if the entrapment is relatively minor, the victims' coworkers will have already extricated them by the time the fire department arrives. Due to the number and complexity of machines used in industrial plants today, it is impossible to have a specific plan for each possible entrapment.

The operations at an industrial entrapment incident should follow a predetermined sequence of events. The process presented previously in the vehicle crash extrication section is applicable at an industrial entrapment with very minor changes.

The following order of procedures works very well and ensures that none of the essential procedures or operations is overlooked:

1. Assess the incident (size-up).
2. Identify on-scene or available experts.
3. Stabilize entrapping machinery.
4. Access the patient.
5. Disentangle the patient.
6. Remove the patient.

Incident assessment should be a predetermined sequence of steps or actions that is used in evaluating an entrapment. This evaluation is usually carried out by the officer. However, at an industrial entrapment incident, there may be a firefighter on the crew with knowledge and/or experience that could be invaluable. A good officer should know this ahead of time and never hesitate to utilize it. As previously stated, while it is important that this assessment be done quickly, it is more important that it be done accurately and completely.

Although identifying on-scene or available experts is shown as the second step it is vitally important that this expert advice be available during the assessment phase if at all possible. It is quite possible that the expert may be able to provide a simple and quick extrication method that is not obvious. It is also possible that the expert can prevent further injury to the victim or injury to the rescuers by preventing the rescuers from taking some action that, although appearing to be logical and effective, would cause the machine to react in an unexpected manner. A good example of this is that while turning off the power to a machine would appear to be the logical thing to do, on some punch press-type machines this would cause them to finish out their cycle.

Personnel conducting an assessment need to take the following into consideration as part of a good incident assessment:

- The apparent extent of injuries
- Current and anticipated hazards to the victim and/or rescuers
- Disentanglement requirements
- Support needs

The results of the assessment will reveal whether any additional assistance is required, such as:

- Additional medical units
- Additional or specialized extrication equipment
- Advanced medical assistance, such as an on-scene doctor
- Other specialized equipment, such as a light unit
- Hazardous materials team

Stabilization of the entrapping machinery should begin with shutting down the power (electrical, hydraulic, and pneumatic) and releasing stored energy (air or hydraulic pressure) to the machine, as soon as it is determined that this will not cause any further movement. Once the power is shut off to the machine, a firefighter should be assigned to stand by the shutoff switch to ensure that it is not turned back on inadvertently.

Generally, the next step in stabilizing the machine is to crib or block the entrapping part so that there will be no further unwanted or unplanned movement. If the machine is in midpoint of a movement, it may be necessary to crib both above and below the moving part. Although in many cases it is quite appropriate and convenient to use the same cribbing materials utilized for stabilizing vehicles, heavier and stronger materials are sometimes required since the machine may be capable of crushing regular cribbing. It may also be possible to use chains or cables to assist in stabilizing the moving part.

Accessing the patient is usually not that difficult in these types of incidents; however, accessing the entrapped appendage may be very difficult or impossible in some cases. It is very beneficial, from the medical point of view, if access can be gained to the entrapped appendage.

The three overriding factors when deciding on the method to be employed to disentangle the patient are (1) the medical condition of the patient, (2) the amount of time required to complete the extrication, and (3) the effect the extrication activities will have on the patient. While the damage to the entrapping machinery should be considered, it has a very low priority when compared to the patient's well-being.

> **Note** Amputation is a viable option in industrial entrapment incidents more than any other particular incident. This is due to the severe damage that is often done to the entrapped appendage and concern for the medical condition of a patient who may not be able to tolerate a complex and lengthy extrication operation.

When it is possible to do so without causing further injury to the patient, the simplest and fastest method of disentanglement is to operate the machinery manually through the rest of its normal cycle or to manually operate the machinery in reverse, backing the entrapped appendage out of the machine.

When these methods are not possible or practical, the entrapping machinery must be disassembled, forced, or cut. Obviously the use of on-scene expert advice and assistance is highly recommended during this phase of the operation. The tools and equipment discussed in the vehicle extrication section may also be very useful in industrial entrapment operations. However, in many instances it might be impossible to force the machinery without causing greater harm to the patient or impossible to cut the machinery components due to their size or strength. In these cases, the machinery must be disassembled piece by piece.

Elevator and Escalator Rescue

When a call is received for an elevator or escalator incident, the responding fire department unit should immediately request that the dispatcher confirm with the calling party that the service company has already been called and a technician is en route; if not, one should be requested immediately. The elevators being installed today and even those that were installed many years ago are very complex. Unless there is a compelling medical emergency requiring that the rescuers access or remove the passengers from the elevator immediately, they should await the arrival of the service technician. In many cases the service technician will be able to repair the problem immediately or at least be able to bring the car to a landing so that the passengers can be removed through normal methods. At incidents where there is no immediate need to access or remove the passengers, the fire department's responsibility is to establish communications with the passengers, ensuring them that they are perfectly safe and help is on the way.

The two basic types of elevator operating systems are hydraulic and electric/cable. Hydraulic elevators are raised and lowered by use of a hydraulic pump connected via a valve system to a multisection ram, on top of which the elevator car sits. A very basic example of this type can be found in many garages where they are utilized to lift vehicles to be worked on. The power unit for a hydraulic elevator is usually found near the lowest floor served by the elevator, **Figure 16-44**. This type of elevator is not generally used in buildings of more than five stories.

Electric/cable elevators are raised and lowered by the use of an electric motor connected to a drum or cable sheave. The cable(s) that is taken up (to raise the car) or let out (to lower the car) by the drum/sheave unit is connected to the frame of the elevator car. The motor drum/sheave unit for this type of elevator is generally located directly above the elevator shaft, **Figure 16-45**, though it can be located at the bottom or in rare cases in an adjacent room. Electric/cable elevators are in use at all high-rise buildings. Extremely tall buildings may have two or more sets of elevators, with one set serving the lower floors and another set serving the upper floors.

Elevators have a variety of doors, openings, and safety interlocks that need to be understood by rescuers. Starting with the **hoistway** door, these doors are locked in the closed position unless the car is at the correct level (the landing) in the hoistway to release the locking devices. These doors can be opened by the use of special hoistway door keys, which come in a wide variety of key shapes and styles. Unless the responding unit has an assortment of these keys, an example of which is shown in **Figure 16-46**, the

Figure 16-44 A hydraulic elevator pump and valve assembly located in the mechanical room, usually on the first floor.

Figure 16-45 An electric/cable elevator motor and drum/cable sheave assembly in the equipment room usually located above the hoistway.

Figure 16-46 An elevator key box carried on responding apparatus shows the wide variety of keys that could be required.

rescuers will have to rely on the building management to have the proper key available when needed. When a hoistway door is opened, an electrical safety interlock will not allow the car to move.

Safety Firefighters should not bet their lives on trusting an interlock switch. They must make sure the power is off.

The elevator car doors themselves do not have a locking mechanism and can be opened from either the inside or the outside of the car by pushing them apart. These doors also have an electrical safety interlock that will not allow the car to move if the slightest opening is detected.

All electric/cable elevator cars have emergency access panels either in the roof or through the side (in multiple car hoistways), **Figure 16-47**. Hydraulic elevators that have an emergency valve that can be used (by an elevator technician) to lower the car to a landing are not required to have these emergency access panels. These emergency panels are equipped with the same type of electrical safety interlock that will not allow the car to move when they are open.

Modern in-car control panels, **Figure 16-48**, generally have the floor selection buttons, an emergency stop button, and a "Fire Service" key slot. Some will have an "Independent Service" switch located behind a locked panel or a separate key slot. Many will also have a telephone or intercom-type system for emergency usage.

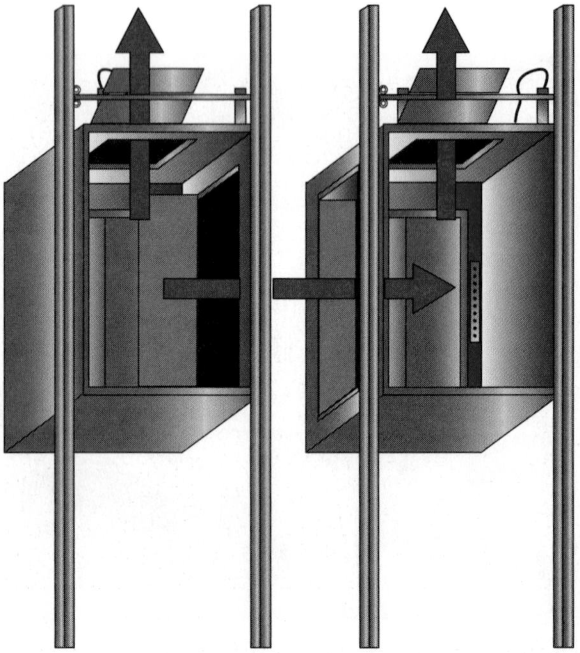

Figure 16-47 Elevator cars may have top emergency access panels, side emergency access panels, or both.

Note Rescuers need to be aware that these telephone or intercom systems may not be answered on site and need to be very cautious relaying information or instructions through an off-site operator.

The Fire Service key slot is for use by firefighters who may want to utilize the elevator car during the investigation of a possible structure fire. When this key switch is operated the car will respond only to commands from the in-car control panel and not to any calls from other floors. The car door will not automatically open on reaching the selected floor; the operator in the car must push and hold the Door Open button.

When the Independent Service switch is operated, the car will respond only to commands from the in-car control panel and will not respond to calls from other floors. However, when the selected floor is reached the door will automatically open.

If it becomes necessary for rescuers to access or remove elevator passengers, this should only be attempted by rescue personnel with special training and experience. This can be a very dangerous process for both rescuers and passengers.

Figure 16-48 Firefighters must know and understand the operation of the various types of elevator control panels that exist within a particular response area.

The first step in accessing the passengers is to determine the location of the car in the hoistway, which can often be a time-consuming process. If rescuers are able to communicate with the passengers, they will be able to assist by telling rescuers at what floor they got on the elevator or possibly the last floor they saw indicated by the in-car floor indicator. If rescuers cannot communicate with passengers or they cannot assist, the quickest method is to go to the elevator pit below the elevator and look up, or to the equipment room above the elevator looking down, and then estimate the car's location.

While personnel are determining the car's location, a firefighter should locate the electrical control panel, which should be in the equipment room, and shut off the power to the elevator. (As with industrial entrapment procedures, this firefighter must maintain control of the electrical power.) Once an estimated location is determined, rescuers open the closest hoistway door to confirm the exact location. An open hoistway door is extremely dangerous to rescuers, passengers, and bystanders, so access to the open hoistway must be controlled. Use of a ladder or other barrier, **Figure 16-49**, is strongly recommended.

If the car is at or very close to a landing, the hoistway doors can be opened as previously discussed, the car doors can be pushed open, and the passengers assisted out of the car. Once the hoistway doors are opened, if enough of the top part of the car doors is accessible from the landing for passengers to be removed through, a ladder needs to placed through the opening down into the car, and a rescuer needs to enter the car to assist the passengers in climbing out, **Figure 16-50**. If the open hoistway doors reveal that enough of the bottom part of the car doors is accessible from the landing, a ladder needs to placed from the landing to the car, and a rescuer needs to climb into the car to assist the passengers, **Figure 16-51**. This is very dangerous; the open hoistway must be blocked.

If the car doors are not accessible from a landing, the rescuers need to go to the landing above and open the hoistway doors so that they can use a ladder to gain access to the top of the car.

> **Safety** Any rescuer entering the hoistway must have a life safety harness and lifeline.

Once on top of the car, the rescuer can open the top emergency panel and gain access to the passengers through it. If the car is close enough for a ladder to be placed from the landing above to the top of the car, the passengers can be removed up the ladder, **Figure 16-52**. If this method is used, passengers need to be placed in a life safety harness and attached to a lifeline as they begin exiting the car.

If the car is not close enough for a ladder to reach it and it is in a common hoistway with other elevator cars, it may be possible to bring one of the other cars alongside and transfer the passengers through the side emergency panel. If this method is chosen, the rescuers will need to build a makeshift bridge between the cars. Short ladders, backboards, wooden planks, and similar items work well for this. Once firefighters are ready to transfer the

Figure 16-49 An open and unprotected hoistway door is extremely dangerous. A ladder can be used to protect the opening.

Figure 16-50 Passengers can be rescued from an elevator with only the top half of the car door opening visible from the landing by extending a ladder down into the car. Firefighters then assist the passengers out.

Figure 16-51 Passengers can be rescued from an elevator with only the bottom half of the car door opening visible from the landing by extending a ladder up into the car. Firefighters then assist the passengers out. Note that the hoistway opening needs to be protected.

Figure 16-53 Passengers can be rescued by using a working elevator stopped alongside of the stuck elevator. Passengers are transferred out through the side emergency access panel. Note the bridging in place and that a firefighter assists the passengers from both cars.

Figure 16-52 Passengers can be rescued from an elevator with the top of the car 6 to 10 feet below the landing by extending a ladder down to the top of the car and a firefighter assisting a passenger out the top emergency access panel. A Class 2 harness and safety lines need to be connected to both firefighter and passenger.

passengers, one rescuer needs to join the passengers in the stuck car in order to assist in the transfer, **Figure 16-53**.

Though not common, incidents involving escalators do occur. They usually involve a passenger getting feet caught in the area where the steps disappear into the **landing plate** or getting fingers caught under the moving hand rail. As with an elevator incident, the responding unit should request that a service technician be dispatched immediately. The other type of incident that can be encountered involving escalators is when a service technician becomes caught in the gear mechanism while working on the unit.

As with all of the other incidents that involve machinery, one of the first things to be done is to shut off and control the power supply. Most modern escalators have emergency shutoff switches located at the bottom landing on one of the end plates adjacent to the hand rail. The motor and drive gear assembly for escalators is usually located under the top landing plate, **Figure 16-54**, or near the top under the escalator itself, behind a removable panel if this area is accessible.

In cases of foot entrapment at the landing plate, the plate itself can be taken up by removing the screws that hold it down. **Figure 16-54** shows one of these plates removed. It is very difficult to

Figure 16-54 The top landing of an escalator with the landing plate removed. The motor/gear assembly can be seen.

reverse the direction of travel of an escalator due to a number of safety devices designed to prevent the escalator from reversing if power is lost. In the case of finger/hand entrapment the firefighters can loosen the wheel that drives the hand rail. It is located in the same area as the motor and drive gear. If more extensive disassembly is required to extricate an entrapped passenger or repair person, it will be necessary to wait for a service technician.

Farm Equipment Rescue

Rescue of victims who are trapped, pinned, impaled, or otherwise injured by farm equipment can be very challenging. While on the surface it may appear that these types of extrications are little different from vehicle extrications, this is not the case. Everything about the farm equipment itself inherently makes extrication a much more difficult process. Farm equipment is built to withstand the rigors of very hard use on a daily basis and not bend, break, or otherwise fail. Therefore, it is routine in extrications involving this type of equipment to have to disassemble the equipment rather than use the other methods discussed in the vehicle extrication section (distortion, displacement, and severing). Couple this with the fact that these events are often located far from paved roads that allow apparatus with all their rescue equipment to be parked close, and rescue becomes all the more difficult.

If a fire department's district has the potential to be required to respond to this type of incident, specialized training should be provided to firefighters. An excellent resource for information and assistance should an incident occur is the local equipment dealer; the department should work on developing a working relationship with dealers prior to an actual incident.

Lessons Learned

This chapter covered a variety of the more common rescue situations that firefighters can expect to be confronted with from the day that they first step onto a unit. However, this does not even scratch the surface of the great variety of real-life rescue situations that firefighters confront on a daily basis. The goal of this chapter was to present a broad spectrum of situations along with the correct methods and procedures to deal with them in a safe, effective, and timely manner.

Note Firefighters need to keep the skills, knowledge, and abilities learned in this chapter in their mental toolbox, building on them throughout their careers and constantly updating them as new technology and equipment is developed and put into use.

KEY TERMS

Asphyxiation Loss of consciousness or death caused by too little oxygen and too much carbon dioxide.

Confined Space A space that is large enough to be entered but is not designed for continuous occupancy.

Cribbing The use of various dimensions of lumber arranged in systematic stacks (pyramid, box, step, etc.) to support an unstable load.

Disassembly The actual taking apart of vehicle components.

Displacement The relocating of major parts (i.e., doors, roof, dash, steering column) of a vehicle.

Distortion The bending of sheet metal or components.

Engulfed To swallow up or overwhelm.

Extricate To set free, release, or disentangle a patient from an entrapment situation.

Ground Pads Sheets of plywood, planks, aluminum sheets, and so on, used to distribute weight over a larger area.

Guideline/Lifeline Rope used as a crew is searching a structure to assist them in finding their way back out.

Hardware Equipment used in conjunction with life safety ropes and harnesses (carabiners, figure eights, rappel racks, etc.).

Harnesses Webbing sewn together to form a belt, seat harness, or seat and chest harness combination.

Hoistway The shaft in which an elevator or a number of elevators travel.

Landing Plate The plate at the top or bottom of an escalator where the steps disappear into the floor.

Packaging The bandaging and preparing of a patient to be moved from the place of injury to a stretcher.

Reel Coil Memory that wire develops from having been placed on a wooden spool as it is being manufactured.

Rescue Those actions that firefighters perform at emergency scenes to remove victims from imminent danger or to extricate them if they are already entrapped.

Severance The cutting off of components (i.e., brake pedal, steering wheel) in a vehicle.

Shoring The use of timbers to support and/or strengthen weakened structural members (roofs, floors, walls, etc.) in order to avoid a secondary collapse during the rescue operation.

Spinal Immobilization The process of protecting a patient against further injury by securing them to a backboard or other rigid device designed to minimize movement.

Suspension System The springs, shock absorbers, tires, and so on, of a vehicle.

Tether Line A rope that is held by a team on shore during a water rescue to be used to haul the rescuer and victim back to shore.

Tunnel Vision The focus of attention on a particular problem without proper regard for possible consequences or alternative approaches.

Tunneling The digging and debris removal accompanied by appropriate shoring to safely move through or under a pile of debris at a structural collapse incident.

Two In/Two Out The procedure of having a crew standing by completely prepared to immediately enter a structure to rescue the interior crew should a problem develop.

Voids Spaces within a collapsed area that are open and may be an area where someone could survive a building collapse.

Webbing Sling Approximately 12 to 15 feet of rescue webbing tied end to end, forming a continuous loop.

REVIEW QUESTIONS

1. What is the biggest danger that firefighters must be aware of when involved in rescue operations?
2. What safety equipment must firefighters always be wearing and have with them prior to entering a burning structure or one in danger of becoming involved?
3. What is meant by the term *two in/two out?*
4. Name at least three of the things firefighters should be looking for when approaching a building they will be searching.
5. Why should firefighters be required to carry a forcible entry tool with them while searching a structure?
6. Describe the difference between the primary search and the secondary search.
7. What carries or drags are recommended in a single rescuer operation?
8. List the sequence of procedures, in order, that should be carried out at a vehicle crash incident.
9. List and describe each of the four disentanglement methods described.
10. List in order and describe the four methods presented for carrying out a water rescue.
11. What are the dangers presented by using a tether line in moving water?

12. Rescuers should not approach too close to a victim who has fallen through ice. Why?
13. List and describe the three types of structural collapses covered.
14. Describe the difference between *shoring* and *cribbing*.
15. Describe the actions and procedures that must be taken to make a trench collapse safe prior to rescuers entering.
16. What is the most common hazard encountered at confined space incidents?
17. Describe the differences between hydraulic and electric/cable elevators.
18. Describe the difference in how an elevator operates when in the "Fire Service" and "Independent Service" modes.

Additional Resources

Bechdel, Less and Slim Ray, *River Rescue.* Appalachian Mountain Club, 1985.

Brown, Michael G., *Engineering Practical Rope Rescue Systems.* Delmar Learning, a part of the Thomson Corporation, Clifton Park, NY, 2000.

Browne, George J. and Gus S. Crist. *Confined Space Rescue.* Delmar Learning, a part of the Thomson Corporation, Clifton Park, NY, 1999.

Downey, Ray, *The Rescue Company.* Fire Engineering Books and Videos, 1992.

Erven, Lawrence, *Emergency Rescue.* Macmillan, New York, 1980.

Frank, James and Jerrold Smith, *Rope Rescue Manual.* California Mountain Company Limited, Santa Barbara, CA, 1987.

Linton, Steven and Damon Rust, *Ice Rescue.* International Association of Dive Rescue Specialists, 1982.

McRae, Max, *Fire Department Operations with Modern Elevators.* Robert J. Brady Company, Bowie, MD, 1977.

Ohio Trade and Industrial Education Service, *Victim Rescue.* Trade and Industrial Education Instructional Materials Laboratory, Columbus, OH, 1976.

Padgett, Alan and Bruce Smith, *On Rope.* National Speleological Society, Huntsville, AL, 1987.

Setnicka, Tim, *Wilderness Search and Rescue.* Appalachian Mountain Club, Boston, MA, 1980.

Technical Rescue Program Development Manual. U.S. Fire Administration, Washington, DC.

CHAPTER 17

FORCIBLE ENTRY

Robert R. Morris, New York City Fire Department

OUTLINE

- Objectives
- Introduction
- Forcible Entry Tools
- Safety with Forcible Entry Tools
- Maintenance of Forcible Entry Tools
- Construction and Forcible Entry
- Methods of Forcible Entry
- Windows
- Breaching Walls and Floors
- Tool Assignments
- Lessons Learned
- Key Terms
- Review Questions

STREET STORY

As a firefighter, many hours are spent training and trying to prepare yourself for any situation you could possibly encounter. After spending twenty-three years in this business, I have found there are some situations you cannot train for, especially when it comes to forcible entry. I would like to share one such incident that happened to me and how I reacted.

In my rookie year as a career firefighter we were dispatched to a structure assignment at one of our mobile home communities. En route to the scene, the dispatcher advised us that she had received a call from one of the neighbors stating that they could see flames coming from the windows, and they were unsure if anyone was home. It was around 9:00 p.m. on Saturday night so we were all hopeful that the home was unoccupied. Upon our arrival we found a mobile home approximately 50 percent involved, and the captain called for a $1\frac{3}{4}$*-inch attack line and instructed us to make entry through the front door. After advancing the line, we discovered the door was locked, and we would have to make forcible entry. My partner retrieved a Halligan tool and after a quick pry the door snapped open. As the door flew open, something immediately hit me in the chest, knocking me backwards and landing on top of me. I opened my eyes and found that it was a severely burned victim who had obviously been trying to escape. After catching my breath and putting my heart back in my chest, we made entry and extinguished the fire.*

When the mop-up crew arrived and started doing their thing, I found myself sitting on the tailboard of the engine with every emotion from fear to sorrow going through my head. My captain, a hardened veteran of the fire service, approached me to talk about all the events that had taken place. He told me the body was that of a seventy-year-old gentleman who had apparently fallen asleep with a cigarette.

This one incident bothered me for several years, and I will honestly admit I even considered leaving the fire service. I still find myself thinking about this fire once in a while, but now I am the hardened veteran because I have seen so much more.

The moral of this story is that forcible entry, like all of firefighting, is not only a physically and mentally challenging task, but can also be emotionally challenging, with no way to prepare. You never know what you will find behind that door.

—Street Story by Rick Townsend, Firefighter/Paramedic, Sierra Vista Fire Department, Sierra Vista, Arizona

OBJECTIVES

After completing this chapter, the reader should be able to:

- Identify forcible entry tools by common name and use.
- Describe the inspection and maintenance procedures for each type of forcible entry tool.
- Describe building features and methods of forcible entry for doors, windows, gates, walls, and floors.
- Identify five types of locks and describe their operation.
- List the steps for the three types of conventional forcible entry.
- Demonstrate conventional forcible entry on a variety of doors.
- Describe or demonstrate the through-the-lock forcible entry method.
- List or describe four construction features of windows and methods of gaining entry.
- List or describe three considerations of breaching walls.
- Explain the three considerations of tool assignments.

INTRODUCTION

According to statistics from the Federal Bureau of Investigation over 2½ million homes are broken into every year, approximately 8,600 each day, or one break-in every 13 seconds. Because of these incidences, residents and owners install a variety of locking devices to protect both their homes and businesses. These devices present a number of challenges to firefighters who must gain access to a structure to conduct firefighting operations.

The ability to gain entry to secured areas and buildings is a primary operation at fires and other emergencies, and **forcible entry** is often one of the first operations conducted at a scene. At structural fires, a delay in gaining entry will reduce the ability to mount an aggressive fire attack, allowing the fire to extend out of control. Also, rescue operations require forcible entry to gain access to areas where victims are located. In addition, the renewed emphasis on rapid intervention teams, for the rescue of trapped firefighters, requires all firefighters to be knowledgeable and proficient with the forcible entry tools, **Figure 17-1**.

Forcible entry is a combination of knowledge and skills used to gain entry to a structure when the windows and doors are locked, blocked, or in some cases do not exist. To perform this essential task, firefighters must be knowledgeable about building construction and the operation of **locking devices** to break, remove, or bypass these elements. In addition, the firefighter must have the skills, gained by training and experience, to apply this knowledge using a variety of tools.

Knowledge

A working knowledge of the many types of locks, hardware, doors, and other assemblies is essential to successful forcible entry operations. Firefighters must be able to "size up" the quickest and easiest way to gain access to buildings, such as the doors

Figure 17-1 A typical assortment of forcible entry tools used by fire departments.

shown in **Figure 17-2**. In addition, the firefighter must know which type of tool to use and the best method to gain access through the door to the building.

Streetsmart Tip As with all firefighting skills, forcible entry must be studied and practiced on a continuing basis. As security concerns increase, the types and number of locks used will also increase. Firefighters should study the locks used in their area during the development of pre-incident plans. They can visit a locksmith to learn of new types of locks or check out the Web sites of lock manufacturers. Forcible entry is more brain power than it is muscle power.

Skill

The element of skill involves a firefighter's ability to apply knowledge of building construction, lock assemblies, tools, and techniques to accomplish the necessary tasks of forcible entry. This means choosing the proper tools and applying the best techniques when using the tools. Again, skills are developed by repeated practice and experimenting with new tools, locks, and techniques.

Experience

Experience is acquired by two means. One is through drills and practice at training sessions and the other is at the scene of actual fires and emergencies. Both are the means by which skill is developed and knowledge is gained as well as reinforced. The most important experience is gained from field operations where firefighters' skills and knowledge are put to the true test.

FORCIBLE ENTRY TOOLS

Firefighters must have an understanding and knowledge of the tools available to conduct forcible entry. The selection and use of the "right" tool are essential if the task of forcible entry is to be completed as quickly as possible. Although any number of tools can be used to accomplish a task, the right tool for the right job is the quickest and easiest way to complete the operation. Because of geography or local tradition, many of the tools shown or listed may have another name or nomenclature. Firefighters must know the tools and their names as used by their particular department. Tools used for forcible entry operations are divided into several families or groups based on their intended uses as shown in **Table 17-1**.

This chapter does not attempt to show or describe *all* of the tools used for forcible entry because some tools are obsolete or used infrequently. The tools described in this chapter are the most common forcible entry tools used today.

Streetsmart Tip **"Try Before You Pry."** The first rule of forcible entry is to attempt to open the door or window and determine if it is even locked. Firefighters should not waste time but should always try to open the door first, as this is part of the size-up of forcible entry. Appropriate eye protection is a necessity when using any and all forcible entry tools. The striking, prying, pulling, or cutting action will create sparks, metal chips, and other debris that can cause eye injury.

Figure 17-2 Firefighters should size-up the way to gain entry, "try before you pry," and think about how they would force these doors.

Striking Tools

The group or family of **striking tools**, **Figure 17-3**, is used to deliver impact to other tools, such as a Halligan tool, in order to drive it into place. Striking tools are used for impact delivery to the lock or the door itself. They may force the door or even break it down.

Flathead Ax

The most common of the striking tools is the flathead ax. Although this tool may be used for cutting, it is generally used to drive the Halligan tool. Together, the ax and Halligan tool form the "**irons**," **Figure 17-4**, and are one of the most important and useful of all forcible entry tools. The flathead ax is available in 6- and 8-pound sizes, the weight being at the head of the ax. The 8-pound ax is best for striking purposes and is commonly known as the forcible entry ax. The handle of the ax can be comprised of wood or fiberglass. Fiberglass is best for all purposes, due to its high strength and ease of maintenance.

Figure 17-3 The group or family of striking tools includes the maul, small hammer, flathead ax, and Denver tools.

Streetsmart Tip A firefighter should never free-swing any of the striking tools. The firefighter using the striking tool, usually to apply force to a prying tool, should hold the tool firmly and use a strong and controlled stroke for power and accuracy.

Maul/Sledge

The maul or sledge is generally found in several sizes with the 10-pound maul being the most common and versatile. Other sizes include the 8-, 12-, and 16-pound versions. The 10-pound maul is sometimes carried with the Halligan tool to again form a set of "irons." The 16-pound maul is considered too heavy to drive other tools such as the Halligan and is sometimes used to break in doors or locks. Once again, fiberglass handles are recommended for strength and safety. Some new tool

Forcible Entry Tools

FAMILY OR GROUP	TOOLS	OPERATION
Striking	Flathead ax, maul, sledgehammer, battering ram, hammer, punch & chisel, lock breaker	Deliver impact force to break lock or drive another tool.
Prying	Crowbar, Halligan tool, hux, claw tool, pry bar, hydraulic	Provide mechanical advantage, leverage.
Cutting	Axes, saws, torches, bolt cutters	Cut material away; cut around locking devices.
Pulling	Hooks, pike poles	Limited for forcible entry; breaks glass, gypsum board, Sheetrock.
Through-the-lock	K tool, A tool, picks & key tool, bam bam, vise grip or channel lock pliers, REX tool	Remove lock cylinder.

TABLE 17-1

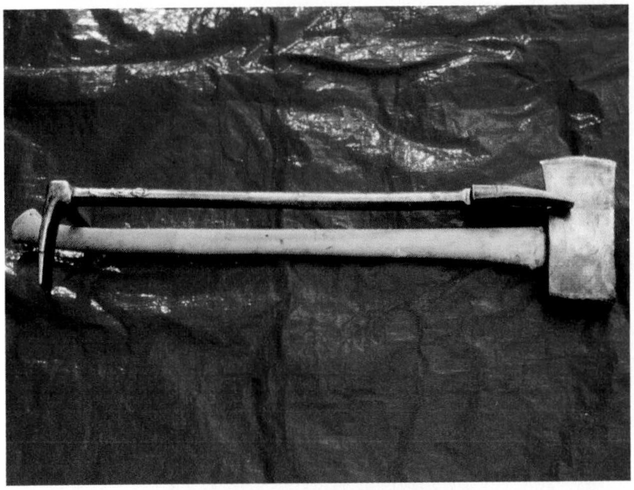

Figure 17-4 The flathead ax and Halligan tool form the tool set known as the *irons.*

styles combine mauls with other tools such as the Denver tool, **Figure 17-5**.

The power a striking tool applies is explained by this equation: *Force* = *Weight* × *Speed.* A tool that is too heavy cannot be moved fast enough to develop proper force. Also, problems with accuracy and control can develop. On the other hand, a tool that is too light will not deliver enough force, even if it is swung hard enough.

Battering Ram

Battering rams, **Figure 17-6**, are used by two or more firefighters to break through a door or wall. The desired result is to break or push the object in. There are several types, but the type most commonly used is the old-style fire department ram. This tool is made of steel and has four handles and a round end for battering and a forked end for breaking and penetrating. Another type of battering ram is smaller and is intended for use by one or two firefighters. This tool weighs approximately 20 to 25 pounds and is used to break doors and breach walls.

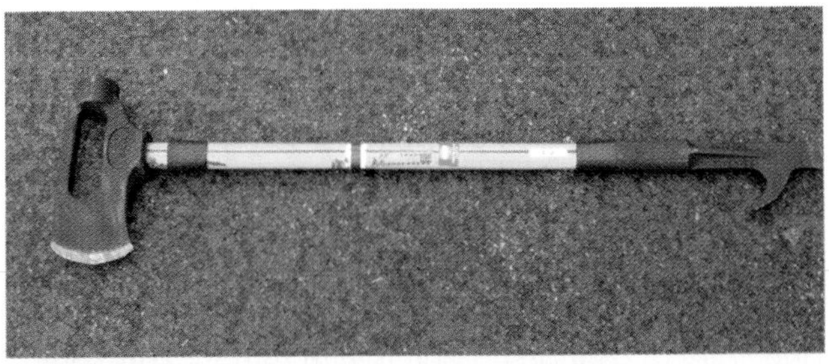

Figure 17-5 The Denver tool is a maul with a cutting edge, prying end, and a pulling hook.

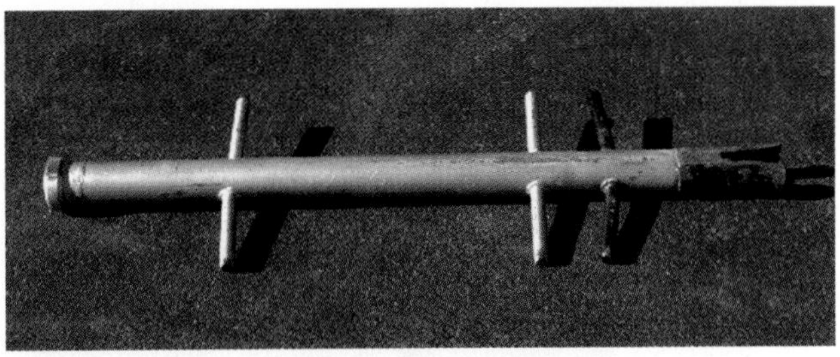

Figure 17-6 The battering ram requires brute strength to break a door, lock, or wall.

Prying and Spreading Tools

The group or family of **prying tools**, **Figure 17-7**, is used to spread apart a door from its jamb, move objects, or expose a locking device.

Halligan Tool

The original **Halligan tool**, **Figure 17-8**, designed by Hugh Halligan of the Fire Department of the City of New York, has proven to be the most important single forcible entry tool used in the fire service. Although the original style tool is not widely used today, the same basic concept with some improvements is capable of many forcible entry tasks.

Halligan-type tools are made in several weights and of different materials. The best type for forcible entry and structural firefighting is a 30-inch forged steel tool. The basic Halligan design is a tool that has three primary parts as shown in **Figure 17-8**. These are the adz end, the pike end, and the fork end. The size and shape of the parts are important because they determine the mechanical advantage the tool will apply to the spreading force. As noted earlier, the Halligan tool is often nested with the flathead ax, forming the "irons."

Claw Tool

The claw tool, a forerunner to the Halligan tool, was the mainstay of forcible entry operations for the previous generation of firefighters. The claw tool had two major parts: the fork end, which had a hollow, curved profile, and the hook end, which was curved and ended with a sharp point. Generally, the tool was made in two sizes. The heavy claw is 42 inches long and the standard claw is 32 inches long. The standard claw was usually carried with a striking tool such as the ax or the kelly tool. The claw tool is obsolete now, having been replaced by the Halligan tool.

Kelly Tool

The Kelly tool is a steel bar, 28 inches long, and has two main features. One end has a large adz, approximately 3 inches wide, and the other end has a large

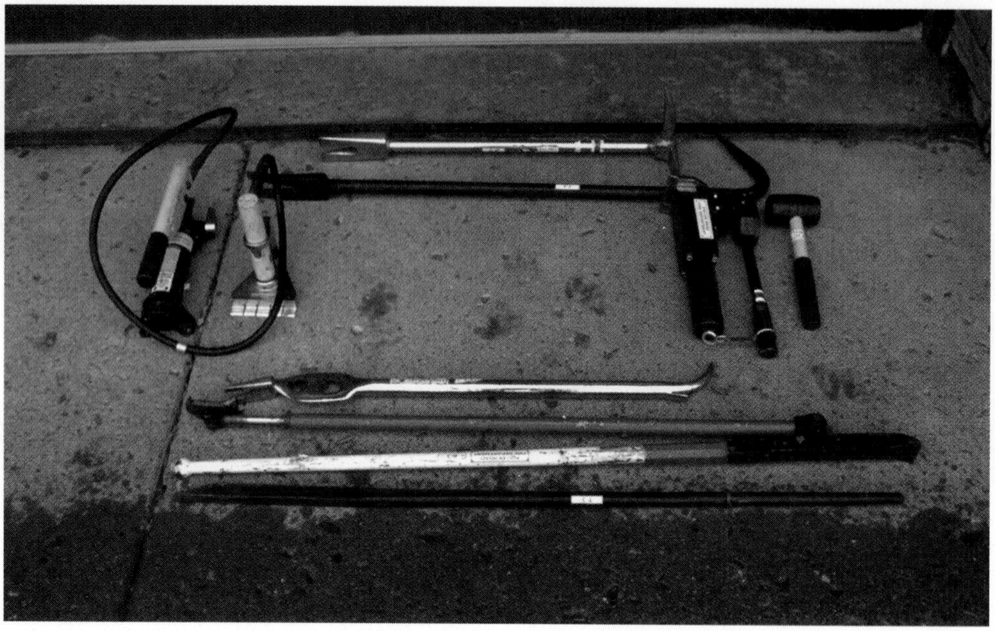

Figure 17-7 The group or family of prying tools includes the Halligan, claw tool, hux bar, Detroit door opener, pry bar, and hydraulic spreaders.

chisel or fork. This is an older tool and has since been replaced by the Halligan tool.

Hydraulic Spreaders

Handheld hydraulic spreaders, **Figure 17-9**, are available in two different styles. One is a hydraulic pump with a spreading device attached by a length of hose and the second is a similar pump with the spreading devices attached directly to it. The spreading end is forced between the door and jamb at the location of the locking device and the pump is operated to spread the door and jamb apart. A tool is used to force the spreaders into the space between the door and jamb.

Power hydraulic spreaders such as a Hurst or Amkus system can also be used to spread objects. Their application is generally for rescue work; however, they may be used to force entry into a building.

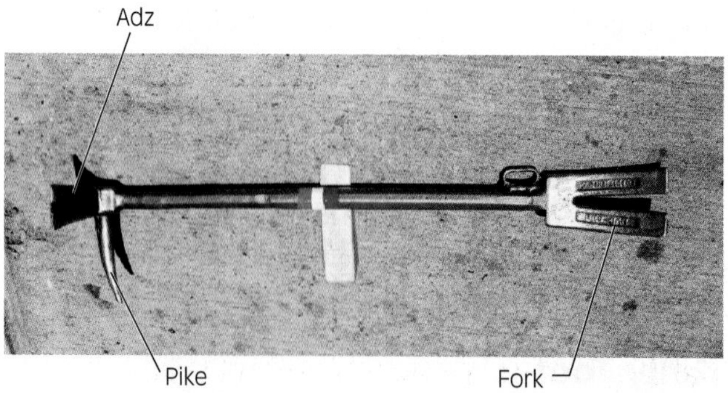

Figure 17-8 The Halligan tool is one of the most useful forcible entry tools.

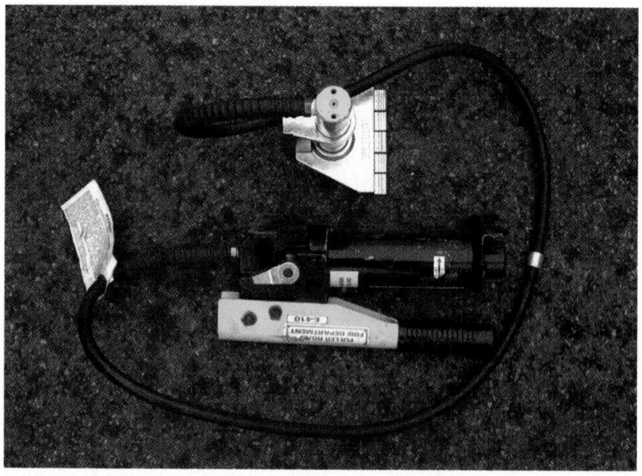

(A)

(B)

Figure 17-9 Handheld hydraulic spreaders are available in two different styles but operate on the same principle. (A) Handheld pump with remote spreader. (B) Handheld unit with integral built-in spreader.

Miscellaneous Prying Tools

There are many other types of prying tools such as the crowbar, pry bar, and others, but most are of limited use. As with any tool its use depends on the type of task to be accomplished and the method of forcible entry.

Cutting Tools

The group or family of **cutting tools**, **Figure 17-10**, is used to cut away materials and expose the locking device or cut through a door or wall to accomplish forcible entry.

Ax

The fire department ax is seen in two basic configurations in **Figure 17-11**: the flathead ax and pick head ax. The flathead ax is most commonly used as a striking tool for forcible entry. In addition to cutting, this tool may be used as a prying tool, especially when equipped with a fiberglass handle. The pike head ax has a point that can be used for puncturing, pulling, and, to a limited extent, prying. It is recommended that all axes be equipped with fiberglass handles for strength and safety.

Handsaws

The most commonly used handsaw is the metal cutting hacksaw. A strong, high-quality frame with a supply of good-quality blades is used for many firefighting operations, especially vehicle rescue. Carpenter saws, both rip and crosscut, may also be carried on fire apparatus for use in areas where power saws are too large, are too noisy, or present a safety hazard.

Bolt Cutters

Bolt cutters are used to cut metal bars, cables, wires, and other hardware. These tools come in a variety of sizes; the most common for forcible entry is 36 inches in length. This size will cut steel up to ⅜ inches thick. Firefighters should avoid using bolt cutters to cut heavy-duty padlocks with case-hardened shackles. The hardness and the thickness of the metal make them difficult to cut and will damage the cutting edge of the tool.

Wire Cutters

Wire cutters are not usually used for forcible entry but are an extremely handy tool to carry. They are used to cut electric wires. They have insulated handles which reduce the risk of electric shock.

Power Cutting Tools—Saws

Power saws, **Figure 17-12**, are the most common power cutting tools used for forcible entry work. They are available in two basic types: the rotary saw with a circular blade and the chain saw. The rotary saw in **Figure 17-12A** has a carbide-tipped blade. **Figure 17-12B** shows a rotary saw with a metal-cutting disc and **Figure 17-12C** shows a chain saw. Rotary and chain saws are usually powered by a two-cycle gasoline engine and require two firefighters for safe operation.

Figure 17-10 The group or family of cutting tools includes axes, saws (both power and manual), and bolt and wire cutters.

(A)

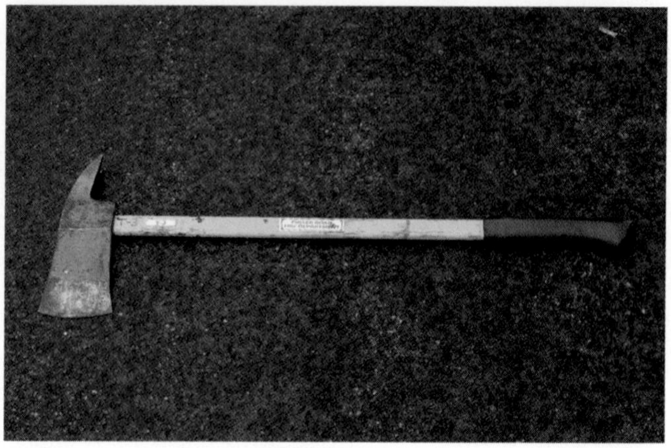

(B)

Figure 17-11 The fire department ax is seen in two basic configurations: (A) flathead and (B) pick head.

(A)

(B)

(C)

Figure 17-12 Gasoline powered saws are available in two types. The rotary saws shown in (A) with the wood cutting blade and (B) with the abrasive disc, for either masonry or metal cutting, or, the chain saw shown in (C). The chain saw is usually used for roof ventilation.

Carbide-Tipped Blades

The rotary power saw is widely used and can be adapted to cut a variety of materials. Wood and composition materials such as roofs and flooring are cut with a carbide-tipped blade, which has 12 to 24 teeth. Light-gauge metal may also be cut with this blade. When cutting with a carbide-tipped blade, it is important to maintain full rpms to avoid having the saw bind in the material being cut.

Metal Cutting Blades

The most common metal cutting blade is an abrasive disc made of aluminum oxide. This blade is used to cut locks, hardware, steel doors, roll-down gates, and other forcible entry applications.

> **Note** With all blades, care must be taken not to bind or bend the blade while cutting. This is particularly important with abrasive discs because they can shatter or disintegrate.

Masonry Cutting Blades

Masonry materials such as concrete, brick, block, and stone can be cut with an abrasive disc made of silicon carbide or a steel with a diamond matrix blade. When cutting with these blades, a spray of water on the blade will cut down on the production of dust and keep the blade cool, thus prolonging the life of the blade.

> **Streetsmart Tip** Gasoline or other hydrocarbon fuels will break down the bonding material used in the manufacture of abrasive disc blades. To avoid this, blades must be stored separately from the fuel can for the saw. One should always refer to the manufacturer's recommendations on use and storage. Checking the fuel supply and starting the engine of gasoline-powered equipment before taking it to the location of the cutting operation will save time and frustration. Some departments start and run the saw each tour to avoid problems.

Chain Saws

Chain saws are used primarily for ventilation purposes; however, they have application as a forcible entry tool depending on the type of building construction and opening required. Several types of saws and chains are available, and many departments use a carbide-tipped chain. The chain saw can also be used to cut through wood siding, wood frame walls, certain doors, and light-gauge metal.

Reciprocating Saws

The reciprocating saw, **Figure 17-13**, is an extremely versatile tool. These saws are powered by electricity, either stationary (house current) or portable (generator). Cordless, battery-powered saws are now available and can be operated remotely from a power source, thus increasing versatility. A variety of blades are available for this type of saw depending on the material to be cut.

Cutting Torch

The cutting torch, **Figure 17-14**, uses a fuel such as acetylene mixed with oxygen to produce a flame to heat metal. A jet of pure oxygen is then applied to intensify the heat, creating a flame temperature in excess of 5,000°F, melting the metal. Many departments use these tools for heavy forcible entry in addition to other cutting tools. Use of cutting torches requires specialized training in addition to following manufacturer's recommendations and department procedures.

> **Caution** Acetylene is an unstable and extremely flammable gas. It is essential to always follow manufacturer's recommendations for storage and use of this material. In addition, cutting torches should never be used in a flammable environment.

Figure 17-13 The reciprocating saw is a versatile tool for forcible entry and rescue work.

Figure 17-14 Cutting torches use a mixture of acetylene and oxygen to generate a high-temperature flame. They cut metal by melting it.

Pulling Tools

The most common type of **pulling tool** is the **hook** or **pike pole**, **Figure 17-15**. These tools are grouped by the type of head and handle length and are used to open up walls and ceilings, to vent windows, and to pull up roof boards or other building materials.

Special Tools

A number of specialized tools are available to assist with or conduct forcible operations. Many of these have been developed by firefighters after years of experimentation and trial and error.

Bam Bam or Dent Puller

Originally designed as an automotive body work tool, the bam bam is used to pull lock cylinders. It consists of a shaft, case-hardened screw, and slide hammer. The screw is turned into the lock cylinder and the slide hammer is operated to pull the cylinder out. As lock technology has advanced, the application of this tool has become limited.

Duck Bill Lock Breaker

The duck bill lock breaker is designed to break open heavy-duty padlocks, **Figure 17-16**. The long, tapered head is placed into the shackle of the lock and driven down with a flathead ax (8-pound minimum) or a maul. This action will pull the shackle out of the body of the lock.

K Tool and Lock Pullers

The "K tool," **Figure 17-17**, is used to perform through-the-lock forcible entry. The K tool is designed to pull out lock cylinders and expose the mechanism in order to open the lock with the various key tools. The back of the tool is shaped like the letter "K" and slides over the lock cylinder. The front of the tool has a loop for the adz of the Halligan tool.

The K tool is placed on the lock cylinder and tapped into place to gain a good purchase or bite on the lock. The Halligan tool is then used to pry/pull

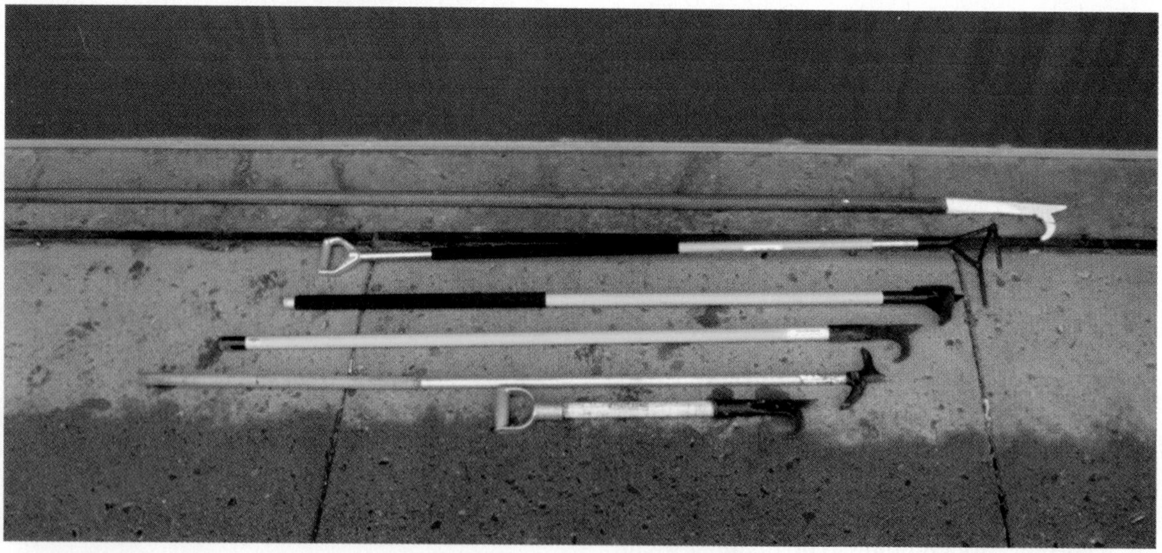

Figure 17-15 The most common type of pulling tool is the hook or pike pole, available in various styles and lengths.

Figure 17-16 Duck bill lock breaker forcing a heavy-duty padlock.

Figure 17-18 The REX tool being used to pull a tubular dead bolt lock.

the cylinder out of the lock mechanism. This exposes the lock mechanism, which is operated with the proper key tool. Key tools include the bent, square, and screwdriver types, as well as a pick to slide open the shutter on some rim locks. A locking-type pliers is a good addition to the K tool kit.

A number of other tools similar to the K tool are available that can be used to perform the same operation of through-the-lock forcible entry. These are the A tool, officer's tool, or REX tool, **Figure 17-18**, to name a few.

Combinations of Tools

To perform the task of forcible entry most of the tools or groups of tools discussed are used in combination with other tools. Striking tools are used with prying tools, such as the flathead ax and Halligan tool, and these tools should be carried or stored together on apparatus. In addition, the firefighter must be able to size up the forcible entry task and choose the right combination of tools to provide adequate leverage or force. Experience will provide the firefighter with the knowledge to choose the right combination and to apply this knowledge to complete the task.

SAFETY WITH FORCIBLE ENTRY TOOLS

As with all tools and equipment, if misused or used for the wrong task, forcible entry tools will create safety hazards. Firefighters will become familiar with the tools, their operation, and their maintenance during training, which should result in safe

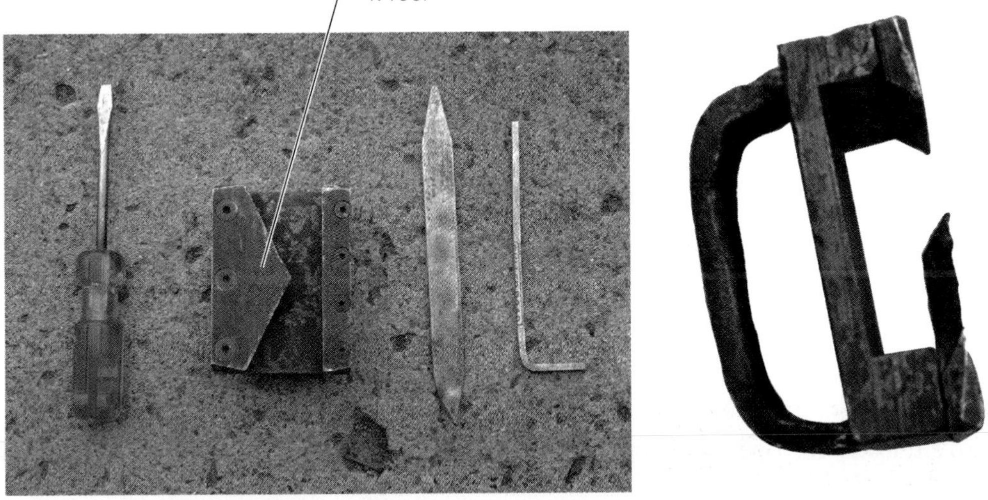

Figure 17-17 The K tool (shown with key tools) is designed to perform "through-the-lock" forcible entry.

operation. A number of general rules apply to all operations as follows:

- Always wear proper personal protective equipment including hand and eye protection.
- Follow manufacturer's guidelines for proper operations.
- Do not attempt to cut material other than that for which a blade or tool was designed for.
- Operate with regard to the safety of others in the immediate work area.
- Make sure tools are in proper operating condition before use.
- Most forcible entry operations require teamwork. Never attempt to use tools alone that require two firefighters.
- When the task is complete and if the tool is no longer needed, secure it to prevent tripping or other hazards.
- Tools should be stored and easily accessible, **Figure 17-19**.

Note The safety guidelines provided are general in nature and firefighters are reminded to read manufacturer's operating instructions for the specific tools used in their department. In addition, they should not use any power tools or other spark-producing equipment in an explosive or flammable atmosphere.

Safety Whenever forcing entry into an unfamiliar area, firefighters must be very cautious as to what they may face when creating the opening. Immediate danger to life and health (IDLH) conditions, extreme fire exposure, drug lab chemicals, booby traps, or other hazards may be present.

Rotary and Chain Saws

As the use of security gates and overhead doors increases, the power saw has become the tool of choice to remove the door or gate. The rotary saw with a metal cutting blade is an effective tool for these operations. These saws present a number of hazards, and firefighters should follow these guidelines to complete the operation safely:

- Always follow the manufacturer's instructions.
- Conduct daily checks for operation and blade condition.
- Check the saw for fuel and proper operation before proceeding to the entry location.

Figure 17-19 These slide-out trays provide easy access to forcible entry tools. Note the mounting brackets that hold the tools in place.

- Equip the saw with a carry strap (standard equipment with some manufacturers).
- Use the right blade for the material being cut, **Figure 17-20**.
- Never carry a running saw up a ladder or through a crowd of firefighters.
- Power saws require two firefighters: the saw operator and a guide firefighter.
- Eye protection is required when running any power equipment, especially power saws.

Carrying Tools

Many forcible entry tools have sharp or pointed ends and must be carried safely from fire apparatus to the fire scene. Firefighters should always be aware of their safety in addition to the safety of other firefighters.

- *Axes.* Carry the ax with the blade away from the body and the pick head covered. Never carry an ax over the shoulder.
- *Prying tools.* Similar to the ax, pointed and sharp edges should be carried away from the body and covered if possible.
- *Hooks or pike poles.* The tool head, the hook end, should be carried down and close to the

(A)

(B)

Figure 17-20 (A) Rotary saws have three types of blades: wood, masonry, and metal. The masonry and wood blades are shown here. (B) This is a new style of wood and roofing blade for the rotary saw.

ground. Depending on handle length, beware of overhead electrical wires and other obstructions. Inside a building, carry the handle close to the body with the hooked end toward the ceiling.

- *Striking tools.* These tools tend to be heavy and the head should be carried close to the ground. When using these tools, do not use a free-swing motion. Firmly grasp the tool with two hands and use a controlled and accurate stroke to move the tool.

Hand Tools

Hand tools used for forcible entry are constructed of metal, wood, fiberglass, or some combination of these materials. All tools must be inspected regularly for condition, cracks in the handles, burrs in the metal, and loose heads.

MAINTENANCE OF FORCIBLE ENTRY TOOLS

Proper tool maintenance is the first step to tool safety. Tools must be inspected and cleaned on a regular basis and checked for wear and damaged parts. Tools should be removed from service or repaired when defects are found.

Metal Heads and Parts

- Remove any dirt or rust with steel wool or emery cloth.
- Use a metal file to maintain the proper profile and cutting edge.
- Sharpen edges and remove burrs with a file.
- Do not keep the blade edge too sharp; this may cause it to chip when in use.
- Do not grind the blade because it can overheat and cause it to lose the temper and become soft.
- Do not paint the metal parts. Keep them lightly oiled if desired.

> **Caution** Oil should not be applied to the striking surface or face of striking tools. When used, an oil coating on the striking surface may cause the tool to slip or glance off, causing injury, and little or no force will be applied to the object being hit.

Fiberglass Handles

- Wash them with soap and water and dry completely.
- Check for damage or cracks.
- Make sure metal parts are secure.

Wood Handles

- Clean with soap and water, rinse, then dry completely.
- Check for damage and sand off any splinters.
- Do not paint or varnish the handles. A coat of boiled linseed oil may be applied if necessary.
- Ensure that the head is securely fastened to the handle.

CONSTRUCTION AND FORCIBLE ENTRY

The type and construction of the many different features of buildings, such as doors, windows, gates, walls, floors, and roofs, must be recognized and understood by firefighters to force entry. A thorough knowledge of the construction of these building features will lead to successful forcible entry operations.

Door Construction

Passage doors are manufactured in many styles as shown in **Figure 17-21** and are manufactured from plastic, wood, metal, and/or glass. The door assembly consists of the door itself, the frame or **jamb**, **mounting hardware**, and locking device mounted in a jamb, which is **rabbeted** or has a stop attached to it, **Figure 17-22**. A rabbeted jamb is formed or milled into the casing that the door closes against to form a seal. The stopped jamb has a piece of molding nailed or attached to the casing for the door to close against. The stopped jamb can be removed, allowing access to the lock assembly and hence easy forcible entry.

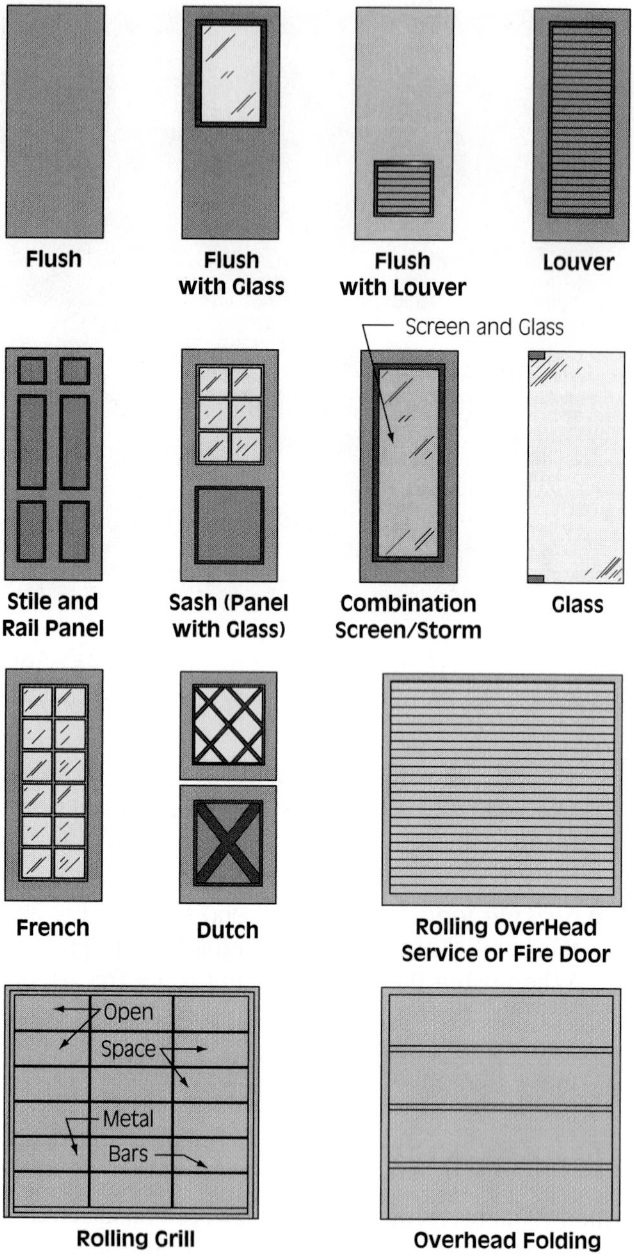

Figure 17-21 Doors are manufactured in a variety of styles and materials.

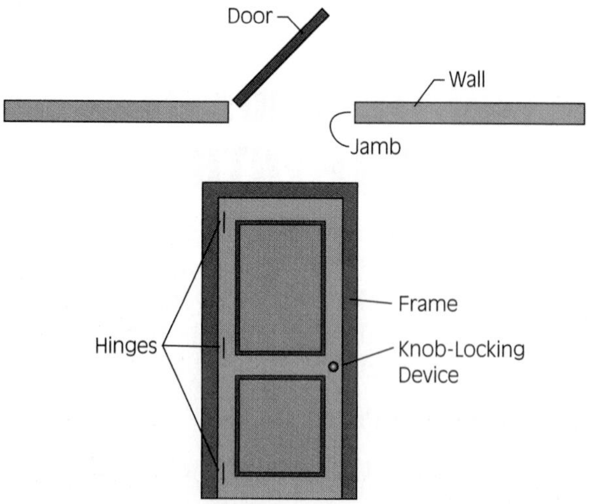

Figure 17-22 The parts of a door assembly.

Types of Doors

Wood Doors

There are three types of wood doors. **Panel doors** have a solid stile and rails with panels made of wood or glass or other materials. **Flush** or **slab doors** are flat or have a smooth surface and may be of either **hollow-core** or **solid-core** construction. Solid-core doors are more resistant to fire and more secure, usually making them more difficult to force. **Ledge doors** are built with solid material, usually individual boards, and are common in barns and warehouses. Wood doors are used primarily in residential construction and are installed on a wood frame or jamb secured to wall framing with nails or screws.

Metal Doors

Metal doors manufactured either as hollow-core or metal clad, **Figure 17-23**, are common in new construction. These are usually installed in metal frames and can be very secure. The metal clad door, which has a steel surface with a wood core, is manufactured in a number of designs and architectural finishes and some may have a fire-resistive rating. Generally metal doors are used in commercial construction and as exterior doors in residential construction. Metal doors are more substantial and secured to the wall construction, **Figure 17-24**. Forcible entry may be

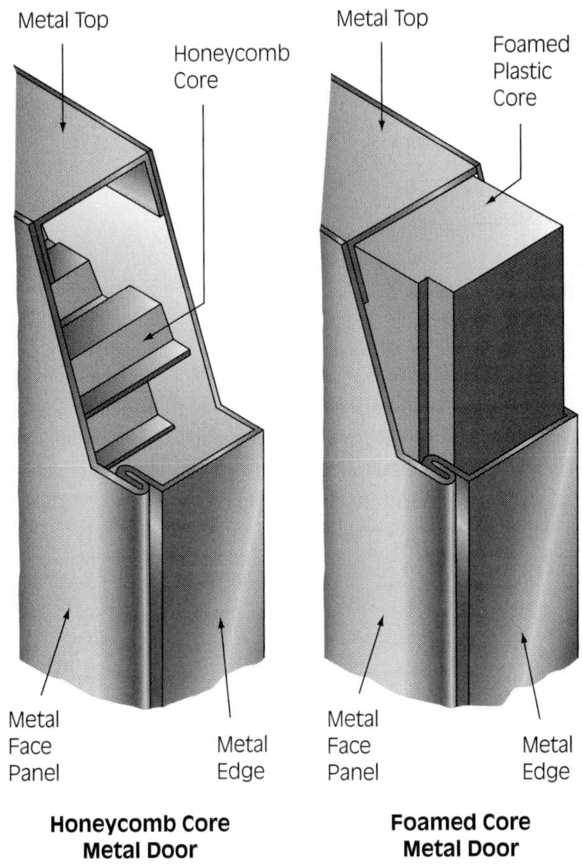

Figure 17-23 Metal doors are of both solid- and hollow-core design.

accomplished in a number of ways depending on installation and number of locks.

Glass Doors

Two main types of glass doors are commonly in use today: the metal or tubular frame glass door and the tempered or frameless glass door, **Figure 17-25**. The metal frame is the most common type of glass door and both styles are generally used in commercial and mercantile occupancies. Most have tempered glass, but a laminated safety glass and polycarbonate glazing (Lexan) are common and can be found in these doors.

Tempered Glass Doors

These doors are made of thick tempered glass, hung on a tubular metal frame, with hinges located at the top and bottom of the door and a locking device set in a metal rail at the bottom of the door. **Tempered glass** is plate glass that has been heat treated to increase its strength. The best method to force this type of door is the through-the-lock technique. If it is necessary to break tempered glass, firefighters should strike a sharp blow with a pointed tool such as the pike of a Halligan tool. Firefighters must be aware that the glass will shatter so they should stand to one side and strike the glass in a lower corner.

Door Swing Direction

Doors are hung in jambs with hinges, and forcible entry is accomplished by working with the direction of swing. The firefighter must determine

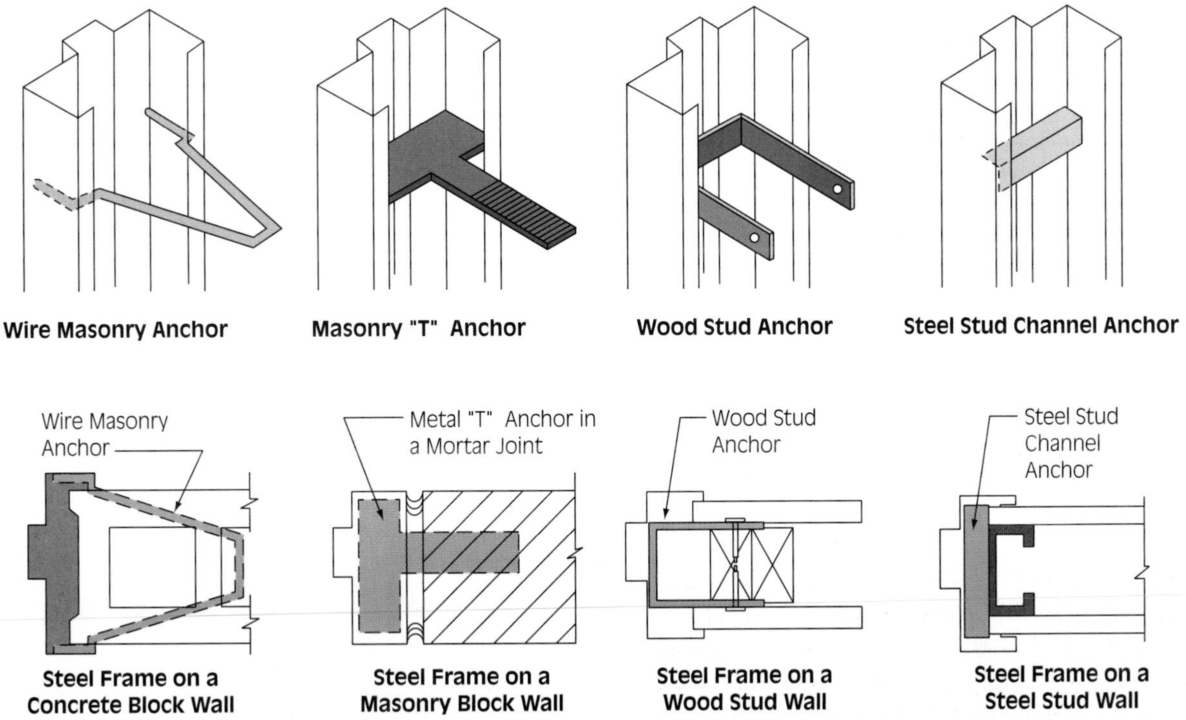

Figure 17-24 Metal doors are secured to walls in a number of ways.

Figure 17-25 Typical commercial glass door.

which way the door swings, and the common way of describing this is in relationship to the forcible entry team. Doors with exposed or visible hinges will swing toward the forcible entry team, **Figure 17-26A**. If no hinges are visible and there is a stop on the door frame, the door will swing away from the team, **Figure 17-26B**.

Sliding Doors

Sliding doors are usually found in residential occupancies and consist of sliding and fixed sections of tempered glass in a wood or metal frame. The locking device is usually very light with a strike screwed into the jamb, **Figure 17-27**. These doors are forced by prying the door away from the jamb with a Halligan tool, breaking the lock striker out of the stationary frame. To prevent entry by intruders, many occupants will place a pipe, broomstick, or other solid object in the track to prevent the door from sliding. With these objects used as additional security devices, breaking the glass may be the only option for entry, but it should be used as a last resort.

Revolving Doors

Revolving doors are made up of four sections (doors) hung on a vertical shaft or hinge that

(A) (B)

Figure 17-26 Hinges indicate the direction of door swing and forcible entry. An outward swinging door is shown in (A) and an inward swinging door in (B).

Figure 17-27 The strike of a residential sliding glass door is secured with two small screws and can be easily forced.

allows the door to rotate. Building and life safety codes require revolving doors to collapse and allow occupants a rapid exit if needed. This is accomplished by releasing the arms that hold the door apart. The "panic-proof" type of revolving door will automatically collapse when two sections are moved in opposite directions. These doors may be locked with key cylinder locks or slide bolts. Generally it is best to avoid using revolving doors as an entry point because of lack of clearance created by the folded door and vertical shaft or hinge.

Overhead Doors

Overhead doors range from the simple garage door to the very secure roll-down steel door, **Figure 17-28**. These doors are built from wood, steel, aluminum, and fiberglass and may have a solid or insulated core. Depending on the occupancy and security requirements, they may have windows of glass or some type of synthetic material.

Residential Garage Doors

Overhead garage doors used in residential construction are typically three- to five-section folding doors of wood or metal construction, **Figure 17-28A**. Older style doors may be one-piece slabs that tilt up into the garage. Overhead doors are installed on tracks with rollers and balance springs to assist in opening. A folding overhead garage door may be forced by any of several methods:

(A)

(B)

Figure 17-28 Typical overhead doors: (A) residential and (B) commercial.

- Break a panel or window, reach in, and unlock the securing device.
- Pull the lock cylinder and utilize through-the-lock forcible entry.
- Automatic openers hold the door in the closed position. To disconnect the opener, break out a panel near the attachment mechanism, reach in with a tool to grab the release cord, and pull as shown in **Figure 17-29**.

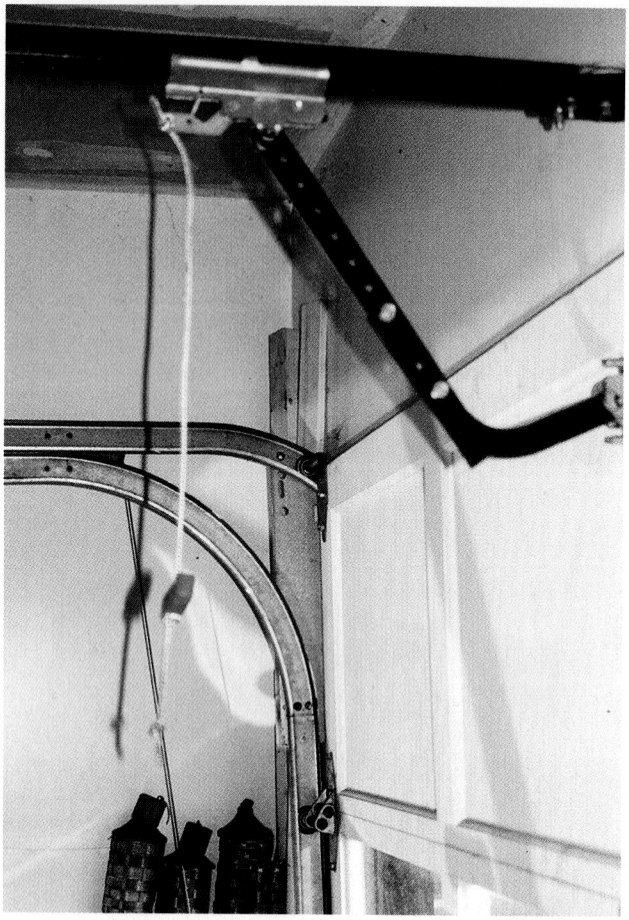

Figure 17-29 To disable a motorized residential overhead door opener, the firefighter can break out a pane of glass or door panel and pull the release rope.

Caution At a recent fire, the initial attack lines advanced through the garage area of a house. The overhead door, which was equipped with an automatic operator, closed either because the balance springs were weakened by fire or the door operator shorted out. This trapped the attack team inside the garage with no means of escape. Once an overhead door is opened, a tool should be placed (a six-foot hook works well) in the track or a pair of locking pliers should be used on the track to prevent the door from closing.

Commercial Garage Doors

Commercial overhead doors are similar in operation and construction to residential doors, **Figure 17-28B**. The exception to this is the type of locking and security devices. These doors may be forced using the same methods mentioned for residential doors or by cutting the door with a rotary saw, **Figure 17-30**. After the door is opened, the same precautions used for residential overhead doors must be taken to ensure that it does not close, thus trapping firefighters.

Roll-Down Steel Doors and Gates

Roll-down steel doors, **Figure 17-31**, are of heavy steel construction to provide a higher level of security or as a rated door in a firewall or separation. Roll-down doors generally use three different methods of opening and closing:

- *Manually operated:* The steel gate is lifted by hand with the assistance of springs.
- *Chain operated:* A chain hoist mechanism is used to lift the door.
- *Electrically powered:* An electric motor connected to a switch is used to raise and lower the door.

The method used to force these doors will depend on the type and number of locks and security devices. Generally the best way to accomplish forcible entry is to:

- Cut or force the locks.
- Attack the hardware.
- Cut through the gate in a manner similar to that shown in **Figure 17-30**.

Security gates used in commercial occupancies provide security while at the same time allow for the display of merchandise. The most common means of securing these gates is to use padlocks. These padlocks can be cut or forced by use of several types of tools; the most reliable method is the power saw with a metal cutting disc. Cutting or otherwise forcing the hardware that secures the gate is a method that can be used with some success. An example of this is cutting the hasp that secures the padlock to the gate.

Locks

Locks are designed and intended to keep unwanted or unwelcome visitors out of a building or occupancy. Firefighters—although not unwelcome visitors—will face the problem of locked door and entry access on a regular basis. As security requirements increase or break-ins occur, additional locks will be placed on a door as shown in **Figure 17-32**.

The most important part of forcible entry is to know the type of lock, how the lock operates, and how to disable it. Knowing this will save valuable time and energy on the fireground and provide quicker access, rescue, and fire extinguishment.

Streetsmart Tip From time to time home improvement and consumer magazines will have articles on locks and security devices. This information is an excellent resource for additional training on new types of locks and methods to force entry.

Figure 17-30 On older commercial overhead doors, a large inverted V-shaped cut is made in the gate with a power saw. A slat on each side is pulled toward the center and removed. This allows all slats to be removed and a large opening created.

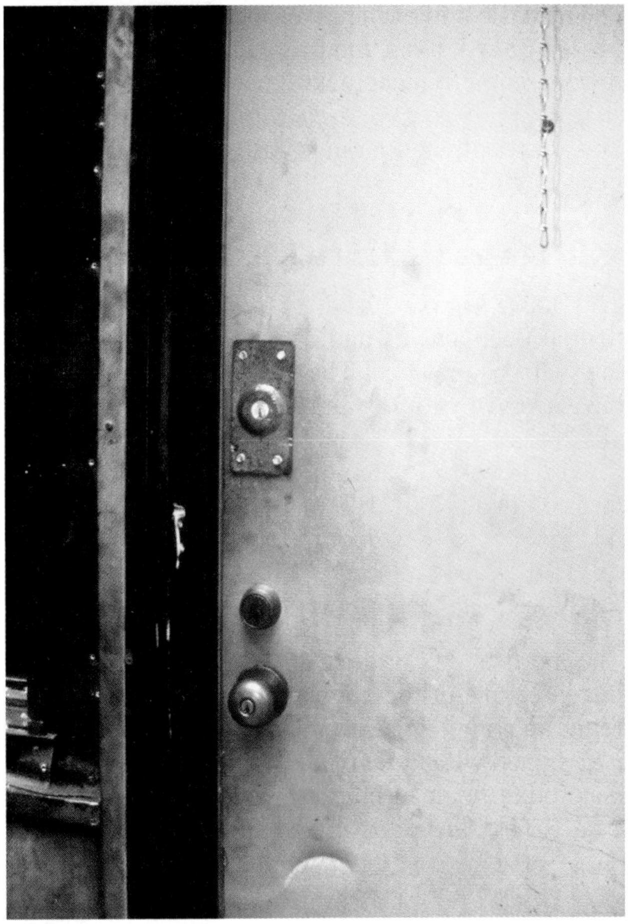

Figure 17-32 To gain access through this door, the firefighter must know how to force these three different styles of locks.

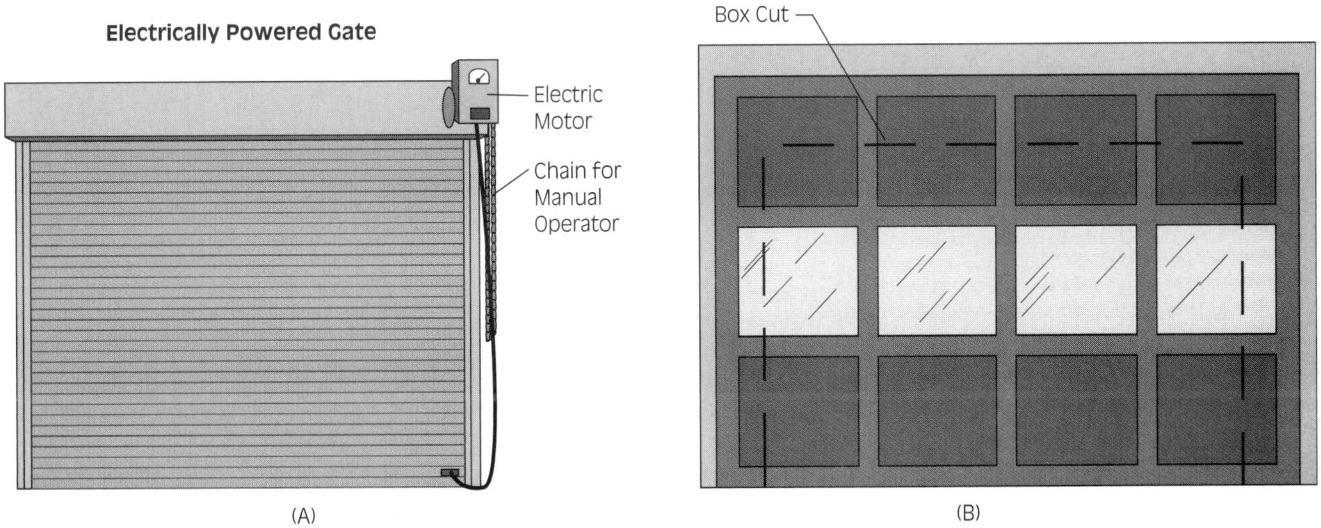

Figure 17-31 (A) Roll-down steel doors are of heavy steel construction to provide a higher level of security or as a rated door in a firewall. (B) Box cut used for this door.

The basic nomenclature of locks as shown in **Figure 17-33** will assist the firefighter in understanding the operation of most locks.

Lock mechanisms used to provide security for doors and other openings fall into the following general categories.

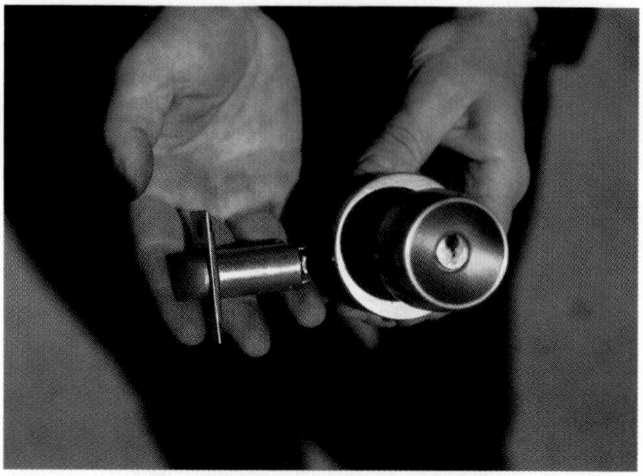

Figure 17-34 A key in the knob-type lock usually found in residential occupancies.

Key in the Knob Lock

This lock, **Figure 17-34**, is most common in residential occupancies and on interior doors in commercial occupancies. The outside of the lock will have a keyway for operation and the inside will usually have a keyway or button. The bolt on this type of lock is either a latch or dead latch bolt type. The latch bolt has a throw of approximately ½ inch, making this an easy door to force.

Mortise Lock

The mortise lock, **Figure 17-35A**, is designed to fit into a cavity in the edge of a door and is usually found in commercial occupancies. The mortise lock is designed with three types of operating latches: the dead bolt, the dead bolt and latch, and the pivoting dead bolt. The pivoting dead bolt may also be known as an Adams Rite lock, which is used in narrow frame/stile aluminum doors. Most mortise locks operate with a key that, when inserted in the keyway and turned, turns a cam, which moves the dead bolt or latch and opens the lock.

- *Dead bolt.* This lock has one sliding bolt and is locked or unlocked by one complete turn of the

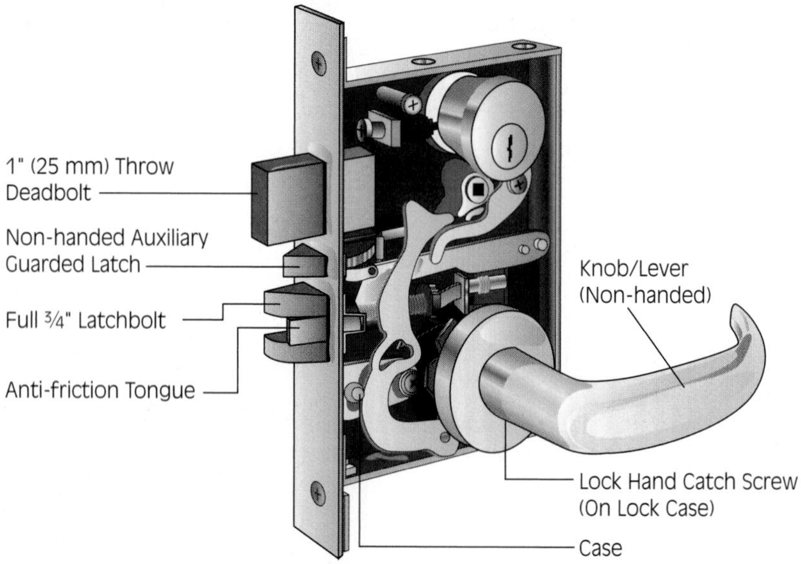

Figure 17-33 The parts of a lock device.

(A) (B) (C)

Figure 17-35 A typical mortise style lock: (A) the exterior, or normal view, (B) the locking mechanism, and (C) the pivoting dead bolt (Adams Rite lock).

key (**Figure 17-35B**) in the cylinder. This will only operate the dead bolt; additional latches or locks are operated separately.

- *Dead bolt and latch.* (**Figure 17-35B**) Similar to the dead bolt, but with an additional latch operated by a doorknob.
- *Pivoting dead bolt.* This lock, **Figure 17-35C**, is used on metal and glass doors and has a bolt that is housed vertically when retracted and pivots up to the horizontal when placed to lock. The **bolt throw** projects approximately 1½ inches into the **strike plate** when locked.

Rim Locks

Rim locks, also known as surface locks, **Figure 17-36**, attach on the inside of a door with the cylinder extending through the door and a keyway visible on the outside. There are many variations of this lock but they are all mounted on the surface of the door.

Tubular Locks

Tubular locks are mounted in a hole that has been bored into the door and are best described as a combination of the key in the knob lock and the mortise lock. However, instead of a knob, the tubular lock uses a cylinder to operate the bolt and is recognized by the cylinder's cover which protrudes about ½ inch on each side of the door. A typical tubular lock is shown in **Figure 17-37**.

Padlocks

Padlocks are portable or detachable locking devices that are manufactured for regular and heavy-duty service. This type of locking device has a movable shackle that locks into the body of the lock and is used to secure a door or gate using a hasp or chain. Padlocks come in a wide variety of shapes and sizes, **Figure 17-38**.

Regular padlocks have a shackle of less than ¼ inch in diameter and are not usually made of hardened steel. These may be cut with a bolt cutter or broken with a lock breaker. Heavy-duty padlocks have shackles larger than ¼ inch in diameter and are made

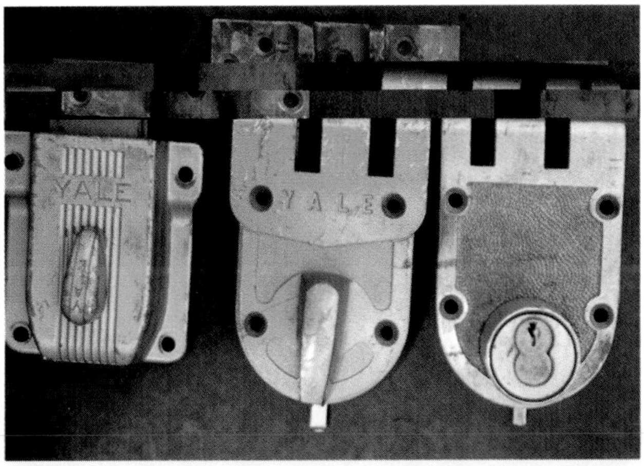

Figure 17-36 Rim locks (from left to right): a dead bolt, a vertical bolt and striker plate, and a vertical bolt with key cylinder.

Figure 17-37 Tubular dead bolts.

of hardened steel, **Figure 17-39**. In addition the shackles on both the heel and toe are locked when the shackle is depressed into the lock case, and both sides of the shackle must be cut to open the lock.

Special Locks

Included here are all devices that do not fit into other categories. Examples of these devices are overhead door locks, magnetic locks, or card key entry systems, **Figures 17-40A** and **B**. Overhead door locks are similar to rim locks and may be forced by gaining access or through the lock. Card key entry systems are regular locking devices operated by an electric actuator or solenoid. Again, conventional or through-the-lock forcible entry is the key to access.

> **Streetsmart Tip** Many office occupancies and hotels/motels are using electronic locks for security. The fire department should contact the facility management people to arrange a procedure for obtaining a master card key if available. This reduces the amount of work for the forcible entry crew and could reduce property damage.

Figure 17-38 Padlocks are available in many shapes and sizes.

Figure 17-39 Cutting a padlock with a power saw. Note that both shackles are being cut at once.

Magnetic locks use a magnetic force to hold a door secure and the system is usually disabled from the interior of the building. In most instances it is quicker and easier to find another means of entry.

Additional Security Devices

As security requirements increase, home and business owners have begun to install many varieties of locks and security devices. These may be as simple as a broom handle in the track of a sliding door or as complicated as a number of additional locks on a door. These additional devices may not provide visible signs or indication of their use to the forcible entry team, **Figure 17-41**. If these types of devices are in use the forcible entry team may need to find an alternate means of entry or use a rotary saw to gain access.

When buildings are being remodeled or demolished, the fire department should check with the owner or contractor to gain permission to salvage unwanted door locks or security devices. These make excellent training aids, especially if similar locks are in use in other buildings in the jurisdiction. In addition, new construction should be surveyed to determine the types of locking devices being installed.

METHODS OF FORCIBLE ENTRY

For fire department operation there are three standard methods of forcible entry:

1. Conventional
2. Through-the-lock
3. Power tools

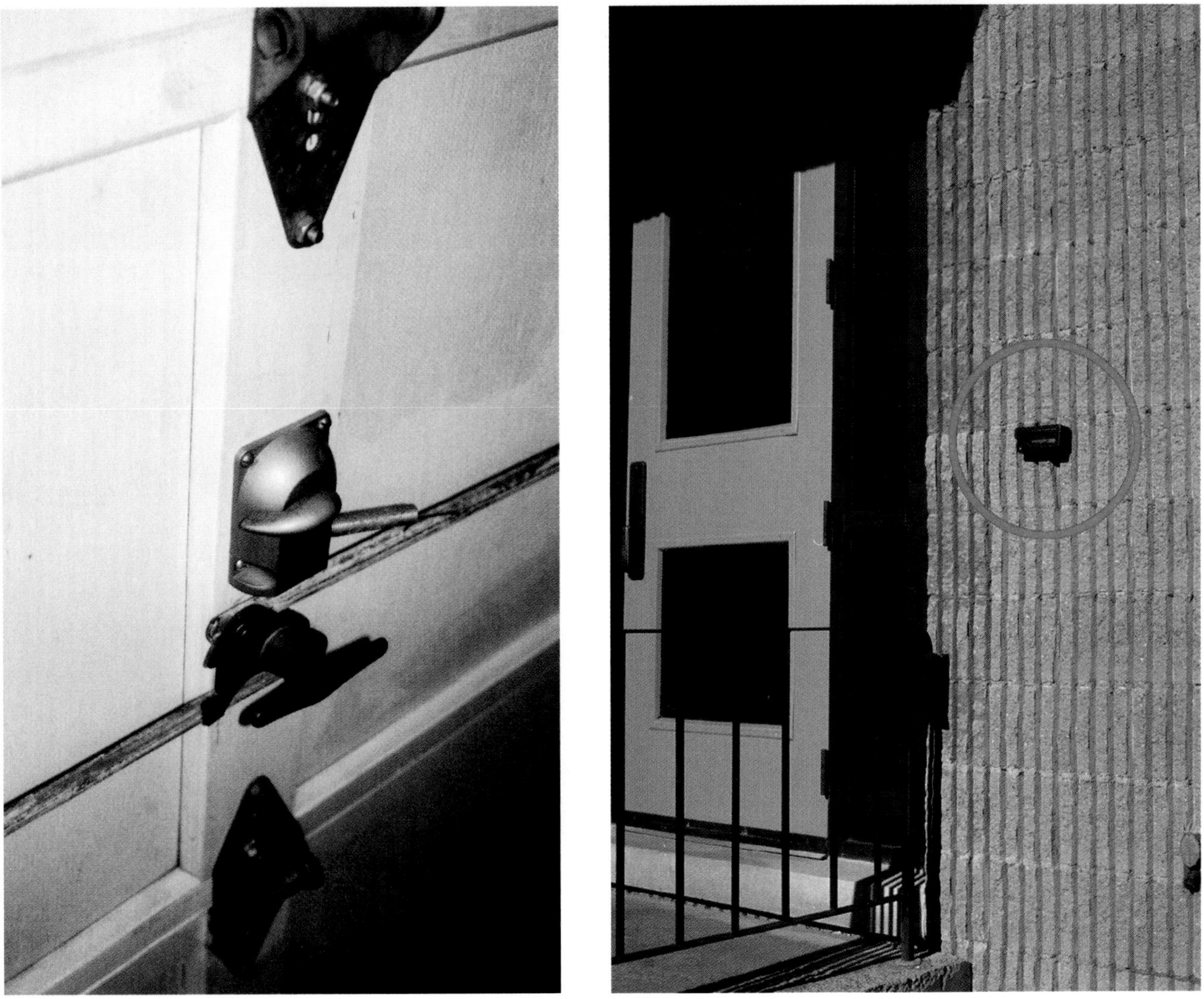

(A) (B)

Figure 17-40 (A) Residential overhead door locks are a type of rim lock and are forced easily using the through-the-lock method. (B) Card key entry systems are standard locks with electric actuators.

Caution All forcible entry operations must be coordinated with fire attack and ventilation. Lack of coordination may result in rapid fire spread or a backdraft.

These methods involve the use of certain tools and the application of many techniques.

Conventional

This is an old and reliable method involving the use of leverage, force, and impact. The primary tools used in this method are the "irons," consisting of the Halligan tool and the flathead ax. This technique requires procedures that will accomplish one or more of the following:

- Force the door away from the jamb, pulling the bolt away and free from the strike plate.
- Break the lock or striker.
- Break the door and/or the frame.
- Force or remove the hinges.
- Breach the wall or door.

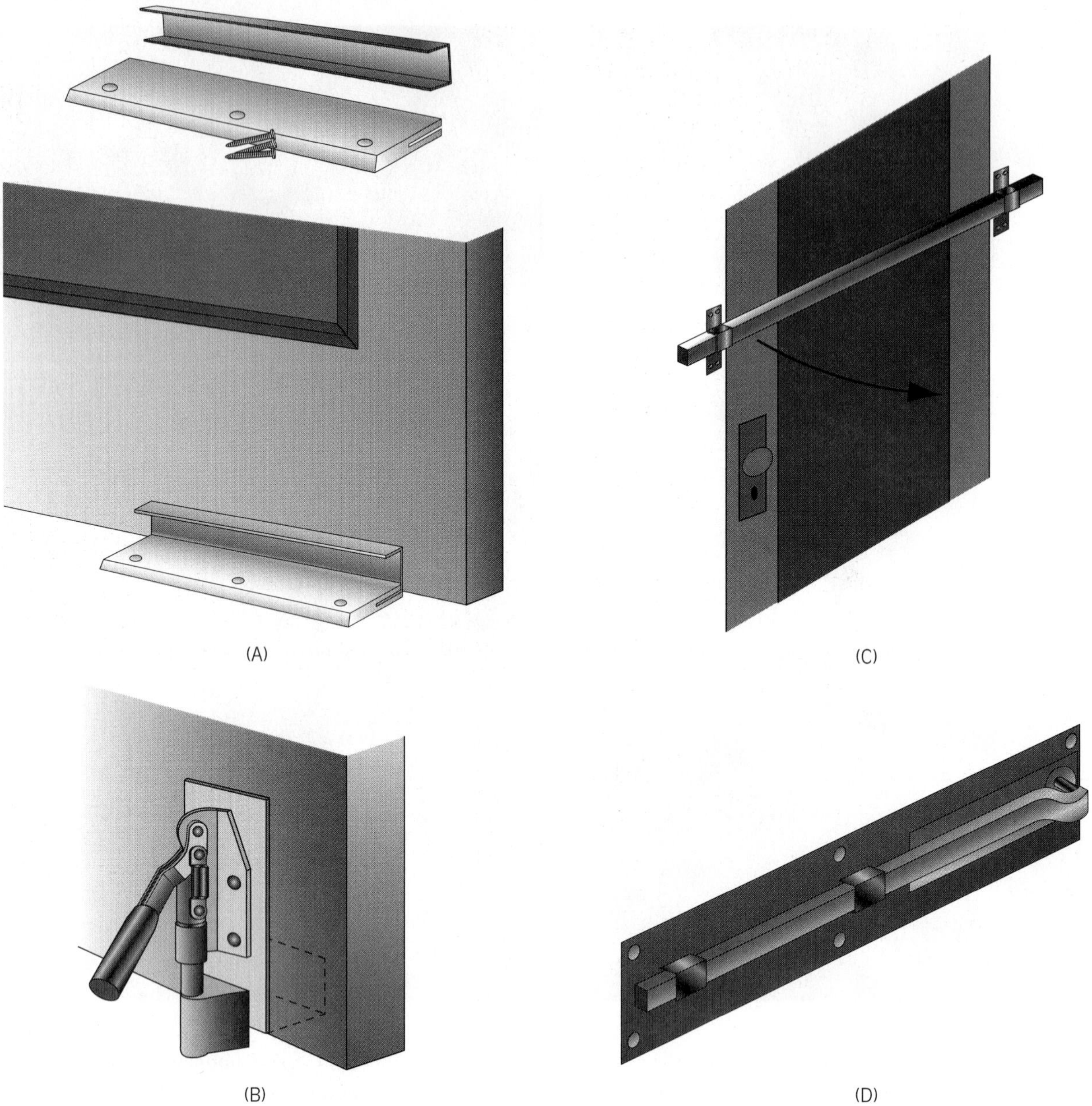

Figure 17-41 These auxiliary locking devices are usually not detected by the forcible entry team: (A) floor-mounted stop plate, (B) door blocker, (C) steel bar and brackets, and (D) sliding bolt. Note that the bar and sliding bolt are often homemade devices.

Streetsmart Tip Always size up the forcible entry task. Regardless of the type of lock, at a quick glance, the fastest way to force entry to the door shown in **Figure 17-42** is to break the glass and unlock the door from the inside. But be aware there may be additional locking devices out of view that may not be easily unlocked. At a fire situation this could magnify the problems associated with heat, smoke, and fire possibly rolling out through the broken glass. On the contrary, once the door in **Figure 17-43** is forced, the firefighter may encounter the household goods stored in front of the door.

In general, conventional forcible entry is quick and reliable and is the primary, preferred method at structural fires. Conventional forcible entry may also involve the use of the hydraulic forcible entry tools.

Conventional Forcible Entry with Door Swinging Away from the Forcible Entry Team (JPR 17-1)

Remember to always try the door to determine if it is locked or secured and forcible entry is actually necessary.

1. Size up the door to determine swing; the number, type, and location of locking devices; and the type of door and frame,

Figure 17-42 Firefighters cannot assume that the quickest way to force entry to this type of door is to break the glass and unlock the door from the inside. Additional locking devices may be out of view.

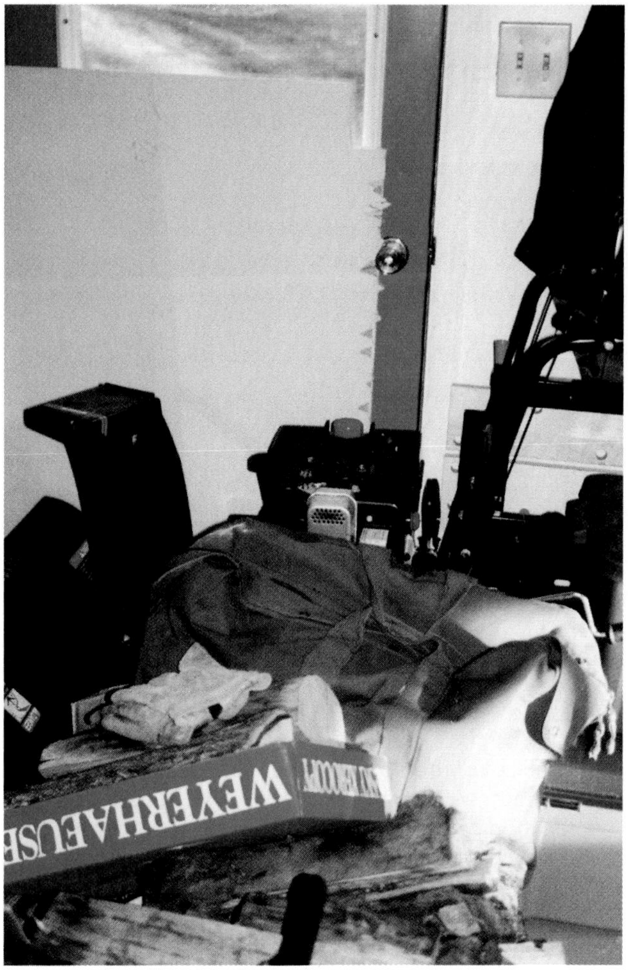

Figure 17-43 Once a door is forced, the firefighter may encounter many obstacles.

JPR 17-1A. With this type of door swing, the hinge pins are not visible. Also check to make sure the door is locked/secured.

2. Gap the door by slipping the adz between the door and the door stop 6 inches above or below center of the lock and push up or down to spread the door from the frame, **JPR 17-1B**. This will allow for placement of the fork end of the Halligan tool. Note that on some doors this may cause the door to open. On door frames with a nailed doorstop, this method will break the stop and allow easy entry of the fork.
3. Set the fork of the Halligan tool into the gapped area with the bevel of the fork toward the door no more than 6 inches above or below the center lock, **JPR 17-1C**. As the Halligan tool is struck with the flathead ax or maul, gradually bring the tool away from and at an approximately 90-degree angle to the door.
4. Set the tool using the flathead ax or maul, driving the Halligan tool in so that the tips are around the door and locked in, **JPR 17-1D**. *If difficulty is experienced, the fork may be inserted with the bevel toward the frame.*
5. Force the door; with the tool set the door is spread from the frame. The Halligan tool is pushed sharply in the direction of the door swing and opens as shown in **JPR 17-1E**. This procedure is repeated as necessary for each lock, **JPR 17-1F.**

On doors that swing away from you, try to control the door as it is forced. Generally, the door is controlled by the firefighter with the Halligan tool. As the door swings in, the firefighter must reach in with the tool and bring the door back under control.

> **Caution** Doors leading into the fire area should be blocked or chocked open when firefighters are conducting a search or attacking the fire in that area in addition to other activities. This will prevent the door from closing behind crews and will allow rapid egress if needed.

JOB PERFORMANCE REQUIREMENT 17-1

Conventional Forcible Entry—Door Swings Away from the Forcible Entry Team

A Size up the door to determine swing; the number, type, and location of locking devices; and the type of door and frame.

B Place the adz between the door and the door stop 6 inches above or below center of the lock and push up or down to spread the door from the frame.

C Place the fork of the Halligan tool into the gapped area with the bevel of the fork toward the door no more than 6 inches above or below the center lock.

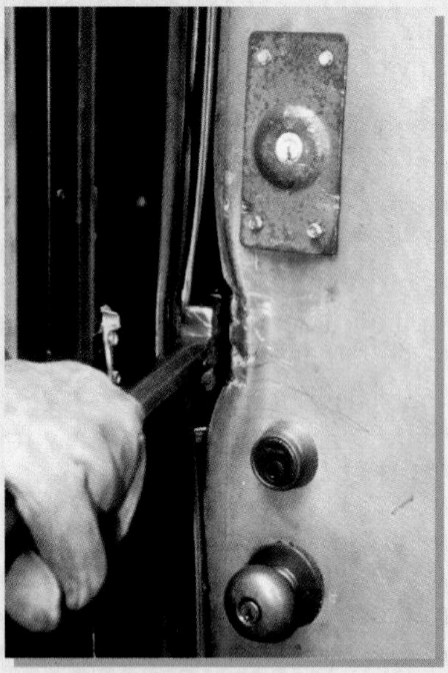

D Set the tool using the flathead ax or maul. Drive the Halligan tool in so that the tips are around the door and locked in.

E If difficulty is experienced in setting the tool, drive fork in with bevel toward frame to allow easier entry.

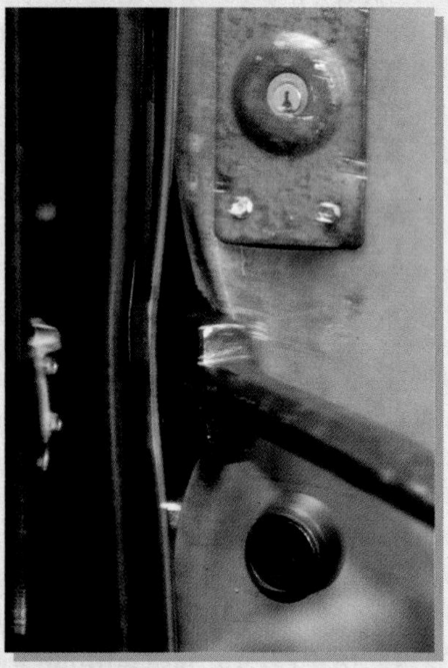

F This procedure is repeated for each lock.

Conventional Forcible Entry with Door Swinging toward the Forcible Entry Team (JPR 17-2)

1. Size up the door to determine swing; the number, type, and location of locking devices; and type of door frame and hinges, **JPR 17-2A**. Also check to make sure the door is locked/secured.
2. Gap the door by driving the adz or fork end of the Halligan tool between the door and frame 6 inches above or below the lock, **JPR 17-2B**. Work the tool in until it contacts the doorstop. The adz end is preferred, but on some doors, especially steel, it will have a tight fit between the door and jamb and the gap may be started by using the fork end, **JPR 17-2C**. If door is tight, work tool up and down to open up gap, **JPR 17-2D**.
3. Set the tool by driving the adz or fork end past the door around the doorstop, **JPR 17-2E**. Drive in the tool until it is "locked in" around the inside of the door. Due to the configuration of the door and wall, using the adz end will allow the tool

JOB PERFORMANCE REQUIREMENT 17-2

Conventional Forcible Entry—Door Swings toward the Forcible Entry Team

A Check the door to determine swing; the number, type, and location of locking devices; and the type of door frame and hinges.

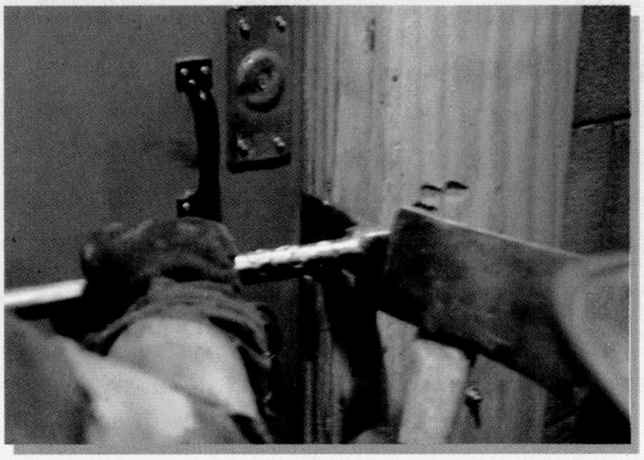

B Create a gap between the door and jamb by driving the adz or fork end of the Halligan tool between the door and frame 6 inches above or below the lock.

C On tight doors, the gap may be started with the fork end.

D If door is tight, work tool up and down to open up gap.

(continued)

JOB PERFORMANCE REQUIREMENT 17-2

Conventional Forcible Entry—Door Swings Toward the Forcible Entry Team (Continued)

E Set the tool by driving the adz or fork end past the door around the doorstop.

F This inside view shows how the adz is set.

G Force the door by pulling the tool away from the door.

to clear the wall. **JPR 17-2F** shows an inside view of how the adz is set.

4. Force the door by pulling the tool away from the door, **JPR 17-2G**. This will cause the bolt to pull out of the strike, the lock to break, or the door frame assembly to fail and the door will open. When dealing with several locks, this procedure may have to be repeated for each lock.

Forcing Doors with Hydraulic Forcible Entry Tools

The hydraulic forcible entry tool (HFT) is primarily used on doors that swing away and have strong metal frames.

1. Place the spreader jaw of the tool between the door and jamb as close as possible to the lock.
2. Set the jaw into the gap to the doorstop using a tool if necessary. On tight-fitting doors, a driving tool may be needed to set the jaw into the doorstop.
3. Pump the tool until the bolt is pulled out of the strike and push the door open.

Streetsmart Tip If there is more than one lock present and they are relatively close together, place the jaw of the tool between the locks.

Through-the-Lock Forcible Entry

The **through-the-lock method** of forcible entry involves attacking the locking mechanism by removing the key cylinder and then operating the lock with alternative means. This method of forcible entry is best applied to mortise, rim, or tubular-type lock cylinders.

In general, the choice of through-the-lock forcible entry is made when entry needs to be gained with damage kept to a minimum. Sometimes it is also the quickest means of entry and is the best method when forcing entry of metal framed glass or all glass doors.

Through-the-Lock Entry by Unscrewing or Wrenching the Locking Cylinder (JPR 17-3)

This method is not as quick as pulling the cylinder with a lock puller and should be used only if time

JOB PERFORMANCE REQUIREMENT 17-3

Through-the-Lock Entry by Unscrewing or Wrenching the Locking Cylinder

A Using locking-type pliers, lock the pliers onto the cylinder. Turn the lock cylinder counterclockwise to unscrew the cylinder.

B Use proper key tool to open lock.

allows. In addition, trim work or walls may not allow the free rotation of the pliers.

1. Size up the lock to determine the type of lock and feasibility of utilizing this method. Lock cylinders with protective collars may not be able to be unscrewed.
2. Using locking-type pliers, lock the pliers onto the cylinder, **JPR 17-3A**.
3. Turn the lock cylinder counterclockwise to unscrew the cylinder.
4. Remove the cylinder and insert the proper end of the key tool as shown in **JPR 17-3B** to operate the locking mechanism as shown later in this chapter.

> **Streetsmart Tip** **Through-the-Lock Technique Modified with Rim Locks:** If the cylinder crumbles and will not pull out or the lock will not unlock, the firefighter can drive the lock off the back of the door with the pike of the Halligan tool or the handle of the lock puller.

Through-the-Lock Entry Using the K Tool (JPR 17-4)

1. Size up the lock to determine the type of lock and feasibility of utilizing this method.
2. The blades of the K tool are forced over the cylinder and decorative ring. Tap the K tool into place until firmly set, **JPR 17-4A**.
3. Place the adz of the Halligan tool into the loop of the K tool, **JPR 17-4B**. If necessary, the Halligan can be tapped to firmly set the K tool.
4. Pull up on the Halligan tool and pull the cylinder, **JPR 17-4C**.

> **Streetsmart Tip** The K tool is designed to be able to pull cylinders that are very close to the bottom or edge of the door. This is done by placing the narrow, straight blade of the K tool toward the bottom of the door to allow room for the tool.

A number of different lock-pulling tools may be used in place of the K tool. The A, officer's, and REX tool, **Figure 17-44**, are designed to be driven behind the lock cylinder to get a "bite" or "purchase" on a substantial part of the cylinder. These tools are most effective on cylinders that are recessed into a door and on odd-shaped cylinders such as tubular locks.

JOB PERFORMANCE REQUIREMENT 17-4

Use of K Tool for Through-the-Lock Forcible Entry

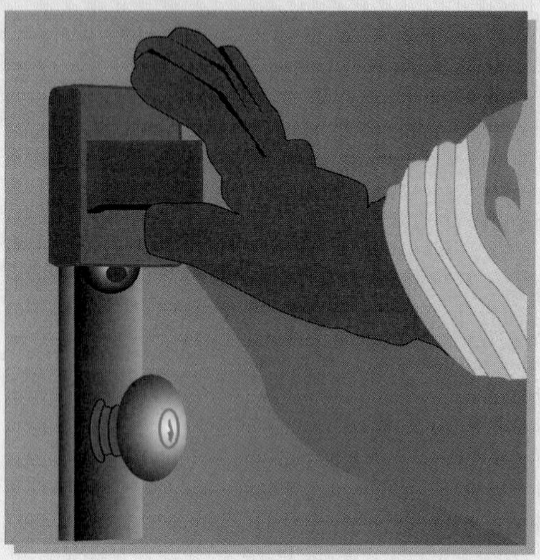

A Force the blades of the K tool over the cylinder.

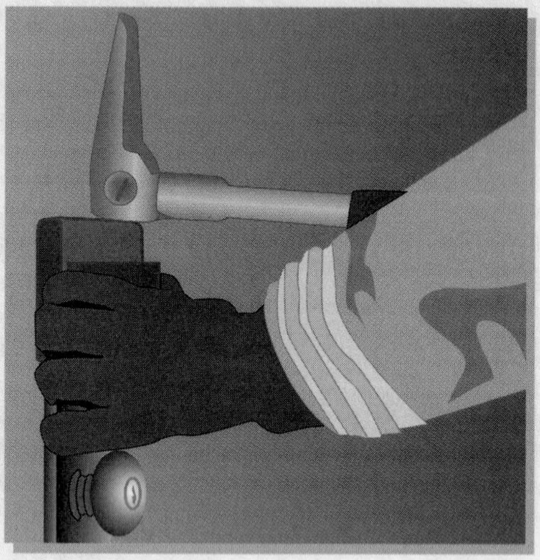

B Place the adz of the Halligan tool into the loop of the K tool.

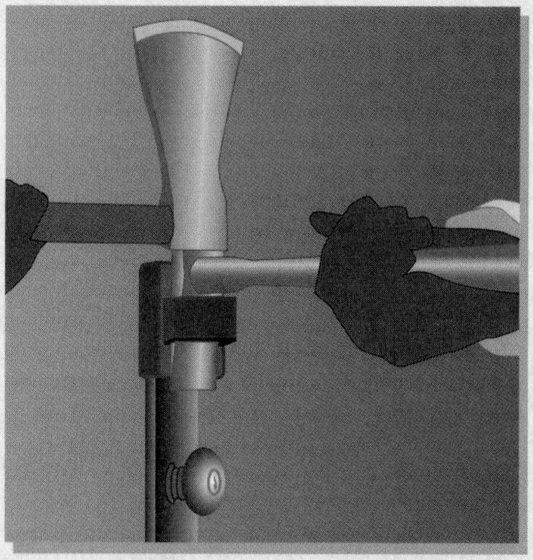

C Pull up on the Halligan tool, and pull the cylinder out.

Operating Lock Mechanisms

The final step using through-the-lock forcible entry is to manipulate the lock by using the proper key tool. Many types of key tools are used, but the most common is the two-sided, flat steel key or square key tool, **Figure 17-45**. This tool has a bent end used for mortise locks and a straight end used for rim and tubular locks.

The correct key tool to operate the locking mechanism is determined by examining the cylinder after it is removed. Lock cylinders fall into two different categories, the mortise having a cam device on the back of the cylinder and the rim lock with a flat or square blade on the back of the cylinder, **Figure 17-46**. The tubular lock has a tailpiece similar to the rim lock.

To open a mortise lock, visualize that the key hole where the cylinder was prior to pulling it out is that of the face of a clock. The keyway of the cylinder prior to removal was at the six o'clock position and the unlocking action will occur within the five to

Figure 17-44 The REX tool is used to pull lock cylinders in the through-the-lock method of forcible entry.

Figure 17-45 The two types of key tools used for through-the-lock forcible entry.

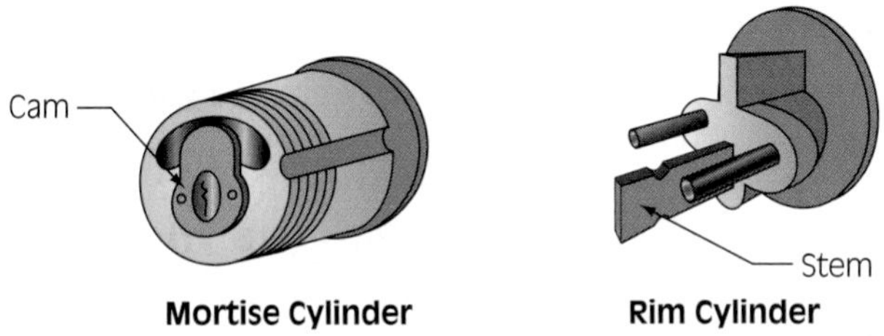

Figure 17-46 To operate the lock, look at the back of the cylinder to determine the right key tool to use.

seven or seven to five o'clock area, **Figure 17-47**. Place the bent end of the key tool into the keyway and move it to the five or seven o'clock position and engage the locking mechanism. Move it to the opposite side to retract the dead bolt. If the lock mechanism has a spring lock, push down with the key tool to depress it and then move it. To unlock the spring latch, move the key tool up to the three o'clock or nine o'clock position and engage the latch mechanism, push, and hold it in place. This will pull back the spring latch and allow the door to open.

To open a rim lock, after removing the cylinder, if a tailpiece is visible, then the lock is a rim lock. Insert the straight blade of the key tool into the slot on the rim lock and rotate the tool to unlock it.

Lock Variations

Shutter Guard

Many locks have a spring-loaded shutter guard that will close when the cylinder is pulled. This will not allow a key tool to open the lock. A shaped tool such as a pick is used to slide the shutter open. Slide the shutter horizontally toward the edge of the door and hold it open until the key tool is in place. Then turn the key tool to open.

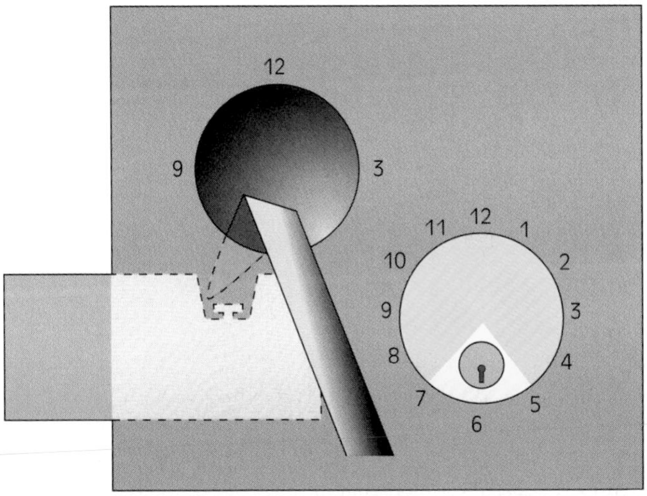

Figure 17-47 The unlocking action will occur within the five to seven or seven to five o'clock area.

Night Latch

A rim lock with the night latch engaged can only be unlocked from the inside.

Through-the-Lock Technique Modified

When the mechanism of a rim lock cannot be operated because of the presence of a night latch or shutter guard, or in any situation where the key tool cannot open the lock, firefighters should use the following procedure. Place the pike of the Halligan tool into the hole and against the back of the *rim lock.* Strike the Halligan tool with an ax and drive the rim lock off the door, **Figure 17-48**.

Tubular Locks

Locks such as the key in the knob and tubular dead bolts are opened in a manner similar to the rim lock. The knob is removed with a lock puller or knocked off with a heavy tool such as the ax or Halligan tool. This will expose the latch mechanism, which can then be operated by the key tool or screwdriver.

Removing the cylinder with a lock puller exposes the tubular dead bolt lock. The adz of the Halligan tool can be used to shear the screws, and the cylinder can sometimes be removed in this manner. The bolt mechanism is then operated with a key tool or screwdriver, depending on the shape of the hub, **Figure 17-49**.

Forcing a Padlock with a Lock Breaker

1. Place the wedge end of the lock breaker through the shackle. Depending on the size of the lock a Halligan tool may also be used for this task.
2. Strike the back or head of the lock breaker with a maul or heavy striking tool.
3. Continue step 2 until both shackles break.

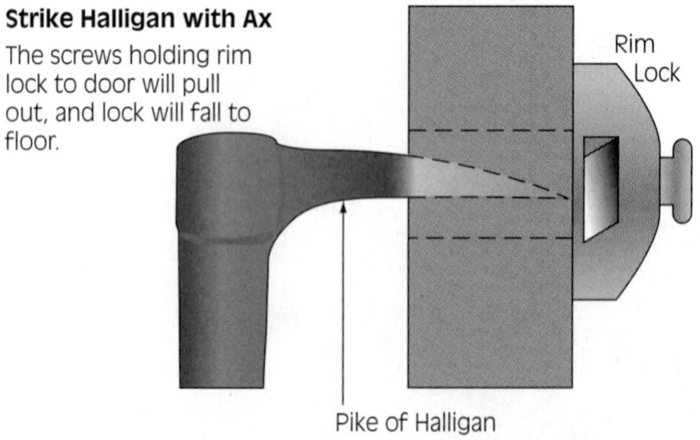

Figure 17-48 Strike the Halligan tool with the ax and drive the lock off the door. This method is used for rim locks only.

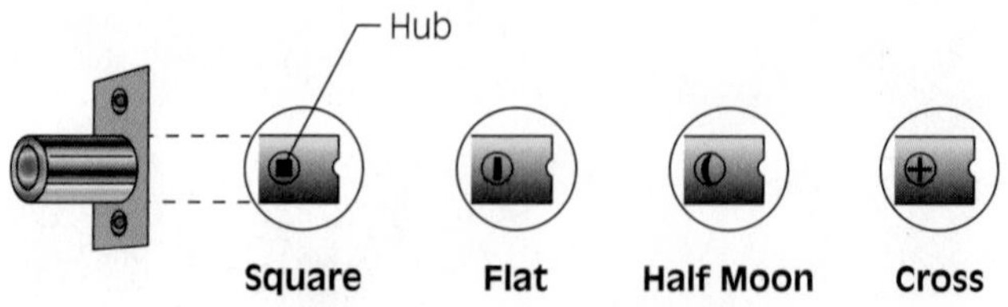

Figure 17-49 The bolt of a tubular lock is operated with a key tool matching one of the shapes shown.

Forcing a Padlock by Cutting with a Rotary Saw

1. Attach locking-type pliers to the lock case and lock the jaws. The locking pliers must have a chain or rope attached so the firefighter can hold the pliers clear of the saw.
2. A rotary saw with a metal cutting blade is used to cut the lock shackles, **Figure 17-50**.

WINDOWS

Windows are found in many types, architectural styles, sizes, and construction as shown in **Figure 17-51**. These are installed in buildings to provide light and natural ventilation. Usually firefighters will force entry into a structure using door openings; however, windows may be used as an alternate means of entry especially for rapid intervention team operations. To successfully conduct forcible entry operations through windows, firefighters must know about the four construction features of windows: glazing, sash, frame, and security devices.

Forcible Entry of Windows

There are many types of windows and each individual one requires a specialized technique to force entry through it. There are two general reasons to force a window: to gain entry and for horizontal ventilation.

> **Streetsmart Tip** When forcing windows, an opening must be created to allow for safe entry and exit with full protective equipment in place. It is important to remove all obstructions such as air-conditioning units, security bars, or child protective gates. The opening created may be needed to make a rapid exit if conditions deteriorate.

To make a large enough opening for entry, it is often necessary to break the glazing and the sash to create the largest possible opening. A common adage heard from experienced firefighters is

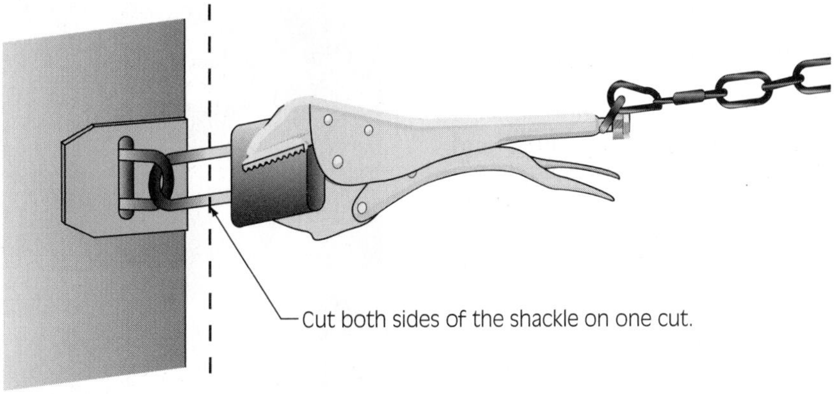

Figure 17-50 The locking pliers must have a chain or rope attached so that the firefighter holding the pliers is clear of the saw.

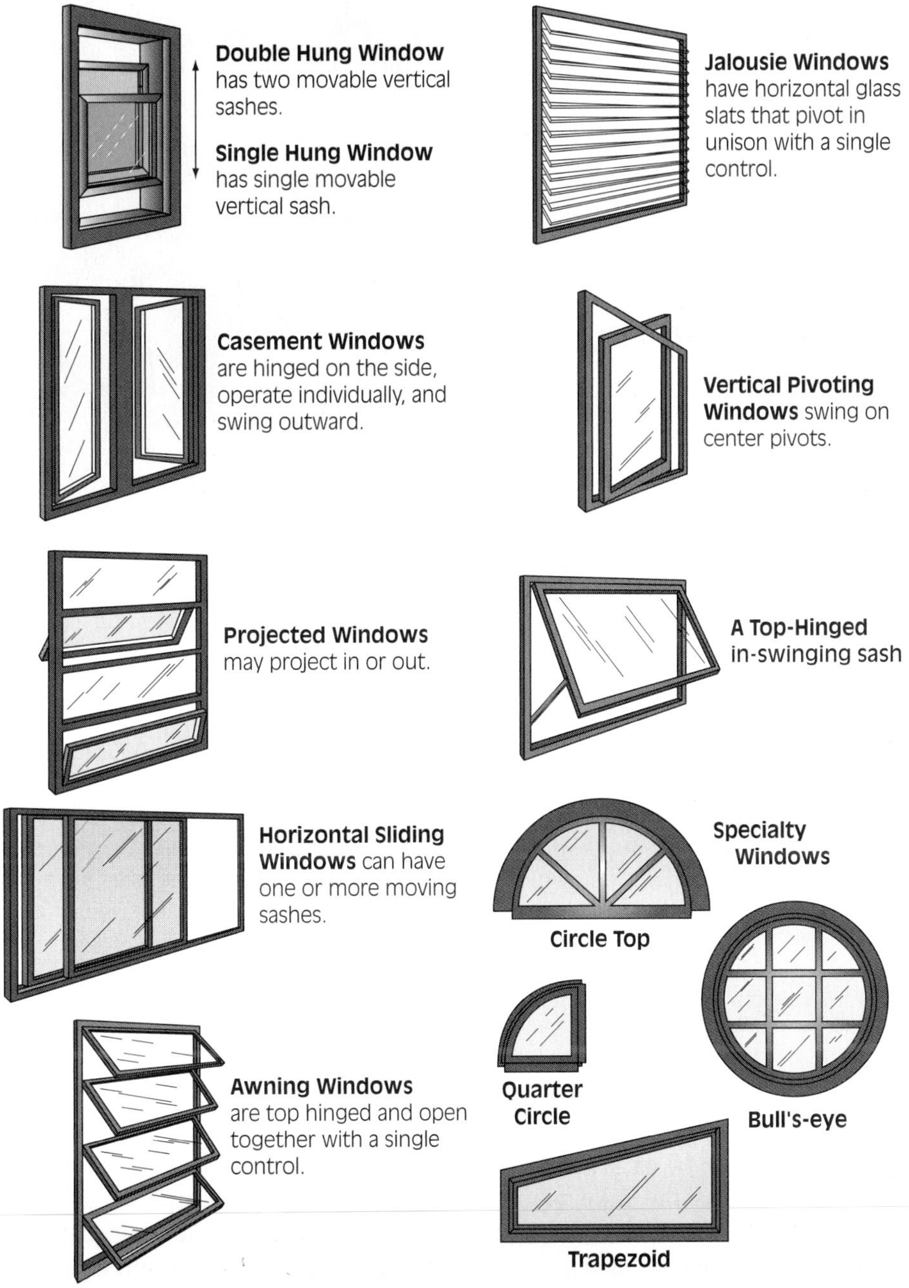

Double Hung Window has two movable vertical sashes.
Single Hung Window has single movable vertical sash.
Jalousie Windows have horizontal glass slats that pivot in unison with a single control.
Casement Windows are hinged on the side, operate individually, and swing outward.
Vertical Pivoting Windows swing on center pivots.
Projected Windows may project in or out.
A Top-Hinged in-swinging sash
Horizontal Sliding Windows can have one or more moving sashes.
Specialty Windows
Circle Top
Quarter Circle
Bull's-eye
Trapezoid
Awning Windows are top hinged and open together with a single control.

Figure 17-51 Windows are found in many types, architectural styles, sizes, and construction.

"make the window a door." At structure fires, the speed of operations is more important than damage done to windows. With this goal in mind, the quickest method is to break the glass and the sash to provide a quick opening for entry, rescue, and ventilation. At the same time, firefighters must remember not to break glass unnecessarily and create a safety hazard.

Glazing

The glass or other clear material portion of the window that allows light to enter is the **glazing**. The most common glazing material is glass. There are different types of glass, of which these are just a few:

- Regular or plate glass
- Tempered glass
- Laminated (safety) glass
- Wire glass

Regular/Plate Glass

Regular glass is relatively weak and easy to break, and when struck with a tool, it breaks into very sharp, knife-like shards. Plate glass is used in larger windows and is generally thicker than regular glass, with thicknesses of ¼ to ¾ inch being the most common. This type of glass will break into large, heavy, and sharp pieces, which can be very dangerous. A long-handled tool such as a hook or pike pole works best, **Figure 17-52**. With both types of this glass, the firefighter should stand to the side and strike the window. All shards of glass that remain must be cleaned out of the window opening.

Figure 17-52 Use a long-handled tool such as a hook or pike pole to break glass out of a window.

Caution Firefighters should always use a tool such as a hook (pike pole) or other long-handled tool to clean all glass out of a window. Care must be taken that glass shards do not slide down the tool's handle.

Streetsmart Tip A unique property of tempered glass is that when it is broken, the whole pane will fracture and fragment into small and relatively dull pieces. Certain panes, even those that have been fractured, may remain in place. These will require the use of a tool to clear out the remaining glass.

Tempered Glass

This material is glass that has been heat treated to give it additional strength. The technique to use when breaking tempered glass is to strike the glass with a pointed tool, such as the pike of the Halligan tool or a pick ax. Best results are obtained when the glass is struck near a corner where the glass is more rigid than at the center of the pane.

Laminated Glass

Laminated glass is also known as *safety glass.* It is commonly composed of two or more sheets of glass with a plastic sheet between them. The purpose of the plastic sheet is to hold the glass together if broken, thus reducing the hazard of flying glass. Laminated glass is most commonly found in automobile windshields; however, it is also used for glazing in windows and doors.

Wire Glass

Wire glass has a wire mesh embedded between two or more layers of glass, **Figure 17-53**. The primary purpose is to give the glass increased fire resistance. When exposed to high heat, the glass will break. The wire will then hold the glass together and allow the glazing to remain intact to stop the horizontal spread of the products of combustion.

To break the wire glass, the firefighter strikes with a tool, an ax or Halligan, and then cleans the glass out of the frame. This is best accomplished by carefully chopping through the wire with the blade of an ax. The firefighter must use caution because the entire piece of wire glass will fall as one complete unit when cut free.

Polycarbonate Glazing

These are plastic products produced to provide additional strength as compared to regular glass. One common type is Lexan. Lexan is approximately 250 times stronger than regular glass and is available in thicknesses up to 4 inches. Polycarbonate glazing can be removed by cutting it with a carbide-tipped circular saw. It can also be removed by striking it near a corner and driving it out. A heavy striking tool is best, and progressively driving it in may work it free from the sash.

Types of Windows

Double Hung/Check Rail Windows

These windows have upper and lower sashes that both slide vertically. The sash may be composed of wood or metal and will have a locking device in place to lock the sashes to one another or to the frame. To force open this type of window, the firefighter should use a tool to force the bottom sash up and break the locking mechanism. Wood sash windows with a single lock are easier to force than those composed of metal sashes/frames or other heavily secured window frames. When this type of window is encountered, breaking the glass and sash may be the most prudent option.

Figure 17-53 Wire glass has a wire mesh embedded between two or more layers of glass.

Energy Efficient Windows

Most new windows and also replacement windows are equipped with double and triple layers of glass. These windows are very well sealed and have a tendency to keep products of combustion trapped inside the structure. From the exterior point of view, they will give little indication of the severe conditions inside. Extra care must be taken when using these windows for entry. Also, energy efficient windows may be more difficult to break.

Casement Windows

Casement windows are hinged on the side of the sash and generally open outward. They commonly have a metal or wood sash and can be secured by one or more latches and a crank for opening and closing. The way to open this type of window is to break the glass with a tool, reach inside, and operate the crank. The screen may have to be cut out of the way to reach inside.

> **Streetsmart Tip** The firefighter needs to size up the window and decide if it is wide enough to enter safely and exit. Often the window sash and center mullion must be removed to provide sufficient room to enter.

Awning Windows

Awning windows are hinged so that they may swing out and may have a crank-operated mechanism to move the sash. They may have a wood or metal sash. The procedure used to open this type of window is similar to that of a casement window. The window must be broken to operate the mechanism. *Many awning windows are too small to enter and the sash will have to be removed or broken out to allow for entry.*

Jalousie Windows

Similar to the awning window, the jalousie window has small sections of tempered glass that are operated with a crank mechanism and overlap when closed.

Projected Windows

This type of window will pivot at the top, bottom, or center and is most commonly found in commercial buildings. Pivoting windows rotate on the top or at the sides.

Fixed Windows

Fixed windows are nonoperable windows used primarily for aesthetics and the introduction of light into a structure. They can be found in inaccessible areas, such as the top of a wall near the roof of a commercial building.

> **Caution** Firefighters must be in full protective equipment before breaking *any* windows. In addition to the hazard of sharp glass and other materials, the firefighter may have to contend with the smoke, heat, and fire that can vent from a broken window.

Bars and Gates

Windows and other openings that require security measures are often fitted with gates or bars, presenting a unique forcible entry situation for firefighters. These bars and gates must be removed or forced out of the way to allow for entry and exit into the structure. The following procedures may be necessary to accomplish this task:

- *Force the locking devices:* Gates that have exposed locks can be opened by forcing or cutting the lock.
- *Attack the fastenings:* Gates and bars can be removed by breaking or cutting through the bolts or attacking the point where the bars or gates are set.
- *Cutting the gates/bars:* The firefighter can use the rotary saw with metal cutting disc (aluminum oxide) or the sawsall to cut the gate or bars.

BREACHING WALLS AND FLOORS

Emergency situations often dictate that the walls of a structure must be opened to allow entry or to remove trapped firefighters or victims. This is especially true in the event of a collapse or blocked exit. There are two main considerations when breaching walls:

- *The type of construction of the building.* Wood-framed buildings with lath and plaster or drywall are usually easy to breach. Solid brick and reinforced concrete buildings are more formidable and will be more difficult to get through and hence require more effort.
- *Tools available.* During the initial stages of an operation, regular hand tools or possibly a power saw may be the only tools available. These tools may be adequate for numerous operations, but solid masonry or reinforced concrete construction will require specialized tools and equipment.

Techniques for Breaching Walls

Breaching Wood-Framed Walls

1. Size up the wall, trying to avoid the area around doors and corners due to narrow stud spaces. *It may be beneficial to create an inspection hole using an ax handle or Halligan bar,* **Figure 17-54**, *to verify the size-up, and check for obstructions, barriers, and fire conditions. The outcome of the inspection will determine whether to move to another location.*
2. Remove the wall covering from your side first to prevent the extension of heat and smoke into the area.
3. If plumbing pipes are encountered, try to bend or break the pipes (plastic and cast iron will shatter, copper will bend).
4. When a large enough hole is made, push in the wall cover over on the other side to complete the hole. If necessary, a stud can be removed by attacking the connection at the sill or plate. If the wall is a bearing wall, do not remove more than one stud. If a larger opening is required, shoring or some other supporting techniques will be necessary for long-term operations.

Breaching Masonry Walls—Block or Brick

To breach these walls without utilizing power tools, the only option may be the 12-pound maul, the Halligan tool, the flathead ax, or a battering ram.

1. Start by removing a single unit of block or brick. Work at the mortar joints because this is usually the weak point.
2. Once the joint is weakened, use the largest striking tool available and break the masonry unit.
3. Proceed by knocking out the surrounding units or release them at the mortar joint.

Figure 17-54 The firefighter uses an ax handle or Halligan tool to poke through a wall to determine if there are any obstructions on the other side.

Streetsmart Tip When making a large opening, a triangular or diamond-shaped hole will help maintain the structural integrity of the wall.

When opening masonry walls with power tools such as a power saw with a masonry disc, cut a triangle by making two angled cuts followed by a cut parallel to the mortar line. It may be quicker to use the maul on the mortar line to break the bricks free.

Breaching Reinforced Concrete Walls and Floors

Solid concrete reinforced walls with steel reinforcing rods are extremely difficult to open up even with the proper tools. The most common tools utilized are the jackhammer, the rotary power saw with masonry blade, and the diamond-tipped chain saw. An oxyacetylene torch may be needed to cut through the reinforcing rods.

Caution Cutting pre- or posttensioned concrete may cause the structural member to lose its strength and collapse. Wall or floor breaching operations in buildings with this type of structural members should be managed by specialized rescue teams with structural engineering assistance.

Breaching Metal Walls

Before any cutting operations are started, the location of heavy structural members, such as columns, should be determined. The tool of choice for this operation will be the rotary saw with a metal cutting disc. The quickest cut is a triangle cut large enough to allow for safe entry, similar to the procedure for overhead doors shown earlier in **Figure 17-30**. Also, depending on the type of metal siding, it may be cut by making two vertical cuts to loosen the wall material. The metal siding can then be pushed in or pulled out depending on the situation. If necessary, the first two cuts can be joined to remove the entire piece. When light-gauge metal is encountered, the wall can be opened up with hand tools (i.e., the blade of the ax).

Techniques for Breaching Floors

Cutting Wood Floors with a Power Saw

The rotary saw with a 12-inch-diameter carbide-tipped blade will cut a maximum depth of 4 inches, **Figure 17-55**. This should be sufficient to cut through most floors in one cut. The hole must be of sufficient size and proper shape. A rectangle,

Figure 17-55 The rotary saw with a 12-inch-diameter carbide-tipped blade will cut a maximum depth of 4 inches.

Figure 17-56 A rectangle, square, or triangle can be cut, followed by the removal of the finished flooring and the subflooring.

square, or triangle can be cut followed by the removal of the finished flooring and the subflooring, **Figure 17-56**. Carpeting and ceramic tiling should be removed before using the power saw.

Cutting Wood Floors with an Ax

When breaching a wood floor using an ax, locate the floor joists and make cuts close to them. Cut along the (parallel) joist and cut the finish floor on a bias (angle). Pull up the finish flooring to expose the subflooring. Cut the subfloor and pull up to expose the area below. Make all cuts on the subfloor first, then pull it up to confine the heat and smoke. Push down any ceiling or other obstructions.

Note these important things to consider when breaching floors:

- Always take precautions to avoid cutting wires, pipes, and conduit when performing breaching operations.
- Maintain the structural stability of the building when opening walls and floors.
- Beware of sparks produced by metal cutting tools.
- Always operate with proper protective equipment including eye, ear, and respiratory protection.

TOOL ASSIGNMENTS

The necessary tools to accomplish the tactics of structural firefighting must be carried in with first on-scene and later arriving units. The timely arrival of primary tools is as important as the placement and operation of the first hoseline and cannot be left to chance.

Tool assignments are based on the occupancy and construction of the building (i.e., multiple dwelling, private dwelling, commercial, etc.), position or task assigned, and department standard operating procedures or policies.

Lessons Learned

Forcible entry is a key tactic in structural firefighting and emergency operations, and firefighters must understand the tools, equipment, and methods used for forcible entry. As with all other fireground tactics, teamwork is an essential element. Failure to conduct effective forcible entry quickly may result in delayed search and rescue operations and unnecessary fire spread. In addition, with the renewed emphasis on rapid intervention teams, the art and skill of forcible entry techniques is a basic element for these operations.

Firefighters must continually size up buildings for firefighting and rescue operations, including how to force entry into a building.

KEY TERMS

Bolt Throw The distance the bolt of a lock travels into the jamb or strike plate. Usually ½ to 1½ inches.

Cutting Tools The group of tools used to cut through or around materials.

Flush or **Slab Doors** Doors that are flat or have a smooth surface and may be of either hollow-core or solid-core construction.

Forcible Entry The fire scene task of gaining entry to a building or secured area by disabling, breaking, or going around locking and security devices.

Glazing The glass or other clear material portion of the window that allows light to enter.

Halligan Tool From the prying group, a 30-inch forged steel tool with three primary parts: the adz end, the pike end, and the fork end.

Hollow-Core Door Any door that is not solid, usually with some type of filler material between face panels.

Hook A tool with a 32-inch to 12-foot handle with a pike and hook on one end. Used for pulling ceilings or separating other materials. Also known as a *pike pole.*

Irons The combination of a Halligan tool and flat-head ax or maul.

Jamb The mounting frame for a door.

Laminated Glass Glass composed of two or more sheets of glass with a plastic sheet between them. The purpose of the plastic sheet is to hold the glass together if broken, thus reducing the hazard of flying glass.

Ledge Door Door built with solid material, usually individual boards, common in barns and warehouses.

Locking Devices A mechanical device or mechanism used to secure a door or window.

Mounting Hardware Hinges, tracks, or other means of attaching a door to the frame or jamb.

Panel Doors Doors with a solid stile and rails with panels made of wood or glass or other materials.

Pike Pole See **hook.**

Prying Tools The group of tools used to separate objects by means of a mechanical advantage.

Pulling Tools The group of tools used to pull away materials.

Rabbeted A door stop that is cut (rabbeted) into the door frame. On metal door frames the stop is an integral part of the frame.

Slab Door See **flush** or **slab door.**

Solid-Core Doors Doors made of solid material such as wood or having a core of solid material between face panels.

Strike Plate The metal piece attached to a door jamb into which the lock bolt slides. Also called a *strike* or *striker.*

Striking Tools The group of tools designed to deliver impact forces to break locks or drive another tool.

Tempered Glass Plate glass that has been heat treated to increase its strength.

Through-the-Lock Method A method of forcible entry in which the lock cylinder is removed by unscrewing or pulling and the internal lock mechanism is operated to open a door. Also, the family of tools used to perform this operation.

Wire Glass Glass with a wire mesh embedded between two or more layers to give increased fire resistance.

REVIEW QUESTIONS

1. Choose an engine or truck company in a fire department and identify five forcible entry tools and describe their use.
2. List the inspection and maintenance procedures for five forcible entry tools.
3. List three different types of doors and describe a method of forcible entry for each.
4. List five types of locks and describe their operation.
5. List the steps for the three types of conventional forcible entry.
6. Demonstrate conventional forcible entry on a variety of doors.
7. Describe or demonstrate the through-the-lock forcible entry method.
8. List four construction features of windows and methods of gaining entry.
9. Describe three considerations when breaching walls.

CHAPTER 18

VENTILATION

Frank J. Miale, Retired Battalion Chief, New York City Fire Department and Lake Carmel Volunteer Fire Department

OUTLINE

- Objectives
- Introduction
- Principles, Advantages, and Effects of Ventilation
- Heat, Smoke, and Toxic Gases
- Considerations for Proper Ventilation
- Fire and Its By-Products
- Flashover
- Backdraft (Smoke Explosion)
- Rollover
- What Needs to Be Vented?
- Air Movement
- Types of Ventilation
- Mechanics of Ventilation
- Ventilation Techniques
- Roof Ventilation
- Safety Considerations
- Obstacles to Ventilation
- Factors Affecting Ventilation
- Lessons Learned
- Key Terms
- Review Questions
- Additional Resources

STREET STORY

On July 4, 1998, we responded as the second truck to a fire on East 4th Street in New York City. As we arrived fire was blowing out five windows on the third floor of a four-story apartment building. The chief ordered me and my members to get above the fire and search the top-floor apartments for trapped occupants. As we arrived on the third floor, the first-due truck was forcing the door to the fire apartment. Thick black smoke was pouring out of the apartment. I told the first-due engine and truck that we were going to the floor above.

As we got to the half landing between the third and fourth floors, I ordered a member of my forcible entry team to totally take out the window on this half landing, knowing that smoke from the fire apartment would escape from this horizontal opening. This would buy us time to search the top floor before the roof bulkhead was opened.

As we were searching the apartments above I heard the familiar "roof bulkheads opened" transmitted from the roof firefighter over my handi talki. I knew that mushrooming would be prevented by this act and that we had a couple of minutes to search. Then I heard a saw operating on the roof and knew additional vertical ventilation was being performed. At this time, however, fire was autoexposing from the third-floor apartments into the fourth-floor apartments we were searching, forcing us to retreat to the hallway, but not before we were able to complete a thorough primary search. Although the fire extended into six apartments and the cockloft, due to the proper immediate horizontal and vertical ventilation, we were given enough time to search the top floor and see that all occupants were removed to safety without injury.

A few months later we responded to a fire in a two-story private dwelling with reports of persons trapped. As we arrived I noticed no fire, but there were black, discolored windows on the second floor. As we entered the fire apartment to search we found there was a tremendous heat condition, but no visible fire. We recognized that this fire was in the smoldering stage. My outside ventilation firefighter asked if he should take out the windows from the exterior. I immediately radioed to all members not to take out any glass until vertical ventilation was performed and the area was cooled down below the gases' explosive range.

Two different fires, two different ventilation strategies. In the first case, immediate horizontal ventilation provided us time to search apartments above the fire for occupants, while in the second case premature horizontal ventilation could possibly have caused a backdraft, seriously injuring us or worse. Proper ventilation is of the utmost importance. Learning the right ventilation operations can truly make a difference. Learn it well.

—Street Story by Mike Gala, Lieutenant, Ladder 148, FDNY

OBJECTIVES

After completing this chapter, the reader should be able to:

- Understand ventilation as a fire service tool.
- Know the principles, advantages, and effects of ventilation.
- Know the origins and effects of heat, smoke, and toxic gases.
- Identify the considerations for proper ventilation.
- Know the effects of fire and its by-products.
- Differentiate between flashover, backdraft (smoke explosion), and rollover.
- Recognize the effects of air movement in ventilation.
- Identify the types of ventilation.
- Identify the mechanics of ventilation.
- Describe ventilation techniques.
- Describe the need for roof ventilation.
- Identify safety considerations when venting operations are in progress.
- Describe the obstacles to ventilation.
- Identify the factors affecting ventilation.

INTRODUCTION

Ventilation can be defined as the planned, methodical, and systematic removal of pressure, heat, smoke, gases, and, in some cases, even flame from an enclosed area through predetermined paths. Ventilation is an essential part of the tactical and strategic objective of modern fire extinguishment. Incorrectly, in some cases ventilation is not employed until all other firefighting attacks have failed. Ventilation is a very complex subject area with many facets. Additional information can be found in books that have been written solely on the subject of ventilation.

PRINCIPLES, ADVANTAGES, AND EFFECTS OF VENTILATION

Ventilation is the relief of the products of combustion from an enclosed area. Combustion of organic material produces heat, smoke, pressure, and fire gases that quickly fill up an enclosed structure. That structure could be a one-room shed, a ten-room private dwelling, or a twenty-story building. In every case, the need to relieve the structure of these products of combustion is a very essential part of the fire suppression effort. The practice provides benefits that go beyond merely suppression to extinguish the fire.

First, by relieving the structure of heat through channeling it into the atmosphere, the fire is deprived of the ability to heat up other parts of the structure. Fire burns when the gases of a combustible substance are liberated. This liberation occurs when heat from an existing fire is applied to the unburned material. The unburned material heats up, liberates gases, and then ignites, permitting the fire to spread. Using ventilation, the heat is exhausted and dissipated into the atmosphere where its ability to spread fire through the structure is reduced.

Secondly, ventilation channels smoke out of the structure. Smoke is a combination of material, mostly unburned hydrocarbons that have a tarry consistency. Because most of these substances are microscopic in size, they are very light and can stay suspended in air. The heat emitted by the fire carries the smoke to all parts of the structure. Because smoke is made up of only partially burned solid bits of microscopic material, it obscures vision, and the more dense the smoke is, the more it will obscure. Light smoke can be like a fog in which shapes can be seen from several feet away. A heavy smoke condition can obscure light so completely that even a powerful light is rendered totally ineffective. Not only does lack of vision seriously hamper firefighting operations, but it also prevents victims from escaping.

Additionally, the unburned tarry hydrocarbons contain substances that irritate the eyes. In the natural function of the eye's protective action, anytime a foreign material is introduced, the tear ducts attempt to flush the eye with body fluids (tears) to rid it of the irritation. These tears blur vision so badly that it is nearly impossible to see.

Smoke contains many other products of combustion that are deadly substances. Many harmful compounds are mixed in with smoke, and a person that is exposed to this material will suffer ill and potentially lethal effects, **Table 18-1**. When people fall victim to smoke or its components, they usually fall into unconsciousness as life slowly ebbs away. The removal of smoke, heat, and toxic gases will add survival time to a potential victim who is unconscious, increasing the chance of successful rescue.

Gases Produced by Fire	
Carbon monoxide	Takes the place of oxygen in the blood.
Carbon dioxide	Overstimulates the rate of breathing.
Hydrogen sulfide	Causes respiratory paralysis.
Sulfur dioxide	Extremely irritating to eyes and respiratory tract.
Ammonia	Extremely irritating to eyes, nose, throat, and lungs.
Hydrogen cyanide	Highly toxic and used commercially as a vermin fumigant.
Hydrogen chloride	Can become hydrochloric acid in mucous membranes.
Nitrogen dioxide	Causes respiratory distress in a delayed reaction.
Acrolein, phosgene	Gases found in certain kinds of fires; are lethal in small doses.

TABLE 18-1

HEAT, SMOKE, AND TOXIC GASES

When fire burns, air heats, expands, becomes lighter, and rises. It also begins to exert pressure on anything that surrounds it. The rising heated air becomes a means by which fire is spread or communicated to surrounding materials by convection, and, to some extent, radiation from the upper levels (see Chapter 4, Fire Behavior). Fire can also spread into other places within a structure by the pressure that is created. Pressure will take the path of least resistance within a confined area as it tries to become equalized. If that path leads to more combustible material, the heated air will spread its heat to that substance and permit it to ignite, thereby spreading the fire. Fire gases consist of many products of combustion that can contribute to the spread of fire, render a living being unconscious leading to death, or contribute the necessary ingredients for an explosion. Many substances found as the by-products of combustion are deadly.

Because of the desire to conserve energy, structures being built today and those being renovated are outfitted with heavy insulation and tight, weatherproof seams. These features make ventilation even more important because the high heat generated by the inability of the heat to escape is turned back into the compartment. These tight construction practices lead to hotter fires, early failure of structural components, and greater incidences of flashover and backdraft. These phenomena are examined in greater detail later in this chapter.

Last, in today's climate-controlled buildings equipped with windows that are unopenable or even windowless, reliance on ventilation is greater in order to move the products of combustion out of the structure. An overview of mechanical ventilation is given later.

CONSIDERATIONS FOR PROPER VENTILATION

In order for ventilation to assist in the fire extinguishment operation, the firefighter must first understand the behavior of fire gases, which include smoke, in a building. Because heat is lighter than air, it tends to rise. Smoke, when mixed with heat, also rises. As the smoke rises, it collects under any vertical obstruction and mushrooms in all directions when it meets that obstruction. Gradually, the smoke and heat fills the structure starting from the highest point in the structure and "banks down" until the entire structure is filled, **Figure 18-1**. To fully understand the concept, imagine a building. Then, pull it out of the ground and turn it upside down. While it is upside down, begin filling the cellar with water. Any channel that the water finds as it drains out of the inverted cellar into the inverted attic will be the same channels that smoke and heat will use to "drain up" out of that cellar, **Figure 18-2**, and, just as water will leak out of small holes in the windows or siding in the inverted structure, so too will smoke in the structure that is right side up. So even though smoke is coming out of the lower floors of a structure, it is only a portion of the smoke and heat that is collecting in the upper part of the structure.

Figure 18-1 Heat, smoke, and fire will follow the path of least resistance and find their way through any available opening.

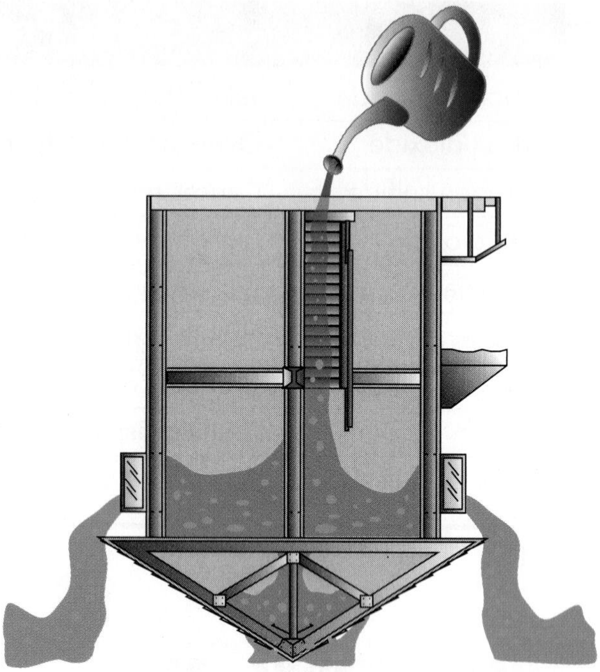

Figure 18-2 Just as water poured into the cellar of an inverted building will fill up any basins and spill over into the next void until it fills, smoke, heat, and fire will behave in a similar fashion in a structure that is right side up.

> **Note** If a large volume of smoke is coming from the windows, it is a good indicator that the smoke and heat are meeting resistance in the attempt to move vertically.

Vertical ventilation is defined as the removal of gases and smoke through vertical channels. This will prevent fire extension by convection from occurring at a remote part of the building. The opening of the structure at the top will permit the fire gases and smoke to exit the building just as punching a hole in the sample inverted building will permit water to drain out of the attic and prevent the building from filling with water.

Another type of ventilation that is often associated with fire suppression is **horizontal ventilation**, which is the channeling of smoke and heat out of the structure through horizontal openings such as windows and doors. There is a distinct difference between the needs and results of vertical and horizontal ventilation. Horizontal ventilation permits the fire's by-products to be pushed out of the structure by the advancement of the fire suppression crews. In areas remote from the fire, it will permit a reduction of smoke, and, to a lesser degree, built-up heat that will aid in a search for victims and any fire extension. Without horizontal ventilation in front of an advancing hose team, the heat, smoke, and now a new element, steam, have nowhere to go. With no outlet for the water to push the smoke, steam, and heat out of, it will be pushed over the hose team at the ceiling level. When water is applied to a fire, the fire is extinguished by the removal of heat from the fire pyramid. When raised to 212°F (100°C) water turns to steam. It takes a tremendous amount of heat to turn water into steam, and, when it does, it expands 1,700 times. That means that when 1 gallon of water is heated to the proper level, it creates 1,700 gallons of steam. If the hose team is about to apply water to a fire in an enclosed room without a ventilation hole opposite their position, the pressure created will push back out through the same opening from which the hose team entered. The fire, heat, smoke, and fire gases will push right over the hose team and superheat the area behind the nozzle team, surrounding them with the heat of the fire.

By opening a channel for the products of combustion to exit as the nozzle team moves in, the heated smoke and steam will take the path of least resistance and vent from the fire room. It is very important for the hose team to move in with a wall of water in front of them and push the fire out the window or another opening, **Figure 18-3**.

Horizontal ventilation requires some forethought before performance. If performed at the wrong time

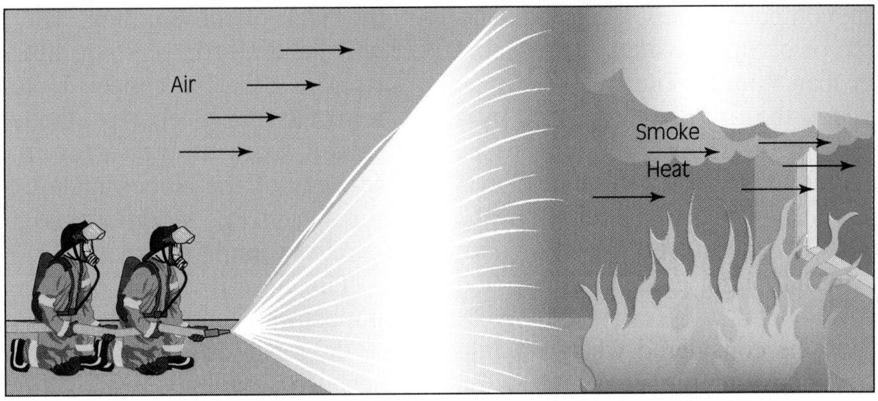

Figure 18-3 Heat, smoke, and fire will always take the path of least resistance. When pressurized by the air created by an advancing hose team, the products of combustion will seek a path that will equalize that pressure. An opening on the opposite side of the advancing nozzle team is essential.

or at the wrong place, it can accelerate fire spread. For example, if the venting is about to be performed in a part of the structure that is two rooms away from the fire room, the air currents that are created could pull the fire into a room that might otherwise not have been involved, **Figure 18-4**, but this will not occur all the time. There are many reasons why this might or might not happen. Some of the factors that will influence the air currents in a structure fire are vertical vent openings, horizontal openings, outside wind direction, the direction the hose attack team is using in relation to the fire, and the room being vented.

When properly performed, the ventilation operation can be as critical as the nozzle team applying water to the fire. It is important to note that in the confusion of a fireground operation, it may not be possible to follow a manual's every line and sentence. The firefighter must learn the conditions that cause events and use every experience to expand that knowledge.

Streetsmart Tip When access to a horizontal ventilation point is impossible, a possible recourse is to use the hose stream to break the windows from inside before mounting the attack.

In rare situations, no ventilation at all is performed at the fire extinguishment operation. Some occupancies are protected by systems that discharge a fire-inhibiting gas such as halon into the room. In this type of occupancy, an identifying placard usually warns that this gas is present. A system such as this might be used where delicate materials are found such as an

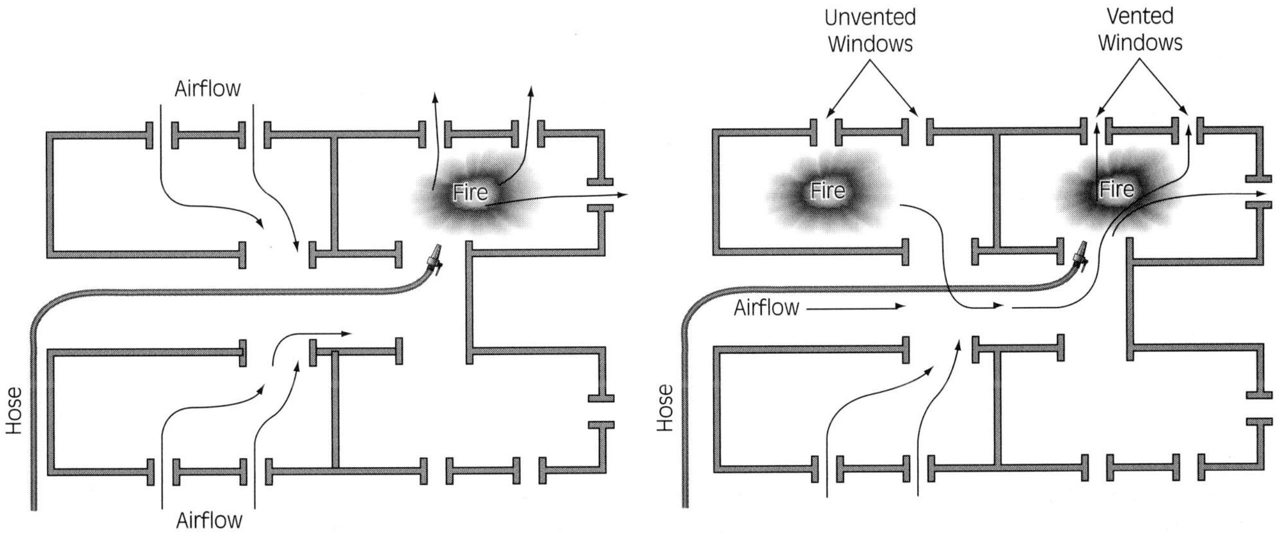

Figure 18-4 Air movement is created by water application. Openings in back of the nozzle team will create airflow from behind in the direction of the hose team. It can be a source of fresh cool air, or it can pull fire to the nozzle from behind. Indiscriminate ventilation can be a liability. Careful assessment and proper timing are important.

archiver, a museum, computer rooms, biological labs, and so on. This type of occupancy might suffer more damage from water than from the fire. With this type of extinguishing agent, the fire is extinguished by a chemical reaction with the flame production (see Chapter 4). In a structure protected by this type of system, ventilation must be avoided until instructed by the incident commander's orders. Venting this room would permit the fire-inhibiting gases to escape and permit reignition.

Many factors must be considered when venting. Some of them are access to the vent site, wind direction, weather conditions, exposures, the material burning, the height of the building, the potential for fire spread, and the escape route. Not every factor will have the same impact at every fire.

Access to the vent site can be tricky. Access to a window to be vented might be impeded by the presence of electrical wires, sloping ground, or sheer height. A window off a porch roof that needs to be vented might require the firefighter to pass windows on that porch roof that are already venting fire. This condition might well prevent an escape route and a second approach might be necessary.

Ventilation duties in the rear of the building might be difficult or impossible because of the presence of locked gates or guard animals. **Guard dogs** are trained for different levels of protection. **Watch dogs** will bark and make a great commotion. Their purpose is to scare an intruder. Guard dogs will bark and assault an intruder. Their function is to warn and injure. Attack dogs are trained not to bark and will attack an intruder with ferocious intent.

A hose stream from a hoseline or even a water extinguisher will send a guard dog running. In all cases, the best approach is to guide the dog into some kind of confined area and leave it there. Occasionally, police might have to be called to deal with the dogs by using tranquilizers or by destroying them.

The presence of electric supply lines to the structure can impede access to a vent site. Unless specifically designed, all fire service ladders, including those made of wood, are electrically conductive, and that presents a great risk when raising or positioning them in the vicinity of electricity. When the window needing ventilation is close to electric power lines, extra care must be employed. Live wires might have had their insulation burned off and energized surrounding metallic substances such as aluminum siding. A ladder resting against that energized siding can result in electrocution. When sizing up the approach, consideration must include the presence of overhead wires and, when necessary, the use of an alternate approach.

Sloping ground will significantly affect the placement of ground ladders. This is especially true if the street or road **frontage** is at a different level than the back of the building. The sides of the building will generally have some kind of slope so that the front and the back meet. Unless the landscaping is terraced with level steps, the ladders will have to be raised from uneven ground. Depending on the severity of the slope, ladders might be unavailable for use in this situation and ventilation of certain windows might be delayed.

The height of the building will also have an impact on access to a window. If the window is out of the reach of ground ladders and not reachable by aerial or tower ladders, venting might have to be performed by using adjoining windows or from windows above or below.

Wind is a factor that can dramatically alter the ventilation. The removal of a window that faces the wind could cause a complete reversal of the air current flow. If the wind is strong enough, the attack hose team can be faced with a blowtorch or wind tunnel effect that will push the fire at them. There are just too many variables to clearly establish a "go/no go" set of rules about whether to leave the glass intact or to remove it under these circumstances. Common sense and experience are factors that will help to solve this judgment call.

Humid, rainy, or foggy weather tends to cool smoke-laden, heated air and prevents the smoke from **lifting** out of the building. If the weather is humid or rainy, the smoke from the chimney does not rise in a column but tends to spread out horizontally. On a clear day, the smoke rises in a crisp column into the upper atmosphere, where it dissipates. Other atmospheric conditions such as pressure or temperature inversions alter natural air movement, affecting ventilation and hindering operations.

Chapter 4 on fire behavior explained how combustion is a process that creates its own life. If the heat produced is not large enough or dissipated quickly, the fire self-extinguishes. Through the use of ventilation, firefighters can remove some of the fire's ability to extend. If the products of combustion, especially heat and smoke, are confined in a structure, two of the three essential ingredients (heat and fuel) are present and ready to be consumed. By venting, those ingredients are removed harmlessly into the atmosphere and permitted to burn where they will do no harm to advancing firefighters.

FIRE AND ITS BY-PRODUCTS

During the pure combustion process, energy is released from an exothermic reaction as heat and light (flame), **Figure 18-5A**. Heat without smoke

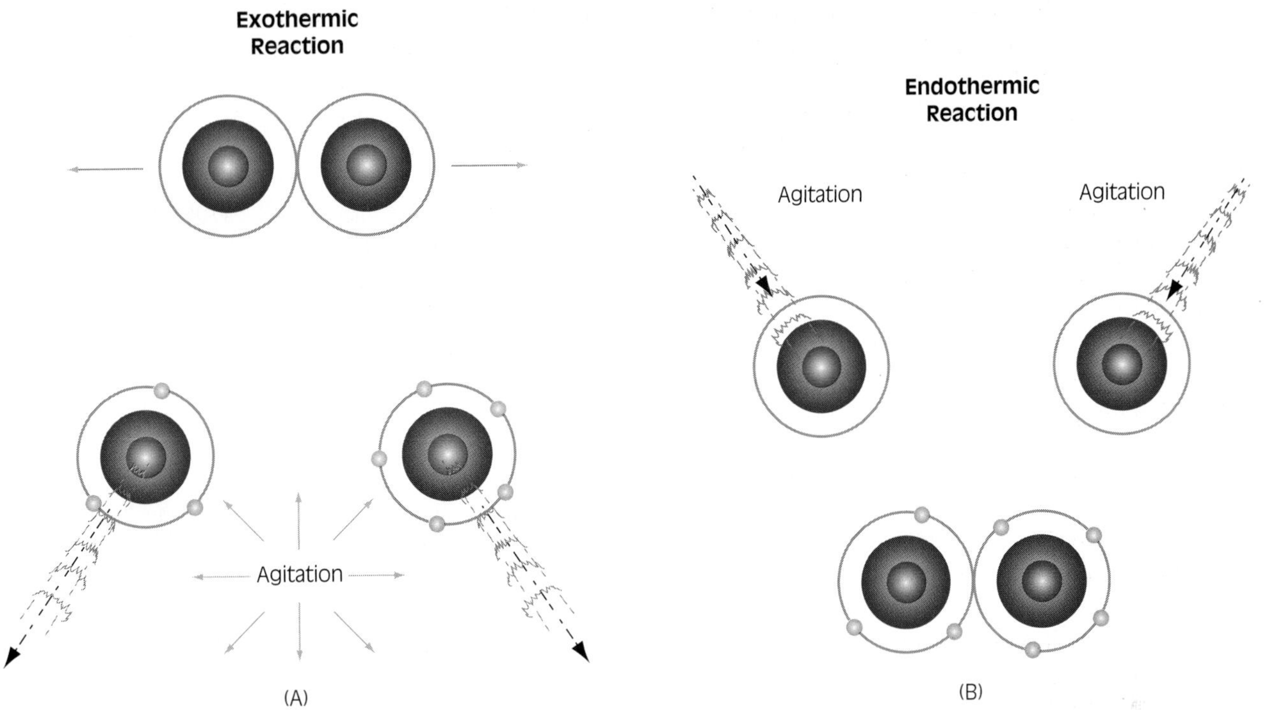

Figure 18-5 (A) In an exothermic reaction, the bond broken between two atoms or molecules creates vibration or agitation, and heat is created by that action. There are times when bonds are broken, heat is created, and the atoms or molecules find stability in another combination minus that bond. In that case, a chemical reaction releases heat, but it stabilizes as another substance. (B) Heat is present when atoms or molecules are agitated or vibrating. Very simply stated, if the loose ends of the atoms or molecules attach to another and become balanced, the agitation ceases and the heat that was being generated is absorbed by the newly created bond.

can be observed as wavy lines emanating from the fire. Other gases that are produced by the combustion process are not visible. The reuniting of molecules to form new substances is part of the endothermic reaction, **Figure 18-5B**. This process actually takes heat from its surroundings and uses it to bond the loose ends of other molecules.

Substances created by fire can be very caustic to a human. The body reacts to the invasion of foreign substances by attempting to rid itself of them. Unburned particulate matter that settles into the mucous membranes of the eye and nasal passages results in a flow of liquid from the tear ducts and nasal passages to naturally flush out the foreign material. When inhaled into the lungs, the body attempts to isolate the foreign material by coating it with phlegm. Then, the body attempts to take the collected phlegm and expel the foreign material by coughing.

When properly performed, ventilation will remove some of the harmful agents associated with the by-products of combustion. Additionally, the presence of certain chemicals and chemical compounds serves to accelerate the human respiration rate. The human body uses two mechanisms to regulate the rate of breathing. One is the level of oxygen. If the oxygen level diminishes, the rate of breathing increases. The oxygen-reading capability of the brain is actually the alternate monitor that regulates the breathing rate. The primary monitor that regulates breathing is the level of carbon dioxide in the blood. Under normal conditions, the body takes in oxygen, uses it, and produces carbon dioxide as a by-product of energy usage by the various muscles of the body. When the brain sensors read an increasing level of carbon dioxide, the body speeds up the breathing rate to increase the oxygen/carbon dioxide exchange in the lungs.

Because the by-product of combustion is carbon dioxide, the body inhales it along with the other poisonous gases produced. As the brain senses the increasing levels of carbon dioxide, it increases the breathing rate to expel the building level of CO_2. In the case of carbon monoxide, another product of combustion, which the blood cells absorb 210 times faster than oxygen, the result is a faster and faster rate of poison absorption.

Firefighter Fact The faster the structure is opened and ventilated, the quicker harmful gases are replaced with fresh air, and, consequently, the better the working environment for the firefighters.

Since all firefighters should be wearing self-contained breathing apparatus, one might ask if ventilation is important since a fresh air supply is available to the wearer. The answer is an emphatic *yes!* Ventilation increases the survival time of a trapped victim who might be overcome or a firefighter who has become disoriented and run out of air.

Ventilation has many other benefits beyond limiting the inhalation of contaminated air. Removal of smoke improves visibility. Victims can be discovered more quickly, and danger from a hostile environment is reduced. The chance of injury will be reduced. Dangers such as holes in floors are easier to spot, and avenues of fire extension become more obvious. With better visibility, the use of tools becomes safer.

If heat is not permitted to linger on a material long enough, it will be unable to liberate the material's combustible gases. By venting the enclosure, the heat level is kept from becoming capable of producing these phenomena:

Flashover: Everything in a confined area ignites at almost the same time.

Backdraft (smoke explosion): Unburned smoke is heated in the absence of oxygen, and, when oxygen is introduced, produces an explosive force.

Rollover: Fire begins to ignite smoke overhead in "fingers of fire" that reach out and begin to consume fuel in the gaseous state.

The mechanics associated with each of these phenomena and prevention techniques are discussed as each relates to ventilation.

FLASHOVER

Light, smoke, and heat are liberated as part of the combustion process. As the trapped heat collects at the ceiling level, its cumulative temperature increases. Across the upper area of the room, heat begins to radiate downward, heating the contents of the room. When the overall temperature reaches the **ignition point** of another substance in the room, a new chain reaction combustion site occurs and additional heat is added beyond the initial source of fire. As each item in the room follows suit, more and more heat is created and more and more items ignite. In a very short time, the entire room and all of its contents are on fire. This happens very rapidly. Slow-motion photography shows that it is not an instantaneous occurrence but a very rapid fire spread. It is important to know the mechanics of a flashover in order to recognize its development. It is even more important to know what course of action to take before, during, and after a flashover.

> **Firefighter Fact** At best, the survival time of a firefighter in bunker gear and breathing apparatus, fully encapsulated with gloves, hood, and helmet flaps down, is estimated to be between ten and fifteen seconds.

The best survival skill here is recognition and avoidance.

BACKDRAFT (SMOKE EXPLOSION)

Backdraft, also sometimes called a smoke explosion, is the rapid ignition of smoke. Sometimes confused with a flashover, it is very different.

When fire burns organic materials in the presence of oxygen, carbonaceous materials are transformed into carbon, carbon dioxide, and water. When there is a shortage of oxygen, incomplete combustion occurs, and, instead of forming carbon dioxide (CO_2), a less stable compound called carbon monoxide (CO) is formed, **Figure 18-6**. During this process, two of the three classic substances for a complete burn are present: fuel and heat.

Incomplete combustion occurs as the oxygen level decreases. Visually, this can be observed by the amount of yellow flame present. With adequate oxygen, the flame is bright yellow. As the oxygen level decreases, the yellow becomes bright orange, then dull orange, then red. As the flame approaches red, the level of oxygen is getting so low that if continued the fire would self-extinguish for lack of oxygen. As discussed in Chapter 4 on fire behavior, carbon monoxide has a very wide range of flammability. When oxygen is introduced into the mix, combustion will again occur to complete the process by converting the CO into CO_2. Under the proper conditions, this can occur with a violent reaction.

Some form of backdraft occurs at almost every fire that remains unvented prior to the arrival of firefighters. Its violence and damage will be directly commensurate with the distance from where the oxygen is introduced and the location of the flames, **Figure 18-7**.

As the fire consumes greater amounts of the oxygen in the room, the production of carbon monoxide increases. With the heat, pressure builds in the confined space. Smoke can be seen puffing from openings, and black smoke condenses on the surface of any

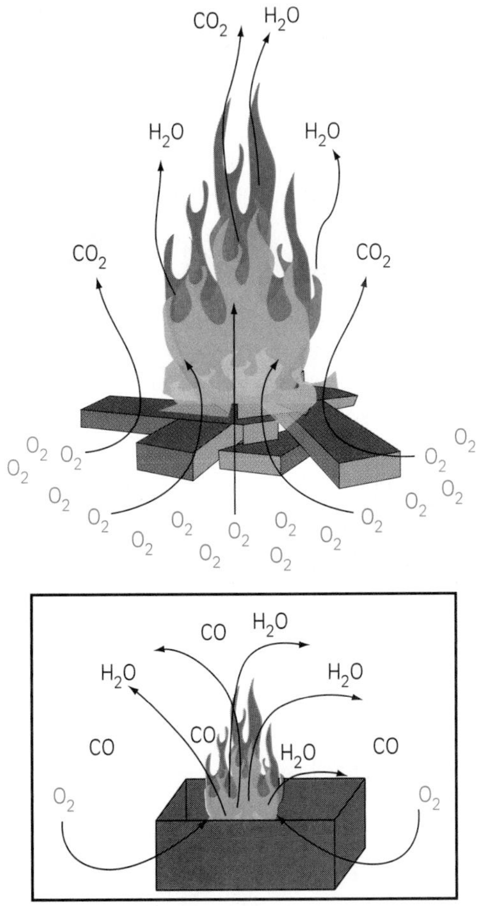

Figure 18-6 When combustion occurs, oxygen is combined with an organic material, and, through a series of bond breaking and recombination processes, carbon dioxide and water are driven off. When oxygen is lacking, instead of two oxygen atoms combining with a carbon atom to form carbon dioxide (*dioxide* meaning "two oxygens"), the carbon atom combines with only one oxygen atom, forming carbon monoxide (*monoxide* meaning "one oxygen"). In the absence of adequate oxygen, carbon monoxide is generated instead of carbon dioxide.

cooler surface, such as glass or a wall. Initially the entire room is filled with heat, carbon monoxide, and other fire gases. When an opening occurs, a billow of smoke escapes as the pressure from inside attempts to equalize. Then, what is described as "sucking" occurs where, once a large majority of the pressure in the space is released, cool outside air is introduced into the room. The cooler air causes the volume of air to contract and pull air in to attempt to equalize again. Being cool, it stays low to the floor and at some **thermal level** a mix of highly concentrated CO is mixed with the fresh, 21 percent oxygen-laden air. As the cool air snakes its way to the source of the fire to replace air that is being heated and driven upward, what could be described as a tunnel of fresh air finds its way to the fire, **Figure 18-8**. An observer might even see the smoke lift off the floor and just before ignition, actually see the fire burning across the expanse of the floor below the smoke level. In the meantime, in terms of the oxygen level on the floor, with each few inches the smoke rises, the CO and O_2 mix more. This mixture of gases at some point will permit the concentration of CO to be in the range of 14.5 to 74 percent. Once that concentration reaches the flame, the three components for combustion are present and primed for ignition. The burn begins at the seat of the fire and, like the fuse on a firecracker, burns back to the opening in the space along the perimeter of the tunnel of oxygen-laden air. This occurs so rapidly that it seems like an explosion. The burn causes the heat, which in turn causes rapid expansion of the surrounding air with such rapidity and force that it can blow out windows, knock down walls, and hurl firefighters out openings and across the street. The degree of force is dependent on several factors.

Signs of a Potential Backdraft

- Smoke-stained windows
- Puffing of smoke at seams and cracks of windows and doors
- Smoke pushing out under pressure
- No visible flame or very dull red flame in the depth of the smoke
- Heavy black smoke
- Tightly sealed building
- Large, open area structure (supermarket, bowling alley, department store)
- Can also be a large, open void (cockloft, between-space of hanging ceiling)
- Extreme heat

Consider these two backdraft conditions and the difference in the results:

- *Short distance to opening.* The atmosphere is laden with carbon monoxide. As the fresh air works in toward the thermal plume, it brings fresh oxygen. The outer perimeter mixes with the CO and achieves the proper flammability range. When the seat of the fire is reached, the tunnel perimeter ignites and follows the path to the opening. Since it is limited and relatively short, the results are unspectacular and appear merely as the fire flaring up.
- *Long distance to opening.* The same initial condition exists. However, in this case the amount of the proper concentration from the opening to the seat of the fire is much greater. When the ignition occurs, it travels a greater distance and amasses much greater force. Although technically it is moving in a chain reaction fashion, the almost

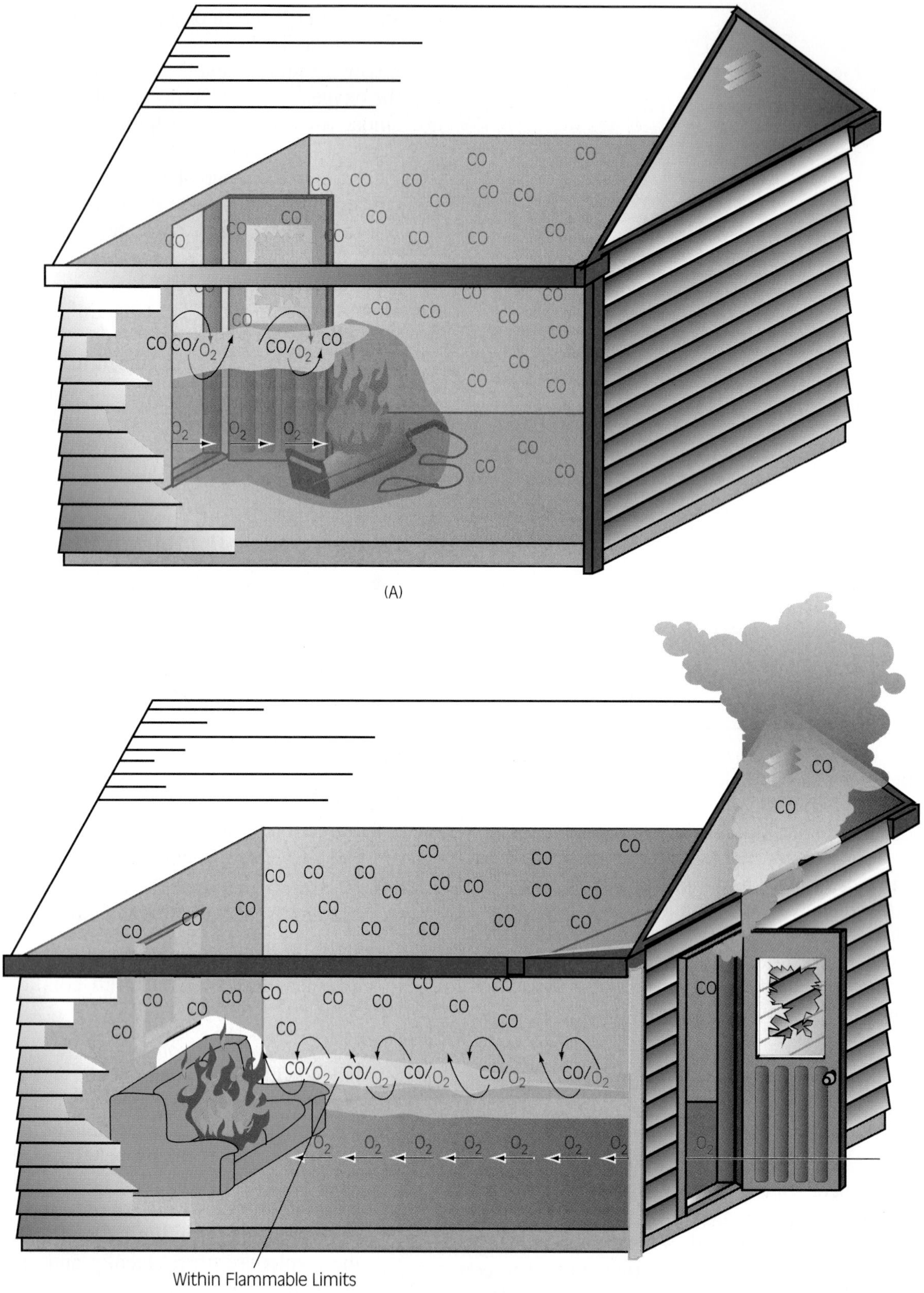

Figure 18-7 (A) When the distance between the opening and the source of ignition is short, there is little opportunity for the gases to circulate and form a mixture that is within carbon monoxide's flammable limits. When the mixture does ignite, it appears more like a flame flare-up than an explosion-like ignition. (B) A large distance between the opening and the location of the fire will permit a greater mixing of carbon monoxide and cool oxygen-laden air. The greater distance will permit the carbon monoxide concentration to be in its flammable limits over a greater area before it reaches an ignition source. A larger instantaneous ignition will result with great force. The greater the distance, the greater will be the resultant ignition force.

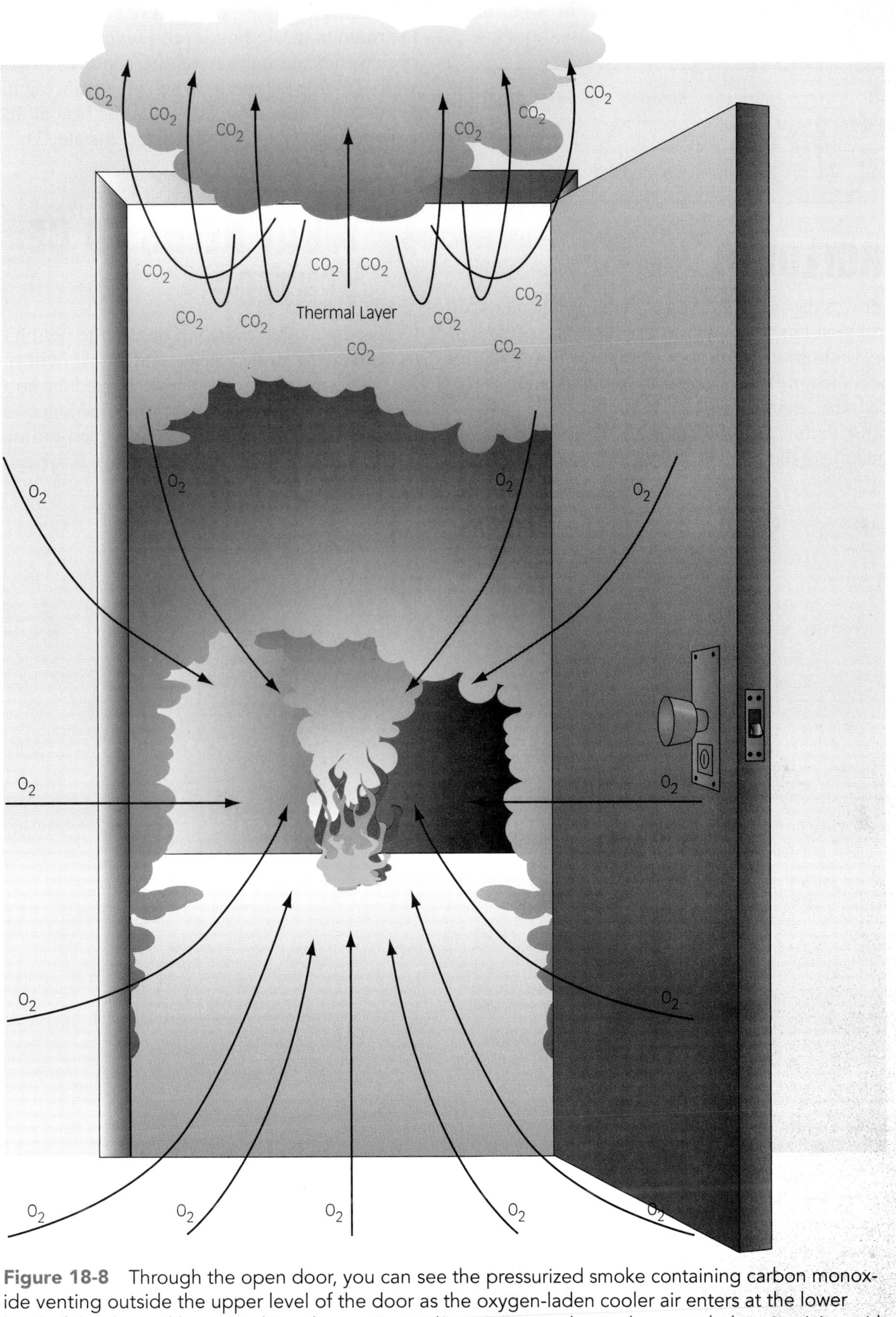

Figure 18-8 Through the open door, you can see the pressurized smoke containing carbon monoxide venting outside the upper level of the door as the oxygen-laden cooler air enters at the lower level of the thermal layer. Looking down a "tunnel" you can see the cool oxygen-laden air mixing with the oxygen-starved carbon monoxide around the perimeter of the tunnel. When it reaches the fire, the mixture will ignite and, in chain reaction fashion, follow the ignitable mixture in the tunnel right out to the opening.

instantaneous chemical reaction is like an explosion.

> **Note** The greater the distance between the smoldering fire and the opening, the greater the explosive mixture involved, and, consequently, the greater the force created.

ROLLOVER

Rollover was described in greater detail in Chapter 4, Fire Behavior, but the subject is briefly revisited here because ventilation is so very important to prevent this phenomenon. When products of combustion are produced, the heat brings them to higher levels, usually accumulating near the ceiling. The heated gases accumulating at the upper levels of the compartments reach their ignition point and begin to spread across the room at the ceiling level. Fingers of flame can be observed reaching across the room followed by a wall of flame behind it. When an advancing hoseline disrupts the upper thermal layer, the heat at that layer is forced down into an unbalanced state. This action can disrupt the rollover phenomenon.

WHAT NEEDS TO BE VENTED?

Unless there is a ventilation opening to permit escape, the expanding heated steam and smoke will roll over the wall of water and drop down behind the hose team, **Figure 18-9**. When ventilation is mentioned, the first image that appears is a fully involved building fire emitting smoke from every opening. It appears to be

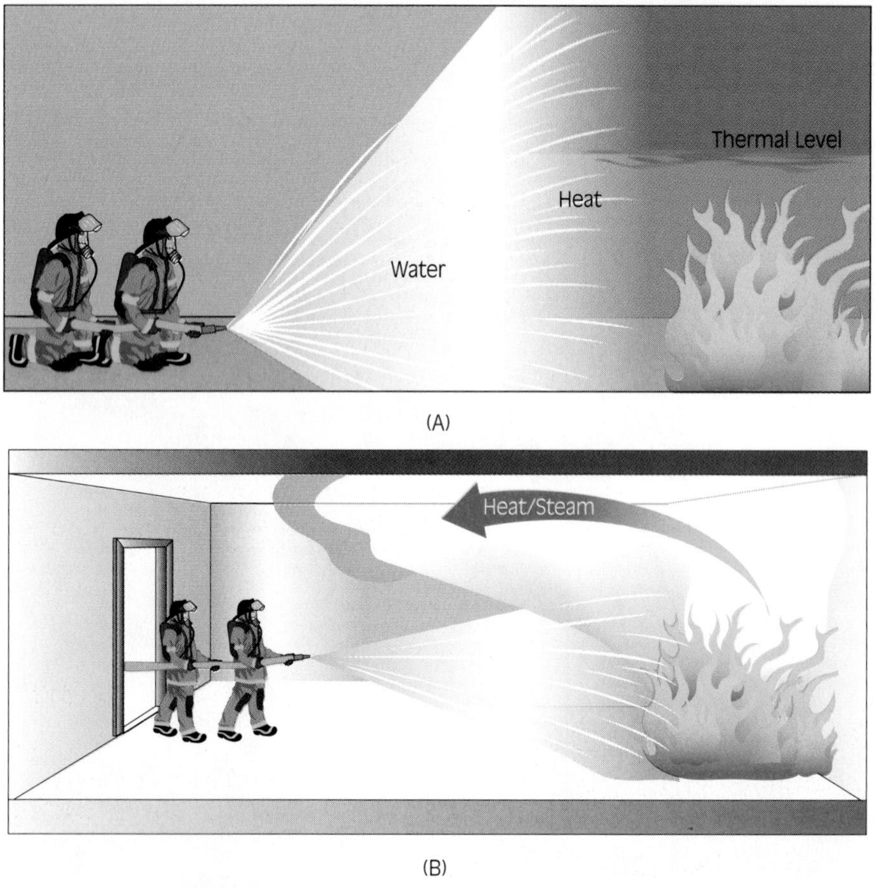

Figure 18-9 (A) Applying water to the upper levels of a thermal layer will cool and disrupt the rollover effect that is apt to occur with the proper conditions. Ventilation is critical when this is done. (B) As a hose team advances into the fire and sprays water in droplet form, it creates a wall of water and disrupts the high-heat thermal layer and cools the upper levels of the compartment. Water absorbs heat as it turns to steam, expanding 1,700 times as it does. If there is no path for the expanding water/steam conversion, it will take the path of least resistance, in this case over the wall of water and the nozzle team. The water movement will then pull any heat from the back of the nozzle team and roll over on top of it.

well vented. However, long before the entire building requires venting, the smaller voids and compartments need to exhaust the increasing pressure and intensifying heat condition. If done in a timely manner, the involvement of the entire building might be avoided.

Voids and Compartments

When examining the subject of ventilation, all compartments must be treated with the same understanding. For example, a residential building is merely a very large compartment with many subcompartments (i.e., apartments). Each subcompartment can be subdivided further (i.e., rooms), then again subdivided (i.e., closets), **Figure 18-10A**. Rounding out the picture, the building might also have eaves, peaks, gables, cocklofts, and other voids, **Figure 18-10B**.

Cocklofts

Between the ceiling of the top floor and the bottom of the roof is a space called the **cockloft**, **Figure 18-10C**,

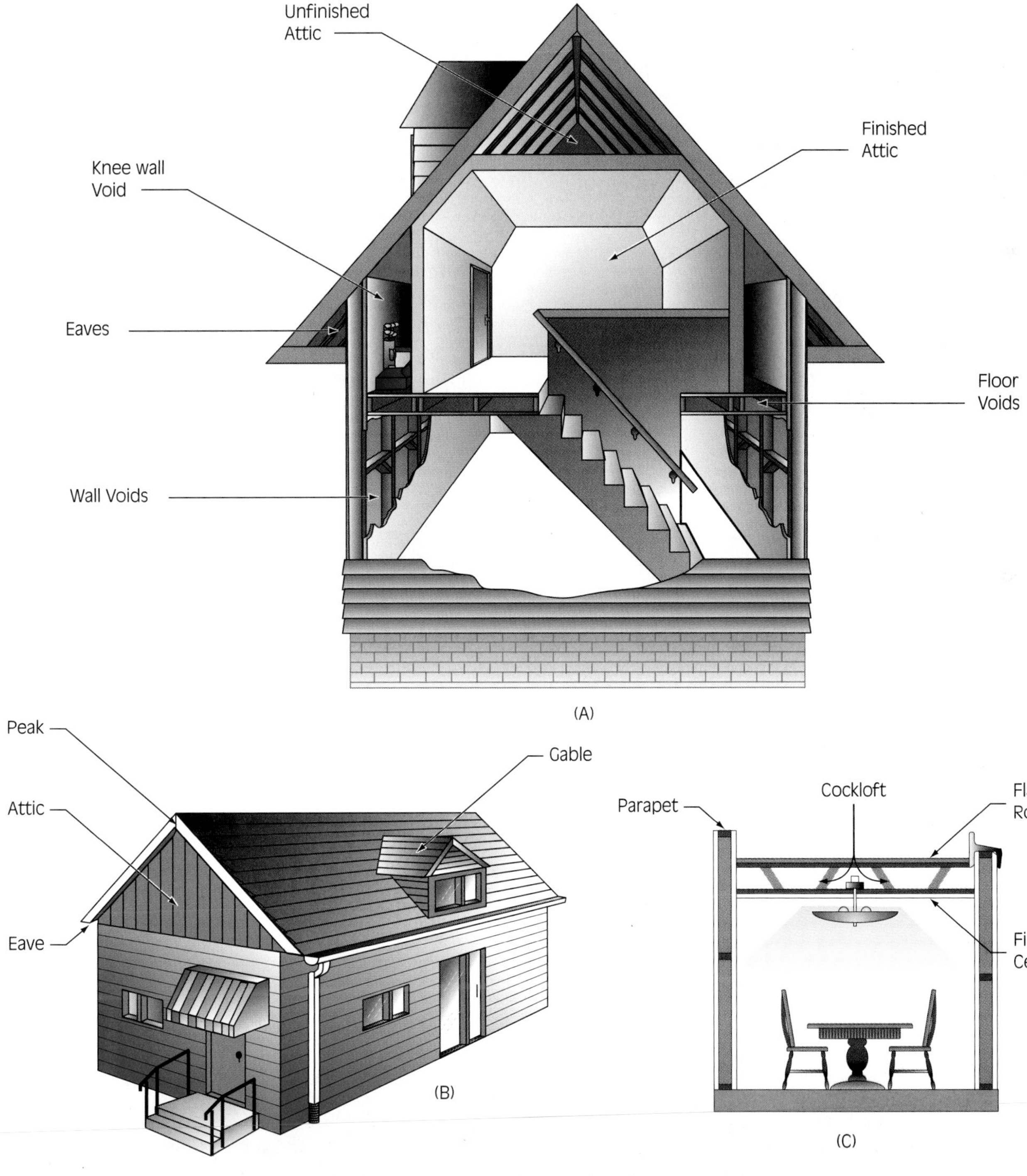

Figure 18-10 Voids in a typical structure that can trap heat and permit fire extension.

which is often the route of major fire extension. Chapter 13 on building construction dealt with this in greater detail. It cannot be overemphasized that this void is a major attack point for a ventilation crew, especially in a top-floor fire or a fire that has extended into that space.

Horizontal and Vertical Voids

Because all heat will follow the path of least resistance, an unobstructed channel could be in the form of a horizontal or vertical void. By following **pipe chases** or electrical wire pathways, heat and fire may extend without being seen. Ventilating these voids by opening them at the highest points can expose any extending fire for extinguishment. In addition, by opening the voids, the trapped heat is diffused into unheated spaces, thus minimizing the chance of heating up a new, uninvolved portion of the building, **Figure 18-11**.

AIR MOVEMENT

Defined and discussed in greater depth in Chapter 4, the attributes of the modes of heat transmission that relate directly to ventilation are revisited here:

Convection. The normal travel path of heat as it follows the convection cycle can be interrupted and challenged by opening a hole in the upper portions of the compartment. Then, instead of traveling up, across, and back down, the heated air will travel up, across, and out of the structure. By providing a path of least resistance in the normal path of heated air, the heat is carried into the atmosphere where it harmlessly dilutes with cooler air.

Conduction. Heat tends to travel along dense material. It can cause remote combustibles hidden from view to ignite. By exposing the heat conductor and ventilating the void where it exists, the trapped heat is provided with another less resistant path, thereby preventing the heat from traveling to uninvolved areas of the structure.

Radiation. In most cases, radiation is an outside function. Under normal conditions, there would be no real need to ventilate. However, in smaller applications, such as an overheated wire, the heat from the wire can radiate to nearby combustible building

Figure 18-11 Ventilation openings are necessary to permit fire gases and steam to vent into the open air. On the left, with no openings provided, the fire is pushed into uninvolved areas of the building, and water is having no effect on the fire in the wall. On the right, an opening provides a path out through the roof to vent heat and fire gases, thus limiting extension.

components. While a combination of the three modes of heat transfer would be at work here, radiation is contributing. Ventilation would have its desired effect as heat is carried away from the recipient of the radiant heat, thereby carrying away and preventing, or at least delaying, the exposed material from reaching its ignition temperature.

TYPES OF VENTILATION

Ventilation can be performed using one of several methods either singularly or in combination with one another. Natural ventilation merely requires opening doors and windows and letting physics take care of the rest, **Figure 18-12**. Among the mechanical means are the use of smoke fans consisting of ejectors and blowers, and the use of water to create air movement.

Natural

Firefighters still use natural methods for evacuating smoke from a building. Merely opening windows and doors will provide natural ventilation. This choice is the most appropriate for incidents where time is not essential and there is no urgency to the emergency, permitting a slower venting operation. Diffusion of smoke and heat is a slow process, but it will eventually be complete when equilibrium is achieved. The natural movement of air currents is the action in this type of ventilation. This method would be appropriate for a light smoke condition where the incident is under control. It is the method of choice for removal of residual smoke from an incident such as food on the stove, overheated motor, leftover smoke from a fire some distance away in the building, or similar situations. A combination of air movement and diffusion is employed in the unassisted natural ventilation process.

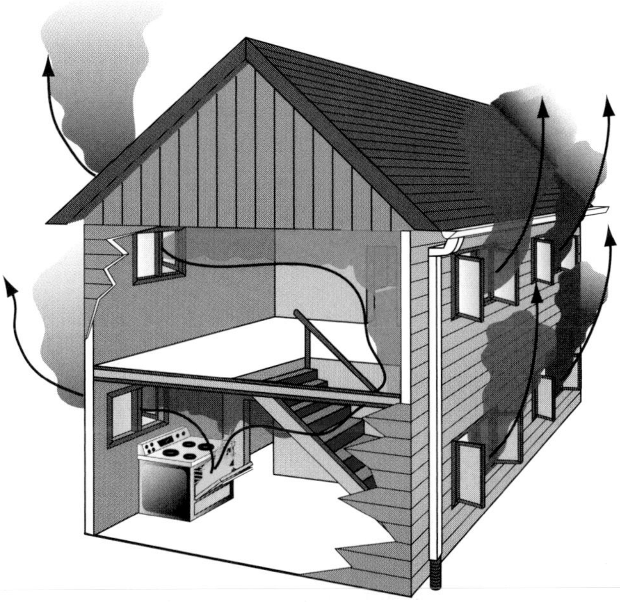

Figure 18-12 Smoke will be carried throughout the building to upper floors by normal air currents mixed in with the heat.

Cutting a hole in the roof or breaking out windows is another type of natural ventilation when greater speed or increased volumes of smoke and heat need to be removed from a structure, **Figure 18-13**. The rising heat will take smoke and fire gases to upper levels following the path of least resistance. A large opening in the roof with an unobstructed path from the lower floor affords an express channel for ventilation using the natural method.

Mechanical

The use of mechanical devices can speed the ventilation process, especially in compartments deep within a structure or under conditions where natural air movement is not satisfactory. Mechanical aids can accelerate the air movement and even reverse the airflow against natural air current movement. Use of fans or blowers in a positive or negative mode and water from a nozzle can provide a large air movement volume.

Heating, Ventilation, and Air Conditioning

Heating, ventilation, and air conditioning (HVAC) systems can be used effectively for ventilation in sealed, climate-controlled buildings. Closing intake dampers that would draw in smoke from the outside robs the fire of fresh air. With the opening and activation of exhaust fans, smoke and heat are removed from the compartment. Uninvolved floors can be vented mechanically, and air movement in the building in general can be controlled.

Smoke Fans

Two types of fans are used for smoke and heat removal. *Ejectors* were among the first fans used by the fire service and were employed primarily to suck out the smoke and heat. Placed in an opening, the fan developed a negative pressure on the structure side of the unit and drew the heated air and smoke into the fan to be ejected, **Figure 18-14**.

The fan is placed in the window with the arrow for airflow leading to the outside and safely secured. The openings that could cause the ejected

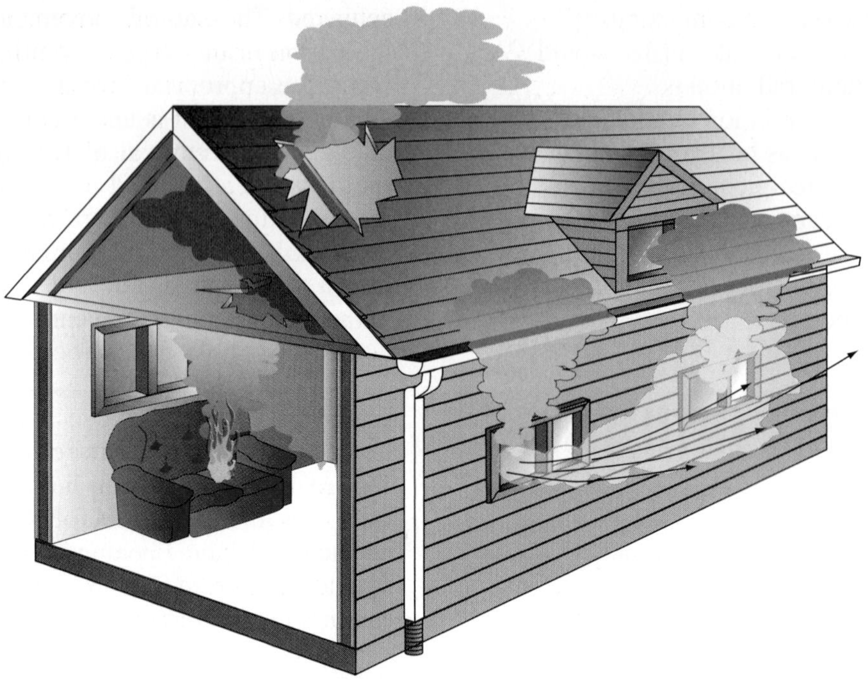

Figure 18-13 The hotter the air, the faster it will circulate throughout the building. Without the roof opening, the smoke will "bank down" and fill the structure. The openings provide a path to the outside for the heat and smoke.

smoke to be pulled back in and recirculated are sealed with tarps or plastic. During this time, establishment of an unobstructed path of airflow from the outside opposite the fan to facilitate fresh air introduction would be beneficial.

The disadvantage of this type of smoke removal is that the firefighting crew positioning the fan has to work exposed to the smoke and heat, **Figure 18-15**. In addition, the smoke must be pulled through the unit, subjecting it to caustic substances and deposits that could eventually cause mechanical difficulties to the motor. Deposits on fan blades can also create inefficiencies. Positioning a fan on upper level floors from

Figure 18-14 A smoke ejector exhaust fan placed in an opening will pull air through the fan as it ejects air out of the structure.

Figure 18-15 Placement of a smoke ejector will subject the firefighters to products of combustion because in order to operate it, the firefighter must stand in the path of the airflow. In addition, it forces the firefighters to work in a hostile environment in which it is difficult to see and breathe.

the exterior is even more hazardous. In these situations, the fan is usually placed on the inside.

Positive Pressure

Positive pressure ventilation (PPV) has been around for over thirty years but only recently has it become an active tool in the arsenal. Air is introduced into the smoke-filled area through the use of PPV fans or blowers. The compartment becomes positively charged with air, and the heat and smoke are displaced through chosen avenues to the exterior, **Figure 18-16**. In addition to being used in the later stages of a fire to clear the smoke to ease overhauling, PPV has been employed with some success during the attack stage of fire suppression. Because of the narrow applicability of the practice during the attack phase, some fire officials are skeptical about introducing the practice to their operations.

> **Streetsmart Tip** Misuse of positive pressure ventilation can severely accelerate fire growth.

However, when used properly, the smoke and heat are totally removed from the access paths of the nozzle team and search team.

A distinct advantage of the PPV blower is ease of setup. The fan merely needs to be positioned at the entrance to the structure with no exposure of the setup crew to dangers from within the structure. Furthermore, some of the blowers are designed for a single-person setup. This is a particularly distinct advantage where personnel are limited. If properly applied, the blower can do the job of several firefighters attempting to open multiple holes with ladders at different locations. While not a cure-all for ventilation problems, PPV can be a material asset to an operation. The placement of the blower will depend on the fan's size and capacity and the size of the opening.

Hydraulic

The last of the mechanical modes of ventilation is to use water. Known as hydraulic ventilation, water is employed to create air movement, **Figure 18-17**. It is a quick method for expulsion of smoke and heat without the need for additional tools. Use of water can cause water damage, but it has some assets, too. If the need to evacuate a smoke-filled room is essential to search for a victim, the time needed to set up an alternative mechanical ventilation tool might be unacceptable. Water can also be used to quickly cool down a heated atmosphere in a structure by sucking and pushing the heated smoke out of the structure.

The nozzle team will quickly feel the benefits as the cooler air from the uninvolved portion of the structure is pulled to them from behind. At the same time, the air movement evacuates the heated upper portion of the compartment, **Figure 18-18**.

This method of ventilation is the most rudimentary. Tests have shown that when properly positioned, the mechanical blowers move a greater amount of cubic feet per minute (cfm) than smoke ejectors, and the use of a nozzle to move air is the most inefficient.

Using a Water Stream to Ventilate. Ventilation by hydraulic means is as simple as directing the hose stream out of the window to remove smoke and heat. Using a fog or spray stream is most effective because the water movement actually pushes and pulls the air surrounding the droplets created by the nozzle. A solid stream nozzle can also be effective by opening the nozzle partially to create a coarse, broken stream.

Whatever type of stream is used, it is placed so that the water is being directed out of the opening. A spray stream setting should be between a 30- to 60-degree angle. Fog and broken solid streams are placed so that the water discharge is entirely outside and not quite touching the sides of the opening. Ideally, the stream should fill 90 percent of the opening. The nozzle stream distance from the opening will depend on the width of the stream. The nozzle team can then move closer or back off

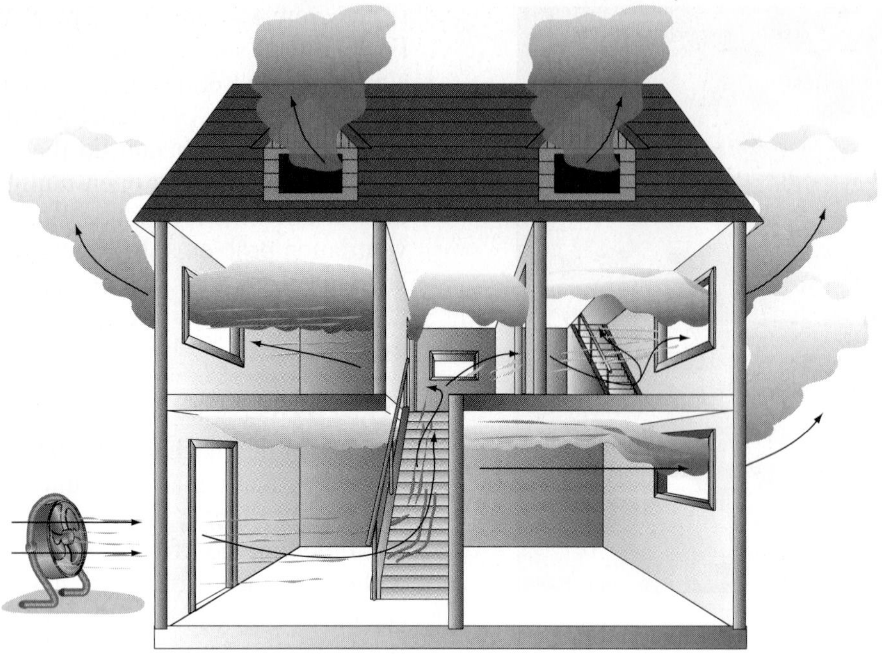

Figure 18-16 Positive pressure literally pressurizes the structure and forces smoke out any path of least resistance. Almost the same effect would occur if a light breeze were blowing directly into the structure from one side and venting out the other side.

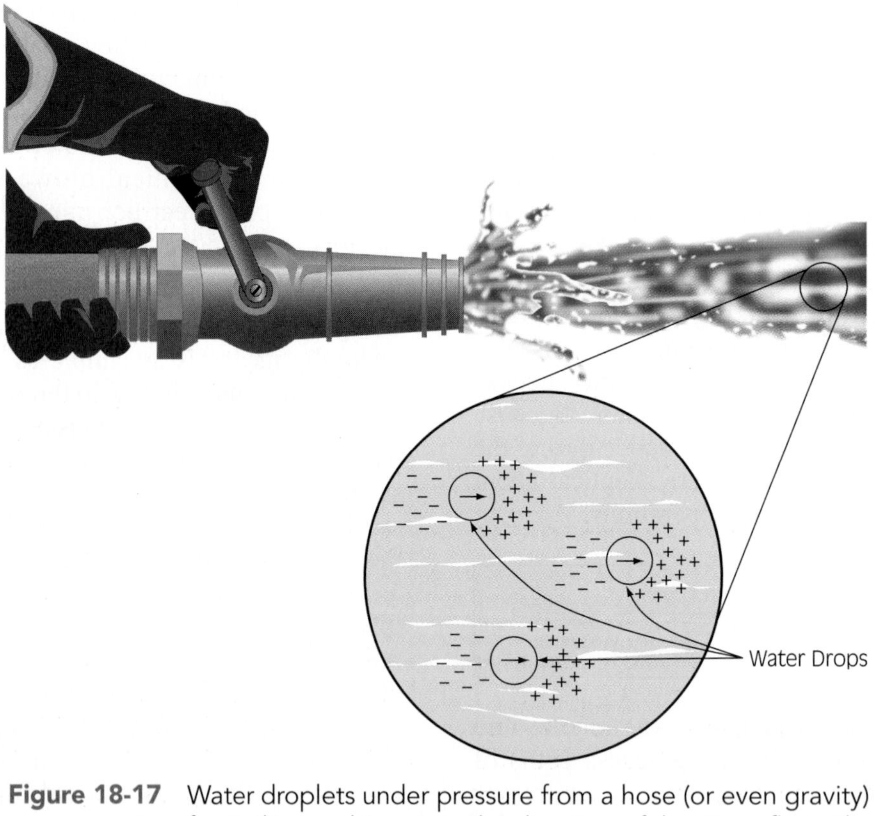

Figure 18-17 Water droplets under pressure from a hose (or even gravity) compress air in front, thus pushing air in the direction of the water flow. The vacuum created behind the droplet pulls the air along with it as the air behind the droplet attempts to fill the low-pressure void.

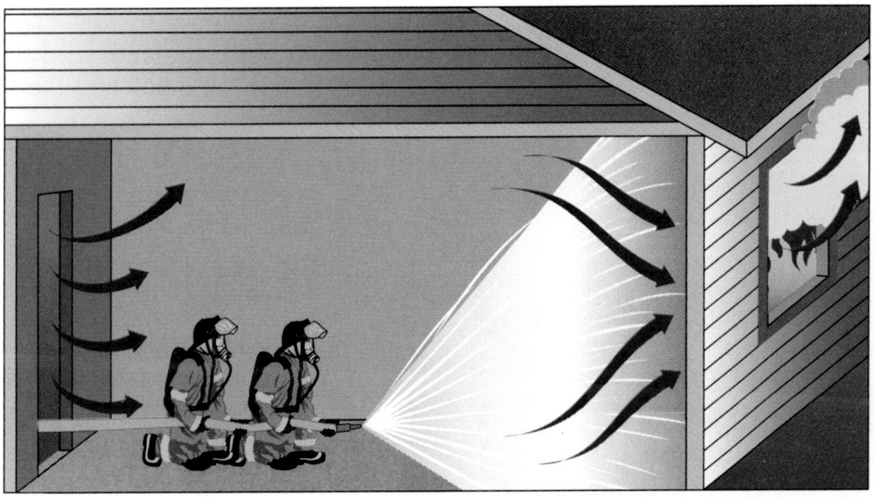

Figure 18-18 A coarse stream from a solid-bore nozzle or a spray or fog stream will create the airflow necessary to move smoke out an opening.

until the maximum beneficial effect is achieved. Because of the many variables that are always present, a certain amount of trial and error is involved with finding the best position for maximum effectiveness.

MECHANICS OF VENTILATION

The entire ventilation process, regardless of how it is achieved, is simply the movement of air from a high-pressure location to a place where lower pressure exists. Whether created artificially or naturally, air will move from a high-pressure area to a low-pressure one. Knowing this natural tendency of air to move will assist the firefighter assigned to the task of ventilation.

Vertical Ventilation

Based on the "heat rises" rule of physics, the collection of heat at the upper levels of a structure will spread fire to those upper levels if the heat level rises to the ignition temperature of the structure. By opening vertical arteries to release any pent-up gases, the odds of fire spreading to involve other parts of the structure or contents are reduced. In addition, the heated air being released will be replaced with cooler air at the lower levels. This improvement of the conditions will assist fire suppression crews attempting to locate the fire, search for possible victims, and position for suppression activities, **Figure 18-19** and **Figure 18-20**.

Horizontal Ventilation

Horizontal ventilation conforms to the same rules of vertical ventilation, which are governed by physics. Both types are actually a form of what is called **diffusion**, a naturally occurring event in which molecules keep spilling excess levels of high concentrations to areas of low concentrations. In addition to the vertical gravitation, the molecules also move laterally. Given a wall, the molecules bounce back into the mix. Given an opening, the molecules will pass into the less concentrated environment, **Figure 18-21**. That is a

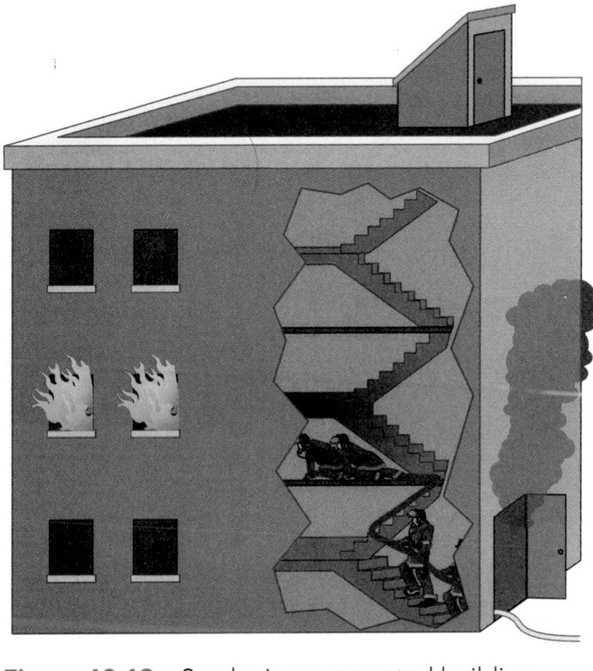

Figure 18-19 Smoke in an unvented building impedes progress. Searches are slowed. Discovery of the fire location and deployment of hoselines are delayed.

scientific description of what happens, but to the firefighter on the fireground, all that is important is that the openings are made and the smoke and heat are channeled out of the structure.

Figure 18-20 While a vented building will still have smoke in it, the difference is profound. The processes of searching, hose deployment, and fire discovery are markedly improved. In addition, the survival time of overcome victims is increased. Last but not least, potential victims might be able to self-evacuate, freeing up personnel for firefighting efforts.

VENTILATION TECHNIQUES

Many techniques are used to effect ventilation. Some are simple and require no special tools; others are more complex, are dangerous to implement, and require sophisticated tools.

Break Glass

The quickest way to open a building for ventilation is to break glass. While not always the technique of choice, it is the best investment of time for results gained if done properly. Many windows can be broken in the time it takes to force open one locked door.

> **Safety** When breaking glass, a fully equipped firefighter wearing proper protective equipment is mandatory. Slivers of broken glass can become lodged in eyes and shards of glass can act as airborne daggers.

Glass can and has penetrated skin deep enough to sever arteries and veins. The wearing of

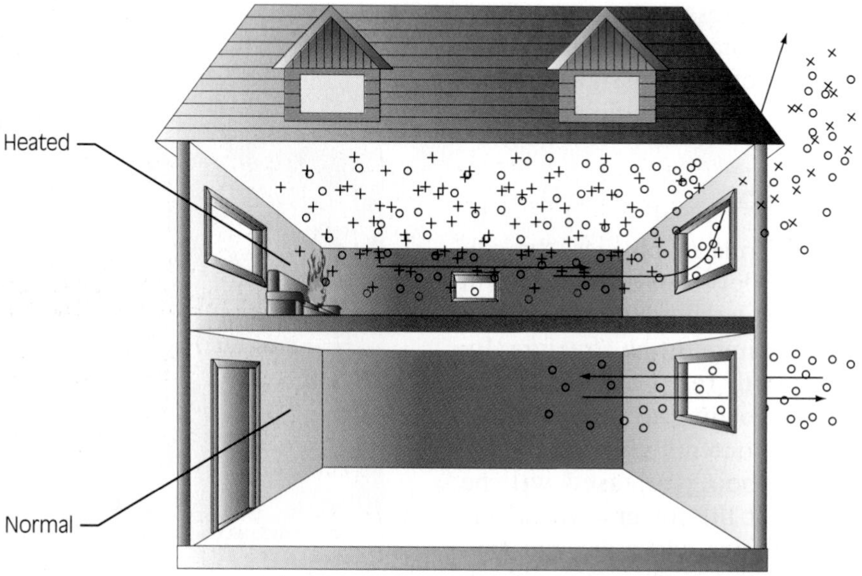

Figure 18-21 Heated air has more agitation in its molecules, causing internal pressure in a compartment. This will, in turn, create greater velocity when air exits an opening. Normal diffusion takes much longer to occur when only natural air movement and currents are employed.

protective equipment will reduce the severity or perhaps even completely eliminate an injury.

Open Doors

Merely opening a door will usually exhaust huge volumes of smoke and heat built up on the other side by pressure from the combustion. Other doors will also be helpful in opening the compartment. Keeping the door on its hinges is a good practice because the need might arise to close that door to limit fire extension or to employ a PPV blower with prechosen ventilation channels.

Effects of Glass Panes

Many windows have several panes of glass separated by wood or aluminum dividers, **Figure 18-22**. When the need to remove glass is evident, the entire window should be removed. An adage exists in the fire service vernacular: "Turn every window into a door." This refers to the possibility that the window may be needed for escape. Removal of the sashes, cross members, and any other obstacle that would impede access or egress is necessary.

When removing glass for ventilation, it is important to remove *all* glass. Shards that remain in the sash will impede airflow and reduce ventilation. A sweep around the perimeter of the sash with a tool will often produce quick satisfactory results.

Using a Tool to Take Out a Window. Regardless of the tool used, basic principles should be followed. Except for special types of glass (i.e., tempered, bulletproof), glass will break regardless of the method employed. What is important is the amount of open area created once the glass is removed.

Once the glass is broken, whatever was on the other side of that glass will be coming out. Therefore, when removing glass, it is imperative that the firefighter not be in the path created by its removal. Firefighters should position themselves off to one side and crouch underneath the window level; this is far more preferable and safer than standing directly in front of the window while removing the glass.

Using the tool to its maximum length to protect the firefighter, the tool is brought down from the upper sash through the window's midsupport and into the lower sash. Usually, this will remove a large portion of the glass. Then, the tool is used to sweep around the perimeter of the window to remove any shards that remain in the sash. Finally, the tool is used to reach into the window and unhitch any shade, blinds, or curtains that will impede airflow. If the firefighter is not able to capture the fastening, the tool can be used to hook the curtains or blinds and pull the entire unit out of the window and down until it releases. These units are held in place by small nails or screws and forcible removal is not difficult. It is essential to use care when dropping window treatments inside the structure. A victim might be lying under the window and could be inadvertently covered up. This area should be thoroughly searched as soon as possible.

The glass removal should be a two-step movement: (1) breaking through the glass from upper to lower sashes and then (2) sweeping the perimeter to remove remaining glass. Of course, not every single piece of glass will disengage. The remaining small shards will not materially affect the airflow. Good judgment must be employed when deciding whether to move on to the next window or continue to work on that opening.

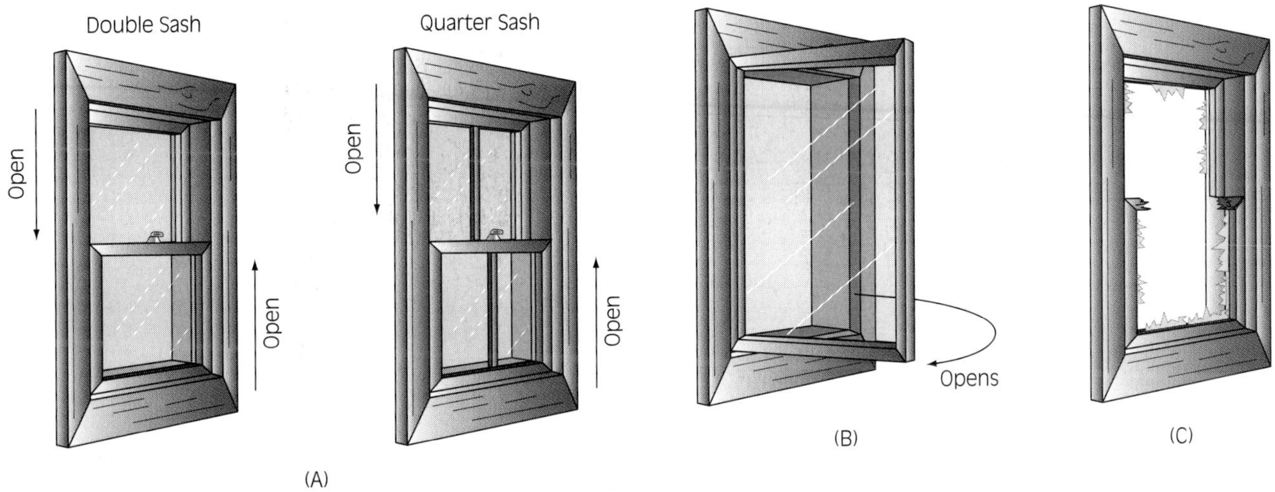

Figure 18-22 (A) Sash windows open by pulling the lower section up or the upper section down. The number of panes of glass will not change the number of sashes. (B) A casement window opens like a door, from right to left or from left to right, usually through the use of a crank or handle. (C) Regardless of the type of window, when it is vented, all the glass, including cross members, should be removed to maximize airflow with a minimum of obstructions.

JOB PERFORMANCE REQUIREMENT 18-1

Using a Tool to Take Out a Window

A Make sure the rope is securely attached to the tool using an approved slip-free knot. The tool is then lowered to the window.

B Wrap a turn of rope around your hand and toss the tool out as far as possible in a horizontal direction. The tool is then swung into the glass and will break out the window.

Note: Before performing this task, make sure there are no firefighters or victims in the building or below the building.

Rope and a Tool

A technique that has produced satisfactory results when operating off a flat roof beyond the reach of a ladder is to use a tool and a rope, **JPR 18-1**. Many windows can be taken out in a short period with this operation. The rope must be secured to the tool so there is no chance it will detach.

Streetsmart Tip Some firefighters have had small rings welded to the tool and use a positive latch rope clip for quick attachment.

Using a Rope and a Tool to Ventilate Upper Floor Windows from Above. The rope is securely attached to the tool using an approved slip-free knot. The tool is then lowered to the window the firefighter desires to ventilate, **JPR 18-1A.** A turn of rope is taken around the firefighter's hand. The tool is then tossed out as far as possible from the building in a horizontal direction. As gravity takes the tool down and the end of the fully extended rope is reached, the tool will swing into the glass at the predetermined length of rope and break out the window, **JPR 18-1B.**

This technique is not the best method because it leaves shards of glass in place and does not remove screens or window shades or curtains, but it is effective when no other approach can be easily made.

Hook or Pike Pole

The use of a hook, also called a pike pole, is also an effective way to take out glass. The length of the pole keeps the firefighter a safe distance from the falling glass and it affords the firefighter the opportunity to stand off to the side of the opening. Pike poles or hooks come in assorted lengths. The use of a longer hook enables access to windows that would otherwise be out of reach without a ladder. A hook can also be used to extend the reach of a firefighter attempting to open or close a door, as a pry tool to remove roof material while standing out of the way of the opening, and for other useful purposes that will extend a firefighter's reach or remove a firefighter from dangerous positions, **Figure 18-23**.

Iron or Halligan

The iron or Halligan tool is brought down diagonally through the glass starting at the upper corner. The tool should be brought through the cross members to create one large opening. Then the tool is used to sweep around the perimeter to clean out any large shards. A disadvantage of this tool is its short length. It places the firefighter in proximity to flying glass. If prying

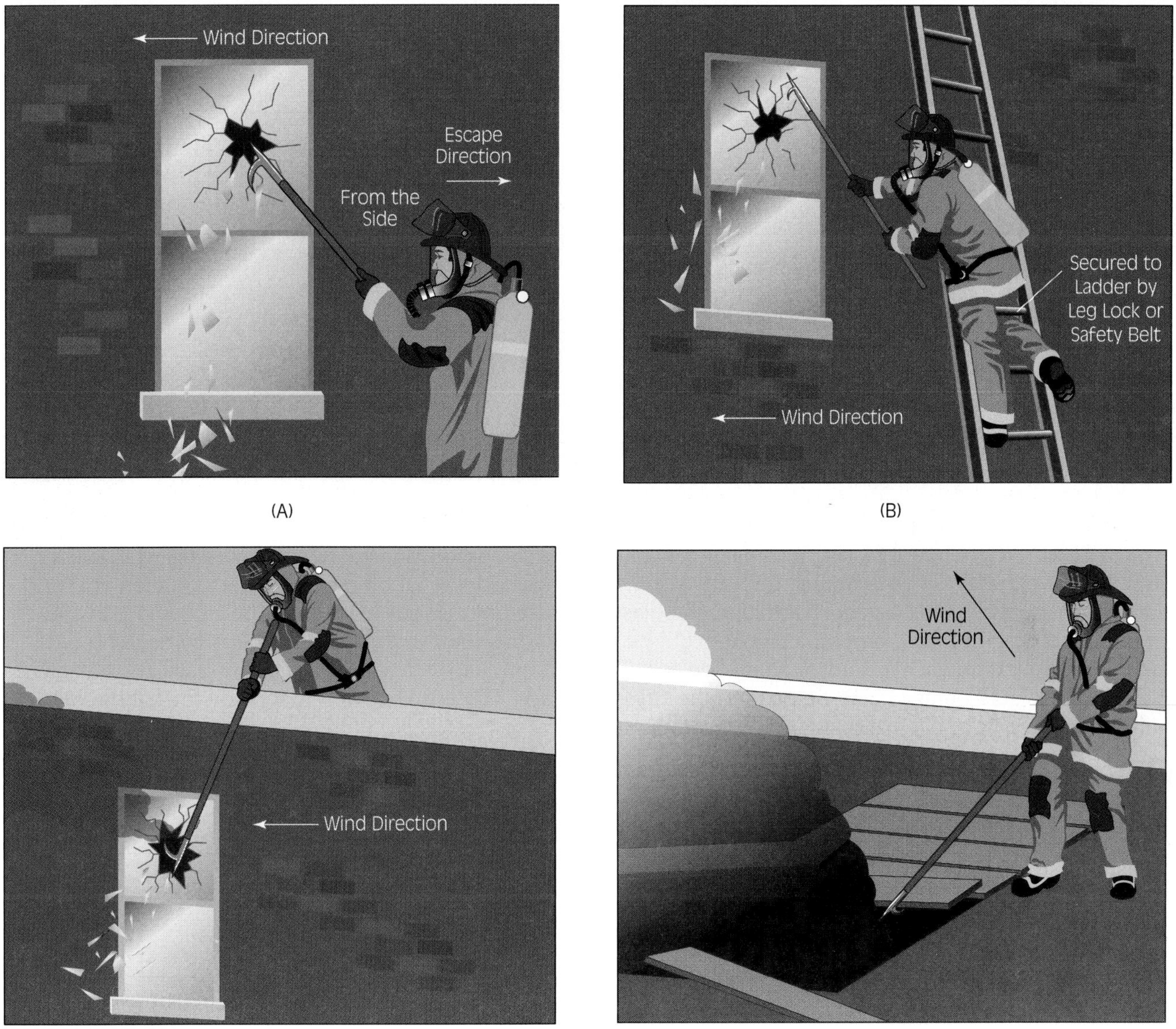

Figure 18-23 Making a ventilation hole requires some preplanning. (A) Firefighters should make the hole so that heat, smoke, and possibly flame do not envelop them. They must not cut off escape routes, and they should work so that the smoke is blowing away from firefighters' positions. (B) When working off a ladder, the same general precautions are necessary. Firefighters must be secured to the ladder before performing any action. This could be through use of a safety belt or a leg lock. (C) When venting from above, firefighters use the wind to their advantage and stand off to one side so that they are not standing in the path of any initial billow of heat. (D) When pulling off roof boards, firefighters should work in the clear air with the wind blowing smoke away, and be careful with roof debris. It will most likely be hidden in the smoke.

open a door or hatch, it places the firefighter almost in the path that the sudden gush of heat and smoke will take. Careful planning on the part of the firefighter must be employed to minimize exposure.

Ax

Similar to the iron or Halligan tool, the ax affords limited reach and places the firefighter in a potentially hazardous location. When using an ax for venting glass, the side of the ax is used to break the glass. Use of the blade portion or the striking head might cause the firefighter to lose a grip on the handle and lose the tool due to the manner in which a hand grasps the handle. This technique will not break tempered glass, for which a sharp pointed tool is required.

Portable Ladder

A very effective method of taking out glass in upper level windows, especially in private dwellings, is by the use of a portable ground ladder. The first thing to

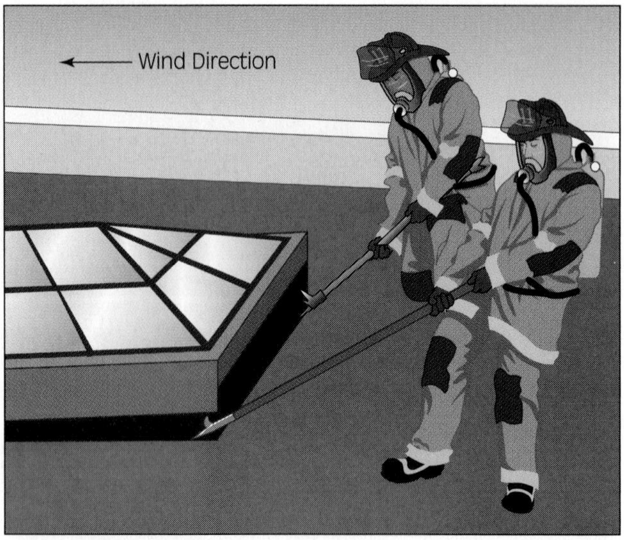

(E)

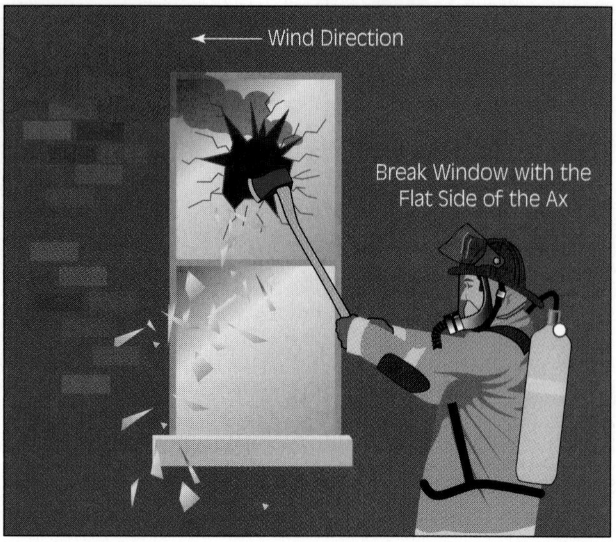

(F)

Figure 18-23 (Continued) Making a ventilation hole requires some preplanning. (E) When removing a skylight, firefighters work with the wind at their backs. It is sometimes less work to lift off the entire housing than to break out each individual pane of glass. In addition, the glass for these units is usually reinforced. Depending on the amount of tar used to waterproof the unit, it might be easier to just remove the panes of glass by bending back the frame and letting each individual pane slide out. (F) When using an ax to remove window glass, the flat side of the ax head should be used, not the point or the striking surface. The position of the hands on the handle will offer better control of the ax as it breaks through the glass.

do is check for overhead obstructions. The ladder is placed to the side of the window against the siding of the house. Then the base (or heel) is arranged and measured so that the tip will fall into the glass at about two-thirds the height of the window. The ladder is then repositioned in front of the window perpendicular to the ground and shoved into the building. The tip of the ladder crashes through the upper pane and often comes down through the sash into the lower pane. The disadvantage of this technique is that sometimes only the top sash window breaks and window hangings remain in place that impede airflow. It is important to know how the interior attack is progressing and the location of the interior team before performing this operation.

Although far from totally effective, it does afford a quick method of getting some ventilation into operation in several upper level windows in a short period of time. When completed with the initial openings, ladders can be repositioned and firefighters can climb them to clean out the rest of the glass, window dressings, blinds, and so on, and to enter the building for a search if necessary.

Safety When using a tool or a ladder to break glass, large glass shards can ride down the tool or the ladder into the firefighter. This shard can shatter in the face of the firefighter or slice through protective equipment into flesh.

Safety When operating on a roof, there should always be a second escape route provided opposite the entry point.

Negative Pressure Ventilation

Areas of high pressure will flow to areas of low pressure. A fan moves air through the use of this physics principle. In ventilation, positive pressure is created by a fan that compresses air through the use of fan blades. That will tend to flow to an area of low pressure in the direction in which the fan is pointed. On the back side of the fan blade, negative pressure is being created. When placed in a window facing the outward flow, the positive pressure is forced out into the area outside the compartment. Along with that flow comes the heat and smoke trapped in the compartment. That is the principle under which a smoke ejector works.

Knowing this principle it is better understood that the closing of all avenues of pressure equalization from the discharge side of the fan to the input side is essential. If not properly sealed, the air will churn or recirculate and merely circle around to the intake side of the fan, **Figure 18-24**. For effective air movement, a barrier must be created that separates the positively charged air from the negatively charged air sectors. In a structure where many windows are broken and there are

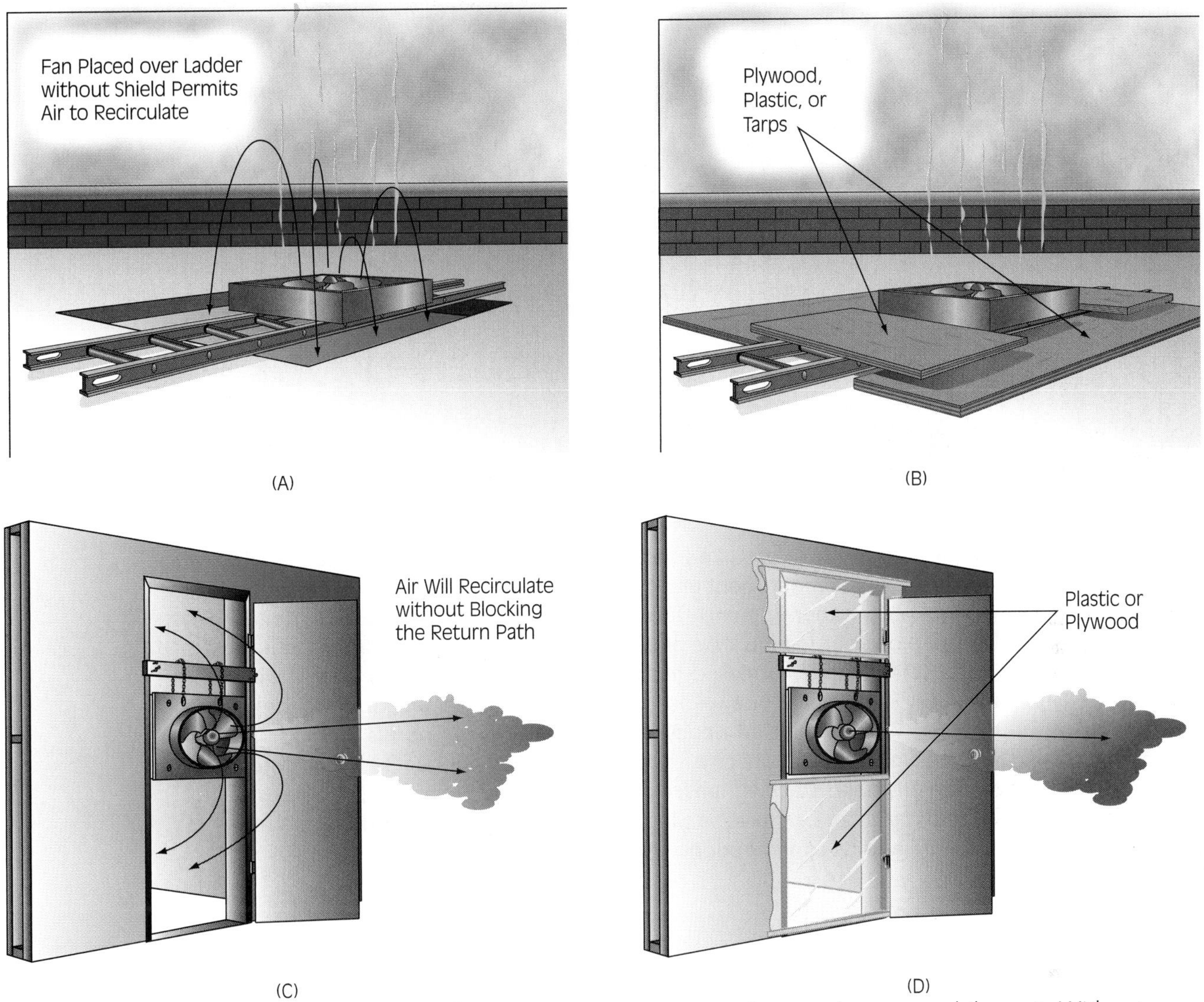

Figure 18-24 (A) When using an exhaust fan, it is important to cover the openings around the unit. Without a means to block the air from returning, it will merely circulate, reducing efficiency. (B) When covering any opening around the exhaust fan, the vacuum necessary to operate efficiently will be created and the exhausted air will not be sucked back around the fan. (C) When using an exhaust fan in a door, the air will circulate from the exhaust side into the intake side if no provision is made to block that flow. (D) Through the use of plastic, tarps, or even a piece of plywood, the air is prevented from being pulled back into the intake side of the fan.

holes in roofs and walls, this is not easily done. Therein lies the major weakness with this technique. Drapes of plastic and tarps can be employed to limit churning.

This type of setup can be very effective in limited access compartments such as cellars or basements where few openings exist. Fans can be placed over the opening in the cellar stairs or in a hole made in the first floor near a window and used to draw out the smoke through suction. This tool is primarily an overhauling aid and should not be used during the attack mode. The fan could bring fire to the location, placing the operators in a dangerous position.

Positive Pressure Ventilation

PPV uses an entirely different approach than negative pressure ventilation, but with the same principle. The positive pressure technique actually injects air into the compartment and pressurizes it. In an attempt to equalize, the smoke and heat are carried out into the areas of lower pressure outside the structure. There are many variations of this practice. The basic principle is to create a cone of air and force it from the outside to the inside, **Figure 18-25**.

Depending on the circumstances, fans can be set up to augment one another. Two fans can be placed

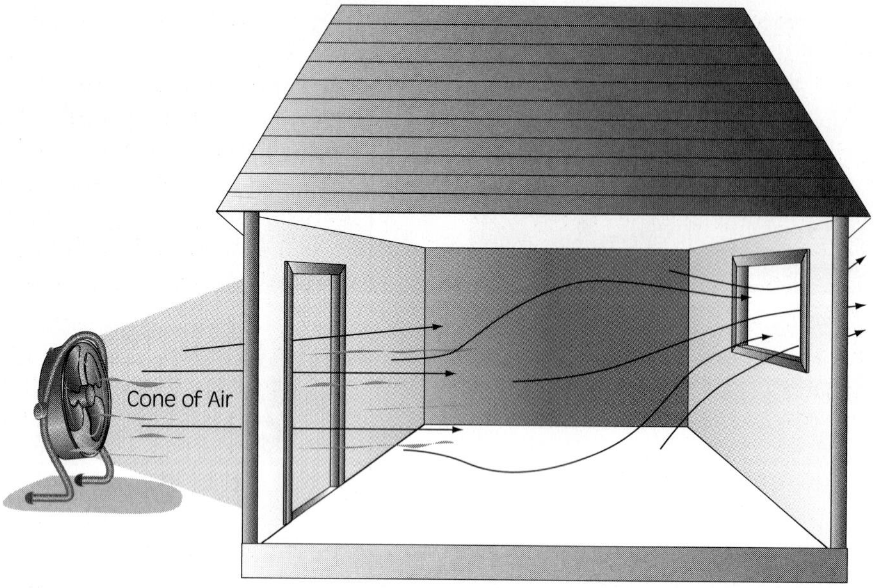

Figure 18-25 With positive pressure ventilation, the theory is to actually pressurize the compartment and then the smoke and heat will actually be pushed out another opening. To be effective, certain actions must be taken. (1) The blower or fan must be placed a short distance from the opening so that a "cone of air" is created that just barely exceeds the opening being used. (2) The exhaust opening should be smaller than the introduction opening for maximum efficiency. That opening size depends on the number of blowers and their capacity. There are many variations where this practice can be effective. Stacking fans, using them in tandem, or placing one behind the other can have varying degrees of success depending on the structure and its configuration.

side by side to cover a wide opening. They may be stacked to create an elongated cone of air. If positioned one behind the other, the rear fan can supercharge the front fan for greater efficiency. The fans can be placed in remote locations to augment the airflow from one part of the structure to another, and three or more fans can be used in conjunction with one another: one or two at the opening and others spaced throughout the structure in upper floors and along long corridors or split in halls that fork. Understanding the principle is all that is needed to create a custom application to fit an unusual situation.

Tests have shown that effective pressures can be generated up to twenty-five stories from the street opening location of a PPV blower. Augmentation beyond that level can still provide effective pressure generation. Intermediate fans will also assist in maintaining the flow.

Generally speaking, for every cubic foot of air injected into a compartment, a cubic foot of air must be ejected. The pressure loss experienced is due to the many openings in any run that will permit some air pressure to escape.

Streetsmart Tip Practical experimentation of positive pressure ventilation can be employed in any structure (even the fire station) through the use of tissue paper. With a fan or blower set up, tissue paper strips are hung in openings throughout the structure. When the fan is activated, the airflow can be demonstrated, redirected to different openings, and observed. The velocity of the air movement can be observed by the degree to which the tissues in the openings are affected.

Positive Pressure Fan/Blower Setup and Operating Principles

To effectuate positive pressure, the blower should be set up outside the structure. The cone of air that is produced must completely cover the opening. This can be achieved by moving the fan closer to or farther away from the building opening. Generally, the size of the exhaust opening must be larger than the air introduction opening. Positive pressure is most efficient when the exhaust opening is ¼ to

1½ times the size of the entrance opening. Communication is essential between the intake side personnel and the exhaust side personnel. All members engaged in a positive pressure operation must be properly trained and understand the principles of the evolution.

Many different combinations of blowers and fans can be utilized. They can be placed in tandem, stacked, side by side, in-line, or in any configuration that accomplishes the goal of pressurizing the compartment. Depending on the size, shape, and number of cubic feet in the structure, many different layouts will be successful.

- Positive pressure evacuation of smoke is successful for distances of up to 1,000 feet.
- For each cubic foot in a building, a cubic foot needs to be exhausted. Multiple floors and multiple rooms can be systematically vented by opening up a single room and window until it is vented, then closing the room off and moving to the next room. This technique is particularly effective in a school classroom-type building, motel, hotel, or similar room layout.

Depending on the size of the blower and the department's practice in using it in overhauling and/or attack modes, the novice firefighter should become familiar with the many different aspects of positive pressure techniques. The basic rules are as follows:

1. The introduction opening should be larger than the exhaust opening.
2. The air can be channeled into different parts of a building by opening and closing windows and doors.
3. A cone larger than the introduction opening is necessary to prevent recirculation of air back out of the structure.
4. Communication at the point of entry and the point of exhaust is important.
5. Practice will provide the understanding of the basic principles that can then be applied to actual fire scene operations.

ROOF VENTILATION

Vertical ventilation can be attained using several methods. One such method is to use the existing building features that allow for removal of heat, smoke, and the by-products of combustion. Those features include penthouse doors, skylights, and ventilation shafts. Care should be taken to ensure that use of these features would not spread the fire, **Figure 18-26.**

Two primary types, offensive and defensive, characterize vertical ventilation openings. The primary offensive vertical ventilation openings are generally referred to as offensive heat holes. These openings are typically placed over the fire or as close to the main body of the fire as reasonable or necessary. The primary defensive vertical ventilation openings are generally considered strip or trench ventilation operations and are broken down into subtypes from there. The goal is to place offensive openings into the structure and evaluate the need for a defensive ventilation opening. Communication with interior forces will assist in the evaluation process for vertical ventilation operations.

With snow, the hot spot might be melted away or slushy compared to the rest of the snow on the roof.

The primary hole should be cut directly over the fire if possible, and several primary holes can be cut. The more direct the heat travel to the outer air, the less potential for extension of the fire. Secondary hole cuts are any cuts that follow the primary hole. Some holes are always secondary cuts, while some types of cuts can be either primary or secondary. An example is a trench cut, also called a strip cut. It is a secondary cut and should never be employed as a primary ventilation hole. Its purpose is to create a line of defense and, as such, should not be ordered until primary holes are completed. Secondary holes are cut away from the immediate fire area and are used for inspection or to cut off fire extension.

Expandable Cut

An expandable cut is the most efficient type of cut for the time expended, **Figure 18-27**. It can produce a hole as large as is needed. Because ventilation is such a necessary operation, the roof must be opened quickly. After planning the cut so that the wind will tend to blow the smoke away from the work area, an initial cut line of about 4 feet long is made. The second cut is a small triangular purchase point. Then a third, fourth, and fifth cut are made. Using a hook (pike pole), the purchase hole is used to hook the roof material and, like a giant lid, the top is lifted off. This type of cut can produce a hole of about 16 square feet.

Care must be taken to prevent damage to support rafters or cross members when cutting a roof. A support rafter can be felt when the saw suddenly begins to slow down as it tries to cut a thickness greater than the roof sheathing. The speed of the saw will increase and the operator will actually be able to feel the difference when the rafter or support beam is passed.

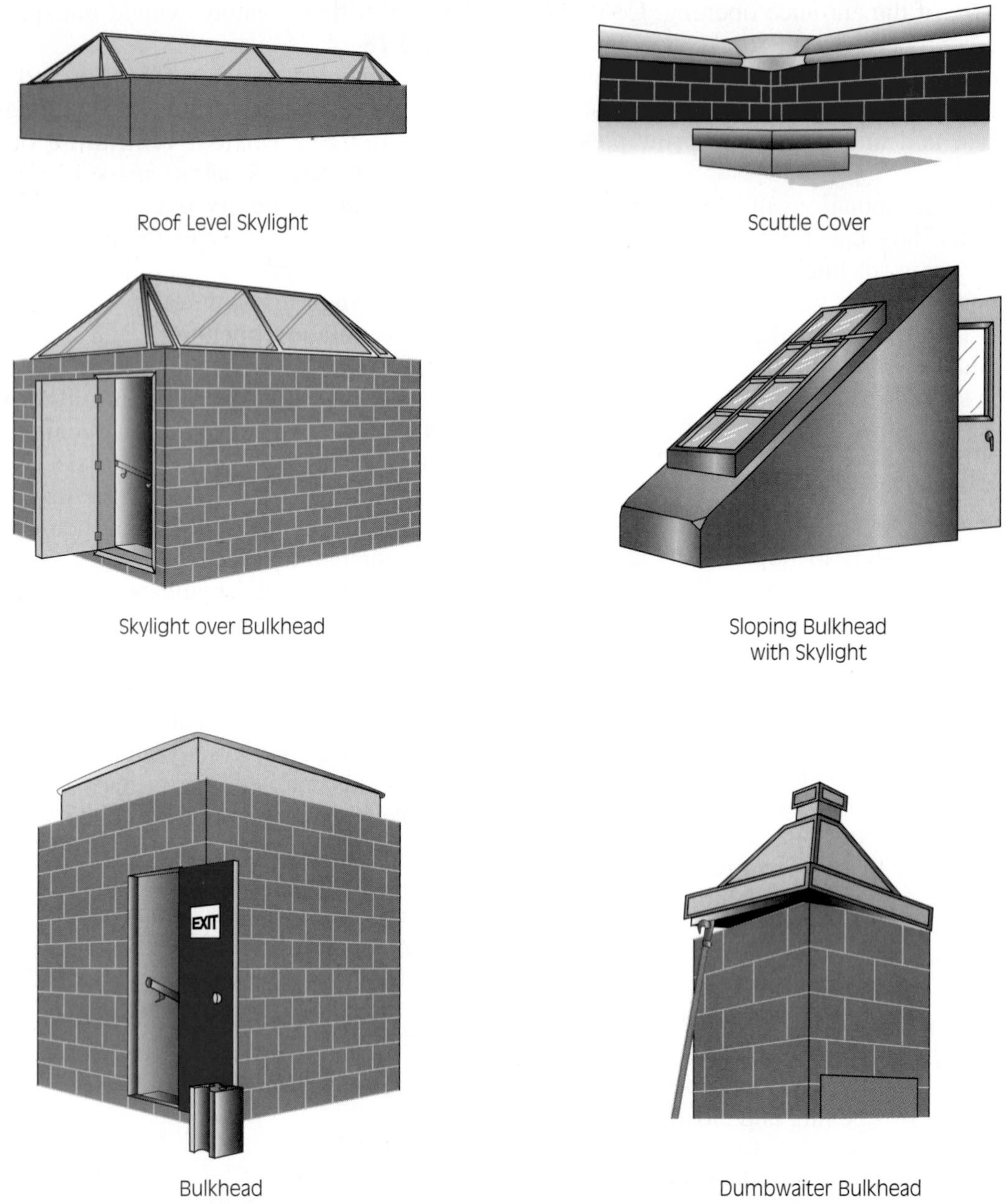

Figure 18-26 Many types of roof structures afford quick and effective initial ventilation. As is always the case, the time spent in creating a hole should be balanced by the amount of efficiency that will be gained. For example, opening a bulkhead door will be quick and effective. It should be performed before a roof cut is initiated.

If a larger opening is necessary, making three additional cuts using one of the sides of the original hole as the first cut expands the hole. With only three more cuts, the size of the hole is doubled to 32 square feet. Again, if an even larger opening is required, another three cuts will increase the hole size by another 16 square feet to a total of 48 square feet. If a larger opening is still required, the large hole can be finished off with only two cuts, **Figure 18-27**.

Occasionally, a close approach to a roof opening is impossible because of the heat condition. In this case, the expanded cut might have to be performed by making as large a hole as possible using whatever part of the hole is approachable, **Figure 18-28** and **Figure 18-29**.

By comparing the numbers, the use of this type of cut will produce a hole that is 64 square feet. In terms of manual labor, if each hole were cut individually, it would take five cuts to produce each hole. A total of four holes would require twenty-five cuts to produce 64 square feet of opening. Using the expandable cut, the same opening is produced with only thirteen cuts. In addition, one large

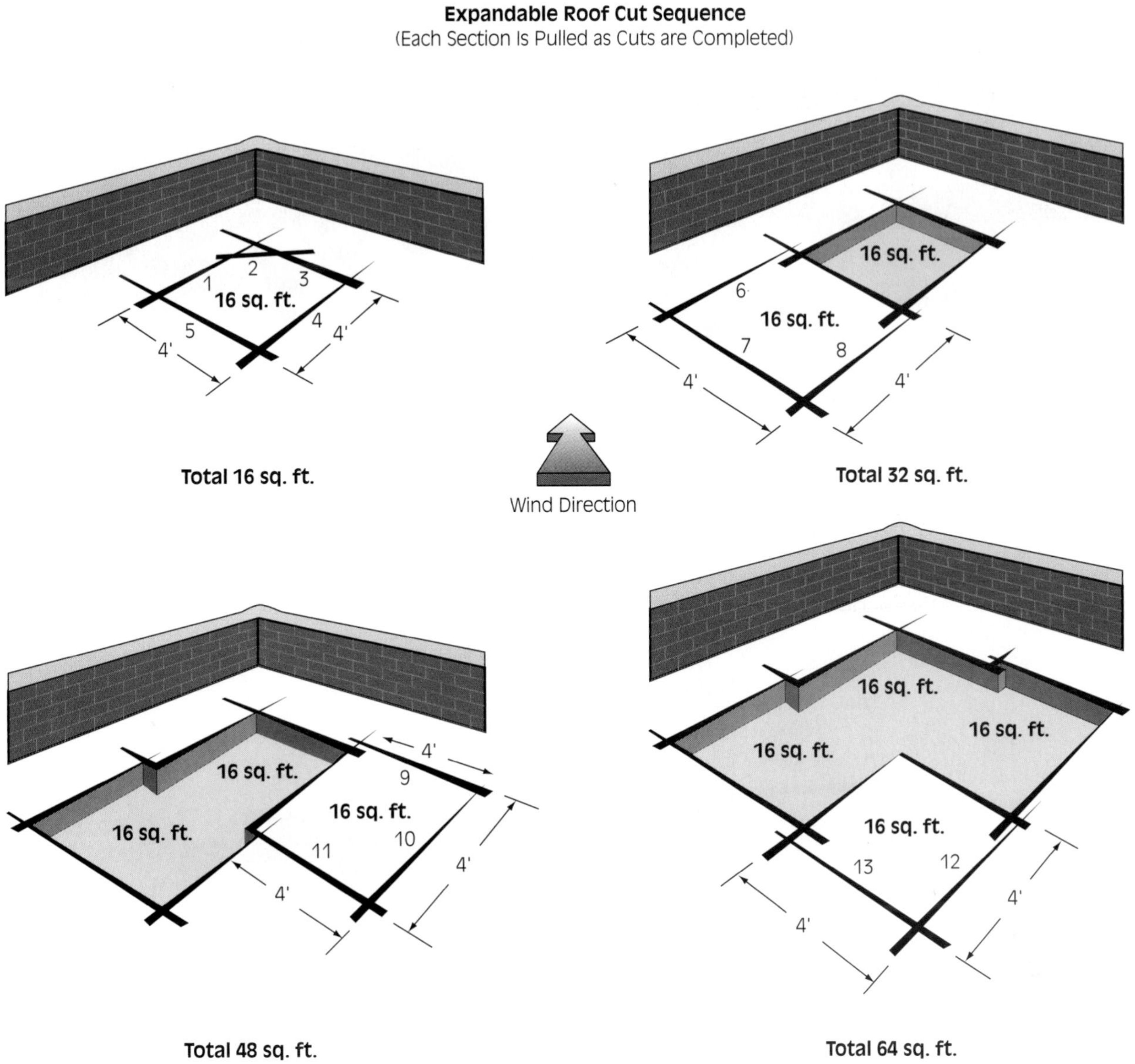

Figure 18-27 When cutting open a roof is necessary, the expandable cut is the most effective, quickest, and most efficient way to get a large opening for vertical ventilation. Each new cut is merely an extension of the previous opening. The overall number of cuts required is dramatically fewer than if each hole were individually cut.

opening produces more airflow than several smaller holes of comparable area.

Among the disadvantages of this type of cut is the amount of debris produced. With each section that is opened, roofing material, roof boards, and wood with protruding nails are left lying about. Additionally, the hole produced is large and a firefighter can fall into the opening, even with the joists that are spanning the opening after the roof material is removed.

Center Rafter Cut (Louver)

The center rafter louver cut is utilized primarily as an offensive heat hole. It is often used on building construction that utilizes plywood sheeting under paper as a roof cover. It can also be used on roofs with 1 × 6 sheeting.

The center rafter louver is based on a series of cuts that begin with an outside cut, **Figure 18-30.** The saw is placed on the roof decking and a cut is made until the roof rafter is felt with the nose of the chain saw bar. The rafter is rolled (raising the saw so as not to cut through the rafter) and the cut is continued. This "head cut" is run as far as the opening is desired, continuing over rafters as they are met.

Then the saw is taken back to the first rafter met, and a lateral cut is made from the head cut. Another cut is made parallel to the head cut until the second rafter is met. The next cut is made to "box" this cut, **Figure 18-30.** After that, the roofperson continues

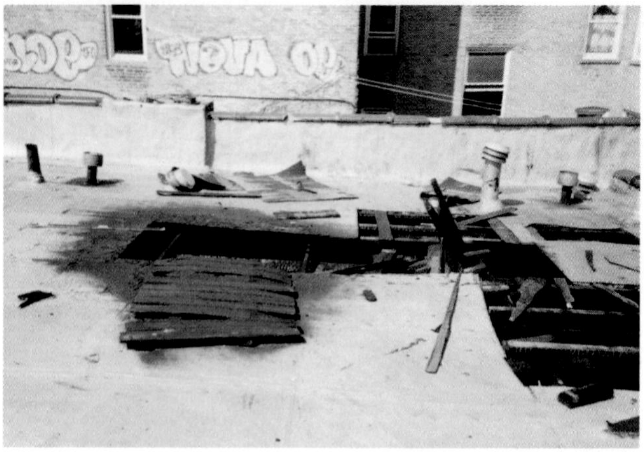

Figure 18-28 Notice here that the opening had to be made in a linear fashion. In this fire, the heat was so intense that a rectangular cut was impossible. However, using the existing hole to make the next series of cuts could expand the opening. Also notice the amount of debris that accumulates. Visible in clear conditions, a severe tripping factor exists under smoke conditions.

Figure 18-29 The initial hole was cut and the "lid" lifted off in one piece. In the foreground, the cuts of the second hole can be seen along the right side of the hole. This type of roof, called an inverted roof, consists of a flimsy under roof support system while the actual weight-carrying members exist at the occupancy ceiling level. This photo illustrates how little support is available.

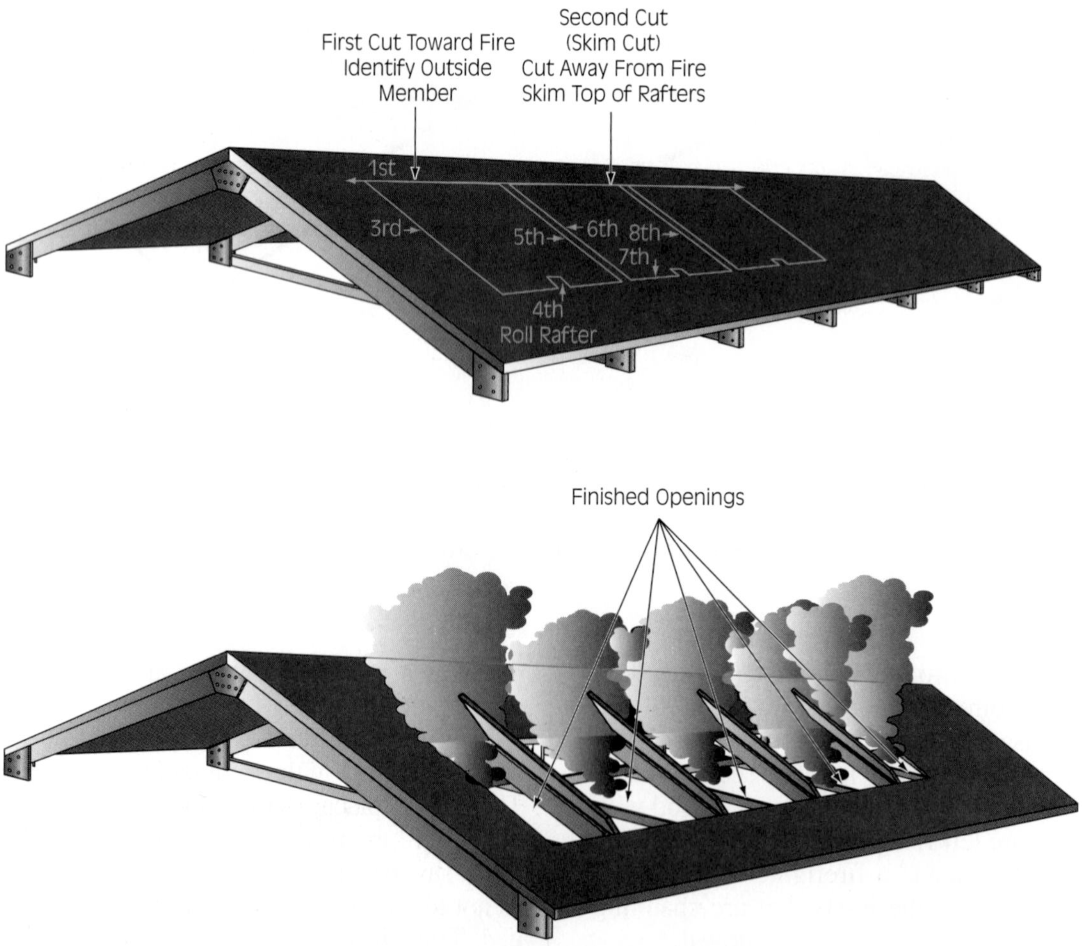

Figure 18-30 A louver opening is made by initially making several cuts parallel to and between the supporting roof joists. Then a connecting cut is made along the end of each cut perpendicular to the first set of cuts. Finally, each section is pushed down on one side and pulled up on the other. The result is a seesaw effect where panels tilt open and provide a hole for airflow. Because it is partially obstructed by each of the panels, it is not as effective as the total removal of the roof material. However, it is safer because there is less chance that a firefighter will fall into the hole. This type of cut is most effective on a plywood-sheathed roof.

duplicating this process until the cut meets the length of the head cut.

Lastly, the roofing pieces are pushed in, creating louvers along the cut and allowing the by-products of combustion to escape.

Triangular Cut

On some types of roofs, the use of the triangular cut technique is favored. A roof supported by an open web bar joist with Q-decking is the best candidate for this type of cut. Because the span of the web bar joist often exceeds 24 inches, the opening can become a funnel if opened using conventional square cuts, **Figure 18-31**.

The use of a triangular cut will help support the underlying Q-decking because it is interlocked. By cutting across the panel on a diagonal, the tendency to sag will be reduced.

> **Streetsmart Tip** Extreme care must be used in open bar truss type roofs because they are prone to failure at the early stages when this type of support system is directly involved.

Of necessity, these holes will be relatively small and might require several cuts to adequately perform the ventilation mission.

Trench Cut or Strip Cut

Purely defensive in design and execution, the trench cut is a roof opening that ventilates the cockloft area of a building where the fire is spreading under the roof, **Figure 18-32**. After the primary roof opening is made, the strategy might call for making a stop at a particular location. It is usually predicated on structural elements inherent in the building itself such as the connecting point of several wings or a place where the roof area narrows due to the presence of structural elements.

With the trench cut, the gases that are expanding horizontally under the roof are vented to the atmosphere and not permitted to pass a chosen defensive line, **Figure 18-33**. When deciding to use a trench cut, it is understood that the building on the fire side of the cut has been determined to be unsavable.

When a trench cut is being properly utilized, a hoseline should be positioned below the cut with ceilings pulled down to expose the opening. Any fire that actually reaches the opening and threatens to jump the gap should be extinguished with the hose from below. Under no circumstance should a hoseline be operated into a hole from the roof side. It will only serve to drive the heat and smoke down into the top floor environment. The purpose of this hole is to permit the laterally extending fire and heat to escape into the atmosphere where it poses little

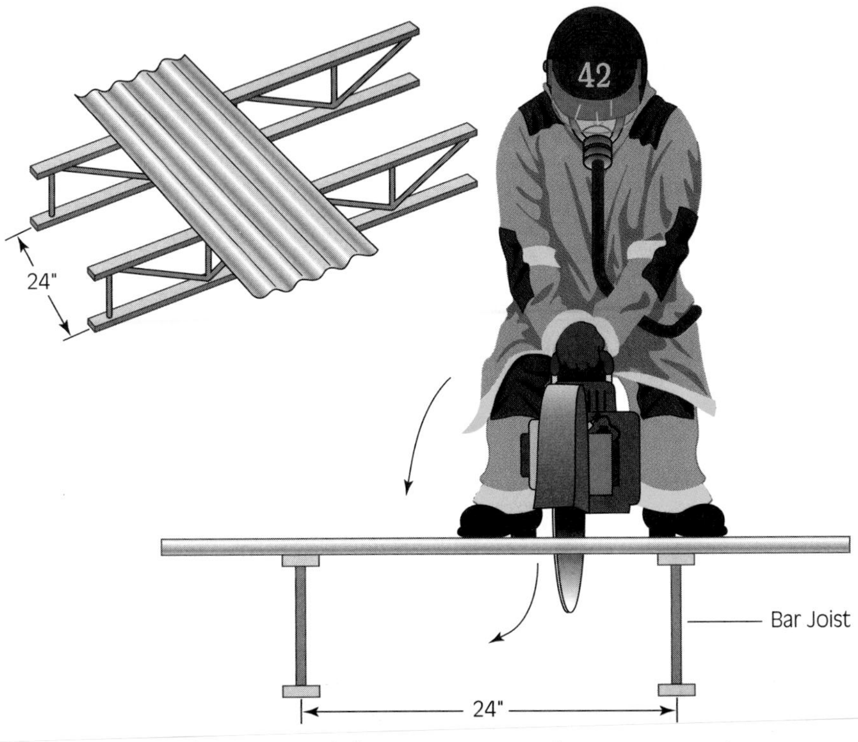

Figure 18-31 The open web bar joist-type roof structure requires extra vigilance. It is possible to cut away the supporting panel on which the firefighter is standing. Because of the large span between the supporting joists, the member can topple into the newly created hole.

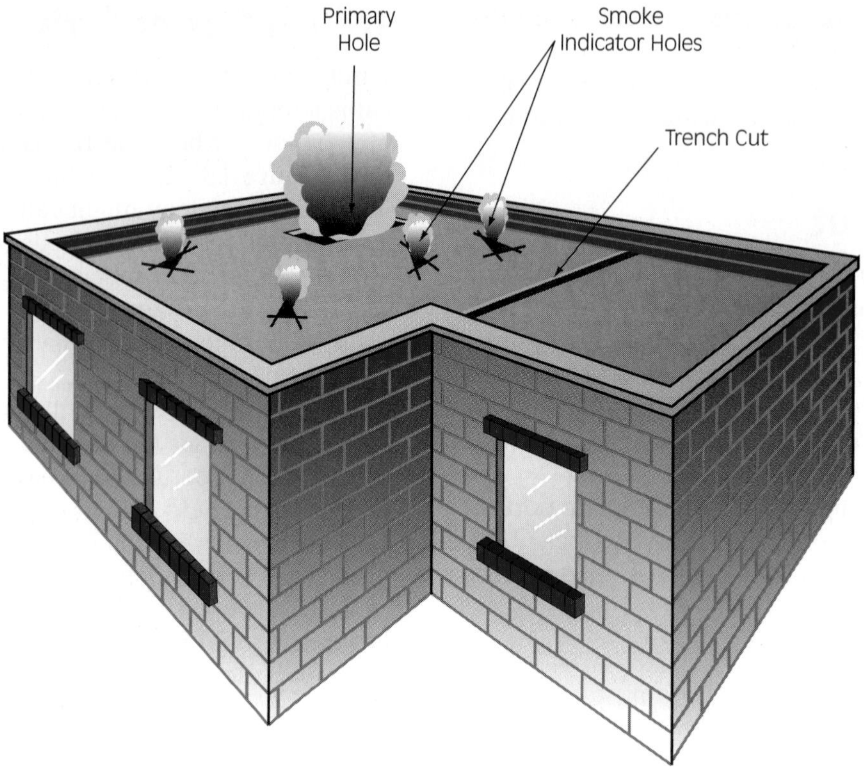

Figure 18-32 A trench cut is used to cut off fire extension. A totally defensive action, its use means that the fire side of the cut is being given up. Once opened, no firefighters should be operating on the fire side of the cut. A trench cut is (1) 2 to 3 feet wide, (2) not opened until the entire cut is complete, (3) not opened until all personnel are on the "safe side" of the cut, and (4) made using any available building features.

Figure 18-33 A trench cut is a defensive move. Ceilings should be pulled below the cut to promote vertical airflow through the trench. Additionally, a handline should be in place below the opening to cut off any horizontal extension.

threat. A hose may be operated on the roof to extinguish a roof fire, but it should never be operated into the ventilation hole.

Inspection Cut

The first operation that needs to be accomplished on a flat roof is the inspection cut. The inspection cut is placed into the roof to determine the following:

1. Roof covering and depth of covering
2. Roof sheeting material
3. Rafter direction
4. Conditions directly below firefighters

The first cut in establishing the inspection cut is made at 45 degrees to a bearing wall, followed by another cut opposite the first. Then the triangular inspection cut is completed with another cut, **Figure 18-34.** The goal is to identify the rafter and roll it when coming in contact with it. The inspection cut is larger than the smoke indicator hole and could pose a danger if placed in a path of access and egress to the ladder.

Smoke Indicator Hole

The smoke indicator hole is the only opening that will adequately determine the conditions directly below firefighters operating on the roof. The smoke indicator hole is a small triangular opening approximately large enough to push the D-handle of a roof hook through but small enough that the firefighters operating on the roof will not put a foot through the roof itself. A smoke indicator hole should be placed into the path of access and egress approximately every 15 to 20 feet of travel.

Figure 18-34 A triangular inspection hole cut is made by dropping the blade of the circular saw through the roof material in three intersecting cuts. If there is no support underneath, the triangle will drop into the cockloft. If there is a supporting member underneath, the roof piece can be tilted up by using a tool to depress one of the triangle points to obtain a purchase point to pull it up for removal.

SAFETY CONSIDERATIONS

When considering the placement of a hole for ventilation, whether it is horizontal or vertical, it is of paramount importance to consider the benefit gained against the possible liability created. In some cases the best choice will be *not* to vent at that particular location, for example, if venting would expose a victim and rescuer on a ladder to danger.

Will Ventilation Permit the Fire to Extend?

There is no justification for permitting a fire to extend in order to complete a task that is assigned. On occasion, an order may be given without full understanding of the consequences of that order because early reconnaissance is incomplete and fact gathering is still taking place. Some action may still be required but the firefighter should consider the possible consequences of that action and report to the incident commander for reevaluation if conditions are changing out of sight of that incident commander. Most incident commanders in the field would gladly countermand an order given when new information shows the original order to be unsound or unsafe.

Will the Escape Route Be Cut Off?

Of paramount importance to firefighters is their individual safety. When performing a ventilation activity, an eye toward escape must always be in the forefront. When venting a series of windows, the firefighter must work toward the escape point, **Figure 18-35**. There should always be two easily recognizable ways off a roof.

Escape routes should be lighted at night if possible and the presence of such escape routes made known to everyone on the roof.

Will Ventilation Endanger Others?

An overview of the operation in progress must always be in place when operating. Careful consideration must be given to what is called **mission vision**, a "blinder" view of the work being

Figure 18-35 Ventilate in the direction of the escape route so escape is not cut off.

performed because the mission of the individual is overriding all other activities. The activity of one firefighter performing a venting task must never endanger the position of another person.

When opening a roof, care must be taken to advise everyone else on the roof where the holes are. In dark, smoky conditions, it is easy to fall into the fire.

In addition, the torn-up roofing material should be cleared away as soon as possible to remove the tripping hazards. As the roof team pulls off the roof, they should pile it up in one place if possible. Thinking ahead of possible safety problems and working to remove them before they present themselves is safety in action.

Work in Teams

A firefighter should never work alone. The presence of a team member will afford a second set of eyes and another mind at work. It is very difficult for any one person to be aware of everything that is taking place at the same time. At times the presence of another firefighter will afford the opportunity to quickly discuss the manner in which to attack the problem at hand. A team member might make the difference between being located when a floor collapses or making a misjudgment resulting in a fall down a shaft. The presence of a team member could make the difference between being able to pry open a bulkhead door or lift a heavy skylight and not being able to do it because it is too difficult for one person.

Proper Supervision

Unfortunately, in some places, the presence of supervision is all too often considered to be a waste of personnel. There should be a person on the scene to make decisions when conflicting options are available. The presence of recognized supervision will ensure that the effort is unified. Supervision also typically brings a certain amount of experience to the activity, because supervisors are usually the more experienced personnel.

A supervisor's presence also helps to prevent team members from becoming too focused on the mission itself, because the nature of a supervisor's job is to oversee the whole operation of the sector and coordinate subordinates' activities to meet the objectives of the teams under the supervisor's command.

OBSTACLES TO VENTILATION

The importance of ventilation cannot be overstated. Because of the unpredictability of fire operations, firefighters will be confronted with many unforeseen circumstances that will delay ventilation activities. Listed next are some of the obstacles that might be encountered.

Access

Access should be one of the initial size-up considerations when arriving at the scene. The firefighter should first assess the needs of the ventilation objective, then determine the route to be employed to reach the location of the job performance task. It might be via the adjoining building, if present, by use of a ladder, or perhaps from the ground. Roof access might be delayed because of surrounding impediments such as overhead wires, elevated trains or roadways, light pole placement, or any other obstruction that might delay placement of a ladder or apparatus, **Figure 18-36**.

Access to a rear yard might be impeded by the presence of high, locked fences or by the presence of watch dogs. Alternative measures must be employed and thought out in advance before the obstacles are encountered. In routine nonemergency responses or activities, it is a good practice to map out an access strategy for a particular building and then assume that such a route would not be available. The constant practice of formulating a second and third plan on a routine basis will permit the ability to do so at a real emergency with greater ease. It is simply a matter of practicing the thinking process routine until it becomes a reflex action.

Security Devices

The presence of security devices can impede ventilation both in access and in timing. Building owners may have barricaded themselves and their possessions behind gates, screens, steel doors, and closed-up windows. In some cases, window openings have been blocked off and sealed, **Figure 18-37**.

In high-crime areas, some shop owners have even removed skylights and replaced the hole in the roof with cheap plywood attached to inferior structural support, all held in place by a few nails. Once covered with roofing, the former hole is no longer discernible and has been the cause of firefighter deaths as members walked on the covered-over holes and fell through.

> **Streetsmart Tip** These unknown alterations mandate a very cautious approach to roof operations and often result in delay of roof ventilation procedures.

Height

From multistory skyscrapers to one-story garages, the firefighter must be alert to the structure's ventilation needs. Sometimes the need to cut a hole in an area that is out of reach of the tool at hand or from a

Figure 18-36 Roof access by ladder can be slowed down when obstacles are encountered. Aerial ladder placement must take into account overhead wires, structures, and presence of utility poles before being set into place.

Figure 18-37 In the interest of security, occupants will place steel security screens or completely close off an opening that might serve as an entry point for a would-be burglar. Notice the structural material behind the steel screen on the window to the left. Any obstructions to a full opening will reduce ventilation.

particular position will be encountered. The need for a longer tool for reach or a rope tied to a tool or a ladder for access should be part of the initial size-up routine for the member assigned to ventilation duties. It must be assumed that reach will be a problem. What appears to be a single-story building from the street side can possess a topographical drop-off in the rear, placing the same floor that is at street level in the front several stories off the ground in the rear. The firefighter must practice thinking in a proactive fashion rather than reactive.

Poor Planning

One great obstacle to ventilation that is addressable is planning. Time is not a luxury. Proper on-site planning must occur without delay. A quick size-up and implementation of the plan are essential for timely ventilation. If ventilation is delayed, the interior team will suffer. Backdrafts or flashovers, decreased survival time of trapped victims, and arduous working conditions will result from delayed ventilation.

Personnel Assignment

A task that is assigned to a shorthanded or inexperienced crew will delay ventilation. If a ladder is needed to reach a setback in order to vent windows, and only one person is assigned to the task, not only will OSHA rules be violated, but objective attainment will be delayed.

Of course, there are times when adequate personnel are not yet on the scene. It is at this time that the ingenuity of the firefighter comes into play. With safety as an overriding concern the heads-up firefighter might be able to effect some ventilation while waiting for reinforcements. In a roof operation, a two-person team can open many openings relatively quickly while awaiting assistance. Structural components such as skylights, doors, roof access hatches, and ventilators can be easily opened or removed by individuals working together. A roof-cutting operation might need additional assistance but at least some openings are being produced to vent. Or two firefighters might be able to tie their tools together to get to a window that is out of the reach of just one tool. However, doing this while five other reachable windows are still not vented would be a poor use of resources.

Unfamiliar Building Layout

Especially in large buildings, the floor layout can be very confusing to the firefighter outside the building attempting to figure out which set of windows serves the fire area. This is particularly true of occupancy floor layouts that form L-shaped or U-shaped configurations in a building that appears to be square on the outside.

The building layout can also be an obstacle to a firefighter attempting to reach the rear of the building. Building wings, fences, lower floor extensions, or unusual configurations of the building's dimensions can confuse the firefighter and delay access. Some occupancies such as contemporary cluster-type construction townhouses can have door entrances on the same floor but serve three different levels. For example, there could be three entrances on the same floor. One door opens to a stairway that leads to the apartment on the floor below, one opens on the same floor as the entrance doors, and the last opens to a stairway that serves the apartment on the floor above. To the firefighter attempting to get to adjoining occupancy on the floor above the fire to vent, the need to force open three different doors might be necessary and will delay ventilation.

Observing structures while under construction, requiring building owners to identify the occupancy served, conducting familiarization drills, entering pertinent information on response ticket critical information (if such a feature is available), or simply entering an uninvolved lower floor occupancy to look at the floor layouts are just some of the ways to address this problem. Often the same floor plan exists throughout a multistory building to reduce construction costs. Utility supply voids will often dictate the "rubber stamp" floor layout of the multistory occupancies.

Ventilation Timing

Venting too early can lead to fire extension; venting too late causes unnecessary punishment to the interior forces and can even prevent forward progress. The possibilities of backdraft and flashover have already been discussed. The timing is dependent on the type of ventilation being performed.

Vertical ventilation of stairways, hallways, and any paths of egress and ingress are paramount and must be effected without delay. If fire is known to be in the area under the roof boards or on a top floor extending into the cockloft, opening of the roof must commence without delay.

> **Note** The importance of early removal of the heat and smoke through vertical channels cannot be overstated.

As with all facets of firefighting, however, nothing can be written in stone. The opening of an entrance door to a fire occupancy on a lower floor that will vent up the stairs and out the roof opening must be carefully planned. There may be situations where escaping building occupants are coming down the stairs or firefighters are moving up the stairs. In such situations, vertical ventilation through this channel must be timed properly and may even have to be delayed. Delay might be necessary when opening a roof cut because the opening might expose an adjoining building to extension. A hoseline might have to be put into place to protect the exposure before the cut can be opened.

Cut a Roof—Open a Roof

Cutting a roof and opening a roof are different processes and this distinction should be clear to firefighters. Cutting a roof is the process of making the necessary cuts to perform the ventilation. Once the cuts are made, the opening of the roof can begin. The opening refers to the actual removal of the roof material. Therefore, a roof can be cut and not yet opened. There might be situations where a delay is necessary in opening the roof because of exposures or line placement or some other reason. This does not necessarily stop the cut. The cuts can be made but the opening delayed until the time is right.

FACTORS AFFECTING VENTILATION

Several factors can affect the attainment of proper ventilation. Partial openings, screens, type of roof material, wind direction, weather, building size, and construction features are some of the more prominent elements.

Partial Openings

Research has demonstrated that one single opening of a given size has greater ventilation capability than several holes that equal the same opening area. Much of this has to do with airflow characteristics. Like water, airflow is not like a chain of molecules that stay in line and move along in single file. Air is made up of many individual molecules that will flow in generally the same direction, but will tend to drift randomly in all directions. Air will move in the direction of less resistance. The eddy currents that can be observed when smoke is released into an unconfined and unchannelled area demonstrate this. If under enough pressure, smoke appears to billow.

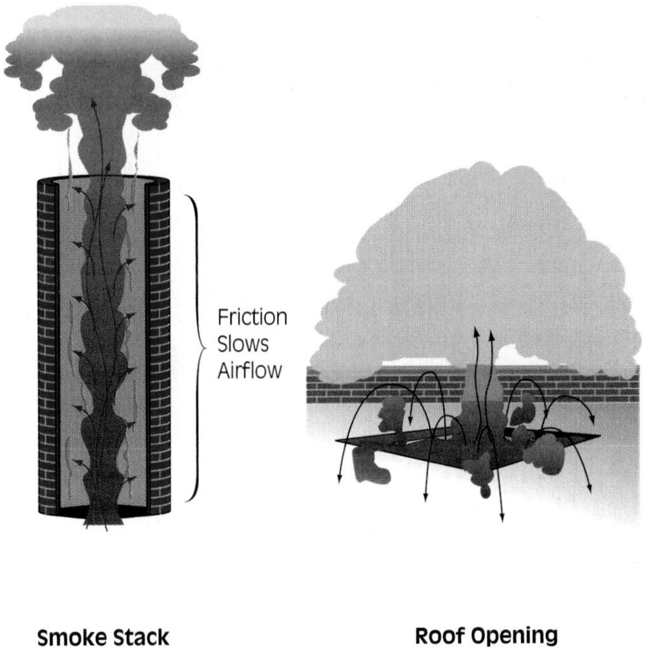

Figure 18-38 Airflow is reduced by friction.

Looking at a cross section of a chimney, it is clear that smoke rubs along the edges but tends to flow in a more uniform manner in the center. The greater the circumference of the vessel (i.e., smokestack) the less friction on the sides and the greater the flow. Once released into the atmosphere, the column of smoke continues to rise until the central core of the column is finally exposed to the reduced outside temperatures and begins to stratify. However, along the outside perimeter there is a constant peeling off of the smoke in eddy currents. The eddy currents will tend to reduce the flow, **Figure 18-38**.

Transforming the large picture of the smokestack into a roof opening, the same principles can be seen at work. With each opening, the outside perimeter of the opening will afford the airflow opportunities to eddy and slow down the overall flow. This is the underlying factor in the principle that many small holes will not be as efficient as one large hole. The large hole will have one perimeter of a given size. The many small holes will have a greater overall length of perimeters. The greater the length of perimeter, the more opportunity to reduce the speed of the airflow.

Partially Broken Windows

The same principle just described comes into play again. If a windowpane that is square is broken and many crescent-shaped shards of glass are left in place, not only is the area of the opening reduced but the presence of the shards creates more perimeter opening distance, **Figure 18-39**.

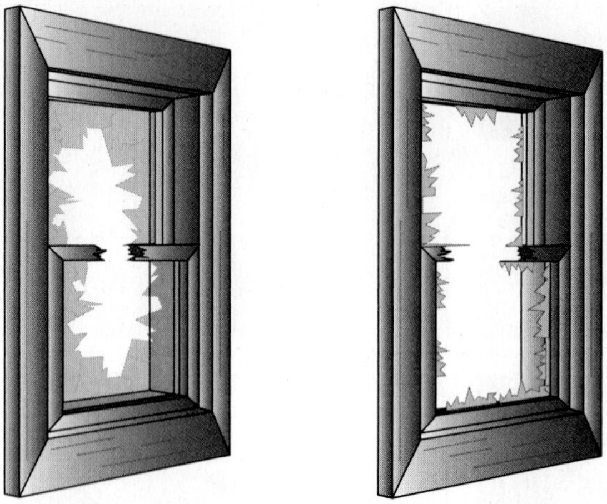

Figure 18-39 Airflow is greatest through a window where glass is fully removed. Screens, shades, curtains, and window cross members should also be removed.

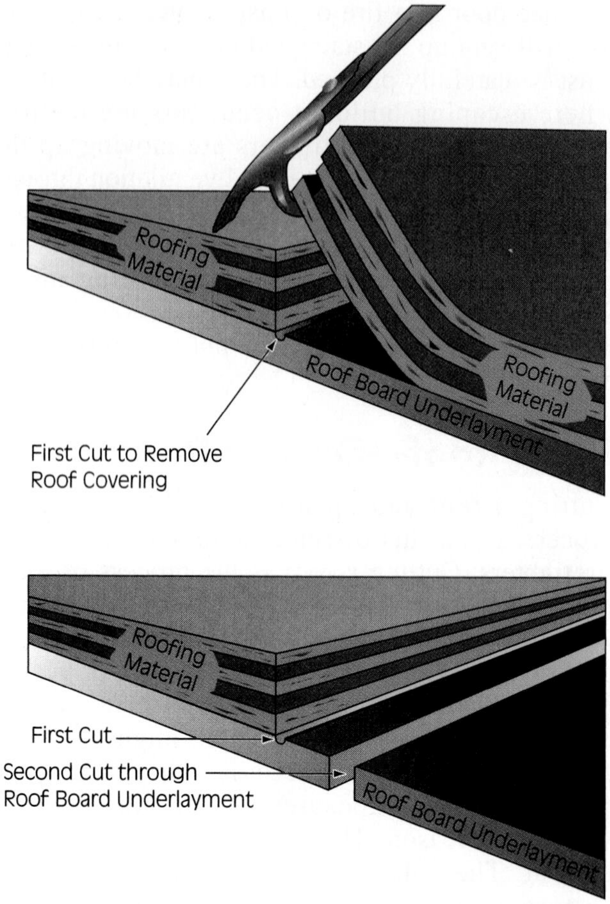

Figure 18-40 If a roof is thick with roof covering, two cuts might have to be made. The first cut is made shallow to remove the covering. It should not pierce the sheathing to prevent smoke from interfering with the completion of the first cut. With the second cut, the roof sheathing is cut and then removed.

Screens

The presence of insect screens in a window that has been broken out is magnified when one considers that a screen is like a solid panel with hundreds of holes.

> **Note** Conservative estimates say that the presence of a screen in a window reduces the airflow by approximately 50 percent.

Put into real-time numbers, failure to remove a screen after taking out the window is like opening 100 windows, then permitting a shutter to close on 44 of them. The effort is wasted and the ventilation is greatly reduced. Removal of the window treatments, curtains, shades, or blinds is also important. Any obstruction in the opening will reduce the airflow.

Roof Material

In some buildings, particularly those of older vintage, the roof material could be several layers thick and several layers might need to be removed before the cut is opened. When planning the cut, the firefighter must be aware that the kerf cut of the saw will usually permit a thin line of smoke to vent. If several layers of roofing must be removed, the original cut should not penetrate the under roof area. The cut should only go as deep as the roofing material that will be removed on the first pass, **Figure 18-40**. The smoke from the kerf cut can be enough to obscure vision and impede the initial roof material removal.

In some types of construction, the roof material might actually prevent the opening of a roof. The purpose of a roof is to shed water. Some materials are well designed for that purpose but are deadly in a fire. In the case of a corrugated metal roof over an open bar joist support structure, the heat from the fire melts the underside of the waterproofing material, usually some form of petrochemical product. Flammable gases are created and trapped below the cooler upper roof and tend to collect in the troughs of the correlated roof. When the heat generates enough pressure, the gases find a way into the underside of the roof through the joints of the metal. The presence of this flammable gas in the cockloft area can be explosive when reached by fire. Additionally, the unsuspecting firefighter might be providing the missing oxygen as the roof is cut. A kerf cut is all that is necessary to open an oxygen supply route. Usually without warning the roof will erupt into fire at the place of least resistance—most often the place where the roof has just been weakened by the saw cuts.

Special care must be employed and constant study of the latest innovations and techniques is essential for the firefighter to maintain and improve a margin of safety in a known hostile environment.

Other types of roofs also have their particular strong and weak points. Any type of truss roof is always suspect because of the nature of its collapse potential. Placement of any personnel on this type of roof must be very carefully considered. The full extent of the fire must be known before commitment of personnel to this type of roof for ventilation is ordered.

In many cases, one roof has been built over another, sometimes with several feet between them. Obviously, the opening of the upper roof will have no effect on the ventilation of any spaces below the under roof. The effects produced by the opening made must be evaluated by the firefighter. If there is only a little smoke and heat being vented from an opening that is over an obviously large body of fire, the incident commander should be notified and an investigation conducted to determine why. The notification lets everyone at the scene know that something is not right and the investigation determines why it is so in order to address the problem and complete the objective.

Dropped or Hanging Ceilings

Like a second roof built over the original, the installation of dropped ceilings creates a similar impediment to the ventilation process. Trapped air pockets will conceal fire and smoke, and the raging fire on the floor will not be able to vent through the opening in the roof. In older buildings, especially of the commercial type that have a regular occupant turnover, several hanging ceilings may be in place.

The space between ceilings is a haven for fire gases to collect. Venting them can be a real challenge because the dropped ceilings are (1) not always obvious, (2) often out of reach of the roof firefighter, and (3) difficult to open from either above or below. The space is usually large and is ripe for a backdraft that will cause the entire ceiling to drop into the firefighting forces below.

Advancing Under a Dropped Ceiling

The airflow in front of a nozzle can work its way into the space above a hanging ceiling, especially if of the grid and insert type. The fire is pushed into the space when the heat lifts the tiles like trap doors. Heat and fire gases fill the space and, when conditions are right, can ignite and blow down the whole grid onto the advancing firefighters, trapping them in a web of metal grid.

Streetsmart Tip When entering an occupancy with a suspended ceiling, the first thing that should be done is to have a tile pushed up from the safety of the doorway into the area to determine the conditions above the ceiling.

If fire is already in the space, the tiles can usually be unseated with the hoseline merely by directing the stream upward and extinguishing fire in the ceiling while advancing.

Building Size

The building size will affect attempts to ventilate. A building's height and width both have profound effects on the ability to move air.

In tall buildings a phenomenon called a *neutral plane* can occur. It is the location where smoke will collect instead of continuing to rise. Fireground experience has found it to inhabit just a couple of floors in some cases, and other experiences have had it collect on up to ten floors many stories above the fire floor. Although certain texts give a particular number of stories above a fire floor where it will occur, the fact is that it will occur where it occurs. There is no good rule of thumb to apply because in most cases, the numbers are laboratory generated or results of carefully structured tests in existing buildings. In a fire, because so many variables are at play, there is just no way to predict accurately where this plane will occur if it occurs at all.

Factors that affect the development of a neutral plane include the presence or absence of HVAC systems and their associated ducts; wind direction around the building; presence of other buildings in the general area and height in relation to the structure on fire; outside temperature and its relative difference to the structure temperature inside; and presence or absence of smoke shafts in the building. The only way to determine what the smoke is doing in the upper floors of a high-rise building is to have reconnaissance teams report on their findings.

In large buildings, air movement can defy normal expectations. Wind direction on the outside of the building can create a negative pressure on one side of the building and cause the smoke to gravitate toward that section. This can become very confusing when smoke is being reported on a side of the building that is remote from the fire. Ventilating for such a fire can be a monumental task, and careful analysis of air movement as reported by reconnaissance teams must be conducted in order to maximize a coordinated ventilation effort.

The smaller the structure, the greater the similarity the air movement will be between buildings of the same type. Light breezes will not usually be much of a factor in a residential 2½-story wood frame building surrounded by trees and other dwellings. However, the same breeze in the same structure that sits high on an unprotected hill might exhibit very different characteristics. The same goes for a building that is facing a wide river or ravine as opposed to a structure that is surrounded by buildings of similar height, even if they are 70, 80, or more feet high. There are no absolutes in this area. So many variables factor into the formula, only experience, observation, and common sense can prepare the firefighter to observe and predict what will most likely be the result of a given situation.

Weather

Some general observation can be made about how weather will affect air movement. On cool dry days, smoke-filled structures will vent quickly. On rainy humid days, the smoke will lift slowly. Snow affects the air movement and a light breeze will tend to "pull" the smoke out of the upper vertical openings. There is no magic to this knowledge. It goes back to fire behavior and what makes smoke rise.

The heated, smoke-filled air rises because the heated air weighs less than the surrounding air sample. In humid environments, the air is actually cooled and that slows the vertical velocity of the plume. While the fire is still pumping smoke out, the column flattens quickly as it cools and might not even leave the structure at the upper levels. That will tend to make it stay low and not lift.

Rain will have the same effect with an added element. The droplets of water will, in addition to cooling the superheated gases, actually impede the vertical flow physically. There are two forces at work: (1) The heat is lifting the smoke particles, and (2) the water droplet is colliding with the airflow as gravity pulls it down resulting in a diminished vertical airflow. The ultimate result is a slowing of the venting process.

Snow affects the rising column of smoke even more profoundly. If the rising air is hot enough, the snow melts, robbing the column of heat and cooling off the air volume. The result is water that further cools down the thermal plume. Snow adds an additional factor of air resistance. Since it has weight, the snowflake will actually force the thermal plume to work harder to rise. The combination of the additional weight to lift, the cooling of the snowmelt, and the water that does further cooling works against the venting process. One can expect difficulty when trying to ventilate vertically in snow.

Horizontal venting, however, is not necessarily affected the same way because gravity or thermal plume generation does not generally affect it. Heated, pressurized air will look for a path of least resistance and will find an opening in a wall (i.e., window or door) regardless of the weather conditions. However, once outside the structure, the horizontal venting turns to a vertical direction and up the side of the structure. At that point the smoke will be affected as described earlier. Once the fire is extinguished and the source of the heat is removed, the ambient atmospheric conditions will tend to move indoors. When the humidity reaches the inside of the structure, the smoke will then tend to stay low and not lift.

Great success can be achieved with positive pressure ventilation or with the use of a hoseline stream to create negative pressure and pull the smoke out. A fog or spray stream can be used to move air by opening the nozzle and directing the stream out the window or door. The droplets of water actually push the air in front of it (positive pressure) and pull the air behind it (negative pressure). The overall effect is the creation of an artificial airflow that can be very successful.

Opening Windows

Opening windows is the simplest way to open a compartment. Not every fire requires the removal of glass. Sometimes areas unaffected by fire can become filled with smoke and merely need to be vented to remove smoke. For many years firefighters were taught to open windows two-thirds from the top and one-third from the bottom to effect ventilation. That concept has changed. Based on what was previously discussed it is now known that one large hole is more efficient than several small holes, so one full sash opening is better than two equal in area. The same holds for windows with some exceptions.

Generally, it is better to open the top sash fully and let the airflow out. The replacement air will come from the open door and the lower levels of the room. There are exceptions to this rule of thumb, **Figure 18-41**.

If the smoke condition from the door opening will make the room conditions worse, then the choice would be to close the door and open the window. If there is one window, the two-thirds/one-third rule is appropriate. The heated upper air will vent from the upper level and the cooler fresh air will enter through the lower opening.

If there are two windows, one should be fully opened at the upper sash and the other opened fully from the bottom sash. Obviously, the greater the opening at the upper level, the larger the heated air volume that can be allowed to escape.

If there are three windows, two can be opened at the top and one at the bottom. If there is a breeze or wind blowing into the outside wall on which there is a window, the window facing the breeze should be

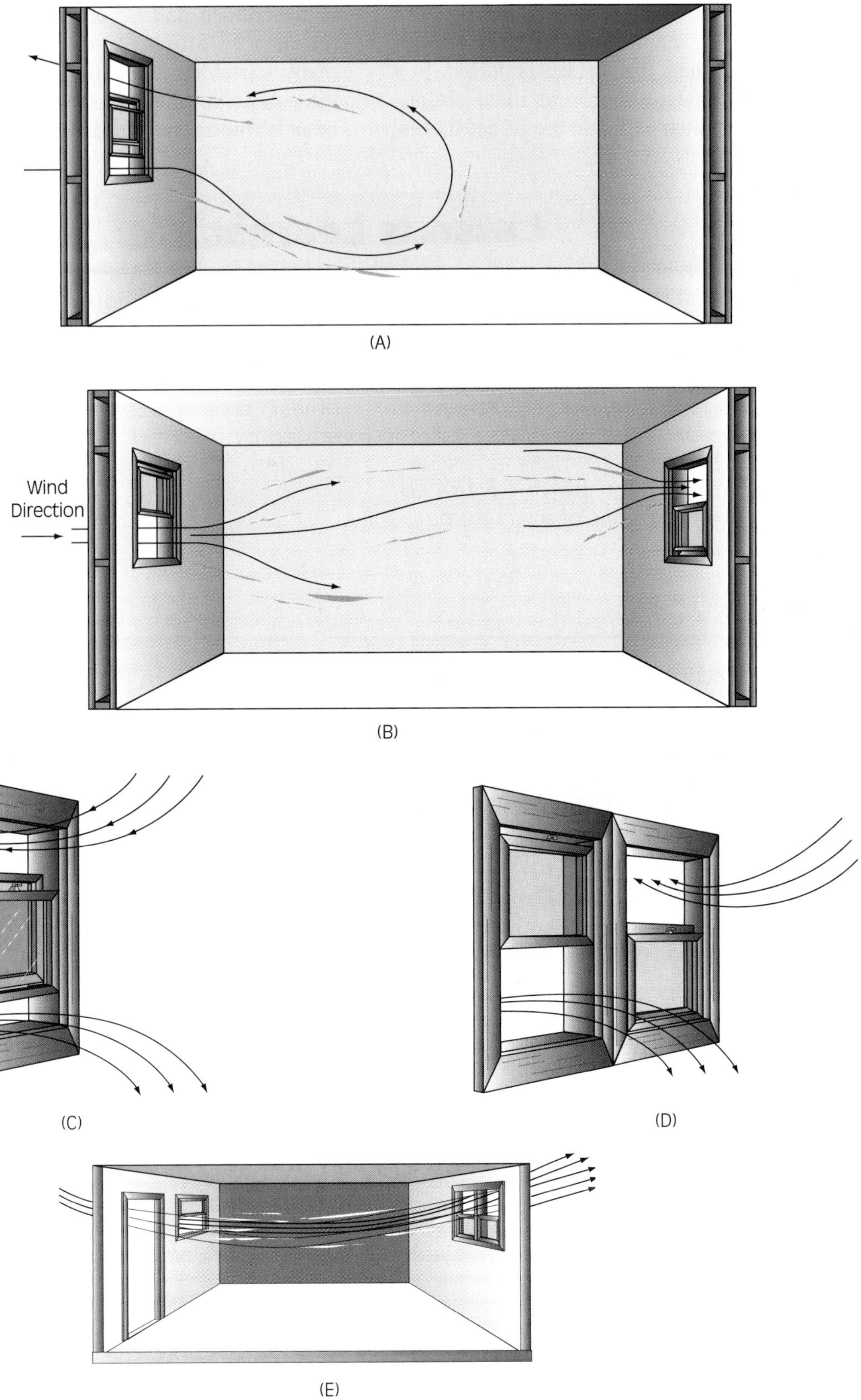

Figure 18-41 Sash windows should be opened to maximize natural airflow characteristics while maintaining the principle that large openings are more efficient than many small openings. Heat rises and cool air sinks. Depending on the number of windows and their placement, various different combinations will be effective. (A) Single window in a room. Open it two-thirds from the top, one-third from the bottom. (B) Two windows in a room. Open bottom sash of window facing wind fully. Open opposite window fully from the top sash. (C) Single window; another view. (D) Double window. Open one sash fully from the top and the other sash fully from the bottom. (E) Cross ventilation. Open window sash facing wind fully from the bottom and the others across the room fully open from the top.

opened from the bottom. The forced air will then actually pressurize the room and the vent/exit opening will be at the upper levels where the heat has collected.

The preceding scenarios represent ideal conditions. The firefighter must evaluate the effect that is being created and be prepared to adjust the plan if the desired effect is not forthcoming. Again, the many variables that tend to be present can change the rules. Nothing can be assumed and everything must be monitored until the incident is over.

Lessons Learned

Ventilation is a tool that is used in firefighting just as any other tool is used. It must be understood, manipulated to the greatest advantage, and used carefully. Its proper use can make the difference between extinguishing a fire and creating a conflagration. Its judicious use can enable a firefighter to enter a structure to make a rescue. Used inappropriately it can permit a fire to extend into uninvolved portions of a structure.

There are no hard and fast rules beyond some very simple truisms. Heat rises and cold air drops. Heated air expands and cooled air contracts. Natural air movement will follow the path of least resistance. Airflow can be artificially generated by mechanical means. Beyond these, the rest depends on the conditions.

KEY TERMS

Cockloft The area between the roof and the ceiling.

Diffusion A naturally occurring event in which molecules travel from levels of high concentration to areas of low concentration.

Frontage The portion of a property that faces and actually touches the street.

Guard Dogs Trained animals that will bark and attack an intruder.

Horizontal Ventilation Channeled pathway for fire ventilation via horizontal openings.

Ignition Point The temperature at which a substance will continue to burn after the source is removed.

Lifting A term used to describe the removal of upper level smoke and heat when cool air replaces the upper level hot air that is escaping.

Mission Vision A term used to describe a condition in which a person becomes so focused on an objective that peripheral conditions are not noticed, as if the person is wearing blinders.

Pipe Chases A construction term used to describe voids designed to house building water supply and waste pipes. The term *electrical chase* is used for wiring.

Thermal Level A layer of air that is of the same approximate temperature.

Vertical Ventilation Channeled pathway for fire ventilation via vertical openings.

Watch Dogs Trained dogs that will bark and create a commotion, but will not attack.

REVIEW QUESTIONS

1. Define ventilation.
2. What are the advantages of venting a structure?
3. Name the two types of ventilation.
4. Define and explain the terms *backdraft, flashover,* and *rollover.*
5. Describe an expandable roof cut.
6. Describe the concept of the two types of forced ventilation, negative and positive.
7. What is the difference between opening a roof and cutting a roof?

Additional Resources

Carter, Harry, *Firefighting Strategy and Tactics.* Fire Protection Publications, Oklahoma State University, Stillwater, OK, 1998.

Casey, James, *Fire Chief's Handbook.* Technical Publishing Company, New York, 1978.

Mittendorf, John, *Ventilation Methods and Techniques.* Fire Technology Services Publishing, 1988.

CHAPTER 19

FIRE SUPPRESSION

Dennis R. Childress, Orange County Fire Authority

OUTLINE

- Objectives
- Introduction
- Elements of Fire Control
- Tactical Considerations
- Lessons Learned
- Key Terms
- Review Questions
- Additional Resources

Photo courtesy of Central Net Fire.

STREET STORY

The Monroeville Fire Department's Rescue 4 arrived at the scene of an apartment building in the early morning hours to find heavy smoke and fire venting out of a basement apartment. The building was a **garden apartment** *structure with two stories in the front and three in the rear because of the topography, since it was built into a hillside. Each floor had four apartments between stairwells. Each stairwell was protected by firewalls or separation walls. The basement apartment was fully involved in the rear. The crew stretched a preconnected 1½-inch hoseline through the front door and down the steps to attack the fire and protect the interior stairwell.*

The other first-alarm units arrived and began normal fireground operations, search and rescue, water supply, ventilation, and secondary fire attack lines. The arriving chief quickly ordered the second alarm.

Upon my arrival, on a second-alarm engine, the fire had **autoextended** *up the rear of the building, entering the cockloft, and was venting through the roof in several areas. The third alarm was sounded. We were assigned the task of stretching a 2½-inch handline up the southern stairwell, pulling the ceiling, and supporting the firewall. By supporting the firewall, I mean we were to ensure that the fire did not get past us into the next set of apartments. We were assisted by a second engine company with a 1½-inch handline. Although an uncommon procedure today, we operated for the duration of our SCBA bottles, exited for fresh bottles twice, and returned.*

After our third SCBA bottle we exited the building and were told to take a break. By this time the fire was under control, forward extension stopped, and the fire would just require overhaul. It was daybreak. The scene in the daylight was impressive. At the height of the fire, three elevated master streams had operated over the main body of fire while six handlines supported interior firewalls, preventing further horizontal fire spread.

What I saw and experienced reinforced everything I had learned from fire academy instructors and through training. For a safe and effective outcome, the suppression of fires requires a coordinated attack, with crews working under a controlled command system and all personnel carrying out the assignments given them. Firefighters must understand the importance of fire suppression techniques and understand the role they play on the fire scene. At this fire, spread was prevented by the incident commander knowing the proper application of fireground tactics and the firefighters understanding suppression methods, application of water, and working as a team.

—Street Story by James Angle, Chief, Palm Harbor Fire Department, Palm Harbor, Florida

OBJECTIVES

After completing this chapter, the reader should be able to:

- Identify structural fire considerations to be made prior to extinguishment.
- Explain the process of fire moving from contents to structure.
- Discuss the resources a fire department considers important in fighting fire.
- Explain the fire tetrahedron in relation to wildland or ground cover firefighting.
- Discuss some of the features of topography to be considered in wildland firefighting.
- Identify some of the automotive structural dangers to firefighters in vehicle fires.
- Explain some of the components of flammable liquids and gases that affect firefighting.
- Define a plan of action for fighting fire regarding attack modes and styles.
- Explain fire stream selection considerations that must be considered in firefighting.
- Explain tactical goals to be considered when fighting fire.
- Discuss the incident management system and how it affects the way fires are controlled.
- Explain the difference between offensive and defensive modes of fire attack.
- Discuss teamwork and its part in firefighting.
- Explain the concept of the two in/two out rule and how it affects structural firefighting procedures.
- Explain why caution must be observed in the overhaul phase of fighting fire.

INTRODUCTION

This chapter gives the reader information that pulls together all of the other information presented to this point. Like a recipe, which lists ingredients and then explains how to blend and cook them, this chapter explains how to put into action the techniques and methods that have already been discussed.

This chapter looks at some of the common types of fire as well as some common techniques of firefighting in different disciplines. It opens with the more common elements of fire control, detailing most of the areas with which average firefighters will be involved during their careers. In this section, seven subjects provide information needed when working in various situations common to the average fire department. A second section details tactical considerations. At the conclusion of this chapter, the reader will have a good foundation of knowledge in fighting fire and tactical situations involving a number of elements common to today's fire service.

ELEMENTS OF FIRE CONTROL

Before firefighters can respond appropriately to an emergency involving fire suppression, they must know the basic principles involved in the processes that create and sustain fire. This section deals with those areas common to the average fire department. It discusses structural firefighting elements, ground cover or wildland firefighting elements, vehicle fires, flammable liquid and gas fires, the process of fire extinguishment, and proper stream selection.

In a previous chapter the fire tetrahedron and its importance in the suppression of fire was discussed. The elements of that tetrahedron are mentioned again here in an effort to instruct firefighters in safe, efficient, expedient fire control. A simple rule is that the earlier the fire department arrives on scene, and the more knowledge they bring with them in the fire suppression process, the lower the losses and the less risk taken.

Structural Fire Components and Considerations

When discussing structural fire components and considerations, it makes sense to look again at the fire tetrahedron, **Figure 19-1**. Most structural firefighting involves the suppression of Class A materials within the structure or as part of the structure itself. These elements are commonly suppressed by removing one side of the fire tetrahedron. The typical fire

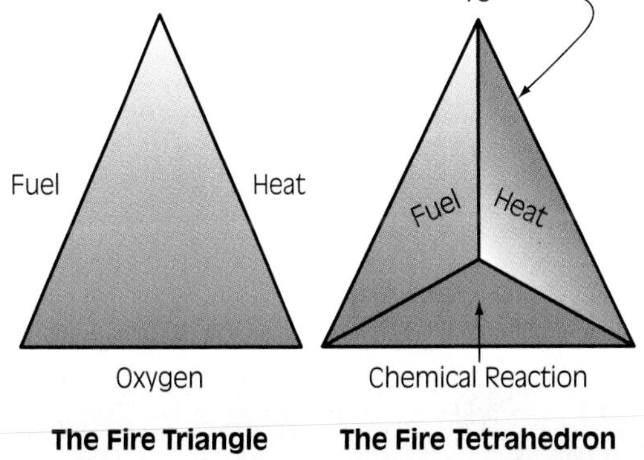

Figure 19-1 The old and new ways of visualizing the combustion process, the fire triangle and the fire tetrahedron.

department will respond with water as an extinguishing agent and suppress the fire by removing the heat, oxygen, or both as quickly as possible. This may not seem difficult in its purest form but can be extremely complicated or hazardous if not done properly and safely. A number of structural firefighting considerations must be made prior to the extinguishment, or a great number of things can go wrong.

Listed in random order are a number of factors that must be taken into consideration.

1. Length of time the fire has been burning
2. Building construction materials
3. Occupancy type and contents
4. Resources available (amount of water, staffing, equipment, etc.)

Taking these factors in order of listing and not in order of importance, recommendations can be made as to how a structure fire should be fought.

The length of time the fire has been burning is an important factor because it can determine the stability of the structure and the amount of fire the responding units will face. It will also determine which stage of involvement the fire is presenting.

As discussed in previous chapters, flashover and backdraft are extremely dangerous fire phenomena that are predicated on the stages of fire in structures. In **Figure 19-2** a flashover is shown as it occurs over the heads of a couple of firefighters. Based on length of burn, the integrity of the structure itself can be in jeopardy.

> **Caution** Most structure fires begin in contents and then spread to the structure itself. If not contained, this structural involvement can create concerns about collapse, both inside as well as outside.

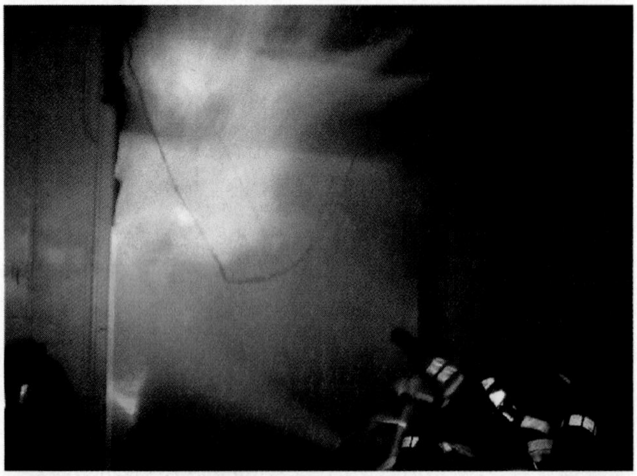

Figure 19-2 A flashover may occur in the early stages of a structure fire. *(Photo courtesy of Phill Queen)*

This brings up the second factor: the building materials themselves. Is the structure made of highly combustible or noncombustible materials? How long will it take for the fire to become involved in the structure itself? **Figure 19-3** shows an apartment fire that began in its contents and then burned into the structure. Chapter 13 details the types of construction materials and demonstrates the dangers involved in structural firefighting.

The third factor listed is the building occupancy and contents. Building occupancy means the proposed use of the structure. Is it designed to be an automobile repair shop or a preschool? Is it a retail store or a board and care facility? Each of these occupancies presents different fire attack protocols. The life hazards, the entrance accesses, and the ventilation components are among many details associated with the occupancy types that have to be taken into consideration by the responding units. Along with the occupancy, the building contents must also be considered. In most cases, the contents will be predicated by the occupancy type. The preschool will have dramatically different contents than the repair garage. These contents will have a bearing on the suppression efforts and control problems encountered by responding units.

Figure 19-3 A contents fire has burned into the structure itself. *(Photo courtesy of Rocco Di Francesco)*

The fourth factor listed is resources available. Fire suppression involves all of the factors listed, as well as the resources available to the firefighting team. Some areas of the country have limited water supplies while others have practically unlimited water. Some departments have large numbers of responding apparatus while others have only one engine with the second responding unit a very long distance away. Some areas of the country have staffing problems while others do not. Some areas have unlimited training opportunities while others do not. These and many other situations are considered when dealing with response resources. These resources—or the lack thereof—have a great bearing on the tactics and strategy used in fighting structure fires.

Structural fire components and considerations are a large part of the responding agency's responsibility. Fire service personnel must be well schooled in all of these factors and situations so as to ensure efficiency, safety, and responsibility.

Ground Cover Fire Components and Considerations

In some areas of the country ground cover fires are a very large part of the fire service's work. Not all departments have jurisdictions with open areas that burn, but those that do find it to be a responsibility that is not taken lightly. In fact, large campaign fires are usually wildland (ground cover) fires. These fires, being so large and involving so many resources, usually cause the first-in departments to utilize other departments from miles around, and in some cases out-of-state departments, to assist in the attack. **Figure 19-4** shows a large wildland fire and its impact on the surrounding area. So, even if a department does not have any open land to burn in its response area, any firefighter in any department could be summoned to help in another agency's wildland area. This being the case, it is a good idea for all firefighters to know the components and considerations involved in ground cover firefighting.

Figure 19-4 The Baker fire in Southern California lays out a large footprint on the area. *(Photo courtesy of Craig Covey)*

In Chapter 4 the fire tetrahedron was discussed. In the previous section of this chapter the same tetrahedron was mentioned, but in the field of ground cover firefighting a fire triangle is used, **Figure 19-5.** In wildland firefighting this triangle reflects a different set of principles related to a different set of conditions when compared to structural firefighting. Each side of the triangle can be broken down into smaller components in order to understand the components and considerations of ground cover or wildland firefighting.

Weather

This component of the wildland fire triangle is the most dynamic of the group and hence is discussed first, **Figure 19-6**. This is because of its dramatic effect on wildland fire conditions. It changes all the time, including any number of times in a given day. Given similar fuels and topography, wildland fires will burn in a radically different way in different weather conditions.

> **Caution** Weather is one of the main causes of firefighter fatalities on wildland fires. In fact, whole teams of firefighters have been caught and killed on wildland fires a number of times over the years because of weather changes that were not recognized in time to change tactics or locations.

> **Caution** Frontal changes (cold and hot), high and low pressures, winds, and storms all have direct and dangerous effects on wildland fires.

Each of these elements can change in very little time on any given day. It is also a fact that some ground cover fires grow so large that they can create their own weather.

A number of safety rules or laws apply to wildland firefighting, based on years of study and practice. A great number of them pertain to the weather and its dangers to the fire environment because this area of wildland firefighting has caused so many injuries and deaths among the firefighter ranks over the years.

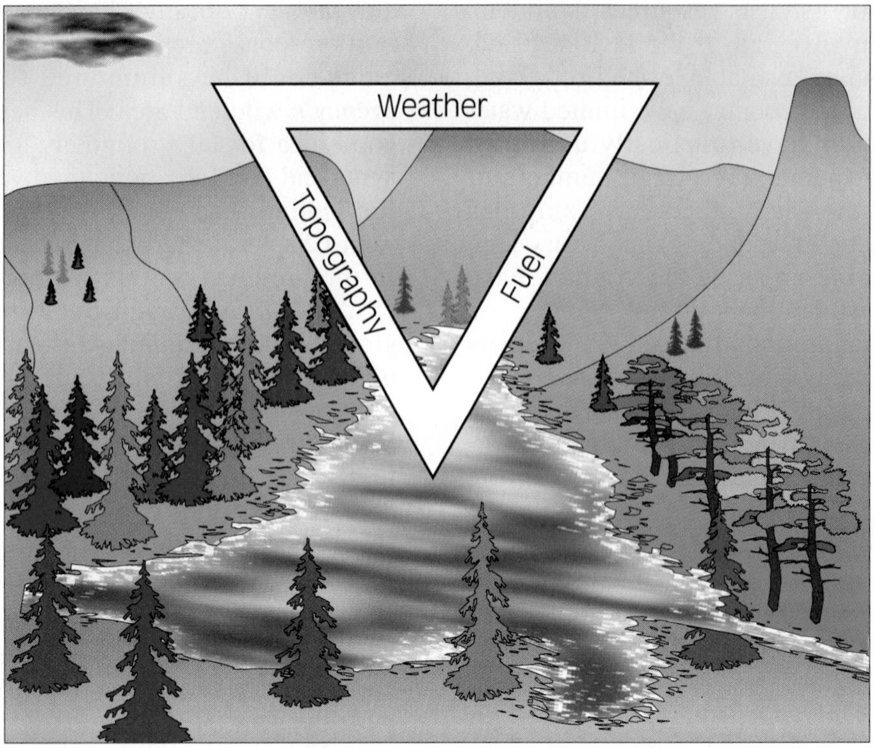

Figure 19-5 The wildland fire triangle differs from the structural fire triangle.

Figure 19-6 The components that make up the weather side of the wildland fire triangle.

Fuels

In the wildland environment, there are a great number of fuel types, **Figure 19-7**. From fine grasses to large timber, the fuels burn differently. There are two main variables in fuels and their effects on fire. These are **rate of spread** and **fire intensity**.

The rate of spread is the speed with which the fire travels across the ground or through the bushes and trees. This rate of spread is affected by a number of variables, all of which are in the wildland fire triangle. In the case of fuels, it can be said that the lighter or finer the fuel, the faster the rate of spread. Given similar weather and topography, fire will run much faster in grasses and light vegetation than in heavy brush or timber. **Figure 19-8** shows a grassy fuel fire moving quickly across the terrain. In fact, this deceptiveness has caused a great number of firefighter deaths and injuries over the years. Numbers of firefighters have been overrun in grassy fuels thinking they could "outrun" the fire when things went bad. It is a sad fact that many firefighters have been caught and burned when running from "light fuels" wildland fires.

Figure 19-8 Grasses and light fuels burn fast and must be respected. *(Photo courtesy of Rocco Di Francesco)*

The second term used here is the *fire intensity*. This is the amount of heat and flame production given off by a given fuel type. Rate of spread has an inverse relationship to fire intensity. Whereas lighter fuels have a higher rate of spread, they have a lower fire intensity. Heavier fuels have a lower rate of spread

Figure 19-7 The components that make up the fuels side of the wildland fire triangle.

and a greater fire intensity. Light grassy fuels burn fast, but not with great flame lengths, while heavy timber burns slower, but with much more intensity, releasing higher flame lengths and Btus. **Figure 19-9** shows a timber fire with long flame heights and high intensity. Flames can run up to 100 feet or more into the air over a timber or heavy brush fire.

Figure 19-9 Heavy timber fires burn with high intensity and flame length. *(Photo courtesy of Phill Queen)*

The last fuel concept discussed here is fuel spacing or continuity. The fuel can be spaced tightly or loosely. Depending on its compactness, the fuel may or may not burn well. If air can circulate around the fuels and they are close enough to each other that they practically touch, then the fire will burn better than if the fuels are so densely packed that no air can circulate in and around them. At the other extreme is fuel that is so sparse that the fire cannot travel from plant to plant. In this case the fire may not run as well. It could even put itself out.

Topography

The last side of the wildland fire triangle is the side called topography, **Figure 19-10**. This term deals with the layout of the land. The **steepness of the slope**, the direction the slope faces, the **drainages** and **river bottoms**, and the **ridges** and **saddles** are all critical to the behavior of fire in the wildland.

It is a fact of nature that fire runs uphill faster than downhill. This is because the flame will run vertically and because the slope places the uphill vegetation close to the flame, **Figure 19-11**. The plants are preheated and burn quicker as the heat travels

Figure 19-10 The components that make up the topography side of the wildland fire triangle.

Figure 19-11 The angle of the terrain will have an effect on a running fire. Note how the flame "leans" into the hill, preheating the fuel above it. If this fire was burning or "backing" down the hill, it would move much slower because the fuel is not preheated.

upward. Conversely, it is said that when a fire is burning down a hill, it is "backing down" because the flame is leaning away from its direction of travel. When the fire backs down a hill, it naturally travels slower because the fuel ahead of it is not being preheated and has to catch fire by conduction rather than convection. It then naturally follows that the steeper the hill, the faster the fire will travel upward and the slower it will travel downward.

> **Streetsmart Tip** A concept the wildland firefighters follow here is that from flat to a 30 percent slope the fire will double its rate of spread and at a 55 percent slope it will double again.

The direction the slope faces is also very important because vegetation grows differently on different **aspects**. In the western United States, for example, the sun warms southern-facing slopes more than it does others. Because of this, the vegetation on the southern-facing slopes is dryer and generally lighter in consistency. This causes the southern-facing slopes to burn with higher rates of spread than the other aspects. Also, with the sun preheating the fuel, the chance of fire activity on these southern-facing slopes is much higher. In other parts of the country a different aspect will be dryer, but the principle is the same. Firefighters should know what aspect gets more sun in an area before planning their strategy in fighting a ground cover fire. **Figure 19-12** shows a team of firefighters entering an area that has multiple topographic features. All must be understood and monitored.

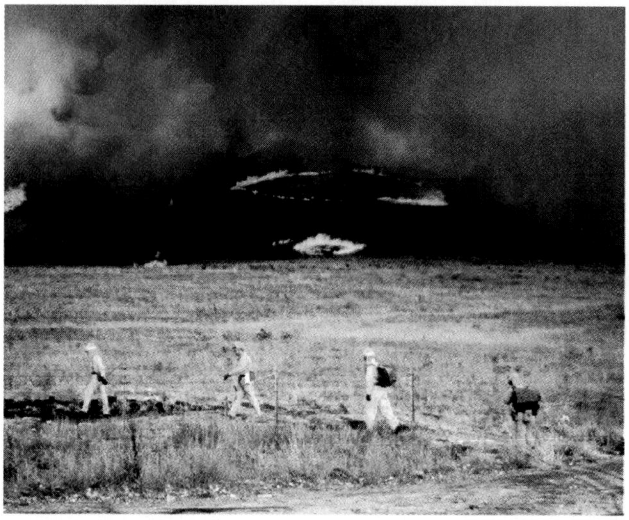

Figure 19-12 A fire crew enters an area with complex topographic features and stops to study the terrain. *(Photo courtesy of Rocco Di Francesco)*

Some more common wildland firefighting terms related to topography are saddle, ridge, drainage or **chimney**, **box canyon**, and **midslope**. These are described in the Key Terms section at the end of this chapter. They all have great relevance in wildland firefighting. Each one has associated dangers when involved in fire. All but the ridge tend to draw fire to themselves. If a ground cover fire is anywhere near these topographic features, the danger level is very high because each will draw the fire to itself in a rate-of-spread increase that will astound the firefighter. These features are responsible for many firefighter deaths in wildland fires.

When discussing ground cover fire components and considerations, it becomes clear that it is a very complex science. Firefighters must understand these components to better ensure the safety and survival of those who fight fires in this environment. The numbers of firefighter deaths and injuries in wildland firefighting are very high, second only to residential firefighting. It is for this reason that ground cover firefighting must be understood and respected by firefighters the world over.

Vehicular Fire Components and Considerations

Studying vehicular fire components and considerations is not a matter of just walking up and applying water. A number of variables can be involved in this practice. This section deals with the more common variables associated with today's automotive industry.

The most logical way to discuss the automobile and its dangers when burning will be to start at the front and work toward the rear. In past years automobile models had a metal bumper held on to the car with brackets. When hit, the bumper nearly always damaged the car. The industry then developed hydraulic supports for the bumpers in order to lessen the chance of car damage when lightly struck. Today's cars have hydraulic supports in both the front and rear to protect the car.

> **Safety** Firefighters should understand that when heated, as in an automobile fire, these hydraulic fluid-filled bumper systems undergo great stress as the fluid expands. This stress can cause the bumper to be propelled off the car, traveling up to 40 feet or more. Persons standing in front of the bumper (front or rear) when this happens can be severely injured.

The point here is to avoid standing in front of any bumper after it is been heated.

The engine compartment has a mass of very hot metal filled with flammable liquids (fuel, oil, and hydraulic fluids). Any fuel leak can ignite quickly because of the heat from the engine. In many cases the fuel fire is extinguished and then keeps reigniting as the heat from the engine contacts new fuel leaking from burned lines or melted parts. These are Class B fires as opposed to Class A fires. In some cases, the engine itself is made of flammable metals such as magnesium. This is a Class D fire. Another danger in this area is the battery and its associated wiring. A battery is a plastic box filled with a hazardous material (sulfuric acid) that creates a flammable gas (hydrogen) as well as electricity. There is the potential for a Class C fire.

The recent addition of electric vehicles to the nation's highways adds another component to the firefighter's information needs. The firefighter must be able to recognize an electric vehicle and be knowledgeable about the differences it presents on the fireground. The first thing to look for will be an insignia of some type, designating electric power. The insignia will be located on the sides or rear of the body. The next step in identification comes when the hood is opened. The firefighter will not find items normally found on a gasoline engine automobile, such as an engine, a transmission, cylinder heads, an alternator, fuel injectors or spark plugs, an air filter, or accessory drive belts. Instead, the firefighter will find high-voltage warning labels on components. Although the chances of finding an electrical (Class C) fire are remote due to numerous manufacturer safeguards, firefighters may have to alter their tactics when fighting fire in this type of vehicle.

The primary dangers in the passenger compartment are the materials used in construction. Most materials found in the passenger compartment are plastic, a form of polyvinyl chloride (PVC). PVC is extremely toxic when burned. A great number of studies have been performed on this material during the past few years and all have come to the same conclusion.

> **Safety** Plastic PVC is many times more toxic than wood and cloth when burned. Firefighters must wear SCBA when working around this material in any stage of combustion. A second danger in the passenger compartment that few consider is that of the hydraulic pistons used to open the hatchback and hoods in some models. These pistons will react to heat in the same way as bumpers do.

Another item that must be mentioned here is the air bag mounted in the dash on either or both sides, and also on the side of the front seat in some models. Great care must be taken when working around these items because of the chance of an accidental discharge. Each model has its own idiosyncrasies that must be studied individually to truly understand the mechanics involved in its design. In most cases, they can be disarmed by disconnecting the battery to the car and waiting some period of time before working around the bag. The side air bag can be disconnected by cutting the power cord running up the side of the seat from the floor to the bag. It is suggested that even after disconnecting the power to air bags, caution should be used when working around them.

The next area to consider is the trunk. At first, a car's trunk might not seem to pose much of a problem but that is not the case. The trunk is used for storage. Firefighters have no idea what is stored in the trunk (without being told by the owner) until it is opened. It could be any type of flammable liquid or gas container, ammunition, explosive commodity, or other dangerous cargo. It has been known to even contain humans.

> **Safety** When opening the trunk, great care must be taken to avoid exposure to potential harm. The firefighter should stand to the side, or crouch very low, and wear full protective clothing with the eyes protected by a helmet shield pulled down or by goggles.

It may be prudent to also have a hoseline charged and ready when accessing this area of the automobile just as is recommended for all other compartments.

The last area of the automobile to consider as dangerous is the underside or fuel storage area. Action movies almost never stage a car crash without the fuel tank rupturing and exploding. In real life this is extremely rare; however, firefighters must approach with caution. What can and usually does happen in car accidents is that the fuel tank will be damaged and be leaking raw fuel on the ground around the vehicle. In this case all precautions must be taken to secure possible ignition sources until the fuel leak is stopped. Dirt or any nonflammable absorbent material will do the job of recovery until something more efficient can be obtained.

When discussing the fuel tank it is important to remember that gasoline may not be the only fuel present. Cars are run on a number of fuels today. Gasoline, diesel, propane, methanol, electricity, and steam are but a few of the fuel systems in use today. Each one carries its own risk and will dictate the tactics used in securing the fuel source.

The last consideration to be mentioned here is that of another type of vehicle, a freight truck. For the purposes of fire components and considerations, trucks (all but the trailer) are simplified as short vehicles. The compartments are the same, only compressed. The materials used are relatively the same and in the same locations. The engine compartment is larger but contains just about the same materials and hazards associated with automobiles. The same precautions are necessary.

The trailer on the truck is another story and can also be dealt with by simplifying it. For the purposes of this chapter, the trailer body is treated as a structure off the ground. It can contain just about anything a structure on the ground contains and can be just as dangerous. All of the principles of caution with structure fires will hold true here.

Studying vehicular fire components and considerations is not an easy task. A vehicle contains nearly all hazards associated with other areas of firefighting, but they are compressed into an area approximately 8 by 20 feet with possible occupants. The risks are there and the firefighter must be aware of every possibility.

Flammable Liquids Fire Components and Considerations

Unlike most Class A fires that either go out when water is applied or continue to burn at a given pace if water is not applied, flammable liquids fires can actually be complicated by improper actions of firefighters. Whereas Class A fires are somewhat predictable and follow given laws of nature when burning or when water is applied to them, flammable liquids fires are not. Each flammable liquid has its own specifications that will affect its burning characteristics and extinguishment.

Chapter 4 introduced a number of terms related to flammable liquids. Knowledge of these terms is critical. Flash point was mentioned, and that characteristic of every flammable liquid will make a very big difference in a material's storage, handling, and extinguishment.

> **Caution** The vapors given off by a flammable liquid can be that material's most hazardous profile, sometimes traveling great distances before finding an ignition source.

For instance, an automobile accident with leaking fuel and no fire can turn into a very dangerous situation if the vapors traveling outward find a distant ignition source bringing fire back to the leaking area.

Another term used in the field of flammable liquids is **solubility**. This is the degree with which a liquid will mix with water. In chemistry, the materials that are soluble in water are called **polar solvents**. Materials that are polar solvent, such as alcohol, will dilute or mix with water quite easily. Hydrocarbon-based liquids (gasoline, oil, etc.) are nonpolar solvents and do not mix with water. They either float on water or sink to the bottom of it. Most flammable liquids are not water soluble, but knowing which ones are can go a long way toward the prevention or extinguishment of a fire involving these materials. Liquids that are not water soluble do not mix with water and either float on water or sink below it. This is termed the liquid's **specific gravity**. Knowing the specific gravity of the liquid in question will tell the firefighter which method of extinguishment should be utilized. If the liquid is heavier than water and sinks, then we know water can be used to extinguish it, if applied properly. If the liquid is lighter than water, and floats on the water, then an alternate means of extinguishment must be used. **Figure 19-13** shows firefighters applying Class B foam to a flammable liquid fire by gently covering its burning surface.

Knowing the material that has spilled, or is burning, and some of its characteristics will greatly benefit the fire company, providing them with a margin of safety and the ability to effectively accomplish their goal of mitigating the hazard.

Figure 19-13 Firefighters gently apply foam to a flammable liquid fire. *(Photo courtesy of Phill Queen)*

Figure 19-14 Firefighters apply water fog to cool a venting propane tank. *(Photo courtesy of Phill Queen)*

Flammable Gas Fire Components and Considerations

Flammable gases, in many cases, can be grouped with the flammable liquids category. In fact, a great number of flammable gases are products of flammable liquids. Often when shipping a flammable gas, the manufacturer will compress the gas in order to ship a larger quantity, turning the gas into liquid under pressure.

As with flammable liquids, it is very important for the firefighter to understand the hazards associated with gases. Gases are typically stored under pressure. Therefore, when a container is ruptured or leaking, the gas escapes immediately, creating a hazard as large as the gas can spread. An example is propane. **Figure 19-14** shows a propane tank fire with firefighters cooling and containing the flames. Propane is stored as a liquid and the gas coming off the liquid is used for a multitude of purposes. Should a propane container rupture, the liquid turns into a gas, expanding over 250 times. That can be a really large problem for the responding firefighter.

The firefighter will want to know the properties of the flammable liquids so a plan of action can be determined and the hazard can be abated. One of those properties will be the gas's **flammable range**. This range will determine whether the fuel is too rich or too lean to burn in the given space it occupies. If the spill or leak is inside a structure with no windows or doors open, then there is a chance that the gas is too thick or rich to burn, as the ratio between oxygen and the gas will not support combustion. Conversely, if the spill is in a structure with open windows and a stiff breeze is blowing through the building, the gas-to-oxygen ratio may be too thin or too low to support the burning of the gas.

Another term common to flammable gases is **vapor density**. This term is very similar to the specific gravity term used with liquids. Basically, it is the weight of the gas as compared to air. Is the gas lighter or heavier than air? Knowing this fact about the material will predicate the plan of action taken because the gas may be hanging low and heavy, looking for an ignition source, or it may be light and dissipate quickly, causing far less concern.

The last term discussed here is possibly the most important: **toxicity**. Simply put, will the product kill a person who encounters it? This contact can be by a means other than breathing. It can be through skin absorption or ingestion also. Knowing this fact will determine what the responders should wear. Will turnout clothing be enough or will a hazardous materials suit have to be worn? And, in some cases, will that hazardous materials suit be enough?

The flammable gas fire components and considerations are many. Just as there are literally thousands of places where these gases can be found, there are a number of ways to deal with them when they are found. The firefighter interested in longevity will pay close attention to the details involved with this firefighting hazard.

Process of Fire Extinguishment

This section begins to develop thought processes based on the previous sections within this and other chapters. The concepts and considerations of the differing situations and classes of fire previously discussed begin to be put into the context of fighting fire.

> **Note** Before firefighters take on the beast of flame, they must first understand that fire is a multisided three-dimensional presentation of heat and chemistry. It must be considered from six sides. It has a top, a bottom, and four sides. It will follow laws of physics and applied science that only the most seasoned firefighter will understand, and even then things will go wrong on occasion. The goal in defeating this beast can only be met by extending an aggressive, fast, well-placed, adequate fire attack.

To accomplish this goal, a number of steps must be taken. The first step is to create a plan of attack. This is done by first locating the fire and determining its extent. Generally speaking, the longer it takes to find the fire, the longer it will take to put it out. Good communication becomes very important here. As Chapter 2 described, this report is built on careful and complete observation of the fire and its development to this point. Each department will usually have SOPs that determine who is first on scene and where each unit should be located. These policies are not set in concrete because each fire is different. However, they have been set up in order to maintain order and efficiency. The attack plan will be built on the observations, training, and opinions of the first-in officer after considering the facts as perceived and predicting a course of action.

The second step is then to apply the plan of action as quickly, efficiently, and safely as possible. Firefighters will take their skills, knowledge, and ability into the face of danger in an effort to mitigate the situation as quickly as possible. Three methods of fire attack are possible: the direct attack, the indirect attack, and the combination attack of fighting fire. The direct attack is quite simply the act of putting water directly onto the seat of the fire, **Figure 19-15**. This method is the most efficient use of water on free-burning fires. A solid or straight stream of water is applied in short bursts directly onto the burning fuels (Class A) until the fire darkens or is extinguished. **Figure 19-16** shows a firefighting team about to enter a burning structure and apply a direct attack on the seat of this fire. If used indoors and done properly, the thermal balance of the room may still hold, giving the firefighting team a clear view of the fire and contents of the structure.

Figure 19-15 A direct attack with firefighters applying a straight stream onto the seat of a fire.

Another method is that of the indirect attack, **Figure 19-17**. This attack is usually utilized when the firefighting team does not see the seat of the fire. A fog stream is applied to the upper areas of the room or above the fire seat in an effort to "steam" the fire out. **Figure 19-18** shows a firefighting team applying

Figure 19-16 A firefighting team prepares for a direct attack on a contents fire. *(Photo courtesy of Phill Queen)*

Figure 19-17 The indirect attack has the firefighters applying a 30-degree fog into the upper heat layer of the fire in order to create a steam that will extinguish the fire.

a fog stream into the upper area of a structure in an indirect attack method. Directing the stream into the superheated upper atmosphere above the fire causes the water to "boil" or "vaporize," steaming the area and causing the fire to die. This method, if used indoors, will usually greatly disturb the thermal balance, causing a loss of visibility in the structure as heat and products of combustion are circulated downward. When using this method, the firefighting team should attempt to shield themselves from the steam vapor by backing out of the area after applying the water. They can reenter as soon as the vapor begins to dissipate.

Caution Note also that the indirect method is not appropriate for occupancies with victims in need of rescue, because the steam created by the attack will burn the victims quickly.

The last method is that of a combination attack. This attack is a blend of both the direct and indirect methods. The straight stream is directed to the seat of the fire and then the nozzle is quickly changed to a fog and the upper areas of the room are covered with a quick shot into the thermal cover. This method will darken the fire and stop the growth of flashover.

In each of these methods of applying water, firefighters must remember that the goal of the fire department is to save property. If water is not applied in proportion to the amount of fire, two things can happen. If not enough water is used, extinguishment will not take place. If too much is used, then the damage by water may exceed that done by fire. The lesson here is to use only what is needed, saving the rest for overhaul if needed.

Proper Stream Selection

Fire streams were discussed at length in Chapter 11 with information regarding a number of things that are mentioned again here. The purpose of this section is to transition from knowledge of the burning characteristics of various occupancies and hazards to actual fireground use of the information.

In the case of fighting fire with water, it is important to understand that in order to be successful, sufficient water must be applied directly to the fire in order to control it. That statement may sound simple until one begins to take into account all of the mechanics of performing that task—not only the physical movement but the mental as well. Success will be based on a number of factors in the selection of the stream alone. Some of those factors, in no particular order, are proper stream type, stream size, stream placement, timing, water supply or quantity of water, stream reach, mobility needs, tactics required, speed of deployment, and personnel available.

Consideration of some of these factors provides a better understanding of the firefighter's role in proper stream selection. The first one was the proper stream type. Is a fog stream the correct choice? If so, which width? **Figure 19-19** shows

Figure 19-18 A team of firefighters applies a fog stream in an indirect attack on a fire. *(Photo courtesy of Central Net Fire)*

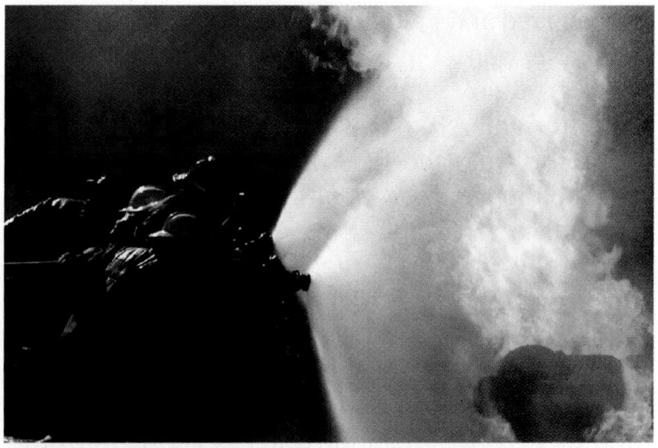

Figure 19-19 Firefighters use a fog pattern to shield themselves from radiant heat. *(Photo courtesy of Phill Queen)*

firefighters using a fog stream to protect themselves from radiant heat. Or is the straight stream a better choice? Each situation calls for differing stream requirements. The most basic answer to stream type may be in the type of attack required: direct, indirect, or combination. Each has its place in given circumstances.

Next to be considered is stream size. The typical response to this question may very well be "Big water, big fire and little water, little fire." It was once said that firefighting is like going to battle. The weapon must match the target. **Figure 19-20** and **Figure 19-21** show examples of the need to have the stream size match the fire size to be effective.

Figure 19-21 If the line selected does not match the size needed, the fire will burn longer and hotter and can jeopardize the operation.

Caution If a large line is needed, then it should be pulled right away so as not to play "catch up" with the fire. Some firefighters think the stream to start with should be the stream that will eventually be used, going smaller later, rather than ever having to go larger later.

The placement of the fire stream can make or break an attack plan. The fireground must be coordinated at all times without exception. This is a battle with an enemy that can injure or kill. When an order is given to place a fire stream in a particular location, it is part of a plan. Without that stream, the plan may fail, so this factor can be very important. An example may be a fire in a dwelling where the first line goes between the trapped occupants and the fire, with the second stream going to the seat of the fire, **Figure 19-22**. This operation will not work without both streams being coordinated and controlled. Care must also be taken to avoid opposing streams, because these streams may drive heat and products of combustion toward the opposing firefighting teams. Still another related factor is that of timing, which will go with speed of deployment. This is self-explanatory. Timing is everything on the fireground. Stopping the progress of the fire as quickly as possible will allow the firefighter to work toward the goals of rescue and property conservation.

Many differing types of nozzles are used in the fire service. One of the commonly used types has the ability to control the amount of water it puts out. In cases of water conservation because of poor sources, a nozzle that conserves water would be advantageous. In such a case, water supply or quantity of water available plays a significant role. A different set of rules applies when the water is limited.

Another factor to consider is the mobility of the stream that will be needed. In some cases, lines are advanced quickly into the fire, whereas in other cases the line is placed in a fixed position holding the fire until another more mobile line can be set up to take over. **Figure 19-23** shows a deck gun attack on a large fire in an effort to bring the fire into handline size by reducing some of its heat. In this same area are the factors of stream reach and tactics. A line may be placed on an exposed structure that is large and

Figure 19-20 A firefighter applies a single straight stream onto a structure in an attempt to reduce some of the fire's heat. *(Photo courtesy of Phill Queen)*

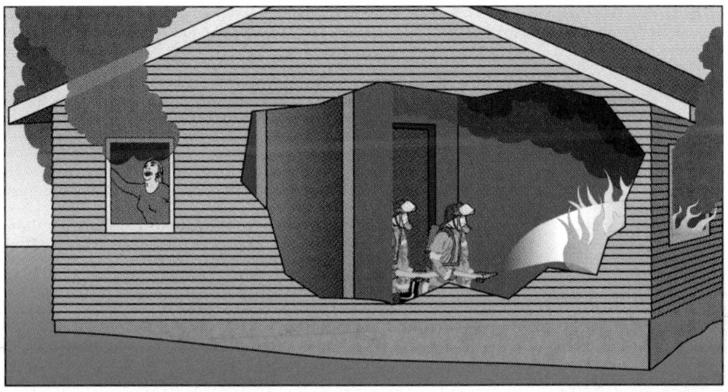

Figure 19-22 The first line pulled should be positioned between the potential victims and the fire.

Figure 19-23 Firefighters use a deck gun to hold the fire until more lines can be put into place. *(Photo courtesy of Rocco Di Francesco)*

Figure 19-24 Large stream appliances at work on a large fire. *(Photo courtesy of Brentt Sporn)*

unmoving, while a smaller, more mobile line is extended into the burning occupancy to fight the fire.

Everything taken to a fire is a resource operated by the fire department. The personnel, water, equipment, and knowledge are all resources. The selection of the proper fire stream is a significant decision. **Figure 19-24** shows the proper use of large streams on a fully involved mercantile fire. **Table 19-1** shows the differing sizes of nozzles and the effort required in order to work properly.

Firefighters carry out their part of the attack plan using water and a good basic knowledge of the requirements being predicted by the fire they are fighting. Knowledge of the factors listed, as well as many others, will arm the firefighter with an arsenal when going into battle against a fire.

TACTICAL CONSIDERATIONS

Earlier in this chapter the elements of fire control were discussed in some detail regarding the most common fire expectations of the average firefighter. Those sections detailed the components and considerations a firefighter must be knowledgeable in to be safe and successful on the fireground. This section deals with those same types of fires in a tactical firefighting capacity. It builds on the information presented in the previous section in an effort to actually use the information in fighting fire.

The actual fighting of fire on the fireground is broken down into tactical objectives. Each of these

Hose Stream Characteristics

TYPE OR SIZE OF LINE	REACH (FT)	MOBILITY	GPM	COMMON USE
1-inch or greater booster or reel line	25–50	Excellent	10–40	Very small nonstructural fires or overhaul
1½ to 1¾ inches	25–50	Good	40–175	Quick attack, one to three rooms
2½ inches	50–100	Fair to poor	125–350	One floor or more, personnel permitting
Master stream	100–200	Poor to none	350–2,000	Large, fully involved structures or exposure protection

TABLE 19-1

objectives must be accomplished in order that the safety of the firefighters, occupants, and others is a primary consideration. If there are no occupants, the overall plan still remains in order, keeping the firefighters safe and efficient. These goals are universal in the fire service, although different departments may use different terms. The priority was first set in place a great many years ago by Lloyd Layman, with the following acronym: **RECEO**—Rescue, Exposures, Confinement, Extinguishment, Overhaul.

Another way of setting this same priority is Rescue, Fire Control, and Property Conservation. Still another acronym is termed REVAS: Rescue, Exposures, Ventilation, Attack, and Salvage. There are a number more, but the point is well made that the fire service has set itself a series of goals that must be accomplished in order that the fireground remains as safe and orderly as possible.

> **Note** A second point that must be made is that the command structure must be solid and well defined in order to accomplish the set objectives in each of the goal areas.

Just as an army employs a rank structure to fight a battle, the fire service uses a rank structure to fight fire. One person will be in charge and appoint people who will command different parts of the operation under that person. **Figure 19-25** shows a battalion chief conferring with his captains in planning his objectives. This system is called the **incident management system (IMS)**. The IMS is basically designed to maintain order in any emergency operation. It can be fire, flood, rescue, first aid, mass casualty, or anything demanding a number of emergency responders. A single person will be in charge of the incident and this person is called the incident commander (IC). This person works from the goals listed earlier utilizing strategy and tactics just as a general would in the military. The IC will delegate assignments to arriving units in an effort to accomplish the goals in a safe and efficient order.

Figure 19-25 A battalion chief meets with his captains to set operational plans into action. *(Photo courtesy of Rocco Di Francesco)*

To explain the system and complexities of fighting a fire, this discussion sets the scene with an average, generic fire, because the methods and operating principles will apply to any fireground operation.

As soon as the fire is reported, and before units ever pull from the station, mental preparation takes place. These are the recall factors of past similar incidents, the recall factors of knowledge about the area where the alarm is reported to be, the recall factors regarding the fire prevention inspections of that area, the recall factors of time of day and route to be taken in response, and the factors regarding the condition of the responding units as to their capabilities. Are they fully staffed today? Are the water tanks full and the pumps working properly? Are the crews fresh and ready to go to work in possibly very dangerous conditions? All of these factors, and many more, will race through the minds of the responders. Each of these factors will eventually dictate the strategies to be considered.

Upon arrival another set of factors must be considered. All of them have to do with the environment of the response. **Table 19-2** demonstrates that the number of factors having to do with the operation can be almost overwhelming. The seasoned firefighter will go through all of these and possibly more in calculating the environment presented at the emergency. And each of them can greatly affect the outcome of the incident depending on how they are handled.

Based on the situation presenting itself to the arriving firefighters, three methods or modes of attack are possible: the **offensive attack**, the **defensive attack**, and the **combination attack**. These modes are predicated on the resources available at the emergency. If the arriving units have adequate resources to handle the situation, then they will fight the fire aggressively and offensively. They will attack the problem head-on and, following department standards, will accomplish their objectives

Fireground Factors

BUILDING		
Size	Construction type	Condition
Age	Openings	Utilities
Concealed spaces	Access	Effect of fire
Extent of fire	Interior fuel load	Exterior fuel load
FIRE		
Size	Location	Direction of travel
Time since ignition	Extent	Materials involved
Material left to burn	Fire load	Stage of involvement
OCCUPANCY		
Type	Value	Fire load
Status (used/vacant)	Hazards of occupancy	Life hazard
Arrangement	Obstructions	

TABLE 19-2

efficiently, effectively, and safely. If they do not have adequate resources to aggressively handle the situation, then they will have to fight the fire in a defensive mode of attack. This mode will be continued until enough resources can be massed to then change to an aggressive, offensive attack. The combination mode is most often utilized when a rescue must be accomplished but not enough resources are on scene to handle the operation entirely. An example would be when a line is pulled to defend potential victims until they are removed, knowing that the line will do nothing to fight fire because it is not large enough or practical for the size of the fire.

As the operation begins to unfold, the IC calls for an attack on the fire. The IC's call will be based on the conditions presented earlier. Hoselines will be pulled, utilizing methods prescribed by the responding department's SOPs. Some common hose evolutions were covered in Chapter 10. The concept of **teamwork** will dominate the scene as each person will be responsible for his or her own assigned details of the operation. This teamwork will continue through the entire operation just as it would in a military operation. Each team member will be given an assignment based on the priority of the incident. As mentioned earlier, rescue will be the first priority, and all actions will be directed toward accomplishing that task. A line may be pulled and placed between the fire and the people needing rescue, or the situation may call for rescue first and no line pulled until it is completed, **Figure 19-26**. Each situation will dictate its own need.

A recent rule by the Occupational Safety and Health Administration sets safety standards on the fireground for rescuing occupants from burning buildings.

> **Note** The two in/two out rule tells the fire department that the teamwork concept must be followed on the fireground at all times.

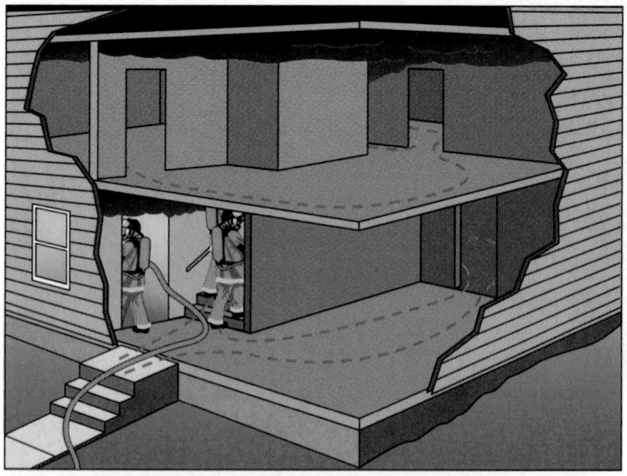

Figure 19-26 If it is not known whether victims exist, then the first line into the structure should assist the search and rescue functions.

This law is designed to protect the firefighter from being injured in rescue attempts where fire conditions are critical.

Firefighter Fact In 1998 the Federal Occupational Safety and Health Administration issued a Final Rule titled *29 CFR Parts 1910 and 1926 Respiratory Protection*. This regulation requires both private industry and the fire service to practice the safe use of self-contained breathing apparatus (SCBA) by mandating a number of rules regarding face piece fit testing, physical requirements for wearers, apparatus requirements, and more. A part of this Final Rule also mandates the number of firefighters required on the fire scene when interior operations are to take place in hazardous situations. The two in/two out rule dictates that during interior structural firefighting in atmospheres that are immediately dangerous to life and health (IDLH), and that do not have lives requiring immediate rescue, at least two firefighters must enter together and remain in visual and voice contact with each other at all times. Two more firefighters must be on standby for the potential rescue of the interior team.

The 1997 NFPA 1500 Standard for Occupational Safety and Health also requires this same two in/two out rule as part of its requirements. This was created in an effort to further follow OSHA laws and protect the safety of firefighters the world over.

If the situation does not call for rescue because there are no dangers to people or animals, then the second priority is placed into action. That priority is exposure. Almost anything in the path of the fire can be called an exposure. The most common exposure is a structure next to the burning building where the fire is heating it to a point where it is close to catching on fire, **Figure 19-27**. Another example would be a car fire that is about to ignite nearby brush or tall grass. If an unburned object (i.e., exposure) is about to catch on fire, it is the responsibility of the responding firefighter to stop the spread of the fire to that object. The logic here is that the item burning has already sustained damage, but the exposed item has not. So firefighters save the exposed item before concentrating on the burning item. Exposure protection is accomplished by either removing the exposed item from the area or wetting it to cool its surface below the point of ignition, **Figure 19-28**. Radiant heat is the main cause of exposure fire, so wetting the exposed item removes that heat at the point of contact.

Figure 19-27 The most common exposure problem is the closest building to the building on fire.

As soon as all exposures are deemed safe, it is time to consider the next step in the firefighting process, which is fire confinement. The fire's parameters are not always clearly defined. Visibility may be difficult at best. The members of the firefighting team must use all of their senses in order to locate the fire. Once located, water or another extinguishing agent is applied in a method that will confine the fire to its area of burn. Stopping the fire's forward progress must be done first, before the fire can be extinguished. It was mentioned earlier that after rescue came fire control. The confinement of the fire is controlling the fire. Some departments will even broadcast a "fire is under control" call to the IC when this step is accomplished. In the wildland, this point becomes even more dominant. An "under control" term may be issued days before actual fire extinguishment is accomplished. When the fire is surrounded and can burn no further outside of this confinement, then it is considered to be under control.

In many cases, to bring the fire under control in structure firefighting another task may have to be added. Ventilation (described in Chapter 18) coordinated with water application may be the only way to achieve the confinement results safely. This will be described more later in the discussion of particular structure fires.

It is common practice to attack most fires from the unburned side, **Figure 19-29**. That is to say that the line will be brought into action between the fire and its intended direction of travel. The opposite of this tactic would be to attack the fire from the rear, pushing the heat and other products of combustion forward into unburned areas, causing far more damage. This is true in almost all applications of

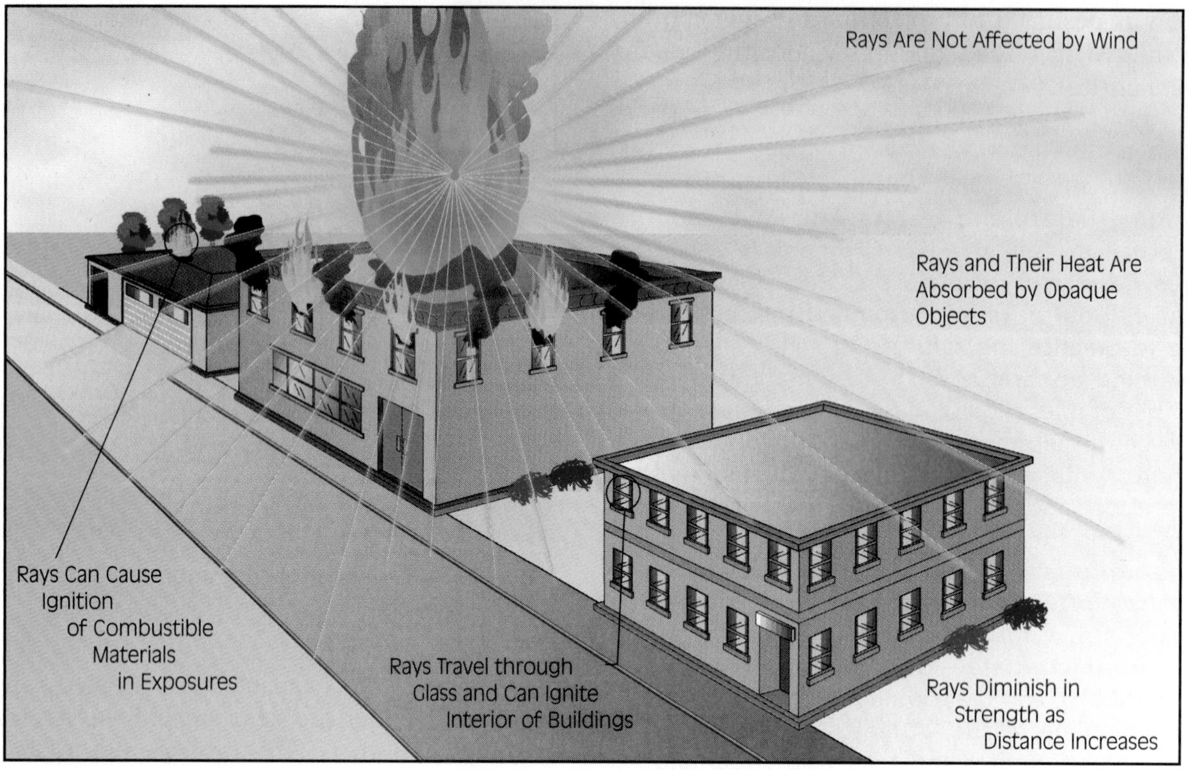

Figure 19-28 Radiant heat is the main cause of exposure fires within short distances. Radiant heat will travel in straight lines from the heat source to nearby objects.

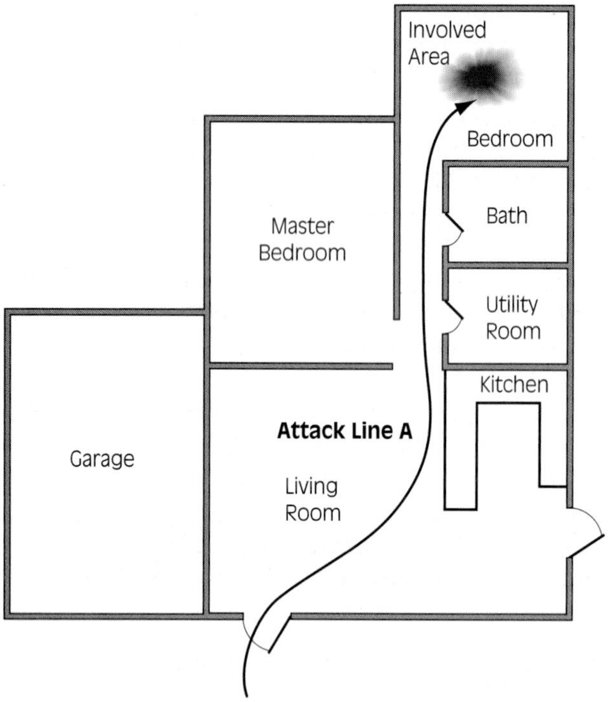

Figure 19-29 Fires most commonly should be attacked from their unburned side, pushing the products of combustion away from unburned materials and areas.

firefighting from structure to wildland or ground cover to flammable liquids and gases.

When the fire is controlled or confined, then the actual extinguishment begins. This is accomplished with direct or indirect attack strategies and resource use. This can also be thought of as property conservation. Saving the property not damaged is a very high priority here. The fire has been confined and now is extinguished utilizing the latest methods and practices in order to save the undamaged property. The use of water is restricted to only as much as is necessary, because water can cause as much and, in some cases, more damage than the fire.

As soon as the fire is extinguished, the overhaul begins. This is a methodical system of ensuring that all embers and chances of reignition are removed. The area is carefully worked by moving every item and wetting all combustibles that could carry new life to the extinguished fire. During this phase of the operation, the cause of the fire is usually determined. By carefully sifting through the fire debris the investigator looks for clues that will lead to the cause of ignition. A closely coordinated effort between these people will result in the call being closed out so the firefighters can return to their respective stations to prepare for the next call.

> **Caution** The firefighter must use extreme caution when in the overhaul phase so that the subtle clues the investigator is working with are not disturbed.

One more point must be made about firefighter safety. That point is about acts of terrorism. The fire service is called upon in nearly every conceivable emergency, representing local authority in the form of a fire department, fire district, or emergency services agency. Unfortunately the firefighter may represent the authority being targeted by an act of terrorism. For this reason it is important that the firefighter be made aware of the possibility of danger from cowardly acts of aggressiveness. Some recognized threats to firefighters are:

Undetonated explosives: secondary devices, low-order explosions, malfunctions

Structural instability: collapsed buildings, falling/thrown debris, live electrical lines, leaking gas mains

Fire

Hazardous materials (poisons, etc.)

Biohazards

Further training in this area is highly recommended for all firefighters. Local jurisdictions, police, FBI, the Department of Justice, the State Fire Marshal, and the National Fire Academy offer this type of training. More information on terrorism is given in Chapter 30.

All of the information given in this section is next applied to tactical considerations in fighting fire in a number of given situations, bringing the information in this chapter into a fine focus for practical use.

Residential Occupancies

Fighting fire in residential occupancies has the greatest impact on national fire fatality figures, for both the public and the fire service. A direct result of this statistic was the passage of the two in/two out ruling by OSHA mentioned earlier and in Chapter 5.

> **Firefighter Fact** More people die in residential fires than in any other type of fire in the United States. This is true for both the public and the firefighters.

For this reason, the fire service must place a much greater emphasis on safety here than almost anywhere else. This subsection deals with firefighting in residential occupancies in both single-family and multifamily dwellings.

Before discussing particular residential occupancy types, it is important to discuss a number of generic considerations that hold for all residential fire emergencies. The best way to do this is to go on scene and discuss each step of consideration for the firefighter. Then, subsequent sections will detail single-family and multifamily dwelling operations.

En route to the alarm, the firefighting teams begin the size-up process discussed earlier. The team considers all factors, both silently and in discussions with each other. This practice prepares the team members for the multiple details that will be thrust on them as soon as they arrive on scene.

On arrival, the members of the firefighting team immediately scan the scene for all details remotely related to the fire. As they complete this scan they have already begun to form an attack plan based on the information gathered. The officer in charge of the team (captain, lieutenant, senior firefighter, or other) will then make a decision based on the input received, and the action will begin.

As discussed earlier, the order of attack must follow guidelines set forth in the department SOPs or training mandates. The first priority will be the rescue of occupants. Information is sought from bystanders and other sources as to the number, location, and condition of potential victims, **Table 19-3**. Determining the stage the fire is in at this time is important because it will have a direct influence on the direction the rescue or

Rescue Factors		
Number of occupants	Location of occupants	Mobility of occupants
Verification by bystanders	Condition of occupants	Firefighting required
EMS/ALS required	Manpower for search	Fire's burn time
Hazards to firefighters	Access to victims	Escape after rescue

TABLE 19-3

attack will take. In the earlier stages, the fire can be attacked aggressively and quickly in an effort to cut it off from the rescue attempt. In later stages that will be much more difficult. The firefighting team must make a decision as to the mode of attack based on the resource needs in making the rescue. Are there adequate resources, that is, enough personnel, equipment, and water to accomplish the order of attack selected? Or are the resources on scene limited, causing the team to hold as it waits for adequate help to arrive?

The two in/two out rule is important here. If a rescue is being considered, but there are no facts to substantiate that one is needed, then the rule applies that adequate manpower must be on scene before entering the structure. If the facts presented show that a definite rescue must be executed, then the rule is relaxed and that rescue can take place. As described in Chapter 16, a primary search for victims is conducted very early, as soon as the amount of personnel on scene will allow.

The sequence of priority of multiple rescues is another factor to consider. The sequence shown in **Figure 19-30**, or order of priority, is most generally as follows:

1. Those closest to the fire
2. Largest grouping of threatened people
3. Anyone else in the fire area
4. Those in the areas that will eventually be exposed

After any necessary rescues have been performed, the next priority will be exposure control. This was discussed earlier when tactical considerations were covered. Following exposure control is the step of confinement of the fire, then the actual firefighting

Figure 19-30 When numerous rescues face the first team they must prioritize their actions based on victim exposure.

itself. As these events begin, a number of simultaneous events must take place to ensure the safety of the firefighting team members. The utilities to the occupancy must be secured and ventilation must be considered. The utilities will consist of the electrical power, the water, and gas supplying the occupancy. A firefighter is typically assigned the task of utility shutoff. This person must find the electrical panel and shut off the breaker switches or unscrew the fuses in older occupancies. Then the gas must be secured by going to the meter or tank and turning off the flow of gas supplying the structure. After that, the company commander will decide if the domestic water to the occupancy should be shut off also. This is usually done if there has been any chance of the water lines breaking during the operation.

Caution The firefighting force does not want to be electrocuted, blown up, or have water running that they do not control. Water can cause as much—or more—damage than the fire.

During this part of the operation, ventilation must be considered. Ventilation is discussed in Chapter 18. A well-coordinated ventilation effort will greatly assist the fire attack team in almost all circumstances. The extinguishment and overhaul parts of the fire attack then will take place bringing the situation to a close.

At this point, many fire departments will offer services to the occupants of the burned structure. These services consist of housing for the night, clothing, food, and, in some cases, counseling. These services are offered by the fire department through a number of volunteer and charity organizations all over the country. Two primary organizations are the American Red Cross and the Salvation Army, but there are a great many others.

This section dealt generically with residential occupancy firefighting. The next topic is the finer details of fighting fire in single-family and multifamily occupancies.

Single-Family Occupancies

The most common element of single-family dwellings is the layout. Granted, floor plans will differ more than porcupines have quills, but there is still commonality. In the one-story home, the bedrooms will generally be on one side of the dwelling, and the living, cooking, and laundry facilities on the other. In the two-story home, the bedrooms will generally be on the upper floor (not always) with possibly one of them downstairs. The other functional areas of the home will be downstairs. In the two-story home it is usually true that there are fewer rooms downstairs than upstairs also, because the functional rooms tend to be larger than the bedrooms. Why is this important? Rescue! In the dark, smoke-filled, hot confines of the fire environment, it is nice to have a little bearing on where the search may be most productive.

Another common consideration of two-story homes is that of heat travel. The heated, toxic products of combustion will tend to travel upward because they are usually lighter than air. They will travel toward the sleeping areas of the common two-story dwelling, many times moving up the open stairways.

It is typically the responsibility of the first-in engine to either take command of the incident and control the operation or to immediately attack the fire. As the engine arrives, the driver pulls to the far side of the structure passing by the front, giving firefighters a view of both sides and the front, and making room for other responding units, **Figure 19-31**. This gives the team three of four sides and quite a lot of information from which to draw an attack plan. Then more often than not, lines are pulled to either attack the fire or protect rescue team members. **Figure 19-32** shows an engine parked with lines pulled back to the structure it passed on arrival. In rural areas the engine may have laid a line from a water source to the fire, or they may have brought water with them in the form of a very large tank on the apparatus or a water tank truck. In metropolitan areas it is common for the first-in unit to attack the fire with the tank water in the unit and the second-in unit to bring water from the hydrant or water source, **Figure 19-33**. Departments' SOPs will differ in this area based on experience and resource availability.

As the second-in team arrives at the fire, they are commonly required to offer backup to the first-in unit. **Figure 19-34** shows two engines at a fire scene where the second-in unit has laid hose and is assisting the first-in unit. Some of the common tasks they are responsible for include these:

- Back up the initial lines laid by the first-in unit
- Protect the access and egress of the first-in company

Figure 19-31 As the first-in unit arrives, it should pull past the building, giving firefighters a good look at up to three different sides or views of the fire.

Figure 19-32 An engine has pulled past the involved structure to pull working lines. *(Photo courtesy of Phill Queen)*

- Assist with attack while the first-in works on exposures
- Assist with exposure protection if the situation warrants more lines
- Back up the initial team in any requests they make

The truck company in most areas will arrive and secure the occupancy's utilities and begin ventilation.

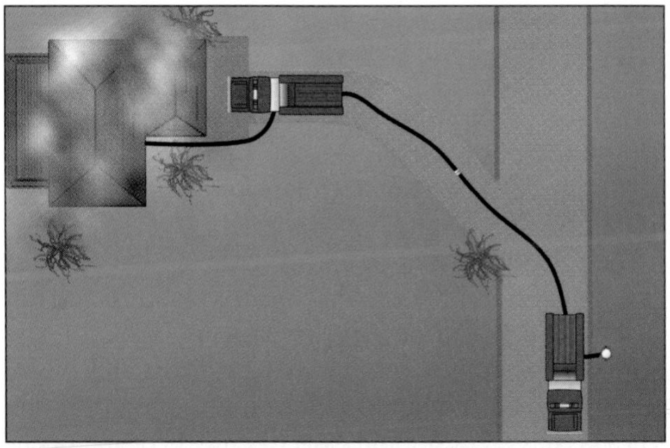

Figure 19-33 The second-due engine may bring water to the first-due unit that went into a "quick attack" using tank water.

Figure 19-34 Here the second-in engine has laid hose from the hydrant and is assisting the first-in unit. *(Photo courtesy of Phill Queen)*

Figure 19-35 A ventilation operation on a single-family dwelling. *(Photo courtesy of Rocco Di Francesco)*

In occupancies that require entry tools not common to engine companies, the truck assists with building entry, and the ladders will be used for a great many purposes as mentioned in Chapter 14. **Figure 19-35** shows a ventilation operation on a single-family residence.

The role of the firefighting teams in single-family dwellings is basic and, as mentioned before, potentially hazardous because of the need to rescue occupants in dangerous situations.

Multifamily Occupancies

Multifamily occupancy firefighting is also potentially dangerous and many times frustrating.

Caution Apartment complexes, condominium developments, upstairs and downstairs tenants, common walls, common attics, poor access, confusing addressing, people or bystanders in the way, parked cars, and many more issues and items make this a hazardous environment for the arriving firefighters. The multifamily occupancy fire is more difficult, dangerous, and demanding on the firefighters than the single-family residence or single-use occupancy.

In large developments, very often the smoke may be the best indicator of the fire's location, giving the firefighters a better mark on which to focus, **Figure 19-36**.

The role of the first-in unit is basically the same as for a single-family occupancy. Size-up and the action must be quick and decisive. It will be based on the factors as perceived by the arriving team and whatever they have gathered from other sources. If the structure has a common means of egress for the occupants, they must be protected by the team, allowing escape. This can be considered a form of rescue, which is the first priority of firefighting (RECEO).

The resources needed at a multifamily occupancy fire are going to have to be upgraded because of the logistics of moving personnel and equipment into areas of need. The distance from the street alone can be great and complex. The greater number of occupants will pose a rescue situation demanding more effort. The structural design lends itself to exposure problems, and confinement will be more difficult due to access problems and larger running areas for the fire to travel. Extinguishment may be difficult because of hidden areas in which the fire will run between floors and units. **Figure 19-37** shows a large multifamily complex on fire during its construction and the size of the pending firefight.

Figure 19-36 In many cases, in larger occupancies or complexes, the smoke may be the best indicator of the fire's location.

Figure 19-37 A large multistory family complex fire will entail a lot more resources than a single-family dwelling fire. *(Photo courtesy of Central Net Fire)*

Figure 19-38 The first-in unit at a mercantile fire should consider parking in front. *(Photo courtesy of Brentt Sporn)*

Business and Mercantile Occupancies

The business or mercantile fire is fought somewhat differently than the residential fire. Here, depending on the time of day, the life hazard potential for fire victims is not as great, and the fire and life hazards to the firefighter can be greater.

In most cases the business having the fire is either closed with no occupants, or open with the occupants either evacuated or fighting the fire. In either case, the chance of having to save lives is far less than for residential firefighting. On the other hand, however, the hazards associated with the occupancy itself can be greater to the firefighter.

> **Caution** Many businesses utilize hazardous materials or dangerous processes, and sometimes storage can be piled high. These can pose a serious risk to the firefighters entering these occupancies under dark, smoky, hot conditions.

Typically, the first-in engine will go to the front of the building in an effort to locate occupants who can assist with information about the fire. They will ask about its location, spread potential (what is burning), lives endangered, time of ignition and source, hazards to watch out for (see Chapter 24 for hazardous materials placarding and recognition), most important items or processes to protect from fire or water, and where access can be made safely. The officer will then decide on an attack plan based on these and other factors known about the area or business. **Figure 19-38** shows the first-in engine in front of the occupancy and the truck behind it utilizing a master stream. If the officer has decided to attack the fire, then the firefighter is told to pull lines from the apparatus following department SOPs in preparation for attack. If the fire can be attacked without endangering crew members, then an attack is made on the fire. If entering the building will put the crew in danger (per the two in/two out rule), the crew will prepare the attack by laying lines and placing equipment, but wait for reinforcements to assist in the interior attack.

The second-in engine has a number of options based on the conditions of the fire, the department's SOPs, or the direction of the IC. Some of these options are as follows:

1. Respond to the rear of the occupancy.
2. Supply the first-in engine with water.
3. Join the first-in unit with manpower for the initial attack.
4. Lay into the sprinkler connection and supply added water to the system.
5. Stage at the street or hydrant waiting for instructions.

If the first-in engine has not made entry due to either the business being closed or not being able to determine the extent or possibility of fire, then the second-in engine will respond to the rear of the occupancy. The second-in engine will often find the fire first, **Figure 19-39**, because many business occupancies have storage in the rear and this is where fires most often begin.

The first-due truck also has a number of options based on SOPs or need. Some of the uses for the first-in truck can be any of the following:

1. Forcible entry
2. Ventilation of structure
3. Search and rescue (primarily multistory ladder work)
4. Ground ladder operations
5. Securing of utilities
6. Salvage
7. Aerial streams

Figure 19-39 Many times the second-in unit will find the greater hazards when they respond to the rear of the structure.

This is only a partial list and if one truck cannot accomplish all of these tasks in a very short time by itself, then a second-in truck company may be called for assistance.

In these types of fires, firefighter injuries and fatalities can occur from falls and building collapse. The falls can occur from roofs either into the building or off the roof on the side of the building. Chapters 4, 13, and 18 describe safety on roofs and collapse potential.

The business or mercantile fire can be a potentially dangerous operation based on the hazards associated with the products manufactured and stored, and the structural integrity during a fire. This type of firefighting must be taken very seriously and studied well in order for the firefighter to be successful and safe.

Multistory Occupancies

The preceding discussions about fighting fires in structures apply in this section as well. Every tactic and factor considered can be utilized in attacking fires in multistory or high-rise structures. In the case of a multistory fire, it must also be expected that the resources needed for this task will be much greater than for any single-story structure. **Figure 19-40** shows a staging area for a multistory fire beginning to fill with responding units. The support structure of the firefighting system will require multiple layers because the fatigue factors alone can overwhelm smaller systems. In downtown Los Angeles, high-rise responses have been known to contain over seventy-five engines, not counting truck companies and other units.

The number of units responding, as well as the workload, will be predicated on the height of the structure, the location of the fire, and local protocol. If the structure is four stories and the fire is on the third floor, not as many units will be required as for a fire on the fifty-sixth floor of a seventy-five-floor building.

This section is intended to give a basic overview of the high-rise fire operation, because an in-depth study is beyond the scope of this book. Common courses in multistory or high-rise firefighting often run up to forty hours in length. Multiple courses are available and a number of books have been written on the subject.

In the typical high-rise response, a greater number of units will respond due to the nature of the work involved in these fires. Common practice is for the first-in unit to go to the **control room**. This room contains items that will assist in the determination of the location and severity of the fire. Some of the items found in a typical control room are:

- Alarm panels giving alarm details as to location, and so on
- Building utility controls
- Keys to all areas of the building including the elevators
- Communications to all parts of the building
- Sprinkler system activation details and controls
- Controls to auxiliary water pumping systems to supplement the building systems
- Maps and diagrams of the building layout and systems

Communications will be extremely important in this operation because the next unit to arrive typically will go to the fire's reported location. As this unit climbs the stairs or accesses the elevators they must be in communication with the first-in unit getting feedback as to that unit's findings regarding location, and so on.

> **Streetsmart Tip** In many jurisdictions, a general rule in high-rise firefighting is not to use the elevator. There is too great a chance that the crew would be taken to the fire floor, and as the door opened, they would be exposed to fire and conditions that could be dangerous to their survival, **Figure 19-41**. Because of the extreme height of many of today's high-rise structures, some jurisdictions may take elevators under certain circumstances. Two of those are if the elevator has firefighter lock-out controls built for fire department access and control or if they are part of a split bank of elevators that does not access the reported fire floor.

If an elevator is utilized, the firefighters will attempt to stop and get off the elevator two or more floors below the reported fire floor and walk the stairs the rest of the way. This way they can remain in the

Figure 19-40 A staging area for a high-rise fire begins to fill with resources. *(Photo courtesy of Dennis Childress)*

stairwell safely before deploying onto the fire floor. This crew will typically take a number of items with them for setting up an attack on the fire. The typical list for an engine or truck company to take with them is shown in **Table 19-4**.

Upon reaching the fire floor the attack team will set up an operation using the standpipe for water supply to

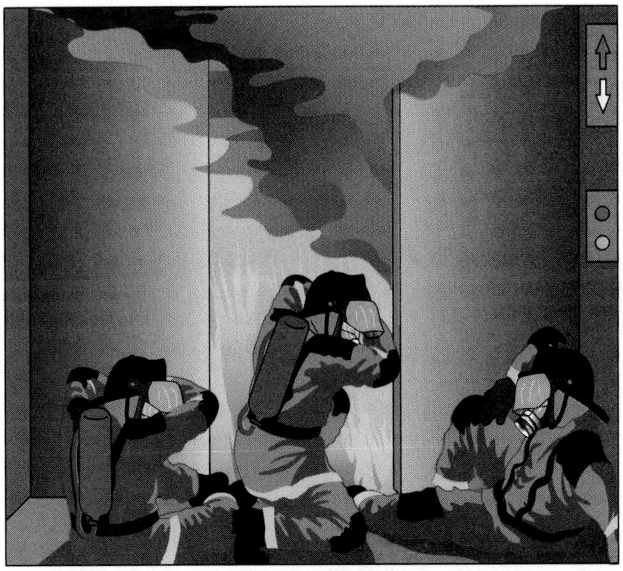

Figure 19-41 Firefighters must only use elevators that are secured so that accidental fire exposure will never occur.

Suggested High-Rise Equipment to Be Carried Aloft
ENGINE COMPANY
One spare SCBA cylinder per member
One large hand lantern or light per company
One standpipe hose bundle to include gated wye and nozzle (see Chapter 12)
Two spanner wrenches and a small pipe wrench or adjustable pliers
One ax
One Halligan tool
One short pike pole
Portable radio
TRUCK COMPANY (FOUR-PERSON UNIT)
One spare SCBA cylinder per member
Two large hand lanterns or lights
Two axes (one flathead)
One Halligan tool
One midlength pike pole
Circular saw with fuel can
One 100-foot drop line (⅜-inch rope if possible)
Portable radio

TABLE 19-4

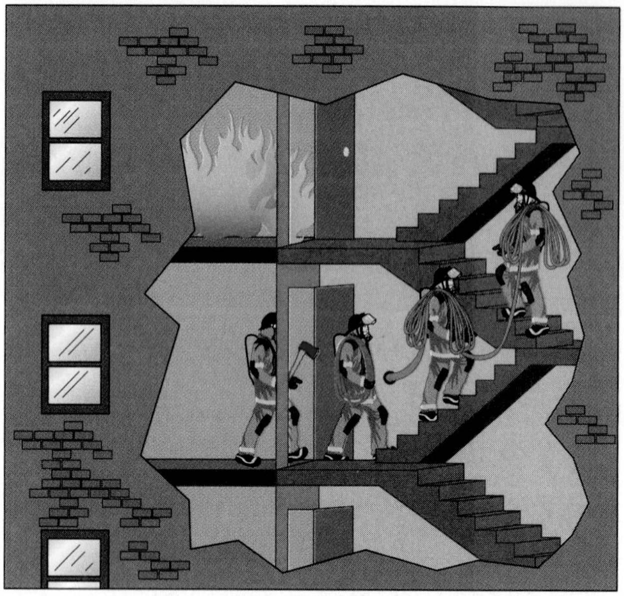

Figure 19-42 The high-rise fire is usually attacked from the stairwells. Laying hose up from the floor below the fire floor will give the firefighting team a safer and more efficient operation.

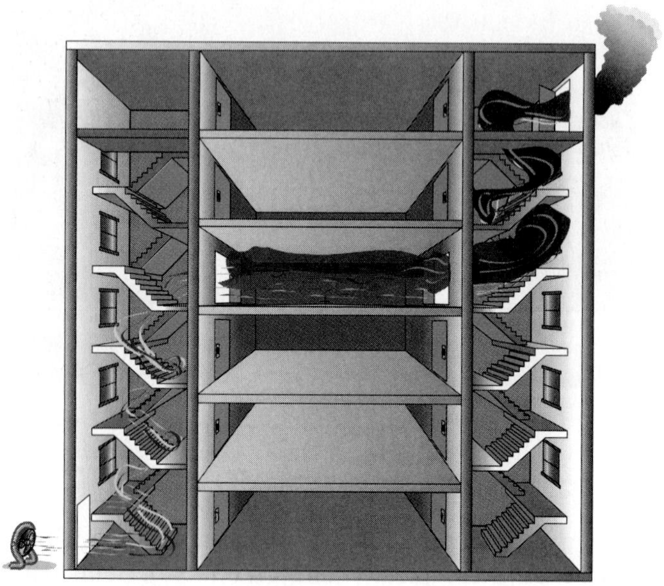

Figure 19-43 Truck company begins ventilation process.

the hose they brought with them. They will go to the floor below the fire and connect the hose to the standpipe, **Figure 19-42**, bringing it up to the fire floor in a layout that will play out as they enter the fire floor to attack the fire. Depending on the size of the fire, they may either attack or wait for a second unit for support. As in residential fire attack, if occupants are on the fire floor, then the first line must go between them and the fire in an effort to create an escape path or at least to contain the fire until sufficient personnel are on scene to focus on the rescue of the people.

The first-in truck will most usually go into a ventilation mode because ventilation will be extremely important in this situation. The products of combustion will be concentrated on the fire floor and possibly others, making rescue and firefighting very difficult at best. Their first task will be to capture the stairwells before the fire does, giving access to the fire to the fire teams as well as exits for the evacuees, **Figure 19-43**.

As other units arrive, they will begin filling in a great number of important operational positions. Some of these assignments could include search, attack team assistance, elevator control, equipment staging two floors below the fire floor (SCBAs, hose, lighting, etc.), ventilation team assistance, water supply to sprinkler or standpipe connections, salvage below the fire floor, and many more.

The multistory or high-rise fire can be one of the most taxing to personnel in the fire service. It will require more personnel, planning, practice, and other resources than almost any other type of operation. The only fire that will exceed it in resource needs will be the wildland conflagration, which is discussed elsewhere in this chapter. The firefighter assigned to a unit with the possibility of responding to a high-rise fire must be prepared for this assignment, because it is dangerous and extremely taxing physically. **Figure 19-44** demonstrates that a multistory fire requires more resources than a smaller single-story structure.

Below-Ground Structures or Basements

Fires below a structure, such as in basements, can be some of the hardest fires to fight due to the harsh conditions the firefighter may have to face. Fire and its combustion products travel upward with heat by convection and radiation. To fight one of these fires, the firefighters must travel down through this superheated air and smoke in order to reach the fire itself, **Figure 19-45**. In just about any other firefighting attack the firefighter can access the fire from below or the side, keeping low and out of the products of combustion. That is not true here.

> **Note** The key to fighting a basement fire is to ventilate as soon as possible, releasing the heat and smoke, making access and attack easier for the firefighting team.

The basement can also be very difficult to ventilate. Chapter 18 tells of a number of ventilation techniques and that information will prove invaluable in a below-ground fire such as a basement.

Figure 19-44 More resources, and many times more water, must be available for the multistory fire. *(Photo courtesy of Rocco Di Francesco)*

The role of the first-in engine company in these fires is to get water onto the fire as quickly as possible. This is primarily because the fire itself will be traveling upward and outward, creating a very poor situation for anything or anybody above the fire in the structure. The second-in unit must quickly assist with the ventilation as well as assist with the line already pulled. A backup line must be pulled as quickly as is practical for the safety of the team that has entered the basement. The attack team must be backed up as soon as possible and this is a situation where a single firefighter will *never* enter the area alone. The team concept must be strictly followed.

In cases where access to the basement cannot be made, it may be possible to punch through the floor above the room on fire and flood the room with water or high expansion foam, **Figure 19-46**. Whatever the attack method, it must be done quickly because everything above the fire is exposed.

Figure 19-45 Firefighters must travel down through superheated gases and toxic products of combustion in order to fight a basement fire.

Structures Equipped with Sprinklers or Standpipes

Fighting fire in occupancies equipped with sprinkler systems or standpipes has been a boon to the fire service over the last decade or two. Before the mandates of these systems, firefighters were not nearly as successful as they are today fighting fires in these types of occupancies.

The sprinklered building creates a unique situation for the firefighter.

> **Firefighter Fact** In over 94 percent of the fires recorded in the recent past, in sprinklered buildings, the sprinklers have either extinguished or confined the fire prior to the arrival of the fire department.

Figure 19-46 If access to the below-ground fire is too dangerous or not practical, then other means must be used to extinguish the fire through the ceiling of the burning room.

The team entering the structure has but to find the seat of the fire and complete extinguishment. This will still not be easy because the smoke and heat will be held down over the combustibles by the cooling action of the water spray from the system. The firefighting team will have to enter the structure very carefully without shutting down the protection afforded by the operating sprinklers, and they will have to find the seat of the fire through very limited visibility to complete the extinguishment of the fire. After this is done, the system can be shut down for a short time during overhaul. Chapter 12 details the operation of sprinkler systems.

The standpipe system is somewhat different. This pipe system will carry water to the firefighting team, easing their dependence on hoselines over great distances. This system is invaluable in fighting fires in high-rise structures. It saves a great amount of time by having the water ready for the team as it arrives at the floor on fire. Team members simply hook the hose carried aloft to the system and then charge it by turning on a valve at the standpipe. Chapter 10 describes this operation in some detail.

Exposure Fires

An **exposure fire** can be described as any combustible item being threatened by something burning in another area. The fire may be carried to the exposure by three ways. Chapter 4 described these as conduction, convection, and radiation.

As stated earlier in this chapter, the second priority after rescue, in most fire scenarios, is exposure protection. This form of property protection is one of the primary focus points in most fires. A number of factors are involved in most exposure fires. In no order of preference they are:

- *Wind:* Convection carrying embers can be a problem.
- *Distance:* The nearer the exposure, the greater chance of fire.
- *Material:* The makeup of the exposed surface determines its combustibility.
- *Intensity of fire:* The intensity of the primary fire factors into its spread.

A couple of basic methods are used to defend an exposure from fire and they are based on the methods of heat transfer. In conditions of conduction, a cooling medium must be applied to the material being heated by the conduction. An example would be a water stream applied to a tank exposed to a fire burning under or on it. **Figure 19-47** shows foam being applied to a tanker that is exposed to a fire next to it. In cases of convection a downwind patrol must be provided so as to ensure that embers do not ignite other materials. An example would be a downwind house with a wood shake roof, near a fire with flying brands being put into the air. Next is radiation as a medium for ignition of an exposure. The best method would be to apply water on to the exposed area to cool it, **Figure 19-48**. An example would be, as a structure is free burning, the first hoseline is placed between the burning structure and the next door unit. The water is applied to the exposed structure until another line can be trained onto the burning structure itself.

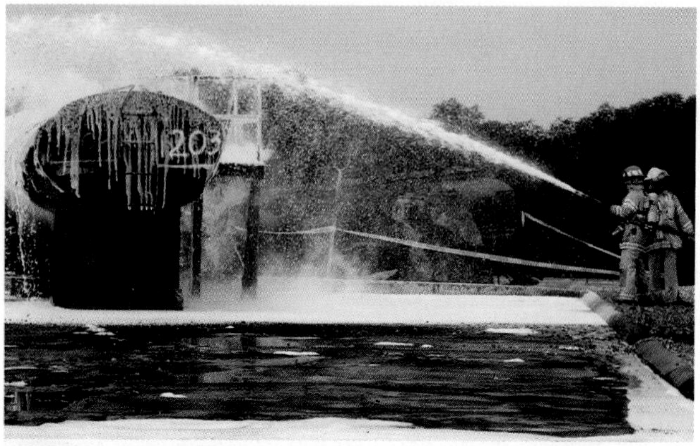

Figure 19-47 Firefighters apply foam to a tank truck exposed to a nearby fire. *(Photo courtesy of Central Net Fire)*

Exposure fires are common and dangerous, especially when the firefighting teams do not see the second fire and are put in danger by it.

> **Safety** It is important to put lookouts in service in order to maintain safety whenever there is any possibility of an exposure fire.

Nonstructural Fires

This subsection discusses fighting fire in nonstructural situations. It opens with wildland or ground cover firefighting, then transitions to vehicle fires and then flammable liquids and gas fires. The section finishes with a short discussion on trash and dumpster fires.

Ground Cover or Wildland Fire

The ground cover or wildland fire can be either a short pickup-type situation or a campaign fire situation lasting days to weeks or more. It can be a short, easy, nonthreatening call or a hard-fought, killer-of-firefighters call. In each of these cases the objectives remain basically the same:

- Confine the spread of the fire by surrounding it with hoselines, removing the fuel, or by another means of stopping its growth.

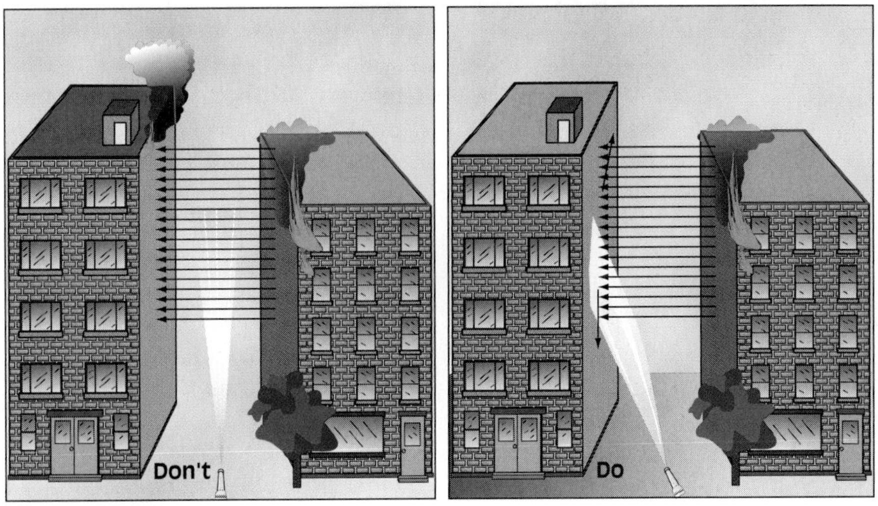

Figure 19-48 Radiation will travel through water or any opaque material. In order for water to be effective, it must be applied to the exposed surface in order to cool it.

- Guide the fire by control measures in order to keep it from burning into areas of higher intensity or faster growth.
- Operate from a position of strength and safety, never giving the fire the advantage of jeopardizing the safety of the fire crews or the citizens being protected.

The initial size-up is very important in that this type of firefighting can be very time consuming and exhausting. A number of basic facts must be considered on arrival that will dictate the attack plan for this type of fire. These facts are based on the wildland fire triangle discussed earlier and are used to determine which attack method will be most successful in fighting this fire. Based on the facts, the officer will ask two questions: What is the fire doing now and what will it be doing in the future? From this, the attack plan will be formed and the fight will begin.

The direct attack is a form of attacking the fire itself, in some fashion. It will usually involve hoselines, but not always. Hand tools or aircraft can be used in many cases as successfully as handlines. A point of attack is made on the fire called an **anchor point**. From there, the team will work up one of the **flanks of the fire** working toward the **head** or progressive end of the fire, **Figure 19-49**. Based on the factors listed in the subsection on ground cover fire components and considerations, the firefighting team will progress up the most active side of the fire until eventually the fire is surrounded and extinguished. **Figure 19-50** shows a firefighting team in direct attack of a grass fire by working up the flank while staying in the burn.

Safety By staying in or near the burn (walking with one foot in the black, or burn, and one foot in the green, or unburned area) a firefighter's safety is greatly enhanced.

If a team is progressing up the opposite flank, that is called a parallel method of attack. Both teams will meet at the head or near there and join lines, controlling the fire. Note that there is a difference between controlling and extinguishing the wildland fire. Controlling is just that. The fire is controlled and no longer burns into threatened areas. Extinguishment is the full extinguishment of all embers and fires contained within the perimeter of the control line.

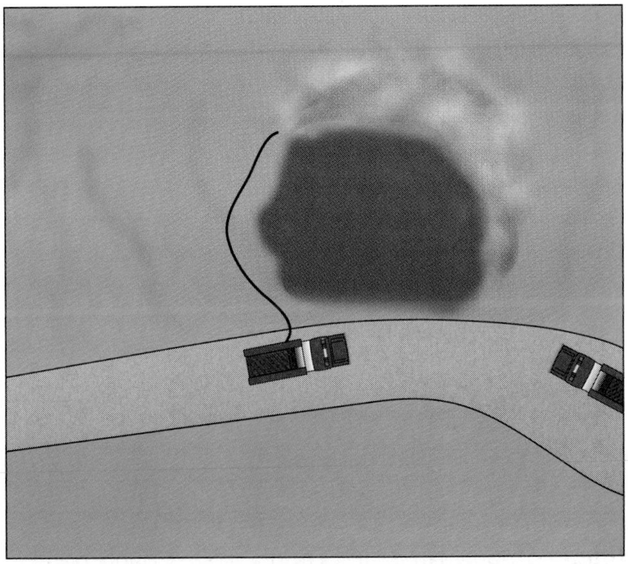

Figure 19-49 A direct attack on a ground cover fire.

Figure 19-50 A fire crew in the direct attack mode works on a grass fire. *(Photo courtesy of Phill Queen)*

Figure 19-52 An indirect attack is accomplished by burning off fuel just before the main fire arrives. *(Photo courtesy of Phill Queen)*

Extinguishment may take days to accomplish after the fire has been controlled.

Once a fire has reached such proportions that a direct attack is impractical, an indirect attack is utilized. This type of attack is common on large campaign fires that sometimes burn for days or even weeks. The indirect attack involves getting ahead of the fire in order to remove fuel from its path, **Figure 19-51**. This can be done by any of several methods. A few of the more common methods are removing fuel by hand or with bulldozers, removing fuel by utilizing natural barriers such as roads and rock outcroppings, or burning off the fuel in controlled conditions. **Figure 19-52** shows the last method as a firefighter burns off fuel before the main body of the fire can reach it. This method of firefighting is very scientific and practiced by only the most experienced fire teams. It requires the ability to read the weather, the fuels, and topography very carefully in order to remain in control of the attack. Consideration of safety and all of the safety lessons and rules must be followed very carefully on this type of firefighting.

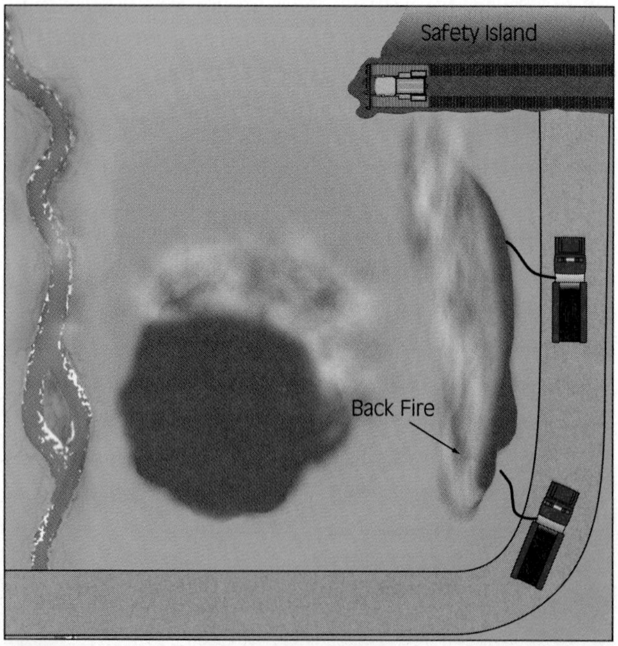

Figure 19-51 An indirect attack on a ground cover fire with two engines and a dozer.

Another type of wildland firefighting is now being utilized quite commonly in the fire service today and that is **interface firefighting**. This type of firefighting involves crews that are often oriented more toward structural firefighting. They are asked to go into the rural areas in an effort to protect structures from wildland fires, **Figure 19-53**. The skills involved with this type of firefighting are very different from urban firefighting. A number of classes and curricula are being promoted in these areas for fire service personnel that are designed to create a much safer and smarter firefighter as society continues to move out of the cities.

Ground cover or wildland firefighting is much different than structural firefighting. It requires different tools, clothing, methods of attack, and knowledge. **Figure 19-54** shows a common tool used in wildland firefighting: a helicopter with a water dropping tank. To be safe, efficient, and effective, the firefighter of today must be educated in this area as well as that of the structural fire world.

Vehicular Firefighting

Vehicle fires are very common in today's fire service. Some of the more common reasons for vehicle fires are:

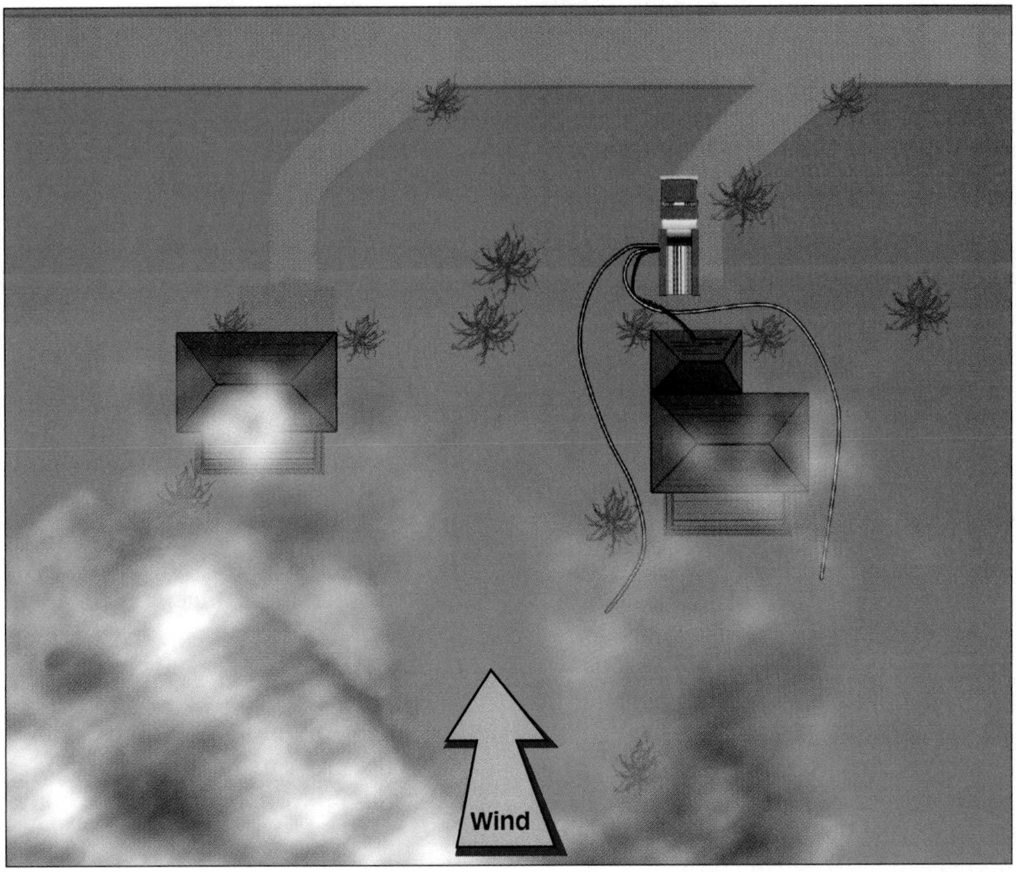

Figure 19-53 Protecting homes in the interface is a growing practice across the country for engine companies used to fighting fire in the cities.

- Collision resulting in leaking fuel and sparks
- Electrical problems causing short circuits
- Overheated components such as tires, engines, or brakes
- Muffler and catalytic converter problems
- Discarded smoking materials
- Arson

The NFPA recommends at least a 1½-inch hoseline for vehicle firefighting and preferably a combination nozzle with a partial fog of approximately a 30-degree pattern initially. The first water should be used to protect the occupants if they are still in the vehicle. **Figure 19-55** shows a firefighting team directing its initial attack on a car's interior into the passenger compartment first. After the occupants

Figure 19-54 A helicopter is readied for wildland firefighting. *(Photo courtesy of Brentt Sporn)*

Figure 19-55 The firefighters' first attack point is the passenger compartment in order to protect life. *(Photo courtesy of Brentt Sporn)*

have been rescued, then the attack on the fire itself may begin. The suggested method of attack is direct. The only time an indirect attack would take place is when the vehicle fire is exposing something else, causing concern on the part of the firefighting team.

Caution As mentioned earlier in the book, backdraft and flashover are two structure fire phenomena that firefighters must be aware of. This is also true with automobile fires. A closed passenger compartment or trunk may be the perfect place for a backdraft to be set up. Or, that same area can see a flashover if the conditions are right. Care must be taken when working in these environments.

The automobile fire can also pose situations in areas not encountered in other fires—closed hood and trunk areas. It will be important to gain entry for a number of reasons. Some of those reasons are the complete extinguishment of the fire, the investigation of the fire's origin, the securing of the vehicle and all of its contents, and possibly the determination of whether the vehicle was carrying illegal materials that factored into why it burned. Gaining entry into the hood or trunk area can be extremely difficult.

In most of today's automobiles a steel cable runs from the interior dash area to the latch at the front of the engine compartment. Unfortunately, this cable burns through quickly and access to the engine area must be by force, as the cable mechanism will be out of service. Entry tools, as discussed in Chapter 27, must be used as well as ingenuity in the entry effort.

The trunk is no different. While there is a cable from the interior driver area to the trunk in most new vehicles, it will be of no use due to the fire. Entry tools will also be needed here.

Safety The attack should be from the upwind and uphill side of the vehicle in order to reduce the firefighter's exposure to heat and the products of combustion. This is also important in situations where leaking fuel is under the vehicle and can run toward the firefighting team members. Consideration must also be applied to chocking the tires on any vehicle burning on unlevel ground. It could be disastrous if the vehicle's brake lines burned through and the vehicle began to roll away or toward the fire engine. Full protective clothing with SCBA must be worn when fighting vehicle fires because the combustibles are toxic and sometimes explosive.

Care must be taken around the bumpers, hatchback struts, and air bag assemblies as discussed in the vehicle fire components and considerations subsection. Care must also be exercised around the tires if they are burning as well as the fuel areas of the vehicle. It is also recommended that the battery to the vehicle be disconnected by removing the cables before any overhaul begins to ensure firefighter safety.

The procedures around electric vehicles may differ due to the nature of the power source and its effect on the automobile itself. The automobile industry has gone to great lengths to educate firefighters in safe methods of fighting electric vehicle fires. The most recommended method to date consists of seven steps:

1. Approach vehicle at an angle to front or rear.
2. Identify vehicle type.
3. Secure and stabilize vehicle.
4. Power off electric vehicles.
5. Make the initial attack on vehicle fires with water and/or foam.
6. Use standard tools and cut-in areas for victim extrication.
7. Do not cut into high-voltage components. These are usually color coded by law to stand out from the rest of the components.

The action taken by an attack team on a vehicle fire in step 5 is to extinguish electric vehicle fires with water and/or foam. If it appears that the fire has extended into the battery pack, the firefighter should attempt to extinguish it following the instructions for the specific battery type. This information changes as technology changes, so it must be obtained from in-service training on an ongoing basis. If the vehicle is in a safe area, under some circumstances, firefighters may allow the fire to self-extinguish. Vehicle manufacturers are now beginning to use a number of alternative fuels in an effort to become more fuel efficient. A firefighter may find compressed natural gas (CNG), liquefied propane gas (LPG), and others. Fighting fire involving these fuels must take on the methods discussed in the next section on flammable liquids and gases fires.

As the firefighters work on the vehicle fire, caution must be considered regarding other traffic in the area, **Figure 19-56**. Unfortunately, a great number of firefighter injuries occur each year at vehicle fires when passing motorists strike firefighters working on the accident scene. Great care must be taken knowing that passing motorists will be looking at the accident more than the people working in the area. It is also important to realize that flashing lights and excitement will attract some drivers to the scene. **Figure 19-57** shows an engine being used to block traffic from the vehicle fire scene. Traffic cones, road flares, or fire engines are to be used to block traffic from the work area in order to better protect firefighters working an accident.

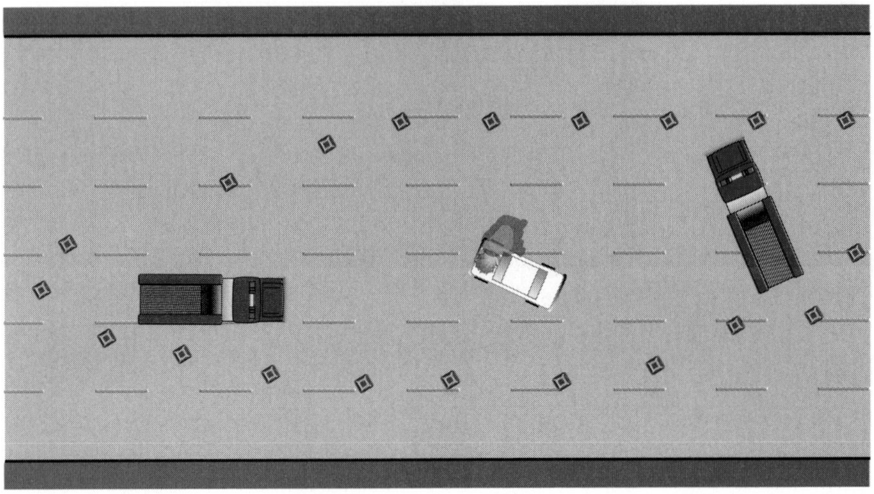

Figure 19-56 To protect firefighters on scene in the street, the engine is typically used to block traffic. Cones are used quite often to assist in this endeavor.

While fighting vehicle fires may at first seem simple, this is far from the truth. Vehicle fires are complex, dangerous situations, requiring skill in fighting each of the four classes of fires. Safety must always be considered in these situations.

Flammable Liquids and Gases Fires

Flammable liquids and gases have a number of things in common that the firefighter can use to control a fire in these commodities. The mechanisms of extinguishment are well known to the fire service. These are some of them:

- Smother the fuel, cutting off the supply of oxygen.
- Starve the fuel by removing unburned material from the area.
- Interrupt the chain reaction process with chemical agents.
- Cool the fuel, reducing vapor pressure.

Figure 19-57 The fire apparatus is used to block traffic from the firefighters' working area. *(Photo courtesy of Phill Queen)*

A great range of possible situations can arise involving combustible and flammable liquids and gases and their burning. As the first-in unit arrives, a number of facts must be obtained in order to handle the situation effectively. Finding answers to the questions asked may be difficult or sometimes impossible. In many situations, hazardous materials teams may be needed in order to identify the materials and the recommended method of handling them. Hazardous materials are discussed in more detail in Chapter 24.

As opposed to most of the other fire scenarios discussed thus far, flammable liquids and gases do not follow the laws of extinguishment so basic to the fire service. Water may not be the choice best suited for firefighting here. **Figure 19-58** shows a crash crew applying foam to a ground fire of flammable liquids.

Figure 19-58 A team of firefighters applies foam to a pool of flammable liquid. *(Photo courtesy of Phill Queen)*

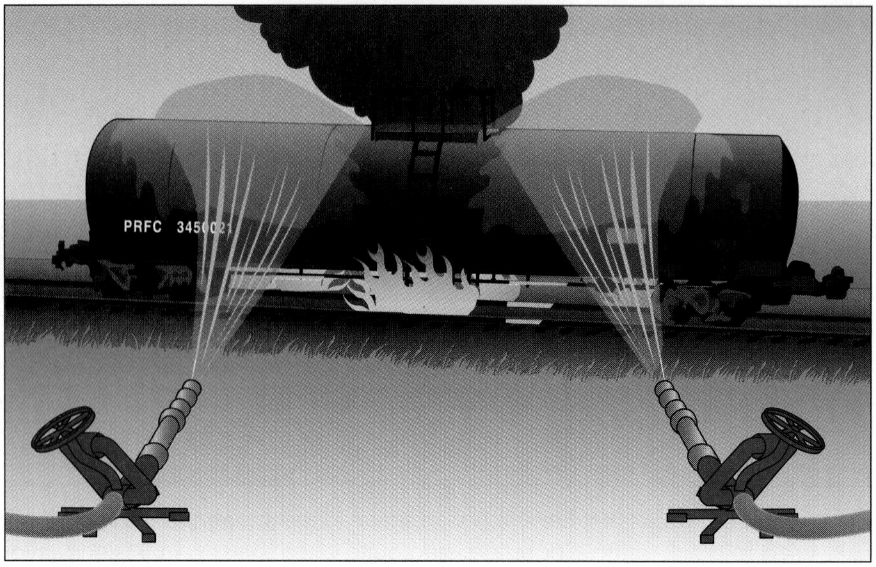

Figure 19-59 Water is applied to the heated metal surface in order to keep it cool, slowing the pressure buildup inside the exposed tank car. This will reduce the possibility of a BLEVE.

> **Caution** In fact, in many cases, water could create more problems than the firefighting team cares to imagine. No one agent applies to all fires in this category, and, in some cases, it may not be the best solution to extinguish the fire at all. It may be better to let it burn itself out as in the case of some pesticides and gases.

Knowing this it is possible to set up some very generic principles for fighting most flammable liquid and gas fires as follows:

1. Identify the material involved and its hazards.
2. Utilize written assistance with the DOT *Emergency Response Guidebook* and other sources.
3. Evaluate the threat to nearby similar commodities such as gas cylinders and tanks. For instance, get water on exposed tanks to reduce internal pressure buildup and prevent the possibility of a BLEVE (see Chapter 4), **Figure 19-59**.
4. Determine which extinguishment agent and principle is best.
   - **a.** For small fires consider CO_2 or dry chemicals.
   - **b.** For larger fires consider foam suited for that particular commodity.
   - **c.** For some fires consider letting the fire burn itself out. This is particularly true in cases where the leak, when burning, is contained, and if extinguished, may spread and reignite, causing greater hazards and situations.
   - **d.** Look for shutoff valves or switches to shut down the flow of material to the fire. **Figure 19-60** shows a group of firefighters approaching a shutoff valve at a flammable liquid fire under the protection of a water fog.

Incidents involving combustible and flammable liquids or gases are very dangerous and have caused the fire service to bring in the concept of hazardous materials teams to deal with these special situations. The number of materials considered hazardous in today's society pose a very great threat to the fire service and its ability to fight fire. Knowledge and safety are the keys in working with these materials.

Trash and Dumpster Fires

A trash fire, whether contained in a bin (dumpster), truck, or spread on the ground, can surprise the most seasoned firefighter with a burning rate or intensity not expected. Trash can be anything discarded by anyone. Trash can produce toxic gases and, in many cases, items discarded can be hazardous and sometimes even explosive. The unprepared firefighter can be injured or killed fighting these seemingly innocent fires.

As the fire engine arrives on scene it is important for the team members to be fully dressed in their

Figure 19-60 A team of firefighters approaches a flammable liquid fire control valve behind a fog pattern. *(Photo courtesy of Phill Queen)*

Figure 19-61 A firefighter approaches a dumpster fire, keeping below the top edge of the container. *(Photo courtesy of Brentt Sporn)*

full protective clothing including SCBA. Water is most usually the agent used in the extinguishment of these typically Class A combustibles.

> **Safety** When first applying water, the firefighter must be protected from the possible explosion of heated containers such as aerosol cans, bottles, and glass items. Using the protection of the side of the bin or the side of the engine should be considered when first attacking a trash fire.

Figure 19-61 shows a firefighter applying water while being protected from exposure.

Many times these fires will deeply seat themselves, not allowing the water to penetrate the materials. In these situations a number of things can be done, including the following:

- Use a tool such as a pike pole or rubbish hook to stir the material as water is applied.
- Use a tractor or dozer to move larger masses of trash in order to penetrate piles.
- If the fire is in a truck, consider having it dumped while water is applied.
- The use of water agents (light water) can be an effective method of achieving deeper penetration into the materials.
- In some cases the bin or dumpster may have to be filled with water to achieve better results.

The trash or rubbish fire can pose extinguishment problems as well as hazards not considered in such nonemergency-type calls. The firefighter must be knowledgeable about the possibilities and proceed with caution when dealing with the unknown contents of these fires.

Outside Stacked or Piled Materials Fires

Many times the firefighter is asked to extinguish a fire in outside storage. These may be lumber piles, used tire piles, stored pallets, bundled paper or cardboard, or just about anything else that is flammable and stored outside. **Figure 19-62** shows a typical yard storage of high piled pallets.

The hose size, stream used, amount of water, appliances for water delivery, and tools for overhaul will either make the job easy or, if not selected properly, make the job difficult.

> **Safety** Working around any high piled storage is dangerous. This danger is magnified if the materials are on fire or even wet after extinguished. Piled storage is held up by the integrity of the items stored, and as that changes with either fire or water damage, those piles may topple. Collapse zones must be maintained whenever working around high piled storage.

Logically, large piles with large fires will require large amounts of water. The delivery of that water will be the key to success. Large-diameter lines with the proper nozzles, usually smooth bore or straight stream capable, will most often be utilized for quick knockdown. As this knockdown is taking place, it will be most important to keep an eye on any exposures that may be threatened.

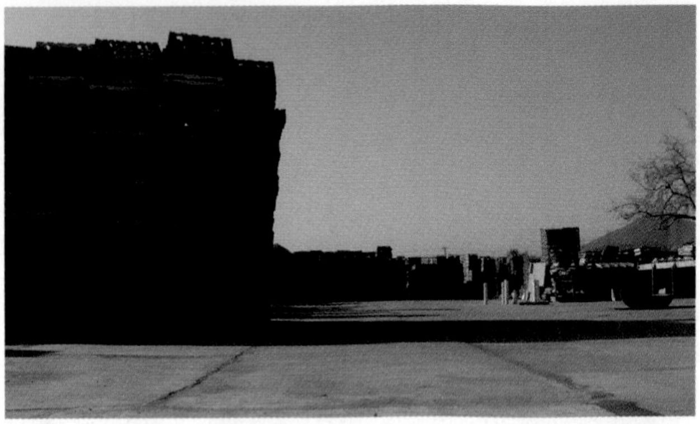

Figure 19-62 Typical high piled storage that can jeopardize firefighter safety.

Those may consist of nearby piles of the same material, small buildings, or even distant exposures threatened by flying fire brands. Then, after initial knockdown, the fog and/or foam nozzles will come into effect. Many times the streams will require additives, such as those described in Chapter 11, in order to add to the penetrating power of the water.

Then the overhaul begins. In most cases, firefighters can expect to have to pull all the materials out of every pile while water is being applied. In the case of used tire storage, it is not uncommon to have to bring in large tractors, cranes with buckets, or other large hauling or moving equipment to break down the piles. Hand tools will also be used as the piles are brought down to manageable sizes. In many cases the overhaul will last far longer than fighting the fire itself.

Caution The products of combustion will continue to develop during the overhaul phase of the work, and it is important to continue to wear SCBA or some type of respiratory protection during this time.

Lessons Learned

The basic principles of firefighting are based on sound scientific laws as well as years of firefighting experience. Every combustible item and occupancy must be carefully studied and understood in order for firefighters to be effective, efficient, and safe in their work.

Just as both synthetic and natural items differ, so will the tactics and strategies necessary to extinguish them differ when they burn. Knowing the basic elements of fire control, and applying them safely, will result in a better environment for all.

KEY TERMS

Anchor Point A safe location from which to begin line construction on a wildland fire.

Aspect The direction a slope faces given in compass directions.

Autoextended When a fire goes out the window on one floor, up the side of the building, which is often noncombustible, and extends through the window or cockloft directly above.

Box Canyon A canyon open on one end and closed on the other. They become very dangerous when wildfire enters them.

Chimney Another term for drainage. Given because of the draw of fire as in heat going up the chimney.

Combination Attack A combined attack based on partial use of both offensive and defensive attack modes.

Control Room A room on the ground floor of a high-rise building where all building systems controls are located.

Defensive Attack A calculated attack on part of a problem or situation in an effort to hold ground until sufficient resources are available to convert to an offensive form of attack.

Drainage A topographic feature on the side of a hill or mountain that naturally collects water runoff, channeling it to the bottom of the rise. Fire is attracted to this feature.

Exposure Fire Any combustible item threatened by something burning nearby that has caught on fire.

Fire Intensity A measurement of Btus produced by a fire. Sometimes measured in flame length in the wildland environment.

Flammable Range Ratio of gas to air that will sustain fire if exposed to flame or spark.

Flanks of the Fire The sides of a wildland fire running from the start point up each side to the end of the fire running into unburned areas.

Garden Apartment A two- or three-story apartment building with common entryways and layouts on each floor, surrounded by greenery and landscaping, sometimes having porches and patios.

Head of the Fire The running top or aggressive end of the fire away from the start point.

Incident Management System (IMS) A management system utilized on the emergency scene that is designed to keep order and follow a sequence of set guidelines.

Interface Firefighting Fighting wildland fire and protecting exposed structures in rural settings.

Midslope An area partway up a slope. Any location not on the bottom or top of a slope, as in a midslope road crossing the slope horizontally.

Offensive Attack An aggressive attack on a situation where resources are adequate and capable of handling the situation.

Polar Solvent A material that will mix with water, diluting itself.

Rate of Spread A ground cover fire's forward movement or spread speed. Usually expressed in chains or acres per hour.

RECEO Acronym coined by Lloyd Layman standing for Rescue, Exposures, Confinement, Extinguishment, and Overhaul.

Ridge The land running between mountain peaks or along a wide peak. A high area separating two drainages running parallel with them.

River Bottom Topographic feature where water runs from higher elevations to lower. Can be dry or wet depending on season or recent rains.

Saddle A pass between two peaks that has a lower elevation than the peaks. Wind will pass through this area faster than over the peaks, so fire is drawn into this feature.

Solubility A liquid's ability to mix with another liquid.

Specific Gravity Weight of a liquid in relation to water. Water is rated 1.

Stairwell An enclosed stairway attached to the side of a high-rise building or in the center core of same.

Steepness of Slope The degree of incline or vertical rise to a given piece of land.

Teamwork A number of persons working together in an effort to reach a common goal.

Toxicity Poisonous level of a substance.

Vapor Density Weight of a gas in relation to air. Air is rated 1.

REVIEW QUESTIONS

1. Explain how water suppresses fire in relation to the fire tetrahedron.
2. Give at least two factors that must be taken into consideration prior to an attack plan being formed and extinguishment of a structure fire.
3. Ground cover firefighting involves a number of principles very different from those used for structural firefighting. Explain how the wildland fire triangle differs from the fire tetrahedron.
4. Vehicle firefighting involves a number of built-in dangers. Name an exterior feature and an interior feature that can injure a firefighter who is not very careful.
5. Explain why firefighting tactics will differ based on a flammable liquid's specific gravity.
6. Why does a basement fire have the potential to be so much more hazardous to a firefighter than a ground-level fire in the same occupancy?
7. OSHA has enacted the two in/two out ruling. Explain this rule and its impact on the fireground during structural operations.
8. Explain the procedures recommended if a firefighter is considering taking an elevator to the fire floor in a high-rise or above-ground fire. What does your jurisdiction recommend?

Additional Resources

Brunacini, Alan, *Fire Command.* National Fire Protection Association, Quincy, MA, 1985.

California State Fire Training, *Fire Command 1A and 1B.* California State Fire Marshal Office.

Coleman, Ronny J., *Management of Fire Service.* Daybury Press, North Scituate, MA, 1978.

Layman, Lloyd, *Fire Tactics and Strategy.* NFPA, Quincy, MA, 1953.

Lowe, Joseph, *Wildland Firefighting Practices.* Delmar Learning, a part of the Thomson Corporation, Clifton Park, NY, 2001.

NFPA 1001: Standard for Fire Fighter Professional Qualifications, 1997 Edition. National Fire Protection Association, Quincy, MA, 1997.

Richman, Harold, *Engine Company Operations.* National Fire Protection Association, Quincy, MA, 1986.

Richman, Harold, *Truck Company Operations.* National Fire Protection Association, Quincy, MA, 1986.

CHAPTER 20

SALVAGE, OVERHAUL, AND FIRE CAUSE DETERMINATION

Geoff Miller, Sacramento Metropolitan Fire District

OUTLINE

- Objectives
- Introduction
- Salvage Tools and Equipment
- Maintenance of Tools and Equipment Used in Salvage
- Salvage Operations
- Salvage Operations in Sprinklered Buildings
- Overhaul Tools and Equipment
- Overhaul Operations
- Fire Cause Determination Concerns
- Securing the Building
- Lessons Learned
- Key Terms
- Review Questions
- Additional Resources

STREET STORY

Many firefighters take more of an interest in going in and attacking the fire. But salvage is important too. Most of the damage to a person's property is caused by smoke and water, not by the actual fire. And those contents can have so much meaning for people—all they want is their private possessions. Beyond that, if there's a possibility of arson, it's your job to preserve the evidence.

I remember receiving a call for a fire at The Museum of Science and Industry—it came in through enhanced 9-1-1. By the time we responded, the museum had white smoke curling from the roof. The initial companies were locating the seat of the fire (not easy since the museum occupies four city blocks). They figured out it was in a void in the roof, and we started doing salvage. I acted as coordinator. In a house fire, we'd be likely to rope the belongings in the center of the room and put salvage covers over everything. But in this case, there were fixed displays, so we couldn't move them, but we still covered everything. We built chutes for the stairwells so that the water could be channeled without damaging the carpets, and we diked the elevators.

If we can prevent water damage and channel out some of the smoke, we'll have a better goodwill image with citizens. In this case, you may not gain the admiration of fellow firefighters, but what we did was save a lot of artifacts that couldn't have been replaced.

—Street Story by Julius Stanley, Lieutenant, Chicago Fire Department, Chicago, Illinois

OBJECTIVES

After completing this chapter, the reader should be able to:

- Explain the purpose of salvage and overhaul operations.
- Explain the importance of salvage from a "customer service" standpoint.
- Identify tools and equipment used in salvage and overhaul operations.
- Identify needed maintenance of salvage and overhaul tools and equipment.
- Perform different identified salvage cover throws, folds, and rolls.
- Arrange a room's contents into a salvageable position.
- Identify sprinkler system shutoff valves.
- Shut off a flowing sprinkler head.
- Explain how fire attack and ventilation assist in the salvage effort.
- Explain where to search buildings for hidden fires.
- Explain how to look for structural stability.
- List debris removal techniques.
- Explain the importance of evidence preservation.
- Identify how to determine the area of origin of a fire.
- Identify how to secure a building after emergency operations are complete.

INTRODUCTION

Salvage and overhaul operations are not often viewed as critical tasks within the fire service. They are not associated with the excitement of fire attack, rescue, or ventilation. This perception leads to a downplaying of importance and an "afterthought" mentality. Fire cause determination is something new firefighters will not be actively involved in, but they need to understand what actions to take so as not to make the cause determination unattainable or more difficult for the fire investigator. This chapter explains the reasons for and importance of salvage, overhaul, and fire cause determination and describes the necessary firefighter skills and abilities needed to effectively perform these operations.

SALVAGE TOOLS AND EQUIPMENT

All fire incidents involve a potential loss of material goods. Whether they are easily replaceable articles of clothing or furniture or irreplaceable photographs, heirlooms, antiques, or memories, they all have meaning to their owner. It is for this reason that firefighters can make a tremendous impact on an incident by aggressively salvaging the occupancy and making sure not to cause more damage than the emergency itself has, **Figures 20-1A** and **B**.

People usually think of the interior of a structure when salvage is considered, but items on the exterior of a house can be just as valuable. While first-in company officers are making their size-up to determine the extent of the involvement, they may be able to *quickly* move some patio furniture or arrange to have vehicle(s) covered in the front of the structure with a salvage cover. It is important that the company officer "triage" the entire scene. If quick action can save property without endangering the operation, then the officer should make an effort to do so. It is possible for any firefighter to get tunnel vision and waste time trying to save one portion of the scene while endangering the integrity of the scene as a whole. Quick thinking and experience are vital traits of a good company officer.

> **Streetsmart Tip** Company officers who are in command of a scene can often perform a small necessary or safety-oriented task if there are no other firefighters in staging and if the task will not interfere with the safety of the on-scene crews or the command of the scene. Commanding the scene is the primary role of that officer.

It is rewarding to report to owners who fear the worst possible loss that salvage crews were able to save a particular special item or memory for them. Special items or memories make a home unique, and firefighters should always consider this while working any fire scene. A quick scan followed by consideration of what the homeowner might want to save can be a great service to a person who suffers the tragedy of a fire.

The basic premise of salvage operations is to remove the harmful atmosphere from the material

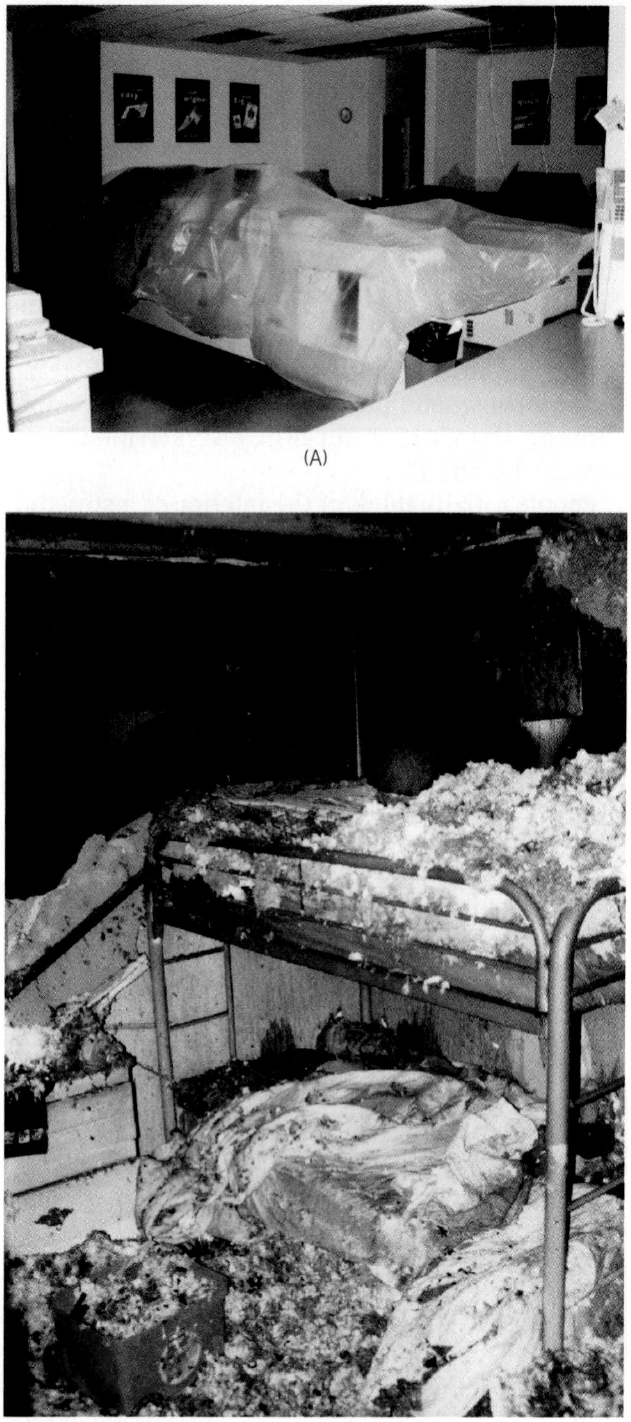

(A)

(B)

Figure 20-1 (A) Good and (B) bad salvage operations.

or to protect the material from the harmful atmosphere. Several techniques are used to accomplish this goal and will be discussed later. Any technique is rendered useless unless the proper tools are used, and there are many tools and different pieces of equipment for salvage work. Some operations might require complex equipment while others demand a hammer and a few nails.

Salvage Covers

The mainstay of salvage is the salvage cover or a variation of the salvage cover. Salvage covers are made out of several different materials, most commonly plastic, canvas, and treated canvas.

The size of the salvage cover varies. It can be anywhere from 10 × 12 feet to 12 × 16 feet. The perimeter of the cover is ringed with grommets spaced at intervals, **Figure 20-2.** The ideal interval is 16 inches to match up with the studs in a building wall. A plastic cover has advantages in weight and water resistance. A canvas cover is durable and, if treated, can compare to the plastic cover for water resistance capability.

Salvage covers are the main materials used for constructing water chutes and catch-alls, which are discussed later.

A new tool that is making the salvage cover a less desirable option is **Visqueen** or black plastic, which can be carried easily by a firefighter, **Figure 20-3**. The Visqueen comes in rolls 120 feet long, with a width, once unfolded, of 20 feet. A roll can also be carried by putting a dowel in the center tube (with a cord attached) and hanging the cord over the neck so the roll is allowed to hang in front of the firefighter. The only additional tool needed is a razor knife or scissors to cut off the desired lengths. In some cases the plastic is precut into various sizes for quick deployment. By precutting some lengths, the weight of the roll is reduced, thus minimizing the stress on the firefighter carrying it.

Floor Runner

Another item used for salvage operations is the floor runner, which is usually around 3 feet wide and about 20 feet long. It is normally made of a lightweight canvas-type material for easy deployment. It is used just like its name implies—to cover the floor

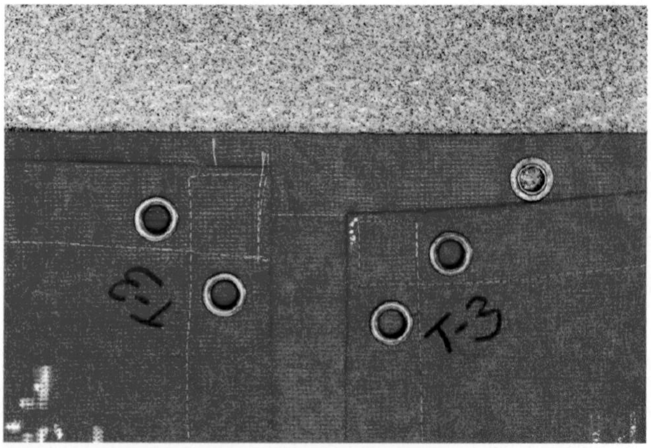

Figure 20-2 Salvage cover grommets.

Figure 20-3 Firefighter carrying Visqueen roll.

Figure 20-4 Floor runner in place.

down a hallway or along a traffic area, **Figure 20-4**. If there is a fire in one area of the building, firefighters should try not to damage the carpet in another. Remember, customer service really makes the difference and is a sign of professionalism.

Water Vacuum

A very useful tool in salvage operations is the water vacuum, **Figure 20-5**, which is available in two basic types. One is worn like a backpack and is used to remove water from areas where access prohibits squeegee use; the other is usually larger and is moved around on wheels. If SCBAs are being worn, the "backpack" water vacuum cannot also be worn. Other methods will have to be considered, or firefighters will have to wait until the air quality is safe and the area is safe to work in without SCBA.

Streetsmart Tip Squeegees are very helpful where the need exists to move large amounts of water on a flat surface. A sprinkler head discharge in a large warehouse can result in many gallons of water on the floor. Once the flow is stopped, this water can be quickly moved from the warehouse by several firefighters working together with squeegees.

The water vacuum can also be used to quickly drain catch-alls so that they do not overflow. The water vacuum usually has a limited capacity of around 5 gallons and thus needs to be emptied. If a large amount of water must be removed by a water vacuum, it is advisable to have the owner contract with a professional water removal company.

Another option carried by some fire agencies is either a submersible pump or "float-a-pump." The submersible pump can be dropped directly into water and used to pump water out through an attached line. This is a very good tool for flooded basements. The float-a-pump simply floats on top of the water and essentially drafts water and pumps it out. The submersible needs electrical power, while the float-a-pump is gasoline-driven.

Figure 20-5 Firefighter using water vacuum.

Figure 20-6 Various fastening tools.

Miscellaneous Salvage Tools

Basic hammers and nails or staple guns also have a place in salvage work. Having these tools available allows firefighters the ability to secure doors, windows, ventilation holes, or other openings to further protect the occupant's property from weather. This is an area where the use of Visqueen comes in handy. It can be left behind on a roof to cover a ventilation hole and does not have to be retrieved.

Doors and windows need to be secured with plywood or some other sturdy material to offer any resistance to vandals. This task is handled by board-up crews and is discussed later.

Sometimes salvage covers need to be used to prevent a leak from damaging an item to be salvaged or an item that is too large to move to a safer location. A hammer and nails can be used to nail the salvage cover to the wall and cover the exposed item. Firefighters also use a small fastening tool called an S-hook that can be hammered into the wall to establish a sturdy place to hang the cover, **Figure 20-6**.

A thorough knowledge of available tools to be used in salvage work will make the task much easier to accomplish. With the right tools, the job will be done correctly and the damage minimized.

MAINTENANCE OF TOOLS AND EQUIPMENT USED IN SALVAGE

After salvage and overhaul tools and equipment are used, they must be cleaned and inspected like any other piece of firefighting equipment to make sure they are ready for the next emergency. A written log should be kept on the maintenance and use of the tools so proper maintenance is done.

Salvage tools and equipment are often exposed to hazardous materials and should be placed out of service if damaged. If through the course of a salvage operation, the equipment suffers damage, the fire department should seek reimbursement from the property owners through their insurance company. Just as the insurance company is liable for the owners' property, it may also be liable for fire department equipment if it is damaged. This is particularly true for salvage equipment where loss is suffered while protecting the insured's items from damage. The department should document the damage thoroughly and then contact the insurance company through the **chain of command** to determine the means to secure reimbursement.

> **Streetsmart Tip** This is one of the many reasons why it is a good idea to carry a camera on the apparatus. It provides an easy way to document items, and it can also capture some great training pictures.

This may not apply in all states, so it is important to contact a local insurance commissioner to determine if this is an option.

When salvage covers are used at an emergency scene they will end up with material or debris on them. This material should not be spilled on clean areas after it has been collected. The best way to keep this material in the cover is to do a "loose fold and roll," **Figure 20-7**. Essentially, with this technique the cover is folded into itself so the debris caught is trapped inside. Once the cover is folded into itself by bringing both sides to the middle once and then once again, the cover is then rolled up. This will keep all of the material inside until it can be disposed of. Once the cover is dumped out and brushed off, it must be washed using a mild soap solution, hung to dry, inspected for tears or holes, repaired, folded, and placed back on the apparatus. (Note that this reveals another plus of Visqueen—no maintenance.)

Drying the cover can prove difficult due to its size. Sometimes hooks are placed along high rooflines to hang the cover to dry. Another method is to rig a system to haul the cover up alongside of a building to dry, **Figure 20-8**. If neither of these is available, it can be laid out on the apparatus floor, but it must be turned over so both sides dry. In any case the cover should be dry before it is folded and returned to service.

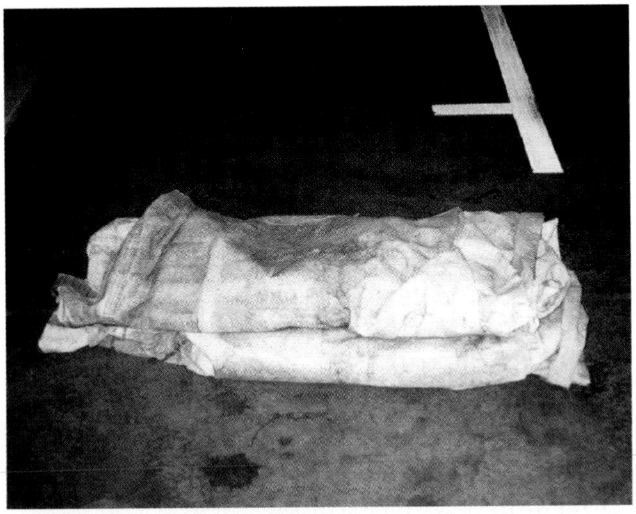

Figure 20-7 Loose fold and roll.

Figure 20-8 Salvage cover drying rack.

Once dry, the cover should be inspected for holes or tears. A simple way to inspect is for firefighters to drape the cover over themsleves and look for light shining through any holes or rips. Holes or tears should be marked with chalk and repaired.

Salvage Cover Folds and Rolls

Salvage covers are either rolled or folded for storage. The available compartment space often dictates which method to use. The roll is accomplished by placing the cover on a large flat surface; if it has a treated side, it is placed up. Sprinkling some baby powder on the cover and spreading it around with a broom will keep the cover from sticking to itself if it is not used for long periods. Two firefighters are needed to roll a cover. Place one firefighter at either end of the cover, **JPR 20-1A**. Both firefighters place one hand on the edge of the cover one-quarter of the way in and the other hand on the outside edge. Each firefighter should be on the same outside edge facing each other. While grasping the cover, with a quick flipping motion, bring the outside edge to the middle, **JPR 20-1B**.

The firefighter's inside hand should be at the fold. Repeat this fold again and again until the fold is at the middle of the cover, **JPR 20-1C.** Fold the opposite end of the cover in the same manner. The cover should now be about 3 feet wide and ready to be rolled, **JPR 20-1D.** The treated side is rolled inside and the cover is ready for deployment.

Many salvage covers are folded for deployment by one firefighter, while others are folded for deployment by two or more firefighters. Folding a

JOB PERFORMANCE REQUIREMENT 20-1

Salvage Cover Roll

A Place one firefighter at either end of the salvage cover. Both firefighters place one hand on the edge of the cover one-quarter of the way in and the other hand on the outer edge.

B While holding on to the cover, with a quick flipping motion, bring the outside edge to the middle.

C Repeat step B with the outside edge, the first fold, and again bring the outside edge to the middle.

D The folded salvage cover should be approximately 3 feet wide and is now ready to be tightly rolled.

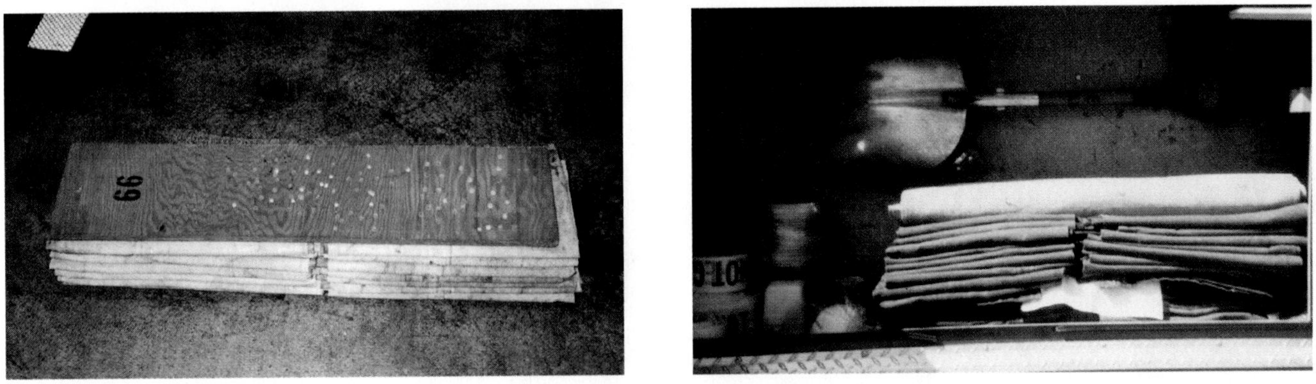

(A) (B)

Figure 20-9 Salvage cover fold. (A) Salvage cover fold template. (B) Paper towel between salvage covers in compartment.

cover is similar to folding a bed sheet. It should be neat and orderly and take up minimal space on the apparatus. Some departments, to keep the folds neat and tidy, use a plywood template to adjust the size, **Figures 20-9A** and **B. JPR 20-2** shows the proper skills to prepare a salvage cover for a one-person deployment. **JPR 20-3** shows the proper skills to prepare a salvage cover for a two-person deployment. When folding a salvage cover, it is vital that the deployment method be considered. These covers often must be deployed quickly.

Other tools must be thoroughly cleaned, dried, and checked for damage. The water vacuum must be rinsed several times to remove all contaminants from the tank. Chain saws will require more than the normal amount of maintenance. They must be taken apart and cleaned thoroughly. They are run in atmospheres they are not designed to run in and used to cut things they are not really designed to cut. They must be inspected and cleaned from top to bottom. When the saw is put back into service, it should start within a few pulls and should run perfectly. Saws are not used all that often, but when they are needed they are *really* needed and are crucial to fire attack, rescue, ventilation, salvage, and overhaul.

Tool maintenance is an important aspect of firefighting. Tools need to be in good working order every time they are used.

SALVAGE OPERATIONS

Safety Considerations

The basic goal behind salvage operations is to eliminate the possibility of damage from a harmful substance to as many belongings as possible in the shortest amount of time. This may entail moving the item out of harm's way, covering the item, or removing the harmful substance from the area.

As with all emergency scene operations, the salvage group must be aware of its surroundings and work as safely as possible. If salvage operations are going on while active suppression operations are taking place—as is often the case—the crew must be appropriately dressed for the conditions; this means dressed the same as the attack crew—full protective clothing with self-contained breathing apparatus. (The individual components of full protective clothing are discussed in Chapter 6.)

The same is true for forcible entry tools. The salvage group is often sent to areas where the fire attack is not taking place, such as the apartment below the involved apartment. In this case it is necessary for group members to carry forcible entry tools with them so they make entry into the apartment. This also applies to interior doors within a business or home that are locked for security reasons.

> **Caution** The crew members must be aware of their surroundings at all times to ensure they do not find themselves in the wrong place at the wrong time. Although the content of a structure is important to its owner, it is not worth injury or a firefighter's life!

One of the most common hazards of salvage work is ceiling collapse. The salvage group may be covering material to protect it from water that is accumulating in the ceiling area due to several possible factors. The roof could have leaks or be under repair, sprinkler heads in the attic could have fused, or the attack crew might not have done a

JOB PERFORMANCE REQUIREMENT 20-2

Preparing a Folded Salvage Cover for a One-Firefighter Spread

A With the cover flat on the ground, fold the outside edge to the center point of the cover. Smooth out wrinkles and align edges.

B Fold the same edge to the center point again. One fold is now beneath the other fold. Smooth out wrinkles and align edges.

C Fold the opposite side in a like manner. Smooth out wrinkles and align edges.

D Fold one end to a point just short of the center point. Smooth out wrinkles and align edges.

E Fold the same end again. One fold is now beneath the other fold. Smooth out wrinkles and align edges.

F Repeat steps to fold the opposite end. Smooth out wrinkles and align edges. The cover is ready for a one-person deployment.

JOB PERFORMANCE REQUIREMENT 20-3

Preparing a Folded Salvage Cover for a Two-Firefighter Spread

A Fold the cover in half and place it on the ground. Smooth out wrinkles and align edges.

B In similar fashion, fold the cover in half (lengthwise) again. Smooth out wrinkles and align edges.

C Fold the cover in half (widthwise). Smooth out wrinkles and align edges.

D Fold the cover end over again (widthwise). Smooth out wrinkles and align edges.

E Fold the cover in half two more times (widthwise). The cover is ready for a two-person deployment.

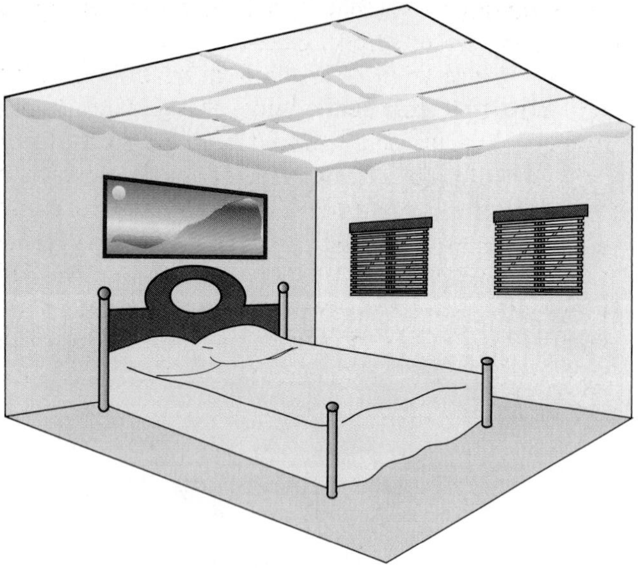

Figure 20-10 Plasterboard (Sheetrock) with water seeping through the seams.

good job of monitoring water flow. For these reasons, the plasterboard (**Sheetrock**) will become weak and sometimes fail. In most cases there will be warning signs. The most common is plasterboard seams begin to show as water seeps into them, **Figure 20-10**. With this sign comes the knowledge that water is building up in the insulation. At a weight of 8⅓ pounds per gallon, the water will make the ceiling fail. Another sign will be plasterboard that sags as if it were not nailed properly.

If these signs appear, the water must be drained before the ceiling collapses. The easiest mitigation is to stop the flow and remove the water. This can be accomplished by quickly salvaging the area below the ceiling and creating a drain hole in the plasterboard to relieve the water buildup. If it is not possible to salvage the area, a quick decision must be made: Is water damage more serious than ceiling collapse? In most cases, water damage will be the lesser of the two evils.

A safe scene involves knowing what other operations are ongoing and where they are taking place. Ventilation crews working above the salvage operation can create a dangerous situation if the cut roof or ceiling drops. Firefighters should be listening for noise and radio traffic and should be aware of crews working above and below them. The incident commander or operations section chief should be aware of all crews and their location in the building and should be contacted if the members of the salvage crew have any questions.

Stopping Water Flowing from Sprinkler Heads

Stopping the flow of water from sprinkler heads is often the main step in a salvage operation. Several types of sprinkler stops are available. The most common is the sprinkler wedge. This is usually a wooden wedge, like a doorstop, that can be jammed into place in the opening of a flowing sprinkler to stop the flow of water. The firefighter must be careful not to break the head off the sprinkler while hammering the wedge in place. This job is not easily accomplished as it is usually done from a ladder while water is flowing under a considerable amount of pressure.

The firefighter assigned this task must wear eye protection and full protective clothing. The firefighter will need to lock into the ladder in order to get some force behind the wedge and eliminate the possibility of falling. Then it is merely a matter of patience. As long as the main drain has been opened and the system is shut down, the firefighter is only working against residual pressure within the system. Getting this assignment is something most firefighters rarely forget!

Other types of sprinkler stops work on more of a mechanical theory: The stop is placed in the opening and either screwed or locked into place by the operator.

Methods of Protecting Material Goods

The "removing items from harm's way" approach can be determined very quickly by asking one simple question: Can it be moved quicker than it can be covered? If the answer is yes, firefighters should go for it! This will usually apply to quickly accessed items like patio furniture and garage items. These are easily moved from harm's way because they usually do not have to be moved very far to get them out of the smoke and heat. Items that are in the way of the fire attack crews should be moved to assist the operation and also accomplish salvage.

Arranging of Furnishings and Salvage Cover Deployment

Covering the item is usually the best choice when the item is either too large to be moved or the movement would be too time consuming. When the decision is made to cover items with salvage covers or Visqueen, the room should be surveyed for the largest item or central piece. All smaller items should be moved to

Figure 20-11 Furniture arranged for salvaging.

this area, **Figure 20-11**. Consider a bedroom for example. Items should be removed from the closets and placed on the bed, dressers should be pushed up close to the bed, and pictures should be removed from the walls and placed on the bed. If dressers are too large to be moved then the drawers can be placed on the bed.

Once everything is ready to be covered, the salvage cover or Visqueen should be deployed. This is done by one or two firefighters. This deployment method is called a *counter payoff.* With one firefighter holding the cover or Visqueen firmly by the bottom fold, the other pulls the top fold or leading edge, if it is rolled, away from the firefighter holding the cover, **Figure 20-12**. It is easier and will save time if the cover is rolled or unfolded directly onto the items needing protection. When the cover is extended, it is then placed on the furnishing and the sides are unfolded. At this point, the reason for having the treated side out is obvious—so that it is facing the ceiling to deflect any water dripping down.

Figure 20-12 Two firefighters deploying a salvage cover.

If there are breakables under the cover, extra care must be taken. Depending on the room being salvaged, a glass piece should be placed on top of the cover after the room is salvaged to designate that valuables are under the cover. This signals other crews that the cover should be removed gently. The salvage cover edges should be "rolled" under so that they won't collect any water or debris. If this is a one-firefighter operation, the cover is just laid on the items and the firefighter plays it out individually.

Ethics Salvage operations can provide temptations to firefighters who would like a souvenir of an incident. Sometimes items are slightly damaged and crews will determine the value to be zero and take the item back to the firehouse. This is stealing and should never be allowed to take place.

Covering items does not just apply to furnishings; it can also apply to carpeting. When attacking an attic fire, the attack crew can bring with them a salvage cover and quickly spread it out prior to pulling ceilings or making attic access cuts. This is also where the floor runner comes into play. This again is just one of those little things that create excellent customer service.

Shoulder Toss

Another way to deploy a salvage cover is to use the *shoulder toss.* This method is done by a single firefighter and is reserved for covering large unbreakable items. Rack storage in a warehouse is one example where this could be used. The firefighter takes a cover that has been folded (not rolled) to the location of deployment. The firefighter then places the cover on one arm so it can be tossed with the firefighter's throwing arm, **JPR 20-4A**. The firefighter grabs the top couple of folds that are nearest and flips them back so that the arm is in a throwing position behind the shoulder. By using a straight arm movement, the cover is tossed over the item being salvaged, **JPR 20-4B**. The cover is then spread out over the item that needs to be covered. This takes practice and some upper body strength, but is quite effective in areas where the item is too high to just spread the cover over it.

JOB PERFORMANCE REQUIREMENT 20-4

Shoulder Toss

A Place a folded salvage cover over one arm.

B Using a straight arm movement, toss the cover over the items being salvaged.

Balloon Toss

If two firefighters are available, the balloon toss is another method of deploying a salvage cover, **JPR 20-5.** This method is used if the objective to be covered is slightly taller than the firefighters. The two firefighters place the unfolded cover on the ground alongside the full length of the base of the objective, **JPR 20-5A.** The firefighters grab the top inside edge of the cover, pick it up to about waist height, and pull with tension on the cover. Both firefighters simultaneously move the cover's edge in an upward motion so that it catches air, and while still holding the end of the cover they move to the other side of the objective while the cover is still in the air, **JPR 20-5B.** The objective is to use the air to help raise the cover over the objective. Both firefighters guide the side of the cover over the objective to the ground on the opposite side. The cover is then adjusted to help water flow off the objective, **JPR 20-5C.**

Working with Visqueen

Working with Visqueen or a rolled salvage cover is essentially the same as working with a folded cover. The limiting factor is the size of the rolled cover. In both cases, the roll is held by one firefighter and the other pays out the remaining portion. Once the material is in place, the sides are folded out and the edges are rolled under. A good example of where Visqueen is much better than a salvage cover is in a supermarket where the long rows of food would need to be covered. By starting at one end of the row, the proper length of Visqueen can be determined and cut. The Visqueen is then placed over the row and everything is covered.

Another use is covering the ventilation holes in a roof. The Visqueen can be nailed to the roof using wood strips. This is normally a task for the board-up crew, and it is discussed later.

Limiting Exposure to Smoke

As discussed in Chapter 18, firefighters should always consider the by-products of combustion—heat, smoke, and toxic gases—as a danger. These by-products should be contained from spreading into undamaged portions of a structure and should be removed from the entire structure using forced air ventilation, **Figure 20-13.** Containment of the by-products can be an action as simple as closing doors after a room has been searched or closing any nearby doors during the fire attack advance. This action can save the contents of rooms from additional smoke, fire, and water damage. Adding positive pressure ventilation not only allows firefighters better visibility and safety from the products of combustion but also affords an additional measure of protection against damaging the contents of the unburned portion of the structure.

Water Removal

Water from fire attack operations can leak through the floor and damage rooms or entire floors below the fire. It is critical to capture, divert, or remove the water to successfully salvage these areas. If there is a

JOB PERFORMANCE REQUIREMENT 20-5

Balloon Toss

A Two firefighters place the cover on the ground next to the item needing protection. The salvage cover is unfolded out of the accordion fold so that the cover is across the base of the item at its full length.

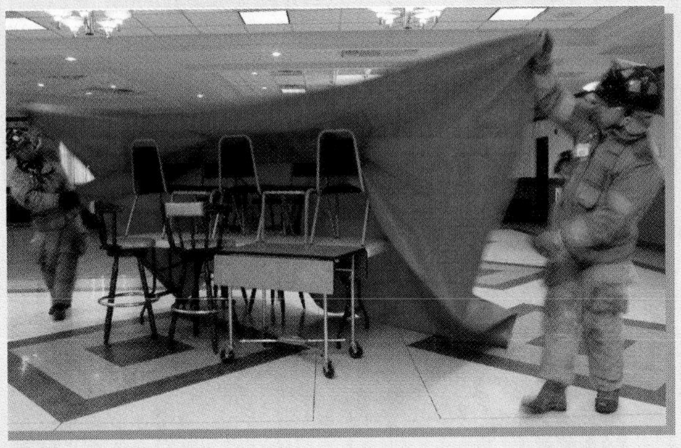

B The firefighters grab the top inside edge of the cover and pick it up to about waist height. Both firefighters simultaneously move the cover's edge in an upward motion so that it catches air, and they then move to the other side of the item while the cover is still in the air.

C The firefighters guide the cover to the floor on the opposite side. Once the items are covered, the firefighters dress the cover by folding the edges under to keep water from pooling and to lessen the chance for the edges of the cover to be disturbed by accident.

floor drain, the operation becomes a little easier. If no drain exists, firefighters should examine their options. If a toilet can be removed, a drain hole can be created. Firefighters can squeegee water down this created drain as they would a floor drain. Firefighters can be creative and use salvage covers to create dikes to channel water into the drain or out of the building. Lining a stairwell with a salvage cover can create a chute to channel the water out of the area.

Catch-All

An example of the need to use a catch-all would be a multiple-story office building. A fire may occur in an office and be extinguished by the sprinkler system. This could cause a major salvage problem on a lower floor.

Catch-alls are used to contain the water dripping through the ceiling until a system is set up to remove the water from the building (chutes, water

Figure 20-13 Positive pressure ventilation in place.

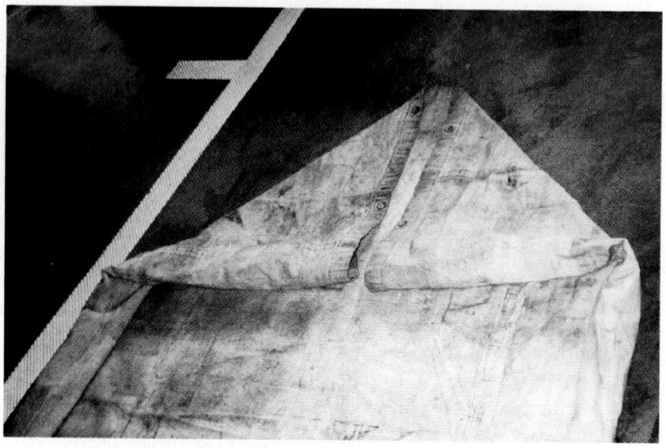
Figure 20-14 Catch-all end folds.

vacuums, squeegees, etc.). Catch-alls have sometimes been constructed using ladders and salvage covers to form a containment area. However, firefighters should consider the weight contained inside this catch-all. At 8.33 pounds per gallon, this weight can soon cause a structural integrity problem in an already damaged building, as the weight is all focused on a small area of the floor. It is much safer to consider a catch-all as a mitigation effort and construct it using a single salvage cover. A system to remove the water from the area should be constructed or devised as quickly as possible.

Catch-alls are easy to make. The firefighter places the cover with the treated side down and rolls the sides in to the desired size of the catch-all. The next step is to fold the ends as shown in **Figure 20-14,** roll the ends in, and tuck them tightly so they do not unroll. The cover is then flipped over, **Figure 20-15.** As the catch-all fills up it will not unroll the sides or ends. As the water depth will only be 2 to 3 inches, a water vacuum may be necessary to adjust the water level until the water can be directed away with a chute or other method.

Water Chute

If the water volume is too much for the catch-all, then a water chute is used so the water can be diverted out of the building through a window or doorway or down a stairwell. The water chutes are a combination of ladders or pike poles and salvage covers. The salvage covers are draped on the ladders so they work like a trough and the water flows down them. Office furnishings or other equipment is used to support the ladders. The salvage cover can also be stretched out and a pike pole placed on either side. The cover is wrapped around the poles and the ends rolled in until the desired width of the chute is attained.

If the length is longer than a single cover, a leakproof seal must join the two, **Figure 20-16**. Once the direction of the flow is determined, the upper cover is put in place with the last 6 inches folded back on itself. The next cover is laid on the next portion of the ladder with its top 6 inches laid on top of the bottom 6 inches of the cover above.

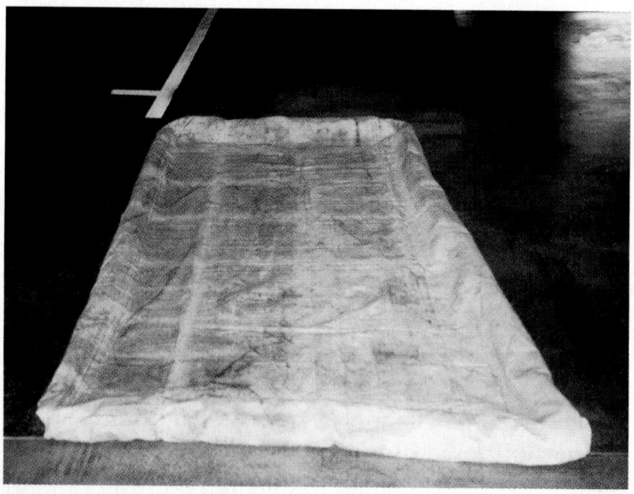
Figure 20-15 Catch-all.

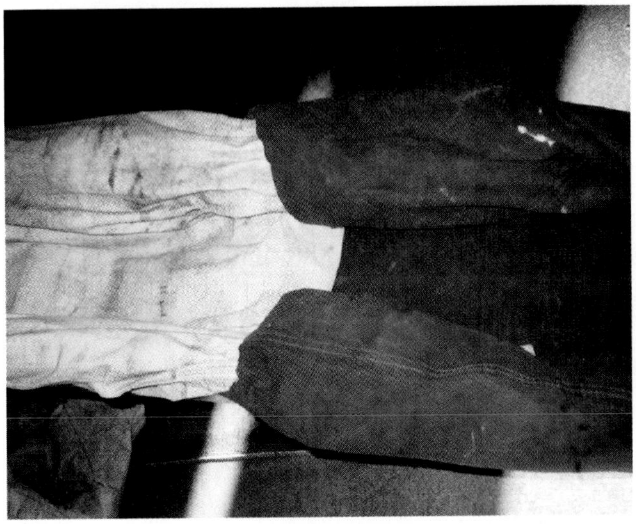

Figure 20-16 Leakproof fold of water chute.

The covers are then rolled together toward the bottom cover and flattened out when there is no more to roll.

SALVAGE OPERATIONS IN SPRINKLERED BUILDINGS

Post Indicator Valve and Outside Screw and Yoke Valve

Many buildings are now equipped with sprinkler systems. These systems are beneficial at stopping or containing a fire but can cause damage to the unburned portions of the buildings if allowed to flow freely. It is important that firefighters know how to shut these systems down to prevent unnecessary damage. The incident commander will determine when the sprinkler is no longer needed for fire suppression.

Essentially two types of valves are used to shut down a sprinkler system: the post indicator valve (PIV), **Figure 20-17**, and the outside screw and yoke (OS&Y) valve, **Figure 20-18**. The PIV is generally located near the sprinkler connection for the building, but it may be just about anywhere around the building. The PIV is about 3 feet tall with a window near the top that states either "Open" or "Shut." This indicates the status of the valve. The PIV should be locked in the open position. Most PIVs are monitored so that if someone attempts to close them, an alarm sounds. If a PIV is found in the closed position it should be noted to the incident commander and turned to the open position if active firefighting is still taking place. To close the valve, the firefighter should break the lock, which is usually a knockoff-type lock, and remove the handle, which is usually locked on the valve.

Figure 20-17 Post indicator valve. Note that the PIV is shut and unlocked. If found in this position at an incident, it should be opened and the condition reported to the incident commander and the fire investigator.

> **Streetsmart Tip** A firefighter should always take bolt cutters when sent to close a sprinkler valve in case it is locked with a regular lock rather than a knockoff lock.

The handle is used to close the valve. It helps to remember, "Righty, tighty—lefty, loosey," or clockwise to tighten and counterclockwise to loosen. Once the PIV is closed, the firefighter should also locate the system drain and open it. This will assist in bleeding the system and reducing the amount of water that will drain through the sprinkler heads.

Figure 20-18 Outside screw and yoke valve.

The OS&Y valve will be located either on an exterior wall of the building or just inside the building. It looks like a steering wheel with a large screw sticking out of the middle of it. If the screw is out then the valve is open; if it is in then it is closed. The wheel may be locked so it cannot be turned to close the valve, or it may be a supervised system that sends an alarm if attempts are made to close the wheel. Bolt cutters are used to cut the lock to shut this valve. Again, if the valve is found in the closed position, the firefighter should notify the incident commander and turn it on if active firefighting is still taking place. If there is any possibility to have the valve reopened due to fire spread then it should not be turned off. The system should only be shut down when *all* active firefighting is completed.

Sprinkler Stops

If no valves can be located, sprinkler stops will have to be used. In some cases, sprinkler stops will need to be put in place to minimize the amount of water that will flow out of the head. An A-frame ladder will reach most ceilings in sprinklered buildings. When placing a sprinkler stop from an A-frame ladder, the ladder is steadied by another firefighter as the firefighter climbs it to provide a good platform. Sometimes the sprinkler head is in a position where it cannot be reached using normal laddering techniques. In this case an auditorium raise can be used. This is a risky operation and those risks must be weighed against the amount of property damage that can be saved. The auditorium raise is discussed in Chapter 14.

Salvage Operations Lessons Learned

Salvage operations can be as important as any actions a fire service agency can perform on an emergency scene. In most cases, a majority of the damage has already been done to the structure and the main job is to make sure the emergency does not get any worse after arrival. This is where salvage operations can be very important to the success of the incident. Firefighters will learn throughout their career that the ability to make something good out of a very bad situation becomes paramount in order to find some value in what they do. With fires, that outcome is in doubt until the end of the incident. Firefighters apply all of their knowledge, skills, and abilities to aggressively attack the problem, but if all that is left is a smoldering parking lot, what have they accomplished? Salvage can immediately give firefighters that sense of purpose. It is no longer acceptable to believe that merely saving the house next door means a successful operation. Firefighters need to preserve as much of the salvageable property as possible. The operation must ensure life safety, exposure protection, and property conservation. With this thought process directing firefighters' efforts, the "customer" will agree that they accomplished what they set out to do.

OVERHAUL TOOLS AND EQUIPMENT

The process of overhaul is as important as the initial extinguishment of the fire. The process of making sure the fire is completely extinguished can be tedious but very worthwhile. The ability to check walls, floors, ceilings, and attics—especially with blown-in insulation and dead air spaces—is a skill that needs to be learned and practiced whenever firefighters attend a working incident. Due to many factors, including successful fire prevention programs, the fire service is responding to fewer working fire incidents. The need to learn good and thorough firefighting practices and make them habits is paramount to successful firefighting.

Tearing into buildings, cutting through floors, pulling ceilings, and searching for hidden fires are tasks that most firefighters enjoy. The tools used for overhaul are built for these tasks, so these tools are among firefighters' favorites. In most cases, firefighting requires precision and attention to details. In overhaul, the mission is simple—leave no area unsearched if there is potential exposure. A firefighter's knowledge of building construction and fire behavior is an asset in property conservation efforts.

Common Tools

Common overhaul tools are pike poles, pitchforks, rubbish hooks, shovels, axes, chain saws, carry-alls, and wheelbarrows, **Figure 20-19**. All of these tools meet the goals of getting into areas to determine if they were exposed to fire and removing necessary items so they will not cause any more fire. Most of these tools are common and their purpose does not need to be explained. The few exceptions are the rubbish hook and the carry-all. The rubbish hook is a heavy tool with a D-handle that is between 4 and 6 feet in length. It has two tongs spaced about 6 inches apart that run perpendicular to the handle: It looks like a rake with only two teeth. The tool is heavy enough to easily break through ceilings or strip shingles from a roof. It is a mainstay for ventilation crews on roof operations. Because of its weight and the D-handle, it is easily used to breach and pull plasterboard.

Carry-All

The other tool needing a little explanation is the carry-all. Its name makes it pretty clear as to its purpose. It is an approximately 6-foot-square piece of heavy canvas with a rope strung through the grommets for handles that is used to carry debris out of the building once it is loaded, **Figure 20-20**. It is used in areas where wheelbarrows cannot be, due to access or other restrictions.

OVERHAUL OPERATIONS

Once the determination has been made that fire investigation concerns have been met, overhaul can commence. Fire investigation concerns are discussed later in this chapter. Overhaul operations consist of breaching walls, floors, ceilings, and dead air spaces in order to confirm that fire is not present in these areas. All areas directly exposed to the fire should be stripped of coverings such as plasterboard or paneling.

Firefighters can look for obvious signs of hidden fire and utilize their senses by feeling the wall for heat, smelling around outlets and other openings for a scent of anything burning, and listening for sounds of items burning. The bottom line is if there is a chance something could be burning behind a wall covering, the wall covering must be removed to make sure it is all clear. It is much better to be safe than sorry in this instance.

Firefighters must visually observe the wooden structural members to make sure there is no smoldering fire behind the coverings. The building structural members should also be visualized in

Figure 20-19 Various overhaul tools.

Figure 20-20 Firefighters carrying debris in a carry-all.

order to confirm that the building's structural integrity has not been compromised by the overhaul activities. Fire can get into these areas through electrical outlets; via conducted heat from metallic pipes or wiring; through breaches in the walls or floors where conduit runs; via heating, ventilation, and air conditioning (HVAC) ductwork; or because of a lack of fire stops in improperly constructed buildings.

Once these areas are breached, all of the insulation, if any is in place, needs to be removed so the entire space can be visualized. This is extremely important in attic spaces. Several different types of blown-in insulation can smolder for many hours prior to igniting, **Figure 20-21**. Some of the more common are cellulose or redwood insulation, which is often found in older homes. If this type of blown-in insulation is found, then the task of removing all of the exposed material becomes paramount. The smallest of embers can travel to remote areas within the attic and smolder for many hours prior to growing to a free burning state. The entire attic must be checked and revisits are recommended, as discussed later in the chapter. Newer homes have rolled-in fiberglass bats that are easily removed and for the most part noncombustible, although the paper backing the fiberglass is attached to will burn.

Overhauling Roofs

Overhauling roofs can be a very long and tedious process. Sometimes there can be multiple roofs under the most visible one and each one needs to be overhauled for possible fire extension. Sometimes flat roofs are left in place while pitched roofs are built over them, thus creating two attic spaces and a real overhaul nightmare. As with all overhaul operations, the material exposed to fire that could smolder for some time needs to be removed. This is why roofs are particularly difficult. During a hot summer day there is nothing worse than working on a roof to overhaul it because of the heat, access problems, the sheer magnitude of the operation, and the inherent safety concerns of working above ground.

Figure 20-21 Blown-in insulation in an attic.

Safety Overhauling roofs is something that needs to be done on all working attic fires and this can turn into a very time-consuming task. This situation provides a good opportunity to bring crews in from outside the area to work a few hours on the roof while the initial attack crews rest or perform other needed tasks prior to resuming their overhaul duties. In overhaul situations, it is important to set up a "rehab" station where crew members can rehydrate and cool off.

Electronic Heat Sensors

A relatively new way to check buildings for hot spots is to use electronic heat sensors or thermal imaging cameras. These sensors determine where heat is higher as compared to the surrounding areas. The sensors' alerting mechanisms differ depending on the sophistication of the equipment. Some simply sound different when they are pointing at an area that is hotter than the surrounding areas, whereas others give the operator the ability to see and pinpoint the exact area that is showing a heat signature. Some of the more sophisticated devices are also used during searches to find people through the smoke, under water, or lost in the wilderness, and for other applications, **Figure 20-22**.

Revisits of the Involved Structure

Even after all areas have been checked, it is important to revisit the scene of the incident. Crews should never leave the scene if there is any active evidence that any part of the building is not completely extinguished. As with all good operations and focusing on customer service, it is always wise to return to make additional checks. The crews are checking for any smoldering areas and to see if the occupants have any questions or need assistance. A good rule is to revisit 2 hours

Figure 20-22 Helmet with thermal imaging camera attached.

after the last company or personnel leaves the scene and then again within 10 hours. This way the building is visited twice within 12 hours of incident termination.

In most cases crews from the next shift can go by the incident for the revisit and also learn from it by looking at the areas of involvement and problems encountered. Volunteer companies will have to work around their normal work schedules. Another option is to have the business owner hire a security company to maintain a fire watch, or local law enforcement may be able to assist if the fire service agency is unable.

Debris Removal

The need to remove the debris while searching for hidden fires is important, but as discussed later in this chapter this needs to be done in cooperation with the fire investigator. All of the loose material damaged in the fire must be removed and further extinguished outside the structure. As walls or ceilings are pulled, the material needs to be removed from the building. This is where carry-alls or wheelbarrows come in handy, **Figure 20-23**. The material needs to be placed in an area outside the structure where cleanup crews can easily access it after the firefighters leave.

The material is placed in a pile. It is *lightly* hosed down to keep from creating a bigger problem down the street with a substantial amount of runoff. Firefighters must be certain the material is extinguished. Debris placed on driveways is the easiest to clean up after the event. Although placing debris in front yards outside a main window may seem convenient for firefighters, customer service should always be considered.

Figure 20-23 Firefighter using wheelbarrow to remove debris.

Streetsmart Tip Sometimes the debris needs to be placed outside a window or other opening, instead of taking it to the driveway or other hard surface. Visqueen should be placed outside the opening to protect any plants or lawn under the plastic. Remember: customer service! This effort will be appreciated by the building owner.

After the overhaul crews have completed the removal of material and are sure it is completely extinguished, they can cover the debris with Visqueen to reduce the eyesore.

Overhaul Operations Lessons Learned

As with salvage operations, the customer service aspects of overhaul are many. The detailed search for hidden fires, the ability to limit the amount of uncovered debris left behind, and the overall ability to leave the building in a state where it is safe for the owners to get in and start looking for lost articles are very important processes. Overhaul work is often the only thing the owners get to see the result of. Making sure overhaul works in concert with the fire investigator is another way to ensure the operation is successful.

FIRE CAUSE DETERMINATION CONCERNS

As the incident progresses through all the different aspects of initial dispatch, response, arrival, rescue, containment, extinguishment, ventilation, salvage, and overhaul, firefighters need to be aware of possible clues as to how the incident started. During dispatch, they should note all of the information available at that time. Information such as weather, time of day, address, type of dispatch (i.e., garage fire versus bedroom fire), and the number of reports all give firefighters a clue as to what the fire was doing prior to arrival, **Figure 20-24**.

During response, additional information may be gathered by answering these types of questions: Is the structure evacuated or still occupied? Are callers reporting explosions or different colored smoke? Were citizens seen leaving the incident at a high rate of speed? Observant firefighters on scene can provide fire investigators with reliable information as to where the fire was, what spectators said about the fire, and how the fire reacted during fire attack. Once inside the structure, firefighters should observe how the fire reacted, where the fire was, how the contents of the building were arranged, if there were signs of a break in, how the electrical breakers looked, if anybody was found in the building and how were they found, how appliances were found, how interior doors were found, and so on. All of these things need to be committed to memory so they can be relayed to the fire investigator, **Figure 20-25**, and also listed in the company officer's report. The smallest of items that seems out of place needs to be reported. Although it may not seem like much, it may be the missing piece of the puzzle for the investigator.

During overhaul, firefighters must make sure the investigator approves of the removal of any part of the structure or any piece of furnishing prior to removing it. Firefighters cannot just throw things out of the building for expediency. Each step has to be taken with knowledge and care so as not to make things worse. As investigators work through a structure, they will be able to release the different rooms for overhaul. Markings on plasterboard from the smoke may mean nothing to a firefighter, but may be that missing puzzle piece for the investigator. Experienced investigators are only as good as the clues they observe. If they show up and all they

Town of Colonie **FIRE/EMS** Incident DATE ________ RUN #________

How Received : [ ] TELEPHONE [ ] RADIO [ ] IN PERSON [ ] 2ND/3RD PARTY

Person Reporting:		Incident/Priority Type:					
Address:		Dept's					
Location of Incident:		Units					
Cross Street/Landmark:							
Callback #:		Rec'd					
Chief Complaint:		Disp'd					
Age:	Sex: Male / Female	Ack'd					
Conscious: Y / N / Unknown		Enrt					
Breathing: Y / N / Unknown		Arrv'd					
Rec'd By:	Disp By:	I/S					
EMS Zone:	Fire Zone:	Amb:		Hosp:	Enrt:	Arrv:	
Fire Signal/Time: /		Amb:		Hosp:	Enrt:	Arrv:	

Priority Change/Time: ______ / ______ Remarks: ____________________

Figure 20-24 Dispatch sheet showing various information. *(Courtesy of Town of Colonie Fire & EMS)*

Mutual Aid Dept/Unit	Disp'd	Ack'd	Enrt	Scene	I/S

Amb:	Hosp:	Enrt:	Arrv:	Remarks:
Amb:	Hosp:	Enrt:	Arrv:	

Sig	Cover Co.	Comm. Post	Fire Inv	Op. Off	Comm. Dir.	Mob. Air	Batt. Co	Car 1
30								
40								

*Signal 20 gas leaks or hazmat incident: page fire investigation (399)

Figure 20-24 Continued

Figure 20-25 Fire investigator working a scene.

have is a bare wood floor and wooden studs for walls, they are at a disadvantage.

Preservation of Evidence

In the process of their work, firefighters may come upon something out of the ordinary and will need to make every effort to preserve it until the fire investigator can observe it, even if this means stopping their entire operation. Firefighters need to leave items where they found them and cover them up with Visqueen or mark off the area with fire line tape to keep everybody out of the affected area. Firefighters do not want to become involved in a big legal matter because they did not do their job correctly. If a room is overhauled and cleaned of all contents and later is found that an incendiary device was located in that room, it could result in a legal matter for the firefighter and the department.

Basics of Point of Origin Determination

The determination of the point or area of origin can be a very scientific pursuit in which exotic testing is done and the exact point is determined—or it can be as simple as looking at a room and knowing where the fire started based on the indicators present. The basic clues a firefighter should watch for are where the fire was first noticed or reported; where the heaviest damage was; or any markings on the walls, ceilings, or floor indicating where the fire started. The basic premise behind fire behavior is that it travels the path of least resistance, **Figure 20-26**. As fires grow, smoke and heat are forced up until they reach an obstruction, usually the ceiling. The smoke and heat will branch out, taking the path of least resistance. In a structure fire this path will be along the ceiling. As

Figure 20-26 "V" pattern in a structure fire.

the smoke and heat go in opposite directions the wall will be marked with a "V." *V pattern* is a common term in the fire investigation field. If the fire moves down a hallway from the front room, the V will start at the bottom nearest the area of origin and proceed up the wall.

Another simple manner to determine the fire's starting point and where it spread is the **depth of char**. Wherever the fire burned the longest is where the fire's depth of char will be the greatest, **Figure 20-27**. This can be illustrated by looking at a log in the fireplace. Usually the center of the log has the most "charring" and it diminishes toward each end. Wherever the fire started would have had the longest period of time to burn the wood members in that area, thus the charring would be greater in that area. The only misleading part of this indicator is when an area of the fire took longer to extinguish than others. This attack crew information can assist the investigator in determining the point of origin.

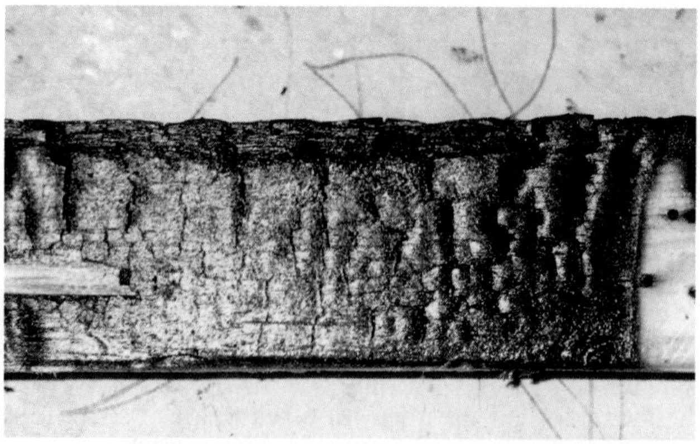

Figure 20-27 Depth of char.

Depth of char is just one of many indicators and can be influenced by many other factors. It is a simple way to determine the point of origin for a simple incident. The key to assisting the fire investigators is not to rush to complete the job and completely mess up theirs. It is imperative that firefighters pay attention to what is going on around them during all aspects of an incident and report items that fall outside the norm.

Fire Cause Determination Lessons Learned

Assisting the fire investigator is a skill that should be second nature to all firefighters. The gathering of information from the time of dispatch throughout the incident should become automatic. Every little detail should be noted and passed along to the investigator. Fire investigators are just another part of the team who are attempting to make this tragedy as tolerable as possible for the owner. Determining the cause and origin of the incident is every firefighter's duty and, like all of the other parts of the job, a very important factor. Firefighters always want to assist an investigation or criminal prosecution.

SECURING THE BUILDING

Firefighters have an obligation to make sure the building is secured after they have completed all of their operations. This would include the revisits discussed earlier in the chapter, boarding up all openings—including roof openings—so the building is secure, and securing the utilities. Some jurisdictions rely on board-up crews to handle this function, **Figure 20-28**, and the insurance carrier will pay those crews. If a building is not insured then the local public works or a local lumberyard might donate some plywood to help the owner defray costs. Again, customer service will leave a lasting positive experience for the homeowner.

These are simple items to consider prior to leaving an incident scene. But what about the displaced owners? Where are they going to stay or do business? What are they going to wear for clothes if everything was destroyed? Do they have the means to contact their insurance company? Do they have insurance? Does the incident commander need to contact the Red Cross or other charitable organization in order to get them a place

Figure 20-28 Building that has been boarded up after a fire.

to stay or some clothing? All of these things should be considered and the fire service should play an active role in solving these issues for the owners.

In many of these cases the owners are insured, have family or friends they can stay with and borrow clothes from, and only need on-scene crews to make sure the building is safe for them to enter so they can get on with their lives and start rebuilding. It is that other small percentage that should be assisted in any way possible to help them get back on their feet. Some fire service agencies have hotel rooms reserved if the need arises to house families that have no place to go; contacts with the local phone company for rerouting the business or homeowner phone lines; and contacts with local churches, the Red Cross, or the Salvation Army for food or clothing. This list could grow as large as the fire service wishes it to. Fire service agencies are taking a much more active role in going above and beyond when it comes to customer service and this will pay off in the long run.

Lessons Learned

This chapter dealt with areas not commonly reinforced through training. Most fire training consists of the vital roles of fire attack, building construction, search and rescue, and so on. Salvage and overhaul duties, while not glamorous, are vital to the functions of fire investigation and property conservation. These duties can often go unnoticed as firefighters are focused on the more exciting aspects of an incident. Most firefighters would rather talk about the great "stop" they made instead of the minimal damage the lower floor of the structure received. However, salvage and overhaul duties do have a direct impact on the success of an incident. They are also a great customer service for the homeowner, even if they are never recognized.

KEY TERMS

Chain of Command Common fire service term that means to always work through one's direct supervisor. The fire service is viewed as a paramilitary organization and because of this all requests for information outside the assigned workplace should go through the supervisor.

Depth of Char A term commonly used by fire investigators to describe the amount of time wooden material had burned. The deeper the char, the longer the material was burning or exposed to direct flame.

Sheetrock A trademark and another name for plasterboard.

Visqueen A trade name for black plastic. It can be used very effectively in salvage and overhaul operations. Many examples are discussed in this chapter.

REVIEW QUESTIONS

1. Why is the exterior of a building important to salvage operations?
2. What are the three types of material that salvage covers are made of?
3. What advantages does Visqueen have over customary salvage covers?
4. How can a fire department be reimbursed for equipment damaged in the course of working the incident?
5. How do firefighters check for holes in the salvage covers?
6. What are the basics of furniture arranging for salvage operations?
7. What are some indicators that a Sheetrock ceiling has a potential to fall?
8. What does PIV stand for?
9. How are overhaul and fire investigation linked?
10. What does a revisit of a structure fire accomplish?
11. What is a "V" pattern?
12. Why is knowing the "depth of char" important?

Additional Resources

NFPA 1921: Guide for Fire and Explosion Investigations. National Fire Protection Association, Battery March Park, MA, 1998.

CHAPTER
21

PREVENTION, PUBLIC EDUCATION, AND PRE-INCIDENT PLANNING

Donald C. Tully, Orange County Fire Authority

OUTLINE

- Objectives
- Introduction
- Administration of the Fire Prevention Division
- Fire Company Inspection Program
- Home Inspections
- Fire and Life Safety Education
- Pre-Incident Management Process
- Lessons Learned
- Key Terms
- Review Questions
- Additional Resources

STREET STORY

As a fire service educator for many years, the one thing I have noticed that sets apart people in prevention from people in suppression is the time it takes to see results. In suppression, the effects of the interventions are seen in minutes, hours, or at the most several days. Prevention is different. The results may not be seen for months, or years, for that matter.

I have been fortunate in my career to see fire prevention pay big dividends. Fire deaths and severe fires have been drastically reduced in Montgomery County, Maryland, while the population and rate of building construction continue to grow. This is true not only here but throughout the country. Safety campaigns, smoke detectors, sprinklers, child-resistant lighters, school programs, and the facts that people are eating out more and smoking less are all contributing factors.

A particular case stands out in which a school fire safety program made the difference between life and death. Alena Tune of Rockville, Maryland, paid attention to the advice she got in school on fire safety and used it to save herself and her four-year-old sister when a fire broke out in their home. It was 3 a.m. on November 17, 1995. Alena woke up to the screams of her sister. She put to practice the survival techniques she learned in school. She got down on her hands and knees, got to her sister, and put her on her back. She crawled 30 feet to the front door and safety. This happened thirteen years after firefighters began teaching fire safety to fourth graders.

So, remember that your prevention message may not pay off today or tomorrow, but armed with lifesaving information, people can use it during their lifetime to save lives and property.

—Street Story by Mary K. Marchone, Program Manager, Montgomery County Fire and Rescue Services, Montgomery County, Maryland

OBJECTIVES

After completing this chapter, the reader should be able to:

- Describe the function of the Fire Prevention Division.
- Understand the purpose and value of a quality fire prevention inspection program.
- Prepare, conduct, and follow up a quality fire prevention inspection of a business occupancy.
- Identify typical violations found in business occupancies and initiate the appropriate corrective action.
- Understand the value and goals of a home inspection program.
- Conduct a fire prevention inspection of a residential occupancy.
- Identify various types and levels of fire service public education programs.
- Prepare and present a fire safety educational program.
- Understand the necessity of pre-incident management for emergencies at target hazards.
- Prepare a pre-incident management plan for a target hazard.

INTRODUCTION

What is the cardinal mission of the fire service in America? Unquestionably, it is the preservation of life and property. The scope of this lofty task is seemingly ever expanding.

Firefighter Fact As recently as thirty years ago firefighters were generally expected to respond only to fires—structure, vehicle, wildland. Today's firefighters are called on to be experts in all aspects of emergency activities: medical emergencies, hazardous materials incidents, urban search and rescue, swift water rescue, high-angle and confined space rescue—and the list goes on.

With all the demands placed on firefighters by the increasing expectations of the public to perform these relatively new areas of technical expertise, it is easy to lose sight of the simplest and most effective method of achieving the goal of the preservation of life and property: prevention. Any physician will attest to the benefits of physical conditioning. Keeping in shape now helps to prevent devastating injury and illness in the future. Why do fire departments do preventive maintenance on their vehicles and equipment? Changing the oil regularly prevents costly engine rebuilds in the future. So it is with fire prevention and education. In an enlightened world, the firefighter's job must be to save lives and property from the effects of fire and other emergencies through the effective application of time-honored and innovative fire prevention and education techniques.

Note Fire prevention officers have long been guided by the adage, "The Three Es of Fire Prevention—Education, Engineering, and Enforcement."

This chapter shows how, through the application of the three Es, the level of fire safety in the community can be enhanced. This information is presented to assist the firefighter/inspector in conducting a quality fire prevention inspection of both residential and nonresidential occupancies. Furthermore, the chapter highlights programs available for instructing communities in proper fire prevention activities and appropriate responses to emergency situations. Finally, a discussion is included on the need and methodology for pre-incident management, or preplanning, by firefighters for emergencies at target hazards.

ADMINISTRATION OF THE FIRE PREVENTION DIVISION

The duties and responsibilities of fire prevention officers are often not well understood by suppression firefighters. These men and women are seen around the station or at headquarters, but little is known about their daily routines. Fire prevention officers play a key role in ensuring that the fire department meets its goal of preserving life and property in the community.

The Fire Prevention Division is responsible for all aspects of the fire and life safety of buildings and occupants prior to an emergency incident. This includes the "engineering" aspect of fire prevention such as interaction with architects and builders, new construction plans review and approval, **Figure 21-1**, installation of fire detection and suppression systems, inspection of newly constructed or remodeled occupancies, **Figure 21-2**, administration of hazardous materials disclosure programs, inspection of high-risk occupancies requiring specialized expertise, and public education and information programs. After a fire or hazardous materials incident has occurred, the Fire Prevention Division is often called on to

Figure 21-1 Fire prevention officer reviewing plans.

conduct the origin and cause investigation, **Figure 21-3**, and coordinate the follow-up activities of other agencies, when appropriate.

These duties may be the responsibility of one fire prevention officer, or specialized assignments may be given to a number of officers, depending on local conditions. Whichever approach is taken, it is certain that the effectiveness of the Fire Prevention Division's efforts is greatly minimized without the cooperation and assistance of suppression firefighters.

FIRE COMPANY INSPECTION PROGRAM

An effective fire company inspection program will reap immediate and future benefits for the community and fire departments. The "enforcement" program should be approached with the following objectives in mind: reduction of fire hazards, opportunity to increase public awareness of the dangers of fire, positive public contact, and building familiarization and **pre-incident management**.

Figure 21-2 Fire prevention inspector walking through new construction.

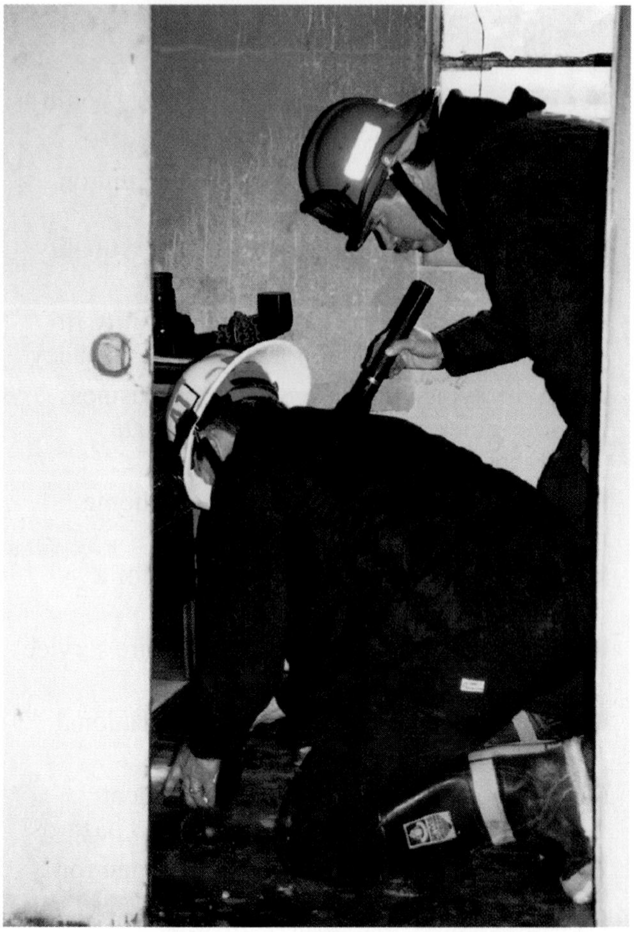

Figure 21-3 Fire prevention officer inspecting a fire scene.

Inspection responsibilities are generally assigned to fire companies by the Fire Prevention Division. For obvious reasons, every effort must be made to ensure that such assignments do not require companies to leave their first-due response areas. The number of inspections assigned to particular fire companies will vary depending on the size and complexity of the occupancies and the level of emergency activity of the fire company.

Equipment

Any job, done right, requires the correct tools and equipment. A minimum complement of tools and equipment for the firefighter/inspector is suggested here:

- Complete standard-issue uniform, clean and pressed
- Occupancy inspection files

- Clipboard with notepad and pencil
- Flashlight
- Inspection forms
- Standardized information bulletins
- Violation notices
- Fire code reference manual

Many company officers find it convenient to keep their forms and equipment organized in a catalog case (large briefcase), **Figure 21-4**.

Preparation for Inspections

With ever-increasing demands for firefighters' time, it is important for company officers to routinely designate specific days and times for conducting fire prevention inspection. Planning and organizing an afternoon's inspections is essential. Preparation should include these steps:

1. Plan the area to be inspected. When practical, all occupancies within a given shopping center, office building, or block should be visited on the same day.
2. Review occupancy files prior to leaving the station. Note past violations, corrections, and recommendations to check for compliance or reoccurrence, **Figure 21-5**.

Figure 21-4 Company officer preparing for inspections.

Figure 21-5 An officer reviews occupancy files before leaving the station.

3. Give consideration to the type of activities conducted at the business relative to the time of day chosen for the inspection. For example, it is usually not good practice to visit a restaurant during lunch hour or a small manufacturing plant after 3:30 p.m. In the case of the restaurant, the staff may be too busy to assist inspectors, and it is not unusual for a small manufacturing plant to shut down after 3:30 p.m.
4. Should the inspector make an appointment with the business owner or arrive unannounced? There is no specific rule in this regard. Generally, fire inspectors do not make "surprise inspections." Many business owners appreciate a telephone call prior to the inspection; others feel it is not necessary. Writing a note in the inspection file indicating the owner's preference is good practice.

   Note, too, that given the unpredictable nature of the firefighting business, keeping an appointment is not always possible. When making an appointment, remember to explain this to the business owner.

Conducting the Inspection

How many firefighter/inspectors are necessary to conduct a fire prevention inspection? Again, there is no magic number. This is a judgment call for the company officer. Remembering that building familiarization is an important aspect of the fire company inspection program, it would seem appropriate to include the entire company. However, consideration must be given to the business owner's perspective. Too many firefighters may be seen as a "show of force" or an intimidation. Too few inspectors may result in a less than thorough or inefficient inspection. Care must also be taken to minimize the disruption to business activities. It certainly does not take three or four inspectors to inspect a small barbershop, for example, but this may be the right number for a large warehouse or manufacturing occupancy.

After arriving at the occupancy to be inspected, the crew should proceed directly to the front office, **Figure 21-6**. They should not linger in front or immediately begin inspecting the exterior of the building. Although the fire department has a statutory right to inspect all businesses, permission must always be obtained from the owners or their representatives prior to beginning any inspection.

Firefighters should introduce themselves to the receptionist, state their business, and request to see the person responsible for general building and employee safety. This may be the business owner, plant safety engineer, operations manager, or a maintenance engineer, **Figure 21-7**. Whoever is designated should have the authority to speak for management and be able to correct any violations noted.

Note The company officer should insist that the crew be escorted during the inspection. The escort will be able to grant access to all areas of the building and answer any questions that may arise. Additionally, the possibility of unfounded questions or impropriety on the part of the inspectors will be eliminated.

Figure 21-6 Upon arrival at the site to be inspected, inspectors should proceed directly to the office.

Figure 21-7 Introduction of a company officer to a business owner.

Occasionally, but fortunately infrequently, a business owner will deny permission to inspect. The company officer should ask for an explanation for the denial. Perhaps it is as simple as the time is not convenient for the business owner, the owner is too busy, or the safety manager is not present. Rarely is it a matter of the owner just refusing to be inspected with no rational explanation. This does occur, though. Under no circumstances should company officers engage in an argument with owners or attempt to persuade them to change their minds through the use of threats of authority or intimidation. A calm and reassuring approach is always the best tactic. "Would another day be more convenient?" or "We are not here to disrupt your business, only to help you maintain a safe working environment for you, your employees, and your customers." If all else fails, the matter should be referred to the Fire Prevention Division. Ultimately, an **administrative warrant** may have to be obtained from the local magistrate.

Note There are several schools of thought regarding the proper methodology for routing through a building when conducting the inspection. Outside to inside, top to bottom, left to right are all acceptable. It is generally of little consequence in what order the occupancy is inspected as long as the method is efficient, systematic, and thorough.

Typical Violations

A primary objective of the fire company inspection program is, of course, the elimination of fire and safety hazards. A **fire hazard** can be defined as any condition, situation, or operation that could lead to the unwanted ignition of combustibles or result in proper combustion becoming uncontrolled. Fire code violations in all occupancies can be categorized in general terms: exiting, fire protection equipment, use and storage of hazardous materials, and electrical and general fire safety.

Exiting

Most fire prevention officials would agree that a building's **means of egress**, **Figure 21-8**, provides the most basic level of life safety to its occupants.

> **Note** Exits are the most often abused and neglected system in a building.

The inspector must verify that no exits are compromised by ensuring the following:

- Clear and unobstructed access is provided to all exits.
- Exits are identified and well lit.
- The proper type of door is used and it opens in the direction of travel.
- Exits are equipped with the proper opening and locking hardware.
- Exit discharges to a public way are clear.

Exit accesses must be kept clear and unobstructed by storage or other materials that may hinder their use. Service corridors in malls and hotels are easy targets for storage of recently delivered merchandise or unused tables and chairs, **Figure 21-9**.

All exit accesses and exits are required to be identified with an exit sign, **Figure 21-10**. An exception is made for the main entrance to an occupancy because it is assumed that it is obvious even to visitors. Generally, exit signs are required to be illuminated only in assembly occupancies or high-rise buildings. When the exit access is not straightforward or is confusing, as in a warehouse or office with many corridors, intermittent directional signs should be used. If the fire inspector becomes lost or misdirected while walking through the building, then it can be assumed that visitors may become lost as well. This situation should be addressed by the use of intermittent directional exit signs.

> **Streetsmart Tip** The use of a "This Is Not An Exit" sign is not considered good practice because people in a hurry to leave the building under emergency circumstances are likely to focus on the word "Exit" and may become disoriented as a result.

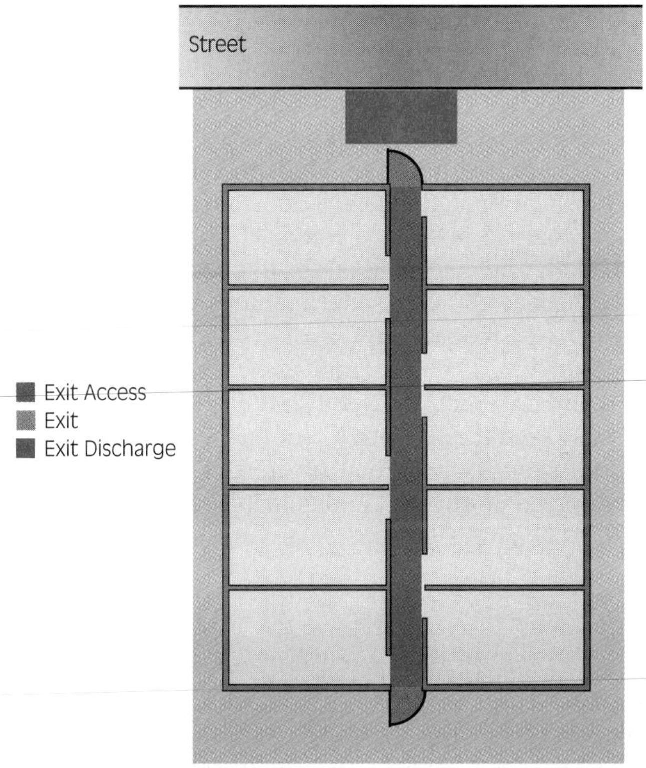

Figure 21-8 Means of egress.

Figure 21-9 Exit corridors must be kept clear of all storage.

Figure 21-10 All exits are required to be clearly identified.

The inspector must ensure that the proper type of door is used as an exit. Of the four classifications of doors—revolving, overhead, sliding, and swinging—only the swinging door is permitted to be used for a required exit. **Figure 21-11** shows the four classifications of doors.

Exits are required to be equipped with specific types of opening and latching devices depending on the occupancy classification. Approved exit hardware can be grouped in three categories: no-knowledge, panic, and special egress devices.

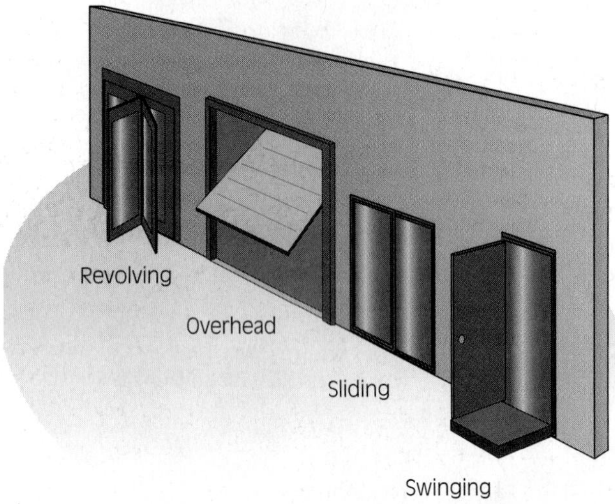

Figure 21-11 Of the four types of doors, only the swinging door can be used for a required exit.

No-knowledge hardware, **Figure 21-12**, is a broad category of locking devices that require no key or special knowledge to operate. Lock sets that combine the latching mechanism with a dead bolt are an example, **Figure 21-13**. The dead bolt is retracted simultaneously with the normal latching device when the knob or lever is turned. This device provides the security of a dead bolt while maintaining simplicity of operation. No-knowledge hardware is required on all exits, except the main entrance, in all occupancies when panic hardware is not required.

> **Note** A double-cylinder lock set is key operated on both sides. This type of hardware is permitted only on the main entrance to most occupancies provided a sign reading "This Door to Remain Unlocked During Business Hours" is displayed above the door. Double-cylinder dead bolts are also permitted in private residential occupancies.

> **Streetsmart Tip** Although double-cylinder dead bolts are permitted in residential occupancies by most building and fire codes, firefighters should strongly discourage their use. During an emergency evacuation of a home, there is no time to fumble for a key.

Panic hardware is required on exits in all assembly, educational, and institutional occupancies. It usually consists of a locking mechanism activated by a bar across the door, **Figure 21-14**. The mechanism must operate and the door must open with no more than 15 pounds of force. This is to allow even small children and frail adults to open the door.

Occasionally, for security purposes, a business owner will request the use of a **special egress control device**. The operators of occupancies such as jewelry or electronic equipment stores, weapons dealers, and the like often feel the need for the additional security provided by locked exits. This, of course, is an obvious violation of the fire code. To

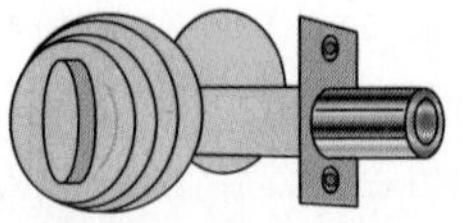

Figure 21-12 No-knowledge hardware is required on all exits not requiring panic hardware, except on the main entrance.

Figure 21-13 Interconnected lock set and dead bolt.

Figure 21-15 Special egress control devices often look like normal panic hardware, but must be identified with a sign.

Figure 21-14 Exit doors should be opened fully to ensure their function.

address the concerns of these owners while ensuring the safety of their employees and customers, fire and building officials have approved the limited use of special egress control devices. This type of device is allowed, provided that an approved automatic fire sprinkler system and an approved automatic smoke detection system protect the building. The door-release will unlock the door in a maximum of fifteen seconds. Pushing on the bar, **Figure 21-15,** activates the device. With the specific approval of the fire marshal, these devices are permitted in the following occupancies: general businesses, factories, institutions, mercantile businesses, storage facilities, and group care facilities and residential care facilities housing clients with various forms of dementia.

Special egress control devices must automatically deactivate whenever the sprinkler or smoke detection systems activate or whenever there is a loss of electrical power to the building or the device. There must also be posted on the door a sign that reads "Keep Pushing. Door Will Open in 15 Seconds. Alarm Will Sound" or similar wording.

Exit discharges must be kept clear to allow for occupants to continue their travel to a safe position away from the structure, **Figure 21-16**. Fire codes

Figure 21-16 Firefighters should inspect the exit discharge to ensure occupants have clear access to the public street or parking area.

usually define a "safe distance" as the nearest public way, that is, a street, alley, sidewalk, or major parking lot.

Fire Protection Equipment

Note Fire protection equipment, whether portable or fixed, is the building occupants' first line of defense against unwanted fire. Firefighters must ensure that this equipment is not only present and maintained in proper working condition, but that the building occupants know when and how to use it.

During an inspection, it is most appropriate for the inspector to question the business representative: "Do your employees know where your fire extinguishers are located and how to use them?" "Do you provide any training in their proper use?" "Do you know how to manually activate your kitchen cooking equipment fire suppression system?" "Do you know where the fire sprinkler control valves are located?" "Do you know how a fire sprinkler system operates?" Many people are still under the misunderstanding that all fire sprinkler heads operate simultaneously or that they are activated by smoke!

Portable Fire Extinguishers. Firefighters should inspect for the presence of fire extinguishers, **Figure 21-17**, and ensure the following:

- Fire extinguishers are located near all exits and in exit accesses.
- The extinguisher has the proper classification and rating for its location.

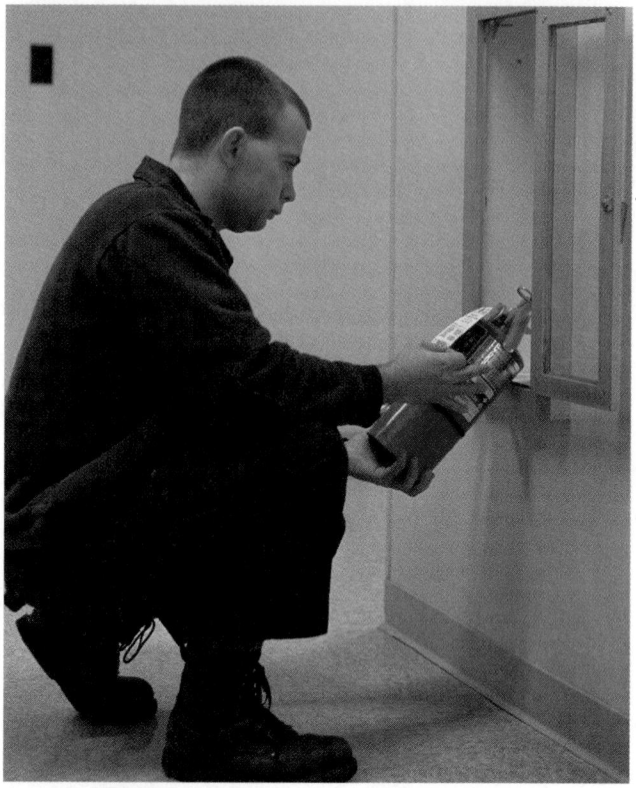

Figure 21-17 Firefighter inspecting a fire extinguisher.

- The fire extinguisher is mounted on the wall and is easily visible and accessible.
- The pressure gauge indicates that the extinguisher is fully charged.
- The tag on the extinguisher indicates that it has been serviced in a time frame conforming to local requirements. (Some jurisdictions require annual servicing, whereas others permit a time span as long as five years between services.)

Automatic Fire Sprinkler Systems. Firefighters should check fire sprinkler systems for the following:

- All water supply valves are open and secured.
- The fire department connection should be free of obstructions and capped, **Figure 21-18**.
- Access should be provided to system risers, valves, and gauges, **Figure 21-19**.
- Pressure gauges on the system riser should show appropriate pressure.
- All areas of the building must be properly protected.
- Sprinkler heads should not be damaged, painted, or corroded.
- The system should be flow tested regularly, preferably quarterly, and records kept, **Figure 21-20**.
- The system should be serviced in a timely manner, usually at five-year intervals.

Figure 21-18 Outside control valves must be open and secured. The fire department connection must be clear of obstructions and capped.

Sprinkler systems in spray paint booths deserve special mention. By the very nature of the operation, sprinkler heads in spray booths quickly become covered with overspray. This condition will not only delay activation of the head but will most likely prevent its operation altogether. Therefore, keeping the heads clean is essential. To protect sprinkler heads that are likely to be ruined by overspray residue, NFPA 25, *Inspection, Testing and Maintenance of Water-Based Fire Protection Systems,* requires that they be covered with thin plastic or small paper bags, as shown in **Figure 21-21**. These bags should be changed daily.

Figure 21-19 Access should be provided to fire sprinkler control valves and gauges.

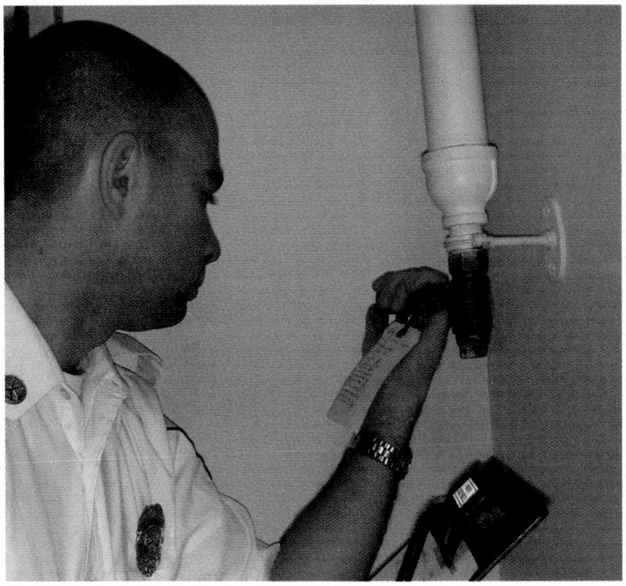

Figure 21-20 The sprinkler system should be flow tested regularly. Some inspectors flow the system themselves, while others ask the business owner to do so.

Restaurant Cooking Surface and Exhaust Hood Suppression Systems. Fixed extinguishing systems protecting the cooking surfaces and exhaust hood and ducts in restaurants are required whenever the cooking process produces grease-laden vapors. Pizza and bread ovens and stoves that are used for heating water or soup would not require a system; however, nearly all other commercial cooking would. Such

Figure 21-21 Many departments allow plastic sandwich-type bags to be placed over sprinkler heads in spray paint booths to prevent the accumulation of overspray on heads.

systems are essentially large fire extinguishers mounted on the wall, **Figure 21-22**. They contain either a wet or dry extinguishing agent.

When inspecting cooking surfaces and hood and duct systems, firefighters should ensure the following:

- The extinguishing agent cylinders are charged and armed.
- Nozzles are free of grease and are capped.
- Fusible links are free of grease.
- Hood filters are clean and in place, **Figure 21-23**. (Filters not only trap grease, but also restrict a surface fire from spreading to the duct.)
- The manual activation control is accessible, **Figure 21-24**.
- The tag on the system indicates that it has been serviced annually.

Similar systems are found in automotive, furniture finishing, and spray paint booths. They should be inspected in the same manner as restaurant systems.

Heat and Smoke Detection Systems. Beginning at the system control panel, **Figure 21-25**, the firefighter should ensure the following:

- All indicators on the panel are in a "normal" condition and not in "trouble" mode.
- The panel is receiving AC power.

Figure 21-23 Extinguishing system nozzles and missing filters.

- All detection devices are present and properly mounted—not hanging from the ceiling.
- Records are provided indicating that the system has been tested periodically.

It is often a good idea for firefighters to request that the system be tested in their presence. While not always practical given the nature of the occupancy, witnessing the activation of the system will not only ensure that it functions but will stress to the owner the importance of the system.

Figure 21-22 Fixed fire extinguishing system for cooking surfaces.

Figure 21-24 Restaurant workers should know the location of the manual release and how to activate the extinguishing system.

Figure 21-25 Firefighter checking fire alarm system control panel.

Fixed Gaseous Extinguishing Systems. These types of extinguishing systems protect building areas where water from a sprinkler system would cause massive amounts of damage to sensitive equipment, such as computers or electrical systems. The extinguishing agent is generally CO_2, halon, and halon replacement. In essence, these protection systems serve the purposes of detection *and* suppression.

Firefighter Fact Because of its danger to the environment, halon has virtually been replaced by other agents that are more environmentally friendly. Existing halon systems are "grandfathered" and can continue operation. Halon is no longer produced, and only remaining stockpiles of halon are available. Recharging systems with halon is generally cost-prohibitive. The result is that most systems are modified to use more environmentally safe extinguishing agents.

Caution The detection system should be inspected as would be any other heat and smoke detection system with the exception of the function test. Function tests should never be done by firefighters because an accidental discharge may result.

The inspection of the suppression portion of the system involves checking the cylinder gauges for proper pressure, ensuring that all discharge nozzles are unobstructed either by foreign objects or storage, and making sure access to the manual discharge control is provided.

Use and Storage of Hazardous Materials

Ensuring the proper use and storage of hazardous materials is a subject that often requires a great deal of knowledge, expertise, training, and experience. Many fire prevention officers dedicate large portions of their careers to such endeavors. However, from the firefighters' perspective, some general principles apply.

Inside Storage and Use

- Storage and use areas must be well ventilated.
- Sources of ignition must be eliminated in any area using or storing flammable or combustible liquids or gases. These include open flames and welding or grinding of metal.
- The maximum quantity of flammable and combustible liquids permitted varies according to the occupancy.
- Quantities of less than 5 gallons must be stored in approved **safety containers**, **Figure 21-26**.

Figure 21-26 Approved flammable liquid safety can.

- Small quantities should be stored in approved metal or wood storage cabinets, **Figure 21-27**.
- Larger quantities must be stored in rooms specially designed for such purposes. These rooms are constructed of fire-resistant material, have raised door sills to prevent leakage, have liquid-tight floors and walls, have explosion-proof electrical fixtures, and are well ventilated.

Figure 21-27 Small quantities of hazardous material should be stored in approved storage cabinets.

- Containers with a capacity greater than 30 gallons may not be stacked.
- Dispensing must be done through an approved pump or self-closing faucet, **Figure 21-28**.
- Dispensing containers must be properly bonded and grounded to protect against the discharge of static electricity, **Figure 21-29**.
- Compressed gas cylinders, whether full or empty, should be stored with protective caps in place and secured with chains or straps to prevent falling, **Figure 21-30**.
- Sources of ignition must be eliminated in any area using, dispensing, or storing flammable or combustible liquids or gases. These include open flames and welding or grinding of metal.

Outside Storage

- The selected storage area should be properly distanced from buildings and property lines to minimize the potential for fire spread or exposure.
- Storage areas must have **secondary containment**. This typically means a 6-inch curb around the storage area to contain spilled liquid, **Figure 21-31A**.

Figure 21-28 Approved self-closing faucet.

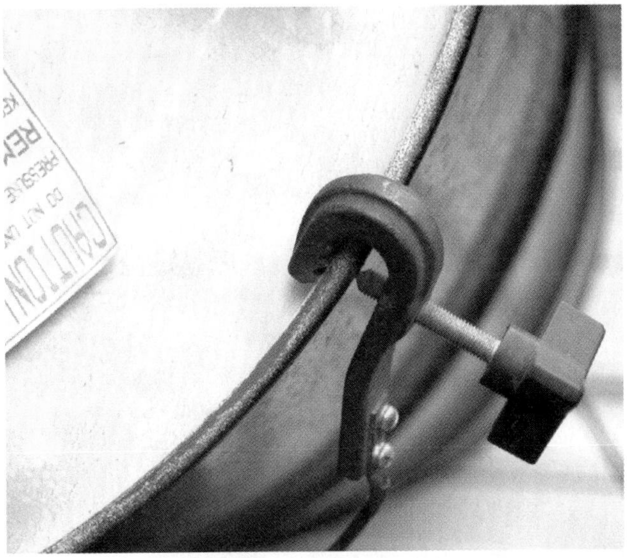

Figure 21-29 Properly bonded and grounded barrels.

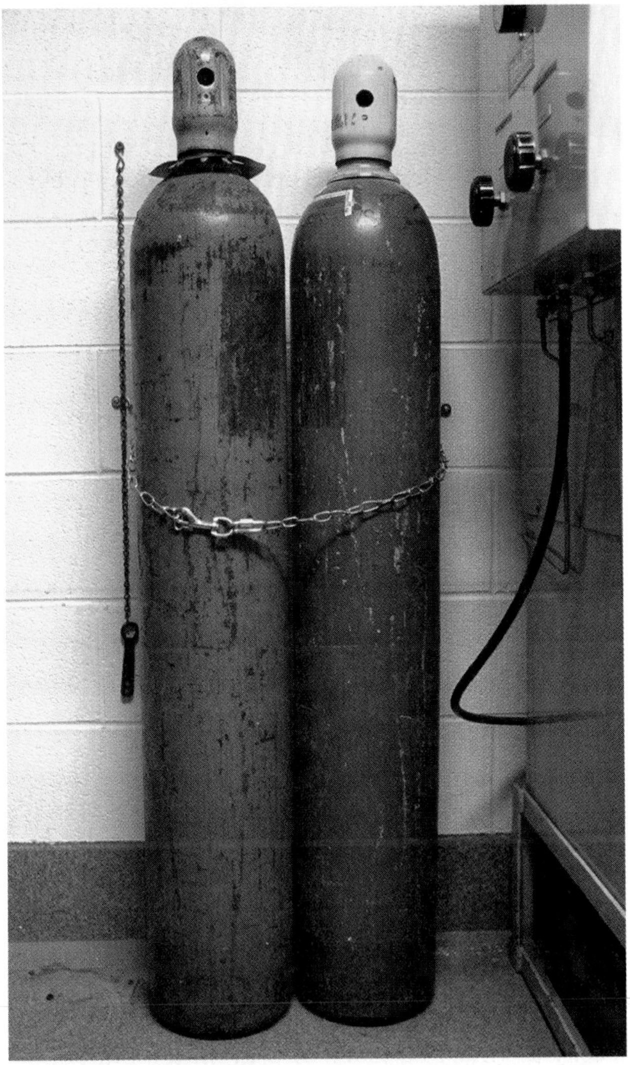

Figure 21-30 Compressed gas cylinders secured with a chain and with caps in place.

(A)

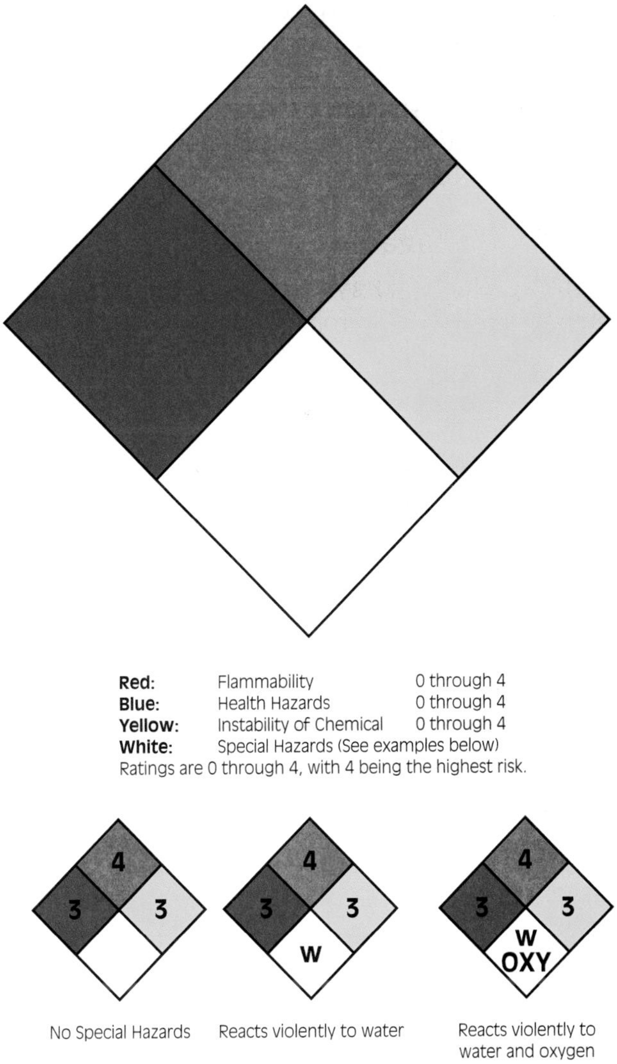

(B)

Figure 21-31 (A) Storage of flammable liquids with curb around storage area. (B) The NFPA 704 placard system identifies hazards in fixed storage facilities. The colors and numbers quickly give first responders vital information.

- Sources of ignition must be eliminated, keeping in mind the potential for the migration of vapors from spilled liquids.

Hazardous materials that are located in fixed storage facilities are not governed by the same regulations as the transportation industry. Local jurisdictions may develop a system of labeling for hazardous materials. Most local jurisdictions choose to adopt the NFPA 704 (*Standard System for the Identification of the Fire Hazards of Materials*) placard system, **Figure 21-31B.** This system provides valuable information to first responders and notifies employees and other persons using the facilities of the contents. The placard system is simple to read and provides information about health hazards, flammability, and reactivity. Firefighters inspecting facilities with these types of placards should speak to the appropriate persons at the facility to review the types of materials contained on the property and the associated hazards. These hazardous materials should be documented in the inspection report and first-due responders should be advised.

Electrical Hazards

Building electrical systems require specialized knowledge to ensure proper design and installation. The firefighter should call on the expertise of a local electrical inspector whenever questionable installation practices are noted. However, a firefighter can be effective in eliminating electrical hazards by observing the following:

- Check all fuse and breaker panels to verify that overcurrent protection devices have not been defeated and that all spaces on the panel are equipped with a breaker switch or blank to cover the opening, **Figure 21-32.**
- Access to the panels should be maintained free of storage and debris.
- All outlets and junction boxes should have an approved cover in place.
- Ensure that outlets, switches, lights, and appliances are approved for the location installed. Remember that ordinary electrical devices are not permitted in rooms designed for the use and storage of flammable or combustible liquids or explosives.
- Extension cords are not permitted to be used in place of permanent wiring. This problem is widespread and becomes more acute when light-duty "zip cords" are utilized. If the installation of additional permanent outlets is not possible, multiplug adapters with built-in circuit breakers are generally an acceptable alternative.
- No exposed wiring, approved or otherwise, should be permitted to run through doorways or under floor coverings or be stapled to a wall.

Figure 21-32 Circuit breakers must never be prevented from tripping by the use of tape or other objects. All spaces on the panel must be covered with either a breaker switch or a blank.

General Fire Safety

> **Note** Poor housekeeping practices, while seemingly a simple issue, are often the root cause of many fire hazards and eventually fires and injuries. Accumulated trash and debris are often found blocking exits or access to fire protection equipment. Haphazard storage of combustibles near open flame or hot surface devices such as water heaters or space heaters is another cause of accidental fires.

Dust accumulation is a common problem in woodworking and textile manufacturing occupancies. Care must be taken to prevent the buildup of dust on all horizontal surfaces, especially those

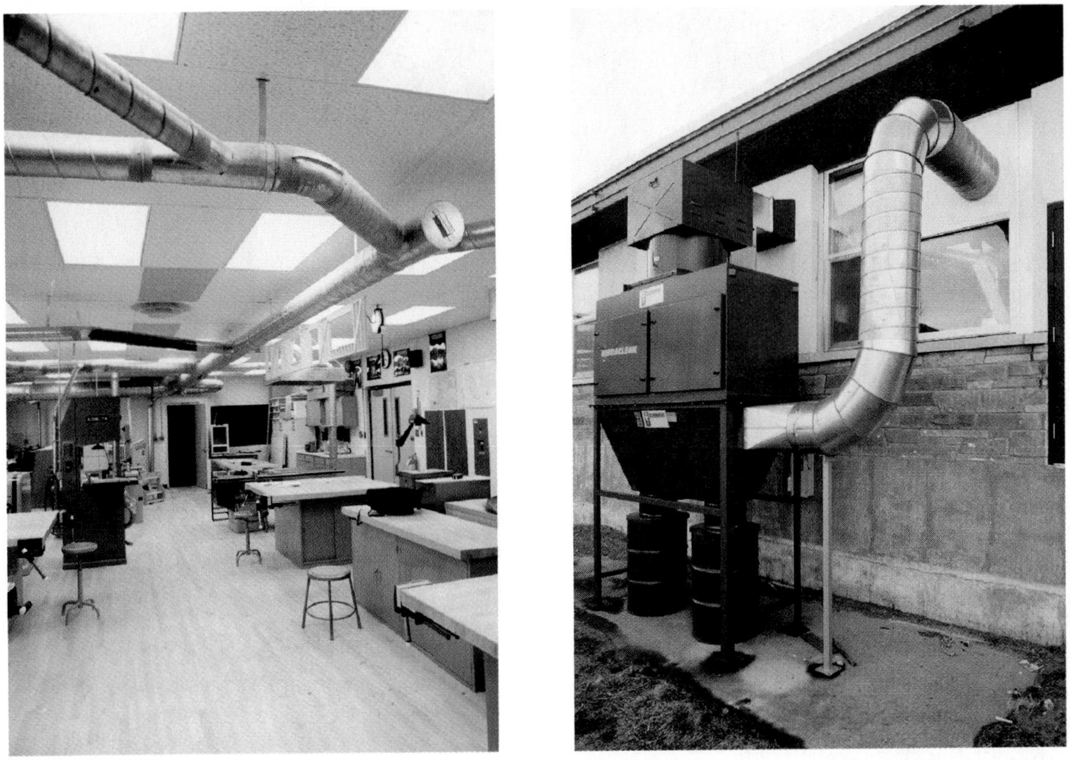

Figure 21-33 Woodworking occupancy with a dust collection system.

overhead. Dust explosions are caused when combustible dust particles suspended in the air contact an ignition source. Fortunately these devastating explosions are an infrequent occurrence. **Figure 21-33** shows the ductwork for a dust collection system.

Proper disposal of rags used with flammable or combustible fuels, solvents, and oils is easily accomplished but so often neglected. Autoignition of such rags due to their casual treatment has resulted in more than a few fires. Oily rags must be stored or discarded in an approved safety can, **Figure 21-34**.

Fire codes restrict smoking in certain occupancies. General smoking in buildings should be discouraged by firefighters. Smoking and nonsmoking areas should be clearly identified and proper containers provided for the disposal of smoking materials.

Building Exterior

No fire prevention inspection is complete until the inspector has walked around the outside of the building. The following observations should be noted:

- Address numerals of sufficient size should be posted where responding emergency personnel can easily see them, **Figure 21-35.**
- All access roads should be considered fire lanes and kept clear of vehicles and storage, **Figures 21-36A** and **B**.
- Secured key boxes (also called Knox Boxes), when provided, should be checked for current keys, **Figure 21-37**. Firefighters should

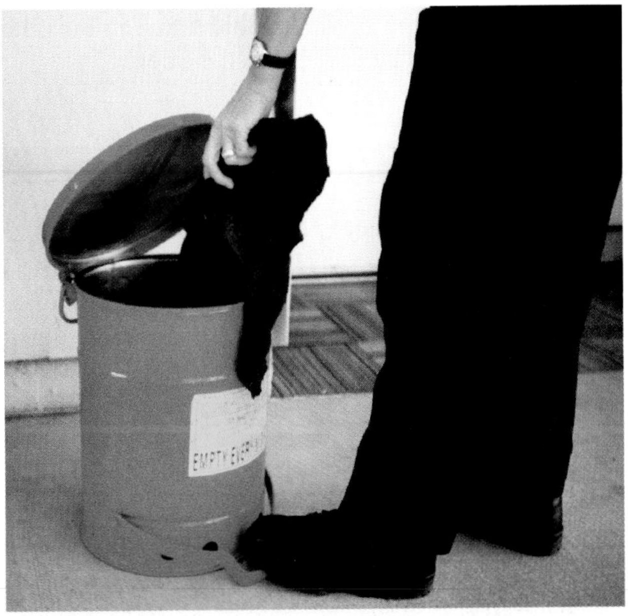

Figure 21-34 Oily rags must be stored in approved safety cans with self-closing lids.

Figure 21-35 All occupancies should have plainly visible address numerals.

encourage, if not require, the use of such boxes for after-hours entry into buildings by emergency personnel.

- Trash disposal areas should be away from buildings and free of debris, **Figure 21-38**.
- Outside storage of flammable and combustible liquids should be as described earlier.

Figure 21-37 Key boxes on the outside of a building allow firefighters access to a locked building. They should be checked for current keys.

- Accumulated dried vegetation near buildings should be eliminated, **Figure 21-39**.

Concluding the Inspection

At the conclusion of the inspection, the firefighter should convey the findings to the building representative. This "wrap-up meeting," as shown in **Figure 21-40**, need not be a formal event. In fact, the efficient inspector will have pointed out any deficiencies or suggestions noted while walking through the business at the time observed. In this way, at the wrap-up, the firefighter can reiterate the

(A) (B)

Figure 21-36 (A) Unblocked fire lane. Concrete blocks under the turf are often used to allow fire apparatus to drive in landscaped areas. (B) Blocked fire lane. Fire lanes must be kept open at all times.

Figure 21-38 Trash storage areas should be free of debris and away from buildings.

findings in summary form prior to leaving the premises. The inspector should remember to thank occupants for their time and praise them for their fire safety efforts. If criticism is necessary, the inspecting officer should exercise tact.

All deficiencies, violations, and suggestions must be documented, even those corrected in the presence of the firefighter.

Figure 21-39 Dried grass at exterior of building should be eliminated.

> **Note** Violations that pose an imminent life hazard, in the opinion of the firefighter, must be corrected before the firefighter leaves the premises. Examples include locked or blocked exits, welding operations in the vicinity of open flammable liquids, and electrical wiring that presents an obvious shock hazard.

Many different styles of notices are available. An example of a Fire and Life Safety Inspection Notice is shown in **Figure 21-41**. At a minimum, the notice should include the name and address of the business, the date of the inspection, the building representative,

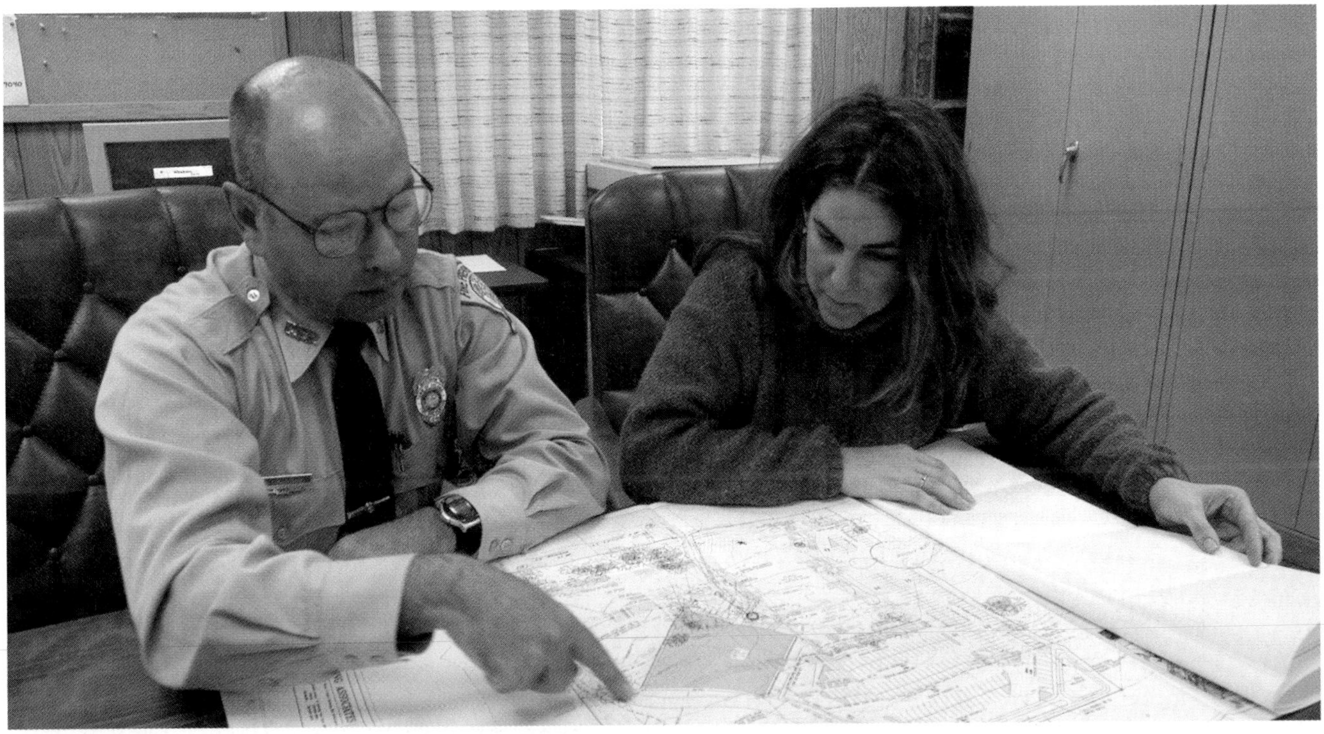

Figure 21-40 Inspector discussing findings with business owner.

FIRE SAFETY CHECKLIST

BUSINESS NAME		PHONE #
STREET ADDRESS	**SUITE**	**ZIP CODE**
BUSINESS OWNER		**HOME PHONE**
OWNER ADDRESS		
EMERGENCY CONTACT		
ADDRESS		**EMERGENCY CONTACT PHONE #**
INSPECTION CONTACT	**INSPECTION DATE**	**INSPECTION COMPLETED**
	#1 #2	( ) YES ( ) NO

#1	#2	
		ARTICLE 1: ADMINISTRATION
☐	☐	0136: No space heaters allowed in occupancy
☐	☐	0137: Notices & orders shall be complied with
		ARTICLE 9: ACCESS
☐	☐	0901: Install approved address identification
☐	☐	0902: Maintain & post fire department access lanes
☐	☐	0903: Lock Box required for access along with current keys
		ARTICLE 10: FIRE PROTECTION SYSTEMS
☐	☐	1001: Maintain fire protection systems in an operable condition
☐	☐	1002: Inspection & test reports to be made available
☐	☐	1004: FDC shall remain clear of obstructions
☐	☐	1005: 3ft. clear space to be maintained around hydrants
☐	☐	1009: All control valves & flow switches shall be monitored
☐	☐	1011: All commercial cooking equipment shall have automatic fire protection
☐	☐	1012: Provide a 40 BC fire extinguisher within 30 feet of cooking equipment
☐	☐	1013: Hood & duct system shall be free of grease
☐	☐	1014: Hood extinguishing system to be serviced and links replaced every 6 months
☐	☐	1015: Fire alarm equipment shall not be obstructed
☐	☐	1017: "Local Alarm Only" signage shall be provided
☐	☐	1018: Maximum travel distance for fire extinguishers exceeded
☐	☐	1019: Properly mount fire extinguisher
☐	☐	1020: Service fire extinguisher annually
		ARTICLE 11: GENERAL
☐	☐	1103: Dumpsters shall be 5 feet from building
☐	☐	1104: Maintain good housekeeping
☐	☐	1105: Maintain storage 18 in. below sprinkler heads
☐	☐	1106: Maintain storage 24 in. below ceiling
☐	☐	1107: No storage in boiler, mechanical, or electrical room
☐	☐	1108: Drapes and other decorative material shall be fire resistive
☐	☐	1109: Aboveground gas meters & piping shall be protected
☐	☐	1110: Maintain clearance from combustibles and heat producing devices
☐	☐	1114: Maintain fire resistive construction
☐	☐	1115: Maintain fire doors and dampers
☐	☐	1117: Remove accumulation of exterior waste

#1	#2	
		ARTICLE 12: EGRESS
☐	☐	1201: Remove all obstructions from path of exit
☐	☐	1202: Exit door shall swing out when occupancy exceeds 50
☐	☐	1203: Remove all unapproved locking devices from exit doors
☐	☐	1205: Emergency power equipment for egress illumination shall be maintained
☐	☐	1206: Exit paths shall be identified by exit signs
☐	☐	1207: Exit signs shall be illuminated
☐	☐	1208: No storage under unprotected stairways
☐	☐	1213: Dead end corridors over 20 feet in length
		ARTICLE 25: PLACES OF ASSEMBLY
☐	☐	2501: Maximum occupant load signs to be posted in occupancies over 50
		ARTICLE 74: COMPRESSED GAS
☐	☐	7401: Compressed gas cylinders shall be secured
		ARTICLE 80: HAZARDOUS MATERIALS
☐	☐	8002: Tanks, piping & valves shall be protected from vehicular damage
		ARTICLE 85: ELECTRICAL EQUIPMENT AND WIRING
☐	☐	8501: Install cover plate on all electrical outlets, switches and junction boxes
☐	☐	8503: Discontinue the use of extension cords
☐	☐	8504: Cords shall not be subject to damage
☐	☐	8505: Cords shall be maintained in good condition
☐	☐	8506: Discontinue the use of multi-plug adapters
☐	☐	8507: Maintain 30 inch clearance around electrical panels
☐	☐	8508: Electrical panels shall be labeled as to area served
☐	☐	8509: Doors into electrical rooms shall be labeled
		MISCELLANEOUS:
☐		**SEE ATTACHED SUPPLEMENTAL**
☐		**COMMENTS:** ______________________

REQUIRED PERMIT	ASSOCIATED FEE	NOT CURRENT	CURRENT

A routine fire inspection has been made of these premises to determine if any life or fire hazard exists. The inspection also was made to determine if any violations of the Uniform Fire Code exist. If violations have been found, this shall serve as your official notice. All violations must be corrected within the time allowed. A re-inspection will be made at a later date to verify the required corrections. Should you need assistance regarding this inspection, contact the Fire Prevention Bureau at 962-2537.

I, the undersigned, am in receipt of a copy of this inspection report and am aware of the hazards noted. I am also aware that this is a routine inspection for correction of violations and **may not encompass every possible violation.**

SIGNATURE OF BUSINESS OWNER, MANAGER, OR RESPONSIBLE PARTY	TITLE	DATE

SIGNATURE OF INSPECTOR

FIRE DEPARTMENT: WHITE   1ST INSPECTION:CANARY/BUSINESS OWNER   2ND INSPECTION:PINK/BUSINESS OWNER

Figure 21-41 Inspection notice should be given to business representatives at the conclusion of the inspection. *(Courtesy of Loveland Fire and Rescue, Loveland, Colorado.)*

Dept. Use
Batt. ______________
Unit ________ Shift ________
Insp. ______________
(Print name)

DELMAR COUNTY FIRE AUTHORITY
INSPECTION NOTICE

INSP. # ______________ Received by ______________________

Violation

Business Name: ______________________ Address: ______________________
As a result of an inspection by the Delmar County Fire Authority on (date ______________
the violation(s) listed below were noted.

☐ Vio. Code 25 1. State law requires that all fire extinguishers be serviced annually CCR TITLE 19
A. Service accomplished by (Co.) ______________________
B. State license number ______________________

☐ Vio. Code 25 2. Provide a type 2A 10BC fire extinguisher. Extinguisher shall be installed on the hanger, conspicuously located, readily accessible in the event of fire. UFC 10.301

☐ Vio. Code 01 3. Address numbers shall be placed on all buildings, plainly visible from the street. Numbers shall contrast with their background. UFC 10.208

☐ Vio. Code 09 4. Discontinue use of extension cords and/or multiplug adapters. UFC 85.106

☐ Vio Code ___ 5. Other: ______________________

I certify under penalty of perjury that the Violations noted above have been corrected.

Signature ______________________ Date corrected ______________

Sign and return this Notice. Failure to correct the identified Violation(s) within 14 days after inspection date will result in reinspection or other appropriate legal action to ensure compliance.

F121-16.2 (R3/90) **FILE COPY**

Figure 21-42 A business owner may affirm that minor code violations have been corrected by signing and returning a self-clearing card.

violations noted with their corresponding code references, the fire department inspector's name, and a compliance date. Most often, notices are completed prior to leaving the building. If this is not possible because further research is required, it is perfectly acceptable to return later with the completed notice. The original should be given to the building occupant and a copy should be filed in the occupancy file. Remember, this is a legal document and must be treated as such.

Reinspections

Few things can negate the validity of a fire inspection faster than the lack of follow-up. All documented violations (and they should *all* be documented) must be reinspected in some manner to emphasize their significance and ensure compliance. *If it is important enough to document, then it is important enough to follow up.* However, not all violations require the time and effort of another trip to the business site. Many departments use a self-clearing card, an example of which is shown in **Figure 21-42**. It may have different names, but the concept is the same. For a limited number of very specific infractions, the business owner affirms that the violation has been corrected by signing a postcard and mailing it to the fire department. Examples of qualifying infractions are address numerals that have been enlarged so they are clearly visible, removal of an extension cord, or recharging a fire extinguisher.

More serious violations require a reinspection. The time allotted to make the corrections will vary with the complexity and seriousness of the violation. However, once a reinspection date has been established and agreed to by the business owner, every effort should be made by the inspector to return on the specified day. Additional compliance time may be granted at the discretion of the inspector. Failure on the part of the business owner to make a reasonable effort to comply with violation notices should be referred to the Fire Prevention Division.

HOME INSPECTIONS

Firefighter Fact National fire statistics prove that year after year roughly 80 percent of all injuries and deaths in America occur in the home. To the credit of the fire service and the public, actual numbers of injuries and deaths have been decreasing in recent years. But the fact remains that eight of ten people killed or injured by fire are in the "safety" of their homes.

The overwhelming majority of fire prevention efforts are concentrated on the business community. The reasons are deep seated and fundamental but are also centered around privacy issues. Americans simply will not allow authorities legislated access to their homes, no matter how noble the reason.

This does not preclude the fire service from making an effort to conduct voluntary home inspection programs. Such programs, whether available to entire jurisdictions or only targeted areas, can be beneficial to the community as well as the fire department. Not only will the threat of fire in family homes be reduced, but much goodwill for the fire department will be fostered.

Voluntary residential inspections can be used to point out fire and life safety hazards, check and install smoke detectors and carbon monoxide detectors, and instruct families in proper emergency preparedness techniques. When granted permission to enter a home, firefighters are cautioned to limit their movement in the home to the "less private" areas. In other words, stay out of the bedrooms.

Typical hazards found in the home are not unlike those discussed previously. Exiting problems are usually centered around double-cylinder dead bolts on doors and security bars on doors and windows, **Figure 21-43**. Electrical problems involve overloaded circuits and extension cords. Storage around furnaces and water heaters is often noted. Combustibles left on top of floor furnaces continue to cause many fires at the start of each heating season. Improperly stored flammables and combustibles in basements and garages are regularly observed.

Attention should be given to issues of safety to small children. Accessible electrical outlets without protective caps, unsecured household chemicals, swimming pools without security fences, and electrical appliances in bathrooms are areas that deserve the serious attention of parents and child care providers.

Figure 21-43 House with security bars.

Smoke and carbon monoxide detectors should be installed, **Figure 21-44**. The importance of these devices cannot be overstated and must be stressed to homeowners. Firefighters should remind residents that their chance for survival in

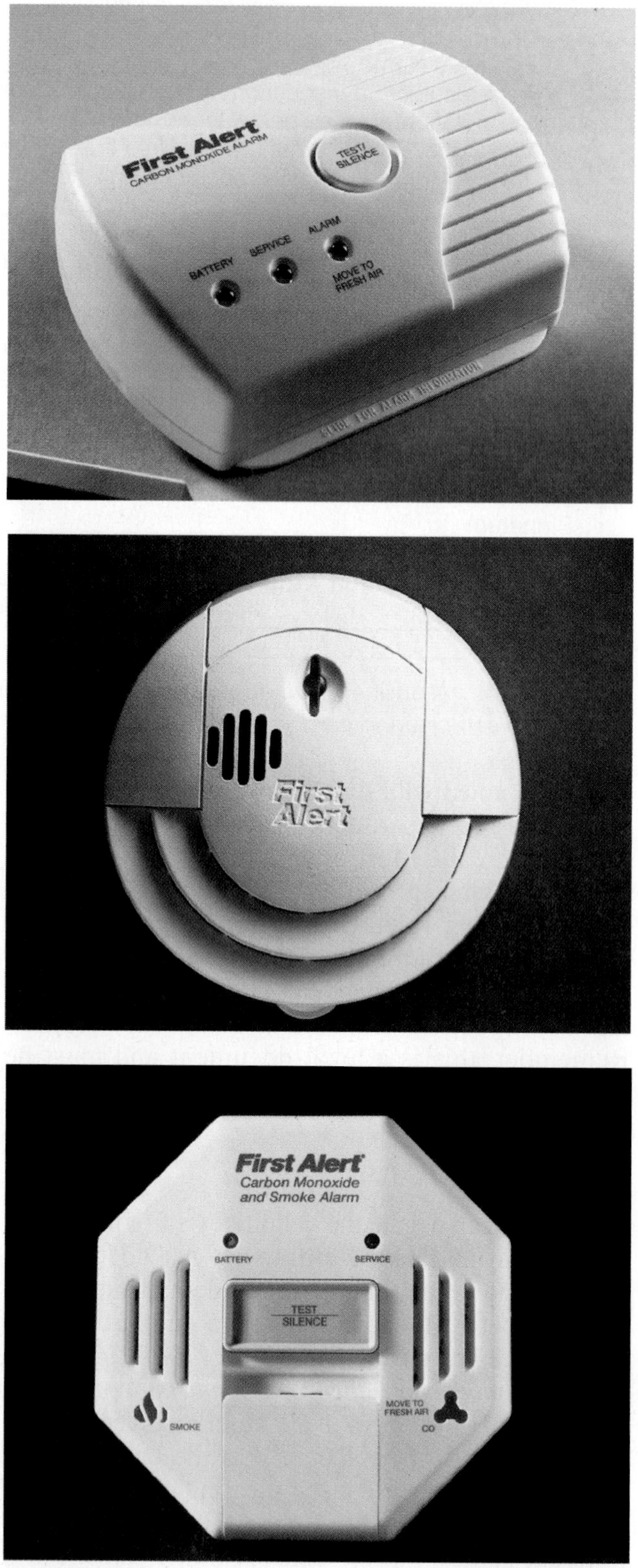

Figure 21-44 Smoke and carbon monoxide detectors are available separately and as one unit. *(Photos courtesy of BRK Brands, Inc. makers of First Alert® branded safety products)*

case of a fire increases by 50 percent if a smoke detector is installed and functioning. However, correct installation is only a good first step. Detectors must be maintained by changing their batteries on a regular basis, **Figure 21-45**. The International Association of Fire Chiefs sponsors the annual "Change Your Clock, Change Your Battery" program to remind residents to change their smoke detectors' batteries when they reset their clocks for Daylight Savings Time.

Figure 21-45 Batteries in smoke detectors should be checked monthly and replaced every six months. Some detectors have lithium batteries that do not require replacement.

Finally, firefighters should not fail to seize the opportunity to discuss emergency preparedness with the home's occupants. This should include first-aid firefighting techniques such as fire extinguisher placement and use, stove top fire extinguishment, **Figure 21-46**, and barbecue fires. Instruction should be given in proper home evacuation in case of fire. But this area could be expanded to include preparedness for natural disasters such as earthquakes, tornadoes, and flooding.

FIRE AND LIFE SAFETY EDUCATION

For most people, the tragedy of losing a home or a loved one to fire is something that happens to someone else in some other neighborhood or town. How many post-incident reports have appeared on television where the neighbor remarks, "I can't believe this happened on our block. This is such a quiet street"? Teaching citizens to recognize life safety hazards and to react

Figure 21-46 Firefighter showing others how to place a lid on a burning pan.

appropriately is clearly a fire department function and responsibility. The problem for fire safety educators is how to make fire and safety personal issues in the community.

> **Note** National and state statistics do little to impress people about the immediacy of a problem or its potential impact on their lives. Citizens must be made aware that fire safety is an issue in their community and be taught what they can personally do to prevent becoming a statistic.

Some departments have personalized the tragedy of fire by conducting tours of burned-out homes where occupants have been injured or killed. Specially built fire safety houses, as shown in **Figure 21-47,** allow more intensive training for the general public. These fire safety houses are designed to simulate fire conditions in a safe and controlled environment. They allow citizens to experience firsthand some of the conditions present during smoke or fire conditions. The citizens learn that a basic evaluation can become a confusing scenario under adverse conditions. Educational efforts must be tailored to meet the specific needs of the community and not be limited strictly to fire-related issues. For example, water safety classes are frequently taught by fire service personnel in communities with a high concentration of backyard swimming pools.

Fire and Life Safety Program Presentations

Firefighters are frequently asked to speak to groups of all ages regarding fire and life safety issues. For a presentation to be worthwhile and effective, it should contain three main components: preparation, presentation, and practice.

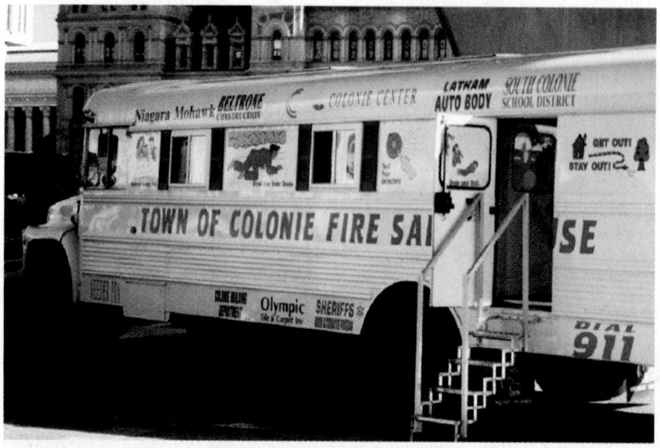

Figure 21-47 This school bus was redesigned and specially equipped to become a fire safety house.

The preparation phase involves the firefighter and the audience. The firefighter must be well versed with current knowledge on the subject matter to be presented and come prepared with appropriate audiovisual equipment, training aids, and literature. Once in front of the audience, the firefighter should prepare the group for the information they are about to receive. An effective method for gaining the audience's attention is to make the subject matter personal to them. For example, when speaking on the topic of drown-proofing to a group of parents, it might be beneficial to recount a recent incident in the community, minus the graphic details, of course. The firefighter should let them know that they are about to receive information they can use to ensure the incident is not repeated at their homes. Getting the audience interested and personally involved is conducive to learning and the goal of the preparation phase.

Now that the audience is primed to learn, it is time to present the information. How the information is transferred from the firefighter to the individuals in the group is left to the creativity of the speaker. Showing a video and answering questions may be easy for the firefighter, but generally not much learning takes place. Professional educators know that the level of learning increases proportionally with the number of senses affected. In other words, the more the audience can see, hear, touch, or smell the subject matter, the more the information will be retained. When speaking to a group of hotel workers about evacuation of their guests during an emergency, it is important to let them hear the fire alarm sounding and see the strobe lights flashing. When speaking to children, use of fire prevention and education specialized equipment helps to keep the children's attention, **Figure 21-48**.

In the final phase of the program, the participants are asked to apply, practice, or demonstrate what they have learned. A group of small children being taught how to report an emergency could practice by using fire station telephones with a firefighter on an extension playing the role of a dispatcher receiving the call. An industrial fire brigade, having been instructed in the proper selection and use of fire extinguishers, could be allowed to extinguish small pan fires. (Appropriate safety equipment is mandatory. Firefighters must always verify local department policy regarding live fire training.) Firefighters should observe this phase closely, ensuring that the information pre-

Figure 21-48 These fire safety characters keep the children focused and help to get the message across.

sented is being applied properly and making correction where necessary.

Forms of Fire and Life Safety Programs

Public education programs usually take three general forms: public service announcements, school programs, and adult/homeowner programs. Fire station tours also provide an excellent opportunity to heighten the public's awareness of fire safety issues.

Public Service Announcements (PSAs)

A simple, inexpensive, and timely method of spreading fire safety messages is to use the media. Radio, television, and newspapers are always willing to assist public agencies with their efforts. Short five- to ten-second television or radio spots or small news articles can be used for delivering safety messages on such seasonal topics as drowning prevention, fireworks safety, smoke detector maintenance, Halloween costumes and candy, and Christmas tree safety.

School Programs

Fire safety and education programs for elementary school children can be valuable and productive, **Figure 21-49**. They are designed to teach students the dangers of fire, how to prevent fires from happening, and what to do if they are caught in a fire. These programs can take many forms: two of the more popular are NFPA's Learn Not to Burn and the Junior Firefighter Program.

Figure 21-49 Popular public education programs are available through the NFPA.

The Learn Not to Burn program is designed for children in preschool through the third grade. It teaches children safety lessons using educational material and technical support provided by NFPA. Teachers attend workshops where they are instructed in course material by fire educators. The teachers then teach the program over a three-month span at their schools. Incorporated into the Learn Not to Burn curriculum are lessons on home evacuation using material from Operation EDITH (Exit Drills In The Home) and the ever-popular Stop, Drop, and Roll.

The Junior Firefighter Program can take many forms. It is generally presented to fifth-grade students in three stages. Phase 1 is usually an oral presentation and tabletop demonstration by a fire education specialist. Using an instructor's manual and student workbooks provided by the fire department, teachers complete phase 2 in about four weeks. Phase 3 consists of a visit to the school by an engine company. Firefighters describe their jobs and demonstrate their equipment. Additionally, they present Junior Firefighter patches or badges to students who have completed the program.

School Evacuation Drills

An effective fire prevention program in schools should be a matter of high priority not only to the fire department and school staff but to the community as a whole. A proper program will include regular and thorough fire safety inspections of the school premises by school staff and the fire department as described throughout this chapter. Additionally, the program must include emergency evacuation drills. Although such drills are usually referred to as "fire" drills, it must be emphasized that emergency evacuations can be necessary for other occurrences as well—bomb threats and natural disasters, to name two. Many states and local communities have adopted statutes requiring evacuation drills to be conducted at designated intervals, generally twice per school year. Where no such mandatory requirement exists, the fire department should work with school administrators to ensure that drills are conducted voluntarily.

Emergency exit drills are necessary to ensure that students and staff can safely, efficiently, and effectively evacuate their buildings in times of stress. Order and control are the key elements that will help to prevent injuries during the evacuation. A good emergency evacuation plan and drill will include the following points:

1. Evacuation drills should be conducted as often as necessary to ensure that all occupants of the building are thoroughly familiar with the process. The fire department should be present whenever possible but at least once per school year.
2. The focus of the drill should be placed on disciplined control and order. Speed, while important, should not be overemphasized. Speed is a by-product of properly planned and supervised evacuation.
3. Specific exits should be assigned to groups of classrooms; for example, Classrooms 1, 2, and 3 should use Exit A and Classrooms 4, 5, and 6 should use Exit B. All building

exits should be utilized to provide for the even distribution of exiting occupants. Alternative exits should also be designated.

4. Students and teachers should proceed to predesignated assembly points outside and away from the building. Playgrounds are appropriate; parking lots are not. Teachers should be required to account for all students in their classes. Taking roll call once assembled outside is the preferred method. They should report the status of their classes to the school administrator in charge. Missing and unaccounted for children must be reported to responding firefighters.
5. Emergency evacuation plans should be drawn in graphic form and posted in each classroom and at various locations throughout the school. **Figure 21-50** shows such a plan.
6. To familiarize students and staff with the sound, the fire alarm system should be used whenever conducting an evacuation drill.

Adult Programs

The types of programs and lessons taught to adults are limited only by the imagination of fire service educators. Common themes include:

- Home fire safety
- Smoke detector placement and maintenance
- Drownproofing
- Fire extinguisher demonstrations and instruction for use
- Operation EDITH (Exit Drills In The Home)
- CPR instruction
- General first aid
- Hazardous materials awareness
- Earthquake, tornado, and flooding preparedness

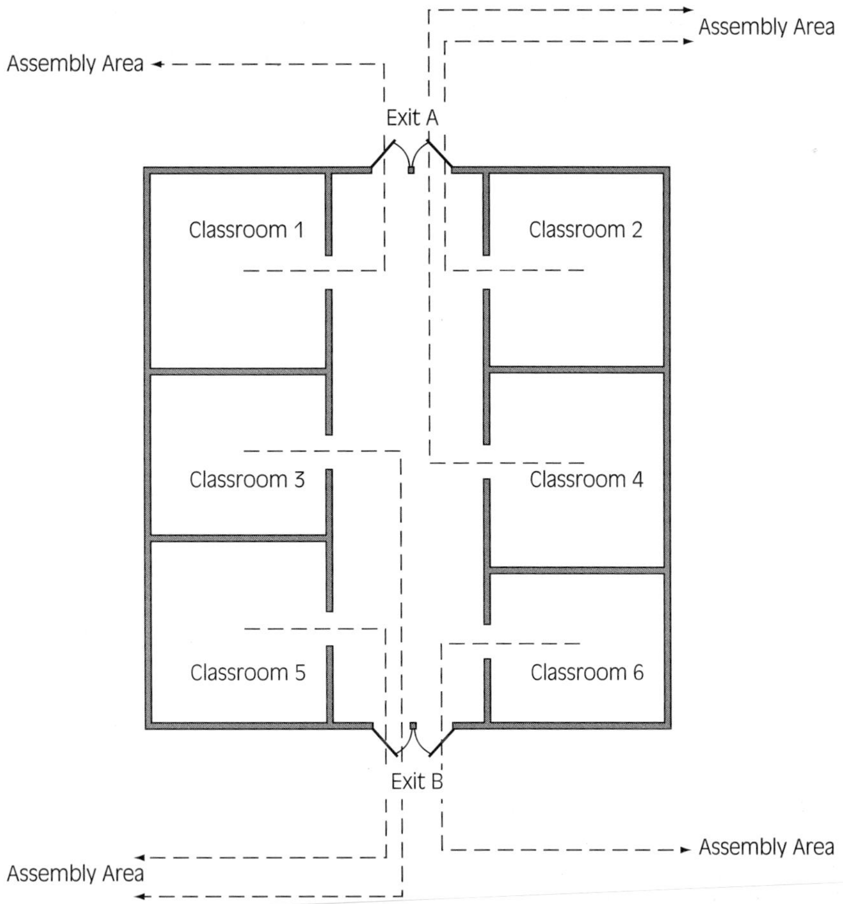

Figure 21-50 A graphic view of the school building showing exits, evacuation routes, and assembly areas should be posted in all classrooms and throughout the building.

Fire Station Tours

Visiting a local fire station has always been, and continues to be, one of the most popular field trips for all types of clubs, organizations, and schools. Station tours not only give the visitors a chance to see firefighting equipment and apparatus up close, but also give them a glimpse into life in a fire station. From the firefighters' perspective, tours provide an excellent opportunity to "spread the word" about fire and life safety.

Fire station tours should be tailored to fit the needs and interests of the visitors. Young children may be interested in trying on a firefighter's safety equipment.

Caution Helmets should not be placed on very small children; neck injuries could result.

Children might also enjoy sitting in the engine's cab or watching a firefighter "slide the pole." Older children and adults may be interested in the operation of a pumper or dispatch and response procedures.

Fire safety messages should be interwoven into the station tour and directed at the appropriate level. Small children should be asked or taught how to call for emergency assistance, to "stop, drop, and roll," or how to properly react to an operating smoke detector and exit their homes safely. Older visitors should be given information regarding such issues as smoke detector and carbon monoxide placement and maintenance, home evacuation planning and drills, and general fire safety. This information should be presented orally during the tour and reinforced with written literature distributed at the end of the visit, **Figure 21-51**. It is generally not good practice to give visitors brochures and flyers during the tour because this will serve as a distraction. Firefighters should also take the opportunity to inform visitors about the specialized educational programs offered by their department. Giving visitors a printed schedule of upcoming events and a telephone number for registration or further information is an excellent method for encouraging participation.

Figure 21-51 Fire safety brochures available through the NFPA.

PRE-INCIDENT MANAGEMENT PROCESS

Fire prevention activities and pre-incident management, often referred to as preplanning, are related and yet different. From a fire prevention perspective, one assumes an incident *can* occur and seeks to take measures to ensure that it does not. Pre-incident management assumes an incident *has* occurred and utilizes tactics, strategies, and coordination of resources to minimize the impact on lives and property. The link between the two is education. Fire service educators strive to teach the public how to prevent fires and other emergencies, while at the same time teaching the proper action to take should an emergency occur.

Pre-incident management can be as simple as company officers deciding from which hydrant they will lay hose for a particular building, or as complex as a coordinated effort between many agencies and many jurisdictions. Regardless of its scope, pre-incident management must be a collaborative effort between all divisions of the fire department and involve other agencies when appropriate.

Deciding to Preplan

In a perfect world, all structures in a given jurisdiction would be preplanned with the information stored in a massive database. That information would be available to emergency responders via mobile data terminals located on all apparatus. While such systems currently exist, they are often out of the financial reach of many fire agencies—at least for the immediate future. Therefore, most data gathered during the pre-incident management process is stored in "hardcopy" form. For practical reasons—storage, staffing, and time constraints—fire agencies must prioritize occupancies to be preplanned. In so doing, consideration should be given to the following:

- Type of occupancy
- Type of incident expected
- Life hazards—civilian and firefighter
- Nature of activities conducted at the occupancy
- Exposure to surrounding areas
- Complexity of firefighting operations
- Resources required

Structures such as high-rise buildings, hotels, malls, large industrial buildings, and multistory and multibuilding apartment complexes should be given high priority. Such occupancies are often referred to as **target hazards**, indicating that a greater than average life hazard or complexity of firefighting operations can be expected.

Site Visit

Generally, after deciding to preplan a structure, the first step is the on-site visit or inspection. The site visit for pre-incident management purposes differs from a fire prevention inspection in that the firefighters' purpose is fact gathering from an operational strategic perspective as opposed to ensuring compliance with fire code regulations. Certainly if code violations are noted, they should not be ignored but rather dealt with in the usual manner.

During the pre-incident management site visit, firefighters should be gathering information that will enable emergency responders to deal effectively with all levels of situations at the site. At a minimum, the following information should be obtained and documented:

- Occupancy classification
- Construction type and method
- Structure size, height, and number of stories
- Exiting systems
- Built-in fire protection
- Access points to the site and interior of the structure
- Exposure problems
- Hazardous materials usage and storage areas
- Personnel—civilian and firefighter—safety issues and features
- General firefighting concerns

Diagrams

An essential element of the pre-incident management process is the diagram of the site, structure, or occupancy. Here is where the information gathered during the site visit is displayed in graphic form. Diagrams of the site and floor plans should be included and should be presented in both **plan** and **sectional views** using a standardized set of symbols, **Figure 21-52**, and drawn to scale.

Diagrams can be hand drawn but the preferable method is to produce them using a personal computer. Sophisticated computer-aided drawing software need not be used. A simple paint or draw application will do the job. Updating the diagram will be a simple process if the original is produced in this manner.

The site plan should include the perimeter of the building, surrounding roadways, access points to the site and structures, fire hydrants, water main sectional valves, fire sprinkler control valves and connections, cross fencing, gates or other obstacles that may impede the movement of emergency vehicles, staging areas, and likely apparatus placement locations.

Floor plans should include a general layout of the interior of the building, floor by floor. Areas of high life hazard, exiting systems, fire protection features, hazardous materials use and storage areas, type of construction, roof openings, stairs and elevators, and other pertinent information can be shown on this drawing.

Seek Input from Others

Structures in need of preplanning have been identified and prioritized, site visits have been conducted, and diagrams have been completed. Now what? As suggested earlier, the pre-incident management process is a collaborative effort involving all levels and divisions of the fire department and often other agencies. All relevant information obtained should be assembled in narrative and graphic form and routed as appropriate to obtain

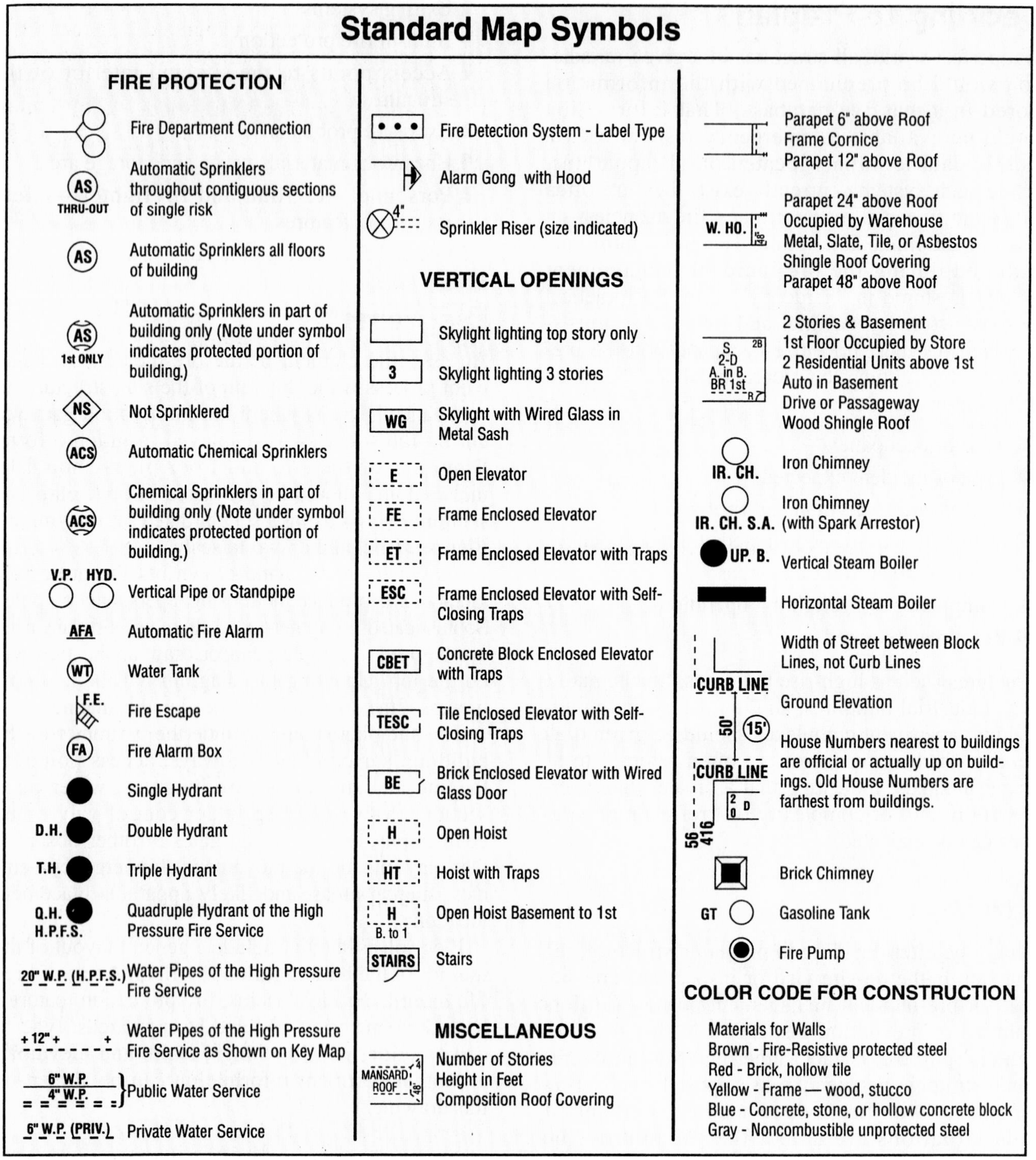

Figure 21-52 Standardized symbols for preplanning.

the perspective, expertise, and input of others. For example, in urban areas, the water department may suggest that, due to anticipated volume of fire streams during a full-scale firefighting operation, the water pressure in city mains be boosted. Law enforcement may have input regarding anticipated traffic problems. The Red Cross or other disaster relief agency may have suggestions regarding evacuation points or temporary shelter locations. In short, the pre-incident management process is not complete until everyone with potential involvement in an incident has had the opportunity for input.

The Finished Document

What should happen to the finished document? Certainly, preplans serve no purpose if they are stored on a hard drive on someone's PC or in a binder on a shelf. They should be carried in command vehicles and on all apparatus. Beyond being carried on vehicles, they should be practiced. Periodic training at specific locations using preplans enables firefighters not only to practice their skills, but to become familiar with facilities in their jurisdictions and to update pre-incident management plans when necessary.

Lessons Learned

It is hard for a firefighter to imagine anything more exhilarating and satisfying than aggressively entering a burning structure, rescuing its occupants, and defeating the sworn enemy—uncontrolled fire. Or perhaps pulling a drowning child from a swimming pool and watching him respond as a firefighter applies lifesaving techniques. Unless it would be responding to a "fire out on arrival" call because a smoke detector that the homeowner had installed at the fire department's recommendation had alerted his sleeping family to the fire downstairs in the kitchen. Or reading in the morning newspaper about the child who escaped serious injury when a fireworks sparkler ignited her clothes and she "stopped, dropped, and rolled" because the firefighters who came to her school had taught her how. There will always be emergencies and disasters. Preventing or lessening the impact of these devastating events must be the firefighter's constant goal.

KEY TERMS

Administrative Warrant An order issued by a magistrate that grants authority for fire personnel to enter private property for the purpose of conducting a fire prevention inspection.

Fire Hazard Any condition, situation, or operation that could lead to the ignition of unwanted combustion or result in proper combustion becoming uncontrolled.

Means of Egress A safe and continuous path of travel from any point in a structure leading to a public way. Composed of three parts: the exit access, the exit, and the exit discharge.

No-Knowledge Hardware Locking devices that require no key or special knowledge to operate.

Panic Hardware Hardware mounted on doors that enable them to be opened by pushing from the inside.

Plan View A drawing or diagram of a building or area as seen from directly overhead. May include a site plan or a floor plan.

Pre-Incident Management Advance planning of firefighting tactics and strategies or other emergency activities that can be anticipated to occur at a particular location. Often referred to as preplanning.

Safety Container A storage can that eliminates vapor release by using a self-closing lid. Also contains a flame arrestor in the dispenser opening.

Secondary Containment Any approved method that will prevent the runoff of spilled hazardous materials and confine it to the storage area.

Sectional View A vertical view of a structure as if it were cut in two pieces. Each piece is a cross section of the structure showing roof, wall, horizontal floor construction, and the location of stairs, balconies, and mezzanines.

Special Egress Control Device Door hardware that will release and unlock the door a maximum of 15 seconds after it has been activated by pushing on the bar.

Target Hazard An occupancy that has been determined to have a greater than average life hazard or complexity of firefighting operations. Such occupancies receive a high priority in the pre-incident management process and often a higher level of first-alarm response assignment.

REVIEW QUESTIONS

1. List the three Es of fire prevention.
2. What are the objectives of a fire company fire prevention inspection program?
3. Identify the first step in conducting a fire prevention inspection after arrival at the occupancy to be inspected.
4. Identify the steps to be taken in the event that permission to inspect is denied by the business owner or representative.
5. Define a fire hazard.
6. Identify the three components of a means of egress.
7. Where is panic hardware required?
8. Discuss the key areas of concern to the fire inspector when inspecting a fire sprinkler system.
9. When are commercial cooking surfaces and equipment required to be protected by a fixed fire extinguishing system?
10. Is it necessary to document a deficiency noted during a fire prevention inspection if it was corrected in the presence of the inspector?
11. When conducting a home fire prevention and safety inspection, the firefighter should be aware of several typical fire hazards found in the home. Identify three.
12. Should fire department public education efforts be tailored to meet the needs of the community and not be limited to fire-related issues?

Additional Resources

Halon Recycling Corporation http://www.halon.org

NFPA Fire Prevention Handbook http://www.NFPAcodesonline.org

Klinoff, Robert, *Introduction to Fire Protection.* Delmar Learning, part of the Thomson Corporation, Clifton Park, NY, 1997.

United States Fire Administration, http://www.usfa.fema.gov

Occupational Safety and Health Administration, http://www.osha-safety-training.net

CHAPTER 22

EMERGENCY MEDICAL SERVICES

Andrea A. Walter, Sterling Park Rescue Squad

OUTLINE

- Objectives
- Introduction
- Roles and Responsibilities of an Emergency Care Provider
- Safety Considerations
- Assessing a Patient
- Cardiopulmonary Resuscitation/AED
- Bleeding Control and Shock Management
- Emergency Care for Common Emergencies
- Lessons Learned
- Key Terms
- Review Questions
- Additional Resources

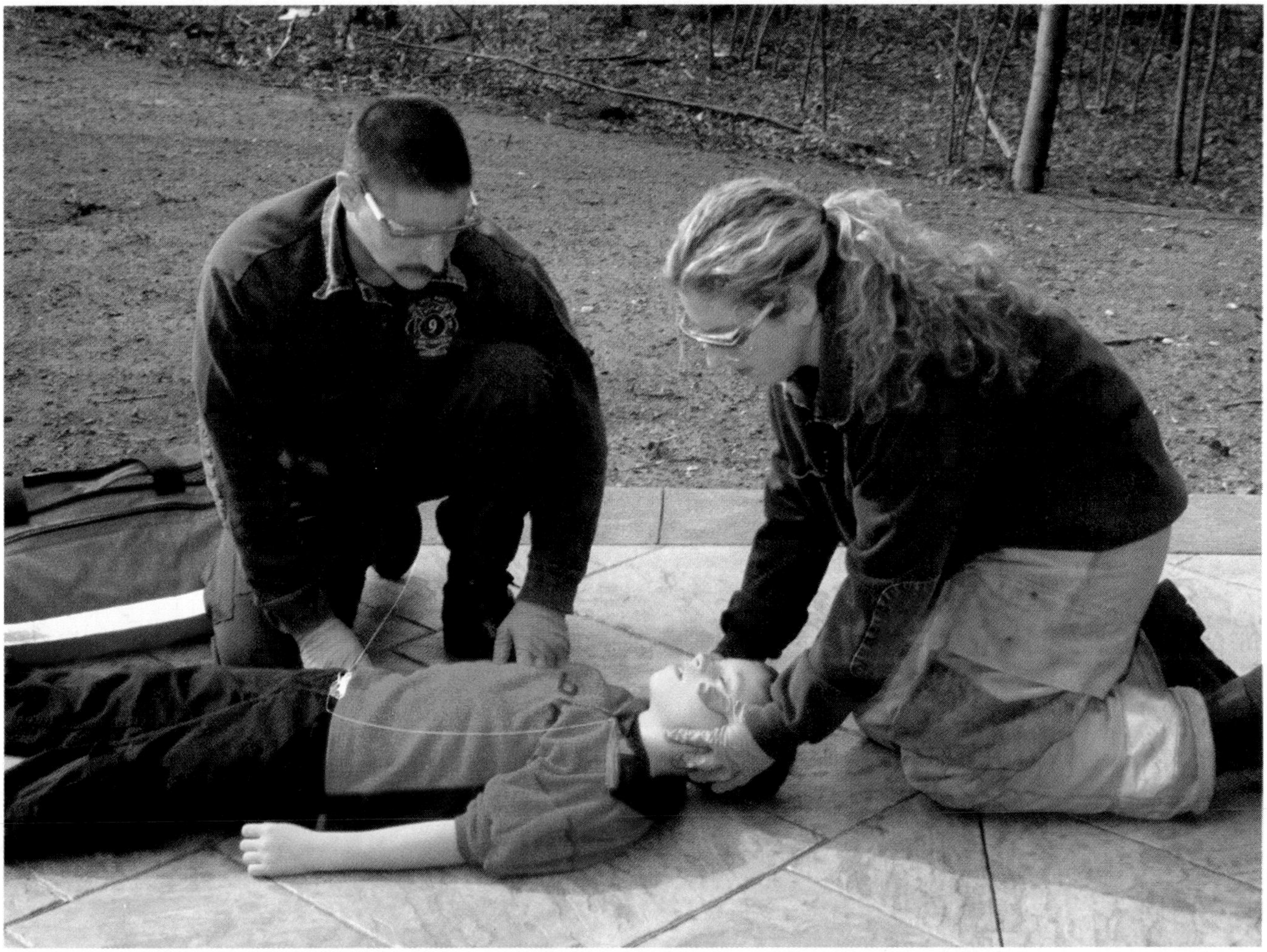

STREET STORY

During the past twenty-two years, I have seen many occasions where EMS training of firefighters has been essential to the success of incidents—especially to the patients involved. There have been many times where an engine I was on arrived at the scene of a collision before an ambulance. There were times when we pulled victims from buildings and no EMS units were available. There have been other rescue situations—in a storm drain, in a steam tunnel, in overturned vehicles—where EMS was on the scene but did not have the proper personal protective equipment to enter the hazardous area to provide care. Still, one incident stands out in my mind as a clear example of why firefighters should have EMS training.

It was a clear weekday afternoon in Fairfax County, Virginia. It had been a typical day with typical calls. It was rush hour—traffic was at a standstill; no one seemed to be moving. A vehicle fire was dispatched, a short distance from Fire Station 34. I knew that Engine 34 and Ambulance 34 were out of service for HAZMAT training and that the next due engine (Fairfax City Engine 33) was about 4 miles away and would have to battle traffic. I was about 3 miles away, just leaving headquarters, but could see the smoke—a billowing, black loom-up. I decided to roll in, just to watch.

Squad 18, a heavy squad with a crew of three, responded from Fire Station 34 and arrived within a couple of minutes. As I recall, their size-up was "a stake body truck well involved." No big deal, I thought. Then, after a minute, they requested an ambulance for a burned subject. Then they requested another, along with two medic units. Then they requested a helicopter. What seemed at the start to be "no big deal" ended up to be a serious situation with five civilians critically burned.

The crew from Squad 18 began treating the patients. I assisted when I arrived, as did the crew from Engine 33. As ambulances and medic units began to arrive, they took over patient care, but the lifesaving measures had already been taken. Yes, the fire was eventually extinguished, but our job as firefighters is life safety first, then property conservation. Thanks to our efforts, all five patients survived.

If we had not been trained to provide EMS care, there could easily have been five fatalities at that scene. Further, with rush hour traffic, there were hundreds (if not thousands) of people who could have perceived that firefighters are only trained to put "wet stuff on red stuff" and that injured victims are of no concern to us. That is not the image the fire service should portray.

The truck was a total loss even before the fire department was called. The five victims, employed by a landscaping company, were riding in the back of the truck when one tried to light a cigarette. Unable to do so because of the wind, he knelt down or bent down and lit it—right over a can of gasoline. The victims were burned from the flash fire that occurred.

Ever since this incident, I have made sure that I have kept my EMS training current, and as a chief officer I have made sure that all personnel under my command have had the opportunity and encouragement to become EMS-trained. Patient care is a key part of what we do as firefighters, and we cannot do it effectively if we are not trained.

—Street Story by Gordon M. Sachs, Chief, Fairfield Fire and EMS, Fairfield, Pennsylvania

OBJECTIVES

After completing this chapter, the reader should be able to:

- Explain the basic elements of an emergency medical system.
- Describe and practice the principles of infection control and body substance isolation for all patients.
- Perform an initial assessment on all patients, obtain vital signs, and conduct a focused history and physical exam for signs of illness and/or injury.
- List the different types of bleeding in patients, demonstrate methods for controlling the bleeding, and be able to treat patients in shock.
- Describe types of burns and ways to treat them.
- Identify the effects of ingested poisonous or controlled substances, how to contact a poison control center, and how to treat patients who have been exposed to a caustic substance.

INTRODUCTION

"When I became a firefighter, I expected to be fighting fires on every shift. I never anticipated that my engine company would respond to more emergency medical calls than fires last year. I am certainly glad that I received training in emergency care in firefighter school, because I have really helped save lives. There is more to being a firefighter than putting water on fire!"

It is true. There is more to being a firefighter than "putting the wet stuff on the red stuff." Firefighters are community defenders and trusted public saviors. When people do not know who else to turn to, they call the fire department. The community relies on firefighters to be skilled at fighting fires and to be creative problem solvers. That means firefighters need to know more than just basic firefighting techniques. That is where emergency medical care comes into the picture, **Figure 22-1**.

Figure 22-1 Emergency medical services star of life.

Emergency medical responses constitute more than 50 percent of total emergency responses for many fire departments all across the country. In some jurisdictions, emergency medical calls make up 75 to 80 percent of the fire department's total emergency responses per year.

This chapter covers very basic material that firefighters will need to know based on the NFPA 1001 standard and also covers standard first-aid practices. A wide variety of training courses in emergency medical services (EMS) are available that allow firefighters to increase their EMS knowledge and skills, and all firefighters are encouraged to continue their EMS training.

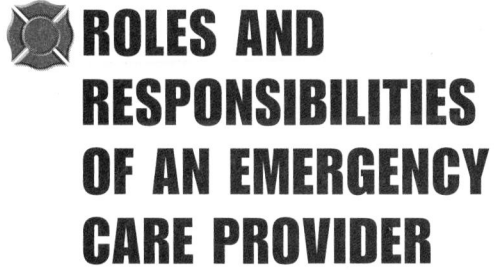

> **Firefighter Fact** Emergency medical care has evolved into an essential part of the fire service.

With the technology available today, firefighters can deliver lifesaving techniques to stabilize patients until emergency medical technicians and paramedics arrive, **Figure 22-2**.

> **Firefighter Fact** Firefighters are a critical, lifesaving link in the emergency response community.

ROLES AND RESPONSIBILITIES OF AN EMERGENCY CARE PROVIDER

As a part of the emergency response system, firefighters often act as providers of emergency care and first aid. Every emergency service delivery system is different, yet many call on firefighters to assist as the first arriving emergency unit on the scene or as a helping hand to the emergency medical technicians and paramedics. In any case, a working knowledge of the basics of emergency medical care is an important part of a firefighter's training.

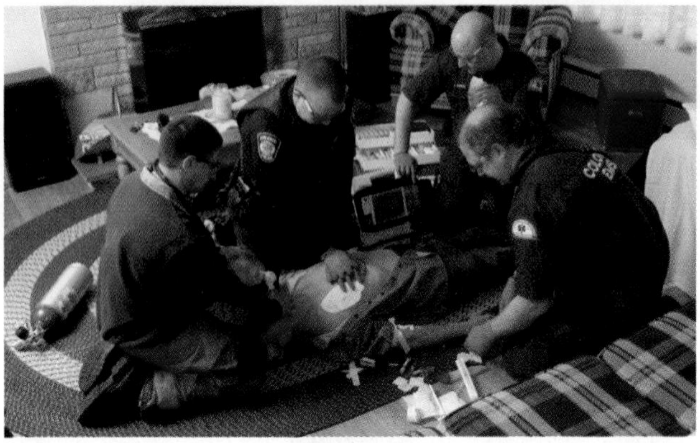

Figure 22-2 Firefighters are often called on to assist EMS crews with patient care.

Key Responsibilities

Firefighters have several key responsibilities when responding to emergency medical incidents:

- To ensure their own safety, the safety of their team, and the safety of the patient.

> **Safety** The most important factor to consider when firefighters respond to emergency medical calls is safety, for themselves, their team members, and their patient.

- To act safely from the minute they step on the fire engine until every piece of equipment is clean or sterilized and placed back on the fire engine, **Figure 22-3A–C**.
- To act in a professional manner at all times. People who have called for assistance are counting on firefighters to help them or their friends or family members. Firefighters must always be respectful and considerate, **Figure 22-4**.
- To never cause a situation to become worse or act beyond the bounds of their training.

> **Ethics** Emergency medical training is not only about what firefighters can do to assist a patient in crisis, it is also about recognizing what is beyond an individual firefighter's training and ability. If firefighters recognize that a situation is beyond their capabilities, and they call for the appropriate assistance and act within the limits of their training, they have done the best job they can for their patient.

- To practice and update emergency care and first-aid skills with training. The emergency medical field is constantly changing. That means firefighters may have to learn new skills,

(A)

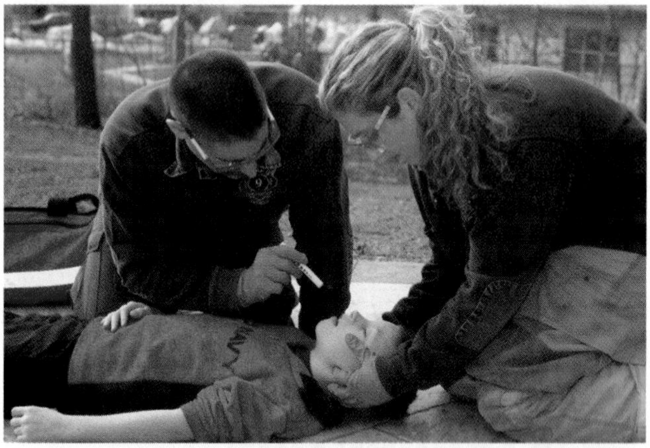

(B)

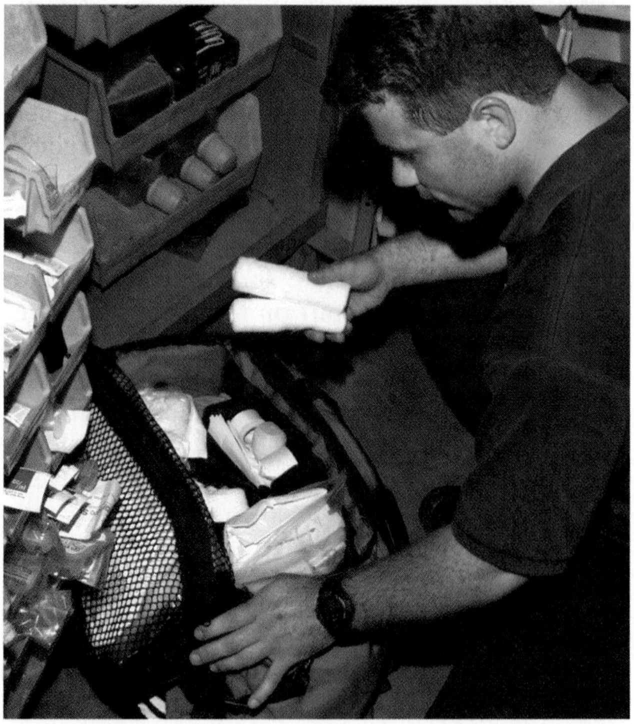

(C)

Figure 22-3 (A) EMS calls begin when firefighters leave the station, (B) continue throughout the time on the scene, and (C) do not end until firefighters have cleaned the equipment and restocked the supplies.

Figure 22-4 Firefighters should always treat patients with respect. They should treat patients as they would want to be treated in the same situation.

Figure 22-5 Firefighters should train on new EMS skills and continue to practice skills they have already learned. Practice makes skilled responders.

as well as practice the old skills, **Figure 22-5**. Training for emergency medical responses and maintaining current knowledge and skills will allow firefighters to be ready for any emergency medical incident that occurs.

- To know the EMS equipment and maintain it properly. While emergency medical care may require the use of a firefighter's brain and hands, it also requires specialized equipment and tools. Firefighters should be familiar with each piece of EMS equipment and all supplies they carry. Also, equipment needs to be maintained properly. It should be cleaned after each use (or discarded properly if it is disposable) and checked at the beginning of each shift, **Figure 22-6**.
- To gather important information. Information about the patient on an emergency medical call

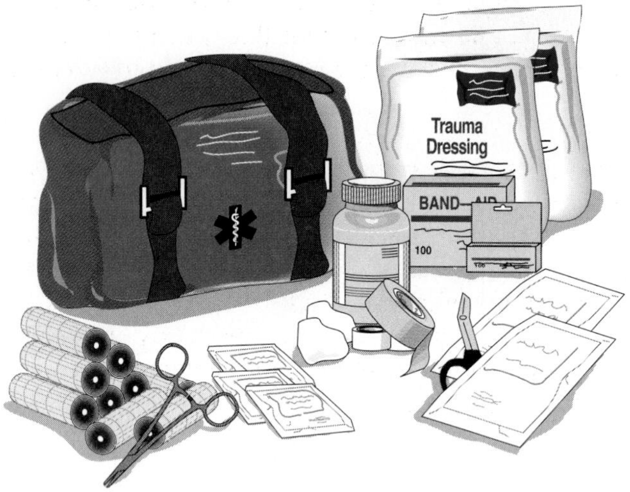

Figure 22-6 Firefighters should check their EMS equipment and supplies at the beginning of each shift.

can be critical for treatment. The firefighter must collect all important information about a patient before deciding on a course of action. Then, when EMS personnel arrive, the firefighter should transfer that information so that the EMS personnel can continue with appropriate care in the ambulance and in the hospital. Collecting patient information is discussed later in this chapter.

Legal Considerations for Emergency Care Providers

Emergency care providers should understand several important legal issues. First is the principle called **standard of care**, which is a legal term that means for every emergency medical incident, an emergency responder should treat the patient in the same manner as would another emergency responder with the same training. In short, this legal principle is what is used in court to demonstrate that the actions of an emergency care provider were adequate or inadequate. This legal issue emphasizes the importance of firefighters acting within the bounds of their emergency care and first-aid training and doing what they have been trained to do, and *only* what they have been trained to do.

Another important term to understand is **consent**. Emergency responders should always attempt to gain consent from the patient before beginning treatment. In the case of children or minors, consent should be obtained from a parent or legal guardian. Emergency responders should not withhold treatment from minors when a parent or guardian is not present. Emergency responders can begin treatment while making every effort to contact the parents or guardians to obtain permission.

Patients who are unconscious when emergency responders arrive are considered to be giving **implied consent** for emergency medical treatment. Patients who are not unconscious, and are mentally capable, have the right to refuse treatment.

Abandonment occurs when an emergency care provider begins treatment of a patient, and then leaves the patient or discontinues treatment prior to the arrival of an equally or higher trained responder.

Caution If firefighters begin patient care, they must remain with the patient until another emergency responder arrives whose training is equal to or higher than theirs.

Interacting with Emergency Medical Services Personnel

As emergency care providers, firefighters interact with emergency medical service personnel while on the scene of an emergency. Emergency medical transport is done in a variety of ways. An ambulance staffed with EMTs and paramedics may come from the firefighters' own fire department, it may come from a separate rescue squad or EMS organization, or it may even come from a private transport company or a hospital.

In some cases, firefighters may be interacting with emergency medical personnel who have arrived by helicopter, commonly referred to as a medi-vac. **Medi-vac** units are staffed by nurses, paramedics, and EMTs who are usually skilled in treating patients with traumatic or serious injuries. A medi-vac unit can be seen in **Figure 22-7**.

It is also important to understand the different levels of care that are provided by EMS crews who may arrive on the scene. An ambulance may provide basic life support (BLS) care, which is the primary level of EMS care, or it may provide advanced life support (ALS) care, which involves much more aggressive treatments for patients and specialized monitoring.

In any case, the arriving EMS personnel will rely on the firefighters on scene to give them critical patient information. What is important patient information? It varies from situation to situation and requires firefighters to use their best judgment, but a few basic pieces of information are important for every call: the patient's age, the patient's medical history, known drug allergies of the patient, medications the patient is taking, the circumstances sur-

Figure 22-7 Medi-vac helicopters are used to transport patients to specialty care hospitals.

rounding the illness or injury the patient is reporting, and the patient's vital signs, **Figure 22-8**.

Many other pieces of information can be helpful, and firefighters should consider gathering such information during an emergency medical call, if feasible: the patient's name and address, what condition the patient was found in, the patient's surroundings (example: condition of an automobile in a car accident), and the treatment that was given to the patient prior to the EMS personnel arriving on the scene.

Figure 22-8 Firefighters should record important information for arriving EMS personnel and for their agency's patient information reports. Effective documentation is critical. *(Photo courtesy of Fred Schall)*

All firefighters need to check with their department about their responsibility for documenting an emergency medical response. Many fire departments require that a patient care report or an emergency call record be completed for every emergency medical response. Thorough and accurate documentation of emergency medical responses will help protect the department in the event of legal action.

SAFETY CONSIDERATIONS

One of the most important considerations when performing emergency care and applying first-aid skills is the safety of the caregiver or individual firefighter performing patient care and the safety of the emergency response team, such as an engine company crew responding on an emergency medical call.

> **Safety** Safety, including **infection control**, should be the primary consideration of every emergency responder.

Safety and infection control should also remain a primary consideration for the duration of the emergency run and even afterwards in the cleanup phase of the incident. *Remember:* Emergency responders who get sick or injured are no longer a part of the solution; they are a part of the problem. Safety is the primary consideration!

Analyzing the Safety of the Emergency Scene

Firefighting training teaches firefighters to observe each and every scene for situations that may be unsafe to the public or to the firefighters present. For example, firefighters must check for overhead obstructions whenever raising a ladder to a structure and check the floor of a structure for stability before stepping on to it. Engine company officers must survey the condition of a burning building to make sure it is not on the verge of collapse before sending firefighters inside for an interior fire attack.

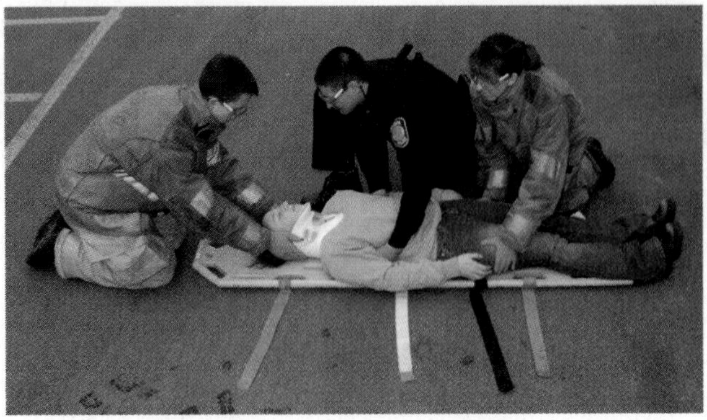

Figure 22-9 The firefighter should pay careful attention to the officer or the lead EMS provider to ensure the team is working together in patient care. Teamwork is important in all aspects of firefighting, including EMS.

While functioning on emergency medical calls, firefighters must also check for scene safety prior to entering. There are a few important considerations for EMS scene safety. Firefighters should listen carefully to all information given prior to arrival on the scene. It is essential to plan actions ahead of time! Also, firefighters should listen to the direction of the officer or the EMS provider in charge of the incident so the team is functioning as one unit, **Figure 22-9**. It may be necessary to stage away from the scene until it can be stabilized by law-enforcement personnel.

Firefighters should carefully observe the emergency scene and people present before entering. Indications of violent behavior, the use or presence of weapons, or signs of controlled substance or alcohol use may all be initial indicators of a safety concern. If after entering the emergency scene a safety concern is discovered, it is acceptable to take the team out of the situation until law-enforcement officers have secured the scene or the dangerous individual or condition has left the scene.

The firefighter should be sure to ask for additional assistance when safety concerns are present, whether in the form of additional fire and emergency medical units or a law-enforcement officer. It is also important to keep the communications center informed of the situation and the team's status to ensure continued safety throughout the incident.

Firefighter Physical and Mental Health

Firefighting is a strenuous occupation, and firefighters are encouraged to maintain a healthy lifestyle to meet the demands of firefighting work.

Emergency medical calls for assistance can also be strenuous work, both physically and mentally.

> **Note** A healthy lifestyle will help firefighters in their ability to respond to emergency medical calls for assistance as well as in typical firefighting activities.

A "healthy lifestyle" is different for each firefighter but generally includes regular exercise, a proper diet, and getting the right amount of sleep. Firefighters should contact their personal physician or department health care provider for assistance with developing a healthy lifestyle that meets their needs.

One of the most common problems for firefighters on emergency medical responses is back injuries. Firefighters should learn and practice proper lifting techniques (lifting with the legs, keeping the weight close to the body, etc.) when moving patients or emergency medical equipment.

Emergency medical calls can also take their toll on the mental health of a firefighter or emergency medical responder. Firefighters should consult their department's policies on critical incident stress debriefing or other intervention techniques available for firefighters who feel adversely affected by emergency medical responses.

Infection Control

Safety for the firefighter responding to emergency medical incidents involves not only scene safety and the prevention of injury but also protection from contracting **communicable diseases** or **infectious diseases**.

Prevention of Exposure to Infectious Diseases

The Centers for Disease Control and Prevention (CDC) is a federal organization that monitors outbreaks of infections and advises agencies on how to handle the situation and control the disease. The CDC, the Occupational Safety and Health Administration (OSHA), and the National Fire Protection Association (NFPA) have all written standards of practice or operational guidelines that firefighters should use when providing medical care.

Many laws and regulations have been enacted on the local, state, and national levels that assist in protecting emergency responders and health care providers from infectious diseases. In addition, first responders should follow any local or organizational policies for infection control.

Safety Firefighters should understand and follow the infection control guidance issued by their departments, their states, and appropriate federal agencies. These guidelines and precautions are developed to protect firefighters and first responders from infectious diseases.

Any time a firefighter treats a sick or injured patient, the risk of **exposure** to disease is present, both for the firefighter and for the patient. Firefighters should protect themselves and the patient from disease. The most effective means of reducing the risk of spreading disease is hand washing, **Figure 22-10.**

Proper hand washing is a necessity in the medical profession, and many types of antibacterial soaps are available for this purpose. Although soap is a great tool, the technique of washing is more important than the type of soap.

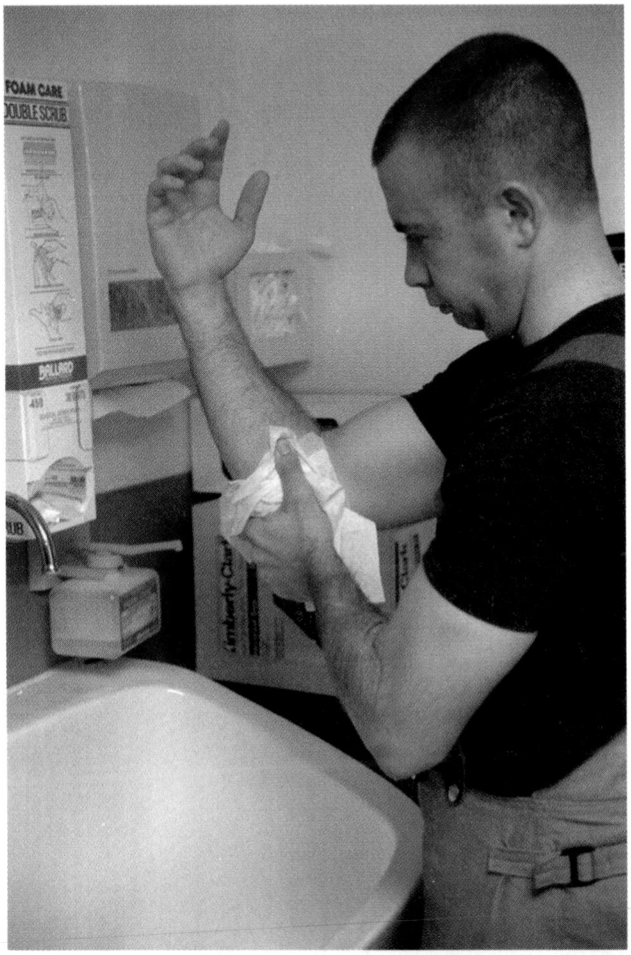

Figure 22-10 The most effective means of reducing the risk of spreading disease is hand washing.

Safety The industry standard for washing hands is vigorous lathering with soap and water for fifteen seconds (or longer) followed by thorough rinsing, **Figure 22-11.** A disposable towel should be used to thoroughly dry the hands, and the towel should be used to shut off nonautomatic faucets to ensure there will be no exposure to the organisms on a dirty faucet handle.

Frequent hand washing, although critical, may have a drying effect on the skin, and a moisturizing lotion may be necessary to combat the discomfort.

Firefighters must also be aware of the possibility of transmission of disease and sickness to their family and friends. Firefighters who do not practice proper infection control procedures can carry disease-causing microorganisms home with them to their families. As a general practice, firefighters should be extra careful never to wear soiled duty uniforms home. Washing soiled items in a separate station machine designed for cleaning contaminated articles can help lower cross-contamination to other washables. It is advisable to shower and change at the station before leaving for home. Likewise, firefighters who respond from a station should keep soiled uniforms and personal protective equipment out of living areas of the station.

Body Substance Isolation

Simple precautions, called **body substance isolation (BSI) precautions**, can reduce the odds of disease transmission significantly. BSI involves wearing proper protective equipment on every call.

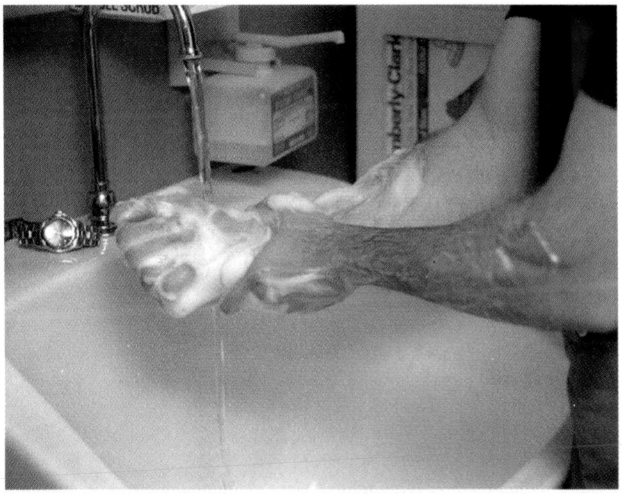

Figure 22-11 Proper hand washing involves vigorous lathering with soap and water for fifteen seconds (or longer) followed by thorough rinsing.

A person who is ill with a communicable disease will not necessarily appear ill and may not want to make the disease known to others. Therefore, firefighters should have an agency-approved face mask available and should wear the minimum of goggles or safety glasses and protective gloves any time a medical scene is entered. There is no exception to this rule!

Safety Appropriate BSI precautions must be taken for every patient.

The mainstay rule of emergency medical care is to treat all body fluids as if they were infectious, **Table 22-1.** If a substance is wet, it has the potential to be an infectious substance. Again, a firefighter should wear a minimum of gloves and eye protection on every call. A face mask can be added, if necessary, for splash potential. A nonabsorbent or plastic gown may be added for medical scenes with gross contamination potential, **Figure 22-12.**

Protective gloves are an important barrier device for obvious reasons but are effective only if they fit and are used properly. Firefighters must don new gloves for each patient contact. They must never use the same pair of gloves on more than one patient. Double gloving can be used for incidents with gross contamination potential. If

Potentially Infectious Body Fluids
Blood
Amniotic fluid
Vaginal discharge
Semen
Cerebrospinal fluid
Pleural fluid
Synovial fluid
Peritoneal fluid
Pericardial fluid
Fluids with little potential to transmit bloodborne diseases:
▮ Tears
▮ Nasal discharge
▮ Vomitus
▮ Sputum
▮ Saliva
▮ Feces
▮ Urine

TABLE 22-1

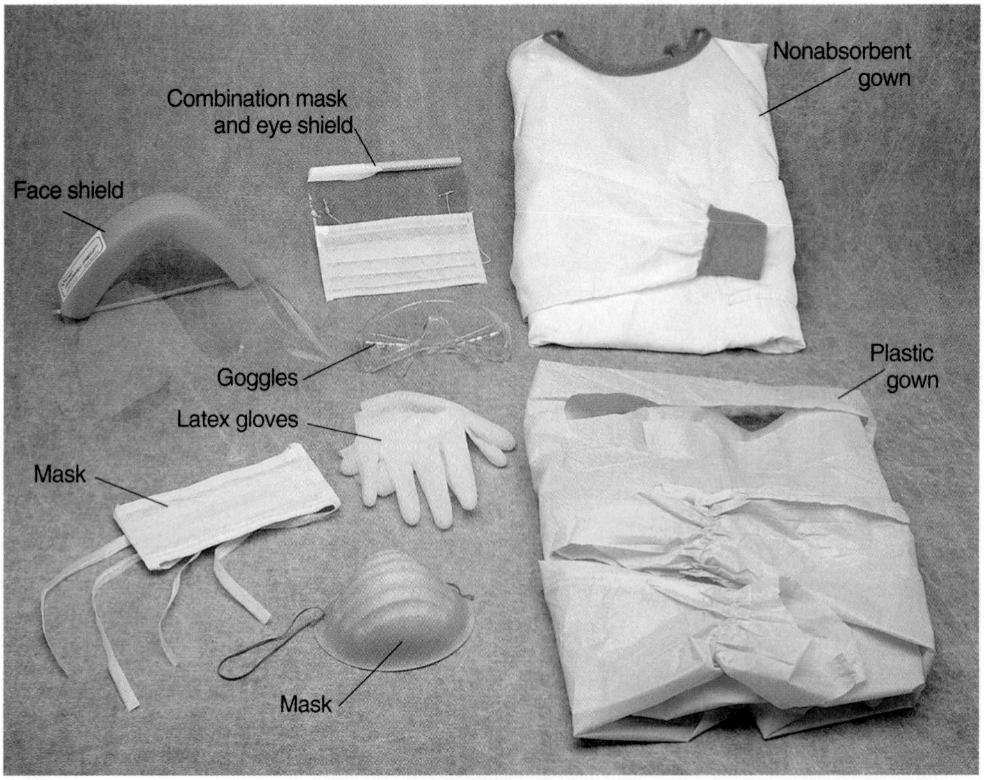

Figure 22-12 Types of personal protective equipment.

gloves fail during use, firefighters should stop immediately, wash their hands, and don another pair of gloves before returning to patient care. Gloves should only be removed after all patient contact possibilities are finished and should be removed without touching the skin with the outer layer of the glove, **JPR 22-1A–C.** Firefighters should dispose of gloves in a container clearly marked with the biohazard label and wash their wash hands after every use of gloves, **JPR 22-1D.**

JOB PERFORMANCE REQUIREMENT 22-1

Removing Gloves

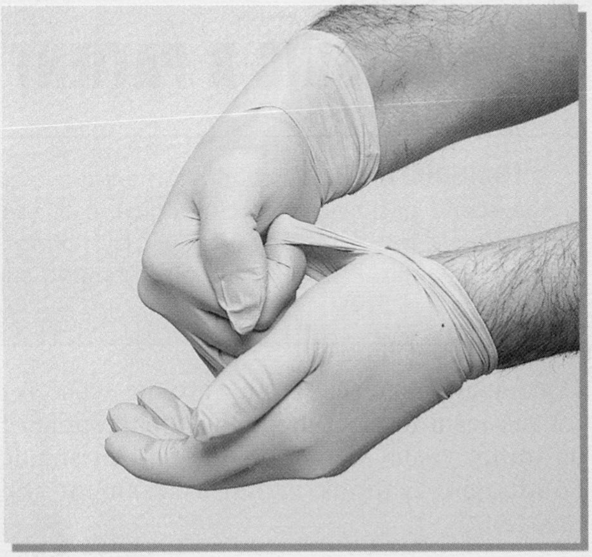

A Grasp the palm or outside cuff of the left glove with the gloved right hand.

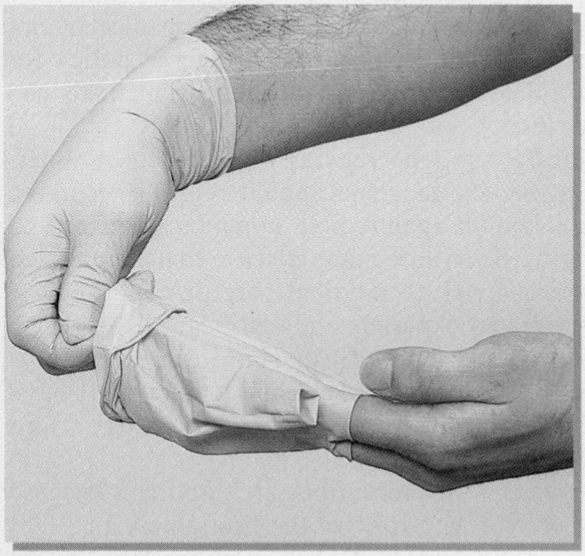

B Pull the left glove toward the fingertips. The glove should turn inside out as it is removed.

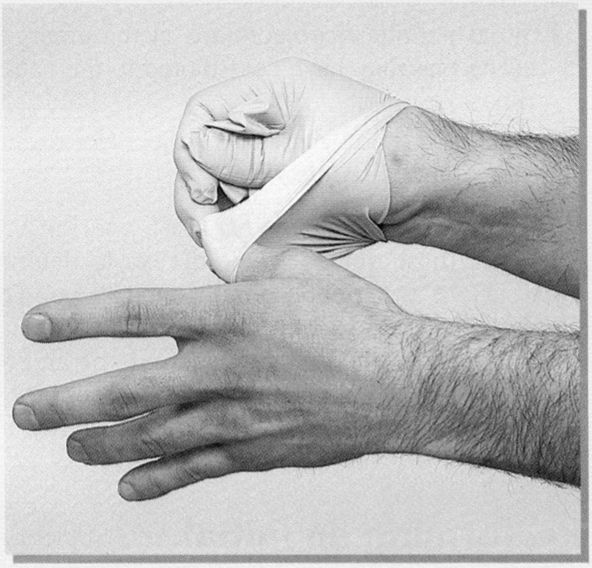

C Hold the removed glove in the still-gloved right hand. Insert the thumb or two fingers of the ungloved left hand under the cuff of the right glove, carefully avoiding any contaminated areas. Pull the right glove toward the fingertips, turning the glove inside out as it is removed. The soiled left glove should remain in the palm of the right glove as it is removed.

D Dispose of the gloves in a container clearly marked with the biohazard level and wash hands thoroughly.

Splashes of body fluid or chemicals can occur on emergency medical scenes, which is the purpose for eye protection in the BSI precautions. The eyes can provide a route for a disease microorganism to enter the firefighter's body. As stated earlier, goggles or safety glasses should also be used on every emergency call.

Any supplies or equipment used for treating a patient should either be disposed of in an appropriate biohazard container or cleansed and disinfected properly, **Figure 22-13.** Firefighters should wear gloves when disinfecting emergency medical equipment and should consult departmental policy for appropriate cleaning and disinfecting methods and procedures.

Firefighters who regularly participate in emergency medical incidents should consider receiving immunization against both common diseases and hepatitis B, an infectious disease that is the most dangerous risk to medical care providers. Firefighters who experience a significant exposure to blood or any body fluid should contact their supervisor immediately. Each agency will have policies concerning infection control and exposures.

In short, BSI precautions are important because they protect the safety of the firefighters and their families and friends. BSI precautions also protect patients with compromised disease-fighting systems from disease-causing microorganisms. A review of precautions for infection control is covered in **Table 22-2.**

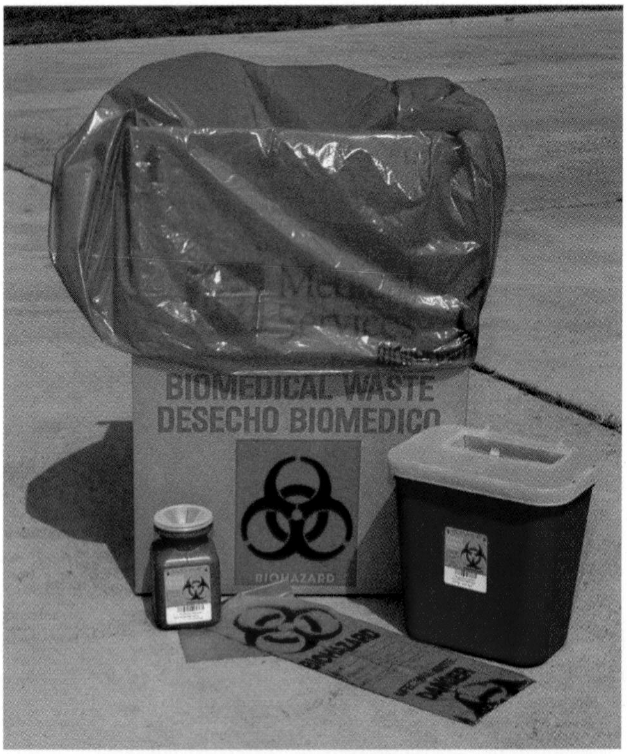

Figure 22-13 Examples of biohazard containers for disposal of infectious waste.

For more information on infection control, BSI, and exposure to infectious diseases, firefighters can contact the CDC or OSHA. Both agencies are a part of the federal government and will provide information as requested. More information is also available on their Web sites at http://www.cdc.gov and http://www.osha.gov.

ASSESSING A PATIENT

Streetsmart Tip When firefighters arrive on the scene of an emergency medical call, they can get a lot of information just by taking a quick look at the patient, the environment, and the situation.

Once a firefighter has made sure the scene is safe for the team to enter and donned appropriate BSI, an initial visual size-up of the patient should be conducted. A quick visual assessment should reveal:

- Is the patient awake?
- Is the patient in a harmful situation or environment?
- What position is the patient in?
- What people or objects are in the emergency scene that may have contributed to the patient's illness or injury?
- What is the skin color of a patient (for example, pale, red, bluish, etc.)?

Firefighters with many years of experience can often use these quick visual surveys to determine how serious a patient's condition may be. These quick visual surveys can provide important patient information in the first few seconds of an emergency responder's arrival, as well as recognize any safety concerns for the patient and the care providers.

Performing an Initial Assessment

After a quick visual survey of the patient and the scene is done, the next step is an **initial assessment**. An initial assessment is the initial investigative action taken by care providers to determine if the patient has the basic signs of life as

Standard Precautions for Infection Control

WASH HANDS (PLAIN SOAP)

Wash after touching **blood, body fluids, secretions, excretions,** and **contaminated items.**

Wash immediately **after gloves are removed** and **between patient contacts.**

Avoid transfer of microorganisms to other patients or environments.

WEAR GLOVES

Wear when touching **blood, body fluids, secretions, excretions,** and **contaminated items.**

Put on **clean** gloves just **before touching mucous membranes** and **nonintact skin.**

Change gloves between tasks and procedures on the same patient after contact with material that may contain high concentrations of microorganisms. Remove gloves promptly after use, before touching noncontaminated items and environmental surfaces, and before going to another patient, and wash hands immediately to avoid transfer of microorganisms to other patients or environments.

WEAR MASK AND EYE PROTECTION OR FACE SHIELD

Protect mucous membranes of the eyes, nose, and mouth during procedures and patient care activities that are likely to generate **splashes** or **sprays** of **blood, body fluids, secretions,** or **excretions.**

WEAR GOWN

Protect skin and prevent soiling of clothing during procedures that are likely to generate **splashes** or **sprays** of **blood, body fluids, secretions,** or **excretions.** Remove a soiled gown as promptly as possible and wash hands to avoid transfer of microorganisms to other patients or environments.

PATIENT CARE EQUIPMENT

Handle used patient care equipment soiled with **blood, body fluids, secretions,** or **excretions** in a manner that prevents skin and mucous membrane exposures, contamination of clothing, and transfer of microorganisms to other patients or environments. Ensure that reusable equipment is not used for the care of another patient until it has been appropriately cleaned and reprocessed and that single-use items are properly discarded.

LINEN

Handle, transport, and process used linen soiled with **blood, body fluids, secretions,** or **excretions** in a manner that prevents exposures and contamination of clothing and avoids transfer of microorganisms to other patients or environments.

Use **resuscitation devices** as an alternative to mouth-to-mouth resuscitation.

TABLE 22-2

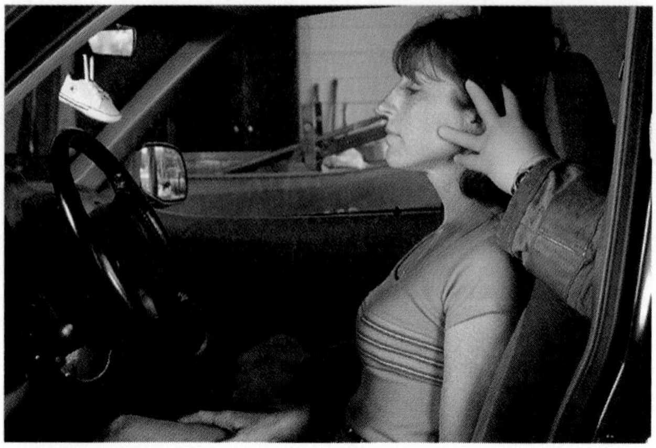

(A)

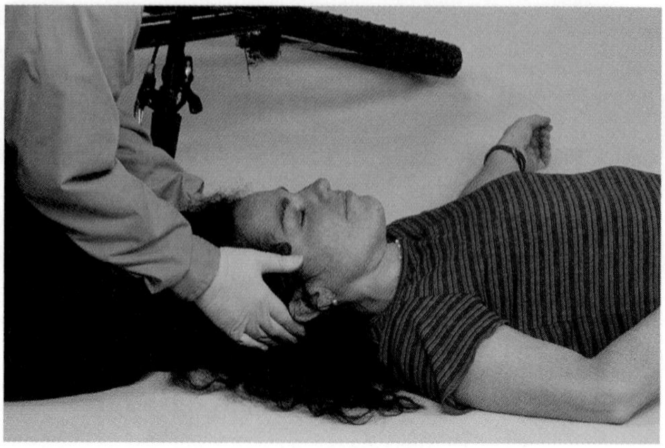

(B)

Figure 22-14 Firefighters stabilizing the head and neck of a (A) seated victim and (B) a victim lying on the ground.

well as any serious, life-threatening injuries. The initial assessment covers the following:

1. Level of consciousness of the patient (Is the patient awake?)
2. Airway (Does the patient have an open airway?)
3. Breathing (Is the patient breathing adequately?)
4. Circulation (Is the patient's heart pumping blood to the body adequately?)
5. Major bleeding (Is there any major bleeding?)

The initial assessment can be done very quickly on a conscious patient who is alert and talking. Patients who are unconscious will require closer examination to conduct the initial assessment.

When performing an initial assessment on a patient, firefighters should consider the following:

Level of Consciousness

Is the patient awake? If not, is the patient responsive to loud verbal commands or painful stimulation? After the firefighter determines the level of consciousness of the patient, another member of the team should maintain stabilization of the head and neck if the patient is either unconscious or has experienced a traumatic injury that may endanger the spine (automobile accident, fall, industrial accident, etc.).

To stabilize the head and neck of the patient, the firefighter places one hand on either side of the patient's head and holds firmly so that the head and neck are in a straight line with the body, **Figures 22-14A** and **B**. The purpose of stabilizing the head and neck is to keep them from moving so that if there is a neck injury, no further damage will be done to the neck or spinal cord.

Airway

The firefighter must ensure that the patient has an open airway. To do this for a patient without traumatic injury, the patient's airway is opened by placing the palm of one hand on the patient's forehead and the fingers of the other hand underneath the chin. The firefighter lifts the chin with the fingers and presses lightly on the forehead to roll the head into the open airway position, **Figures 22-15A** and **B.**

If the patient has a traumatic injury and there could be damage to the patient's neck, the firefighter will need to use a different method for opening the airway. To do this, one hand is placed on either side of the patient's head, placing the fingers along the curve in the jaw bone near the ear. Using the fingers, the jaw bone is pressed out to open the airway, **Figures 22-16A** and **B**.

The act of breathing brings air into the lungs where oxygen is passed into the bloodstream. Oxygen is vital to sustaining life. Many fire departments carry oxygen on their apparatus. Many patients experiencing difficulty breathing may benefit from the use of supplemental oxygen from cylinders that firefighters and EMS providers carry on their equipment. Note, however, that in very rare circumstances, the use of supplemental oxygen may worsen the patient's condition. Firefighters should consult their department's policies on the use of supplemental oxygen. Oxygen administration and the devices used to deliver oxygen to a patient are part of the curriculum of most higher level emergency medical training courses.

Breathing

When it is established that the patient has an open airway, or a firefighter has taken measures to open the airway, the next step is to check to see if the patient is

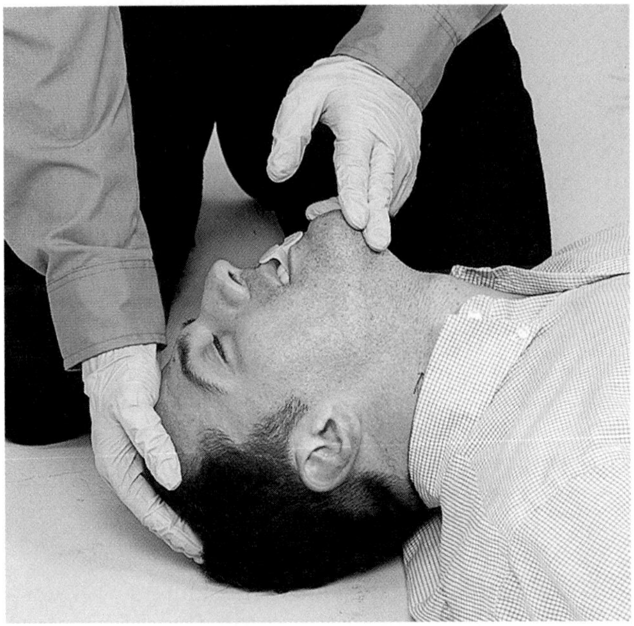
(A)

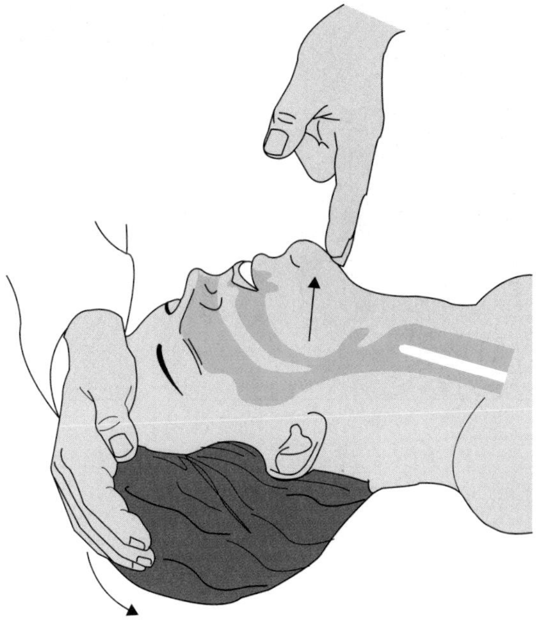
(B)

Figure 22-15 Opening the airway of a patient without traumatic injury using the head-tilt, chin-lift method.

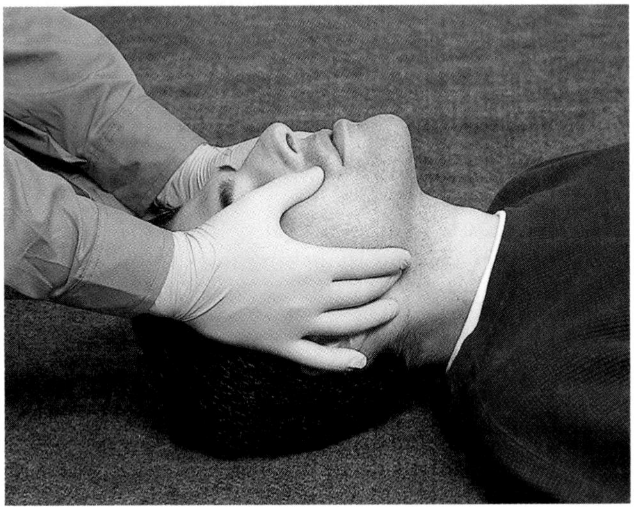
(A)

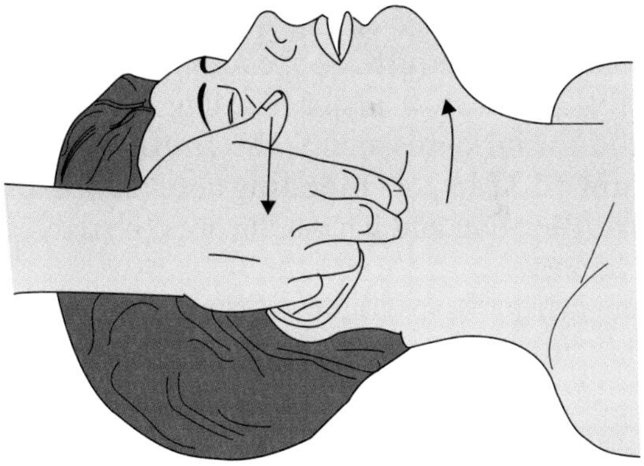
(B)

Figure 22-16 Opening the airway of a patient with traumatic injury using a jaw thrust.

breathing. If there is no breathing, the firefighter should provide rescue breathing for the patient.

Circulation

After the patient has an open airway and is breathing (or someone is breathing for the patient) the firefighter should check for a pulse in the patient. For an unconscious victim, the firefighter should locate the **carotid pulse** in the neck. On a conscious patient, the firefighter can use the **radial pulse** in the wrist to determine the pulse rate and quality. For a small child or infant, the best place to locate the pulse rate is the **brachial artery** in the inside of the upper arm. The pulse should be checked for 3 to 5 seconds to determine if it is present. The locations of these pulse points are shown in **Figure 22-17**.

> **Streetsmart Tip** Firefighters should never use a thumb to check a patient's pulse. The thumb has its own pulse, and firefighters may confuse their own pulse with the one they are trying to locate in a patient. It is best to use the index and middle fingers together to locate a patient's pulse.

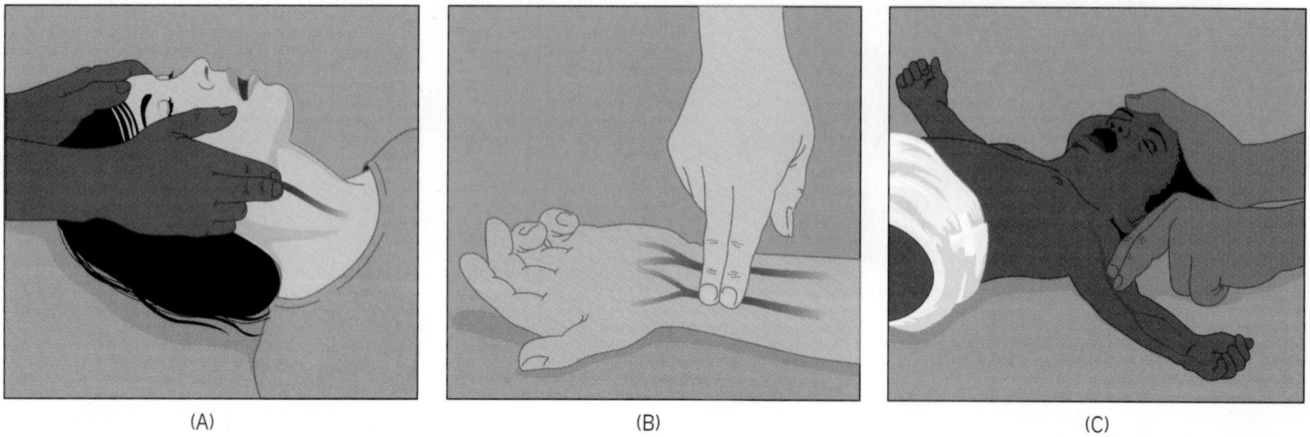

Figure 22-17 Locating pulses in (A) the carotid artery in the neck, (B) the radial artery in the wrist, and (C) the brachial artery in an infant's upper arm.

Major Bleeding

After determining that the patient has a pulse, the firefighter checks the patient from head to toe, front and back, for major bleeding. A sweeping motion is used with both hands along the entire length of the patient's body to check for major bleeding. If the patient is lying down, it is important to check underneath the patient's body. This can be done without moving the patient by simply reaching under the patient's body when doing the sweeping motion, and checking the gloves frequently for signs of blood. If a source of major bleeding is located during the initial assessment, the firefighter should attempt to stop the bleeding and bandage the wound. Bleeding and bandaging are covered later in this chapter.

The initial assessment allows firefighters to quickly identify major problems and start treatment of patients. In a mass casualty situation, where there are many patients, a triage system may be used. **Triage** is a quick and systematic method of identifying which patients are in serious condition and which patients are not, so that the more seriously injured patients can be treated first. Many triage systems are based on the same principles in the initial assessment. Firefighters can learn more about triage systems in higher levels of EMS training.

Vital Signs and the Focused History and Physical Exam

After the initial assessment of the patient is complete, the firefighter continues on to a focused history and physical exam, which is a thorough examination of the patient. This consists of three parts: (1) patient fact-finding, (2) vital signs, and (3) a head-to-toe survey of the patient.

If the patient is conscious, firefighters will need to go on a fact-finding mission for information about the patient. This includes trying to determine the age of the patient, important medical history, and allergies to medications. Firefighters should also try to establish what the patient's current situation is. Has the patient had chest pain for the past two hours? Does the patient remember exactly what happened in the automobile accident? Also, firefighters should always check patients for a bracelet, necklace, or watch that may contain vital medical information, **Figure 22-18**.

If the patient is unconscious, the firefighter can try to do fact-finding by asking questions of family members, friends, or bystanders present on the scene. If no one else is present or no one can provide medical facts about the patient, the firefighter may be able to find clues on the emergency scene to help establish some information about the patient.

After the fact-finding is done, the next step is to get a set of vital signs on the patient. For firefighters providing emergency first aid, the set of vital signs should include a pulse rate, a respiratory rate, and the color and temperature of the skin. **Figure 22-19** shows average vital sign ranges by age of the patient.

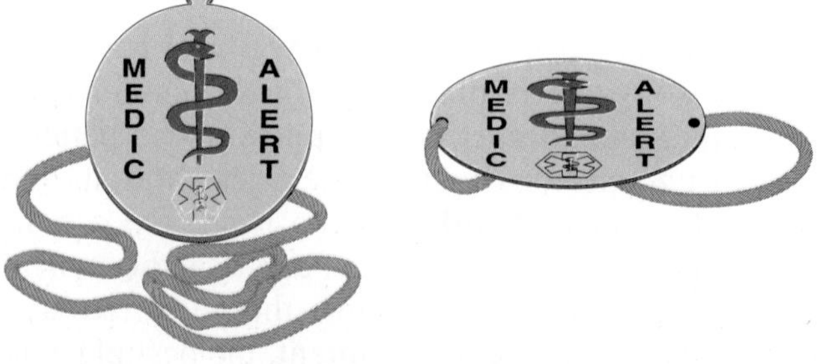

Figure 22-18 Examples of medical alert tags.

AGE	PULSE	RESPIRATIONS
Newborn	120–160	40–60
1 year	80–140	30–40
3 years	80–120	25–30
5 years	70–115	20–25
7 years	70–115	20–25
10 years	70–115	15–20
15 years	70–90	15–20
Adult	60–80	12–20

Figure 22-19 Average vital sign ranges by age.

Pulse Rate

The pulse rate is the number of times the heart is beating per minute. To establish the pulse rate, the firefighter should first locate the carotid pulse in the neck or the radial pulse in the wrist. Using the second hand on a watch or clock, the firefighter should count the number of beats for a minute to get the patient's pulse rate. With experience and practice, firefighters may choose to count the beats over a thirty-second period and multiply the number by two for the pulse rate per minute.

Respiratory Rate

The respiratory rate, like the pulse rate, is the number of times the patient is breathing in one minute. To get the respiratory rate, the firefighter observes the patient's chest and abdomen for the characteristic rise and fall that occur with each breath for one minute.

> **Streetsmart Tip** If the patient is wearing heavy clothing, or it is difficult to see the chest and abdomen rise and fall, the firefighter may lightly place one hand on the upper abdomen to feel the rise and fall of breathing.

The firefighter should also try to observe the quality of the respirations (labored, shallow, normal).

Color and Temperature of the Skin

The firefighter should carefully observe the skin color and temperature of the patient, because this can be an important indicator of the patient's condition. The skin color may be normal, ashen (gray in color), blue or cyanotic, red, yellow, or pale (white). The skin temperature may be normal, cool and clammy, hot and dry, or hot and moist. These are important signs, and should be noted and recorded on all patients.

After the patient's vital signs are recorded, it is time to do a head-to-toe examination of the patient. It is important to check the patient's vital signs periodically. A good practice to follow is obtaining vital signs every three to five minutes. To conduct a head-to-toe survey of the patient, the firefighter should follow the procedure outlined here:

Head and Neck. Gently palpate the head and neck to check for any structural damage to the bones in the head, face, and neck.

> **Caution** In the initial assessment, one of the first steps after checking the scene for safety purposes and donning BSI is to establish and maintain head and neck stabilization if the patient is unconscious or a traumatic injury is suspected. If head and neck stabilization have been initiated on a patient, the firefighter should *not* move the head or neck when performing a head-to-toe survey.

Face and Eyes. Gently palpate the bones in the face to check for swelling or deformities. Examine the mouth, nose, and ears for the presence of foreign objects, bleeding, or discharge of fluid.

Carefully examine the patient's eyes and note the size of the pupil: normal, **constricted,** or **dilated**. Also, check to make sure that the pupils are equal in size for both the right and left eye. Using a small flashlight, quickly shine the light in each eye briefly and note whether or not the pupil is "reactive" to light. If the pupil is reactive, it will shrink with the presence of light and grow larger with the absence of the light source. Refer to **Figure 22-20** for examples of different pupil sizes.

Upper Torso and Chest. Check the bones of the rib cage and breastbone (sternum) for structural integrity. Also, check to see if the chest rise and fall from breathing is equal on both the right and left sides. An unequal chest rise and fall may indicate a serious breathing problem, such as a collapsed lung.

Lower Torso, Abdomen, and Pelvis. Check the abdomen and pelvic area for any obvious signs of traumatic injury, like bruising, swelling, or pulsating masses. A pulsating mass is a sign of a rupture, or impending rupture, of the body's largest artery, the aorta, and is a very serious condition. Using one hand on top of the other, press down in each of the four quadrants of the abdomen to check for pain, tenderness, or abdominal rigidity. Next, place one hand on each hip on the side of the body, squeeze together gently to check the stability of the pelvis, then press down on the top of the hips gently to continue to check the structural integrity of the pelvis.

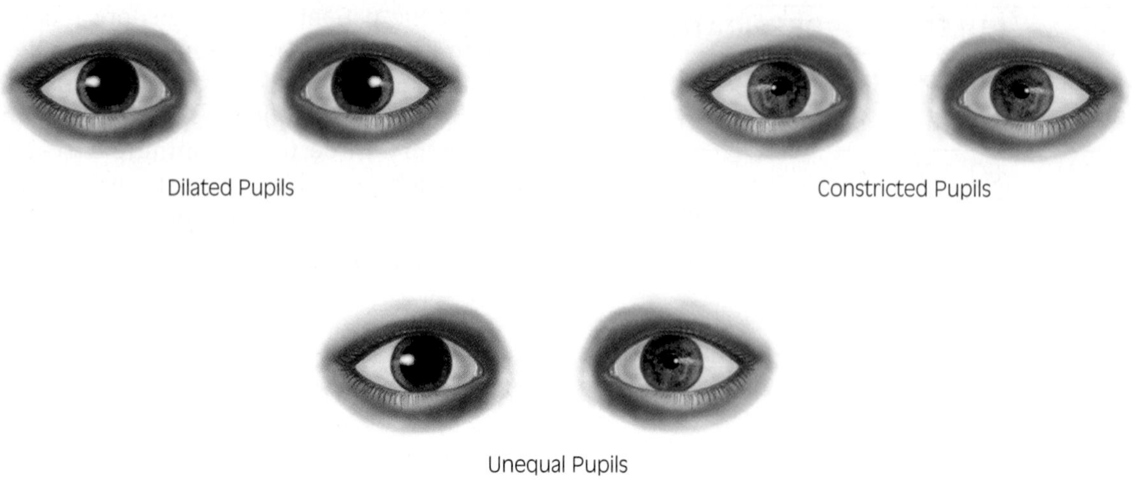

Figure 22-20 Examples of pupil size.

Arms and Legs. Gently feel the arms and legs of the patient to check for any possible damage to the extremities. Also, look carefully at the arms and legs in comparison to one another. Is one leg, or arm, longer or shorter than the other? If the patient is awake, check for movability of the fingers and toes to make sure the neurological system (the brain and spinal cord) is functioning. You can also ask a patient to grip your hand and squeeze, or place your hands at the bottom of the patient's feet and ask the patient to push against your hands. Check for the pulse in both arms and both legs to be sure that the heart is properly circulating blood to the extremities. The pulse in the foot can be felt on the top of the foot and the inside of the ankle underneath the protruding bone structure in the ankle, **Figure 22-21**.

Another very effective way of checking for circulation in the arms and legs of children under six years of age is by checking capillary refill. To do this, pinch the end of the fingers, or toes, until the skin underneath the nailbed becomes white. When you

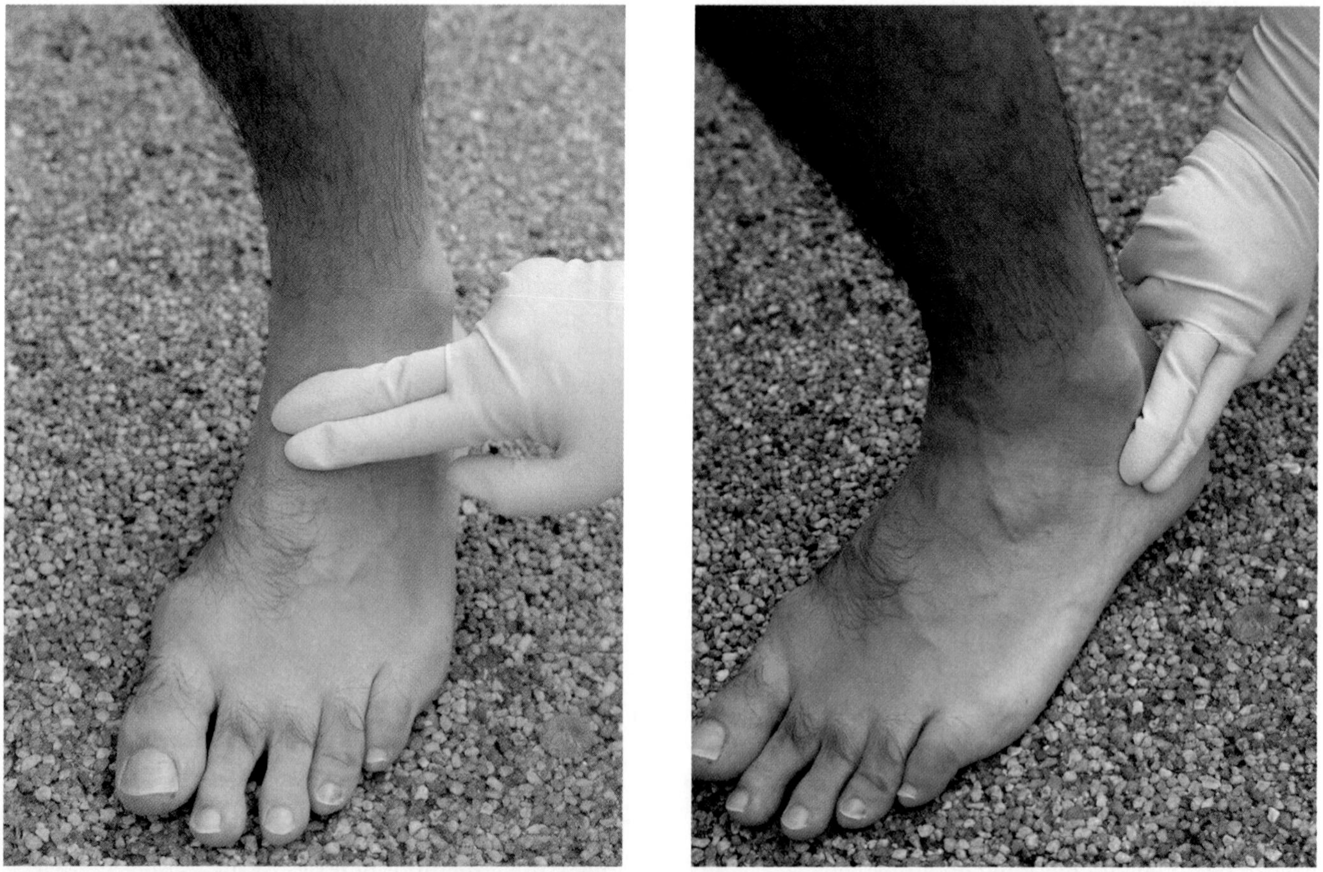

Figure 22-21 Locating pulses in the foot.

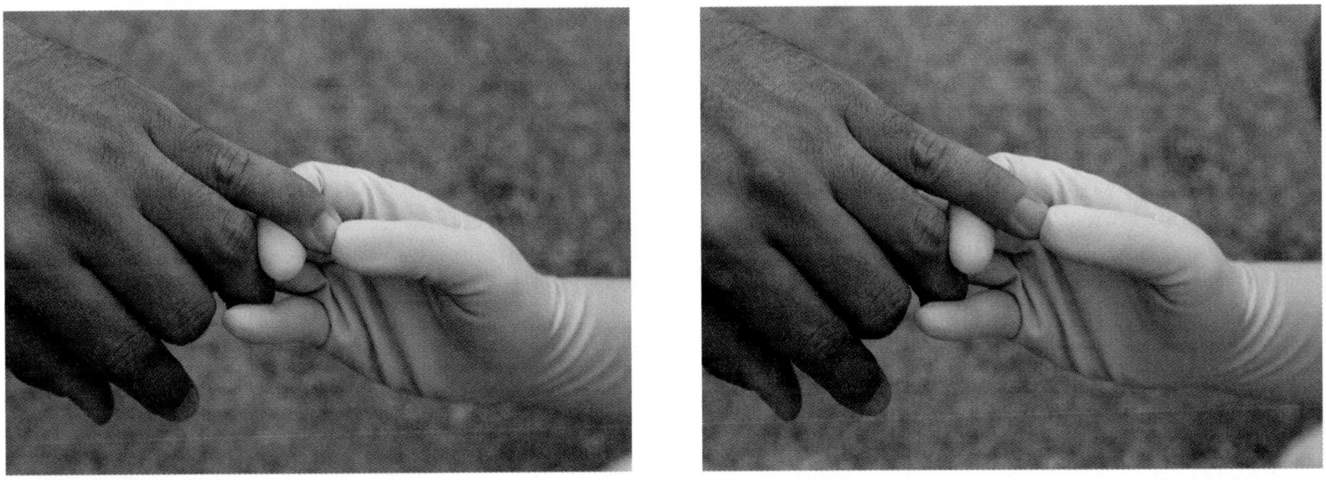

Figure 22-22 Capillary refill.

release the pressure, watch to see how quickly the skin becomes pink or normal again. Normally, the skin will return to its original color right away. If it takes longer than a second or two for the skin under the nail to return to its original color, there may be a circulation problem in the patient. **Figure 22-22** demonstrates the capillary refill technique.

Patient Findings

If the fire department requires that firefighters complete a call record for emergency medical patients, they should be sure to document the findings of the assessment and the patient's vital signs.

> **Note** Documenting patient information is very important, and firefighters should be familiar with their department's policies concerning the recording of patient and emergency response information.

Firefighters may have found something in the survey that is important to the treatment of the patient, both by an ambulance crew and in the hospital. Everything that is found must be passed on to the emergency medical provider taking over care of the patient. That includes the initial assessment, the focused history and physical exam, the patient's vital signs, and all patient fact-finding that was done while on the scene.

CARDIOPULMONARY RESUSCITATION/AED

One of the most basic and widely learned emergency response skills is cardiopulmonary resuscitation, or CPR. This skill is learned by all types of people, from lifeguards to physicians. CPR is a critical, lifesaving skill for firefighters and emergency responders to learn and practice, **Figure 22-23**.

> **Note** CPR and techniques for helping choking victims are important knowledge to have not only while on duty with the fire department but also anywhere a firefighter may go.

Figure 22-23 Firefighters learning CPR. *(Photo courtesy of Fred Schall)*

Coronary heart disease and cardiovascular diseases are the leading killers in the United States. Heart disease, while more prevalent in the elderly, can affect anyone. Emergency care providers will most certainly be faced with situations where cardiovascular disease has caused a sudden respiratory arrest (a patient who is not breathing) or cardiac arrest (a patient who is not breathing and has no pulse). In these situations, CPR is a basic lifesaving skill.

Several different organizations support CPR training, education, and research. The two largest organizations are the American Heart Association (AHA) and the American Red Cross. CPR training is widely available in the United States. If a particular department or agency does not have CPR classes available, firefighters can contact a local chapter of the AHA or the American Red Cross and ask them about CPR training.

Safety CPR is a critical skill for firefighters to know and practice. They need this skill not only to provide care to patients on emergency medical calls but also to be prepared to help fellow firefighters on the scene of a fire or emergency incident. Heart disease is a leading killer of firefighters operating on emergency scenes every year. Firefighters should have the skills necessary to help the members of their team in need, including CPR and first-aid training.

As research on cardiovascular disease and cardiac arrest procedures continues, the skills and information regarding CPR may be changed or revised. Firefighters should follow the most current standards set forth by the AHA or the American Red Cross for education, practice, and performance of CPR. Firefighters should also be aware of recommendations concerning retraining on CPR skills, which should be done every one to two years.

Automated external defibrillators (AEDs), Figure 22-24, have become an important part of the provision of emergency medical care. Many types of first response organizations, including fire departments, use AEDs as a lifesaving tool. AEDs are becoming more common in nonemergency settings as well, including shopping malls, casinos, airplanes, and even private homes. AEDs are becoming more common because they dramatically improve the outcome of cardiac arrest.

An AED is a small machine with an internal computer that analyzes a patient's heart rhythm and delivers electrical shocks, or defibrillations, as necessary to correct a heart with a malfunctioning rhythm. It requires some basic training to place the AED on a patient and use it appropriately. Many first response agencies are training firefighters to use AEDs at the same time they are training them on CPR techniques. Firefighters should consult their agency's policies and explore training courses for AED use by firefighters.

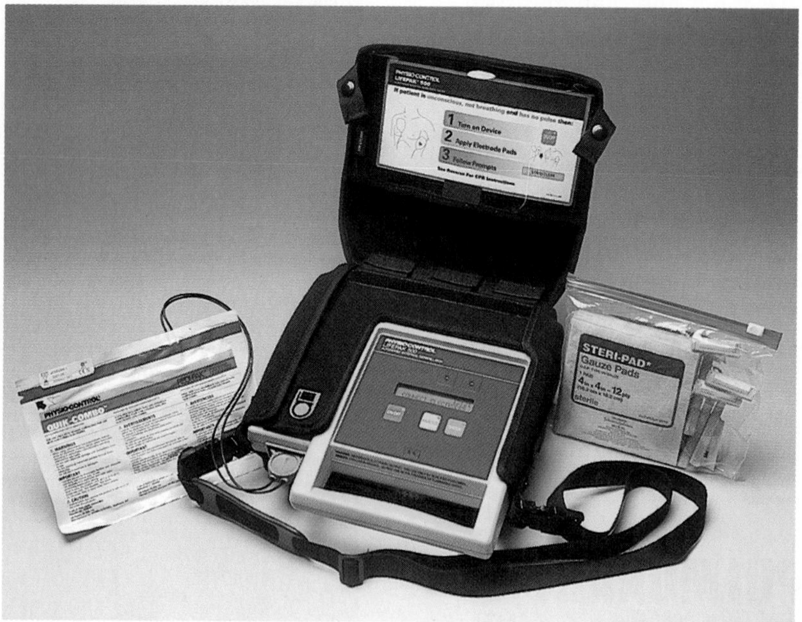

Figure 22-24 An AED is a small machine with a computer that analyzes a patient's heart rhythm and delivers electrical shocks, or defibrillations, as necessary to correct a heart with a malfunctioning rhythm.

BLEEDING CONTROL AND SHOCK MANAGEMENT

The blood in the body brings oxygen and nutrients to the cells in the body in order for the cells to survive, grow, and reproduce. The heart and the system of tubes through which blood travels (**arteries**, **veins**, and **capillaries**) are very complex and vital systems, **Figure 22-25**. The **cardiovascular system** is a closed system with a pump (the heart) and specialized tubing (the blood vessels) that provide the body with blood containing the elements humans need to live. Damage to the heart or the blood vessels may result in a lower volume of blood or an ineffective pumping system, which are serious, if not life-threatening, conditions that can cause a patient to go into shock, or hypoperfusion (a lack of oxygen and nutrients to the tissues).

Caution Bleeding control and shock (hypoperfusion) management are important skills to learn in first-aid training. The firefighter should use BSI, as with all patients, but especially so in this situation because exposure to blood and body fluids is very likely.

Internal and External Bleeding

Internal bleeding occurs when there is bleeding within the body and no visible open wound is present. Internal bleeding can occur because of trauma to the body or because of illness. Internal bleeding can be a very serious condition for a patient.

Firefighters should look for the following signs, which may indicate internal bleeding:

- Bruising of the skin
- Pale skin
- Cold and clammy skin

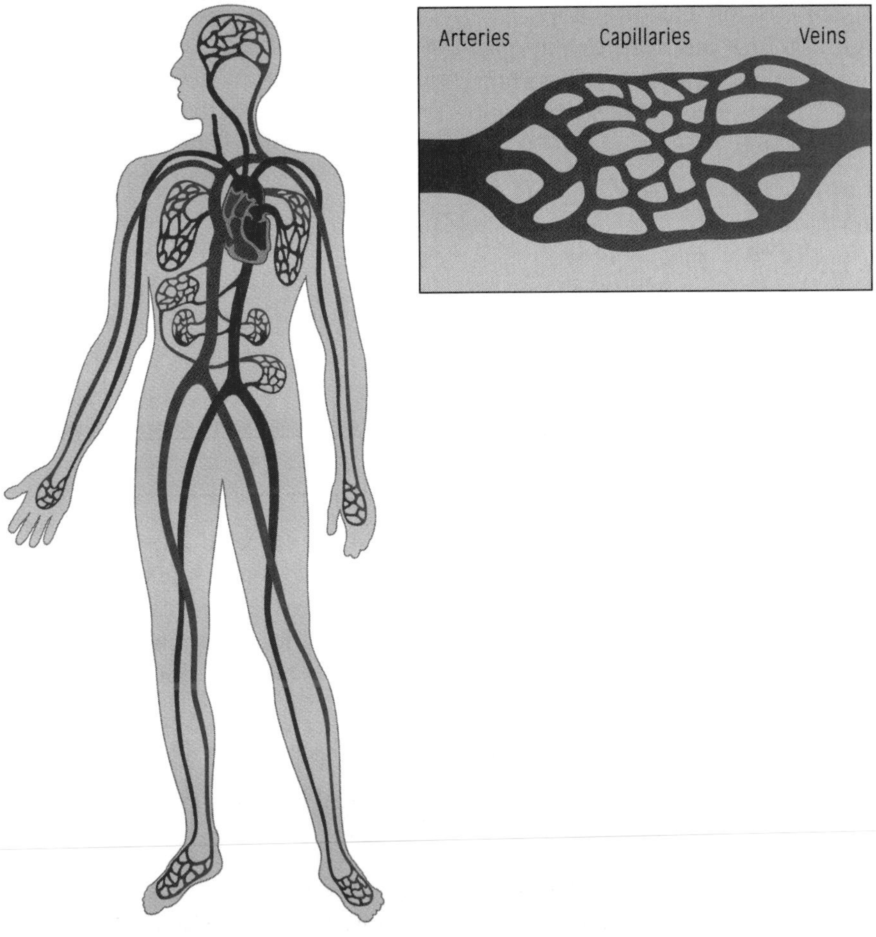

Figure 22-25 The blood vessel system in the body, which is composed of arteries, veins, and capillaries.

- Dilated pupils
- Obvious deformities to major bones (like the pelvis, the upper leg, etc.)
- A rigid and tender abdomen
- Blood in the urine or from the rectum
- Blood from the mouth or nose, or blood in vomitus

Patients who the firefighter suspects are suffering from internal bleeding should be treated for shock (hypoperfusion), as covered later in this chapter.

External bleeding is bleeding that is coming from an open wound on the body. The three types of external bleeding are discussed next.

Arterial Bleeding

Arterial bleeding is bleeding from arteries, which are the blood vessels that carry blood that has been oxygenated away from the heart to the cells in the body. Arterial blood, because it has just been supercharged with oxygen, is bright red in color. Bleeding from the arteries can be very serious. In order for the oxygenated blood to get to the cells, the heart pushes it through the arteries under pressure. When an artery is severed or damaged, this pressure can force a large quantity of blood out of the blood vessel system, depleting the system and possibly causing shock.

Firefighters providing first-aid care to patients can recognize arterial bleeding by looking for bright red blood from the wound site or blood spurting or pulsating from the wound site.

Venous Bleeding

Venous bleeding is bleeding from a vein. Veins are the blood vessels that carry blood from the cells in the body, after the oxygen and nutrients are used, back to the heart to be reoxygenated in the lungs. Venous blood, because it has been stripped of the oxygen it was carrying, is bluish red in color. Bleeding from the veins can also be very serious, especially if the bleeding is from a large vein. Unlike the arteries, however, the pressure on the blood in the veins is not as great, but can still be life threatening.

Firefighters providing first-aid care to patients can recognize venous bleeding by looking for bluish red blood from the wound site or blood flowing steadily from the wound site.

Capillary Bleeding

Capillary bleeding is bleeding from a capillary. Capillaries are the very small blood vessels connecting the arteries and the veins. These small blood vessels filter the oxygenated blood from the arteries to the cells and then take the used blood back from the cells and into the veins to return to the heart and lungs. Bleeding from the capillaries is what a person sees when he or she cuts a finger or skins a knee. It is usually not serious; however, it should be treated by the care provider.

Firefighters providing first-aid care to patients can recognize capillary bleeding by looking for blood slowly oozing from the wound site.

Caring for Patients with Internal Bleeding

Once firefighters have established what kind of bleeding is occurring in a patient, they can take measures to get the bleeding under control. Internal bleeding cannot be brought under control, however, by first-aid actions. For patients exhibiting signs of internal bleeding or shock, there are other actions a firefighter can take to provide assistance.

For minor internal bleeding or, in other words, bruising, the best treatment firefighters can give the patient is to apply a cold pack or ice pack to the affected site. The purpose of applying cold to the bruised area is to attempt to slow down the blood flow seeping into the skin by narrowing the blood vessels. The cold will also help with alleviating some of the pain of bruising as well. After applying cold to the injury site, the firefighter should attempt to elevate the part of the body that is injured, unless there is a chance there is a fracture. For example, if bruising and pain are found on the lower leg, and there appears to be no deformity of the leg structure or signs of a bone fracture, cold should be applied to the site and the leg raised.

> **Streetsmart Tip** When applying cold packs or ice packs to a patient, the firefighter should place a towel, cloth, or gauze pad between the cold pack and the skin to prevent freezing of the tissues.

Major internal bleeding is a serious problem and requires rapid medical attention. The loss of a large volume of blood can cause a patient to go into shock. The treatment for patients in shock is covered in a later section.

Caring for Patients with External Bleeding

Three methods are used when dealing with a patient experiencing external bleeding, as explained next. Note that, before acting, the firefighter should remember to wear protective gloves, since exposure to blood and body fluids is likely.

Pressure on the Site of the Bleeding, or Direct Pressure

Capillary and some venous bleeding can usually be controlled by direct pressure. Direct pressure can also help in the case of arterial bleeding. Firefighters controlling bleeding with direct pressure should first locate the wound and source of the bleeding. (They must look carefully, because a large volume of blood may be coming from a small cut or puncture.) The firefighter should place a sterile dressing or bandage over the site of the wound and apply pressure by pressing a hand against the wound site. Using a roll of gauze (preferably sterile) or a roll of bandage material, the firefighter wraps the site with the material, pulling it tight to create moderate pressure to the wound site. It is important to use a piece of medical or first-aid tape to secure the bandage and keep pressure on the wound site. If the bleeding soaks through the bandaging, it should not be removed to redo the bandage. Once a bandage is placed on a patient, it should not be removed. Instead, the firefighter should place another sterile dressing on the bandaged wound and wrap it again with a roll or gauze, this time applying a little more pressure with the bandaging, **Figure 22-26**.

Elevate the Site of the Bleeding

After applying direct pressure and a bandage to the wound that is bleeding, the firefighter should elevate the extremity or area that is bleeding. If direct pressure stops the bleeding, elevation may not be necessary. Elevation will make the blood flow to the site slower because it must travel "uphill" and, therefore, may slow down the bleeding. If the bleeding is from an arm or leg, it should be simply propped or raised above the level of the heart. If the bleeding is from the head or face, the firefighter should make sure the patient sits upright.

Use Pressure Points

The blood vessel system in the body has some main vessels, or tubes, that supply certain areas of the body. If direct pressure and elevation do not

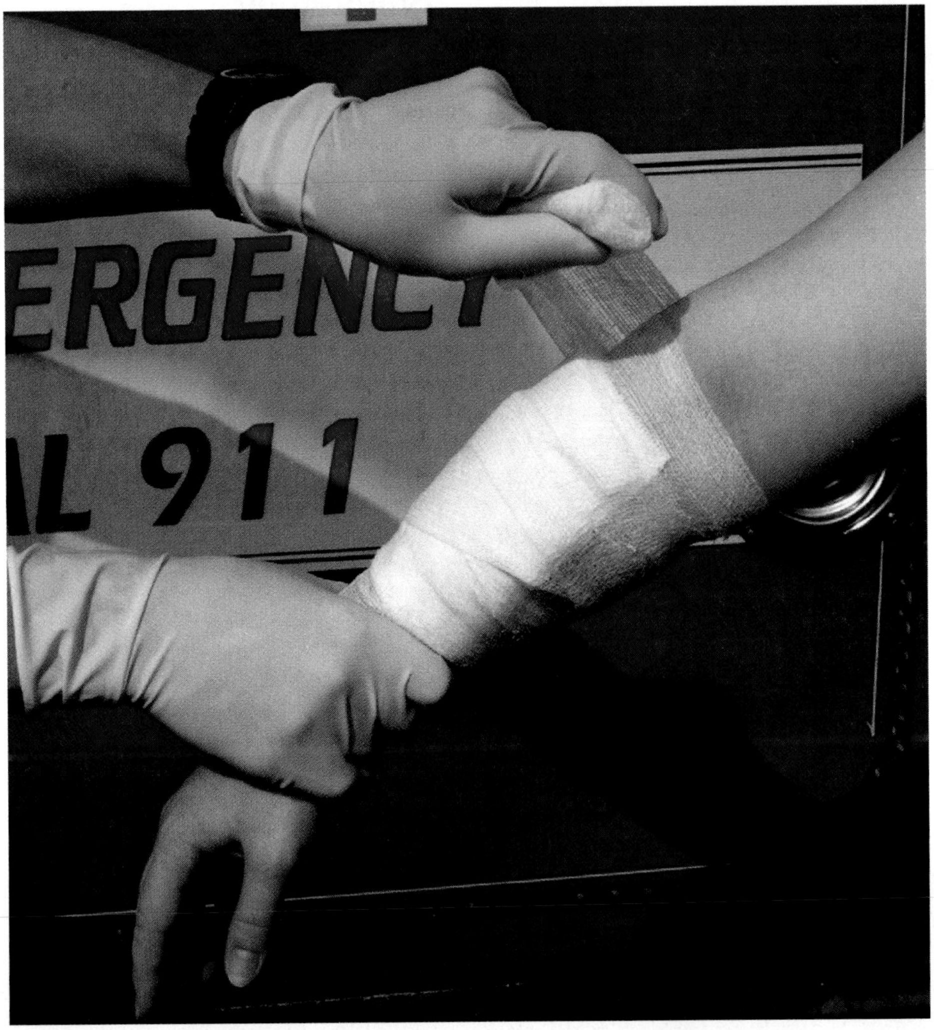

Figure 22-26 Applying a pressure dressing to a bleeding wound.

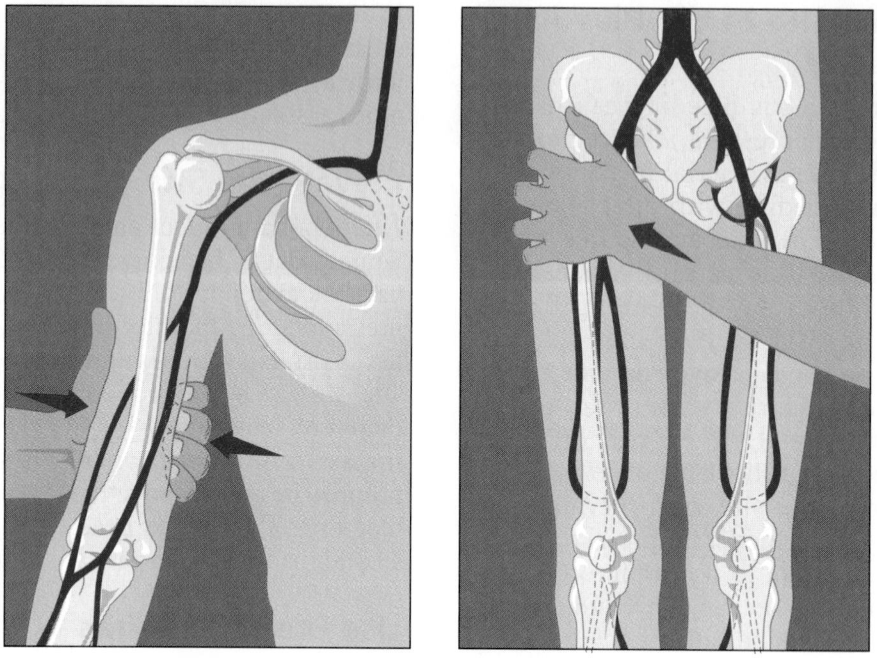

Figure 22-27 Using pressure points in the upper arm and the leg to decrease blood flow to a wound.

help stop or slow the bleeding, applying pressure to these "pressure points" may slow the flow of blood to the body area, thus slowing the flow of blood from the wound, **Figure 22-27**. The best pressure points are in the upper arm (brachial artery) and in the hip/pelvic area (**femoral artery**). In both of these areas, the major artery is near a bone where it can be compressed to slow blood flow.

Types of Wounds Requiring First Aid

There are several types of wounds that firefighters providing first aid should be able to recognize and treat.

Abrasion

An **abrasion** is a scrape or brush of the skin usually making it reddish in color and resulting in minor capillary bleeding. An abrasion is what results when someone "skins a knee," a common event in childhood. Abrasions should be treated by applying sterile bandages and, if the bleeding has not stopped, direct pressure.

Avulsion

An **avulsion** is an injury where a part of the skin is torn away, but still attached, leaving a flap or loose area hanging. Avulsions should be treated by applying sterile bandages and, if the bleeding has not stopped, direct pressure.

Amputation

An **amputation** occurs when a part of the body is severed completely as a result of an injury. This can happen to fingers and toes and even arms and legs. Firefighters should first apply sterile bandages to the site where the amputation occurred with direct pressure, using elevation or pressure points as needed. Firefighters should then locate the severed body part and wrap it completely with a sterile dressing. Once it is wrapped in the dressing, the body part should be placed into a plastic bag which is then sealed tightly. The severed body part must be kept cold and transported to the hospital with the patient. Firefighters can place the severed body part that is sealed in a plastic bag into a second plastic bag filled with ice to keep the body part cold for transport.

Laceration or Incision

A **laceration** is a cut to the skin and underlying tissues that leaves an irregular, uneven pattern. An **incision** is a cut to the skin that leaves a straight, even pattern. In both cases, the firefighter should apply sterile bandages to the wound and, if the bleeding has not stopped, direct pressure. Elevation and pressure points may be necessary for a large or deep laceration or incision.

Puncture

A **puncture** is caused by an object that has stabbed the body. If the object is no longer in the wound, firefighters can treat the puncture wound by applying a sterile bandage and, if necessary, direct pressure, elevation, or pressure points. If the object causing the puncture remains in the body, firefighters should pack the sterile dressings around the wound site, leaving the object in the body. They should not attempt to remove an object imbedded in a wound, because that could cause more harm to the patient. The object must be secured so that it will not move around and cause more damage to the skin and tissues.

What Is Shock? (Hypoperfusion)

Shock, or **hypoperfusion**, is a condition caused by a problem with or failure of the circulatory system that results in a decrease of oxygen and vital nutrients to the body's tissues. In other words, it is a lack of blood to the organs and tissues of the body. The body requires a regular and constant supply of oxygen and nutrients to function properly.

> **Caution** Shock (hypoperfusion) is a very serious condition and can cause death.

Shock (hypoperfusion) can result from a problem with the heart, which pumps the blood to the tissues. This can happen when a patient is experiencing a heart attack, and the heart has lost its ability to pump the blood adequately. Shock can also result from a failure in the blood vessels that carry the blood to and from the heart. This can be caused by a severe allergic reaction, which causes the blood vessels to expand rapidly, or by a trauma to the head or spinal cord, also causing a rapid expansion, or dilation, of the blood vessels.

Lastly, shock can result from a problem within the blood vessels, namely, a drop in the quantity of blood contained in the vessels. This can occur when the patient has an injury that has caused major internal or external bleeding. As the blood is lost from the body, the vessels and the heart have less oxygen carrying material to work with, resulting in the shock state. This type of shock can also occur in a patient who has experienced an illness and is experiencing **dehydration**. A severely dehydrated state can cause shock.

Recognizing the Signs and Symptoms of Shock (Hypoperfusion)

There are many signs and symptoms of shock, or hypoperfusion, that firefighters should look for in a patient, **Figure 22-28**:

- Pale color to the skin, or a bluish tint to the skin, especially around the lips and in the nailbeds
- A cool temperature of the skin

Signs of Shock

- Pale or Bluish Tint to Skin Color
- Cool Skin Temperature
- Moist or Sweating Skin
- Dilated Pupils
- Rapid, Shallow Breathing
- Rapid, Weak Pulse
- Nausea, Thirst, or Vomiting
- Unconsciousness or an Altered Level of Consciousness

Figure 22-28 Signs of shock.

- Sweating, or moist skin
- Pupils larger than normal, or dilated
- Rapid, shallow breathing
- A rapid, weak pulse
- Complaints of nausea, thirst, or vomiting
- Unconsciousness, or an altered level of consciousness

Firefighters should look for signs of shock (hypoperfusion) in all patients. Many different traumatic injuries or severe illnesses can result in shock, so firefighters should always be prepared for a shock state in a patient.

Caring for Patients in Shock

Firefighters who find a patient with the signs and symptoms of shock (hypoperfusion) should follow the basic treatment outlined here. Firefighters should also treat patients in this manner if they suspect that they could go into shock.

Treating a Patient for Shock (Hypoperfusion)

1. Ensure scene safety and BSI precautions.
2. Assess the patient's level of consciousness and, if needed, maintain head and neck stabilization.
3. Make sure the airway of the victim is open. This is especially important in unconscious patients.
4. Ensure that the patient is breathing and has a pulse. If either is absent, follow the procedures for cardiopulmonary resuscitation and rescue breathing.
5. Treat the injuries present and control any major bleeding.
6. Keep the patient warm. Place some blankets or a covering over the patient to prevent

any body heat loss or exposure to cold in the environment.

7. Position the patient properly. The best possible position for a patient in shock is lying on the back. If head and neck stabilization is initiated during the initial assessment it should be continued throughout patient treatment. Placing a patient on the back may not be possible if the patient is having difficulty breathing, so the patient can be placed in a reclined position. A patient who is conscious and vomiting is best placed on the side so that the vomitus does not obstruct the airway.
8. Raise the legs of the patient to allow more blood to reach the heart. The legs should be raised together and no more than a foot above the level of the heart. Do not raise the legs of a patient who has a traumatic injury to the legs or pelvis.
9. Provide reassurance to the patient, and monitor vital signs every few minutes.

EMERGENCY CARE FOR COMMON EMERGENCIES

No two emergency medical calls are the same. Firefighters must learn the basics of first-aid treatment and then adapt them to each emergency. This section lists some common emergencies, things to look out for, and first-aid treatment.

Trouble Breathing

Trouble breathing is a very common emergency that firefighters may be called on for first-aid care. There are many reasons why a patient may be experiencing trouble breathing, anything from anxiety to a major respiratory disease. Common signs and symptoms of trouble breathing include wheezing, gasping for air, shallow breathing, pale or bluish skin color, anxiety, or even unconsciousness if the patient is having serious difficulty breathing.

Care Guidelines

1. After assessing scene safety and donning BSI, perform an initial assessment and attend to any major problems found with the airway, breathing, or circulation.
2. Reassure the patient, keep the patient calm, and monitor vital signs frequently.
3. Recognize that this is a serious emergency and emergency medical care is needed right away.

Chest Pain

Chest pain is another common emergency firefighters may encounter, as shown in **Figure 22-29**. Chest pain related to a heart problem or heart attack may be described as a tightness or squeezing feeling in the chest, and the pain may be felt not only in the chest but also in the jaw, abdomen, or arm (often the left arm). Chest pain can also result from some respiratory disorders, traumatic injuries, or even stomach problems such as indigestion.

Care Guidelines

1. After assessing scene safety and donning BSI, perform an initial assessment and attend to any major problems found with the airway, breathing, or circulation.
2. Reassure the patient, keep the patient calm, and monitor vital signs frequently.
3. Recognize that this is a serious emergency and emergency medical care is needed right away.
4. Because chest pain may be a sign of a problem with the heart, there is a possibility the patient could go into shock. Look for signs and symptoms of shock and treat the patient accordingly.

Medical Illnesses

A firefigher may be called on to assist a patient with a medical illness, which can range from stomach problems like nausea and vomiting to general illnesses like fevers, the flu, or the common cold.

Care Guidelines

1. After assessing scene safety and donning BSI, perform an initial assessment and attend to any major problems found with the airway, breathing, or circulation.
2. Reassure the patient, keep the patient calm, and monitor vital signs frequently.
3. Recognize that this could be a serious emergency and emergency medical care is needed right away.
4. Because a medical illness may cause dehydration (lack of fluids) in a patient, there is a possibility the patient could go into shock. Look for signs and symptoms of shock and treat the patient accordingly.

Allergic Reactions

Allergic reactions result from the body's reaction to a substance to which there is an allergy. Many people experience allergy problems associated with hay fever or bee stings. In some cases, allergic reactions

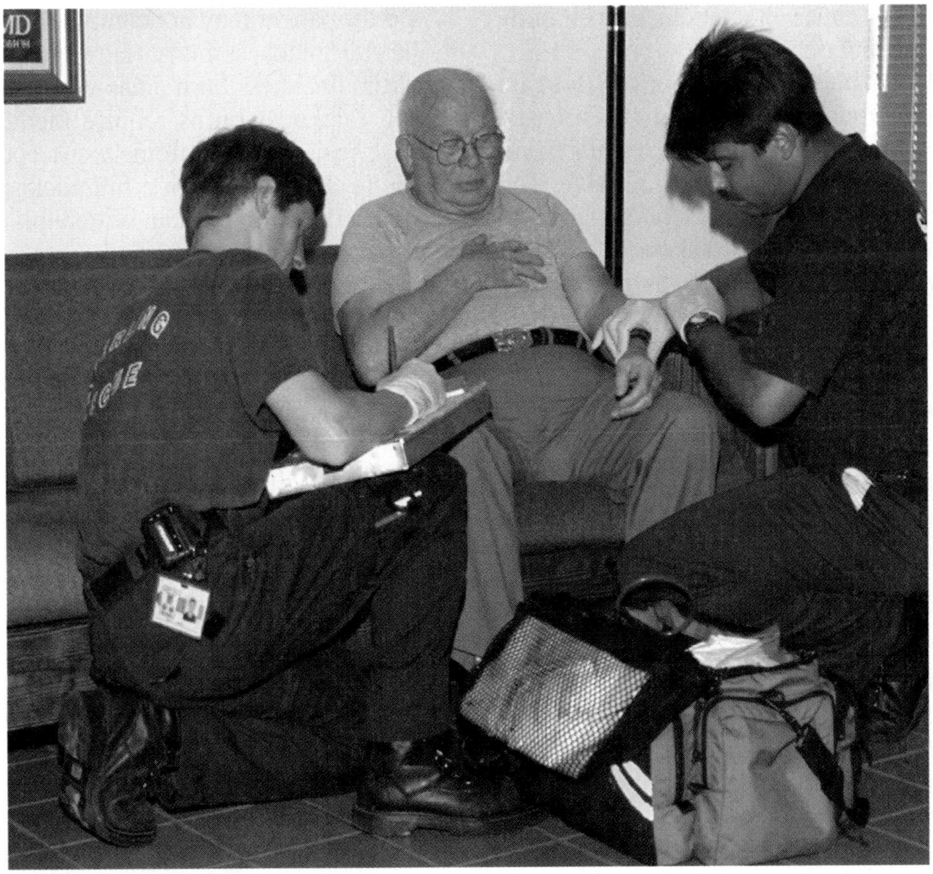

Figure 22-29 Firefighters treating a patient complaining of chest pain. *(Photo courtesy of Fred Schall)*

can be very severe, even life threatening. Allergic reactions can result from a variety of substances, such as types of food, things in the environment, bites and stings, and medications. Allergic reactions may be localized at the part of the body where the substance was introduced, and present as localized swelling, redness, and itching. Allergic reactions involving the entire body system may include trouble breathing, rapid pulse, shallow breathing, hives (red splotches or bumps on the body, typically concentrated on the torso), and anxiety, **Figure 22-30**.

Care Guidelines

1. After assessing scene safety and donning BSI, perform an initial assessment and attend to any major problems found with the airway, breathing, or circulation. Severe allergic reactions can compromise the airway and breathing of a patient because the reaction tightens the breathing passages in the lungs and airway. Be prepared to breathe for the patient if necessary.
2. Reassure the patient, keep the patient calm, and monitor vital signs frequently. Also, if the patient is still in contact with or near the substance causing the allergic reaction, remove the patient from the area or isolate and dispose of the substance.
3. Recognize that in an allergic reaction involving the body system, this is a serious emergency and emergency medical care is needed right away.
4. Because severe allergic reactions often cause the blood vessels to widen, there is a possibility the patient could go into shock. Look for signs and symptoms of shock and treat the patient accordingly.

Thermal Burns

Thermal burns, or burns caused by heat, can be as simple as a sunburn or as life threatening as severe burns across much of the body. Thermal burns are

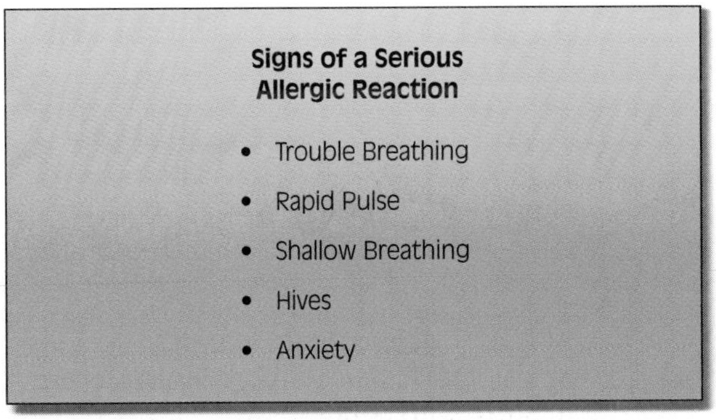

Figure 22-30 Signs of a serious allergic reaction.

very painful and can cause the patient to lose body heat and fluids, causing dehydration.

Burns are divided into three categories based on their severity. **Superficial burns**, sometimes referred to as first-degree burns, occur when the outer layer or layers of skin are burned, **Figure 22-31A.** The skin will turn red and feel hot to the touch. Depending on the severity of the superficial burn, swelling may result. These burns are often painful to the patient but will generally heal in about a week, even if left untreated. Sunburn is an example of a superficial burn.

In **partial thickness burns**, sometimes called second-degree burns, additional layers of skin are burned, but the deeper layers of the skin remain undamaged, **Figure 22-31B.** The skin will be very red, and blisters will develop. These burns can cause the patient intense pain because the nerves under the skin are sometimes affected. The major sign of a partial thickness burn is the blistering of the skin; however, this blistering does not always occur immediately. It may take several hours before blisters develop. These types of burns require medical attention, and with treatment they generally heal within a few weeks.

The last category of burn is a **full thickness burn**, often referred to as a third-degree burn, **Figure 22-31C.** In full thickness burns, all layers of the skin are burned. These burns will have the characteristics of partial thickness burns, but will also have areas of charred or blackened skin, which is the definitive sign of a full thickness burn. These burns are the most serious of the three types because the entire area of skin has been damaged, and the body is no longer able to hold fluids inside or keep bacteria and dirt outside. Many times nerves are also damaged or destroyed, and the patient may not complain of pain in the area of the full thickness burn. (However, the accompanying partial thickness burn areas will still be very painful.) Full thickness burns require medical attention, and these patients are extremely susceptible to infections. The healing process of a full thickness burn is usually long and painful and can be debilitating.

Care Guidelines

1. After assessing scene safety and donning BSI, perform an initial assessment and attend to any major problems found with the airway, breathing, or circulation. Be sure to remove the patient from the environment to stop the burning process.
2. Reassure the patient, keep the patient calm, and monitor vital signs frequently.
3. Cover the burned area with dry, sterile dressings to prevent infection. (Firefighters should consult their agency for specific treatment for burns. Some fire departments may have a different treatment for burns.)
4. Monitor the patient's airway and breathing constantly. Burn victims may also have inhaled smoke or hot gases, which can cause damage in the lungs and create airway or breathing problems.
5. Also, monitor the patient for signs and symptoms of shock and treat accordingly.

Chemical Burns

Chemical burns are caused by chemical substances that come into contact with the skin or tissues of the body, creating a caustic reaction. A variety of chemical substances can cause a chemical burn. They can

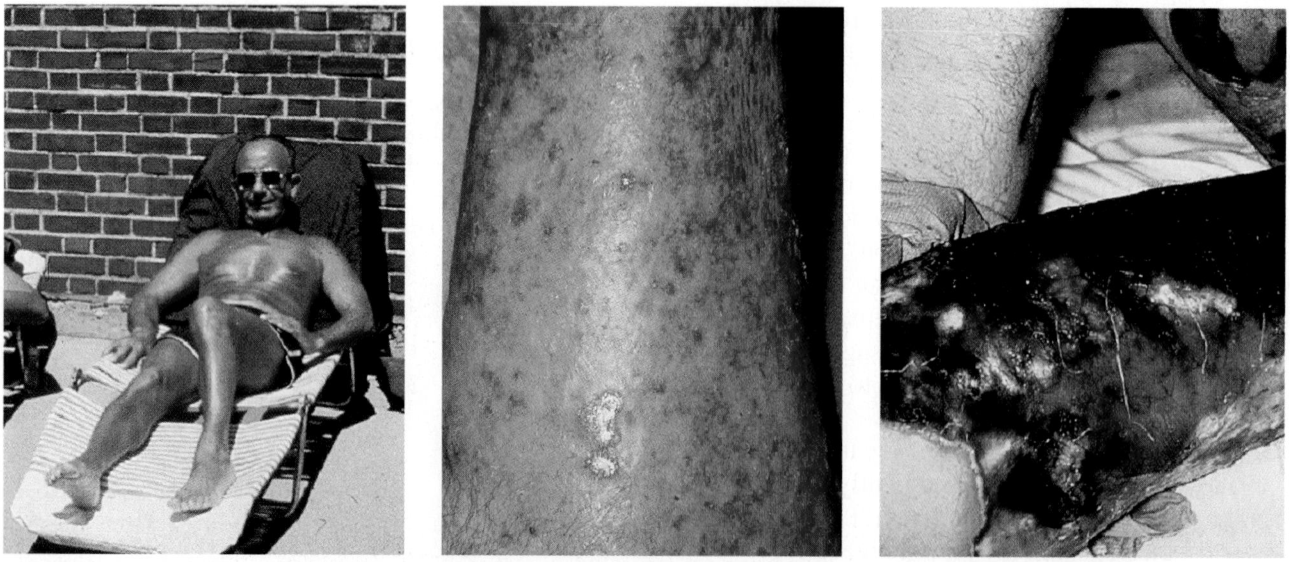

(A) (B) (C)

Figure 22-31 (A) Superficial burns, (B) partial thickness burns, and (C) full thickness burns.

be found in many different locations, from large, industrial plants to family residences.

Care Guidelines

1. After assessing scene safety and donning BSI, perform an initial assessment and attend to any major problems found with the airway, breathing, or circulation.
2. Reassure the patient, keep the patient calm, and monitor vital signs frequently.
3. If the chemical substance is dry or in powder form, brush off as much of the substance as you can from the patient, and remove the patient's clothing, watches, and jewelry. (Firefighters should make sure they protect themselves when removing the chemical from the patient. Do not use bare hands or brush the substance into the wind where it may become airborne and harm others. Consider additional protective gear, such as masks and gowns.)
4. If the chemical substance is not in dry or powder form (liquid chemicals, acids, etc.), flush the substance off the patient with large volumes of gently flowing water. Use a garden hose without a pressure nozzle, bottles of water, or water from a faucet to flush the substance off the patient. Remove all clothing, watches, and jewelry from the patient.
5. Monitor the patient's airway and breathing constantly. Chemical burn victims may also have inhaled a substance that can cause damage to the lungs and create airway or breathing problems.
6. If the chemical has gotten into the patient's eyes, flush the eyes with large amounts of gently flowing water for at least fifteen to twenty minutes.

Poisoning

The ingestion or inhalation of a caustic substance is considered poisoning. Poisoning can also result from the ingestion of a large quantity of a normally harmless substance, like over-the-counter medications or household substances as shown in **Figure 22-32**.

Care Guidelines

1. After assessing scene safety and donning BSI, perform an initial assessment and attend to any major problems found with the airway, breathing, or circulation.

Figure 22-32 Many types of substances and materials can be harmful to people and cause poisoning. When caring for a patient who has possible poisoning, the firefighter should make every attempt to find out exactly what substance was ingested or inhaled and how much.

2. Reassure the patient, keep the patient calm, and monitor vital signs frequently.
3. Do some investigative work to find out *exactly* what substance was ingested or inhaled. Keep that information handy.
4. *Do not* give the patient any liquid or food to ingest.
5. Contact the local poison control center. Check with your department or agency and find out what your local poison control center is and the telephone number for the center. Tell the poison control center all essential information about the patient, such as age, sex, medical history, and symptoms. Also, tell the poison control center *exactly* what substance was ingested or inhaled, and how much of it was ingested or inhaled. The poison control center will be able to suggest additional treatment for the patient before the EMS unit arrives.

Streetsmart Tip Local poison control centers are very valuable and helpful resources. Even if the exact name or dosage of the substance is unknown, they may be able to help the emergency responder figure out what it is.

Poison control centers have special references to assist them in figuring out "mystery" substances. For example, if a pill is found, a firefighter can describe it to the poison control center, and they may be able to find out what the name of the drug is and even the dosage.

Fractures and Sprains

A **fracture** is a medical term for a broken bone. A **sprain** is an injury to the ligaments that hold joints in the body together and allow them to move. Firefighters may notice swelling and tenderness to the area of a fracture or sprain. In the case of a fracture, there may be deformity to the bone and body structure.

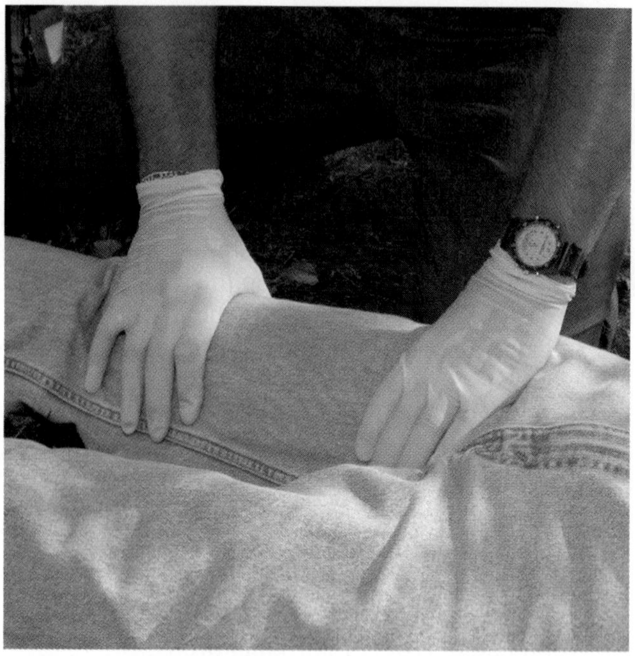

Figure 22-33 Firefighter stabilizing a leg injury until the EMS unit arrives.

Care Guidelines

1. After assessing scene safety and donning BSI, perform an initial assessment and attend to any major problems found with the airway, breathing, or circulation.
2. Reassure the patient, keep the patient calm, and monitor vital signs frequently.
3. Do not move the patient, and protect the cervical spine area in cases where the patient has fallen or been involved in a moving vehicle accident.
4. Do not move the injured part of the body. Do not attempt to straighten an arm or leg that is deformed. Carefully protect the injured area and be sure it does not move or shift, **Figure 22-33**.
5. As needed, apply cold packs or ice packs to the injured area to help relieve some of the pain and swelling.

Lessons Learned

Emergency medical care is a major part of most firefighters' responsibilities, and the treatment a firefighter gives a patient may make the difference between life and death. As the field of emergency medicine has progressed, the fire service has played an increasing role in providing first response emergency care. This chapter has provided some very basic information about first aid and emergency medical treatment, but there is a lot more to be learned.

The most important tools used to provide emergency medical first aid do not come from a trauma bag or a first-aid kit. The most important tool is the firefighter's brain. The job of a first-aid provider is a lot like that of a detective. Firefighters should use their senses to discover all that they can about a patient and the environment, and then treat the patient accordingly. A life could be saved in the process!

> **Streetsmart Tip** All firefighters should consider furthering their education to become emergency medical technicians and even paramedics.

KEY TERMS

Abandonment Abandonment occurs when an emergency responder begins treatment of a patient, and then leaves the patient or discontinues treatment prior to the arrival of an equally or higher trained responder.

Abrasion A scrape or brush of the skin usually making it reddish in color and resulting in minor capillary bleeding.

Allergic Reaction The body's reaction to a substance to which there is an allergy.

Amputation Occurs when part of the body is severed completely as a result of an injury.

Arterial Bleeding Bleeding from an artery.

Arteries The blood vessels, or tubes, within the body that carry blood rich with oxygen and nutrients away from the heart.

Automated External Defibrillator (AED) A portable computer-driven device that analyzes a patient's heart rhythm and delivers defibrillation shocks when necessary.

Avulsion An injury where a part of the skin is torn away, but still attached, leaving a flap or loose area hanging.

Body Substance Isolation (BSI) Precautions A set of precautions for emergency responders designed to prevent exposure to any body fluid or substance.

Brachial Artery A major artery in the inside of the upper arm that supplies blood to the arm. Can be used as a pressure point for controlling bleeding and for locating a pulse on an infant.

Capillaries The very small blood vessels in the body that connect arteries and veins and filter the oxygen and nutrients from the blood into the tissues of the body.

Capillary Bleeding Bleeding from a capillary.

Cardiovascular System The heart, blood vessels, and blood within the body.

Carotid Pulse The pulse located on either side of the neck.

Chemical Burns Burns caused by chemical substances that come into contact with the skin or tissues of the body, creating a caustic reaction.

Communicable Disease A disease that can be transmitted from one person to another.

Consent The acceptance of emergency medical treatment by a patient or victim.

Constricted A condition of the pupils where they are much smaller than normal and may appear almost like a "pinpoint."

Dehydration A loss of water and vital fluids in the body.

Dilated A condition of the pupils where they are much larger than normal and can take up almost the whole colored portion of the eye.

Exposure A contact with a potentially disease-producing organism; the contact does not necessarily produce the disease in the exposed individual.

External Bleeding Bleeding that is coming from an open wound on the body.

Femoral Artery A major artery in the lower body near the groin that supplies the leg with blood. Can be used as a pressure point for controlling bleeding in the lower extremities.

Fracture A medical term for a broken or cracked bone in the body.

Full Thickness Burns Burns affecting not only the skin structure but the tissues and muscles underneath. Full thickness burns may be red, white, or charred in color and will appear dry because the blood vessels in the skin are damaged extensively and are not supplying fluids to the area.

Hypoperfusion A serious condition caused by a problem or failure of the circulatory system that results in a decrease of oxygen and vital nutrients to the body's tissues. Also known as shock.

Implied Consent The assumption of acceptance of emergency medical treatment by an unconscious patient or a child with no parents or legal guardians present.

Incision A cut to the skin that leaves a straight, even pattern.

Infection Control Procedures and practices for firefighters and emergency medical care providers to follow to prevent the transmission of diseases and germs from a patient to themselves or other patients.

Infectious Disease See *communicable disease.*

Initial Assessment The initial investigative action taken by care providers to determine if the patient has the basic signs of life as well as any serious, life-threatening injuries.

Internal Bleeding Bleeding within the body when no visible open wound is present.

Laceration A cut to the skin and underlying tissues that leaves an irregular, even pattern.

Medi-vac An ambulance that transports patients by air. Typically, medi-vac units are helicopters with highly trained EMS personnel and nurses.

Partial Thickness Burns Burns affecting the entire skin structure that lies over the top of the fatty tissues and muscles causing skin to turn red and blistering of the skin.

Puncture An injury caused by an object that has stabbed the body.

Radial Pulse The pulse located in either wrist.

Shock A serious condition caused by a problem or failure of the circulatory system that results in a decrease of oxygen and vital nutrients to the body's tissues. Also known as hypoperfusion.

Sprain Injury to the ligaments that hold the body's joints together and allow them to move.

Standard of Care A legal term that means for every emergency medical incident, an emergency responder should treat the patient in the same manner as another emergency responder with the same training.

Superficial Burns Burns affecting the outermost layer of skin, which typically cause redness of the skin, swelling, and pain.

Thermal Burns Burns caused by heat or hot objects.

Triage A quick and systematic method of identifying which patients are in serious condition and which patients are not, so that the more seriously injured patients can be treated first.

Veins The blood vessels, or tubes, within the body that carry blood lacking oxygen and nutrients back to the heart.

Venous Bleeding Bleeding from a vein.

REVIEW QUESTIONS

1. What are the basic elements of an emergency medical system? What is the firefighter's role in the emergency medical system?
2. Describe the principles of BSI. Discuss how you would use BSI as a firefighter providing emergency medical care to a patient. When are BSI precautions needed?
3. What are the five major parts of an initial assessment?
4. What are the three types of external bleeding and the characteristics of each?
5. What are the three types of thermal burns, and how would a firefighter treat a patient with thermal burns? What is the difference in treatment between a patient with thermal burns and a patient with chemical burns?
6. How do you contact your local poison control center? What information do you give to the poison control center when contacting them for assistance?

Additional Resources

American Association of Poison Control Centers, http://www.aapcc.org

American Heart Association, http://www.americanheart.org

American Red Cross, http://www.redcross.org

Beebe, Richard, Med, RN, NREMT-P; and Deborah Funk, MD, NREMT-P, *Fundamentals of Emergency Care,* Delmar Learning, a part of the Thomson Corporation, Clifton Park, NY, 2001.

BLS for Healthcare Providers, American Heart Association, Dallas, Texas, 2002.

Burn Resource Center, http://www.burnsurvivor.com

Centers for Disease Control, http://www.cdc.gov

Fundamentals of BLS for Healthcare Providers, American Heart Association, Dallas, Texas, 2002.

Fundamentals of BLS for Healthcare Providers (Video), American Heart Association, Dallas, Texas, 2002.

Occupational Safety and Health Administration, http://www.osha.gov

Walter, Andrea; Edgar, Chris; and Marty Rutledge, *First Responder Handbook,* Delmar Learning, a part of the Thomson Corporation, Clifton Park, NY, 2003.

CHAPTER 23

FIREFIGHTER SURVIVAL

David W. Dodson, Lead Instructor, Response Solutions, Colorado

OUTLINE

- Objectives
- Introduction
- Incident Readiness
- Safety at Incidents
- Firefighter Emergencies
- Lessons Learned
- Key Terms
- Review Questions
- Endnotes
- Additional Resources

STREET STORY

It is one thing to talk about the life and death nature of firefighting—yet another when you are actually experiencing a situation that is about to kill you. A million thoughts go through your head and time seems to slow down. You rely on training, instinct, awareness, and even luck to get you through. My own experience happened one cold winter night—the kind of night where you didn't want to go outside. I was the second-arriving company officer to a rural, single-story farmhouse with a well-involved attic fire. The older home had a sturdy, high-mass roof structure so we elected an aggressive interior attic—that is, pulling ceilings and hitting it from underneath. Conditions inside the home were clear, although we knew there was significant fire over our heads in the attic space. In short time, we had two lines operating inside and making good progress. I noticed that the ceiling was sagging in the large "great-room" area of the house—and advised all crews to stay near walls and doorways and away from any open-span areas in case the roof collapsed. Seeing this potential increase, I thought we could simply force the collapse by pulling down the sagging area. I told my crew to take cover behind me in a doorway while I yanked down the ceiling. All was "safe" in my mind because we were eliminating a hazard and making access to the burning attic easier. Finding a good leverage point became troublesome so I moved into the room just a few feet. The roof then let go behind me.

As I turned to escape out the doorway, the roof hit me and flipped my 230 pounds over a couch. The first thought to cross my mind was dying. The second thought was a question: Where did my crew go? Training then kicked in. I took inventory of my environment: I was on my back (with SCBA—ouch), still breathing, legs working, no flames, light smoke, and a ceiling/roof structure pinning me to the floor and couch. I called for my crew—no reply. I called out for anyone—no reply. I couldn't get to my radio or PASS device manual activator. I thought about lying motionless to help activate the device but blew off that idea and went into self-help mode. I reasoned that I should try to self-extricate and get to my crew somehow. My instinct told me to move. I knew the fire was right above me with only drywall and framing separating me from it. With an energy burst, I was able to spin myself over to my side and began to crawl through a slight void created by the furniture in the room. Earlier, I was aware of three exits to the area. The way I came in was now blocked but I recalled a side door and a rear sliding glass door. Luckily, the void led to the sliding glass door. I exited and went to the entry point of my crew. They were shielded from the collapse by the doorway and they simply exited to begin looking for another way in to find me. What seemed like ten minutes under that roof was more like seconds according to my crew. They didn't even have time to report the collapse before they were out and I met them.

Training, instinct, awareness, and luck figured into this survival. Use this chapter to help build the training, instinct, and awareness to help you survive. Don't rely on the luck.

—Street Story by Dave Dodson, Lead Instructor, Response Solutions, Colorado

OBJECTIVES

After completing this chapter, the reader should be able to:

- List the three main components that lead to incident readiness.
- Define the four key checks to ensure that PPE is ready for response.
- List three types of personal accountability systems.
- Define personal size-up.
- Describe the three components that lead to "fitness for duty."
- Name three practices that lead to team continuity.
- Define risk/benefit.
- List and briefly describe the three components of rehabilitation.
- Describe the procedures that should be taken to establish and prepare for the assignment of a rapid intervention team.
- List the five steps that can lead to an organized rapid escape.
- List the three steps that should be taken when entrapment occurs.
- Compare and contrast post-incident thought patterns and critical incident stress.

INTRODUCTION

The excited radio transmission of "Mayday, mayday, firefighters injured and trapped" will present perhaps the most stressful circumstances that a firefighter will ever face. While no firefighter or fire officer expects this to happen, it would be negligence if training efforts failed to address firefighter survival and the handling of situations in which firefighters are injured, trapped, missing, disoriented, or imperiled—all firefighter emergencies.

Obviously, it makes sense to attempt to prevent a firefighter emergency from happening. Firefighter survival is accomplished through significant proficiency training and education. Daily, firefighters can help prevent firefighter emergencies through incident readiness—that is, the efforts to ensure that all firefighters are mentally and physically ready to respond. Personal protective equipment (PPE), task accountability, and fitness for duty are all issues that help prevent firefighter emergencies. At an incident, firefighter actions, inaction, and attention to hazards will help prevent a firefighter emergency. Likewise, attention to team continuity, orders and communication, rapid intervention planning, and rehabilitation all help prevent firefighter emergencies.

If a firefighter emergency were to occur, a planned, systematic process of rescue would have to be established to avoid compounding the seriousness of the emergency. Having procedures in place for rapid escape and self-extrication, accountability recall, and post-incident emotional processing is essential. This chapter discusses all of these areas and gives firefighters survival tools to help prevent firefighter emergencies and to assist them when such an unfortunate event does occur.

INCIDENT READINESS

Unlike most professionals, firefighters must perform at peak mental and physical levels with little or no notice. Constant readiness is imperative. Preparing for incident response involves more than firefighters merely making sure they know where they are going. Incident readiness is a mental process that answers a few questions:

1. Am I in a position to respond?
2. Is my protective gear available?
3. What is my relationship to the response?
4. Physically, can I respond?
5. Mentally, can I check out of my current thoughts and focus on response?

A negative answer to any of these questions may set a firefighter up for an injury. To avoid this risk, firefighters must take steps to ensure that their "system" is assembled and ready for response. Some readiness system items are addressed before an alarm is received, while others are addressed on receipt of the alarm. This "system" is the key to survival and includes some important components.

> **Note** Namely, the readiness system includes PPE, personal accountability, and fit-for-duty status.

Personal Protective Equipment

The personal protective ensemble can be considered the first step in one's ability to focus on incident handling.

> **Safety** One fire service adage claims that PPE is the first and last means of defense from risks.

Specifically, PPE is the first thing firefighters put on and the last thing they want to be left with when the incident is over, **Figure 23-1**. If firefighters rely on PPE to protect them from some hazard, they have in effect used up all other protective measures. As an example, if firefighters find that their PPE prevented a serious burn, that means the PPE became the last means of prevention. Firefighting streams, ventilation, zoning, and access/egress efforts have likely failed if the firefighter got to the point that PPE made the difference between injury and no injury.

Caution At the beginning of a duty tour or shift, firefighters must ensure that their protective equipment is dry, serviceable, and ready for quick donning.

In many departments, firefighters must check *all* their various ensembles. Firefighters working at Fire Station 4 in Loveland, Colorado, must prepare their structural ensemble, aluminized proximity suits, wildland gear, and (EMS) PPE each and every shift. This readiness check ensures that the firefighters are ready for any type of incident that is likely to occur within Station 4's response district, which includes an airport, commercial and residential structures, rural areas and fields, and EMS responsibility for 10,000 people.

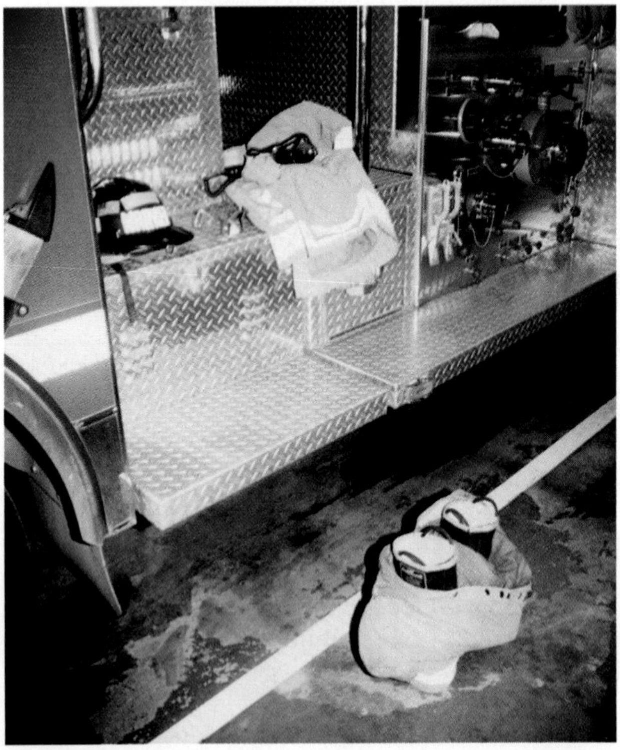

Figure 23-1 PPE is part of the survival system. Is it ready?

When preparing PPE for readiness, firefighters should ensure the following:

1. All clothing materials are dry. Wet clothing reduces the protective insulation of the clothing and may lead to steam burns in heated environments.
2. All PPE is present and positioned so that it may be rapidly donned. This is important for station-assigned firefighters and those who respond to a station to staff an apparatus. Firefighters who respond from home directly to an incident must pack their dry gear in such a way that it can be donned quickly and appropriately.
3. Essential "pocket tools" are available and in working order. It is not uncommon for firefighters to carry an assortment of personal tools that they feel are important. Small flashlights, trauma shears, doorstops, nylon webbing, carabiners, grease pencils, chalk, sprinkler wedges, multiple-tool pliers, knives, foam earplugs, self-escape rope, and two-way radios are all common pocket tools found in firefighters' PPE.
4. Alternative PPE items are appropriately packed and ready for use. As stated earlier, many firefighters must use (EMS) PPE, wildland PPE, proximity PPE, or lightweight rescue PPE. These should also be checked and made ready.

When an alarm is received, the time spent donning gear *before* arrival at the incident is time well invested. People waiting until incident arrival to don PPE are setting themselves up to "shortcut" their protective ensemble. These shortcuts are caused by firefighters who are concerned and distracted by the situation and the actions that need to be started rather than by completion of their ensemble. Donning structural PPE prior to incident response is preferable; however, driving in full PPE can introduce some of its own safety issues. While large fire apparatus allows some room for the bulky structural PPE, driving a personally owned vehicle (POV) or a small support apparatus may be more cumbersome. If a driver makes the choice to don PPE after driving to an incident, strict discipline *must* be exercised to fully don PPE before engaging in the incident.

Caution Firefighters should not find themselves at an incident without their PPE fully donned; they must practice strict self-discipline to complete the ensemble. They must not let the urgency of the situation override prudent judgment!

Personal Accountability

Personal accountability is an essential part of firefighter readiness. Personal accountability refers not only to the established accountability system used by a particular fire department, but also to the firefighter's relationship to the response and ability to perform as trained.

Accountability System

The accountability system used by a department can take on many forms. Typically, the system will fall into one of three general types:

1. *Passport:* This is a crew-card system that is tracked on a status board by a monitor. Members of a crew give a name chip to the company officer or team leader, who places all of the team names on a card or **passport**. When given an assignment, the passport is given to the monitor, who tracks all crews. Team leaders must report any changes in their teams' location or assignment.
2. *Tag:* In this system, individual firefighters report to staging and give an identification tag to the staging manager. The staging manager groups the tags into teams and tracks the progress of each. First arriving crews and personnel will not have a staging/accountability manager to collect their tags. In these cases, individual tags are usually hooked on a ring of the first-due apparatus, and the staging/accountability manager, once assigned, will collect these.
3. *Company officer:* This system is perhaps the oldest and most used system. Here, the company officer, team leader, or other supervisor is responsible for keeping track of the crew. This system is the least preferred system in that many firefighters have been lost and many crews broken up due to the lack of formality of the system.

Regardless of the system used, each firefighter must be aware of how the system works and follow its specific guidelines. Failure to be accounted for on the incident is akin to **freelancing**—that is, performing a task that has not been assigned or performing a task alone—both are fire service taboos, **Figure 23-2**.

> **Streetsmart Tip** Not only can firefighters endanger themselves through freelancing, but may also indirectly injure others when they attempt to find the freelancer.

Figure 23-2 Accountability systems take on many forms—firefighters must know how to check in!

Worse, if freelancers become lost or trapped, nobody will come to assist if they do not know freelancers are on scene and working.

Relationship to the Response

Another part of the accountability equation is attention to each individual's relationship to the response. Specifically, firefighters' relationship to the response includes their assignments and their own personal size-ups.

- *Assignment.* In most cases, firefighters are given readiness assignments—that is, being assigned as given crew members for given positions on the apparatus. With these assignments come responsibilities to ensure that they perform preassigned or reactive duties. In cases where firefighters have no preassigned duty, firefighters must be prepared to perform a host of duties. The readiness elements entail a mental preparation to perform the tasks of that position. If firefighters are assigned positions that they have not performed recently, they may not be ready to act—leaving an invitation to injury. In some departments, a list of crew duties for a given seated position on the apparatus is listed on a plastic card—this is a great reminder. Firefighters should take the time to reacquaint themselves with the expected tasks, tools, and procedures required. This improves readiness and reduces injury potential.
- *Personal Size-Up.* **Personal size-up** can be defined as a continuous mental evaluation of firefighters' immediate environments, facts, and probabilities. This mental evaluation should have them evaluating the weather, time of day, current

chain of command, and likely assignment. Once again, they are mentally preparing themselves—thereby reducing injury potential.

Work as Trained

Streetsmart Tip Personal accountability is achieved when all firefighters and fire officers are able to perform assigned essential tasks—tasks they have been trained for—and keep the chain of command advised of their progress.

Of significant importance here is everyone's ability to perform as trained. A common thread discovered in post-incident investigations is the fact that many of the injured were performing tasks they had never experienced before. In some cases, the training had been given by the individual's department, but the individual either forgot *how* to accomplish the task or chose a method that was inappropriate for the situation. How can this be prevented? Listed next are two key points to assist firefighters.

Firefighters should perform as trained. This sounds easier than it is. Most departments invest considerable cost to train firefighters, so firefighters should do their part and constantly practice and perfect their skills. For career firefighters, this should be no challenge—evenings around the station are perfect times for doing this—it does, however, take motivation. Volunteers have a bigger challenge. They should take the initiative to go to the station and practice, and get others to join in the effort.

Firefighters need to know their strengths and weaknesses. The achievement of a firefighter certificate does not mean a firefighter has all of the necessary skills to be an effective emergency responder. It is merely a minimum. Training never ends for a firefighter—ever! Firefighters must constantly test themselves and try to improve those areas where they discover weaknesses. As a matter of fact, firefighters should strive for *mastery* of their assigned tasks, **Figure 23-3**. Mastery can be defined as the ability to achieve 90 percent of an objective 90 percent of the time. Further, firefighters should expand their abilities.

Caution Firefighters should not let the incident scene become an avenue for discovery and on-the-job training.

Fitness for Duty

Fitness for duty is more than just being "in good shape."

Figure 23-3 Firefighters achieve mastery of tasks through repeated training. Mastery reduces the chance of injury.

Note Fitness for duty includes mental fitness, physical fitness, wellness, energy, and rest, **Figure 23-4.**

If just one of these is less than adequate, the window to being injured opens.

Mental Fitness

Being mentally ready to respond to an incident at all times rarely happens—firefighters' lives are complex, busy, and full of mental and, in some cases, emotional challenges. As has been stated

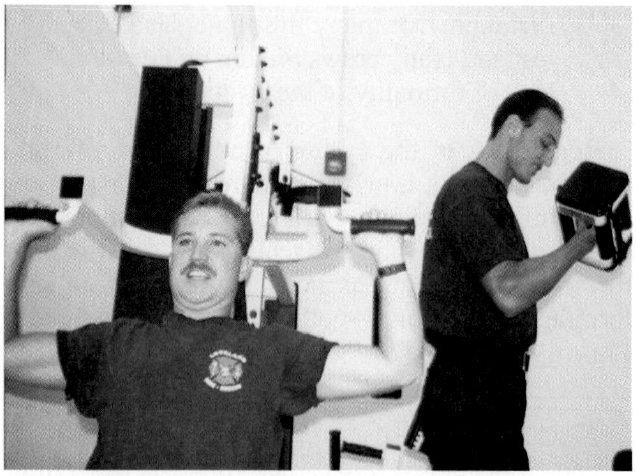

Figure 23-4 Fitness for duty includes more than just being "in shape." It also includes wellness, rest, energy, and mental fitness.

often, a firefighter's size-up begins on receipt of the incident. If a firefighter is involved in an emotional event during this phase, key size-up information may be missed. The key is for the firefighter to "check out" of the environment and "check in" to size-up. The firefighter's non-firefighting acquaintances may not understand the rapid mental departure—it is common, however, for them to forgive the firefighter.

Once again, it is easier to say a firefighter should "check in" to the incident than to actually do it. Many firefighters have a mental ritual to help them achieve the check-in state every time. They create a ritual—a physical or mental act that reminds them to check out of their emotional or current mental state. For some, it comes when they buckle their seat belt and respond. Others take a deep breath and force the exhalation. Still others perform a rapid mental checklist: What is the call? Where is the call? Is my gear ready? What is my relationship to the response—first due, second due?

Physical Fitness and Wellness

There is no doubt that fit firefighters are less injury prone than those who are unfit. Fitness is a lifestyle commitment. On receipt of the alarm, physical fitness becomes more than an issue of how much firefighters "work out." Do they have injuries? Are they recovering from injuries? Are they recovering from a cold or the flu? Any of these conditions can affect performance and fitness. The simple act of going from a resting heart rate at alarm time to a maximal heart rate on arrival at a working fire can stress the body to the point of collapse if it is not capable of accommodating the spike. A firefighter who is in "great shape" can collapse quickly at a fire if fighting off the infection of a flu. Likewise, persons taking prescription drugs may experience rapid loss of strength or fatigue easier. The lasting effects of alcohol can also impair the firefighter well after consumption.

Physical fitness and wellness are highly individual—that is, firefighters are all different and have varied fitness needs based on genetics, gender, metabolic rate, and personal history (injuries, tolerance levels, etc.). However, firefighting contains some universal demands that require a certain level of fitness and wellness. Specifically, firefighters need to create a fitness and wellness lifestyle that includes the following:

- *Cardiovascular conditioning.* Without a doubt, this is critical. Firefighters seldom operate at incidents with a resting heartbeat. Repeated conditioning at elevated heart/breathing rates strengthens the heart/lung relationship and reduces the chance of overexertion at incidents.
- *Core strengthening.* The human core can be defined as the muscle structure that supports the hips, back, and thoracic cavity. This "power center" is used in virtually all physical activity at an incident. Sit-ups, crunches, and balance-oriented exercises (using a fitness ball) help strengthen this core.
- *Flexibility improvement.* The stress on joints, tendons, and ligaments can be brutal in the firefighting world—strains and sprains are the leading firefighter injury. Following a brief cardiovascular warm-up, firefighters should apply stretching and flexing routines to help condition joints, tendons, and ligaments.
- *Resistance training.* Firefighting is labor intensive. Resistance training (weight lifting) helps build muscle that makes the firefighting labor demand easier on the body. While many strategies exist, resistance training that systematically works *all* major muscle groups is best and will help keep postural balance.
- *Nutritional balance.* Perhaps the most neglected area of fitness and wellness of firefighters is the daily intake of food. Although this highly personal topic is fodder for much debate, most fitness experts agree that the key is to eat less but eat more often. Eating a balance of carbohydrates, protein, and fats along with a programmed intake of supplements helps the firefighter find nutritional balance. Firefighters working a shift should stay well hydrated by drinking plenty of water throughout the shift. Firefighters who are on call or in volunteer systems should keep drinking water close and hydrate en route to the fire station or incident.

Energy and Rest

Safety is directly affected by the energy potential and rest a firefighter has stored. Obviously, if the firefighter is short on sleep, the physical and mental capacities are reduced. Likewise, if it has been more than three hours since a meal or the food (energy) in the stomach has been used up, the firefighter will reach a fatigued state quicker.

> **Caution** For most firefighters, the worst time to respond to an emergency is just after midnight, when they have not received much sleep after a long day, it has been five or six hours since the last meal, and the mind and body are in repair mode.

What is the solution? The answer lies in the firefighter's awareness. The firefighter should be attentive to energy and rest levels and plan to communicate an early rehabilitation need to an officer during operations.

> **Streetsmart Tip** One simple suggestion: Firefighters should keep a balanced energy bar (protein, carbohydrate, and fat) in the apparatus or car and eat part of it during their response. A half a bar of protein, carbohydrate, and fat will serve the body well if it has been more than three hours since the last meal. Hydrating prior to incident engagement, however, is essential. Drinking some water while en route helps maintain heart stroke volume when sudden work demands are required.

SAFETY AT INCIDENTS

With roughly half of all duty-related injuries and deaths happening at the incident scene, it makes sense for each firefighter to develop a system to follow to minimize risk of injury or death. It is important to note here that *nobody* tries to get injured at an incident—the circumstances of the incident have created the potential. In most cases, the individual receiving the injury or the other firefighters and fire officers have failed to proactively "see" the events that led to the injury. Additionally, injuries and death occur when firefighters allow the urgency of the situation to override prudent judgment. What is being said here is valid—injuries can be prevented. At the individual level, a firefighter can prevent injuries and even death through a series of mental and physical actions. This section shows these mental and physical actions.

Team Continuity

> **Note** The success of all incident handling is dependent on trained, assembled teams accomplishing organized tasks, **Figure 23-5**.

Figure 23-5 Freelancing is eliminated and incident success is gained when specific tasks are assigned to teams of two or more firefighters operating from a single incident action plan.

Departments that allow well-meaning firefighters to perform as each sees fit will ultimately experience a duty death or significant injury. The fire service has coined the term *freelancing* for this type of action. Freelancing is dangerous. Countless investigative reports cite freelancing as a significant contributing event to death or injury. To eliminate freelancing, each and every firefighter must be assigned to a team of two or more and be given specific assignments. On completion of a task, the team must report that the task is done and also report any update information that could be useful to the incident commander. The incident commander will then reassign the team or have the team report to staging or rehabilitation.

Additionally, the team must exercise guarded judgment when completing a task. For example, a team performing overhaul may decide that an additional tool is needed. The simple decision is to send a lone firefighter to retrieve the tool, but this sets up a situation that breaks team continuity. Rather than sending a lone firefighter, it may be prudent to send two, leaving two to continue task completion. In the case of a three-person crew, the team leader should weigh the potential of breaking the team. In most cases, it is best to keep the team together and ask for a shuttle team to bring the needed tool or to take a momentary break and have the team exit the area together, get the tool (and perhaps a drink of water), and then return. This process may seem inefficient, but it underscores a guarded, safe approach to team continuity. **Figure 23-6** outlines practices that contribute to team continuity.

Keys to Team Continuity

- Know who is on your team and who is the team leader.
- Keep your team in *visual* contact at all times. If visibility is obscured, be within touch or voice distance of your partner.
- Communicate your needs and observations to the team leader.
- Rotate to rehab and staging as a team.
- Watch your team members—practice a strong "buddy-care" approach.

Figure 23-6 Team continuity reduces the chance for injuries.

Orders/Communication

The incident commander is responsible for assembling an **incident action plan (IAP)** that is implemented by teams performing tasks. These tasks are assigned to organized teams in the form of orders. Each team is responsible for carrying out the order and providing updates on a regular basis. Additionally, the team must relay information about any pertinent hazards or conditions that may be important to the overall IAP.

Safety As individuals, firefighters must keep their team leader advised of conditions and hazards they find as work is performed.

The discovery of a weak stairway, of a significant weight load in an attic or on the roof, or of fire in a void space are all important facts that should be relayed to the team leader and subsequently up the chain of command of the incident management system (IMS).

Occasionally, a team performing an order is given a different order by someone else in the IMS. In this case, the team leader needs to inform the person giving the new order that they are already under orders. The person giving the new order then must either countermand the original order or find someone else to perform the new order. In these cases, it is *imperative* that both the team and the person giving the new order communicate any changes to the person giving the original order.

Typically, the first arriving group of teams and apparatus performs a prescribed set of tasks or orders based on a preplanned or standard set of procedures. In these cases, the teams should know not only the tasks,

Figure 23-7 Late-arriving teams can benefit from a quick, formal safety briefing prior to performing tasks.

but also the tools required and any safety considerations. An official safety briefing is usually not held and teams carry on their task. This does not mean that team members should not vocalize safety thoughts among themselves—in fact, this should be encouraged. Building construction considerations, fire behavior and smoke observations, and access/egress routes should be discussed as the team approaches its task.

Later-arriving teams are given assignments based on the IAP. These teams should ascertain essential information and hold a quick, formal safety briefing based on the conditions present and the hazards that are faced, **Figure 23-7**. This practice is not new—wildland fire crews are *required* to perform a crew briefing prior to task accomplishment. Most proactive fire departments are initiating these safety briefings as a matter of habit. **Figure 23-8** lists good habits regarding orders and communications.

Good Habits Regarding Orders and Communications

- Clearly understand any order; ask questions and seek clarity.
- Assemble tools and openly discuss procedure as you approach tasks.
- Relay all pertinent hazards found during the course of task completion.
- Update crew status on a regular basis.
- Advise when tasks are complete.
- Communicate rotation to rehab and/or staging.

Figure 23-8 Practicing good reporting habits enhances communications.

Risk/Benefit

Another key to firefighter survival is the concept of risk/benefit. Simply stated, **risk/benefit** is an evaluation of the potential benefit that a task will accomplish in relationship to the hazards that will be faced while completing the task. As an example, the task of vertically ventilating a residential structure fire will benefit the fire suppression and lifesaving effort by increasing visibility, reducing heat buildup, and reducing overall damage. That benefit, however, is at the risk of roof collapse, operating on a potentially steep incline, or operating without quick egress. The team operating on the roof is often in the best place to judge this relationship. Signs of collapse, difficult footing, and obscured visibility may present a significant risk to the team. The risk/benefit evaluation in this example may cause the team leader to withdraw the team.

Risk/benefit evaluations take place at different levels within an IMS. At the command level, an overall evaluation takes place, as reflected in the IAP. At the team level, crews evaluate their immediate environment and report this back to the person giving them their orders, **Figure 23-9**. Regardless of the level of evaluation, some basic guidelines can be used to help make risk/benefit decisions:

- Firefighters will take a significant risk to save a known life.
- Firefighters will take a calculated risk, and provide for additional safety, to save valuable property or reduce the potential for civilian and firefighter injuries.
- Firefighters will take no risk to their safety to save what is already lost.

Figure 23-9 Solid risk/benefit analysis means taking no risk for that which is already lost. *(Photo courtesy of Richard W. Davis)*

These guidelines are common within the fire service and are more clearly spelled out in NFPA Standard 1500, *Fire Department Occupational Safety and Health Program.*

Personal Size-Up

As mentioned several times in this book, firefighters must perform a personal size-up. A size-up is a continuous mental evaluation process. Specifically, firefighters should continually evaluate the safety of their environment by staying aware of the following:

- *Established work areas.* Firefighters have to be aware of collapse zones, barriers to traffic, and hazard isolation zones.
- *Hazardous energy.* Electrical equipment, pressure vessels, chemicals, and even springs and cables can suddenly release—causing injury and death to firefighters. It is important to *scan* the environment and mentally log all forms of hazardous energy.
- *Smoke conditions.* Smoke is fuel waiting to ignite. Firefighters crawling in low-visibility, hot smoke are flirting with an explosive environment. Cooling and ventilation are key.
- *Escape routes.* The firefighter who continually plans two or three escape routes while performing tasks is investing in survivability if the established work area breaks down. The firefighter should count doors and windows, scan for traffic barriers, and plot pathways to "safe havens."

Rehabilitation

> **Note** A study of firefighter duty-related injury and death statistics shows that stress and overexertion consistently rank as leading causes.

Firefighting is hot, arduous work performed in PPE that does not allow body heat to evaporate through sweat. A key health and safety concern is controlling the heat stress that occurs when a firefighter's internal core temperature rises above its normal level during incident activities. Core temperatures that have risen above normal can lead to heat- and heart-related injuries and death. Proper rehabilitation efforts (*rehab* for short) will help keep firefighters safe by reducing heat stress. Additionally, good rehab practices will provide the rest, hydration, and nourishment needed for sustained operations. NFPA addresses specific practices for rehab in its document, NFPA 1584, *Rehabilitation of Members Operating at Incident Scene Operations and Training Exercises.*

> **Note** The key elements of rehab include rest, active cooling, hydration, and nourishment.

Rest

At an incident, rest is achieved during crew rotation. When a firefighter is rotated to rehabilitation, the firefighter should maximize the opportunity to rest by sitting down, by allowing medical personnel (which should be assigned) to do vital sign check, **Figure 23-10**, and by mentally disengaging from the event (which rests the mind). The firefighter who achieves these three important steps can actually work longer at any given incident—it allows a brief recovery from physical and mental stress and allows other rehabilitation elements (hydration and nourishment) to repair and prepare the firefighter for further work. The medical evaluation by a trained medical provider is an important part of rest. Many signs and symptoms indicate the need for further rest—or removal from the incident scene.

One additional point can be made here. Some firefighters do not feel the need for rehabilitation and may be reluctant to spend time in rehabilitation while others are aggressively engaged in the incident—these firefighters are potential candidates for overexertion. Remember, the time to rehabilitate is *before* a firefighter gets tired. "Rehab early and often" should be the fire service creed, especially if the incident duration will span hours, **Figure 23-11**.

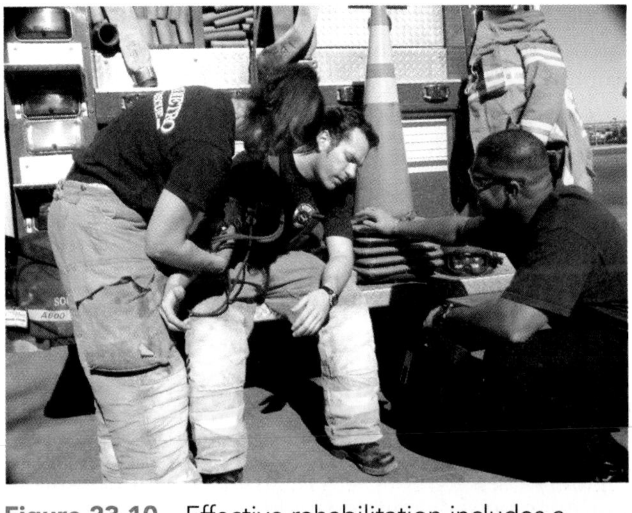

Figure 23-10 Effective rehabilitation includes a medical evaluation by BLS or ALS personnel.

Active Cooling

As mentioned at the start of this section, increased core temperatures can cause heat- and heart-related injuries and death. During rehab, firefighters should engage in active cooling to reduce their core body temperature. Typically, firefighters have used passive cooling as a strategy to reduce core temperatures. Passive cooling includes the use of shade, air movement, and rest to bring down core temperatures. Medical personnel have typically monitored heart rate and a firefighter's perceived comfort to determine when a person has recovered sufficiently to don PPE and resume firefighting operations. New studies have shown that this strategy is *not* adequate and that heart rate and perceived comfort are not good indicators of sufficient core temperature cooling.[1] Effective core temperature cooling is achieved using an "active cooling" strategy. Active cooling is best achieved using a forearm immersion technique. Simply put, firefighters should doff their coats and submerge their hands and forearms in a basin of cool water. This technique is more effective than misting fans and has a reduced tendency to cause the chills that can happen when firefighters go immediately from hot environments to cold environments.

Figure 23-11 Rehabilitation should start well before a firefighter is thirsty or tired. Failure to rehabilitate "early and often" opens the door to injury.

Hydration

Hydration cannot be overemphasized in any incident environment, whether it is one of heat stress, cold stress, heavy workload, light workload, mentally taxing events, or long-duration events. Water is vital to the peak operation of virtually every body system from transport of nutrients, to blood flow, to waste removal, to temperature regulation. When the body becomes dehydrated, these systems start to shut down in order to protect themselves. With this shutdown comes fatigue, reduced mental ability, and, in extreme cases, medical emergencies such as renal (kidney) failure, shock, and death. The working firefighter must accommodate for the wearing of heavy clothing that does not allow evaporation of sweat. Additionally, the firefighter must account for strenuous physical activity under stressful situations. Hydration of firefighters should become paramount—even to the point of excess. Firefighters must not *wait until they are thirsty to drink water!*

Safety As a rule, firefighters should strive to drink a quart of water an hour during periods of work—this is best delivered in 8-ounce increments spread over the hour.[2]

Substituting carbonated and/or sugared beverages or other liquids for water can slow the absorption of water into the system. For this reason, just water should be given for the first hour. For activities lasting longer than an hour, some consideration can be given to adding essential electrolytes and nutrients along with the water. Many sports drinks are available that can achieve this. These sports drinks are best diluted 50 percent with water in order to speed their absorption into the system.

Nourishment

While dehydration and thermal stress can lead to energy depletion, most firefighters associate energy depletion with the need for food. Nutrition or "fueling" of the firefighter can have the effect of rejuvenating the firefighter or putting the firefighter to sleep. Too often, rehabilitation-feeding efforts accomplish the latter. A firefighter properly nourished will work smarter and safer. The firefighter who is improperly fed will not only want to "crash," but will likely make sluggish mental calculations leading to injury.

So what is the proper way to nourish firefighters? A brief study of essential nourishment theory can help answer this question. In basic terms, maximizing energy from the human machine takes a balance of four essential elements: oxygen, water, blood sugar (from food), and insulin. When this balance is present, other essential hormones and enzymes are created that make for a well-running human machine. The element that is often misprescribed is food—the foundation for building balance. True, *all* food is fuel for the firefighter. That is, all food is converted to glucose (blood sugar) for the muscles to use as fuel. Insulin is released into the bloodstream to help convert the blood sugar into energy for muscle use. Some foods are digested faster than others though. Complex carbohydrates, such as breads, pasta, and beans, require tremendous amounts of blood, and time, for digestion. This explains why some firefighters want to go to sleep after eating—blood is needed for digestion—and less energy is available for thinking and physical tasks.

The key to providing quick energy to the firefighter is to find a balance of protein, fat, and carbohydrate. Ideally, this balance should be a 30/30/40 mix—that is, 30 percent protein, 30 percent fat, and 40 percent carbohydrate.[3] This balance provides essential elements from three food groups. A balanced approach achieves a few benefits. One, the balance will stabilize insulin release into the bloodstream, helping to reduce the rollercoaster of blood sugar levels that often lead to sporadic activity, chemical imbalance, and fatigue. Second, the balance approach stimulates the release of hormones and enzymes that optimize human performance—both physical and mental.

Choosing the best protein, carbohydrate, and fat also promotes steady, sustained performance. Protein is best derived from low-fat meats such as turkey, chicken, and fish. Eggs and cheese also offer protein. Fats should be of the monounsaturated type like olive oil, nuts, and peanut butter. Often, carbohydrates are dangerous in that so many of the foods typically found on the fire scene are rich in unfavorable carbohydrates. Candy, bread, potatoes, and bananas are all carbohydrates that are quite rich and have a tendency to slow down the worker. Good carbohydrates include green vegetables, apples, tomatoes, and oatmeal.

A good example of a quick, balanced rehabilitation meal would include water, sliced turkey sandwiches on thin-sliced bread with mustard or ketchup, apples, and a handful of peanuts. Low-sugar energy bars can be stocked on apparatus and used as a meal replacement. It is best to use energy bars that are balanced 30/30/40.

Rapid Intervention Teams

Rapid intervention teams (RIT) are being formed by many fire departments as a proactive practice to rescue lost or trapped firefighters. NFPA 1500, *Standard for Fire Department Occupational Safety and Health Pro-*

gram, and NFPA 1561, *Standard for Emergency Services Incident Management System,* both require the formation of a RIT early in the incident if firefighters are exposed to dangerous environments or tasks.

As mentioned in Chapter 19 (two in/two out rule), the initial RIT includes the two "out" team members that support the two "in" members. This requirement is specific to interior fire attack but should be conceptually used for all incidents where firefighters are taking risks. As more resources arrive, the RIT becomes a designated team for a defined task. This RIT should have no other assignment than to prepare for the rapid deployment in support of a search and/or rescue.

Streetsmart Tip Firefighters assigned to the RIT should be prepared to act immediately on orders to rescue lost or trapped firefighters. This means the crew assigned the RIT task should be equipped and prepared for the task.

When preparing for the RIT assignment, firefighters should gather appropriate tools and equipment such as SCBA (donned and in the ready position), forcible entry tools (including a power saw), hand lanterns, radios, and a life rope. Additionally, the RIT crew should start preplanning a rescue operation. Information about building construction, layout, and entry/egress routes should be shared. The team should also inquire as to the IAP and current crew location and assignments and should monitor radio traffic. As a rule, the RIT should position near the main entry point of the incident hot zone and be ready for immediate deployment.

The radio transmission of "Mayday" or "Emergency traffic" should cause the RIT to size up the situation and mask up. Typically, it is best for the RIT to await orders from the incident commander to initiate the rescue/search, unless the IC has given the RIT the responsibility to self-initiate actions. Regardless, the RIT needs to communicate its actions on a frequent basis. **Figure 23-12** can serve as a RIT checklist for the firefighter assigned to a rapid intervention team.

Rapid Intervention Team Checklist

- ☐ Don PPE/SCBA
- ☐ Assemble team tools including:
  - Hand lanterns
  - Radio
  - Prying tools/ax
  - Lifeline rope
  - Thermal Imaging Camera
  - Lifting Equipment
  - Spare Air Supply
- ☐ Inventory building/environment
  - Construction type
  - Egress/entrance
  - Weak links
  - Layout/hazards
- ☐ Receive briefing on incident action plan and crew assignments/locations.
- ☐ Be ready by listening to the radio and staging near primary access point.

Figure 23-12 Firefighters assigned to a rapid intervention team should increase their readiness using a "checklist" approach.

FIREFIGHTER EMERGENCIES

So far, this chapter has focused on preventing injuries and deaths through proactive actions. Occasionally, preventive measures do not address every circumstance or the evolution of the incident presents the working firefighters with an emergency that threatens their lives. In these cases, each firefighter must rely on instincts and training to help survive the situation. To help understand the actions to be taken during an actual or potential firefighter emergency, the firefighter must study procedures for rapid escape and declaring a **Mayday** for lost and trapped situations. Further, the firefighter must understand that survival includes processes for the rescue of trapped and lost firefighters as well as long-term mental survival.

Rapid Escape

During the course of task completion, firefighting teams may discover a situation that requires immediate escape or may be ordered to do so via radio and fireground warnings. The firefighting teams must then take immediate steps to exit the building or environment.

Caution The two most common situations that will lead to this rapid escape need are rapid fire spread and building collapse. An event such as flashover, backdraft, or partial collapse of a building can lead to rapid fire spread, potentially trapping or endangering firefighters, **Figure 23-13**.

Figure 23-13 Rapid fire spread and partial collapse are likely to trigger the need for rapid escape.

A building being attacked by fire is always subject to collapse. Once this attack degrades the load-bearing portions of the structure, a collapse will occur. Teams that witness these events must take steps to initiate an evacuation. Teams that hear an established evacuation signal must also immediately egress. Some typical evacuation signals include a repeated series of air horn blasts from apparatus or a special radio tone followed by the evacuation order. The next section discusses some important steps that will assist in achieving immediate escape.

Rapid Escape Steps

1. *Preplan the escape.* All firefighters operating in a hazardous environment should continually look for multiple escape routes. Even in routine or nonthreatening environments, the firefighter should be constantly evaluating escape routes. This process develops good habits that will serve the firefighter well when a threatening situation does develop.
2. *Immediately report the need for rapid evacuation.* Actual collapse or signs of impending collapse can be communicated via radio while exiting the area. A delay in reporting these conditions could endanger other crews. Likewise, the witnessing of a flashover (or signs of an impending flashover or other hostile fire event) should be cause for immediate reporting.
3. *Acknowledge rapid evacuation or escape signals.* In cases where the team has not initiated the rapid escape, the team must acknowledge that it has heard the signal. Signals vary from department to department but typically consist of an audible fireground signal (a defined air horn signal or a barrage of sirens and air horns) as well as a radio broadcast (an alert tone followed by the evacuation order).
4. *Rapidly escape.* Rapid escape means just that—firefighters should leave the area immediately. They should leave hoselines and heavy tools and escape. Trying to bring these types of items with them has resulted in injuries to many firefighters. Often, it is faster for the team to use a way out that is different from the way it entered. This is where the team's escape preplan observations pay off. At times, firefighters will be faced with a situation that requires a temporary escape—in these cases, the priority is to protect the crew and plan for further escape. Here are some examples:

- *Room escape.* It may be possible to enter an uninvolved room, shut the door, and use a window for further escape. If the window is above the first floor, the firefighter may have to clear the window and wait for a ladder. These actions should be communicated if it is not obvious to crews outside. If the room has no window, the firefighter should think through the options. Often, a simple wall breach (drywall over wood studs) can get the team to rooms with alternative escape routes.

- *Roof/balcony escape.* When fire or collapse has cut off stairwells and hallways, it may be easier to get to the roof or balcony and await ladders. In these cases, it is best to close doors or create barriers that separate the team from the fire. It is essential to communicate the team's needs if the firefighters assisting the team are not in visual contact.

- *Self-rescue.* In some extreme cases, it might be possible to escape through a system of self-rescue. Many firefighters carry a personal rescue rope or strap so they can lower themselves to another floor. Firefighters equipped and trained to do this need to establish a barrier to fire spread, secure a reliable anchor point, have a device or method to descend the rope, and have a method to break windows on lower floors.

- *Ladder escape.* When a ladder is in place and available for rapid escape, firefighters should use safe and common methods to mount and descend the ladder. In hurried situations (like collapse and flashover conditions), firefighters may opt for dangerous headfirst mounting and ladder slides. It is important to note that these techniques can lead to serious injuries or death. Locally, fire departments may choose to instruct firefighters on these techniques. These skills are highly technical and require close supervision and instruction for practice. A firefighter should *not* attempt to practice these without an experienced instructor who uses protective systems.

5. *Report successful escape.* Once the team has escaped the building or area, members should report that the team is safe and accounted for. This process is usually accomplished with a fireground **personnel accountability report (PAR)**. A PAR is an organized approach to accounting for everyone on scene, **Figure 23-14**. Typically, the accountability manager runs through all the assignments given thus far and asks (usually via radio) for each crew to report its status. From this, unaccounted teams are identified and rapid intervention is initiated. Many departments perform a PAR at given intervals, such as when dispatch announces time intervals (every 10 to 30 minutes). Most departments perform a PAR after an emergency or when the incident operation changes modes or strategies (for instance, switches from offensive to defensive operations).

Lost, Trapped, and Injured Firefighters

Firefighting operations place crews in environments in which they have potentially never been—large buildings with unusual floorplans, small confined areas with minimal escape routes, and a whole host of stock, furniture, and arrangement obstacles. With this comes the potential to be trapped or lost during fire suppression, search, or rescue assignments. While most firefighters will agree that an entrapment is a true firefighter emergency, fewer believe that being lost is an emergency. They believe that, eventually, they will find a way out. This thinking needs to be altered.

Figure 23-14 A PAR is a personnel accountability report organized to check the status of all crews working an incident. PARs should take place every half-hour or after an evacuation or any firefighter emergency.

Caution A firefighter or crew that has lost spatial bearings in an IDLH atmosphere is experiencing a firefighter emergency!

With entrapments and lost situations, it is important to have a mental process to follow to avoid complicating the situation.

Entrapments

The first step a firefighter should take in an entrapment is to get assistance. Activation of a PASS device is warranted and the declaration of a "Mayday" should be made over the radio. Some radios are equipped with an emergency assistance button. It should be activated if the radio is so equipped. The Mayday will be followed up with radio procedures and communications to get assistance to the trapped firefighter—often via the rapid intervention team. The firefighter should follow this up with other noise-making efforts. Banging on a pipe or throwing debris may be helpful, but the firefighter must be careful not to use up energy or excessive air in doing this. Using visual signals such as a flashlight may also prove helpful. The second step in dealing with the entrapment is to size up the situation and develop a plan. Some key questions to answer include:

- What exactly is causing the entrapment?
- What is the exposure to fire/smoke/further collapse?
- How much breathing air is left in the SCBA bottle?
- What are the extent of injuries?
- Is there anything that can be done to self-extricate?
- Is there any self-first-aid that can be performed?
- How can air be preserved?

The trapped firefighter should attempt self-extrication. This must be planned and systematic as opposed to reckless and panic driven. All of these steps accomplish two important points. First, a process is created to help with rescue. Second, the firefighter's mind is kept active to help ward off panic.

Lost Firefighters

It cannot be overemphasized that a firefighter or team lost in an IDLH atmosphere is in fact experiencing a firefighter emergency. The steps to overcome this emergency are simple and can result in a quick resolution prior to an injury. First, the firefighter or team must report the fact they are lost. This is also a Mayday situation and should be transmitted as such over the radio. The Mayday will lead to radio traffic trying to ascertain last known position and any clues that might help a RIT crew locate the firefighters. Second, the lost firefighters need to manually activate their PASS devices. Finally, the firefighters need to stop and take inventory of their surroundings. From this, they can establish direction, door and window locations, and potential paths. As with firefighter entrapments, some questions should be answered:

- What is the exposure to fire/smoke/further collapse?
- How much breathing air is left in the SCBA bottle?
- How can air be preserved?
- Are there other options that have not been explored?

It is important to maintain radio contact with the RIT members and the incident commander. A team that is lost needs to help each other remain calm and preserve air. Any member that appears to panic can have a negative effect on the others—the team should exercise steps to keep minds open.

Injured Firefighters

Perhaps the most stressful situation that will ever face a firefighter is the realization that a fellow firefighter has become seriously injured. Regardless of what caused the injury, firefighters will typically drop whatever task they have been assigned and rush to aid their fellow firefighter, **Figure 23-15**. In some cases, this rapid focus on assisting a coworker becomes heroic due to extreme dangers faced by the now-rescuers and the firefighter victim. It should also be said that the potential to injure or kill many firefighters could be risked during firefighter rescues. One example of this unnecessary risk is the all-too-often scenario where firefighters rush to a debris pile to start rescue of firefighters caught under a collapse. A secondary collapse occurs and compounds the event. To minimize this risk, firefighters must trust RITs. The RIT is designed to attempt the rescue and remove the victim firefighter from the hazard so appropriate medical care can begin.

Once a firefighter is found to be trapped or injured, the RIT should be activated. It is natural for another team that is in proximity of the trapped firefighter or crew to engage in the rescue—this must be communicated! Collapse, exposure to the fire, and smoky conditions may exist and make the rescue difficult. Furthermore, the trapped firefighter or crew will be in an area that is hard to access by rescuers. Practicing the removal of firefighters from tight enclosed areas or through small windows is an investment that can save firefighters' lives.

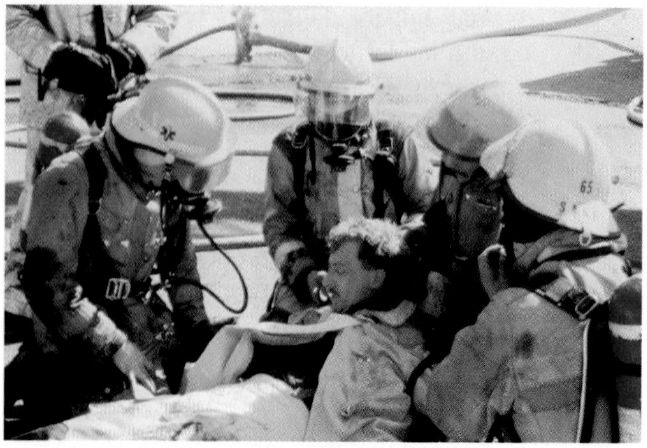

Figure 23-15 A serious firefighter injury or fatality will cause significant incident stress. Focus and use of RITs will minimize unnecessary risk during firefighter rescue and help maintain incident control.

Firefighters who are not on the RIT or in the immediate vicinity of the rescue should resist the urge to rush in and help. History shows that most firefighter rescues require the assignment of many teams. Firefighters should let the RIT take the lead and wait for their assignment. In this way, they are part of the plan—not the problem.

Post-Incident Survival

Once an incident has reached the point where crews are starting to clean up and be released, a relaxed phenomenon encroaches on the minds of the firefighters. This phenomenon has been called **post-incident thought patterns** and is responsible for many injuries and deaths. Common post-incident injuries include strains, sprains, and being struck by objects.

> **Note** The causes of these post-incident injuries seem almost ironic in a profession where aggressive and calculated risk taking is a hallmark.

Incidents that are especially gruesome or involve significant human tragedy can easily impact firefighters in an emotional sense, leading to long-term mental and health issues. This is called *critical incident stress* (CIS). It is important that firefighters address post-incident thought patterns as well as CIS.

Post-Incident Thought Patterns

One cause of post-incident injuries has to do with the little-studied notion of post-incident thought patterns. In essence, this is inattentiveness. In cases of especially difficult, unusually spectacular, or particularly challenging incidents, firefighters will tend to reflect on their actions. The replay of the incident starts almost instantly when the order is given to "pick up." This introspection is normal. The switch from activities requiring brainpower and physical energy to an activity that is so routine so as to be dull is a hard one to adapt to. Herein lies the problem. Some signs that post-incident thought patterns are affecting crews include faraway stares, firefighters wanting to be alone, and firefighters who stop and look about as if they have forgotten their task.

> **Note** Simple safety reminders or jocularity can help firefighters regain focus and reduce injury potential.

One method to reduce the impact of these thought patterns is to take a time-out and have everyone gather for a quick incident summary and safety reminder, **Figure 23-16**. These huddles can be effective for everyone or just small groups.

Another factor to consider that leads to inattentiveness is chemical imbalance. Even the most successful rehabilitation program cannot prevent firefighters from experiencing fatigue and mental drain. With the end of an incident, especially one requiring major physical effort, comes the relaxation of the firefighter's mind, which, in turn, starts shutting down protective chemicals that stimulate performance. The adrenaline "rush" is over and the firefighter's metabolism will return to a "repair" state. This causes a mental slowdown that can lead to unclear thinking and result in injuries.

Figure 23-16 A quick huddle during post-incident activities can serve to remind firefighters to continue to work safe and *smart* during pickup activities.

Another way to see this chemical and mind imbalance is to look at a firefighter's tools from a layperson's point of view. Most laypersons would literally have to concentrate on carrying an ax, pike pole, chain saw, roof ladder, or hose length in order to not hurt themselves or anyone around. Consider also doing these tasks in bulky, restrictive clothing and heavy boots. The firefighter does these things after incredible energy bursts under frightening conditions. Familiarity plays a certain role here, but concentration is still required. If the mind has been taxed, the body has been fatigued, and the signal to "relax" has been given—yet the concentration required to do the task has remained the same—the potential for injury rises.

Whether the issue is chemical imbalance or post-incident thought patterns, the firefighter needs to stay alert and try to pick up signs of potential injury and take steps to "survive" without injury.

Critical Incident Stress

Firefighters are expected to tolerate a certain level of incident stress given the nature of the environment that firefighting brings. Dealing with community hazards, injuries, and death can even be classified as normal for the profession. Some events, however, can trigger a significant emotional response from responders. These emotional responses may not always be external. In fact, many firefighters harbor the reaction internally. Events that typically induce an emotional response include:

- Death or serious injury to a coworker or a person known to the responder
- Mass casualty incidents
- Death or serious injury to a child, especially from an act of violence or crime
- A prolonged rescue attempt that results in a death

Firefighters will exhibit signs of CIS in many ways. **Figure 23-17** outlines a few of these signs and symptoms. It is important to note that none of these signs is abnormal—emergency responders will continue to feel the stress of critical incidents. What is important is to "survive" the CIS through a process of critical incident stress management (CISM).

CISM can take on many forms, both formal and informal. Formal CISM is usually accomplished following an incident of significant magnitude—one affecting numerous responders. Typically, formal CISM involves a critical incident stress debriefing (CISD) process that is led by peers and health pro-

Selected Signs and Symptoms of Critical Incident Stress

Behavioral:	Increased irritation, aggressiveness, withdrawal, flashbacks, inattentiveness, alcohol/drug abuse, and memory loss
Physical:	Loss of appetite, insomnia, fatigue, headaches, muscle stiffness, and hypertension
Psychological:	Guilt, sadness, depression, career introspection, claimed burnout, fear, and unsociability

Figure 23-17 Signs of critical incident stress are natural following "trigger" incidents.

fessionals. The CISD should be mandatory for all responders and is designed to help everyone with immediate and long-term mental health needs. An informal process usually includes a "defusing" that allows persons to deal with an ongoing rescue or to voice concerns and thoughts of the incident in a peer environment. Defusing should take place just prior to releasing firefighters from the scene, or back at the station just after the incident, **Figure 23-18**.

To help manage critical incident stress as well as daily incident stress, the firefighter should use CIS processes and tested stress-reduction principles such as exercise, adequate and appropriate nutrition, deep breathing, deep relaxation (massage, meditation, etc.), hobby development, and positive social interaction. With these steps, the firefighter can survive post-incident stress in the long term.

(A)

(B)

Figure 23-18 Critical incident stress management sessions can be (A) informal such as a "defusing" or (B) formal such as a process that includes peer support and mental health professionals.

Lessons Learned

Firefighter survival is dependent on many proactive and preventive actions. These actions include incident readiness, safe operations at incidents, and appropriate preparation and response to firefighter emergencies. Incident readiness includes those efforts to prepare the firefighter's personal protective equipment, personal accountability, and individual fitness for duty. Of particular note, fitness for duty relies on the firefighter's mental and physical wellness as well as the firefighter's attention to energy and rest.

Safe operations and, therefore, survival are dependent on team continuity and the elimination of freelancing. Additionally, safe incident operations are achieved when teams perform orders with attention to communications and timely updates. While operations are under way, individual teams must practice solid risk/benefit analysis for their specific job. During the incident, crews must take advantage of the rest, hydration, and nourishment offered through established rehabilitation processes.

Rapid intervention teams are formed for immediate deployment should a firefighter emergency take place. To meet this challenge, RITs must assemble tools and be briefed on essential information that will assist in a more rapid reaction should an emergency occur. Firefighter emergencies require the individual firefighter to practice a clear and concise approach to dealing with the emergency. Using the term "Mayday" can signal everyone of the need for firefighter rescue. Whether trapped or lost, the firefighter must remain calm and follow a defined series of steps to mitigate the situation.

Finally, firefighters must survive long term through an understanding of post-incident thought patterns and the effects of critical incident stress. Understanding the factors that lead to each is a useful tool in dealing with the reality of emergency responder stress and can help reduce the potential for firefighter injury or death.

KEY TERMS

Freelancing The act of working alone or performing a task for which the firefighter has not been assigned.

Incident Action Plan (IAP) A strategic and tactical plan developed by the incident commander.

Mayday A universal call for help. A Mayday indicates that an individual or team is in extreme danger.

Passport A term given to a specific accountability system where crews are tracked using a card (passport) with all members listed. An accountability manager tracks the passports on an accountability board.

Personal Size-Up A continuous mental evaluation of an individual's immediate environment, facts, and probabilities.

Personnel Accountability Report (PAR) This is an organized roll call of all units assigned to an incident.

Post-Incident Thought Patterns A phenomenon that describes an individual's inattentiveness following a significant incident. Post-incident thought patterns can lead to injuries or even death.

Risk/Benefit An evaluation of the potential benefit that a task will accomplish in relationship to the hazards that will be faced while completing the task.

REVIEW QUESTIONS

1. List the three main components that lead to incident readiness.
2. Define the four key checks to ensure that individual personal protective equipment is ready for response.
3. Describe the advantages and disadvantages of three kinds of personnel accountability systems.
4. Define personal size-up.
5. Briefly describe the three components that lead to fitness for duty.
6. Name three practices that lead to team continuity. Describe how each can increase firefighter survival.
7. Define risk/benefit.
8. List and briefly describe the three components of on-scene rehabilitation.
9. Describe the procedures that should be taken to establish and prepare for the assignment of a rapid intervention team.
10. List the five steps that can lead to an organized rapid escape.
11. Describe the reasons why a lost firefighter is considered an emergency.
12. List the three steps that should be taken when entrapment occurs.
13. Compare and contrast post-incident thought patterns and critical incident stress.

Endnotes

1. McLellan, Thomas, PhD, *Safe Work Limits While Wearing Firefighting Protective Clothing*. Defense Research and Development Council, Toronto Fire Department Grant Study Report, Toronto, Ontario, Canada, 2002.
2. U.S. Fire Administration, *Emergency Incident Rehabilitation*, FA-114. USFA Publications, Washington, DC, July 1992.
3. Sears, Barry, *Enter the Zone*. HarperCollins Publishers, New York, 1995.

Additional Resources

Angle, James S., *Safety in the Emergency Services.* Delmar Learning, a part of the Thomson Corporation, Clifton Park, NY, 1999.

Dodson, David W., *Fire Department Incident Safety Officer.* Delmar Learning, a part of the Thomson Corporation, Clifton Park, NY, 1999.

Emergency Incident Rehabilitation, FA-114. U.S. Fire Administration, Washington, DC, 1996.

EMS Safety: Techniques and Applications, FA-144. U.S. Fire Administration, Washington, DC, 1996.

Firefighter Fatality Investigations, http://www.cdc.gov/niosh

LeCuyer, John, *Designing the Fitness Program—A Guide for Public Safety Organizations,* Fire Engineering Books & Videos, a Division of PennWell Corporation, Saddlebrook, NJ, 2001.

NFPA 1500: Standard on Fire Department Occupational Safety and Health Program, 1997 Edition. National Fire Protection Association, Quincy, MA, 1997.

NFPA 1561: Standard on Emergency Services Incident Management System, 2000 Edition. National Fire Protection Association, Quincy, MA, 2000.

NFPA 1584: Rehabilitation of Members Operating at Incident Scene Operations and Training Exercises, 2003 Edition. National Fire Protection Association, Quincy, MA, 2003.

Risk Management Practices for the Fire Service, FA-166. U.S. Fire Administration, Washington, DC, 1996.

CHAPTER 24

HAZARDOUS MATERIALS: LAWS, REGULATIONS, AND STANDARDS

Christopher Hawley, Baltimore County Fire Department (Ret.)

OUTLINE

- Objectives
- Introduction
- Laws, Regulations, and Standards
- Emergency Planning
- OSHA HAZWOPER Regulation
- Standards
- Additional Laws, Regulations, and Standards
- Lessons Learned
- Key Terms
- Review Questions
- Endnotes
- Additional Resources

STREET STORY

We arrived on scene to find the patient lying beside the hot dusty road, baking in the summer sun. He was a male in his mid-thirties, wearing torn blue jeans, old tennis shoes, and no shirt. A local farmer was on the scene. He was driving home when he noticed the man beside the road. It was obvious the man was in distress, so the farmer called 9-1-1 from a cellular phone.

When we got off the fire engine, it was apparent that the man was struggling to breathe. He was sunburned, sweaty, and unable to respond to any of our questions. He was shaking slightly but it just did not look like the typical epileptic seizure. We put the man on high-flow oxygen and began to assess his vital signs. It was the size of his pupils that tipped us off. In addition to the pinpoint pupils, the patient was incontinent and producing volumes of frothy sputum. We stripped off his clothes, decontaminated him with water, and alerted the incoming ambulance by radio that we might have an organophosphate poisoning. Fortunately, we had recently had some hazardous materials training on pesticide exposures. Had we not been armed with that information, we may not have picked up on the clues so quickly.

The transport ambulance arrived what seemed like hours later, and by that time we were assisting his respirations with a bag valve mask and suctioning his airway. An IV was started, and atropine was administered in large doses to counteract the effects of a suspected pesticide. Unfortunately, we were never able to identify the substance he was exposed to or even know the duration of the exposure.

Sometimes, despite the greatest efforts and the best technology available, people do not survive. He fought hard to breathe and even harder to live, but in the end, neither we nor the hospital staff were able to save his life. He died an hour or so after we first saw him, probably never knowing why he became so sick.

I think about him from time to time, wondering if the call could have gone smoother or we could have done something different, but I always arrive at the same conclusion: You do your best, learn from any mistakes or situations you find yourself in, and just hope you have a better outcome the next time around.

We could have suffered a secondary exposure because we were not properly protected, but we did not. We could have had an inhalation exposure from his contaminated clothing, but we did not. We could all have ended up as patients, but we did not. It was only luck that brought us through without harm. Would I do it differently the next time? Absolutely. Our first responsibility as firefighters is to not make the problem worse when we show up! You are no good to anyone if you become part of the problem or end up being a victim yourself. Use common sense, wear the right PPE for the right situation, and, above all, think before you act. Do not count on luck to bring you back to the station.

—Street Story by Rob Schnepp, Captain Paramedic, Alameda County Fire Department, Alameda County, California

OBJECTIVES

After completing this chapter, the reader should be able to:

- Explain the local emergency response plan and standard operating guidelines.*
- Explain the student's role within these documents at the awareness level.
- Explain the notification process to request assistance.
- Explain the role of the local emergency planning committee.
- Explain the role of the SARA Title III regulation and emergency response.
- Explain other regulations that have an effect on fire department activities.

INTRODUCTION

Hazardous materials response, as shown in **Figure 24-1**, is one of the specialty fields within the fire service. When the fire department is called to a hazardous materials release scene, the actual handling of chemical releases is usually done by a well-trained team whose function it is to handle such emergencies, but in almost all cases firefighters are required to assist in the effort.

> **Note** Many times at hazardous materials incidents the actions that the first-in firefighters take in the first five minutes determine how the next five hours/days/months will go.

Firefighters and EMS providers are bombarded with exposures to hazardous materials each and every day in their regular fire suppression and EMS duties. A large number of firefighters and EMS providers are exposed to bloodborne pathogen materials each day, sometimes suffering fatal effects from this exposure. By the nature of their jobs they are exposed to numerous toxic and cancer-causing agents, many times in non-emergency situations. When confronted with emergency situations, the exposure and risks increase as well. This chapter covers some hazardous materials risks and some fundamental response profiles that will help keep firefighters safe. As the fire service furthers its efforts to improve firefighter health and safety, the response to chemical accidents is one area in which extra precautions are needed.

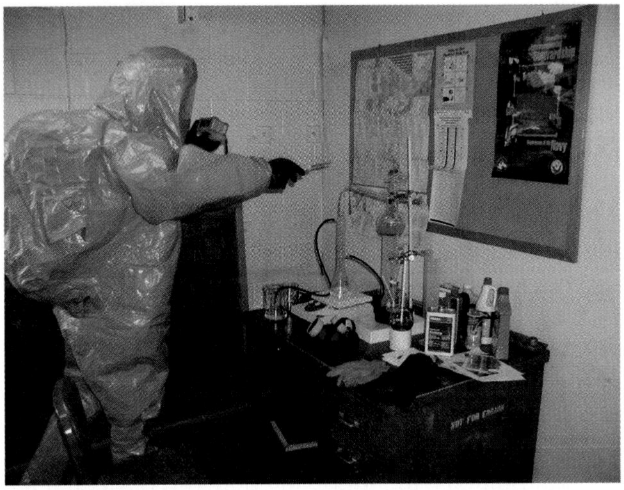

Figure 24-1 A hazardous material team member surveys a chemical agent lab using air monitors.

> **Safety** In a fire situation, the injuries can be acute, for instance, if a building collapses during firefighting operations. A chemical exposure can also kill a firefighter immediately with a deadly dose, or it can cause serious illness or injury that prolongs death for ten to twenty years.

As society has become more environmentally aware, so has the fire service. Hazardous materials response is still a very young field, and most communities have only had fire department–based hazardous materials response teams for sixteen years or less. The Jacksonville, Florida, team is the oldest hazardous materials response team in the country, and it is only twenty-three years old. This field is constantly changing and dynamic, and the response to hazardous materials emergencies has undergone three distinct changes in response methodology. Tactics used to handle incidents ten years ago certainly were appropriate then but might bring criticism if used today. Technology is rapidly changing as well. **Air monitoring devices** that were used in the past strictly by hazardous materials response teams are now common on engine, truck, medic, and squad companies. On the hazardous materials response team side, technology has also increased to a level that requires a high degree of training and expertise. Response to chemical accidents is a rapidly changing field and brings new experiences every day.

Reader's Note: In the hazardous materials section, NFPA 1001 requires that the student receive hazardous materials training at either the Awareness level or the Operations level. The information in this text covers both levels and in some cases exceeds the Operations level. All of this information is important for firefighters' survival. See the discussion on NFPA standards and OSHA's HAZWOPER for more information on Awareness and Operations level training.

From building fires, tank truck fires, **clandestine drug labs**, shipboard incidents, pressurized cylinder leaks, and overturned tank trucks to the emerging incidents of terrorism, hazardous materials response teams face new challenges on every response. This chapter provides firefighters with a practical framework that can help keep them and their crews alive in situations that involve hazardous materials.

The next few sections describe in detail the many aspects of hazardous materials and appropriate actions when working those incidents. To establish the groundwork for these sections, it is necessary to provide a definition for hazardous materials. The text that follows provides these standard definitions. Out of these definitions, the following may be considered the best definition: A hazardous material is any substance that jumps out of its container when something goes wrong, and hurts and harms the things it touches. Ludwig Benner, Jr., who worked for the National Transportation Safety Board, provided this early definition. It is a perfect definition for this field.

Caution Almost everything is hazardous to human health if it is used improperly or escapes its container. This is demonstrated in **Figure 24-2**.

Figure 24-2 The material shown here is an example of one that ignites when it escapes its container and comes in contact with the air. A material that is air reactive is known as *pyrophoric*.

Even the most toxic material is not hazardous as long as it is used correctly and stays in its intended container. The other principle that is important here is that even if the material escapes its container, it must have the ability to cause harm to humans, property, or the environment. Every day hazardous materials are packaged to be transported and are intended to stay in that package. When firefighters become involved in an incident that involves a hazardous material, an increased level of concern results, but as long as the material remains packaged, and the firefighters do not open the package, the risk to them is very minimal. The material may be toxic, flammable, corrosive, or reactive, and these properties remain no matter what happens—but to reach out and harm people it must escape or be forced from the container. The following are some standard definitions:

DOT hazardous material: Any substance or material in any form or quantity that poses an unreasonable risk to the safety, health, and property when transported in commerce. The DOT also provides more exacting definitions with each hazard class.

EPA hazardous material: A chemical released into the environment that could be potentially harmful to the public's health or welfare.

OSHA hazardous chemical: Those chemicals that would be a risk to employees if exposed in the workplace.

LAWS, REGULATIONS, AND STANDARDS

It is important for the first responder to have a basic understanding of the legislative history of hazardous materials. It was in this area that fire departments first became regulated as to how they would respond and how they would be trained. Many environmental and safety regulations affect how firefighters respond to emergencies. This section provides an overview of the major federal laws, regulations, and standards. Readers should consult their local environmental and OSHA offices to learn about state and local laws and regulations.

Development Process

It is important to understand the differences among laws, regulations, and standards. **Laws** result from legislation passed by Congress and signed by the president. **Regulations**, on the other hand, are developed by government agencies, like OSHA or the EPA, and have the weight of law but are not passed by Congress nor signed by the president. In some cases laws require the development of regulations and provide a framework for the government agency to follow. **Standards** are developed and reviewed by a nongovernmental consensus committee, such as the NFPA. These do not have the weight of law but could be applied by a regulating agency or in court.

> **Note** As time goes on, standards are being applied with the weight of law, typically by OSHA.

As citizens and as members of an industry, it is important for firefighters to participate in the development and review of laws, regulations, and standards.

EMERGENCY PLANNING

The first law that regulated how fire departments respond to emergencies was the **Superfund Amendments and Reauthorization Act**, commonly referred to as **SARA**. This law was passed in 1986 for the protection of emergency responders and the community. Its intent was to inform emergency responders of chemical hazards within their community. The main component of the law is called the **Emergency Planning and Community Right to Know Act (EPCRA)**. It is divided into two sections: planning for emergencies and providing a mechanism to get chemical storage information to emergency responders.

State and Local Emergency Response Committees

The planning portion established a requirement to provide a **State Emergency Response Committee (SERC)** in each state. It is the responsibility of the SERC to ensure the state has the resources necessary to respond safely and effectively to chemical releases. The SERC also had to provide the framework to implement **Local Emergency Planning Committees** known as **LEPCs**. The LEPC is a group composed of the representatives of the community, emergency responders, industry, hospitals, media, and other government agencies. Most LEPCs are set up on a county basis,[1] although larger communities may have their own. The LEPC is responsible for the development of a hazardous materials emergency plan and its annual revision.

Local Emergency Response Plans

The emergency plan is an important component to a successful response. The emergency plan should outline emergency contacts and procedures. Target facilities such as those that store extremely hazardous substances are to be included, along with specific information and response tactics. A decision tree for evacuation and in-place sheltering is usually included. If the decision is made to evacuate citizens, the plan provides the mechanism to shelter and take care of these evacuees. It is important for personnel to have an understanding of this plan and know how to access its resources during an emergency.

The LEPC is also responsible for ensuring that local resources are adequate to handle a chemical release in the community. Ensuring that local responders receive the proper amount of training and that the emergency plan is evaluated annually by conducting an exercise (drill), is also part of their responsibility. It is important for the emergency services to be an integral player in the LEPC. Decisions regarding a fire department's response to chemical incidents may be planned at the LEPC level. Many of the facilities using hazardous materials will have representatives at these meetings, and before an emergency incident is a better time to meet these representatives. Funding for training and planning is also available through some LEPCs, since most federal HAZMAT grants are provided through the LEPC.

Chemical Inventory Reporting

The other section of EPCRA, usually referred to as SARA Title III, is the chemical reporting portion of the act. It requires some facilities to report chemical information to the state, LEPC, and local fire department. Failure to report this information on an annual basis can result in fines of $25,000 a day.

To qualify as a reporting facility, the facility has two methods of reaching a reporting threshold. Most facilities meet the reporting threshold of storing more than 10,000 pounds of a chemical. The 10,000-pound number is for a single chemical and is not the total of all chemicals on site. A facility that has more than 1,300 gallons of acetone is required to report because the amount represents more than 10,000 pounds of acetone. A facility that stores drums of ammonia and water is only required to report the ammonia, not the total weight of the drum. They do have to account for all of the ammonia at the facility, regardless of the location or type of container.

Another method of meeting the threshold of reporting is to store one of 366 chemicals that the EPA considers an **extremely hazardous substance (EHS)**. The EHSs have separate reporting requirements and lower reporting thresholds. Some must be reported at 100 pounds. If these chemicals are released, then the facility has a whole host of responsibilities to comply with, including immediate notification of the LEPC, usually by dialing 9-1-1 or some other listed emergency contact number as determined by the LEPC. The common EHSs are chlorine, sulfur dioxide, anhydrous ammonia, and sulfuric acid. Because EHS materials can present an extreme threat to the community, it is important for emergency responders to become familiar with the facilities that use them and to use caution when responding to potential releases involving these materials.

Facilities that are required to report as SARA facilities are required to submit either a Tier 1 or Tier 2 Chemical Inventory Report. The Tier 2 report, **Figure 24-3**, is the most common and lists the chemical name, storage amount in a range, storage location and information, and emergency contact information, including twenty-four hour phone numbers. The facility is also required to submit a list of chemicals or **Material Safety Data Sheets (MSDS)**. A site plan may also be required that outlines the storage areas. EHS facilities are also required to submit a copy of their emergency plan. Depending on the state or locality, reporting requirements may be more stringent, although all have to meet the minimum listed. Note, however, that the information contained on a Tier 2 form regarding the chemicals that are located on site is for the *prior* year. Facilities report from January 1 to March 1 for the prior year.

The basis for this reporting is to inform the emergency responders of the hazard of responding to a SARA facility. It also allows the emergency responder to make informed decisions as to site entry, tactical decisions, and potential evacuation. MSDS are a valuable tool in obtaining chemical information, and SARA[2] requires their availability to emergency responders.

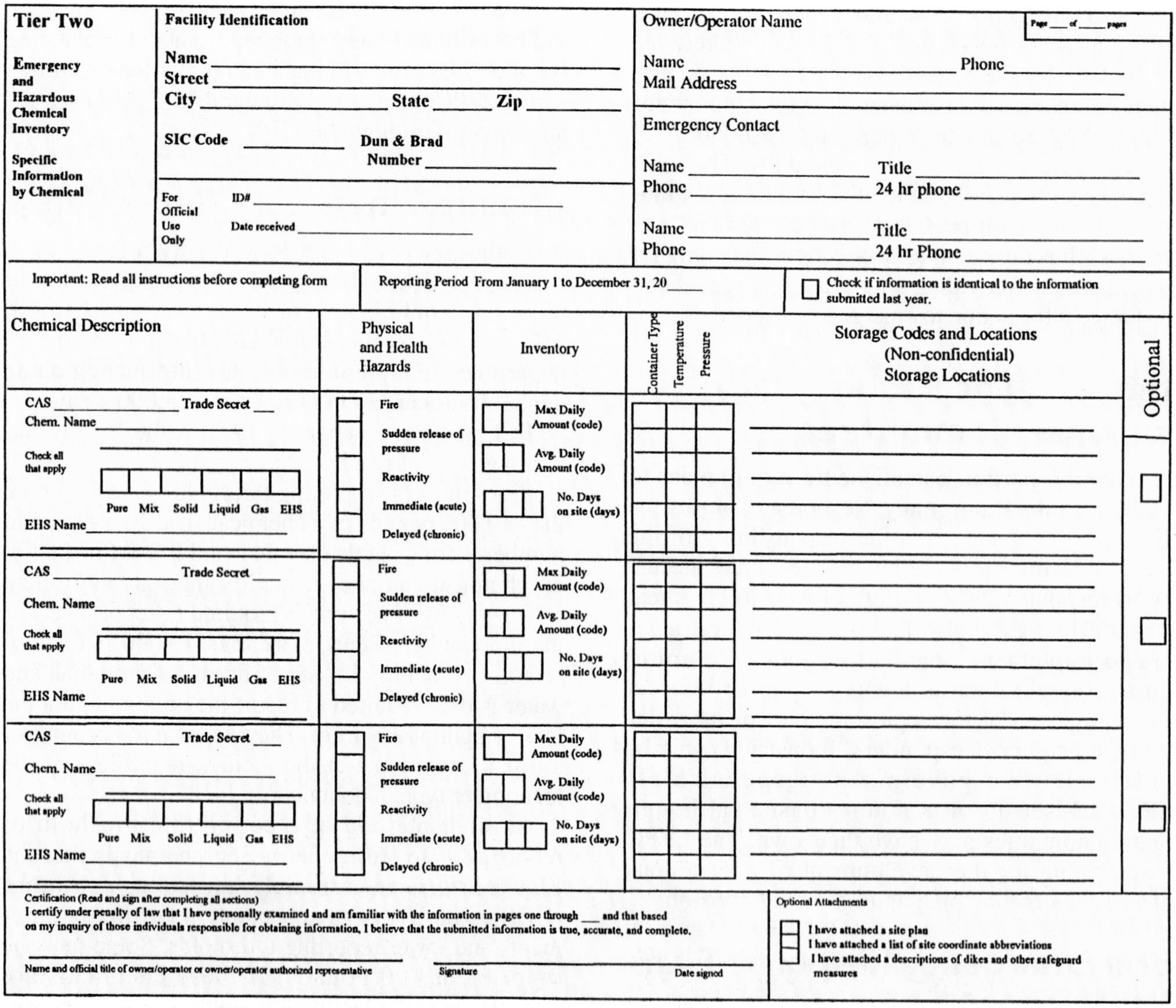
Tier Two

Emergency and Hazardous Chemical Inventory

Specific Information by Chemical

Facility Identification

Name ____
Street ____
City ____ State ____ Zip ____

SIC Code ____ Dun & Brad Number ____

For Official Use Only — ID# ____ Date received ____

Owner/Operator Name — Page ___ of ___ pages

Name ____ Phone ____
Mail Address ____

Emergency Contact

Name ____ Title ____
Phone ____ 24 hr phone ____

Name ____ Title ____
Phone ____ 24 hr Phone ____

Important: Read all instructions before completing form

Reporting Period From January 1 to December 31, 20___

☐ Check if information is identical to the information submitted last year.

Chemical Description | Physical and Health Hazards | Inventory | Container Type | Temperature | Pressure | Storage Codes and Locations (Non-confidential) Storage Locations | Optional

CAS ____ Trade Secret ___
Chem. Name ____
Check all that apply: Pure Mix Solid Liquid Gas EHS
EHS Name ____

Fire / Sudden release of pressure / Reactivity / Immediate (acute) / Delayed (chronic)

Max Daily Amount (code) / Avg. Daily Amount (code) / No. Days on site (days)

CAS ____ Trade Secret ___
Chem. Name ____
Check all that apply: Pure Mix Solid Liquid Gas EHS
EHS Name ____

Fire / Sudden release of pressure / Reactivity / Immediate (acute) / Delayed (chronic)

Max Daily Amount (code) / Avg. Daily Amount (code) / No. Days on site (days)

CAS ____ Trade Secret ___
Chem. Name ____
Check all that apply: Pure Mix Solid Liquid Gas EHS
EHS Name ____

Fire / Sudden release of pressure / Reactivity / Immediate (acute) / Delayed (chronic)

Max Daily Amount (code) / Avg. Daily Amount (code) / No. Days on site (days)

Certification (Read and sign after completing all sections)
I certify under penalty of law that I have personally examined and am familiar with the information in pages one through ___ and that based on my inquiry of those individuals responsible for obtaining information, I believe that the submitted information is true, accurate, and complete.

Name and official title of owner/operator or owner/operator authorized representative — Signature — Date signed

Optional Attachments
☐ I have attached a site plan
☐ I have attached a list of site coordinate abbreviations
☐ I have attached a descriptions of dikes and other safeguard measures

Figure 24-3 This form is an example of what facilities are required to submit to the fire department and the Local Emergency Planning Committee on an annual basis.

Tier Two

Emergency and Hazardous Chemical Inventory

Specific Information by Chemical

Facility Identification

Name ____
Street ____
City ____ State ____ Zip ____

SIC Code ____ Dun & Brad Number ____

For Official Use Only ID# ____ Date received ____

Owner/Operator Name

Page ___ of ___ pages

Name ____ Phone ____
Mail Address ____

Emergency Contact

Name ____ Title ____
Phone ____ 24 hr phone ____

Name ____ Title ____
Phone ____ 24 hr Phone ____

Important: Read all instructions before completing form

Reporting Period From January 1 to December 31, 20____

☐ Check if information is identical to the information submitted last year.

Confidential Location Information Sheet

	Container Type	Temperature	Pressure	Storage Codes and Locations (Confidential) Storage Locations	Optional
CAS # ____ Chemical Name ____					☐
CAS # ____ Chemical Name ____					☐
CAS # ____ Chemical Name ____					☐

Certification (Read and sign after completing all sections)
I certify under penalty of law that I have personally examined and am familiar with the information in pages one through ___ and that based on my inquiry of those individuals responsible for obtaining information, I believe that the submitted information is true, accurate, and complete.

Name and official title of owner/operator or owner/operator authorized representative ____ Signature ____ Date signed ____

Optional Attachments
☐ I have attached a site plan
☐ I have attached a list of site coordinate abbreviations
☐ I have attached a descriptions of dikes and other safeguard measures

Figure 24-3 Continued

OSHA HAZWOPER REGULATION

Another part of SARA was the requirement for OSHA to develop a regulation covering activities that involve hazardous materials. The regulation, known as the **Hazardous Waste Operations and Emergency Response**, became final on March 6, 1989. It is referred to as **HAZWOPER** or by its identification number[3] of 29 CFR 1910.120. There was considerable discussion within the fire service when this regulation was established as to whether or not it applied to all of the fire service. The primary unknown was whether this regulation applied to volunteer firefighters. In some states HAZWOPER only affected the private sector and not career firefighters, while in other states only the public sector was subject to this regulation. In many cases volunteer firefighters were determined to be employees of the local government since they received workers' compensation benefits or received safety equipment from the employer, and so this regulation applied. To end this argument and confusion, the EPA adopted the same regulation and issued it on the EPA's behalf.

> **Note** Because employment is not a concern of the EPA, as it is with OSHA, all persons are covered by this regulation regardless of their employment status.

The EPA regulation is also called HAZWOPER and is referenced by 40 CFR 311. It is interesting to note that OSHA now puts a provision in some of its regulations that specifically mandates the coverage of volunteer employees.

This regulation has had far-reaching effects for the fire service. It has required that certain training be provided, requires the development of standard operating procedures, and mandates certain requirements when handling chemical releases. It only allows persons trained in the handling of chemical releases to respond to chemical incidents. Only certain levels allow operation in and around chemical releases. Even just to be present at a chemical accident requires training, and as firefighters' activity levels increase, so does their need for further training.

Paragraph q

The majority of this OSHA regulation covers employers' responsibilities at hazardous waste sites. The last section, **paragraph q**, covers emergency response and applies to the fire service. It established five levels of training, as well as a requirement for annual refresher training. It requires that an incident command system be used and that there be an incident commander at chemical releases. At larger incidents a safety officer must be appointed. As firefighters progress up the training levels the requirements become more detailed. The two in/two out rule, as discussed in other chapters, originated from this OSHA regulation.

The five training levels established by OSHA and the NFPA and their basic responsibilities are as follows:

Awareness: Persons trained at the Awareness level respond to possible chemical releases, identify the potential for a chemical release, call for assistance, and stand by isolating the area and denying entry to other persons. Persons trained at the Awareness level cannot take any action beyond this. This level is intended for police officers, public works employees, and other government employees.

Operations: Persons trained at the operations level can act in defensive fashion to chemical spills. Acting defensively means that the responder does not enter a hazardous area, but can set up dikes, dams, and other confinement measures. Training at the operations level allows the firefighter to assist technicians in setting up the various activities that are required at an incident. Operations training can be expanded to include specialized activities such as decontamination, so that persons trained at this level can assist with this activity. This level is intended for the fire and EMS service.

Incident Commander: This level is designed for a person who has received operations level training and has received training in incident command procedures. This person will be in charge of the incident. To be the incident commander does not mean that this person has the highest level of chemical response training but is the senior response official. The IC must rely on the expertise of the other responders, such as the HAZMAT team, facility officials, or other technical specialists, to make strategic and tactical decisions.

Technician: This is the level at which offensive activities can be completed in the hazard area. Other than some specific restrictions outlined in HAZWOPER, there are no general restrictions as to the activities technicians can perform, as long as the activities fit within the scope of their training. This is the level at which leaks can be stopped and mitigation of the incident can be completed. HAZMAT technicians are expected to mitigate or stop the incident from progressing.

Specialist: This level is identified only in HAZWOPER and is intended to be a level that has received a higher level of training above a technician, or may be someone who specializes in a specific chemical or area of expertise. The training concentrates on chemistry and the identification of unknown materials. In some instances, the specialist supervises the technicians. The NFPA removed this level in 1992.

Medical Monitoring

Other sections of this regulation include a requirement for medical monitoring of certain employees. An individual's responsibilities mandate whether the employer must provide that individual with an annual physical. An annual physical is required if an individual meets any of these requirements:

- Was exposed to a chemical above the permissible exposure limit.
- Wears a respirator or is covered by the OSHA respiratory regulation (29 CFR 1910.134).
- Was injured due to a chemical exposure.
- Is a member of a HAZMAT team.

The physician determines the extent of the exam and can establish that the exams be given every two years if that time period is determined to be acceptable. For the fire service the only persons required to be given a physical are the members of a HAZMAT team, although OSHA 1910.134 (Respiratory Protection) has a requirement that a medical survey questionnaire be answered by every firefighter. Depending on the answers to the questions, a physical may or may not be required. If a firefighter is exposed to a hazardous material above the OSHA-specified level, then the department is required to provide a physical exam to the firefighter. The physician determines the extent of the exam and any other future visit or tests.

OSHA requires that the medical records be kept by the employer for a period of thirty years past the last date of employment. This applies to the HAZWOPER regulation as well as other OSHA regulations. In simple terms any medical record generated by the employer, or for the employer, must be kept and must be available for employees if they request it.

STANDARDS

The group that establishes most of the standards for the fire service is the National Fire Protection Association. The NFPA establishes a variety of committees that develop standards, which are then made available for review and comment by the public. After the review process, each standard is voted on by the NFPA committee and, if passed, then goes to the next public meeting where it is voted on by the group in attendance. Once passed it then is sent to the Standards Council, where it is voted on again. If it passes there, it becomes a standard. As with emergency medical services in which an acceptable standard of care applies, typically based on a national or regional standard, the NFPA establishes a standard of care. A person cannot be held criminally liable for violating an NFPA standard, but may be held civilly liable.

One of the areas in which the NFPA standards have been applied as having the weight of a regulation is in the hazardous materials arena.

> **Note** OSHA has used what is known as the *general duty clause*, which means that all employers have a general duty or an obligation to provide a workplace free from hazards.

OSHA has used this clause to cite employers for violating an NFPA standard.

NFPA 471

Three primary NFPA standards apply to hazardous materials response and training, although there are others that cover the storage and use of chemicals. The first standard is NFPA 471, *Recommended Practice for Responding to Hazardous Materials Incidents.* This standard, which provides detailed methods and operational procedures, goes beyond the requirements established by the HAZWOPER regulation.

NFPA 472

The second standard is NFPA 472, *Professional Competence of Responders to Hazardous Materials Incidents.* This is the listing of objectives required to meet the training levels established by the NFPA and OSHA. OSHA established only basic criteria to meet in order for the employer to certify their employees as to their ability to respond to hazardous materials incidents. This NFPA document expands the requirements in order for the employer to certify their employees. The levels mirror those of the HAZWOPER regulation, but in 1992 the NFPA removed the specialist category. In the 1997 edition they added the following competencies: private sector specialist employee, hazardous materials branch officer, hazardous materials branch safety officer, tank car specialist, cargo tank specialist, and inter-modal specialist.

In the 2002 edition the NFPA added objectives related to terrorism response. These changes were part of the 1997 edition but were listed as tentative interim amendments[4] (TIAs). The 2002 version adopts them as part of the standard, and they are covered by this text.

NFPA 473

The third standard is NFPA 473, *Competencies for EMS Personnel Responding to Hazardous Materials Incidents.* This standard adds some additional competencies above NFPA 472 with regard to EMS issues. It provides EMS Level I and Level II training levels. The competencies for Level I allow EMS providers to perform patient care in the cold sector and provide a higher level of patient care information. At Level II, EMS providers can perform patient decontamination and work in the warm sector. The level of training is that of a technician with additional competencies with regard to patient care. The book *Emergency Medical Response to Hazardous Materials Incidents* offers more information on this topic.

Standard of Care

As mentioned before, emergency responders have to abide by a standard of care. In years past it was deemed acceptable to wash spilled gasoline down a storm drain. To do that now is illegal, and personnel could face state and federal charges for violating the Clean Water Act. This standard of care is composed of the laws, regulations, standards, local protocols, and experience detailed earlier. Violations of this standard of care are based on three theories: **liability**, **negligence**, and **gross negligence**.

When operating in and around hazardous materials, liability becomes a very real issue. Other than EMS responses like those shown in **Figure 24-4**, firefighters at a chemical release are likely to find themselves personally liable for any wrongdoing. Liability is being responsible for personal actions.

> **Note** Environmental and safety regulations are designed to place blame on the person who made the decision, not on the organization itself, unless the organization has policies that violate these regulations.

In the example of the individual who made the decision to wash gasoline down a storm drain, that individual could end up in jail or have to pay a substantial fine.

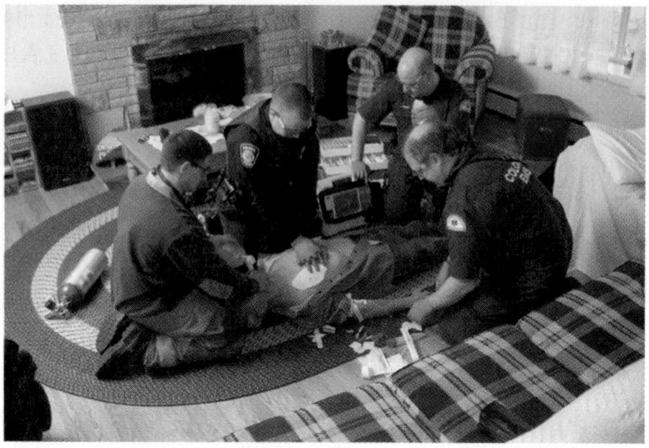

Figure 24-4 Just as EMS responders have to follow a standard of care so that the patient is provided an appropriate level of care, HAZMAT response has a similar standard of care.

Negligence is not following the standard of care or an accepted practice. Gross negligence is the willful disregard for the standard of care. In other words, a conscious decision was made not to follow the standard of care. To avoid becoming liable, firefighters should receive the training that is required for their positions and act only to that level of training. If firefighters are not sure what the next step should be, they should consult with their local HAZMAT team or other local or state environmental protection agency.

ADDITIONAL LAWS, REGULATIONS, AND STANDARDS

The items discussed next are ones that firefighters should be aware of, because they are commonly encountered or applied in chemical releases.

Hazard Communication

The hazard communication regulation issued by OSHA (29 CFR 1910.1200) requires that employers provide an MSDS for all chemicals located at a facility at quantities above "household quantities." The term *household quantities* allows for small amounts, such as those that would be purchased for home use, such that MSDS are not required. An example would be a gallon of household ammonia for use in an office. This would not require an MSDS, because this is a normal household quantity. If, however, a case of household ammonia were brought in, an MSDS would be required because this is not a normal household quantity.

The employer must provide a list of these MSDS materials and when they arrived on site. The employer must provide training on these MSDS materials and the hazard communication program. One of the important components of this program is the requirement for the facility to make these available for emergency responders. Fire departments[5] themselves are responsible for following this regulation.

Superfund Act

The Comprehensive Environmental Response, Compensation and Liability Act (CERCLA) is commonly referred to as the Superfund law. This law was established for the cleanup of toxic waste sites around the country. It was the first law to set the groundwork for the regulating of the fire service with regard to response to chemical emergencies. When responding to a Superfund site, some additional concerns and requirements must be followed.

The site has an existing emergency response plan, of which the local fire department should have a copy. The site should have its access limited, hazard zones established, and, depending on the nature of the work involved, may have a decontamination corridor created. Prior to work beginning on a Superfund site, the local fire department should meet with the site supervisor to learn the hazards of the site and any protection measures that may be in place. Because Superfund sites vary from old landfills to mercury dumping sites, it is not possible to provide here a breakdown of the potential site hazards and precautions that the site may hold. Prior to becoming a Superfund site, the area may be listed as a National Priority List (NPL) site. Firefighters should contact their state EPA or federal regional EPA office to learn of any Superfund or NPL sites in their area.

Clean Air Act

The Clean Air Act Amendments (CAAA), passed in 1990, have some language that affects the fire service. It requires that certain facilities file additional planning documents and that the LEPC and the local fire service be involved in training and exercises. As of June 1999, facilities were required to submit emergency plans, many of which must be coordinated with the local fire department. Other requirements, such as exercises and worst-case scenarios, may also involve the fire department.

Respiratory Protection

OSHA's respiratory protection regulation (29 CFR 1910.134) also impacts the fire service, because it adds additional requirements that were previously exempted. The biggest changes are the inclusion of the two in/two out rule, which was in place for chemical incidents but now also includes structural fire situations. The fire service is now required to fit test all firefighters and provide them with a medical survey or a physical exam.

Record keeping has also changed. Specific records must be kept by the fire department regarding the employees and the respiratory protection program. Records include the listing of personnel training, daily equipment checks, periodic maintenance, routine service, review of the respiratory protection program, and other items.

Firefighter Safety

NFPA Standard 1500, *Fire Department Occupational Safety and Health Program,* is sometimes referred to when discussing hazardous materials issues. This standard is a "broad-based" program for providing a safe workplace for firefighters. It has been applied by OSHA using the general duty clause, specifically regarding the two in/two out requirement.

NFPA Chemical Protective Clothing

NFPA Standards 1991, 1992, and 1993 are standards for chemical protective clothing, including encapsulated suits, splash-protective suits, and support garments. They establish design and use requirements.

NFPA 1994, *Standard on Protective Ensembles for Chemical/Biological Terrorism Incidents,* has three levels of protective equipment that can be used in the event of a chemical or biological attack.

Lessons Learned

The maze of laws, regulations, and standards can be confusing and sometimes overwhelming. Most laws and regulations are not easy to read and it can take years to comprehend their true intent. They are subject to interpretation and change frequently. Emergency responders must keep abreast of the ones that affect their everyday jobs. In larger departments, staff may be assigned to monitor health, safety, and environmental laws and regulations. In smaller departments, this may not be missed, but the fire service is an industry the same as the chemical plant or gas station in any community. The fire service has to follow the same health, safety, and environmental laws and regulations. For everyone's health and safety, firefighters should keep abreast of these items, so they can help make communities safer places to live and work.

KEY TERMS

Air Monitoring Devices Used to determine oxygen, explosive, or toxic levels of gases in air.

Awareness Level The basic level of training for emergency response to a chemical accident, the basis of which is the ability to recognize a hazardous situation and call for assistance.

Clandestine Drug Labs Illegal labs set up to manufacture street drugs.

Emergency Planning and Community Right to Know Act (EPCRA) The portion of SARA that specifically outlines how industries report their chemical inventory to the community.

Extremely Hazardous Substances (EHS) A list of 366 substances that the EPA has determined present an extreme risk to the community if released.

Gross Negligence Occurs when an individual disregards training and continues to act in a manner without regard for others.

Hazardous Waste Operations and Emergency Response (HAZWOPER) The OSHA regulation that covers safety and health issues at hazardous waste sites, as well as response to chemical incidents.

Incident Commander Level A training level that encompasses the operations level with the addition of incident command training. Intended to be the person who may command a chemical incident.

Laws Legislation that is passed by the House and Senate, and signed by the president.

Liability The possibility of being held responsible for individual actions.

Local Emergency Planning Committee (LEPC) A group composed of members of the community, industry, and emergency responders to plan for a chemical incident, and to ensure that local resources are adequate to handle an incident.

Material Safety Data Sheet (MSDS) Information sheet for employees that provides specific information about a chemical, with attention to health effects, handling, and emergency procedures.

Negligence Acting in an irresponsible manner, or different from the way in which someone was trained; that is, differing from the standard of care.

Operations Level The next level of training above awareness that provides the foundation which allows for the responder to perform defensive activities at a chemical incident.

Paragraph q The paragraph within HAZWOPER that outlines the regulations that govern emergency response to chemical incidents.

Regulations Developed and issued by a governmental agency and have the weight of law.

Specialist Level A level of training that provides for a specific type of training, such as railcar specialist; someone who has a higher level of training than a technician.

Standards Usually developed by consensus groups establishing a recommended practice or standard to follow.

State Emergency Response Committee (SERC) A group that ensures that the state has adequate training and resources to respond to a chemical incident.

Superfund Amendments and Reauthorization Act (SARA) A law that regulates a number of environmental issues, but is primarily for chemical inventory reporting by industry to the local community.

Technician Level A high level of training that allows specific offensive activities to take place, to stop or handle a chemical incident.

REVIEW QUESTIONS

1. How does the response to hazardous materials differ from a structural fire response?
2. How can hazardous materials be easily defined?
3. Who is to receive the chemical reporting required under SARA Title III?
4. Who is responsible for ensuring the proper response to a hazardous materials emergency?
5. Which regulation covers all emergency responders who respond to chemical accidents?
6. What is the difference between negligence and gross negligence?
7. Which regulation requires employers to maintain Material Safety Data Sheets?

Endnotes

1. Some LEPCs are established on a city basis; some states may only have one. The format varies from state to state.
2. The OSHA hazard communication regulation (29 CFR 1910.1200) also makes this requirement.
3. Federal regulations are published in the *Federal Register,* which comes out daily. Annually, they are published in a text called the *Code of Federal Regulations* (CFR). Each federal agency is identified by a two-digit number (29 for OSHA, 40 for EPA, 49 for DOT), and each regulation is given a number to identify it.
4. A TIA is tentative because it has been established outside of the normal NFPA process, in between revisions. When the standard comes up for revision it will be subject to the normal review process.
5. The applicability varies from state to state, depending on OSHA's jurisdiction. Some states may have more stringent regulations.

Additional Resources

Bevelacqua, Armando S., *Hazardous Materials Chemistry.* Delmar Learning, a part of the Thomson Corporation, Clifton Park, NY, 2001.

Bevelacqua, Armando S. and Richard H. Stilp, *Hazardous Materials Field Guide,* Delmar Learning, a part of the Thomson Corporation, Clifton Park, NY, 1998.

Bevelacqua, Armando S. and Richard H. Stilp, *Terrorism Handbook for Operational Responders,* 2nd ed. Delmar Learning, a part of the Thomson Corporation, Clifton Park, NY, 2004.

Hawley, Chris, *HazMat Air Monitoring & Detection Devices.* Delmar Learning, a part of the Thomson Corporation, Clifton Park, NY, 2002.

Hawley, Chris, *Hazardous Materials Incidents,* 2nd ed. Delmar Learning, a part of the Thomson Corporation, Clifton Park, NY, 2004.

Henry, Timothy V. *Decontamination for Hazardous Materials Emergencies,* Delmar Learning, a part of the Thomson Corporation, Clifton Park, NY, 1999.

Laughlin, Jerry and David Trebisacci, editors, *Hazardous Materials Response Handbook,* 4th ed. National Fire Protection Association, Quincy, MA, 2002.

Lesak, David, *Hazardous Materials Strategies and Tactics.* Prentice Hall, 1998.

Noll, Gregory, Michael Hildebrand, and James Yvorra, *Hazardous Materials Managing the Incident.* Fire Protection Publications, Oklahoma University, 1995.

Schnepp, Rob and Paul Gantt, *Hazardous Materials: Regulations, Response & Site Operations.* Delmar Learning, a part of the Thomson Corporation, Clifton Park, Clifton Park, NY, 1999.

Stilp, Richard and Armando Bevelacqua, *Emergency Medical Response to Hazardous Materials Incidents.* Delmar Learning, a part of the Thomson Corporation, Clifton Park, NY, 1997.

CHAPTER

25

HAZARDOUS MATERIALS: RECOGNITION AND IDENTIFICATION

Christopher Hawley, Baltimore County Fire Department (Ret.)

OUTLINE

- Objectives
- Introduction
- Location and Occupancy
- Placards, Labels, and Markings
- Other Identification Systems
- Containers
- Senses
- Chemical and Physical Properties
- Lessons Learned
- Key Terms
- Review Questions
- Endnotes
- Additional Resources

Photo courtesy of Baltimore County Fire Department

STREET STORY

While the news media tends to focus on the large and spectacular, the reality is that every day the fire service responds to literally hundreds of hazardous materials–related incidents. These incidents usually do not get much attention because they are small in scope and are handled with a minimal amount of fire department resources. Common examples include natural gas leaks and flammable and combustible liquid spills.

We have an old saying in the HAZMAT community that the initial ten minutes of an incident will dictate the tone for the first hour of an incident. Clearly, the actions—or inactions—of the first responders will set the tone for an incident. I think back to two incidents in my career that reinforce this point.

The first incident involved an engine company being sent to investigate a call regarding an unknown liquid spilled along a highway. Upon arrival, the engine company officer found a very viscous red liquid spilled along a long distance of a road, apparently from the rear of a tractor-trailer. Unsure of the identity and the potential hazard of the liquid, the officer requested a HAZMAT unit to respond to the scene and provide assistance. Using their monitoring and detection equipment, responders were eventually able to determine that the liquid posed no hazard to the community. In fact, the unknown liquid was eventually identified as strawberry syrup! Although the officer took some ribbing from his peers, he clearly made the proper decisions to ensure the safety of both his personnel and the community.

In the second incident, a police officer was called to provide assistance to a public works road crew that had discovered what appeared to be a 5-pound portable fire extinguisher wrapped in duct tape with a fuse on the top. Although he believed it to be a hoax, the officer still requested the response of the fire department bomb squad. After conducting a thorough risk assessment and requesting the additional assistance of a HAZMAT response team, the bomb squad was able to determine that the perceived hoax was, in fact, an actual explosive device. When the device was disrupted (i.e., blown up) by the bomb squad, it made a lasting impression on the police officer!

In both of these instances, recognition and identification by the first responders set the tone for the incident. If you do not know what to do, isolate the area, deny entry, and call for help.

—Street Story by Gregory G. Noll, Emergency Planning and Response Consultant, Hildebrand and Noll Associates, Lancaster, Pennsylvania

OBJECTIVES

After completing this chapter, the reader should be able to:

- Identify the nine hazard classes as defined by DOT.*
- Identify the hazards associated with each hazard class.
- Identify the standard occupancies where hazardous materials may be used or stored.
- Identify the standard container shapes and sizes and common products.
- Identify both facility- and transportation-related markings and warning signs.
- Identify the standard transportation types for highway and rail.
- Explain the use of the NFPA 704 system.
- Explain the use of transportation containers in identifying possible contents.
- Explain the location of emergency shutoffs on highway containers.
- Explain the importance of understanding chemical and physical properties of hazardous materials.

INTRODUCTION

This chapter is of primary importance to the emergency responder.

> **Safety** It is through recognition and identification (R&I) that firefighters can impact their ability to stay alive.

The inability to recognize the potential for chemicals to be present and the inability to identify the chemical hazard can place firefighters in severe danger. Firefighting is inherently dangerous, but the response to a hazardous materials release creates an additional risk. Not only can there be immediate effects from some materials, but multiple exposures can have far-reaching effects. As depicted in the opening photo, firefighters are using proper PPE but taking a considerable risk to rescue a live victim. More information on this rescue and risk is provided in the case study in Chapter 28. Although fires have killed hundreds before, these types of fires are rare. Hazardous materials incidents have killed thousands and injured countless more. In 1984 a release of methyl isocyanate in Bhopal, India, killed more than 2,000 and injured thousands more. This incident was the basis for the Emergency Planning and Community Right to Know Act (EPCRA) because several facilities in the United States use this material.

Four basic clues to recognition and identification are (1) location and occupancy; (2) placards, labels, and markings; (3) container types; and (4) the senses. A mere suspicion in any of these areas should be enough to place a first responder on guard for the possibilities of a chemical release and its associated hazards, **Figure 25-1**.

LOCATION AND OCCUPANCY

> **Caution** The size of the community does not impact the potential for hazardous materials; every community has hazardous materials.

Figure 25-1 The four basic clues to recognition and identification are location and occupancy; placards, labels, and markings; container types; and senses.

Reader's Note: In the hazardous materials section, NFPA 1001 requires that the student receive hazardous materials training at either the Awareness level or the Operations level. The information in this text covers both levels and in some cases exceeds the Operations level. All of this information is important for firefighters' survival. See the discussion on NFPA standards and OSHA's HAZWOPER for more information on Awareness and Operations level training.

Figure 25-2 Agricultural supply stores have a large quantity of hazardous materials, including pesticides, herbicides, and fertilizers. In many cases they also have fuels, including propane.

Most communities have a gas station or a hardware store. The average home has a large amount of hazardous materials that can cause enormous problems during a response. In rural communities, farms present unique risks due to the storage of pesticides and fertilizers, **Figure 25-2**. All of these locations and occupancies provide the potential for the storage of hazardous materials. In general, the more industrialized a community is, the more hazardous materials the community will contain. Communities adjacent to industrialized areas or along major transportation corridors (interstate highways, rail, water) may also have the same hazards because these materials can travel through the community, **Figure 25-3**. Buildings that typically store hazardous materials include hardware stores, hospitals, auto part supply stores, dry cleaners, manufacturing facilities, print shops, doctors' offices, photo labs, agricultural supply stores, semiconductor manufacturing facilities, electronics manufacturing facilities, light to heavy industrial facilities, marine terminals, rail yards, airport terminals and fueling areas, pool chemical stores, paint stores, hotels, swimming pools, and food manufacturing facilities.

Figure 25-3 If a community has a road, the potential for a hazardous materials incident exists. One of the most common chemical releases is a gasoline spill.

PLACARDS, LABELS, AND MARKINGS

This section examines the first concrete evidence of the presence of hazardous materials. A number of systems are used to mark hazardous materials containers, buildings, and transportation vehicles. The systems result from laws, regulations, and standards, and in some cases from a combination of the three. As an example the **Building Officials Conference Association (BOCA)** code, which has been adopted as a regulation in local communities, requires the use of the NFPA 704 marking system, **Figure 25-4**, for certain occupancies.

Placards

The most commonly seen item for identifying the location of hazardous materials is the placard. The Department of Transportation (DOT) regulates the movement of hazardous materials ("dangerous goods" in Canada) by air, rail, water, roadway, and pipeline by means of 49 CFR 170-180. After meeting certain guidelines a shipper must placard a vehicle to warn of the storage of chemicals on the vehicle, an example of which is shown in **Figure 25-5**.

> **Note** The quantity of hazardous materials that must be carried in order to require placarding is 1,001 pounds, unless it is one of five classes of materials that require placarding at any amount.

Figure 25-5 The DOT requires some shippers of hazardous materials to provide placards to warn responders of chemicals that may be on the truck.

Table 25-1 and **Table 25-2** provide further explanations of the placarding system.

The DOT has established a system of nine hazard classes that uses more than twenty-seven placards to identify a shipment. The idea behind these hazard classes is to provide a general grouping to a shipment and to provide some basic information regarding the potential hazards. The placards, like the one shown in **Figure 25-6**, are 10¾ inches by 10¾ inches

Figure 25-4 The NFPA has developed a system for identifying potential chemical hazards in a building. The system is known as NFPA 704.

Materials That Require Placarding at Any Amount (DOT Table 1)

HAZARD CLASS OR DIVISION	PLACARD TYPE
1.1	Explosives 1.1
1.2	Explosives 1.2
1.3	Explosives 1.3
2.3	Poison gas
4.3	Dangerous when wet
5.2 (Organic peroxide, type B, liquid or solid, temperature controlled)	Organic peroxide
6.1 (Inhalation hazard Zone A or B)	Poison inhalation hazard
7 (radioactive label III only)	Radioactive

TABLE 25-1

Materials That Require Placarding at 1,001 Pounds (DOT Table 2)

CLASS OR DIVISION	PLACARD TYPE
1.4	Explosives 1.4
1.5	Explosives 1.5
1.6	Explosives 1.6
2.1	Flammable gas
2.2	Nonflammable gas
3	Flammable
COMBUSTIBLE LIQUID	**COMBUSTIBLE**
4.1	Flammable solid
4.2	Spontaneously combustible
5.1	Oxidizer
5.2 (Other than organic peroxide, type B, liquid or organic peroxide solid, temperature controlled)	Organic peroxide
6.1 (Other than inhalation)	Poison
6.1 (PG III)	Keep away from food
6.2	None
8	Corrosive
9	Class 9
ORM-D	None

TABLE 25-2

Figure 25-6 "Corrosive" placard. Not to scale.

and are to be placed on four sides of the vehicle. Labels are $3\frac{9}{10}$ inches by $3\frac{9}{10}$ inches and are affixed near the shipping name on the container. Labels, for the most part, are smaller versions of placards and are designed to provide warnings about the package contents. There are labels for some materials that do not have or require placards.

The system is designed so that materials designated by the DOT as potentially harmful to the environment, humans, and animals are easily identified. Such a material has to present an unreasonable risk to the safety, health, or property upon contact. The DOT has two placarding tables, which are called Table 1 and Table 2. The materials that are on Table 1 (as shown in **Table 25-1**) are those that are most hazardous and require the use of a placard no matter what the quantity being shipped. The materials in DOT Table 2 (as shown in **Table 25-2**) are those that require placarding at 1,001 pounds. The criteria establishing this 1,001-pound rule are not clearly defined, but responders should be aware that a spill of 999 pounds of a material can be as hazardous as 1,001 pounds. The shipper uses the hazardous materials tables (49 CFR 172.504) to determine which labels and placards are required. The tables may also list a packing group for the material, which indicates the danger associated with the material being transported.

- *Packaging Group I:* Greatest danger
- *Packaging Group II:* Medium danger
- *Packaging Group III:* Minor danger

Packaging groups are only assigned to classes 1, 3 through 6, 8, and 9 as shown in **Table 25-3** through **Table 25-6**. These are determined based on flash points, boiling points, and toxicity.

> **Note** Placards are useful because they take advantage of four ways to communicate the hazard class they represent.

Packing Groups

FLAMMABILITY—CLASS 3 PACKING GROUP (PG)	FLASH POINT (°F)	BOILING POINT (°F)
I		≤ 95
II	< 73	> 95
III	≥ 73 and ≤ 141	> 35

TABLE 25-3

Division 2.3, Poisonous Gas

HAZARD ZONE	INHALATION TOXICITY—LETHAL CONCENTRATION (LC_{50})*
Hazard Zone A	LC_{50} less than or equal to 200 ppm
Hazard Zone B	LC_{50} greater than 200 ppm and less than or equal to 1,000 ppm
Hazard Zone C	LC_{50} greater than 1,000 ppm and less than or equal to 3,000 ppm
Hazard Zone D	LC_{50} greater than 3,000 ppm and less than or equal to 5,000 ppm

*LC_{50} is the lethal concentration to 50 percent of an exposed population (gases).

TABLE 25-4

Division 6.1, Packing Materials That Are Toxic by a Route Other Than Inhalation

PACKING GROUP	ORAL TOXICITY—LETHAL DOSE (LD_{50})* (MG/KG)	DERMAL TOXICITY—LETHAL DOSE (LD_{50}) (MG/KG)	INHALATION BY DUSTS AND MISTS LC_{50} (MG/L)
I	≤ 5	≤ 40	≤ 0.5
II	$<5, \leq 50$	$>40, \geq 200$	$>0.5, \leq 2$
III	Solids: $>50, \leq 200$ Liquids: $>50, \leq 500$	$>200, \leq 1{,}000$	$>2, \leq 10$

*(LD_{50}) Lethal dose to 50 percent of the exposed population (solids and liquids).

TABLE 25-5

Division 6.1, Toxic Materials Poisonous by Inhalation

PACKING GROUP	VAPOR CONCENTRATION AND TOXICITY
I (Hazard Zone A)	$V \geq 500\ LC_{50}$ and LC_{50} 200 mL/m^3

TABLE 25-6

They have distinct colors, they have a picture at the top of the triangle depicting a representation of the hazard, they state the hazard class in the middle of the placard, and they display the hazard class and division number in the bottom triangle.

As an example the placard shown in **Figure 25-6** is black and white, which represents the corrosive class. The top of the placard has a picture of a hand and a steel bar being eaten away by the corrosive material being poured on it, the middle of the placard states corrosive, and at the bottom the class number 8 shows.

The DOT also requires the addition of a four-digit number, known as the United Nations/North America (UN/NA) identification number, either on a placard or on an adjacent orange strip. This identifies a bulk shipment of over 119 gallons and provides an identity to the material. A tank truck carrying gasoline, which would be considered a bulk shipment, would display a flammable placard[1] with the number 1203 either in the middle of the placard or on an orange strip adjacent to the placard as shown in **Figure 25-7**. This provides an additional bit of information to the responder, because without the UN/NA number, the only information provided would be that a flammable liquid was on board.

The nine hazard classes and subdivisions[2] are discussed next.

Class 1, Explosives (Figure 25-8)

- *Division 1.1.* Mass explosion hazard, such as black powder, dynamite, ammonium perchlorate, detonators for blasting, and RDX explosives.
- *Division 1.2.* Projectile hazard, such as aerial flares, detonating cord, detonators for ammunition, and power device cartridges.
- *Division 1.3.* Fire hazard or minor blast hazard. Examples include liquid-fuel rocket motors and propellant explosives.
- *Division 1.4.* Minor explosion hazard, which includes line throwing rockets, practice ammunition, detonation cord, and signal cartridges.
- *Division 1.5.* Very insensitive explosives, which do have mass explosion potential but during nor-

Figure 25-7 There are three ways to signify the four-digit ID number for bulk shipments. The most common is for the four-digit DOT identification number to be placed in the middle of the placard.

mal shipping would not present a risk. Ammonium nitrate and fuel oil (ANFO) mixtures are an example of this division.

- *Division 1.6.* Also very insensitive explosives that do not have mass explosion potential. Materials that present an unlikely chance of ignition are part of this grouping.

Safety Incidents involving explosives can be very dangerous, especially when involved in fire.

Making a tactical decision to attack a fire involving explosives can endanger the responders, especially if the fire has reached the cargo area of the vehicle. The recommendations in the **Emergency Response Guidebook (ERG)** should be followed, paying particular attention to the isolation and evacuation distances. When explosives are involved in traffic accidents not involving fire, the actual threat is minimized depending on the circumstances. As long as the explosives were transported legally and as they were intended to be transported, the responders should face little danger. Some cities require an escort and that the explosives be transported at nonpeak hours. Spilled explosive materials may present a health hazard if inhaled or absorbed through the skin.

Figure 25-8 "Explosive" placards and labels.

ANFO Explosion On November 29, 1988, an engine company from the Kansas City, Missouri, Fire Department was dispatched to a reported pickup truck fire. While en route to the incident the engine company was told to use caution because explosives were reportedly involved. When they arrived they found that they had two separate fires, one in the pickup truck and the other in a trailer. The first engine began to extinguish the fire in the pickup truck and requested a second engine for assistance as well as the district battalion chief. They attempted to contact the second engine to warn them of the explosives on fire on top of the hill. The second engine arrived and began to attack the fire on top of the hill. They requested the assistance of the first engine and also requested a squad for water. From the radio communications with the battalion chief the crews thought that the explosives had already detonated.

The battalion chief was a quarter of a mile away where he had stopped to talk with the security guards when the explosives detonated. The explosion moved the chief's car 50 feet and blew in the windshield. The blast was heard for 60 miles and damaged homes within 15 miles. The chief requested additional assistance and staged the responding companies. There was a report that there were more explosives on the hill that had not yet detonated. Luckily the chief did not let any other responders into the scene because shortly after pulling back, a second explosion went off, reportedly larger than the first one.

The next morning a team of investigators went to the site to begin the investigation. They discovered that six firefighters were killed in the blast and the explosion was very devastating. Only one engine was recognizable; the other was reduced to the frame rail. The first blast was from 17,000 pounds of an ammonium nitrate, fuel oil, and aluminum mixture. It also had 3,500 pounds of ANFO. The second explosion involved 30,000 pounds of the mixture. The use of ANFO is common throughout the United States for blasting purposes. Ammonium nitrate is commonly used as a fertilizer in the agricultural business and is used on residential lawns. The Oklahoma City Alfred P. Murrah Federal Building explosive was devised of ammonium nitrate and nitromethane, very similar to ANFO. The World Trade Center bombing in 1993 used an explosive made up primarily of urea nitrate and three cylinders of hydrogen. Both explosives are comparable to each other, and when they are used improperly the results can be devastating.

Class 2, Gases (Figure 25-9)

- *Division 2.1.* Flammable gases that are ignitable at 14.7 psi in a mixture of 13 percent or less in air, or have a flammable range with air of at least 12 percent regardless of the lower explosive limit (LEL). Propane and isobutylene are examples of this division.
- *Division 2.2.* Nonflammable, nonpoisonous, compressed gas, including liquefied gas, pressurized **cryogenic gas**, and compressed gas in solution. Carbon dioxide, liquid argon, and nitrogen are examples.
- *Division 2.3.* Poisonous gases that are known to be toxic to humans and would pose a threat during transportation. Chlorine and liquid cyanogen are common examples of this division. Gases assigned to this division are also assigned a letter code identifying the material's toxicity levels. These levels are discussed further in the section on toxicology. The hazard zones associated with this division are:

  Hazard Zone A: LC_{50} less than or equal to 200 ppm
  Hazard Zone B: LC_{50} greater than 200 ppm and less than or equal to 1,000 ppm
  Hazard Zone C: LC_{50} greater than 1,000 ppm and less than or equal to 3,000 ppm
  Hazard Zone D: LC_{50} greater than 3,000 ppm and less than or equal to 5,000 ppm

The hazard zones are a quick way to determine how toxic a material is. Hazard Zone A materials are more toxic than Hazard Zone B materials, and so forth. The shipping papers will identify these by the addition of Poison Inhalation Hazard (PIH) Zone A, B, C, or D.

Class 3, Flammable Liquids (Figure 25-10)

- Flammable liquids have a flash point of less than 141°F. Gasoline, acetone, and methyl alcohol are examples.
- Combustible liquids are those with flash points above 100°F and below 200°F. The DOT allows liquids with a flash point of 100°F to be shipped as a combustible liquid. Examples include diesel fuel, kerosene, and various oils.

> **Caution** The difference between a flammable liquid and a combustible liquid is based on the material's flash point. The flash point is the temperature of the liquid at which there could be a flash fire if an ignition source is present. The fire service usually refers to a flammable liquid as one that has a flash point of 100°F or lower. The DOT classifies any liquid with a flash point of 141°F or lower as being a flammable liquid. Any liquid with a flash point greater than 141°F is considered combustible by the DOT. It can be confusing when using the terms *flammable* and *combustible* to describe a liquid material. Most references, with the exception of the DOT, use 100°F as the criteria for flammable and combustible.

Figure 25-9 "Gas" placards and labels.

Figure 25-10 "Flammable" and "Combustible" placards and label.

Figure 25-11 Class 4 placards and labels.

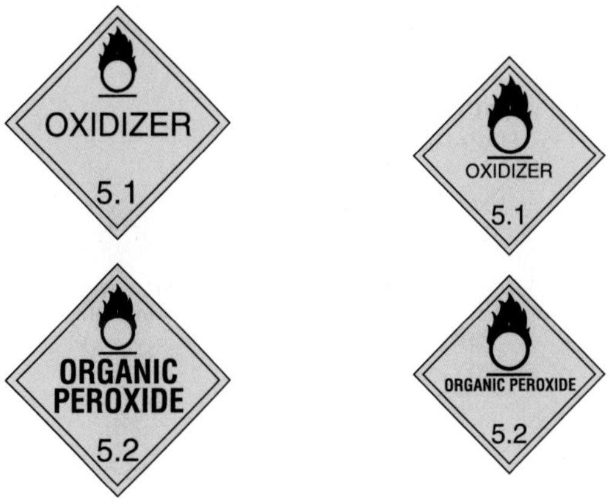

Figure 25-12 Class 5 placards and labels.

Class 4, Flammable Solids (Figure 25-11)

- *Division 4.1.* Includes wetted explosives, self-reactive materials, and readily combustible solids. Examples include magnesium ribbons, picric acid, explosives wetted with water or alcohol, or plasticized explosives.
- *Division 4.2.* Composed of spontaneously combustible materials including pyrophoric materials or self-heating materials. An example is zirconium powder.
- *Division 4.3.* Dangerous-when-wet materials are those that when in contact with water can ignite or give off flammable or toxic gas. Calcium carbide when mixed with water makes acetylene gas, which is very flammable as well as unstable in this form. Sodium is another example of a material that when wet can ignite explosively. Lithium and magnesium are not as explosive as sodium but will react with water. Magnesium if on fire will react violently if water is used in an attempt to extinguish the fire.

Class 5, Oxidizers and Organic Peroxides (Figure 25-12)

- *Division 5.1.* The class assigned to materials that have the ability to produce oxygen, which in turn increases the chance of fire, and during fires make the fire burn more intensely. Ammonium nitrate and calcium hypochlorite are examples.
- *Division 5.2.* The organic peroxides, which have the ability to explode or polymerize, which if contained is an explosive reaction. These are further subdivided into seven types:

Type A: Can explode upon packaging. These are DOT forbidden, which means they cannot be transported and must instead be produced on site.
Type B: Can thermally explode, considered a very slow explosion.
Type C: Neither detonates nor **deflagrates** rapidly, and will not thermally explode.
Type D: Only detonates partially or deflagrates slowly, and has medium or no effect when heated and confined.
Type E: Shows low or no effect when heated and confined.
Type F: Shows low or no effect when heated and confined, and has low or no explosive power.
Type G: Is thermally stable and is desensitized.

Class 6, Poisonous Materials (Figure 25-13)

- *Division 6.1.* Materials that are so toxic to humans that they would present a risk during transportation. Examples include arsenic and aniline.
- *Division 6.2.* Composed of microorganisms or their toxins, which can cause disease to humans or animals. Anthrax, rabies, tetanus, and botulism are examples.

Hazard zones are associated with Class 6 materials:

- *Hazard Zone A:* LC_{50} less than or equal to 200 ppm
- *Hazard Zone B:* LC_{50} greater than 200 ppm and less than or equal to 1,000 ppm.

Class 7, Radioactive Materials (Figure 25-14)

- Those materials determined to have radioactive activity at certain levels.
- Although there is only one placard, the labels shown in **Figure 25-14** are further subdivided

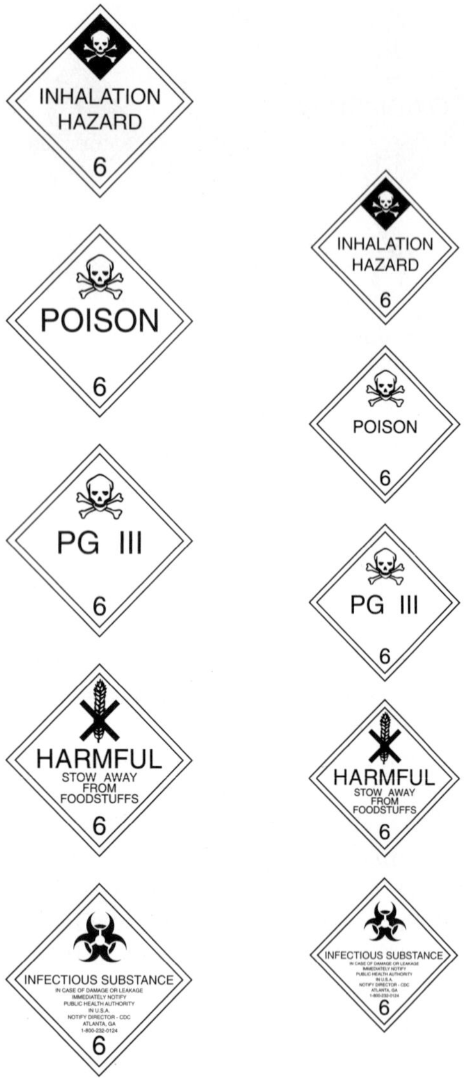

Figure 25-13 Class 6 placards and labels.

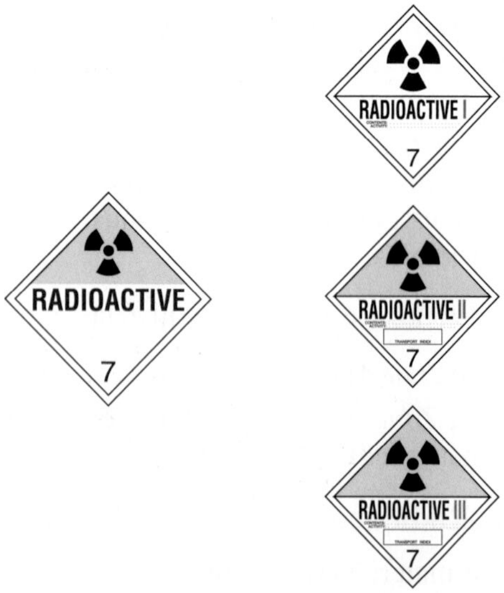

Figure 25-14 Class 7 placards and labels.

Figure 25-15 Class 8 placards and label.

into Radioactive I, II, and III, with level III being the highest hazard. The designation I, II, or III is dependent on two criteria: the transport index and the radiation level coming from the package. The transport index is the degree of control the shipper is to use and is based on a calculation of the radiation threat the package presents. Responders should understand that a package labeled radioactive may be emitting radiation. These emissions of radiation are legal within certain guidelines:

- Radioactive I label—less than 0.005 mSv/hr (0.5 mR/hr)
- Radioactive II label—more than Radioactive I and less than 0.5 mSv/hr (50 mR/hr)
- Radioactive III label—more than Radioactive II and less than 2 mSv/hr (200 mR/hr)

Packages emitting more than 2 mSv/hr require special handling and transportation and are subject to additional regulations.

Class 8, Corrosives (Figure 25-15)

- Includes both acids and bases, and is described by the DOT as a material capable of causing visible destruction in skin or corroding steel or aluminum. Examples include sulfuric acid and sodium hydroxide.

Class 9, Miscellaneous Hazardous Materials (Figure 25-16)

- A general grouping that is composed of mostly hazardous waste. Dry ice, molten sulfur, and polymeric beads are examples that would use a Class 9 placard.

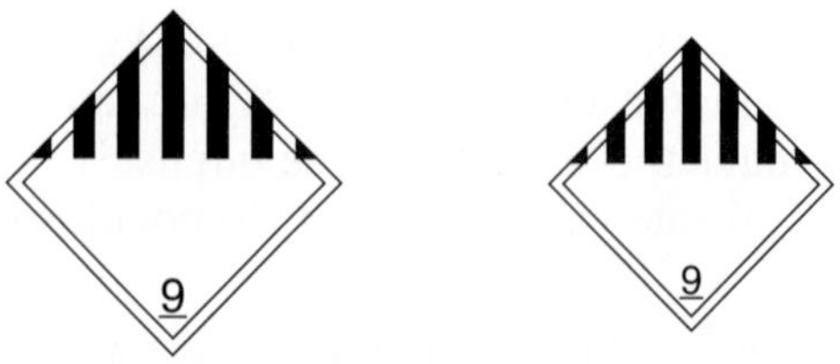

Figure 25-16 Class 9 placards and label.

Figure 25-17 "Dangerous" placard.

- This is known as a catch-all category. If a substance does not fit into any other category and presents a risk during transportation then it becomes Class 9.

Other Placards and Labels

- The Dangerous placard is used when the shipper is sending a mixed load of hazardous materials. If the shipper sends 2,000 pounds of corrosives and 2,000 pounds of a flammable liquid, then instead of displaying two placards the shipper can display a Dangerous placard, as shown in **Figure 25-17**. If any of the items exceeds 2,205 pounds and is picked up at one location, then in addition to the Dangerous placard the shipper is to display the placard for the material that exceeds the 2,205 pounds.
- The Stow Away from Foodstuffs placard indicates that a poisonous material is being transported, but it is not poisonous enough to meet the rules to be placarded as a poison; most are PG III. Chloroform is an example that would use a placard like the one shown in **Figure 25-18**.
- "Other Regulated Material—Class D" (ORM-D) is a classification that is left over from a previous DOT regulation. It is a subdivision that includes ammunition and consumer commodities, such as cases of hair spray. The package will have the printing "ORM-D" on the outside of the package. The previous regulation used to have ORM-A through ORM-E, but these are now grouped together in Class 9, Miscellaneous Hazardous Materials.

Figure 25-18 "Harmful" placard.

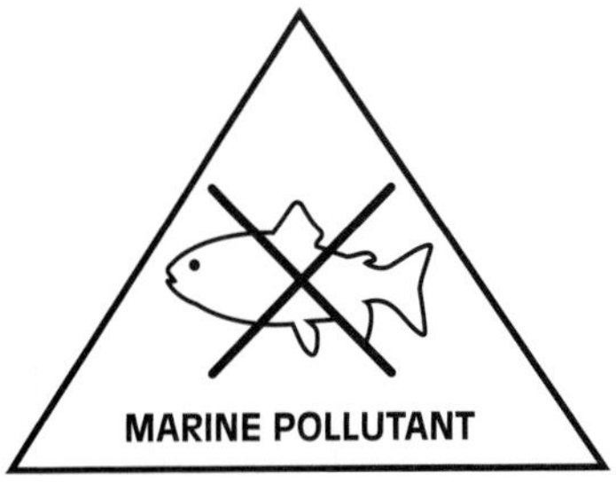

Figure 25-19 "Marine Pollutant" marking.

- "Marine Pollutant" is displayed on shipments that, if the material were released into a waterway, would damage the marine life, **Figure 25-19**.
- Elevated temperature material will have a "HOT" label either to the side of or on the placard, as shown in **Figure 25-20**, if it meets one of the following criteria:
  - Is a liquid above 212°F.
  - Is a liquid that is intentionally heated and has a flash point above 100°F.
  - Is a solid at 464°F or above.
  - "Infectious Substances" is a label like the one shown in **Figure 25-21** that is sometimes used on the outside of trucks. It is not required by the DOT but may be required by other agencies such as Health and Human Services or a state agency.
  - The Fumigated placard, like the one shown in **Figure 25-22**, is used when a trailer or railcar has been fumigated with a poisonous material. This placard is commonly found near ports where containers are frequently fumigated after arriving from a foreign port.

A white square background as shown in **Figure 25-23** is used in the following situations:

- On the highway for controlled Radioactive III shipments

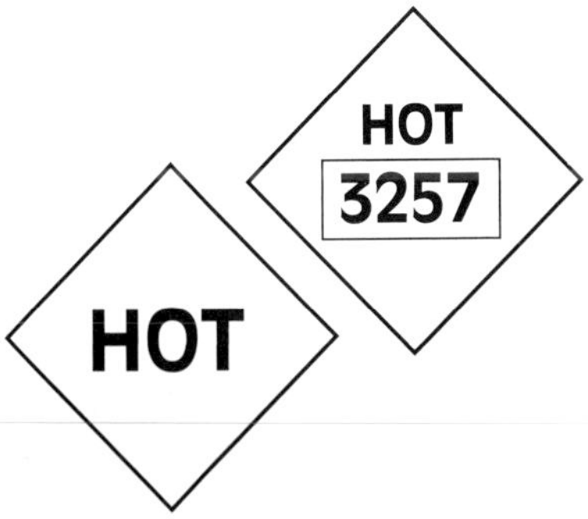

Figure 25-20 "Hot" placard.

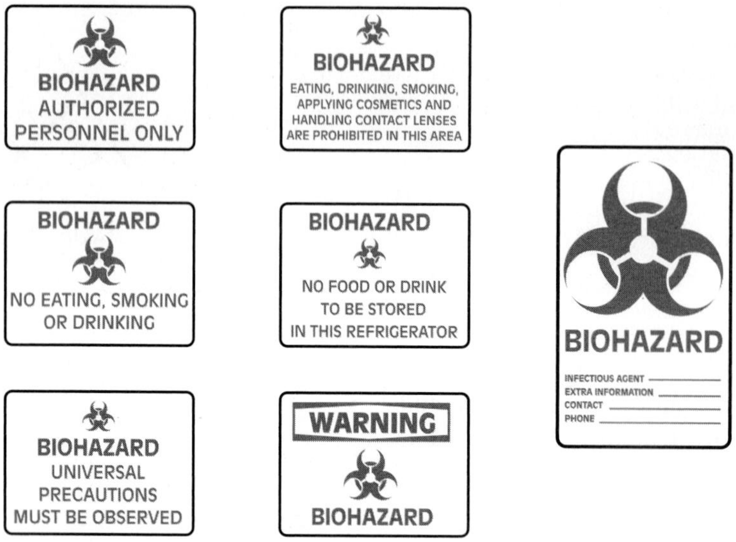

Figure 25-21 "Biohazard" labels.

Figure 25-22 "Fumigation" marking.

Figure 25-23 White-squared "Flammable Gas" placard.

- On rail for:
  - Explosives 1.1 or 1.2
  - Division 2.3 Hazard Zone A (poison gas) materials
  - Division 6.1 Packing Group I Hazard Zone A (poison)
  - Division 2.1 (flammable gas) in a DOT 113 tank car
  - Division 1.1 or 1.2, which is chemical ammunition that also meets the definition of a material that is poisonous by inhalation

Problems with the Placarding System

The placarding system relies on a human to determine the extent of the load, determine the appropriate hazard classes, and interpret difficult regulations to determine if a placard is required. The placard then must be affixed to all four sides of the vehicle before shipment. Placards are only required for shipments that exceed 1,001 pounds, except for materials listed in **Table 25-1**, which require placarding at any amount. It is suspected that 10 to 20 percent of the trucks traveling the highway are not placarded at all or not placarded correctly.

> **Caution** Given the restrictions of many cities, bridges, and tunnels where hazardous materials are not allowed, many trucks are probably not carrying the proper designations.

A placard can come off during transport and legally may not have to be immediately replaced. The fact that an incident involves a truck or train should alert the first responder to the potential for hazardous materials, and when a placard is involved extra precautions should be taken.

Streetsmart Tip A tractor trailer was involved in an accident and had jackknifed in Baltimore County, Maryland. The HAZMAT team was called to assist with the fuel spill. Upon arrival the HAZMAT team inquired as to the contents of the truck and was told by the first responders that the truck was carrying rocking chairs. While working to control the fuel leak the HAZMAT team noticed a greenish blue liquid coming from the front of the trailer, where it had been damaged in the accident. They located the driver and asked him about the contents of the trailer, which did not display a placard. The driver stated he was carrying rocking chairs, which is what the shipping papers showed the load to be.

The crews opened the back of the trailer to inspect the cargo and found rocking chairs. Upon closer inspection of the front of the truck, several drums were noticed in the very front of the trailer. When the driver was confronted by the HAZMAT team and a state trooper, the driver confessed he had picked up a load of unknown waste from his first stop. He did not know the contents of the drums and it took several days to determine the exact contents of the drums. Beware—it is unlikely that someone trying to evade the law will mark the shipment appropriately.

Labels

Labeling and Marking Specifics

Package markings must include the shipping name of the material, the UN/NA identification number, and the shipping and receiving companies' names and addresses. Packages that contain more than a **reportable quantity (RQ)** of a material must also be marked with an RQ near the shipping name. Packages that are listed as ORM-D materials should be marked as such. Some packages with liquids in them must use orientation arrows. Materials that pose inhalation hazards must affix an Inhalation Hazard label next to the shipping name, as shown in **Figure 25-24**. Hazardous wastes will be marked "Waste" or will use the EPA labeling system to identify these packages.

Labels are identical to placards, other than their size.

Note Materials that have more than one hazard may be required to display a primary hazard label and a subsidiary label.

The primary label will have the class and division number in the bottom triangle, while the subsidiary label will not have the number at all, as shown in **Figure 25-25**. As an example, the material

Figure 25-24 The DOT adds the "Poison—Inhalation Hazard" label to those materials that present severe toxic hazards.

Figure 25-25 Primary and subsidiary placards.

acrylonitrile, inhibited, is required to be labeled "Flammable" with a subsidiary label of "Poison."

OTHER IDENTIFICATION SYSTEMS

There are several other identification systems that are used in private industry to mark facilities and containers. Military shipments and pipelines are also marked to provide a warning as to the potential for hazardous materials. Much like the transportation system the warnings are a clue to the potential presence of hazardous materials that could cause harm to the responders.

NFPA 704 System

One of the other more common systems used to identify the presence of hazardous materials is the NFPA 704 system. This system is designed for buildings, not transportation, and alerts the first responders to the potential hazards in and around a facility. The system is much like the placarding system and relies on a triangular sign that is divided into four areas, as shown in **Figure 25-26**. The four areas are divided by color as well, and use a ranking system to identify severity. The four areas and colors are:

- Health hazard—blue
- Fire hazard—red
- Reactivity hazard—yellow
- Special hazards—white

Figure 25-26 NFPA 704 system placard.

The system uses a ranking of 0–4 with 0 presenting no risk and a ranking of 4 indicating severe risk. The specific listings are discussed next.

Health

This listing is based on a limited exposure to the materials using standard firefighting protective clothing as the protective clothing for the exposure.

4—Severe health hazard
3—Serious health hazard
2—Moderate hazard
1—Slight hazard
0—No hazard

Flammability

This listing pertains to the ability of the material to burn or be ignited.

4—Flammable gases, volatile liquids, pyrophoric materials
3—Ignites at room temperature
2—Ignites when slightly heated
1—Needs to be preheated to burn
0—Will not burn

Reactivity

This listing is based on the material's ability to react, especially when shocked or placed under pressure.

4—Can detonate or explode at normal conditions
3—Can detonate or explode if strong initiating source is used
2—Violent chemical change if temperature and pressure are elevated
1—Unstable if heated
0—Normally stable

Special Hazards

This listing is used to indicate water reactivity and oxidizers, which are included in the NFPA 704 system. In some cases other symbols may be used such as the tri-foil for radiation hazards, "ALK" for alkalis, and "CORR" for corrosives. In the presence of the slashed W there is also an accompanying ranking structure for water reactivity in addition to the hazards listed in the other triangles:

4—Not used with the slashed W and a reactivity ranking of 4
3—Can react explosively with water
2—May react with water or form explosive mixtures with water
1—May react vigorously with water
0—Slashed W is not used with a reactivity ranking of 0

Some potential problems exist with the NFPA 704 system, because it groups all of the chemical hazards listed in a building into one sign. If the sign is placed on a tank that contains one material, the system does a good job of warning about the contents of the tank, but does not provide the name of the product. For a facility that has hundreds—if not thousands—of materials, the system will only warn of the worst-case scenario. As an example, dramatically different tactics are used to handle a flammable gas incident versus a flammable liquids incident, but both can be classified as fire hazard 4. The system is best used to alert the first responder to the presence of hazardous materials and to warn of the worst-case scenario.

Hazardous Materials Information System

Commonly referred to as HMIS, the Hazardous Materials Information System was designed to provide a mechanism to comply with the federal hazard communication regulation, which requires that all containers be marked with the appropriate hazard warnings and the ingredients be provided on the label, **Figure 25-27**. Many products that come into the workplace are missing adequate warning labels. The HMIS is not a uniform system. It can be developed by the facility or by the manufacturer of the labels, so one system may vary from another. Most systems are similar to the NFPA 704 system and use blue, red, and yellow colors with a numbering system that provides an indication of hazard. The colors may be used in a triangle format or, in most cases, as stacked bars. The numbers are usually 0–4, the same as the NFPA system, but in rare cases may differ from the NFPA system. The facility manager or other representative should have the key to the symbols, or a chart should be provided somewhere in the facility indicating what the symbols and the warning levels are. The chart is usually stored with the MSDS. In most cases, a central location should be chosen for the MSDS and other hazard communication information.

In some systems, a picture is provided of the level of PPE required for the substance. Each HMIS is different, and responders should not assume any particular hazard level until the warning levels can be determined.

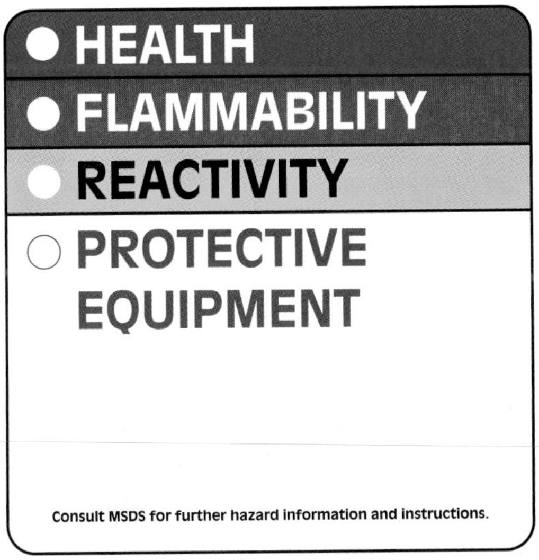

HEALTH
FLAMMABILITY
REACTIVITY
PROTECTIVE EQUIPMENT
Consult MSDS for further hazard information and instructions.

Figure 25-27 HMIS label.

Military Warning System

The military uses the DOT placarding system when possible, but in some cases may use its own system. Within the DOT's *Emergency Response Guidebook,* an emergency contact number is given when responding to an incident involving a military shipment.

In most cases, for extremely hazardous materials, arms, explosives, or secret shipments, firefighters can assume prior to their arrival that the military is already aware of the incident and probably already responding. The higher the hazard the more likely there will be an escort for the shipment. There may be shipments in which the driver of the truck is not allowed to leave the cab of the truck and may provide warnings to stay away from the truck.

> **Safety** For high-security shipments the driver is armed as are the personnel in the escort vehicle. If an incident occurs involving one of these vehicles, firefighters must obey the commands of the escorting personnel and determine if they have made the appropriate notifications.

If the driver and escort crew are killed or seriously injured in the accident, it would be advisable to notify the military about the incident, although with satellite tracking help is probably already on the way. The phone number to contact the military is in the DOT ERG, along with **Chemtrec's** and other emergency contact numbers.

Other incidents involving fuels, food, or military equipment may require notification of the military.

The military typically uses its own marking system at its facilities to mark the buildings. The military uses a series of symbols and a numerical ranking system as shown in **Figure 25-28**.

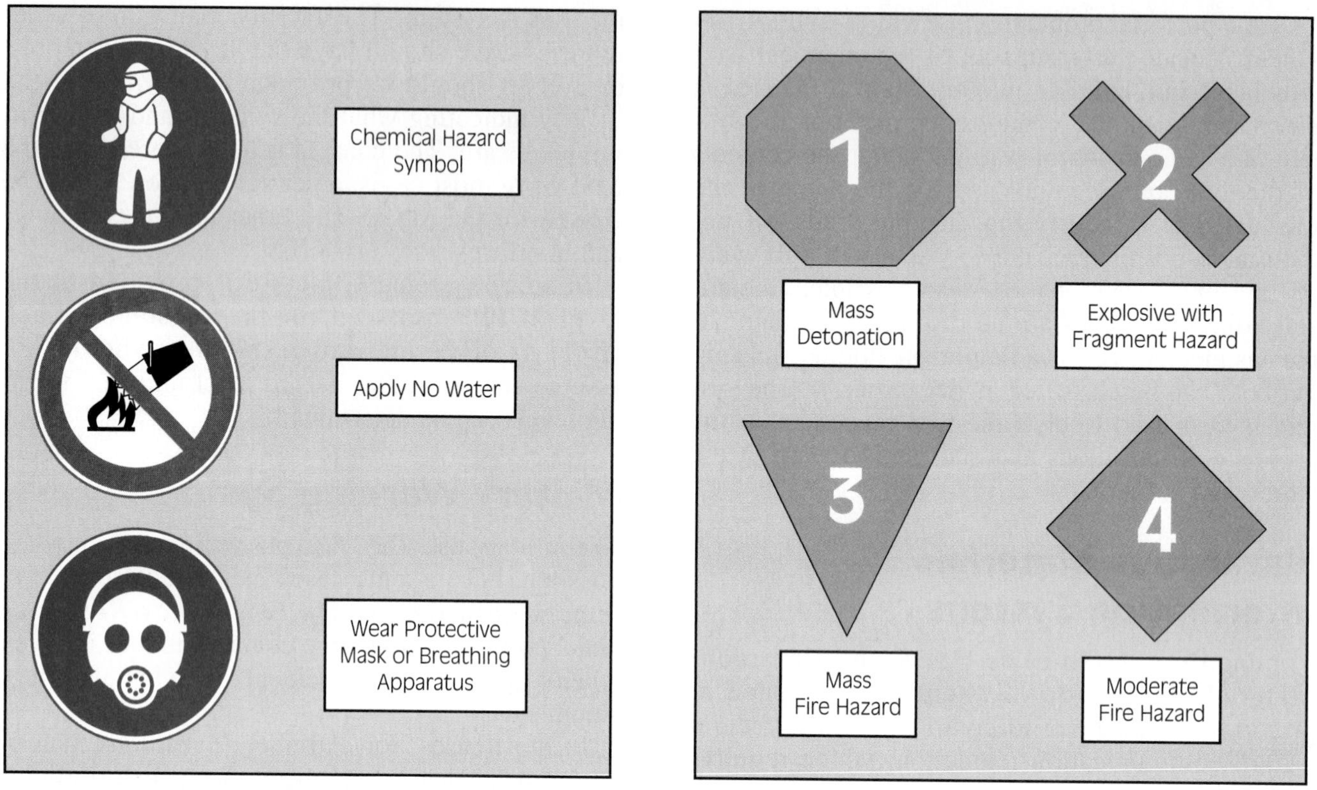

Figure 25-28 Military placards.

Pipeline Markings

Any place an underground pipeline crosses a mode of transportation the pipeline owner is required to place a sign like the one shown in **Figure 25-29** that indicates the pipeline contents, owner, and emergency contact number. The pipeline contents may be general, as in "petroleum products," because the same pipeline may be used to ship fuel oil, gasoline, motor oil, or other products. Dedicated pipelines that carry only one product will be marked with the specific product that it carries. The pipeline should be buried a minimum of 3 feet and should be adequately marked.

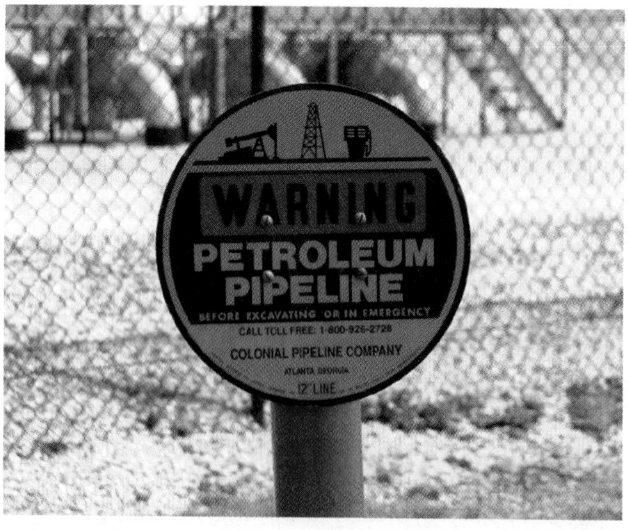

Figure 25-29 The owner of a pipeline is required to provide the contents of the pipe, the owner's name, and an emergency contact number.

Many of the larger pipelines, such as the Colonial pipeline that originates in Texas and ends in New York, are 26 inches in diameter along the main pipeline. In the event of a release involving the pipeline, the line will be shut down immediately. Even with this immediate shutdown, the potential exists to lose several hundred thousand gallons of hazardous materials because the distance between the shutoffs is substantial.

An incident involving a pipeline can be a serious event—firefighters should not underestimate the need for considerable local, state, and federal resources. Within the fire department alone, considerable resources may be required such as command staff, logistic support, communications, and tactical units. A fuel oil pipeline rupture in Reston, Virginia, resulted in the loss of more than 400,000 gallons of fuel, requiring considerable resources from several states to control the spill. The resources included emergency response organizations; local, state, and federal assistance; and a considerable number of private cleanup companies.

> **Note** Some pipelines move one type of product, while others move several different types each day.

The product in the pipelines varies from liquefied gases and petroleum products to slurried material. Pipeline companies are required to conduct in-service training and tours for the emergency responders in the communities their pipelines transverse. When firefighters have an incident on or near any pipeline, it is advisable to notify the pipeline owner of the incident, even if they are pretty sure the pipeline was not damaged. A train derailment in California caused a pipeline to shift, and the pipeline did not release any product until several days after the original derailment occurred. Most pipeline operators would like to have the opportunity to check the line as opposed to having a catastrophic release several days later because the line was not checked.

Container Markings

Most containers such as drums are marked with the contents of the drum, **Figure 25-30**, while cylinders have the name of the product stenciled on the side of the cylinder. In bulk shipments the bulk container will have the name of the product stenciled on the side.

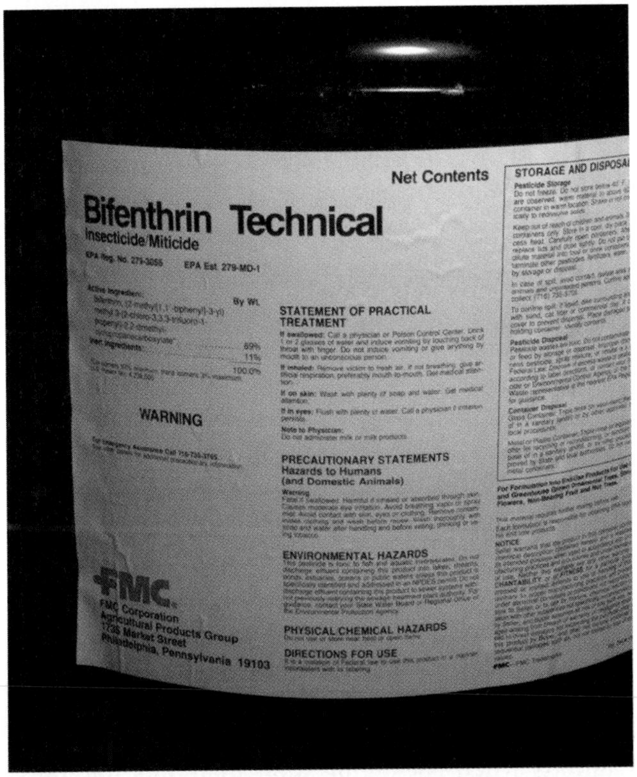

Figure 25-30 The label describes the contents of the drum.

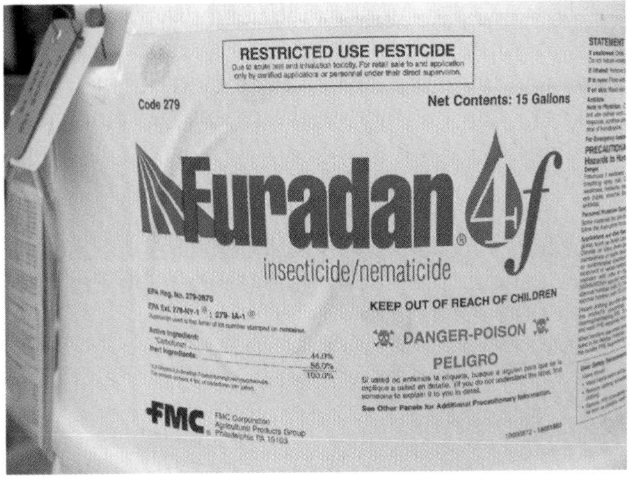

Figure 25-31 "Pesticide" labels.

Trucks that are dedicated haulers will also stencil or mark the product name on two sides of the vehicle.

Pesticide Container Markings

Due to their toxicity, pesticides are regulated by the EPA as to how they are to be marked. The label on a pesticide container, such as the one shown in **Figure 25-31**, will have the manufacturer's name for the pesticide, which is not usually the chemical name for the product. The label will also contain a signal word such as "Danger," "Warning," or "Caution."

In the United States the EPA issues an EPA registration number and in Canada the label will have a pest control number. The label will also include a precautionary statement and a hazard statement, examples of which include "Keep from Waterways," and "Keep Away from Children." The active ingredients will be listed by name and percentage; in most cases the active ingredients are usually a small percentage of the product. Inert ingredients are also listed but not specifically named. For liquid pesticides the "inert" ingredient is usually kerosene or diesel fuel, an item not normally considered "inert," except by the EPA.

CONTAINERS

Hazardous materials come in a variety of containers of many shapes and sizes, from 1-ounce bottles and larger bags to tanks and ships carrying hundreds of thousands of gallons. A survey of the materials in the average home will reveal a wide variety of storage containers. Compressed gas cylinders hold propane; steel containers hold flammable and combustible liquids; bottles, jars, and small drums hold various products. Plastic-lined cardboard boxes, and bags of various types are also used for chemical storage.

The type of material and the end use for the product determine the packaging. Packaging used to store household or consumer commodities is usually different than the industrial version. In some cases the industrial version may be full-strength undiluted product, whereas the household version is only a small percentage of that strength mixed with a less hazardous substance such as water. The type of container usually provides a good clue as to the contents of the package.

Streetsmart Tip The more substantial, durable, and fortified a container is, like the container shown in **Figure 25-32**, the more likely the material inside is dangerous.

On the other hand, materials transported in fiberboard drums usually have no significance with regard to human health, although they may pose a risk to the environment.

Figure 25-32 The type of container can provide some clues as to the contents of the container. Because this drum is reinforced, it has a high likelihood of containing an extremely hazardous material.

When looking at the recognition and identification process, **first responders** should be alert for anything unusual when arriving at an incident. When on an EMS call to a residential home, it would be unusual to find a 55-gallon drum in a bedroom along with glassware associated with a lab environment. These types of recognition and identification clues should alert the first responders that additional assistance may be required. Arriving at an auto repair garage and finding 55-gallon drums and compressed gas cylinders should not be unexpected, however.

General

Containers come in a variety of sizes and shapes and the general category of containers is not exceptional. Most of the general containers are designed for household use but will be carried in large quantities when moved in transportation. When moving to bags and into drums and cylinders the move is made from household to industrial use. All of these types can be used in the home, but a super sack which can hold thousands of pounds of materials is not usually considered household.

Cardboard Boxes

With the popularity of shopping clubs and discount warehouses, more and more homeowners are buying materials by the case, when in the past they bought in much smaller quantities. Cardboard boxes are used to ship and contain hazardous materials. They can hold glass, metal, or plastic bottles. In some cases they may have a plastic lining, such as the box shown in **Figure 25-33**, which holds sulfuric acid. Many household pesticides, insecticides, and fertilizers are contained in cardboard boxes. With the exception of these products and sulfuric acid, most products contained in boxes are usually not extremely toxic to humans, but may present an environmental threat. Materials in transport to suppliers may be transported in larger cardboard boxes and then broken down at the retail level. Responders should note any labels on these packages, but the absence of any labels does not indicate that hazardous materials are not present.

Bottles

From 1-ounce bottles to 1-gallon bottles, the variety of containers is endless and the types of products contained in them too numerous to mention. In recent times manufacturers have begun to take precautions when packaging their materials for transport and use, especially when glass bottles are used. Nowadays, when chemicals are shipped in glass

Figure 25-33 Typically, chemicals that can cause harm are not packaged in cardboard. This sulfuric acid is one example of a material that can cause harm, but unfortunately is packaged in cardboard.

bottles, the bottles are usually packed in cardboard boxes and insulated from potential damage. One-gallon glass containers are usually shipped in what is known as carboys, like the one shown in **Figure 25-34**. Carboys provide a protective cover to

Figure 25-34 To protect the glass bottle, which has a corrosive in it, a carboy is used in case the bottle is dropped.

protect against potential damage during transportation. If the container is dropped, the bottle should survive the fall. Carboys are usually seen in laboratories and in smaller chemical production facilities.

Ensuring the material's compatibility with the container it will be stored in is important, but the one area that usually results in a release is the use of an improper cap. The chemical must not only be compatible with the glass, it must also be compatible with the material the lid is composed of. A variety of materials are used in the manufacturing of lids. Many new glass containers, like the one shown in **Figure 25-35**, are coated with plastic to avoid the bottle being broken when dropped. Even if the bottle is cracked, the contents are supposed to remain sealed within the plastic coating.

Bags

Bags are also commonly used as containers for chemicals. Bags can be as simple as paper bags or plastic-lined paper bags to fiber bags, plastic bags, and the reinforced super sacks or tote bags. It might be a surprise to open the back of a trailer and find four super sacks like those shown in **Figure 25-36** carrying a material that is classified as a poison. Bags

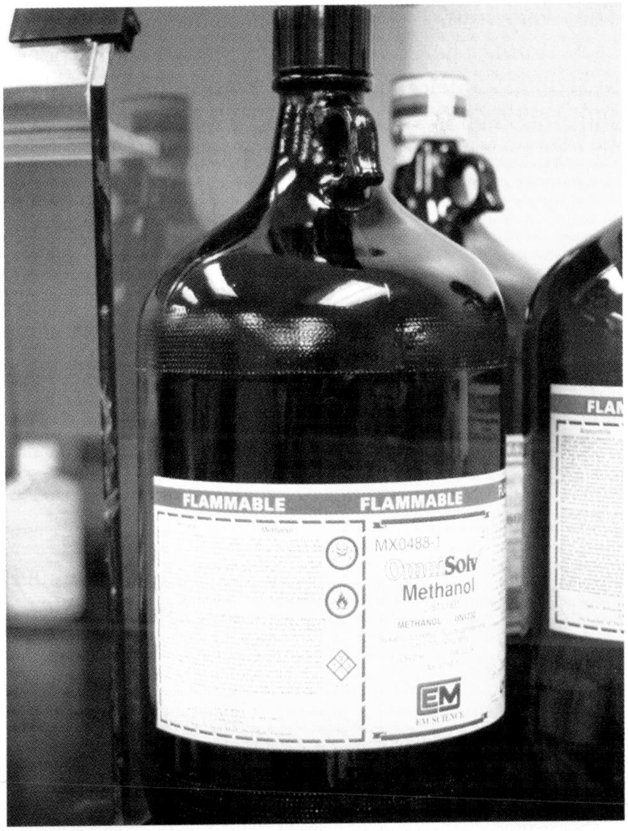

Figure 25-35 This glass jar is coated with a plastic coating that will not allow the liquid to spill out if the glass is broken or dropped.

Figure 25-36 It can be quite surprising to open the back of a tractor trailer and find these super sacks. They can hold solid materials, some of which can be toxic.

carry anything from food items to poisonous pesticides, and the method of transportation varies widely.

Drums

When discussing hazardous materials, drums are the containers with which most responders are familiar. They vary from 1-gallon sizes up to a 95-gallon overpack drum. The construction varies from fiberboard to stainless steel. The typical drum holds 55 gallons and weighs 400 to 1,000 pounds. It is possible to get an idea of what a drum may contain by the construction of the drum. **Table 25-7** provides an indication of potential drum contents, but this is not an absolute listing; contents can and do vary from drum to drum.

Cylinders

Cylinders, like those shown in **Figure 25-37**, come in 1-pound sizes up to several thousand pounds and carry a variety of chemical products. The product and its chemical and physical properties will determine in what type of container the product is stored.

> **Caution** Other than the hazard of the chemical itself, the big hazard of all cylinders is that they are pressurized.

The pressures range from a low of 200 psi to a high of 5,000 psi. One of the most common cylinders firefighters run across is the propane tank, which range from 1 pound up to millions of gallons. In residential homes, firefighters will find everything from the 20-gallon cylinder for barbecues to the 100- to 250-pound cylinders used as a fuel source for the home. In some areas it is not uncommon to find 1,000-pound cylinders and often they are buried underground.

Specialized cylinders that hold cryogenic gases (extremely cold) appear to be high pressure, but in reality are low pressure. The bulkiness of the cylinders is a result of the large amount of insulation required to keep the material cold. Cylinders usually have **relief valves** or **frangible disks** in the event they are overpressurized or are involved in a fire. Most communities, regardless of their size, have

Drum Contents

TYPE OF DRUM	POSSIBLE CONTENTS (IN ORDER OF LIKELIHOOD)
Fiberboard (cardboard), unlined	Dry, granular material such as floor sweep, sawdust, fertilizers, plastic pellets, grain, etc.
Fiberboard, plastic lined	Wetted material, slurries, foodstuffs, material that may affect cardboard or could permeate the cardboard
Plastic (poly)	Corrosives such as hydrochloric acid and sodium hydroxide, some combustibles, foodstuffs such as pig intestines
Steel	Flammable materials such as methyl alcohol, combustible materials such as fuel oil, motor oil, mild corrosives, and liquid materials used in food production
Stainless steel	More hazardous corrosives such as oleum (concentrated sulfuric acid)
Aluminum	Pesticides or materials that react with steel and cannot be shipped in a poly drum

TABLE 25-7

Figure 25-37 Cylinders present additional risks to responders because not only can the contents be hazardous, but if the cylinder is involved in a fire it may explode.

cylinders of chlorine and sulfur dioxide used in water treatment. These cylinders come in 100- to 150-pound and 1-ton cylinders, which could create a major incident if they were ruptured or suffered a release.

Note The area affected by a 100-pound chlorine cylinder release can be several miles, **Figure 25-38**, causing serious injuries if not fatal effects.

Totes and Bulk Tanks

Both **totes** and **bulk tanks** are becoming more common, sized as they are between drums and tank trucks, and are used for a variety of purposes. Used in industrial and food applications, they hold flammable, combustible, toxic, and corrosive materials. They are constructed of steel, aluminum, stainless steel, lined materials, poly tanks, and other products, **Figure 25-39**. They can carry up to 500 gallons, but the usual capacity is 300 gallons. They are transported on flatbed tractor trailers or in box-type tractor trailers.

A common incident with totes can occur during offloading. Tanks are offloaded from the bottom through a swinging valve, such as the one shown in **Figure 25-40**. It is a common occurrence for this valve to swing out during transport and get knocked off during movement.

One unusual tote is made to transport calcium carbide, a material that when it gets wet forms acetylene gas, which is reactive and very flammable.

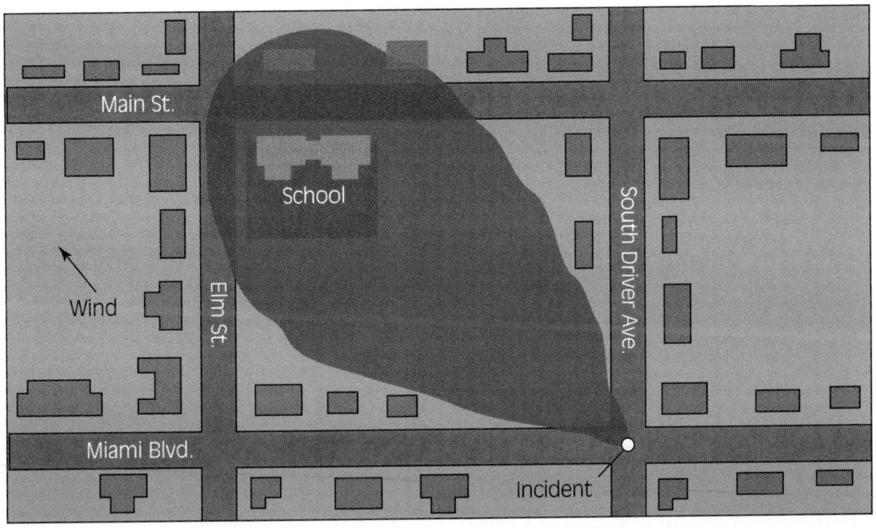

Figure 25-38 The type of vapor cloud commonly referred to as a plume varies with the terrain and buildings in the vicinity. The plume here represents one of the most common types.

Figure 25-39 The use of 55-gallon drums is decreasing and the use of these portable bulk tanks is increasing. Like the super sacks these can be hidden away in the back of a trailer.

Figure 25-40 The most common type of spill occurs when a valve is knocked off, releasing the contents.

Pipelines

Pipelines vary in size and pressure, but can be sized between 1/2 inch and more than 6 feet. They are commonly buried underground. The most common products they carry are natural gas, propane, and assorted liquid petroleum products. The larger petroleum pipelines originate in Texas and Louisiana and then proceed up throughout the East Coast. The West Coast also has its share of large pipelines, with Alaska having a majority. Pipelines can originate from any bulk storage facility and can cross many states, and some type of pipeline system is found in every state. Because the amount in the pipelines varies it is important that first responders know the location of the pipelines and emergency contact names and phone numbers so if there is a suspected problem they can notify the pipeline owner immediately.

Highway Transportation Containers

The type of vehicle provides some important clues as to the possible contents of the vehicle. The most common truck is a tractor trailer or a box truck. There are four basic tank truck types that carry hazardous materials, with some additional specialized containers. Tractor trailers carry the whole variety of hazardous materials and portable containers. They can carry loose material that is not contained in any fashion other than by the truck itself. They can carry portable tanks that hold 500 gallons or bulk bags that weigh several tons.

Caution When dealing with tractor trailers the rule is to expect the unexpected.

Nothing is routine. Until the driver has been interviewed, the shipping papers looked at, and the cargo actually examined, a firefighter cannot confirm or deny the presence of hazardous materials. Sometimes the signage on the trailer is an indication of the possible contents, and a trailer that has several placard holders is a likely candidate for hazardous materials transport. If a tractor trailer is refrigerated like the one shown in **Figure 25-41** and is carrying hazardous materials, extra precautions must be taken because the materials may require the cold temperature to remain stable.

Leakage is often found in containers known as **intermodal containers** or, more commonly, **sea containers**. These types of containers are typically used on ships, then offloaded onto a tractor trailer or loaded directly onto a flatbed railcar. These containers come from all over the world and can contain any imaginable commodity.

Note The types of containers that are shipped in these trailers vary from bags, boxes, and drums to bulk tanks and cylinders.

Although the driver is supposed to have the shipping papers for the contents, on occasion the paperwork is missing or is sealed in the back of the

Figure 25-41 Although in most cases refrigerated trailers are carrying food, there exists the possibility that they may have chemicals that require refrigeration to remain stable.

trailer. Determining the contents of a trailer can be very difficult and frustrating.

Tank trucks carry several hundred gallons up to a maximum of 10,000 gallons. The DOT allows maximum loads by weight not by gallons, so the actual capacity varies state to state. The most common tank truck is the gasoline tank truck, which usually carries 5,000 to 10,000 gallons. In September 1995 the DOT changed the regulations covering tank trucks, so two systems are used for identifying tank trucks, **Figure 25-42**. In the past the DOT wrote specifications as to how the manufacturer should build a tank truck. Today, they have established performance-based standards for the construction. The DOT allows trucks that were manufactured before the new regulation to remain on the road, as long as they meet the applicable inspection requirements. The four basic types of tank trucks are:

- DOT-406/MC-306 gasoline tank truck
- DOT-407/MC-307 chemical hauler
- DOT-412/MC-312 corrosive tanker
- MC-331 pressurized tanker

Figure 25-42 Many of the differences between a 306 and a 406 tank truck are internal. The biggest difference is that the dome covers on a 406 are less likely to open during a rollover, although the skin of the tank is thinner. *(Photo courtesy of Maryland Department of the Environment)*

The DOT numbers are the more recently manufactured tanks, and the MC (motor carrier) numbers identify those tanks manufactured prior to September 1995. There are some differences in construction of the two types—some of which favor the emergency responders, others favor the shipping company—but overall the newer tanks hold up better during accidents and rollovers. If unsure of the type of tank, all tank trucks have **specification (spec) plates**, which list all the pertinent information regarding that tank. The spec plate in many cases is located on the passenger side of the tank near the front of the tank. In some cases, shipping papers or MSDS are located in a paper holder as shown in **Figure 25-43** (tube).

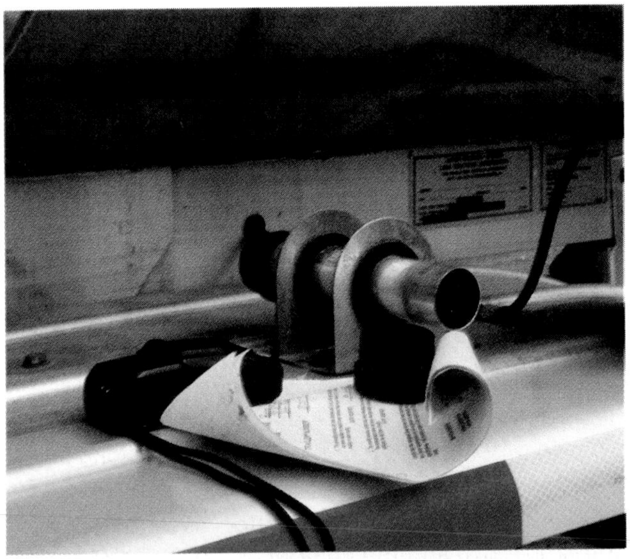

Figure 25-43 In most cases the shipping papers will be with the driver in the cab of the truck, but they may be in a special tube located on the trailer.

Figure 25-44 The DOT-406/MC-306 is the most common truck on the highway today and is referred to as a gasoline tanker. It can and frequently does carry other types of flammable and combustible liquids. *(Photo courtesy of Maryland Department of the Environment)*

DOT-406/MC-306

This is the most common tank truck on the road today, **Figure 25-44**, and for that reason, along with the large number of shipments, suffers the most accidents. Although the most common products carried on these trucks are gasoline and diesel fuel, almost any flammable or combustible liquid can be found on these types of trucks. This truck is known for its elliptical shape and is usually made of aluminum. Older style tanks used to be made of steel, which presented an explosive situation known as a BLEVE. The tanks generally have three or four separate compartments, but two to five compartments are not uncommon. Newer style tanks have considerable vapor recovery systems as well as rollover protection, and in some cases these features are combined. The valving and piping are contained on the bottom of the tank, as shown in **Figure 25-45**, and the number of pots is indicated by the number of outlets as well as the number of manhole assemblies located on top of the tank. The maximum pressure that this truck can hold is 3 psi. The compartments are separated by bulkheads, which usually fail during a rollover situation. Although in most states the shipper is not allowed to ship a mixed load of flammable and combustible materials, widely differing loads can be encountered from compartment to compartment.

Figure 25-45 Most 406/306 trucks have more than one tank (pots) and the number of tanks is indicated by the number of valves or the number of dome covers on top of the truck. *(Photo courtesy of Maryland Department of the Environment)*

> **Caution** During a rollover, the bulkheads may shift, allowing all products to mix.

On initial examination it may appear that only one tank has ruptured and is leaking, but it is possible to lose the entire contents of the tank through a leak in one compartment, like that shown in **Figure 25-46**. In the majority of rollovers experience has shown that at least one bulkhead has separated in almost every accident, resulting in product mixing. In most cases this is not a big problem because it may only be the mixing of different grades of gasoline. Within the individual tanks themselves, baffles limit the movement of the product within the tank. The emergency shutoff valves, **Figure 25-47**, are located on the driver's side near the front of the tank and near the piping on the passenger side.

DOT-407/MC-307

The DOT-407/MC-307 tankers are the workhorses of the chemical industry, **Figure 25-48**. They carry a variety of materials including flammable, combustible, corrosive, poisonous, and food products.

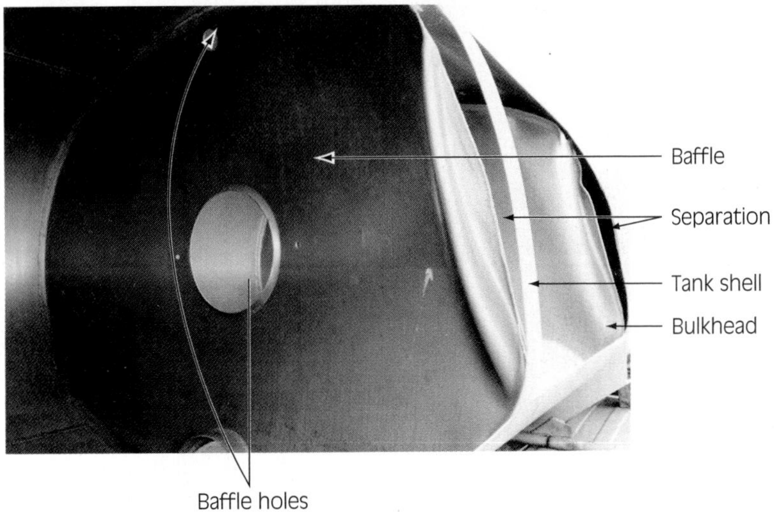

Figure 25-46 When a truck rolls over on its side, the internal baffles and bulkheads may shift. The internal baffles reduce the amount of sloshing the liquid will do when the truck starts and stops. The baffles will allow product to move through the holes in the baffle wall. A bulkhead separates the compartments and does not allow the products to mix. Shown in the photograph is an MC-306 that rolled on its side. The side wall of the tank has been cut away, revealing a baffle and a bulkhead. Also shown is the separation that took place between the tank shell and the baffles/bulkhead. When the tank wall was cut away, strips of the tank shell were left where they should have connected to a baffle or bulkhead.

The two basic types of chemical tanks are insulated and uninsulated. The insulated tank can have a number of additional concerns that do not apply to an uninsulated tank. These tanks usually hold 2,000 to 7,000 gallons, lower amounts than the 406/306 because most of the products they carry are heavier than petroleum products. The average amount found in these tanks is 5,000 gallons.

The uninsulated tank is round and has stiffening rings around the tank. The offloading piping is located on the bottom or off the rear of the tank. These tanks are composed of only one compartment, and its loading piping and manhole are usually on the top in the middle. These tanks generally do not hold up as well during rollovers

Figure 25-47 On most trucks there is a minimum of one emergency shutoff, and with most there are two. The most common location is near the driver's door; the other is usually located near the valve area.

Figure 25-48 On the right is an insulated MC-307/DOT-407, and on the left is an uninsulated MC-307/DOT-407. Both trucks hold comparable products, but the insulated one holds products that are heated or may require heating to offload.

Figure 25-49 This insulated version is identical to the uninsulated tank but has an aluminum cover and insulation. Note the differences in these two trucks. The one on the right has safety railings around the manhole. Although not an absolute rule, the truck on the right would carry more dangerous products and would have other added safety features. Most of these items are not required but were added by the trucking company.

and accidents as the insulated version. The shell is made of stainless steel and can hold pressures up to 40 psi.

The insulated tank, **Figure 25-49**, is a covered version of the uninsulated tank, although in some cases it has a slightly smaller inner tank. The inner tank is made of stainless steel with about 6 inches of insulation, and the outer shell is made of aluminum. The inner tank may also be lined with a fiberglass or other liner depending on the chemical that is carried. Due to the aluminum and insulation these tanks hold up remarkably well during rollovers.

Safety One of the major problems with this insulated tank is that in the event of a leak, the location where the material leaks out of the outer shell is usually nowhere near the leak on the inner shell.

Within the insulation there can be heating and cooling lines depending on the product being carried, **Figure 25-50**. Products such as paint are shipped at 170°F and need to be heated to that temperature for offloading. Some products need to remain at certain temperatures to remain stable, and first responders need to be aware of any special requirements.

In general, both types of tanks have rollover protection, similar piping, and relief valves that serve two purposes: overpressurization and vacuum protection. The emergency shutoffs are located near the front of the tank on the driver's side and near the offloading piping.

Figure 25-50 Products carried in an insulated 407/307 need to remain either heated or cooled. Some products may need to be heated for offloading. The heater coils that run around the tank heat the product up, allowing it to be offloaded.

DOT-412/MC-312

These tankers, **Figure 25-51**, carry a wide variety of corrosives, both acids and bases. These tankers are round and are smaller than the 306s and 307s due to the weight of the corrosives they carry. Most petroleum products weigh about 8 pounds per gallon while some corrosives weigh up to 15 pounds per gallon. Because of the weight, the stiffening rings used are generally bulkier than the ones used on DOT-407 tanks. These tankers are constructed of a single tank that carries up to 7,000 gallons, with most tanks holding 5,000 or fewer gallons. The tanks are made of stainless steel and are usually lined to protect against corrosion. The piping can be on top of the tank located in the middle, but

Figure 25-51 The DOT-412/MC-312 is designed to carry corrosives and is similar in design to the uninsulated 307, although smaller. The inner tank may be lined with a variety of materials to prevent the corrosive from attacking the tank.

Figure 25-52 The black coating around the manhole indicates that a corrosive is being carried. It is used to protect the tank from spillage. It is not required nor will it be found on all corrosive tanks.

is usually located on the end of the tanker. The piping is usually contained within a housing that includes the manhole and offload piping. This housing protects the piping in the event of a rollover, **Figure 25-52**. The area around the manhole is usually coated with a material that resists the chemical being carried and is usually a black, tar-like coating.

MC-331

MC-331 tanks look like bullets and are noted for their rounded ends and smooth exterior, as shown in **Figure 25-53**. They carry liquefied gases that are liquefied by pressure. One of the most common products carried in this type of tank is propane. They also carry ammonia, butane, and other flammable and corrosive gases. These tanks carry up to 11,500 gallons and have a general pressure of 200 psi, although it can be as high as 500 psi. The relief valves are located on top of the tank at the rear of the trailer, and they sometimes malfunction during a rollover. The tanks are made of steel, are uninsulated, and are heavily fortified with heavy bolts used in the piping and manholes. The tanks are usually painted white or silver to reduce the potential heating by sunlight.

The tanks contain a liquid along with a certain amount of vapor. The most liquid the tank is supposed to have is 80 percent to allow for expansion when heated. The liquid in the tanks is at atmospheric temperature but on release can go below 0°F (propane is −90°F) and could cause frostbite upon contact. The pressure in these tanks is of concern when responding to incidents involving these tanks.

Figure 25-53 MC-331 tanks carry liquefied compressed gases such as propane and ammonia. They are made of steel and are designed to carry a variety of products.

Firefighter Fact Temperatures and Pressurized Container When a propane tank is emptied and, hence, the pressure reduced, the temperature of the propane drops below 0°F. Temperature, pressure, and volume are interrelated. Think of an SCBA bottle. When it gets filled it becomes hot because the pressure and volume are increasing. When the SCBA bottle is used, it becomes cold because it is losing pressure and volume. Any time one of the parameters is changed, there is a corresponding change in the other properties. When a pressurized gas is pressurized to a point that it becomes a liquid, as is the case with propane, it allows for a lot of propane to be stored in a small container. When released, however, the temperature will drop because the pressure is decreasing in the container. This is known as autorefrigeration.

Boiling Liquid Expanding Vapor Explosion (BLEVE)

Safety If the pressure increases at a rate higher than the relief valve can handle, the tank will explode. These explosions have been known to send pieces of the tank up to a mile, with the ends of the tank typically traveling the farthest, although any part is subject to become a projectile.

When tanks, trucks, tank cars, or other containers are involved in a fire situation there are a number of hazards. One very deadly hazard is known as a boiling liquid expanding vapor explosion (BLEVE), **Figure 25-54.** A large number of firefighters have been killed by propane tank BLEVEs, and when a BLEVE

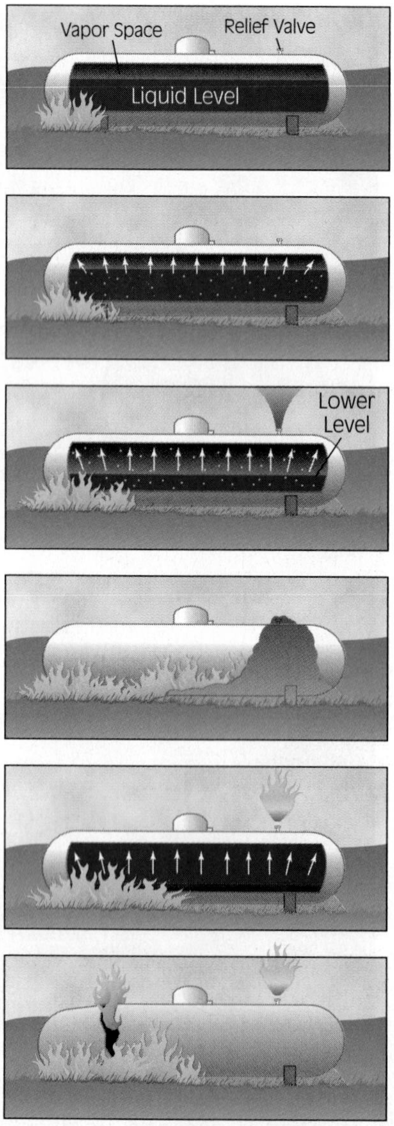

Fire impinging on a propane tank.

As heat increases inside of tank, the pressure also increases. The liquid will begin to boil.

As the pressure increases, the relief valve will open, releasing propane. The propane, being heavier than air, will sink.

The vapor will reach the fire and ignite.

The relief valve will ignite, also causing heat to be on the tank by that flame. The pressure will increase in the tank.

As the pressure in the tank is increasing, the tank may discolor and the pitch of the relief valve will get higher. Eventually the tank will rupture. This is known as a BLEVE.

Figure 25-54 Diagram of a BLEVE.

occurs it usually results in more than one firefighter being killed at a single incident. The type of container and the product within the container will dictate how severe a BLEVE may be. The basis of a BLEVE is the fact that the pressure inside the container increases and cannot be held by the container. The contents are violently released, and if the material is flammable, an explosion or large fireball occurs. In the recent past there have been several incidents involving BLEVEs that resulted in emergency responder deaths and injuries, thus emphasizing the need to recognize and prevent this event before it occurs.

Another phenomenon that transpires with containers is known as violent tank rupture (VTR), which occurs with nonflammable materials. The concept is the same as a BLEVE, but there is not a characteristic fireball and explosion. With both a BLEVE and a VTR there is some form of heat increase inside the container, typically from a nearby fire. The fire heats the container, which heats the contents. The contents will boil, which creates expanding vapors, which in turn increase the pressure inside the container. In some containers the relief valve will activate, relieving the pressure. In some cases the pressure inside the tank is greater than the relief valve can handle and the pressure continues to increase. One of two possibilities can occur with a BLEVE or a VTR. One possibility is that the relief valve will not be able to handle the increase in pressure and the tank will fail. The other possibility is that the fire or heat source that is creating the problem will weaken the container shell, and the resulting increasing pressure will vent at this weakened portion of the container. If the heat source is a fire and the container product is flammable, when the relief valve activates the raw product coming out of the container typically ignites. This may increase the temperature of the tank, increasing the pressure. It is never advisable to extinguish the fire coming from a relief valve, as that is a safety mechanism. It is possible to cool the top of the tank near the relief valve with an unstaffed hose stream. The difference between a BLEVE and a VTR occurs when the container fails. A BLEVE occurs with flammable liquids, such as propane. When the container fails, the vapors of the flammable liquid ignite, creating the explosion. How the container ruptures and the amount of product released will determine how severe the explosion will be. With a VTR the container will fail. Since the product is nonflammable, the product will not ignite. The only event is a rupture of the container, spilling its contents. The container can still rocket, and the resulting release of pressure can be violent. A VTR can occur with a container of water or other "nonhazardous" material. As an example a 55-gallon steel drum of water, when heated, can travel several hundred feet depending on where the release point is on the container. In most cases the bungs (screw-top caps) will release, flying a considerable distance, and the container will remain mostly intact.

The failure of the container when impinged from a fire usually occurs as the fire is impacting the tank in the vapor space. When heat is applied to a section of the tank that has no internal mechanism to provide cooling, failure of the steel can occur. When fire impinges on the liquid portion of the tank, the liquid will distribute the heat spread internally and the steel tank typically will not weaken. The problem is that responding firefighters do not usually know the liquid level, and one cannot easily predict when a tank may fail. Any fire impingement on a tank is a serious problem, and withdrawing from the scene may be the best course of action. This brings up another interesting note: A casual observation of BLEVEs indicates that most BLEVEs and firefighter deaths occur within the first few minutes of arrival. The clock starts ticking on the BLEVE time bomb from the first minute heat is applied to the tank. The clock does not start with the 9-1-1 call or the arrival of firefighters. The critical time for safety may have already passed before firefighters arrive on the scene. If firefighters arrive at an incident involving a propane tank on fire or being impinged by fire, they are in extreme danger. If the relief valve is not operating, the danger is even more pronounced. Operating in close proximity to a tank in this situation can be a fatal mistake. Firefighters should follow this risk/benefit analysis: Risk a lot to save a lot, and risk a little to save a little.

One recent event involved the death of two firefighters and injuries to seven other emergency responders. The location was a farm, at which a propane tank was on fire. The relief valve on the tank was operating and the vapors from the relief valve were on fire. About eight minutes after the firefighters had arrived, the tank exploded into four separate parts. The four parts went in four different directions. The two firefighters who died were 105 feet away and were struck by one piece of the tank, dying instantly.

In two other farm incidents, firefighters lost their lives. In 1993 a fire and BLEVE in Ste. Elisabeth de Warwick, Quebec, Canada, killed three firefighters. In 1997 in Burnside, Illinois, two firefighters lost their lives in a fire and BLEVE. In all three of these fatal BLEVE events, the relief valves were operating upon arrival of the fire department.

One other event worth noting is the 1984 PEMEX LPG Terminal fire in Mexico City, Mexico. An 8-inch pipeline broke while filling a tank at the terminal. The resulting vapor cloud, which was 650 feet by 572 feet by 8 feet, ignited. This resulted in an explosion, which included a ground shock, and a major fire. The terminal had two large spheres, four small spheres, and forty-eight horizontal tanks. About fifteen minutes into the fire the first BLEVE occurred, and for the next ninety minutes tanks continued to BLEVE. Tanks and liquid propane rained down on the adjacent community. The death toll exceeded 500 people, and thousands were injured.

In 1983 five Buffalo, New York, firefighters lost their lives when a three-story building collapsed. Nine other firefighters were injured in the massive explosion. The force of the explosion caused the first arriving apparatus, including a ladder truck, to be blown across the street. A propane tank inside the building had been struck and was leaking. An unknown ignition source sparked the explosion minutes after fire crews arrived.

In January 2003 a propane truck (MC-331) went over a guardrail and fell to the ground 35 feet below. The tank ruptured and the leaking propane ignited, resulting in a 600-foot-high fireball. The resulting explosion moved the truck several hundred feet away from the first impact area. The driver of the truck was killed in the accident.

As can be noted in the stories above, BLEVEs and VTRs can result in injuries and fatalities. Any time containers are under stress, such as during a fire, they can fail. Many times they fail violently and with severe consequences. Some containers have relief valves, while others do not. Materials that are highly poisonous such as chlorine will not have a relief valve; they will have a fusible metal plug that vents the pressure of the tank. This fusible metal plug is not like a relief valve that shuts off when the pressure inside the tank is decreased. Once a fusible metal plug melts, the contents of the tank come out, no matter the pressure. When a tank is on fire or is being impinged by fire, the tank is being weakened and the contents are being heated, creating increased pressure. Some of the general rules of firefighting and propane tanks (or other containers under pressure) are as follows:

- Firefighters should withdraw immediately in the case of rising sound from venting relief valves or discoloration of the tank.
- Fire must be fought from a distance with unstaffed hose holders or monitor nozzles.
- The tank should be cooled with flooding quantities long after the fire is out. A minimum of 500 gpm at the point of flame impingement is recommended by the NFPA.
  - If the water is vaporizing on contact, firefighters are not putting enough water on the tank. Water should be running off the tank if it is being cooled.
- Firefighters should not direct water at relief valves or safety devices, as icing may occur. Icing would block the venting material, which could cause an increase in pressure inside the tank.
- The tank may fail from any direction, but firefighters should avoid the ends of the tank.
- For massive fire, it is recommended to use unmanned hose holders or monitor nozzles. If this is impossible, firefighters should withdraw from the area and let the fire burn.
- The dangers associated with a BLEVE are:
  - The fireball can engulf responders and exposures.
  - Metal parts of the tank can fly considerable distances.
  - Liquid propane can be released into the surrounding area and be ignited.
  - The shock wave, air blast, or flying metal parts created by a BLEVE can collapse buildings or move responders and equipment.

Specialized Tank Trucks

These types of tank trucks are used to carry unique chemicals or chemicals that have to be transported in a certain fashion. When gases are transported, they are transported as liquefied gases, as was described with the MC-331 tank trucks. They can also be transported as refrigerated gases or as compressed gases as will be described in this section. Other trucks are dry bulk which can carry a variety of products from grain to explosives. Materials that required high temperatures are transported in special vehicles to keep them hot. The intermodal series of tanks are cousins to their full size highway tanks but carry the chemicals in a comparable fashion.

MC-338 Cryogenic Tankers

MC-338 tankers are uniquely constructed like the one shown in **Figure 25-55**. They have a tank with an outer shell. The inner container is steel or nickel, with a substantial layer of insulation; the exterior is aluminum or mild steel. The space between the shells is placed under a vacuum to assist in the cooling process. The ends of the tank are flat, and the piping is contained usually at the end of the tanker in a double door box. Relief valves are located on top of the tank, to the rear of the tank. The best way to describe this tanker is to compare it to a Thermos bottle on wheels. Cryogenic materials are gases that have been refrigerated to a temperature that converts them to liquids. Unlike liquefied gases, which use pressure to reduce them to liquids, these are cooled to the point of becoming liquid. To remain

Figure 25-55 The MC-338 carries cynogenic liquefied gases. The tank resembles a rolling Thermos bottle.

liquids, the material must be kept cool. To be a cryogenic material the liquid has to be at least −150°F and can be as cold as −456°F.

The most common products are nitrogen, carbon dioxide, oxygen (liquid oxygen or LOX), argon, and hydrogen. The material inside is kept liquid by vacuum, and when on the road the tank will have a maximum of 25 psi. As the truck travels the sun will heat the material, and the pressure will increase. As the pressure increases the relief valve will open up, relieving any access pressure into the atmosphere.

> **Note** It is not uncommon to see a white vapor cloud like that shown in **Figure 25-56** coming from the relief valves while the truck is traveling on the highway or sitting alongside the road. This is a normal occurrence and is not cause for alarm.

When the truck makes a delivery, the pressure needs to be increased to push the liquid out of the tank. The driver opens the piping and allows the material to flow into an evaporator which is located just in front of the rear wheels of the tank. Once in the evaporator the material will heat up, increasing the pressure in the tank and forcing the liquid into the receiving tank. During transportation and offloading, a large amount of ice will build up on the piping and valves.

Tube Trailers

Tube trailers, **Figure 25-57**, contain several pressurized vessels, constructed much like the MC-331 tank. They are constructed of steel and have pressures ranging from 2,000 to 6,000 psi. They hold pressurized gases such as air, helium, and oxygen. The piping and controls are usually located on the

Figure 25-56 When transporting liquids it is not uncommon to see vapors coming from the relief valves. The truck can only travel with the tank at 25 psi or less. As it heats, the pressure increases, triggering the relief valve. This is a normal situation, and the tank will vent until below the 25 psi. When offloading the pressure can be increased to assist with the offloading procedure.

Figure 25-57 The tubes on this trailer contain pressurized gases. The pressure can vary from 2,000 to 6,000 psi.

Figure 25-58 Dry bulk tanks carry a variety of products. Some examples include fertilizers, explosives, and concrete.

rear of the trailer, but could be in the front. The typical delivery mode is that the driver will drop a full trailer off at a facility and pick up an empty one for refilling. Although they are not subject to BLEVEs because they only contain a gas, if involved in a fire, tube trailers can experience a VTR and rocket in the same fashion as a BLEVE.

Dry Bulk Tanks

Dry bulk tanks resemble large uninsulated MC-307s in shape with bottom hoppers to unload the product, as shown in **Figure 25-58**. The tanks hold dry products and sometimes a slurry, like concrete. The most common products are fertilizers, lime, flour, grain, and other food products. The potential hazard when dealing with these tankers is predominantly environmental but at times these tankers contain toxic materials. They are usually offloaded using air pressure either from the truck itself or at the facility.

Hot Materials Tanker

Hot materials tankers vary in that they can be modified MC-306s, 307s, or dry bulk containers, **Figure 25-59**. They may have a mechanism to keep the material hot or it may be loaded hot. It may require heating prior to offloading. Common products are tar, asphalt, and molten sulfur, fuel oil #7 and #8.

Safety The major problem with these tankers is the heated material itself. Anyone coming into contact with the material could be seriously burned. The molten material could ignite the truck or other combustibles.

Figure 25-59 These trucks carry molten products and can be heating the product while driving. This practice is illegal but is found on occasion. The fuels used to heat the product are either diesel/kerosene or propane.

IMO Containers

CONTAINER	MAXIMUM CAPACITY (GAL)	PRESSURES (PSIG)	PRODUCTS
IM-101 or IMO Type 1	6,340	25.4–100	Nonflammable liquids, mild corrosives, foods, and other products
IM-102 or IMO Type 2	6,340	14.5–25.4	Flammable materials, corrosives, and other industrial materials
Spec 51 or IMO Type 5	6,418	100–500	Liquefied gases, much like an MC-331
Specialized tanks	Varies	Varies	Includes cryogenic tanks and tube banks (small tube trailers) that carry the same products as their highway counterparts

TABLE 25-8

If the material is allowed to cool, it can cause problems for the responders or the shipping company. Tar trucks may be transported with a propane or fuel oil flame ignited to heat the product en route to the job site, although this is illegal in most states.

Intermodal Tanks

Intermodal tanks are increasing in use and carry the same types of products as their highway and rail companions, **Table 25-8** and **Figures 25-60, 25-61,** and **25-62**. They are called *intermodal* (IM) because they can be used on ships, railways, or highways.

Note Smaller intermodal tanks may be found inside of box-type tractor trailers.

Figure 25-60 This is an IMO-101 tank. Like the totes, these are bulk tanks capable of carrying a large quantity of product. These are normally placed on ships, then delivered locally by a truck, although trains can also be used.

Figure 25-61 This is an intermodal tank, commonly referred to as an IM or IMO. This is a bulk tank that carries an average of 3,000 to 5,000 gallons. This tank is an IMO-101, which it is an atmospheric tank. The orange panel indicates that it has a United Nations (UN) hazard code of 60 and a UN number of 2572. The code of 60 indicates that the product is a toxic material. The UN number indicates that the product is phenylhydrazine. The name is also stenciled on the sides of the container, and there are toxic placards.

Figure 25-62 This is another IMO-101, and the photo shows that the orange panel indicates a hazard code of 66 and a UN number of 2810. The hazard code indicates that the product is a highly toxic material, and there are fifty-two different materials listed for UN 2810. One would have to look at the shipping papers to figure out what is being carried in this tank. The tank is marked "foodstuff only," which means that it is holding a product used in food. It is interesting to note that many of the chemicals listed for UN 2810 are chemical warfare agents such as sarin nerve agent and comparable materials. The most likely candidate, considering the "foodstuff only" label, would be a medical-type product.

Intermodals follow three basic types: nonpressurized, pressurized, and highly pressurized. They are built in two ways. In one, they sit inside a steel frame, called a box-type framework; in the other, the tank is part of the framework, called a beam-type intermodal. Like tank trucks they are assigned specification numbers, IM-101, IM-102, Spec 51 (specification 51). When used internationally they are called IMOs. They are made to be dropped off at a facility and when empty picked up for refilling.

Rail Transportation

As with highway transportation there are only a few types of railcars, and they are similar to their highway counterparts. The piping and shape may be the same but that is the extent of the similarities.

Note In rail transportation the quantities are greatly increased—up to 30,000 gallons for hazardous materials and up to 45,000 gallons for nonhazardous materials.

Rail incidents usually involve multiple railcars, whereas highway incidents usually involve one or two trucks. The incidents may occur in rural areas, away from water supplies and easy access. Rail incidents will involve multiple agencies, and the local community can expect a large contingent of assistance coming from the state and federal government, which in itself will be difficult to manage.

Railcars come in three basic types: nonpressurized, pressurized, and specialized cars. Although they are categorized in this fashion, the commodities they carry will determine the ultimate use of the car. As with highway transportation there are dedicated railcars that will be marked with the products they carry.

Caution The term *nonpressurized car* is actually misapplied because it can have up to 100 psi in the tank.

A nonpressurized car carries chemicals, combustible and flammable liquids, corrosives, and slurries. The easiest way to determine if the car is nonpressurized, such as the one shown in **Figure 25-63**, is to observe whether the valves, piping, and other appliances are located outside of a protective housing. In some cases, the car will be bottom unloaded, with the ability to load the car through the top, or the car may be top loaded and unloaded. The cars will usually have a small dome cover, but the relief valves and other piping are located outside of this dome. Most of the piping is located on the catwalk on top of the car.

Some cars have unique paint schemes such as those shown in **Figure 25-64.** Nonpressurized cars can be insulated and may have heating and cooling lines around the tank. Prior to offloading, the tank may need to be heated to ease the offloading process. Some nonpressurized railcars have an expansion dome. The piping valves and fittings sit on top of this dome, which was constructed to hold

Figure 25-63 The indication that this is a nonpressurized railcar is given by the fact that all of the valves are on the outside of the tank and not contained within any protective housing.

Figure 25-64 This railcar is white with a red stripe, which is used to indicate hydrogen cyanide but on occasion is used to indicate dangerous materials. This car holds hydrogen fluoride anhydrous, which is a severe inhalation hazard and is corrosive. The term *anhydrous* indicates "no water" and means the hydrogen fluoride is pure.

any potential expansion. These cars are not in regular service but may be seen in larger industrial facilities that have their own railroad service on site.

Pressurized tank cars, **Figure 25-65**, also carry a wide variety of products, including flammable gases like propane, and poisonous gases like chlorine and sulfur dioxide. The pressurized cars will have pressures in excess of 100 psi up to 600 psi. Most pressure cars that carry flammable materials will be insulated with a spray-on insulation or may be thermally insulated. This insulation is 1 to 2 inches thick and helps reduce the chance of a BLEVE during a fire situation. To determine if the railcar is a pressurized car the firefighter can look at the spec plate, which will have all of the valves, piping, and fittings located under the protective housing, on top

Figure 25-65 The pressurized car is indicated by the valves being contained within the protective housing of the railcar.

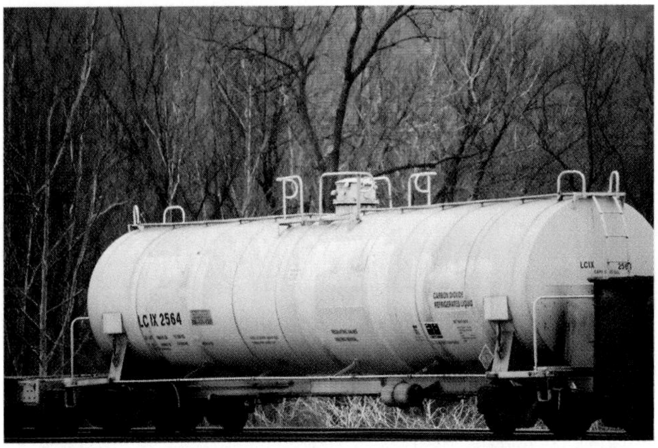

Figure 25-66 This specialized railcar holds liquefied carbon dioxide, a cryogenic material.

of the railcar. A catwalk around the protective housing provides relatively easy access.

Specialized Railcars

Specialized railcars have the same characteristics as highway vehicles, and, in fact, in some cases highway box trailers are loaded onto flatbed railcars. When transported in this fashion they are referred to as trailers on flat cars (TOFC). Regular box trailers as well as refrigerated trailers can be found on flat cars. Much like highway trailers, there are freight boxcars that haul the same products as their highway trailers, with the exception of carrying much larger quantities. Examples of these cars are shown in **Figure 25-66**. The boxcars can carry boxes, cylinders, bulk tanks, totes, and super sacks. Refrigerated railcars are similar to highway and intermodal boxes and are referred to as reefers. They contain their own fuel source. Dry bulk closed railcars are common, and open hopper cars can also be found. Tube trailers for rail are also found, but they are rarely used; the most common is a highway tube trailer set on a flat car. Cryogenics are also carried in railcars and have the same low pressure (25 psi) as in highway transportation, with the characteristic venting when the pressure increases.

Markings on Railcars

Railroads use the same placarding system used on highways, with the exceptions noted in the placard section. Railcars, however, are usually marked better than their counterparts on the highway; at the very least, the information is printed larger. In addition to a placard, the name of the hazardous material is stenciled on two sides of the car. In addition, by looking at the sides of a railcar, a firefighter can learn the specification type, maximum quantities, test pressures, relief valve settings, and other pertinent information. These types of markings are shown in **Figure 25-67**.

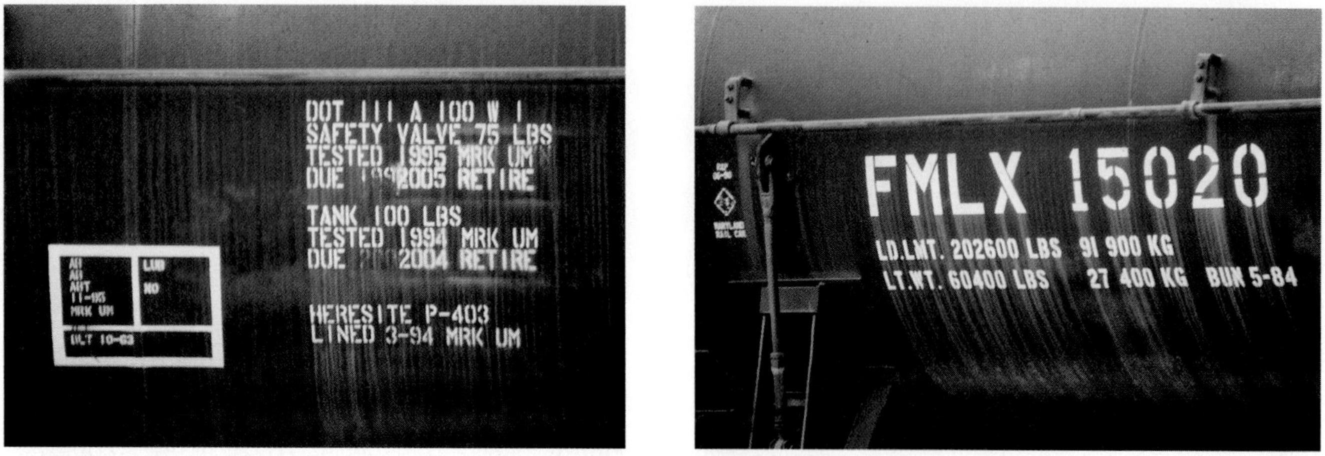

Figure 25-67 A railcar has a lot of information stenciled on the side. The FMLX indentifies who owns the car. The X at the end indicates that the car is privately owned, that is, the railroad does not own the car. The 15020 is the car number and can be used to cross-reference the shipping papers. The photo on the left is the certification and testing data. The top line DOT 111 A 100 W 1 provides most of the information. The tank is a DOT 111 specification, which means that it is a non-pressure car. The A indicates the type of couplers, and the 100 is the maximum pressure for the car. The W and the 1 indicate the type of welds and other tank construction information. The last line is also important, and it shows that the tank was lined on 3-94, which means that it has an internal lining. In a derailment the lining could separate from the tank and the chemical could react with the tank metal.

Note Certain railcars may be painted in a certain configuration to identify their hazardous loads.

It used to be common practice to paint a car carrying hydrogen cyanide white with a red stripe running around the middle, hence the name "candy-striper."

Bulk Storage Tanks

Bulk storage tanks range in size from 250 gallons up to millions of gallons and store a variety of products, the most common being petroleum products. These tanks are seen in residential homes and rural areas, and are common at an industrial facility, such as the one shown in **Figure 25-68**. In residential homes the most common tank is a 250-gallon home heating oil tank, and some homes or small businesses may have gasoline or diesel fuel tanks that vary from 250 to 500 gallons.

The two basic groupings of tanks are in-ground and aboveground. Since the passage of the EPA underground storage regulations, there has been considerable movement to remove **underground storage tanks (USTs)** and replace them with **aboveground storage tanks (ASTs)**. If the facility

Figure 25-68 The sizes of tanks vary from a few hundred gallons up to several million gallons. A catastrophic failure of the tank may overwhelm the responders' ability to handle the incident.

owner elects to keep its tanks underground then additional requirements are placed on them for testing, containment, and leak prevention.

> **Caution** Leaks from these types of tanks can be overwhelming because the leak may not be detected for a period of time, giving it a chance to spread throughout the area, contaminating a large area.

A substantial release from a million-gallon tank of gasoline that escapes a facility would require an enormous response from the emergency responders and environmental contractors.

Underground tanks are usually constructed of steel and are coated with an anticorrosion material. New tanks are usually double walled, that is, a tank within a tank, to prevent any spillage into the environment. The piping comes up through the ground and its use will determine its route. The offload piping for an UST at a gasoline station comes up through the pump, which **Figure 25-69** demonstrates. The piping is manufactured so that in the event that a car knocks off the pump it will shut off the flow of gasoline and will snap the piping off at or near the ground level. The only spillage of gasoline would be from the amount in the piping aboveground and in the hose, if everything works properly.

The loading piping is located separately and all of the fill pipes are located in the same area for easy transfer from a tank truck. The fill pipes are usually color coded and marked so that the driver can differentiate between unleaded, unleaded super, and diesel fuel, although the color coding is not standard across the country and varies from company to company. At some location on the property there will be vent pipes for the tanks, generally away from the pumps and near the property line. There will also be other manhole covers approximately 6 to 8 inches in diameter that have a triangle on the top of the cover. These are inspection wells and typically surround the tank. The holes are drilled at various depths so that leaks can be easily detected by air monitoring of the well, or if water is in the well, by taking a sample. If a facility has had a leak, there may be a large number of these wells on and around the property. Most gas station tanks are 10,000 to 25,000 gallons in size and, if not properly monitored, can slowly release a substantial amount of product over a short period of time.

> **Note** It is not uncommon to find gasoline bubbling up in a basement miles away from the gas station and to find that the release occurred several years ago.

Another common problem arises when farms are redeveloped into housing developments. If unknown USTs are located on the property, they may or may not be discovered during the construction and can eventually leak, causing problems. The tank and environmental industry refers to a **leaking underground storage tank** as a **LUST**.

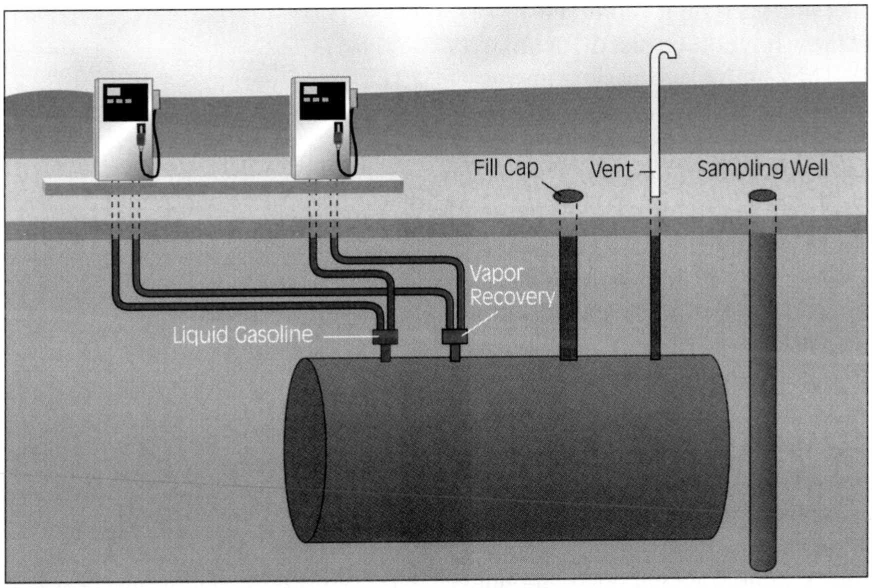

Figure 25-69 Piping system of a gas station.

Safety A recent incident in Baltimore County, Maryland, demonstrates that regardless of how many protection systems may be in place, a release off the property is still possible.

One afternoon a gas station was getting a tank filled with fuel, as per normal procedure, and the tank filling went as expected with no problems. The gas station had installed a state-of-the-art alarm system that would indicate a fault in the system and alert the owners to any potential releases. The system goes through a system check after each sale, if no pump is operated for a period of thirty seconds. If a pump is on, or the period between sales is less than thirty seconds, then the system check does not begin to work. During a busy afternoon, it could be estimated that the system check would not function for a long period of time, because at least one of the pumps is operating all the time.

In addition to the system check of each of the storage tank locations, some piping and the pumps all have electronic monitors that detect the presence of a liquid. When a liquid is detected, an alarm activates—the same type of alarm as if the system check fails the system. One of the potential problems with many self-service gas stations is that there is only one attendant, who for many reasons cannot leave the work area to either check on a problem outside or, depending on the alarm panel location, cannot check any alarms. In some gas stations the alarms are in other rooms and cannot be seen or heard by the only attendant. Another problem is that the liquid alarms will activate when it rains, tripping for rainwater.

The pumps that sit on top of the storage tanks have the ability to supply upward of 80 psi of pressure to all pump station nozzles, although this is governed down to an actual working pressure of 10 psi at all pumps. In this particular incident, a flange that comes out of the tank pump failed and separated from the tank pump. From the pump discharge there is a 2-inch-diameter opening, which at 10 to 80 psi could pump a considerable amount of fuel in a short time. One theory is that a customer may have set a nozzle down after not being able to pump fuel, or the amount of fuel being pumped was inadequate. This action would have resulted in the tank pump continuing to pump. A liquid alarm sounded for a number of areas at the gas station. The attendant notified the pump repair company as per protocol, but a response was delayed due to a number of extenuating circumstances.

Once the repair company arrived they found the tank empty; it was at this time the clerk informed them of the recent delivery. It was later found that an estimated 4,500 gallons were lost and were causing flammable vapor readings in a several-block area, although all the gasoline was later recovered. Within the gas delivery system, this station had the best protection in place, and had just retrofitted the station with a new piping system, but it shows that no matter how many protection systems are in place, it is still possible to have a release.

Aboveground tanks are becoming more and more common, although they have been used for many years, **Figure 25-70**. They are of two basic construction types—upright and horizontal—and some are nonpressurized or atmospheric tanks and some are pressurized tanks. ASTs hold a wider variety of chemicals as opposed to their UST counterparts, but petroleum products are still a leading commodity stored in these type of tanks. They vary in size from the 250-gallon home heating oil tank to the several million gallon oil tank. In industrial and commercial applications, a containment area is required around the tank. The containment area must be able to hold the contents of the largest tank within the containment. Regular inspection of these areas is required to ensure that rainwater, snow, or ice does not cause a buildup of liquid in the containment area, which would reduce the containment's ability to hold its intended amount. Depending on the weather conditions, the containment area's gate valve may be left in an open position,

Figure 25-70 Due to increased environmental concerns, many tanks are being placed aboveground and are called aboveground storage tanks or ASTs.

which leaves the facility at risk for product to escape the facility through the open drain or gate valve.

Upright storage tanks come in three basic construction types: **ordinary tank**, **external floating roof tank**, and **internal floating roof tank**. The ordinary tank, **Figure 25-71**, is constructed of steel and typically has a sloped or cone roof to shed rainwater, snow, and ice. The roof and tank shell seam is purposely made weak so that in the event of an explosion, only the top of the tank is relocated. One of the major problems with this type of tank is that when the tank is not full, there is room for vapors to accumulate.

Caution Any time vapors are allowed to accumulate, the potential exists for a fire or explosion.

Some ordinary tanks are purposely constructed without a roof in place. These are typically used in safety vent situations and water treatment areas where a roof may cause additional problems. It is important to identify these tanks during pre-incident surveys so as not to misjudge the severity of an incident. In many chemical processes it is not uncommon to have a storage tank with piping into the tank just to catch overflow or the contents of a system in the event of overpressurization or failure. The materials can be hot and produce large amount of vapors, which, if a roof were present, would allow a buildup of pressure, causing a catastrophic failure of the tank and roof.

External floating roof tanks are used to eliminate the buildup of vapors. They ride on top of the liquid, **Figure 25-72**, and since there is no space between the roof and the liquid, vapors cannot accumulate and create problems. External floaters can be seen from the top of the tanks, and a ladder is affixed to the roof to allow access. A common incident with these types of tanks results from a lightning strike, which can cause a fire in the roof/shell interface area. *This type of fire is difficult to extinguish and can result in roof/tank failure if not controlled quickly.* It is also possible that water used for firefighting could also sink the roof, causing further problems.

Internal floaters are constructed in the same manner as external floaters but have an additional roof over the top of the tank. An example is shown in **Figure 25-73**. The type of roof varies, but is usually a slightly coned roof with vent holes along the outer edge of the tank shell, or a geodesic type of roof that is usually made of fiberglass. The internal floater suffers from the same type of fire problem as the external, although it is a reduced risk. If a fire were to start in the roof/shell interface, the roof makes it more difficult to extinguish.

Figure 25-71 This is a cone roof tank. It has a weak roof-to-shell seam so that in the event of an explosion the roof will come off, but the tank should remain intact.

Figure 25-72 This is an open floating roof tank, in which the roof floats on top of the product. This reduces the release of vapors, as there is no vapor space, and reduces the fire potential.

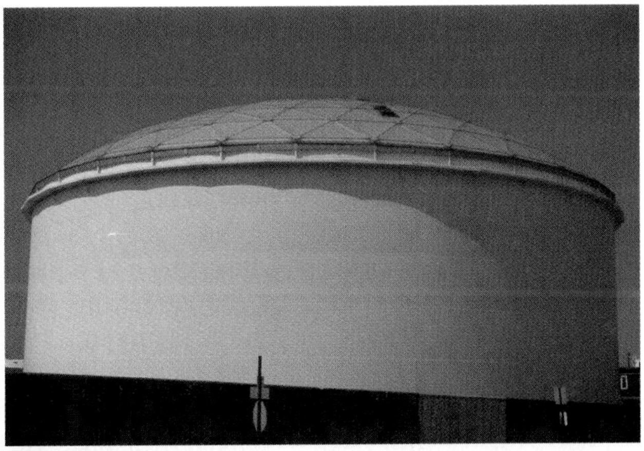

Figure 25-73 This is a covered floating roof tank, which is the same as an open floating roof tank but has a cover to keep out snow, rain, and debris. Another term for this type of tank is *geodesic domed tank*.

Figure 25-74 The specialized tank such as the propane tank shown here has some of the same properties as its transportation equivalents.

Figure 25-75 This tank holds cryogenic liquid oxygen and is typical of a cryogenic upright tank.

Specialized Tanks

Specialized tanks are a combination of the tank types discussed earlier in this section and include pressurized tanks and cryogenic tanks. The larger pressurized vessels are divided into two categories: low-pressure and high-pressure tanks. The low-pressure tanks hold flammable liquids, corrosive liquids, and some gases, up to 15 psig. Common high-pressure commodities are liquefied propane, liquefied natural gas, or other gaseous or liquefied petroleum gases. An acid like hydrochloric acid, which has a high vapor pressure, would not be uncommon in a tank like this.

These types of tanks may have an external cover, which appears to be a tank within a tank. The pressurized tanks are larger versions of the propane tanks discussed earlier and can have a capacity of up to 9 million gallons although less than 40,000 gallons is typical. Liquefied petroleum gases are not only stored in pressurized tanks such as the one shown in **Figure 25-74**; these types of gases are also stored in other locations such as caves carved out of mountains, although this is rare. Because these facilities are not required to report under the SARA Title III regulations, firefighters will have to contact their local gas suppliers to find out how they store their gas products.

Upright cryogenic storage tanks, **Figure 25-75**, are located in almost every community, especially if the community has a hospital or medical center. Most hospitals are supplied with liquid oxygen (LOX) through the cryogenic tank. Facilities that sell or distribute compressed gases are likely to have cryogenic tanks. Fast-food restaurants and convenience stores are now using cryogenic tanks to supply carbon dioxide to the soda dispensing machines.

SENSES

Vision and hearing are acceptable senses to use while investigating potential chemical releases. The use of touch, smell, and taste, however, is a dangerous use of the senses.

> **Caution** The lack of an odor cannot be equated with a lack of toxicity. Many severely toxic materials are colorless and odorless.

Sensory clues gained from other persons are useful to help identify the spilled material. The smell of materials is an important clue when trying to identify a spilled material, so if bystanders or evacuated persons can assist by describing an odor, firefighters should take advantage of this clue, but they should not endanger their lives to determine an odor. Many toxic materials can be harmful if touched because the material can be absorbed through the skin. Anytime a person handles a chemical, appropriate chemical protective clothing should be worn.

CHEMICAL AND PHYSICAL PROPERTIES

Although not intended to be a full chemistry lesson, the material in this section is a key component of safety. The chemical and physical properties outlined here are appropriate for a firefighter's level of response. The identification and use of these key terms can determine the outcome of an incident and firefighter well-being. As a firefighter progresses up through the response levels the need for additional chemistry also increases. When in doubt, the firefighter should consult with a HAZMAT team or other resources such as Chemtrec or a local specialist. A lot of the terms to be discussed next can be applied throughout the entire firefighter text. The basis of a fire is a chemical reaction. The better that firefighters understand this chemical reaction, the better off they will be, which will also benefit the citizens of their communities.

States of Matter

The basic chemical and physical properties that are important to understand are the **states of matter**: **solid, liquid**, and **gas** as shown in **Figure 25-76**. The severity of an incident can be determined by knowing if the material is a solid chunk of stuff, a pool of liquid, or an invisible gas.

Note The level of concern rises with each change of state; relatively speaking, a release of a solid material is much easier to handle than a liquid release, and it is nearly impossible to control a gas.

The control methodology for each state increases in difficulty from simple controls with a solid, to difficult with a gas. Evacuations have to be larger for releases involving gases, whereas a minimal evacuation would be required for most solid materials.

How the chemicals can hurt someone also varies with the state of material. Solids usually can only enter the body through contact or ingestion, although inhalation of dusts is possible. Liquids can be ingested, absorbed through the skin, and, if evaporating, inhaled. Gases on the other hand can be absorbed through the skin, and inhaled, and to some extent ingested.

Adjoining the states of matter are melting point, freezing point, **Figure 25-77**, boiling point, **Figure 25-78**, and condensation point. All of these are related because they are the points at which a material changes its state. The **melting point** is the temperature at which solid must be heated to transform the solid to the liquid state. Ice, for instance, has a melting point of 32°F. The **freezing point** is the temperature of a liquid when it is transformed into a solid. For water, the freezing point is 32°F. The actual temperatures vary by the tenths of a degree, but are very close. The **boiling point** is reached when the liquid is heated to the point at which evaporization takes place, that is, the liquid

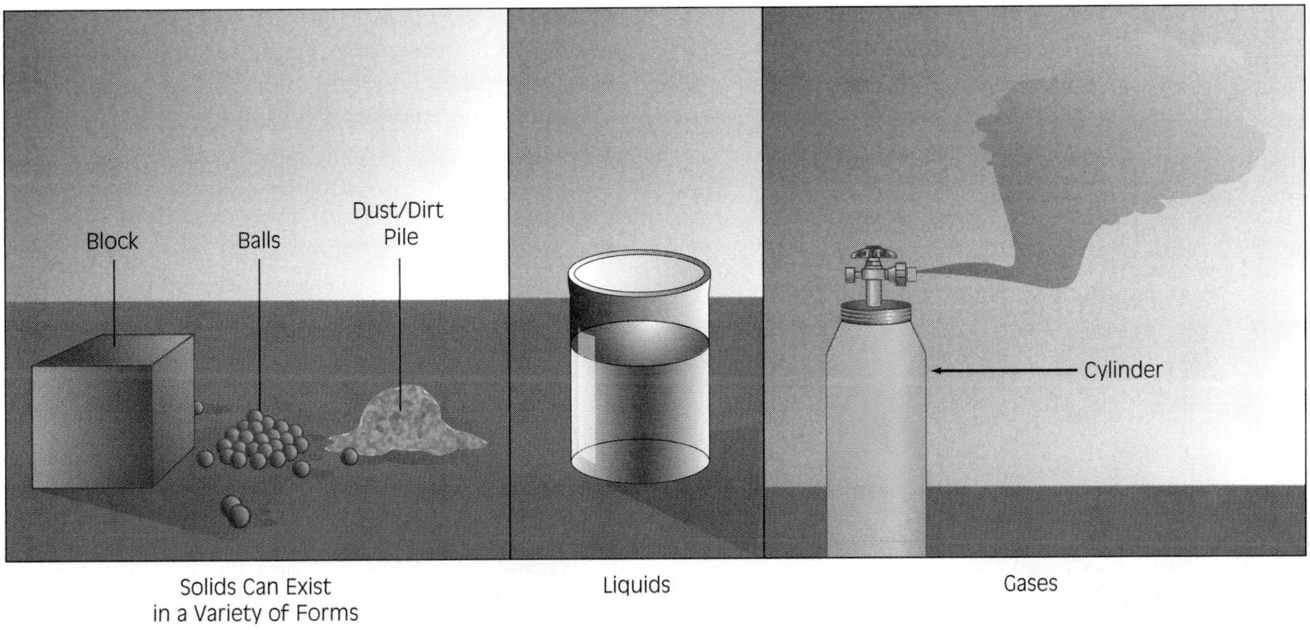

Figure 25-76 States of matter.

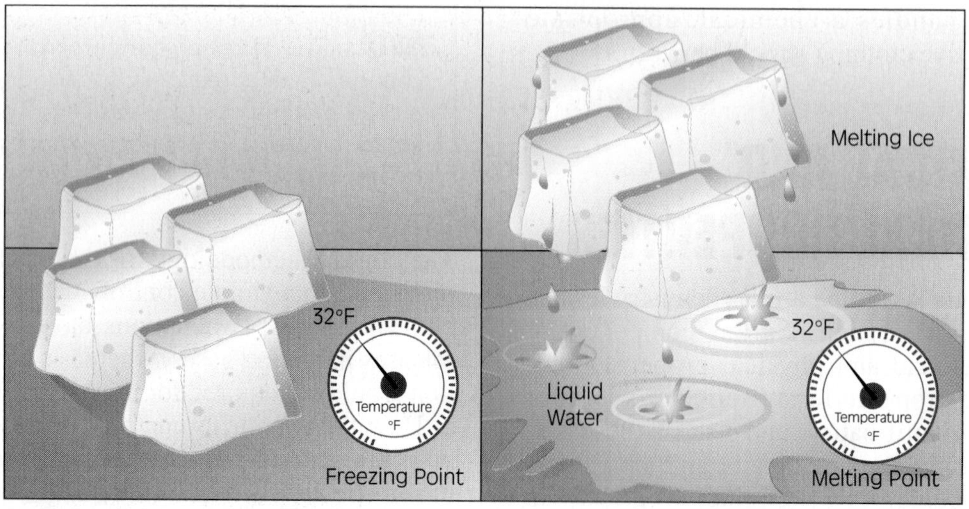

Figure 25-77 Melting and freezing points.

is being changed into a gas. Another way of defining boiling point is the temperature of a liquid when the vapor pressure exceeds the atmospheric pressure and a gas is produced. Water boils at 212°F and changes into a gaseous state. The important thing to remember about boiling points is the fact that when the liquid approaches this temperature, vapors are being produced that can cause larger problems.

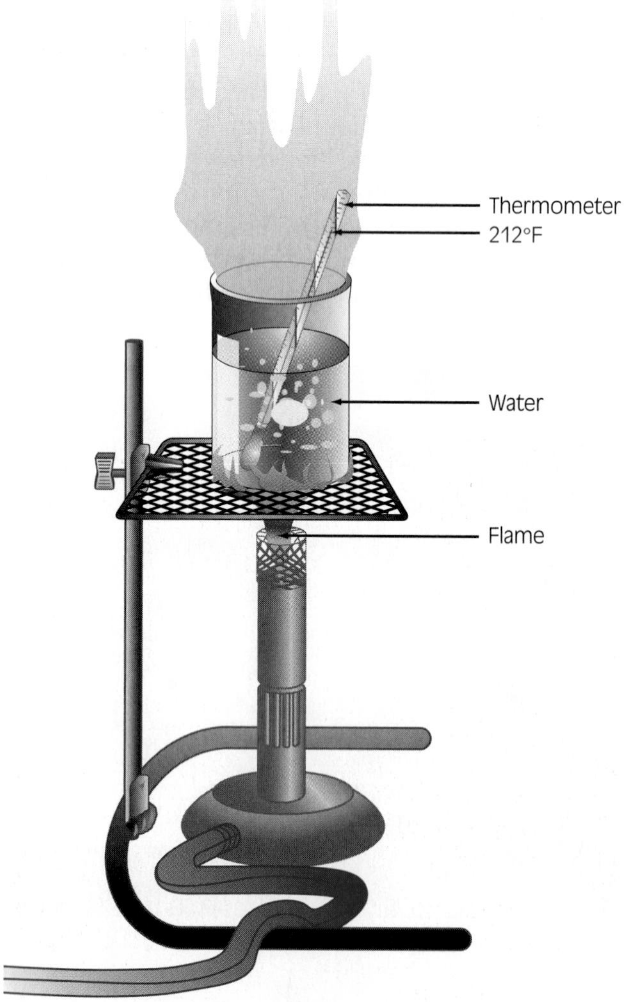

Figure 25-78 Boiling point.

Vapor Pressure

Out of all of the chemical and physical properties, vapor pressure is one of the most important to a hazardous materials responder. If a product has a high **vapor pressure**, it can be very dangerous. A material with a low vapor pressure is typically not a major concern. Vapor pressure has to do with the amount of vapors released from a liquid or a solid, **Figure 25-79**. The true definition is the pressure that is exerted on a closed container by the vapors coming from the liquid or solid. Vapor pressure can be related with the ability of a material to evaporate, in that the material is not really disappearing, it is just moving to another state of matter. Chemicals like gasoline, acetone, and alcohol all have high vapor pressures, whereas diesel fuel, motor oil, and water all have low vapor pressures.

Vapor pressures are measured in three ways: millimeters of mercury (mm Hg), pounds per square inch (psi), and atmospheres (atm). Normal or aver-

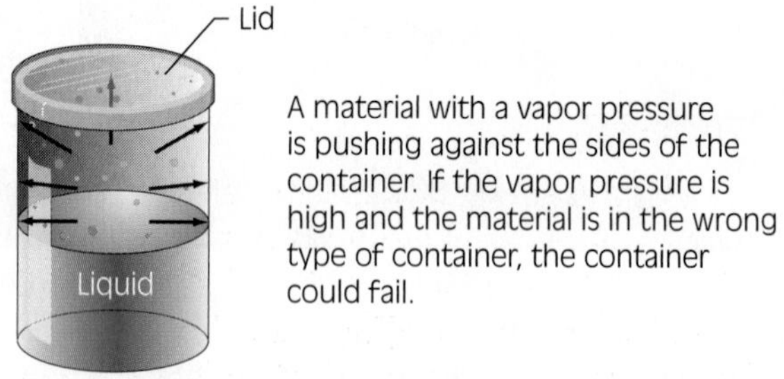

Lid
Liquid
A material with a vapor pressure is pushing against the sides of the container. If the vapor pressure is high and the material is in the wrong type of container, the container could fail.

Figure 25-79 Vapor pressure.

Common Products and Their Vapor Pressures

NAME	VAPOR PRESSURE @ 68°F (MM HG)	BOILING POINT (°F)
Water	25*	212
Acetone	180	134
Gasoline	300–400	399
Diesel fuel	2–5	304–574
Methyl alcohol	100	149
Ethion (pesticide)	0.0000015	304 (decomposes)
Sarin nerve agent	2.1	316

*The vapor pressure of water has been reported in various texts as between 17 and 25 mm Hg. This text uses 25, as it is the highest reported vapor pressure for water.

TABLE 25-9

age vapor pressures are 760 mm Hg, 14.7 psi, and 1 atm. Although these figures are used to describe normal vapor pressure, they best describe atmospheric pressure. The temperature which is used is 68°F (20°C), which is the standard temperature. If a temperature is not provided, then it can be assumed it is 68°F. Chemicals that have a true vapor hazard are those in excess of 40 mm Hg, and they are considered volatile. Chemicals with a vapor pressure of less than 40 mm Hg do not present much inhalation hazard even though they can still present extreme toxicity through skin absorption. Chemicals with a vapor pressure above 40 mm Hg can be considered inhalation hazards in addition to any other route of exposure they may possess. **Table 25-9** lists vapor pressures for some common products.

A unique chemical phenomenon called **sublimation** occurs when a chemical goes from a solid state of matter to a gas. The material never enters the liquid phase. The sublimation ability means that some solids have a vapor pressure and can move directly to the gaseous stage. Some solids such as dry ice (carbon dioxide) move quickly to the gaseous stage, while others, such as mothballs (naphthalene or paradichlorobenzene), move slowly.

Vapor Density

It is easy to confuse vapor pressure with vapor density, but they have two entirely different meanings. Vapor density determines whether the vapors will rise or fall, **Figure 25-80**. When deciding on potential evacuations and other tactical objectives (e.g., sampling and monitoring) this is an important consideration. One of the primary reasons that natural gas leaks do not ignite or flash back more often is the vapor density of natural gas. Air is given a value of 1, and all other gases are compared to air. Gases that have a vapor density of less than 1 will rise in air and will dissipate, whereas gases with a vapor density greater than 1 will stay low to the ground. Natural gas has a vapor density of 0.5, while propane has a vapor density of 1.56, which causes it to seek out any low spots, like gullies or sewers. Propane is more likely to ignite or flash back because it has a greater potential of finding an ignition source.

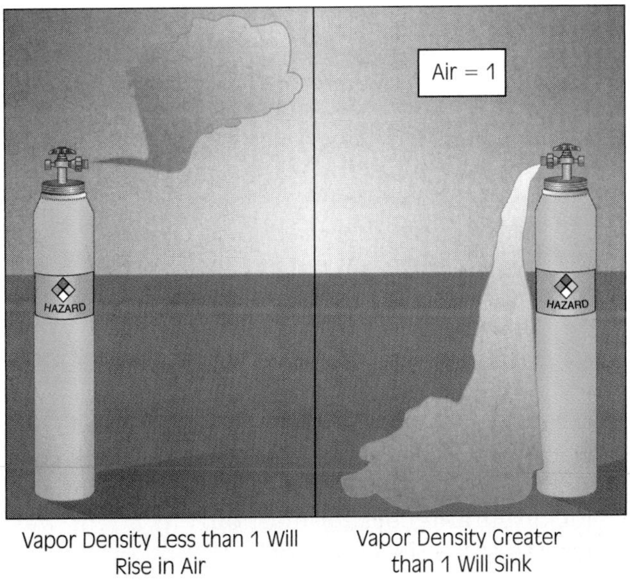
Air = 1
HAZARD
HAZARD
Vapor Density Less than 1 Will Rise in Air
Vapor Density Greater than 1 Will Sink

Figure 25-80 Vapor density.

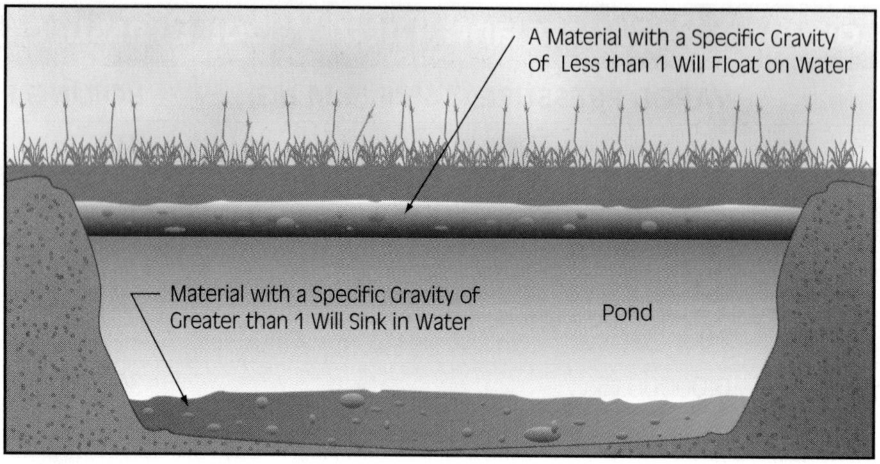

Figure 25-81 Specific gravity.

Specific Gravity

This chemical property is similar to vapor density in that it determines whether a material sinks or floats in water, **Figure 25-81**. Specific gravity is of prime concern when efforts are being taken to limit the spread of a spill by the use of booms or absorbent material. Water is given a value of 1, and chemicals that have a specific gravity of less than 1 will float on water. Fuels, oils, and other common hydrocarbons (chemicals composed of hydrogen and carbon, typically combustible and flammable liquids) have a specific gravity of less than 1 and, hence, will float on water.

Materials that have a specific gravity of greater than 1 will sink in water. It is much easier to recover materials floating on top of the water as opposed to those underwater. Carbon disulfide and 1,1,1-trichloroethane are two common materials that sink in water. Any material that sinks in water is especially troublesome if it has the potential to reach any groundwater, because remediation (cleanup) efforts are difficult and expensive.

Also of concern are materials that are water soluble, that is, have the ability to mix with water, not sink or float. Corrosives and many poisons are water soluble and are difficult to remove from a water source. Examples are provided in the following Firefighter Fact. Alcohol is also water soluble. Materials that are water soluble are difficult to extinguish.

Corrosivity

This text has already discussed tanks and containers that hold corrosives, as well as the concerns of dealing with corrosives. *Corrosive* is a term that is

Firefighter Fact When dealing with water-soluble materials it is important to protect bodies of water, because the cleanup can be difficult or impossible if the material enters the water. In one incident more than 10,000 gallons of sodium hydroxide, a very corrosive material, entered a small stream. The creek had to be dammed, and a neutralizing agent was put into the stream. Any life existing in this mile of stream was killed by the sodium hydroxide, but the environmental damage could have spread for many more miles if it had not been neutralized. Once neutral, the water was released, which eventually led to a larger body of water, where no damage occurred. This took considerable resources and time to accomplish, and luckily the necessary neutralizing agent was located at the site of the spill.

In another incident a very small amount of a pesticide was sprayed near a pond. After a rainfall the pesticide ran into the pond, killing all of the fish. The only method of removing this pesticide is the addition of other chemicals that are also hazardous. It will be years before the pond is able to sustain life.

In another incident in Baltimore City a railcar was involved in an accident and released 18,000 gallons of hydrochloric acid, which eventually ran into a local stream. The addition of the acid actually raised the pH level of the stream to near acceptable levels, because this stream was already heavily contaminated from a number of other sources.

Note that nothing should be flushed into any body of water without the express permission of the local or state environmental agencies.

pH of Common Materials

MATERIAL	pH	MATERIAL	pH
Water*	7	Sulfuric acid	1
Stomach acid	2	Gasoline	7
Orange juice†	2	Hydrochloric acid	0
Drain cleaner**	14	Pepsi‡	2
Potassium hydroxide	14	Household ammonia***	14

*Tap water is usually a 7, while rainwater in the Northeast can be 3–6 (acid rain).

†Citric acid is the main ingredient.

**The main ingredient is sodium hydroxide (lye).

‡Phosphoric acid is the main ingredient.

***This is a 5 percent solution of ammonium hydroxide and water.

TABLE 25-10

applied to both acids and bases, and is used to describe a material that has the potential to corrode or eat away skin or metal. Some examples are shown in **Table 25-10.** People deal with corrosives every day in that the human body is naturally acidic and they use many corrosive materials in their everyday lives. Acids are sometimes referred to as corrosives, whereas bases are also known as alkalis or caustics. The accurate way to describe a corrosive is to identify the material's pH, which provides some measure of corrosiveness. pH is an abbreviation for positive hydronium ions. It is used to designate the corrosive nature of a material. Acids have hydronium ions, and bases have hydroxide ions. It is the percent (ratio) of these ions that makes up the pH number. Materials having a pH of 0 to 6.9 are considered acids, and materials with a pH of 7.1 to 14 are considered bases. A material having a pH of 7 is considered neutral. The pH scale is a logarithmic scale, meaning the movement from 0 to 1 is an increase of 10. The movement from 0 to 2 is an increase of 100.

When dealing with a corrosive response, one of the common methods to mitigate the release is to neutralize the corrosive. One thought is that water can be used to dilute and thereby neutralize the spill. This presents two major issues: one, the mixing of water (chemical reaction), and two, the runoff from the reaction. Corrosives and water can be a dangerous combination. One should never add a corrosive to water, as it presents great risk. There may be some spattering and heat generation.

If 1 gallon of an acid had a pH of 0, it would take 10 gallons of water to move the material to a pH of 1. To move it to a pH of 2 would take 100 gallons of water, and to change the pH to 6 would require 1 million gallons of water—all for a 1-gallon spill. Chemically neutralizing a corrosive spill is the better choice, but even that can present some issues. When neutralizing a strong acid, a weak base should be used to perform the neutralization. A street method of calculation for neutralization is that it takes more than 8,800 pounds of potash to bring 1,000 gallons of 50 percent sulfuric acid to neutral, which is more realistic than controlling 1 million gallons of runoff. These examples are provided for information only; the neutralization of corrosives is a technician-level skill and can be very dangerous.

If the skin or eyes are exposed to a corrosive material, they should be immediately flushed with large quantities of water. This flushing should continue for at least twenty minutes uninterrupted. Some corrosive materials will cause immediate blindness and skin burns, and water should be used to prevent further injury.

Chemical Reactivity

Chemicals when they mix will have one of three types of reactions: exothermic, endothermic, or no reaction. The most common is the **exothermic reaction**, meaning the release of heat.

As discussed in Chapter 4, fire is a rapid oxidation reaction (exothermic) accompanied by heat and light. When most chemicals mix and provide an exothermic reaction there usually is not a lot of light but there can be substantial heat. By mixing one tablespoon each of vinegar (acetic acid) and ammonia (base) at the same temperature, an immediate rise in temperature of 10°F will occur. When handling oleum (concentrated sulfuric acid) and applying water to a spill, the resulting mixture will bubble,

Flash Points of Some Common Materials

MATERIAL	FLASH POINT (°F)	MATERIAL	FLASH POINT (°F)
Gasoline	−45	Diesel fuel	>100
Isopropyl alcohol	53	Motor oil	300–450
Acetone	−4	Xylene	90

TABLE 25-11

fume, boil, and heat to over 300°F the instant the water hits the acid. An **endothermic reaction** is one in which the energy created by the reaction is absorbed and cooling occurs.

Flash Point

A flash point is described as the temperature of a liquid at which, when heated by an ignition source, a flash fire occurs, **Table 25-11**. This resulting flash fire will ignite just the vapors and self-extinguish once those vapors are burned up. The liquid itself does not burn; it is the mixture of vapors and air that ignites. Following closely behind the flash point is the fire point of a liquid. The fire point is the temperature of the liquid at which, when heated, vapors are produced that when ignited will sustain burning. In the laboratory, scientists can replicate flash points and fire points; on the street, however, they are usually one and the same.

Autoignition Temperature

Sometimes referred to as **ignition temperature**, the autoignition temperature is the temperature at which a material will ignite on its own without an ignition source. Ignition temperatures are much higher than flash points and represent a potential hazard level depending on the temperature. Depending on the context, the term **self-accelerating decomposition temperature (SADT)** may be used, which is essentially the same thing as the autoignition temperature. Regardless of what it is called, it is a level to avoid.

Flammable Range

The two main areas within a flammable range are the lower explosive limit and upper explosive limit. The flammable range is the difference between the two extremes. A fire or explosion needs an ignition source, a fuel (vapors), and air. The proper vapor-to-air ratio is also required or ignition cannot occur. The **lower explosive** (flammable) **limit (LEL)** is the lowest amount of vapor mixed with air that can provide the proper mixture for a fire or explosion. The air monitor that is used to detect combustible gases is designed to read this level. The **upper explosive** (flammable) **limit (UEL)** is the highest amount of vapor mixed with air that will sustain a fire or explosion.

Each year many natural gas explosions occur throughout the United States. The LEL of natural gas is 5 percent, so if it is mixed with 95 percent air with an ignition source present an explosion or fire is possible, as shown in **Figure 25-82**. The UEL for natural gas is 15 percent, so 85 percent air is required to result in a fire or explosion. The flammable range for natural gas is 5 to 15 percent, and a fire or explosion can occur at any point in that range, as long as there is enough air to complete the mixture. With 4 percent methane and 96 percent air, there would not be any fire or explosion nor would there be at 16 percent natural gas and 84 percent air. Acetylene, which is a common gas, has a LEL of 2.5 percent and UEL of 100 percent, which means that when 2.5 percent LEL is reached, the potential for a fire exists. Having less than the LEL is referred to as *too lean,* and amounts in excess of the UEL are called *too rich.* Levels less than the LEL are much safer than levels above the UEL.

> **Safety** When confronted with a situation that has a level higher than the UEL, the situation is very dangerous.

Ventilation should be carried out using non-spark-inducing devices, and great care should be taken to minimize any potential electrical arcs such as not using light switches or doorbells.

Toxic Products of Combustion

This is the one area in which firefighters suffer considerable chemical exposures. Any time a person is in smoke or breathes smoke, the body is being bombarded with toxic chemicals. Many toxic chemicals, such as carbon monoxide, carbon dioxide, hydrogen cyanide, hydrochloric acid, and phosgene, are produced in a fire. The worst type

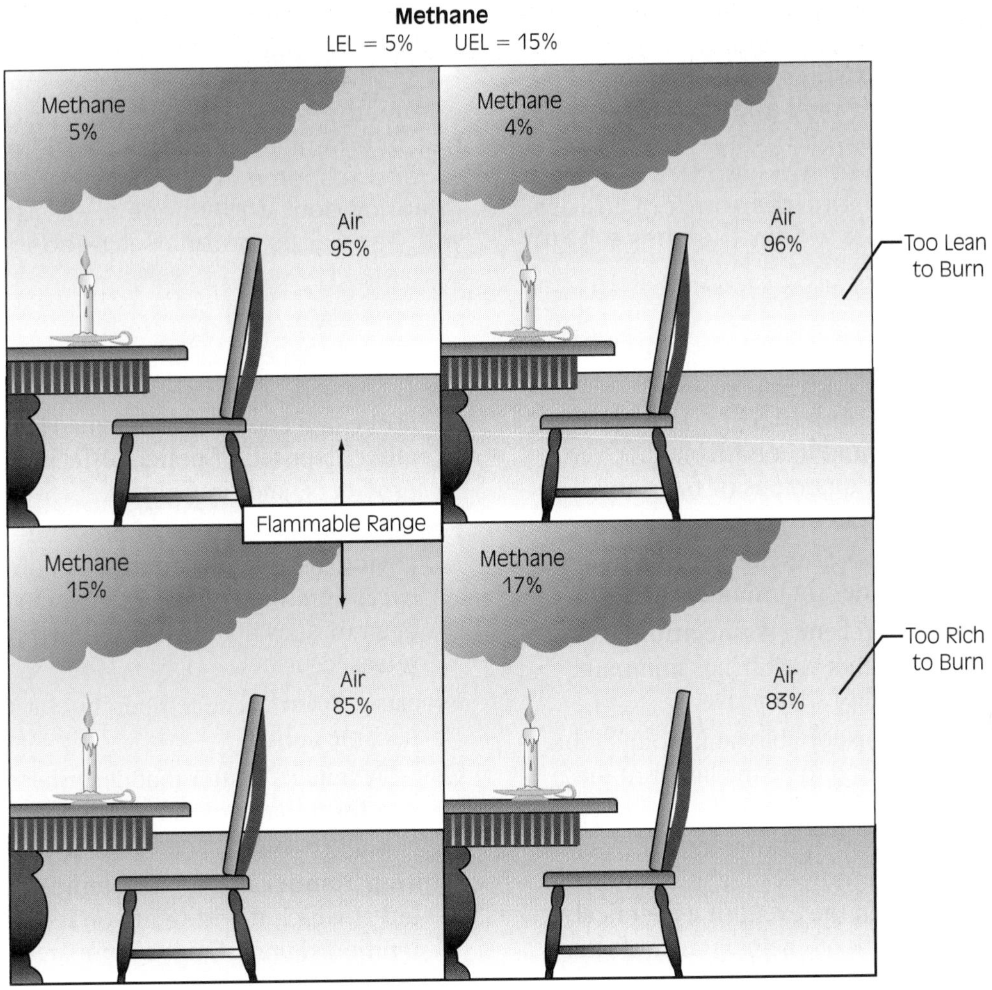
Methane
LEL = 5% UEL = 15%
Methane 5%
Air 95%
Methane 4%
Air 96%
Too Lean to Burn
Flammable Range
Methane 15%
Air 85%
Methane 17%
Air 83%
Too Rich to Burn
To have a fire or explosion the lower explosive limit must be reached. Each gas has a flammable range in which there can be a fire or explosion. Below the LEL or above the UEL means there cannot be a fire.

Figure 25-82 Flammable range.

of chemical accident a firefighter can respond to is a house, car, or Dumpster fire. Even brush fires are not exempt from the toxic products of combustion, because the field may have been sprayed with pesticides, herbicides, or other chemicals. Even burning wool or hay produces extremely toxic gases. Due to this constant exposure to these and other materials, it is important for firefighters to wear all of their protective clothing, especially the SCBA.

Lessons Learned

The ability to recognize and identify the potential for hazardous materials to be present at an incident is important for the first responder. The numbers of tank trucks, tank cars, and containers can easily overwhelm the beginning student. At any incident there is always a factor that relates to the recognition and identification process, whether it is the location, placards, container type, or physical senses. It could even be that sixth sense that alerts people to a potential problem. When that occurs, it is important to proceed with caution.

One of the important lessons for responders to remember is that they do not have to commit everything to memory, but they should know where to access hazardous materials information. It is not expected that each community have a railcar expert in their fire department. What is expected is that each fire department have a contact person available around the clock to obtain that resource. It is possible that every material on this earth has the ability to cause harm in some fashion, but the chemical and physical properties play a factor in the type of harm that can be caused.

Materials that have low vapor pressures present little risk to the responders unless touched or eaten. Materials that have high vapor pressures do present a great risk to responders and to the community and should be treated with caution. Vapor pressure is only one of the terms with which responders should become familiar. Local HAZMAT responders are a good resource and should be contacted early in an incident and whenever assistance dealing with hazardous materials is needed.

KEY TERMS

Aboveground Storage Tank (AST) Tank that is stored above the ground in a horizontal or vertical position. Smaller quantities of fuels are often stored in this fashion.

Boiling Point The temperature to which a liquid must be heated in order to turn into a gas.

Building Officials Conference Association (BOCA) A group that establishes minimum building and fire safety standards.

Bulk Tank A large transportable tank, comparable to a tote, but considered to be the larger of the two.

Chemtrec The Chemical Transportation Emergency Center, which provides technical assistance and guidance in the event of a chemical emergency; a network of chemical manufacturers that provide emergency information and response teams if necessary.

Cryogenic Gas Any gas that exists as a liquid at a very cold temperature, always below −150°F.

Deflagrates Rapid burning, which in reality with regard to explosions can be considered a slow explosion, but is traveling at a lesser speed than a detonation.

Emergency Response Guidebook (ERG) Book provided by the DOT that assists the first responder in making decisions at a transportation-related chemical incident.

Endothermic Reaction A chemical reaction in which heat is absorbed, and the resulting mixture is cold.

Exothermic Reaction A chemical reaction that releases heat, such as when two chemicals are mixed and the resulting mixture is hot.

External Floating Roof Tank Tank with the roof exposed on the outside that covers the liquid within the tank. The roof floats on the top of the liquid, which does not allow for vapors to build up.

First Responders A group designated by the community as those who may be the first to arrive at a chemical incident. This group is usually composed of police officers, EMS providers, and firefighters.

Frangible Disk A type of pressure-relieving device that actually ruptures in order to vent the excess pressure. Once opened the disk remains open; it does not close after the pressure is released.

Freezing Point The temperature at which liquids become solids.

Gas A state of matter that describes the material in a form that moves freely about and is difficult to control. Steam is an example.

Ignition Temperature The temperature of a liquid at which it will ignite on its own without an ignition source. Can be compared to SADT.

Intermodal Containers These are constructed in a fashion so that they can be transported by highway, rail, or ship. Intermodal containers exist for solids, liquids, and gases.

Internal Floating Roof Tank Tank with a roof that floats on the surface of the stored liquid, but also has a cover on top of the tank, so as to protect the top of the floating roof.

Leaking Underground Storage Tank (LUST) Describes a leaking tank that is underground.

Liquid A state of matter that implies fluidity, which means a material has the ability to move as water would. There are varying states of being a liquid from moving very quickly to very slowly. Water is an example.

Lower Explosive Limit (LEL) The lower part of the flammable range, and is the minimum required to have a fire or explosion.

Melting Point The temperature at which solids become liquids.

Ordinary Tank A horizontal or vertical tank that usually contains combustible or other less hazardous chemicals. Flammable materials and other hazardous chemicals may be stored in smaller quantities in these types of tanks.

Relief Valve A device designed to vent pressure in a tank, so that the tank itself does not rupture due to an increase in pressure. In most cases these devices are spring loaded so that when the pressure decreases the valve shuts, keeping the chemical inside the tank.

Reportable Quantity (RQ) Both the EPA and DOT use the term. It is a quantity of chemicals that may require some type of action, such as reporting an inventory or reporting an accident involving a certain amount of the chemical.

Sea Containers Shipping boxes that were designed to be stacked on a ship, then placed onto a truck or railcar.

Self-Accelerating Decomposition Temperature (SADT) Temperature at which a material will ignite itself without an ignition source present. Can be compared to ignition temperature.

Solid A state of matter that describes materials that may exist in chunks, blocks, chips, crystals, powders, dusts, and other types. Ice is an example.

Specification (Spec) Plates All trucks and tanks have a specification plate that outlines the type of tank, capacity, construction, and testing information.

States of Matter Describe in what form matter exists, such as solids, liquids, or gases.

Sublimation The ability of a solid to go to the gas phase without being liquid.

Tote A large tank usually 250 to 500 gallons, constructed to be transported to a facility and dropped for use.

Underground Storage Tank (UST) Tank that is buried under the ground. The most common are gasoline and other fuel tanks.

Upper Explosive Limit (UEL) The upper part of the flammable range. Above the UEL, fire or an explosion cannot occur because there is too much fuel and not enough oxygen.

Vapor Pressure The amount of force that is pushing vapors from a liquid. The higher the force the more vapors (gas) being put into the air.

REVIEW QUESTIONS

1. What are the nine hazard classes as defined by DOT?
2. An explosive placard is what color?
3. What three things on a placard indicate the potential hazards?
4. A DOT-406/MC-306 tank truck commonly carries what product?
5. A DOT-406 tank truck has a characteristic shape from the rear. What is it?
6. An MC-331 carries what type of gases?
7. What does the blue section of the NFPA 704 system refer to?
8. What are the locations of emergency shutoffs on an MC-331?
9. What are the four basic clues to recognition and identification?
10. The **Table 25-1** placards are required for what quantities?
11. A tractor trailer carrying Division 1.1 materials is well involved in fire. What should be the firefighter's initial tactics?
12. When propane tanks are involved in a fire, what is a potential consequence?

Endnotes

1. A tank truck placarded for gasoline could also haul diesel fuel or fuel oil without having to change placards.
2. The requirements for the subdivisions are presented in general. Other requirements may need to be met for a material to be assigned to a subdivision. Refer to 49 CFR 170–180 for more information.

Additional Resources

Bevelacqua, Armando, *Hazardous Materials Chemistry.* Delmar Learning, a part of the Thomson Corporation, Clifton Park, NY, 2001.

Bevelacqua, Armando and Richard Stilp, *Hazardous Materials Field Guide.* Delmar Learning, a part of the Thomson Corporation, Clifton Park, NY, 1998.

Bevelacqua, Armando S. and Richard H. Stilp, *Terrorism Handbook for Operational Responders,* 2nd ed. Delmar Learning, a part of the Thomson Corporation, Clifton Park, NY, 2004.

Hawley, Chris, *Hadardous Materials Incidents,* 2nd ed. Delmar Learning, a part of the Thomson Corporation, Clifton Park, NY, 2004.

Hawley, Chris, *HazMat Air Monitoring & Detection Devices.* Delmar Learning, a part of the Thomson Corporation, Clifton Park, NY, 2002.

Henry, Timothy V. *Decontamination for Hazardous Materials Emergencies,* Delmar Learning, a part of the Thomson Corporation, Clifton Park, NY, 1999.

Lesak, David, *Hazardous Materials Strategies and Tactics.* Prentice Hall, 1998.

Noll, Gregory, Michael Hildebrand, and James Yvorra, *Hazardous Materials Managing the Incident.* Fire Protection Publications, Oklahoma University, 1995.

Schnepp, Rob and Paul Gantt, *Hazardous Materials: Regulations, Response & Site Operations.* Delmar Learning, a part of the Thomson Corporation, Clifton Park, NY, 1999.

Stilp, Richard and Armando Bevelacqua, *Emergency Medical Response to Hazardous Materials Incidents.* Delmar Learning, a part of the Thomson Corporation, Clifton Park, NY, 1997.

CHAPTER
26

HAZARDOUS MATERIALS: INFORMATION RESOURCES

Christopher Hawley, Baltimore County Fire Department (Ret.)

OUTLINE

- Objectives
- Introduction
- Emergency Response Guidebook
- Material Safety Data Sheets
- Shipping Papers
- Facility Documents
- Computer Resources
- Chemtrec
- Reference and Information Texts
- Industrial Technical Assistance
- Lessons Learned
- Key Terms
- Review Questions
- Additional Resources

STREET STORY

I had been on the job for several years, and there was nothing unusual about the call when it first came in—nothing remarkable that would lead to any heightened sense of awareness. "Engine 29, check out the smell of smoke in the building at 211 Mason Street." It would be a simple food on the stove call or something to that effect.

When we arrived on the scene the local police department had already made initial entry to the facility (which turned out to be a restaurant) and reported a slight haze with a strange pungent odor but no visible fire. We made entry into the vacant facility and immediately noticed the haze and pungent odor, but because it was a restaurant we thought it was simply something to do with the food preparation process. As we made our way through the facility, we began to notice a strange taste in our mouths and several of those present began to experience irritation to the eyes and throat, yet still nothing clicked. It was, after all, simply a restaurant, perhaps some burnt wiring or bad cooking, no big deal. As we continued our investigation, several of those present began to complain of tightness in the chest, headaches, and nausea. I too was feeling strange but did not want to throw in the towel so another responder and I continued on.

Then suddenly as the haze began to clear via the doors and windows that had been opened for ventilation, we began to notice an abundance of dead insects and rodents throughout the facility. And now on the radio we heard the call for an ambulance as one of the responders who had been on scene for some time began to experience more than simply mild irritations. We began to realize that the haze was in fact some type of aerosolized insecticide/pesticide and that we had been exposed and contaminated. Now and only now did we begin to comprehend the error of our ways, so to speak, and implement the proper response and protocols.

As the weeks followed, I replayed the incident over and over and came to realize all of the mistakes that we had made, and how imperative it was to seek the proper training to ensure that I never put myself or those with me in this position again. Two of the responders present were unable to return to work for a period of time (in one case several months). They had to be treated for the effects of exposure to organophosphates. Both are fine today, but much has been written about the possible long-term effects of exposure to organophosphates, and one can only wonder what course these individuals' health could have taken had they been further exposed. And that I do, all too often.

Speaking only to my errors, there was no attempt to obtain the proper information from the appropriate resources or the owners. (After we realized we had no fire, we should have consulted the owners before further entry, because there was no life hazard.) Nor was subsequent contact made with the other tenants of this row structure who may have been able to provide information, or to even take basic actions to ensure the health and safety of those involved. The failure to properly research and gather information on this call could have resulted in much more severe health problems for all involved had the concentrations been higher or ventilation unsuccessful. But the fact that two responders were unnecessarily exposed still remains with me today.

—Street Story by Tom Creamer, Special Operations Coordinator, City of Worcester Fire Department, Worcester, Massachusetts

OBJECTIVES

After completing this chapter, the reader should be able to:

- Explain the terms used on Material Safety Data Sheets.
- Tell where MSDS are located.
- Identify the standard information available on an MSDS.
- Use the *Emergency Response Guidebook* (ERG).
- List the types of assistance that can be provided by Chemtrec.
- Explain the methods used to contact a shipper or emergency contact.
- Describe other resources that may be available in the community.

INTRODUCTION

Caution Specific and current chemical information is essential if the first responder is going to protect lives and property.

Chemical information is available through a variety of sources, including those carried on emergency apparatus and information sources that can be reached via telephone, fax, or computer. The shipper and the facility are required to maintain certain documents that will assist first responders in determining the chemical hazards they may face. Knowing what information is available and how to interpret this information is a valuable tool. This chapter discusses the most common sources of information.

EMERGENCY RESPONSE GUIDEBOOK

The DOT's ERG is a well-known book for emergency responders, **Figure 26-1**. The DOT makes one copy available for every emergency response apparatus in the country.

The ERG is commonly referred to as the DOT book or the orange book. This book, which is published almost every three years, contains information regarding the most commonly transported chemicals as regulated by the DOT. It is intended as a guidebook for first responders during the initial phase of a hazardous materials incident. It provides information regarding the potential hazards of responding to these materials and is one of the only books that provides specific evacuation recommendations. Although an excellent book for first responders, it does have limitations for the more advanced responder, who requires more specific chemical information.

Figure 26-1 The DOT *Emergency Response Guidebook* should be found in every emergency vehicle in the United States. It provides chemical emergency response information that is valuable to the first responder.

The book consists of these major sections:

- Placard information
- Listing by DOT identification number
- Alphabetical listing by shipping name
- Response guides
- Table of initial isolation and protective action distances
- List of dangerous water-reactive materials

On the inside cover, the book provides an example of a shipping document used in truck transportation and provides information on how the shipping document should be written. It also provides an example of a placard with an identification number panel. The first page, which is shown in **Figure 26-2**, is an essential page for responder safety because it outlines all of the actions the first responder should take. It provides a decision tree process by taking the reader through the incident step by step. It also has an important listing of the guides for explosives.

The next few pages of the ERG provide information on safety precautions and whom to call for assistance. The first responder should already know where the closest local assistance will be coming from and where to contact state assistance if required. The DOT book provides a contact number for federal assistance, although responders should proceed by requesting local, state, and *then* federal

RESIST RUSHING IN !
APPROACH INCIDENT FROM UPWIND
STAY CLEAR OF ALL SPILLS, VAPORS, FUMES AND SMOKE

HOW TO USE THIS GUIDEBOOK DURING AN INCIDENT INVOLVING DANGEROUS GOODS

ONE **IDENTIFY THE MATERIAL** BY FINDING ANY **ONE** OF THE FOLLOWING:

THE 4-DIGIT ID NUMBER ON A PLACARD OR ORANGE PANEL

THE 4-DIGIT ID NUMBER (after UN/NA) ON A SHIPPING DOCUMENT OR PACKAGE

THE NAME OF THE MATERIAL ON A SHIPPING DOCUMENT, PLACARD OR PACKAGE

IF AN **ID NUMBER** OR THE **NAME OF THE MATERIAL** CANNOT BE FOUND, SKIP TO THE NOTES BELOW.

TWO **LOOK UP THE MATERIAL'S 3-DIGIT GUIDE NUMBER** IN EITHER:

THE ID NUMBER INDEX..(the yellow-bordered pages of the guidebook)

THE NAME OF MATERIAL INDEX..(the blue-bordered pages of the guidebook)

If the guide number is supplemented with the letter "**P**", it indicates that the material may undergo violent polymerization if subjected to heat or contamination.

If the index entry is highlighted (in either yellow or blue), it is a TIH (Toxic Inhalation Hazard) material or a Dangerous Water Reactive Material (produces toxic gas upon contact with water). **LOOK FOR THE ID NUMBER AND NAME OF THE MATERIAL** IN THE TABLE OF INITIAL ISOLATION AND PROTECTIVE ACTION DISTANCES (the green-bordered pages). Then, if necessary, **BEGIN PROTECTIVE ACTIONS IMMEDIATELY** (see Protective Actions on page 314). If protective action is not required, use the information jointly with the 3-digit guide.

USE GUIDE 112 FOR ALL EXPLOSIVES EXCEPT FOR EXPLOSIVES 1.4 (EXPLOSIVES C) WHERE GUIDE 114 IS TO BE CONSULTED.

THREE **TURN TO THE NUMBERED GUIDE** (the orange-bordered pages) **AND READ CAREFULLY.**

NOTES **IF A NUMBERED GUIDE CANNOT BE OBTAINED BY FOLLOWING THE ABOVE STEPS**, AND A PLACARD CAN BE SEEN, LOCATE THE PLACARD IN THE TABLE OF PLACARDS (pages 16-17), THEN GO TO THE 3-DIGIT GUIDE SHOWN NEXT TO THE SAMPLE PLACARD.

IF A REFERENCE TO A GUIDE CANNOT BE FOUND AND THIS INCIDENT IS BELIEVED TO INVOLVE DANGEROUS GOODS, TURN TO **GUIDE 111** NOW, AND USE IT UNTIL ADDITIONAL INFORMATION BECOMES AVAILABLE. If the shipping document lists an emergency response telephone number, call that number. If the shipping document is not available, or no emergency response telephone number is listed, IMMEDIATELY CALL the appropriate **emergency response agency listed on the inside back cover of this guidebook.** Provide as much information as possible, such as the name of the carrier (trucking company or railroad) and vehicle number. AS A LAST RESORT, CONSULT THE TABLE OF RAIL CAR AND ROAD TRAILER IDENTIFICATION CHART (pages 18-19). IF THE CONTAINER CAN BE IDENTIFIED, REMEMBER THAT THE INFORMATION ASSOCIATED WITH THESE CONTAINERS IS FOR THE WORST CASE POSSIBLE.

Figure 26-2 The first page of the ERG is a step-by-step listing of how to use the book. When not sure how to proceed, the responder should turn to this page.

assistance. The contact numbers and agencies are divided among Canada, the United States, and Mexico. It is important for the first responder to read the DOT book prior to an incident because it provides a large amount of background material that could not be effectively read during an emergency.

The book also provides a listing of the hazard class system, **Figure 26-3**, and provides a reference point for the hazard classes that may be listed on placards or shipping papers. This listing precedes the placard section, which is valuable if the only information available is the placard. The placard section, **Figure 26-4**, provides information about how to proceed at an incident where the only information is a placard. All of the possible placards are listed with their pictures, and the reader is referred to the accompanying guide page.

HAZARD CLASSIFICATION SYSTEM

The hazard class of dangerous goods is indicated either by its class (or division) number or name. For a placard corresponding to the primary hazard class of a material, the hazard class or division number must be displayed in the lower corner of the placard. However, no hazard class or division number may be displayed on a placard representing the subsidiary hazard of a material. For other than Class 7 or the OXYGEN placard, text indicating a hazard (for example, "CORROSIVE") is not required. Text is shown only in the U.S. The hazard class or division number must appear on the shipping document after each shipping name.

Class 1 - Explosives

Division 1.1	Explosives with a mass explosion hazard
Division 1.2	Explosives with a projection hazard
Division 1.3	Explosives with predominantly a fire hazard
Division 1.4	Explosives with no significant blast hazard
Division 1.5	Very insensitive explosives; blasting agents
Division 1.6	Extremely insensitive detonating articles

Class 2 - Gases

Division 2.1	Flammable gases
Division 2.2	Non-flammable, non-toxic* compressed gases
Division 2.3	Gases toxic* by inhalation
Division 2.4	Corrosive gases (Canada)

Class 3 - Flammable liquids (and Combustible liquids [U.S.])

Class 4 - Flammable solids; Spontaneously combustible materials; and Dangerous when wet materials

Division 4.1	Flammable solids
Division 4.2	Spontaneously combustible materials
Division 4.3	Dangerous when wet materials

Class 5 - Oxidizers and Organic peroxides

Division 5.1	Oxidizers
Division 5.2	Organic peroxides

Class 6 - Toxic* materials and Infectious substances

Division 6.1	Toxic* materials
Division 6.2	Infectious substances

Class 7 - Radioactive materials

Class 8 - Corrosive materials

Class 9 - Miscellaneous dangerous goods

Division 9.1	Miscellaneous dangerous goods (Canada)
Division 9.2	Environmentally hazardous substances (Canada)
Division 9.3	Dangerous wastes (Canada)

*** The words "poison" or "poisonous" are synonymous with the word "toxic".**

Figure 26-3 The DOT uses nine classes of hazardous materials, as described in Chapter 25, that are listed in the ERG. Occasionally the only information that is available is the hazard class.

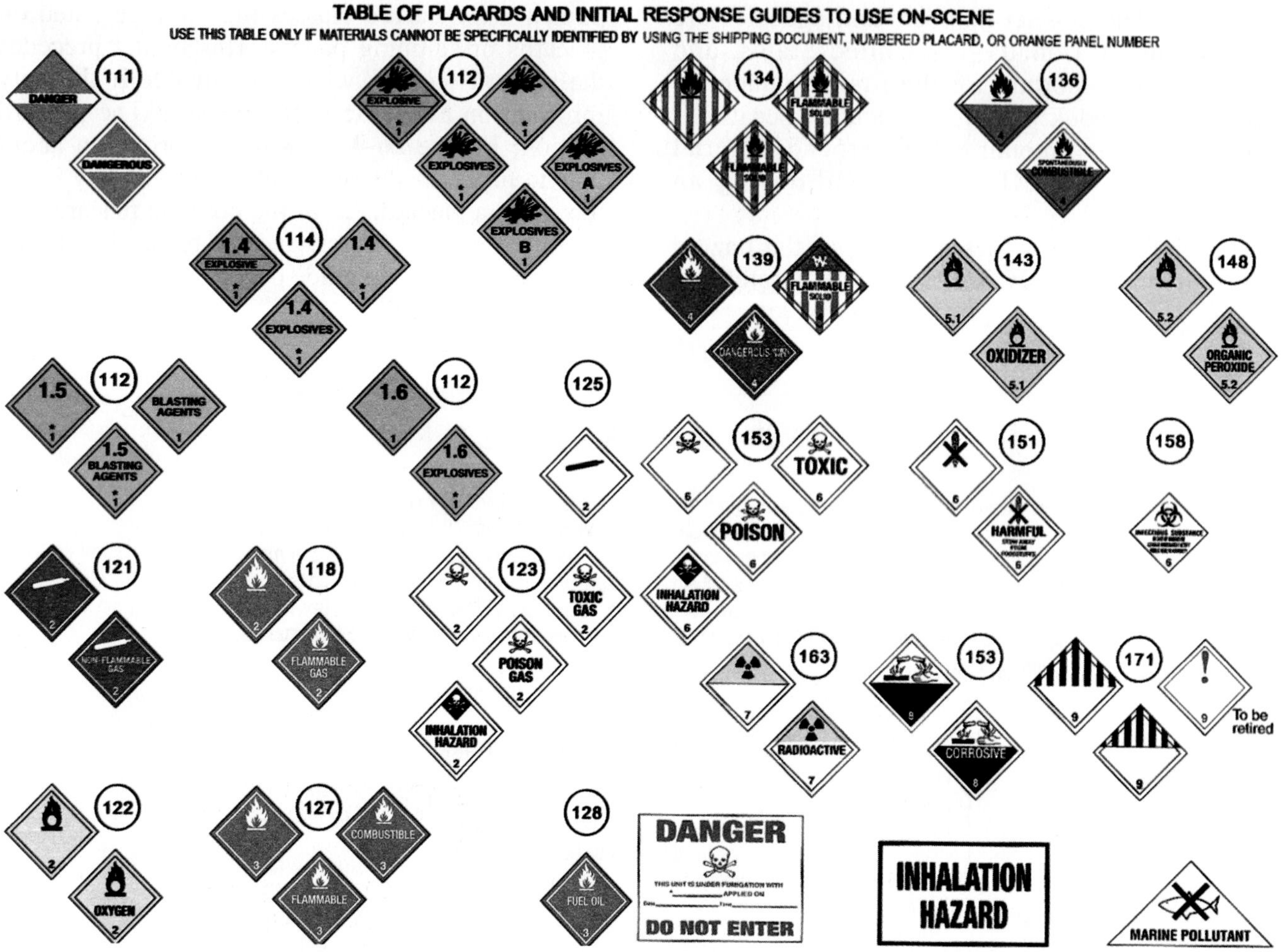

Figure 26-4 If the only information that is available is a placard, the ERG can provide assistance. Listed below each placard is a guide number that provides response information based on the placard information.

The yellow section, **Figure 26-5**, is a numerical listing by the identification, or ID, number. This is a number assigned by the DOT that identifies a material being shipped. This number takes two forms, a North America (NA) number or a United Nations (UN) number, but it is always a four-digit number. Although one would think that this number would specifically identify a material, in some cases it may not. Depending on the material it may be lumped into a general category such as Hazardous Waste, Liquid, or NOS (Not Otherwise Specified), but in most cases it will identify a specific substance. The yellow section lists the ID number, shipping name, and guide reference number. If the material is highlighted as shown in **Figure 26-6**, then the reader must refer to the guide page listed as well as the Table of Initial Isolation and Protective Action Distances, located at the back of the book in the green section.

The following abbreviations are used in the DOT ERG:

- NOS—not otherwise specified
- PIH—poison by inhalation
- P—polymerization hazard
- SCA—surface contaminated articles (radiation)
- LSA—low specific activity (radiation)
- PG I, II, or III—Packing Group I, II, or III
- ORM—other regulated material (listed as ORM A-E)
- LC_{50}—lethal concentration to 50 percent of the population
- LD_{50}—lethal dose to 50 percent of the population

The blue section of the book mirrors the yellow section except that it is alphabetical by shipping

ID No.	Guide No.	Name of Material
1014	**122**	Carbon dioxide and Oxygen mixture, compressed
1014	**122**	Oxygen and Carbon dioxide mixture
1014	**122**	Oxygen and Carbon dioxide mixture, compressed
1015	**126**	Carbon dioxide and Nitrous oxide mixture
1015	**126**	Nitrous oxide and Carbon dioxide mixture
1016	**119**	Carbon monoxide
1016	**119**	Carbon monoxide, compressed
1017	**124**	Chlorine
1018	**126**	Chlorodifluoromethane
1018	**126**	Refrigerant gas R-22
1020	**126**	Chloropentafluoroethane
1020	**126**	Refrigerant gas R-115
1021	**126**	1-Chloro-1,2,2,2-tetrafluoroethane
1021	**126**	Chlorotetrafluoroethane
1021	**126**	Refrigerant gas R-124
1022	**126**	Chlorotrifluoromethane
1022	**126**	Refrigerant gas R-13
1023	**119**	Coal gas
1023	**119**	Coal gas, compressed
1026	**119**	Cyanogen
1026	**119**	Cyanogen, liquefied
1026	**119**	Cyanogen gas
1027	**115**	Cyclopropane
1027	**115**	Cyclopropane, liquefied
1028	**126**	Dichlorodifluoromethane
1028	**126**	Refrigerant gas R-12
1029	**126**	Dichlorofluoromethane
1029	**126**	Refrigerant gas R-21
1030	**115**	1,1-Difluoroethane
1030	**115**	Difluoroethane
1030	**115**	Refrigerant gas R-152a
1032	**118**	Dimethylamine, anhydrous
1033	**115**	Dimethyl ether
1035	**115**	Ethane
1035	**115**	Ethane, compressed
1036	**118**	Ethylamine
1037	**115**	Ethyl chloride
1038	**115**	Ethylene, refrigerated liquid (cryogenic liquid)
1039	**115**	Ethyl methyl ether
1039	**115**	Methyl ethyl ether
1040	**119P**	Ethylene oxide
1040	**119P**	Ethylene oxide with Nitrogen
1041	**115**	Carbon dioxide and Ethylene oxide mixture, with more than 9% but not more than 87% Ethylene oxide
1041	**115**	Carbon dioxide and Ethylene oxide mixtures, with more than 6% Ethylene oxide
1041	**115**	Ethylene oxide and Carbon dioxide mixture, with more than 9% but not more than 87% Ethylene oxide
1041	**115**	Ethylene oxide and Carbon dioxide mixtures, with more than 6 % Ethylene oxide
1043	**125**	Fertilizer, ammoniating solution, with free Ammonia
1044	**126**	Fire extinguishers with compressed gas
1044	**126**	Fire extinguishers with liquefied gas
1045	**124**	Fluorine

Figure 26-5 The yellow pages are a numeric listing of the chemicals regulated by the DOT. This listing is by four-digit United Nations/North America identification number and also provides the shipping name and the emergency response guide page information.

ID No.	Guide No.	Name of Material
2465	**140**	Sodium dichloroisocyanurate
2465	**140**	Sodium dichloro-s-triazinetrione
2466	**143**	Potassium superoxide
2467	**140**	Sodium percarbonates
2468	**140**	Trichloroisocyanuric acid, dry
2468	**140**	Trichloro-s-triazinetrione, dry
2468	**140**	(mono)-(Trichloro)-tetra-(monopotassium dichloro)-penta-s-triazinetrione, dry
2469	**140**	Zinc bromate
2470	**152**	Phenylacetonitrile, liquid
2471	**154**	Osmium tetroxide
2473	**154**	Sodium arsanilate
2474	**157**	Thiophosgene
2475	**157**	Vanadium trichloride
2477	**131**	Methyl isothiocyanate
2478	**155**	Isocyanate solution, flammable, poisonous, n.o.s.
2478	**155**	Isocyanate solution, flammable, toxic, n.o.s.
2478	**155**	Isocyanate solutions, n.o.s.
2478	**155**	Isocyanates, flammable, poisonous, n.o.s.
2478	**155**	Isocyanates, flammable, toxic, n.o.s.
2478	**155**	Isocyanates, n.o.s.
2480	**155**	Methyl isocyanate
2481	**155**	Ethyl isocyanate
2482	**155**	n-Propyl isocyanate
2483	**155**	Isopropyl isocyanate
2484	**155**	tert-Butyl isocyanate
2485	**155**	n-Butyl isocyanate
2486	**155**	Isobutyl isocyanate
2487	**155**	Phenyl isocyanate

Figure 26-6 If the listing in the yellow pages or the blue pages is highlighted, as is the listing of ID number 2481, Ethyl isocyanate, then the reader must turn to the orange guide and to the green section. The green section is the Table of Initial Isolation and Protective Action Distances.

name, **Figure 26-7**. These shipping names are assigned by the DOT and for the most part are identical to the chemical names, but in some cases they can vary. A lot of names are assigned to chemicals, including the actual chemical name, synonyms, trade names, and shipping names.

Safety Some chemicals can have more than sixty synonyms, which can create a confusing situation. **Table 26-1** provides one example.

The chemical known by the DOT as chlordane is also called by the names given in **Table 26-1**. This section lists the ID number, shipping name, and guide reference number. If the material is highlighted in blue, the reader should refer to the guide page listed as well as the Table of Initial Isolation and Protective Action Distances, located at the back of the book in the green section.

The middle of the book, the orange section, makes up the actual guide pages, **Figure 26-8**. A total of sixty-one guides is given for the more than 4,000 chemicals listed by the DOT. It is this generalization that makes this book valuable to the initial responder but less so to an advanced responder, such as a HAZMAT technician. A HAZMAT technician needs more specific information, which the ERG does not provide.

Each guide takes up two pages and is divided into three major sections: potential hazards, public safety, and emergency response. The guide may also provide some additional information about a specific chemical. It is important for the responder to read the entire guide, and, if referred to the Table of Initial Isolation and Protective Action Distances, to read that section also.

The potential hazards section lists the predominant hazard on the top line. If fire is the major concern, then it will be listed on top; if health is the major concern, then it will be listed on top. The information regarding the health effects provides the route of exposure and any major symptoms of exposure. It also details other potential health and environmental concerns with fire products and runoff. The fire section is not as detailed as the health section but does provide some assistance in identifying potential hazards. It provides information related to the flammability of the product and will state whether the vapors will stay low to the ground or will rise in air. Depending on the product, it may also describe the material's ability to float on water. Products shipped at an elevated temperature as well as those materials that have the ability to **polymerize** will have a notation.

One of the problems with the ERG is the fact that gasoline, which is highly flammable, is in the same guide that is provided for diesel fuel, which to the fire service is combustible. Although the DOT uses 141°F as the difference between flammable and combustible, the fire service still uses 100°F as the

Name of Material	Guide No.	ID No.
Alkyl sulphonic acids, solid, with more than 5% free Sulphuric acid	153	2583
Alkyl sulphonic acids, solid, with not more than 5% free Sulphuric acid	153	2585
Alkylsulphuric acids	156	2571
Allethrin	151	2902
Allyl acetate	131	2333
Allyl alcohol	131	1098
Allylamine	131	2334
Allyl bromide	131	1099
Allyl chloride	131	1100
Allyl chlorocarbonate	155	1722
Allyl chloroformate	155	1722
Allyl ethyl ether	131	2335
Allyl formate	131	2336
Allyl glycidyl ether	129	2219
Allyl iodide	132	1723
Allyl isothiocyanate, inhibited	155	1545
Allyl isothiocyanate, stabilized	155	1545
Allyltrichlorosilane, stabilized	155	1724
Aluminum, molten	169	9260
Aluminum alkyl halides	135	3052
Aluminum alkyl hydrides	138	3076
Aluminum alkyls	135	3051
Aluminum borohydride	135	2870
Aluminum borohydride in devices	135	2870
Aluminum bromide, anhydrous	137	1725
Aluminum bromide, solution	154	2580
Aluminum carbide	138	1394
Aluminum chloride, anhydrous	137	1726
Aluminum chloride, solution	154	2581

Name of Material	Guide No.	ID No.
Aluminum dross	138	3170
Aluminum ferrosilicon powder	139	1395
Aluminum hydride	138	2463
Aluminum nitrate	140	1438
Aluminum phosphate, solution	154	1760
Aluminum phosphide	139	1397
Aluminum phosphide pesticide	157	3048
Aluminum powder, coated	170	1309
Aluminum powder, pyrophoric	135	1383
Aluminum powder, uncoated	138	1396
Aluminum processing by-products	138	3170
Aluminum remelting by-products	138	3170
Aluminum resinate	133	2715
Aluminum silicon powder, uncoated	138	1398
Aluminum smelting by-products	138	3170
Aluminum sulfate, solid	171	9078
Aluminum sulfate, solution	154	1760
Aluminum sulphate, solid	171	9078
Aluminum sulphate, solution	154	1760
Amines, flammable, corrosive, n.o.s.	132	2733
Amines, liquid, corrosive, flammable, n.o.s.	132	2734
Amines, liquid, corrosive, n.o.s.	153	2735
Amines, solid, corrosive, n.o.s.	154	3259
2-Amino-4-chlorophenol	151	2673
2-Amino-5-diethylaminopentane	153	2946
2-Amino-4,6-dinitrophenol, wetted with not less than 20% water	113	3317
2-(2-Aminoethoxy)ethanol	154	1760
2-(2-Aminoethoxy)ethanol	154	3055

Figure 26-7 The blue pages are alphabetical by shipping name. These listings also provide the UN/NA number and the emergency response guide page.

difference. A major difference is seen when handling a gasoline spill as opposed to a diesel fuel spill under normal conditions. By only using the guide as intended, the reader has no way of differentiating between the two products. Although taking this route might be safer, because it assumes a worst-case scenario, there are times when this action is inappropriate and can lead to other problems, such as an unnecessary evacuation for a diesel fuel spill.

The public safety section provides information for the initial public protection options and key issues for the safety of the responders.

> **Note** Only the public safety section of the ERG provides recommendations about how far to isolate the scene and provides some initial strategic goals for the first arriving company.

Chemical Names

Aspon-Chlordane	Octachlorodihydrodicyclopentadiene
Belt	1,2,4,5,6,7,8,8-Octachloro-2,3,3a,4,7,7a-hexahydro-4,7-methanoindene
CD 68	1,2,4,5,6,7,8,8-Octachloro-2,3,3a,4,7,7a-hexahydro-4,7-methano-1H-indene
Chloordaan	1,2,4,5,6,7,8,8-Octachloro-3a,4,7,7a-hexahydro-4,7-methylene indane
Chlordan	Octachloro-4,7-methanohydroindane
g-Chlordan	Octachloro-4,7-methanotetrahydroindane
Chlorindan	1,2,4,5,6,7,8,8-Octachloro-4,7-methano-3a,4,7,7a-tetrahydroindane
Chlor kil	1,2,4,5,6,7,8,8-Octachloro-3a,4,7,7a-tetrahydro-4,7-methanoindan
Chlorodane	1,2,4,5,6,7,8,8-Octachloro-4,7,7a-Tetrahydro-4,7-methanoindane
Chlortox	1,2,4,5,6,7,8,8-Octachlor-3a,4,7,7a-Tetrahydro-4,7-endo-methano-indan (German)
Clordan	Octa-Klor
Clorodane	Oktaterr
Cortilan-Neu	Ortho-Klor
Dichlorochlordene	1,2,4,5,6,7,8,8-Ottochloro-3A,4,7,7A-Tetraidro-4,7-endo-methano-indano (Italian)
Dowchlor	RCRA Waste Number U036
ENT 9,932	SD 5532
ENT 25,552-X	Shell SD-5532
HCS 3260	Synklor
Kypchlor	Tat Chlor 4
M 140	Topichlor 20
NCI-C00099	Topiclor
Niran	Topiclor 20
1,2,4,5,6,7,8,8-Octachloor-3a,4,7,7a-tetrahydro-4,7-endo-methano-indaan (Dutch)	Toxichlor
1,2,4,5,6,7,10,10-Octachloro-4,7,8,9-tetrahydro-4,7-methyleneindane	Velsicol 1068

Excerpted from Richard J. Lewis, Sr. *Dangerous Properties of Industrial Materials,* 8th ed. Van Nostrand Reinhold Co., New York, 1994.

TABLE 26-1

GUIDE 124 GASES - TOXIC AND/OR CORROSIVE - OXIDIZING ERG2000

POTENTIAL HAZARDS

HEALTH
- **TOXIC; may be fatal if inhaled or absorbed through skin.**
- Fire will produce irritating, corrosive and/or toxic gases.
- Contact with gas or liquefied gas may cause burns, severe injury and/or frostbite.
- Runoff from fire control may cause pollution.

FIRE OR EXPLOSION
- Substance does not burn but will support combustion.
- Vapors from liquefied gas are initially heavier than air and spread along ground.
- These are strong oxidizers and will react vigorously or explosively with many materials including fuels.
- May ignite combustibles (wood, paper, oil, clothing, etc.).
- Some will react violently with air, moist air and/or water.
- Containers may explode when heated.
- Ruptured cylinders may rocket.

PUBLIC SAFETY
- **CALL Emergency Response Telephone Number on Shipping Paper first. If Shipping Paper not available or no answer, refer to appropriate telephone number listed on the inside back cover.**
- Isolate spill or leak area immediately for at least 100 to 200 meters (330 to 660 feet) in all directions.
- Keep unauthorized personnel away.
- Stay upwind.
- Many gases are heavier than air and will spread along ground and collect in low or confined areas (sewers, basements, tanks).
- Keep out of low areas.
- Ventilate closed spaces before entering.

PROTECTIVE CLOTHING
- Wear positive pressure self-contained breathing apparatus (SCBA).
- Wear chemical protective clothing which is specifically recommended by the manufacturer. It may provide little or no thermal protection.
- Structural firefighters' protective clothing provides limited protection in fire situations ONLY; it is not effective in spill situations.

EVACUATION

Spill
- See the Table of Initial Isolation and Protective Action Distances for highlighted substances. For non-highlighted substances, increase, in the downwind direction, as necessary, the isolation distance shown under "PUBLIC SAFETY".

Fire
- If tank, rail car or tank truck is involved in a fire, ISOLATE for 800 meters (1/2 mile) in all directions; also, consider initial evacuation for 800 meters (1/2 mile) in all directions.

ERG2000 GASES - TOXIC AND/OR CORROSIVE - OXIDIZING **GUIDE 124**

EMERGENCY RESPONSE

FIRE

Small Fires: Water only; no dry chemical, CO_2 or Halon®.
- Contain fire and let burn. If fire must be fought, water spray or fog is recommended.
- Do not get water inside containers.
- Move containers from fire area if you can do it without risk.
- Damaged cylinders should be handled only by specialists.

Fire involving Tanks
- Fight fire from maximum distance or use unmanned hose holders or monitor nozzles.
- Cool containers with flooding quantities of water until well after fire is out.
- Do not direct water at source of leak or safety devices; icing may occur.
- Withdraw immediately in case of rising sound from venting safety devices or discoloration of tank.
- ALWAYS stay away from tanks engulfed in fire.
- For massive fire, use unmanned hose holders or monitor nozzles; if this is impossible, withdraw from area and let fire burn.

SPILL OR LEAK
- Fully encapsulating, vapor protective clothing should be worn for spills and leaks with no fire.
- Do not touch or walk through spilled material.
- Keep combustibles (wood, paper, oil, etc.) away from spilled material.
- Stop leak if you can do it without risk.
- Use water spray to reduce vapors or divert vapor cloud drift. Avoid allowing water runoff to contact spilled material.
- Do not direct water at spill or source of leak.
- If possible, turn leaking containers so that gas escapes rather than liquid.
- Prevent entry into waterways, sewers, basements or confined areas.
- Isolate area until gas has dispersed.
- Ventilate the area.

FIRST AID
- Move victim to fresh air. • Call 911 or emergency medical service.
- Apply artificial respiration if victim is not breathing.
- Do not use mouth-to-mouth method if victim ingested or inhaled the substance; induce artificial respiration with the aid of a pocket mask equipped with a one-way valve or other proper respiratory medical device.
- Administer oxygen if breathing is difficult.
- Clothing frozen to the skin should be thawed before being removed.
- Remove and isolate contaminated clothing and shoes.
- In case of contact with substance, immediately flush skin or eyes with running water for at least 20 minutes.
- Keep victim warm and quiet. • Keep victim under observation.
- Effects of contact or inhalation may be delayed.
- Ensure that medical personnel are aware of the material(s) involved, and take precautions to protect themselves.

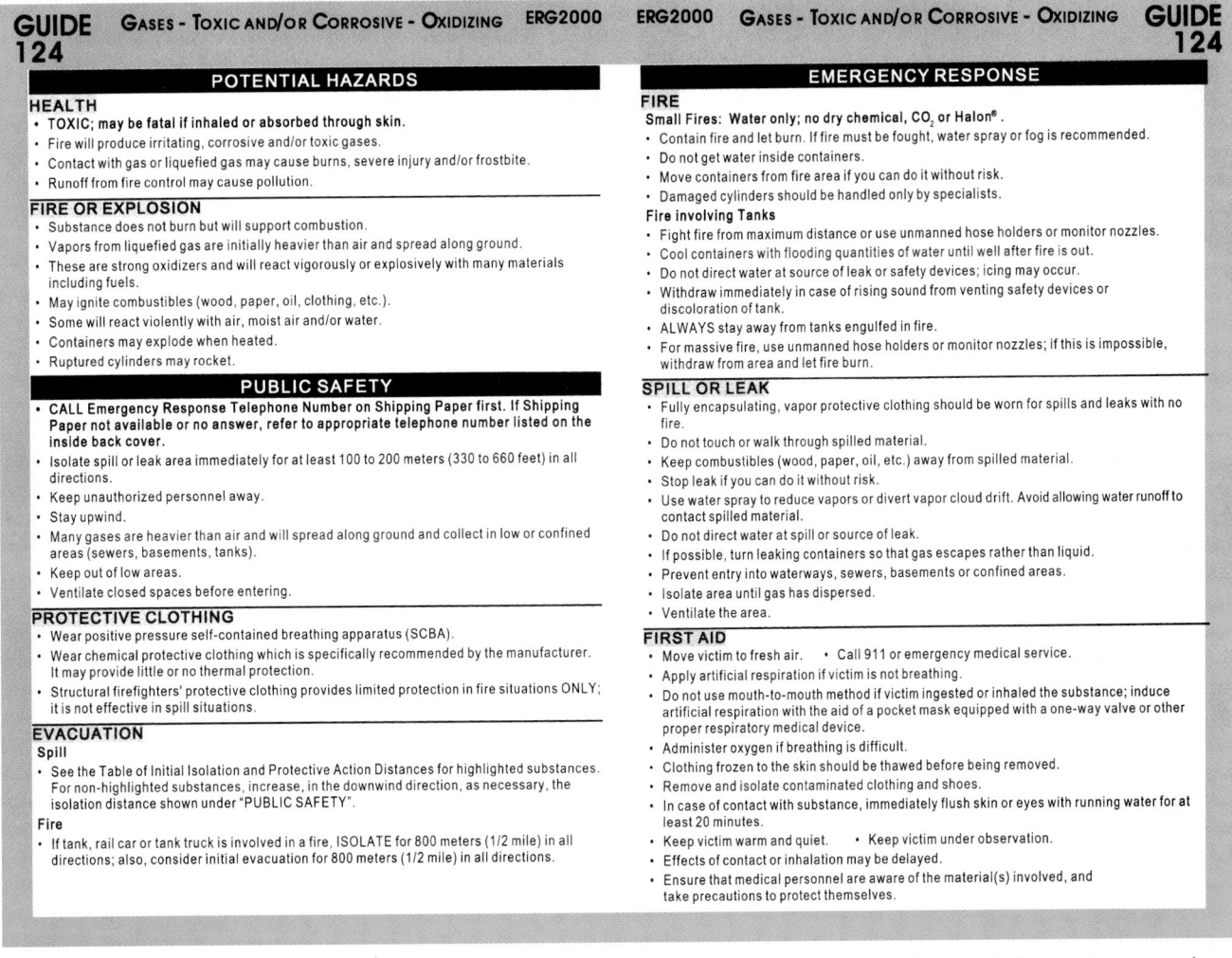

Figure 26-8 The response guide in the orange section provides the responder with basic information regarding the health, fire, and public safety issues. Each part needs to be read completely before taking any action.

It is this section as well as the Table of Initial Isolation and Protective Action Distances, **Figure 26-9**, that make this book one of the necessary tools in the mitigation of a chemical release. It also recommends that responders contact the emergency contact provided on the shipping papers or, if these are not available, then to use one of the contacts listed on the inside back cover. It provides some tactical objectives for handling radioactive substances and some additional considerations regarding these types of incidents.

The public safety section also lists personal protective equipment (PPE) recommendations and provides four basic PPE scenarios. It always recommends the use of positive-pressure SCBA, which is always a good recommendation when responding to chemical releases. The next level in severity is one in which chemical protective equipment is required because turnout gear will provide limited protection. In some cases it will state that turnout gear will provide limited protection, without any recommendation for other PPE. PPE is discussed in Chapter 27 of this text, and many of these same issues will be covered in that chapter. The last recommendation is the suggestion that in some cases chemical protective equipment is a good idea, but that turnout gear is acceptable for fire situations.

This section also provides information regarding initial evacuation distances for large spills and for fires. If the material is listed in the Table of Initial Isolation and Protective Action Distances, the reader will be referred to this section, located at the back of the book in the green section. The isolation distances listed in the DOT book vary from a minimum of 30 feet to a maximum of 800 feet. The minimums are derived for materials that are solid or present little risk to the responder. The higher the toxicity or the risk, the greater the isolation distance. The average distance is 330 feet, which is what should be used for the establishment

TABLE OF INITIAL ISOLATION AND PROTECTIVE ACTION DISTANCES

		SMALL SPILLS (From a small package or small leak from a large package)			LARGE SPILLS (From a large package or from many small packages)		
		First ISOLATE in all Directions	Then PROTECT persons Downwind during-		First ISOLATE in all Directions	Then PROTECT persons Downwind during-	
ID No.	NAME OF MATERIAL	Meters (Feet)	DAY Kilometers (Miles)	NIGHT Kilometers (Miles)	Meters (Feet)	DAY Kilometers (Miles)	NIGHT Kilometers (Miles)
2232 2232	Chloroacetaldehyde 2-Chloroethanal	30 m (100 ft)	0.2 km (0.1 mi)	0.5 km (0.3 mi)	60 m (200 ft)	0.6 km (0.4 mi)	1.6 km (1.0 mi)
2334	Allylamine	30 m (100 ft)	0.2 km (0.1 mi)	0.5 km (0.3 mi)	95 m (300 ft)	1.0 km (0.6 mi)	2.4 km (1.5 mi)
2337	Phenyl mercaptan	30 m (100 ft)	0.2 km (0.1 mi)	0.2 km (0.1 mi)	30 m (100 ft)	0.3 km (0.2 mi)	0.6 km (0.4 mi)
2382 2382	1,2-Dimethylhydrazine Dimethylhydrazine, symmetrical	30 m (100 ft)	0.2 km (0.1 mi)	0.3 km (0.2 mi)	60 m (200 ft)	0.5 km (0.3 mi)	1.1 km (0.7 mi)
2407	Isopropyl chloroformate	30 m (100 ft)	0.2 km (0.1 mi)	0.3 km (0.2 mi)	95 m (300 ft)	0.8 km (0.5 mi)	1.9 km (1.2 mi)
2417 2417	Carbonyl fluoride Carbonyl fluoride, compressed	30 m (100 ft)	0.2 km (0.1 mi)	1.1 km (0.7 mi)	125 m (400 ft)	1.0 km (0.6 mi)	3.1 km (1.9 mi)
2418 2418	Sulfur tetrafluoride Sulphur tetrafluoride	60 m (200 ft)	0.5 km (0.3 mi)	1.9 km (1.2 mi)	305 m (1000 ft)	2.9 km (1.8 mi)	6.9 km (4.3 mi)
2420	Hexafluoroacetone	30 m (100 ft)	0.3 km (0.2 mi)	1.4 km (0.9 mi)	365 m (1200 ft)	3.7 km (2.3 mi)	8.5 km (5.3 mi)
2421	Nitrogen trioxide	30 m (100 ft)	0.2 km (0.1 mi)	0.2 km (0.1 mi)	155 m (500 ft)	0.6 km (0.4 mi)	2.1 km (1.3 mi)
2438	Trimethylacetyl chloride	30 m (100 ft)	0.2 km (0.1 mi)	0.2 km (0.1 mi)	30 m (100 ft)	0.3 km (0.2 mi)	0.8 km (0.5 mi)
2442	Trichloroacetyl chloride (when spilled on land)	30 m (100 ft)	0.2 km (0.1 mi)	0.3 km (0.2 mi)	60 m (200 ft)	0.6 km (0.4 mi)	1.4 km (0.9 mi)
2442	Trichloroacetyl chloride (when spilled in water)	30 m (100 ft)	0.2 km (0.1 mi)	0.2 km (0.1 mi)	30 m (100 ft)	0.3 km (0.2 mi)	1.3 km (0.8 mi)
2474	Thiophosgene	60 m (200 ft)	0.6 km (0.4 mi)	1.8 km (1.1 mi)	275 m (900 ft)	2.6 km (1.6 mi)	5.0 km (3.1 mi)
2477	Methyl isothiocyanate	30 m (100 ft)	0.2 km (0.1 mi)	0.3 km (0.2 mi)	60 m (200 ft)	0.5 km (0.3 mi)	1.1 km (0.7 mi)
2480	Methyl isocyanate	95 m (300 ft)	0.8 km (0.5 mi)	2.7 km (1.7 mi)	490 m (1600 ft)	4.8 km (3.0 mi)	9.8 km (6.1 mi)
2481	Ethyl isocyanate	215 m (700 ft)	1.9 km (1.2 mi)	4.3 km (2.7 mi)	915 m (3000 ft)	11.0+ km (7.0+ mi)	11.0+ km (7.0+ mi)

Figure 26-9 The Initial Isolation and Protective Action Distances table provides recommended isolation distances for the materials that were highlighted in the yellow or blue sections. The table is divided into small spills and large spills.

of an isolation zone for an unknown material, a distance that in an urban setting can present some significant problems. As more information becomes available, this isolation distance can increase or decrease as needed. When dealing with explosives the minimum distance should be 800 feet for the initial isolation distance. The isolation distances for persons downwind are expanded from the initial isolation distances and vary from 0.1 to 7 miles.

The emergency response section provides information regarding fires, spills, and first aid. The fire section is divided between small fires and large fires, and potential tactics that can be used for both. For both types of fires, there is a listing of what type of extinguishing agent may be needed such as water, foam, dry chemical, halon, or CO_2. Specific materials may also list special agents that may be needed. Listed under the large fire section are some tactical considerations as well as some recommendations for specific substances. If the material is carried by tank truck or railcar, the book provides some additional information on fighting those types of fires. In this section some helpful hints are given that apply in many cases involving fires and these types of containers. The hints include the following:

- Fight fire from a distance using unstaffed monitors.
- Withdraw immediately if the sound level from the venting safety device rises or the tank begins to discolor.
- Cool containers with flooding quantities of water until well after the fire is out.

The spill or leak section lists some general tactical objectives and provides some specific information on some substances. It is important to be aware that this section could lead responders into an action that may be beyond their level of training if not used correctly. As an example, most of the guides recommend that the reader can stop a leak if it can be done without risk. The intention is for a responder *trained to the operations level* to be able to close a remote shutoff valve if available. Most readers may incorrectly follow the recommendation and try to stop the leak regardless of the method, a tactic that requires a HAZMAT technician to accomplish safely. Another area of concern is the suggestion that a water spray be used to knock down vapors. If necessary to save lives, using a water spray is an acceptable tactic.

> **Safety** The use of water indiscriminately to knock down "vapors" can create many other problems.

These problems include increasing the spill size, creating runoff problems, possible reactions, and adding an additional hazard to other responders who will try to mitigate the leak. If necessary, as in the case of knocking down anhydrous ammonia vapors that may affect a neighborhood, it is an acceptable tactic. To flow water into the back of a trailer that has a leaking drum, in which the contents have not been identified, is inappropriate and could cause further problems.

The first-aid section provides basic medical treatment and some basic decontamination recommendations for chemical burns. The information is basic and if confronted with a patient, the responder should contact the local hospital or Chemtrec for further assistance.

The Table of Initial Isolation and Protective Action Distances is the green section in the back of the book, **Figure 26-9**. It provides specific isolation and evacuation distances for the materials that were highlighted in the yellow or blue section. This section is further subdivided between small spills and large spills, and both are divided between day and night distances. The criteria used to establish these distances used the following information: the DOT incident database (HMRIS), typical package size, typical flow rate from a ruptured package, and the release rate of vapors from a spill. The DOT chose the average day as being warm and sunny with a temperature of 95°F. They chose sixty-one cities and did a five-year study of the weather to establish a pattern. It was during this study that it was determined that materials will travel farther at night than during the day. They also chose to use an evacuation distance such that in 90 percent of the incidents, the distance used would be too large. But using this scenario, in 10 percent of the incidents the distance would be too small. The distances are a guide for the first thirty minutes of an incident and use a typical day as defined by the DOT. The ERG is a guide to get an evacuation started, and as soon as possible air monitoring needs to be established to definitively define the evacuation distances. In addition an emergency response software package known as CAMEO uses a vapor cloud modeling program known as ALOHA. This modeling program is generally referred to as a plume dispersion model, an example of which is shown in **Figure 26-10**, and can assist with evaluating potential downwind evacuations, using real-time weather and data.

The DOT defines a small spill as a leaking container smaller than a 55-gallon drum or a leak from a small cylinder, whereas a large spill is defined as coming from a container larger than 55 gallons or a large leak from a cylinder. These are further explained in that a leak from several small containers may be considered a large leak. A small leak from a large container may also be considered a small leak.

The last section included in the green section is the List of Dangerous Water-Reactive Materials, **Figure 26-11**. This section provides the evacuation distances for these materials if they contact water. The distances vary from 0.3 to 6 miles. Another helpful item that this section provides is the chemical that this material makes when it contacts water, an unusual piece of information for this type of text.

The last pages are filled with definitions, glossary, and explanations. The inside back cover provides a listing of additional emergency contacts for the United States, Canada, and Mexico. Further discussion regarding these agencies is found later in this chapter.

The firefighter should consider these general reminders when using the DOT ERG:

- It is an excellent source for evacuation and isolation distances.
- It is an excellent source for water-reactive hazards.
- It is an excellent first responder document, one that will keep the responders safe.
- Is a great starting point for the first thirty minutes of the incident.
- The specific chemical information is limited, and in some cases too general for a HAZMAT team, although it is well suited for the first responder level.
- Specific identification of the product is essential to responder and citizen safety.

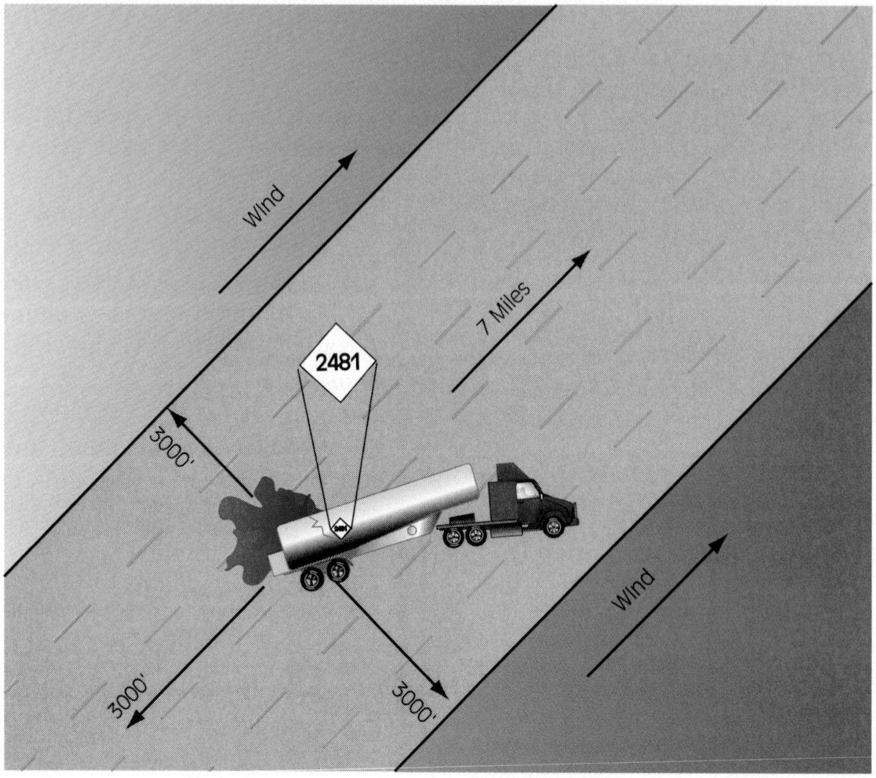

Figure 26-10 A large spill of ethyl isocyanate (ID 2481) is listed as having an isolation distance of 1,700 feet in all directions. In such a case, responders need to protect those persons downwind for 7+ miles.

MATERIAL SAFETY DATA SHEETS

Material Safety Data Sheets, or MSDS, as they are commonly called, are a result of the hazard communication standard, which is OSHA regulation 29 CFR 1910.1200, which became effective in 1994. As stated in Chapter 24, this regulation requires employers who use chemicals above the household quantity to create MSDS. They are also required to develop a hazard communication plan, label all chemical containers, and provide training to employees on an annual basis. This training is based on the chemical hazards the employees will face in the workplace. An MSDS is required to have a variety of information. The amount, quality, and order of information is determined by the manufacturer. The original intent of the MSDS was to protect employees working at the facility, not emergency responders. Although many of the sheets do have applicability and serve a useful purpose, in some cases they may not have the necessary emergency response information. The information that may be found on an MSDS is listed in **Table 26-2**.

The quality of information varies from MSDS to MSDS and from manufacturer to manufacturer. The early MSDS had a lot of good information; however, as litigation has increased, the typical MSDS provides a worst-case scenario, which results in an extremely conservative approach to the handling of the chemical.

Given the choice of using the DOT ERG or an MSDS, the firefighter should rely more on the technical information on the MSDS. The DOT ERG places many chemicals into a few general categories. Although it may be conservative, the MSDS is specific to the chemical. It is important through the preplanning process that responders learn about the chemicals from a facility representative. Responders should work through the MSDS and the ERG and determine prior to the incident which information is most beneficial.

Although many attempts have been made to modify the format of the MSDS and to improve the qual-

TABLE OF WATER-REACTIVE MATERIALS WHICH PRODUCE TOXIC GASES

Materials Which Produce Large Amounts of Toxic-by-Inhalation (TIH) Gas(es)
When Spilled in Water

ID No.	Guide No.	Name of Material	TIH Gas(es) Produced
1726	137	Aluminum chloride, anhydrous	HCl
1728	155	Amyltrichlorosilane	HCl
1732	157	Antimony pentafluoride	HF
1736	137	Benzoyl chloride	HCl
1745	144	Bromine pentafluoride	HF HBr Br_2
1746	144	Bromine trifluoride	HF HBr Br_2
1747	155	Butyltrichlorosilane	HCl
1752	156	Chloroacetyl chloride	HCl
1754	137	Chlorosulfonic acid	HCl
1754	137	Chlorosulfonic acid and Sulfur trioxide mixture	HCl
1754	137	Chlorosulphonic acid	HCl
1754	137	Chlorosulphonic acid and Sulphur trioxide mixture	HCl
1754	137	Sulfur trioxide and Chlorosulfonic acid	HCl
1754	137	Sulphur trioxide and Chlorosulphonic acid	HCl
1758	137	Chromium oxychloride	HCl
1777	137	Fluorosulfonic acid	HF
1777	137	Fluorosulphonic acid	HF
1801	156	Octyltrichlorosilane	HCl
1806	137	Phosphorus pentachloride	HCl
1809	137	Phosphorus trichloride	HCl
1810	137	Phosphorus oxychloride	HCl
1818	157	Silicon tetrachloride	HCl
1828	137	Sulfur chlorides	HCl SO_2 H_2S
1828	137	Sulphur chlorides	HCl SO_2 H_2S

Chemical Symbols for TIH Gases:

Br_2	Bromine	HF	Hydrogen fluoride	PH_3	Phosphine
Cl_2	Chlorine	HI	Hydrogen iodide	SO_2	Sulfur dioxide
HBr	Hydrogen bromide	H_2S	Hydrogen sulfide	SO_2	Sulphur dioxide
HCl	Hydrogen chloride	H_2S	Hydrogen sulphide	SO_3	Sulfur trioxide
HCN	Hydrogen cyanide	NH_3	Ammonia	SO_3	Sulphur trioxide

Use this list only when material is spilled in water.

Figure 26-11 List of Dangerous Water-Reactive Materials from the green section of the ERG (1 of 4 pages).

ity of information, the MSDS has remained the same since inception. There are a couple of recommended formats, and a new format exists for the European chemical industry. It can be anticipated that the United States will have to conform to this format. The information provided in **Table 26-2** is from this format. One of the improvements with this new MSDS format is a section specifically on emergency response procedures and considerations for firefighters. A sample MSDS is provided in **Figure 26-12**.

Information Included on an MSDS

INFORMATION	NOTE
Chemical product and company identification	The name the chemical company uses to identify the product is used near the top of the MSDS. Information related to the manufacturer, such as the address and other contact information, is provided near the top of the first sheet. The MSDS is required to have a twenty-four-hour emergency contact number, although in most cases this number is Chemtrec's.
Chemical composition	The ingredients of the chemical are listed here. If the mixture is a trade secret, there is a provision to exclude this information. If this information is needed for medical reasons, the manufacturer is to provide this information to the physician. In some cases the true identity of the material may never be known, and only the hazards may be provided.
Hazards identification	In this section an emergency overview is provided, for both an employer and emergency responders. This section is lengthy and includes potential health effects, first aid, and firefighting measures. The exposure levels are typically included with the health effects, but may be listed separately. New to MSDS will be a section on accidental releases. This section will be beneficial to emergency responders. Handling and storage considerations, as well as engineering controls including proper PPE, are also provided in this section.
Physical and chemical properties	Specific chemical information such as boiling points, vapor pressures, flash points, and many other specific items are included in this section.
Stability and reactivity	Some chemicals become unstable after a period of time or as a result of poor storage conditions, or they may react with other chemicals. Any information regarding this type of information is provided here.
Toxicology information	The long-term health effects and other concerns regarding acute and chronic exposures are provided in this section.
Ecological information	Information regarding any potential environmental effects is listed here.
Disposal considerations	Although in many cases this section states "follow local regulations," this is the section that outlines any regulatory requirements for disposal.
Transport information	Information regarding the DOT regulations is listed here, typically the hazard class and UN identification number.
Regulatory information	Any other regulations that apply to the use, storage, or disposal of the chemical are listed here. If the chemical is covered by any of the other regulations, this information is listed in this section.

TABLE 26-2

AIR PRODUCTS

Material Safety Data Sheet

Version 1.2
Revision Date 08/12/2003
MSDS Number 300000000099
Print Date 10/01/2003

1. PRODUCT AND COMPANY IDENTIFICATION

Product name : Nitrogen

Chemical formula : N2

Synonyms : Nitrogen, Nitrogen gas, Gaseous Nitrogen, GAN

Product Use Description : General Industrial

Company : Air Products and Chemicals,Inc
7201 Hamilton Blvd.
Allentown, PA 18195-1501

Telephone : 800-345-3148

Emergency telephone number : 800-523-9374 USA
01-610-481-7711 International

2. COMPOSITION/INFORMATION ON INGREDIENTS

Components	CAS Number	Concentration (Volume)
Nitrogen	7727-37-9	100 %

Concentration is nominal. For the exact product composition, please refer to Air Products technical specifications.

3. HAZARDS IDENTIFICATION

Emergency Overview

High pressure gas.
Can cause rapid suffocation.
Self contained breathing apparatus (SCBA) may be required.

Potential Health Effects

Inhalation : In high concentrations may cause asphyxiation. Asphyxiation may bring about unconsciousness without warning and so rapidly that victim may be unable to protect themselves.

Eye contact : No adverse effect.

Skin contact : No adverse effect.

Ingestion : Ingestion is not considered a potential route of exposure.

Chronic Health Hazard : Not applicable.

Exposure Guidelines

1/7

Air Products and Chemicals,Inc — Nitrogen

Material Safety Data Sheet

Version 1.2
Revision Date 08/12/2003
MSDS Number 300000000099
Print Date 10/01/2003

Primary Routes of Entry : Inhalation

Target Organs : None known.

Symptoms : Exposure to oxygen deficient atmosphere may cause the following symptoms: Dizziness. Salivation. Nausea. Vomiting. Loss of mobility/consciousness.

Aggravated Medical Condition

None.

Environmental Effects

Not harmful.

4. FIRST AID MEASURES

General advice : Remove victim to uncontaminated area wearing self contained breathing apparatus. Keep victim warm and rested. Call a doctor. Apply artificial respiration if breathing stopped.

Eye contact : Not applicable.

Skin contact : Not applicable.

Ingestion : Ingestion is not considered a potential route of exposure.

Inhalation : Remove to fresh air. If breathing is irregular or stopped, administer artificial respiration. In case of shortness of breath, give oxygen.

5. FIRE-FIGHTING MEASURES

Suitable extinguishing media : All known extinguishing media can be used.

Specific hazards : Upon exposure to intense heat or flame, cylinder will vent rapidly and or rupture violently. Product is nonflammable and does not support combustion. Move away from container and cool with water from a protected position. Keep containers and surroundings cool with water spray.

Special protective equipment for fire-fighters : Wear self contained breathing apparatus for fire fighting if necessary.

6. ACCIDENTAL RELEASE MEASURES

Personal precautions : Evacuate personnel to safe areas. Wear self-contained breathing apparatus when entering area unless atmosphere is proved to be safe. Monitor oxygen level. Ventilate the area.

Environmental precautions : Do not discharge into any place where its accumulation could be dangerous. Prevent further leakage or spillage if safe to do so.

2/7

Air Products and Chemicals,Inc — Nitrogen

Material Safety Data Sheet

Version 1.2
Revision Date 08/12/2003
MSDS Number 300000000099
Print Date 10/01/2003

Methods for cleaning up : Ventilate the area.

Additional advice : If possible, stop flow of product. Increase ventilation to the release area and monitor oxygen level. If leak is from cylinder or cylinder valve, call the Air Products emergency telephone number. If the leak is in the user's system, close the cylinder valve, safely vent the pressure, and purge with an inert gas before attempting repairs.

7. HANDLING AND STORAGE

Handling

Protect cylinders from physical damage; do not drag, roll, slide or drop. Do not allow storage area temperature to exceed 50°C (122°F). Only experienced and properly instructed persons should handle compressed gases. Before using the product, determine its identity by reading the label. Know and understand the properties and hazards of the product before use. When doubt exists as to the correct handling procedure for a particular gas, contact the supplier. Do not remove or deface labels provided by the supplier for the identification of the cylinder contents. When moving cylinders, even for short distances, use a cart (trolley, hand truck, etc.) designed to transport cylinders. Leave valve protection caps in place until the container has been secured against either a wall or bench or placed in a container stand and is ready for use. Use an adjustable strap wrench to remove over-tight or rusted caps. Before connecting the container, check the complete gas system for suitability, particularly for pressure rating and materials. Before connecting the container for use, ensure that back feed from the system into the container is prevented. Ensure the complete gas system is compatible for pressure rating and materials of construction. Ensure the complete gas system has been checked for leaks before use. Employ suitable pressure regulating devices on all containers when the gas is being emitted to systems with lower pressure rating than that of the container. Never insert an object (e.g. wrench, screwdriver, pry bar, etc.) into valve cap openings. Doing so may damage valve, causing a leak to occur. Open valve slowly. If user experiences any difficulty operating cylinder valve discontinue use and contact supplier. Close container valve after each use and when empty, even if still connected to equipment. Never attempt to repair or modify container valves or safety relief devices. Damaged valves should be reported immediately to the supplier. Close valve after each use and when empty. Replace outlet caps or plugs and container caps as soon as container is disconnected from equipment. Do not subject containers to abnormal mechanical shocks which may cause damage to their valve or safety devices. Never attempt to lift a cylinder by its valve protection cap or guard. Do not use containers as rollers or supports or for any other purpose than to contain the gas as supplied. Never strike an arc on a compressed gas cylinder or make a cylinder a part of an electrical circuit. Do not smoke while handling product or cylinders. Never re-compress a gas or a gas mixture without first consulting the supplier. Never attempt to transfer gases from one cylinder/container to another. Always use backflow protective device in piping. When returning cylinder install valve outlet cap or plug leak tight. Never use direct flame or electrical heating devices to raise the pressure of a container. Containers should not be subjected to temperatures above 50°C (122°F). Prolonged periods of cold temperature below -30°C (-20°F) should be avoided.

Storage

Full containers should be stored so that oldest stock is used first. Containers should be stored in a purpose build compound which should be well ventilated, preferably in the open air. Stored containers should be periodically checked for general condition and leakage. Observe all regulations and local requirements regarding storage of containers. Protect containers stored in the open against rusting and extremes of weather. Containers should not be stored in conditions likely to encourage corrosion. Containers should be stored in the vertical position and properly secured to prevent toppling. The container valves should be tightly closed and where appropriate valve outlets should be capped or plugged. Container valve guards or caps should be in place. Keep containers tightly closed in a cool, well-ventilated place. Store containers in location free from fire risk and away from sources of heat and ignition. Full and empty cylinders should be segregated. Do not allow storage temperature to exceed 50°C (122°F). Return empty containers in a timely manner.

3/7

Air Products and Chemicals,Inc — Nitrogen

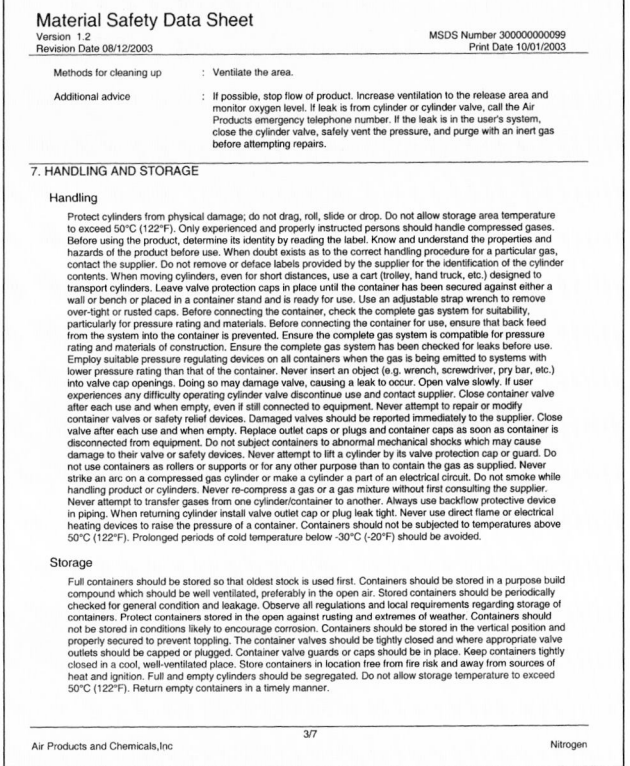

Figure 26-12 An example of an MSDS. *(Courtesy of Air Products and Chemicals, Inc. Presented for illustrative purposes only)*

Material Safety Data Sheet

Version 1.2 — MSDS Number 300000000099
Revision Date 08/12/2003 — Print Date 10/01/2003

Technical measures/Precautions

Containers should be segregated in the storage area according to the various categories (e.g. flammable, toxic, etc.) and in accordance with local regulations. Keep away from combustible material.

8. EXPOSURE CONTROLS / PERSONAL PROTECTION

Personal protective equipment

Respiratory protection	: Self contained breathing apparatus (SCBA) or positive pressure airline with mask are to be used in oxygen-deficient atmosphere. Air purifying respirators will not provide protection. Users of breathing apparatus must be trained.
Hand protection	: Sturdy work gloves are recommended for handling cylinders. The breakthrough time of the selected glove(s) must be greater than the intended use period.
Eye protection	: Safety glasses recommended when handling cylinders.
Skin and body protection	: Safety shoes are recommended when handling cylinders.
Special instructions for protection and hygiene	: Ensure adequate ventilation, especially in confined areas.
Remarks	: Simple asphyxiant.

9. PHYSICAL AND CHEMICAL PROPERTIES

Form	: Compressed gas.
Color	: Colorless gas
Odor	: No odor warning properties.
Molecular Weight	: 28 g/mol
Relative vapor density	: 0.97 (air = 1)
Density at 70 °F (21 °C)	: 0.075 lb/ft3 (0.0012 g/cm3) Note: (as vapor)
Specific Volume at 70 °F (21 °C)	: 13.80 ft3/lb (0.8615 m3/kg)
Boiling point/range	: -320.8 °F (-196 °C)
Critical temperature	: -232.6 °F (-147 °C)
Melting point/range	: -346.0 °F (-210 °C)
Water solubility	: 0.02 g/l

Air Products and Chemicals,Inc — 4/7 — Nitrogen

Material Safety Data Sheet

Version 1.2 — MSDS Number 300000000099
Revision Date 08/12/2003 — Print Date 10/01/2003

10. STABILITY AND REACTIVITY

Stability	: Stable under normal conditions.
Hazardous decomposition products	: None.

11. TOXICOLOGICAL INFORMATION

No known toxicological effects from this product.

12. ECOLOGICAL INFORMATION

Ecotoxicity effects

Aquatic toxicity	: No data available.
Toxicity to other organisms	: No data available.
Mobility	: No data available.
Bioaccumulation	: No data available.

Further information

No ecological damage caused by this product.

13. DISPOSAL CONSIDERATIONS

Waste from residues / unused products	: Contact supplier if guidance is required. Return unused product in orginal cylinder to supplier.
Contaminated packaging	: Return cylinder to supplier.

14. TRANSPORT INFORMATION

CFR

Proper shipping name	: Nitrogen, compressed
Class	: 2.2
UN/ID No.	: UN1066

IATA

Proper shipping name	: Nitrogen, compressed
Class	: 2.2
UN/ID No.	: UN1066

IMDG

Proper shipping name	: NITROGEN, COMPRESSED
Class	: 2.2
UN/ID No.	: UN1066

Air Products and Chemicals,Inc — 5/7 — Nitrogen

Material Safety Data Sheet

Version 1.2 — MSDS Number 300000000099
Revision Date 08/12/2003 — Print Date 10/01/2003

CTC

Proper shipping name	: NITROGEN, COMPRESSED
Class	: 2.2
UN/ID No.	: UN1066

Further Information

Avoid transport on vehicles where the load space is not separated from the driver's compartment. Ensure vehicle driver is aware of the potential hazards of the load and knows what to do in the event of an accident or an emergency.

15. REGULATORY INFORMATION

OSHA Hazard Communication Standard (29 CFR 1910.1200) Hazard Class(es)
Compressed Gas

Country	Regulatory list	Notification
USA	TSCA	Included on Inventory.
EU	EINECS	Included on Inventory.
Canada	DSL	Included on Inventory.
Australia	AICS	Included on Inventory.
South Korea	ECL	Included on Inventory.
China	SEPA	Included on Inventory.
Philippines	PICCS	Included on Inventory.
Japan	ENCS	Included on Inventory.

EPA SARA Title III Section 312 (40 CFR 370) Hazard Classification:
Sudden Release of Pressure Hazard

US. California Safe Drinking Water & Toxic Enforcement Act (Proposition 65)
This product does not contain any chemicals known to State of California to cause cancer, birth defects or any other harm.

16. OTHER INFORMATION

NFPA Rating

Health	: 0
Fire	: 0
Instability	: 0
Special	: SA

HMIS Rating

Health	: 0
Flammability	: 0
Physical hazard	: 3

Prepared by : Air Products and Chemicals, Inc. Global EH&S Product Safety Department

For additional information, please visit our Product Stewardship web site at

http//www.airproducts.com/productstewardship/

Air Products and Chemicals,Inc — 6/7 — Nitrogen

Figure 26-12 (Continued) *(Courtesy of Air Products and Chemicals, Inc. Presented for illustrative purposes only)*

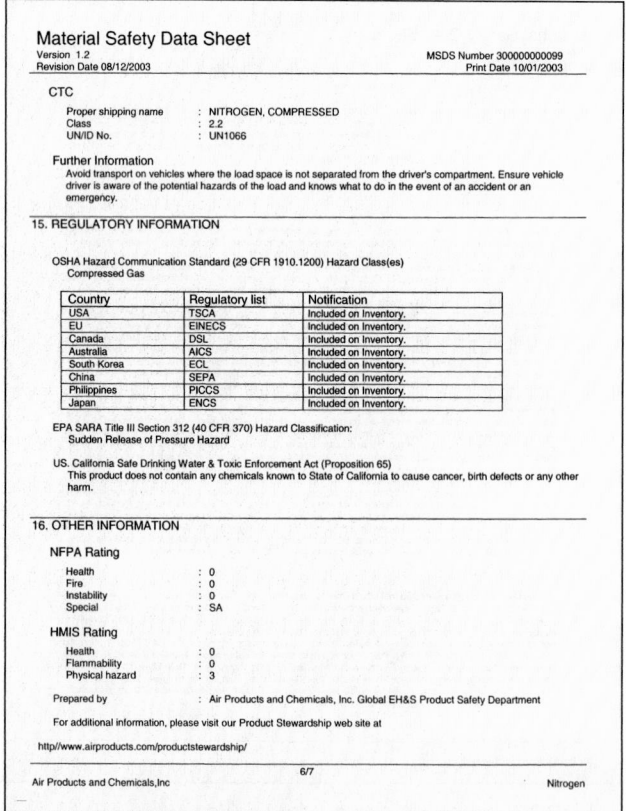

Using the MSDS Wisely

It is always recommended that responders use more than one source of information. The EPA issued a safety alert in June 1999 on the use of the MSDS, and the majority of that bulletin is reproduced here. "PROBLEM: A critical consideration when choosing a response strategy is the safety of emergency responders. Adequate information about on-site chemicals can make a big difference when choosing a safe response strategy. This information must include name, toxicity, physical and chemical characteristics, fire and reactivity hazards, emergency response procedures, spill control, and protective equipment. Generally, responders rely primarily on Material Safety Data Sheets (MSDS) maintained at the facility. However, MSDS may not provide sufficient information to effectively and safely respond to accidental releases. This *Alert* is designed to increase awareness of MSDS limitations, so that first responders can take proper precautions, and identify additional sources of chemical information, which could help prevent death or injury."

Accidents and How the MSDS Relates

In May 1997 a massive explosion and fire occurred at an agricultural chemical packaging facility in eastern Arkansas. Prior to the explosion, employees observed smoke in a back warehouse and evacuated. The facility called local responders and asked for help to control smoldering inside a pesticide container. The local fire department rapidly responded and reviewed the smoldering product's MSDS. The MSDS lacked information on decomposition temperatures or explosion hazards. The firefighters decided to investigate the building. While they were approaching, a violent explosion occurred.

Fragments from a collapsing cinder block wall killed three firefighters and seriously injured a fourth. In April 1995, an explosion and fire at a manufacturing facility in Lodi, New Jersey, caused the death of five responders. The explosion occurred while the company was blending aluminum powder, sodium hydrosulfite, and other ingredients. Even though the material was water reactive, the MSDS for the product advised the use of a "water spray . . . to extinguish fire." The recommendation in the MSDS for "small fires" was to flood with water; however, "small fire" was not defined, the amount of water necessary was not specified, and no information dealt with how to respond to large fires (which can occur during blending processes).

The MSDS only described the hazards associated with the product. In this case, responders needed information on the hazards associated with the reactivity during the blending process (which was significantly different from the product).

Emergency responders should note that the chemical information provided on an MSDS usually presents the hazards associated with that particular product. Once the product is placed in a process some factors may change, resulting in the increase, decrease, or elimination of hazards. These factors may include reactions with other chemicals and changes in temperature, pressure, and physical/chemical characteristics.

MSDS in the Workplace

In 1988 the Occupational Safety and Health Administration (OSHA) required facilities storing or using hazardous chemicals to comply with the Hazard Communication Standard. This standard requires employers to provide employees with an MSDS for every hazardous chemical present on site and to train those employees to properly recognize the hazards of the chemicals and to handle them safely. An MSDS normally provides information on the physical/chemical characteristics and first-aid procedures. This information is valuable for employees to safely work with the chemical. However, the content for the MSDS on emergency response procedures, fire, and reactive hazards may be insufficient for local responder use in an emergency situation. Vagueness, technical jargon, understandability, product versus process concerns, and missing information on an MSDS may increase the risk to emergency responders. MSDS are provided by manufacturers, importers, and/or distributors. MSDS chemical hazard information can vary substantially depending on the provider. Sometimes this discrepancy is due to different testing procedures. However, whoever prepared the MSDS is responsible for ensuring the accuracy of the hazard information. **Table 26-3** summarizes information from various MSDS for the chemical *azinphos methyl* and illustrates how different sources can provide varied and conflicting information.

Comparison of MSDS Data for Azinphos Methyl—AZM

	MSDS–A	MSDS–B	MSDS–C	MSDS–D	CAMEO
Hazard Rating	Health—2 Flammable—0 Reactivity—0	None listed	Health—3 Flammable—2 Reactivity—2	Health—4 Flammable—0 Reactivity—0	Health—3 (extremely hazardous) Flammable—2 (ignites when moderately heated) Reactivity—2 (violent chemical change possible)
Reactivity Hazards	Stable under normal conditions. Hazardous polymerization will not occur.	Depends on characteristics of dust. Decomposes under influence of acids and base.	Stable material. Unstable above 100°F sustained temperature. Hazardous polymerization will not occur.	Releases toxic, corrosive, flammable, or explosive gases. Polymerization will not occur.	Will decompose.
Incompatibility	High temperatures, oxidizers, alkaline substances	Acids and bases	Heat, moisture	Heat, flames, sparks, and other ignition sources	Heat, UV light
Fire Hazards	Vapors from fire may be hazardous.	Combustible, gives off irritating toxic fumes (or gases) in a fire.	Decomposes above 130°F with gas evolution and dense smoke. Dust explosion hazard for large dust cloud.	Containers may rupture or explode if exposed to heat.	Decomposes giving off ammonia, hydrogen, and CO.

TABLE 26-3

SHIPPING PAPERS

Besides the use of placards when chemicals are transported the carrier is required to provide shipping papers for the cargo. The shipping papers generally provide the following information:

- Who is shipping the packages
- Where the packages are going
- Emergency contact information
- The number and weight of the packages
- The name of the materials or at least the hazard class of the materials
- Special notations for hazardous materials

If the vehicle is not carrying hazardous materials, there is no requirement to have any specific information, which at times can be confusing for first responders. When carrying hazardous materials the shipping papers may include a packing group (PG) number listed as a I, II, or III. The lower the number, the worse the chemical is, and it may have special shipping requirements.

Other information may include a reportable quantity (RQ) for a hazardous material. Some chemicals have a threshold of reporting when spilled much like in storage situations and SARA section 304. The RQ will be listed on the shipping papers. If the material is spilled and exceeds the quantity listed, the driver must report the spill to the **National Response Center (NRC)**.

The driver/operator of the vehicle is supposed to keep the shipping papers with the vehicle at all times and should be able to provide them upon request.

> **Caution** Problems occur in accidents where there are multiple shipping papers from a variety of pickups. If the driver has kept all of the previous shipping papers from other days, it can be confusing sorting out what the actual cargo is.

If the driver provides information regarding the cargo, and the shipping papers agree with that information, the results should be confirmed with the shipping company. If all three are in agreement as to what is being shipped, it is probably accurate, but this is not guaranteed. A driver who picked up some illegal waste is not going to put it on a shipping paper, nor placard the vehicle, and probably will not admit to having it. It is obvious that the shipping company will not know about it, so until visually confirmed with appropriate levels of PPE, one must never assume what a cargo consists of.

Mode of Transportation

In highway transportation the shipping papers are called a bill of lading, or most commonly just shipping papers. The papers are supposed to be within arm's reach of the driver, most commonly in a pouch on the driver's door. On tank trucks a duplicate set of papers may be located in a tube attached near the landing gear on the driver's side. In most cases the driver will leave the papers in the truck, and they usually need to be retrieved. If the driver had many pickup stops there will be multiple papers, and some companies put each individual item on a ticket. The hazardous materials are sometimes color coded, as in the case of United Parcel Service (UPS), which uses red tabs to identify HAZMAT packages.

For rail the shipping papers are called the **consist** or **waybill** and are in the control of the engineer, or the conductor if the train has one. There may be two sets of papers, one listing the contents of each car and one that is by car number, starting with the first car past the engine being number 1 (in some cases car number 1 is the last car). In a derailment this numbering system usually goes out the window but in some cases may be useful. Most railcars are identified well with car numbers, and as long as they are upright can usually be identified.

Rail uses a number known as a **Standard Transportation Commodity Code (STCC)**, usually referred to as a "stick" number. It is a seven-digit number, and if it starts with a "49" the material is hazardous. Out of all of the modes of transportation, in rail incidents the engineer is not likely to surrender the papers and will want to accompany the papers anywhere they go. Rail rules require that the engineer be in possession of the papers at all times. If they are lost, engineers face serious personal penalties. There is a computer system in place called Operation Respond that tracks rail shipments. If the car number is known, the computer can identify what the cargo is in a particular car. The system is being slowly implemented in several cities and only involves a handful of carriers at this time.

On a ship the papers are called the **dangerous cargo manifest**, or DCM for short. These papers are in the control of the ship's captain or the master who is second in command of the ship. Regardless of the location of the ship and the crew complement, there is always someone in charge of the ship. There may be a considerable number of crew aboard a ship, but there may be only one person who speaks English. The most difficult time to obtain information regarding cargo is when the ship is loading and unloading, especially for container ships. Although the papers may be on board for all of the containers, it can be some time before the crew actually knows where each container is and what the actual cargo is. One of the major problems is determining what is in each individual container, and this information may be limited or not available at all. Luckily each container is well marked with identification information and after some time the source can be traced to get information.

It is interesting to note that one of the biggest hazards from ships comes from the fuel on board. Even a cruise ship can have up to a million gallons of fuel on board. When dealing with ships, special training is required, because they are extremely hazardous to operate on, and responders can get easily lost and/or fall into deep holds in the ship.

In air shipping the papers are called **air bills** and are in the control of the pilot and usually stored in the cockpit. The type of aircraft (passenger or cargo) will determine the type and amount of materials that can be sent via air. The most common aircraft involved with chemical spills are cargo aircraft, and in many cases exact identification is very time consuming and difficult.

Figure 26-13 Most HAZMAT teams and many fire departments have chemical information on computer and can quickly and easily access this information.

FACILITY DOCUMENTS

Each facility that has chemicals above consumer quantities is supposed to have MSDS as described earlier. In SARA Title III, facilities that are covered by these requirements are also required to submit a Tier 2 form, a listing of all chemicals on site that have an MSDS, and a site plan. If they are using extremely hazardous substances (EHS) they are also required to have an emergency plan. Facilities may also be covered by other regulations and may be required to have other plans in place. Each facility should be able to provide upon request, relatively quickly and without any hassles, an MSDS for a given material. Many facilities will leave a binder of MSDS at the gate or with the security guard. Twenty-four-hour staff should have full access to information as well as emergency contact numbers. The size of the facility will determine the amount of staff available to track this process. The SARA reports and information are updated annually and must be submitted by March 1 of each year. The MSDS may not be revised annually but the facility must ensure their accuracy. Responders should meet with the facility to review these documents and to view the facility.

COMPUTER RESOURCES

Many of the chemical information texts are also available electronically, **Figure 26-13**, typically as a CD-ROM disk. Many first responders use the **Computer-Aided Management for Emergency Operations (CAMEO) program** for chemical information. CAMEO is a software package that combines chemical response information with emergency planning capability. The information within CAMEO is easily accessed and can be used by first responders. CAMEO also has the ability to provide a vapor cloud model, known as plume dispersion, as shown in **Figure 26-13**. When the local data and leak information are input, a program known as Aerial Location of Hazardous Atmospheres (ALOHA) can determine the worst-case scenario for the vapor cloud travel.

There are also CDs that have MSDS on them, and many companies, especially chemical distributors, and universities have their MSDS on the Internet. A simple Internet search of "Material Safety Data Sheets" provides several thousand results of possible MSDS locations. Other programs provide chemical information, but in most cases they are specifically designed for a HAZMAT team and may be above the first responder level. One of the greatest advantages of computer software is the ability to search for a chemical by its name and synonyms quickly.

CHEMTREC

The Chemical Transportation Emergency Center, or Chemtrec, as it is called, is an information service provided by the American Chemistry Council (ACC). A group of chemical manufacturers established the association for several purposes, but one of the outgrowths is the Chemtrec service, which is a free service to emergency responders, paid for by the chemical manufacturers through an annual fee. Although other services also provide chemical information, Chemtrec is the largest and has been providing this service longer than any other company. When a company joins the Chemtrec system, it provides MSDS to Chemtrec as well as emergency contact information. Chemtrec can be considered a large phone book, so if a responder has a company name or chemical, Chemtrec can provide a contact name and number. Many shippers use Chemtrec as their emergency contact point.

If Chemtrec does not have an MSDS on file, they do have other chemical information databases. One of the big advantages is their ability to contact the manufacturer directly, and they will conference call with a responder so that accurate information can be obtained right from the product specialist. If Chemtrec does not have a specific manufacturer's name, they can connect the responder with other specialists who may be able to provide technical assistance. Chemtrec is well connected when dealing with chemical injuries and exposures and can provide medical information as well as provide a contact for further information. One thing Chemtrec does not do is make any regulatory notifications—that is the responsibility of the shipper. When calling Chemtrec, the responder should have the following information available:

- *Caller's name and phone number.* Most responders provide the dispatch phone number, because the on-scene cell phone may be tied up.
- *Name of the shipper or manufacturer.* If this information is not known, Chemtrec will contact a manufacturer for assistance, until the manufacturer is identified.
- *Shipping paper information.* This includes truck or railcar number, the carrier name, the consignee (receiver) name, type of incident, and local conditions.

The Canadian equivalent is called CANUTEC and stands for Canadian Transportation Emergency Center and provides the same services as Chemtrec. In Mexico the Emergency Transportation System for the Chemical Industry is known as SETIQ and provides the same service as Chemtrec and CANUTEC. All three services' phone numbers and other emergency contact numbers are provided in the DOT ERG.

REFERENCE AND INFORMATION TEXTS

Many texts are available from a variety of sources that provide chemical information. In the Additional Resources section at the end of the chapter there is a listing of common texts used by HAZMAT teams, like those shown in **Figure 26-14**. One

Figure 26-14 Many texts provide chemical information. No one text covers all chemicals, and responders should use several references to make sure the information is available and correct.

Information Sources

TEXT	DEVELOPED BY	GENERAL INFORMATION
Chemical Hazards Risk Information System (CHRIS)	Coast Guard	Chemicals discussed are common in water transportation, and it provides water-based response information. This should not be considered a detriment because most of these materials are also common on the highway, and it is a great source of chemical and physical properties.
NIOSH Pocket Guide to Chemical Hazards	National Institute for Occupational Safety and Health (NIOSH)	This is an easy-to-use source of technical information and is one of the best sources of chemical information. It covers the most common industrial materials used today.
Dangerous Properties of Industrial Materials	Sax and Lewis	This is one of the best sources; it has information on chemicals used in industrial applications.
Emergency Handling of Hazardous Materials in Surface Transportation	American Association of Railroads	This text covers materials commonly transported by rail, which also covers most of the items on the highway as well.
Farm Chemical Handbook	Meister Publishing	This is the best source for pesticide, herbicide, and insecticide information; no other text is as up to date or as in depth.

TABLE 26-4

thing to consider is that every piece of apparatus should carry several reference sources. If a HAZMAT team is available in a community, responders probably do not have to have any additional sources of information, because contacting the HAZMAT team by radio or phone is quick and easy. If responders are in an area where the HAZMAT team is not immediately available or it faces considerable travel time, then additional sources of information would be recommended.

The DOT ERG is fairly easy to use, but many of the other reference sources can be complicated and may require some knowledge of chemistry terms, many of which may be above the first responder level. With some study of the preliminary information contained in each text, most responders can readily access the information they need. The other difference is that these texts are chemical specific and the information is for that material only. The reference texts are also slanted toward the group that develops the text. Some examples are included in **Table 26-4**.

A variety of texts is recommended because responders cannot always anticipate the type of material that they may run across. If a responder is in an area where the HAZMAT team has an extended response time, it is recommended that several persons in that department be trained to the technician level, so as to be able to provide some technical assistance to the incident commander. To function as a HAZMAT team, several persons must be trained, but one trained person could provide chemical-specific information and act as a liaison after the arrival of the HAZMAT team.

INDUSTRIAL TECHNICAL ASSISTANCE

Each community usually has a technical specialist in a given field. As an example, in a facility that ships and receives railcars, there is usually someone within that facility who is knowledgeable regarding railcars. In the event of an emergency in that community, it may be wise to use the resources of that person. Almost every town that has some type of industrial facility has a technical specialist within their community.

Many areas of the country, including Baltimore, Houston, and Baton Rouge, have industrial mutual aid groups that are designed to assist each other and the community in the event of a chemical release. In Baltimore this group is called the South Baltimore Industrial Mutual Aid Plan (SBIMAP) and provides technical specialists who assist on the scene and also provides specialized equipment throughout a three-state area. It has been in existence for many years, and one HAZMAT team uses the services of a chemist each time it responds.

Each industrial facility usually has a person responsible for safety and health, and that person is usually a good technical resource. Within the facility there may be chemists or chemical engineers who can provide chemical information. When dealing with chemical exposures and toxicology, many facilities have industrial hygienists who work with those issues daily. These people should be contacted prior to a large incident, to find out where they live and what their availability is. When approached, many of these professionals are more than willing to assist their community.

Lessons Learned

Note Accurate and quick information is essential to every HAZMAT incident. The more information that can be obtained, the easier the incident is to resolve.

First responders should be starting the information process by trying to obtain as much information as possible, so they can make safe and accurate tactical decisions. If waiting for a HAZMAT team, the more information that can be obtained prior to the arrival of the team will make the task easier and less stressful. As essential as this information is, first responders should not take any risks attempting to get this information, such as entering a hazard area to get shipping papers.

KEY TERMS

Air Bill The term used to describe the shipping papers used in air transportation.

Computer-Aided Management for Emergency Operations (CAMEO) Program A computer program that combines a chemical information database with emergency planning software. It is commonly used by HAZMAT teams to determine chemical information.

Consist The shipping papers that list the cargo of a train. The listing is by railcar, and the consist lists all of the cars.

Dangerous Cargo Manifest (DCM) The shipping papers for a ship, which list the hazardous materials on board.

National Response Center (NRC) The location that must be called to report a spill if it is in excess of the reportable quantity.

Polymerize A chain reaction in which the material quickly duplicates itself and, if contained, can be very explosive.

Standard Transportation Commodity Code (STCC) A number assigned to chemicals that travel by rail.

Waybill A term that may be used in conjunction with consist, but is a description of what is on a specific railcar.

REVIEW QUESTIONS

1. What are three common methods of obtaining chemical information?
2. The DOT ERG is intended to be useful for how long at an incident?
3. If given a four-digit UN identification number, which resource provides quick initial information?
4. What mode of transportation uses STCC numbers?
5. Are MSDS required in a home for household quantities of hazardous materials?
6. Can Chemtrec be used for medical information?
7. Which section of the DOT book is used if you have a shipping name?
8. Which reference book provides information related to suggested isolation distances?
9. What advantage does an industrial contact provide at a chemical spill?
10. What cargo probably will not be listed on the shipping papers?

Additional Resources

Bevelacqua, Armando S., *Hazardous Materials Chemistry.* Delmar Learning, a part of the Thomson Corporation, Clifton Park, NY, 2001.

Bevelacqua, Armando S. and Richard H. Stilp, *Terrorism Handbook for Operational Responders,* 2nd ed., Delmar Learning, a part of the Thomson Corporation, Clifton Park, NY, 2004.

Bevelacqua, Armando S. and Stilp, Richard, *Hazardous Materials Field Guide.* Delmar Learning, a part of the Thomson Corporation, Clifton Park, NY, 1998.

Chemical Hazard Risk Information System (CHRIS) U.S. DOT/U.S. Coast Guard. www.chrismanual.com

Emergency Handling of Hazardous Materials in Surface Transportation, American Association of Railroads, 1998.

Farm Chemicals Handbook, Meister Publishing, 1998.

Hawley, Christopher, *Hazardous Materials Air Monitoring/Detection Devices.* Delmar Learning, a part of the Thomson Corporation, Clifton Park, NY, 2002.

Hawley, Christopher, *Hazardous Materials Incidents,* 2nd ed., Delmar Learning, a part of the Thomson Corporation, Clifton Park, NY, 2004.

Hawley, Christopher, *Hazardous Materials Response and Operations.* Delmar Learning, a part of the Thomson Corporation, Clifton Park, NY, 2001.

Henry, Timothy, *Emergency Decontamination for Hazardous Materials Responders.* Delmar Learning, a part of the Thomson Corporation, Clifton Park, NY, 1999.

Lesak, David, *Hazardous Materials Strategies and Tactics.* Prentice Hall, 1998.

Lewis, Richard J., Sr., editor, *Dangerous Properties of Industrial Materials.* Van Nostrand Reinhold, 1994.

Noll, Gregory, Michael Hildebrand, and James Yvorra, *Hazardous Materials Managing the Incident.* Fire Protection Publications, Oklahoma University, 1995.

Schnepp, Rob and Paul Gantt, *Hazardous Materials: Regulations, Response & Site Operations.* Delmar Learning, a part of the Thomson Corporation, Clifton Park, NY, 1999.

Stilp, Richard and Bevelacqua, Armando S., *Emergency Medical Response to Hazardous Materials Incidents.* Delmar Learning, a part of the Thomson Corporation, Clifton Park, NY, 1997.

CHAPTER 27

HAZARDOUS MATERIALS: PERSONAL PROTECTIVE EQUIPMENT

Christopher Hawley, Baltimore County Fire Department (Ret.)

OUTLINE

- Objectives
- Introduction
- Health Hazards
- Exposure Levels
- Types of Personal Protective Equipment
- Lessons Learned
- Key Terms
- Review Questions
- Endnote
- Additional Resources

STREET STORY

Chemical protective clothing has come a long way since I was introduced to chemical hazards one evening in 1974. I had responded to a dumpster fire early one morning. We arrived to see a thick, red-brown cloud issuing from the waste container. Obviously, but unfortunately not to us, we had a hazardous materials situation, not a fire. Undaunted, we went about our tasks, and, after half an hour, we eventually extinguished the fire. However, we all had burning eyes, noses, throats, and severe irritation to our wrists and necks. We called for the assistant chief on duty. He suggested our ailments were from the diesel fumes issuing from the fire truck. We thought otherwise but went back to the station, finished our tours, and went home.

The next morning I was visited by a man from the chemical company who wanted to make sure we were all right. He explained to me that the dumpster had contained discarded nitric acid bottles, and that some had still been full. They had started the fire. He was more concerned about our exposure, however, since we had not worn chemical protective clothing or even SCBA. (Hey, the fire was outside! Remember those days?) He went on to explain that nitric acid had "a delayed reaction to any acute exposure." This, he told me, would have affected the lungs. They might have developed fluid or edema, creating chemical pneumonia. In addition, my throat might have swollen and made it difficult to breathe. I remember saying to him, "You're probably glad I answered the door!" He said that, actually, he was.

Over the years I have looked back on that incident, and I know it is what got me into HAZMAT. It was the driving force behind my signing up for one of the first National Fire Academy Chemistry of Hazardous Materials courses. I know now that chemical protective clothing, and at the very least SCBA, should always be worn in any smoke, vapor, or cloud situation, but hindsight is the best sight, I guess.

A final note for anyone who might be concerned about me. Do not worry; everything works out in the end. A few months later I went to an acetylene-tank fire. I was acting officer and had decided to extinguish the fire with a sodium-bicarbonate extinguisher. Upon closer examination I noticed the fire was out. I turned to my engine crew to tell them not to charge the extinguisher. Sadly, one overanxious fellow discharged the unit right into my face—and me, once again, without SCBA! With my mouth wide open, I got a lung full of baking soda. So we can all relax, because I am balanced—chemically neutral—again!

—Street Story by Mike Callan, President, Callan and Company, Middlefield, Connecticut

OBJECTIVES

After completing this chapter, the reader should be able to:

- Describe the causes of harm.
- Explain the health hazards associated with chemical releases.
- Discuss various chemical-related health terms.
- Identify the various levels of PPE.
- Demonstrate the use of SCBA and other respiratory protection at chemical releases.
- Demonstrate the use of firefighting protective clothing at chemical releases.
- Explain the signs and symptoms of heat stress.

INTRODUCTION

Personal protective equipment, or PPE as it is commonly called, takes on many different shapes and versions. One example is provided in **Figure 27-1**. Even standard firefighter turnout gear (TOG) has many variations and ensembles. When dealing with hazardous materials these configurations are endless.

> **Note** The use of PPE is essential to the health and safety of the first responders.

The failure to use PPE or to use it properly may cause an injury or have fatal effects.

> **Safety** Many firefighter injuries can be prevented by the proper use of PPE, most of which only takes a few seconds to put in place.

The best protection for the first responder above all else is to use the self-contained breathing apparatus (SCBA), which is described in detail in Chapter 7. SCBA offers a substantial amount of chemical protection and should be considered the minimum protection against chemical spills. Firefighter turnout gear offers limited protection against some chemicals, but it is not intended to be used for chemical spill response. Firefighter turnout gear is not tested or approved for chemical spills, and, although it may offer some protection in some chemical environments, it should only be used for immediate life-threatening rescue situations where there will be limited time in the hazard area.

Figure 27-1 PPE is used to protect the wearer from a variety of hazards, but no one type of PPE protects the wearer from all types of hazards. This Level A fully encapsulated suit protects the wearer from chemical hazards but offers little protection from heat, flame, or mechanical hazards. All PPE adds heat stress, and this garment adds significant heat stress on the wearer. Decisions on which PPE to wear are difficult to make and require additional training.

As the need to be in the hazard area increases, so does the need for the proper PPE. This chapter outlines specialized types of PPE, but specific hands-on training is required prior to the use of this PPE. In general, HAZMAT technicians and specialists wear chemical-protective clothing, but persons trained to the operations level may be required to don chemical-protective clothing to perform decontamination operations or other patient-related activities after specific PPE training is provided.

HEALTH HAZARDS

Health issues are a serious concern with hazardous materials emergencies. They can affect every responder and may even be carried home. They can affect the responder immediately, or it may be years before they take a toll. Protecting the body from hazardous materials is easily accomplished by wearing some form of protective clothing and having a basic understanding of toxicology and how chemicals cause harm.

Toxicology

Toxicology is the study of poisons and their effect on the body, and people who study the effect of poisons on the body are known as toxicologists. Although toxicologists are typically found in the medical community, most industrial facilities have industrial hygienists on staff whose responsibility is to protect the workers' health and safety. Their primary focus is on the chemical hazards that exist within the facility. These people have extensive training in toxicology and chemical exposures and are great resources to the emergency services. Because the world of toxic exposures can be complicated and in emergency situations information is needed quickly, a quick consultation with an industrial hygienist may make the incident easier to resolve.

Types of Exposures

There are two types of exposures, **acute** and **chronic**, both of which can have serious health effects.

> **Note** An acute exposure is a quick, one-time exposure to a chemical.

Typically, little damage or effect is noted after an acute exposure. One exception would be exposure to some unusual acids in which a short exposure could cause minor burns. Overall, the human body does well with short-duration exposures and recovers from them, but all chemical exposures should be avoided. A simple example of an acute exposure is that of a nonsmoker who decides to smoke one cigarette. This one-time exposure for most people is not harmful, nor would it cause any long-term health effects.

An example of a chronic exposure, however, is that of the person who smokes three packs of cigarettes a day for twenty years. This person has received doses of cigarette smoke several times a day for a long period of time and is likely to have health problems associated with this chronic exposure. (Note that abnormalities do exist: The person who smoked one cigarette may develop lung problems from that one acute exposure, and the person who chain-smokes for twenty years may never develop health problems associated with the chronic exposure.)

Types of Hazards

As with the IMS, several methods are used to identify possible hazards at a chemical release. The most common one in use today is known by the acronym **TRACEM**, which stands for thermal, radiation, asphyxiation, chemical, **etiological**, and mechanical hazards. Each of the individual hazards has additional hazards that fit within that classification. Much like the **risk-based response** theory, the use of TRACEM assigns a chemical to a risk category so that tactical decisions can be based on that classification. The subcategories within TRACEM are given in **Table 27-1**.

Categories of Health Hazards

Within the TRACEM categories, there are terms that responders should understand, because MSDSs or industrial contacts may describe some chemicals as fitting into one or more of these categories. One of the most commonly used terms is **carcinogen**, which refers to a material with cancer-causing potential. There are two classifications of carcinogens, known and suspected, with the majority being suspected carcinogens. According to the American Chemical Society (ACS), a group that tracks chemicals, there are 21 million chemicals in existence today. Each day the ACS adds 4,000 more chemicals to their listing. The National Toxicology Program under the Public Health Service of the U.S. Health and Human Services Administration issues an annual report on carcinogens. The tenth annual report lists 52 chemicals that are known to cause cancer and 176 chemicals that are suspected of causing cancer. When dealing with chemical spill response, the risk of getting cancer always causes great fear. Firefighters are exposed to a large number of chemicals, many of them known cancer-causing agents, but if firefighters wear their SCBA these exposures are unlikely to cause problems. For older firefighters who worked in earlier years when SCBA was not used as extensively as it is today, cancer is still a leading cause of death, and many retired firefighters have not had the chance to enjoy retirement due to an early death.

> **Safety** Wearing SCBA and avoiding other off-duty activities that involve cancer-causing materials can prevent the development of cancer in most persons.

Another term that is commonly used is **irritant**, which is self-explanatory. An irritant is not corrosive but mimics the effect of a corrosive material in

Subcategories within TRACEM	
Thermal	Both heat and cold hazards fit into this category. If a flammable liquid ignites, it is classified as a thermal hazard. If liquefied oxygen contacts a person's skin, it could cause frostbite and a thermal (cold) burn.
Radiation	Any of the types of radiation, such as alpha, beta, gamma, and neutron, fit into this category.
Asphyxiation	Both simple and chemical asphyxiants fit into this category.
Chemical	This category has a number of subgroups, including poisons and corrosives. The poisons may be called by another name—toxic—and some chemicals are highly toxic. There are specific levels of exposure that determine into which category a chemical fits. Also within this category are convulsants, irritants, sensitizers, and allergens. The reaction a person has to a particular chemical varies from person to person, much in the way some people are allergic to bee stings, while others are not. Chemicals affect some people and not others.
Etiological	Bloodborne pathogens and biological materials exist within this category.
Mechanical	Although not chemical in nature, mechanical hazards exist within the hazard area of a chemical spill, including the standard slip, trip, and fall hazards that one should always be concerned about. Other examples of mechanical hazards would be getting hit from blast particles, such as from a bomb or BLEVE, or a drum falling on a responder.

TABLE 27-1

that it can cause irritation of the eyes and possibly the respiratory tract. One notable characteristic of an exposure to irritants is that the effects are easily reversed by exposure to fresh air. Mace and pepper spray are classified as irritants and may be called incapacitating agents by the military.

Sensitizer is a term that is used to describe a chemical that causes an effect that is in reality an allergic reaction. In most cases an employee can work with a chemical for years and suffer no effects and then one day suffer a severe reaction to the material. Skin reddening, hives, itching, and difficulty breathing are possible symptoms when dealing with a sensitizing agent. Some persons, however, can become sensitive to a chemical after one exposure.

Some chemicals only affect one or more organs and are described as target organ hazards, or they may affect a body system, such as the central nervous system. The effects depend on the individual, dose, concentration, and length of exposure. Some of the target organ descriptions are provided in **Table 27-2**.

Routes of Exposure

The four primary routes of exposure are respiratory, absorption, ingestion, and injection, **Figure 27-2**. The route that is the most commonly associated with causing health effects, both acute and chronic, is the respiratory system route, as shown in **Figure 27-3**.

> **Safety** In almost all cases the respiratory system requires some type of protection. For emergency services workers this is the easiest system to protect because they have easy access to SCBA, should be familiar with it, and are comfortable with its use.

The respiratory system can be affected by gases, vapors, and solid materials such as dust and other particles. In many cases the chemicals may not have any effect on the respiratory system itself, but may enter the body through the respiratory system and affect other organs or body systems. When dealing with respiratory hazards, there are two categories of asphyxiants, simple and chemical. Although the end result is usually the same, the manner in which a person is killed does differ significantly. *Simple asphyxiants simply exclude the oxygen in the air and push it out of the area.* It is not an adverse chemical reaction but simply a matter of something other than oxygen occupying the space in the body where the oxygen should be. Normal oxygen levels are 20.9 percent in air, and the body starts to develop difficulty breathing

Target Organs and Systems

NAME	TARGET ORGAN OR SYSTEM
Central nervous system (CNS) chemicals	Affect the central nervous system and can cause short-term or long-term effects. Commonly short-term memory is lost after exposure to a CNS hazard material. Many of the hydrocarbons cause CNS effects, and people sometimes purposely expose themselves to a CNS agent to receive a "high" from the exposure. In the long term, the brain cells are damaged, never to recover. Neurotoxins essentially cause the same effects.
Peripheral nervous system (PNS) chemicals	Much like CNS chemicals, the PNS chemicals affect the body's ability to move in a coordinated fashion. Exposure to a PNS chemical causes a disruption of the brain's ability to move messages to the other body systems.
Hepatoxins	These types of materials affect the liver and if the exposure is high enough can cause severe damage to the liver.
Nephrotoxins	These materials adversely affect the kidneys.
Reproductive toxins	These toxins affect the ability to reproduce and can cause birth defects. These types of toxins can stay within the body, so they can have adverse effects on a pregnancy even if the exposure occurred a considerable time before the pregnancy.
Mutagens	An exposure to a mutagen may not cause any harm to the people who received the exposure, but the effect could be transmitted to their offspring. Mutagens cause damage to the genetic system and can cause mutations that can become hereditary.
Teratogens	These materials can affect an unborn child. The effects, however, do not happen at a cellular level and would not be passed along from generation to generation.

TABLE 27-2

at less than 19 percent. A person will start to have serious problems at less than 16.5 percent oxygen. Gases such as nitrogen, halon, and carbon dioxide (CO_2) will move oxygen out of the area and cause people in the area to have difficulty breathing. If the concentrations are high enough, death could result. Chemical asphyxiants work in a different manner. They cause a chemical reaction within the body and will not allow it to use the readily available oxygen. The most common chemical asphyxiant is carbon monoxide (CO). When CO is in the air in sufficient quantities, it enters the bloodstream through the lungs. It binds with the hemoglobin in the blood, forming carboxyhemoglobin. Because hemoglobin has an attraction for CO about 225 times that of oxygen, it will not allow the oxygen molecules to bind with the blood, which causes severe health problems and often death.

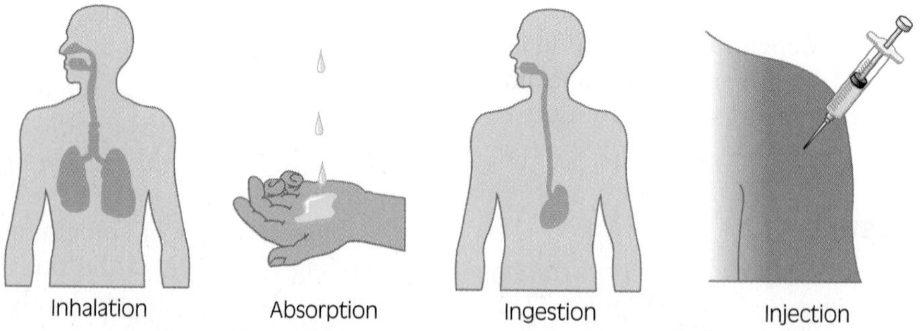

Figure 27-2 Routes of exposure.

Figure 27-3 Respiratory system route of exposure.

The other common route of entry is via skin absorption, because the skin is the body's largest organ. Although some chemicals can cause damage to the skin and may irritate it, this does not mean that it is toxic by skin absorption. The number of chemicals that are toxic by skin absorption is relatively low, but precautions should be taken to minimize contact, and, if at all possible, have no skin contact with chemicals. Skin contact with chemicals can cause burns, rashes, or drying of the skin. The only way to provide skin protection is to wear proper protective clothing that will not allow the chemicals to get onto the skin. Firefighter turnout gear will slow the process down but will not prevent the eventual migration of the chemical to the skin. Effective decontamination is required to ensure that all of the chemical is cleaned from the skin.

The other route of entry is ingestion, which is more common than one would think.

> **Caution** After fighting a fire, responders are typically covered in soot and other debris. Most firefighters do not decontaminate themselves prior to eating or drinking any refreshments that may be available to them at the scene. Without proper cleaning, they will typically ingest some of these products of combustion, none of which is healthy to eat.

The use of SCBA generally prevents this route of exposure, at least until the rehabilitation phase of the incident.

One other route of exposure is injection, although it is not considered to be one of the major routes. Many emergency services workers are exposed to hazardous materials via this route, and for this population it is probably one of the leading routes, after inhalation. The most common material that emergency services workers are exposed to is body fluids or what is referred to as bloodborne pathogens. Other methods of injection are through being near a high-pressure line when it breaks, such as a hydraulic rescue tool fluid line that would inject hydraulic fluid into a person's body. Other than standard infection control practices, there is little protection for these types of exposures except to properly wear full PPE when working in and around situations where exposure to these fluids is possible.

Factors that affect the rate of exposure, regardless of the route, are basic items such as temperature, pulse, and respiratory rate. The higher each of these items is in an individual, the more likely it is that the chemical will have some effect. The damage that chemicals have is based on the equation *Effect = Dose + Concentration + Time.* This equation relates to acute and chronic exposures. A small dose, that is, a small concentration, for a small period of time is not likely to have an adverse effect on a normal human. A large dose at a high concentration over a long period of time will in most cases have an effect. People exhibiting normal vital signs who are exposed to a chemical may not have any effects. But if they jog around the block prior to being exposed to a material, they may be affected.

> **Note** The increased temperature of the body will allow for faster absorption into the body, and the accompanying increased respiratory rate will cause more chemicals to enter the respiratory system.

The increased pulse rate allows for the chemicals to be spread throughout the body faster. In some confined space incidents, where chemicals may have played a factor in injuries and deaths, increased vital signs play a role. In most cases the first victim may be unconscious and therefore the body's system has slowed down, and in some cases the victim went down due to a lack of oxygen. When a person recognizes that a coworker has gone down, the vital signs increase and when trying to perform a rescue the person may actually be exposed to a higher level of the chemical than the coworker he or she is trying to rescue. This is one of the reasons why in some cases the rescuer dies, and the original victim ends up surviving.

EXPOSURE LEVELS

In industry, monitoring for exposures is commonplace and is usually a preventive action, but in the emergency services it can be an afterthought, typically after an incident has occurred. Several different types of exposure values have been issued by a variety of agencies, some of which can be very confusing. Exposure values have been established for the commonly used chemicals and for a variety of situations. The key to preventing exposures is to monitor for hazardous materials and to wear appropriate PPE. The one key agency involved with exposure values is the Occupational Safety and Health Administration (OSHA). The exposure values that they set are the ones that must be followed by all industries, including the fire service because it too is considered an industry.[1] Another organization that issues exposure values is the American Conference of Govenmental Industrial Hygienists (ACGIH), a group that advocates worker safety and conducts a lot of studies regarding chemical exposures. The National Institute of Occupational Safety and Health (NIOSH), a research arm of OSHA, issues recommendations for exposure levels. OSHA is the only agency that provides legally binding exposure values; all of the others are recommendations. In some OSHA regulations the employer is required to use the lowest published exposure values, which in many cases are not OSHA's own values. When dealing with emergency situations, it is always recommended that responders follow the lowest published values.

The exposure values are based on an average male and are for an industrial application. The exposure values are typically based on an eight-hour day, forty-hour workweek with a sixteen-hour break between exposures. The values that are issued for the various substances are typically conservative; in a given population it is not atypical to find someone who is sensitive to a chemical at a lower value than the rest of the group. Each value has an extra margin for safety built in, ranging from 1 to 10,000 times the actual value. The exposure values are typically listed in parts per million (ppm) or as milligrams per meter cubed (mg/m^3). Explanations of these terms are provided in **Figure 27-4**.

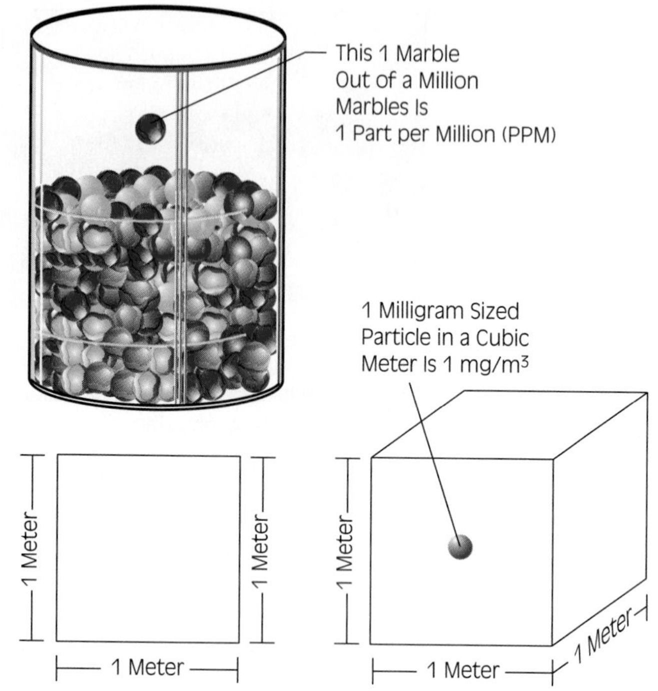
This 1 Marble
Out of a Million
Marbles Is
1 Part per Million (PPM)
1 Milligram Sized
Particle in a Cubic
Meter Is 1 mg/m^3
1 Meter
1 Meter
1 Meter
1 Meter
1 Meter
1 Meter

Figure 27-4 Explanation of units of measure.

The most common exposure values are expressed in ppm, but for some materials (typically solids) the values may be expressed as mg/m^3. The values are generally for a period of time, usually eight hours. OSHA refers to this eight-hour exposure value as the **permissible exposure limit (PEL)**. The ACGIH refers to this eight-hour exposure as the **threshold limit value (TLV)**. Both are average exposures over the eight-hour period. A worker can be exposed to more than the PEL or TLV as long as at the end of the day the exposure value is less than the PEL or TLV. In some cases these exposure values are called time weighted averages (TWAs), and they may be expressed as the OSHA-TWA or the ACGIH-TWA, which is an eight-hour daily average exposure. NIOSH issues **recommended exposure limits (RELs)**, which are for a ten-hour day, forty-hour workweek. These exposure values are outlined in graphic form in **Figure 27-5**.

Other values that may be listed for chemicals are the **ceiling levels**, generally referred to as PEL-C or TLV-C. These provide an amount that is the highest level to which an employee can be exposed. When figuring an average, there are times when employees are going to be exposed to chemicals at a level higher than the PEL or TLV, but there are also times when they will be exposed to less than those levels. As long as their exposure average is less than the PEL or TLV, they are acceptable. Certain chemicals will cause effects at levels that may be obtained through worker exposure and would not be considered safe, but the overall exposure would fall below

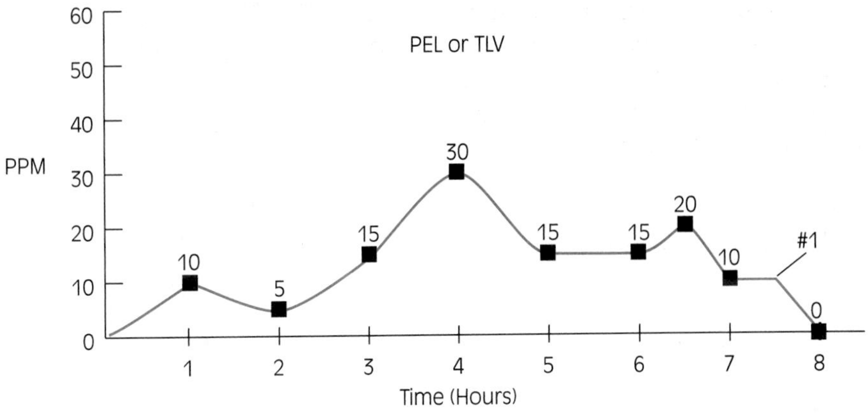

#1

PEL or TWA

Hour	Exposure
1	10
2	5
3	15
4	30
5	15
6	15
7	20
8	10
Total 8	120 Avg Exposure 15 ppm

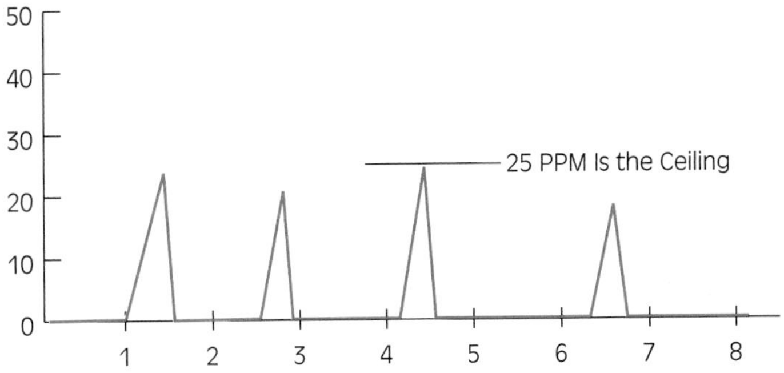

STEL: Allows four exposures of 15 minutes in length, with a minimum of an hour break in between. The ceiling is the highest exposure, no matter the time.

Figure 27-5 Exposure values.

the PEL or TLV. To avoid unsafe levels, the safety organizations may attach a ceiling level to an exposure value, and that is the highest level that the employee can be exposed to, regardless of what the end average is.

Another exposure value is known as the **short-term exposure limit (STEL)**. This value is assigned to a fifteen-minute exposure. An employee can be exposed at this level for fifteen minutes and then is required to take an hour break from the exposure. The employee can do this four times a day without any adverse effects. NIOSH is also using an excursion value that is coupled with a time limit, five to thirty minutes typically. At this level, an employee can enter an environment one time and not suffer any effects.

The last value can be confusing because it is called the immediately dangerous to life or health (IDLH) value. One would think that being exposed to a chemical at the IDLH level would mean that death may be imminent. In reality this is a value that is the maximum airborne concentration that an individual could escape and not suffer any adverse effects. The actual definition does not match the legal definition. At the IDLH level, emergency responders need to be using SCBA as an absolute minimum, and if the chemical is toxic by skin absorption then a fully encapsulated gastight suit (Level A) must be used.

Other values that a responder may see are called **lethal doses (LD_{50})** or **lethal concentrations (LC_{50})**. The LD is for solids and liquids, and the LC is for gases. In most cases animal studies provide these values, but some are derived from human studies, suicides, and murders. The 50 attached to the LD or LC means 50 percent of the exposed population. In the studies a certain number of test subjects were exposed to a low level of chemicals. After a period of time, the test subjects were studied for any adverse effects. Another group was exposed to a higher level of the chemical. When the subjects were exposed to a certain level and 50 percent of the test subjects died, this established the LD_{50} or LC_{50}. With terrorism, emergency responders are having to use some military data, much of which is based on LD_{50} or LC_{50} type studies. The military values are generally expressed as **LCt_{50}** or lethal concentration to 50 percent of the population, with "t" representing time, usually expressed in minutes. The military also uses **ICt_{50}**, which is the incapacitating concentration to 50 percent of the population in a certain amount of time. **Table 27-3** lists exposure values for some chemicals.

Although the exposure values can be confusing it is important to know what each of the values means, and emergency responders should be aware of the exposures they receive during incidents.

> **Safety** Air monitoring of these exposure levels is a good way to ensure responder safety.

It is generally recommended for emergency responders to use the TLVs or PELs as the point at which SCBA should be utilized. Emergency responders are faced with a lot of chemical exposures, and at these levels their safety can be ensured if proper PPE is worn.

TYPES OF PERSONAL PROTECTIVE EQUIPMENT

The most common type of PPE for the firefighter is turnout gear. Full turnout gear is defined as helmet, hood, coat, pants, boots, gloves, PASS, and SCBA, as shown in **Figure 27-6**. To be fully protected, all of this equipment must be in use. All of the snaps, zippers, and closures must be used to offer optimal protection. Firefighter TOG offers protection against heat and water. The exact amount of protection depends on the type of gear used. Firefighter TOG is not certified for chemical contact, nor should it be used for chemical protection. Some of the latest style gear is certified to protect against bloodborne pathogens, but to be sure, the firefighter should check for the NFPA certification.

Self-Contained Breathing Apparatus

With regard to chemical exposures, SCBA offers a protection factor of 10,000. *What this means is that a person wearing SCBA has a survivability rating 10,000 times greater than a person who is not wearing SCBA. To offer this protection, the SCBA must be fitted properly to the person and must be a positive pressure device.* If these two factors are not followed then the protection factor can be consider-

Exposure Values

CHEMICAL	PEL (PPM)	IDLH (PPM)	LCT_{50} (PPM)	ICT_{50} (PPM)
Sarin	0.000017	0.03	12	8
Mustard	0.0005	0.0005	231	21.5
Acetone	1000	2500	N/A	N/A
Acrolein	0.1	2	N/A	N/A
Ammonia	50 (ST)	300	N/A	N/A
Chlorine	1	10	N/A	N/A
Ethion	0.02	N/A	N/A	N/A
Hydrogen sulfide	20 (ceiling)	100	N/A	N/A
Carbon monoxide	50 (35 ppm NIOSH)	200 (ceiling)	N/A	N/A

N/A, not applicable.

TABLE 27-3

ably less than the 10,000. A positive pressure SCBA has an airflow in the face piece all the time, and it maintains positive air pressure inside the mask to keep contaminants out. For firefighting or chemical spill response, it is imperative that the SCBA be a positive pressure unit and that it be activated automatically without any intervention on the part of the wearer.

Safety A firefighter's chance for survival at a hazardous materials incident is dramatically improved when wearing SCBA, and it should be considered the minimum when dealing with chemical spills.

Although various types of SCBA equipment are available, the most common for chemical spill response is a sixty-minute type, which on an average allows twenty to thirty minutes of work time for a HAZMAT team member. When determining the use of SCBA, one has to consider the time it takes to enter the hazard area, working time, time to leave the area, and time to be decontaminated and undressed. In general, thirty- and forty-five-minute air supplies are inadequate for spill response.

Figure 27-6 The use of SCBA offers tremendous protection against heat and chemical hazards. Responders who enter any environment that may have a chemical present should always use SCBA.

Some teams and persons who deal with waste sites may use **supplied air respirators (SARs)**, which have some advantages over SCBA. They do not have the weight of the SCBA, but do restrict movement somewhat due to the hoseline. Some logistical issues are associated with the use of SARs, but for long-term incidents they are of assistance.

Some response teams use **air-purifying respirators (APRs)** for minor spills, as shown in **Figure 27-7**. Although they offer some advantages and are common in industry they are not commonly used within the emergency services. With terrorism and bloodborne pathogen issues becoming more commonplace, however, the use of APRs will increase.

APRs also require **fit testing** just like the SCBA to ensure that the respirator fits and will offer the wearer protection. If a responder chooses to use an APR at an incident, a decision flowchart must be followed. In general, the responder must know the

Figure 27-7 Some HAZMAT teams use air-purifying respirators that filter out possible contaminants. However, they do not offer as much protection as SCBA.

chemical, and it must have good warning properties, such as a characteristic odor, and the responder has to identify the amount in the air. The amount of oxygen in the area must be verified and good levels must be maintained. In addition, a variety of cartridges can be used, and when dealing with a variety of chemicals a large stock of these cartridges must be maintained. With the exception of waste sites, minor releases, and possibly some decontamination work, SCBA is much easier to use.

Problems with SCBA include extra weight, fatigue, lack of full visibility, lack of mobility, contribution to heat stress, need to refill air supply, and other limiting factors. It is for this reason that many HAZMAT teams choose the other types of respiratory protection. The type of SCBA used will determine the amount of weight added to the responder, and in general the longer the work time the higher the weight of the apparatus. This added weight adds to the overall heat stress on the user. When using SCBA the wearer has limited vision because the face piece does not allow for a full spectrum of vision. When a chemical protective suit is added this limited field of vision gets even smaller. The use of SCBA in some applications such as confined spaces limits wearers as to how they can move about the confined space. For people who may be slightly claustrophobic with SCBA, there is an additional layer of stress placed on them when donning chemical protective clothing. The users of SCBA and chemical protective clothing need to be medically cleared to function with this type of PPE and should receive periodical medical exams, according to the applicable regulations.

Chemical Protective Clothing

There are four basic levels of chemical protective clothing, but these levels are broken down into further components. The levels are assigned letters to signify their protection levels. Both OSHA and the EPA use Level A, B, C, and D with Level A being the highest level of chemical protection.

Since the establishment of these levels, many changes have been made to PPE styles and types. When HAZMAT teams first started they had reusable protective clothing. It was used, cleaned, tested, and then reused. But the integrity of the suit became questionable after each use, so the suit manufacturers decided to switch to disposable one-time-use garments. Today, most teams use disposable suits, although some teams maintain at least one type of reusable garment, usually a Teflon fabric suit.

All chemical protective clothing must be checked prior to use for compatibility with the chemical that has been spilled.

Caution No one suit fits all chemicals.

Some of the suits that are NFPA certified do come close, but there are a couple of chemicals that they are not to be used with. Chemical compatibility is based on the **permeation** of a chemical through the fabric of the suit. Permeation is the movement of a chemical through the fabric on the molecular level, **Figure 27-8**. In an acceptable suit, no damage to the fabric should result from a permeation test, nor should anything be seen visually. Imagine wrapping garlic in plastic wrap; initially there is no smell, but after a few hours the scent of garlic has permeated the refrigerator as if it had not been wrapped at all. Compatibility charts are provided by the manufacturer for the chemicals against which they have tested the fabric.

Firefighter Fact Permeation is based on an 8-hour day or 480 minutes. When a department's inventory has more than one suit fabric, it is best to choose a fabric that is given a value of >8 hours, because that means that in tests the fabric showed no breakthrough in 8 hours. After the fabric reaches 8 hours, the test is usually halted, unless specific times are indicated by the manufacturer.

Most fabrics are tested against a battery of chemicals as specified by the American Society for Testing and Materials (ASTM) or the NFPA. This battery of chemicals represents the majority of the chemical families that responders are most likely to encounter.

Two other areas that are of concern when dealing with PPE are degradation and penetration. The degradation of a suit involves the physical destruction of the suit, leaving a hole or damage. Penetration is the movement of chemicals through natural openings such as zippers, glove/suit interface points, and other areas of the suit. It does not involve damage but is an area where chemicals may enter the suit.

Level A Protective Clothing

Level A protective clothing, as shown in **Figure 27-9**, is thought of as providing the highest level of protection against chemical exposure.

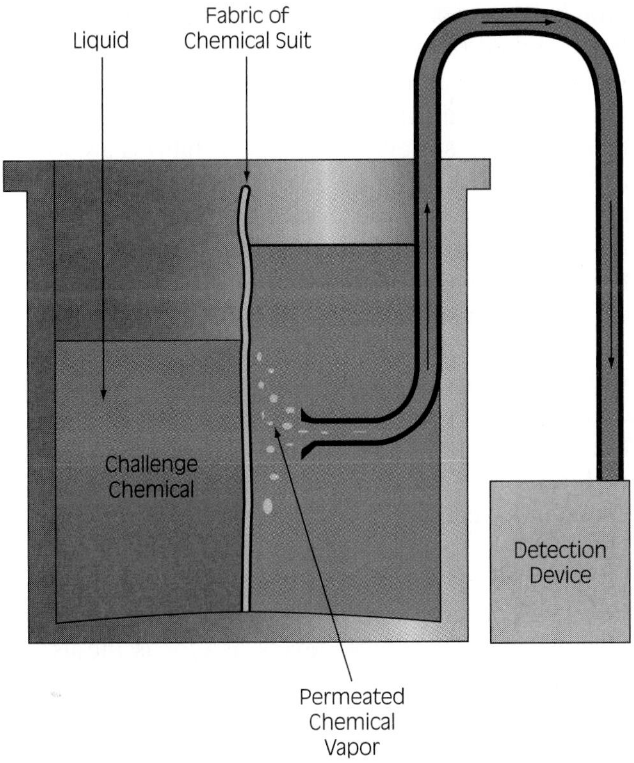

Figure 27-8 When conducting a permeation test, the fabric splits a test container, and a measurement device is used to see if the chemical goes through the fabric.

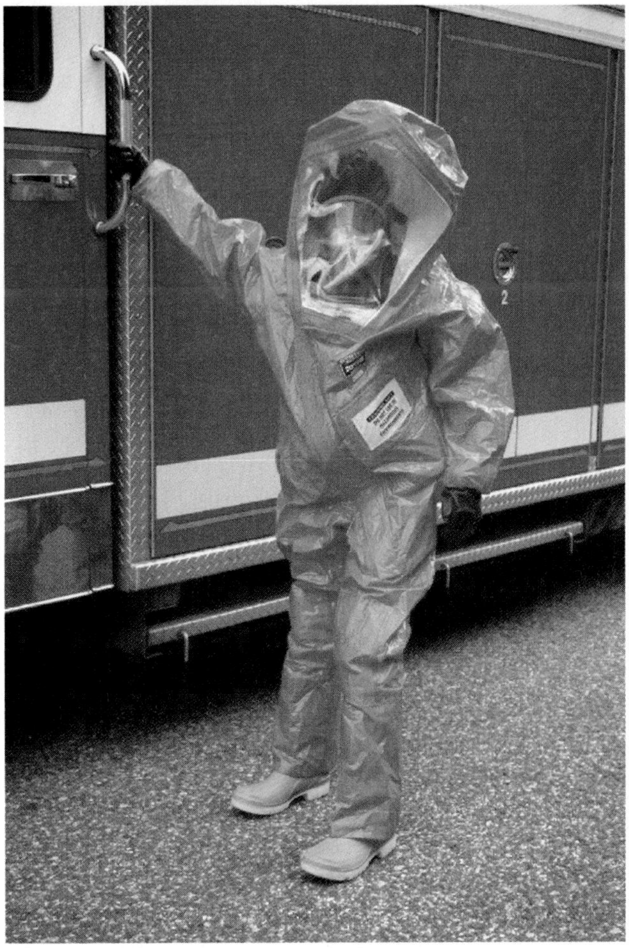

Figure 27-9 The Level A suit offers good protection from chemicals. It is a vapor-tight suit and is permeation resistant to most chemicals.

Note While it is true that Level A clothing offers the maximum level of chemical protection, it is also the leader as far as causing heat stress and physical and psychological stress on the responder wearing the garment.

It is a fully **encapsulated suit** and sometimes is called an encapsulated suit instead of a Level A suit.

To be considered a Level A the suit must have attached gloves and attached boots, and the zipper must be gastight. The suit is designed not to allow any gases to penetrate the garment, and by accomplishing this, it becomes liquid tight as well. Because materials cannot get inside the suit, including air, and gases cannot escape, the person wearing the suit needs to have an SCBA on the whole time. The suit does have relief valves that vent the exhaled air after a certain pressure buildup.

The requirement to use Level A suits within the HAZWOPER regulation is when a firefighter is entering an atmosphere above the IDLH value and the chemical is toxic by skin absorption. There are several occasions, however, where the use of Level A clothing would be recommended for some chemicals that do not meet that definition. On occasions where a responder may be potentially covered with a toxic or corrosive material over the whole body, a Level A suit would be advisable.

Streetsmart Tip When wearing a Level A suit prehydration is highly recommended, and dressing should take place in a cool quiet area, preferably in the shade.

During emergency operations all members must be monitored. Lack of full visibility and heat stress are major concerns when using Level A protective clothing. The Level A garment typically consists of the following components:

- Encapsulated suit with attached gastight gloves and boots
- Inner and outer gloves

- Hard hat
- Communication system
- Cooling system
- SCBA
- PBI/Nomex coveralls
- Overboots

To be an NFPA 1991 (Encapsulated Suit Specifications) certified Level A suit, the suit must have some flash fire resistance. To assist in compliance, newer suits use a blended fabric that offers chemical resistance as well as the flash resistance. Older style suits use a flash suit overgarment that is made of aluminized PBI/Kevlar fabric. This flash protection is not intended for firefighting, but it offers three to thirteen seconds of protection when involved in a flash fire.

Level B Suits

Within the **Level B protective clothing** family, there is a lot of variety in suit types, as shown in **Figure 27-10**. Although the EPA and OSHA acknowledge two basic types, a large number of styles are available. The two basic types are coverall style and encapsulated, but even these have subvarieties. The encapsulated style of Level B suits is similar to the Level A style, but does not have attached gloves, nor is it vapor-tight. It typically has attached booties, and the SCBA is worn on the inside. Some manufacturers provide glove ring assemblies that allow for the gloves to be preattached during storage. A variety of fabrics are available for the Level B style of suits, and compatibility is important. *The Level B encapsulated suit is the workhorse of HAZMAT teams; it is the most common suit used.*

The encapsulated Level B suit is sometimes referred to as a Bubble B or a B plus suit. These suits have some of the same heat stress issues associated with them as do the Level A suits, even though they are lighter. Other styles of Level B suits include a two-piece garment consisting of a jacket and pants, usually bib-overall style. The coverall style Level B suit may have attached booties or a hood, but there are a number of styles for a variety of uses. The one item that makes the Level B suit different from the lower levels of PPE is the use of SCBA. A Level A suit is a gastight suit, whereas the Level B is intended for splash protection and the SCBA offers the respiratory protection.

A Level B suit ensemble consists of these components:

- Level B suit
- Hard hat

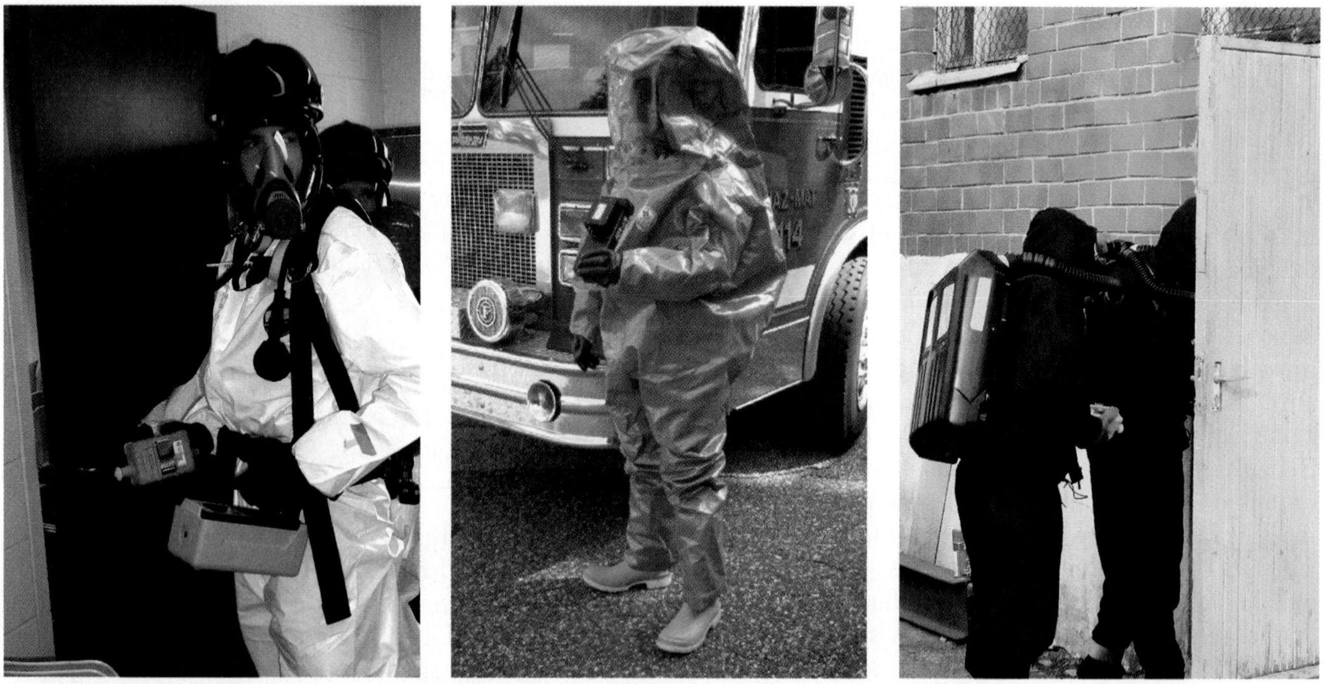

(A) (B) (C)

Figure 27-10 The photos here all represent Level B suits: (A) a coverall style for law-enforcement officers; (B) an encapsulated style, which is not gastight; and (C) a military-designed two-piece garment worn by tactical officers. The respiratory protection is a rebreather style, which provides a four-hour air supply.

- Inner and outer gloves
- SCBA
- Communication system
- Outer boots
- Nomex/PBI coveralls

Level C Suits

A Level C suit, as shown in **Figure 27-11**, incorporates the use of an air-purifying respirator within the ensemble. For obvious reasons an APR cannot be used within an encapsulated suit, but it can be used with the other styles. A Level C suit can be a coverall or a two-piece garment. *A Level C suit is used where splashes may occur, but where respiratory hazards are minimal and are covered by the use of an APR.*

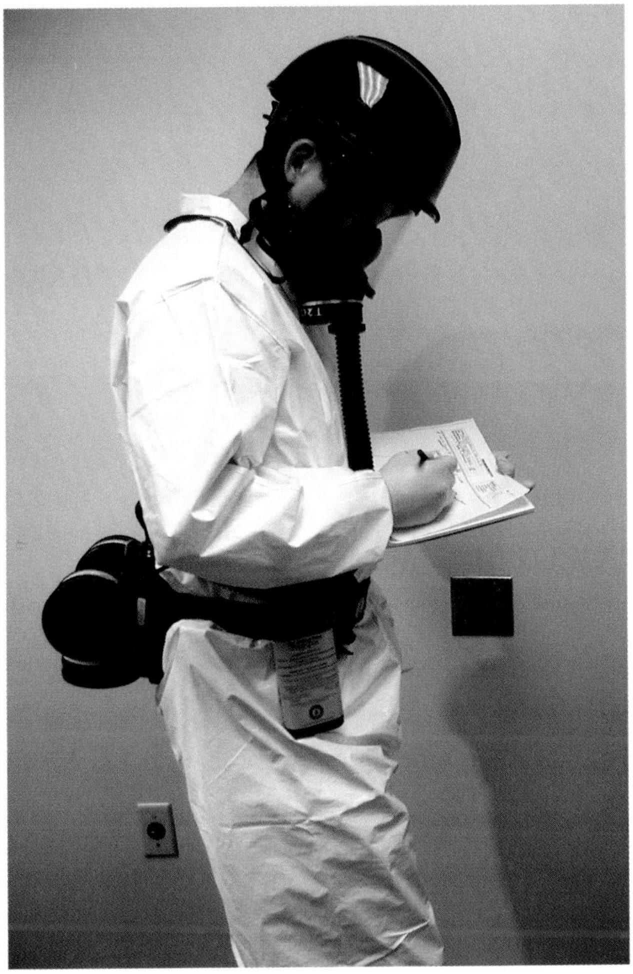

Figure 27-11 The difference between Level C and Level B suits is the respirator. When using an APR or a powered APR (PAPR) as shown, the suit is a Level C; when using an SCBA it is a Level B. An encapsulated suit and an APR cannot be used together because the wearer of an APR needs sufficient oxygen, which may be lacking in an encapsulated suit.

When using an APR, an extensive listing of requirements must be met.

A Level C garment consists of these components:

- Level C suit
- Air-purifying respirator
- Hard hat
- Inner and outer gloves
- Outer boots
- Nomex/PBI coveralls

Level D suit

A Level D suit, shown in **Figure 27-12**, is actually regular work clothing. It is used when respiratory protection is not required and splashes are not a concern. The Level D suit provides no chemical protection, but does offer protection against other workplace hazards. Level D protective clothing consists of these items:

- Work clothes
- Hard hat
- Chemical/work gloves
- Safety glasses
- Safety shoes/boots

High-Temperature Clothing

The two basic types of high-temperature clothing are proximity and fire entry gear. A set of proximity gear is shown in **Figure 27-13**. *The most common use for this type of gear is in airport firefighting and flammable liquids firefighting.* The entry suits may also be used in high-temperature applications within industry, such as steel making.

High-temperature gear is usually identified by its characteristic aluminized outer shell. This shell is usually attached to a PBI/Kevlar fabric that offers a higher heat resistance than normal structural firefighting TOG. This proximity gear is named for its ability to allow the wearer to get close to the burning liquid. It offers protection for temperatures up to 300° to 400°F.

Fire entry gear is designed to allow the wearer to enter a fully involved fire area for a period of thirty to sixty seconds over the life of the suit. The thought process was that it would allow for the rescue of trapped victims, which firefighters now know would not be alive if that type of garment is needed. There are certain applications for this gear involving industrial incidents where high temperatures would be found, but these are rare and most departments no longer carry this type of

Figure 27-12 Level D is standard work clothing and may include chemical gloves, goggles, and boots.

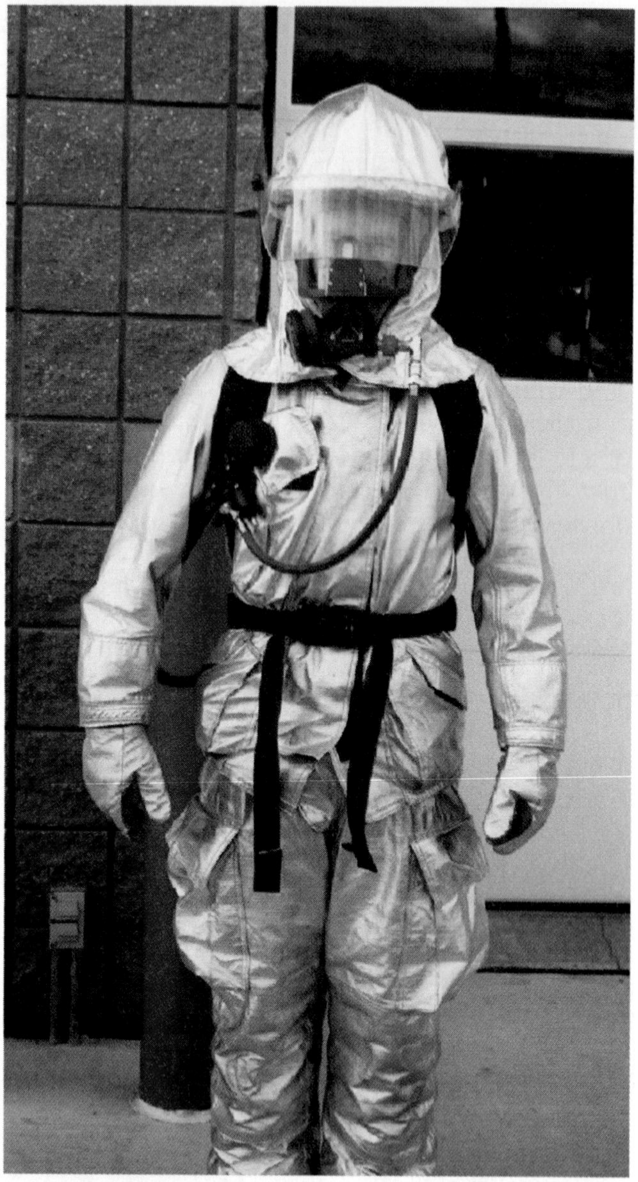

Figure 27-13 Proximity gear is used for firefighting of flammable liquids and is commonly used by airport firefighters and industrial firefighters.

gear. Fire entry gear can be used in temperatures ranging up to 2000°F.

Low-Temperature Clothing

When dealing with cryogenics the responder must wear protective clothing that protects the wearer against very cold temperatures. Propane, once it is released from a cylinder to the atmosphere, can reach temperatures well below 0°F. Cryogenics by their definition are at least −150°F and are usually colder. Standard protective clothing is not effective against these types of materials. Responders should take precautions against cold stress, and prevent hypothermia. Although no full protective clothing ensemble is available for dealing with cold materials, various types of gloves, **Figure 27-14**, and gauntlets and aprons are available. Layering of clothing can offer some additional protection. Unfortunately in addition to the cold, cryogenics usually present other hazards such as fire, corrosion, toxicity, and asphyxiation.

Limitations of Personal Protective Equipment

There are four basic limitations to protective clothing, and they apply across the board from EMS infec-

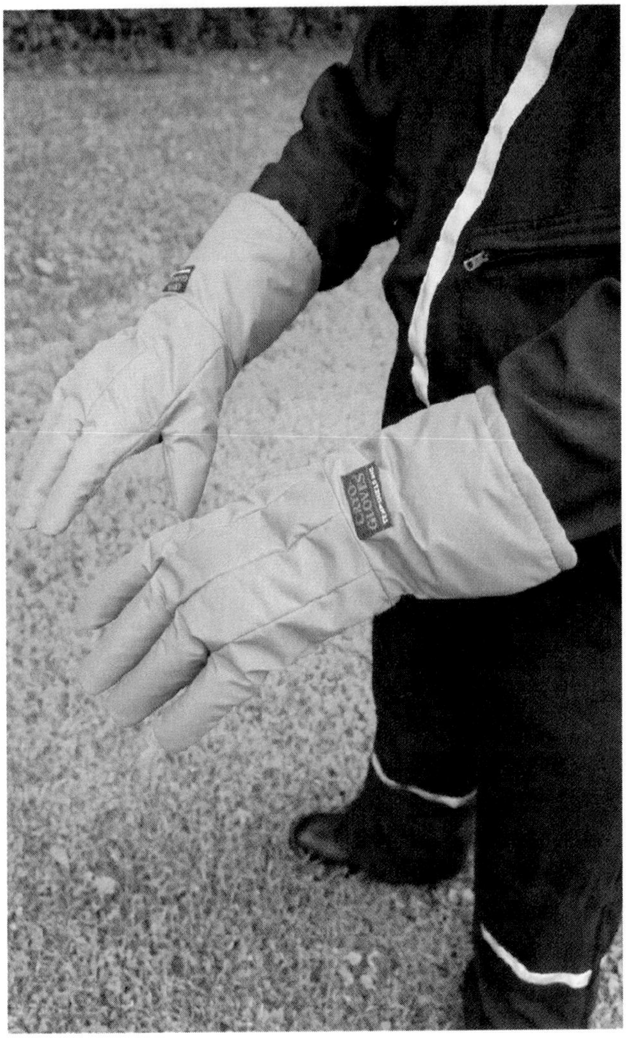

Figure 27-14 Cryogenic gloves are used when dealing with very cold refrigerated liquids. They offer protection from a range of −150° to −450°F.

Figure 27-15 Firefighter turnout gear and chemical protective clothing all place considerable stress on the responder. Steps should be taken to minimize the heat stress.

tion control gear, to firefighter TOG, to the fully encapsulated Level A suit. Most of these issues are only thought of when dealing with chemical protective clothing, but more thought and emphasis should be placed on these issues when using any type of protective clothing. The four major issues are heat stress, mobility, visibility, and communications problems.

> **Streetsmart Tip** Stress is a leading killer of firefighters, and heat stress plays a major factor in many of these deaths.

Turnout gear, although getting lighter, places tremendous stress on the wearer. The type of vapor barrier used in the TOG will determine the amount of heat stress a person will encounter, and not all vapor barriers are created equal. It is important for emergency responders to be able to recognize heat stress and its degrees of seriousness. Heat stress can lead to heat stroke, a condition that is almost always fatal. When wearing a Level A suit the temperature inside the suit can easily exceed 100°F, even on a cold day. When using PPE of any type, responders should take care to prehydrate and take their time when doffing their PPE—immediate removal of their PPE can shock the body, **Figure 27-15**, and cause serious health issues. This is important to remember when removing PPE from a HAZMAT responder, who is usually zipped up inside a very warm environment and cannot control the undressing progress.

Caution Dehydration is a major factor when operating at a fire or a chemical spill. Frequent hydration and frequent rest breaks are important.

The progression of heat stress is dependent on the amount of work being performed and the physical ability of the responders. To best combat any heat-related emergency, responders should hydrate, **Figure 27-16**, take frequent breaks, and until acclimated take extra precautions. When an incident occurs on a day in which the weather conditions would not be considered normal, personnel should pay extra attention to their activity levels. Early recognition is important for the health and safety of the responders. The levels and their warning signs are listed in **Table 27-4**.

Figure 27-16 Hydration is important for any emergency incident. At chemical spills, it is important that the entry team hydrate prior to entering the hot sector.

Health Hazards and Common Warning Signs

Heat exhaustion	This is the real first step toward dehydration, and it is fairly common. The root cause is excess sweating, resulting in a loss of body fluid. Symptoms include dizziness, headache, nausea, diarrhea, and vomiting. Heat exhaustion affects not only HAZMAT responders, but firefighters as well. Responders should have had enough fluid that they have the feeling they need to urinate. Heat exhaustion that is not treated can lead to heat stroke, which is a very serious and often fatal condition.
Heat stroke	This is a serious emergency and can be fatal to upward of 80 percent of the patients. In simple terms it means the body's ability to regulate its temperature has failed and is no longer functioning. The symptoms include unconsciousness, hot and dry skin, seizures, confusion, and disorientation. Being dehydrated when approaching heat stroke levels only complicates the process and actually speeds up the patient's deterioration. Heat stroke should be prevented because once the body shuts down, the end result is usually fatal.

TABLE 27-4

Lessons Learned

The use of protective clothing is important for the various hazards that responders may face. The absolute minimum should be SCBA, which offers a high level of protection. If a responder cannot confirm the absence of hazardous materials through the use of air monitors, then a basic level of protective clothing should be used. The decision to use chemical protective clothing is not an easy one because many factors must be examined prior to its use. With the ensuing heat stress that accompanies the wearing of chemical protective clothing, cases can arise in which the protective clothing may be more dangerous than the chemical hazard. It is for this reason the responders must use effective risk assessment to determine the true hazard—and then dress for that hazard.

KEY TERMS

Acute A quick one-time exposure to a chemical.

Air-Purifying Respirator (APR) Respiratory protection that filters contaminants out of the air, using filter cartridges. Requires the atmosphere to have sufficient oxygen, in addition to other regulatory requirements.

Carcinogen A material that is capable of causing cancer in humans.

Ceiling Level The highest exposure a person can receive without suffering any ill effects. It is combined with the PEL, TLV, or REL as a maximum exposure.

Chronic A continual or repeated exposure to a hazardous material.

Encapsulated Suit A chemical suit that covers the responder, including the breathing apparatus. Usually associated with Level A clothing, that is gas- and liquid tight, but there are some Level B styles that are fully encapsulated, but not gas- or liquid tight.

Etiological A form of a hazard that includes biological, viral, and other disease-causing materials.

Fit Testing A test that ensures the respiratory protection fits the face and offers maximum protection.

ICt_{50} The incapacitating level for time to 50 percent of the exposed group. It is a military term that is often used in conjunction with LCt_{50}.

Irritant A material that is irritating to humans, but usually does not cause any long-term adverse health effects.

LCt_{50} The lethal concentration for time to 50 percent of the group. Same as the LC_{50}, but adds the element of time. It is a military term.

Lethal Concentration (LC_{50}) A value for gases that provides the amount of chemical that could kill 50 percent of the exposed group.

Lethal Dose (LD_{50}) A value for solids and liquids that provides the amount of a chemical that could kill 50 percent of an exposed group.

Level A Protective Clothing Fully encapsulated chemical protective clothing. It is gas- and liquid tight and offers protection against chemical attack.

Level B Protective Clothing A level of protective clothing that is usually associated with splash protection. Level B requires the use of SCBA. Various clothing styles are considered Level B.

Permeation The movement of chemicals through chemical protective clothing on a molecular level; does not cause visual damage to the clothing.

Permissible Exposure Limit (PEL) An OSHA value that regulates the amount of a chemical that a person can be exposed to during an eight-hour day.

Recommended Exposure Limit (REL) An exposure value established by NIOSH for a ten-hour day, forty-hour workweek. Similar to the PEL and TLV.

Risk-Based Response An approach to responding to a chemical incident by categorizing a chemical into a fire, corrosive, or toxic risk. Use of a risk-based approach can assist the responder in making tactical, evacuation, and PPE decisions.

Sensitizer A chemical that after repeated exposures may cause an allergic-type effect on some people.

Short-Term Exposure Limit (STEL) A fifteen-minute exposure to a chemical followed by a one-hour break between exposures. Only allowed four times a day.

Supplied Air Respirator (SAR) Respiratory protection that provides a face mask, air hose connected to a large air supply, and an escape bottle. Typically used for waste sites or confined spaces.

Threshold Limit Value (TLV) An exposure value that is similar to the PEL, but is issued by the ACGIH. It is based on an eight-hour day.

TRACEM An acronym for the types of hazards that exist at a chemical incident: thermal, radiation, asphyxiation, chemical, etiological, and mechanical.

REVIEW QUESTIONS

1. What route of entry is the easiest to protect against?
2. What material are emergency responders commonly exposed to through injection?
3. After twenty years of smoking what type of exposure has a person received?
4. Which type of exposure is a one-time event?
5. What are the three common routes of exposure?
6. What six hazards potentially exist at a chemical release?
7. Which adverse medical effect mentioned in the chapter can become hereditary?
8. Nitrogen is what type of asphyxiant?
9. Which exposure value uses a fifteen-minute time limit?
10. What is the term *ceiling* used for with regard to exposure values?
11. Which level of chemical protective clothing offers a high level of protection against chemicals that are toxic through skin absorption?
12. What is a major concern as the levels of protective clothing increase?
13. Which type of respiratory protection offers the highest level of protection?

Endnote

1. Not all emergency services workers are covered by the Federal OSHA regulations. Some are covered by their state OSHA or are not covered at all by OSHA regulations. This varies from state to state.

Additional Resources

Bevelacqua, Armando S., *Hazardous Materials Chemistry.* Delmar Learning, a part of the Thomson Corporation, Clifton Park, NY, 2001.

Bevelacqua, Armando S., and Stilp, Richard H., *Hazardous Materials Field Guide,* Delmar Learning, a part of the Thomson Corporation, Clifton Park, NY, 1998.

Bevelacqua, Armando S., and Stilp, Richard H., *Terrorism Handbook for Operational Responders,* 2nd ed., Delmar Learning, a part of the Thomson Corporation, Clifton Park, NY, 2004.

Hawley, Chris, *Hazardous Materials Incidents,* 2nd ed., Delmar Learning, a part of the Thomson Corporation, Clifton Park, NY, 2004.

Hawley, Chris, *HazMat Air Monitoring & Detection Devices.* Delmar Learning, a part of the Thomson Corporation, Clifton Park, NY, 2002.

Henry, Timothy V., *Decontamination for Hazardous Materials Emergencies,* Delmar Learning, a division of the Thomson Corporation, Clifton Park, NY, 1999.

Lesak, David, *Hazardous Materials Strategies and Tactics.* Prentice Hall, 1998.

Noll, Gregory, Michael Hildebrand, and James Yvorra, *Hazardous Materials Managing the Incident.* Fire Protection Publications, Oklahoma University, 1995.

Schnepp, Rob and Paul Gantt, *Hazardous Materials: Regulations, Response & Site Operations.* Delmar Learning, a part of the Thomson Corporation, Clifton Park, NY, 1999.

Stilp, Richard and Armando Bevelacqua, *Emergency Medical Response to Hazardous Materials Incidents.* Delmar Learning, a part of the Thomson Corporation, Clifton Park, NY, 1997.

CHAPTER 28

HAZARDOUS MATERIALS: PROTECTIVE ACTIONS

Christopher Hawley, Baltimore County Fire Department (Ret.)

OUTLINE

- Objectives
- Introduction
- Hazardous Materials Management Processes
- Common Incidents
- Decontamination
- Methods of Decontamination
- Lessons Learned
- Key Terms
- Review Questions
- Endnotes
- Additional Resources

Courtesy of Baltimore County Fire Department

STREET STORY

It was the start of what we call a normal day on the job, until later in the morning when an alarm for a chemical leak inside a beverage warehouse was sounded. The dispatch consisted of what we call a HAZMAT box: four engine companies, one truck company, a rescue squad, basic life support unit, HAZMAT company, and command officer. I was working at the HAZMAT company that day. While en route, a radio transmission by the first arriving company advised the Emergency Communications Center of a major ammonia leak inside the warehouse storage area and that they were taking protective actions. This area contained multiple storage of beverages and boxes within an enclosed and secured area that also contained valves and piping for the anhydrous ammonia refrigeration system.

Upon our arrival, the warehouse had been evacuated and a strong odor of ammonia had already consumed the entire area surrounding the warehouse. Once we had performed a hazard risk assessment and ensured that the first responders had taken appropriate protective actions, we then selected our level of protection. Three hazardous materials technicians and I entered the release area to shut off the valve to the leaking pipe. After locating the release area we located the valve and made an attempt to close the valve. While closing the valve a sudden release of gaseous and liquid ammonia covered the personnel working at and around the valve. Visibility was taken from us almost instantaneously because of the gaseous release and communications were lost between all four technicians. I was able to find my way out and noticed that my personnel were still inside the release area. Prior to making another entry to locate my personnel, I noticed a white smoke coming from my chemical boots.

After further investigation I realized that the oil-based paint from the concrete floor was causing a chemical reaction under the soles of my boots. I reentered the release area, located my personnel, and immediately withdrew from the release area to the decontamination area. Once we were refreshed, a second entry attempt was made into the release area, where we were able to locate another sectional valve and stop the leak. The hardest part of the second attempt was removing our SCBA inside our suits to squeeze past piping and valves to get to the right one.

Prior to leaving the scene we finally determined that prior to our arrival a firefighter had entered the release area and closed the valve without notifying command and/or hazardous materials personnel. When hazardous materials personnel entered the release area thinking that the valve was not closed, they actually reopened it, which caused the valve to freeze in the open position. In this event, a number of factors affected our response. First-arriving crews needed to address isolation and evacuation issues, and the type of release, but one firefighter was endangered by not taking appropriate protective actions. No personnel were injured or exposed to the ammonia, but the incident proved to be very dangerous as a result of personnel freelancing and the lack of training present at an emergency scene.

—Street Story by Gregory L. Socks, Captain, Montgomery County, MD Hazmat Team

OBJECTIVES

After completing this chapter, the reader should be able to:

- Discuss the various incident management systems.
- Discuss the various methods of container breaching.
- Explain the four methods of vapor cloud movement.
- Explain the methods used to make isolation and evacuation decisions.
- Describe the use of hot, warm, and cold zones.
- Identify the use of air monitors in the determination of zones.
- Discuss the use of incident levels to describe the severity of the incident.
- Describe common incidents within each hazard class.
- Discuss methods of emergency decontamination.
- Describe the five types of decontamination.
- Describe the various methods of accomplishing decontamination.

INTRODUCTION

This chapter provides a myriad of topics for the responder and focuses on some general tactics that should be followed at a hazardous materials incident. The tactical considerations provided here are for general situations and may not apply to specific situations because each chemical spill is different, and for each spill there may be another way of handling that release.

> **Note** The basic concept for first responders is one of isolation—first responders should not allow other people to become part of the incident, and they should protect those involved with the incident or those who may become part of the incident in a short time.

A lot of the information in this section may not apply to specific cases. For firefighters beginning their training, it is unlikely they will be making community evacuation decisions for the next couple of years, but the material in this chapter should be kept in mind for the time when they will be making these types of decisions.

HAZARDOUS MATERIALS MANAGEMENT PROCESSES

Several different management processes exist that can be used for hazardous materials incidents, many of which have been in use for many years and offer well-proven methods of organizing an incident. All of the systems have been adapted from fire service systems to fit the needs of a chemical release. The cores of all of these systems are basically the same, but they do differ in some areas. The core to all systems is the protection of life, property, and the environment.

One of the systems developed is the **8-Step Process**, which was devised by Mike Hildebrand, Greg Noll, and Jim Yvorra. Dave Lesak developed another system called the **GEDAPER process** of hazardous materials management. Another system developed early on by Ludwig Benner, Jr., is the **DECIDE process**, which is listed along with the other systems in **Table 28-1**. Regardless of which system a department chooses, it is important to choose a system that everyone understands and can use. OSHA's HAZWOPER regulation requires the use of an IMS but does not state which type of system is required. The reality is that for the overall incident an incident management team (IMS) will be in place, and the group of responders responsible for HAZMAT will fit into that IMS using one of the processes mentioned or a combination, depending on the situation.

Isolation and Protection

> **Note** When arriving at a suspected chemical release, it is important to isolate the area from other people who may inadvertently wander into a hazardous environment.

Methods of isolation can be as simple as barrier tape, **Figure 28-1**, to the use of law enforcement at traffic control points. With a chemical hazards incident, it is important to control the incident quickly. The more people entering the suspected hazard area, the more people who may later need to be rescued or, depending on the situation, may need decontamination.

The protection of the people in a hazard area can best be accomplished by evacuation of the immediate area. This does not imply that everyone is simply told to leave, because they may need to be

Hazardous Materials Management Systems

DECIDE	8-STEP PROCESS	GEDAPER	HAZMAT STRATEGIC GOALS
Detect the presence of the hazardous materials	Site management and control	Gather information	Isolation
Estimate the likely harm	Identify the problem	Estimate potential course and harm	Evacuation
Choose a response objective	Hazard and risk identification	Determine strategic goals	Notification
Identify the action	Select personal protective clothing and equipment	Assess tactical options and resources	Product identification
Do the best possible	Information management and resource coordination	Plan and implement chosen actions	Determination of appropriate personal protective equipment
Evaluate your progress	Implement response objectives	Evaluate	Decontamination
	Decontamination	Review	Spill and leak control
	Terminate the incident		Termination

Sources: The information on the 8-Step Process® is from *Managing the Incident,* Gregory Noll, Michael Hildebrand, Jim Yvorra, Fire Protection Publications, Oklahoma University, 1995. The information on the GEDAPER® is from *Hazardous Materials Strategies and Tactics,* David Lesak, Prentice Hall, 1998. Both systems reproduced with permission of the authors.

TABLE 28-1

decontaminated or at least medically evaluated depending on the situation. A plan must be established for the holding of these people until a determination can be made as to their status. The HAZMAT team is usually the only group that can make this determination. When dealing with people in the suspected hazard area, frequent communication with the HAZMAT team is important. The other use of protection at a chemical spill is usually associated with the adjacent community evacuation or sheltering in place. Both of these issues are further discussed later in this chapter.

Figure 28-1 One of the first priorities should be to isolate the area so as to prevent other people from becoming involved with the incident.

Rescue

When discussing isolation and protection, rescue is a topic that naturally follows those two important issues. The rescue of victims from a suspected hazard area can be extremely controversial. The decision to make a rescue is a personal one, because it may involve substantial risk to the rescuer. Local protocol and SOPs must be considered.

Safety Firefighting is inherently risky. As much as firefighters would like to eliminate all risk from their occupation, it is impossible to do so. Their best hope is to safely manage that risk and use methods to identify that risk.

In reality, if responders arrive at the incident safe and sound, their chance for survival dramatically increases. One of the most dangerous parts of firefighting is responding to the incident—each year 20 to 30 percent of firefighter deaths occur while going to and from an incident.

With specific regard to turnout gear (TOG), scientific data published in August 1999 provide incident commanders with some information as to the ability to make rescues in hazardous situations. The Soldiers Biological and Chemical Command (SBCCOM), located at the Aberdeen Proving Grounds in Maryland, performed several studies involving firefighters' protective clothing. The lead study is known as the 3/30 Rule, and researchers found that firefighters' protective clothing can offer protection where significant hazards exist to someone without protective clothing. They tested military chemical nerve and blister agents, both of which present toxicity hazards; the nerve agent is extremely toxic. In situations where a nerve agent has been released and there are both live and dead victims, firefighters in full PPE and SCBA can enter this environment and make rescues with no ill effects. In a situation where all the victims are dead, firefighters can enter this severely toxic environment for three minutes with no ill effects. There are a number of factors that go into entering this type of toxic environment. The study outlines the best way for firefighters to protect themselves with TOG. There are methods discussed that use tape to add additional protection, as well as some other unique suggestions. The discussion on entering a terrorism event is controversial, and this study only offers some science behind the decision-making process. It goes without saying that in these cases emergency decontamination is a must and the gear should probably be destroyed afterward.

Once on the scene, firefighters need to evaluate the incident. Is it a rescue situation? This information should be confirmed as best it can; in some cases this information cannot be verified. What are conditions at the incident? Is there a confirmed chemical spill? If there is a release, can a person wearing no protection survive? In many cases the fire department is called to suspected chemical releases, and, after investigation, it turns out there really was not a chemical release. If the people who need to be rescued are alive, then the actual risk to the firefighters making the rescue attempt wearing full PPE is minimal. If, on the other hand, the first responders arrive at a local mall and are told that twenty people are unconscious and appear to be dead in the hardware store, the situation is different. When the responders approach the store entrance and see the twenty people lying on the floor, not moving, the responders may not want to enter that environment.

Other considerations need to be taken into account when making a rescue decision, such as response and notification time. The fact that the people in the hazardous environment have been there for a period of five to fifteen minutes, depending on the response time of the department, is critical to the decision-making process. They are not wearing any PPE and have been exposed to the material for a considerable amount of time. If they are alive when responders arrive, the risk to the rescuers is minimal. A method of emergency decontamination should be set up, and a backup crew should be standing by. The responding HAZMAT team should be consulted prior to entry so as to verify the chemical information. When rescuing the victims there should be no delay in their evacuation—a swoop-and-scoop technique is the order of the day. Stokes baskets or other methods of quick evacuation should be employed. Firefighters should not take vitals, ask medical histories, or perform medical procedures; instead, they should move the victims to a safe area.

Once out of the area, decontamination should be performed if required on both victims and rescuers, and they should be isolated from the remainder of the responders. Once decontaminated, EMS personnel can begin to work on the patients using appropriate levels of PPE. After decontamination, the rescue team should remove their PPE and bag them. After evaluation by EMS providers and consultation with the HAZMAT team the rescuers should be sent to rehabilitation. Their turnout gear and SCBA will need to be evaluated and possibly sent for cleaning or replacement, depending on the possible contamination and the material in question.

Safety The rescue of victims is made using simple risk/benefit analysis: a lot of risk is taken when a lot is at stake. No risk is taken when the benefit is little.

When handling hazardous materials incidents, encountering trapped victims is an unusual incident, but procedures should be in place to cover this contingency. The most likely scenario is a traffic accident, in which persons are trapped and chemicals are involved. This type of scenario is common if gasoline is introduced into the picture. Firefighters are used to extricating trapped victims with gasoline leaking from one or more vehicles. Protection lines are established, and a higher level of PPE is typically used by the rescue crews. The rescue crews limit the number of people in the hazard area, and a quick extrication is usually performed. If other chemicals are involved the scenario may change slightly. As HAZMAT companies arrive, personnel can be replaced by those with better chemical protection, and the HAZMAT company can work to control or eliminate the chemical hazard.

Making the decision to enter a hazardous environment takes training and experience. This decision should be made in direct consultation with the HAZMAT company. To make that decision, some HAZMAT teams use an approach known as risk-based response. By the heavy use of air monitors, an unknown chemical can be placed into one of three risk categories: fire, corrosive, or toxic hazard. In examining chemical and physical properties, a chemical can be placed into one or more of these categories. The major factor of concern in a potentially hazardous situation is a chemical's vapor pressure. If the chemical does not have a high vapor pressure, the risk of entering an environment containing this spilled material is very low unless a responder touches or falls into the material. One of the other items that a HAZMAT team can use to its advantage is the fact that the majority of incidents they respond to involve flammable and combustible liquids or gases. Although not designed as such, firefighter turnout gear offers ample protection for a rescue situation. HAZMAT teams and now many first responders have very capable detection devices for these types of materials and can determine the true risk during an operation. Included here is a listing of the top ten chemicals spilled in this country every year. This is a HAZMAT team's bread and butter, and they should be comfortable with the handling of these materials. First responders should know the locations in which these materials are stored and used. The response to an incident involving these materials should be no different than a response to a bedroom fire. The top ten chemicals spilled are:[1]

1. Sulfuric acid
2. Hydrochloric acid
3. Chlorine
4. Oil
5. Ammonia
6. Sodium hydroxide
7. Gasoline
8. Propane
9. Natural gas
10. Methanol

Risk-Based Response A risk-based response philosophy is based on the use of air monitors to guide responders safely. The monitors guide the responders as to which situations are safe to proceed into and when the use of PPE is appropriate. Unfortunately, many first responders do not have all of the required air monitors for a high level of protection. When confronted with rescue situations in potentially hazardous environments, first responders should use full protective equipment, including SCBA, as well as air monitors. They should also consult with the HAZMAT team, who can help guide them to the appropriate level of action. All emergency response involves some risk; steps need to be taken to manage (and thereby minimize) that risk.

In one situation, a truck carrying food rear-ended another tractor-trailer stopped in the middle lane of the highway as shown in **Figures 28-2, 28-3,** and **28-4.** It is estimated that the food truck was traveling at 65 mph when it hit the stopped truck. The truck it hit was carrying a mixed load consisting of mostly 55-gallon drums. The driver of the stopped truck was not hurt and was able to remove his shipping papers. The driver of the food truck was still alive but was extremely entangled in the wreckage. The driver was conscious and alert, and initial vital signs were stable. The trucks were tangled together, and the door to the truck carrying the drums was torn away from the truck. Drums had shifted onto the food truck and were laid over the front of that truck.

The first responders saw the placards on the first truck and requested a HAZMAT assignment. When they approached the food truck to evaluate the driver, they saw the drums in a precarious position in the back of the truck. They already had turnout gear on but had also donned SCBA. They obtained the shipping papers from the driver and consulted with the HAZMAT company, which was about twenty minutes away. The contents of the drums were mostly flammable liquids and some combustible liquids. The first responders were advised to use a combustible gas indicator, continue with full PPE, establish foam lines, and begin the rescue.

Upon arrival, the HAZMAT company team members met with the incident commander (IC) to evaluate the scene. They confirmed the monitoring being done by the first responders and began to evaluate the other parts of the load. It was determined that whenever the rescue companies moved a part of the dash of the truck,

Figure 28-2 The truck to the right rear-ended the front truck trapping the driver in the truck on the right. In the trailer to the left are 55-gallon drums, some of which are leaking. Using PPE, air monitoring devices, and protection lines, crews continued with the rescue. *(Photo courtesy of the Baltimore County Fire Department)*

one of the leaking drums would increase its flow. The HAZMAT companies secured the leak, moved the drum, and secured the remainder of the drums. They examined the rest of the load to make sure there were no other problems and continued to monitor the atmosphere. Once the victim was removed and the rescue companies had moved away, the HAZMAT team **over-packed** and pumped the contents of the other drums into other containers.

The risk category was fire, and the PPE chosen was appropriate for the risk category. At no time were any flammable readings indicated during the rescue, although some were encountered during the transfer operation. This was an example of the various disciplines working together to rescue a victim in a hazardous situation.

Site Management

The management of a hazardous materials incident can be very difficult even for a seasoned incident commander. A number of incident management strategies are available to the IC, some of which were outlined in previous chapters. The hazardous materials specific positions are outlined in **Table 28-2**. A fire department IC on a normal fire-type incident deals predominantly with resources within the IC's own agency. Occasions may arise when the IC will discuss items with the police department, and depending on the location a police officer may be in the command area. The IC may also deal with other municipal agencies such as the health department or social services. Other outside agencies such as the Red Cross may also be involved and work under the direction of the IC. Depending on the size of the community, the media may play a factor in the management of the incident. But in the overall scheme of the incident the IC deals mostly with the IC's own agency and coworkers.

Figure 28-3 Rescue crews are extricating the driver of a second vehicle involved in a rear-end collision. The driver remained conscious throughout the extrication and was severely entangled. Complicating the rescue was the presence of a truck full of hazardous materials containers. *(Photo courtesy of the Baltimore County Fire Department)*

Figure 28-4 Leaking drums were found during this rescue operation. Rescue crews had a foam line standing by and were air monitoring for flammable levels. When HAZMAT crews arrived, they took over air monitoring, quickly removed the leaking drums, and secured the other drums. When the dirver was extricated, HAZMAT crews took care of the remaining drums. *(Photo courtesy of the Baltimore County Fire Department)*

HAZMAT Branch Positions

- *Backup:* Personnel assigned to rescue the entry team if necessary. They are dressed in the same suits as the entry team and are fully prepared to make an entry, with the exception of being on air. The number of backup personnel is a minimum of two, but can be expanded to more personnel depending on the incident. The backup team must be trained to the minimum of the Technician level.
- *Decontamination:* A person is assigned to oversee the setup and operation of the decontamination area, sometimes referred to as the decon area, or the contamination reduction corridor. Other personnel will also be operating within the area performing the decon or PPE removal. People operating in this area are usually trained to the Technician level, but people trained to the Operations level may also be operating in this area. If Operations level personnel are used, they should have received specific training in this job function.
- *Entry:* A minimum of a two-person team will enter the hazard area or hot zone. The type of chemical will determine the type of PPE that will be used. Prior to entry the IC must brief the personnel who will be working in the hazard area. To be on the entry team, responders must be a technician or a specialist.
- *Hazardous materials branch management:* This person is responsible for handling the tactical objectives, as assigned by the incident commander. This person is in charge of the HAZMAT team and therefore coordinates its efforts. May be referred to as HAZMAT operations.
- *Hazardous materials branch safety:* This person assists the overall safety officer, but is concerned mostly with HAZMAT-specific issues such as PPE selection and use. This HAZMAT safety position is usually a technician or above. The overall safety officer is concerned with all personnel at the incident and usually focuses on other hazards not related to the chemical (e.g., slip, trip, and fall hazards).
- *Information/research:* This position provides information regarding the chemical and physical properties to the HAZMAT branch officer and the IC. After completion of these duties, these personnel may then assume responsibility for documenting the incident.
- *Reconnaissance:* Personnel use binoculars or may even enter the hazard area to determine incident severity, potential identification of the hazardous material, and gather any additional information.
- *Resources:* These personnel are responsible for gathering and maintaining supplies required for the incident. May be referred to as logistics.

TABLE 28-2

When involved in a chemical release, especially one of major proportions, many more agencies get involved. In some communities, the HAZMAT team may not be associated with the local fire department; instead it may be from a county mutual aid group or from an adjoining community. In many incidents, major road closures are necessary, which increases the presence of police officers, and depending on the size of the road may bring command level officers to the scene. If the incident involves a state road or highway or has the potential to impact these roads, the state highway department may attend the incident. County or state environmental representatives or responders may also arrive at the incident to assist. In some states the Department of Natural Resources (DNR) has jurisdiction in chemical spills and, hence, may respond to an incident.

On a large spill the media will be a much larger group and may be from outside the immediate area, such as at the incident shown in **Figure 28-5**. A chemical release incident usually requires the services of a cleanup contractor, who will be arriving at the incident to assist in the cleanup effort. Depending on the incident, it is not uncommon to see insurance adjusters arriving within a few hours of the spill. On larger spills or spills that may endanger a waterway, the EPA or Coast Guard may arrive to assist with the spill.

Although this listing of agencies is nowhere near complete, it gives the reader an idea of the number of different agencies that may assist with the mitigation of the incident. In some cases the IC may have several alarms worth of equipment of his own to manage, not to mention these other agencies. Until the incident is moved from the emergency phase to the nonemergency cleanup phase, the fire department IC is usually still in charge.[2] In some cases the IC will have a difficult time and can end up handling the "assistance" rather than actually managing the incident. In this situation the department's public information officer (PIO) and liaison officer will be of great assistance.

Firefighter Fact At a hazardous materials incident, the incident commander is responsible for a number of tasks. Regardless of the local SOP, the IC has many legal obligations under the HAZWOPER regulation. The IC is responsible for the overall actions at the incident, regardless of who is performing the tasks. In most cases, the HAZMAT team performs the mitigation of the incident, and depending on the interaction of the IC and the HAZMAT team there may not be much verbal communication.

In the perfect world of IMS the IC makes all of the decisions, but the reality is that the technical group provides the suggested options. In some cases there may be only one option. If the incident goes to court it will be the IC who has to answer for the HAZMAT team's actions, so the IC must stay informed of the incident action plan and be given the opportunity to choose an appropriate response.

At a chemical release the IC is also responsible for the pre-entry briefing, which is when the entry crew is informed of the hazards that exist and what actions are expected of them. If there are no predesignated emergency signals, these must be decided on prior to entry. The IC must use a system of monitoring the progress of the incident and make changes to the system as needed. The mitigation of an incident takes the cooperation of many persons and agencies, all working toward a common goal under the guidance of the IC.

The management scenario painted did not mention victims, hospitals, or an evacuation—all of which further complicate the incident.

Note The use of an incident management system is not only mandated by OSHA but is a good idea so that personnel can be tracked and the outside agencies can be managed effectively.

Figure 28-5 The media will play a role in the handling of an incident, from evacuation instructions to minimizing any hysteria that may have occurred during the incident. A good flow of public information is essential to keeping the surrounding community at ease. *(Photo courtesy of Baltimore County Fire Department)*

Regardless of the IMS system, only a small component of the overall system is typically used during routine fire operations, such as a finance branch being established during a house fire. It is likely, however, that in a major chemical release a finance branch would be established and have several persons assigned to assist in this function.

In some fashion a liaison must be established with all of the responding agencies, and their specific roles also have to function within the IMS. A HAZMAT incident usually requires a minimum of two safety officers, one for overall safety and one assigned to HAZMAT specific issues.

Note OSHA requires the use of a safety officer, and requires that the safety officer be knowledgeable about the tasks the worker is going to perform.

A person trained to the operations level is not adequate to be the safety officer for the HAZMAT operation. A HAZMAT safety officer should be trained to the Technician or Specialist level to make sure that the mitigation efforts are carried out effectively and safely. Some specific HAZMAT branch functions that may be utilized during an incident are listed in **Table 28-2**.

Establishment of Zones

Some jurisdictions refer to **zones** and some to **sectors,** but whatever the term, they are referring to areas established to identify the various isolation points, such as those shown in **Figure 28-6.** These

Figure 28-6 Proper isolation and work zones are needed to ensure the savety of responders.

zones are referred to as the hot, warm, and cold zones. The hot zone may also be referred to as the exclusion zone, isolation area, hazard area, or a similar term. The warm zone may be known as the contamination reduction area and the cold zone as the support area. In many cases these areas are identified through colored barriers, cones, or other markings. Hot is usually identified by the color red, warm by the color yellow, and the cold zone by the color green, **Figure 28-7**.

> **Safety** The most important zone that needs to be established is the **isolation area**, and this needs to be established by the first arriving responder.

This area normally becomes the hot zone after the arrival of the HAZMAT team. This isolation or hot zone is the area immediately around the release and is an area that requires the use of the proper PPE. Entering the hazard area could expose personnel to the substance and possible contamination.

The minimum zone that should be established is this isolation area. The distances for this area can be determined by the use of the DOT (ERG). The minimum distance established by this book is 330 feet, which should be the absolute minimum for an unknown material. The distance for this area is obviously determined by the local situation, so it is difficult to make blanket statements about recommended distances. If it is convenient and easy to isolate a block then it makes sense to do so, but if isolating the block will create havoc in the community, then 330 feet may be recommended. This distance is only valid if there are no indicators of an actual release. Any indications of a flowing spill or

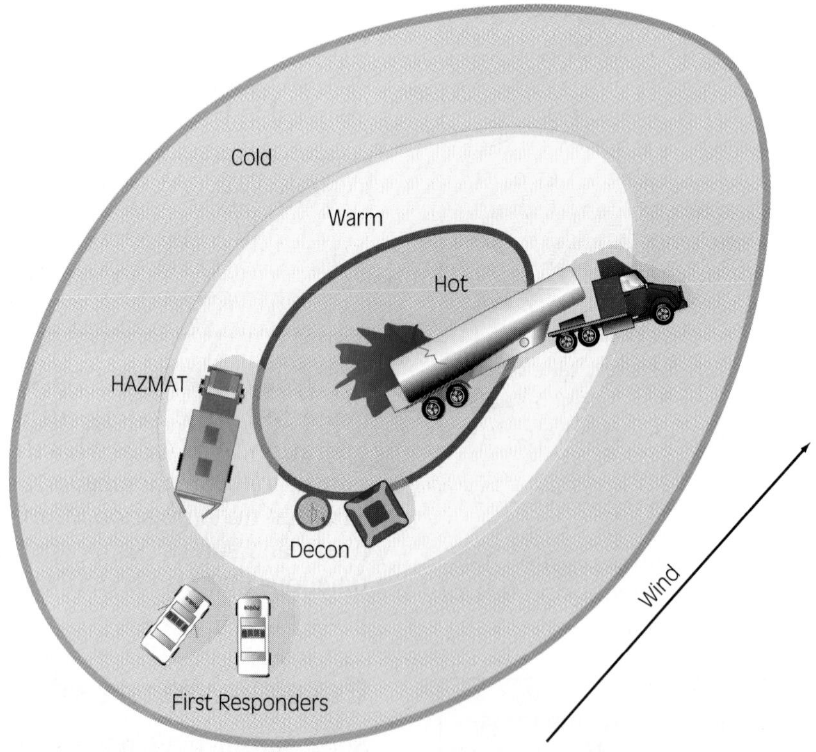

Figure 28-7 The establishment of zones is usually based on the types of hazards that may be present. For general chemical spills, the zones established are referred to as the hot, warm, and cold zones.

a vapor cloud for an unknown substance will dramatically increase this distance. If time permits, and resources are available, then an attempt should be made to set up the warm and cold zones. These zones are usually established by the HAZMAT team after arrival and setup.

First responders should position themselves in an upwind and uphill position, as shown in **Figure 28-8**. If operating at a water-based spill, they should operate upwind and upstream. The first arriving apparatus should have set up in that position when it arrived, but if not, the vehicle should be moved to that type of position as quickly as possible. Although it cannot always be accomplished, being in an upwind position is the first priority, and uphill the second priority. If responders are upwind, but forced to set up in the downhill position, they must extend the distance away. One of the common mistakes made is that the first arriving apparatus does not communicate the on-scene wind directions and the best route of travel for other apparatus to follow to arrive and stage at the upwind and uphill position. All apparatus should be positioned so that they are pointed away from the incident so in the event of a catastrophic release or other emergency, the apparatus can be moved without turning around or backing.

Other items to consider for the hot zone are topography, accessibility for responding units and other resources, weather conditions, and bodies of water, including those for drinking water. Other items to consider are setup areas for the HAZMAT team including decontamination and dress-out locations. Other exposures must be examined, and part of the isolation process is the limiting of possible ignition sources.

Public exposure potential must be considered, and this can be broken down into specific time frames such as short term (minutes and hours), medium term (days, weeks, and months), and long term (years and generations). Emergency responders usually deal only with short-term and medium-term types of incidents, but the actions that first responders take can result in long-term exposures, for both the responders and the public. Sewer lines, both sanitary and storm sewers, must be taken into account as well as other utilities such as cable and phone, both aboveground and belowground. Transportation corridors such as highways, rail lines, ports, and airports must be considered because the incident may well involve or affect these areas. For buildings the internal areas can be broken down into zones, but items that should be considered are floor drains, ventilation ducts, and air returns.

The warm zone is set up after the arrival of the HAZMAT team and is usually where the decontamination area is established. In some cases the warm zone is also extended to allow for the HAZMAT team setup. In unusual situations the warm zone is an area where some type of PPE may be required, and depending on the circumstances may be an area that can be affected by wind shifts or a catastrophic failure. The establishment of these zones is arbitrary although it is based on the best judgment of the IC or the HAZMAT team. There is no magic line of hot and warm zones, and nor could one be established. The cold zone is the area where the incident command

Figure 28-8 The best position for first responders is uphill and upwind from the release.

post is established and is where the first responding companies will be positioned. All support operations such as medical and rehabilitation are set up in this area, and movement between the zones is controlled at access control points. No PPE is required for the cold zone because there should be no chance for chemical exposure in this area. A person should be assigned to act as security for each of these zones to ensure that only authorized personnel enter these areas.

The Importance of Isolating the Scene

The first arriving responders can make a hazardous materials release easier to manage if they begin the isolation process. A bus accident often causes responders to act quickly to isolate the area and deny entry. In many cities the people on a bus who have been involved in an accident will receive a check in exchange for a release from further damage claims against the bus company. Failure to control access points to the bus will result in a dramatic increase in the number of occupants, many of whom may want to become "patients." What holds true with bus accidents is now true with chemical releases. In many cases, checks are being written or people plan to sue for possible damages, making isolation difficult. The mention of chemical exposure may result in people attempting to enter a potentially hazardous area. One of the most difficult areas to attempt to control is a mall or a shopping center where hundreds of people may be involved. A simple discharge of pepper spray will result in a number of real victims and usually generate a few more suspect victims. Everyone must be treated and possibly transported, which can create a strain on the EMS system. Early isolation and evacuation can help reduce the number of potential victims.

Some HAZMAT teams use monitoring devices to help establish the zones. The use of air monitors to establish these isolation points is crucial, because the distances provided in the DOT ERG are very conservative and are based on the worst-case scenario. Real-time, on-scene air monitoring cannot be replaced by plume projections, estimations, or models depicted in a text.

The zone distances are based on theories established by the risk-based response system. The materials that require the greatest isolation are those that have high vapor pressure or exist as gases in their natural state. A material such as sodium hydroxide (sometimes referred to as lye or caustic) requires an isolation distance of only a few feet. Although very corrosive, and contact with skin would cause some irritation and burns if not washed off immediately, it has a very low vapor pressure. The vapor pressure of sodium hydroxide is 1 mm Hg at 1,390°F, which means that the spill would have to be heated to 1,390°F in order to produce a very small amount of vapor, considerably less than the vapor pressure of water (25 mm Hg). With this material, it is unlikely that a large isolation distance would have to be established, nor would any evacuations probably be necessary.

A chlorine release from a railcar, on the other hand, does present an extreme risk to the community because once in the atmosphere chlorine rapidly changes to a gas. Chlorine is not only poisonous but also an oxidizer and a corrosive material with a very high vapor pressure of 5,168 mm Hg. A substantial evacuation area is required for chlorine. One advantage though is that chlorine is easily detected, so a true hazard area can be established using air monitoring.

Evacuations and Sheltering in Place

Note Making a decision to conduct an **evacuation** or to **shelter in place** can be one of the most difficult decisions for an emergency responder. Regardless of the decision, there are usually political ramifications, right, wrong, or indifferent.

The IC will be bombarded by the public, media, and coworkers about the decision. To a first responder, unfortunately, not much assistance is readily available. The only resource is the DOT ERG, and that book is conservative and may not apply in a specific situation. The best way to determine whether to evacuate or shelter in place is to conduct real-time air monitoring that can determine the exact hazard area. Consultation with the HAZMAT team and the local conditions guided by the DOT ERG can establish a starting point. If the incident is at a SARA Title III facility, the jurisdiction's emergency plan should have some recommendations regarding evacuation. Some plans provide a checklist to follow to assist with the evacuation decision-making process. Close coordination with the local emergency management agency is essential to making the outcome successful.

If a decision is made to evacuate, a suitable location needs to be found, transportation may be required, and accommodations need to be established for the evacuees. This type of assistance is usually provided by the emergency management coordinator, who can be a good point of contact. The HAZMAT team can also run a plume projection, **Figure 28-9**, using the CAMEO computer pro-

EXIT

Figure 28-9 Most HAZMAT teams use computer software to provide plume projection, which could provide the travel of a vapor cloud.

gram, but that is also conservative and may cause a larger evacuation than necessary.

Chemical vapor plumes have characteristic shapes as determined by computer models, such as the ALOHA model used by CAMEO. These standard plumes are to be used for worst-case scenarios and may not apply to a specific locality. Although the plumes are computed for a variety of topographies and types of weather, each local area differs, and only local weather conditions and air monitoring results can provide truly accurate results. The standard shapes for plumes are provided in **Figure 28-10**.

Some studies have shown that in most cases sheltering in place is safer than evacuation. Just imagine evacuating the most populated area of a community; how would a responder notify *all* of the citizens? In

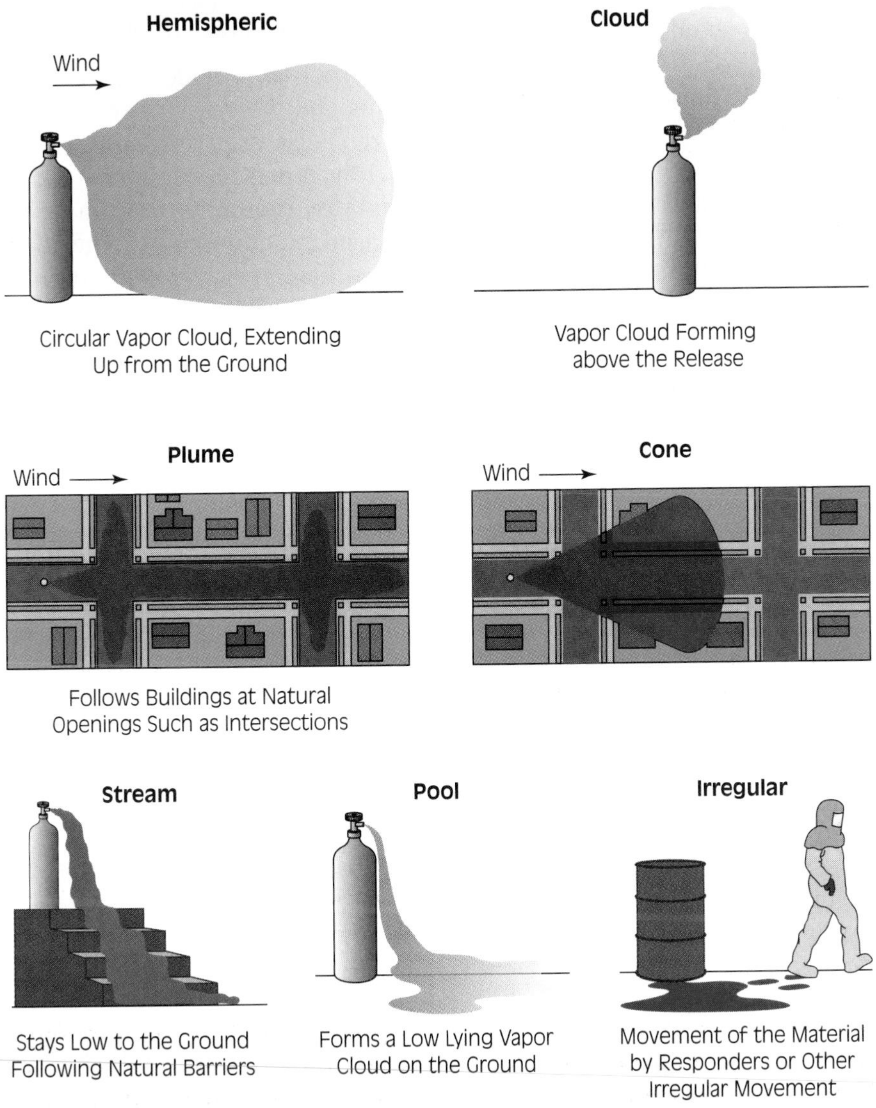
Hemispheric
Wind
Circular Vapor Cloud, Extending Up from the Ground
Cloud
Vapor Cloud Forming above the Release
Plume
Wind
Follows Buildings at Natural Openings Such as Intersections
Cone
Wind
Stream
Stays Low to the Ground Following Natural Barriers
Pool
Forms a Low Lying Vapor Cloud on the Ground
Irregular
Movement of the Material by Responders or Other Irregular Movement

Figure 28-10 Standard shapes for plumes or vapor clouds may form after a gas is released. The exact type varies with the topography and the buildings in the area.

most communities the police department may have this responsibility, but do they have the resources to accomplish this task? Is there an emergency alerting system for TV or radio in the community? In urban or metropolitan areas evacuating just a few blocks can affect thousands of people. The worst-case scenario using the DOT ERG is an evacuation distance of 4,000 feet by 7+ miles, which in a city could be 25,000 to 100,000 evacuees. It would be nearly impossible to evacuate that many people, not to mention the panic and chaos that such action would bring.

> **Note** Injuries and fatalities often result from an evacuation, even when the evacuation is announced days ahead of time.

There are certain times when evacuation is required such as when an explosion is probable, when explosives are involved, when a container may suffer a BLEVE or rupture with violent consequences, or when the release will continue for more than a few hours.

Sometimes an evacuation may not be recommended, such as for a hospital, nursing home, jail, or other facility in which rapid removal of the occupants is not practical. In these types of occupancies air monitoring and control of HVAC are advisable, in addition to having a liaison remain at the facility who is in radio contact with the IC. The chemical properties of the released material will have an effect on the decision, because some materials will rise up in the air and dissipate quickly, whereas others may stay low to the ground causing evacuation problems. The type of leak must be considered: Can it be quickly and easily controlled or is there a probability that it cannot be controlled by the HAZMAT team? Will evacuating the citizens subject them to a higher level of the chemical than keeping them in place?

ACGIH has provided some planning levels for the chemicals that require planning under the Clean Air Act Amendments (CAAA). These planning levels are on three tiers and provide actual levels that can be used for emergency planning. Because the permissible exposure limits and threshold limit values were established for an eight-hour exposure during normal conditions, they do not have much applicability during an emergency release. These **emergency response planning (ERP)** levels are designed to assist with the emergency planners' preparation of the community emergency plan, and would be useful in determining the evacuation zone. When making the decision to evacuate or shelter in place, the flow of information to the public and the media is essential, because a lack of information can bring disastrous results.

When sheltering in place the citizens should shut all windows and doors, shut off air handling systems, and stay tuned to a TV or radio station. If a continual flow of information is not available, people will become frustrated and may make attempts to go find the information for themselves. When dealing with larger facilities such as high-rises, hospitals, nursing homes, schools, or jails, it may be best to station an emergency responder at that location who acts as a direct link to the incident, so that any questions can be immediately answered and fears alleviated. The emergency management office can also establish a rumor control hotline, which can be used to answer questions about the incident from concerned citizens and family members.

COMMON INCIDENTS

This section provides an overview of common incidents and the types of releases in each of the DOT hazard classes. This listing is far from complete; it merely represents some of the most common incidents or incidents that have a great potential to impact the community. The recommendations provided here are only suggestions; local policies and procedures should be followed.

Types of Releases

When talking about chemical incidents it is important to classify how the chemicals are released from their container. In many cases the manner in which the release occurred can help provide clues to the successful mitigation of the incident. The type of release can be classified as a breach in a container or as a release within a containment system. The NFPA provides two distinct categories for both breaches and releases.

There are several ways of looking at the potential release of a chemical. Either the chemical itself is stressed or the container is stressed. In an incident involving a gasoline tanker that is rolled over at 55 mph the container is stressed, and the contents are likely to come out of the container. In an incident in which a paint waste is reacting within a drum, the material is stressed, but if the drum is sealed tight, the container will be stressed due to the chemical reaction creating pressure. The three general types of stress are thermal stress, mechanical stress, and chemical stress. Thermal stress is the addition of heat or cold to a container, and on opposite sides cold stress can be as damaging as heat stress. Placing a metal drum into a pool of a liquefied gas could result in the drum becoming brittle and easily cracked. Dropping a drum is an example of mechanical stress, and putting a corrosive in a metal drum is an example of chemical stress.

Both pressurized and nonpressurized containers can breach in several different ways. **Figure 28-11** provides examples of each type of container breach. The most common container breaches are punctures and closures that open up. It is uncommon to find a container that has disintegrated or has runaway cracks, splits, or tears. Many factors are involved with container breaches, but the most common breaches are nail punctures in drums, forklift punctures, dropped drums, and in many cases closures not used or not in place. Many incidents have been quickly mitigated by the use of a drum lid. On tank trucks a common breach point is the frangible disk, which may rupture due to overfilling or a quick stop causing the liquid to slosh and rupture the disk, releasing some of the contents. The incident is quickly handled by the replacement of the disk.

The methods in which chemicals can be released through a containment system are detonation, violent rupture, rapid relief, and a spill or leak. When these methods result in a chemical release, the action can be violent and can have catastrophic consequences. When a container such as a propane tank, **Figure 28-12**, detonates, it can travel up to a mile. If a container ruptures violently, it means the container was under pressure, causing the violent release. Much like a BLEVE, a violent tank rupture can send the tank or portions thereof a considerable distance.

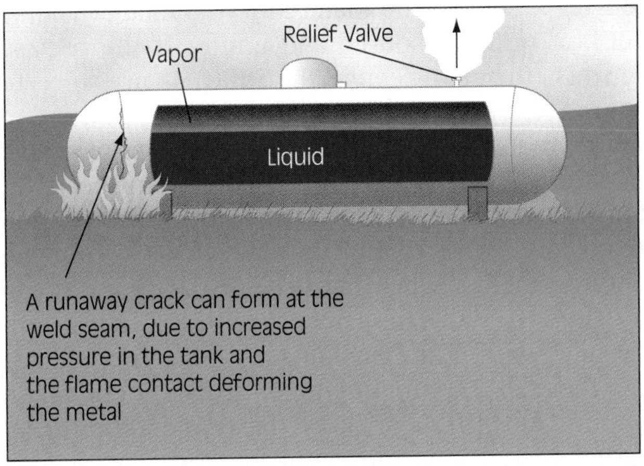

Figure 28-12 Propane tank detonation.

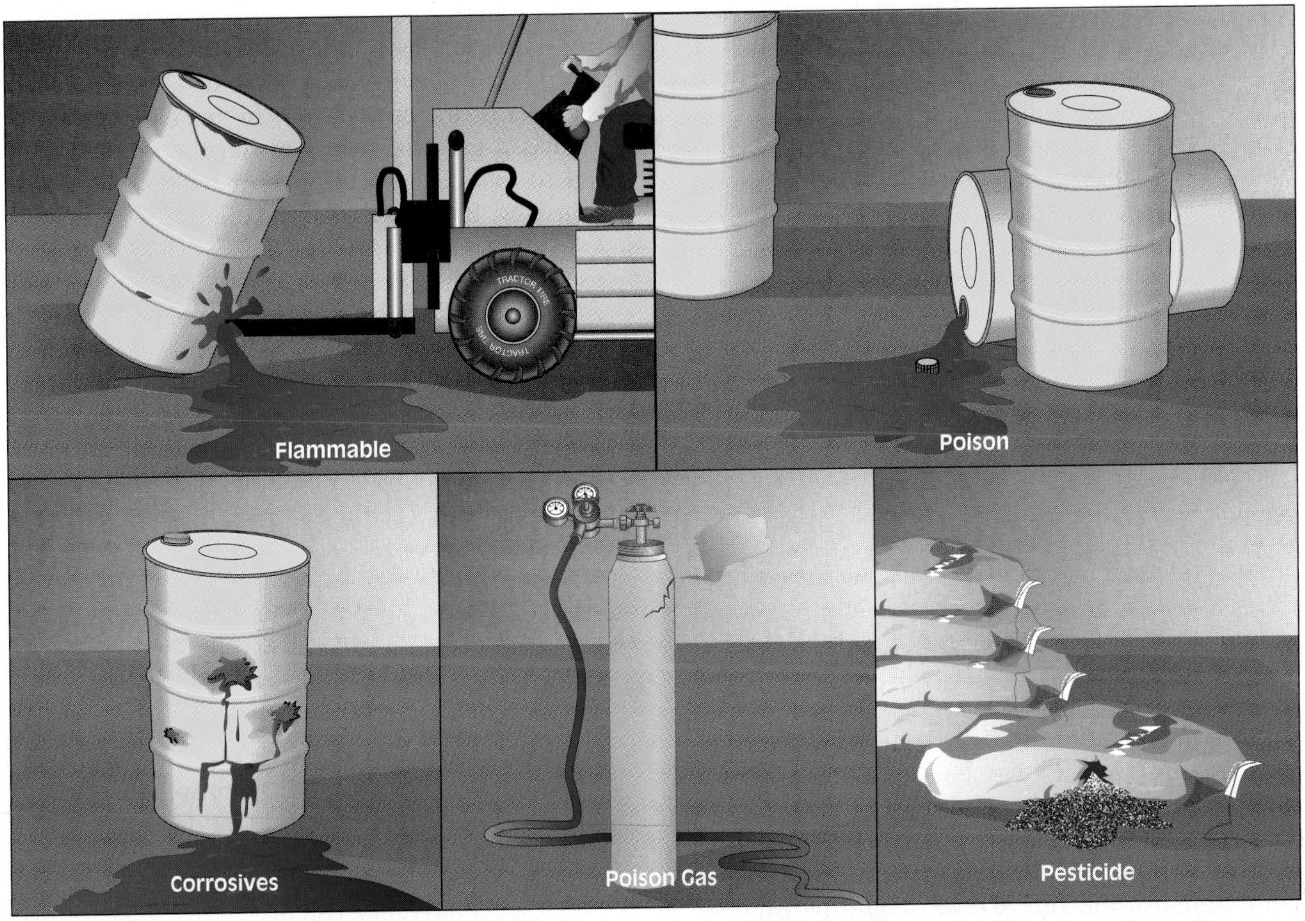

Figure 28-11 Types of container breaches.

A container that has a rapid relief valve should be able to release enough pressure to bring a margin of safety to the incident. This is only true if the cause of the increased pressure is removed and the pressure does not continue to climb within the container. If a relief valve is operating, yet still not able to relieve building pressure, a true emergency condition exists and failure to bring the pressure under control can have catastrophic consequences. An operational relief valve is allowing the product to escape, and if the material is flammable it may find an ignition source and ignite. If the release ignites, this may cause the internal pressure to increase even more, causing more product to ignite. It is never recommended that a relief valve be stopped, nor that a fire coming from the relief valve be extinguished. The only time a fire coming from any valve (relief or otherwise) should be extinguished is when the flow of the material can be stopped with 100 percent assurance that it will be successful. Extinguishing any other fire, including those that are impinging on a tank or container, should be a high priority. Hoselines should be directed such that they cool the container, which will reduce the pressure and allow the relief valve to stop operating.

Spills and leaks both result in product being lost. A spill is defined as a loss of product from a naturally occurring opening such as a bung on a drum or a leaking valve. A leak is defined as a release from an unnatural opening such as a puncture in the side of a drum. The end result for a spill or a leak is that product is released and can create problems for the responders.

Explosives

> **Safety** The general rule with explosives is that if the fire is near or is affecting the explosives, then emergency responders should isolate the area and back away. Life safety is the top priority.

All persons must be removed from the area, and a defensive operation should be established.

Many other considerations come into play if the fire is not directly impacting the explosives. A brake fire on a truck carrying explosives is one example; as long as the fire has not reached the cargo area, firefighters may be able to extinguish it without incident. When making the decision to fight a fire in this situation, water must be applied quickly and in large quantities to cool the cargo and then extinguish the fire. Crews should be limited, and adjoining areas evacuated. The recommendations in the DOT ERG, **Figure 28-13**, are a good starting point for isolation and evacuation.

Another incident that involves explosives arises when the fire department assists a bomb squad with a suspected explosive device. Close coordination with the bomb squad is required to make sure that the operation is conducted safely. Some example standoff distances are provided in **Table 28-3.** Although the distance varies with some jurisdictions, radios, cell phones, or other electronic devices should not be used within 500 to 1,000 feet of a suspected device. Many bomb squads detonate suspected devices in place, which may start a fire or cause a potential structural collapse. Other techniques are to "disrupt" the device or render it safe by the use of a water cannon. These techniques could cause the device to detonate. The time to discuss these scenarios and plans of action is *prior* to an incident, not *during* it. At incidents where there is a suspected device, firefighters should be aware of the potential for a secondary device, one designed to injure the responders. When operating at the scene of an explosion, it is important to have the bomb squad search the scene for secondary devices, **Figure 28-14**. The terrorism chapter, Chapter 30, has more information on secondary devices.

Other incidents may involve a shipment of explosives that has been involved in an accident. In most cases these incidents present little risk to the responder or the community. In this case the bomb squad is a great technical resource and should be used to evaluate the scene prior to moving any explosives. In some cases an explosives truck, such as the one shown in **Figure 28-15**, carrying ammonium nitrate may be involved in an accident and if it has fuel oil on board, the mixture known as **ANFO** could be created. Without an initiation charge or other substantial energy source, the material is unlikely to explode, but the ammonium nitrate does present a toxicity hazard and care should be taken around spills of this material. In incidents of this type the bomb squad should also be consulted.

Chapter 30 discusses some possible terrorism scenarios, and explosives are used in almost all of the incidents. HAZMAT and bomb squad duties are becoming more and more interlaced.

> **Safety** It is not uncommon for well-meaning citizens to bring explosives to the fire station. If the explosives are outside or in the citizen's vehicle, the firefighter should leave them there and call for the bomb squad.

RESIST RUSHING IN !
APPROACH INCIDENT FROM UPWIND
STAY CLEAR OF ALL SPILLS, VAPORS, FUMES AND SMOKE

HOW TO USE THIS GUIDEBOOK DURING AN INCIDENT INVOLVING DANGEROUS GOODS

ONE **IDENTIFY THE MATERIAL** BY FINDING ANY **ONE** OF THE FOLLOWING:

THE 4-DIGIT ID NUMBER ON A PLACARD OR ORANGE PANEL

THE 4-DIGIT ID NUMBER (after UN/NA) ON A SHIPPING DOCUMENT OR PACKAGE

THE NAME OF THE MATERIAL ON A SHIPPING DOCUMENT, PLACARD OR PACKAGE

IF AN **ID NUMBER** OR THE **NAME OF THE MATERIAL** CANNOT BE FOUND, SKIP TO THE NOTES BELOW.

TWO **LOOK UP THE MATERIAL'S 3-DIGIT GUIDE NUMBER** IN EITHER:

THE ID NUMBER INDEX..(the yellow-bordered pages of the guidebook)

THE NAME OF MATERIAL INDEX..(the blue-bordered pages of the guidebook)

If the guide number is supplemented with the letter "**P**", it indicates that the material may undergo violent polymerization if subjected to heat or contamination.

If the index entry is highlighted (in either yellow or blue), it is a TIH (Toxic Inhalation Hazard) material or a Dangerous Water Reactive Material (produces toxic gas upon contact with water). **LOOK FOR THE ID NUMBER AND NAME OF THE MATERIAL** IN THE TABLE OF INITIAL ISOLATION AND PROTECTIVE ACTION DISTANCES (the green-bordered pages). Then, if necessary, **BEGIN PROTECTIVE ACTIONS IMMEDIATELY** (see Protective Actions on page 314). If protective action is not required, use the information jointly with the 3-digit guide.

USE GUIDE 112 FOR ALL EXPLOSIVES EXCEPT FOR EXPLOSIVES 1.4 (EXPLOSIVES C) WHERE GUIDE 114 IS TO BE CONSULTED.

THREE **TURN TO THE NUMBERED GUIDE** (the orange-bordered pages) **AND READ CAREFULLY.**

NOTES **IF A NUMBERED GUIDE CANNOT BE OBTAINED BY FOLLOWING THE ABOVE STEPS,** AND A PLACARD CAN BE SEEN, LOCATE THE PLACARD IN THE TABLE OF PLACARDS (pages 16-17), THEN GO TO THE 3-DIGIT GUIDE SHOWN NEXT TO THE SAMPLE PLACARD.

IF A REFERENCE TO A GUIDE CANNOT BE FOUND AND THIS INCIDENT IS BELIEVED TO INVOLVE DANGEROUS GOODS, TURN TO **GUIDE 111** NOW, AND USE IT UNTIL ADDITIONAL INFORMATION BECOMES AVAILABLE. If the shipping document lists an emergency response telephone number, call that number. If the shipping document is not available, or no emergency response telephone number is listed, IMMEDIATELY CALL the appropriate **emergency response agency listed on the inside back cover of this guidebook.** Provide as much information as possible, such as the name of the carrier (trucking company or railroad) and vehicle number. AS A LAST RESORT, CONSULT THE TABLE OF RAIL CAR AND ROAD TRAILER IDENTIFICATION CHART (pages 18-19). IF THE CONTAINER CAN BE IDENTIFIED, REMEMBER THAT THE INFORMATION ASSOCIATED WITH THESE CONTAINERS IS FOR THE WORST CASE POSSIBLE.

Figure 28-13 The DOT ERG provides some basic explosives information that provides a good margin of safety when dealing with these types of incidents.

Everything from distress flares and ammunition to hand grenades, pipe bombs, dynamite, and blasting caps is commonly brought to fire stations. Firefighters should not be handling these types of devices; the bomb squad should handle these types of situations. Depending on the type of device and the size, isolation and some evacuation may be necessary.

Safety It is a good general practice not to bring unknown materials into the fire station.

Citizens may also bring in old chemicals, some of which may be explosive and shock sensitive. With picric acid just the simple removal of the cap or vibration of the container is enough to initiate the explosion of the container, with enough force to cause fatal injuries to the person who attempts to open or set down the container. Another common item that may be brought in is containers of old ether (ethyl ether), which is outdated after a year in storage. The opening of this container may also bring fatal consequences. The job of handling these containers usually falls to the bomb squad with a HAZMAT team interface.

Bomb Threat Standoff Distances

Threat description	Explosive capacity	Lethal air blast range	Mandatory evacuation distance	Desired evacuation distance
Pipe bomb	5 lbs.	25 ft.	70 ft.	850 ft.
Briefcase or suitcase	50 lbs.	40 ft.	150 ft.	1,850 ft.
Compact sedan	220 lbs.	60 ft.	240 ft.	915 ft.
Sedan	500 lbs.	100 ft.	320 ft.	1,050 ft.
Van	1,000 lbs.	125 ft.	400 ft.	1,200 ft.
Moving van or delivery truck	4,000 lbs.	200 ft.	640 ft.	1,750 ft.
Semi-trailer	40,000 lbs.	450 ft.	1,400 ft.	3,500 ft.

Explosive capacity—based on the maximum volume or weight of explosives (TNT equivalent) that could reasonably be hidden in the package or vehicle.

Lethal air blast range—the minimum distance personnel in the open are expected to survive from blast effects. It is based on severe lung damage or fatal impact injury from body translation.

Mandatory evacuation distance—the range to which all buildings must be evacuated. From this range to the desired evacuation distance, personnel may remain in the building (with some risk) but should move to a safe area in the interior of the building away from windows and exterior walls. Evacuated personnel must move to the desired evacuation distance.

Desired evacuation distance—the range to which personnel in the open must be evacuated and the preferred range for building evacuation. This is the maximum range of the threat from flying shrapnel/debris or flying glass from window breakage.

Source: Developed by the ATF, with technical assistance from the U.S. Corps of Engineers. Supported by the Technical Support Working Group (TSWG), a research and development arm of the National Security Council Interagency working group.

TABLE 28-3

Figure 28-14 The handling of explosives requires extensive training and equipment.

In some cases toxic materials may be brought in, which can contaminate the person who accepts the package and cause severe health problems for the firefighters in the station. Depending on state regulations, if fire department personnel accept a package from a citizen, the department may end up owning the package and may be required to pay for the proper disposal of the item—sometimes a costly good deed.

Gases

Incidents involving gases include both flammable and nonflammable gases, with the most commonly released gases being flammable. Also in this category are poisonous gases. The poisons section provides more information on poisonous gases.

Luckily for many first responders, many departments carry gas detection devices that will warn of potentially explosive atmospheres involving flammable gases. Depending on the setup the detector may warn of oxygen-deficient atmospheres, which may be caused by the release of a nonflammable gas. The two most commonly released flammable

Figure 28-15 This truck has separate tanks containing ammonium, nitrate, and fuel oil. In the event of a crash, they may mix, but without an initiating charge it is unlikely to detonate. *(Photo courtesy of the Maryland Department of the Environment)*

gases are natural gas and propane. Common propane bottles are shown in **Figure 28-16**. Although used for the same purposes, propane and natural gas do have differing characteristics that can affect a response to a gas leak. Both gases are odorless when they exist in their natural state, so when moved through a distribution system an odorant is added. In major interstate pipelines, however, the gas may be transported without an odorant added.

> **Note** The largest difference between natural gas and propane is the vapor density. Propane is heavier than air and will stay low to the ground trying to find an ignition source. Natural gas, on the other hand, will rise in air and should dissipate quickly.

This is weather dependent, and certain weather conditions can affect these characteristics. Under some conditions the gas may travel the length of a pipe, following the path of least resistance, usually coming up inside a building. The leak may not be detected for a few days or weeks until it surfaces. This is most common in areas of the country where the ground freezes and limits the upward movement of the gas. Many states offer training on natural gas and propane emergencies; this training is often provided by the gas companies themselves.

Incidents involving these two gases are commonplace, and for natural gas the most common incident

Figure 28-16 These 20-pound cylinders, found in most homes, can create large fireballs and can explode with considerable force if involved in fire.

involves a ruptured gas main. The gas system is set up on a grid, with both large distribution pipes and smaller delivery pipes. The pipes leading into a home are typically ½ to 1 inch in diameter. Distribution pipes may be 2 to 36 inches depending on the region. First responder actions at these types of incidents generally involve isolation and protection, and then the team stands by until the line can be shut off. Gas detection devices should be employed to determine the true hazard area, and adjacent buildings should be checked for gas. When checking buildings, sampling should be continuous and the highest point should be checked prior to declaring a building "safe."

> **Caution** At no point should first responders jump into a hole to attempt to shut off a leak, nor should underground valves be shut in an attempt to stop the leak.

Gas line valves may actually be keeping the amount of gas reduced, and moving the valve may increase the amount of gas escaping. If the leak is on the outtake site of the meter, it is acceptable for the valve on the meter to be shut off. The meter should be locked out and tagged out, and only the gas company should turn the flow of gas back on. Some departments carry special tags that mark the system as being out of service, and some actually lock the system in the off position. When the gas company arrives, they repair the leak and place the system back in service. When dealing with pipes and electrical systems, an OSHA regulation prescribes the procedure for shutting off a system and marking it so that the system is not accidentally turned back on prematurely. This regulation is known as the "lock out, tag out" regulation and applies to many fire service situations.

Propane releases generally involve cylinders, but leaks can also happen in a pipeline. A number of the pipelines running across many states carry propane as well as natural gas. Other than its flammability, the fact that propane sinks and stays low to the ground creates an additional hazard. If a propane cylinder is releasing liquid propane, the liquid will be very cold, sometimes −90°F. A common incident involving propane cylinders, especially those 20-pound cylinders designed for home barbecue grills, involves the overfilling of the cylinder. Propane cylinders are only supposed to be filled to 80 percent of the capacity of the cylinder to allow for expansion of the gas. If the cylinder is filled more than the 80 percent and the temperature increases, the gas may escape the relief valve or frangible disk.

For fires involving propane tanks, there is the potential for a BLEVE, an event that can be catastrophic to the responders. A BLEVE can occur with any flammable gas storage tank, but is usually associated with propane tanks. When pressurized tanks explode, they can travel for a considerable distance. Although no one can predict exactly when a tank will fail, some indicators that a BLEVE may be forthcoming are increased flame height, or the appearance that the flames coming from the relief valve are under high pressure. The sound of a relief valve is deafening, but when the pressure increases the pitch of the sound will get higher and may become louder.

> **Safety** If the tank becomes discolored, distended, or loses shape, or the sound from the relief valve increases, an immediate withdrawal is indicated.

When making the decision to fight a propane tank fire, a large quantity of water needs to be applied quickly and continuously. The vapor space of the tank should be concentrated on, and the fire should not be extinguished unless the responder is certain the flow of gas can be stopped. When preplanning, first responders should plan to establish a flow of water on the tank in excess of 500 gpm within a few minutes of arrival. Each year a number of firefighters are killed by BLEVEs; in fact, in many cases whole alarm assignments have been killed and seriously injured during the BLEVE. More information in BLEVEs is found in Chapter 24 and Chapter 25.

Some cars now are powered by natural gas or propane; the most common are cars that are part of a fleet of vehicles, including government and utility company vehicles. To determine if a vehicle is powered by one of these gases, the firefighter should look for a sticker that reads CNG (compressed natural gas), LP (liquefied petroleum), propane powered, or natural gas powered. Sample stickers are shown in **Figures 28-17A** and **B.** Other vehicles that may have compressed gas cylinders are recreational vehicles (motor homes) and work utility vehicles.

Other common gas releases involve carbon dioxide, chlorine, and ammonia. Although carbon dioxide (CO_2) is a common fire extinguishing agent, it is also used for the distribution of beverages and is commonly found in restaurants, bars, and convenience stores. An increase of CO_2 in a building can make people sick, and it is becoming a common **sick building chemical**. Although not commonly looked for, when dealing with emer-

(A)

(B)

Figure 28-17 Vehicles that use alternative fuels such as natural gas, propane, or electric are marked to indicate the fuel source. Photo A shows a CNG sticker on the rear bumper, indicating that the car is fueled by compressed natural gas. The sticker in photo B is beside the front driver's quarter panel and indicates this vehicle is environmentally friendly alternatively fueled. Cars with alternative fuels typically still have a gasoline tank, and there may be gasoline present in emergencies.

gency response to a **sick building**, CO_2 should be considered as a possible source of the problem. Both chlorine and ammonia have good warning properties. Their distinct odors are often easily identified by the people in the building. In most cases the odor is so irritating that people will self-evacuate the building prior to hazardous levels being built up.

Figure 28-18 Responding to this emergency presented several challenges, due mainly to the final resting place of the overturned truck. Although HAZMAT response prefers an uphill position, the best place for the offload truck was downhill. Extra distance and additional protection lines were necessary to compensate for the disadvantaged position. *(Photo courtesy of the Maryland Department of the Environment)*

Flammable and Combustible Liquids

By far this is the leading category for the most common type of releases, because it includes gasoline and diesel fuel, **Figure 28-18**. Firefighters respond to these types of incidents thousands of times a day.

> **Safety** Gasoline is the leading chemical when it comes to chemical accident fatalities, and transportation leads in those fatalities. Unfortunately, due to its familiarity, many emergency responders do not adequately protect themselves when responding to incidents of this type.

Gasoline has between 1 and 5 percent benzene, which is a confirmed human cancer-causing agent (carcinogen) to which firefighters are commonly exposed. A person who smells gasoline is receiving an exposure to benzene, an exposure that is repeated on a regular basis. When dealing with small spills firefighters may have a tendency to use little or no protective clothing, but their bodies are continually being exposed to this toxic material. In addition to its toxicity, gasoline is also very flammable. It can even be ignited by the static electricity generated by clothing as a responder approaches an incident.

Although of less concern than gasoline, diesel fuel, fuel oil, and kerosene do have some toxicity concerns. This lower level of concern can catch a responder unaware. Although in normal circumstances it is difficult to ignite these materials, if their temperature increases they easily ignite. On a day with a temperature below 70°F the potential for ignition is low, but a spill onto blacktop when it is 100°F outside adds considerable risk for a fire because a larger quantity of vapors is going to be produced on the pavement, which may have temperatures in excess of 110°F.

In years past, spills of flammable or combustible liquids were commonly flushed down storm drains. This practice for the most part has been discontinued because it is severely environmentally damaging, not to mention that it only moves the problem to another location. Many areas of the country collect the spilled material and dispose of it in an environmentally sound manner. The discharging of oil (includes fuels) is a violation of the federal Clean Water Act, and emergency responders could face serious fines if the material were flushed into a waterway.[3]

Streetsmart Tip **To Fight or Not to Fight?** One common incident involves overturned or burning tanker trucks. The product found most often in these trucks is gasoline, and when one of these trucks is involved in an accident, a fire usually results.

When first responding companies arrive they have a tendency to try to extinguish the fire. Without large quantities of firefighting foam it is very difficult to extinguish a tanker fire. If, on arrival, the truck is well involved and is not near any exposures and is not impacting an adjacent community, it is usually best to let the fire continue to burn because when attempting to extinguish this type of fire there is considerable runoff, which usually contains both the gasoline and firefighting foam, both of which are best kept out of waterways. Also, if the fire is extinguished, a large amount of gasoline at an elevated temperature may remain so foam must be reapplied every few minutes to ensure that the fire does not reignite, **Figure 28-19.** It is then necessary to remove the hot gasoline from the truck and pump it into another truck, a dangerous proposition. If, on the other hand, the fire is allowed to burn, nothing would remain that could cause harm to responders or the environment. The resulting black smoke, although it looks horrible, is in reality less damaging to the environment than a liquid spill. The smoke is predominantly carbon, which provides the thick black smoke. However, if the truck is impacting a community or adjacent structure, then all attempts should be made to extinguish the fire. The truck may be wrecked under a bridge, which is usually considered a critical structure in a community. The cost of a bridge can be in the millions, so responders should try to calculate the financial impact in terms of hard dollars and inconvenience if the bridge were destroyed and it took a year to rebuild. If units arrive and are confronted with a large fire, the initial water should be directed toward the bridge. When sufficient foam and water have arrived, the truck fire should be extinguished, while continually applying foam.

The application of foam at a non-fire incident usually causes friction between firefighters and environmental agencies because the foam can be damaging to the environment. Even if an "environmentally safe" foam is used, the breakdown products of the fuel are not environmentally safe, so environmental agencies prefer the use of minimal foam.

Another concern arises when a truck is overturned and it must be drilled and pumped out prior to righting the truck. The application of foam makes the situation very slippery, hence creating additional hazards. The use of air monitors to determine when the foam blanket is breaking down is recommended, and the use of as little foam as possible is recommended.

Flammable Solids, Water Reactives, and Spontaneously Combustible Materials

When dealing with materials in these categories, a specific identity and emergency response information are crucial. Consultation with the HAZMAT team is important, because using the wrong tactic can be devastating to the community. Most emergency responders have some experience with flammable solids, because road flares are classified as flammable solids. In most cases flammable solids are difficult to ignite, but once ignited they burn vigorously and are difficult to extinguish.

Figure 28-19 Although a necessary evil when dealing with flammable liquids, responders should minimize the amount of foam applied at a spill. It can cause further environmental damage, not to mention the slip, trip, and fall hazards. *(Photo courtesy of Baltimore County Fire Department)*

The water-reactive group can be defined in two ways. When a water-reactive material gets wet, a violent reaction, such as a fire or explosion, can result. This is what most people think about when they hear the term *water reactive.* The water-reactive category also includes reactive metals that, once ignited, can create a problem, such as a violent reaction, if water is applied. Magnesium is one such reactive metal. Once ignited, magnesium burns vigorously and when water is applied it may explode, as shown in **Figure 28-20**.

Some materials such as calcium carbide are shipped as water-reactive materials. When water is applied to calcium carbide, a little bubbling occurs, which would not be considered very violent. The gas that is being released by the bubbles, however, is acetylene gas. When pure acetylene gas is produced in this manner it is unstable and reactive. If there is an ignition source nearby there will be a fire or explosion. So the actual application of water in itself does not produce a significant problem, but the action of the water may create additional concerns.

Materials that are spontaneously combustible are usually transported in a manner designed to keep them stable. To remain stable, they may require storage at a specific temperature or have a material added to them. White phosphorus, which is usually covered with water, is an example of a spontaneously combustible material, because it ignites in the presence of air. Communities with facilities that use flammable metals should have a stockpile of metal extinguishing agents such as Metal-X or Lith-X, commonly referred to as Class D extinguishing powder. Sand that is known to be dry (e.g., it has been stored inside in a closed container) can also be used, but most sand has moisture contained within it.

Figure 28-20 Magnesium is water reactive, and when on fire is very reactive to the water being used to attempt extinguishment.

Safety The use of water on flammable metals is not recommended and can cause severe injuries to the hose crew, because the metal has a tendency to explode, sending hot metal and other fragments flying.

The water will break down and release hydrogen gas, which will increase the amount of heat and flame being produced. Most HAZMAT teams have a limited supply of this type of extinguishing agent and usually rely on a local industrial resource for more.

Oxidizers and Organic Peroxides

Like the previous group, this is another class of materials for which help should be requested early. Oxidizers and organic peroxides can have explosive characteristics, and attempting to deal with them can lead to fatal mistakes. Organic peroxides can react explosively even if the responders have taken no action.

One of the best known oxidizers is ammonium nitrate. By itself it is relatively harmless, although it does have some toxicity associated with it. It cannot explode unless mixed with a fuel and an initiating charge added. Although all three items should not be transported together, on occasion ammonium nitrate, fuel, and an initiating charge might be on the same truck. In other situations the oxidizer does not require any oxygen to start a fire, because it will provide its own. Oxygen is placarded with a specific oxygen placard or it can be shipped with an oxidizer placard. Oxygen itself cannot be ignited, but it will greatly intensify a fire when it is involved in amounts greater than the 20.9 percent that is found in air.

Liquefied oxygen (LOX), which is a cryogenic, presents even more hazards in addition to supporting combustion. On asphalt (or other hydrocarbon material) it is shock sensitive and can detonate if compressed, such as by a firefighter walking on it. Another concern is the absorption of oxygen from a LOX release by a firefighter's turnout gear, which can present a flammability problem. Although caution should be used, and the exposed gear should be given time to air out, this would be a very rare occurrence. Most hospitals have large upright cryogenic storage tanks of liquid oxygen, as do many nursing homes, **Figure 28-21**. Smaller in-home versions are available as well, presenting a large fire risk in a residential home, not to mention a freezing and contact hazard if knocked over.

Figure 28-21 A leak of liquid oxygen on asphalt can present a shock-sensitivity problem in addition to the increased risk of a fire.

Another commonplace situation involving oxidizers is encountered when dealing with pool chemicals. Most pool chemicals are oxidizers or may react with oxidizers and can be involved in a chemical reaction or fire. If the materials do not ignite, a very dangerous situation can develop, because a large vapor cloud can be created from just a handful of pool chemicals—one that can have disastrous effects on a neighborhood. Specific advice from the HAZMAT team is required prior to handling or disposing of any of these types of chemicals. There is a lessened level of concern with these types of chemicals because they are considered "household," but they do present a large hazard to the community.

Poisons

Although Class 6 is poison liquids, included here are poisonous gases from Class 2 (gases). These types of materials, which include the "Stow Away from Foodstuffs" and "Marine Pollutant" materials, are toxic to both humans and the environment. Although they are toxic in varying degrees, for the first responder they are poisonous and should be treated as such. The materials labeled as poisonous are very much so, because the DOT does not include many of the materials that are toxic in this category. When choosing a hazard class for a material, the predominant hazard determines its placement into one or another category. Some improvements have been made in the subsidiary placarding and labeling category, but there is room for more. Poisonous gases may be accompanied by a label on a bulk container that reads "Poisonous by Inhalation" and would be listed on the shipped papers as PIH. By definition, these materials present a risk to humans and the community and require extra precautions.

The most common incidents with these types of materials result from pesticides and agricultural chemicals. Although most household materials are of lesser concentrations, on occasion higher strength materials have been used. There have been several incidents in which bug-spraying companies (or individuals) have used full-strength agricultural products in residential homes, requiring evacuations lasting a few weeks and removal of some homes due to the contamination.

Safety Technical-grade pesticides are very dangerous and should be handled with care.

Incidents involving commercial home fertilizing trucks are common, but in most cases this material is diluted and ready for application. Fertilizers do not present much risk to the responders or the environment unless found in large quantities. Pesticides and insecticides do, however, present a risk to the responders and the environment, and responders should consult with a HAZMAT team prior to taking any action. Many home pest control vehicles carry a mixed load of premixed and undiluted materials. If such a vehicle is involved in an auto accident, the use of full gear is recommended, and anyone who comes in contact with any of the liquid or solid materials should be decontaminated.

Radioactive Materials

Although considerable emphasis was placed on radiation in the early days of HAZMAT, due to a low number of incidents involving these materials, some of the emphasis has been redirected to other areas. It is true that incidents involving these types of materials are rare, but they do have a significant impact on the well-being of the community, and with the potential now for terrorism, radiation is being emphasized again.

Radioactive materials are commonly used in the community, in smoke detectors, in ground imaging equipment, and in the medical community. Although many people express concern when around radioactive materials, most radioactive materials are not harmful even for a lengthy exposure.

Radiation is divided into two categories: ionizing and nonionizing radiation. Radioactive materials that emit alpha, beta, gamma, and neutron forms of radiation are ionizing types. The ionizing forms of radiation are able to make changes in atoms, which can lead to health problems with humans. Nonionizing radiation takes the form of sunlight (visible light), microwaves, and radio waves. These forms of energy are not as strong as their ionizing counterparts, but high amounts or repeated exposure (e.g., to sunlight) can cause health problems.

Safety Keeping the material in the container that it is intended to be in is paramount to reducing the exposure levels of the responders.

Some radioactive materials, such as those coming to and from a nuclear power–generating facility, can present some risk to the community. Radioactive materials that present a risk to the community are shipped in high-strength containers designed to withstand a substantial crash and to be involved in a fire situation without any release of radioactive materials. These shipments are usually well tracked and, when involved in an incident, specialized help is usually already on the way before local responders call for it. Knowing who is the local contact for radiation emergencies is key to an effective response when dealing with radioactive materials.

When dealing with potential radiation hazards, the adage "time, distance, and shielding" should always be followed. This basic principle of radiation exposure should be applied to *all* chemical exposures. Firefighters must be careful to limit their time around a radiation source, keep their distance, and wear some type of shielding; then they can be protected against most forms of radiation. Most radiation exposure guidelines are based on a one-hour exposure time. By reducing their time, firefighters reduce their exposure. By doubling their distance, they can reduce their radiation exposure to one-fourth of the original exposure. Protective clothing and SCBA can provide some shielding for some forms of radiation, while more substantial forms of shielding such as lead or concrete may be needed for more dangerous types of radiation. Persons who have been exposed to a radiation source are not radioactive. They may suffer some significant health problems, but they do not present a risk to others. Persons who have radioactive material on them, such as a radioactive powder, are contaminated and may present a risk to others. Simple decontamination with soap and water is needed to reduce the damage to the person. When dealing with radioactive materials it is best to try to contain the runoff so as not to further contaminate the incident scene. The best protection against radiation is to keep the radiation source in its protective container.

Corrosives

After the flammable and combustible categories, the next most likely HAZMAT incident will probably involve a corrosive. The most common incidents occur with sulfuric acid, hydrochloric (muriatic) acid, and sodium hydroxide. Both of the acids are transported as liquids, but the sodium hydroxide may be transported wet or dry. Both sulfuric acid and sodium hydroxide have little or no vapor pressure, so they present little risk outside of the immediate spill area. Hydrochloric acid, on the other hand, does have a high vapor pressure and may require some additional isolation and evacuation.

To handle any of these materials, chemical protective clothing is required, because after some time they can eat a hole in turnout gear. A responder who is splashed with any of these three materials should wash the material off as soon as possible. Note that the actual burns will take a couple of minutes to become evident, and will progress in severity if not quickly washed off. This applies to these three materials only, because some of the less common acids, such as oleum or sulfur trioxide, will cause burns immediately upon contact, but it is still a matter of time before disfiguring injuries occur.

Chemical neutralization may be the best choice for handling a corrosive spill. This type of operation needs to be performed by a HAZMAT technician who has donned appropriate levels of PPE. When neutralizing chemicals, heat and violent reactions may occur, so neutralization should take place only after consultation with a chemist.

Other Incidents

It is impossible to outline each specific action that first responders should take at a chemical release. The first rule if a responder suspects that hazardous materials are involved is to isolate the area and evacuate other persons in the immediate area until further information is available. Request the assistance of the closest HAZMAT team because their technical expertise and equipment will be needed. First responders are not trained nor equipped to provide detection or characterization of unknown materials, and to use their senses, such as smell, is to play Russian roulette.

Safety The absence of a vapor cloud or odor does not mean that a potentially deadly material is not present.

Many toxic materials are odorless and colorless and thus will provide little warning prior to causing lethal effects. It is not the job of first responders to seek the cause of a chemical release nor to provide an identification. If the information is presented to them or can be obtained without risk, then it is of benefit, but otherwise it is the job of a HAZMAT team to perform those tasks.

Common incidents that first responders may get involved with are sick buildings, or odor complaints that can occur in buildings such as the one shown in **Figure 28-22**. The use of air monitors is the key to survival with these types of incidents,

Figure 28-22 In "sick buildings," the occupants become symptomatic from a variety of potential sources.

but the sense of smell is not sufficient to determine immediately dangerous levels. When involved in sick buildings or odor complaints, the best course of action for first responders is to determine the validity of the complaint. Once it is determined that the call is valid and not related to a sunny Friday afternoon, the first responders should don SCBA and request the services of a HAZMAT team. The occupants of the building should be removed and the windows and doors shut. The HVAC system should be shut down so as not to remove the unknown material. These steps are crucial to the success of the HAZMAT team's attempts to find the source of the problem.

First responders may also get called to gas leaks inside a building, and their ultimate safety is determined by a suitable air monitor. It is impossible to determine the level of a gas in a building based on smell. Once in an environment that has a gas odor, the body will desensitize itself to the odor and responders will not be able to smell the gas anymore, even though the amount has been increasing, moving closer to a potential explosion.

Gas grills are prone to propane leaks, and if the leak reaches an ignition source can cause severe damage to buildings. Many HAZMAT teams carry pipe flares that burn off the gas, preventing an explosion. First responder actions should be limited to isolation and evacuation and to preparing standby hoselines. If the vapors are traveling toward other homes, then the use of hoselines with fog nozzles to move the vapors is appropriate.

DECONTAMINATION

The task of **decontamination** may fall to a person trained at the first responder Operations level. If a first responder is expected to perform decontamination, then specific training in that area is required, as well as training in the use of chemical PPE. The definition of *decontamination,* or *decon* for short, is the physical removal of contaminants from people, equipment, and the environment. It is important to note that the concepts provided here may be for one or all three of these areas of concern. Just because one type of decontaminating solution is effective for a tool does not mean that it can be used on humans. Prior to performing any decontamination procedure on humans, first responders should consult with the HAZMAT team or a chemist. Some decon solutions use very dangerous materials, especially if they come in contact with skin. For equipment it may be acceptable to use large quantities of sodium hydroxide mixed with detergent, but this same solution would cause severe burns if placed on the skin. When using decon solutions on humans, caution and consultation is required.

The two ways to become contaminated are via direct contact or secondary contamination. If first responders go into a spill area and put their hands into a drum of blue paint, direct contamination has occurred. When they leave the area and shake hands with another responder, they have secondarily contaminated the other responder with the blue paint.

Types of Decontamination

There are five general types of decontamination levels: **emergency decon, gross decon, formal decon, fine decon**, and **mass decon**. In a large full-scale incident, all five types may be used, but in reality the majority of incidents require only minimal decon. There are two large categories of decontamination, wet and dry decontamination, and both categories have the five levels with some modifications. The two categories should be self-explanatory: One requires the use of water or other liquid solution, whereas the other does not involve a wet process.

The process of decontamination is chemical specific so there are no absolute rules. First responders should as a minimum have a good understanding of the emergency decon procedures and develop a local procedure to handle this type of problem. When confronted with a contaminated patient who is in need of decon, there will be no time to develop the procedure. With potential terrorism in mind, responders should think about the emergency decon of not one person, but thousands of people.

Emergency Decontamination

Emergency decontamination can be as simple as taking a hoseline and washing someone off, or it can be complicated by an unconscious patient on a backboard or in a Stokes basket. For most chemicals, plain water washing is sufficient to decontaminate the person, as long as the person's clothes are removed. Removal of the patient's clothes removes 60 to 90 percent of the contamination. As toxic as nerve agents are, if the patient has been contaminated with an aerosol, then just having the person stay in fresh air for fifteen minutes can be an effective decon method. Between fresh air and clothing removal, effective emergency decon can be accomplished.

The common methods of emergency decon involve the use of a water line and some method of runoff control, such as that shown in **Figure 28-23**.

Figure 28-23 One of the simplest forms of emergency decon is the use of a hoseline.

Note In reality when people's lives are on the line, runoff is not a large concern, but when not under those types of conditions, attempts should be made to recover any runoff.

An inflatable children's pool, which can be easily carried on a fire truck, can be used to contain the runoff. Another method involves the use of a tarp or plastic that is spread between two ladders. If none of these methods is possible, then using a tarp or plastic to catch as much runoff as possible is advised. If all other methods are not available then it is best to use natural areas that would collect the runoff.

Decon should not be performed over storm drains or other environmental concerns. An industrial plant may have safety showers, and it is acceptable to use these safety showers. Prior to using the showers it is important to ask the people from the facility where the runoff goes. In most cases the runoff will go to the facility's own water treatment area and can be taken care of by that system. Caution should be used when the safety shower drains straight to the sanitary system; prior to using such a shower, the local sewer authority should be contacted.

One of the other general rules with regard to emergency decon involves the removal of corrosives from a victim. Large quantities of water should be used, because if small amounts of water are mixed with the corrosive there is the potential for a heat increase. Massive amounts of water will eliminate this potential for further injury. Besides using large quantities of water on corrosives, it is necessary to continue the flushing for a minimum of twenty minutes. No matter how bad the temptation is for EMS providers to remove the patient, the best treatment for corrosive burns is the water flush, especially for burns caused by a base, such as sodium hydroxide.

Figure 28-24 Gross decontamination is used to remove the majority of the contamination.

Gross Decontamination

The process of gross decon is the removal of the majority of the contaminants, and is usually the first rinsing station within a full decon setup, as shown in **Figure 28-24**. In many cases gross decon is just part of the whole process, but it can be the only step in a minor setup. Some HAZMAT teams use shower setups for gross decon, whereas others may use hoselines or pump tanks. Because this step is the most likely location for contaminants to collect, attempts should be made to recover any of the runoff. When part of a multistep process, the entry team usually performs this step on themselves without any assistance.

When dry decon is used, overgarments such as booties or extra gloves may have been used that, when leaving the hot sector, are placed in a designated container. For many materials dry decon is an acceptable use of PPE and resources.

Figure 28-25 Formal decontamination is used to remove any further contamination that may remain after gross decontamination.

When using any type of decon, the bottom line is to make it safe for the responder to get out of the PPE and safe for the responders assisting with the PPE removal. When using reusable PPE, decon needs to be more formal, but even extensive cleaning can be done to the PPE after the incident and removal of the responder.

Formal Decontamination

The process of formal decon is one in which some actual scrubbing and cleaning take place to remove any residual contaminants, **Figure 28-25**. It is usually done after the gross decon and usually has a couple of steps. Attempts to recover any runoff should be used, but this is not as critical as for the gross decon step.

It is during the formal decon procedure that decon solutions are typically added to the process. These solutions are chemical specific. This process may involve showers, hoselines, or pump sprayers, and one or two people dressed in a lower level of PPE may perform this activity. When doing formal decon it is important to pay particular attention to the areas most likely to be contaminated, the hands and feet.

Fine Decontamination

This form of decon is not performed by first responders, but by hospital-based personnel. It is a cleaning that removes all of the contaminants from the body. It would be nearly impossible to clean eyes, ears, fingernails, and other areas of the body in the street; these tasks are better accomplished in a hospital, as shown in **Figure 28-26**. According to OSHA the staff must have proper training to perform fine decon, and they must be trained to the Operations level, although some training to the Technician level is preferred. Some hospitals have designated decon areas with a separate entrance, separate ventilation system, and a holding tank for runoff control.

Through the Local Emergency Planning Committee (LEPC) plan, hospitals are required to identify themselves as being able to accomplish decontamination of patients. Every hospital should have a policy to cover a walk-in contaminated patient, and emergency responders should test these plans on a regular basis. Regular training and interfacing are required to keep this type of system functioning well within a community. Responders should ensure that this capability is well run and that sufficient training has taken place, because they are likely candidates to become patients someday!

Mass Decontamination

Most HAZMAT decontamination setups are designed primarily to decontaminate the members of the HAZMAT team. They are designed to handle two to four people dressed in chemical protective clothing. Although most of these systems can handle civilians, they are designed to decontaminate responders' PPE. When decontaminating civilians, several issues arise: clothing, privacy, valuables, weather impact, and inconvenience. Civilians may not be willing to undergo decontamination or to separate from their relatives. They may not want to remove their clothes or valuables. The thought for mass decontamination is that plans must be developed to handle decon of thousands of people. Most of the mass decontamination issues arise when terrorism events are discussed. It is best to have the victims decontaminated before they get to the hospital, because contaminated victims will create additional problems at the hospital. More information on possible terrorism scenarios is discussed in Chapter 30. If victims are contaminated with military

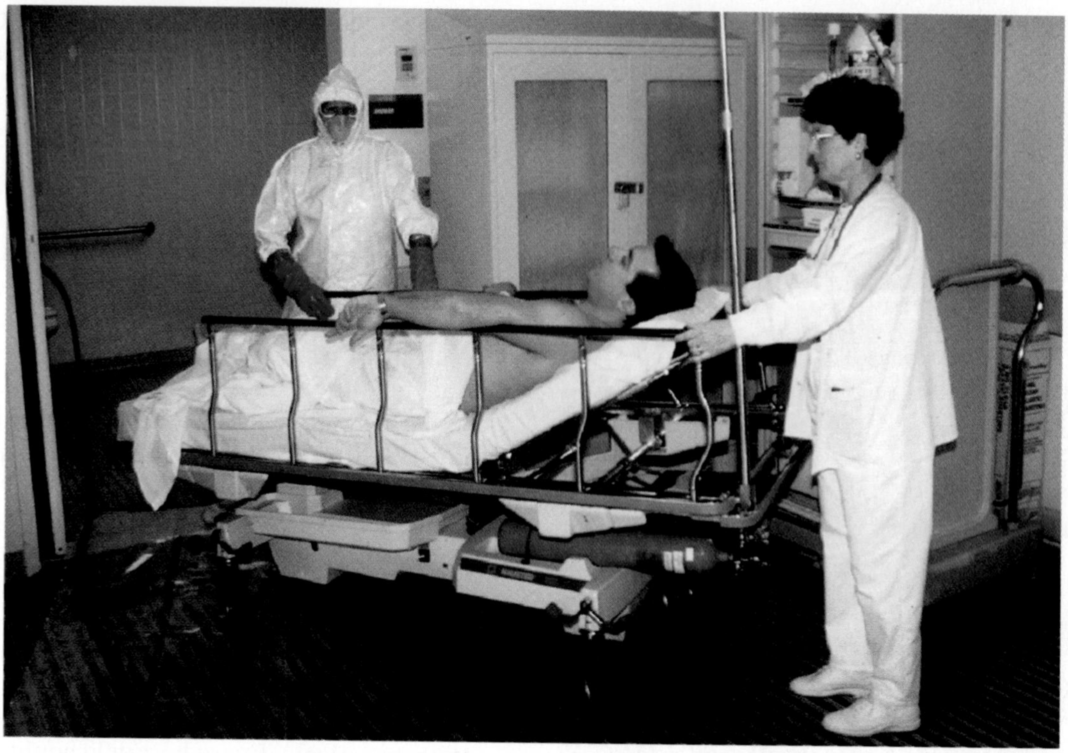

Figure 28-26 Fine decontamination occurs at a hospital and is used to clean the fingernails, ears, and other hard-to-reach areas. This drill victim is headed out from the decon room to the regular emergency department. *(Photo courtesy of the Baltimore County Fire Department)*

nerve or blister agents, speed is of the essence. If decontamination is not performed on symptomatic contaminated victims, their chance for survival may be slim. The hardest part of this process is identifying those who need physical decontamination and those who need **psychological decontamination**. If victims are symptomatic and there is a credible threat presented, then decontamination should be accomplished quickly. First arriving units need to establish this decontamination process. The best decontamination solution is water, followed by soap.

Safety The use of a bleach and water solution in terrorism decontamination is no longer considered an acceptable practice.

Studies from the military have shown that the use of bleach can actually cause more deaths of and injuries to contaminated victims. Decontamination is chemical specific, and unless first responders can identify the specific chemical compound they should use a simple soap and water solution. A hoseline can be used to accomplish emergency victim decontamination, and as more assistance arrives the system can become more elaborate. When setting up the decontamination process, it is important to avoid victim backup and if there are delays, to make sure that the victims wait in water. After the victims exit the decontamination area they should be guided to holding areas. It is important to keep symptomatic victims from nonsymptomatic victims. If these victims are mixed, the nonsymptomatic victims will tend to pick up the symptoms due to psychological reasons. Examples of mass decontamination setup are provided in **Figure 28-27** and **Figure 28-28.**

Figure 28-27 Mass decon should have the victims in water for as long as possible. If time permits, soap and water should be used to better the decontamination process.

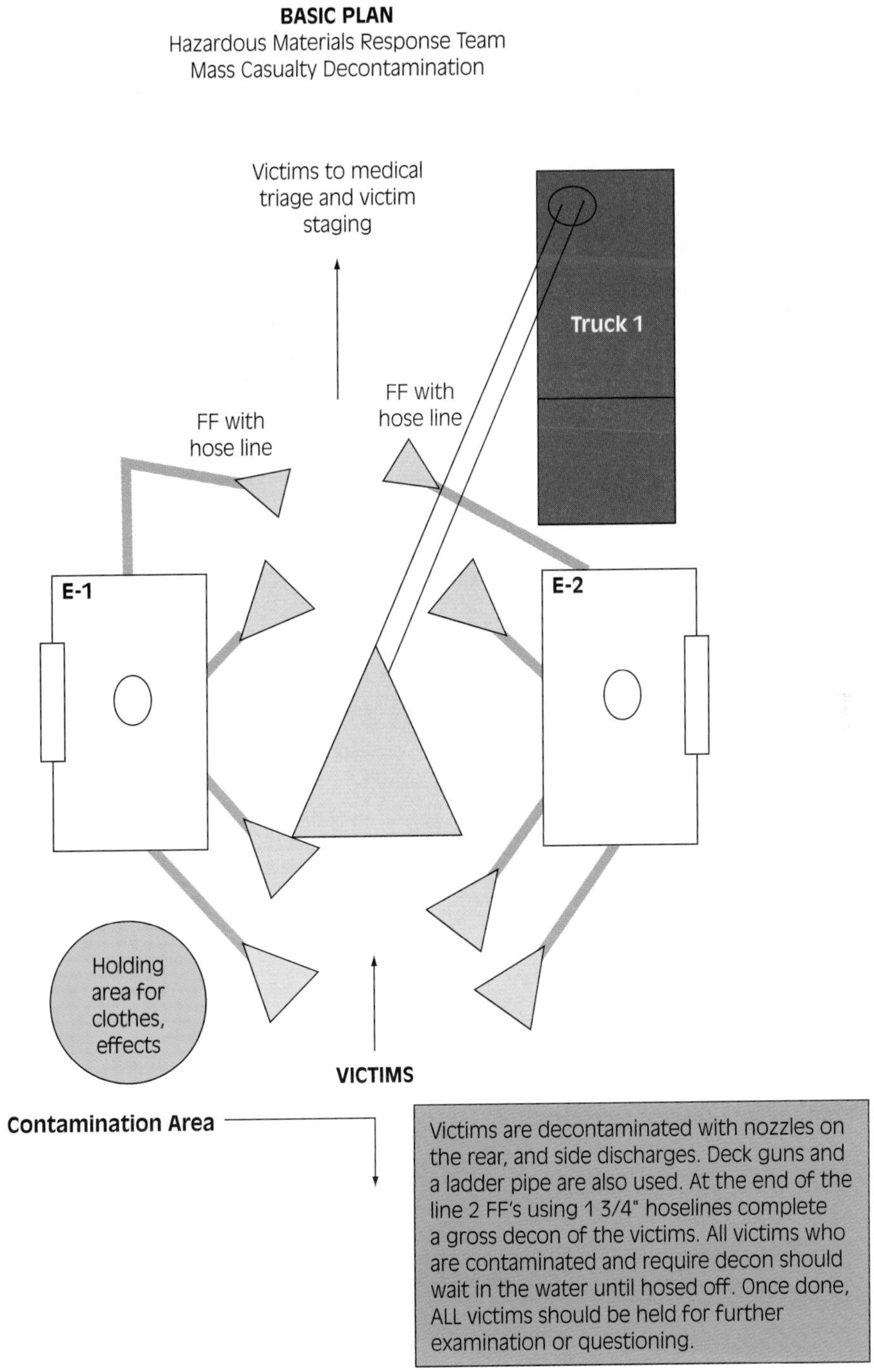
BASIC PLAN
Hazardous Materials Response Team
Mass Casualty Decontamination
Victims to medical triage and victim staging
Truck 1
FF with hose line
FF with hose line
E-1
E-2
Holding area for clothes, effects
VICTIMS
Contamination Area
Victims are decontaminated with nozzles on the rear, and side discharges. Deck guns and a ladder pipe are also used. At the end of the line 2 FF's using 1 3/4" hoselines complete a gross decon of the victims. All victims who are contaminated and require decon should wait in the water until hosed off. Once done, ALL victims should be held for further examination or questioning.

Figure 28-28 Mass decon setups. *(Courtesy of Baltimore County Fire Department)*

Decontamination Process

There are several variations to a decontamination process, but in the overall scheme of things they are basically similar. **JPR 28-1A** through **H** show the basic process and outline the following steps: tool drop, gross decon, formal decon, PPE removal, SCBA removal, clothing removal, body wash (emergency only), dry off, and medical evaluation, including rehydration. The variations include the use of shower systems, tents, number and location of steps, and personnel. The method of runoff collection varies team to team as well, and the pictures depict the use of a 3-inch hoseline that is cut and coupled to provide a rectangular setup, and includes a shower wand manifold. For most situations that involve solids and liquids a minimum type of decon should be set up.

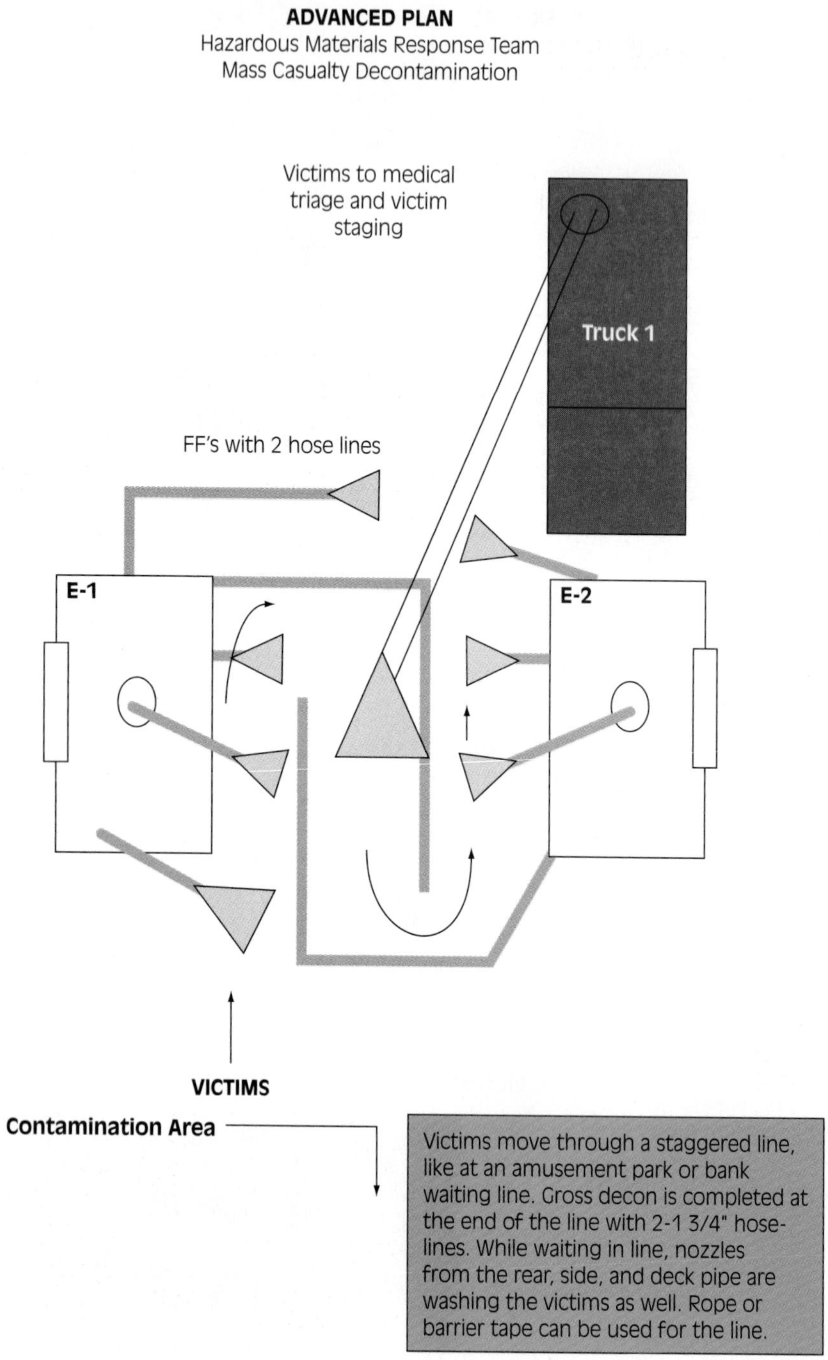
ADVANCED PLAN
Hazardous Materials Response Team
Mass Casualty Decontamination
Victims to medical triage and victim staging
Truck 1
FF's with 2 hose lines
E-1
E-2
VICTIMS
Contamination Area
Victims move through a staggered line, like at an amusement park or bank waiting line. Gross decon is completed at the end of the line with 2-1 3/4" hose-lines. While waiting in line, nozzles from the rear, side, and deck pipe are washing the victims as well. Rope or barrier tape can be used for the line.

Figure 28-28 (Continued) *(Courtesy of Baltimore County Fire Department)*

The HAZWOPER regulation requires that the IC have a plan for decontamination, which can vary from no setup through a multistep process as depicted in the decontamination photo series. For a material such as a technical-grade pesticide, a full setup with all stations is recommended. For gases, in reality no decon is required; however, a minimum setup should be established prior to entry in the event of an emergency and for psychological purposes. Regardless of the type or size of the setup used, it should be set up and ready prior to any entry into a hazard area.

JOB PERFORMANCE REQUIREMENT 28-1

Steps in the Decontamination Process

A The tool drop is usually the first stage in the decon setup. The tools are placed for the use of another entry team or may be collected here for later decontamination.

B In most systems gross decon is the next step. Some response teams use showers or hoselines to accomplish this task.

C The next step is formal decon and may involve two stations in which the responder is rinsed, scrubbed, and rinsed again. The solution used is chemical dependent and varies depending on the contaminant. This is the first step in which other responders may assist.

D After the wet portion of the decon process, the entry team removes their PPE, with the assistance of other responders.

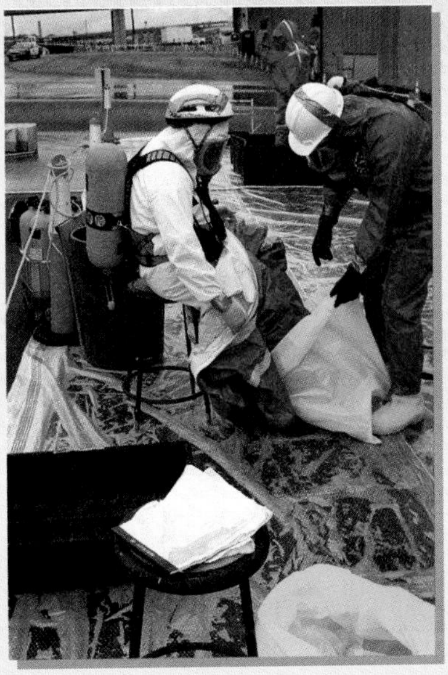

E After the removal of the PPE, the entry team will remove their SCBA. It is usually placed into some type of containment system for later cleaning.

(continued)

JOB PERFORMANCE REQUIREMENT 28-1

Steps in the Decontamination Process (Continued)

F In some systems a fourth washing area is established for body washing. Some teams may set up a tent for this purpose.

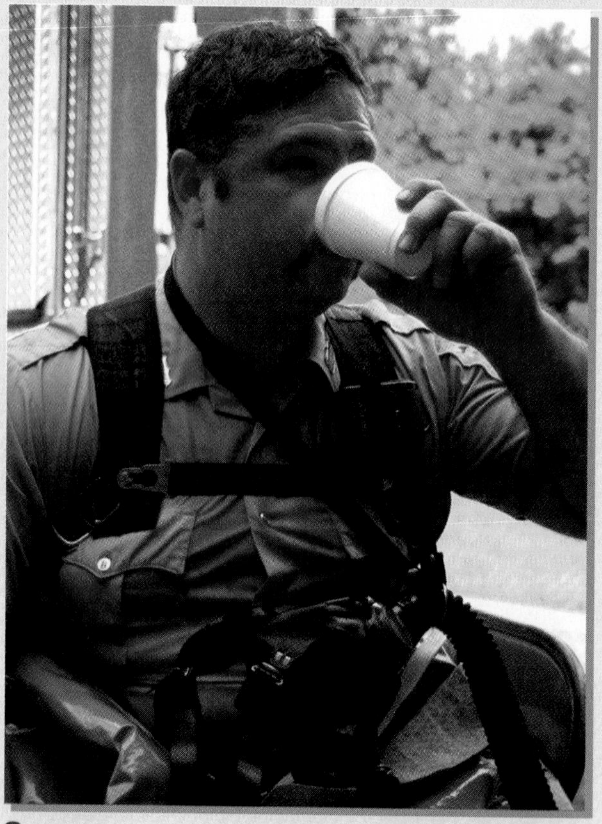

G The last steps are hydration and medical evaluation.

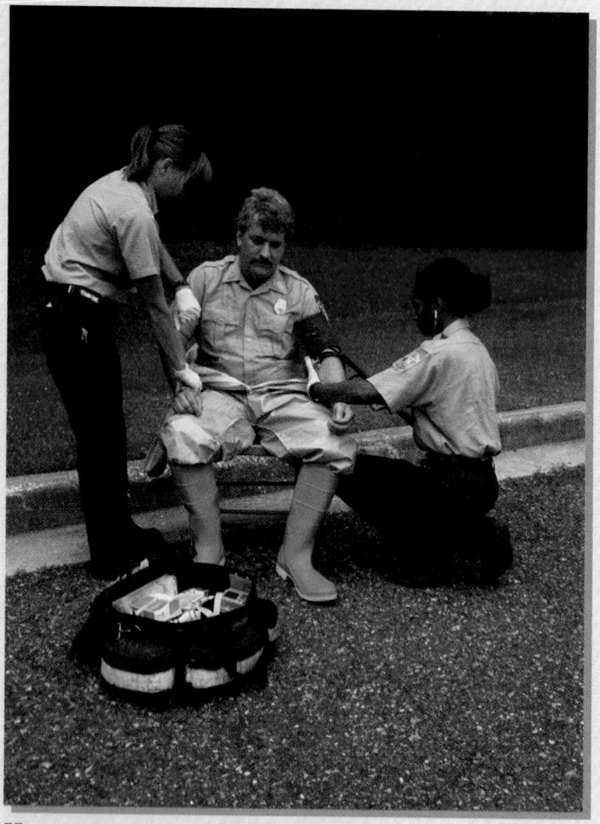

H Medical evaluation after entry is very important. HAZMAT suit work is very stressful on the body.

Methods of Decontamination

TYPE	USED ON HUMANS	USED ON EQUIPMENT	USED ON ENVIRONMENT
Absorption	Yes	Yes	Yes
Adsorption	Yes	Yes	Yes
Covering	No	Yes	Yes
Dilution	Yes	Yes	Yes
Disinfection	Yes	Yes	Yes
Disposal	No	Yes	Yes
Emulsification	Yes	Yes	Yes
Neutralization	Under specific conditions	Yes	Yes
Overpacking	No	Yes	Yes
Removal	Yes	Yes	Yes
Solidification	No	Yes	Yes
Vacuuming	Yes	Yes	Yes
Vapor Dispersion	Yes	Yes	Yes

TABLE 28-4

METHODS OF DECONTAMINATION

There are general methods of decontamination that apply to humans, equipment, and the environment, **Table 28-4**. Each method may not apply directly to all three categories. It is important to consult with the HAZMAT team or a chemist prior to using any of these methods on a human, because they may be more dangerous than the contaminant.

Absorption

The spilled material is picked up by the absorbent material, which acts like a sponge. Common absorbent materials include ground-up newspaper, clay, kitty litter, sawdust, charcoal, and poly fiber. The absorbed material does not change, and the volume of material may increase. The whole mixture will have to be treated as waste and will not change any of the characteristics such as flammability. Compatibility needs to be researched prior to using these materials, because some materials could cause an adverse reaction.

Adsorption

The material to be picked up bonds to the outside of the adsorption medium. Activated carbon and sand are the most common adsorbents and should be available locally at a chemical facility. Many chemical facilities have drums, bags, or totes of activated carbon set up to receive liquids within a process, and may have some available in the event of an emergency.

Covering

Solid materials can often be covered as a way of decontaminating them, but covering could also be used for liquids. As long as the material is compatible with the covering material and will not permeate the material, it is useful to cover a spill. Once the spill area is covered, a method of securing the seal must be used to prevent the cover from blowing away.

Dilution

The ability to dilute a contaminant is dependent on the chemical structure of the spilled material, and it does not work on all contaminants. Imagine getting coated with used motor oil; flushing with water will not remove all of the contamination. When dealing with corrosives, large quantities of water are required, sometimes into the hundreds of thousands of gallons. In recent studies, it has been determined that plain water is as effective as some of the available bleach and water solutions. But with some types of neutralization, dilution is a possible solution.

Disinfection

When dealing with humans a 0.5 percent bleach and water solution can be used for some etiological contaminants. It was initially thought that bleach solutions would also be effective for biological contaminants, but it has been determined that plain water is just as effective for humans. For equipment or the environment, disinfection can be accomplished using a higher percentage of bleach or other chemical.

Disposal

In limited cases, a tool cannot be decontaminated, or a more likely scenario is that it is not cost effective to decontaminate a particular type of tool. In these cases, disposal is the chosen decon method. Contaminated soil is often dug up and disposed of either at an incinerator or hazardous waste landfill.

Emulsification

Emulsification is a form of neutralization and is the chemical altering of a contaminant. Many emulsifiers are sold for fuel and oil products, because they chemically break down the fuels into other components. The individual components may still be harmful to the environment, and the use of emulsifiers may be illegal in some states, except under extreme conditions. Emulsifiers are useful when the contaminant is in an unusual area, such as in a sump pump in a residential basement, that would be difficult to remove otherwise.

Neutralization

Neutralization is usually reserved for corrosive materials, but is also used here to describe the procedure that reduces the toxicity of a poisonous material. Responders are urged to consult with a chemist prior to performing this type of activity. The chemicals used to reduce this risk are in themselves hazardous—but in a much lesser fashion than the contaminant. An emergency responder should avoid mixing a neutralization solution using sodium hydroxide because it is very corrosive, but with some chemicals it is extremely useful in small amounts in reducing the hazard of some types of contaminants. In general neutralization it is exothermic and can produce violent reactions, although these are extremely rare.

Overpacking

This objective is dependent on the material involved and the size of the container. Overpacking is the placing of the material into another container, usually a larger drum. Contaminated clothes may be placed into a 55-gallon drum known as an overpack drum, hence the name.

Removal

Removal is an option reserved for soil in most cases, although equipment or tools could also be removed from the area eliminating the contamination. In most cases the contaminant is removed and overpacked.

Solidification

Solidification is another method that, depending on the solidification agent used, may alter the suspect agent. In some cases, however, the solidification agent will not have any effect on the agent and will reduce the hazard and enable sampling to occur. Chemical compatibility is an item that needs to be confirmed prior to the use of a solidification material.

Vacuuming

For solid materials such as dusts or fibers, a vacuum would certainly reduce the hazard. An ordinary shop vac is not recommended for this task, however, because they are not filtered well enough to keep the exhaust from blowing the agent back into the air. Instead it is recommended to use a vacuum that is equipped with a high-efficiency particulate air (HEPA) filter. Special vacuums are made exclusively for picking up mercury without the mercury vapors being released into the air.

Vapor Dispersion

General ventilation can be used to decontaminate or clear a building. When humans are involved, fresh air can be a decontaminating solution. The use of PPV fans can assist in the removal of contaminants in a building, but the levels of carbon monoxide must be watched so that those levels do

Incident Levels

LEVEL	INCIDENT SCALE
1	Small-scale incident can usually be handled by the first responders, but may also require minimal additional resources. Notifications are usually local and may only be internal to the fire department. The highest level of PPE required is firefighter turnout gear and SCBA, and even that may not be necessary. The material is not toxic by skin absorption, although it may present an inhalation hazard. The size of the spill is usually small and will have minimal environmental impact. Incidents of this type include natural gas or propane leaks and small fuel spills.
2	Incidents at this level usually require additional assistance, such as from a HAZMAT team. The incident may require additional notifications to environmental or emergency management agencies on a local or state basis. The amount of material may be larger or more hazardous than in a Level 1 incident. The type of PPE will probably be chemical protective clothing or other clothing not carried by the first responders. The first responders may have switched from an active role to a support role. A Level 2 incident may require a small evacuation and possibly a large isolation. The incident will also probably impact the ability of the jurisdiction to respond to other emergencies, depending on the resources available. An example incident would be an overturned gasoline tanker or a leaking propane tanker. A leaking drum in the back of a tractor-trailer would be classed as a Level 2 incident.
3	A Level 3 incident requires substantial local resources and requires the assistance of other agencies, local and state. Resources on the federal level may also be required, or are at a minimum notified. The incident will require the evacuation of the affected area and a substantial isolation area. The release is large or the material is extremely toxic. Examples of a Level 3 incident include a train derailment with leaking chlorine railcars. A substantial leak from an ammonia tank truck would also be an example of a Level 3 incident.

TABLE 28-5

not become dangerous. It is possible to add a vapor dispersant into a building to speed up the breakdown of the contaminant, but consultation with a chemist is recommended.

Some response teams and emergency plans outline incident levels to provide quick notifications. **Table 28-5** lists the incident levels and their associated concerns.

Lessons Learned

Protective actions are used for a variety of purposes, the most important of which is to protect the public and in many cases the responder. Management of a chemical release incident is not an easy task, because there are multiagency jurisdictions and a variety of tasks may need to be accomplished at the same time. Unlike most fires or accidents, political pressure is usually associated with a chemical release and there is the potential that thousands of people may require evacuation and relocation.

The determination for decontamination can also be a difficult decision, because most departments do not have the ability to decontaminate large numbers of victims. Choosing the method of decontamination can be difficult as well because of the many possibilities. The best options for protection are to provide prompt isolation and prevent further escalation of the exposed population. For those people who are contaminated, a water-based decontamination method is usually recommended.

KEY TERMS

8-Step Process A management system used to organize the response to a chemical incident. The elements are site management and control, identifying the problem, hazard and risk evaluation, selecting PPE and equipment, information management and resource coordination, implementing response objectives, decon and cleanup operations, and terminating the incident.

ANFO The acronym that is used for ammonium nitrate fuel oil mixture, which is a common explosive. ANFO was used in the Oklahoma City bombing incident.

DECIDE Process A management system used to organize the response to a chemical incident. The factors of DECIDE are detect, estimate, choose, identify, do the best, and evaluate.

Decontamination The physical removal of contaminants (chemicals) from people, equipment, or the environment. Most often used to describe the process of cleaning to remove chemicals from a person.

Emergency Decon The rapid removal of a material from a person when that person (or responder) has become contaminated and needs immediate cleaning. Most emergency decon setups use a single hoseline to perform a quick gross decon of a person with water.

Emergency Response Planning (ERP) Levels that are used for planning purposes and are usually associated with the preplanning for evacuation zones.

Evacuation The movement of people from an area, usually their homes, to another area that is considered to be safe. People are evacuated when they are no longer safe in their current area.

Fine Decon The most detailed of the types of decontamination. Usually performed at a hospital that has trained staff and is equipped to perform fine decon procedures.

Formal Decon The washing and scrubbing portion of the decontamination process. The process is usually repeated and is performed by a decon crew.

GEDAPER Process A management system used to organize the response to a chemical incident. The factors are gather information, estimate potential, determine goals, assess tactical options, plan, evaluate, and review.

Gross Decon The portion of the decontamination process that removes the majority of the chemicals through a flushing process. The gross washing is done using large amounts of water, and is usually done by the individual or the individual's partner.

Isolation Area An area that is set up by responders and is intended to keep people, both citizens and responders, out. May later become the hot zone/sector as the incident evolves. Is the minimum area that should be established at any chemical spill.

Mass Decon A process used to decontaminate large numbers of contaminated victims.

Overpacked A response action that involves the placing of a leaking drum (or container) into another drum. There are drums made specifically to be used as overpack drums in that they are oversized to handle a normal size drum.

Psychological Decon The process performed when persons who have been involved in a situation think they have been contaminated and want to be decontaminated. Responders who have identified that the persons have *not* been contaminated should still consider what can be done to make them feel better.

Sector An area established and identified for a specific reason, typically because a hazard exists within the sector. The sectors are usually referred to as hot, warm, and cold sectors and provide an indication of the expected hazard in each sector. Sometimes referred to as a zone.

Shelter in Place A form of isolation that provides a level of protection while leaving people in place, usually in their homes. People are usually sheltered in place when they may be placed in further danger by an evacuation.

Sick Building A term that is associated with indoor air quality. A building that has an air quality problem is referred to as a *sick building.* In a sick building, occupants become ill as a result of chemicals in and around the building.

Sick Building Chemical When a building is referred to as a *sick building,* certain chemicals exist within that cause health problems for the occupants. These chemicals are referred to as *sick building chemicals.*

Zone An area established and identified for a specific reason, typically because a hazard exists within the zone. The zones are usually referred to as hot, warm, and cold zones and provide an indication of the expected hazard in each zone. Sometimes referred to as a sector.

REVIEW QUESTIONS

1. What is the first priority of any incident management system?
2. What zone or sector should be established first?
3. When dealing with a flammable liquid spill, what could be used to establish the hot zone?
4. What level of incident would be used to describe a chemical release in which 2,000 people were evacuated?
5. What is the minimum amount of decontamination that should be set up for every incident?
6. What are the four types of decontamination?
7. Which decon step are hospitals involved in?
8. The decision to attempt a rescue in a hazardous materials situation depends on what information?
9. What causes the most common breach in a 55-gallon drum?
10. If fire has reached the cargo portion of a truck carrying DOT Class 1.1 materials, what tactics are used to fight the fire?
11. How can air monitors be used in the determination of hazard zones?

Endnotes

1. These top ten (which are in random order) are derived from the EPA, OSHA, and DOT by examining the chemical release records. Because each agency tracks releases in a different fashion, we calculated each agency's top ten and determined an average, the result of which is the list given.
2. In most states the fire department IC is in charge of the incident until command is relinquished to another agency. Check local and state regulations for more information. Although most fire departments will leave when the cleanup contractor begins to work, a problem exists if the contractor's crew is working in a hazardous environment. As long as they are, the FD should maintain a presence because the FD is solely responsible for public safety, a responsibility that cannot be delegated.
3. Any location such as a storm drain, ditch, or culvert that eventually leads to a waterway is defined as a waterway.

Additional Resources

Bevelacqua, Armando S., *Hazardous Materials Chemistry.* Delmar Learning, a part of the Thomson Corporation, Clifton Park, NY, 2001.

Bevelacqua, Armando S. and Stilp, Richard, *Hazardous Materials Field Guide.* Delmar Learning, a part of the Thomson Corporation, Clifton Park, NY, 1998.

Hawley, Christopher, *Hazardous Materials Air Monitoring/Detection Devices.* Delmar Learning, a part of the Thomson Corporation, Clifton Park, NY, 2002.

Hawley, Christopher, *Hazardous Materials Incidents,* 2nd ed., Delmar Learning, a part of the Thomson Corporation, Clifton Park, NY, 2004.

Hawley, Christopher, *Hazardous Materials Response and Operations.* Delmar Learning, a part of the Thomson Corporation, Clifton Park, NY, 2001.

Henry, Timothy, *Emergency Decontamination for Hazardous Materials Responders.* Delmar Learning, a part of the Thomson Corporation, Clifton Park, NY, 1999.

Lesak, David, *Hazardous Materials Strategies and Tactics.* Prentice Hall, 1998.

Noll, Gregory, Michael Hildebrand, and James Yvorra, *Hazardous Materials Managing the Incident.* Fire Protection Publications, Oklahoma University, 1995.

Schnepp, Rob and Gantt, Paul, *Hazardous Meterials: Regulations, Response, and Site Operations.* Delmar Learning, a part of the Thomson Corporation, Clifton Park, NY, 1999.

Stilp, Richard and Armando Bevelacqua, *Emergency Medical Response to Hazardous Materials Incidents.* Delmar Learning, a part of the Thomson Corporation, Clifton Park, NY, 1997.

Stilp, Richard and Bevelacqua, Armando S., *Emergency Medical Response to Hazardous Materials Incidents.* Delmar Learning, a part of the Thomson Corporation, Clifton Park, NY, 1997.

CHAPTER 29

PRODUCT CONTROL AND AIR MONITORING

Christopher Hawley, Baltimore County Fire Department (Ret.)

OUTLINE

- Objectives
- Introduction
- Defensive Operations
- Air Monitoring at the First Responder Operations Level
- Meter Terminology
- Carbon Monoxide Incidents
- Lessons Learned
- Key Terms
- Review Questions
- Additional Resources

STREET STORY

In January 1983 I was faced with the realization of how little I knew about chemical emergencies. It became a reality all too quickly and was a hard lesson to learn. I was called at my home around 4 a.m. to respond to the downtown area of my city for a fire that involved chemicals. It was reported that several firefighters had been injured. En route I was in radio contact with several officers on the scene and was informed that there had been an explosion and that four firefighters had been burned.

Once on the scene, I learned that a firefighter who had driven a car in my wedding party was being sent to a burn center in New York City. Now, not only was I dealing with a difficult incident, but it also now had an emotional aspect tied to it. Even with my limited knowledge, I was given the job of informing and advising the incident commander of the potential hazards involved with several of the chemicals they were able to identify. The fire was caused by the overpressurization of a reactor vessel. Because of the explosion no product control was initiated, and to compound the problem, the property was adjacent to Long Island Sound. This incident was causing environmental damage to the land, water, and air, not to mention the four firefighters who had already been injured. My cadre of tools included two reference books and sixteen hours of hazardous materials training. After several difficult hours trying to identify, evaluate, and provide useful information to both the incident commander and the hospital, I came to the conscious decision that I simply did not have the right tools nor the required knowledge to be very helpful.

This one incident changed the way I would respond to chemical emergencies for the rest of my life. This incident taught me the importance of good identification skills, how important it is to understand every term as it relates to the hazards of the product, and the absolute need for air monitoring devices. To keep our troops safe, being armed with knowledge is the only way to face this type, or any type, of emergency again.

—Street Story by Frank Docimo, Special Operations Officer, Turn of River Fire Department, Turn of River, Connecticut

OBJECTIVES

After completing this chapter, the reader should be able to:

- Identify equipment that can determine hazardous environments and isolation areas.
- Describe available defensive operations for a release.
- Describe the various methods of damming, diking, diverting, and other defensive operations.
- Explain what types of detection equipment are available to the first responder and equipment that a HAZMAT team might carry that could be utilized at a chemical release.

INTRODUCTION

The use of product control techniques can provide a quick reduction in the damage that is done to a community and the environment in the event of a spill. The reduction of the surface area over which a product has spread provides a direct reduction in the danger to responders and the community. By implementing these types of measures, first responders can provide an extreme advantage to the overall mitigation of the incident. The use of air monitoring devices is becoming more commonplace, and many first responders have some type of air monitoring device available to them. This chapter provides a brief overview of this complex subject and concentrates on the knowledge that the first responder should have. To ensure firefighter safety, basic air monitoring must be accomplished, even for basic firefighting, natural gas leaks, gasoline spills, or any other incident in which firefighters may be exposed to hazardous materials.

DEFENSIVE OPERATIONS

According to the NFPA, two large tasks fit into defensive operations at the first responder level: containment and confinement. Both of these can be combined into the task of product control, which is the one task that persons trained to the Operations level can perform as long as they perform these duties away from the hazard area. To perform defensive operations, the responders do not really get close to the actual spill itself. The activities that fit into defensive operations are absorption, diking, damming, diverting, retention, dilution, vapor dispersion, vapor suppression, and the use of remote shutoffs. The type and level of PPE required for each of these tasks will vary with the chemical involved, but a minimum would be turnout gear and SCBA. If making a dam a considerable distance away from the chemical release—and no contact with the material is possible—then a lesser level could be used. The tasks discussed next are common tasks assigned to responders trained to the Operations level, and they are critical tasks that can protect the community and the environment. Although it is not necessary for first responders to carry all of this type of equipment, they should know its location in the event it is needed or know how to improvise, adapt, and overcome if the equipment is not available.

Absorption

First responders are often asked to clean up a spilled material, and on a daily basis, they probably use the **absorption** technique on a large quantity of fuel and oil products using a product similar to the one shown in **Figure 29-1**. First responders should know the type of absorbent materials carried on their apparatus. Some absorbent materials will not pick up water, while others absorb any liquid with which they come

Figure 29-1 A common defensive action is the application of absorbent to a spilled material.

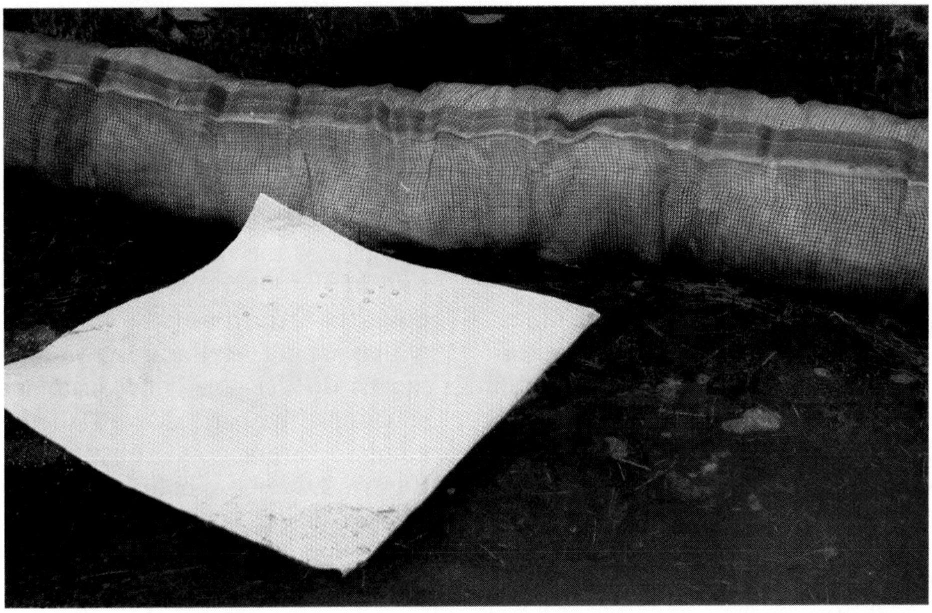

Figure 29-2 Absorbent pads will pick up oil from the water, but will not absorb the water, allowing them to float.

in contact. When trying to pick up fuels off water, having an absorbent that does not pick up water is important. Absorbents vary from clay kitty litter, ground-up newspaper, and corn husks to synthetic fibers. Their weight and absorbent capabilities vary, and not all can be used interchangeably. They may be available in loose bulk form, as pads, **Figure 29-2**, as large rolls, or in boom style.

To hold the absorbent material, construction of a filter fence such as the one shown in **Figure 29-3** is recommended. The filter fence can be made of "rat wire" or other small-weave fencing material. Wooden stakes can be used to hold the fence in place, unless the soil is rocky, in which case metal stakes are recommended.

Also available on the market are solidification agents that encapsulate the spill, as well as microbiological oil-eating bugs. Both of these agents are expensive and although sold as absorption materials, are not generally used by emergency services due to their cost and the quantity required. It is best to carry a variety of absorbent products, as well as several different styles.

Figure 29-3 A screen fence can be used to collect the absorbent material floating on the water after the absorbent has collected the spilled material. The water flows through the fence below the absorbent.

Diking and Damming

The quantity of material spilled will determine to what extent **diking** and **damming** operations are performed. The actions required at a 50-gallon spill differ from those of a 500,000-gallon release. Diking, damming, and diverting are techniques to consider using when a spill occurs on the ground or on a waterway. A spill on the ground is much easier to control than one on the water, but the principles are the same. When creating a dike or a dam, the first responder is either stopping the flow of the material or keeping it from a specific area. Except for small streams, it is nearly impossible to stop a body of water from flowing. If the spilled material is soluble with water, and the waterway has a good flow, it will be impossible to collect the spilled material, unless heavy equipment and a large area are available that could be used to create a very large dam.

First responders often use earth, sand, or rocks for dikes or dams depending on what is available. Having local contacts available around the clock that can supply these items is important, and some jurisdictions have emergency contacts for dump truck loads of sand or dirt. Local or state highway departments are a good place to start when these items are required. Heavy equipment such as front-end loaders or track hoes can also be used to create dams or dikes. When constructing a dam or a dike, it is best to construct three setups. Responders should start at the farthest point away from the spill and work back toward the spill. It may be necessary to establish the first containment miles away from the spill. If the barrier is started near the spill, responders may be overtaken by the spill and be forced to play catch-up, not to mention the need for additional PPE.

Two basic types of dams can be created: overflow and underflow dams, **Figure 29-4**. The specific gravity of the spilled material dictates whether the material will float on top of the water or travel on the bottom, which determines what type of dam needs to be constructed. An overflow dam allows the water to flow over the dam and contain the spilled material at the base of the dam. An underflow dam allows the water to flow under the dam and collect the material on top of the water. Most spilled materials float on top of the water and hence require the use of an underflow dam.

The materials required to construct either type of dam are shovels, dirt or sand, and pipes. Pipes should be at least 4 inches in diameter, but larger is better. The higher the flow, the bigger the pipe and the greater the number of pipes required. A good rule of thumb to use is to have enough pipes to cover two-thirds of the width of the waterway, which would be the minimum amount required. Standard PVC pipe is adequate and, for emergency situations, the hard sleeves (suction) on the engine company are a good substitute. **Figure 29-5** and **Figure 29-6** show both styles of dams and the use of hard sleeves, respectively. For an overflow dam, first responders tend to establish a dam, place the pipes on top, and consider the dam finished. It is best, however, to set the pipes on top of the dam, and then continue with a layer of dirt or sand on top of the pipes, because there is a tendency for the water to erode the dam, and the containment will collapse without this extra layer of dirt or sand.

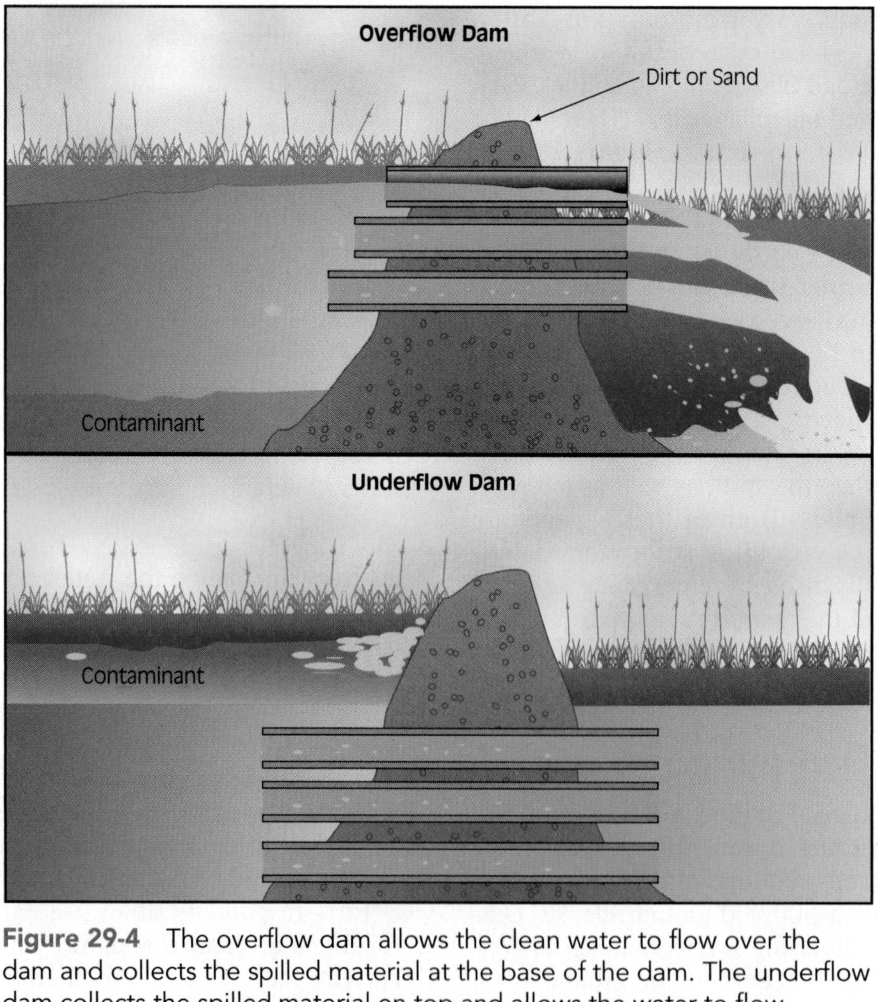

Figure 29-4 The overflow dam allows the clean water to flow over the dam and collects the spilled material at the base of the dam. The underflow dam collects the spilled material on top and allows the water to flow through the bottom of the dam.

(A)

(B)

Figure 29-5 (A) Overflow dam and (B) underflow dam.

Figure 29-6 First responders have equipment such as hard sleeves to establish overflow and underflow dams.

Diverting

Almost anything can be used, such as dirt, sand, absorbent material, tarps, and hoselines, to divert a running spill, **Figure 29-7**. The most common use of the **diverting** technique is to keep a running

Figure 29-7 If a spill cannot be controlled, it may be better to divert the spill around items such as sewers and allow the spill to continue down the street.

Figure 29-8 Shown is an example of using a harbor boom. The boom should have been angled so that the material being collected would be near a shoreline. In this case, the boom was stretched straight across the waterway causing the material to be collected in the middle.

spill from entering a storm drain. A ring of dirt around the storm drain will keep the spilled material out, but does not stop the flow of the product. There are times when, due to the large quantity of spilled material, stopping the flow would be impossible, but it would be possible to keep it from the storm drain.

The diversion tactic can also be used for spills on water. By using a solid boom, such as a harbor boom, **Figure 29-8**, material can be diverted into another area or even contained. Another method of diversion is to dig a trench alongside the waterway to collect

Figure 29-9 Floating boom collects the spilled material that is floating on top of the water. Like the pads, the boom does not absorb water and should stay afloat even when saturated with the spilled material.

the spilled material off the top of the water. This is effective for large spills if there is a collection vehicle such as a vacuum truck standing by to collect the spilled material as it runs into the diversion. These types of vehicles are available from hazardous waste cleanup contractors. The most common method is to use a floating absorbent boom that absorbs the material off the water as shown in **Figure 29-9**.

Retention

The most common method of **retention** is the digging of a hole, either by hand or by machine, such as the track hoe shown in **Figure 29-10**. To catch a running

Figure 29-10 Small retention areas can be hand dug, but larger spills require larger equipment. Responders should have established contacts for accessing such equipment.

spill, the best method is to dig a hole large enough to collect the spill and then divert the spill into it.

Having the ability to create a large enough containment area is paramount to success. If unable to build a large enough containment area, then first responders should concentrate on building several in a row to collect the material. As with dikes and dams, it is best to start construction of the containment area farthest from the spill and then create several other areas closer to the spill.

Streetsmart Tip When digging retention areas, it is important to be aware of other underground utilities such as water, electric, sewer, cable TV, phone, or natural gas lines.

Figure 29-11 The use of a water spray is not necessary unless the vapor cloud is impacting the public. In some cases it creates more hazards than it eliminates.

Dilution

Much like **dilution** when conducting decontamination processes, this technique is not always the solution to pollution. If a water-soluble material is in a waterway, then for all intents and purposes it is being diluted. If the waterway is small, the fire department may need to add some water to the spilled materials. Simple flushing of a material into a waterway is not an acceptable tactic anymore and should not be considered. Items such as fuels, oils, or other hydrocarbons are not usually water soluble and cannot be diluted by water. For some specific chemicals, dilution is possible only when combined with some of the other containment tactics.

Vapor Dispersion

The topic of **vapor dispersion** can create some confusion for responders, because by their nature firefighters like to use water, usually in large quantities, and on some occasions the application of water will make spill situations safer. For instance, if a severe life threat exists because an adjacent nursing home cannot be evacuated, then using a water spray to disperse vapors would be a good idea. If, on the other hand, the release is occurring in a remote area away from any population, then a water spray to disperse vapors such as the one shown in **Figure 29-11** is not necessary. To be effective, the material that is being released must be water soluble or the vapor cloud must be able to be moved by the water streams. Although the DOT ERG states for many types of spills "use water spray to reduce vapors or divert vapor cloud," water is not recommended unless a life threat scenario exists.

Some departments use water sprays at natural gas leaks, which is not necessary and further complicates repair of the leak. Many of the leaks occur in a hole that was created by a backhoe. Water can fill the hole and make any attempted repair very dangerous and complicated. Also the use of water may actually knock down vapors that normally rise up and dissipate quickly.

Use of a water spray can also result in the creation of another substance. When anhydrous ammonia is released into the air, and a water spray is applied, the resulting runoff is ammonium hydroxide, a very corrosive liquid that itself would require containment and cleanup. Both natural gas and propane are not water soluble and the use of water vapor will just relocate the vapors to another area.

Note It is important to know what material is leaking prior to the application of water, because the water spray may cause more problems. It is very important to consult with the HAZMAT team prior to utilizing this tactic.

Vapor Suppression

With the use of firefighting foams, vapor suppression is another tactic that first responders can use. The type of material spilled dictates the type of foam to be used, because not all foams will work on all products. See previous chapters for more information regarding foam and its application. The use of foam to suppress vapors is generally limited to flammable liquid spills, but some chemical foams are available for other types of materials.

Before the application of foam the responders should ensure that the material is contained, and that the application of foam will not cause any further problems and is compatible with the spilled material. As when using foam to extinguish a fire, responders need to make sure they have enough foam stockpiled to keep a layer of foam on the spilled material.

Remote Shutoffs

The shutting of valves is not usually a tactic used by first responders, but there are some exceptions to this rule. On most tank trucks and at some fixed facilities, there are **remote shutoffs** that could be operated by first responders. Each type of truck is different but in general an emergency shutoff is located behind the driver's side of the cab and another near the control valves. The shutoffs are both mechanically operated and in most cases are self-operating in the event of a fire. They are usually well marked as emergency shutoffs and in an easy-to-find location. The two most common locations are behind the driver's door at the tank or at the valve controls. At some facilities, typically those with loading racks, a remote shutoff may be located near the entrance so that in the event of an emergency the first responders can shut off the flow of product.

AIR MONITORING AT THE FIRST RESPONDER OPERATIONS LEVEL

The use of air monitoring is one of the most important tasks to accomplish when responding to chemical incidents. Air monitoring can keep responders alive. With many departments purchasing detectors to assist with carbon monoxide alarms, first responders are experiencing associated benefits. Depending on the type of alarm purchased, it may also be used to detect flammable gases, oxygen levels, and one or two toxic gases, **Figure 29-12**. Many departments use these instruments for flammable gas leaks and for confined space entries.

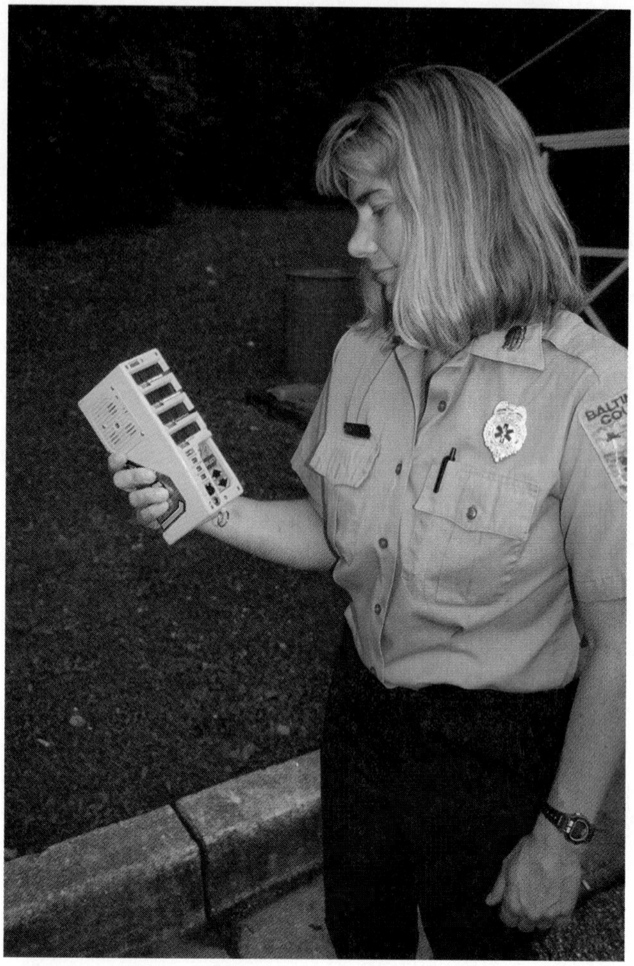

Figure 29-12 Air monitors, especially those called three- and four-gas instruments, are being used more commonly by first responders.

When responding to situations that involve unknown materials, HAZMAT responders need pH detection for corrosives, a combustible gas detector for the fire risks, and a **photo-ionization detector (PID)** for the toxic risks. Unfortunately, many responders only rely on one or two of these detectors, and the most common detector is a three-, four-, or five-gas instrument. Other instruments are not available or not used. By not using a method of detecting toxic materials, responders (and the public) can be lulled into a false sense of security. Combustible gas detectors are not made or designed to measure toxic gases in air; they only read those types of gases when they become fire hazards. Many flammable gases are very toxic at considerably lower levels than their lower explosive limit. Many first responders have purchased air monitors, but may not have a full understanding of how they work.

> **Note** The use of air monitoring and sampling equipment can be complicated and requires practice.

Most fire service responders have a reasonable understanding of combustible gas indicators but even that knowledge can be limited. Society is becoming much more sophisticated and the citizens (the fire service's customers) expect more. This section presents some concepts on air monitoring, monitoring strategies, and information on how the monitors work and their uses. The field of air monitoring technology is ever changing and new technology emerges each year, although the basic principles of safe decision making remain the same. When purchasing air monitors it is important to understand the basic features of the instruments so that the purchase benefits the organization making the purchase.

Note The general rule when purchasing an air monitor is to figure one-quarter of the purchase price into the department's annual budget for upkeep and repairs. Even though the purchaser may have been told differently, it is necessary to keep the instrument calibrated and maintained on a regular basis.

Regulations and Standards

The Occupational Safety and Health Administration (OSHA) HAZWOPER regulation (29 CFR 1910.120) does not provide a lot of specific requirements for air monitoring, even for the HAZMAT technician, but air monitoring is the principal safety element throughout the document. OSHA wants the incident commander (IC) to identify and classify the hazards that are present on a site. The use of air monitoring is the primary key to fulfilling this obligation, but in order to fulfill this obligation a HAZMAT technician must be present along with other detection devices.

The NFPA 472 document also has some requirements for air monitoring but, like the OSHA regulation, they are fairly generic. There are no requirements at the Operations level for the use of air monitors.

It is important for first responders to understand how monitors work and what their deficiencies are because they may be placing their lives on the line depending on the readings received on an air monitor. First responders are more affected by the confined space regulation with regard to air monitors than they are by the HAZMAT regulation.

It is unfortunate, but it can be predicted that a lack of understanding and/or maintenance will play a factor in future firefighter fatalities and injuries based on the large number of departments using these instruments.

Air Monitor Configurations

Most departments purchase an instrument that is known as a three-, four-, or five-gas instrument, commonly referred to as a **multigas detector**, such as the one shown in **Figure 29-13**. In other words it samples for the presence of three to five different gases. Normal units sample for oxygen levels, a lower explosive limit (LEL), and two or three toxic gases. The normal "toxics" are carbon monoxide (CO) and hydrogen sulfide (H_2S). If a third gas can be sampled for, most departments choose chlorine or sulfur dioxide. For most departments, though, sampling for CO and H_2S is more than sufficient.

Figure 29-13 These instruments are four-gas instruments and measure LEL, oxygen, carbon monoxide, and hydrogen sulfide.

Although these instruments are considered expensive, in the grand scheme of things they are inexpensive compared to other detection devices used by a HAZMAT team. Many departments may purchase a device with little or no training, which can be detrimental to the successful use of the instrument. When buying a detector it is essential to purchase several maintenance items designed to keep the instrument functioning. Typically these items are included in a special kit price and include a battery-powered sample pump, calibration gas, and hardware. The calibration gas usually has an expiration date and will need to be replaced in six months to a year. The sensors in the detector need regular calibration, and the instrument should be turned on and used on a regular basis for a period specified by the manufacturer. Depending on the charging system, rechargeable batteries may need to be replaced on occasion.

The sensors also need replacement. Although most manufacturers provide two-year warranties for the sensors, in general, the LEL sensor will last four to five years, the oxygen sensor twelve to eighteen months, and the CO and H_2S sensors eighteen to twenty-four months. Other toxic sensors may last six to eighteen months depending on the type.

Most of the perceived problems with instruments are created by a lack of maintenance and adequate training. Regardless of what a salesperson says, no device exists that can conclusively identify an unknown material for a first responder.

METER TERMINOLOGY

To comprehend fully the use of air monitors, the responder must understand the basic terminology that is generic to all monitors and applies across the board. More detailed information related to the specific monitors is provided in later sections.

Bump Test

Two terms that need further explanation are bump test and calibration. A **bump test**, also known as a field test, exposes a monitor to known gases, allowing the monitor to go into alarm mode, and then removes the gas, **Figure 29-14**. By exposing the monitor to a known gas, a person can verify the monitor's response to that gas. Most manufacturers provide bump gas cylinders, which contain the gases required to check their instrument. When bump testing, firefighters should follow the manufacturer's recommendations.

In most cases, the bump test is used to ensure that the alarms function as intended and the instrument is reading. By regularly calibrating and bump testing, the user can determine how accurate the bump test will be in the field. Some instruments react very well to bump testing and will display the levels as provided on the bump gas cylinder, while others will be off slightly. When responding to confined spaces, responders are required to bump test the instrument prior to entry into the confined space.

Calibration

Calibration is used to determine if a monitor responds accurately to exposure to a known quantity of gas. When new sensors are installed, they will usually read higher than intended; calibration electronically changes the sensor to read the intended value. As the sensor gets older, it becomes less sensitive and calibration electronically raises the value that the sensor displays. When a sensor fails, it cannot be electronically brought up to the correct value. Regular calibration gives a picture of the expected life of a sensor, which will deteriorate over time.

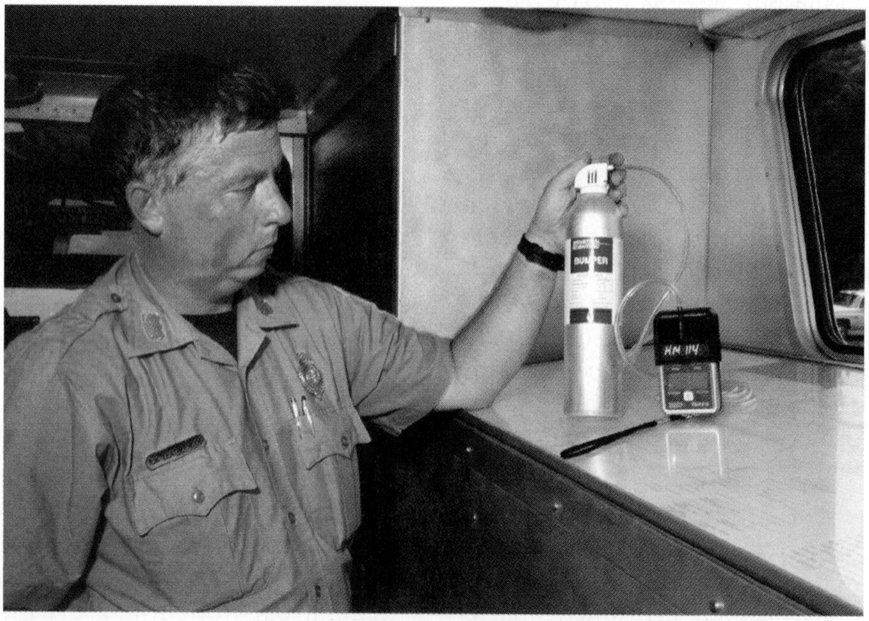

Figure 29-14 A bump test ensures that the instrument responds to known gases and should reach alarm levels to verify that the alarms are functioning.

The regularity of calibration is subject to great debate. Some departments calibrate daily, others every six months. The only item found in the regulations (for anything that requires air monitoring) is in the confined space regulation, which requires calibration according to the manufacturer's recommendations. Most of the written instruction guides from the manufacturers require calibration before each use. The definition of a *calibration* at this point is also subject to debate, because the department can verify a monitor's accuracy by exposing it to a known quantity of gas, but not perform a "full" calibration. Most response teams establish a regular schedule of calibration (weekly/monthly) and then perform "bump" or "field" checks during an emergency response. It is essential to check with manufacturers as to what calibration/bump test policy they recommend because they are all different.

Reaction Time

All monitors have a lag time or, as it is better known, reaction time. The use of a pump or not will vary the reaction time. Monitors operating without a pump are in what is known as the *diffusion mode* and will generally have a fifteen- to thirty-second lag time. Monitors operating with a pump have a typical reaction time of three to five seconds, but this can vary from manufacturer to manufacturer. Hand-aspirated pumps usually require ten to fifteen pumps to draw in an appropriate sample. Hand-aspirated pumps are not recommended, because the readings they provide will be too varied depending on how well the user operates the hand pump. The goal is to provide a given amount of volume across the sensors. It is important to follow the manufacturer's recommended lengths of hose to ensure that the pump operates correctly.

> **Note** When using sampling tubing, one to two seconds of lag time should be added for each foot of hose.

Recovery Time

Monitors also have a recovery time. Recovery time is the amount of time it takes the monitor to clear itself of the air sample. This time is affected by the chemical and physical properties of the sample, the amount of sampling hose, and the amount absorbed by the monitor. Some monitors take an extended period of time to clear if they are exposed to a large quantity of a gas. In some cases the instrument must be taken out of the environment and shut off, then started again. The reaction time will affect the overall recovery time.

Relative Response

When a gas monitor is purchased, it is set to read a specific type of gas, such as methane. If any other gas is sampled, the instrument will usually read the gas but at different values. The term *relative response* is used to describe the way the monitor reacts to a gas other than the one it was calibrated for, and is a term that is not commonly used by emergency responders.

To maintain a high level of safety, each person operating an air monitor must have a basic understanding of relative response so as not to let the monitor lead them into dangerous situations. Each detector has what is known as a relative response curve that compensates for different types of gases, **Table 29-1**. The monitor's manufacturer has tested the monitor against other gases and has provided a factor (relative response factor) that can be used to determine the amount of gas present when sampling. The user multiplies the displayed reading by the factor to arrive at the reading for the gas that is actually present. For instance, consider a responder at a xylene spill who is using an ISC TMX-412 calibrated for pentane. The detector is reading 50 percent of the LEL. According to **Table 29-1**, the response curve factor for xylene is 1.3, which is multiplied by the LEL reading:

$$\text{Detector reading} \times \text{Response curve factor} = \text{Actual LEL reading}$$

$$50 \times 1.3 = 65$$

So the actual LEL is 65 percent, a number higher than what the instrument was providing. On the other hand, if the responder used the same instrument for a spill of propane, the response curve factor is 0.8. Using the same scenario—that the instrument was reading 50 percent of the LEL—the actual reading is 40 percent ($50 \times 0.8 = 40$), a safer situation than reported by the detector.

Oxygen Monitors

Oxygen is one of the most important things to sample for. Humans need it to survive, and the other instruments need it to function correctly. Normal air contains 20.9 percent oxygen; below 19.5 percent is considered oxygen deficient and is considered to be a health risk, and above 23.5 percent is considered a fire risk. If an oxygen drop is noted on the monitor, one or possibly more than one contaminants are present, causing the reduced oxygen levels, and

Response Curve Factors*		
GAS BEING SAMPLED	ISC TMX-412 FACTORS	MSA 261 FACTORS
Hydrogen	0.5	0.6
Methane	0.5	0.6
Acetylene	0.7	0.8
Ethylene	0.7	0.8
Ethane	0.7	0.7
Methanol	0.6	0.7
Propane	0.8	0.9
Ethanol	0.8	N/A
Acetone	0.9	1
Butane	0.9	1
Isopropanol	1	1.1
Pentane	1	1
Benzene	1	1.1
Hexane	1.2	1.3
Toluene	1.1	1.2
Styrene	1.1	N/A
Xylene	1.3	N/A

*Always use the response factors supplied by the manufacturer, but keep in mind that they are laboratory estimates. ISC factors are for Industrial Scientific Corporation LEL sensor 1704A1856-200 calibrated with pentane. The MSA factors are for the Mine Safety Appliances sensor MSA 360/361 calibrated with pentane.

TABLE 29-1

another material (i.e., toxic, flammable, corrosive, or inert) is causing the oxygen-deficient atmosphere. When in oxygen-deficient atmospheres, any combustible gas readings will also be deficient and cannot be relied on.

If in an oxygen-enriched atmosphere, then the combustible gas readings will be increased and not accurate. Many oxygen-enriched atmospheres result from a chemical reaction involving oxidizers and typically present a dangerous situation.

Oxygen Monitor Limitations

Most oxygen sensors last only a year since they are always working. An oxygen sensor is an electrochemical sensor that has two electrodes within a gel-like material. When oxygen passes through the sensor, it causes a chemical reaction, hence creating an electrical charge and causing a readout to be provided on the monitor. As long as the sensor is exposed to O_2, it will cause a reaction within the electrolyte solution sealed within the sensor. This is the most commonly replaced sensor and usually needs to be replaced annually, although some may last longer. Chemicals that hurt O_2 sensors are those with lots of oxygen in their molecular structure such as carbon dioxide (CO_2) and strong oxidizing materials such as chlorine and ozone. The problem with CO_2 is that it is always present in the air, and the higher the percentage the faster the sensor will deteriorate.

The optimal temperature for operation is between 32° and 120°F. Between 0° and 32°F the sensor slows down (electrolyte is like a slushy), and temperatures below 0°F can permanently damage the sensor. Operation depends on absolute atmospheric pressure, and calibration is required at the atmospheric pressure at which the user will be sampling. The sensor should also be calibrated for the temperature and weather conditions for the area being sampled.

Flammable Gas Indicators

> **Note** *Flammable gas indicator (FGI)* is a new term. The old terminology for these detectors described them as combustible gas indicators (CGIs), which is a misapplied term. These devices only detect the presence of flammable gases, not combustible gases. The term *combustible gases* is derived from their history and development in mining safety. Another term that could be used to describe these units is *LEL sensors,* as they are used to determine the percent of the lower explosive limit (LEL) of the material present. This text uses FGI, but LEL sensor would also apply.

All of these types of sensors work, some better than others in different situations. It is important to understand how each of these sensors works, because budgetary considerations may dictate the purchase of only one type. (Note, however, that to be able to detect a wide variety of chemicals effectively, it may be necessary to have more than one type of sensor.)

Most of the new FGIs are used to measure the LEL of the gas for which they are calibrated. The majority of FGIs are calibrated for methane (natural gas) using pentane gas. When calibrated for methane, the FGI sensor will read up to the LEL, and with some of the new units will in fact shut off the sensor when the atmosphere exceeds the LEL. This is an important consideration, because the longer the sensor is exposed to an atmosphere above the LEL, the quicker it loses its life.

The FGI reads up to 100 percent of the LEL, so that if a FGI is calibrated for methane and the user is sampling methane, the FGI will read 100 percent, but the actual concentration in that area will be 5 percent for methane (the LEL for methane is 5 percent). If the FGI displays a reading of 50 percent, then the concentration for methane is 2.5 percent. Any flammable gas sample that passes over the sensor will cause a reaction; how much of a reaction depends on the gas. Each FGI comes with a relative response curve for other gases, and the exact number of other gases referenced varies from manufacturer to manufacturer. To further complicate this issue, the reading on the FGI can either be higher or lower depending on the gas. If the reading is below the actual percentage the user is safe, but if the actual percentage is above what the FGI is reading then the user may no longer be in a safe atmosphere because the LEL may have been exceeded.

The Environmental Protection Agency (EPA) has established action levels that provide a safe layer expressly because of the relative response curve problem. They have factored the layer in and provided guidelines for users to follow, as provided in **Table 29-2**.

The basic principle of most FGIs is that a stream of sampled air passes through the sensor housing, causing a heat increase and, conversely, creating an electric charge and causing a reading on the instrument. The three combustible gas sensor types are shown in **Figure 29-15**. When purchasing or using a monitor, responders should be aware of the different sensor

EPA Air Monitoring Guidelines

ATMOSPHERE	LEVEL %	ACTION
Flammable gas	<10 LEL	Continue to monitor with caution.
	>10	Evacuate space (OSHA).
	10–25 LEL	Continue to monitor, but use extreme caution especially as higher levels are found (EPA).
Oxygen	<19.5	Monitor with SCBA. FGI values are not valid (EPA). OSHA requires SCBA.
	19.5–25 (OSHA, 23.5)	Continue monitoring with caution. SCBA not needed based on O_2 content only. OSHA requires evacuation of the space.
	>25 (EPA)	Explosion hazard; withdraw immediately.

TABLE 29-2

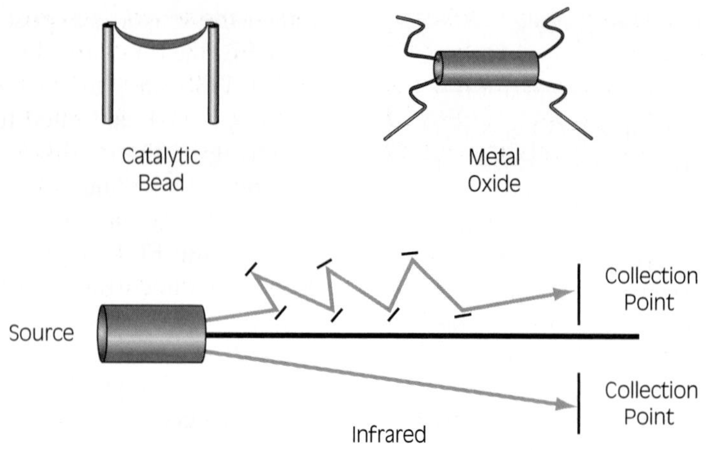

Figure 29-15 Examples of combustible gas sensors: catalytic bead, metal oxide, and infrared.

types. Readings can and do vary among the three, and the safety of responders is in the hands of that instrument so they must understand how it works.

Catalytic Bead

The **catalytic bead** sensor is the most common sensor in use today. A catalytic bead sensor comes in two configurations: a bowl-shaped piece of metal with a solid bead in the center, or a wire between two poles with a solid bead of metal in the center. In both cases the beads are coated with a catalytic material that aids the gas sample to be burned off efficiently. The sensor has two sets of these bowls. One burns the gas, and the other is used to compare the heat increase or decrease. These sensors replaced the **wheatstone bridge sensor,** which was a coiled piece of platinum wire between two poles.

Metal Oxide Sensor

The **metal oxide sensor** is commonly referred to as an **MOS**. Its very nature makes it a very sensitive sensor, which causes all of its "perceived" problems. If used correctly (and interpreted correctly) this sensor can provide many clues and answers at chemical incidents.

The MOS is a semiconductor in a sealed unit that has a wheatstone bridge sensor in it surrounded by a coating of a metal oxide. Heater coils provide a constant temperature. When the sample gas passes over the heated bridge, it combines with a pocket of oxygen created from the metal oxide. This reaction causes an electrical change, which causes the FGI to provide a reading. This sensor is very sensitive and will pick up almost anything that crosses it, which can be confusing to the user. It requires regular maintenance and calibration, more so than the other types of sensors. Most MOS FGIs do not provide a readout of the percentage; most provide only an audible warning or provide a number from within a range. The range extends from 0 to 50,000 units and is a relative scale. If a tablespoon of baby oil is spilled on a table and an MOS sensor passed over it, the reading will be in the range of 5 to 30. If an MOS sensor is taken into a room with 5 percent methane, the reading will be within the range for a flammable gas, probably near 45,000 to 50,000 units. Some monitors allow the MOS to read in percentages of the LEL, in addition to a general sensing range of 0 to 50,000 units. The MOS reacts to tiny amounts, which is an outstanding feature of the monitor—most monitors are not nearly as sensitive. Note, however, that if a department can afford only one combustible gas sensor, a catalytic bead sensor is preferred because the MOS can be erratic. The MOS sensor does have great applicability for a HAZMAT team or for a department that already has another type of LEL sensor.

Infrared Sensors

The **infrared sensor** is new to the emergency response field. It uses a hot wire to produce a broad range of wavelengths, a filter to obtain the desired wavelength, and a detection device on the other side of the sensor housing. The light that is emitted from the hot wire is split, one wavelength going through the filter, the other to the detection device to be used as a reference source. When a gas is sent into the sample chamber, the gas molecules will absorb some of the infrared light and will not reach the detection device, which will read the amount of light reaching the detector. The amount of light reaching the detection device is compared and a reading is provided. The big advantage of infrared is that it does not require oxygen in which to function, that is, it can take readings in oxygen-deficient atmospheres. The device also is not affected by temperature, nor is it easily poisoned by high exposures. The disadvantages include cost and its many cross sensitivities. This is another sensor that should not be purchased as a stand-alone unit, but should be considered after the purchase of a catalytic bead or wheatstone bridge sensor.

Toxic Gas Monitors

This section describes the sensors that are commonly used in three-, four-, and five-gas units (LEL, O_2, toxic 1, toxic 2, toxic 3) and are usually used to measure carbon monoxide (CO) and hydrogen sulfide (H_2S). Toxic sensors are available for a variety of gases, however, including chlorine, sulfur dioxide, hydrogen chloride, hydrogen cyanide, and nitrogen dioxide. The most common unit sold today by far is a four-gas unit that measures LEL, O_2, CO, and H_2S. This combination is a direct result of the confined space regulation issued by OSHA. Many people understand why the first three sensors were chosen, but do not understand why H_2S was chosen because OSHA did not specify H_2S. Most confined space entries (and consequently the most deaths and injuries) take place in sewers and manholes, and H_2S is commonly found in sewers and anyplace else things are rotting. So when choosing one of the optional gases to sample for, it is best to choose H_2S unless the department has a specialized need for some of the other gases.

Many response teams consider choosing one of the other toxic gases mentioned in the preceding paragraph as their fourth or fifth gas, but they must weigh the cost of doing so. More cost-effective methods of detecting toxic gases such as chlorine or sulfur dioxide are available. The average cost for the sensor is $400 to $500 and is usually guaranteed for one year. The calibration gas is $300 to $400 and is only good for six months before it expires. The other problem is that if the instrument is taken into an environment with more than 20 ppm chlorine, it will be ruined.

Most toxic sensors are electrochemical sensors that have electrodes (two or more) and a chemical mixture sealed in a sensor housing. The gases pass over the sensor, causing a chemical reaction within it, and an electrical charge is created, which causes a readout to be displayed. All toxic sensors display in parts per million. Some toxic sensors use metal oxide technology and react in the same fashion as FGI metal oxide sensors.

Other Detectors

In most cases the detectors discussed in the following subsections will be used by a HAZMAT team, but the first responder should be aware of the devices and their capabilities. In areas without a HAZMAT team or where the HAZMAT team response times are unusually long, the first responder may want to consider the purchase of these items. As is true of the monitoring devices already discussed, these instruments are complicated, and this is not by far a complete listing of the instruments that should be carried by a HAZMAT team. First responders are capable of beginning the detection process but in almost all cases further detection capability is required.

Photo-Ionization Detectors

The photo-ionization detector (PID) is sometimes referred to as a total vapor survey instrument, and an example is shown in **Figure 29-16**. Because of their ability to detect a wide variety of gases in small amounts, PIDs are becoming essential tools of HAZMAT response teams. The PID does not indicate what materials are present, much the same as the FGI will not identify the specific material that is present, but when used as a general survey instrument, the user can identify potential areas of concern and possible leaks/contamination. These are the instruments that look for toxic materials in the air, and they are essential to determining exposures to many toxic materials. The combustible gas detector will start reading toxic materials in a range of 50 to 500 ppm, which for some gases is extremely high and could have immediate effects on a responder. The PID starts to read at levels near 0.1 ppm, which is very sensitive. Because of their sensitive nature, they can detect small amounts of hydrocarbons in the soil. Sick building calls are on the increase, and the PID is a valuable tool in identifying possible hot spots within the building.

Figure 29-16 The photo-ionization detector is used by HAZMAT teams to look for toxic gases in the air. It is a very sensitive instrument that can measure small amounts of gases in the air.

Figure 29-17 A standard colorimetric tube is a glass tube with a crystalline material in it that changes color when a certain type of gas is present. These tubes can be used to identify unknown gases and can provide a reading for a known gas.

Figure 29-18 The chip measurement system uses a chip to analyze a gas and provides a digital readout. Standard colorimetric sampling requires an interpretation of the color change and the length of the change. The CMS takes the guesswork out of the system by providing the reading in ppm.

Colorimetric Sampling

Colorimetric tubes are used for the detection of known and unknown vapors. With sick building calls becoming more and more frequent, the use of colorimetric tubes by HAZMAT teams has greatly increased. Colorimetric sampling consists of taking a glass tube filled with a reagent (usually a powder or crystal). The reagent is placed into a pumping mechanism that causes air to pass over the reagent. A colorimetric sampling device is shown in **Figure 29-17**. If the gas reacts with the reagent, then a color change should occur, indicating a response to the gas sample. Detection tubes are made for a wide variety of gases and generally follow the chemical family lines (i.e., hydrocarbons, halogenated hydrocarbons, acid gases, amines, etc.). Although the tubes may be marked for a specific gas, they usually have cross sensitivities (react to other materials), which at times is the most valuable aspect of colorimetric sampling.

One system that has been developed involves the use of bar-coded sampling chips. This system, called the **chip measurement system (CMS)**, involves the insertion of a sampling chip into a pump, **Figure 29-18**. The pump recognizes the chip in use and provides the correct amount of sample through the reagent. The pump, by means of optics and light transfer system reflective measurement, provides an accurate reading of the gas that may be present.

Other instruments are also used by a HAZMAT team, some of which include pH detection, flame-ionization detectors, warfare agent detection, mace/pepper spray detection, and characterization sample kits. The detection of unknown materials is complicated even for some HAZMAT teams, and when first responders are confronted with a possible chemical release, no matter how insignificant they think it is, they should request or as a minimum consult their closest HAZMAT team.

CARBON MONOXIDE INCIDENTS

The fire service is responding to an increasing number of calls involving carbon monoxide detector alarms. In 1995 the city of Chicago experienced several thousand carbon monoxide detector alarms in one day, due to an inversion that kept the smog, pollution, and carbon monoxide at a low elevation within the city. Because carbon monoxide is colorless, odorless, and very toxic it is important that first responders understand the characteristics of carbon monoxide and how the detectors in the home work. As with other chemicals, carbon monoxide (CO) can be an acute or chronic toxicity hazard. It is only acutely toxic at high levels, typically at levels in excess of 100 ppm. At levels of less than 100 ppm, the hazard comes from a chronic exposure, which can be hazardous. Because of carbon monoxide's properties, it can only be detected by a CO detector. In extremely high concentrations it can be explosive. Exposure to CO causes flu-like symptoms, headache, nausea, dizziness, confusion, and irritability. Exposure to high levels can cause vomiting, chest pain, shortness of breath, loss of consciousness, brain damage, and death.

Although confronted with levels of CO in a house, the residents may not be exhibiting any signs or symptoms.

> **Note** If CO is found in a home, its residents should seek medical attention because signs and symptoms of CO poisoning may be delayed for twenty-four to seventy-two hours.

Levels over 100 ppm are extremely dangerous and the residents should be medically evaluated. Monitoring with a CO monitor is essential to determine the possible exposure to CO. Persons who may only show minor effects of CO poisoning and who would normally be transported to the closest hospital need to have the residence monitored. If high levels are found using a monitor, then the exposed residents will need alternative treatment such as the **hyperbaric chamber** and such treatment should not be delayed. An oxygen saturation monitor will often be used incorrectly to determine the O_2 level in a patient. Patients who have been exposed to CO will cause an oxygen saturation monitor to read 100 percent because the monitor reads the oxygen molecule in CO as being O_2. The elderly, children, or women who are pregnant are especially susceptible to CO and may have had a serious exposure without showing any effects.

If the first arriving units do not have a CO monitor and victims may still be in the residence, personnel are to have SCBA on and functioning when searching the residence. After determining no victims are present, crews are to ensure that the house is closed up and then wait outside for a CO monitor. Crews should not enter an area with a CO detector that is activated without the use of SCBA. When using a monitor, if crews find levels that exceed 35 ppm, crews should use SCBA to continue the investigation. Crews should be suspicious when responding to reports of an unconscious person or reports of "several people down" and should not enter an area without SCBA if it is possible that CO (or other toxic gases) may be present. An air monitor will ensure responder safety with regard to the gases for which it samples.

It is possible that people may be found unconscious due to a natural gas leak, but this is very unlikely. Natural gas has a distinctive odor, is nontoxic, and will only asphyxiate a person by pushing oxygen out of an area. The only sign of this exposure is unconsciousness or death; any of the flu-like symptoms are due to CO poisoning, not natural gas. If the level of natural gas is high enough to cause unconsciousness, then a very severe explosion hazard is present; in fact, an explosion would be imminent.

When home CO detectors are activated, it is possible that a standard fire department air monitor will not pick up any CO when first responders arrive. This is because the CO detectors purchased for the homes are made to detect small amounts of CO over a long period of time, but fire service detectors provide "instant" readings and only pick up 1 ppm or more. The fact that firefighters may not pick up any readings, however, does not mean the residence's detector is defective. Many factors may cause fire service monitors not to get any readings, including if low amounts of CO are present or a momentary high level of CO activated the alarm but then dissipated prior to fire department arrival. The amount of time the residence is open will also dramatically affect fire service readings. Crews are reminded to keep the residence closed up so that the air monitor has a chance to monitor the level of CO. As a reminder, any time units respond to unknown odors or sick building calls, responders should remove any people from the building and keep it closed up. Because the amounts of toxic gases in sick building incidents are usually small, keeping them contained is very important. A patient cannot be treated for toxic gas exposure unless the source and type of exposure are known. For the patient's long-term health as well as the responder's, quick, reliable gas samples are a necessity.

In some parts of the country, CO detectors are required much in the same fashion as smoke detectors. The brand and type of device will determine how well the device actually performs. The three basic sensing technologies are **biomimetic**, metal oxide, and electrochemical, each having advantages and disadvantages. Location, weather conditions, and the type of sensor will determine the types of readings that can be expected from a particular brand of detector.

Biomimetic. A biomimetic is a gel-like material that is designed to operate in the same fashion as the human body does when exposed to CO. This type of sensor is prone to false alarms, because it cannot reset itself unless it is placed in an environment free of CO, which in most homes is impossible. The sensor may need twenty-four to forty-eight hours to clear itself after an exposure to CO. The actual concentration of CO at the time the detector sounds may be low, but the exposure may have been enough to send the detector over the alarming threshold. If responding to an incident in which one of these detectors has activated, it will need to be placed in a CO-free environment for twenty-four to forty-eight hours to allow itself to clear. It is important for some type of detection device to be left in place for the residents until their detector clears because it is not advisable to leave them unprotected.

Metal Oxide. This is the same type of sensor that is used in the combustible gas detector, but it is designed to read carbon monoxide. How successful this design is in reading only CO is subject to debate, and it is known that this detector—although superior to the biomimetic sensor—does have some cross sensitivities and will react to other gases.

Although it is hoped that responders would be using a three- to five-gas detection device to check a home, it is possible for this type of sensor to alarm for propane. Responders using only a CO instrument may find themselves walking into a flammable atmosphere. This type of detector can usually be identified by the use of a power cord because the sensor requires a lot of energy and, in most cases, provides a digital readout. Once activated this sensor needs some time to clear itself, but this is usually less than twenty-four hours.

Electrochemical. An electrochemical sensor is also referred to as an instant detection and response (IDR) sensor, and it is the same type of electrochemical sensor found in three- to five-gas instruments. It has a sensor housing with two charged poles in a chemical slurry. When CO goes across the sensor it causes a chemical reaction, which changes the resistance within the housing. If the amount is high enough, it will cause an alarm. It provides an instant reading of CO and does not require a buildup of CO to activate. It has an internal mechanism that checks the sensor to make sure it is functioning, which is a unique feature. Out of the three types of residential detectors, based on sensor technology, the electrochemical sensor would provide the best sensing capability.

Common sources of CO include furnaces (oil and gas), hot water heaters (oil and gas), fireplaces (wood, coal, and gas), kerosene heaters or other fueled heaters, gasoline engines running inside (basements or garages), barbecue grills burning near the residence (garage or porch), and faulty flues or exhaust pipes.

Lessons Learned

The use of defensive product control methods is a key component for the protection of a community and the environment. In most cases first responders have the equipment necessary to handle these tasks. With some modification or adaptation, first responders can accomplish the control of many spills. The limiting of spills will mitigate the incident sooner and prevent its spread. If first responders cannot stockpile the necessary equipment, then contacts should be made with those facilities that may have the materials. (Under the requirements of the Oil Pollution Act (OPA) of 1990, certain facilities are required to maintain stockpiles of emergency equipment.)

First responders are also becoming more involved with air monitoring and are becoming more aware of the hazards chemicals present. When using air monitors first responders are reminded that they are not all encompassing and their use requires training and experience. They require testing and regular maintenance and cannot be expected to function properly without this upkeep. To determine true levels, first responders must use a range of instruments, which may be above their level. When in doubt, first responders should consult with the local HAZMAT team.

KEY TERMS

Absorption A defensive method of controlling a spill by applying a material that absorbs the spilled chemical.

Biomimetic A form of gas sensor that is used to determine levels of carbon monoxide. It is of the type of sensors used in home CO detectors. It closely re-creates the body's reaction to CO and activates an alarm.

Bump Test Used to determine if an air monitor is working. It will alarm if a toxic gas is present. It is a quick check to make sure the instrument responds to a sample of gas.

Calibration Used to set the air monitor and to ensure that it reads correctly. When calibrating a monitor, it is exposed to a known quantity of gas to make sure it reads the values correctly.

Catalytic Bead The most common type of combustible gas sensor that uses two heated beads of metal to determine the presence of flammable gases.

Chip Measuring System (CMS) A form of colorimetric air sampling in which the gas sample passes through a tube. If the correct color change occurs, the monitor interprets the amount of change and indicates a level of the gas on an LCD screen.

Colorimetric Tubes Crystal-filled tubes that change colors in the presence of the intended gases. These tubes are made for the detection of known and unknown gases.

Damming The stopping of a body of water, which at the same time stops the spread of the spilled material.

Diking A defensive method of stopping a spill. A common dike is constructed of dirt or sand and is used to hold a spilled product. In some facilities, a dike may be preconstructed such as around a tank farm.

Dilution The addition of a material to the spilled material to make it less hazardous. In most cases water is used to dilute a spilled material, although other chemicals could be used.

Diverting Using materials to divert a spill around an item. For instance, several shovels full of dirt can be used to divert a running spill around a storm drain.

Hyperbaric Chamber A chamber that is usually used to treat scuba divers who ascended too quickly and need extra oxygen to survive. The chamber re-creates the high-pressure atmosphere of diving and forces oxygen into the body. It is also successful in the treating of carbon monoxide poisoning and smoke inhalation, because both of these problems require high amounts of oxygen to assist with the patient's recovery.

Infrared Sensor A sensor that uses infrared light to determine the presence of flammable gases. The light is emitted in the sensor housing and the gas passes through the light. If it is flammable the sensor will indicate the presence of the gas.

Metal Oxide Sensor (MOS) A coiled piece of wire that is heated to determine the presence of flammable gases.

Multigas Detector A term used to describe an air monitor that measures oxygen levels, explosive (flammable) levels, and one or two toxic gases such as carbon monoxide or hydrogen sulfide.

Photo-ionization Detector (PID) An air monitoring device used by HAZMAT teams to determine the amount of toxic materials in the air.

Remote Shutoffs Valves that can be used to shut off the flow of a chemical. The term *remote* is used to denote valves that are located away from the spill.

Retention The digging of a hole in which to collect a spill. Can be used to contain a running spill or collect a spill from the water.

Vapor Dispersion The intentional movement of vapors to another area, usually by the use of master streams or hoselines.

Wheatstone Bridge Sensor A type of combustible gas sensor that uses a heated coil of wire to determine the presence of flammable gases.

REVIEW QUESTIONS

1. What type of dam would be required for a fuel spill in which the fuel has a specific gravity of less than 1?

2. What type of dam is required for a chemical spill in which the material has a specific gravity of greater than 1?

3. Describe the two key items needed to construct an underflow or overflow dam.
4. Who should be consulted prior to using vapor dispersion techniques?
5. Describe the normal configuration for a multi-gas detector.
6. If a detector is calibrated for methane, and a person responds to a propane release, describe whether the instrument will detect propane and, if it will, how it does so.
7. Explain why and how often detectors should be calibrated.
8. With regard to safety and the ability to detect various toxic gases, how would you rate the use of a multigas detector with this combination: LEL, O_2, CO, H_2S?

Additional Resources

Bevelacqua, Armando S., *Hazardous Materials Chemistry.* Delmar Learning, a part of the Thomson Corporation, Clifton Park, NY, 2001.

Bevelacqua, Armando S. and Stilp, Richard, *Hazardous Materials Field Guide.* Delmar Learning, a part of the Thomson Corporation, Clifton Park, NY, 1998.

Bevelacqua, Armando S. and Stilp, Richard H., *Terrorism Handbook for Operational Responders,* 2nd ed. Delmar Learning, a part of the Thomson Corporation, Clifton Park, NY, 2004.

Hawley, Christopher, *Hazardous Material Air Monitoring and Detection Devices.* Delmar Learning, a part of the Thomson Corporation, Clifton Park, NY, 2002.

Hawley, Christopher, *Hazardous Materials Response and Operations.* Delmar Learning, a part of the Thomson Corporation, Clifton Park, NY, 2001.

Henry, Timothy, *Emergency Decontamination for Hazardous Materials Responders.* Delmar Learning, a part of the Thomson Corporation, Clifton Park, NY, 1999.

Lesak, David, *Hazardous Materials Strategies and Tactics.* Prentice Hall, 1998.

Maslansky, Carol J. and Steven P. Maslansky, *Air Monitoring Instrumentation.* Van Nostrand, Reinhold, 1993.

Noll, Gregory, Michael Hildebrand, and James Yvorra, *Hazardous Materials: Managing the Incident.* Fire Protection Publications, Oklahoma University, 1995.

Schnepp, Rob and Paul W. Gantt, *Hazardous Materials: Regulations, Response and Site Operations.* Delmar Learning, a part of the Thomson Corporation, Clifton Park, NY, 1999.

Stilp, Richard and Bevelacqua, Armando S., *Emergency Medical Response to Hazardous Materials Incidents.* Delmar Learning, a part of the Thomson Corporation, Clifton Park, NY, 1997.

CHAPTER 30

TERRORISM AWARENESS

Christopher Hawley, Baltimore County Fire Department (Ret.)

Courtesy of FEMA

OUTLINE

- Objectives
- Introduction
- Types of Terrorism
- Potential Targets
- Indicators of Terrorism
- HAZMAT Crimes
- Incident Actions
- General Groupings of Warfare Agents
- Detection of Terrorism Agents
- Federal Assistance
- Lessons Learned
- Key Terms
- Review Questions
- Endnote
- Additional Resources

STREET STORY

The fire service is made up of a large, tight-knit family, and the special bond we have is carried over into the other emergency services.

On September 11, 2001, our family and our nation were attacked in a manner that was unprecedented. After the attacks in New York and Arlington, Virginia, and the plane crash in Shanksville, Pennsylvania, America was at war. Terrorists had struck, and in a few precious minutes we suffered a grievous loss. America changed that day, and the emergency services took a severe hit, but we persevered. During the first few minutes of the fire in the World Trade Center my thoughts were that it was a large fire, and that it would be a tough battle, but one we could win. When the towers collapsed, my heart sank. I knew that I had just witnessed hundreds of my family members being struck down. My thoughts were immediately for my good friends who work in FDNY and for all of the other emergency responders at that scene.

The events of that day caused the deaths of many emergency responders and an enormous amount of emotional damage to the survivors, family members, and population of the United States. But on that day, when citizens across the country dialed 9-1-1, emergency responders showed up . . . we proved that we were not defeated. We are emergency responders and the keystones of our communities. We are the first line of defense against terrorism in this country. We must remain focused on our mission, which is to protect the citizens in our community. With that said, we as emergency responders must look at the real threat that terrorism presents to our communities.

There are many events that occur in this country that present substantial risks to you as a responder and to the community. The events discussed in this chapter are events that can have catastrophic consequences for your community. You must prepare for these events and become tuned in to the potential for terrorism or other criminal behavior in your community. Every community has the potential to be attacked or to be a base of terrorist operations. Read the listing of the case histories, and you will learn that all communities are at risk. As you respond in your community, remember those who served on September 11, 2001, and dedicate your career to being the most knowledgeable and prepared firefighter that you can be. . . You never know what the future may bring.

—Street Story by Christopher Hawley, Baltimore County Fire Department (Ret.)

OBJECTIVES

After completing this chapter, the reader should be able to:

- Discuss potential target locations.
- Discuss indicators of potential terrorist activity.
- Describe incident actions to be taken at a terrorist attack.
- Describe additional hazards at a terrorist attack.
- Describe other specialized resources to assist with a terrorist attack.
- Describe methods of requesting federal assistance.
- Identify common agents that may be used in a terrorist attack.

Figure 30-1 In 1993 a van was used to carry explosives to an underground parking garage in the World Trade Center. Six people were killed and more than 1,000 were injured. There were 50,000 people in the building, and the goal of the terrorist was to collapse the building into the adjacent tower.

INTRODUCTION

It is unfortunate that a chapter on terrorism needs to be included in this firefighting text, and until recent times this would not have been necessary. Although everyone has seen terrorism on the evening news, it used to occur in places such as Northern Ireland, Beirut, Israel, or somewhere other than the United States. Until recently, the United States remained for the most part immune to the reign of terrorists. This changed in February 1993 when the World Trade Center was bombed in New York City, **Figure 30-1**. Even when this bombing occurred, the fire service did not pay much attention, because the bombing was looked at as the type of incident that happened only in big cities. In addition, the persons who were found to be responsible for the bombing were controlled by an influence from outside the United States, so the thought was that it was an isolated foreign attack.

However, when the Alfred P. Murrah Federal Building in Oklahoma City was devastated by a bomb on April 19, 1995, the United States fire service took notice. The attack on the federal building, **Figure 30-2**, was brought on by a person who did not have ties to another country and was a natural born citizen of the United States. It was perceived as an attack on America from one of its own, not from some unknown citizen from a foreign nation.

And perhaps the most devastating attack on the United States occurred on September 11, 2001, when the World Trade Center and the Pentagon were hit with a total of three airplanes. Then, in Shanksville, Pennsylvania, a hijacked plane crashed into a field. As a result of the impact of the planes and the resulting fire, the 110-story twin towers of the World Trade Center collapsed. The death toll from all three sites was more than 2,600 civilians and 346 emergency responders.

Shortly following three horrific crashes, in October 2001 a flood of noncredible anthrax letters hit the country. Mixed in with the thousands of noncredible letters were several letters containing real anthrax. These letters were responsible for five deaths and up to thirteen other illnesses. The letters were sent to news media outlets and to members of Congress.

Regardless of the origin, the potential for terrorism is here in this country and has to remain in firefighters' thoughts as they respond to any incident. Other incidents are occurring on a regular basis, and, although they do not fit the exact definition of terrorism, they involve the use of large-caliber weapons or a large number of weapons. This chapter looks at terrorism, **HAZMAT crimes**, and other potentially dangerous criminal situations that are sometimes just as deadly as terrorism. Firefighters have to think outside the box in regard to terrorism. There are many events that occur in this country that place responders at risk. Terrorism, HAZMAT crimes, murders, and other criminal events all present a risk to firefighters. Someone using ricin to kill another person is committing murder, but the weapon is as deadly to the responders as it is to the intended victim. A booby-trapped drug lab is not terrorism, but it presents significant risk to responders. Many criminals are protecting themselves with

Figure 30-2 A truck bomb caused the devastation in the Oklahoma City bombing in which 167 people were killed and 759 injured. The damage extended several blocks in each direction, and 300 buildings were damaged. Fatalities occurred in 14 separate buildings. *(Photo courtesy of John O'Connell)*

body armor and fortified vehicles. Many crimes such as bank robberies involve the use of explosives and sophisticated weaponry. Street violence that can be associated with gangs is increasingly violent, and when people are killed or injured the fire service is called into action. Crimes such as assaults and murders are increasing in the school system. In recent times there have been deadly riots in Los Angeles and in St. Petersburg, Florida, usually after a sporting event. In Chicago, twenty-one people were killed and fifty were injured in a nightclub after a fight broke out. These deaths may have been caused by a lack of exits and the use of pepper spray by security personnel, which panicked the crowd. The workplace is also becoming an increasingly dangerous place to be. A firefighter in Jackson, Mississippi, entered the headquarters fire station intent on killing the fire chief. Several persons were killed, not to mention the emotional damage that occurred.

Crimes are becoming more violent, and it is perceived that this trend will continue to increase. Emergency responders are immediately placed into dangerous situations, and can get caught—literally and figuratively—in the cross fire.

History of Modern Terrorism and Hazardous Materials Crimes

As a result of the terrorist attacks on September 11, 2001, the previous bombing of the World Trade Center in 1993, the Alfred P. Murrah Federal Building bombing in Oklahoma City, and other incidents, the fire service's response to some types of incidents will be forever changed. In the late 1990s there were two other bombings that are fairly well known to the fire service: the Atlanta Olympic Park bombing in 1996 and the Atlanta abortion clinic bombing in 1997. That particular abortion clinic bombing is significant because a secondary device was used. It is thought that it was strategically placed with the sole intention of harming the responders. The device was placed near the location where the incident command post was set up. Luckily, only minor injuries were received by the responders who were near the blast. In 1998 there was another bombing at a Birmingham, Alabama, abortion clinic in which an off-duty police officer employed as a security guard was killed by a device that was more accurately placed to target the responders. This secondary device was activated by the person responsible as he watched the victim approach the radio-controlled device. A brief overview of some of the recent acts of terrorism, HAZMAT crimes, use of terrorism materials, and criminal acts that presented risk to responders follows:

1980s A series of bombings targeted primarily at the Internal Revenue Service (IRS) included an attempt at a chemical bomb using ammonia and bleach. Another attempt included the use of a hot water heater as a very large pipe bomb, but the vehicle carrying the bomb caught fire while the terrorist was driving the vehicle.

1984 Dr. Michael Swango had a long history of suspicious deaths while he moved throughout the United States and abroad. He was stripped of his license to work as a doctor and got a job as a paramedic. Although his crimes do not fit the pattern of standard terrorism, he was finally arrested for attempting to poison his paramedic coworkers with chemicals.

1984 The Rajneesh Foundation was responsible for poisoning 715 people with *Salmonella* bacteria. They poisoned the salad bars in ten restaurants. The group had used other biological agents and had targeted a water supply for attack. They had previously used raw sewage and dead rodents in an attempt to poison the system. They used nursing home patients as their test targets for some of the biological attacks. The attacks were an attempt to effect a change in a local election.

1985 Members of a militant group developed a plot to poison a water supply. They were going to use 35 gallons of cyanide, which they had in their possession. Other members of the group were arrested for church arsons and attempting to blow up gas pipelines.

1993 The World Trade Center was damaged by a van bomb. Six people were killed.

1995 A man was arrested after manufacturing **ricin**, an extremely deadly toxin, with the intention of killing someone he was jealous of.

1995 The Alfred P. Murrah Federal Building in Oklahoma City was the target of a truck bomb. Several hundred people were injured and 186 people were killed in the blast.

1995 The members of a militia group known as the Patriots Council were arrested for the manufacture of ricin in the attempted assassination of a U.S. marshal.

1995 Two separate ricin incidents involved two doctors. Dr. Deborah Green killed two of her three children by burning her house, and she attempted to kill her husband three times with ricin. In Virginia, Dr. Ray Mettetal attempted to murder his boss with a syringe filled with boric acid and saline. He also possessed ricin and a number of other materials.

1995 (and throughout the 1990s) The Aum Shinrikyo are the cult group that is best known for a 1995 sarin nerve agent attack in the Tokyo subway. For a long period of time the group carried off chemical and biological weapons attacks and went to great lengths to develop their weapons program. An examination of this group yields many valuable lessons on chemical and biological attacks. The group had assets of more than $300 million. The group was also home to more than a hundred scientists whose sole function was to develop chemical and biological weapons. The group used, or explored the use of, *Clostridium botulinum, Bacillus anthracis* (anthrax), Q fever, Ebola virus, and viral hemorrhagic fever. They carried off ten chemical attacks and nine biological attacks, not to mention numerous small and large full-scale tests. They killed seven people and injured 200 in a sarin nerve agent attack in Matsumoto in 1994. The subway attack in 1995 killed thirteen but injured more than 5,000. The attack consisted of placing sarin in three subway lines, with a total of eleven sarin bags that were pierced with umbrellas. The deaths occurred to those who came in contact with liquid or were in small confined spaces. Over 4,000 people went on their own to 278 hospitals. The Tokyo Fire Department ambulances transported 688 patients. The breakdown of the injured was seventeen critical and thirty-seven severe. There were 948 patients suffering from miosis (pinpoint pupils). All of the moderately injured patients were treated and released within six hours after they arrived at the hospital. Although 85 percent of the patients did not require any treatment, they still flooded the hospitals.

1995 A train was derailed near Hyder, Arizona, and a terrorist group claimed responsibility, although no one has been charged in the attack. One person was killed and the derailment seriously injured twelve people.

1995 The Anaheim Fire Department in California was made aware of a potential sarin nerve agent attack at Disneyland. The chief was notified at midnight to be at the command post four hours later. Police agencies, federal law enforcement, and the military had known about the threat for five days. Up until notifying the fire chief and assembling the resources prior to the event, no other planning had occurred. The fire department was placed in charge of the incident and quickly developed a plan of action. Disneyland did not close and 30,000 to 40,000 people visited the park during the threat period. Luckily, the threat never materialized.

1996 Stuart Adelman, Buffalo, New York, used his employer's Nuclear Regulatory Commission (NRC) license to order several radioactive sources. The purpose of the purchase is not known, but it is suspected that foul play was being planned.

1996 Members of a paramilitary group who had access to preplanning information obtained through the fire department were arrested when they attempted to blow up a Department of Justice complex in West Virginia.

1996 and 1997 The Atlanta area was besieged with bombings, including one at the Olympic Park that killed one person. At an abortion clinic in Birmingham, Alabama, a booby-trapped device killed an off-duty police officer and wounded a nurse.

1996 Theodore Kaczynski, known as the Unabomber, was responsible for a string of bombings that lasted eighteen years. He sent sixteen bombs, which killed three people and injured twenty-three. He wrote a manifesto that appeared in the *New York Times* and was recognized by his brother, who turned him in to the Federal Bureau of Investigation (FBI).

1996 A disgruntled firefighter in Jacksonville, Mississippi, went to his department's headquarters with the intent of killing the fire chief. The fire chief was not injured, but several other fire department personnel were killed.

1996 A lab worker brought in pastries that were poisoned with *Shigella dysenterie* to coworkers. She made twelve people ill and was found guilty on five felony assault charges. Police also learned that she had attempted on more than one occasion to poison a former boyfriend with *Shigella* and other biological agents.

1997 EMS and police responded to a shooting incident, and the suspect was found to have ricin, *E. coli,* and a mixture of nicotine and dimethyl sulfoxide (DMSO).

1997 Four members of the Ku Klux Klan were arrested for plotting to blow up a hydrogen sulfide tank in order to create a diversion for an armored car robbery.

1998 In Nairobi, Kenya and Dar-es-Salaam, Tanzania, bombs exploded in two U.S. embassies, killing 224 people.

1998 Kathryn Schoonover attempted to mail 100 envelopes with cyanide disguised as diet powder all over the United States.

1998 Two men were arrested on the suspicion of **anthrax** possession. They were later released, since it was determined that they only possessed a possible anthrax vaccine. Larry Wayne Harris had previously been arrested in 1995 for the possession of plague, a biological toxin that he had ordered through the mail. He was a previous member of the Aryan Nation but was deemed too radical for the group.

1998 In Charlotte, North Carolina, a man with an explosive device held some hostages in a government building. The explosive device was thought to also contain some type of chemical agent. It was later determined that the filling agent was harmless, although the explosive was live.

1998 Abortion clinics in Florida, Louisiana, and Texas were affected by attacks using **butyric acid**, a material with a horrible, irritating odor.

1998 In Lafayette, Indiana, a pickup truck rammed into the courthouse. The bed of the pickup had flammable and combustible materials as well as several explosives.

1999 In Colorado, two students who were armed with an array of guns and explosives attacked their own high school. The suicide attack resulted in fifteen deaths and spurred a rash of bomb hoaxes throughout the country.

1999 The FBI investigated hundreds of anthrax hoaxes, none of which involved the actual use of the biological agent. Abortion clinics were the targets in most of the cases.

1999 Ahmed Ressam was arrested in Port Angeles when he crossed over the border into the United States. He was responsible for plotting the millennium "border bomb" and had the components to a large bomb in his truck. His target was the Los Angeles Airport (LAX).

2000 Police and fire officials scrambled after more than eighty vials filled with an unknown liquid were found throughout Milwaukee and its surrounding suburbs.

2000 Dr. Larry Ford's house in Irvine, California, was searched after he attempted to murder his partner. Dr. Ford committed suicide, and his house was found to have held numerous chemicals and chemical agents. The search took several weeks, and chemicals, guns, and biological materials were found.

2000 During a genetics conference in Minneapolis, Minnesota, protestors attacked three restaurants with hydrogen cyanide.

2000 Members of a militia group in Sacramento, California, were arrested for plotting to blow up a 24-million-gallon propane tank.

2001 A Chevron employee in the El Segundo, California, refinery threatened to blow up the facility. The facility produces more than 4 million gallons of jet fuel a day.

2001 Over 2,600 civilians and 365 emergency responders lost their lives at the World Trade Center in New York City; the Pentagon in Arlington, Virginia; and a field in Shanksville, Pennsylvania. Four planes were used in the attack, two hitting the World Trade Center and one hitting the Pentagon. The fourth plane crashed into a field. Members of the al Qaeda group have been held responsible for the attack.

2001 In October there was one death as a result of an anthrax-laden letter. Later, there was a total of seven deaths. The emergency services across the country were deluged with calls about white powder. A very small percentage of the calls was investigated by the FBI, but they opened more than 14,000 cases regarding the white powder events. Only five cases involved real anthrax, which was sent to members of Congress, to New York City, and to Boca Raton, Florida. In December, Clayton Lee Waagner was arrested and admitted to sending more than 500 noncredible anthrax letters to women's reproductive health centers (WRHCs). He was a member of the Army of God, a group known for pursuit of the right to life cause.

2001 The Animal Liberation Front (ALF) and the Earth Liberation Front (ELF) committed a number of terrorist acts throughout the year. The ALF admitted to 137 illegal actions at a variety of locations. Both groups target buildings and businesses that have any connection to the use of animals or perceived violation of the environment. The acts typically create inconveniences, such as glued locks, but on occasion do include arson and other violent destruction of property. For the most part these groups take great care to avoid any potential injuries to humans, but responders could be killed or injured while responding to or handling these acts.

2001 Two Jewish Defense League (JDL) members were arrested for plotting to blow up a mosque and a congressman's office. Other members of the JDL have been quoted as having a desire to continue the militant work of the two men arrested.

2001 Richard Reid was arrested on a Paris to Miami flight for attempting to detonate a PETN explosive that was located in his tennis shoes. PETN is one of the more powerful explosives. He was overpowered by other passengers and the flight crew while trying to light the fuse. He has ties to al Qaeda and is supportive of their cause.

2002 Two different bombers created fear in the United States, one in Philadelphia and the other in the Midwest. Preston Lit set pipe bombs off in U.S. Postal Service mailboxes in the Philadelphia area. Lucas Helder, a college student, set off pipe bombs in residential home mailboxes in five states. Helder's intention was to set off bombs so that the explosions when drawn on a map would form a smiley face. His bombings injured six people. He set a total of eighteen pipe bombs, six of which exploded.

2002 A fifteen-year-old stole a small plane and flew it into a high-rise building in Tampa, Florida, killing himself. His suicide note stated that he wanted to be just like Tim (McVeigh), Eric (Rudolph), and Osama (bin Laden).

2002 Joseph Konopka, a member of the Realm of Chaos group, had stashed potassium and hydrogen cyanide in a Chicago subway tunnel. The group is known for wanting to destroy public utility, water, sewage, and telecommunications systems.

2002 The leaders of two militia groups in Kentucky and Pennsylvania were arrested for possession of weapons and explosives.

2002 The FBI arrested Abdullah al Muhajir, who is associated with al Qaeda, in Chicago. He was researching the use of, and looking for materials to detonate, a radiological dispersion device. He had traveled to Pakistan and had studied methods to pull off such an attack with al Qaeda operatives. In late 2002 the FBI was still looking for 100 al Qaeda members and investigating 150 persons and groups who may have al Qaeda ties. They made two large arrests in Detroit and Buffalo, as well as other arrests throughout the United States, apprehending a number of suspects who had al Qaeda ties.

The following sections provide some information about the type of terrorism agents that currently exist. Some are unique ideas and thoughts, and some represent possible scenarios. There exists some thought within the fire service that this type of information should not be published. All of this information is readily available in the public domain and in training programs and texts throughout the United States. Most of the exact recipes and "how-to" instructions for these and any other device are easily obtained through printed texts and the Internet. This text does not provide any information that cannot be easily obtained elsewhere. If an ordinary citizen is able to obtain it in an easy, normal, and legal fashion, it is certain that the terrorist already has it. Responders need to be aware of the various chemicals and devices that someone may design to kill them and others in their community. The best defenses are education and trying to stay one step ahead of the terrorist. By being informed as to how a terrorist may operate or what some of the devices may involve, responders will be alert to a potentially fatal situation.

TYPES OF TERRORISM

The types of terrorism are divided into two distinct areas: foreign based and domestic. Until the Murrah building bombing, the fear of terrorism was aimed at a foreign source, and it was thought that any terrorist attack would be from a foreign country. To be foreign based, the motivation or supervision must come from a foreign country. Domestic terrorism originates from within the United States and is not influenced by any foreign party. From the list of terrorist acts provided earlier and from other statistics, it is clear that the largest percentage of terrorism is domestic in origin.

The FBI defines terrorism as a violent act or an act dangerous to human life in violation of the criminal laws of the United States or any segment to intimidate or coerce a government, the civilian population, or any segment thereof, in furtherance of political or social objectives. The key to this definition is the intimidation of the government or the civilian population. A militant group trying to influence the local political process sprayed *Salmonella* bacteria on a fast-food restaurant salad bar and was successful in making more than 600 people sick. The Tokyo subway sarin attack was an attempt to destroy a good portion of the police department in an effort to prevent a raid on the terrorist compound.

> **Safety** The fire service will be called to many incidents that will not fit the exact definition of terrorism, but the hazards from a pipe bomb are the same regardless of the motivation of the builder.

Many responses that would have been routinely handled in the past must now be treated much differently, and responders must always be on their guard for terrorist-style devices or potential acts of terrorism.

The terrorist's motivation is to produce fear that may be aimed at the general public or the government. Fear can be provoked by large-scale actions such as the acts on September 11, 2001, the original bombing of the World Trade Center in 1993, or the bombing of the Alfred P. Murrah Federal Building in Oklahoma City. Even acts that are not terrorism can create terror in the community. The Chicago nightclub incident, which was caused by pepper spray, and the Rhode Island nightclub fire have sparked fear and concern about safety in nightclubs. A terrorist can also incite fear just by planting the thought of potential terrorism or by devising a hoax. The latter scenario is the more likely and can be very difficult to handle from an emergency service perspective.

The thought process for determining if a threat is credible or not has five elements. If the person known or thought to be responsible is determined to have several of these capabilities, it increases the credibility factor:

1. The first of the five qualifiers involves the potential terrorist's educational ability to make a device or agent that, unlike explosives or ricin, is very difficult to attain. To truly make a biological pathogen agent, in most cases, one needs an advanced knowledge of biology. Ricin, a biological toxin, does not require any advanced knowledge compared with weaponized anthrax. Some of the threats with letters or packages have misstated the origin of the material, such as calling anthrax a virus, or have misspelled the agent's name. If the terrorists do not know the true origin of the material or

cannot spell the material correctly, they probably do not have the education necessary to make the material they state they produced. This does not take into account a person who may purchase the material.

2. The next qualifier is a person's ability to obtain the raw materials necessary to make the agents. Many of the materials necessary to make chemical warfare agents are banned for sale. Others appear on hot lists, which means they are only sold to legitimate businesses. This would not preclude someone from buying the raw materials on the black market or stealing them from a legitimate business.
3. The third qualifier is the ability to manufacture the devices or machinery required to make the agent. To manufacture chemical warfare agents requires the use of a reactor vessel, which requires about a 10-foot by 10-foot space to produce less than a gallon. There are some agents that could be produced in a bathtub using backyard chemistry, but these are not the high-end agents that attract much attention. Many people who have attempted to make agents in less than ideal conditions have died during the production process. Many criminals do not take the time to follow standard industrial safety precautions.
4. One qualifier that is often overlooked is the ability to disseminate these agents. The military conducted many tests on chemical and biological warfare agents, and although they have some good methods of dissemination, even they lack a 100 percent effective method of dissemination. The Aum Shinrikyo cult in Japan is a perfect example, as they were a group with millions of dollars in assets and full chemical and biological lab and production facilities. They employed the services of 235 scientists to develop and manufacture chemical and biological agents. The Aum Shinrikyo abandoned their biological weapons program after a full-scale release of anthrax that failed. They used sarin nerve agent twice, the first time in Matsumoto, Japan, in which seven people were killed and 200 injured. The dissemination method used in the Matsumoto attack was much more effective than the one used in the Tokyo subway attack. If they had used the same dissemination method they would have greatly altered the course of events. They would have been limited by the amount of agent that could have been produced in a short period of time.
5. The last qualifier, which is the most important, is whether the person or group has the motivation to pull off the attack. The intentional killing of one person takes significant motivation, and the intentional killing of hundreds requires a whole lot of motivation. There is always the potential for an attack, but it takes considerable education, raw materials, manufacturing, and dissemination ability to pull off a chemical or biological terrorist attack. On the other hand, explosives are easy to manufacture and do not require much education, only simple tools, and the materials required are easy to assemble. It is for this reason that explosives are used in the majority of cases and are quite successful in completing an attack. In many cases the terrorist can be successful because of the hysteria associated with a potential terrorism incident. A balance must be struck between a cautious approach and one that does not allow the terrorist to win by crippling a community and causing hysteria.

Another consideration, as we have seen in England and Ireland, is the disturbing trend of viewing emergency service personnel as targets. One theory currently under examination is that the second explosion at one of the Atlanta abortion clinic bombings was aimed at the emergency responders.

> **Safety** Responders must always be alert to the potential of a secondary device.

POTENTIAL TARGETS

Potential targets exist throughout every community in the nation and can be commercial buildings, high-rise buildings, and even residential homes. Although some incidents do not fit the definition of terrorism, the materials used are the same as those a terrorist might use. Whether the objective is murder or terrorism, the danger to the responder is the same. When looking at terrorism, potential targets can be grouped into several categories: public assembly such as the area shown in **Figure 30-3**; federal, state, and local public buildings; mass transit systems; high economic impact areas; telecommunication facilities; and historical or symbolic locations.

While obviously not an exclusive list, buildings that could be targeted include the Federal Bureau of

Figure 30-3 Any location is a potential target for a terrorist. Any location where large numbers of people are present, such as a mall or sports event, is a prime target.

Investigation (FBI); the Bureau of Alcohol, Tobacco, and Firearms (ATF); the Internal Revenue Service (IRS); military installations, Social Security buildings; transportation areas; city or county buildings, including fire and police stations; abortion clinics and Planned Parenthood offices; fur stores; laboratories; colleges; cosmetic production/testing facilities; banks; utility buildings; churches; and chemical storage facilities. Transportation facilities such as airports and train, bus, or subway stations are high on the potential list of targets, given the number of people who may be potential targets and the relative ease of escape. In the southeastern United States, a large number of churches have been subjected to arson fires, and in some cases explosive devices have been used. A large number of abortion clinics have been the subject of bombings, attacks, and other threats. Any incident in or near one of these facilities should be approached with caution. Responders should know the location of these facilities in their jurisdiction. Preplans for these facilities should be thought out by the company officers, but it is not recommended that these plans be committed to paper. As the battle between pro-life and pro-choice groups continues to rage, incidents at these facilities can only be expected to rise, with emergency responders caught in the cross fire.

Many of the potential targets of terrorism have not been buildings at all but events where large numbers of people are present. The Atlanta Olympic Park bombing is an example. Other scenarios involve sports stadiums, such as the one pictured in **Figure 30-4**, public assembly locations, transportation hubs, and fairs and festivals. First responders should have some preplans for these types of locations. One possible scenario for a stadium, devised by Captain Richard Brooks of the Baltimore County Fire Department, describes the first-in medic unit arriving at a stadium where in Section 300 there are 40 people projectile vomiting. After five minutes 200 people are projectile vomiting, and as time goes on the number increases. What happens to the responders when confronted with a situation of this nature? How many responders would be affected by this massive amount of people vomiting? How many responders would be needed to handle this incident? This act of terrorism could be accomplished by putting syrup of ipecac in the ketchup container beside one of the hot dog vendors, an easy task. Imagine the hysteria if a note was found stating that a biological agent was distributed in that section. What impact would that have on the remaining 50,000 people in the stadium if that information got out? Planners and responders involved on the national level in trying to develop response profiles to terrorism are grappling with how to plan for incidents involving 100 people, 1,000 people, 10,000 people, and 50,000 people. Terrorism incidents can very quickly overwhelm the responders and their whole emergency response system.

When dignitaries visit locations, a lengthy planning process typically takes place in which the fire department should be involved. When the Pope visited Baltimore in 1997 the planning process took

Figure 30-4 Other than special events, the most common location where large numbers of people are together is at sporting events. At this stadium, if an incident were to occur, more than 50,000 people could become part of the incident.

more than eight months. Planning for such a large event takes the cooperation of local, state, and federal agencies. Even when dignitaries visit locations such as New York City or Chicago, advance planning occurs. Other events such as political conventions or other large political gatherings all bring the potential for an incident. When one of these events comes to a community, local responders need to be prepared for not only the people arriving to the event, but also the massive federal response that may be pre-positioned. For many special events, whole task forces of federal resources may be hidden away just in case of an incident.

Certain dates have significance to several militant groups. The date of April 19 is the anniversary of the Waco, Texas, incident in which the ATF stormed a compound that housed the Branch Davidians, a group thought to be a militia group. April 19 was also the date of the Oklahoma City bombing, and the date was chosen by the bomber as a way to retaliate for the Waco incident. Other dates provoke the potential for terrorist acts. For instance, the anniversary of *Roe v. Wade,*[1] January 22, could incite a strike by antiabortion groups. Within the United States, forty states are suspected to have members of militia, patriot, or constitutionalist groups. Membership counts vary from fifty people in some states to several thousand members in other states. Groups of concern include anarchist groups and white supremacy groups such as the Ku Klux Klan, Zionist Occupation Government, skinheads, and neo-Nazi groups, including the Aryan Nation. Other groups suspected of activity or thought to have terrorism potential include patriots, New World Order militia, constitutionalists, and tax protesters. To learn which groups are active, it is relatively simple to use a Web browser and search the Internet for many of these groups, because most have Web sites.

INDICATORS OF TERRORISM

An explosion or explosive device is the most common tool of the terrorist, and police across the country have made several arrests of persons for making or storing large quantities of explosives.

> **Firefighter Fact** According to an FBI source, more than 93 percent of terrorism incidents use explosives as the weapon of choice.

The most common device is a pipe bomb such as the one shown in **Figure 30-5.** Any incident where an explosion has occurred or first responders believe that an explosion has occurred should be suspected of being a terrorist incident.

> **Safety** If one explosion has occurred, responders should always assume that there is a second device awaiting their arrival.

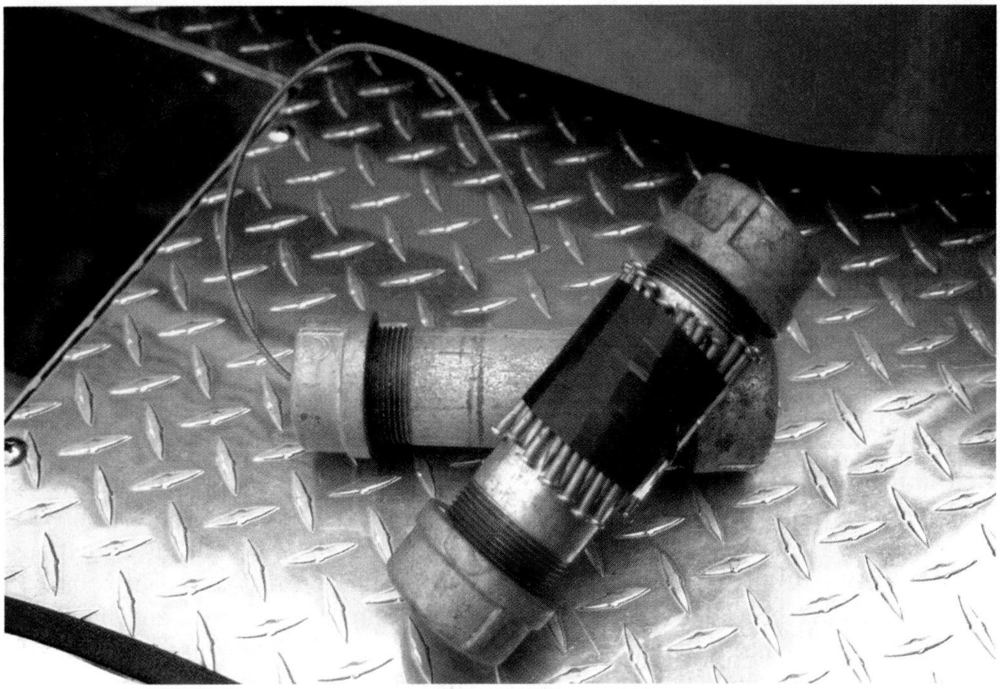

Figure 30-5 The most common explosive device is a pipe bomb, and it is very effective. It is a very dangerous device, not only for responders but for the builder as well.

In this day and age any suspicious package should be suspected of containing explosives and should be dealt with by a bomb technician. Firefighters should never assume that they can handle the package or remove it from the area. Just like EMS and HAZMAT, the handling of bombs or suspicious packages is a very specialized field and should only be done by a person trained to do that type of job.

The presence of chemicals or lab equipment in an unusual location, such as a home or apartment, is an indication of possible illegal activity. When looking at chemicals or a lab there are three possibilities: drug making, bomb making, or terrorism agent production (chemical or biological). The most likely, based on statistics, is drug making, followed by bomb making. Although very common in some parts of the country, predominantly the West Coast and the Northwest, drug production labs are not commonly located throughout the whole United States. Currently the number of drug labs is increasing in the Midwest, with responders taking down more than twenty a month in some areas. This trend is slowly moving from the West Coast to the East Coast. The most likely scenario is locating someone making explosive devices, as many individuals like to make homemade devices. It is possible, but very unlikely, that firefighters would locate a facility attempting to make a terrorism agent such as sarin. The exception to the biological agents would be for the production of ricin, as the items required to make ricin are easily obtained and the production is just as easy. Fortunately, ricin does not have the potential to easily kill large numbers of people. It is primarily an injection hazard, although it is still very toxic through inhalation or ingestion. A responder in a small community or a rural setting is the most likely to run across any of these types of production areas. Several persons a year are arrested for the possession of ricin with the intent to use it for some type of criminal act. Most of the arrests are occurring in small towns throughout the United States. No matter what agent is located, the responder should immediately isolate the area and call for assistance. The call should go simultaneously to the police, the HAZMAT team, and the bomb squad. This request would apply to all three types of labs—drug, bomb, or warfare agent—as any of these labs should be handled as a cooperative effort.

Another indicator of potential terrorism is the intentional release of chemicals into a building or the environment. Finding a chlorine cylinder in a courthouse would be unusual and should put the responders on alert to the fact that there is a high probability that a terrorist may be at work. Finding chemical containers such as bottles, bags, cylinders, or other containers in unusual locations would also be suspect. In an industrial facility in which chemicals may be common, responders may find that there is the intentional release of a chemical.

Note One of the best indicators of potential terrorism will be a pattern of unexplained illness or injury.

A response to a mall for a seizure patient is not unusual. However, a response to a mall for six people having seizures is very unlikely and could involve a chemical release from a terrorist attack. Seizures, twitching, tightness in the chest, pinpoint pupils, runny nose, nausea, and vomiting are all signs and symptoms of a warfare agent attack. EMS providers will probably be the first group to identify the use of a chemical agent. Imagine arriving at an explosion where there is a large amount of debris and twenty victims. To most the injuries would appear to be blast injuries, as they would in most cases be visible during a quick survey of a patient. Other signs and symptoms, primarily pinpoint pupils, probably would not be identified until the patient is given a more thorough exam. When confronted with victims that are unconscious or dead and now have outward signs of death, such as trauma, the responder must face the possibility that they may have been the victims of a chemical attack. This obviously has to be put in perspective. If responders in a metropolitan area of the Northeast are called to an apartment building in the winter because there are twelve unconscious people, and this happens several times each winter, it is probably carbon monoxide poisoning. However, if responders are called to a mall in the summertime because there are twelve unconscious people, it is probably not due to carbon monoxide but to a chemical release of some type. When distributing a warfare agent the explosive device will usually not have a destructive effect on the building or the surrounding area. In some cases the larger the device the more likely it is that the detonation will consume the agent. Most of the victims will not display signs or symptoms of a blast, although the one closest to the device may suffer some of those types of injuries.

Smelling unusual odors or seeing a vapor cloud may be an indicator that chemicals have been released. As may be seen with other toxic gases such as chlorine, arriving and finding dead birds or other animals should alert the responder not only to a chemical hazard but also to the potential for terrorism. It would also be unusual to respond to an office building or a house and notice security measures of the type that one would not expect in that occupancy. Items such as extra locks, bars on windows, surveillance cameras, fortified doors, guards, and other unusual protection devices may provide a clue to the responder that something out of the ordinary is at play.

HAZMAT CRIMES

Although this chapter covers terrorism, emergency responders may respond to other criminal-related events. Whether the incident is terrorism related or not, the effects on responders can be the same. Criminals are using more weapons than just guns today, and the use of clandestine labs for illicit production is on the rise. These incidents are referred to as HAZMAT crimes, since chemicals are being used in an illegal fashion. Occasionally, there will be a robbery of a convenience store where the weapon is a chemical. The robbery suspect might throw a corrosive liquid on the clerk in order to rob the cash register. There have been other robberies, attacks, or attempted murders using corrosive liquids. Whether or not the law-enforcement community or the prosecutor decides the incident is an act of terrorism, emergency responders need to recognize potentially hazardous situations and have the ability to protect themselves.

One big exposure issue for emergency responders is drug related. Many drug addicts use chemicals as part of the process to get high, and the drugs themselves are usually toxic and may present other hazards. Drug users may use a combination of chemicals to get high, and many of these items are toxic and flammable. Ether is used to assist in the heating process of several types of drugs. This material in pure form is extremely flammable, and the container may eventually become a shock-sensitive explosive. Most drugs are in solid form, which means that they present little risk unless eaten or touched with bare hands. In some cases the drugs may be stored or used with flammable and toxic liquids.

The most common situation in which emergency responders could be directly affected by drug use is when a person is huffing. While people are huffing they typically use a toxic and/or flammable material. Many common household items such as paints, glues, hair spray, and solvents provide the high that some people desire. From a HAZMAT point of view, these materials are toxic and flammable and airborne. In most cases, users will spray the material into a bag to concentrate the vapors, but when they are done the vapors remain in the room of use.

Clandestine Labs

There are several types of clandestine labs that emergency responders are likely to encounter. The most common is the drug lab, but other possible labs include explosives labs, chemical labs, and biological weapons labs.

> **Safety** All labs have inherent dangers for responders, and all are very dangerous locations to occupy. All may be booby-trapped or be set up to harm responders, and a booby trap does not know if the person coming through the door is a police officer, a firefighter, or an EMT.

Figure 30-6 Example of a possible biological lab. Shown is a microscope and an incubator; both would be used in the production of biological materials.

For the most part a biological weapons lab, shown in **Figure 30-6,** can run unattended without any major concern and may be shut down in any number of ways without major consequence. A drug, explosives, or chemical lab, on the other hand, should only be shut down by someone trained to do so, as these labs are especially dangerous.

Drug Labs

There are many types of drug labs. Just for methamphetamines, there are more than six common methods of production. In 1973 the Drug Enforcement Administration (DEA) discovered 41 labs; in 1999 they discovered 2,155 labs, and in 2001 they raided 12,715 methamphetamine labs. The map shown in **Figure 30-7** provides proof that the labs are moving eastward at a rapid rate; soon all of the United States will have to deal with this problem. Due to its popularity, meth production is becoming common but it does involve a dangerous process. The production of drugs requires the use of many chemicals. These chemicals can be purchased outright, stolen, or manufactured using other chemicals. As many drug-producing chemicals are hot listed or cannot be purchased, the producer must resort to innovative methods to produce the chemicals. The "nazi" method of meth production involves the use of anhydrous ammonia, which is usually stolen from a chemical facility. If a 150-pound cylinder of anhydrous ammonia developed a leak or catastrophically failed, people for a considerable distance downwind could be affected, and those in the immediate area would be in grave danger.

Drug labs can be very complicated setups and can be found in any number of locations such as homes, barns, hotels, storage units, and even trucks. Emergency responders routinely encounter these labs inadvertently through other responses. Responders who encounter a drug lab should notify their HAZMAT team, the bomb squad, and the local office of the DEA. The shutting down of a drug lab is a complicated and very dangerous process.

> **Safety** When chemicals are being heated or cooled there is a chance for a violent reaction.

It is this heating and cooling process that indicates the type of lab that may be present. When responders see glassware that is distilling chemicals—in other words, evaporating a certain component—and then rehydrating or condensing another portion of the original chemical, this is indicative of possible drug production. The end result of the process is a solid form, usually a powder. There will be a production line-type formation to the glassware, with some solutions being heated while others are cooled. At some point, gas cylinders may be present and the gas is allowed to mix with some part of the process. Many of the chemicals involved in the production of drugs are flammable, and although many are also toxic the predominant hazard is flammability.

> **Safety** When choosing protective clothing to enter a lab, it is important to protect responders against the predominant hazard.

The materials that are toxic cannot harm the responder as long as SCBA is worn and the materials are not touched with bare hands or eaten. It is highly recommended that personnel from the police, fire, and EMS departments receive training in drug lab awareness, with some receiving specialized training in this area.

Explosives Labs

Although not common, it is possible that emergency responders might encounter an explosives lab, which is the predominant weapon of choice for a terrorist.

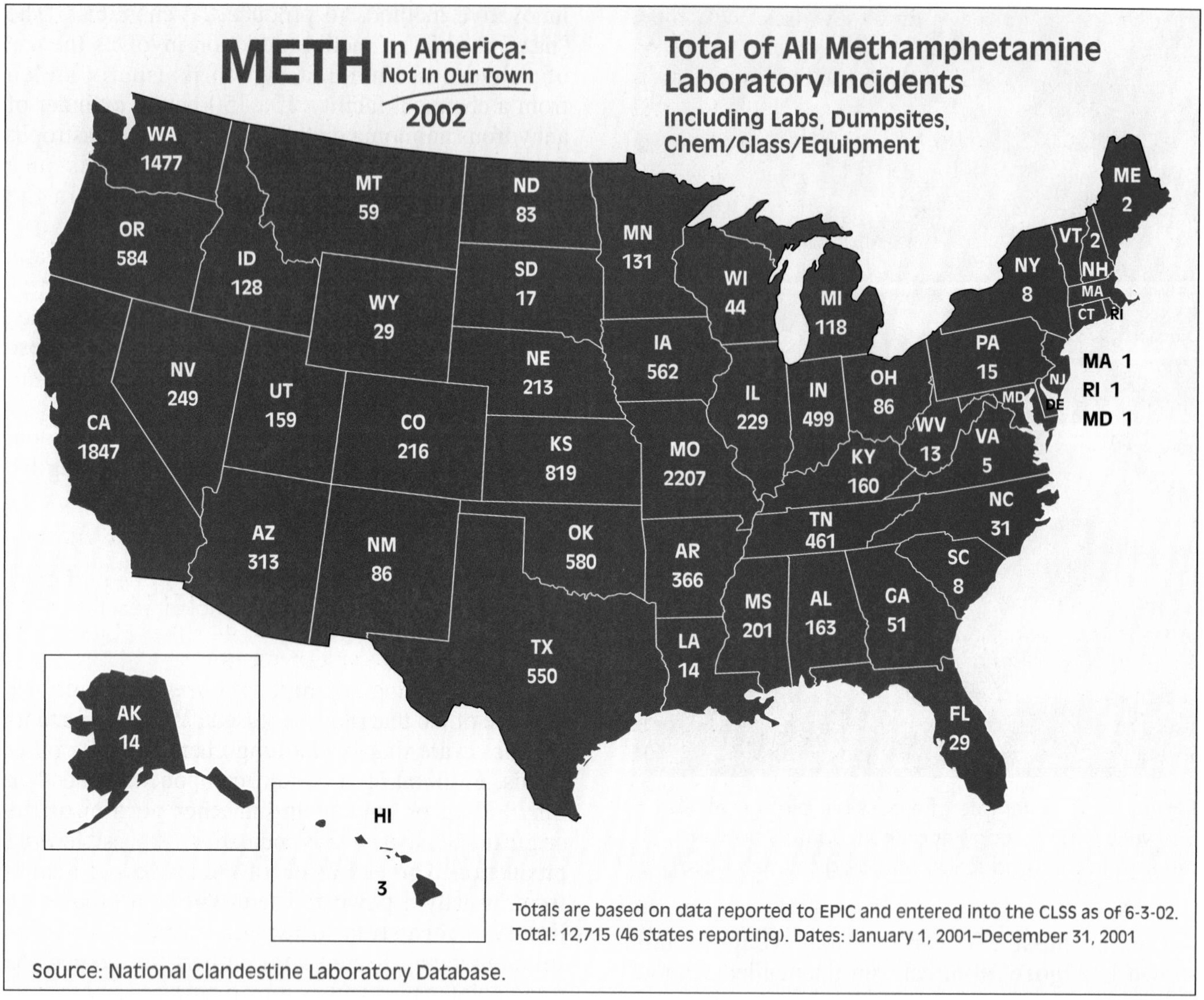

Figure 30-7 Methamphetamine lab seizures across the United States. Note the high number of labs in the Midwest. In years past, the largest numbers of lab seizures occurred in the West. The prevalence of methamphetamine labs is moving Eastward at a fast pace.

An explosives lab can be anything from a workbench pipe bomb builder to a full chemical production facility. A person building a pipe bomb does not need much equipment other than some simple hand tools, pipes, caps, and powder. A person making cyclotrimethylenetrinitramine, commonly referred to as RDX, or another more sophisticated explosive will need some more equipment. Depending on the availability of some materials, the bomb maker may have to make some chemical components as opposed to purchasing them. It is when materials are made at home that the danger increases for a responder. The processes to make the chemical components for explosives are very dangerous and present a significant risk to the builder and responders. An explosives lab differs from a drug lab in that most of the processes do not involve heating, cooling, condensing, or distilling. However, some processes used to make the chemicals do perform some of these functions, so there are no black-and-white rules for identification. The major work at an explosives lab is usually mixing of materials, in most cases solids and liquids. If gases are used, they are usually coupled with the explosive device and may be used to increase the heat of the explosion, boosting its efficiency. People making explosives will usually have a large amount of powders in their house.

Some of the other indicators of an explosives lab are the presence of ignition devices, boosting charges, or blasting caps. Many people who manufacture explosives are doing so to make fireworks and will have cardboard tubes for the explosives to be placed into. In one case, it was originally thought that an explosives maker was just making fireworks, but some other explosive devices were found to have BBs and nails taped and glued to the outside of the explosive. Neither of these have an effect on the display ability of the explosive device and are only designed to kill or maim. When an explosives lab is discovered, the local bomb squad and the local HAZMAT team should be called in to assess the materials. As the stability of many of the materials cannot be ensured, it is usually necessary to remotely destroy many of the found materials.

Terrorism Agent Labs

These are the least likely labs to be encountered in emergency response, but responders must be aware of their existence and some of the unique features of these types of labs. The two types of terrorism agent, or **weapons of mass destruction (WMD)**, labs are chemical weapons or biological weapons labs. Statistically the most likely lab is a biological toxin lab, which may be used to manufacture ricin, which is shown in **Figure 30-8.** The FBI, on average, arrests a few people a year for the possession of ricin, usually accompanied by a threat. Biological labs may be set up to attempt to make other biological materials. Other than ricin, botulinum toxin, and a few other biological materials, the manufacture of biological weapons is very difficult. The manufacture of ricin is a simple process that only requires a few items, such as castor beans and some readily available chemicals. The process to make some of the more advanced biological agents such as anthrax is more difficult and requires a higher level of education, sophisticated equipment, and access to raw materials. The development of biological agents involves culturing the material, usually in a petri dish. These petri or culturing dishes may be placed in an incubator or oven-like device to keep the material warm and at a constant temperature. Depending on the type of agent, there may be grinders, dryers, and sieves present to finish off the product. The major route of entry for many of these products is through touch or ingestion. The only inhalation concern would be during the grinding process, but a simple high-efficiency particulate air (HEPA) mask offers more than enough protection for a responder. Some of the materials used as part of the process may be flammable and have some toxicity, but the chemicals used to clean glassware and tools are usually highly corrosive and in most cases will be sodium hydroxide (lye) and/or bleach. A bioweapons lab will involve some chemicals, but may resemble more of a greenhouse than a chemical lab. Some biological materials are sensitive to light and will be kept in the dark, and the characteristic distilling and condensing glass apparatus will be missing.

Chemical weapons labs use two methods of production: the development of new chemical agents through standard production methods and the synthesis of existing materials. The development of a new chemical agent is the more difficult of the two processes and is nearly impossible except for someone with a chemistry background and access to some chemistry equipment. The recipes for chemical weapons are fairly sophisticated and require access to many raw materials that are hot listed. There have only been two arrests of persons in the United States for attempting to make a chemical weapon such as sarin nerve agent. Both were arrested for ordering the raw materials, and it is thought that the two individuals did not have the educational capacity to manufacture the agent. The production of these agents can be very risky to the producer and, without safety precautions, may result in death. Some of the off gases from the production of chemical warfare agents are extremely dangerous.

Figure 30-8 Shown is a lab that could be used to produce a biological toxin such as ricin.

The most probable scenario for the development of a chemical warfare agent is the synthesis of an existing product. The criminal would take existing materials, which are usually in diluted form, and synthesize them or reduce them down to a concentrated product. Luckily most, if not all, of the existing products do not present much risk to humans, as they are strictly engineered to harm only insects. This would not stop a terrorist from attempting to use one of these products in an illegal manner, however. There are some pesticides on the market that are applicable to this type of scenario. The standard household pesticide usually has 0.05 to 0.5 percent of pure product mixed with an inert ingredient. A mixture that is used by a farmer may be on the order of 40 to 50 percent pure form and is then diluted in the farmer's tank. In order to be more harmful to humans the pesticide must be concentrated and not diluted. The criminal must devise a method of removing one or more of the inert ingredients. As the inert ingredient is usually flammable or combustible, this is not a difficult task. A mechanism must be devised to off-gas the inert ingredient and then capture the pesticide and collect it.

One form of pesticides, known as technical grade pesticides, is already concentrated. These technical grade pesticides are in pure form and are not diluted. These materials are not in common use but can be found in and around the United States.

INCIDENT ACTIONS

A terrorist incident combines four types of emergency response into a large incident.

> **Note** These types of incidents have these four characteristics until proven otherwise: Mass casualty EMS incident + hazardous materials release and/or explosive devices + crime scene considerations = incident management challenges.

The first-in companies at these types of incidents can easily be overwhelmed and are going to be committed to basic actions, such as life preservation. The IC will have enormous responsibilities dealing with all of the required actions. All of the components present are difficult to handle individually, and now in this type of incident they are combined and must be handled simultaneously. The handling of a 100-person **mass casualty** incident is difficult and has the potential to overload the IMS, and that may only be one-quarter of the result of a terrorist incident. Responders should examine their response systems and determine how many patients present a concern. Some systems define a mass casualty as an incident involving five victims, while in other systems it may be ten to twelve. Obviously in a system with one EMS unit, more than one patient begins to cause system problems and may involve delays in treatment and transportation.

Imagine responding to an explosion at a mall, with 100 people injured. Such a mass casualty incident (MCI) will have EMS playing a predominant role. The police department, on the other hand, will be concerned with evidence preservation and crime scene considerations. The possibility exists that the perpetrator(s) also used a chemical or biological weapon, and the explosion was the means of distribution. This situation would be an MCI, a crime scene, and a potential chemical release situation with some, if not most, of the patients contaminated. On top of this scenario, add the potential for a secondary explosive device, one that is aimed at the responders! If responders can eliminate the chemical and secondary device issues, then all they have to deal with is the simple 100-person MCI and the crime scene issues. EMS will handle the patients and the police department would handle the criminal element.

The other aspect of a suspected terrorism incident will be the massive response from the federal government, even if not requested. Later in this chapter more information is provided about federal resources that are available to respond. In most cases, the minimum response to a suspected terrorism incident would come from the FBI, initially from the local agent. Field offices are located across the country, and almost every major city has a field office. Agent(s) usually arrive one to two hours into the incident. If terrorism is suspected the FBI is the lead agency of the incident as provided by a Presidential Decision Directive known as **PDD 39**. As with hazardous materials incidents it is important to know all involved players before an incident. Knowing who the local FBI agent is can be very important, because meeting for the first time in front of an incident is not conducive to effective scene management.

Working out the "who's in charge" concerns prior to an incident is important. The fire department should liaison with its local police, EMS, and emergency management agency prior to incidents. These and many other agencies are going to be involved and have a variety of responsibilities at an incident. In general, a unified command is recommended, and although this does not mean command by consensus, input from the various agencies should be considered. The agency with the majority of the tasks to do is generally in charge. In the initial stages the fire/EMS authority would be in charge while rescuing victims, but after the victims are removed and evidence recovery becomes the next priority, the command may switch to the police department.

The magnitude of response to a terrorism incident can be the most difficult part of the incident to manage. The IC will be overtaken by a large number of federal agencies' representatives, who on a regular basis will be replaced by a later arriving supervisor. Response groups consisting of two to seventy responders may arrive uninvited, all trying to assist with the incident. Another group that can also overwhelm the system is the media. In most cases the local media will react as normal, and for the most part will cooperate. Members of the

> **Note** If an incident occurs that involves federal responders, the national media will follow close behind.

national media do not know, nor do they follow, local protocol, and they will require information. A lack of accurate information can be disastrous to an incident by causing a deterioration of the incident and creating more hysteria than is already present. Media relations are very important in these types of events.

One of the most important issues that will arise other than the life safety hazard is that of evidence preservation. As much care as possible should be taken to preserve evidence and make sure it is taken care of appropriately. The collection of evidence is primarily a law-enforcement responsibility. Unless properly trained, fire service personnel should not collect evidence but should alert the police of its presence. A whole host of issues goes along with the collection of evidence, including the chain of custody. This chain of custody, or the paper trail that follows any evidence, is crucial to the successful prosecution of the persons responsible. The failure to follow proper procedure or document the travel of evidence can result in a case being dismissed, regardless of any other evidence. There have been cases, such as the Murrah bombing, in which fire service personnel have collected evidence, but that is a very unusual occurrence. Firefighters who are assigned to collect evidence should be fire investigators or fire marshals because they are typically trained in evidence collection. Another alternative is to double up and use a firefighter and a police officer to collect evidence.

A cooperative effort is needed to combat a terrorist attack. The primary functions are rescue/life safety by fire and EMS personnel, hazard identification by the HAZMAT team, identification of possible secondary devices by a bomb technician, and incident management. It is important to communicate the hazards to all personnel, and to limit the response to essential personnel. Instead of having the whole alarm assignment report to the front of the building, it is preferable to use one or two companies to investigate while staging the other companies away from dumpsters, mailboxes, or dead-end streets. When dealing with victims it is essential to isolate them until the cause is identified. The victims can have a large amount of information and should be questioned quickly. Questions to ask include these: What did you see, hear, or smell? Was this coming from one area or was it throughout the building? Did you see anything else suspicious? What type of signs and symptoms do you have? In addition, the police will need to conduct interviews. Documentation and preservation of the evidence are essential to the successful prosecution of the terrorist.

GENERAL GROUPINGS OF WARFARE AGENTS

Terrorists could use any of a number of possible warfare agents. They are classified into three broad areas. Weapons of mass destruction are commonly used by the military. Some of the regulations that prohibit the making, storing, or using of terrorism agents are called WMD laws or regulations. Any item that has the potential to cause significant harm or damage to a community or a large group of people is considered a WMD. The other two classifications are nuclear, biological, and chemical (NBC) or biological, nuclear, incendiary, chemical, and explosive (BNICE). Both of these are descriptions of the types of materials that could be used in a terrorism attack. Although there are slight variations, they are all used to describe the various types of agents that a terrorist could use. Most of the language differences come from funding legislation or a specific federal agency.

The military has devised a naming system for many of these agents, many of which are listed in **Table 30-1**. Responders should become familiar with these names because much of the literature and help guides refer to these agents by these names. For instance, when using a military detection device, the military name is used. When dealing with terrorism, firefighters are entering another world that has its own language. The fire service has to adopt this new language to survive in this new world. The three groupings mentioned earlier are further subdivided into the categories discussed in the following subsections.

Nerve Agents

Nerve agents are related to organophosphorus pesticides and include tabun, sarin, soman, and V agent. They were designed for one purpose and that is to kill people. Although very toxic, their ability to kill large numbers of people requires that the dissemination device function correctly and a number of other critical factors be in place to be truly effective. Although several gallons of sarin agent were used in the Tokyo subway attack, the distribution method was ineffective, so out of the twelve people who died, the only people killed by the sarin

Military Designations for Agents

NAME	MILITARY DESIGNATION	UN/DOT HAZARD CLASS	NAME	MILITARY DESIGNATION	UN/DOT HAZARD CLASS
Tabun	GA	6.1	Sarin	GB	6.1
Soman	GD	6.1	Thickened Soman	TGD	6.1
V agent	VX	6.1	Mustard	H	6.1
Distilled mustard	HD	6.1	Nitrogen mustard	HN	6.1
Lewisite	L	6.1	Hydrogen cyanide	AC	6.1
Cyanogen chloride	CK	2.3	Chlorine	CL	2.3
Phosgene	CG	2.3	Tear gas	CS	6.1
Mace	CN	6.1	Pepper spray	OC	2.2 and 6.1

TABLE 30-1

itself were the two people who actually touched the liquid. The chemical and physical properties of these agents hinder their ability to be effective as a stand-alone killer. To best produce the desired effect, the agents must touch people in liquid form or be breathed in while the materials are in aerosol form. The materials will not stay in aerosol form for very long, and they have a very low vapor pressure and thus will not create vapors as standing liquid. All of the military warfare agents have a vapor pressure less than water, which means they do not evaporate quickly and unless the liquid is touched or placed on the skin it does not present a large hazard.

Signs and Symptoms of Nerve Agents

All of the nerve agents present the same types of signs and symptoms as organophosphorus pesticides, and in reality the difference is minor. Nerve agents are pesticides for humans and are a stronger, more concentrated version of commercially available pesticides. The signs and symptoms can be generally described using the acronym SLUDGEM, which stands for:

S alivation—excessive drooling
L acrimation—tearing of the eyes
U rination—loss of bladder control
D efecation—loss of bowel control (diarrhea)
G astrointestinal—nausea and vomiting
E mesis—vomiting
M iosis—pinpoint pupils

The term SLUDGEM describes all of the symptoms from the minor ones to the extreme signs and symptoms. A slight exposure to any of the nerve agents will cause pinpoint pupils, a runny nose, and difficulty breathing. A person who has come in contact with the liquid will be experiencing all the SLUDGEM signs in addition to convulsions. A person who is in convulsions needs immediate decontamination and medical treatment in order to survive. This treatment sequence has to occur in less than five to ten minutes. In addition, there must be sufficient medication available on scene to accomplish the treatment. Most paramedic units carry enough medication to treat one or two patients who have severe symptoms.

Incendiary Agents

For the sake of classification, **incendiary agents** are placed into the chemical classification, because chemicals are used in these devices. The most commonly used chemicals are flammable and combustible liquids. The standard Molotov cocktail is an example of an incendiary device that could be used by a terrorist. In some cases arsonists have used a mixture of chemicals, usually oxidizers, to create very fast high-temperature fires.

Blister (Vesicants)

The category of **vesicants**, or, as they are more commonly called, **blister agents,** includes chemical compounds called mustard, distilled mustard, nitrogen mustard, and lewisite. These materials were never designed to kill. They were designed to incapacitate the enemy so that if one person was affected by one of these agents several more would be needed to care for the affected person. Although at high concentrations these materials can be toxic, their biggest threat is from skin contact which causes severe irritation and blistering. Their chemical and physical properties make them less of a hazard than the nerve agents. One large concern with these agents is the fact that the effects from an exposure can be delayed from fifteen minutes to several hours. Quick identification of a blister agent is key to keeping the victims safe.

Signs and Symptoms of Blister Agent Exposure One of the biggest risks with the blister agents is that they may present delayed effects. If not detected this could result in victims being released only to later have problems. In general, blister agents are not designed to kill; they were designed to incapacitate the enemy, resulting in troops being assigned to assist with the wounded. It is possible to create scenarios in which fatalities could occur, but these would be unusual cases. The effects from blister agents include irritation of the eyes, burning of the skin, and difficulty in breathing. The more severe exposure results in blisters, which may be delayed. The only real street treatment for these signs and symptoms is decontamination and supportive measures. It is important to have anyone with liquid contact blot the liquid off the skin and avoid spreading the agent. Fortunately, the chemical and physical properties of these agents make them difficult to disseminate, and coming in contact with the liquid would be the primary means of injury.

Blood and Choking Agents

The four chemicals discussed here in addition to being terrorism agents are also common industrial chemicals. The first category is **blood agents** and includes hydrogen cyanide and cyanogen chloride. Both of these materials are gases that disrupt the body's ability to use the oxygen within the bloodstream. They are also referred to as chemical asphyxiates. The **choking agents**, chlorine and phosgene, are very common in industry. Chlorine is present in almost every town, because it is used for water treatment processes and in swimming pools. Any community that has a water system or swimming pool has some form of chlorine. Chlorine comes in cylinders of 150 pounds to 90-ton railcars. It also comes in the tablet form (HTH) that is typically used in residential pools. The release of chlorine from a 90-ton railcar could result in several hundred thousand injuries and possibly an equal number of deaths, especially in an urban area. A small amount of chlorine can be very deadly or at a minimum create substantial panic in a community.

Signs and Symptoms of Blood and Choking Agents Many of the blood agents are commonly found in industry and may be found at normal chemical facilities. The signs and symptoms of slight exposure to blood agents include dizziness, difficulty in breathing, nausea, and general weakness. With cyanides, the breathing initially will be rapid and deep, followed by respiratory depression and usually death. The two most common choking agents are very common in industrial use, and chlorine can be found in most communities. Signs and symptoms include difficulty in breathing and respiratory distress, eye irritation, and, in higher amounts, skin irritation. Phosgene may present delayed effects, while chlorine's effects are immediate.

Irritants (Riot Control)

The most commonly used materials that are classified as potential terrorism agents are irritants and include mace, pepper spray, and tear gas. An incident that uses an irritant often impacts a large number of people, because the usual target is a school, mall, or other large place of assembly. The use of one small container can affect large numbers of people and make them immediately symptomatic. Luckily these materials are not extremely toxic—although they are extremely irritating—and the symptoms will usually disappear after fifteen to twenty minutes of exposure to fresh air. The response to one of these incidents is difficult because patients with real medical problems need treatment and the source of the irritant is often difficult to identify.

Signs and Symptoms of Irritants The signs and symptoms for a slight exposure up to a high dose are the same, with the exception of increasing severity. The signs and symptoms are eye and respiratory irritation. There is no real treatment except removal to fresh air; in fifteen to twenty minutes the symptoms will begin to disappear. Supportive care can be provided.

Biological Agents and Toxins

The most likely agents to be used in a terrorism scenario are **biological agents** and **toxins**. Some of the materials in this grouping include anthrax, mycotoxins, smallpox, plague, tularemia, and ricin. Out of all of the agents for a terrorist to make, this grouping is the easiest, especially ricin. The fatal route of entry differs with each of the agents and could occur via skin contact, inhalation, or injection. These agents are difficult to distribute effectively, and in some cases exposure to sunlight may neutralize many of these agents. Specific information about the two most popular agents is provided next.

Anthrax

Anthrax is a naturally occurring bacterial disease that is commonly found in dead sheep. It is contagious through skin contact or by inhalation of the anthrax spores. Although relatively easy to obtain, it is more difficult to culture and grow the proper grade of anthrax. To produce fatal effects, the type of anthrax required is called weapons-grade anthrax and is very difficult to produce. Even if developed it must be distributed effectively and under the right conditions.

Anthrax Scare 2001 The FBI is still investigating the anthrax attacks of October 2001 that started in Boca Raton, Florida, at the American Media building. It is suspected that a letter was sent and opened at this facility, which publishes the *National Enquirer.* One person died and several others fell ill at this building, but a letter was never recovered from this facility. The letters that were sent targeted the media and members of Congress. The NBC studios, the *New York Post*, and ABC all received letters. Two members of Congress, Senator Daschle and Senator Leahy, received letters. The letters killed five people and made eighteen others ill. Some of the deaths and illnesses involved people at various post offices. It is thought that the mail handling process enabled anthrax to aerosolize and get into the air. The letters contained small amounts of anthrax, which has been tested to be the Ames strain of anthrax that was produced in the United States. A lot has been learned about anthrax and its ability to be used as a weapon. The theoretical dose of anthrax in each of the recovered letters was estimated to be able to kill several hundred thousand people. The reality is that each letter typically killed one person. The other major factor in the deaths was a delay in treatment and in some cases misdiagnoses. The persons who did not seek quick treatment or who were misdiagnosed had difficult or unsuccessful recoveries. Those who sought quick medical treatment, during which the agent was recognized as anthrax, survived. The Centers for Disease Control (CDC) reported that there were ten cases of inhalational anthrax and twelve confirmed or suspected cases of cutaneous anthrax. Of the ten inhalational cases, seven occurred to postal workers in New Jersey and Washington, D.C., at mail sorting facilities. In the American Media case, one person who received the letter died, and the person sorting the mail fell ill. Six of the ten individuals with inhalational cases of anthrax survived after treatment, which is higher than originally anticipated.

The signs and symptoms of inhalational anthrax are a one- to four-day period of malaise, fatigue, fever, muscle tenderness, and a nonproductive cough followed by a rapid onset of respiratory distress, cyanosis, and sweating. The recent cases also had profound, often drenching, sweating, along with nausea and vomiting.

Ricin

Although ricin is easy to make, the required distribution method leaves a lot to be desired because it must be injected to be truly effective. It is 10,000 times more toxic than the nerve agent sarin. A small amount such as one milligram, about the size of a pinhead, can be fatal. Death usually occurs several days after injection. By other routes of entry, such as inhalation or ingestion, the most likely consequence is that many people would get sick but would eventually recover. After explosives, ricin is the leading choice of domestic terrorists, and several times a year someone is arrested for possession of ricin.

Response to Anthrax Hoaxes With anthrax hoaxes becoming a daily occurrence and dramatically peaking in late 2001 and early 2002, emergency responders should establish a plan for response. From the outset it would be difficult for someone to manufacture and successfully pull off an anthrax threat. The use of real anthrax is most likely an isolated case. One cannot say with absolute certainty, but a new attack is not likely and is not likely to kill or injure mass numbers of people. In the HAZ-

MAT and WMD business one should avoid the use of the words *never, always,* and *best.* The credibility factor and the technological difficulty make weaponized anthrax an unlikely candidate for use as a weapon. In many of the hoax incidents, letters proclaimed that the recipients had been exposed to anthrax. In some of these cases the only thing present in the envelope was a letter, which meant that there was no other material present. Anthrax is not invisible, and in order to improve the likelihood that the attack may be successful a quantity of the material must be present. In order to cause health effects the material must be inhaled, and in order to be inhaled the material needs to be distributed to put it into the air. The material, since it is a solid, requires a device to put it into the air. If this device is not present the risk to any persons in a building is very small. The only action required is to double- or triple-bag the envelope, package, or material in bags suitable for evidence collection. Procedures for the collection of evidence should be followed, and the local police as well as the FBI should be consulted. The persons who touched the material should be instructed to wash their hands with soap and water. They do not require full body decontamination nor are special solutions required such as a bleach and water mixture. The military advises that water is more than sufficient for humans. Anyone who was in the immediate vicinity (i.e., several feet) of the person does not require decontamination. The only reason for a full body wash would be if the material was splashed on a person from head to toe, but it is not required to be done immediately. The person can be taken to a shower or be provided privacy to take a shower. The person who did open the envelope should be entered into the health care system, advised of the signs and symptoms of exposure, and provided with emergency contact information. There is no need to start prophylactic antibiotics just because a person opened a letter with a powder in it. There is sufficient time to do lab analysis on the material before medical treatment is required. The threat from a solid material (no matter how toxic) is from inhaling the dust or touching the material. Gloves and SCBA are adequate protection for the collection of evidence. The FBI has a number of labs around the country set up to assist with the identification of WMD agents, and the local FBI office should be contacted for assistance.

Radioactive Agents

Nuclear agents, unfortunately, have to be put back on the list of possibilities that a terrorist could use. There are two types of radiation events, nuclear detonation and **radiological dispersion devices (RDDs)**. The use of an actual nuclear detonation device is very unlikely given the security these materials have. The amount required and the specific type also make the use unlikely. Although there is current speculation that there are some small nuclear devices missing from Russia, this has never been substantiated. There is the potential, however, for some nuclear material that could be used in an RDD coming from Eastern Europe. An RDD is a device that disperses radiological materials, usually through an explosive device. One example would be the use of a pharmaceutical grade radioactive material attached to a pipe bomb, which could cause a large amount of radioactive material to be distributed. The strength of the radiation source dictates how harmful an RDD would be. In many cases the RDD would present more of an investigative issue than a significant health issue. There are some radiation sources that would present a risk, but these are not in common use. The other factor is the explosive device and its limitations. Technologically an RDD is viable for a small area such as one-half acre of land. Attempting to distribute a significant amount of radioactive material takes a significant explosive device, not to mention the radiation source. The larger the RDD the more danger there is to the terrorist who has to assemble, transport, and detonate the device. There is an advantage to a noncredible RDD or a small RDD, and that is the public's reaction. The perception to the public and to many responders is that this would be a radioactive disaster. Radiation meters would indicate that there were radioactive materials spread around. The reality is that the amount of radiation would not be dangerous. As time passed, the danger would lessen as the radioactive material became less hazardous. Still, there would likely be tremendous concern in the community. Radiation causes fear because it is a big unknown, making it a prime weapon for a terrorist. Education on the hazards of radiation and the effective use of radiation monitors can ease this fear and allow responders to make an informed response.

Other Terrorism Agents

Although considerable emphasis is placed on warfare agents, in reality many other common industrial or household materials can be just as deadly, if not more so.

> **Note** Remember that a terrorist wants to create panic, not necessarily kill or injure massive numbers of people.

The most likely scenario is a pipe bomb, followed by the use of ricin. Responders should not be lulled into a false sense of security if they do not find a "warfare" agent or if the device is only a small pipe bomb and not a moving truck filled full of ammonium nitrate and fuel oil. Nerve agents (or any other chemical warfare agent) have never been used in the United States, nor manufactured by anyone other than the military. The FBI Bomb Data Center reports that out of the 3,000 bombings that occur each year, there have been only two large bombings, but these "small" bombs kill an average of 32 and injure 277 people each year.

DETECTION OF TERRORISM AGENTS

The detection of terrorism agents is difficult, and given the potential circumstances is done under severe conditions. The response to terrorism incidents has changed how many HAZMAT teams operate and has increased their capability to handle other situations. The confirmation that terrorist agents have been used is difficult because they are extremely hard to detect. The exceptions are the standard industrial materials such as chlorine, which is easy to detect and confirm its presence.

The detection of terrorism agents addresses three major categories of hazards: chemical agents, radiological materials, and biological agents. There are a number of devices that detect chemical warfare agents. They range from inexpensive paper strip tests (M-8 and M-9) to sophisticated electronic devices costing more than $125,000. The most common are direct reading devices, which detect nerve and blister agents, **Figure 30-9.** Some devices will also detect drugs, explosives, mace and pepper spray, and industrial chemicals. The radiological detection field has dosimeters that track the dose of radiation that one is receiving to handheld instruments. Radiation pagers and pager/dosimeters are very small. Pager-sized devices are available and can detect levels of radiation and alert the user to potential danger. Typical radiation monitors detect the dose a human is absorbing, while some newer models will determine and identify the exact type of radiation source present. The detection of biological agents is more difficult, and there are limited choices. One detection device that is new to the emergency response world is a polymerase chain reaction (PCR) unit, which detects the DNA of the sample and compares it to other DNA samples loaded in the library. It is the same method that is used in the laboratory to detect biological agents. The only other choice is handheld bioassays, which, depending on the brand, function with varying degrees of accuracy. Some work well and have accuracy rates in the 90th percentile, while others have accuracy rates in the 30th percentile. The major issue with any biological agent detection is that the sample collection method has to follow exacting standards without any deviation. The detection of any terrorism agent is difficult, and the HAZMAT team should take responsibility for doing the testing.

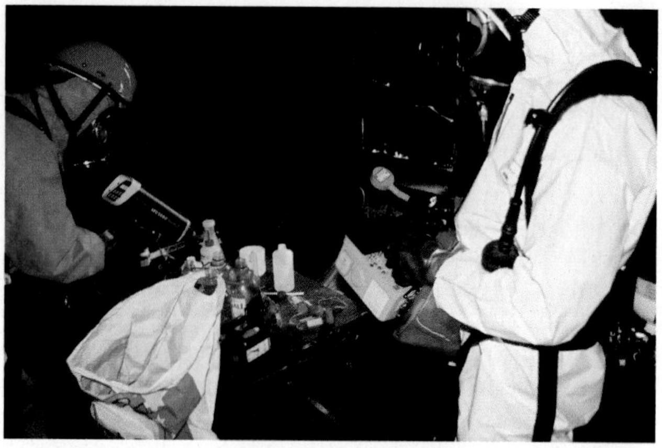

Figure 30-9 This air monitor device detects chemical warfare agents, gamma radiation, and irritants such as mace and pepper spray.

In some cases standard civilian detection devices such as photo-ionization detectors also play a role in the detection of terrorism agents. Other tests are used by HAZMAT teams that have applicability in the detection of terrorism agents. Due to the many mitigating factors the detection of these agents is going to be difficult, and to be conclusive will probably require lab tests.

FEDERAL ASSISTANCE

The federal government has established roles and responsibilities in the event an act of terrorism occurs, and these roles are provided in PDD 39. Per this document, the FBI is designated the lead agency during the emergency (crisis management) stage of an incident. The Federal Emergency Management Agency (FEMA) becomes the lead when the incident is no longer in the emergency phase (consequence management). The FBI has a Hazardous Materials Response Unit (HMRU) that provides identification, mitigation assistance, and evidence collection for potential terrorist incidents. The HMRU is a multifaceted group that not only responds to terrorism incidents, but also responds to incidents involving explosives, drug labs/incidents, and environmental crimes. A response from FEMA will vary with the incident, as will the number of FEMA personnel, but the response will be similar to any other disaster; FEMA assists in restoration and recovery issues.

The Urban Search and Rescue (USAR) teams fall under FEMA, and are activated by following the emergency management chain from the local level to the state level and then to FEMA for activation. At this time there are twenty-eight USAR teams across the country. They provide expertise in heavy rescue operations such as building collapses. This team of sixty-two people is composed of rescue specialists, dog search teams, medical specialists, communications specialists, and an engineering and rigging component. The majority of victims are rescued by the first responders and other local specialized resources. But in some cases trapped victims may be alive for many days and may require the expertise and equipment that a USAR team has available. The unfortunate thing is that the USAR teams have a delayed response, because it takes them several hours to become airborne, not to mention travel time. If there is a possibility they may be needed, their assistance should be requested early. Like many of the teams, it is not uncommon for an advance party to arrive hours prior to the arrival of the remainder of the team.

A number of other agencies may be involved in the event of an incident. The military has a couple of units that have terrorism response capabilities and responsibilities. The army has the Technical Escort Unit (TEU), headquartered at the Aberdeen Proving Grounds, Maryland. There are other units at Dugway Proving Grounds, Utah; Pine Bluff Arsenal, Arkansas; and Fort Belvoir, Virginia. The TEU is assigned to provide escort service for warfare agents and to be the troubleshooting group in the event of an incident involving warfare agents or explosives. When such an incident occurs, the TEU will respond to assist with identification and the mitigation of the incident. Team members handle chemical, biological, and explosive materials as well as other hazardous materials. TEU is a self-contained unit and has the lab resources of the Soldiers Biological and Chemical Command (SBCCOM) of the Aberdeen Proving Grounds available to assist them.

Basic Incident Priorities When dealing with an incident that involves terrorism, first responders should follow these guidelines:

- To rescue live victims, use full protective clothing including SCBA. All of the agents listed in this chapter are predominantly hazardous through inhalation. Avoid touching any unknown liquids or solids, because most are also toxic through skin contact.
- Use a quick in/quick out approach: Do not treat victims, remove them from the area. Keep in mind the terrorist may be among the injured. Watch for secondary devices.
- Request HAZMAT and the police bomb squad. The sooner responders eliminate the potential for chemical agents or a secondary device, the better off they will be. The HAZMAT team may have the ability to detect the agents listed in this chapter, and most HAZMAT teams are working with their bomb squads on a more frequent basis.
- Limit personnel operating in the hazard area.
- Establish multiple staging areas, out of the line of sight.
- Notify the local emergency management agency so that they can mobilize the state and federal resources.
- If a building has collapsed or there is potential for a building collapse, request assistance from a tactical rescue team or a USAR team.
- Isolate all victims, separating contaminated from clean victims.
- Establish a safe triage, treatment, and transport area away from the impact or hot zone.
- Notify all area hospitals of the incident.
- Remember that the incident is a crime scene and make provisions to preserve as much evidence as possible.
- If there is reason to suspect the presence of chemical agents, use the DOT ERG or other reference sources such as the *Medical Management of Chemical Casualties Handbook* to suggest safety precautions and patient treatments.

The marines have a unit known as the Chemical and Biological Incident Response Force (CBIRF), which comes from Indian Head, Maryland. This unit responds to acts of terrorism across the world and has three main components: decontamination, medical, and security. It is a self-contained unit and requires only a water source for conducting long-term operations. CBIRF is able to respond nationwide upon request to terrorist incidents and can provide detection and mitigation as well. All of the federal resources including CBIRF and TEU can be and have been pre-positioned for certain events, such as the Olympics, political conventions, and visits by dignitaries. These resources integrate within the local system, usually hidden away from the public, and are immediately available.

Other federal programs include training under the Nunn-Lugar-Domenici Legislation, which was passed in September 1996 (P.L. 104-201). This law, known as the Domestic Preparedness Training Initiative, mandated that the Department of Defense provide training to the 157 major cities and counties across the United States. It is intended to enhance the capability of the local, state, and federal response to incidents involving NBC materials. Part of this process is an assessment to determine the cities' capabilities after the training. A second component of the law provides for thirty teams to be fielded by the National Guard across the United States. These teams known as Weapons of Mass Destruction Civil Support Teams (WMD-CSTs) are being established to provide a National Guard component to the response to terrorists. Also set up around the country are Metropolitan Medical Response Teams (MMRTs), which are a group of 129 people on each team. They are trained and equipped to handle the medical component of a terrorism event. The plan is to have them in the 120 largest metropolitan areas in the country.

Lessons Learned

The response to a potential terrorism incident can be very challenging, and every responder must be alert to the possibility of such an incident. To be safe, responders should wear all PPE, not linger in the environment, and relocate to a safe area once the live victims are out. It is important to be aware of the potential for secondary devices and request HAZMAT and bomb squad assistance quickly. One to thousands of victims may be injured or killed. Scenarios involve tremendous loss of life, including a large number of responders. There exists the possibility that responders may lose and the terrorist will win, a situation that can be avoided by training, planning, and preparing for such an incident.

KEY TERMS

Anthrax A biological material that is naturally occurring and is severely toxic to humans. It is commonly used in hoax incidents.

Biological Agents Microorganisms that cause disease in humans, plants, and animals; they also cause the victims' health to deteriorate. Biological agents have been designed for warfare purposes.

Blister Agents A group of chemical agents that cause blistering and irritation of the skin. Sometimes referred to as vesicants.

Blood Agents Chemicals that affect the body's ability to use oxygen. If they prevent the body from using oxygen, fatalities result.

Butyric Acid A fairly common lab acid that has been used in many attacks on abortion clinics. Although not extremely hazardous, it has a characteristic stench that permeates the entire area where it is spilled.

Choking Agents Agents that cause a person to cough and have difficulty breathing. The terrorism agents that are considered choking agents are chlorine and phosgene, both very toxic gases.

HAZMAT Crime A criminal act that uses or threatens the use of chemicals as a weapon.

Incendiary Agents Chemicals that are used to start fires, the most common being a Molotov cocktail.

Mass Casualty An incident in which the number of patients exceeds the capability of the EMS to manage the incident effectively. In some jurisdictions this can be two patients, while in others it may take ten to make the incident a mass casualty.

Nerve Agents Chemicals that are designed to kill humans, specifically in warfare. They are chemically similar to organophosphorus pesticides and cause the same medical reaction in humans.

PDD 39 Presidential Decision Directive 39, which established the FBI as the lead agency in terrorism incidents responsible for crisis management. It also established FEMA as the lead for consequence management.

Radiological Dispersion Device (RDD) An explosive device that spreads radioactive material throughout an area.

Ricin A biological toxin that can be used by a terrorist or other person attempting to kill or injure someone. It is the easiest terrorist agent to produce and one of the most common.

Toxins Disease-causing materials that are extremely toxic and in some cases more toxic than other warfare agents such as nerve agents.

Vesicants A group of chemical agents that cause blistering and irritation of the skin. Commonly referred to as blister agents.

Weapons of Mass Destruction (WMD) A term that is used to describe explosive, chemical, biological, and radiological weapons used for terrorism and mass destruction.

REVIEW QUESTIONS

1. Describe four potential targets of terrorism.
2. Describe four indicators of potential terrorist activity.
3. What are the most readily available agents that could be used in a terrorist attack?
4. Which three main local/regional agencies or groups should be notified immediately of a suspected terrorist attack?
5. Describe the process of requesting federal assistance.
6. Explain which of the BNICE agents is the most likely to be found at an incident.
7. Describe the second most likely agent.
8. Describe which agent is designed to kill immediately.
9. What are the immediate signs and symptoms of sarin exposure?

Endnote

1. The *Roe v. Wade* decision was the Supreme Court case that allowed legalized abortions, and is the case that has created a lot of the turmoil between pro-choice and pro-life forces.

Additional Resources

Bevelacqua, Armando S., *Hazardous Materials Chemistry.* Delmar Learning, a part of the Thomson Corporation, Clifton Park, NY, 2001.

Bevelacqua, Armando S. and Stilp, Richard H., *Hazardous Materials Field Guide.* Delmar Learning, a part of the Thomson Corporation, Clifton Park, NY, 1998.

Bevelacqua, Armando S., and Stilp, Richard, *Terrorism Handbook for Operational Responders,* 2nd ed., Delmar Learning, a part of the Thomson Corporation, Clifton Park, NY, 2004.

Buck, George, *Preparing for Terrorism: An Emergency Services Guide.* Delmar Learning, a part of the Thomson Corporation, Clifton Park, NY, 1998.

Buck, George, *Preparing for Biological Terrorism: An Emergency Services Guide.* Delmar Learning, a part of the Thomson Corporation, Clifton Park, NY, 2002.

Buck, George, *Preparing for Terrorism: An Emergency Services Guide.* Delmar Learning, a part of the Thomson Corporation, Clifton Park, NY, 1998.

Buck, George, Buck, Lori, and Mogil, Barry, *Preparing for Terrorism: The Public Safety Communicator's Guide.* Delmar Learning, a part of the Thomson Corporation, Clifton Park, NY, 2003.

Hawley, Chris, *Hazardous Materials Incidents.* 2nd ed., Delmar Learning, Clifton Park, NY, 2004.

Hawley, Christopher, *Hazardous Material Air Monitoring and Detection Devices.* Delmar Learning, a part of the Thomson Corporation, Clifton Park, NY, 2002.

Henry, Timothy V., *Decontamination for Hazardous Materials Emergencies.* Delmar Learning, a part of the Thomson Corporation, Clifton Park, NY, 1999.

Medical Management of Biological Casualties. U.S. Army Medical Research Institute of Infectious Diseases, Ft. Dietrick, Fredrick, MD, 1996.

Medical Management of Chemical Casualties Handbook. U.S. Army Chemical Casualty Care Office, Medical Research Institute of Chemical Defense, Aberdeen Proving Ground, MD, Sept. 1995.

Pickett, Mike, *Explosives Identification Guide.* Delmar Learning, a part of the Thomson Corporation, Clifton Park, NY, 1999.

Scnepp, Rob and Gantt, Paul W., *Hazardous Materials, Regulations, Response & Site Operations.* Delmar Learning, a part of the Thomson Corporation, Clifton Park, NY, 1999.

Smelby Jr., L. Charles, ed., *Hazardous Materials Response Handbook,* 3rd ed. National Fire Protection Association, Quincy, MA, 1997.

Stilp, Richard and Bevelacqua, Armando. *Citizen's Guide to Terrorism Preparedness.* Delmar Learning, a part of the Thomson Corporation, Clifton Park, NY, 2003.

Stilp, Richard and Armando Bevelacqua, *Emergency Medical Response to Hazardous Materials Incidents.* Delmar Learning, a part of the Thomson Corporation, Clifton Park,NY, 1997.

GLOSSARY

Abandonment Abandonment occurs when an emergency responder begins treatment of a patient and then leaves the patient or discontinues treatment prior to the arrival of an equally or higher trained responder.

Aboveground Storage Tank (AST) Tank that is stored above the ground in a horizontal or vertical position. Smaller quantities of fuels are often stored in this fashion.

Abrasion A scrape or brush of the skin usually making it reddish in color and resulting in minor capillary bleeding.

Absolute Pressure The measurement of pressure, including atmospheric pressure. Measured in pounds per square inch absolute.

Absorption A defensive method of controlling a spill by applying a material that absorbs the spilled chemical.

Accelerator A device to speed the operation of the dry pipe valve by detecting the decrease in air pressure. It pipes air pressure below the clapper valve, speeding its opening.

Accident The result of a series of events and conditions that lead to an unsafe situation resulting in injury and/or property damage.

Accident Chain A series of events and conditions that can lead to or have led to an accident. These events and conditions are typically classified into five areas: environment, human factors, equipment, events, and injury.

Acclimation The act of becoming accustomed or used to something. Typically achieved through repeated practice within a given set of conditions.

Acute A quick one-time exposure to a chemical.

Adapter Device that adapts or changes one type of hose thread to another, allowing connection of two different lines. Adapters have a male end on one side and a female on the other with each side being a different thread type, for example, an iron pipe to national standard adapter.

Administrative Warrant An order issued by a magistrate that grants authority for fire personnel to enter private property for the purpose of conducting a fire prevention inspection.

Aerial Apparatus Fire apparatus using mounted ladders and other devices for reaching areas beyond the length of ground ladders.

Air Bill The term used to describe the shipping papers used in air transportation.

Aircraft Rescue and Firefighting (ARFF) Of or pertaining to firefighting operations involving fixed or rotary wing aircraft.

Air Monitoring Devices Used to determine oxygen, explosive, or toxic levels of gases in air.

Air-Purifying Respirators (APR) Respiratory protection that filters contaminants out of the air, using filter cartridges. Requires the atmosphere to have sufficient oxygen, in addition to other regulatory requirements.

Allergic Reaction The body's reaction to a substance to which there is an allergy.

Americans with Disabilities Act Public law that bars discrimination on the basis of disability in state and local services. Enacted in 1990.

Amputation Occurs when part of the body is severed completely as a result of an injury.

Anchor Point A safe location from which to begin line construction on a wildland fire.

ANFO The acronym that is used for ammonium nitrate fuel oil mixture, which is a common explosive. ANFO was used in the Oklahoma City bombing incident.

Anthrax A biological material that is naturally occurring and is severely toxic to humans. It is commonly used in hoax incidents.

Application Rate Amount of foam or foam solution needed to extinguish a fire. Usually expressed in gallons per minute per square foot or liters per minute per square meter.

Aqueous Film-Forming Foam (AFFF) A synthetic foam that as it breaks down forms an aqueous layer or film over a flammable liquid.

Aquifer A formation of permeable rock, gravel, or sand holding water or allowing water to flow through it.

Arson A malicious fire or fires set intentionally by humans for vengeance or profit.

Arterial Bleeding Bleeding from an artery.

Arteries The blood vessels, or tubes, within the body that carry blood rich with oxygen and nutrients away from the heart.

Articulating Boom Ladder An apparatus with a series of booms and a platform on the end. It is maneuvered into position by adjusting the various boom sections into place to position the platform at the desired location.

Aspect The direction a slope faces given in compass directions.

Asphyxiation Condition that causes death due to lack of oxygen and an excessive amount of carbon monoxide or other gases in the blood.

Association of Public Safety Communications Officials-Int., Inc. (APCO) International not-for-profit organization dedicated to the advancement of public safety communications. Membership is made up of public safety professionals from around the world.

Atmospheric Pressure The pressure exerted by the atmosphere, which for Earth is 14.7 pounds per square inch at sea level.

Atomization The separation of atoms and molecules into an unconnected state where they are in suspension rather than in liquid form.

Attack Hose Small- to large-diameter hose used to supply nozzles and other applicators or protective system for fire attack. Attack hose commonly means handheld hoselines from 1½ to 2½ inches (38 or 63 mm) in diameter.

Authority Having Jurisdiction (AHJ) The responsible governing organization or body having legal jurisdiction.

Autoextended When a fire goes out the window on one floor, up the side of the building, which is often noncombustible, and extends through the window or cockloft directly above.

Automated External Defibrillator (AED) A portable computer-driven device that analyzes a patient's heart rhythm and delivers defibrillation shocks when necessary.

Automatic or **Constant Pressure Nozzle** Nozzle with a spring mechanism built in that reacts to pressure changes and adjusts the flow and resultant reach of the nozzle.

Automatic Sprinkler System A system of devices that will activate when exposed to fire, connected to a piping system that will supply water to control the fire. Typically, an automatic sprinkler system is also supported by firefighters when they arrive on the scene.

Auxiliary Appliances Another term for protective devices, particularly sprinkler and standpipe systems.

Available Flow Amount of water that can be moved to extinguish the fire. Depends on the water supply, pump(s) and their capabilities, and the size and length of hose.

Avulsion An injury where a part of the skin is torn away, but still attached, leaving a flap or loose area hanging.

Awareness Level The basic level of training for emergency response to a chemical accident, the basis of which is the ability to recognize a hazardous situation and call for assistance.

Axial Load A load passing through the center of the mass of the supporting element, perpendicular to its cross section.

Backdraft A sudden, violent reignition of the contents of a closed container fire that has consumed the oxygen within the space when a new source of oxygen is introduced.

Backflow Preventers A check valve or set of valves used to prevent a backflow of water from one system into another. Required where a building water or fire protection system connects with the public water system. Backflow preventers are being required for environmental and health reasons.

Backstretch or **Flying Stretch** An attack line lay where the engine is at the hydrant and the line is stretched back from the engine to the fire. The flying stretch is a version of the backstretch where the engine stops in front of the fire, the attack portion is removed, and the engine proceeds to the hydrant.

Balloon Frame A style of wood frame construction in which studs are continuous for the full height of a building.

Bank Down A condition in which the heat, smoke, and fire gases have reached the uppermost level in a compartment and, instead of continuing up, begin to push down from the ceiling toward the floor.

Base Radio Radio station that contains all of the antennas, receivers, and transmitters necessary to transmit and receive messages.

Basic 9-1-1 Telephone system that automatically connects a person dialing the digits "9-1-1" to a predetermined answering point through normal telephone service facilities. Number and location information is not normally provided in *basic* systems.

Beam A structural member subjected to loads perpendicular to its length.

Bed Ladder The nonextending part of an extension ladder.

Bevel The outside curve of the fork end of the Halligan tool.

Bight A doubled section of rope, usually made along the standing part, that forms a U-turn in the rope that does not cross itself.

Biological Agents Microorganisms that cause disease in humans, plants, and animals; they also cause the victim's health to deteriorate. Biological agents have been designed for warfare purposes.

Biomimetic A form of gas sensor that is used to determine levels of carbon monoxide. It is of the type of sensors used in home CO detectors. It closely re-creates the body's reaction to CO and activates an alarm.

Blister Agents A group of chemical agents that cause blistering and irritation of the skin. Sometimes referred to as vesicants.

Blood Agents Chemicals that affect the body's ability to use oxygen. If they prevent the body from using oxygen, fatalities result.

Body Substance Isolation Precautions A set of precautions for emergency responders designed to prevent exposure to any body fluid or substance.

Boiling Liquid Expanding Vapor Explosion (BLEVE) Describes the rupture of a container when a confined liquid boils and creates a vapor pressure that exceeds the container's ability to hold it.

Boiling Point The temperature at which liquids must be heated in order to turn into a gas.

Bolt Throw The distance the bolt of a lock travels into the jamb or strike plate. Usually ½ to 1½ inches.

Bond A substance or an agent that causes two or more objects or parts to bind.

Booster Hose Smaller diameter, flexible hard-rubber-coated hose of ¾- or 1-inch (19- to 25-mm) size usually mounted on a reel that can be used for small trash and grass fires or overhaul operations after the fire is out.

Bourdon Gauge The type of gauge found on most fire apparatus that operates by pressure in a curved tube moving an indicating needle.

Box Canyon A canyon open on one end and closed on the other. They become very dangerous when wildfire enters them.

Brachial Artery A major artery in the inside of the upper arm that supplies blood to the arm. Can be used as a pressure point for controlling bleeding and for locating a pulse on an infant.

Branch The command designation established to maintain span of control over a number of divisions, sectors, or groups.

Bresnan Distributors Has six or nine solid tips or broken stream openings designed to rotate in a circular spray pattern. Used to fight fire in basements or cellars when firefighters cannot make a direct attack on the fire.

British Thermal Unit (BTU) A measurement of heat that describes the amount of heat required to raise 1 pound of water 1°F.

Brush Gear Another term for a wildland personal protective ensemble.

Building Officials Conference Association (BOCA) A group that establishes minimum building and fire safety standards.

Bulk Tank A large transportable tank, comparable to a tote, but considered to be the larger of the two.

Bump Test Used to determine if an air monitor is working. It will alarm if a toxic gas is present. It is a quick check to make sure the instrument responds to a sample of gas.

Bunkers A slang term that is used mostly to describe the components of a structural firefighting ensemble. The original use of the term *bunkers* referred only to the pant/boot combination that firefighters wore at night and placed next to their "bunks" for rapid donning.

Butyric Acid A fairly common lab acid that has been used in many attacks on abortion clinics. Although not extremely hazardous, it has a characteristic stench that permeates the entire area where it is spilled.

Bypass Eductor Eductor with two waterways and a valve that allows plain water to pass by the venturi or through the venturi to create foam solution.

Calibration Used to set the air monitor and to ensure that it reads correctly. When calibrating a monitor, it is exposed to a known quantity of gas to make sure it reads the values correctly.

Cantilever Beam A beam that is supported at only one end.

Capillaries The very small blood vessels in the body that connect arteries and veins and filter the oxygen and nutrients from the blood into the tissues of the body.

Capillary Bleeding Bleeding from a capillary.

Carbon Dioxide (CO_2) An inert colorless and odorless gas that is stored under pressure as a liquid that is capable of being self-expelled and is effective in smothering Class B and C fires.

Carbon Monoxide Colorless, odorless, poisonous gas that when inhaled combines with the red blood cells excluding oxygen.

Carcinogen A material that is capable of causing cancer in humans.

Cardiovascular System The heart, blood vessels, and blood within the body.

Carotid Pulse The pulse located on either side of the neck.

Catalytic Bead The most common type of combustible gas sensor that uses two heated beads of metal to determine the presence of flammable gases.

Ceiling Level The highest exposure a person can receive without suffering any ill effects. It is combined with the PEL, TLV, or REL as a maximum exposure.

Cellar Nozzles Has four spray nozzles designed to rotate in a circular spray pattern for fighting fires in basements or cellars when firefighters cannot make a direct attack on the fire.

Chain of Command Common fire service term that means to always work through one's direct supervisor. The fire service is viewed as a paramilitary organization and because of this all requests for information outside the assigned workplace should go through the supervisor.

Check Valves Valves installed to control water flow in one direction, typically when different systems are interconnected.

Chemical Burns Burns caused by chemical substances that come into contact with the skin or tissues of the body, creating a caustic reaction.

Chemtrec The Chemical Transportation Emergency Center, which provides technical assistance and guidance in the event of a chemical emergency; a network of chemical manufacturers who provide emergency information and response teams if necessary.

Chimney Another term for drainage. Given because of the draw of fire as in heat going up the chimney.

Chip Measuring System (CMS) A form of colorimetric air sampling in which the gas sample passes through a tube. If the correct color change occurs, the monitor interprets the amount of change and indicates a level of the gas on an LCD screen.

Choking Agents Agents that cause a person to cough and have difficulty breathing. The terrorism agents that are considered choking agents are chlorine and phosgene, both very toxic gases.

Chord The top and bottom components of a beam or truss. The top chord is subjected to compressive force; the bottom chord is subjected to tensile force.

Chronic A continual or repeated exposure to a hazardous material.

Cistern An underground water tank made from natural rock or concrete. Cisterns store large quantities of water—30,000 gallons or more—in areas without other water supplies or as a backup supply.

Clandestine Drug Labs Illegal labs set up to manufacture street drugs.

Class A Classification of fire involving ordinary combustibles such as wood, paper, cloth, plastics, and rubber.

Class B Classification of fire involving flammable and combustible liquids, gases, and greases. Common products are gasoline, oils, alcohol, propane, and cooking oils.

Class C Classification of fire involving energized electrical equipment, which eliminates using water-based agents.

Class D Classification of fire involving combustible metals and alloys such as magnesium, sodium, lithium, and potassium.

Class K A new classification of fire as of 1998 that involves fires in combustible cooking fuels such as vegetable or animal oils and fats.

Clipping Term associated with the use of two-way radios that is used to describe instances when either the first part of a message or the last part of a message is cut off as the result of either speaking before pressing the transmit key or releasing the transmit key prior to the end of a transmission.

Closed-Circuit SCBA A type of SCBA unit in which the exhaled air remains in the system to be filtered and mixed with oxygen for reuse.

Cockloft The area between the roof and the ceiling.

Code of Federal Regulations (CFR) The documents that include federally promulgated regulations for all federal agencies.

Collapse Zone The area around a building where debris will land when it falls. As an absolute minimum this distance must be at least 1½ times the height of the building.

Colorimetric Tubes Crystal-filled tubes that change colors in the presence of the intended gases. These tubes are made for the detection of known and unknown gases.

Column A structural element that is subjected to compressive forces—typically a vertical member.

Combination Attack A combined attack based on partial use of both offensive and defensive attack modes.

Combination Fire Attack A blend of the direct and indirect fire attack methods, with firefighters applying water to both the fuel and the atmosphere of the room.

Combination Nozzle A spray nozzle that is capable of providing straight stream and spray patterns, which are adjustable or variable by the operator. Most fog nozzles used today are combination nozzles.

Combustion The chemical action in which heat and light are produced and the heat is used to maintain the chemical chain reaction to continue the process.

Command Vehicle Typically used by operations chief officers in the fire service.

Common Terminology The designation of a term that is the same throughout an IMS.

Communicable Disease A disease that can be transmitted from one person to another.

Communications Sending, giving, or exchanging of information.

Company A team of firefighters with apparatus assigned to perform a specific function in a designated response area.

Compound A combination of substances joined in a chemical bond that exists in a proportional amount and cannot be separated without chemical interaction.

Compressed Air Foam System (CAFS) A foam system where compressed air is injected into the foam solution prior to entering any hoselines. The fluffy foam created needs no further aspiration of air by the nozzle.

Compression A force that tends to push materials together.

Computer-Aided Dispatch Computer-based automated system that assists the telecommunicator in assessing dispatch information and recommends responses.

Computer-Aided Management for Emergency Operations (CAMEO) Program A computer program that combines a chemical information database with emergency planning software. It is commonly used by HAZMAT teams to determine chemical information.

Concentrated Load A load applied to a small area.

Confined Space A space that is large enough to be entered but is not designed for continuous occupancy.

Conflagration A large and destructive fire.

Consent The acceptance of emergency medical treatment by a patient or victim.

Consist The shipping papers that list the cargo of a train. The listing is by railcar, and the consist lists all of the cars.

Consolidated Incident Action Plan The strategic goals to eliminate the hazard or control the incident.

Constant or **Set Volume Nozzle** Nozzle with one set volume at a set pressure. For example, 60 gpm at 100 psi (227 L/min 690 kPa). The only adjustment is the pattern.

Constricted A condition of the pupils where they are much smaller than normal and may appear almost like a "pinpoint."

Continuous Beam A beam that is supported in three or more places.

Control Room A room on the ground floor of a high-rise building where all building systems controls are located.

Cribbing The use of various dimensions of lumber arranged in systematic stacks (pyramid, box, step, etc.) to support an unstable load.

Critical Incident Stress Debriefing (CISD) A formal gathering of incident responders to help defuse and address stress from a given incident.

Critical Incident Stress Management (CISM) A process for managing the short- and long-term effects of critical incident stress reactions.

Cryogenic gas Any gas that exists as a liquid at a very cold temperature, always below −150°F.

Cutting Tools The group of tools used to cut through or around materials.

Damming The stopping of a body of water, which at the same time stops the spread of the spilled material.

Dangerous Cargo Manifest (DCM) The shipping papers for a ship, which lists the hazardous materials on board.

Database Organized collection of similar facts.

Dead Load The weight of the building materials and any part of the building permanently attached or built-in.

DECIDE Process A management system used to organize the response to a chemical incident. The factors of DECIDE are detect, estimate, choose, identify, do the best, and evaluate.

Decontamination The physical removal of contaminants (chemicals) from people, equipment, or the environment. Most often used to describe the process of cleaning to remove chemicals from a person.

Defensive Attack A calculated attack on part of a problem or situation in an effort to hold ground until sufficient resources are available to convert to an offensive form of attack.

Deflagrates Rapid burning, which in reality with regard to explosions can be considered a slow explosion, but is traveling at a lesser speed than a detonation.

Dehydration A loss of water and vital fluids in the body.

Deluge Systems Designed to protect areas that may have a fast-spreading fire engulfing the entire area. All of its sprinkler heads are already open, and the piping contains atmospheric air. When the system operates, water flows to all heads, allowing total coverage. The system uses a deluge valve that opens when a separate fire detection system senses the fire and signals to trip the valve open.

Density The mass per unit volume of a substance under specified conditions of pressure and temperature.

Deployment Plan Predetermined response plan of apparatus and personnel for specific types of incidents and specific locations.

Depth of Char A term commonly used by fire investigators to describe the amount of time wooden material had burned. The deeper the char, the longer the material was burning or exposed to direct flame.

Design Load A load the engineer planned for or anticipated in the structural design.

Detergent-Type Foams Use synthetic surfactants to break down the surface tension of water to create a foaming blanket.

Diffusion A naturally occurring event in which molecules travel from levels of high concentration to areas of low concentration.

Diking A defensive method of stopping a spill. A common dike is constructed of dirt or sand and is used to hold a spilled product. In some facilities, a dike may be preconstructed such as around a tank farm.

Dilated A condition of the pupils where they are much larger than normal and can take up almost the whole colored portion of the eye.

Dilution The addition of a material to the spilled material to make it less hazardous. In most cases water is used to dilute a spilled material, although other chemicals could be used.

Direct Fire Attack An attack on the fire made by aiming the flow of water directly at the material on fire.

Disassembly The actual taking apart of vehicle components.

Discharge Flow Total amount of water flowing from the discharge side of the pump.

Displacement The relocating of major parts (i.e., doors, roof, dash, steering column) of a vehicle.

Distortion The bending of sheet metal or components.

Distributed Load A load applied equally over a broad area.

Distributor Pipe or **Extension Pipe** Devices that allow a nozzle or other device to be directed into holes to reach basements, attic, and floors that cannot be accessed by personnel. The distributor pipe has self-supporting brackets that help hold it into place when in use.

Diverting Using materials to divert a spill around an item. For instance, several shovels full of dirt can be used to divert a running spill around a storm drain.

Division Command designation responsible for operations within an assigned geographic area.

Double Female Allows the two male ends of hose to be connected.

Double Male Used to connect two female thread couplings.

Drafting The pumping of water from a static source by taking advantage of atmospheric pressure to force water from the source into the pump.

Drainage A topographic feature on the side of a hill or mountain that naturally collects water runoff, channeling it to the bottom of the rise. Fire is attracted to this feature.

Dressing The practice of making sure that all parts of a knot are lying in the proper orientation to the other parts and look exactly as the pictures herein indicate.

Dry Chemicals Dry extinguishing agents divided into two categories. Regular dry chemicals work on Class B and C fires; multipurpose dry chemicals work on Class A, B, and C fires.

Dry Hydrant A piping system for drafting from a static water source with a fire department connection at one end and a strainer at the water end.

Dry Pipe Systems Air under pressure replaces the water in the system to protect against freezing temperatures. The sprinkler control valve uses a dry pipe valve to keep pressurized air maintained above with the supply water under pressure below the valve.

Dry Powders Extinguishing agents for Class D fires.

Dump Site The area where tenders are unloaded or their load dumped.

Dutchman A short fold of hose or a reverse fold that is used when loading hose and a coupling comes at a point where a fold should take place or when two sets of couplings end up on top of or next to each other. The dutchman moves the coupling to another point in the load.

Dynamic A rope having a high degree of elongation (10 to 15 percent) at normal safe working loads.

Ears Elongated folds or flaps at the ends of a layer of hose to assist in pulling that layer.

Eccentric Load A load perpendicular to the cross section of the supporting element that does not pass through the center of mass.

Eductor Device that siphons a liquid from a container into a moving stream.

8-Step Process A management system used to organize the response to a chemical incident. The elements are site management and control, identifying the problem, hazard and risk evaluation, selecting PPE and equipment, information management and resource coordination, implementing response objectives, decon and cleanup operations, and terminating the incident.

Electrical Conductor Any material that will permit electricity to flow through it.

Emergency Call Box System of telephones connected by private line telephone, radio-frequency, or cellular technology usually located in remote areas and used to report emergency situations.

Emergency Communications Center Facility either wholly or partially dedicated to being able to receive emergency and, in some instances, nonemergency reports from citizens. Centers such as these are sometimes referred to as fire alarm, headquarters, dispatch, or a public safety answering point (PSAP).

Emergency Decon The rapid removal of a material from a person when that person (or responder) has become contaminated and needs immediate cleaning. Most emergency decon setups use a single hoseline to perform a quick gross decon of a person with water.

Emergency Medical Dispatch System designed for use by telecommunicators to assist them in evaluating patient symptoms using predetermined criteria and responses.

Emergency Medical Services The delivery of prehospital medical treatment.

Emergency Medical Technician (EMT) An individual trained and certified to provide basic life support emergency medical care.

Emergency Planning and Community Right to Know Act (EPCRA) The portion of SARA that specifically outlines how industries report their chemical inventory to the community.

Emergency Response Guidebook (ERG) Book provided by the DOT that assists the first responder in making decisions at a transportation-related chemical incident.

Emergency Response Planning (ERP) Levels that are used for planning purposes and are usually associated with the preplanning for evacuation zones.

Employee Assistance Program (EAP) A defined program that offers professional mental health and other health services to employees.

Encapsulated Suit A chemical suit that covers the responder, including the breathing apparatus. Usually associated with Level A clothing, that is gas- and liquid tight, but there are some Level B styles that are fully encapsulated, but not gas- or liquid tight.

Encoder Device that converts an "entered" code into paging codes, which in turn activate a variety of paging devices.

Endothermic Reaction A chemical reaction in which heat is absorbed, and the resulting mixture is cold.

Engine Company The unit designation of a group of firefighters assigned to a piece of apparatus designed to deliver water to the fire scene.

Engulfed To swallow up or overwhelm.

Enhanced 9-1-1 Similar in nature to basic 9-1-1 but with the capability to provide the caller's telephone number and address.

Equilibrium When referring to gas or liquids, a state where a balance has occurred in mixture or weight.

Etiological A form of a hazard that includes biological, viral, and other disease-causing materials.

Evacuation The movement of people from an area, usually their homes, to another area that is considered to be safe. People are evacuated when they are no longer safe in their current area.

Evaporation A process in which the molecules of a liquid are liberated into the atmosphere at a rate greater than the rate at which the molecules return to the liquid. Ultimately the liquid becomes fully airborne in a gaseous state.

Exhauster A device to speed the operation of the dry pipe valve by detecting the decrease in air pressure. It helps bleed off air.

Exit Drills in the Home (EDITH) A fire survival program to encourage people to practice fire drills from their home or residence.

Exothermic Reaction A chemical reaction that releases heat, such as when two chemicals are mixed and the resulting mixture is hot.

Explosive Limits A concentration of a gas or liquid that is not too rich or too lean to ignite with force.

Exposure A contact with a potentially disease-producing organism; the contact does not necessarily produce the disease in the exposed individual.

Exposure Fire Any combustible item threatened by something burning nearby that has caught on fire.

Extension Ladder A ladder consisting of two or more sections that has the ability to be extended to a desired height through the use of a halyard.

External Bleeding Bleeding that is coming from an open wound on the body.

External Floating Roof Tank Tank with the roof exposed on the outside that covers the liquid within the tank. The roof floats on the top of the liquid, which does not allow for vapors to build up.

Extremely Hazardous Substances (EHS) A list of 366 substances that the EPA has determined present an extreme risk to the community if released.

Extricate To set free, release, or disentangle a patient from an entrapment situation.

Federal Communications Commission Government agency charged with administering the provisions of the Communications Act of 1934 and the revised Telecommunications Act of 1996 and is responsible for nonfederal radio-frequency users.

Femoral Artery A major artery in the lower body near the groin that supplies the leg with blood. Can be used as a pressure point for controlling bleeding in the lower extremities.

Fill Site The area where tenders are filled or get their water.

Fine Decon The most detailed of the types of decontamination. Usually performed at a hospital that has trained staff and is equipped to perform fine decon procedures.

Fire Alarm Notification to the fire department that a fire or other related emergency is in progess, which results in a response.

Fire Engineering The study of fire, fire behavior, fire extinguishment, and suppression.

Fire Flow Capacity The amount of water available or amount that the water distribution system is capable of flowing.

Fire Flow Requirement A measure comparing the amount of heat the fire is capable of generating versus the amount of water required for cooling the fuels below their ignition temperature.

Fire Hazard Any condition, situation, or operation that could lead to the ignition of unwanted combustion or result in proper combustion becoming uncontrolled.

Fire Hose A flexible conduit used to convey water or other agent from a water source to the fire.

Fire Hydraulics The principles associated with the storage and transfer of water in firefighting activities.

Fire Intensity A measurement of Btus produced by a fire. Sometimes measured in flame length in the wildland environment.

Fire Load The amount of heat energy released when combustibles burn in a given area or building—expressed in British thermal units (Btus).

Fire Resistive The capacity of a material to withstand the effects of fire.

Fire-Resistive Rating The time in hours that a material or assembly can withstand fire exposure. Fire-resistive ratings are usually provided for testing organizations. The ratings are expressed in a time frame, usually hours or portions thereof.

Fire Shelter A last-resort protective device for wildland firefighters caught or trapped in an environment where a firestorm or blowup is imminent.

Fire Societies Groups of people who voluntarily banded together to deal with a community's fire problems.

Fire Station Alerting System System used to transmit emergency response information to fire station personnel via voice and/or digital transmissions.

Fire Stopping Pieces of material, usually wood or masonry, placed in stud or joist channels to slow the extension of fire.

Fire Stream The water or other agent as it leaves the hose and nozzle toward its objective, usually the fire.

Fire Tetrahedron Four-sided pyramid-like figure describing the heat, fuel, oxygen, and chemical reaction necessary for combustion.

Fire Wardens Designated community individuals who walked the streets at night looking for fire and carrying large wooden rattles with which to signify a found fire.

Fire Watch An organized patrol of a protected property when the sprinkler or other protection system is down for maintenance. Personnel from the property regularly check to make sure a fire has not started and assist in evacuation and prompt notification of the fire department.

Firefighter Assist and Search Team (FAST) A company designated to search for and rescue trapped or lost firefighters. May also be called a rapid intervention team (RIT).

Firemark Signs on sheets of metal telling firefighters which company held the insurance policy on a home or building.

First Responders A group designated by the community as those who may be the first to arrive at a chemical incident. This group is usually composed of police officers, EMS providers, and firefighters.

Fit Testing A test that ensures the respiratory protection fits the face and offers maximum protection.

Flammable Limits The concentration level of a substance at which it will burn.

Flammable Range Ratio of gas to air that will sustain fire if exposed to flame or spark.

Flanks of the Fire The sides of a wildland fire running from the start point up each side to the end of the fire running into unburned areas.

Flash Point The temperature at which a liquid will liberate a flammable gas.

Flashover A sudden event that occurs when all the contents of a container reach their ignition temperature simultaneously.

Flow The rate or quantity of water delivered, usually measured in gallons per minute or liters per minute (1 gpm = 3.785 L/min).

Fluoroprotein Film–Forming Foam (FFFP) Combines protein with the film-forming fluorinated surfactants of AFFF to improve on the qualities of both types of foam.

Fluoroprotein Foam Designed as an improved protein foam with a fluorinated surfactant added.

Flush or **Slab Doors** Doors that are flat or have a smooth surface and may be of either hollow-core or solid-core construction.

Fly Ladder That portion of a ladder that extends out from the bed ladder. Also called *fly section.*

Foam An aggregate of gas-filled bubbles formed from aqueous solutions of specially formulated concentrated liquid foaming agents.

Fog Nozzle Delivers either a fixed spray pattern or variable combination of straight stream and spray patterns.

Forcible Entry The fire scene task of gaining entry to a building or secured area by disabling, breaking, or going around locking and security devices.

Foreman Individual designated as the leader of an early fire company; a predecessor to the modern title of fire chief.

Forestry Hose Specially designed hose for use in forestry and wildland firefighting. It comes in 1- and 1½-inch (25- and 38-mm) sizes and should meet U.S. Forestry Service specifications.

Formal Decon The washing and scrubbing portion of the decontamination process. The process is usually repeated and is performed by a decon crew.

Fracture A medical term for a broken or cracked bone in the body.

Frangible Disk A type of pressure-relieving device that actually ruptures in order to vent the excess pressure. Once opened the disk remains open; it does not close after the pressure is released.

Freelancing The act of working alone or performing a task for which the firefighter has not been assigned.

Freezing point The temperature at which liquids become solids.

Friction Caused by the rubbing of materials against each other while in movement and converts or robs some of the movement energy into heat energy.

Friction Loss Measurement of friction in a system such as a hoseline.

Frontage The portion of a property that faces and actually touches the street.

Fuel Resistance Ability to tolerate the fuel and to avoid being saturated by or picking up the fuel.

Full Thickness Burns Burns affecting not only the skin structure but the tissues and muscles underneath. Full thickness burns may be red, white, or charred in color, and will appear dry because the blood vessels in the skin are damaged extensively and are not supplying fluids to the area.

Garden Apartment A two- or three-story apartment building with common entryways and layouts on each floor, surrounded by greenery and landscaping, sometimes having porches and patios.

Gas A state of matter that describes the material in a form that moves freely about and is difficult to control. Steam is an example.

Gate Valves Indicating and nonindicating valves that are opened and closed to control water flow.

Gauge Pressure Measures pressure without atmospheric pressure. Normally fire department gauges do not measure atmospheric pressure. Gauge pressure is measured in psi or psig.

GEDAPER Process A management system used to organize the response to a chemical incident. The factors are gather information, estimate potential, determine goals, assess tactical options, plan, evaluate, and review.

Girder A large structural member used to support beams or joists—that is, a beam that supports beams.

Glazing The glass or other clear material portion of the window that allows light to enter.

Gross Decon The portion of the decontamination process that removes the majority of the chemicals through a flushing process. The gross washing is done using large amounts of water and is usually done by the individual or the individual's partner.

Gross Negligence Occurs when an individual disregards training and continues to act in a manner without regard for others.

Ground Pads Sheets of plywood, planks, aluminum sheets, and so on, used to distribute weight over a larger area.

Guard Dogs Trained animals that will attack an intruder.

Guideline/Lifeline Rope used as a crew is searching a structure to assist them in finding their way back out.

Gusset Plate A connecting plate used in truss construction. In steel trusses, these plates are flat steel stock. In wood trusses, the plates are either light-gauge metal or plywood.

Halligan Tool From the prying group, a 30-inch forged steel tool with three primary parts: the adz end, the pike end, and the fork end.

Halyard A rope or cable that is used to raise the fly ladders of an extension ladder.

Hard Suction Hose A special type of hose that does not collapse when used for drafting.

Hardware Equipment used in conjunction with life safety ropes and harnesses (carabiners, figure eights, rappel racks, etc.).

Harnesses Webbing sewn together to form a belt, seat harness, or seat and chest harness combination.

Hazardous Materials Chemicals that are flammable, explosive, or otherwise capable of causing death or destruction when improperly handled or released.

Hazardous Materials Technician An individual trained to meet the requirements of CFR OSHA 1910.120, *Technician Level for Hazardous Materials Response.*

Hazardous Waste Operations and Emergency Response (HAZWOPER) The OSHA regulation that covers safety and health issues at hazardous waste sites, as well as response to chemical incidents.

HAZMAT Crime A criminal act that uses or threatens the use of chemicals as a weapon.

Head of the Fire The running top or aggressive end of the fire away from the start point.

Head Pressure Measures the pressure of a column of water in feet (meters). Head pressure gain or loss results when water is being pumped above or below the level of the pump. A head of 2.31 feet (0.7 m) would equal 1 psi (6.895 kPa).

Heat Resistance The ability of foam to stand up to the heat of the fire or to hot surfaces near the fire.

Heat Sink The term used to denote a place where heat is drained away from a source.

Helix The metal or plastic bands or rings used in hard suction hose to prevent its collapse under drafting conditions.

Higbee Cut The blunt ending of the threads of fire hose couplings that allows the threads to be properly matched, avoiding cross-threading.

Hoistway The shaft in which an elevator or a number of elevators travel.

Hollow-Core Door Any door that is not solid, usually with some type of filler material between face panels.

Home Alerting Devices Emergency alerting devices primarily used by volunteer department personnel to receive reports of emergency incidents.

Hook A tool with a 32-inch to 12-foot handle with a pike and hook on one end. Used for pulling ceilings or separating other materials. Also known as a *pike pole.*

Horizontal Ventilation Channeled pathway for fire ventilation via horizontal openings.

Hose Bed The portion or compartment of fire apparatus that carries the hose.

Hose Bridges Devices that allow vehicles to pass over a section of hose without damaging it.

Hose Cap Does not allow water to flow through it. Instead, it caps the end of a hoseline or appliance to prevent water flow.

Hose Cart A handcart or flat cart modified to be able to carry hose and other equipment around large buildings. Some departments use them for high-rise situations.

Hose Clamp A device to control the flow of water by squeezing or clamping the hose shut. Some work by pushing a lever that closes the jaws of the device and others have a screw mechanism or hydraulic pump that closes the jaws.

Hose Jackets Metal or leather devices used for stopping leaks without shutting down the line that is fitted over the leaking area and either clamped or strapped together to control the leak.

Hose Roller or **Hoist** A metal frame, with a securing rope, shaped to fit over a windowsill or edge of a roof with two rollers to allow the hose to roll over the edge, preventing chafe.

Hose Strap A short strap with a forged handle and cinch clip attached. Used to help maneuver hose and attach hose to ladders and stair rails.

HVAC Acronym for heating, ventilation, and air-conditioning unit. HVACs are typically a rooftop unit on commercial buildings. Buildings may have one or dozens of these units.

Hydrant Valves or **Switch Valves** Valve used on a hydrant that allows an engine to connect and charge its supply line immediately but also allows an additional engine to connect to the same hydrant without shutting down the hydrant, and increases the flow of the hydrant.

Hydrant Wrenches Tools used to operate the valves on a hydrant. May also be used as a spanner wrench. Some are plain wrenches and others have a ratchet feature to speed the operation of the valve.

Hydraulic Pistons Mechanical rams that operate by pressure exerted through the use of a liquid, usually some form of oil.

Hydraulics The study of fluids at rest and in motion.

Hydrocarbon Any of numerous organic compounds, such as benzene and methane, that contain only carbon and hydrogen.

Hyperbaric Chamber A chamber that is usually used to treat scuba divers who ascended too quickly and need extra oxygen to survive. The chamber re-creates the high-pressure atmosphere of diving and forces oxygen into the body. It is also successful in the treating of carbon monoxide poisoning and smoke inhalation, because both of these problems require high amounts of oxygen to assist with the patient's recovery.

Hypoperfusion A serious condition caused by a problem or failure of the circulatory system that results in a decrease of oxygen and vital nutrients to the body's tissues. Also known as shock.

Hypoxia A deficiency of oxygen.

ICt_{50} The incapacitating level for time to 50 percent of the exposed group. It is a military term that is often used in conjunction with LCt_{50}.

Ignition The point at which the need for outside heat application ceases and a material sustains combustion based on its own generation of heat.

Ignition Point The temperature at which a substance will continue to burn after the source is removed.

Ignition Temperature The temperature of a liquid at which it will ignite on its own without an ignition source. Can be compared to SADT.

Immediately Dangerous to Life and Health (IDLH) The maximum level of danger one could be exposed to and still escape without experiencing any effects that may impair escape or cause irreversible health effects.

Impact Load A load that is in motion when it is applied.

Implied Consent The assumption of acceptance of emergency medical treatment by an unconscious patient or a child with no parents or legal guardians present.

Incendiary Agents Chemicals that are used to start fires, the most common being a Molotov cocktail.

Incident Action Plan (IAP) A strategic and tactical plan developed by the incident commander.

Incident Commander Level A training level that encompasses the operations level with the addition of incident command training. Intended to be the person who may command a chemical incident.

Incident Management System (IMS) A management system utilized on the emergency scene that is designed to keep order and follow a sequence of set guidelines.

Incision A cut to the skin that leaves a straight, even pattern.

Increaser Used to connect a smaller hose to a larger one.

Indirect Fire Attack An attack made on interior fires by applying a fog stream into a closed room or compartment, thus converting the water into steam to extinguish the fire.

Infection Control Procedures and practices for firefighters and emergency medical care providers to follow to prevent the transmission of diseases and germs from a patient to themselves or other patients.

Infectious Disease See **Communicable Disease.**

Infrared Sensor A sensor that uses infrared light to determine the presence of flammable gases. The light is emitted in the sensor housing and the gas passes through the light. If it is flammable the sensor will indicate the presence of the gas.

Initial Assessment The initial investigative action taken by care providers to determine if the patient has the basic signs of life as well as any serious, life-threatening injuries.

In-Line Eductor Eductor in which the waterway is always piped through a venturi.

Inorganic A substance that is not of any living organism.

Intake Relief Valve Required on large-diameter hose at the receiving engine that functions as a combined overpressurization relief valve, a gate valve, and an air bleed-off.

Integrated Communications The ability of all units or agencies to communicate at an incident.

Interface Firefighting Fighting wildland fire and protecting exposed structures in rural settings.

Intermodal Containers These are constructed in a fashion so that they can be transported by highway, rail, or ship. Intermodal containers exist for solids, liquids, and gases.

Internal Bleeding Bleeding within the body when no visible open wound is present.

Internal Floating Roof Tank Tank with a roof that floats on the surface of the stored liquid, but also has a cover on top of the tank, so as to protect the top of the floating roof.

Intervention The act of intervening; to come between as an influencing force. Typically a reactive action.

Irons The combination of a Halligan tool and flathead ax or maul.

Irritant A material that is irritating to humans, but usually does not cause any long-term adverse health effects.

Isolation Area An area that is set up by responders and is intended to keep people, both citizens and responders, out. May later become the hot zone/sector as the incident evolves. Is the minimum area that should be established at any chemical spill.

Jacket The outer part of the hose, often a woven cloth or rubberized material, which protects the hose from mechanical and other damage.

Jamb The mounting frame for a door.

Jet Dump A device that speeds the process of dumping a load of water from a tanker/tender.

Jet Siphon A device that speeds the process of transferring water from one tank to another.

Joist A wood framing member that supports floor or roof decking.

Kerf Cut A quick and easily made examination hole. It is created by letting the spinning blade of a power saw cut through the material to be cut and pulling it out, leaving only a slit-like cut measuring approximately 12 inches long and only as wide as the cutting blade.

Kern A derivative of the term *kernel,* which is defined as "the central, most important part of something; core; essence."

Knockdown Speed Speed with which foam spreads across the surface of a fuel.

Laceration A cut to the skin and underlying tissues that leaves an irregular, even pattern.

Ladder Pipe An appliance that is attached to the underside of an aerial ladder for an elevated water application.

Laminated Glass Glass composed of two or more sheets of glass with a plastic sheet between them. The purpose of the plastic sheet is to hold the glass together if broken, thus reducing the hazard of flying glass.

Landing Plate The plate at the top or bottom of an escalator where the steps disappear into the floor.

Laws Legislation that is passed by the House and Senate and signed by the president.

LCt_{50} The lethal concentration for time to 50 percent of the group. Same as the LC_{50}, but adds the element of time. It is a military term.

Leaking Underground Storage Tank (LUST) Describes a leaking tank that is underground.

Ledge Door Door built with solid material, usually individual boards, common in barns and warehouses.

Lethal Concentration (LC_{50}) A value for gases that provides the amount of chemical that could kill 50 percent of the exposed group.

Lethal Dose (LD_{50}) A value for solids and liquids that provides the amount of a chemical that could kill 50 percent of an exposed group.

Level A Protective Clothing Fully encapsulated chemical protective clothing. It is gas and liquid tight and offers protection against chemical attack.

Level B Protective Clothing A level of protective clothing that is usually associated with splash protection. Level B requires the use of SCBA. Various clothing styles are considered Level B.

Liability The possibility of being held responsible for individual actions.

Life Safety Term applied to the fire protection concept in which buildings are designed to allow for the escape of building occupants without injuries. Life safety usually makes the building more fire resistant, but this is not the main goal.

Life Safety Line According to NFPA 1983, rope dedicated solely to the purpose of constructing lines for supporting people during rescue, firefighting, or other emergency operations, or during training evolutions.

Lifting A term used to describe the removal of upper-level smoke and heat when cool air replaces the upper-level hot air that is escaping.

Liner The inner layer of fire hose, usually made of rubber or a plastic material, that keeps the water in the tubing of the hose.

Lintel A beam that spans an opening in a load-bearing masonry wall.

Liquid A state of matter that implies fluidity, which means a material has the ability to move as water would. There are varying states of being a liquid from moving very quickly to moving very slowly. Water is an example.

Live Load The weight of all materials and people associated with but not part of a structure.

Load-Bearing Wall Any wall that supports other walls, floors, or roofs.

Loaded Stream Combats the water freezing problem by adding an alkali salt as an antifreezing agent.

Loading The weight of building materials or objects in a building.

Local Application System Designed to protect only a certain or local portion of the building, usually directly where the hazard will occur or spread.

Local Emergency Planning Committee (LEPC) A group composed of members of the community, industry, and emergency responders to plan for a chemical incident, and to ensure that local resources are adequate to handle an incident.

Locking Devices A mechanical device or mechanism used to secure a door or window.

Loop A turn in the standing part that crosses itself and results in the standing part continuing on in the original direction of travel.

Lower Explosive Limit (LEL) The lower part of the flammable range, and is the minimum required to have a fire or explosion.

Mantle Anything that cloaks, envelops, covers, or conceals.

Mask Confidence or "Smoke Divers" Training Training courses designed to develop a firefighter's skills and confidence for using SCBA.

Mass Casualty An incident in which the number of patients exceeds the capability of the EMS to manage the incident effectively. In some jurisdictions this can be two patients, while in others it may take ten to make the incident a mass casualty.

Master Stream or **Heavy Appliances** Non-handheld water applicator capable of flowing over 350 gallons of water per minute (1325 L/min).

Mastery The concept that an individual can achieve 90 percent of an objective 90 percent of the time.

Material Safety Data Sheet (MSDS) Information sheet for employees that provides specific information about a chemical, with attention to health effects, handling, and emergency procedures.

Matter Something that occupies space and can be perceived by one or more senses; a physical body, a physical substance, or the universe as a whole. Something that has mass and exists as a solid, liquid, or gas.

Mayday A universal call for help. A Mayday indicates that an individual or team is in extreme danger.

Means of Egress A safe and continuous path of travel from any point in a structure leading to a public way. Composed of three parts: the exit access, the exit, and the exit discharge.

Medium-Diameter Hose (MDH) Either 2½- or 3-inch (63- or 75-mm) hose.

Medi-Vac An ambulance that transports patients by air. Typically, medi-vac units are helicopters with highly trained EMS personnel and nurses.

Melting Point The temperature at which solids become liquids.

Metal Oxide Sensor (MOS) A coiled piece of wire that is heated to determine the presence of flammable gases.

Midslope An area partway up a slope. Any location not on the bottom or top of a slope, as in a midslope road crossing the slope horizontally.

Miscible Having the ability to mix with water.

Mission Statement A written declaration by a fire agency describing the things that it intends to do to protect its citizenry or customers.

Mission Vision A term used to describe a condition in which a person becomes so focused on an objective that peripheral conditions are not noticed, as if the person is wearing blinders.

Mitigation Actions taken to eliminate a hazard or make a hazard less severe or less likely to cause harm. Typically a proactive action.

Mobile Data Computer Communications device that, unlike the mobile data terminal, does have information processing capabilities.

Mobile Data Terminal Communications device that in most cases has no information processing capabilities.

Mobile Radio Complete receiver/transmitter unit that is designed for use in a vehicle.

Mobile Support Vehicle Vehicle designed exclusively for use as an on-scene communication center and command post.

Modular Organization The ability to start small and expand if an incident becomes more complex.

Molecule The smallest particle into which an element or a compound can be divided without changing its chemical and physical properties; a group of like or different atoms held together by chemical forces.

Mortar Mixture of sand, lime, and portland cement used as a bonding material in masonry construction.

Mounting Hardware Hinges, tracks, or other means of attaching a door to the frame or jamb.

Multigas Detector A term used to describe an air monitor that measures oxygen levels, explosive (flammable) levels, and one or two toxic gases such as carbon monoxide or hydrogen sulfide.

Multiple-Alarm Incident Involves the response of additional personnel.

Mutual Aid or **Assistance Agreements** Prearranged written agreements of the type and amount of assistance one jurisdiction will provide to another in the event of a large-scale fire or disaster. The key to understanding mutual aid is that it is a reciprocal agreement.

National Emergency Number Association Not-for-profit organization founded in 1982 and made up of more than 6,000 members. The association fosters technical advancement, availability, and implementation of a universal emergency telephone number system.

National Fire Protection Association (NFPA) A not-for-profit membership organization that uses a consensus process to develop model fire prevention codes and firefighting training standards.

National Institute for Occupational Safety and Health (NIOSH) A federal institute tasked with investigating firefighter fatalities and making recommendations to prevent reoccurrence.

National Response Center (NRC) The location that must be called to report a spill if it is in excess of the reportable quantity.

Needed or **Required Flow** Estimate of the amount of water required to extinguish a fire in a certain type period. Based on the type and amount of fuel burning.

Negligence Acting in an irresponsible manner or different from the way in which someone was trained; that is, differing from the standard of care.

Nerve Agents Chemicals that are designed to kill humans, specifically in warfare. They are chemically similar to organophosphorus pesticides and cause the same medical reaction in humans.

Nested The state when all the ladders of an extension ladder are unextended.

NFPA 1001 *Standard for Fire Fighter Professional Qualifications,* a national consensus training standard establishing the job performance requirements of tasks to be performed by firefighters.

NFPA 1404 National Fire Protection Association standard created by the Fire Service Training Committee detailing the requirements for fire service SCBA programs, including training and maintenance procedures.

NFPA 1500 National Fire Protection Association standard created by the Technical Committee on Fire Service Occupational Safety and Health that addresses a number of issues concerning protective equipment.

NFPA 1981 National Fire Protection Association standard specific to open-circuit SCBA for fire service use that contains additional requirements above the NIOSH certification.

NFPA 72 National Fire Alarm Code.

NFPA Standard 1931 The standard issued by the National Fire Protection Association that governs fire service ladder testing and certification.

9-1-1 Emergency telephone number that provides access to the public safety services in the community, region, and, ultimately, nation.

NIOSH National Institute for Occupational Safety and Health, 42 CFR Part 84, sole responsibility for testing and certification of respiratory protection including fire service SCBA.

No-Knowledge Hardware Locking devices that require no key or special knowledge to operate.

Nozzle A tapered or constricted tube used to increase the speed or change the direction of water or other fluids.

Nozzle Flow The amount or volume of water that a nozzle will provide. Flow is measured in gallons per minute or liters per minute.

Nozzle Pressure The pressure required to effectively operate a nozzle. Pressure is measured in pounds per square inch or kilopascals.

Nozzle Reach The distance the water will travel after leaving the nozzle. Reach is a function of the pressure, which is converted to velocity or speed of the water leaving the nozzle.

Nozzle Reaction The force of nature that makes the nozzle move in the opposite direction of the water flow. The nozzle operator must counteract the thrust exerted by the nozzle to maintain control.

Occupancy Classifications The use for which a building or structure is designed.

Occupant Use Hose Hose that is used in standpipe systems for building occupants to fight incipient fires. It is usually 1½-inch (38-mm) single-jacket hose similar to attack hose.

Occupational Safety and Health Administration (OSHA) The federal agency, under the Department of Labor, that is responsible for employee occupational safety.

Offensive Attack An aggressive attack on a situation where resources are adequate and capable of handling the situation.

One- or **Two-Person Rope** According to NFPA 1983, a one-person rope requires a minimum tensile strength of 4,500 pounds, and a two-person rope requires a minimum tensile strength of 9,000 pounds.

Open-Circuit SCBA A type of SCBA unit in which the exhaled air is vented to the outside atmosphere.

Operational Period The time frames for operations at an incident. At large-scale or complex incidents these will usually be eight- to twelve-hour time frames.

Operations Level The next level of training above awareness that provides the foundation which allows for the responder to perform defensive activities at a chemical incident.

Ordinary Tank A horizontal or vertical tank that usually contains combustible or other less hazardous chemicals. Flammable materials and other hazardous chemicals may be stored in smaller quantities in these types of tanks.

Organic A substance derived from living organisms.

OSHA 29 CFR 1910.134 Standard establishing minimum medical, training, and equipment levels for respiratory protection programs.

Outside Stem and Yoke (OS&Y) Valve Has a wheel on a stem housed in a yoke or housing. When the stem is exposed or outside, the valve is open. Also called an outside screw and yoke valve.

Overpacked A response action that involves the placing of a leaking drum (or container) into another drum. There are drums made specifically to be used as overpack drums in that they are oversized to handle a normal size drum.

Oxidizer A catalyst in the breakdown of molecules.

Oxygen Deficient Atmosphere An atmosphere with an oxygen content below 19.5 percent by volume.

Packaging The bandaging and preparing of a patient to be moved from the place of injury to a stretcher.

Panel Doors Doors with a solid stile and rails with panels made of wood or glass or other materials.

Panic Hardware Hardware mounted on doors that enable them to be opened by pushing from the inside.

Paragraph q The paragraph within HAZWOPER that outlines the regulations that govern emergency response to chemical incidents.

Paramedic (EMT-P) An individual trained and certified to provide advanced life support emergency medical care, including drug therapy.

Parapet The projection of a wall above the roofline of a building.

Partial Thickness Burns Burns affecting the entire skin structure that lies over the top of the fatty tissues and muscles causing skin to turn red and blistering of the skin.

Passport A term given to a specific accountability system where crews are tracked using a card (passport) with all members listed. An accountability manager tracks the passports on an accountability board.

PDD 39 Presidential Decision Directive 39, which established the FBI as the lead agency in terrorism incidents responsible for crisis management. It also established FEMA as the lead for consequence management.

Permeation The movement of chemicals through chemical protective clothing on a molecular level; does not cause visual damage to the clothing.

Permissible Exposure Limit (PEL) An OSHA value that regulates the amount of a chemical that a person can be exposed to during an eight-hour day.

Personal Alert Safety System (PASS) A device that emits a loud alert or warning that the wearer is motionless.

Personal Size-Up A continuous mental evaluation of an individual's immediate environment, facts, and probabilities.

Personnel Accountability Report (PAR) This is an organized roll call of all units assigned to an incident.

Photo-ionization Detector (PID) An air monitoring device used by HAZMAT teams to determine the amount of toxic materials in the air.

Piercing Nozzles Originally designed to penetrate the skin of aircraft and now have been modified to pierce through building walls and floors.

Pike Pole See **Hook.**

Pipe Chases A construction term used to describe voids designed to house building water supply and waste pipes. The term *electrical chase* is used for wiring.

Pitot Gauge A device with an opening in its blade-shaped section that allows water to flow to a Bourdon gauge and registers the flowing discharge pressure of an orifice.

Plan View A drawing or diagram of a building or area as seen from directly overhead. May include a site plan or a floor plan.

Platform Framing A style of wood frame construction in which each story is built on a platform, providing fire stopping at each level.

Polar Solvent A material that will mix with water, diluting itself.

Polar Solvent Type of Foam or Alcohol-Resistant Foam Foam that is compatible with alcohol and/or polar solvents by creating a polymeric barrier between the water in the foam and the polar solvent.

Polymeric Barrier A separation barrier made up of polymer or a chain of molecules linked in a series of long strands. This separates a polar solvent from an ATC foam blanket.

Polymerize A chain reaction in which the material quickly duplicates itself and, if contained, can be very explosive.

Portable Hydrant or Manifold Like a large water thief and may have one or more intakes and numerous outlets to allow multiple hoselines to be utilized with or without a pumper at the fire location.

Portable Water Tanks Collapsible or inflatable temporary tanks for the storage of water that is dumped from tankers or tenders. Usually carried by the tender to set up a dump site.

Positive Pressure A feature of SCBA providing a continuous supply of air, delivered by the regulator to the face piece, keeping toxic gases from entering. This pressure (1½ to 2 psi, depending on the manufacturer) is slightly above atmospheric pressure.

Post Indicator Valve (PIV) A control valve that is mounted on a post case with a small window, reading either "OPEN" or "SHUT."

Post-Incident Thought Patterns A phenomenon that describes an individual's inattentiveness following a significant incident. Post-incident thought patterns can lead to injuries or even death.

Preaction System Similar to the dry pipe and deluge systems. The system has closed piping and heads with air under no or little pressure, but the water does not flow until signaled open from a separate fire detection system. The preaction valve then opens and allows water to flow through the system to the closed heads. When an individual head is heat activated, it opens and water attacks the fire. Usually used when water can cause a large dollar loss.

Prearrival Instructions Self-help instructions intended to enhance the overall safety of the citizen until first responders arrive on the scene.

Pre-Incident Management Advance planning of firefighting tactics and strategies or other emergency activities that can be anticipated to occur at a particular location. Often referred to as preplanning.

Pressure The force, or weight, of a substance, usually water, measured over an area.

Pressure-Regulating Device Designed to control the head pressure at the outlet of a standpipe system to prevent excessive nozzle pressures in hoselines.

Primary Hole Ventilation term used to describe the first holes to be cut in a roof. They must be located as close to directly over the fire as possible to prevent laterally drawing the fire across unburned areas.

Protein Foam Made from chemically broken down natural protein materials, such as animal blood, that have metallic salts added for foaming.

Prying Tools The group of tools used to separate objects by means of a mechanical advantage.

Psychological Decon The process performed when persons who have been involved in a situation think they have been contaminated and want to be decontaminated. Responders who have identified that the persons have *not* been contaminated should still consider what can be done to make them feel better.

Pulling Tools The group of tools used to pull away materials.

Pulmonary Edema Fluid filling the lungs causing death by drowning.

Pump Operator A generic term to describe the person responsible for operating a fire apparatus pump. Other commonly used titles include motor pump operator, engineer, technician, chauffeur, and driver/operator.

Puncture An injury caused by an object that has stabbed the body.

Purlins A series of wood beams placed perpendicular to steel trusses to help support roof decking.

Pyrolysis Decomposition or transformation of a compound caused by heat.

Quint A combination fire service apparatus with components of both engine company and a truck company.

Rabbeted A door stop that is cut (rabbeted) into the door frame. On metal door frames the stop is an integral part of the frame.

Radial Pulse The pulse located in either wrist.

Radiological Dispersion Device (RDD) An explosive device that spreads radioactive material throughout an area.

Rafter A wood joist that is attached to a ridge board to help form a peak.

Rapid Intervention Crew (RIC) See **Rapid Intervention Team.**

Rapid Intervention Team (RIT) A company designated to search for and rescue trapped or lost firefighters. Depending on location, may also be called a FAST.

Rate of Spread A ground cover fire's forward movement or spread speed. Usually expressed in chains or acres per hour.

RECEO Acronym coined by Lloyd Layman standing for Rescue, Exposures, Confinement, Extinguishment, and Overhaul.

Recommended Exposure Limit (REL) An exposure value established by NIOSH for a ten-hour day, forty-hour workweek. Similar to the PEL and TLV.

Reducers Used to connect a larger hose to a smaller one.

Reel Coil Memory that wire develops from having been placed on a wooden spool as it is being manufactured.

Regulations Developed and issued by a governmental agency and have the weight of law.

Rehab A shortened word meaning *rehabilitation.* Rehab typically consists of rest, medical evaluation, hydration, and nourishment.

Relief Valve A device designed to vent pressure in a tank, so that the tank itself does not rupture due to an increase in pressure. In most cases these devices are spring loaded so that when the pressure decreases the valve shuts, keeping the chemical inside the tank.

Remote Shutoffs Valves that can be used to shut off the flow of a chemical. The term *remote* is used to denote valves that are located away from the spill.

Reportable Quantity (RQ) Both the EPA and DOT use the term. It is a quantity of chemicals that may require some type of action, such as reporting an inventory or reporting an accident involving a certain amount of the chemical.

Rescue Those actions that firefighters perform at emergency scenes to remove victims from imminent danger or to extricate them if they are already entrapped.

Rescue Company The unit designation of a group of firefighters assigned to perform specialized rescue work and/or tactics and functions such as forcible entry, search and rescue, ventilation, and so on.

Rescue Specialist A firefighter with specialized training and experience in areas such as high angle rope rescue, confined space, trench, or structural collapse rescue.

Residential Sprinkler System Smaller and more affordable version of a wet or dry pipe sprinkler system designed to control the level of fire involvement such that residents can escape.

Residual Pressure The pressure in a system after water has begun flowing.

Respiratory Protection Programs Management programs designed to ensure employee respiratory protection as required by OSHA 29 CFR 1910.134 and NFPA 1500.

Respiratory System The system of the human body that exchanges oxygen and waste gases to and from the circulatory system.

Retard Chamber Acts to prevent false alarms from a sudden pressure surge in the water supply by collecting a small volume of water before allowing a continued flow to the alarm device. The water from a surge is drained from a small hole in the bottom of the collection chamber.

Retention The digging of a hole in which to collect a spill. Can be used to contain a running spill or collect a spill from the water.

Ricin A biological toxin that can be used by a terrorist or other person attempting to kill or injure someone. It is the easiest terrorist agent to produce and one of the most common.

Ridge The land running between mountain peaks or along a wide peak. A high area separating two drainages running parallel with them.

Ringdown Circuits Telephone connection between two points. Going "off-hook" on one end of the circuit causes the telephone on the other end of the circuit to "ring" without having to dial a number.

Risk The chance of injury, damage, or loss; hazard.

Risk-Based Response An approach to responding to a chemical incident by categorizing a chemical into a fire, corrosive, or toxic risk. Use of a risk-based approach can assist the responder in making tactical, evacuation, and PPE decisions.

Risk/Benefit An evaluation of the potential benefit that a task will accomplish in relationship to the hazards that will be faced while completing the task.

Risk Management The process of minimizing the chance, degree, or probability of damage, loss, or injury.

River Bottom Topographic feature where water runs from higher elevations to lower. Can be dry or wet depending on season or recent rains.

Rollover A phenomenon where the burning of superheated gases from fire extends into the top areas of the compartment in the upper thermal layers.

Rope Hose Tool About 6 feet (2 m) of ½-inch (13-mm) rope spliced into a loop with a large metal hook at one end and a 2-inch (50-mm) ring at the other. Used to tie in hose and ladders, carry hose, and perform many other tasks requiring a short piece of rope.

Round Turn Formed by continuing the loop on around until the sections of the standing part on either side of the round turn are parallel to one another.

Run Card System System of cards or other form of documentation that provides specific information on what apparatus and personnel respond to specific areas of a jurisdiction.

Running End End of the rope that is not rigged or tied off.

Saddle A pass between two peaks that has a lower elevation than the peaks. Wind will pass through this area faster than over the peaks, so fire is drawn into this feature.

Safety Container A storage can that eliminates vapor release by using a self-closing lid. Also contains a flame arrestor in the dispenser opening.

Sea Containers Shipping boxes that were designed to be stacked on a ship, then placed onto a truck or railcar.

Search and Rescue Attempts by fire and emergency service personnel to coordinate and implement a search for a missing person and then effect a rescue.

Secondary Containment Any approved method that will prevent the runoff of spilled hazardous materials and confine it to the storage area.

Secondary Hole A ventilation hole that is opened only after the primary holes have been opened. It complements the primary holes.

Sectional View A vertical view of a structure as if it were cut in two pieces. Each piece is a cross section of the structure showing roof, wall, horizontal floor construction, and the location of stairs, balconies, and mezzanines.

Sector An area established and identified for a specific reason, typically because a hazard exists within the sector. The sectors are usually referred to as hot, warm, and cold sectors and provide an indication of the expected hazard in each sector. Sometimes referred to as a zone.

Self-Accelerating Decomposition Temperature (SADT) Temperature at which a material will ignite itself without an ignition source present. Can be compared to ignition temperature.

Self-Contained Breathing Apparatus (SCBA) A type of respiratory protection in which a self-contained air supply and related equipment are worn or attached to the user. Fire service SCBA is required to be of the positive pressure type.

Sensitizer A chemical that after repeated exposures may cause an allergic-type effect on some people.

Setting The finishing step, making sure that the knot is snug in all directions of pull.

Severance The cutting off of components (i.e., brake pedal, steering wheel) in a vehicle.

Shear A force that tends to tear a material by causing its molecules to slide past each other.

Sheetrock A trademark and another name for plasterboard.

Shelter in Place A form of isolation that provides a level of protection while leaving people in place, usually in their homes. People are usually sheltered in place when they may be placed in further danger by an evacuation.

Shock A serious condition caused by a problem or failure of the circulatory system that results in a decrease of oxygen and vital nutrients to the body's tissues. Also known as hypoperfusion.

Shock Load A load or impact being transferred to a rope suddenly and all at one time.

Shoring The use of timbers to support and/or strengthen weakened structural members (roofs, floors, walls, etc.) in order to avoid a secondary collapse during the rescue operation.

Short-Term Exposure Limit (STEL) A fifteen-minute exposure to a chemical followed by a one-hour break between exposures. Only allowed four times a day.

Shoulder Load Hose load designed to be carried on the shoulders of firefighters.

Shuttle Operation The cycle in which mobile water supply apparatus is dumped, moves to a fill site for refilling, and is returned to the dump site.

Siamese A device that connects two or more hoselines into one line with either a clapper valve or gate valve to prevent loss of water if only one line is connected.

Sick Building A term that is associated with indoor air quality. A building that has an air quality problem is referred to as a *sick building.* In a sick building, occupants become ill as a result of chemicals in and around the building.

Sick Building Chemical When a building is referred to as a *sick building,* certain chemicals exist within that cause health problems for the occupants. These chemicals are referred to as *sick building chemicals.*

Simple Beam A beam supported at the two points near its end.

Slab Door See **Flush** or **Slab Doors.**

Slot Loads Narrow section of a hose bed where hose is flat loaded in the slot.

Small Lines or **Small-Diameter Hose** Hose less than 2½ inches (63 mm) in diameter.

Soft Suction Hose Large-diameter woven hose used to connect a pumper to a hydrant. Also known as a soft sleeve.

Solid-Core Doors Doors made of solid material such as wood or having a core of solid material between face panels.

Solid Stream Nozzles Type of nozzle that delivers an unbroken or solid stream of water to the fire. Also called solid tip, straight bore, or smooth bore.

Solid A state of matter that describes materials that may exist in chunks, blocks, chips, crystals, powders, dusts, and other types. Ice is an example.

Solubility A liquid's ability to mix with another liquid.

Spalling Deterioration of concrete by the loss of surface material due to the expansion of moisture when exposed to heat.

Span of Control The ability of one individual to supervise a number of other people or units. The normal range is three to seven units or individuals, with the ideal being five.

Spanner Wrenches Used to tighten or loosen couplings. They may also be useful as a pry bar, door chock, gas valve control, and so on.

Speaking Trumpet Trumpet used by a foreman or crew boss to shout orders above the noise of firefighting activities.

Special Egress Control Device Door hardware that will release and unlock the door a maximum of fifteen seconds after it has been activated by pushing on the bar.

Specialist Level A level of training that provides for a specific type of training, such as railcar specialist; someone who has a higher level of training than a technician.

Specific Gravity Weight of a liquid in relation to water. Water is rated 1.

Specification (Spec) Plates All trucks and tanks have a specification plate that outlines the type of tank, capacity, construction, and testing information.

Spinal Immobilization The process of protecting patients against further injury by securing them to a backboard or other rigid device designed to minimize movement.

Sprain Injury to the ligaments that hold the body's joints together and allow them to move.

Sprinkler Systems Designed to automatically distribute water through sprinklers placed at set intervals on a system of piping, usually in the ceiling area, to extinguish or control the spread of fires.

Staging Part of the operations section where apparatus and personnel assigned to the incident are available for deployment within three minutes.

Stairwell An enclosed stairway attached to the side of a high-rise building or in the center core of same.

Standard of Care A legal term that means for every emergency medical incident, an emergency responder should treat the patient in the same manner as another emergency responder with the same training.

Standard Operating Procedure (SOP) Specific information and instruction on how a task or assignment is to be accomplished.

Standard Transportation Commodity Code (STCC) A number assigned to chemicals that travel by rail.

Standards Usually developed by consensus groups establishing a recommended practice or standard to follow.

Standing Part The part of a rope that is not used to tie off.

Standpipe Systems Piping systems that allow for the manual application of water in large buildings.

State Emergency Response Committee (SERC) A group that ensures that the state has adequate training and resources to respond to a chemical incident.

States of Matter Describes in what form matter exists, such as solids, liquids, or gases.

Static A rope having very little (less than 2 percent) elongation at normal safe working loads.

Static Pressure The pressure in the system with no hydrants or water flowing.

Staypoles The stabilizer poles attached to the sides of Bangor ladders that are used to assist in the raising of this type of ladder. Once raised, they are not used to support the extended ladder.

Steepness of Slope The degree of incline or vertical rise to a given piece of land.

Storz Couplings The most popular of the nonthreaded hose couplings.

Straight Stream A nozzle pattern that creates a hollow stream, similar in shape to the solid stream pattern, but the straight stream pattern must pass around the baffle of the nozzle. Newer fog nozzle designs, especially the automatic nozzles, only have this hollow effect from the tip on and, hence, create a solid stream with good reach and penetration abilities, some better than solid stream nozzles.

Strainers Placed over the end of a suction hose to prevent debris from being sucked into the pump. Some strainers have a float attached to keep them at or near the water's surface. A different style of strainer or screen is located on each intake of a pump.

Strategic Goals The overall plan developed and used to control an incident.

Stream Shape The arrangement or configuration of the water or other agent droplets as they leave the nozzle.

Stream Straighter A metal tube, commonly with metal vanes inside it, between a master stream appliance and its solid nozzle tip. The purpose is to reduce any turbulence in the stream, allowing it to flow straighter.

Strike Plate The metal piece attached to a door jamb into which the lock bolt slides. Also called a *strike* or *striker.*

Striking Tools The group of tools designed to deliver impact forces to break locks or drive another tool.

Sublimation The ability of a solid to go to the gas phase without being liquid.

Superficial Burns Burns affecting the outermost layer of skin, which typically cause redness of the skin, swelling, and pain.

Superfund Amendments and Reauthorization Act (SARA) A law that regulates a number of environmental issues, but is primarily for chemical inventory reporting by industry to the local community.

Supplied Air Respirator (SAR) A type of SCBA in which the self-contained air supply is remote from the user, and the air is supplied by means of air hoses.

Supply Hose or **Large-Diameter Hose (LDH)** Larger hose [3½ inches (90 mm) or bigger] used to move water from the water source to attack units. Common sizes are 4 and 5 inches (100 to 125 mm).

Surface-to-Mass Ratio Exposed exterior surface area of a material divided by its weight.

Suspension System The springs, shock absorbers, tires, and so on, of a vehicle.

Tactics The specific operations performed to satisfy the strategic goals for an incident.

Tactilely Using the sense of touch to feel for any differences or abnormality.

Tag/Guide Lines Tag lines are ropes held and controlled by firefighters on the ground or lower elevations in order to keep items being hoisted from banging against or getting caught on the structure as they are being hoisted.

Tanker The term given to aircraft capable of carrying and dropping water or fire retardant. Some departments still use the term to describe land-based water apparatus.

Target Hazard An occupancy that has been determined to have a greater than average life hazard or complexity of firefighting operations. Such occupancies receive a high priority in the pre-incident management process and often a higher level of first-alarm response assignment.

TDD Device that allows citizens to communicate with the telecommunicator through the use of a keyboard over telephone circuits instead of voice communications.

Teamwork A number of persons working together in an effort to reach a common goal.

Technician Level A high level of training that allows specific offensive activities to take place, to stop or handle a chemical incident.

Telecommunicator Individual whose primary responsibility is to receive emergency requests from citizens, evaluate the need for a response, and ultimately sound the alarm that sends first responders to the scene of an emergency.

Tempered Glass Plate glass that has been heat treated to increase its strength.

Tender The abbreviated term for *water tender.* A water tender is defined as a land-based mobile water supply apparatus. Some departments still use the term *tender* to describe a hose-carrying support apparatus.

Tensile Strength Breaking strength of a rope when a load is applied along the direction of the length, generally measured in pounds per square inch.

Tension A force that pulls materials apart.

Terra Cotta Tiles composed of clay and sand that are kiln fired. May be structural or decorative.

Terrorism Acts of violence that are arbitrarily committed against lives or property and intended to create fear and anxiety.

Tether Line A rope that is held by a team on shore during a water rescue to be used to haul the rescuer and victim back to shore.

Thermal Burns Burns caused by heat or hot objects.

Thermal Layering The stratification of gases produced by fire into layers based on their temperature.

Thermal Level A layer of air that is of the same approximate temperature.

Thermal Plume A column of heat rising from a heat source. A fully formed plume will resemble a mushroom as the upper level of the heat plume cools, stratifies, and begins to drop outside the rising column.

Thermal Protective Performance (TPP) A rating level, expressed in seconds, used to characterize the protective qualities of a PPE component before serious injury is experienced by the wearer.

Threshold Limit Value (TLV) An exposure value that is similar to the PEL, but is issued by the ACGIH. It is based on an eight-hour day.

Through-the-Lock Method A method of forcible entry in which the lock cylinder is removed by unscrewing or pulling and the internal lock mechanism is operated to open a door. Also, the family of tools used to perform this operation.

Tidal Changes The rising and falling of the surface water levels due to the gravitational effects between the Earth and the moon. In some areas, these changes are insignificant but in others there is more than 40 feet of difference between high and low tide.

Tip Arc The path that a ladder's tip will take while being raised.

Torsion Load A load parallel to the cross section of the supporting member that does not pass through the long axis. A torsion load tries to "twist" a structural element.

Total Flooding System Used to protect an entire area, room, or building by discharging an extinguishing agent that completely fills or floods the area with the extinguishing agent to smother or cool the fire or break the chain reaction.

Tote A large tank usually 250 to 500 gallons, constructed to be transported to a facility and dropped for use.

Tower Ladder An apparatus with a telescopic boom that has a platform on the end of the boom or ladder. It can be extended or retracted and rotated like an aerial ladder.

Toxicity Poisonous level of a substance.

Toxins Disease-causing materials that are extremely toxic and in some cases more toxic than other warfare agents such as nerve agents.

TRACEM An acronym for the types of hazards that exist at a chemical incident: thermal, radiation, asphyxiation, chemical, etiological, and mechanical.

Triage A quick and systematic method of identifying which patients are in serious condition and which patients are not, so that the more seriously injured patients can be treated first.

Triple Combination Engine Company Fire apparatus that can carry water, pump water, and carry hose and equipment.

Truck Company The unit designation of a group of firefighters assigned to perform tactics and functions such as forcible entry, search and rescue, ventilation, and so on.

Truss A rigid framework using the triangle as its basic shape.

Tunnel Vision The focus of attention on a particular problem without proper regard for possible consequences or alternative approaches.

Tunneling The digging and debris removal accompanied by appropriate shoring to safely move through or under a pile of debris at a structural collapse incident.

Turntable The rotating platform of a ladder that affords an elevating ladder device the ability to turn to any target from a fixed position.

Two In/Two Out The procedure of having a crew standing by completely prepared to immediately enter a structure to rescue the interior crew should a problem develop.

Type A Reporting System System in which an alarm from a fire alarm box is received and retransmitted to fire stations either manually or automatically.

Type B Reporting System System in which an alarm from a fire alarm box is automatically transmitted to fire stations and, if used, to outside alerting devices.

Type I, Fire-Resistive Construction Type in which the structural members, including walls, columns, beams, girders, trusses, arches, floors, and roofs, are of approved noncombustible or limited combustible materials with sufficient fire-resistive rating to withstand the effects of fire and prevent its spread from story to story.

Type II, Noncombustible Construction Type not qualifying as Type I construction, in which the structural members, including walls, columns, beams, girders, trusses, arches, floors, and roofs, are of approved noncombustible or limited combustible materials with sufficient fire-resistive rating to withstand the effects of fire and prevent its spread from story to story.

Type III, Ordinary Construction Type in which the exterior walls and structural members that are portions of exterior walls are of approved noncombustible or limited combustible materials, and interior structural members, including walls, columns, beams, girders, trusses, arches, floors, and roofs, are entirely or partially of wood of smaller dimension than required for Type IV construction or of approved noncombustible or limited combustible materials.

Type IV, Heavy Timber Construction Type in which exterior and interior walls and structural members that are portions of such walls are of approved noncombustible or limited combustible materials. Other interior structural members, including columns, beams, girders, trusses, arches, floors, and roofs, shall be of solid or laminated wood without concealed spaces.

Type V, Wood Frame Construction Type in which the exterior walls, bearing walls, columns, beams, girders, trusses, arches, floors, and roofs are entirely or partially of wood or other approved combustible material smaller than the material required for Type IV construction.

Underground Storage Tank (UST) Tank that is buried under the ground. The most common are gasoline and other fuel tanks.

Undesigned Load A load not planned for or anticipated.

Unified Command The structure used to manage an incident involving multiple response agencies or when multiple jurisdictions have responsibility for control of an incident.

Unity of Command One designated leader or officer to command an incident.

Upper Explosive Limit (UEL) The upper part of the flammable range. Above the UEL, fire or an explosion cannot occur because there is too much fuel and not enough oxygen.

Utility Rope Rope used for utility purposes only. Some of the tasks utility ropes are used for in most every fire department are hoisting tools and equipment, cordoning off areas, and stabilizing objects. Also used as ladder halyards.

Vacuum (Negative) Pressure The measurement of the pressure less than atmospheric pressure, which is usually read in inches of mercury (in. Hg or mm Hg) on a compound gauge.

Vapor Density Weight of a gas in relation to air. Air is rated 1.

Vapor Dispersion The intentional movement of vapors to another area, usually by the use of master streams or hoselines.

Vapor Pressure The amount of force that is pushing vapors from a liquid. The higher the force the more vapors (gas) being put into the air.

Vapor Suppression Ability to contain or control the production of fuel vapors.

Vaporization The process in which liquids are converted to a gas or vapor.

Variable, Adjustable, or **Selectable Gallonage Nozzle** Nozzle that allows the nozzleperson to select the flow, with usually two or three choices, and the pattern.

Veins The blood vessels, or tubes, within the body that carry blood lacking oxygen and nutrients back to the heart.

Velocity Pressure The forward pressure of water as it leaves an opening.

Veneer A covering or facing, not a load-bearing wall, usually with brick or stone.

Venous Bleeding Bleeding from a vein.

Venturi Principle A process that creates a low-pressure area in the induction chamber of the eductor and allows the foam concentrate to be drawn into and mix with the water stream.

Vertical Ventilation Channeled pathway for fire ventilation via vertical openings.

Vesicants A group of chemical agents that cause blistering and irritation of the skin. Commonly referred to as blister agents.

Vicarious Experience A shared experience by imagined participation in another's experience.

Visqueen A trade name for black plastic. It can be used very effectively in salvage and overhaul operations.

Voice Inflection Change of tone or pitch of voice.

Voids Spaces within a collapsed area that are open and may be an area where someone could survive a building collapse.

Wall Indicator Valve (WIV) A control valve that is mounted on a wall in a metal case with a small window, reading either "OPEN" or "SHUT."

Watch Dogs Trained dogs that will bark and create a commotion, but will not attack.

Water Columning A condition in a dry pipe sprinkler system in which the weight of the water column in the riser prevents the operation of the dry pipe valve.

Water Curtain Nozzle Designed to spray water to protect exposures against heat by wetting the exposure's surface.

Water Hammer A sudden surge of pressure created by the quick opening or closing of valves in a water system. The surge is capable of damaging piping and valves.

Water Table The level of groundwater under the surface.

Water Tender The term given to land-based water supply apparatus.

Water Thief A variation of the wye that has one inlet and one outlet of the same size plus two smaller outlets with all of the outlets being gated. The standard water thief usually has a 2½-inch (65-mm) inlet with one 2½-inch (65-mm) and two 1½-inch (38-mm) outlets.

Waybill A term that may be used in conjunction with consist, but is a description of what is on a specific railcar.

Weapon of Mass Destruction (WMD) A term that is used to describe explosive, chemical, biological, and radiological weapons used for terrorism and mass destruction.

Web The vertical portion of a truss or I beam that connects the top chord with the bottom chord.

Web Gear The term given to a whole host of personal items carried on a belt/harness arrangement worn by wildland firefighters. Items include water bottles, a fire shelter, radio, and day sack.

Webbing Nylon strapping, available in tubular and flat construction methods.

Webbing Sling Approximately 12 to 15 feet of rescue webbing tied end to end, forming a continuous loop.

Western or **Platform Framing** A style of wood frame construction in which each story is built on a platform, providing fire stopping at each level.

Wet Chemicals Extinguishing agents that are water-based solutions of potassium carbonate–based chemicals, potassium acetate–based chemicals, or potassium citrate–based chemicals, or a combination.

Wet Pipe Sprinkler System Has automatic sprinklers attached to pipes with water under pressure all the time.

Wheatstone Bridge Sensor A type of combustible gas sensor that uses a heated coil of wire to determine the presence of flammable gases.

Wire Glass Glass with a wire mesh embedded between two or more layers to give increased fire resistance.

Work Hardening A phrase given to the effort and physical training designed to prepare an individual to better perform the physical tasks that are expected of the individual. Work hardening is key in preventing injuries resulting from typical firefighting tasks.

Working End The end of the rope that is utilized to secure/tie off the rope.

Working Length The length of the ladder that spans the distance from the ground to the point of contact with the structure. This does not include any distance the ladder might go beyond the point of contact as would be the case when the tip extends beyond the roof.

Wye A device that divides one hoseline into two or more. The wye lines may be the same size or smaller size and the wye may or may not have gate control valves to control the water flow.

Zone An area established and identified for a specific reason, typically because a hazard exists within the zone. The zones are usually referred to as hot, warm, and cold zones and provide an indication of the expected hazard in each zone. Sometimes referred to as a sector.

Zoning A term given to the establishment of specific hazard zones; that is, hot zone, warm zone, cold zone. Also collapse zones.

ACRONYMS

ACGIH American Conference of Governmental Industrial Hygienists
ACS American Chemical Society
AFFF aqueous film–forming foam
AHA American Heart Association
ALOHA aerial location of hazardous atmospheres
ALS advanced life support
ANFO ammonium nitrate and fuel oil
APCO Association of Public Safety Communications Officials—International, Inc.
APR air-purifying respirators
ARFF aircraft rescue and firefighting
AST aboveground storage tanks
ASTM American Society for Testing and Materials
ATC alcohol-type concentrates
ATF Alcohol, Tobacco, and Firearms
BLEVE boiling liquid expanding vapor explosion
BLS basic life support
BNICE biological, nuclear, incendiary, chemical, and explosive
BOCA Building Officials Conference Association
BSI body substance isolation
Btu British thermal unit
CAAA Clean Air Act Amendment
CAD computer-aided drawing; computer-aided dispatch
CAFS compressed air foam systems
CAMEO Computer-Aided Management for Emergency Operations
CANUTEC Canadian Transportation Emergency Center
CBDCOM Chemical and Biological Defense Command
CBIRF Chemical and Biological Incident Response Force (Marines)
CERCLA Comprehensive Environmental Response, Compensation, and Liability Act
CFR *Code of Federal Regulations*
CGI combustible gas indicators
CHEMTREC Chemical Transportation Emergency Center
CHRIS Chemical Hazards Risk Information System
CIS critical incident stress
CISD critical incident stress debriefing
CISM critical incident stress management
CMA Chemical Manufacturers Association
CMS chip measurement system
CNS central nervous system
CPR cardiopulmonary resuscitation
DCM dangerous cargo manifest
DECIDE detect, estimate, choose, identify, do the best, evaluate
DNR Department of Natural Resources
DOT Department of Transportation (U.S.)
EAP Employee Assistance Program
EDITH Exit Drills In The Home
EHS extremely hazardous substance
EMS emergency medical services
EMT emergency medical technician
EMT-P emergency medical technician-paramedic
EPA Environmental Protection Agency
EPCRA Emergency Planning and Community Right to Know Act
ERG Emergency Response Guidebook
ERP emergency response planning
FAST firefighter assist and search team
FBI Federal Bureau of Investigation
FCC Federal Communications Commission
FD fire department
FEMA Federal Emergency Management Agency
FFFP fluoroprotein film–forming foam
GEDAPER gather information, estimate potential, determine goals, assess tactical options, plan, evaluate, review
gpm gallons per minute
HAZMAT hazardous materials
HAZWOPER hazardous waste operations and emergency response
HCS hazard communication standard
HFT hydraulic forcible entry tools
HIV human immunodeficiency virus
HMIS Hazardous Materials Information System
HMRU Hazardous Materials Response Unit (FBI)
HVAC heating, ventilation, and air conditioning
IAFF International Association of Fire Fighters
IAP incident action plan
IC incident commander
ICt_{50} incapacitating concentration to 50 percent of the population, with "t" representing time, usually expressed in minutes
IDLH immediately dangerous to life and health
IDR instant detection and response
IFSAC International Fire Service Accreditation Congress
IM intermodal
IMS Incident Management System
IR infrared

IRS Internal Revenue Service
ISC Industrial Scientific Corporation
kPa kilopascals
L/min liters per minute
LC_{50} lethal concentration to 50 percent of the population
LCt_{50} lethal concentration to 50 percent of the population, with "t" representing time, usually expressed in minutes
LD_{50} lethal dose to 50 percent of the population
LEL lower explosive (flammable) limit
LEPC Local Emergency Planning Committee
LOX liquid oxygen
LSA low specific activity (radiation)
LUST leaking underground storage tank
MCI mass casualty incident
MOS metal oxide sensor
mph miles per hour
MSA mine safety appliances
MSDS Material Safety Data Sheet
MSV mobile support vehicle
NBC nuclear, biological, and chemical
NBFU National Board of Fire Underwriters
NENA National Emergency Number Association
NFPA National Fire Protection Association
NHT national hose thread
NIOSH National Institute for Occupational Safety and Health
NOS not otherwise specified
NP nozzle pressure
NPL national priority list
NPQB National Professional Qualifications Board
NR nozzle reaction
NRC National Response Center
NST national standard thread
OPA Oil Pollution Act
ORM-D other regulated material, Class D
OS&Y outside stem and yoke valve
OSHA Occupational Safety and Health Administration
P polymerization hazard
PAR personnel accountability report
PASS personal alert safety system
PDD 39 Presidential Decision Directive 39
PEL permissible exposure limit
PFD personal flotation device
PG I, II, or III packing group I, II, or III
PID photo-ionization detector
PIH poison inhalation hazard
PIO public information officer
PNS peripheral nervous system
PPE personal protective equipment
ppm parts per million
PPV positive pressure ventilation
PSA public service announcements
PSAP public safety answering point
psi pounds per square inch
psia pounds per square inch absolute
psig pounds per square inch gauge
PVC polyvinyl chloride
R&I recognition and identification
RAID Rapid Assessment and Initial Detection (team)
RECEO rescue, exposures, confinement, extinguishment, overhaul
REL recommended exposure limit
REVAS rescue, exposures, ventilation, attack, and salvage
RIT rapid intervention team
RQ reportable quantity
SADT self-accelerating decomposition temperature
SAR supplied air respirator
SARA Superfund Amendments and Reauthorization Act
SBIMAP South Baltimore Industrial Mutual Aid Plan
SCA surface contaminated articles (radiation)
SCBA self-contained breathing apparatus
SERC State Emergency Response Committee
SETIQ Emergency Transportation System for the Chemical Industry
SOG standard operating guideline
SOP standard operating procedure
SRS supplemental restraint systems
SSP standard sprinkler pendent
SSU standard sprinkler upright
STCC Standard Transportation Commodity Code
STEL short-term exposure limit
TC Transportation Canada
TEU Technical Escort Unit
TIA Tentative Interim Amendment
TLV threshold limit value
TOFC trailers on flat cars
TOG turnout gear
TPP thermal protective performance
TRACEM thermal, radiation, asphyxiation, chemical, etiological, and mechanical hazards
UEL upper explosive (flammable) limit
UN/NA United Nations/North America
UPS uninterruptible power supply
USAR Urban Search and Rescue (team)
UST underground storage tank
UV ultraviolet
VTR violent tank rupture
WIV wall indicator valve
WMD weapons of mass destruction

OTHER FIRE SCIENCE TITLES FROM DELMAR

Fundamentals and Suppression

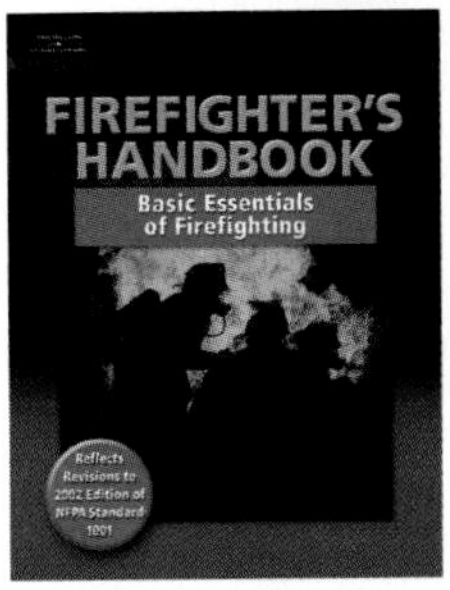

Delmar's Firefighter's Handbook: Basic Essentials of Firefighting

This briefer version of Delmar Learning's *Firefighter's Handbook* features important and up-to-date information about today's fire service without delving into the related topic of hazardous materials. It covers the critical details that apply to the job of today's firefighter, including the basic requirements of Firefighter I and II. The book is ideal for fire departments, academics, and schools in which hazardous materials are taught in a separate course with separate learning materials.

Order #: 1-4018-3582-1

Delmar's Basic Firefighting Video Series and CD-ROM Courseware

The perfect complement to *The Firefighter's Handbook,* this set of four videos leads viewers, step by step, through processes and procedures used every day by highly trained firefighters on the job. The series uses a unique blend of simulated action sequences, professional quality animations, and helpful graphics to assist viewers in acquiring vitally important technical knowledge and hands-on skills—safely and efficiently.

Order #: Videos: 0-7668-4099-9,
CD-ROM Courseware: 0-7668-4104-9

***Introduction to Fire Protection,* 2nd ed./Klinoff**

This book offers a complete introduction to the field of fire protection, technology, and the wide range of services provided by both public and private fire departments of today. It covers fighting fires and the provisions of other emergency services, hazardous materials control, fire prevention, and public education.

Order #: 0-7668-4958-9

***Principles of Fire Behavior*/Quintiere**

While explaining the science of fire with a precision found nowhere else, this text applies science to fire safety design and investigation. Using a quantitative approach, it presents an ideal introduction to the scientific principles behind fire behavior.

Order #: 0-8273-7732-0

***Hydraulics for Firefighting*/Crapo**

Hydraulics for Firefighting leads readers throughout the principles, theory, and practical application of fire service hydraulics. This book is written in a format that will help guide the new firefighter through even the most technical hydraulic principles and complex laws of physics. The author takes care to explain theories in ways that are easily understood by anyone with knowledge of basic algebra.

Order #: 0-7668-1905-1

***Introduction to Fire Pump Operations*/Sturtevant**

Here's the book that offers students and professional fire pump operators the updated knowledge required to efficiently, effectively, and safely operate and maintain fire pumps. With an emphasis on NFPA standards and safety, the book is logically presented in three sections: Pump Construction/Peripherals, Pump Procedures, and Water Flow Calculations.

Order #: 0-8273-7366-X

***Firefighting Strategies and Tactics*/Angle, Gala, Harlow, Lombardo, and Maciuba**

This book is a complete source for learning firefighting strategies and tactics, from standard company responsibilities and assignments to specialized situational strategies and tactics. The reader will progress from basic concepts to the application of tactics and situational strategies for particular occupancies or types of fires. This book was written in an easy-to-follow manner. It is presented in a fashion that can be universally applied in all areas of the country, rural to urban.

Order #: 0-7668-1344-4

***Explosives Identification Guide*/Pickett**

This is a reference guide to explosives for emergency responders such as firefighters, police officers, and EMS staff as well as security personnel. Through color photographs and short descriptions, the student can identify explosives by general type and learn the appropriate way to treat each of them. Written in a general, nontechnical style, the book is a fast and easy guide for those with little or no knowledge of, or experience with, explosives.

Order #: 0-7668-0490-9

***Wildland Firefighting Practices*/Lowe**

The reader will learn in detail all aspects of wildland firefighting with this new, well-illustrated text. Written in a clear, how-to style by a seasoned wildland fire officer, it provides a comprehensive explanation of all the skills a firefighter needs to operate effectively against any type of wildland blaze.

Order #: 0-7668-0147-0

Terrorism

***Preparing for Terrorism: An Emergency Services Guide*/Buck**

This text helps the reader develop the skills for dealing with terrorism on many levels: preparing and planning for a terrorist attack, mitigating its effects, proper emergency response, and recovery from terrorism disasters. It is an essential guide to the planning and implementation of antiterrorist response and operations for the overall safety of the first responder.

Order #: 0-8273-8397-5

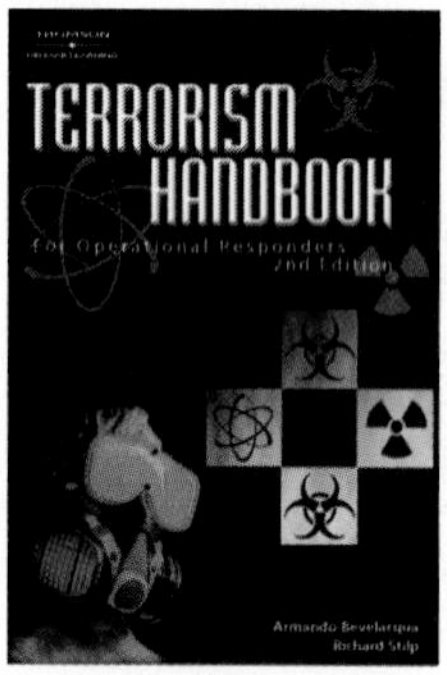

***Terrorism Handbook for Operational Responders*, 2nd ed./Bevelacqua and Stilp**

This updated book is a guide into the most significant points that surround the emergency response processes needed to cope with terrorism incidents. It highlights new equipment and strategies that can enhance a responder's detection, monitoring, and protection capabilities against chemical and biological agents. First responders are provided with the knowledge they need to prepare for and combat acts of terrorism.

Order #: 1-4018-5065-0

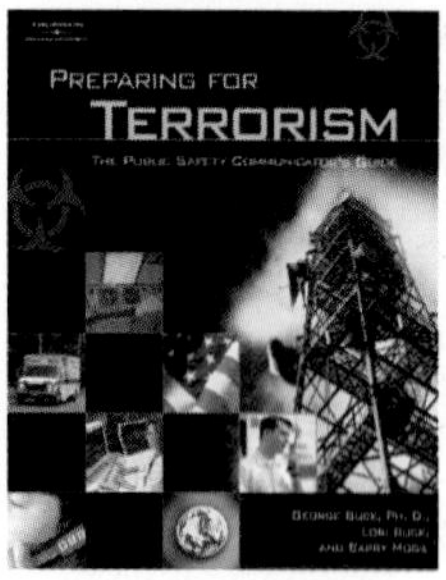

***Preparing for Terrorism: The Public Safety Communicator's Guide*/Buck, Buck, and Mogil**

This book starts with an overview of national and international terrorism, while emphasis throughout the book is on how to prepare communications center staff and their families for a terrorist event by providing them with a well-thought-out employee emergency plans and contingencies. Solutions to communications problems, such as cellular and landline telephone overload situations, are addressed as well.

Order #: 1-4018-7131-3

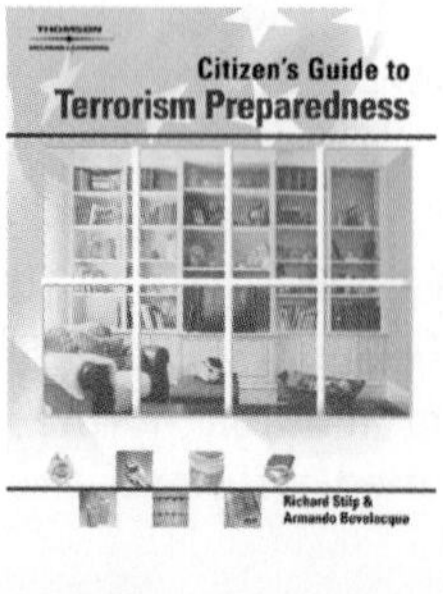

***Citizen's Guide to Terrorism Preparedness*/Stilp and Bevelacqua**

This book provides readers with facts, figures, and practical guidelines to follow as they go about their daily lives. It is designed specifically for average citizens who want to take all of the steps they can to prepare themselves for a terrorist act in their state, city, or neighborhood.

Order #: 1-4018-1474-3

***Preparing for Biological Terrorism: An Emergency Services Guide*/Buck**

This book contains vitally important information to guide local agencies in their efforts to secure and coordinate the influx of state and federal resources before, during, and after an attack. This resource walks through the fundamental concepts of emergency planning. Subsequent chapters enable readers to immerse themselves thoroughly in specific elements of successful emergency planning.

Order #: 1-4018-0987-1

***Explosives Identification Guide*/Pickett**

This book is a reference guide to explosives for emergency responders such as firefighters, police officers, and EMS staff as well as security personnel. Through color photographs and short descriptions, the student can identify explosives by general type and learn the appropriate way to treat each of them.

Order #: 0-7668-0490-9

HAZMAT

***Hazardous Materials Incidents*, 2nd ed./Hawley**

Hazardous Materials Incidents is an invaluable procedural manual and all-inclusive information resource for emergency services professionals. Easy-to-read and perfect for use in HAZMAT awareness, operations, and technician-level training courses, this "Operations Plus" book begins by acquainting readers with current laws and regulations, including those governing emergency planning and workplace safety.

Order #: 1-4018-5758-2

***Hazardous Materials Air Monitoring and Detection Devices*/Hawley**

This book provides HAZMAT teams with a thorough guide to effective air monitoring in emergency response situations. Each type of air monitoring devices available for emergency services is described in detail, including operating guidelines and sampling strategies. A special section discusses the latest trend in HAZMAT detection and monitoring of hazardous agents used for terrorism.

Order #: 0-7668-0727-4

***Hazardous Materials Response and Operations*/Hawley**

While presenting an in-depth look at the response to hazardous materials releases, this book covers each class of hazardous materials and provides recommenda-

tions and guidelines for the protection of responders and victims. The text focuses on a cutting-edge response profile called Risk-Based Response, known for its progressive and aggressive approach to dealing with chemical spills. Also included is a detailed discussion of response strategies to terrorism, preparing students to be leaders in the newest area of the HAZMAT field.

Order #: 0-7668-1150-6

Hazardous Materials Chemistry/Bevelacqua

Hazardous Materials Chemistry covers the basic concepts of chemistry, emphasizing the decision-making process so that appropriate strategies and tactics will be chosen.

Order #: 0-7668-1434-3

Hazardous Materials Field Guide/Bevelacqua and Stilp

Whether the incident involves hazardous materials, a clandestine laboratory, terrorism, or a confined space operation, this user-friendly resource includes information that is consistent with the mission of all agencies. The guidebook's easy access format allows rapid identification of placards, labels, silhouettes, and common commodities that move on roadways and railways. Medical considerations are described throughout the text, identifying potential needs of an affected community.

Order #: 0-7668-0155-1

Hazardous Materials: Regulations, Response, and Site Operations/Schnepp and Gantt

This essential guide provides the student with a practical approach to the concepts of handling hazardous materials. Based on OSHA "HAZWOPER" regulations, this invaluable text addresses the specific competencies required of persons responding to a hazardous materials emergency.

Order #: 0-8273-7999-4

Emergency Medical Response to Hazardous Materials Incidents/Stilp and Bevelacqua

Medical aspects of hazardous materials response including the initial response, chemical and toxicological information, and effects on body systems—injury and treatment, physiology, and treatments of common poisonings—are explained in this book. The student will learn how to make decisions based on a scale of risk versus gain.

Order #: 0-8273-7829-7

Emergency Decontamination for Hazardous Materials Responders/Henry

This one-of-a-kind book focuses entirely on decontamination, a crucial aspect of hazardous materials emergency response. The book brings together facts about chemical contamination gathered over the last ten years and presents them in a simple, streetwise way.

Order #: 0-7668-0693-6

Hazardous Materials Air Monitoring/Detection Devices/Hawley

This text provides HAZMAT students with a thorough guide to effective air monitoring in emergency response situations. The key component to safely responding to hazardous materials is the use of a variety of detection devices for effective air monitoring. The most up-to-date information is provided along with discussion of future trends and rapidly changing technology in this field.

Order #: 0-7668-0727-4

Hazardous Materials Response and Operations/Hawley

This text presents an in-depth look at the response to hazardous materials releases, covering each class of hazardous materials, and providing recommendations and guidelines for the protection of responders and victims. The text focuses on a cutting-edge response profile called risk-based response, known for its progressive and aggressive approach to dealing with chemical spills.

Order #: 0-7668-1150-6

Inspection, Investigation, and Fire Law

Fire Prevention: Inspection and Code Enforcement, 2nd ed./Diamantes

This is a vital resource for the application of building and fire prevention codes in the inspection of buildings and facilities and for compliance through the code enforcement process. Issues such as enforcement authority, determining inspection priorities, maintenance of rated assemblies, fire protection systems, storage occupancies, detonation and deflagration hazards, and hazardous materials storage and processing are covered in depth in this comprehensive guide.

Order #: 0-7668-5285-7

Fire and Emergency Law Casebook/Schneid

Fire and emergency personnel are provided with the information about potential legal liabilities encountered every day. Actual cases are presented in detail and followed by explanations that identify the most important legal issues facing fire departments, EMS, and related organizations.

Order #: 0-8273-7342-2

Rescue

Confined Space Rescue/Browne and Crist

Confined space rescues present unique hazards to victims and emergency service workers. This new, clearly written book identifies the problems, showing users how to address them and how to rescue a victim. With this book, firefighters, police officers, EMS personnel, emergency medical staff, and industrial rescue teams can learn a simple set of skills that will provide a foundation for growth into advanced rescue operations.

Order #: 0-8273-8559-5

Engineering Practical Rope Rescue Systems/Brown

Beginners and advanced rescue technicians will learn from and enjoy this entertaining and educational book. A practical look at rope rescue systems from the point of view of an experienced professional, each chapter features exciting stories and real-life situations. The book provides a complete review of team integrity and development issues as well as team efficiency concepts that create a superior survival profile.

Order #: 0-7668-0197-7

Fire Protection Systems

Design of Water-Based Fire Protection Systems/Gagnon

A vital reference for every inspector and designer of fire protection, sprinkler, architectural, or engineering systems, this book is a must. Hydraulic calculations for the most commonly encountered water-based fire protection systems are covered in detail. Manual hydraulic calculations are thoroughly covered, and a computer disk is included to provide the reader with the opportunity to calculate a wide variety of systems.

Order #: 0-8273-7883-1

Design of Special Hazard and Fire Alarm Systems/Gagnon

As the most current guide to the design of state-of-the-art special hazard and fire protection systems, this book is essential to architects, engineers, layout technicians, plumbers, mechanical contractors, and sprinkler firms. Using the most up-to-date NFPA standards and reference data, it guides the reader through the steps needed to design a wide variety of simple and complex systems.

Order #: 0-8273-8293-6

Officer

Coming Soon! Second Edition, Fall 2004 (Order #: 1-4018-2605-9)

Company Officer/Smoke

Any firefighter that wants to gain certification as a Fire Officer will find this practical guide an excellent resource. Based on the latest information and requirements outlined in NFPA 1021, the Standard for Fire Officer Professional Qualifications, the book gives the user the information necessary to meet NFPA Standard competencies for certification as Officer I, the first-line supervisor.

Order #: 0-8273-8472-6

Going for the Gold/Coleman

Author Ronny Coleman, Chief Deputy Director of the California Department of Forestry and Fire Protection, offers a unique, must-have resource for the thousands of individuals who hope to carry the fire chief's badge. It covers the specifics of how a person actually applies leadership and decision-making concepts on a daily basis. The book provides a realistic appraisal of what it takes to aspire for, achieve, and then succeed as fire chief.

Order #: 0-7668-0868-8

Occupational Health and Safety in the Emergency Services/Angle

A comprehensive approach to program management for fire and emergency service occupational safety and health is provided in this new, practical book. Safety officers and fire department and EMS managers will make good use of this one-stop resource.

Order #: 0-8273-8359-2

Fire Department Incident Safety Officer/Dodson

This is the only book that provides a clear, focused, and detailed approach to making a difference as an incident safety officer. Company officers, battalion chiefs, safety officers, and incident commanders will benefit from the foundation material and the incident safety officer action model presented in this book.

Order #: 0-7668-0362-7

Reference

Firefighter Exam, 2nd ed./Learning Express

This new edition has been completely revised and updated to reflect changes in national firefighter tests. With this book, firefighter candidates will be able to sharpen their skills with four practice tests. Instructional chapters target the basic skills tested on firefighter exams—math, reading comprehension, grammar, vocabulary, memory, judgment, and recall.

Order #: 1-57685-440-X

A Practical Guide to Teaching in the Fire Service/Morse

Designed for the person without formal training or a degree in education who is found teaching in front of a classroom, this how-to book is the perfect resource. Its no-nonsense approach covers the day-to-day information that is needed to conduct a successful class in fire service.

Order #: 0-7668-0432-1

***Practical Problems in Mathematics for Emergency Services*/Sturtevant**

This is the only math-related text specifically written for the emergency service field. Today, most certifications in the emergency service field require written exams that include related math problems. Designed for such exams, this book may be used as a preparation for certification and promotional exams, as well as a quick reference for the seasoned professional.

Order #: 0-7668-0420-8

Codes

2003 International Fire Code

The 2003 International Fire Code references national standards to comprehensively address fire safety in new and existing buildings. Topics addressed include fire department access, fire hydrants, automatic sprinkler systems, fire alarm systems, hazardous material storage, and fire-safety requirements for buildings.

Order #: 1-4018-5074-X

2003 International Building Code

The 2003 International Building Code addresses design and installation of building systems with requirements that emphasize performance. The IBC is coordinated with all the International Codes including structural, fire, and life-safety provisions covering seismic, wind, accessibility, egress, occupancy, roofs, and more.

Order #: 1-4018-5073-1

Emergency Medical Services

***First Responder Handbook: Fire Service Edition*/Walter, Rutledge, and Edgar**

This book covers safety for first responders, patient care principles, emergency response tactics, and first responder actions for incidents involving terrorism. This street-smart book fulfills the U.S. Department of Transportation's requirements for first responder certification while offering firefighters valuable emergency medical training.

Order #: 0-7668-3919-2

***First Responder Handbook: Law Enforcement Edition*/Walter, Rutledge, Edgar, and Davis**

This book is for anyone in law enforcement and provides valuable emergency medical training to potential first responders to an emergency scene.

Order #: 0-7668-4191-X

INDEX

A-frame hoist, 389
A-frame ladder, 379–380
Abandonment, 694
Aboveground storage tank (AST), 792
Abrasion, 712
Absolute pressure, 290
Absorption, 885, 892–893
Accelerator, 320
Access, 583
Accident, 110
Accident chain, 110
Accident prevention, 110, 111
Acclimation, 137
Accordion load, 237, 238, 239
Accountability system, 725
Acetane, 91
Acetone, 797, 800, 840
Acetylene, 517
ACGIH, 838
ACGIH-TWA, 838
Acrolein, 551, 840
Active cooling, 731
Acute exposure, 834
Adams, Samuel, 9
Adams Rite lock, 529
Adapter, 228
Adjustable gallonage nozzle, 283
Administrative warrant, 662
Adsorption, 885
Advanced life support (ALS) care, 694
Advancing flat load, 245, 246
Advancing hoseline at entry point of building, 260
Advancing hoselines, 251–261
 into structures, 253
 standpipe system, 255–257
 up/down stairs, 254–255
 working hose off ladders, 257–261
AED, 708
Aerial fire apparatus, 30
Aerial ladder, 373–375
AFFF, 296
AGCGIH-TWA, 838
Agricultural chemicals, 875
Agricultural supply stores, 756
Air bills, 826
Air chisels, 485
Air consumption, 169
Air cylinders, 153–155. *See also* Self–contained breathing apparatus (SCBA)
Air monitor configurations, 899–900
Air monitoring, 898–906
 bump test, 900
 calibration, 900–901
 catalytic bead, 904
 colorimetric sampling, 906
 EPA guideline, 903
 FGI, 903–904
 infrared sensor, 904–905
 ISC TMX-412 factors, 901
 MOS, 904
 oxygen monitors, 901–902
 PID, 905
 reaction time, 901
 recovery time, 901
 regulations/standards, 899
 relative response, 901
 toxic gas monitor, 905
Air movement, 562–563
Air pressure, 87
Air-purifying respirator (APR), 841
Alarm and monitoring systems, 331
Alarms. *See* Communications and alarms
Allergic reactions, 714–715
Allied agencies and organizations, 35–36
ALOHA, 826
ALS care, 694
Aluminum cylinders, 154
Ambulance stretchers, 480–481
America Burning, 18
American Conference of Governmental Industrial Hygienists (ACGIH), 838
American fire helmet, 11
Ammonia, 551, 840
Ammonium nitrate, 874
Ammonium nitrate and fuel oil (ANFO), 761, 866
Amputation, 500, 712
Ancient history, 5–6
ANFO, 866
ANFO explosion, 761
Angle, James, 46, 592
Anthrax, 930
Anthrax hoaxes, 930–931
Anthrax scare (2001), 930
APCO, 48
Appliance friction loss, 294
Application rate, 295
APR, 841
Aqueous film forming foam (AFFF), 296
Aquifer, 203
Army stretcher, 480
Arrival reports, 67–68
Articulating boom ladder, 376
Artillery, 271
Asphyxiants, 835–836
Asphyxiation, 145
Aspon-Chlordane, 814
Assessing the patient, 700–707. *See also* Medical examination
Assistant chief, 32
Association of Public Safety Communications Officials - International (APCO), 48
AST, 792
Atmospheric pressure, 212, 213
Atomization, 98
Atoms, 79, 80
Attack dogs, 554
Attack hose, 222
Attack line, 244
Attic ladder, 379, 380
Authority having jurisdiction, 148
Autoignition temperature, 800
Automated external defibrillators (AED), 708
Automatic alarm systems, 56–57
Automatic hose washer, 225
Automatic nozzles, 283
Automobile, extrication from. *See* Extrication from motor vehicles
Automobile fire, 599–601, 622–625
Autorefrigeration, 782
Auxiliary appliances, 309
Auxiliary locking devices, 532
Auxiliary protective signaling system, 56
Available flow, 291
Avulsion, 712
Awareness level, 748
Awning windows, 541, 543
Ax, 515, 516, 571
Axial load, 345

Backboard, 478–480
Backdraft, 100, 556–560
Backflow preventers, 209
Backpack/harness assembly, 152, 153
Backpack pump tank fire extinguisher, 189, 190
Backstretch, 244
Bagging a rope, 447–449
Bags, 773–774
Baker fire (Southern California), 595
Balanced pressure demand-type foam proportioner system, 300
Balloon frame, 356–357
Balloon toss, 644, 645
Bam bam, 518
Bangor ladders, 377, 379
Bank-back technique, 303
Bank-in technique, 303
Bar joist, 361
Bars and gates (windows), 544
Base radio, 58
Basement fire, 618–619
Basic 9-1-1, 54
Basic apparatus requirements, 30
Basic figure eight knot, 433, 434
Basic life support (BLS) care, 694
Battalion/district chief, 32
Battering rams, 513
Battery-powered reciprocating saw, 486
Beam, 346, 347
Beam raise, 400, 402, 403–404
Becket bend, 429–430
Bed ladder, 372, 373
Below-grade rescue, 497–498

Below-ground fire (basement), 618–619
Belt, 814
Benner, Ludwig, Jr., 853
Bight, 426
Biohazard containers, 700
Biohazard labels, 766
Biological, nuclear, incendiary, chemical, and explosive (BNICE), 927
Biological agents, 930
Biological weapons lab, 923, 925
Biomimetic, 908
Blanket drag, 469–472
Bleeding control, 709–712
BLEVE, 88–89, 626, 782–784, 870
Blister agents, 929
Blood agents, 929
Blood vessel system, 709
BLS care, 694
BNICE, 927
BOCA, 757
Body fluids, 698
Body substance isolation (BSI) precautions, 697–700
Boiling liquid expanding vapor explosion (BLEVE), 88–89, 626, 782–784, 870
Boiling point, 89, 795, 796
Bolt cutters, 515, 648
Bomb threat standoff distances, 868
Bond, 80
Booster hose, 222
Boots, 129
Boston Vendome hotel fire (1972), 343
Bottles, 772–773
Bounce-off technique, 302
Bourdon gauge, 214
Bowline, 431–432
Box trucks, 776. *See also* Tank trucks
Brachial artery, 703, 704
Braid-on-braid rope, 423
Braided rope, 422–423
Brannigan, Francis L., 348
Breaching floors, 545–546
Breaching walls, 544–545
Break-apart nozzle, 265, 267
Breaking glass, 568–569
Breathing problems, 714
Bresnan distributor, 286
Brick, 349
British thermal unit (Btu), 77
Brush gear, 130
BSI precautions, 697–700
Btu, 77
Buddy breathing attachment, 171
Buddy system, 164
Building collapse
  collapse signs, 364
  collapse zone, 365
  factors to consider, 346
  famous collapses, 343–344
  masonry walls collapse, 350
  preparing for collapse, 364–365
  rescue, 494–497
  types, 495–496
  typical hazards, 360
Building construction, 339–368
  beams, 346
  building types, 352–357
  buildings under construction, 364
  collapse. *See* Building collapse
  columns, 346
  composites, 350–351
  concrete, 349–350
  connections, 347
  doors, 522–527
  fire effects, 347–352
  forces, 345
  historically significant building collapses, 343–344
  loads, 342–345
  locks, 526–530
  masonry, 350
  occupancy use, 359–360
  other construction types, 357–359
  parapet walls, 363
  roof structures, 362–363
  stairs, 363
  steel, 349
  trusses, 361
  void spaces, 362
  walls, 347
  wood, 348–349
Building loading, 342
Building Officials Conference Association (BOCA), 757
Building size, 587
Building types, 352–357
Buildings under construction, 364
Bulk storage tanks, 790–794
Bulk tanks, 775
Bump test, 900
Bunkers, 127
Burning process, 91–93
Burning tanker trucks, 872
Burns, 715–717
Burst hose, 268–269
Bus accident, 862
Business/mercantile fire, 615–616
Butane, 91
Butterfly roof, 362
Bylaws, 35
Bypass eductor, 298
Bypass eductor bypass valve, 299

CAD systems, 51
CAFS, 299, 301
Calcium carbide, 873
Calibration, 900–901
Call-taking process. *See* Receiving reports of emergencies
Callan, Mike, 832
CAMEO, 826
Camera, 637
Candy-striper, 790
Cantilever beam, 346
CANUTEC, 827
Capillary bleeding, 710
Capillary refill, 707
Carbide-tipped blade, 517
Carbon dioxide (CO_2), 146, 188, 333, 551, 555, 870
Carbon dioxide detectors, 907
Carbon dioxide extinguisher, 192
Carbon monoxide (CO), 91, 146–148, 551, 555, 836, 840, 899
Carbon monoxide (CO) detectors, 313, 678, 907
Carbon monoxide incidents, 906–908
Carbon monoxide poisoning, 147
Carboys, 773
Carcinogen, 834
Card key entry systems, 531
Cardboard boxes, 772
Cardiopulmonary resuscitation (CPR), 707–708
Cardiovascular conditioning, 727
Cardiovascular system, 709
Carotid pulse, 703, 704
Carries, 468–469
Carry-all, 649
Carrying ladders, 395–399
Cartridge-operated dry chemical extinguisher, 193
Cartridge-operated extinguishers, 189–190
Cascade system, 156, 177–179
Casement windows, 541, 543, 569
Cast iron, 349
Catalytic bead sensor, 904
Catch-all, 645–646
CBIRF, 934
CD 68, 814
CDC, 696
Ceiling levels, 838
Cellar nozzle, 286
Cellulose, 79
Celsius scale, 78
Center rafter louver, 577–579
Centers for Disease Control and Prevention (CDC), 696
Centimeter, 78
Central nervous system (CNS) chemicals, 836
Central station protective signaling system, 56
CERCLA, 750
CFR, 109
Chain saws, 517
Charring, 654
Check rail windows, 543
Check valves, 209
Chemical and Biological Incident Response Force (CBIRF), 934
Chemical and physical properties - hazardous materials, 795–801
  autoignition temperature, 800
  chemical reactivity, 799–800
  corrosivity, 798–799
  flammable range, 800
  flash point, 800
  specific gravity, 798
  states of matter, 795–796
  toxic products of combustion, 800–801
  vapor density, 797
  vapor pressure, 796–797
Chemical asphyxiants, 836

Chemical burns, 716–717
Chemical inventory report, 746–747
Chemical names, 812, 814
Chemical protective clothing, 842–846
Chemical reaction, 83
Chemical reactivity, 799–800
Chemical release/breach, 864–866
Chemical stress, 865
Chemical weapons labs, 925
Chemtrec, 826–827
Chernobyl nuclear disaster, 85
Chest pain, 714, 715
Chief of department, 32
Chief officers, 32
Childress, Dennis R., 591
Chimney, 585
Chip measurement system (CMS), 905
Chloordaan, 814
Chlor kil, 814
Chlordan, 814
Chloridan, 814
Chlorine, 82, 840, 862
Chlorodane, 814
Chloroform, 765
Chlortox, 814
Choking agents, 929
Chord, 346
Chronic exposure, 834
Circuit breaker, 672
CISD, 737–738
CISM, 116, 737–738
Cistern, 205
Civil War, 13–14
Clandestine labs, 922–923
Class 1 harness, 491
Class 2 harness, 491
Class 3 harness, 491, 492
Class A extinguishers, 193–194
Class A fire, 100, 101, 185
Class A foam, 296–297
Class B extinguishers, 194
Class B fire, 100, 101, 185–186
Class B foam, 296, 297
Class C extinguishers, 194
Class C fire, 100, 101, 186
Class D extinguishing powder, 873
Class D fire, 186
Class D heavy metal fire, 100, 101
Class I standpipe system, 329, 330
Class II standpipe system, 329, 330
Class III standpipe system, 329
Class K fire, 100, 186–187
Claw tool, 513
Clean Air Act Amendments (CAAA), 750
Climbing path, 391
Clipping, 65
Clordan, 814
Clorodane, 814
Closed-circuit SCBA, 158
Closet ladder, 379, 380
Clothing drag, 471, 473
Clove hitch, 427–429
CMS, 905
CNG sticker, 871
CNS chemicals, 836
CO, 91, 146–148, 551, 555, 836, 840, 899
CO detectors, 313, 678, 907
CO_2, 146, 188, 333, 551, 555, 870
Coat method (donning SCBA), 160, 162, 163
Coats, 128
Cockloft, 561–562
Code of Federal Regulations (CFR), 109
Cohortes Vigilum, 6
Coiling a rope, 445–447
Cold zone, 860
Coleman, Ronny J., 2
Collapse signs, 364
Collapse zone, 365
Colorimetric sampling, 906
Colorimetric tubes, 905
Column, 346, 347
Combination A-frame ladder, 379–381
Combination attack, 608
Combination fire attack, 290
Combination fog nozzle, 283, 288
Combination headsets, 134
Combination tools, 483, 484
Combination units, 31
Combined gravity-pumped water distribution system, 206
Combustible heavy metals, 101, 102
Combustible liquid, 762
Combustion, 85, 86
Command staff positions, 41–42
Commercial cooking surfaces, 667–668
Commercial garage doors, 526
Common incidents - hazardous materials, 864–877
  chemical release/breach, 864–866
  corrosives, 876
  explosives, 866–868
  flammable/combustible liquids, 871–872
  flammable solids, 872
  gas leaks (buildings), 877
  gases, 868–871
  gasoline, 871–872
  oxidizers/organic peroxides, 874
  poisonous gases, 874–875
  reactive materials, 875–876
  spontaneously combustible materials, 873
  water reactives, 873
Common terminology, 37
Communications and alarms, 45–73
  arrival reports, 67–68
  communications facility, 49–51
  communications personnel, 48–49
  computers, 51
  emergency services deployment, 58–62
  MSVs, 68
  radio systems and procedures, 63–67
  receiving reports of emergencies, 51–58. *See also* Receiving reports of emergencies
  records, 68–70
  traffic control systems, 62–63
Communications facility, 49–51
Communications personnel, 48–49
Communications process, 47
Company, 28–29
Company officer, 28, 29
Compartments/subcompartments, 561
Composite cylinders, 154
Composite truss, 351
Composites, 350–351
Compound, 79
Comprehensive Environment Response, Compensation and Liability Act (CERCLA), 750
Compressed air foam systems (CAFS), 299, 301
Compression, 345
Compressor/purifier system, 177, 179–180
Computer-aided dispatch (CAD) systems, 51
Computer resources, 826
Computers, 51
Concentrated load, 342
Concrete, 349–350
Conduction, 94–95
Conductor, 84
Cone roof tank, 793
Confined space, 498
Confined space rescue, 498
Connections, 347
Consent, 694
Consist, 825
Consolidated incident action plans, 38
Constant pressure nozzle, 283
Constant volume nozzle, 283
Constricted pupils, 706
Container breaches, 865
Container markings, 771
Containers, 771–794
  bags, 773–774
  bottles, 772–773
  bulk storage tanks, 790–794
  bulk tanks, 775
  cardboard boxes, 772
  cylinders, 774–775
  drums, 774, 776
  highway transportation containers, 776–778. *See also* Tank trucks
  pipelines, 776
  rail transportation, 788–790
  temperature, 782
  totes, 775
Contamination reduction area (warm zone), 860, 861
Contents fire, 594
Continuous beam, 346
Convection, 95–96
Conventional forcible entry, 531–536
Core strengthening, 727
Corrosive placards/labels, 764
Corrosives, 876
Corrosivity, 798–799
Cortilan-Neu, 814
Cotton fiber rope, 420
Counter payoff, 643
Coupling/uncoupling hose, 228
Couplings, 225–227
Covered floating roof tank, 793
Covering, 885

CPR, 707–708
Creamer, Tom, 806
Cribbing, 488, 496
Critical incident stress (CIS), 737–738
Critical incident stress debriefing (CISD), 737–738
Critical incident stress management (CISM), 116, 737–738
Cross-arm carry, 469
Cryogenic gloves, 847
Cryogenic materials, 784
Cryogenic upright storage tanks, 794
Cutters, 483
Cutting a roof, 585
Cutting tools, 515–517
Cutting torch, 517
Cyanogen chloride, 928
Cylinder replacement procedure, 172–175
Cylinders, 774–775
Cylinders, air, 153–155. *See also* Self–contained breathing apparatus (SCBA)

Dalmatian, 13, 14
Damaged rope, 440
Damming, 893–894
Dangerous cargo manifest (DCM), 825
Dangerous placard, 765
Dangerous water-reactive materials, 817, 819
Dangerous-when-wet materials, 763
DCM, 825
De Francesco, Joseph, 419
Dead bolt, 528–529
Dead bolt and latch, 529
Dead-end water mains, 206
Dead loads, 342
Debris removal, 652
Decay stage, 92–93
DECIDE, 854
Deck gun, 271
Decontamination, 877–887
- defined, 877
- emergency, 877–878
- fine, 879, 880
- formal, 879
- gross, 878–879
- mass, 879–881
- methods of, 885–887

Decontamination process, 881–884
Deep wells, 204
Defense attack, 608
Defensive operations
- absorption, 892–893
- diking/damming, 893–894
- dilution, 897
- diverting, 895–896
- remote shutoff, 898
- retention, 896–897
- vapor dispersion, 897
- vapor suppression, 897–898

Dehydration, 848
Deluge set, 271–273
Deluge systems, 320–321
Density, 84, 88
Dent puller, 518
Denver tool, 513
Department of Transport (DOT), 757
Deployment plan, 59
Deployment tables, 60
Depth of char, 654
Deputy chief, 32
Design load, 344
Designated incident facilities, 38
Detection systems, 309–314
- flame detectors, 313–314
- gas detectors, 313
- heat detectors, 310–311
- people or manual systems, 309–310
- smoke detectors, 311–312

Detector activation operations, 335
Detergent-type foams, 296
Diagram of the site, 685
Diagrammatical guides. *See* Job performance requirements
Dichlorochlordene, 814
Diesel fuel, 797
Diffusion, 87, 567
Digital logging recorders, 69
Diking/damming, 893–894
Dilated pupils, 706
Dilution, 885–886, 897
Direct attack, 289, 603, 621
Direct pump water distribution system, 206
Disassembly, 489
Discharge flow, 292
Disentanglement, 489
Disinfection, 885, 886
Displacement, 489
Disposal, 885, 886
Distilled mustard, 928
Distortion, 489
Distributed load, 342
Distributor pipe, 228
Diverting, 895–896
Division chief, 32
Docimo, Frank, 891
Dodson, Dave, 106, 123, 339, 343, 348, 721
Domestic Preparedness Training Initiative, 934
Door blocker, 532
Door swing direction, 523–524
Doors, 522–527
Dormers, 363
DOT, 757
DOT-406/MC-306 gasoline tank trucks, 778
DOT-407/MC-307 chemical haulers, 778–780
DOT-412/MC-312 corrosive tankers, 780–781
DOT book. *See Emergency Response Guidebook* (ERG)
DOT hazardous material, 744
DOT identification number, 760
DOT placarding tables, 758
Double becket bend, 430–431
Double-cylinder lock set, 664
Double-donut roll, 231, 233
Double female, 228
Double gloving, 698
Double hung window, 541
Double male, 228
Dowchlor, 814
Drafting operation, 264–265, 266
Drags, 469–476
Drain and carry, 234
Drain cleaner, 799
Dressing a knot, 426
Dropped ceilings, 587
Drug labs, 923
Drums, 774, 776
Dry barrel hydrants, 207, 208
Dry bulk tanks, 786
Dry chemical extinguishing agents, 188
Dry chemical total flooding systems, 333
Dry chemicals, 185
Dry hydrant, 205, 207, 209
Dry pipe systems, 319–320
Dry powders, 186
Duck bill lock breaker, 518, 519
Duck's foot, 228
Dump site, 211
Dumpster fire, 626–627
Duplex radio system, 64
Dust explosions, 673
Dutchman, 237
Duty training standards, 26–28

EAP, 116
Early American history, 8–13
Ears, 245, 246
Eccentric load, 345
EDITH, 33
Educating the public, 679–684. *See also* Fire and life safety education
Eductor, 297–299
EHS, 745
8-Step Process, 854
Electric/cable elevators, 501, 502
Electrical conductor, 84
Electrical hazards, 672–673
Electrical power, 83–84
Electrical situations, rescue from, 498–499
Electricity, 84
Electrochemical sensor, 908
Electronic heat sensors, 650
Electrons, 80, 84
Elevated temperatures, 145
Elevator control panels, 502
Elevator rescue, 500–504
Emergency call boxes, 55
Emergency communications centers, 49–51
Emergency decontamination, 877–878
Emergency evacuation plan, 682
Emergency evacuation signals, 67
Emergency medical dispatch, 53
Emergency Medical Responses to Hazardous Materials Incidents, 749
Emergency medical services (EMS), 689–720
- AED, 708
- allergic reactions, 714–715

Emergency medical services (EMS), (*Cont.*)
assessing the patient, 700–707. *See also* Medical examination
bleeding control, 709–712
breathing problems, 714
BSI presentations, 697–701
burns, 715–717
chest pain, 714
common emergencies, 714–718
CPR, 707–708
first aid, 712–713
fractures/sprains, 718
infection control, 696–697, 701
interacting with EMS personnel, 694–695
legal issues, 694
medical illness, 714
poisoning, 717–718
roles/responsibilities, 691–694
safety considerations, 695–700
shock, 713–714
Emergency Planning and Community Right to Know Act (EPCRA), 745
Emergency Response Guidebook (ERG)
abbreviations, 810
blue section, 810, 812, 813
chemical names, 812, 814
contact numbers, 809
emergency response section, 816
first-aid section, 817
first pages, 808
general reminders, 817
green section, 817, 819
guide pages, 812, 815
hazard classes, 809
inside back cover, 807–817
inside front cover, 808
last pages, 817
orange section, 812, 815
placard section, 809, 810
potential hazards section, 815–816
public safety section, 813–816
sections, listed, 807
shipping names, 812, 813
spill or leak section, 816–817
yellow section, 810, 811
Emergency response planning (ERP), 864
Emergency services deployment, 58–62
Employee assistance program (EAP), 116
EMS. *See* Emergency medical services (EMS)
Emulsification, 885, 886
Encapsulated suit, 842–845
Encoders, 61
Endothermic reaction, 80, 81, 555, 800
Energy bar, 728, 732
Energy efficient windows, 543
Engine company, 29
Engineered wood, 349
English measurement system, 77, 78
Enhanced 9-1-1, 54, 59
Ensembles, 127–133
ENT 9,932, 814
ENT 25,552-X, 814
Entrapment, 735
Environmental and safety regulations, 749
EPA air monitoring guidelines, 903
EPA hazardous material, 744
EPCRA, 745
Equilibrium, 87
Equipment
extrication from motor vehicles, 482–486
fire extinguishers. *See* Portable fire extinguisher
fire hose. *See* Fire hose and appliances
fire inspectors, 660–661
forcible entry, 511–521
high-rise firefighting, 617
ladder. *See* Ladder
nozzle. *See* Nozzle
overhaul operations, 648–649
PPE. *See* Personal protective equipment (PPE)
rope. *See* Rope
safety, 112–114
salvage operations, 633–639
SCBA. *See* Self-contained breathing apparatus (SCBA)
standpipe operations, 334
ERP, 864
Escalator rescue, 504–505
Ethion, 797, 840
Evacuation, 862–864
Evacuation distance, 817, 868
Evacuation drills, 682
Evaporation, 89
Exclusion area, 860
Exhaust fan, 573
Exhauster, 320
Exit doors, 665
Exit drills in the home (EDITH), 33
Exothermic reaction, 80, 81, 555, 799
Expandable cut, 575–577
Explosive limits, 90
Explosive placards/labels, 761
Explosives, 866–868
Explosives lab, 923–925
Exposure levels, 838–840
Exposure to hazardous materials, 834–838
Extended aerial ladder, 374
Extending hoseline, 265–268
Extension ladder, 377–378
Extension pipe, 228
External bleeding, 710–712
External floating roof tank, 793
Extinguishing agents, 187–188
Extinguishing fires. *See* Fire suppression
Extremely hazardous substance (EHS), 745
Extremity carry, 469, 470
Extrication from motor vehicles, 482–490
disentanglement, 489
fluid/parts cleanup, 490
hazards zoning, 487
patient access, 488
patient removal, 489–490
scene assessment, 486
scene stabilization, 490
tools and equipment, 482–486
traffic barrier, 486
traffic calming, 487
vehicle recovery, 490
work areas, 486
Eye protection, 134

Face shield, 134
Facility reports, 826
Fahrenheit scale, 78
Fans, 573–575
Farm equipment rescue, 505
FCC, 63
FDNY, 114
Federal Communications Commission (FCC), 63
Federal Emergency Management Agency (FEMA), 932–933
FFFP, 296
FGI, 903–904
Fiberglass composite cylinders, 154
Fiberglass cylinders, 154
Field test, 900
Figure eight, 432–436
Figure eight on a bight, 433, 436
Fill site, 211
Fine decontamination, 879, 880
Finish loads, 243–245
Fire
by-products, 554–556
gases, 551
Fire and life safety education, 679–684
adult programs, 683
fire drills, 682–683
fire station tours, 684
PSAs, 681
school programs, 681–682
Fire attack, 289
Fire behavior, 74–105
backdraft, 100
BLEVE, 88–89
boiling point, 89
burning process, 91–93
classes of fire, 100–102
combustion, 85, 86
conduction, 94–95
convection, 95–96
fire tetrahedron, 77
flammable/explosive limits, 90–91
flashover, 99
heat transfer, 94–97
measurements, 77–78
oxygen, 86
physics of fire, 78–82
radiation, 96–97
rollover, 99
sources of heat, 82–85
states of matter, 98
structural fire, 102–103
thermal conductivity, 97
thermal layering, 99
vapor pressure/density, 86–88

Fire cause determination concerns, 652–654
Fire classification, 100–102, 185–187
Fire company inspection program, 660–679
building exterior, 673–674
concluding the inspection, 674
conducting the inspection, 661–662
electrical hazards, 672–673
equipment, 660–661
exiting, 663–666
fire protection equipment, 666–669
hazardous materials, 669–673
home inspections, 677–679
preparation, 661
reinspections, 677
restaurants, 667–668
violations, 662–669
wrap-up meeting, 674
Fire drills, 682–683
Fire entry gear, 845–846
Fire escapes, 386–387
Fire extinguisher classification symbols, 102
Fire extinguishers. *See* Portable fire extinguishers
Fire extinguishment. *See* Fire suppression
Fire flow capacity, 203
Fire flow rate, 212
Fire flow requirement, 203
Fire hose and appliances, 219–278
advancing flat load, 245, 246
advancing hoselines, 251–261. *See also* Advancing hoselines
burst hose, 268–269
care/maintenance, 223–224
coupling/uncoupling hose, 228
couplings, 225–227
extending hoseline, 265–268
finish loads/preconnected loads, 243–245
fire stream, 289–295
friction loss coefficients, 294
hoisting, 451–454
hose carries, 234–236
hose lay procedures, 269–271
hose loads, 236–242
hose rolls, 230–233
hose tools/appliances, 227–228
master stream devices, 271–274
minuteman load, 246–248
nozzle. *See* Nozzle
operating, 288–289
service testing of hose, 274–275
stored hose loads/packs, 248
triple-layer load, 247–248
types of hoses, 222–223, 288–289
water supply connection, 261–265
wildland firefighting hose loads, 248–250
Fire hydrants, 206–209, 213–215
Fire intensity, 597
Fire lane, 674
Fire load, 344
Fire plume, 96
Fire Prevention Division, 659–660
Fire prevention inspections, 32, 33
Fire prevention office, 32
Fire protection valves, 325
Fire resistive, 353
Fire-resistive construction, 353–354
Fire resistive ratings, 353
Fire safety brochures, 684
Fire safety characters, 681
Fire service key slot, 502
Fire service knots. *See* Knot
Fire shelter, 132
Fire societies, 8
Fire sprinkler systems. *See* Sprinkler systems
Fire station alerting system, 60
Fire station tours, 684
Fire stopping, 357
Fire stream, 281, 289–295, 604–606
Fire suppression, 591–630
below-ground structure (basements), 618–619
business/mercantile occupancies, 615–616
combination attack, 608
defense attack, 608
direct attack, 603
dumpster fire, 626–627
flammable liquid/gas considerations, 601–603
flammable liquid/gas fires, 625–626
ground cover considerations, 595–599
groundcover fire, 620–622
IMS, 607–609
indirect attack, 603–604
multistory occupancies, 616–618
nonstructural fires, 620
offensive attack, 607–608
outside stacked materials fire, 627–628
piled materials fire, 627–628
RECEO, 607
residential occupancies, 611–612
REVAS, 607
single family occupancies, 612–614
sprinkler system, 619–620
standpipe system, 620
stream selection, 604–606
structural fire considerations, 593–595
tactical considerations, 606
trash fire, 626–627
vehicular fire, 622–625
vehicular fire considerations, 599–601
wildland fire, 620–622
wildland fire triangle, 599–601
Fire survival programs, 33
Fire tetrahedron, 77
Fire triangle, 77
Fire wardens, 8
Fire watch, 329
Firefighter, 26
Firefighter deaths, 108, 109
Firefighter emergencies, 733–738
Firefighter fitness and health, 115–116
Firefighter injuries, 109
Firefighter readiness, 119
Firefighter safety, 106–122. *See also* Safety
accident prevention, 110, 111
equipment, 112–114
firefighter responsibilities, 117–120
injury/death, 108–109
personnel, 114–117
procedures, 111–112. *See also* SOPs
safety attitude, 116–117
safety standards and regulations, 109–110
Firefighter safety responsibilities, 117–120
Firefighter survival, 721–740
active cooling, 731
energy/rest, 727–728, 731
entrapment, 735
firefighter emergencies, 733–738
fitness for duty, 726
hydration, 732
incident readiness, 723–728
injured firefighters, 736
lost firefighters, 736
mental fitness, 726–727
nutrition, 727, 732
on-scene rehabilitation, 730–731
orders/communication, 729
personal accountability, 725–726
personal size-up, 730
physical fitness/wellness, 727
post-incident survival, 736–738
PPE, 723–724
rapid escape, 733–735
risk/benefit analysis, 730
RITs, 732–733
safety at incidents, 728–733
team continuity, 728
Firefighter's carry, 468–469
Firefighter's drag, 476
Fireground factors, 608
Firemarks, 8
FiRP, 350
First aid, 712–713
Fishing line, 421
Fitness for duty, 726
Fixed gaseous extinguishing systems, 669
Fixed temperature heat detector, 310, 311
Flame detectors, 313–314
Flameover, 99
Flammable and combustible liquids, 871–872
Flammable and combustible placards/labels, 762
Flammable gas indicator (FGI), 903–904
Flammable limits, 90–91
Flammable liquid/gas fires, 601–603, 625–626
Flammable liquid safety can, 670
Flammable range, 800, 801
Flammable solids, 872
Flammable solids placards/labels, 763
Flash point, 98, 601, 762, 800
Flashover, 92, 93, 97, 99, 102, 556, 594

Flat carry, 396–398
Flat load, 237–238, 240, 241
Flathead ax, 513, 515
Flexibility improvement, 727
Floor-mounted stop plate, 532
Floor plans, 685
Floor runner, 634–635
Flow, 282, 291–292
Flow meters, 292
Fluoroprotein film forming foam (FFFP), 296
Fluoroprotein foam, 296
Flush doors, 522
Flush-type hydrant, 209
Fly extension raise, 402, 405
Fly ladder, 372, 373
Flying stretch, 244
Foam, 295–303
  application, 297–303
  characteristics, 295–296
  classes, 296–297
  venturi principle, 297–298
Foam eductor, 298
Foam flooding systems, 333
Foam nozzles, 300, 302
Foamed core metal door, 523
Fog nozzles, 283–285
Folding ladder, 379, 380
Follow-through figure eight knot, 433, 435
Foot-tilt method, 228, 229
Footwear, 129
Force, 345
Forcible entry, 508–547
  breaching floors, 545–546
  breaching walls, 544–545
  conventional entry, 531–536
  doors, 522–527
  glazing, 542–543
  knowledge, skill, experience, 508–509
  lock variations, 539–540
  locks, 526–530
  operating lock mechanisms, 538–539
  through-the-lock entry, 536–538
  tool assignments, 546
  tools, 511–521
  windows, 540–544
Forcing a padlock, 540
Forestry hose, 223
Formal decontamination, 879
Formal procedures, 111
40 CFR 311, 747
42 CFR Part 84, 149
49 CFR 170.180, 757
49 CFR 172.504, 757
Forward lay, 270
Four-digit UN/NA number, 760, 811
Four-gas instruments, 898
Four-person flat carry, 398
Fracture, 718
Franklin, Benjamin, 9
Freelancing, 119, 725, 728
Freezing point, 795, 796
Freshwater, 203
Friction, 83
Friction loss, 292–294
Friction loss coefficients, 294
Fuel resistance, 296
Full escape procedure, 170
Full thickness burn, 716
Full-wrapped fiberglass cylinders, 154
Fully developed stage, 92
Fumigated placard, 765, 766

g-Chlordan, 814
Gable roof, 362
Gala, Mike, 184, 549
Gambrel roof, 362
Garage doors, 525–526
Gas, 98, 795, 868–871
Gas detectors, 313
Gas leaks (buildings), 877
Gas placards/labels, 762
Gasoline, 91, 797, 799, 800, 871–872
Gasoline engine-powered hydraulic pumps, 482
Gasoline-powered saws, 516
Gate valves, 209
Gauge pressure, 290
GEDAPER, 854
General duty clause, 749
Girder, 346
Glass doors, 523
Glazing, 542
Gloves, 129, 698–700, 847
Go method, 493
Goldfeder, Billy, 202
Gratacap, Andrew, 11
Gravity connections, 347
Gravity-fed water distribution system, 206
Great Chicago Fire, 14
Great Fire of London, 7
Grid (looped) system, 207
Grooved and butterfly valve, 325
Grooved check valve, 325
Gross decontamination, 878–879
Gross negligence, 750
Ground cover fire fighting, 595–599, 620–622
Ground ladder, 376–381
Ground pads, 497
Groundwater, 203–204
Growth stage, 91–92
Guard dogs, 554
Guideline/lifeline, 464
Gusset plates, 361

Half hitch, 427
Halligan tool, 513, 570
Halon, 333, 669
Halyard, 372
Hancock, John, 9
Hancock, Robert F., 417, 459
Hand-aspirated pump, 901
Hand in Hand, 10
Hand protection, 129
Hand washing, 697, 701
Handheld hydraulic spreaders, 514
Handline-operated medium expansion foam generator, 301
Handsaw, 515
Hanging ceilings, 587
Harbor boom, 896
Hard laid, 422
Hard sleeve hydrant connection, 265
Hard sleeves, 895
Hard suction hose, 222, 223
Harmful placard, 765
Harness, 491–492
Hawley, Christopher, 741, 753, 805, 831, 851, 890, 911, 912
Hayes, Daniel, 14
Hazard area, 860
Hazard classes
  corrosives, 764
  ERG, 809
  explosives, 760–761
  flammable liquids, 762
  flammable solids, 763
  gases, 762
  miscellaneous materials, 764–765
  oxidizers and organic peroxides, 763
  poisonous materials, 763
  radioactive materials, 763–764
Hazardous materials
  chemical and physical properties, 795–801. *See also* Chemical and physical properties – hazardous materials
  common incidents, 864–877. *See also* Common incidents – hazardous materials
  container markings, 771
  containers, 771–794. *See also* Containers
  decontamination, 887–887. *See also* Decontamination
  defined, 744
  electrical hazards, 672–673
  exposure, 834–838
  exposure levels, 838–840
  health hazards, 833–838, 848
  HMIS, 769
  incident levels, 887
  information sources, 805–830. *See also* Information sources – hazardous materials
  labels, 767–768
  laws, 745, 750
  location and occupancy, 755–756
  medical monitoring, 748
  military warning system, 769–770
  MSDS, 818–824
  NFPA 704 system, 768–769
  OSHA HAZWOPER regulation, 747–748
  pesticide container markings, 771
  pipeline markings, 770–771
  placards, 757–768. *See also* Placards
  PPE, 840–848. *See also* Personal protective equipment (PPE)
  protective actions, 851–889. *See also* Protective actions – hazardous materials
  recognition and identification, 753–804
  reporting requirements, 745–747
  senses, 794–795

standards, 749–751
storage, 669–672
TRACEM, 834, 835
training levels, 748
Hazardous materials containers. *See* Containers
Hazardous materials information system (HMIS), 769
Hazardous materials management systems, 854
Hazardous Materials Response Unit (HMRU), 932–933
Hazardous Waster Operations and Emergency Response (HAZWOPER), 747
Hazards zoning, 487
HAZMAT branch positions, 858
HAZMAT crimes, 922–927
HAZWOPER, 747
HCS 3260, 814
Head pressure, 291
Head-tilt, chin-lift method, 703
Header, 346
Head's up display (HUD), 168
Health hazards, 833–838, 848
Hearing protection, 134
Heat detectors, 57, 310–311
Heat exhaustion, 848
Heat release rate, 94
Heat resistance, 296
Heat sink, 94
Heat stroke, 848
Heat transfer, 94–97
Heating, ventilation, and air conditioning (HVAC) systems, 563
Heavy-duty padlocks, 529
Heavy timber construction, 355–356
Heavy timber dimensions, 356
Heavy timber fires, 598
Heeling a ladder, 410, 411
Helix, 222
Helmet, 128–129
Hepatoxins, 836
Hero of Alexandria, 6
HFT, 536
Hibernia Fire Company, 10
Higbee cut, 225, 226
Higbee indicator, 226
High expansion foam generator, 302, 333
High-piled storage fire, 627–628
High-pressure air bag set, 485
High-pressure air chisel kit, 485
High-pressure hydrants, 209
High-pressure tanks, 794
High-rise firefighting, 616–618
High-rise (standpipe) hose packs, 248, 251
High-temperature clothing, 845–846
Highway transportation containers, 776–778. *See also* Tank trucks
Hildebrand, Mike, 853
Hip roof, 362
Historical overview, 4–17
ancient beliefs, 5–6
building collapses, 343–344
Civil War, 13–14
early American history, 8–13
Industrial Revolution, 14
Middle Ages, 7
Roman Empire, 6
terrorism, 914–917
World War II, 17
HMIS, 769
HMRU, 932–933
Hoist, 227
Hoisting, 449–454
ax, 450
hoseline, 451–454
ladders, 412, 414, 451, 454, 455
pike pole, 450, 451
small equipment, 451, 454
Hoistway door, 503
Hollow-core doors, 522
Home alerting devices, 60
Home evacuation, 682
Home inspections, 677–679
Honeycomb core metal door, 523
Hood systems, 332
Hoods, 129, 130
Hook, 518, 570
Hook ladder, 378–379
Horizontal sliding windows, 541
Horizontal ventilation, 552, 567–568
Horizontal voids, 562
Horseshoe load, 239, 242, 243
Hose. *See* Fire hose and appliances
Hose bed, 223
Hose bridge, 227
Hose cap, 228
Hose carries, 234–236
Hose cart, 227
Hose clamp, 227, 268
Hose couplings, 225–227
Hose drag, 236
Hose jacket, 227
Hose lay procedures, 269–271
Hose loads, 236–242
Hose packs, 248
Hose roller, 227
Hose rolls, 230–233, 248
Hose storage rack, 226
Hose strap, 227
Hose stream characteristics, 606
Hose testing, 274–275
Hose threads, 226
Hose washer on hydrant, 224
HOT label, 765
Hot materials tankers, 786
Hot zone, 860–861
Household ammonia, 799
Household quantities, 750
Housekeeping practices, 672
HUD, 168
Human chain technique, 463
Humid environments, 588
HVAC systems, 563
Hydrant connections, 262–265
Hydrant valves, 228
Hydrant wrench, 227
Hydration, 137, 732, 848
Hydraulic elevators, 501
Hydraulic forcible entry tool (HFT), 536
Hydraulic spreaders, 514
Hydraulic ventilation, 565–567
Hydraulics, 290
Hydrocarbon, 79, 80, 297
Hydrochloic acid, 799
Hydrochloric (muriatic) acid, 876
Hydrogen chloride, 146, 551
Hydrogen cyanide, 143, 146, 551
Hydrogen sulfide (H2S), 551, 840, 849
Hyperbaric chamber, 907
Hypoperfusion, 713–714
Hypoxia, 145, 147

IAP, 729
IC, 607
Ice rescue, 494
Ice rescue ensemble, 132, 133
ICF, 357
ICt_{50}, 840
IDLH atmosphere, 125, 143, 839
IDR sensor, 908
Ignition, 91
Ignition stage, 91
Ignition temperature, 800
IM/IMO tanks, 787–788
Immediate danger to life and health (IDLH), 125, 143, 839
IMO 101, 787, 788
Impact load, 342
Implied consent, 694
Imposition of loads, 344–345
IMS. *See* Incident management system (IMS)
In-line eductor, 298
Incendiary agents, 928
Incident action plan (IAP), 729
Incident command designations, 41–42
Incident commander (IC), 607
Incident commander level, 748
Incident engagement checklist, 120
Incident levels, 887
Incident management, 36–37
Incident management system (IMS), 37–42
command staff positions, 41–42
components, 37–38
fire suppression, 607–609
firefighter survival, 729
functions, 39–41
safety, 118
unified command, 42
Incision, 712
Increaser, 228
Indirect attack, 603–604, 622
Indirect fire attack, 289–290
Induction heat, 84
Industrial entrapment rescue, 499–500
Industrial Revolution, 14
Industrial technical assistance, 829
Infectious substances labels, 765, 766
Informal procedures, 111
Information half-life, 17–18
Information sources-hazardous materials, 805–830
Chemtrec, 826–827
computer resources, 826

Information sources-hazardous materials, (*Cont.*)
ERG, 807–819. See also *Emergency Response Guidebook* (ERG)
facility documents, 826
industrial technical assistance, 829
MSDS, 818–824
reference and information texts, 827–829
shipping papers, 825–826
Infrared flame detectors, 314
Infrared sensor, 904–905
Inhalation Hazard label, 767
Initial assessment, 700, 702
Initial Isolation and Protective Action Disturbances table, 816
Injured firefighters, 736
Inorganic, 80
Insect screens, 586
Insecticides, 875
Inspection cut, 581
Inspection notice, 676
Inspections. *See* Fire company inspection program
Instant detection and response (IDR) sensor, 908
Insulated MC-307/DOT-407, 779–780
Intake relief valve, 228
Integrated communications, 38
Interconnected lock set and dead bolt, 665
Interface firefighting, 622
Intermodal (IM) tanks, 787–788
Internal bleeding, 709–710
Internal floating roof tank, 793
Intersecting roof, 362
Intervention, 110
Ionization-type smoke detector, 311–312
IR detectors, 314
Irons, 513
Irritants, 834, 929
ISC TMX-412 factors, 902
Isolation and protection, 853–854
Isolation area (hot zone), 860–861
Isolation distance, 816
Isopropyl alcohol, 800

Jacket, 221
Jalousie windows, 541, 543
Jamb, 522
Jaw thrust, 703
Jefferson, Thomas, 9
Jet siphon valve, 211
Job performance requirements
carries, 468–471
coupling/uncoupling hoses, 229–230
decontamination process, 883–884
drags, 472–476
forcible entry, 534–538
hoisting, 450–455
hose carries, 234–236
hose loads, 238–243
hose rolls, 231–233
knots, 427–432
ladders, 395–414. *See also* Ladder skills
removing gloves, 699
returning sprinkler system to service, 327–328
reverse horseshoe load, 244
rope, 446–449
salvage operations, 638–645
SCBA, 159–180. *See also* Job performance requirements – SCBA
taking out a window, 570
victim removal, 468–477
Job performance requirements - SCBA
changing out a cylinder (other firefighter), 175–176
cylinder replacement procedure, 172–175
daily inspection, 172, 173
donning- coat method, 160, 162, 163
donning- face piece, 166, 167
donning- over the head method, 159–160, 161
donning- seat-mounted apparatus, 164, 165
servicing cylinder (cascade system), 177–179
servicing cylinder (compressor/purifier system), 179–180
Joist, 346
JPR. *See* Job performance requirements
Junior firefighter program, 682

K tool, 518–519, 537, 538
Kelleher, Mike, 75
Kelly tool, 513–514
Kernmantle rope, 423
Kerosene, 91
Kertzie, Peter F., 370
Key in the knob lock, 528
Key tools, 539
Kilometer, 78
Knee-press one-person uncoupling method, 230, 231
Knights of Malta, 7
Knockdown speed, 296
Knot, 424–439
becket bend, 429–430
bowline, 431–432
clover hitch, 427–429
double becket bend, 430–431
elements of, 425
figure eight, 432–436
half-hitch, 427
overhand, 427
rescue, 437–438
water, 438–439
Knox boxes, 673–674
Koonce, Ric, 183, 201, 219, 307
Kypchlor, 814

Labels, 767–768
Laceration, 712
Lach, Bernard, 142
Ladder, 369–416
A-frame, 379–381
aerial, 373–375
articulating boom, 376
breaking glass, 571–572
certification/testing procedures, 395
extension, 377–378
fire escapes, and, 386–387
folding, 378, 379
ground, 376–381
hoisting, 412, 414, 451, 454, 455
hook, 378–379
maintenance, cleaning, inspection, 381–383
painting, 394–395
parking, 393–394
parts, 371–373
Pompier, 381
roof level, and, 386
safety, 383, 390–393
selection criteria, 384–387
skills. *See* Ladder skills
special uses, 387–390
storage, 393
straight, 376–377
terminology, 374, 375
tower, 375–376
truck-mounted, 373–376
uses, 383–384, 387–390
windows, and, 385–386
working off, 392–393
Ladder belt, 491
Ladder commands, 400
Ladder company, 29
Ladder escape, 734–735
Ladder load, 391–392
Ladder painting, 394–395
Ladder pipe, 271
Ladder pipe operations, 274
Ladder positioning terminology, 374
Ladder raising terminology, 375
Ladder selection, 384–387
Ladder skills, 395–414
carrying ladders, 395–399
carrying tools, 408, 409
engaging hook on hook ladder, 412, 413
flat carry, 396–398
heeling a ladder, 410, 411
hoisting ladders by rope, 412, 414
leg lock, 408
lower ladder into building, 404, 406
mounting/dismounting, 410–412
raising ladders, 399–408
roof ladder deployment, 412
shoulder carry, 396, 397
single carry, 398–399
suitcase carry, 395, 396
window dismount, 410, 412
Ladder truck, 272
Ladder uses, 383–384, 387–390
Laid rope, 422
Laminated glass, 542
Landing plate, 504
Large-diameter hose (LDH), 222
Large spill, 817
Layman, Lloyd, 17
LC_{50} 839–840
LCt_{50}, 840
LD_{50}, 839–840
LDH, 222

Leaking underground storage tank (LUST), 791
Lean-to collapse, 495
Learn Not to Burn program, 682
Leather lace-up boot, 129
Leather pull-up boot, 129
Ledge doors, 522
Left-hand search, 463
Leg lock, 408
LEL, 800
LEPC, 745
Lesak, Dave, 853
Lethal concentrations (LC_{50}), 839–840
Lethal dose (LD_{50}), 839–840
Level 1 incident, 887
Level 2 incident, 887
Level 3 incident, 887
Level A protective clothing, 842–844
Level B encapsulated suit, 844–845
Level C suits, 845
Level D suits, 845
Lewisite, 928
Liability, 749
Liaison officer, 41
Life safety rope, 424, 425
Light obscuration detector, 312, 313
Light obscuration photoelectric smoke detector, 313
Light-scattering detector, 312, 313
Light-scattering photoelectric smoke detector, 313
Lightweight floor truss systems, 362
Lightweight wooden I beams, 350, 351
Line-of-duty deaths, 26
Line-type heat detector, 311
Lintel, 346
Liquefied oxygen (LOX), 874
Liquid, 86, 98, 795
List of dangerous water-reactive materials, 817, 819
Lith-X, 873
Live loads, 342
Load, 342–345
Load-bearing wall, 347
Loaded stream, 187
Local application system, 332
Local Emergency Planning Committee (LEPC), 745
Local emergency response plans, 745
Local protective signaling system, 56
Lock cylinders, 538, 539
Lock variations, 539–540
Locks, 526–530
Logging recorders, 69
Logistics, 40
Long spine boards, 478
Loop, 426
Loose fold and roll, 637
Lost firefighters, 736
Louver opening, 577–579
Low/medium-pressure bag, 485
Low-pressure carbon dioxide extinguishing system, 333
Low-pressure tanks, 794
Low-temperature clothing, 846
Lower explosive limit (LEL), 800
LOX, 874
Lug style butterfly valve, 325
LUST, 791

M 140, 814
Mace, 928
Magnesium, 873
Magnetic tape logging recorders, 69
Manifold, 228
Manila rope, 420, 421, 442
Mansard roof, 362
Manual incident card, 70
Manual pull box, 57
Manufactured buildings, 358
Marchone, Mary K., 658
Marine Pollutant placard, 765
Mask rule, 143
Masonry, 349, 350
Masonry cutting blades, 517
Masonry wall collapse, 350
Mass casualty incident (MCI), 926
Mass decontamination, 879–881
Master streams, 271–274, 288
Material safety data sheet (MSDS), 818–824
Matter, 78
Maul, 512
MC-331 pressurized tankers, 781
MC-338 cryogenic tankers, 784–785
MCI, 926
MDH, 222
Means of egress, 663
Measurements, 77–78
Mechanical stress, 865
Mechanical ventilation, 563
Medi-vac, 694, 695
Medical alert tags, 704
Medical examination, 700–707
  airway, 702
  bleeding, 704
  breathing, 702
  circulation, 703
  color/temperature of skin, 705–707
  fact-finding, 704
  level of consciousness, 702
  vital signs, 705
Medical illnesses, 714
Medium-diameter handlines, 289
Medium-diameter hose (MDH), 222
Meltdown, 85
Melting point, 795, 796
Mental fitness, 726–727
Mental health, 116
Mercantile fire, 615–616
Metal cutting blade, 517
Metal doors, 522–523
Metal oxide sensor (MOS), 904, 908
Metal-X, 873
Meter, 78
Methamphetamine laboratory incidents, 924
Methane, 82, 88
Methyl alcohol, 797
Metric system, 77, 78
Metropolitan medical response team (MMRT), 934
Miale,
Micropho
Microwave
Middle Ages,
Military placard
Military warning
Mill construction, 3
Miller, Geoff, 631
Minuteman load, 246–2
Miscellaneous hazardous placards, 764
Mission statement, 4, 24
Mission vision, 581
Mitchell, David, 722
Mitigation, 110
MMRT, 934
Mobile data computers, 63
Mobile data terminals, 62
Mobile microphone, 66
Mobile radio, 58
Mobile support vehicles (MSVs), 68, 69
Mobile water supply apparatus, 30, 31, 204–205
Model Procedures Guide for Structural Firefighting, 204
Modified Gasner bar pack, 250, 251, 252
Modular organization, 38
Molecules, 79
Monitor pipe, 271, 272
Montagna, Frand, 280
Morris, Robert R., 508
Mortar, 350
Mortise lock, 528–529, 538
MOS, 904, 908
Motor vehicle, extrication from. *See* Extrication from motor vehicles
Motor vehicular firefighting, 599–601, 622–625
Mounting hardware, 522
MSDS, 818–824
MSVs, 68, 69
Multifamily occupancy firefighting, 614
Multigas detector, 899
Multiple ladder removal technique, 387
Multipurpose dry chemical agent, 188
Multisite trunked radio systems, 64, 65
Multistory occupancy firefighting, 616
Multivalved wet barrel hydrant, 207
Municipal fire alarm systems, 55
Mustard, 840
Mutagens, 836
Mutual aid or assistance agreements, 36

National Board of Fire Underwriters (NBFU), 15
National Emergency Number Association (NENA), 54
National Fire Protection Association (NFPA), 26, 749
National Institute of Occupational Safety and Health (NIOSH), 110, 838
National priority list (NPL) site, 750
Natural gas, 91
Natural ventilation, 563

NBC, 927
NCI-C00099, 81
Needed flow, 2
Negatively ch
Negligence,
Nephrotox
Nerve ag
Neutral
Neutra
Neut
Ne
2
749
2, 749, 899
473, 749
PA 704, 671, 672
NFPA 704 placard system, 671, 768–769
NFPA 1001, 26, 27, 743n
NFPA 1021, 28, 32
NFPA 1031, 33
NFPA 1041, 34
NFPA 1061, 48
NFPA 1221, 49
NFPA 1404, 144, 148, 149
NFPA 1500, 109, 126, 144, 148, 609, 730, 732, 751
NFPA 1500 series, 110
NFPA 1561, 733
NFPA 1581, 136
NFPA 1901, 28, 29, 30
NFPA 1931, 381, 395
NFPA 1932, 382
NFPA 1975, 136
NFPA 1981, 148, 149, 1506
NFPA 1983, 128, 419, 424, 439, 491
NFPA 1991, 844
NFPA standards, 749
Night latch, 540
9-1-1, 54
NIOSH, 110, 838
Niran, 814
Nitrogen, 333
Nitrogen dioxide, 146, 551
No-knowledge hardware, 664
Noll, Gregory G., 754, 853
Nomex, 86
Non-load-bearing wall, 347
Noncombustible construction, 354
Nonpressurized railcar, 788
Nourishment theory, 732
Nozzle, 281–288
  cellar, 286
  defined, 281
  fire stream, 289–295
  foam, 300–302
  fog, 283–285
  operations, 287–288
  piercing, 286
  playpipes, 286, 287
  pressure, 293
  shutoff, 286, 287
  solid tip, 282–283
  straight steam, 285
Nozzle flow, 282
Nozzle pressure, 282, 293
Nozzle reach, 282
Nozzle reaction, 282, 293
NPL site, 750
Nuclear, biological, and chemical (NBC), 927
Nuclear energy and heat, 85
Nucleus, 80
Nunn-Lugar-Domenici Legislation, 934
Nutrition, 115, 727, 732
Nylon ropes, 420–421

Occupancy classifications, 359–360
Occupant use hose, 223
Occupational Safety and Health Administration (OSHA), 109, 838
Octa-Klor, 814
Offensive attack, 607–608
Offensive heat holes, 575
Oily rags, 673
Old-style sprinklers, 315, 316
On-scene rehabilitation, 730–731
One-person foot-tilt method, 228, 229
One-person over-the-hip method, 229
One-person raise, 406–408
One-person rope, 424
Open-circuit SCBA, 151, 152
Open floating roof tank, 793
Open hoistway door, 503
Open web steel joists, 362
Opening a roof, 585
Opening doors, 569
Opening windows, 588–590
Operating lock mechanisms, 538–539
Operation Respond, 825
Operations level, 748
Orange book. *See Emergency Response Guidebook* (ERG)
Orange juice, 799
Order of Saint John of Hospitaliers, 7
Ordinary construction, 354–355
Ordinary tanks, 793
Organic peroxides, 874
Organic substances, 79, 81
Organizational structure, 24–25
Oriented strand board (OSB), 351
ORM-D, 765
Ortho-Klor, 814
OSB, 351
OSHA 109, 838
OSHA 29 CFR 1910.134, 148, 149, 748, 750
OSHA CFRs, 109
OSHA hazardous material, 744
OSHA HAZWOPER regulation, 747–748
OSHA-TWA, 838
OS&Y valve, 318, 319, 324, 325, 648
Outside stacked materials fire, 627–628
Outside stem and yoke (OS&Y) valve, 318, 319, 324, 325, 648
Over the head method (donning SCBA), 159–160, 161
Over-the-hip method, 229
Overflow dam, 894, 895
Overhand safety knot, 427
Overhaul, 610, 611
Overhaul operations, 648–652
Overhauling roofs, 650
Overhead doors, 525
Overhead garage door, 525–527
Overpacking, 885, 886
Overturned tanker trucks, 872
Oxidation process, 81–82
Oxidizer, 81
Oxidizers, 874
Oxidizers/organic peroxides placards, 763
Oxygen, 86, 90
Oxygen-deficient atmosphere, 145
Oxygen monitors, 901–902

Packing groups, 759–760
Padlock, 529–530, 540
Pagers, 61
Pancake collapse, 495
Panel doors, 522
Panic hardware, 664
Pant, 128
PAR, 735
Paragraph q, 748
Parallel chord truss, 361
Parallel method of attack, 621
Parapet, 363
Parmalee, Henry, 14
Partial openings, 585
Partial thickness burns, 716
Partially broken windows, 585, 586
PASS devices, 135–136, 144
Passport, 725
Patient, examining the. *See* Medical examination
PCR, 932
PDD 39, 926
PEL, 838
PEL-C, 838
PEMEX LPG Terminal fire (Mexico City), 783
Pendent sprinkler, 316, 317
Pepsi, 799
Percent drop, 215
Peripheral nervous system (PNS) chemicals, 836
Permeation, 842
Permissible exposure limit (PEL), 838
Personal accountability, 725–726
Personal alert safety system (PASS), 135–136
Personal flotation device (PFD), 492, 493
Personal protective equipment (PPE), 123–140
  APRs, 841
  care/maintenance, 136
  chemical protective clothing, 842–846
  cryogenic gloves, 847
  eye protection, 134
  firefighter survival, 723–724
  hazardous materials, 840–848
  hearing protection, 134
  high-temperature clothing, 845–846

ice rescue ensemble, 132, 133
Level A-D suits, 842–845
low-temperature clothing, 846
PASS devices, 135–136
proximity ensemble, 129–130, 131
SARs, 841
standards/regulations, 126
street smarts, 137–138
structural ensemble, 127–129
swift-water ensemble, 133
technical rescue ensemble, 133
turnout gear (TOG), 847, 855
wildland ensemble, 130–132
work uniform, 136
Personal size-up, 725–726, 730
Personnel accountability report (PAR), 735
Peshtigo (Wisconsin) forest fire, 14
Pesticide container markings, 771
Pesticides, 875, 925–926
PFD, 492, 493
Philadelphia Contributionship, 9, 10
Phoenix from the ashes, 5, 6
Phosgene, 146, 551, 928
Photo-ionization detector (PID), 905
Photoelectric detectors, 312
Physical fitness/wellness, 727
Physics of fire, 78–82
Pick head ax, 515, 516
PID, 905
Piercing nozzle, 286
Pike pole, 518, 570
Piled materials fire, 627–628
Pinned connections, 347
PIO, 41
Pipe bomb, 931, 932
Pipeline markings, 770–771
Pipelines, 776
Piping system (gas station), 791
Pitot gauge, 214
PIV, 324–325, 647
Pivoting dead bolt, 529
Placards, 757–766
DOT placarding tables, 758
ERG, 809, 810
hazard classes, 760–765. *See also* Hazard classes
miscellaneous, 765–766
packing groups, 759–760
problems, 766
UN/NA number, 760
Plastic PVC, 600
Plate glass, 542
Platform framing, 357
Playpipes, 286, 287
Plug, 206
Plume, 775, 863
Plume dispersion model, 817
Plywood, 349
PNS chemicals, 836
Point of origin determination, 653–654
Poisoning, 717–718
Poisonous gases, 874–875
Poisonous materials placards, 763, 764
Polar solvents, 297, 601
Pole ladders, 377, 379
Policies, 34
Polycarbonate glazing, 543
Polyester rope, 421
Polyethylene rope, 421–422
Polymerase chain reaction (PCR), 932
Polymeric barrier, 297
Polypropylene rope, 421
Polyvinyl chloride (PVC), 600
Pompier ladder, 381
Ponds, 205
Pool chemicals, 874
Popcorn, 89
Portable deluge sets, 272–273
Portable fire extinguishers
care/maintenance, 197
classes, 193–194
extinguishing agents, 187–188
fire classification, 185–187
inspection requirements, 197–198
kinds, 188–193
limitations, 194–195
operation, 195–197
rating system, 193–194
Portable hydrant, 228
Portable ladders, 376–381
Portable radio, 66
Portable water tanks, 210–211
Positive pressure ventilation (PPV), 565, 566, 573–575
Positively charged, 84
Post and beam, 347
Post-incident survival, 736–738
Post-incident thought patterns, 737
Post indicator valve (PIV), 324–325, 647
Potassium hydroxide, 799
Power hydraulic combination tools, 484
Power hydraulic cutters, 483
Power hydraulic rams, 483
Power hydraulic spreaders, 483, 514
Power hydraulic tools, 482–484
Power saws, 515, 516
PPE. *See* Personal protective equipment (PPE)
PPE ensemble, 127–133
PPV, 565, 566, 573–575
Pre-incident management process, 684–687
Pre-incident management site visit, 685
Preaction systems, 321, 322
Prearrival instructions, 48
Precast concrete, 350
Preconnected combination loads, 245
Preconnected loads, 243–245
Pressure
absolute, 290
atmospheric, 212, 213
defined, 86
gauge, 290
head, 291
nozzle, 282, 293
pump discharge, 292
residual, 213
static, 214
units of measure, 282
vacuum, 291
vapor,
velocity,
water dis
Pressure dress
Pressure points,
Pressure-regulatin
Primary label, 767,
Primary search, 465
Printers, 62
Procedures, 35, 111. *See a*
performance requireme
Product control, 892–898. *See*
Defensive operations
Projected windows, 541, 544
Propane, 88
Propane tank detonation, 865
Proprietary protective signaling system, 56
Protection plates, 373
Protective actions - hazardous materials, 851–864
evacuation/sheltering in place, 862–864
isolation and protection, 853–854
rescue, 854–856
risk-based response, 856–857
site management, 857–859
zones, 862–864
Protective gloves, 698–700
Protective hoods, 129, 130
Protective systems, 307–338
alarm and monitoring systems, 331
detection systems, 309–314
fire department operations, 333–336
local application and hood systems, 332
sprinkler systems, 314–329
standpipe systems, 329–331
total flooding systems, 332
Protein foam, 296
Protons, 80
Proximity gear, 845, 846
Proximity PPE, 129–130, 131
Prying and spreading tools, 513–515
PSA, 681
Psychological decontamination, 880
Public education, 679–684. *See also* Fire and life safety education
Public fire/life safety education, 32
Public information officer (PIO), 41
Public service announcement (PSA), 681
Pulling tools, 518
Pulmonary edema, 145
Pulse rate, 703, 704, 705, 707
Pump discharge pressure, 292
Pump operator, 281
Pump panel gauges, 214
Pump pressure gauges, 291
Pump-type extinguishers, 188–190
Pumper, 30
Puncture, 712–713
Pupil size, 706
Purlin, 346
PVC, 600
Pyrolysis, 83
Pyrophoric, 744

Quarter-turn t
Quick respo
Quint, 30, 3
Rabbete 63–67
Radial
Radiacards, 763–764
Radi device
Ra
R
F
ion, 788–790

technique, 303
ladders, 399–408
s, 483
apid escape, 733–735
Rapid intervention crew universal air connection (RIC/UAC), 168
Rapid intervention team (RIT), 732–733
Rate-of-rise heat detector, 310
Rate of spread, 597
Rattlewatch, 8, 9
RCRA Waste Number U036, 814
RDD, 931
RDX, 924
Reach, 282
Reach method, 492
Reaction time, 901
Reactive materials, 875–876
Readiness assignments, 725
Reading buildings, 360
Reading smoke, 102–103
Rearmount aerial ladder, 373
Receiving reports of emergencies, 51–58
  automatic alarm systems, 56–57
  cellular telephone, 54–55
  municipal fire alarm systems, 55
  still alarm/walk-ups, 58
  TDD, 58
  telephone, 54
RECEO, 607
Reciprocating saw, 485, 486, 517
Recommended exposure limit (REL), 838
Records, 68–70
Recovery time, 901
Red Rover, 11
Reducer, 228
Reel coil, 499
Reference and information texts, 827–829
Reflective trim, 128
Regular glass, 542
Regulations, 34
Rehabilitation, 730–731
Reinforced concrete, 349
Reinspections, 677
REL, 838
Relative response, 901
Remote shutoff, 898
Remote station protective signaling system, 56
emoval, 885, 886
portable quantity (RQ), 767, 825
porting facility, 745, 746
eproductive toxins, 836
Required flow, 291
Rescue, 437–438
Rescue company, 29
Rescue procedures, 459–507
  backboards, 478–480
  below-grade rescue, 497–498
  carries, 468–469
  confined space rescue, 498
  drags, 469–476
  electrical situations, 498–499
  elevator rescue, 500–504
  escalator rescue, 504–505
  extrication from motor vehicles, 482–490. *See also* Extrication from motor vehicles
  farm equipment rescue, 505
  hazardous materials, 854–856
  ice rescue, 494
  industrial entrapment rescue, 499–500
  rescue of firefighter wearing SCBA, 476–477
  searching burning structures, 461–467
  stretchers, 480–481
  structural collapse rescue, 494–497
  TICs, 466–467
  trench rescue, 497–498
  tunnel vision, 461
  vertical rescue, 491–492
  victim removal, 467–476
  water rescue, 492–494
Residential garage doors, 525, 526
Residential occupancy firefighting, 611–612
Residential overhead door locks, 531
Residential sprinkler systems, 321, 322
Residual pressure, 213
Resistance training, 727
Resource management, 38
Respiratory hazards, 144–148
Respiratory protection policy, 143
Respiratory protection regulation, 750–751
Respiratory rate, 705
Respiratory route of exposure, 837
Rest, 727–728, 731
Restaurants, 667–668
Restricted openings, 170
Retard chamber, 318, 319
Retention, 896–897
REVAS, 607
Reverse horseshoe load, 244
Reverse lay, 270
Revolving doors, 524–525
REX tool, 519, 539
RIC/UAC, 168
Ricin, 930
Riding the fire apparatus, 120
Right-hand search, 463
Rigid connections, 347
Rim lock, 529, 539
Ringdown circuits, 58
Risk-based response, 856–857
Risk/benefit analysis, 730
Risk/benefit philosophy, 118
Risk management, 108
RIT, 732–733
RIT checklist, 733
Roll-down steel doors, 526, 527
Rollover, 99, 560
Roman Empire, 6
Roof/balcony escape, 734
Roof ladder, 378–379
Roof ladder deployment, 412
Roof structures, 362–363
Roof ventilation, 575–581, 586–587
Room escape, 734
Rope, 417–458. *See also* Knot
  cleaning, 441–444
  coiling, 445–447
  construction methods, 422–423
  drying, 443–444
  hoisting, 449–455
  inspection, 439–441
  lagging, 447–449
  maintenance, 441
  parts, 425
  storage, 444
  strength, 421
  tying, between two objects, 454–457
  types, 420–423
  uses, 423–424
  washing, 442–443
Rope hose tool, 227
Rope strengths, 421
Rope washer, 442, 443
Rotary power saw, 517
Rotary saws, 520, 521
Round turn, 426
Routes of exposure, 835–838
Row method, 492, 493
Rowdies and Rum era, 11
RQ, 767, 825
Rubber boot, 129
Rubber-coated fire hose, 222
Run card system, 59
Rung raise, 400–402
Rungs, 373
Running end, 425
Rural water supply, 210–212
"Rushing to the Conflict," 12
Rusting metal, 82

Sachs, Gordon M., 690
SADT, 800
Safety
  air bags, 484
  air chisels, 486
  ambulance stretchers, 480
  EMS, 695–700
  equipment, 112–114
  firefighters, of, 721–740. *See also* Firefighter survival
  forcible entry tools, 511–521
  generally, 106–122. *See also* Firefighter safety
  ladders, 383, 390–393
  personnel, 114–117
  power hydraulic tools, 484

reciprocating electric saws, 486
salvage operations, 639, 642
SCBA, 168–169
scene assessment, 486
water rescue, 492
Safety attitude, 116–117
Safety glass (building), 542
Safety glasses (eyewear), 134
Safety officer, 41
Safety standards and regulations, 109–110
Safety triad, 111
Salvage cover, 634
Salvage cover fold, 639
Salvage cover roll, 637, 638
Salvage operations, 633–648
balloon toss, 644, 645
catch-all, 645–646
ethics, 643
limiting exposure to smoke, 644
protecting material goods, 642–643
safety, 639, 642
shoulder toss, 643, 644
sprinkled buildings, 647–648
tools and equipment, 633–639
Visqueen, 644
water chute, 646–647
water removal, 644–645
SAR, 158, 841
SARA, 745
SARA facility, 745, 746
SARA reports, 826
SARA Title III, 745
Sarin, 797, 840
Sash windows, 569, 589
Saws, 515–517
Scaling ladder, 381
SCBA. *See* Self-contained breathing apparatus (SCBA)
SCBA cylinders, 153–155
SCBA face piece, 157
SCBA field maintenance sheet, 174
Scheerer, Randy, 124
Schnepp, Rob, 742
School evacuation drills, 682–683
Scott, Ralph, 15
Screens, 586
SD 5532, 814
Sea (IMO) containers, 776, 787–788
Seat carry, 469, 471
Second-degree burns, 716
Secondary label, 767, 768
Secondary search, 465
Sectors, 859
Secured key boxes, 673–674
Securing the building, 654–655
Security bars, 678
Security devices, 583
Selectable gallonage nozzles, 283
Self-accelerating decomposition temperature (SADT), 800
Self-clearing card, 677
Self-closing faucet, 670
Self-contained breathing apparatus (SCBA), 141–182
air quality, 155–156
backpack/harness assembly, 152, 153
changing cylinders, 172–176
closed-circuit SCBA, 158
cylinders, 153–155
daily maintenance checks, 159, 172, 173
design/size, 150
donning/doffing, 158–168
emergency procedures, 171
face piece assembly, 157
hazardous materials, 840–842
hostile environment, 170
JPRs. *See* Job performance requirements - SCBA
legal requirements, 148, 149
limitations, 148–150
maintenance, 159, 172, 173
new designs, 168
open-circuit SCBA, 152
regulator assembly, 156
respiratory hazards, 145–146
restricted openings, 170
safe use, 168–169
SARs, 158
seat-mounted apparatus, 162, 164, 165
servicing cylinders, 176–180
toxic gases, 146–148
types, 151–158
Self-rescue, 734
Senses, 794–795
Sensitizer, 835
Sensor label, 372
SERC, 745
Set volume nozzle, 283
SETIQ, 827
Setting a knot, 426
Severance, 489
Shallow wells, 204
Shear (force), 345
Shears (tool), 483
Shed roof, 362
Sheet bend, 429
Shell SD-5532, 814
Sheltering in place, 862–864
Shipping names, 812, 813
Shipping papers, 825–826
Shock, 713–714
Shock load, 419
Shoring, 496–497
Short-term exposure limit (STEL), 839
Shoulder carry, 396, 397
Shoulder load, 237
Shoulder loop carry, 234–235
Shoulder toss, 643, 644
Shouldis, William, 308
Shutoff, 286, 287
Shutter guard, 539
Siamese, 228
Sick building syndrome, 870–871, 876
Silencing activated detector, 335
Simple asphyxiants, 835–836
Simple beam, 346
Simplex radio system, 64
Single-donut hose roll, 231, 233
Single hung window, 541
Single occupancy firefighting, 612–614
Single-p
Single-pe
Single-secti
SIP, 351
Sisal rope, 420
Sit and drag meth
Site plan, 685
Skeletal frame, 347
Slab doors, 522
Sledge, 512
Sliding bolt, 532
Sliding doors, 524, 525
Slot load, 246–248
SLUDGEM, 928
Small-diameter headlines, 288
Small-diameter house, 222
Small lines, 222
Small spill, 817
Smith, James P., 340
Smith, Mike, 22
Smoke, 145
Smoke color, 103
Smoke density, 103
Smoke detectors, 57, 311–313, 678, 679
Smoke ejector, 563–565
Smoke explosion, 556–560
Smoke fans, 563–565
Smoke indicator hole, 581
Smoke velocity, 102
Smoke volume, 102
Smokestack, 585
Smooth bore nozzles, 282–283
Snorkel basket, 376
Snorkel ladder, 376
Snow, 588
Snowflake technique, 302
Socks, Gregory L., 852
Sodium bicarbonate, 188
Sodium chlorate, 82
Sodium chloride, 82
Sodium chlorite, 82
Sodium hydroxide, 876
Soft sleeve, 223
Soft sleeve hydrant connection, 264
Soft suction hose, 223
SOG, 35
Solid, 98, 795
Solid-core doors, 522
Solid stream nozzle, 282–283
Solid tip nozzle, 282–283
Solidification, 885, 886
Solubility, 601
Soman, 928
SOPs, 35, 112, 113
Sources of heat, 82–85
Spalling, 350
Span of control, 36, 38
Spanner wrenches, 227
Special appliance friction loss, 294
Special egress control device, 664, 665
Special locks, 530
Special-purpose nozzles, 286
Specialist level, 748
Specialized ensembles, 133
Specialized railcars, 789
Specialized tank trucks, 784

Specialized t
Specialty h
Specialty
Specific
Specific
Split l
Spon
Sp
S

...am, 323
...2, 648
...n operations, 335
...tems, 314–329
...ions/piping, 321–324
...ol devices, 324–325
...uge systems, 320–321
dry pipe systems, 319–320
fire department operations, 335
fire protection valves, 325
fire suppression, 619–620
preaction systems, 321
residential systems, 321
returning system to service, 325
sprinkler heads, 315–317
sprinkler stops, 642, 648
water pipe systems, 318–319
Sprinkler tongs/wedges, 326
Sprinkler wedge, 642
Squad, 29
Squad company, 29
Squeegees, 635
SRS devices, 488
SSP, 315
SSU, 315
Stages of matter, 795–796
Staging, 39
Stairs, 363
Standard map symbols, 686
Standard of care, 694, 749
Standard operating guideline (SOG), 35
Standard operating procedures (SOPs), 35, 112, 113
Standard response sprinkler, 316
Standard Transportation Commodity Code (STCC), 825
Standing part, 425
Standoff distances, 866, 868
Standpipe hose packs, 248, 251
Standpipe operations, 334–335
Standpipe pack equipment, 334
Standpipe system, 255–257, 329–331
Standpipe valves, 331
Stanley, Julius, 632
State Emergency Response Committee (SERC), 745
States of matter, 98
Static electricity, 84
Static pressure, 214
Status reports, 68
Staypoles, 377
STC, 825
Steel, 349
...eel cylinders, 153
...el trusses, 361
...le, Michelle, 220
...L, 839
...tep-by-step procedures. *See* Job performance requirements
Stick number, 825
Stiff-arm method, 230, 231
Still alarm, 58
Stomach acid, 799
Stone, 349
Stops, 373
Storage hose roll, 230, 232
Stored pressure AFFF/FFFP extinguisher, 191
Stored pressure dry chemical extinguisher, 191, 192
Stored pressure foam extinguisher, 191
Stored pressure water extinguisher, 190, 191
Storz coupling, 225, 226
Stow Away from Foodstuffs placard, 765
Straight bore nozzle, 282–283
Straight finish load, 243
Straight hose lay, 270
Straight (storage) hose roll, 230, 232
Straight ladder, 376–377
Straight stream nozzle, 285
Strainer, 228
Stream application, 289–295
Stream pattern, 282
Stream shape, 282
Stream straighter, 273
Stress, 737–738
Stretchers, 480–481
Striking tools, 512–513
Strip cut, 579–581
Structural collapse. *See* Building collapse
Structural collapse rescue, 494–497
Structural fire, 102–103
Structural firefighting ensemble, 127–129
Structural fires, 102–103
Structural insulated panel (SIP), 351
Structural steel, 349
Suitcase carry, 395, 396
Suitcase ladder, 379, 380
Sulfur dioxide, 551
Sulfuric acid, 799, 876
Superficial burns, 716
Superfund Act, 750
Superfund Amendments and Reauthorization Act (SARA), 745
Superfund site, 750
Supplemental restraint system (SRS), 488
Supplied air respirator (SAR), 158, 841
Supply hose, 222–223
Support area (cold zone), 860
Surface locks, 529
Surface-to-mass ratio, 346
Surface water, 204
Survival, firefighter. *See* Firefighter survival
Suspended ceiling, 587
Swift-water ensemble, 133
Swift-water rescue, 493
Switch valves, 228
Symbols
fire extinguisher classification, 102
labels. *See* Labels
placards. *See* Placards
preplanning, 686
Synklor, 814

Tabun, 928
Tactical worksheets, 40
Tag, 725
Tag/guide line, 449, 450
Tank trucks, 776–788
BLEVE, 782–784
DOT 406/MC-306, 778
DOT 407/MC-307, 778–780
DOT 412/MC-312, 780–781
dry bulk tanks, 786
hot materials tankers, 786
intermodal tanks, 787–788
MC-331, 781
MC-338 cryogenic tankers, 784–785
spec plates, 777
specialized trucks, 784
tar trucks, 787
tube trailers, 785–786
Tanker, 204
Target hazards, 685
Tat Chlor 4, 814
TDD, 58
Team continuity, 728
Tear and run printers, 62
Technical Escort Unit (TEU), 933
Technical grade pesticides, 926
Technical rescue ensemble, 133
Technician level, 748
Technological obsolescence, 18
Telecommunicators, 48, 51, 52
Temperatures and pressurized container, 782
Tempered glass, 542
Tempered glass doors, 523
Tender shuttle operation, 211–212
Tension, 345
Tentative interim amendment (TIA), 749, 752n
Teratogens, 836
Terrorism, 611, 911–936
clandestine labs, 922–923
credibility of threat, 917–918
detection of terrorism agents, 932
drug labs, 923
explosives labs, 923–925
federal assistance, 932–934
groupings of terrorism agents, 927–932
HAZMAT crimes, 922–927
historical overview, 914–917
incident actions, 926–927
incident priorities, 933
indicators of, 920–922
potential targets, 918–920
terrorism agent labs, 925–926

Terrorism agent labs, 925–926
TEU, 933
Theory of catastrophic reform, 15
Thermal burns, 715–716
Thermal conductivity, 97
Thermal imaging camera (TIC), 466–467, 650–651
Thermal layering, 99
Thermal plume, 95
Thermal protective performance (TPP), 128
Thermal stress, 864
Third-degree burn, 716
Threaded couplings, 225, 226
Three Es of fire prevention, 659
Three-gas instruments, 898
Three-person flat carry, 396, 398
3/30 rule, 855
Threshold limit value (TLV), 838
Through-the-lock forcible entry, 536–538
Through-the-lock technique modified, 540
Throw method, 492, 493
TIA, 749, 752n
TIC, 650–651
Tidal changes, 204
Tie-rods, 373
Tilt-up construction, 350
Time weighted average (TWA), 838
Tipiclor, 814
Tipiclor 20, 814
TLV, 838
TLV-C, 838
TOFC, 789
TOG, 847, 855
Too lean, 800
Too rich, 800
Tools. *See* Equipment
Topichlor, 814
Topography, 598–599
Tormentor poles, 377
Torsion load, 345
Total flooding systems, 332
Totes, 775
Tower ladder, 375–376
Townsend, Rick, 509
Toxic gas monitor, 905
Toxic gases, 146–148
Toxic products of combustion, 800–801
Toxichlor, 814
Toxicity, 602
Toxicology, 834
Toxins, 930
TPP, 128
TRACEM, 834, 835
Tractor trailers, 776. *See also* Tank trucks
Traffic barrier, 486
Traffic calming, 487
Traffic cones, 487
Traffic control systems, 62–63
Trailers on flat cars (TOFC), 789
Training, 114–115
Training division, 33–34
Transfer of command, 36–37
Trash storage areas, 675
Trench cut, 579–581
Trench rescue, 497–498
Triage, 633, 704
Triangle Shirtwaist Factory fire, 314
Triangular cut, 579
Triangular inspection hole cut, 581
Triple combination engine company, 16
Triple-layer load, 247–250
Trousers, 128
Truck company, 29
Truck-mounted ladders, 373–376
Truss, 346, 361
Tube trailers, 785–786
Tubular locks, 529–530, 540
Tully, Donald C., 657
Tunnel vision, 461
Tunneling, 497
Turnout gear (TOG), 847, 855
Turntable, 374
TWA, 838
29 CFR 1910.120, 747, 750, 899
29 CFR 1910.134, 148, 149, 166
29 CFR 1910.156, 149
29 CFR Parts 1910 and 1926 Respiratory Protection, 609
Twin donut roll, 231, 233
Two in/two out rule, 110, 462, 609, 612
Two-person beam raise, 402, 403–404
Two-person coupling method, 229, 230
Two-person rope, 424
Two-person rung raise, 400–402
Two-person stiff-arm method, 230, 231
Two-way radios, 67
Type 1 manila rope, 420
Type A reporting system, 56
Type B reporting system, 56, 57
Type I fire-resistive construction, 353–354
Type II noncombustible construction, 354
Type III ordinary construction, 354–355
Type IV heavy timber construction, 355–356
Type V wood frame construction, 356–357

UEL, 800
Ultraviolet flame detectors, 313–314
Ultraviolet-infrared flame detectors, 314
UN/NA number, 760, 811
Uncoupling hose with spanners, 230
Underflow dam, 894, 895
Underground storage tank (UST), 791
Underwriters' playpipe, 286, 287
Undesigned load, 344
Unequal pupils, 706
Uninsulated MC-307/DOT-407, 779–780
Uninterruptible power supply (UPS), 50
Units of measurement, 77–78
Unity of command, 36
Upper explosive limit (UEL), 800
Upright cryogenic storage tanks, 794
Upright sprinkler, 316, 317
Upright storage tanks, 793
UP
Urba
te
UST, 791
UV detecto
UV-IR detec

V agent, 928
V pattern, 654
V-type collapse, 495
Vacuum (negative) pre
Vacuuming, 885, 886
Vapor cloud (plume), 775,
Vapor density, 87–88, 602, 7
Vapor dispersion, 885, 886–88
Vapor pressure, 86–88, 796–797
Vapor suppression, 296, 897–898
Variable combination fog nozzle, 28
Variable gallonage nozzle, 283
Vehicular firefighting, 599–601, 622–625
Velocity pressure, 291
Velsicol 1068, 814
Veneer, 350
Venous bleeding, 710
Vent for life strategy, 465
Ventilation, 548–590
  air movement, 562–563
  backdraft, 556–560
  benefits, 550
  breaking glass, 568–569
  building size, 587–588
  defined, 550
  dropped/hanging ceilings, 587
  factors to consider, 551–554
  flashover, 556
  horizontal, 552, 567–568
  obstacles to, 583–585
  opening doors, 569
  partial openings, 585
  partially broken windows, 585
  PPV, 573–575
  rollover, 560
  roof, 575–581
  roof materials, 586–587
  safety, 581–582
  screens, 586
  types, 563–567
  vertical, 552, 567
  weather, 588
  when needed, 560–562
  windows, 569–572, 585, 588–590
Ventilation timing, 584
Venturi principle, 297, 298
Vertical indicator post, 325
Vertical pivoting windows, 541
Vertical rescue, 491–492
Vertical ventilation, 552, 567
Vertical voids, 562
Vesicants, 929
Vesta, 6
Vicarious experience, 116
Victim removal, 467–476
Vigiles, 6
Violent tank rupture (VTR), 783–784
Visqueen, 634, 635, 644

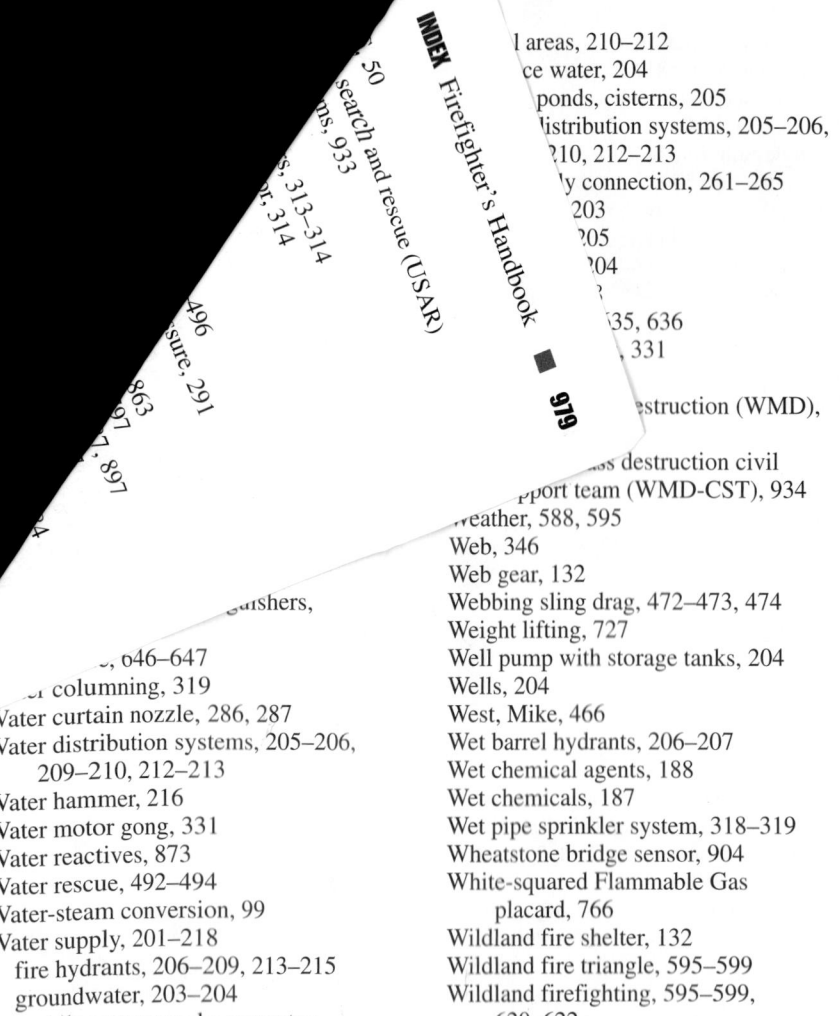

...uishers,
..., 646–647
...columning, 319
Water curtain nozzle, 286, 287
Water distribution systems, 205–206, 209–210, 212–213
Water hammer, 216
Water motor gong, 331
Water reactives, 873
Water rescue, 492–494
Water-steam conversion, 99
Water supply, 201–218
  fire hydrants, 206–209, 213–215
  groundwater, 203–204
  mobile water supply apparatus, 204–205
  obstructions, 215–216
  percent drop, 215
  pressure, 212–213, 214
...l areas, 210–212
...ce water, 204
...ponds, cisterns, 205
...distribution systems, 205–206, ...210, 212–213
...ly connection, 261–265
...203
...205
...204
...35, 636
..., 331
...estruction (WMD),
...ss destruction civil ...pport team (WMD-CST), 934
...eather, 588, 595
Web, 346
Web gear, 132
Webbing sling drag, 472–473, 474
Weight lifting, 727
Well pump with storage tanks, 204
Wells, 204
West, Mike, 466
Wet barrel hydrants, 206–207
Wet chemical agents, 188
Wet chemicals, 187
Wet pipe sprinkler system, 318–319
Wheatstone bridge sensor, 904
White-squared Flammable Gas placard, 766
Wildland fire shelter, 132
Wildland fire triangle, 595–599
Wildland firefighting, 595–599, 620–622
Wildland firefighting hose loads, 248–250
Wildland hose advancing and extension, 269
Wildland PPE, 130–132
Wildland web gear, 132
William Stieger Act, 109
Wind, 554
Windisch, F. C. (Fred), 107
Window dismount, 410, 412
Window search, 465
Windows
  bars and gates, 544
  forcible entry, 540–544
  glazing, 542–543
  opening, 588–590
  partially broken, 585
  types/styles, 541, 543–544
  ventilation, 569–572, 585, 588–590
Wire cutters, 171, 515
Wire glass, 543
Wisko, Mike, 460
WIV, 324–325
WMD, 925, 927
WMD-CST, 934
WMD laws, 927
Wood, 348–349
Wood doors, 522
Wood frame construction, 356–357
Wood sash windows, 543
Wood trusses, 361
Work hardening, 115, 116
Work uniform, 136
Working end, 425
Working length, 399, 400
World War II, 17
Wound requiring first aid, 712–713
Wutz, Thomas J., 141
Wye, 228

Yvorra, Jim, 853

Zones, 862–864

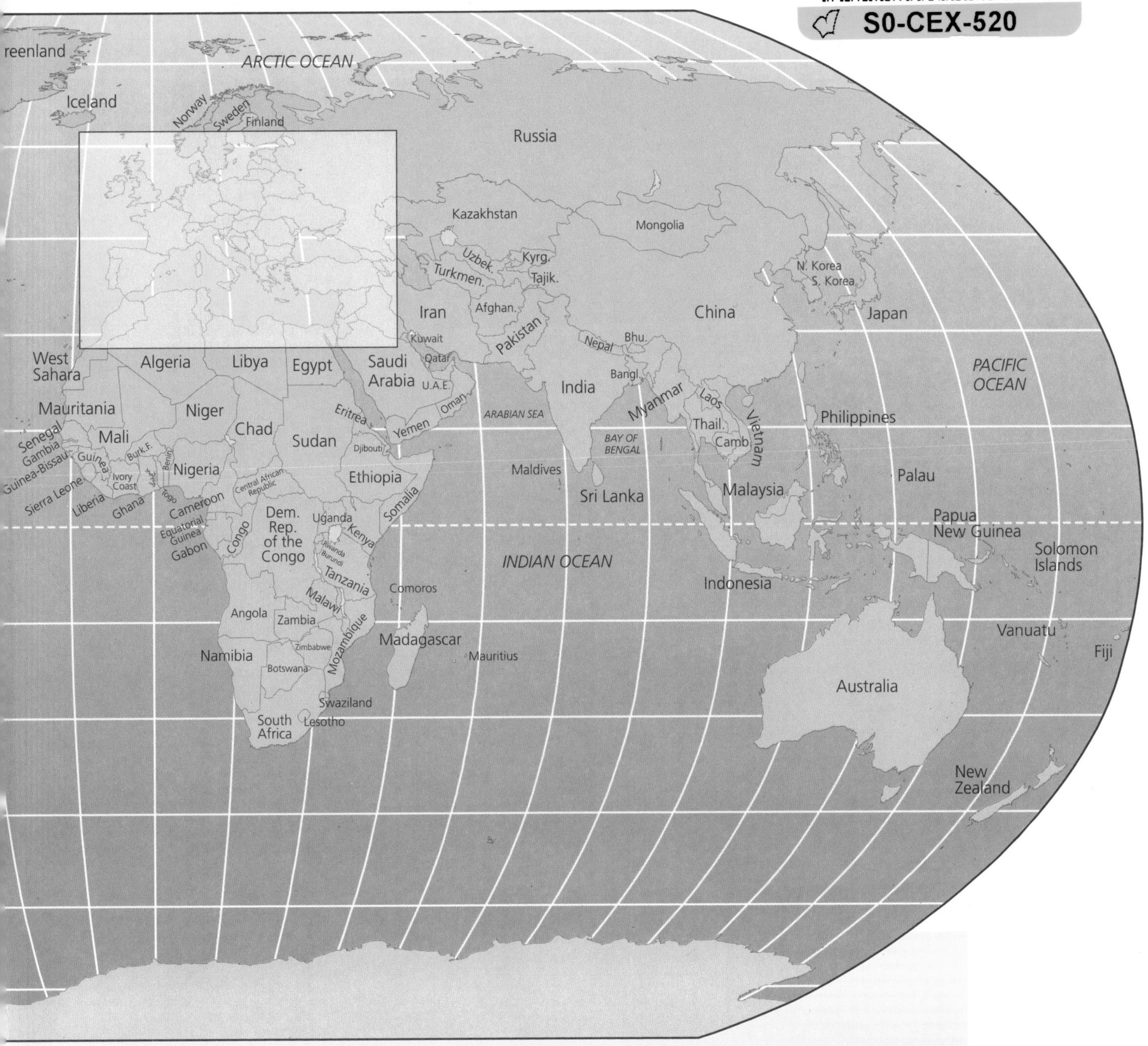
reenland
ARCTIC OCEAN
Iceland
Norway
Sweden
Finland
Russia
Kazakhstan
Mongolia
Uzbek.
Kyrg.
Turkmen.
Tajik.
N. Korea
S. Korea
Iran
Afghan.
China
Japan
Kuwait
Pakistan
Nepal
Bhu.
West Sahara
Algeria
Libya
Egypt
Saudi Arabia
Qatar
U.A.E.
India
Bangl.
PACIFIC OCEAN
Mauritania
Niger
Eritrea
Oman
ARABIAN SEA
Myanmar
Laos
Philippines
Senegal
Gambia
Guinea-Bissau
Mali
Chad
Sudan
Yemen
Thail.
Vietnam
Guinea
Burk.F.
Benin
Nigeria
Djibouti
BAY OF BENGAL
Camb.
Sierra Leone
Liberia
Ivory Coast
Ghana
Togo
Central African Republic
Ethiopia
Maldives
Palau
Cameroon
Somalia
Sri Lanka
Malaysia
Equatorial Guinea
Gabon
Congo
Dem. Rep. of the Congo
Uganda
Kenya
Rwanda
Burundi
Papua New Guinea
INDIAN OCEAN
Solomon Islands
Tanzania
Indonesia
Comoros
Malawi
Angola
Zambia
Vanuatu
Zimbabwe
Mozambique
Madagascar
Mauritius
Fiji
Namibia
Botswana
Australia
Swaziland
South Africa
Lesotho
New Zealand

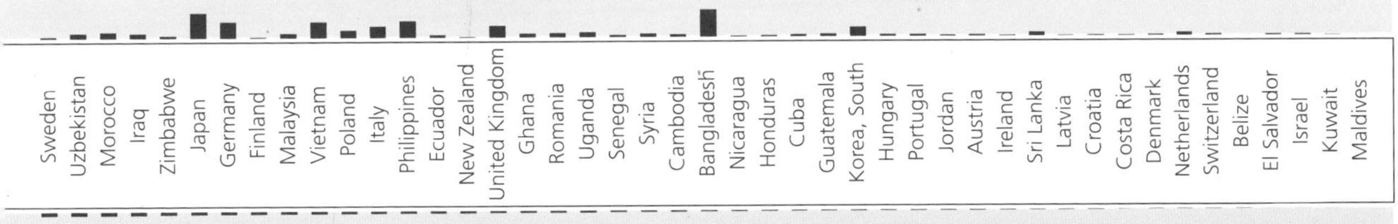
Sweden
Uzbekistan
Morocco
Iraq
Zimbabwe
Japan
Germany
Finland
Malaysia
Vietnam
Poland
Italy
Philippines
Ecuador
New Zealand
United Kingdom
Ghana
Romania
Uganda
Senegal
Syria
Cambodia
Bangladesh
Nicaragua
Honduras
Cuba
Guatemala
Korea, South
Hungary
Portugal
Jordan
Austria
Ireland
Sri Lanka
Latvia
Croatia
Costa Rica
Denmark
Netherlands
Switzerland
Belize
El Salvador
Israel
Kuwait
Maldives

ENVIRONMENT

THE SCIENCE BEHIND THE STORIES

SECOND EDITION

JAY WITHGOTT

SCOTT BRENNAN

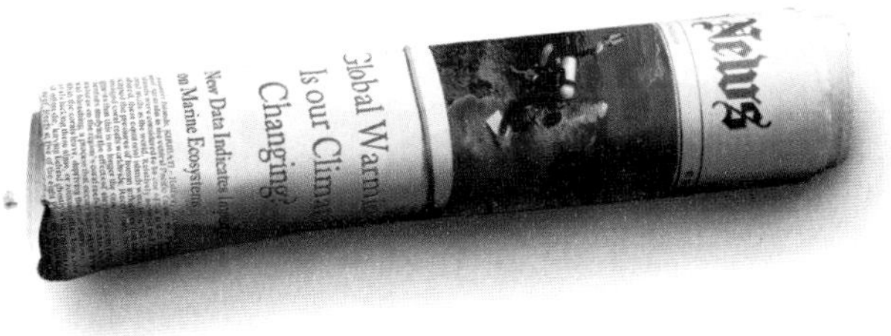

PEARSON

Benjamin Cummings

San Francisco • Boston • New York
Cape Town • Hong Kong • London • Madrid • Mexico City
Montreal • Munich • Paris • Singapore • Sydney • Tokyo • Toronto

Senior Acquisitions Editor: Chalon Bridges
Senior Project Editor: Mary Ann Murray
Executive Managing Editor: Erin Gregg
Managing Editor: Michael Early
Manufacturing Buyer: Stacy Wong
Production Supervisor: Lori Newman
Art Development: Russell Chun
Director, Media Development: Lauren Fogel
Media Producer: Ziki Dekel
Editorial Assistant: Haig MacGregor
Photo Production Manager: Travis Amos
Photo Researcher: Kristin Piljay
Marketing Manager: Jeff Hester
Production Supervisor, Media: Jennifer Mattson
Composition: TechBooks/GTS
Project Manager: Christine Knapp
Illustrations: Dragonfly Media Group
Copyeditor: Sally Peyrefitte
Proofreader: William Heckman
Design Manager: Mark Ong
Text Design: Gary Hespenheide
Cover Design: Yvo Riezebos
Cover Printer: Phoenix Color Corporation
Text Printer: Quebecor World Dubuque

Cover photograph: Phoenix Islands, Kiribati, Polynesia. Divers mapping the area find fragile table coral in Kanton Lagoon. Paul Nickelen/National Geographic/Getty Images.

Photo credits continue following the glossary.

Printed using soy-based ink. Paper is recycled containing at least 20% post-consumer waste.

ISBN 0-8053-4467-5 [Student Text Component]
ISBN 0-8053-8204-6 [Instructor]
ISBN 0-13-134642-3 [High School]

Library of Congress Cataloging-in-Publication Data

Withgott, Jay.
Environment: the science behind the stories.–2nd ed. / Jay Withgott, Scott Brennan.
p. cm.
Brennan's name appears first on the previous ed.
ISBN 0-8053-8203-8
1. Environmental sciences. I. Brennan, Scott R. II. Title.
GE105.B74 2006
363.7—dc22 2005036344

PEARSON
Benjamin Cummings

www.aw-bc.com

1 2 3 4 5 6 7 8 9 10 -QWD-08 07 06

About the Authors

Jay H. Withgott is a science and environmental writer with a background in scientific research and teaching. He holds degrees from Yale University, the University of Arkansas, and the University of Arizona. As a researcher, he has published scientific papers on topics in ecology, evolution, animal behavior, and conservation biology in a variety of journals including *Proceedings of the National Academy of Sciences, Proceedings of the Royal Society of London B, Evolution,* and *Animal Behavior.* He has taught university-level laboratory courses in ecology, ornithology, vertebrate diversity, anatomy, and general biology.

As a science writer, Jay has authored articles for a variety of journals and magazines including *Science, New Scientist, BioScience, Current Biology, Conservation in Practice,* and *Natural History.* He combines his scientific expertise with his past experience as a reporter and editor for daily newspapers to make science accessible and engaging for general audiences.

Jay lives with his wife, biologist Susan Masta, in Portland, Oregon, and takes every opportunity he can to explore the diverse landscapes of Oregon and the American West.

Scott Brennan has taught environmental science, ecology, resource policy, and journalism at Western Washington University and at Walla Walla Community College. He has also worked as a journalist, photographer, and consultant.

Scott has cultivated his expertise in environmental science and public policy by serving as Campaign Director of Alaskans for Responsible Mining, as Executive Conservation Fellow of the National Parks Conservation Association in Washington, D.C., and as a consultant to the U.S. Department of Defense Environmental Security Office at the Pentagon.

When not at work, Scott is likely to be found exploring the Chugach Mountains and the Bristol Bay drainages in southwest Alaska. He lives with his wife, Angela, and their dogs Raven and Hatcher, in south central Alaska's Chester Creek Watershed.

How do environmental issues affect

Integrated Central Case Studies begin each chapter and are further developed throughout the chapter text. These highlight real people and real places to bring environmental issues to life, making general concepts more understandable and interesting to learn.

149

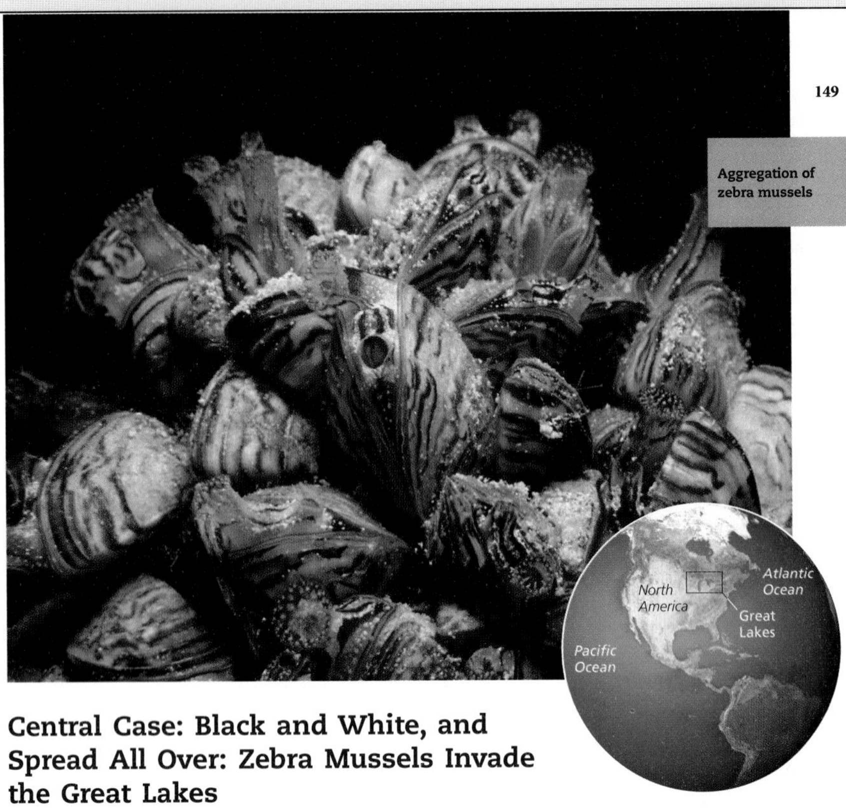

Aggregation of zebra mussels

Central Case: Black and White, and Spread All Over: Zebra Mussels Invade the Great Lakes

"We are seeing changes in the Great Lakes that are more rapid and more destructive than any time in the history of the Great Lakes."
—Andy Buchsbaum, Director of the National Wildlife Federation's Great Lakes Office

"When you tear away the bottom of the food chain, everything that is above it is going to be disrupted."
—Tom Nalepa, National Oceanic and Atmospheric Administration research biologist

As if the Great Lakes hadn't been through enough already, the last thing they needed was the zebra mussel. The pollution-fouled waters of Lake Erie and the other Great Lakes shared by Canada and the United States had become gradually cleaner in the years following the Clean Water Act of 1970. As government regulation brought industrial discharges under control, people once again began to use the lakes for recreation, and populations of fish rebounded.

Then the zebra mussel arrived. Black-and-white-striped shellfish the size of a dime, zebra mussels attach to hard surfaces and open their paired shells, feeding on algae by filtering water through their gills. This mollusc, given the scientific name *Dreissena polymorpha*, is native to the Caspian Sea, Black Sea, and Azov Sea in western Asia and eastern Europe. It made its North American debut in 1988 when it was discovered in Canadian waters at Lake St. Clair, which connects Lake Erie with Lake Huron. Evidently ships arriving from Europe had discharged ballast water containing the mussels or their larvae into the Great Lakes.

Within just two years of their discovery in Lake St. Clair, zebra mussels had reached all five of the Great Lakes. The next year, these invaders entered New York's Hudson River to the east, and the Illinois River at Chicago to the west. From the Illinois River and its canals, they soon reached the Mississippi River, giving them access to a vast watershed covering 40% of the

people and places?

INVESTIGATE it! on the Withgott/Brennan Companion Website provides an additional 120 case studies beyond those presented in the text. Browse by topic or geographic region to access 100 recent articles from The New York Times and 20 abc NEWS clips that explore environmental issues that are in the news today.

INVESTIGATE it!

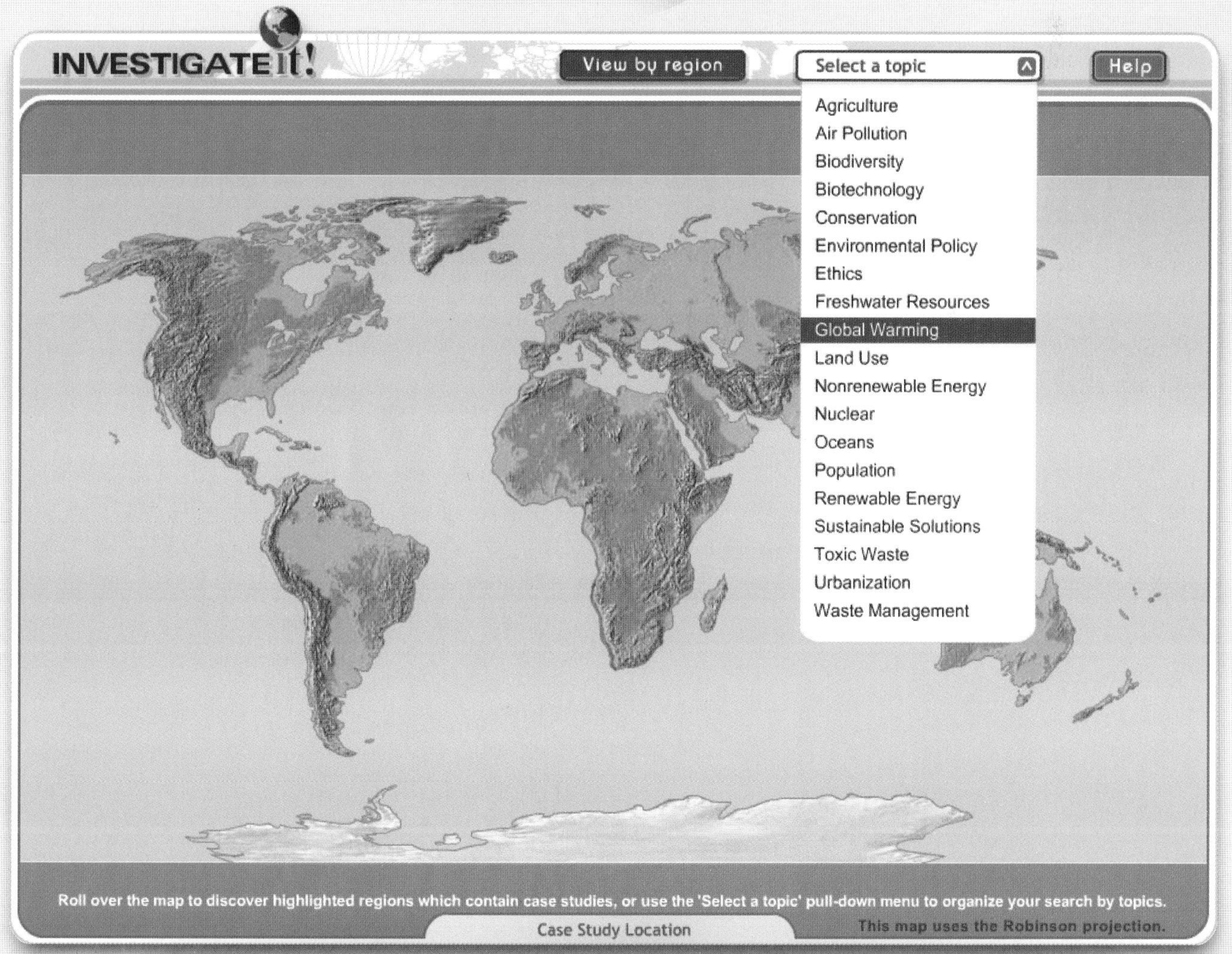

Do you understand the science behind

The Science behind the Story highlights how scientists develop hypotheses, test predictions, and analyze and interpret data. Each *Science behind the Story* carefully walks you through the scientific process—not only *what* scientists have discovered, but *how* they discovered it. These engaging accounts help you understand "how we know what we know" about environmental issues.

The Science behind the Story

Inferring Zebra Mussels' Impacts on Fish Communities

Food webs are complicated systems, and disentangling them to infer the effects of any one species is fraught with difficulty. When zebra mussels appeared in the Great Lakes, people feared for sport fisheries, and estimated that fish population declines could cost billions of dollars. The mussels would deplete the phytoplankton and zooplankton that fish depended on, people reasoned, and many fewer fish would survive. Yet even after 15 years, there was no solid evidence of widespread harm to fish populations.

So, aquatic biologist David Strayer of the Institute of Ecosystem Studies in Millbrook, New York, joined Kathyrn Hattala and Andrew Kahnle of New York State's Department of Environmental Conservation (DEC). They mined datasets on fish populations in the Hudson River, which zebra mussels had invaded in 1991.

Strayer and others had already been studying effects of zebra mussels on aspects of the community for years. Their data showed that since the species' introduction to the Hudson:

- Biomass of phytoplankton fell 80%.
- Biomass of small zooplankton fell 76%.
- Biomass of large zooplankton fell 52%.

Zebra mussels increased filter-feeding in the community 30-fold, thereby depleting the phytoplankton and small zooplankton, and leaving all sizes of zooplankton with less phytoplankton to eat. Overall, the zooplankton and invertebrate animals of the open water that are eaten by open-water fish declined by 70%.

However, Strayer's work had also found that *benthic,* or bottom-dwelling, invertebrates in shallow water (especially in the nearshore, or *littoral,* zone) had increased by 10%, and likely much more, because the mussels' shells provide habitat structure, and their feces provide nutrients.

These contrasting trends in the benthic shallows and the open deep water led Strayer's team to hypothesize that zebra mussels would harm open-water fish that ate plankton but would help littoral-feeding fish. They predicted that after zebra mussel introduction, larvae and juveniles of six common open-water fish species would decline in number, decline in growth rate, and shift downriver toward saltier water, where mussels are absent. Conversely, they predicted that larvae and juveniles of 10 littoral fish species would increase in number, increase in growth rate, and shift upriver to regions of greatest zebra mussel density.

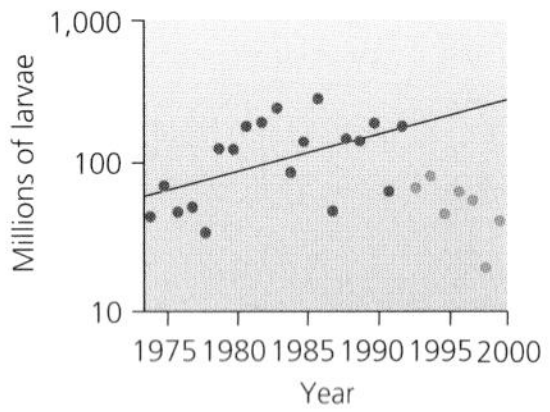

(a) American shad

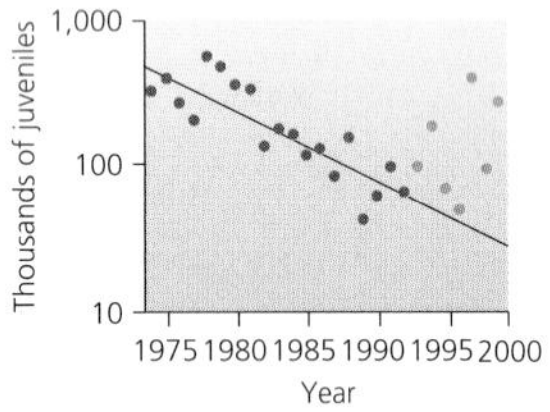

(b) Tessellated darter

Larvae of American shad **(a)**, an open-water fish, had been increasing in abundance before zebra mussels were introduced (red points and trend line). After zebra mussel introduction, shad larvae decreased in abundance (orange points). Juveniles of the tessellated darter **(b)**, a littoral zone fish, had been decreasing in abundance before zebra mussels were introduced (red points and trend line). After zebra mussel introduction, they increased in abundance (orange points). *Source:* Strayer, D., et al. 2004. Effects of an invasive bivalve (*Dreissena polymorpha*) on fish in the Hudson River estuary. *Can. J. Fish. Aquat. Sci.* 61: 924–941.

To test their predictions, the researchers analyzed data from three

community, removal of a keystone species will have substantial ripple effects and will alter a large portion of the food web.

Often, large-bodied secondary or tertiary consumers near the tops of food chains are considered keystone species. Top predators control populations of herbivores, which would otherwise multiply and could, through increased herbivory, greatly modify the plant community. In the United States, government bounties promoted the hunting of wolves and mountain lions, which were largely exterminated by the middle of the 20th century. In the absence of these predators, unnaturally dense deer

the news stories?

20 PART ONE Foundations of Environmental Science

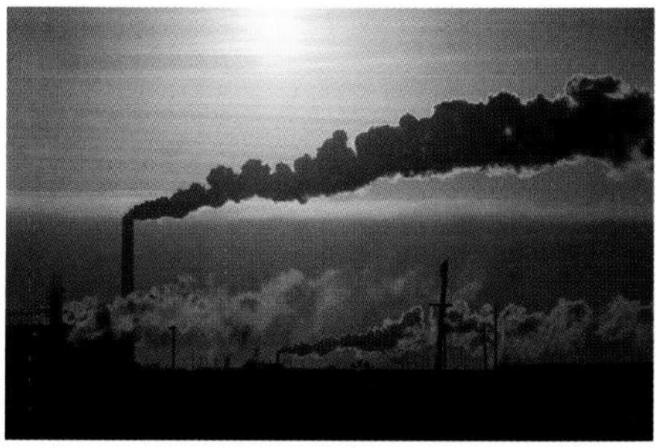

FIGURE 1.14 Indoor and outdoor air pollution contribute to millions of premature deaths each year, and environmental scientists and policymakers are working to reduce this problem in a variety of ways.

toxicologists are chronicling the impacts on people and wildlife of the many synthetic chemicals and other pollutants we emit into the environment (Chapter 14). Our most pressing pollution challenge may be to address the looming specter of global climate change (Chapter 18). Scientists have firmly concluded that human activity is altering the composition of the atmosphere and that these changes are affecting Earth's climate. Since the start of the industrial revolution, atmospheric carbon dioxide concentrations have risen by 31%, to a level not present in at least 420,000 years. This increase results from our reliance on burning fossil fuels to power our civilization. Carbon dioxide and several other gases absorb heat and warm Earth's surface, which is likely responsible for glacial melting, sea-level rise, impacts on wildlife and crops, and increased episodes of destructive weather.

The combined impact of human actions such as climate change, overharvesting, pollution, the introduction of non-native species, and particularly habitat alteration, has driven many aquatic and terrestrial species out of large parts of their ranges and toward the brink of extinction (Chapter 11). Today Earth's biological diversity, or **biodiversity,** the cumulative number and diversity of living things, is declining dramatically. Many biologists say we are already at the outset of a mass extinction event comparable to only five others documented in all of Earth's history. Biologist Edward O. Wilson has warned that the loss of biodiversity is our most serious and threatening environmental dilemma, because it is not the kind of problem that responsible human action can remedy. Rather, the extinction of species is irreversible; once a species has become extinct, it is lost forever.

Solutions to environmental problems must be global and sustainable

The nature of virtually all of these environmental issues is being changed by the set of ongoing phenomena commonly dubbed *globalization.* Our increased global interconnectedness in trade, politics, and the movement of people and of other species poses many challenging problems, but it also sets the stage for novel and effective solutions.

The most comprehensive scientific assessment of the present condition of the world's ecological systems and their ability to continue supporting our civilization was completed in 2005. In this year, over 2,000 of the world's leading environmental scientists from nearly 100 nations completed the **Millennium Ecosystem Assessment.** The four main findings of this exhaustive project are summarized in Table 1.1. The Assessment makes clear that our degradation of the world's environmental systems is having negative impacts on all of us, but that with care and diligence we can still turn many of these trends around.

Table 1.1 Main Findings of the Millennium Ecosystem Assessment

- Over the past 50 years, humans have changed ecosystems more rapidly and extensively than in any comparable period of time in human history, largely to meet rapidly growing demands for food, freshwater, timber, fiber, and fuel. This has resulted in a substantial and largely irreversible loss in the diversity of life on Earth.
- The changes made to ecosystems have contributed to substantial net gains in human well-being and economic development, but these gains have been achieved at growing costs. These costs include the degradation of ecosystems and the services they provide for us, and the exacerbation of poverty for some groups of people.
- This degradation could grow significantly worse during the first half of this century.
- The challenge of reversing the degradation of ecosystems while meeting increasing demands for their services can be partially overcome, but doing so will involve significantly changing many policies, institutions, and practices.

Adapted from *Millennium Ecosystem Assessment, Synthesis Report,* 2005.

Environmental issues change quickly, so Withgott/Brennan uses the most current data available.

References are clearly cited so you can trace the source of the information presented.

Do you know how to interpret graphs

Interpreting Graphs and Data activities at the end of each chapter give you hands-on experience in working with graphs, so you can develop the skills you'll need to understand scientific information when you see it in the news.

114 PART ONE Foundations of Environmental Science

INTERPRETING GRAPHS AND DATA

In phytoremediation, plants are used to clean up soil or water contaminated by heavy metals such as lead (Pb), arsenic (As), zinc (Zn), and cadmium (Cd). For plants to absorb these metals from soil, the metals must be dissolved in soil water. For any given instance, all metal can be accounted for as either remaining bound to soil particles, being dissolved in soil water, or being stored in the plant.

In a study on the effectiveness of alpine penny-cress (*Thlaspi caerulescens*) for phytoremediation, Enzo Lombi and his colleagues grew crops of this small perennial plant for approximately one year in pots of soil from contaminated sites. They then measured the amount of zinc and cadmium in the soil and in the plants when they were harvested.

1. What were the zinc and cadmium concentrations in the soil prior to phytoremediation? What were the zinc and cadmium concentrations in the soil after one year of phytoremediation?

2. How much zinc and cadmium were removed from the soil? If the plants continued to remove zinc and cadmium from the soil at the rates shown above, approximately how long would it take to remove all the zinc and cadmium?

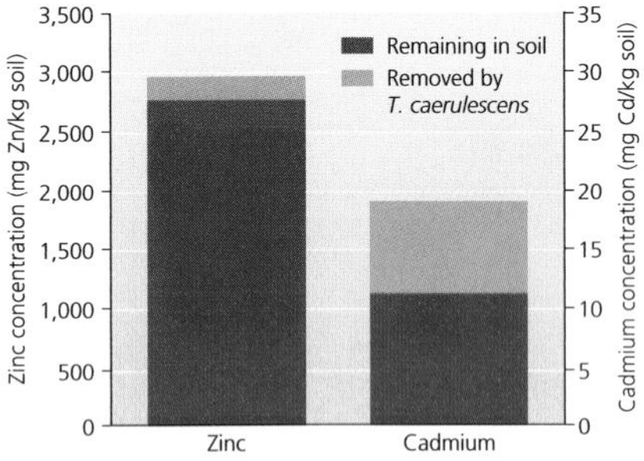

Removal of zinc and cadmium from contaminated soil by alpine penny-cress, *Thlaspi caerulescens*. Data from Lombi, E., et al. 2001. Phytoremediation of heavy metal-contaminated soils: natural hyperaccumulation versus chemically enhanced phytoextraction. *Journal of Environmental Quality* 30: 1919–1926.

3. Alpine penny-cress produces natural chelating agents (see "The Science behind the Story," pp. 92–93) that increase the solubility of metals in soil water. If these dissolved metals are not taken up by the plants, what may be an unintended consequence of having increased their solubility?

GRAPHit! exercises on the Withgott/Brennan Companion Website help you to better understand how to work with and interpret graphs.

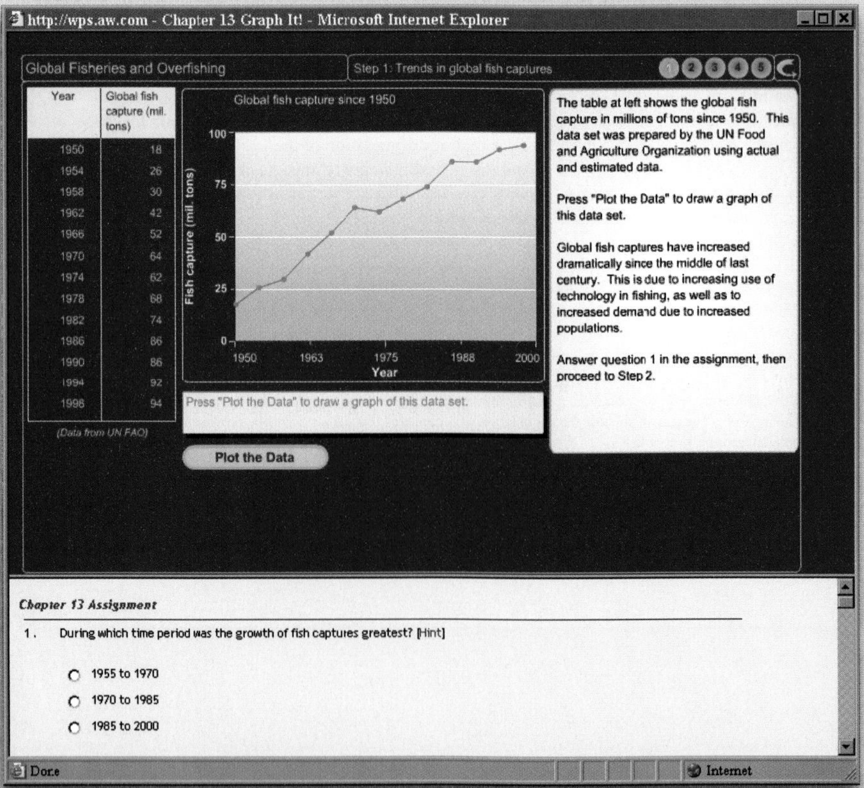

and data?

CALCULATING ECOLOGICAL FOOTPRINTS

In ecological systems, a rough rule of thumb is that when energy is transferred from plants to plant-eaters or from prey to predator, the efficiency is only about 10%. Much of this inefficiency is a consequence of the second law of thermodynamics. Another way to think of this is that eating 10 calories of plant material is the ecological equivalent of eating 1 calorie of material from an animal.

Humans are considered omnivores because we can eat both plants and animals. The choices we make about what to eat have significant ecological impacts. With this in mind, calculate the ecological energy requirements for four different diets, each of which provides a total of 2,000 dietary calories per day.

Diet	Source of calories	Number of calories consumed by source	Ecologically equivalent calories by source	Total ecologically equivalent calories per day
100% plant	Plant			
0% animal	Animal			
90% plant	Plant	1,800	1,800	3,800
10% animal	Animal	200	2,000	
50% plant	Plant			
50% animal	Animal			
0% plant	Plant			
100% animal	Animal			

1. How many ecologically equivalent calories would it take to support you for a year on each of the four diets listed?
2. What is the relative ecological impact of including as little as 10% of your calories from animal sources (e.g., milk, dairy products, eggs, and meat)? What is the ecological impact of a strictly carnivorous diet compared with a strict vegetarian diet?
3. What percentages of the calories in your own diet do you think come from plant versus animal sources? Estimate the ecological impact of your diet, relative to a strictly vegetarian one.
4. Describe some challenges of providing food for the growing human population, especially as people in many poorer nations develop a taste for an American-style diet rich in animal protein and fat.

Calculating Ecological Footprints activities at the end of each chapter let you work with numbers to evaluate the impact of actions—including your own—on a local and global scale.

Are you prepared to make informed

The *Viewpoints* feature in each chapter presents two opposing views on an environmental issue related to the chapter's central theme, allowing you to consider multiple sides of the story.

112 PART ONE Foundations of Environmental Science

VIEWPOINTS

Bioremediation

Naturally occurring microbes were put to use cleaning up beaches following the *Exxon Valdez* oil spill. **Was bioremediation a success in Prince William Sound? Can this technique address many of our society's problems with chemical contamination, or does it suffer from too many limitations?**

Bioremediation: An Effective Solution

Bioremediation was definitely a success in Prince William Sound. It was important to remove the spilled oil that reached the shorelines in an environmentally responsible way. Washing it into the sea and collecting it by skimming was the first step, but what to do about the residual oil on the beach? Bioremediation seemed a likely technology.

Oil seeps have released oil into the sea for millions of years, feeding a diverse group of microorganisms. Oil is thus very biodegradable, though unusual in that although it is rich in energy, it contains none of the inorganic nutrients required for microbial growth. The arrival of oil on a shoreline causes a dramatic increase in the number of oil-degrading microorganisms, but their growth is soon limited by the low levels of bioavailable nitrogen and phosphorus in seawater.

Our approach in Alaska was to add oleophilic (oil-adhering) and slow-release fertilizers to provide enough of these limiting nutrients over many tidal cycles so that biodegradation would be stimulated, but not so much that they would cause environmental harm. Rigorous testing, in collaboration with the State of Alaska and U.S. EPA, showed that the approach stimulated the natural rate of biodegradation some twofold to fivefold, with no detectable adverse environmental impact. Bioremediation was used on about 74 miles of shoreline—by far the largest use of this technology to date—and most shorelines were oil-free within 3 years instead of the predicted decades.

Can bioremediation address other environmental problems? Yes. There are many situations where stimulating the biodegradation of fuels, explosives, chlorinated solvents, transformer fluids, pesticides, and other substances will be an environmentally benign and effective treatment. Although the best ways of stimulating biodegradation will not be the same in all situations, working to stimulate natural processes without causing adverse effects will be an excellent way of cleaning many contaminated sites.

Roger Prince is a senior research associate at ExxonMobil's Corporate Strategic Research Laboratory in Annandale, New Jersey. He was Exxon's lead scientist in the monitoring of the successful bioremediation of shorelines oiled during the *Exxon Valdez* spill in Alaska, and he continues to work on the bioremediation of marine oil spills, including a recent multinational collaboration in the Arctic.

Biostimulation Can Sometimes Enhance Environmental Cleanup

The *Exxon Valdez* oil spill, which led to the enactment of the Oil Pollution Act of 1990, gave rise to the largest bioremediation field trial ever attempted. A research study was conducted by EPA in 1989 and 1990 to develop data to support the recommendation to go forward with a full-scale cleanup. Unfortunately, the data generated were equivocal and did not provide sufficient evidence to prove the success of the treatment. The study was equivocal because its experimental design did not provide sufficient replication or randomization of plots to allow for the calculation of experimental error, or to account for the high variability of oil contamination on the beaches.

A later field study in Delaware did provide the unequivocal evidence needed. Data generated in this and other EPA-funded field studies provided sufficient, statistically sound evidence that biostimulation of indigenous microorganisms can accelerate the disappearance of hydrocarbons at a spill site.

However, two other studies I have been involved with, both done on Canadian wetlands, have showed that biostimulation may not always be appropriate. One needs to determine the background levels of nutrients already present and make sure the affected environment is aerobic in nature. Only if nutrients are limiting in concentration, and if dissolved oxygen in the pore space is high enough to support microbial growth on the hydrocarbons, is biostimulation appropriate.

Based on various studies such as these, EPA has published two guidance documents on bioremediation on marine shorelines, freshwater wetlands, and salt marshes.

Although it is not appropriate in every case, biostimulation can play a key role in the environmental cleanup of oil spills. It is a tool to be seriously considered when contemplating how to restore a contaminated environment.

Albert D. Venosa is a senior research scientist with the U.S. Environmental Protection Agency, National Risk Management Research Laboratory, in Cincinnati, Ohio. He currently heads EPA's oil spill research program in the Office of Research and Development. He has published over 100 works in many aspects of wastewater treatment and hydrocarbon biodegradation.

Explore this issue further by accessing **Viewpoints** at www.aw-bc.com/withgott.

Reach your own conclusion on *Viewpoints* questions by accessing the *Viewpoints* link on the companion website. There you'll find questions to consider when exploring the issues further, and links to websites that support each opinion.

decisions on environmental issues?

Issues in environmental science often lack black-and-white answers, so critical thinking skills help you navigate the gray areas. ***Weighing the Issues*** questions throughout each chapter encourage you to grapple with questions about science, policy, and ethics.

Weighing the Issues:
Ecosystems Where You Live

Think about the area where you live. How would you describe that area's ecosystems? How do these systems interact with one another? If one ecosystem were greatly disturbed (say, if a wetland or forest were replaced by a shopping mall), what impacts might that have on nearby natural systems?

You Decide activities on the Withgott/Brennan Companion Website allow you to play the role of decision maker as you study the data, then form your own plan for saving endangered grizzlies or stopping global warming.

Preface

We live in extraordinary times. Human impact on our environment has never been so intensive or so far-reaching. The future of Earth's systems and of our society depends more critically than ever on the way we interact with the natural systems around us. Fundamental aspects of nutrient cycling, biological diversity, atmospheric composition, and climate are changing at dizzying speeds. Yet thanks to environmental science, we now understand better than ever how our planet's systems function and how we influence these systems. Understanding environmental science helps us to characterize human-induced problems and also illuminates the tremendous opportunities we have before us for effecting positive change.

The field of environmental science captures the very essence of this unique moment in history. An interdisciplinary field, environmental science integrates the natural sciences with the social sciences, studying both the workings of our planet and the workings of our own species. Environmental science draws upon the methods and findings of numerous established academic disciplines, from ecology to geology to chemistry to economics to political science to ethics. This interdisciplinary pursuit stands at the vanguard of the current need to synthesize academic disciplines and to incorporate their contributions into a big-picture understanding of the world and our place within it.

We wrote this book because we feel that the vital importance of environmental science in today's world makes it imperative to engage, educate, and inspire a broad audience of today's students—the citizens and leaders of tomorrow. We have therefore tried to implement the very best in modern teaching approaches and to clarify how the scientific process can inform human efforts. We also have aimed to maintain a balanced approach and to encourage critical thinking as we flesh out the social debate over many environmental issues. Finally, we have resolved to avoid gloom and doom and instead provide hope and solutions.

These several aims guided our crafting of the second edition of this text, as they had guided the first. Moreover, the second edition is significantly improved in many ways. We revisited every word of the text, incorporating the most current information from this fast-moving field and streamlining our presentation to make learning more straightforward. Dozens of new figures and an enhanced art style enable us to educate more effectively; the biogeochemical cycle figures in Chapter 7 are just one example. We expanded our coverage of ecology, energy, urbanization, and other topics that many instructors deem especially important or that are of growing interest. We also enhanced our focus on sustainability and on the ecological footprint concept, making them themes of the text, because of their central importance in environmental science. The new section on campus sustainability in our final chapter showcases examples of sustainable solutions that students are enacting on campuses worldwide.

Students often desire help with study skills, comprehending graphs and data, and understanding the impacts of their environmental choices in a concrete quantitative way. To address these needs, we added three new features to the end of each chapter. "Reviewing Objectives" summarizes each chapter's main points and relates them to the learning objectives presented at the opening of the chapter, enabling students to confirm that they have understood the most crucial ideas and to review concepts by turning to specified page numbers. "Interpreting Graphs and Data" uses figures from recent scientific studies to help students build quantitative and analytical skills in reading graphs and making sense of data. "Calculating Ecological Footprints" enables students to calculate the environmental impacts of their own choices and then see how individual impacts scale up to impacts at the societal level.

We have also retained the major features that made the first edition of our book unique and that are proving so successful in classrooms across North America:

- **Integrated Central Case Studies.** Our teaching experiences, together with feedback from colleagues across the continent, clearly reveal that telling compelling stories about real people and real places is the best way to capture students' interest. Providing narratives with concrete detail also helps teach abstract concepts, because it gives students a tangible framework with which to incorporate new ideas. Whereas many textbooks these days serve up case studies in isolated boxes, we have chosen to integrate each chapter's central case study into the main text, weaving information and elaboration throughout the chapter. In this way, we use the concrete realities of the people and places of the central case study to help illustrate the topics we cover. We are gratified that students and instructors using our first edition have consistently applauded this approach, and we hope it can help bring about a new level of effectiveness in environmental science education.

- **The Science behind the Story.** Our goal is not simply to present students with facts, but to engage them in the scientific process of testing and discovery. To do this, we discuss the scientific method and the social context of science in our opening chapter, and we describe hundreds of real-life studies throughout the main text. We also feature in each chapter "The Science behind the Story" boxes, which elaborate on particular studies important to the chapter topic, guiding readers through the details of the research. In this way we show not merely *what* scientists discovered, but also *how* they discovered it. Instructors using our first edition have confirmed that this feature enhances student comprehension of each chapter's material and deepens understanding of the scientific process itself—a key component of effective citizenship in today's science-driven world.

- **Viewpoints.** In our text we have striven to present a balanced picture of environmental issues, informed by the best science that bears upon them. Yet we all know that sometimes intelligent people can examine the same data and come to dramatically different conclusions. To ensure that students are exposed to a diversity of interpretations on key issues, we include in each chapter the *Viewpoints* feature, which consists of paired essays authored by invited experts who present different points of view on particular questions of importance. The essays provide students a taste of informed arguments directly from individuals who are actively involved in work—and debate—on environmental issues. To encourage critical thinking, we refer students to an online resource at the book's website that presents questions they can use to critically examine and discuss the ideas in these essays and that provides links to websites that support the contributors' viewpoints.

- **Weighing the Issues.** Because the multifaceted issues in environmental science often lack black-and white answers, students need critical-thinking skills to help navigate the gray areas at the juncture of science, policy, and ethics. We have aimed to help develop these skills with our end-of-chapter questions and with our "Weighing the Issues" feature. Several "Weighing the Issues" questions are dispersed throughout each chapter, serving as stopping points for students to absorb and reflect upon what they have read and wrestle with some of the complex dilemmas in environmental science.

- **An emphasis on solutions.** The complaint we most frequently hear from students in environmental science courses is that the deluge of environmental problems can seem overwhelming. In the face of so many problems, students often come to feel that there is no hope or that there is little they can personally do to make a difference. We have aimed to counter this impression by drawing out innovative solutions that have worked, are being implemented, or can be tried in the future. While we do not paint an unrealistically rosy picture of the challenges that lie ahead, we portray dilemmas as opportunities and we try to instill hope and encourage action. Indeed, for every problem that human carelessness has managed to create, human ingenuity can devise one—and likely multiple—solutions.

Environment: The Science behind the Stories has grown directly from our professional experiences in teaching, research, and writing. Jay Withgott has synthesized and presented science to a wide readership. His experience in distilling and making accessible the fruits of scientific inquiry has shaped our book's content and the presentation of its material. Scott Brennan has taught environmental science to thousands of undergraduates and has developed an intimate feeling for what works in the classroom. His knowledge and experience have shaped the pedagogical approaches we have taken in this book.

We have also been guided in our efforts by extensive input from our professional colleagues and from hundreds of instructors from around North America who have served as reviewers for our chapters and as advisors in focus group meetings arranged by Benjamin Cummings. The participation of so many learned and thoughtful experts has improved this volume in countless ways.

We sincerely hope that our efforts will come close to being worthy of the immense importance of our subject matter. We invite you, students and instructors alike, to let us know how well we have achieved our goals and where you feel we have fallen short. We are committed to continual improvement, and value your feedback. Please write the authors in care of Chalon Bridges (chalon.bridges @aw.com), Benjamin Cummings Publishing, 1301 Sansome Street, San Francisco, California, 94111.

At this most historic time to study environmental science, we are honored to serve as your guides in the quest to better understand our world and ourselves.

Jay Withgott and Scott Brennan

INSTRUCTOR SUPPLEMENTS

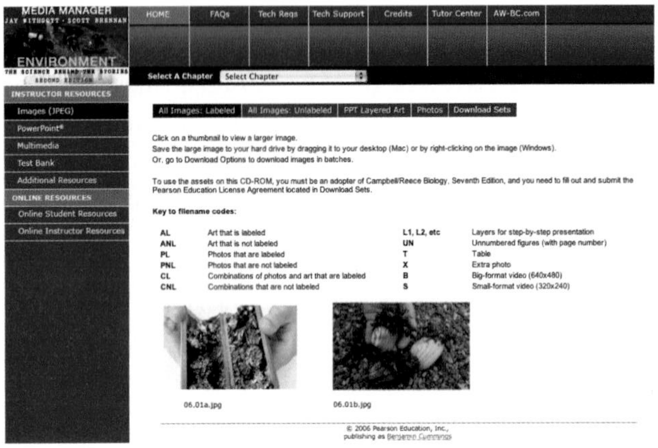

The Withgott/Brennan Media Manager
0-8053-8111-2

This powerful media package is organized chapter-by-chapter and includes all teaching resources in one convenient location. You'll find 5-minute *ABC News* Lecture Launcher videos, PowerPoint presentations, active lecture questions to facilitate class discussions (for use with or without clickers), and an image library that includes all art and tables from the text.

Instructor's Guide and Test Bank
0-8053-4468-3

This comprehensive resource provides chapter outlines, key terms, a listing of website and media resources, and teaching tips for lecture and classroom activities. A printed version of the Test Bank is conveniently included in the manual, offering hundreds of multiple-choice, short-answer, and essay questions to use on tests and quizzes. New to this edition are graphing and scenario-based questions to test students' critical-thinking abilities.

Computerized Test Bank
0-8053-8110-4

Hundreds of multiple-choice, short-answer, essay, graphing, and scenario-based questions on a cross-platform CD-ROM. Categorized by chapter objective for instructor ease in searching for question types.

Transparency Acetates
0-8053-8112-0

Includes 300 full-color acetates of all the art and tables from the text.

CourseCompass for Environment
0-8053-8198-8

This nationally hosted, easy-to-use course management tool allows professors to combine their own material with the material in the Environmental Science Place to create dynamic, online learning environments. Professors can post their syllabus, assign tutorials, customize quizzes, automatically grade them, and track the results instantly in the grade book. Go to **www.coursecompass.com**

Blackboard Premium for Environment
0-8053-8197-X

Blackboard Open Access
0-8053-8195-3

WebCT Premium for Environment
0-8053-8196-1

WebCT Open Access
0-8053-8194-5

Acknowledgments

A textbook is the product of *many* more minds and hearts than one might guess from the names on the cover. The two of us are exceedingly fortunate to be supported and guided by the tremendous staff at Benjamin Cummings and by a small army of experts in environmental science who have generously shared their time and expertise. Although we alone, as authors, bear responsibility for any inaccuracies, the strengths of this book result from the collective labor and dedication of innumerable people.

We would first like to thank our acquisitions editor, Chalon Bridges. Chalon's commitment and unremitting enthusiasm have inspired our entire team to relish the challenge of taking a successful and well-received first edition and making it still better. Moreover, her extensive interaction with instructors across North America has helped us define and refine our pedagogy and our innovative features. The approach, design, and essence of this book owe a great deal to Chalon's astute guidance and vision.

Senior project editor Mary Ann Murray worked hard every step of the way to help create this second edition, and her skill and devotion have touched every page of this book. Mary Ann's sharp eye and sound judgment improved the text in countless ways, even while she coordinated the Herculean logistics of our many features, reviews, supplements, and contributors. We are particularly grateful for her cheery optimism in the face of her authors' seemingly interminable delays.

We are excited by our book's newly enhanced art program. Senior art producer Russell Chun's ability to transform our nebulous suggestions into clear and insightful illustrations gave life and light to key points from the text. Photo researcher Kristin Piljay and photo manager Travis Amos helped provide the other half of our visual impact, reliably coming up with arresting photographic images.

We also are thrilled with two new features, "Interpreting Graphs and Data" and "Calculating Ecological Footprints." For each chapter, these were imaginatively conceived and ably authored by Jonathan Frye of McPherson College. Our thanks also go to Ned Knight of Linfield College for his insightful reviews of these features. Significant contributions from the book's first edition by April Lynch and Etienne Benson were retained in the second. We also would like to thank the authors of our Viewpoints essays, each of whom is credited along with his or her essay.

Editorial assistant Haig MacGregor was there when we needed him, coordinating reviews and a dozen other tasks. Copyeditor Sally Peyrefitte again provided thorough and meticulous examination of our text. Once the manuscript was ready, production supervisor Lori Newman saw it through to production, along with managing editors Erin Gregg and Michael Early. We thank project manager Christine Knapp and the rest of the staff at The GTS Companies for terrific work putting the whole thing together.

Finally, we would like to acknowledge the authors of our Instructor's Guide and Test Bank supplements. JodyLee Estrada Duek of Pima Community College, Debra Socci of Seminole Community College, and Steven Uyeda of Pima Community College performed extensive yet careful, quality work on a tight schedule. Kristy Manning thoroughly revised the PowerPoint slides for the main text. We also wish to thank Ziki Dekel for the development and production of the innovative media that accompanies this book.

Of course, none of this has any impact on education without the marketing staff to get the book into your hands. Marketing manager Jeff Hester dedicated his talent and enthusiasm to the book's promotion and distribution. And last but surely not least, the many field representatives who help communicate our vision and deliver our product to instructors are absolutely vital, and we deeply appreciate their work and commitment.

In the list that follows, we acknowledge the many instructors and outside experts who helped us maximize the quality and accuracy of our presentation through their chapter reviews, feature reviews, class tests, or other services. If the thoughtfulness and thoroughness of these reviewers are any indication, we feel confident that the teaching of environmental science is in excellent hands!

Lastly, Jay gives loving thanks to his wife Susan Masta, who endured the book's writing and revision with tremendous patience and sacrifice and provided support and sustenance throughout. Scott would like to thank Angela, Sean, Jonathan, Korby, Karl and Jess, and Jodi and Andy. We dedicate this book to today's students, who will shape tomorrow's world.

Jay Withgott and Scott Brennan

Reviewers

Jeffrey Albert, *Watson Institute of International Studies;* John V. Aliff, *Georgia Perimeter College;* Mary E. Allen, *Hartwick College;* Dula Amarasiriwardena, *Hampshire College;* Corey Andries, *Albuquerque Technical Vocational Institute;* David M. Armstrong, *University of* Colorado–*Boulder;* David L. Arnold, *Ball State University;* Joseph Arruda, *Pittsbur State University;* Thomas W. H. Backman, *Linfield College;* Marilynn Bartels, *Black Hawk College;* David Bass, *University of Central Oklahoma;* Christy Bazan, *Illinois State University;* Christopher Beals, *Volunteer State Community College;* Hans T. Beck, *Northern Illinois University;* Timothy Bell, *Chicago State University;* Terrence Bensel, *Allegheny College;* Gary Beluzo, *Holyoke Community College;* Bob Bennett, *University of Arkansas;* William B. N. Berry, *University of California, Berkeley;* Grady Price Blount, *Texas A & M University–Corpus Christi;* Marsha Bollinger, *Winthrop University;* Richard D. Bowden, Allegheny College; Frederick J. Brenner, *Grove City College;* Hugh Brown, *Ball State University;* J. Christopher Brown, *University of Kansas;* Dan Buresh, *Sitting Bull College;* John S. Campbell, *Northwestern College;* Mike Carney, *Jenks High School;* Kelly S. Cartwright, *College of Lake County;* Michelle Cawthorn, *Georgia Southern University;* Brad S. Chandler, *Palo Alto College;* Paul Chandler, *Ball State University;* David A. Charlet, *Community College Southern Nevada;* Kenneth E. Clifton, *Lewis and Clark College;* John E. Cochran, *Columbia Basin College;* Thomas L. Crisman, *University of Florida;* Jessica Crowe, *South Georgia College;* Gregory A. Dahlem, *Northern Kentucky University;* Mary E. Davis, *University of Massachusetts, Boston;* Thomas A. Davis, *Loras College;* Lola M. Deets, *Pennsylvania State University–Erie;* Roger del Moral, *University of Washington;* Craig Diamond, *Florida State University;* Darren Divine, *Community College of Southern Nevada;* Toby Dogwiler, *Winona State University;* Jeffrey Dorale, *University of Iowa;* Tracey Dosch, *Waubonsee Community College;* JodyLee Estrada Duek, *Pima Community College;* Jeffrey R. Dunk, *Humboldt State University;* Jean W. Dupon, *Menlo College;* Robert M. East, Jr., *Washington & Jefferson College;* Thomas R. Embich, *Harrisburg Area Community College;* Kenneth Engelbrecht, *Metropolitan State College of Denver;* Bill Epperly, *Robert Morris College;* Bonnie Fancher, *Switzerland County High School;* Francette Fey, *Macomb Community College;* Steve Fields, *Winthrop University;* Brad Fiero, *Pima Community College;* David G. Fisher, *Maharishi University of Management;* Linda Fitzhugh, *Gulf Coast Community College;* Laura Furlong, *Northwestern College;* Steven Frankel, *Northeastern Illinois University;* Arthur Fredeen, *University of Northern British Columbia;* Sandi B. Gardner, *Triton College;* Kristen S. Genet, *Anoka Ramsey Community College;* Marcia Gillette, *Indiana University–Kokomo;* Thad Godish, *Ball State University;* Michele Goldsmith, *Emerson College;* Amy R. Gregory, *University of Cincinnati;* Carol Griffin, *Grand Valley State University;* Judy Guinan, *Radford University;* Gian Gupta, *University of Maryland, Eastern Shore;* Mark Gustafson, *Texas Lutheran University;* Greg Haenel, *Elon University;* Grace Hanners, *Huntingtown High School;* Alton Harestad, *Simon Fraser University;* Barbara Harvey, *Kirkwood Community College;* Jill Haukos, *South Plains College;* Keith Hench, *Kirkwood Community College;* George Hinman, *Washington State University;* Joseph Hobbs, *University of Missouri;* Jason Hoeksema, *Cabrillo College;* Curtis Hollabaugh, *University of West Georgia;* David Hong, *Diamond Bar High School;* Debra Howell, *Chabot College;* April Huff, *North Seattle Community College;* Pamela Davey Huggins, *Fairmont State University;* Barbara Hunnicutt, *Seminole Community College;* Jon E. Hutchins, *Buena Vista University;* Daniel Hyke, *Alhambra High School;* Juana Ibáñez, *University of New Orleans;* Walter Illman, *University of Iowa;* Daniel Ippolito, *Anderson University;* Bonnie Jacobs, *Southern Methodist University;* Nan Jenks-Jay, *Middlebury College;* Stephen R. Johnson, *William Penn University;* Gina Johnston, *California State University, Chico;* Richard R. Jurin, *University of Northern Colorado;* Thomas M. Justice, *McLennan Community College;* Stanley S. Kabala, *Duquesne University;* Steve Kahl, *Plymouth State University;* Richard R. Keenan, *Providence Senior High School;* Dawn G. Keller, *Hawkeye Community College;* Ned J. Knight, *Linfield College;* Penelope M. Koines, *University of Maryland;* Alexander Kolovos, *University of North Carolina–Chapel Hill;* Erica Kosal, *North Carolina Wesleyan College;* Steven Kosztya, *Baldwin Wallace College;* John C. Kinworthy, *Concordia University;* Robert J. Koester, *Ball State University;* Jim Krest, *University of South Florida–South Florida;* Sushma Krishnamurthy, *Texas A&M International University;* James Kubicki, *Penn State University;* Diane M. LaCole, *Georgia Perimeter College;* Troy A. Ladine, *East Texas Baptist University;* Vic Landrum, *Washburn University;* Tom Langen, *Clarkson University;* Andrew Lapinski, *Reading Area Community College;* Kim D. B. Largen, *George Mason University;* Lissa Leege,

Georgia Southern University; John F. Looney, Jr., *University of Massachusetts–Boston;* Linda Lusby, *Acadia University;* Les M. Lynn, *Bergen Community College;* Richard A. Lutz, *Rutgers University;* Timothy F. Lyon, *Ball State University;* Sue Ellen Lyons, *Holy Cross School;* James G. March, *Washington & Jefferson College;* Blasé Maffia, *University of Miami;* Keith Malmos, *Valencia Community College East;* Anthony J.M. Marcattilio, *St. Cloud State University;* Patrick S. Market, *University of Missouri–Columbia;* Allan Matthias, *University of Arizona;* Jake McDonald, *University of New Mexico;* Dan McNally, *Bryant University;* Steven J. Meyer, *University of Wisconsin–Green Bay;* Kiran Misra, *Edinboro University of Pennsylvania;* Paul Montagna, *University of Texas–Austin;* Brian W. Moores, *Randolph-Macon College;* James T. Morris, *University of South Carolina;* Sherri Morris, *Bradley University;* Mary Murphy, *Penn State Abington;* Carla S. Murray, *Carl Sandburg College;* Richard A. Niesenbaum, *Muhlenberg College;* Mark P. Oemke, *Alma College;* Bruce Olszewski, *San Jose State University;* Nancy Ostiguy, *Penn State University;* David R. Ownby, *Stephen F. Austin State University;* Philip Parker, *University of Wisconsin–Platteville;* Brian D. Peer, *Simpson College;* Clayton Penniman, *Central Connecticut State;* Donald J. Perkey, *University of Alabama–Huntsville;* Raymond Pierotti, *University of Kansas;* Craig D. Phelps, *Rutgers University;* Frank X. Phillips, *McNeese State University;* Thomas E. Pliske, *Florida International University;* Avram G. Primack, *Miami University of Ohio;* Barbara Reynolds, *University of North Carolina–Asheville;* Samuel K. Riffell, *Mississippi State University;* Tom Robertson, *Portland Community College, Rock Creek Campus;* Mark Robson, *University of Medicine and Dentistry of New Jersey;* Angel M. Rodriguez, *Broward Community College;* Steven Rudnick, *University of Massachusetts–Boston;* Deanne Roquet, *Lake Superior College;* George E. Rough, *South Puget Sound Community College;* John Rueter, *Portland State University;* Shamili A. Sandiford; *College of Dupage;* Robert Sanford, *University of Southern Maine;* Ronald Sass, *Rice University;* Carl Schafer, *University of Connecticut;* Jeffery A. Schneider, *State University of New York–Oswego;* Mark Schwartz, *University of California–Davis;* Jennifer Scrafford, *Loyola College;* Richard Seigel, *Towson University.* Maureen Sevigny, *Oregon Institute of Technology;* Rebecca Sheesley, *University of Wisconsin–Madison;* William Shockner, *Community College of Baltimore County;* Christian V. Shorey, *University of Iowa;* Robert Sidorsky, *Northfield Mt. Hermon High School;* Cynthia Simon, *University of New England;* Michael Singer, *Wesleyan University;* Mark Smith, *Chaffey College;* Debra Socci, *Seminole Community College;* Ravi Srinivas, *University of St. Thomas;* Bruce Stallsmith, *University of Alabama–Huntsville;* Jeff Steinmetz, *Queens University of Charlotte;* Robert Strikwerda, *Indiana University–Kokomo;* Andrew Suarez, *University of Illinois;* Keith S. Summerville, *Drake University;* Mark L. Taper, *Montana State University;* Julienne Thomas, *Robert Morris College;* Jamey Thompson, *Hudson Valley Community College;* Todd Tracy, *Northwestern College;* Frederick R. Troeh, *Iowa State University;* Virginia Turner, *Robert Morris College;* Michael Tveten, *Pima Community College;* G. Peter van Walsum, *Baylor University;* Michael Vorwerk, *Westfield State College;* Daniel W. Ward, *Waubonsee Community College;* Caryl Waggett, *Allegheny College;* Lisa Weasel, *Portland State University;* John F. Weishampel, *University of Central Florida;* James W. C. White, *University of Colorado;* Donald L. Williams, *Park University;* Ray E. Williams, *Rio Hondo College;* Dwina Willis, *Freed-Hardeman University;* James Winebrake, *Rochester Institute of Technology;* Danielle Wirth, *Des Moines Area Community College;* Marjorie Wonham, *University of Alberta;* Wes Wood, *Auburn University;* Joan G. Wright, *Truckee Meadows Community College;* Michael Wright, *Truckee Meadows Community College;* S. Rebecca Yeomans, *South Georgia College;* Lynne Zeman, *Kirkwood Community College.*

Viewpoints Essayists

Frank Ackerman, *Tufts University;* Jock Anderson, *World Bank;* Thomas M. Bonnicksen, *University of California–Davis;* Lester Brown, *Earth Island Institute;* James T. Carlton, *Williams College;* Ignacio Chapela, *University of California–Berkeley;* Timothy L. Cline, *Population Connection;* Thomas Flint, *AgFARMation;* Pete Geddes, *Foundation for Research on Economics and the Environment;* Eban Goodstein, *Lewis and Clark College;* Karl Grossman, *State University of New York–College at Old Westbury;* Adrian Herrera, *Arctic Power;* Susan Hock, *National Renewable Energy Laboratory;* Nan Jenks-Jay, *Middlebury College;* Peter Kareiva, *The Nature Conservancy;* Debra Knopman, *RAND Corporation;* Matthew Koehler, *Native Forest Network;* Michael Leech, *International Game Fish Association;* Jane Lubchenco, *Oregon State University;* David McIntosh, *Natural Resources Defense Council;* Norman Myers, *Oxford University;* Nalini M. Nadkarni, *The Evergreen State College;* Sara Nicolas, *American Rivers;* Randal O'Toole, *American Dream Coalition;* Warren Porter, *University of Wisconsin–Madison;* Daryl Prigmore, *University of Colorado–Colorado Springs;* Roger Prince, *ExxonMobil;* John Ritch, *World Nuclear Association;* Terry Roberts, *Potash & Phosphate Institute;* Lori Saldaña, *California State Assembly;* Gavin Schmidt, *NASA/Goddard Institute for Space Studies;* Jane S. Shaw, *Property and Environment Research Center;* Daniel Simberloff, *University of Tennessee;* S. Fred Singer, *Science and Environmental Policy Project;* Gary Sirota, *Coast Law Group;* Jeff Speck, *National Endowment for the Arts;* Marian K. Stanley, *American Chemistry Council;* Douglas Sylva, *Catholic Family and Human Rights Institute;* Paul H.

Templet, *Louisiana State University;* Frederick Troeh, *Iowa State University;* Indra K. Vasil, *University of Florida;* Albert D. Venosa, *U.S. Environmental Protection Agency;* Karen Wayland, *Natural Resources Defense Council;* Nathaniel Wheelwright, *Bowdoin College.*

First Edition Reviewers

Mary E. Allen, *Hartwick College;* Dulasiri Amarasiriwardena, *Hampshire College;* Gary I. Anderson, *Santa Rosa Junior College;* Joseph A. Arruda, *Pittsburg State University;* Timothy J. Bailey, *Pittsburg State University;* Stokes Baker, *University of Detroit;* David Bass, *University of Central Oklahoma;* Timothy J. Bell, *Chicago State University;* William Berry, *University of California–Berkeley;* Kristina Beuning, *University of Wisconsin–Eau Claire;* Richard Drew Bowden, *Allegheny College;* Nancy Broshot, *Linfield College;* David Brown, *California State University–Chico;* Lee Burras, *Iowa State University;* Charles E. Button, *University of Cincinnati Clermont College;* Jon Cawley, *Roanoke College;* Linda Chalker-Scott, *University of Washington;* Sudip Chattopadhyay, *San Francisco State University;* Luke W. Cole, *Center on Race, Poverty, & the Environment;* Darren Divine, *Community College of Southern Nevada;* Iver W. Duedall, *Florida Institute of Technology;* Margaret L. Edwards-Wilson, *Ferris State University;* Anne H. Ehrlich, *Stanford University;* Thomas R. Embich, *Harrisburg Area Community College;* W. F. J. Evans, *Trent University;* Jiasong Fang, *Iowa State University;* M. Siobhan Fennessy, *Kenyon College;* Linda Mueller Fitzhugh, *Gulf Coast Community College;* Doug Flournoy, *Indian Hills Community College–Ottumwa;* Johanna Foster, *Johnson County Community College;* Chris Fox, *Catonsville Community College;* Nancy Frank, *University of Wisconsin–Milwaukee;* Robert Frye, *University of Arizona;* Sandi Gardner, *Triton College;* Marcia L. Gillette, *Indiana University–Kokomo;* Jeffrey J. Gordon, *Bowling Green State University;* John G. Graveel, *Purdue University;* Cheryl Greengrove, *University of Washington;* Amy R. Gregory, *University of Cincinnati Clermont College;* Sherri Gross, *Ithaca College;* David E. Grunklee, *Hawkeye Community College;* Mark Gustafson, *Texas Lutheran University;* Daniel Guthrie, *Claremont College;* David Hacker, *New Mexico Highlands University;* Greg Haenel, *Elon University;* David Hassenzahl, *University of Nevada Las Vegas;* Joseph Hobbs, *University of Missouri–Columbia;* Catherine Hooey, *Pittsburgh State University;* Jonathan E. Hutchins, *Buena Vista University;* Juana Ibáñez, *University of New Orleans;* Walter Illman, *University of Iowa;* Gina Johnston, *California State University–Chico;* Stanley Kabala, *Duquesne University;* Carol Kearns, *University of Colorado–Boulder;* Dawn Keller, *Hawkeye Community College;* Tom Kozel, *Anderson College;* Frank T. Kuserk, *Moravian College;* William R. Lammela, *Nazareth College;* Michael T. Lares, *University of Mary;* John Latto, *University of California–Berkeley;* Joseph Luczkovich, *East Carolina University;* Jennifer Lyman, *Rocky Mountain College;* Ian R. MacDonald, *Texas A & M University;* Kenneth Mantai, *State University of New York–Fredonia;* Patrick S. Market, *University of Missouri–Columbia;* Steven R. Martin, *Humboldt State University;* John Mathwig, *College of Lake County;* Allan Matthias, *University of Arizona;* Robert Mauck, *Kenyon College;* Debbie McClinton, *Brevard Community College;* Paul McDaniel, *University of Idaho;* Gregory McIsaac, *Cornell University;* Dan McNally, *Bryant College;* Richard McNeil, *Cornell University;* Mike L. Meyer, *New Mexico Highlands University;* Patrick Michaels, *Cato Institute;* Chris Migliaccio, *Miami Dade Community College;* Kiran Misra, *Edinboro University of Pennsylvania;* Mark Mitch, *New England College;* Brian W. Moores, *Randolph-Macon College;* James T. Morris, *University of South Carolina;* Sherri Morris, *Bradley University;* William M. Murphy, *California State University–Chico;* Rao Mylavarapu, *University of Florida;* Jane Nadel-Klein, *Trinity College;* Muthena Naseri, *Moorpark College;* Michael J. Neilson, *University of Alabama–Birmingham;* Moti Nissani, *Wayne State University;* Richard B. Norgaard, *University of California–Berkeley;* Niamh O'Leary, *Wells College;* Brian O'Neill, *Brown University;* Eric Pallant, *Allegheny College;* Phillip J. Parker, *University of Wisconsin–Platteville;* Daryl Prigmore, *University of Colorado;* Loren A. Raymond, *Appalachian State University;* Barbara C. Reynolds, *University of North Carolina–Asheville;* Thomas J. Rice, *California Polytechnic State University;* Gary Ritchison, *Eastern Kentucky University;* Mark G. Robson, *University of Medicine and Dentistry of New Jersey;* Carlton Lee Rockett, *Bowling Green State University;* Armin Rosencranz, *Stanford University;* Robert E. Roth, *The Ohio State University;* Christopher T. Ruhland, *Minnesota State University;* Ronald L. Sass, *Rice University;* Richard A. Seigel, *Towson University;* Wendy E. Sera, *NDAA's National Ocean Service;* Maureen Sevigny, *Oregon Institute of Technology;* Linda Sigismondi, *University of Rio Grande;* Jeffrey Simmons, *West Virginia Wesleyan College;* Jan Simpkin, *College of Southern Idaho;* Patricia L. Smith, *Valencia Community College;* Douglas J. Spieles, *Denison University;* Bruce Stallsmith, *University of Alabama at Huntsville;* Richard J. Strange, *University of Tennessee;* Robert A. Strikwerda, *Indiana University at Kokomo;* Richard Stringer, *Harrisburg Area Community College;* Ronald Sundell, *Northern Michigan University;* Bruce Sundrud, *Harrisburg Area Community College;* Max R. Terman, *Tabor College;* Adrian Treves, *Wildlife Conservation Society;* Charles Umbanhowar, *St. Olaf College;* G. Peter van Walsum, *Baylor University;* Callie A. Vanderbilt, *San Juan College;* Elichia A. Venso, *Salisbury University;* Rob Viens, *Bellevue Community College;* Maud M. Walsh, *Louisiana State University;* Phillip L. Watson, *Ferris State University;* Richard

D. Wilk, *Union College;* James J. Winebrake, *Rochester Institute of Technology;* Jeffrey S. Wooters, *Pensacola Junior College;* Zhihong Zhang, *Chatham College.*

First Edition Class Testers

David Aborne, *University of Tennessee–Chattanooga;* Reuben Barret, *Prairie State College;* Morgan Barrows, *Saddleback College;* Henry Bart, *LaSalle University;* James Bartalome, *University of California–Berkeley;* Christy Bazan, *Illinois State University;* Richard Beckwitt, *Framingham State College;* Elizabeth Bell, *Santa Clara University;* Peter Biesmeyer, *North Country Community College;* Donna Bivans, *Pitt Community College;* Evert Brown, *Casper College;* Christina Buttington, *University of Wisconsin, Milwaukee;* Tait Chirenje, *Richard Stockton College;* Reggie Cobb, *Nash Community College;* Ann Cutter, *Randolph Community College;* Lola Deets, *Pennsylvania State University–Erie;* Ed DeGrauw, *Portland Community College;* Mrs. Dockstader, *Monroe Community College;* Dee Eggers, *University of North Carolina–Asheville;* Jane Ellis, *Presbyterian College;* Paul Fader, *Freed Hardeman University;* Joseph Fail, *Johnson C. Smith University;* Brad Fiero, *Pima Community College, West Campus;* Dane Fisher, *Pfeiffer University;* Chad Freed, *Widener University;* Sue Glenn, *Gloucester County College;* Sue Habeck, *Tacoma Community College;* Mark Hammer, *Wayne State University;* Michael Hanson, *Bellevue Community College;* David Hassenzahl, *Oakland Community College;* Kathleen Hornberger, *Widener University;* Paul Jurena, *University of Texas–San Antonio;* Dawn Keller, *Hawkeye Community College;* David Knowles, *East Carolina University;* Erica Kosal, *Wesleyan College; John Logue, University of Southern Carolina Sumter;* Keith Malmos, *Valencia Community College;* Nancy Markee, *University of Nevada–Reno;* Bill Mautz, *University of New Hampshire;* Julie Meents, *Columbia College;* Mr. Getchell, *Mohawk Valley Community College;* Lori Moore, *Northwest Iowa Community College;* Elizabeth Pixley, *Monroe Community College;* John Novak, *Colgate University;* Brian Peck, *Simpson College;* Sarah Quast, *Middlesex Community College;* Roger Robbins, *East Carolina University;* Mark Schwartz, *University of California–Davis;* Julie Seiter, *University of Nevada–Las Vegas;* Brian Shmaefsky, *Kingwood College;* Diane Sklensky, *Le Moyne College;* Mark Smith, *Fullerton College;* Patricia Smith, *Valencia Community College East;* Sherilyn Smith, *Le Moyne College;* Jim Swan, *Albuquerque Technical Vocational Institute;* Amy Treonis, *Creighton University;* Darrell Watson, *The University of Mary Hardin Baylor;* Barry Welch, *San Antonio College;* Susan Whitehead, *Becker College;* Roberta Williams, *University of Nevada–Las Vegas;* Justin Williams, *Sam Houston University;* Tom Wilson, *University of Arizona.*

Brief Contents

PART ONE
Foundations of Environmental Science

1 An Introduction to Environmental Science 2
2 Environmental Ethics and Economics: Values and Choices 26
3 Environmental Policy: Decision Making and Problem Solving 57
4 From Chemistry to Energy to Life 88
5 Evolution, Biodiversity, and Population Ecology 116
6 Species Interactions and Community Ecology 148
7 Environmental Systems and Ecosystem Ecology 182

PART TWO
Environmental Issues and the Search for Solutions

8 Human Population 216
9 Soil and Agriculture 244
10 Agriculture, Biotechnology, and the Future of Food 276
11 Biodiversity and Conservation Biology 308
12 Resource Management, Forestry, Land Use, and Protected Areas 342
13 Urbanization and Creating Livable Cities 372
14 Environmental Health and Toxicology 400
15 Freshwater Resources: Natural Systems, Human Impact, and Conservation 432
16 The Oceans: Natural Systems, Human Use, and Marine Conservation 465
17 Atmospheric Science and Air Pollution 496
18 Global Climate Change 528
19 Fossil Fuels: Energy and Impacts 556
20 Conventional Energy Alternatives 589
21 New Renewable Energy Alternatives 619
22 Waste Management 646
23 Sustainable Solutions 674

Appendix A: Some Basics on Graphs A-1
Appendix B: Metric System B-1
Appendix C: Periodic Table of the Elements C-1
Appendix D: Geologic Time Scale D-1

Detailed Contents

PART ONE
Foundations of Environmental Science

1 An Introduction to Environmental Science 2

Our Island, Earth 3

The Nature of Environmental Science 10

The Nature of Science 12

Sustainability and the Future of Our World 19

The Science behind the Story: *The Lesson of Easter Island* 8

2 Environmental Ethics and Economics: Values and Choices 26

CENTRAL CASE: *The Mirrar Clan Confronts the Jabiluka Uranium Mine* 27

Culture, Worldview, and the Environment 28

Environmental Ethics 30

Economics: Approaches and Environmental Implications 38

The Science behind the Story: *Gross Domestic Product versus Genuine Progress Indicator* 46

The Science behind the Story: *Calculating the Economic Value of Earth's Ecosystems* 50

Viewpoints: *Environment versus Economy?* 52

3 Environmental Policy: Decision Making and Problem Solving 57

CENTRAL CASE: *San Diego and Tijuana's Sewage Pollution Problems and Policy Solutions* 58

Environmental Policy: An Introduction 59

U.S. Environmental Policy: An Overview 62

Approaches to Environmental Policy 70

The Environmental Policy Process 75

International Environmental Policy 81

The Science behind the Story: *Spotting Sewage by Satellite* 76

The Science behind the Story: *Assessing the Environmental Impact of Treating Transboundary Sewage* 78

Viewpoints: *Public versus Private: In Whose Best Interest?* 84

4 From Chemistry to Energy to Life 88

CENTRAL CASE:
Bioremediation of the Exxon Valdez *Oil Spill* 89

Chemistry and the Environment 90

Energy Fundamentals 103

The Origin of Life 108

The Science behind the Story: *Student Chemist Lets Plants Do the Dirty Work* 92

The Science behind the Story: *How Isotopes Reveal Secrets of Earth and Life* 94

Viewpoints: *Bioremediation* 112

5 Evolution, Biodiversity, and Population Ecology 116

CENTRAL CASE:
Striking Gold in a Costa Rican Cloud Forest 117

Evolution as the Wellspring of Earth's Biodiversity 118

Levels of Ecological Organization 129

Population Ecology 131

The Conservation of Biodiversity 141

The Science behind the Story: *The K-T Mass Extinction* 128

The Science behind the Story: *Climate Change and Its Effects on Monteverde* 140

Viewpoints: *Conservation at Monteverde* 144

6 Species Interactions and Community Ecology 148

CENTRAL CASE:
Black and White, and Spread All Over: Zebra Mussels Invade the Great Lakes 149

Species Interactions 150

Ecological Communities 156

Earth's Biomes 170

The Science behind the Story: *Inferring Zebra Mussels' Impacts on Fish Communities* 160

The Science behind the Story: *Otters, Urchins, Kelp, and a Whale of a Chain Reaction* 164

Viewpoints: *Invasive Species* 169

7 Environmental Systems and Ecosystem Ecology 182

CENTRAL CASE:
The Gulf of Mexico's "Dead Zone" 183

Earth's Environmental Systems 184

Ecosystems 189

Biogeochemical Cycles 195

Geological Systems: How Earth Works 206

The Science behind the Story: *Biosphere 2* 190

The Science behind the Story: *Hypoxia and the Gulf of Mexico's "Dead Zone"* 202

Viewpoints: *The Dead Zone* 210

PART TWO
Environmental Issues and the Search for Solutions

8 Human Population 216

CENTRAL CASE:
China's One-Child Policy 217

Human Population Growth: Baby 6 Billion and Beyond 218

Demography 222

Population and Society 229

The Science behind the Story: *Causes of Fertility Decline in Bangladesh* 230

The Science behind the Story: *AIDS Resistance Genes and the Black Death: An Unexpected Connection?* 238

Viewpoints: *Population Control* 234

9 Soil and Agriculture 244

CENTRAL CASE:
No-Till Agriculture in Southern Brazil 245

Soil: The Foundation for Feeding a Growing Population 246

Soil as a System 250

Soil Degradation: Problems and Solutions 255

The Science behind the Story: *Measuring Erosion* 256

The Science behind the Story: *Overgrazing and Fire Suppression in the Malpai Borderlands* 270

Viewpoints: *Soil Conservation* 269

10 Agriculture, Biotechnology, and the Future of Food 276

CENTRAL CASE:
Possible Transgenic Maize in Oaxaca, Mexico 277

The Race to Feed the World 278

Pests and Pollinators 282

Genetic Modification of Food 287

Preserving Crop Diversity 295

Feedlot Agriculture: Livestock and Poultry 296

Aquaculture 299

Sustainable Agriculture 300

The Science behind the Story: *Transgenic Contamination of Native Maize?* 292

The Science behind the Story: *Organic Farming* 302

Viewpoints: *Genetically Modified Foods* 294

11 Biodiversity and Conservation Biology 308

CENTRAL CASE:
Saving the Siberian Tiger 309

Our Planet of Life 310

Biodiversity Loss and Species Extinction 316

Benefits of Biodiversity 324

Conservation Biology: The Search for Solutions 329

The Science behind the Story: *Amphibian Diversity and Amphibian Declines* 322

The Science behind the Story: *Testing and Applying Island Biogeography Theory* 332

Viewpoints: *Biodiversity* 338

12 Resource Management, Forestry, Land Use, and Protected Areas 342

CENTRAL CASE: ***Battling over the Last Big Trees at Clayoquot Sound*** 343

Resource Management 344

Forest Management 346

Agricultural Land Use 361

Parks and Reserves 364

The Science behind the Story: *Adaptive Management and the Northwest Forest Plan* 348

The Science behind the Story: *Using Geographic Information Systems to Survey Earth's Forests* 352

Viewpoints: *Managing Forests* 359

13 Urbanization and Creating Livable Cities 372

CENTRAL CASE: ***Managing Growth in Portland, Oregon*** 373

Our Urbanizing World 374

Sprawl 378

Creating Livable Cities 384

Urban Sustainability 393

The Science behind the Story: *Measuring the Impacts of Sprawl* 382

The Science behind the Story: *Assessing Benefits of Rail Transit* 390

Viewpoints: *Suburban Sprawl* 381

14 Environmental Health and Toxicology 400

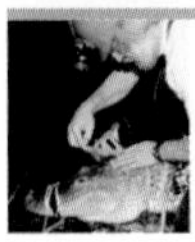

CENTRAL CASE: ***Alligators and Endocrine Disruptors at Lake Apopka, Florida*** 401

Environmental Health 402

Toxic Agents in the Environment 407

Studying Effects of Hazards 417

Risk Assessment and Risk Management 422

Philosophical and Policy Approaches 424

The Science behind the Story: *Bisphenol-A* 412

The Science behind the Story: *Pesticides and Child Development in Mexico's Yaqui Valley* 418

Viewpoints: *Chemical Product Testing: Industry or Government?* 427

15 Freshwater Resources: Natural Systems, Human Impact, and Conservation 432

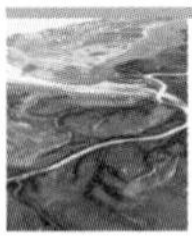

CENTRAL CASE: ***Plumbing the Colorado River*** 433

Freshwater Systems 434

How We Use Water 440

Solutions to Freshwater Depletion 449

Freshwater Pollution and Its Control 451

Wastewater and Its Treatment 458

The Science behind the Story: *Arsenic in the Waters of Bangladesh* 456

The Science behind the Story: *Using Nature to Treat Our Wastewater* 460

Viewpoints: *Dam Removal* 448

16 The Oceans: Natural Systems, Human Use, and Marine Conservation 465

CENTRAL CASE: *Seeding the Seas with Marine Reserves* 466

Oceanography 467

Marine Ecosystems 473

Human Use and Impact 478

Emptying the Oceans 482

Marine Conservation Biology 486

The Science behind the Story: *China's Fisheries Data* 486

The Science behind the Story: *Do Marine Reserves Work?* 490

Viewpoints: *Marine Reserves* 489

17 Atmospheric Science and Air Pollution 496

CENTRAL CASE: *The 1952 "Killer Smog" of London* 497

Atmospheric Science 498

Outdoor Air Pollution 505

Indoor Air Pollution 519

The Science behind the Story: *Identifying CFCs as the Main Cause of Ozone Depletion* 516

The Science behind the Story: *Acid Rain at Hubbard Brook Research Forest* 520

Viewpoints: *Air Pollution* 512

18 Global Climate Change 528

CENTRAL CASE: *Rising Temperatures and Seas May Take the Maldives Under* 529

Earth's Hospitable Climate 530

Methods of Studying Climate Change 534

Climate Change Estimates and Predictions 540

Debate Over Climate Change 546

Strategies for Reducing Emissions 548

The Science behind the Story: *Understanding El Niño and La Niña* 536

The Science behind the Story: *Scientists Use Pollen to Study Past Climate* 538

Viewpoints: *Global Climate Change* 551

19 Fossil Fuels: Energy and Impacts 556

CENTRAL CASE: *Oil or Wilderness on Alaska's North Slope?* 557

Sources of Energy 559

Coal 564

Oil 565

Natural Gas 574

Environmental Impacts of Fossil Fuel Use 575

Political, Social, and Economic Impacts of Fossil Fuel Use 579

Energy Conservation 582

The Science behind the Story: *How Electricity is Generated* 566

The Science behind the Story: *How Crude Oil is Refined* 572

Viewpoints: *Drilling for Oil in ANWR* 584

20 Conventional Energy Alternatives 589

CENTRAL CASE: *Sweden's Search for Alternative Energy* 590

Alternatives to Fossil Fuels 591

Nuclear Power 592

Biomass Energy 607

Hydroelectric Power 612

The Science behind the Story: *Assessing Emissions from Power Sources* 598

The Science behind the Story: *Health Impacts of Chernobyl* 604

Viewpoints: *Nuclear Power* 606

21 New Renewable Energy Alternatives 619

CENTRAL CASE:
Iceland Moves toward a Hydrogen Economy 620

"New" Renewable Energy Sources 621

Solar Energy 623

Wind Energy 627

Geothermal Energy 633

Ocean Energy Sources 635

Hydrogen 637

The Science behind the Story: *Idaho's Wind Prospectors* 632

The Science behind the Story: *Algae as a Hydrogen Fuel Source* 638

Viewpoints: *Hydrogen and Renewable Energy* 641

22 Waste Management 646

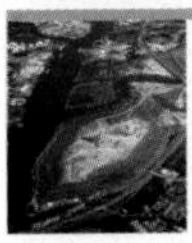

CENTRAL CASE:
Transforming New York's Fresh Kills Landfill 647

Approaches to Waste Management 648

Municipal Solid Waste 649

Industrial Solid Waste 663

Hazardous Waste 664

The Science behind the Story: *Digging Garbage: The Archaeology of Solid Waste* 654

The Science behind the Story: *Testing the Toxicity of "E-Waste"* 668

Viewpoints: *Recycling* 661

23 Sustainable Solutions 674

CENTRAL CASE:
Ball State University Aims for Campus Sustainability 675

Sustainability on Campus 676

Sustainability and Sustainable Development 682

Strategies for Sustainability 685

Precious Time 695

The Science behind the Story: *Assessing Costs and Benefits of Environmental Regulations* 686

Viewpoints: *Sustainability* 694

Appendix A: Some Basics on Graphs A-1

Appendix B: Metric System B-1

Appendix C: Periodic Table of the Elements C-1

Appendix D: Geologic Time Scale D-1

Glossary G-1

Credits CR-1

Selected Sources R-1

Index I-1

PART ONE

Foundations of Environmental Science

Researcher studying eucalyptus forest, Australia

1 An Introduction to Environmental Science

Our island, Earth

Upon successfully completing this chapter, you will be able to:

- Define the term *environment*
- Describe natural resources and explain their importance to human life
- Characterize the interdisciplinary nature of environmental science
- Understand the scientific method and how science operates
- Diagnose and illustrate some of the pressures on the global environment
- Evaluate the concepts of sustainability and sustainable development

Our Island, Earth

Viewed from space, our home planet resembles a small blue marble suspended against a vast inky-black backdrop. Although few of us will ever get to witness that sight directly, photographs taken by astronauts convey a sense that Earth is small, isolated, and fragile. It may seem vast to us as we go about our lives on its surface, but from the astronaut's perspective it is apparent that Earth and its natural systems are not unlimited. From this perspective, it becomes clear that as our population, our technological powers, and our consumption of resources increase, so do our abilities to alter our planet and damage the very systems that keep us alive.

Our environment is the sum total of our surroundings

A photograph of Earth reveals a great deal, but it does not convey the complexity of our environment. Our **environment** (a term that comes from the French *environner,* "to surround") is more than water, land, and air; it is the sum total of our surroundings. It includes all of the **biotic factors,** or living things, with which we interact. It also includes the **abiotic factors,** or nonliving things, with which we interact. Our environment includes the continents, oceans, clouds, and ice caps you can see in the photo of Earth from space, as well as the animals, plants, forests, and farms that comprise the landscapes around us. In a more inclusive sense, it also encompasses our built environment, the structures, urban centers, and living spaces humans have created. In its most inclusive sense, our environment also includes the complex webs of scientific, ethical, political, economic, and social relationships and institutions that shape our daily lives.

From day to day, people most commonly use the term *environment* in the first, narrow sense—of a nonhuman or "natural" world apart from human society. This connotation is unfortunate, because it masks the very important fact that humans exist within the environment and are a part of nature. As one of many species of animals on Earth, we share with others the same dependence on a healthy functioning planet. The limitations of language make it all too easy to speak of "people and nature," or "human society and the environment," as though they are separate and do not interact. However, the fundamental insight of environmental science is that we are part of the natural world and that our interactions with other parts of it matter a great deal.

Environmental science explores interactions between humans and our environment

Appreciating how we interact with our environment is crucial for a well-informed view of our place in the world and for a mature awareness that we are one species among many on a planet full of life. Understanding our relationship with the environment is also vital because we are altering the very natural systems we need, in ways we do not yet fully comprehend.

We depend utterly on our environment for air, water, food, shelter, and everything else essential for living. However, our actions modify our environment, whether we intend them to or not. Many of these actions have enriched our lives, bringing us longer life spans, better health, and greater material wealth, mobility, and leisure time. However, these improvements have often degraded the natural systems that sustain us. Impacts such as air and water pollution, soil erosion, and species extinction can compromise human well-being, pose risks to human life, and threaten our ability to build a society that will survive and thrive in the long term. The elements of our environment were functioning long before the human species appeared, and we would be wise to realize that we need to keep these elements in place.

Environmental science is the study of how the natural world works, how our environment affects us, and how we affect our environment. We need to understand our interactions with our environment because such knowledge is the essential first step toward devising solutions to our most pressing environmental problems. Many environmental scientists are taking this next step, trying to apply their knowledge to develop solutions to the many environmental challenges we face.

It can be daunting to reflect on the sheer magnitude of environmental dilemmas that confront us today, but with these problems also come countless opportunities for devising creative solutions. The topics studied by environmental scientists are the most centrally important issues to our world and its future. Right now, global conditions are changing more quickly than ever. Right now, through science, we as a civilization are gaining knowledge more rapidly than ever. And right now, the window of opportunity for acting to solve problems is still open. With such bountiful challenges and opportunities, this particular moment in history is indeed an exciting time to be studying environmental science.

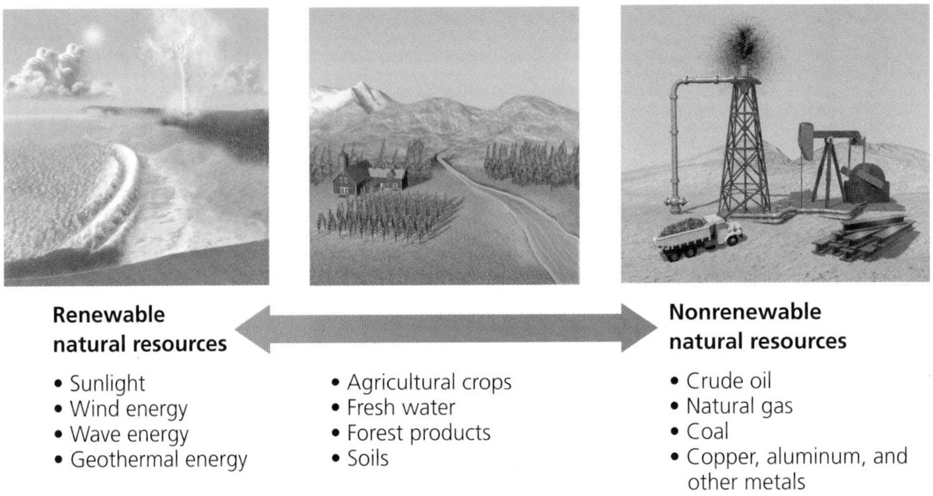

FIGURE 1.1 Natural resources lie along a continuum from perpetually renewable to nonrenewable. Perpetually renewable resources, such as sunlight, will always be there for us. Nonrenewable resources, such as oil and coal, exist in limited amounts that could one day be gone. Other resources, such as timber, soils, and food crops, can be renewed on intermediate time scales, if we are careful not to deplete them.

Natural resources are vital to our survival

An island by definition is finite and bounded, and its inhabitants must cope with limitations in the materials they need. On our island, Earth, human beings, like all living things, ultimately face environmental constraints. Specifically, there are limits to many of our **natural resources,** the various substances and energy sources we need to survive. Natural resources that are virtually unlimited or that are replenished over short periods are known as **renewable natural resources.** Some renewable resources, such as sunlight, wind, and wave energy, are perpetually available. Others, such as timber, food crops, water, and soil, renew themselves over months, years, or decades, if we are careful not to use them up too quickly or destructively. In contrast, resources such as mineral ores and crude oil are in finite supply and are formed much more slowly than we use them. These are known as **nonrenewable natural resources.** Once we use them up, they are no longer available.

We can view the renewability of natural resources as a continuum (Figure 1.1). Some renewable resources may turn nonrenewable if we overuse them. For example, overpumping groundwater can deplete underground aquifers and turn a lush landscape into a desert. Populations of animals and plants we harvest from the wild may be renewable if we do not overharvest them but may vanish if we do. In recent years, our consumption of natural resources has increased greatly, driven by rising affluence and the growth of the largest human population in history.

Human population growth has shaped our relationship with natural resources

For nearly all of human history, only a few million people populated Earth at any one time. Although past populations cannot be calculated precisely, Figure 1.2 gives some idea of just how recently and suddenly our population has grown beyond 6 billion people.

Two phenomena triggered remarkable increases in population size. The first was our transition from a hunter-gatherer lifestyle to an agricultural way of life. This change began to occur around 10,000 years ago and is known as the **agricultural revolution.** As people began to grow their own crops, raise domestic animals, and live sedentary lives in villages, they found it easier to meet their nutritional needs. As a result, they began to live longer and to produce more children who survived to adulthood. The second notable phenomenon, known as the **industrial revolution,** began in the mid-1700s. It entailed a shift from rural life, animal-powered agriculture, and manufacturing by craftsmen, to an urban society powered by **fossil fuels** (nonrenewable energy sources, such as oil, coal, and natural gas, produced by the decomposition and fossilization of ancient life). The industrial revolution introduced improvements in sanitation and medical technology, and it enhanced agricultural production with fossil-fuel-powered equipment and synthetic fertilizer (▸ pp. 278–282).

Thomas Malthus and population growth At the outset of the industrial revolution in England, population growth was regarded as a good thing. For parents, high birth rates meant more children to support them in old age. For society, it meant a greater pool of labor for factory work.

British economist **Thomas Malthus** (1766–1834) had a different opinion. Malthus claimed that unless population growth were controlled by laws or other social strictures, the number of people would outgrow the available food

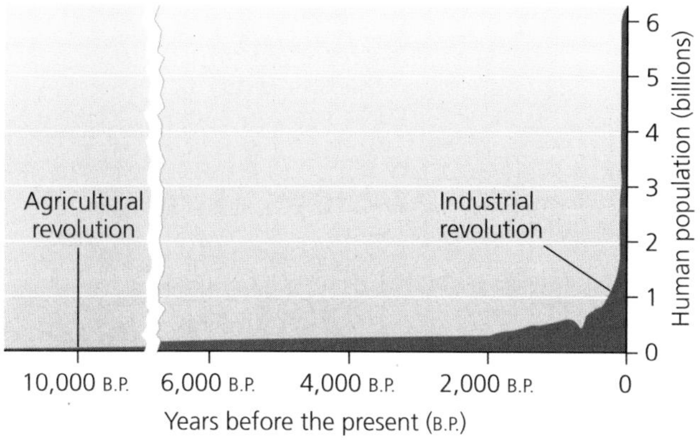

(a) World population growth

(b) Urban society

FIGURE 1.2 For almost all of human history, our population was low and relatively stable. It increased significantly as a result first of the agricultural revolution and then of the industrial revolution **(a)**. Our skyrocketing population has given rise to congested urban areas, such as this city in Java, Indonesia **(b)**.

supply until starvation, war, or disease arose and reduced the population (Figure 1.3). Malthus's most influential work, *An Essay on the Principle of Population,* published in 1798, argued that a growing population would eventually be checked either by limits on births or increases in deaths. If limits on births (such as abstinence and contraception) were not implemented soon enough, Malthus wrote, deaths would increase through famine, plague, and war.

Malthus's thinking was shaped by the rapid urbanization and industrialization he witnessed during the early years of the industrial revolution, but debates over his views continue today. As we will see in Chapter 8 and throughout this book, global population growth has indeed helped spawn famine, disease, and social and political conflict. However, increasing material prosperity has also helped bring down birth rates—something Malthus did not foresee.

Paul Ehrlich and the "population bomb" In our day, biologist Paul Ehrlich of Stanford University has been called a "neo-Malthusian" because he too has warned that population growth will have disastrous effects on human welfare. In his 1968 book, *The Population Bomb,* Ehrlich predicted that the rapidly increasing human population would unleash widespread famine and conflict that would consume civilization by the end of the 20th century. Like Malthus, Ehrlich argued that population was growing much faster than our ability to produce and distribute food, and he maintained that population control was the only way to prevent massive starvation and civil strife.

Although human population nearly quadrupled in the past 100 years—the fastest it has ever grown (see Figure 1.2a)—Ehrlich's predictions have not materialized on the scale he predicted. This is due, in part, to agricultural advances made in recent decades (► pp. 278–279). As

(a) 18th-century London, England

(b) Thomas Malthus

FIGURE 1.3 The England of Thomas Malthus's era (1766–1834), shown in this engraving **(a)**, favored population growth as society industrialized. Malthus **(b)** argued that population growth could lead to disaster.

a result, Ehrlich and other neo-Malthusians have revised their predictions accordingly and now warn of a postponed, but still impending, global crisis.

Resource consumption exerts social and environmental impacts

Population growth affects resource availability and is unquestionably at the root of many environmental problems. However, the growth in consumption is also to blame. The industrial revolution enhanced the material affluence of many of the world's people by considerably increasing our consumption of natural resources and manufactured goods.

Garrett Hardin and the "tragedy of the commons"

The late Garrett Hardin of the University of California, Santa Barbara, disputed the economic theory that unfettered exercise of individual self-interest will serve the public interest. According to Hardin's best-known essay, "The Tragedy of the Commons," published in the journal *Science* in 1968, resources that are open to unregulated exploitation will eventually be depleted.

Hardin based his argument on a scenario described in a pamphlet published in 1833. In a public pasture, or "common," that is open to unregulated grazing, Hardin argued, each person who grazes animals will be motivated to increase the number of his or her animals in the pasture. Ultimately, overgrazing will cause the pasture's food production to collapse (Figure 1.4). Because no single person owns the pasture, no one has incentive to expend effort taking care of it, and everyone takes what he or she can until the resource is depleted.

Some have argued that private ownership can address this problem. Others point to cases in which people sharing a common resource have voluntarily organized and cooperated in enforcing its responsible use. Still others maintain that the dilemma justifies government regulation of the use of resources held in common by the public, from forests to clean air to clean water.

FIGURE 1.4 Unregulated areas that offer limited resources freely to the public are prone to be depleted by the process that Garrett Hardin dubbed "the tragedy of the commons."

Weighing the Issues: The Tragedy of the Commons

Imagine you make your living fishing for lobster. You are free to boat anywhere and set out as many traps as you like. Your harvests have been good, and nothing is stopping you from increasing the number of your traps. However, all the other lobster fishers are thinking the same thing, and the fishing grounds are getting crowded. Catches decline year by year, until one year the fishery crashes, leaving you and all the others with catches too meager to support your families. Some of your fellow fishers call for dividing the waters and selling access to individuals plot-by-plot. Others urge the fishers to team up, set quotas among themselves, and prevent newcomers from entering the market. Still others are imploring the government to get involved and pass laws regulating how much fishers can catch. What do you think is the best way to combat this tragedy of the commons and restore the fishery? Why?

Wackernagel, Rees, and the ecological footprint

As global affluence has increased, human society has consumed more and more of the planet's limited resources. We can quantify resource consumption using the concept of the "ecological footprint," developed in the 1990s by environmental scientists Mathis Wackernagel and William Rees. The **ecological footprint** expresses the environmental impact of an individual or population in terms of the cumulative amount of land and water required to provide the

FIGURE 1.5 The "ecological footprint" represents the total area of land and water needed to produce the resources a given person or population uses, together with the total amount of land and water needed to dispose of their waste. The footprints of the urbanized and affluent societies of today's developed nations tend to be much larger than the geographic areas these societies take up directly. Adapted from Wackernagel, M., and W. Rees. 1996. *Our ecological footprint: Reducing human impact on the Earth.* Gabriola Island, British Columbia: New Society Publishers.

raw materials the person or population consumes and to dispose of or recycle the waste the person or population produces (Figure 1.5). It measures the total amount of Earth's surface "used" by a given person or population, once all direct and indirect impacts are totaled up.

For humanity as a whole, Wackernagel and Rees have calculated that our species is using 30% more resources than are available on a sustainable basis from all the land on the planet. That is, we are depleting renewable resources 30% faster than they are being replenished—like drawing the principal out of a bank account rather than living off the interest. Furthermore, people from wealthy nations have much larger ecological footprints than do people from poorer nations. If all the world's people consumed resources at the rate of North Americans, these researchers concluded, we would need the equivalent of two additional planet Earths.

Environmental science can help us avoid mistakes made by past civilizations

It remains to be seen whether the direst predictions of Malthus, Ehrlich, and others will come to pass for today's global society, but we already have historical evidence that civilizations can crumble when pressures from population and consumption overwhelm resource availability. Easter Island is the classic case (see "The Science behind the Story," ▸ pp. 8–9), but it is not the only example. Many great civilizations have fallen after depleting resources from their environments, and each has left devastated landscapes in its wake. The Greek and Roman empires show evidence of such a trajectory, as do the Maya, the Anasazi, and other civilizations of the New World. Plato wrote of the deforestation and environmental degradation accompanying ancient Greek cities, and today further evidence is accumulating from research by archaeologists, historians, and paleoecologists who study past societies and landscapes. The arid deserts of today's Middle Eastern countries were far more vegetated when the great ancient civilizations thrived there; at that time these regions were lush enough to support the very origin of agriculture. While deforestation created deserts in temperate regions, in more tropical climates, the ancient cities of fallen civilizations became overgrown by jungle. The gigantic stone monuments of the Angkor civilization in Southeast Asia, like those of the Maya in Mexico and Central America, remained unknown to Westerners until the

The Science behind the Story

The Lesson of Easter Island

Easter Island is one of the most remote spots on the globe, located in the Pacific Ocean 3,750 km (2,325 mi) from South America and 2,250 km (1,395 mi) from the nearest inhabited island. When the first European explorers reached the island (today called Rapa Nui) in 1722, they found a barren landscape populated by fewer than 2,000 people, who lived in caves and eked out a marginal existence from a few meager crops. However, explorers also noted that the desolate island featured hundreds of gigantic statues of carved stone, evidence that a sophisticated civilization had once inhabited the island.

The haunting statues of Easter Island were erected by a sophisticated civilization that collapsed after depleting its resource base and devastating its island environment.

Historians and anthropologists long wondered how people without wheels or ropes, on an island without trees, could have moved statues 10 m (33 ft) high weighing 90 metric tons (99 tons) as far as 10 km (6.2 mi) from the quarries where they were chiseled to the coastal sites where they were erected. The explanation, scientists have discovered, lay in the fact that the island did not always lack trees, and its people were not always without rope.

Indeed, scientific research tells us that the island had once been lushly forested, with all the appeal of a South Pacific paradise, and had supported a prosperous society with a population of 6,000 to 30,000 people. Tragically, this once-flourishing civilization overused its resources and cut down all its trees, destroying itself in a downward spiral of starvation and conflict. Today Easter Island stands as a parable and a warning for what can happen when a population grows too large and consumes too much of the limited resources that support it.

To solve the mystery of Easter Island's past, scientists have used various methods. Some, such as British scientist John Flenley, have excavated sediments from the bottom of the island's volcanic crater lakes, drilling cores deep into the mud and examining ancient grains of pollen preserved there. Because pollen grains vary from one plant species to another, scientists, by identifying specific pollen grains, can reconstruct, layer by layer, the history of vegetation in a region through time. By analyzing pollen grains under scanning electron microscopes, Flenley and other researchers found that when Polynesian people arrived (likely between A.D. 300 and A.D. 900), the island was covered with a species of palm tree related to the Chilean wine palm, a tall and thick-trunked tree. Archaeologists located ancient palm nut casings in caves and crevices, and a geologist found carbon-lined channels in the soil that matched root channels typical of the Chilean wine palm. Furthermore, scientists deciphering the island people's script on stone tablets discerned characters etched in the form of palm trees.

By studying pollen and the remains of wood from charcoal, scientists such as French archaeologist Catherine Orliac have found that at least 21 other species of plants, many of them trees, had also been common, and are now completely gone. The island had clearly supported a diverse forest. However, starting around A.D. 750, tree populations declined and ferns and grasses became more common, according to pollen analysis from one lake site. By A.D. 950, the trees were largely gone, and around A.D. 1400 overall pollen levels plummeted, indicating a dearth of vegetation. The same sequence of events occurred about two centuries later at the other two lake sites, which were higher and more remote from village areas. Researchers first hypothesized that the forest loss was due to climate change, but evidence instead supported the hypothesis that the people had gradually denuded their own island.

The palms and other trees provided fuelwood, building material for houses and canoes, fruit to eat and fiber for clothing, and presumably, logs to move the stone statues. Several anthropologists in recent years have experimentally tested hypotheses about how the islanders moved their monoliths down from the quarries, by hiring groups of men to recreate the feat. The methods that have worked involve using numerous tree trunks as rollers or sleds, as well as great quantities of rope. The only likely source of rope on the island would have been the fibrous inner bark of the hauhau tree, a species that today is near extinction.

With the trees gone, soil would have eroded away—a phenomenon confirmed by data from the bottom of Easter Island lakes, where large quantities of sediment accumulated. Faster runoff of rainwater would have meant less fresh water available for drinking. Runoff and erosion would have degraded the islanders' agricultural land, lowering yields of crops, such as bananas, sugar cane, and sweet potatoes. Reduced agricultural production would have led to starvation and subsequent population decline.

Archaeological evidence supports the scenario of environmental degradation and civilization decline. Analysis of 6,500 bones by archaeologist David Steadman has shown that at least 6 species of land birds and 25 species of seabirds nested on Easter Island and were eaten by islanders. Today no native land birds and only 1 seabird are left. Remains from charcoal fires can be aged by radiocarbon dating (▸ pp. 94–95), and show that islanders' diets shifted over the years. Besides their crops and the island's birds, early islanders feasted on the bounty of the sea, including porpoises, fish, sharks, turtles, octopus, and shellfish. Analysis of islanders' diets in the later years indicated that little seafood was consumed. With the trees gone, the islanders could no longer build the great double-canoes their proud Polynesian ancestors had used for centuries to fish and travel among islands. Indeed, the Europeans who visited Easter Island in the 1700s observed only a few old small canoes and flimsy rafts made of reeds. As resources declined, the islanders' main domesticated food animal, the chicken, became more valuable. Archaeologists found that later islanders kept their chickens in stone fortresses with entrances designed to prevent theft. The once prosperous and peaceful civilization fell into clan warfare, as revealed by unearthed skeletons, skulls with head wounds, and artifacts of weapons made of obsidian, a hard volcanic rock.

Is the story of Easter Island as unique and isolated as the island itself, or does it hold lessons for our world today? Like the Easter Islanders, we are all stranded together on an island with limited resources. Earth may be vastly larger and richer in resources than was Easter Island, but Earth's human population is also much greater. The Easter Islanders must have seen that they were depleting their resources, but it seems that they could not stop. Whether we can learn from the history of Easter Island and act more wisely to conserve the resources on our island, Earth, is entirely up to us.

19th century, and most of these cities remain covered by rainforest. Researchers have learned enough by now, however, that scientist and author Jared Diamond in his 2005 book, *Collapse,* could synthesize this information and formulate sets of reasons why civilizations succeed and persist, or fail and collapse. Success and persistence, it turns out, depend largely on how societies interact with their environments.

Today we are confronted with news and predictions of environmental catastrophes on a regular basis, but it can be difficult to assess the reliability of such reports. It is even harder to evaluate the causes and effects of environmental change. Perhaps most difficult is to devise solutions to environmental problems. Studying environmental science will outfit you with the tools that can help you evaluate information on environmental change and think critically and creatively about possible actions to take in response. Let us examine this broad field we call environmental science, and then explore the process and methods of science in general.

The Nature of Environmental Science

Environmental scientists aim to comprehend how Earth's natural systems function, how humans are influenced by those systems, and how we are influencing those systems. In addition, many environmental scientists are motivated by a desire to develop solutions to environmental quandaries. The solutions themselves (such as new technologies, policy decisions, or resource management strategies) are applications of environmental science. However, the study of such applications and their consequences is, in turn, also part of environmental science.

People vary in their perception of environmental problems

Environmental science arose in the latter half of the 20th century as people sought to better understand environmental problems and their origins. An *environmental problem,* stated simply, is any undesirable change in the environment. However, the perception of what constitutes an undesirable change may vary from one person or group of people to another, or from one context or situation to another. A person's age, gender, class, race, nationality, employment, and educational background can all affect whether he or she considers a given environmental change to be a "problem."

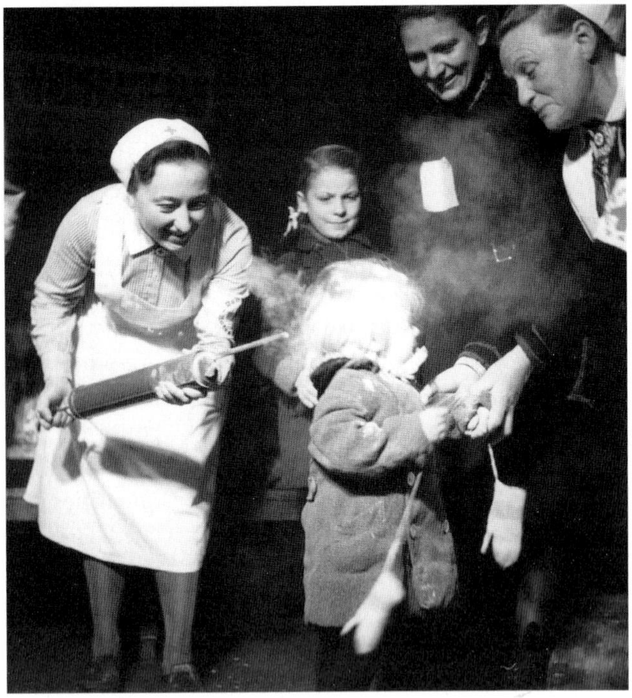

FIGURE 1.6 How a person or a society defines an environmental problem can vary with time and circumstance. In Germany in 1945, health hazards of the pesticide DDT were not yet known, so children were doused with the chemical to treat head lice. Today, knowing of its toxicity to people and wildlife, many developed nations have banned DDT. However, in some developing countries where malaria is a threat, DDT is welcomed to combat mosquitoes that transmit the disease.

For instance, today's industrial societies are more likely to view the spraying of the pesticide DDT as a problem than those societies viewed it in the 1950s, because today more is known about the health risks of pesticides (Figure 1.6). At the same time, a person living today in a malaria-infested village in Africa or India may welcome the use of DDT if it kills mosquitoes that transmit malaria, because malaria is viewed as a more immediate health threat. Thus an African and an American who have each knowledgeably assessed the pros and cons may, because of differences in their circumstances, differ in their judgment of DDT's severity as an environmental problem.

Different types of people may also vary in their awareness of problems. For example, in many cultures women are responsible for collecting water and fuelwood. As a result, they are often the first to perceive environmental degradation affecting these resources, whereas men in the same area simply might not "see" the problem. As another example, in most societies information about environmental health risks tends to reach wealthy people more readily than poor people. Thus, who you are, where you live, and what you do can have a huge effect on how you perceive your environment, how you perceive and react to change, and what impact those changes may have on how you live your life. In Chapter 2, we will examine the diversity of human values and philosophies and consider their effects on how we define environmental problems.

Environmental science provides interdisciplinary solutions

Studying and addressing environmental problems is a complex endeavor that requires expertise from many disciplines, including ecology, earth science, chemistry, biology, economics, political science, demography, ethics, and others. Environmental science is thus an **interdisciplinary** field—one that borrows techniques from numerous disciplines and brings research results from these disciplines together into a broad synthesis (Figure 1.7). Traditional established disciplines are valuable because their scholars delve deeply into topics, uncovering new knowledge and developing expertise in particular areas. Interdisciplinary fields are valuable because their practitioners take specialized knowledge from different disciplines, consolidate it, synthesize it, and make sense of it in a broad context to better serve the multifaceted interests of society.

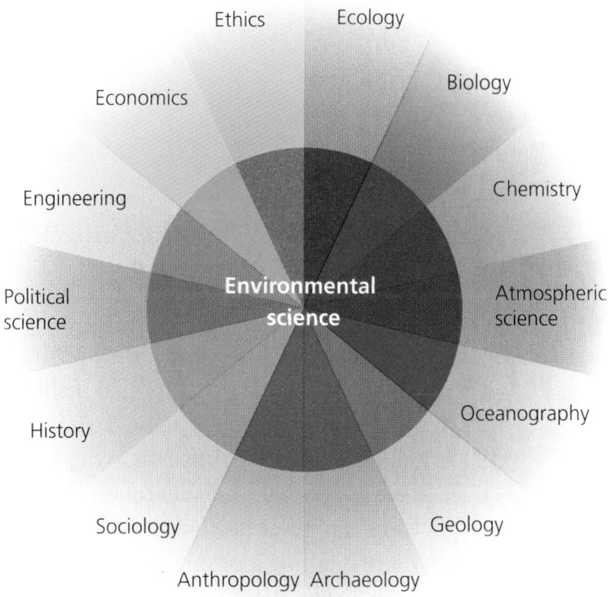

FIGURE 1.7 Environmental science is a highly interdisciplinary pursuit, involving input from many different established fields of study across the natural sciences and social sciences.

Environmental science is especially broad because it encompasses not only the **natural sciences** (disciplines that study the natural world), but also the **social sciences** (disciplines that study human interactions and institutions). The natural sciences provide us the means to gain accurate information about our environment and to interpret it reasonably. Addressing environmental problems, however, also involves weighing values and understanding human behavior, and this requires the social sciences. Most environmental science programs focus predominantly on the natural sciences as they pertain to environmental issues. In contrast, programs incorporating the social sciences heavily often prefer using the term **environmental studies** to describe their academic umbrella. Whichever approach one takes, these fields reflect many diverse perspectives and sources of knowledge.

Just as an interdisciplinary approach to studying issues can help us better understand them, an integrated approach to addressing problems can produce effective and lasting solutions. One example is the dramatic improvement in one aspect of air quality in the United States over the past few decades. Ever since automobiles were invented, lead had been added to gasoline to make cars run more smoothly, even though medical professionals knew that lead emissions from tailpipes could cause health problems, including brain damage and premature death. In 1970 air pollution was severe, and motor vehicles accounted for 78% of U.S. lead emissions. But over the following years, engineers, physicians, atmospheric scientists, and politicians all merged their knowledge and skills into a process that eventually resulted in a ban on leaded gasoline. By 1996 all gasoline sold in the United States was unleaded, and the nation's largest source of atmospheric lead emissions had been completely eliminated.

Environmental science is not the same as environmentalism

Although many environmental scientists are interested in solving problems, it would be incorrect to confuse environmental science with environmentalism, or environmental activism. They are *not* the same. Environmental science is the pursuit of knowledge about the workings of the environment and our interactions with it. **Environmentalism** is a social movement dedicated to protecting the natural world—and, by extension, humans—from undesirable changes brought about by human choices (Figure 1.8). Although environmental scientists may study many of the same issues environmentalists care about, as scientists

FIGURE 1.8 Environmental scientists and environmental activists play very different roles. Some scientists have become activists to promote particular solutions to environmental problems. However, most have not, and those who have generally try hard to keep their advocacy separate from their pursuit of objective scientific work.

they attempt to maintain an objective approach in their work. Remaining free from personal or ideological bias, and open to whatever conclusions the data demand, is a hallmark of the effective scientist. We will now proceed with a brief overview of how science works and how scientists go about this enterprise that brings our society so much valuable knowledge.

The Nature of Science

Modern scientists describe **science** (from the Latin *scire,* "to know") as a systematic process for learning about the world and testing our understanding of it. The term *science* is also commonly used to refer to the accumulated body of knowledge that arises from this dynamic process of observation, testing, and discovery.

Knowledge gained from science can be applied to address societal problems. Among the applications of science are its use in developing technology and its use in informing policy and management decisions (Figure 1.9). These pragmatic applications in themselves are not science, but they must be informed by science in order to be effective. Many scientists are motivated simply by a desire to know how the world works, and others are motivated by the potential for developing useful applications.

Environmental science is a dynamic yet systematic way of studying the world, and it is also the body of knowledge accumulated from this process. Like science in general, environmental science informs its practical applications and often is motivated by them.

Why does science matter? The late astronomer and author Carl Sagan wrote the following in his 1995 treatise, *The Demon Haunted World: Science as a Candle in the Dark:*

> We've arranged a global civilization in which the most crucial elements—transportation, communications, and all other industries; agriculture, medicine, education, entertainment, protecting the environment; and even the key democratic institution of voting—profoundly depend on science and technology. We have also arranged things so that almost no one understands science and technology. This is a prescription for disaster. We might get away with it for a while, but sooner or later this combustible mixture of ignorance and power is going to blow up in our faces. . . . Science is an attempt, largely successful, to understand the world, to get a grip on things, to get hold of ourselves, to steer a safe course.

Sagan and many other thinkers before and since have argued that science is essential if we hope to sort fact from fiction and develop solutions to the problems—environmental and otherwise—that we face today.

Scientists test ideas by weighing evidence

How can we tell whether warnings of impending environmental catastrophes—or any other claims, for that matter—are based on scientific thinking? Scientists examine ideas about how the world works by designing tests to determine whether these ideas are supported by evidence. Ideas can be refuted by evidence but can never be absolutely proven, so, strictly speaking, scientific testing amounts to attempting to disprove ideas. If a particular statement or explanation is testable and resists repeated attempts to disprove it, scientists are likely to accept it as a useful and true explanation. Scientific inquiry thus consists of an incremental approach to the truth.

(a) Prescribed burning

(b) Methanol-powered fuel-cell car

FIGURE 1.9 Scientific knowledge can be applied in policy and management decisions and in technology. Prescribed burning, shown here in the Ouachita National Forest, Arkansas (**a**), is a management practice to restore healthy forests, and is informed by scientific research into forest ecology. Energy-efficient automobiles, like this methanol-powered fuel-cell car from Daimler–Chrysler (**b**), are technological advances made possible by materials and energy research.

The scientific method is the key element of science

Scientists generally follow a process called the **scientific method.** A technique for testing ideas with observations, it involves several assumptions and a series of interrelated steps. There is nothing mysterious about the scientific method; it is merely a formalized version of the procedure any of us might naturally take, using common sense, to resolve a question.

The scientific method is a theme with variations, however, and scientists pursue their work in many different ways. Because science is an active, creative, imaginative process, an innovative scientist may find good reason to stray from the traditional scientific method when a particular situation demands it. Moreover, scientists from different fields approach their work differently because they deal with dissimilar types of information. A natural scientist, such as a chemist, will conduct research quite differently from a social scientist, such as a sociologist. Because environmental science includes both natural and social sciences, in our discussion here we use the term *science* in its broad sense, to include both. Despite their many differences, scientists of all persuasions broadly agree on fundamental elements of the process of scientific inquiry.

The scientific method relies on the following assumptions:

- The universe functions in accordance with fixed natural laws that do not change from time to time or from place to place.
- All events arise from some cause or causes and, in turn, cause other events.
- We can use our senses and reasoning abilities to detect and describe natural laws that underlie the cause-and-effect relationships we observe in nature.

As practiced by individual researchers or research teams, the scientific method (Figure 1.10) typically consists of the steps outlined below.

Make observations Advances in science typically begin with the observation of some phenomenon that the scientist wishes to explain. Observations set the scientific method in motion and also function throughout the process.

Ask questions Curiosity is a fundamental human characteristic. This is evident to anyone who has observed the explorations of a young child in a new environment. Babies want to touch, taste, watch, and listen to anything that catches their attention, and as soon as they can speak,

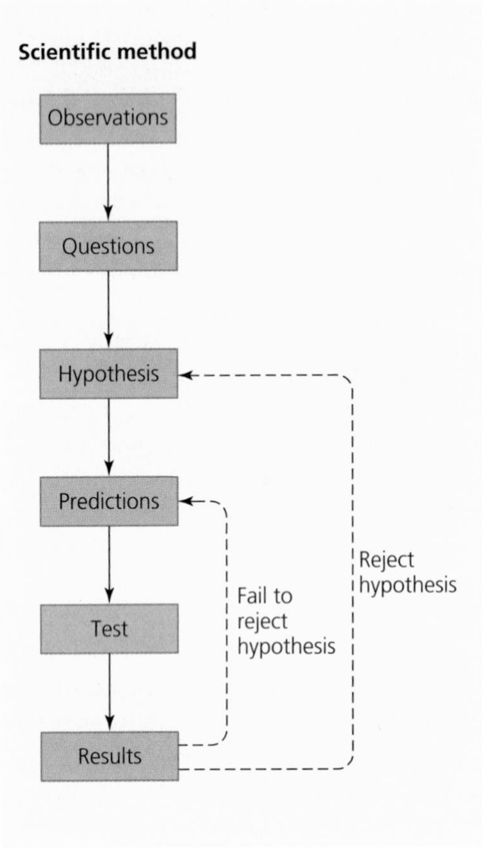

FIGURE 1.10 The scientific method is the observation-based hypothesis-testing approach that scientists use to learn how the world works. This diagram is a simplified generalization that, although useful for instructive purposes, cannot convey the true dynamic and creative nature of science. Moreover, researchers from different disciplines may pursue their work in ways that legitimately vary from this model.

they begin asking questions. Scientists, in this respect, are kids at heart. Why are certain plants or animals less common today than they once were? Why are storms becoming more severe and flooding more frequent? What is causing excessive growth of algae in local ponds? Do pesticide impacts on fish or frogs indicate that people may be affected in the same ways? All of these are questions environmental scientists have asked and attempted to answer.

Develop a hypothesis Scientists attempt to answer their questions by devising explanations that they can test. A **hypothesis** is an educated guess that explains a phenomenon or answers a scientific question. For example, a scientist investigating the question of why algae are growing excessively in local ponds might observe chemical fertilizers being applied on farm fields nearby. The scientist might then state a hypothesis as follows: "Agricultural fertilizers running into ponds cause the amount of algae in the ponds to increase."

Make predictions The scientist next uses the hypothesis to generate **predictions,** which are specific statements that can be directly and unequivocally tested. In our algae example, a prediction might be: "If agricultural fertilizers are added to a pond, the quantity of algae in the pond will increase."

Test the predictions Predictions are tested one at a time by gathering evidence that could potentially refute the prediction and thus refute the hypothesis. The strongest form of evidence comes from experimentation. An **experiment** is an activity designed to test the validity of a hypothesis. It involves manipulating **variables,** or conditions that can change. For example, a scientist could test the hypothesis linking algal growth to fertilizer by selecting two identical ponds and adding fertilizer to one while leaving the other in its natural state. In this example, fertilizer input is an **independent variable,** a variable the scientist manipulates, whereas the quantity of algae that results is the **dependent variable,** one that depends on the fertilizer input. If the two ponds are identical except for a single independent variable (fertilizer input), then any differences that arise between the ponds can be attributed to that variable. Such an experiment is known as a **controlled experiment** because the scientist controls for the effects of all variables except the one whose effect he or she is testing. In our example, the pond left unfertilized serves as a **control,** an unmanipulated point of comparison for the manipulated **treatment** pond. Whenever possible, it is best to *replicate* one's experiment, that is, to stage multiple tests of the same comparison of control and treatment. Our scientist could perform a replicated experiment on, say, 10 pairs of ponds, adding fertilizer to one of each pair.

Experiments can establish causal relationships, showing that changes in an independent variable cause changes in a dependent variable. However, experiments are not the only way of testing a hypothesis. Sometimes a hypothesis can be convincingly addressed through **correlation,** searching for relationships among variables. Let's suppose our scientist surveys 50 ponds, 20 of which happen to be fed by fertilizer runoff from nearby farm fields and 30 of which are not. Let's also say he or she finds seven times as much algal growth in the fertilized ponds as in the unfertilized ponds. The scientist would conclude that algal growth is correlated with fertilizer input; that is, that one tends to increase along with the other. Although this type of evidence is weaker than the causal demonstration that controlled experiments can provide, sometimes it is the best approach, or the only feasible one. For example, in studying the effects of global climate change (Chapter 18), we could hardly run an experiment adding carbon dioxide to 10 treatment planets and comparing the result to 10 control planets.

Analyze and interpret results Scientists record **data,** or information, from their studies. They particularly value *quantitative* data, which is information expressed using numbers, because numbers provide precision and are easy to compare. The scientist running the fertilization experiment, for instance, might quantify the area of water surface covered by algae in each pond or might measure the dry weight of algae in a certain volume of water taken from each.

However, even with the precision that numbers provide, a scientist's results may not be clear-cut. Data from treatments and controls may vary only slightly, or different replicates may yield different results. The scientist must therefore analyze the data using statistical tests. With these mathematical methods, scientists can determine objectively and precisely the strength and reliability of patterns they find.

Some research, especially in the social sciences, involves data that is *qualitative,* or not expressible in terms of numbers. Research involving historical texts, personal interviews, surveys, detailed examination of case studies, or descriptive observation of behavior can include qualitative data on which statistical analyses may not be possible. Such studies are still scientific in the broad sense, because their data can be interpreted systematically using other accepted methods of analysis.

Weighing the Issues: Replicates and Data Analysis

Let's say our scientist who is testing for the effects of agricultural fertilizer on algal growth uses experimental replicates, testing 10 pairs of ponds. If 5 of the 10 treatments grow more algae than controls while 5 grow less, what do you think the scientist should conclude? What if all 10 treatments grow more algae than do controls? What if 8 do? If 8 treatment ponds show 10% more growth than the control ponds they are paired with, but the remaining 2 control ponds show 300% more growth than their paired treatments, then what should the scientist conclude? Such cases require statistical analysis so that we can judge levels of confidence to assign to our conclusions. Given possibilities like this, can you explain why scientists believe replicates are important?

If experiments refute a hypothesis, the scientist will reject it and may develop a new hypothesis to replace it. If experiments fail to reject the hypothesis, this outcome lends support to the hypothesis but does not *prove* it is correct. The scientist may choose to generate new predictions to test the hypothesis in a different way and further assess its likelihood of being true. Thus, the scientific method loops back on itself, often giving rise to repeated rounds of hypothesis-revision and new experimentation (see Figure 1.10).

If repeated tests fail to reject a particular hypothesis, evidence in favor of it accumulates, and the researcher may eventually conclude that the idea is well supported. One would ideally also want to test different potential explanations for the question of interest. For instance, our scientist might propose an additional hypothesis that algae increase in fertilized ponds because numbers of fish or invertebrate animals that eat algae decrease. It is possible, of course, that both hypotheses could be correct and that each may explain some portion of the initial observation that local ponds were experiencing algal blooms.

There are different ways to test hypotheses

An experiment in which the researcher actively chooses and manipulates the independent variable is known as a **manipulative experiment** (Figure 1.11a). A manipulative experiment provides the strongest type of evidence a scientist can obtain. In practice, however, some modes of scientific inquiry are more amenable to manipulative experimentation than others. Physics and chemistry tend to involve manipulative experiments, but many other fields deal with entities less easily manipulated than physical forces and chemical reagents. This is true of *historical sciences* such as cosmology, which deals with the history of the universe, and paleontology, which explores the history of past life. It is difficult to manipulate experimentally a star thousands of light years away, or the fossil tooth from a mastodon. Moreover, many of the most interesting questions in these fields center on the causes and consequences

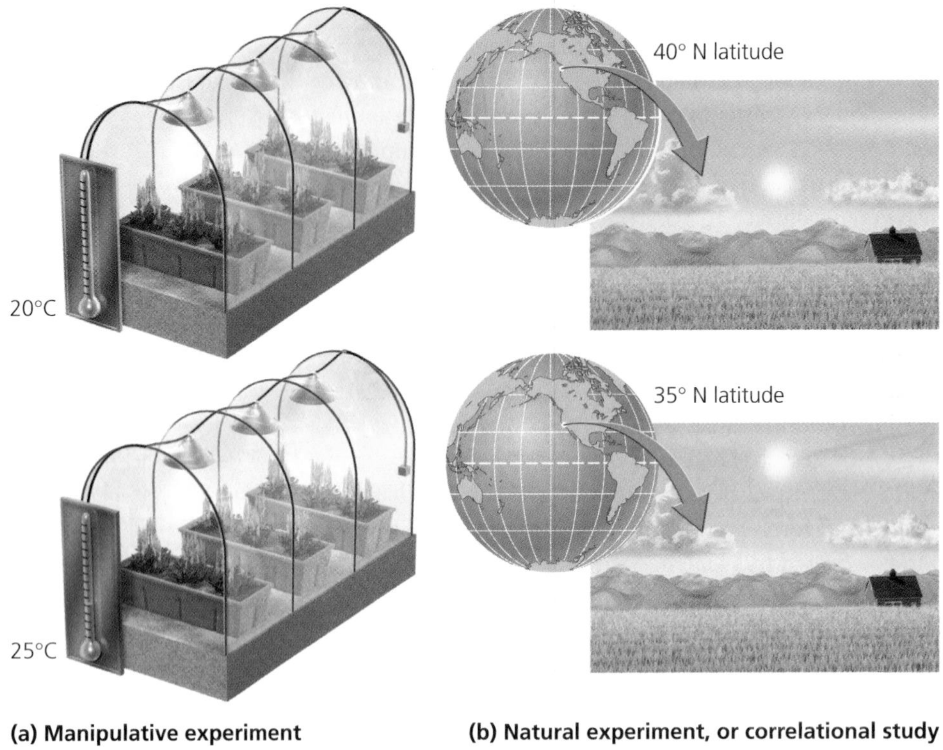

FIGURE 1.11 A researcher wishing to test how temperature affects the growth of wheat might run a manipulative experiment in which wheat is grown in two identical greenhouses, one kept at 20°C (68°F) and the other kept at 25°C (77°F) **(a)**. Alternatively, the researcher might run a "natural experiment" in which he or she compares the growth of wheat in two fields at different latitudes, a cool northerly location and a warm southerly one **(b)**. Because it would be difficult to hold all variables besides temperature constant, the researcher might want to collect data on a number of northern and southern fields and correlate temperature and wheat growth using statistical methods.

of particular historical events, rather than the behavior of general constants.

Disciplines that do not quite fit the so-called physics model of science sometimes rely on **natural experiments** rather than manipulative ones (Figure 1.11b). For instance, an evolutionary biologist might want to test whether animal species isolated on oceanic islands tend to evolve large body size over time. The biologist cannot run a manipulative experiment by placing animals on islands and continents and waiting long enough for evolution to do its work. However, this is exactly what nature has already done. The biologist might test the idea by comparing pairs of closely related species, in which one of each pair lives on an island and the other on a continental mainland. The experiment has in essence been conducted naturally, and it is up to the scientist to interpret the results.

In ecology, both manipulative and natural experimentation is used. The science of **ecology** deals with the distribution and abundance of organisms (living things), the interactions among them, and the interactions between organisms and their abiotic environments. When possible, ecologists try to run manipulative experiments. An ecologist wanting to measure the importance of a certain insect in pollinating the flowers of a given crop plant might, for example, fit some flowers with a device to keep the insects out while leaving other flowers accessible, and later measure the fruit output of each group. Other questions that involve large spatial scales or long time scales may instead require natural experiments.

The social sciences generally involve less experimentation than the natural sciences, depending more on careful observation and interpretation of patterns in data. A sociologist studying how people from different cultures conceive the notion of wilderness might conduct a survey and analyze responses to its questions, looking for similarities and differences among respondents. Such analyses may be either quantitative or qualitative, depending on the nature of the data and the researchers' particular questions and approaches.

Descriptive observational studies and natural experiments can show correlation between variables, but they cannot demonstrate that one variable *causes* change in another, as manipulative experiments can. Not all variables are controlled for in a natural experiment, so a single result could give rise to several interpretations. However, correlative studies, when done well, can make for very convincing science, and they preserve the real-world complexity that manipulative experiments often sacrifice. Moreover, sometimes correlation is all we have. Because manipulations are difficult at large scales, some of the most important questions in environmental science tend to be addressed with correlative data. The large scale and complexity of many questions in environmental science also mean that few studies, manipulative or correlative, come up with neat and clean results. As such, scientists are not always able to give policymakers and society black-and-white answers to questions.

The scientific process does not stop with the scientific method

Individual researchers or teams of researchers follow the scientific method as they investigate questions that interest them. However, scientific work takes place within the context of a community of peers, and to have any impact, a researcher's work must be published and made accessible to this community. Thus, the scientific method is embedded within a larger process that takes place at the level of the scientific community as a whole (Figure 1.12).

Peer review When a researcher's work is done and the results analyzed, he or she writes up the findings and submits them to a journal for publication. Several other scientists specializing in the topic of the paper examine the manuscript, provide comments and criticism (generally anonymously), and judge whether the work merits publication in the journal. This procedure, known as **peer review,** is an essential part of the scientific process. Peer review is a valuable guard against faulty science contaminating the literature on which all scientists rely. However, because scientists are human and may have their own personal biases and agendas, politics can sometimes creep into the review process. Fortunately, just as individual scientists strive their best to remain objective in conducting their research, the scientific community does its best to ensure fair review of all work. Winston Churchill once called democracy the worst form of government, except for all the others that had been tried. The same might be said about peer review; it is an imperfect system, yet no one has come up with a better one.

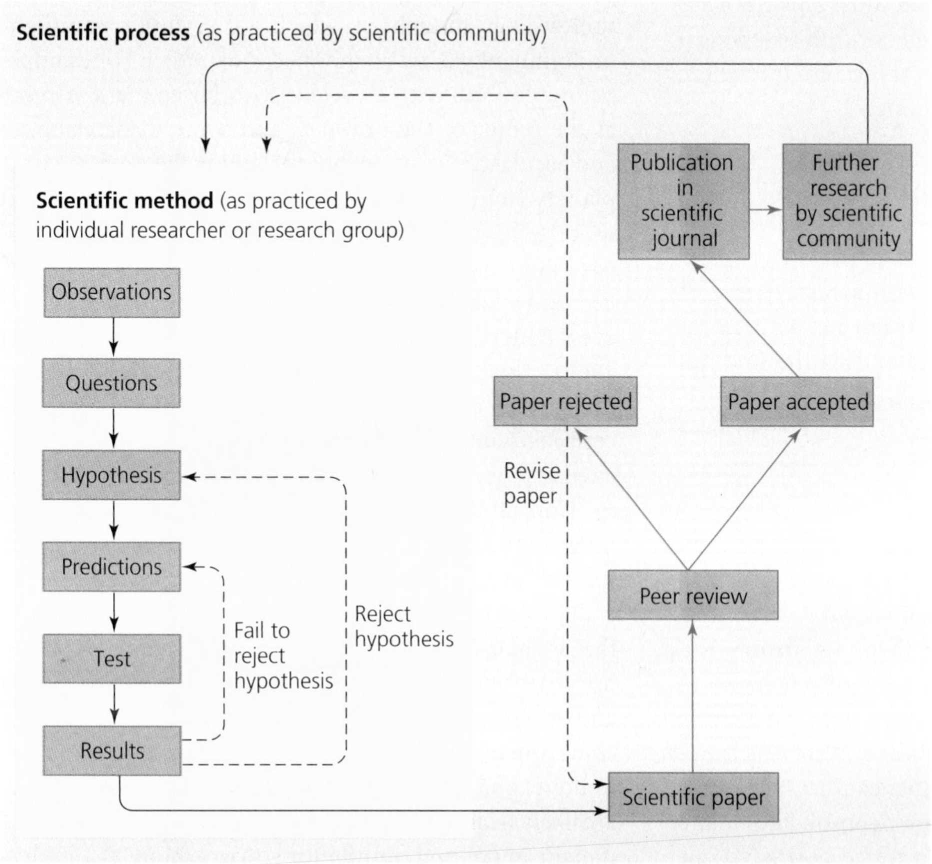

FIGURE 1.12 The scientific method (inner box) followed by individual researchers or research teams exists within the context of the overall process of science at the level of the scientific community (outer box). This process includes peer review and publication of research, acquisition of funding, and the development of theory through the cumulative work of many researchers.

Conference presentations Scientists frequently present their work at professional conferences, where they interact with colleagues and often receive informal comments on their research. When research has not yet been published, feedback from colleagues can help improve the quality of a scientist's work before it is submitted for publication.

Grants and funding Research scientists spend large portions of their time writing grant applications requesting money to fund their research from private foundations or government agencies such as the National Science Foundation. Grant applications undergo peer review just as scientific papers do, and competition for funding is often intense. Scientists' reliance on funding sources can also lead to potential conflicts of interest. A scientist who obtains data showing his or her funding source in an unfavorable light may be reluctant to publish the results for fear of losing funding—or worse yet, may be tempted to doctor the results. This situation can arise, for instance, when an industry funds research to test its products for safety or environmental impact. Most scientists do not succumb to these temptations, but some funding sources have been known to pressure their scientists for certain results. This is why as a student or informed citizen, when critically assessing a scientific study, you should always try to find out where the researchers obtained their funding.

Repeatability Sound science is based on doubt rather than certainty and on repeatability rather than one-time occurrence. Even when a hypothesis appears to explain observed phenomena, scientists are inherently wary of accepting it. The careful scientist may test a hypothesis repeatedly in various ways before submitting the findings for publication. Following publication, other scientists may attempt to reproduce the results in their own experiments and analyses.

Theories If a hypothesis survives repeated testing by numerous research teams and continues to predict experimental outcomes and observations accurately, it may potentially be incorporated into a theory. A **theory** is a widely accepted, well-tested explanation of one or more cause-and-effect relationships that has been extensively validated by a great amount of research. Whereas a hypothesis is a simple explanatory statement that may be refuted by a single experiment, a theory consolidates many related hypotheses that have been tested and have not been refuted.

Note that scientific use of the word *theory* differs from popular usage of the word. In everyday language when we say something is "just a theory," we are suggesting it is a speculative idea without much substance. Scientists, however, mean just the opposite when they use the term; to them, a theory is a conceptual framework that effectively explains a phenomenon and has undergone extensive and rigorous testing, such that confidence in it is extremely strong. For example, Darwin's theory of evolution by natural selection (▸pp. 118–121) has been supported and elaborated by many thousands of studies over 150 years of intensive research. Such research has shown repeatedly and in great detail how plants and animals change over generations, or evolve, to express characteristics that best promote survival and reproduction. Because of its strong support and explanatory power, evolutionary theory is the central unifying principle of modern biology.

Science may go through "paradigm shifts"

It is crucial to realize that results obtained by the scientific method may sometimes later be reinterpreted to show that earlier interpretations were incorrect. Thomas Kuhn's 1962 book *The Structure of Scientific Revolutions* argued that science goes through periodic revolutions, dramatic upheavals in thought, in which one scientific **paradigm,** or dominant view, is abandoned for another. For example, before the 16th century, scientists believed that Earth was at the center of the universe, and some made elaborate and accurate measurements explaining the movements of planets from that viewpoint. Their data fit the theory quite well, yet the theory eventually was disproved by Nicolaus Copernicus, who showed that placing the sun at the center of the universe explained the planetary data even better. A similar paradigm shift occurred in the 1960s, when geologists accepted the theory of plate tectonics (▸pp. 207–209), once evidence for the movement of continents and the action of tectonic plates had accumulated and become overwhelmingly convincing.

Understanding how science works is vital to assessing how scientific ideas and interpretations change through time as new information accrues. This process is especially relevant in environmental science, a young field that is changing rapidly as we learn vast amounts of new information, as human impacts on the planet multiply, and as lessons from the consequences of our actions become apparent. Because so much remains unstudied and undone, and because so many issues we cannot foresee are likely to arise in the future, environmental science will remain an exciting frontier for you to explore as a student and as an informed citizen throughout your life.

Sustainability and the Future of Our World

Throughout this book you will see examples of environmental scientists asking questions, developing hypotheses, conducting experiments, gathering and analyzing data, and drawing conclusions about environmental processes and the causes and consequences of environmental change. Environmental scientists who aim to understand the condition of our environment and the consequences of our impacts are studying the most centrally important issues of our time.

Population and consumption lie at the root of many environmental changes

We modify our environment in diverse ways, but the steep and sudden rise in human population has amplified nearly all of our impacts (Chapter 8). Our numbers have nearly quadrupled in the past 100 years, passing 6 billion in 1999 and 6.5 billion in 2006. We add about 78 million people to the planet each year—that's over 200,000 per day. Today, the rate of population growth is slowing, but our absolute numbers continue to increase and to shape our interactions with one another and with our environment.

Our consumption of resources has risen even faster than our population growth. The rise in affluence has been a positive development for humanity, and our conversion of the planet's natural capital has made life more pleasant for us so far. However, like rising population, rising per capita consumption amplifies the demands we make on our environment. Moreover, affluence and consumption have not grown equally for all the world's citizens. Today the 20 wealthiest nations boast 40 times the income of the 20 poorest nations—twice the gap that existed four decades earlier. The ecological footprint of the average citizen of a developed nation such as the United States is considerably larger than that of the average resident of a developing country (Figure 1.13). Within the United States, the richest fifth of people claim nearly half the income, whereas the poorest fifth receive only 5%.

We face challenges in agriculture, pollution, energy, and biodiversity

The dramatic growth in human population and consumption is due in part to our successful efforts to expand and intensify the production of food (Chapters 9 and 10).

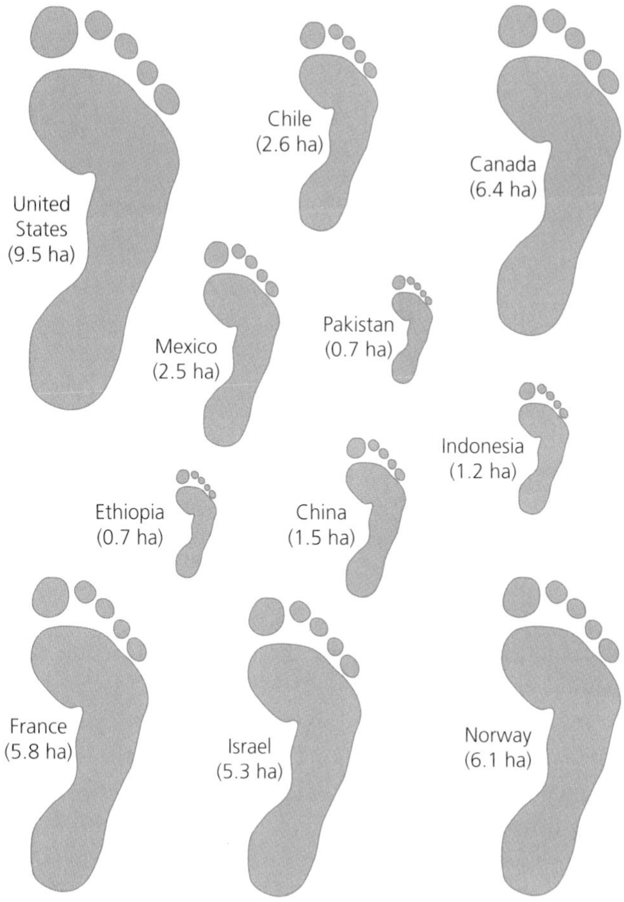

FIGURE 1.13 The citizens of some nations have larger ecological footprints than the citizens of others. U.S. residents consume more resources—and thus use more land—than residents of any other nation. Shown here are ecological footprints for average citizens of several developed and developing nations, as of 2001. Data from Global Footprint Network, 2005.

Since the agricultural revolution, new technologies have enabled us to grow increasingly more food per unit of land. These advances in agriculture must be counted as one of humanity's great achievements, but they have come at some cost. We have converted nearly half the planet's land surface for agriculture; our extensive use of chemical fertilizers and pesticides poisons organisms and alters natural systems; and erosion, climate change, and poorly managed irrigation are destroying 5–7 million hectares (ha; 12.5–17.5 million acres) of productive cropland each year.

Meanwhile, pollution from our farms, industries, households, and individual actions dirties our land, water, and air (Figure 1.14). Outdoor air pollution, indoor air pollution, and water pollution contribute to the deaths of millions of people each year (Chapters 15–17). Environmental

FIGURE 1.14 Indoor and outdoor air pollution contribute to millions of premature deaths each year, and environmental scientists and policymakers are working to reduce this problem in a variety of ways.

toxicologists are chronicling the impacts on people and wildlife of the many synthetic chemicals and other pollutants we emit into the environment (Chapter 14). Our most pressing pollution challenge may be to address the looming specter of global climate change (Chapter 18). Scientists have firmly concluded that human activity is altering the composition of the atmosphere and that these changes are affecting Earth's climate. Since the start of the industrial revolution, atmospheric carbon dioxide concentrations have risen by 31%, to a level not present in at least 420,000 years. This increase results from our reliance on burning fossil fuels to power our civilization. Carbon dioxide and several other gases absorb heat and warm Earth's surface, which is likely responsible for glacial melting, sea-level rise, impacts on wildlife and crops, and increased episodes of destructive weather.

The combined impact of human actions such as climate change, overharvesting, pollution, the introduction of non-native species, and particularly habitat alteration, has driven many aquatic and terrestrial species out of large parts of their ranges and toward the brink of extinction (Chapter 11). Today Earth's biological diversity, or **biodiversity,** the cumulative number and diversity of living things, is declining dramatically. Many biologists say we are already at the outset of a mass extinction event comparable to only five others documented in all of Earth's history. Biologist Edward O. Wilson has warned that the loss of biodiversity is our most serious and threatening environmental dilemma, because it is not the kind of problem that responsible human action can remedy. Rather, the extinction of species is irreversible; once a species has become extinct, it is lost forever.

Solutions to environmental problems must be global and sustainable

The nature of virtually all of these environmental issues is being changed by the set of ongoing phenomena commonly dubbed *globalization*. Our increased global interconnectedness in trade, politics, and the movement of people and of other species poses many challenging problems, but it also sets the stage for novel and effective solutions.

The most comprehensive scientific assessment of the present condition of the world's ecological systems and their ability to continue supporting our civilization was completed in 2005. In this year, over 2,000 of the world's leading environmental scientists from nearly 100 nations completed the **Millennium Ecosystem Assessment.** The four main findings of this exhaustive project are summarized in Table 1.1. The Assessment makes clear that our degradation of the world's environmental systems is having negative impacts on all of us, but that with care and diligence we can still turn many of these trends around.

Table 1.1 Main Findings of the Millennium Ecosystem Assessment

- Over the past 50 years, humans have changed ecosystems more rapidly and extensively than in any comparable period of time in human history, largely to meet rapidly growing demands for food, freshwater, timber, fiber, and fuel. This has resulted in a substantial and largely irreversible loss in the diversity of life on Earth.
- The changes made to ecosystems have contributed to substantial net gains in human well-being and economic development, but these gains have been achieved at growing costs. These costs include the degradation of ecosystems and the services they provide for us, and the exacerbation of poverty for some groups of people.
- This degradation could grow significantly worse during the first half of this century.
- The challenge of reversing the degradation of ecosystems while meeting increasing demands for their services can be partially overcome, but doing so will involve significantly changing many policies, institutions, and practices.

Adapted from *Millennium Ecosystem Assessment, Synthesis Report,* 2005.

FIGURE 1.15 Human activities are pushing many organisms, including the panda, toward extinction. Efforts to save endangered species and reduce biodiversity loss include many approaches, but all require that adequate areas of appropriate habitat be preserved in the wild.

Fortunately, potential solutions abound

We cannot, of course, live without exerting any impact on Earth's systems. We face trade-offs with many environmental issues, and the challenge is to develop solutions that further our quality of life while minimizing harm to the environment that supports us. Fortunately, there are many workable solutions at hand, and many more potential solutions we can achieve with further effort.

In response to agricultural problems, scientists and others have developed and promoted soil conservation, high-efficiency irrigation, and organic agriculture. Technological advances and new laws have greatly reduced the pollution emitted by industry and automobiles in wealthier countries. Although the U.S. government has resisted international efforts to rein in pollutants to halt climate change, American scientists have been at the forefront of climate change science, and other nations are beginning to address the problem, as are the governments of some U.S. states. Amid ample reasons for concern about the state of global biodiversity, advances in conservation biology are enabling scientists and policymakers in many cases to work together to protect habitat, slow extinction, and safeguard endangered species (Figure 1.15). Recycling is helping relieve our waste disposal problems, and alternative renewable energy sources are being developed to take the place of fossil fuels (Figure 1.16). These are but a few of the many solutions we will explore in the course of this book.

Are things getting better or worse?

Despite the myriad challenges we review in this book, many people maintain that the general conditions of human life and the environment are in fact getting better, not worse. A recent proponent of this view, Danish statistician Bjorn Lomborg, wrote in his book *The Skeptical Environmentalist:*

> We are not running out of energy or natural resources. There will be more and more food per head of the world's population. Fewer and fewer people are starving. In 1900 we lived for an average of 30 years; today we live for 67. . . . The air and water around us are becoming less and less polluted. Mankind's lot has actually improved in terms of practically every measurable indicator.

Furthermore, some people maintain that we will find ways to make Earth's natural resources meet all of our needs indefinitely and that human ingenuity will see us through any difficulty. Such views are sometimes characterized as *Cornucopian*. In Greek mythology, *cornucopia*—literally "horn of plenty"—is the name for a magical goat's horn that overflowed with grain, fruit, and flowers. In contrast, people who predict doom and disaster for the world because of our impact upon it have been called *Cassandras*, after the mythical princess of Troy with the gift of prophecy whose dire predictions were not believed.

At least three questions are worth asking each time you are confronted with seemingly conflicting statements from Cassandras and Cornucopians. One question is whether the impacts being debated pertain only to humans or also to other organisms and natural systems. The second question is whether the debaters are thinking in the short term or the long term. The third question is whether they are considering all costs and

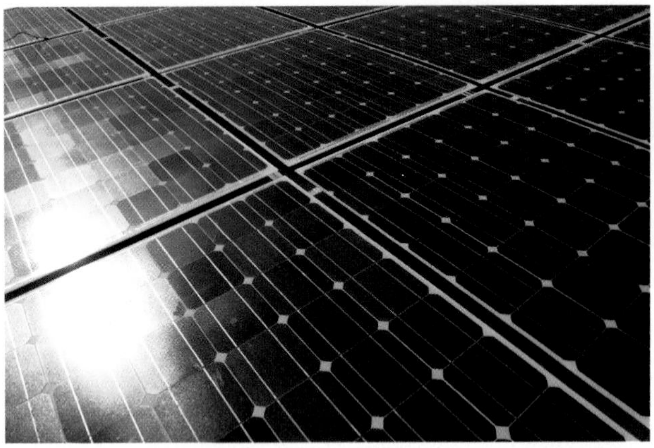

FIGURE 1.16 Our dependence on fossil fuels has caused a wide array of environmental impacts. Although fossil fuels have powered our civilization since the industrial revolution, many renewable energy sources exist, such as solar energy that can be collected with panels like these. Such alternative energy sources could be further developed for sustainable use now and in the future.

benefits relevant for the question at hand, or only some. As you proceed through this book and encounter countless contentious issues, consider how one's perception of them may be influenced by these three factors.

Sustainability is a goal for the future

The primary challenge in our increasingly populated world is how to live within our planet's means, such that Earth and its resources can sustain us and the rest of Earth's biota for the foreseeable future. This is the challenge of **sustainability,** a guiding principle of modern environmental science. Sustainability means leaving our children and grandchildren a world as rich and full as the world we live in now. It means not depleting Earth's natural capital, so that after we are gone our descendants will enjoy the use of resources as we have. It means developing solutions that are able to work in the long term. Sustainability requires maintaining fully functioning ecological systems, because we cannot sustain human civilization without sustaining the natural systems that nourish it.

Sustainability is a concept you will encounter throughout this book. Our final chapter (Chapter 23) takes a wide-ranging look at emerging sustainable solutions—on college and university campuses and in the world at large.

Sustainability need not require great sacrifice of us. We will naturally always desire to enhance our quality of life, and as we will see, there are many ways we can do so while also encouraging a more sustainable lifestyle. Economists employ the term *development* to describe the use of natural resources for economic advancement (as opposed to simple subsistence, or survival). Logging, farming, mining, and building homes and factories are all types of development, and each of them affects the environment and gives rise to changes that environmental scientists study. **Sustainable development** is the use of renewable and nonrenewable resources in a manner that satisfies our current needs without compromising future availability of resources. The United Nations defines sustainable development as development that ". . . meets the needs of the present without sacrificing the ability of future generations to meet theirs." Answering a simple question—"Can this activity continue forever?"—indicates whether a particular activity is sustainable.

Sustainability depends, in large part, on the ability of the current human population to limit its environmental impact. Doing so will require us to make an ethical commitment, while also applying information we gain from the sciences. Science can help us devise ways to limit our impact and maintain the functioning of the environmental systems on which we depend.

Conclusion

Finding effective ways of living peacefully, healthfully, and sustainably on our diverse and complex planet will require a thorough scientific understanding of both natural and social systems. Environmental science helps us understand our intricate relationship with the environment and informs our attempts to solve and prevent environmental problems.

It is important to keep in mind that identifying a problem is the first step in devising a solution to it. Many of the trends detailed in this book may cause us worry, but others give us reason to hope. One often-heard criticism of environmental science courses and textbooks is that too often they emphasize the negative. Recognizing the validity of this criticism, in this book we attempt to balance the discussion of environmental problems with a corresponding focus on potential solutions. Solving environmental problems can move us toward health, longevity, peace, and prosperity. Science in general, and environmental science in particular, can aid us in our efforts to develop balanced and workable solutions to the many environmental dilemmas we face today and to create a better world for ourselves and our children.

REVIEWING OBJECTIVES

You should now be able to:

Define the term *environment*

- Our environment consists of everything around us, including living and nonliving things. (p. 3)
- Humans are a part of the environment and are not separate from nature. (p. 3)

Describe natural resources and explain their importance to human life

- Resources from nature are essential to human life and civilization. (p. 4)
- Some resources are perpetually renewable, others are nonrenewable, and still others are renewable if we are careful not to exploit them at too fast a rate. (p. 4)
- Malthus and Ehrlich pointed out risks of human population growth, while Hardin and Wackernagel and Rees pioneered important concepts in resource consumption. (pp. 4–7)

Characterize the interdisciplinary nature of environmental science

- Environmental science uses the approaches and insights of numerous disciplines from the natural sciences and the social sciences. (p. 11)

Understand the scientific method and how science operates

- Science is a process of using observations to test ideas. (p. 12)
- The scientific method consists of a series of steps, including making observations, formulating questions, stating a hypothesis, generating predictions, testing predictions, and analyzing the results obtained from the tests. (pp. 13–15)
- The scientific method is not always followed strictly, and there are different ways to test questions scientifically. (pp. 15–17)
- Scientific research occurs within a larger process that includes peer review of work, journal publication, and interaction with colleagues. (pp. 17–18)

Diagnose and illustrate some of the pressures on the global environment

- Increasing human population and increasing per capita consumption exacerbate human impacts on the environment. (p. 19)
- Human activities such as industrial agriculture and the use of fossil fuels for energy are having diverse environmental impacts, including resource depletion, air and water pollution, habitat destruction, and the diminishment of biodiversity. (pp. 19–20)

Evaluate the concepts of sustainability and sustainable development

- Sustainability means living within the planet's means, such that Earth's resources can sustain us—and other species—for the foreseeable future. (pp. 20–22)
- Sustainable development is possible; we need not decrease our quality of life to establish sustainable lifestyles. (p. 22)

TESTING YOUR COMPREHENSION

1. What do renewable resources and nonrenewable resources have in common? How are they different? Identify two renewable and two nonrenewable resources.
2. How did the agricultural revolution affect human population size? How did the industrial revolution affect human population size? Explain your answers.
3. What is "the tragedy of the commons"? Explain how the concept might apply to an unregulated industry that is a source of water pollution.
4. What is environmental science? Name several disciplines involved in environmental science.
5. What are the two meanings of *science?* Name three applications of science.
6. Describe the scientific method. What is the typical sequence of steps?
7. Explain the difference between a manipulative experiment and a natural experiment.
8. What needs to occur before a researcher's results are published? Why is this important?
9. Give examples of three major environmental problems in the world today, along with their causes.
10. What is sustainable development?

SEEKING SOLUTIONS

1. Many resources are renewable if we use them in moderation but can become nonrenewable if we overexploit them. Order the following resources on a continuum of renewability (see Figure 1.1), from most renewable to least renewable: soils, timber, fresh water, food crops, and biodiversity. What factors influenced your decision? For each of these resources, what might constitute overexploitation, and what might constitute sustainable use?
2. Why do you think the Easter Islanders did not or could not stop themselves from stripping their island of all its trees? Do you see similarities between the history of the Easter Islanders and the modern history of our society? Why or why not?
3. What environmental problem do *you* feel most acutely yourself? Do you think there are people in the world who do not view your issue as an environmental problem? Who might they be, and why might they take a different view?
4. Name an environmental problem you would like to see solved or mitigated. Describe the scientific research you think would need to be completed so that workable solutions to this problem can be developed. Would more than science be needed?
5. If the human population were to stabilize tomorrow and never surpass 7 billion people, would that solve our environmental problems? Which types of problems might be alleviated, and which might continue to become worse?
6. Consider the historic expansion of agriculture and our ability to feed increasing numbers of people, as described in this chapter. Now ask yourself, "Are things getting better or worse?" Ask this question from four points of view: (1) from the human perspective, (2) from the perspective of other organisms, (3) from a short-term perspective, and (4) from a long-term perspective. Do your answers to this question change? If so, how?

INTERPRETING GRAPHS AND DATA

Environmental scientists study phenomena that range in size from individual molecules (Chapter 4) to the entire Earth (Chapter 7), and that occur over time periods lasting from fractions of a second to billions of years. To simultaneously and meaningfully represent data covering so many orders of magnitude, scientists have devised a variety of mathematical and graphical techniques, such as exponential notation and logarithmic scales. Below are two graphical representations *of the same data,* representing the growth of a hypothetical population from an initial size of 10 individuals at a rate of increase of approximately 2.3% per generation. The graph in part (a) uses a conventional linear scale for the population size; the graph in part (b) uses a logarithmic scale.

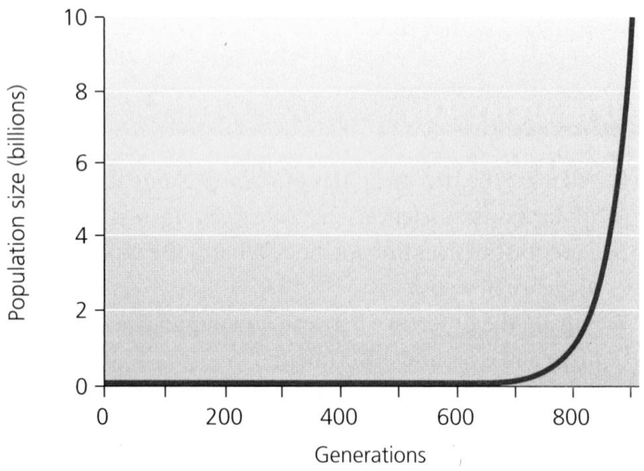

(a) Linear scale

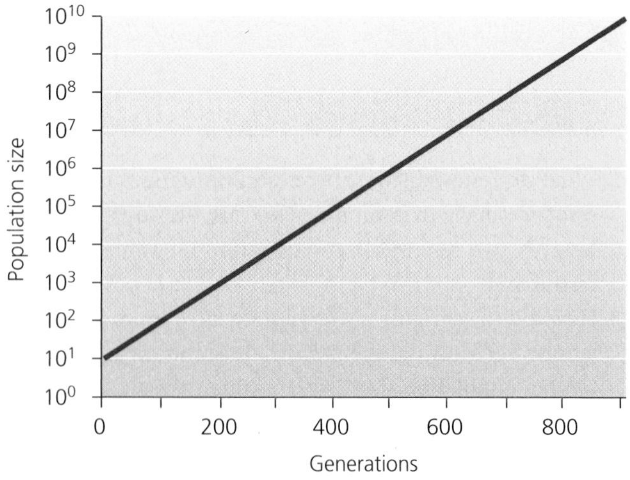

(b) Logarithmic scale

Hypothetical population growth curves, assuming an initial size of 10 and a constant rate of increase of approximately 2.3% per generation.

1. Using the graph in part (a), what would you say was the population size after 200 generations? After 400? After 600? After 800? How would you answer the same questions using the graph in part (b)? What impression does the graph in part (a) give about population change for the first 600 generations? What impression does the graph in part (b) give?
2. Compare these graphs to Figure 1.2a in the text. What does the human population appear to be doing between 10,000 B.P. and 2,000 B.P.?
3. The size of a population that is growing by a constant rate of increase will plot as a straight line on a logarithmically scaled graph like the one in part (b) above, but if the annual rate of increase changes, the line will curve. Do you think the data for the human population over the past 12,000 years would plot as a straight line on a logarithmically scaled graph? If not, when and why do you think the line would bend?

CALCULATING ECOLOGICAL FOOTPRINTS

Mathis Wackernagel and his colleagues have continued to refine the method of calculating ecological footprints—the amount of land and water required to produce the energy and natural resources we consume. In a 1999 paper, they applied their method to 52 nations that together account for 80% of the world's population and 95% of the World Domestic Product. According to their study, there are 4.9 acres available for every person in the world.

Compare the ecological footprints of each of the countries listed in the table below. Calculate their proportional relationships to the world population's average ecological footprint and to the land available globally to meet our ecological demands.

Country	Ecological footprint (acres per person)	Proportion relative to world average footprint	Proportion relative to world land available
Bangladesh	1.2		
Colombia	4.9		1.0 (4.9/4.9)
Mexico	6.4		
Sweden	14.6		
Thailand	6.9		
United States	25.4		
World Average	6.9	1.0 (6.9/6.9)	1.4 (6.9/4.9)

Data from Wackernagel, M., et al. 1999. National natural capital accounting with the ecological footprint concept. *Ecological Economics* 29: 375–390.

1. Why is the ecological footprint for people in Bangladesh so low?
2. Why is it so high in the United States?
3. The population of the United States is expected to grow to 349 million by 2025. What impact, if any, do you think this growth will have on the average global ecological footprint?
4. Based on the data in the table, what impacts do you think average family income has on ecological footprints?

Take It Further

Go to www.aw-bc.com/withgott or the student CD-ROM where you'll find:

- Suggested answers to end-of-chapter questions
- Quizzes, animations, and flashcards to help you study
- *Research Navigator*™ database of credible and reliable sources to assist you with your research projects
- **GRAPHit!** tutorials to help you master how to interpret graphs
- **INVESTIGATEit!** current news articles that link the topics that you study to case studies from your region to around the world

2 Environmental Ethics and Economics: Values and Choices

Kakadu National Park, Australia

Upon successfully completing this chapter, you will be able to:

- Characterize the influences of culture and worldview on the choices people make
- Outline the nature, evolution, and expansion of environmental ethics in Western cultures
- Describe precepts of classical and neoclassical economic theory, and summarize their implications for the environment
- Compare the concepts of economic growth, economic health, and sustainability
- Explain the fundamentals of environmental economics and ecological economics

Protestors against proposed Jabiluka uranium mine

Central Case: The Mirrar Clan Confronts the Jabiluka Uranium Mine

"For some people, what they are is not finished at the skin, but continues with the reach of the senses out into the land. If the land is summarily disfigured or reorganized, it causes them psychological pain."
—Barry Lopez, American nature writer

"The Jabiluka uranium mine will improve the quality of the environment. The uranium resource there has been polluting the river system naturally, probably for thousands of years. . . . With the uranium resource removed and put to good use, the level of radioactivity will fall."
—Michael Darby, Australian political commentator

The remote Kakadu region of Australia's Northern Territory is home to several groups of Australian Aborigines, native people who lived there long before the British colonization of Australia. The region also features Kakadu National Park, a World Heritage Site recognized by the United Nations for its irreplaceable natural and cultural resources. In addition, the region's land holds uranium, the naturally occurring radioactive metal valued for its use in nuclear power plants, nuclear weapons, and various medical and industrial tools. Uranium mining is a key contributor to the Australian national economy, accounting for 7% of Australia's economic output.

Many of Australia's uranium deposits occur on Aboriginal lands, giving rise to conflict between corporations seeking to develop mining operations and Aboriginal people trying to maintain their traditional culture. One such group is the Mirrar Clan, an extended family of 25 Kakadu-area Aborigines. The Mirrar have been living with the region's first uranium mine, the Ranger Mine, on their land since the Australian government approved its development in 1978.

In recent years, the Mirrar have been fighting the proposed development of a second mine, Jabiluka, nearby on their land. The Mirrar see Jabiluka as a threat to their culture and religion, which are deeply tied to the landscape. Like other Aborigines, they hold the landscape to be sacred, and they depend on its resources for their daily needs. The proposed mine site is near traditional

Mirrar hunting and gathering sites and is in the floodplain of a river that provides the clan food and water.

The Mirrar also view Jabiluka as a threat to their health and to the integrity of their environment, particularly given repeated radioactive spills at the Ranger mine. Many Mirrar fear that contaminated water could be released into area creeks and that radioactive radon gas would emanate from stored waste materials. Moreover, mindful of geological faults that exist in the area, the Mirrar worry that dams holding mine waste could fail catastrophically in an earthquake. Environmental activists worldwide have joined the Mirrar's struggle; in 1998 nearly 3,000 people traveled to the Kakadu region to protest Jabiluka.

In late 2002 their efforts appeared to have succeeded. Sir Robert Wilson, chief executive officer of the corporation holding rights to the Jabiluka ore body, announced the cancellation of mining plans at Jabiluka, citing economic factors (declining world uranium prices) and ethical factors (concerns about developing the mine without Mirrar consent). Wilson added that the company planned to rehabilitate the site and restore damage done there during exploration and assessment. However, a formal agreement was never signed, and since that time the price of uranium has risen on the world market. The corporation's plans are now in a holding pattern as it waits and hopes that the Mirrar will one day give their consent.

The Mirrar oppose mining despite the economic benefits the mining company has promised them in the form of jobs, income, development, and a higher material standard of living. The decision to bypass these economic incentives was not easy, and indeed, a number of other Aboriginal groups in the Kakadu region support mine development. In formulating their approaches to the mining proposal, the Mirrar and other Australians weighed economic, social, cultural, and philosophical questions as well as scientific ones. The story of mining and the Mirrar exemplifies some of the ways in which values, beliefs, and traditions interact with economic interests to influence the choices all of us make about how to live within our environment.

Culture, Worldview, and the Environment

The Mirrar faced difficult choices. They were offered substantial economic benefits, but they also felt that mine development ran counter to their ethical respect for their land. Such trade-offs between economic benefits and ethical concerns crop up frequently in environmental issues.

Ethics and economics involve values

As we saw in Chapter 1, environmental science examines Earth's natural systems, the ways they affect humans, and the ways humans affect them. Thus, environmental science entails a firm understanding of the natural sciences. To address environmental problems, however, it is also necessary to understand how people perceive their environment, how they relate to it philosophically and pragmatically, and how they value its elements. Ethics and economics are two quite different disciplines, but each deals with questions of what we value and how those values influence our decisions and actions. Anyone trying to address an environmental problem must try to understand not only how natural systems work, but also how values shape human behavior.

Culture and worldview influence one's perception of the environment

Almost every action we take affects our environment. Growing food requires soil, cultivation, and often irrigation. Building homes requires land, lumber, and metal. Manufacturing and fueling vehicles require metal, plastic, glass, and petroleum. From nutrition to housing to transportation, we meet our needs by altering our surroundings. Our decisions about how we manipulate and exploit our environment to meet our needs depend in part on rational assessments of costs and benefits. However, our decisions are also heavily influenced by the particular culture of which we are a part and by our particular worldview. **Culture** can be defined as the ensemble of knowledge, beliefs, values, and learned ways of life shared by a group of people. Culture, together with personal experience, influences each person's perception of the world and his or her place within it, something described as the person's **worldview.** A worldview reflects a person's (or group's) beliefs about the meaning, operation, and essence of the world (Figure 2.1).

People with different worldviews can study the same situation and review identical data yet draw dramatically different conclusions. For example, many well-meaning people have supported the Jabiluka mine while many other well-meaning people have opposed it. The officers, employees, and shareholders of the mining company, and government officials who support the mine, view uranium mining as a source of jobs, income, energy, and economic growth. They believe mining will benefit Australia in general and the Mirrar Clan in particular. Mine opponents, in contrast, foresee environmental problems, injustice, and negative social consequences. They recognize that uranium mining disturbs the landscape, pollutes air

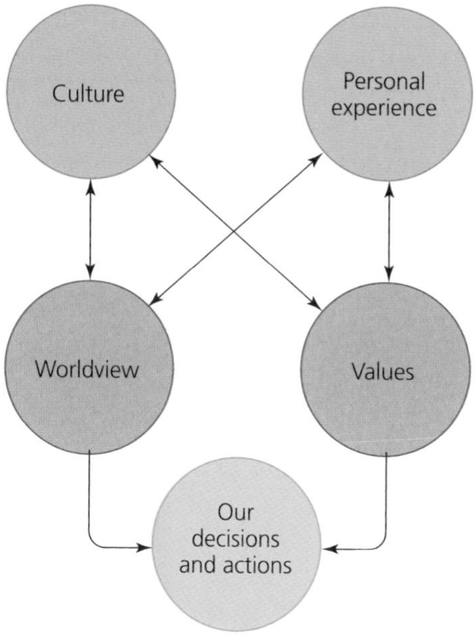

FIGURE 2.1 Culture and personal experience influence a person's worldview and values, which in turn influence his or her actions and decisions. The disciplines of ethics and economics each examine (in very different ways) factors that influence our values and guide human behavior. Ethics, economics, and the behavior and preferences of individuals, together with information from the natural sciences, all inform the making of policy.

(a) Ranger mine

(b) Proposed Jabiluka mine site

FIGURE 2.2 The Ranger mine (**a**), located on Aboriginal lands amid sacred sites, has caused enough environmental impacts to spark the Mirrar Clan's opposition to the proposed Jabiluka mine (**b**), whose preparation had begun nearby.

and water, and can expose miners to radiation, while community disruption, alcoholism, and crime frequently accompany mining booms.

Many factors shape our worldviews and perception of the environment

The traditional culture and worldview of the Mirrar Clan have played large roles in its response to the proposed Jabiluka mine. Australian Aborigines view the landscape around them as the physical embodiment of stories that express the beliefs and values central to their culture. The Australian landscape to them is a sacred text, analogous to the Bible in Christianity, the Koran in Islam, or the Torah in Judaism. Aborigines believe that spirit ancestors possessing human and animal features traveled routes called "dreaming tracks," leaving signs and lessons in the landscape. Modern Aborigines still engage in "walkabouts," long walks that retrace the dreaming tracks. By explaining the origins of specific landscape features, dreaming-track stories assign meaning to notable landmarks and help Aborigines construct detailed mental maps of their surroundings. The stories also teach lessons concerning family relations, hunting, food gathering, and conflict resolution. Dreaming-track stories passed from one generation to the next help maintain Aboriginal culture. The Mirrar who oppose Jabiluka believe the mine would desecrate sacred sites and compromise their culture (Figure 2.2).

Weighing the Issues: Uranium Mining in Bethlehem

Suppose a mining company discovered uranium beneath the site in Bethlehem believed to be the birthplace of Jesus—or near the Wailing Wall in Jerusalem, or the mosque at Mecca. What do you think would happen if the company announced plans to develop a mine there, assuring the public that environmental impacts would be minimal and that the mine would create jobs and stimulate economic growth? What aspects of this unlikely situation resemble that of the Mirrar case, and what factors are different? Explain your answers.

Religion is one of many factors that can shape people's worldviews and perception of the environment. A community may also share a particular view of its environment if its members have lived through similar experiences. For example, early European settlers in both Australia and North America viewed their environment as a hostile force because inclement weather, wild animals, and other natural forces frequently destroyed crops, killed livestock, and took settlers' lives. Such experiences were shared in stories and in songs and helped shape prevailing social attitudes in many frontier communities. The view of nature as a hostile force and an adversary to be overcome has passed from one generation to the next and still influences the way many North Americans and Australians view their surroundings.

A person's political ideology can also shape his or her attitude toward the environment. For instance, one's view of the proper role of government will influence whether or not one wants government to intervene in a market economy to protect environmental quality. Economic factors also sway how people perceive their environment and make decisions. An individual with a strong interest in the outcome of a decision that may result in his or her private gain or loss is said to have a *vested interest.* Mining company executives have a vested interest in a decision to open an area to mining because a new mine can increase profits, to which executive compensation is frequently tied. Likewise, a company's shareholders have a vested interest in such a decision because the value of the shares they hold increases with profits. In each case, vested interests may lead people to view a proposed mine such as Jabiluka primarily as a source of economic gain.

Throughout this book you will encounter scientific data regarding the environmental impacts of our choices (where to make our homes, how to make a living, what to wear, what to eat, how to travel, how to spend our leisure time, and so on). You will see that culture, worldviews, and values play critical roles in such choices and even can influence the interpretation of scientific data. Thus, acquiring scientific understanding is only one part of the search for solutions to environmental problems. Attention to ethics and economics helps us understand why and how we value those things we value.

Environmental Ethics

The field of **ethics** is a branch of philosophy that involves the study of good and bad, of right and wrong. The term *ethics* can also refer to the set of moral principles or values held by a person or a society. Ethicists help clarify how people judge right from wrong by elucidating the criteria, standards, or rules that people use in making these judgments. Such criteria are grounded in values—for instance, promoting human welfare, maximizing individual freedom, or minimizing pain and suffering.

People of different cultures or with different worldviews may differ in their values and thus may differ in the specific actions they consider to be right or wrong. This is why some ethicists are **relativists;** that is, they believe that ethics do and should vary with social context. However, different human societies show a remarkable extent of agreement on what moral standards are appropriate. Thus many ethicists are **universalists;** that is, they maintain that there exist objective notions of right and wrong that hold across cultures and situations. For both relativists and universalists, ethics is a *normative* or *prescriptive* pursuit; it tells us how we *ought to* behave.

Ethical standards are the criteria that help differentiate right from wrong. One classic ethical standard is *virtue,* which, as the ancient Greek philosopher Aristotle held, involves the personal achievement of moral excellence in character through reasoning and moderation. Another ethical standard is the *categorical imperative* proposed by Immanuel Kant, which roughly approximates Christianity's "golden rule": to treat others as you would prefer to be treated yourself. A third standard is the principle of *utility,* elaborated by British philosophers Jeremy Bentham and John Stuart Mill. The utilitarian principle holds that something is right when it produces the greatest practical benefits for the most people. We employ such ethical standards as tools for decision making, consciously or unconsciously, to situations in everyday life.

Environmental ethics pertains to humans and the environment

The application of ethical standards to relationships between humans and nonhuman entities is known as **environmental ethics.** This relatively new branch of ethics arose once people began to perceive environmental changes brought about by industrialization. Human interactions with the environment frequently give rise to ethical questions that can be difficult to resolve. Consider some examples:

- Does the present generation have an obligation to conserve resources for future generations? If so, how should this influence our decision making, and how much are we obligated to sacrifice?
- Are there situations that justify exposing some communities to a disproportionate share of pollution? If not, what actions are warranted in preventing this problem?

- Are humans justified in driving species to extinction? Are we justified in causing other changes in ecological systems? If destroying a forest would drive extinct an insect species few people have heard of but would create jobs for 10,000 people, would that action be ethically admissible? What if it were an owl species? What if only 100 jobs would be created?

We have extended ethical consideration to more entities through time

Answers to questions like those above depend partly on what ethical standard(s) a person chooses to use. They also depend on the breadth and inclusiveness of the person's domain of ethical concern. A person who feels responsibility for the welfare of insects would answer the third question very differently from a person whose domain of ethical concern ends with humans. Most of us feel moral obligations to some entities in the world, but by no means to all.

Throughout Western history, people have gradually enlarged the array of entities they feel deserve ethical consideration. The enslavement of human beings by other human beings was common in many societies until recently, for instance. Women in the United States were not allowed to vote until 1920, and they still face lower pay for equal work. Consider, too, how little ethical consideration citizens of one nation generally extend to those of another on which their government has declared war. Human societies are only now beginning to embrace the principle that all people be granted equal ethical consideration.

Our expanding domain of ethical concern has begun to include nonhuman entities as well. Concern for the welfare of domesticated animals is evident in humane societies and in the lengths many people go to provide for their pets. Animal-rights activists voice concern for animals that are hunted, eaten, or used in laboratory testing. A great many people now accept that wild animals (at least obviously sentient animals, such as large vertebrates, with which we share similarities) merit ethical consideration. Moreover, today many environmentalists are concerned not only with certain animals but also with the well-being of whole natural communities. Some people have gone still further, suggesting that all of nature—living things and nonliving things, even rocks—should be ethically represented. The historian Roderick Nash illustrated this historical expansion of ethics in his 1989 book, *The Rights of Nature* (Figure 2.3).

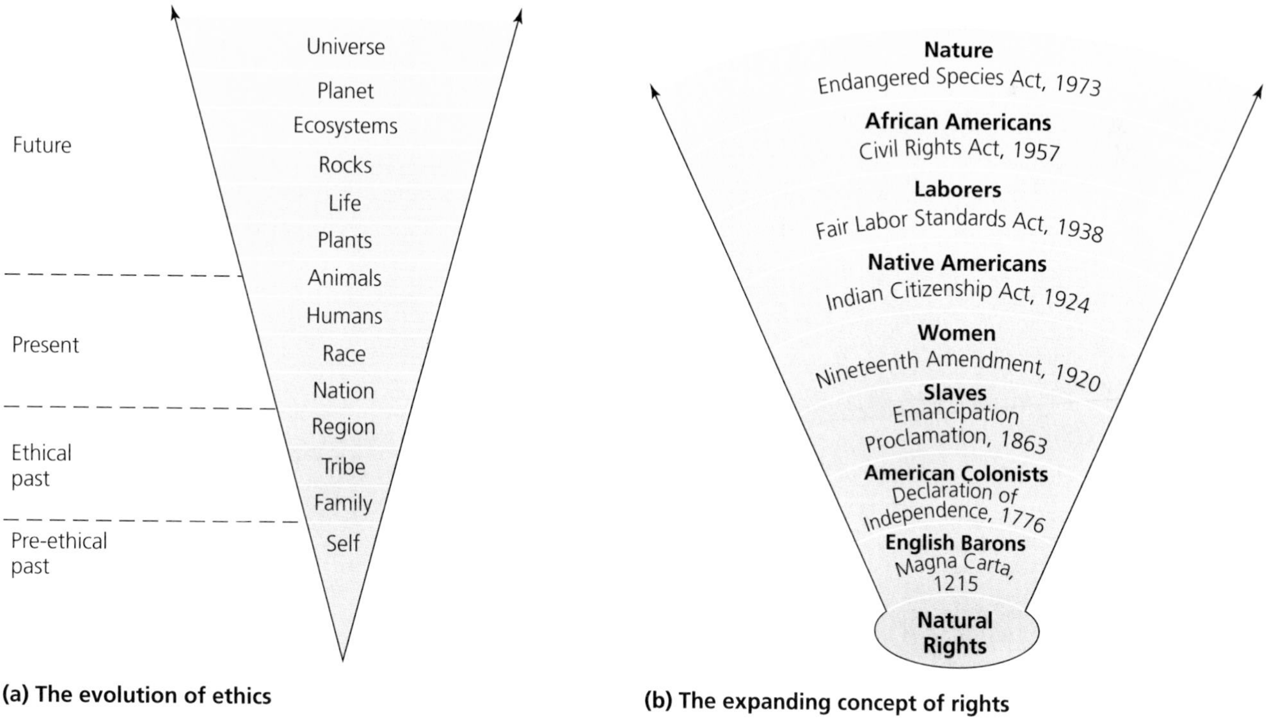

FIGURE 2.3 Through time, people in Western cultures have broadened the scope of their ethical consideration for others. We can view ethics progressing through time in a generalized way outward from the self (**a**). This historical expansion of ethics is reflected by key legal milestones in the expansion of rights granted by Britain and then the United States (**b**). *Source:* Nash, R. F. 1989. *The rights of nature.* University of Wisconsin Press.

What is behind this ongoing expansion? Rising economic prosperity in Western cultures, as people gained more leisure time and became less anxious about their day-to-day survival, has helped enlarge our ethical domain. Science has also played a role. Ecology, as it has developed over the past 75 years, has made clear that all organisms are interconnected and that what affects plants, animals, and ecosystems can in turn affect humans. Evolutionary biology over the past 150 years has shown that humans are merely one species out of millions and have evolved subject to the same pressures as other organisms. Ecology and evolution have demonstrated scientifically that humans do not stand apart from nature, but rather are part of it.

For many non-Western cultures, expanded ethical domains are nothing new. Many traditional hunter-gatherer cultures have long granted nonhuman entities ethical standing. The Mirrar, who view their landscape as sacred and alive, are a case in point. However, it is worthwhile to examine Western ethical expansion because it is tied to so many of our society's beliefs and actions regarding the environment. People often simplify the continuum Nash portrayed by dividing it into three ethical perspectives: anthropocentrism, biocentrism, and ecocentrism.

Anthropocentrism **Anthropocentrism** describes a human-centered view of our relationship with the environment. An anthropocentrist denies or ignores the notion that nonhuman entities can have rights. An anthropocentrist also measures the costs and benefits of actions solely according to their impact on people (Figure 2.4). To evaluate a human action that affects the environment, an anthropocentrist might use criteria such as impacts on human health, economic costs and benefits, and aesthetic concerns. For example, if the Jabiluka mine provided a net economic benefit while doing no harm to human health and having little aesthetic impact, the anthropocentrist would conclude it was a worthwhile venture, even if it might drive some native species extinct. If protecting the area would provide spiritual, economic, or other benefits to humans now or in the future, an anthropocentrist might favor its protection. In the anthropocentric perspective, anything not providing benefit to people is considered to be of negligible value.

Biocentrism In contrast to anthropocentrism, **biocentrism** ascribes values to actions, entities, or properties on the basis of their effects on all living things or on the integrity of the biotic realm in general (see Figure 2.4). In

FIGURE 2.4 We can categorize people's ethical perspectives as anthropocentric, biocentric, or ecocentric. An anthropocentrist extends ethical standing only to humans and judges actions in terms of their effects on humans. A biocentrist values and considers all living things, human and otherwise. An ecocentrist extends ethical consideration to living and nonliving components of the environment. The ecocentrist also takes a holistic view of the connections among these components, valuing the larger functional systems of which they are a part.

this perspective, all life has ethical standing. A biocentrist evaluates actions in terms of their overall impact on living things, including—but not exclusively focusing on—human beings. In the case of the Jabiluka mine proposal, a biocentrist might oppose the mine if it posed a serious threat to the abundance and variety of living things in the area, even if it would create jobs, generate economic growth, and pose no threat to human health. Some biocentrists advocate equal consideration of all living things, whereas others advocate that some types of organisms should receive more than others.

Ecocentrism **Ecocentrism** judges actions in terms of their benefit or harm to the integrity of whole ecological systems, which consist of biotic and abiotic elements and the relationships among them (see Figure 2.4). For an ecocentrist, the well-being of an individual organism—human or otherwise—is less important than the well-being of a larger integrated ecological system. An ecocentrist might approve of an action that harmed human health, caused economic loss, or took a number of lives, if such impacts were necessary to protect an entire species, community, or ecosystem (we will study species, communities, and ecosystems in Chapters 5 through 7). Ecocentrism is a more holistic perspective than biocentrism or anthropocentrism. It not only includes a wider variety of entities, but also stresses the need to preserve the connections that tie the entities together into functional systems.

Environmental ethics has ancient roots

Environmental ethics arose as a distinct academic discipline in the early 1970s, but people have contemplated our relationship with, and possible responsibilities toward, nature for thousands of years. Ancient Aboriginal dreaming-track stories treat the environment as a source of sacred teachings, so boulders, caves, patches of lichen, and other entities are felt to have moral significance worthy of contemplation and protection. In the Western tradition, the ancient Greek philosopher Plato expressed what he considered humans' moral obligation to the environment, writing, "The land is our ancestral home and we must cherish it even more than children cherish their mother."

Some ethicists and theologians have pointed to the religious traditions of Christianity, Judaism, and Islam as sources of anthropocentric hostility toward the environment. They point out biblical passages such as, "Be fruitful and multiply, and fill the earth and subdue it; and have dominion over the fish of the sea and over the birds of the air and over every living thing that moves upon the earth." Such wording has justified and encouraged an animosity toward nature that has characterized Western culture over the centuries, some scholars say. Others interpret sacred texts of these religions to encourage benevolent human stewardship over nature. Consider the directive, "You shall not defile the land in which you live. . . ." In fact, a 2003 poll showed that 56% of Americans supported environmental protection because they considered the environment to be "God's creation." Although people have held differing views of their ethical relationship with their environment for millennia, environmental impacts that became apparent during the industrial revolution intensified debate about our species' relationship with its environment.

The industrial revolution inspired environmental philosophers

As the industrial revolution spread in the 19th century from Great Britain throughout Europe and to North America and elsewhere, it amplified human impacts on the environment. In this period of social and economic transformation, agricultural economies became industrial ones, machines enhanced or replaced human and animal labor, and much of the rural population moved into cities. Consumption of natural resources accelerated rapidly, and pollution increased dramatically as coal combustion fueled railroads, steamships, ironworks, and factories.

Many British writers and philosophers of the time criticized the drawbacks of industrialization. Critic **John Ruskin** (1819–1900) called cities "little more than laboratories for the distillation into heaven of venomous smokes and smells." Ruskin also complained that people prized the material benefits that nature could provide but no longer appreciated its spiritual and aesthetic benefits. Motivated by similar concerns, a number of citizens' groups sprang up in 19th-century England that could be considered some of the first environmental organizations (Table 2.1).

In the United States during the 1840s, a philosophical movement called *transcendentalism* flourished, espoused in New England by the American philosophers **Ralph Waldo Emerson** and **Henry David Thoreau** and by poet **Walt Whitman.** The transcendentalists viewed nature as a direct manifestation of the divine, emphasizing the soul's oneness with nature and God. Like Ruskin, the transcendentalists objected to what they saw as their fellow citizens' obsession with material things, and through their writing they promoted their holistic view of nature. The transcendentalist worldview resembled that of the Mirrar in some respects. Both traditions identify a need to experience wild nature, and both view natural entities as symbols or messengers of some deeper truth. Although Thoreau viewed nature as

Table 2.1 Early Environmental Organizations in 19th-century Great Britain

Organization	Year established	Purpose
Scottish Rights of Way Society	1843	Protect walking paths in and near cities
Commons Preservation Society	1865	Preserve forests and other landscapes
Society for the Protection of Ancient Buildings	1877	Protect the built environment, especially historic buildings
Selborne League	1885	Protect rare birds, plants, and landscapes
Coal Smoke Abatement Society	1898	Improve urban air quality

divine, he also observed the natural world closely and came to understand it in the manner of a scientist; he was in many ways one of the first ecologists. His book *Walden,* in which he recorded his observations and thoughts while he lived at Walden Pond away from the bustle of urban Massachusetts, remains a classic of American literature.

Conservation and preservation arose around the turn of the 20th century

One admirer of Emerson and Thoreau was **John Muir** (1838–1914), a Scottish immigrant to the United States who eventually settled in California and made the Yosemite Valley his wilderness home. Although Muir chose to live in isolation in his beloved Sierra Nevada for long stretches of time, he nonetheless became politically active and won fame as a tireless advocate for the preservation of wilderness (Figure 2.5). Muir was motivated by the rapid deforestation and environmental degradation he witnessed throughout North America and by his belief that the natural world should be treated with the same respect that cathedrals receive. Today he is associated with the **preservation** ethic, which holds that we should protect the natural environment in a pristine, unaltered state. Muir argued that nature deserved protection for its own inherent value (an ecocentrist argument), but he also maintained that nature played a large role in human happiness and fulfillment (an anthropocentrist argument). "Everybody needs beauty as well as bread," he wrote in 1912, "Places to play in and pray in, where nature may heal and give strength to body and soul alike."

Some of the same factors that motivated Muir also inspired the first professionally trained American forester, **Gifford Pinchot** (1865–1946; Figure 2.6). Both men opposed the rapid deforestation and unregulated economic development of North American lands that occurred during their lifetimes. However, Pinchot took a more anthropocentric view of how and why nature should be valued. He is today the person most closely associated with the **conservation** ethic, which holds that humans should put natural resources to use but also that we have a responsibility to manage them wisely. Whereas preservation aims to preserve nature for its own sake and for the aesthetic, spiritual, symbolic, and recreational benefit of people, conservation promotes the prudent, efficient, and sustainable extraction and use of natural resources for the benefit of present and future generations. The conservation ethic uses a utilitarian standard, stating that in using resources, humans should attempt to provide the greatest good to the greatest number of people for the longest time.

Pinchot and Muir came to represent different branches of the American environmental movement, and their contrasting ethical approaches often pitted them against

FIGURE 2.5 A pioneering advocate of the preservation ethic, John Muir is also remembered for his efforts to protect the Sierra Nevada from development and for his role in founding the Sierra Club, a leading environmental organization. Here Muir (right) is shown with President Theodore Roosevelt in Yosemite National Park. After his 1903 wilderness camping trip with Muir, the president instructed his interior secretary to increase protected areas in the Sierra Nevada.

FIGURE 2.6 Gifford Pinchot, the first chief of what would become the U.S. Forest Service, was a leading proponent of the conservation ethic. The conservation ethic holds that humans should use natural resources, but strive to ensure the greatest good for the greatest number for the longest time.

one another on policy issues of the day. Nonetheless, they both represented reactions against a prevailing "development ethic," which holds that humans are and should be masters of nature and which promotes economic development without regard to its negative consequences. Pinchot eventually founded what would become the U.S. Forest Service and served as its chief in Theodore Roosevelt's administration. Both Pinchot and Muir left legacies that reverberate today in the different ethical approaches to environmentalism.

Weighing the Issues: Preservation and Conservation

With which ethic do you most identify—preservation or conservation? Think of a forest or other important natural resource in your region. Give an example of a situation in which you might adopt a preservation ethic and an example of one in which you might adopt a conservation ethic. Are there conditions under which you'd follow neither, but instead adopt a "development ethic"?

Aldo Leopold's land ethic arose from the conservation and preservation ethics

As a young forester and wildlife manager, **Aldo Leopold** (1887–1949; Figure 2.7) began his career fully in the conservationist camp, having graduated from Yale Forestry School, which Pinchot had helped found just as Roosevelt and Pinchot were advancing conservation on the national stage. As a forest manager in Arizona and New Mexico, Leopold embraced the government policy of shooting predators, such as wolves, to increase populations of deer and other game animals. At the same time, Leopold followed the development of ecological science. He eventually ceased to view certain species as "good" or "bad" and instead came to see that healthy ecological systems depend on the protection of all their interacting parts, including predators as well as prey. Drawing an analogy to mechanical maintenance, he wrote, "to keep every cog and wheel is the first precaution of intelligent tinkering."

It was more than science that pulled Leopold from an anthropocentric perspective toward a more holistic one. One day he shot a wolf, and when he reached the animal, Leopold was transfixed by "a fierce green fire dying in her eyes." The experience remained with him for the rest of his life and helped lead him to a more ecocentric ethical outlook. Years later, as a University of Wisconsin professor,

FIGURE 2.7 Aldo Leopold, a wildlife manager and pioneering environmental philosopher, articulated a new relationship between people and the environment. In his essay "The Land Ethic," he called on people to include the environment in their ethical framework.

Leopold argued that humans should view themselves and "the land" as members of the same community, and that people are obliged to treat the land in an ethical manner. In his 1949 essay "The Land Ethic," he wrote:

> All ethics so far evolved rest upon a single premise: that the individual is a member of a community of interdependent parts. . . . The land ethic simply enlarges the boundaries of the community to include soils, waters, plants, and animals, or collectively: the land. . . . A land ethic changes the role of *Homo sapiens* from conqueror of the land-community to plain member and citizen of it. . . . It implies respect for his fellow-members, and also respect for the community as such.

Leopold intended that the land ethic would help guide decision making. "A thing is right," he wrote, "when it tends to preserve the integrity, stability, and beauty of the biotic community. It is wrong when it tends otherwise." Leopold died before seeing "The Land Ethic" and his best-known book, *A Sand County Almanac,* in print, but today many view him as the most eloquent and important philosopher of environmental ethics.

Deep ecology extends environmental ethics

One philosophical perspective that goes beyond even Leopold's ecocentrism is **deep ecology,** established in the 1970s. Proponents of deep ecology describe the movement as resting on principles of "self-realization" and biocentric equality. They define self-realization as the awareness that humans are inseparable from nature and that the air we breathe, the water we drink, and the foods we consume are both products of the environment and integral parts of us. Biocentric equality is the precept that all living beings have equal value and that because we are truly inseparable from our environment, we should protect all other living things as we would protect ourselves.

Ecofeminism equates male attitudes toward nature and toward women

As deep ecology and mainstream environmentalism were extending people's ethical domains outward during the 1960s and 1970s, major social movements, such as the civil rights movement and the feminist movement, were gaining prominence. A number of feminist scholars saw parallels in human behavior toward nature and men's behavior toward women. The degradation of nature and the social oppression of women shared common roots, these scholars asserted.

Ecological feminism, or **ecofeminism,** argues that the patriarchal (male-dominated) structure of society—which traditionally grants more power and prestige to men than to women—is a root cause of both social and environmental problems. Ecofeminists hold that a worldview traditionally associated with women, which interprets the world in terms of interrelationships and cooperation, is more compatible with nature than a worldview traditionally associated with men, which interprets the world in terms of hierarchies and competition. Ecofeminists maintain that a male tendency to try to dominate and conquer what men hate, fear, or do not understand has historically been exercised against both women and the natural environment.

Environmental justice seeks equal treatment for all races and classes

Our society's domain of ethical concern has been expanding from rich to poor and from majority races and ethnic groups to minority ones. This ethical expansion involves applying a standard of fairness and equality and has given rise to the environmental justice movement. The U.S. Environmental Protection Agency (EPA) defines **environmental justice** as "the fair treatment and meaningful involvement of all people regardless of race, color, national origin, or income with respect to the development, implementation, and enforcement of environmental laws, regulations, and policies."

The environmental justice movement was fueled by the perception that poor people and minorities tend to be exposed to a greater share of pollution, hazards, and environmental degradation than are richer people and whites. A protest in the early 1980s by African Americans in Warren County, North Carolina, against a toxic waste dump in their community is widely seen as the beginning of the movement (Figure 2.8). The state had chosen to site the dump in the county with the highest percentage of African Americans, prompting Warren County residents to suspect "environmental racism." Environmental justice grew to prominence in the early 1990s as more people across North America began fighting environmental hazards in their communities. This movement—in contrast to earlier environmental movements—was made up largely of low-income people and minorities.

In 1983, a U.S. General Accounting Office (GAO) study found that three of four toxic waste landfills in the southeastern United States were located in communities where the population of minorities exceeded that of

FIGURE 2.8 Communities of poor people and people of color have suffered more than their share of environmental problems, a situation that has given rise to the environmental justice movement. The movement gained prominence with this protest of a toxic waste dump in Warren County, North Carolina.

whites, and the fourth landfill was located in a community that was 38% African American. In contrast, minorities made up only 20% of the region's population. In 1987, the United Church of Christ Commission for Racial Justice found that the percentage of minorities in areas with toxic waste sites was twice that of areas without toxic waste sites. Researchers studying air pollution, lead poisoning, pesticide exposure, and workplace hazards have found similar patterns. Today the environmental justice movement has broadened to encompass equity in transportation options, redevelopment of abandoned urban sites, worker health and safety, and access to parklands.

Weighing the Issues: Environmental Justice

Consider the place where you grew up. Where were the factories, waste dumps, and polluting facilities located, and who lived closest to them? Who lives nearest them in the town or city that hosts your campus? Do you think the concerns of environmental justice advocates are justified? If so, what could be done to ensure that poor communities are no more polluted than wealthy ones?

The attempts of the predominantly white Australian government and uranium mining companies to open mines on traditional lands of the Mirrar (Figure 2.9) have been characterized by critics as violations of environmental

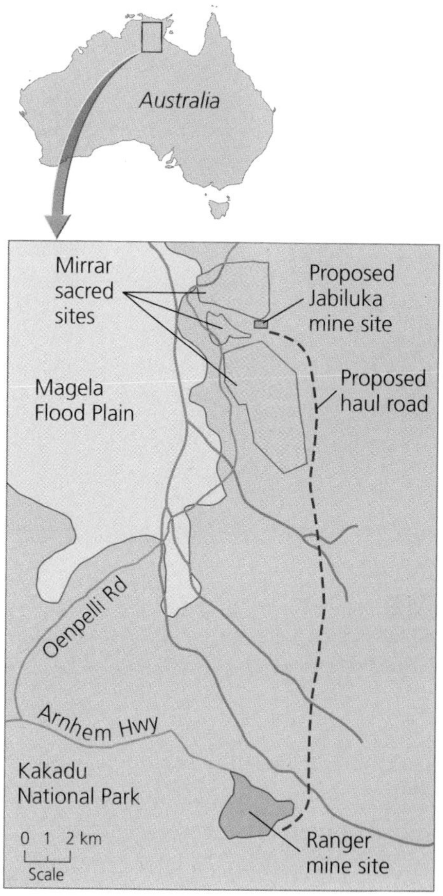

FIGURE 2.9 The Ranger mine and the site of the proposed Jabiluka mine lie amid Aboriginal lands and sacred sites.

justice. In North America, uranium mining has been a focus of environmental justice concerns as well. Native Americans of the Dene Nation, in Canada's Northwest Territories and Saskatchewan, have suffered health effects from working in uranium mines with minimal safeguards. In the southwestern United States from 1948 through the late 1960s, uranium mines employed many Native Americans, including those of the Navajo nation. Although uranium mining had already been linked to health problems and premature death, the miners had practically no awareness of radiation and its risks. The Navajo language did not even have a word for *radiation,* and for nearly two decades neither the mining industry nor the U.S. government provided information or safeguards to the miners. Many Navajo families built homes and bread-baking ovens from the abundant waste rock produced as a by-product of the mining process, not realizing it was radioactive (Figure 2.10).

Cases of lung cancer began to appear among Navajo miners in the early 1960s, but scientific studies of radiation's

FIGURE 2.10 Native Americans employed as uranium miners in Canada and the United States, such as the Navajo miner shown here, have suffered from adverse effects of mining.

effects on miners at the time excluded Native American workers. The decision to include only white miners in the studies was attributed to the researchers' desire to study a "homogeneous population." A later generation of Americans perceived this as negligence and discrimination, and their desire for justice gave rise to the Radiation Exposure Compensation Act of 1990, a federal law that compensated Navajo miners who suffered health effects from unprotected work in the mines. Such developments illustrate the interplay between changing ethical values and resultant policymaking, which we will examine further in Chapter 3. First, we will explore economics, which, like ethics, addresses people's values and widely informs policy.

Economics: Approaches and Environmental Implications

People who oppose the Jabiluka mine do so largely on the basis of ethical concerns and worries over environmental impacts. Few have challenged the mining plan on economic grounds; mine opponents generally recognize uranium as a lucrative resource that generates jobs, income, and electricity, and they do not dispute the contribution of uranium exports to the Australian economy. Support for the mine is based primarily on economic factors. Such conflict between ethical and economic motivations is a recurrent theme in environmental issues worldwide.

Is there a trade-off between economics and the environment?

Although measures to safeguard the environment may frequently mesh well with ethical considerations, we often hear it said that environmental protection works in opposition to economic health. But is this necessarily the case? Growing numbers of economists assert that there need be no such trade-off—that in fact, environmental protection can be *good* for the economy. As we will see, the view one takes often depends on whether one thinks in the short term or the long term and whether one holds to traditional economic schools of thought or to newer ones that view human economies as coupled to the natural environment.

Economics studies the allocation of scarce resources

Like ethics, economics examines factors that guide human behavior. **Economics** is the study of how people decide to use scarce resources to provide goods and services in the face of demand for them. By this definition, environmental problems are also economic problems that can intensify as population and per capita resource consumption increase. For example, pollution may be viewed as depletion of the scarce resources of clean air, water, or soil. Indeed, the word *economics* and the word *ecology* come from the same Greek root, *oikos,* meaning "household." In its broadest context, the human "household" is Earth itself. Economists traditionally have studied the household of human society, and ecologists the broader household of all life.

Several types of economies exist today

An **economy** is a social system that converts resources into **goods,** material commodities manufactured for and bought by individuals and businesses; and **services,** work done for others as a form of business. The oldest type of economy is the **subsistence economy.** People in subsistence economies—who still comprise much of the human population—meet most or all of their daily needs directly from nature and do not purchase or trade for most of life's necessities.

A second type of economy is the **capitalist market economy.** In this system, buyers and sellers interact to determine which goods and services to produce, how much to produce, and how these should be produced and distributed. Capitalist economies are often contrasted with state socialist economies, or **centrally planned economies,** in which government determines in a top-down manner how to allocate resources. In today's world, capitalism predominates over socialism. However, a pure market economy would operate without government intervention. In reality, all capitalist market economies today, including that of the United States, have borrowed much from state socialism and are in fact hybrid systems. In modern market economies, as we will see in Chapter 3, governments typically intervene for several reasons: (1) to eliminate unfair advantages held by single buyers or sellers; (2) to provide social services, such as national defense, medical care, and education; (3) to provide "safety nets" (for the elderly, victims of natural disasters, and so on); (4) to manage the commons (▸ p. 6); and (5) to mitigate pollution.

Many societies function somewhere along a continuum between a pure subsistence economy and a capitalist market economy. For example, the Mirrar acquire essential food and water directly from their environment, but they purchase many other necessities.

Environment and economy are intricately linked

All human economies exist within the larger environment and depend on it in important ways. Economies receive inputs from the environment, process them in complex ways that enable human society to function, then discharge outputs of waste from this process into the environment. Economies are thus *open systems* (▸ p. 189) integrated with the larger environmental system of which they are a part. Earth, in turn, is a materially *closed system,* so the inputs Earth can provide to economies are ultimately limited.

Although these interactions between human economies and the nonhuman environment are readily apparent, traditional economic schools of thought have long overlooked the importance of these connections. Indeed, most conventional economists today still adhere to a worldview that largely ignores the environment (Figure 2.11a), and this worldview continues to drive most policy decisions. However, modern economists belonging to the fast-growing fields of environmental economics and ecological economics (▸ p. 45) explicitly accept that human economies are subsets of the environment and depend crucially on the environment (Figure 2.11b).

Economic activity uses resources from the environment. Natural resources (▸ p. 4) are the various substances and forces we need to survive: the sun's energy, the fresh water we drink, the trees that provide our lumber, the rocks that provide our metals, and the fossil fuels that power our machines and produce our plastics. We can think of natural resources as "goods" produced by nature. Without Earth's natural resources, there would be no human economies and, in fact, no human beings.

Environmental systems also naturally function in a manner that supports economies. Earth's ecological systems purify air and water, cycle nutrients, provide for the pollination of plants by animals, and serve as receptacles and recycling systems for the waste generated by our economic activity. Such essential services, often called **ecosystem services** (Table 2.2), support the life that makes our economic activity possible. Some ecosystem services represent the very nuts-and-bolts of our survival, and others enhance our quality of life.

While the environment enables economic activity by providing ecosystem goods and services, economic activity can affect the environment in return. When we deplete natural resources and produce too much pollution, we can degrade the ability of ecological systems to function. In fact, the Millennium Ecosystem Assessment (▸ p. 20) concluded in 2005 that 15 of 24 ecosystem services its scientists surveyed globally were being degraded or used unsustainably. The degradation of ecosystem services can in turn negatively affect economies. Currently, ecological degradation is harming poor people more than wealthy people, the Millennium Ecosystem Assessment found. As a result, restoring ecosystem services stands as a prime avenue for alleviating poverty in much of the world. These interrelationships among economic and environmental conditions have only recently become widely recognized, however. Let's briefly examine how economic thought has changed over the years, tracing the path that is now beginning to lead economies to become more compatible with natural systems.

Adam Smith and other philosophers founded classical economics

Economics shares a common intellectual heritage with ethics, and practitioners of both have long been interested in the relationship between individual action and societal well-being. Some philosophers argued that individuals acting in their own self-interest would harm society. Others believed that such behavior could benefit society, as long as the behavior was constrained by the rule of law and private property rights and operated within fairly competitive markets. The latter view was articulated by Scottish philosopher **Adam Smith** (1723–1790). Known

FIGURE 2.11 Modern environmental and ecological economists view economic activity differently from economists of the more conventional neoclassical school. Standard neoclassical economics focuses on processes of production and consumption between households and businesses (**a**), viewing the environment only as a "factor of production" that helps enable the production of goods. Environmental and ecological economists view the human economy as existing within the natural environment (**b**), receiving resources from it, discharging waste into it, and interacting with it through various ecosystem services.

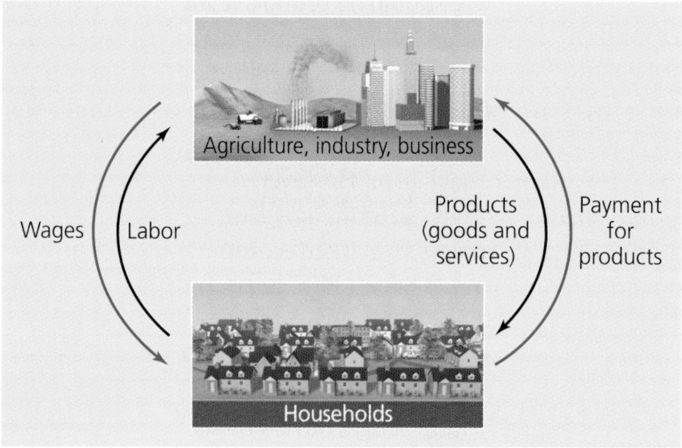

(a) Conventional view of economic activity

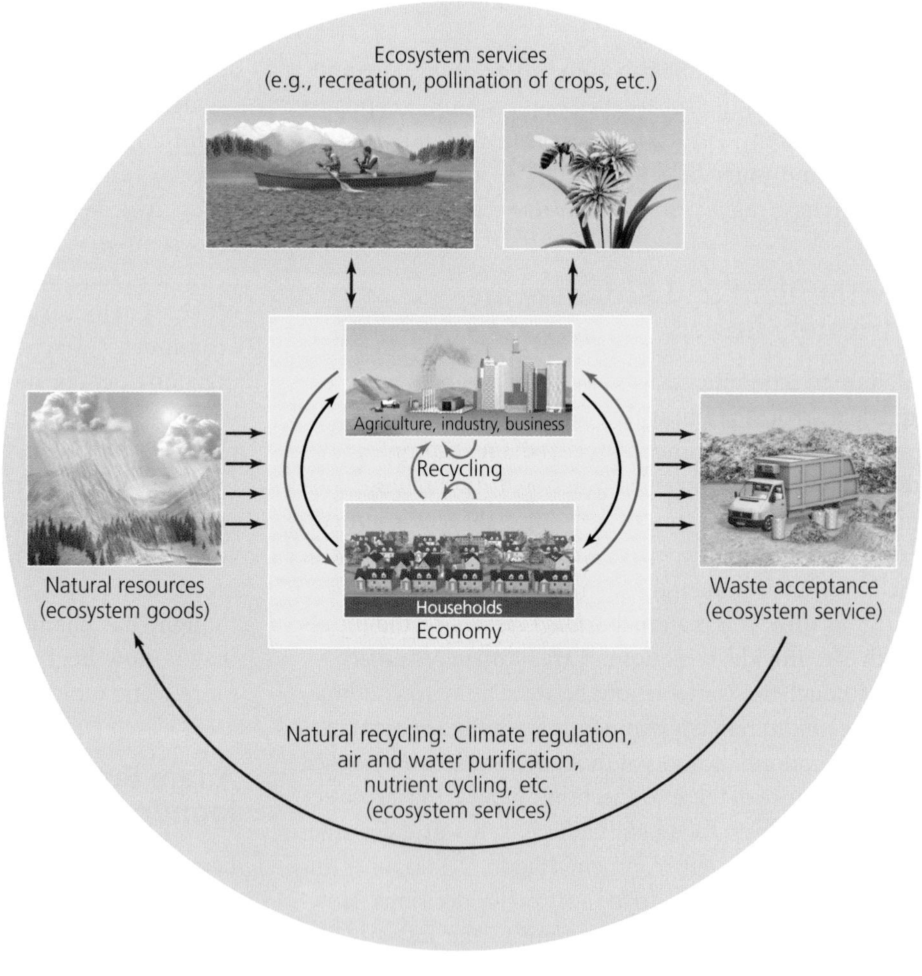

(b) Economic activity as viewed by environmental and ecological economists

Table 2.2 Ecosystem Services

Type of ecosystem service*	Example(s)
Regulating atmospheric gases	Maintaining the ozone layer; balancing oxygen, carbon dioxide, and other gases
Regulating climate	Controlling global temperature and precipitation through oceanic and atmospheric currents, greenhouse gases, cloud formation, and so on
Damping impacts from disturbance	Providing storm protection, flood control, and drought recovery, mainly through vegetation structure
Regulating water flow	Providing water for agriculture, industry, transportation
Storing and retaining water	Providing water through watersheds, reservoirs, aquifers
Controlling erosion and promoting soil retention	Preventing soil loss from wind or runoff; storing silt in lakes and wetlands
Forming soil	Weathering rock; accumulating organic material
Cycling nutrients	Cycling carbon, nitrogen, phosphorus, sulfur, and other nutrients through ecosystems
Treating waste	Removing toxins, recovering nutrients, controlling pollution
Pollinating plants	Transporting floral gametes by wind or pollinating animals, enabling crops and wild plants to reproduce
Controlling populations biologically	Controlling prey with predators; controlling hosts with parasites; controlling herbivory on crops with predators and parasites
Providing habitat	Providing ecological settings in which creatures can breed, feed, rest, migrate, winter
Providing food	Producing fish, game, crops, nuts, and fruits that humans obtain by hunting, gathering, fishing, subsistence farming
Supplying raw materials	Producing lumber, fuel, metals, fodder
Furnishing genetic resources	Providing unique biological sources for medicine, materials science, genes for resistance to plant pathogens and crop pests, ornamental species (pets and horticultural plant varieties)
Providing recreational opportunities	Ecotourism, sport-fishing, hiking, birding, kayaking, other outdoor recreation
Providing cultural or noncommercial uses and goods	Aesthetic, artistic, educational, spiritual, and/or scientific values of ecosystems

*Ecosystem "goods" are here included in ecosystem services.
Adapted with permission from Costanza, R., et al. 1997. The value of the world's ecosystem services and natural capital. *Nature* 387: 253–260.

today as the father of **classical economics,** Smith believed that when people are free to pursue their own economic self-interest in a competitive marketplace, the marketplace will behave as if guided by "an invisible hand" that ensures their actions will benefit society as a whole. In his 1776 book *Inquiry into the Nature and Causes of the Wealth of Nations,* Smith wrote:

> It is not from the benevolence of the butcher, the brewer, or the baker that we expect our dinner, but from their regard to their own self-interest. [Each individual] intends only his own security, only his own gain. And he is led in this by an invisible hand to promote an end which was no part of his intention. By pursuing his own interests he frequently promotes that of society more effectually than when he really intends to.

Smith's philosophy remains a pillar of free-market thought today, and many credit it for the tremendous gains in material prosperity that industrialized nations have experienced in the past few centuries. The *laissez-faire* policies that free-market thought has spawned, however, have also been widely criticized by those who feel that market capitalism exacerbates inequalities between rich and poor and contributes to environmental degradation. Market capitalism, these critics assert, should be constrained and regulated by democratic government.

Neoclassical economics incorporates human psychology

Economists subsequently took more quantitative a proaches and incorporated human psychology int work. Modern **neoclassical economics** exam psychological factors underlying consumer choices, explaining market prices in terms of consumer preferences for units of particular commodities. In neoclassical economic theory, buyers desire the lowest possible price, whereas sellers desire the highest possible price. This conflict between buyers and sellers results in a compromise price being reached and the "right" quantity of commodities being bought and sold. This is often phrased in terms of *supply,* the amount of a product offered for sale at a given price, and *demand,* the amount of a product people will buy at a given price if free to do so. Theoretically, when prices go up, demand drops and supply increases; and when prices fall, demand rises and supply decreases. The market automatically moves toward an equilibrium point, a price at which supply equals demand (Figure 2.12a). Similar reasoning can be applied to environmental issues, such that economists can determine "optimal" levels of resource use or pollution control (Figure 2.12b).

Cost–benefit analysis is a widespread tool

A method commonly used by neoclassical economists is **cost–benefit analysis.** In this approach, estimated costs for a proposed action are totaled up and compared to the sum of benefits estimated to result from the action. If total benefits exceed costs, the action should be pursued; if costs exceed benefits, it should not. When choosing among multiple alternative actions, the one with the greatest excess of benefits over costs should be chosen. This reasoning seems eminently logical, and because the analysis aims to be quantitative and precise, it would seem likely to result in clear recommendations for policy.

However, problems often crop up because not all costs and benefits are easily quantified, or even identified or defined. It may be quite feasible to quantify wages paid to uranium miners, the market value of uranium extracted from a mine, or the cost of measures to minimize health risks for miners. But it is difficult to assess the cost of a valued landscape being scarred by mine development, or the cost of radioactive contamination of a stream. Because some costs and benefits cannot easily be assigned monetary values—and because it is difficult to identify and agree on all costs and benefits—cost–benefit analysis is often controversial. Moreover, because economic bene- ... ily quantified than environmental ... ts tend to be overrepresented in ... analyses. As a result, environmen- ... these analyses are biased in favor ... ent and against environmental

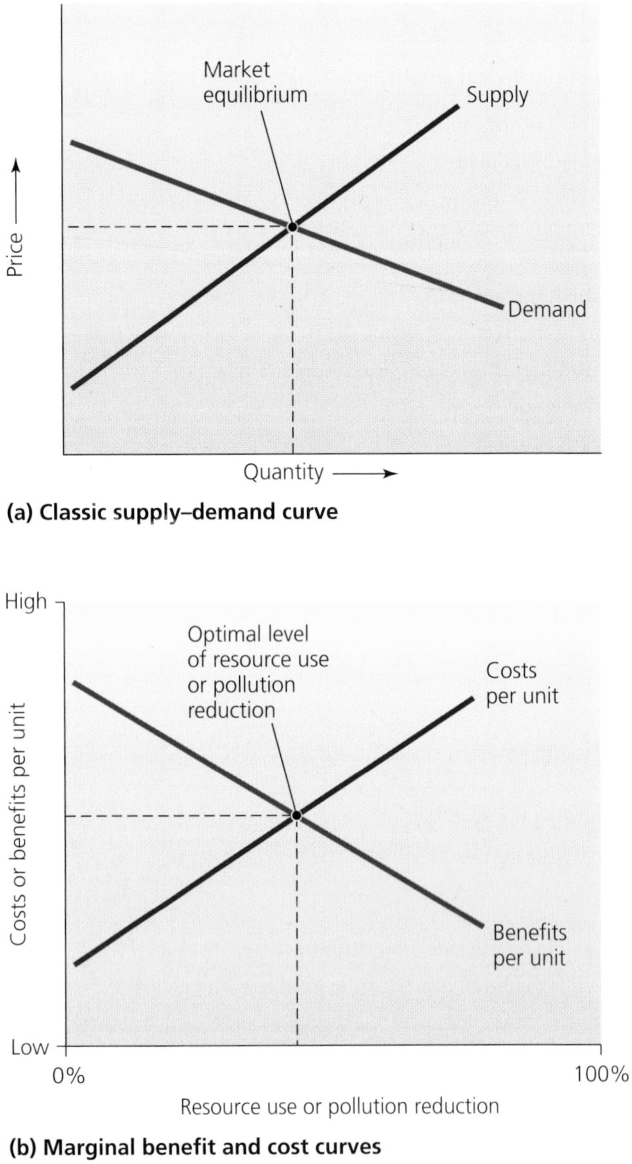

FIGURE 2.12 This basic supply-and-demand curve (**a**) illustrates the relationship between supply, demand, and market equilibrium, the "balance point" at which demand is equal to supply. We can use a similar graph (**b**) to determine an "optimal" level of resource use or pollution mitigation. In this graph, the cost per unit of resource use or pollution cleanup (blue line) rises as the resource use or pollution cleanup proceeds and it becomes expensive to extract or clean up the remaining amounts. Meanwhile, the benefits per unit of resource use or pollution cleanup (red line) decrease. The point where the lines intersect gives the optimal level.

Aspects of neoclassical economics have profound implications for the environment

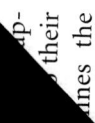

Today's capitalist market systems operate largely in accord with the precepts of neoclassical economics. These systems have generated unprecedented material wealth, employment, and other desirable outcomes, but they have also contributed to environmental problems. Four fundamental assumptions of neoclassical economics have implications for the environment:

- Resources are infinite or substitutable.
- Long-term effects should be discounted.
- Costs and benefits are internal.
- Growth is good.

Resources are infinite or substitutable Neoclassical economic models generally treat workers and other resources as being either infinite or largely "substitutable and interchangeable." This implies that once we have depleted a resource—natural, human, or otherwise—we should be able to find a replacement for it. Human resources can substitute for financial resources, for instance, or manufactured resources can substitute for natural resources. Theory allows that the substituted resource may be less efficient or more costly, but some degree of substitutability is generally assumed.

Certainly it is true that many resources can be replaced; our societies have transitioned from manual labor to animal labor to steam-driven power to fossil-fuel power, and they may yet transition to renewable power sources, such as solar energy. However, Earth's material resources are ultimately limited. Nonrenewable resources such as fossil fuels can be depleted, and many renewable resources can be used up as well if we exploit them faster than they can be replenished. This is what happened to the Easter Islanders (▸ pp. 8–9) who harvested their forests faster than the forests could regrow.

Weighing the Issues: Substitutability and the Environment

Can you think of a natural resource that might be difficult to replace with a substitute? What problems might arise from the assumption that all resources, including clean air and water, are substitutable and interchangeable? Can you think of examples that violate any of the other three assumptions of neoclassical economics? Explain your answers.

Long-term effects should be discounted Although few people would dispute that resources are *ultimately* limited, many assume that their depletion will take place so far in the future that there is no need for current generations to worry. For economists in the neoclassical tradition, an event far in the future counts much less than one in the present; in economic terminology, future effects are "discounted." In discounting, short-term costs and benefits are granted more importance than long-term costs and benefits, causing policy to play down long-term consequences of decisions we make today. Some governments and businesses use a 10% annual discount rate for decisions on resource use. This means that the long-term value of a stand of ancient trees worth $500,000 would drop by 10% each year and, after 10 years of discounting, would be worth only $174,339.22. By this logic, the more quickly the trees are cut, the more they are worth.

Costs and benefits are internal A third assumption of neoclassical economics is that all costs and benefits associated with a particular exchange of goods or services are borne by individuals engaging directly in the transaction. In other words, it is assumed that the costs and benefits of a transaction are "internal" to the transaction, experienced by the buyer and seller alone, and do not affect other members of society.

However, in many situations this is simply not the case. Pollution from a factory can harm people living nearby. In such cases, someone—often taxpayers not involved in producing the pollution—ends up paying the costs of alleviating it. Market prices do not take the social, environmental, or economic costs of this pollution into account. Costs or benefits of a transaction that involve people other than the buyer or seller are known as **externalities.** A positive externality is a benefit enjoyed by someone not involved in a transaction, and a negative externality, or **external cost,** is a cost borne by someone not involved in a transaction (Figure 2.13). Negative externalities often harm groups of people or society as a whole, while allowing certain individuals private gain. External costs commonly include the following:

- Human health problems
- Property damage
- Declines in desirable elements of the environment, such as fewer fish in a stream
- Aesthetic damage, such as that resulting from air pollution or clear-cutting
- Stress and anxiety experienced by people downstream or downwind from a pollution source
- Declining real estate values resulting from these problems

The Mirrar have experienced external costs in the form of pollution from the Ranger mine. In March 2002, the Australian Broadcasting Corporation reported that radioactive material from the Ranger mine had contaminated a stream on Mirrar land with uranium concentrations

FIGURE 2.13 An Indonesian boy wading in a polluted river suffers external costs: costs that are not borne by the buyer or seller. External costs may include water pollution, aesthetic harm, human health problems, property damage, harm to aquatic life, aesthetic degradation, declining real estate values, and other impacts.

4,000 times higher than allowed by law. According to an Aboriginal representative, this was the fourth such violation in less than three months.

By ignoring external costs, economies create a false idea of the true and complete costs of particular choices and unjustly subject people to the consequences of transactions in which they did not participate. External costs comprise one reason governments develop environmental legislation and regulations (▸ p. 63). Unfortunately, external costs are difficult to account for and eliminate. It is tough to assign a monetary value to illness, premature death, or degradation of an aesthetically or spiritually significant site.

Growth is good A fourth assumption of the neoclassical economic approach is that economic growth is required to keep employment high and maintain social order. The argument goes something like this: If the poor view the wealthy as the source of their suffering, they may revolt. Promoting economic growth can defuse this situation by creating opportunities for the poor to become wealthier themselves. By making the overall economic pie larger, everyone's slice becomes larger, even if some people still have much smaller slices than others.

The idea that economic growth is good has been encouraged over the centuries by the concept of material progress, espoused by Western cultures since the Enlightenment. The modern-day United States may represent the clearest example of the worldview that "more and bigger" is always better. We hear constantly in business news of increases in an industry's output or percentage growth in a country's economy, with increases touted as good news and decreases, stability, or even a minor drop in the *rate* of growth presented as bad news. Economic growth has become the quantitative yardstick by which progress is measured.

Is the growth paradigm good for us?

The rate of economic growth in recent decades is unprecedented in human history, and the world economy is seven times the size it was half a century ago. All measures of economic activity—trade, rates of production, amount and value of goods manufactured—are higher than they have ever been and are still increasing. This growth has brought many people much greater material wealth (although not equitably, and gaps between rich and poor remain immense).

To the extent that economic growth is a means to an end—a tool with which we can achieve greater human happiness—it can be a good thing. However, many observers today worry that growth has become an end in itself and is no longer necessarily the best tool with which to pursue happiness. Critics of the growth paradigm often note that runaway growth resembles the multiplication of cancer cells, which eventually overwhelm and destroy the organism in which they grow. These critics fear that runaway economic growth will likewise destroy the economic system on which we all depend. Resources for growth are ultimately limited, they argue, so nonstop growth is not sustainable and will fail as a long-term strategy.

Defenders of traditional economic approaches reply that Cassandras (▸ p. 22) have been saying for decades that limited resources would doom growth-oriented economies, yet most of these economies are still expanding dramatically. Thomas Malthus (▸ pp. 4–5) warned two centuries ago that increasing populations would bump into limits on resources. British classical economist David Ricardo (1772–1823) reasoned that as more high-quality farmland became occupied, profits from the lower-quality farmland that was left would progressively decrease, eventually halting growth. German philosopher Karl Marx (1818–1883) held that limits on economic growth would occur in the form of working-class revolutions against the capitalist leisure class. So why then, 150 to 200 years after such predictions were made, are we witnessing the most rapid growth of material wealth in human history?

FIGURE 2.14 Advances in technology have enabled people to push back the natural limits on growth and continue expanding their economies. Innovations in machinery for petroleum extraction, shown here, have allowed us to extract more fossil fuels than ever before and to push our economic productivity far beyond what was possible a century or two ago.

One prime reason is technological innovation. In case after case, improved technology has enabled us to push back the limits on growth. More powerful technology for extracting minerals, fossil fuels, and groundwater has expanded the amounts of these natural resources available to us (Figure 2.14). Technological developments such as automated farm machinery, fertilizers, and chemical pesticides have allowed us to grow more food per unit area of land, boosting agricultural output. Faster, more powerful machines in our factories have enabled us to translate our enhanced resource extraction and agricultural production into faster rates of manufacturing. Such increases in scale and efficiency explain why many of the dire predictions of those concerned about resource limitations have only partly come to pass.

Economists disagree on whether economic growth is sustainable

Can we conclude, then, that endless improvements in technology are possible and that we will never run into shortages of resources? At one end of the spectrum are those Cornucopians (▸ p. 21), economists, businesspeople, and policymakers who believe technology can solve everything—a philosophy that has greatly influenced economic policy in market economies over the past century.

At the other end of the spectrum, **ecological economists** argue that a couple of centuries is not a very long period of time and that history suggests that civilizations do not, in the long run, overcome their environmental limitations. Ecological economics, which has emerged as a discipline only in the past decade or two, applies principles of ecology and systems science (Chapters 5 through 7) to the analysis of economic systems. Earth's natural systems generally operate in self-renewing cycles, not in a linear or progressive manner. Ecological economists advocate sustainability in economies and see natural systems as good models. To evaluate an economy's sustainability, ecological economists take a long-term perspective and ask, "Could we continue this activity forever and be happy with the outcome?" Most ecological economists argue that the growth paradigm will eventually fail and that if nothing is done to rein in population growth and resource consumption, depleted natural systems could plunge our economies into ruin. Many of these economists advocate economies that do not grow and do not shrink, but rather are stable. Such **steady-state economies** are intended to mirror natural ecological systems.

In the middle of the spectrum are **environmental economists.** They tend to agree with ecological economists that economies are unsustainable if population growth is not reduced and resource use is not made more efficient. Environmental economists, however, maintain that we can accomplish these changes and attain sustainability within our current economic systems. By retaining the principles of neoclassical economics but modifying them to address environmental challenges, environmental economists argue that we can keep our economies growing and that technology can continue to improve efficiency. Environmental economists were the first to develop ways to tackle the problems of external costs and discounting. In the 1950s and 1960s, a nonprofit, nonpartisan organization called Resources for the Future pioneered ways to weigh the true costs and benefits associated with resource use. These environmental economists blazed a trail, and ecological economists then went further, proposing that sustainability requires far-reaching changes leading to a steady-state economy. Thus, whereas environmental economists implemented reform, ecological economists call for revolution.

A steady-state economy is a revolutionary alternative to growth

The idea of a steady-state economy did not originate with the rise of ecological economics. Back in the 19th century, British economist **John Stuart Mill** (1806–1873) hypothesized

The Science behind the Story

Gross Domestic Product versus Genuine Progress Indicator

For decades, economists have assessed the economic robustness of a nation by calculating its **Gross Domestic Product (GDP),** the total monetary value of final goods and services produced in the country each year.

However, there are problems with using this measure of economic activity to represent a nation's economic well-being. For one, GDP does not account for the nonmarket values we discuss in this chapter. For another, GDP is not necessarily an expression of *desirable* economic activity. In fact, GDP can increase whether the economic activities driving it help or hurt the environment or society. For example, a large oil spill in a U.S. national park could increase the U.S. GDP because oil spills require cleanups, which cost money and, as a result, increase the production of goods and services. A radiation leak at a uranium mine on Aboriginal homelands would likely add to the Australian GDP because of the many monetary transactions required for cleanup and medical care.

Some economists have attempted to develop economic indicators that differentiate between desirable and undesirable economic activity. Such indicators can function as more accurate guides to nations' welfare. One alternative to the GDP is the **Genuine Progress Indicator (GPI),** introduced in 1995 by Redefining Progress, a nonprofit organization that develops economic and policy tools to promote accurate market prices and sustainability. The GPI has not yet gained widespread acceptance, but it has generated a

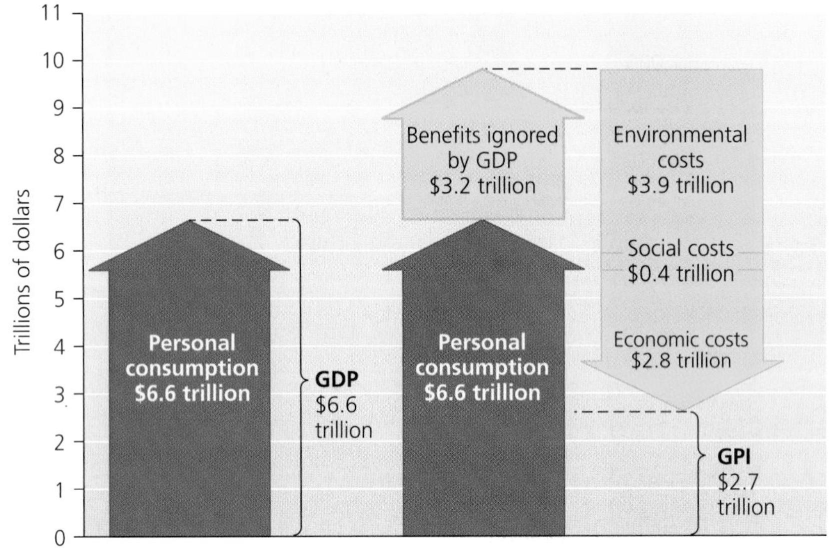

The Gross Domestic Product (GDP; red arrow) sums together all economic activity, whether good or bad. The Genuine Progress Indicator (GPI) adds to the GDP benefits such as volunteering and parenting (upward pointing gold arrow). The GPI also subtracts external environmental costs such as pollution, social costs such as divorce and crime, and economic costs such as borrowing and the gap between rich and poor (downward pointing gold arrow). Shown are values for GDP and GPI for the United States in the year 2002. Data from Venetoulis, J. and C. Cobb. 2004. *The genuine progress indicator, 1950–2002 (2004 update).* Redefining Progress.

that as resources became harder to find and extract, economic growth would slow and eventually stabilize. Economies would carry on in a state in which individuals and society subsist on steady flows of natural resources and on savings accrued during occasional productive but finite periods of growth. Such a model appears to match the economies of Australian Aborigines and many other traditional societies throughout the world before the global expansion of European culture.

Modern proponents of a steady-state global economy, such as American economist Herman Daly, are not so sanguine that a steady state will evolve on its own from a capitalist market system. Instead, most believe we will need to rethink our assumptions and fundamentally change the way we conduct economic transactions if we are to achieve a steady state.

Those resisting such notions often assume that an end to growth will mean an end to a rising quality of life. Ecological

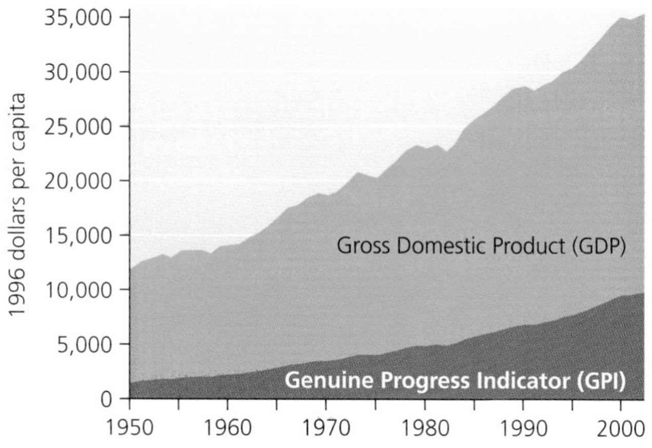

Although the GDP of the United States has increased dramatically since 1950, the GPI has risen more slowly. GPI advocates suggest that this discrepancy means that we are spending more money than ever but that our lives are not that much better. Data from Venetoulis, J. and C. Cobb. 2004. *The genuine progress indicator, 1950–2002 (2004 update).* Redefining Progress.

great deal of discussion that has drawn attention to the weaknesses of the GDP.

To calculate GPI, economists begin with conventional economic activity and then add to it all those positive contributions to the economy that do not have to be paid for with money, such as volunteer work and parenting. They then subtract negative impacts, such as crime, pollution, gaps between rich and poor, and other detrimental social, environmental, and economic factors (see the first figure). The GPI thereby summarizes many more forms of economic activity than does the GDP, and it differentiates between economic activity that increases societal well-being and economic activity that decreases it.

Thus, whereas the GDP increases when fossil fuel use increases, the GPI declines because of the adverse environmental and social impacts of such consumption, including air and water pollution, increased road congestion and traffic accidents, and global climate change. The second figure compares changes in per capita GPI and GDP in the United States from 1950 to 2002. The country's GDP has increased greatly as a result of increased economic activity. Its GPI has also increased, but not nearly as rapidly.

The GPI is not the only alternative to the GDP. The United Nations uses a tool of its own called the Human Development Index, which is calculated by assessing a nation's standard of living, life expectancy, and education. Another alternative to the GDP is the Index of Sustainable Economic Welfare (ISEW), developed by ecological economists Herman Daly and John Cobb in 1989. The ISEW is based on income, wealth distribution, natural resource depletion, benefits associated with volunteerism, and environmental degradation. Any one of these indices, ecological economists maintain, should give a more accurate portrait of a nation's welfare than the GDP, which policymakers currently use so widely.

economists, however, argue that quality of life can continue to rise under a steady-state economy and, in fact, may be more likely to do so. Technological advances will not cease just because growth stabilizes, they argue, and neither will behavioral changes (such as greater use of recycling) that enhance sustainability. Instead, wealth and human happiness can continue to rise after economic growth has leveled off.

Attaining sustainability will certainly require the reforms that environmental economists advocate and may well require the fundamental shifts in thinking, values, and behavior that ecological economists say is necessary (see "The Science behind the Story," above). How can these goals be attained in a world whose economic policies are still largely swayed by a Cornucopian worldview that barely takes the environment into account? While keeping in mind that ecological and environmental economic approaches are still actively being developed, we will now survey a few strategies for sustainability that have been offered so far.

We can give ecosystem goods and services monetary values

As we have noted, economies receive from the environment vital resources and ecosystem services. However, any survey of environmental problems in the world today—deforestation, biodiversity loss, pollution, collapsed fisheries, climate change, and so on—makes it immediately apparent that our society often mistreats the very systems that keep it alive and healthy. Why is this? From the economist's perspective, humans exploit natural resources and systems in large part because the market assigns these entities no quantitative monetary value or, at best, assigns values that underestimate their true worth.

Think for a minute about the nature of some of these services. The aesthetic and recreational pleasure we obtain from natural landscapes, whether wildernesses or city parks, is something of real value. Yet this value is hard to quantify and appears in no traditional measures of economic worth. Or consider Earth's water cycle (▸ pp. 206–207), by which rain fills our reservoirs with drinking water, rivers give us hydropower and flush away our waste, and water evaporates, purifying itself of contaminants and readying itself to fall again as rain. This natural cycle is absolutely vital to our existence, yet because its value is not quantified, markets impose no financial penalties when we interfere with it. Ecosystem services are said to have **nonmarket values,** values not usually included in the price of a good or service (Table 2.3 and Figure 2.15). Because the market does not assign value to ecosystem services, debates such as that over the Jabiluka mine often involve comparing apples and oranges—in this case, the intangible cultural, ecological, and spiritual arguments of the Mirrar versus the hard numbers of mine proponents.

To resolve this dilemma, environmental and ecological economists have sought ways to assign values to ecosystem services. One technique, **contingent valuation,** uses surveys to determine how much people are willing to pay to protect a resource or to restore it after damage has been done. Such an exercise was conducted with a mining proposal in the Kakadu region in the early 1990s that preceded the Jabiluka proposal. The Kakadu Conservation Zone, a government-owned 50-km^2 (19-mi^2) plot of land surrounded by Kakadu National Park, was either to be developed for mining or preserved and added to the park. To determine the degree of public support for environmental protection versus mining, a government commission sponsored a contingent valuation study to determine how much Australian citizens valued keeping the Kakadu Conservation Zone preserved and undeveloped. Researchers interviewed 2,034 citizens, asking them how much money they would be willing to pay to stop mine development.

The interviewers presented two scenarios: (1) a "major impact" scenario based on predictions of environmentalists who held that mining would cause great harm, and (2) a "minor impact" scenario based on predictions of mining executives who held that development would have few downsides. After presenting both scenarios in detail, complete with photographs, the interviewers asked the respondents how much their households would pay if each scenario, in turn, was correct. Respondents on average said their households would pay \$80 per year to prevent the minor-impact scenario and \$143 per year to prevent the major-impact scenario. Multiplying these figures by the number of households in Australia (5.4 million at the time), the researchers found that preservation was "worth" \$435 million annually to the Australian population under the minor-impact scenario, and \$777 million under the major-impact scenario. Because both of these numbers exceeded the \$102 million in annual economic benefits expected from mine development, the researchers concluded that preserving the land undeveloped was worth more than mining it.

Table 2.3 Values That Modern Market Economies Generally Do Not Address

Nonmarket value	Is the worth we ascribe to things . . .
Use value	that we use directly
Option value	that we do not use now but might use later
Aesthetic value	for their beauty or emotional appeal
Cultural value	that sustain or help define our culture
Scientific value	that may be the subject of scientific research
Educational value	that may teach us about ourselves and the world
Existence value	simply because they exist, even though we may never experience them directly (e.g., an endangered species in a far-off place)

(a) Existence values
(b) Use values
(c) Option values
(d) Aesthetic values
(e) Scientific values
(f) Educational values
(g) Cultural values

FIGURE 2.15 Accounting for nonmarket values such as those shown here may help us to make better environmental and economic decisions.

Because contingent valuation relies on survey questions and not actual expenditures, critics complain that in such cases people will volunteer idealistic (inflated) values rather than realistic ones, knowing that they will not actually have to pay the price they name. In part because of such concerns, the Australian government commission decided not to use the Kakadu contingent valuation study's results. (The mine was stopped, but as a result of Aboriginal opposition.)

Whereas contingent valuation measures people's *expressed* preferences, other methods aim to measure people's *revealed* preferences—preferences as revealed by data on actual behavior. For example, the amount of money, time, or effort people expend to travel to parks for recreation has been used to measure the value people place on parks. Economists have also analyzed housing prices, comparing homes with similar characteristics but different environmental settings, to infer the dollar value of landscapes, views, and peace and quiet. Another approach assigns environmental amenities value by measuring the cost required to restore natural systems that have been damaged or to mitigate harm from pollution.

In 1997 one research team reviewed studies using various valuation methods in an effort to calculate the overall global economic value of all the services that ecosystems provide (see "The Science behind the Story," ▸ pp. 50–51). The researchers came up with a figure of $33 trillion per year ($40 trillion in 2005 dollars)—greater than the combined gross domestic products of all nations in the world. A follow-up study in 2002 concluded that the economic benefits of preserving the world's remaining natural areas outweighed the benefits of exploiting them by a factor of 100 to 1.

Markets can fail

When they do not reflect the full costs and benefits of actions, markets are said to fail. **Market failure** occurs when markets do not take into account the environment's positive effects on economies (such as ecosystem services) or when they do not reflect the negative effects of economic activity on the environment or on people (external costs). Traditionally, market failure has been countered by government intervention. Governments

The Science behind the Story

Calculating the Economic Value of Earth's Ecosystems

To Robert Costanza, the problem was like an elephant in the living room, which economists had ignored for decades: Earth's ecosystems provide essential life-support services, including arable soil, waste treatment, clean water, and clean air. However, economists had failed to account for how much those services contribute economically to human welfare, as is routinely done for conventional goods and services.

So Costanza, an environmental economist at the University of Maryland, joined with 12 colleagues in 1996 at the National Center for Ecological Analysis and Synthesis in Santa Barbara, California, and combed the scientific literature. The team identified more than 100 studies that estimated the worth of such ecosystem services as water purification, greenhouse gas regulation, plant pollination, and pollution cleanup.

The studies the team reviewed had estimated values of particular ecosystem services in several ways. Methods such as contingent valuation had frequently been used because people clearly value such aspects of natural systems as biodiversity and aesthetics, even though no one pays actual money specifically for them. After poring over studies that examined the value of 17 services provided by oceans, forests, wetlands, and other ecosystems, Costanza and his colleagues synthesized the results to provide the first comprehensive quantitative estimate of the global value of ecosystem services.

To estimate the worth of the services more accurately, Costanza and his team reevaluated the data from the earlier studies using alternative valuation techniques. One method was to calculate the cost of replacing ecosystem services with technology. For example, marshes protect people from floods and filter out water pollutants. If a marsh were destroyed, the researchers would calculate the value of the services it had provided by measuring the cost of the levees and water-purification technology that would be needed to assume those tasks. The researchers then calculated the global monetary value of such wetlands by multiplying those totals by the global area occupied by the ecosystem. By calculating similar totals from coral reefs, deserts, tundra, and other ecosystems, they arrived at a global value for ecosystem services.

By their calculation, the biosphere provides at least $33 trillion ($40 trillion in 2005 dollars) worth of ecosystem services each year—greater than the gross domestic product (GDP) of all nations combined. Published in the journal *Nature* in 1997, their research paper ignited a firestorm of controversy.

Some environmental advocates and ethicists argued that it was a bad idea to put a dollar figure on priceless services such as clean air and clean water. The value of these services cannot be calculated, they held, because we would all perish if they disappeared. Environmental ethicist Timothy Weiskel of Harvard Divinity School contended that to make a commodity out of biodiversity confused "sacred space with the market place."

Some economists, meanwhile, disparaged the methods by which the researchers calculated values. Replacement costs were not a legitimate way of determining value,

can dictate limits on corporate behavior through laws and regulations. They can institute *green taxes,* which penalize environmentally harmful activities. Or, they can design economic incentives that put market mechanisms to work to promote fairness, resource conservation, and economic sustainability. We will examine legislation, regulation, green taxation, and market incentives in Chapter 3 as part of our policy discussion, but we will briefly note a few examples of alternative methods here.

Ecolabeling and permit-trading address market failure

Legislation, regulation, or taxation by government may not always effectively address market failure, so economists have contrived ways to use aspects of the market to counteract market failure. In one strategy, manufacturers of certain products are required or encouraged to designate on their labels how the products were grown, harvested, or manufactured. This method, called **ecolabeling,** they held, because nature's services, like other economic goods and services, are worth only what people will demonstrably pay for them. Critics also argued that combining the value of ecosystem services from various small tracts of land is meaningless because people decide whether to preserve or exploit resources based on particular local considerations, not on generalized global ones.

To address economists' criticisms, Costanza joined Andrew Balmford of Cambridge University and 17 other colleagues to conduct another analysis. They compared the benefits and costs of preserving natural systems intact with those of converting wild lands for agriculture, logging, or fish farming. Again, they synthesized studies on ecosystem services, this time focusing on just five ecosystems: west African and Malaysian tropical forests, Thai mangrove swamps, Canadian wetlands, and Philippine coral reefs.

In their paper, published in the journal *Science* in 2002, the team reported that a global network of nature reserves covering 15% of Earth's land surface and 30% of the ocean would be worth between \$4.4 and \$5.2 trillion. This amount is about 100 times the value of the same area were it to be converted for direct exploitative human use. That 100:1 benefit–cost ratio, they wrote, demonstrates clearly that "conservation in reserves represents a strikingly good bargain."

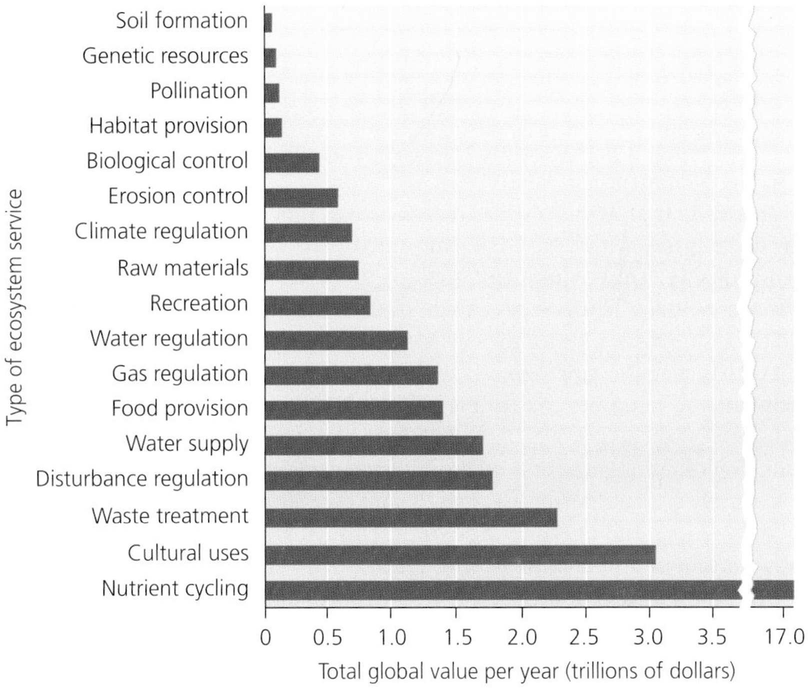

In 1997, Robert Costanza and colleagues estimated the total value of the world's ecosystem services at approximately \$33 trillion. Shown are subtotals for each major class of ecosystem service. The \$33 trillion figure does not include values from some ecosystems, such as deserts and tundra, for which adequate data were unavailable. Data from Costanza, R., et al. 1997. The value of the world's ecosystem services and natural capital. *Nature* 387: 253–260.

serves to tell consumers which brands use environmentally benign processes (Figure 2.16). By preferentially buying ecolabeled products, consumers can provide businesses a powerful incentive to switch to more environmentally friendly processes. The best-known example of this has been labeling cans of tuna as "dolphin-safe," indicating that the methods used to catch the tuna avoid the accidental capture of dolphins. Other examples include labeling recycled paper, organically grown foods, genetically modified foods, and lumber harvested through sustainable forestry. We will come across such instances of ecolabeling in future chapters.

Another technique devised to counteract market failure is to create markets in permits for environmentally harmful activities. Under such a system, companies are allowed to buy and sell the right to conduct environmentally harmful activities. For instance, utilities and industries might trade permits to emit air pollution. In such a scheme, the government first determines an acceptable level of pollution, then issues permits to pollute.

VIEWPOINTS

Environment versus Economy?

Is environmental protection economically costly? How can we best balance the demands of economic development with the demands of environmental protection?

Economic Progress a Prerequisite for Environmental Quality

The Greek word *oikos,* meaning "household," is the root of both *economics* and *ecology.* It suggests complementarities. Protecting the environment is not free, but environmental quality and economic development are not mutually exclusive. The real enemy of the environment is poverty, not affluence.

If economic growth really destroys the environment, why do countries with large GDPs enjoy high environmental quality? The environment is cleaner in developed countries because their citizens have both the inclination and the resources to care for the environment.

Are we, as some critics assert, like the man falling from a ten-story building and concluding as he passes the second story, "so far so good"? I think not. In the long term, technological improvements and productivity gains allow us to use fewer material inputs—and to emit ever fewer pollutants—per unit of economic output. This reduces both our economic and ecological footprint.

For example, genetically modified (GM) crops and synthetic chemicals allow us greater yields on the same amount of land. As a result, the United States can afford to place 50 million acres of farmland into conservation reserves while remaining a major food exporter. The poorest countries are beneficiaries as they adopt our efficient and less environmentally damaging technologies—shortcutting the road to environmental quality.

The question we face is, in what combination and in what amounts should we seek the things we want? We value clean water and the preservation of other species. But we also value fresh produce in winter and fast and convenient transportation. Not all good things go together.

Regardless of claims, environmental quality is only one of several competing values people seek. Scarcity—the fact that virtually no resources are abundant enough to satisfy all human demands at zero cost—dictates that choices must be made among competing values. It is intellectually and ethically irresponsible to pretend away the necessity of such choices.

Pete Geddes is program director of the Foundation for Research on Economics and the Environment (FREE) and Gallatin Writers. Both are based in Bozeman, Montana.

The Trade-off Myth

Reducing pollution and protecting resources costs money. As a nation we spend over $200 billion each year, or more than 2% of GDP, on environmental protection. However, it is often contested how much environmental protection will cost in any particular case. Forecasting the costs of compliance is difficult because economists have an equally difficult time estimating future technological responses that may lower costs. For example, credible industry estimates for sulfur dioxide reduction from power plants under the 1990 Clean Air Act Amendments were eight times too high; the EPA overestimated costs by a factor of 2 to 4.

One "cost" of environmental protection that is blown out of proportion is job loss. Contrary to popular belief, there is no net "jobs–environment trade-off" in the economy, only a steady shift of jobs to cleanup work. On one hand, about 2,000 workers in the U.S. lose their jobs each year for environment-related reasons, which is less than 1/10 of 1% of all layoffs. On the other hand, as we spend more on the environment, more jobs are created. Given the industrial nature of much cleanup work, these jobs are also heavily weighted toward manufacturing and construction. Finally, and again contrary to folk wisdom, very few manufacturing plants flee the industrial countries to escape onerous environmental regulation. Plants do leave, but the overwhelming reason is the cost of labor.

What is the best way to balance costs against the benefits of environmental cleanup? Formal benefit–cost analysis is one approach, but it is of limited value when the benefits of environmental protection are highly uncertain (which is often the case), or when the costs of resource degradation are borne by a relatively small group. In these situations, the best approach is to define a health or ecological standard for cleanup and to trust democratic processes to ensure that the costs of cleanup do not rise too high.

Eban Goodstein is professor of economics at Lewis and Clark College in Portland, Oregon. He is the author of the textbook *Economics and the Environment* (John Wiley and Sons, 2004), as well as *The Trade-off Myth: Fact and Fiction about Jobs and the Environment* (Island Press, 1999).

Explore this issue further by accessing **Viewpoints** at www.aw-bc.com/withgott.

FIGURE 2.16 Ecolabeling allows businesses to promote products that have low environmental impact. Organic juices and produce are examples of ecolabeled products becoming widely available in the mass market.

Each party may emit its permitted amount, but if it is able to reduce its pollution, it receives credit for the amount it did not emit and can sell this credit to other parties that want to expand operations or are less able to control their emissions. Creating markets in permits provides companies an economic incentive to find ways to reduce emissions because the sale of permits brings in money. We will discuss **permit-trading** further in Chapter 3 (▸ p. 73). The method is not perfect. Large firms can hoard permits, deterring smaller new firms from entering the market, an action that suppresses competition. Nevertheless, permit-trading markets have shown great promise for reducing environmental degradation while granting industries the flexibility to lessen their impacts in ways that are economically palatable to them.

Conclusion

Permit-trading, ecolabeling, and valuation of ecosystem services are some recent developments that have brought economic approaches to bear on environmental protection and resource conservation. As economics becomes more environmentally friendly, it renews some of its historic ties to ethics. Environmental ethics has expanded people's sphere of ethical consideration outward to encompass other societies and cultures, other creatures, and even nonliving entities that were formerly outside the realm of ethical concern. This ethical expansion involves the concept of distributional equity, or equal treatment, which is the aim of environmental justice. One type of distributional equity is equity among generations. Such concern by current generations for the welfare of future generations is the basis for the notion of sustainability.

Is sustainability a pragmatic pursuit for us? The answer largely depends on whether we believe that economic well-being and environmental well-being are opposed to one another or whether we accept that they can work in tandem. Equating economic well-being with economic growth, as most economists traditionally have, suggests that economic welfare entails a trade-off with environmental quality. However, if economic welfare can be enhanced in the absence of growth, we can envision economies and environmental quality benefiting from one another.

REVIEWING OBJECTIVES

You should now be able to:

Characterize the influences of culture and worldview on the choices people make

- A person's culture strongly influences his or her worldview. Factors such as religion and political ideology are especially influential. (pp. 28–30)

Outline the nature, evolution, and expansion of environmental ethics in Western cultures

- Our society's domain of ethical concern has been expanding, such that we have granted more and more entities ethical consideration. (pp. 31–32)
- Anthropocentrism values humans above all else, whereas biocentrism values all life and ecocentrism values ecological systems. (pp. 32–33)
- The preservation ethic (preserving natural systems intact) and the conservation ethic (promoting responsible long-term use of resources) have guided branches of the environmental movement during the past century. (pp. 34–35)
- The environmental justice movement, seeking equal treatment for people of all races and income levels, is one of several recent outgrowths of environmental ethics. (pp. 36–38)

Describe precepts of classical and neoclassical economic theory, and summarize their implications for the environment

- Classical economic theory proposes that individuals acting for their own economic good can benefit society as a whole. This view has provided a philosophical basis for free-market capitalism. (pp. 39–41)
- Neoclassical economics focuses on consumer behavior and supply and demand as forces that drive economic activity. (pp. 41–42)
- Several assumptions of neoclassical economic theory contribute to environmental impact. (pp. 42–44)

Compare the concepts of economic growth, economic health, and sustainability

- Conventional economic theory has promoted never-ending economic growth, with little regard to possible environmental impact. (p. 44)
- Economic growth is not necessarily required for economic well-being. (pp. 44–47, 52)
- In the long run, some economists believe that a steady-state economy will be necessary to achieve sustainability. (pp. 45–47)

Explain the fundamentals of environmental economics and ecological economics

- Environmental economists advocate reforming economic practices to promote sustainability. Key approaches are to identify external costs, assign value to nonmonetary items, and attempt to make market prices reflect real costs and benefits. (pp. 45–51)
- Ecological economists support these efforts and others. Many support developing a steady-state economy. (pp. 45–47)

TESTING YOUR COMPREHENSION

1. What does the study of ethics encompass? Describe the three classic ethical standards. What is environmental ethics?
2. Why in Western cultures have ethical considerations expanded to include nonhuman entities?
3. Describe the philosophical perspectives of anthropocentrism, biocentrism, and ecocentrism. How would you characterize the perspective of the Mirrar Clan?
4. Differentiate between the preservation ethic and the conservation ethic. Explain the contributions of John Muir and Gifford Pinchot in the history of environmental ethics.
5. Describe Aldo Leopold's "land ethic." How did Leopold define the "community" to which ethical standards should be applied?
6. Name four key contributions the environment makes to the economy.
7. Describe Adam Smith's metaphor of the "invisible hand." How did neoclassical economists refine classical economics?
8. Describe four ways in which critics hold that neoclassical economic approaches can negatively affect the environment.
9. Compare and contrast the views of neoclassical economists, environmental economists, and ecological economists.
10. What is contingent valuation, and what is one of its weaknesses? Describe an alternative method that addresses this weakness.

SEEKING SOLUTIONS

1. Do you feel that an introduction to environmental ethics and worldviews is an important part of a course in environmental science? Should ethics and worldviews be a component of other science courses? Explain your answers.
2. Describe your worldview as it pertains to your relationship with the environment. How do you think your culture has influenced your worldview? How do you think your personal experience has influenced it? Do you feel that you fit into any particular category discussed in this chapter? Why or why not?
3. How would you analyze the case of the Mirrar Clan and the proposed Jabiluka uranium mine from each of the following perspectives? In your description, list three questions that a person of each perspective would likely ask when attempting to decide whether the mine should be developed. Be as specific as possible, and be sure to identify similarities and differences in approaches:
   - Preservationist
   - Conservationist
   - Deep ecologist

- Environmental justice advocate
- Neoclassical economist
- Ecological economist

4. What is a steady-state economy? Do you think this model is a practical alternative to the growth paradigm? Why or why not?
5. Do you think we should attempt to quantify and assign market values to ecosystem services and other entities that have only nonmarket values? Why or why not?
6. A manufacturing facility on a river near your home provides jobs for 200 people in your community and pays $2 million in taxes to the local government each year. Sales taxes from purchases made by plant employees and their families contribute an additional $1 million to local government coffers. However, a recent peer-reviewed study in a well-respected scientific journal revealed that the plant has been discharging large amounts of waste into the river, causing a 25% increase in cancer rates, a 30% reduction in riverfront property values, and a 75% decrease in native fish populations.

   The plant owner says the facility can stay in business only because there are no regulations mandating expensive treatment of waste from the plant. If any such regulations were imposed, he says he would close the plant, lay off its employees, and relocate to a more business-friendly community.

   How would you recommend resolving this situation? What further information would you want to know before making a recommendation? In arriving at your recommendation, how did you weigh the costs and benefits associated with each of the plant's impacts?

INTERPRETING GRAPHS AND DATA

As described in "The Science behind the Story" on ▸ pp. 46–47, economists use various indicators of economic well-being. One that has been used for decades is the Gross Domestic Product (GDP), the total monetary value of final goods and services produced each year. An alternative measure called the Genuine Progress Indicator (GPI) is calculated as follows:

GDP + (Benefits ignored by GDP) − (Environmental costs) − (Social and economic costs)

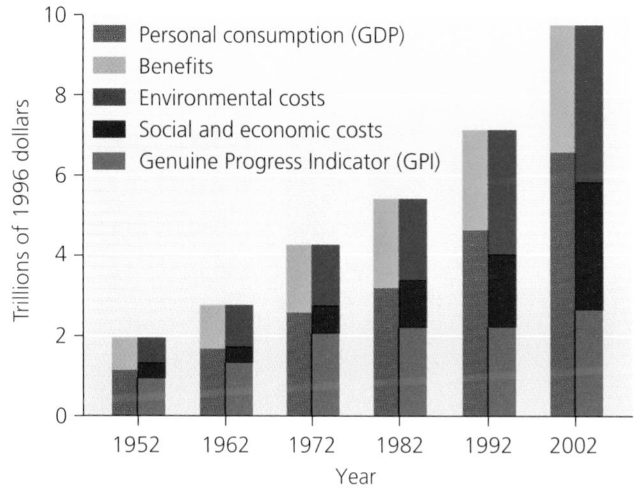

Components of GPI and GDP for the United States, 1952–2002.
Data from Venetoulis, J., and C. Cobb. 2004. *The genuine progress indicator, 1950–2002 (2004 update)*. Redefining Progress.

Benefits include such things as the value of parenting and volunteer work. Environmental costs include the costs of water, air and noise pollution, loss of wetlands, depletion of nonrenewable resources, and other environmental damage. Social and economic costs include investment, lending, and borrowing costs, as well as the costs of crime, family breakdown, underemployment, commuting, pollution abatement, automobile accidents, and loss of leisure time.

1. Describe economic growth as measured by GDP for the United States from 1952 to 2002. Now describe economic growth as measured by GPI over the same time period. To what factors would you attribute the growing difference between these measures?
2. For GPI to grow significantly, one or more things must happen: Either GDP must grow faster, benefits must grow faster, or social, economic, and environmental costs must shrink relative to the other terms. Which of these scenarios do you think is most likely? Which would you prefer? How do the data in the graph support your answer?
3. Even with domestic regulations for air and water pollution control, hazardous waste disposal, solid waste management, forestry practices, and species protection, environmental costs continue to increase. Why do you suppose the trend is still in that direction?

CALCULATING ECOLOGICAL FOOTPRINTS

Although the Gross Domestic Product (GDP) of the United States grew impressively between 1952 and 2002 (see the graph in the "Interpreting Graphs and Data" section), so did the U.S. population. According to the U.S. Census Bureau, the midyear population of the nation was 157,552,740 in 1952, and 288,368,698 in 2002—an 83% increase.

Estimate the values of the components of the Genuine Progress Indicator (GPI) for the United States in 1952 and 2002 from the graph, and enter your estimates into the table below. Then, using the population figures, calculate the per capita values for each component in 1952 and 2002.

Components of GPI	U.S. total in 1952 (trillions of dollars)	Per capita in 1952 (thousands of dollars)	U.S. total in 2002 (trillions of dollars)	Per capita in 2002 (thousands of dollars)
GDP	1.2	7.6	6.6	22.9
Benefits				
Environmental costs				
Social and economic costs				
GPI				

1. Consider your own life. What would you estimate is the value of the benefits in which you participate? Compare your personal estimate with the national average value in 2002.
2. What would you estimate are the values of the environmental, social, and economic costs for which you are personally responsible? How do they compare with the national average values in 2002?
3. In 2002, social and economic costs were proportionally larger relative to GDP than they were in 1952, and environmental costs were roughly the same in proportion to GDP. How would you account for these trends? What could you do to help improve these trends in your own personal accounting?

Take It Further

Go to www.aw-bc.com/withgott or the student CD-ROM where you'll find:

- Suggested answers to end-of-chapter questions
- Quizzes, animations, and flashcards to help you study
- *Research Navigator*™ database of credible and reliable sources to assist you with your research projects
- GRAPHit! tutorials to help you master how to interpret graphs
- INVESTIGATEit! current news articles that link the topics that you study to case studies from your region to around the world

Environmental Policy: Decision Making and Problem Solving

3

Border of Tijuana, Mexico (left), and San Diego, California (right)

Upon successfully completing this chapter, you will be able to:

- Describe environmental policy and assess its societal context
- Identify the institutions important to U.S. environmental policy and recognize major U.S. environmental laws
- Categorize the different approaches to environmental policy
- Delineate the steps of the environmental policy process and evaluate its effectiveness
- List the institutions involved with international environmental policy and describe how nations handle transboundary issues

Closed San Diego beach

Central Case: San Diego and Tijuana's Sewage Pollution Problems and Policy Solutions

"Ignorance is an evil weed, which dictators may cultivate among their dupes, but which no democracy can afford among its citizens."
—William Beveridge, British economist

"It is the continuing policy of the Federal Government . . . to create and maintain conditions under which man and nature can exist in productive harmony and fulfill the social, economic, and other requirements of present and future generations of Americans."
—National Environment Policy Act, 1969

On November 23, 1996, officials closed all the public beaches in San Diego, California. Stormwater runoff following heavy rains had washed pollutants into local rivers and coastal waters. This also occurred across the border in the Mexican city of Tijuana, whose aging sewer system became clogged with debris, causing raw sewage to overflow into streets and onto beaches. Such incidents, called "rogue flows," take place when heavy rain overwhelms the ability of sewage treatment plants to process wastewater. Rogue flows had become so common in San Diego and Tijuana that local surfers and swimmers casually referred to the initial one of each rainy season as the "first flush."

The Tijuana River winds northwestward through the arid landscape of northern Baja California, Mexico, crossing the U.S. border south of San Diego. A river's **watershed** consists of all the land from which water drains into the river, and the Tijuana River's watershed covers 4,500 km^2 (1,750 mi^2) and is home to 2 million people of two nations. The Tijuana River watershed is a *transboundary* watershed (so named because it crosses a political boundary—in this case, a national border), with approximately 70% of its area in Mexico (Figure 3.1). On the Mexican side of the border, the river and the arroyos, or creeks, that flow into it are lined with farms, apartments, shanties, and factories, as well as leaky sewage treatment plants and toxic dump sites. Many of

FIGURE 3.1 The Tijuana River winds northwestward from Mexico into California just south of San Diego, draining 4,500 km^2 (1,750 mi^2) of land in its watershed. Pollution entering the river affects Mexican residents of the watershed and U.S. citizens on San Diego County beaches. During high-water episodes that overwhelm treatment facilities, millions of gallons of raw sewage have flushed into the river and the Pacific Ocean.

these sources release pollutants, which heavy rains wash through the arroyos into the Tijuana River and eventually onto U.S. and Mexican beaches.

Although pollution has flowed in the Tijuana River for at least 70 years, the problem has grown worse in recent years as the region's population has boomed, outstripping the capacity of sewage treatment facilities. Rogue flows have caused thousands of beach closures and pollution advisories in recent years. Garbage carried by the flows also litters the beaches. "Every day I find broken glass, balloons, or can pop-tops. I've even found hypodermic needles. It's really sad," one resident of Imperial Beach told her local newspaper in 2005.

The problem is worse on the Mexican side because most Mexican residents of the Tijuana River watershed live in poverty relative to their U.S. neighbors. In poor neighborhoods such as Loma Taurina, pollution of the river directly affects people's day-to-day lives. The rise of U.S.-owned factories, or *maquiladoras,* on the Mexican side of the border has contributed to the river's pollution, both through direct disposal of industrial waste and by attracting thousands of new workers to the already crowded region.

As impacts increased, people in the San Diego and Tijuana areas, from coastal residents to grassroots activists to businesspeople, pressed policymakers to do something. As we explore environmental policy in this chapter, we will periodically return to the Tijuana River watershed and see how citizens and policymakers together have tried to address these problems.

Environmental Policy: An Introduction

When a society reaches broad agreement that a problem exists, it may persuade its leaders to try to resolve the problem through the making of policy. **Policy** consists of a formal set of general plans and principles intended to address problems and guide decision making in specific instances. **Public policy** is policy made by governments, including those at the local, state, federal, and international levels. Public policy consists of laws, regulations, orders, incentives, and practices intended to advance societal welfare. **Environmental policy** is policy that pertains

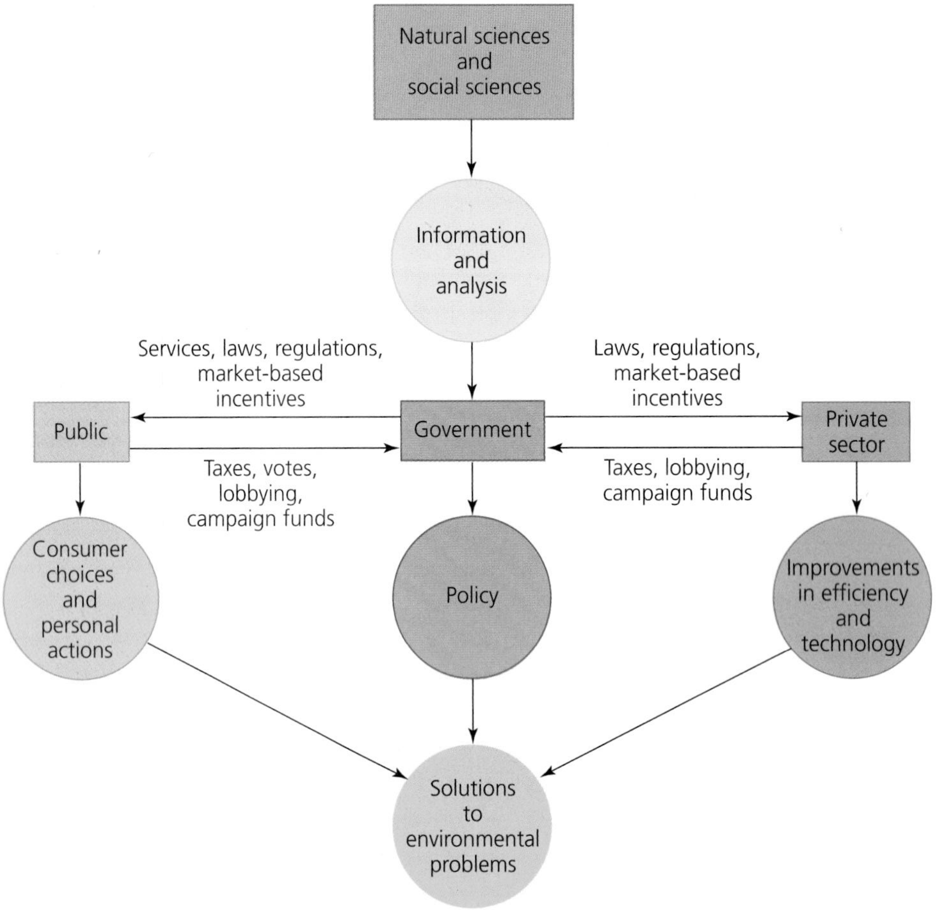

FIGURE 3.2 Policy plays a central role in how we as a society address environmental problems. This process begins when research from the academic disciplines involved in environmental science leads to some understanding of a given environmental problem, its potential consequences, and its possible solutions. Government may use this information in formulating policy, together with ethical and economic considerations and with input from citizens and the private sector. Government policy includes laws, regulations, and market-based incentives that aim, for example, to restrict resource exploitation or ensure fairness in resource use. Along with improvements in technology and efficiency from the private sector and consumer choices exercised by citizens in the marketplace, public policy can produce lasting solutions to environmental problems.

to human interactions with the environment. It generally aims to regulate resource use or reduce pollution to promote human welfare and/or protect natural systems.

Forging effective policy requires input from science, ethics, and economics. Science (Chapter 1) provides the information and analysis needed to identify and understand environmental problems and devise potential solutions to them. Ethics and economics (Chapter 2) offer criteria to assess the extent and nature of problems, and help clarify how society might like to see them addressed. Government interacts with individual citizens, organizations, and the private sector in a variety of ways to formulate policy, which is one of our major tools for devising lasting solutions to environmental problems (Figure 3.2).

Policymaking can be vital in addressing dilemmas like those of the Tijuana River watershed. Sewage-tainted rivers and beaches pose several threats. Scientific research tells us that raw sewage carries pathogens, organisms that can cause illness in humans and other animals (Table 3.1). Science also reveals how untreated sewage can radically alter conditions for aquatic and marine life, lowering concentrations of dissolved oxygen and increasing mortality for many species. Economically, pollution and beach closures cause financial harm by reducing recreation, tourism, and other economic activity associated with clean coastal areas. This is a major consideration both in Mexico and in southern California, whose beaches each year host 175 million visitors who spend over $1.5 billion. Ethically, water pollution also poses problems, because in most cases pollution from upstream users degrades water quality for downstream users. Many phenomena that come to be viewed as environmental problems share this combination of impacts—harming human health, altering ecological systems, inflicting economic damage, and creating inequities among people.

Environmental policy addresses issues of equity and resource use

People have identified environmental problems and asked government officials to develop solutions since at least the first century A.D., when Roman citizens complained to their leaders about air pollution from wood smoke. Today, our population growth and resource consumption have generated environmental problems on an unprecedented scale. Furthermore, the market capitalist economic systems of

Table 3.1 Pathogens and Health Effects Associated with Untreated Sewage

Pathogen	Human health effect
Cryptosporidium	Diarrhea, vomiting, cramps, dehydration
Giardia lamblia	Diarrhea, vomiting, cramps, dehydration
Salmonella typhi	Typhoid fever
Vibrio cholerae	Cholera
Poliovirus	Poliomyelitis
Shigella spp.	Dysentery
Entamoeba histolytica	Dysentery
Hepatitis A virus	Infectious hepatitis

modern constitutional democracies are driven by incentives for short-term economic gain rather than long-term social and environmental stability. Market capitalism provides little incentive for businesses or individuals to behave in ways that minimize environmental impact or equalize costs and benefits among parties. Such *market failure* (▸ pp. 49–51) has traditionally been viewed as justification for government intervention. Thus, environmental policy aims to protect environmental quality and the natural resources people use, and also to promote equity in people's use of resources.

The tragedy of the commons Policy to protect resources held and used in common by the public is intended to safeguard these resources from depletion or degradation. As Garrett Hardin explained in his essay, "The Tragedy of the Commons" (▸ p. 6), a resource held in common that is unregulated will eventually become overused and degraded. Therefore, he argued, it is in our best interest to develop guidelines for the use of common resources. In Hardin's illustrative example of a common pasture, such guidelines might limit the number of animals each individual can graze or might require pasture users to pay to restore and manage the shared resource. These two concepts, restriction of use and active management, are central to environmental policy today.

The tragedy of the commons does not always play itself out as Hardin predicted. Some traditional societies have devised safeguards against exploitation, and in modern Western societies resource users have occasionally cooperated to prevent overexploitation. Moreover, many cases do not meet Hardin's starting assumptions; resources on public lands may not be equally accessible to everyone, but may instead be more accessible to wealthier or more established resource extraction industries. Nonetheless, the threat of overexploiting public resources is always real and has been a driving force behind much environmental policy.

Free riders Another reason to develop policy for publicly held resources is the **free rider** predicament. Let's say a community on a river suffers from water pollution that emanates from 10 different factories. The problem could in theory be solved if every factory voluntarily agreed to reduce its own pollution. However, once they all begin reducing their pollution, it becomes tempting for any one of them to stop doing so. Such a factory, by avoiding the sacrifices others are making, would in essence get a "free ride" on the efforts of others. If enough factories take a free ride, the whole effort will collapse. Because of the free rider problem, private voluntary efforts are often less effective than efforts mandated by public policy. Public policy can prevent the free rider problem and ensure that all parties sacrifice equitably by enforcing compliance with laws and regulations or by taxing parties to attain funds to pursue societal goals.

External costs Environmental policy is also developed to ensure that some parties do not use resources in ways that harm others. One way to promote fairness is by dealing with *negative externalities,* or *external costs* (▸ pp. 43–44), harmful impacts that result from market transactions but are borne by people not involved in the transactions. For example, a factory that discharges waste freely into a river may reap greater profits by avoiding paying for waste disposal or recycling. Its actions, however, impose external costs (water pollution, decreased fish populations, aesthetic degradation, or other problems) on downstream users of the river. U.S.-owned *maquiladoras* in the Tijuana River watershed dump waste that affects downstream users, such as Mexican families who use river water for washing (Figure 3.3). Likewise, the detergents these families use for washing further pollute the river, creating external costs for families farther downstream and for beachgoers in Mexico and California.

These goals of environmental policy—to protect resources against the tragedy of the commons and to promote equity by eliminating free riders and dealing with external costs—will become apparent in case after case explored in this book.

Many factors can hinder implementation of environmental policy

If the goals of environmental policy are seemingly so noble, why is it that environmental laws are so often challenged, environmental regulations frequently derided, and the ideas of environmental activists repeatedly ignored or rejected by citizens and policymakers?

FIGURE 3.3 River pollution raises many issues that have been viewed as justification for environmental policy. This woman washing clothes in the river may suffer upstream pollution from factories, and her use of detergents may cause further pollution for people living downstream.

In the United States, most environmental policy has come in the form of regulations handed down from the federal government or from state or local governments. Businesses and individuals sometimes view these regulations as overly restrictive, bureaucratic, or unresponsive to human needs. For instance, many landowners fear that zoning regulations or protections for endangered species will impose restrictions on the use of their land. Developers complain of time and money lost to bureaucracy in obtaining permits; reviews by government agencies; surveys for endangered species; and required environmental controls, monitoring, and mitigation. In the eyes of such property owners and businesspeople, environmental regulation all too often has meant inconvenience and/or economic loss.

Weighing the Issues:
Do We Really Need Environmental Policy?

Many free-market advocates maintain that environmental laws and regulations are an unnecessary government intrusion into private affairs. As you may recall from Chapter 2 (pp . 39, 41), Adam Smith argued that individuals will benefit both themselves and society by pursuing their own self-interest. Do you agree? Can you describe a situation in which an individual acting in his or her self-interest could harm society by causing an environmental problem? Can you describe how environmental policy might rectify the situation? What are some advantages and disadvantages of instituting environmental laws and regulations, versus allowing unfettered exchange of materials and services?

Another reason people sometimes resent environmental policy stems from the very nature of environmental problems, which often develop slowly and gradually. The degradation of ecosystems and public health due to human impact on the environment are long-term processes. Human behavior, however, is generally geared toward addressing short-term needs—even if they conflict with long-term needs—and this tendency is reflected in our social institutions. Businesses usually opt for short-term economic gain over long-term considerations. The news media have a short attention span traditionally based on the daily news cycle, whereby new and sudden one-time events are given more coverage than slowly developing long-term trends. Politicians most often act out of short-term interest because they depend on reelection every few years. For all these reasons, many environmental policy goals that seem admirable and that attract wide public support in theory may be obstructed in their practical implementation.

U.S. Environmental Policy: An Overview

The United States provides a good focus for understanding environmental policy in constitutional democracies worldwide. First, the United States historically pioneered innovative environmental policy. Second, U.S. policies have served as models—both of success and failure—for many other nations and international government bodies. Third, the United States exerts a great deal of influence on the affairs of other nations. Finally, understanding U.S. environmental policy on the federal level can enable you to better understand environmental policy at local, state, and international levels.

U.S. policy arises from the three branches of government

Environmental policy, like all U.S. policy, results from actions of the three branches of government established under the U.S. Constitution (Figure 3.4). Statutory law, or **legislation,** is created by the **legislative branch,** or Congress, which consists of the Senate and the House of Representatives. For instance, Congress passed the Tijuana River Valley Estuary and Beach Sewage Cleanup Act in 2000 to fund the treatment of sewage flowing into the United States in the Tijuana River.

Legislation is enacted (approved) or vetoed (rejected) by the president, who heads the **executive branch.** The president may also issue *executive orders,* specific legal instructions for government agencies. Once statutory laws are enacted, their implementation and enforcement is assigned to the appropriate administrative agency within the executive branch. Administrative agencies (Figure 3.5), which may be established by Congress or by presidential order, are sometimes nicknamed the "fourth branch" of government because they are the source of a great deal of policy, in the form of regulations. **Regulations** are specific rules based on the more broadly written statutory law. Besides issuing regulations, administrative agencies monitor compliance with laws and regulations and enforce them when individuals or corporations violate them. Administrative agencies have come to be the source of most U.S. environmental policy.

The **judicial branch,** or judiciary, consisting of the Supreme Court and various lower courts, is charged with interpreting law. This is necessary because of changing social factors and technologies and because Congress must write laws broadly to ensure that they apply to varied circumstances throughout the nation. Decisions rendered by the courts make up a body of law known as *case law.* Previous rulings serve as *precedents,* or legal guides, for

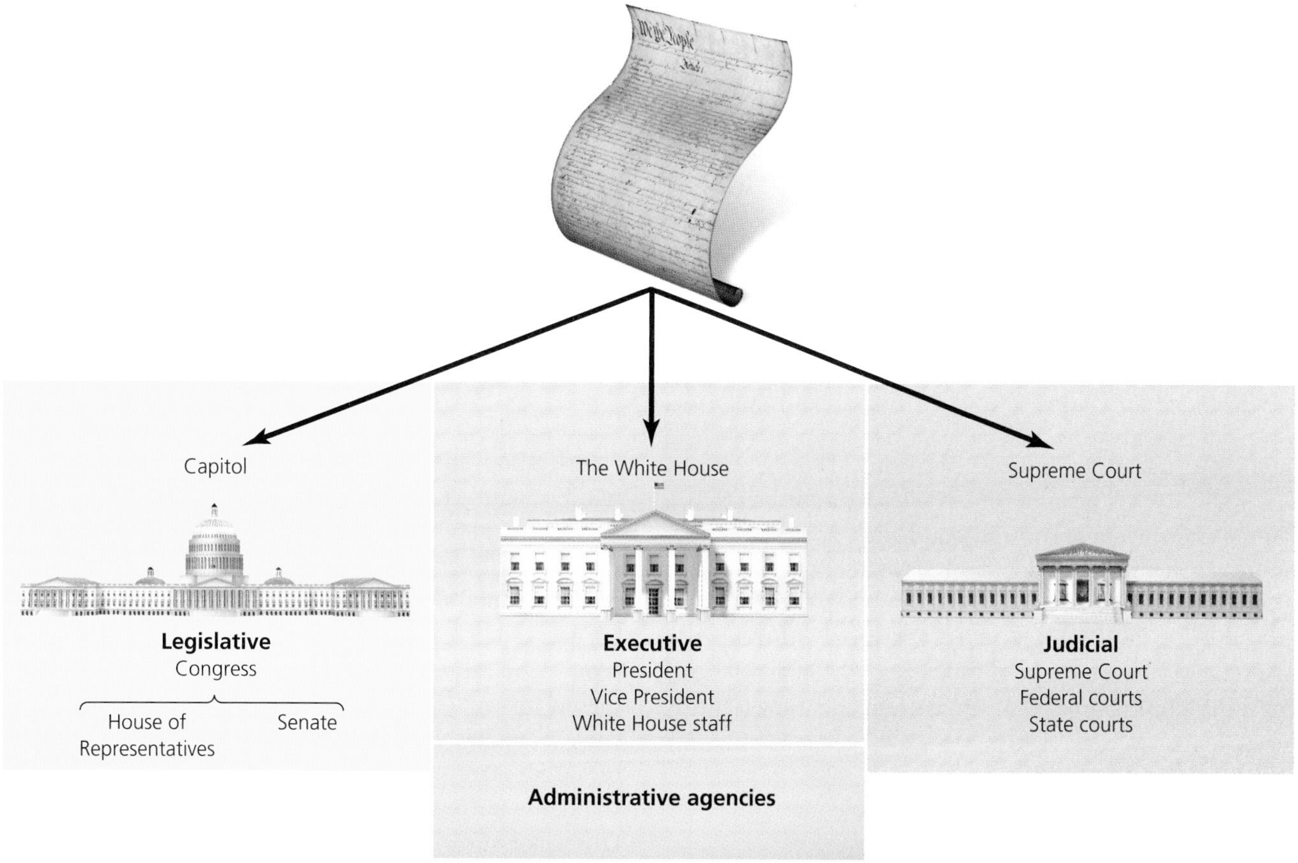

FIGURE 3.4 In the United States, federal powers are shared among the legislative, executive, and judicial branches of government. All three branches (including the administrative agencies of the executive branch) are responsible for aspects of developing and implementing environmental policy.

Federal Agencies and Departments of the Executive Branch that Affect Environmental Policy

- Environmental Protection Agency (EPA)
- Office of Management and Budget (OMB)
- U.S. Trade Representative
- Tennessee Valley Authority (TVA)
- Nuclear Regulatory Commission (NRC)
- Council on Environmental Quality (CEQ)
- Department of Treasury
  - Community Development Financial Institutions Fund
- Department of Commerce
  - National Institute of Standards and Technology
  - National Oceanic and Atmospheric Administration (NOAA)
- Department of Defense (DOD)
  - Army Corps of Engineers
- Department of Energy (DOE)
  - Office of Environment, Safety, and Health
  - Energy Efficiency and Renewable Energy Network
  - Office of Civilian Radioactive Waste Management
  - Office of Fossil Energy
  - Office of Health and Environmental Research
  - Office of Environmental Management
- Department of Health and Human Services
  - Public Health Service
  - Agency for Toxic Substances and Disease Registry
  - Centers for Disease Control and Prevention (CDC)
  - Food and Drug Administration (FDA)
- Department of Housing and Urban Development (HUD)
  - Community Planning and Development
  - Healthy Homes and Lead Hazard Control
  - Policy Development and Research
- Department of Interior
  - Bureau of Indian Affairs
  - Division of Energy and Mineral Resources
  - Bureau of Land Management (BLM)
  - Minerals Management Service
  - National Park Service (NPS)
  - Office of Ethics
  - Fish and Wildlife Service (FWS)
  - U.S. Geological Survey (USGS)
  - Water Resources Division
- Department of Justice
  - Environmental and Natural Resources Division
  - Office of Legal Policy
  - Office of Public Affairs
- Department of Labor
  - Mine Safety and Health Organization
- Department of State
  - U.S. Agency for International Development (USAID)
- Department of Transportation
  - Office of Pipeline Safety
  - Office of Environment and Energy
- Department of Agriculture (USDA)
  - Farm Service Agency
  - Forest Service (USFS)
  - Natural Resources Conservation Service (NRCS)
  - National Rural Development Council
  - Food and Consumer Service

FIGURE 3.5 The many administrative agencies of the executive branch are the source of most U.S. environmental policy. *Source:* U.S. General Services Administration, Washington, D.C.

later cases, steering judicial decisions through time. The judiciary has been an important arena for environmental policy. Grassroots environmental advocates and nongovernmental organizations have used lawsuits as tools to help level the playing field with better-funded large corporations and agencies. Conversely, the courts have allowed businesses and individuals to challenge the constitutional validity of environmental laws they feel to be infringing on their rights.

State and local policy also affects environmental issues

The structure of the federal government is mirrored at the state level with governors, legislatures, judiciaries, and agencies. State laws cannot violate principles of the U.S. Constitution, and if state and federal laws are found to be in direct conflict, federal laws take precedence.

Many states with dense urban populations, such as California, New York, and Massachusetts, have strong environmental laws and well-funded environmental agencies, whereas many less-populous states, such as those of the interior West, put less priority on environmental protection. To safeguard public health in locations such as San Diego's beaches, California legislators in 1997 required state environmental health officials to set standards for testing waters for bacterial contamination. Officials issue an advisory, or warning, when bacterial concentrations in nearshore waters exceed health limits established by California law. In 1999, further legislation passed by the State Assembly mandated another state agency to develop methods for locating sources of contamination. As we proceed through our discussion of the federal government, keep in mind that important environmental policy is also created at the state and local levels.

Some constitutional amendments bear on environmental law

The U.S. Constitution laid out several principles that have come to be especially relevant to environmental policy. One of these is the clause from the Fourteenth Amendment prohibiting a state from denying "equal protection of its laws" to any person. This amendment provides the constitutional basis for the environmental justice movement (▸ pp. 36–37).

The Fifth Amendment ensures, in part, that private property shall not "be taken for public use without just compensation." Courts have interpreted this clause, known as the *takings clause*, to ban not only the literal taking of private property but also what is known as regulatory taking. A **regulatory taking** occurs when the government, by means of a law or regulation, deprives a property owner of some or all economic uses of that property. Many people cite the takings clause in arguing against environmental regulations that restrict land use and development on privately owned land. For example, some would contend that zoning regulations (▸ pp. 385–386) that prohibit a landowner from opening a hazardous waste dump in a residential neighborhood deprive the landowner of an economically valuable use of the land and, therefore, violate the Fifth Amendment.

In a landmark court case in 1992, the Supreme Court ruled that a state land use law intended to "prevent serious public harm" violated the takings clause. The case, known as *Lucas v. South Carolina Coastal Council,* involved a developer named Lucas who in 1986 purchased beachfront property in South Carolina for $975,000. In 1988, before Lucas began to build, South Carolina's legislature passed a law banning construction on eroding beaches. A state agency classified the Lucas property as an eroding beach and prohibited residential construction there. Believing that this amounted to a regulatory taking, Lucas asked a state court to overrule the new law and allow him to build homes on his property. Agreeing with Lucas, the state court ruled that he was entitled to $1.2 million in compensation. South Carolina appealed the case to the state Supreme Court, which overturned the lower court's decision. Lucas then appealed to the U.S. Supreme Court, which overturned the state Supreme Court's ruling and sided with the lower court, declaring that the state law deprived Lucas of all economically beneficial uses of his land. Today, homes stand on the land, and regulatory takings remains a contentious area of law (Figure 3.6).

FIGURE 3.6 Beaches subject to erosion and hurricane damage like this one in South Carolina have given rise to environmental policy debates. Should the government be able to prevent development in areas where erosion, storms, and flooding pose risks to property and human life? If so, does such prohibition constitute a taking of private property rights, and should the property owner be compensated? These were the questions addressed in *Lucas v. South Carolina Coastal Council.*

Weighing the Issues: Regulatory Takings

Imagine you have purchased land and plan to make money by clearing its forest and building condominiums on it. Now suppose an environmental group concerned about forest loss in the region points out that an endangered plant species occurs on the property and petitions the government to prevent development of the land. Use the takings clause to argue why you should be allowed to build or be financially compensated if you are not allowed to build.

Now put yourself in the shoes of the environmentalists, and make a case why the landowner should not be allowed to build nor be financially compensated.

Which argument do you feel has greater merit? Would it make a difference to you if the land in question had been zoned as developable or nondevelopable when the purchase was made? What do you think is the best way to balance private property rights with protection of the public good in cases like this?

FIGURE 3.7 Settlers such as these (**a**) took advantage of the federal government's early environmental policies, including the Homestead Act of 1862. Early mining activities on public lands were largely unregulated (**b**). The Mineral Lands Act of 1866 required that mining could occur "subject . . . to the local customs or rules of miners," rather than in a way that would protect the environment.

(a) Settlers in Custer County, Nebraska, circa 1860

The first U.S. environmental policy addressed public land management

The laws that comprise U.S. environmental policy were created largely in three periods. Laws enacted during the first period, from the 1780s to the late 1800s, dealt primarily with the management of public lands and accompanied the westward expansion of the nation. Environmental laws of this period were intended mainly to promote settlement and the extraction and use of the West's abundant natural resources. Among these early laws were the *General Land Ordinances of 1785 and 1787,* which gave the federal government the right to manage Western lands and created a grid system for surveying them and readying them for private ownership. Between 1785 and the 1870s, the federal government promoted settlement in the West on lands it had appropriated from Native Americans by doling out as many of these lands as possible to its citizens. Western settlement provided these citizens with means to achieve prosperity, while relieving crowding in Eastern cities. It expanded the nation's geographical reach at a time when the young United States was still jostling with European powers for control of the continent. It also wholly displaced the millions of Native Americans who had inhabited these lands. U.S. environmental policy of this era reflected the public perception that Western lands were practically infinite, and inexhaustible in natural resources. The following are a few laws typical of this era:

- The Homestead Act of 1862 allowed any citizen to claim 65 ha (160 acres) of public land by living there for 5 years and cultivating the land or building a home, for a $16 fee (Figure 3.7a). A waiting period of only 14 months was available to those who could pay $176 for the land.
- The Mineral Lands Act of 1866 provided land for $5 per acre to promote mining and settlement. It allowed

(b) Nineteenth-century mining operation, Lynx Creek, Alaska

(c) Loggers felling an old-growth tree, Washington

FIGURE 3.7 *cont.* (**c**) Early forest policy promoted rapid clearing of forests and offered no incentives for replanting trees, conserving timber, or restoring habitat.

mining to occur subject to local customs, with no government oversight (Figure 3.7b).

- The Timber Culture Act of 1873 granted 65 ha (160 acres) to any citizen promising to cultivate trees on one-quarter of that area (Figure 3.7c).

Such laws encouraged settlers, entrepreneurs, and land speculators to move west, hastening the closing of the frontier.

The second wave of U.S. environmental policy addressed impacts of the first

In the late 1800s, as the West became more populated and its resources were increasingly exploited, public perception and government policy toward natural resources began to shift. Laws of this period aimed to mitigate some of the environmental problems associated with westward expansion. In 1872 Congress designated Yellowstone as the world's first national park. In 1891 Congress, to prevent overharvesting and protect forested watersheds, passed a law authorizing the president to create "forest reserves" off-limits to logging. In 1903, President Theodore Roosevelt created the first national wildlife refuge. These acts enabled the creation, over the next few decades, of a national park system, national forest system, and national wildlife refuge system that still stand as global models (► pp. 364–365). These developments reflected a new understanding that the West's resources are not inexhaustible, but instead require legal protection.

Land management policies continued through the 20th century, targeting soil conservation in the Dust Bowl years (► p. 260) and extending through the Wilderness Act of 1964 (► pp. 365–366), which sought to preserve still-pristine lands "untrammeled by man, where man himself is a visitor who does not remain."

The third wave of U.S. environmental policy responded largely to pollution

Further social changes in the mid- to late 20th century gave rise to the third major period of U.S. environmental policy. In a more densely populated country driven by technology, heavy industry, and intensive resource consumption, Americans found themselves better off economically but living amid dirtier air, dirtier water, and more waste and toxic chemicals. During the 1960s and 1970s, several events triggered increased awareness of environmental problems and brought about a shift in public priorities and important changes in public policy.

A landmark event was the 1962 publication of *Silent Spring*, a book by American scientist and writer Rachel Carson (Figure 3.8). *Silent Spring* awakened the public to the negative ecological and health effects of pesticides and industrial chemicals. (The book's title refers to Carson's warning that pesticides might kill so many birds that few would be left to sing in springtime.)

FIGURE 3.8 Scientist, writer, and citizen activist Rachel Carson illuminated the problem of pollution from DDT and other pesticides in her 1962 book, *Silent Spring*.

FIGURE 3.9 In a spectacular display of the need for better control over water pollution, Ohio's Cuyahoga River caught fire several times in the 1950s and 1960s. The Cuyahoga was so polluted with oil and industrial waste that the river would burn for days at a time.

The Cuyahoga River (Figure 3.9) also did its part to raise attention to the hazards of pollution. The Cuyahoga was so polluted with oil and industrial waste that the river actually caught fire near Cleveland, Ohio, more than half a dozen times during the 1950s and 1960s. This spectacle, coupled with an enormous oil spill off the Pacific coast near Santa Barbara, California, in 1969, moved the public to prompt Congress and the president to do more to protect the environment.

Today, largely because of environmental policies enacted since the 1960s, pesticides are more strictly regulated, and the nation's air and water are considerably cleaner. The public enthusiasm for environmental protection that spurred such advances remains strong today. Polls repeatedly show that an overwhelming majority of Americans favor environmental protection. Such support is evident each year in April, when millions of people worldwide celebrate Earth Day in thousands of locally based events featuring speeches, lectures, demonstrations, hikes, bird-walks, and more. In the more than 30 years since the first Earth Day, celebrated on April 22, 1970, participation in this event has grown markedly and has spread to nearly every country in the world (Figure 3.10).

NEPA guarantees citizens input into environmental policy decisions

Besides Earth Day, two federal actions marked 1970 as the dawn of the modern era of environmental policy in the United States. On January 1, 1970, President Richard Nixon signed the **National Environmental Policy Act (NEPA)** into law (Figure 3.11). NEPA created an agency called the Council on Environmental Quality and required that an **environmental impact statement (EIS)** be prepared for any major federal action that might significantly affect environmental quality. An EIS is a report of results from detailed studies that assess the potential impacts on the environment that would likely result from development projects undertaken or funded by the federal government.

NEPA's effects have been far-reaching. The EIS process forces government agencies and any businesses that contract with them to evaluate impacts on the environment before proceeding with a new dam, highway, or building project. Although the EIS process generally does not halt such projects, it can serve as an incentive to lessen the environmental damage done by a development or activity. NEPA also grants ordinary citizens input in the policy process by requiring that environmental impact statements be made publicly available and that public comment on them be solicited and considered.

The creation of the EPA marked a shift in federal environmental policy

Six months after signing NEPA into law, Nixon issued an executive order calling for a new integrated approach to environmental policy based on the understanding that environmental problems are interrelated. "The Government's environmentally-related activities have grown up piecemeal over the years," the order stated. "The time has come to organize them rationally and systematically." Nixon's order moved elements of agencies regulating water quality, air pollution, solid waste, and other issues into a newly created agency, the **Environmental Protection Agency (EPA).** The order charged the EPA with conducting and evaluating research, monitoring environmental quality, setting standards for pollution levels, enforcing those standards, assisting the states in meeting standards and goals, and educating the public.

(a) The first Earth Day, Washington, D.C., 1970

Other prominent laws followed

Ongoing public demand for a cleaner environment during this period resulted in a number of key laws that remain the linchpins of U.S. environmental policy (Figure 3.12). Today there are thousands of federal, state, and local environmental laws in the United States, and thousands more abroad. For problems like the Tijuana River's pollution, a crucial law has been the Clean Water Act of 1977. Throughout much of U.S. history, water policy was left largely to local and state governments. (An early exception was the federal Rivers and Harbors Act of 1899, which restricted dumping and discharges into navigable waters.) Prior to passage of federal legislation such as the Clean Water Act, pollution problems were subject primarily to *tort law* (law addressing harm caused to one entity by another), and individuals suffering external costs from pollution were limited to seeking redress through lawsuits. Reaction to the ruling in the case *Boomer v. Atlantic Cement Company* in 1970 effectively ended tort law's viability as a tool for preventing pollution. The court ruled that even though residents of Albany, New York, were suffering pollution from a neighborhood cement plant, the pollution could continue because the economic costs of controlling the pollution were greater than the economic costs of the damage the pollution caused.

(b) Schoolchildren celebrating Earth Day, Katmandu, Nepal, 2002

FIGURE 3.10 April 22, 1970, marked the first Earth Day celebration. (**a**) This public outpouring of support for environmental protection catalyzed the third wave of environmental policy in the United States. Three decades later, Earth Day is celebrated by millions of people across the globe (**b**), as shown here in Nepal.

The flaming waters of the Cuyahoga, however, indicated to many people that tough legislation was needed. Thanks to restrictions on pollutants by the Federal Water Pollution Control Acts of 1965 and 1972, and then the Clean Water Act, U.S. waterways finally began to recover.

FIGURE 3.11 Richard Nixon served as president during the period when modern environmental policy took shape. Here he signs the National Environmental Policy Act (NEPA) into law on January 1, 1970.

FIGURE 3.12 Most major laws in modern U.S. environmental policy were enacted in the 1960s and 1970s.

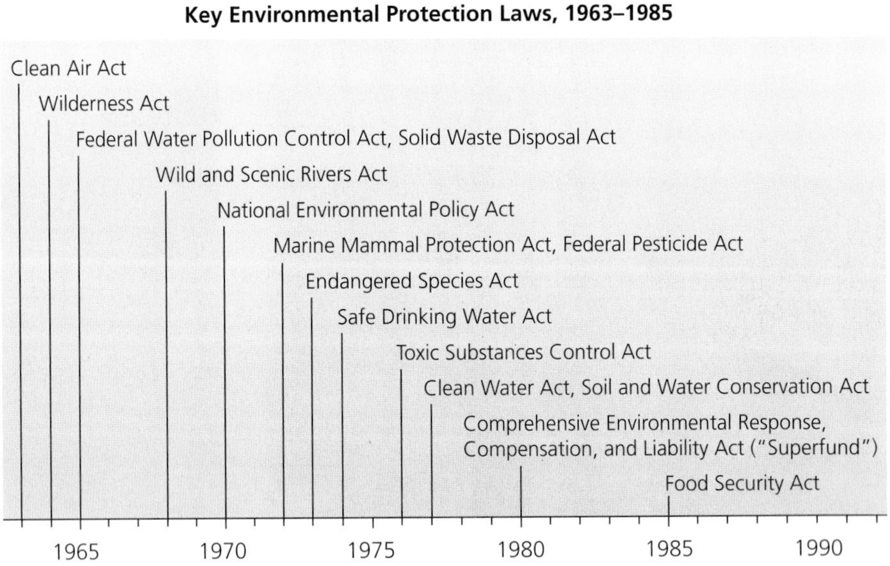

These laws regulated the discharge of wastes, especially from industry, into rivers and streams. The Clean Water Act also aimed to protect wildlife and establish a system for granting permits for the discharge of pollutants.

Weighing the Issues: Judging Costs in Tort Law

Imagine you are the judge presiding over a lawsuit brought by citizens who live next to an industrial plant. Pollution from the plant regularly covers their homes, yards, and cars with dust and soot, making them cough and wheeze. In weighing whether to require the plant to clean up its emissions, will you consider in your ruling the financial costs the factory would need to incur? Or will you ignore economic aspects and base your ruling solely on evidence of harm to health? How easy or difficult do you think it would be to calculate and compare the costs to human health of ongoing pollution versus the costs of mandating emission controls or shutting down the plant? How could one estimate these costs?

The social context for environmental policy changes over time

Historians of environmental policy have suggested that accomplishments such as the Federal Water Pollution Control Acts and the Clean Water Act occurred in the 1960s and 1970s because three factors converged. First, evidence of environmental problems became widely and readily apparent. Second, people could visualize policies to deal with the problems. Third, the political climate was ripe, with a supportive public and leaders who were willing to act.

By the 1980s, the political climate in the United States had changed. Although public support for the goals of environmental protection remained high, many citizens and policy experts began to feel that the legislative and regulatory means used to achieve environmental policy goals too often imposed economic burdens on businesses and personal burdens on individuals. Since 1980, numerous efforts have been made at the federal level to roll back or reform environmental laws.

As the United States retreated from its leadership in environmental policy during the past two decades, other nations were increasing their political attention to environmental issues. The 1992 Earth Summit at Río de Janeiro, Brazil, was the largest international diplomatic conference ever held, drawing representatives from 179 nations and unifying these leaders around the idea of sustainable development (▸ p. 22 and ▸ pp. 682–683). Indeed, we may now be embarking on a fourth wave of environmental policy, one focused on sustainability. This new policy approach tries to find ways to safeguard the functionality of natural systems while raising living standards for the world's poorer people (Figure 3.13). As the world's nations continue to feel the social, economic, and ecological effects of environmental degradation, environmental policy will without doubt become a more central part of governance and everyday life in all nations in the years ahead.

Approaches to Environmental Policy

Criticism of environmental legislation and regulation, and the means by which they often have been implemented, has spurred political scientists, economists, and

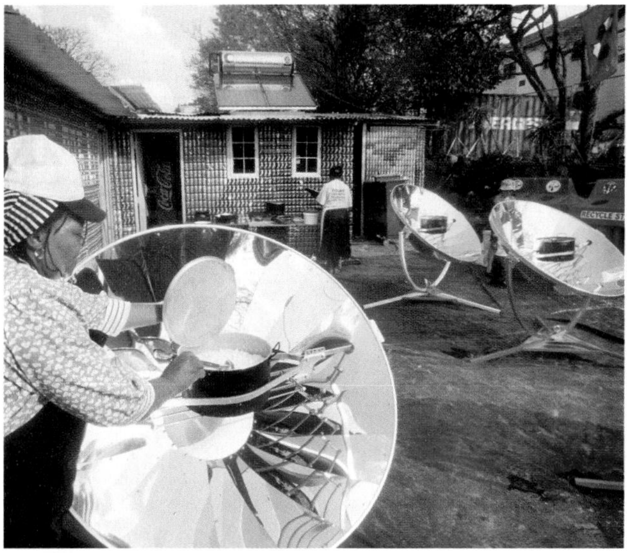

FIGURE 3.13 Many nations are shifting policies toward supporting sustainable development efforts, trying to increase standards of living while safeguarding the environment. Here, a woman stirs rice on a solar-powered oven at a restaurant made of recycled drink cans, showcased at the Ubunto Village at the U.N. World Summit on Sustainable Development in Johannesburg, South Africa, in 2002.

others to come up with alternative approaches for attaining environmental policy goals. The most widely developed alternatives involve the creative use of economic incentives to encourage desired outcomes, discourage undesired outcomes, and set market dynamics in motion to achieve goals in an economically efficient manner.

Top-down policy approaches are not always effective

Many environmental laws and regulations aiming to reduce pollution have simply set strict legal limits and threatened punishment for violating these limits, in what is sometimes called a **command-and-control** approach. The command-and-control approach has resulted in some major successes, as evidenced by the cleaner air and water U.S. residents enjoy today. Without doubt, our environment would be in far worse shape were it not for government regulatory intervention. However, many people have grown disenchanted with the top-down, sometimes heavy-handed nature of the command-and-control approach, and it is clear that government intervention sometimes fails every bit as badly as markets can fail.

Sometimes government actions are well intentioned but not well enough informed, so they can lead to unforeseen consequences. The predator-control policies that Aldo Leopold first supported, and later opposed, are a classic example (▸ p. 35). Killing wolves and mountain lions may have boosted deer populations for hunters and saved a few livestock for ranchers, but we now have enough ecological knowledge to recognize that it also helped cause a population explosion of deer that has altered the vegetative structure of many North American forests.

Policy can also fail if a government does not live up to its responsibilities to protect its citizens or treat them equitably. This may occur when leaders allow themselves to be unduly influenced by *interest groups,* small groups of people seeking private gain that work against the larger public interest. Finally, the command-and-control approach can fail in that it may generate opposition among citizens to government policy. If citizens view laws and regulations primarily as restrictions on their freedom, those policies will not last long in a constitutional democracy.

Subsidies are a widespread economic policy tool

Shortcomings of the command-and-control approach have led many economists to advocate the use of economic tools to compensate for market failure. One set of economic policy tools aims to encourage industries or activities that are deemed desirable. Governments may give *tax breaks* to certain types of businesses or individuals, for instance. Relieving the tax burden lowers costs for the business or individual, thus assisting the desirable industry or activity.

Probably the most widespread economic policy tool is the **subsidy,** a government giveaway of cash or publicly owned resources that is intended to encourage a particular activity. National governments commonly provide subsidies to industries they judge to benefit the nation in some way. Subsidies can be used to promote environmentally sustainable activities, but all too often they have been used to prop up unsustainable ones. Subsidies judged to be harmful to the environment and to the economy total roughly $1.45 *trillion* yearly across the globe, according to British environmental scientist Norman Myers—an amount larger than the economies of all but five nations.

The average U.S. taxpayer pays $2,000 per year in environmentally harmful subsidies, plus $2,000 more through increased prices for goods and through degradation of ecosystem services, Myers estimates. U.S. subsidies for industries and activities promoting automobile transportation alone amount to $1,700 per taxpayer per year, and the nation's heavily subsidized gasoline is generally cheaper than bottled water. The most recent Green Scissors

Report estimates that in 2003, $58 billion of U.S. taxpayers' money was budgeted for 68 environmentally harmful subsidies. The Green Scissors Report is a project of 22 nongovernmental organizations such as Friends of the Earth, Taxpayers for Common Sense, and the U.S. Public Interest Research Group. Among the subsidies highlighted in recent reports are the following:

- *General Mining Law of 1872.* Each year, mining companies extract $500 million to $1 billion in minerals from U.S. public lands without paying a penny in royalties to the taxpayers who own these lands. Since this law was enacted, the U.S. government has given away over $245 billion of mineral resources, and mining activities have polluted more than 40% of the watersheds in the West. The 130-year-old act still allows mining companies to buy public lands for $5 or less per acre.
- *Coal subsidies.* Since 1984, Congress has made $1.8 billion available to the coal industry through the Clean Coal Technology Program, paying for industry research that may lead to technologies to reduce air pollution from coal combustion. An additional $800 million per year goes to the coal industry for further research and development. These amounts dwarf subsidies granted to less-polluting renewable energy sources.
- *Forest Service road-building subsidies.* The U.S. Forest Service manages taxpayer-owned forests for various uses, including timber harvesting. Although many people assume that the corporations that harvest trees from public forests cover the costs of roads required to get the logs out, this is not the case. From 1992 to 1997 the Forest Service spent more than $387 million in tax dollars for road construction for timber companies (Figure 3.14). Despite legislative reform, subsidies are currently estimated at $170 million over 5 years.

FIGURE 3.14 When companies cut timber in U.S. national forests, roads must be built and maintained to enable access. The costs of these roads are paid by taxpayers.

Advocates of sustainable resource use have long urged governments to subsidize environmentally sustainable activities instead. To some extent, this is being done. For instance, subsidies for renewable energy sources totaled nearly $1.1 billion in the United States in 1999, according to the U.S. Department of Energy. But this amount falls short of the $2.8 billion that went to nonrenewable energy sources in that same year.

Green taxes discourage undesirable activities

Another economic policy tool—taxation—can be used to discourage undesirable activities. Taxing undesirable activities helps to "internalize" external costs by making them part of the overall cost of doing business. Taxes on environmentally harmful activities and products are called **green taxes.** By taxing activities and products that cause undesirable environmental impacts, a tax becomes a tool for policy as well as simply a way to fund government.

Green taxes have yet to gain widespread support in the United States, although similar "sin taxes" on cigarettes and alcohol are tools of U.S. social policy. Taxes on pollution have been widely instituted, however, in Europe, where many nations have adopted the *polluter pays principle.* This principle specifies that the price of a good or service should include all its costs, including costs of environmental degradation that would otherwise be passed on as external costs.

Under green taxation, a factory that pollutes a waterway would pay taxes based on the amount of pollution it discharges. The idea is to give companies a financial incentive to reduce pollution, while allowing the polluter the freedom to decide how best to minimize its expenses. One polluter might choose to invest in technologies to reduce its pollution if doing so is less costly than paying the taxes. Another polluter might find abating its pollution more costly, and could choose to pay the taxes instead—funds the government might then apply toward mitigating pollution in some other way. Green taxation provides incentive for industry to lower emissions not merely to a level specified in a regulation, but to still-lower levels. However, green taxes do have disadvantages. One is that businesses will most likely pass on their tax expenses to consumers and the increased costs may affect low-income consumers disproportionately more than high-income ones.

Markets in permits can save money and produce results

A still more creative market-based approach is for government to sell or give to companies the right to pollute, by establishing markets in tradable pollution permits. After determining the overall amount of pollution it will allow an entire industry to produce, the government can issue permits to individual polluters that allow them each to emit a certain fraction of that amount. Polluters are then allowed to buy, sell, and trade these permits with other polluters. With such **marketable emissions permits,** governments create incentives for firms to reduce their pollution in a way that is compatible with market capitalism.

Suppose, for example, you are a plant owner with permits to release 10 units of pollution, but you find that you can become more efficient and release only 5 units of pollution instead. You then have a surplus of permits, which might be very valuable to some other plant that is having trouble reducing its pollution, or to one that wants to expand production. In such a case, you can sell your extra permits. Doing so meets the needs of the other plant and generates income for you while preventing any increase in the total amount of pollution. Moreover, environmental organizations can buy up surplus permits and "retire" them, thus reducing the overall amount of pollution produced.

Such a system of marketable emissions permits has been in place in the United States. It was established by 1990 amendments to the Clean Air Act, which mandated reduced emissions of sulfur dioxide (Figure 3.15), a major contributor to acidic deposition (▸ pp. 514–518). Starting in 1995, permits were issued to power plants, allowing fewer emissions gradually year by year. Los Angeles had success in the 1990s with a similar program to reduce its smog. Marketable emissions permits generally end up costing both industry and government much less than a conventional regulatory system. Savings from the Clean Air Act permits have been estimated to add up to several billion dollars per year. Although such "cap-and-trade" programs can succeed in reducing the overall amount of pollution, they do have the drawback of allowing hotspots of pollution to occur around plants that buy permits to pollute more.

FIGURE 3.15 Markets in emissions permits have worked effectively in the United States to decrease the sulfur dioxide pollution that contributes to acid rain. Currently, nations and companies are developing markets in carbon emissions trading, following ratification of the Kyoto Protocol.

A global market in carbon emissions is currently developing as a result of the Kyoto Protocol to address climate change (▸ pp. 550, 552). Under the Protocol, nations have targets for reducing their carbon emissions from power plants, automobiles, and other sources that are driving climate change. Nations can gain carbon credits by investing in renewable energy or reforesting landscapes, as well as by reducing emissions, and any excess permits to emit carbon can be sold to other nations. European nations and major energy companies have begun such a market, and within a few years it should be clear whether this will develop into the multibillion-dollar market that many analysts predict.

Market incentives are being tried widely on the local level

At a more local scale, you may well have already taken part in transactions involving financial incentives as policy tools. Many municipalities charge residents for waste disposal according to the amount of waste they generate.

VIEWPOINTS

Public versus Private: In Whose Best Interest?

Transboundary pollution problems frequently give rise to debates over the best management approach—for example, public versus private, or some combination thereof. **In your view, what are the appropriate roles of the public and the private sectors in resolving the sewage pollution problems of the San Diego–Tijuana region?**

Clean, Safe Water Should Not Be a For-Profit Enterprise

The persistent problem of water pollution near the U.S.–Mexico border is partly the result of breakdowns in engineering and technology, including inadequately designed sewage treatment plants and missing or broken collection pipelines. However, it also reflects a shared history of water mismanagement and a lack of cooperative binational planning efforts.

Water is vitally important to this arid and rapidly growing region, yet over 90% is imported, used once, then dumped into the sea via sewage outfalls. Renegade sewage flows from these pipelines often pollute the river and beaches. Ultimately, to have sufficient clean water for basic health and future development, both countries will need to reconsider this pattern of use and begin conserving, reusing, and reclaiming water.

The best solutions for protecting water quality and ensuring adequate supply and sufficient treatment will be based on international cooperation and long-term planning projects. Japan is already investing in water and wastewater projects in Baja California, and the United States is beginning to help by using EPA grants to develop long-term master plans for improving water quality in the region.

However, changes in attitude and water use require support from governmental, research, and community organizations. Border residents need to understand not only their rights to the water they use, but also their role in keeping it clean and using it responsibly.

Relying on private for-profit companies to solve water problems—especially if they do not involve the community in the planning and decision-making process—can result in higher costs and fees that many families cannot afford, and will do little to change the public mindset. Public access to clean, safe water is a basic human right and is needed to protect public health. If companies fail to make a profit, who will step in to ensure that safe drinking water is available and sewage is treated?

Lori Saldaña represents District 76 in California's State Assembly. Previously she was an environmental policy researcher, writer, and community activist, specializing in water issues along the U.S.–Mexico border.

Public-Private Partnerships Mean Faster and More Efficient Environmental Change

A policy that excludes private participation in public service projects is bad policy if the government is unwilling or unable to fund project implementation.

It is a philosophical debate, not a technical question: "Should the private sector, motivated by profit, be able to participate in providing public services that are traditionally the exclusive realm of the public sector?" The private sector is more innovative, efficient, and flexible. The public-private partnership model (PPV) also provides for full disclosure, transparency, public comment, oversight, and regulatory control. The private sector can more often implement projects faster, and at a lower cost, than a lumbering government. PPV models incorporate a benefit unavailable in public sector projects: the ability to regulate by contract (as well as by statute). Desired environmental benefits are attained, while profit invigorates the market.

The Bajagua Project, a proposed PPV development model wastewater treatment plant to be built in Tijuana, Mexico, is such a project. It can be built faster and for less than any public facility. Bajagua will use technology selected by the U.S. EPA. It must comply with NEPA and all U.S., Mexican, and Californian environmental laws. Bajagua must also satisfy all contractual obligations *before* payment is due from the government, more closely emulating the "pay for services rendered" transaction model that we use every day. A policy where the public gets the environmental benefit it desires before it pays is a good policy.

In 1993, Congress capped the amount to be spent on the International Wastewater Treatment Plant. The federal agency that built the primary module in 1999 exhausted its funding and now cannot build the secondary module. Yet San Diego's beaches remain polluted, local economies remain depressed, and the federal agency remains in violation of the Clean Water Act. Clearly, PPV offers the only real possibility of success for positive environmental change.

Gary L. Sirota is a business and environmental litigator, consultant in international PPV infrastructure development, and a specialist in legislative finance and policy analysis.

Explore this issue further by accessing **Viewpoints** at www.aw-bc.com/withgott.

Other cities place taxes or disposal fees on items that require costly safe disposal, such as tires and motor oil. Still others give rebates to residents who buy water-efficient toilets and appliances, because this can cost the city less than upgrading its sewage treatment system. Likewise, power companies sometimes offer discounts to customers who buy high-efficiency lightbulbs and appliances, because doing so is cheaper for the utilities than expanding the generating capacity of their plants.

At all levels, from the local to the international, market-based incentives can reduce environmental impact while minimizing overall costs to industry and easing concerns about the intrusiveness of government regulation. Command-and-control policy is straightforward to implement, easy to monitor, and frequently works. Market-based approaches can be more complicated, but if they work, they can lessen environmental impact at a lower overall cost.

Weighing The Issues:
Environmental Policy Approaches

Imagine a factory is polluting a river near where you live. You organize your neighbors and complain to your Congresswoman, who promises to try to help solve the problem. What approach to environmental policy would you urge her to try, and why?

The Environmental Policy Process

Anyone can become involved in helping ideas become public policy. In the U.S. system, it is true that each and every person has a political voice and can make a difference. Unfortunately, it is also true that money wields influence and that some people and organizations are far more politically connected and influential than others. We will explore some of the ways people make themselves influential as we examine the main steps of the policymaking process. Our discussion pertains both to citizens at the grassroots level and to large organizations and corporations.

The environmental policy process begins when a problem is identified

The first step in the environmental policy process is to identify an environmental problem (Figure 3.16). Identifying a problem requires curiosity, observation, record keeping,

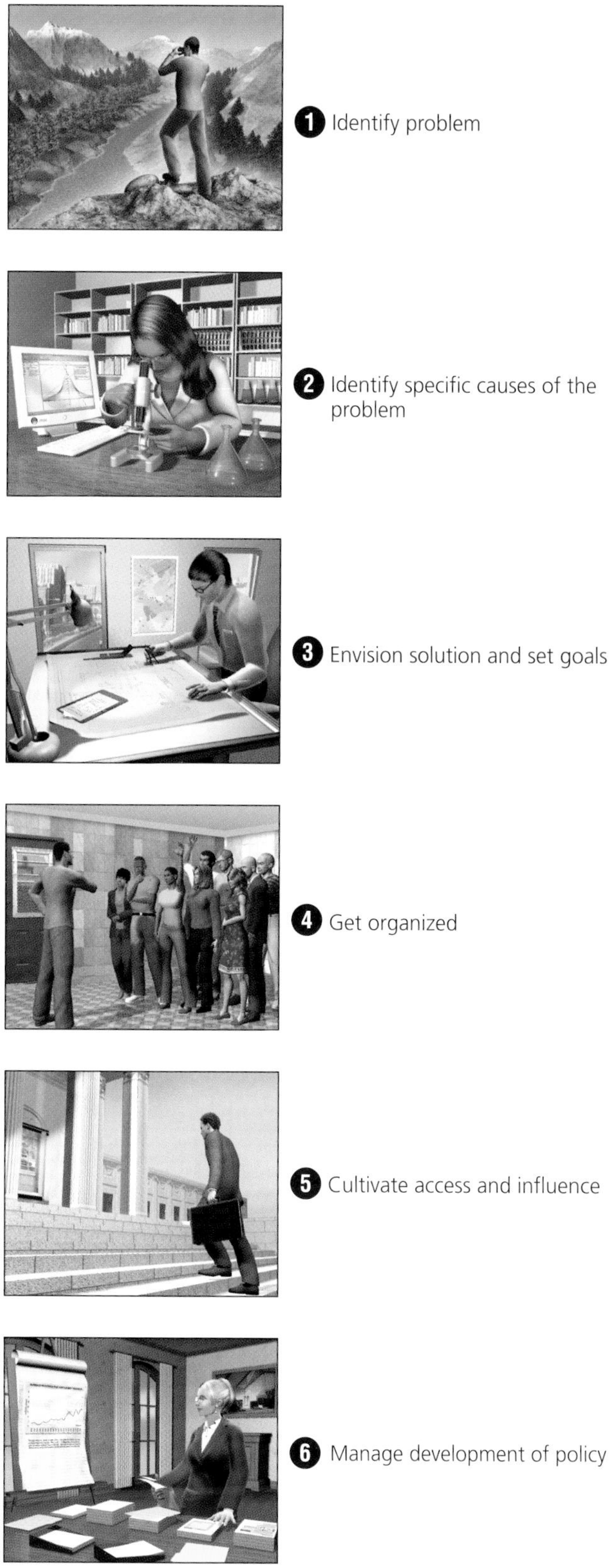

FIGURE 3.16 Understanding the steps of the policy process is an essential element of solving environmental problems.

The Science behind the Story

Spotting Sewage by Satellite

For decades, San Diego's sewage-contaminated coast remained a stubborn mystery—until scientists turned to the sky.

Water quality along San Diego's southern beaches had begun steadily declining in the 1960s, but it was supposed to be improving for surfers and swimmers in the late 1990s. Some key sewage sources had been pinpointed along the U.S.–Mexico border, and the U.S. government had spent $260 million to build a new treatment plant in the Tijuana River Valley and clean up coastal water.

Yet bacteria levels in coastal waters remained troublesome. Between 2000 and 2003, authorities had to close various San Diego County beaches because of sewage pollution 168 times for a total of 687 beach-days. During this time period, pollution advisories were issued 920 times over a total of 5,658 beach-days. The city of Imperial Beach, near the border, saw its beach closed 161 days in 1998 because of sewage contamination. Researchers and policymakers wondered what the earlier cleanup efforts had missed. Were existing sewage treatment plants not operating as well as planned? Or had past tracking efforts overlooked some pollution sources?

In 1999, local water quality officials got together with environmental scientists at Ocean Imaging, Inc., a company based near San Diego, and came up with a hypothesis—and a plan. Undetected pollution sources likely existed along the U.S.–Mexico border, they surmised, and those sources must be large enough to cause widespread contamination. They thought such pollution flows might be traceable using an innovative type of airborne imaging technology, known as synthetic aperture radar, or SAR.

SAR was originally developed as a military tracking tool, but through a joint effort between Ocean Imaging and NASA, the technology is now used to study environmental problems. Installed aboard an airplane or satellite, SAR instruments bounce radar signals off Earth's surface and record their echoes. The echoes change in frequency, based on the material being measured. For example, clean ocean water will return one type of signal, whereas large bodies of spilled sewage or oil suspended in ocean water will usually return a different type.

By analyzing the signals received, researchers can create detailed images of the surfaces they are studying. Marine sewage flows or oil spills will often show up on SAR images as dark plumes or patches in ocean water (see the figure). SAR can function through clouds and at night, making it invaluable in tracking something as mobile and unpredictable as a sewage spill.

Using SAR to scan the coast along the U.S.–Mexico border, scientists quickly found what they had predicted—a little-known source of almost entirely raw

and an awareness of our relationship with the environment. For example, assessing the contamination of San Diego- and Tijuana-area beaches required understanding the ecological and health impacts of untreated sewage. It also required being able to detect contamination on beaches and understanding water flow dynamics among the beaches, the Pacific Ocean, and the Tijuana River watershed (see "The Science behind the Story," above).

Identifying causes of the problem is the second step in the policy process

Once an individual or group has defined a particular environmental problem, discovering specific causes of the problem is next on the agenda. A person seeking causes for pollution in the Tijuana River watershed might notice that transboundary sewage spills took on a more toxic and industrial nature during the mid-1960s, when U.S.-based companies began opening *maquiladoras* on the Mexican side of the border. Advocates of the *maquiladora* system argue that these factories provide much-needed jobs south of the border while keeping companies' costs low by paying Mexican workers far less than U.S. workers. Critics argue that the factories are waste-generating, water-guzzling polluters whose transboundary nature makes them particularly difficult to regulate.

Identifying problems and their causes requires that scientific research play a major role in the policy

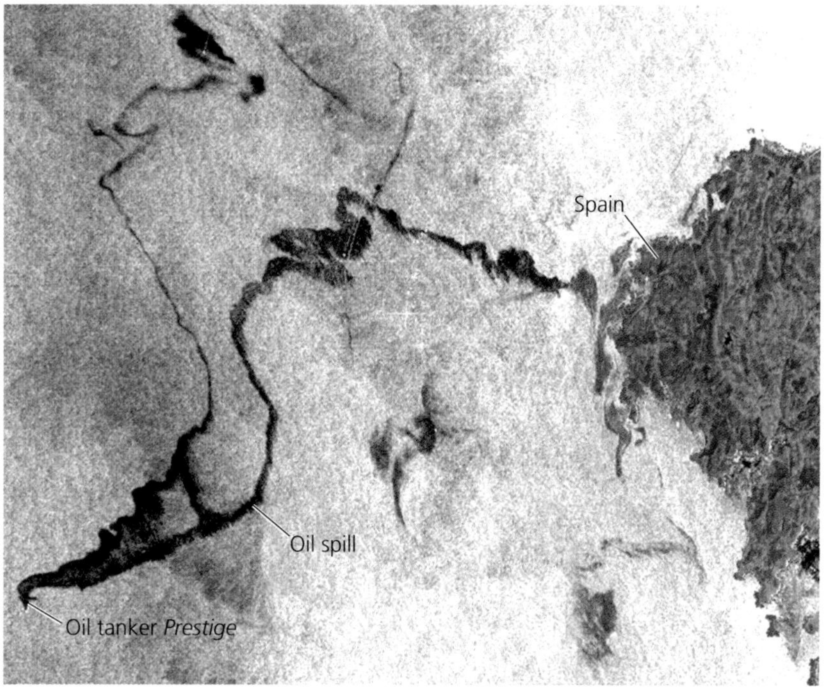

The same radar technology that has mapped San Diego's and Tijuana's sewage flows has also been used to track oil spills at sea. Here, a synthetic aperture radar image shows the sinuous trail of oil spilling from the tanker *Prestige*, which broke apart off the coast of Spain in November 2002.

sewage dumping into the surf in northern Mexico. The sewage, most likely from residential areas, was flowing through a contaminated creek at a rate of 95–133 million L (25–35 million gal) per day. Once in the ocean, the sewage flow showed up in the satellite image as a dark plume of disturbed water that, carried by currents, stretched north along the San Diego coast. Water samples taken at the same time from boats on the water confirmed that high levels of bacteria followed the plume's path.

These findings, made public in 2000, gave San Diego badly needed answers to its pollution puzzle. Everyone from environmental scientists to surfers could now "visualize what is going on," as a San Diego congressman said when the first SAR images were shown. Local policymakers used the report to lobby for more sewage tracking and cleanup efforts along the border. The SAR research gave a push to the Tijuana River Valley Estuary and Beach Sewage Cleanup Act, as well as the cross-border master plan to improve water quality in Tijuana. By unveiling an unknown source of a persistent problem, radar-based pollution tracking allowed San Diego and Tijuana to improve the cleanup of their polluted beaches.

process. Much of this work takes place in the arena of *risk assessment* (▸ pp. 422–423), in which scientists evaluate the extent and nature of problems and judge the risks that they pose to public health or environmental quality.

The third step is envisioning a solution

The better one can identify specific causes of a problem, the more effectively one will be able to envision solutions to it and argue for implementing those solutions. Science plays a role here too, through the process of *risk management* (▸ pp. 423–424), in which scientists develop strategies to minimize risk. Frequently, however, solutions will involve social or political action. In San Diego, citizen activists wanted Tijuana to enforce its own pollution laws more effectively—something that, once visualized, started to happen when San Diego city employees began training and working with their Mexican counterparts to keep hazardous wastes out of the sewage treatment system.

Getting organized is the fourth step

When it comes to gaining the ear of elected officials and influencing policy, organizations are generally more effective than lone individuals. The sole critic or crusader is easily dismissed as a crackpot or troublemaker, but a

The Science behind the Story

Assessing the Environmental Impact of Treating Transboundary Sewage

In 1990 the United States and Mexico formally agreed to construct a water treatment plant in the U.S. portion of the Tijuana River Valley. Before the South Bay International Water Treatment Plant (IWTP) could be built, however, the U.S. government was legally required by the National Environmental Policy Act (NEPA) to assess its environmental impact.

NEPA's environmental impact statement (EIS) process provides the framework for environmental impact assessment, but it is not the only law the IWTP's planners had to take into account. The federal Clean Water Act requires that harmful toxins and bacteria be removed from wastewater discharged into U.S. rivers, lakes, and oceans, and the Endangered Species Act protects species such as the Pacific pocket mouse, an inhabitant of the Tijuana River Valley.

International treaties also constrained the IWTP. For example, one 1989 agreement between the United States and Mexico required that the plant be funded by the EPA, be built on U.S. territory, and treat at least 1,095 L/sec (25 million gal/day). Finally, state and local laws regulated the impact of the plant on nearby communities and on the quality of California's coastal waters. Before construction could begin, all these constraints had to be addressed by the EPA and the U.S. section of the International Boundary and Water Commission (IBWC), the organization that would own and operate the plant.

The IWTP gives sewage primary treatment, but still needs to fund facilities to provide secondary treatment. Currently, treated waste is discharged into the Pacific Ocean at the South Bay Ocean Outfall, shown here during its construction.

In 1991 a draft EIS for the IWTP was released for public comment. The draft provided a preliminary assessment of the plant's impact on biological and cultural resources, public health and safety, scenic and recreational values, water quality, and other environmental factors. Three years later, after extensive research and public discussion, a final EIS was released. In it, the EPA and the IBWC endorsed the plant as the

group of hundreds or thousands of individuals is not as easily dismissed. Furthermore, organizations are more effective at raising funds, which by U.S. law they are permitted to contribute to political campaigns.

As effective as large organizations can be, it is important to remember that small coalitions and even individual citizens who are motivated, informed, and organized can solve environmental problems. As renowned anthropologist Margaret Meade once remarked, "Never doubt that a small group of thoughtful, committed citizens can change the world—indeed it is the only thing that ever has." San Diego-area resident Lori Saldaña provides an example. Concerned about the Tijuana River's pollution, Saldaña reviewed plans for the international wastewater treatment plant that the U.S. government proposed to build (see "The Science behind the Story," above). She concluded that it would merely shift pollution from the river to the ocean, where sewage would be released 5.6 km (3.5 mi) offshore. Working with her local Sierra Club chapter, Saldaña protested the plant's design and participated in a lawsuit that forced the government to conduct further studies. The EPA finally agreed to a design change, although funding for it has stalled in Congress. For her efforts, Saldaña received awards and was "preferred alternative," and the EPA confirmed the endorsement with an official Record of Decision.

In its Record of Decision, the EPA made several choices about the IWTP's design and operation that became controversial. One choice concerned opening the plant in phases. In the first phase, expected to last 2 years, large solids and some suspended particles would be filtered out of the wastewater using an advanced primary treatment process. However, other pollutants—including toxic metals and bacteria—would remain untreated until facilities were constructed for a secondary treatment process known as activated sludge, in which microorganisms convert highly toxic waste into a less toxic substance (see ▸ pp. 458–461 for more information on wastewater treatment). As long as those facilities remained uncompleted, the plant would be releasing polluted water into the ocean, in violation of the Clean Water Act.

In response to the EIS and Record of Decision, two environmental groups sued the EPA and the IBWC. In their lawsuit, the groups argued that the EIS had failed to consider a type of secondary treatment facility known as a completely mixed aerated pond system. The suit was eventually settled out of court, but it helped spur the EPA and IBWC to conduct a supplemental EIS in which they considered seven different secondary treatment alternatives, including activated sludge and the pond system. Compared to activated sludge, the pond system would produce a smaller volume of toxic by-products and would better absorb spikes of high toxicity. Based on these considerations, the EPA eventually decided to endorse the pond system.

In the supplemental EIS, the EPA also reaffirmed its decision to release treated wastewater through the South Bay Ocean Outfall, a 5.8-km (3.6-mi) underwater tunnel with outlets 29 m (95 ft) below the ocean surface (see figure). The decision was based on a computer model of ocean currents and pollution levels off the California coast, which indicated that wastewater released through the outfall—once treated to the secondary level—would be sufficiently diluted to meet federal and state pollution standards.

Even though the formal EIS process has now largely been completed, scientific findings continue to shape the future of the IWTP. Tests conducted in 1997 and 1998 indicated that wastewater treated to the advanced primary level was still acutely toxic to fish and other marine life. The findings suggested that Tijuana's sewage would require more aggressive treatment than most wastewater generated in the United States. In 1999, local activist (and now State Assemblywoman) Lori Saldaña and oceanographer Tim Baumgartner studied water quality above the outfall. They found a noticeable decrease in water quality compared to nearby areas. Saldaña and Baumgartner posted their results on the Internet and urged policymakers to speed construction of the IWTP's secondary treatment facilities, now scheduled to be completed by 2007. The federal government is also funding a more formal monitoring process to ensure that the plant's outflow does not damage California's coastal environment.

appointed to a commission on border environmental issues by President Bill Clinton. After a decade of activism, Saldaña ran for the California State Assembly in 2004 and won. She is now the representative from California's 76th district.

Gaining access to political powerbrokers is the fifth step

The fifth step in the policy process entails gaining access to policymakers who have the clout to help enact the desired changes. People gain access and influence through lobbying, campaign contributions, and the revolving door.

Lobbying Anyone who spends time or money trying to change an elected official's mind is engaged in **lobbying.** The term was originally used to describe the activities of corporate representatives who loitered in the lobbies of Washington, D.C., establishments for opportunities to talk with members of Congress. Although anyone can lobby, it is much more difficult for an ordinary citizen than for the thousands of full-time professional lobbyists employed by the many businesses and organizations seeking a voice in Washington politics. Opponents of environmental advocacy often paint large environmental organizations, such as the National Audubon Society, as interest groups with inordinate lobbying power in Washington. But according to a 2002 report by *Fortune* magazine, these organizations are surprisingly weak. The highest-ranking environmental organization on *Fortune*'s list, the Sierra Club, ranked only at number 52. This ranking placed the Sierra Club among such groups as the Distilled Spirits Council of the United States and the National Association of Letter Carriers, far back from such heavyweights as the National Rifle Association (NRA) and the American Association of Retired Persons (AARP). Indeed, the American Petroleum Institute spends on lobbying nearly as much as the entire budgets of the top five U.S. environmental advocacy groups combined.

Campaign contributions For those of us who can't spend our time hanging around Washington lobbies, supporting a candidate's reelection efforts with money is another way to make our voices heard. Because environmental policy often regulates the activities of corporations, they have a strong interest in shaping it. Although corporations and industries may not legally make direct campaign contributions, they are allowed to establish *political action committees (PACs)* for that purpose. Generally affiliated with industries, environmental groups, and other organizations with an interest in election outcomes, PACs raise money and distribute it to political campaigns, helping like-minded candidates win elections, in hope of gaining access to those individuals once they are elected.

The revolving door Some individuals employed by government-regulated industries gain political influence when they take jobs with the very government agencies responsible for regulating their industry. Businesses also often hire former government bureaucrats who regulated their industries. Such movement of individuals between the private sector and government agencies is known as the **revolving door.** For example, the George W. Bush Administration has included a commerce secretary who was CEO of a petroleum company, a transportation secretary who worked for a leading corporation in the transportation industry, and an agriculture secretary who served on the board of the first company in the nation to market genetically engineered crops. Proponents of the revolving door system assert that corporate executives who take government jobs regulating their own industry bring with them an intimate knowledge of that industry. This inside experience makes them highly qualified and likely to benefit the nation with especially well-informed policy, proponents maintain. Critics of the system contend that taking a job regulating your former employer is a clear conflict of interest that undermines the effectiveness of the regulatory process. Regulators from the private sector will be biased toward assisting their industry, critics say, and may fail to enforce regulations, thus acting against the interests of the taxpayers who pay their salaries.

Weighing the Issues: The Revolving Door

What do you think of each of the arguments for and against the revolving door system? Would the citizens of the United States be better off with or without the revolving door? How could we encourage it or discourage it?

Shepherding a solution into law is the sixth step in the policy process

Whether you're a corporate lobbyist or a grassroots activist, once your organization has the access and clout to influence policymakers, the better-known parts of the policy process come into play. Having gained access to elected officials and convinced them to hear your requests, you may be asked to prepare a bill, or draft law, that embodies the solutions you seek. Anyone can draft a bill. The hard part is finding members of the House and Senate willing to introduce the bill and shepherd it from subcommittee through full committee and on to passage by the full Congress. Lobbying and media attention intensify as the bill progresses through this process (Figure 3.17). If it passes through all of these steps, the bill may become law, but it can die in countless fashions along the way.

Of course, the policy process does not end with the enactment of legislation. Following a law's enactment, administrative agencies implement regulations. Policymakers also evaluate the policy's successes and failures and may revise the policy as necessary. Moreover, the judicial branch interprets law in response to suits in the courts, and much environmental policy has lived and died by judicial interpretation. The full policy process is long and often cumbersome, but it has resulted in a great deal of effective governance in constitutional democracies in the United States and many other nations.

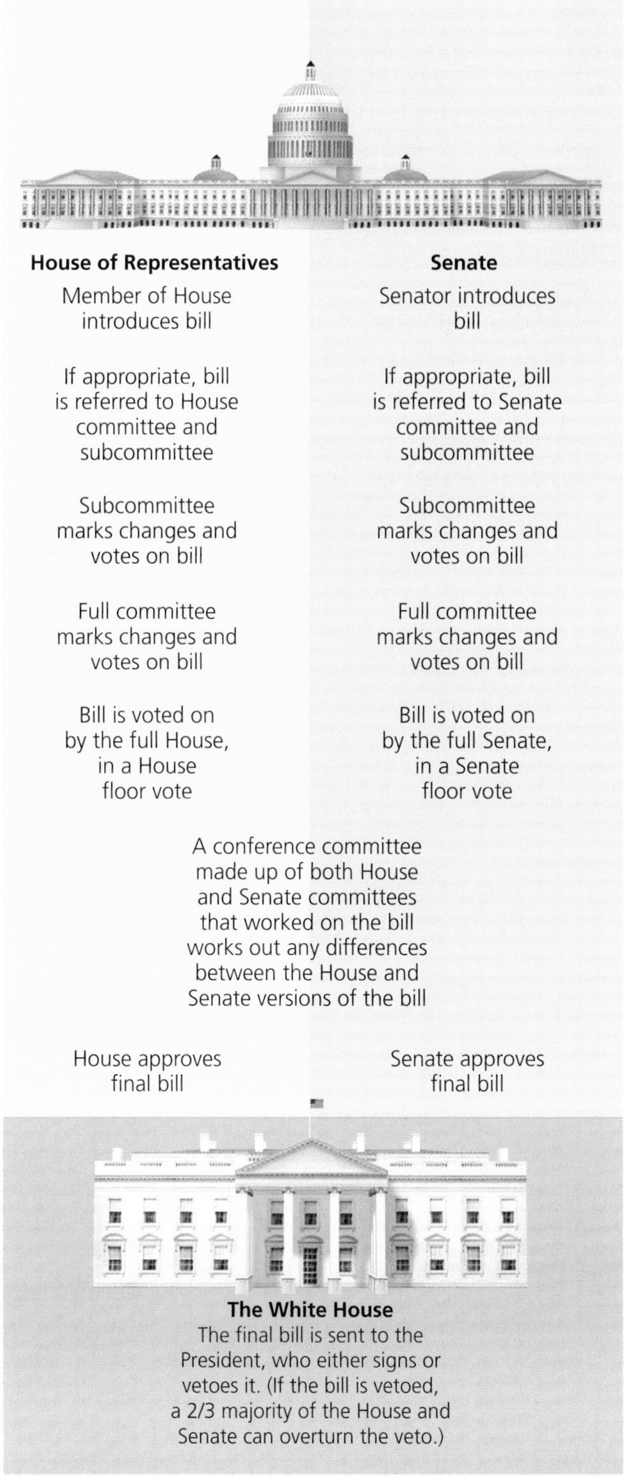
House of Representatives
Member of House introduces bill
If appropriate, bill is referred to House committee and subcommittee
Subcommittee marks changes and votes on bill
Full committee marks changes and votes on bill
Bill is voted on by the full House, in a House floor vote
Senate
Senator introduces bill
If appropriate, bill is referred to Senate committee and subcommittee
Subcommittee marks changes and votes on bill
Full committee marks changes and votes on bill
Bill is voted on by the full Senate, in a Senate floor vote
A conference committee made up of both House and Senate committees that worked on the bill works out any differences between the House and Senate versions of the bill
House approves final bill
Senate approves final bill
The White House
The final bill is sent to the President, who either signs or vetoes it. (If the bill is vetoed, a 2/3 majority of the House and Senate can overturn the veto.)

FIGURE 3.17 Before a bill becomes a law, it must clear a number of hurdles in both legislative bodies. If the bill passes the House and Senate, a conference committee must work out any differences between the House and Senate versions before the bill is sent to the president. The president may then sign or veto the bill.

International Environmental Policy

Environmental systems pay no heed to political boundaries, so environmental problems often are not restricted to the confines of particular countries. For instance, because most of the world's major rivers cross international borders, problems like those along the Tijuana River are frequently international in nature. Because U.S. law has no authority in Mexico or any other nation outside the United States, international law is vital to solving transboundary problems.

Mexico and the United States are working together to manage water

Often countries make progress on international issues not through legislation, but through creative bilateral or multilateral agreements hammered out after a lot of hard work and diplomacy. Such was the case with the successful effort to develop a long-term plan to manage drinking water and wastewater in the Tijuana metropolitan area. This master planning process, funded by the U.S. EPA under Congressional direction, began in January 2002. It involved the Comisión Estatal de Servicios Publicos de Tijuana (CESPT), the city agency that manages water and wastewater in Tijuana; the Mexican National Water Commission; the State Water Commission for Baja California; and the North American Development Bank. The resulting Tijuana Master Plan for Water and Wastewater Infrastructure aims to address a shortage of drinking water and possibilities for its reuse; water infrastructure; wastewater collection and transport; and wastewater treatment.

In 2002 the U.S. Congress also provided funding and authority for the EPA to work with CESPT to upgrade Tijuana's sewer system. The Tijuana Sewer Rehabilitation Project, known as Tijuana Sana ("Healthy Tijuana") was approved in 2001. This cooperative transboundary pollution prevention program will attempt to repair leaky sewer pipes in Tijuana. The 4-year, $43 million project will replace 131 km (81 mi), or 7.5%, of Tijuana's sewer pipes. If it succeeds, the project should reduce or eliminate the most severe sewage spills into the Tijuana River.

International law includes conventional law and customary law

Because solving transboundary dilemmas requires international cooperation, several principles of international law and a number of international organizations have arisen. Whereas U.S. law arises from the Constitution and the Bill of Rights, international environmental law is more nebulous in its origins and authorities.

International law known as **conventional law** arises from *conventions,* or *treaties,* into which nations enter. One example is the Montreal Protocol, a 1987 accord among more than 160 nations to reduce the emission of airborne chemicals that deplete the ozone layer (▸ pp. 513–514). Another example is the Kyoto Protocol to reduce fossil-fuel emissions that contribute to global climate change (▸ p. 530). This agreement took effect in February 2005 without United States participation, after a quorum of nations had ratified the accord. In 1990 Mexico and the United States signed a treaty and agreed to build an international wastewater treatment plant to handle excess sewage from Tijuana (Figure 3.18). In this case the treaty process worked well, but the results fell short of expectations. The facility reached its capacity within 3 years and then began discharging material to the ocean that did not meet safety standards established by U.S. law.

Other international law arises from long-standing practices, or customs, held in common by most cultures. This is known as **customary law.** Four principles underlie customary law as it applies to environmental issues:

FIGURE 3.18 In July of 1990, the United States and Mexico entered into a treaty and agreed to build the International Wastewater Treatment Plant (IWTP) to handle excess sewage from Tijuana. The IWTP began operating just north of the border in 1997 and treats up to 95 million L (25 million gal) of Mexican sewage each day. The IWTP has provided great benefits, but it reached its capacity within 3 years of opening.

- *Good neighborliness.* No nation should use its natural resources in a way that adversely affects other nations.
- *Due diligence.* Every nation should respect and protect the rights of other nations through the prevention and reduction of pollution.
- *Equitable resource use.* No nation should use more than its share of a natural resource.
- *The principle of information and cooperation.* Nations should provide notice and information to other nations when their actions might affect the interests and affairs of those nations.

Several organizations shape international environmental policy

Although there is no real mechanism for enforcing international environmental law, a number of international organizations regularly act to influence the behavior of nations by providing funding, applying peer pressure, and/or directing media attention. These organizations include the United Nations, the World Bank, the World Trade Organization, and the European Union and other multinational consortia, as well as a wide variety of nongovernmental organizations (NGOs).

The United Nations sponsors environmental agencies

In 1945, representatives of 50 countries founded the **United Nations (U.N.).** Headquartered in New York City, this organization's purpose is "to maintain international peace and security; to develop friendly relations among nations; to cooperate in solving international economic, social, cultural and humanitarian problems and in promoting respect for human rights and fundamental freedoms; and to be a centre for harmonizing the actions of nations in attaining these ends."

The United Nations has taken an active role in shaping international environmental policy (Figure 3.19). Of several agencies within it that influence environmental policy, most notable is the United Nations Environment Programme (UNEP), created in 1972, which helps nations understand and solve environmental problems. Based in Nairobi, Kenya, its mission is sustainability, enabling countries and their citizens "to improve their quality of life without compromising that of future generations." UNEP's extensive research and outreach activities provide a wealth of information useful to policymakers and scientists throughout the world and have provided a good deal of the data cited throughout this book.

FIGURE 3.19 The United Nations is active in international environmental policymaking. For instance, it sponsored the 2002 Earth Summit in Johannesburg, South Africa.

The World Bank holds the purse strings for development

The United Nations can encourage cooperation and promote awareness of environmental problems, but the World Bank holds the purse strings. Established in 1944 and based in Washington, D.C., the **World Bank** is one of the globe's largest sources of funding for economic development. This institution can shape environmental policy through its funding of major development projects, including dams, irrigation infrastructure, and other undertakings. In 2004 the World Bank provided over $20.1 billion in loans for projects designed to benefit the poorest people in the poorest countries around the world.

Despite its admirable mission, the World Bank has frequently been criticized for funding unsustainable projects that cause more environmental problems than they solve. In 1987 the World Bank reorganized itself and established an environmental office. Since then, the Bank has given higher priority to assessing the environmental impacts of its projects. In 2001 the World Bank issued its "Environmental Strategy," a document intended as a guide to improve quality of life, promote sustainability, and protect regional and global commons. Providing for the needs of growing human populations in poor nations while minimizing damage to the environmental systems on which people depend can be a tough balancing act. Environmental scientists today agree that the concept of sustainable development must be the guiding principle for such efforts.

The European Union is active in environmental affairs

Like the World Bank and the United Nations, the **European Union (EU)** was not established primarily with environmental problem solving in mind. However, the treaty that created it held as one of its goals the promotion of solutions to environmental problems. Formed after World War II, the EU as of early 2005 contained 25 member nations. It seeks to promote Europe's unity and its economic and social progress (including environmental protection) and to "assert Europe's role in the world." The EU can sign binding treaties on behalf of its members and can enact regulations that have the same authority as national laws in each member nation. It can also issue *directives,* which are more advisory in nature. The EU's European Environment Agency works to address waste management, noise pollution, water pollution, air pollution, habitat degradation, and natural hazards. The EU also seeks to remove trade barriers among member nations. It has classified some nations' environmental regulations as barriers to trade because some northern European nations have traditionally had more stringent environmental laws that prevent the import and sale of environmentally harmful products from other member nations.

The World Trade Organization has recently attained surprising power

Based in Geneva, Switzerland, the **World Trade Organization (WTO)** was established in 1995, having grown from a 50-year-old international trade agreement. The WTO represents multinational corporations and promotes free trade by reducing obstacles to international commerce and enforcing fairness among nations in trading practices. Whereas the United Nations and the European Union have limited influence over nations' internal affairs, the WTO has real authority to impose financial penalties on nations that do not comply with its directives. These penalties can on occasion play major roles in shaping environmental policy.

Like the EU, the WTO has interpreted some national environmental laws as unfair barriers to trade. For instance, in 1995 the U.S. EPA issued regulations requiring cleaner-burning gasoline in U.S. cities, following Congress's amendments of the Clean Air Act. Brazil and Venezuela filed a complaint with the WTO, saying the new rules unfairly discriminated against the petroleum they exported to the United States, which did not burn as cleanly. The WTO agreed, ruling that even though the South American gasoline posed a threat to human health in the United States, the EPA rules represented an illegal trade barrier. The ruling forced the United States to alter its approach to regulating gasoline. Not surprisingly, critics have frequently charged that the WTO aggravates environmental problems.

Weighing the issues: Trade Barriers and Environmental Protection

If Nation A has stricter laws for environmental protection than Nation B, and if these laws restrict the ability of Nation B to export its goods to Nation A, then by the policy of the WTO and the EU, Nation A's environmental protection laws could be overruled in the name of free trade. Do you think this is right? What if Nation A is a wealthy industrialized country and Nation B is a poor developing country that needs every economic boost it can get?

Nongovernmental organizations also exert influence

A number of NGOs have grown to become international in scope and exert influence over international environmental policy (Figure 3.20). The nature of these advocacy groups is diverse. Some, such as The Nature Conservancy, focus on accomplishing conservation objectives on the ground (in its case, purchasing and managing land and habitat for rare species) without becoming politically involved. Other groups, including Conservation International, the World Wide Fund for Nature, Greenpeace, Population Connection, and many others, attempt to shape policy directly or indirectly through research, education, lobbying, or protest. NGOs apply more funding and expertise to environmental problems, and conduct more research intended to solve them, than do many national governments.

FIGURE 3.20 Pursuing different visions of what makes for international environmental progress, nongovernmental organizations, such as the environmental advocacy group Greenpeace, sometimes clash with international institutions, such as the World Bank and the World Trade Organization.

International institutions and dynamics become more important in a globalizing world

As globalization proceeds, our world is becoming ever more interconnected. As a result, both human societies and Earth's ecological systems are being altered at unprecedented rates. Trade and technology have expanded the global reach of all societies, especially those such as the United States, which consume resources from across the world. Highly consumptive nations that import goods and resources from far and wide exert extensive impacts on the planet's environmental systems. Multinational corporations operate outside the reach of national laws and all too often have little incentive to conserve resources or conduct their business sustainably in the nations where they operate. For all these reasons, in today's globalizing world the organizations and institutions that influence international policy are becoming increasingly vital.

Conclusion

Environmental policy is a problem-solving tool that makes use of science, ethics, and economics, and requires an astute understanding of the political process. Conventional command-and-control approaches of legislation and regulation are the most common approaches to policymaking, but various innovative economic policy tools have also been developed. As we have seen in the case of the Tijuana River, environmental issues often overlap political boundaries and require international cooperation. Through the hard work of concerned citizens interacting with their government representatives, the political process eventually produced promising solutions in the Tijuana River Valley Estuary and Beach Sewage Cleanup Act (renewed in 2004) and in binational agreements and management plans. We will draw on the fundamentals of environmental policy introduced in this chapter throughout the remainder of this book. By understanding these fundamentals, you will be well equipped to develop your own creative solutions to many of the challenging problems we will encounter.

REVIEWING OBJECTIVES

You should now be able to:

Describe environmental policy and assess its societal context

- Policy is a tool for decision making and problem solving that makes use of information from science and values from ethics and economics. (pp. 59–60)
- Environmental policy is designed to protect natural resources and environmental amenities from degradation or depletion, and to promote equitable treatment of people. (pp. 60–61)

Identify the institutions important to U.S. environmental policy and recognize major U.S. environmental laws

- The legislative, executive, and judicial branches, together with administrative agencies, all play roles in U. S. environmental policy. (pp. 63–65)
- U. S. environmental policy came in three waves. The first encouraged frontier expansion and resource extraction. The second aimed to mitigate impacts of the first. The third targeted pollution and gave us many of today's major environmental laws. (pp. 66–70)
- Some major U.S. laws include the National Environmental Policy Act, the Clean Air Act, and the Clean Water Act. (pp. 68–70)

Categorize the different approaches to environmental policy

- Legislation from Congress and regulations from administrative agencies make up most federal policy. These top-down approaches are referred to as "command-and-control." (p. 71)
- Economic approaches include subsidies, green taxation, and market-based permit trading. (pp. 71–73, 75)

Delineate the steps of the environmental policy process and evaluate its effectiveness

- The policy process entails several steps: (1) identifying the problem, (2) identifying causes of the problem, (3) envisioning solutions, (4) getting organized, (5) gaining access to power, and (6) guiding a solution into law. (pp. 75–80)
- In a democracy, anyone can use the policy process, although corporations and organizations with money and resources tend to have the most clout. (pp. 77–80)

List the institutions involved with international environmental policy and describe how nations handle transboundary issues

- Many environmental problems cross political boundaries and thus must be addressed internationally. (pp. 81–84)
- International policy includes conventional law (law by treaty) and customary law (law by shared traditional custom). (pp. 81–82)
- Institutions such as the United Nations, European Union, World Bank, World Trade Organization, and nongovernmental organizations all play roles in international policy. (pp. 82–84)

TESTING YOUR COMPREHENSION

1. Describe and critique two common justifications for environmental policy. Explain the concept of external costs, and state why it is relevant to environmental policy.
2. Outline the primary responsibilities of the legislative, executive, and judicial branches of the U.S. government. What is the "fourth branch" of the U.S. government?
3. What is meant by a *regulatory taking*?
4. Summarize the differences between the first, second, and third waves of environmental policy in U.S. history.
5. What did the National Environmental Policy Act accomplish? Briefly describe the origin and mission of the U.S. Environmental Protection Agency.
6. Differentiate between a green tax, a subsidy, a tax break, and a marketable emissions permit.
7. Describe the environmental policy process, from identification of a problem through enactment of a federal law.
8. What kinds of things can an individual citizen do to become influential in the policymaking process?
9. What is the difference between conventional law and customary law? What special difficulties do transboundary environmental problems present?
10. Why are environmental regulations sometimes considered to be unfair barriers to trade?

SEEKING SOLUTIONS

1. Do we need environmental policy? Why or why not?
2. Imagine that you live in the San Diego area and cannot safely use beaches in your neighborhood because of water pollution that originated in Mexico. Who do you think should pay to prevent the pollution of your beaches? You? The Mexican government? The state of California or the U.S. government? *Maquiladoras?* What are the pros and cons of each of these potential funding sources? Now imagine that you live in Mexico in the Tijuana River watershed and depend on its water for your drinking, washing, and the irrigation of your garden. Who do you think should pay to prevent the pollution of your water supply?
3. Compare the main approaches to environmental policy—command-and-control, tort law, and economic or market-based approaches. Can you name an advantage and disadvantage of each? Do you think any one approach is most effective? Could we do with just one approach, or does it help to have more than one?
4. Reflect on the causes for the transitions in U.S. history from one type of environmental policy to another. Now peer into the future, and think about how life might be different in 25, 50, or 100 years. What would you speculate about the environmental policy of the future? What issues might it address? Do you predict we will have more or less environmental policy?
5. Think of one environmental issue that you would like to see solved through legislation. From what you've learned about the policymaking process, how do you think you could best shepherd your ideas through the process?
6. Compare the roles of the United Nations, the European Union, the World Bank, the World Trade Organization, and nongovernmental organizations. If you could gain the support of just one of these institutions for a policy you favored, which would you choose? Why?

INTERPRETING GRAPHS AND DATA

The Clean Air Act legislation of 1970, 1977, and 1990 was designed to improve air quality in the United States by monitoring and reducing the emissions of air pollutants judged to pose threats to human health, such as carbon monoxide, nitrogen dioxide, sulfur dioxide, ozone, particulate matter, and lead (▸ pp. 507–508). The main source of lead emissions in 1970 was the exhaust of vehicles burning gasoline to which tetra-ethyl lead had been added to improve combustion. By 1985, leaded gasoline was phased out of use, although airplanes and racecars were exempted.

The 1990 amendments addressed the growing problem of urban smog by requiring the use of reformulated gas (RFG) in cities with the worst smog problems. One of the RFG requirements specifies 2% oxygen content in fuel, which has been met by adding either ethanol (▸ pp. 611–612) or methyl tert-butyl ether (MTBE). Although it burns cleanly, MTBE is water-soluble and may cause cancer, so groundwater contamination from fuel spills is a concern. Sixteen states that collectively account for 45% of U.S. consumption of MTBE have now passed legislation banning or restricting its use. The following graph shows trends in U.S. lead emissions and MTBE consumption since 1970.

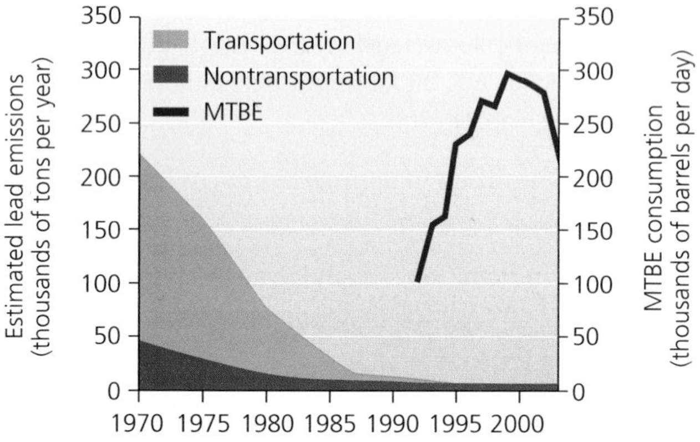

Estimated lead emissions from transportation and nontransportation sources, and consumption of MTBE in the United States. Data from U.S. Department of Transportation, Bureau of Transportation Statistics

1. Did policy resulting from Clean Air Act legislation succeed in reducing the public health risk from exposure to lead? Use data provided in the graph to support your answer.
2. In the 1990s, use of MTBE in RFG increased rapidly on the east and west coasts of the United States. Ethanol blends were used mostly in the corn-growing states of the central part of the country because ethanol is made from

corn, and so provides a market for in-state farmers while low transportation costs make it cheaper for in-state consumers. As MTBE use is being phased out in many states, use of ethanol is increasing. Name three ways in which these trends illustrate how state government actions may differ from federal government actions.

3. Do you think it is better policy to regulate air and water quality separately (e.g., under the Clean Air Act and Clean Water Act) or together under a single, comprehensive act to prevent and control pollution (as some European nations do)? How do the data in the graph support your reasoning?

CALCULATING ECOLOGICAL FOOTPRINTS

How many of us think about the destination of our waste when we flush the toilet? Some nutrient pollution from sewage generally ends up in a local waterway, but the U.S. Environmental Protection Agency establishes strict discharge standards following amendments to the U.S. Clean Water Act. One measure used is total suspended solids (TSS). To comply with federal regulations, wastewater treatment facilities in the United States must discharge water with a monthly average TSS no greater than 30 mg per liter.

Assuming that the average toilet flush is 3 gallons and occurs 4 times per person per day, calculate the total discharge of TSS in wastewater with the maximum permissible content. Note that 1 gal = 3.7 L, 1,000 mg = 1 g, and 1,000 g = 2.2 lbs.

	TSS discharged per day	TSS discharged per month	TSS discharged per year
You	0.440 g	13.4 g	160.6 g
Your class			
Your hometown			
United States			

1. Assuming that water in an average toilet flush contains 100 g TSS, calculate the amount of TSS removed from the nation's wastewater each day, month, and year, if the EPA standard is met.
2. Do you think the standards should be stricter? What would be at least one advantage and one disadvantage of stricter standards?

Take It Further

Go to www.aw-bc.com/withgott or the student CD-ROM where you'll find:

- Suggested answers to end-of-chapter questions
- Quizzes, animations, and flashcards to help you study
- *Research Navigator*™ database of credible and reliable sources to assist you with your research projects
- GRAPHIt! tutorials to help you master how to interpret graphs
- INVESTIGATEit! current news articles that link the topics that you study to case studies from your region to around the world

4 From Chemistry to Energy to Life

Exxon Valdez oil tanker in Prince William Sound, Alaska

Upon successfully completing this chapter you will be able to:

- Explain the fundamentals of environmental chemistry and apply them to real-world situations
- Describe the molecular building blocks of living organisms
- Differentiate among the types of energy and recite the basics of energy flow
- Distinguish photosynthesis, respiration, and chemosynthesis, and summarize their importance to living things
- Itemize and evaluate the major hypotheses for the origin of life on Earth
- Outline our knowledge regarding early life and give supporting evidence for each major concept

Workers spraying fertilizer on an oil-coated Alaska beach

Central Case: Bioremediation of the *Exxon Valdez* Oil Spill

"There is a dramatic difference. . . . It really cleaned the oil off the rock. It looked like someone brought in new rock."

—EPA PROGRAM MANAGER CHUCK COSTA, DESCRIBING EXPERIMENTAL BIOREMEDIATION RESULTS IN 1989

"The rush to bioremediation in Alaska was a function of the size of the problem and limited availability of options. . . . It did not turn out to be the silver bullet that many hoped it would be."

—ALASKA DEPARTMENT OF ENVIRONMENTAL CONSERVATION REPORT, 1993

On March 24, 1989, the tanker *Exxon Valdez* struck a reef in Alaska's Prince William Sound and spilled 42 million L (11 million gal) of crude oil, which eventually coated 2,100 km (1,300 mi) of Alaskan coastline. The largest oil spill in U.S. history, it killed an estimated 100,000–400,000 seabirds, 2,600–5,500 sea otters, 200–300 harbor seals, and countless fish. The oil smothered intertidal plants and animals and defiled the area's relatively pristine environment. The local economy took a nosedive as hundreds of fishermen were thrown out of work and tourism plummeted.

The massive spill was met with a massive response. Thousands of workers employed by Exxon (now ExxonMobil) and by government agencies, together with local volunteers, launched a cleanup effort of unprecedented scope. The cleanup crews corralled the oil with booms, skimmed it from the water, soaked it up with absorbent materials, and dispersed it with chemicals. They pressure-washed the beaches, removed contaminated sand with backhoes and tractors, and even tried burning the oil.

Scientists also used the opportunity to test a new cleanup strategy that enlisted nature to help take care of the mess. They stimulated naturally occurring bacteria to biodegrade, or break down, the oil. About 5% of the single-celled microbes present on Alaskan beaches feed on chemical compounds called *hydrocarbons* that are produced by the region's conifer trees. Hydrocarbons from conifers are chemically similar to the hydrocarbons that make up crude oil, so scientists predicted that the microbes might also be able to degrade oil. Scientists from the EPA and Exxon decided to put the bacteria to work in a process called **bioremediation,** the attempt to clean up pollution by enhancing natural processes of biodegradation by living organisms.

Although the bacteria were presented with an abundant new food source in the form of oil washing up on the beaches, they were not immediately able to consume it. The oil contained plenty of carbon, but not enough nitrogen and phosphorus. To remedy this imbalance of nutrients, scientists applied a fertilizing mixture containing nitrogen and phosphorus to several beaches, leaving other areas of the shore as untreated controls. The fertilizing treatment seemed to work; bacterial numbers increased, and oil residues decreased visibly. Encouraged, scientists put the program into full swing, and by the end of the year workers had treated more than 113 km (70 mi) of contaminated beach. They expanded the applications over the next 2 years.

Because there were many complicating factors, experts have interpreted the results of the study differently, and they still debate how much the treatments increased chemical breakdown of the oil. Some say degradation was sped up fivefold, whereas others think it made no difference. However, the well-publicized *Valdez* operation served as a model effort, and today bioremediation is actively researched and increasingly applied in many situations. Bioremediation has many practical limitations, but, when feasible, it can accomplish much good with a minimum of expense and environmental disturbance.

FIGURE 4.1 The *Exxon Valdez* spill coated hundreds of thousands of seabirds with oil, impairing their ability to insulate themselves with their feathers, and bringing on fatal hypothermia. Rescue workers labored tirelessly to clean oil from those birds they could capture and treat.

Chemistry and the Environment

The *Exxon Valdez* oil spill ignited a wide array of ecological, economic, political, and ethical concerns. Today many wildlife populations have recovered, but some have not, and pockets of oil remain throughout the region. Lawsuits against the company (some still pending) testify to the concerns of fishermen whose livelihoods were wrecked and of cleanup workers who say their health was affected. The U.S. Congress in the year following the spill passed the Oil Pollution Control Act, which required the Coast Guard and the U.S. Environmental Protection Agency (EPA) to strengthen regulations on tankers and their operators.

At the root of all these diverse impacts is the chemical makeup of the oil. It is the chemistry of crude oil that causes it to gum up birds' feathers and mammals' fur, impairing their insulating abilities and bringing on hypothermia (Figure 4.1). It is oil's chemistry that causes it to float on water and accumulate on beaches. It is certain hydrocarbons from oil that, mixed in the water column or volatile in the air, are harmful to wildlife and carcinogenic to humans. Yet the chemistry of crude oil also provides the energy that powers our remarkable civilization and modern way of life—a way of life that allows us the luxury to study, reflect on, and act to address these very issues.

Examine any environmental issue, and you will likely discover chemistry playing a central role. Chemistry is crucial to understanding how gases such as carbon dioxide and methane contribute to global climate change, how pollutants such as sulfur dioxide and nitric oxide cause acid rain, and how pesticides and other artificial compounds we release into the environment affect the health of wildlife and people. Chemistry is central, too, in understanding water pollution and sewage treatment, atmospheric ozone depletion, hazardous waste and its disposal, and just about any energy issue.

Chemistry is also central to developing solutions to environmental problems. Bioremediation is one clear illustration of this, and organisms are now used to clean up pollution in a variety of situations. Hydrocarbon-consuming bacteria and fungi are used to clean up soil beneath leaky gasoline tanks that threaten drinking water supplies. Other kinds of microbes are used to degrade pesticide residues in soil. Plants such as wheat, tobacco, water hyacinth, and cattails have been employed to clean

Table 4.1 Earth's Most Abundant Chemical Elements, by Mass

Earth's crust	Oceans	Air	Organisms
Oxygen (O), 49.5%	Oxygen (O), 88.3%	Nitrogen (N), 78.1%	Oxygen (O), 65.0%
Silicon (Si), 25.7%	Hydrogen (H), 11.0%	Oxygen (O), 21.0%	Carbon (C), 18.5%
Aluminum (Al), 7.4%	Chlorine (Cl), 1.9%	Argon (Ar), 0.9%	Hydrogen (H), 9.5%
Iron (Fe), 4.7%	Sodium (Na), 1.1%	Other, <0.1%	Nitrogen (N), 3.3%
Calcium (Ca), 3.6%	Magnesium (Mg), 0.1%		Calcium (Ca), 1.5%
Sodium (Na), 2.8%	Sulfur (S), 0.1%		Phosphorus (P), 1.0%
Potassium (K), 2.6%	Calcium (Ca), <0.1%		Potassium (K), 0.4%
Magnesium (Mg), 2.1%	Potassium (K), <0.1%		Sulfur (S), 0.3%
Other, 1.6%	Bromine (Br), <0.1%		Other, 0.5%

up toxic waste sites by letting them draw up heavy metals, such as lead and cadmium, through their roots (see "The Science behind the Story," pp. 92–93).

Sometimes suitable plants or bacterial cultures are introduced to a site. Sometimes naturally existing ones are fertilized, as was done at Prince William Sound. And sometimes the best way to mitigate pollution is simply to monitor naturally occurring organisms as they do their work, without disrupting the system. Bioremediation can be far less expensive, less environmentally intrusive, and more effective than conventional methods for cleaning up pollution. However, bioremediation does not always work, and it can sometimes be very slow, leave a job uncompleted, or introduce new problems. Scientists today are seeking ways to take bioremediation to the next level, by genetically engineering microbes and plants to become more efficient at the specific metabolic tasks we ask of them. Environmental chemists are excited about the countless future applications of chemistry that may help us address environmental problems.

Atoms and elements are chemical building blocks

To appreciate the complex chemistry involved in environmental science we must begin with a grasp of the fundamentals. The carbon, nitrogen, and phosphorus that played such key roles in the bioremediation of the oil spill in Prince William Sound are each elements. An **element** is a fundamental type of matter, a chemical substance with a given set of properties, which cannot be broken down into substances with other properties. Chemists currently recognize 92 elements occurring in nature, as well as more than 20 others that have been artificially created. Besides carbon and nitrogen, elements especially abundant in living organisms include hydrogen and oxygen (Table 4.1). Each element is assigned an abbreviation, or chemical symbol. The *periodic table of the elements* (see Appendix B) summarizes information on the elements in a comprehensive and elegant way.

Elements are composed of **atoms,** the smallest components that maintain the chemical properties of the element (Figure 4.2). Every atom has a nucleus of **protons** (positively charged particles) and **neutrons** (particles lacking electric charge). The atoms of each element have a defined number of protons, called the *atomic number.* (Elemental carbon, for instance, has six protons in its nucleus; thus, its atomic number is 6.) An atom's nucleus is surrounded by negatively

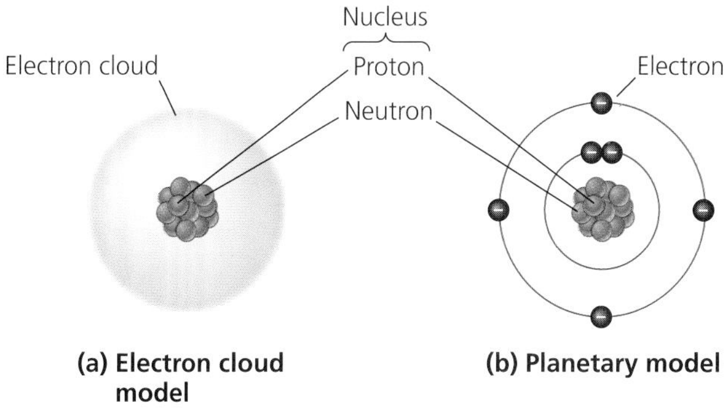

FIGURE 4.2 In an atom, protons and neutrons are held in the nucleus, and electrons move through space around the nucleus. Atoms, such as the carbon atom shown here, can be envisioned in different ways. In **(a)**, electrons are represented by a cloud, indicating the space within which they may likely occur at any given time. In **(b)**, electrons are shown orbiting the nucleus in concentric rings like planets orbit the sun. The diagram in **(a)** is more realistic, but we will use diagrams such as that in **(b)** to compare numbers of protons, neutrons, and electrons in our next figure.

The Science behind the Story

Student Chemist Lets Plants Do the Dirty Work

Bacteria that break down oil are just one example of organisms that scientists are putting to work to clean up environmental pollutants. Green plants can help, too.

When soil is contaminated with heavy metals from mining, manufacturing, oil extraction, or military facilities, the standard solution is to dig up tons of soil and pile it into a hazardous waste dump. Bulldozing so much dirt, however, can release toxic chemicals into the air and cost up to $2.5–7.5 million per hectare ($1–3 million per acre). As an alternative, scientists are developing methods of *phytoremediation*, using plants (*phyto* means "plant") to remediate, or detoxify, contaminated soils.

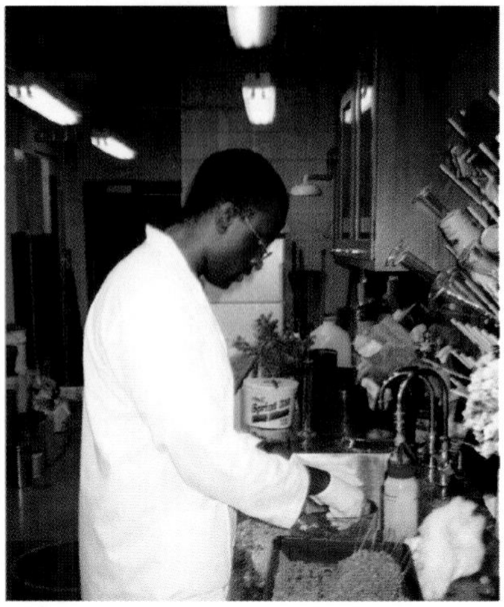

Rice University student Marc Burrell conducted research to find new ways to induce plants to clean up contaminated soil.

One researcher making advances in phytoremediation is Marc Burrell, now a senior at Rice University in Houston, Texas. While still in high school in Wisconsin, Burrell, with phytoremediation expert Peter Goldsbrough of Purdue University in Indiana and other mentors, began researching how to coax plants to remove toxic lead from soil.

Working one summer with Greg and Maria Begonia at Jackson State University in Mississippi, Burrell ran lab experiments to test how wheat could be made to draw up lead from soil. Normally, lead is not accessible to plants, because it is tied up in compounds such as lead carbonate and lead oxide, which do not dissolve in water. But chemicals called *chelating agents* can bind to lead and make it water-soluble so plant roots can draw it up. Burrell's greenhouse experiments showed that adding the chelating agent EDTA to the soil increased wheat's uptake of lead by about 300,000 times.

Burrell hypothesized that adding an acid would enhance the effect, because an acid's hydrogen ions would help break apart lead

charged particles known as **electrons,** which balance the positive charge of the protons (Figure 4.3).

Isotopes Although all atoms of a given element contain the same number of protons, they do not necessarily contain the same number of neutrons. Atoms with differing numbers of neutrons are referred to as **isotopes** (Figure 4.4a). Isotopes are denoted by their elemental symbol preceded by the *mass number,* or combined number of protons and neutrons in the atom. For example, ^{14}C (carbon-14) is an isotope of carbon with 8 neutrons (and 6 protons) in the nucleus rather than the normal 6 neutrons of ^{12}C (carbon-12).

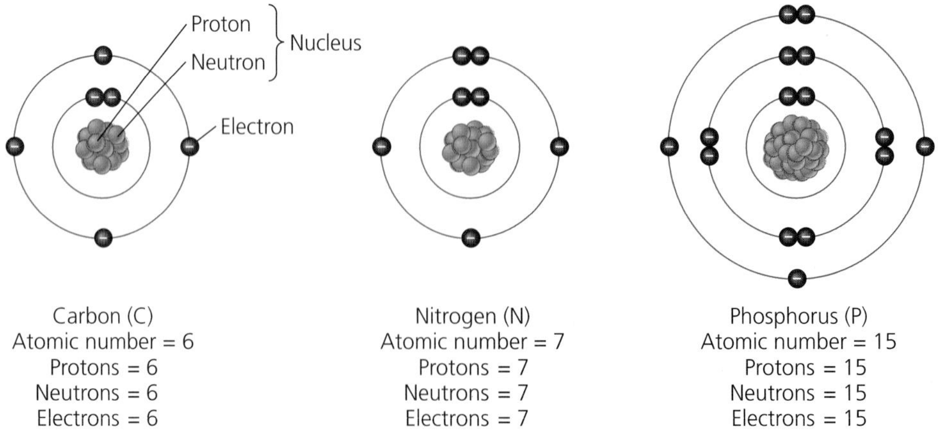

FIGURE 4.3 Each chemical element has a different number of protons, neutrons, and electrons. Carbon possesses 6 of each, nitrogen 7, and phosphorus 15.

compounds, freeing more lead ions to bind to EDTA. His experiments supported this hypothesis; when Burrell added acetic acid, the plants took up three times more EDTA.

Next, Burrell wanted to better understand how genes and proteins affect lead uptake. To deal with toxic metals that they take up by accident, many plants produce chelating agents called *phytochelatins* to drag the metals to vacuoles, or empty cellular sacs, where they can be stashed away without harm. Phytochelatins were thought to work with many metals, but no one had tested them with lead. Working under Heather Owen at the University of Wisconsin–Milwaukee, Burrell experimented with mustard plants engineered by geneticists who had knocked out the gene responsible for producing phytochelatins. He grew these engineered plants alongside normal plants in lead-contaminated soil. He found that the engineered plants that could not produce phytochelatins were more susceptible to lead poisoning and died sooner. This suggested that phytochelatins do, in fact, squirrel away lead into vacuoles.

Such research results are beginning to be applied at contaminated sites. Once plants have accumulated metals, they can be harvested and put through a smelting procedure to recover the metals. Alternatively, the plants can be dried and disposed of at a hazardous waste site.

Phytoremediation is a new pursuit, and it faces some hurdles. One is time; individual plants can take up only so much of a substance, and 5 to 20 years of repeated plantings may be required to reduce a soil's metal content to an acceptable level. Another is that metals need to be in a water-soluble form. In addition, cleanup is limited to the depth of soil that plants' roots reach. Finally, plants that accumulate toxins can potentially harm insects that eat the plants, and in turn, animals that eat the insects.

Despite such obstacles, phytoremediation is catching on. At military bases in Iowa, Tennessee, and Nebraska, the U.S. Army Corps of Engineers is using vegetation in artificial wetlands to minimize contamination of groundwater by ammunition. A Virginia company, Edenspace, has used plants to extract lead from residential sites, arsenic from military and energy facilities, zinc and cadmium at EPA Superfund sites (▸ pp. 670–671), and tungsten from abandoned mines. Such efforts are at the forefront of "green chemistry" today, thanks in part to the endeavors of bright and hardworking young researchers like Marc Burrell.

Because they differ slightly in mass, isotopes of an element differ slightly in their behavior. This fact has turned out to be very useful for researchers. Scientists have been able to use isotopes to study a number of phenomena that help illuminate the history of Earth's physical environment. Researchers also have used them to study the flow of nutrients within and among organisms, and the movement of organisms from one geographic location to another (see "The Science behind the Story," ▸ pp. 94–95).

Some isotopes are radioactive and "decay," changing their chemical identity as they shed subatomic particles and emit high-energy radiation. **Radioisotopes** decay into lighter and lighter radioisotopes, until they become *stable isotopes,* isotopes that are not radioactive. Each radioisotope decays at a rate determined by that isotope's **half-life,** the amount of time it takes for one-half the atoms to give off radiation and decay. Different radioisotopes have very different half-lives, ranging from fractions of a second to billions of years. The radioisotope uranium-235 (^{235}U) is our society's source of energy for

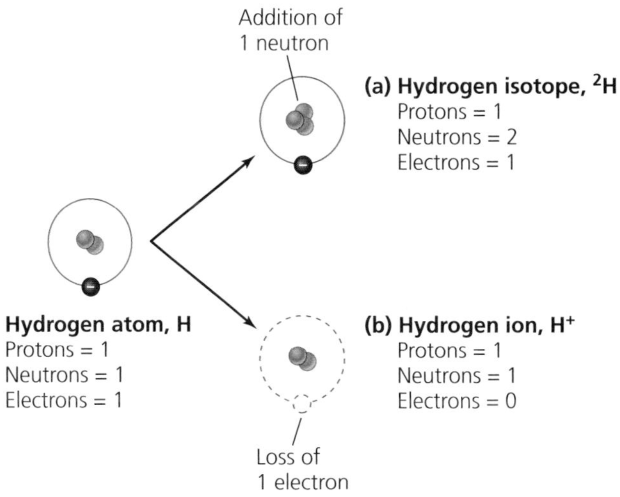

FIGURE 4.4 Atoms of an element such as hydrogen can become chemically altered to form isotopes and ions. Shown in **(a)** is an isotope of hydrogen, hydrogen-2 (^{2}H), or deuterium. This isotope contains two neutrons rather than one, and thus it contains greater mass than a typical hydrogen atom. Shown in **(b)** is the hydrogen ion, H^+. By losing its electron, it gains a positive charge.

The Science behind the Story

How Isotopes Reveal Secrets of Earth and Life

Isotopes, those alternate versions of chemical elements, have become one of the most powerful instruments in the environmental scientist's toolkit. They have enabled scientists interested in the past to date ancient materials, reconstruct the climate of past ages, and study the lifestyles of prehistoric humans. They have allowed researchers focused on the here-and-now to work out photosynthetic pathways, measure animals' diets and health, and trace nutrient flows through organisms and ecosystems.

For researchers studying the past, *radiocarbon dating* is one informative approach. Carbon's most abundant isotope is ^{12}C, but ^{13}C and ^{14}C also occur. Carbon-14 is radioactive and occurs in organisms at the same low concentration that it occurs in the atmosphere. Once an organism dies, no new ^{14}C is incorporated into its structure, and the radioactive decay process (▸ p. 93) gradually reduces its store of ^{14}C, converting these atoms to ^{14}N (nitrogen-14).

The decay is slow and steady enough to act as a kind of clock, so that scientists can date ancient organic materials by measuring the percent of carbon that is ^{14}C and matching this value against the clocklike progression of decay. In this way, archaeologists and paleontologists have dated prehistoric human remains; charcoal, grain, and shells found at ancient campfires; and bones and frozen tissues of recently extinct animals, such as mammoths. Scientists can also estimate the age of a fossil by radiocarbon dating the rock, peat, or sediment that surrounds it. The most recent ice age has been dated from ^{14}C analysis of trees overrun by glacial ice sheets. Ice drilled from glaciers today can be aged by measuring the ^{14}C in air bubbles trapped during its formation.

The half-life (time it takes for one-half of a sample to decay) of ^{14}C is 5,730 years. Thus, radiocarbon dating is not useful for items over 50,000 years old, because too little ^{14}C remains to permit accurate analysis. For dating older items, other isotopes can be used. Uranium-238 (with a half-life of 4.5 billion years) has been used to age very early fossils. For dating geological formations, potassium-argon dating is useful (potassium-40 decays to argon-40). Oxygen-18 has been widely used to infer changes in climate and sea level.

Researchers interested in present-day ecology can use *stable isotopes.* Unlike radioactive isotopes, stable isotopes occur in nature in constant ratios. For instance, nitrogen occurs as 99.63% nitrogen-14 and 0.37% nitrogen-15. Ratios of isotopes are called *isotopic signatures,* and by analyzing these signatures scientists can gain valuable information. For example, organisms tend to retain ^{15}N in their tissues but readily excrete ^{14}N. As a result, animals higher in the food chain show isotopic signatures biased toward ^{15}N, as do animals that are starving. Keith Hobson, an ecologist with Environment Canada and the University of Saskatchewan, has used nitrogen signatures to analyze the diets of seabirds and marine mammals, to show that geese fast while nesting, and to trace artificial contaminants in food chains.

Hobson and other scientists have also used stable carbon isotopes for ecological studies. Plants produce food through one of three photosynthetic pathways, and the isotopic signature of carbon in plants varies among these pathways. Grasses have higher ratios of ^{13}C to ^{12}C than oak trees do, for instance, whereas cacti have intermediate ratios. When animals eat plants, they incorporate the plants' isotopic signatures into their own tissues, and

commercial nuclear power (▸ pp. 594–595). It decays into a series of daughter isotopes, eventually forming lead-207 (^{207}Pb), and has a half-life of about 700 million years.

Ions Atoms may also gain or lose electrons to become **ions,** electrically charged atoms or combinations of atoms (Figure 4.4b). Ions are denoted by their elemental symbol followed by their ionic charge. For instance, a common ion used by mussels and clams to form shells is Ca^{2+}, a calcium atom that has lost two electrons, and so has a charge of positive 2.

Atoms bond to form molecules and compounds

Atoms can bond together and form **molecules,** combinations of two or more atoms. Molecules may contain one this signal passes up the food chain. As a result, carbon isotope studies can tell ecologists what an animal has been eating. Archaeologists have used isotopic signatures in human bone to determine when ancient people switched from a hunter-gatherer diet to an agricultural one.

Carbon signature data can even tell a scientist where an animal has been. For example, nectar-feeding bats have been shown to move seasonally between communities dominated by cacti to communities dominated by trees. Such movements have been inferred for migrating warblers, for elephants hunted for ivory, and for the oceanic movements of seals and salmon.

Recently, researchers have used isotopes to track movements of birds and other animals that migrate thousands of miles. This is possible because the isotopic signature of hydrogen in rainfall varies systematically across large geographic regions. This signature gets passed from rainwater to plants, and from plants to animals, leaving a fingerprint of geographic origin in an animal's tissues. Hobson and his colleagues in 1998 used a combination of isotopic data from hydrogen and carbon to pinpoint the geographic origins of monarch butterflies that had migrated from throughout the United States and Canada to communal roosts in Mexico—providing important information for their conservation (see figure).

Other elements show similar standing patterns of natural variation that have not yet been used or even discovered, researchers say, so there remains much more we can learn from the use of these subtle chemical clues.

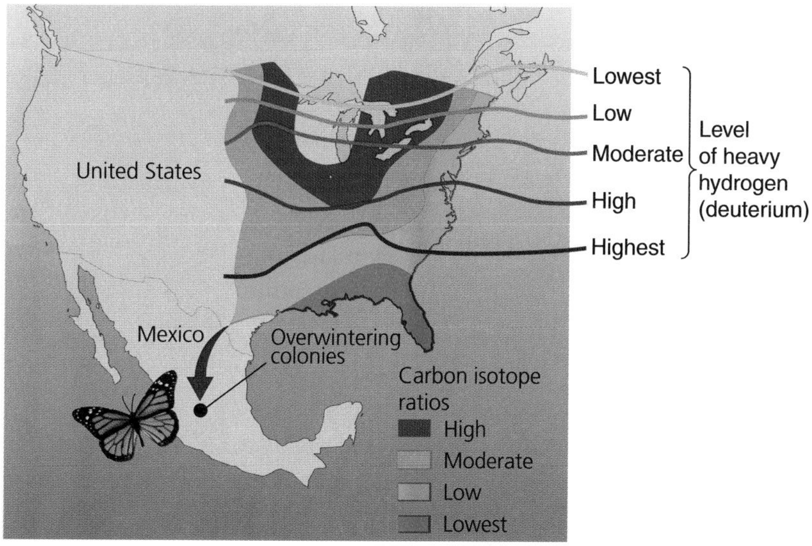

Plants in different geographic areas show different isotopic ratios for elements such as carbon and hydrogen. Caterpillars of monarch butterflies incorporate into their tissues carbon and hydrogen in the isotopic ratios present in the plants they eat. When these caterpillars metamorphose into butterflies and migrate, they carry these isotopic signals with them, providing scientists clues to their origin. Shown is a map of isotopic ratios across eastern North America produced from measurements of monarchs in the summer. The four colored bands show decreasing ratios of ^{13}C to ^{12}C from north to south. The five gray lines show increasing ratios of ^{2}H (heavy hydrogen or deuterium) to ^{1}H from north to south. By measuring carbon and hydrogen isotope ratios in monarchs wintering in Mexico, and matching the combination of these numbers against this map, researchers were able to pinpoint the geographic origin of many of the butterflies. *Source*: Wassenaar, L. I., and K. A. Hobson 1998. *Proceedings of the National Academy of Sciences of the USA* 95:15436–15439.

element or several. Common molecules containing only a single element include those of oxygen gas (O_2) and nitrogen gas (N_2), both of which are abundant in air. A molecule composed of atoms of two or more different elements is called a **compound.** Water is a compound; composed of two hydrogen atoms bonded to one oxygen atom, it is denoted by the chemical formula H_2O. Another compound is carbon dioxide, consisting of one carbon atom bonded to two oxygen atoms; its chemical formula is CO_2.

Atoms bond together because of an attraction for one another's electrons. Because the strength of this attraction varies among elements, atoms may be held together in different ways, according to whether and how they share or transfer electrons. When atoms in a molecule share electrons, they generate a **covalent bond.** For instance, two atoms of hydrogen bond to form hydrogen gas, H_2, by sharing electrons equally. Atoms in a covalent bond can also share electrons unequally, with one atom exerting a greater pull. Such is the case with water, in which oxygen attracts electrons more strongly than hydrogen, forming what are termed *polar* covalent bonds. In contrast, if the strength of attraction is unequal enough, an electron may be transferred from one atom to another. Such a transfer creates oppositely charged ions that are said to form **ionic bonds.** These associations are not considered molecules, but instead are called **ionic compounds,** or **salts.** Table salt (NaCl) contains ionic bonds between positively charged sodium ions (Na^+), each of which donated an electron, and negatively charged chloride ions (Cl^-), each of which received an electron.

Elements, molecules, and compounds can also come together in mixtures without chemically bonding. Homogenous mixtures of substances are called *solutions,* a term most often applied to liquids, but also applicable to some gases and solids. Air in the atmosphere is a solution formed of constituents such as nitrogen, oxygen, water, carbon dioxide, methane (CH_4), and ozone (O_3). Human blood, ocean water, plant sap, and metal alloys such as brass are all solutions. Crude oil at high pressure may carry natural gas in solution and often contains other substances distributed unevenly. It is a heavy liquid mixture of many kinds of molecules consisting primarily of carbon and hydrogen atoms. Its physical properties vary with its temperature, pressure, and composition.

The chemical structure of the water molecule facilitates life

Water dominates Earth's surface, covering over 70% of the globe, and its abundance is a primary reason Earth is hospitable to life. Scientists think life originated in water and stayed there for 3 billion years before moving onto land. Today every land-dwelling creature remains critically tied to water for its existence.

The water molecule's amazing capacity to support life results from its unique chemical properties. A water molecule's single oxygen atom bonds to its two hydrogen atoms at a 105-degree angle. As just mentioned, the oxygen atom attracts electrons more strongly, resulting in a polar molecule in which the oxygen end of the molecule has a partial negative charge and the hydrogen end has a partial positive charge. Because of this configuration, water molecules can adhere to one another in a special type of interaction called a *hydrogen bond,* in which the oxygen atom of one water molecule is weakly attracted to one or two hydrogen atoms of another (Figure 4.5). The weak electrical attraction of hydrogen bonding can also occur between hydrogen and certain other atoms, such as nitrogen. In water, hydrogen bonds are most stable in ice, somewhat stable in liquid water, and broken in water vapor.

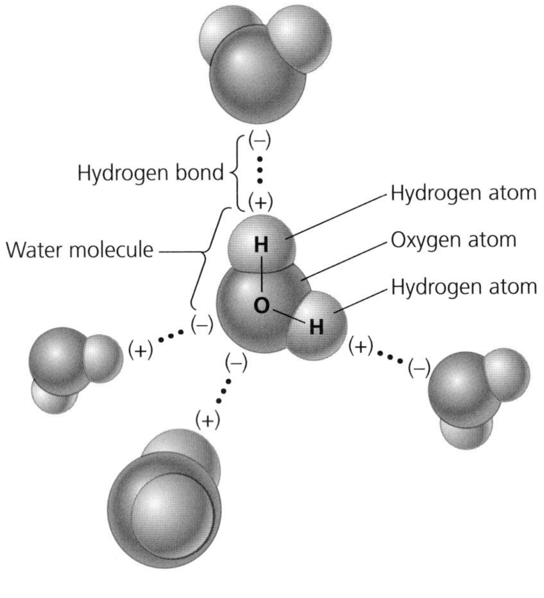

FIGURE 4.5 Water is a unique compound that has several properties crucial for life. Hydrogen bonds give water cohesion by enabling water molecules to adhere loosely to one another.

These loose connections among molecules give water several properties important in supporting life and stabilizing Earth's climate:

- Water exhibits strong cohesion. (Think of how water holds together in drops, and how drops on a surface join together when you touch them to one another.) This cohesion facilitates the transport of chemicals, such as nutrients and waste, in plants and animals.
- Hydrogen bonding provides water molecules a capacity to resist temperature change. Initial heating weakens hydrogen bonds between molecules but does not speed molecular motion. As a result, water can absorb a large amount of heat with only small changes in its temperature. This quality helps stabilize systems against change, whether those systems are organisms, ponds, lakes, or climate systems.
- Water molecules in ice are farther apart than in liquid form (Figure 4.6a). As a result, the solid form of water is less dense than the liquid—the reverse pattern of most other compounds, which become denser as they freeze.

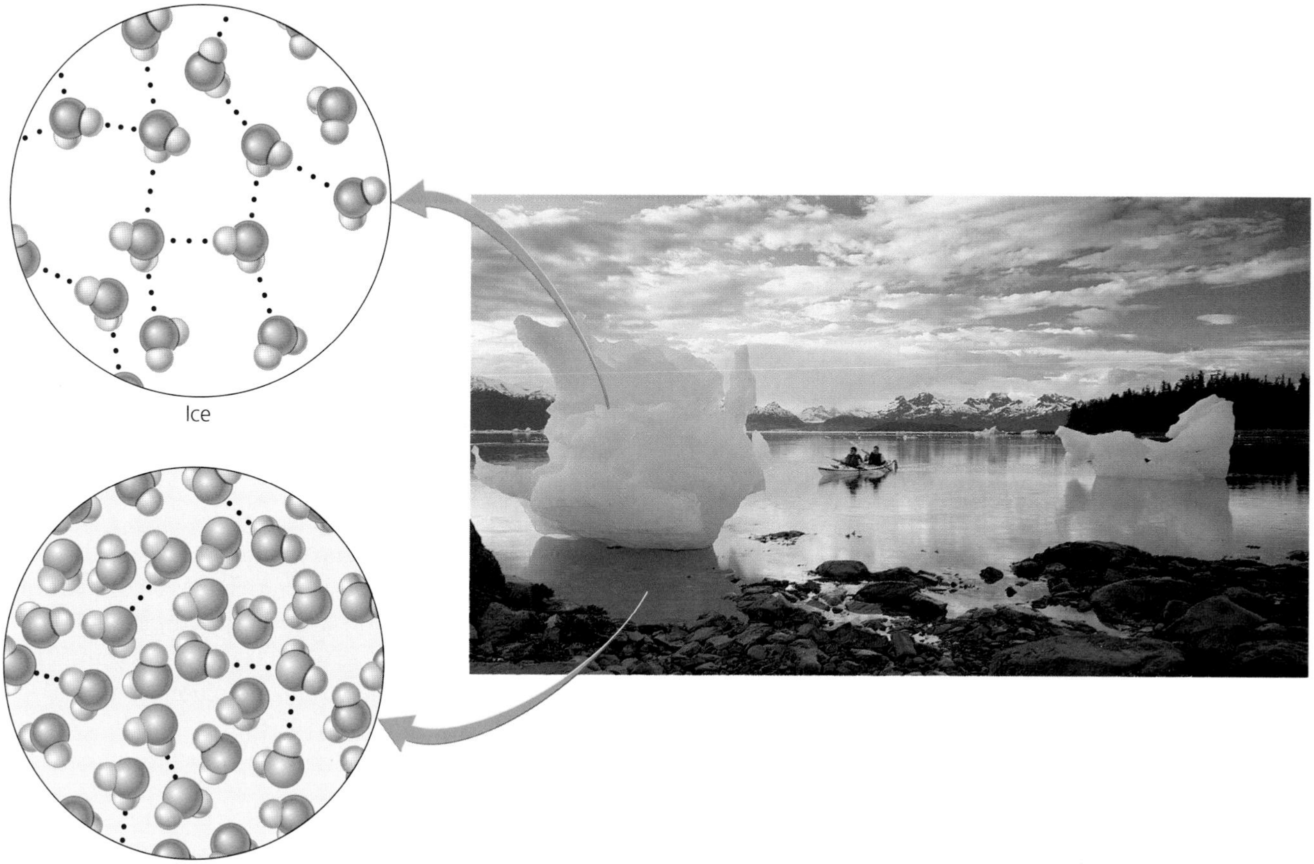

(a) Why ice floats on water

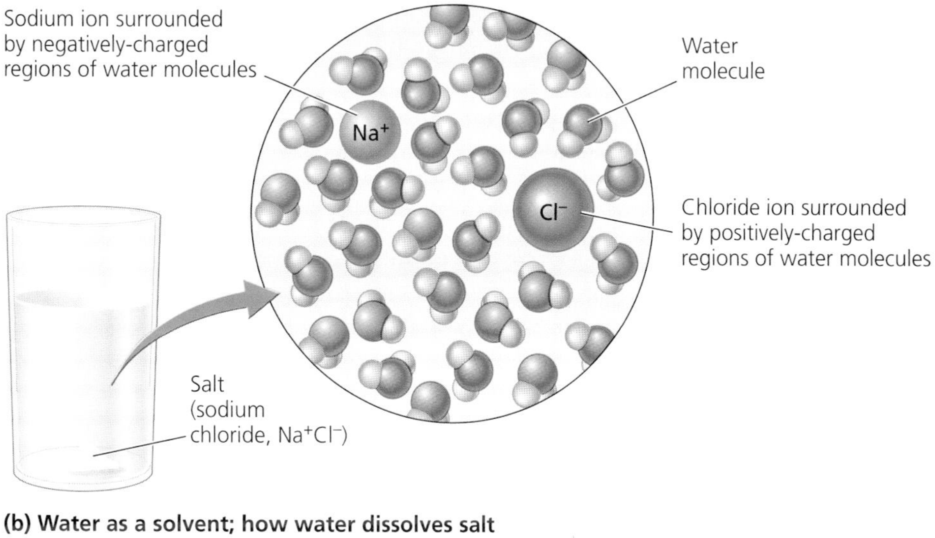

(b) Water as a solvent; how water dissolves salt

FIGURE 4.6 (a) Ice floats in liquid water because ice is less dense. Each molecule is connected to neighboring molecules by stable hydrogen bonds, forming a spacious crystal lattice. In liquid water, hydrogen bonds frequently break and re-form, and the molecules are closer together and less well organized. (b) Water is often called the "universal solvent" because it can dissolve so many chemicals, especially polar and ionic compounds. Seawater holds sodium and chloride ions, among others, in solution.

This is why ice floats on liquid water. Floating ice has an insulating effect that can prevent water bodies from freezing solid in winter.

- The polar nature of water molecules allows them to bond well with other polar molecules, because the positive end of one molecule bonds readily to the negative end of another. As a result, water can hold in solution, or dissolve, many other molecules, including chemicals necessary for life (Figure 4.6b). It follows that most biologically important solutions involve water.

Weighing the Issues: Water's Properties for Life

Water has several special properties that make it accommodating to life. Can you generate examples of specific ways in which each of these properties might help a particular organism, such as a fish in a pond? Can you think of ways in which any of water's properties might also bring harm to the fish?

Hydrogen ions determine acidity

In any aqueous solution, a small number of water molecules dissociate, each forming a hydrogen ion (H^+) and a hydroxide ion (OH^-). The product of hydrogen and hydroxide ion concentrations is always 10^{-14}. As the concentration of one increases, the concentration of the other decreases, and the product of their concentrations remains constant. Pure water contains equal numbers of these ions, each at a concentration of 10^{-7}, and we say that this water is neutral. Most aqueous solutions, however, contain different concentrations of these two ions. Solutions in which the H^+ concentration is greater than the OH^- concentration are **acidic,** whereas solutions in which the OH^- concentration is greater than the H^+ concentration are **basic.**

The **pH** scale (Figure 4.7) was devised to quantify the acidity or basicity of solutions. It runs from 0 to 14, because these numbers reflect the negative logarithm of the hydrogen ion concentration ("pH" comes from "potential Hydrogen"). Thus pure water has a pH of 7, because its hydrogen ion concentration is 10^{-7}. Seawater has a greater concentration of hydroxide ions, close to 10^{-6}. This means that its hydrogen ion concentration is about 10^{-8}, and thus its pH is close to 8. Solutions with pH less than 7 are acidic, those with pH greater than 7 are basic, and those with pH of 7 are neutral. Because the pH scale is logarithmic, each step on the scale represents a tenfold difference in hydrogen ion concentration. Thus, a substance with pH of 6 contains 10 times as many hydrogen ions as a substance with pH of 7, and a substance with pH of 5 contains 100 times as many hydrogen ions as one with pH of 7. Figure 4.7 shows pH for a number of common substances. Industrial air pollution has intensified the acidity of precipitation (► pp. 514–518), and rain in parts of the northeastern and midwestern United States now frequently dips to pH of 4 or lower.

Matter is composed of organic and inorganic compounds

Beyond their need for water, living things also depend on organic compounds, which they create and of which they are created. **Organic compounds** consist of carbon atoms (and generally hydrogen atoms) joined by covalent bonds, and may include other elements, such as nitrogen, oxygen, sulfur, and phosphorus. Carbon's unusual ability to build elaborate molecules has resulted in millions of different organic compounds that show various degrees of complexity. Because of the diversity of organic compounds and their importance in living organisms, chemists differentiate organic compounds from inorganic compounds, which lack carbon–carbon bonds.

Crude oil and petroleum products are made up of organic compounds called hydrocarbons. **Hydrocarbons** consist solely of atoms of carbon and hydrogen. The simplest hydrocarbon is methane (CH_4), the key component of natural gas; it has one carbon atom bonded to four hydrogen atoms (Figure 4.8a). Adding another carbon atom and two more hydrogen atoms gives us ethane

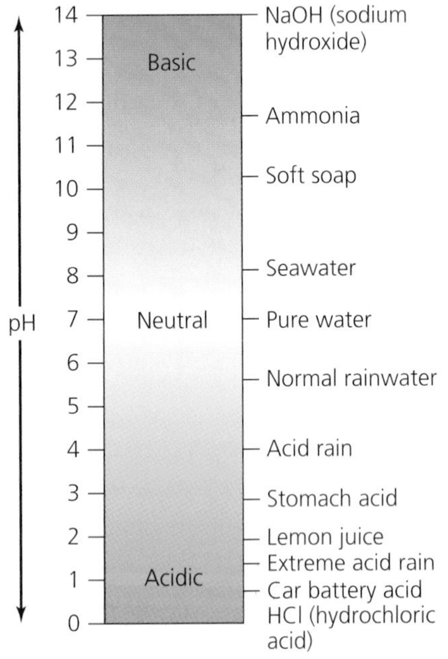

FIGURE 4.7 The pH scale measures how acidic or basic a solution is. The pH of pure water is 7, the midpoint of the scale. Acidic solutions have higher hydrogen ion concentrations and lower pH, whereas basic solutions have lower hydrogen ion concentrations and higher pH.

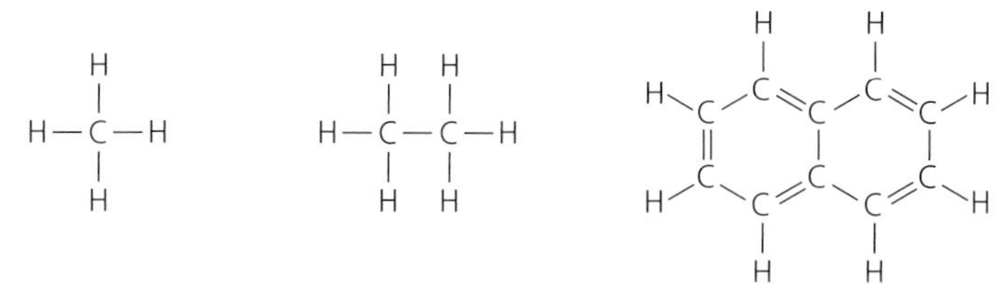

(a) Methane, CH_4 **(b) Ethane, C_2H_6** **(c) Naphthalene, $C_{10}H_8$ (a polycyclic aromatic hydrocarbon)**

FIGURE 4.8 Hydrocarbons are a major class of organic compound, and mixtures of them make up fossil fuels such as crude oil. The simplest hydrocarbon is methane **(a).** Many hydrocarbons consist of linear chains of carbon atoms with hydrogen atoms attached; the shortest of these is ethane **(b).** Volatile hydrocarbons with multiple rings, such as naphthalene **(c),** are called polycyclic aromatic hydrocarbons (PAHs).

(C_2H_6), the next-simplest hydrocarbon (Figure 4.8b). The smallest (and therefore lightest-weight) hydrocarbons (those consisting of four or fewer carbon atoms) exist in a gaseous state at normal temperatures and pressures. Larger (therefore heavier) hydrocarbons are liquids, and those consisting of over 20 carbon atoms long are normally solids. Some hydrocarbons from petroleum are known to pose health hazards to wildlife and people. For example, polycyclic aromatic hydrocarbons, or PAHs (Figure 4.8c), which are volatile molecules with a structure of multiple carbon rings, can evaporate from spilled oil and gasoline and can mix with water. The eggs and young of fish and other aquatic creatures are often most at risk. PAHs also occur in particulate form in various combustion products, including cigarette smoke, wood smoke, and charred meat.

Bacteria used in the bioremediation of petroleum spills do not actually consume the entire hydrocarbon molecules they attack. Rather, the bacteria, facilitated by oxygen, degrade complex hydrocarbon structures into simpler ones, or into their simplest components, hydrogen and carbon. Often this involves sequences of many chemical reactions. For example, bacterial degradation of the PAH naphthalene shown in Figure 4.8c involves 11 steps, during which atoms are added, removed, and rearranged, eventually producing the simpler products pyruvate and acetaldehyde (Figure 4.9).

Macromolecules are building blocks of life

Just as the carbon atoms in hydrocarbons may be strung together in chains, other organic compounds can sometimes combine to form long chains of repeated molecules. Some of these chains, called **polymers,** play key roles as building blocks of life. Three types of polymers are essential to life: proteins, nucleic acids, and carbohydrates. Lipids are not considered polymers but are also fundamental to life. These four types of molecules are referred to as **macromolecules** because of their large size.

Proteins Amino acids are organic molecules consisting of a central carbon linked to a hydrogen atom, an acidic carboxyl group (—COOH), a basic amine group (—NH_2), and an organic side chain unique to each type of amino acid (Figure 4.10a). Organisms combine up to 20 different types of amino acids into long chains to build **proteins** (Figure 4.10b). A protein's identity is determined by its particular sequence of amino acids and by the shape the protein molecule assumes as it folds. Protein molecules typically have highly convoluted shapes, with certain parts of the chain exposed and others hidden inside the folds (Figure 4.10c). A protein's folding pattern affects its function, because the position of each chemical group helps determine how it interacts with cell surfaces and with other molecules.

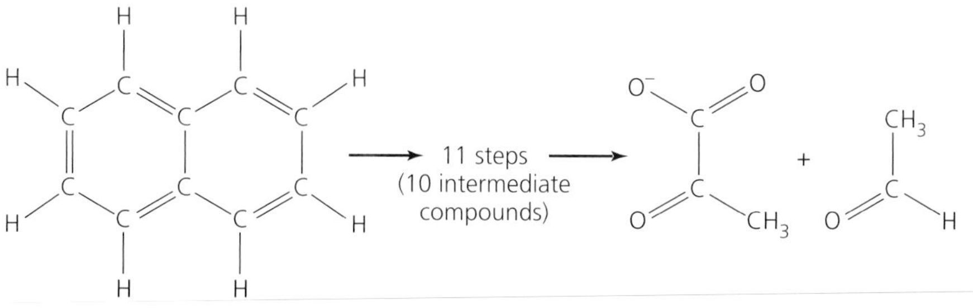

Naphthalene ($C_{10}H_8$) Pyruvate ($C_3H_3O_3^-$) Acetaldehyde (C_2H_4O)

FIGURE 4.9 Bacterial degradation of naphthalene involves 11 steps and results in the simpler organic compounds pyruvate and acetaldehyde. The many chemicals involved in this complicated process are omitted from this simplified diagram.

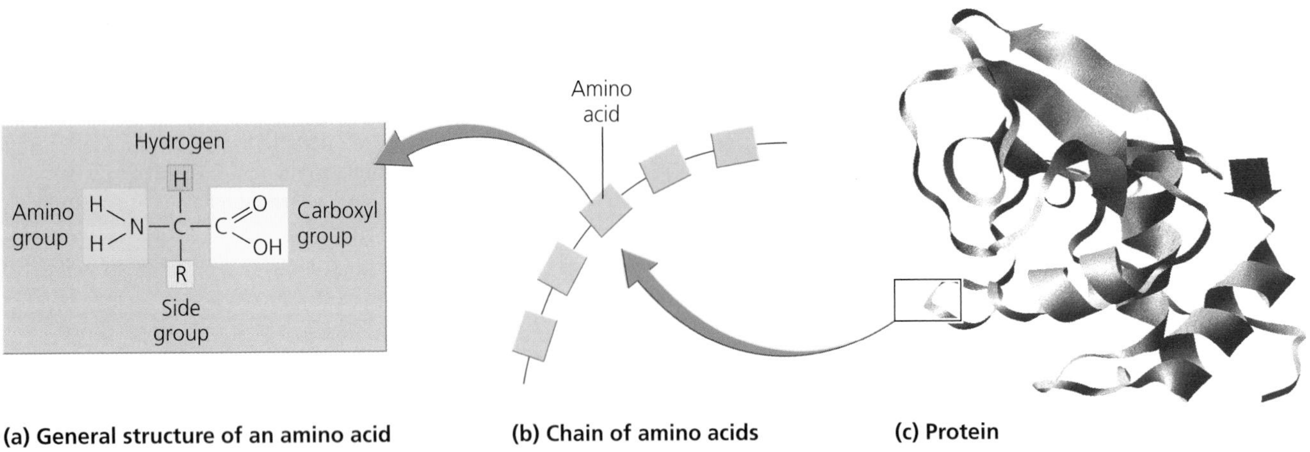

FIGURE 4.10 Proteins are polymers vital for life. They are made up of long chains of amino acids (**a, b**) and fold up into complex convoluted shapes (**c**) that help determine their functions.

Proteins serve many functions. Some help produce tissues and provide structural support for the organism. For example, animals use proteins to generate skin, hair, muscles, and tendons. Some proteins help store energy, and others transport substances. Some function as components of the immune system, defending the organism against foreign attackers. Still others act as hormones, molecules that serve as chemical messengers within an organism. Finally, proteins can serve as enzymes, molecules that catalyze, or promote, certain chemical reactions. For example, bacteria used for bioremediation use specialized enzymes to break down hydrocarbons, just as we use enzymes to digest our food.

Nucleic acids Protein production is directed by **nucleic acids.** The two nucleic acids—**deoxyribonucleic acid (DNA)** and **ribonucleic acid (RNA)**—carry the hereditary information for organisms and are responsible for passing traits from parents to offspring. Nucleic acids are composed of series of nucleotides, each of which contains a sugar molecule, a phosphate group, and a nitrogenous base (Figure 4.11a). DNA includes four types of nucleotides, each with a different nitrogenous base: adenine (A), guanine (G), cytosine (C), and thymine (T). RNA is similar to DNA in structure, except that its sugar group is ribose (instead of deoxyribose), thymine is replaced by uracil (U), and RNA is generally single-stranded whereas DNA is double-stranded. Within DNA or RNA, nucleotides are linked together to form extremely long chains, with a sugar and phosphate backbone and nitrogenous base pairs. Adenine (A) pairs with thymine (T), and cytosine (C) pairs with guanine (G). In DNA, the paired base chains can be pictured as rungs of a ladder, with the ladder twisted into a spiral, giving the entire molecule a shape called a double helix (Figure 4.11b).

In the process of *transcription,* the hereditary information in the nucleotide sequence of DNA is rewritten to a molecule of RNA. Then, during the process of *translation,* RNA directs the order in which amino acids assemble to build proteins (Figure 4.12). Proteins go on to influence the structure and maintenance of the organism. Genetic information from DNA is passed from one generation to another as the strands replicate during cell division

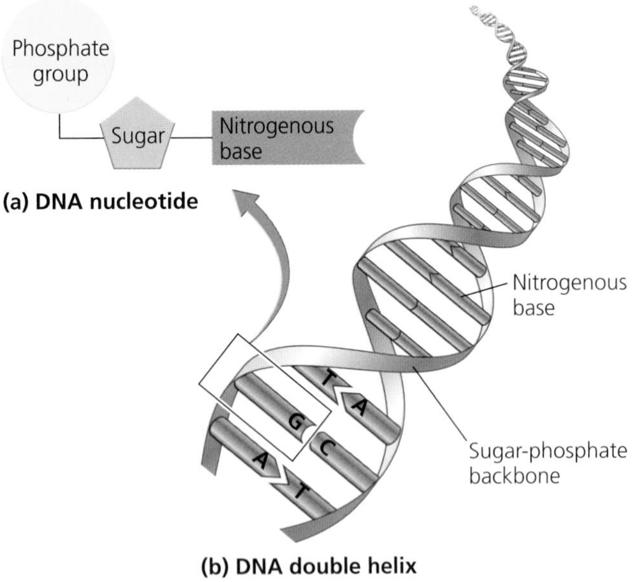

FIGURE 4.11 DNA is the molecule that carries genetic information from parent to offspring across the generations. The information is coded in the sequence of nucleotides (**a**), small molecules that pair together like rungs of a ladder that twist into the shape of a double helix (**b**).

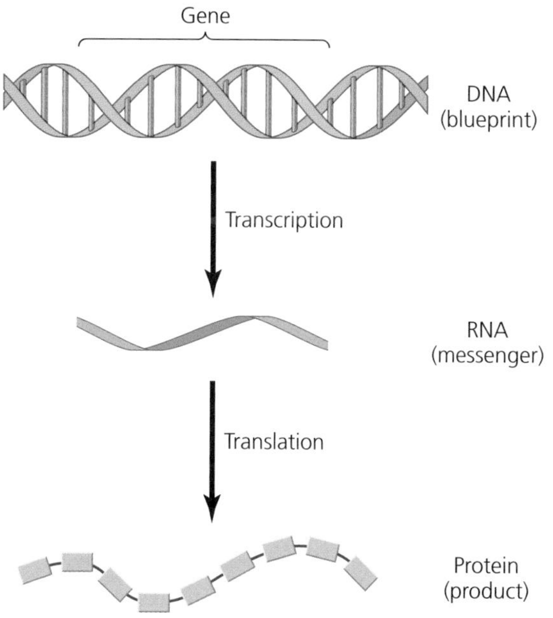

FIGURE 4.12 DNA serves as the blueprint for synthesizing the proteins that help build and maintain organisms. The genetic information in DNA (functional regions of which are called *genes*) instructs the creation of RNA through a process called *transcription*. Messenger RNA then directs the synthesis of proteins through a process called *translation*.

and egg or sperm formation. Regions of DNA coding for particular proteins that perform particular functions are called **genes.** In most organisms, the *genome*—the set of all an organism's genes—is divided into chromosomes. Different types of organisms have different numbers of genes and chromosomes. Most bacteria have a single circular chromosome, for instance, whereas humans have 46 linear ones.

Carbohydrates **Carbohydrates** constitute a third class of biologically vital polymer. These organic compounds consist of atoms of carbon, hydrogen, and oxygen (Figure 4.13). Simple carbohydrates, called sugars or monosaccharides, have structures, or skeletons, that are three to seven carbon atoms long, and formulas that are some multiple of CH_2O. Glucose ($C_6H_{12}O_6$) is one of the most common and important sugars, providing energy that fuels plant and animal cells. Glucose also serves as a building block for complex carbohydrates, or polysaccharides. Plants use starch, a glucose-based polysaccharide, to store energy, and animals eat plants to acquire starch.

In addition, both plants and animals use complex carbohydrates to build structure. Insects and crustaceans form hard shells from the carbohydrate chitin. Cellulose, the most abundant organic compound on Earth, is a complex carbohydrate found in the cell walls of leaves, bark, stems, and roots. Cellulose, like starch, is composed of glucose molecules, bound together in a different way.

Lipids A fourth type of macromolecule includes a chemically diverse group of compounds that are classified together because they do not dissolve in water. These **lipids** include fats, phospholipids, waxes, and steroids:

- *Fats and oils* are convenient forms of energy storage, especially for mobile animals. Their hydrocarbon structures somewhat resemble gasoline, a similarity echoed in their function: to effectively store energy and release it when burned.
- *Phospholipids* are similar to fats but consist of one water-repellant side and one water-attracting side. This characteristic allows them, when arranged in a double

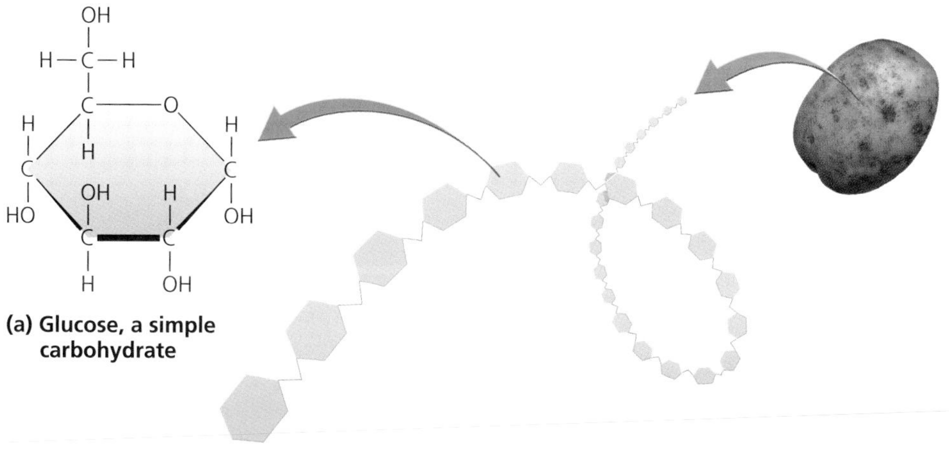

(a) Glucose, a simple carbohydrate

(b) Starch, a polysaccharide

FIGURE 4.13 The monosaccharide glucose (**a**) is the simplest and most abundant carbohydrate and is a vital energy source for organisms. Linked glucose molecules form starch (**b**), an important polysaccharide.

layer, to make up the primary component of animal cell membranes.

- *Waxes* are lipids that are digestible by some but not all organisms, and they can play structural roles (for instance, beeswax in bees' hives).
- *Steroids* are used in animal cell membranes and in production of hormones, including the sex hormones estrogen and androgen, vital to sexual maturation.

Organisms use cells to compartmentalize macromolecules

All living things are composed of **cells,** the most basic unit of organismal organization. Organisms range in complexity from single-celled bacteria to plants and animals that contain millions of cells. Cells vary greatly in size, shape, and function.

Biologists classify organisms into two groups based on the structure of their cells. **Eukaryotes** include plants, animals, fungi, and protists. The cells of **eukaryotes** (Figure 4.14a) consist of an outer membrane of lipids and an inner fluid-filled chamber containing **organelles,** internal structures that perform specific functions. These internal structures include (among others) ribosomes, which are organelles that synthesize proteins, and mitochondria, where the last step in the extraction of energy from sugars and fats occurs. Eukaryotes also have within each of their cells a membrane-enclosed nucleus that houses DNA. Eukaryotic organisms generally have many cells.

Prokaryotic organisms are much simpler. Prokaryotes are generally single-celled, and their cells lack membrane-bound organelles and a nucleus (Figure 4.14b). All bacteria are prokaryotes, as are the lesser-known microorganisms called archaea. Bacteria are diverse and are ubiquitous in the environment, and of course they do far more than attack oil spills. Many types of bacteria perform functions

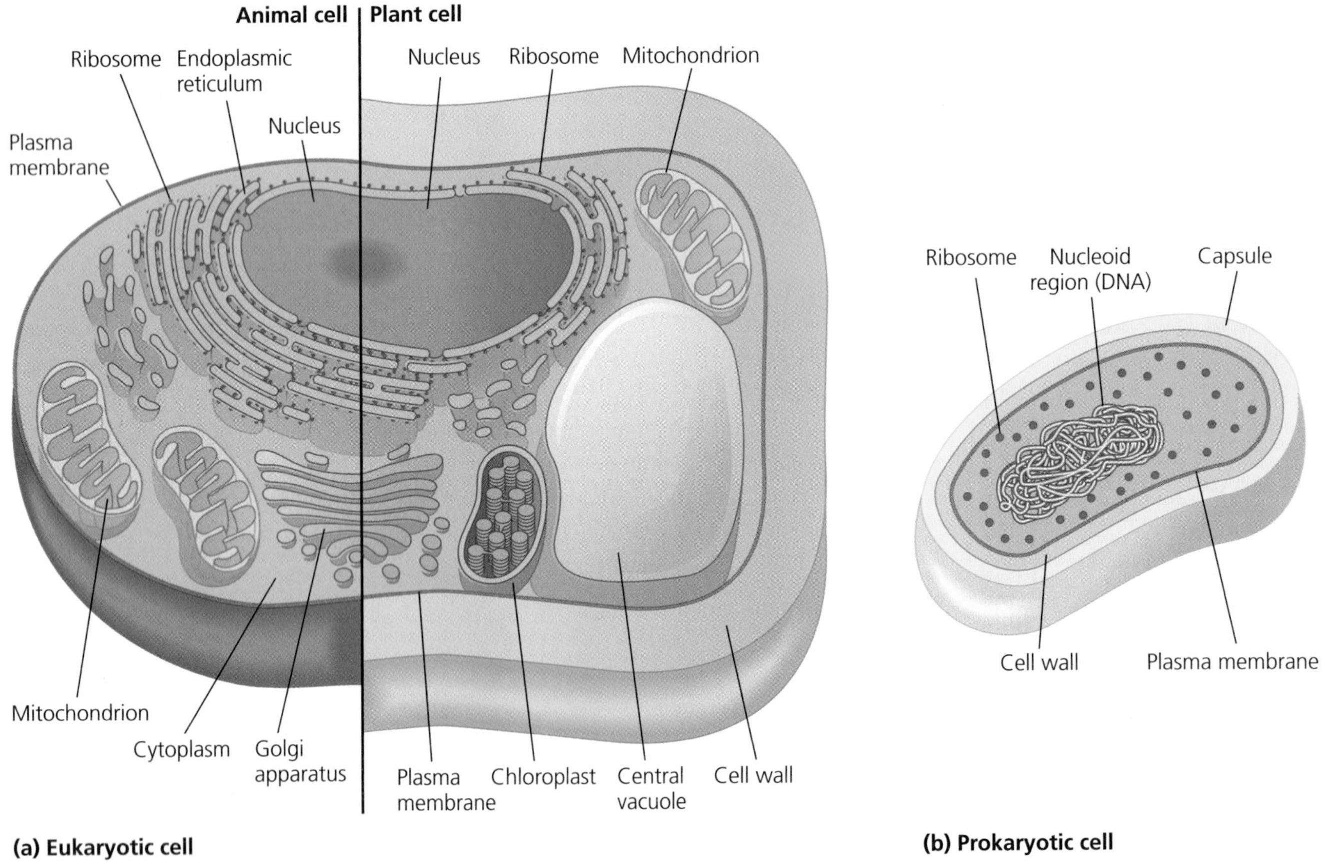

FIGURE 4.14 Cells are the smallest unit of life that can function independently. Eukaryotic cells (**a**) contain organelles, such as mitochondria and chloroplasts, as well as a membrane-enclosed nucleus that contains DNA. Plant cells (right) have rigid cell walls of cellulose, whereas animal cells (left) have more flexible cell membranes. Prokaryotic cells (**b**) are simpler, lacking membrane-bound organelles and an enclosed nucleus.

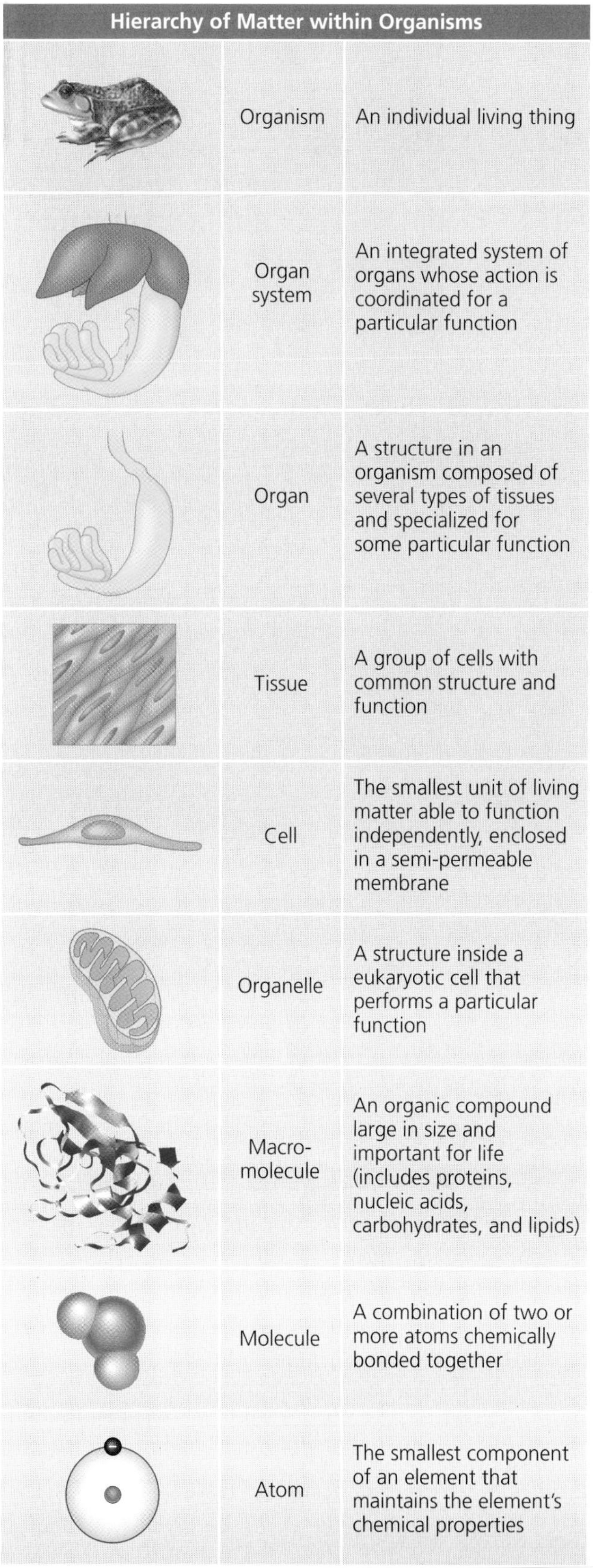

Hierarchy of Matter within Organisms
Organism
An individual living thing
Organ system
An integrated system of organs whose action is coordinated for a particular function
Organ
A structure in an organism composed of several types of tissues and specialized for some particular function
Tissue
A group of cells with common structure and function
Cell
The smallest unit of living matter able to function independently, enclosed in a semi-permeable membrane
Organelle
A structure inside a eukaryotic cell that performs a particular function
Macro-molecule
An organic compound large in size and important for life (includes proteins, nucleic acids, carbohydrates, and lipids)
Molecule
A combination of two or more atoms chemically bonded together
Atom
The smallest component of an element that maintains the element's chemical properties

FIGURE 4.15 Within an organism, we can view matter as being organized in a hierarchy of levels. Atoms (with their protons, neutrons, and electrons) form the base of the hierarchy, and they build the next level, molecules. Macromolecules make up portions of cells, including (in eukaryotes) cell organelles. Cells of similar types function together in tissues, tissues make up organs, and organs make up organ systems, which in turn collectively comprise the organism.

vital to human life—for instance, aiding in digestion and preventing the buildup of harmful wastes.

In eukaryotes, cells specialize in different roles and are organized into collections of cells performing the same function, called *tissues.* Tissues make up *organs,* and organisms are composed of *organ systems.* We have now completed a (very quick!) review of the hierarchy in which matter is organized in living things on Earth (Figure 4.15). Over the next three chapters, we will explore the levels of this hierarchy above the organismal level, as we study the science of ecology. But first we will examine energy, something that underlies every process in environmental science.

Energy Fundamentals

Creating and maintaining organized complexity, whether of a cell or an organism or an ecological system, requires energy. Energy is needed to power the geological forces that shape our planet, to organize matter into complex forms such as biological polymers, to build and maintain cellular structure, and to power the interactions that take place among species. Indeed, energy is somehow involved in nearly every biological, chemical, and physical event.

But what, exactly, is energy? An intangible phenomenon, **energy** is that which can change the position, physical composition, or temperature of matter. A sparrow in flight expends energy to propel its body through the air. When the sparrow lays an egg, its body uses energy to create the calcium-based eggshell and color it with pigment. The sparrow sitting on its nest transfers energy from its body in heating the developing chicks inside its eggs. Some of the most dramatic releases of energy in nature do not involve living things; think of volcanoes erupting or tornadoes sweeping across the plains.

Scientists differentiate between two types of energy: **potential energy,** energy of position; and **kinetic energy,** energy of motion. Consider river water held behind a dam. By preventing water from moving downstream, the dam causes the water to accumulate potential energy. When the dam gates are opened, the potential energy is

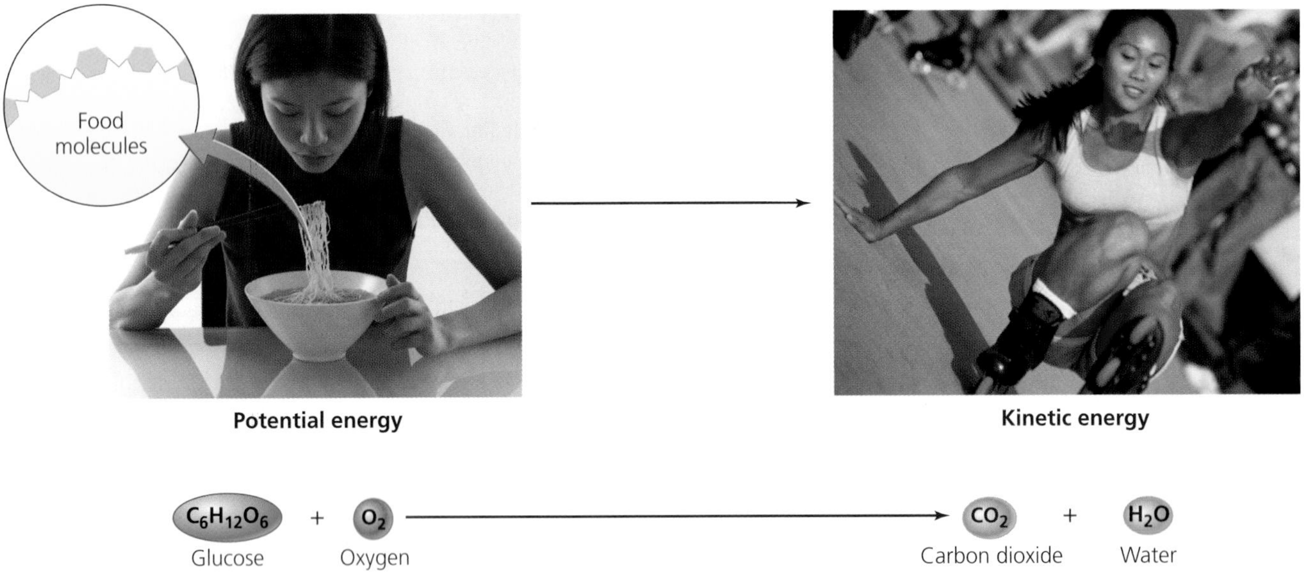

FIGURE 4.16 Energy is released when potential energy is converted to kinetic energy. Potential energy stored in sugars, such as glucose, in the food we eat **(a)**, combined with oxygen, becomes kinetic energy when we exercise **(b)**, releasing carbon dioxide and water as by-products.

converted to kinetic energy, in the form of water's motion as it rushes downstream.

Such energy transfers take place at the atomic level every time a chemical bond is broken or formed. **Chemical energy** is potential energy held in the bonds between atoms. Bonds differ in their amounts of chemical energy, depending on the atoms they hold together. Converting a molecule with high-energy bonds (such as the carbon–carbon bonds of petroleum products) into molecules with lower-energy bonds (such as the bonds in water or carbon dioxide) releases energy by changing potential energy into kinetic energy and produces motion, action, or heat. Just as our automobile engines split the hydrocarbons of gasoline to release chemical energy and generate movement, our bodies split glucose molecules in our food for the same purpose (Figure 4.16).

Energy is always conserved . . .

Although energy can change from one form to another, it cannot be created or lost. The total energy in the universe remains constant and thus is said to be conserved. Scientists have dubbed this principle the **first law of thermodynamics.** The potential energy of the water behind a dam will equal the kinetic energy of its eventual movement down the riverbed. Similarly, burning converts the potential energy in a log of firewood to an equal amount of energy produced as heat and light. We obtain energy from the food we eat, which we expend in exercise, put toward the body's maintenance, or store as fat. We do not somehow create additional energy or end up with less than the food gives us. Any individual system can temporarily increase or decrease in energy, but the total amount in the universe remains constant.

. . . But energy changes in quality

Although the first law of thermodynamics requires that the overall amount of energy be conserved in any process of energy transfer, the **second law of thermodynamics** states that the nature of energy will change from a more-ordered state to a less-ordered state, if no force counteracts this tendency. The degree of disorder in a substance, system, or process is called **entropy,** and the second law of thermodynamics holds that systems tend to move toward increasing entropy. For instance, after death every organism undergoes decomposition and loses its structure. A log of firewood—the highly organized and structurally complex product of many years of slow tree growth—transforms in the campfire to a residue of carbon ash, smoke, and gases such as carbon dioxide and water vapor, as well as the light and the heat of the flame (Figure 4.17). With the help of oxygen, the complex biological polymers making up the wood are converted into a disorganized assortment of rudimentary molecules and heat and light energy.

The nature of any given energy source helps determine how easily humans can harness it. Sources such as

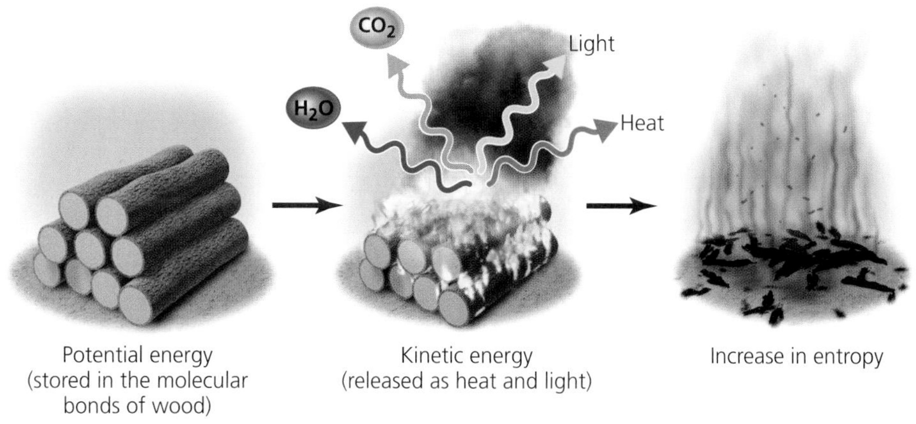

FIGURE 4.17 The burning of firewood demonstrates energy transfer leading from a more-ordered to a less-ordered state. This increase in entropy reflects the second law of thermodynamics.

petroleum products and high-voltage electricity contain concentrated energy that is easily released. It is relatively easy for us to gain large amounts of energy efficiently from these high-quality sources. In contrast, sunlight and the heat stored in ocean water are considered low-quality energy sources. Each and every day the world's oceans absorb heat energy from the sun equivalent to that of 250 billion barrels of oil—more than 3,000 times as much as our global society uses in a year. But because this energy is spread out across such vast spaces, it is diffuse and difficult to harness.

Weighing the Issues: Energy Quality and Energy Policy

Contrast the ease of harnessing high-quality energy, such as that of petroleum, with the ease of harnessing low-quality energy, such as that of heat from the oceans. How do you think this difference has affected our society's energy policy and energy sources through the years?

In every transfer of energy, some portion usable to us is lost. The inefficiency of some of the most common energy conversions that power our society can be surprising. When we burn gasoline in an automobile engine, only about 16% of the energy released is used to power the automobile's movement. The rest of the energy is converted to heat. Incandescent light bulbs are worse; only 5% of their energy is converted to the light that we use them for, while the rest escapes as heat. Viewed in this context, the 15% efficiency of much current solar energy technology does not look bad at all.

Although the second law of thermodynamics specifies that systems tend to move toward disorder, the order of an object or system can be increased through the input of additional energy from outside the system. This is precisely what living organisms do. Organisms maintain their structure and function by consuming energy. They represent a constant struggle to maintain order and combat the natural tendency toward disorder.

Light energy from the sun powers most living systems

The energy that powers Earth's organisms and that flows through ecological systems comes primarily from the sun. The sun releases radiation from large portions of the electromagnetic spectrum, although our atmosphere filters much of this out, and we can see only some of this radiation as visible light (Figure 4.18). Most of the sun's energy is reflected, or else absorbed and re-emitted, by the atmosphere, land, or water. Solar energy drives our weather and climate patterns, including winds and ocean currents. A small amount (less than 1% of the total) powers plant growth, and a still smaller amount flows from plants into the organisms that eat them and the organisms that decompose dead organic matter. A minuscule amount of energy, relatively speaking, is eventually deposited below ground in the form of the chemical bonds in fossil fuels.

The sun's light energy is used directly by some organisms to produce their own food. Such organisms, called **autotrophs** or primary **producers,** include green plants, algae, and cyanobacteria. (Cyanobacteria are a type of bacteria named for their characteristic blue-green, or cyan, color.) Autotrophs turn light energy from the sun into chemical energy in a process called photosynthesis (Figure 4.19). In **photosynthesis,** sunlight powers a series of chemical reactions that convert carbon dioxide and water into sugars, transforming low-quality energy from the sun into high-quality energy the

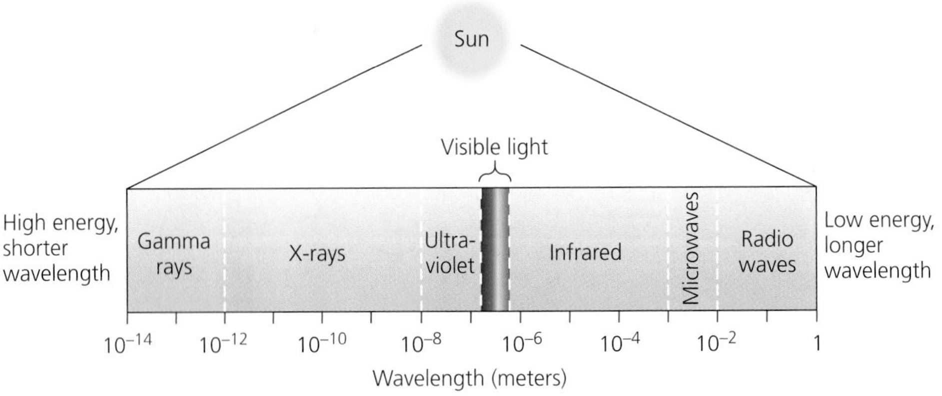

FIGURE 4.18 The sun emits radiation from many portions of the electromagnetic spectrum, and visible light makes up only a small proportion of this energy. Some radiation that reaches our planet is reflected back, some is absorbed by air, land, and water, and a small amount powers photosynthesis.

organism can use. It is an example of moving toward a state of lower entropy, and as such it requires a substantial input of outside energy.

Photosynthesis produces food for plants and animals

Photosynthesis occurs within cell organelles called *chloroplasts,* where the light-absorbing pigment *chlorophyll* (which is what makes plants green) uses solar energy to initiate a series of chemical reactions called *light-dependent reactions.* During these reactions, water molecules are split, and they react to form hydrogen ions (H^+) and molecular oxygen (O_2), thus creating the oxygen that we breathe. The light-dependent reactions also produce small, high-energy molecules that are used to fuel a set of *light-independent reactions.* In these reactions, carbon atoms from carbon dioxide are linked together to manufacture sugars. Photosynthesis is a complex process, but the overall reaction can be summarized with the following equation:

$$6CO_2 + 12H_2O + \text{energy from the sun} \longrightarrow C_6H_{12}O_6 \text{ (sugar)} + 6O_2 + 6H_2O$$

The numbers preceding each molecular formula indicate how many of each molecule are involved in the reaction. Note that the sum of the numbers on each side of the equation for each element are equal; that is, there are 6 C, 24 H, and 24 O on each side. This illustrates how chemical equations are balanced, with each atom recycled and matter conserved. No atoms are lost; they are simply rearranged among molecules. Note also that water appears on both sides of the equation. The reason is that for every 12 water molecules that are input and dissociated in the process, 6 water molecules are newly created. We can streamline the photosynthesis equation by showing only the net loss of 6 water molecules:

$$6CO_2 + 6H_2O + \text{energy from the sun} \longrightarrow C_6H_{12}O_6 \text{ (sugar)} + 6O_2$$

Thus in photosynthesis, water, carbon dioxide, and light energy from the sun are transformed to produce sugar (glucose) and oxygen. To accomplish this, green plants draw up water from the ground through their roots, suck in carbon dioxide from the air through their

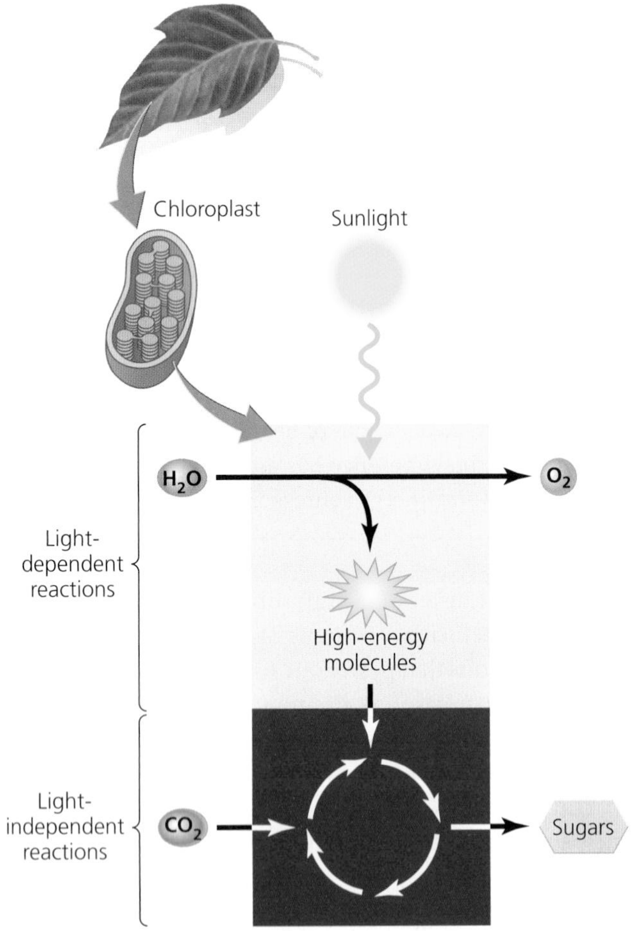

FIGURE 4.19 Autotrophs including plants, algae, and cyanobacteria use sunlight to convert carbon dioxide and water into sugars and oxygen in photosynthesis. Autotrophs provide themselves and the many heterotrophs that eat them with energy for life.

leaves, and harness sunlight. With these ingredients, they create sugars for their growth and maintenance, and release oxygen as a by-product. Animals, in turn, depend on these two outputs of photosynthesis. Animals survive by eating plants, or by eating animals that have eaten plants, and by taking in oxygen. In fact, it is thought that animals appeared on Earth's surface only after the planet's atmosphere had been supplied with oxygen by the earliest autotrophs, the cyanobacteria.

Cellular respiration releases chemical energy

The chemical energy created by photosynthesis can later be used by organisms in a process called **cellular respiration.** To release the chemical energy of glucose, cells use the reactivity of oxygen to convert glucose back into its original starting materials, water and carbon dioxide. The energy released during this process is used to form chemical bonds or to perform other tasks within cells. The net equation for cellular respiration is the exact opposite of that for photosynthesis:

$$C_6H_{12}O_6 \text{ (sugar)} + 6O_2 \rightarrow 6CO_2 + 6H_2O + \text{energy}$$

However, the energy gained per glucose molecule in respiration is only two-thirds of the energy input per glucose molecule in photosynthesis—a prime example of the second law of thermodynamics in action. The extraction of energy from glucose through respiration occurs in the autotrophs that created the glucose and also in the animals that gain glucose by consuming autotrophs. Organisms that consume autotrophs are called **consumers,** or **heterotrophs** (*hetero* means "different"), and include most animals, as well as the fungi and microbes that decompose organic matter. In most ecological systems, plants, algae, or cyanobacteria form the base of a food chain through which energy passes to heterotrophs (▸ pp. 156–158).

Geothermal energy also powers Earth's systems

Although the sun is Earth's primary power source, it is not the only one. A minor additional source is the gravitational pull of the moon, which causes ocean tides. This low-quality energy is weak and diffuse in comparison to solar energy. A more significant additional energy source is radiation emanating from inside Earth, powered by radioactivity. When we think of radioactivity, nuclear power plants and atomic weapons may come to mind, but radioactivity is a natural phenomenon. As discussed earlier (▸ p. 93), it consists of the gradual release of high-energy

FIGURE 4.20 This geyser in the Black Rock Desert of Nevada propels scalding water into the air, powered by geothermal energy from deep below ground. The bright colors of the rocks are from colonies of bacteria that thrive in the hot mineral-laden water.

rays or particles by radioisotopes as their nuclei decay. Radiation from radioisotopes deep inside Earth heats the inside of the planet, and this heat gradually makes its way to the surface. There it drives plate tectonics (▸ pp. 207–209), heats magma that erupts from volcanoes, and warms groundwater, which in some regions of the planet shoots out of the ground in the form of geysers (Figure 4.20). Called **geothermal energy,** this heating from deep within the planet is now being harnessed for commercial power (▸ pp. 633–635).

Long before humans came along, however, geothermal energy was powering other biological communities. On the floor of the ocean, jets of geothermally heated water—essentially underwater geysers—gush into the icy-cold depths. One of the amazing scientific discoveries of recent decades was the realization that these *hydrothermal vents* can host entire communities of organisms that thrive in the extreme high-temperature, high-pressure conditions. Gigantic clams, immense tubeworms, and odd mussels, shrimps, crabs, and fish all flourish in the seemingly hostile environment near scalding water that shoots out of tall chimneys of encrusted minerals (Figure 4.21).

These locations are so deep underwater that they completely lack sunlight, so their communities cannot fuel themselves through photosynthesis. Instead, bacteria in deep-sea vents use the chemical-bond energy of hydrogen sulfide (H_2S) to transform inorganic carbon into organic carbon compounds in a process called **chemosynthesis:**

$$6CO_2 + 6H_2O + 3H_2S \rightarrow C_6H_{12}O_6 \text{ (sugar)} + 3\ H_2SO_4$$

(a) Hydrothermal vent

(b) Giant tubeworms

FIGURE 4.21 Hydrothermal vents on the ocean floor (**a**) send spouts of hot mineral-rich water into the cold blackness of the deep sea. Amazingly, specialized biological communities thrive in these unusual conditions. Odd creatures such as these giant tubeworms (**b**) survive thanks to bacteria that produce food from hydrogen sulfide.

There are many different types of chemosynthesis, but note how this particular reaction for chemosynthesis closely resembles the photosynthesis reaction. These two processes use different energy sources, but each uses water and carbon dioxide to produce sugar and a by-product, and each produces potential energy that is later released during respiration. Energy from chemosynthesis passes through the deep-sea-vent animal community as heterotrophs such as clams, mussels, and shrimp gain nutrition from chemoautotrophic bacteria. Hydrothermal vent communities excited scientists not only because they were novel and unexpected, but also because some researchers believe they may help us understand how life itself originated.

The Origin of Life

How and where life originated is one of the most centrally important—and intensely debated—questions in modern science. In searching for the answer, scientists have learned a great deal about the history of life on Earth and about what early Earth was like.

Early Earth was a very different place

Earth formed about 4.5 billion years ago in the same way as the other planets of our solar system; dispersed bits of material whirling through space around our sun were drawn by gravity into one another, coalescing into a series of spheres. For several hundred million years after the planets formed, there remained enough stray material in the solar system that Earth and the other young planets were regularly bombarded by large chunks of debris in the form of asteroids, meteorites, and comets. The largest impacts were probably so explosive that they vaporized our planet's newly formed oceans. Add to this the severe volcanic and tectonic activity and the intense ultraviolet radiation, and it is clear that early Earth was a pretty hostile place (Figure 4.22). Any life that got under way during this "bombardment stage" might easily have been killed off. Only after most debris was cleared from the solar system was life able to gain a foothold and keep it.

FIGURE 4.22 The young Earth on which life originated was a very different place from our planet today. Microbial life first evolved amid sulfur-spewing volcanoes, intense ultraviolet radiation, frequent extraterrestrial impacts, and an atmosphere containing ammonia.

Earth's early atmosphere was probably very different from our atmosphere today. For instance, oxygen was largely lacking until photosynthesizing microbes started producing it. Whereas today's atmosphere is dominated by nitrogen and oxygen (see Table 4.1), many scientists in the 20th century thought Earth's early atmosphere contained large amounts of hydrogen, ammonia (NH_3), methane, and water vapor. More recently, opinion has shifted, and most scientists now infer that the atmosphere contained less hydrogen but more carbon dioxide, nitrogen, and carbon monoxide (CO) than previously thought.

Several hypotheses have been proposed to explain life's origin

Most biochemists interested in life's origin think that life must have begun when inorganic chemicals linked themselves into small molecules and formed organic compounds. Some of these compounds gained the ability to replicate, or reproduce themselves, whereas others found ways to group together into proto-cells. There is much debate and ongoing research, however, on the details of this process, especially concerning the location of the first chemical reactions and the energy source(s) that powered them.

Primordial soup: The heterotrophic hypothesis In the 1930s, scientists J. B. S. Haldane and Aleksandr Oparin independently advanced the idea that life evolved from a primordial soup of simple inorganic chemicals—carbon dioxide, oxygen, and nitrogen—dissolved in the ocean's surface waters or tidal shallows. They suggested how simple amino acids might have formed under these conditions and how more complex organic compounds could have eventually followed, including simple ribonucleic acids that could replicate themselves. This hypothesis is termed *heterotrophic* because it proposes that the first life forms used organic compounds from their environment as an energy source.

This hypothesis has traditionally been favored, and lab experiments have provided evidence that the proposed process can work. In 1953, biochemists Stanley Miller and Harold Urey passed electricity through a mixture of water vapor, hydrogen, ammonia, and methane, which was believed at that time to represent the early atmosphere, and were readily able to produce impressive amounts of organic compounds, including amino acids. Subsequent experiments confirmed Miller's and Urey's findings, but now that most scientists believe early atmospheric conditions were different, this hypothesis seems somewhat less likely to represent what actually happened.

Seeds from space: The extraterrestrial hypothesis A modification of the heterotrophic hypothesis proposes that early chemical reactions on Earth may have received help from outer space. In the early 1900s, Swedish chemist Svante Arrhenius proposed that microbes from space might have traveled on meteorites that crashed to Earth, seeding our planet with life. Scientists largely rejected this idea, also called the *panspermia hypothesis*, believing that even if amino acids or bacteria were to exist in space, they could not be transported to Earth because the high temperatures that comets and meteors attain as they enter our atmosphere should destroy all biological compounds.

However, the Murchison meteorite, which fell in Australia in 1969, was found to contain many amino acids, suggesting that amino acids within rock can survive impact. Since then, experiments simulating impact conditions have shown that organic compounds and some bacteria can withstand a surprising amount of abuse. Furthermore, planetary scientists have shown that large asteroid impacts on one planet (such as Mars) can throw up so much material that some eventually may make its way to other planets (such as Earth). And recent astrobiology research has made a case that comets have brought large amounts of water, and possibly organic compounds, to Earth throughout its history. As a result of such findings, long-distance travel of microbes through space and into our atmosphere now seems more plausible than previously thought.

Life from the depths: The chemoautotrophic hypothesis In the 1970s and 1980s, several scientists, among them Jack Corliss, a discoverer of deep-sea hydrothermal vents, proposed that life may have emanated from the deep sea. The chemoautotrophic hypothesis proposes that life originated at scalding-hot deep-sea vent systems, where sulfur was abundant. In this scenario, the first organisms were chemoautotrophs, creating their own food from hydrogen sulfide.

Genetic analysis of the relationships of present-day organisms (p. 111) suggests that some of the most ancient ancestors of today's life forms lived in extremely hot, wet environments. Additionally, the extreme heat of hydrothermal vents could act to speed up chemical reactions that link atoms together into long molecules, a necessary early step in life's formation. Scientists have shown experimentally that it is possible to form amino acids and begin a chain of steps that might potentially lead to the formation of life under high-temperature, high-pressure conditions similar to those of hydrothermal vents.

Weighing the Issues: Hypotheses on Life's Origin

Which lines of evidence in the debate over the origin of life strike you as the most convincing, and why? Which strike you as the least convincing, and why? Can you think of any further scientific research that could be done to address the question of how life originated?

Self-replication and cell formation were crucial steps

Whether they formed in deep-sea vents, tidal pools, or comet craters, early organic polymers had to develop the ability to self-replicate before life as we know it could commence. This involves a chicken-or-egg paradox. To replicate biological polymers, nucleic acids, typically DNA, are needed—but to create DNA, enzymes, a type of biological polymer, are required. In order to get one, you would seem to need the other. This is why scientists think the first self-replicating molecule may have been RNA, because RNA can act as both enzyme and information carrier. Early RNA molecules may have acted as their own enzymes, replicating without the help of other polymers. Over time, scientists hypothesize, the replicating process of early RNA was modified, and DNA, which is more stable, came to dominate as the information carrier in modern cells.

The formation of cells must have been another key step in life's origin. Cells have semipermeable membranes and/or walls, which allow them to maintain an internal environment different from their surroundings. Chance aggregations of polymers may have formed cell-like structures, and structures that happened to contain RNA may have replicated. Such cell-like structures have formed spontaneously in lab experiments, have been able to divide, and—if enzymes are provided—have been able to absorb substances from their surroundings, as cells do.

The fossil record has taught us much about life's history

The earliest evidence of life on Earth comes from rocks about 3.5 billion years old. Although these earliest traces are controversial, there is ample evidence that simple forms of life, such as single-celled bacteria, were present on Earth well over 3 billion years ago. Remains of these microscopic life forms (or their chemical by-products) have been preserved just as have later, much larger, creatures, such as dinosaurs—by fossilization.

As organisms die, some are buried by sediment. Under the right conditions, the hard parts of their bodies—such as bones, shells, and teeth—may be preserved as the sediments are compressed into rock (▸ pp. 195–196). Minerals replace the organic material, leaving behind a **fossil,** an imprint in stone of the dead organism (Figure 4.23). In thousands of locations around the world, geological processes over millions of years have buried sedimentary rock layers and later brought them to the surface, revealing assemblages of fossils representing plants and animals from different time periods. The cumulative body of fossils worldwide is known as the **fossil record.** Paleontologists study the fossil record to infer the history of past life on Earth.

The fossil record clearly shows that:

- The species living today are but a tiny fraction of all the species that ever lived; the vast majority of Earth's species are long extinct.

FIGURE 4.23 The fossil record has helped reveal the history of life on Earth. The numerous fossils of trilobites suggest that these animals, now extinct, were abundant in the oceans from roughly 540 million to 250 million years ago.

- Earlier types of organisms changed, or evolved, into later ones.
- The number of species existing at any one time has increased through history.
- There have been several episodes of *mass extinction,* or simultaneous loss of great numbers of species (▸ pp. 127–128).
- Many organisms present early in history were smaller and simpler than modern organisms.

The fossil record also tells us that for most of life's history, microbes, like the bacteria that consume hydrocarbons or the cyanobacteria that produce oxygen, were the only life on Earth. It was not until about 600 million years ago that large and complex organisms such as animals, land plants, and fungi appeared.

The crude oil with which we began this chapter is itself a kind of fossil. Plant and animal matter can be preserved as they sink to the seafloor and are buried in the absence of oxygen; eventually they become compressed and turn into the amorphous mixes of hydrocarbons we call fossil

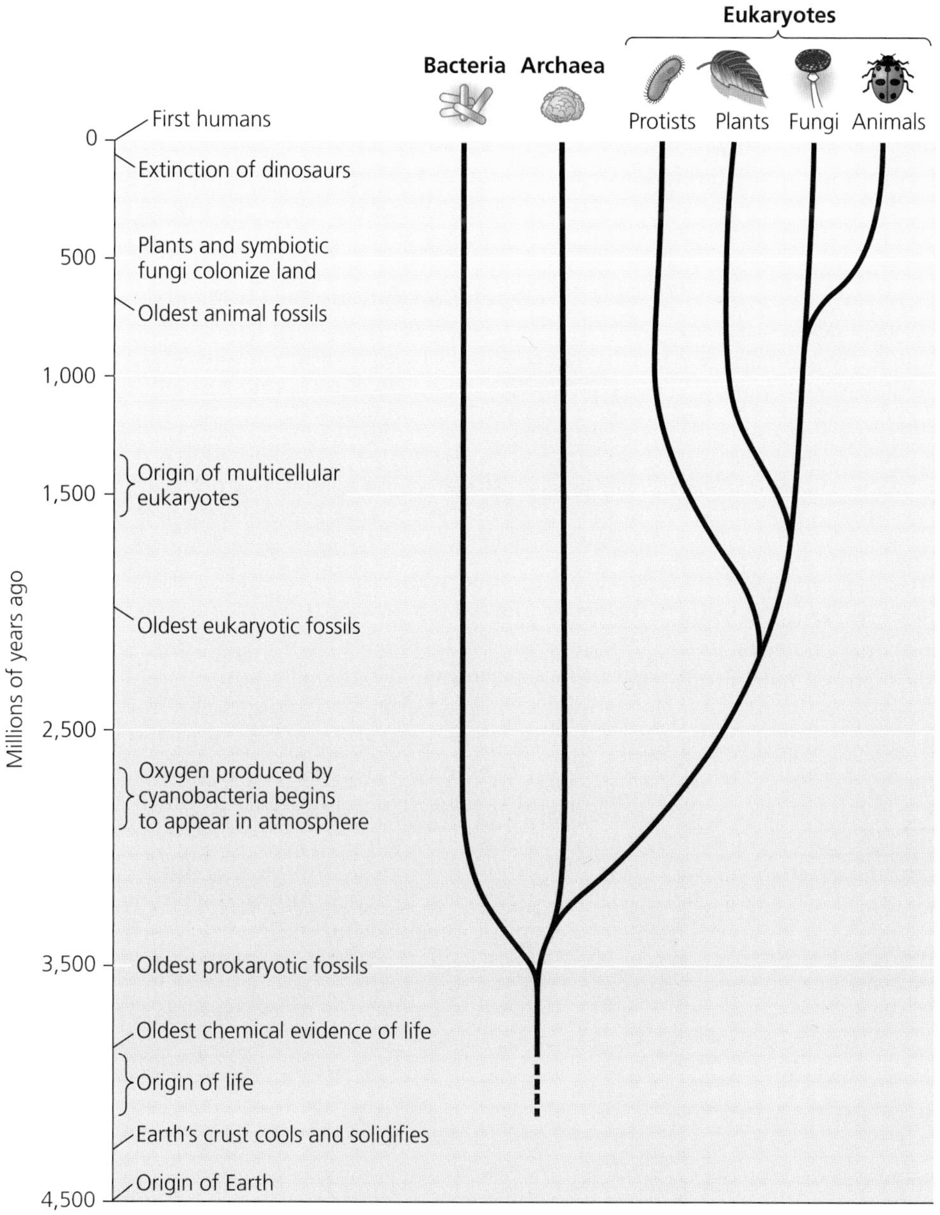

FIGURE 4.24 Microbes have lived on the planet for most of its 4.5-billion-year history. Large complex eukaryotes such as vertebrates, in contrast, are relative newcomers. The fossil record and the analysis of present-day organisms and their genes allow scientists to reconstruct evolutionary relationships among organisms and to build a "tree of life." As you progress upward from the trunk of the tree to the tips of its branches, you are moving forward in time. Each fork denotes the divergence of major groups of organisms, each group of which in this greatly simplified diagram includes many thousands of species. The tree of life as understood by scientists today consists of three main groups—the bacteria, the recently discovered archaea, and the diverse eukaryotes. Protists and ancestral eukaryotes are poorly known, and their future study may well produce discoveries that further revise our understanding of life's history.

fuels (▸ p. 559). Coal, oil, and natural gas are the fossil fuels we use to power our civilization. When we drive a car, ignite a stove, or flick on a light switch, we are using energy from life buried millions of years ago.

Present-day organisms and their genes also help us decipher life's history

Besides fossils, biologists also use present-day organisms to infer how evolution proceeded in the past. By comparing the genes and/or the external characteristics of organisms, scientists can create branching trees, similar to family genealogies, that show the relationships among organisms and thus their history of divergence through time. As you follow such a tree from its trunk to the tips of its branches, you proceed forward through time, tracing the history of life. A major advance was made in recent years as scientists discovered an entire new domain of life, the archaea, single-celled prokaryotes that are genetically very different from bacteria. Today most biologists view the tree of life as a three-pronged edifice consisting of the bacteria, the archaea, and the eukaryotes (Figure 4.24). We will examine the flowering of the diversity of life on our planet in Chapter 5 and further in Chapter 11. The relationships among organisms, and those between organisms and their environments, form the basis for ecology, a discipline of primary importance to environmental science.

VIEWPOINTS

Bioremediation

Naturally occurring microbes were put to use cleaning up beaches following the *Exxon Valdez* oil spill. **Was bioremediation a success in Prince William Sound? Can this technique address many of our society's problems with chemical contamination, or does it suffer from too many limitations?**

Bioremediation: An Effective Solution

Bioremediation was definitely a success in Prince William Sound. It was important to remove the spilled oil that reached the shorelines in an environmentally responsible way. Washing it into the sea and collecting it by skimming was the first step, but what to do about the residual oil on the beach? Bioremediation seemed a likely technology.

Oil seeps have released oil into the sea for millions of years, feeding a diverse group of microorganisms. Oil is thus very biodegradable, though unusual in that although it is rich in energy, it contains none of the inorganic nutrients required for microbial growth. The arrival of oil on a shoreline causes a dramatic increase in the number of oil-degrading microorganisms, but their growth is soon limited by the low levels of bioavailable nitrogen and phosphorus in seawater.

Our approach in Alaska was to add oleophilic (oil-adhering) and slow-release fertilizers to provide enough of these limiting nutrients over many tidal cycles so that biodegradation would be stimulated, but not so much that they would cause environmental harm. Rigorous testing, in collaboration with the State of Alaska and U.S. EPA, showed that the approach stimulated the natural rate of biodegradation some twofold to fivefold, with no detectable adverse environmental impact. Bioremediation was used on about 74 miles of shoreline—by far the largest use of this technology to date—and most shorelines were oil-free within 3 years instead of the predicted decades.

Can bioremediation address other environmental problems? Yes. There are many situations where stimulating the biodegradation of fuels, explosives, chlorinated solvents, transformer fluids, pesticides, and other substances will be an environmentally benign and effective treatment. Although the best ways of stimulating biodegradation will not be the same in all situations, working to stimulate natural processes without causing adverse effects will be an excellent way of cleaning many contaminated sites.

Roger Prince is a senior research associate at ExxonMobil's Corporate Strategic Research Laboratory in Annandale, New Jersey. He was Exxon's lead scientist in the monitoring of the successful bioremediation of shorelines oiled during the *Exxon Valdez* spill in Alaska, and he continues to work on the bioremediation of marine oil spills, including a recent multinational collaboration in the Arctic.

Biostimulation Can Sometimes Enhance Environmental Cleanup

The *Exxon Valdez* oil spill, which led to the enactment of the Oil Pollution Act of 1990, gave rise to the largest bioremediation field trial ever attempted. A research study was conducted by EPA in 1989 and 1990 to develop data to support the recommendation to go forward with a full-scale cleanup. Unfortunately, the data generated were equivocal and did not provide sufficient evidence to prove the success of the treatment. The study was equivocal because its experimental design did not provide sufficient replication or randomization of plots to allow for the calculation of experimental error, or to account for the high variability of oil contamination on the beaches.

A later field study in Delaware did provide the unequivocal evidence needed. Data generated in this and other EPA-funded field studies provided sufficient, statistically sound evidence that biostimulation of indigenous microorganisms can accelerate the disappearance of hydrocarbons at a spill site.

However, two other studies I have been involved with, both done on Canadian wetlands, have showed that biostimulation may not always be appropriate. One needs to determine the background levels of nutrients already present and make sure the affected environment is aerobic in nature. Only if nutrients are limiting in concentration, and if dissolved oxygen in the pore space is high enough to support microbial growth on the hydrocarbons, is biostimulation appropriate.

Based on various studies such as these, EPA has published two guidance documents on bioremediation on marine shorelines, freshwater wetlands, and salt marshes.

Although it is not appropriate in every case, biostimulation can play a key role in the environmental cleanup of oil spills. It is a tool to be seriously considered when contemplating how to restore a contaminated environment.

Albert D. Venosa is a senior research scientist with the U.S. Environmental Protection Agency, National Risk Management Research Laboratory, in Cincinnati, Ohio. He currently heads EPA's oil spill research program in the Office of Research and Development. He has published over 100 works in many aspects of wastewater treatment and hydrocarbon biodegradation.

Explore this issue further by accessing **Viewpoints** at www.aw-bc.com/withgott.

Conclusion

Carbon-based life has flourished on Earth for over 3 billion years, stemming from an origin that scientists are eagerly attempting to understand. Deciphering how life originated depends in part on understanding energy, energy flow, and chemistry. Knowledge in these areas also enhances our understanding of how present-day organisms interact with one another, how they relate to their nonliving environment, and how environmental systems function. Energy and chemistry are in some way tied to nearly every significant process involved in environmental science.

Chemistry can also be a tool for finding solutions to environmental problems. Cleaning up chemical pollution through bioremediation with microbes or plants is just one example. Knowledge of chemistry can be a powerful ally, whether you are interested in analyzing agricultural practices, managing water resources, reforming energy policy, conducting toxicological studies, or finding ways to mitigate global climate change.

REVIEWING OBJECTIVES

You should now be able to:

Explain the fundamentals of environmental chemistry and apply them to real-world situations

- Understanding chemistry provides a powerful tool for developing solutions to many environmental problems. (pp. 90–91)
- Atoms form molecules, and changes at the atomic level can result in alternative forms of elements, such as ions and isotopes. (pp. 91–96)
- Characteristics of the water molecule help facilitate life. (pp. 96–98)
- Living things depend on organic compounds, which are carbon-based. (pp. 98–99)

Describe the molecular building blocks of living organisms

- Macromolecules, including proteins, nucleic acids, carbohydrates, and lipids, are key building blocks of life. (pp. 99–102)
- Organisms use cells to compartmentalize macromolecules. (pp. 102–103)

Differentiate among the types of energy and recite the basics of energy flow

- Energy can change in its nature between potential and kinetic energy. Chemical energy is potential energy in the bonds between atoms. (pp. 103–104)
- The total amount of energy in the universe is conserved; it cannot be created or lost. (p. 104)
- Systems tend to increase in entropy, or disorder, unless energy is added to build or maintain order and complexity. (pp. 104–105)
- Earth's systems are powered by radiation from the sun and by geothermal heating from the planet's core. (pp. 105–108)

Distinguish photosynthesis, respiration, and chemosynthesis, and summarize their importance to living things

- In photosynthesis, autotrophs use carbon dioxide, water, and solar energy to produce the sugars they need, as well as oxygen. (pp. 105–107)
- In respiration, organisms extract energy from sugars by converting them in the presence of oxygen into carbon dioxide and water. (p. 107)
- In chemosynthesis, specialized autotrophs use carbon dioxide, water, and chemical energy from minerals to produce the sugars they need. (pp. 107–108)

Itemize and evaluate the major hypotheses for the origin of life on Earth

- The heterotrophic hypothesis proposes that life arose from chemical reactions in surface or shallow waters of the ocean. (p. 109)
- The panspermia hypothesis proposes that substances needed for life's origin on Earth arrived from space. (p. 109)
- The chemoautotrophic hypothesis proposes that life arose from chemical reactions near deep-sea hydrothermal vents. (p. 109)

Outline our knowledge regarding early life, and give supporting evidence for each major concept

- The fossil record has revealed many patterns in the history of life, including that species evolve, most species are extinct, and species numbers on Earth have increased. (pp. 110–111)
- By comparing modern-day organisms scientists can infer genetic relationships among them and understand their evolutionary history. (p. 111)

TESTING YOUR COMPREHENSION

1. What are the basic building blocks of matter? Provide examples using chemicals common in living organisms.
2. Name four ways in which the chemical nature of the water molecule facilitates life.
3. What are the three classes of biological polymer, and what are their functions?
4. Describe the two major forms of energy, and give examples of each.
5. State the first law of thermodynamics, and describe some of its implications.
6. What is the second law of thermodynamics, and how might it affect our interactions with the environment?
7. What are the two major sources of energy that power Earth's environmental systems?
8. What substances are produced by photosynthesis? By cellular respiration? By chemosynthesis?
9. Compare and contrast three competing hypotheses for the origin of life.
10. Name three things scientists have learned from the fossil record.

SEEKING SOLUTIONS

1. Under what types of conditions might bioremediation be a successful strategy, and when might it not be?
2. Can you think of an example of an environmental problem not mentioned in this chapter that a good knowledge of chemistry could help us solve?
3. Describe an example of energy transformation from one form to another that is not mentioned in this chapter.
4. Give three examples of ways in which the input of energy can resist the tendency toward disorder that the second law of thermodynamics describes.
5. Referring to the chemical reactions for photosynthesis and respiration, provide an argument for why increasing amounts of carbon dioxide in the atmosphere due to global climate change might potentially increase amounts of oxygen in the atmosphere. Give an argument for why it might potentially decrease amounts of atmospheric oxygen. What would you need to know to determine which of these two outcomes might occur?
6. The debate over the origin of life is heated and ongoing. For each of the main hypotheses, what evidence would convince you of its validity over the others?

INTERPRETING GRAPHS AND DATA

In phytoremediation, plants are used to clean up soil or water contaminated by heavy metals such as lead (Pb), arsenic (As), zinc (Zn), and cadmium (Cd). For plants to absorb these metals from soil, the metals must be dissolved in soil water. For any given instance, all metal can be accounted for as either remaining bound to soil particles, being dissolved in soil water, or being stored in the plant.

In a study on the effectiveness of alpine penny-cress (*Thlaspi caerulescens*) for phytoremediation, Enzo Lombi and his colleagues grew crops of this small perennial plant for approximately one year in pots of soil from contaminated sites. They then measured the amount of zinc and cadmium in the soil and in the plants when they were harvested.

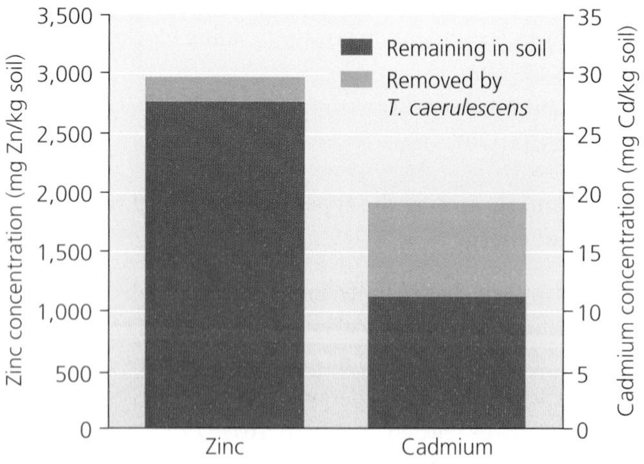

Removal of zinc and cadmium from contaminated soil by alpine penny-cress, *Thlaspi caerulescens*. Data from Lombi, E., et al. 2001. Phytoremediation of heavy metal-contaminated soils: natural hyperaccumulation versus chemically enhanced phytoextraction. *Journal of Environmental Quality* 30: 1919–1926.

1. What were the zinc and cadmium concentrations in the soil prior to phytoremediation? What were the zinc and cadmium concentrations in the soil after one year of phytoremediation?

2. How much zinc and cadmium were removed from the soil? If the plants continued to remove zinc and cadmium from the soil at the rates shown above, approximately how long would it take to remove all the zinc and cadmium?

3. Alpine penny-cress produces natural chelating agents (see "The Science behind the Story," pp. 92–93) that increase the solubility of metals in soil water. If these dissolved metals are not taken up by the plants, what may be an unintended consequence of having increased their solubility?

CALCULATING ECOLOGICAL FOOTPRINTS

In ecological systems, a rough rule of thumb is that when energy is transferred from plants to plant-eaters or from prey to predator, the efficiency is only about 10%. Much of this inefficiency is a consequence of the second law of thermodynamics. Another way to think of this is that eating 10 calories of plant material is the ecological equivalent of eating 1 calorie of material from an animal.

Humans are considered omnivores because we can eat both plants and animals. The choices we make about what to eat have significant ecological impacts. With this in mind, calculate the ecological energy requirements for four different diets, each of which provides a total of 2,000 dietary calories per day.

Diet	Source of calories	Number of calories consumed by source	Ecologically equivalent calories by source	Total ecologically equivalent calories per day
100% plant	Plant			
0% animal	Animal			
90% plant	Plant	1,800	1,800	3,800
10% animal	Animal	200	2,000	
50% plant	Plant			
50% animal	Animal			
0% plant	Plant			
100% animal	Animal			

1. How many ecologically equivalent calories would it take to support you for a year on each of the four diets listed?
2. What is the relative ecological impact of including as little as 10% of your calories from animal sources (e.g., milk, dairy products, eggs, and meat)? What is the ecological impact of a strictly carnivorous diet compared with a strict vegetarian diet?
3. What percentages of the calories in your own diet do you think come from plant versus animal sources? Estimate the ecological impact of your diet, relative to a strictly vegetarian one.
4. Describe some challenges of providing food for the growing human population, especially as people in many poorer nations develop a taste for an American-style diet rich in animal protein and fat.

Take It Further

Go to www.aw-bc.com/withgott or the student CD-ROM where you'll find:

- Suggested answers to end-of-chapter questions
- Quizzes, animations, and flashcards to help you study
- *Research Navigator*™ database of credible and reliable sources to assist you with your research projects
- **GRAPHIt!** tutorials to help you master how to interpret graphs
- **INVESTIGATEIt!** current news articles that link the topics that you study to case studies from your region to around the world

5 Evolution, Biodiversity, and Population Ecology

Monteverde cloud forest, Costa Rica

Upon successfully completing this chapter, you will be able to:

- Explain the process of natural selection, and cite evidence for this process
- Describe the ways in which evolution results in biodiversity
- Discuss reasons for species extinction and mass extinction events
- List the levels of ecological organization
- Outline the characteristics of populations that help predict population growth
- Assess logistic growth, carrying capacity, limiting factors, and other fundamental concepts of population ecology
- Identify efforts and challenges involved in the conservation of biodiversity

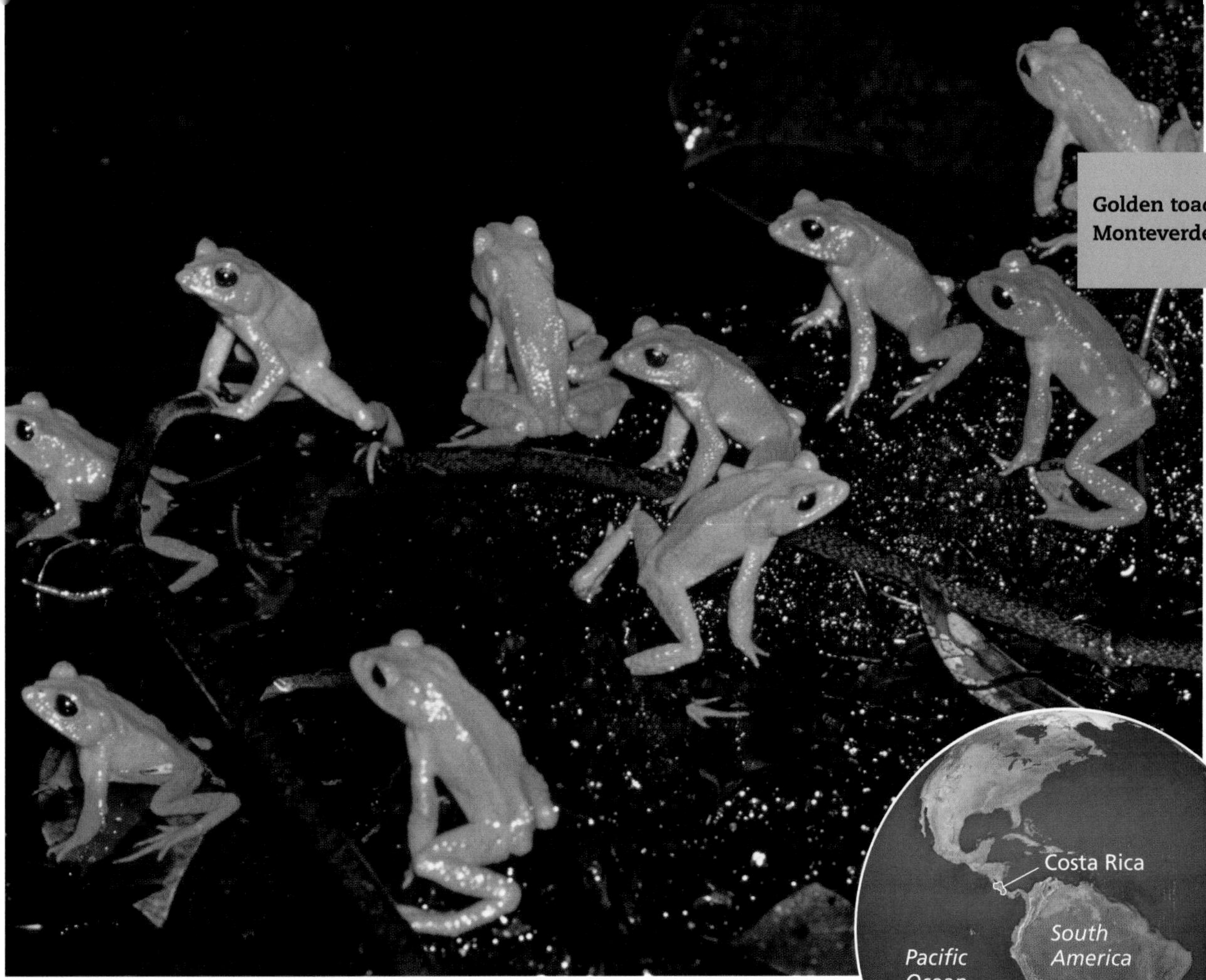

Golden toads at Monteverde

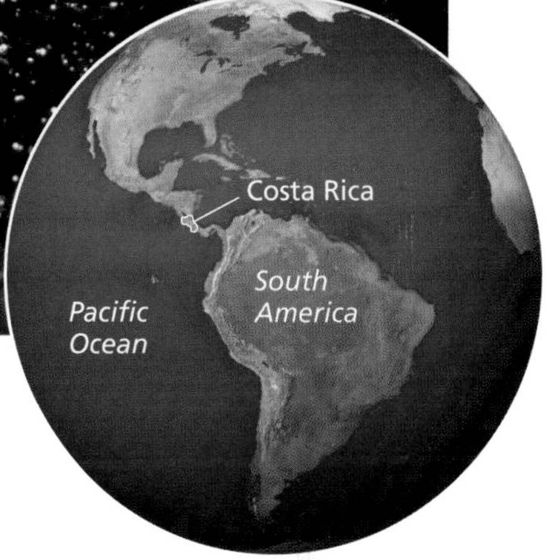

Central Case: Striking Gold in a Costa Rican Cloud Forest

"I must confess that my initial response when I saw them was one of disbelief and suspicion that someone had dipped the examples in enamel paint."
—Dr. Jay M. Savage, describing the golden toad in 1966

"What a terrible feeling to realize that within my own lifetime, a species of such unusual beauty, one that I had discovered, should disappear from our planet."
—Dr. Jay M. Savage, describing the golden toad in 1998

During a 1963 visit to Central America, biologist Jay M. Savage heard rumors of a previously undocumented toad living in Costa Rica's mountainous Monteverde region. The elusive amphibian, according to local residents, was best known for its color: a brilliant golden yellow-orange. Savage was told the toad was hard to find because it appeared only during the early part of the region's rainy season.

Monteverde means "green mountain" in Spanish, and the name couldn't be more appropriate. The village of Monteverde sits beneath the verdant slopes of the Cordillera de Tilarán, mountains that receive over 400 cm (157 in) of annual rainfall. Some of the lush forests above Monteverde, which begin around 1,600 m (5,249 ft, just under a mile high), are known as *lower montane rainforests*. They are also known as *cloud forests* because much of the moisture they receive arrives in the form of low-moving clouds that blow inland from the Caribbean Sea. Monteverde's cloud forest was not fully explored at the time of Savage's first visit, and researchers who had been there described the area as pristine, with a rich bounty of ferns, liverworts, mosses, clinging vines, orchids, and other organisms that thrive in cool, misty environments. Savage knew that such conditions create ideal habitat for many toads and other amphibians.

In May of 1964, Savage organized an expedition into the muddy mountains above Monteverde to try to document the existence of the previously unknown toad species in its natural habitat. Late in the afternoon of May 14, he and his colleagues found what they were looking for. Approaching the mountain's crest, they spotted bright orange patches on the forest's black floor.

In one area that was only 5 m (16.4 ft) in diameter, they counted 200 golden toads.

The discovery received international attention, making a celebrity of the tiny toad—which Savage named *Bufo periglenes* (literally, "the brilliant toad")—and making a travel destination of its mountain home. At the time, no one knew that the Monteverde ecosystem was about to be transformed. No one foresaw that the oceans and atmosphere would begin warming due to global climate change (Chapter 18) and cause Monteverde's moisture-bearing clouds to rise, drying the forest. No one could guess that this newly discovered species of toad would become extinct in less than 25 years.

Evolution as the Wellspring of Earth's Biodiversity

The golden toad was new to science, and countless species still await discovery, but scientists understand quite well how the world became populated with the remarkable diversity of organisms we see today. We know that the process of biological evolution has brought us from a stark planet inhabited solely by microbes to a lush world of 1.5 million (and likely millions more) species (Figure 5.1).

Perceiving how organisms adapt to their environments and change over time is crucial for understanding the history of life. Understanding evolution is also vital for appreciating ecology, a central component of environmental science. Evolutionary processes are relevant to many aspects of environmental science, including pesticide resistance, agriculture, medicine, and environmental health.

The term *evolution* in the broad sense means change over time, but scientists most often use the term to refer specifically to biological evolution. Biological **evolution** consists of genetic change in organisms across generations. This genetic change often leads to modifications in the appearance, functioning, or behavior of organisms through time. Biological evolution results from random genetic changes, and may proceed randomly or be directed by natural selection.

Natural selection is the process by which traits that enhance survival and reproduction are passed on more frequently to future generations than those that do not, altering the genetic makeup of populations through time. The theory of evolution by natural selection is one of the best-supported and most illuminating concepts in all of science, yet it has remained socially controversial among some nonscientists who fear it may threaten their religious beliefs. Although scientists sometimes disagree about the specific mechanisms thought to drive evolution in particular cases, or about the time scales on which it takes place, this routine scientific debate should not be equated with the socially driven opposition of some nonscientists. From a scientific standpoint, evolutionary theory is indispensable, because it is the foundation of modern biology.

(a) Resplendent quetzal

(b) Puffball mushroom

(c) Harlequin frog

(d) Scutellerid bug

FIGURE 5.1 Much of our planet's biological diversity resides in tropical rainforests. Monteverde's cloud-forest community includes organisms such as this **(a)** resplendent quetzal *(Pharomachrus mocinno)*, **(b)** puffball mushroom *(Calostoma cinnabarina)*, **(c)** harlequin frog *(Atelopus varius)*, and **(d)** scutellerid bug *(Pachycoris torridus)*.

Natural selection shapes organisms and diversity

In 1858, **Charles Darwin** and **Alfred Russell Wallace** each proposed the concept of natural selection as a mechanism for evolution and as a way to explain the great variety of living things. Darwin and Wallace were exceptionally keen naturalists from England who had studied plants and animals in such exotic locales as the Galapagos Islands and the Malay Archipelago. Both men recognized that organisms face a constant struggle to gain sufficient resources to survive and reproduce. Both were influenced by the writings of Thomas Malthus (▸pp. 4–5), who feared that human population growth would outstrip resource availability and lead to widespread death and social upheaval. Darwin and Wallace observed that organisms produce more offspring than can possibly survive, and they realized that some offspring may be more likely than others to survive and reproduce. Furthermore, they recognized that whichever characteristics give certain individuals an advantage in surviving and reproducing might be inherited by their offspring. These characteristics, Darwin and Wallace realized, would tend to become more prevalent in the population in future generations.

Natural selection is a simple concept that offers an astonishingly powerful explanation for patterns apparent in nature. The idea of natural selection follows logically from a few straightforward premises (Table 5.1). One is that individuals of the same species vary in their characteristics. Although not known in Darwin and Wallace's time, we now know that variation is due to differences in genes, the environments within which genes are expressed, and the interactions between genes and environment. As a result of this variation, some individuals within a species will happen to be better suited to their environment than others and thus will be able to survive longer and/or reproduce more.

Many characteristics are passed from parent to offspring through the genes, and a parent that is long-lived, robust, and produces many offspring will pass on genes to more offspring than a weaker, shorter-lived individual that produces only a few offspring. In the next generation, therefore, the genes of better-adapted individuals will be more prevalent than those of less well-adapted individuals. From one generation to another through time, species will evolve to possess characteristics that lead to better and better success in a given environment. A trait that promotes success is called an **adaptive trait,** or an **adaptation.** A trait that reduces success is *maladaptive.*

Table 5.1 The Logic of Natural Selection

- Organisms produce more offspring than can survive
- Individuals vary in their characteristics
- Many characteristics are inherited by offspring from parents

Therefore,

- Some individuals will be better suited to their environment than other individuals
- By producing more offspring and/or offspring of higher quality, better-suited individuals transmit more genes to future generations than poorly suited individuals
- Future generations will contain more genes, and thus more characteristics, of the better-suited individuals; as a result, characteristics evolve across generations through time

Natural selection acts on genetic variation

For an organism to pass a trait along to future generations—that is, for the trait to be *heritable*—genes in the organism's DNA (▸pp. 100–101) must code for the trait. Accidental alterations that arise during DNA replication give rise to genetic variation among individuals. In an organism's lifetime, its DNA will be copied millions of times by millions of cells. In all this copying and recopying, sometimes a mistake is made. Accidental changes in DNA, called **mutations,** can range in magnitude from the addition, deletion, or substitution of single nucleotides (▸p. 100) to the insertion or deletion of large sections of DNA. If a mutation occurs in a sperm or egg cell, it may be passed on to the next generation. Although most mutations have little effect, some can be deadly, whereas others can be beneficial. Those that are not lethal provide the genetic variation on which natural selection acts.

Sexual reproduction also generates variation. In sexual organisms, genetic material is mixed, or recombined, so that a portion of each parent's genome is included in the genome of the offspring. This process of *recombination* produces novel combinations of genes, generating variation among individuals.

Genetic variation can lead to variation in organismal-level traits. We can visualize how traits vary using distribution graphs, which help us see that selection can alter the characteristics of organisms through time in three main ways (Figure 5.2). Selection that drives a feature in one direction rather than another—for example, toward larger or smaller, faster or slower—is called *directional selection.* In contrast, *stabilizing selection* produces intermediate traits, in essence preserving the status quo. Under *disruptive selection,* traits diverge from their starting condition in two or more directions.

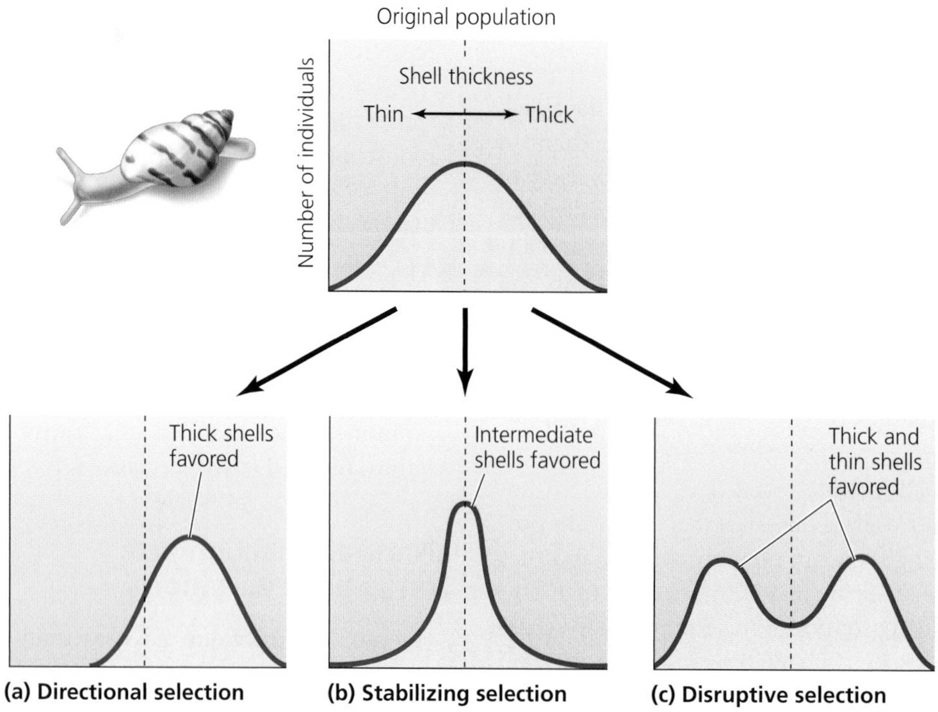

FIGURE 5.2 Selection can act in three ways. Consider snails living in tropical cloud forest, and assume that we begin with a population of snails with shells of different thicknesses (top graph). Because shells protect snails against predators, snails with thick shells may be favored over those with thin shells, through *directional selection* **(a)**. Alternatively, suppose that a shell that is too thin breaks easily, whereas a shell that is too thick wastes resources that are better used for feeding or reproduction. In such as case, *stabilizing selection* **(b)** could act to favor snails with shells that are neither too thick nor too thin. Under *disruptive selection* **(c)**, extreme traits are favored. For example, perhaps thin-shelled snails are so resource-efficient that they can outreproduce intermediate-shelled snails, whereas thick-shelled snails are so well protected from predators that they also outreproduce intermediate-shelled snails. In such a case, each of the "extreme" strategies works more effectively than a compromise between the two, and natural selection increases the relative numbers of thin- *and* thick-shelled snails, while reducing the number of intermediate-shelled ones.

An organism's environment determines what pressures natural selection will exert on the organism. However, environments change, and organisms may move to new places and encounter new conditions. In either case, a trait that is adaptive in one location or season may prove maladaptive in another. Golden toads that had adapted to the moist conditions of Monteverde's cloud forest would not have survived in drier forests, and apparently did not persist after Monteverde's climate became drier starting 25 years ago. Differences in environmental conditions in time and space make adaptation a moving target for organisms.

In all these ways, variable genes and variable environments interact as organisms engage in a perpetual process of adapting to the changing conditions around them. During this process, natural selection does not simply weed out unfit individuals. It also helps to elaborate and diversify traits that in the long term may help lead to the formation of new species and whole new types of organisms. In this way, natural selection has helped bring about the wondrous flowering of life on our planet.

Evidence of natural selection is all around us

The results of natural selection are all around us, visible in every adaptation of every organism (Figure 5.3). In addition, countless lab experiments (mostly with fast-reproducing organisms, such as bacteria and fruit flies) have demonstrated rapid evolution of traits. The evidence for selection that may be most familiar to us is that which Darwin himself cited prominently in his work 150 years ago: our breeding of domesticated animals. In our dogs, our cats, and our livestock, we have conducted selection under our own direction. We have chosen animals with traits we like and bred them together, while not breeding those with variants we do not like. Through such selective

FIGURE 5.3 Natural selection has produced tremendous diversity among organisms in the wild. In the group of birds known as Hawaiian honeycreepers, closely related species have adapted to different food resources, habitats, or ways of life, as indicated by the diversity in their plumage colors and the shapes of their bills. Such a burst of species formation due to natural selection is known as an *adaptive radiation.*

breeding, we have been able to exaggerate particular traits we prefer. Consider the great diversity of dog breeds (Figure 5.4a), all of which comprise variations on a single species. From Great Dane to Chihuahua, they can interbreed freely and produce viable offspring, yet breeders maintain striking differences between them by allowing only like individuals to breed with like. This process of selection conducted under human direction is termed **artificial selection.**

Artificial selection has also given us the many crop plants we depend on for food, all of which were domesticated from wild ancestors and carefully bred over years, centuries, or millennia (Figure 5.4b). Through selective breeding, we have created corn with larger sweeter kernels; wheat and rice with larger and more numerous grains; and apples, pears, and oranges with better taste. We have diversified single types into many, for instance, breeding variants of the plant *Brassica oleracea* to create broccoli, cauliflower, cabbage, and Brussels sprouts. Our entire agricultural system is based on artificial selection.

Weighing the Issues: Artificial Selection

Consider some of the pets and farm animals that humans have domesticated through artificial selection, such as horses, dogs, and cats. In what ways do these domesticated animals differ from their wild relatives? What kinds of characteristics do people prefer in pets and farm animals? Do the evolved traits of the domesticated animals match these preferences?

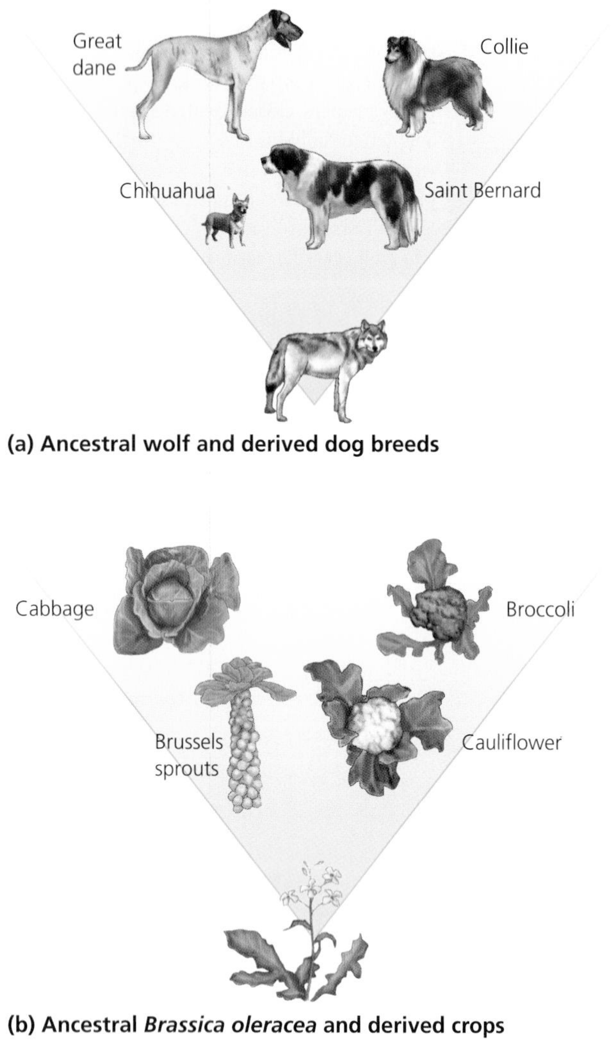

FIGURE 5.4 Selection imposed by humans (selective breeding, or artificial selection) has resulted in the numerous breeds of dogs. By starting with the gray wolf *(Canis lupus)* as the ancestral wild species, and by breeding like with like and selecting for the traits we prefer, we have evolved breeds as different as Great Danes and Chihuahuas (**a**). By this same process we have created the immense variety of crop plants we depend on for sustenance. Cabbage, Brussels sprouts, broccoli, and cauliflower were all evolved from a single ancestral species, *Brassica oleracea* (**b**).

Evolution generates biological diversity

When Charles Darwin wrote about the wonders of a world full of diverse animals and plants, he conjured up the vision of a "tangled bank" of vegetation harboring all kinds of creatures. Such a vision fits well with the arching vines, dripping leaves, and mossy slopes of the tropical cloud forest of Monteverde. Indeed, tropical forests worldwide teem with life and harbor immense biological diversity (see Figure 5.1).

Biological diversity, or **biodiversity** for short, refers to the sum total of all organisms in an area, taking into account the diversity of species, their genes, their populations, and their communities. A **species** is a particular type of organism or, more precisely, a population or group of populations whose members share certain characteristics and can freely breed with one another and produce fertile offspring. A **population** is a group of individuals of a particular species that live in the same area. We have already discussed genes (▸p. 101), and we will introduce communities shortly (▸p. 130; Chapter 6).

Scientists have described between 1.5 million and 1.8 million species, but many more remain undiscovered or unnamed. Estimates for the total number of species in the world range up to 100 million, with many of them thought to occur in tropical forests. In this light, the discovery of a new toad species in Costa Rica in 1964 seems far less surprising. Although Costa Rica covers a tiny fraction (0.01%) of Earth's surface area, it is home to 5–6% of all species known to scientists. And of the 500,000 species scientists estimate exist in the country, only 87,000 (17.4%) have been inventoried and described.

Tropical rainforests such as Costa Rica's, however, are by no means the only places rich in biodiversity. Step outside anywhere on Earth, even in a major city, and you will find numerous species within easy reach. They may not always be large and conspicuous like Yellowstone's bears or Africa's elephants, but they will be there. Plants poke up from cracks in asphalt in every city in the world, and even Antarctic ice harbors microbes. In a handful of backyard soil there may exist an entire miniature world of life, including several insect species, several types of mites, a millipede or two, many nematode worms, a few plant seeds, countless fungi, and millions upon millions of bacteria. We will examine Earth's biodiversity in detail in Chapter 11.

Speciation produces new types of organisms

How did Earth come to have so many species? Whether there are 1.5 million or 100 million, such large numbers require scientific explanation. The process by which new species are generated is termed **speciation.** Speciation can occur in a number of ways, but most biologists consider the main mode of species formation to be *allopatric speciation,* species formation due to the physical separation of populations over some geographic distance. To understand allopatric speciation, begin by picturing a population of organisms. Individuals within the population

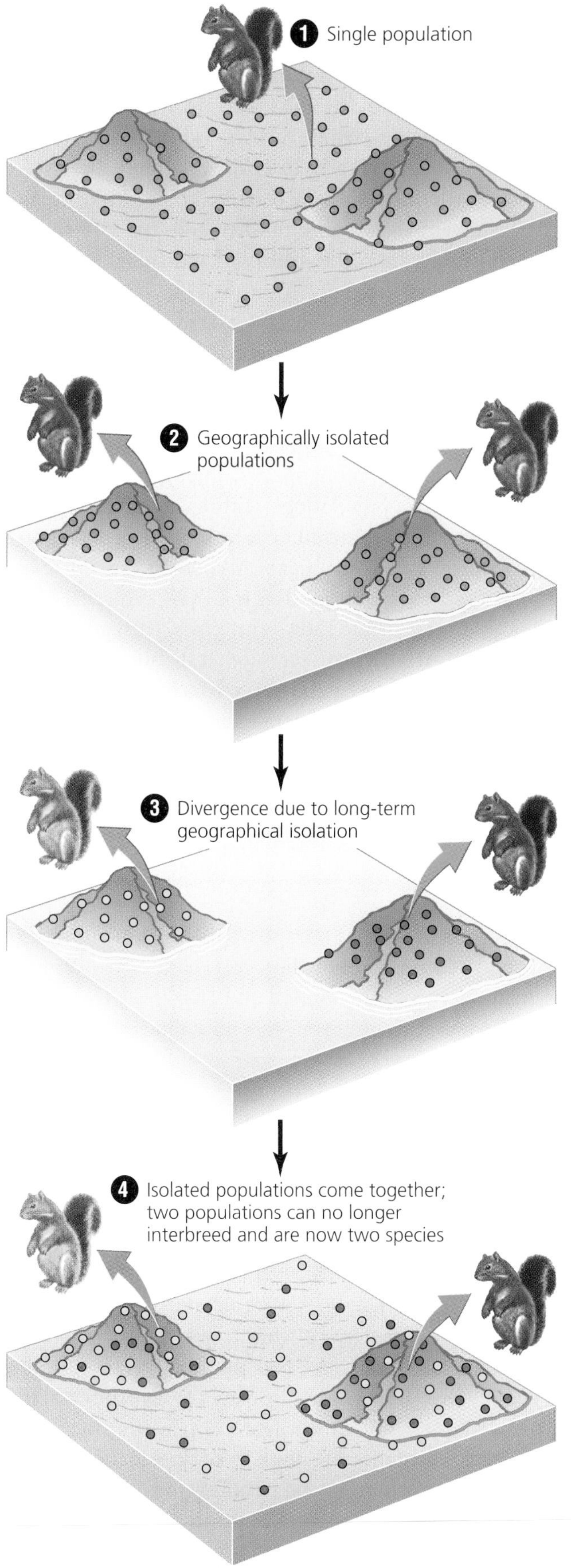
1 Single population
2 Geographically isolated populations
3 Divergence due to long-term geographical isolation
4 Isolated populations come together; two populations can no longer interbreed and are now two species

possess many similarities that unify them as a species, because they are able to reproduce with one another and share genetic information. However, if the population is broken up into two or more populations that become isolated from one another, individuals from one population cannot reproduce with individuals from the others.

When a mutation arises in the DNA of an organism in one of these isolated populations, it cannot spread to the other populations. Over time, each population will independently accumulate its own set of mutations. Eventually, the populations may diverge, or grow different enough, that their members can no longer mate with one another. Perhaps they no longer recognize one another as being the same species because they have diverged so much in appearance or behavior. Once this has happened, there is no going back; the two populations cannot interbreed, and they have embarked on their own independent evolutionary trajectories as separate species (Figure 5.5). The populations will continue diverging in their characteristics as chance mutations accumulate that confer traits causing the populations to become different in random ways. If environmental conditions happen to be different for the two populations, then natural selection may accelerate the divergence. Through the speciation process, single species can generate multiple species, each of which can in turn generate more.

Populations can be separated in many ways

The long-term geographic isolation of populations that can lead to allopatric speciation can occur in various ways (Table 5.2). Glacial ice sheets may move across continents during ice ages and split populations in two. Major rivers may change course and do the same. Mountain ranges may rise and divide regions and their organisms. Drying climate may partially evaporate lakes, subdividing them into multiple smaller bodies of water. Warming or cooling temperatures may cause whole plant communities to move northward or southward, or upslope or downslope, creating new patterns of plant and animal distribution. Regardless of the mechanism of separation, in

FIGURE 5.5 Allopatric speciation has generated much of Earth's diversity. In this process, some geographical barrier splits a population. In this diagram, two mountaintops (1) are turned into islands by rising sea level (2), isolating populations of squirrels. Each isolated population accumulates its own independent set of genetic changes over time, until individuals become genetically distinct and unable to breed with individuals from the other population (3). The two populations now represent separate species and will remain so even if the geographical barrier is removed and the new species intermix (4).

Table 5.2 Mechanisms of Population Isolation That Can Give Rise to Allopatric Speciation

- Glacial ice sheets advance
- Mountain chains are uplifted
- Major rivers change course
- Sea level rises, creating islands (see Figure 5.5)
- Climate warms, pushing vegetation up mountain slopes and fragmenting it
- Climate dries, dividing large single lakes into multiple smaller lakes
- Ocean current patterns shift
- Islands are formed in the sea by volcanism

order for speciation to occur, populations must remain isolated for a long time, generally thousands of generations.

If the geological or climatic process that has isolated populations reverses itself—if the glacier recedes, or the river returns to its old course, or warm temperatures turn cool again—then the populations can come back together. This is the moment of truth for speciation. If the populations have not diverged enough, their members will begin interbreeding and reestablish gene flow, mixing those mutations that each population accrued while isolated. However, if the populations have diverged sufficiently, they will not interbreed, and two species will have been formed, each fated to continue on its own evolutionary path.

Although allopatric speciation has long been considered the main mode of species formation, speciation appears to occur in other ways as well. *Sympatric speciation* occurs when species form from populations that become reproductively isolated within the same geographic area. For example, populations of some insects may become isolated by feeding and mating exclusively on different types of plants. Or they may mate during different seasons, isolating themselves in time rather than space. In some plants, speciation apparently has occurred as a result of hybridization between species. In others, it seems to have resulted from mutations that changed the numbers of chromosomes, creating plants that could not mate with plants with the original number of chromosomes. Garnering solid evidence for speciation mechanisms is difficult, so biologists still actively debate the relative prevalence of each of these modes of speciation.

Life's diversification results from numerous speciation events

Repeated speciation events have generated complex patterns of diversity at levels above the species level. Such patterns are studied by evolutionary biologists, who examine how groups of organisms arose and how they evolved the characteristics they show. For instance, how did we end up with plants as different as mosses, palm trees, daisies, and redwoods? Why do fish swim, snakes slither, and sparrows sing? How and why did the ability to fly evolve independently in birds, bats, and insects? To address such questions, one needs to know how the major groups diverged from one another, and this pattern ultimately results from the history of individual speciation events.

We saw in Chapter 4 how the history of divergence can be represented in a treelike diagram (Figure 4.24, ▸p. 111). Such branching diagrams, called cladograms, or **phylogenetic trees,** illustrate scientists' hypotheses as to how divergence took place (Figure 5.6). Phylogenetic trees can show relationships among species, among major groups of species, among populations within a species, or even among individuals. In addition, by mapping traits onto a tree according to which organisms possess them, one can trace how the traits themselves may have evolved. For instance, the tree of life shows that birds, bats, and insects are distantly related, with many other flightless groups between them. So, it is far simpler to conclude that the three groups evolved flight independently than to conclude that the many flightless groups all lost an ancestral ability to fly. Because phylogenetic trees help biologists make such inferences about so many traits, they have become one of the modern biologist's most powerful tools.

Life's history, as revealed by phylogenetic trees and by the fossil record (▸pp. 110–111), is complex indeed, but a few big-picture trends are apparent. As we mentioned in Chapter 4, life in its 3.5 billion years has evolved complex structures from simple ones, and large sizes from small ones. However, these are only generalizations. Many organisms have evolved to become simpler or smaller when natural selection favored it. Many very complex life forms have disappeared (Figure 5.7), and it is easy to argue that Earth still belongs to the bacteria and other microbes, some of them little changed over eons.

Even fans of microbes, however, must marvel at some of the exquisite adaptations that animals, plants, and fungi have evolved: The heart that beats so reliably for an animal's entire lifetime that we take it for granted. The complex organ system of which the heart is a part. The stunning plumage of a peacock in full display. The ability of each and every plant on the planet to lift water and nutrients from the soil, gather light from the sun, and turn it into food. The staggering diversity of beetles and other insects. The human brain and its ability to reason. All these and more have resulted from the process of evolution as it has generated new species and whole new branches on the tree of life.

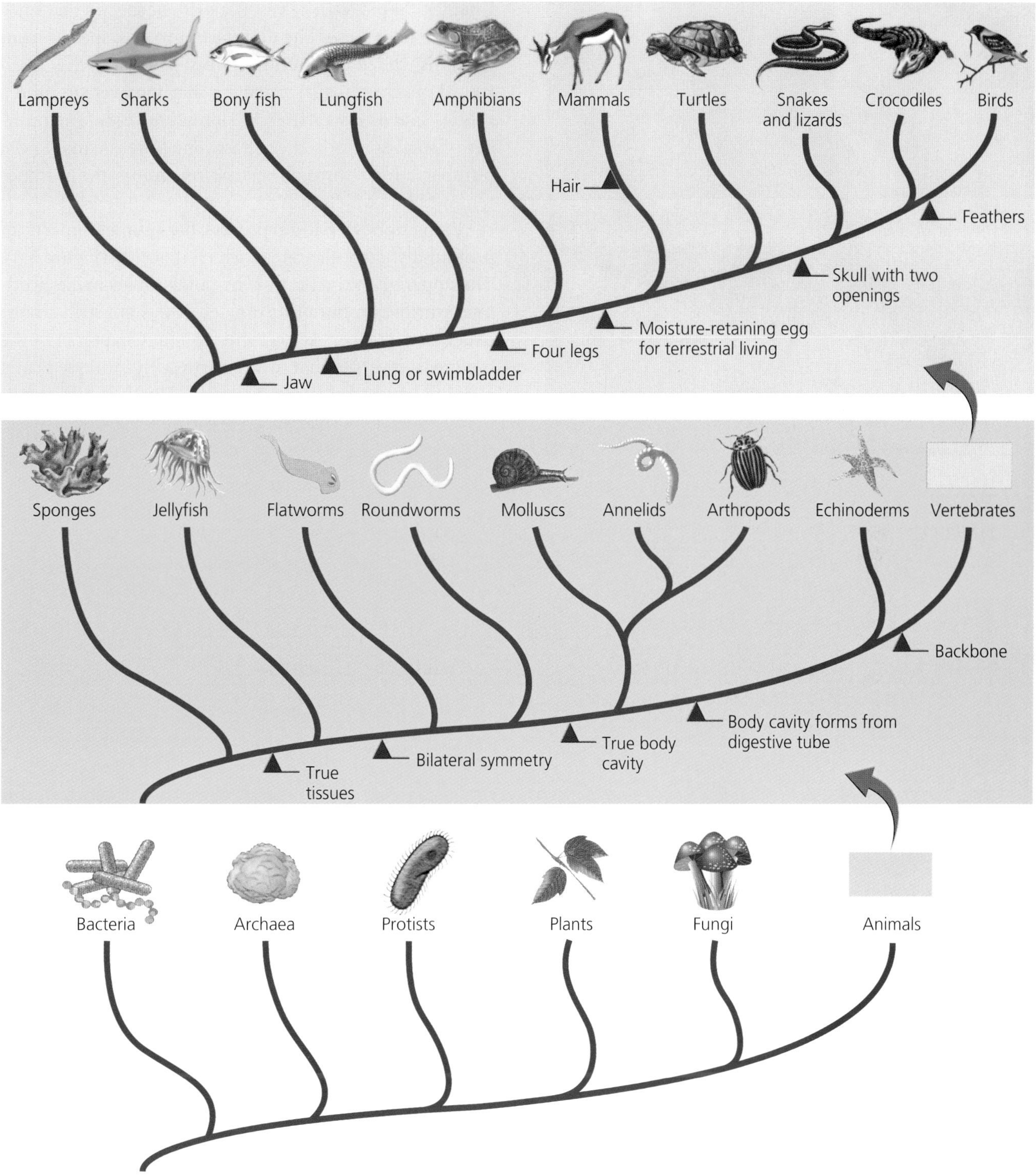

FIGURE 5.6 Phylogenetic trees show the history of life's divergence. Similar to family genealogies, these trees illustrate relationships among groups of organisms, as inferred from the study of similarities and differences among present-day creatures. The diagram here is a greatly simplified representation of relationships among a few major groups—one small portion of the huge and complex "tree of life." Each branch results from a speciation event, and time proceeds upward from bottom to top. By mapping traits onto phylogenetic trees, biologists can study how traits have evolved over time. In this diagram, several major traits are mapped, using triangular arrows indicating the point at which they originated. For instance, all vertebrates "above" the point at which jaws are indicated have jaws, whereas lampreys diverged before jaws originated and thus lack them.

FIGURE 5.7 Life has not always progressed from simple to complex during evolution. Many complex organisms have gone extinct, taking their designs and innovations with them. For example, the strange creatures portrayed in this painting were found fossilized in the Burgess Shale of the Canadian Rockies in British Columbia. They lived in marine environments 530 million years ago and vanished without leaving descendants. Painting by Marella J. Sibbick. *Source:* The National History Museum, London.

Speciation and extinction together determine earth's biodiversity

Although speciation generates Earth's biodiversity, it is only one part of the equation—for, as you will recall from Chapter 4 (▸p. 110), the vast majority of species that once lived are now gone. The disappearance of a species from Earth is called **extinction.** From studying the fossil record, paleontologists calculate that the average time a species spends on Earth is 1–10 million years. The number of species in existence at any one time is equal to the number added through speciation minus the number removed by extinction.

Extinction is a natural process, but human impact can profoundly affect the rate at which it occurs (Figure 5.8). The apparent extinction of the golden toad made headlines worldwide, but unfortunately it was not such an unusual occurrence. As we will see in Chapter 11, the biological diversity that makes Earth such a unique planet is being lost at an astounding rate. This loss affects humans directly, because other organisms provide us with life's necessities—food, fiber, medicine, and ecosystem services (▸p. 39). Species extinction brought about by human impact may well be the single biggest environmental problem we face, because the loss of a species is irreversible.

Some species are more vulnerable to extinction than others

In general, extinction occurs when environmental conditions change rapidly or severely enough that a species cannot adapt genetically to the change; natural selection simply does not have enough time to work. All manner of environmental events can cause extinction, from climate

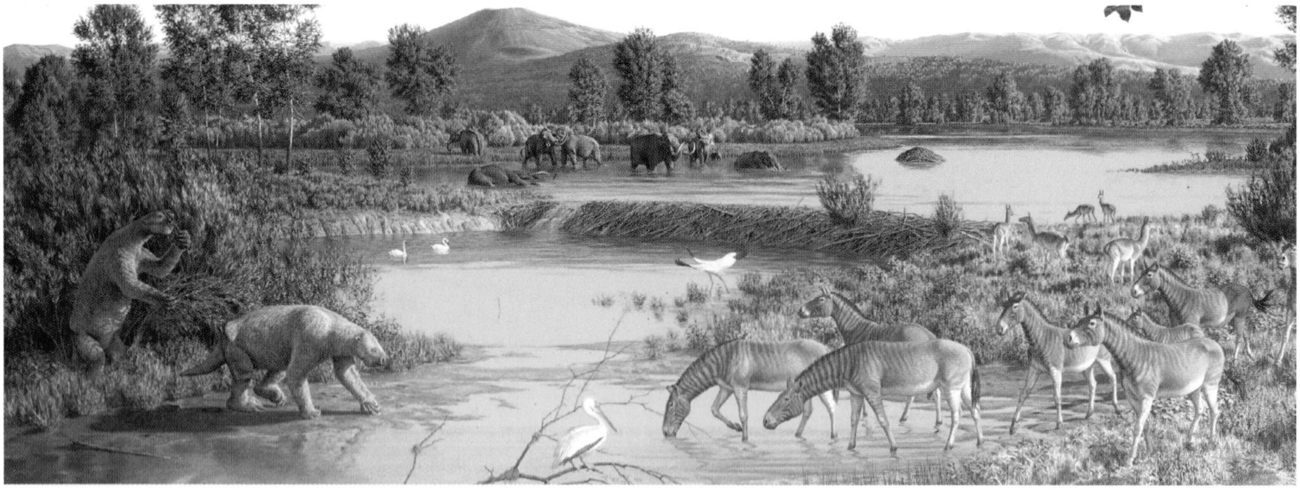

FIGURE 5.8 Until 10,000 years ago, the North American continent teemed with a variety of large mammals, including mammoths, camels, giant ground sloths, lions, saber-toothed cats, and various types of horses, antelope, bears, and others. Nearly all of this megafauna went extinct suddenly about the time that humans first arrived on the continent. Similar extinctions occurred in other areas simultaneously with human arrival, suggesting to many scientists that overhunting or other human impacts were responsible. *Source:* National Museum of Natural History, Smithsonian Institution.

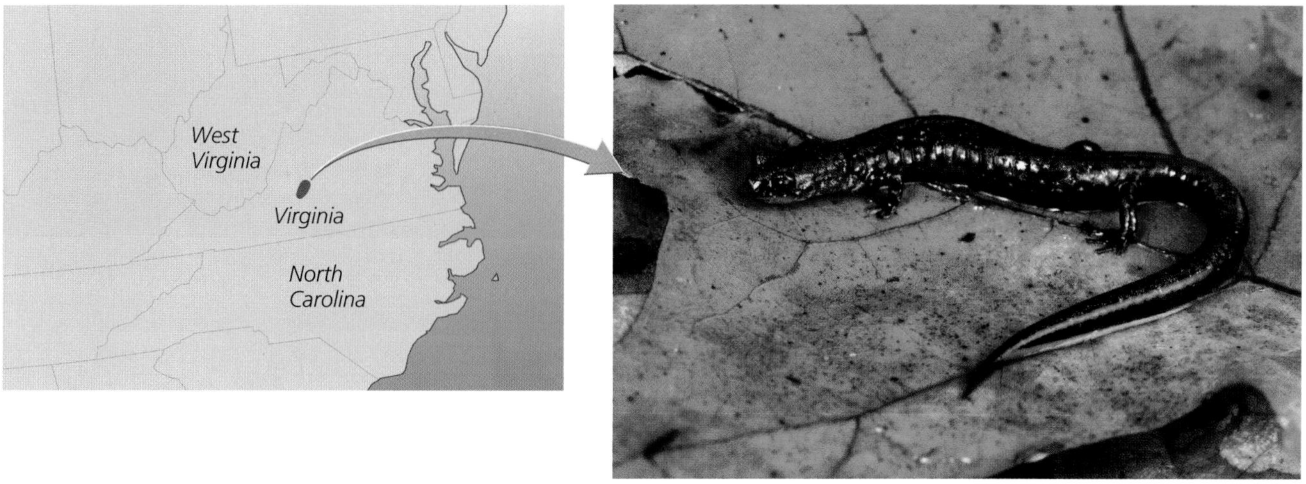

FIGURE 5.9 Forty salamander species in the United States are restricted in range to areas the size of a typical county, and some of these live atop single mountains. Such small range sizes leave these creatures vulnerable to extinction if severe local environmental changes occur. The Peaks of Otter salamander *(Plethodon hubrichti)* pictured here lives on only a few mountains in Virginia's Blue Ridge Mountains.

change to the rise and fall of sea level, to the arrival of new harmful species, to severe weather events such as extended droughts. In general, small populations and species narrowly specialized on some particular resource or way of life are most vulnerable to extinction from environmental change.

The golden toad was a prime example of a vulnerable species. It was **endemic** to the Monteverde cloud forest, meaning that it occurred nowhere else on the planet. Endemic species face relatively high risks of extinction because all their members belong to a single, sometimes small, population. At the time of its discovery, the golden toad was known from only a 4-km^2 (988-acre) area of Monteverde. It also required very specific conditions to breed successfully. During the spring at Monteverde, water collects in shallow pools within the network of roots that span the cloud forest's floor. The golden toad gathered to breed in these root-bound reservoirs, and it was here that Jay Savage and his companions collected their specimens in 1964. Monteverde provided ideal habitat for the golden toad, but the minuscule extent of that habitat meant that any environmental stresses that deprived the toad of the resources it needed to survive might doom the entire world population of the species.

In the United States, a number of amphibians are limited to very small ranges and thus are vulnerable to extinction. The Yosemite toad is restricted to a small region of the Sierra Nevada in California, the Houston toad occupies just a few areas of Texas woodland, and the Florida bog frog lives in a tiny region of Florida wetland. Fully 40 salamander species in the United States are restricted to areas the size of a typical county, and some of these live atop single mountains (Figure 5.9).

Earth has seen several episodes of mass extinction

Most extinction occurs gradually, one species at a time. The rate at which this type of extinction occurs is referred to as the *background extinction rate.* However, Earth has seen five events of staggering proportions that killed off massive numbers of species at once. These episodes, called **mass extinction events,** have occurred at widely spaced intervals in Earth history and have wiped out half to 95% of our planet's species each time. The best-known mass extinction occurred 65 million years ago and brought an end to the dinosaurs (although birds are modern representatives of dinosaurs). Evidence suggests that the impact of a gigantic asteroid caused this event, called the Cretaceous-Tertiary, or K-T, event (see "The Science behind the Story," ▸pp. 128–129).

The K-T event, as massive as it was, was moderate compared to the mass extinction at the end of the Permian period 250 million years ago (see Appendix D for geologic periods). Paleontologists estimate that 75–95% of all species on Earth may have perished during this event. Precisely what caused the event scientists don't know. The evidence for extraterrestrial impact is much weaker than it is for the K-T event, and other ideas

The Science behind the Story

The K-T Mass Extinction

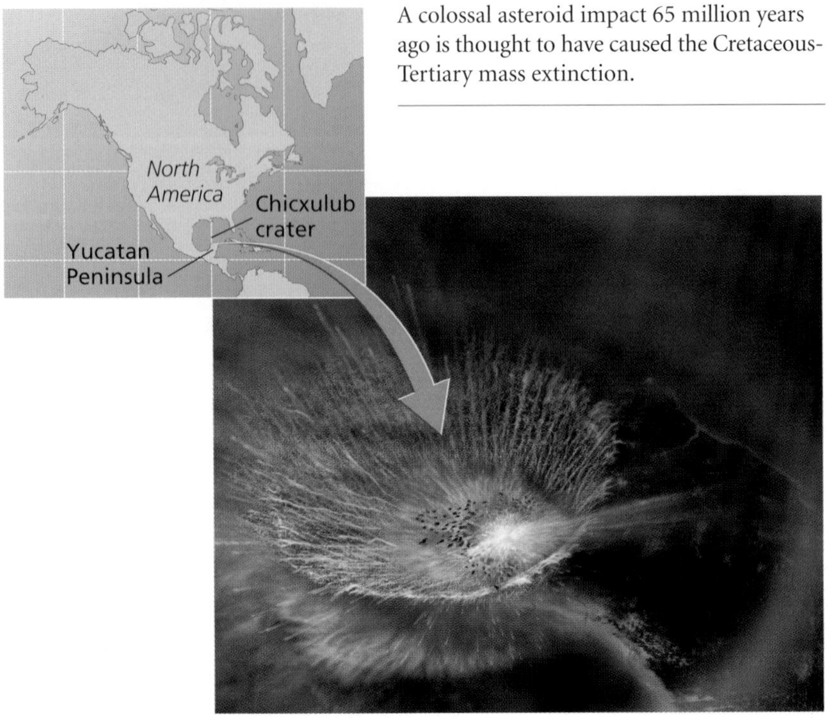

A colossal asteroid impact 65 million years ago is thought to have caused the Cretaceous-Tertiary mass extinction.

On five occasions in the history of life, huge numbers of species went extinct in a geologic instant. The last mass extinction occurred 65 million years ago, at the dividing line between the Cretaceous and Tertiary periods, called the *K-T boundary*. About 70% of the species then living, including the dinosaurs, disappeared.

When he first started working at Bottaccione Gorge in Italy, American geologist Walter Alvarez had no idea he would soon help discover what killed off the dinosaurs. Alvarez was developing a new method to determine the age of sedimentary rocks, and he'd chosen Bottaccione Gorge because it formed an ideal geological archive. Its 400-m (1,300-ft) walls are stacked like layer cake with beds of rose-colored limestone that formed between 100 million and 50 million years ago from dust that had settled to the bottom of an ancient sea. While analyzing these layers, Alvarez noticed a band of reddish clay 1 cm (0.4 in.) thick sandwiched between two layers of limestone. The older layer just below it was packed with fossils of globotruncana, a sand-sized animal that lived in the late Cretaceous period. The newer layer just above it contained just a few scattered fossils of a cousin of globotruncana, typical of sedimentary rock formed in the early Tertiary period. What Alvarez found interesting was that the intermediate clay layer, which had formed just as dinosaurs were going extinct, had no fossils at all.

To see how quickly the K-T mass extinction event occurred,

abound. The hypothesis with the most support so far is that massive volcanism threw into the atmosphere a global blanket of soot and sulfur, smothering the planet, reducing sunlight, and inducing severe climate gyrations.

The sixth mass extinction is upon us

Many biologists have concluded that Earth is currently entering its sixth mass extinction event—and that we are the cause. The changes to Earth's natural systems set in motion by human population growth, development, and resource depletion have driven many species extinct and are threatening countless more. The alteration and outright destruction of natural habitats, the hunting and harvesting of species, and the introduction of invasive species from one place to another where they can harm native species—these processes and many more have combined to threaten Earth's biodiversity.

When we look around us, it may not appear as though a human version of an asteroid impact is taking place, but we cannot judge such things on a human timescale. On the geological timescale, extinction over 100 years or over 10,000 years appears every bit as instantaneous as extinction over a few days.

Walter and his father, physicist Luis Alvarez, analyzed the intermediate clay layer to see how long it had taken to form, using the rare metal iridium as a kind of clock. Almost all of the iridium on Earth's surface comes from dust from tiny meteorites (shooting stars) that burn up in the atmosphere. Because the same amount of meteorite dust rains down each year, the Alvarezes measured iridium levels in the clay layer to determine how many years' worth of iridium had accumulated.

Iridium levels in the limestone were typical for sedimentary rocks, about 0.3 parts per billion. In the clay layer, however, the Alvarezes were surprised to find levels 30 times higher. To make sure the finding was not unique to Bottaccione Gorge, they checked the K-T clay layer at a Danish sea cliff. It had 160 times more iridium than the surrounding rock. The Alvarezes hypothesized that the excess iridium had come from a massive asteroid that smashed into Earth, causing a global environmental catastrophe that had wiped out the dinosaurs.

To convince themselves and the science community that an asteroid impact did in fact cause the K-T event, the Alvarezes had to rule out other possible explanations. For example, the extra iridium could have come from seawater. Calculations, however, proved that seawater did not contain enough iridium to account for the high levels in the clay layers.

An asteroid 10 km (6.2 mi) wide strikes Earth, on average, every 100 million years. Such an impact would unleash an explosion 1,000 times more forceful than the 1883 eruption of the Indonesian volcano Krakatau, which scattered so much dust around the world that sunsets were intense for 2 years afterward. An asteroid impact 65 million years ago, they suggested, kicked up enough soot to blot out the sun for several years. This would have inhibited photosynthesis, causing plants to die off, food webs to collapse, and most animals, including dinosaurs, to die of starvation. Only a few smaller animals survived, feeding on rotting vegetation. When sunlight returned, plants sprouted from dormant seeds, and a long recovery began.

Published in the journal *Science* in 1980, the Alvarezes' explanation was immediately attacked by other geologists, who claimed that spectacular volcanic eruptions more likely explained the high iridium levels in Bottaccione Gorge. But throughout the 1980s, scientists kept finding evidence supporting the asteroid-impact hypothesis. Iridium-enriched clay turned up at K-T layers around the world, as did bits of minerals called shocked quartz and stishovite, which form only under the extreme pressure of thermonuclear explosions and asteroid impacts. The Alvarezes' hypothesis became widely accepted after 1991, once scientists pinpointed a 65-million-year-old crater of the expected size in the ocean off the coast of Mexico. Today, scientists broadly agree that an asteroid impact 65 million years ago caused our planet's most recent mass extinction.

Weighing the Issues:
Should We Care about Extinction?

Although many scientists say biodiversity loss is our biggest environmental problem, some critics say we should not be concerned about biodiversity loss or a new mass extinction. What do you think can account for such a disparity of viewpoints? Thinking back to our discussion of ethics and economics in Chapter 2, can you elaborate reasons why we should be concerned about loss of biodiversity and extinction of species?

Levels of Ecological Organization

The extinction of species, their generation through speciation, and other evolutionary mechanisms and patterns have substantial influence on ecology. Moreover, it's often said that ecology provides the stage on which the play of evolution unfolds. The two, it's clear, are tightly intertwined in many ways. As we discussed in Chapter 1, ecology is the study of interactions among organisms and between organisms and their environments.

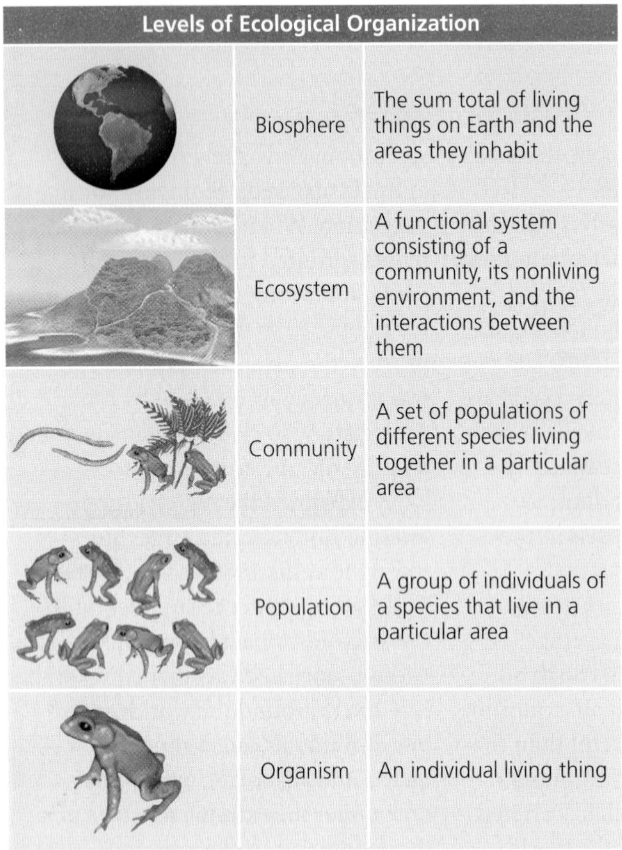

Levels of Ecological Organization
Biosphere
The sum total of living things on Earth and the areas they inhabit
Ecosystem
A functional system consisting of a community, its nonliving environment, and the interactions between them
Community
A set of populations of different species living together in a particular area
Population
A group of individuals of a species that live in a particular area
Organism
An individual living thing

FIGURE 5.10 Life exists in a hierarchy of levels. Ecology includes the study of the organismal, population, community, and ecosystem levels and, increasingly, the level of the biosphere. Levels below the organismal level were illustrated in Figure 4.15, ▸p. 103.

Ecology is studied at several levels

Life occurs in a hierarchy of levels. The atoms, molecules, and cells we reviewed in Chapter 4 represent the lowest levels in this hierarchy (see Figure 4.15, ▸p. 103). Aggregations of cells of particular types form tissues, and tissues form organs, all housed within an individual living organism. Ecologists study relationships on the higher levels of this hierarchy (Figure 5.10), namely on the organismal, population, community, and ecosystem levels. **Communities** are made up of multiple interacting species that live in the same area. A population of golden toads, a population of resplendent quetzals, populations of ferns and mosses, together with all of the other interacting plant, animal, fungal, and microbial populations in the Monteverde cloud forest, would be considered a community. **Ecosystems** encompass communities and the abiotic (nonliving) material and forces with which their members interact. Monteverde's cloud-forest ecosystem consists of the community plus the air, water, soil, nutrients, and energy the community's organisms use.

At the organismal level, the science of ecology describes relationships between organisms and their physical environments. It helps us understand, for example, what aspects of the golden toad's environment were important to it, and why. **Population ecology** investigates the quantitative dynamics of how individuals within a species interact with one another. It helps us understand why populations of some species (such as the golden toad) decline while populations of others (such as ourselves) increase. **Community ecology** focuses on interactions among species, from one-to-one interactions to complex interrelationships involving entire communities. In the case of Monteverde, it allows us to study how the golden toad and many other species of its cloud-forest community interact. Finally, **ecosystem ecology** reveals patterns, such as energy and nutrient flow, by studying living and nonliving components of systems in conjunction. As we will see, changing climate has had a strong influence on the organisms of Monteverde's cloud-forest ecosystem.

As improving technologies allow scientists to learn more about the complex operations of natural systems on a global scale, ecologists are increasingly expanding their horizons beyond ecosystems to the biosphere as a whole. In this chapter we explore ecology up through the population level. In Chapter 6 we examine the community level, and in Chapter 7 we explore the ecosystem and biosphere levels.

Habitat, niche, and degree of specialization are important in organismal ecology

On the organismal level, each organism relates to its environment in ways that tend to maximize its survival and reproduction. One key relationship involves the specific environment in which an organism lives—its **habitat.** A species' habitat consists of both living and nonliving elements—of rock, soil, leaf litter, and humidity, as well as the other organisms around it. The golden toad lived in a habitat of cloud forest— more specifically, on the moist forest floor, using seasonal pools for breeding and burrows for shelter. The plants known as epiphytes use other plants as habitat; they grow on trees for physical support, obtaining water from the air and nutrients from organic debris that collects among their leaves. Epiphytes thrive in cloud forests because they require a habitat with high humidity, and Monteverde hosts more than 330 species of epiphytes, mostly ferns, orchids, and bromeliads (pineapple relatives). By collecting pools of rainwater and pockets of leaf litter, epiphytes create habitat for many other organisms, including many invertebrates and even frogs that lay their eggs in the rainwater pools.

Habitats are scale-dependent. A tiny soil mite may perceive its habitat as a mere square meter of soil. A vulture, in contrast, may view its habitat in terms of miles upon miles of hills and valleys that it easily traverses by air.

Each organism thrives in certain habitats and not in others, leading to nonrandom patterns of **habitat use.** Mobile organisms actively select habitats in which to live from among the range of options they encounter, a process called *habitat selection.* In the case of plants and sessile animals, whose progeny disperse passively, patterns of habitat use result from success in some habitats and failure in others. The criteria by which organisms favor some habitats over others can vary greatly. The soil mite may judge available habitats in terms of the chemistry, moisture, and compactness of the soil and the percentage and type of organic matter. The vulture may ignore not only soil but also topography and vegetation, focusing solely on the abundance of dead animals in the area that it scavenges for food. Every species judges habitats differently because every species has different needs.

Habitat selection is important in environmental science because the availability and quality of habitat are crucial to an organism's well-being. Indeed, because habitats provide everything an organism needs, including nutrition, shelter, breeding sites, and mates, the organism's very survival depends on the availability of suitable habitats. Often this engenders conflict with people who want to alter or develop a habitat for their own purposes.

Another way in which an organism relates to its environment is through its niche. A species' **niche** reflects its use of resources and its functional role in a community. This includes its habitat use, its consumption of certain foods, its role in the flow of energy and matter, and its interactions with other organisms. The niche is a multidimensional concept, a kind of summary of everything an organism does. We will examine the niche concept further in Chapter 6 (▸pp. 151–152).

Organisms vary in the breadth of their niche. Species with narrow breadth, and thus very specific requirements, are said to be **specialists.** Those with broad tolerances, able to use a wide array of habitats or resources, are **generalists.** For example, in a study of eight Costa Rican bird species that feed from epiphytes, ornithologist T. Scott Sillett found that four were generalists. The other four were specialists on the insect resources the epiphytes provided and spent more than 75% of their foraging efforts feeding from ephiphytes. Specialist and generalist strategies each have advantages and disadvantages. Specialists can be successful over evolutionary time by being extremely good at the things they do, but they are vulnerable when conditions change and threaten the habitat or resource on which they have specialized. Generalists meet with success by being able to live in many different places and weather variable conditions, but they may not thrive in any one situation to the degree that a specialist does. An organism's habitat, niche, and degree of specialization each reflect the adaptations of the species and are products of natural selection.

Population Ecology

Individuals of the same species inhabiting a particular area make up a population. Species may consist of multiple populations that are geographically isolated from one another. This is the case with a species characteristic of Monteverde—the resplendent quetzal *(Pharomachrus mocinno),* considered one of the world's most spectacular birds (see Figure 5.1a). Although it ranges from southernmost Mexico to Panama, the resplendent quetzal lives only in high-elevation tropical forest and is absent from low-elevation areas. Moreover, human development has destroyed much of its forest habitat. Thus, the species today exists in many separate populations scattered across Central America.

In contrast, humans have become more mobile than any other species and have spread into nearly every corner of the planet. As a result, it is difficult to define a distinct human population on anything less than the global scale. Some would maintain that in the ecological sense of the word, all 6.5 billion of us comprise one population.

Populations exhibit characteristics that help predict their dynamics

Whether one is considering humans or quetzals or golden toads, all populations show characteristics that help population ecologists predict the future dynamics of that population. Attributes such as density, distribution, sex ratio, age structure, and birth and death rates all help the ecologist understand how a population may grow or decline. The ability to predict growth or decline is especially useful in monitoring and managing threatened and endangered species (▸pp. 331–335). It is also vital in applying to human populations (Chapter 8). Understanding human population dynamics, their causes, and their consequences is one of the central elements of environmental

(b) 19th-century lithograph of pigeon hunting in Iowa

(a) Passenger pigeon

FIGURE 5.11 The passenger pigeon (**a**) was once North America's most numerous bird, and its flocks literally darkened the skies when millions of birds passed overhead (**b**). However, human cutting of forests and hunting drove the species to extinction within a few decades.

science and one of the prime challenges for our society today.

Population size Expressed as the number of individual organisms present at a given time, **population size** may increase, decrease, undergo cyclical change, or remain the same over time. Extinctions are generally preceded by population declines. As late as 1987, scientists documented a golden toad population at Monteverde in excess of 1,500 individuals, but in 1988 and 1989 scientists sighted only a single toad. By 1990, the species had disappeared.

The passenger pigeon *(Ectopistes migratorius),* also now extinct, illustrates the extremes of population size (Figure 5.11). It was once the most abundant bird in North America; flocks of passenger pigeons literally darkened the skies. In the early 1800s, ornithologist Alexander Wilson wrote of watching a flock of 2 billion birds 390 km (240 mi) long that took 5 hours to fly over and sounded like a tornado. Passenger pigeons nested in gigantic colonies in the forests of the upper Midwest and southern Canada. Once people began cutting the forests, however, the birds' great concentrations made them easy targets for market hunters, who gunned down thousands at a time and shipped them to market by the wagonload. By the end of the 19th century, the passenger pigeon population had declined to such a low number that they could not form the large colonies they apparently needed to breed effectively. In 1914, the last passenger pigeon on Earth died in the Cincinnati Zoo, bringing the continent's most numerous bird species to extinction within just a few decades.

Population density The flocks and breeding colonies of passenger pigeons showed high population density, another attribute that ecologists assess to better understand populations. **Population density** describes the number of individuals within a population per unit area. For instance, the 1,500 golden toads counted in 1987 within 4 km^2 (988 acres) indicated a density of 375 toads/km^2. In general, larger organisms have lower population densities because they require more resources to survive.

High population density can make it easier for organisms to group together and find mates, but it can also lead to conflict in the form of competition if space, food, or mates are in limited supply. Overcrowded organisms may also become more vulnerable to the predators that feed on them, and close contact among individuals can increase the transmission of infectious disease. For these reasons, organisms sometimes leave an area when densities become too high. In contrast, at low population densities, organisms benefit from more space and resources but may find it harder to locate mates and companions.

Overcrowding at high population densities is thought to have doomed Monteverde's harlequin frog (*Atelopus varius*; see Figure 5.1c), an amphibian that disappeared from the cloud forest at the same time as the golden toad. The harlequin frog is a habitat specialist, favoring "splash zones," areas alongside rivers and streams that receive spray from waterfalls and rapids. As Monteverde's climate grew warmer and drier in the 1980s and 1990s, water flow decreased, and many streams dried up. Splash zones grew smaller and fewer, and harlequin frogs were forced to cluster together in what remained of the splash-zone habitat. Researchers J. Alan Pounds and Martha Crump recorded frog population densities up to 4.4 times higher than normal, with more than 2 frogs per meter (3.3 ft) of stream. Such overcrowding likely made the frogs more vulnerable to disease transmission, predator attack, and assault from parasitic flies. From their field research, during which Pounds and Crump witnessed 40 frogs dead or dying, the researchers concluded that these factors led to the harlequin frog's disappearance from Monteverde.

Thankfully, a new population of harlequin frogs was found in 2003 on a private reserve elsewhere in Costa Rica, so there is still hope that the species may survive. The frog was rediscovered by University of Delaware student Justin Yeager, who was doing field research during his study abroad trip in Costa Rica that summer.

Population distribution It was not simply the harlequin frog's density, but also its distribution in space that led to its demise at Monteverde. **Population distribution,** or **population dispersion,** describes the spatial arrangement of organisms within an area. Ecologists define three distribution types: random, uniform, and clumped (Figure 5.12). In a *random distribution*, individuals are located haphazardly in space in no particular pattern. This type of distribution can occur when the resources an organism needs are found throughout an area and other organisms do not strongly influence where members of a population settle.

A *uniform distribution* is one in which individuals are evenly spaced. This can occur when individuals hold territories or otherwise compete for space. For instance, in a desert where there is little water, each plant may need a certain amount of space for its roots to gather adequate moisture. As a result, each individual plant may be equidistant from others.

FIGURE 5.12 Individuals in a population can be spatially distributed over a landscape in three fundamental ways. In a random distribution (**a**), organisms are dispersed at random through the environment. In a uniform distribution (**b**), individuals are spaced evenly, at equal distances from one another. Territoriality can result in such a pattern. In a clumped distribution (**c**), individuals occur in patches, concentrated more heavily in some areas than in others. Habitat selection or flocking to avoid predators can result in such a pattern.

In a *clumped distribution,* the pattern most common in nature, organisms arrange themselves according to the availability of the resources they need to survive. Many desert plants grow in patches around isolated springs or along arroyos that flow with water after rainstorms. During their mating season, golden toads were found clumped at seasonal breeding pools. Humans, too, exhibit clumped distribution; people frequently aggregate together in urban centers. Clumped distributions often indicate habitat selection. Distributions can depend on the scale at which one measures them. At very large scales, all organisms show clumped or patchy distributions, because some parts of the total area they inhabit are bound to be more hospitable than others.

Sex ratios For organisms that reproduce sexually and have distinct male and female individuals, the sex ratio of a population can help determine whether it will increase or decrease in size over time. A population's **sex ratio** is its proportion of males to females. In monogamous species (in which each sex takes a single mate), a 50/50 sex ratio maximizes population growth, whereas an unbalanced ratio leaves many individuals of one sex without mates.

Age structure Populations most often consist of individuals of different ages. **Age distribution,** or **age structure,** describes the relative numbers of organisms of each age within a population. Like sex ratio, age distribution can have strong effects on rates of population growth or decline. A population made up mostly of individuals past reproductive age will tend to decline over time. In contrast, a population with many individuals of reproductive age or soon to be of reproductive age is likely to increase. A population with an even age distribution will likely remain stable as births keep pace with deaths.

Age structure diagrams, often called *age pyramids,* are visual tools scientists use to show the age structure of populations (Figure 5.13). The width of each horizontal bar represents the relative size of each age class. A pyramid with a wide base has a relatively large age class that has not yet reached its reproductive stage, indicating a population much more capable of rapid growth. In this respect, a wide base of an age pyramid is like an oversized engine on a rocket—the bigger the booster, the faster the increase. We will examine age pyramids further in Chapter 8 (▸pp. 224–226) in reference to human populations.

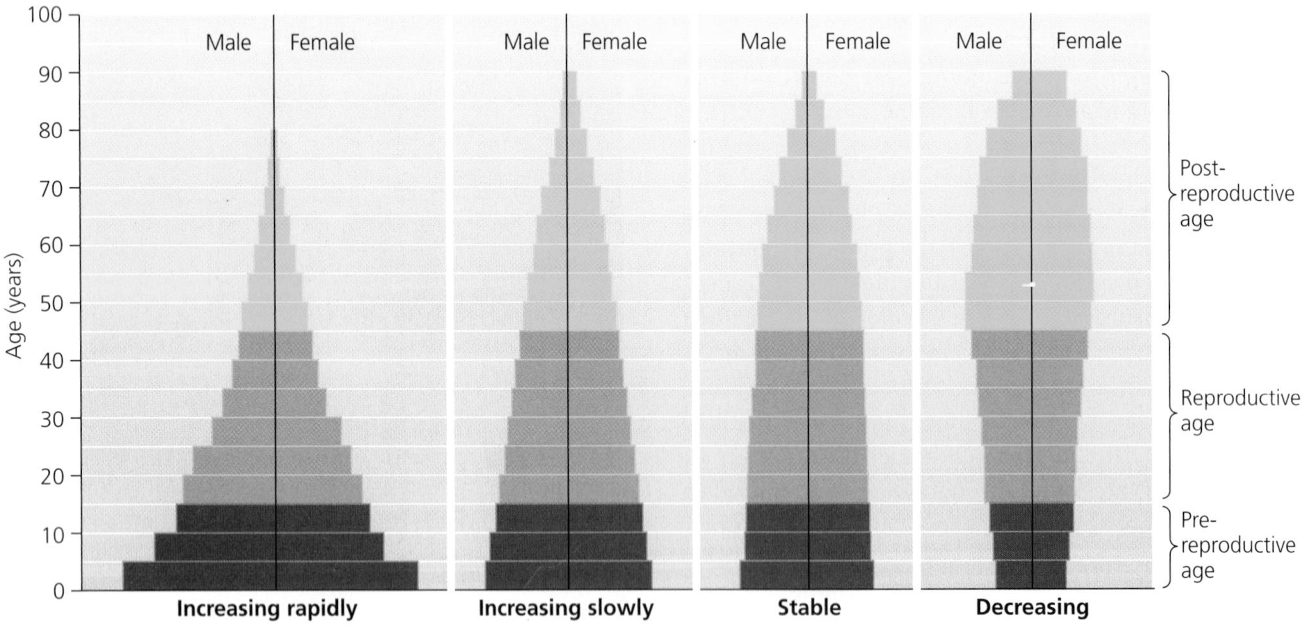

FIGURE 5.13 Age structure diagrams show the relative frequencies of individuals of different age classes in a population. In this example for humans, populations heavily weighted toward young age classes (at left) grow most quickly, whereas those weighted heavily toward old age classes (at right) decline.

Birth and death rates All the preceding factors can influence the rates at which individuals within a population are born and die. A convenient way to express birth and death rates is to measure the number of births and deaths per 1,000 individuals for a given time period. Such a rate is termed a *crude birth rate* or *crude death rate.*

Just as individuals of different ages have different abilities to reproduce, individuals of different ages show different probabilities of dying. For instance, people are more likely to die at old ages than young ages; if you were to follow 1,000 10-year-olds and 1,000 80-year-olds for a year, you would find that at year's end more 80-year-olds had died than 10-year-olds. However, this pattern does not hold for all organisms. Amphibians such as the golden toad produce large numbers of young, which suffer high death rates. For a toad, death is less likely (and survival more likely) at an older age than at a very young age. To show how the likelihood of death can vary with age, ecologists use graphs called **survivorship curves** (Figure 5.14). There are three fundamental types of survivorship curves. Humans, with higher death rates at older ages, show a *type I* survivorship curve. Toads, with highest death rates at young ages, show a *type III* survivorship curve. A *type II* survivorship curve is intermediate and indicates equal rates of death at all ages. Many birds are thought to show type II curves.

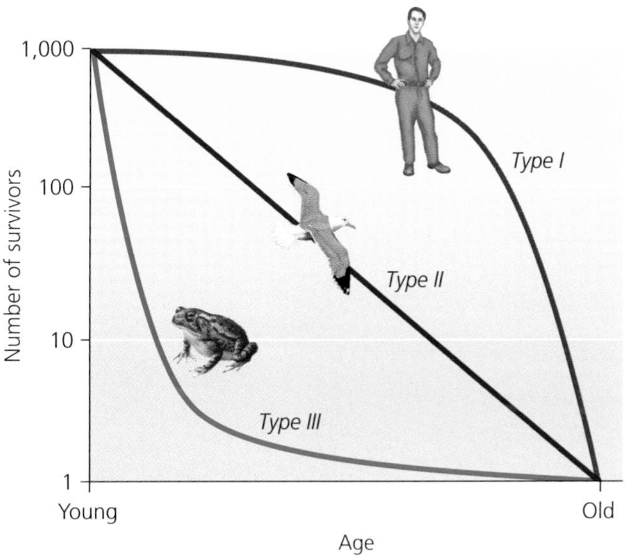

FIGURE 5.14 In a type I survivorship curve, survival rates are high when organisms are young and decrease sharply when organisms are old. In a type II survivorship curve, survival rates are equivalent regardless of an organism's age. In a type III survivorship curve, most mortality takes place at young ages, and survival rates are greater at older ages. Some examples include humans (type I), birds (type II), and amphibians (type III).

Populations may grow, shrink, or remain stable

Now that we have outlined some key attributes of populations, we are ready to take a quantitative view of population change by examining some simple mathematical concepts used by population ecologists and *demographers* (those who study human populations). Population growth, or decline, is determined by four factors:

1. Births within the population *(natality)*
2. Deaths within the population *(mortality)*
3. **Immigration** (arrival of individuals from outside the population)
4. **Emigration** (departure of individuals from the population)

To understand how a population changes, we measure its **growth rate,** which can be calculated as the crude birth rate plus the immigration rate, minus the crude death rate plus the emigration rate, each expressed as the number per 1,000 individuals per year:

$$(\text{Crude birth rate} + \text{immigration rate}) - (\text{Crude death rate} + \text{emigration rate}) = \text{Growth rate}$$

The resulting number tells us the net change in a population's size per 1,000 individuals. For example, a population with a crude birth rate of 18 per 1,000, a crude death rate of 10 per 1,000, an immigration rate of 5 per 1,000, and an emigration rate of 7 per 1,000 would have a growth rate of 6 per 1,000:

$$(18/1{,}000 + 5/1{,}000) - (10/1{,}000 + 7/1{,}000) = 6/1{,}000$$

Thus, a population of 1,000 in one year will reach 1,006 in the next. If the population is 1,000,000, it will reach 1,006,000 the next year. These population increases are often expressed as percentages, which we can calculate using the formula:

$$\text{Growth rate} \times 100\%$$

Thus, a growth rate of 6/1,000 would be expressed as:

$$6/1{,}000 \times 100\% = 0.6\%$$

By measuring population growth in terms of percentages, scientists can compare increases and decreases in species that have far different population sizes. They can also project changes that will occur in the population over longer periods, much like you might calculate the amount of interest your savings account will earn over time.

Table 5.3 Exponential Growth in a Savings Account with 5% Annual Compound Interest

Age (in years)	Principal
0 (birth)	$1,000
10	$1,629
20	$2,653
30	$4,322
40	$7,040
50	$11,467
60	$18,679
70	$30,426
80	$49,561

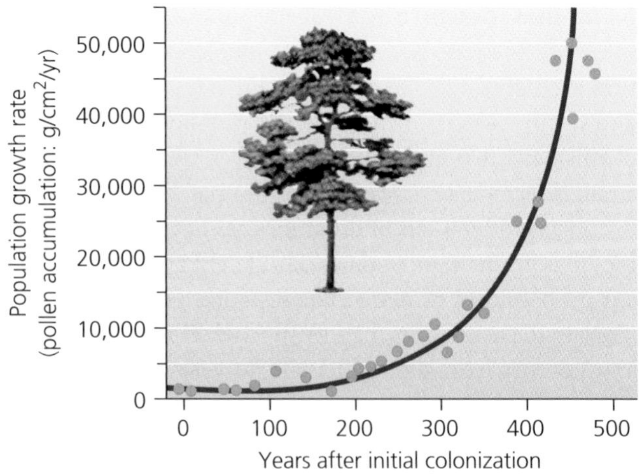

FIGURE 5.15 Although no species can maintain exponential growth indefinitely, some may grow exponentially for a time when colonizing an unoccupied environment or exploiting an unused resource. Scientists have used pollen records to determine that the Scots pine *(Pinus sylvestris)* increased exponentially after the retreat of glaciers following the last ice age around 9,500 years ago. Go to **GRAPHit!** at www.aw-bc.com/withgott or on the student CD-ROM. Data from Bennett, K. D. 1983. Postglacial population expansion of forest trees in Norfolk, U.K. *Nature* 303: 164–167.

Unregulated populations increase by exponential growth

When a population, or anything else, increases by a fixed percentage each year, it is said to undergo **exponential growth.** A savings account is a familiar frame of reference for describing exponential growth. If at the time of your birth your parents had invested $1,000 in a savings account earning 5% interest compounded each year, you would have only $1,629 by age 10, and $2,653 by age 20, but you would have over $30,000 when you turn 70. If you could wait just 10 years more, that figure would rise to nearly $50,000 (Table 5.3). Only $629 was added during your first decade, but approximately $19,000 was added during the decade between ages 70 and 80. The reason is that a fixed percentage of a small number makes for a small increase, but that same percentage of a large number produces a large increase. Thus, as savings accounts (or populations) become larger, each incremental increase likewise gets larger. Such acceleration is a characteristic of exponential growth.

We can visualize changes in population size by using population growth curves. The J-shaped curve in Figure 5.15 shows exponential increase. As Thomas Malthus realized, populations of all organisms increase exponentially unless they meet constraints. Each organism reproduces by a certain amount, and as populations get larger, more individuals reproduce by that amount. If there are no external limits on growth, ecologists theoretically expect exponential growth.

Exponential growth usually occurs in nature when a population is small and environmental conditions are ideal for the organism in question. Most often, these conditions occur when organisms are introduced to a new environment. Mold growing on a piece of bread or fruit, or bacteria colonizing a recently dead animal, are cases in point. But species of any size may show exponential growth under the right conditions. A population of the Scots pine, *Pinus sylvestris,* grew exponentially when it began colonizing the British Isles after the end of the last ice age (see Figure 5.15). Receding glaciers had left conditions ideal for its exponential expansion.

Limiting factors restrain population growth

However, exponential growth rarely lasts long. If even a single species in Earth's history had increased exponentially for very many generations, it would have blanketed the planet's surface, and nothing else could have survived. Instead, every population eventually is constrained by **limiting factors,** physical, chemical, and biological characteristics of the environment that restrain population growth. The interaction of these factors determines the **carrying capacity,** the maximum population size of a species that a given environment can sustain.

Ecologists use the curve in Figure 5.16 to show how an initial exponential increase is slowed and finally brought to a standstill by limiting factors. Called the **logistic**

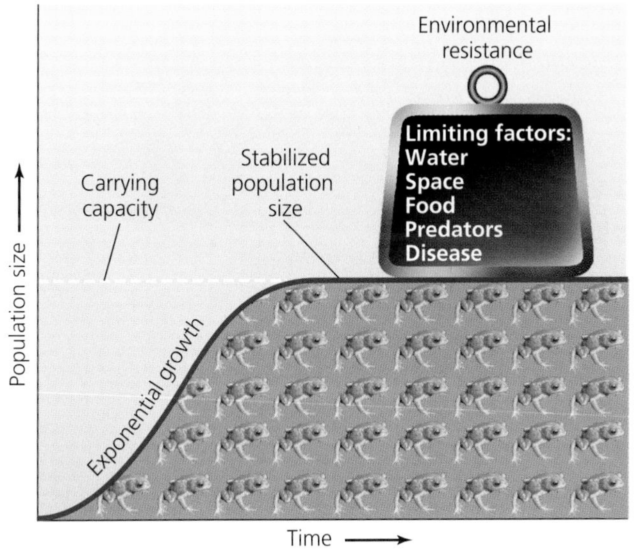

FIGURE 5.16 The logistic growth curve shows how population size may increase rapidly at first, then grow more slowly, and finally stabilize at a carrying capacity. Carrying capacity is determined both by the biotic potential of the organism and by various external limiting factors, collectively termed *environmental resistance*.

growth curve, it rises sharply at first but then begins to level off as the effects of limiting factors become stronger. Eventually the force of these factors—which taken together are termed *environmental resistance*—stabilizes the population size at its carrying capacity.

The logistic curve is a simplified model, and real populations can behave differently. Some may cycle indefinitely above and below the carrying capacity. Some may show cycles that become less extreme and approach the carrying capacity. Others may overshoot the carrying capacity and then crash, fated either for extinction or recovery (Figure 5.17).

Many factors contribute to environmental resistance and influence a population's growth rate and carrying capacity. Space is one factor that limits the number of individuals a given environment can support; if there is no physical room for additional individuals, they are unlikely to survive. Other limiting factors for animals in a terrestrial environment include the availability of food, water, mates, shelter, and suitable breeding sites; temperature extremes; prevalence of disease; and abundance of predators. Plants are often limited by amounts of sunlight and moisture and the type of soil chemistry, in addition to disease and attack from plant-eating animals. In aquatic systems, limiting factors include salinity, sunlight, temperature, dissolved oxygen, fertilizers, and pollutants.

Sometimes one limiting factor may outweigh all others and restrict population growth. For example, scientists hypothesize that Monteverde's population of golden toads had plenty of space, food, and shelter, but that it lacked adequate moisture. If moisture were the primary limiting factor, then increasing moisture would have increased the carrying capacity of the habitat for the toads. Indeed, to determine limiting factors, ecologists often conduct experiments in which they increase or decrease a hypothesized limiting factor to observe its effects on population size. Unfortunately in the case of the golden toad, such experiments could not be conducted before its disappearance.

Carrying capacities can change

Because limiting factors can be numerous, and because environments are complex and ever-changing, carrying capacity can vary constantly. The human species illustrates another reason that carrying capacity is not necessarily a fixed entity. Although all organisms are subject to environmental resistance, they may be capable of altering their environment to reduce this resistance. Our own species has proved particularly effective at this. When our ancestors began to build shelters and use fire for heating and cooking, they reduced the environmental resistance of areas with cold climates and were able to expand into new territory. As limiting factors are overcome (either through the development of new technologies or through natural environmental change), the carrying capacity for a species increases. We humans have managed so far to increase the planet's carrying capacity for ourselves, but unfortunately for the golden toad, the limiting factors on its population growth seemed to exert ever-increasing pressure during the late 1980s.

Weighing the Issues: Carrying Capacity and Human Population Growth

As we saw in Chapter 1, the global human population has risen from fewer than 1 billion 200 years ago to 6.5 billion today, and we have far exceeded our historic carrying capacity. What factors increased Earth's carrying capacity for people? Do you think there are limiting factors for the human population? What might they be? Do you think we can keep raising our carrying capacity in the future? Might Earth's carrying capacity for us decrease?

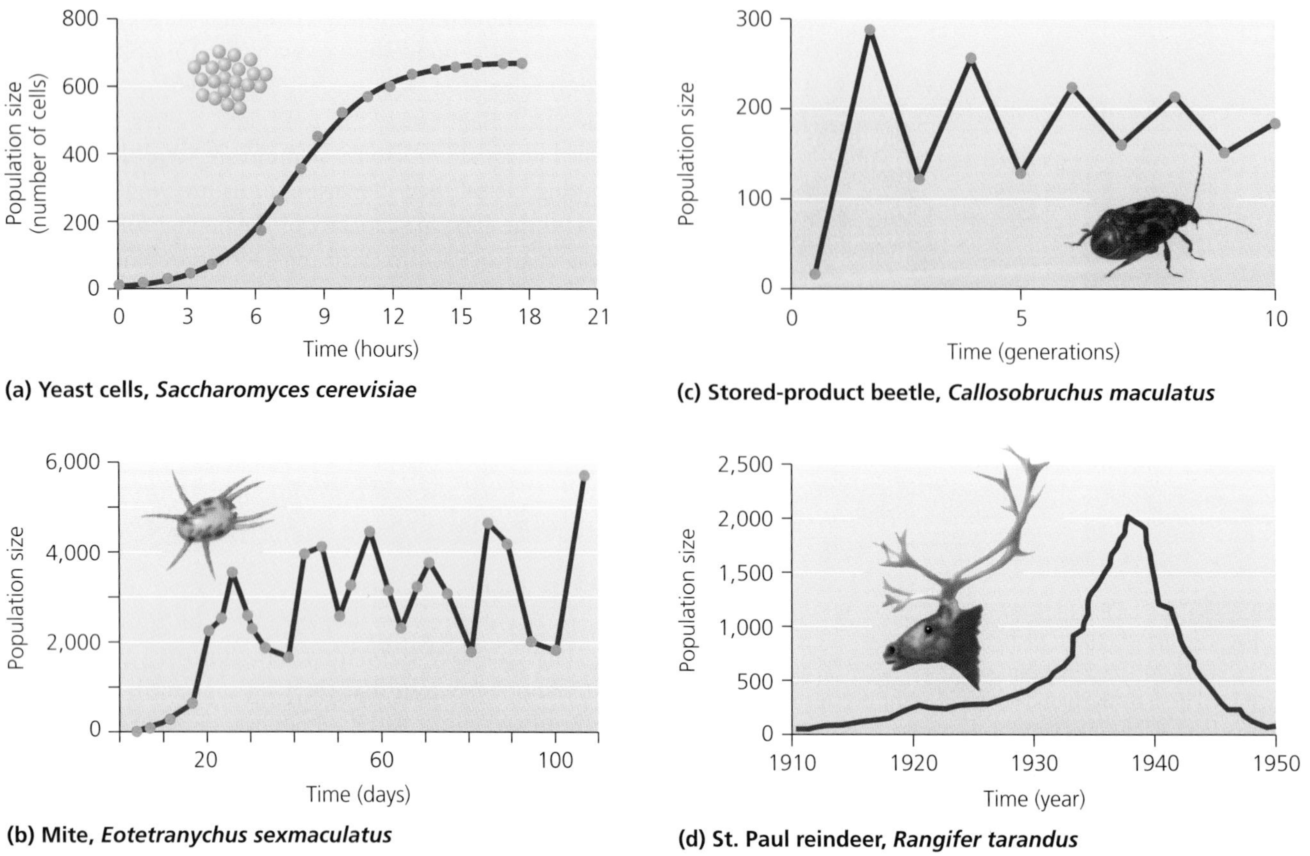

FIGURE 5.17 Population growth in nature often departs from the stereotypical logistic growth curve, and it can do so in several fundamental ways. Yeast cells from an early lab experiment show logistic growth **(a)** that, like the Scots pine in Figure 5.15, closely matches the theoretical model. Some organisms, like the mite shown here, show cycles in which population fluctuates indefinitely above and below the carrying capacity **(b)**. Population oscillations can also dampen, lessening in intensity and eventually stabilizing at carrying capacity **(c)**, as in a lab experiment with the stored-product beetle. Populations that rise too fast and deplete resources may crash just as suddenly **(d)**, like the population of reindeer introduced to the Bering Sea island of St. Paul. Data from Pearl, R. 1927. The growth of populations. *Quarterly Review of Biology* 2: 532–548, (a); Huffaker, C. B. 1958. Experimental studies on predation: Dispersion factors and predator-prey oscillations, *Hilgardia* 27: 343–383, (b); Utida, S. 1967. Damped oscillation of population density at equilibrium, *Researches on Population Ecology* 9: 1–9, (c); Scheffer, V. C. 1951. Rise and fall of a reindeer herd, *Scientific Monthly* 73: 356–362, (d).

The influence of some factors depends on population density

Just as carrying capacity is not a fixed entity, the influence of limiting factors can vary with changing conditions. In particular, the density of a population can increase or decrease the impact of certain factors on that population. Recall that high population density can help organisms find mates but can also increase competition and the risk of predation and disease. Such factors are said to be **density-dependent** factors, because their influence waxes and wanes according to population density. The logistic growth curve in Figure 5.16 represents the effects of density dependence. The more population size rises, the more environmental resistance kicks in.

Density-independent factors are limiting factors whose influence is not affected by population density. Temperature extremes and catastrophic events such as floods, fires, and landslides are examples of density-independent factors, because they can eliminate large numbers of individuals without regard to their density.

Biotic potential and reproductive strategies vary from species to species

Limiting factors from an organism's environment provide only half the story of population regulation. The other half comes from the attributes of the organism itself. For example, organisms differ in their *biotic potential,* or ability to produce offspring. A fish with a short gestation period that lays thousands of eggs at a time has high biotic potential, whereas a whale with a long gestation period that gives birth to a single calf at a time has low biotic potential. The interaction between an organism's biotic potential and the environmental resistance to its population growth helps determine the fate of its population.

Giraffes, elephants, humans, and other large animals with low biotic potential produce a relatively small number of offspring and take a long time to gestate and raise each of their young. Species that take this approach to reproduction compensate by devoting large amounts of energy and resources to caring for and protecting the relatively few offspring they produce during their lifetimes. Such species are said to be **K-selected.** K-selected species are so named because their populations tend to stabilize over time at or near their carrying capacity, and *K* is an abbreviation for carrying capacity. Because their populations stay close to carrying capacity, these organisms must be good competitors, able to hold their own in a crowded world. Thus in these species, natural selection favors individuals that invest in producing offspring of high quality that can be good competitors.

In contrast, species that are **r-selected** focus on quantity, not quality. Species considered to be r-selected have high biotic potential and devote their energy and resources to producing as many offspring as possible in a relatively short time. Their offspring do not require parental care after birth, so r-strategists simply leave their survival to chance. The abbreviation *r* denotes the rate at which a population increases in the absence of limiting factors. Populations of r-selected species fluctuate greatly, such that they are often well below carrying capacity. This is why natural selection in these species favors traits that lead to rapid population growth. Many fish, plants, frogs, insects, and others are r-selected. The golden toad is one example. Each adult female laid 200–400 eggs, and its tadpoles spent 5 weeks unsupervised in the breeding pools metamorphosing into adults.

Table 5.4 summarizes stereotypical traits of r-selected and K-selected species. However, it is important to note that these are two extremes on a continuum and that most species fall somewhere between these endpoints. Moreover, some organisms show combinations of traits that do not clearly correspond to a place on the continuum. A redwood tree *(Sequoia sempervirens),* for instance, is large and long-lived, yet it produces many small seeds and offers no parental care.

Table 5.4 Traits of r-selected and K-selected species

r-selected species	K-selected species
Small size	Large size
Fast development	Slow development
Short-lived	Long-lived
Reproduction early in life	Reproduction later in life
Many small offspring	Few large offspring
Fast population growth rate	Slow population growth rate
No parental care	Parental care
Weak competitive ability	Strong competitive ability
Variable population size, often well below carrying capacity	Constant population size, close to carrying capacity
Variable and unpredictable mortality	More constant and predictable mortality

Populations of K-selected species are generally regulated by density-dependent factors such as disease, predation, and food limitation. In contrast, density-independent factors tend to regulate populations of r-selected species, whose success or failure is often determined by large-scale environmental change. Many r-selected species frequently experience large swings in population size, such as rapid increases during the breeding season and rapid declines soon after, when unfit and unlucky young are removed from the population. For this reason, scientists often have difficulty determining whether steep population declines are a part of natural cycles or a sign of serious trouble. For years, scientists debated the golden toad's apparent extinction. Now that it has failed to reappear for over 15 years, most agree that the toad's population crash was not part of a normal, repeating cycle.

Changes in populations influence the composition of communities

In the late 1980s, the golden toad and the harlequin frog were the most diligently studied species affected by changing environmental conditions in the Costa Rican cloud forest. However, once scientists began looking at populations of other species at Monteverde, they began to notice more troubling changes. By the early 1990s, not only had golden toads, harlequin frogs, and other organisms been pushed from their cloud-forest habitat into apparent extinction, but many species from lower, drier habitats had also begun to appear at Monteverde. These immigrants included species tolerant of drier conditions,

The Science behind the Story

Climate Change and Its Effects on Monteverde

Soon after the golden toad's disappearance, scientists began to investigate the potential role of climate change in driving cloud-forest species toward extinction. They had noted that the period from July 1986 to June 1987 was the driest on record in Monteverde, with unusually high temperatures and record-low stream flows. These conditions had caused the golden toad's breeding pools to dry up shortly after they filled in the spring of 1987, likely killing nearly all of the eggs and tadpoles present in the pools.

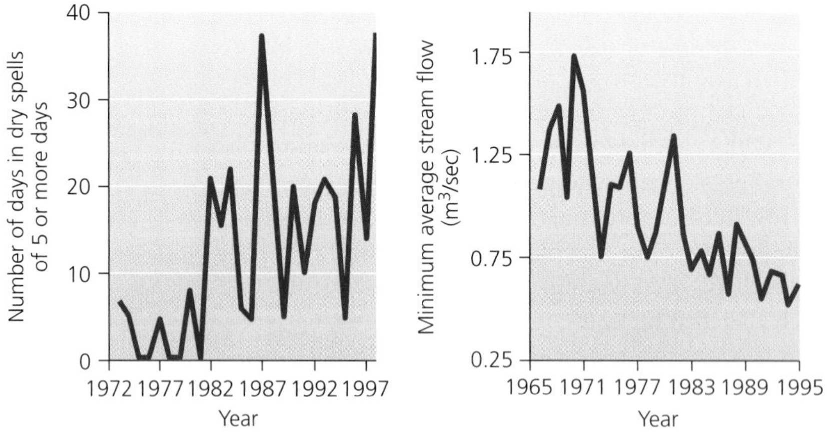

(a) Increase in dry spells **(b) Decrease in stream flow**

Warming and drying trends in Monteverde's climate may have contributed to the region's amphibian declines. Evidence gathered over 25 years shows **(a)** an increase in the annual number of dry days and **(b)** a decrease in the amount of annual stream flow. Data from Pounds, J. A., et al. 1999. Biological response to climate change on a tropical mountain. *Nature* 398: 611–615.

Scientists began reviewing reams of weather data and eventually found that the number of dry days and dry periods each winter in the Monteverde region had increased between 1973 and 1998. Biologists knew that such local climate trends were bad news for amphibians like the golden toad and harlequin frog. Because amphibians breathe and absorb moisture through their skin, they are susceptible to dry conditions, high temperatures, acid rain, and pollutants concentrated by reduced water levels. Based on these facts, herpetologists J. Alan Pounds and Martha Crump in 1994 hypothesized that hot, dry conditions were to blame for increased adult mortality and breeding problems among golden toads and other amphibians.

Throughout this period, scientists worldwide were realizing that the oceans and atmosphere were warming because of human release of carbon dioxide and other gases into the atmosphere. Global climate change (Chapter 18), experts were learning, could produce varying effects on climate at regional and local levels. With this in mind, Pounds and others concerned about Monteverde's changing conditions reviewed the scientific literature on ocean and atmospheric science to analyze the effects on Monteverde's local climate of warming patterns in the ocean regions around Costa Rica.

By 1997 these researchers had determined that Monteverde's cloud forest was becoming drier because the clouds that had given the forest its name and much of its moisture now passed by at higher elevations, where they were no longer in contact with the trees. The primary factor determining the clouds' altitude, the researchers found, is nearby ocean temperatures; as ocean temperatures increase, clouds pass over Monteverde at higher elevations. Once the cloud forest's water supply was pushed upward, out of reach of the mountaintops, the cloud forest began to dry out.

such as blue-crowned motmots *(Momotus momota)* and brown jays *(Cyanocorax morio)*. By the year 2000, 15 dry-forest species had moved into the cloud forest and begun to breed. Meanwhile, population sizes of several cloud-forest bird species had declined. After 1987, 20 of 50 frog species vanished from one part of Monteverde, and ecologists later reported more disappearances, including those of two lizards native to the cloud forest. Scientists hypothesized that the warming, drying trends that researchers were documenting (see "The Science behind the Story," above) were causing population fluctuations and unleashing changes in the composition of the community.

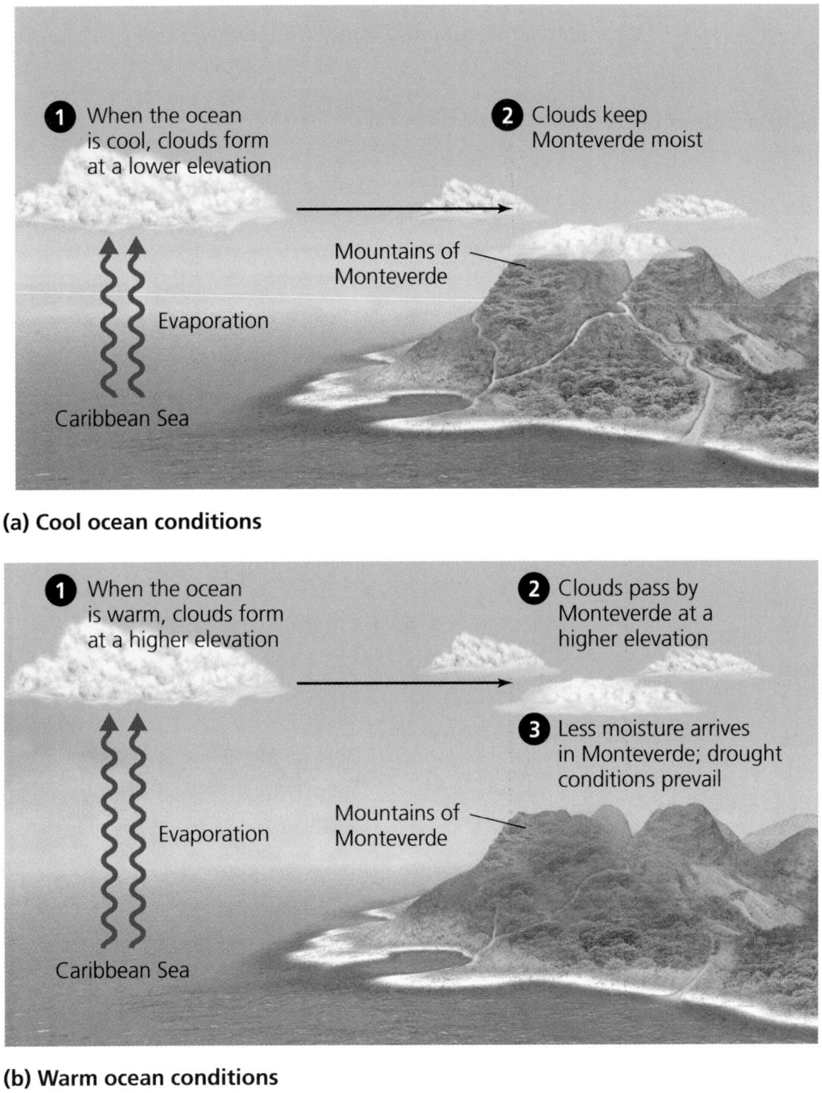

Monteverde's cloud forest gets its name and life-giving moisture from clouds that sweep inland from the oceans. (**a**) When ocean temperatures are cool, the clouds keep Monteverde moist. (**b**) Warmer ocean conditions resulting from global climate change cause clouds to form at higher elevations and pass over the mountains, drying the cloud forest.

In a 1999 paper in the journal *Nature,* Pounds and two colleagues reported these findings. Their conclusion—that broad-scale climate modification was causing local changes at the species, population, and community levels—explained a great number of events occurring at Monteverde. Rising cloud levels and decreasing moisture could explain not only the disappearance of the golden toad and harlequin frog, but also the concurrent population crashes in 1987 and subsequent disappearance of 20 species of frogs and toads from the Monteverde region. Amphibians that survived underwent population crashes in each of the region's three driest years.

Pounds and his co-workers further described "a constellation of demographic changes that have altered communities of birds, reptiles and amphibians" in the area as likely additional consequences of this shift in moisture availability. As these mountaintop forests dried out, dry-tolerant species crept in, and moisture-dependent species were stranded at the mountaintops by a rising tide of dryness. Although organisms may in general be driven from one area to another by changing environmental conditions, if a species has nowhere to go, then extinction may result.

The Conservation of Biodiversity

Changes in populations and communities have been taking place naturally as long as life has existed, but today human development, resource extraction, and population pressure are speeding the rate of change and altering the types of change. The ways we modify our environment cannot be fully understood in a scientific vacuum, however. The actions that threaten biodiversity have complex social, economic, and political roots, and environmental scientists appreciate that we must understand these aspects if we are to develop solutions.

Fortunately, there are things people can do to forestall population declines of species threatened with extinction. Millions of people around the world are already taking action to safeguard the biodiversity and ecological and evolutionary processes that make Earth such a unique place (Chapter 11). Costa Ricans have been confronting the challenges to their nation's biodiversity, and their actions so far show what even a small country of modest means can do.

Social and economic factors affect species and communities

Many of the threats to Costa Rica's species and ecological communities result from past economic and social forces whose influences are still evident. European immigrants and their descendants viewed Costa Rica's lush forests as an obstacle to agricultural development, and timber companies saw them simply as a source of wood products. Costa Rica's leading agricultural products have long included beef and bananas, whose production and cultivation require extensive environmental modification. Between 1945 and 1995, the country's population grew from 860,000 to 3.34 million, and the percentage of land devoted to pasture increased from 12% to 33%. With much of the formerly forested land converted to agriculture, the proportion of the country covered by forest decreased from 80% to 25%. In 1991, Costa Rica was losing its forests faster than any other country in the world—nearly 140 ha (350 acres) per day. As a result, populations of innumerable species were declining, and some were becoming endangered (Figure 5.18). As had occurred in the history of the United States, few people foresaw the need to conserve biological resources until it became clear that they were being rapidly lost.

Costa Rica took steps to protect its environment

During the 1950s a group of Quakers, Christian pacifists who opposed the U.S. military draft, emigrated from Alabama to Costa Rica and founded the village of Monteverde.

(a) Golden-cheeked warbler

(b) Green sea turtle

(c) Red-backed squirrel monkey

FIGURE 5.18 Costa Rica is home to a number of species classified as globally threatened or endangered. The golden-cheeked warbler, *Dendroica chrysoparia* (**a**), winters in Central America but breeds in the Texas hill country, where its habitat is being rapidly lost to housing development. The green sea turtle, *Chelonia mydas* (**b**), is widely distributed throughout the world's oceans and lays eggs on beaches in Costa Rica and elsewhere, but it has undergone steep population declines. The red-backed squirrel monkey, *Saimiri oerstedii* (**c**), is endemic to a tiny area in Costa Rica and is judged vulnerable to forest loss because of its small geographic range. These vertebrate species receive attention from scientists and the media and are the focus of recovery efforts, but many more plants, insects, and other less-celebrated species are declining in number while most of us go about our days unaware of their existence.

FIGURE 5.19 Costa Rica has protected a wide array of its diverse natural areas. This protection has stimulated the nation's economy through ecotourism. Here, visitors experience a walkway through the forest canopy in one of the nation's parks.

The Quakers relied on milk and cheese for much of their economic activity, but they also set aside one-third of their land for conservation. The Quakers' efforts, along with contributions from international conservation organizations, provided the beginnings of what is today the Monteverde Cloud Forest Biological Reserve. This privately managed 10,500-ha (26,000-acre) reserve was established in 1972 to protect the forest and its populations of 2,500 plant species, 400 bird species, 500 butterfly species, 100 mammal species, and 120 reptile and amphibian species, including the golden toad.

In 1970, the Costa Rican government and international representatives came together to create the country's first national parks and protected areas. The first parks centered on areas of spectacular scenery, such as the Poas Volcano National Park. Santa Rosa National Park encompassed valuable tropical dry forest, Tortuguero National Park contained essential nesting beaches for the green turtle (*Chelonia mydas;* see Figure 5.18b), and Cahuita National Park was meant to protect a prominent coral reef system. Initially the government gave the parks little real support. According to Costa Rican conservationist Mario Boza, in their early years the parks were granted only five guards, one vehicle, and no funding.

Today government support is greater. Fully 12% of the nation's area is contained in national parks, and a further 16% is devoted to other types of wildlife and conservation reserves. Costa Ricans, along with international biologists, are working to protect endangered species and recover their populations. Costa Rica and its citizens are now reaping the benefits of their conservation efforts—not only ecological benefits, but also economic ones. Because of its parks and its reputation for conservation, tourists from around the world now visit Costa Rica, a phenomenon called **ecotourism** (Figure 5.19). The ecotourism industry draws more than 1 million visitors to Costa Rica each year, provides thousands of jobs to Costa Ricans, and is a major contributor to the country's economy. Today's Costa Rican economy is fueled in large part by commerce and tourism, whose contributions (40%) outweigh those of industry (22%) and agriculture (13%) combined.

Weighing the Issues: How Best to Conserve Biodiversity?

Most people view national parks and ecotourism as excellent ways to help keep ecological systems intact. Yet the golden toad went extinct despite living within a reserve established to protect it, and climate change does not pay attention to park boundaries. What lessons can we take from this about the conservation of biodiversity?

It remains to be seen how effectively ecotourism can help preserve natural systems in Costa Rica in the long term. As forests outside the parks disappear, the parks are beginning to suffer from illegal hunting and timber extraction. Conservationists like Boza say the parks are still underprotected and underfunded. Ecotourism will likely need to generate still more money to preserve habitat, protect endangered species, and restore altered communities to their former condition. Restoration is being carried out in Costa Rica's Guanacaste Province, for instance, where scientists are restoring dry tropical forest from grazed pasture. Restoration of ecological communities is one phenomenon we will examine in our next chapter, as we move from populations to communities.

Conclusion

The golden toad and other organisms of the Monteverde cloud forest have helped illuminate the fundamentals of evolution and population ecology that are integral to environmental science. The evolutionary processes of natural selection, speciation, and extinction help determine Earth's biodiversity. Understanding how ecological processes work at the population level is crucial to protecting biodiversity threatened by the mass extinction event that many biologists maintain is already under way.

VIEWPOINTS

Conservation at Monteverde

What lessons, if any, can we learn from ecological changes that have occurred in the Monteverde region since the golden toad's discovery in 1964?

Lessons from the Green Mountain: Changes in Ecological and Human Communities

Monteverde—the "Green Mountain"—is one of the best-studied cloud forests in the world. Over the years, researchers have documented negative effects of human activities on the diverse biota. These include forest fragmentation and its isolation of plant and animal populations; lengthened dry seasons due to regional deforestation; and the upward shift of lower-elevation animal and plant populations due to global climate change.

One complex aspect of increasing human presence in Monteverde is ecotourism. The annual influx of nearly 80,000 visitors has had negative effects on the human community. Consumerism has increased. Television has replaced square-dancing. Cell telephones have replaced the single community party line used in previous decades. Guides rely on having cars rather than feet to show visitors around.

However, the presence of outside visitors who are deeply interested in nature has also affected Monteverde in positive ways. Some Monteverde families have abandoned dairy farming as their source of income, and have been able to turn to natural history guiding or managing small hotels. This has led to the reforestation of pastures, and the coalescence of formerly fragmented forest patches.

In addition, each visitor to Monteverde becomes a potential conservation activist. By seeing the unique wildlife and habitats—and learning about the causes that threaten it—visitors can take political and economic action to promote conservation when they return home. Grass roots conservation and educational organizations such as the Monteverde Conservation League and the Cloud Forest School have done much to promote conservation. These organizations rely on the contributions of ecotourists who have been touched by their experiences of seeing a quetzal or flowering orchid, and who become moved to preserve the habitat of the region.

Nalini M. Nadkarni is a member of the faculty at The Evergreen State College in Olympia, Washington. Her research focuses on the ecological interactions that occur in tropical and temperate rainforest canopies. The co-founder and president of the International Canopy Network, she is co-author of the book *Monteverde: Ecology and Conservation of a Tropical Cloud Forest.*

Conservation Successes in the Shadow of the Golden Toad

In the late 1970s, the world began to discover the pristine montane forests and peaceful farming community of Monteverde, Costa Rica. Overnight, it seemed, ecotourism boomed, new houses, hotels, and restaurants sprung up, and tens of thousands of visitors arrived annually, eager to see resplendent quetzals and epiphyte-laden cloud forests. But a decade later, equally suddenly, the golden toad, a stunning amphibian found nowhere else on Earth, vanished. Shortly afterward, 19 other species of frogs and salamanders—40% of all the amphibian species that were present when I began my dissertation research there in 1979—disappeared from the area. Fortunately, none but the golden toad was endemic to Monteverde.

Yet, surprisingly, there have also been impressive achievements in conservation at Monteverde over this same time period. In 1979, poaching of large mammals and birds was commonplace. Now the very people who hunted scarce animals with rifles use binoculars instead as they lead natural history tours. Tapirs and guans are more common today than they have been for more than half a century. As the Monteverde Cloud Forest Preserve has expanded tenfold, clearings on the Atlantic slope have reverted to lush forest. The Guacimal River, formerly rancid because of waste dumped by the community dairy plant, is much cleaner now.

Thus, although ecotourism can have a negative effect on local populations of some plants and animals, it can also have enduring positive impacts on conservation—not just locally by increasing economic opportunities and incentives for land preservation, but also globally by educating the public about environmental values. Monteverde has inspired visitors to appreciate tropical forests and, through that experience, their own natural heritage—and that may help protect threatened habitats worldwide.

Nathaniel Wheelwright is Professor of Biology at Bowdoin College in Brunswick, Maine, director of the Bowdoin Scientific Station on Kent Island, New Brunswick, and co-editor of *Monteverde: Ecology and Conservation of a Tropical Cloud Forest.*

Explore this issue further by accessing **Viewpoints** at www.aw-bc.com/withgott.

REVIEWING OBJECTIVES

You should now be able to:

Explain the process of natural selection and cite evidence for this process

- Because organisms produce excess young, individuals vary in their traits, and many traits are inherited, some individuals will prove better at surviving and reproducing. Their genes will be passed on and become more prominent in future generations. (pp. 118–119)
- Mutations and recombination provide the genetic variation for natural selection. (pp. 119–120)
- We have produced our pets, farm animals, and crop plants through artificial selection. (pp. 120–122)

Describe the ways in which evolution results in biodiversity

- Natural selection can act as a diversifying force as organisms adapt to their environments in myriad ways. (pp. 120–122, 124)
- Speciation (by geographic isolation and other means) produces new species. (pp. 122–124)
- Once they have diverged, lineages continue diverging, a process represented in a phylogenetic tree. (pp. 124–125)

Discuss reasons for species extinction and mass extinction events

- Extinction often occurs when species that are highly specialized or that have small populations encounter rapid environmental change. (pp. 126–127)
- Earth's life has experienced five known episodes of mass extinction, due to an asteroid impact and possibly volcanism and other factors. (pp. 127–129)

List the levels of ecological organization

- Ecologists study phenomena on the organismal, population, community, and ecosystem levels—and, increasingly, on the biosphere level. (pp. 129–130)

Outline the characteristics of populations that help predict population growth

- Populations are characterized by population size, population density, population distribution, sex ratio, age structure, and birth and death rates. (pp. 131–135)
- Immigration and emigration, as well as birth and death rates, determine how a population will grow or decline. (p. 135)

Assess logistic growth, carrying capacity, limiting factors, and other fundamental concepts of population ecology

- Populations unrestrained by limiting factors will undergo exponential growth until they meet environmental resistance. (p. 136)
- Logistic growth describes the effects of density dependence; exponential growth slows as population size increases, and population size levels off at a carrying capacity. (pp. 136–138)
- K-selection and r-selection describe theoretical extremes in how organisms can allocate growth and reproduction. (pp. 138–139)

Identify efforts and challenges involved in the conservation of biodiversity

- Social and economic factors influence our impacts on natural systems. (pp. 141–144)
- Extensive efforts to protect and restore species and habitats will be needed to prevent further erosion of biodiversity. (pp. 141–144)

TESTING YOUR COMPREHENSION

1. Explain the premises and logic that supports the concept of natural selection.
2. How does allopatric speciation occur?
3. Name two examples of evidence for natural selection.
4. Name three organisms that have gone extinct, and give a probable reason for each extinction.
5. What is the difference between a species and a population? Between a population and a community?
6. Contrast the concepts of habitat and niche.
7. List and describe each of the five major population characteristics discussed in this chapter. Explain how each shapes population dynamics.
8. Could any species undergo exponential growth forever? Explain your answer.
9. Describe how limiting factors relate to carrying capacity.
10. Explain the difference between K-selected species and r-selected species. Can you think of examples of each that were not mentioned in the chapter?

SEEKING SOLUTIONS

1. In what ways has artificial selection changed people's quality of life? Give examples. Can you imagine a way in which artificial selection could be used to improve our quality of life further? Can you imagine a way it could be used to lessen our environmental impact?
2. What types of species are most vulnerable to extinction, and what kinds of factors threaten them? Can you think of any species in your region that are threatened with extinction today? What reasons lie behind their endangerment?
3. Do you think the human species can continue raising its global carrying capacity? How so, or why not? Do you think we *should* try to keep raising our carrying capacity? Why or why not?
4. Describe the evidence suggesting that changes in temperature and precipitation led to the extinction of the golden toad and to population crashes for other amphibians at Monteverde. What do you think could be done to help make future such declines less likely?
5. What are the advantages of ecotourism for a country like Costa Rica? Can you think of any disadvantages? What would you recommend that Costa Rica do to prevent the loss of its biodiversity?
6. As Monteverde changed and some species disappeared, scientists reported that others moved in from lower, drier areas. If this is true, should we be concerned about the extinction of the golden toad and disappearance of other species from Monteverde? Explain your answer.

INTERPRETING GRAPHS AND DATA

Amphibians are sensitive biological indicators of climate change because their reproduction and survival are so closely tied to water. One way in which drier conditions may affect amphibians is by reducing the depth of the pools of water in which their eggs develop. Shallower pools offer less protection from UV-B (ultraviolet) radiation, which some scientists maintain may kill embryos directly or make them more susceptible to disease.

Herpetologist Joseph Kiesecker and colleagues conducted a field study of the relationships among water depth, UV-B radiation, and survivorship of western toad *(Bufo boreas)* embryos in the Pacific Northwest. In manipulative experiments, the researchers placed toad embryos in mesh enclosures at three different depths of water. The researchers placed protective filters that blocked all UV-B radiation over some of these embryos, while leaving other embryos unprotected without the filters. Some of the study's results are presented in the accompanying graph.

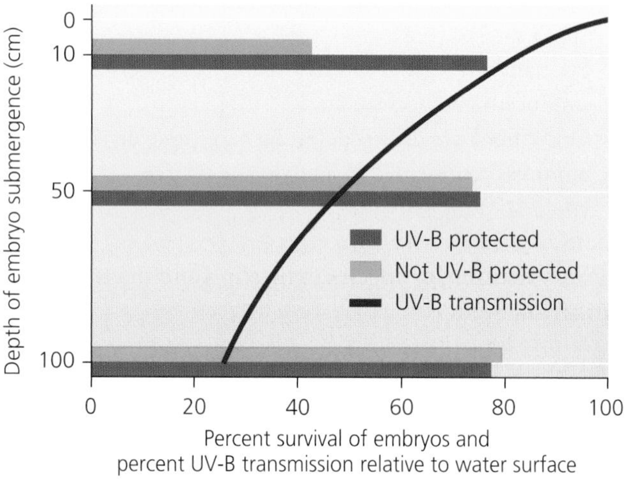

Embryo survivorship in western toads *(Bufo boreas)* at different water depths and UV-B light intensities. Red bars indicate embryos protected under a filter that blocked UV-B light; orange bars indicate unprotected embryos. The blue line indicates the amount of UV-B light reaching different depths in the water column, expressed as a percentage of the UV-B radiation at the water surface. Data from Kiesecker, J. M., et al. 2001. Complex causes of amphibian population declines. *Nature* 410: 681–684.

1. If the UV-B radiation at the surface has an intensity of 0.27 watts/m^2, approximately what is its intensity at depths of 10 cm, 50 cm, and 100 cm?
2. Approximately how much did survival rates at the 10-cm depth differ between the protected and unprotected treatments? Why do you think survival rates differed significantly at the 10-cm depth but not at the other depths?
3. What do you think would be the effect of drier-than-average years on the western toad population, if the average depth of pools available for toad spawning dropped? How do the data above address your hypothesis? Do they support cause-and-effect relationships among water depth, UV-B exposure, disease, and toad mortality?

CALCULATING ECOLOGICAL FOOTPRINTS

In 2004, coffee consumption in the United States topped 2.7 billion pounds (out of 14.8 billion pounds produced globally). Next to petroleum, coffee is the most valuable commodity on the world market, and the United States is its leading importer. Most coffee is produced in large tropical plantations, where coffee is the only tree species and is grown in full sun. However, approximately 2% of coffee is produced in small groves where coffee trees and other species are intermingled. These *shade-grown* coffee forests maintain greater habitat diversity for tropical rainforest wildlife. Given the information above, estimate the coffee consumption rates in the table below.

	Population	Pounds of coffee per day	Pounds of coffee per year
You (or the average American)	1	0.025	9
Your class			
Your hometown			
Your state			
United States			

Data from O'Brien, T. G. and M. F. Kinnaird. 2003. Caffeine and conservation. *Science* 300: 587; and International Coffee Organization.

1. What percentage of global coffee production is consumed in the United States? If only shade-grown coffee were consumed in the United States, how much would shade-grown production need to increase to meet that demand?
2. How much extra would you be willing to pay for a pound of shade-grown coffee, if you knew that your money would help to prevent habitat loss or extinction for animals such as Sumatran tigers, rhinoceroses, and the many songbirds that migrate between Latin America and North America each year?
3. If everyone in the United States were willing to pay as much extra per pound for shade-grown coffee as you are, how much additional money would that provide for conservation of biodiversity in the tropics each year?

Take It Further

Go to www.aw-bc.com/withgott or the student CD-ROM where you'll find:

- Suggested answers to end-of-chapter questions
- Quizzes, animations, and flashcards to help you study
- *Research Navigator*™ database of credible and reliable sources to assist you with your research projects
- **GRAPHit!** tutorials to help you master how to interpret graphs
- **INVESTIGATEit!** current news articles that link the topics that you study to case studies from your region to around the world

6 Species Interactions and Community Ecology

Marsh community along the shore of Lake Ontario

Upon successfully completing this chapter, you will be able to:

- Compare and contrast the major types of species interactions
- Characterize feeding relationships and energy flow, using them to construct trophic levels and food webs
- Distinguish characteristics of a keystone species
- Characterize the process of succession and the debate over the nature of communities
- Perceive and predict the potential impacts of invasive species in communities
- Explain the goals and methods of ecological restoration
- Describe and illustrate the terrestrial biomes of the world

Aggregation of zebra mussels

Central Case: Black and White, and Spread All Over: Zebra Mussels Invade the Great Lakes

"We are seeing changes in the Great Lakes that are more rapid and more destructive than any time in the history of the Great Lakes."

—Andy Buchsbaum, director of the National Wildlife Federation's Great Lakes Office

"When you tear away the bottom of the food chain, everything that is above it is going to be disrupted."

—Tom Nalepa, National Oceanic and Atmospheric Administration research biologist

As if the Great Lakes hadn't been through enough already, the last thing they needed was the zebra mussel. The pollution-fouled waters of Lake Erie and the other Great Lakes shared by Canada and the United States had become gradually cleaner in the years following the Clean Water Act of 1970. As government regulation brought industrial discharges under control, people once again began to use the lakes for recreation, and populations of fish rebounded.

Then the zebra mussel arrived. Black-and-white-striped shellfish the size of a dime, zebra mussels attach to hard surfaces and open their paired shells, feeding on algae by filtering water through their gills. This mollusc, given the scientific name *Dreissena polymorpha,* is native to the Caspian Sea, Black Sea, and Azov Sea in western Asia and eastern Europe. It made its North American debut in 1988 when it was discovered in Canadian waters at Lake St. Clair, which connects Lake Erie with Lake Huron. Evidently ships arriving from Europe had discharged ballast water containing the mussels or their larvae into the Great Lakes.

Within just two years of their discovery in Lake St. Clair, zebra mussels had reached all five of the Great Lakes. The next year, these invaders entered New York's Hudson River to the east, and the Illinois River at Chicago to the west. From the Illinois River and its canals, they soon reached the Mississippi River, giving them access to a vast watershed covering 40% of the

United States. By 1992, zebra mussels had reached the Ohio, Arkansas, and Tennessee Rivers, and by 1994 they had colonized waters in 19 U.S. states and two Canadian provinces.

How could a mussel spread so quickly? The zebra mussel's larval stage is well adapted for long-distance dispersal. Its tiny larvae drift freely for several weeks, traveling as far as the currents take them. Adults that attach themselves to boats and ships may be transported from one place to another, even to small isolated lakes and ponds well away from major rivers. Moreover, in North America the mussels encountered none of the particular species of predators, competitors, and parasites that had evolved to limit their population growth in the Old World.

But why all the fuss? Zebra mussels are best known for clogging up water intake pipes at factories, power plants, municipal water supplies, and wastewater treatment facilities (Figure 6.1a). At one Michigan power plant, workers counted 700,000 mussels per square meter of pipe surface. Great densities of these organisms can damage boat engines, degrade docks, foul fishing gear, and sink buoys that ships use for navigation. Through such impacts, it is estimated that zebra mussels cost the U.S. economy hundreds of millions of dollars each year.

Zebra mussels also have severe impacts on the ecological systems they invade. They eat primarily phytoplankton, microscopic algae that drift in open water. Because each mussel filters a liter or more of water every day, they consume so much phytoplankton that they can deplete populations. Phytoplankton is the foundation of the Great Lakes food web, so its depletion is bad news for zooplankton, the tiny aquatic animals that eat phytoplankton—and for the fish that eat both. Water bodies with zebra mussels have fewer zooplankton and open-water fish than water bodies without them, researchers are finding. Zebra mussels also interfere with native molluscs, suffocating them by attaching to their shells (Figure 6.1b).

However, zebra mussels also benefit some bottom-feeding invertebrates and fish. By filtering algae and organic matter from open water and depositing nutrients in their feces, they shift the community's nutrient balance to the bottom and benefit the species that feed there. Once they have cleared the water, sunlight penetrates more deeply, spurring the growth of large-leafed underwater plants and algae. Such changes have further ripple effects throughout the community that scientists are only beginning to understand.

(a) Clogging a pipe

(b) Suffocating native clams

FIGURE 6.1 Zebra mussels clog up water intake pipes (**a**) of power plants and industrial facilities. They also starve and suffocate native clams by adhering to their shells en masse and sealing them shut (**b**).

Species Interactions

By interacting with many species in a variety of ways, zebra mussels have set in motion an array of changes in the ecological communities they have invaded. Interactions among species are the threads in the fabric of communities, holding them together and determining their nature. Ecologists have organized species interactions into several fundamental categories. Most prominent are competition, predation, parasitism, herbivory, and mutualism. Table 6.1 summarizes the positive and negative impacts of each type of interaction for each participant.

Table 6.1 Effects of Species Interactions on Their Participants

Type of interaction	Effect on species 1	Effect on species 2
Mutualism	+	+
Commensalism	+	0
Predation, parasitism, herbivory	+	–
Neutralism	0	0
Amensalism	–	0
Competition	–	–

"+" denotes a positive effect; "–" denotes a negative effect; "0" denotes no effect.

Competition can occur when resources are limited

When multiple organisms seek the same limited resource, their relationship is said to be one of **competition.** Competing organisms do not usually fight with one another directly and physically. Competition is generally more subtle, involving the consequences of one organism's ability to match or outdo others in procuring resources. The resources for which organisms compete can include just about anything an organism might need to survive, including food, water, space, shelter, mates, sunlight, and more. Competitive interactions can take place among members of the same species *(intraspecific competition)* or among members of two or more different species *(interspecific competition).*

We have already discussed intraspecific competition in Chapter 5, without naming it as such. Recall that density dependence (▸ p. 138) limits the growth of a population; individuals of the same species compete with one another for limited resources, such that competition is more acute when there are more individuals per unit area (denser populations). Thus, intraspecific competition is really a population-level phenomenon. Interspecific competition, however, can have substantial effects on the composition of communities.

Interspecific competition can give rise to different types of outcomes. If one species is a very effective competitor, it may exclude another species from resource use entirely. This outcome, called *competitive exclusion,* occurred in Lake St. Clair and western Lake Erie as the zebra mussel outcompeted a native mussel species.

Alternatively, if neither competing species fully excludes the other, the species may live side by side at a certain ratio of population sizes. This result, called *species coexistence,* may produce a stable point of equilibrium, in which the population size of each remains fairly constant through time.

Coexisting species that use the same resources tend to adjust to their competitors to minimize competition with them. Individuals can do this by changing their behavior so as to use only a portion of the total array of resources they are capable of using. In such cases, individuals are not fulfilling their entire *niche,* or ecological role (▸p. 131). The full niche of a species is called its **fundamental niche** (Figure 6.2a). An individual that plays only part of its role because of competition or other species interactions is said to be displaying a **realized niche** (Figure 6.2b), the portion of its fundamental niche that is actually filled, or realized.

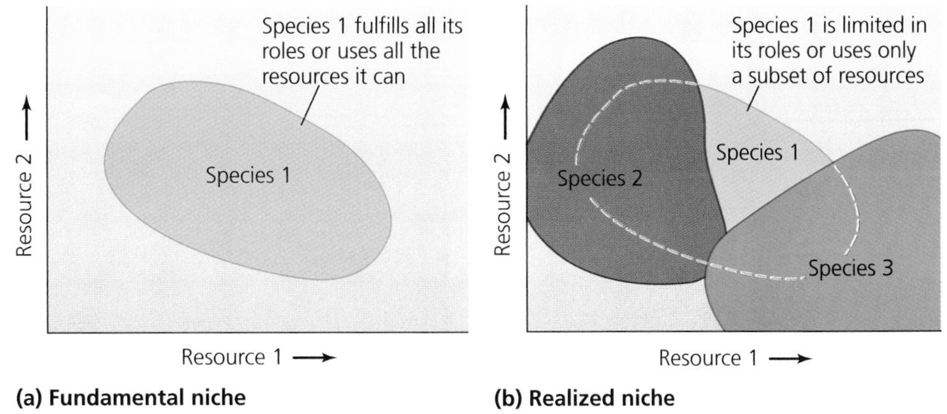

FIGURE 6.2 An organism facing competition may be forced to play a lesser ecological role or use fewer resources than it would in the absence of its competitor. With no competitors, an organism can exploit its full fundamental niche **(a).** But when competitors restrict what an organism can do or what resources it can use, the organism is limited to a realized niche **(b),** which covers only a subset of its fundamental niche. In considering niches, ecologists have traditionally focused on competition, but they now recognize that other species interactions also are influential.

Yellow-bellied sapsucker drills rows of holes in trunk and consumes sap and insects that get stuck in sap
White-breasted nuthatch climbs downward along trunk looking for insects
Pileated woodpecker digs deeply into wood to find large insects
Brown creeper climbs upward along trunk looking for tiny insects

FIGURE 6.3 When species compete, they tend to partition resources, each specializing on a slightly different resource or way of attaining a shared resource. A number of types of birds—including the woodpeckers, creeper, and nuthatch shown here—feed on insects from tree trunks, but they use different portions of the trunk, seeking different foods in different ways.

Species make similar adjustments over evolutionary time. They adapt to competition by evolving to use slightly different resources or to use their shared resources in different ways. If two bird species eat the same type of seeds, one might come to specialize on larger seeds and the other to specialize on smaller seeds. Or one bird might become more active in the morning and the other more active in the evening, thus avoiding direct interference. This process is called **resource partitioning,** because the species divide, or partition, the resource they use in common by specializing in different ways (Figure 6.3). Resource partitioning can lead to *character displacement,* which occurs when competing species evolve physical characteristics that reflect their reliance on the portion of the resource they use. By becoming more different from one another, two species reduce their competition. Through natural selection (▸pp. 118–121), birds that specialize on larger seeds may evolve larger bills that enable them to make best use of the resource, whereas birds specializing on smaller seeds may evolve smaller bills. This is precisely what extensive research has revealed about the finches from the Galapagos Islands that were first described by Charles Darwin.

Several types of interactions are exploitative

In competitive interactions, each participant has a negative effect on other participants, because each takes resources the others could have used. This is reflected in the two minus signs shown for competition in Table 6.1. In other types of interactions, some participants benefit while others are harmed (note the +/− interactions in Table 6.1). We can think of interactions in which one member exploits another for its own gain as exploitative interactions, or *exploitation.* Such interactions include predation, parasitism, herbivory, and related concepts, as outlined below.

FIGURE 6.4 Predator-prey interactions have ecological and evolutionary consequences for both prey and predator. Here, a fire-bellied snake *(Liophis epinephalus)* devours a frog in the Monteverde cloud forest we studied in Chapter 5.

Predators kill and consume prey

Every living thing needs to procure food, and for most animals, that means eating other living organisms. **Predation** is the process by which individuals of one species, a *predator,* hunt, capture, kill, and consume individuals of another species, its *prey* (Figure 6.4). Along with competition, predation has traditionally been viewed as one of the primary organizing forces in community ecology. Interactions between predators and prey structure the food webs that we will examine shortly, and they influence community composition by helping determine the relative abundance of predators and prey.

Zebra mussel predation on phytoplankton has reduced phytoplankton populations by up to 90%, according to many studies in the Great Lakes and Hudson River. Zebra mussels also consume the smaller types of zooplankton. This predation, combined with the competition mentioned above, has caused zooplankton population sizes and biomass to decline by up to 70% in Lake Erie and the Hudson River since zebra mussels arrived. Meanwhile, the mussels do not eat some cyanobacteria (▸p. 105), so concentrations of these cyanobacteria rise in lakes with zebra mussels. Most predators are also prey, however, and zebra mussels have become a food source for a number of North American species since their introduction. These include diving ducks, muskrats, crayfish, flounder, sturgeon, eels, and several types of fish with grinding teeth, such as carp and freshwater drum.

Predation can sometimes drive population dynamics by causing cycles in population sizes. An increase in the population size of prey creates more food for predators, which may survive and reproduce more effectively as a result. As the predator population rises, additional predation drives down the population of prey. Fewer prey in turn causes some predators to starve, so that the predator population declines. This allows the prey population to begin rising again, starting the cycle anew. Most natural systems involve so many factors that such cycles don't last long, but in some cases we see extended cycles (Figure 6.5).

Predation also has evolutionary ramifications. Individual predators that are more adept at capturing prey will likely live longer, healthier lives and be better able to provide for their offspring than will less adept individuals. Thus, natural selection on individuals within a predator species leads to the evolution of adaptations that make them better hunters. Prey face an even stronger selective pressure—the risk of immediate death. For this reason, predation pressure has caused organisms to evolve an elaborate array of defenses against being eaten (Figure 6.6).

Parasites exploit living hosts

Organisms can exploit other organisms without killing them. **Parasitism** is a relationship in which one organism, the *parasite,* depends on another, the *host,* for nourishment or some other benefit while simultaneously doing

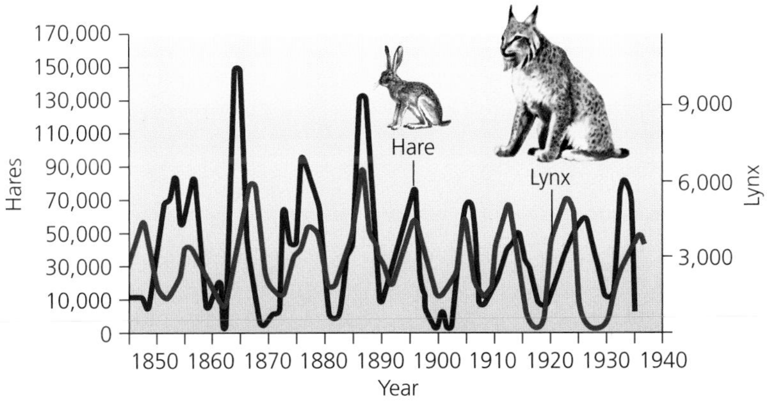

FIGURE 6.5 Predator-prey systems sometimes show paired cycles, in which increases and decreases in one organism apparently drive increases and decreases in the other. Although such cycles are predicted by theory and are seen in lab experiments, they are very difficult to document conclusively in natural systems. Data from MacLulich, D.A. 1937. Fluctuation in the numbers of varying hare *(Lepus americanus). Univ. Toronto Stud. Biol. Ser. 43,* Toronto: University of Toronto Press.

(a) Cryptic coloration (b) Warning coloration (c) Mimicry

FIGURE 6.6 Natural selection to avoid predation has resulted in many fabulous adaptations. Some prey hide from predators by *crypsis*, or camouflage, such as this gecko on tree bark **(a)**. Other prey are brightly colored to warn predators that they are toxic or distasteful, such as this poison dart frog **(b)**. Still others fool predators with mimicry. Some, like walking sticks imitating twigs, mimic for crypsis. Others mimic toxic, distasteful, or dangerous organisms, like this caterpillar **(c)**; when it is disturbed, the caterpillar swells and curves its tail end and shows eyespots, to look like a snake's head.

the host harm. Unlike predation, parasitism usually does not result in an organism's immediate death, although it sometimes contributes to the host's eventual death.

Many parasites live in close contact with their hosts. These parasites include disease pathogens, such as the protists that cause malaria and dysentery, as well as animals, such as tapeworms, that live in the digestive tracts of their hosts. Other parasites live on the exterior of their hosts, such as the ticks that attach themselves to their hosts' skin, and the sea lamprey *(Petromyzon marinus)*, another invader of the Great Lakes (Figure 6.7). Sea lampreys are tube-shaped vertebrates that grasp the bodies of fish with a suction-cup mouth and a rasping tongue, sucking blood from the fish for days or weeks. Sea lampreys invaded the Great Lakes from the Atlantic Ocean after people dug canals to connect the lakes for shipping, and the lampreys soon devastated economically important fisheries of chubs, lake herring, whitefish, and lake trout. Since the 1950s, U.S. and Canadian fisheries managers have reduced lamprey populations by applying chemicals that selectively kill lamprey larvae.

FIGURE 6.7 Parasites harm their host organism in some way. With its suction-like mouth and rasping tongue, the sea lamprey *(Petromyzon marinus)* attaches itself to fish and sucks their blood for days or weeks, sometimes killing the fish. Sea lampreys wreaked havoc on Great Lakes fisheries after entering the lakes through human-built canals.

Other types of parasites are free-living and come into contact with their hosts only infrequently. For example, the cuckoos of Eurasia and the cowbirds of the Americas parasitize other birds by laying eggs in their nests and letting the host bird raise the parasite's young.

Some parasites cause little harm, but others may kill their hosts. Many insects parasitize other insects, often killing them in the process, and are called *parasitoids.* Various species of parasitoid wasps lay eggs on caterpillars. When the eggs hatch, the wasp larvae burrow into the caterpillar's tissues and slowly consume them. The wasp larvae metamorphose into adults and fly from the body of the dying caterpillar.

Just as predators and prey evolve in response to one another, so do parasites and hosts, in a process termed *coevolution.* Hosts and parasites can become locked in a duel of escalating adaptations, a situation sometimes referred to as an *evolutionary arms race.* Like rival nations racing to stay ahead of one another in military technology, host and parasite may repeatedly evolve new responses to the other's latest advance. In the long run, though, it may not be in a

FIGURE 6.8 Herbivory is a common way to make a living. The world holds many thousands, and perhaps millions, of species of plant-eating insects, such as this larva (caterpillar) of the death's head hawk moth *(Acherontia atropos)* from western Europe.

parasite's best interest to become too harmful to its host. Instead, a parasite might leave more offspring in the next generation—and thus be favored by natural selection—if it allows its host to live a longer time, or even to thrive.

Herbivores exploit plants

One of the most common types of exploitation is **herbivory,** which occurs when animals feed on the tissues of plants. Insects that feed on plants are the most widespread type of herbivore; just about every plant in the world is attacked by some type of insect (Figure 6.8). In most cases, herbivory does not kill a plant outright, but may affect its growth and reproduction.

Like animal prey, plants have evolved a wide array of defenses against the animals that feed on them. Many plants produce chemicals that are toxic or distasteful to herbivores. Others arm themselves with thorns, spines, or irritating hairs. In response, herbivores may evolve ways to overcome these defenses, and the plant and the animal may embark on an evolutionary arms race.

Some plants go a step further and recruit certain animals as allies to assist in their defense. Many such plants encourage ants to take up residence by providing thorns or swelled stems for the ants to nest in or nectar-bearing structures for the ants to feed from. These ants protect the plant in return by attacking other insects that land or crawl on it. Other plants respond to herbivory by releasing volatile chemicals when they are bitten or pierced. The airborne chemicals attract predatory insects that may attack the herbivore. Such cooperative strategies, trading defense for food, are examples of our next type of species interaction, mutualism.

Mutualists help one another

Mutualism is a relationship in which two or more species benefit from interaction with one another. Generally each partner provides some resource or service that the other needs.

Many mutualistic relationships—like many parasitic relationships—occur between organisms that live in close physical contact. (Indeed, biologists hypothesize that many mutualistic associations evolved from parasitic ones.) Such physically close association is called **symbiosis.** Thousands of terrestrial plant species depend on mutualisms with fungi; plant roots and some fungi together form symbiotic associations called mycorrhizae. In these symbioses, the plant provides energy and protection to the fungus, while the fungus assists the plant in absorbing nutrients from the soil. In the ocean, coral polyps, the tiny animals that build coral reefs, share beneficial arrangements with algae known as zooxanthellae. The coral provide housing and nutrients for the algae in exchange for a steady supply of food—90% of their nutritional requirements.

You, too, are part of a symbiotic association. Your digestive tract is filled with microbes that help you digest food—microbes for which you are providing a place to live. Indeed, we may owe our very existence to symbiotic mutualisms. It is now widely accepted that the eukaryotic cell (▸p. 102) originated after certain prokaryotic cells engulfed other prokaryotic cells and established mutualistic symbioses. Scientists have inferred that some of the engulfed cells eventually evolved into cell organelles.

Not all mutualists live in close proximity. One of the most important mutualisms in environmental science involves free-living organisms that may encounter each other only once in their lifetimes. This is *pollination* (Figure 6.9), an interaction of key significance to agriculture and our food supply (▸pp. 285–286). Bees, birds, bats, and other creatures transfer pollen (male sex cells) from one flower to ova (female cells) of another, fertilizing the female egg, which subsequently grows into a fruit. The

FIGURE 6.9 In mutualism, organisms of different species benefit one another. An important mutualistic interaction for environmental science is pollination. This hummingbird visits flowers to gather nectar and in the process transfers pollen between flowers, helping the plant reproduce. Pollination is of key importance to agriculture, ensuring the reproduction of many crop plants.

pollinating animals visit flowers for their nectar, a reward the plant uses to entice them. The pollinators receive food, and the plants are pollinated and reproduce. Various types of bees alone pollinate 73% of our crops, one expert has estimated—from soybeans to potatoes to tomatoes to beans to cabbage to oranges.

Some interactions have no effect on some participants

Two other types of species interaction get far less attention. **Amensalism** is a relationship in which one organism is harmed and the other is unaffected. In **commensalism**, one species benefits and the other is unaffected. Amensalism has been difficult to pin down, because it is hard to prove that the organism doing the harm is not in fact besting a competitor for a resource. For instance, some plants release poisonous chemicals that harm nearby plants (a phenomenon called *allelopathy*), and some experts have suggested that this is an example of amensalism. However, allelopathy can also be viewed as one plant investing in chemicals to outcompete others for space.

One association commonly cited as an example of commensalism is when the conditions created by one plant happen to make it easier for another plant to establish and grow. For instance, palo verde trees in the Sonoran Desert create shade and leaf litter that allow the soil beneath them to hold moisture longer, creating an area that is cooler and moister than the surrounding sunbaked ground. Young plants find it easier to germinate and grow in these conditions, so seedling cacti and other desert plants generally grow up directly beneath "nurse" trees such as palo verde. This phenomenon, called *facilitation,* influences the structure and composition of communities and how they change through time.

Ecological Communities

In Chapter 5 we defined a *community* as a group of populations of organisms that live in the same place at the same time. The members of a community interact with one another in the ways described above, and the direct interactions among species often have indirect effects that ripple outward to affect other community members. The strength of interactions also varies, and together species interactions determine the species composition, structure, and function of communities. *Community ecologists* are interested in what species coexist, how they relate to one another, how communities change through time, and why these patterns exist.

Energy passes among trophic levels

The interactions among members of a community are many and varied, but some of the most important involve who eats whom. As we saw in Chapter 4 (▸pp. 105–106), the energy that drives such interactions in most systems comes ultimately from the sun via photosynthesis. As organisms feed on one another, this energy moves through the community, from one rank in the feeding hierarchy, or **trophic level,** to another (Figure 6.10).

Producers Producers, or autotrophs ("self-feeders"), comprise the first trophic level, as we saw in Chapter 4 (▸pp. 105–108). Terrestrial green plants, cyanobacteria, and algae capture solar energy and use photosynthesis to produce sugars. The chemosynthetic bacteria of hot springs and deep-sea hydrothermal vents use geothermal energy in a similar way to produce food.

Consumers Organisms that consume producers are known as *primary consumers* and comprise the second

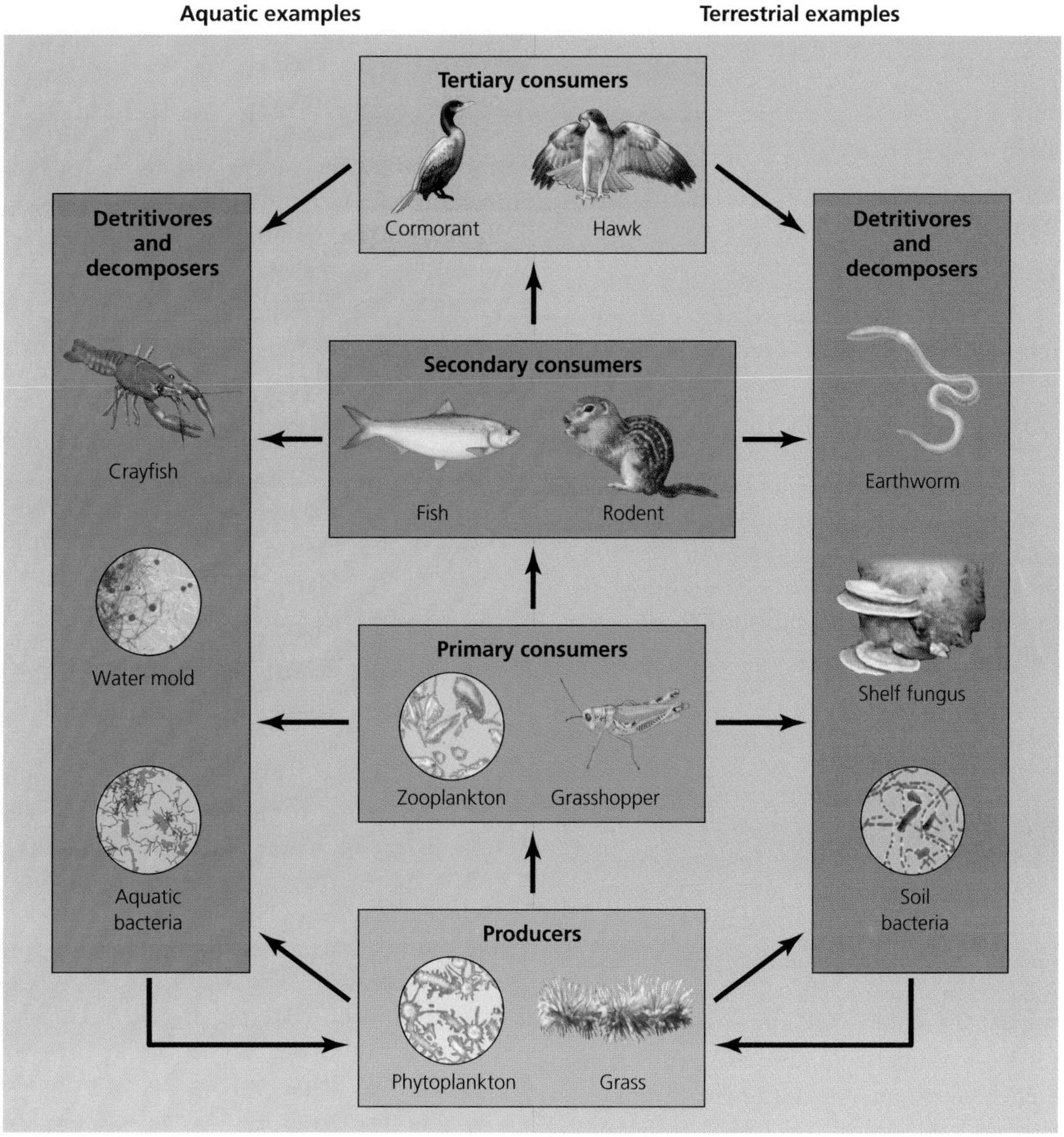

FIGURE 6.10 Ecologists organize species hierarchically by their feeding rank, or trophic level. The diagram shows aquatic (left) and terrestrial (right) examples at each level. Arrows indicate the direction of energy flow. Producers produce food by photosynthesis, primary consumers (herbivores) feed on producers, secondary consumers eat primary consumers, and tertiary consumers eat secondary consumers. Communities can have more or fewer trophic levels than in this example. Detritivores and decomposers feed on nonliving organic matter and the remains of dead organisms from all trophic levels, and they "close the loop" by returning nutrients to the soil or the water column for use by producers.

trophic level. Grazing animals, such as deer and grasshoppers, are primary consumers. The third trophic level consists of *secondary consumers,* which prey on primary consumers. Wolves that prey on deer are considered secondary consumers, as are rodents and birds that prey on grasshoppers. Predators that feed at even higher trophic levels are known as *tertiary consumers.* Examples of tertiary consumers include hawks and owls that eat rodents that have eaten grasshoppers. Note that most primary consumers are *herbivores* because they consume plants, whereas secondary and tertiary consumers are *carnivores* because they eat animals. Animals that eat both plant and animal food are referred to as *omnivores.*

Detritivores and decomposers *Detritivores* and *decomposers* consume nonliving organic matter. Detritivores, such as millipedes and soil insects, scavenge the waste products or the dead bodies of other community members. Decomposers, such as fungi and bacteria, break down leaf litter and other nonliving matter further into simpler constituents that can then be taken up and used by plants. These organisms play an essential role as the community's recyclers, making nutrients from organic matter available for reuse by living members of the community.

In Great Lakes communities, phytoplankton are the main producers, floating freely and photosynthesizing with sunlight that penetrates the upper layer of the water. Zooplankton are primary consumers, feeding on the phytoplankton. Phytoplankton-eating fish are primary consumers, and zooplankton-eating fish are secondary consumers. At higher trophic levels are tertiary consumers such as larger fish and birds that feed on plankton-eating fish. Zebra mussels, by eating both phytoplankton and zooplankton, function on multiple trophic levels. When any of these organisms dies and sinks to the bottom, detritivores scavenge its tissues and microbial decomposers recycle its nutrients.

Energy, biomass, and numbers decrease at higher trophic levels

At each trophic level, most of the energy that organisms use is lost through respiration. Only a small amount of the energy is transferred to the next trophic level through predation, herbivory, or parasitism. The first trophic level (producers) contains a large amount of energy, but the second (primary consumers) contains less energy—only that amount gained from consuming producers. The third trophic level (secondary consumers) contains still less energy, and higher trophic levels (tertiary consumers) contain the least. A general rule of thumb is that each trophic level contains just 10% of the energy of the trophic level below it, although the actual proportion can vary greatly.

This pattern, which can be visualized as a pyramid, generally also holds for the numbers of organisms at each trophic level. Generally, fewer organisms exist at higher trophic levels than at lower trophic levels. A grasshopper eats many plants in its lifetime, a rodent eats many grasshoppers, and a hawk eats many rodents. Thus, for every hawk in a community there must be many rodents, still more grasshoppers, and an immense number of plants. Because the difference in numbers of organisms among trophic levels tends to be large, the same pyramid-like relationship also often holds true for biomass. Even though rodents are larger than grasshoppers, and hawks larger than rodents, the sheer number of prey relative to the predators means that prey biomass will likely be greater overall.

Food webs show feeding relationships and energy flow

As energy is transferred from species on one trophic level to species on other trophic levels, it is said to pass up a *food chain.* Plant, grasshopper, rodent, and hawk make up a food chain, a linear series of feeding relationships. Thinking in terms of food chains is conceptually useful, but in reality ecological systems are far more complex than simple linear chains. A more accurate representation of the feeding relationships in a community is a **food web,** a visual map of feeding relationships and energy flow, showing the many paths by which energy passes among organisms as they consume one another.

Figure 6.11 shows a food web from a temperate deciduous forest of eastern North America. Like virtually all diagrams of ecological systems, it is greatly simplified, leaving out the vast majority of species and interactions that occur. Note, however, that even within this simplified diagram, we can pick out a number of different food chains involving different sets of species.

A Great Lakes food web would involve the phytoplankton and cyanobacteria that photosynthesize near the water's surface, the zooplankton that eat them, the fish that eat all these, the larger fish that eat the smaller fish, and the lampreys that parasitize the fish. It would include a number of native mussels and clams and, since 1988, the zebra mussel that is crowding them out. It would include diving ducks that used to feed on native bivalves and now are preying on zebra mussels. This food web would also show that an array of bottom-dwelling invertebrates feed from the refuse of zebra mussels. These waste products promote bacterial growth and disease pathogens that harm native bivalves, but they also provide nutrients that nourish crayfish and many smaller *benthic* (bottom-dwelling) invertebrate animals. Finally, the food web would include underwater plants and macroscopic algae, whose growth is promoted by zebra mussels. The mussels clarify the water by filtering out phytoplankton, and sunlight penetrates more deeply into the water column, spurring photosynthesis and plant growth. Overall, zebra mussels alter this food web essentially by shifting productivity from the open-water regions to the benthic and *littoral* (nearshore) regions.

The many direct interactions in this food web also create indirect interactions. For example, zebra mussels affect fish indirectly, helping benthic and littoral fishes and making life harder for open-water fishes (see "The Science behind the Story," ▸ pp. 160–161).

FIGURE 6.11 Food webs are conceptual representations of feeding relationships in a community. This food web pertains to eastern North America's temperate deciduous forest and includes organisms on several trophic levels. In a food web diagram, arrows are drawn from one organism to another to indicate the direction of energy flow as a result of predation, parasitism, or herbivory. For example, an arrow leads from the grass to the cottontail rabbit to indicate that cottontails consume grasses. The arrow from the cottontail to the tick indicates that parasitic ticks derive nourishment from cottontails. Communities include so many species and are complex enough, however, that most food web diagrams are bound to be gross simplifications.

Some organisms play bigger roles in communities than others

"Some animals are more equal than others," George Orwell wrote in his 1945 book *Animal Farm.* Although Orwell was making wry sociopolitical commentary, his remark hints at a truth in ecology. In communities, ecologists have found, some species exert greater influence than do others. A species that has particularly strong or far-reaching impact is often called a **keystone species.** A keystone is the wedge-shaped stone at the top of an arch that is vital for holding the structure together; remove the keystone, and the arch will collapse. In an ecological

The Science behind the Story

Inferring Zebra Mussels' Impacts on Fish Communities

Food webs are complicated systems, and disentangling them to infer the effects of any one species is fraught with difficulty. When zebra mussels appeared in the Great Lakes, people feared for sport fisheries, and estimated that fish population declines could cost billions of dollars. The mussels would deplete the phytoplankton and zooplankton that fish depended on, people reasoned, and many fewer fish would survive. Yet even after 15 years, there was no solid evidence of widespread harm to fish populations.

So, aquatic biologist David Strayer of the Institute of Ecosystem Studies in Millbrook, New York, joined Kathyrn Hattala and Andrew Kahnle of New York State's Department of Environmental Conservation (DEC). They mined datasets on fish populations in the Hudson River, which zebra mussels had invaded in 1991.

Strayer and others had already been studying effects of zebra mussels on aspects of the community for years. Their data showed that since the species' introduction to the Hudson:

- Biomass of phytoplankton fell 80%.
- Biomass of small zooplankton fell 76%.
- Biomass of large zooplankton fell 52%.

Zebra mussels increased filter-feeding in the community 30-fold, thereby depleting the phytoplankton and small zooplankton, and leaving all sizes of zooplankton with less phytoplankton to eat. Overall, the zooplankton and invertebrate animals of the open water that are eaten by open-water fish declined by 70%.

However, Strayer's work had also found that *benthic,* or bottom-dwelling, invertebrates in shallow water (especially in the nearshore, or *littoral,* zone) had increased by 10%, and likely much more, because the mussels' shells provide habitat structure, and their feces provide nutrients.

These contrasting trends in the benthic shallows and the open deep water led Strayer's team to hypothesize that zebra mussels would harm open-water fish that ate plankton but would help littoral-feeding fish. They predicted that after zebra mussel introduction, larvae and juveniles of six common open-water fish species would decline in number, decline in growth rate, and shift downriver toward saltier water, where mussels are absent. Conversely, they predicted that larvae and juveniles of 10 littoral fish species would increase in number, increase in growth rate, and shift upriver to regions of greatest zebra mussel density.

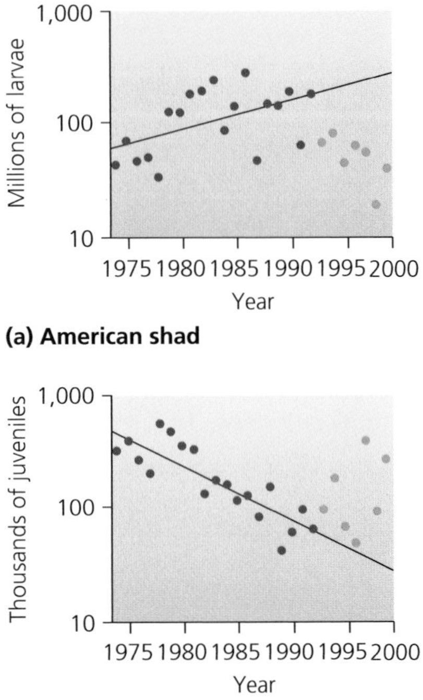

(b) Tessellated darter

Larvae of American shad (**a**), an open-water fish, had been increasing in abundance before zebra mussels were introduced (red points and trend line). After zebra mussel introduction, shad larvae decreased in abundance (orange points). Juveniles of the tessellated darter (**b**), a littoral zone fish, had been decreasing in abundance before zebra mussels were introduced (red points and trend line). After zebra mussel introduction, they increased in abundance (orange points). *Source:* Strayer, D., et al. 2004. Effects of an invasive bivalve (*Dreissena polymorpha*) on fish in the Hudson River estuary. *Can. J. Fish. Aquat. Sci.* 61: 924–941.

To test their predictions, the researchers analyzed data from three

community, removal of a keystone species will have substantial ripple effects and will alter a large portion of the food web.

Often, large-bodied secondary or tertiary consumers near the tops of food chains are considered keystone species. Top predators control populations of herbivores, which would otherwise multiply and could, through increased herbivory, greatly modify the plant community. In the United States, government bounties promoted the hunting of wolves and mountain lions, which were largely exterminated by the middle of the 20th century. In the absence of these predators, unnaturally dense deer types of fish surveys carried out over 26 years spanning periods before and after the zebra mussel's arrival. One data set came from surveys conducted from 1985 to 1999 by the DEC. The other two came from surveys conducted from 1974 to 1999 by biologists hired by electric utilities, which in New York are required to monitor fish populations in return for using the Hudson's water for cooling at their power plants.

The researchers compared values for abundance, growth, and distribution for young fish before 1991 with those after 1991. The results supported their predictions. Larvae and juveniles of open-water fish, such as American shad, blueback herring, and alewife, tended to decline in abundance in the years after zebra mussel introduction (first figure, part (a)). Those of littoral fish, such as tessellated darter, bluegill, and largemouth bass, tended to increase (first figure, part (b)). Growth rates showed the same trend: Open-water fish grew more slowly after zebra mussel introduction, whereas littoral fish grew more quickly. In terms of distribution in the 248-km (154-mi) stretch of river studied, open-water fish shifted downstream toward areas with fewer zebra mussels, whereas littoral fish shifted upstream toward areas with more zebra mussels (second figure). The results were published in 2004 in the *Canadian Journal of Fisheries and Aquatic Sciences.*

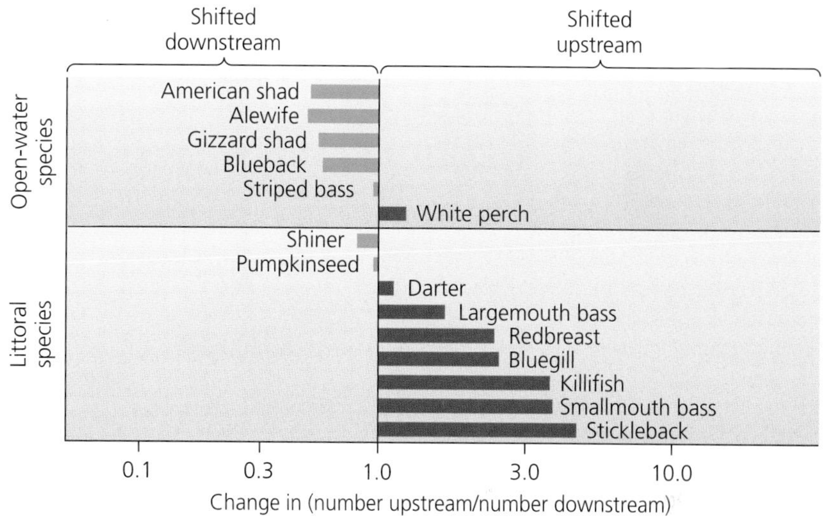

Young of open-water fish, such as American shad, blueback herring, and alewife, tended to shift downstream toward areas with fewer zebra mussels in the years following zebra mussel arrival. Young of littoral fish, such as killifish, bluegill, and largemouth bass, tended to shift upstream toward areas with more zebra mussels. *Source:* Strayer, D., et al. 2004. Effects of an invasive bivalve (*Dreissena polymorpha*) on fish in the Hudson River estuary. *Can. J. Fish. Aquat. Sci.* 61: 924–941.

Overall, the results supported the hypothesis that the fish community would respond to changes in its food resources caused by zebra mussels. The results are correlative, and correlation does not prove causation. However, previous attempts to address these questions experimentally were so limited in time and scale that they could not reflect true effects in natural systems.

Research such as this helps illuminate the often obscure connections and impacts that particular species interactions have on communities as a whole. In this case, the research may also help fisheries managers predict changes in commercially and recreationally important fish populations. With this knowledge, biologists may be able to manage fisheries more effectively in the Hudson and other areas invaded by zebra mussels.

populations have overgrazed forest-floor vegetation and eliminated whole cohorts of tree seedlings, causing major changes in forest structure.

The removal of top predators in the United States was an uncontrolled large-scale experiment with unintended consequences. But ecologists have verified the keystone species concept in controlled experiments. Classic work by marine biologist Robert Paine established that the predatory starfish *Pisaster ochraceus* has great influence on the community composition of intertidal organisms on the Pacific coast of North America. When *Pisaster* is present in this community, species diversity is high, with several

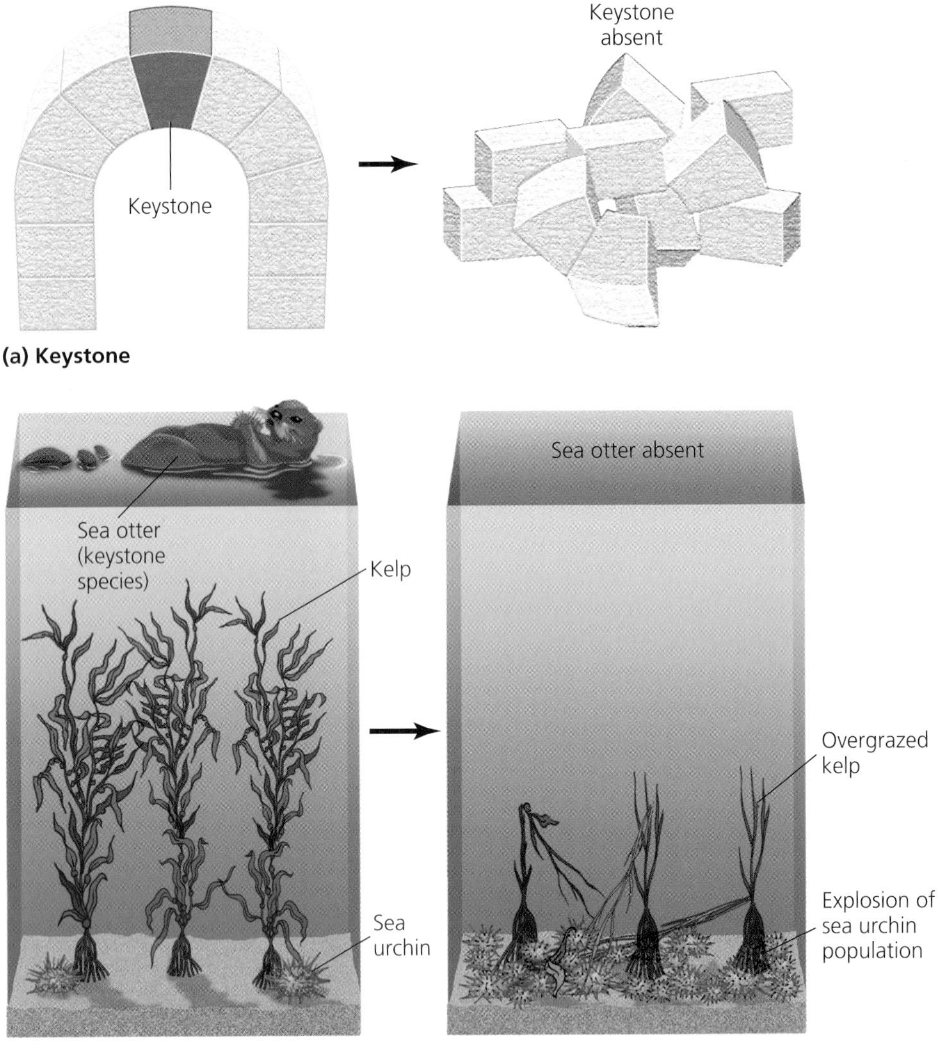

URE 6.12 A keystone is the edge-shaped stone at the top of an arch that holds its structure together (**a**). A keystone species, such as the sea otter, is one that exerts great influence on a community's composition and structure (**b**). Sea otters consume sea urchins that eat kelp in marine nearshore environments of the Pacific. When otters are present, they keep urchin numbers down, which allows lush underwater forests of kelp to grow and provide habitat for many other species. When otters are absent, urchin populations increase and the kelp is devoured, destroying habitat and depressing species diversity. See "The Science behind the Story," ▸pp. 164–165.

types of barnacles, mussels, and algae. When *Pisaster* is removed, the mussels it preys on become numerous and displace other species, suppressing species diversity.

Animals at high trophic levels, such as wolves, starfish, and sea otters (Figure 6.12), are most often seen as keystone species. But other species attain keystone status as "ecosystem engineers" by physically modifying the environment shared by community members. Beavers build dams and turn streams into ponds, flooding acres of dry land and turning them to swamp. Prairie dogs dig burrows that aerate the soil and serve as homes for other animals. Less conspicuous organisms and those toward the bottoms of food chains can potentially be viewed as keystone species, too. Remove the fungi that decompose dead matter, or the insects that control plant growth, or the phytoplankton that are the base of the marine food chain, and a community may change very rapidly indeed. Because there are usually more species at lower trophic levels, however, it is less likely that any one of them alone might have wide influence; if one species is removed, other species that remain may be able to perform many of its functions.

Identifying keystone species is no simple task, and there is no cut-and-dried definition of the term to help us. Community dynamics are complex, species interactions differ in their strength, and the strength of species interactions can vary through time and space. "The Science behind the Story" (▸pp. 164–165) gives an idea of the surprises that are sometimes in store for ecologists studying these interactions.

Weighing the Issues: Keystone Species and Conservation

Imagine the government is funding a development project in your town and is gathering citizen input on three options. The environmental impact statement (EIS) (▸p. 68) states that option 1 would likely result in the

FIGURE 6.13 Secondary succession in a terrestrial sett… occurs after a disturbance, such … fire, landslides, or farming, remov… most vegetation from an area. Here is shown a typical series of changes in a plant community of eastern North America following the abandonment of a farmed field.

extermination of bobcats, a tertiary consumer in the community. Option 2, the EIS predicts, would probably kill off a species of pocket mouse, a primary consumer that is common in the community. Option 3 would likely eliminate a species of lupine, a plant that covers a large percentage of the ground in the present community.

You are a citizen desiring minimal change in the natural community so that your children grow up in an area like the one in which you grew up. What questions would you ask of an ecologist about the bobcat, pocket mouse, and lupine so that you could decide which might most likely be a keystone species? If you instead had to provide input without further information, what would you advise the government?

Communities respond to disturbance in different ways

The removal of a keystone species and the spread of an invasive species are just two of many types of disturbance that can modify the composition, structure, or function of an ecological community. Over time, any given community may experience natural disturbances ranging from gradual phenomena such as climate change to sudden events such as hurricanes, floods, or avalanches.

Communities are dynamic systems and may respond to disturbance in several ways. A community that resists change and remains stable despite disturbance is said to show **resistance** to the disturbance. Alternatively, a community may show **resilience,** meaning that it changes in response to disturbance but later returns to its original state. Or, a community may be modified by disturbance permanently and may never return to its original state.

Succession follows severe disturbance

If a disturbance is severe enough to eliminate all or most of the species in a community, the affected site will undergo a somewhat predictable series of changes that ecologists call **succession.** In the traditional view of this process, ecologists described two types of succession. **Primary succession** follows a disturbance so severe that no vegetation or soil life remains from the community that occupied the site. In primary succession, a biotic community is built essentially from scratch. In contrast, **secondary succession** begins when a disturbance dramatically alters an existing community but does not destroy all living things or all organic matter in the soil. In secondary succession (Figure 6.13), vestiges of the previous community remain, and these building blocks help shape the process.

At terrestrial sites, primary succession takes place after a bare expanse of rock, sand, or sediment becomes newly exposed to the atmosphere. This can occur when glaciers retreat, lakes dry up, or volcanic lava flows spread across the landscape. Species that arrive first and colonize the new substrate are referred to as **pioneer species.** Pioneer species are well adapted for colonization, having traits such as spores or seeds that can travel long distances. The pioneers best suited to colonizing bare rock are the mutualistic aggregates of fungi and algae known as *lichens.* Lichens succeed because their algal component provides food and energy via photosynthesis while the fungal component takes a firm hold on rock and captures the moisture that both organisms need to survive. As lichens grow, they secrete acids that break down the rock surface. The resulting waste material forms the beginnings of soil, and once soil begins to form, small plants, insects, and worms find the rocky outcrops more hospitable. As new organisms arrive,

The Science behind the Story

Otters, Urchins, Kelp, and a Whale of a Chain Reaction

Ecologists required years of careful study to comprehend the relationship among sea otters, sea urchins, and kelp forests diagrammed in Figure 6.12. And even after they thought they understood it, some surprises were in store.

Sea otters live in coastal waters of the Pacific Ocean. These mammals float on their backs amid the waves, feasting on sea urchins that they pry from the bottom and bring to the surface. Once abundant, sea otters were hunted nearly to extinction for their fur. Their protection by international treaty in 1911 allowed their numbers to regrow. Otters returned to high densities in some regions (such as off Alaska and the Aleutian Islands), but failed to return at all in others.

Biologists noted that regions with otters hosted dense "forests" of kelp. Kelp is a brown alga (seaweed) that anchors to the seafloor and reaches toward sunlit surface waters, growing up to 60 m (200 ft) high. Kelp forests provide complex physical structure in which diverse communities of fish and invertebrates find shelter and food.

In regions without sea otters, scientists found kelp forests absent. Urchins in such areas become so numerous that they may eat every last bit of kelp, creating empty seafloors called "urchin barrens" that are relatively devoid of life.

Ecologists observed that once areas with urchin barrens were recolonized by sea otters, urchin numbers declined and kelp forests returned. Through comparative research of this kind, ecologists determined that otters were largely responsible for the presence of the kelp forest community, simply by keeping urchin numbers in check through predation. This research—mostly by James Estes of the University of California at Santa Cruz and his colleagues—established sea otters as a prime example of a keystone species.

But the story did not end there. In the 1990s, otter populations dropped precipitously in areas off Alaska and the Aleutians. No one knew why. So, Estes and his coworkers placed radio tags on Aleutian otters and studied them at sea. Their first hypothesis was that fertility rates had dropped, but the radio-tracking observations showed that females were raising pups without problem. They then tested a second hypothesis, that otters were simply moving to other locations. But the radio tracking showed no unusual dispersal, and the population declines were taking place evenly over large areas. They were left with one viable hypothesis: increased mortality.

Biologists were finding no evidence of disease or starvation, however. Then one day in 1991, Estes's team witnessed something never seen before. They watched as a sea otter was killed and eaten by an orca, or killer whale. These striking black-and-white predators grow up to 10 m (32 ft) long, hunt in groups, and had always attacked larger prey. A sea otter to them is a mere snack. Yet over the following years, Estes's team saw nine more cases of orca predation on otters. Could killer whales be killing off the otters?

The researchers calculated the hours they had spent studying otters at sea before and after 1991 to assess the likelihood that they had simply missed seeing predation by orcas earlier. Their statistical analysis convinced them the rate of attacks had indeed increased. Then they compared a bay where otters were vulnerable to orcas with a lagoon where they were protected. Otter numbers in the lagoon remained stable over 4 years, whereas those in the bay dropped by 76%. Radio tracking showed no movement between these locations.

Finally, using data on otter birth rates, death rates, and population age structure (▸p. 134), they estimated that to account for the otter decline, 6,788 orca attacks per year would have had to occur in their study area. This produced an *expected* rate of observed attacks that matched their *actual* number of observed attacks well. These lines of evidence led the researchers to propose that predation by orcas was eliminating otters.

As otters declined in the Aleutians, urchins increased, and kelp

they provide more nutrients and habitat for future arrivals. As time passes, larger plants establish themselves, the amount of vegetation increases, and species diversity rises.

Secondary succession on land begins when a fire, a hurricane, logging, or farming removes much of the biotic community. Consider a farmed field in eastern North America that has been abandoned. In the first few years after farming ends, the site will be colonized by pioneer species of grasses, herbs, and forbs that were already in the vicinity and that disperse effectively. As density fell dramatically (see the figure). These changes supported the idea that otters were keystone species, but now it seemed that one keystone species was being controlled by another.

Why had orcas suddenly started eating otters? A possible answer came in 2003, after Alan Springer, Estes, and others determined that sea otters were only the latest in a series of population crashes in the northern Pacific. Harbor seals had declined by more than 90% in the late 1970s and 1980s. Fur seals had fallen by 60% since the 1970s. Sea lions crashed by 80% in the 1980s. And preceding these declines were the collapses of the great whales—gray whales, blue whales, humpback whales, and others.

Historically, most orcas specialized on eating great whales, which they would kill in groups, like a wolf pack taking down an elk. But industrial whaling by ships from Japan, Russia, and other nations since the 1950s caused populations of great whales to plummet by 99% between 1965 and 1973 in the northern Pacific. Once human hunting decimated the great whales, the orcas had to turn elsewhere. Springer, Estes, and their colleagues argued that they had shifted their diet to smaller, less-favored seals and sea lions. When their predation had depleted those populations, the chain reaction continued, and they turned to smaller prey still—sea otters.

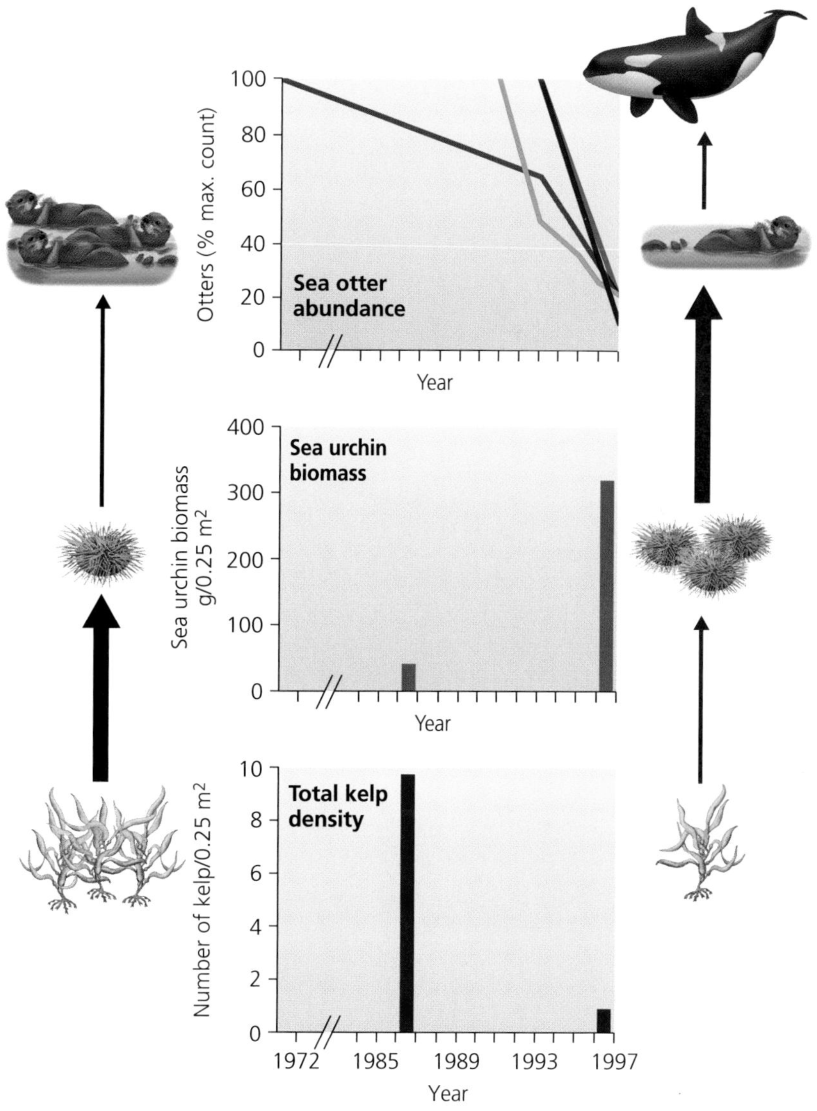

Before the 1990s (left side of figure), otters kept urchin numbers in check, allowing kelp forests to grow. By the end of the 1990s (right side of figure), orcas (killer whales), deprived of their usual food sources, were eating otters. This set off a chain reaction across several trophic levels: Otters decreased, sea urchins increased, and kelp decreased. The lines in the top graph indicate trends in otter populations from four different Aleutian islands. Width of arrows indicates strength of interaction. *Source:* Estes, J., et al. 1998. Killer whale predation on sea otters linking oceanic and nearshore ecosystems. *Science* 282: 473–476.

time passes, shrubs and fast-growing trees such as aspens rise from the field. Pine trees subsequently rise above the aspens and shrubs, forming a pine-dominated forest. This pine forest develops an understory of hardwood trees, because pine seedlings do not grow well under mature pines, whereas some hardwood seedlings do. Eventually the hardwoods outgrow the pines, creating a hardwood forest (see Figure 6.13).

Succession also occurs in aquatic systems. A lake or pond that originates as nothing but water on a lifeless

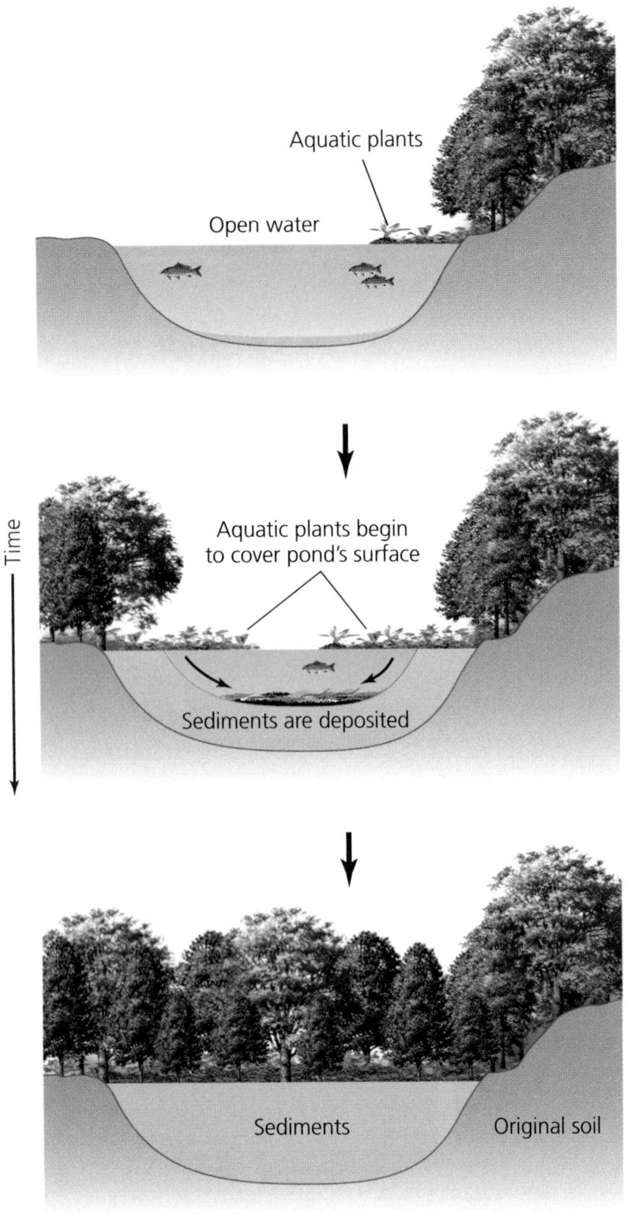

FIGURE 6.14 Primary aquatic succession occurs when plant growth gradually fills in a pond or lake and converts an aquatic system to a wet meadow and ultimately to a terrestrial system. Increased nutrient input can accelerate this process.

substrate begins to undergo succession as it is colonized by algae, microbes, plants, and zooplankton. As these organisms grow, reproduce, and die, the water body slowly fills with organic matter. The lake or pond acquires further organic matter and sediments from the water it receives from rivers, streams, and surface runoff. Eventually, the water body fills in and undergoes a gradual transition to a terrestrial system (Figure 6.14).

In this traditional view of succession that we have described, the transitions between stages of succession eventually lead to a *climax community*, which remains in place, with little modification, until some disturbance restarts succession. Early ecologists felt that each region had its own characteristic climax community, determined by the region's climate.

Today, ecologists recognize that succession is far more variable and less predictable than originally thought. The trajectory of succession can vary greatly according to chance factors, such as which particular species happen to gain an early foothold. The stages of succession blur into one another and vary from place to place, and some stages may sometimes be skipped completely. In addition, climax communities are not predetermined solely by climate, but may vary with other conditions from one time or place to another. Once a climax community is disturbed and succession is set in motion, there is no guarantee that the community will ever return to that climax state. Many communities disturbed by human impact have not returned to their former conditions. This is the case with vast areas of the Middle East that once were fertile enough to support productive farming but now are deserts.

How cohesive are communities?

Ecologists who have studied how communities change in response to disturbance have conceptualized communities in different ways. Early in the 20th century, botanist Frederick Clements promoted the view that communities are cohesive entities whose members remain associated over time and space. Communities, he argued, are discrete units with integrated parts, much like organisms. Clements's view implied that the many varied members of a community share similar limiting factors and evolutionary histories.

Henry Gleason disagreed. Gleason, also a botanist, maintained that each species responds independently to its own limiting factors and that species can join or leave communities without greatly altering their composition. Communities, Gleason argued, are not cohesive units, but temporary associations of individual species that can reassemble into different combinations.

Today ecologists side largely with Gleason, although most see validity in aspects of both men's ideas. Indeed, many ecologists still find it useful to refer to communities by names that highlight certain key plants (such as "oak-hickory forest," "tallgrass prairie," and "pine-bluestem community"). Ecologists find labeling communities as though they were cohesive units to be a pragmatic tool, even though they know the associations could change radically decades or centuries hence.

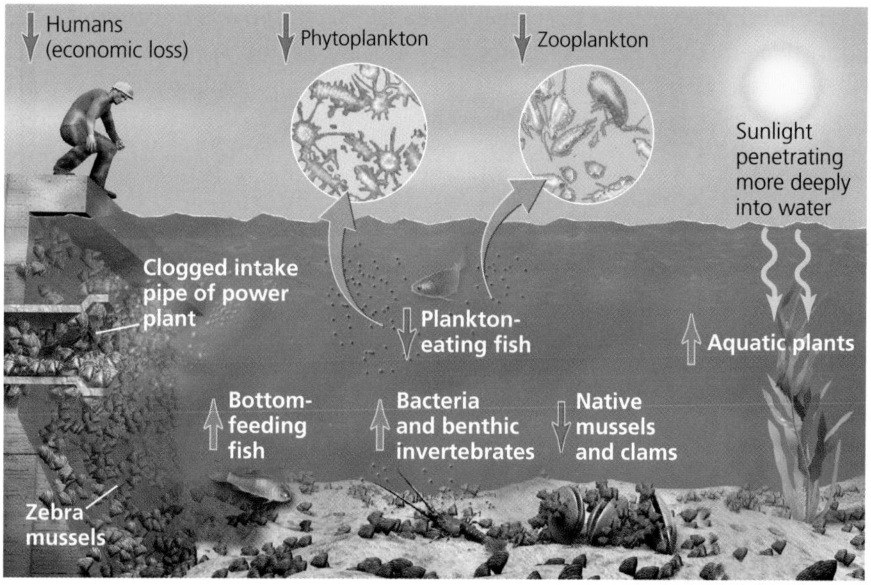

(a) Impacts of zebra mussels on members of a Great Lakes nearshore community

FIGURE 6.15 The zebra mussel is a example of a biological invader that has modified an ecological community. By filtering phytoplankton and small zooplankton from open water, it generates a number of impacts on other species, both negative (red downward arrows) and positive (green upward arrows) **(a)**. This map **(b)** shows the range of the zebra mussel in the United States as of 2005. In less than two decades it has spread from the Great Lakes east to Vermont and Connecticut; west to Nebraska and Kansas; and south to Louisiana and Mississippi (shaded states and red dots), and it has been transported by people to other states as well (blue dots). *Source:* (b) U.S. Geological Survey.

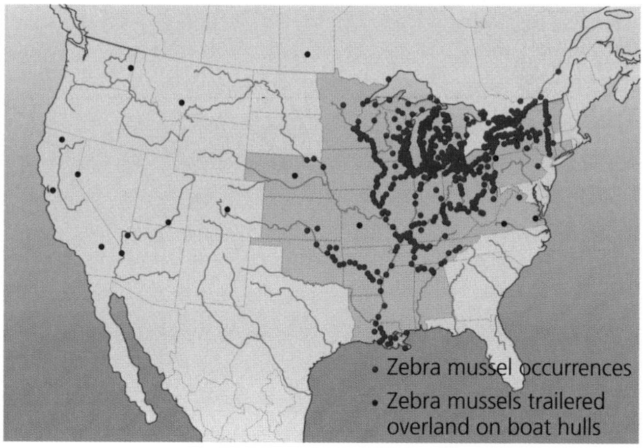

(b) Occurrence of zebra mussels in the United States, 2005

Invasive species pose new threats to community stability

Traditional concepts of community cohesion and of succession involve sets of organisms understood to be native to an area. But what if a new organism arrives from elsewhere? And what if this non-native organism turns *invasive,* spreading widely and becoming dominant in a community? Such **invasive species** can potentially alter a community substantially and are one of the central ecological forces in today's world.

Most often, invasive species are non-native species that people have introduced, intentionally or by accident, from elsewhere in the world. Species become invasive when limiting factors that regulate their population growth are removed. Many ecologists have suggested that plants and animals brought to one area from another may leave their predators, parasites, and competitors behind and be freed from these constraints on their population growth. If there happen to be few organisms in the new environment that can act as predators, parasites, or competitors, the introduced species can do very well. As it proliferates, it may exert diverse influences on its fellow community members (Figure 6.15).

An example is the chestnut blight, an Asian fungus that killed nearly every mature American chestnut *(Castanea dentata),* the dominant tree species of many forests of eastern North America, in the quarter-century preceding 1930. Asian trees had evolved defenses against the fungus over long millennia of coevolution, but the American chestnut had not. A different fungus caused Dutch elm disease, destroying most of the American elms *(Ulmus americana)* that once gracefully lined the streets of many U.S. cities. Grasses introduced in the American West by ranchers have overrun entire regions, pushing out native vegetation. Fish introduced into streams for sport compete with and exclude native fish. Hundreds of island-dwelling animals and plants worldwide have been driven extinct by the goats, pigs, and rats introduced by human colonists. We will examine more examples in our discussion of biodiversity in Chapter 11 (▸pp. 320–322). The impact of invasive species on native species and ecological communities is severe already, and it is growing year by year with the increasing mobility of humans and the globalization of our society.

Our global trade helped spread zebra mussels, which were unintentionally transported in the ballast water of cargo ships. To maintain stability at sea, ships take water into their hulls as they begin their voyage and discharge that water at their destination. Decades of unregulated exchange of ballast water have ferried hundreds of species across the oceans.

In North America, zebra mussels—and the media attention they generated—helped put invasive species on the map as a major environmental and economic problem. Scientific research into introduced species proliferated, and many ecologists came to view invasive species as the second-greatest threat to species and natural systems, behind only habitat destruction. In 1990 the U.S. Congress passed the Nonindigenous Aquatic Nuisance Prevention and Control Act, which became the National Invasive Species Act of 1996. Among other things, this law directed the Coast Guard to ensure that ships dump their freshwater ballast at sea and exchange it with saltwater before entering the Great Lakes.

Since then, funding has become widely available for the control and eradication of invasive species. Managers have been trying a wide variety of techniques to control the spread of zebra mussels—removing them manually, applying toxic chemicals, drying them out, depriving them of oxygen, introducing predators and diseases, and stressing them with heat, sound, electricity, carbon dioxide, and ultraviolet light. However, most of these are localized and short-term fixes that are not capable of making a dent in the huge populations at large in the environment. In case after case, managers are finding that controlling and eradicating invasive species are so difficult and expensive that preventive measures (such as ballast water regulations) represent a much better investment.

Weighing the Issues: Are Invasive Species All Bad?

Some ethicists have questioned the notion that all invasive species should automatically be considered bad. If we introduce a non-native species to a community and it greatly modifies the community, do you think that is a bad thing? What if it drives another species extinct? What if the invasive species arrived on its own, rather than through human intervention? What ethical standard(s) (▸p. 30) would you apply to determine whether an invasive species should be battled or accepted?

Altered communities can be restored to their former condition

Invasive species are adding to the tremendous transformations that humans have already forced on natural landscapes and communities through habitat alteration, deforestation, hunting of keystone species, pollution, and other activities. With so much of Earth's landscape altered by human impact, it is impossible to find areas that are truly pristine. This realization has given rise to the conservation effort known as **ecological restoration.** The practice of ecological restoration is informed by the science of **restoration ecology.** Restoration ecologists research the historical conditions of ecological communities as they existed before our industrialized civilization altered them. They then try to devise ways to restore some of these areas to an earlier condition, often to a natural "presettlement" condition.

For instance, in the United States nearly every last scrap of tallgrass prairie that once covered the eastern Great Plains and parts of the Midwest was converted to agriculture in the 19th century. Now a number of efforts are underway to restore small patches of prairie by planting native prairie plants, weeding out invaders and competitors, and introducing controlled fire to mimic the fires that historically maintained this community. Illinois boasts several of the largest prairie restoration projects so far. The Morton Arboretum in Lisle has a 40-ha (100-acre) prairie, and one at the Forest Glen Preserve includes more than 120 species of native prairie plants. A 184-ha (455-acre) area inside the massive ring of the Fermilab nuclear accelerator in Batavia is being restored, and plans are underway for a larger project at an old U.S. Army ammunition facility near Joliet.

Perhaps the world's largest restoration project is the ongoing effort to restore the Florida Everglades. The Everglades, a 7,500-km^2 (4,700-mi^2) ecosystem of marshes and seasonally flooded grasslands, has been drying out for decades because the water that feeds it has been managed for flood control and overdrawn for irrigation and development. Populations of wading birds have dropped 90–95%, and economically important fisheries have suffered greatly as a result. The 30-year, $7.8-billion restoration project intends to restore water by undoing damming and diversions of 1,600 km (1,000 mi) of canals, 1,150 km (720 mi) of levees, and 200 water control structures.

One of the most intriguing new restoration efforts is the drive to restore the Mesopotamian marshes between the Tigris and Euphrates Rivers in Iraq, formerly one of the world's greatest wetlands. The government of Saddam Hussein diverted water from the marshes in an effort to debilitate the minority peoples who had lived here for thousands of years. Following the U.S. occupation of Iraq, ecologists from many nations want to help restore the marshes and give their human residents a place to live again in their traditional lifestyle. Whether the funds, resources, and access are granted to allow this project to succeed remains to be seen. Regardless, the more our population grows and development spreads, the more ecological restoration will become a vital conservation strategy for the future.

VIEWPOINTS

Invasive Species

Invasive species can substantially modify the ecological communities they invade. **How should we as a society respond to invasive species?**

Prevention Is the Best Strategy

Non-native species have been introduced globally for centuries for food and other human uses, but many human-mediated invasions have caused widespread impacts in freshwater, marine, and terrestrial ecosystems. Environmental impacts include displacement of native fauna and flora, modification of nutrient and chemical cycles and food webs, reduction of groundwater levels, and erosion. Economic problems include extensive impacts on industry, recreation, and fishing. Numerous diseases and pathogenic organisms have been introduced as well.

All of these considerations have led to efforts to *prevent* further invasions and *control* those exotic species that have become established. *Prevention* is one of the most important tools of invasion management. Public education, starting in elementary school, is the long-term foundation of prevention, such that individuals are aware of the concern about introduced species, and take responsibility for their personal roles in accidentally or intentionally moving non-native species. For terrestrial species, preventing the arrival of non-native species includes thorough examination of imported goods, shipping containers, or luggage at airports, as well as chemical treatment of imported wood and other products. Non-native aquatic species may arrive by the movement of fishing bait, live seafood, aquaculture species, and particularly ships' ballast water. Prevention measures for ballast water include filtering, heating, ultraviolet treatment, oxygen deprivation, and other techniques.

Controlling invaders that have become established involves a broad range of tools (such as chemical treatments, barriers, and physical removal) focused on limiting the spread and abundance of the target species. Eradication is sometimes possible for terrestrial plants and animals, but is generally impossible in aquatic environments, where populations often become too widespread to locate and remove all individuals.

Stopping the arrival of unwanted species in the first place is the best long-term management strategy: If it's raining, close the windows.

James T. Carlton is Professor of Marine Sciences at Williams College in Williamstown, Massachusetts, and Director of Williams-Mystic, The Maritime Studies Program of Williams College and Mystic Seaport (Mystic, Connecticut). His research is on global marine bioinvasions and marine extinctions in modern times. He is the founding Editor-in-Chief of the international journal *Biological Invasions*.

Deal with Invasions on a Case-By-Case Basis

Certain introduced species invade natural ecosystems and damage human enterprises such as agriculture and fisheries. However, these harmful invaders are a minority of introduced species. Other introduced species are useful as food crops, pets, or in other ways. Controversy therefore arises over how to deal with introduced species: Which ones should we target for management, and what sort of management?

It is more efficient to keep introduced species out in the first place than to try to eradicate or manage established populations. For deliberately introduced species, most nations now use risk assessment methods to try to quantify the likelihood that a species will become a pest and then to decide whether to permit introduction. For species that arrive on their own, biologists and policymakers now focus on restricting frequently used pathways, such as ballast water and untreated wood.

Once a species establishes, it is often possible to eradicate it, especially if it has not spread far. Because of early failed eradication attempts that wasted much money and harmed non-target species, many managers hesitate to attempt eradication. However, technologies have improved to the point where there are now many successes.

If eradication fails or is not attempted, there are three general approaches to managing introduced species. Mechanical or physical control entails pulling out weeds by hand or machine, shooting vertebrates, hand-collecting snails, and the like. Chemical control involves using pesticides, but is controversial because these chemicals can have non-target impacts. Biological control uses natural enemies—predators, parasites, and pathogens—imported from the native range of the introduced pest, but is controversial because it entails introducing new species, some of which have become invasive pests themselves.

Pronouncements that chemicals are dangerous or biological control is dangerous are far too sweeping, and what is needed is case-by-case examination of all available technologies to deal with established invaders.

Daniel Simberloff is the Nancy Gore Hunger Professor of Environmental Studies at the University of Tennessee. His research is on ecology, biogeography, and evolution, and often relates to the causes and consequences of species associations in communities. Much of his research has focused on conservation issues, especially the impacts of introduced species.

Explore this issue further by accessing **Viewpoints** at www.aw-bc.com/withgott.

Weighing the Issues:

Restoring "Natural" Communities

Practitioners of ecological restoration in North America aim to restore communities to their natural state. But what is meant by "natural"? Does it mean the state of the community before industrialization? Before Europeans came to the New World? Before any people laid eyes on the community? Let's say Native Americans altered a forest community 8,000 years ago by burning the underbrush regularly to improve hunting, and continued doing so until Europeans arrived 400 years ago and cut down the forest for farming. Today the area's inhabitants want to restore the land to its "natural" forested state. Should restorationists try to recreate the forest of the Native Americans, or the forest that existed before Native Americans arrived? What are some advantages and disadvantages of each approach?

Earth's Biomes

Across the world, each portion of each continent has different sets of species, leading to endless variety in community composition. However, communities in far-flung places often share strong similarities in their structure and function. This allows us to classify communities into broad types. A **biome** is a major regional complex of similar communities—a large ecological unit recognized primarily by its dominant plant type and vegetation structure. The world contains a number of biomes, each covering large contiguous geographic areas (Figure 6.16).

Biomes are groupings of communities that cover large geographic areas

Which biome covers any particular portion of the planet depends on a variety of abiotic factors, including

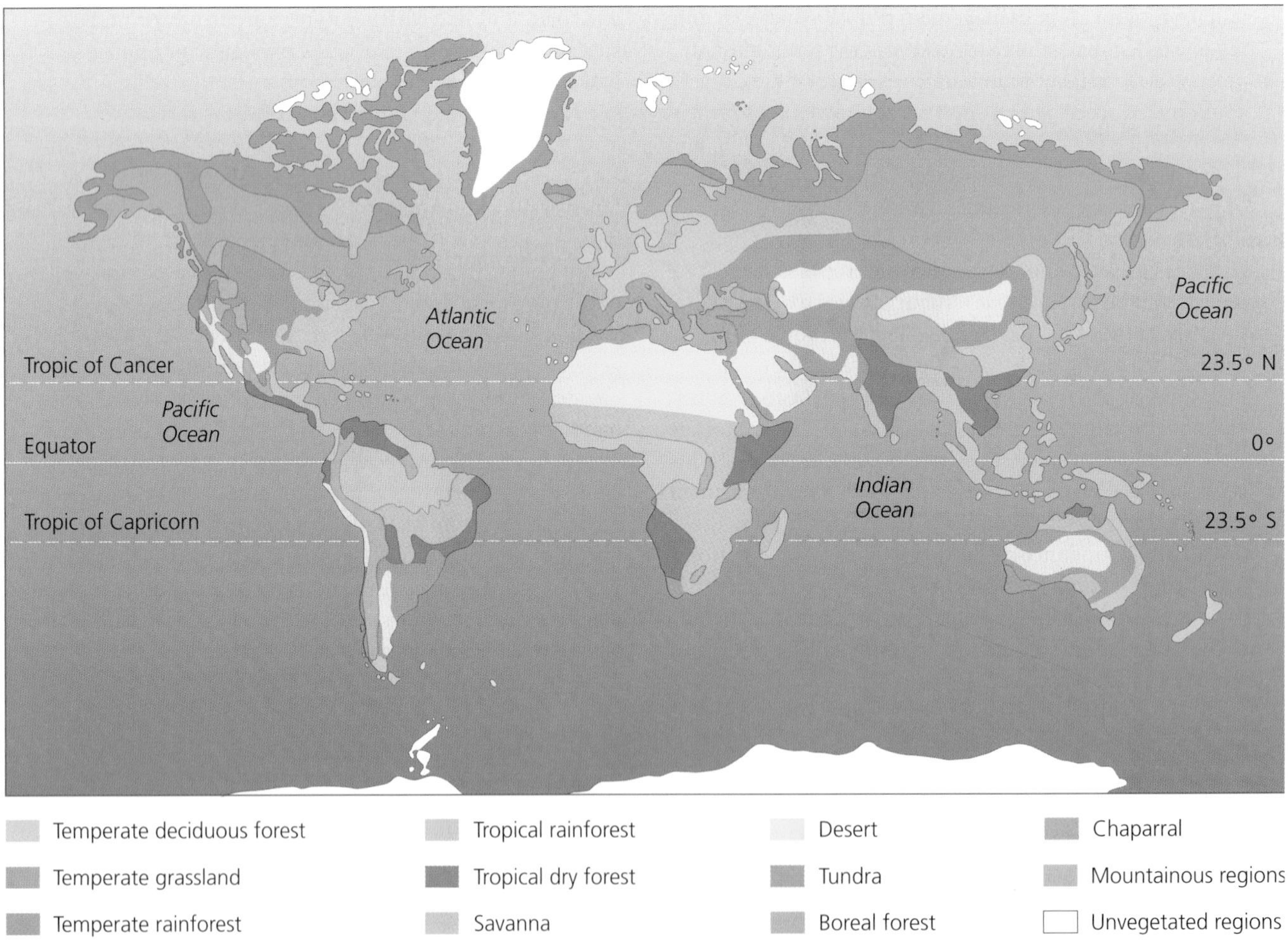

FIGURE 6.16 Biomes are distributed around the world according to temperature, precipitation, atmospheric and oceanic circulation patterns, and other factors.

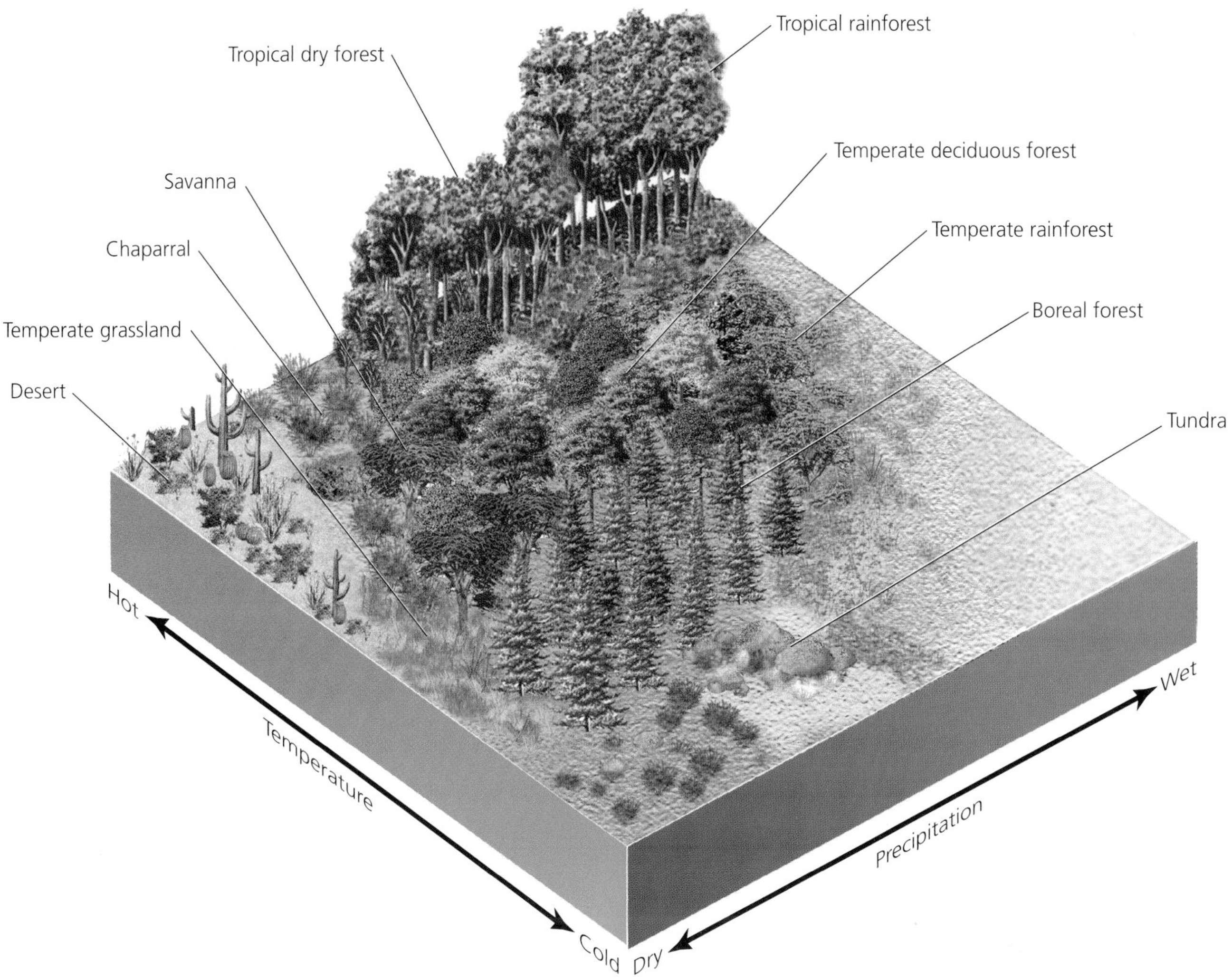

FIGURE 6.17 As precipitation increases, vegetation generally becomes taller and more luxuriant. As temperature increases, types of plant communities change. Together, temperature and precipitation are the main factors determining which biome occurs in a given area. For instance, deserts occur in dry regions, tropical rainforests occur in warm wet regions, and tundra occurs in the coldest regions.

temperature, precipitation, atmospheric circulation, and soil characteristics. Among these factors, temperature and precipitation exert the greatest influence (Figure 6.17). Because biome type is largely a function of climate, and because average monthly temperature and precipitation are among the best indicators of an area's climate, scientists often use climate diagrams, or **climatographs,** to depict such information. Global climate patterns cause biomes to occur in large patches in different parts of the world. For instance, temperate deciduous forest occurs in eastern North America, north-central Europe, and eastern China. Note in Figure 6.16 how patches representing the same biome tend to occur at similar latitudes. This is due to the north-south gradient in temperature and to atmospheric circulation patterns (▸p. 504).

Each biome encompasses a variety of communities that share similarities. For example, the eastern United States supports part of the temperate deciduous forest biome. From New Hampshire to the Great Lakes to eastern Texas, precipitation and temperature are similar enough that most of the region's natural plant cover consists of broad-leafed trees that lose their leaves in winter. Within this region, however, there exist many different types of temperate deciduous forest, such as oak-hickory, beech-maple, and pine-oak forests, each sufficiently different to be designated a separate community.

(a) Temperate deciduous forest

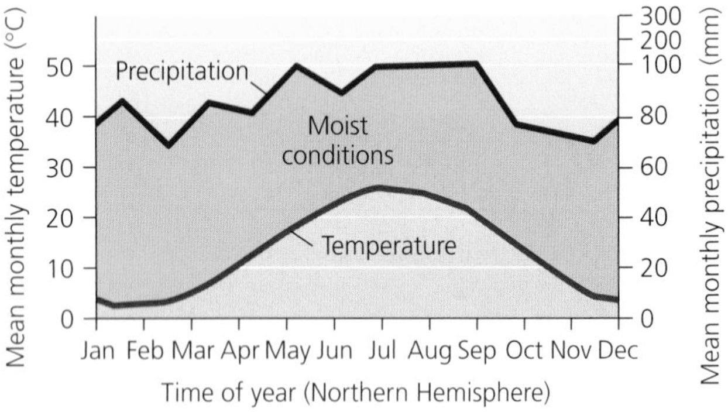

(b) Washington, D.C., USA

FIGURE 6.18 Temperate deciduous forests experience relatively stable seasonal precipitation but more variation in seasonal temperatures. Scientists use climate diagrams to illustrate an area's average monthly precipitation and temperature. Typically in these diagrams, the *x* axis marks months of the year (beginning in January for regions in the Northern Hemisphere and in July for regions in the Southern Hemisphere). Paired *y* axes denote average monthly temperature and average monthly precipitation. The twin curves plotted on a climate diagram indicate trends in precipitation (blue) and in temperature (red) from month to month. When the precipitation curve lies above the temperature curve, as is the case throughout the year in the temperate deciduous forest biome around Washington, D.C., the region experiences relatively "moist" conditions, which we indicate with green coloration. Climatograph adapted from Breckle, S.W., 1999. *Walter's vegetation of the Earth: The ecological systems of the geo-biosphere*, 4th ed. Berlin: Springer-Verlag.

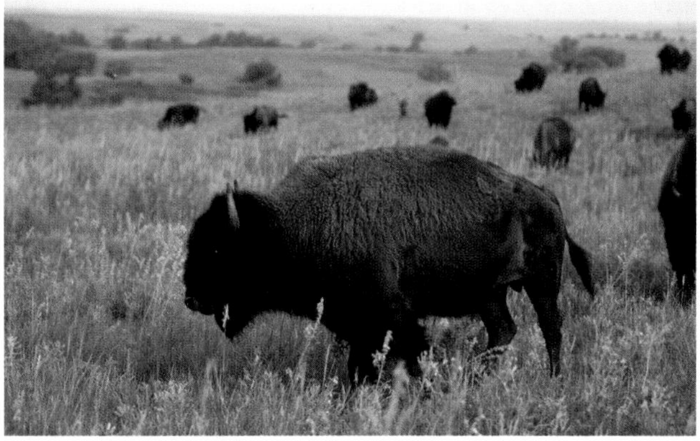

(a) Temperate grassland

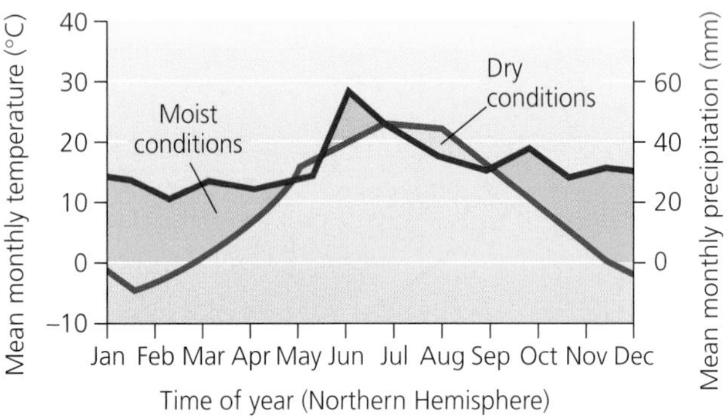

(b) Odessa, Ukraine

FIGURE 6.19 Temperate grasslands experience temperature variations throughout the year and too little precipitation for many trees to grow. Constructed for Odessa, Ukraine, this climatograph indicates both "moist" (green) and "dry" (yellow) climate conditions. When the temperature curve is above the precipitation curve, as is the case in May and mid-June through September, the climate conditions are "dry." Climatograph adapted from Breckle, S.W., 1999.

We can divide the world into roughly ten terrestrial biomes

Temperate deciduous forest The **temperate deciduous forest** (Figure 6.18) that dominates the landscape around the central and southern Great Lakes is characterized by broad-leafed trees that are *deciduous,* meaning that they lose their leaves each fall and remain dormant during winter, when hard freezes would endanger leaves. These mid-latitude forests occur in much of Europe and eastern China as well as in eastern North America—all areas where precipitation is spread relatively evenly throughout the year. Although soils of the temperate deciduous forest are relatively fertile, the biome generally consists of far fewer tree species than are found in tropical rainforests. Oaks, beeches, and maples are a few of the most abundant types of trees in these forests. A sampling of typical animals of the temperate deciduous forest of eastern North America is shown in Figure 6.11.

Temperate grassland Moving westward from the Great Lakes, we find **temperate grasslands** (Figure 6.19).

This is because temperature differences between winter and summer become more extreme, and rainfall diminishes. The limited amount of precipitation in the Great Plains region west of the Mississippi River can support grasses more easily than trees. Also known as steppe or prairie, temperate grasslands were once widespread throughout parts of North and South America and much of central Asia. Today people have converted most of the world's grasslands for agriculture, greatly reducing the abundance of native plants and animals. Characteristic vertebrate animals of the North American grasslands include American bison *(Bison bison)*, prairie dogs, pronghorn antelope *(Antilocapra americana)*, and ground-nesting birds such as meadowlarks.

Temperate rainforest Moving further west in North America, the topography becomes more varied, and biome types are intermixed. The coastal Pacific Northwest region, with its heavy rainfall, features **temperate rainforest** (Figure 6.20), a forest type known for its potential to produce large volumes of commercially important forest products, such as lumber and paper. Coniferous trees, such as cedars, spruces, hemlocks, and Douglas fir *(Pseudotsuga menziesii)* grow very tall in the temperate rainforest, and the forest interior is shaded and damp. In the Pacific Northwest, moisture-loving animals such as the bright yellow banana slug *(Ariolimax columbianus)* are common, and old-growth stands hold the endangered spotted owl *(Strix occidentalis)*. The soils of temperate rainforests are usually quite fertile but are susceptible to landslides and erosion if forests are cleared. Temperate rainforests have been the focus of controversy in the Pacific Northwest, where overharvesting has driven some species toward extinction and pushed many forest-dependent human communities toward economic stagnation.

(a) Temperate rainforest

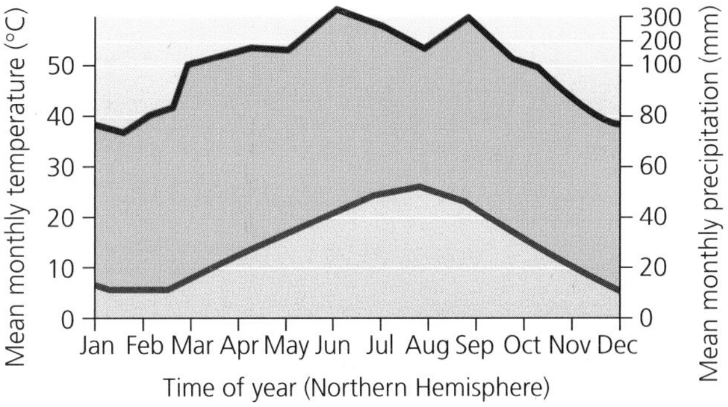

(b) Nagasaki, Japan

FIGURE 6.20 Temperate rainforests receive a great deal of precipitation and feature moist, mossy interiors. Climatograph adapted from Breckle, S.W., 1999.

Tropical rainforest In tropical regions we see the same pattern found in temperate regions: Areas of high rainfall grow rainforests, areas of intermediate rainfall host dry or deciduous forests, and areas of lower rainfall become dominated by grasses. However, tropical biomes differ from their temperate counterparts in other ways because they are closer to the equator and therefore warmer on average year-round. For one thing, they hold far greater biodiversity.

Tropical rainforest (Figure 6.21) is found in Central America, South America, southeast Asia, west Africa, and other tropical regions and is characterized by year-round rain and uniformly warm temperatures. Tropical rainforests have dark, damp interiors, lush vegetation, and highly diverse biotic communities, with greater numbers of species of insects, birds, amphibians, and various other animals than any other biome. These forests are not dominated by single species of trees, as are forests closer to the poles, but instead consist of very high numbers of tree species intermixed, each at a low density. Any given tree may be draped with vines, enveloped by strangler figs, and loaded with epiphytes (orchids and other plants that grow in trees), such that trees occasionally collapse under the weight of all the life they support. Despite this profusion of life, tropical rainforests have very poor, acidic soils that are low in organic matter. Nearly all nutrients present in this biome are contained in the trees, vines, and other plants—not in the soil. An unfortunate consequence is that once tropical rainforests are cleared, the nutrient-poor soil can support agriculture for only a short time. As a result, farmed areas are abandoned quickly, and the soil and forest vegetation recover slowly.

(a) Tropical rainforest

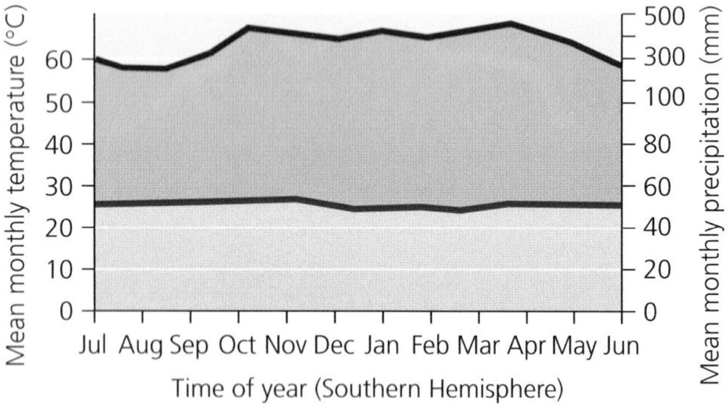

(b) Bogor, Java, Indonesia

FIGURE 6.21 Tropical rainforests, famed for their biodiversity, grow under constant, warm temperatures and a great deal of rain. Climatograph adapted from Breckle, S.W., 1999.

(a) Tropical dry forest

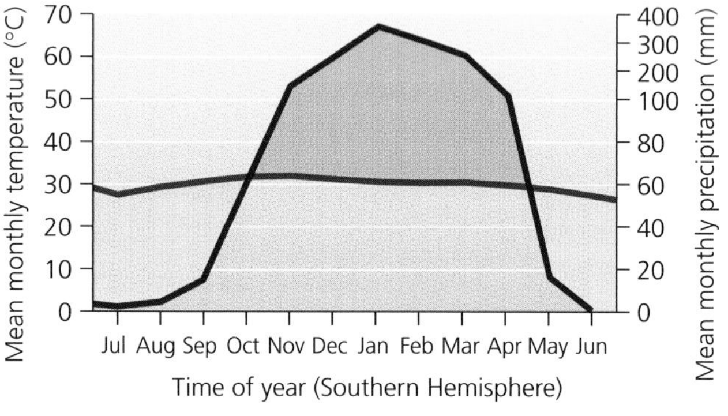

(b) Darwin, Australia

FIGURE 6.22 Tropical dry forests experience significant seasonal variations in precipitation and relatively stable warm temperatures. Climatograph adapted from Breckle, S.W., 1999.

Tropical dry forest Tropical areas that are warm year-round but where rainfall is lower overall and highly seasonal give rise to **tropical dry forest,** or tropical deciduous forest (Figure 6.22), a biome widespread in India, Africa, South America, and northern Australia. Wet and dry seasons each span about half a year in tropical dry forest. Rains during the wet season can be extremely heavy and, coupled with erosion-prone soils, can lead to severe soil loss when forest clearing occurs over large areas. Across the globe, much tropical dry forest has been converted to agriculture. Clearing for farming or ranching is made easier by the fact that vegetation heights are much lower and canopies less dense than in tropical rainforest. Organisms that inhabit tropical dry forest have adapted to seasonal fluctuations in precipitation and temperature. For instance, plants are deciduous and often leaf out and grow profusely with the rains, then drop their leaves during the driest times of year.

Savanna Drier tropical regions give rise to **savanna** (Figure 6.23), tropical grassland interspersed with clusters of acacias or other trees. The savanna biome is found today across stretches of Africa (the ancestral home of our species), South America, Australia, India, and other dry tropical regions. Precipitation in savannas usually arrives during distinct rainy seasons and concentrates grazing animals near widely spaced water holes. Common herbivores on the African savanna include zebras, gazelles, and giraffes, and the predators of these grazers include lions, hyenas, and other highly mobile carnivores.

Desert Where rainfall is very sparse, **desert** (Figure 6.24) forms. This is the driest biome on Earth; most deserts receive well under 25 cm (9.8 in.) of precipitation per year, much of it during isolated storms months or years apart. Depending on rainfall, deserts vary greatly in the

(a) Savanna

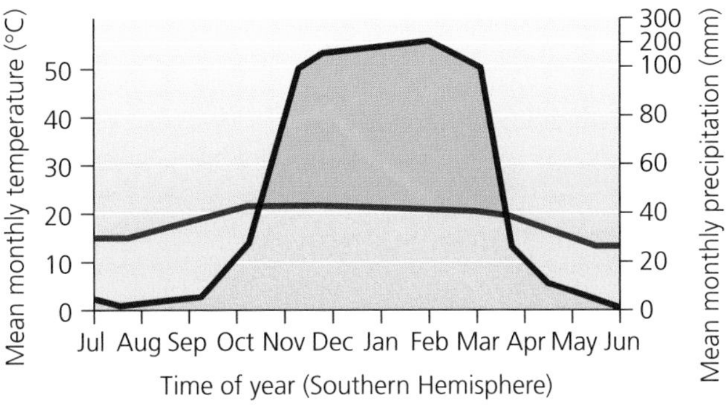

(b) Harare, Zimbabwe

FIGURE 6.23 Savannas are grasslands with clusters of trees. They experience slight seasonal variation in temperature but significant variation in rainfall. Climatograph adapted from Breckle, S.W., 1999.

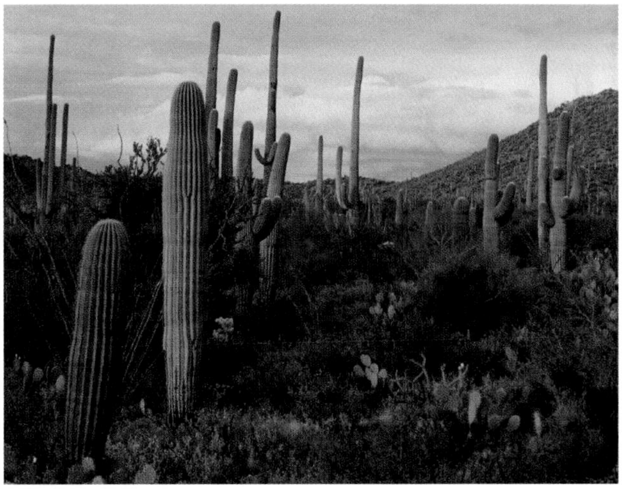

(a) Desert

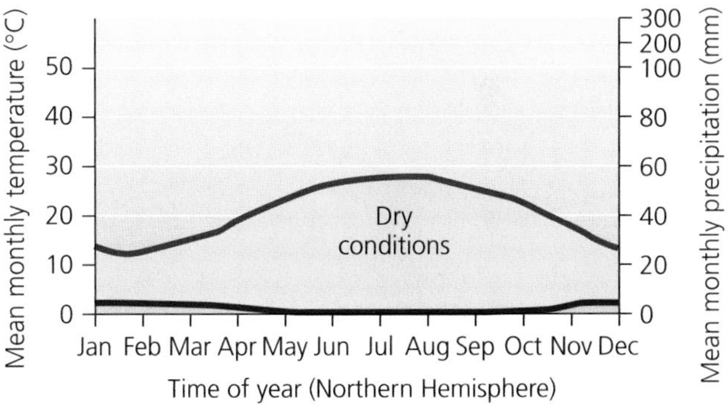

(b) Cairo, Egypt

FIGURE 6.24 Deserts are dry year-round, but they are not always hot. Precipitation can arrive in intense, widely spaced storm events. The temperature curve is consistently above the precipitation curve in this climatograph of Cairo, Egypt, indicating that the region experiences "dry" conditions all year. The photograph, from the Sonoran Desert in Arizona, shows the maximum amount of vegetation a desert can support. Climatograph adapted from Breckle, S.W., 1999.

amount of vegetation they support. Some, like the Sahara and Namib deserts of Africa, are mostly bare sand dunes; others, like the Sonoran Desert of Arizona and northwest Mexico, are quite heavily vegetated. Deserts are not always hot; the high desert of the western United States is one example. Because deserts have low humidity and relatively little vegetation to insulate them from temperature extremes, sunlight readily heats them in the daytime, but daytime heat is quickly lost at night. As a result, temperatures vary widely from day to night and across seasons of the year. Desert soils can often be quite saline and are sometimes known as lithosols, or stone soils, for their high mineral and low organic-matter content.

Desert animals and plants show many adaptations to deal with a harsh climate. Most reptiles and mammals, such as rattlesnakes and kangaroo mice, are active in the cool of night, and many Australian desert birds are nomadic, wandering long distances to find areas of recent rainfall and plant growth. Many desert plants have thick leathery leaves to reduce water loss, or green trunks so that the plant can photosynthesize without leaves, which would lose water. The spines of cacti and many other desert plants guard those plants from being eaten by herbivores desperate for the precious water they hold.

Tundra **Tundra** (Figure 6.25) is nearly as dry as desert but is located at very high latitudes along the northern edges of Russia, Canada, and Scandinavia. Extremely cold winters with little daylight and moderately cool summers with lengthy days characterize this landscape of

(a) Tundra

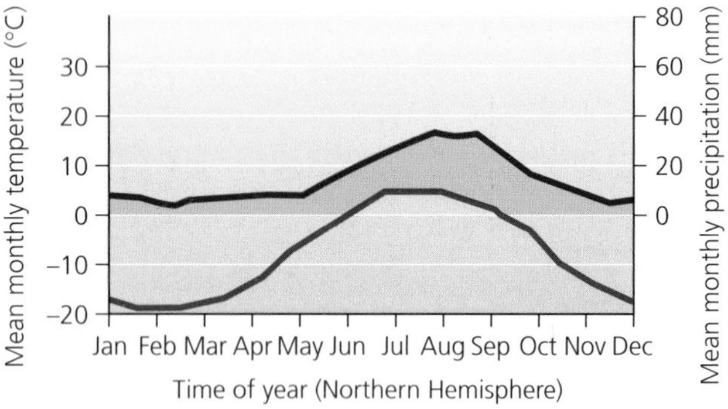

(b) Vaigach, Russia

FIGURE 6.25 Tundra is a cold, dry biome found near the poles and atop high mountains at lower latitudes. Climatograph adapted from Breckle, S.W., 1999.

(a) Boreal forest

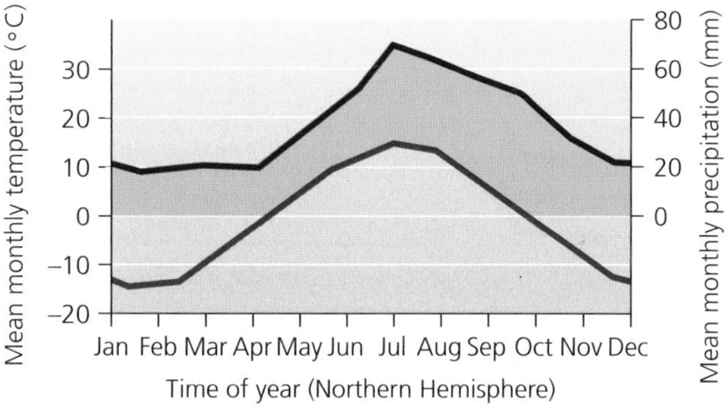

(b) Archangelsk, Russia

FIGURE 6.26 Boreal forest is defined by long, cold winters, relatively cool summers, and moderate precipitation. Climatograph adapted from Breckle, S.W., 1999.

lichens and low, scrubby vegetation without trees. The great seasonal variation in temperature and day length results from this biome's position close to the poles, which are angled toward the sun in the summer and away from the sun in the winter. Because of the cold climate, underground soil remains more or less permanently frozen, and is called *permafrost.* During the long, cold winters, the surface soils freeze as well; then, when the weather warms, they melt and produce seasonal accumulations of surface water that make ideal habitat for mosquitoes and other biting insects. The swarms of insects benefit bird species that migrate long distances to breed during the brief but productive summer. Caribou also migrate to the tundra to breed, and then leave for the winter. Only a few animals, such as polar bears *(Ursus maritimus)* and musk oxen *(Ovibos moschatus),* can survive year-round in this extreme climate. Tundra also occurs as *alpine tundra* at the tops of high mountains in temperate and tropical regions.

Boreal forest The northern coniferous forest, or **boreal forest,** often called *taiga* (Figure 6.26), stretches in a broad band across much of Canada, Alaska, Russia, and Scandinavia. It consists of a few species of evergreen trees, such as black spruce *(Picea mariana),* that dominate large stretches of forests interspersed with occasional bogs and lakes. The boreal forest's uniformity over huge areas reflects the climate common to this latitudinal band of the globe: These forests develop in cooler, drier regions than do temperate forests, and they experience long, cold winters and short, cool summers. Soils are typically nutrient-poor and somewhat acidic. As a result of the strong

(a) Chaparral

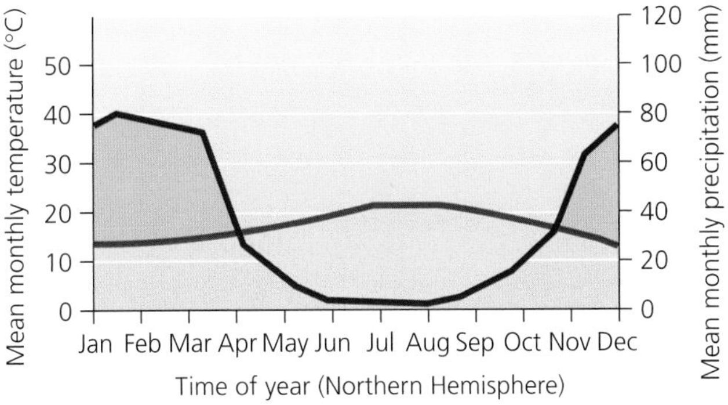

(b) Los Angeles, California, USA

FIGURE 6.27 Chaparral is a highly seasonal biome dominated by shrubs, influenced by marine weather, and dependent on fire. Climatograph adapted from Breckle, S.W., 1999.

seasonal variation in day length, temperature, and precipitation, many organisms compress a year's worth of feeding, breeding, and rearing of young into a few warm, wet months. Year-round residents of boreal forest include mammals such as moose *(Alces alces)*, wolves *(Canis lupus)*, bears, lynx *(Felis lynx)*, and many burrowing rodents. This biome also hosts many insect-eating birds that migrate from the tropics to breed during the brief, intensely productive, summer season.

Chaparral In contrast to the boreal forest's broad, continuous distribution, **chaparral** (Figure 6.27) is limited to fairly small patches widely flung around the globe. Chaparral consists mostly of evergreen shrubs and is densely thicketed. This biome is also highly seasonal, with mild, wet winters and warm, dry summers. This type of climate is induced by oceanic influences and is often termed "Mediterranean." In addition to ringing the Mediterranean Sea, chaparral occurs along the coasts of California, Chile, and southern Australia. Chaparral communities experience frequent fire, and their plant species are adapted to resist fire or even to depend on it for germination of their seeds.

Altitude creates patterns analogous to latitude

As any hiker or skier knows, climbing in elevation causes a much more rapid change in climate than moving the same distance toward the poles. Vegetative communities change along mountain slopes in correspondence with this small-scale climate variation (Figure 6.28). It is often said that hiking up a mountain in the southwestern United States is like walking from Mexico to Canada. A hiker ascending one of southern Arizona's higher mountains would begin in Sonoran Desert or desert grassland and proceed through oak woodland, pine forest, and finally spruce-fir forest—the equivalent of passing through several biomes. A hiker scaling one of the great peaks of the Andes in Ecuador could begin in tropical rainforest and end amid glaciers in alpine tundra.

Aquatic systems also show biome-like patterns

In our discussion of biomes, we have focused exclusively on terrestrial systems, because the biome concept, as traditionally developed and applied, has been limited to terrestrial systems. Areas equivalent to biomes also exist in the oceans, but their geographic shapes would look very different from those of terrestrial biomes if plotted on a world map. One might consider the thin strips along the world's coastlines to represent one aquatic system, the continental shelves another, and the open ocean, the deep sea, coral reefs, and kelp forests as still other distinct sets of communities. There are also many coastal systems that straddle the line between terrestrial and aquatic, such as salt marshes, rocky intertidal communities, mangrove forests, and estuaries. And of course there are freshwater systems such as those of the Great Lakes.

Unlike terrestrial biomes, aquatic systems are shaped not by air temperature and precipitation, but by factors such as water temperature, salinity, dissolved nutrients, wave action, currents, depth, and type of substrate (e.g., sandy, muddy, or rocky bottom). Marine communities are also more clearly delineated by their animal life than by their plant life. We will examine freshwater and marine systems in the greater detail they deserve in Chapters 15 and 16.

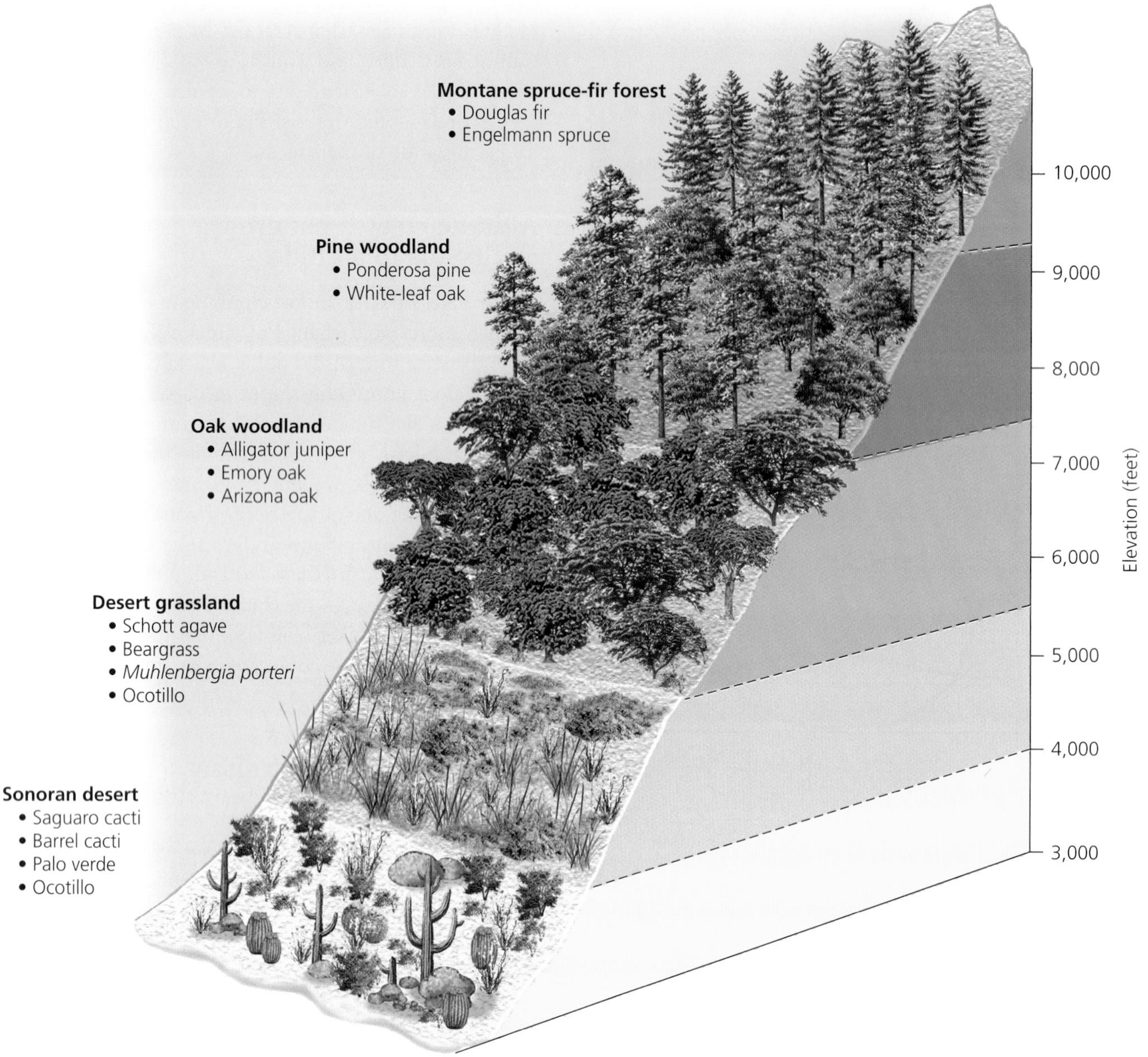

FIGURE 6.28 As altitude increases, vegetation changes in ways similar to how it changes as one moves toward the poles. Climbing a mountain in southern Arizona, as pictured here, can be likened to traveling from Mexico to Canada, taking the hiker through the local equivalent of several biomes.

Conclusion

The natural world is so complex that we can visualize it in many ways and at various scales. Dividing the world's communities into major types, or biomes, is informative at the broadest geographic scales. Understanding how communities function at more local scales requires understanding how species interact with one another. Species interactions such as predation, parasitism, competition, and mutualism give rise to effects both weak and strong, direct and indirect. Feeding relationships can be represented by the concepts of trophic levels and food webs, and particularly influential species are sometimes called keystone species. Increasingly humans are altering communities, in part by introducing non-native species that may turn invasive. But increasingly, through ecological restoration, we are also attempting to undo the changes we have caused.

REVIEWING OBJECTIVES

You should now be able to:

Compare and contrast the major types of species interactions

- Competition results when individuals or species vie for limited resources. It can occur within or among species and can result in coexistence or exclusion. It also can lead to realized niches, resource partitioning, and character displacement. (pp. 151–152)
- In predation, one species kills and consumes another. It is the basis of food webs and can influence population dynamics and community composition. (p. 153)
- In parasitism, one species derives benefit by harming (but usually not killing) another. (pp. 153–155)
- Herbivory is an exploitative interaction whereby an animal feeds on a plant. (p. 155)
- In mutualism, species benefit from one another. Some mutualisms are symbiotic, whereas other mutualists are free-living. (pp. 155–156)

Characterize feeding relationships and energy flow, using them to construct trophic levels and food webs

- Energy is transferred in food chains among trophic levels. (pp. 156–157)
- Lower trophic levels generally contain more energy, biomass, and numbers of individuals than higher trophic levels. (p. 158)
- Food webs illustrate feeding relationships and energy flow among species in a community. (pp. 158–159)

Distinguish characteristics of a keystone species

- Keystone species have impacts on communities that are far out of proportion to their abundance. (pp. 159–162)
- Top predators are frequently considered keystone species, but other organisms may be thought of as keystones for other reasons. (p. 162)

Characterize the process of succession and the debate over the nature of communities

- Succession is a stereotypical pattern of change within a community through time. (p. 163)
- Primary succession begins with an area devoid of life. Secondary succession begins with an area that has been severely disturbed (pp. 163–166)
- Clements held that communities are discrete, cohesive units. His view has largely been replaced by that of Gleason, who held that species may be added to and deleted from communities through time. (p. 166)

Perceive and predict the potential impacts of invasive species in communities

- Invasive species such as the zebra mussel have altered the composition, structure, and function of communities. (p. 167)
- Humans are the cause of most modern species invasions, but we can also respond to invasions with prevention and control measures. (pp. 167–168)

Explain the goals and methods of ecological restoration

- Ecological restoration aims to restore communities to a more "natural" state, variously defined as before human or industrial interference. (p. 168)
- Restoration efforts in the field are informed by the growing science of restoration ecology. (p. 168)

Describe and illustrate the terrestrial biomes of the world

- Biomes represent major classes of communities spanning large geographic areas. (pp. 170–171)
- The distribution of biomes is determined by temperature, precipitation, and other factors. (pp. 170–177)
- The biome concept by tradition refers to terrestrial systems. Aquatic systems can be classified in similar ways, determined by different factors. (p. 177)

TESTING YOUR COMPREHENSION

1. How does competition lead to a realized niche? How does it promote resource partitioning?
2. Contrast the several types of exploitation. How do predation, parasitism, and herbivory differ?
3. Give examples of symbiotic and nonsymbiotic mutualisms. Describe at least one way in which mutualisms affect your daily life.
4. Explain how trophic levels, food chains, and food webs are related.
5. Name several ways in which a species could be considered a keystone species.
6. Explain and contrast primary and secondary terrestrial succession.

7. Explain and contrast Clements's and Gleason's views of ecological communities.
8. Name five changes to Great Lakes communities that have occurred since the invasion of the zebra mussel.
9. What factors most strongly influence the type of biome that forms in a particular place on land? What factors determine the type of aquatic system that may form in a given location?
10. Draw climate diagrams for a tropical rainforest and for a desert. Label all parts of the diagram, and describe all of the types of information an ecologist could glean from such a diagram.

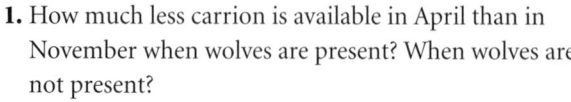

SEEKING SOLUTIONS

1. Imagine that you spot two species of birds feeding side by side, eating seeds from the same plant, and that you begin to wonder whether competition is at work. Describe how you might design scientific research to address this question. What observations would you try to make at the outset? Would you try to manipulate the system to test your hypothesis that the two birds are competing? If so, how?
2. Spend some time outside on your campus or in your yard or in the nearest park or natural area. Find at least 10 species of organisms, and observe them long enough to watch them feed or to make an educated guess about what they feed on. Now, using Figure 6.11 as a model, draw a simple food web involving all the organisms you observed.
3. Can you think of one organism not mentioned in this chapter as a keystone species that you believe may be a keystone species? For what reasons do you suspect this? How could you experimentally test whether an organism is a keystone species?
4. Why do scientists consider invasive species to be a problem? What makes a species "invasive," and what ecological effects can invasive species have?
5. From year to year, biomes are stable entities, and our map of world biomes appears to be a permanent record of patterns across the planet. But are the locations and identities of biomes permanent, or could they change over time? Provide reasons for your answers.
6. Can you devise possible responses to the zebra mussel invasion? What strategies would you consider if you were put in charge of the effort to control this species' spread and reduce its impacts? Name some advantages of each of your ideas, and identify some obstacles it might face in being implemented.

INTERPRETING GRAPHS AND DATA

The gray wolf *(Canis lupus)* is a keystone species in Yellowstone National Park's ecosystem. Wolf packs hunt elk, gorge themselves on the kill, and leave the carcass as carrion for scavenger species such as ravens, magpies, eagles, coyotes, and bears. As the global climate has warmed, winters in Yellowstone have become shorter over the past 55 years. Fewer elk weaken and die in milder weather, and so less carrion is available to scavengers during warmer, shorter winters. Biologists Christopher Wilmers and Wayne Getz studied the links among climate change, wolves, elk, and scavenger populations in Yellowstone. They used empirical field data on wolf predation rates and elk carrion availability recorded over 55 years to develop a model that estimated carrion availability with and without wolves for each winter month. Some of their findings are presented in the graph.

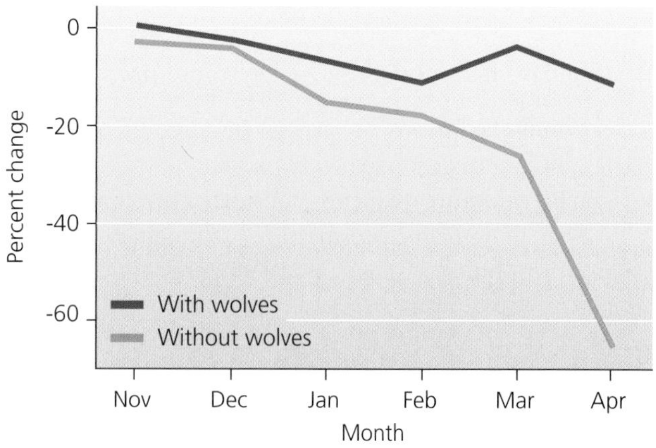

Changes in amount of winter carrion available to scavengers in Yellowstone National Park, with and without wolves, according to the model. Differences in March and April are statistically significant. Data from Wilmers, C. C., and W. M. Getz, 2005. Gray wolves as climate change buffers in Yellowstone. *PLoS Biology* 3 (4): e92.

1. How much less carrion is available in April than in November when wolves are present? When wolves are not present?

2. Wolves were hunted nearly to extinction in the 1930s and were only reintroduced to Yellowstone in 1995. How, would you suspect, has their reintroduction affected scavenger populations since then? Why?

3. What effect would you predict that continued shorter, warmer winters would have on scavenger populations? Why? Are the predicted effects of wolf reintroduction and of climate change compounded, or do they tend to cancel one another out?

CALCULATING ECOLOGICAL FOOTPRINTS

Species appearing in a new area are generally called "invasive" if they increase markedly in population and also increase in their impacts on the biotic communities and landscapes around them. By these measures, are human beings an invasive species? The table below shows human population and per capita energy consumption for the United States across the time period of roughly two generations, from 1950 to 1975, and from 1975 to 2000. Total energy consumption gives us a very rough measure of total environmental impact. Calculate total energy consumption in the table by multiplying the population and per capita consumption values.

Year	Human population (U.S., in millions)	Per capita energy consumption (U.S., in million BTU)	Total energy consumption (U.S., in quadrillion BTU)
1950	151	229	
1975	216	334	
2000	281	352	

Data source: U.S. Census Bureau and U.S. Energy Information Administration.

1. In percentage terms, how much did total energy consumption, as calculated here, increase between 1950 and 1975? Between 1975 and 2000?

2. What effects on biotic communities do you think this increase in total energy consumption has had? Speculate on overall effects, and give several specific known or likely examples.

3. After considering these data, do you consider yourself to be a member of an invasive species? Why or why not?

Take It Further

Go to www.aw-bc.com/withgott or the student CD-ROM where you'll find:

- Suggested answers to end-of-chapter questions
- Quizzes, animations, and flashcards to help you study
- *Research Navigator*™ database of credible and reliable sources to assist you with your research projects
- GRAPHit! tutorials to help you master how to interpret graphs
- INVESTIGATEit! current news articles that link the topics that you study to case studies from your region to around the world

7 Environmental Systems and Ecosystem Ecology

The Mississippi River as it enters the Gulf of Mexico

Upon successfully completing this chapter, you will be able to:

- Describe the nature of environmental systems
- Define ecosystems and evaluate how living and nonliving entities interact in ecosystem-level ecology
- Compare and contrast how carbon, phosphorus, nitrogen, and water cycle through the environment
- Explain how plate tectonics and the rock cycle shape the earth beneath our feet

Fisherman in the Gulf of Mexico's "dead zone"

Central Case: The Gulf of Mexico's "Dead Zone"

"In nature there is no 'above' or 'below,' and there are no hierarchies. There are only networks nesting within other networks."
—Fritjof Capra, theoretical physicist

"Let's say you put Saran Wrap over south Louisiana and suck the oxygen out. Where would all the people go?"
—Nancy Rabalais, biologist for the Louisiana Universities Marine Consortium

Louisiana fishermen have long hauled in more seafood than those of any other U.S. state except Alaska. Each year they have plied the rich waters of the northern Gulf of Mexico and have sent nearly 600 million kg (1.3 billion lb) of shrimp, fish, and shellfish to our dinner tables. Then in 2005, Hurricane Katrina and Hurricane Rita pummeled the Gulf Coast and left Louisiana's fisheries in ruin. Boats, docks, marinas, and fueling stations were destroyed, and thousands were suddenly out of work. Today, fishermen are struggling to reestablish their livelihoods as the industry continues to rebuild itself.

But for years before the hurricanes hit, fishing had become increasingly difficult. In the words of longtime Louisiana fisherman Johnny Glover, it was "getting harder and harder to make a living." The reason? Each year billions of organisms were suffocating in the Gulf's "dead zone," a region of water so depleted of oxygen that marine organisms are killed or driven away.

The low concentrations of dissolved oxygen in the bottom waters of this region represent a condition called **hypoxia** (see "The Science behind the Story," ▸ pp. 202–203 and Figure 7.5, ▸ p. 189). Aquatic animals obtain oxygen by respiring through their gills, and, like us, these animals will asphyxiate if deprived of oxygen. Fully oxygenated water contains up to 10 parts per million (ppm) of oxygen, but when concentrations drop below 2 ppm, creatures that can leave an affected area will do so. Below 1.5 ppm, most marine organisms die. In the Gulf's hypoxic zone, oxygen concentrations frequently drop well below these levels.

The dead zone appears each spring and grows through the summer and fall, starting near the mouths of the Mississippi and Atchafalaya Rivers off the Louisiana coast. In 2002 the dead zone reached a record 22,000 km^2 (8,500 mi^2)—an area larger than New Jersey. Shrimp boats came up with nets nearly empty. One shrimper derided his meager catch as "cat food." Others, ironically, said they hoped a hurricane would strike and stir some oxygen into the Gulf's stagnant waters.

What's starving these waters of oxygen? Scientists studying the dead zone have identified modern Midwestern farm practices and other human impacts hundreds of kilometers away. The Gulf, they say, is being over-enriched by nitrogen and phosphorus flushed down the Mississippi River. This nutrient pollution comes from fertilizers used on farms far upstream in the Mississippi River basin, as well as from other sources, including urban runoff, industrial discharges, atmospheric deposition from fossil fuel combustion, and municipal sewage outflow.

The U.S. government has acted on these findings, proposing that farmers in states such as Ohio, Iowa, and Illinois cut down on fertilizer use. Farmers' advocates protest that farmers are being singled out while urban pollution sources are being ignored. Meanwhile, coastal dead zones have appeared in 150 other areas throughout the world, from Chesapeake Bay to Oregon to Denmark to the Black Sea. The story of how scientists have determined the causes of these dead zones involves understanding environmental systems and the often complex behavior they exhibit.

Earth's Environmental Systems

Our planet's environment consists of complex networks of interlinked systems. In the realm of community ecology (Chapter 6), these systems include the ecological webs of relationships among species. At the ecosystem level, they include the interaction of living species with the nonliving entities around them. Earth's systems also include cycles that guide the flow of key chemical elements and compounds that support life, regulate climate, and control other aspects of Earth's functioning. We depend on these systems for our very survival.

Assessing questions holistically by taking a "systems approach" is helpful in environmental science, in which so many issues are multifaceted and complex. Such an approach poses a challenge, however, because systems often show behavior that is difficult to understand and predict. The scientific method operates best when researchers isolate and manipulate small parts of complex systems, focusing in depth on manageable components. However, environmental scientists increasingly are accepting the challenge of studying systems broadly and are beginning to find solutions to problems such as the Gulf of Mexico's hypoxic zone.

Systems show several defining properties

A **system** is a network of relationships among parts, elements, or components that interact with and influence one another through the exchange of energy, matter, or information. Systems receive inputs of energy, matter, or information, process these inputs, and produce outputs. Energy inputs to Earth's environmental systems include solar radiation as well as heat released by geothermal activity, organismal metabolism, and human activities such as fossil fuel combustion. Information inputs can come in the form of sensory cues from visual, olfactory (chemical), magnetic, or thermal signals. Inputs of matter occur when chemicals or physical material moves among systems, such as when seeds are dispersed long distances, migratory animals deposit waste far from where they consumed food, or plants convert carbon in the air to living tissue by photosynthesis. As a system, the Gulf of Mexico receives inputs of freshwater, sediments, nutrients, and pollutants from the Mississippi and other rivers. Shrimpers harvest some of the Gulf system's output: matter and energy in the form of shrimp. This output subsequently becomes input to the human economic system and to the digestive systems of the many people who consume the shrimp. Even greater than the output of shrimp is that of menhaden, a fish that supplies oil for many consumer products and is used to feed farmed animals.

Sometimes a system's output can serve as input to that same system, a circular process described as a **feedback loop.** Feedback loops are of two types, negative and positive. In a **negative feedback loop** (Figure 7.1a), output that results from a system moving in one direction acts as input that moves the system in the other direction. Input and output essentially neutralize one another's effects, stabilizing the system. A thermostat, for instance, stabilizes a room's temperature by turning the furnace on when the room gets cold and shutting it off when the room gets hot. Similarly, negative feedback regulates our body temperature. If we get too hot, our sweat glands pump out moisture that evaporates to cool us down, or we may move from sun to shade. If we get too cold, we shiver, creating heat, or we move into the sun or put on more clothing. Another example of negative feedback is a predator-prey system in which predator and prey populations rise and fall in response to one another (see Figure 6.5, ▸p. 153). Most systems in nature involve negative feedback

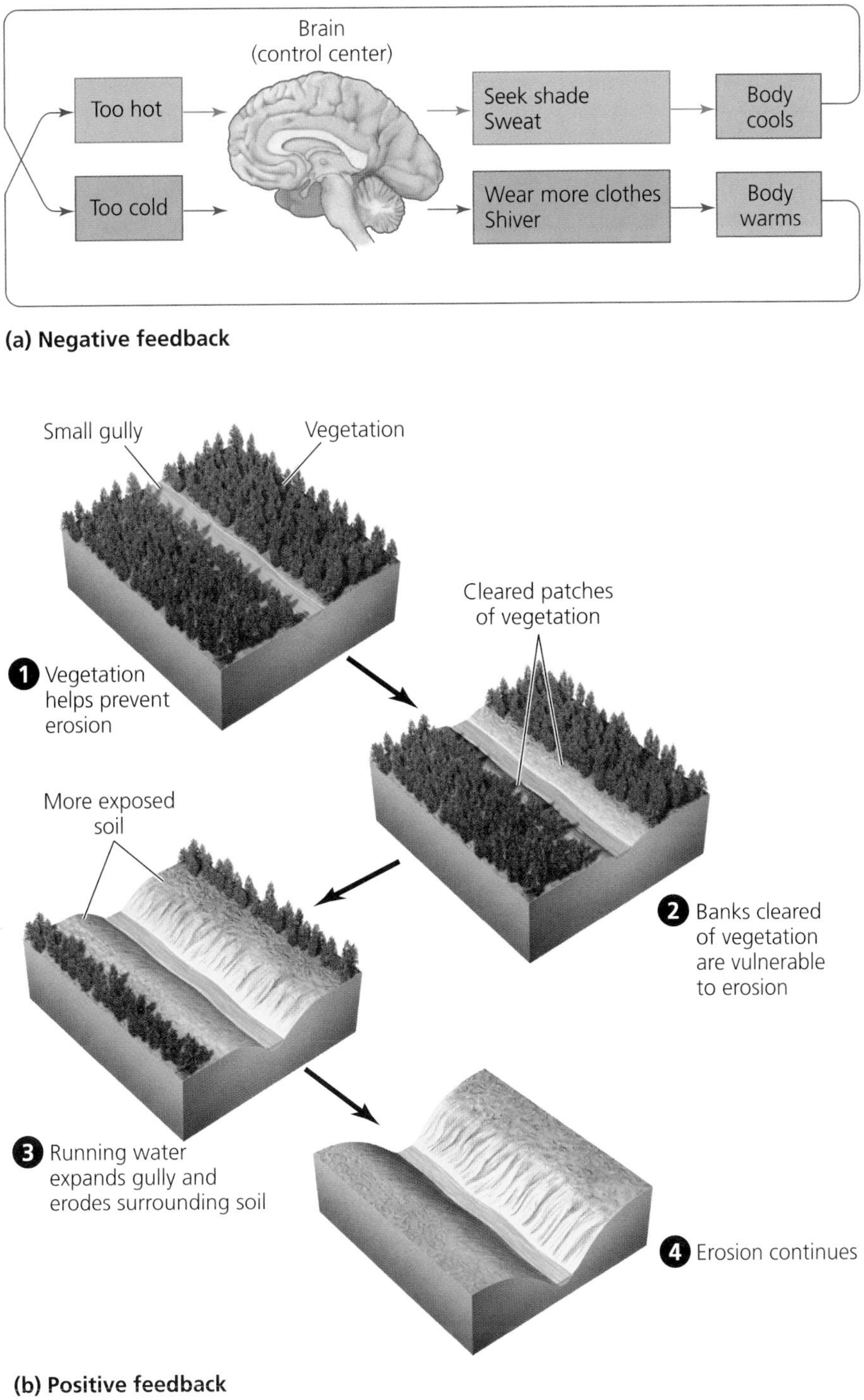

FIGURE 7.1 Negative feedback loops (**a**) exert a stabilizing influence on systems and are common in nature. The human body's response to heat and cold involves a negative feedback loop. Positive feedback loops (**b**) have a destabilizing effect on systems and push them toward extremes. Rare in nature, they are common in natural systems altered by human impact. The clearing of forested land, for instance, can lead to a runaway process of soil erosion.

loops. Negative feedback loops enhance stability, and in the long run, only those systems that are stable will persist.

Positive feedback loops have the opposite effect. Rather than stabilizing a system, they drive it further toward one extreme or another. Exponential growth in human population (▸ pp. 218–219) provides an example. The more people are born, the more there are to give birth to further people; increased output leads to increased input, leading to further increased output. Another example is the spread of cancer; as cells multiply out of control, the process is self-accelerating. Positive feedback can also occur with the process of erosion, the removal of soil by water or wind (▸ pp. 257–258). Once vegetation has been cleared to expose soil, erosion may become progressively more severe if the forces of water or wind surpass the rate of vegetative regrowth. Water flowing through an eroded gully may expand the gully and lead to further erosion (Figure 7.1b). Positive feedback can alter a system substantially. Positive feedback loops are rare in nature, but they are common in natural systems altered by human impact.

The inputs and outputs of complex natural systems usually occur simultaneously, keeping the system constantly active. Earth's climate system, for instance, does not ever stop. When processes within a system move in opposing directions at equivalent rates so that their effects balance out, the process is said to be in **dynamic equilibrium.** Processes in dynamic equilibrium can contribute to **homeostasis,** the tendency of a system to maintain constant or stable internal conditions. When homeostasis exists, organisms and other systems can keep their internal conditions within a range that allows them to function. Homeostatic systems are often thought of as being in a steady state; however, the steady state itself may change slowly over time while the system maintains its ability to stabilize conditions internally. For instance, organisms grow and mature. Similarly, Earth has experienced a gradual increase in atmospheric oxygen over its history (▸p. 108), yet life has adapted, and Earth remains, by most definitions, a homeostatic system.

Often it is difficult to understand systems fully by focusing on their individual components because systems can show **emergent properties,** characteristics not evident in the components alone. Stating that systems possess emergent properties is a lot like saying, "The whole is more than the sum of its parts." For example, if you were to reduce a tree to its component parts (leaves, branches, trunk, bark, roots, fruit, and so on) you would not be able to predict the whole tree's emergent properties, which include the role the tree plays as habitat for birds, insects, parasitic vines, and other organisms (Figure 7.2). You could analyze the tree's chloroplasts (photosynthetic cell organelles), diagram its branch structure, and evaluate its fruit's nutritional content, but you would still be unable to understand the tree as habitat, as part of a forest landscape, or as a reservoir for carbon storage.

Systems seldom have well-defined boundaries, so deciding where one system ends and another begins can be difficult. Consider a desktop computer system. It is

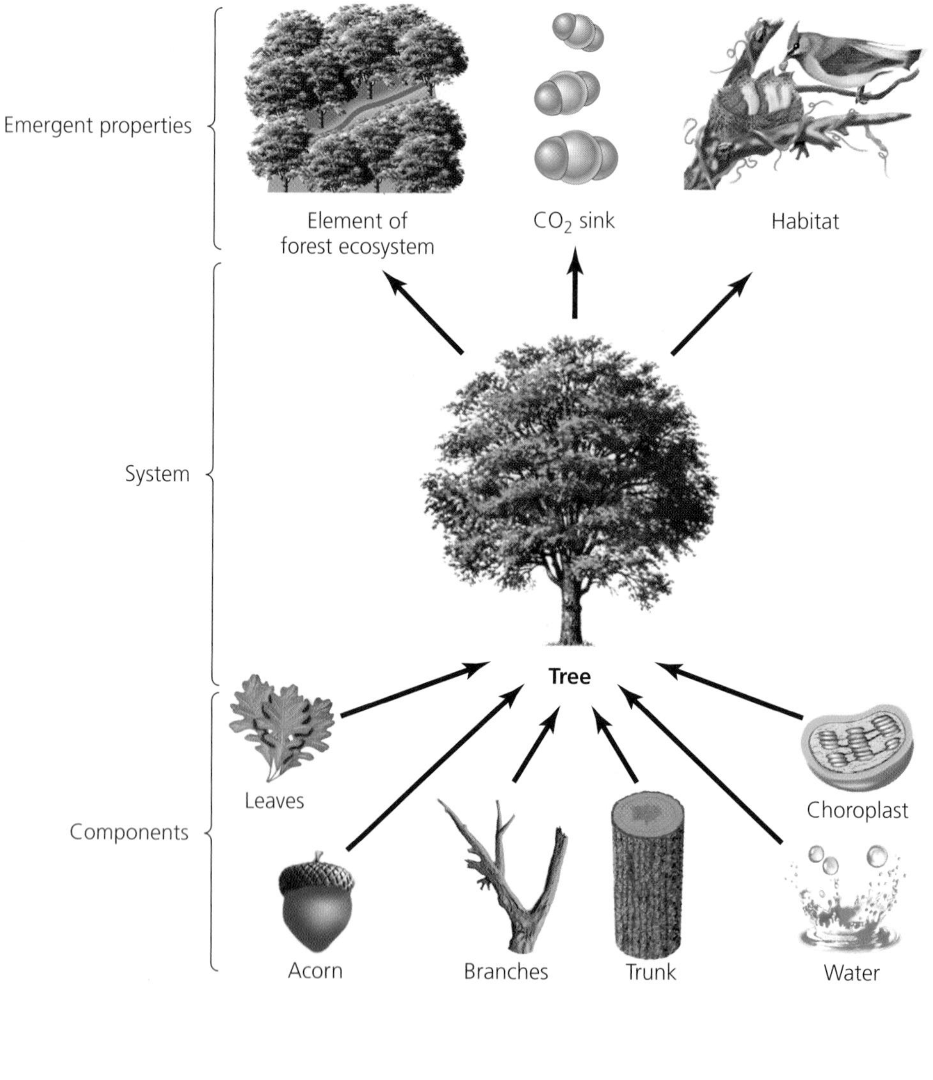

FIGURE 7.2 A system's emergent properties are not evident when we break the system down into its component parts. For example, a tree serves as wildlife habitat and plays roles in forest ecology and global climate regulation, but you would not know that from considering the tree only as a collection of leaves, branches, and chloroplasts. If we try to understand systems solely by breaking them into component parts, we will miss much of what makes them important.

certainly a network of parts that interact and exchange energy and information, but what are its boundaries? Is the system what arrives in a packing crate and sits on top of your desk? Or does it include the network you connect it to at school, home, or work? What about the energy grid you plug it into, with its distant power plants and transmission lines? And what of the Internet? Browsing the Web, you are drawing in digitized text, light, and sound from around the world. No matter how we attempt to isolate or define a system, we soon see that it has many connections to systems larger and smaller than itself. Systems overlap, and one may be contained within others, so where we draw boundaries may depend on the spatial or temporal scale on which we choose to focus.

Scientists will often treat a system as if it is a **closed system,** one that is isolated and self-contained, for the purpose of simplifying some problem with which they are grappling. However, no matter how closed a system might seem, if we look closely enough or wait long enough, we will detect interactions with other systems. Thus, when viewed in context, all systems are **open systems,** exchanging energy, matter, and information with other systems. Our planet is clearly an open system in regard to energy because it receives sunlight continuously and emanates heat into space. In regard to physical matter, Earth is largely a closed system, but comets, asteroids, and meteors do introduce small amounts of material every day.

Understanding the dead zone requires considering Mississippi River and Gulf of Mexico systems together

The Gulf of Mexico and the Mississippi River are systems that interact with one another. On a map, the Mississippi River appears as a branched and braided network of water channels. But where are this system's boundaries? You might argue that the Mississippi consists primarily of water, originates in Minnesota, and ends in the Gulf of Mexico near New Orleans. But what about the rivers that feed it and the farms, cities, and forests that line its banks (Figure 7.3)? Major rivers such as the Missouri, Arkansas, and Ohio flow into the Mississippi. Hundreds of smaller tributaries drain vast expanses of farmland, woodland, fields, cities, towns, and industrial areas before their water joins the Mississippi's. These waterways carry with them millions of tons of sediment, hundreds of species of plants and animals, and numerous pollutants.

For an environmental scientist interested in runoff and the flow of water, sediment, or pollutants, it may make best sense to view the Mississippi River's watershed as a system. As with the Tijuana River in Chapter 3, one must consider the entire area of land a river drains to comprehend and solve problems of river pollution. However, for a scientist interested in the Gulf of Mexico's dead zone, it may make best sense to view the Mississippi River watershed together with the Gulf as the

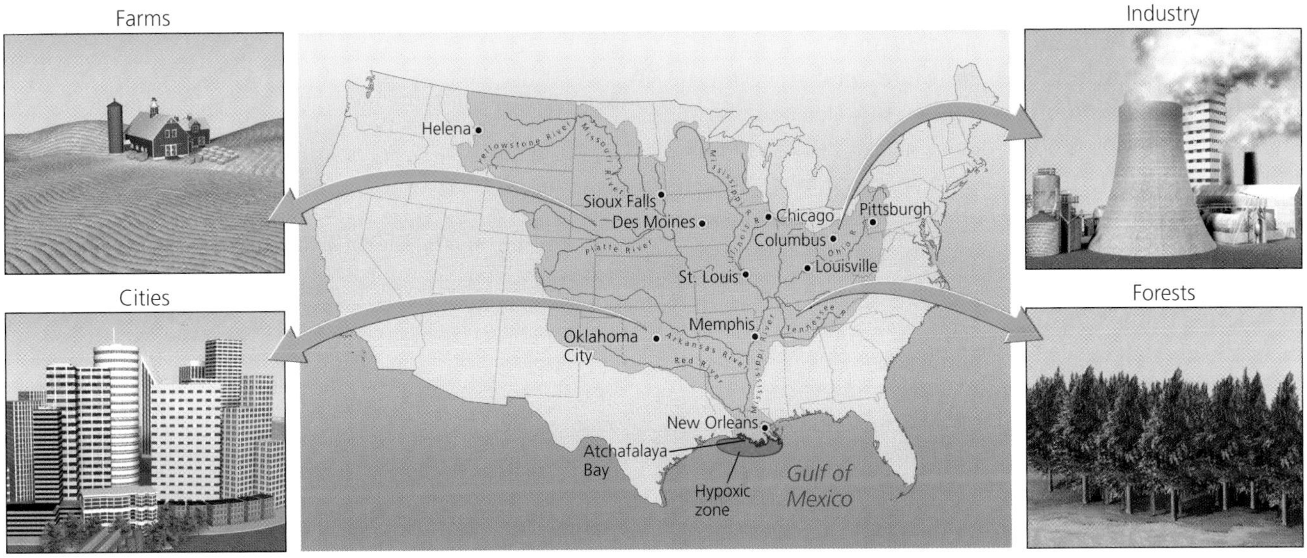

FIGURE 7.3 The Mississippi River system is the largest in North America. The river's watershed encompasses 3.2 million km^2 (1.2 million mi^2), or 41% of the area of the lower 48 U.S. states. The river carries water, sediment, and pollutants from a variety of sources downriver to the Gulf of Mexico, where nutrient pollution has given rise to a hypoxic zone.

system of interest because their interaction is central to the problem. In environmental science, one's delineation of a system can and should depend on the questions one is addressing.

The reason for the Gulf of Mexico's dangerously low levels of oxygen, scientists have concluded, is abnormally high levels of nutrients such as nitrogen and phosphorus. The excess nutrients originate with sources in the Mississippi River watershed, particularly nitrogen- and phosphorus-rich fertilizers applied to crops. Inorganic nitrogen fertilizers used on Midwestern farms account for roughly 30% of total nitrogen contributions to the river, an amount equal to that from natural soil decomposition. The remainder comes from various sources, including animal manure, nitrogen-fixing crops, sewage treatment facilities, street runoff, and industrial and automobile emissions. Altogether, agricultural sources are thought to contribute 74% of the nitrate and 65% of the total nitrogen carried in the river. Much of the nitrate originates in the upper portions of the watershed (Figure 7.4a) from farms growing corn and soybeans in Iowa, Illinois, Indiana, Minnesota, and Ohio. Nitrogen fertilizer input to farmland in the Mississippi River watershed has increased dramatically since 1950 (Figure 7.4b). Since 1980, the Mississippi River and the Atchafalaya River (which drains a third of the Mississippi's water through a second delta) have pumped about 1 million metric tons of nitrate into the Gulf of Mexico each year, three times as much as during the 1960s.

The enhanced nitrogen input to the Gulf boosts the growth of *phytoplankton* (▸ p. 150), microscopic photosynthetic algae, protists, and cyanobacteria that drift near the surface. Phytoplankton ordinarily are limited in their growth by scarcity of nutrients such as nitrogen and phosphorus. As phytoplankton flourish at the surface and as zooplankton (▸ p. 150) consume them, more dead phytoplankton and waste products of phytoplankton and zooplankton drift to the bottom, providing food for organisms (mainly bacteria) that decompose them. The result is a population explosion of bacteria. These decomposers consume enough oxygen to cause oxygen concentrations in bottom waters to plummet, suffocating shrimp and fish that live at the bottom, and creating the dead zone. The fresh water from the river remains naturally stratified in a layer at the surface that mixes only very slowly with the denser salty ocean water, so that oxygenated surface water does not make its way down to the bottom-dwelling life that needs it. This process of nutrient over-enrichment, blooms of algae, increased production of organic matter, and subsequent ecosystem degradation is known as **eutrophication** (Figure 7.5). As we will see in

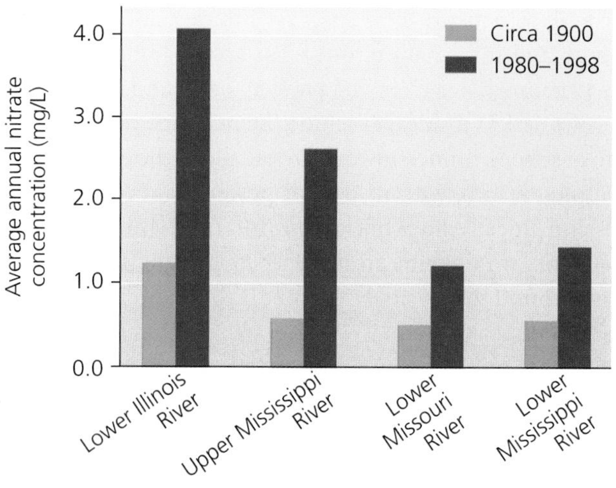

(a) Nitrate concentrations in portions of the Mississippi River watershed

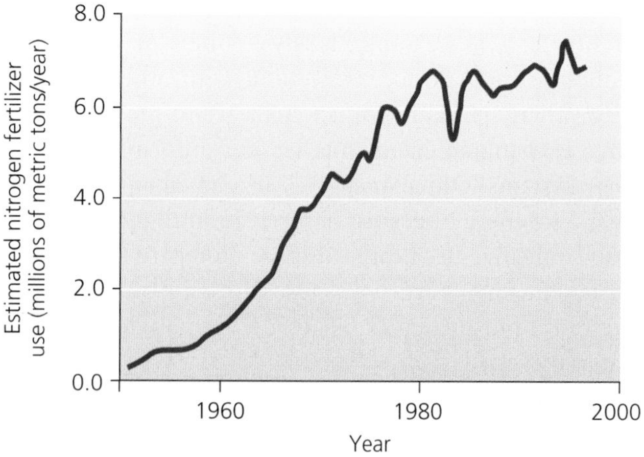

(b) Nitrogen use in the Mississippi-Atchafalaya River basin

FIGURE 7.4 Concentrations of nitrogen in the Mississippi and its tributaries rose during the 20th century, especially in the upper portion of the watershed **(a)**. Orange bars show average concentrations from circa 1900 (based on more recent analysis of sediments), and red bars show average concentrations measured from 1980 to 1998. Usage of nitrogen-based fertilizer in the Mississippi-Atchafalaya River basin skyrocketed after 1950 **(b)**. About 15% of nitrogen from fertilizer applied to farms eventually reaches rivers that flow to the Gulf. Data from National Science and Technology Council: Committee on Environment and Natural Resources. May 2000. Hypoxia: An integrated assessment in the northern Gulf of Mexico.

Chapters 15 and 16, eutrophication can take place in both freshwater and saltwater systems.

Moderate amounts of additional nutrients may increase the productivity of fisheries, but at higher concentrations, this fertilizing effect is offset by hypoxia, and fishery yields decline.

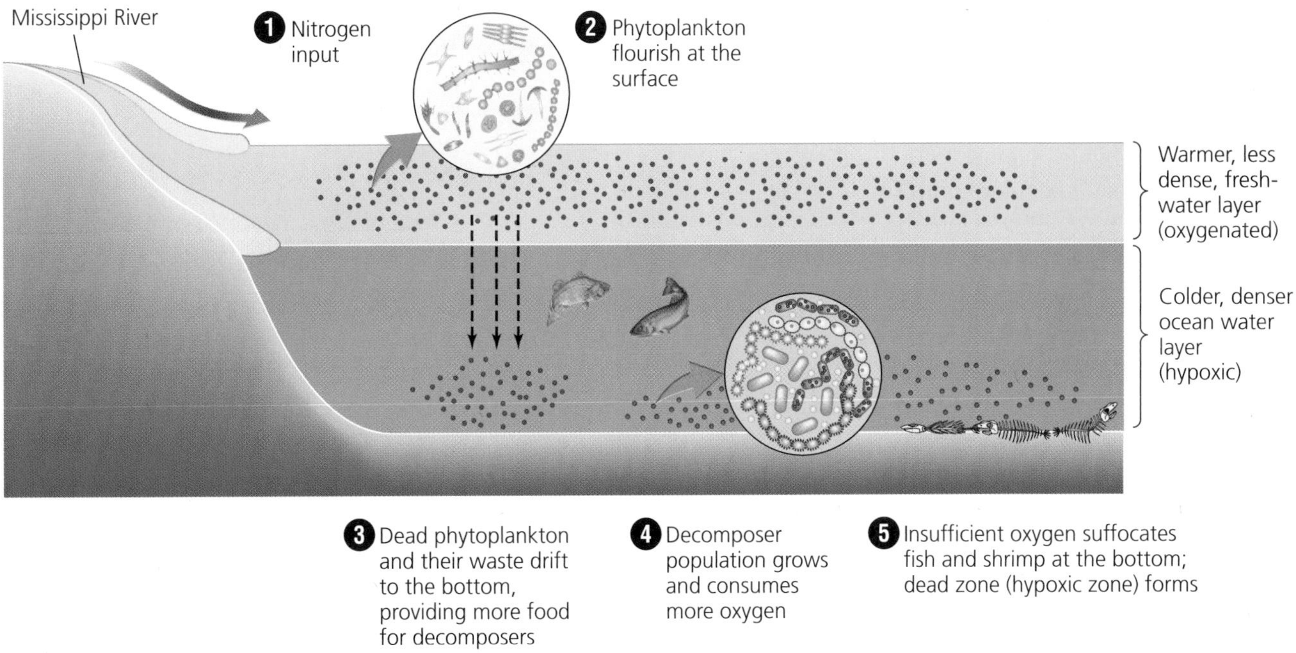
Mississippi River
1 Nitrogen input
2 Phytoplankton flourish at the surface
Warmer, less dense, fresh-water layer (oxygenated)
Colder, denser ocean water layer (hypoxic)
3 Dead phytoplankton and their waste drift to the bottom, providing more food for decomposers
4 Decomposer population grows and consumes more oxygen
5 Insufficient oxygen suffocates fish and shrimp at the bottom; dead zone (hypoxic zone) forms

FIGURE 7.5 Excess nitrogen causes eutrophication in coastal marine systems such as the Gulf of Mexico. Coupled with stratification (layering) of water, eutrophication can severely deplete dissolved oxygen. Nitrogen from river water (1) boosts growth of phytoplankton (2), which die and are decomposed at the bottom by bacteria (3). Stability of the surface layer prevents deeper water from absorbing oxygen to replace that consumed by decomposers (4), and the oxygen depletion suffocates or drives away bottom-dwelling marine life (5). This process gives rise to hypoxic zones like that of the Gulf of Mexico.

Environmental systems may be perceived in various ways

There are many ways to delineate natural systems, and your choice will depend on the particular issues in which you are interested. Categorizing environmental systems can help us make Earth's dazzling complexity comprehensible to the human brain and accessible to problem solving. However, a fundamental insight of environmental science is that systems overlap and interact.

For instance, scientists sometimes divide Earth's components into structural spheres. The **lithosphere** is the rock and sediment beneath our feet, in the planet's uppermost layers. The **atmosphere** is composed of the air surrounding our planet. The **hydrosphere** encompasses all water—salt or fresh, liquid, ice, or vapor—in surface bodies, underground, and in the atmosphere. The **biosphere** consists of the total of all the planet's living organisms and the abiotic (nonliving) portions of the environment with which they interact. We will examine the workings of various environmental systems within these structural spheres in more detail in later chapters.

Although these categories can be useful, their boundaries overlap, so the systems interact. Picture a robin plucking an earthworm from the ground after a rain. You are witnessing an organism (the robin) consuming another organism (the earthworm) by removing it from part of the lithosphere (soil) that the earthworm had been modifying—all this made possible because rain (from the hydrosphere) recently wet the ground. The robin might then fly through the air (the atmosphere) to a tree (an organism), in the process respiring (combining oxygen from the atmosphere with glucose from an organism, and adding water to the hydrosphere and carbon dioxide and heat to the atmosphere). Finally, the bird might defecate, adding nutrients from an organism to the lithosphere below. The study of such interactions among living and nonliving things is a key part of ecology at the ecosystem level. As scientists become more inclined to approach systems holistically, ecology at the ecosystem level and beyond has become increasingly important.

Ecosystems

An **ecosystem** consists of all organisms and nonliving entities that occur and interact in a particular area at the same time. The ecosystem concept builds on the idea of

The Science behind the Story

Biosphere 2

In September 1991, eight people and nearly 3,800 species of plants and animals were sealed within Biosphere 2, a collection of airtight, interconnected domes spanning more than 1.2 hectares (3.0 acres) in the Arizona desert (see figure). The goal of the biospherians, as they were known, was to survive for two years within a self-contained ecosystem that might someday be used to colonize other planets, while also learning about environmental processes on Earth. Only nine months later, however, oxygen levels in Biosphere 2's artificial atmosphere began to drop at an alarming rate. Within 18 months, the biospherians were literally gasping for breath.

Within 18 months of being sealed off from Earth's atmosphere, Biosphere 2 had oxygen levels of less than 14%—close to the lowest levels at which humans can survive.

The near-failure of Biosphere 2's life-support system was not due to a lack of data. Scattered throughout Biosphere 2's ocean, rainforest, savanna, and desert biomes were more than 1,000 sensors that tracked day-to-day changes in oxygen, carbon dioxide, temperature, pH, and other environmental variables. Nonetheless, despite a sophisticated computer monitoring system and the help of external advisers, the biospherians, only some of whom were scientists, were unable to locate the missing oxygen. In desperation, they called on Wallace S. Broecker, a geochemist at Columbia University's Lamont-Doherty Earth Observatory.

Broecker and graduate student Jeff Severinghaus quickly ruled out the possibility that the biospherians themselves were consuming the oxygen. The amount disappearing—roughly 450 kg (1,000 lb) of oxygen per month—was far larger than they, or any of the project's large animals,

the biological community (Chapter 6), but ecosystems include abiotic components as well as biotic ones. In ecosystems, energy flows and matter cycles among these living and nonliving components.

Ecosystems are systems of interacting living and nonliving entities

The idea of ecosystems originated early last century with people such as British ecologist Arthur Tansley, who saw that biological entities are tightly intertwined with chemical and physical entities. Tansley and others felt that there was so much interaction and feedback between organisms and their abiotic environments that it made most sense to view living and nonliving elements together. For instance, the input of moisture from clouds plays a key role in the Monteverde cloud forest ecosystem we visited in Chapter 5. The flow of water, sediment, and nutrients from the Mississippi River plays a key role in the nearshore ecosystem of the northern Gulf of Mexico. And as we saw in our discussion of biomes in Chapter 6 (▸pp. 170–177), abiotic factors such as temperature, precipitation, latitude, and elevation have substantial influence over which species and communities exist in a given locality.

Ecologists soon began analyzing ecosystems as an engineer might analyze the operation of a machine. In this view, ecosystems are systems that receive inputs of energy, process and transform that energy while cycling matter internally, and produce a variety of outputs (such as heat, water flow, and animal waste products) that can move could be using. So, they focused their investigation on Biosphere 2's 30,000 tons of soil, which had an extraordinarily high percentage—nearly 30%—of organic matter. They determined that microbes in the soil were converting unexpectedly large amounts of oxygen to carbon dioxide. Still, one question remained: If the microbes were responsible for the severe decrease in oxygen, then CO_2 levels in the atmosphere should have skyrocketed, when instead the opposite had occurred. Broecker and Severinghaus suspected that the missing carbon dioxide was being stored in the soil itself.

On close examination, however, that proved not to be the case, so the researchers turned their attention to another possible culprit: the exposed concrete supporting the building's glass and metal shell. In Earth's atmosphere, they knew, concrete can react with carbon dioxide to form a solid substance called calcium carbonate. Usually, the reaction takes place only in a thin outer layer of exposed concrete, which is what the designers of Biosphere 2 had expected. Concentrations of CO_2 in Biosphere 2's atmosphere, however, ranged 3–10 times higher than Earth's, and the researchers found that carbon dioxide, converted to calcium carbonate, had been deposited up to 15 cm (6 in.) deep in the concrete walls.

The most effective solution would have been to replace all 30,000 tons of soil, but that would have cost millions of dollars and set the project back several years. Instead, Biosphere 2's management team settled on two stopgap measures. First, to address the urgent need for oxygen, they injected more than 23 tons of pure oxygen gas. Then, to minimize calcium carbonate deposits, they covered the exposed concrete with a layer of paint. With the most pressing problem solved, at least temporarily, the biospherians were able to focus on other issues, such as invasive species. Aquatic fire worms were preying on coral, hundred-foot-long morning glory vines were overwhelming the rainforest, and crazy ants—an aggressive species unintentionally included in the project—had spread throughout Biosphere 2 and driven most of its other insect species extinct.

Nonscientific problems also plagued the project. By the time the biospherians emerged from their two-year seclusion in September 1993, the project had become embroiled in accusations of mismanagement and scientific fraud. In 1994, its financial backer, Texas oil billionaire Edward P. Bass, dismissed the project's management team and invited a panel of independent scientists to assess its future.

In 1996, Columbia University began the difficult process of transforming Biosphere 2 into a center for research and education on global climate change. Despite some successes, the project remained troubled, and Columbia eventually backed out. Whatever the future of Biosphere 2 may hold, the lesson of its first mission is clear: Creating self-contained ecosystems from scratch is no easy task. We're better off preserving the ones we have.

into other ecosystems. Energy flows in one direction through ecosystems; most arrives as radiation from the sun, powers the system, and exits in the form of heat. Matter, in contrast, is generally recycled within ecosystems. We saw in Chapter 6 (▸pp. 156–159) how energy and matter are passed from one organism to another through food web relationships, from producers to primary consumers to secondary consumers. Matter is recycled because when organisms die and decay their nutrients remain in the system. In contrast, most energy that organisms gain is lost during their lifetimes through respiration.

Ecosystems can be conceptualized at different scales. An ecosystem can be as small as an ephemeral puddle of water where brine shrimp and tadpoles feed on algae and detritus with mad abandon as the pool dries up. Or an ecosystem might be as large as a lake or a forest. For some purposes, scientists even view the entire biosphere as a single all-encompassing ecosystem (see The Science behind the Story, above). The term is most often used, however, to refer to systems of moderate geographic extent that are somewhat self-contained. For example, Monteverde's mountaintop system is delineated by drier forests and cleared agricultural land on its lower slopes.

Landscape ecologists study geographic areas with multiple ecosystems

Ecosystems are open systems, and those that physically abut one another may interact extensively, no matter how distinctly different they may appear. For instance, a pond ecosystem is very different from a forest ecosystem that surrounds it, but the two interact. Salamanders that develop in the pond live their adult lives under logs on the forest floor until returning to the pond to breed. Rainwater that nourishes forest plants may eventually make its way to the pond, carrying with it nutrients from the forest's leaf litter. Similarly, coastal dunes, the ocean, and the lagoon or salt marsh between them all interact, as do forests and prairie where they converge. Areas where ecosystems meet may consist of transitional zones called **ecotones,** in which elements of each ecosystem mix.

Because of this mixing, sometimes ecologists find it useful to view these systems at a larger landscape scale that focuses on geographic areas that include multiple ecosystems. For instance, if you are interested in large mammals, such as black bears, that move seasonally from mountains to valleys or move between mountain ranges, you had better consider the landscape scale that includes all these habitats. Such a broad-scale approach, often called **landscape ecology,** is important in studying birds that migrate between continents, or fish, such as salmon, that move between marine and freshwater ecosystems. Some conservation groups, such as the Nature Conservancy, have begun applying the landscape ecology approach widely in their land management strategies.

Weighing the Issues: Ecosystems Where You Live

Think about the area where you live. How would you describe that area's ecosystems? How do these systems interact with one another? If one ecosystem were greatly disturbed (say, if a wetland or forest were replaced by a shopping mall), what impacts might that have on nearby natural systems?

Energy is converted to biomass

Regardless of scale, the energy flow in most ecosystems begins with radiation from the sun. In Chapter 4 (▸pp. 105–107) we explored how autotrophs, such as green plants and phytoplankton, use photosynthesis to capture the sun's energy and produce food. We then saw in Chapter 6 (▸pp. 156–159) how organisms at higher trophic levels consume organisms at lower trophic levels, transferring energy.

As autotrophs convert solar energy to the energy of chemical bonds in sugars, they perform *primary production.* Specifically, the assimilation of energy by autotrophs is termed **gross primary production.** Autotrophs must use a portion of this production to power their own metabolism by respiration (▸p. 107). The energy that remains after respiration and is used to generate biomass ecologists call **net primary production.** Thus, net primary production equals gross primary production minus respiration. Net primary production can be measured by the energy or the organic matter stored by plants after they have metabolized enough for their own maintenance.

Another way to think of net primary production is that it represents the energy or biomass available for consumption by heterotrophs. Plant matter not eaten by herbivores becomes fodder for detritivores and decomposers. Heterotrophs use the energy they gain from plants for their own metabolism, growth, and reproduction. The total biomass that heterotrophs generate by consuming autotrophs is termed *secondary production.*

Ecosystems vary in the rate at which plants convert energy to biomass. The rate at which production occurs is termed **productivity,** and ecosystems whose plants convert solar energy to biomass rapidly are said to have high **net primary productivity.** Freshwater wetlands, tropical forests, and coral reefs and algal beds tend to have the highest net primary productivities, whereas deserts, tundra, and open ocean tend to have the lowest (Figure 7.6a). Variation among ecosystems and among biomes in net primary productivity results in geographic patterns of variation across the globe (Figure 7.6b). In terrestrial ecosystems, net primary productivity tends to increase with temperature and precipitation. In aquatic ecosystems, net primary productivity tends to rise with light and the availability of nutrients.

Nutrients can limit ecosystem productivity

Nutrients are elements and compounds that organisms consume and require for survival. Organisms need several dozen naturally occurring chemical elements to survive. Elements and compounds required in relatively large amounts are called *macronutrients* and include such nutrients as nitrogen, carbon, and phosphorus. Nutrients needed in small amounts are called *micronutrients.*

Nutrients stimulate production by plants, and the lack of nutrients can limit production. The availability of nitrogen or phosphorus frequently is a limiting factor (▸p. 136) for plant growth. When these nutrients are added to a system, plants show the greatest response to whichever nutrient has been in shortest supply. Phosphorus tends to be limiting in freshwater systems, and nitrogen in marine systems. Thus the Gulf of Mexico's hypoxic zone is thought to result primarily from excess nitrogen, whereas ponds and lakes in the Mississippi

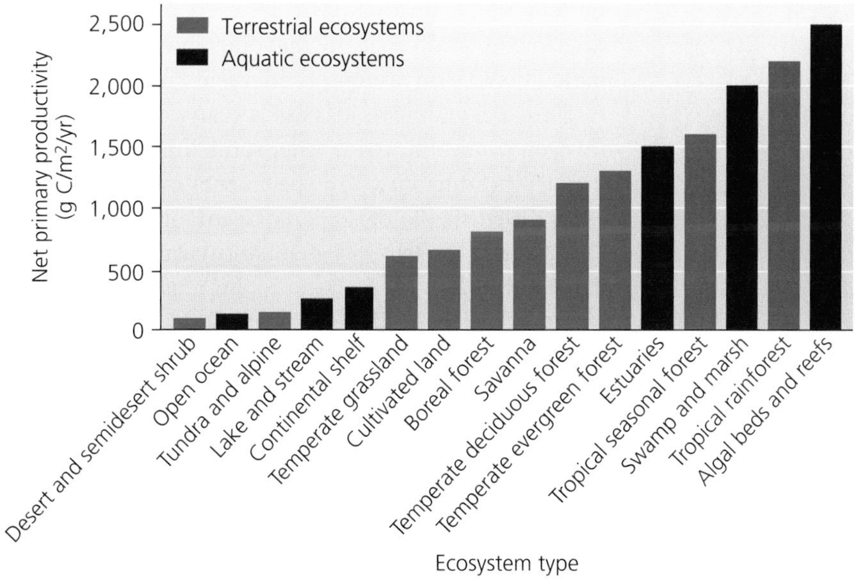

(a) Net primary productivity for major ecosystem types

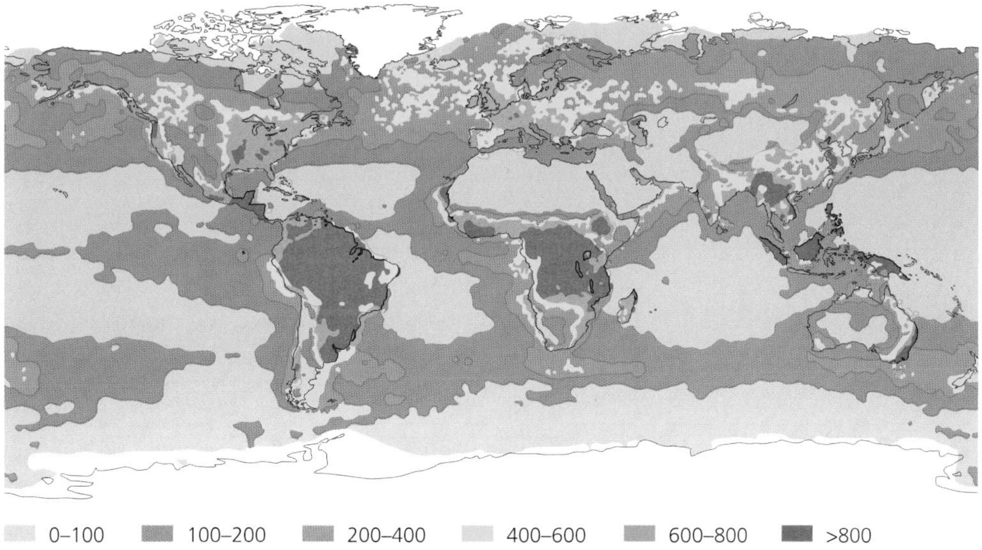

(b) Global map of net primary productivity

FIGURE 7.6 (**a**) Freshwater wetlands, tropical forests, coral reefs, and algal beds show high net primary productivities on average, whereas deserts, tundra, and the open ocean show low values. (**b**) On land, net primary production varies geographically with temperature and precipitation, which also influence the locations of biomes (▶pp. 170–178). In the world's oceans, net primary production is highest around the margins of continents, where nutrients (of both natural and human origin) run off from land. Data in (a) from Whittaker, R. H. 1975. *Communities and ecosystems*, 2nd ed., New York: MacMillan. Map in (b) from satellite data presented by Field, C. B., et al. 1998. Primary production of the biosphere: Integrating terrestrial and oceanic components. *Science* 281: 237–240.

River Valley tend to suffer eutrophication when they contain too much phosphorus.

Canadian ecologist David Schindler and others demonstrated the effects of phosphorus on freshwater systems in the 1970s by experimentally manipulating entire lakes. In one experiment, the researchers divided a 16-ha (40-acre) lake in Ontario in half with a plastic barrier. To one half the researchers added carbon, nitrate, and phosphate; to the other they added only carbon and nitrate. Soon after the experiment began, they witnessed a

FIGURE 7.7 A portion of this lake in Ontario was experimentally treated with the addition of phosphate. This treated portion experienced an immediate, dramatic, and prolonged algal bloom, visible in the opaque water in the topmost part of this photo. Photo by David W. Schindler.

dramatic increase in algae in the half of the lake that received phosphate, whereas the other half hosted algal levels typical for lakes in the region (Figure 7.7). This difference held until shortly after they stopped fertilizing seven years later, when algae decreased to normal levels in the half that had previously received phosphate. Such experiments showed clearly that phosphorus addition can markedly increase primary productivity in lakes.

Similar experiments in coastal ocean waters have shown nitrogen to be the more important limiting factor for primary productivity. In experiments in the 1980s and 1990s, Swedish ecologist Edna Granéli took samples of ocean water from sites in the Baltic Sea and added phosphate, nitrate, or nothing. Chlorophyll and phytoplankton increased greatly in the flasks with nitrate, whereas those with phosphate did not differ from the controls. Seawater experiments in Long Island Sound by other researchers have shown similar results. For open ocean waters far from shore, research has shown that iron is a highly effective nutrient.

Because nutrients run off from land in the Baltic, Long Island Sound, the Gulf of Mexico, and worldwide, primary productivity in the oceans tends to be greatest in nearshore waters, which receive nutrient runoff from land, and lowest in open ocean areas far from land (see Figure 7.6b). Satellite imaging technology that reveals phytoplankton densities has given scientists an improved view of productivity at regional and global scales, and has helped them track blooms of algae that contribute to coastal hypoxic zones (Figure 7.8).

The number of known dead zones is increasing globally, with about 150 documented so far. Most are located off the coasts of Europe and the eastern United States. Specific causes vary from place to place, but most result from rising nutrient pollution from farms, cities, and industry.

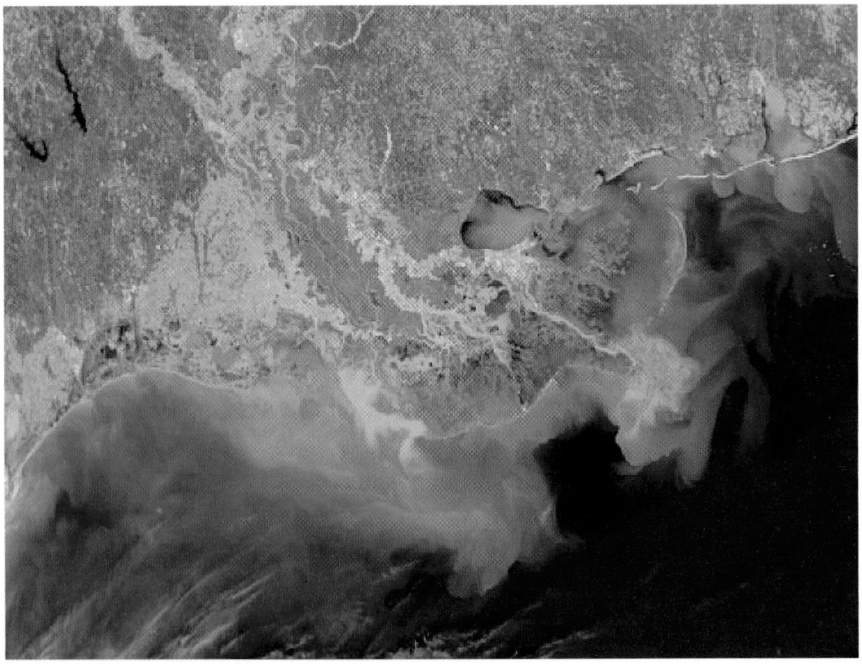

FIGURE 7.8 Satellite images like this one have helped scientists track runoff and phytoplankton blooms. Shown is the Louisiana coast on December 30, 1997. Greenish brown at the top is land, and deep blue at the bottom right is water of the Gulf of Mexico. Along the coast, two light brown plumes of sediment are visible expanding into the Gulf's waters from the mouths of the Mississippi and Atchafalaya Rivers. The shades of turquoise represent phytoplankton blooms that result from nutrient inputs and that foster the dead zone. Image from the SeaWiFS Project, NASA/Goddard Space Flight Center, and ORBIMAGE.

In North America, Chesapeake Bay may be the most severely affected area besides the Gulf of Mexico. Decades of pollution and human impact here have devastated fisheries and greatly altered the bay's ecology. Good news of a sort comes from the Black Sea, which borders Ukraine, Russia, Turkey, and eastern Europe. This immense inland sea had long suffered one of the world's worst hypoxic zones, and many of its valuable fisheries and seagrass beds were destroyed. Then in the 1990s, after the Soviet Union collapsed, industrial agriculture in the region declined drastically. With fewer fertilizers running off into the sea, this water body began to recover, and today fish and mussel beds are reviving. However, agricultural collapse is not a strategy anyone would choose to alleviate hypoxia. Rather, scientists are proposing a variety of innovative and economically acceptable ways to reduce nutrient runoff.

Biogeochemical Cycles

Just as nitrogen and phosphorus from Minnesota soybean fields end up in Gulf shrimp on our dinner plates, all nutrients move through the environment in complex and fascinating ways. Whereas energy enters an ecosystem from the sun, flows from one organism to another, and is dissipated to the atmosphere as heat, the physical matter of an ecosystem is circulated over and over again.

Nutrients circulate through ecosystems in biogeochemical cycles

Nutrients move through ecosystems in **nutrient cycles** or **biogeochemical cycles.** In these cycles they travel through the atmosphere, hydrosphere, and lithosphere, and from one organism to another, in dynamic equilibrium. A carbon atom in your fingernail today might have helped comprise the muscle of a cow a year earlier, may have resided in a blade of grass a month before that, and may have been part of a dinosaur's tooth 100 million years ago. After we die, the nutrients in our bodies will spread widely through the environment, eventually being incorporated by an untold number of organisms far into the future.

Nutrients move from one *pool*, or *reservoir*, to another, remaining for varying amounts of time *(residence time)* in each. The dinosaur, the grass, the cow, and you are each reservoirs for carbon atoms. The movement of nutrients among pools is termed *flux*, and the rates of flux between any given pair of pools can change over time. Human activity has influenced certain flux rates. For instance, it has increased the flux of nitrogen from the atmosphere to pools on the Earth's surface and has shifted the flux of carbon in the opposite direction. Reservoirs that release more nutrients than they accept are called *sources*, and reservoirs accepting more nutrients than they are releasing are called *sinks*. As we discuss biogeochemical cycles, think about how they involve negative feedback loops that promote dynamic equilibrium, and about how some human actions can generate positive feedback loops.

The carbon cycle circulates a vital organic nutrient

As the definitive component of organic molecules (▸pp. 98–99), carbon is an ingredient in carbohydrates, fats, and proteins, and in the bones, cartilage, and shells of all living things. From fossil fuels to DNA, from plastics to pharmaceuticals, carbon (C) atoms are everywhere. The **carbon cycle** describes the routes that carbon atoms take through the environment (Figure 7.9).

Photosynthesis, respiration, and food webs Producers, including terrestrial and aquatic plants, algae, and cyanobacteria, pull carbon dioxide out of the atmosphere and out of surface water to use in photosynthesis. Photosynthesis (▸pp. 105–107) breaks the bonds in carbon dioxide (CO_2) and water (H_2O) to produce oxygen (O_2) and carbohydrates (e.g., glucose, $C_6H_{12}O_6$). Autotrophs use some of the carbohydrates to fuel their own respiration, thereby releasing some of the carbon back into the atmosphere and oceans as CO_2. When producers are eaten by primary consumers, who in turn are eaten by secondary and tertiary consumers, more carbohydrates are broken down in respiration, producing carbon dioxide and water. The same process occurs when decomposers consume waste and dead organic matter. Respiration from all these organisms releases carbon back into the atmosphere and oceans.

All organisms use carbon for structural growth, so a portion of the carbon an organism takes in becomes incorporated into its tissues. The abundance of plants and the fact that they take in so much carbon dioxide for photosynthesis makes plants a major reservoir for carbon. Because CO_2 is a greenhouse gas of primary concern (▸pp. 530–532), much research on global climate change has been directed toward measuring the amount of CO_2 that plants tie up. Scientists are working toward understanding exactly how much this portion of the carbon cycle influences Earth's climate.

Sediment storage of carbon As organisms die, their remains may settle in sediments in ocean basins

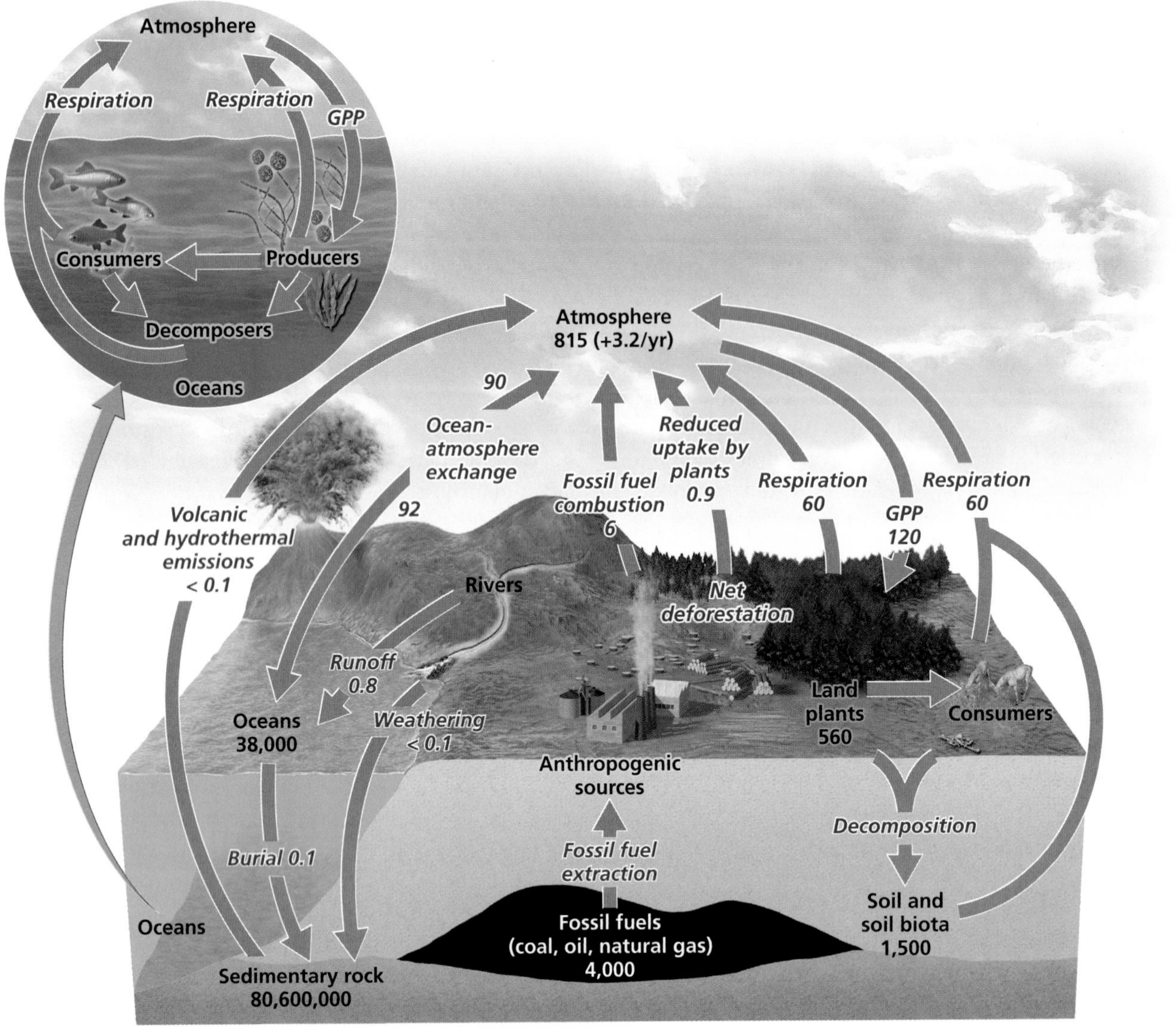

FIGURE 7.9 The carbon cycle summarizes the many routes that carbon atoms take as they move through the environment. Gray arrows represent fluxes among reservoirs, or pools, for carbon. In the carbon cycle, plants use carbon dioxide from the atmosphere for photosynthesis (gross primary production, or "GPP" in the figure). Carbon dioxide is returned to the atmosphere through respiration by plants, their consumers, and decomposers. The oceans sequester carbon in their water and in deep sediments. The vast majority of the planet's carbon is stored in sedimentary rock. In the figure, pool names are printed in plain bold type, and numbers in plain bold type represent pool sizes, expressed in units of 10^{15} g C. Processes printed in italic bold red type give rise to fluxes, printed in italic bold red type and expressed in units of 10^{15} g C per year. Data from Schlesinger, W. H. 1997. *Biogeochemistry: An analysis of global change*. 2nd ed. London: Academic Press.

or in freshwater wetlands. As layers of sediment accumulate, older layers are buried more deeply, experiencing high pressure over long periods of time. These conditions can convert soft tissues into fossil fuels—coal, oil, and natural gas—and shells and skeletons into sedimentary rock, such as limestone. Sedimentary rock comprises the largest single reservoir in the carbon cycle. Although any given carbon atom spends a relatively short time in the atmosphere, carbon trapped in sedimentary rock may reside there for hundreds of millions of years.

Carbon trapped in sediments and fossil fuel deposits may eventually be released into the oceans or atmosphere by geological processes such as uplift, erosion, and volcanic eruptions. It also reenters the atmosphere when we extract and burn fossil fuels.

The oceans The world's oceans are the second-largest reservoir in the carbon cycle. They absorb carbon-containing compounds from the atmosphere, from terrestrial runoff, from undersea volcanoes, and from the waste products and detritus of marine organisms. Some carbon atoms absorbed by the oceans—in the form of carbon dioxide, carbonate ions (CO_3^{2-}), and bicarbonate ions (HCO_3^-)—combine with calcium ions (Ca^{2+}) to form calcium carbonate ($CaCO_3$), an essential ingredient in the skeletons and shells of microscopic marine organisms. As these organisms die, their calcium carbonate shells sink to the ocean floor and begin to form sedimentary rock. The rates at which the oceans absorb and release carbon depend on many factors, including temperature and the numbers of marine organisms converting CO_2 into carbohydrates and carbonates.

Humans are shifting carbon from the lithosphere to the atmosphere

By mining fossil fuel deposits, we are essentially removing carbon from an underground reservoir with a residence time of millions of years. By combusting fossil fuels in our automobiles, homes, and industries, we release carbon dioxide and greatly increase the flux of carbon from the ground to the air. Since the mid-18th century, our fossil fuel combustion has added 246 billion metric tons (271 billion tons) of carbon to the atmosphere. Meanwhile, the movement of CO_2 from the atmosphere back to the hydrosphere, lithosphere, and biosphere has not kept pace.

In addition, cutting down forests and burning fields removes carbon from the pool of vegetation and releases it to the air. And if less vegetation is left on the surface, there are fewer plants to draw CO_2 back out of the atmosphere.

As a result, scientists estimate that today's atmospheric carbon dioxide reservoir is the largest that Earth has experienced in the past 420,000 years, and perhaps in the past 20 million years. The ongoing flux of carbon out of the fossil-fuel reservoir and into the atmosphere is a driving force behind global climate change (Chapter 18).

The phosphorus cycle involves mainly lithosphere and ocean

The element phosphorus (P) is a key component of DNA, RNA, and cell membranes (▸pp. 100–102). Organisms also use phosphorus to build two important biochemical compounds, adenosine diphosphate (ADP) and adenosine triphosphate (ATP). Cells use ADP and ATP to transfer and convert energy from one form to another during metabolism, and these compounds play roles in processes involving DNA and RNA. Although phosphorus is vital for life in these ways, the amount of phosphorus in organisms is dwarfed by the vast amounts in rocks, soil, sediments, and the oceans. Unlike the carbon and nitrogen cycles, the **phosphorus cycle** (Figure 7.10) has no appreciable atmospheric component besides the transport of tiny amounts of windblown dust and seaspray.

Geology and phosphorus availability Most of Earth's phosphorus is contained within rocks and is released only by weathering (▸pp. 251–252), an extremely slow process. The weathering of rocks releases phosphate (PO_4^{3-}) ions into water. Phosphates dissolved in lakes or in the oceans precipitate into solid form, settle to the bottom, and reenter the lithosphere's phosphorus reservoir in sediments. Because most phosphorus is bound up in rock and only slowly released, environmental concentrations of phosphorus available to organisms tend to be very low. This relative rarity explains why phosphorus is frequently a limiting factor for plant growth and why an artificial influx of phosphorus can produce immediate and dramatic effects.

Food webs Plants can take up phosphorus through their roots only when phosphate is dissolved in water. Primary consumers acquire phosphorus from water and plants and pass it on to secondary and tertiary consumers. Consumers also pass phosphorus to the soil through the excretion of waste. Decomposers break down phosphorus-rich organisms and their wastes and, in so doing, return phosphorus to the soil.

We affect the phosphorus cycle

Humans influence the phosphorus cycle in several ways. We mine rocks containing phosphorus to extract this nutrient for the inorganic fertilizers we use on crops and lawns. Treated and untreated sewage discharge also tends to be rich in phosphates. Those phosphates that run off into waterways can boost algal growth and cause eutrophication, leading to murkier waters and altering the structure and function of aquatic ecosystems. Phosphates are also present in detergents, and one way each of us can reduce phosphorus input into the environment is to purchase phosphate-free detergents.

The nitrogen cycle involves specialized bacteria

Nitrogen (N) makes up 78% of our atmosphere by mass, and is the sixth most abundant element on Earth. It is an essential ingredient in the proteins, DNA, and RNA that

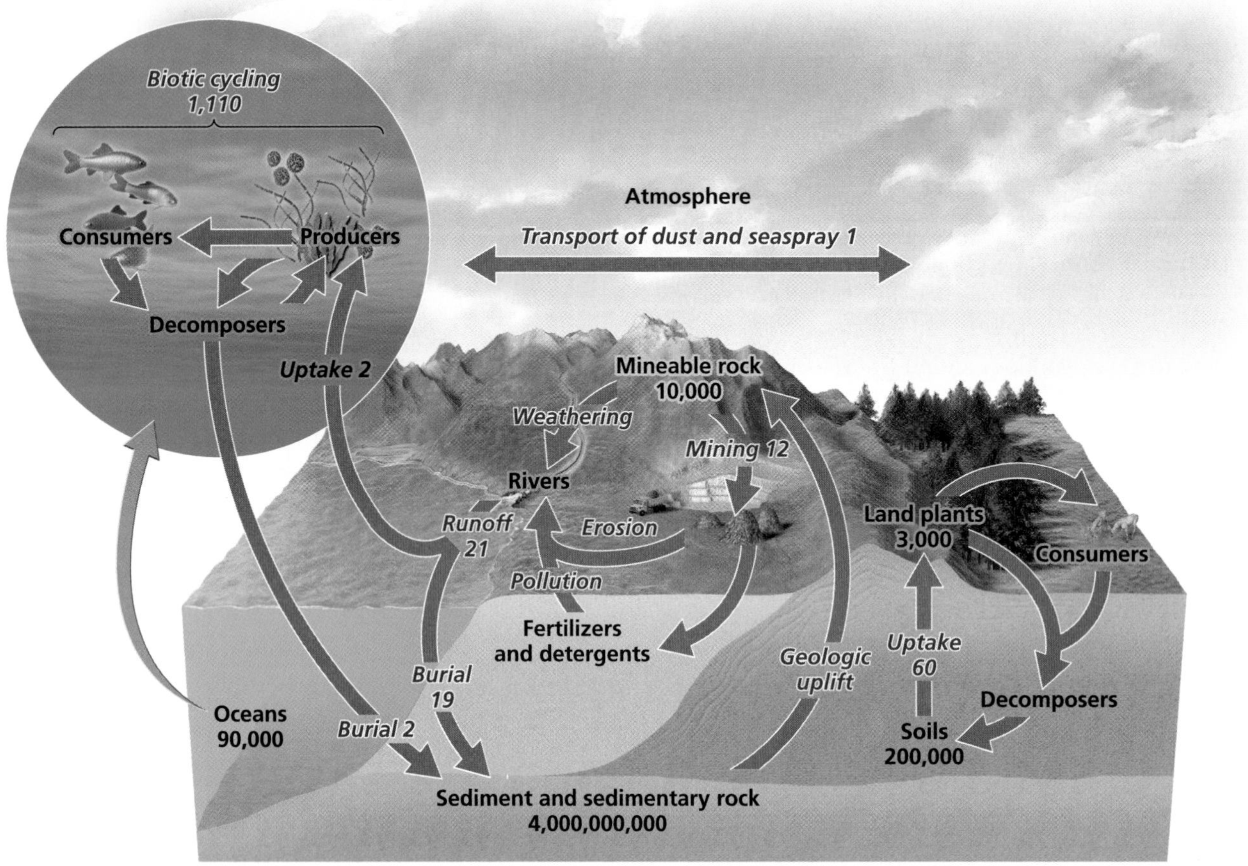

FIGURE 7.10 The phosphorus cycle summarizes the many routes that phosphorus atoms take as they move through the environment. Gray arrows represent fluxes among reservoirs, or pools, for phosphorus. Most phosphorus resides underground in rock and sediment, but the phosphorus cycle moves this element through the soil, the oceans, and freshwater and terrestrial ecosystems. Rocks containing phosphorus are uplifted geologically and weathered away in this slow process, and small amounts of phosphorus cycle through food webs, where this nutrient is often a limiting factor for plant growth. In the figure, pool names are printed in plain bold type, and numbers in plain bold type represent pool sizes, expressed in units of 10^{12} g P. Processes printed in italic bold red type give rise to fluxes, printed in italic bold red type and expressed in units of 10^{12} g P per year. Data from Schlesinger, W. H. 1997. *Biogeochemistry: An analysis of global change.* 2nd ed. London: Academic Press.

build our bodies and, like phosphorus, is an essential nutrient for plant growth. Thus the **nitrogen cycle** (Figure 7.11) is of vital importance to us and to all other organisms. Despite its abundance in the air, nitrogen gas (N_2) is chemically inert and cannot cycle out of the atmosphere and into living organisms without assistance from lightning, highly specialized bacteria, or human intervention. For this reason, the element is relatively scarce in the lithosphere and hydrosphere and in organisms. However, once nitrogen undergoes the right kind of chemical change, it becomes biologically active and available to the organisms that need it, and it can act as a potent fertilizer. Its scarcity makes biologically active nitrogen a limiting factor for plant growth.

Nitrogen fixation To become biologically available, inert nitrogen gas (N_2) must be "fixed," or combined with hydrogen in nature to form ammonia (NH_3), whose water-soluble ions of ammonium (NH_4^+) can be taken up by plants. **Nitrogen fixation** can be accomplished in two ways: by the intense energy of lightning strikes, or when air in the top layer of soil comes in contact with particular types of **nitrogen-fixing bacteria.** These bacteria live in a mutualistic relationship (▸pp. 155–156) with many types

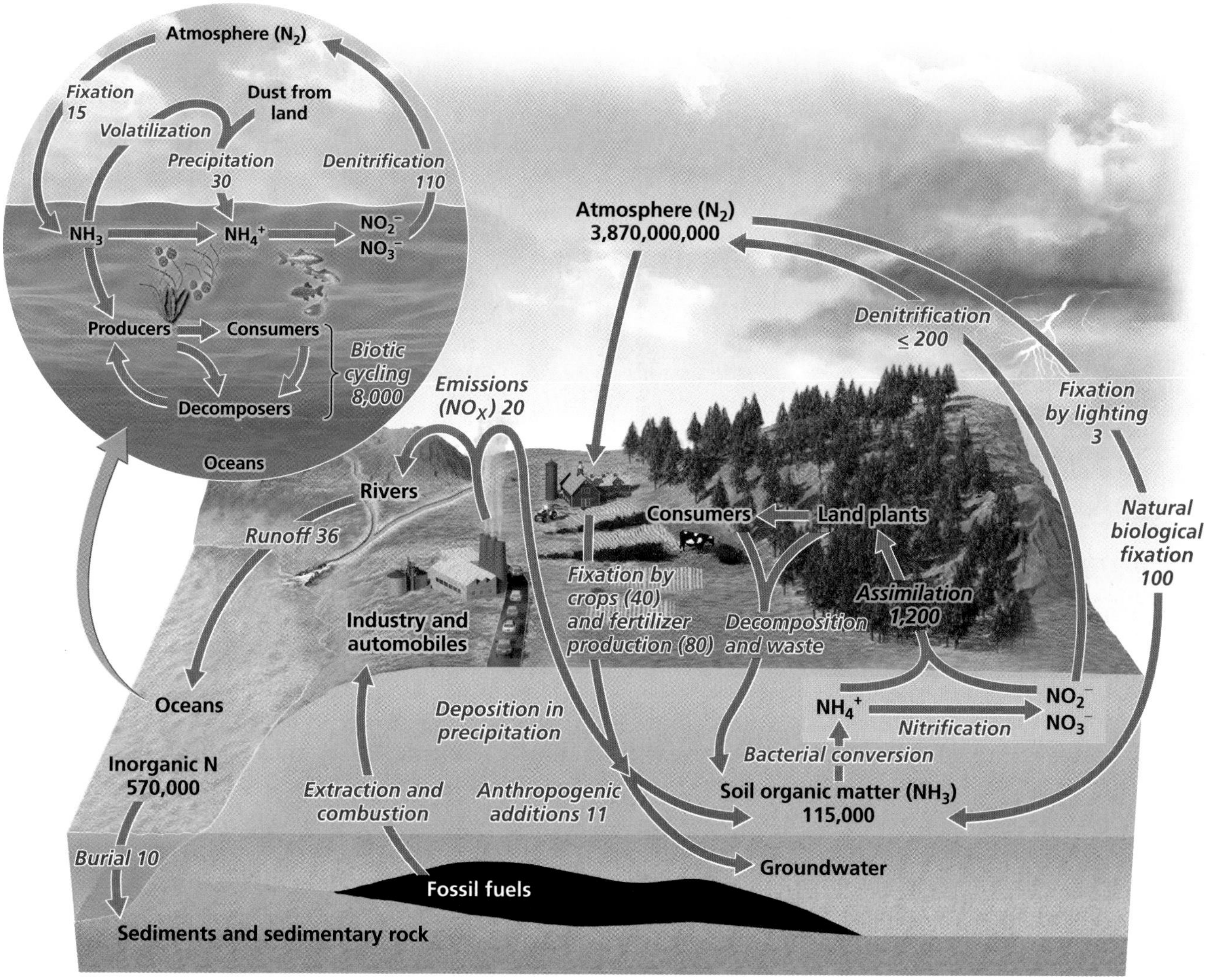

FIGURE 7.11 The nitrogen cycle summarizes the many routes that nitrogen atoms take as they move through the environment. Gray arrows represent fluxes among reservoirs, or pools, for nitrogen. In the nitrogen cycle, specialized bacteria play key roles in "fixing" atmospheric nitrogen and converting it to chemical forms that plants can use, while other types of bacteria convert nitrogen compounds back to the atmospheric gas N_2. In the oceans, inorganic nitrogen is buried in sediments while nitrogen compounds are cycled through food webs as they are on land. In the figure, pool names are printed in plain bold type, and numbers in plain bold type represent pool sizes, expressed in units of 10^{12} g N. Processes printed in italic bold red type give rise to fluxes, printed in italic bold red type and expressed in units of 10^{12} g N per year. Data from Schlesinger, W. H. 1997. *Biogeochemistry: An analysis of global change*. 2nd ed. London: Academic Press.

of plants, including soybeans and other legumes, providing them nutrients by converting nitrogen to a usable form. As we will see in Chapter 9, farmers have long nourished their soils by planting crops that host nitrogen-fixing bacteria among their roots (Figure 7.12).

Nitrification and denitrification Other types of specialized bacteria then perform a process known as **nitrification.** In this process, ammonium ions are first converted into nitrite ions (NO_2^-), then into nitrate ions (NO_3^-). Plants can take up these ions, which also become available after atmospheric deposition on soils or in water or after application of nitrate-based fertilizer.

Animals obtain the nitrogen they need by consuming plants or other animals. Recall from Chapter 4 ("The Science behind the Story," ▸pp. 94–95) how scientists use stable isotopes of nitrogen to study the trophic level and nutritional condition of animals. Decomposers obtain

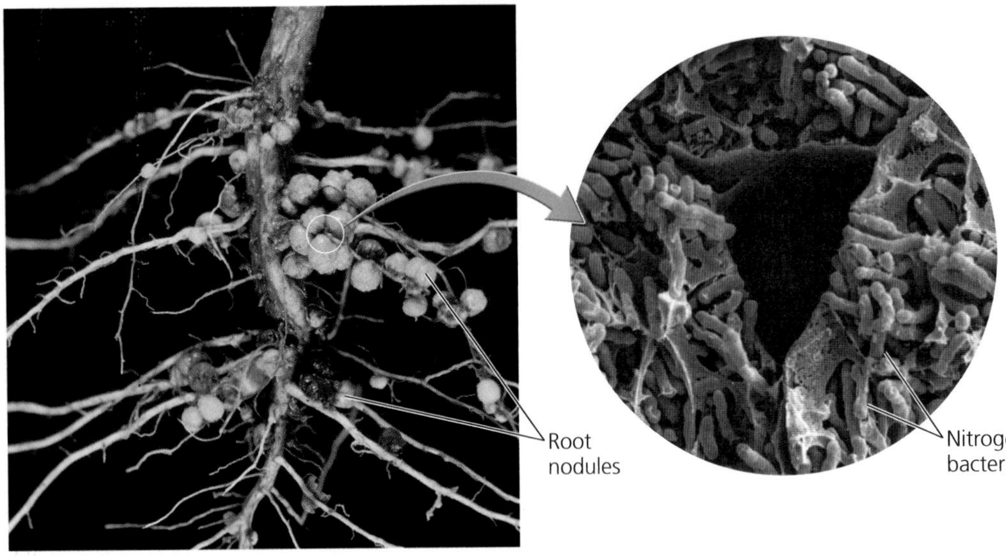

FIGURE 7.12 Specialized bacteria live in nodules on the roots of this legume plant. In the process of nitrogen fixation, the bacteria convert nitrogen to a form that the plant can take up into its roots.

nitrogen from dead and decaying plant and animal matter and from the urine and feces of animals. Once decomposers process the nitrogen-rich compounds they take in, they release ammonium ions, making these available to nitrifying bacteria to convert again to nitrates and nitrites.

The next step in the nitrogen cycle occurs when **denitrifying bacteria** convert nitrates in soil or water to gaseous nitrogen via a multistep process. Denitrification thereby completes the cycle by releasing nitrogen back into the atmosphere as a gas.

Humans have greatly influenced the nitrogen cycle

The impacts of excess nitrogen from agriculture and other human activities in the Mississippi River watershed have become painfully evident to shrimpers and scientists with an interest in the Gulf of Mexico (see "The Science behind the Story" on ▸pp. 202–203). But hypoxia in the Gulf and other coastal locations around the world is hardly the only problem resulting from human manipulation of the nitrogen cycle.

Historically, nitrogen fixation was a *bottleneck*, a step that limited the flux of nitrogen out of the atmosphere. This changed when the research of two German chemists enabled us to fix nitrogen on an industrial scale. Fritz Haber worked in the German army's chemical weapons program during World War I. Shortly before the war, Haber found a way to combine nitrogen and hydrogen gases to synthesize ammonia, a key ingredient in modern explosives and agricultural fertilizers. Several years later, Carl Bosch built on Haber's work and devised methods to produce ammonia on an industrial scale. The work of these two scientists enabled people to overcome the limits on productivity long imposed by nitrogen scarcity in nature. The widespread application of their findings has enhanced agriculture and thereby contributed to the enormous increase in human population over the past 90 years. Farmers, golf course managers, and homeowners have all taken advantage of the fertilizers made possible by the **Haber-Bosch process.** These developments have also led to a dramatic alteration of the nitrogen cycle, however. Today, using the Haber-Bosch process, our species is fixing at least as much nitrogen artificially as is being fixed naturally. We have effectively doubled the natural rate of nitrogen fixation on Earth (Figure 7.13).

By fixing atmospheric nitrogen, we increase its flux out of the atmosphere and into other reservoirs. In addition, we have affected fluxes in other parts of the cycle. When we burn forests and fields, we force nitrogen out of soils and vegetation and into the atmosphere. When we burn fossil fuels, we increase the rate at which nitric oxide (NO) enters the atmosphere and reacts to form nitrogen dioxide (NO_2). This compound is a precursor to nitric acid (HNO_3), a key component of acid precipitation (▸pp. 514–518). We introduce another nitrogen-containing gas, nitrous oxide (N_2O), by allowing anaerobic bacteria to break down the tremendous volume of animal waste produced in agricultural feedlots (▸pp. 296–297). We have also accelerated the introduction of nitrogen-rich compounds into terrestrial

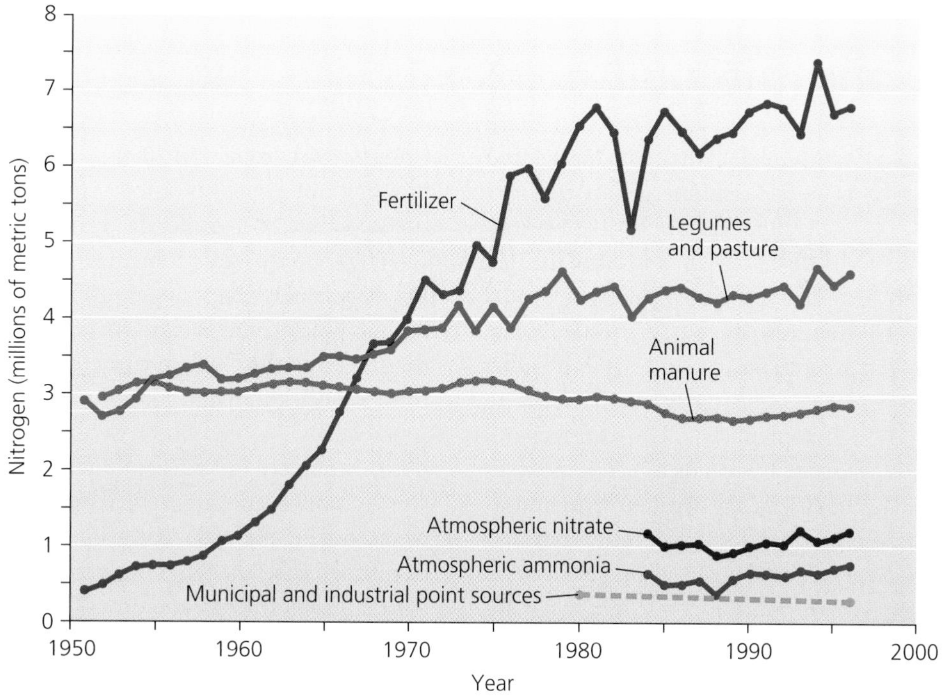

FIGURE 7.13 In the past half century, human inputs of nitrogen into the environment have greatly increased, such that today fully half of the nitrogen entering the environment is of human origin. Agricultural fertilizer has for the past 35 years been the leading source of all nitrogen inputs, natural and artificial. Data from National Science and Technology Council: Committee on Environment and Natural Resources. May 2000. *Hypoxia: An integrated assessment in the northern Gulf of Mexico.*

and aquatic systems by destroying wetlands and cultivating more legume crops that host nitrogen-fixing bacteria in their roots.

These activities increase amounts of nitrogen available to aquatic plants and algae, boosting their growth. Algal populations soon outstrip the availability of other required nutrients and begin to die and decompose. As in the Gulf of Mexico, this large-scale decomposition can rob other aquatic organisms of oxygen, leading to shellfish die-offs and other significant impacts on ecosystems. Increased nitrate pollution, such as that found in groundwater in the Mississippi River basin, can also lead to human health effects.

In 1997, a team of scientists led by Peter Vitousek of Stanford University summarized the dramatic changes humans have caused in the global nitrogen cycle. Each of these alterations can reverberate, through feedback loops, to produce unintended outcomes and unpredictable results. Human alterations of the nitrogen cycle, according to the Vitousek team's report, have:

- Doubled the rate that fixed nitrogen enters terrestrial ecosystems (and the rate is still increasing).
- Increased atmospheric concentrations of the greenhouse gas N_2O and of other oxides of nitrogen that produce smog.
- Depleted essential nutrients, such as calcium and potassium, from soils, because fertilizer helps flush them out.
- Acidified surface water and soils.
- Greatly increased transfer of nitrogen from rivers to oceans.
- Encouraged plant growth, causing more carbon to be stored within terrestrial ecosystems.
- Reduced biological diversity, especially plants adapted to low nitrogen concentrations.
- Changed the composition and function of estuaries and coastal ecosystems.
- Harmed many coastal marine fisheries.

Weighing the Issues: Nitrogen Pollution and Its Financial Impacts

Most nitrate that enters the Gulf of Mexico originates from farms and other sources in the upper Midwest, yet many of its negative impacts are borne by downstream users, such as fishermen and shrimpers along the Gulf Coast. Who do you believe should be responsible for addressing this problem? Should environmental policies on this particular issue be developed and enforced by state governments, the federal government, both, or neither? Explain the reasons for your answer.

In 1998, the U.S. Congress passed the Harmful Algal Bloom and Hypoxia Research and Control Act. This law called for an "integrated assessment" of hypoxia in the northern Gulf to address the extent, nature, and causes of the dead zone, as well as its ecological and economic

The Science behind the Story

Hypoxia and the Gulf of Mexico's "Dead Zone"

She was prone to seasickness, but Nancy Rabalais cared too much about the Gulf of Mexico to let that stop her. Leaning over the side of an open boat idling miles from shore, she hauled a water sample aboard—and helped launch the long effort to breathe life back into the Gulf's "dead zone."

Since that expedition in 1985, Rabalais, her colleague and husband Eugene Turner, and fellow scientists at the Louisiana Universities Marine Consortium (LUMCON) and Louisiana State University have made great progress in unraveling the mysteries of the region's hypoxia—and in getting it on the political radar screen.

A few scientists had noticed abnormally low oxygen levels in parts of the Gulf in the 1970s. But such observations gained urgency in the 1980s as fishing trawlers had to head farther and farther offshore to find anything to catch. Such early signs of trouble led to bigger questions for Rabalais and her colleagues: How widespread was the hypoxia? What was causing it? How much worse might it get?

Rabalais and other researchers started tracking oxygen levels at nine sites in the Gulf every month, and continued those measurements for five years. At dozens of other spots near the shore and in deep water, they took less frequent oxygen readings. For some of this work, the researchers have relied on mobile oxygen probes. Sensors, as they are lowered into the water, measure oxygen levels and send continuous readings back to a shipboard computer. Further data have come from fixed, submerged oxygen meters that continuously measure dissolved oxygen and store the data.

The team also collected hundreds of coastal and Gulf water samples, using lab tests to measure levels of nitrogen, salt, bacteria, and phytoplankton. LUMCON scientists logged hundreds of miles in their boats, regularly monitoring more than 70 sites in the Gulf. They also donned scuba gear to view firsthand the condition of shrimp, fish, and other sea life. Such a range of long-term data allowed the scientists to build a "map" of the dead zone, tracking its location and effects.

In 1991, Rabalais made that map public, earning immediate headlines. That year, her group mapped the size of the zone at more than 10,000 km^2 (about 4,000 mi^2). Bottom-dwelling shrimp were stretching out of their burrows, straining for oxygen. Many fish had fled. The bottom waters, infused with sulfur from bacterial decomposition, smelled of rotten eggs.

The group's years of continuous tracking also explained the dead zone's predictable emergence. As rivers rose each spring (and as fertilizers were applied in the Midwestern farm states), oxygen would start to disappear in the northern Gulf. The hypoxia would last through the summer or fall, until seasonal storms mixed oxygen into hypoxic areas. Over time, monitoring linked the dead zone's size to the volume of river flow and its nutrient load; the 1993 flooding of the Mississippi created a zone much larger than the year before. Conversely, a drought in 2000 brought lower river flows, lower nutrient loads, and a smaller dead zone.

The source of the problem, Rabalais said, lay back on land. The Mississippi and Atchafalaya Rivers draining into the Gulf were polluted from runoff, and that pollution spurred algal blooms whose decomposition snuffed out oxygen in wide stretches of ocean water. The rivers were carrying high concentrations of nitrates and other chemicals from agricultural fertilizers and other sources. As outlined in this chapter, excess nitrogen sets off a chain reaction in marine waters that eventually drains oxygen in deeper waters.

Many Midwestern farming advocates and some scientists, such as Derek Winstanley, chief of the Illinois State Water Survey, challenged the findings. They argued that the Mississippi naturally carries high loads of nitrogen from the rich prairie soil and that the Rabalais team had not ruled out upwelling in the Gulf as a source of nutrients. But sediment analyses showed that Mississippi River mud was much lower in nitrates early in the century, and Rabalais and Turner found that silica residue from phytoplankton blooms had increased in Gulf sediments between 1970 and 1989, paralleling rising nitrogen levels. In 2000, the federal integrative assessment involving dozens of scientists laid the blame for the dead zone on nutrients from fertilizers and other sources.

Then in 2004, while representatives of farmers and fishermen bickered over political fixes, EPA water quality scientist Howard Marshall suggested that phosphorus

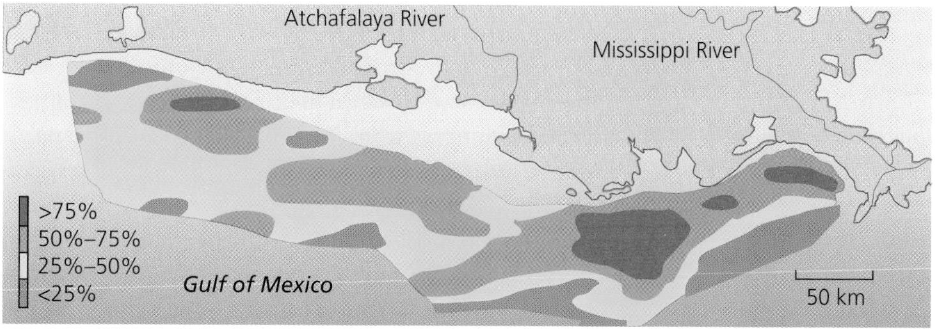

(a) Frequency of hypoxia, 1985–1999

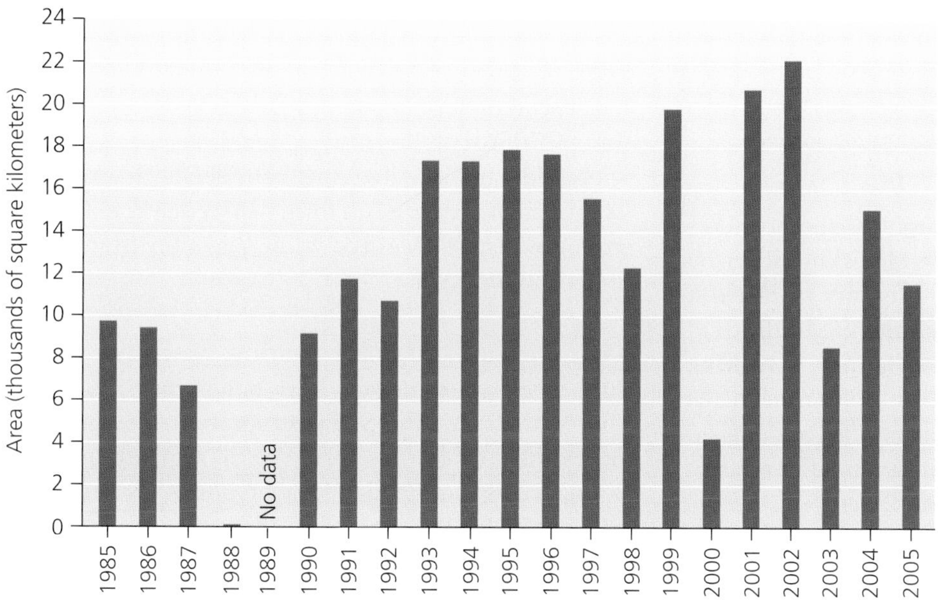

(b) Area of hypoxic zone in the northern Gulf of Mexico

Some parts of the Gulf of Mexico suffer from hypoxia more frequently than others (**a**). Areas in red have experienced oxygen concentrations below 2 ppm in more than 75% of surveys; those in orange experienced hypoxia in 50–75% of surveys; those in yellow were affected in 25–50% of surveys; and those in green were affected in less than 25% of surveys. The size of the Gulf's hypoxic zone varies (**b**) as a result of several factors. These include floods, which increase its size by bringing additional runoff (as with the Mississippi River floods of 1993), and tropical storms, which decrease its size by mixing oxygen-rich water into the dead zone (as in 2003). Between 1985 and 2005, the hypoxic zone averaged 12,700 km^2 in size and ranged up to 22,000 km^2. Data in (a) from Rabalais, N., et al. 2002. Beyond science into policy: Gulf of Mexico hypoxia and the Mississippi. *BioScience* 52: 129–142. Data in (b) from Nancy Rabalais, LUMCON.

pollution from industry and sewage treatment might instead be at fault. His reasoning: the ratio of nitrogen to phosphorus in the Gulf was so overwhelmingly biased toward nitrogen that phosphorus had become the limiting factor. Because phytoplankton need both, they now had more nitrogen than they could use but not enough phosphorus. Thus, Marshall suggested, we'd be best off reducing phosphorus if we want to alleviate the dead zone.

Other scientists are giving this idea a mixed reception, and many have proposed that nitrogen and phosphorus should be managed jointly. Further suggestions have come from still more research. One is that the federally mandated 30% reduction in nitrogen in the river will not be enough. Another is that large-scale restoration of wetlands along the river and at the river's delta would best filter pollutants before they reached the Gulf. All this research is guiding a federal plan to reduce farm runoff, clean up the Mississippi, restore wetlands, and shrink the Gulf's dead zone—and it has also led to a better understanding of hypoxic zones around the world.

"What people do 800 miles away from the Gulf of Mexico directly affects the Gulf of Mexico," Rabalais said when she accepted a major environmental award in 1999. "It's hard for many people to realize that."

impacts. The assessment report published two years later also outlined potential solutions to the problem, along with estimates of the social and economic costs associated with these solutions. The report proposed that the federal government work with Gulf Coast and Midwestern communities to:

- Reduce nitrogen fertilizer use on Midwestern farms.
- Change the timing of fertilizer application to minimize rainy-season runoff.
- Use alternative crops.
- Manage nitrogen-rich livestock manure more effectively.
- Restore nitrogen-absorbing wetlands in the Mississippi River basin.
- Use artificial wetlands to filter farm runoff.
- Install better nitrogen-removing technologies in sewage treatment plants.
- Restore frequently flooded lands to reduce runoff.
- Restore wetland ecosystems near the Mississippi River's mouth to enhance nitrogen-absorbing ability.
- Continue evaluating which of these approaches work and which do not.

A combined state and federal task force proposed reducing nitrogen flowing down the Mississippi River by 30% by 2015, although some scientists later estimated it will require cuts of 50% to decrease the size of the dead zone. Farmers' advocates argued that farmers were being unfairly singled out, when many other artificial and natural sources contribute to the problem. They also argued that severe restrictions on fertilizer use would hurt farmers economically and decrease crop yields. As people debated the science and the proposed remedies, Congress in 2003 reauthorized the Harmful Algal Bloom and Hypoxia Research and Control Act for an additional five years, promising funding for further research and the development of solutions.

Many scientists and farmers are now searching for innovative solutions that would alleviate pollution while not hurting agriculture. One proposal offers farmers insurance and economic incentives for not using excess fertilizer. Another program is testing new farming strategies to see whether any can maintain yields while decreasing fertilizer use. A third approach involves planting cover crops in the off-season to reduce runoff from bare fields. Yet another encourages farmers to maintain artificial wetlands on their lands that serve as natural buffers against pollution. Wetland plants host denitrifying bacteria that convert nitrates to nitrogen gas, so wetlands can effectively clean up a large amount of nitrogen pollution. Many Midwestern farmers have already taken part in one or more of these strategies.

The hydrologic cycle influences all other cycles

Water is so integral to life that we frequently take it for granted. Water is the essential medium for all manner of biochemical reactions (▸ pp. 96–98). It plays key roles in nearly every environmental system, including each of the nutrient cycles we have just discussed. Water carries nutrients and sediments from the continents to the oceans via rivers, streams, and surface runoff, and it distributes sediments onward in ocean currents. Increasingly, water also distributes artificial pollutants. The water cycle, or **hydrologic cycle** (Figure 7.14), summarizes how water—in liquid, gaseous, and solid forms—flows through our environment. Our brief introduction to the hydrologic cycle here sets the stage for our more in-depth discussion of freshwater and marine systems in Chapters 15 and 16.

The oceans are the main reservoir in the hydrologic cycle, holding 97% of all water on Earth. The freshwater we depend on for our survival accounts for less than 3%, and two-thirds of this small amount is tied up in glaciers, snowfields, and ice caps. Thus, considerably less than 1% of the planet's water is in a form that we can readily use—groundwater, surface freshwater, and rain from atmospheric water vapor.

Evaporation and transpiration Water moves from surface reservoirs—oceans, lakes, ponds, rivers, and moist soil—into the atmosphere by **evaporation,** the conversion of a liquid to gaseous form. Warm temperatures and strong winds speed rates of evaporation. A greater degree of exposure has the same effect; an area logged of its forest or converted to agriculture or residential use will lose water more readily than a comparable area that remains vegetated. Water also enters the atmosphere by **transpiration,** the release of water vapor by plants through their leaves. Transpiration and evaporation act as natural processes of distillation, effectively creating pure water by filtering out minerals carried in solution.

Precipitation, runoff, and surface water Water returns from the atmosphere to Earth's surface as **precipitation** when water vapor condenses and falls as rain or snow. Precipitation may be taken up by plants and used by animals, but much of it flows as **runoff** into streams, rivers, lakes, ponds, and oceans. Amounts of precipitation vary greatly from region to region globally, helping give rise to the variety of biomes.

Groundwater Some precipitation and surface water soaks down through soil and rock to recharge under-

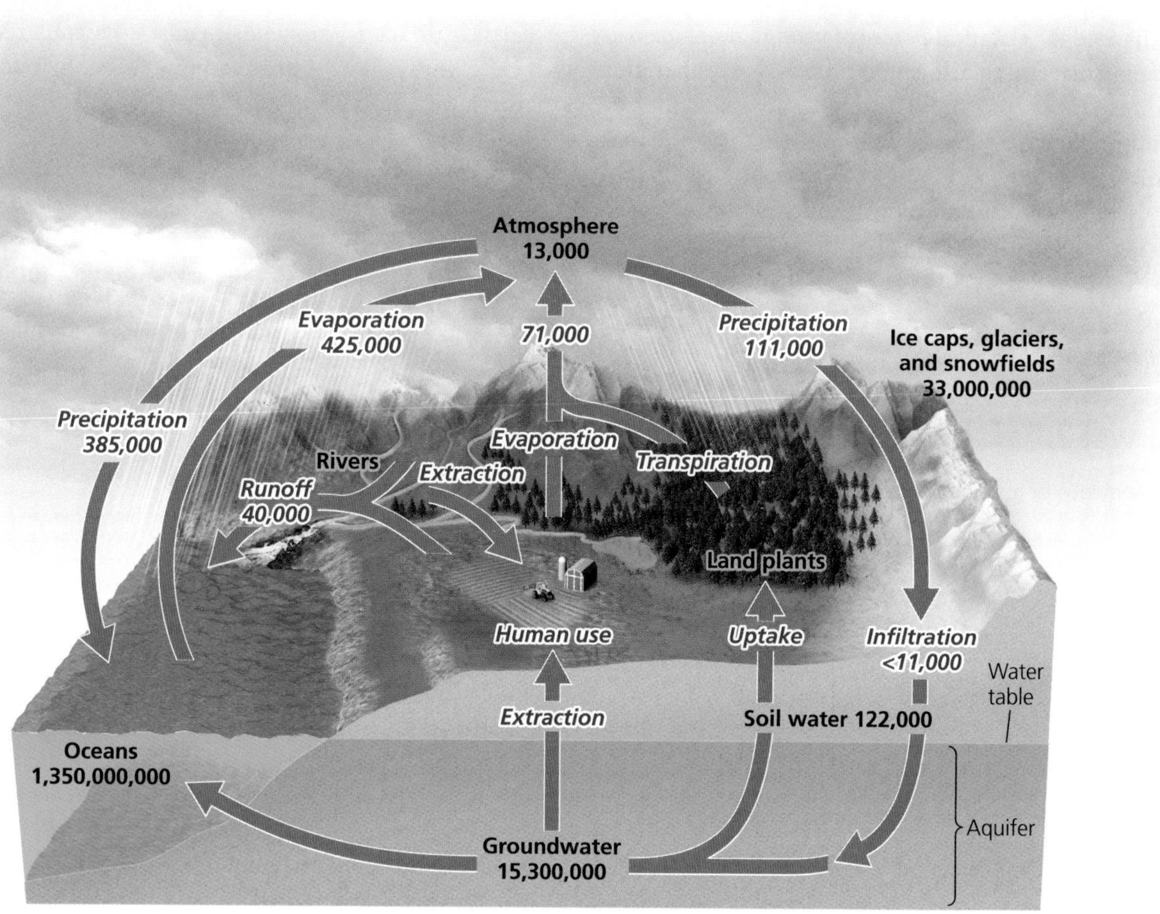

FIGURE 7.14 The hydrologic cycle summarizes the many routes that water molecules take as they move through the environment. Gray arrows represent fluxes among reservoirs, or pools, for water. The hydrologic cycle is a system unto itself but also plays key roles in other biogeochemical cycles. Oceans hold 97% of our planet's water, while most freshwater resides in groundwater and icecaps. Water vapor in the atmosphere condenses and falls to the surface as precipitation, then evaporates from land and transpires from plants to return to the atmosphere. Water flows downhill into rivers, eventually reaching the oceans. In the figure, pool names are printed in plain bold type, and numbers in plain bold type represent pool sizes, expressed in units of km^3. Processes printed in italic bold red type give rise to fluxes, printed in italic bold red type and expressed in units of km^3 per year. Data from Schlesinger, W. H. 1997. *Biogeochemistry: An analysis of global change,* 2nd ed. London: Academic Press.

ground water reservoirs known as **aquifers.** Aquifers are spongelike regions of rock and soil that hold **groundwater,** water found underground beneath layers of soil. The upper limit of groundwater held in an aquifer is referred to as the **water table.** Aquifers may hold groundwater for long periods of time, so the water may be quite ancient. In some cases groundwater can take hundreds or even thousands of years to fully recharge after being depleted. Groundwater becomes exposed to the air where the water table reaches the surface, and the exposed water can run off toward the ocean or evaporate into the atmosphere.

Our impacts on the hydrologic cycle are extensive

Human activity has affected every aspect of the water cycle. By damming rivers to create reservoirs, we have increased evaporation and, in some cases, infiltration of surface water into aquifers. By altering Earth's surface and its vegetation, we have increased surface runoff and erosion. By spreading water on agricultural fields, we have depleted rivers, lakes, and streams, and have increased evaporation. By removing forests and other vegetation, we have reduced transpiration and have lowered water tables. And by emitting into the atmosphere pollutants that dissolve in water droplets, we have changed the chemical nature of precipitation, in effect sabotaging the natural distillation process that evaporation and transpiration provide. Perhaps most threatening to our future, we have drawn groundwater to the surface for drinking, irrigation, and industrial use, and have thereby begun to deplete groundwater resources (Figure 7.15). Water shortages have already given rise to numerous conflicts worldwide, from the Middle East to the American West (▸p. 444).

Weighing the Issues: Your Water

Are you aware of any evidence of water shortages or conflict over water use in your region? What about pollution and water quality? In what ways might these issues affect the hydrologic cycle in your area?

Geological Systems: How Earth Works

Biogeochemical cycles are not the planet's only cyclical environmental systems. Physical processes of geology determine Earth's landscape and form the foundation for the biotic patterns that overlay the landscape.

The rock cycle is a fundamental environmental system

We tend to think of rock as pretty solid stuff; this is clear when we say something is "hard as rock." Yet in the long run, over geological time, rocks do change. Rocks and the minerals that comprise them are heated, melted, cooled, broken down, and reassembled in a very slow process called the **rock cycle** (Figure 7.15). The type of rock in a given region helps determine soil chemistry and thereby influences the biotic components of the region's ecosystems.

Igneous rock All rocks can melt. At high enough temperatures, rock will enter a molten, liquid state called **magma.** If magma is released from the lithosphere (as in a volcanic eruption), it may flow or spatter across Earth's surface as **lava.** Rock that forms when magma cools is called **igneous** (from the Latin *ignis,* meaning "fire") **rock** (Figure 7.15a). Igneous rock comes in several different types, because there are different ways in which magma can solidify. Magma that cools slowly while it is well below Earth's surface is known as *intrusive* igneous rock. Half Dome and many other famous rock formations at Yosemite National Park in California were formed in this way and later exposed above the surface. Granite is the best-known type of intrusive rock. A slow cooling process allows minerals of different types to segregate from one another and aggregate with minerals of their own type, forming the crystals that give granite its multicolored, coarse-grained appearance. In contrast, when magma is ejected from a volcano, it cools quickly, so minerals have little time to differentiate into clusters. This kind of igneous rock is called *extrusive* igneous rock, and its most common representative is basalt.

Sedimentary rock All rock weathers away with time. The relentless forces of wind, water, freezing, and thawing eat away at rocks, stripping off one tiny grain (or large chunk) at a time. Particles of rock blown by wind or washed away by water finally come to rest somewhere, where they help to form **sediments.** These eroded remains of rocks usually are deposited very slowly, but floods can accelerate the process. The floods that sweep down the Mississippi River deposit sediments and nutrients at the river's mouth, building up its delta. They also spread them across the floodplain along the river's banks, where soils are enriched as a result. Sediments collect downhill, downstream, or downwind from their sources and are deposited in layers. These layers accumulate over time, causing the weight and pressure of overlying layers to increase. **Sedimentary rock** (Figure 7.15b) is formed when dissolved minerals seep through sediment layers and act as a kind of glue, crystallizing and binding sediment particles together. The formation of rock through these processes of compaction, binding, and crystallization is termed **lithification.** The fossils of organisms—and the fossil fuels we use for energy—are created by similar processes of physical compaction and chemical transformation (▸pp. 561–562).

Like igneous rock, the several types of sedimentary rock are classified by the way they form and the size of particles they contain. Rock such as limestone and rock salt forms by chemical means when rocks dissolve and their components crystallize to form new rock. A second type of sedimentary rock forms when layers of sediment are compressed and physically bonded to one another. Examples include conglomerate, made up of large particles that give it the appearance of nougat; sandstone, made of cemented sand particles; and shale, composed of still smaller mud particles.

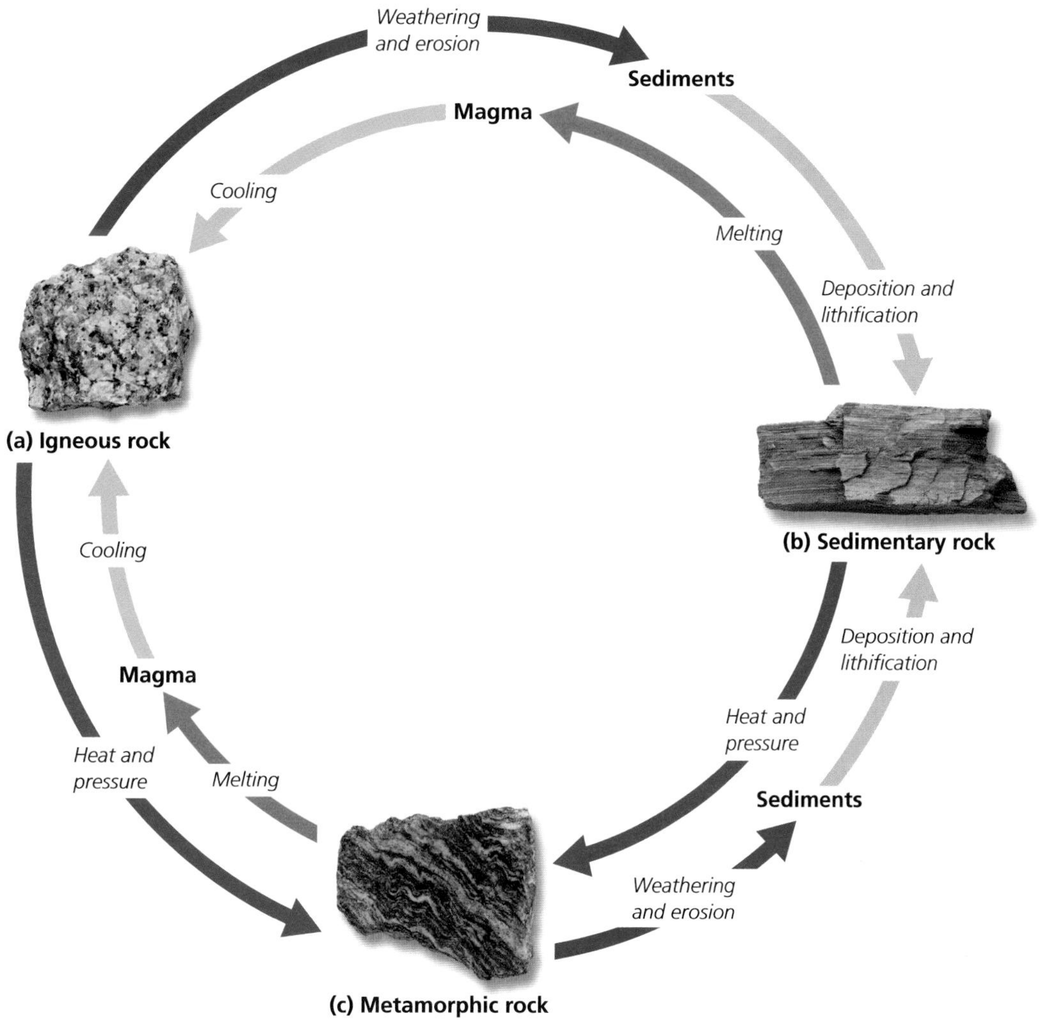

FIGURE 7.15 The rock cycle is a slow cycle that shapes Earth's crust and affects Earth's nutrient cycles. Igneous rock (**a**) is formed when rock melts to form magma, and the magma then cools. Sedimentary rock (**b**) is formed when rock is weathered and eroded, and the resulting sediments are compressed to form new rock. Metamorphic rock (**c**) is formed when rock is subjected to intense heat and pressure underground. Through these several processes (shown by differently colored arrows in the figure), each type of rock can be converted into either of the other two types.

Metamorphic rock Geological forces may bend, uplift, compress, or stretch rock. When great heat or pressure is exerted on rock, the rock may change its form, becoming **metamorphic** (from the Greek for "changed form" or "changed shape") **rock** (Figure 7.15c). The forces that metamorphose rock occur at temperatures lower than the rock's melting point but high enough to reshape crystals within the rock and change its appearance and physical properties. Metamorphic rock may resemble the rock from which it was created or may be quite different. Common types of metamorphic rock include marble, formed when limestone is heated and pressurized, strengthening its structure; and slate, formed when shale is heated and metamorphosed.

The changes that occur as rocks are altered from one type to another may proceed in any direction. Understanding the transition of rocks from one stage in the rock cycle to another enables us to appreciate more clearly the formation and conservation of soils, mineral resources, fossil fuels, and other natural resources.

Plate tectonics shapes Earth's geography

The rock cycle takes place within the broader context of **plate tectonics,** a process that underlies earthquakes and volcanoes and that determines the geography of the Earth's surface. Earth's surface consists of a lightweight

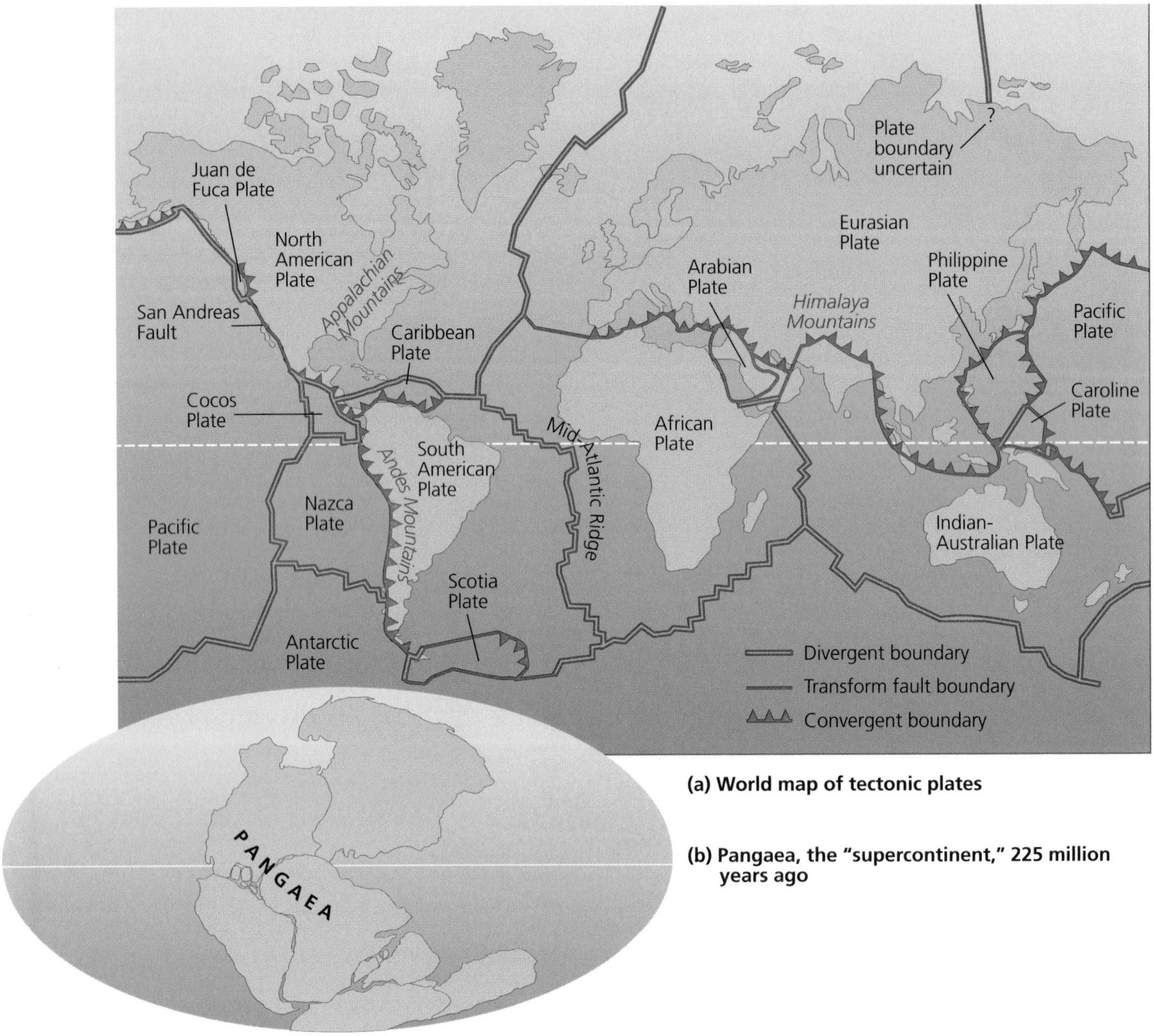

(a) World map of tectonic plates

(b) Pangaea, the "supercontinent," 225 million years ago

FIGURE 7.16 Earth's crust consists of roughly 15 major plates (**a**) that move through time by the process of plate tectonics. Today's continents were joined together in the "supercontinent" Pangaea (**b**) about 225 million years ago.

thin **crust** of rock floating atop a malleable **mantle,** which in turn surrounds a molten heavy **core** made mostly of iron. Earth's internal heat drives convection currents that flow in loops in the mantle, pushing the mantle's soft rock cyclically upward (as it warms) and downward (as it cools), like a gigantic conveyor belt. As the mantle material moves, it drags large plates of crust along its surface edge.

Earth's surface consists of about 15 major tectonic plates, most including some combination of ocean and continent (Figure 7.16a). Imagine peeling an orange and putting the pieces of peel back onto the fruit; the ragged pieces of peel are like the plates of crust riding atop Earth's surface. These plates move at rates of roughly 2–15 cm (1–6 in.) per year. This movement has influenced Earth's climate and life's evolution throughout our planet's history as the continents combined, separated, and recombined in various configurations. By studying ancient rock formations throughout the world, geologists have determined that at least twice, all landmasses were joined together in a supercontinent scientists have dubbed *Pangaea* (Figure 7.16b).

At **divergent plate boundaries,** magma surging upward to the surface divides plates and pushes them apart, creating new crust as it cools and spreads (Figure 7.17a). A

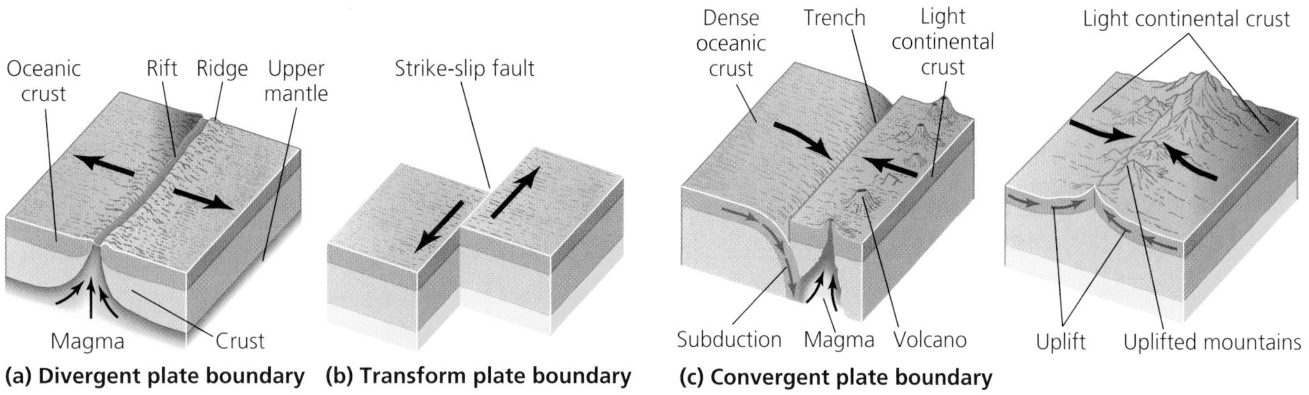

FIGURE 7.17 Different types of boundaries between tectonic plates result in different geologic processes. At a divergent plate boundary, such as a mid-ocean ridge on the seafloor **(a),** magma extrudes from beneath the crust, and the two plates move gradually away from the boundary in the manner of conveyor belts. At a transform plate boundary **(b),** two plates slide alongside one another, creating friction that leads to earthquakes. Where plates collide at a convergent plate boundary **(c),** one plate may be subducted beneath another, leading to volcanism, or both plates may be uplifted, leading to the formation of mountain ranges.

prime example is the Mid-Atlantic Ridge, part of a 74,000-km (46,000-mi) system of magmatic extrusion cutting across the seafloor. Plates expanding outward from divergent plate boundaries at mid-ocean ridges bump against other plates, creating different types of plate boundaries.

When two plates meet, they may slip and grind alongside one another, forming a **transform plate boundary** (Figure 7.17b). The friction at these boundaries spawns earthquakes along slipstrike faults. This is the case with the Pacific Plate and the North American Plate, which rub against each other along California's San Andreas Fault. Southern California is slowly inching its way toward northern California along this fault, and Los Angeles will eventually reach San Francisco.

When plates collide at **convergent plate boundaries,** either of two consequences may result (Figure 7.17c). First, one plate of crust may slide beneath another in a process called **subduction.** The subducted crust is heated as it dives into the mantle, and it may send up magma that erupts through the surface in volcanoes. Mount Saint Helens in Washington, which erupted violently in 1980 and renewed its activity in 2004, is fueled by magma from subduction. When denser ocean crust slides beneath lighter continental crust, volcanic mountain ranges are formed that parallel coastlines. Examples are the Cascades, which include Mount Saint Helens, and South America's Andes Mountains, where the Nazca Plate slides below the South American Plate. When one plate of oceanic crust is subducted beneath another, the resultant volcanism may form arcs of islands, such as Japan and the Aleutians. In addition, deep trenches may be created, such as the Mariana Trench, the planet's deepest abyss.

Alternatively, two colliding plates of continental crust may slowly lift material from both plates. The Himalayas, the world's highest mountains, are the result of the Indian-Australian Plate's collision with the Eurasian Plate 40–50 million years ago, and these mountains are still being uplifted today. The Appalachian Mountains of the eastern United States, once the world's highest mountains themselves, resulted from a much earlier collision with the edge of what is today Africa.

Weighing the Issues: Geology and Topography

Think about the topography in your region. Can you infer what geological processes produced it? Now recall the most recent earthquake or volcano in the news. With what type of plate boundary do you suspect this event was associated?

Amazingly, this environmental system of such fundamental importance was totally unknown to us just half a century ago. Our civilization was sending humans to the moon by the time our geologists were explaining the movement of land under our very feet. It is humbling to reflect on this; what other fundamental systems might we not yet appreciate or understand while our technology—and our ability to affect Earth's processes—continues racing ahead?

VIEWPOINTS

The Dead Zone

Scientific research has indicated that nitrogen fertilizers from the Mississippi River watershed are contributing to hypoxia in the Gulf of Mexico. **Do you agree this is occurring? Why or why not? If so, what steps should be taken to solve the problem?**

Evidence Not Conclusive

Scientific evidence that nitrogen (N) fertilizer is polluting the northern Gulf of Mexico is not conclusive. Hypoxia in the Gulf has been recognized since 1935, long before fertilizer use became widespread in the 1960s. Nitrogen fertilizer use in the Mississippi River basin has remained relatively stable in the last two decades, but the size of the hypoxic zone has fluctuated markedly, especially since 1988.

According to the U.S. Geological Survey, the annual discharge of N from the Mississippi River has tripled in the last 30 years, with most of the increase occurring from 1970 to 1983. However, since 1980, river N discharge has changed very little, whereas N fertilizer use has grown by almost 10%. From 1980 to 1999, N fertilizer sales in the Mississippi River basin explained 8% of the variation in river nitrate-N discharge, whereas the Mississippi River's average annual flow explained nearly 80% of its variation.

Recent data suggest that the molar ratio of inorganic N to inorganic phosphorus (P) is the principal indicator of phytoplankton blooms that cause hypoxia, and that the N:P ratio in freshwater entering the Gulf is high enough that any increase in inorganic P loading will increase the hypoxic zone. The blame is shifting from N to P fertilizer. However, P fertilizer sales in the Mississippi River Basin have declined by 17% since 1980.

Can hypoxia be blamed on sales of N or P fertilizer? Numerous nutrient sources contribute to Gulf loading. Atmospheric deposition, decomposition of crop residue and soil organic matter, legumes, animal manure, municipal sewage sludge and effluent, and composted household wastes all contribute nutrients to the Gulf.

Hypoxia results from a complex interaction of chemical, biological, and physical factors. Fertilizer is a potential pollutant if used improperly, but used correctly, it increases food production and helps protect the environment.

Act Now to Save These Resources

The springtime area of low-oxygen (anoxic) water in the Gulf of Mexico, known as the dead zone, is driven by a massive influx of nutrients into a system no longer able to process them. Eutrophication begins when nutrients from farmlands in the floodplain states wash to the sea. These nutrients (nitrogen fertilizers) now present in the water lead to plankton blooms, which in turn reduce dissolved oxygen in the water and eventually kill fish.

Taking a system view is slightly more complicated, but understanding the system is important for the most effective long-term management. Before people built levees all along the delta, the Mississippi River flooded each spring, and the waters of the river covered the extensive wetlands. This important renewal process deposited sediment on the wetlands to build up the soil base while the plants of the wetlands made use of the nutrient pollutants in the water. The result was cleaner water, richer wetlands, and a sustained environment. Levees now prevent the flooding, the dead zone emerges, and the wetlands are lost as they sink below sea level. Rises in sea level speed the loss.

Loss of wetlands is serious. The wetlands are the base of the fisheries of the Gulf of Mexico, and their loss is irreversible. Saving the wetlands and reversing the dead zone requires a twofold approach. First, reduce the amount of fertilizer so that it is used more efficiently by plants and so that less enters streams. This has the added benefit of saving money and reducing energy consumption (making fertilizers is energy intensive). Second, reinstate the flooding of the wetlands.

Should we wait to act? No. We know enough now to design strategies that can sustain these resources, and new information is unlikely to change what we know. The precautionary principle, which environmental managers use, says that even if information is imperfect, it is important to act before the resource is lost entirely and while any possible cost of error is small and manageable. We need to act now to save these resources.

Terry L. Roberts is vice president of the Potash & Phosphate Institute (PPI) and vice president of the Foundation for Agronomic Research (FAR), located in Norcross, Georgia. Dr. Roberts is a Certified Crop Adviser and a Fellow of the American Society of Agronomy.

Explore this issue further by accessing **Viewpoints** at www.aw-bc.com/withgott.

Paul Templet is a professor at the Institute for Environmental Studies at Louisiana State University. He organized the first Earth Day at LSU in 1970 and served as the secretary of the Louisiana Department of Environmental Quality from 1988 to 1992.

Conclusion

Earth hosts many interacting systems, and the way one perceives them depends on the questions in which one is interested. Physical systems and processes such as the hydrologic cycle, the rock cycle, and plate tectonics lay the groundwork for the ways in which life spreads itself across the planet. Life interacts with its abiotic environment in ecosystems, systems through which energy flows and materials are recycled. Understanding the biogeochemical cycles that describe the movement of nutrients within and among ecosystems is crucial, because human activities are causing significant changes in the ways those cycles operate.

Thinking in terms of systems is important in understanding how Earth works, so that we may learn how to avoid disrupting its processes and how to mitigate any disruptions we cause. By studying the environment from a systems perspective and by integrating scientific findings with the policy process, people who care about the Mississippi River and the Gulf of Mexico are working today to solve their pressing problems. Their model is one that we can adapt to many other issues in environmental science.

We might also consider adopting other models more generally. Think again about unperturbed ecosystems, their use of renewable solar energy, their recycling of nutrients, and the extent to which they exhibit dynamic equilibrium and involve negative feedback loops. The environmental systems we see on Earth today are those that have survived the test of time. Our industrialized civilization is young in comparison. Might we not be able to take a few lessons about sustainability from a careful look at the natural systems of our planet?

REVIEWING OBJECTIVES

You should now be able to:

Describe the nature of environmental systems

- Systems are networks of interacting components that generally involve feedback loops, show dynamic equilibrium, and result in emergent properties. (pp. 184–187)
- Because Earth's natural systems are so complex, environmental scientists often take a holistic approach to studying environmental systems. (pp. 184, 187–188)
- Because environmental systems interact and overlap, one's delineation of systems depends on the questions in which one is interested. (p. 189)

Define ecosystems and evaluate how living and nonliving entities interact in ecosystem-level ecology

- Ecosystems consist of all organisms and nonliving entities that occur and interact in a particular area at the same time. (pp. 189–190)
- Energy flows in one direction through ecosystems, whereas matter is recycled. (p. 191)
- Energy is converted to biomass, and ecosystems vary in their productivity. (pp. 192–193)
- Input of nutrients can boost productivity, but an excess of nutrients can alter ecosystems in ways that cause severe ecological and economic consequences. (pp. 192–195)

Compare and contrast how carbon, phosphorus, nitrogen, and water cycle through the environment

- Most carbon is contained in sedimentary rock. Substantial amounts also occur in the oceans and in soil. Carbon flux between organisms and the atmosphere occurs via photosynthesis and respiration. (pp. 195–197)
- Phosphorus, like carbon, is most abundant in sedimentary rock, with substantial amounts in soil and the oceans. Phosphorus has no appreciable atmospheric pool, however. It is a key nutrient for plant growth. (pp. 197–198)
- Nitrogen, like phosphorus, is a vital nutrient for plant growth. Most nitrogen is in the atmosphere, however, and must be "fixed" by specialized bacteria or lightning before plants can use it. (pp. 197–200)
- Water moves widely through the environment in the hydrologic cycle. (pp. 204–205)
- Humans are causing substantial impacts to Earth's biogeochemical cycles. These impacts include shifting carbon from fossil fuel reservoirs into the atmosphere, shifting nitrogen from the atmosphere to the planet's surface, and depleting groundwater supplies, among many others. (pp. 197, 200–206)

Explain how plate tectonics and the rock cycle shape the earth beneath our feet

- Matter is cycled within the lithosphere, and rocks transform from one type to another. (pp. 206–207)
- Plate tectonics is a fundamental system that produces earthquakes and volcanoes and guides Earth's physical geography. (pp. 207–209)

TESTING YOUR COMPREHENSION

1. Which type of feedback loop is most common in nature, and which more commonly results from human action? How might the emergence of a positive feedback loop affect a system in homeostasis?
2. Describe how hypoxic conditions can develop in coastal marine ecosystems such as the northern Gulf of Mexico.
3. What is the difference between an ecosystem and a community?
4. Describe the typical movement of energy through an ecosystem. Describe the typical movement of matter through an ecosystem.
5. What role do each of the following play in the carbon cycle?
    - Cars
    - Photosynthesis
    - The oceans
    - Earth's crust
6. Describe the difference between the function performed by nitrogen-fixing bacteria and that performed by denitrifying bacteria.
7. How has human activity altered the carbon cycle? The phosphorus cycle? The nitrogen cycle? To what environmental problems have these changes given rise?
8. What is the difference between evaporation and transpiration? Give examples of how the hydrologic cycle interacts with the carbon, phosphorus, and nitrogen cycles.
9. Name the three main types of rocks, and describe how each type may be converted to the others via the rock cycle.
10. How does plate tectonics account for mountains? For volcanoes? For earthquakes? Why do you think it took so long for scientists to discover such a fundamental environmental system as plate tectonics?

SEEKING SOLUTIONS

1. In this chapter we discussed how system boundaries can be difficult to determine. Can you think of a truly closed system whose boundaries are easily defined? If so, try to describe such a system and its boundaries.
2. Consider the ecosystem(s) that surround(s) your campus. How do some of the principles from our discussion on ecosystems apply to the ecosystem(s) around your campus?
3. A simple change in the flux rate between just two reservoirs in a single nutrient cycle can potentially have major consequences for ecosystems and, indeed, for the globe. Explain how this can be, using one example from the carbon cycle and one example from the nitrogen cycle.
4. How do you think we might solve the problem of eutrophication in the Gulf of Mexico? Assess several possible solutions, your reasons for believing they might work, and the likely hurdles we might face. Explain who should be responsible for implementing solutions, and why.
5. Imagine that you are a shrimper on the Louisiana coast and that your income is decreasing year by year because the dead zone is making it harder and harder to catch shrimp. One day your senator comes to town, and you have a one-minute audience with her. What steps would you urge her to take in Washington, D.C., to try to help alleviate the dead zone and bring back the shrimp fishery?
6. Now imagine that you are an Iowa farmer and that you have learned that the federal government is insisting that you use 30% less fertilizer on your crops each year. You know that in good growing years you could do without that fertilizer, and you'd be glad not to have to pay for it. But in bad growing years, you need the fertilizer to ensure a harvest so that you can continue making a living. And you must apply the fertilizer each spring before you know whether it will be a good or bad year. What would you tell your senator when she comes to town?

INTERPRETING GRAPHS AND DATA

Scientists are debating what effects global climate change (Chapter 18) may have on nutrient cycles. As soil becomes warmer, especially at far northern latitudes, nutrients in the soil should become more available to plants, stimulating plant growth. One hypothesis is that more carbon will end up stored in the soil as a result, because plants will pull carbon from the atmosphere and transfer it to the soil reservoir as they shed leaves or die.[1] Under this hypothesis, increased flux of carbon from the atmosphere to the soil would act as negative feedback counteracting climate warming, because less carbon in the atmosphere would lead to less warming.

To test whether the carbon flux actually changes in this way when nutrients are made more available in a tundra ecosystem, researchers are conducting a long-term study in Alaska.[2] For 20 years, they have added fertilizer to treatment plots while leaving control plots unfertilized. Recently, they estimated amounts of carbon by measuring biomass aboveground and belowground in both sets of plots. Aboveground biomass consists of living plant material, whereas belowground biomass consists mostly of nonliving organic material stored in the soil and not yet decomposed. Some of the research team's results are presented in the graph below.

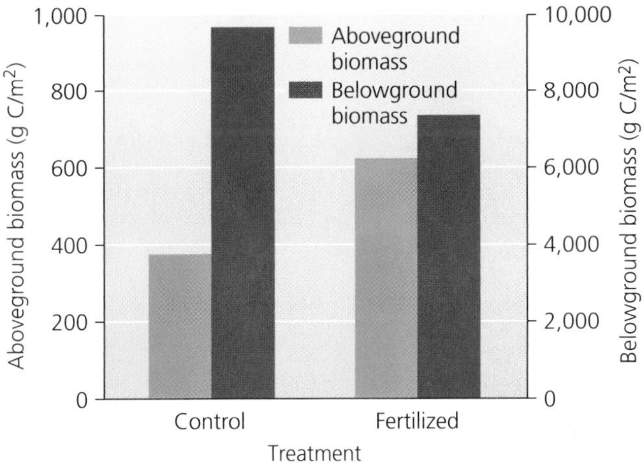

Effects of 20 years of fertilization (10 g N and 5 g P/m^2/yr) on C pools in tundra near Toolik Lake, Alaska. Differences are statistically significant for the aboveground, belowground, and total carbon pools.

1. Calculate the sizes of the aboveground, belowground, and total carbon pools (in g C/m^2) for the Control and Fertilized treatment groups.
2. What do the aboveground data indicate about the effect of fertilizer on plant growth? What do the belowground data indicate about the effect of fertilizer on organic material stored in the soil?
3. Do the data support the hypothesis that the net effect of increased nutrient availability will be to remove carbon from the atmosphere and store it in the soil? Using Figure 7.9 as a reference, can you suggest a different hypothesis that might explain the data better? Based on this data, would you predict that the warming of tundra soil will decrease the atmospheric concentration of CO_2 and act as negative feedback to climate change, or increase it and act as positive feedback?

[1]Hobbie, S. E., et al. 2002. A synthesis: The role of nutrients as constraints on carbon balances in boreal and arctic regions. *Plant Soil* 242: 163–170.
[2]Mack, M. C., et al. 2004. Ecosystem carbon storage in arctic tundra reduced by long-term nitrogen fertilization. *Nature* 431: 440–443.

CALCULATING ECOLOGICAL FOOTPRINTS

In the United States, a common dream is to own your own home, surrounded by a weed-free, green lawn. Nationwide there are about 20 million acres of lawn grass, making it our largest single crop!

Assuming that all of the populations indicated in the following table have lawns and will fertilize them at a typical fertilizer application rate of 45 lb per acre, calculate the total amount of nitrogen that will be applied to their lawns. When estimating the number of lawns, assume that the typical household includes three people.

Fertilizer application	Number of lawns	Pounds of nitrogen
To your 1/3-acre lawn	1	15
To the lawns of your classmates		
To the lawns of all your schoolmates		
To all the lawns in your hometown		
To all the lawns in your state		
To all the lawns in the United States	60,000,000	

1. Where does all of this nitrogen go? Where does it come from?
2. What other environmental impacts are caused by fertilizer production, transport, and application?

Take It Further

Go to www.aw-bc.com/withgott or the student CD-ROM where you'll find:

- Suggested answers to end-of-chapter questions
- Quizzes, animations, and flashcards to help you study
- *Research Navigator*™ database of credible and reliable sources to assist you with your research projects
- GRAPHit! tutorials to help you master how to interpret graphs
- INVESTIGATEit! current news articles that link the topics that you study to case studies from your region to around the world

PART TWO

Environmental Issues and the Search for Solutions

Canal Street, New Orleans, after Hurricane Katrina

8 Human Population

Hong Kong, China

Upon successfully completing this chapter, you will be able to:

- Assess the scope of human population growth
- Evaluate how human population, affluence, and technology affect the environment
- Explain and apply the fundamentals of demography
- Outline and assess the concept of demographic transition
- Describe how wealth and poverty, the status of women, and family planning programs affect population growth
- Characterize the dimensions of the HIV/AIDS epidemic

Billboard in Chengdu, China, promoting One-Child Policy

Central Case: China's One-Child Policy

"Population growth is analogous to a plague of locusts. What we have on this earth today is a plague of people."
—Ted Turner, media magnate and supporter of the United Nations Population Fund

"There is no population problem."
—Sheldon Richman, senior editor, Cato Institute

The People's Republic of China is the world's most populous nation, home to one-fifth of the 6.5 billion people living on Earth at the start of 2006. When Mao Zedong founded the country's current regime 57 years earlier, roughly 540 million people lived in a mostly rural, war-torn, impoverished nation. Mao believed population growth was desirable, and under his leadership China grew and changed. By 1970, improvements in food production, food distribution, and public health allowed China's population to swell to approximately 790 million people. At that time, the average Chinese woman gave birth to 5.8 children in her lifetime.

Unfortunately, the country's burgeoning population and its industrial and agricultural development were eroding the nation's soils, depleting its water, leveling its forests, and polluting its air. Chinese leaders realized that the nation might not be able to feed its people if their numbers grew much larger. They saw that continued population growth could exhaust resources and threaten the stability and economic progress of Chinese society. The government decided to institute a population-control program that precluded large numbers of Chinese couples from having more than one child.

The program began with education and outreach efforts encouraging people to marry later and have fewer children. Along with these efforts, the Chinese government increased the accessibility of contraceptives and abortion. By 1975, China's annual population growth rate had dropped from 2.8% to 1.8%. To further decrease birthrates, in 1979 the government took the more drastic step of instituting a system of rewards and punishments to enforce a one-child limit. One-child families received better access to schools, medical care, housing, and government jobs, and mothers with only one child were given longer maternity leaves. Families with more than one child, meanwhile, were subjected to social scorn and ridicule, employment discrimination, and monetary fines. In some cases, the fines exceeded half the offending couple's annual income.

In enforcing these policies, China has, in effect, been conducting one of the largest and most controversial social experiments in history. In purely quantitative terms, the experiment has been a major success; the nation's growth rate is now down to 0.6%, making it easier for the country to deal with its many social, economic, and environmental challenges. However, China's population control policies have also produced unintended consequences, such as widespread killing of female infants, an unbalanced sex ratio, and a black-market trade in teen-aged girls. Moreover, the policies have elicited intense criticism from those who oppose government intrusion into personal reproductive choices.

China embarked on its policy because its leaders felt it necessary. As other nations become more and more crowded, might their governments also feel forced to turn to drastic policies that restrict individual freedoms? In this chapter, we examine human population dynamics worldwide, consider their causes, and assess their consequences for the environment and our society.

Human Population Growth: Baby 6 Billion and Beyond

While China was working to slow its population growth and speed its economic growth, on the other side of the Eurasian continent, a milestone was reached in 1999. On the morning of October 12 of that year, the first cries of a newborn baby in Sarajevo, Bosnia-Herzegovina, marked the arrival of the six-billionth human being on our planet (Figure 8.1). At least that was how the milestone was symbolically marked by the United Nations, which monitors human population growth, among other global trends.

Just how much is 6 billion? We often have trouble conceptualizing the scale of huge numbers like a billion. Keep in mind that a billion is 1,000 times greater than a million. If you were to count once each second without ever sleeping, it would take over 30 years to reach a billion. In order to put a billion miles on your car, you would need to drive from New York to Los Angeles more than 350,000 times.

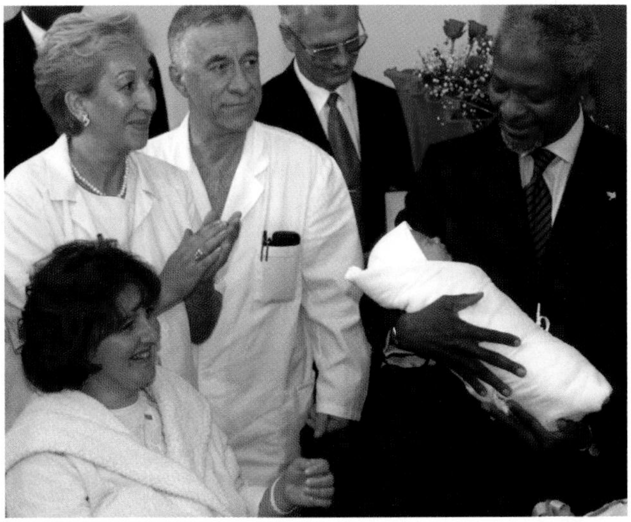

FIGURE 8.1 U.N. Secretary-General Kofi Annan recognized the newborn son of Fatima Nevic and her husband, Jasminko, as our six-billionth neighbor. Although it is impossible to know the precise moment—or day, week, or even month—the world's population reached 6 billion, U.N. population experts pinpointed October 12, 1999, as the best approximation to make the symbolic declaration. Many observers interpreted the selection of a child born in war-ravaged Sarajevo as a harbinger of the hard times that could face future generations as population grows and competition for scarce resources increases.

The human population is growing nearly as fast as ever

As we saw in Chapter 1 (▸pp. 4–5), the human population has been growing at a tremendous rate. Our population has doubled just since 1964 and is growing by roughly 78 million people annually (nearly 2.5 people every *second*). This is the equivalent of adding all the people of California, Texas, and New York to the world each year. It took until after 1800, virtually all of human history, for our population to reach 1 billion. Yet we reached 2 billion by 1930, and 3 billion in just 30 more years, in 1960. Our population added its next billion in just 15 years (1975), its next billion in a mere 12 years (1987), and its most recent billion in another 12 years (Figure 8.2). Think about when you were born and how many people have been added to the planet just since that time. This unprecedented growth means that today's generations are in circumstances that previous generations never experienced. Our grandparents never had to deal with the number of people that crowd our planet today.

How and why has our growth accelerated? We saw in Chapter 5 (▸p. 136) how exponential growth—the increase in a quantity by a *fixed percentage* per unit time—accelerates the absolute increase of population size over time, just as compound interest accrues in a savings

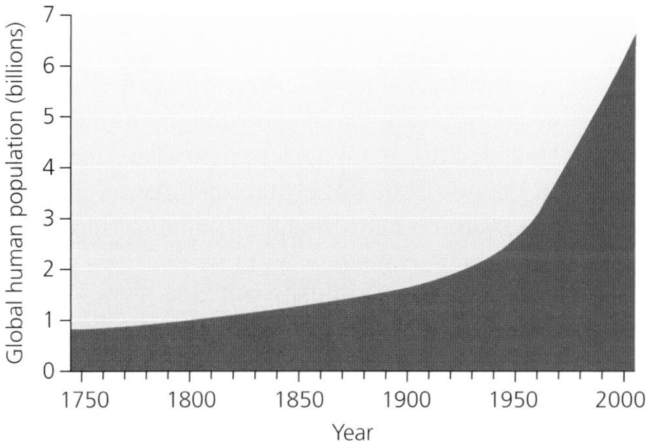

FIGURE 8.2 The global human population has grown exponentially, rising from less than 1 billion in 1800 to over 6.5 billion today. Data from U.S. Bureau of the Census.

account. The reason, you will recall, is that a given percentage of a large number is a greater quantity than the same percentage of a small number. Thus, even if the growth rate remains steady, population size will increase by greater increments with each successive generation.

In fact, our growth rate has not remained steady. Instead, for much of the 20th century, the growth rate of the human population actually rose from year to year. It peaked at 2.1% during the 1960s and has declined to 1.2% since then. Although 1.2% may sound small, exponential growth endows small numbers with large consequences. For instance, a hypothetical population starting with 1 man and 1 woman that grows at 1.2% gives rise to a population of 2,939 after 40 generations and 112,695 after 60 generations. In today's world, rates of annual growth vary greatly from region to region. Figure 8.3 maps this variation.

At a 2.1% annual growth rate, a population doubles in size in only 33 years. For low rates of increase, we can estimate doubling times with a handy rule-of-thumb. Just take the number 70, and divide it by the annual percentage growth rate: $70 \div 2.1 = 33.3$. Had China not instituted its one-child policy—that is, had its growth rate remained unchecked at 2.8%—it would have taken only 25 years to double in size. Had population growth continued at this rate, China's population would have surpassed 2 billion people in 2004.

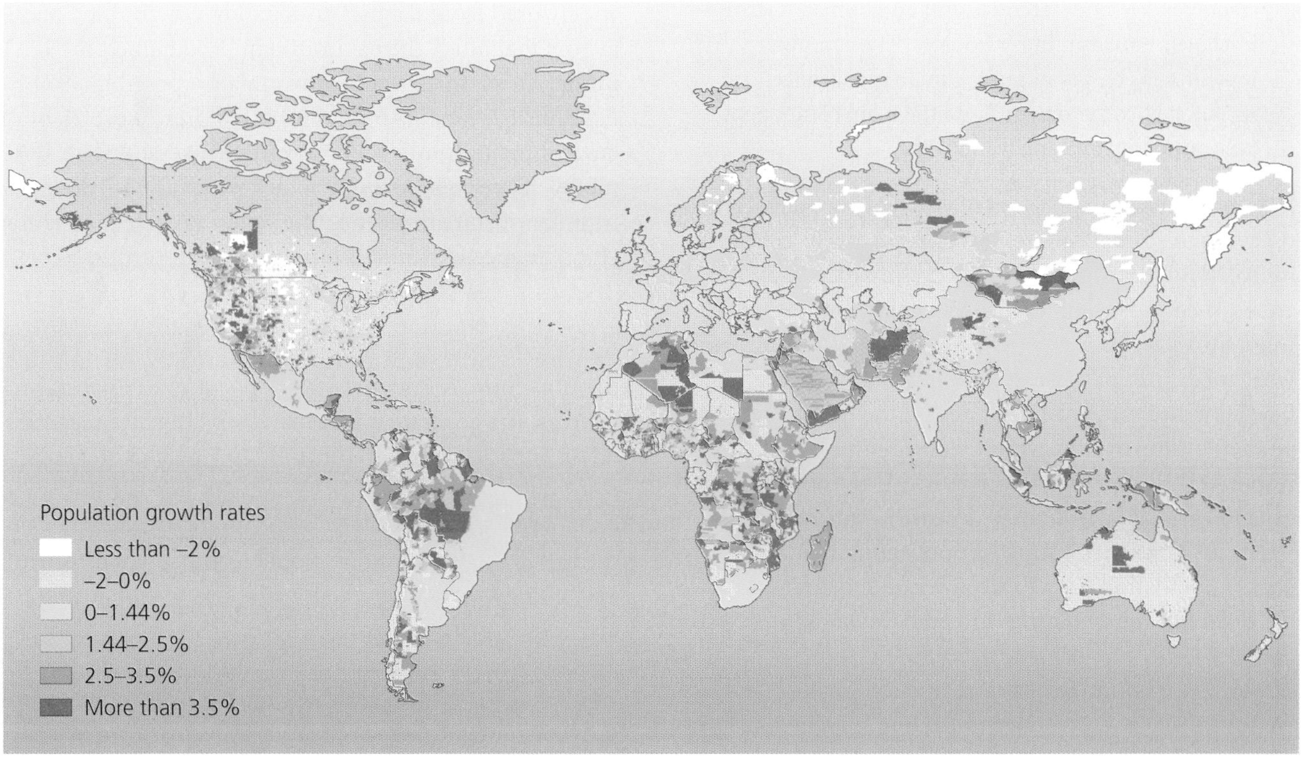

FIGURE 8.3 A map of population growth rates from the period 1990–1995 shows great variation from place to place. Population is growing fastest in tropical regions and in some desert and rainforest areas that have historically been sparsely populated. Data from Center for International Earth Science Information Network (CIESIN), Columbia University; and Harrison, P., and F. Pearce. 2000. *AAAS atlas of population and environment.* Berkeley, CA: University of California Press.

Is population growth really a "problem"?

Our ongoing population growth has resulted largely from technological innovations, improved sanitation, better medical care, increased agricultural output, and other factors that have led to a decline in death rates, particularly a drop in rates of infant mortality. Birth rates have not declined as much, so births have outpaced deaths for many years now. Thus, the so-called population problem actually arises from a very good thing—our ability to keep more of our fellow humans alive longer.

Indeed, just as the mainstream view in the day of Thomas Malthus (▸pp. 4–5) held that population increase was a good thing, there are many people today who argue that population growth poses no problems. Under the Cornucopian view that many economists hold, resource depletion due to population increases is not a problem if new resources can be found to replace depleted resources (▸p. 43). Libertarian writer Sheldon Richman expressed this view at the time the six-billionth baby was born:

> The idea of carrying capacity doesn't apply to the human world because humans aren't passive with respect to their environment. Human beings create resources. We find potential stuff and human intelligence turns it into resources. The computer revolution is based on sand; human intelligence turned that common stuff into the main component [silicon] of an amazing technology.

In contrast to Richman's point of view, environmental scientists recognize that few resources are actually created by humans and that not all resources can be replaced once they are depleted. For example, once species have gone extinct, we cannot replicate their exact function in ecosystems, or know what medicines or other practical applications we might have obtained from them, or regain the educational and aesthetic value of observing them. Another irreplaceable resource is land, that is, space in which to live; we cannot expand Earth like a balloon to increase its surface area.

Even if resource substitution could hypothetically enable population growth to continue indefinitely, could we maintain the *quality* of life that we would desire for ourselves and our descendants? Surely some of today's resources are bound to be easier or cheaper to use, and less environmentally destructive to harvest or mine, than any resources that can replace them. Replacing such resources might make our lives more difficult or less pleasant. In any case, unless resource availability keeps pace with population growth, the average person in the future will have less space in which to live, less food to eat, and less material wealth than the average person does today. Thus population increases are indeed a problem if they create stress on resources, social systems, or the natural environment, such that our quality of life declines.

Despite these considerations—and despite the fact that in today's world population growth is correlated with poverty, not wealth—many governments have found it difficult to let go of the notion that population growth increases a nation's economic, political, or military strength. Many national governments, even those that view global population increase as a problem, still offer financial and social incentives that encourage their own citizens to produce more children. Governments of countries currently experiencing population declines (such as many in Europe) feel especially uneasy. According to the Population Reference Bureau, more than 3 of every 5 European national governments now take the view that their birth rates are too low, and none state that theirs is too high. However, outside Europe, 56% of national governments feel their birth rates are too high, and only 8% feel they are too low.

Population is one of several factors that affect the environment

The extent to which population increase can be considered a problem involves more than just numbers of people. One widely used formula gives us a handy way to think about factors that affect the environment. Nicknamed the **IPAT model,** it is a variation of a formula proposed in 1974 by Paul Ehrlich (▸pp. 5–6) and John Holdren, a professor of environmental policy at Harvard University. The IPAT model represents how our total impact (I) on the environment results from the interaction among population (P), affluence (A), and technology (T):

$$I = P \times A \times T$$

Increased population intensifies impact on the environment as more individuals take up space, use natural resources, and generate waste. Increased affluence magnifies environmental impact through the greater per capita resource consumption that generally has accompanied enhanced wealth. Changes in technology may either decrease or increase human impact on the environment. Technology that enhances our abilities to exploit minerals, fossil fuels, old-growth forests, or ocean fisheries generally increases impact, but technology to reduce smokestack emissions, harness renewable energy, or improve manufacturing efficiency can decrease impact.

We might also add a sensitivity factor (S) to the equation to denote how sensitive a given environment is to human pressures:

$$I = P \times A \times T \times S$$

For instance, the arid lands of western China are more sensitive to human disturbance than the moist regions of southeastern China. Plants grow more slowly in the arid west, making deforestation and soil degradation more likely. Thus, adding an additional person to western China should have more environmental impact than adding one to southeastern China.

We could refine the IPAT equation further by adding terms for the effects of social institutions, such as education, laws and their enforcement, stable and cohesive societies, and ethical standards that promote environmental well-being. Factors like these all affect how population, affluence, and technology translate into environmental impact.

Impact can be thought of in various ways, but it can generally be boiled down to either pollution or resource consumption. Pollution became a problem in the modern world once our population grew large enough that we produced great quantities. The depletion of resources by larger and hungrier populations has been a focus of scientists and philosophers since Malthus's time. Recall how on Easter Island (▸ pp. 8–9), islanders brought down their own civilization by depleting their most important limited resource, trees. History offers other cases in which resource depletion helped end civilizations, from the Mayans to the Mesopotamians. Some environmental scientists have predicted similar problems for our global society in the near future if we do not manage to embark on a path toward sustainability (Figure 8.4).

However, as we noted in Chapter 1, Malthus and his "neo-Malthusian" followers have not yet seen their direst predictions come true. The reason is that we have developed technology—the T in the IPAT equation—time and again to alleviate our strain on resources and allow us to further expand our population. For instance, we have employed technological advances to increase global agricultural production faster than our population has risen (▸ pp. 278–279).

Modern-day China shows how all elements of the IPAT formula can combine to cause tremendous environmental impact in very little time. The world's fastest-growing economy over the past two decades, China is "demonstrating what happens when large numbers of poor people rapidly become more affluent," in the words of Earth Policy Institute president Lester Brown. While millions of Chinese are increasing their material wealth and their consumption of resources, the country is battling unprecedented environmental challenges brought

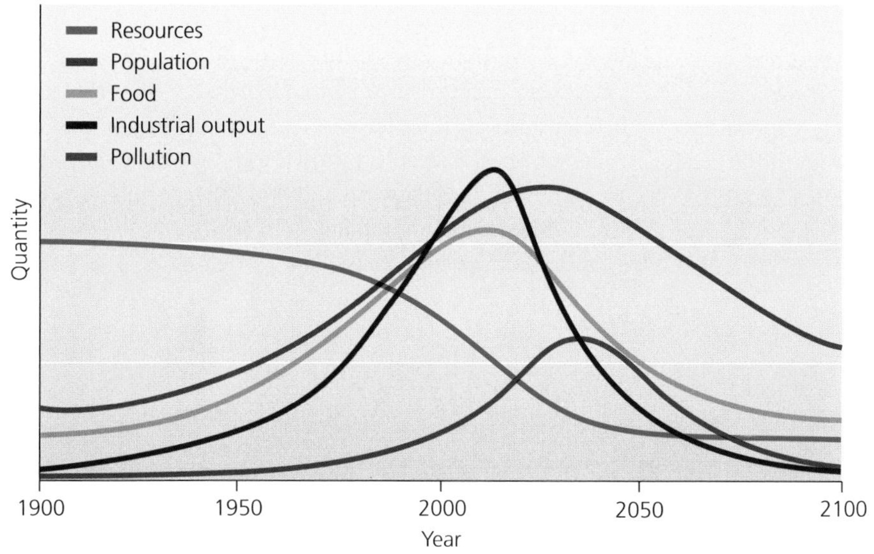

FIGURE 8.4 Environmental scientists Donella Meadows, Jorgen Randers, and Dennis Meadows used computer simulations to generate a series of projections of trends in human population, resource availability, food production, industrial output, and pollution. Their projections for trends over the coming century are based on data from the past century and current scientific understanding of the environment's biophysical limits. Shown here is their projection for a world in which "society proceeds in a traditional manner without any major deviation from the policies pursued during most of the twentieth century." In this projection, population and production increase until declining nonrenewable resources make further growth impossible, causing population and production to decline rather suddenly. The researchers also ran their simulations with different parameters to examine possible alternative futures. Under a scenario with policies aimed at sustainability, population leveled off at 8 billion, production and resource availability leveled off at medium-high levels, and pollution declined to low levels. Data from Meadows, D., et al. 2004. *Limits to growth: The 30-year update.* White River Junction, VT: Chelsea Green Publishing.

about by its pell-mell economic development. Intensive agriculture has expanded westward out of the country's historic moist rice-growing areas, causing farmland to erode and literally blow away, much like the Dust Bowl tragedy that befell the U.S. agricultural heartland in the 1930s (▸ p. 260). China has overpumped many of its aquifers and has drawn so much water for irrigation from the Yellow River that the once-mighty waterway now dries up in many stretches. Although China has been reducing its air pollution from industry and charcoal-burning homes, the country faces new urban pollution and congestion threats from rapidly increasing numbers of automobiles. As the world's developing countries try to attain the level of material prosperity that industrialized nations enjoy, China is a window on what much of the rest of the world could soon become.

Weighing the Issues: Population Growth and Reproductive Freedom

It is often suggested that if human population growth remains unchecked, everyone will eventually suffer a poorer quality of life. Would you be willing to make this sacrifice if it meant that people in other countries (such as China) could avoid government-imposed limitations on their reproductive freedom? If your own government ever implemented a strict reproductive policy, how would you feel? Would you rather have the government limit your reproductive freedom or your consumption?

Demography

As we have seen, it is a fallacy to think of people as being somehow outside nature. Humans exist within their environment as one species out of many. As such, all the principles of population ecology we outlined in Chapter 5 that apply to toads, frogs, and passenger pigeons apply to humans as well. Environmental factors set limits on our population growth, and the environment has a carrying capacity (▸ pp. 136–137) for our species, just as it does for every other.

We happen to be a particularly successful organism, however—one that has repeatedly raised its carrying capacity by developing technology to overcome the natural limits on its growth. We did so with the agricultural and the industrial revolutions (▸ p. 4) and likely before that with our invention of tools (Figure 8.5).

Environmental scientists who have tried to pin a number to the human carrying capacity have come up with wildly differing estimates. The most rigorous estimates range from 1–2 billion people living prosperously in a healthy environment to 33 billion living in extreme poverty in a degraded world of intensive cultivation without natural areas. As our population climbs toward 7 billion and beyond, we may yet continue to find ways to raise our carrying capacity. Given our knowledge of population ecology, however, we have no reason to presume that human numbers can go on growing indefinitely. Indeed, as we have seen (see Figure 5.17d, ▸ p. 138), populations that exceed their carrying capacity can crash.

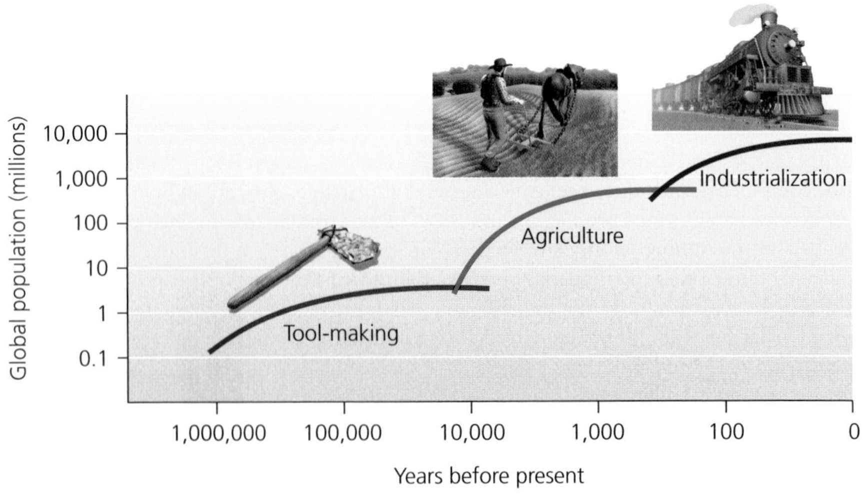

FIGURE 8.5 Tool making, the advent of agriculture, and industrialization each allowed our species to raise its global carrying capacity. The logarithmic scale of the axes makes it easier to visualize this pattern. Data from Goudie, A. 2000. *The human impact*. Cambridge, MA: MIT Press.

Demography is the study of human population

The application of population ecology principles to the study of statistical change in human populations is the focus of the social science of **demography.** The field of demography developed along with and partly preceding population ecology, and the disciplines have influenced and borrowed from one another. Data gathered by demographers help us understand how differences in population characteristics and related phenomena (for instance, decisions about reproduction) affect human communities and their environments. Demographers study population size, density, distribution, age structure, sex ratio, and rates of birth, death, immigration, and emigration of humans, just as population ecologists study these characteristics in other organisms. Each of these characteristics is useful for predicting population dynamics and potential environmental impacts.

Population size The global human population of more than 6.5 billion consists of well over 200 nations with populations ranging from China's 1.3 billion, India's 1.1 billion, and the 300 million of the United States down to a number of island nations with populations below 100,000 (Figure 8.6). The size that our global population will eventually reach remains to be seen (Figure 8.7). However, population size alone—the absolute number of individuals—doesn't tell the whole story. Rather, a population's environmental impact depends on its density, distribution, and composition (as well as on affluence, technology, and other factors outlined earlier).

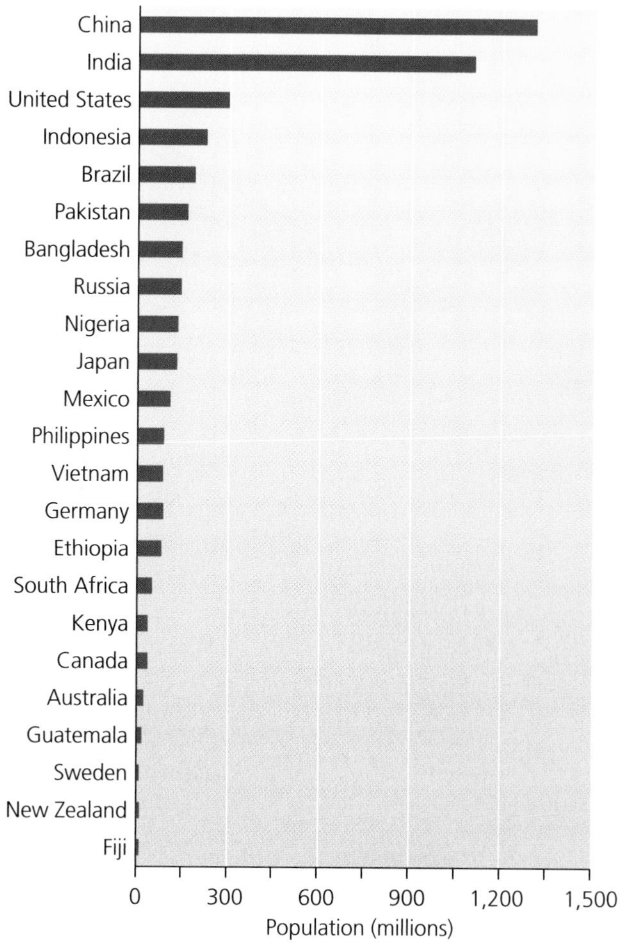

FIGURE 8.6 The world's nations range in human population from several thousand (on some South Pacific islands) up to China's 1.3 billion. Shown here are the 2005 populations for the world's most populous 15 countries, followed by a selection of other countries. Data from Population Reference Bureau. 2005. *2005 world population data sheet.*

Population density and distribution People are distributed very unevenly over the globe. In ecological terms, our distribution is clumped (▸ pp. 133–134) at all spatial scales. At the largest scales (Figure 8.8), population density is high in regions with temperate, subtropical, and tropical climates, such as China, Europe, Mexico, southern Africa, and India. Population density is low in regions with extreme-climate biomes, such as desert, deep rainforest, and tundra. Dense along seacoasts and rivers, human population is less dense at locations far from water. At intermediate scales, we cluster together in cities and suburbs and are spread more sparsely across rural areas. At small scales, we cluster in certain neighborhoods and in individual households.

This uneven distribution means that certain areas bear far more environmental impact than others. Just as the Yellow River has experienced intense pressure from millions of Chinese farmers, the world's other major rivers, from the Nile to the Danube to the Ganges to the Mississippi, have all received more than their share of human impact. The urban way of life entails the packaging and transport of goods, intensive fossil fuel consumption, and hotspots of pollution. However, people's concentration in cities relieves pressure on ecosystems in less-populated areas by releasing some of them from direct human development (▸ p. 395).

At the same time, areas with low population density are often vulnerable to environmental impacts, because the reason they have low populations in the first place is that they are sensitive and cannot support many people (a high S value in our revised IPAT model). Deserts, for instance, are easily affected by development that commandeers a substantial share of available water. Grasslands can be turned to deserts if they are farmed too intensively, as has happened across vast stretches of the Sahel region

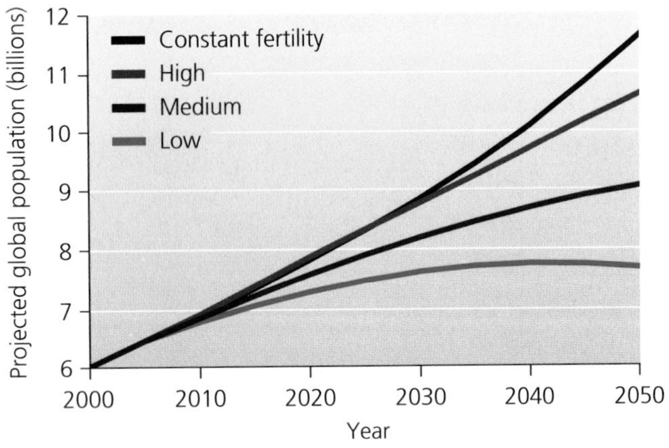

FIGURE 8.7 The United Nations predicts trajectories of world population growth, presenting its estimates in several scenarios based on different assumptions of fertility rates. In this 2004 projection, population is estimated to reach 11.7 billion in the year 2050 if fertility rates remain constant at 2004 levels (top line in graph). However, U.N. demographers expect fertility rates to continue falling, so they arrived at a best guess (*medium* scenario) of 9.1 billion for the human population in 2050. In the *high* scenario, if women on average have 0.5 child more than in the medium scenario, population will reach 10.6 billion in 2050. In the *low* scenario, if women have 0.5 child less than in the medium scenario, the world will contain 7.7 billion people in 2050. Data from United Nations Population Division. 2004. *World population prospects: The 2004 revision.*

bordering Africa's Sahara Desert, in the Middle East, and in parts of China and the United States.

Age structure Data on the age structure or age distribution of human populations are especially valuable to demographers trying to predict future dynamics of populations. As we saw in Chapter 5 (▸p. 134), large proportions of individuals in young age groups portend a great deal of reproduction and, thus, rapid population growth. Examine age pyramids for the nations of Canada and Madagascar (Figure 8.9). Not surprisingly, it is Madagascar that has the greater population growth rate. In fact, its annual growth rate, 2.7%, is 9 times that of Canada's 0.3%.

By causing dramatic reductions in the number of children born since 1970, China virtually guaranteed that its population age structure would change. Indeed, in 1995 the median age in China was 27; by 2030 it will be 39. In 1997 there were 125 children under age 5 for every 100 people 65 or older in China, but by 2030 there will be only 32. The number of people older than 65 will rise from 100 million in 2005 to 236 million in 2030 (Figure 8.10). This dramatic shift in age structure will challenge China's economy, health care systems, families, and military forces because fewer working-age people will be available to support social programs that assist the increasing number of older people. However, the shift in age structure also reduces the proportion of dependent children. The reduced

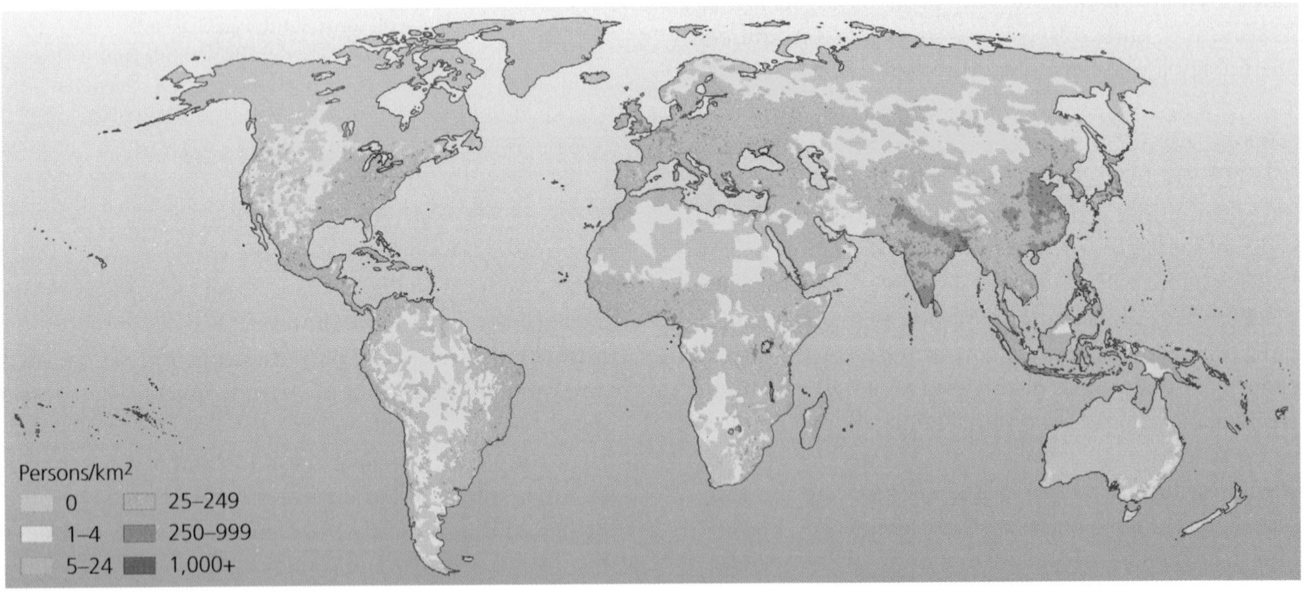

FIGURE 8.8 Human population density varies tremendously from one region to another. Arctic and desert regions have the lowest population densities, whereas areas of India, Bangladesh, and eastern China have the densest populations. Data are for 2000, from Center for International Earth Science Information Network (CIESIN), Columbia University; and Centro Internacional de Agricultura Tropical (CIAT), 2004.

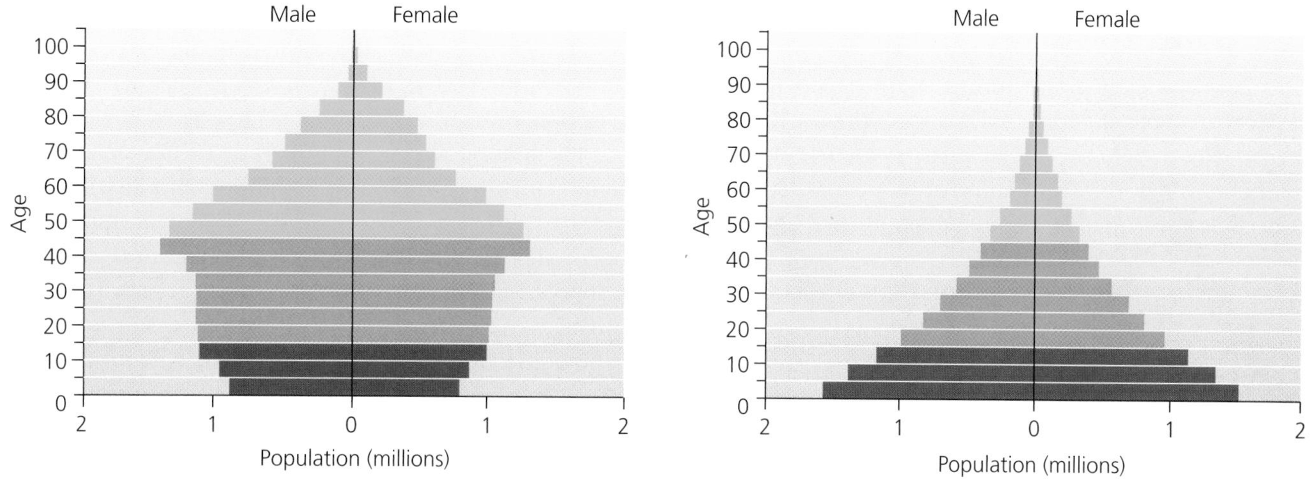

(a) Age pyramid of Canada in 2005

(b) Age pyramid of Madagascar in 2005

FIGURE 8.9 Canada (**a**) shows a balanced age structure, with relatively even numbers of individuals in various age classes. Madagascar (**b**) shows an age distribution heavily weighted toward young people. Madagascar's population growth rate is 9 times that of Canada's. Go to GRAPHit! at www.aw-bc.com/withgott or on the student CD-ROM. Data from U.N. Population Division.

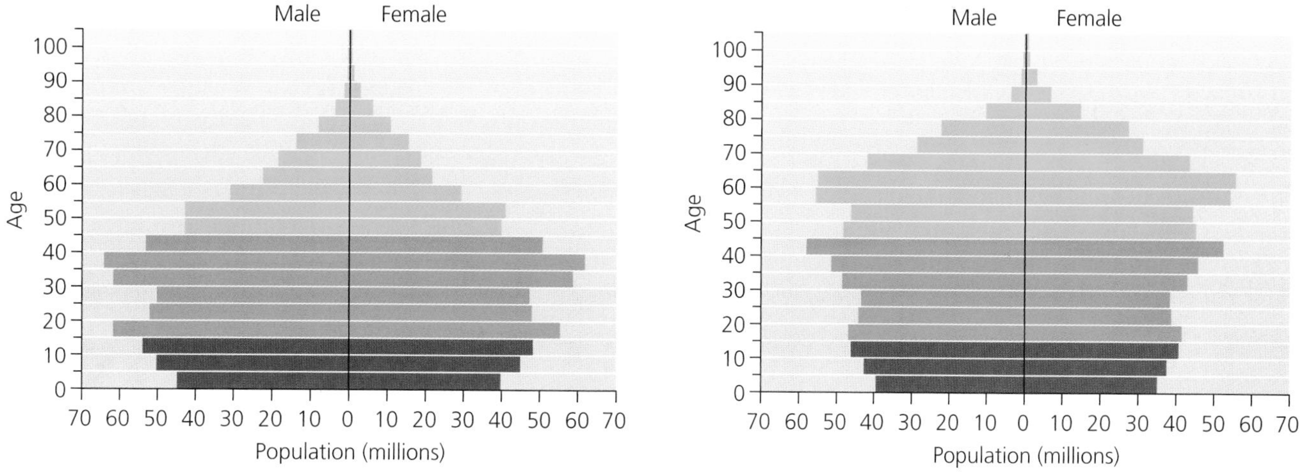

(a) Age pyramid of China in 2005

(b) Projected age pyramid of China in 2030

FIGURE 8.10 As China's population ages, older people will outnumber the young. Age pyramids show the predicted graying of the Chinese population between 2005 (**a**) and 2030 (**b**). Today's children may, as working-age adults (**c**), face pressures to support greater numbers of older citizens than has any previous generation. Data from U.N. Population Division.

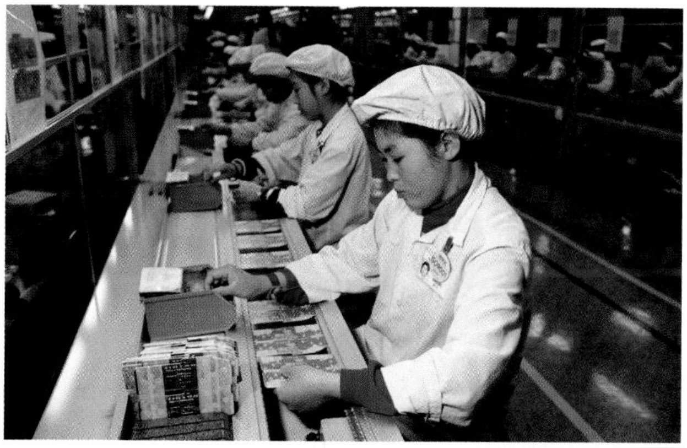

(c) Young female factory workers in Hong Kong

number of young adults may mean a decrease in the crime rate. Moreover, older people are often productive members of society, contributing volunteer activities and services to their children and grandchildren. Clearly, in terms of both benefits and drawbacks, life in China will continue to be profoundly affected by the particular approach its government has taken to population control.

Weighing the Issues: China's Reproductive Policy

Consider the benefits as well as the problems associated with a reproductive policy such as China's. Do you think a government should be able to enforce strict penalties for citizens who fail to abide by such a policy? If you disagree with China's policy, what alternatives can you suggest for dealing with the resource demands of a quickly growing population?

This pattern of aging in the population is occurring in many countries, including the United States (Figure 8.11). Older populations will present new challenges for many nations, as increasing numbers of older people require the care and financial assistance of relatively fewer working-age citizens.

Sex ratios The ratio of males to females also can affect population dynamics. Imagine two islands, one populated by 99 men and 1 woman and the other by 50 men and 50 women. Where would we be likely to see the greatest population increase over time? Of course, the island with an equal number of men and women would have a greater number of potential mothers and thus a greater potential for population growth.

The naturally occurring sex ratio in human populations at birth features a slight preponderance of males; for every 100 female infants born, 105 to 106 male infants are born. This phenomenon may be an evolutionary adaptation to the fact that males are slightly more prone to death during any given year of life. It usually ensures that the ratio of men to women is approximately equal at the time people reach reproductive age. Thus, a slightly uneven sex ratio at birth may be beneficial. However, a greatly distorted ratio can lead to problems.

In recent years, demographers have witnessed an unsettling trend in China: The ratio of newborn boys to girls has become strongly skewed. In the 2000 census, 120 boys were reported born for every 100 girls. Some provinces reported sex ratios as high as 138 boys for every 100 girls. A leading hypothesis for these unusual sex ratios is that many parents, having learned the sex of their fetuses by ultrasound, are selectively aborting female fetuses. Traditionally, Chinese culture has valued sons because they can carry on the family name, assist with farm labor in rural areas, and care for aging parents. Daughters, in contrast, will most likely marry and leave their parents, as the culture dictates. As a result, they will not provide the same benefits to their parents as will sons. Sociologists hold that this cultural gender preference, combined with the government's one-child policy, has led some couples to selectively abort female fetuses or to abandon or kill female infants.

China's skewed sex ratio may have the effect of further lowering population growth rates. However, it has proved tragic for the "missing girls." It is also beginning to have the undesirable social consequence of leaving many Chinese men single. This, in turn, has resulted in a grim new phenomenon. In parts of rural China, teen-aged girls are being kidnapped and sold to families in other parts of the country as brides for single men.

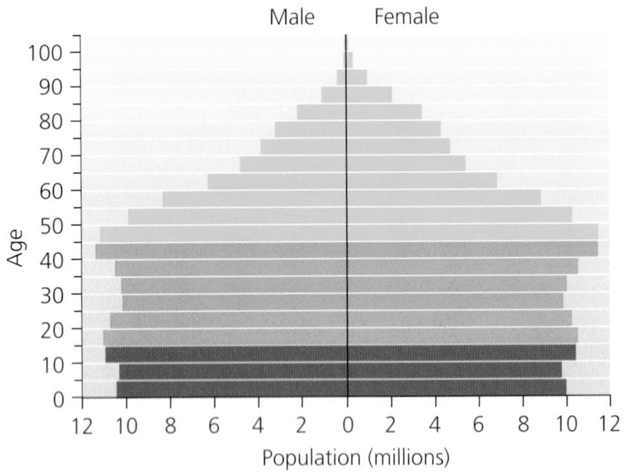

FIGURE 8.11 The "baby boom" is visible in the 2005 age pyramid for the United States, in the age brackets between 40 and 50. In future years the nation will experience an aging population as baby-boomers grow older. Data from U.N. Population Division.

Population growth depends on rates of birth, death, immigration, and emigration

Just as they do for other organisms, rates of birth, death, immigration, and emigration help determine whether a human population grows, shrinks, or remains stable. The formula for measuring population growth that we used in Chapter 5 (▸p. 135) also pertains to humans: birth and immigration add individuals to a population, whereas death and emigration remove individuals. Technological advances have led to a dramatic decline in human death rates, widening the gap between birth rates and death rates and resulting in the global human population expansion.

In today's ever-more-crowded world, immigration and emigration are playing increasingly large roles. Refugees, people forced to flee their home country or region, have become more numerous in recent decades as a result of war, civil strife, and environmental degradation. The United Nations puts the number of refugees who flee to escape poor environmental conditions at 25 million per year and possibly many more. Often the movement of refugees causes environmental problems in the receiving region as these desperate victims try to eke out an existence with no livelihood and with no cultural or economic attachment to the land or incentive to conserve its resources. The millions who fled Rwanda following the genocide there in the mid-1990s, for example, inadvertently destroyed large areas of forest while trying to obtain fuelwood, food, and shelter to stay alive once they reached the Democratic Republic of Congo (Figure 8.12).

For most of the past 2,000 years, China's population has been relatively stable. The first significant increases resulted from enhanced agricultural production and a powerful government during the Qing, or Manchu, Dynasty in the 1800s. Population growth began to outstrip food supplies by the mid-1850s, and quality of life for the average Chinese peasant began to decline. From the mid-1800s, an era of increased European intervention in China, until 1949, China's population grew very slowly, at about 0.3% per year. This slow population growth was due, in part, to food shortages and political instability. As we have seen, population growth rates rose again following Mao's establishment of the People's Republic, and they have declined since the establishment of the one-child policy (Table 8.1).

Table 8.1 Trends in China's Population Growth

Measure	1950	1970	1990	2005
Total fertility rate	5.8	5.8	2.2	1.6
Rate of natural population increase (% per year)	1.9	2.6	1.4	0.6
Doubling time (years)	37	27	49	117

Data from China Population Information and Research Center, 2005; and Population Reference Bureau. 2005. *2005 World population data sheet.*

Since 1970, growth rates in many countries have been declining, even without population control policies, and the global growth rate has declined (Figure 8.13). This decline has come about, in part, from a steep drop in birth rates.

A population's total fertility rate influences population growth

One key statistic demographers calculate to examine a population's potential for growth is the **total fertility rate (TFR),** or the average number of children born per female member of a population during her lifetime. **Replacement fertility** is the TFR that keeps the size of a population stable. For humans, replacement fertility is equal to a TFR of 2.1. When the TFR drops below 2.1, population size, in the absence of immigration, will shrink.

FIGURE 8.12 The flight of refugees from Rwanda into the Democratic Republic of Congo in 1994 following the Rwandan genocide caused tremendous hardship for the refugees and tremendous stress on the environment into which they moved.

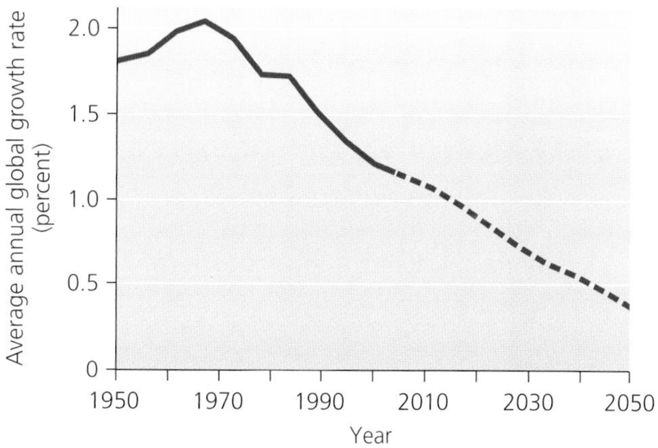

FIGURE 8.13 The annual growth rate of the global human population peaked in the 1960s and has been declining since then. The dashed line indicates projected future trends. Data from United Nations Population Division. 2004. *World population prospects: The 2004 revision.*

Table 8.2 Total Fertility Rates for Major Continental Regions

Region	Total fertility rate (TFR)
Africa	5.1
Latin America and the Caribbean	2.6
Asia	2.5
Oceania	2.1
North America	2.0
Europe	1.4

Data from Population Reference Bureau. 2005. *2005 World population data sheet.*

Various factors influence TFR and have acted to drive it downward in many countries in recent years. Historically, people tended to conceive many children, which helped ensure that at least some would survive, but lower infant mortality rates have made this less necessary. Increasing urbanization has also driven TFR down; whereas rural families need children to contribute to farm labor, in urban areas children are usually excluded from the labor market, are required to go to school, and impose economic costs on their families. If a government provides some form of social security, as most do these days, parents need fewer children to support them in their old age when they can no longer work. Finally, with greater education and changing roles in society, women tend to shift into the labor force, putting less emphasis on child rearing.

All these factors have come together in Europe, where TFR has dropped from 2.6 to 1.4 in the past half-century. Every European nation now has a fertility rate below the replacement level, and populations are declining in 18 of 43 European nations. In 2005, Europe's overall annual **natural rate of population change** (change due to birth and death rates alone, excluding migration) was –0.1%. Worldwide by 2005, a total of 71 countries had fallen below the replacement fertility of 2.1. These countries made up roughly 45% of the world's population and included China (with a TFR of 1.6). Table 8.2 shows the TFRs of major continental regions.

Weighing the Issues: Consequences of Low Fertility?

In the United States, Canada, and every European nation, the total fertility rate has now dipped below the replacement fertility rate. What economic or social consequences do you think might result from below-replacement fertility rates?

Some nations have experienced a change called the demographic transition

Many nations that have lowered their birth rates and TFRs have been going through a similar set of interrelated changes. In countries with good sanitation, good health care, and reliable food supplies, more people than ever before are living long lives. As a result, over the past 50 years the life expectancy for the average person has increased from 46 to 67 years as the global crude death rate has dropped from 20 deaths per 1,000 people to 9 deaths per 1,000 people. Strictly speaking, **life expectancy** is the average number of years that an individual in a particular age group is likely to continue to live, but often people use this term to refer to the average number of years a person can expect to live from birth. Much of the increase in life expectancy is due to reduced rates of infant mortality. Societies going through these changes are mostly the ones that have undergone urbanization and industrialization and have been able to generate personal wealth for their citizens.

To make sense of these trends, demographers developed a concept called the **demographic transition.** This is a model of economic and cultural change proposed in the 1940s and 1950s by demographer Frank Notestein and elaborated on by others to explain the declining death rates and birth rates that have occurred in Western nations as they became industrialized. Notestein believed nations moved from a stable pre-industrial state of high birth and death rates to a stable post-industrial state of low birth and death rates. Industrialization, he proposed, caused these rates to fall naturally by first decreasing mortality and then lessening the need for large families. Parents would thereafter choose to invest in quality of life rather than quantity of children. Because death rates fall before birth rates fall, a period of net population growth results. Thus, under the demographic transition model, population growth is seen as a temporary phenomenon that occurs as societies move from one condition to another.

The pre-industrial stage Notestein's demographic model describing the population impacts of industrialization proceeds in several stages (Figure 8.14). The first is the **pre-industrial stage,** characterized by conditions that have defined most of human history. In pre-industrial societies, both death rates and birth rates are high. Death rates are high because disease is widespread, medical care rudimentary, and food supplies unreliable and difficult to obtain. Birth rates are high because people must compensate for high mortality rates in infants and young children by having several children. In this stage, children are

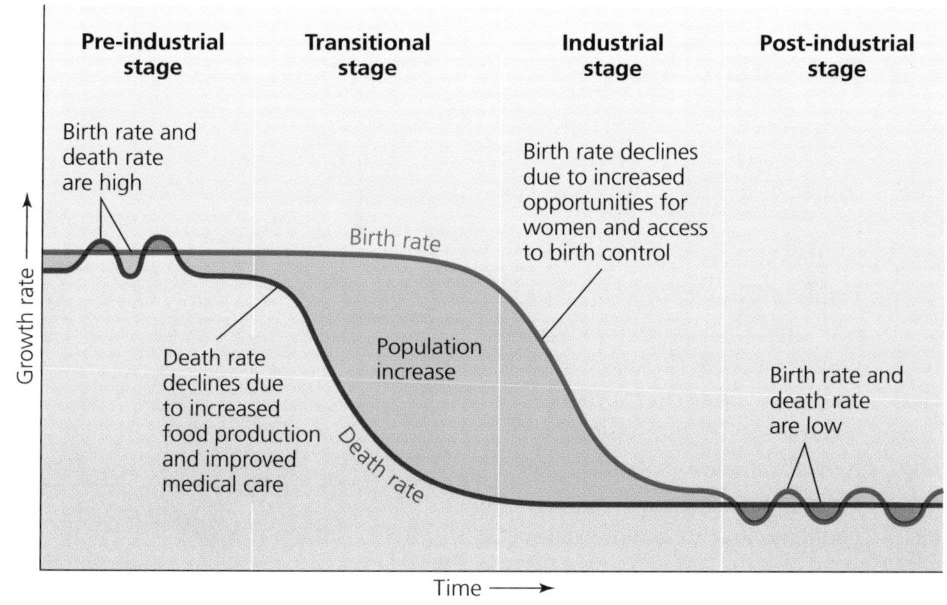
Pre-industrial stage
Transitional stage
Industrial stage
Post-industrial stage
Birth rate and death rate are high
Birth rate
Birth rate declines due to increased opportunities for women and access to birth control
Growth rate
Death rate declines due to increased food production and improved medical care
Population increase
Death rate
Birth rate and death rate are low
Time

FIGURE 8.14 The demographic transition is an idealized process that has taken some populations from a pre-industrial state of high birth rates and high death rates to a post-industrial state of low birth rates and low death rates. In this diagram, the wide green area between the two curves illustrates the gap between birth and death rates that causes rapid population growth during the middle portion of this process. Data from Kent, M. M. and K. A. Crews. 1990. *World population: Fundamentals of growth*. Population Reference Bureau.

valuable as additional workers who can help meet a family's basic needs. Populations within the pre-industrial stage are not likely to experience much growth, which is why the human population was relatively stable from Neolithic times until the industrial revolution.

Industrialization and falling death rates Industrialization initiates the second stage of the demographic transition, known as the **transitional stage.** This transition from the pre-industrial stage to the industrial stage is generally characterized by declining death rates due to increased food production and improved medical care. Birth rates in the transitional stage remain high, however, because people have not yet grown used to the new economic and social conditions. As a result, population growth surges.

The industrial stage and falling birth rates The third stage in the demographic transition is the **industrial stage.** Industrialization increases opportunities for employment outside the home, particularly for women. Children become less valuable, in economic terms, because they do not help meet family food needs as they did in the pre-industrial stage. If couples are aware of this, and if they have access to birth control, they may choose to have fewer children. Birth rates fall, closing the gap with death rates and reducing the rate of population growth.

The post-industrial stage In the final stage, the **post-industrial stage,** both birth and death rates have fallen to low and stable levels. Population sizes stabilize or decline slightly. The society enjoys the fruits of industrialization without the threat of runaway population growth.

Is the demographic transition a universal process?

The demographic transition has occurred in many European countries, the United States, Canada, Japan, and several other developed nations over the past 200 to 300 years. Nonetheless, it is a model that may not apply to all developing nations as they industrialize now and in the future. Some social scientists doubt that it will apply; they point out that population dynamics may be different for developing nations that adopt the Western world's industrial model rather than devising their own. Some demographers assert that the transition will fail in cultures that place greater value on childbirth or grant women fewer freedoms.

Moreover, natural scientists warn that there are not enough resources in the world to enable all countries to attain the standard of living that developed countries now enjoy. It has been estimated that for all nations to enjoy the quality of life that United States citizens enjoy, we would need the natural resources of two more planet Earths. Whether developing nations, which include the vast majority of the planet's people, pass through the demographic transition as developed nations have is one of the most important and far-reaching questions for the future of our civilization and Earth's environment.

Population and Society

Demographic transition theory links the statistical study of human populations with various societal factors that influence, and are influenced by, population dynamics. Let's now examine a few of these major societal factors more closely.

The Science behind the Story

Causes of Fertility Decline in Bangladesh

Research in developing countries indicates that poverty and overpopulation can create a vicious cycle, in which poverty encourages high fertility and high fertility obstructs economic development. Are there policy steps that such countries can take to bring down fertility rates? Scientific analysis of family-planning programs in the South Asian nation of Bangladesh suggests that the answer is yes.

Bangladesh is one of the poorest, most densely populated countries on the planet. Its 145 million people live in an area about the size of Wisconsin, and 45% of them live below the poverty line. With few natural resources and 1,000 people per km^2 (over $2,500/mi^2$—more than twice the population density of New Jersey), limiting population growth is critically important. As Bangladeshi president Ziaur Rahman declared in 1976, "If we cannot do something about population, nothing else that we accomplish will matter much."

Fortunately, Bangladesh has made striking progress in controlling population growth in the past three decades. Despite stagnant economic development, low literacy rates, poor health care, and limited rights for women, the nation's total fertility rate (TFR) has dropped markedly. In the 1970s, the average woman in Bangladesh gave birth to more than six children over the course of her life. Today, the TFR is 3.0.

Researchers hypothesized that family-planning programs were responsible for Bangladesh's rapid reduction in TFR. Because conducting an experiment to test such a hypothesis is difficult, some researchers took advantage of a natural experiment. By comparing Bangladesh to countries that are socioeconomically similar but have had less success in lowering TFR, such as Pakistan, researchers concluded that Bangladesh succeeded because of aggressive, well-funded outreach efforts that were sensitive to the values of its traditional society.

However, because no two countries are identical, it is difficult to draw firm conclusions from such broad-scale studies. This is why the Matlab Family Planning and Health Services Project, in the isolated rural area of Matlab, Bangladesh, has become one of the best-known experiments in family planning in developing countries. The Matlab Project was an intensive outreach program run collaboratively by the Bangladeshi government and international aid organizations. Each household in the project area received biweekly visits from local women offering counseling, education, and free contraceptives. Compared to a similar government-run program in a nearby area, the Matlab Project featured more training, more services, and more frequent visits. In both areas, a highly organized health surveillance system gave researchers detailed information about births, deaths, and health-related behaviors such as contraceptive use. The result was an experiment comparing the Matlab Project with the government-run area.

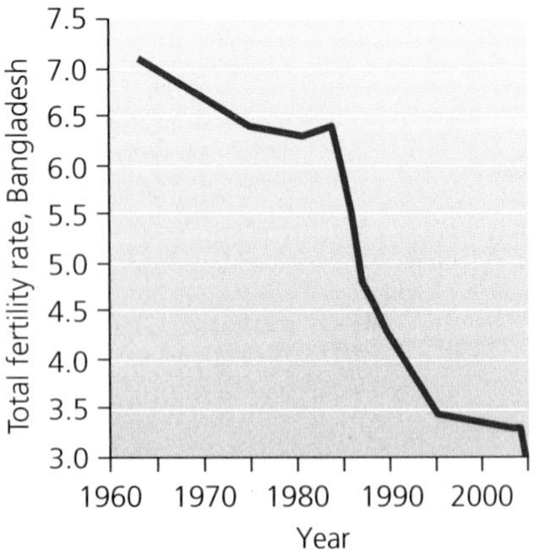

Total fertility rate has declined markedly in Bangladesh in the past 40 years, in part because of the enhanced availability of contraceptives. However, TFR has leveled off in recent years, suggesting that other societal changes are needed to lower it further. Data from Bangladesh Bureau of Statistics; Bangladesh Fertility Survey; The Global Reproductive Health Forum at Harvard; International Centre for Diarrhoeal Disease Research, Bangladesh; National Family Planning and Fertility Survey; United States Agency for International Development.

When Matlab Project director James Phillips and his colleagues

Women's empowerment greatly affects population growth rates

Many demographers had long believed that fertility rates were influenced largely by degrees of wealth or poverty. However, affluence alone cannot determine TFR, because a number of developing countries now have fertility rates lower than that of the United States. Instead, recent research is highlighting factors pertaining to the social empowerment of women. Drops in TFR have reviewed a decade's worth of data in 1988, they found that fertility rates had declined in both areas. The decline appeared to be due almost entirely to a rise in contraceptive use, because other factors—such as the average age of marriage—remained the same. Phillips and his colleagues also found that the declines had been significantly greater in the Matlab area than in the government-run area. These findings suggested that high-intensity outreach efforts can affect fertility rates even in the absence of significant improvements in women's status, education, or economic development.

But why exactly was the outreach program successful? One hypothesis was that visits from health care workers had helped convince local women that small families are desirable. However, in 1999, Mary Arends-Kuenning, a graduate student in economics at the University of Michigan, and her colleagues reported that there was no relationship between women's perception of the ideal family size and the number of visits made by outreach workers, either in Matlab or nearby comparison areas. Ideal family size declined equally in all areas. Instead of creating new demand for birth control, the Matlab Project appears to have helped women convert an already-existing desire for fewer children into behaviors, such as contraceptive use, that reduce fertility.

In the Matlab Project, Bangladeshi households received visits from local women offering counseling, education, and free contraceptives.

Bangladesh's ability to rein in fertility rates despite unfavorable social and economic conditions bodes well for impoverished nations facing explosive population growth. However, significant challenges remain. Since the 1990s, Bangladesh's TFR has appeared to level off at slightly more than 3 children per woman. If rates fail to decline further, the country's population could double to 290 million—nearly the size of today's U.S. population—within 30 years. Scientific research has helped illuminate the impact of family-planning programs on fertility, but further reductions may require fundamental social, political, and economic changes that are difficult to implement in traditional, resource-strapped countries such as Bangladesh. Nonetheless, the scientific evidence collected at Matlab since the 1970s has played an important role in informing population control efforts in Bangladesh and elsewhere.

been most noticeable in countries where women have gained improved access to contraceptives and education, particularly family-planning education (see "The Science behind the Story," above; also see Figure 8.15 and Figure 8.16).

In 2005, 53% of married women worldwide (ages 15–49) reported using some modern method of contraception to plan or prevent pregnancy. China, at 86%, had the highest rate of contraceptive use of any nation. Six western European nations showed rates of contraceptive

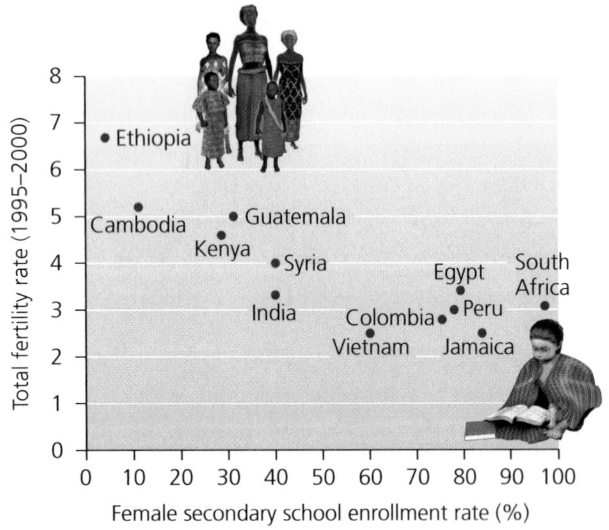

FIGURE 8.15 Increasing female literacy is strongly associated with reduced birth rates in many nations. Data from McDonald, M., and D. Nierenberg. 2003. Linking population, women, and biodiversity. *State of the World 2003.* Washington, D.C.: Worldwatch Institute.

use above 70%, as did Costa Rica, Cuba, New Zealand, Canada, Brazil, and Thailand (the U.S. rate was 68%). At the other end of the spectrum, 23 African nations had rates below 10%. These low rates of contraceptive use contribute to high fertility rates in sub-Saharan Africa, where the region's TFR is 5.6 children per woman. By comparison, in Asia, where the TFR in 1950 was 5.9, today it is 2.5—in part a legacy of the population control policies of China and some other Asian countries.

These data clearly indicate that in societies where women have little power, substantial numbers of pregnancies are unintended. Unfortunately, many women still lack the information and personal freedom of choice to allow them to make their own decisions about when to have children and how many to have. Today, many social scientists and policymakers recognize that for population growth to slow and stabilize, women need to achieve equal power with men in societies worldwide. Studies show that in societies in which women are freer to decide whether and when to have children, fertility rates have fallen, and the resulting children are better cared for, healthier, and better educated.

Unfortunately, we are still a long way from achieving gender equality. Over two-thirds of the world's people who cannot read, and 60% of those living in poverty, are women. Violence against women remains shockingly common. In many societies, by tradition men restrict women's decision-making abilities, including decisions as to how many children they will bear. The gap between the power held by men and the power held by women is just as obvious at the highest levels of government. Worldwide, only 13% of elected government officials in national legislatures are women (Figure 8.17). The United States

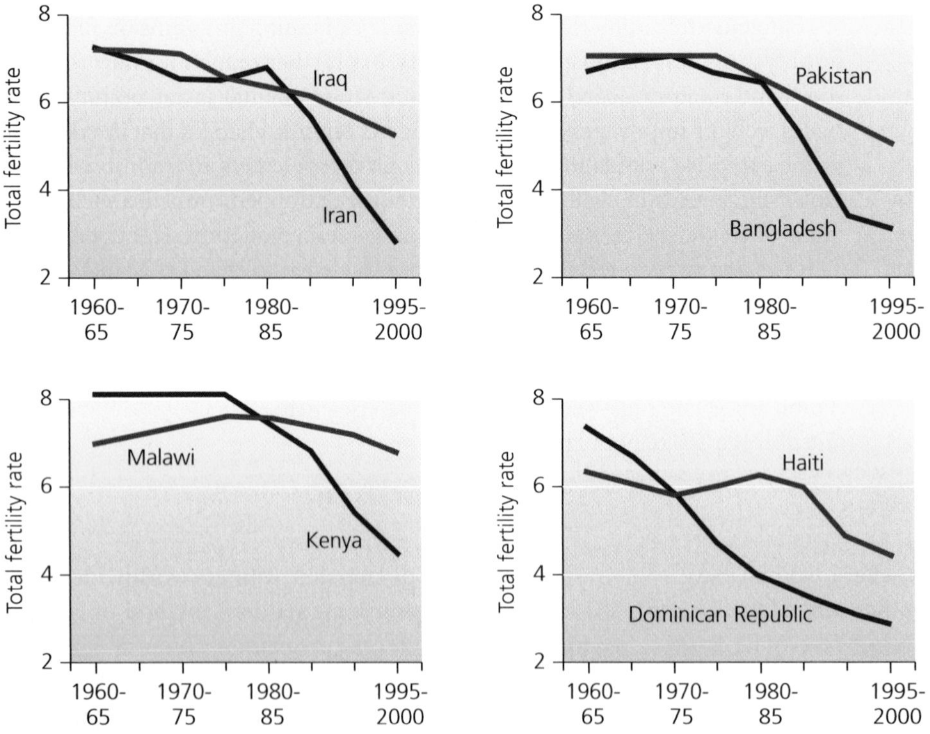

FIGURE 8.16 Data from four pairs of neighboring countries demonstrate the effectiveness of family planning in reducing fertility rates. In each case, the nation that invested in family planning and (in some cases) made other reproductive rights, education, and health care more available to women (blue lines) reduced its total fertility rate (TFR) far more dramatically than its neighbor (red lines). Data from U.N. Population Division; and Harrison, P., and F. Pearce. 2000. *AAAS atlas of population and environment* Berkeley, CA: University of California Press.

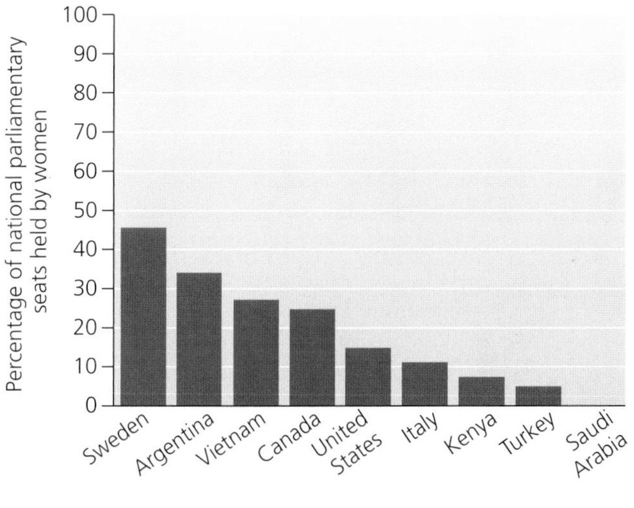

FIGURE 8.17 Although women make up more than half of the world's population, they are vastly underrepresented in positions of political power. Measured by the percentages of seats held by women in national legislatures, women in some regions fare better than those in others. Note that the United States may not rank quite as highly as you might guess. Data from Inter-Parliamentary Union, Women in national parliaments. (Aug. 2005).

lags behind not only Europe but also many developing nations in the proportion of women in positions of power in its government. As more women win positions of power, perhaps gender equality will become a more tangible reality. Such equality would have environmental consequences, for when women have economic and political power and access to education, they gain the option, and often the motivation, to limit the number of children they bear.

Population policies and family-planning programs are working around the globe

Data show that funding and policies that encourage family planning have been effective in lowering population growth rates in all types of nations, even those that are least industrialized. No nation has pursued a population control program as extreme as China's, but other rapidly growing nations have implemented less-restrictive programs.

The government of Thailand has relied on an education-based approach to family planning that has reduced birth rates and slowed population growth. In the 1960s, Thailand's growth rate was 2.3%, but by 2005 it had declined to 0.7%. This decline was achieved without a one-child policy. It has resulted, in large part, from government-sponsored programs devoted to family-planning education and increased availability of contraceptives.

India has had long-standing policies, but many observers think they are too weak. Unless it strengthens its efforts to slow population growth, India seems set to overtake China in population soon, and its population is projected to exceed China's by 200 million people in the year 2050. Brazil, Mexico, Iran, Cuba, and many other developing countries have instituted active programs to reduce their population growth. These programs entail setting targets and providing incentives, education, contraception, and reproductive health care.

Many of these programs are working. The data shown in Figure 8.16 are not the only cases in which family-planning programs have helped lower fertility rates. One study in 2000 examined four different pairs of nations located in the same parts of the world, with one country in each pair having a stronger program: Thailand and the Philippines, Pakistan and Bangladesh, Tunisia and Algeria, and Zimbabwe and Zambia. The demographers concluded that in all four cases, the country with the stronger program (Thailand, Bangladesh, Tunisia, and Zimbabwe) initiated or accelerated a decline in fertility with its policies. In the case of Thailand and the Philippines, the researchers also concluded that the Catholic Church's strong presence in the Philippines held back the success of family planning there.

In 1994, the United Nations hosted a milestone conference on population and development in Cairo, Egypt, at which 179 nations endorsed a platform calling on all governments to offer universal access to reproductive health care within 20 years. The conference marked a turn away from older notions of command-and-control population policy geared toward pushing contraception and lowering population to preset targets. Instead, it urged governments to offer better education and health care and to address social needs that bear indirectly on population (such as alleviating poverty, disease, and sexism).

Despite the successes of family planning internationally, the United States has often declined to fund family-planning efforts by the United Nations. Canceling this funding, for example, was one of George W. Bush's first acts on becoming U.S. president in 2001.

Weighing the Issues: U.S. Involvement in International Family Planning

From 1998 to 2001, the U.S. government provided $46.5 million to the United Nations Population Fund (UNFPA), whose programs provide education in family planning, HIV/AIDS prevention, and teen pregnancy prevention in many nations, including China. Since then,

VIEWPOINTS

Population Control

Debate over human population growth and environmental problems is often contentious. **Do you believe that national governments should implement policies, subsidies, or other programs to reduce birth rates?**

Implement ICPD Program of Action

Access to reproductive health care, including family planning, is a basic human right. To exercise this right, men and women need to be informed about family planning. They also need to have access to safe, effective, affordable, and acceptable methods of family planning of their choice.

All national governments should adopt policies, subsidies, and other programs to help implement the Program of Action, which the United Nations agreed to at the International Conference on Population and Development (ICPD), held in Cairo in 1994.

The ICPD is based on principles virtually every country in the world agreed to, one of which is as follows:

> . . . States should take all appropriate measures to ensure, on a basis of equality of men and women, universal access to health-care services, including those related to reproductive health care, which includes family planning and sexual health. Reproductive health-care programs should provide the widest range of services without any form of coercion. All couples and individuals have the basic right to decide freely and responsibly the number and spacing of their children and to have the information, education and means to do so.

In the same way that democratic nations are obligated to assist emerging democracies in holding fair and free elections, developed nations ought to assist developing nations in implementing the ICPD Program of Action. In every society where the principles of the ICPD have been implemented, birth rates have gone down, infant survival rates have gone up, maternal mortality has declined, and the quality of life has improved.

Timothy Cline is director of communication at Population Connection, where he has held a variety of positions, including publications manager. He has also served as chief of advocacy and policy research at the Johns Hopkins University Center for Communication Programs. Before joining Population Connection, Cline served as operations manager for the Washington Regional Alliance and was a senior editor at Ecomedia.

Population Control: A Bad Idea for Governments

To address this issue, we must ask three questions. First, is population growing at an unsustainable rate? Second, are the world's problems—poverty, hunger, ecological degradation, etc.—caused by population growth? And third, do governments possess rightful authority to engage in population control?

1. The world is experiencing historically unprecedented fertility decline. In about half the countries on Earth—including rich and poor, developed and developing—fertility is at or below replacement rate, meaning that in many countries women are not having enough babies to maintain current population levels. Some demographers now worry that entire economies and social welfare systems may face profound difficulties because of this decline.

2. The population growth that has occurred over the past century (from roughly 2 billion to 6 billion) is not the direct cause of severe development problems. For instance, according to the U.N. Population Division, "Even for those environmental problems that are concentrated in countries with rapid population growth, it is not necessarily the case that population increase is the main cause, nor that slowing population growth would make an important contribution to resolving the problem." Persistent development problems are more likely linked to types of economic and political organization (nondemocratic countries tend to have worse environmental records, for instance) and should be addressed through political reform.

3. The quest for government control over such an intimate matter as family size seems prone to abuse, which has occurred in such diverse nations as India, Peru, and China. In China, where the "one child policy" has often been touted as history's most successful population control program, over 100 million women have been forced to abort unborn babies or to be sterilized.

For these practical and ethical reasons, the world's limited development funds could be better spent than in encouraging further fertility decline.

Douglas A. Sylva is senior fellow of the Catholic Family and Human Rights Institute, a think tank and lobbying group that consults governments on international social policy. He is also a regular columnist for thefactis.org, focusing on international affairs and development. He received a Ph.D. in political science from Columbia University.

Explore this issue further by accessing **Viewpoints** at www.aw-bc.com/withgott.

Table 8.3 Per Capita Wealth, with Rates of Fertility, Population Growth, and Contraceptive Use, for Selected Nations

Nation	Per Capita GNI PPP (U.S. $)*	Population increase (% per year)	Children born per woman (TFR)	Population density (per mi^2)	Infant mortality (per 1,000)	Percentage of couples using birth control
Ethiopia	810	2.5	5.9	182	100	6
Niger	830	3.4	8.0	29	153	4
Haiti	1,680	1.9	4.7	774	80	22
Cameroon	2,090	2.3	5.0	89	74	13
Pakistan	2,160	2.4	4.8	528	85	20
India	3,100	1.7	3.0	869	60	43
Nicaragua	3,300	2.7	3.8	115	36	66
Syria	3,550	2.7	3.7	257	22	35
China	5,530	0.6	1.6	353	27	86
Brazil	8,020	1.4	2.4	56	27	70
Mexico	9,590	1.9	2.6	142	25	59
Czech Republic	18,400	−0.1	1.2	335	4	58
Spain	25,070	0.1	1.3	223	4	53
United Kingdom	31,460	0.2	1.7	635	5	79
Japan	30,040	0.1	1.3	876	3	48
Canada	30,660	0.3	1.5	8	5	73
United States	39,710	0.6	2.0	80	7	68

*GNI PPP is "gross national income in purchasing power parity," a measure that standardizes income and makes it comparable among nations, by converting income to "international" dollars using a conversion factor. International dollars indicate the amount of goods and services one could buy in the United States with a given amount of money. Data from Population Reference Bureau. 2005. *2005 World population data sheet.*

the Bush administration has withheld funds, pointing out that U.S. law prohibits funding any organization that "supports or participates in the management of a program of coercive abortion or involuntary sterilization," and claiming that the Chinese government has been implicated in both these activities. Many nations and organizations criticized the U.S. decision, and the European Union offered additional funding to UNFPA to offset the loss of U.S. contributions. What do you think of the U.S. decision? Should the United States fund family planning efforts in other nations? What conditions, if any, should it place on the use of such funds?

Poverty is strongly correlated with population growth

The alleviation of poverty, one target of the Cairo conference, has been linked to population because poorer societies tend to show higher population growth rates than do wealthier societies. This pattern is consistent with demographic transition theory. Note in Table 8.3 how poorer nations tend to have higher fertility and growth rates, along with higher birth and infant mortality rates and lower rates of contraceptive use.

Trends such as these have affected the distribution of people on the planet. In 1960, 70% of all people lived in developing nations. By 2005, 81% of the world's population was living in these countries. Moreover, fully 98% of the next billion people to be added to the global population will be born in these poor, less developed regions (Figure 8.18). This is unfortunate from a social standpoint, because these people will be added to the countries that are least able to provide for them. It is also unfortunate from an environmental standpoint, because poverty often results in environmental degradation. People dependent on agriculture in an area of poor farmland, for instance, may need to try to farm even if doing so degrades the soil and is not sustainable. This is largely why Africa's once-productive Sahel region, like many regions of western China, is turning to desert (Figure 8.19). Poverty also drives the hunting of many large mammals in Africa's forests, including the great apes that are now disappearing as local settlers and miners kill them for their "bush meat."

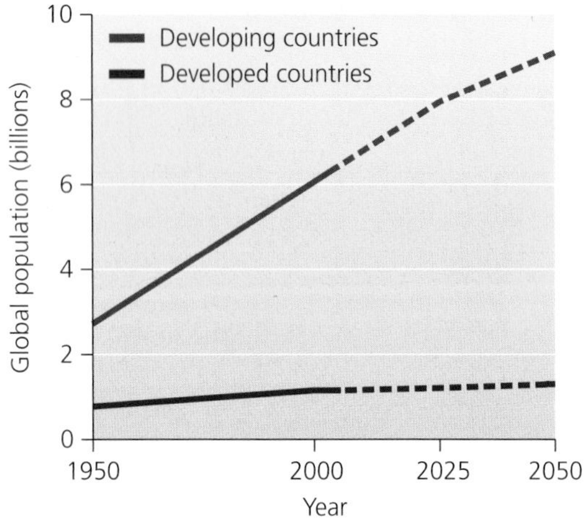

FIGURE 8.18 Nearly 98% of the next 1 billion people added to Earth's human population will reside in the less developed, poorer parts of the world. The dashed line indicates projected future trends. Data from U.N. Population Division; and Harrison, P., and F. Pearce. 2000. *AAAS atlas of population and environment.* 2000. Berkeley, CA: University of California Press.

Consumption from affluence creates environmental impact

Poverty can lead people into environmentally destructive behavior, but wealth can produce even more severe and far-reaching environmental impacts. The affluence that characterizes a society such as the United States, Japan, or the Netherlands is built on massive and unprecedented levels of resource consumption. Much of this chapter has dealt with numbers of people rather than on the amount of resources each member of the population consumes or the amount of waste each member produces. The environmental impact of human activities, however, depends not only on the number of people involved but also on the way those people live. Recall the A for affluence in the IPAT equation. Patterns of affluence and consumption are spread unevenly across the world, and affluent societies generally consume resources from other societies as well as from their own.

In Chapter 1 (▸pp. 6–7, 19, and 25), we introduced the concept of the *ecological footprint,* the cumulative amount of Earth's surface area required to provide the raw materials a person or population consumes and to dispose of or recycle the waste that they produce. Individuals from affluent societies leave a considerably larger per capita ecological footprint (see Figure 1.13, ▸p. 19). This fact should remind us that the "population problem" does not lie entirely with the developing world. Just as population is rising, so is consumption, and environmental scientists have calculated that we are already living beyond the planet's means to support us sustainably. One recent analysis concluded that humanity's global ecological footprint surpassed Earth's capacity to support us in 1987 and that our species is now living more than 20% beyond its means (Figure 8.20).

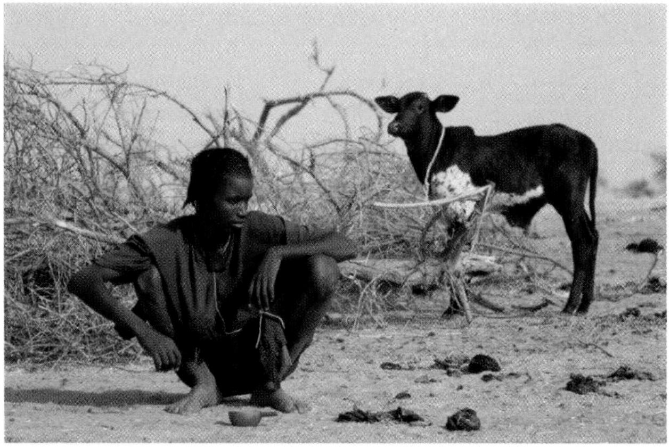

FIGURE 8.19 In the semi-arid Sahel region of Africa, where population is increasing beyond the land's ability to handle it, dependence on grazing agriculture has led to environmental degradation.

The wealth gap and population growth contribute to violent conflict

The stark contrast between affluent and poor societies in today's world is, of course, the cause of social as well as

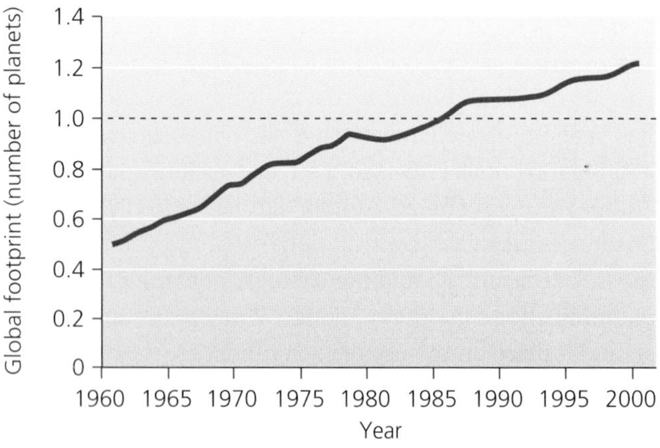

FIGURE 8.20 The global ecological footprint of the human population is 2.5 times larger than it was in 1961 and now exceeds what Earth can bear in the long run, scientists have calculated. The estimate shown here indicates that we have already overshot our carrying capacity by at least 20%; that is, we are using renewable natural resources 20% faster then they are being replenished. Data from WWF-World Wide Fund for Nature, 2004. *Living planet report.* Gland, Switzerland: WWF.

(a) A family living in the United States

(b) A family living in Egypt

FIGURE 8.21 A typical U.S. family (**a**) may own a large house, keep numerous material possessions, and have enough money to afford luxuries such as vacation travel. A typical family in a developing nation such as Egypt (**b**) may live in a small, sparsely furnished dwelling with few material possessions and little money or time for luxuries. The ubiquity of television sets, even among poor families of the developing world, means that the world's poor see representations (both real and exaggerated) of wealth in the United States as depicted on American TV shows. Many sociologists hold that this has increased the poor's awareness of the global wealth gap and has spurred aspirations for consumption among the poor of developing nations.

environmental stress. Over half the world's people live below the internationally defined poverty line of U.S. $2 per day. The richest one-fifth of the world's people possesses over 80 times the income of the poorest one-fifth (Figure 8.21). The richest one-fifth also uses 86% of the world's resources. That leaves only 14% of global resources—energy, food, water, and other essentials—for the remaining four-fifths of the world's population to share. As the gap between rich and poor grows wider and as the sheer numbers of those living in poverty continue to increase, it seems reasonable to predict increasing tensions between the "haves" and the "have-nots." This is why the inequitable distribution of wealth is one of the key factors the U.S. Departments of Defense and State take into account when assessing the potential for armed conflict around the world, whether it be conventional warfare or terrorism.

HIV/AIDS is a major influence on populations in parts of the world

The rising material wealth and falling fertility rates of many industrialized nations today is slowing population growth in accordance with the demographic transition model. Some other nations, however, are not following Notestein's script. Instead, in these countries mortality is beginning to increase, presenting a scenario more akin to Malthus's fears. This is especially the case in countries where the HIV/AIDS epidemic has taken hold (Figure 8.22). African nations are being hit hardest. Of the 38 million people around the world infected with HIV/AIDS as of 2004, 25 million live in the nations of sub-Saharan Africa. The low rate of use of contraceptives, which contributes to this region's high fertility rate, also fuels the expansion of AIDS. One in every 13 people aged 15 to 49 in sub-Saharan Africa is infected with HIV, and for southern African nations, the figure is more than one in five.

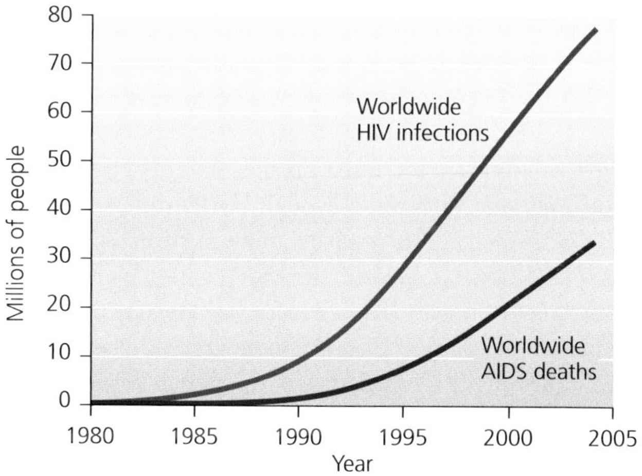

FIGURE 8.22 AIDS cases are increasing rapidly in much of the world. As of 2004, total cumulative HIV infections since 1980 were estimated at nearly 78 million, and 34 million people are estimated to have died from the disease so far. Data from UNAIDS; and *Vital signs 2005*. Washington, D.C.: Worldwatch Institute.

Science behind the Story

AIDS Resistance Genes and the Black Death: An Unexpected Connection?

When HIV was first identified as the cause of AIDS in the early 1980s, some scientists predicted that a cure for the disease would be found within a few years. In the two decades that followed, however, progress toward a cure or vaccine remained slow, even as the number infected grew to more that 60 million. Thus, the discovery in 1996 of a genetic mutation that appeared to confer resistance to AIDS was an important breakthrough. Still more surprising was a possible connection between AIDS and the Black Death, a disease that decimated Europe in the mid-14th century.

Beginning in 1984, geneticists Stephen J. O'Brien, Michael Dean, and colleagues at the National Cancer Institute (NCI) began searching for individuals who were naturally resistant to HIV. They gathered genetic samples from people with and without the virus, trying to identify mutations unique to people who had been exposed to HIV but remained uninfected.

Until the mid-1990s, the search was largely unsuccessful. Then in 1995, NCI researchers identified molecules called chemokines that carry signals from one part of the immune system to another. Other researchers discovered that HIV uses the immune system's own chemokine receptors to penetrate the cell membrane. Together, these advances revealed how HIV gains entry into human hosts. They also gave O'Brien, Dean, and other researchers a promising place to look for resistance genes.

The search quickly produced exciting results. In 1996, three separate groups reported the discovery of a mutation conferring resistance to HIV. The mutation—a deletion of 32 base-pairs of DNA—affected the chemokine receptor CCR5, which is expressed on the surface of macrophages, the first cells HIV attacks when it enters a new host. People with two copies of the mutation appeared to be immune to HIV. People with one copy developed AIDS, but slowly.

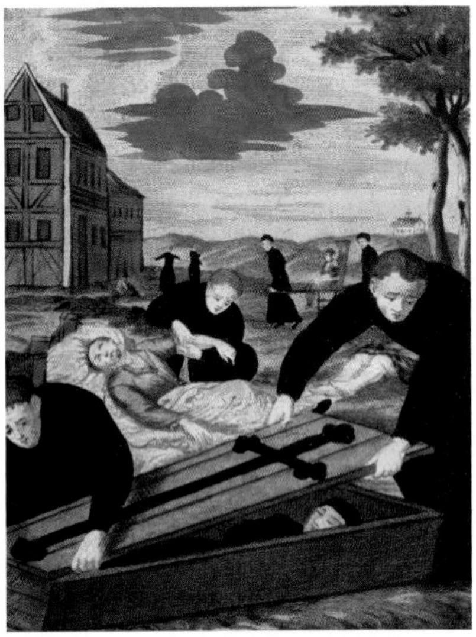

The Black Death killed one of every three Europeans in the 14th century but may have conferred on the descendants of its survivors some resistance to AIDS.

The AIDS epidemic is having the greatest impact on human populations of any disease since the Black Death killed roughly one of three people in 14th-century Europe, and since smallpox brought by Europeans to the New World wiped out perhaps millions of native people (see "The Science behind the Story," above). As AIDS takes roughly 6,000 lives in Africa every day, the epidemic is unleashing a variety of demographic changes. Infant mortality in sub-Saharan Africa has risen to 96 deaths out of 1,000 live births—14 times the rate in the developed world. The high numbers of infant deaths and premature deaths of young adults has caused life expectancy in parts of southern Africa to fall from a high of close to 59 years in the early 1990s back down to less than 40 years, where it stood in the early 1950s. AIDS is also leaving behind millions of orphans. As of 2002, 14 million children under the age of 15 worldwide had lost one or both parents to the disease.

Weighing the Issues: HIV/AIDS and Population

What sorts of problems would you predict might occur in the surviving population after a major disease such as AIDS kills a high percentage of the population?

Severe demographic changes have social, political, and economic repercussions

Beyond its demographic effects, AIDS is having immediate social, economic, and political consequences. Everywhere in sub-Saharan Africa, AIDS is undermining the

Surprisingly, the mutation was unevenly distributed across human populations. One research team, led by Belgian scientists Michael Samson and Marc Parmentier, found that about 9% of the Caucasians they tested had at least one copy of the mutation, but none of the Japanese or Africans they tested had the mutation. Other groups found similar results, and it soon became clear that the mutation was present only in Europeans and their descendants.

A key step toward answering why the mutation was unique to Europeans occurred when O'Brien and Dean determined when the mutation had first arisen. To do so, they analyzed the mutation's association with distinctive genetic variants, or alleles, on the chromosome. In each generation, the association, or linkage, between the mutation and its neighbors is weakened by recombination, the chromosomal reshuffling that takes place during reproduction. Scientists can thus use the strength of the linkage to identify the mutation's approximate age. When O'Brien and Dean used this technique, they found that the CCR5 mutation was about 700 years old.

That put the mutation's origin at about the time when the Black Death, which most scientists have identified as bubonic plague, was ravaging Europe. Like the AIDS virus, the bacterium responsible for bubonic plague *(Yersinia pestis)* initially attacks macrophages, using them as a refuge before spreading to the rest of the body. Thus a mutation in CCR5 could potentially have conferred resistance to bubonic plague, just as it now confers resistance to AIDS. Under the strong selection pressure of the Black Death, which killed a third of Europe's population, the prevalence of the mutation could have risen greatly. Thus, the researchers suggested, the plague had one positive legacy: it offered the descendants of Europeans who survived it some degree of immunity to AIDS.

As with any exciting hypothesis, other researchers set out to confirm or refute it by testing the idea in different ways. This time, two labs taking different approaches cast doubt on the idea. Joan Mecsas and colleagues at Stanford University experimented with lab mice and found that mice with the CCR5 mutation were *not* protected against bubonic plague.

Shortly before the Mecsas team published its results in February 2004, two researchers from University of California–Berkeley conducted a modeling study that considered the population dynamics of diseases, and found that plague was not the most likely candidate for causing natural selection for the CCR5 mutation. Instead, they suggested that smallpox—another disease that killed millions in the Middle Ages—played this role.

ability of developing countries to make the transition to modern technologies because it is removing many of the youngest and most productive members of society. For example, in 1999 Zambia lost 600 teachers to AIDS, and only 300 new teachers graduated to replace them. In Rwanda, more than one in three college-educated residents of the city of Kigali are infected with the virus. South Africa loses an estimated $7 billion per year to declines in its labor force as AIDS patients fill the nation's hospitals (Figure 8.23). The loss of productive household members to AIDS causes families and communities to break down as income and food production decline while medical expenses and debt skyrocket.

These problems are hitting many countries at a time when their governments are already experiencing what has been called *demographic fatigue.* Demographically

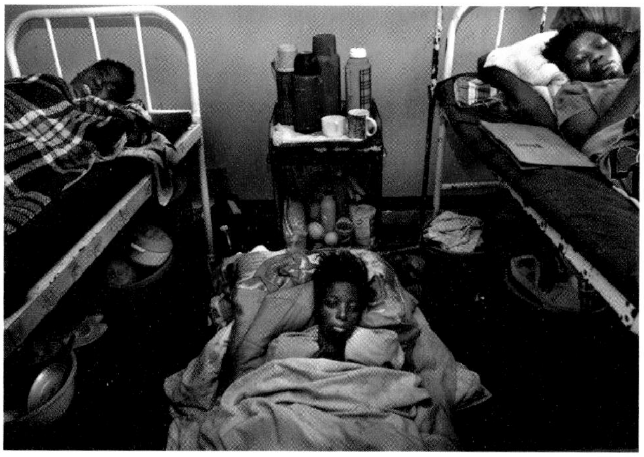

FIGURE 8.23 AIDS patients occupy 60% of South Africa's hospital beds. By 2010, AIDS in Africa may orphan an estimated 40 million children.

fatigued governments face overwhelming challenges related to population growth, including educating and finding jobs for their swelling ranks of young people. With the added stress of HIV/AIDS, these governments face so many demands that they are stretched beyond their capabilities to address problems. As a result, the problems grow worse, and citizens lose faith in their governments' abilities to help them.

If nations in sub-Saharan Africa—and other regions where the disease is spreading fast, such as India and southeast Asia—do not take aggressive steps soon, and if the rest of the world does not step in to help, these countries could fail to advance through the demographic transition. Instead, their rising death rates could push birth rates back up, potentially causing these countries to fall back to the pre-industrial stage of the demographic transition model. Such an outcome would lead to greater population growth while economic and social conditions worsen. It would be a profoundly negative outcome, both for human welfare and for the well-being of the environment.

If one of humanity's goals is to generate a high standard of living and quality of life for all the world's people, then developing nations must find ways to reduce their population growth. However, those of us living in the industrialized world must also be willing to reduce our consumption. Earth does not hold enough resources to sustain all 6.5 billion of us at the current North American standard of living, nor can we go out and find extra planets; so, we must make the best of the one place that supports us all.

Conclusion

Today, several years after welcoming its six-billionth member, the human population is larger than at any time in the past. Our growing population, as well as our growing consumption, affects the environment and our ability to meet the needs of all the world's people. Approximately 90% of children born today are likely to live their lives in conditions far less healthy and prosperous than most of us in the industrialized world are accustomed to.

However, there are at least two major reasons to be encouraged. First, although global population is still rising, the *rate* of growth has decreased nearly everywhere, and some countries are even seeing population declines. Most developed nations have passed through the demographic transition, showing that it is possible to lower death rates while stabilizing population and creating more prosperous societies. A second reason to feel encouraged is the progress in expanding rights for women worldwide. Although there is still a long way to go, women are slowly being treated more fairly, receiving better education, obtaining more economic independence, and gaining more ability to control their reproductive decisions. Aside from the clear ethical progress these developments entail, they are helping slow population growth.

Human population cannot continue to rise forever. The question is how it will stop rising: through the gentle and benign process of the demographic transition, through restrictive governmental intervention such as China's one-child policy, or through the miserable Malthusian checks of disease and social conflict caused by overcrowding and competition for scarce resources. Moreover, sustainability demands a further challenge—that we stabilize our population size in time to avoid destroying the natural systems that support our economies and societies. We are indeed a special species. We are the only one to come to such dominance as to change fundamentally so much of Earth's landscape, and even its climate system. We are also the only species with the intelligence needed to turn around an increase in our own numbers before we destroy the very systems on which we depend.

REVIEWING OBJECTIVES

You should now be able to:

Assess the scope of human population growth

- Our global population of 6.5 billion people adds about 78 million people per year (2.5 people every second). (p. 218)
- Our growth rate peaked at 2.1% in the 1960s and now stands at 1.2%. Growth rates vary among regions of the world. (p. 219)
- Rising population is a problem to the extent that it depletes resources, intensifies pollution, stresses social systems, or degrades ecosystems, such that the natural environment or our quality of life decline. (p. 220)

Evaluate how human population, affluence, and technology affect the environment

- The IPAT model summarizes how environmental impact (I) results from interactions among population size (P), affluence (A), and technology (T). (pp. 220–221)
- Rising population and rising affluence (leading to greater consumption) each increase environmental impact. Technological advances have frequently

exacerbated environmental degradation, but they can also help mitigate our impact. (pp. 220–222)

Explain and apply the fundamentals of demography

- ▶ Demography applies principles of population ecology to the statistical study of human populations. (p. 223)
- ▶ Demographers study size, density, distribution, age structure, and sex ratios of populations, as well as rates of birth, death, immigration, and emigration. (pp. 223–227)
- ▶ Total fertility rate (TFR) contributes greatly to change in a population's size. (pp. 227–228)

Outline and assess the concept of demographic transition

- ▶ The demographic transition model explains why population growth has slowed in industrialized nations. Industrialization and urbanization have reduced the economic need for children, while education and the empowerment of women have decreased unwanted pregnancies. Parents in developed nations choose to invest in quality of life rather than quantity of children. (pp. 228–229)
- ▶ The demographic transition may or may not proceed to completion in all of today's developing nations. Whether it does is of immense importance for the quest for sustainability. (p. 229)

Describe how wealth and poverty, the status of women, and family planning programs affect population growth

- ▶ When women are empowered and achieve equality with men, fertility rates fall, and children tend to be better cared for, healthier, and better educated. (pp. 230–233)
- ▶ Family-planning programs and reproductive education have successfully reduced population growth in many nations. (pp. 230–233)
- ▶ Poorer societies tend to have higher population growth rates than do wealthier societies. (p. 235)
- ▶ The high consumption rates of affluent societies may make their ecological impact greater than that of poorer nations with larger populations. (pp. 236–237)

Characterize the dimensions of the HIV/AIDS epidemic

- ▶ About 38 million people worldwide are infected with HIV/AIDS, of which 25 million live in sub-Saharan Africa. (pp. 237–238)
- ▶ Epidemics that claim large numbers of young and productive members of society influence population dynamics and can have severe social and political ramifications. (pp. 238–240)

TESTING YOUR COMPREHENSION

1. What is the approximate current human global population? How many people are being added to the population each day?
2. Why has the human population continued to grow in spite of environmental limitations?
3. Contrast the views of environmental scientists with those of the libertarian writer Sheldon Richman and similar-thinking economists over whether population growth is a problem. Why does Richman think the concept of carrying capacity does not apply to human populations?
4. Explain the IPAT model. How can technology either increase or decrease environmental impact? Provide at least two examples.
5. What characteristics and measures do demographers use to study human populations? Which of these help determine the impact of human population on the environment?
6. What is the total fertility rate (TFR)? Can you explain why the replacement fertility for humans is approximately 2.1? How is Europe's TFR affecting its natural rate of population change?
7. Why have fertility rates fallen in many countries?
8. In the demographic transition model, why is the pre-industrial stage characterized by high birth and death rates, and the industrial stage by falling birth and death rates?
9. How does the demographic transition model explain the increase in population growth rates in recent centuries? How does it explain the decrease in population growth rates in recent decades?
10. Why do poorer societies have higher population growth rates than wealthier societies? How does poverty affect the environment? How does affluence affect the environment?

SEEKING SOLUTIONS

1. China's reduction in birth rates is leading to significant change in the nation's age structure. Review Figure 8.10, which portrays the projected change. You can see that the population is growing older, based on the top-heavy age pyramid for the year 2030. What sorts of effects might this ultimately have on Chinese society? Explain your answer.
2. The World Bank estimates that half the world's people survive on less than the equivalent of two dollars per day. What effect would you expect this situation to have on the political stability of the world? Explain your answer.
3. Apply the IPAT model to the example of China provided in the chapter. How do population, affluence, technology, and ecological sensitivity affect China's environment? Now consider your own country, or your own state. How do population, affluence, technology, and ecological sensitivity affect your environment? How can we regulate the relationship between population and its effects on the environment?
4. Do you think that all of today's developing nations will complete the demographic transition and come to enjoy a permanent state of low birth and death rates? Why or why not? What steps might we as a global society take to help ensure that they do? Now think about developed nations like the United States and Canada. Do you think these nations will continue to lower and stabilize their birth and death rates in a state of prosperity? What factors might affect whether they do so?
5. Imagine that India's prime minister puts you in charge of that nation's population policy. India has a population growth rate of 1.7% per year, a TFR of 3.0, a 43% rate of contraceptive use, and a population that is 72% rural. What policy steps would you recommend, and why?
6. Now imagine that you have been tapped to design population policy for Germany. Germany is losing population at an annual rate of 0.1%, has a TFR of 1.3, a 72% rate of contraceptive use, and a population that is 88% urban. What policy steps would you recommend, and why?

INTERPRETING GRAPHS AND DATA

Below are graphed data representing the economic condition of the world's population. The *y* axis indicates the per capita income for each country or region expressed as purchasing power in U.S. dollars (termed *gross national income in purchasing power parity*, or *GNI PPP*; see Table 8.3). The *x* axis indicates the cumulative percentage of the world population whose per capita GNI PPP is equal to or greater than that country's or region's per capita GNI PPP. The horizontal dotted line indicates the global average per capita GNI PPP.

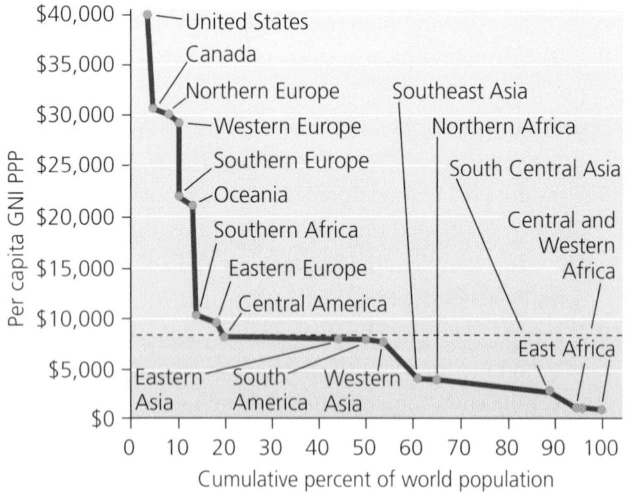

Percent of world population at various income levels. Data source: Population Reference Bureau. 2005. *World population data sheet 2005.*

1. What percentage of the world population lives at or below the global average per capita GNI PPP? What percentage lives at or below one half of the global average per capita GNI PPP? What percentage lives at or above twice the global average per capita GNI PPP?
2. Given a global average per capita GNI PPP of $8,540 and a world population of 6,477,000,000 people (as of mid-2005), what is the total global GNI PPP? What would the global GNI PPP be if everyone lived at the level of affluence of the United States?
3. How do you personally resolve the ethical conflict between the desirable goal of raising the standard of living of the billions of desperately poor people in the world and the likelihood that increasing their affluence (A in the equation I = PAT) will have a negative impact on the environment?

CALCULATING ECOLOGICAL FOOTPRINTS

The equation I = PAT (Impact = Population × Affluence × Technology) suggests that a population's size and affluence are not the only determinants of its ecological impact; its technological choices also have an effect. Technologies can be either efficient or wasteful. One way of gauging the relative value of T is to calculate a per capita value of I/A (equivalent to I divided by A divided by P). The table presents per capita values of I (estimated ecological footprints) and A (income). Calculate the relative values of T by completing the blank column.

Country	Impact (ecological footprint, in acres per capita)	Affluence (per capita income, in GNI PPP)	Technology (I/A) (footprint per $1,000 income)
Bangladesh	1.2	$1,980	0.61
Colombia	4.9	$6,820	
Mexico	6.4	$9,590	
Sweden	14.6	$29,770	
Thailand	6.9	$8,020	
United States	25.4	$39,710	
World average	6.9	$8,540	

Data sources: Population Reference Bureau. 2005. *World population data sheet 2005*; and Wackernagel, M., et al. 1999. National natural capital accounting with the ecological footprint concept. *Ecological Economics* 29: 375–390.

1. If the world average value of T were decreased (improved) to that of the United States, what per capita GNI PPP could be supported at the current average per capita ecological footprint of 6.9 acres?

2. What value of T would enable the world's population to live at its current affluence within the 4.9 acres per capita that Wackernagel et al. estimate are available? Do you think this is achievable?

3. Which country's technological choices would you choose to study if you were interested in learning how to maximize your standard of living while minimizing your ecological impact? Using the value of T for this country and the mid-2005 world population of 6,477,000,000, calculate the following:

- **(a)** The number of people the world could support at the current per capita impact of 6.9 acres and affluence of $8,540
- **(b)** The number of people the world could support sustainably on the available 4.9 acres per capita and at an affluence of $8,540
- **(c)** The per capita GNI PPP that the world's current population could achieve on 6.9 acres per capita
- **(d)** The per capita GNI PPP that the world's current population could achieve on 4.9 acres per capita

Take It Further

Go to www.aw-bc.com/withgott or the student CD-ROM where you'll find:

- Suggested answers to end-of-chapter questions
- Quizzes, animations, and flashcards to help you study
- *Research Navigator*™ database of credible and reliable sources to assist you with your research projects
- GRAPHIt! tutorials to help you master how to interpret graphs
- INVESTIGATEIt! current news articles that link the topics that you study to case studies from your region to around the world

9 Soil and Agriculture

Wheat fields in Paraná, Brazil

Upon successfully completing this chapter, you will be able to:

- Explain the importance of soils to agriculture, and describe the impacts of agriculture on soils
- Outline major historical developments in agriculture
- Delineate the fundamentals of soil science, including soil formation and the properties of soil
- State the causes and predict the consequences of soil erosion and soil degradation
- Recite the history and explain the principles of soil conservation

No-till farm in Ceara, Brazil

Central Case: No-Till Agriculture in Southern Brazil

"The nation that destroys its soil destroys itself."
— U.S. President Franklin D. Roosevelt

"There are two spiritual dangers in not owning a farm. One is the danger of supposing that breakfast comes from the grocery, and the other that heat comes from the furnace."
— Conservationist and philosopher Aldo Leopold

In southernmost Brazil, hundreds of thousands of people make their living farming. The warm climate and rich soils of this region's rolling highlands and coastal plains have historically made for bountiful harvests. However, repeated cycles of plowing and planting over many decades diminished the productivity of the soil. More and more topsoil—the valuable surface layer of soil richest in organic matter and nutrients—was being eroded away by water and wind. Meanwhile, the synthetic fertilizers used to restore nutrients were polluting area waterways. Yields were falling, and by 1990 farmers were looking for help.

As a result, many of southern Brazil's farmers abandoned the conventional practice of tilling the soil after harvests. In its place, they turned to *no-tillage* farming, otherwise known as *zero-tillage, no-till,* or in Brazil, *plantio direto*. Turning the earth by tilling (plowing, disking, harrowing, or chiseling) aerates the soil and works weeds and old crop residue into the soil to nourish it. Tilling, however, also leaves the surface bare of vegetation for a period of time, during which erosion by wind and water can remove precious topsoil. Although tilling historically boosted the productivity of agriculture in Europe, many experts now think it is less appropriate for soils in subtropical regions such as southern Brazil. The reason is that the heavy rainfall of tropical and subtropical regions results in greater rates of erosion, causing tilled soils to lose organic matter and nutrients.

Working with agricultural scientists and government extension agents, southern Brazil's farmers began leaving crop residues on their fields after harvesting, and planting "cover crops" to keep soil protected during periods when they weren't raising a commercial crop.

When they went to plant the next crop, they merely cut a thin, shallow groove into the soil surface, dropped in seeds, and covered them. They did not invert the soil as they had when tilling, and the soil stayed covered with plants or their residues at all times, reducing erosion by 90%.

With less soil eroding away, and more organic material being added to it, the soil held more water and was better able to support crops. The improved soil quality meant better plant growth and greater crop production. In the state of Santa Catarina, maize yields per hectare increased by 47% between 1991 and 1999, wheat yields rose by 82%, and soybean yields by 83%, according to local farmers, extension agents, and international scientists. In the states of Paraná and Rio Grande do Sul, maize yields were up 67% over 10 years, and soybean yields were up 68%.

Besides boosting yields, no-till farming methods reduced farmers' costs, because farmers now used less labor and less fuel. No-till agriculture spread quickly in the region as farmers saw their neighbors' successes and traded information through "Friends of the Land" clubs organized on local, municipal, regional, and statewide levels. In Paraná and Rio Grande do Sul, the area being farmed with no-till methods shot up from 700,000 ha (1.7 million acres) in 1990 to 10.5 million ha (25.9 million acres) in 1999, when it involved 200,000 farmers. In Santa Catarina, where farms are generally smaller, over 100,000 farmers now apply no-till methods to 880,000 ha (2.2 million acres) of farmland. No-till farming is now spreading northward into Brazil's tropical regions and to other parts of Latin America.

By enhancing soil conditions and reducing erosion, no-till techniques have benefited southern Brazil's society and environment as well; its air, waterways, and ecosystems are less polluted. Similar effects are being felt in parts of the United States and elsewhere in the world where no-till and reduced-tillage methods are increasingly being applied.

Reduced tillage is certainly not a panacea for all areas of the world. In general, tropical areas benefit more than temperate regions, because erosion is greater in the tropics and hot weather can overheat tilled soil. The benefits and drawbacks of different tillage approaches vary with location, soil characteristics, and type of crop. In regions suitable for reduced tillage, proponents say these approaches can help make agriculture sustainable. We will need sustainable agriculture if we are to feed the world's human population while protecting the natural environment, including the soils that vitally support our production of food.

Soil: The Foundation for Feeding a Growing Population

As the human population has increased, so have the amounts of land and resources we devote to agriculture, which currently covers 38% of Earth's land surface. We can define **agriculture** as the practice of raising crops and livestock for human use and consumption. We obtain most of our food and fiber from **cropland,** land used to raise plants for human use, and **rangeland** or pasture, land used for grazing livestock.

Healthy soil is vital for agriculture, for forestry (Chapter 12), and for the functioning of Earth's natural systems. **Soil** is not merely lifeless dirt; it is a complex plant-supporting system consisting of disintegrated rock, organic matter, water, gases, nutrients, and microorganisms. Each of these components can be altered by the way we treat soil. Productive soil is a renewable resource, but if we abuse it through careless or uninformed practices, we can greatly reduce its productivity.

As population and consumption increase, soils are being degraded

If we are to feed the world's rising human population, we will need to change our diet patterns or increase agricultural production—and do so sustainably, without degrading the environment and reducing its ability to support agriculture. However, we cannot simply keep expanding agriculture into new areas, because land suitable and available for farming is running out. Instead, we must find ways to improve the efficiency of food production in areas that are already in agricultural use.

Today many lands unsuitable for farming are being farmed, causing considerable environmental damage. Mismanaged agriculture has turned grasslands into deserts and has removed ecologically precious forests. It has extracted nutrients from soils and added them to water bodies, harming both systems. It has diminished biodiversity; encouraged invasive species; and polluted soil, air, and water with toxic chemicals. Poor agricultural practices have allowed countless tons of fertile soil to be blown and washed away.

As our planet gains over 70 million people each year, we lose 5–7 million ha (12–17 million acres) of productive cropland annually. Throughout the world, especially in drier regions, it has gotten more difficult to raise crops and graze livestock as soils have become eroded and

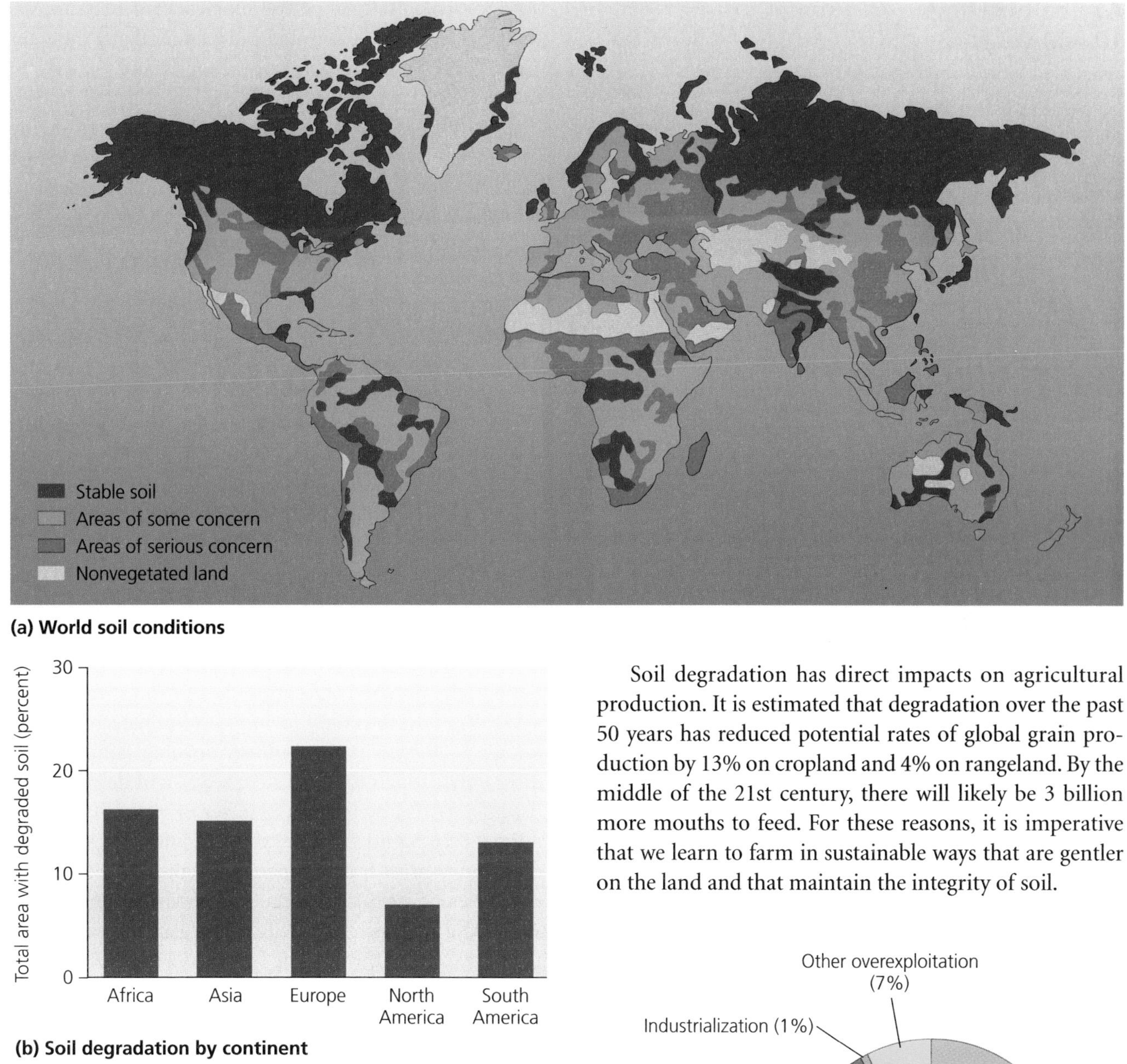

FIGURE 9.1 Soils are becoming degraded in many areas worldwide (**a**). Europe currently has a higher proportion of degraded land than other continents (**b**) because of its long history of intensive agriculture, but degradation is rising quickly in developing countries in Africa and Asia. Go to GRAPHIt! at www.aw-bc.com/withgott or on the student CD–ROM. Data from International Soil Reference and Information Centre (ISRIC) and United Nations Environment Programme (UNEP), 1996. *Human–induced soil degradation.* Rome: ISRIC, UNEP, and U.N. Food and Agriculture Organization (FAO) (a); UNEP 2002. *Global environmental outlook 3.* London: UNEP and Earthscan Publ. (b).

degraded (Figure 9.1). Soil degradation around the globe has resulted from roughly equal parts forest removal, cropland agriculture, and overgrazing of livestock (Figure 9.2).

Soil degradation has direct impacts on agricultural production. It is estimated that degradation over the past 50 years has reduced potential rates of global grain production by 13% on cropland and 4% on rangeland. By the middle of the 21st century, there will likely be 3 billion more mouths to feed. For these reasons, it is imperative that we learn to farm in sustainable ways that are gentler on the land and that maintain the integrity of soil.

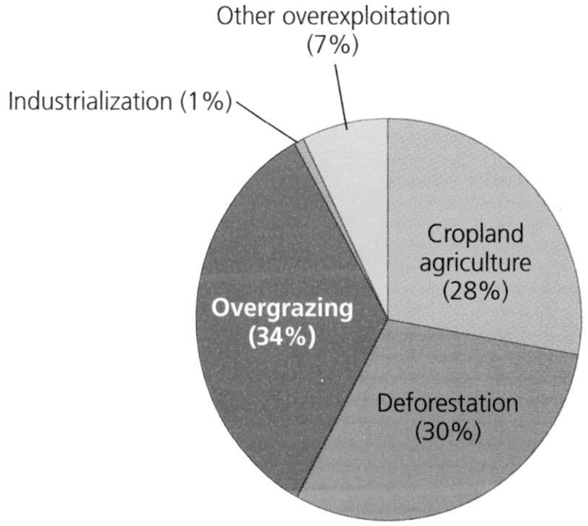

FIGURE 9.2 The great majority of the world's soil degradation results from cropland agriculture, overgrazing by livestock, and deforestation. Data from Wali, M. K. et al., 1999. Assessing terrestrial ecosystem sustainability: Usefulness of regional carbon and nitrogen models. *Nature and Resources* 35: 21–33.

Agriculture began to appear around 10,000 years ago

Agriculture is a relatively new approach for meeting our nutritional needs; on the scale of human history, there was no such thing as a farm until very recently. During most of our species' 160,000-year existence, we were hunter-gatherers, depending on wild plants and animals. Then about 10,000 years ago, as the climate warmed following a period of glaciation, people in some cultures began to raise plants from seed and to domesticate animals.

How and why did agriculture begin? The most plausible hypothesis is that it began as hunter-gatherers brought back to their encampments wild fruits, grains, and nuts. Some of these foods fell to the ground, were thrown away, or were eaten and survived passage through the digestive system. The plants that grew from these seeds near human encampments likely produced fruits that were on average larger and tastier than those in the wild, because they sprang from seeds of fruits selected by people because they were especially large and delicious. As these plants bred with others nearby that shared their characteristics, they gave rise to subsequent generations of plants with large and flavorful fruits. Eventually, people realized that they could guide this selective process through conscious effort, and our ancestors began intentionally planting seeds from the plants whose produce was most desirable. This is, of course, artificial selection at work (▸pp. 120–122). This practice of selective breeding continues to the present day and has produced the many hundreds of crops we enjoy, all of which are artificially selected versions of wild plants. People followed the same process of selective breeding with animals, creating livestock from wild species.

Once our ancestors learned to cultivate crops and raise animals, they began to settle in more permanent camps and villages, often near water sources. Agriculture and a sedentary lifestyle likely reinforced one another in a positive feedback cycle (▸p. 185). The need to harvest crops kept people sedentary, and once they were sedentary, it made sense to plant more crops. Population increase resulted from these developments and further promoted them. Moreover, the ability to grow excess farm produce enabled some people to leave farming and live off the food that others produced. This led to the development of professional specialties, commerce, technology, densely populated urban centers, social stratification, and politically powerful elites. For better or worse, the advent of agriculture eventually brought us the civilization we have today.

Archaeological and paleoecological evidence suggests that agriculture was invented independently by different cultures in at least five areas of the world and possibly 10 or more (Figure 9.3). The earliest widely accepted archaeological evidence for plant domestication is from the "Fertile Crescent" region of the Middle East about 10,500 years ago, and the earliest evidence for animal domestication also is from that region, just 500 years later. Crop remains have been dated using radiocarbon dating (▸p. 94) and similar methods. Wheat and barley originated in the Fertile Crescent, as did rye, peas, lentils, onions, garlic, carrots, grapes, and other food plants familiar to us today. The people of this region also domesticated goats and sheep. Meanwhile, in China, domestication began as early as 9,500 years ago, leading eventually to the rice, millet, and pigs we know today. Agriculture in Africa (coffee, yams, sorghum, and more) and the Americas (corn, beans, squash, potatoes, llamas, and more) developed later in several areas, 4,500–7,000 years ago.

For most of these thousands of years, the work of cultivating, harvesting, storing, and distributing crops was performed by human and animal muscle power, along with hand tools and simple machines (Figure 9.4). This biologically powered agriculture is known as **traditional agriculture.** In the oldest form of traditional agriculture, known as *subsistence agriculture,* farming families produce only enough food for themselves and do not make use of large-scale irrigation, fertilizer, or teams of laboring animals. In contrast, *intensive traditional agriculture* sometimes uses draft animals and employs significant quantities of irrigation water and fertilizer, but it stops short of using fossil fuels. This type of agriculture aims to produce food for the farming family, as well as excess food to sell in the market.

Industrialized agriculture is newer still

The industrial revolution introduced large-scale mechanization and fossil fuel combustion to agriculture just as it did to industry, enabling farmers to replace horses and oxen with faster and more powerful means of cultivating, harvesting, transporting, and processing crops. Other advances facilitated irrigation and fertilizing, while the invention of chemical pesticides reduced competition from weeds and herbivory by insects and other crop pests. To be efficient, however, **industrialized agriculture** demands that vast fields be planted with single types of crops. The uniform planting of a single crop, termed **monoculture,** is distinct from the *polyculture* approach of much traditional agriculture, such as Native American

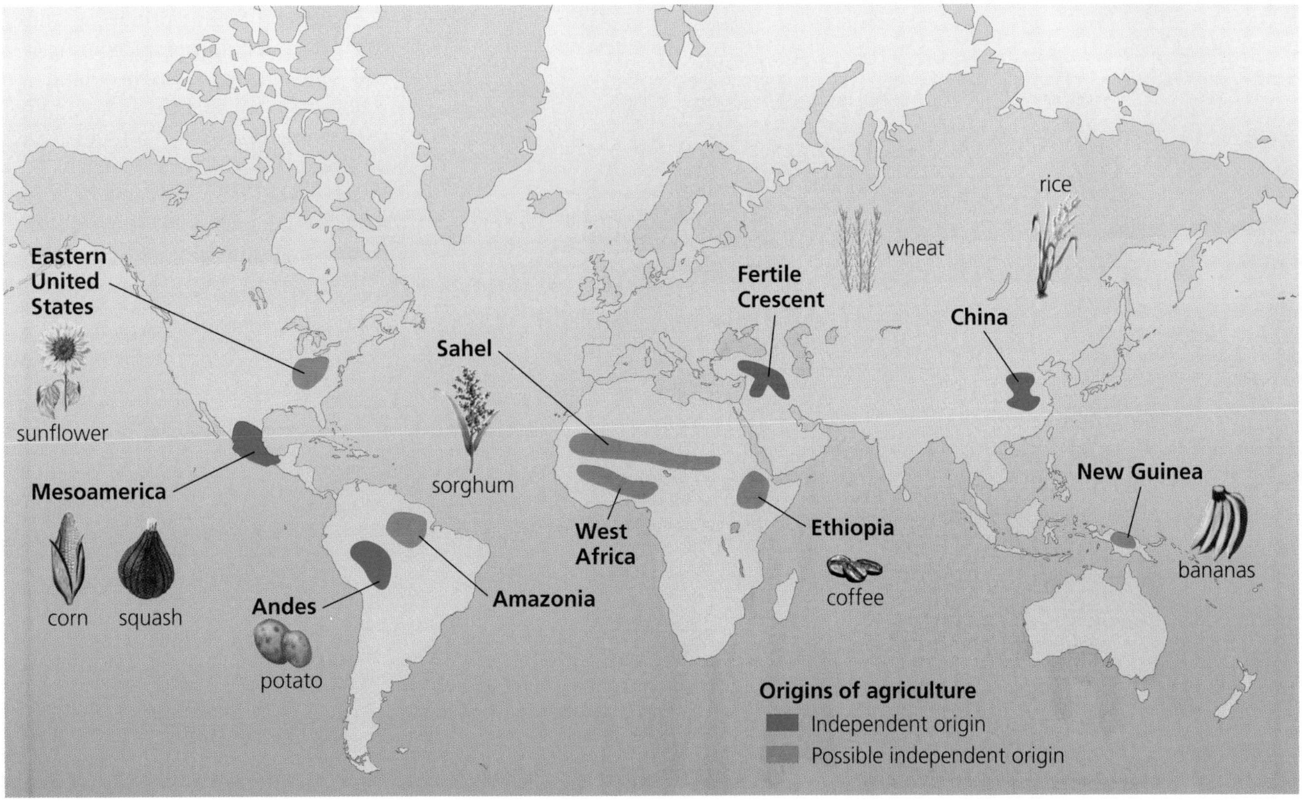

FIGURE 9.3 Agriculture appears to have originated independently in multiple locations throughout the world, as different cultures domesticated certain plants and animals from wild species living in their environments. This depiction summarizes conclusions from diverse sources of research on evidence for early agriculture. Areas where people are thought to have independently invented agriculture are colored green. (China may represent two independent origins.) Areas colored blue represent regions where people either invented agriculture independently or obtained the idea from cultures of other regions. A few of the many crop plants domesticated in each region are shown. Data from syntheses in Diamond, J., 1997. *Guns, germs, and steel.* New York: W.W. Norton; and Goudie, A. 2000. *The human impact,* 5th ed. Cambridge, MA: MIT Press.

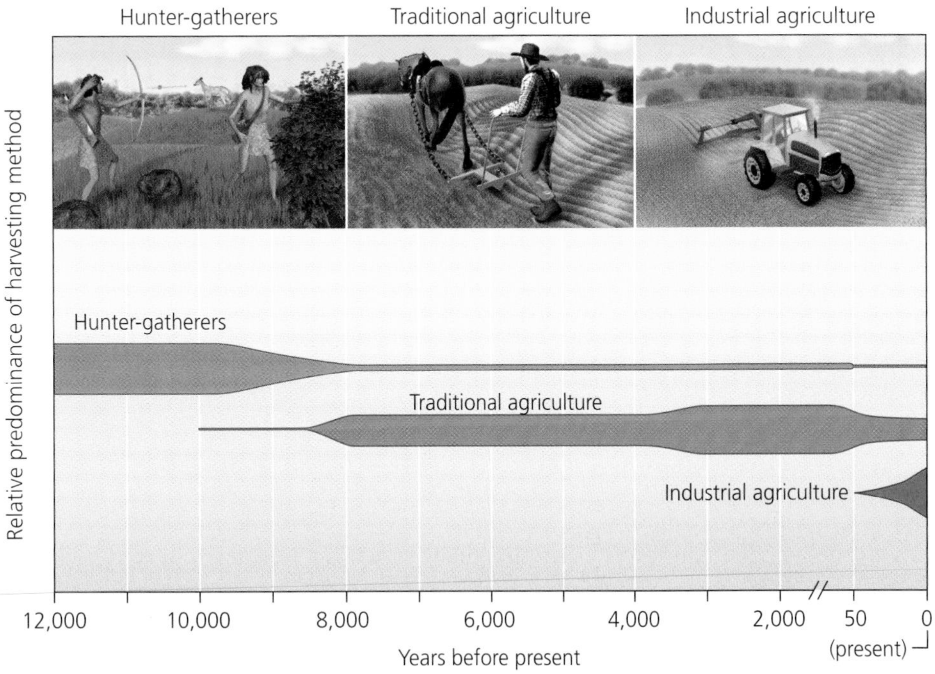

FIGURE 9.4 Hunting and gathering was the predominant human lifestyle until the onset of agriculture and sedentary living, which centered around farms, villages, and cities, beginning nearly 10,000 years ago. Over the millennia, societies practicing traditional agriculture gradually replaced hunter-gatherer cultures. Only within the past century has industrialized agriculture spread, replacing much traditional agriculture.

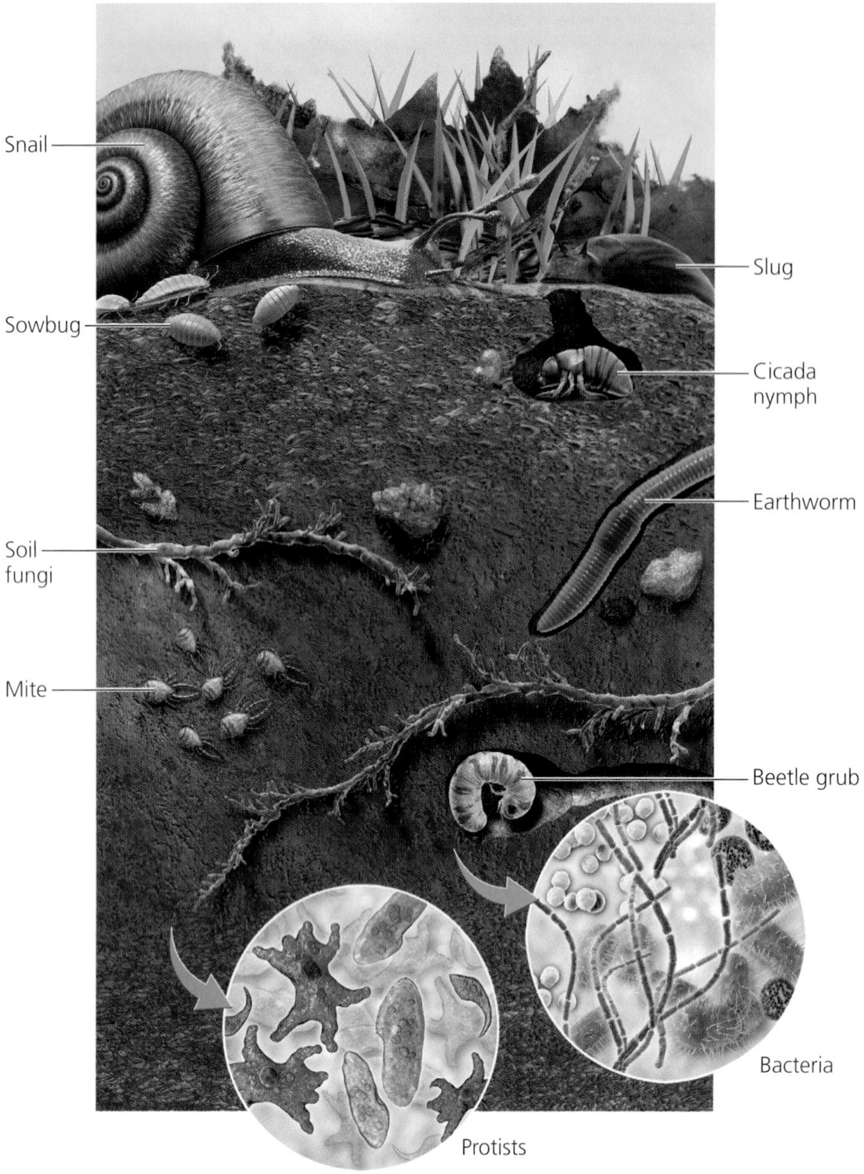

FIGURE 9.5 Soil is a complex mixture of organic and inorganic components and is full of living organisms whose actions help keep it fertile. In fact, entire ecosystems exist in soil. Most soil organisms, from bacteria to fungi to insects to earthworms, decompose organic matter. Many, such as earthworms, also help to aerate the soil.

farming systems that mixed maize, beans, squash, and peppers in the same fields. Today, industrialized agriculture occupies about 25% of the world's cropland.

Industrialized agriculture spread from developed nations to developing nations with the advent of the **green revolution,** a phenomenon we will explore in Chapter 10 (▸pp. 279–282). Beginning around 1950, the green revolution introduced new technology, crop varieties, and farming practices to the developing world. These advances dramatically increased yields per acre of cropland, and helped millions avoid starvation. But despite its successes, the green revolution is exacting a high price. The intensive cultivation of farmland is creating new problems and exacerbating old ones. Many of these problems pertain to the integrity of soil, which is the very foundation of our terrestrial food supply.

Soil as a System

We generally overlook the startling complexity of soils. We tend to equate the word *soil* with the word *dirt,* which connotes something useless or undesirable. Soil, however, is much more. It is not merely loose material derived from rock; it also contains a large biotic component, is molded by life, and is capable of supporting plant growth (Figure 9.5).

By volume, soil consists very roughly of half mineral matter and up to 5% organic matter. The rest consists of pore space taken up by air or water. The organic matter in soil includes living and dead microorganisms as well as decaying material derived from plants and animals. Most of us tend to think of soil as inert and lifeless, but a single teaspoon of soil can contain 100 million bacteria, 500,000 fungi, 100,000 algae, and 50,000 protists. Soil also provides habitat for earthworms, insects, mites, millipedes, centipedes, nematodes, sow bugs, and other invertebrates, as well as burrowing mammals, amphibians, and reptiles. The composition and quality of a region's soil can have as much influence on the region's ecosystems as do the climate, latitude, and elevation. In fact, because soil is composed of living and nonliving components that interact in complex ways, soil itself meets the definition of an ecosystem (▸pp. 189–191).

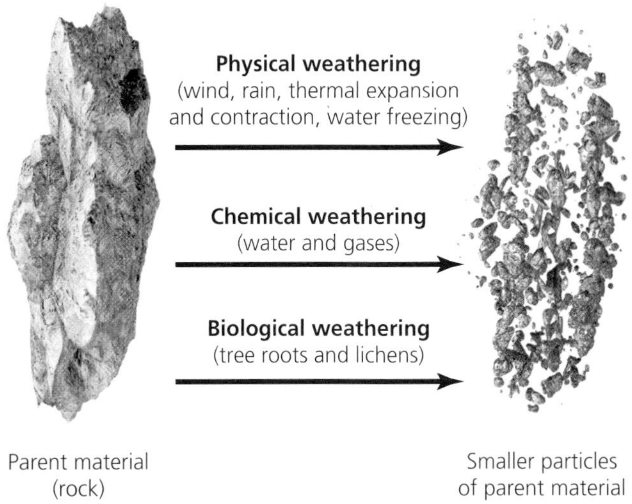

FIGURE 9.6 The weathering of parent material is the first step in soil formation. Rock is broken down into finer particles by physical, chemical, or biological means.

Soil formation is slow and complex

The formation of soil plays a key role in terrestrial primary succession (▸p. 163), which begins when the lithosphere's parent material is exposed to the effects of the atmosphere, hydrosphere, and biosphere. **Parent material** is the base geological material in a particular location. It can include lava or volcanic ash; rock or sediment deposited by glaciers; wind-blown dunes; sediments deposited by rivers, in lakes, or in the ocean; or **bedrock,** the continuous mass of solid rock that makes up Earth's crust.

The processes most responsible for soil formation are weathering, erosion, and the deposition and decomposition of organic matter. **Weathering** describes the physical, chemical, and biological processes that break down rocks and minerals, turning large particles into smaller particles (Figure 9.6).

Physical or *mechanical weathering* breaks rocks down without triggering a chemical change in the parent material. Wind and rain are two main forces of physical weathering. Daily and seasonal temperature variation aids their action by causing the thermal expansion and contraction of parent material. Areas with extreme temperature fluctuations experience rapid rates of physical weathering. Water freezing and expanding in cracks in rock also causes physical weathering.

Chemical weathering results when water or other substances chemically interact with parent material. Warm, wet conditions usually accelerate chemical weathering.

Biological weathering occurs when living things break down parent material by physical or chemical means. For instance, lichens initiate primary terrestrial succession by producing acid, which chemically weathers rock. A tree may accelerate weathering through the physical action of its roots as they grow and rub against rock. It may also accelerate weathering chemically through the decomposition of its leaves and branches or with chemicals it releases from its roots.

Weathering produces fine particles, and is the first step in soil formation. Another process often involved is **erosion,** the movement of soil from one area to another. Erosion may sometimes help form soil in one locality by depositing material it has depleted from another. Erosion is particularly prevalent when soil is denuded of vegetation, leaving the surface exposed to water and wind that may wash or blow it away. Although erosion can sometimes help build new soil in the long term, on the timescale of human lifetimes and for the natural systems on which we depend, erosion is generally perceived as a destructive process that reduces the amount of life that a given area of land can support.

Biological activity contributes to soil formation through the deposition, decomposition, and accumulation of organic matter. As plants, animals, and microbes die or deposit waste, this material is incorporated into the substrate, mixing with minerals. The deciduous trees of temperate forests, for example, drop their leaves each fall, making leaf litter available to the detritivores and decomposers (▸pp. 157–158) that break it down and incorporate its nutrients into the soil. In decomposition, complex organic molecules are broken down into simpler ones, including those that plants can take up through their roots. Partial decomposition of organic matter creates *humus,* a

Table 9.1 Five Factors That Influence Soil Formation

Factor	Effects
Climate	Soil forms faster in warm, wet climates. Heat speeds chemical reactions and accelerates weathering, decomposition, and biological growth. Moisture is required for many biological processes and can speed weathering.
Organisms	Earthworms and other burrowing animals mix and aerate soil, add organic matter, and facilitate microbial decomposition. Plants add organic matter and affect a soil's composition and structure.
Topographical relief	Hills and valleys affect exposure to sun, wind, and water, and they influence where and how soil moves. Steeper slopes result in more runoff and erosion and in less leaching, accumulation of organic matter, and differentiation of soil layers.
Parent material	Chemical and physical attributes of the parent material influence properties of the resulting soil.
Time	Soil formation takes decades, centuries, or millennia. The four factors above change over time, so the soil we see today may be the result of multiple sets of factors.

Adapted from: Jenny, H. 1941. *Factors of soil formation: A system of quantitative pedology.* New York: McGraw-Hill, Inc. Reprinted 1994 by Dover Publications, Mineola, New York.

dark, spongy, crumbly mass of material made up of complex organic compounds. Soils with high humus content hold moisture well and are productive for plant life.

Weathering, erosion, the accumulation and transformation of organic matter, and other processes that contribute to soil formation are all influenced by outside factors. Soil scientists cite five primary factors that influence the formation of soil (Table 9.1).

Weighing the Issues: Earth's Soil Resources

It can take 500 to 1,000 years to produce 1 inch of natural topsoil. Is soil a renewable resource? How do you think soil's long renewal time should influence its management? What types of practices encourage the formation of new topsoil?

A soil profile consists of distinct layers known as horizons

Once weathering has produced an abundance of small particles between the parent material and the atmosphere, then wind, water, and organisms begin to move and sort them. Eventually, distinct layers develop. Each layer of soil is known as a **horizon,** and the cross-section as a whole, from surface to bedrock, is known as a **soil profile.**

The simplest way to categorize soil horizons is to recognize A, B, and C horizons corresponding to topsoil, subsoil, and parent material. However, soil scientists often find it useful to subdivide the layers more finely, by their characteristics and the processes that take place within them. For our purposes we will discuss six major horizons, known as the O, A, E, B, C, and R horizons (Figure 9.7). Soils from different locations vary, and few soil profiles contain all six of these horizons, but any given soil contains at least some of them. Generally, the degree of weathering and the concentration of organic matter decrease as one moves downward in the soil profile.

Many soil profiles include an uppermost layer consisting mostly of organic matter, such as decomposing branches, leaves, and animal waste. This thin layer is designated the **O horizon** (O for *organic*) or litter layer. Just below the O horizon lies the **A horizon,** consisting of inorganic mineral components such as weathered substrate, with organic matter and humus from above mixed in. The A horizon is often referred to as **topsoil,** that portion of the soil that is most nutritive for plants and therefore most vital to ecosystems and agriculture. Topsoil takes its loose texture and dark coloration from its humus content. The O and A horizons are home to most of the countless organisms that give life to soil.

Beneath the A horizon in some soils lies the **E horizon.** E refers to *eluviation,* meaning loss, and the E horizon is characterized by the loss of some minerals and organic matter through leaching. **Leaching** is the process whereby solid particles suspended or dissolved in liquid are transported to another location. Generally in soils, the solvent is water, and leaching carries minerals downward. Soil that undergoes leaching is a bit like coffee grounds in a drip filter. When it rains, water infiltrates the soil (just as it infiltrates coffee grounds), dissolves

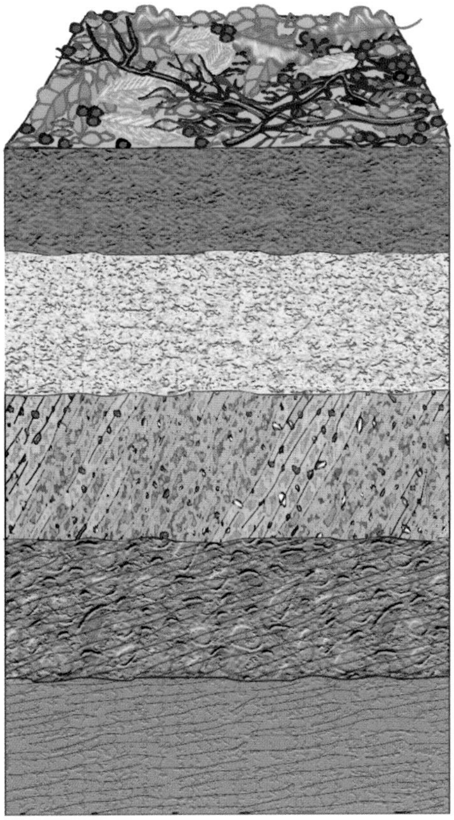

FIGURE 9.7 Mature soil consists of layers, or horizons, that have different compositions and characteristics. The number and depth of horizons vary from place to place and from soil type to soil type, producing different soil profiles. In general, organic matter and the degree of weathering decrease as one moves downward in a soil profile. The O horizon consists mostly of organic matter deposited by organisms. The A horizon, or topsoil, consists of some organic material mixed with mineral components. Minerals tend to leach out of the E horizon down into the B horizon. The C horizon consists largely of weathered parent material, which may overlie an R horizon of pure parent material.

some of its components, and carries them downward into the deeper horizons. Minerals commonly leached from the E horizon include iron, aluminum, and silicate clay. In some soils, minerals may be leached so rapidly that plants are deprived of nutrients. Minerals that leach rapidly from soils may be carried into groundwater and can pose human health threats when the water is extracted.

Minerals leaching from the A and E horizons move into the layer beneath them, the **B horizon,** or subsoil. This horizon collects and accumulates minerals from above. Often called the *illuviation horizon, zone of accumulation,* or *zone of deposition,* the B horizon contains a greater concentration of minerals and organic acids leached from above than does the E horizon.

The **C horizon,** if present, is located below the B horizon and consists of parent material unaltered or only slightly altered by the processes of soil formation. It therefore contains rock particles that are larger and less weathered than the layers above. The C horizon sits directly above the **R horizon,** or parent material.

Soil can be characterized by color, texture, structure, and pH

The six horizons presented above depict an idealized, "typical" soil, but soils display great variety. U.S. soil scientists classify soils into 12 major groups, based largely on the processes thought to form them. Within these 12 "orders," there are dozens of "suborders," hundreds of "great groups," and thousands of soils belonging to lower categories, all arranged in a hierarchical system. Scientists classify soils using properties such as color, texture, structure, and pH.

Soil color The color of soil (Figure 9.8) can indicate soil composition and sometimes soil fertility. Black or dark brown soils are usually rich in organic matter, whereas a

FIGURE 9.8 The color of soil may vary drastically from one location to another. A soil's composition affects its color. For instance, soils high in organic matter tend to be dark brown or black.

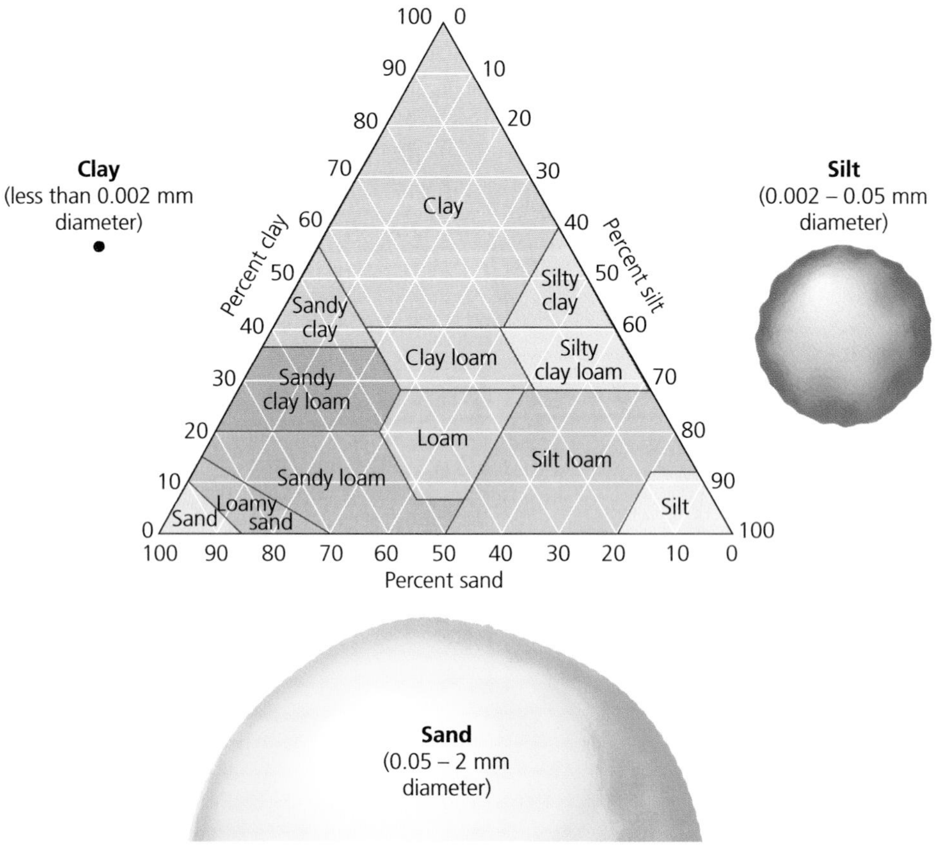

FIGURE 9.9 The texture of soil depends on its mix of particle sizes. Using the triangular diagram shown, scientists classify soil texture according to the relative proportions of sand, silt, and clay. After measuring the percentage of each type of particle size in a soil sample, a scientist can trace the appropriate white lines extending inward from each side of the triangle to determine what type of soil texture that particular combination of values creates. Loam is generally the best for plant growth, although some types of plants grow better in other textures of soil.

pale gray to white color often indicates leaching or low organic content. This color variation occurs among soil horizons in any given location and also among soils from different geographic locations. Long before modern analytical tests of soil content were developed, the color of topsoil provided farmers and ranchers with information about a region's potential to support crops and provide forage for livestock.

Soil texture Soil texture is determined by the size of particles and is the basis on which the United States Department of Agriculture (USDA) assigns soils to one of three general categories (Figure 9.9). **Clay** consists of particles less than 0.002 mm in diameter, **silt** of particles 0.002–0.05 mm, and **sand** of particles 0.05–2 mm. Sand grains, as any beachgoer knows, are large enough to see individually and do not adhere to one another. Clay particles, in contrast, readily adhere to one another and give clay a sticky feeling when moist. Soil with a relatively even mixture of the three particle sizes is known as **loam.**

For the farmer, soil texture influences a soil's "workability," its relative ease or difficulty of cultivation. Soil texture also influences *soil porosity,* a measure of the size of spaces between particles. In general, the finer the particles, the smaller the spaces between them. The smaller the spaces, the harder it is for water and air to travel through the soil, slowing infiltration and reducing the amount of oxygen available to soil biota. Conversely, soils with large particles allow water to pass through (and beyond the reach of plants' roots) too quickly. Thus, crops planted in sandy soils require frequent irrigation. For this reason, silty soils with medium-sized pores, or loamy soils with mixtures of pore sizes, are generally best for plant growth and crop agriculture.

Soil structure Soil structure is a measure of the "clumpiness" of soil. Some degree of structure encourages soil productivity, and biological activity helps promote this structure. However, soil clumps that are too large can discourage plant roots from establishing if soil particles are compacted too tightly together. Repeated tilling can compact soil and make it less able to absorb water. When farmers repeatedly till the same field at the same depth, they may end up forming *plowpan,* a hard layer that resists the infiltration of water and the penetration of roots.

Soil pH The degree of acidity or alkalinity (▸p. 98) influences a soil's ability to support plant growth. Plants can die in soils that are too acidic or alkaline, but moderate variation can influence the availability of nutrients for

FIGURE 9.10 In tropical forested areas, the traditional form of farming is *swidden* agriculture, as seen here in Surinam. In this practice, forest is cut, the plot is farmed for one to a few years, and the farmer then moves on to clear another plot, leaving the first to regrow into forest. This frequent movement is necessary because tropical soils are nutrient-poor, with nearly all nutrients held in the vegetation. Burning the cut vegetation adds nutrients to the soil, which is why this practice is often called "slash-and-burn" agriculture. At low population densities, this form of farming had little large-scale impact on forests, but at today's high population densities, it is a leading cause of deforestation.

plants' roots. During leaching, for instance, acids from organic matter may remove some nutrients from the sites of exchange between plant roots and soil particles, and water carries these nutrients deeper.

Regional differences in soil traits can affect agriculture

The characteristics of soil and soil profiles can vary from place to place. One example that bears on agriculture is the difference between soils of tropical rainforests and those of temperate grasslands. Although rainforest ecosystems have high primary productivity (▸ pp. 192–193), most of their nutrients are tied up in plant tissues and not in the soil. The soil of Amazonian rainforest in northern Brazil is in fact much less productive than the soil of grassland in Kansas.

To understand how this can be, consider the main differences between the two regions: temperature and rainfall. The enormous amount of rain that falls in the Amazon readily leaches minerals and nutrients out of the topsoil and E horizon. Those not captured by plants are taken quickly down to the water table, out of reach of most plants' roots. High temperatures speed the decomposition of leaf litter and the uptake of nutrients by plants, so amounts of humus remain small, and the topsoil layer remains thin.

Thus when forest is cleared for farming, cultivation quickly depletes the soil's fertility. This is why the traditional form of agriculture in tropical forested areas is *swidden* agriculture, in which the farmer cultivates a plot for one to a few years and then moves on to clear another plot, leaving the first to grow back to forest (Figure 9.10). This method may work well at low population densities, but with today's high human populations, soils may not be allowed enough time to regenerate. As a result, intensive agriculture has ruined the soils and forests of many tropical areas.

In temperate grassland areas such as the Kansas prairie, in contrast, rainfall is low enough that leaching is reduced and nutrients remain high in the soil profile, within reach of plants' roots. Plants take up nutrients and then return them to the topsoil when they die; this cycle maintains the soil's fertility. The thick, rich topsoil of temperate grasslands can be farmed repeatedly with minimal loss of fertility if proper farming techniques are used. However, growing and harvesting crops without returning adequate organic matter to the soil gradually depletes organic material, and leaving soil exposed to the elements increases erosion of topsoil. It is such consequences that farmers in southern Brazil, the U.S. Midwest, and other locations have sought to forestall through the use of reduced tillage.

Soil Degradation: Problems and Solutions

Scientists' studies of soil and the practical experience of farmers have shown that the most desirable soil for agriculture is a loamy mixture with a pH close to neutral that is workable and capable of holding nutrients. Many soils

The Science behind the Story

Measuring Erosion

Can a hedge of grass help stop soil erosion? Grass hedges are widely used, especially in the tropics, to trap eroding soil by slowing runoff from rain. But Jerry Ritchie, a researcher with the U.S. Department of Agriculture, wanted to measure how well they actually work. Going beyond the visible signs of soil loss, he used techniques ranging from simple measuring pins to complex radiation detectors to document erosion and calculate just how much soil has moved.

In the 1990s, Ritchie and his team of researchers began by measuring erosion around hedges planted near a set of gullies in Maryland. They relied on cheap, simple tools known as erosion pins (see the figure), which were developed in the 1960s and 1970s by scientists working for the U.N. Food and Agriculture Organization. Erosion pins are spikes that can be made from almost anything, including bamboo stakes or pieces of plastic pipe. The pins, each cut to a uniform length, are driven into the soil until their tops are level with the ground's surface. Over time, if soil in the area is eroding, the soil surface will recede, and the erosion pins will be increasingly exposed.

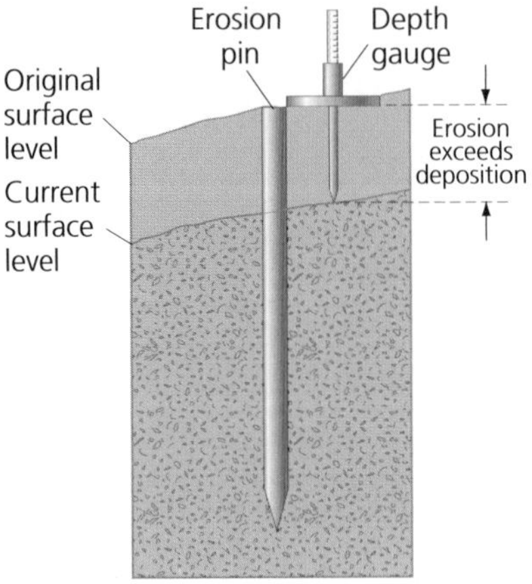

As soil erodes around an erosion pin, more of the pin is exposed, enabling the soil scientist to measure the amount of soil loss.

By using many pins over a wide area and averaging their readings, scientists can determine an overall erosion rate for the area.

Researchers often dig expansive holes, called catchpits, nearby. These pits are lined with plastic and serve as collection sites for eroding soil. Researchers measure the volume of soil that accumulates in the catchpits and compare that data with the extent of exposure on erosion pins. In the Maryland experiment, Ritchie used such techniques and found that 1–2 cm (0.4–0.8 in.) of soil was accumulating upslope from the hedges per year, indicating that the grass was trapping soil as it moved.

Erosion pins and catchpits work well in one spot but are impractical over a large region. To get the best evidence of erosion on a wide scale, scientists have turned to measuring a modern-day leftover rarely considered an environmental benefit—nuclear fallout from atomic

deviate from this ideal and prevent land from being arable or limit the productivity of arable land. Increasingly, limits to productivity are being set by human impact that has degraded many once-excellent soils. Common problems affecting soil productivity include erosion, desertification, salinization, waterlogging, nutrient depletion, structural breakdown, and pollution.

Erosion can degrade ecosystems and agriculture

Erosion, as we have noted, is the removal of material from one place and its transport toward another by the action of wind or water. *Deposition* is the arrival of eroded material at its new location. Erosion and deposition are natural processes that in the long run can help create soil. Flowing water can deposit eroded sediment in river valleys and deltas, producing rich and productive soils. This is why floodplains are excellent for farming and why flood-control measures can decrease the productivity of agriculture in the long run.

However, erosion often becomes a problem locally for ecosystems and agriculture because it nearly always takes place much more quickly than soil is formed. Furthermore, erosion tends to remove topsoil, the most valuable soil layer for living things. People have increased the vulnerability of fertile lands to erosion through three widespread practices:

1. Overcultivating fields through poor planning or excessive plowing, disking, or harrowing
2. Overgrazing rangelands with more livestock than the land can support
3. Clearing forested areas on steep slopes or with large clear-cuts (► pp. 355–357)

weapons testing. In 1945, the United States exploded the world's first nuclear bomb, and since then more than 2,000 nuclear devices have been tested by the United States, the former Soviet Union, and other nations. Nuclear weapons testing has spread radioactive material through the atmosphere worldwide, and fallout from the atmosphere has covered Earth's surface with minuscule but measurable amounts of nuclear debris.

Fallout includes cesium-137, a radioactive isotope of the element cesium. As discussed in Chapter 4 (▸pp. 94–95), isotopes of chemical elements are often tracked and measured by environmental scientists. Cesium-137, a product of nuclear fuel and weapons reactions, has a half-life of 30 years—a duration that enables soil scientists to use the isotope as a universal environmental tracer for erosion and sediment deposits. Soil tends to absorb cesium-137 quickly and evenly, so if soil in an area hasn't moved or been heavily disturbed, testing will show fairly uniform levels of the isotope. But if such measurements are not uniform—with some areas showing lower concentrations of cesium-137 and others showing higher concentrations—then erosion may be at work. Ritchie decided to use cesium-137 tests on the area around the hedges and compare the results against his physical soil measurements.

In studies involving cesium-137, soil samples are tested using a gamma spectrometer. This device measures gamma rays, which serve as unique signatures of energy emitted by chemical elements in soil or rock. As they are emitted, gamma rays show up as sharp emission lines on a spectrum. The energy represented in these emissions reflects which elements are present, and the intensity of the lines reveals the concentration of the element. Thus it is possible to calculate the amount of an isotope such as cesium-137 in a test sample. In erosion tests, each sample from the study area is measured for cesium-137, and those levels are compared to baseline levels for the region. By pinpointing places in the study area with lower or higher levels of accumulated cesium, scientists can detect where soil has moved and how much has shifted.

In Maryland, the radioactive testing helped Ritchie determine that hedges may offer only partial help against erosion. Although his team's physical measurements of soil accumulation showed soil being deposited near the hedges, the cesium-137 tests revealed that the area around the hedges had nonetheless undergone a net loss of soil over a period of four decades. Grass hedges can help, Ritchie wrote when releasing his findings for the Federal Agricultural Research Service in 2000, but they "should not be seen as a panacea."

Erosion can be gradual and hard to detect. For example, an erosion rate of 12 tons/ha (5 tons/acre) removes only a penny's thickness of soil. In many parts of the world, scientists, farmers, and extension agents are measuring erosion rates in hopes of identifying areas in danger of serious degradation before they become too badly damaged (see "The Science behind the Story," above).

Soil erodes by several mechanisms

Grasslands, forests, and other plant communities protect soil from wind and water erosion. Vegetation breaks the wind and slows water flow, while plant roots hold soil in place and take up water. Removing plant cover will nearly always accelerate erosion. Several types of erosion can occur, including wind erosion and four principal kinds of water erosion (Figure 9.11).

Splash erosion (Figure 9.11a) occurs when rain striking the soil surface breaks aggregates into smaller sizes. Soil particles are released and fill in gaps between the remaining clumps, decreasing a soil's ability to absorb water. In *sheet erosion*, or overland flow (Figure 9.11b), surface water flows downhill, washing topsoil away in relatively uniform layers. *Rill erosion* (Figure 9.11c) takes place when water runs along small furrows on the surface of the topsoil, gradually deepening and widening the furrows into rills, or small channels. Rills can merge to form larger and larger channels and eventually gullies. *Gully erosion* (Figure 9.11d) is least common but causes the most dramatic and visible changes in the landscape.

Research indicates that rill erosion has the greatest potential to move topsoil, followed by sheet erosion and splash erosion, respectively. All types of water erosion—particularly gully erosion—are more likely to occur where

(a) Splash erosion

(b) Sheet erosion

(c) Rill erosion

(d) Gully erosion

FIGURE 9.11 The erosion of soil by water can be classified into at least four categories. Splash erosion (**a**) occurs as raindrops strike the ground with enough force to dislodge small amounts of soil. Sheet erosion (**b**) results when thin layers of water traverse broad expanses of sloping land. Rill erosion (**c**) leaves small pathways along the surface where water has carried topsoil away. Gully erosion (**d**) cuts deep into soil, leaving large gullies that can expand as erosion proceeds.

slopes are steeper. In general, steeper slopes, greater precipitation intensities, and sparser vegetative cover all lead to greater water erosion.

One study conducted in the early 1990s determined that at erosion rates typical for the United States, U.S. croplands lose about 2.5 cm (1 in.) of topsoil every 15–30 years, reducing corn yields by 4.7–8.7% and wheat yields by 2.2–9.5%. According to U.S. government figures, erosion rates in the United States declined from 9.1 tons/ha (3.7 tons/acre) in 1982 to 5.9 tons/ha (2.4 tons/acre) in 2001, thanks to soil conservation measures discussed below. Yet in spite of these measures, U.S. farmlands still lose 6 tons of soil for every ton of grain harvested.

Soil erosion is a global problem

Erosion has become a major problem in many areas of the world, including Australia, sub-Saharan Africa, central Asia, India, the Middle East, and parts of South America, Central America, Europe, and the United States. In total, more than 19 billion ha (47 billion acres) of the world's croplands suffer from erosion and other forms of soil degradation resulting from human activities. Between 1957 and 1990, China lost as much arable farmland as exists in Denmark, France, Germany, and the Netherlands combined. In Kazakhstan, central Asia's largest nation, industrial cropland agriculture imposed on land better suited for grazing caused tens of millions of hectares to be degraded by wind erosion. For Africa, projections indicate that soil degradation over the next 40 years could reduce crop yields by half. Couple these declines in soil quality and crop yields with the rapid population growth occurring in many of these areas, and we begin to see why some observers describe the future of agriculture as a crisis situation.

In today's world, humans are the primary cause of erosion, and we have accelerated it to unnaturally high rates. A 2004 study by geologist Bruce Wilkinson analyzed prehistoric erosion rates from the geologic record and compared these with modern rates. Wilkinson concluded that humans are over 10 times more influential at moving soil than are all other natural processes on the surface of the planet combined.

Arid land may lose productivity by desertification

Much of the world's population lives and farms in arid environments, where **desertification** is a concern. This term describes a loss of more than 10% productivity due to erosion, soil compaction, forest removal, overgrazing, drought, salinization, climate change, depletion of water sources, and other factors. Severe desertification can result in the expansion of desert areas or creation of new ones in areas that once supported fertile land. This process has occurred in many areas of the Middle East that have been inhabited, farmed, and grazed for long periods of time. To appreciate the cumulative impact of centuries of traditional agriculture, we need only look at the present desertified state of that portion of the Middle East where agriculture originated, nicknamed the "Fertile Crescent." These arid lands—in present-day Iraq, Syria, Turkey, Lebanon, and Israel—are not so fertile anymore.

Arid and semiarid lands are prone to desertification because their precipitation is too meager to meet the demand for water from growing human populations. According to the United Nations Environment Programme (UNEP), 40% of Earth's land surface can be classified as drylands, arid areas that are particularly subject to degradation. Declines of soil quality in these areas have endangered the food supply or well-being of more than 1 billion people around the world. Of the affected lands, most degradation results from wind and water erosion (Figure 9.12).

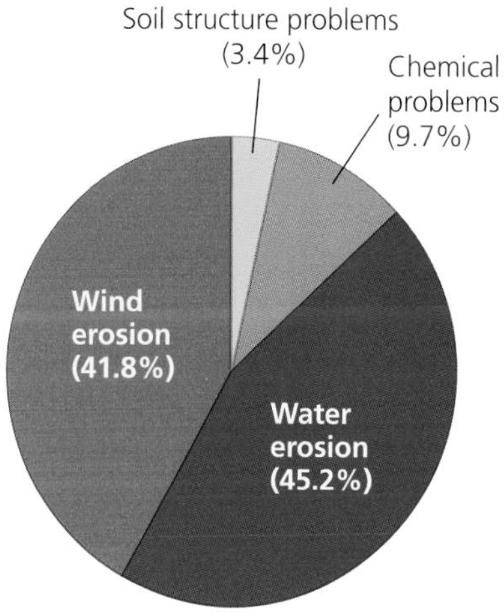

FIGURE 9.12 Soil degradation on drylands is due primarily to erosion by wind and water. Data from United Nations Environment Programme. 2002. *Tackling land degradation and desertification.* Washington and Rome: Global Environment Facility and International Fund for Agricultural Development.

It has been estimated that desertification affects fully one-third of the planet's land area, impinging on people in 110 countries. Desertification cost the world's people at least $300–600 billion in income just in the period 1978–1991, UNEP estimates. China alone loses $6.5 billion annually from desertification. In its western reaches, desert areas are expanding and joining one another because of overgrazing from over 400 million goats, sheep, and cattle. In the Sistan Basin along the border of Iran and Afghanistan, an oasis that supported a million livestock recently turned barren in just 5 years, and windblown sand buried over 100 villages. In Africa, the continent's most populous nation, Nigeria, loses an amount of land equal to half the state of Delaware each year to the expanding Sahara Desert. In Kenya, overgrazing and deforestation fueled by rapid population growth has left 80% of its land vulnerable to desertification. In a positive feedback cycle, the soil degradation forces ranchers to crowd onto more marginal land and farmers to reduce fallow periods, both of which further exacerbate soil degradation.

As a result of desertification, in recent years gigantic dust storms from denuded land in China have blown across the Pacific Ocean to North America, and dust storms from Africa's Sahara Desert have blown across the

Atlantic Ocean to the Caribbean Sea. Such massive dust storms occurred in the United States during the Dust Bowl days of the early 20th century, when desertification shook American agriculture and society to their very roots.

The Dust Bowl was a monumental event in the United States

Prior to large-scale cultivation of the southern Great Plains of the United States, native prairie grasses of this temperate grassland region held erosion-prone soils in place. In the late 19th and early 20th centuries, however, many homesteading settlers arrived in Oklahoma, Texas, Kansas, New Mexico, and Colorado with hopes of making a living there as farmers. Between 1879 and 1929, cultivated area in the region soared from around 5 million ha (12 million acres) to 40 million ha (100 million acres). Farmers grew abundant wheat, and ranchers grazed many thousands of cattle, sometimes expanding onto unsuitable land. Both types of agriculture contributed to erosion by removing native grasses and breaking down soil structure.

During the early 1930s, a drought in the region exacerbated the ongoing human impacts on the soil. The region's strong winds began to carry away millions of tons of topsoil. Dust storms traveled up to 2,000 km (1,250 mi), blackening rain and snow as far away as New York and Vermont. Some areas in the affected states lost as much as 10 cm (4 in.) of topsoil in a few short years (Figure 9.13). The affected region in the Great Plains became known as the **Dust Bowl,** a term now also used for the historical event itself. The "black blizzards" of the Dust Bowl destroyed livelihoods and caused many people to suffer a type of chronic lung irritation and degradation known as dust pneumonia, similar to the silicosis that afflicts coal miners exposed to high concentrations of coal dust. Large numbers of farmers were forced off their land, and many who remained had to rely on government assistance programs to survive.

The Soil Conservation Service pioneered measures to slow soil degradation

In response to the devastation in the Dust Bowl, the U.S. government, along with state and local governments, increased its support of research into soil conservation measures. The U.S. Congress also passed the Soil Conservation Act of 1935. This act described soil erosion as a threat to the nation's well-being and established the Soil Conservation Service (SCS) to address the problem. The new agency worked closely with farmers to develop conservation plans for individual farms, following several aims and principles:

- Assess the land's resources, its problems, and opportunities for conservation.
- Draw on science to prepare an integrated plan for the property.
- Work closely with land users to ensure that conservation plans harmonize with the users' objectives.
- Implement conservation measures on individual properties to contribute to the overall quality of life in the watershed or region.

The early teams that the SCS formed to combat erosion typically included soil scientists, forestry experts, engineers, economists, and biologists. These teams were among the earliest examples of interdisciplinary approaches to environmental problem solving. The first director of the SCS, Hugh Hammond Bennett, was an innovator and

FIGURE 9.13 Drought combined with poor agricultural practices brought devastation and despair to millions of U.S. farmers in the 1930s, especially in the Dust Bowl region of the southern Great Plains. The tragedy spurred the development of soil conservation practices that have since been put into place in the United States and around the world.

evangelist for soil conservation. Under his leadership, the agency promoted soil-conservation practices through county-based **conservation districts.** These districts operate with federal direction, authorization, and funding, but they are organized by state law. The districts implement soil conservation programs locally and aim to empower local residents to plan and set priorities in their home areas. In 1994 the SCS was renamed the *Natural Resources Conservation Service,* and its responsibilities were expanded to include water quality protection and pollution control.

The SCS served as a model for similar efforts elsewhere in the world (Figure 9.14). Southern Brazil's no-till movement came about through local grass-roots organization by farmers, with the help of agronomists and government extension agents who provided them information and resources. In this model of collaboration between local farmers and trained experts, 8,000 Friends of the Land clubs now exist in Paraná and Rio Grande do Sul, and 7,700 in Santa Catarina. Many of these groups are delineated by the boundaries of the more than 3,000 small-scale watersheds *(microbacias)* in which they farm.

No-till agriculture as practiced in southern Brazil is one of many approaches to soil conservation. Hugh Hammond Bennett advocated a complex approach, combining techniques such as crop rotation, contour farming, strip-cropping, terracing, grazing management, and reforestation, as well as wildlife management. Such measures have been widely applied in many places around the world.

FIGURE 9.14 Government agricultural extension agents assist farmers by providing information on the newest research and techniques that can help them farm productively while minimizing damage to the land. Such specialists have helped U.S. farmers since the Dust Bowl and now assist farmers worldwide. Here, an extension agent from Colombia's Instituto Colombiano Agropecuario inspects yuca plants grown by farmer Pedro Gomez on a farm in Valle del Cauca.

Farmers can protect soil against degradation in various ways

Several farming techniques can reduce the impacts of conventional cultivation on soils (Figure 9.15). Some of these have been promoted by the SCS since the Dust Bowl. Some, like no-till farming in Brazil, are finding popularity more recently. Others have been practiced by certain cultures for centuries.

Crop rotation The practice of alternating the kind of crop grown in a particular field from one season or year to the next is **crop rotation** (Figure 9.15a). Rotating crops can return nutrients to the soil, break cycles of disease associated with continuous cropping, and minimize the erosion that can come from letting fields lie fallow. Many U.S. farmers rotate their fields between wheat or corn and soybeans from one year to the next. Soybeans are legumes, plants that have specialized bacteria on their roots that can fix nitrogen (▸pp. 198–200). Soybeans revitalize soil that the previous crop had partially depleted of nutrients. Crop rotation also reduces insect pests; if an insect is adapted to feed and lay eggs on one particular crop, planting a different crop will leave its offspring with nothing to eat.

In a practice similar to crop rotation, southern Brazil's farmers plant "cover crops" designed to prevent erosion and forestall nitrogen loss from leaching during times of the year when the main crops are not growing. Santa Catarina's extension agents have worked with farmers to test over 60 species as cover crops, including both legumes and other plants, such as oats and turnips.

Contour farming Water running down a hillside can easily carry soil away, particularly if there is too little vegetative cover to hold the soil in place. Thus, sloped agricultural land is especially vulnerable to erosion. Several methods have been developed for farming on slopes. **Contour farming**

FIGURE 9.15 The world's farmers have adopted various strategies to conserve soil. Rotating crops such as soybeans and corn (**a**) helps restore soil nutrients and reduce impacts of crop pests. Contour farming (**b**) reduces erosion on hillsides. Intercropping (**c**) can reduce soil loss while maintaining soil fertility. Terracing (**d**) minimizes erosion in steep mountainous areas. Shelterbelts (**e**) protect against wind erosion. In (**f**), corn grows up from amid the remnants of a "cover crop" used in no-till agriculture.

(Figure 9.15b) consists of plowing furrows sideways across a hillside, perpendicular to its slope, to help prevent formation of rills and gullies. The technique is so named because the furrows follow the natural contours of the land. In contour farming, the downhill side of each furrow acts as a small dam that slows runoff and catches soil before it is carried away. Contour farming is most effective on gradually sloping land with crops that grow well in rows. Extension agents from Santa Catarina in Brazil have helped those farmers who still plow to switch to contour plowing and to plant barriers of grass along their contours.

Intercropping Farmers may also gain protection against erosion by **intercropping,** planting different types of crops in alternating bands or other spatially mixed arrangements (Figure 9.15c). Intercropping helps slow erosion by providing more complete ground cover than does a single crop. Like crop rotation, intercropping offers the additional benefits of reducing vulnerability to insect and disease incidence, and, when a nitrogen-fixing legume is one of the crops, of replenishing the soil. Some southern Brazilian farmers intercrop food crops with cover crops. The cover crops are physically mixed with primary crops, which include maize, soybeans, wheat, onions, cassava, grapes, tomatoes, tobacco, and orchard fruit.

Terracing On extremely steep terrain, terracing (Figure 9.15d) is the most effective method for preventing erosion. Terraces are level platforms, sometimes with raised edges, that are cut into steep hillsides to contain water from irrigation and precipitation. **Terracing** transforms slopes into series of steps like a staircase, enabling farmers to cultivate hilly land without losing huge amounts of soil to water erosion. Terracing is common in ruggedly mountainous regions, such as the foothills of the Himalayas and the Andes, and has been used for centuries by farming communities in such areas. Terracing is labor-intensive to establish but in the long term is likely the only sustainable way to farm in mountainous terrain.

Shelterbelts A widespread technique to reduce erosion from wind is to establish **shelterbelts** or *windbreaks* (Figure 9.15e). These are rows of trees or other tall, perennial plants that are planted along the edges of fields to slow the wind. Shelterbelts have been widely planted across the U.S. Great Plains, where fast-growing species such as poplars are often used. Shelterbelts have also been combined with intercropping in a practice known as *agroforestry,* or *alley cropping.* In this approach, fields planted in rows of mixed crops are surrounded by or interspersed with rows of trees that provide fruit, wood, or protection from wind. Such methods have been used in India, Africa, and in Brazil, where coffee growers near a national conservation area have established farming systems combining farming and forestry.

Table 9.2 No-Till Farming in Brazil

Benefits of no-till farming

- Conserves biodiversity in soil and in terrestrial and aquatic ecosystems
- Produces sustainable, high crop yields
- Heightens environmental awareness among farmers
- Provides shelter and winter food for animals
- Reduces irrigation demands by 10–20%
- Crop residues act as a sink for carbon (1 metric ton/ha)
- Reduces fossil fuel use by 40–70%
- Enhances food security by increasing drought resistance
- Reduces erosion by 90%

Other benefits arising from the reduction in erosion

- Reduces silt deposition in reservoirs
- Reduces water pollution from chemicals
- Increases groundwater recharge and lessens flooding
- Increases sustained crop yields and lowers food prices
- Lowers costs of treating drinking water
- Reduces costs of maintaining dirt roads
- Eliminates dust storms in towns and cities
- Increases efficiency in use of fertilizer and machinery

Modified from Shaxson, T. F. 1999. The roots of sustainability: Concepts and practice: Zero tillage in Brazil, *ABLH Newsletter ENABLE; World Association for Soil and Water Conservation (WASWC) Newsletter.*

Reduced tillage To plant using the zero-tillage method (Figure 9.15f), a tractor pulls a "no-till drill" that cuts long furrows through the O horizon of dead weeds and crop residue and the upper levels of the A horizon. The device drops seeds into the furrow and closes the furrow over the seeds. Often a localized dose of fertilizer is added to the soil along with the seed. Reduced-tillage agriculture disturbs the soil surface more than no-tillage does, but less than conventional cultivation does. By increasing organic matter and soil biota while reducing erosion, no-till and reduced tillage farming can build soil up, restore it, and improve it. Based on the Brazilian experience, proponents of no-till farming have claimed that the practice offers a number of benefits (Table 9.2).

No-till and reduced tillage methods were pioneered in the United States and United Kingdom, where no-till is still rare but where reduced tillage has been slowly spreading for decades. Today nearly half of U.S. acreage is farmed with reduced-tillage methods. As the appeal of no-till farming spread in Brazil, it also spread in neighboring Argentina and Paraguay. In Argentina, the area under no-till farming exploded from 100,000 ha (247,000 acres) in 1990 to 7.3 million ha (18.0 million acres) in 1999, covering 30% of all arable land in the country. The results there parallel those in Brazil: increased crop yields, reduced erosion, enhanced soils, and a healthier environment. Maize yields grew by 37% and soybean yields by 11%, while costs to farmers fell by 40–57%. Erosion, pesticide use, and water pollution declined. As in Brazil, the techniques spread largely because of the actions of farmers themselves and their national no-till farmers' organization.

Critics of no-till and reduced-tillage farming in the United States have noted that these techniques often require substantial use of chemical herbicides (because weeds are not physically removed from fields) and synthetic fertilizer (because other plants take up a significant portion of the soil's nutrients). In many industrialized countries, this has indeed been the case. Proponents of the Brazilian program, however, assert that it does not always need to be so. Southern Brazil's farmers have departed somewhat from the industrialized model by relying more heavily on *green manures* (dead plants as fertilizer) and by rotating fields with cover crops, including nitrogen-fixing legumes. The manures and legumes nourish the soil, and cover crops also reduce weeds by taking up space the weeds might occupy. Critics maintain, however, that green manures are generally not practical for large-scale intensive agriculture. Certainly, reduced tillage methods work well in some areas but not in others, and they work better with some crops than with others. Farmers will do best by educating themselves on the options and doing what is best for their particular crops on their own land.

The methods we have described to combat soil degradation can be used in combination. When they have been, the results have sometimes been dramatic. One town in the Guatemalan highlands that established shelterbelts, crop rotation, and cover crops with the help of a U.S.-based nonprofit organization improved its corn production from 0.4 tons/ha (1.0 tons/acre) in 1972 to 2.5 tons/ha (6.2 tons/acre) in 1979. It went on to improve production to 4.5 tons/ha (11.1 tons/acre) by 1994.

Weighing the Issues:
How Would You Farm?

You are a farmer owning land on both sides of a steep ridge. You want to plant a sun-loving crop on the sunny, but very windy, south slope of the ridge and a crop that needs a great deal of irrigation on the north slope. What type of farming techniques might maximize conservation of your soil? What other factors might you want to know about before you decide to commit to one or more methods?

FIGURE 9.16 Vast swathes of countryside in western China have been planted with fast-growing poplar trees. These "reforestation" efforts do not create ecologically functional forests—the plantations are too biologically simple—but they do greatly slow soil erosion.

Erosion-control practices protect and restore plant cover

Farming methods to control erosion make use of the general principle that maximizing vegetative cover will protect soils, and this principle has been applied widely beyond farming. It is common throughout the developed world to stabilize eroding banks along creeks and roadsides by planting plants to anchor the soil. In areas with severe and widespread erosion, some nations have planted vast plantations of fast-growing trees. China has embarked on the world's largest tree-planting program to slow its soil loss (Figure 9.16). Although such "reforestation" efforts do help slow erosion, they do not at the same time produce ecologically functioning forests, because tree species are selected only for their fast growth and are planted in monocultures.

Irrigation has boosted productivity but has also caused long-term soil problems

Erosion is not the only threat to the health and integrity of soils. Soil degradation can result from other factors as well, such as impacts caused by our application of water to crops. The artificial provision of water to support agriculture is known as **irrigation.** Some crops, such as rice and cotton, require large amounts of water, whereas others, such as beans and wheat, require relatively little. Other factors influencing the amount of water required for growth include the rate of evaporation, as determined by climate, and the soil's ability to hold water and make it available to plant roots. If the climate is too dry or too much water evaporates or runs off before it can be absorbed into the soil, crops may require irrigation. By irrigating crops, people have managed to turn previously dry and unproductive regions into fertile farmland. Seventy percent of all freshwater withdrawn by people is used for irrigation. Irrigated acreage has increased dramatically around the world, reaching 276 million ha (683 million acres) in 2002, greater than the entire area of Mexico and Central America. We will examine irrigation further in Chapter 15 (▸ pp. 440–441).

If some water is good for plants and soil, it might seem that more must be better. But this is not necessarily the case; there is indeed such a thing as too much water. Overirrigation in poorly drained areas can cause or exacerbate certain soil problems. Soils too saturated with water may become waterlogged. When **waterlogging** occurs, the water table is raised to the point that water bathes plant roots, depriving them of access to gases and essentially suffocating them. If it lasts long enough, waterlogging can damage or kill plants.

An even more frequent problem is **salinization,** the buildup of salts in surface soil layers. In dryland areas where precipitation is minimal and evaporation rates are high, water evaporating from the soil's A horizon may pull water from lower horizons upward by capillary action. As this water rises through the soil, it carries dissolved salts, and when it evaporates at the surface, those salts precipitate and are left at the surface. Irrigation in arid areas generally hastens salinization, because it provides repeated doses of moderate amounts of water, which dissolve salts in the soil and gradually raise them to the surface. Moreover, because irrigation water often contains some dissolved salt in the first place, irrigation introduces new sources of salt to the soil. Overirrigation and waterlogging can worsen salinization problems, and in many areas of farmland, soil is turning whitish with encrusted salt. Salinization now inhibits agricultural production on one-fifth of all irrigated cropland globally, costing more than $11 billion annually.

Salinization is easier to prevent than to correct

The remedies for mitigating salinization once it has occurred are more expensive and difficult to implement than the techniques for preventing it in the first place. The best way to prevent salinization is to avoid planting crops that require a great deal of water in areas that are prone to the problem. A second way is to irrigate with water that is as low as possible in salt content. A third way is to irrigate efficiently, supplying no more water than the crop requires, thus minimizing the amount of water that evaporates and hence the amount of salt that accumulates in the topsoil. Currently, irrigation efficiency worldwide is low; only 43% of the water applied actually gets used by plants. Drip irrigation systems (Figure 9.17) that target water directly to plants are one solution to the problem. These systems allow more control over where water is aimed and waste far less water. Once considered expensive to install, they are becoming cheaper, such that more farmers in developing countries will be able to afford them.

(a) Conventional irrigation

(b) Drip irrigation

FIGURE 9.17 Currently, less than half the water we apply in irrigation actually gets taken up by plants. Conventional methods that lose a great deal of water to evaporation (**a**) are now being replaced by more efficient ones in which water is more precisely targeted to plants. In drip irrigation systems, such as this one watering grape vines in California (**b**), hoses are arranged so that water drips from holes in the hoses directly onto the plants that need the water.

If salinization has occurred, one potential way to mitigate it would be to stop irrigating and wait for rain to flush salts from the soil. However, this solution is unrealistic because salinization generally becomes a problem in dryland areas where precipitation is never adequate to flush soils. A better option may be to plant salt-tolerant plants, such as barley, that can be used as food or pasture. A third option is to bring in large quantities of less-saline water with which to flush the soil. However, using too much water may cause waterlogging. As is the case with many environmental problems, preventing salinization is easier than correcting it after the fact.

Weighing the Issues: Measuring and Regulating Soil Quality

The U.S. EPA has adopted measures of air quality and water quality and has set legal standards for allowable levels of various pollutants in air and water. Could such standards be developed for soil quality? If so, what properties should be measured to inform such standards? Should such standards be developed? Why or why not?

Agricultural fertilizers boost crop yields but can be over-applied

Salinization is not the only source of chemical damage to soil. Overapplying fertilizers can also chemically damage soils. Plants grow through photosynthesis, requiring sunlight, water, and carbon dioxide, but they also require nitrogen, phosphorus, and potassium, as well as smaller amounts of over a dozen other nutrients. Plants remove these nutrients from soil as they grow, and leaching likewise removes nutrients. If agricultural soils come to contain too few nutrients, crop yields decline. Therefore, a great deal of effort has aimed to enhance nutrient-limited soils by adding **fertilizer,** any of various substances that contain essential nutrients (Figure 9.18).

There are two main types of fertilizers. **Inorganic fertilizers** are mined or synthetically manufactured mineral supplements. **Organic fertilizers** consist of natural materials (largely the remains or wastes of organisms) and include animal manure; crop residues; fresh vegetation, known as *green manure;* and *compost,* a mixture produced when decomposers break down organic matter, including food and crop waste, in a controlled environment. Organic fertilizers can provide some benefits that inorganic fertilizers cannot. The proper use of compost improves soil structure, nutrient retention, and water-retaining capacity, helping to prevent erosion. As a form of recycling, composting reduces the amount of waste consigned to landfills

FIGURE 9.18 Farmers often add nutrients to soils with fertilizers. Organic fertilizers such as manure and vegetation may be used, or synthetically manufactured chemicals may be applied to supply nitrogen, phosphorus, and other nutrients, as this North Dakota farmer is doing.

and incinerators (▸ p. 658). However, organic fertilizers are no panacea. For instance, manure, when applied in amounts needed to supply sufficient nitrogen for a crop, may introduce excess phosphorus that can run off into waterways. Inorganic fertilizers are generally more susceptible than are organic fertilizers to leaching and runoff, and they are somewhat more likely to cause unintended off-site impacts. Inorganic and organic fertilizer use is growing globally (Figure 9.19). Unfortunately, its mismanagement is causing increasingly severe pollution problems.

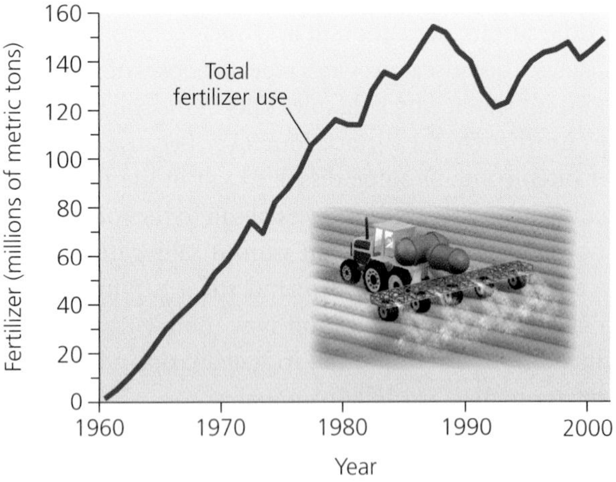

FIGURE 9.19 Use of synthetic fertilizers has risen sharply over the past half-century and now stands at over 140 million metric tons. (The temporary drop during the early 1990s was due to economic decline in countries of the former Soviet Union following that nation's dissolution.) Data from Food and Agriculture Organization of the United Nations (FAO); and Worldwatch Institute, 2001. *Vital signs 2001.*

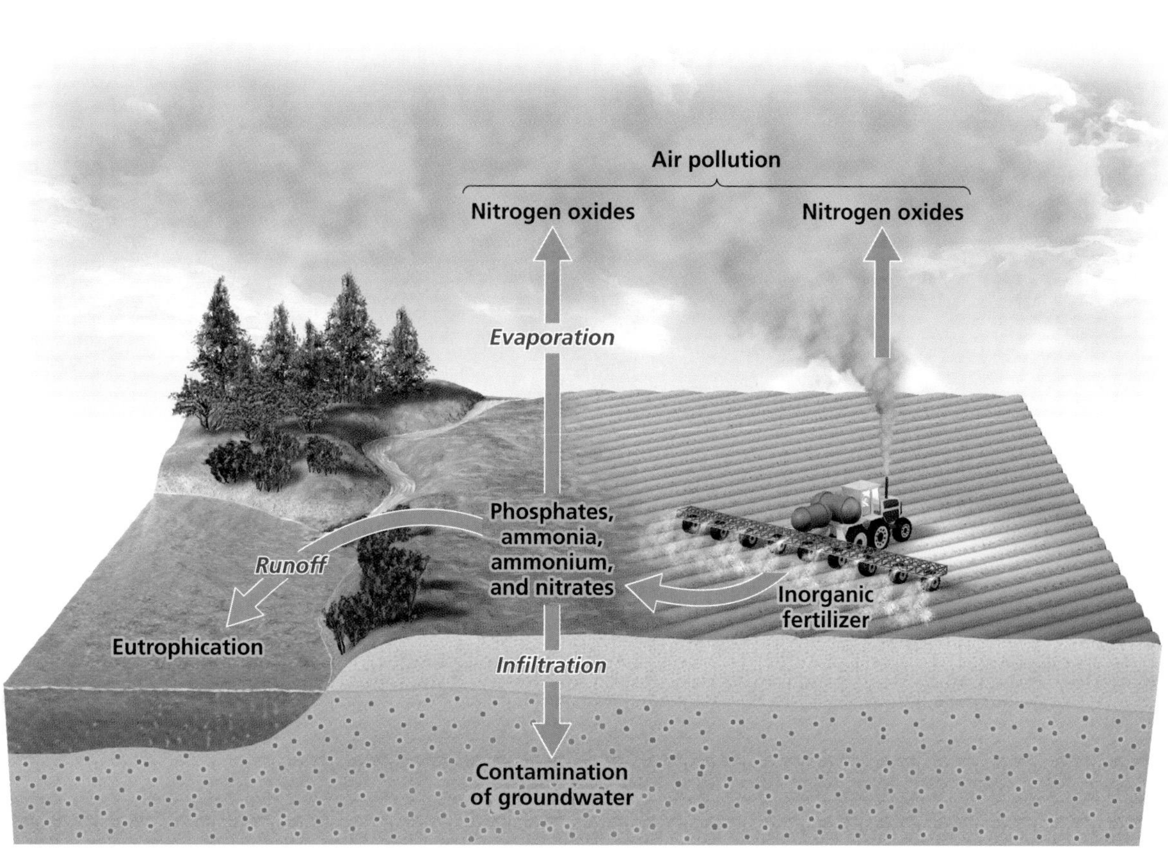

FIGURE 9.20 The overapplication of inorganic (or organic) fertilizers can have effects beyond the farm field, because nutrients that are not taken up by plants may end up in other places. Nitrates can leach into groundwater, where they can pose a threat to human health in drinking water. Phosphates and some nitrogen compounds can run off into surface waterways and alter the ecology of streams, rivers, ponds, and lakes through eutrophication. Some compounds like nitrogen oxides can even enter and pollute the air. Anthropogenic inputs of nitrogen have greatly modified the nitrogen cycle (▸ pp. 197–199), and now account for one-half the total nitrogen flux on Earth.

Applying substantial amounts of fertilizer to croplands has impacts far beyond the boundaries of the fields (Figure 9.20). We saw in Chapter 7 one impact of excess fertilizer use. Nitrogen and phosphorus runoff from farms and other sources in the Mississippi River basin each year spurs phytoplankton blooms in the Gulf of Mexico and creates an oxygen-depleted "dead zone" that kills fish and shrimp. Such eutrophication occurs at many river mouths, lakes, and ponds throughout the world. Moreover, nitrates readily leach through soil and contaminate groundwater, and components of some nitrogen fertilizers can even volatilize (evaporate) into the air. Through these processes, unnatural amounts of nitrates and phosphates spread through ecosystems and pose human health risks, including cancer and methemoglobinemia, or blue-baby disease, which can asphyxiate and kill infants. The U.S. Environmental Protection Agency (EPA) has determined that nitrate concentrations in excess of 10 mg/L for adults and 5 mg/L for infants in drinking water are unsafe, yet many sources around the world exceed even the looser standard of 50 mg/L set by the World Health Organization.

Grazing practices and policies can contribute to soil degradation

We have focused in this chapter largely on the cultivation of crops as a source of impacts on soils and ecosystems, but raising livestock also has such impacts. When sheep, goats, cattle, or other livestock graze on the open range, they feed primarily on grasses. As long as livestock populations do not exceed a range's carrying capacity (▸ pp. 136–137) and do not consume grasses faster than grasses can be replaced, grazing may be sustainable. However, when too many animals eat too much of the plant cover, impeding plant regrowth and

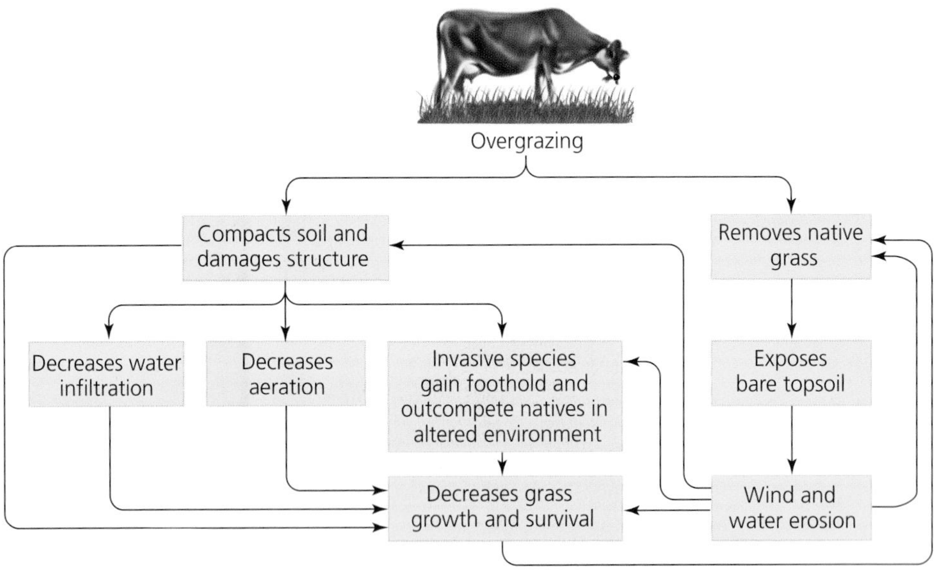

FIGURE 9.21 When grazing by livestock exceeds the carrying capacity of rangelands and their soil, overgrazing can set in motion a series of consequences and positive feedback loops that degrade soils and grassland ecosystems.

preventing the replacement of biomass, the result is **overgrazing.**

Rangeland scientists have shown that overgrazing causes a number of impacts, some of which give rise to positive feedback cycles that exacerbate damage to soils, natural communities, and the land's productivity for grazing (Figure 9.21). When livestock remove too much of an area's plant cover, more soil surface is exposed and made vulnerable to erosion. Soil erosion makes it difficult for vegetation to regrow, perpetuating the lack of cover and giving rise to more erosion. Moreover, nonnative weedy plants may invade denuded soils (Figure 9.22). These invasive plants are usually less palatable to livestock and can outcompete native vegetation in the new, modified environment, further decreasing native plant cover.

In addition, overgrazing can compact soils and alter their structure. Soil compaction makes it harder for water to infiltrate, harder for soils to be aerated, harder for plants' roots to expand, and harder for roots to conduct cellular respiration (▸ p. 107). All of these effects further decrease the growth and survival of native plants.

Overgrazing is a serious problem worldwide. As a cause of soil degradation, it is equal to cropland agriculture, and it is a greater cause of desertification. Humans keep a total of 3.3 billion cattle, sheep, and goats. Rangeland classified as degraded now adds up to 680 million ha (1.7 billion acres), five times the area of U.S. cropland, although some estimates put the number as high as 2.4 billion ha (5.9 billion acres), fully 70% of the world's rangeland area. Rangeland degradation is estimated to cost $23.3 billion per year. Grazing exceeds the sustainable supply of grass in India by 30% and in parts of China by up to 50%. To relieve pressure on rangelands, both nations are now beginning to feed crop residues to livestock.

Range managers in the United States do their best to assess the carrying capacity of rangelands and inform livestock owners of these limits, so that herds are rotated from site to site as needed to conserve grass cover and soil integrity. Managers also can establish and enforce limits on grazing on publicly owned land when necessary. U.S.

FIGURE 9.22 The effects of overgrazing can be striking, as shown in this photo along a fence line separating a grazed plot (right) from an ungrazed plot (left) in the Konza Prairie Reserve in Kansas. The overgrazed plot has lost much of its native grass and has been invaded by weedy plants that compete more effectively in conditions of degraded soil.

VIEWPOINTS

Soil Conservation

Productive farming depends on fertile soil, but our farming practices have all too often eroded and degraded soil. **How much erosion constitutes a problem? How can we best protect the condition of soil while we farm?**

Land Policy Is Necessary for Soil Conservation

Claiming that soil erosion is a major problem in the more-developed parts of the world is environmental nonsense. Pierre Crosson of Resources for the Future, Washington, D.C., has looked carefully at the U.S. data over many years and concluded that soil erosion does *not* constitute a significant problem for contemporary agriculture and its sustainability.

There are surely many cases of local erosive effects, but these usually amount to small quantities of soil shifting around the agricultural landscape. Erosion can be beneficial in some cases, such as in the creation of new farmland in Bangladesh, which formed when materials washed down from Nepal and other parts of the Himalayas.

Soil conservation is a problem that must be approached from all sides. The most important factor will be to secure legally protected property rights for resource users. Farmers must have the right to own their own land resource. Ethiopia is usually regarded as one of the most eroded countries in the world. The state owns all farmland and has a history of moving people against their will to different areas. If a farmer thinks he will not be able to continue managing his farm for the next several years, he will not invest his scarce resources in erosion-reducing practices such as terracing, stone building, maintaining vegetative contour strips, and reducing grazing pressure. For resource custody, and hence the treatment of soil, land policy really matters.

Land policy also affects public investment in infrastructure such as rural roads, which reduce the cost of farmers' transactions. Farmers must haul supplies such as fertilizer to their farms and must transport their product to market. Inadequate access to markets can significantly influence the incentives to employ resource-consuming soil conservation practices. Thus, social science has key roles to play along with physical and biological sciences in addressing soil management.

Jock R. Anderson joined the World Bank in its Agriculture and Natural Resources Department, where he served variously as adviser of agricultural technology policy and, more recently, as adviser of strategy and policy for agriculture and rural development. He is a fellow of the Australian Institute of Agricultural Science and of the American Agricultural Economics Association.

Sustainable Soil Productivity Requires Good Land Stewards

How much soil erosion is tolerable? The standard answer to this question ranges from 1 to 5 tons of soil per acre annually, depending on soil depth and other characteristics that affect sustainability. However, this answer is admittedly subjective and debatable.

Some say these values are too low. They argue that land has been used for centuries with higher rates of erosion, and it is difficult or impossible to reduce erosion on some land without drastic reductions in high-value crops. Others say the standard tolerable erosion values are too high. They argue that the average annual rate of soil formation is less than 1 ton per acre, so a greater erosion rate will inevitably make the soil shallower and less productive. Most agree that excessive erosion is undesirable and a major cause of land abandonment.

Tillage was fundamental to the development of civilization, but it also exposed the soil to erosion—a problem that became more serious as population increased. Tillage can now be reduced or even eliminated by using modern no-till equipment that can cut through crop residues and plant seeds in a narrow slot. Weeds, insects, and plant diseases are controlled by integrated pest management using crop rotations, biological methods, and chemicals. We can now grow crops with much less erosion and still produce good yields.

Conserving water and planting drought-tolerant crops in semi-arid climates is another important advance in areas where erosion due to summer fallow has been common. These practices are effective ways to conserve soil.

The most important requirement for sustainable productivity is a fervent desire to be good land stewards—people who use the best soil conservation practices and are alert for detecting problem spots.

Frederick R. Troeh grew up in Idaho, surveyed soils for the Soil Conservation Service, earned his Ph.D. from Cornell University, and taught soil science and researched for Iowa State University. He has worked internationally in Uruguay, Argentina, and Morocco, and he is lead author of the textbooks *Soils and Soil Fertility* and *Soil and Water Conservation.*

Explore this issue further by accessing **Viewpoints** at www.aw-bc.com/withgott.

The Science behind the Story

Overgrazing and Fire Suppression in the Malpai Borderlands

In the high desert of southern Arizona and New Mexico, scientists trying to heal the scars left by decades of cattle ranching and overgrazing found they had to contend with a creature even more damaging than a hungry steer: Smokey the Bear.

Wildfires might seem a natural enemy of grasslands, but in the Malpai Borderlands, researchers realized that people's efforts to suppress fire had done far more harm. Before large numbers of Anglo settlers and ranchers arrived more than a century ago, this rocky stretch of arid Western landscape thrived under an ecological cycle common to many grasslands. Scrubby trees such as mesquite grew near creeks. Hardy grasses such as black grama covered the drier plains. Periodic wildfires, usually sparked by lightning, burned back shrubs and trees and kept grasslands open. Native grazing animals, such as deer, rabbits, and bighorn sheep, ate the grasses, but rarely ate enough to deplete the range. Fed by seasonal rains, new grasses sprouted without being overeaten or crowded out by larger plants.

To restore native grasslands, the Malpai Borderlands Group reinstated fire as a natural landscape process, conducting controlled burns on over 100,000 ha since 1994. Monitoring indicates that restoring fire has improved ecological conditions in the region.

By the early 1990s, however, those grasslands were increasingly scarce. Ranchers had brought large cattle herds to the area in the late 1800s. The cows chewed through vast expanses of grass, trampled soil, and scattered mesquite seed into areas where grasses had dominated. Ranchers fought wildfires to keep their herds safe, and the federal government joined in the firefighting efforts. Gradually, the Malpai's ranching families found themselves struggling to feed their cattle and make a living from the land. Decades of photos taken by ranchers showed how the area's soil had eroded, and how trees and brush had overgrown the grass. The ranchers knew their cattle were part of the problem, but they also suspected that firefighting efforts might be to blame.

In 1993, a group of ranchers launched an innovative plan. They formed the Malpai Borderlands

ranchers have traditionally had little incentive to conserve rangelands because most grazing has taken place on public lands leased from the government, not on lands privately owned by ranchers. In addition, the U.S. government has heavily subsidized grazing. As a result, overgrazing has been extensive and has caused many environmental problems in the American West. Today increasing numbers of ranchers are working cooperatively with government agencies, environmental scientists, and even environmental advocates to find ways to ranch more sustainably and safeguard the health of the land (see "The Science behind the Story," above).

Forestry, too, has impacts on soil

Farming and grazing are agricultural practices that help feed human populations, that depend on healthy soils, and that affect the conditions of those soils. Forestry, the

Group, designating about 325,000 ha (800,000 acres) of land as an area for protection and study. Through study of the region, ranchers joined government agencies, environmentalists, and scientists to bring back grasses, restore native animal species, and return periodic fires to the landscape (see the figure).

The group's research efforts have centered on the Gray Ranch, a 121,000-ha (300,000-acre) parcel in the heart of the borderlands. Scientists led by researcher Charles Curtin created study sites by dividing pastureland on the ranch into four study areas of about 890 ha (2,200 acres) each. Each area is further divided into four different "treatments," or areas with varying land management techniques:

- In Treatment 1, fire is not suppressed, and grazing by cattle and native animals is permitted.
- In Treatment 2, fire is not suppressed, and grazing is not allowed.
- In Treatment 3, animals can graze, but fire is suppressed.
- In Treatment 4, fire is suppressed, and grazing is not permitted.

Treatments 1 and 3, which allow grazing, also feature small fenced-off areas that prevent animals from eating grass. These exclosures allow scientists to make precise side-by-side comparisons of how grazing affects grasses. Scientists measure rainfall in each area and monitor soils for degradation and erosion. Teams of wildlife and vegetation specialists monitor each treatment for the distribution and abundance of birds, insects, animals, and vegetation.

By comparing areas where fire is suppressed to those where fire can burn, researchers have documented how the suppression of fire leads to more brush and trees and less grass. When fire burns an area, woody plants such as mesquite are damaged and their seed production disrupted, and the flames are often followed by a return of grass. Such benefits follow both natural fires and carefully monitored, deliberately set controlled burns.

Cattle don't do heavy damage to grass if they are managed carefully, the researchers have found. The scientists have helped ranchers develop a cycle of grassbanking, in which herds are allowed to graze on shared plots of land while other areas recover. And ranchers must work with the weather. Scientists have found that controlled burns or grassbanking should track with cycles of rain and drought to bring back grass.

Because of such research, controlled burns are now a regular part of the Malpai landscape. More than 100,000 ha (250,000 acres) have been burned since 1994. Natural fires are often allowed to run their course with little or no intervention. Damaged areas have been reseeded with native grasses. Scientists increasingly believe that ranching, if managed properly, can help bring damaged grazing areas back to life. "We cannot assume rangelands will recover on their own," Curtin wrote in a recent study on the Malpai Borderlands. "Conservation of grazed lands requires restoring and sustaining natural processes."

cultivation of trees, is a similar practice that we will discuss in Chapter 12. Forestry practices can have substantial impacts on soils, just as farming and ranching can. As with farming and ranching techniques, forestry practices have been modified over the years to try to minimize damage to soils. Some practices, such as clear-cutting—the removal of all trees from an area at once—can lead to severe erosion. This is particularly the case on steep slopes (Figure 9.23). Alternative timbering methods that remove fewer trees over longer periods of time are more successful in minimizing erosion.

A number of U.S. and international programs promote soil conservation

In recent years, the U.S. Congress has enacted a number of laws promoting soil conservation. The Food Security Act of 1985 required farmers to adopt soil conservation

FIGURE 9.23 Deforestation, discussed further in Chapter 12, can be a major cause of erosion, particularly when trees are clear-cut for timber on steep slopes.

plans and practices as a prerequisite for receiving price supports and other government benefits. The Conservation Reserve Program, also enacted in 1985, pays farmers to stop cultivating highly erodible cropland and instead place it in conservation reserves planted with grasses and trees. The USDA estimates that for an annual cost of $1 billion, this program saves 700 million metric tons (771 million tons) of topsoil each year. Besides reducing erosion, the Conservation Reserve Program has generated income for farmers and has provided habitat for native wildlife. In 1996, Congress extended the program by passing the Federal Agricultural Improvement and Reform Act. Also known as the "Freedom to Farm Act," this law aimed to reduce subsidies and government influence over many farm products. It also created the Environmental Quality Incentive Program and the Natural Resources Conservation Foundation to promote and pay for the adoption of conservation practices in agriculture. In 1998, the USDA initiated the Low-Input Sustainable Agriculture Program to provide funding for individual farmers to develop and practice sustainable agriculture.

Internationally, the United Nations promotes soil conservation and sustainable agriculture through a variety of programs of its Food and Agriculture Organization (FAO). The FAO's Farmer-Centered Agricultural Resource Management Program (FAR) is a project undertaken in partnership by China, Thailand, Vietnam, Indonesia, Sri Lanka, Nepal, the Philippines, and India to support innovative approaches to resource management and sustainable agriculture. This program studies agricultural success stories and tries to help other farmers duplicate the successful efforts. Rather than following a top-down, government-controlled approach, the FAR program calls on the creativity of local communities to educate and encourage farmers throughout Asia to conserve soils and secure their food supply.

Conclusion

Many of the policies enacted and the practices developed to combat soil degradation in the United States and worldwide have been quite successful, particularly in reducing the erosion of topsoil. Despite these successes, however, soil is still being degraded at a rate that calls into question the sustainability of industrial agriculture.

Our species has enjoyed a 10,000-year history with agriculture, yet despite all we have learned about soil degradation and conservation, many challenges remain. It is clear that even the best-conceived soil conservation programs require research, education, funding, and commitment from both farmers and governments if they are to fulfill their potential. In light of continued population growth, we will likely need better technology and wider adoption of soil conservation techniques to avoid an eventual food crisis. Increasingly, it seems relevant to consider whether the universal adoption of Aldo Leopold's land ethic (▸ pp. 35–36) will also be required if we are to feed the 9 billion people expected to crowd our planet in mid-century.

REVIEWING OBJECTIVES

You should now be able to:

Explain the importance of soils to agriculture, and describe the impacts of agriculture on soils

- Successful agriculture requires healthy soil. (p. 246)
- As the human population grows and consumption increases, pressures from agriculture are degrading Earth's soil, and we are losing 5–7 million ha (12–17 million acres) of productive cropland annually. (pp. 246–247)

Outline major historical developments in agriculture

- Beginning about 10,000 years ago, people began breeding crop plants and domesticating animals. (p. 248)
- Domestication took place through the process of selective breeding, or artificial selection. (p. 248)
- Agriculture is thought to have originated multiple times independently in different cultures across the world. (pp. 248–249)
- Industrial agriculture is gradually replacing traditional agriculture, which largely replaced hunting and gathering. (pp. 248–250)

Delineate the fundamentals of soil science, including soil formation and the properties of soil

- Soil includes diverse biotic communities that decompose organic matter. (pp. 250–251)
- Climate, organisms, relief, parent material, and time are factors influencing soil formation. (pp. 251–252)
- Soil profiles consist of distinct horizons with characteristic properties. (pp. 252–253)
- Soil can be characterized by color, texture, structure, and pH. (pp. 253–255)
- Soil properties affect the potential for plant growth and agriculture in any given location. (p. 255)

State the causes and predict the consequences of soil erosion and soil degradation

- Some agricultural practices have resulted in high rates of erosion across the world, lowering crop yields. (pp. 256–258)
- Desertification affects a large portion of the world's soils, especially those in arid regions. (pp. 259–260)
- Overirrigation can cause salinization and waterlogging, which lower crop yields and are difficult to mitigate. (pp. 264–266)
- Overapplying fertilizers can cause pollution problems that affect ecosystems and human health. (pp. 266–267)
- Overgrazing can cause soil degradation on grasslands, as well as diverse impacts to native ecosystems. (pp. 267–270)
- Careless forestry practices, such as deforesting steep slopes, are a major cause of erosion. (pp. 270–271)

Recite the history and explain the principles of soil conservation

- The Dust Bowl in the United States and similar events elsewhere have encouraged scientists and farmers to develop ways of better protecting and conserving topsoil. (pp. 260–261)
- Farming techniques such as crop rotation, contour farming, intercropping, terracing, shelterbelts, and reduced tillage enable farmers to reduce soil erosion and boost crop yields. (pp. 261–264)
- In the United States and across the world, governments are devising innovative policies and programs to deal with the problems of soil degradation. (pp. 271–272)

TESTING YOUR COMPREHENSION

1. How did the practices of selective breeding and human agriculture begin roughly 10,000 years ago? Summarize the influence of agriculture on the development and organization of human communities.
2. Describe the methods used in traditional agriculture, and contrast subsistence agriculture with intensive traditional agriculture. What makes industrialized agriculture different from traditional agriculture?
3. What processes are most responsible for the formation of soil? Describe the three types of weathering that may contribute to the process of soil formation.
4. Name the five primary factors thought to influence soil formation, and describe one effect of each.
5. How are soil horizons created? What is the general pattern of distribution of organic matter in a typical soil profile?
6. Why is erosion generally considered a destructive process? Name three human activities that can promote

soil erosion. Describe four kinds of soil erosion by water. What factors affect the intensity of water erosion?

7. List innovations in soil conservation introduced by Hugh Hammond Bennett, first director of the SCS. What other farming techniques can help reduce the risk of erosion due to conventional cultivation methods?
8. How does terracing effectively turn very steep and mountainous areas into arable land? Explain the method of no-till farming. Why does this method reduce soil erosion?
9. How do fertilizers boost crop growth? How can large amounts of fertilizer added to soil also end up in water supplies and the atmosphere?
10. What policies can be linked to the practice of overgrazing? Describe the effects of overgrazing on soil. What conditions characterize sustainable grazing practices?

SEEKING SOLUTIONS

1. How do you think a farmer can best help to conserve soil? How do you think a scientist can best help to conserve soil? How do you think a national government can best help to conserve soil?
2. How and why might actual soils differ from the idealized six-horizon soil profile presented in the chapter? How might departures from the idealized profile indicate the impact of human activities? Provide at least three examples.
3. What method of farming would you choose to employ on a gradual slope with the threat of natural erosion? What kinds of plants might you use to prevent erosion, and why?
4. Discuss how the methods of no-till or reduced tillage farming, as described in this chapter, can enhance soil quality. What drawbacks or negative effects might no-till or reduced-tillage practices have on soil quality, and how might these be prevented?
5. Discuss how methods of locally based sustainable agriculture described in this chapter are promoting the science of soil conservation. In reference to Aldo Leopold's land ethic (▸pp. 35–36), how are humans and the soil members of the same community?
6. Imagine that you are the head of an international granting agency that assists farmers with soil conservation and sustainable agriculture. You have $10 million to disburse. Your agency's staff has decided that the funding should go to (1) farmers in an arid area of Africa prone to salinization, (2) farmers in a fast-growing area of Indonesia where swidden agriculture is practiced, (3) farmers in southern Brazil practicing no-till agriculture, and (4) farmers in a dryland area of Mongolia undergoing desertification. What types of projects would you recommend funding in each of these areas, how would you apportion your funding among them, and why?

INTERPRETING GRAPHS AND DATA

Kishor Atreya and his colleagues at Kathmandu University in Nepal conducted a field experiment to test the effects of reduced tillage versus conventional tillage on erosion and nutrient loss in the Himalayan Mountains in central Nepal. The region in which they worked has extremely steep terrain (with an average slope of 18%), and receives over 138 cm (55 in.) of rain per year, with 90% of it coming during the monsoon season from May to September. Atreya's team measured the amounts of soil, organic carbon, and nitrogen lost from the research plots (which were unterraced) over the course of a year. Some of their results are presented in the graph.

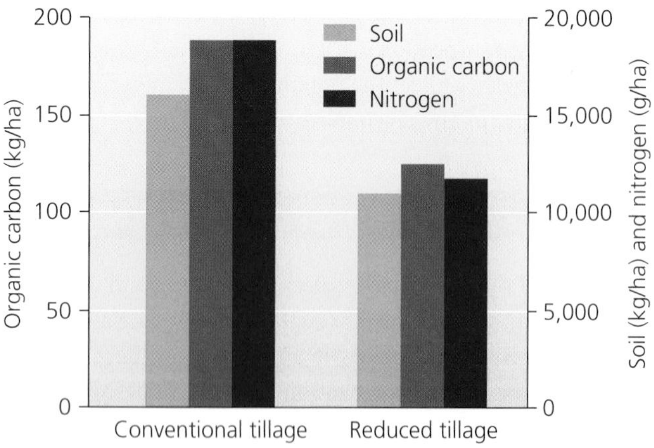

Annual soil and nutrient losses in plots under conventional and reduced tillage systems. All reduced tillage values are significantly different than their conventional tillage counterparts. Data from Atreya, K., et al. 2005. Applications of reduced tillage in hills of central Nepal. *Soil & Tillage Research*, in press.

1. Under the conditions of the study reported above, how much soil, organic carbon, and nitrogen would be saved annually in fields with reduced tillage relative to fields with conventional tillage? Express your answers both in absolute units and as percentages.

2. Given that annual crop yields in the study plots were approximately 4 metric tons/ha, what is the ratio of soil lost to crop yield under conventional tillage? Under reduced tillage?
3. Is reduced tillage a sustainable management practice for Nepalese farmers? If so, what data from the study above would you cite in support of your answer? If not, or if you cannot say, then what concerns raised by the data above would still need to be addressed, or what additional data would be needed, to answer the question?

CALCULATING ECOLOGICAL FOOTPRINTS

As you learned in this chapter, rates of soil loss due to erosion can be high. Even in the United States, approximately 6 pounds of topsoil are lost for every 1 pound of grain harvested. Erosion rates vary greatly with soil type, topography, tillage method, and crop type. For simplicity let us assume that the 6:1 ratio applies to all plant crops and that a typical diet includes 1 pound of plant material or its derived products (sugar, for example) per day. In the first two columns of the table, calculate the annual topsoil losses associated with growing this food for you and for other groups, assuming the same diet.

	Plant products consumed (lb)	Soil loss at 6:1 ratio (lb)	Soil loss at 4:1 ratio (lb)	Reduced soil loss at 4:1 relative to 6:1 ratio (lb)
You	365	2,190	1,460	730
Your class				
Your state				
United States				

1. Improved soil conservation measures reduced erosion by approximately one-third from 1982 to 1997. If additional measures were again able to reduce the current rate of soil loss by a third, the ratio of soil lost to grain harvested would fall from 6:1 to 4:1. Calculate the soil losses associated with food production at a 4:1 ratio, and record your answers in the third column of the table.
2. Calculate the amount of topsoil hypothetically saved by the additional conservation measures in Question 1, and record your answers in the fourth column of the table.
3. Define a "sustainable" rate of soil loss. Describe how you might determine if a given farm was practicing sustainable use of soil.

Take It Further

Go to www.aw-bc.com/withgott or the student CD-ROM where you'll find:

- Suggested answers to end-of-chapter questions
- Quizzes, animations, and flashcards to help you study
- *Research Navigator*™ database of credible and reliable sources to assist you with your research projects
- GRAPHIt! tutorials to help you master how to interpret graphs
- INVESTIGATEIt! current news articles that link the topics that you study to case studies from your region to around the world

10 Agriculture, Biotechnology, and the Future of Food

Organic vegetable farm in Whatcom County, Washington

Upon successfully completing this chapter, you will be able to:

- Explain the challenge of feeding a growing human population
- Identify the goals, methods, and environmental impacts of the "green revolution"
- Categorize the strategies of pest management
- Discuss the importance of pollination
- Describe the science behind genetically modified food
- Evaluate controversies and the debate over genetically modified food
- Ascertain approaches for preserving crop diversity
- Assess feedlot agriculture for livestock and poultry
- Weigh approaches in aquaculture
- Evaluate sustainable agriculture

Farmer in maize field in Oaxaca, Mexico

Central Case: Possible Transgenic Maize in Oaxaca, Mexico

"Worrying about starving future generations won't feed them. Food biotechnology will. . . . At Monsanto, we now believe food biotechnology is a better way forward."
—Advertising campaign of the Monsanto Company

"Industrial agriculture has not produced more food. It has destroyed diverse sources of food, and it has stolen food from other species . . . using huge quantities of fossil fuels and water and toxic chemicals in the process."
—Vandana Shiva, director of the Research Foundation for Science, Technology, and Natural Resource Policy, India

Corn is a staple grain of the world's food supply. We can trace its ancestry back roughly 5,500 years, when people in the highland valleys of what is now the state of Oaxaca in southern Mexico first domesticated that region's wild maize plants. The corn we eat today arose from some of the many varieties that evolved from the early selective crop breeding conducted by the people of this region.

Today Oaxaca remains a world center of biodiversity for maize, with many native varieties, or *cultivars,* growing in the rich, well-watered soil. Preserving such varieties of crops in their ancestral homelands is important for securing the future of our food supply, scientists maintain, because these varieties serve as reservoirs of genetic diversity—reservoirs we may need to draw on to sustain or advance our agriculture.

Thus, it caused global consternation when, in 2001, Mexican government scientists conducting routine genetic tests of Oaxacan farmers' maize announced that they had turned up DNA that matched genes from genetically modified (GM) corn. GM corn was widely grown in the United States, but Mexico had banned its cultivation in 1998. Corn is one of many crops that scientists have genetically engineered to express desirable traits such as large size, fast growth, and resistance to insect pests. To genetically engineer crops, scientists extract genes from the DNA of one organism and transfer them into the DNA of another. The aim is to improve crop performance and feed the world's hungry, but many people worry that the **transgenes** from these **transgenic** plants could cause unpredictable harm. One concern is that transgenic crops might crossbreed with native crops and thereby "contaminate" the genetic makeup of native crops.

Two researchers at the University of California at Berkeley, Ignacio Chapela and his postdoctoral associate David Quist, shared these concerns, and they ventured to Oaxaca to test samples of native maize. Their analyses seemed to confirm the government scientists' findings, revealing what they argued were traces of DNA from genetically engineered corn in the genes of native maize plants. They also suggested that the invading genes had split up and spread throughout the maize genome. Quist and Chapela published their findings in the scientific journal *Nature* in November 2001. Activists opposed to GM food trumpeted the news and urged a ban on imports of transgenic crops from producer countries such as the United States into developing nations. The agrobiotech industry defended the safety of its crops and questioned the validity of the research—as did many of Quist and Chapela's peers. Responding to criticisms from researchers, *Nature* took the unprecedented step of stating that Quist and Chapela's paper should never have been published—a move that created a firestorm of controversy (see "The Science behind the Story," ▸ pp. 292–293).

Since that time, further research by the Mexican government has confirmed that transgenes exist in Mexican maize. These findings have not yet (as of late 2005) been published in a scientific journal, but they were accepted by a special commission of experts convened under the North American Free Trade Agreement (NAFTA) to study the issue. In November 2004 the commission reported that in 11 communities, 3–13% of maize contained transgenes, and in four other localities, 20–60% of maize contained transgenes. It also reported that 37% of maize grains distributed by the government food distribution agency, Diconsa, were transgenic. The commission reasoned that corn imported from the United States was the source. U.S. corn shipments contain an undifferentiated mix of GM and non-GM grain. Once the transgenes were inside the country, they would have easily spread by wind pollination and by interbreeding with native cultivars of maize.

A harder question to answer is how genetically modified crops may affect people and the environment. In this chapter, we take a wide-ranging view of the ways people have devised to increase agricultural output, the environmental effects of these efforts, and the food of the future.

The Race to Feed the World

Although human population growth has slowed, we can still expect our numbers to swell to 9 billion by the middle of this century. For every two people living today, there will be three in 2050. Feeding 50% more mouths half a century from now while protecting the integrity of soil, water, and ecosystems will require sustainable agriculture. This could involve approaches as diverse as organic farming and the genetically modified crops that are eliciting so much controversy in Oaxaca and elsewhere.

Agricultural production has outpaced population increase so far

Over the past half century, our ability to produce food has grown even faster than global population (Figure 10.1). However, largely because of political obstacles and inefficiencies in distribution, today 850 million people in developing countries do not have enough to eat. Every 5 seconds, somewhere in the world, a child starves to death. Agricultural scientists and policymakers pursue a goal of **food security,** the guarantee of an adequate, reliable, and available food supply to all people at all times. Making a food supply sustainable depends on maintaining healthy soil, water, and biodiversity. As we saw in Chapter 9, careless expansion of agriculture can have devastating effects on the environment and the long-term ability of the world's soils to continue supporting crops and livestock.

Starting in the 1960s, a number of scientists including Paul Ehrlich (▸ pp. 5–6) predicted widespread starvation and a catastrophic failure of agricultural systems, arguing that the human population could not continue to grow without outstripping its food supply. However, the human population has continued to increase well past their predictions. Although it is tragic that 850 million people are hungry today, this number is smaller than the 960 million who lacked reliable and sufficient food in 1970. In percentage terms, we have reduced hunger by half, from 26% of the population in 1970 to 13% today.

We have achieved these advances in part by increasing our ability to produce food (see Figure 10.1). We have increased food production by devoting more energy (especially fossil fuel energy) to agriculture; by planting and harvesting more frequently; by increasing the use of irrigation, fertilizer, and pesticides; by increasing the amount of cultivated land; and by developing (through crossbreeding and genetic engineering) more productive crop and livestock varieties.

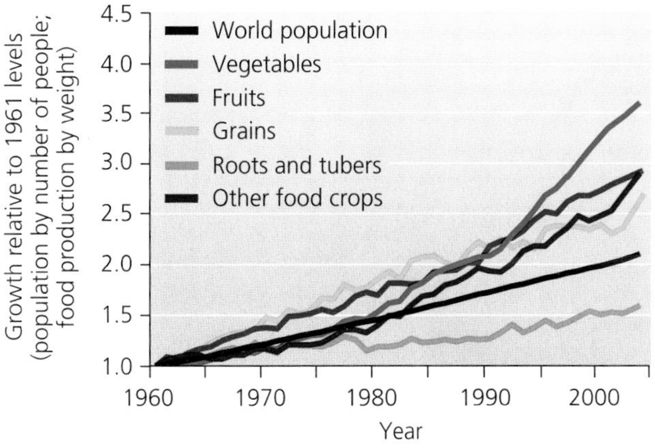

FIGURE 10.1 Global agricultural food production rose by over two-and-a-half times in the past four decades, growing at a faster rate than world population. Production of all types of foods, particularly vegetables, increased from 1961 to 2004. Trend lines show cumulative increases relative to 1961 levels. Data from Food and Agriculture Organization of the United Nations.

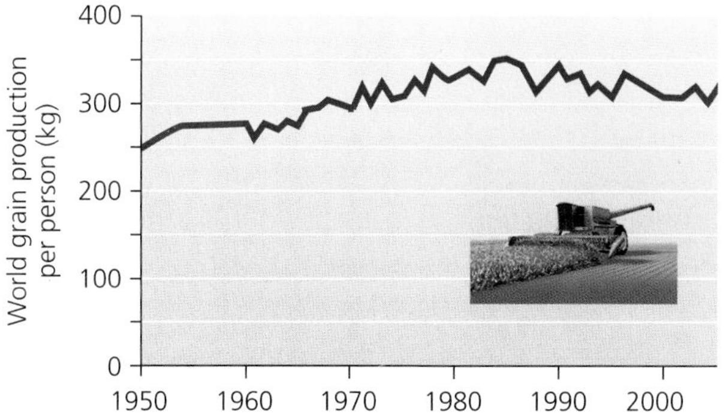

FIGURE 10.2 World per capita production of grain rose until 1985. Since then it has declined slightly as grain production has grown more slowly than global population. Some have pointed to this trend as a sign that our agriculture is no longer able to cope with the deteriorating soil and environmental conditions it has caused. Others propose that the decline is due to two factors: (1) populations in developed nations, where people may be near their limit of grain consumption already, have stabilized; and (2) population growth has been greatest in developing nations, where people consume far less than in developed nations. As a result, a larger share of the world's people lives in societies in which people consume less, thus dragging down the global average demand for grain production. Data from UN FAO and U.S. Department of Agriculture.

However, with grain crops, the world's staple foods, we are producing less food per person each year. World grain production per person peaked in 1985 and has since slowly fallen (Figure 10.2). Moreover, the world's soils are in decline, and nearly all the planet's arable land has already been claimed. Simply because agricultural production has outpaced population growth so far, there is no guarantee that it will continue to do so.

We face undernourishment, overnutrition, and malnourishment

Although many people lack access to adequate food, others are affluent enough to consume more than is healthy. People who are **undernourished,** receiving less than 90% of their daily caloric needs, live mostly in the developing world. Meanwhile, in the developed world, many people suffer from **overnutrition,** receiving too many calories each day. In the United States, where food is available in abundance and people tend to lead sedentary lives with little exercise, more than three out of five adults are technically overweight, and over one out of four are obese. For most people who are undernourished, the reasons are economic. One-fifth of the world's people live on less than $1 per day, and over half live on less than $2 per day, the World Bank estimates. Hunger is a problem even in the United States, where the U.S. Department of Agriculture (USDA) has classified 31 million Americans as "food insecure," lacking the income required to procure sufficient food at all times.

Just as the *quantity* of food a person eats is important for health, so is the *quality* of food. **Malnutrition,** a shortage of nutrients the body needs, including a complete complement of vitamins and minerals, can occur in both undernourished and overnourished individuals. Malnutrition can lead to disease (Figure 10.3). When people eat a high-starch diet but not enough protein or essential amino acids (▸pp. 99–100), then *kwashiorkor* results. Children who have recently stopped breast-feeding are most at risk for developing kwashiorkor, which causes bloating of the abdomen, deterioration and discoloration of hair, mental disability, immune suppression, developmental delays, anemia, and reduced growth. Protein deficiency together with a lack of calories can lead to *marasmus*, which causes wasting or shriveling among millions of children in the developing world.

The "green revolution" led to dramatic increases in agricultural production

The desire for greater quantity and quality of food for the growing human population led in the mid- and late 20th century to the **green revolution** (first introduced in Chapter 9, ▸p. 250). Realizing that farmers could not go on indefinitely cultivating more and more land to increase crop output, agricultural scientists created methods and technologies to increase crop output per unit area of existing cultivated land. Industrialized nations had been

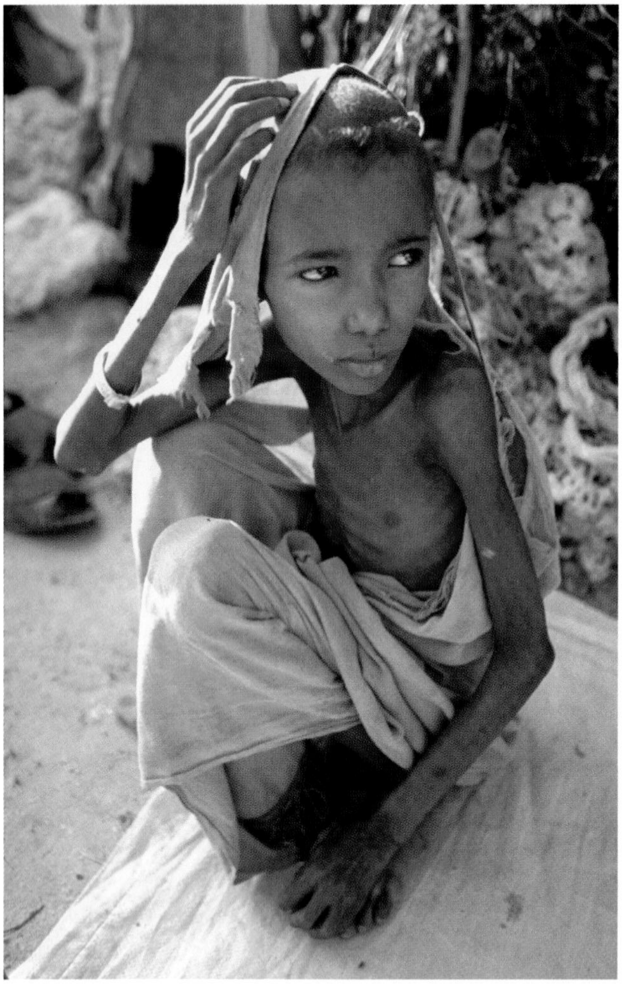

FIGURE 10.3 Millions of children, including this child in Somalia, suffer from forms of malnutrition, such as kwashiorkor and marasmus.

FIGURE 10.4 Norman Borlaug holds examples of the wheat variety he bred that helped launch the green revolution. The high-yielding disease-resistant wheat helped increase agricultural productivity in many developing countries.

dramatically increasing their per-area yields; the average hectare of U.S. cornfield during the 20th century, for instance, upped its corn output fivefold. Many people saw such growth in production and efficiency as key to ending starvation in developing nations.

The transfer of technology to the developing world that marked the green revolution began in the 1940s, when U.S. agricultural scientist Norman Borlaug introduced Mexico's farmers to a specially bred type of wheat (Figure 10.4). This strain of wheat produced large seed heads, was short in stature to resist wind, was resistant to diseases, and produced high yields. Within two decades of planting and harvesting this specially bred crop, Mexico tripled its wheat production and began exporting wheat. The stunning success of this program inspired others. Borlaug—who won the Nobel Peace Prize for his work—took his wheat to India and Pakistan and helped transform agriculture there. Soon many developing countries were increasing their crop yields using selectively bred strains of wheat, rice, corn, and other crops from developed nations. Some varieties yielded three or four times as much per acre as did their predecessors.

The green revolution has caused the environment both benefit and harm

Along with the new grains, developing nations imported the methods of industrialized agriculture. They began applying large amounts of synthetic fertilizers and chemical pesticides on their fields, irrigating crops with generous amounts of water, and using heavy equipment powered by fossil fuels. From 1900 to 2000, humans expanded the world's total cultivated area by 33%, yet increased energy inputs into agriculture by 80 *times*.

This high-input agriculture succeeded dramatically in allowing farmers to harvest more corn, wheat, rice, and soybeans from each hectare of land. Intensive agriculture saved millions in India from starvation in the 1970s and eventually turned that nation into a net exporter of grain (Figure 10.5). However, it had mixed effects on the environment. On the positive side, the intensified use of already-cultivated land reduced pressures to convert additional natural lands for new cultivation. Between 1961 and 2002, food production rose 150% and population rose 100%, while area converted for agriculture increased only 10% (see "Interpreting Graphs and Data," ▸pp. 306–307). For this reason, the green revolution prevented some degree

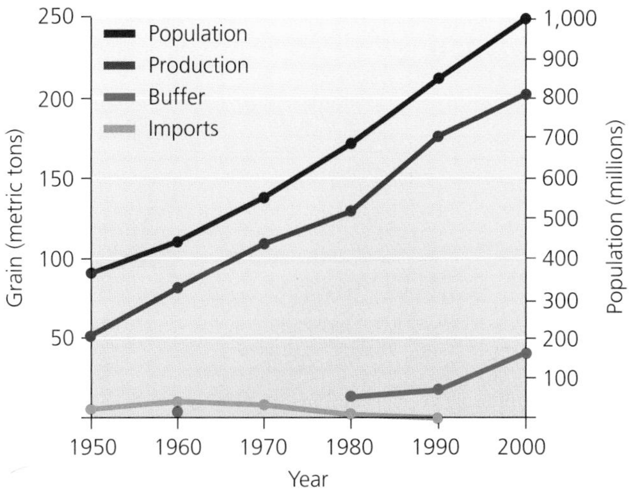

FIGURE 10.5 Thanks to green revolution technology, India's grain production has kept pace with its rapid population growth. This has enabled India to stop importing grain and to maintain extra grain in reserve as a buffer against food shortages.

Weighing the Issues: The Green Revolution and Population

In the 1960s, India's population was growing at an unprecedented rate, and its traditional agriculture was not producing enough food to support the growth. By adopting green revolution agriculture, India sidestepped mass starvation. In the years since intensifying its agriculture, India has added several hundred million more people and continues to suffer widespread poverty and hunger. Norman Borlaug called his green revolution methods "a temporary success in man's war against hunger and deprivation," something to give us breathing room in which to deal with what he called "the Population Monster."

Do you think we can call the green revolution a success? Do you think the green revolution has solved problems, or delayed our resolution of problems, or created new ones? How sustainable are green revolution approaches? Consider our discussion of demographic transition theory from Chapter 8 (▸ pp. 228–229). Have we been dealing with the "Population Monster" during the breathing room that the green revolution has bought for us?

of deforestation and habitat conversion in many countries at the very time those countries were experiencing their fastest population growth rates. In this sense, the green revolution was a boon for the preservation of biodiversity and natural ecosystems. However, the intensive use of water, fossil fuels, and chemical fertilizers and pesticides had extensive negative impacts on the environment in terms of pollution, salinization, and desertification (Chapter 9).

One key aspect of green revolution techniques has had negative consequences for biodiversity and mixed consequences for crop yields. The planting of crops in *monocultures,* large expanses of single crop types (Figure 10.6a), has made planting and harvesting more

(a) Wheat monoculture

(b) Armyworm

FIGURE 10.6 Most agricultural production in industrialized countries comes from monocultures—large stands of single types of crop plant, such as this wheat field in Washington (**a**). Clustering crop types in uniform fields on large scales greatly improves the efficiency of planting and harvesting, but it also decreases biodiversity and makes crops susceptible to outbreaks of pests that specialize on particular crops. Armyworms (**b**) are major agricultural pests whose outbreaks can substantially reduce crop yields. The caterpillars of these moths defoliate a wide variety of crops, including wheat, corn, cotton, alfalfa, and beets.

efficient and has thereby increased output. However, monocultural planting has reduced biodiversity over huge areas, because many fewer wild organisms are able to live in monocultures than in native habitats or in traditional small-scale polycultures. Moreover, when all plants in a field are genetically similar, as in monocultures, all will be equally susceptible to viral diseases, fungal pathogens, or insect pests that can spread quickly from plant to plant (Figure 10.6b). For this reason, monocultures bring significant risks of catastrophic failure.

Monocultures have also contributed to a narrowing of the human diet. Globally, 90% of the food we consume now comes from only 15 crop species and eight livestock species—a drastic reduction in diversity from earlier times. The nutritional dangers of such dietary restriction have been alleviated by the fact that expanded global trade has provided many people access to a wider diversity of foods from around the world. However, this effect has benefited wealthy people far more than poor people. One reason farmers and scientists were so concerned about transgenic contamination of Oaxaca's native maize is that Oaxacan maize varieties serve as a valuable source of genetic variation in a world where so much variation is being lost to monocultural practices.

Table 10.1 Most Commonly Used Pesticides* in Agriculture in the United States

Active ingredient†	Type of pesticide	Millions of kg applied per year
Glyphosate	Herbicide	39–41
Atrazine	Herbicide	34–36
Metam sodium	Fumigant	26–28
Acetochlor	Herbicide	14–16
2,4-D	Herbicide	13–15
Malathion	Insecticide	9–11
Methyl bromide	Fumigant	9–11
Dichloropropene	Fumigant	9–11
Metolachlor-s	Herbicide	9–11
Metolachlor	Herbicide	7–10

*Includes only "conventional pesticides" used in agriculture. Does not include many other types of pesticides, such as disinfectants and wood preservatives.

†Includes only active ingredients, not ingredients such as oil, sulfur, and sulfuric acid.

Data from Kiely, T., et al. 2004. *Pesticides industry sales and usage: 2000 and 2001 market estimates.* Washington, DC: U.S. Environmental Protection Agency.

Pests and Pollinators

Throughout the history of agriculture, the insects, fungi, viruses, rats, and weeds that eat or compete with our crop plants have taken advantage of the ways we cluster food plants into agricultural fields. These organisms, in making a living for themselves, cut crop yields and make it harder for farmers to make a living. As just one example of thousands, various species of moth caterpillars known as armyworms (see Figure 10.6b) lower yields of everything from beets to sorghum to millet to canola to pasture grasses. Pests and weeds have always posed problems for traditional agriculture, and they pose an even greater threat to monocultures, where a pest adapted to specialize on the crop can easily move from one individual plant to many others of the same type.

What humans term a *pest* is any organism that damages crops that are valuable to us. What we term a *weed* is any plant that competes with our crops. These are subjective categories that we define entirely by our own economic interests. There is nothing inherently malevolent in the behavior of a pest or a weed. These organisms are simply trying to survive and reproduce. From the viewpoint of an insect that happens to be adapted to feed on corn, grapes, or apples, a grain field, vineyard, or orchard represents an endless buffet.

Many thousands of chemical pesticides have been developed

To prevent pest outbreaks and to limit competition with weeds, people have developed thousands of artificial chemicals to kill insects *(insecticides)*, plants *(herbicides)*, and fungi *(fungicides)*. Such poisons that target pest organisms are collectively termed **pesticides.** Table 10.1 shows the 10 most widely used pesticides in U.S. agriculture. All told, roughly 400 million kg (900 million lb) of active ingredients from conventional pesticides are applied in the United States each year. Three-quarters of this total is applied on agricultural land. The remainder is used by industry and municipalities (13%) and applied by homeowners to homes, lawns, and gardens (11%). Since 1960, pesticide use has risen fourfold worldwide. Usage in the United States and the rest of the industrialized world has leveled off in the past two decades, but it continues to increase in the developing world. Today more than $32 billion is expended annually on pesticides, with one-third

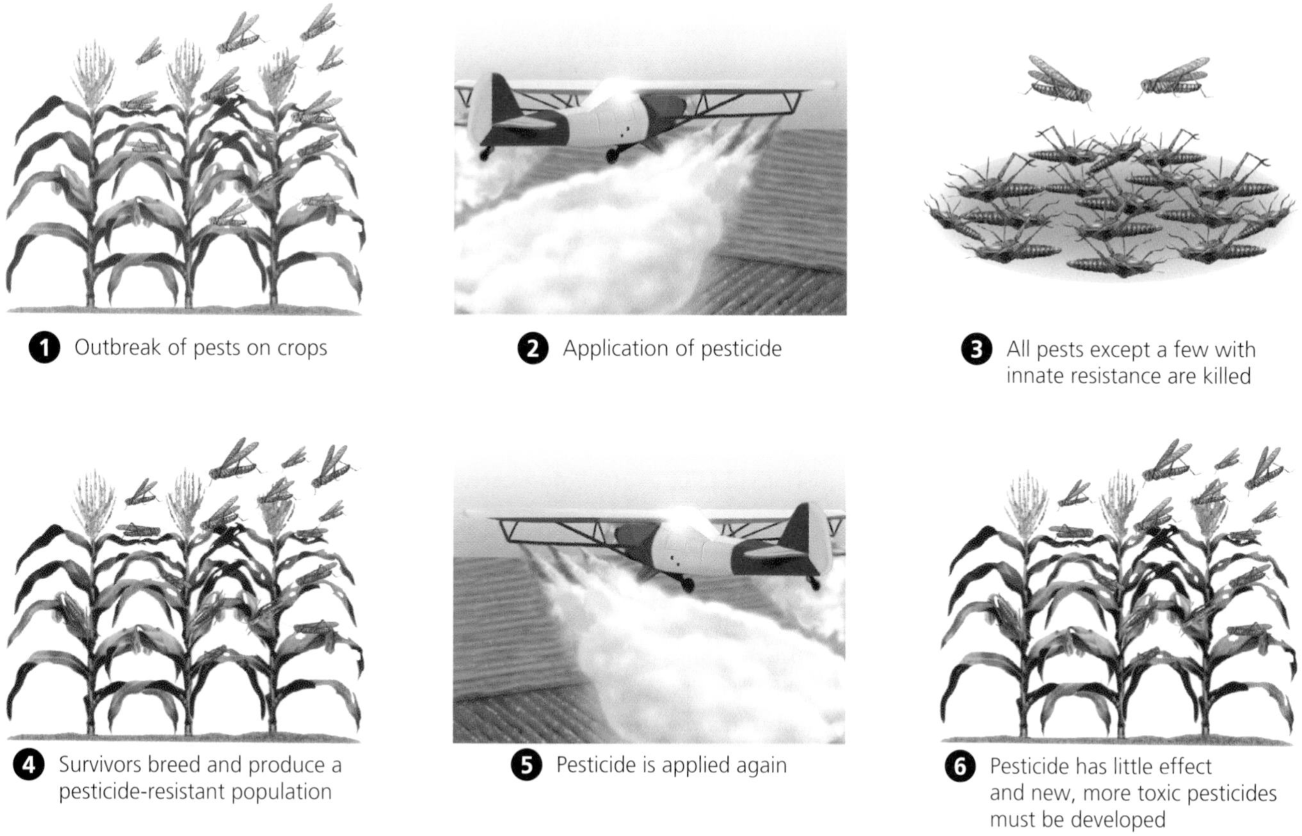
1 Outbreak of pests on crops
2 Application of pesticide
3 All pests except a few with innate resistance are killed
4 Survivors breed and produce a pesticide-resistant population
5 Pesticide is applied again
6 Pesticide has little effect and new, more toxic pesticides must be developed

FIGURE 10.7 Through the process of natural selection, crop pests frequently evolve resistance to the poisons we apply to kill them. This simplified diagram shows that when a pesticide is applied to an outbreak of insect pests, it may kill virtually all individuals except those few with an innate immunity to the poison. Those surviving individuals may found a population with genes for resistance to the poison. Future applications of the pesticide may then be ineffective, forcing us to develop a more potent poison or an alternative means of pest control.

of that total spent in the United States. We will address the health consequences of synthetic pesticides for humans and other organisms in Chapter 14.

Pests evolve resistance to pesticides

Despite the toxicity of these chemicals (▸pp. 406–407), their usefulness tends to decline with time as pests evolve resistance to them. Recall from our discussion of natural selection (▸pp. 118–121) that organisms within populations vary in their traits. Because most insects and microbes occur in huge numbers, it is likely that a small fraction of individuals may by chance have genes that confer some degree of immunity to a given pesticide. Even if a pesticide application kills 99.99% of the insects in a field, 1 in 10,000 survives. If an insect survives by being genetically resistant to a pesticide, and if it mates with other resistant individuals of the same species, the insect population may grow. This new population will consist of individuals that are genetically resistant to the pesticide. As a result, pesticide applications will cease to be effective (Figure 10.7).

In many cases, industrial chemists are caught up in an "evolutionary arms race" with the pests they battle, racing to increase or retarget the toxicity of their chemicals while the armies of pests evolve ever-stronger resistance to their efforts. The number of species known to have evolved resistance to pesticides has grown over the decades. As of 2000, there were nearly 2,700 known cases of resistance by 540 species to over 300 pesticides. Several insects, such as the green peach aphid, Colorado potato beetle, and diamondback moth, have evolved resistance to multiple insecticides. Resistant pests can take a significant economic toll on crops. As just one example, gummy stem blight destroyed two-thirds of Texas's melon crop in 1997 after the blight evolved resistance to the pesticide Benlate.

Biological control pits one organism against another

Because of pesticide resistance and the health risks of some synthetic chemicals, agricultural scientists increasingly battle pests and weeds with organisms that eat or infect them. This strategy, called **biological control,** or **biocontrol** for short, operates on the principle that "the enemy of one's enemy is one's friend." For example, parasitoid wasps (▸p. 154) are natural enemies of many caterpillars. These wasps lay eggs on a caterpillar, and the larvae that hatch from the eggs feed on the caterpillar, eventually killing it. Parasitoid wasps have been used as biocontrol agents in many situations. Some such efforts have succeeded at pest control and have led to steep reductions in chemical pesticide use.

One classic case of successful biological control is the introduction of the cactus moth, *Cactoblastis cactorum*, from Argentina to Australia in the 1920s to control invasive prickly pear cactus that was overrunning rangeland (Figure 10.8). Within just a few years, the moth managed to free millions of hectares of rangeland from the cactus.

A widespread modern biocontrol effort has been the use of ***Bacillus thuringiensis*** **(Bt),** a naturally occurring soil bacterium that produces a protein that kills many caterpillars and the larvae of some flies and beetles. Farmers have used the natural pesticidal activity of this bacterium to their advantage by spraying spores of this bacterium on their crops. If used correctly, Bt can protect crops from pest-related losses.

(a) Before cactus moth introduction

(b) After cactus moth introduction

FIGURE 10.8 In one of the classic cases of biocontrol, larvae of the cactus moth, *Cactoblastis cactorum,* were used to clear non-native prickly pear cactus from millions of hectares of rangeland in Queensland, Australia. These photos from the 1920s show an Australian ranch before (**a**) and after (**b**) introduction of the moth.

Biological control agents themselves may become pests

In most cases, biological control involves introducing an animal or microbe into a foreign ecosystem, often on another continent. Such relocation helps ensure that the target pest has not already evolved ways to deal with the biocontrol agent, but it also introduces risks. In some cases, biocontrol has produced unintended consequences once the biocontrol agent became invasive and began affecting nontarget organisms. Following the cactus moth's success in Australia, for example, it was introduced in other countries to control prickly pear. Moths introduced to Caribbean islands spread to Florida on their own and are now eating their way through rare native cacti in Florida and spreading to other states. If these moths reach Mexico and the southwestern United States, they could decimate many native and economically important species of prickly pear there.

One recent study revealed the extent to which some biocontrol agents in Hawaii have missed their targets. Wasps and flies have been introduced to control agricultural pests in Hawaii at least 122 times over the past century, and biologists Laurie Henneman and Jane Memmott suspected that some of these might be adversely affecting native Hawaiian caterpillars that were not pests. They sampled parasitoid wasp larvae from 2,000 caterpillars of various species in a remote mountain swamp far from farmland. In this area, designated as a wilderness preserve, they found that fully 83% of the parasitoids were biocontrol agents that had been intended to combat lowland agricultural pests.

Scientists debate the relative benefits and risks of biocontrol measures. If biocontrol works as planned, it can be a permanent solution that requires no further maintenance and is environmentally benign. However, if the agent has nontarget effects, the harm done may also become permanent, because removing the agent from the system once it is established is far more difficult than simply stopping a chemical pesticide application. One noted skeptic of biocontrol, ecologist Daniel Simberloff (▸p. 169, and ▸pp. 332–333), has remarked that biocontrol "should be used with our eyes wide open—and as a last resort." However, two British scientists reviewing cases as of the year 2000 concluded that only a small percentage of efforts have resulted in demonstrable nontarget effects, and perhaps less than 10% of these effects were substantial. Because of concerns about unintended impacts, researchers now study biocontrol proposals carefully before putting them into action, and government regulators must approve these efforts. However, there will never be a sure-fire way of knowing in advance whether a given biocontrol program will work as planned.

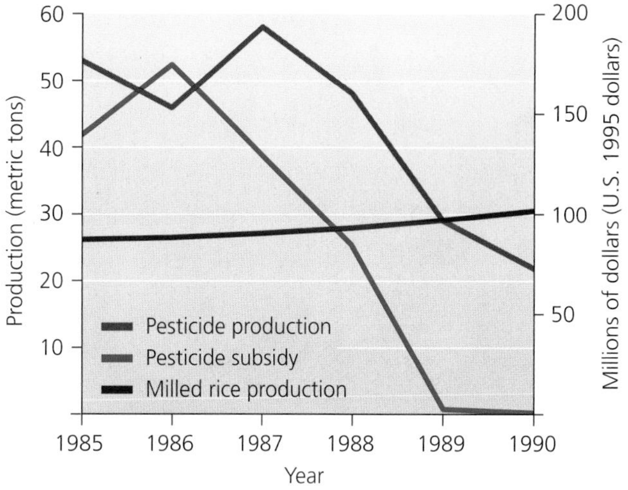

FIGURE 10.9 The Indonesian government threw its weight behind integrated pest management starting in 1986. Within just a few years, pesticide production and pesticide imports were down drastically, pesticide subsidies were phased out, and yields of rice increased slightly.

Integrated pest management combines biocontrol and chemical methods

As it became clear that both chemical and biocontrol approaches have their drawbacks, many agricultural scientists and farmers developed a more sophisticated strategy, trying to combine the best attributes of each approach. In **integrated pest management (IPM),** numerous techniques are integrated to achieve long-term suppression of pests, including biocontrol, use of chemicals, close monitoring of populations, habitat alteration, crop rotation, transgenic crops, alternative tillage methods, and mechanical pest removal. IPM is broadly enough defined that it encompasses a wide variety of strategies.

In recent decades, IPM has become popular in many parts of the world. Indonesia stands as an exemplary case (Figure 10.9). The nation had subsidized pesticide use heavily for years, but its scientists came to understand that pesticides were actually making pest problems worse. They were killing the natural enemies of the brown planthopper, which began to devastate rice fields as its populations exploded. Concluding that pesticide subsidies were costing money, causing pollution, and apparently decreasing yields, the Indonesian government in 1986 banned the importation of 57 pesticides, slashed pesticide subsidies, and encouraged IPM. Within 4 years, pesticide production fell to below half its 1986 level, imports fell to one-third, and subsidies were phased out (saving $179 million annually). Rice yields rose 13%.

We depend on insects to pollinate crops

Managing insect pests is such a major issue in agriculture that many people fall into a habit of thinking of all insects as somehow bad or threatening. But in fact, most insects are harmless to agriculture, and some are absolutely essential. The insects that pollinate agricultural crops are one of the most vital, yet least understood and least appreciated, factors in cropland agriculture. Pollinators are the unsung heroes of agriculture.

Pollination is the process by which male sex cells of a plant (pollen) fertilize female sex cells of a plant; it is the botanical version of sexual intercourse. Without pollination, no plants could reproduce sexually, and no plant species would persist for long. Plants such as ferns, conifer trees, and grasses achieve pollination by the wind. Millions of minuscule pollen grains are blown long distances, and by chance a small number land on the female parts of other plants of their species. The many kinds of plants that sport showy flowers, however, typically are pollinated by animals, such as hummingbirds, bats, and insects (Figure 10.10). Flowers are, in fact, evolutionary adaptations that function to attract pollinators. The sugary nectar and protein-rich pollen in flowers serve as rewards to lure these sexual intermediaries, and the sweet smells and bright colors of flowers are signals to advertise these rewards.

Although our staple grain crops are derived from grasses and are wind-pollinated, many other crops

FIGURE 10.10 Many agricultural crops depend on insects or other animals to pollinate them. Our food supply, therefore, depends partly on conservation of these vital organisms. These apple blossoms are being visited by a European honeybee. Flowers use colors and sweet smells to advertise nectar and pollen, enticements that attract pollinators.

FIGURE 10.11 European honeybees are widely used to pollinate crop plants, and beekeepers transport hives of bees to crops when it is time for flowers to be pollinated. However, honeybees have recently suffered devastating epidemics of parasitism, making it increasingly important for us to conserve native species of pollinators.

depend on insects for pollination. The most complete survey to date, by tropical bee biologist Dave Roubik, documented 800 species of cultivated plants that rely on bees and other insects for pollination. An estimated 73% of cultivars are pollinated, at least in part, by bees; 19% by flies; 5% by wasps; 5% by beetles; and 4% by moths and butterflies. In addition, bats pollinate 6.5% and birds 4%. As one of many examples, alfalfa has long been a major cash crop in the U.S. Great Basin states and is pollinated mostly by native alkali bees that live in the soil as larvae. In the 1940s to 1960s, farmers began plowing the land and increasing pesticide use in an effort to boost yields. These measures killed vast numbers of the soil-dwelling bees, and alfalfa production declined. By 1990, only 15% of U.S. alfalfa-growing lands were inhabited by alkali bees and native leaf-cutter bees. Despite their decline in numbers, bees and their pollination services help these lands produce fully 85% of the U.S. alfalfa crop.

Conservation of pollinators is vital

Preserving the biodiversity of native pollinators is especially important today because the domesticated workhorse of pollination, the European honeybee *(Apis apis)*, is being devastated by parasites. North American farmers regularly hire beekeepers to bring colonies of this introduced honeybee to their fields when it is time to pollinate crops (Figure 10.11). In recent years, certain parasitic mites have swept through honeybee populations, decimating hives and pushing many beekeepers toward financial ruin. Moreover, research indicates that honeybees are sometimes less effective pollinators than many native species, and often outcompete them, keeping the native species away from the plants.

Farmers and homeowners alike can help maintain populations of pollinating insects by reducing or eliminating pesticide use. All insect pollinators, including honeybees, are vulnerable to the vast arsenal of insecticides that modern industrial agriculture applies to crops and that many homeowners apply to lawns and gardens. Some insecticides are designed to specifically target certain types of insects, but many are not. Without full and detailed information on the effects of pesticides, farmers and homeowners trying to control the "bad" bugs that threaten the plants they value all too often kill the "good" insects as well.

Homeowners, even in the middle of a large city, can encourage populations of pollinating insects by planting gardens of flowering plants that nourish pollinating insects and by providing nesting sites for bees. These can be simple contraptions made of wood that have holes and plastic straws. And by allowing noncrop flowering plants (such as clover) to grow around the edges of their fields, farmers can maintain a diverse community of insects—some of which will pollinate their crops.

Genetic Modification of Food

The green revolution enabled us to feed a greater number and proportion of the world's people, but relentless population growth demands still more. A new set of potential solutions began to arise in the 1980s and 1990s as advances in genetics enabled scientists to directly alter the genes of organisms, including crop plants and livestock. The genetic modification of organisms that provide us food holds promise for increasing nutrition and the efficiency of agriculture while lessening the impacts of agriculture on the planet's environmental systems. However, genetic modification may also pose risks that are not yet well understood, which has given rise to protest around the globe from consumer advocates, small farmers, opponents of big business, and environmental activists. Because genetically modified foods have generated so much emotion and controversy, it is vital at the outset to clear up the terminology and clarify exactly what the techniques involve.

Genetic modification of organisms depends on recombinant DNA

The genetic modification of crops and livestock is one type of genetic engineering. **Genetic engineering** is any process whereby scientists directly manipulate an organism's genetic material in the lab, by adding, deleting, or changing segments of its DNA. **Genetically modified (GM) organisms** are organisms that have been genetically engineered using a technique called recombinant DNA technology. **Recombinant DNA** is DNA that has been patched together from the DNA of multiple organisms. In this process, scientists break up DNA from multiple organisms and then splice segments together, trying to place genes that produce certain proteins and code for certain desirable traits (such as rapid growth, disease and pest resistance, or higher nutritional content) into the genomes of organisms lacking those traits.

Recombinant DNA technology was developed in the 1970s by scientists studying the bacterium *Escherichia coli*. As shown in Figure 10.12, scientists first isolate plasmids, small, circular DNA molecules, from a bacterial culture. At the same time, DNA containing a gene of interest is removed from the cells of another organism. Scientists insert the gene of interest into the plasmid to form recombinant DNA. This recombinant DNA enters new bacteria, which then reproduce, generating many copies of the desired gene.

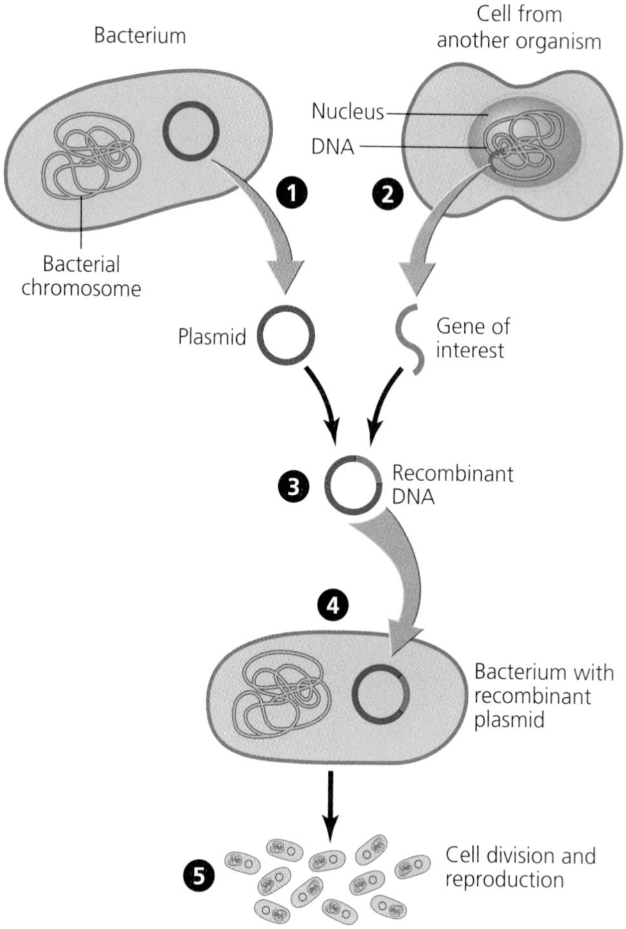

FIGURE 10.12 The creation of recombinant DNA is a key to genetically modifying organisms. In this process, a gene of interest is excised from the DNA of one type of organism and is inserted into a stretch of bacterial DNA called a plasmid. The plasmid is then introduced into cells of the organism to be modified. If all goes as planned, the new gene will be expressed in the GM organism as a desirable trait, such as rapid growth or high nutritional content in a food crop.

When scientists use recombinant DNA technology to develop new varieties of crops, they can often introduce the recombinant DNA directly into a plant cell and regenerate an entire plant from that single cell. Some plants, including many grains, are not receptive to plasmids, in which case scientists may use a "gene gun" to shoot DNA directly into plant cells. An organism that contains DNA from another species is called a *transgenic* organism, and the genes that have moved between them are called *transgenes*. The creation of transgenic organisms is one type of **biotechnology,** the material application of biological science to create products derived from organisms. Recombinant DNA and other types of biotechnology have helped us develop medicines, clean up pollution,

Several Notable Examples of Genetically Modified Food Technology	
Food	Development
Golden rice 	Millions of people in the developing world get too little vitamin A in their diets, causing diarrhea, blindness, immune suppression, and even death. The problem is worst with children in east Asia, where the staple grain, white rice, contains no vitamin A. Researchers took genes from plants that produce vitamin A and spliced the genes into rice DNA to create more-nutritious "golden rice" (the vitamin precursor gives it a golden color). Critics charged that biotech companies hyped their product, which contains only small amounts of the nutrient and may not be the best way to combat vitamin A deficiency. India's foremost critic of GM food, Vandana Shiva, charged that "vitamin A rice is a hoax... a very effective strategy for corporate takeover of rice production, using the public sector as a Trojan horse." Backers of the technology counter that the nutritive value can be further improved and could enhance the health of millions of people.
Flavr Savr tomato 	By reversing the function of a normal tomato gene, the Calgene Corporation created the Flavr Savr tomato, which Calgene maintained would ripen longer on the vine, taste better, stay firm during shipping, and last longer in the produce department. The U.S. Food and Drug Administration approved the Flavr Savr tomato for sale in the United States in 1994. Calgene stopped selling the Flavr Savr in 1996, however, for several reasons, including problems with the technique and public safety concerns.
Ice-minus strawberries 	University of California–Berkeley researcher Steven Lindow removed a gene that facilitated the formation of ice crystals from the DNA of a particular bacterium, *Pseudomonas syringae*. The modified, frost-resistant bacteria could then serve as a kind of antifreeze when sprayed on the surface of frost-sensitive crops such as strawberries. The multiplying bacteria would coat the berries, protecting them from frost damage. However, early news coverage of this technique showed scientists spraying plants while wearing face masks and protective clothing, an image that caused public alarm.
Bt crops 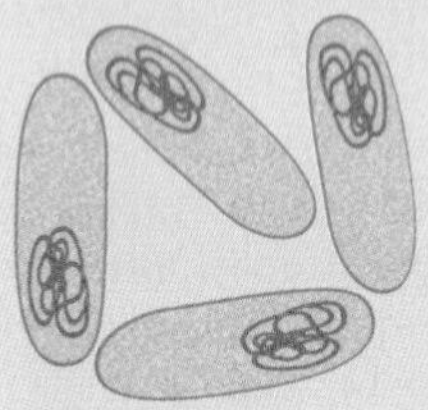	By equipping plants with the ability to produce their own pesticides, scientists hoped to boost crop yields by reducing losses to insects. By the late 1980s, scientists working with *Bacillus thuringiensis* (Bt) had pinpointed the genes responsible for producing that bacterium's toxic effects on insects, and had managed to insert the genes into the DNA of crops. The USDA and EPA approved Bt versions of 18 crops for field testing, from apples to broccoli to cranberries. Corn and cotton are the most widely planted Bt crops today. Proponents say Bt crops reduce the need for chemical pesticides. However, critics worry that the continuous presence of Bt in the environment will induce insects to evolve resistance to the toxins and that Bt crops might cause allergic reactions in humans. Another concern is that the crops may harm nontarget species. A 1999 study reported that pollen from Bt corn can kill the larvae of monarch butterflies, a nontarget species, when corn pollen drifts onto milkweed plants monarchs eat. Another study that year showed the Bt toxin could leach from corn roots and poison the soil.

FIGURE 10.13 The early development of genetically modified foods has been marked by a number of cases in which these products ran into trouble in the marketplace or were opposed by activists. A selection of these cases serves to illustrate some of the issues that proponents and opponents of GM foods have being debating.

understand the causes of cancer and other diseases, dissolve blood clots after heart attacks, and make better beer and cheese. Figure 10.13 details several of the most notable developments in GM foods. These examples and the stories behind them illustrate both the promises and pitfalls of food biotechnology.

Genetic engineering is like, and unlike, traditional agricultural breeding

The genetic alteration of plants and animals by humans is nothing new; we have been influencing the genetic makeup of our livestock and crop plants for thousands of years. As we saw in Chapter 9 (▸p. 248), our ancestors altered the gene pools of our domesticated plants and animals through selective breeding by preferentially mating individuals with favored traits so that offspring would inherit those traits. Early farmers selected plants and animals that grew faster, were more resistant to disease and drought, and produced large amounts of fruit, grain, or meat.

Proponents of GM crops often stress this continuity with our past and say there is little reason to expect that today's GM food will be any less safe than selectively bred

Several Notable Examples of Genetically Modified Food Technology	
Food	Development
StarLink corn 	StarLink corn, a variety of Bt corn, had been approved and used in the United States for animal feed but not for human consumption. In 2000, StarLink corn DNA was discovered in taco shells and other corn products, causing fears that the corn might cause allergic reactions. No such health effects have been confirmed, but the corn's French manufacturer, Aventis CropScience, chose to withdraw the product from the market. Although StarLink corn was grown on only a tiny portion of U.S. farmland, its transgene apparently spread widely to other corn through cross-pollination. This episode cost U.S. taxpayers, because the U.S. government spent $20 million to purchase contaminated corn and remove it from the food supply.
Sunflowers and superweeds 	Sunflowers have also been engineered to express the Bt toxin. Research on Bt sunflowers suggests that their transgenes might spread to other plants and turn them into vigorous weeds that compete with the crop. This is most likely to happen with crops like squash, canola, and sunflowers that can breed with their wild relatives. In 2002, Ohio State University researcher Alison Snow and colleagues bred wild sunflowers with Bt sunflowers and found that hybrids with the Bt gene produced more seeds and suffered less herbivory than hybrids without it. They concluded that if Bt sunflowers were planted commercially, the Bt gene would spread into wild sunflowers, potentially turning them into superweeds. Researcher Norman Ellstrand of the University of California–Riverside, meanwhile, had found that transgenes from radishes were transferred to wild relatives 1 km (0.6 mi) away and that hybrids produced more seeds, so the gene could be expected to spread in wild populations. He found the same results with sorghum and its weedy relative, johnsongrass. Such results suggest that transgenic crops can potentially create superweeds that can compete with crops and harm nontarget organisms.
Roundup Ready crops 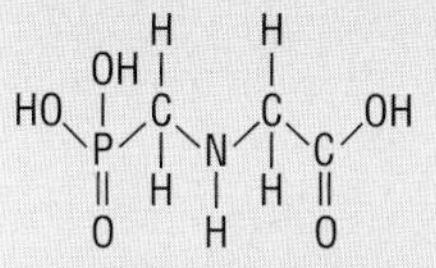	The Monsanto Company manufactures a widely used herbicide called Roundup. Roundup kills weeds, but kills crops too, so farmers must apply it carefully. Thus, Monsanto engineered Roundup Ready crops, including soybeans, corn, cotton, and canola, that are immune to the effects of its herbicide. With these variants, farmers can spray Roundup on their fields without killing their crops, in theory making the farmer's life easier. Of course, this also creates an incentive for farmers to use Monsanto's Roundup herbicide rather than a competing brand. Unfortunately, Roundup is not completely benign; its active ingredient, glyphosate, is the third-leading cause of illness for California farm workers. It also harms nitrogen-fixing bacteria and desirable fungi in soils that are essential for crop production. Biotech proponents have argued that GM crops are good for the environment because they reduce pesticide use. This may often be the case, but some studies have shown that farmers apply more herbicide when they use Roundup Ready crops.
Terminator seeds 	In the late 1990s the USDA worked with Delta and Pine Land Company to engineer a line of crop plants that can kill their own seeds. This so-called "terminator" technology would ensure that farmers buy seeds from seed companies every year rather than planting seeds saved from the previous year's harvest. Because GM crops require a great deal of research and development, seed companies reason that they need to charge farmers annually for seeds in order to recoup their investment. Critics worried that pollen from terminator plants might fertilize normal plants, damaging the crops of farmers who save seeds from year to year. Some nations, like India and Zimbabwe, banned terminator seeds. These countries saw the efforts of biotech seed companies to sell them terminator seeds as a ploy to make poor farmers dependent on multinational corporations for seeds. In the face of this opposition, in 1999 agrobiotech companies Monsanto and AstraZeneca announced that they would not bring their terminator technologies to market.

food. Dan Glickman, head of the USDA from 1995 to 2001, remarked:

> Biotechnology's been around almost since the beginning of time. It's cavemen saving seeds of a high-yielding plant. It's Gregor Mendel, the father of genetics, cross-pollinating his garden peas. It's a diabetic's insulin, and the enzymes in your yogurt. . . . Without exception, the biotech products on our shelves have proven safe.

However, as biotech critics are quick to point out, the techniques geneticists use to create GM organisms do differ from traditional selective breeding in several ways. For one, selective breeding generally mixes genes of individuals of the same species, whereas with recombinant DNA technology, scientists mix genes of different ones, even species as different as viruses and crops, or spiders and goats. For another, selective breeding deals with whole organisms living in the field, whereas genetic engineering involves lab experiments dealing with genetic material apart from the organism. And whereas traditional breeding selects from among combinations of genes that come together on their own, genetic engineering creates the novel combinations directly. Thus, traditional breeding changes organisms through the process of selection, whereas genetic engineering is more akin to the process of mutation.

Biotechnology is transforming the products around us

In just three decades, GM foods have gone from science fiction to big business. As recombinant DNA technology first developed in the 1970s, scientists debated among themselves whether the new methods were safe. They collectively regulated and monitored their own research until most scientists were satisfied that reassembling genes in bacteria did not create dangerous superbacteria. Once the scientific community declared itself confident that the technique was safe in the 1980s, industry leaped at the chance to develop hundreds of applications, from improved medicines (such as hepatitis B vaccine and insulin for diabetes) to designer plants and animals. Traits engineered into crops, such as built-in pest resistance and herbicide resistance, made it efficient, and in some cases more economical, for large-scale commercial farmers to do their jobs. As a result, sales of GM seeds to these farmers in the United States and other countries rose quickly.

Today well over two-thirds of the U.S. harvests of soybeans, corn, and cotton consists of genetically modified strains. Worldwide, more than half the soybean harvest is transgenic, as is one of every four cotton plants, one of every five canola plants, and one of every six corn plants. Globally in 2004, it was estimated that GM crops grew on 81 million ha (200 million acres) of farmland, an area the size of Kansas, Nebraska, and California combined.

Of the 17 nations growing GM crops in 2004, five (the United States, Argentina, Canada, Brazil, and China) accounted for 96% of GM crops, with the United States alone accounting for three-fifths of the global total (Figure 10.14). Because these nations are major food exporters, much of the produce on the world market is transgenic for crops such as soybeans, corn, cotton, and canola. The global area planted in GM crops has jumped by more than 10% annually every year since 1996, and over half the world's people now live in nations in which GM crops are grown. The market value of GM crops in 2004 was estimated at $4.7 billion.

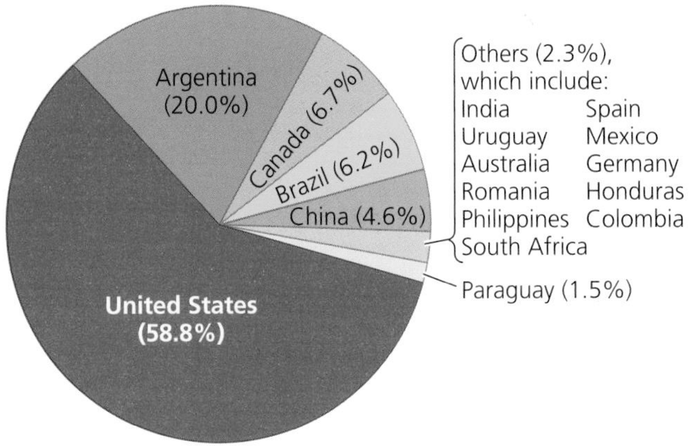

FIGURE 10.14 At 58.8% of the global total, the United States leads the world in land area dedicated to genetically modified crops. Following far behind the United States is Argentina at 20.0%. Data from International Service for the Acquisition of Agri-Biotech Application (ISAAA), 2004.

What are the impacts of GM crops?

As GM crops were adopted, as research proceeded, and as biotech business expanded, many citizens, scientists, and policymakers became concerned. Some feared the new foods might be dangerous for people to eat. Others were concerned that transgenes might escape and pollute ecosystems and damage nontarget organisms. Still others worried that pests would evolve resistance to the supercrops and become "superpests" or that transgenes would be transferred from crops to other plants and turn them into "superweeds." Some, like Quist and Chapela, worried that transgenes might ruin the integrity of native ancestral races of crops.

Because the technology is new and its large-scale introduction into the environment is newer still, there remains a lot scientists don't know about how transgenic crops behave in the field. Certainly, millions of Americans eat GM foods every day without outwardly obvious signs of harm, and evidence for negative ecological effects is limited so far. However, it is still too early to dismiss the concerns discussed above without further scientific research. Therefore, critics argue that we should proceed with caution, adopting the **precautionary principle,** the idea that one should not undertake a new action until the ramifications of that action are well understood.

The best efforts so far to test the ramifications of GM crops produced results in 2003 through 2005. The British government, in considering whether to allow the planting of GM crops, commissioned three large-scale studies. The first study, on economics, found that GM crops could produce long-term financial benefits for Britain, although short-term benefits would be minor. The second study addressed health risks and found little to no evidence of harm to human health, but noted that effects on wildlife and ecosystems should be tested before crops are approved. The third study (involving 19 researchers, 200 sites, and $8 million in funding) tested effects on bird and invertebrate populations from four GM crops modified for herbicide resistance. Results showed that fields of GM beets and GM spring oilseed rape supported less biodiversity than fields of their non-GM counterparts.

Fields of GM maize supported more, however, and fields of winter oilseed rape showed mixed results. Policymakers had hoped that the biodiversity study would end the debate, but the science showed that the impacts of GM crops are complex.

Debate over GM foods involves more than science

Much more than science has been involved in the debate over GM foods. Ethical issues have played a large role. For many people, the idea of "tinkering" with the food supply seems dangerous or morally wrong. Even though our agricultural produce is the highly artificial product of thousands of years of selective breeding, people tend to think of food as natural. Furthermore, because every person relies on food for survival and cannot choose *not* to eat, the genetic modification of dietary staples such as corn, wheat, and rice essentially forces people to consume GM products or to go to special effort to avoid them.

The perceived lack of control over one's own food has driven widespread concern about domination of the global food supply by a few large businesses. Gigantic agrobiotech companies, among them Monsanto, Syngenta, Bayer CropScience, Dow, DuPont, and BASF, create GM technologies. Many activists say these multinational corporations threaten the independence and well-being of the small farmer. This perceived loss of democratic local control is a driving force in the opposition to GM foods, especially in Europe and the developing world. Critics of biotechnology also voice concern that much of the research into the safety of GM organisms is funded, overseen, or conducted by the corporations that stand to profit if their transgenic crops are approved for human consumption, animal feed, or ingredients in other products.

So far, GM crops have not lived up to their promise of feeding the world's hungry. Nearly all commercially available GM crops have been engineered to express either pesticidal properties (e.g., Bt crops) or herbicide tolerance. Often, these GM crops are tolerant to herbicides that the same company manufactures and profits from (e.g., Monsanto's Roundup Ready crops). Crops with traits that might benefit poor small-scale farmers of developing countries (such as increased nutrition, drought tolerance, and salinity tolerance) have not been widely commercialized, perhaps because corporations have less economic incentive to do so. Whereas the green revolution was a largely public venture, the "gene revolution" promised by GM crops has been largely driven by market considerations of companies selling proprietary products.

FIGURE 10.15 Saskatchewan farmer Percy Schmeiser was accused by the Monsanto Company of planting its patented Roundup Ready soybeans in his field without a contract with the company. Schmeiser said his non-GM plants were contaminated with Monsanto's transgenes from neighboring farms. Monsanto won the David-and-Goliath case in Canada's Supreme Court, but Schmeiser became a hero to small farmers and anti-GM food activists worldwide.

When the U.S.-based Monsanto Company began developing GM products in the mid-1980s, it foresaw public anxiety and worked hard to inform, reassure, and work with environmental and consumer advocates, whom the company feared would otherwise oppose the technology. Monsanto even lobbied the U.S. government to regulate the industry so the public would feel safer about it. These efforts were undermined, however, when the company's first GM product to market, a growth hormone to spur milk production in cows, alarmed consumers concerned about children's health. Then, when the company went through a leadership change, its new head changed tactics and pushed new products aggressively without first reaching out to opponents. Opposition built, and the company lost the public's trust, especially in Europe and in the developing world.

In Canada, Monsanto has been engaged in a high-publicity battle with a third-generation Saskatchewan farmer named Percy Schmeiser (Figure 10.15). Schmeiser maintained that pollen from Monsanto's Roundup Ready canola (see Figure 10.13) used by his neighbors blew onto his land and pollinated his non-GM canola. Schmeiser had never purchased Monsanto's patented

The Science behind the Story

Transgenic Contamination of Native Maize?

David Quist and Ignacio Chapela's *Nature* paper reporting DNA from genetically engineered corn in native Mexican maize was controversial from the moment it was published. The idea that a transgene had entered native maize in remote areas of Oaxaca was provocative enough. But their claim that the transgene also had been jumping throughout the genome came under particular fire.

Soon after publication, several geneticists pointed out flaws in the methods Quist and Chapela used to determine the location of the transgene. Interpreting results of an experimental technique known as inverse polymerase chain reaction (i-PCR), Quist and Chapela had reported that the transgene was surrounded by essentially random sequences of DNA. Critics argued that i-PCR was unreliable and that Quist and Chapela had used insufficient controls. Their results, the critics suggested, could have arisen from similarities between the transgene and stretches of maize DNA.

Ignacio Chapela (left) and David Quist (right) ignited a firestorm of controversy with their scientific paper reporting transgenic corn DNA in Mexican maize.

Several geneticists also attacked Quist and Chapela's more fundamental claim that transgenes had been integrated into the genome of native maize. Matthew Metz of the University of Washington and Johannes Fütterer of the Institute of Plant Sciences in Zürich, Switzerland, argued that the low levels of transgenic DNA detected indicated, at most, first-generation hybrids between local cultivars and transgenic varieties. Other critics, such as Paul Christou, editor of the journal *Transgenic Research,* noted that PCR was a highly sensitive technique and that all of Quist and Chapela's findings could be unreliable if laboratory practices had been careless.

On April 11, 2002, *Nature* published two letters from scientists critical of the study, along with an editorial note concluding that "the evidence available is not sufficient to justify the publication of the original paper." Quist and Chapela responded by acknowledging that some of their initial findings, particularly those based on the i-PCR technique, were probably invalid. But they also

seed and said he did not want the crossbreeding. Monsanto investigators took seed samples from his plants and charged him with violating Canada's law that makes it illegal for farmers to reuse patented seed or grow the seed without a contract with the company. Monsanto sued Schmeiser, and the court sided with Monsanto, ordering the farmer to pay the corporation roughly $238,000. Schmeiser appealed the case to the Supreme Court of Canada, which in 2004 ruled by a 5-4 vote that Schmeiser had violated Monsanto's patent (although it spared the 74-year-old farmer from having to make payments to the company). Schmeiser received wide public support, a government committee called for revising the patent law, and the National Farmers Union of Canada called for a moratorium on GM food. Meanwhile, Monsanto continues to demand that small farmers in Canada and the United States heed the Canadian and U.S. patent laws.

Weighing the Issues: Early Hurdles for GM Foods

As the vignettes in Figure 10.13 illustrate, a number of GM foods have run into difficulties. Do you think this reveals an underlying problem with the approach, or are these simply examples of unavoidable glitches that are bound to occur during the early development of any new technology? Can you take any lessons from the stories in Figure 10.13? Do you expect that debate between proponents and opponents of GM organisms will subside in the future, or not?

Given such developments, the future of GM foods seems likely to hinge on social, economic, legal, and political factors as well as scientific ones. European consumers have expressed widespread unease about possible presented new analyses to support their fundamental claim and pointed to a Mexican government study that also found high rates of transgenic contamination. However, this did little to convince skeptics, especially because the Mexican government's results remained unpublished.

The correspondence published in *Nature* was only part of a broader debate that took place on the editorial pages of journals such as *Nature Biotechnology* and *Transgenic Research,* in newspapers and magazines, over the Internet, and within the halls of academe. The debate sometimes turned personal. GM supporters pointed out that Chapela and Quist had long opposed transgenic crops and biotechnology corporations. Chapela had spoken out against his university's plan to enter into a $25 million partnership with the biotechnology firm Novartis (now Syngenta), sparking a debate that created divisions among Berkeley faculty. GM opponents countered that Quist and Chapela's most vocal critics received funding from biotechnology corporations. As the controversy progressed, it became clear that few participants in the debate could be considered entirely objective. Even *Nature,* whose impartiality is critical to its reputation as a first-tier science journal, was engaged in commercial partnerships with biotechnology corporations.

Chapela was denied tenure by Berkeley's administration, even though his department and dean had recommended him. Alleging that he had received unfair treatment, Chapela waged a 3-year fight to win tenure and gained a following that included 230 academics who signed a letter of support. In May 2005, a new chancellor awarded him tenure.

The twists and turns of the still-unresolved debate illustrate that there can be more to the scientific process than merely the scientific method. Almost overlooked, however, was the fact that nearly all researchers agreed that gene flow between transgenic corn and Mexico's native landraces was bound to happen at some point.

Researchers anxious about transgenic crops emphasize that such gene flow could decrease genetic diversity and negatively affect agriculture and the environment. Those in favor of transgenic foods argue that gene flow is natural and that the genetic diversity of Mexican maize has persisted despite thousands of years of cross-pollination. The paucity of adequate scientific data—and the high economic and environmental stakes surrounding Mexico's ban on transgenic corn—go a long way toward explaining why the debate has been so heated.

risks of GM technologies, whereas U.S. consumers have largely accepted the GM crops approved by U.S. agencies. Opposition in nations of the European Union resulted in a de facto moratorium on GM foods from 1998 to 2003, blocking the importation of hundreds of millions of dollars in U.S. agricultural products. This prompted the United States to bring a case before the World Trade Organization in 2003, complaining that Europe's resistance was hindering free trade. Europeans now demand that GM foods be labeled and criticize the United States for not joining 100 other nations in signing the Cartagena Protocol on Biosafety, a treaty that lays out guidelines for open information about exported crops.

Transnational spats between Europe and the United States will surely affect the future direction of agriculture, but the world's developing nations could exert the most influence in the end. Recent decisions by the governments of India and Brazil to approve GM crops (following long and divisive debate) are already adding greatly to the world's transgenic agriculture. Moreover, China is aggressively expanding its use of transgenic crops.

A counterexample is Zambia, one of several African nations that refused U.S. food aid meant to relieve starvation during a drought in late 2002. The governments of these nations worried that their farmers would plant some of the GM corn seed meant to be eaten and that GM corn would thereby establish itself in their countries. They viewed this as undesirable because African economies depend on exporting food to Europe, which has put severe restrictions on GM food. In the end, Zambia's neighbors accepted the grain after it had been milled (so none could be planted), but Zambia held out. Citing health and environmental risks, uncertain science, and the precautionary principle, the Zambian government declined the aid, despite the fact that 2–3 million of its people were at risk

VIEWPOINTS

Genetically Modified Foods

Proponents of GM foods say these products can alleviate hunger and malnutrition while posing no known threats to human health or the environment. Opponents say that these foods help only the large corporations that sell them and that they do pose risks to human health, wild organisms, and ecosystems. **What do you think? Should we encourage the continued development of GM foods? If so, what, if any, restrictions should we put on their dissemination?**

A Global Experiment Without Controls

Genetic engineering, specifically transgenesis, gives us the unprecedented capacity to move DNA. In so doing, this technology breaches boundaries established through millions of years of evolution. As such, we should expect fundamental alterations in ecosystems with the release of transgenic crops, fish, insects, microbes, and so forth into uncontrolled areas. These alterations are similar in nature to those caused by the introduction of exotic species into new environments. Both processes are unpredictable and could have serious consequences.

Yet because of political and short-term economic imperatives, releases of transgenic organisms have continued unabated for at least a decade. Science has barely started to imagine the ecological and evolutionary consequences of releasing transgenic crops. Not only are we experiencing a global experiment without controls, we don't have the tools to document it. Serious research, although extremely scarce, has already confirmed some of the theoretical fears concerning transgenesis aired by scientists a quarter century ago.

Today, we have cataclysmic world hunger paired with food surpluses. The claim that transgenesis can solve this problem is merely a diversion tailored to conceal how transgenesis manipulates the biosphere. Molecular biology might one day become part of the solution to world hunger, but it is certainly not the science most relevant to address the problem today.

What checks should be placed on the release of transgenic organisms? Every check. Through the unaccountable releases so far, we have seen enough, and possibly caused enough, environmental insult for me to say today that we should stop. We need to take stock of the consequences of transgenesis and continue researching under strictly regulated conditions.

Ignacio H. Chapela is associate professor (microbial ecology) in the Department of Environmental Science, Policy, and Management at the University of California, Berkeley. He helped found the Mycological Facility: Oaxaca, Mexico, where he serves as scientific director.

The United States Should Begin a Phased Deregulation of Biotech Crops

During the past two decades the international scientific community, biotechnology industry, and regulatory agencies in many countries have accumulated and critically evaluated a wealth of information about the production and use of biotech crops and products. Biotech crops have been planted since 1996 on more than 1 billion acres of farmland in nearly 20 countries. More than 1 billion humans and hundreds of millions of farm animals have consumed biotech foods and products. Yet there is not a single instance in which biotech crops and foods have been shown to cause illness in humans or animals or to damage the environment.

In spite of this exemplary safety record, a small but well-organized, well-financed, and vocal antibiotechnology lobby has alleged that biotech crops and products are unsafe for humans and a danger to the environment, demanding a moratorium or outright ban on biotech crops. The rhetoric of the antibiotechnology groups is alarming, confusing, and frightening to the public, but it is devoid of any substance, because they have never provided any credible scientific evidence to support their allegations.

Any further delay in combining the power of biotechnology with conventional breeding will seriously endanger future food security, political and economic stability, and the environment. Plant biotechnology is still the best hope for meeting the food needs of the ever-growing world population. Biotech crops are already helping to conserve valuable natural resources, reduce the use of harmful agro-chemicals, produce more nutritious foods, and promote economic development.

Twenty years ago, the United States set the precedent by developing regulations for the development and use of biotech crops. Now, as the world leader in plant biotechnology, it is imperative that it lead again by phasing out these redundant regulations in an organized and responsible manner.

Indra K. Vasil is graduate research professor emeritus at the University of Florida (Gainesville, Florida). His research focuses on the biotechnology of cereal crops, and he has been recognized as one of the world's most highly cited authors in the plant and animal sciences.

Explore this issue further by accessing **Viewpoints** at www.aw-bc.com/withgott.

FIGURE 10.16 Debate over GM foods reached a dramatic climax in Zambia in 2002, when the government refused U.S. shipments of GM corn that were intended to relieve starvation due to drought. Here a Zambian mother with her child waits in a line for food assistance.

of starvation (Figure 10.16). Intense debate followed within the country and around the world. Eventually the United Nations delivered non-GM grain, and in April 2003 the Zambian government announced a plan to coordinate a comprehensive long-term policy on GM foods.

The Zambian experience demonstrates some of the ethical, economic, and political dilemmas modern nations face. The corporate manufacturers of GM crops naturally aim to maximize their profits, but they also aim to develop products that can boost yields, increase food security, and reduce hunger. Although industry, activists, policymakers, and scientists all agree that hunger and malnutrition are problems and that agriculture should be made environmentally safer, they often disagree about the solutions to these dilemmas and the risks that each proposed solution presents.

Preserving Crop Diversity

As the excitement over Quist and Chapela's findings in Oaxaca demonstrate, one concern many people harbor about transgenic crops is that transgenes might move, by pollination, into local native races of crop plants.

Crop diversity provides insurance against failure

Preserving the integrity of native variants gives us a bulwark against commercial crop failure. The regions where crops first were domesticated generally remain important repositories of crop biodiversity. Although modern industrial agriculture relies on a small number of plant types, its foundation lies in the diverse varieties that still exist in places like Oaxaca. These varieties contain genes that, through conventional crossbreeding or genetic engineering, might confer resistance to disease, pests, inbreeding, and other pressures that challenge modern agriculture. Monocultures essentially place all our eggs in one basket, such that any single catastrophic cause could potentially wipe out entire crops. Having available the wild relatives of crop plants or the domesticated varieties, or cultivars, of crop plants gives us the genetic diversity that may include ready-made solutions to unforeseen problems. Because accidental interbreeding can decrease the diversity of local variants, many scientists argue that we need to protect areas like Oaxaca. For this reason, the Mexican government helped create the Sierra de Manantlan Biosphere Reserve around an area harboring the localized plant thought to be the direct ancestor of maize. For this reason, too, it imposed a national moratorium in 1998 on the planting of transgenic corn.

We have lost a great deal of genetic diversity in our crop plants already. The number of wheat varieties in China is estimated to have dropped from 10,000 in 1949 to 1,000 by the 1970s, and Mexico's famed maize varieties now number only 30% of what was extant in the 1930s. In the United States, many fruit and vegetable crops have decreased in diversity by 90% in less than a century. A primary cause of this loss of diversity is that market forces have discouraged diversity in the appearance of fruits and vegetables. Commercial food processors prefer items to be similar in size and shape, for convenience. Consumers, for their part, have shown preferences for uniform, standardized food products over the years. Now that local organic agriculture is growing in affluent societies, however, consumer preferences for diversity are increasing.

Seed banks are living museums for seeds

Protecting areas with high crop diversity is one way to preserve genetic assets for our agricultural systems. Another is to collect and store seeds from crop varieties and periodically plant and harvest them to maintain a diversity of cultivars. This is the work of **seed banks** or **gene banks,** institutions that preserve seed types as a kind of living museum of genetic diversity (Figure 10.17). In total, these facilities hold roughly 6 million seed samples, keeping them in cold, dry conditions to encourage long-term viability. The $300 million in global funding for these facilities is not adequate for proper storage and for the labor of growing out the seed periodically to renew

(a) Traditional food plants of the Desert Southwest

(b) Pollination by hand

FIGURE 10.17 Seed banks preserve genetic diversity of traditional crop plants. Native Seeds/SEARCH of Tucson, Arizona, preserves seeds of food plants important to traditional diets of Native Americans of Arizona, New Mexico, and northwestern Mexico. Beans, chiles, squashes, gourds, maize, cotton, and lentils are all in its collections, as well as less-known plants such as amaranth, lemon basil, and devil's claw (**a**). Traditional foods such as mesquite flour, prickly pear pads, chia seeds, tepary beans, and cholla cactus buds help fight the diabetes that Native Americans frequently suffer from having adopted a Western diet. At the farm where seeds are grown, care is taken to pollinate varieties by hand (**b**) to protect their genetic distinctiveness.

the stocks. Therefore, it is questionable how many of these 6 million seeds are actually preserved. Major efforts include large seed banks such as the U.S. National Seed Storage Laboratory at Colorado State University, the Royal Botanic Garden's Millennium Seed Bank in Britain, Seed Savers Exchange in Iowa, and the Wheat and Maize Improvement Center (CIMMYT) in Mexico.

Feedlot Agriculture: Livestock and Poultry

Food from cropland agriculture makes up a large portion of the human diet, but most people also eat animal products. People don't *need* to eat meat or other animal products to live full, active, healthy lives, but for most people it is difficult to obtain a balanced diet without incorporating animal products. Most of us do eat animal products, and this choice has significant environmental, social, agricultural, and economic impacts.

Consumption of animal products is growing

As wealth and global commerce have increased, so has our consumption of meat, milk, eggs, and other animal products (Figure 10.18). The world population of domesticated animals raised for food rose from 7.3 billion animals to 20.6 billion animals between 1961 and 2000. Most of these animals are chickens. Global meat production has increased fivefold since 1950, and per capita meat consumption has nearly doubled. The most-eaten meat per unit weight is pork.

High consumption has led to feedlot agriculture

In traditional agriculture, livestock were kept by farming families near their homes or were grazed on open grasslands by nomadic herders or sedentary ranchers. These traditions have survived, but the advent of industrial agriculture has

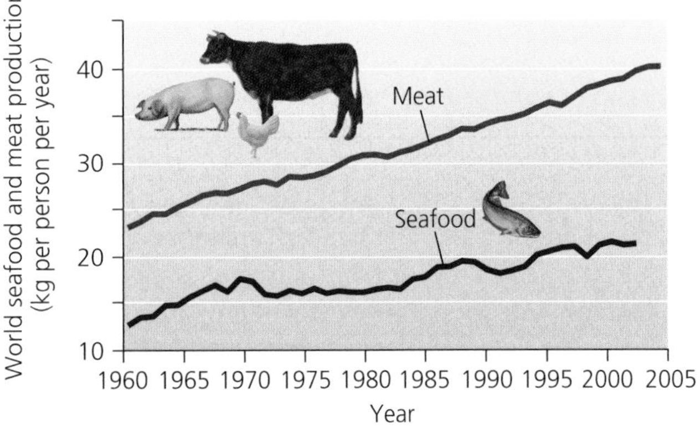

FIGURE 10.18 Per capita consumption of meat from farm animals has increased steadily worldwide over the past few decades, as has per capita consumption of seafood (marine and freshwater, harvested and farmed). Data from Food and Agriculture Organization of the United Nations.

FIGURE 10.19 These chickens at a Pennsylvania factory farm are housed several to a cage and have been "debeaked," the tips of their beaks cut off to prevent them from pecking one another. The hens cannot leave the cages and essentially spend their lives eating, defecating, and laying eggs, which roll down slanted floors to collection trays. The largest U.S. chicken farms house hundreds of thousands of individuals.

added a new method. **Feedlots,** also known as *factory farms* or *concentrated animal feeding operations (CAFOs),* are essentially huge warehouses or pens designed to deliver energy-rich food to animals living at extremely high densities (Figure 10.19). Today over half of the world's pork and poultry come from feedlots, as does much of its beef.

Feedlot operations allow for greater production of food and are probably necessary for a country with a level of meat consumption like that of the United States. Feedlots have one overarching benefit for environmental quality: Taking cattle, sheep, goats, and other livestock off the land and concentrating them in feedlots reduces the impact they would otherwise exert on large portions of the landscape. In Chapter 9 (▸ pp. 267–268, and 270) we saw how overgrazing can degrade soils and vegetation and that hundreds of millions of hectares of land are considered overgrazed. Animals that are densely concentrated in feedlots will not contribute to overgrazing and soil degradation.

Of course, feedlots are not without impact, and many environmental advocates have attacked them for their contributions to water and air pollution. Waste from feedlots can emit strong odors and can pollute surface water and groundwater, because livestock produce prodigious amounts of feces and urine. One dairy cow can produce about 20,400 kg (44,975 lb) of waste in a single year. Greeley, Colorado, is home to North America's largest meat-packing plant and two adjacent feedlots, all of which are owned by the agribusiness firm ConAgra. Each feedlot has room for 100,000 cattle that are fed surplus grain and injected with anabolic steroids to stimulate growth. During its stay at the feedlot, a typical steer will eat 1,360 kg (3,000 lb) of grain, gain 180 kg (400 lb) in body weight, and generate 23 kg (50 lb) of manure each day. The amount of manure that 200,000 such animals generate exceeds the amount of waste produced by all the human residents of Atlanta, St. Louis, Boston, and Denver combined. Poor waste containment practices at feedlots in North Carolina, Maryland, and other states have been linked to outbreaks of disease, including virulent strains of *Pfiesteria,* a microbe that poisons fish. The crowded and dirty conditions under which animals are often kept necessitates heavy use of antibiotics to control disease. These chemicals can be transferred up the food chain, and their overuse can cause microbes to evolve resistance to them.

Feedlot impacts can be minimized when properly managed, and both the EPA and the states regulate U.S. feedlots. Most feedlot manure is applied to farm fields as fertilizer, reducing the need for chemical fertilizers. Manure in liquid form can be injected into the ground where plants need it, and farmers can conduct tests to determine amounts that are appropriate to apply.

Weighing the Issues: Feedlots and Animal Rights

Animal rights activists decry factory farming because they say it mistreats animals. Chickens, pigs, and cattle are kept crowded together in small pens their entire lives, fattened up, and slaughtered. Chickens are often "debeaked." Do you think animal rights concerns should be given weight as we determine how best to raise our food? Do you think these are as important as the environmental issues? Are conditions at feedlots a good reason for being vegetarian?

Our food choices are also energy choices

What we choose to eat has ramifications for how we use energy and the land that supports agriculture. Recall our discussions of thermodynamics and trophic levels (▸ pp. 104–105 and ▸ pp. 156–157). Every time energy moves from one trophic level to the next, as much as 90% is lost. For example, if we feed grain to a cow and then eat beef from the cow, we lose a great deal of the grain's energy to the cow's digestion and metabolism. Energy is used up

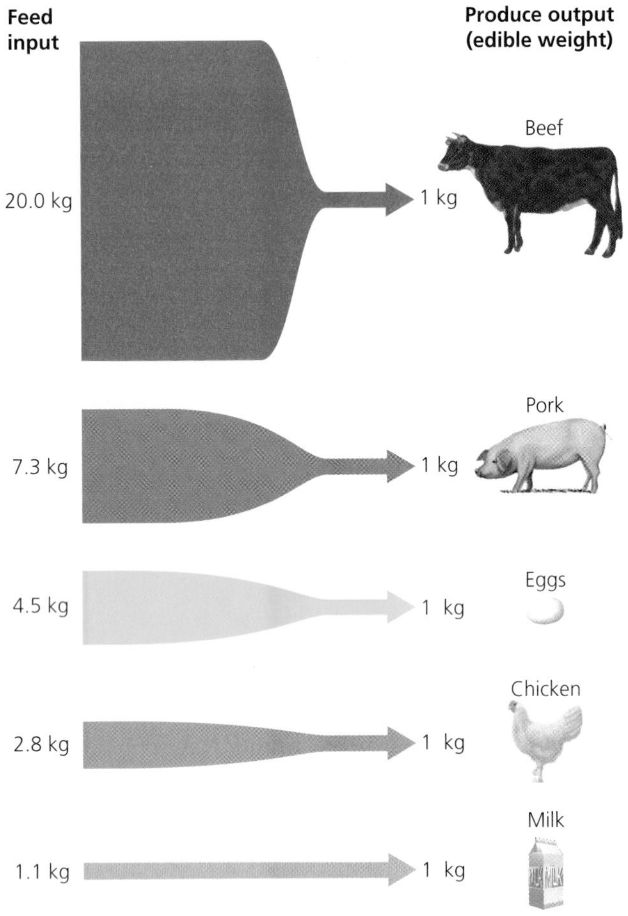

FIGURE 10.20 Different animal food products require different amounts of input of animal feed. Chickens must be fed 2.8 kg of feed for each 1 kg of resulting chicken meat, for instance, whereas 20 kg of feed must be provided to cattle to produce 1 kg of beef. Go to GRAPHIt! at www.aw-bc.com/withgott or on the student CD-ROM. Data from Smil, V. 2001. *Feeding the world: A challenge for the twenty-first century.* Cambridge, MA: MIT Press.

when the cow converts the grain to tissue as it grows, and as the cow uses its muscle mass on a daily basis to maintain itself. For this reason, eating meat is far less energy-efficient than relying on a vegetarian diet. The lower in the food chain from which we take our food sources, the greater the proportion of the sun's energy we put to use as food, and the more people Earth can support.

Some animals convert grain feed into milk, eggs, or meat more efficiently than others (Figure 10.20). Scientists have calculated relative energy-conversion efficiencies for different types of animals. Such energy efficiencies have ramifications for land use—land and water are required to raise food for the animals, and some animals require more than others. Figure 10.21 shows the area of land and weight of water required to produce 1 kg (2.2 lb) of food protein for milk, eggs, chicken, pork, and beef. Producing eggs and chicken meat requires the least space and water, whereas producing beef requires the most. Such differences make clear that when we choose what to eat, we are also indirectly choosing how to make use of resources such as land and water.

In 1900 we fed about 10% of global grain production to animals. In 1950 this number had reached 20%, and by the beginning of the 21st century we were feeding 45% of global grain production to animals. Although much of the grain fed to animals is not of a quality suitable for human consumption, the resources required to grow it could have instead been applied toward growing food for people. One partial solution is to feed livestock crop residues, plant matter such as stems and stalks that we would not consume anyway, and this is increasingly being done.

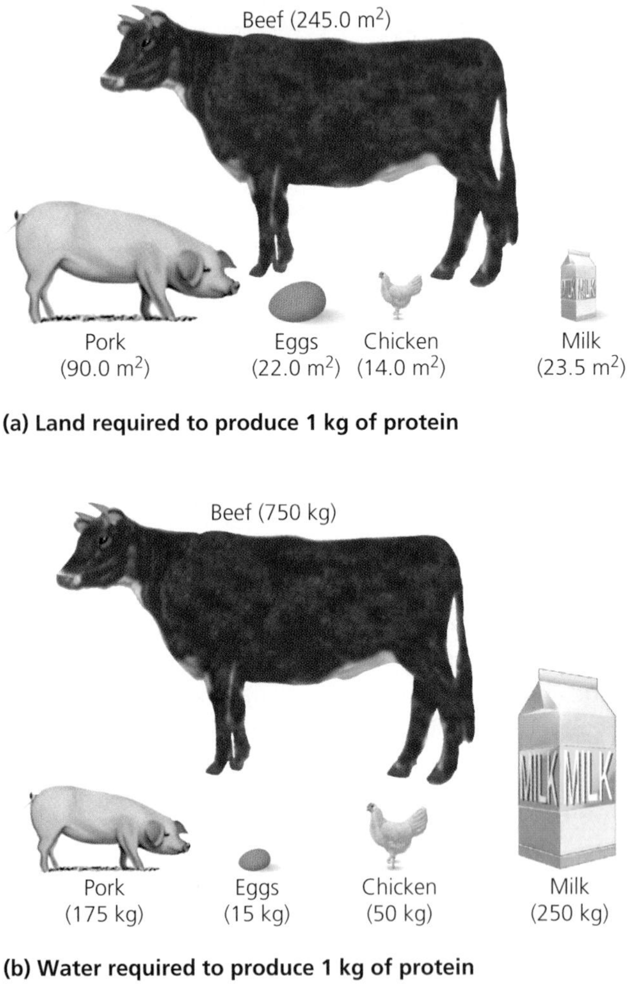

FIGURE 10.21 Producing different types of animal products requires different amounts of land and water. Raising cattle for beef requires by far the most land and water of all animal products. Go to GRAPHIt! at www.aw-bc.com/withgott or on the student CD–ROM. Data from Smil, V. 2001. *Feeding the world: A challenge for the twenty-first century,* Cambridge, MA: MIT Press.

Aquaculture

In addition to plants grown in croplands and animals raised on rangelands and in feedlots, we rely on aquatic organisms for food. Wild fish populations are plummeting throughout the world's oceans as increased demand and new technologies have led us to overharvest most marine fisheries (▸pp. 482–485). This means that raising fish and shellfish on "fish farms" may be the only way to meet the growing demand for these foods.

We call the raising of aquatic organisms for food in controlled environments **aquaculture.** Many aquatic species are grown in open water in large, floating net-pens. Others are raised in land-based ponds or holding tanks. People pursue both freshwater and marine aquaculture. Aquaculture is the fastest-growing type of food production; in the past 20 years, global output has increased sevenfold (Figure 10.22). Aquaculture today provides a third of the world's fish for human consumption, is most common in Asia, and involves over 220 species. Some, such as carp, are grown for local consumption, whereas others, such as salmon and shrimp, are exported to affluent countries.

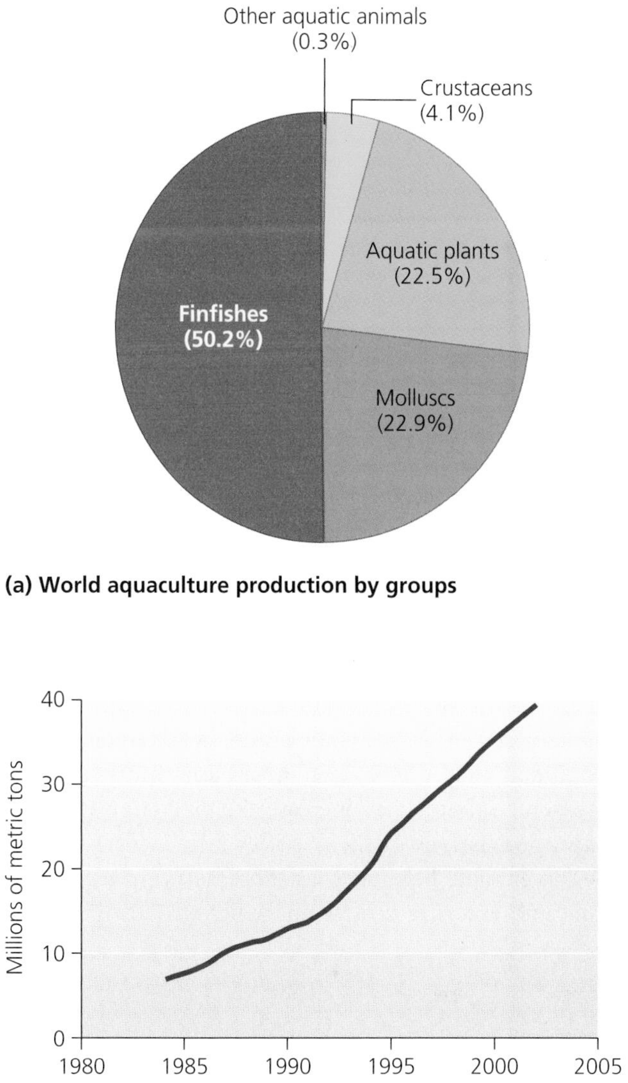

(a) World aquaculture production by groups

(b) World aquaculture production

FIGURE 10.22 Aquaculture involves a wide diversity of marine and freshwater organisms **(a).** Global production of meat (all species other than plants) from aquaculture has risen steeply in the past two decades **(b).** Data in (a) from FAO 2004. *The state of world fisheries and aquaculture, 2004;* and (b) FAO *Aquaculture production statistics, 1984–1993* and *Fishery statistics: Aquaculture production.*

Aquaculture brings a number of benefits

When conducted on a small scale by families or villages, as in China and much of the developing world, aquaculture helps ensure people a reliable protein source. This type of small-scale aquaculture can be sustainable, and it is compatible with other activities. For instance, uneaten fish scraps make excellent fertilizers for crops. Aquaculture on larger scales can help improve a region's or nation's food security by increasing overall amounts of fish available. Aquaculture on any scale also has the benefit of reducing fishing pressure on overharvested and declining wild stocks. Reducing fishing pressure also reduces the *by-catch* (▸p. 485; the unintended catch of nontarget organisms) that results from commercial fishing. Furthermore, aquaculture relies far less on fossil fuels than do fishing vessels and provides a safer work environment than does commercial fishing. Fish farming can also be remarkably energy-efficient, producing as much as 10 times more fish per unit area than is harvested from oceanic waters on the continental shelf and up to 1,000 times as much as is harvested from the open ocean.

Aquaculture has negative impacts

Along with its benefits, aquaculture has disadvantages. Dense concentrations of farmed animals can increase the incidence of disease, which reduces food security, necessitates antibiotic treatment, and results in additional expense. A virus outbreak wiped out half a billion dollars in shrimp in Ecuador in 1999, for instance. Aquaculture can also produce prodigious amounts of waste, from the farmed organisms and from the feed that goes uneaten and decomposes in the water column. Farmed fish often are fed grain, and as we have discussed, growing grain to feed animals that we then eat reduces the energy efficiency of food production and consumption. In other cases, farmed fish are fed fish meal made from wild ocean fish such as herring and anchovies, whose harvest may place additional stress on wild fish populations.

If farmed aquatic organisms escape into ecosystems where they are not native, they may spread disease to native stocks or may outcompete native organisms for food or habitat. The possibility of competition also arises when the farmed animals have been genetically modified. Like the transgenic corn that has influenced Mexican maize, transgenic fish have become a part of the food production system in recent years. Genetic engineering of Pacific salmon has produced transgenic fish that weigh up to 11 times more than nontransgenic ones. Transgenic Atlantic salmon raised in Scotland have been engineered to grow to 5–50 times the normal size for their species (Figure 10.23). GM fish such as these may outcompete their non-GM wild cousins while also spreading disease to them. They may also interbreed with native and hatchery-raised fish and weaken already troubled stocks. Researchers have concluded that under certain circumstances, escaped transgenic salmon may increase the extinction risk that native populations of their species face, in part because the larger male fish (such as those carrying a gene for rapid and excessive growth) have better odds of mating successfully.

FIGURE 10.23 Efforts to genetically modify important food fish have resulted in the creation of transgenic salmon (top), which can be considerably larger than wild salmon of the same species.

Sustainable Agriculture

Industrialized agriculture involves many adverse environmental impacts, from the degradation of soils (Chapter 9) to reliance on fossil fuels (Chapter 19) to problems arising from pesticide use, genetic modification, and intensive feedlot and aquaculture operations. Although many of these developments in intensive commercial agriculture have alleviated some environmental pressures, they have often exacerbated others. Industrial agriculture in some form seems necessary to feed our planet's 6.5 billion people, but many feel we will be better off in the long run by practicing less-intensive methods of raising animals and crops.

Farmers and researchers have made great advances toward sustainable agriculture in recent years. **Sustainable agriculture** is agriculture that does not deplete soils faster than they form. It is farming and ranching that does not reduce the amount of healthy soil, clean water, and genetic diversity essential to long-term crop and livestock production. It is, simply, agriculture that can be practiced in the same way far into the future. For example, the no-till agriculture practiced in southern Brazil that we examined in Chapter 9 appears to fit the notion of sustainable agriculture, as does the traditional Chinese practice of carp aquaculture in small ponds. Sustainable agriculture is closely related to *low-input agriculture,* agriculture that uses smaller amounts of pesticides, fertilizers, growth hormones, water, and fossil fuel energy than are currently used in industrial agriculture. Food-growing practices that use no synthetic fertilizers, insecticides, fungicides, or herbicides—but instead rely on biological approaches such as composting and biocontrol—are termed **organic agriculture.**

Organic agriculture is on the increase

Citizens, government officials, farmers, and agricultural industry representatives have debated the meaning of the word *organic* for many years. Experimental organic gardens began to appear in the United States in the 1940s, but it was decades before the U.S. government developed a clear definition of what it meant to be organic. In 1990, Congress passed the Organic Food Production Act. This law established national standards for organic products and facilitated the sale of organic food. As required by this act, the USDA in 2000 issued criteria by which crops and livestock could be officially certified as organic (Table 10.2). California legislation passed the same year established stricter state guidelines for labeling foods organic, and Washington and Texas followed with still

Table 10.2 USDA Criteria for Certifying Crops and Livestock as Organic

For crops to be considered organic

- The land where they are grown must be free of prohibited substances for at least 3 years.
- They must not be genetically engineered.
- They must not be treated with ionizing radiation (a means of eliminating bacteria in packaged food).
- The use of sewage sludge is prohibited.
- They must be produced without fertilizer containing synthetic ingredients.
- Fertility and crop nutrients must be achieved through crop rotations, cover crops, animal and crop wastes, or synthetic materials approved by the National Organic Standards Board.
- Use of most conventional pesticides is prohibited.
- Use of organic seeds and other planting stock is preferred.
- Crop pests, weeds, and diseases should be controlled through physical, mechanical, and biological management practices or with synthetic substances approved by the National Organic Standards Board.

For livestock to be considered organic

- Mammals must be raised under organic management from the last third of gestation; poultry, no later than the second day of life.
- Producers must feed livestock 100% organic agricultural feed; however, vitamin and mineral supplements are allowed.
- Producers of existing dairy herds must provide 80% organically produced feed for 9 months, followed by 3 months of 100% organically produced feed.
- Use of hormones or antibiotics is prohibited, although vaccines are permitted.
- Animals must have access to the outdoors.

Data from The National Organic Program. 2002. *Organic production and handling standards.* U.S. Department of Agriculture.

stricter guidelines. Today 17 U.S. states have laws spelling out standards for organic products.

Weighing the Issues: Do You Want Your Food Labeled?

The USDA issues labels to certify that produce claiming to be organic has met the government's organic standards. Increasingly, critics of GM products want them to be labeled as well. Given that 70% of processed food currently contains GM ingredients, labeling would cause added—and many people think, unnecessary—costs. But the European Union currently labels such foods. Do you want your food to be labeled? Would you choose among foods based on whether they are organic or genetically modified? Do you feel your food choices have environmental impacts, good or bad? Is purchasing power an effective way to make your views heard?

Long viewed as a small niche market, the market for organic foods is on the increase. Although it accounts for only 1% of food expenditures in the United States and Canada, sales of organic products increased 20% annually from 1989 to 2002, when global sales of organic products reached $25 billion. In 2001, 3–5% (close to $10 billion) of Europe's food market was organic. Although 3–5% may not seem like much, organic agriculture expanded by a factor of 35 between 1985 and 2001 in Europe, representing an annual growth rate of 30%.

Production is increasing along with demand. Although organic agriculture takes up less than 1% of cultivated land worldwide (24 million ha [59 million acres] in 2004), this area is rapidly expanding. In the United States and Canada, the amount of land used in organic agriculture has recently increased 15–20% each year. As of 2001, 550,000 ha (1.36 million acres) of U.S. farmland and 1 million ha (2.47 million acres) of Canadian farmland were in organic production. Today farmers in more than 130 nations practice organic farming commercially to some extent.

Two motivating forces have fueled these trends. Many consumers favor organic products because of concern that consuming produce grown with the use of pesticides may pose risks to their health. Consumers also buy organic produce out of a desire to improve environmental quality by reducing chemical pollution and soil degradation (see "The Science behind the Story," ▸ p. 302). Many other consumers, however, will not buy organic produce because it usually is more expensive and often looks less uniform and aesthetically appealing in the supermarket aisle compared to the standard produce of high-input agriculture. Overall, enough consumers are willing to pay more for organic meat, fruit, and vegetables that businesses are making such foods more widely available. In early 2000, one of Britain's largest supermarket brands announced that it would sell only organic food—and that the new organic products would cost their customers no more than had nonorganic products. In addition to food products, many textile makers (among them The Gap, Levi's, and Patagonia) are increasing their use of organic cotton.

Government initiatives have also spurred the growth of organic farming. For example, several million hectares of land has undergone conversion from conventional to

The Science behind the Story

Organic Farming

Fields of wheat and potatoes, some grown organically and some cultivated with the synthetic chemicals favored by industrialized agriculture, stand side by side on an experimental farm in Switzerland. Although conventionally farmed fields receive up to 50% more fertilizer, they produce only 20% more food than organically farmed fields. How are organic fields able to produce decent yields without synthetic agricultural chemicals? The answer, scientists have found, lies in the soil.

Researchers have long documented that organic farming is better for the environment, but farmers have long found that synthetic chemicals help them fight pests and increase crop yields. Although organic farming might put fewer synthetic chemicals into the air and water, some have contended that it will never contribute much to the world's food supply.

To address concerns about crop yields, Swiss researchers at the Research Institute of Organic Agriculture have been comparing organic and conventional fields since 1978, using a series of growing areas that feature four different farming systems. One group of plots mirrors conventional farms, in which large amounts of chemical pesticides, herbicides, and fertilizer are applied to soil and plants. Another set of fields is treated with a mixed approach of conventional and organic practices, including chemical additives, synthetic sprays, and livestock manure as fertilizer. Organic plots use only manure, mechanical weeding machines, and plant extracts to control pests. A fourth group of

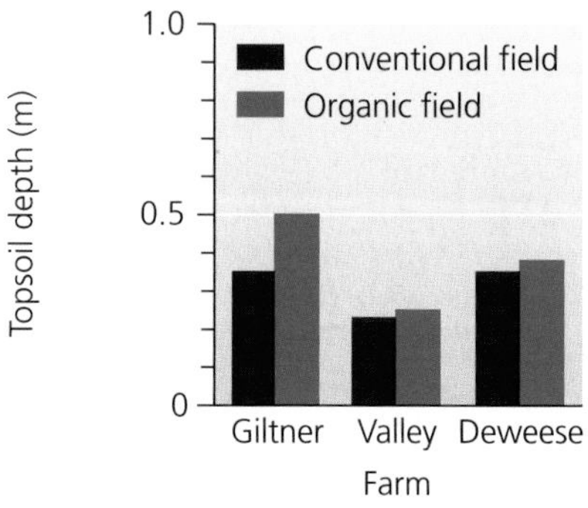

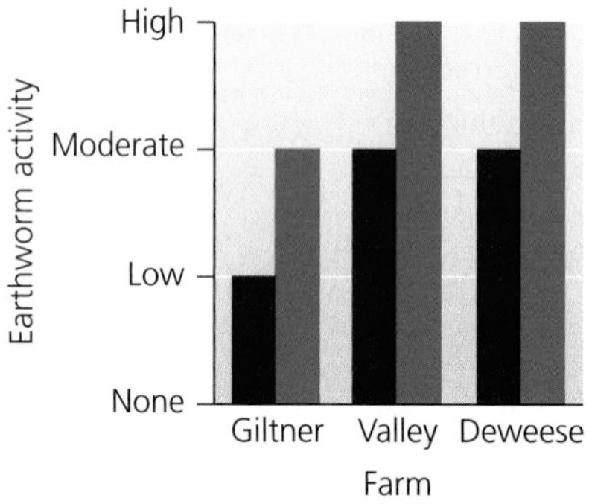

Researchers in Nebraska and North Dakota demonstrated that organic farming at three sites increased topsoil depth moderately and activity of earthworms dramatically. Data from Liebig, M. A., and J. W. Doran, 1999. Impact of organic production practices on soil quality indicators, *Journal of Environmental Quality* 28: 1601–1609.

plots follows similar organic practices but also uses extra natural boosts, such as adding herbal extracts to compost. The two organic plots receive about 35–50% less fertilizer than the conventional fields and 97% fewer pesticides.

Over more than 20 years of monitoring, the organic fields yielded 80% of what the conventional fields produced, researchers reported in the journal *Science* in 2002. Organic crops of winter wheat yielded about 90% of the conventional wheat crop yield. Organic potato crops averaged about 68% of the conventional potato yields. The comparatively low potato yield was due to nutrient deficiency and a fungus-caused potato blight.

Scientists have hypothesized that organic farms keep their yields high because organic agricultural practices better conserve soil quality, keeping soil fertile over the long term. In one study that appears to back this hypothesis, U.S. researchers compared five pairs of organic and conventional farms in the mid-1990s in North Dakota and Nebraska. After extracting soil samples from each of the farms, the researchers analyzed the soil for water-holding ability, microbial biomass, and nutrients such as carbon and nitrogen. Organic farming, they found, produced soils that contained more naturally occurring nutrients, held greater quantities of water, and had higher concentrations of microbial life than conventionally farmed soil. Organic farms also had deeper nutrient-rich topsoil and greater earthworm activity—all signs of soil healthy enough to produce impressive crops without help from synthetic chemicals.

Scientists at the Swiss research project have found similar signs of soil fertility on their organic research plots. Increasingly, researchers are concluding that organically managed soil supports a more diverse range of microbial and plant life, which translates into increased biodiversity, self-sustaining fields, and strong crop yields. Such findings may be pivotal as large growers increasingly debate whether to turn to organic farming.

organic farming in Europe since the European Union adopted a policy in 1993 to support farmers financially during the first years of conversion. The United States offers no such support, which may be why the U.S. organic market lags behind that of Europe. Such support is important, because conversion often means a temporary loss in income for farmers. More and more studies, however, suggest that reduced inputs and higher market prices can, in the long run, make organic farming more profitable for the farmer than conventional methods.

Locally supported agriculture is growing

Increasing numbers of farmers and consumers are also supporting local small-scale agriculture. Farmers' markets (Figure 10.24) are becoming more numerous throughout North America as consumers rediscover the joys of fresh, locally grown produce. The average food product sold in U.S. supermarkets travels at least 2,300 km (1,400 mi) between the farm and the shelf, and supermarket produce is often chemically treated to preserve freshness and color. At farmers' markets, consumers can buy fresh produce in season from local farmers and often have a wide choice of organic items and unique local varieties.

Some consumers are even partnering with local farmers in a phenomenon called *community-supported agriculture.* In this practice, consumers pay farmers in advance for a share of their yield, usually in the form of weekly deliveries of produce. Consumers get fresh seasonal produce, while farmers get a guaranteed income stream up front to invest in their crops—an alternative to taking out loans and being at the mercy of the weather. As of 2005, about 1,700 U.S. farms were supplying 340,000 families per week in community-supported agriculture programs.

Cuba has embraced organic agriculture

Perhaps no other nation has implemented local organic farming to the extent that Cuba has. Long a close ally of the former Soviet Union, Cuba suffered economic and agricultural upheaval following the Soviet Union's dissolution. In 1989, as the USSR was breaking up, Cuba lost 75% of its total imports, 53% of its oil imports, and 80% of its fertilizer and pesticide imports. Faced with such losses, Cuba's farmers had little choice but to "go organic."

Because far less oil was available to fuel Cuba's transportation system, farmers began growing food closer to cities and even within them. By 1998 the Cuban government's Urban Agriculture Department had encouraged the development of more than 8,000 gardens in the capital city of Havana (Figure 10.25). Over 30,000 people, including farmers, government workers, and private citizens, worked in these gardens, which covered 30% of the

FIGURE 10.24 Farmers' markets, like this one in San Francisco, have become more widespread as consumers have rediscovered the benefits of buying fresh, locally grown produce.

FIGURE 10.25 Organic gardening takes place within the city limits of Havana, Cuba, out of necessity. With little money to pay for the large amounts of fertilizers and pesticides required for industrialized agriculture, Cubans get much of their food from local agriculture without these inputs.

city's available land. Cuba has also taken steps to compensate for the loss of fossil fuels, fertilizers, and pesticides by, for example, using oxen instead of tractors, using integrated pest management, encouraging people to live outside urban areas and to remain involved in agriculture, and establishing centers to breed organisms for biological pest control.

Cuba's agriculture likely requires more human labor per unit output than do intensive commercial farms of developed nations, and Cuba's economic and agricultural policies are guided by tight top-down control in a rigid state socialist system. Nevertheless, Cuba's lowinput farming has produced some positive achievements. The practices have led to the complete control of the sweet-potato borer, a significant pest insect, and in the 1996–1997 growing season the Cuban people produced record yields for 10 crops. Although Cuba's move toward organic agriculture was involuntary, its response to its economic and agricultural crisis illustrates how other nations might, by choice, begin to farm in ways that rely less on enormous inputs of fossil fuels and synthetic chemicals.

Organic and sustainable agriculture will likely need to play a large role in our future

Organic agriculture succeeds in part because it alleviates many problems introduced by high-input agriculture, even while passing up many of the benefits. For instance, although in many cases more insect pests attack organic crops because of the lack of chemical pesticides, biocontrol methods can often keep these pests in check. Moreover, the lack of synthetic chemicals maintains soil quality and encourages helpful pollinating insects. In the end, consumer choice will determine the future of organic agriculture. Falling prices and wider availability suggest that organic agriculture will continue to increase. In addition, sustainable agriculture, whether organic or not, will sooner or later need to become the rule rather than the exception.

Conclusion

Many of the intensive commercial agricultural practices we have discussed have substantial negative environmental impacts. At the same time, it is important to realize that many aspects of industrialized agriculture have had positive environmental effects by relieving certain pressures on land or resources. Whether Earth's natural systems would be under more pressure from 6.5 billion people practicing traditional agriculture or from 6.5 billion people living under the industrialized agriculture model is a very complicated question.

What is certain is that if our planet is to support 9 billion people by mid-century without further degradation of the soil, water, pollinators, and other ecosystem services that support our food production, we must find ways to shift to sustainable agriculture. Approaches such as biological pest control, organic agriculture, pollinator conservation, preservation of native crop diversity, sustainable aquaculture, and likely some degree of careful and responsible genetic modification of food may all be parts of the game plan we will need to set in motion. What remains to be seen is the extent to which individuals, governments, and corporations will be able to put their own interests and agendas in perspective to work together toward a sustainable future.

REVIEWING OBJECTIVES

You should now be able to:

Explain the challenge of feeding a growing human population

- Our food production has outpaced the growth of our population, yet there are still 850 million hungry people in the world. (pp. 278–293)

Identify the goals, methods, and environmental impacts of the "green revolution"

- The goal of the green revolution was to increase agricultural productivity per unit area of land to feed the world's hungry without further degrading natural lands. (pp. 279–280)
- Agricultural scientists used selective breeding to develop strains of crops that grew quickly, were more nutritious, or were resistant to disease or drought. (p. 280)
- The expanded use of fossil fuels and chemical fertilizers and pesticides has increased pollution. However, the increased efficiency of production has reduced the amount of natural land converted for farming. (pp. 280–281)

Categorize the strategies of pest management

- Most "pests" and "weeds" are killed with synthetic chemicals that also can pollute the environment and pose health hazards. (pp. 282–283)
- Pests tend to evolve resistance to chemical pesticides, forcing chemists to design ever more toxic poisons. (p. 283)
- Natural enemies of pests can be employed against them in the practice of biological control. (pp. 284–285)
- Integrated pest management includes a combination of techniques, and attempts to minimize use of synthetic chemicals. (p. 285)

Discuss the importance of pollination

- Insects and other organisms are essential for ensuring the reproduction of many of our crop plants. (pp. 285–286)
- Conservation of native pollinating insects is vitally important to our food supply. (p. 286)

Describe the science behind genetically modified food

- Genetic modification depends on the technology of recombinant DNA. Genes containing desirable traits are moved from one type of organism into another. (pp. 287–288)
- Modification through genetic engineering is both like and unlike traditional selective breeding. (pp. 288–289)
- GM crops may have ecological impacts, including the spread of transgenes, the creation of "superweeds," and indirect impacts on biodiversity. More research is needed to determine how widespread or severe these impacts may be. (pp. 288–291)

Evaluate controversies and the debate over genetically modified food

- Little evidence exists so far for human health impacts from GM foods, but anxiety over health impacts inspires wide opposition to GM foods. (p. 290–291)
- Many people have ethical qualms about altering the food we eat through genetic engineering. (p. 291)
- Opponents of GM foods view multinational biotechnology corporations as a threat to the independence of small farmers. (pp. 291–292)

Ascertain approaches for preserving crop diversity

- Protecting regions of diversity of native crop varieties, such as Oaxaca, can provide insurance against failure of major commercial crops. (p. 295)
- Seed banks preserve rare and local varieties of seed, acting as storehouses for genetic diversity. (pp. 295–296)

Assess feedlot agriculture for livestock and poultry

- Increased consumption of animal products has driven the development of high-density feedlots. (pp. 296–297)
- Feedlots create tremendous amounts of waste and other environmental impacts, but they also relieve pressure on lands that could otherwise be overgrazed. (pp. 297–298)

Weigh approaches in aquaculture

- Aquaculture provides economic benefits and food security, can relieve pressures on wild fish stocks, and can be sustainable. (p. 299)
- Aquaculture also creates pollution, habitat loss, and other environmental impacts. (pp. 299–300)

Evaluate sustainable agriculture

- Organic agriculture has fewer environmental impacts than industrial agriculture. It is a small part of the market but is growing rapidly. (pp. 300–304)
- Locally supported agriculture, as shown by farmers' markets and community-supported agriculture, is also growing. (pp. 303–304)

TESTING YOUR COMPREHENSION

1. What kinds of techniques have people employed to increase agricultural food production? How did agricultural scientist Norman Borlaug help inaugurate the green revolution?
2. Explain how pesticide resistance occurs.
3. Explain the concept of biocontrol. List several components of a system of integrated pest management (IPM).
4. About how many and what types of cultivated plants are known to rely on insects for pollination? Why is it important to preserve the biodiversity of native pollinators?
5. What is recombinant DNA? How is a transgenic organism created? How is genetic engineering different from traditional agricultural breeding? How is it similar?
6. Describe several reasons why many people support the development of genetically modified organisms, and name several uses of such organisms that have been developed so far.
7. Describe the scientific concerns of those opposed to genetically modified crops. Describe some of the other concerns.
8. Name several positive and negative environmental effects of feedlot operations. Why is beef an inefficient food from the perspective of energy consumption?
9. What are some economic benefits of aquaculture? What are some negative environmental impacts?
10. What are the objectives of sustainable agriculture? What factors are causing organic agriculture to expand?

SEEKING SOLUTIONS

1. Assess several ways in which high-input agriculture can be beneficial for the environment and several ways in which it can be detrimental to the environment. Now suggest several ways in which we might modify industrial agriculture to lessen its environmental impact.
2. What factors make for an effective biological control strategy of pest management? What risks are involved in biocontrol? If you had to decide whether to use biocontrol against a particular pest, what questions would you want to have answered before you decide?
3. From what you have learned in this chapter about the staple crop corn, how would you choose to farm corn, if you had to do so for a living? Would you choose to grow genetically modified corn? How would you manage pests and weeds? Would you grow corn for people, livestock, or both? Think of the various ways corn is grown, purchased, and valued in different places—such as the United States, Europe, Oaxaca, and Zambia—as you formulate your answer.
4. Those who view GM foods as solutions to world hunger and pesticide overuse often want to speed their development and approval. Others adhere to the precautionary principle and want extensive testing for health and environmental safety. How much caution do you think is warranted before a new GM crop is introduced?
5. Imagine it is your job to make the regulatory decision as to whether to allow the planting of a new genetically modified strain of cabbage that produces its own pesticide and has twice the vitamin content of regular cabbage. What questions would you ask of scientists before deciding whether to approve the new crop? What scientific data would you want to see, and how much would be enough? Would you also consult nonscientists or take ethical, economic, and social factors into consideration?
6. Cuba adopted low-input organic agriculture out of necessity. If the country were to become economically prosperous once more, do you think Cubans would maintain this form of agriculture, or do you think they would turn to intensive, high-input farming instead? What path do you think they should pursue, and why?

INTERPRETING GRAPHS AND DATA

In the year 2000, over 80 million metric tons of nitrogen fertilizer was used in producing food for the world's 6 billion people. Food production, use of nitrogen fertilizers, and world population all had grown over the preceding 40 years, but at somewhat different rates. Food production grew slightly faster than population while relatively little additional land was converted to agricultural use during this time. Fertilizer use grew most rapidly.

1. Express the year 2002 values of the four graphed indices as percentages of the value of each index in 1961.
2. Calculate the ratio of the food production index to the nitrogen fertilizer use index in 1961 and in 2002. What does comparing these two ratios tell you about how the efficiency of nitrogen use in agriculture has changed? Is this an example of the law of diminishing returns?

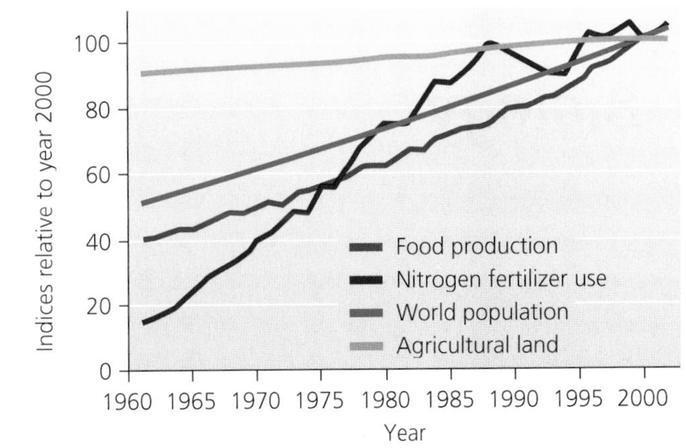

Global food production, nitrogen fertilizer use, human population, and land converted to agriculture, 1961–2002, relative to 2000 levels (2000 = 100). Data from Food and Agriculture Organization of the United Nations.

3. As world population has grown, so has the demand for food, yet little additional land has been devoted to food production. Calculate the ratio of the agricultural land index to the population index for 1961 and 2002. What does comparing these two ratios tell you about how the per capita demand on agricultural land has changed over the years? To what factors can you attribute this change?

CALCULATING ECOLOGICAL FOOTPRINTS

As food production became more industrialized during the 20th century, several trends emerged. One trend, documented in this chapter, was a loss in the number of varieties of crops grown. A second trend was the increasing amount of energy expended to store food and ship it to market. In the United States today, food travels an average of 1,400 miles from the field to your table. The price you pay for the food covers the cost of this long-distance transportation, which in 2004 was approximately one dollar per ton per mile. Assuming that the average person eats 2 pounds of food per day, calculate the food transportation costs for each category in the table below.

Consumer	Daily Cost	Annual Cost
You	$1.40	$511
Your class		
Your hometown		
Your state		
United States		

1. What specific challenges to environmental sustainability are imposed by a food production and distribution system that relies on long-range transportation to bring food to market?
2. A study by Pirog and Benjamin* noted that locally produced food traveled only 50 miles or so to market, thus saving 96% of the transportation costs. Locally grown foods may be fresher and cause less environmental impact as they are brought to market, but what are the disadvantages to you as a consumer in relying on local food production? Do you think the advantages outweigh those disadvantages?
3. What has happened to gasoline prices recently? Would future increases in the price of gas affect your answers to the preceding questions?

*Data from Pirog, R., and A. Benjamin. 2003. Checking the food odometer: Comparing food miles for local versus conventional produce sales to Iowa institutions. Ames, IA: Leopold Center for Sustainable Agriculture, Iowa State University.

Take It Further

Go to www.aw-bc.com/withgott or the student CD-ROM where you'll find:

- Suggested answers to end-of-chapter questions
- Quizzes, animations, and flashcards to help you study
- *Research Navigator*™ database of credible and reliable sources to assist you with your research projects
- GRAPHit! tutorials to help you master how to interpret graphs
- INVESTIGATEit! current news articles that link the topics that you study to case studies from your region to around the world

11 Biodiversity and Conservation Biology

The Sikhote-Alin Mountains meet the Pacific Ocean

Upon successfully completing this chapter, you will be able to:

- Characterize the scope of biodiversity on Earth
- Describe ways to measure biodiversity
- Contrast background extinction rates and periods of mass extinction
- Evaluate the primary causes of biodiversity loss
- Specify the benefits of biodiversity
- Assess conservation biology and its practice
- Explain island biogeography theory and its application to conservation biology
- Compare and contrast traditional and more innovative biodiversity conservation efforts

A Siberian tiger in the Sikhote-Alin Mountains

Central Case: Saving the Siberian Tiger

"Future generations would be truly saddened that this century had so little foresight, so little compassion, such lack of generosity of spirit for the future that it would eliminate one of the most dramatic and beautiful animals this world has ever seen."

—George Schaller, wildlife biologist, on the tiger

"Except in pockets of ignorance and malice, there is no longer an ideological war between conservationists and developers. Both share the perception that health and prosperity decline in a deteriorating environment. They also understand that useful products cannot be harvested from extinct species."

—Edward O. Wilson, Harvard University biodiversity expert

Historically, tigers roamed widely across Asia from Turkey to northeast Russia to Indonesia. Within the past 200 years, however, people have driven the majestic striped cats from most of their historic range. Today, tigers are exceedingly rare and are creeping toward extinction.

Of the tigers that still survive, those of the subspecies known as the Siberian tiger are the largest cats in the world. Males reach 3.66 m (12 ft) in length and weigh up to 363 kg (800 lb). Also named Amur tigers for the watershed they occupied along the Amur River, which divides Siberian Russia from Manchurian China, these cats now find their last refuge in the forests of the remote Sikhote-Alin Mountains of the Russian Far East.

For thousands of years the Siberian tiger coexisted with the region's native people and held a prominent place in native language and lore. These people referred to the tiger as "Old Man" or "Grandfather" and equated it with royalty or viewed it as a guardian of the mountains and forests. Indigenous people of the region rarely killed a tiger unless it had preyed on a person.

The Russians who moved into the region and exerted control in the early 20th century had no such cultural traditions. They hunted tigers for sport and hides, and some Russians reported killing as many as 10 tigers in a single hunt. In addition, poachers began killing tigers to sell their body parts to China and other Asian countries, where they are used in traditional medicine and as aphrodisiacs. Meanwhile, road building, logging, and agriculture began to fragment tiger habitat and provide easy access for well-armed hunters. The tiger population dipped to perhaps 20–30 animals.

International conservation groups began to get involved, working with Russian biologists to try to save the dwindling tiger population. One such group was the Hornocker Wildlife Institute, now part of the Wildlife Conservation Society. In 1991 the group helped launch the Siberian Tiger Project, devoted to studying the tiger and its habitat. The team put together a plan to protect the tiger, began educating people regarding the tiger's importance and value, and worked closely with those who live in proximity to the big cats.

Thanks to such efforts by conservation biologists, today Siberian tigers in the wild number roughly 330–370, and about 600 more survive in zoos and captive breeding programs around the world. The outlook for the species' survival still looks challenging, but many people are trying to save these endangered animals. It is one of many efforts around the world today to stem the loss of our planet's priceless biological diversity.

Our Planet of Life

Growing human population and resource consumption are putting ever-greater pressure on the flora and fauna of the planet, from tigers to tiger beetles. We are diminishing Earth's diversity of life, the very quality that makes our planet so special. In Chapter 5 we introduced the concept of **biological diversity,** or **biodiversity,** as the sum total of all organisms in an area, taking into account the diversity of species, their genes, their populations, and their communities. In this chapter we will refine this definition and examine current biodiversity trends and their relevance to our lives. We will then explore science-based solutions to biodiversity loss.

Biodiversity encompasses several levels

Biodiversity is a concept as multifaceted as life itself, and definitions of the term are plentiful. As sociologist of science David Takacs explains in his 1996 book, *The Idea of Biodiversity,* different biologists employ different working definitions according to their own aims, interests, and values. Nonetheless, there is broad agreement that the concept applies across several major levels in the organization of life (Figure 11.1). The level that is easiest to visualize and most commonly used is species diversity.

FIGURE 11.1 The concept of biodiversity encompasses several levels in the hierarchy of life. Species diversity (middle frame of figure) refers to the number or variety of species. Genetic diversity (bottom frame) refers to variation in DNA composition among individuals within a species. Ecosystem diversity (top frame) and related concepts refer to variety at levels above the species level, such as ecosystems, communities, habitats, or landscapes.

Species diversity As you recall from Chapter 5 (▸p. 122), a *species* is a distinct type of organism, a set of individuals that uniquely share certain characteristics and can breed with one another and produce fertile offspring. Biologists may use somewhat differing criteria to delineate species boundaries; some emphasize characteristics shared because of common ancestry, whereas others emphasize ability to interbreed. In practice, however, scientists broadly agree on species identities. We can express **species diversity** in terms of the number or variety of species in the world or in a particular region. One component of species diversity is *species richness,* the number of species. Another is *evenness* or *relative abundance,* the extent to which numbers of individuals of different species are equal or skewed.

As we saw in Chapter 5 (▸pp. 122–126), speciation generates new species, adding to species richness, whereas extinction decreases species richness. Although immigration, emigration, and local extinction may increase or decrease species richness locally, only speciation and extinction change it globally.

Taxonomists, the scientists who classify species, use an organism's physical appearance and genetic makeup to determine its species. Taxonomists also group species by their similarity into a hierarchy of categories meant to reflect evolutionary relationships. Related species are grouped together into *genera* (singular, *genus*), related genera are grouped into families, and so on (Figure 11.2). Every species is given a two-part Latin or Latinized scientific name denoting its genus and species. The tiger, *Panthera tigris,* differs from the world's other species of large cats such as the jaguar *(Panthera onca),* the leopard *(Panthera pardus),* and the African lion *(Panthera leo).* These four species are closely related in evolutionary terms, as indicated by the genus name they share, *Panthera.* They are more distantly related to cats in other genera such as the cheetah *(Acinonyx jubatus)* and the bobcat *(Felis rufus),* although all cats are classified together in the family Felidae.

Biodiversity exists below the species level in the form of *subspecies,* populations of a species that occur in different geographic areas and differ from one another in some characteristics. Subspecies are formed by the same processes that drive speciation, but result when divergence does not proceed far enough to create separate species. Scientists denote subspecies with a third part of the scientific name. The Siberian tiger, *Panthera tigris altaica,* is one of five subspecies of tiger still surviving (Figure 11.3). Tiger subspecies differ in color, coat thickness, stripe patterns, and size. For example, *Panthera tigris altaica* is 5–10 cm (2–4 in.) taller at the shoulder than the Bengal tiger *(Panthera tigris tigris)* of India and Nepal, and it has a thicker coat and larger paws.

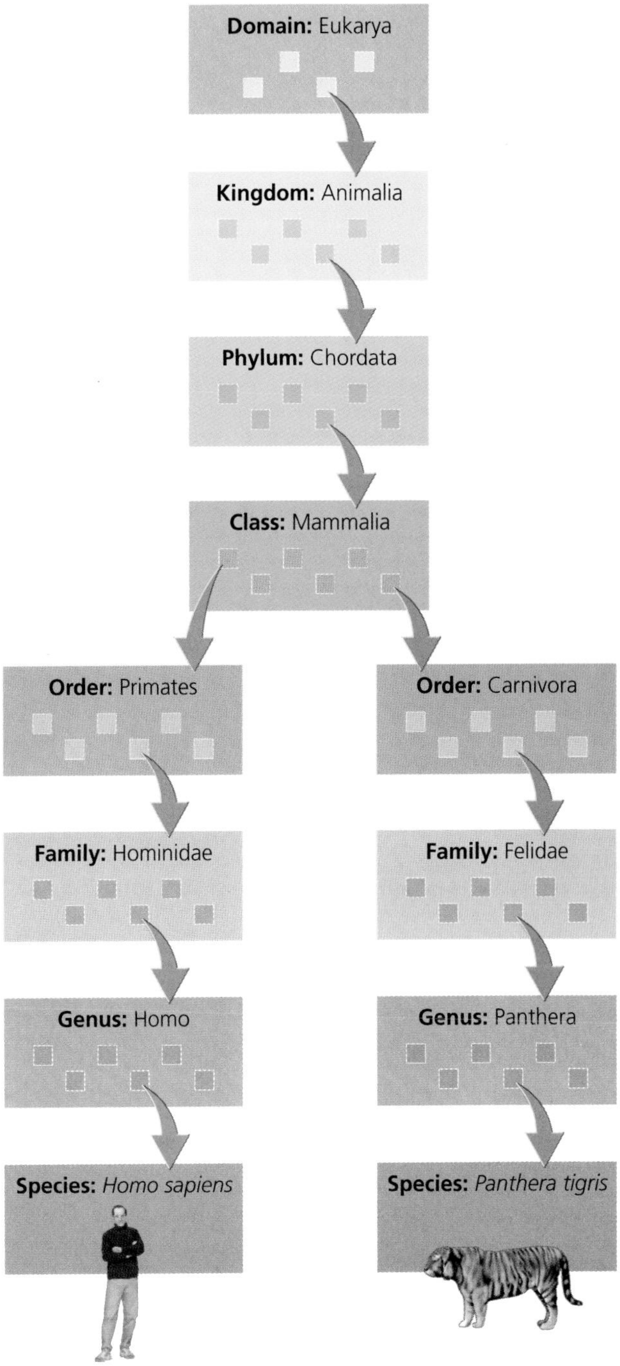

FIGURE 11.2 Taxonomists classify organisms using a hierarchical system meant to reflect evolutionary relationships. Species that are similar in their appearance, behavior, and genetics (because of recent common ancestry) are placed in the same genus. Organisms of similar genera are placed within the same family. Families are placed within orders, orders within classes, classes within phyla, phyla within kingdoms, and kingdoms within domains. For instance, humans (*Homo sapiens,* a species in the genus *Homo*) and tigers (*Panthera tigris,* a species in the genus *Panthera*) are both within the class Mammalia. However, the differences between our two species, which have evolved over millions of years, are great enough that we are placed in different orders and families.

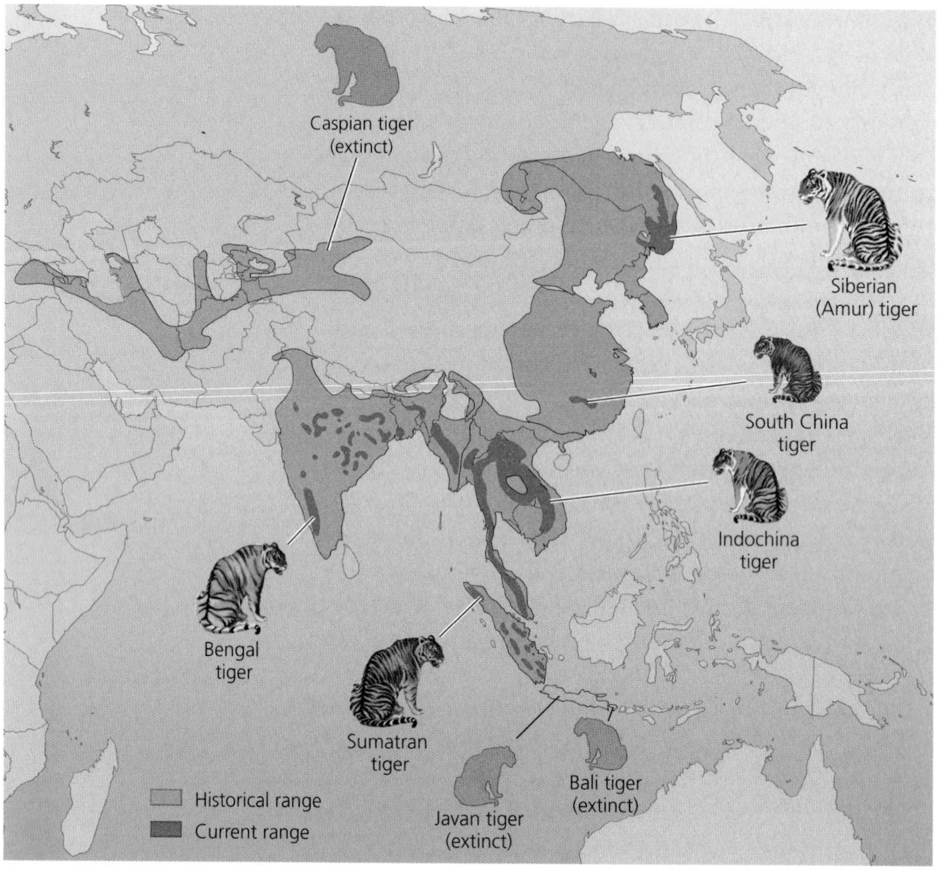

FIGURE 11.3 Three of the eight subspecies of tiger became extinct during the 20th century. The Bali, Javan, and Caspian tigers are extinct. Today only the Siberian (Amur), Bengal, Indochina, Sumatran, and South China tigers persist, and the Chinese government estimates that less than 30 individuals of the South China tiger remain. Deforestation, hunting, and other pressures from people have caused tigers of all subspecies to disappear from most of the geographic range they historically occupied. This map contrasts the ranges of the eight subspecies in the years 1800 (orange) and 2000 (red). Data from the Tiger Information Center.

Genetic diversity Scientists designate subspecies when they recognize substantial genetically based differences among individuals from different populations of a species. However, all species consist of individuals that vary genetically from one another to some degree, and this genetic diversity is an important component of biodiversity. **Genetic diversity** encompasses the differences in DNA composition among individuals within species and populations.

Genetic diversity provides the raw material for adaptation to local conditions. A diversity of genes for coat thickness in tigers allowed natural selection to favor genes for thin coats of fur in Bengal tigers living in warm regions, and genes for thick coats of fur for Siberian tigers living in cold regions. In the long term, populations with more genetic diversity may stand better chances of persisting, because their variation better enables them to cope with environmental change. Populations with little genetic diversity are vulnerable to environmental change for which they are not genetically prepared. Populations with depressed genetic diversity may also be more vulnerable to disease and may suffer *inbreeding depression,* which occurs when genetically similar parents mate and produce weak or defective offspring. Scientists have sounded warnings over low genetic diversity in species that have dropped to low population sizes in the past, including cheetahs, bison, and elephant seals, but the full consequences of reduced diversity in these species remain to be seen. Diminishing genetic diversity in our crop plants also is a prime concern to humanity, as we saw in Chapter 10 (▸pp. 295–296).

Ecosystem diversity Biodiversity also encompasses levels above the species level. *Ecosystem diversity* refers to the number and variety of ecosystems, but biologists may also refer to the diversity of biotic community types or habitats within some specified area. If the area is large, scientists may also consider the geographic arrangement of habitats, communities, or ecosystems at the landscape level, including the sizes, shapes, and interconnectedness of patches of these entities. Under any of these concepts, a seashore of rocky and sandy beaches, forested cliffs, offshore coral reefs, and ocean waters would hold far more biodiversity than the same acreage of a monocultural cornfield. A mountain slope whose vegetation changes from desert to hardwood forest to coniferous forest to alpine meadow would hold more biodiversity than an area the same size consisting of only desert, forest, or meadow.

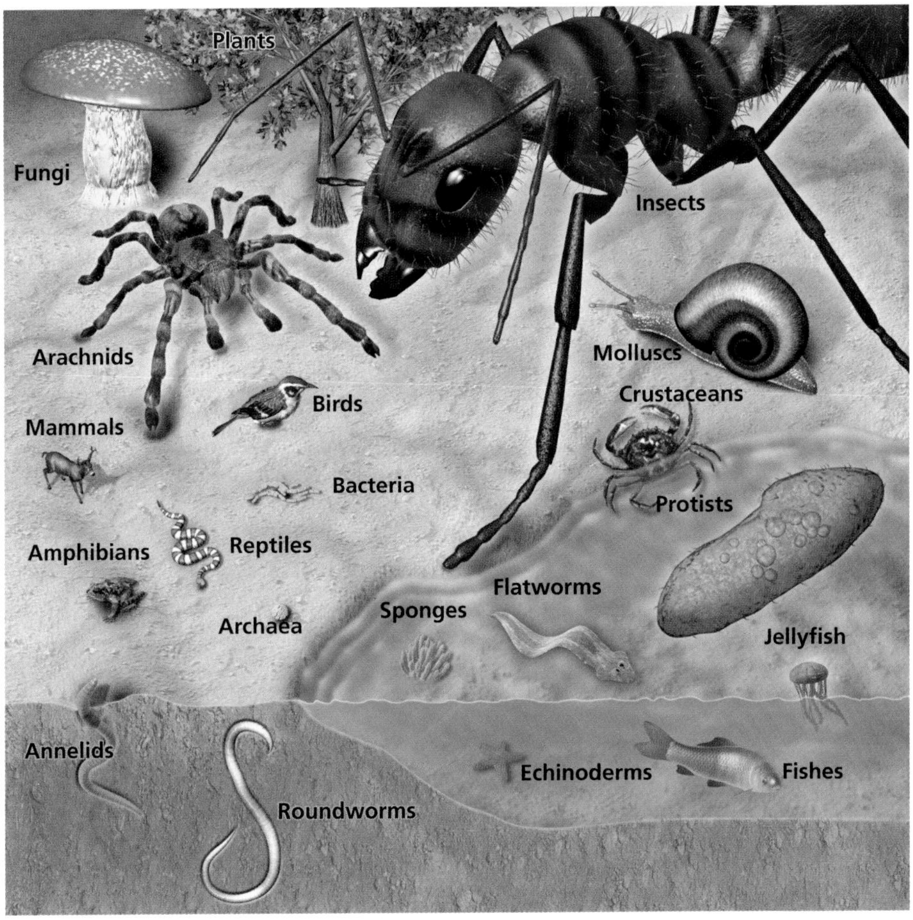

FIGURE 11.4 This illustration shows organisms scaled in size to the number of species known so far from each major taxonomic group, giving a visual sense of the disparity in species richness among groups. However, because most species are not yet discovered or described, some groups (such as bacteria, archaea, insects, nematodes, protists, fungi, and others) may contain far more species than we now know of. Data from Groombridge, B., and M. D. Jenkins. 2002. *Global biodiversity: Earth's living resources in the 21st century*. UNEP-World Conservation Monitoring Centre. Cambridge, U.K.: Hoechst Foundation.

Measuring biodiversity is not easy

Coming up with precise quantitative measurements to express a region's biodiversity is difficult. This is partly why scientists often express biodiversity in terms of its most easily measured component, species diversity, and in particular, species richness. Species richness is a good gauge for overall biodiversity, but we still are profoundly ignorant of the number of species that exist worldwide. So far, scientists have identified and described 1.7–2.0 million species of plants, animals, and microorganisms. However, estimates for the total number that actually exist range from 3 million to 100 million, with our best educated guesses ranging from 5 million to 30 million.

Species are not evenly distributed among taxonomic groups. In terms of number of species, insects show a staggering predominance over all other forms of life (Figures 11.4 and 11.5). Within insects, about 40% are beetles. Beetles outnumber all noninsect animals and all plants. No wonder the 20th-century British biologist J. B. S. Haldane famously quipped that God must have had "an inordinate fondness for beetles."

Our knowledge of species numbers is incomplete for several reasons. First, some areas of Earth remain little explored. We have barely sampled the ocean depths, hydrothermal vents (►pp. 107–108), or the tree canopies and soils of tropical forests. Second, many species are tiny and easily overlooked. These inconspicuous organisms include bacteria, nematodes (roundworms), fungi, protists, and soil-dwelling arthropods. Third, many organisms are so difficult to identify that ones thought to be identical sometimes turn out, once biologists look more closely, to be multiple species. This is frequently the case with microbes, fungi, and small insects, but also sometimes with organisms as large as birds, trees, and whales.

Smithsonian Institution entomologist Terry Erwin pioneered one method of estimating species numbers. In 1982, Erwin's crews fogged rainforest trees in Central America with clouds of insecticide and then collected insects, spiders, and other arthropods as they died and fell from the treetops. Erwin concluded that 163 beetle species specialized on the tree species *Luehea seemannii*. If this were typical, he figured, then the world's 50,000 tropical tree species would hold 8,150,000 beetle species and—since beetles represent 40% of all arthropods—20 million arthropod species. If canopies hold two-thirds of all arthropods, then arthropod species in tropical forests alone would number 30 million. Many assumptions were

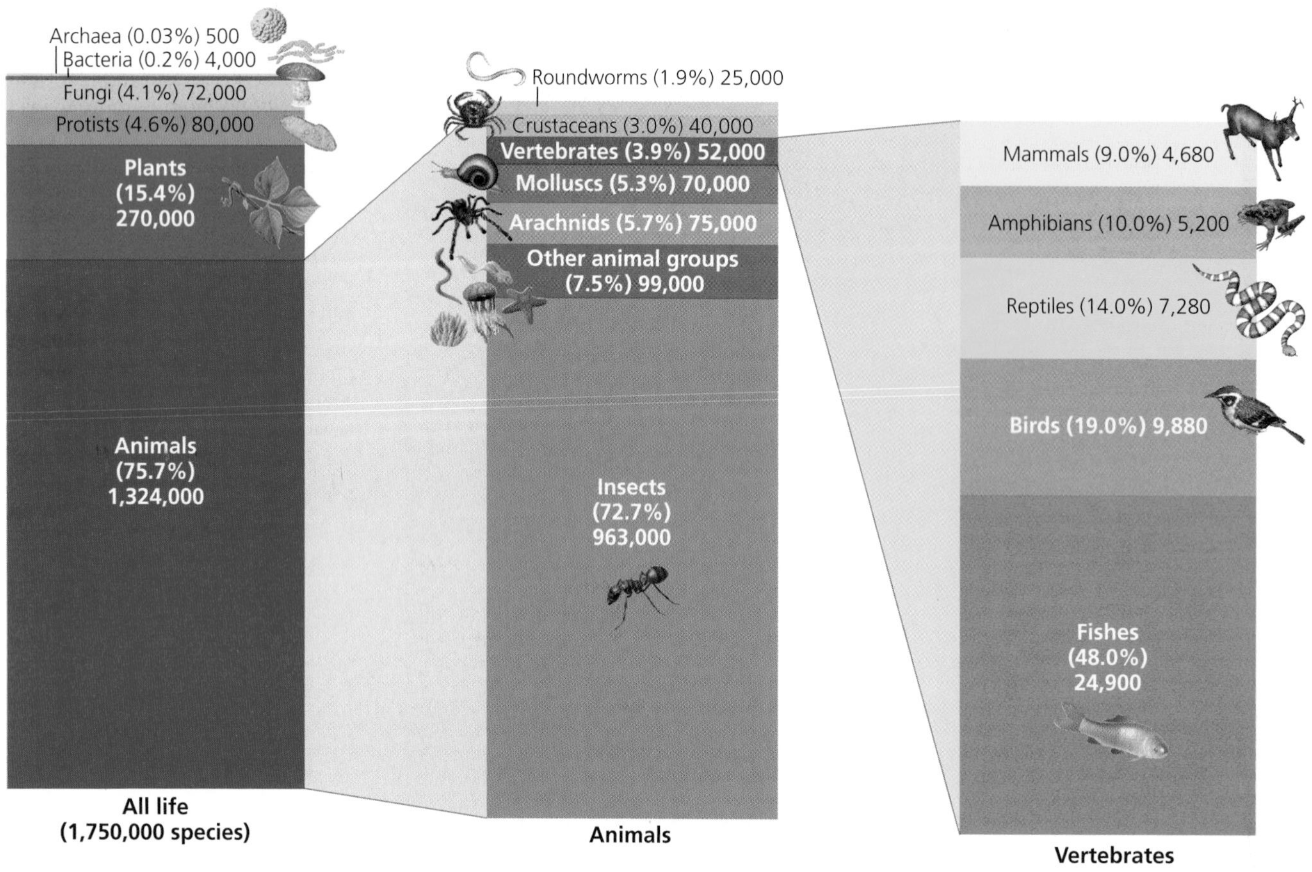

FIGURE 11.5 In the left portion of the figure, we see that three-quarters of known species are animals. The central portion subdivides animals, revealing that nearly three-quarters of animals are insects and that vertebrates comprise only 3.9% of animals. Among vertebrates (right portion of figure), nearly half are fishes, and mammals comprise only 9%. As noted, most species are not yet discovered or described, so some groups may contain far more species than we now know of. Data from Groombridge, B., and M.D. Jenkins. 2002. *Global biodiversity: Earth's living resources in the 21st century.* UNEP-World Conservation Monitoring Centre. Cambridge, U.K.: Hoechst Foundation.

involved in this calculation, and several follow-up studies have revised Erwin's estimate downward.

Biodiversity is unevenly distributed

Numbers of species tell only part of the story of Earth's biodiversity. Living things are distributed across our planet unevenly, and scientists have long sought to explain the distributional patterns they see. For example, as we have noted, some groups of organisms include only one or a few species, whereas other groups contain many. Some groups have given rise to many species in a relatively short period of time through the process of adaptive radiation (see Figure 5.3, ▸p. 121). Species diversity also varies according to biome. Tropical dry forests and rainforests tend to support more species than tundra and boreal forests, for instance. The variation in diversity by biome is related to one of the planet's most striking patterns of species diversity: the fact that species richness generally increases as one approaches the equator (Figure 11.6). This pattern of variation with latitude, called the *latitudinal gradient,* has been one of the most obvious patterns in ecology, but it also has been one of the most difficult ones for scientists to explain.

Hypotheses abound for the cause of the latitudinal gradient in species richness, but it seems likely that plant productivity and climate stability play key roles in the phenomenon (Figure 11.7). Greater amounts of solar energy, heat, and humidity at tropical latitudes lead to more plant growth, making areas nearer the equator more productive and able to support larger numbers of animals. In addition, the relatively stable climates of equatorial regions—their similar temperatures and rainfall from day to day and season to season—help ensure that single species won't dominate ecosystems, but that instead numerous species can coexist. Whereas varying environmental

FIGURE 11.6 For many types of organisms, number of species per unit area tends to increase as one moves toward the equator. This trend, the latitudinal gradient in species richness, is one of the most readily apparent—yet least understood—patterns in ecology. One example is bird species in North and Central America: In any one spot in arctic Canada and Alaska, 30 to 100 species can be counted; in areas of Costa Rica and Panama, the number rises to over 600. Adapted from Cook, R. E. 1969. Variation in species density in North American birds. *Systematic Zoology* 18: 63–84.

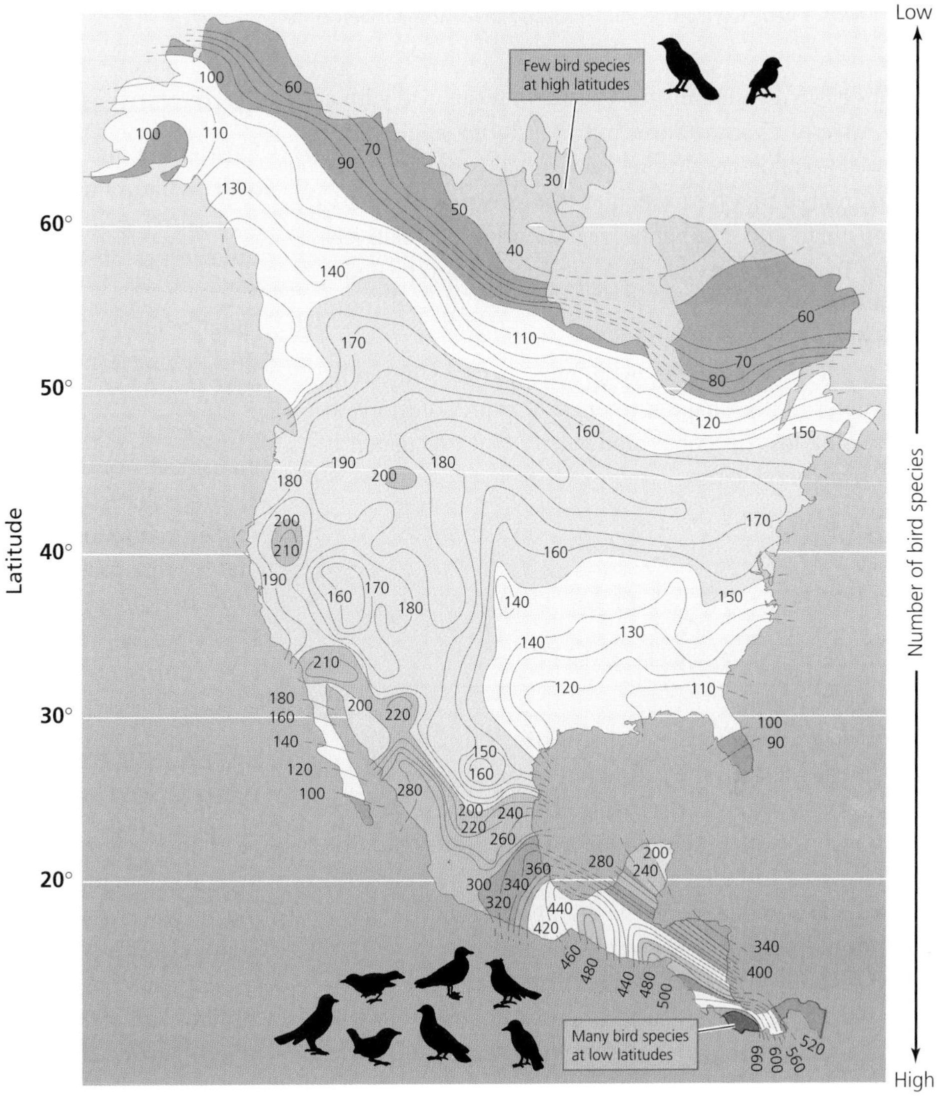

FIGURE 11.7 Ecologists have offered many hypotheses for the latitudinal gradient in species richness, and one set of ideas is summarized here. The variable climates (across days, seasons, and years) of polar and temperate latitudes favor organisms that can survive a wide range of conditions. Such generalist species have expansive niches; they can do many things well enough to survive, and they spread over large areas. In tropical latitudes, the abundant solar energy, heat, and humidity induce greater plant growth, which supports more organisms. The stable climates of equatorial regions favor specialist species, which have restricted niches but do certain things very well. Together these factors are thought to promote greater species richness in the tropics.

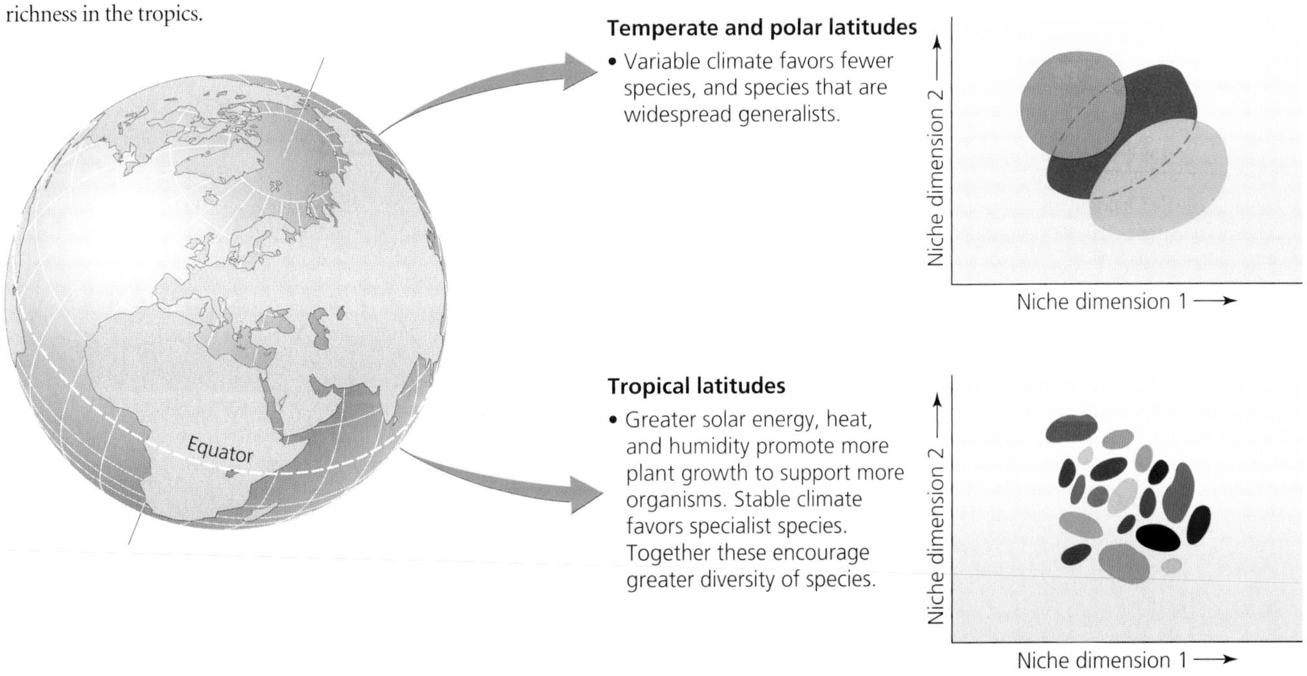

conditions favor generalists—species that can deal with a wide range of circumstances but that do no single thing very well—stable conditions favor organisms with specialized niches that do particular things very well. In addition, polar and temperate regions may be relatively lacking in species because glaciation events repeatedly forced organisms out of these regions and toward more tropical latitudes.

We will discuss further geographic patterns in biodiversity later in this chapter, when we explore solutions to the ongoing loss of global biodiversity that our planet is currently experiencing.

Biodiversity Loss and Species Extinction

Biodiversity at all levels is being lost to human impact, most irretrievably in the extinction of species. Once vanished, a species can never return. *Extinction* (►p. 126) occurs when the last member of a species dies and the species ceases to exist, as apparently was the case with Monteverde's golden toad. The disappearance of a particular population from a given area, but not the entire species globally, is referred to as **extirpation.** The tiger has been extirpated from most of its historic range, but it is not yet extinct. Although a species that is extirpated from one place may still exist in others, extirpation is an erosive process that can, over time, lead to extinction.

Extinction is a natural process

Extirpation and extinction occur naturally. If organisms did not naturally go extinct, we would be up to our ears in dinosaurs, trilobites, ammonites, and the millions of other types of creatures that vanished from Earth long before humans appeared. Paleontologists estimate that roughly 99% of all species that have ever lived are now extinct. This means that the wealth of species on our planet today comprises only about 1% of all species that ever lived. Most extinctions preceding the appearance of humans have occurred one by one for independent reasons, at a rate that paleontologists refer to as the **background rate of extinction.** For example, the fossil record indicates that for mammals and marine animals, one species out of 1,000 would typically become extinct every 1,000 to 10,000 years. This translates to an annual rate of one extinction per 1 to 10 million species.

Earth has experienced five previous mass extinction episodes

Extinction rates have risen far above this background rate during several mass extinction events in Earth's history. In the past 440 million years, our planet has experienced five major episodes of **mass extinction** (Figure 11.8). Each of these events has eliminated more than one-fifth of life's families and at least half its species (Table 11.1). The most severe episode occurred at the end of the Permian period, 248 million years ago, when close to 54% of all families, 90% of all species, and 95% of marine species went extinct.

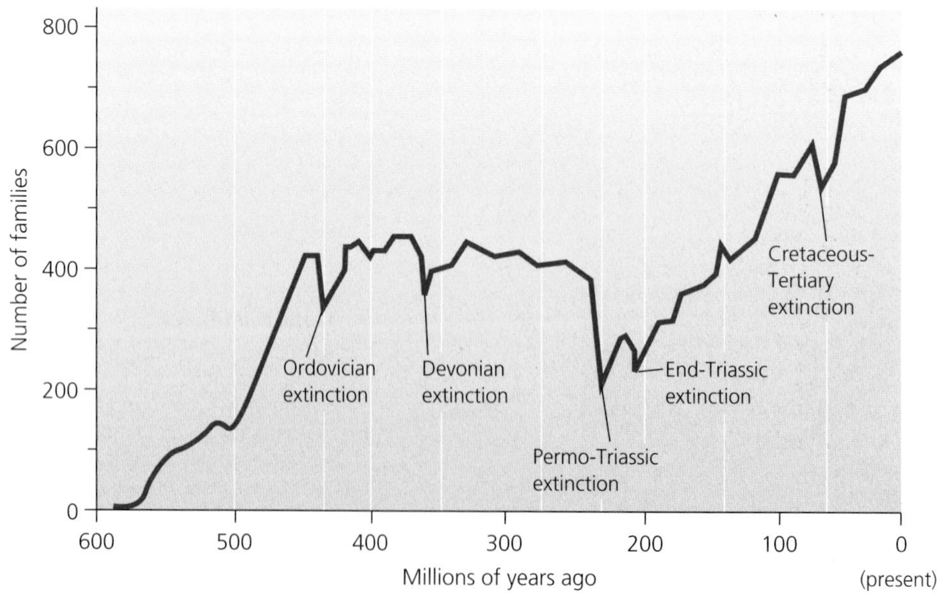

FIGURE 11.8 The fossil record shows evidence of five episodes of mass extinction during the past half-billion years of Earth history. At the end of the Ordovician, Devonian, Permian, Triassic, and Cretaceous periods, 50–95% of the world's species appear to have gone extinct. Each time, biodiversity later rebounded to equal or higher levels, but the rebound required millions of years in each case. Data from Raup, D. M., and J. J. Sepkoski. 1982. Mass extinctions in the marine fossil record. *Science* 215: 1501–1503.

Table 11.1 Mass Extinctions

Event	Date (millions of years ago)	Cause	Types of life most affected	Percentage of life depleted
Ordovician	440 mya	Unknown	Marine organisms; terrestrial record is unknown	>20% of families
Devonian	370 mya	Unknown	Marine organisms; terrestrial record is unknown	>20% of families
Permo-Triassic	250 mya	Possibly volcanism	Marine organisms; terrestrial record is less known	>50% of families; 80–95% of species
End-Triassic	202 mya	Unknown	Marine organisms; terrestrial record is less known	20% of families; 50% of genera
Cretaceous-Tertiary	65 mya	Asteroid impact	Marine and terrestrial organisms, including dinosaurs	15% of families; >50% of species
Current	Beginning 0.01 mya	Human impact, through habitat destruction and other means	Large animals, specialized organisms, island organisms, and organisms hunted or harvested by humans	Ongoing

The best-known episode occurred at the end of the Cretaceous period, 65 million years ago, when an apparent asteroid impact brought an end to the dinosaurs and many other groups (▸pp. 128–129). In addition, there is evidence for further mass extinctions in the Cambrian period and earlier, more than half a billion years ago.

If current trends continue, the modern era, known as the Quaternary period, may see the extinction of more than half of all species. Although similar in scale to previous mass extinctions, today's ongoing mass extinction is different in two primary respects. First, humans are causing it. Second, humans will suffer as a result of it.

Humans set the sixth mass extinction in motion years ago

We have recorded many instances of human-induced species extinction over the past few hundred years. Sailors documented the extinction of the dodo on the Indian Ocean island of Mauritius in the 17th century, and we still have a few of the dodo's body parts in museums. Among North American birds in the past two centuries, we have driven into extinction the Carolina parakeet, great auk, Labrador duck, and passenger pigeon (▸p. 132), and probably the Bachman's warbler and Eskimo curlew. Several more species, including the whooping crane, California condor, Kirtland's warbler, and the ivory-billed woodpecker, recently rediscovered in the wooded swamps of Arkansas, teeter on the brink of extinction.

However, species extinctions caused by humans precede written history. Indeed, people may have been hunting species to extinction for thousands of years. Archaeological evidence shows that in case after case, a wave of extinctions followed close on the heels of human arrival on islands and continents (Figure 11.9). After Polynesians reached Hawaii, half its birds went extinct. Birds, mammals, and reptiles vanished following human arrival on many other oceanic islands, including large island masses such as New Zealand and Madagascar. The pattern appears to hold for at least two continents, as well. Dozens of species of large vertebrates died off in Australia after Aborigines arrived roughly 50,000 years ago, and North America lost 33 genera of large mammals after people arrived on the continent at least 10,000 years ago.

Current extinction rates are much higher than normal

Today, species loss is accelerating as our population growth and resource consumption put increasing strain on habitats and wildlife. A decade ago, 1,500 of the world's leading scientists reported to the United Nations in their Global Biodiversity Assessment that in the preceding 400 years, 484 animal and 654 plant species were known to have become extinct, and that more than 30,000 plant and animal species currently faced extinction. In 2005, scientists with the Millennium Ecosystem Assessment (▸p. 20) calculated that the current global extinction rate is 100 to 1,000 times greater than the background rate. Moreover, they projected that the rate would increase tenfold or more in future decades.

To keep track of the current status of endangered species, the World Conservation Union maintains the

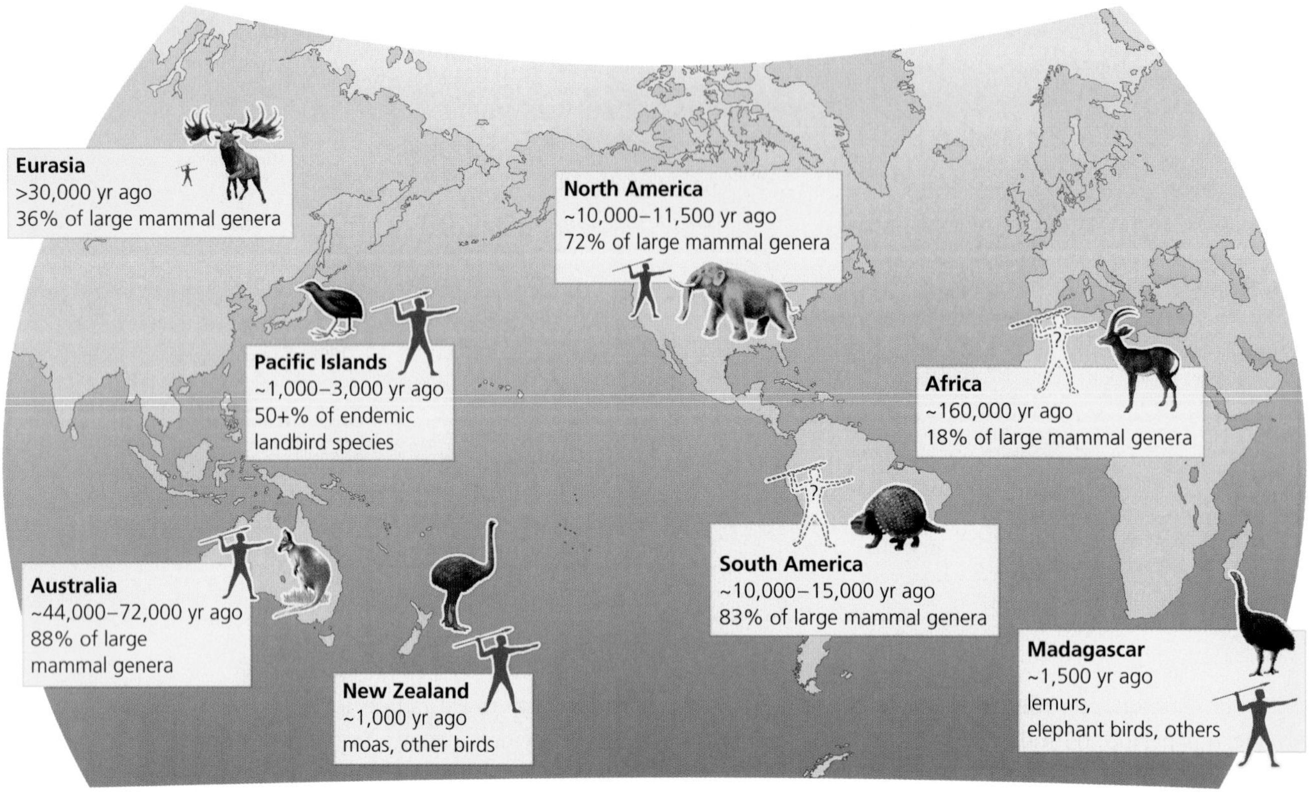

FIGURE 11.9 This map shows for each region the time of human arrival and the extent of the recent extinction wave. Illustrated are representative extinct megafauna from each region. The human hunter icons are sized according to the degree of evidence that human hunting was a cause of extinctions; larger icons indicate more certainty that humans (as opposed to climate change or other forces) were the cause. Data for South America and Africa are so far too sparse to be conclusive, and future archaeological and paleontological research could well alter these interpretations. Adapted from Barnosky, A. D., et al. 2004. Assessing the causes of late Pleistocene extinctions on the continents. *Science* 306: 70–75; and Wilson, E. O. 1992. *The diversity of life.* Cambridge, MA: Belknap Press.

Red List, an updated list of species facing high risks of extinction. The 2004 Red List reported that 23% (1,101) of mammal species and 12% (1,213) of bird species are threatened with extinction. Among other major groups (for which assessments are not fully complete), estimates of the percentage of species threatened ranged from 31% to 86%. Since 1996, the total number of vertebrate animals listed as threatened climbed by more than 6%. Since 1970, at least 58 fish species, 9 bird species, and 1 mammal species have become extinct, and in the United States alone over the past 500 years, 236 animals and 17 plants are confirmed to have gone extinct. For all of these figures, the *actual* numbers of species extinct and threatened, like the actual number of total species in the world, are doubtless greater than the *known* numbers.

Among the 1,101 mammals facing possible extinction on the Red List is the tiger, which despite—or perhaps because of—its tremendous size and reputation as a fierce predator, is one of the most endangered large animals on the planet. In 1950, eight tiger subspecies existed (see Figure 11.3). Today, three are extinct. The Bali tiger, *Panthera tigris balica,* went extinct in the 1940s; the Caspian tiger, *Panthera tigris virgata,* during the 1970s; and the Javan tiger, *Panthera tigris sondaica,* during the 1980s.

Biodiversity loss involves more than extinction

Statistics on extinction tell only part of the story of biodiversity loss. The larger part of the story is the decline in population sizes of many organisms. Declines in numbers are accompanied by shrinkage of species' geographic ranges. Thus, many species today are less numerous and occupy less area than they once did. These patterns mean that genetic diversity and ecosystem diversity, as well as species diversity, are being lost. To measure and quantify

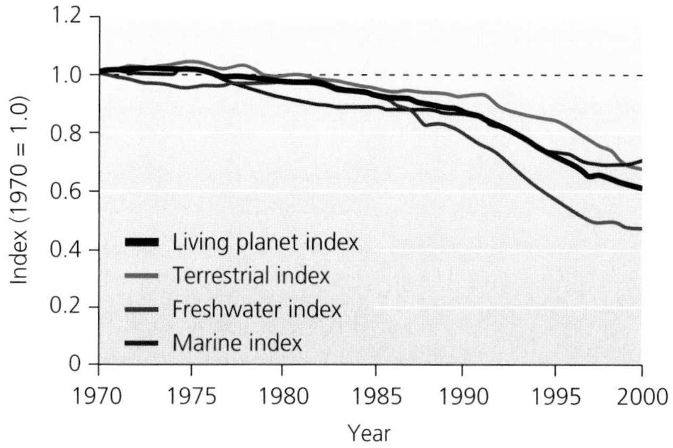

FIGURE 11.10 The Living Planet Index serves as an indicator of the state of global biodiversity. Index values summarize population trends for 1,145 species. Between 1970 and 2000, the Living Planet Index fell by roughly 40%. The indices for terrestrial and marine species fell 30%, and the index for freshwater species fell 50%. Data from World Wide Fund for Nature and U.N. Environment Programme. 2004. *The Living Planet Report, 2004*. Gland, Switzerland: WWF.

this degradation, scientists at the World Wildlife Fund and the United Nations Environment Programme (UNEP) developed a metric called the *Living Planet Index.* This index summarizes trends in the populations of 555 terrestrial species, 323 freshwater species, and 267 marine species that are well enough monitored to provide reliable data. Between 1970 and 2000, the Living Planet Index fell by roughly 40% (Figure 11.10).

There are several major causes of biodiversity loss

Reasons for the decline of any given species are often multifaceted and complex, so they can be difficult to determine. The current precipitous decline in populations of amphibians throughout the world provides an example. Frogs, toads, and salamanders worldwide are decreasing drastically in abundance. Several have already gone extinct, and scientists are struggling to explain why. Recent studies have implicated a wide array of factors, and most scientists now suspect that such factors may be interacting synergistically (see "The Science behind the Story," ▸pp. 322–323).

Nonetheless, overall, scientists have identified four primary causes of population decline and species extinction: habitat alteration, invasive species, pollution, and overharvesting. Global climate change (Chapter 18) now threatens to become the fifth. Each of these factors is exacerbated by human population growth and by our increase in per capita consumption of resources.

Habitat alteration Nearly every human activity can alter the habitat of the organisms around us. Farming replaces diverse natural communities with simplified ones of only one or a few plant species. Grazing modifies the structure and species composition of grasslands. Either type of agriculture can lead to desertification. Clearing forests removes the food, shelter, and other resources that forest-dwelling organisms need to survive. Hydroelectric dams turn rivers into reservoirs upstream and thereby affect water conditions and floodplain communities downstream. Urbanization and suburban sprawl supplant diverse natural communities with simplified human-made ones, driving many species from their homes.

Such changes in habitat generally have negative effects. Organisms are already adapted to the habitats in which they live, so any change is likely to render the habitat less suitable for them. Of course, human-induced habitat change may benefit some species. Animals such as starlings, house sparrows, pigeons, and gray squirrels do very well in urban and suburban environments and benefit from our modification of natural habitats. However, the species that benefit are relatively few; for every species that gains, more lose. Furthermore, the species that do well in our midst tend to be weedy, cosmopolitan generalists that are in little danger of disappearing any time soon.

Habitat alteration is by far the greatest cause of biodiversity loss today. It is the primary source of population declines for 83% of threatened mammals and 85% of threatened birds, according to UNEP data. As just one example of thousands, the prairies native to North America's Great Plains have been almost entirely converted to agriculture. The area of prairie habitat has been reduced by more than 99%. As a result, grassland bird populations have declined by an estimated 82–99%. Many grassland species have been extirpated from large areas, and the two species of prairie chickens still persisting in pockets of the Great Plains could soon go extinct. Habitat destruction has occurred widely in nearly every biome. Estimates by UNEP in 2002 reported that 45% of Earth's forests, 50% of its mangrove ecosystems, and 10% of its coral reefs had been destroyed by recent human activity.

Invasive species Human introduction of non-native species to new environments, where some may become invasive (Figure 11.11), has also pushed native species toward extinction. Some introductions have been accidental. Examples include aquatic organisms (such as zebra mussels; Chapter 6) transported among continents in the ballast water of ships, animals that have escaped from the pet trade, and the weed seeds that cling to our socks as we travel from place to place. Other introductions have been

Invasive Species			
Species	**Native to...**	**Invasive in...**	**Effects**
Mosquito fish *(Gambusia affinis)* 	North America	Africa, Asia, Europe, and Australia	Introduced to control mosquito populations, the mosquito fish outcompetes native fish, eats their eggs, and does no better than native species in controlling mosquitoes.
Zebra mussels *(Dreissenna polymorpha)* 	Caspian Sea	Freshwater ecosystems including the Great Lakes of Canada and the United States	Zebra mussels (Chapter 6) most likely made their way from their home by traveling in ballast water taken on by cargo ships. They compete with native species and clog water treatment facilities and power plant cooling systems.
Kudzu *(Pueraria montana)* 	Japan	Southeastern United States	A vine that can grow 30 m (100 ft) in a single season. The U.S. Soil Conservation Service introduced kudzu in the 1930s to help control erosion. Adaptable and extraordinarily fast-growing, kudzu has taken over thousands of hectares of forests, fields, and roadsides in the southeastern United States.
Asian long-horned beetles *(Anoplophora glabripennis)* 	Asia	United States	Having first arrived in the United States in imported lumber in the 1990s, these beetles burrow into hardwood trees and interfere with the trees' ability to absorb and process water and nutrients. They may wipe out the majority of hardwood trees in an area. Several U.S. cities, including Chicago in 1999 and Seattle in 2002, have cleared thousands of trees after detecting these invaders.
Rosy wolfsnail *(Euglandina rosea)* 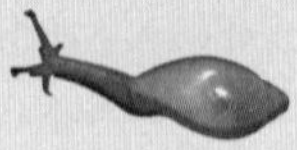	Southeastern United States and Latin America	Hawaii	In the 1950s, well-meaning scientists introduced the rosy wolfsnail to Hawaii to prey upon and reduce the population of another invasive species, the giant African land snail (*Achatina fulica*), which had been introduced early in the 20th century as an ornamental garden animal. Within a few decades, however, the carnivorous rosy wolfsnail had instead driven more than half of Hawaii's native species of banded tree snails to extinction.
Cane toad *(Bufo marinus)* 	Southern United States to tropical South America	Northern Australia and other locations	Since being introduced 70 years ago to control insects in sugarcane fields, the cane toad has wreaked havoc across northern Australia (and other locations). The skin of this tropical American toad can kill its predators, and the cane toad outcompetes native amphibians.
Bullfrog *(Rana catesbiana)* 	Eastern North America	Western North America	The bullfrog is contributing to amphibian and reptile declines in western North America. Bullfrog tadpoles grow large and can outcompete and prey on other tadpoles, but need to grow a long time in permanent water to do so. Historically most water bodies in the arid West dried up part of the year, making it impossible for bullfrogs to live there, but artificial impoundments—dams, farm ponds, canals—gave the bullfrogs bases from which they could spread.

FIGURE 11.11 Invasive species are species that thrive in areas where they are introduced, outcompeting, preying on, or otherwise harming native species. Of the many thousands of invasive species, this chart shows a few of the best known.

intentional. People have brought with them food crops, domesticated animals, and other organisms as they colonized new places, generally unaware of the ecological consequences that could result. Species native to islands are especially vulnerable to disruption from introduced species because the native species have been in isolation for so long with relatively few parasites, predators, and competitors. As a result, they have not evolved the

Invasive Species

Species	Native to...	Invasive in...	Effects
Gypsy moth *(Lymantria dispar)* 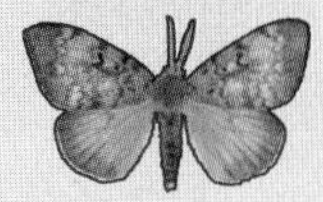	Eurasia	Northeastern United States	In the 1860s, a scientist introduced the gypsy moth to Massachusetts in the mistaken belief that it might be bred with others to produce a commercial-quality silk. The gypsy moth failed to start a silk industry, and instead spread through the northeastern United States and beyond, where its outbreaks defoliate trees over large regions every few years.
European starling *(Sturnus vulgaris)* 	Europe	North America	The bird was first introduced to New York City in the late 19th century by Shakespeare devotees intent on bringing every bird mentioned in Shakespeare's plays to the new continent. It only took 75 years for the birds to spread to the Pacific coast, Alaska, and Mexico, becoming one of the most abundant birds on the continent. Starlings are thought to outcompete native birds for nest sites.
Indian mongoose *(Herpestes auropunctatus)* 	Southeast Asia	Hawaii	Rats that had invaded the Hawaiian islands from ships in the 17th century were damaging sugarcane fields, so in 1883 the Indian mongoose was introduced to control rat populations. Unfortunately, the rats were active at night and the mongooses fed during the day, so the plan didn't work. Instead mongooses began preying on native species like ground-nesting seabirds and the now-endangered Nene or Hawaiian goose (*Branta sandvicensis*).
A green alga *(Caulerpa taxifolia)* 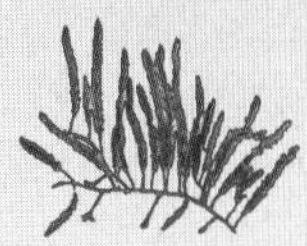	Tropical oceans and seas	Mediterranean Sea	Dubbed the "killer algae," *Caulerpa taxifolia* has spread along the coasts of several Mediterranean countries since it apparently escaped from Monaco's aquarium in 1984. Creeping underwater over the sand and mud like a green shag carpet, it crowds out other plants, is inedible to most animals, and tangles boat propellers. It has been the focus of intense eradication efforts since arriving recently in Australia and California.
Cheatgrass *(Bromus tectorum)* 	Eurasia	Western United States	In just 30 years after its introduction to Washington state in the 1890s, cheatgrass has spread across much of the western United States. Its secret: fire. Its thick patches that choke out other plants and use up the soil's nitrogen burn readily. Fire kills many of the native plants, but not cheatgrass, which grows back even stronger amid the lack of competition.
Brown tree snake *(Boiga irregularis)* 	Southeast Asia	Guam	Nearly all native forest bird species on the South Pacific island of Guam have disappeared. The culprit is the brown tree snake. The snakes were likely brought to the island inadvertantly as stowaways in cargo bays of military planes in World War II. Guam's birds had not evolved with tree snakes, and so had no defenses against the snake's nighttime predation. The snakes also cause numerous power outages each year on Guam and have spread to other islands where they are repeating their ecological devastation. The arrival of this snake is the greatest fear of conservation biologists in Hawaii.

defenses necessary to resist invaders that are better adapted to these pressures.

Most organisms introduced to new areas perish, but the few types that survive may do very well, especially if they find themselves without the predators and parasites that attacked them back home or without the competitors that had limited their access to resources. Once released from the limiting factors of predation, parasitism, and

The Science behind the Story

Amphibian Diversity and Amphibian Declines

Amphibians illustrate the two most salient aspects of Earth's biodiversity today. Scientists are discovering more and more species while more and more populations and species are vanishing.

New species of most classes of vertebrates are discovered at a rate of only one or a few per year, but the number of known amphibian species (which include frogs, salamanders, and others)—about 5,800 as of 2005—has jumped nearly 42% just since 1985.

At the same time, however, over 200 species are in steep decline. Researchers feel that they may be naming some species just before they go extinct and losing others before they are even discovered. At least 32 species of frogs, toads, and salamanders studied just years or decades ago, including the golden toad (Chapter 5), are now altogether gone.

These losses are especially worrying because amphibians are widely regarded as "biological indicators" that indicate whether an ecosystem is in good shape or is degraded. Amphibians rely on both aquatic and terrestrial environments and may breathe and absorb water through their skin, so they are sensitive to pollution and other environmental stresses. The link between amphibians and environmental quality suggests that studying the reasons for their declines can tell us much about the state of our environment.

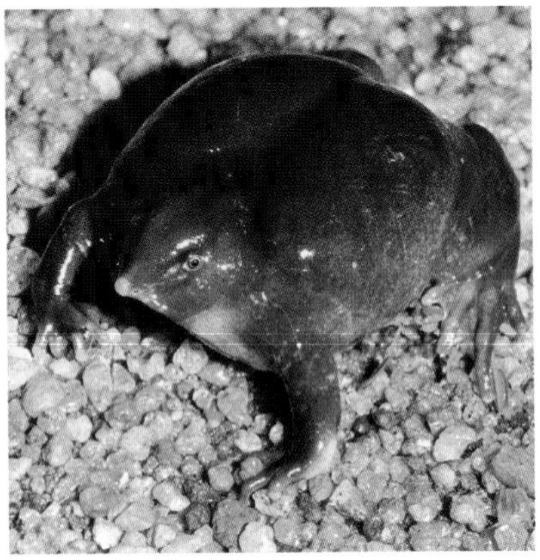
The odd-looking purplish frog *Nasikabatrachus sahyadrensis*, of India, is one of many new amphibian species recently discovered.

In Sri Lanka and other countries, scientific scrutiny and improved technology have revealed amphibian "hot spots." In the 1990s, an international team of scientists set out to determine whether Sri Lanka, a large tropical island off the coast of India, held more than the 40 frog species that were already known. Researcher Madhava Meegaskumbura and his team combed through trees, rivers, ponds, and leaf litter for 8 years, collecting more than 1,400 frogs at 300 study sites. The scientists analyzed the frogs' physical appearance, habitat use, and vocalizations. They also examined the frogs' genes by obtaining sequences of nucleotides in several regions of their DNA. They then compared these genetic, physical, and behavioral characteristics to those of known species of frogs.

The team found that the DNA from many of their frogs didn't match that of known species. And they found that many of their frogs looked different, sounded different, or behaved differently from known species. Clearly, they had discovered new species of frogs unknown to science. Some of these novel species live on rocks and sport leg fringes and markings that help disguise them as clumps of moss. Others are tree frogs that lay their eggs in baskets they construct. In all, more

competition, an introduced species may increase rapidly, spread, and displace native species. Moreover, invasive species cause billions of dollars in economic damage each year.

Pollution Pollution can negatively affect organisms in many ways. Air pollution (Chapter 17) can degrade forest ecosystems. Water pollution (Chapter 15) can adversely affect fish and amphibians. Agricultural runoff (including fertilizers, pesticides, and sediments; Chapters 7, 9, and 10) can harm many terrestrial and aquatic species. Heavy metals, PCBs, endocrine-disrupting compounds, and various other toxic chemicals can poison people and wildlife (Chapter 14), and the effects of oil and chemical spills on wildlife are dramatic and well known. However, although pollution is a substantial threat, it tends to be less significant than public perception holds it to be. The damage to wildlife and ecosystems caused by pollution can be severe, than 100 new species of amphibians were discovered—all on an island only slightly larger than the state of West Virginia! When reported in the journal *Science* in 2002, the study caught the attention of conservation biologists worldwide.

Such promising discoveries, however, come against a backdrop of distressing declines. Observed numbers of amphibians are down around the globe. Scientists are racing to pin down the causes, and have found evidence for causes as varied as habitat destruction, chemical pollution, disease, invasive species, and climate change. Most worrisome is that many populations are vanishing even when no direct damage, such as habitat loss, is apparent. In some cases, researchers surmise that a combination of factors may be at work.

In one study, researchers Rick Relyea and Nathan Mills presented young frogs with two common dangers—pesticides and predators—to see how the mix affected their survival. The team collected 10 pairs of tree frogs from a wildlife area in Missouri and placed their eggs in clean water. When tadpoles emerged from the eggs, groups of 10 were each put in different tubs of water. Some tubs contained pure water, others contained varying levels of the pesticide carbaryl, and others contained the harmless solvent acetone as a control. To some of each of these three types of tubs, the researchers added a hungry predator—a young salamander. The salamander was caged and couldn't reach the tadpoles, but the tadpoles were aware of its presence. In a series of experiments, over up to 16 days, researchers watched to see how many tadpoles survived the different combinations of stress factors.

Their results, published in the *Proceedings of the National Academy of Sciences* in 2001, revealed that tadpoles that withstood one type of stress might not survive two. As expected, all tadpoles in clean water with no predators survived, and all tadpoles exposed to high concentrations of carbaryl died within several days, regardless of predator presence. But when carbaryl levels were lower, the presence of the salamander made a noticeable difference. In one trial, about 75% of tadpoles survived the pesticide if no predator was present, but in the presence of the salamander, survival rates dropped to about 25%. Thus, when both stresses were present (a condition likely in the tadpoles' natural habitat), death rates increased by two to four times.

One year later, a study published in the same journal by herpetologist Joseph Kiesecker found similar results with pathogens and pesticides. His field and lab experiments revealed that wood frogs were more vulnerable to parasitic infections that cause limb deformities when they were exposed to water containing pesticides.

As scientists learn more about how such factors combine to threaten amphibians, they are gaining a clearer picture of how the fate of these creatures may foreshadow the future for other organisms. "Amphibians have been around for 300 million years. They're tough, and yet they're checking out all around us," says David Wake, a biologist at the University of California at Berkeley, who was among the first to note the creatures' decline. "We really do see amphibians as biodiversity bellwethers."

but it tends to be less than the damage caused by habitat alteration or invasive species.

Overharvesting For most species, a high intensity of hunting or harvesting by humans will not *in itself* pose a threat of extinction, but for some species it can. The Siberian tiger is one such species. Large in size, few in number, long-lived, and raising few young in its lifetime—a classic K-selected species (▸p. 139)—the Siberian tiger is just the type of animal to be vulnerable to population reduction by hunting. The advent of Russian hunting nearly drove the animal extinct, whereas decreased hunting during and after World War II contributed to a population increase. By the mid-1980s, the Siberian tiger population was likely up to 250 individuals. The political freedom that came with the Soviet Union's breakup in 1989, however, brought with it a freedom to harvest Siberia's natural resources, the tiger included, without

FIGURE 11.12 Body parts from tigers have long been used as medicines or aphrodisiacs in some traditional Asian cultures. Hunters and poachers have illegally killed countless tigers through the years to satisfy market demand for these items. Here a street vendor in northern China displays tiger penises and other body parts for sale.

regulations or rules. This coincided with an economic expansion in many Asian countries, where tiger penises are traditionally used to try to boost human sexual performance and where tiger bones, claws, whiskers, and other body parts are used to treat a wide variety of maladies (Figure 11.12). Thus, the early 1990s brought a boom in poaching (poachers killed at least 180 Siberian tigers between 1991 and 1996), as well as a dramatic increase in logging of the Korean pine forests on which the tigers and their prey depend.

Over the past century, hunting has led to steep declines in the populations of many other K-selected animals. The Atlantic gray whale has gone extinct, and several other whales remain threatened or endangered. Gorillas and other primates that are killed for their meat may be facing extinction soon. Thousands of sharks are killed each year simply for their fins, which are used in soup. Today the oceans contain only 10% of the large animals they once did (▸ p. 483).

Climate change The preceding four types of human impacts affect biodiversity in discrete places and times. In contrast, our manipulation of Earth's climate system (Chapter 18) is beginning to have global impacts on biodiversity. As we will explore in Chapter 18, our emissions of carbon dioxide and other "greenhouse gases" that trap heat in the atmosphere are causing average temperatures to warm worldwide, modifying global weather patterns and increasing the frequency of extreme weather events. Scientists foresee that these effects, together termed *global climate change*, will accelerate and become more severe in the years ahead until we find ways to reduce our emissions from fossil fuels.

Climate change is beginning to exert effects on plants and animals. Extreme weather events such as droughts put increased stress on populations, and warming temperatures are forcing species to move toward the poles and higher in altitude. Some species will be able to adapt, but others will not. Consider the cloud-forest fauna from Monteverde that we examined in Chapter 5. Mountaintop organisms cannot move further upslope to escape warming temperatures, so they will likely perish. Trees may not be able to move poleward fast enough. Animals and plants may find themselves among different communities of prey, predators, and parasites to which they are not adapted. The impacts of climate change will likely play a large role in shaping the future world that we and our children will inhabit.

All five of these avenues are influenced by human population growth and rising per capita consumption. More people and more consumption mean more habitat alteration, more invasive species, more pollution, more overharvesting, and more climate change. Growth in population and growth in consumption are the ultimate reasons behind the proximate threats to biodiversity.

Benefits of Biodiversity

Scientists worldwide are presenting us with data that confirm what any naturalist who has watched the habitat change in his or her hometown already knows: From amphibians to tigers, biodiversity is being lost rapidly and visibly within our lifetimes. This suggests the question, "Does it matter?" There are many ways to answer this question, but we can begin by considering the ways that biodiversity benefits people. Scientists have offered a number of tangible, pragmatic reasons for preserving biodiversity, showing how biodiversity directly or indirectly supports human society. In addition, many people feel that organisms have an intrinsic right to exist and that ethical and aesthetic dimensions to biodiversity preservation cannot be ignored.

Biodiversity provides ecosystem services free of charge

Contrary to popular opinion, some things in life can indeed be free, as long as we choose to protect the living systems that provide them. Intact forests provide clean air and buffer hydrologic systems against flooding and drought. Native crop varieties provide insurance against disease and drought. Abundant wildlife can attract tourists and boost the economies of developing nations. Intact ecosystems provide these and other valuable processes, known as *ecosystem services* (▸pp. 39–41, 48–51), for all of us, free of charge.

Maintaining these ecosystem services is one clear benefit of protecting biodiversity. According to UNEP, biodiversity:

- Provides food, fuel, and fiber
- Provides shelter and building materials
- Purifies air and water
- Detoxifies and decomposes wastes
- Stabilizes and moderates Earth's climate
- Moderates floods, droughts, wind, and temperature extremes
- Generates and renews soil fertility and cycles nutrients
- Pollinates plants, including many crops
- Controls pests and diseases
- Maintains genetic resources as inputs to crop varieties, livestock breeds, and medicines
- Provides cultural and aesthetic benefits
- Gives us the means to adapt to change

Organisms and ecosystems support a vast number of vital processes that humans could not replicate or would need to pay for if nature did not provide them. As we saw in Chapter 2, the annual value of just 17 of these ecosystem services may be in the neighborhood of \$16–54 trillion per year.

Biodiversity helps maintain ecosystem function

Even if functioning ecosystems are important, however, does biodiversity really help them maintain their function? Ecologists have found that the answer appears to be yes. Research has demonstrated that high levels of biodiversity tend to increase the *stability* of communities and ecosystems. Research has also found that high biodiversity tends to increase the *resilience* of ecological systems—their ability to weather disturbance, bounce back from stresses, or adapt to change. Most of this research has dealt with species diversity, but new work is finding similar effects for genetic diversity. Thus, a decrease in biodiversity could diminish a natural system's ability to function and to provide services to our society.

What about the extinction of selected species, however? Skeptics have asked whether the loss of a few endangered species will really make much difference in an ecosystem's ability to function. Ecological research suggests that the answer to this question depends on which species are removed. Removing a species that can be functionally replaced by others may make little difference. Recall, however, from Chapter 6 our discussion of keystone species (▸pp. 159–162). Like the keystone that holds together an arch, a keystone species is one whose removal results in significant changes in an ecological system. If a keystone species is extirpated or driven extinct, other species may disappear or experience significant population changes as a result.

Often top predators, such as tigers, are considered keystone species. A single top predator may prey on many other carnivores, each of which may prey on many herbivores, each of which may consume many plants. Thus the removal of a single individual at the top of a food chain can have impacts that multiply as they cascade down the food chain. Moreover, top predators such as tigers, wolves, and grizzly bears are among the species most vulnerable to human impact. Large animals are frequently hunted, and also need large areas of habitat, making them susceptible to habitat loss and fragmentation. In addition, top predators are vulnerable to the buildup of toxic pollutants in their tissues through the process of biomagnification (▸p. 416).

Top predators are not the only species that exert far-reaching influence over their ecosystems. The influence of other species, including "ecosystem engineers" such as ants and earthworms, can be equally significant. Ecosystems are complex, and it is difficult to predict which particular species may be important. Thus, many people prefer to apply the precautionary principle in the spirit of Aldo Leopold (▸pp. 35–36), who advised, "To keep every cog and wheel is the first precaution of intelligent tinkering."

Biodiversity enhances food security

Biodiversity benefits our agriculture, as well. As our discussion of native landraces of corn in Oaxaca, Mexico, in Chapter 10 showed, genetic diversity within crop species and their ancestors is enormously valuable. In 1995, Turkey's wheat crops received at least $50 billion worth of disease resistance from wild wheat strains. California's barley crops annually receive $160 million in disease resistance benefits from Ethiopian strains of barley. During the 1970s a researcher discovered a maize species in Mexico known as *Zea diploperennis.* This maize is highly resistant to disease, and it is a perennial, meaning it will grow back year after year without being replanted. At the time of its discovery, its entire range was limited to a 10-ha (25-acre) plot of land in the mountains of the Mexican state of Jalisco.

Other potentially important food crops await utilization (Figure 11.13). The babassu palm *(Orbignya phalerata)* of the Amazon produces more vegetable oil than any other plant. The serendipity berry *(Dioscoreophyllum cumminsii)* produces a sweetener that is 3,000 times sweeter than table sugar. Several species of salt-tolerant grasses and trees are so hardy that farmers can irrigate them with saltwater. These same plants also produce animal feed, a substitute for conventional vegetable oil, and other economically important products. Such species could be immeasurably beneficial to areas undergoing soil salinization due to poorly managed irrigation (▸pp. 265–266).

Biodiversity provides drugs and medicines

People have made medicines from plants for centuries, and many of today's widely used drugs were discovered by studying chemical compounds present in wild plants, animals, and microbes (Figure 11.14). Each year pharmaceutical products owing their origin to wild species generate up to $150 billion in sales.

It can truly be argued that every species that goes extinct represents one lost opportunity to find a cure for cancer or AIDS. The rosy periwinkle *(Catharanthus roseus)* produces compounds that treat Hodgkin's disease and a particularly deadly form of leukemia. Had this native plant of Madagascar become extinct prior to its discovery by medical researchers, two deadly diseases would have claimed far more victims than they have to date. In Australia, where the government has placed high priority on research into products from rare and endangered species, a rare species of cork, *Duboisia leichhardtii,* now provides hyoscine, a compound that physicians use to treat cancer, stomach disorders, and motion sickness. Another Australian plant, *Tylophora,* provides a drug that treats lymphoid leukemia. Researchers are now exploring the potential of the compound prostaglandin E2

Food Security and Biodiversity: Potential new food sources

Species	Native to...	Potential uses and benefits
Amaranths (three species of *Amaranthus*) 	Tropical and Andean America	Grain and leafy vegetable; livestock feed; rapid growth, drought resistant
Buriti palm *(Mauritia flexuosa)* 	Amazon lowlands	"Tree of life" to Amerindians; vitamin-rich fruit; pith as source for bread; palm heart from shoots
Maca *(Lepidium meyenii)* 	Andes Mountains	Cold-resistant root vegetable resembling radish, with distinctive flavor; near extinction
Tree tomato *(Cyphomandra betacea)* 	South America	Elongated fruit with sweet taste
Babirusa *(Babyrousa babyrussa)* 	Indonesia: Moluccas and Sulawesi	A deep-forest pig; thrives on vegetation high in cellulose and hence less dependent on grain
Capybara *(Hydrochoeris hydrochoeris)* 	South America	World's largest rodent; meat esteemed; easily ranched in open habitats near water
Vicuna *(Lama vicugna)* 	Central Andes	Threatened species related to llama; valuable source of meat, fur, and hides; can be profitably ranched
Chachalacas (*Ortalis,* many species) 	South and Central America	Birds, potentially tropical chickens; thrive in dense populations; adaptable to human habitations; fast-growing
Sand grouse *(Pterocles,* many species) 	Deserts of Africa and Asia	Pigeon-like birds adapted to harshest deserts; domestication a possibility

FIGURE 11.13 By protecting biodiversity, we can enhance food security. The wild species shown here are a tiny fraction of the many plants and animals that could someday supplement our food supply. Adapted from Wilson, E. O. 1992. *The diversity of life.* Cambridge, MA: Belknap Press.

Medicines and Biodiversity: Natural sources of pharmaceuticals			
Plant		Drug	Medical application
Pineapple *(Ananas comosus)*		Bromelain	Controls tissue inflammation
Autumn crocus *(Colchicum autumnale)*		Colchicine	Anticancer agent
Yellow cinchona *(Cinchona ledgeriana)*		Quinine	Antimalarial
Common thyme *(Thymus vulgaris)*		Thymol	Cures fungal infection
Pacific yew *(Taxus brevifolia)*		Taxol	Anticancer (especially ovarian cancer)
Velvet bean *(Mucuna deeringiana)*		L-Dopa	Parkinson's disease suppressant
Common foxglove *(Digitalis purpurea)*		Digitoxin	Cardiac stimulant

FIGURE 11.14 By protecting biodiversity, we can enhance our ability to treat illness. Shown here are just a few of the plants that have so far been found to provide chemical compounds of medical benefit. Adapted from Wilson, E. O. 1992. *The diversity of life.* Cambridge, MA: Belknap Press.

in treating gastric ulcers. This compound was first discovered in two frog species unique to the rainforest of Queensland, Australia. Scientists believe that both species are now extinct.

Weighing the Issues: Bioprospecting in Costa Rica

Bioprospectors search for organisms that can provide new drugs, foods, or other valuable products. Scientists working for pharmaceutical companies, for instance, scour biodiversity-rich countries for potential drugs and medicines. Many have been criticized for harvesting indigenous species to create commercial products that do not benefit the country of origin. To make sure it would not lose the benefits of its own biodiversity, Costa Rica reached an agreement with the Merck pharmaceutical company in 1991. The nonprofit National Biodiversity Institute of Costa Rica (INBio) allowed Merck to evaluate a limited number of Costa Rica's species for their commercial potential in return for $1 million, plus equipment and training for Costa Rican scientists.

Do you think that both sides win in this agreement? What if Merck discovers a compound that could be turned into a billion-dollar drug? Does this provide a good model for other countries? For other companies?

Biodiversity provides economic benefits through tourism and recreation

Besides providing for our food and health, biodiversity can represent a direct source of income through tourism, particularly for developing countries in the tropics that have impressive species diversity. As we saw in Chapter 5 with Costa Rica, many people like to travel to experience protected natural areas, and in so doing they create economic opportunity for residents living near those natural areas. Visitors spend money at local businesses, hire local people as guides, and support the parks that employ local residents. Ecotourism thus can bring jobs and income to areas that otherwise might be poverty-stricken.

Ecotourism has become a vital source of income for nations such as Costa Rica, with its rainforests; Australia, with its Great Barrier Reef; Belize, with its reefs, caves, and rainforests; and Kenya and Tanzania, with their savanna wildlife. The United States, too, benefits from ecotourism; its national parks draw millions of visitors domestically and from around the world. Ecotourism serves as a powerful financial incentive for nations, states, and local communities to preserve natural areas and reduce impacts on the landscape and on native species.

As ecotourism increases in popularity, however, critics have warned that too many visitors to natural areas can degrade the outdoor experience and disturb wildlife. Anyone who has been to Yosemite, the Grand Canyon, or the Great Smokies on a crowded summer weekend can attest to this. Ecotourism's effects on species living in parks and reserves are much debated, and likely they vary enormously from one case to the next. As ecotourism continues to increase, so will debate over its costs and benefits for local communities and for biodiversity.

People value and seek out connections with nature

Not all of the benefits of biodiversity to humans can be expressed in the hard numbers of economics or the day-to-day practicalities of food and medicine. Some scientists and philosophers argue that there is a deeper

FIGURE 11.15 Edward O. Wilson is the world's most recognized authority on biodiversity and its conservation and has inspired many people who study our planet's life. A Harvard professor and world-renowned expert on ants, Wilson has written over 20 books and has won two Pulitzer prizes. His books *The Diversity of Life* and *The Future of Life* address the value of biodiversity and its outlook for the future.

importance to biodiversity. E. O. Wilson (Figure 11.15) has described a phenomenon he calls **biophilia,** "the connections that human beings subconsciously seek with the rest of life." Wilson and others have cited as evidence of biophilia our affinity for parks and wildlife, our keeping of pets, the high value of real estate with a view of natural landscapes, and our interest—despite being far removed from a hunter-gatherer lifestyle—in hiking, bird-watching, fishing, hunting, backpacking, and similar outdoor pursuits.

In a 2005 book, writer Richard Louv adds that as today's children are increasingly deprived of outdoor experiences and direct contact with wild organisms, they suffer what he calls "nature-deficit disorder." Although it is not a medical condition, this alienation from biodiversity and the natural environment, Louv argues, may damage childhood development and lie behind many of the emotional and physical problems young people in developed nations face today.

Weighing the Issues: Biophilia and Nature-Deficit Disorder

What do you think of the concepts of biophilia and "nature-deficit disorder"? Have you ever felt a connection to other living things that you couldn't explain in scientific or economic terms? Do you think that an affinity for other living things is innately human? How could you determine whether or not most people in your community feel this way?

Do we have ethical obligations toward other species?

If Wilson, Louv, and others are right, then biophilia not only may affect ecotourism and real estate prices, but also may influence our ethics. When Maurice Hornocker and his associates first established the Siberian Tiger Project, he wrote: "Saving the most magnificent of all the cat species and one of the most endangered should be a global responsibility. . . . If they aren't worthy of saving, then what are we all about? What is worth saving?"

On one hand, we humans are part of nature, and like any other animal we need to use resources and consume other organisms to survive. In that sense, there is nothing immoral about our doing so. On the other hand, we have conscious reasoning ability and are able to control our actions and make conscious decisions. Our ethical sense has developed from this intelligence and ability to choose. As our society's sphere of ethical consideration has widened over time, and as more of us take up biocentric or ecocentric worldviews (▸pp. 32–33), more people have come to feel that other organisms have intrinsic value and an inherent right to exist.

Despite our ethical convictions, however, and despite biodiversity's many benefits—from the pragmatic and economic to the philosophical and spiritual—the future of biodiversity is far from secure. Even our protected areas and national parks are not big enough or protected well enough to ensure that biodiversity is fully safeguarded within their borders. The search for solutions to today's biodiversity crisis is an exciting and active one, and scientists are playing a leading role in developing innovative approaches to maintaining the diversity of life on Earth.

Conservation Biology: The Search for Solutions

Today, more and more scientists and citizens perceive a need to do something to stem the loss of biodiversity. In his 1994 autobiography, *Naturalist,* E. O. Wilson wrote:

> When the [20th] century began, people still thought of the planet as infinite in its bounty. The highest mountains were still unclimbed, the ocean depths never visited, and vast wildernesses stretched across the equatorial continents. . . . In one lifetime exploding human populations have reduced wildernesses to threatened nature reserves. Ecosystems and species are vanishing at the fastest rate in 65 million years. Troubled by what we have wrought, we have begun to turn in our role from local conqueror to global steward.

Conservation biology arose in response to biodiversity loss

The urge to act as responsible stewards of natural systems, and to use science as a tool in that endeavor, helped spark the rise of conservation biology. **Conservation biology** is a scientific discipline devoted to understanding the factors, forces, and processes that influence the loss, protection, and restoration of biological diversity. It arose as biologists became increasingly alarmed at the degradation of the natural systems they had spent their lives studying.

Conservation biologists choose questions and pursue research with the aim of developing solutions to such problems as habitat degradation and species loss. Conservation biology is thus an applied and goal-oriented science, with implicit values and ethical standards. This perceived element of advocacy sparked some criticism of conservation biology in its early years. However, as scientists have come to recognize the scope of human impact on the planet, more of them have directed their work to address environmental problems. Today conservation biology is a thriving pursuit that is central to environmental science and to achieving a sustainable society.

Conservation biologists work at multiple levels

Conservation biologists integrate an understanding of evolution and extinction with ecology and the dynamic nature of environmental systems. They use field data, lab data, theory, and experiments to study the impacts of humans on other organisms. They also attempt to design, test, and implement ways to mitigate human impact.

These researchers address the challenges facing biological diversity at all levels, from genetic diversity to species diversity to ecosystem diversity. At the genetic level, *conservation geneticists* study genetic attributes of organisms, generally to infer the status of their populations. If two populations of a species are found to be genetically distinct enough to be considered subspecies, they may have different ecological needs and may require different types of management. In addition, as a population dwindles, genetic variation is lost from the gene pool. Conservation geneticists ask how small a population can become and how much genetic variation it can lose before running into problems such as inbreeding depression. By determining a *minimum viable population size* for a given population, conservation geneticists and population biologists provide wildlife managers with an indication of how vital it may be to increase the population.

Problems for populations and subspecies spell problems for species, because declines and local extirpation generally precede range-wide endangerment and extinction. As we will soon see, it is at the species level that much of the funding and resources for conservation biology exist. However, many efforts also revolve around habitats, communities, and ecosystems.

Island biogeography theory is a key component of conservation biology

Safeguarding habitat for species and conserving communities and ecosystems requires thinking and working at the landscape level. One key conceptual tool for doing so is the **equilibrium theory of island biogeography.** This theory, introduced by E. O. Wilson and ecologist Robert MacArthur in 1963, explains how species come to be distributed among oceanic islands. Since then, researchers have also applied it to "habitat islands"—patches of one habitat type isolated within "seas" of others. The Sikhote-Alin Mountains, last refuge of the Siberian tiger, are a habitat island, isolated from other mountains by deforested regions, a seacoast, and populated lowlands.

Island biogeography theory predicts the number of species on an island based on the island's size and its distance from the nearest mainland. The number of species on an island results from a balance between the number added by immigration and the number lost through extinction (or more precisely, extirpation from the particular island). Immigration and extinction are ongoing

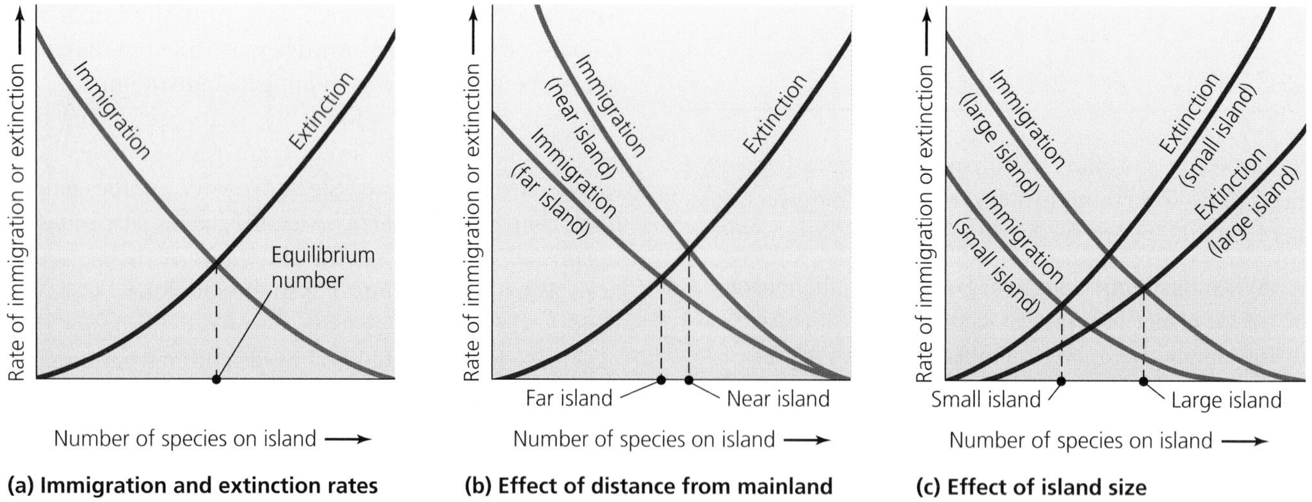

FIGURE 11.16 Island biogeography theory explains species richness on islands as a function of immigration and extinction rates interacting with island size and distance from the mainland. In **(a)**, immigration rates are highest for islands with few species (because each new immigrant species represents a high proportion of the island's total species). Extinction rates are greatest on islands with many species (because interspecific competition for resources keeps population sizes small). These two trends set the stage for examining the effects of island size and distance. In **(b)**, islands closer to a mainland tend to hold more species, because immigration rates are higher, whereas extinction rates are not affected by distance. In **(c)**, immigration is greater on large islands because large islands are easier for dispersing organisms to discover, while extinction rates are lower on large islands, because more area allows populations to grow larger. Together these trends give large islands a higher number of species at equilibrium.

dynamic processes, so the balance between them represents an equilibrium state (Figure 11.16a).

The farther an island is located from a continent, the fewer species tend to find and colonize it. Thus, remote islands host fewer species because of lower immigration rates (Figure 11.16b). Large islands have higher immigration rates because they present fatter targets for wandering or dispersing organisms to encounter. Large islands have lower extinction rates because more space allows for larger populations, which are less vulnerable to dropping to zero by chance. Together, these trends give large islands more species at equilibrium than small islands (Figure 11.16c). Very roughly, the number of species on an island is expected to double as island size increases tenfold. This effect can be illustrated with *species-area curves* (Figure 11.17). Large islands also tend to contain more species because they generally possess more habitats than smaller islands, providing suitable environments for a wider variety of arriving species.

These theoretical patterns have been widely supported by empirical data from the study of species on islands (see "The Science behind the Story," ▸pp. 332–333). The patterns hold up for terrestrial habitat islands, such as forests fragmented by logging and road building (Figure 11.18). Small islands of forest lose their diversity fastest, starting with large species that were few in number to begin with. In a landscape of fragmented habitat, species requiring the habitat will gradually disappear, winking out from one fragment after another over time. Fragmentation of forests and other habitats constitutes one of the prime threats to biodiversity. In response to habitat fragmentation, conservation biologists have designed landscape-level strategies to try to optimize the arrangement of areas to be preserved. We will examine a few of these strategies in our discussion of parks and preserves in Chapter 12 (▸pp. 364–369).

Weighing the Issues: Fragmentation and Biodiversity

Suppose a critic of conservation tells you that human development increases biodiversity, pointing out that when a forest is fragmented, new habitats, such as grassy lots and gardens, may be introduced to an area and allow additional species to live there. How would you respond?

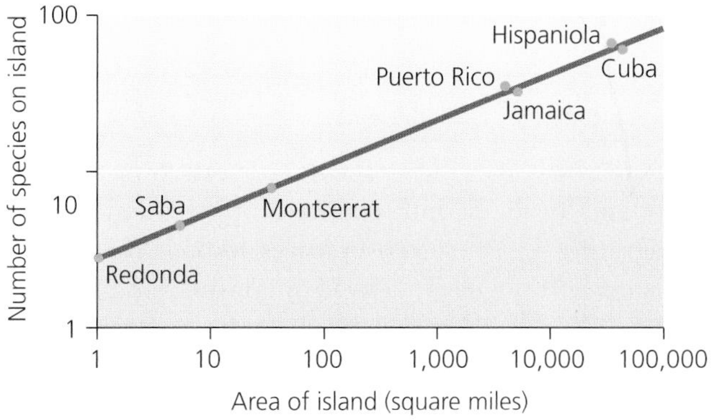

FIGURE 11.17 The larger the island, the greater the number of species—a prediction borne out by data from around the world. By plotting the number of amphibians and reptile species on Caribbean islands as a function of the areas of these islands, the species-area curve shows that species richness increases with area. The increase is not linear, but logarithmic; note the scales of the axes. Go to GRAPHIt! at www.aw-bc.com/withgott or on the student CD-ROM. Data from MacArthur, R. H., and E. O. Wilson. 1967. *The theory of island biogeography.* Princeton University Press.

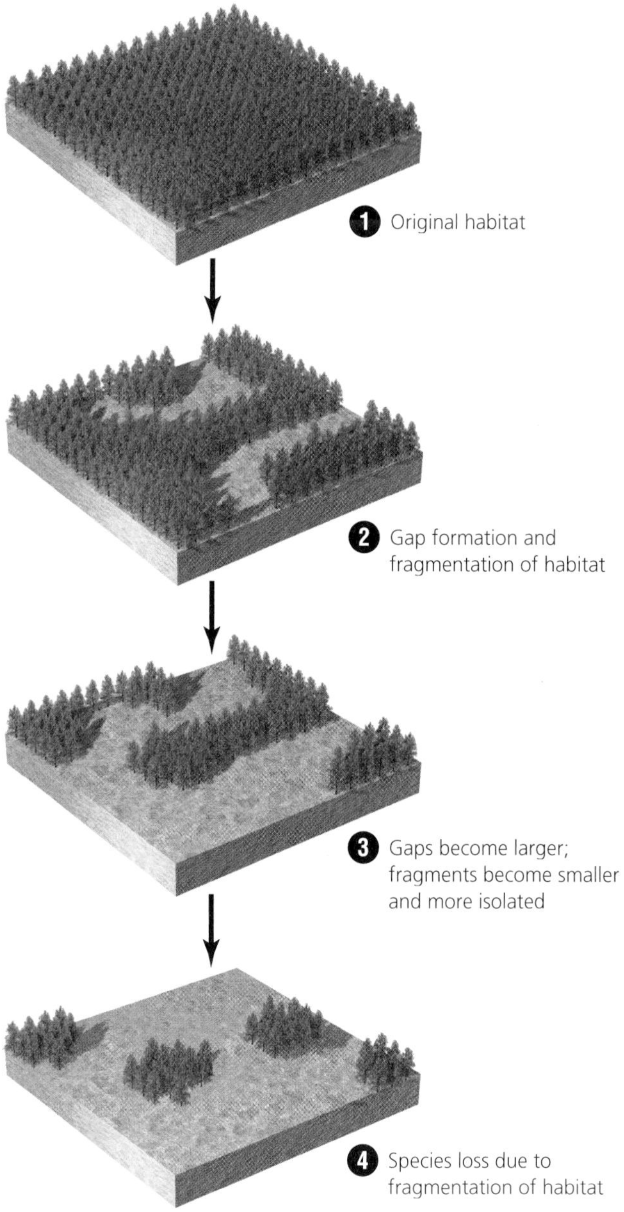

FIGURE 11.18 Forest clearing, farming, road building, and other types of human land use and development can fragment natural habitats. Habitat fragmentation usually begins when gaps are created within a natural habitat. As development proceeds, these gaps expand, join together, and eventually dominate the landscape, stranding islands of habitat in their midst. As habitat becomes fragmented, fewer populations can persist, and numbers of species in the fragments decrease with time.

Should endangered species be the focus of conservation efforts?

The primary legislation for protecting biodiversity in the United States is the **Endangered Species Act (ESA).** Passed in 1973, the ESA forbids the government and private citizens from taking actions (such as developing land) that destroy endangered species or their habitats. The ESA also forbids trade in products made from endangered species. The aim is to prevent extinctions, stabilize declining populations, and enable populations to recover. As of 2005, there were 1,264 species in the United States listed as "endangered" or as "threatened," the latter status considered one notch less severe than endangered.

The ESA has had a number of notable successes. Following the ban on the pesticide DDT and years of intensive effort by wildlife managers, the peregrine falcon, brown pelican, bald eagle, and other birds have recovered and have been removed from the endangered list (▸p. 416). Intensive management programs with other species, such as the red-cockaded woodpecker, have held formerly declining populations steady in the face of continued pressure on habitat. In fact, roughly 40% of declining populations have been held stable.

This success comes despite the fact that the U.S. Fish and Wildlife Service and the National Marine Fisheries Service, the agencies responsible for upholding the ESA, have been perennially underfunded for the job. These agencies have faced repeated budgetary shortfalls for endangered species protection. Moreover, efforts to reauthorize the ESA have faced stiff opposition in the U.S. Congress since the 1990s. As this book went to press, Republican Congressman Richard Pombo was leading efforts to weaken the ESA, including stripping it of its ability to safeguard habitat.

The Science behind the Story

Testing and Applying Island Biogeography Theory

The researchers who first experimentally tested the equilibrium theory of island biogeography and applied it to conservation biology leaned on their own resourcefulness, along with some plastic, some pesticides, and a small station wagon.

Robert MacArthur and Edward O. Wilson had originally developed island biogeography theory by using observational data from oceanic islands, correlating numbers of species found on islands with island size and distance between landmasses. Yet as of 1966, no one had tested its precepts in the field with a manipulative experiment. Wilson decided to remedy that.

Wilson began looking for islands in the United States where he could run an experimental test: to remove all animal life from islands and then observe and measure recolonization. To be suitable, the islands would need to be small, contain few forms of life, and be situated close to the mainland to ensure an influx of immigrating species. Wilson found his research sites off the tip of Florida: six small mangrove islands 11–18 m (36–59 ft) in diameter, home only to trees and a few dozen species of insects, spiders, centipedes, and other arthropods.

Daniel Simberloff, Wilson's graduate student at the time, painstakingly counted each island's arthropods, breaking up bark and poking under branches to find every mite, midge, and millipede. Then with the help of professional exterminators, Wilson and Simberloff wrapped each island in a plastic tarpaulin and gassed the interior with the pesticide methyl bromide. After removing the tarpaulins, Simberloff checked to make sure no creatures were left alive. The researchers then waited to see how life on the islands would return.

Over the next 2 years, Simberloff scrambled up trees and turned over leaves, looking for newly arrived organisms. His monitoring showed that life recovered on most islands within a year, regaining about the same number of species and total number of arthropods the islands had sheltered originally. Larger islands once again became home to a greater number of species than smaller islands. Outlying islands recovered more slowly and reached lower species diversities than did islands near the mainland. These results provided the first evidence from a manipulative experiment for the predictions of island biogeography theory.

Published in the journal *Ecology* in 1969 and 1970, Simberloff and Wilson's research gave new empirical rigor to a set of ideas that was increasingly helping scientists understand geographic patterns of biodiversity. Their research also fueled a question of

Daniel Simberloff and E. O. Wilson used mangrove islands in the Florida Keys to test island biogeography theory.

pressing concern for conservation biology: Could island biogeography theory also be applied to isolated "islands" of habitat on continents? At the University of Michigan, biology graduate student William Newmark set out to address this question.

Newmark had learned that many North American national parks kept records that documented sightings of wildlife over the course of the parks' existence. He surmised that by examining these historical records, he could infer which species had vanished from parks, which were new arrivals, and roughly when these changes occurred. The parks, increasingly surrounded by development, were islands of natural habitat isolated by farms, roads, towns, and cities, so Newmark hypothesized that island biogeography theory would apply.

In 1983, Newmark drove his Toyota station wagon to 24 parks in the western United States and Canada. At each park, he studied the wildlife sighting records, focusing on larger mammals such as bear, lynx, and river otter (but not species, such as wolves, that had been deliberately eradicated by hunting). Newmark found a few species missing from many parks, and they added up to a troubling total. Forty-two species had disappeared in all, and not as a result of direct human action. The red fox and river otter had vanished from Sequoia and Kings Canyon National Parks, for example, and the white-tailed jackrabbit and spotted skunk no longer lived in Bryce Canyon National Park. As theory predicted, the smallest parks showed the greatest number of losses, and the largest parks retained a greater number of species.

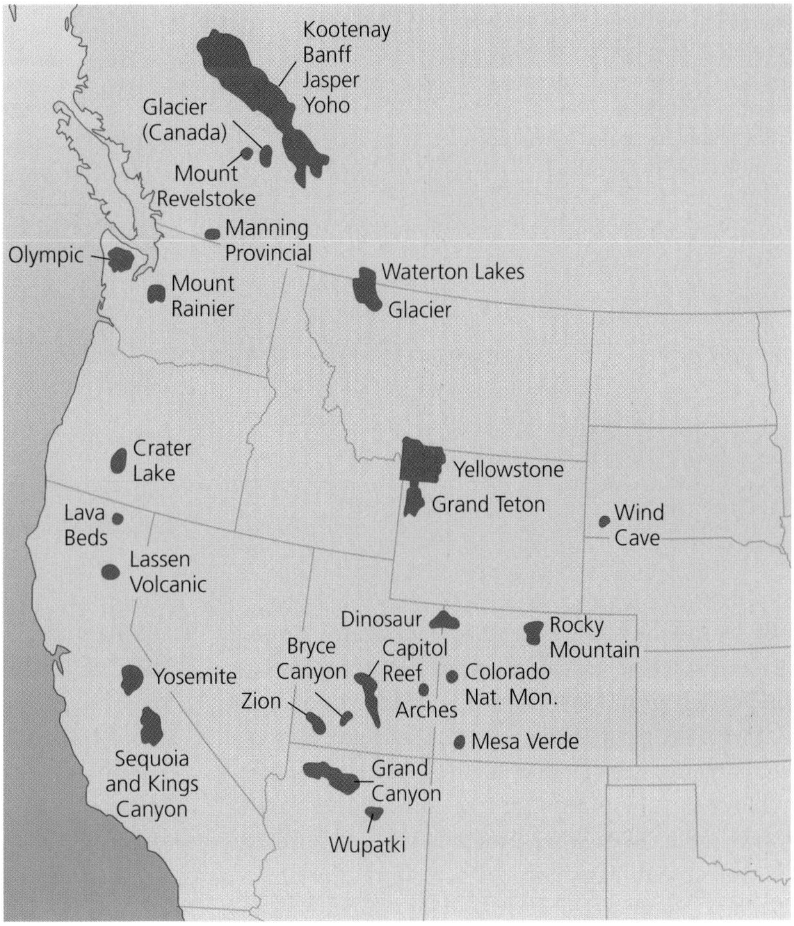

William Newmark visualized national parks of western North America as islands of natural habitat in a sea of development. His data showed that mammal species were disappearing from these recently created terrestrial "islands," in accordance with island biogeography theory. Adapted from Quammen, D. 1997. *The song of the dodo.* New York: Simon & Schuster.

These species disappeared because the parks, Newmark concluded, were too small to sustain their populations in the long term. Moreover, the parks had become island habitats too isolated to be recolonized by new arrivals. Newmark's findings, published in the journal *Nature* in 1987, placed island biogeography theory squarely on the mainland. Today, this theory helps inform national park policy and biodiversity conservation plans around the world.

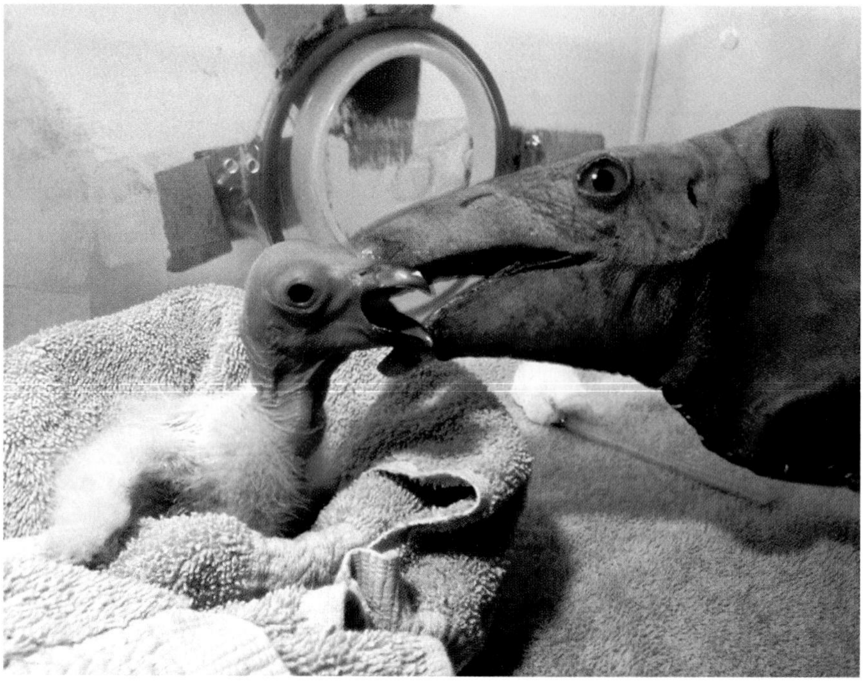

FIGURE 11.19 In efforts to save the California condor *(Gymnogyps californianus)* from extinction, biologists have raised hundreds of chicks in captivity with the help of hand puppets designed to look and feel like the heads of adult condors. Using these puppets, biologists feed the growing chicks in an enclosure and shield them from all contact with humans, so that when the chick is grown it does not feel an attachment to people.

Polls repeatedly show that most Americans support the idea of protecting endangered species. Yet some oppose and resent provisions of the ESA. Many opponents feel that the ESA places more value on the life of an endangered organism than it does on the livelihood of a person. This was a common perception in the Pacific Northwest in the 1990s, when protection for the northern spotted owl slowed logging in old-growth rainforest and caused many loggers to fear for their jobs. Resentment toward the ESA also comes from landowners worried that federal officials will restrict the use of private land if threatened or endangered species are found on it. This has led in many cases to a practice described as "Shoot, shovel, and shut up," among landowners who want to conceal the presence of such species on their land.

ESA supporters maintain that such fears are overblown, pointing out that the ESA has stopped few development projects. Moreover, a number of provisions of the ESA and its amendments promote cooperation with landowners. *Habitat conservation plans* and *safe harbor agreements* allow landowners to harm species in some ways if they voluntarily improve habitat for the species in others.

Debate over the U.S. law has influenced other nations' approaches to species protection. When Canada enacted its long-awaited endangered species law in 2002, the *Species at Risk Act (SARA),* the Canadian government was careful to stress cooperation with landowners and provincial governments, rather than presenting the law as a decree from the national government. Canada's environment minister, David Anderson, wanted none of the hostility the U.S. act had unleashed. Environmentalists and many scientists, however, protested that SARA was too weak and failed to protect habitat adequately. Today a number of nations have laws protecting species. In Russia, the government issued Decree 795 in 1995, creating a Siberian tiger conservation program and declaring the tiger one of the nation's most important natural and national treasures. Time will tell how effective this decree will be.

Captive breeding, reintroduction, and cloning are single-species approaches

In the effort to save threatened and endangered species, biologists are going to impressive lengths. Zoos and botanical gardens have become centers for the **captive breeding** of these species, so that individuals can be raised and reintroduced into the wild. One example is the program to save the California condor, North America's largest bird (Figure 11.19). Condors were persecuted in the early 20th century, collided with electrical wires, and succumbed to lead poisoning from scavenging carcasses of animals killed with lead shot. By 1982, only 22 condors remained, and biologists decided to take all the birds into captivity, in hopes of boosting their numbers and then releasing them. The ongoing program is succeeding. So far, over 100 of the 250 birds raised in captivity have been released into the wild at sites in California and Arizona, where a few pairs have begun nesting.

Other reintroduction programs have been more controversial. Reintroducing wolves to Yellowstone National Park has proven popular with the public, but reintroducing them to sites in Arizona and New Mexico has met stiff resistance from ranchers who fear the wolves will attack their livestock. The program is making slow headway; several of the wolves have been shot.

The newest idea for saving species from extinction is to create more individuals by cloning them. In this technique, DNA from an endangered species is inserted into a cultured egg without a nucleus, and the egg is implanted into a closely related species that can act as a surrogate mother. So far two Eurasian mammals have been cloned in this way. With future genetic technology, some scientists even talk of recreating extinct species from DNA recovered from preserved body parts. However, even if cloning can succeed from a technical standpoint, most biologists agree that such efforts are not an adequate response to biodiversity loss. Without ample habitat and protection in the wild, having cloned animals in a zoo does little good.

Some species act as "umbrellas" for protecting habitat and communities

Protecting habitat and conserving communities, ecosystems, and landscapes are the goals of many conservation biologists. Often, particular species are essentially used as tools to conserve communities and ecosystems. This is because the ESA provides legal justification and resources for species conservation, but no such law exists for communities or ecosystems. Large species that roam great distances, such as the Siberian tiger, require large areas of habitat. Meeting the habitat needs of these so-called *umbrella species* automatically helps meet those of thousands of less charismatic animals, plants, and fungi that would never elicit as much public interest.

Environmental advocacy organizations have found that using large and charismatic vertebrates as spearheads for biodiversity conservation has been an effective strategy. This approach of promoting particular *flagship species* is evident in the longtime symbol of the World Wide Fund for Nature (World Wildlife Fund in North America), the panda. The panda is a large endangered animal requiring sizeable stands of undisturbed bamboo forest. Its lovable appearance has made it a favorite with the public—and an effective tool for soliciting funding for conservation efforts that protect far more than just the panda. At the same time, many conservation organizations today are moving beyond the single-species approach. The Nature Conservancy, for instance, has in recent years focused more on whole communities and landscapes. The most ambitious effort may be the Wildlands Project, a group proposing to restore huge amounts of North America's land to its presettlement state.

Weighing the Issues: Single-Species Conservation?

What would you say are some advantages of focusing on conserving single species, versus trying to conserve broader communities, ecosystems, or landscapes? What might be some of the disadvantages? Which do you think is the better approach, or should we use both?

International conservation efforts include widely signed treaties

At the international level, biodiversity protection has been pursued in a variety of ways. Most effective so far have been several treaties facilitated by the United Nations. The 1973 **Convention on International Trade in Endangered Species of Wild Fauna and Flora (CITES)** protects endangered species by banning the international transport of their body parts. When nations enforce it, CITES can protect the tiger and other rare species whose body parts are traded internationally.

In 1992, leaders of many nations agreed to the **Convention on Biological Diversity.** This treaty embodies three goals: to conserve biodiversity, to use biodiversity in a sustainable manner, and to ensure the fair distribution of biodiversity's benefits. The Convention addresses a number of topics, including the following:

- Providing incentives for biodiversity conservation
- Managing access to and use of genetic resources
- Transferring technology, including biotechnology
- Promoting scientific cooperation
- Assessing the effects of human actions on biodiversity
- Promoting biodiversity education and awareness
- Providing funding for critical activities
- Encouraging every nation to report regularly on their biodiversity conservation efforts

Since the treaty was proposed, UNEP has identified a number of accomplishments, such as ensuring that Ugandan people share in the economic benefits of wildlife preserves, increasing global markets for "shade-grown" coffee and other crops grown without removing forests, and replacing pesticide-intensive farming practices with sustainable ones in some rice-producing Asian nations. As of 2005, 188 nations had become parties to the

FIGURE 11.20 The golden lion tamarin *(Leontopithecus rosalia)*, a species endemic to Brazil's Atlantic rainforest, is one of the world's most endangered primates. Captive breeding programs have produced roughly 500 individuals in zoos, but the tamarin's habitat is fast disappearing.

Convention on Biological Diversity. Those choosing *not* to do so include Iraq, Somalia, the Vatican, and the United States. This decision is just one example of why the U.S. government is no longer widely regarded as a leader in biodiversity conservation efforts.

Biodiversity hotspots pinpoint areas of high diversity

One international approach oriented around geographic regions, rather than single species, has been the effort to map **biodiversity hotspots.** The concept of biodiversity hotspots was introduced in 1988 by British ecologist Norman Myers (▸p. 338) as a way to prioritize regions that are most important globally for biodiversity conservation. A hotspot is an area that supports an especially great number of species that are **endemic** to the area, that is, found nowhere else in the world (Figure 11.20). To qualify as a hotspot, a location must harbor at least 1,500 endemic plant species, or 0.5% of the world total. In addition, a hotspot must have already lost 70% of its habitat as a result of human impact and be in danger of losing more.

The nonprofit group Conservation International maintains a list of 34 biodiversity hotspots (Figure 11.21). The ecosystems of these areas together once covered 15.7% of the planet's land surface, but today, because of habitat loss, cover only 2.3%. This small amount of land is the exclusive home for 50% of the world's plant species and 42% of all terrestrial vertebrate species. The hotspot concept gives incentive to focus on these areas of endemism, where the greatest number of unique species can be protected with the least amount of effort.

Community-based conservation is increasingly popular

Taking a global perspective and prioritizing optimal locations to set aside as parks and reserves makes good sense. However, setting aside land for preservation affects the people that live in and near these areas. In past decades, many conservationists from developed nations, in their zeal to preserve ecosystems in other nations, too often neglected the needs of people in the areas they wanted to protect. Many developing nations came to view this international environmentalism as a kind of neocolonialism.

Today this has largely changed, and many conservation biologists actively engage local people in efforts to protect land and wildlife in their own backyards, in an approach sometimes called **community-based conservation.** Setting aside land for preservation deprives local people of access to natural resources, but it can also guarantee that these resources will not be used up or sold to foreign corporations and can instead be sustainably managed. Moreover, parks and reserves draw ecotourism (▸p. 143), which can support local economies.

In the small Central American country of Belize, conservation biologist Robert Horwich and his Wisconsin-based group Community Conservation, Inc., have helped start a number of community-based conservation projects. The Community Baboon Sanctuary consists of tracts of riparian forest that farmers have agreed to leave intact, to serve as homes and traveling corridors for the black howler monkey, a centerpiece of ecotourism. The fact that the reserve uses the local nickname for the monkey signals respect for residents, and today a local women's cooperative is running the project. A museum was built, and residents receive income for guiding and housing visiting researchers and tourists. Some other projects have not turned out as well, however. Nearby on the Belizean coast, efforts to create a locally run reserve for the manatee have

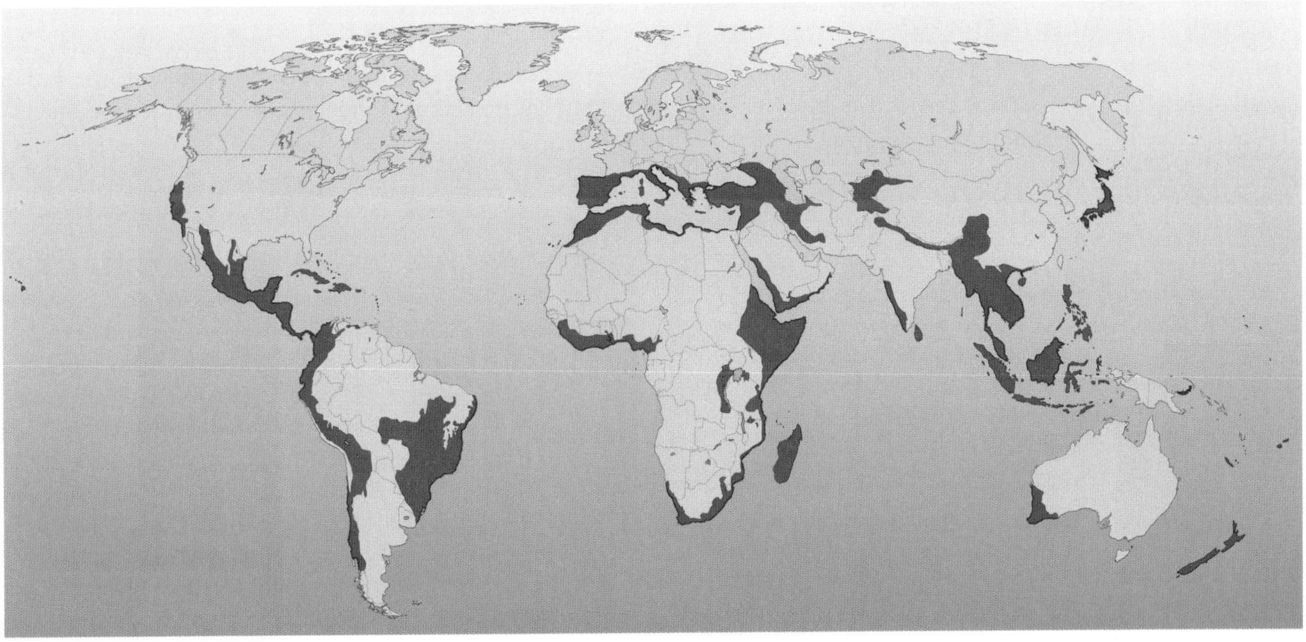

FIGURE 11.21 Some areas of the world possess exceptionally high numbers of species found nowhere else. Some conservation biologists have suggested prioritizing habitat preservation in these areas, dubbed *biodiversity hotspots*. Shown in red are the 34 biodiversity hotspots mapped by Conservation International in 2005. Data from Conservation International, 2005.

struggled because of shortfalls in funding. Community-based conservation has not always been successful, but in a world of increasing human population, locally based management that meets people's needs sustainably will likely be essential.

Innovative economic strategies are being employed

As conservation moves from single-species approaches to the hotspot approach to community-based conservation, innovative economic strategies are also being attempted. One strategy is the *debt-for-nature swap*. In such a swap, a conservation organization raises money and offers to pay off a portion of a developing country's international debt in exchange for a promise by the country to set aside reserves, fund environmental education, and better manage protected areas.

A newer strategy that Conservation International has pioneered is the *conservation concession*. Nations often sell concessions to foreign multinational corporations, allowing them to extract resources from the nation's land. A nation can, for instance, earn money by selling to an international logging company the right to log its forests. Conservation International has stepped in and paid nations for concessions for conservation rather than resource extraction. The nation gets the money *and* keeps its natural resources intact. The South American country of Surinam, which still has extensive areas of pristine rainforest, entered into such an agreement and has virtually halted logging while pulling in $15 million. It remains to be seen how large a role such strategies will play in the future protection of biodiversity.

Conclusion

The erosion of biological diversity on our planet threatens to result in a mass extinction event equivalent to the mass extinctions of the geological past. Human-induced habitat alteration, invasive species, pollution, and overharvesting of biotic resources are the primary causes of biodiversity loss. This loss matters, because human society could not function without biodiversity's pragmatic benefits. As a result, conservation biologists are rising to the challenge of conducting science aimed at saving endangered species, preserving their habitats, restoring populations, and keeping natural ecosystems intact. The innovative strategies of these scientists hold promise to slow the erosion of biodiversity that threatens life on Earth.

VIEWPOINTS

Biodiversity

Biodiversity is being lost worldwide at an accelerating pace. **How should we respond to biodiversity loss? What solutions should we seek, and what strategies should we prioritize?**

Mainstreaming Conservation through Ecosystem Services

Parks and nature reserves are the jewels of conservation, but they are not, and never will be, enough. Even the most ambitious conservation plans aspire to having no more than 20–30% of the world's lands protected as nature reserves, and the reality is that closer to 5–10% of land is currently under protection. If we put a fence around our parks to keep humans out, our nature preserves will still end up deteriorating because of how we treat the remaining 70–90% of Earth. Nature cannot be sequestered in some small portion of the globe while the bulk of land and water are left at the mercy of humans. Conservation will never be attainable unless humans learn to live in, work in, exploit, and harvest nature in a manner that does not ravage biodiversity.

An additional 3 billion people are expected in the next 50 years, bringing the total population to 9 billion. Many of the most impoverished people in the world depend on nature for their livelihood. All of us depend on nature for flood control, protection against storm damage, and other regulating services. To protect biodiversity and meet human needs, we must align biodiversity protection with these ecosystem services. As we make clear the links between ecosystem services and biodiversity protection, institutions and business should increasingly be willing to pay for nature's protection.

The disastrous tsunami of 2004 did minimal damage in coastal areas sheltered by natural mangrove forests. If only more of the Asia-Pacific region had been willing to invest in the insurance provided by mangroves, we would have saved human lives and biodiversity. The solution to biodiversity loss lies in clearly recognizing and valuing ecosystem services and in setting priorities that focus simultaneously on biodiversity and ecosystem services.

Peter Kareiva is lead scientist for The Nature Conservancy and an adjunct professor at the University of California at Santa Barbara and at Santa Clara University. Dr. Kareiva has also served as director of the Division for Conservation Biology at NOAA Fisheries, at the Northwest Fisheries Science Center, Seattle. His research has concerned organisms as diverse as whales, owls, Antarctic seabirds, ladybug beetles, butterflies, wildebeest, and genetically engineered microbes and crops.

Parks: The Best Way to Protect Biodiversity?

Although protected areas cover 12% of Earth's land surface, we need many more of them. Consider 34 "biodiversity hotspots" containing the last habitats of 50% of Earth's vascular plant species and 42% of vertebrate species (excluding fish). Once covering 16% of Earth's land surface, their habitats have since lost 86% of their expanse and now cover just 2.3%. If we could safeguard these relatively small areas, we could reduce the number of extinct species by at least one-third. Protection of the hotspots (parks, reserves, and so on) would cost roughly $1.5 billion per year for 5 years—just one-seventh of all conservation funding worldwide. What a massive need and massive opportunity for protected areas!

In the tropics, however, where most biodiversity is found and where it is most threatened, no island is an island, so to speak. The Kruger Park in South Africa is drying out because rivers arising outside the park are losing their waters to ranches and other development works. The park is also being overtaken by acid rain from South Africa's main industrial region upwind.

Moreover, global warming will shift temperature bands, and consequently vegetation communities, away from the equator and toward the poles. Many plants and animals in Hawaii will have little place to go but into the sea, as will those in the southern tip of Africa, northern Philippines, and dozens of other unfortunate locales. Even if they were to be turned into giant parks and perfectly protected on the ground, they would still be vulnerable to global warming—half of which is caused by carbon dioxide emissions from fossil fuels.

Biodiversity enthusiasts, here's a key question for you: Which country has less than one-twentieth of the world's population but causes one-quarter of the world's carbon dioxide emissions?

Norman Myers is a Fellow of Oxford University and a member of the U.S. National Academy of Sciences. He works as an independent scientist, advising the United Nations, the World Bank, the World Conservation Union, the World Wildlife Fund, and dozens of other conservation bodies around the world.

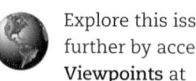

Explore this issue further by accessing **Viewpoints** at www.aw-bc.com/withgott.

REVIEWING OBJECTIVES

You should now be able to:

Characterize the scope of biodiversity on Earth

- Biodiversity can be thought of at three levels, commonly called species diversity, genetic diversity, and ecosystem diversity. (pp. 310–312)
- Roughly 1.7–2.0 million species have been described so far, but scientists agree that the world holds millions more. (pp. 313–314)
- Some taxonomic groups (such as insects) hold far more diversity than others. (pp. 313–314)
- Diversity is unevenly spread across different habitats and areas of the world. (pp. 314–316)

Describe ways to measure biodiversity

- Global estimates of biodiversity are based on extrapolations from scientific assessments in local areas and certain taxonomic groups. (pp. 313–314)

Contrast background extinction rates and periods of mass extinction

- Species have gone extinct at a background rate of roughly one species per 1 to 10 million species each year. Most species that have ever lived are now extinct. (p. 316)
- Earth's life has experienced five mass extinction events in the past 440 million years. (pp. 316–317)
- Human impact is presently causing the beginnings of a sixth mass extinction. (pp. 317–319)

Evaluate the primary causes of biodiversity loss

- Habitat alteration is the main cause of current biodiversity loss. Invasive species, pollution, and overharvesting are also important causes. Climate change threatens to become a major cause very soon. (pp. 319–324)

Specify the benefits of biodiversity

- Biodiversity is vital for functioning ecosystems and the services they provide us. (p. 325)
- Wild species are sources of food, medicine, and economic development. (pp. 325–327)
- Many people feel humans have a psychological need to connect with the natural world. (pp. 327–328)

Assess conservation biology and its practice

- Conservation biology is an applied science that studies biodiversity loss and seeks ways to protect and restore biodiversity at all its levels. (p. 329)

Explain island biogeography theory and its application to conservation biology

- Island biogeography theory explains how size and distance influence the number of species occurring on islands. (pp. 329–330)
- The theory applies to terrestrial islands of habitat in fragmented landscapes. (pp. 330–331)

Compare and contrast traditional and more innovative biodiversity conservation efforts

- Most conservation efforts and laws so far have focused on threatened and endangered species. Efforts include captive breeding and reintroduction programs. (pp. 331, 334–335)
- Species that are charismatic and well known are often used as tools to conserve habitats and ecosystems. Increasingly, landscape-level conservation is being pursued in its own right. (p. 335)
- International conservation approaches include treaties, biodiversity hotspots, community-based conservation, debt-for-nature swaps, and conservation concessions. (pp. 335–337)

TESTING YOUR COMPREHENSION

1. What is biodiversity? List and describe three levels of biodiversity.
2. What are the five primary causes of biodiversity loss? Can you give a specific example of each?
3. List and describe five invasive species and the adverse effects they have had.
4. Define the term *ecosystem services.* Give five examples of ecosystem services that humans would have a hard time replacing if their natural sources were eliminated.
5. What is the relationship between biodiversity and food security? Between biodiversity and pharmaceuticals? Give three examples of potential benefits of biodiversity conservation for food security and medicine.
6. Describe four reasons why people suggest biodiversity conservation is important.
7. What is the difference between an umbrella species and a keystone species? Could one species be both an umbrella species and a keystone species?
8. Explain the theory of island biogeography. Use the example of the Siberian tiger to describe how this theory can be applied to fragmented terrestrial landscapes.

9. Name two successful accomplishments of the U.S. Endangered Species Act. Now name two reasons some people have criticized it.
10. What is a biodiversity hotspot? Describe community-based conservation.

SEEKING SOLUTIONS

1. In one of the quotes that open this chapter, biologist E. O. Wilson argues that "except in pockets of ignorance and malice, there is no longer an ideological war between conservationists and developers. Both share the perception that health and prosperity decline in a deteriorating environment." Do you agree or disagree? How do people in your community view biodiversity?
2. Many arguments have been advanced for the importance of preserving biodiversity. Which argument do you think is most compelling, and why? Which argument do you think is least compelling, and why?
3. Some people argue that we shouldn't worry about endangered species because extinction has always occurred. How would you respond to this view?
4. Imagine that you are an influential legislator in a country that has no endangered species act and that you want to introduce legislation to protect your country's vanishing biodiversity. Consider the U.S. Endangered Species Act and the Canadian Species At Risk Act, as well as international efforts such as CITES and the Convention on Biological Diversity. What strategies would you write into your legislation? How would your law be similar to and different from the U.S., Canadian, and international efforts?
5. Environmental advocates from developed nations who want to preserve biodiversity globally have long argued for setting aside land in biodiversity-rich regions of developing nations. Many leaders of developing nations have responded by accusing the advocates of neocolonialism. "Your nations attained their prosperity and power by overexploiting their environments decades or centuries ago," these leaders asked, "so why should we now sacrifice our own development by setting aside our land and resources?" What would you say to these leaders of developing countries? What would you say to the environmental advocates? Do you see ways that both preservation and development goals might be reached?
6. Compare the biodiversity hotspot approach to the approach of community-based conservation. What are the advantages and disadvantages of each? Can we—and should we—follow both approaches?

INTERPRETING GRAPHS AND DATA

Habitat alteration is the primary cause of present-day biodiversity loss. Of all human activities, the one that has resulted in the most habitat alteration is agriculture. Between 1850 and 2000, 95% of the native grasslands of the Midwestern United States were converted to agricultural use. As a result, conventional farming practices replaced diverse natural communities with greatly simplified ones. The vast monocultures of industrialized agriculture produce bountiful harvests, but at substantial costs in lost ecosystem services.

Data from a recent study reviewing the scientific literature on the effects of organic farming versus conventional farming practices on biodiversity are shown in the graph.

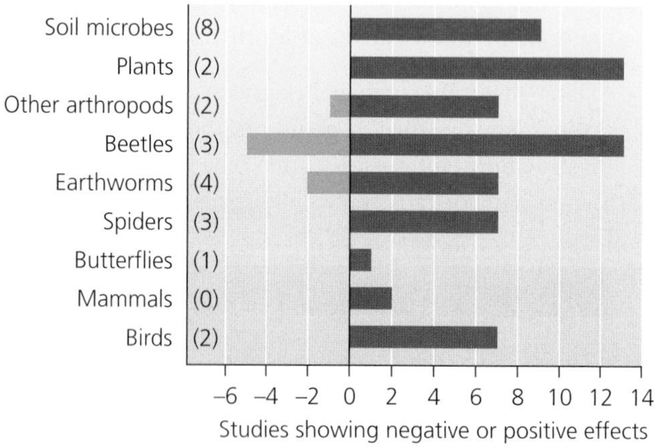

Numbers of scientific studies reporting negative or positive effects on biodiversity of organic agriculture versus conventional farming practices. In parentheses are numbers of studies reporting no effect. Data from Hole, D., et al. 2005. Does organic farming benefit biodiversity? *Biological Conservation* 122: 113–130.

1. How many studies showed a positive effect of organic farming on biodiversity, relative to conventional farming? How many studies reported a negative effect? How many studies reported no effect?
2. For which group or groups of organisms is evidence of positive effects the strongest? Reference the numbers to support your choice(s).
3. Recall the ecosystem services provided by biodiversity (▸p. 325). What services do the groups you chose in Question 2 provide?

CALCULATING ECOLOGICAL FOOTPRINTS

Of the five major causes of biodiversity loss discussed in this chapter, habitat alteration arguably has the greatest impact. In their 1996 book introducing the ecological footprint concept, authors Mathis Wackernagel and William Rees present a consumption/land-use matrix for an average North American. Each cell in the matrix lists the number of hectares of land of that type required to provide for the different categories of a person's consumption (food, housing, transportation, consumer goods, and services). Of the 4.27 hectares required to support this average person, 0.59 hectares are forest, with most (0.40 hectares) being used to meet the housing demand. Using this information, calculate the missing values in the table.

	Hectares of forest used for housing	Total forest hectares used
You	0.40	0.59
Your class		
Your state		
United States		

Data from Wackernagel, M., and W. Rees. 1996. *Our ecological footprint: reducing human impact on the earth.* British Columbia, Canada: New Society Publishers.

1. Approximately two-thirds of the forests' productivity is consumed for housing. To what use(s) would you speculate that most of the other third is put?
2. If the harvesting of forest products exceeds the sustainable harvest rate, what will be the likely consequence for the forest?
3. What will be the impact of deforestation, or of the loss of old-growth forests and their replacement with plantations of young trees, on the species diversity of the forest community? In your answer, discuss the possibilities of both extirpation and extinction.

Take It Further

Go to www.aw-bc.com/withgott or the student CD-ROM where you'll find:

- Suggested answers to end-of-chapter questions
- Quizzes, animations, and flashcards to help you study
- *Research Navigator*™ database of credible and reliable sources to assist you with your research projects
- GRAPHIt! tutorials to help you master how to interpret graphs
- INVESTIGATEit! current news articles that link the topics that you study to case studies from your region to around the world

12 Resource Management, Forestry, Land Use, and Protected Areas

Clearcuts on timber industry land adjacent to Gifford Pinchot National Forest, Washington

Upon successfully completing this chapter, you will be able to:

- Identify the principles, goals, and approaches of resource management
- Summarize the ecological roles and economic contributions of forests, and outline the history and scale of forest loss.
- Explain the fundamentals of forest management, and describe the major methods of harvesting timber
- Analyze the scale and impacts of agricultural land use
- Identify major federal land management agencies and the lands they manage
- Recognize types of parks and reserves, and evaluate issues involved in their design

Anti-logging protest at Clayoquot Sound

Central Case: Battling Over the Last Big Trees at Clayoquot Sound

> **"What we have is nothing less than an ecological Holocaust occurring right now in British Columbia."**
> —Mark Wareing, Western Canada Wilderness Committee, 1990

> **"Clear-cutting . . . may be either desirable or undesirable, acceptable or unacceptable, according to the type of forest and the management objectives for the forest."**
> —Dr. Hamish Kimmins, University of British Columbia, 1992

It was the largest act of civil disobedience in Canadian history, and it played out along a seacoast of stunning majestic beauty, at the foot of some of the world's biggest trees. Protestors blocked logging trucks, preventing them from entering stands of ancient temperate rainforest, the forest that had once carpeted most of Vancouver Island, British Columbia.

The activists were opposing **clear-cutting,** the logging practice that removes all trees from an area. Most of Canada's old-growth temperate rainforest had already been cut, and the forests of Clayoquot Sound on the western coast of Vancouver Island were among the largest undisturbed stands of temperate rainforest left on the planet.

The protestors chanted slogans, sang songs, and chained themselves to trees. The loggers complained that the protestors were keeping them from doing their jobs and making a living. In the end, 850 of the 12,000 protestors were arrested in full view of the global media, and this remote, mist-enshrouded land of cedars and hemlocks became ground zero in the global debate over how we manage forests.

That was in 1993. Timber from old-growth forests had powered British Columbia's economy since the province's early days. Historically, one in five jobs in British Columbia depended on its \$13 billion timber industry, and many small towns would have gone under without it. By 1993, however, the timber industry was cutting thousands of jobs a year because of mechanization, and the looming depletion of old growth threatened to slow the industry.

Meanwhile, half a world away in Great Britain, the environmental group Greenpeace was busy trying to convince British companies and customers to boycott forest products made from British Columbia trees that had been clear-cut by multinational timber company MacMillan Bloedel. Soon two British corporations cancelled contracts worth several million dollars with the beleaguered company, and British Columbia's premier found himself touring European nations trying to persuade them not to boycott his province's main export.

Then in 1995, the provincial government called for an end to clear-cutting at Clayoquot Sound, after its appointed scientific panel of experts submitted a new forestry plan for the region. The plan recommended reducing harvests, retaining 15–70% of old-growth trees in each stand, decreasing the logging road network, designating forest reserve areas, and managing riparian zones. Two years later the provincial government reversed many of these regulations on logging and a new premier pronounced forest activists "enemies of British Columbia," but the progressive management objectives for Clayoquot Sound have largely survived to the present day.

The antagonists called a truce and struck a deal; wilderness advocates and MacMillan Bloedel agreed to logging of old growth in limited areas, using more environmentally friendly practices. In 1998, Native people of the region formed a timber company, Iisaak, in agreement with MacMillan Bloedel's successor, Weyerhaeuser, and began logging in a more environmentally sensitive manner.

Meanwhile, leaving most of the trees standing had accomplished just what forest advocates had predicted: People from all over the world—1 million each year—were now visiting Clayoquot Sound for its natural beauty, and kayaking and whale-watching in its waters. Ecotourism (along with fishing and aquaculture) surpassed logging as the driver of local economies. The United Nations designated the site as an international biosphere reserve, encouraging land protection and sustainable development. The trees appeared to be worth more left standing than cut down.

Tensions continue today, however. Another timber company, Interfor, is harvesting areas near park and biosphere reserve boundaries. Local forest advocates worry that the provincial government's new Working Forest Policy will increase logging in the region. The town of Tofino has petitioned the province to exempt Clayoquot Sound's forests from the new policy so that pristine valleys can be preserved and the town's ecotourism economy can be maintained. As long as our demand for lumber, paper, and forest products keeps increasing, pressures will keep building on the remaining forests on Vancouver Island and around the world.

Resource Management

Debates over forest resources epitomize the broader questions of how to manage natural resources in general. We need to manage the resources we take from the natural world because many of them are limited. In Chapter 1, we saw how some resources, such as fossil fuels, are nonrenewable on human time scales, whereas other resources, such as the sun's energy, are perpetually renewable. Between these extremes lie resources such as timber, which are renewable if they are not exploited too rapidly or carelessly. **Resource management** is the practice of harvesting potentially renewable resources in ways that do not deplete them. Resource managers are guided in their decision making by available research in the natural sciences, as well as by political, economic, and social factors.

A key question in managing resources is whether to focus narrowly on the resource of interest or to look more broadly at the environmental system of which the resource is a part. Taking a broader view can often help avoid damaging the system and can thereby help sustain the availability of the resource in the long term.

Several natural resources are vital to us

Besides timber, several other types of natural resources are vital to our civilization. These include soils, freshwater, wildlife and fisheries, rangeland, and minerals (Figure 12.1). All natural resources also serve functions in the ecosystems of which they are a part.

Soils Soil resources, particularly topsoil, are of direct importance to us because they support the plants we grow for food and fiber and thus play a central role in agriculture. As we saw in Chapter 9 (▸ pp. 261–264), certain farm practices, such as terracing and use of windbreaks, can help guard against loss of topsoil to erosion, and other farm practices, such as planting nitrogen-fixing crops, can help maintain soil quality. Healthy soils also support forests and other natural communities, serving as a site for decomposition and a reservoir for nutrients.

Freshwater Each of us depends directly on freshwater, so ensuring a dependable supply of drinking water is a life-or-death issue. Freshwater also is necessary for agriculture; indeed, we use most freshwater not for drinking but for irrigating crops. In addition, waterways and wetlands are crucial for wildlife and properly functioning ecosystems, so those who manage water resources try to maintain supplies for all these reasons. Water managers also try to protect the

FIGURE 12.1 Our civilization depends on a number of natural resources, including timber, soils, freshwater, wildlife and fisheries, rangeland, and minerals. Minerals are a nonrenewable resource that is mined, but the others can be kept renewable through responsible resource management.

quality of these supplies by guarding against pollution. We will examine freshwater resources in depth in Chapter 15.

Wildlife and fisheries Humans have long hunted animals for food. Managers regulate the hunting of game animals, such as deer and quail, for food and sport in an effort to maintain populations of these animals at desired levels. As populations of many other terrestrial species decline from habitat loss and other causes (▸ pp. 319–324), management for nongame species is becoming increasingly important. Marine animals are also harvested, and despite extensive management of fisheries, many stocks of fish and shellfish have declined throughout the world's oceans from overharvesting, as we will see in Chapter 16. Besides their use as food, living organisms provide humans with many other benefits, from materials to medicines to aesthetic appreciation (▸ pp. 325–328). Organisms are also, of course, vital components of the ecosystems that sustain our world.

Rangeland Most of our food from animals today comes not from wild animals but from domesticated ones that are farmed. Most cattle in North America are raised in crowded feedlots, but they have traditionally been raised by grazing on open grassland. As we saw in Chapter 9 (▸ pp. 267–268, 270), grazing can be sustainable, but overgrazing results in damage to soils, waterways, and vegetative communities. Range managers are responsible for regulating ranching on public lands, and they advise ranchers on sustainable grazing practices on private lands.

Minerals Our civilization depends on numerous minerals, and until we achieve a closed-loop economy (▸ pp. 664–665), we will rely on the mining of minerals. Iron is mined and processed to make steel. Copper is used in pipes, electrical wires, and a variety of other applications. Aluminum is extracted via bauxite ore and used in packaging and other end products. Lead is used in batteries, to shield medical patients from radiation, and in many other ways. Zinc, tungsten, phosphate, uranium, gold, silver—the list goes on and on.

Although we rely on these mineral resources, we do not manage their extraction as we do with the aforementioned resources. Like fossil fuels, minerals are nonrenewable resources that are mined rather than harvested. Therefore, the mining industry has no built-in incentive to conserve. Instead, it benefits by extracting as much as it can as fast as it can and then, once extraction has become too inefficient to be profitable, moving on to new sites.

The process of mining also has historically degraded surrounding environments, sometimes on massive scales. Mining may directly remove vegetation from large areas, cause erosion, and produce acidic runoff that poisons area waterways. In addition, the smelting of metals following mining can create severe air pollution. We touch on some of the environmental and social costs of coal mining in Chapter 19 (▸ pp. 576–577), but these issues exist with all types of mining. Improved technology and more environmentally sensitive extraction procedures can help minimize environmental impacts. Such advances come as a result of public pressure, government legislation, or economic savings. Industries extracting potentially renewable resources, in contrast, have an added incentive to reduce ecological impacts: doing so may help sustain their ability to harvest resources—and this is where resource management comes in.

Managers have tried to achieve maximum sustainable yield

One guiding principle in resource management has been **maximum sustainable yield.** The aim is to achieve the maximum amount of resource extraction possible, without depleting the resource from one harvest to the next. At first this goal may sound ideal, but in reality it may sometimes harm the ecosystems from which the resource is derived. If those ecosystems cease to function effectively as a result, then in time this may decrease the availability of the resource.

For instance, recall the logistic growth curve (Figure 5.16, ▸ p. 137), which shows that a population grows most quickly when it is at an intermediate size. A fisheries manager aiming for maximum sustainable yield will therefore prefer to keep fish populations at intermediate levels so that they rebound quickly after each harvest. Doing so should

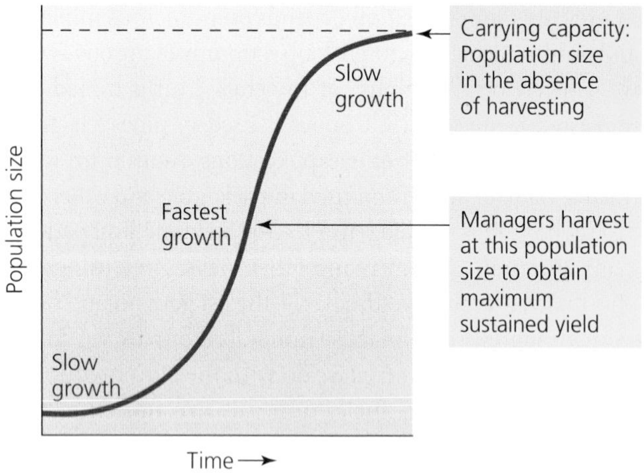

FIGURE 12.2 Using the concept of maximum sustainable yield, resource managers attempt to maximize the amount of resource harvested, as long as the harvest is sustainable in perpetuity. In the case of a wildlife population or fisheries stock that grows according to a logistic growth curve, managers aim to keep the population at an intermediate level, well below the carrying capacity, because populations grow fastest at intermediate sizes.

result in the greatest amount of fish harvested over time, while the population sustains itself (Figure 12.2). However, this management approach keeps the fish population at only about half its carrying capacity (▸ p. 136)—well below the level it would attain in the absence of fishing. Reducing one population in this way will likely have effects on other species and on the food web dynamics of the community. From an ecological point of view, management for maximum sustainable yield may thereby set in motion complex and significant ecological changes.

In forest management, maximum sustainable yield argues for cutting trees shortly after they have gone through their fastest stage of growth, and trees often grow most quickly at intermediate ages. Thus, trees may be cut long before they have grown as large as they would in the absence of harvesting. Although this practice may maximize timber production over time, it can cause drastic changes in the ecology of a forest by eliminating habitat for species that depend on mature trees.

Today many managers pursue ecosystem-based management

Because of these dilemmas, increasing numbers of managers today espouse ecosystem-based management. **Ecosystem-based management** attempts to manage the harvesting of resources in ways that minimize impact on the ecosystems and ecological processes that provide the resource. The plan proposed in 1995 by the Scientific Panel for Sustainable Forest Practices on Clayoquot Sound and approved by British Columbia's government was essentially a plan for ecosystem-based management. By carefully managing ecologically important areas such as riparian corridors, by considering patterns at the landscape level, and by affording protection to some forested areas, the plan aimed to allow continued timber harvesting at reduced levels while preserving the functional integrity of the ecosystem.

Although ecosystem-based management has gained a great deal of support in recent years, it is challenging for managers to determine how best to implement this type of management. Ecosystems are complex, and our understanding of how they operate is limited. Thus, ecosystem-based management has often come to mean different things to different people.

Adaptive management evolves and improves

Some management actions will succeed, and some will fail. A wise manager will try new approaches if old ones are not effective. In recent years, many resource managers have taken this a step further by implementing adaptive management. **Adaptive management** involves systematically testing different management approaches with the aim of improving methods as time goes on. It calls for changing practices midstream if necessary, as managers learn which work best. This approach represents a true fusion of science and management, because hypotheses about how best to manage resources are explicitly tested.

Adaptive management can be time-consuming and complicated. It entails monitoring the results of one's practices and continually adjusting them as needed, based on what is learned. It has posed a challenge for many managers, because those who adopt new approaches must often overcome inertia and resistance to change from proponents of established practices. Adaptive management is beginning to be used in forestry (see "The Science behind the Story," ▸ pp. 348–349). The management of timber and other forest resources is a clear and representative example of resource management.

Forest Management

Forests cover much of Earth's land surface, provide habitat for countless organisms, and play key roles in our planet's biogeochemical cycles (▸ pp. 195–206). Forests have also long provided humanity with wood for fuel, construction, paper production, and more. Foresters, those professionals who manage forests through the practice of **forestry**, today must balance the central importance

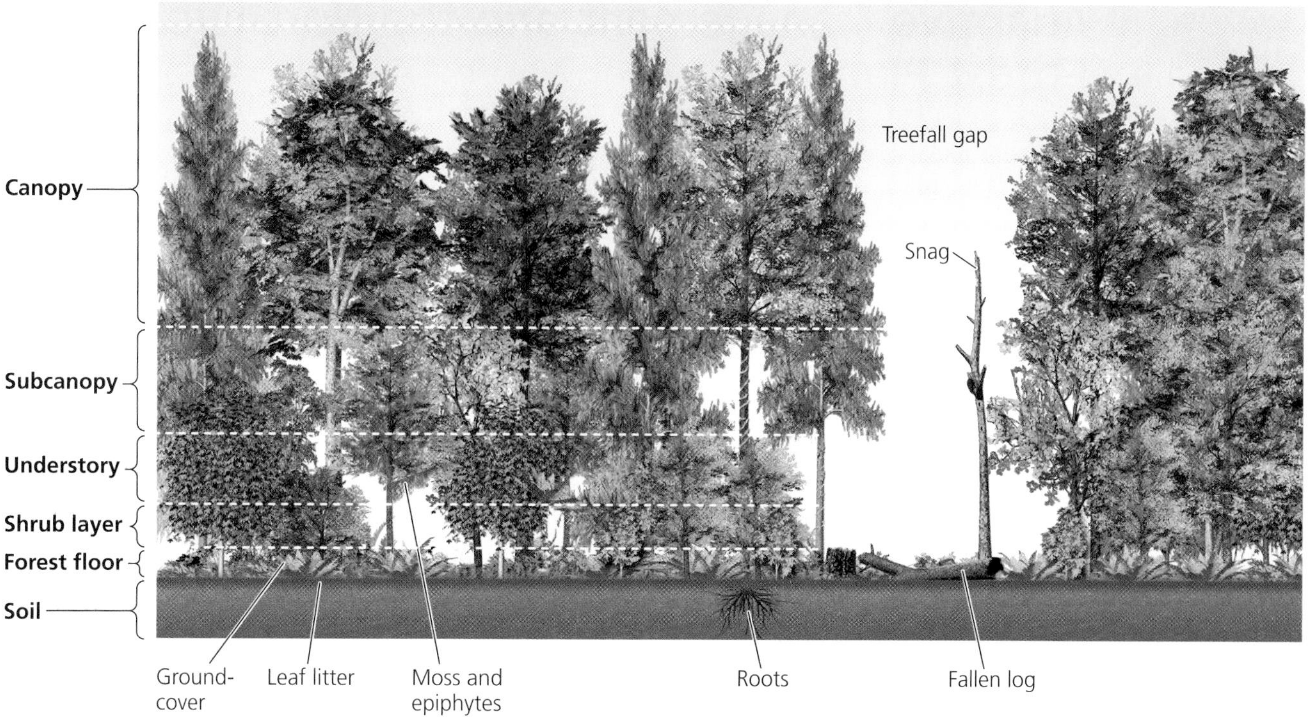

FIGURE 12.3 Mature forests are complex ecosystems. In this cross-section of a generalized mature forest, the crowns of the largest trees form the canopy, and smaller trees beneath them form the shaded subcanopy and understory. Shrubs and groundcover grow just above the forest floor, which may be covered in leaf litter rich with invertebrate animals. Vines, mosses, lichens, and epiphytes cover portions of trees and the forest floor. Snags (standing dead trees), whose wood can be easily hollowed out, provide food and nesting and roosting sites for woodpeckers and other animals. Fallen logs nourish the soil and young plants and provide habitat for countless invertebrates as the logs decompose. Treefall gaps caused by fallen trees let light through the canopy and create small openings in the forest, allowing early successional plants to grow in patches within the mature forest.

of forests as ecosystems with civilization's demand for wood products.

Forests are ecologically valuable

Most of the world's forests occur as boreal forest, the biome (▸ pp. 170, 176–177) that stretches across much of Canada, Russia, and Scandinavia, or as tropical rainforest, the biome that occurs in South and Central America, equatorial Africa, and Indonesia and southeast Asia. Temperate forests cover less area globally, in part because so many have already been cleared by people.

Because of their structural complexity and their ability to provide many niches for organisms, forests comprise some of the richest ecosystems for biodiversity (Figure 12.3). Trees furnish food and shelter for an immense diversity of vertebrate and invertebrate animals. Countless insects, birds, mammals, and other organisms subsist on the leaves, fruits, and seeds that trees produce. Some animals are adapted for living in dense treetop canopies. Here beetles, caterpillars, and other leaf-eating insects abound, providing food for birds such as tanagers and warblers, while arboreal mammals from squirrels to sloths to monkeys consume fruit and leaves. Other animals specialize on the subcanopies of trees, and still others utilize the bark, branches, and trunks. Cavities in trunks provide nest and shelter sites for a wide variety of vertebrates. Dead and dying trees are valuable for many species; these snags are decayed by insects that are eaten by woodpeckers and other animals.

Understory shrubs and groundcover plants give a forest structural complexity and provide habitat for still more organisms. Moreover, the leaves, stems, and roots of forest plants are colonized by an extensive array of fungi and microbes, in both parasitic and mutualistic relationships (▸ pp. 153–155). And much of a forest's diversity resides in the forest floor, where the soil is generally nourished by leaf litter. As we saw in Chapter 11, myriad

The Science behind the Story

Adaptive Management and the Northwest Forest Plan

While environmental advocates and the timber industry were battling one another at Clayoquot Sound, similar debates were occurring in the United States, particularly in the Pacific Northwest. Here loggers were closing in on some of the nation's last stands of old-growth conifers, and preservationists were staging protests to protect the stands. Environmentalists also were winning lawsuits to force the U.S. Forest Service and other agencies to enforce provisions to protect the northern spotted owl (and thus its habitat, old-growth forest) under the Endangered Species Act (► pp. 331–333). As a result, logging was being significantly restricted.

In 1993, the Clinton administration waded into the impasse and helped direct the development of a new plan to manage the forests of western Washington, Oregon, and northwestern California so that consensus could be reached and logging could continue with adequate protections for species and ecosystems. The resulting **Northwest Forest Plan,** approved in February 1994, used science extensively to guide management.

Interstate 90 at Snoqualmie Pass, Washington, is a barrier to many animal species traveling through the forested areas on either side of it.

Research conducted over the past decade since the plan's inception has fallen into several categories: (1) wildlife conservation and population viability, (2) aquatic conservation, (3) socioeconomic research, (4) ecological processes and function, (5) spatial and landscape studies, (6) tree growth studies, (7) adaptive management concepts, and (8) adaptive management areas.

The Northwest Forest Plan represented one of the first large-scale applications of adaptive management. Ten adaptive management areas (AMAs) were established throughout the region. These areas ranged from 37,000 to 200,000 ha (92,000 to 500,000 acres), and were located in regions where reduced timber harvests were affecting local economies.

Several studies were carried out at one AMA, Snoqualmie Pass in Washington's Cascade Mountains (see figure). Some of this research found that understanding plant associations at the landscape scale could help predict disturbance from fire, insects, and disease. Another study addressed habitat fragmentation and its effects on wildlife. Land ownership at Snoqualmie Pass in 1994 was like a checkerboard, with alternating blocks of land owned by various government agencies, timber companies, and individuals. Landscape ecologists thought the area might

soil organisms help decompose plant material and cycle nutrients.

In general, forests with a greater diversity of plants host a greater diversity of organisms overall. And, in general, fully mature forests, such as the undisturbed old-growth forests remaining at Clayoquot Sound, contain more biodiversity than younger forests, because older forests contain more structural diversity and thus more microhabitats and resources for more species.

The complex systems we call forests provide all manner of vital ecosystem services (► pp. 39–41). Forest vegetation acts to stabilize soil and prevent erosion. Forest plants also help regulate the hydrologic cycle, slowing runoff, lessening flooding, and purifying water as they take it in from the soil and release it to the atmosphere. Forest plants also store carbon, release oxygen, and moderate climate. By performing such ecological functions, forests are indispensable for our survival and help make our planet's environment what it is.

Forest products are economically valued

Forests also provide people with economically valuable wood products. For millennia, wood from forests has fueled our fires, keeping people warm and well fed. It has serve as a vital north–south corridor for wildlife to travel, so land was transacted to try to consolidate ownership to form corridors of publicly managed land. But an outstanding question was whether wildlife would be able to cross Interstate 90, which runs east–west over the pass.

Forest Service researchers led by Peter Singleton and John Lehmkuhl spent 2 years studying the movements of wildlife in relation to a 48-km (30-mi) stretch of I-90 to find out how much of a barrier the highway represented.

First, they used GIS methods (see the next "Science behind the Story" on ▸pp. 352–353) to predict how accessible areas were to animal species with different dispersal abilities and habitat preferences.

They also set up cameras that took pictures of animals that passed in front of heat and motion detectors. From data at 115 stations, they found no difference between sites near and far from the interstate. However, they did find that some animals, for example, bobcats, skunks, hares, and bears, appeared near certain portions of the highway more than other portions.

Third, the researchers mapped all road-killed animals discovered along 86 km (54 mi) of the interstate by Washington Department of Transportation employees. They found that the 450 deer and elk killed in the 2-year period were concentrated along four particular stretches of interstate.

Fourth, they surveyed the interstate in winter for tracks left in snow by animals crossing the highway. Twenty-three of 37 crossings by coyotes, bobcats, and raccoons took place in a single 2-mile stretch of road.

Fifth, they monitored underpasses and overpasses, as well as drainage culverts crossing beneath the interstate, to determine how animals were using these structures. Using cameras and plates of soot or sand that recorded footprints, the researchers detected 15 mammal species using culverts to cross the interstate, with deer mice the most frequent crossers. Different species used different types of culverts, but dry ones 0.4–1.1 m in diameter were regularly used.

With these data, Singleton and Lehmkuhl could pinpoint locations preferred by wildlife and could recommend that opportunities for safely crossing the interstate be provided at those locations. In particular, they wrote, providing dry drainage culverts near areas of mature forest could help promote the ecological functions that small mammals provide, such as dispersing plant seeds and fungal spores. They also urged that habitat be restored and maintained on the landscape level in patterns that facilitated animal movement.

Key for adaptive management is that data continue to be collected, to assess whether actions that are taken succeed or fail. In this respect, the AMAs under the Northwest Forest Plan have not yet lived up to their potential. Doing adaptive management right requires time, money, and a rare combination of flexibility and commitment. However, science and management under the Northwest Forest Plan is ongoing, and the Plan's proponents are committed to making it work.

housed people, keeping us sheltered. It built the ships that carried people and cultures from one continent to another. It allowed us to produce paper, the medium of the first information revolution.

In recent decades, industrial harvesting has allowed the extraction of more timber than ever before, supplying all these needs of a rapidly growing human population and its expanding economy. The exploitation of forest resources has been instrumental in helping our society achieve the standard of living we enjoy today. Indeed, without industrial timber harvesting, you would not be reading this book.

Most commercial logging today takes place in Canada, Russia, and other nations that hold large expanses of boreal forest, and in tropical countries with large amounts of rainforest, such as Brazil. In the United States, most logging takes place on land both private and public, primarily in the conifer forests of the West and the pine plantations of the South.

Demand for wood has led to deforestation

We all depend in some way on wood, from the subsistence herder in Nepal cutting trees for firewood to the

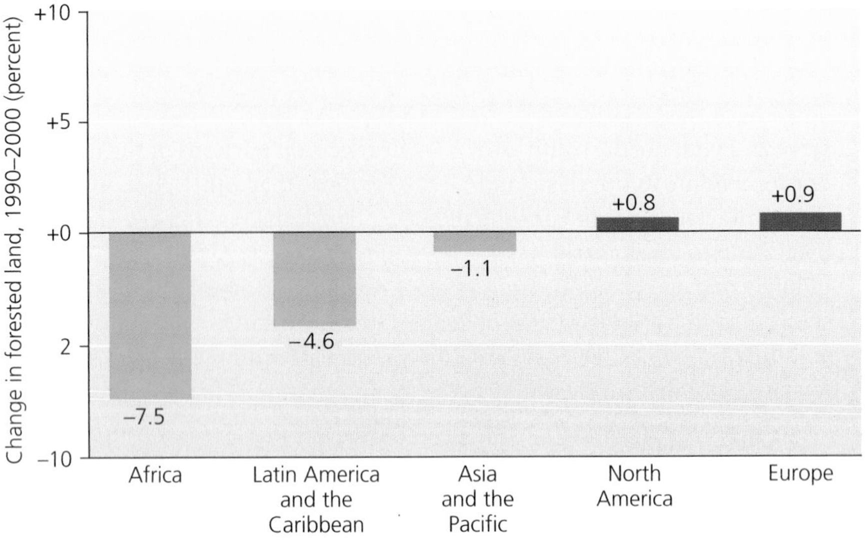

FIGURE 12.4 Nations in Africa and in Latin America are experiencing rapid deforestation as they attempt to develop, extract resources, and provide new agricultural land for their growing populations. In parts of Europe and North America, meanwhile, forested area is slowly increasing as some former farmed areas are abandoned and allowed to grow back into forest. Go to GRAPHIt! at www.aw-bc.com/withgott or on the student CD-ROM. Data from United Nations Environmental Programme. 2002. *Global environmental outlook 3*. London: Earthscan Publications.

American student consuming reams upon reams of paper in the course of getting a degree. For such reasons, people have cleared forests for millennia to exploit timber resources. Still more forest has been cleared to make way for agriculture.

Deforestation, the clearing and loss of forests, has altered the landscapes and ecosystems of much of the planet. It has caused soil degradation, population declines, and species extinctions, and, as we saw with Easter Island (▸ pp. 8–9), has in some cases helped bring whole civilizations to ruin. Continued deforestation threatens dire ecological and economic consequences. Impacts are greatest in tropical areas, because of the potentially massive loss of biodiversity, and in arid regions, because of the vulnerability to desertification (▸ pp. 259–260). In addition, deforestation adds carbon dioxide (CO_2) to the atmosphere: CO_2 is released when plant matter is burned or decomposed, and thereafter less vegetation remains to soak up CO_2. Deforestation is thereby one contributor to global climate change (▸ p. 530).

All continents have experienced deforestation, but forests are being felled at the fastest rates today in the tropical rainforests of Latin America and Africa (Figure 12.4). Developing countries in these regions are striving to expand areas of settlement for their burgeoning populations and to boost their economies by extracting natural resources and selling them abroad. Meanwhile, areas of Europe and eastern North America are slowly gaining forest cover as they recover from severe deforestation of past decades and centuries. In total, depending on one's definition, one-fifth to one-third of Earth's land area is currently covered by forest (see "The Science behind the Story," ▸ pp. 352–353).

The growth of the United States and Canada was fed by deforestation

Deforestation for timber and farmland propelled the growth of the United States and Canada throughout their phenomenal expansion westward across the North American continent over the past 400 years. The vast deciduous forests of the East were virtually stripped of their trees by the mid-19th century, making way for countless small farms. Timber from these forests built the cities of the Atlantic seaboard. Later, cities such as Chicago were constructed with timber felled in the vast pine and hardwood forests of Wisconsin and Michigan. As a farming economy shifted to an industrial one, wood was used to stoke the furnaces of industry. Logging operations moved south to the Ozarks of Missouri and Arkansas, while the pine woodlands of the South were logged and converted to pine plantations. Once most mature trees were removed from these areas, timber companies moved west, cutting the continent's biggest trees in the Rocky Mountains, the Sierra Nevada, the Cascade Mountains, and the Pacific Coast ranges (Figure 12.5).

By the early 20th century, very little virgin timber was left in the lower 48 U.S. states (Figure 12.6). Today, the largest oaks and maples found in eastern North America, and even most redwoods of the California coast, are merely **second-growth** trees, trees that have sprouted and grown to partial maturity after old-growth timber has been cut. The size of the gargantuan trees they replaced can be seen in the enormous stumps that remain in the more recently logged areas of the Pacific coast. The scarcity of old-growth trees on the North American continent

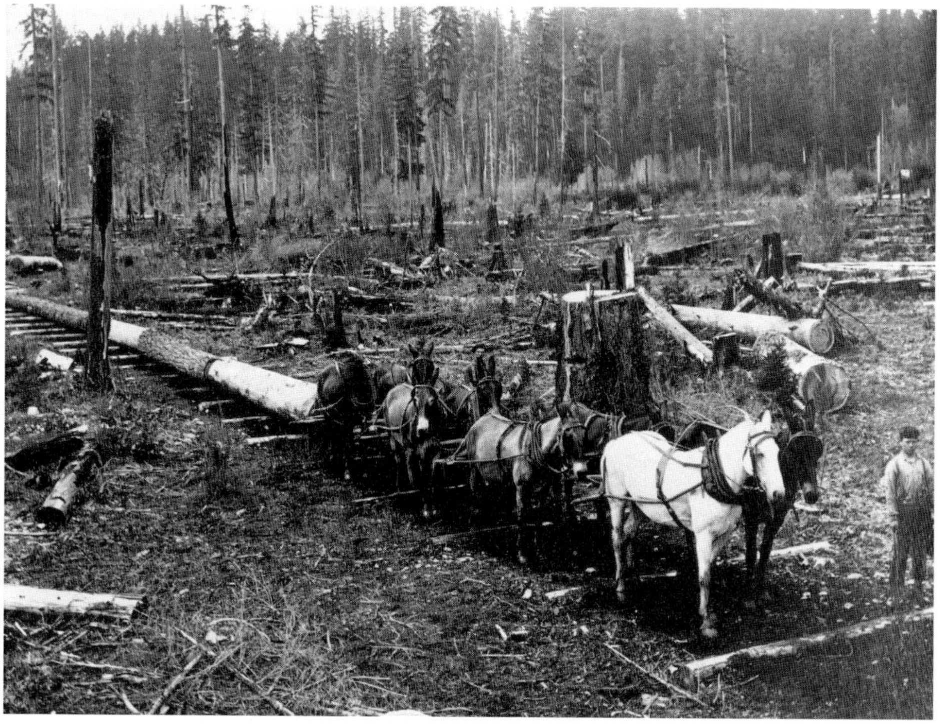

FIGURE 12.5 A mule team drags logs of Douglas fir from a clear-cut in Mason County, Washington, in 1901. Early timber harvesting practices in North America caused significant environmental impacts and removed virtually all the virgin timber from one region after another.

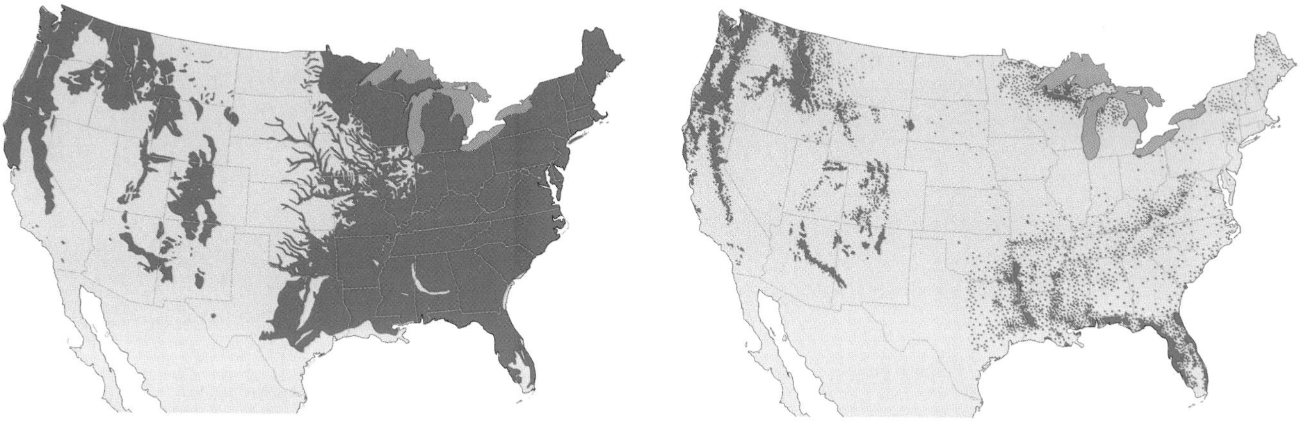

(a) Areas of natural forest, 1620

(b) Areas of natural forest, 1920

FIGURE 12.6 When Europeans first were colonizing North America, the entire eastern half of the continent and substantial portions of the western half were covered in forest (**a**). By the early 20th century, the vast majority of this forest had been cut, replaced mostly by agriculture and other human land uses (**b**). Since that time, most of the remaining original forest has been cut, but much of the landscape has also become reforested with second-growth forest. *Source:* Williams, M. 1989. *Americans and their forests.* Cambridge: Cambridge University Press. As adapted by Goudie, A., 2000. *The human impact.* Cambridge, MA: MIT Press.

today explains the passion with which environmental advocates have fought for the preservation of ancient forests in areas such as Clayoquot Sound.

The fortunes of loggers have risen and fallen with the availability of big trees. As each region was denuded of timber, the industry declined there and the timber companies moved on, while local loggers lost their jobs. If the remaining ancient trees of North America—most in British Columbia and Alaska—are cut, many U.S. and Canadian loggers will likely be out of jobs once again. Meanwhile, their employers will move on to nations of the developing world, as many already have.

The Science behind the Story

Using Geographic Information Systems to Survey Earth's Forests

One map is washed in splashes of green, showing broad stretches of largely undisturbed forest. Another outlines political boundaries. A third map marks human population centers, denoting their density with different colors. A fourth lays out parks, reserves, and other land protected from development. On its own, each map shows only one physical characteristic of Earth. But layered together through the use of a powerful technology, these maps reveal something entirely different: the condition of the world's last intact forests.

Used in a groundbreaking study published by the United Nations Environment Programme (UNEP) in 2001, these maps, in conjunction with a *geographic information system (GIS),* have become an essential tool for environmental planners and resource managers around the world.

A GIS consists of software that builds layered maps, using different sets of data to compile various views of landscapes (see figure). Environmental researchers most often use data on the status and uses of natural resources. In the case of the UNEP study, researchers wanted a comprehensive look at global forests—where they are, which countries govern them, which forests face the most pressure from encroaching human settlements, and which are most protected. The researchers especially wanted to know the status of "closed" forests: forests whose canopies cover more than 40% of their area. Conducting such a survey on the ground, going from forest to forest and country to country, would have been neither practical nor accurate. Synthesizing the findings from such research into a comprehensive paper map would have been almost impossible. With a GIS, however, researchers pull data such as satellite imagery and computerized maps into an all-inclusive view of the area they are studying.

Some of the researchers' most critical data are gathered via a sensor, an advanced very high resolution radiometer (AVHRR), orbiting the Earth by satellite. The rotating mirror, telescope, and internal electronics of the AVHRR measure wavelengths of energy rising from Earth's surface. Different types of substrate release different wavelengths, which the AVHRR translates into color images. Different colors in the satellite imagery thus denote whether an area is covered by water, rock, houses, or vegetation.

For more than a decade, scientists have taken advantage of AVHHR to produce the Normalized Difference Vegetation Index (NDVI). NDVI researchers survey forests and plants and gauge changes to vegetation. Using the NDVI's catalog of worldwide vegetation images, UNEP researchers loaded forest data into their GIS. They then added three other sets of electronic data into the GIS: the distribution of the world's population, global political boundaries, and land protected against development.

Once compiled, the data layers offered UNEP researchers an unprecedented global view of forests. About 20% of Earth's surface remains covered by closed forest, the study indicated. Moreover, the study confirmed previous findings that forests in densely populated countries, such as India and Indonesia, are under pressure from expanding human settlement. Such

Deforestation is proceeding rapidly in many developing nations

Uncut tropical forests still remain in many developing countries, and these nations are in the position the United States and Canada faced a century or two ago: having a vast frontier that they can develop for human use. Today's advanced technology, however, has allowed these countries to exploit their resources and push back their frontiers even faster than occurred in North America. As a result, deforestation is rapid in places such as Brazil and Indonesia.

Developing nations are often desperate enough for economic development that they impose few or no restrictions on logging. Often their timber is extracted by foreign multinational corporations, which have paid fees to the developing nation's government for a *concession,* or right to extract the resource. In such cases, the foreign corporation has little or no incentive to manage forest resources sustainably. Many of the short-term economic benefits are reaped not by local residents but by the corporations that log the timber and export it elsewhere.

countries require stronger and immediate conservation efforts, researchers determined. The team also found that more than 80% of the world's intact forests were concentrated in just 15 countries and that 88% of those forests were sparsely inhabited by people. These findings could prove critical in starting conservation efforts in regions at risk, to help prevent further forest degradation.

GIS is not foolproof; GIS maps are only as good as the data that go into them. Some of the UNEP data—those on population size and protected areas—may not be reliable, researchers warn, because the countries providing the information may not always keep accurate records. Some satellite images used in the UNEP study were ambiguous, as was the case with the maps indicating forest cover; the team couldn't be sure every tree farm or plantation had been removed from that set of data before loading into the GIS.

However, the UNEP scientists stressed that without GIS such a major forest assessment would never have been possible. With it, forest planners can focus conservation priorities on areas with the best prospects for continued existence.

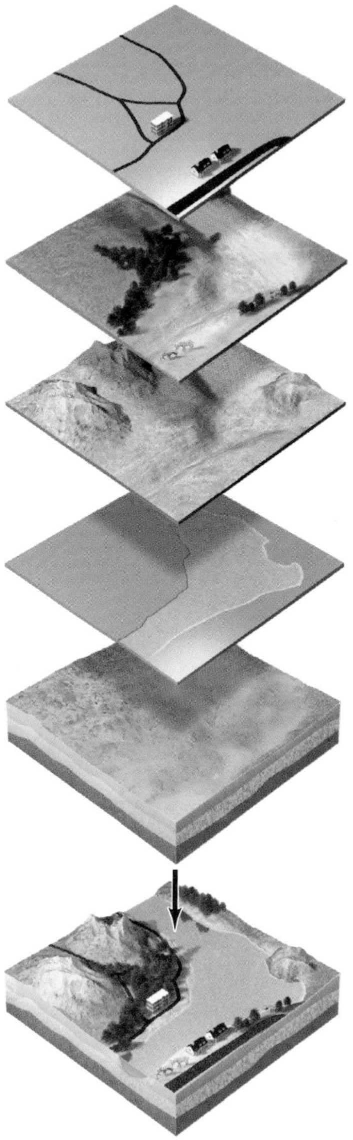

Geographic information systems (GIS) allow the layering of different types of data on natural landscape features and human land uses so as to produce maps integrating this information. GIS can then be used to explore informative correlations between these data sets.

Local people may or may not receive temporary employment from the corporation, but once the timber is harvested they no longer have the forest and the ecosystem services it once provided.

In Sarawak, the Malaysian portion of the island of Borneo, foreign corporations that were granted logging concessions have deforested several million hectares of tropical rainforest since 1963. The clearing of this forest—one of the world's richest, hosting such organisms as orangutans and the world's largest flower, *Rafflesia arnoldii*—has had direct impacts on the 22 tribes of people who live as hunter-gatherers in Sarawak's rainforest. The Malaysian government did not consult the tribes about the logging, which decreased the wild game on which these people depended. Oil palm agriculture was established afterward, leading to pesticide and fertilizer runoff that killed fish in local streams. The tribes protested peacefully and finally began blockading logging roads. The government, which at first jailed them, now is negotiating, but it insists on converting the tribes to a farming way of life.

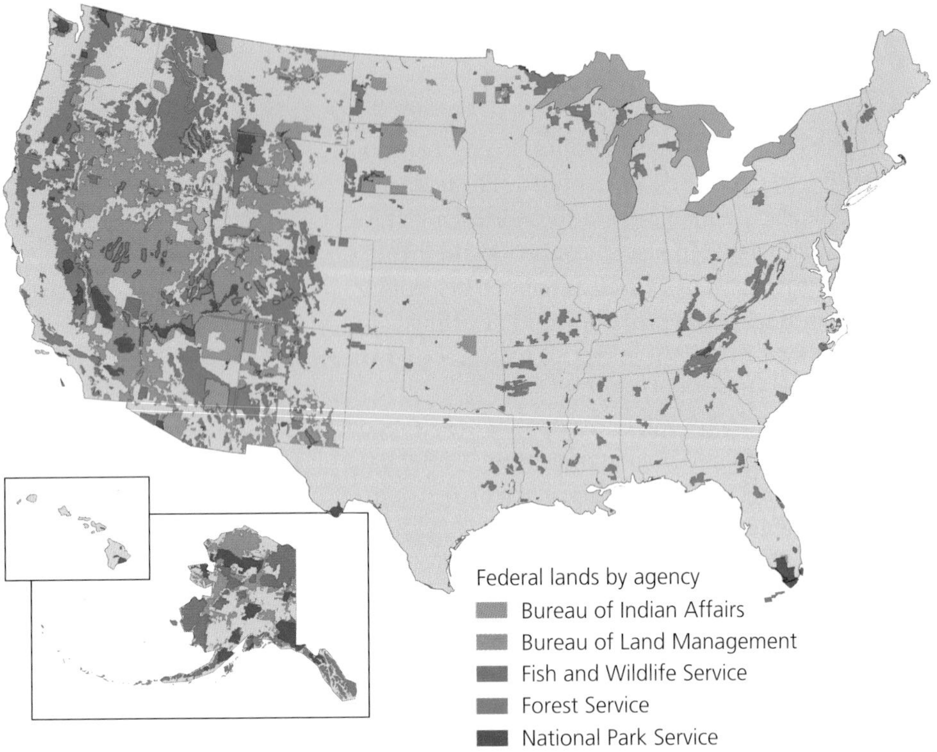

FIGURE 12.7 Federal agencies own and manage well over 250 million ha (600 million acres) of land in the United States, particularly in the western states. These include national forests, national parks, national wildlife refuges, Native American reservations, and Bureau of Land Management lands. Data from United States Geological Survey.

Weighing the Issues: Logging Here or There

Imagine you are an environmental activist protesting a logging operation that is cutting old-growth trees near your hometown. Now let's say you know that if the protest is successful, the company will move to a developing country and cut its virgin timber instead. Would you still protest the logging in your hometown? Would you pursue any other approaches?

Fear of a "timber famine" spurred establishment of national forests

In the United States, the depletion of the eastern forests and widespread fear of a "timber famine" spurred the formation of a system of forest reserves: public lands set aside to grow trees, produce timber, protect watersheds, and serve as insurance against future scarcities of lumber. Today the U.S. **national forest** system consists of 77 million ha (191 million acres) spread across all but a few states (Figure 12.7). The system, managed by the U.S. Forest Service, covers over 8% of the nation's land area.

The U.S. Forest Service was established in 1905 under the leadership of Gifford Pinchot (▸ pp. 34–35). Pinchot and others developed the concepts of resource management and conservation during the Progressive Era, a time of social reform when people urged that science and education be applied to public policy to improve society. In line with Pinchot's conservation ethic, the Forest Service aimed to manage the forests for "the greatest good of the greatest number in the long run." Pinchot believed the nation should extract and use resources from its public lands, so timber harvesting was, from the start, a goal behind establishing the national forests. But conservation meant planting trees as well as harvesting them, and the Forest Service intended to pursue wise management of timber resources.

Management goals are similar for the Canadian Forest Service and the provincial forestry ministries. Of Canada's 310 million ha (765 million acres) of forested land, 77% belongs to the provinces, and only 16% is federally owned and 7% privately owned. About one-third of the nation's forests are in British Columbia, and 38% are in Quebec and Ontario.

Timber is extracted from public and private lands

Timber is extracted from publicly held forests in the United States and Canada not by the governments of these nations, but by private timber companies. In the United States, Forest Service employees plan and manage timber sales and build roads to provide access for logging companies, which sell the timber they harvest for profit.

However, most timber harvesting in the United States these days takes place on private land, including land owned by timber companies. In 2001, the most recent year for which good data are available, timber companies extracted 31.7 million m^3 (340 million ft^3) of live timber from national forests. Although this is a large amount, it is considerably less than the amount cut from private lands and other public forests (Figure 12.8). Timber harvesting declined on national forests during the 1980s and 1990s, and in 2001 tree regrowth outpaced tree removal on these lands by nearly 12 to 1. Overall, timber production in the United States and other developed countries has remained roughly stable for the past 40 years. Meanwhile, it has more than doubled in developing countries.

The equivalent rates of growth and removal on private lands in Figure 12.8 reflect attempts by timber companies to manage their resources in accordance with the maximum sustainable yield approach, so that they can obtain maximal profits over many years for their owners and investors. On public lands, rates of growth and removal reflect not only economic forces but social and political ones as well, and these have changed over time. From the U.S. national forests, private timber extraction began to increase in the 1950s as the country experienced a postwar economic boom, consumption of paper products rose, and the population expanded into newly built suburban homes. In more recent decades, harvests from national forests decreased as economic trends shifted, public concern over clear-cutting grew, and forest management philosophy evolved.

Plantation forestry has grown

Today the North American timber industry focuses most on maximizing production from tree plantations in the Northwest and the South. These plantations feature stands of fast-growing tree species, and are single-species monocultures (▸ pp. 281–282). Because all trees in a given stand are planted at the same time, the stands are **even-aged,** with all trees the same age (Figure 12.9). Stands are cut after a certain number of years (called the *rotation time*), and the land is replanted with seedlings. Most ecologists and foresters view these plantations more as crop agriculture than as ecologically functional forests. Because there are few tree species and little variation in tree age, plantations do not offer many forest organisms the habitat they need. However, some harvesting methods aim to maintain **uneven-aged** stands, where a mix of ages (and often a mix of tree species) makes the stand more similar to a natural forest.

Timber is harvested by several methods

When they harvest trees, timber companies use any of several methods. From the 1950s through the 1970s, many timber harvests were conducted using the clear-cutting method, in which all trees in an area are cut, leaving only stumps. Clear-cutting is generally the most cost-efficient method in the short term, but it has the greatest impacts

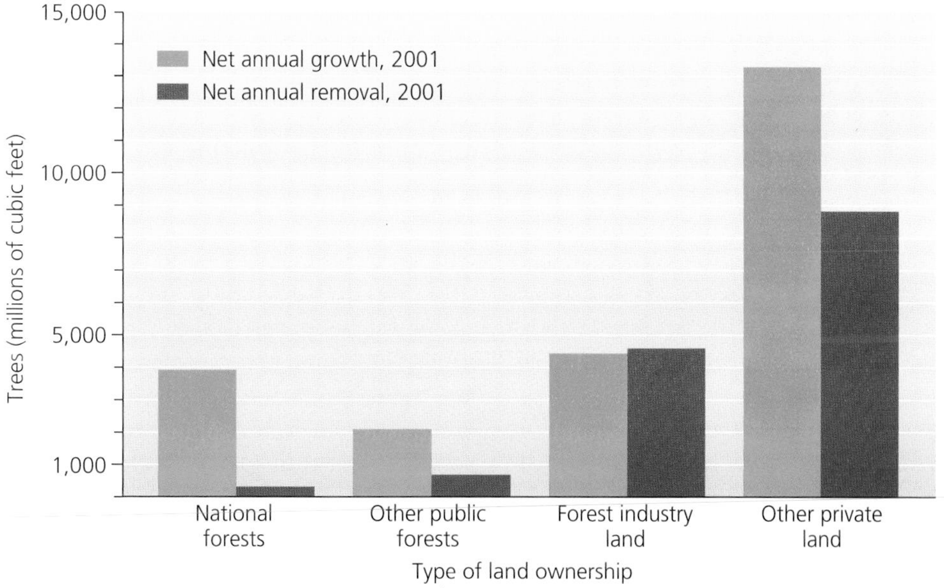

FIGURE 12.8 Forest Service data indicate that in the United States, trees (measured in cubic feet of wood biomass) are growing at a faster rate than they are being removed. The exception is on land privately owned by the timber industry, where extraction is narrowly outpacing growth. Go to GRAPHIt! at www.aw-bc.com/withgott or on the student CD-ROM. Data are for 2001, from United States Forest Service, 2001.

FIGURE 12.9 Even-aged management is practiced on tree plantations where all trees are of equal age, as seen in the stand in the foreground that is regrowing after clear-cutting. In uneven-aged management, harvests are designed so as to maintain a mix of tree ages, as seen in the more mature forest in the background. The increased structural diversity of uneven-aged stands provides superior habitat for most wild species and makes these stands more akin to ecologically functional forests.

on forest ecosystems (Figure 12.10). In the best-case scenario, clear-cutting may mimic natural disturbance events such as fires, tornadoes, or windstorms that knock down trees across large areas. In the worst-case scenario, entire communities of organisms are destroyed or displaced, soil erodes, and the penetration of sunlight to ground level changes microclimatic conditions such that new types of plants replace those that dominated the native forest. Essentially, an artificially driven process of succession (▸ p. 163) is set in motion, in which the resulting

FIGURE 12.10 Clear-cutting is the most cost-efficient method for timber companies, but it can have severe ecological consequences, including soil erosion and species turnover. Although certain species do use clear-cuts as they regrow, most people find these areas aesthetically unappealing, and public reaction to clear-cutting has driven changes in forestry methods.

climax community may turn out to be quite different from the original climax community.

Widespread clear-cutting occurred across North America at a time when public awareness of environmental problems was blossoming. The combination produced public outrage toward the timber industry and public forest managers. Eventually the industry integrated other harvesting methods (Figure 12.11). Clear-cutting (Figure 12.11a) is still widely practiced, but other methods involve cutting some trees and leaving some standing. In the *seed-tree* approach (Figure 12.11b), small numbers of mature and vigorous seed-producing trees are left standing so that they can reseed the logged area. In the *shelterwood* approach (also Figure 12.11b), small numbers of mature trees are left in place to provide shelter for seedlings as they grow. These three methods all lead to even-aged stands of trees.

Selection systems, in contrast, allow uneven-aged stand management. In selection systems (Figure 12.11c), only some trees in a forest are cut at any one time. The stand's overall rotation time may be the same as in an even-aged approach, because multiple harvests are made, but the stand remains mostly intact between harvests. Selection systems include single-tree selection, in which widely spaced trees are cut one at a time, and group selection, in which small patches of trees are cut.

It was a form of selection harvesting that MacMillan Bloedel and other timber companies pursued at Clayoquot Sound, after old-growth advocates applied pressure and the scientific panel published its guidelines. Not wanting to bring a complete end to logging when so many local people depended on the industry for work, these activists and scientists instead promoted what they considered a

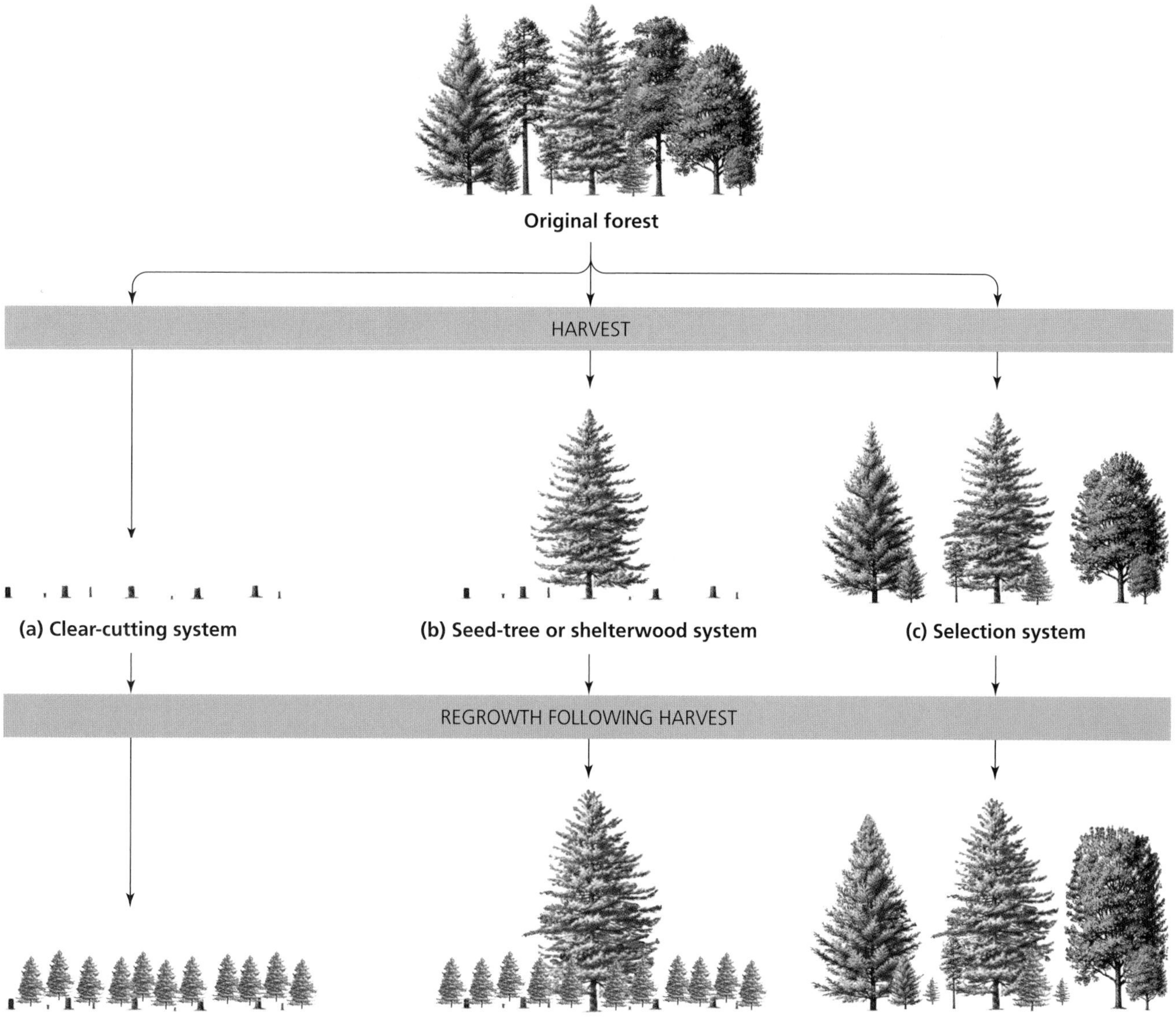

FIGURE 12.11 Foresters and timber companies have devised various methods of harvesting timber from forests. In clear-cutting (**a**), all trees in an area are cut, extracting a great deal of timber inexpensively but leaving a vastly altered landscape. In seed-tree systems and shelterwood systems (**b**), small numbers of large trees are left in clearcuts, to help re-seed the area or provide shelter for growing seedlings. In selection systems (**c**), a minority of trees is removed at any one time, while most are left standing. These latter methods involve less environmental impact than clear-cutting, but all methods can cause significant changes to the structure and function of natural forest communities.

more environmentally friendly method of timber removal. However, selection systems are by no means ecologically harmless. Moving trucks and machinery over an extensive network of roads and trails to access individual trees compacts the soil and disturbs much of the forest floor. Selection methods are also unpopular with timber companies because they are expensive, and loggers dislike them because they are more dangerous than clear-cutting.

All methods of logging result in habitat disturbance, which invariably affects the plants and animals inhabiting an area. All methods change forest structure and composition. Most methods increase soil erosion, leading to siltation of waterways, which can degrade habitat and affect drinking water quality. Most methods also speed runoff, sometimes causing flooding. In extreme cases, as when steep hillsides are clear-cut, landslides can result.

Public forests may be managed for recreation and ecosystems

In recent decades, increased awareness of these problems has prompted many citizens to protest the way public forests are managed in the United States and Canada. These citizens have urged that the national, state, and provincial forests be managed for recreation, wildlife, and ecosystem integrity, rather than for timber. They want forests managed as ecologically functional entities, not as cropland for trees.

Critics of the U.S. Forest Service have also protested the fact that taxpayers' money is used to subsidize the extraction of publicly held resources by private corporations. Scientists who have analyzed government subsidies have concluded that the U.S. Forest Service loses at least $100 million of taxpayers' money each year by selling timber well below its costs for marketing and administering the harvest and for building access roads. Subsidies also inflate harvest levels beyond what would occur in a free market.

In one sense, the U.S. Forest Service has long had a policy of attending to interests beside timber production. For the past half century, forest management has nominally been guided by the policy of **multiple use,** meaning that the national forests were to be managed for recreation, wildlife habitat, mineral extraction, and various other uses. In reality, however, timber production was most often the primary use.

In 1976 the U.S. Congress passed the **National Forest Management Act,** which mandated that plans for renewable resource management be drawn up for every national forest. These plans were to be explicitly based on the concepts of multiple use and sustained yield, and they were to be subject to broad public participation under the National Environmental Policy Act (▸ p. 68). Guidelines specified that these plans:

- Consider both economic and environmental factors.
- Provide for diversity of plant and animal communities and preserve the regional diversity of tree species.
- Ensure research and monitoring of management practices.
- Permit increases in harvest levels only if sustainable.
- Ensure that timber is harvested only where soils and wetlands will not be irreversibly damaged, lands can be restocked quickly, and economic return alone does not guide the choice of harvest method.
- Ensure that logging is conducted only where impacts have been assessed; cuts are shaped to the terrain; maximum size limits are established; and "cuts are carried out in a manner consistent with the protection of soil, watershed, fish, wildlife, recreation, and aesthetic resources, and the regeneration of the timber resource."

Over the years following passage of the National Forest Management Act, the U.S. Forest Service began responding to the increasing public demand that national forests be managed for uses other than timber. It developed new programs to manage wildlife, nongame animals, and endangered species. It pushed for ecosystem-based management and even ran extensive programs of ecological restoration, attempting to recover whole plant and animal communities that had been lost or degraded. Moreover, timber harvesting methods were brought more in line with ecosystem-based management goals. A set of approaches dubbed **new forestry** called for timber cuts that came closer to mimicking natural disturbances. For instance, "sloppy clear-cuts" that leave a variety of trees standing were intended to mimic the changes a forest might experience if hit by a severe windstorm.

In late 2004, however, the George W. Bush administration issued new regulations that bucked these trends. These new rules freed local forest managers from requirements imposed by the National Forest Management Act, granting them more flexibility in managing forests, but loosening environmental protections and restricting public oversight.

Then in 2005, the Bush administration repealed the Clinton administration's roadless rule, by which 23.7 million ha (58.5 million acres)—31% of national forest land and 2% of total U.S. land—were in 2001 put off limits to further road construction or maintenance. Although the roadless rule had been supported by a record 4.2 million public comments, the Bush administration overturned it and required state governors to petition the federal government if they want to keep areas in their states roadless. The states of California, Oregon, and New Mexico responded by suing the federal government, asking that the roadless rule be reinstated.

Fire policy has also stirred controversy

Some ecosystem management efforts, ironically, run counter to the U.S. Forest Service's best-known symbol, Smokey the Bear. The cartoon bear wearing a ranger's hat who advises us to fight forest fires is widely recognized, but unfortunately Smokey's message is badly outdated, and many scientists assert that it has done great harm to American forests.

For over a century, the Forest Service and other land management agencies suppressed fire whenever and wherever it broke out. Yet ecological research now clearly shows that many ecosystems depend on fire. Certain plants have seeds that germinate only in response to fire, and researchers studying tree rings have documented that many ecosystems historically experienced frequent fire. Burn marks in a tree's rings reveal past fires, giving scientists

VIEWPOINTS

Managing Forests

The U.S. Forest Service attempts to manage the national forests for multiple uses, including timber production, wildlife habitat, and recreation. **How well is the agency doing in balancing its multiple goals? What changes, if any, do you think are needed in the way the national forests are managed?**

Restoring America's National Forests

The U.S. Forest Service completed the 20th century with a celebrated and criticized record. Its challenge is that laws direct the Forest Service to provide society with often conflicting values from the national forests, including wood, water, wildlife, fish, forage, and recreation.

The Forest Service also must "preserve and protect" forests as required in the Organic Act of 1897. Therefore, conservation guides management because it integrates the need to use and protect national forests.

For much of the last century, the Forest Service protected forests by putting out fires, without realizing that frequent, gentle fires kept forests thinned and healthy. Then they stopped timber harvests and active management on many national forests, so forests grew even thicker and less healthy.

Today's forests are 10 to 20 times denser than historic forests. Consequently, between the years 2000 and 2003, unnatural wildfires destroyed 9.7 million ha (24 million acres) of forest, many human lives, and thousands of homes. Insects devoured millions more hectares of unhealthy forest.

Restoration should guide the Forest Service in the 21st century. Restoration means using active management to restore ecologically and economically sustainable forests that represent natural historic landscapes. Restored forests provide diverse public values, from scenery and recreation to lumber and safety.

However, we don't have the billions of tax dollars needed to manage our forests. That means we must form a partnership with the private sector to make restoration economically feasible.

We have a choice. Adopt a "hands-off" policy and let the harsh indifference of unnatural wildfires and mindless insects determine the fate of national forests, or shape their destiny through restoration forestry.

Thomas M. Bonnicksen is a historian of North American forests and originator of the concept of restoration forestry. He is professor emeritus of forest science at Texas A&M University, visiting professor at the University of California at Davis, and the author of *America's Ancient Forests* (John Wiley, 2000.)

Protect Our Forests from Logging

Most Americans are shocked when they learn that over the past century the Forest Service's management has largely emphasized logging, road building, and other forms of resource extraction.

Unfortunately, because the Forest Service's budgets are still tied to logging and resource extraction—not to forest protection and restoration—the public's clean water, wildlife habitat, wildlands, and recreational opportunities continue to be sacrificed. Just consider these facts:

- There are 716,500 km (445,000 mi) of roads on national forests—enough to circle Earth 18 times.
- An estimated 50% of riparian areas on national forests require restoration because of impacts from logging, road building, grazing, mining, and offroad vehicles.
- Taxpayers spend over $1 billion annually subsidizing private logging companies to cut down national forests—all to supply less than 2% of our nation's wood products.
- Although less than 5% of America's ancient, old-growth forests remain, these heritage forests continue to be logged, with over 160 km^2 (100 mi^2) of the public's ancient forests currently on the chopping block in the Northwest alone.

Fully protecting and restoring our national forests will take a heroic effort. The first step in the restoration process is to prevent further ecological degradation by protecting our national forests from logging and other forms of resource extraction. Next, we need to redirect taxpayer subsidies toward ecologically based restoration projects—such as road removal and watershed restoration—with the goal of restoring natural processes and reestablishing fully functioning ecosystems.

Only once this happens will we see the Forest Service's management of national forests in step with the desires of an American public that wants to see our forests protected and restored.

Matthew Koehler is executive director of the Native Forest Network and a co-founder of the National Forest Protection Alliance, a national network of 130 grassroots forest protection organizations working to protect and restore national forests.

Explore this issue further by accessing **Viewpoints** at www.aw-bc.com/withgott.

FIGURE 12.12 Forest fires are natural phenomena to which many plants are well adapted and which maintain many ecosystems. The suppression of fires by humans over the past century has led to a buildup of leaf litter and young trees, which serve as fuel to increase the severity of fires when they do occur. As a result, catastrophic forest fires (such as this one in Yellowstone National Park in 1988) have become more common in recent years. These unnaturally severe fires can do great damage to ecosystems and human communities. The best solution, most fire ecologists agree, is to forego suppressing natural fires as much as possible and to institute controlled burning to reduce fuel loads and restore forest ecosystems.

an accurate history of fire events extending back hundreds or even thousands of years. Researchers have found that North America's grasslands and open pine woodlands burned regularly. Ecosystems dependent on fire are adversely affected by its suppression; pine woodlands become cluttered with hardwood understory that ordinarily would be cleared away by fire, for instance, and animal diversity and abundance decline in such cluttered habitats.

In the long term, fire suppression can lead to catastrophic fires that truly do damage forests, destroy human property, and threaten human lives. Fire suppression allows limbs, logs, sticks, and leaf litter to accumulate on the forest floor over the years, effectively producing kindling for a catastrophic fire. Such fuel buildup helped cause the 1988 fires in Yellowstone National Park (Figure 12.12), the 2003 fires in southern California, the 2003 fires in British Columbia, and thousands of other fires across the continent. Fire suppression and fuel buildup have made catastrophic fires significantly greater problems than they were in the past. At the same time, increasing residential development on the edges of forested land is placing more homes in fire-prone situations.

To reduce fuel load and improve the health and safety of forests, the Forest Service and other land management agencies have in recent years been burning areas of forest under carefully controlled conditions. These **prescribed burns** or **controlled burns** have worked effectively, but have been implemented on only a relatively small amount of land. And every once in a while, a prescribed burn may get out of control, as happened in 2000 when homes and government labs were destroyed at Los Alamos, New Mexico. All too often, these worthy efforts have been impeded by public misunderstanding and by interference from politicians who have not taken time to understand the science behind the approach.

In the wake of the 2003 California fires, the U.S. Congress, intending to make forests less fire-prone, passed the Bush administration's Healthy Forests Restoration Act. Although this legislation encourages some prescribed burning, it primarily promotes the physical removal of small trees, underbrush, and dead trees by timber companies. The removal of dead trees, or snags, following a natural disturbance is called **salvage logging.** From an economic standpoint, salvage logging may seem to make good sense. However, ecologically, snags have immense value; the insects that decay them provide food for wildlife, and many birds, mammals, and reptiles depend on holes in snags for nesting and roosting sites. Conducting timber removal operations on recently burned land can also cause severe erosion and soil damage. Many scientists and environmental advocates have criticized the Healthy Forests Restoration Act, saying it increases commercial logging in national forests while doing little to reduce catastrophic fires near populated areas. By streamlining procedures for timber removal on public lands, the law also decreases oversight and public participation in enforcing environmental regulations, critics contend.

Weighing the Issues: How to Handle Fire?

A century of fire suppression has left millions of hectares of forested lands in North America in danger of catastrophic wildfires. Yet we will probably never have adequate resources to conduct careful prescribed burning over all these lands. Can you suggest any possible solutions that might help protect people's homes near forests while improving the ecological condition of some forested lands?

Sustainable forestry is gaining ground

Any company can claim that its timber harvesting practices are sustainable, but how is the purchaser of wood products to know whether they really are? In the last several years, a consumer movement has grown that is making informed consumer choice possible. Several organizations now examine the practices of timber companies and offer **sustainable forestry certification** to products produced using methods they consider sustainable (Figure 12.13).

Organizations such as the International Organization for Standardization (ISO), the Sustainable Forestry Initiative (SFI) program, and the Forest Stewardship Council (FSC) have varying standards for certification. Consumers can look for the logos of these organizations on forest products they purchase. The FSC is widely perceived to have the strictest certification standards. In 2001, Iisaak, the Native-run timber company at Clayoquot Sound, became the first tree farm license holder in British Columbia to receive FSC certification.

Consumer demand for sustainable wood has been great enough that several major retail businesses have announced that they will sell only sustainable wood. Home Depot in 2002 began selling only FSC-certified lumber, and the company says it is doing its best to keep prices as low as possible. B&Q, a major British retailer similar to Home Depot, also switched to sustainable wood, and the company's head said he was "taken aback" by the favorable public response. The decisions of such retailers are influencing the logging practices of many timber companies. In British Columbia, 70% of the province's annual harvest now is certified or meets ISO requirements.

Sustainable forestry is more costly for the timber industry, but if certification standards can be kept adequately strong, then consumer choice in the marketplace can be a powerful driver for good forestry practices for the future.

FIGURE 12.13 A Brazilian woodcutter taking inventory marks timber harvested from a forest certified for sustainable management in Amazonian Brazil. A consumer movement centered on independent certification of sustainable wood products is allowing consumer choice to promote sustainable forestry practices.

Agricultural Land Use

Having replaced many forests, agriculture now covers more of the planet's surface than does forest. Thirty-eight percent of Earth's terrestrial surface is devoted to agriculture—more than the area of North America and Africa combined. Of this land, 26% supports pasture, and 12% consists of crops and arable land. Agriculture is the most widespread type of human land use, and causes tremendous impacts on land and ecosystems. Although agricultural methods such as organic farming and no-till farming can be sustainable, the majority of the world's cropland hosts intensive traditional agriculture and monocultural industrial agriculture, involving heavy use of fertilizers, pesticides, and irrigation, and often producing soil erosion, salinization, and desertification (Chapters 9 and 10).

In theory, the marketplace should discourage people from farming with intensive methods that degrade land they own if such practices are not profitable. But agriculture in many countries is supported by massive subsidies. Governments of 30 developed nations handed out $311 billion in farm subsidies in 2001, averaging $12,000 per farmer. Roughly one-fifth of the income of the average U.S. or Canadian farmer comes from subsidies. Proponents of such subsidies stress that the vagaries of weather make profits and losses from farming unpredictable from year to year. To persist in the long term,

FIGURE 12.14 Cropland agriculture exerts dramatic effects over a large portion of Earth's landscape. The huge green circles visible during any airplane trip over the Great Plains are created by center-pivot irrigation. Immense sprinklers pivot around a central point, watering a circular area. Shown are fields in eastern Oregon.

these proponents say, an agricultural system needs some way to compensate farmers for bad years. Opponents of subsidies argue that farmers can buy insurance to protect themselves against crop failures and that subsidizing environmentally destructive agricultural practices is unsustainable (Figure 12.14).

Wetlands have been drained for farming

Many of today's crops grow on the sites of former wetlands (▸ pp. 435–436)—swamps, marshes, bogs, and river floodplains—that were drained and filled in (Figure 12.15). Throughout recent history, governments have encouraged laborious efforts to drain wetlands. To promote settlement and farming, the United States passed a series of laws known as the Swamp Lands Acts in 1849, 1850, and 1860, which encouraged draining wetlands for agriculture. The government transferred over 24 million ha (60 million acres) of wetlands to state ownership (and eventually to private hands) to stimulate drainage, conversion, and flood control.

In the Mississippi River valley, the Midwest, and a handful of states from Florida to Oregon, these transfers eradicated malaria (because mosquito vectors breed in wetlands) and created over 10 million ha (25 million acres) of new farmland. A U.S. Department of Agriculture (USDA) program in 1940 provided monetary aid and technical assistance to farmers draining wetlands on their property, resulting in the conversion of almost 23 million ha (57 million acres).

Today, less than half the original wetlands in the lower 48 U.S. states and southern Canada remain. However, many people now have a new view of wetlands. Rather than viewing them as worthless swamps, science has made clear that they are valuable ecosystems. This scientific knowledge, along with a preservation ethic, has induced policymakers to develop regulations to safeguard remaining wetlands. However, because of loopholes, differing state laws, development pressures—and even debate over the legal definition of wetlands—many of these valuable ecosystems are still being lost.

Financial incentives are also being used as a tool to influence agricultural land use in the United States. The Conservation Reserve Program begun in 1985 was a landmark initiative that provided farmers with a different kind of subsidy—it paid them to take highly erodible lands out of production and instead encouraged them to make the areas more habitable for wildlife. Now, under the Wetland Reserve Program, the U.S. government is offering subsidies to landowners who refrain from developing wetland areas.

Weighing the Issues: Subsidies, Soil, and Wetlands

Do you think that subsidy programs such as the Conservation Reserve Program and the Wetland Reserve Program are a good use of taxpayers' money? Are financial incentives such as these better tools than government regulation for promoting certain land use goals?

FIGURE 12.15 Most of North America's wetlands have been drained and filled, and the land converted to agricultural use. The northern Great Plains region of Canada and the United States was pockmarked with thousands of "prairie potholes," water-filled depressions that served as nesting sites for most of the continent's waterfowl. Today many of these wetlands have been lost; shown are farmlands encroaching on prairie potholes in North Dakota.

Livestock graze one-fourth of Earth's land

Cropland agriculture uses less than half the land taken up by livestock grazing, which covers a quarter of the world's land surface. Human use of rangeland, however, does not necessarily exclude its use by wildlife or its continued functioning as a grassland ecosystem. Grazing can be sustainable if done carefully and at low intensity. In the American West, ranching proponents claim that cattle are merely taking the place of the vast herds of bison that once roamed the plains. Indeed, most of the world's grasslands have historically been home to large herds of grass-eating mammals, and grasses have adapted to herbivory. Nonetheless, poorly managed grazing, as we saw in Chapter 9, can have adverse impacts on soil and grassland ecosystems.

Most U.S. rangelands are federally owned and managed by the **Bureau of Land Management (BLM).** The BLM is the nation's single largest landowner; its 106 million ha (261 million acres) are spread across 12 western states (see Figure 12.7). Ranchers are allowed to graze cattle on BLM lands for inexpensive fees, a practice that many public lands advocates say encourages overgrazing. Thus ranchers and environmentalists have traditionally been at loggerheads. In the past several years, however, ranchers and environmentalists have been finding common ground, teaming up to preserve ranchland against what each of them views in common as a threat—the encroaching housing developments of suburban sprawl (▸ pp. 378–379). Although developers often pay high prices for ranchland, many ranchers do not want to see the loss of the wide open spaces and the ranching lifestyle that they cherish.

Land use in the American West might have been better managed

Land uses such as grazing, farming, and timber harvesting need not have strongly adverse impacts. It is not these activities per se that cause environmental problems, but rather the overexploitation of resources beyond what ecosystems can handle. In the American West, a great deal of damage was done to the land by poor farming practices, overgrazing, and attempts to farm arid lands that were more suitable for grazing or preservation.

Most land to the west of the 100th meridian, the longitudinal line slicing through the Great Plains from Manitoba south through Texas, receives less than 50 cm (20 in.) of rain per year, making it too arid for unirrigated agriculture. One man in U.S. history recognized this fact and attempted to reorient policy so the West could be settled in a way that allowed farmers to succeed. John Wesley Powell, an extraordinary individual who explored the raging Colorado River by boat despite having lost an arm in the Civil War, undertook vast surveys of the Western lands for the U.S. government in the late 19th century (Figure 12.16). A pioneer in calling for government

FIGURE 12.16 John Wesley Powell, Civil War hero and 19th-century Western explorer, tried to shift U.S. land use policy to take account of the aridity of Western lands.

agencies to base their policies on science, Powell maintained that lands beyond the 100th meridian were too dry to support farming on the 65-ha (160-acre) plots the government parceled out through the Homestead Act (▸ p. 66). Plots in the West, he said, would have to be 16 times as large and would require irrigation. Moreover, to prevent individuals or corporations from monopolizing scarce water resources, he urged the government to organize farmers into cooperative irrigation districts, with each district encompassing a watershed.

Powell's ideas, based on science and close study of the land, were too revolutionary for the entrenched political interests and prevailing misconceptions of his time, which held that the West was a utopia for frontier settlement. The ideas in Powell's 1878 *Report on the Lands of the Arid Region of the United States* were, for the most part, never implemented. Instead, existing land use policies contributed to failures such as the Dust Bowl of the 1930s (▸ p. 260).

For agriculture and forestry alike, debates continue today over how best to use land and manage resources. Resource extraction from public lands in the United States and Canada has helped propel the economies of these countries. But as resources dwindle, as forests and soils are degraded, and as the landscape fills with more people, the arguments for conservation of resources—for their sustainable use—have grown stronger. Also growing stronger is the argument for preservation of land—setting aside tracts of relatively undisturbed land intended to remain forever undeveloped.

Parks and Reserves

Preservation has been part of the American psyche ever since John Muir rallied support for saving scenic lands in the Sierras (▸ p. 34). For ethical reasons as well as pragmatic ecological and economic ones, U.S. citizens and many other people worldwide have chosen to set aside tracts of land in perpetuity to be preserved and protected from development.

Why have we created parks and reserves?

What specifically in this mix of ethics, ecology, and economics has driven so many cultures to refrain voluntarily from exploiting land for material resources? The historian Alfred Runte has cited four traditional reasons that parks and protected areas have been established:

1. Enormous, beautiful, or unusual features such as the Grand Canyon, Mount Rainier, or Yosemite Valley inspire people to protect them—an impulse termed *monumentalism* (Figure 12.17).
2. Protected areas offer recreational value to tourists, hikers, fishers, hunters, and others.
3. Protected areas offer utilitarian benefits. For example, undeveloped watersheds provide cities with clean drinking water and a buffer against floods.
4. Parks make use of sites lacking economically valuable material resources; land that holds little monetary value is easy to set aside because no one wants to buy it.

To these four traditional reasons, a fifth has been added in recent years: the preservation of biodiversity. As we saw in Chapter 11, human impact alters habitats and has led to countless population declines and species extinctions. A park or reserve is widely viewed as a kind of Noah's Ark, an island of habitat that can, scientists hope, maintain species that might otherwise disappear.

Federal parks and reserves began in the United States

The striking scenery of the American West impelled the U.S. government to create the world's first **national parks,** publicly held lands protected from resource extraction and development but open to nature appreciation and various forms of recreation. In 1872, Yellowstone National Park was established as "a public park or pleasuring-ground for the benefit and enjoyment of the people." Yosemite, General Grant, Sequoia, and Mount Rainier National Parks followed after 1890. The Antiquities Act of 1906 gave the president authority to declare selected

FIGURE 12.17 The awe-inspiring beauty of some regions of the western United States was one reason for the establishment of national parks. Images of scenic vistas such as this one of Bridal Veil Falls in Yosemite National Park, portrayed by the landscape painter Albert Bierstadt, have inspired millions of people from North America and abroad to visit these parks.

public lands as national monuments, which can be an interim step to national park status. Presidents from Theodore Roosevelt to Bill Clinton have used this authority to expand the nation's system of protected lands.

The National Park Service (NPS) was created in 1916 to administer the growing system of parks and monuments, which today numbers 388 sites totaling 32 million ha (79 million acres) and includes national historic sites, national recreation areas, national wild and scenic rivers, and other types of areas (see Figure 12.7). This most widely used park system in the world received 277 million reported recreation visits in 2004—about as many visits as there are U.S. residents. The high visitation rates signal the success of the park system, but they also create overcrowded conditions at some parks. Many observers have therefore suggested that the parks' popularity indicates a pressing need to expand the system.

Many other nations now have national park systems. Canada's system covers 26.5 million ha (65.5 million acres) and receives 16 million visits yearly. At Clayoquot Sound, Pacific Rim National Park Reserve is a protected area designated for future national park status. The Clayoquot Sound region also encompasses several provincial parks. Provincial parks in Canada cover more area than national parks. In the United States, state parks are numerous and tend to be more oriented toward recreation than are national parks.

Another type of federal protected area in the United States is the **national wildlife refuge.** The system of national wildlife refuges, begun in 1903 by President Theodore Roosevelt, now totals 37 million ha (91 million acres) spread over 541 sites (see Figure 12.7). The U.S. Fish and Wildlife Service administers the refuges with management ranging "from preservation to active manipulation of habitats and populations." Indeed, these refuges not only serve as havens for wildlife, but also in many cases encourage hunting, fishing, wildlife observation, photography, environmental education, and other public uses. Some wildlife advocates find it ironic that hunting is allowed at many refuges, but hunters have long been in the forefront of the conservation movement and have traditionally supplied the bulk of funding for land acquisition and habitat management for the refuges. Many refuges are managed for waterfowl, but the FWS increasingly considers nongame species as well as game species. The FWS manages at the habitat and ecosystem levels, engaging in ecological restoration of marshes and grasslands, for example.

Wilderness areas have been established on various federal lands

In response to the public's desire for undeveloped areas of land, in 1964 the U.S. Congress passed the Wilderness Act, which allowed areas of existing federal lands to be designated as **wilderness areas.** These areas are off-limits to development of any kind, but they are open to public recreation such as hiking, nature study, and other activities that have minimal impact on the land. Congress declared that wilderness areas were necessary "to assure that an increasing population, accompanied by expanding settlement and growing mechanization, does not occupy and modify all areas within the United States and its possessions, leaving no lands designated for preservation and protection in their natural condition." Wilderness areas have been established within portions of national forests, national parks, national wildlife refuges, and BLM land, and they are overseen by the agencies that

FIGURE 12.18 Wilderness areas were designated on various federally managed lands in the United States following the 1964 Wilderness Act. These include areas little disturbed by human activities, such as the Selway-Bitterroot Wilderness in Idaho, shown here.

administer those areas (Figure 12.18). Some preexisting extractive land uses, such as grazing and mining, were "grandfathered in," or allowed to continue, within wilderness areas as a political compromise so the act could be passed.

Not everyone supports land set-asides

The restriction of activities in wilderness areas has helped generate opposition to U.S. land protection policies. Sources of such opposition include the governments of some western states, where large portions of land are federally owned. When those states came into existence, the federal government retained ownership of much of the acreage inside their borders. Idaho, Oregon, and Utah own less than 50% of the land within their borders, and in Nevada 80% of the land is federally owned. Western state governments have traditionally sought to obtain land from the federal government and encourage resource extraction and development on it.

The drive to extract more resources, secure local control of lands, and expand recreational access to public lands is epitomized by the **wise-use movement,** a loose confederation of individuals and groups that coalesced in the 1980s and 1990s in response to the increasing success of environmental advocacy. Wise-use advocates are dedicated to protecting private property rights; opposing government regulation; transferring federal lands to state, local, or private hands; and promoting more motorized recreation on public lands. Wise-use advocates include farmers, ranchers, trappers, and mineral prospectors at the grassroots level who live off the land, as well as groups representing the large corporations of industries that extract timber, mineral, and fossil fuel resources.

Debate between mainstream environmental groups and wise-use spokespeople has been vitriolic. Each side claims to represent the will of the people and paints the other as the oppressive establishment. Wise-use advocates have played key roles in ongoing debates over national park policy, such as whether recreational activities that disturb wildlife, such as snowmobiles and jet-skis, should be allowed. Under the Bush administration, wilderness protection policies have been weakened, and federal agencies have generally shifted policies and enforcement away from preservation and conservation, and toward recreation and extractive uses.

Nonfederal entities also protect land

Efforts to set aside land—and the debates over such efforts—at the federal level are paralleled at regional and local levels. Each U.S. state and Canadian province has agencies that manage resources on state or provincial lands, as do many counties and municipalities. As just one example, New York State in the 19th century created Adirondack State Park in a mountainous area whose streams converge to form the headwaters of the Hudson River, which flows south past Albany to New York City. Seeing the need for river water to power industries, keep canals filled, and provide drinking water, the state set the land aside, a farsighted decision that has paid dividends through the years.

Private nonprofit groups also preserve land. **Land trusts** are local or regional organizations that preserve lands valued by their members. In most cases, land trusts purchase land outright with the aim of preserving it in its natural condition. The Nature Conservancy can be considered the world's largest land trust, but smaller ones are springing up throughout North America. By one estimate, there are 900 local and regional land trusts in the United States that together own 177,000 ha (437,000 acres) and have helped preserve an additional 930,000 ha (2.3 million acres), including well-known scenic areas such as Big Sur on the California coast, Jackson Hole in Wyoming, and Maine's Mount Desert Island.

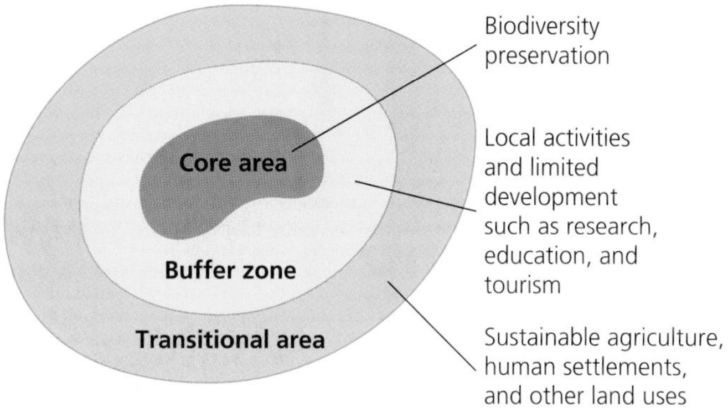

FIGURE 12.19 Biosphere reserves are international efforts that couple preservation with sustainable development to benefit local residents. Each reserve includes a core area that preserves biodiversity, a buffer zone that allows limited development, and a transition zone that permits various uses.

Parks and reserves are increasing internationally

Many nations have established national park systems and are benefiting from ecotourism as a result—from Costa Rica (Chapter 5) to Ecuador to Thailand to Tanzania. The total worldwide area in protected parks and reserves increased more than fourfold from 1970 to 2000, and in 2003 the world's 38,536 protected areas covered 1.3 billion ha (3.2 billion acres), or 9.6% of the planet's land area. However, parks in developing countries do not always receive the funding they need to manage resources, provide for recreation, and protect wildlife from poaching and timber from logging. Thus many of the world's protected areas are merely *paper parks*—protected on paper but not in reality.

Some types of protected areas fall under national sovereignty but are designated or partly managed internationally by the United Nations. *World heritage sites* are an example; currently over 560 sites across 125 countries are listed for their cultural value and nearly 150 for their natural value. One such site is Australia's Kakadu National Park, discussed in Chapter 2. Another is the mountain gorilla reserve shared by three African countries. The gorilla reserve, which integrates national parklands of Rwanda, Uganda, and the Democratic Republic of Congo, is also an example of a *transboundary park,* an area of protected land overlapping national borders. Transboundary parks can be quite large, and they account for 10% of protected areas worldwide, involving 113 countries. A North American example is Waterton-Glacier National Parks on the Canadian–U.S. border. Some transboundary reserves function as *peace parks,* helping ease tensions by acting as buffers between nations that have quarreled over boundary disputes. This is the case with Peru and Ecuador as well as Costa Rica and Panama, and many people hope that peace parks can also help resolve conflicts between Israel and its neighbors.

Biosphere reserves are tracts of land with exceptional biodiversity that couple preservation with sustainable development to benefit local people. They are designated by UNESCO (the United Nations Educational, Scientific, and Cultural Organization) following application by local stakeholders. Each biosphere reserve consists of (1) a core area that preserves biodiversity, (2) a buffer zone that allows local activities and limited development that do not hinder the core area's function, and (3) an outer transition zone in which agriculture, human settlement, and other land uses can be pursued in a sustainable way (Figure 12.19).

Clayoquot Sound was designated as Canada's 12th biosphere reserve in 2000, in an attempt to help build cooperation among environmentalists, timber companies, Native peoples, and local residents and businesses. The core area consists of provincial parks and Pacific Rim National Park Reserve. Environmentalists hoped the designation would help promote stronger land preservation efforts. Local residents supported it because outside money was being offered for local development efforts. The timber industry did not stand in the way once it was clear that harvesting operations would not be affected. The designation has brought Clayoquot Sound more international attention, but it has not created new protected areas and has not altered land use policies.

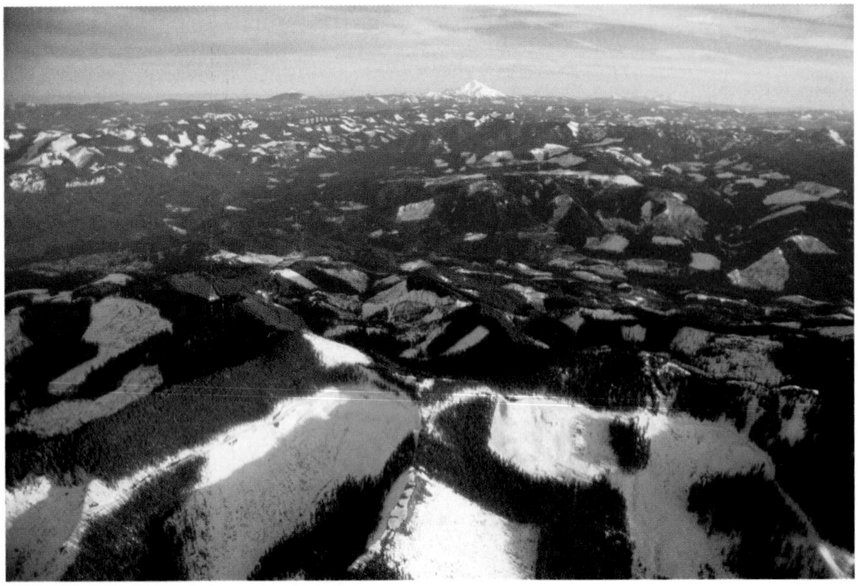

(a) Mount Hood National Forest, Oregon

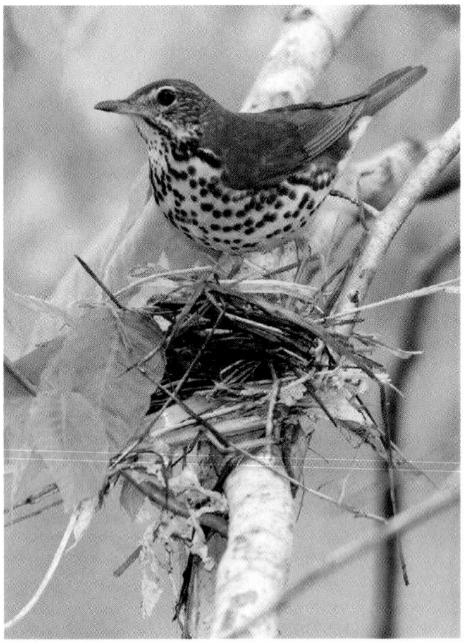

(b) Wood thrush

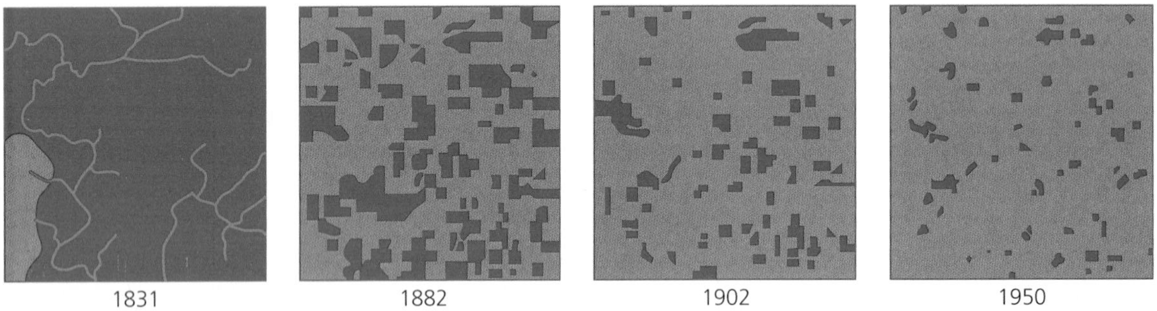

(c) Fragmentation of wooded area (green) in Cadiz Township, Wisconsin

FIGURE 12.20 As human populations have grown and human impacts have increased, most large expanses of natural habitat have become fragmented into smaller disconnected areas. Forest fragmentation from timber harvesting, for example, is evident on the Mount Hood National Forest, Oregon **(a)**. Fragmentation has significant impacts on forest-dwelling species such as the wood thrush, *Hylocichla mustelina* **(b)**, a migrant songbird of eastern North America. In forest fragments, wood thrush nests are parasitized by cowbirds that thrive in open country and edge habitats. Forest fragmentation has been extreme in the eastern and midwestern United States; shown in **(c)** are historical changes in forested area in Cadiz Township, Wisconsin, between 1831 and 1950. *Source for (c):* Curtis, J. T. 1956. The modification of mid-latitude grasslands and forests by man. In Thomas, W. L. Jr., ed., *Man's role in changing the face of the earth.* Chicago: Univ. of Chicago Press.

The design of parks and reserves has consequences for biodiversity

Often it is not outright destruction of habitat that threatens species, but rather the fragmentation of habitat (▸ pp. 330–331). Expanding agriculture, spreading cities, highways, logging, and many other impacts have chopped up large contiguous expanses of habitat into small disconnected ones (Figure 12.20a and 12.20c). When this happens, many species suffer. Bears, mountain lions, and other animals that need large ranges in which to roam may disappear. Bird species that thrive in the interior of forests may fail to reproduce when forced near the edge of a fragment (Figure 12.20b). Their nests often are attacked by predators and parasites that favor open habitats surrounding the fragment or that travel along habitat edges. Avian ecologists judge forest fragmentation to be a main reason why populations of many songbirds of eastern North America are declining.

Because habitat fragmentation is such a central issue in biodiversity conservation, and because there are limits on how much land can be set aside, conservation biologists have argued heatedly about whether it is better to make reserves large in size and few in number, or many in number but small in size. Nicknamed the **SLOSS dilemma,** for "**s**ingle **l**arge **o**r **s**everal **s**mall," this debate is ongoing and complex, but it seems clear that large species that roam great distances, such as the Siberian tiger (Chapter 11), benefit more from the "single large" approach to reserve design. In contrast, creatures such as insects that live as larvae in small areas may do just fine in a number of small isolated reserves, if they can disperse as adults by flying from one reserve to another.

A related issue is whether **corridors** of protected land are important for allowing animals to travel between islands of protected habitat. In theory, connections between fragments provide animals access to more habitat and help enable gene flow to maintain populations in the long term. Many land management agencies and environmental groups try, when possible, to join new reserves to existing reserves for these reasons. It is clear that we will need to think on the landscape level if we are to preserve a great deal of our natural heritage.

Conclusion

Managing natural resources is necessary for resources such as timber, which can be either responsibly and sustainably managed or carelessly exploited and overharvested. The United States, Canada, and other nations have established various federal and regional agencies to oversee and manage publicly held land and the natural resources that are extracted from public land.

Forest management in North America reflects trends in land and resource management in general. Early emphasis on resource extraction evolved into policies on sustained yield and multiple use, a shift that occurred as land and resource availability declined and as the public became more aware of environmental degradation. Public forests today are managed not only for timber production, but also for recreation, wildlife habitat, and ecosystem integrity.

Meanwhile, public support for preservation of natural lands has resulted in parks, wilderness areas, and other reserves, both in North America and abroad. These trends are positive ones, because the preservation and conservation of land and resources is essential if we wish our society to be sustainable and to thrive in the future.

REVIEWING OBJECTIVES

You should now be able to:

Identify the principles, goals, and approaches of resource management

- Resource management enables us to sustain natural resources that are renewable if we are careful not to deplete them. (pp. 344–345)
- Resource managers have increasingly focused not only on extraction, but also on sustaining the ecological systems that make resources available. (pp. 344–346)
- Resource managers have long managed for maximum sustainable yield and are beginning to implement ecosystem-based management and adaptive management. (pp. 345–346)

Summarize the ecological roles and economic contributions of forests, and outline the history and scale of forest loss

- Forests provide us economically important timber, but also support biodiversity and contribute ecosystem services. (pp. 347–349)
- Developed nations deforested much of their land as settlement, farming, and industrialization proceeded. Today deforestation is taking place most rapidly in developing nations. (pp. 350–354)

Explain the fundamentals of forest management, and describe the major methods of harvesting timber

- The U.S. national forests were established to conserve timber for the nation and allow for its sustainable extraction. (p. 354)
- Harvesting methods include clear-cutting and other even-aged techniques, as well as selection strategies that maintain uneven-aged stands that more closely resemble natural forest. (pp. 355–357)
- Foresters have responded to public demand by beginning to manage for recreation, wildlife habitat, and ecosystem integrity, as well as timber production. (p. 358)
- Fire policy has been politically controversial, but scientists agree that we need to take steps to reverse the impacts of a century of fire suppression. (pp. 358, 360)
- Certification of sustainable forest products is allowing consumer choice in the marketplace to influence forestry techniques. (p. 361)

Analyze the scale and impacts of agricultural land use

- Agriculture has contributed greatly to deforestation, and has had enormous impacts on landscapes and ecosystems worldwide. (pp. 361–364)

Identify major federal land management agencies and the lands they manage

- The U.S. Forest Service, National Park Service, Fish and Wildlife Service, and Bureau of Land Management manage U.S. national forests, national parks, national wildlife refuges, and BLM land. (pp. 354, 364–366)

Recognize types of parks and reserves, and evaluate issues involved in their design

- Public demand for preservation and recreation has led to the creation of parks, reserves, and wilderness areas in North America and across the world. (pp. 364, 367)
- Biosphere reserves are one of several types of internationally managed protected lands. (p. 367)
- Because habitat fragmentation threatens wildlife, and landscape patterns matter, conservation biologists are working on how best to design parks and reserves. (pp. 368–369)

TESTING YOUR COMPREHENSION

1. How do minerals differ from timber when it comes to resource management?
2. Compare and contrast maximum sustainable yield, adaptive management, and ecosystem management. Why may pursuing maximum sustainable yield sometimes conflict with what is ecologically desirable?
3. Name several major causes of deforestation. Where is deforestation most severe today?
4. Describe the major methods of timber harvesting.
5. What are some ecological effects of logging? What has been the U.S. Forest Service's response to public concern over the ecological effects of logging?
6. Are forest fires a bad thing? Explain your answer.
7. Approximately what percentage of Earth's land is used for agriculture? What policies have caused conversion of wetlands for agriculture in the United States?
8. Name five reasons that parks and reserves have been created. Why did the U.S. Congress determine in 1964 that wilderness areas were necessary? How do these areas differ from national parks and national wildlife refuges?
9. Why do some people in the United States oppose federal land protection?
10. Roughly what percentage of Earth's land is protected? What types of protected areas have been established in countries outside the United States?

SEEKING SOLUTIONS

1. Do you think maximum sustained yield represents an appropriate policy for resource managers to follow?
2. Consider the economic importance of timber and the ecological roles that forests play. How would you manage the public forests of your country, if you were in charge?
3. People in developed countries are fond of warning people in developing countries to stop destroying rainforest. People of developing countries often respond that this is hypocritical, because the developed nations became wealthy by deforesting their land and exploiting its resources in the past. What would you say to the president of a developing nation, such as Brazil, that is seeking to clear much of its forest?
4. Can you think of a land use conflict that has occurred in your region? How was it resolved, or is it unresolved?
5. What are some ecological effects of agricultural subsidies? Propose arguments for and against subsidies from an ecological point of view.
6. Imagine you have just been elected mayor of a town on Clayoquot Sound. A timber company that employs 20% of your town's residents wants to log a hillside above the town, and the provincial government is supportive of the harvest. But owners of ecotourism businesses that run whale-watching excursions and rent kayaks to out-of-town visitors are complaining that the logging would destroy the area's aesthetic appeal and devastate their businesses—and these businesses provide 40% of the tax base for your town. Greenpeace is organizing a demonstration in your town soon, and news reporters are beginning to call your office, asking what you will do. How will you proceed?

INTERPRETING GRAPHS AND DATA

The invention of the moveable-type printing press by Johannes Gutenberg in 1450 stimulated a demand for paper that has only increased as the world population has grown. The 20th-century invention of the xerographic printing process used in photocopiers and laser printers has accelerated our demand for paper, with most of the raw fiber for paper production coming from wood pulp from forest trees.

1. Approximately how many millions of tons of paper and paperboard were consumed worldwide in 1970? 1980? 1990? 2000?
2. By what percentage did worldwide consumption of paper and paperboard increase from 1970 to 1980? From 1980 to 1990? From 1990 to 2000?
3. If consumption continues to increase at current rates, what do you predict the worldwide consumption of paper and paperboard will be in 2010?

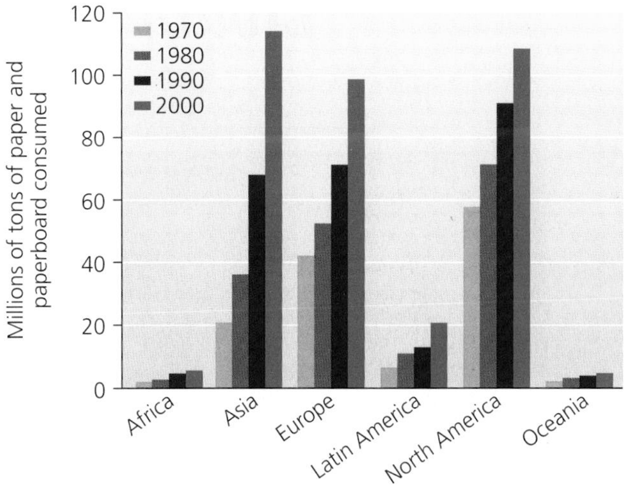

Global consumption of paper and paperboard, 1970–2000. Data from the Food and Agriculture Organization of the United Nations.

CALCULATING ECOLOGICAL FOOTPRINTS

	Population (millions)*	Total paper consumed in 2000 (millions of tons)	Per capita paper consumed in 2000 (pounds)
Africa	840	6	14
Asia	3,766		
Europe	728		
Latin America	531		
North America	319		
Oceania	32		
World	6,216		~114

**Data source:* Population Reference Bureau.

How much paper do you think you use? You may be surprised to learn that the average North American uses over 300 kg (660 lb) of paper and paperboard per year. Using the estimates of paper and paperboard consumption for each region of the world for the year 2000—as shown in the table in the "Interpreting Graphs and Data" section—calculate the per capita consumption of paper and paperboard for each region of the world using the population data in the table.

1. How much paper would North Americans save each year if we consumed paper at the rate of Europeans?
2. How much paper would be consumed if everyone in the world used as much paper as the average European? As the average North American?
3. Why do you think people in other regions consume less paper, per capita, than North Americans? Name three things you could do to reduce your paper consumption.

Take It Further

Go to www.aw-bc.com/withgott or the student CD-ROM where you'll find:

- Suggested answers to end-of-chapter questions
- Quizzes, animations, and flashcards to help you study
- *Research Navigator*™ database of credible and reliable sources to assist you with your research projects
- GRAPHIT! tutorials to help you master how to interpret graphs
- INVESTIGATEIT! current news articles that link the topics that you study to case studies from your region to around the world

13 Urbanization and Creating Livable Cities

Intersection near Shibuya station, Tokyo, Japan

Upon successfully completing this chapter, you will be able to:

- Describe the scale of urbanization
- Assess urban and suburban sprawl
- Outline city and regional planning and land use strategies
- Evaluate transportation options
- Describe the roles of urban parks
- Analyze environmental impacts and advantages of urban centers
- Assess the pursuit of sustainable cities

Pedestrians and joggers use Portland, Oregon's Tom McCall Waterfront Park

Central Case: Managing Growth in Portland, Oregon

"Sagebrush subdivisions, coastal condomania, and the ravenous rampage of suburbia in the Willamette Valley all threaten to mock Oregon's status as the environmental model for the nation."
—Oregon governor Tom McCall, 1973

"We have planning boards. We have zoning regulations. We have urban growth boundaries and 'smart growth' and sprawl conferences. And we still have sprawl."
—Environmental scientist Donella Meadows, 1999

With the fighting words above, Oregon governor Tom McCall challenged his state's legislature in 1973 to take action against runaway urbanization, which many Oregon residents feared would ruin the communities and landscapes they had come to love. McCall echoed the growing concerns of state residents that farms, forests, and open space were being gobbled up for development, including housing for people moving in from California and elsewhere. Foreseeing a future of subdivisions, strip malls, and traffic jams engulfing the pastoral Willamette Valley, Oregon acted. With Senate Bill 100, the state legislature in 1973 passed a sweeping land use law that would become the focus of acclaim, criticism, and careful study for years afterward by other states and communities trying to manage their own urban and suburban growth.

Oregon's law required every city and county to draw up a comprehensive land use plan, in line with statewide guidelines that had gained popular support from the state's electorate. As part of each land use plan, each metropolitan area had to establish an **urban growth boundary (UGB)**, a line on a map intended to separate areas desired to be urban from areas desired to remain rural. Development for housing, commerce, and industry would be encouraged within these urban growth boundaries, but severely restricted beyond them. The intent was to revitalize city centers, prevent suburban sprawl, and protect farmland, forests, and open landscapes around the edges of urbanized areas.

At the time, Oregon was taking many pioneering environmental steps. The legislature passed the world's first bottle bill (▸ pp. 660, 662), while policymakers and citizen volunteers enhanced state parks and began cleaning up the Willamette River. In Portland, the state's largest city, voters turned down a proposal to construct a new freeway, then chose to rip out an old highway along the riverfront and replace it with a public park. Portland-area residents also established a new regional entity to help plan how land would be apportioned in their region. The Metropolitan Service District, or Metro, represents 25 municipalities and three counties.

Metro adopted the Portland-area urban growth boundary in 1979 and has tried to focus growth on existing urban centers and to build communities where people can walk or take mass transit between home, work, and shopping. These policies have largely worked as intended; Portland's downtown and older neighborhoods have thrived, regional urban centers are becoming denser and more community oriented, mass transit has expanded, and development has been limited on land beyond the UGB. Portland began attracting international attention for its "livability."

To many Portlanders today, the UGB remains the key to maintaining quality of life in city and countryside alike. In the view of its critics, however, the "Great Wall of Portland" is an elitist and intrusive government regulatory tool. Ironically, the Portland area's successes may one day prove its undoing. A continuing influx of people has meant rapid development and rising housing prices. Still, most citizens have supported Oregon's land use rules for the past 30 years, and the system has survived three state referenda and many legal challenges.

In November 2004, however, Oregon voters approved a ballot measure that threatens to eviscerate the very land use reforms they had backed for three decades. Ballot Measure 37 requires the state to compensate landowners if government regulation has decreased the value of their land. For example, regulations prevent landowners outside UGBs from subdividing their lots and selling them for housing development. Under Ballot Measure 37, the state now has to pay these landowners to make up for theoretically lost income, or else allow them to ignore the regulations. Because the state does not have enough money to pay such claims, the measure could effectively gut Oregon's zoning, planning, and land use rules.

A county judge struck down the measure in late 2005, and its fate now seems destined to be decided in the courts. Events in Oregon over the next few years could tell us much about how our cities and landscapes may change in the future.

Our Urbanizing World

We live at a turning point. Beginning about the year 2007, for the first time in human history, more people will live in urban areas than in rural areas. This shift from the countryside into towns and cities, or **urbanization,** is arguably the single greatest change our society has undergone since its transition from a nomadic hunter-gatherer lifestyle to a sedentary agricultural one.

Industrialization has driven the move to urban centers

Agricultural harvests that produced surplus food freed a proportion of citizens from farm life and allowed the rise of specialized manufacturing professions, class structure, political hierarchies, and urban centers (▸p. 248). Technological innovations spawned by the industrial revolution (▸p. 4) created jobs and opportunities in urban centers for people no longer needed on farms. Industrialization and urbanization bred further technological advances that increased production efficiencies, both on the farm and in the city. This process of positive feedback continues today.

Worldwide, the proportion of population that is urban rose from 30% half a century ago to 48% today. Since 1950, the world's urban population has quadrupled, growing faster than population overall. Between 1975 and 2000, the global urban population increased by 2.53% each year, while the rural population rose only by 0.92% annually. From 2000 to 2030, the United Nations projects that the urban population will grow by 1.83% annually, while the rural population will decline by 0.03% each year.

Trends differ between developed and developing nations, however. In developed nations such as the United States and Canada, urbanization has slowed, because roughly three of every four people already live in cities, towns, and **suburbs,** the smaller communities that ring cities. In 1850, the U.S. Census Bureau classified only 15% of U.S. citizens as urban dwellers. That percentage passed 50% shortly before 1920 and now stands at 80%. Most U.S. urban dwellers reside in suburbs; fully 50% of the U.S. population today is suburban.

In contrast, today's developing nations, where many people still reside on farms, are urbanizing rapidly (Figure 13.1). In nations such as China, India, and Nigeria, rural people are streaming to cities in search of jobs and urban lifestyles. U.N. demographers estimate that virtually all the world's population growth over the next 25 years will be absorbed by urban areas of developing nations.

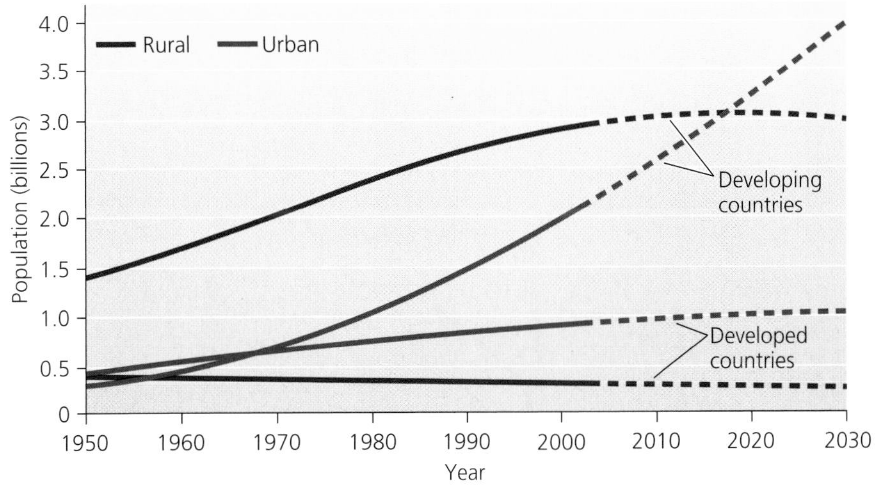

FIGURE 13.1 In developing countries today, urban populations are growing quickly, whereas rural populations are leveling off and may soon begin to decline. Developed countries are already largely urbanized, so in these countries urban populations are growing more slowly, whereas rural populations are falling. Solid lines in the graph indicate past data, and dashed lines indicate projections of future trends. Data from United Nations Population Division. 2004. *World urbanization prospects: The 2003 revision.* New York: UNPD.

Today's urban centers are unprecedented in scale

Cities in themselves are nothing novel. Urban centers where population—and political power—is concentrated have been part of human culture for several thousand years. Ancient Mediterranean civilizations, the great Chinese dynasties, and the Mayan and Incan empires all featured sophisticated and powerful urban centers.

What is new is the sheer scale of today's metropolitan areas. Human population growth (Chapter 8) has placed greater numbers of people in towns and cities than ever before. Today, 46 cities hold more than 5 million people, and 20 cities are home to over 10 million residents (Table 13.1). The metropolitan area of the world's most populous city, Tokyo, Japan, is home to 35 million people. North America's largest metropolises, Mexico City and New York City, each hold about 18 million. However, less than 5% of urban dwellers live in cities of greater than 10 million. Rather, most of them live in smaller cities, such as Portland, Omaha, Winnipeg, Raleigh, Austin, and their still-smaller suburbs.

Urban growth has often been rapid

Urban populations are growing for two reasons: (1) more people are moving from farms to cities than are moving from cities to farms, and (2) the human population overall is growing. Most cities are influenced by both of these trends, but the particular reasons for growth (or decline) vary from one city to another. Portland got its start in the mid-19th century as pioneers arriving by the Oregon Trail settled where the Willamette River flowed into the Columbia River. Situated at the juncture of these two major rivers, Portland had a strategic advantage in trade. The city grew as it shipped farm products from the fertile Willamette Valley overseas and accepted products shipped in from other North American ports and from Asia. Like many U.S. cities, Portland's population growth stalled in the 1950s to 1970s, as crowding and deteriorating economic conditions caused

Table 13.1 Metropolitan Areas with 10 Million Inhabitants or More, as of 2003

City, Country	Millions of people
Tokyo, Japan	35.0
Mexico City, Mexico	18.7
New York, United States	18.3
Sao Paulo, Brazil	17.9
Mumbai (Bombay), India	17.4
Delhi, India	14.1
Calcutta, India	13.8
Buenos Aires, Argentina	13.0
Shanghai, China	12.8
Jakarta, Indonesia	12.3
Los Angeles, United States	12.0
Dhaka, Bangladesh	11.6
Osaka-Kobe, Japan	11.2
Rio de Janeiro, Brazil	11.2
Karachi, Pakistan	11.1
Beijing, China	10.8
Cairo, Egypt	10.8
Moscow, Russian Federation	10.5
Metro Manila, Philippines	10.4
Lagos, Nigeria	10.1

Source: United Nations Population Division. 2004. *World urbanization prospects: The 2003 revision.* New York: UNPD.

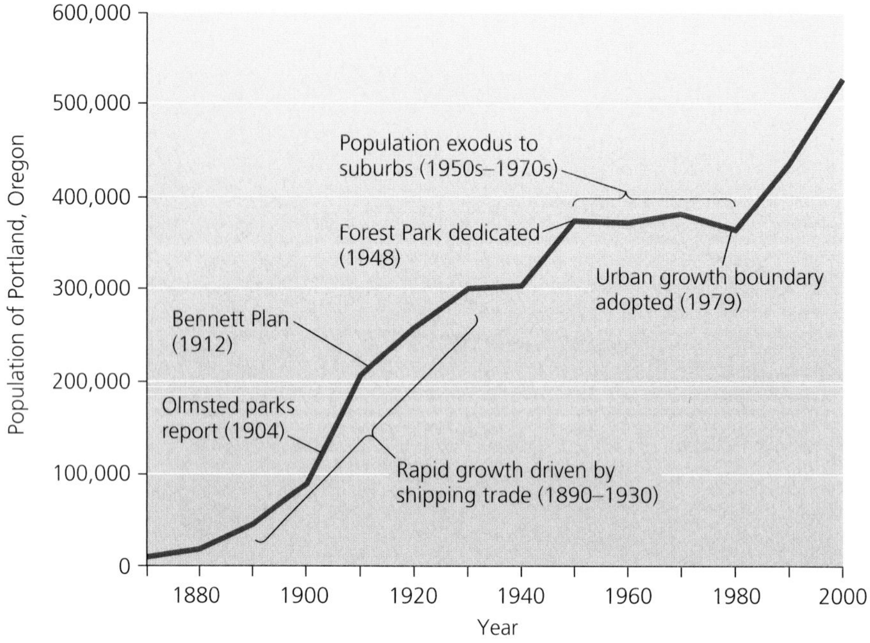

FIGURE 13.2 Once Portland established its position as a strategically located port, international shipping trade provided jobs and boosted its economy, and the city's population grew rapidly. City residents began leaving for the suburbs in the 1950s to 1970s, but policies designed to make the city center more attractive then restarted the city's growth.

many city residents to move outward into the growing suburbs. Unlike many U.S. cities, however, policies undertaken to improve the city center's attractiveness helped restart Portland's growth (Figure 13.2).

Some American cities have grown faster than Portland. For instance, Chicago was incorporated in 1833 with a population of 350, but this frontier town grew with extraordinary speed as railroads funneled through it the resources of the vast lands to the west on their way to the cities and consumers to the east. Chicago became a center for grain processing, livestock slaughtering, meatpacking, and much else. With ample employment opportunities, the city's population soared to 500,000 in 1880, 1.1 million in 1890, and 3.5 million during World War II. In recent years, many cities in the southern and western United States have gone through similar growth spurts, as people (particularly retirees) from northern and eastern states have moved south and west in search of warmer weather or more space. Between 1990 and 2000, for instance, the population of the Atlanta metropolitan area grew by 39%, while that of the Phoenix region grew by 45%, and that of the Las Vegas metropolitan area grew by 83%.

Internationally, most fast-growing cities today are in the developing world, because industrialization is decreasing the need for farm labor and is increasing commerce and jobs in cities. Sadly, another reason is because wars, conflict, and ecological degradation are driving millions of people out of the countryside and into cities. Cities like Mumbai (Bombay), India; Lagos, Nigeria; and Cairo, Egypt are growing in population even faster than American cities did. All too often, they are doing so without the economic growth to match their population growth. As a result, many of these cities are facing overcrowding, pollution, and poverty. Nearly three of every four governments of developing nations have by now enacted policies to discourage the movement of people from the countryside into cities.

Various factors influence the geography of urban areas

Real estate agents have long used the saying, "Location, location, location," to stress how much a home's whereabouts determines its value—and location is vitally important for urban centers, as well. Successful cities tend to be located in places that give them economic advantages. Think of any major city, and chances are it's situated along a major river, seacoast, railroad, or highway—some corridor for trade that has driven economic growth. Environmental variables such as climate, topography, and the configuration of waterways go a long way toward determining whether a small settlement will become a large city. Many well-located cities, like Portland and Chicago, have acted as linchpins in trading networks, funneling in resources from agricultural regions, processing them and manufacturing products, and shipping those products to other markets.

In fact, all cities, from ancient times to the present day, have supported themselves by drawing in resources from outlying rural areas through trade, persuasion, or conquest. In turn, cities have historically influenced how people use land in surrounding areas (Figure 13.3). Although city life and country life may seem very different, cities and the rural regions surrounding them have always been linked by tight economic relationships.

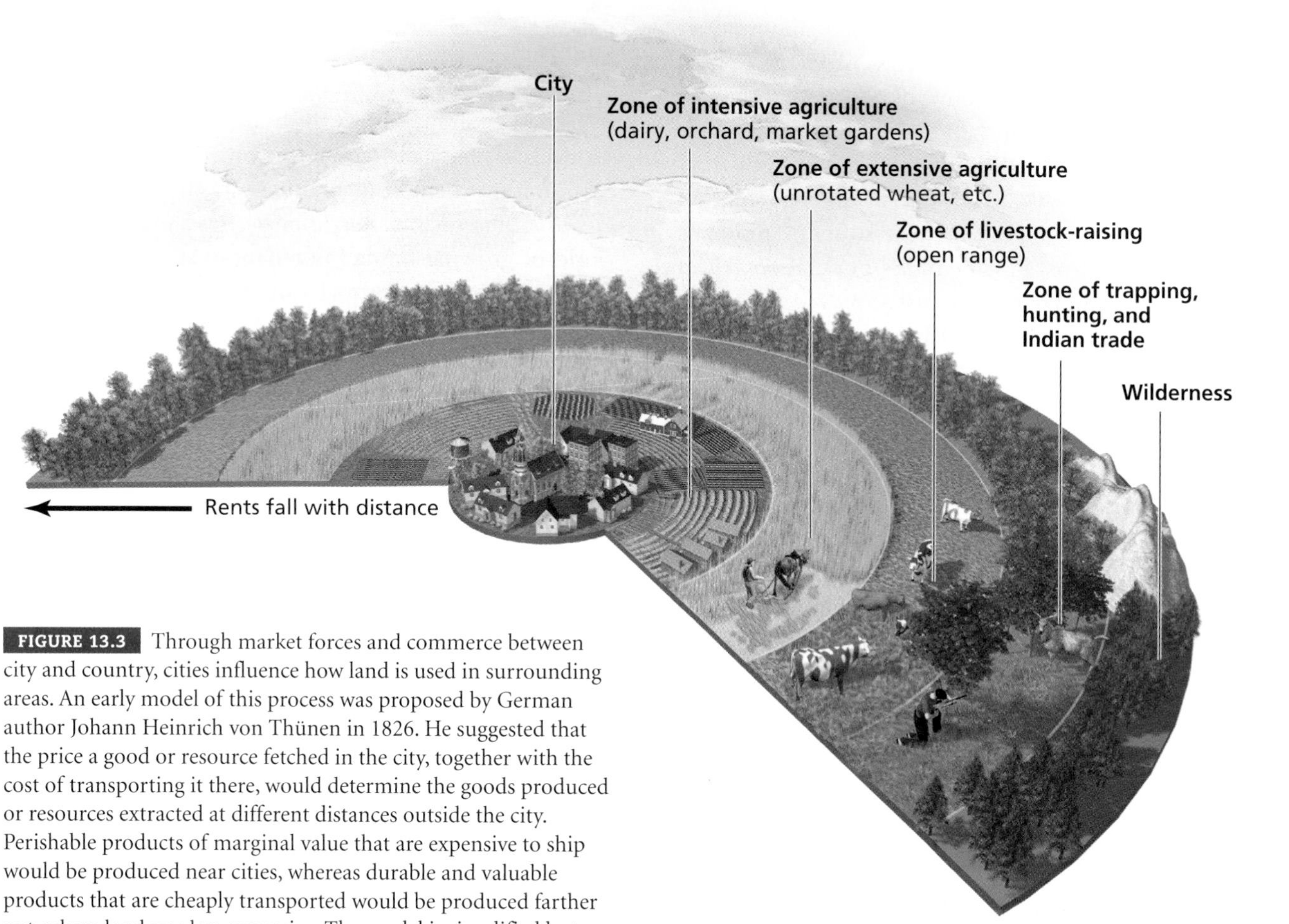

FIGURE 13.3 Through market forces and commerce between city and country, cities influence how land is used in surrounding areas. An early model of this process was proposed by German author Johann Heinrich von Thünen in 1826. He suggested that the price a good or resource fetched in the city, together with the cost of transporting it there, would determine the goods produced or resources extracted at different distances outside the city. Perishable products of marginal value that are expensive to ship would be produced near cities, whereas durable and valuable products that are cheaply transported would be produced farther out, where land was less expensive. The model is simplified but suggests how commerce between urban and rural areas helps structure land use far beyond a city's borders. Today, suburbs have proliferated between the city and agricultural zones. *Source:* von Thünen, J. H., as adapted by W. Cronon. 1991. *Nature's metropolis: Chicago and the Great West*. New York: W.W. Norton.

Weighing the Issues: What Made Your City?

Consider the town or city in which you live, or the major urban center located nearest you. Why do you think it developed into an urban area? What physical, social, or environmental factors may have aided its growth?

Spatial patterns of urbanization can change, however, with changing times. Today several factors are enabling population centers to decentralize in developed nations. For one thing, people now are globally interconnected to an unprecedented degree. Being located on a river or seacoast is no longer as vital to a city's success in our age of global commerce, jet travel, diplomacy, television, cell phones, and the Internet. Globalization has connected distant societies, and businesses and individuals can more easily communicate from locations away from major city centers. Moreover, fossil fuels have enabled the outward spread of cities. By easing long-distance transport, fossil fuels have made it easier to commute into and out of cities and to import and export resources, goods, and waste. Such factors have enabled a shift of population from cities to suburbs, particularly in the United States and Canada.

People have moved to suburbs

By the mid-20th century, many cities in the United States, Canada, and other developed nations had accumulated more people than these cities had jobs to offer. Unemployment rose, and crowded inner-city areas began to suffer increasing poverty and crime. As inner cities declined economically from the 1950s onward, many affluent city dwellers chose to move outward to the cleaner, less crowded, and more parklike suburban communities beginning to surround the cities. These people were pursuing more space, better economic opportunities, cheaper real estate, less crime, and better schools for their children. The exodus to the suburbs further hastened the economic decline of central cities. Chicago's population declined to 80% of its peak because so many residents moved to its suburbs. Philadelphia's population fell to 76% of its peak, and Detroit's to just 55%.

In most ways, suburbs have delivered the qualities people sought in them. The wide spacing of houses, with each house on its own plot of land, gives families room and privacy. However, by allotting more space to each person, suburban growth has spread human impact across the landscape. Natural areas have disappeared as housing developments are constructed. We have built extensive road networks to ease travel, but suburbanites now find themselves needing to drive everywhere. They commute longer distances to work and spend more time commuting in more congested traffic. The expanding rings of suburbs surrounding cities have grown larger than the cities themselves, and towns are running into one another. These aspects of suburban growth have inspired a new term: *sprawl.*

Sprawl

The term *sprawl* has become laden with meanings and connotes different things to different people. To some, sprawl is aesthetically ugly, environmentally harmful, and economically inefficient. To others, it is the collective outgrowth of reasonable individual desires and decisions in a world of growing human population. We can begin our discussion by giving **sprawl** a simple nonjudgmental definition: the spread of low-density urban or suburban development outward from an urban center.

Today's urban areas spread outward

As urban and suburban areas have grown in population, they have also grown spatially. This growth is obvious from maps and satellite images of rapidly spreading cities such as Las Vegas (Figure 13.4). Houses and roads supplant over 1 million ha (2.5 million acres) of U.S. rural land each year—over 2,700 ha (6,700 acres) every day.

Because suburban growth entails allotting more space per person than in cities, in most cases this outward spatial growth across the landscape has outpaced the growth in

(a) Las Vegas, Nevada, 1972 **(b) Las Vegas, Nevada, 2002**

FIGURE 13.4 Satellite images show the type of rapid urban and suburban expansion that many people have dubbed *sprawl*. Las Vegas, Nevada, is currently one of the fastest-growing cities in North America. Between 1972 (**a**) and 2002 (**b**), the population increased more than fivefold, and the developed area has risen more than threefold.

(a) Uncentered commercial strip development

(b) Low-density single-use development

(c) Scattered, or leapfrog, development

(d) Sparse street network

FIGURE 13.5 Several standard approaches to development can result in sprawl. In uncentered commercial strip development **(a)**, businesses are arrayed in a long strip along a roadway, and no attempt is made to create a centralized community with easy access for consumers. In low-density, single-use residential development **(b)**, homes are located on large lots in residential tracts far away from commercial amenities. In scattered or leapfrog development **(c)**, developments are created at great distances from a city center and are not integrated. In developments with a sparse street network **(d)**, roads are far enough apart that moderate-sized areas go undeveloped, but not far enough apart for these areas to function as natural areas or sites for recreation. All these development approaches necessitate frequent automobile use.

numbers of people. In fact, many researchers define *sprawl* as the physical spread of development at a rate greater than the rate of population growth. For instance, Phoenix grew from 105,000 residents spread over 44 km^2 (17 mi^2) in 1950 to 1.3 million residents spread over 1,200 km^2 (470 mi^2) in 2002; its land area grew 27 times larger, while the population grew 12 times larger. Between 1950 and 1990, the population of 58 major U.S. metropolitan areas rose by 80%, but the land area they covered rose by 305%. Even in 11 metro areas where population declined between 1970 and 1990, the amount of land covered increased.

Several types of development approaches can lead to sprawl (Figure 13.5). As a result, Chicago's metropolitan area now consists of more than 9 million people spread over 23,000 km^2 (9,000 mi^2)—an area 40 times the size of the city proper. Each person in the suburban region takes up an average of 11 times as much space as do residents of the city proper. As for Portland, today its growing suburbs hold as many people as does the city proper.

Sprawl has several causes

There are two main components of sprawl. One is human population growth—there are simply more people alive each year. The other is per capita land consumption—each person takes up more land. The amount of sprawl is a function of the number of people added to an area times the amount of land the average person occupies.

A study of the 100 major metropolitan areas of the United States between 1970 and 1990 found that, on average, each of these two factors contributes about equally to sprawl. Cities varied, however, in which factor was more important. The Los Angeles metro area increased in population density by 9% between 1970 and 1990, becoming the nation's most densely populated metro area. Increasing density should be a good recipe for preventing sprawl. Yet L.A. still grew in size by a whopping 1,021 km^2 (394 mi^2). This spatial growth, despite the increase in density, clearly resulted from an overwhelming influx of new people.

Detroit provides a contrasting example. The Detroit metro area lost 7% of its population between 1970 and 1990, yet it expanded in area by 28%. Clearly population growth was not the issue here; rather, sprawl was caused solely by increased per capita land consumption.

We discussed reasons for human population growth in general in Chapter 8. As for the increase in per capita land consumption, there are numerous reasons. Technologies such as telecommunications and the Internet have fostered movement away from city centers. Because of these technologies, many businesses no longer need the infrastructure a major city provides, and many workers have greater flexibility to live wherever they desire. The primary reasons for greater per capita land consumption, however, are that most people simply like having some space and privacy and dislike congestion. Furthermore, in the consumption-oriented American lifestyle that promotes bigger houses, bigger cars, and bigger TVs, having more space to house one's possessions becomes important. Unless there are overriding economic or social disadvantages, most people prefer living in a cleaner, more spacious, more affluent community.

Economists, politicians, and city boosters have almost universally encouraged the unbridled spatial expansion of cities and suburbs. The conventional assumption has been that growth is good and that attracting business, industry, and residents will unfailingly increase a community's economic well-being, political power, and cultural influence. Today, however, this assumption is increasingly being challenged. As the negative effects of sprawl on citizens' lifestyles accumulate, growing numbers of people have begun to question the mantra that all growth is good.

What is wrong with sprawl?

Sprawl means different things to different people. To some, the word evokes strip malls, homogenous commercial development, and tracts of cookie-cutter houses encroaching on farmland and ranchland. It may suggest traffic jams, destruction of wildlife habitat, and loss of natural land around cities. However, for other people, sprawl represents the collective result of choices made by millions of well-meaning individuals trying to make a better life for themselves and their families. In this view, those who decry sprawl are elitist, show disdain for the popular will, and fail to appreciate the good things about suburban life. Let us try, then, to leave the emotional debate aside and assess the impacts of sprawl (see "The Science behind the Story," ▸ pp. 382–383).

Transportation Most studies show that sprawl constrains transportation options, essentially forcing people to drive cars. These constraints include the need to own a vehicle and to drive it most places, the need to drive greater distances or to spend more time in vehicles, a lack of mass transit options, and more traffic accidents. Across the United States, during the 1980s and 1990s the average length of work trips rose by 36%, and total vehicle miles driven increased at three times the rate of population growth. An automobile-oriented culture also increases dependence on nonrenewable petroleum, with the attendant economic and environmental consequences (▸ pp. 575–580).

Pollution Sprawl's effects on transportation give rise to increased pollution. Carbon dioxide emissions from vehicles exacerbate global climate change (Chapter 18) while nitrogen- and sulfur-containing air pollutants contribute to tropospheric ozone, urban smog, and acid precipitation (▸ pp. 514–518). Waterways are polluted by substances such as motor oil and road salt from roads and parking lots. Runoff of polluted water from paved areas is estimated to be about 16 times greater than from naturally vegetated areas. Such air and water pollution has been shown to degrade natural environments and pose risks to human health.

Health Besides the health impacts of pollution, some recent studies suggest that sprawl promotes physical inactivity because driving cars largely takes the place of walking during daily errands. This research has suggested that physical inactivity has increased obesity and high blood pressure, which can in turn lead to other ailments. A 2003 study found that people from the most-sprawling U.S. counties weigh 2.7 kg (6 lb) more for their height than people from the least-sprawling U.S.

VIEWPOINTS

Suburban Sprawl

Do you see problems with urban and suburban sprawl? **If so, what are the best solutions? What strategies should we pursue in making our communities more livable?**

The Myth of Urban Sprawl

Russians say Americans have no real problems, so they make them up. Urban sprawl is one of those made-up problems. Yet the proposed remedy to sprawl—sometimes called "smart growth," though it is anything but smart—will cause far more problems than it solves.

The U.S. Department of Agriculture says that urban development does not threaten American farm productivity. Nor is it a threat to rural open space: All of the cities, suburbs, and towns in the United States occupy less than 4% of the nation's land area.

University of Southern California planning professor Peter Gordon points out that low-density development is a remedy for, not the cause of, congestion, air pollution, and many other problems. Studies claiming that suburbs cause obesity, crime, and other social problems are little more than junk science, being based on inadequate data with little statistical significance and usually confusing cause and effect.

So-called smart growth says more people should live in high-density, mixed-use developments. These developments can be attractive to some people, mainly young adults with no children. But the market for them is limited. Most Americans still find a single-family home with a large yard to be their American dream.

Attempting to impose smart growth on more people has many unfortunate effects. Because it doesn't significantly reduce the miles people drive, it increases congestion; and because cars pollute more in stop-and-go traffic, it increases air pollution. Smart growth makes housing unaffordable to low- and even middle-income families, and makes neighborhoods more vulnerable to crime.

Instead of attempting to impose their lifestyle preferences on others, city officials should simply ensure that people pay the full costs of whatever lifestyle they prefer. Once that happens, people can be free to choose to live in high densities or low, and to drive, walk, bicycle, or ride transit.

Randal O'Toole is an economist and the director of the American Dream Coalition, which seeks to solve urban problems without reducing people's personal freedom. He is the author of *The Vanishing Automobile and Other Urban Myths: How Smart Growth Will Harm American Cities.* He has taught environmental economics at Yale, the University of California at Berkeley, and Utah State University.

The Real Problem with Sprawl

The most visible problem associated with sprawl is one of livability—sprawl's effect on our quality of life. However, the bigger but less visible problem is sprawl's contribution to global warming.

Until the mid-20th century, every village, town, and city in the world was made up of mixed-use, pedestrian-friendly neighborhoods. But after World War II, the time-tested *neighborhood* was discarded in favor of an untested invention, now known as *sprawl.*

Sprawl is based on two simple premises: first, that each land use be separated from every other; and second, that these now distant land uses be connected by a massive automotive infrastructure. In sprawl, walking not only serves limited purposes but also can often be dangerous. For this reason, a typical suburban household generates more than 12 one-way car trips per day.

The ecological impact is profound. Motor vehicles, the lifeblood of sprawl, are the single greatest contributor to global warming. Over the past 60 years, we have created a built environment that requires most adults to own a car and drive it every day. Unless we quickly make the change to nonpolluting vehicles, global warming will remain the strongest argument against sprawl.

But what if cars did not pollute; would sprawl then offer a satisfactory solution? This is where quality of life enters the picture. In a society in which cars are a prerequisite to social and economic viability, those who can't drive—one-third of the population—become second-class citizens. And because everyone who can drive must drive, we spend inordinate amounts of time stuck in traffic, time that would be better spent in less stressful pursuits. The frustration with this situation—the hours that we spend trying to reconnect our artificially disassociated lives—is the reason why most people who argue against sprawl cite concerns over quality of life.

Jeff Speck is director of design at the National Endowment for the Arts. Prior to joining the Endowment in 2003, he spent 10 years as Director of Town Planning at the firm of Duany Plater-Zyberk and Co., Architects and Town Planners (DPZ). With Andres Duany and Elizabeth Plater-Zyberk, he is the co-author of the book *Suburban Nation: The Rise of Sprawl and the Decline of the American Dream,* published in March 2000 by North Point/Farrar Straus Giroux.

Explore this issue further by accessing **Viewpoints** at www.aw-bc.com/withgott.

The Science behind the Story

Measuring the Impacts of Sprawl

Critics of sprawl have blamed it for so many societal ills that it can make a person feel guilty just for being born in the suburbs or shopping at a mall. But what does scientific research tell us are the actual consequences of sprawl?

When Reid Ewing of Rutgers University and his team set out to measure the impacts of sprawl, they discovered that researchers studying sprawl have been hard-pressed even to agree on a definition of the term or on how to measure it. Surveying the literature, Ewing's team found that researchers using different criteria ranked cities in very different ways. For instance, in most studies Los Angeles was deemed more sprawling than Portland, but in some, Portland was judged to suffer worse sprawl than L.A.

So Ewing, Rolf Pendall of Cornell University, and Don Chen of the nonprofit group Smart Growth America tried to define *sprawl* in terms as simple as possible, without mixing sprawl's consequences into the definition. They decided that sprawl occurs when the spread of development across the landscape far outpaces population growth.

Ewing, Pendall, and Chen then devised four criteria by which to rank 83 of the largest U.S. metropolitan areas in terms of sprawl. Sprawling cities would show:

Traffic congestion is one of the most recognized impacts of sprawl.

- Low residential density
- Distant separation of homes, employment, shopping, and schools
- Lack of "centeredness," that is, lack of activity in community centers and downtown areas
- Street networks that make many streets hard to access

For each criterion, Ewing's team measured multiple factors—22 variables in all. They then devised a way to analyze the variables and arrive at a cumulative index of sprawl. Finally, they obtained data from municipalities throughout the country.

Ewing's team's rankings showed that the most sprawling area in the nation was the Riverside–San Bernardino, California, region, which has expanded quickly in recent decades. The area with the least sprawl was New York City, whose historically dense population and vibrant neighborhoods kept it geographically compact, relative to its number of inhabitants. The 10 most- and least-sprawling areas are listed in the accompanying tables.

With their sprawl scores in hand, the researchers next correlated those scores with a number of transportation variables. They found that people in the 10 most-sprawling metros owned more cars (180 per 100 households) than people in the 10 least-sprawling metros (162 per 100 households). They also found that residents of the most-sprawling metros drove an average of 43 km (27 mi) per day, whereas those of the least-sprawling metros drove only 34 km (21 mi). In addition, people in the most-sprawling metros used public transit far less and suffered 67% more traffic fatalities than those in the least-sprawling metros.

Strikingly, the study found no significant difference in commute time from home to work for people of sprawling versus less-sprawling metro areas. Critics of sprawl have long blamed sprawl for traffic congestion and commute delays. Advocates of suburban spread have argued that more streets ease commutes and that regions can sprawl their way out of congestion. The Ewing team's results seem to suggest that each side may have a point and that the effects may cancel one another out.

counties. Over 25% of people from the most-sprawling counties showed hypertension (high blood pressure), whereas fewer than 23% of people from the least-sprawling U.S. counties showed this condition.

Land use The spread of low-density development over large areas of land means that more land is developed while less is left as forests, fields, farmland, or ranchland. Of the estimated 1 million ha (2.5 million acres) of U.S.

Because vehicle emissions cause air pollution, the researchers also measured levels of tropospheric ozone (▸p. 507). Sprawling areas had worse air pollution, they found; ozone levels were 40% higher in the most-sprawling metros.

None of these results prove that sprawl *causes* these impacts, because statistical correlation alone does not imply causation. However, taken together, the results suggest that spatial patterns of development may influence people's options, impacts, and behavior relative to transportation. The results were published in 2003 in the *Transportation Research Record*, and in a 2002 report published by Smart Growth America.

Ewing and his colleagues are continuing to examine other impacts of sprawl. For instance, a study in 2004 went beyond ozone to look at other health correlates of sprawl. Because other studies have found that residents of sprawling areas depend more on cars and walk less, the researchers hypothesized that they would find more obesity and poorer health in people living in sprawling areas. That is exactly what they found: People from sprawling metros are on average heavier for their height and show increased instances of high blood pressure (although not more diabetes or cardiovascular disease).

The 10 Most-Sprawling American Urban Areas

Rank	Metropolitan Region
1	Riverside–San Bernardino, CA
2	Greensboro–Winston-Salem–High Point, NC
3	Raleigh–Durham, NC
4	Atlanta, GA
5	Greenville–Spartanburg, SC
6	West Palm Beach–Boca Raton–Delray Beach, FL
7	Bridgeport–Stamford–Norwalk–Danbury, CT
8	Knoxville, TN
9	Oxnard–Ventura, CA
10	Fort Worth–Arlington, TX

1 = most-sprawling, 10 = less-sprawling.

The 10 Least-Sprawling American Urban Areas

Rank	Metropolitan Region
83	New York, NY
82	Jersey City, NJ
81	Providence–Pawtucket–Woonsocket, RI
80	San Francisco, CA
79	Honolulu, HI
78	Omaha, NE-IA
77	Boston–Lawrence–Salem–Lowell–Brockton, MA
76	Portland, OR
75	Miami–Hialeah, FL
74	New Orleans, LA

83 = least-sprawling, 74 = more-sprawling. *Source:* Ewing, R., et al. 2002. *Measuring sprawl and its impact.* Washington, D.C.: Smart Growth America.

As more studies on the impacts of sprawl accumulate, we will have a better idea of the benefits and costs of our choices in urban design.

land converted each year, roughly 60% is agricultural land and 40% is forest. These types of open space provide vital resource production, aesthetic beauty, habitat for wildlife, cleansing of water, places for recreation, and many other ecosystem services (▸p. 39, 41). Sprawl generally diminishes all these amenities.

Economics Sprawl drains tax dollars from existing communities and funnels them into infrastructure for new development on the fringes of those communities. Money that could be spent maintaining and improving downtown centers is instead spent on extending the road system, water and sewer system, electricity grid, and telephone lines to distant developments, and extending police and fire service, schools, and libraries. For instance, one study calculated that sprawling development at Virginia Beach, Virginia, would require 81% more in infrastructure costs and would drain 3.7 times more from the community's general fund each year than compact urban development. Advocates for sprawling development argue that taxes on new development eventually pay back the investment made in infrastructure, but studies have found that in most cases taxpayers continue to subsidize new development, especially if municipalities do not pass on infrastructure costs to developers.

Weighing the Issues:
Sprawl Near You

Is there sprawl in the area where you live? Are you bothered by it, or not? Has development in your area had any of the impacts described above? Do you think your city or town should use its resources to encourage outward growth?

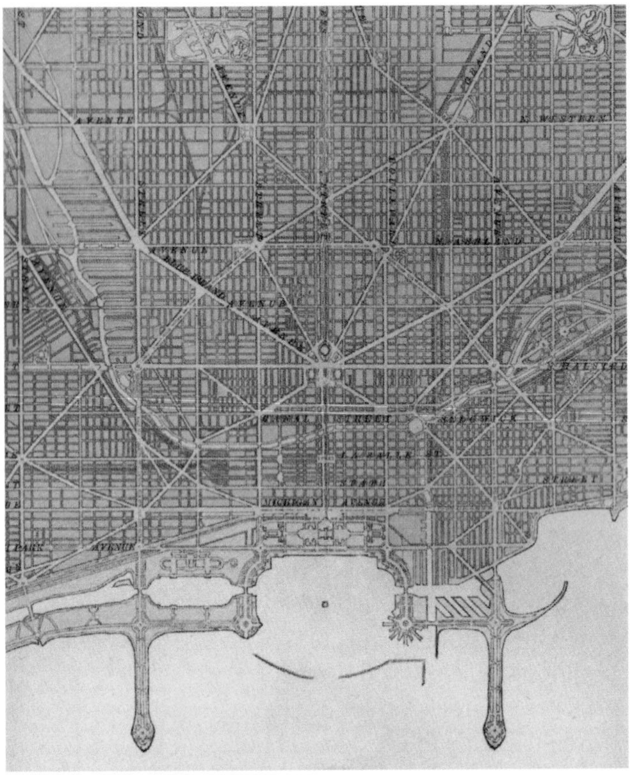

FIGURE 13.6 Daniel Burnham's 1909 *Plan of Chicago* included parks and greenways, efficient transportation routes, and increased access to the lakefront for city residents. The plan has served as a model for city planners ever since.

Creating Livable Cities

Architects, planners, developers, and policymakers across North America today are trying to restore the vitality of city centers and respond to the challenges that suburban sprawl presents. But efforts to design and restore cities are as old as cities themselves.

City and regional planning are means for creating livable urban areas

City planning is the professional pursuit that attempts to design cities so as to maximize their efficiency, functionality, and beauty. City planners advise policymakers on where different types of development should be allowed, transportation needs, public parks, and other matters.

City planning in North America came into its own in the early 20th century. Landscape architect Daniel Burnham's 1909 *Plan of Chicago* (Figure 13.6) represented the first thorough plan for an American city, and it was largely implemented over the following years and decades. This plan expanded city parks and playgrounds, improved neighborhood living conditions, streamlined traffic systems, and cleared industry and railroads from the shore of Lake Michigan to provide public access to the lake.

Portland gained its own comprehensive plan just 3 years later. Edward Bennett's *Greater Portland Plan* recommended rebuilding the harbor; dredging the river channel; constructing new docks, bridges, tunnels, and a waterfront railroad; superimposing wide radial boulevards on the old city street grid; establishing civic centers downtown; and greatly expanding the number of parks. Voters approved the plan by a 2:1 margin, but they defeated a bond issue that would have paid for park development. As the century progressed, several other major planning efforts were conducted, and some ideas, such as establishing a downtown public square, came to fruition.

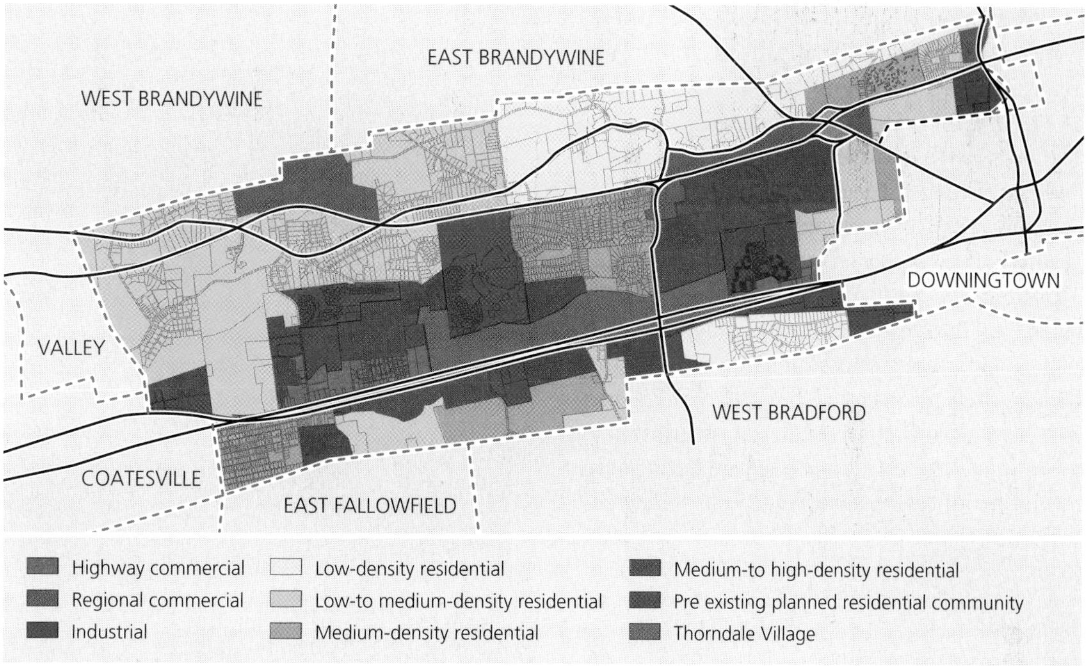

FIGURE 13.7 By zoning areas of a city for different uses, planners can guide how an urban area develops. Zoning puts restrictions on what private landowners can do with their land, but it is intended to maximize prosperity, efficiency, and quality of life for the community. This zoning map for Caln Township, Pennsylvania, shows several patterns common to modern zoning practice. Public and institutional uses are clustered together in a downtown area. Industrial uses are clustered together, away from most residential areas. Commercial uses are clustered along major roadways, and residential zones generally are higher in density toward the center of town.

City planning grew in importance throughout the 20th century as urban populations expanded, inner cities decayed, and wealthier residents fled to the suburbs. In today's world of sprawling metropolitan areas, **regional planning** has become just as important. Regional planners deal with the same issues as city planners, but they work on broader geographic scales and must coordinate their work with multiple municipal governments. In some places, regional planning has been institutionalized in formal governmental bodies; the Portland area's Metro is the epitome of such a regional planning entity.

Weighing the Issues: Your Urban Area

Think of your favorite parts of the city you know best. What aspects do you like about them? What do you dislike about some of your least favorite parts of the city? What could this city do to improve the quality of life for its inhabitants?

Zoning is a key tool for planning

One tool planners use is **zoning,** the practice of classifying areas for different types of development and land use (Figure 13.7). Industrial plants may be kept out of districts zoned for residential use to preserve the cleanliness and tranquility of residential neighborhoods, for instance. The specification of zones for different types of development gives planners a powerful means of guiding what gets built where. Zoning can restrict areas to just one use, as is often done with suburban residential tracts in so-called bedroom communities. Or, zoning can allow the type of mixed use—residential and commercial, for instance—that some planners say can reinvigorate urban neighborhoods. Zoning also gives home buyers and business owners security; they know in advance what types of development can and cannot be located nearby.

Zoning involves government restriction of the use of private land and represents a top-down constraint on personal property rights. For this reason, some people consider zoning a regulatory taking (▸ p. 65) that violates individual freedoms. Others defend zoning, saying that government has a proper role in setting certain limitations on property rights for the good of the community. Similar debates arise with endangered species management (▸ pp. 331–334) and other environmental issues.

Oregon voters sided with private property rights when in 2004 they passed Ballot Measure 37, which shackled government's ability to enforce zoning regulations with landowners who had owned their land before the regulations were enacted. However, opponents of the measure insisted that the state's voters had not understood the complicated terms of the proposal. They predicted that many will change their minds once they begin seeing new development they do not wish to occur. For the most part, people have supported zoning over the years because the common good it produces for communities is widely felt to outweigh the restrictions on private use.

Weighing the Issues: Zoning and Development

Imagine you own a 10-acre parcel of land that you want to sell for housing development—but the local zoning board rezones the land so as to prohibit the development. How would you respond?

Now imagine that you live next to someone else's undeveloped 10-acre parcel, and you enjoy the privacy it provides—but the local zoning board rezones the land so that it can be developed into a dense housing subdivision. How would you respond?

What factors do you think members of a zoning board should take into consideration when deciding how to zone or rezone land in a community?

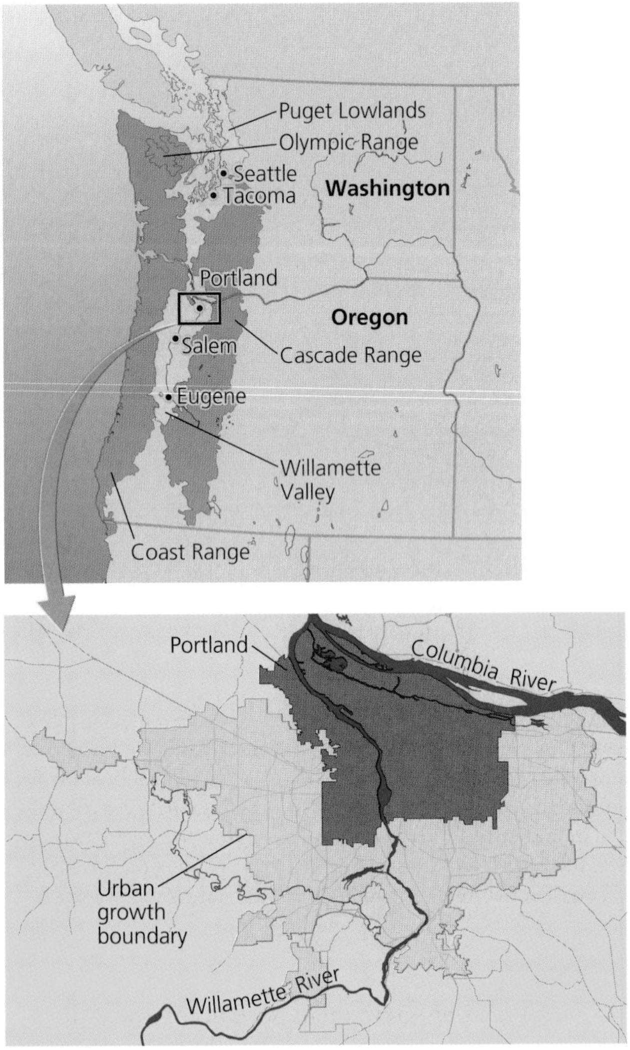

FIGURE 13.8 Oregon's urban growth boundaries (UGBs) were a response to fears that suburban sprawl might one day stretch in a great megalopolis from Eugene, Oregon, up the Willamette Valley to Portland, and up the Puget Lowlands to Seattle, Washington. The Portland area's 956-km^2 (369-mi^2) UGB encompasses Portland (dark gray) and portions of 24 other communities and three counties (light gray). Its jagged edge separates areas planners have earmarked for urban development from areas they have chosen to protect from urban development.

Urban growth boundaries have become popular

Planners intended Oregon's urban growth boundaries to limit sprawl by containing future growth largely within existing urbanized areas. The UGBs aimed to revitalize downtowns; protect farms, forests, and their industries; and ensure urban dwellers some access to open space near cities. Among Willamette Valley towns, the long-term goal was to head off the growth of a potential megalopolis stretching from Eugene to Seattle (Figure 13.8).

Since Oregon began its experiment, a number of other states, regions, and cities have adopted UGBs—from Boulder, Colorado, to Lancaster, Pennsylvania, to many California communities. In their own ways, all the UGBs aim to concentrate development, prevent sprawl, and preserve working farms, orchards, ranches, and forests. UGBs unquestionably help maintain farms and natural areas. UGBs also appear to reduce the amounts municipalities have to pay for infrastructure, compared to sprawl. The best estimate nationally is that UGBs save taxpayers about 20% on infrastructure costs. However, UGBs also seem to increase housing prices within their boundaries, although precisely how much is difficult to assess; the best estimate for Portland is that its UGB adds roughly $10,000 to the average home price.

Although housing is becoming less affordable, in most other ways the Portland-area UGB (see Figure 13.8) is working as intended. It has lowered prices for land outside the UGB while increasing prices within it. It has restricted development outside the UGB. It has increased the density of new housing inside the UGB by over 50% as homes are built on smaller lots and as multistory apartments fulfill a vision of "building up, not out." Downtown employment rose by 73% between 1970 and 1995 as businesses and residents alike invested in the central city. And Portland has been able to absorb considerable immigration while avoiding rampant sprawl. However, urbanized area still did increase by 101 km^2 (39 mi^2) in the decade after the UGB was established, because 146,000 people were added to the population. This fact suggests that relentless population growth may thwart even the best antisprawl efforts.

When Oregon's lawmakers passed Senate Bill 100, they understood that the UGBs would likely need to be enlarged from time to time to accommodate population growth. Indeed, Metro has enlarged the Portland-area UGB three dozen times, expanding it by over 10% between 1998 and 2004. Population projections for the region suggest there will be pressure for still more expansion. Many other locations that have instituted UGBs have chosen to expand them later, as well. For instance, a group of 48 local governments from cities, towns, and counties along the Colorado Front Range in 1997 agreed to a UGB that allowed expansion of the urbanized area to 1,130 km^2 (700 mi^2) by the year 2020, a 30% increase in size. Soon, however, this limit was raised three times, finally to 1,369 km^2 (850 mi^2) by the year 2030. Given the reality of repeated UGB expansions, it must be asked whether UGBs will truly limit sprawl in the long run.

"Smart growth" aims to counter sprawl

As more people have begun to feel negative effects of sprawl on their everyday lives, efforts to control growth have sprung up throughout North America. Oregon's Senate Bill 100 was one of the first, and since then dozens of states, regions, and cities have adopted similar land use policies. Urban growth boundaries and many other ideas from these policies have coalesced under the concept of **smart growth**.

Proponents of smart growth want communities to manage their growth in ways that maintain or improve quality of life for residents. To them, this means guiding the rate, placement, and style of development such that it serves the environment, the economy, and the community. Smart growth tries to promote healthy neighborhoods and communities, jobs and economic development, transportation options, and environmental quality. It aims to counter sprawl and to rejuvenate the older existing communities that so often are drained and impoverished by sprawl. Smart growth often means "building up, not out"—focusing development and economic investment in existing urban centers and favoring multistory shop-houses and high-rises over one-story homes spreading outward. Ten principles of smart growth are listed in Table 13.2.

Table 13.2 Ten Principles of "Smart Growth"

- Mix land uses
- Take advantage of compact building design
- Create a range of housing opportunities and choices
- Create walkable neighborhoods
- Foster distinctive, attractive communities with a strong sense of place
- Preserve open space, farmland, natural beauty, and critical environmental areas
- Strengthen and direct development towards existing communities
- Provide a variety of transportation choices
- Make development decisions predictable, fair, and cost-effective
- Encourage community and stakeholder collaboration in development decisions

Source: U.S. Environmental Protection Agency, 2005.

People have pursued the goals of smart growth in various ways. Several hundred referenda and ballot measures in dozens of U.S. states in the past decade have proposed various solutions to slow sprawl and rechannel development into more community-friendly paths. Many of these ballot measures have won, and a number of experiments are going on right now across North America.

The "new urbanism" is now in vogue

Much of what Portland and cities like it have been trying to accomplish is compatible with a school of thought called the **new urbanism,** currently in vogue among many architects, planners, and developers. This approach seeks to design neighborhoods on a walkable scale, with homes, businesses, schools, and other amenities all close together for convenience. The aim is to create functional neighborhoods in which most of a family's needs can be met close to home without the use of a car. Green spaces, trees, a mix of architectural styles,

FIGURE 13.9 This plaza at Mizner Park, Boca Raton, Florida, is part of a planned community built in the style of the "new urbanism." Homes, schools, and businesses are mixed close together in a centered neighborhood so that most amenities are within walking distance.

and creative street layouts add to the visual interest and pleasantness of new urbanist developments (Figure 13.9). In these ways, these developments mimic the traditional urban neighborhoods that existed until the advent of suburbs.

Over 600 communities in the new urbanist style across North America are in various stages of planning or construction, and many are now complete. Seaside, Florida, was the first, and others have followed, such as Kentlands in Gaithersburg, Maryland; Addison Circle in Addison, Texas; Mashpee Commons in Mashpee, Massachusetts; Harbor Town in Memphis, Tennessee; Celebration in Orlando, Florida; and Orenco Station, west of Portland.

New urbanist neighborhoods are generally connected to public transit systems. In a related approach called *transit-oriented development,* compact communities in the new urbanist style are arrayed around the stops on a major rail transit line, enabling people to travel most places they need to go by train and foot alone. Several lines of the Washington, D.C. Metro system have been successfully developed in this manner.

Transportation options are vital to livable cities

A key ingredient in any planner's recipe for improving the quality of urban life is making multiple transportation options available to citizens. These options include public buses, trains and subways, and light rail (smaller rail systems powered by overhead electric wires). As long as an urban center is large enough to support the infrastructure necessary, these mass transit options are cheaper, more energy-efficient, and cleaner than roadways choked with cars (Figure 13.10). They also ease traffic congestion by carrying passengers who would otherwise be driving cars. An average Portland bus is estimated to keep about 250 cars off the road each day. Compared to mass transit rail systems, road networks take up more space, and cars emit more pollution. The fuel and productivity lost to traffic jams have been estimated to cost the U.S. economy $74 billion each year.

The nation's most-used train systems have been the extensive heavy rail systems in America's largest cities, such as New York's subways, Washington, D.C.'s Metro, the T in Boston, and the San Francisco Bay area's BART, each of which carries more than one-fourth of each city's daily commuters. In Canada, rail systems in Montreal and Toronto enjoy similarly heavy use. Major cities internationally from Moscow to Beijing to Paris to Tokyo also have large and heavily utilized subway systems. Some cities with severe traffic problems, such as Bangkok, Thailand, and Athens, Greece, have recently opened new rail systems that carry hundreds of thousands of commuters a day. Light rail use is increasing in Europe, and U.S. ridership is now rising faster than is the rate of new car drivers. Most countries have bus systems that are far more accessible to citizens than are those of the United States.

One city famous for its efficient transportation system is Curitiba, Brazil. Faced with a heavy influx of immigrants from outlying farms in the 1970s, visionary city leaders led by Mayor Jaime Lerner decided to pursue an

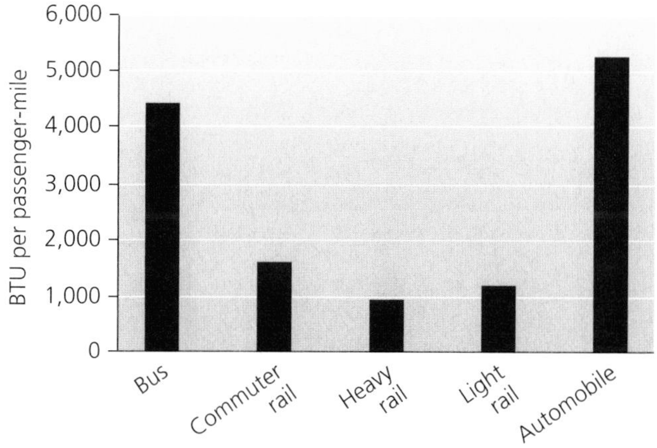

(a) Energy consumption for different modes of transit

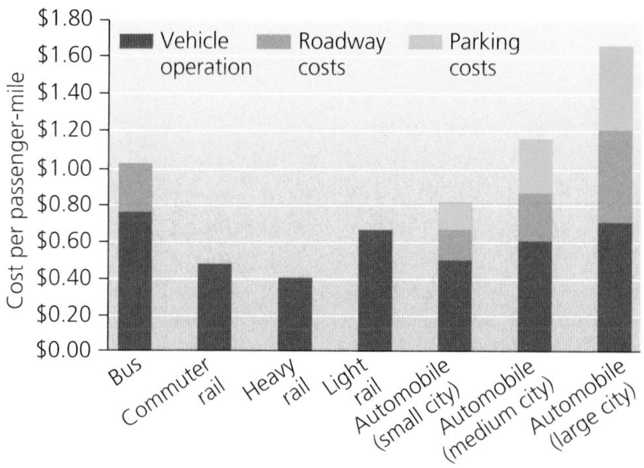

(b) Operating costs for different modes of transit

FIGURE 13.10 Rail transit tends to be cheaper and more energy-efficient than road-based transportation. Rail transit consumes far less energy per passenger mile (**a**) than bus or automobile transit. Rail transit involves fewer costs per passenger mile (**b**) than bus or automobile transit. *Source:* Litman, T. 2005. *Rail transit in America: A comprehensive evaluation of benefits.* Victoria, B.C.: Victoria Transport Policy Institute.

(a) MAX light rail train

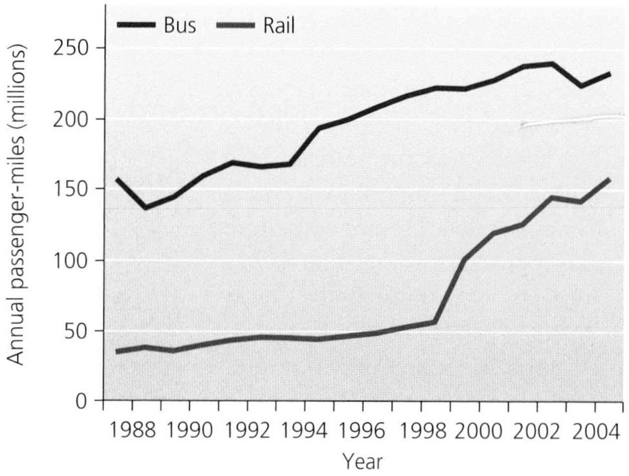

(b) Portland transit ridership trends

FIGURE 13.11 Portland's light rail system (**a**) is one component of an urban planning strategy that has helped make it one of North America's most livable cities. In Portland, bus ridership has been increasing slowly, whereas ridership on the MAX light rail system is growing more quickly (**b**).

aggressive planning process so that they could direct growth rather than being overwhelmed by it. They reconfigured Curitiba's road system to maximize the efficiency of a large fleet of public buses. Today this metropolis of 2.5 million people has an outstanding bus system that is used each day by three-quarters of the population. The 340 bus routes, 250 terminals, and 1,900 buses accompany measures to encourage bicycles and pedestrians. All of this has resulted in a steep drop in car usage, despite the city's rapidly growing population.

Portland's well-organized bus system carries 63 million riders per year. In 1986 Portland introduced a light rail system called MAX, and Metro policy encouraged the development of self-sufficient neighborhood communities in the new urbanist style along the rail lines. Light rail ridership has steadily increased as the system has expanded (Figure 13.11).

Establishing mass transit is not always easy, however. Planners and zoning boards face many constraints, and one of the biggest is the road-based transportation system that encourages individuals to drive their own automobiles. Once a road system has been developed and businesses and homes are built alongside roads, it can be difficult and expensive to replace or complement the road system with a mass transit system. In addition, modes of mass transit differ in their effectiveness (see "The Science behind the Story", ▸ pp. 390–391), depending on city size, size of the transit system, and other factors.

The Science Behind the Story

Assessing Benefits of Rail Transit

Most urban planning experts see benefits in public mass transit and in making transit systems readily accessible to citizens. But building a mass transit system—and then maintaining and expanding it—can be a hugely expensive undertaking. Thus, planners and policymakers greatly value quantitative information on the costs and benefits of different types of transit systems and on how extensive a given transit system should be to produce the benefits they desire.

For this reason, researcher Todd Litman of the nonprofit Victoria Transport Policy Institute conducted a comprehensive evaluation of the benefits of rail transit in the United States, published in 2005 by the Victoria Transport Policy Institute, with support from the American Public Transport Association.

Litman first divided American cities into three categories. "Bus-only" cities have bus service but no rail service. "Small-rail" cities have rail systems serving fewer than 12% of daily commuters; these include 16 cities ranging from Atlanta to St. Louis to Denver to Salt Lake City. "Large-rail" cities feature rail service

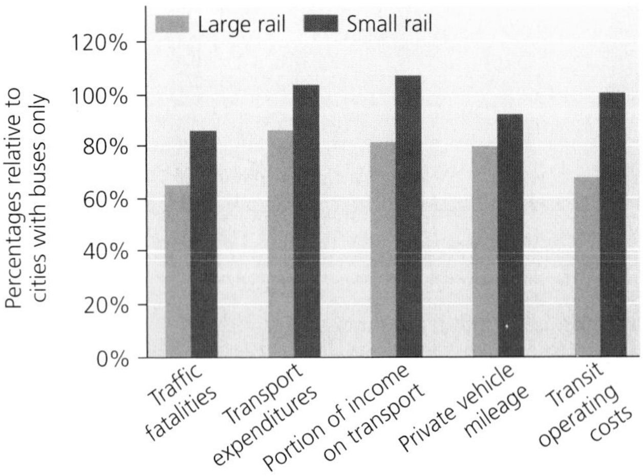

(a) Benefits of rail transport per capita

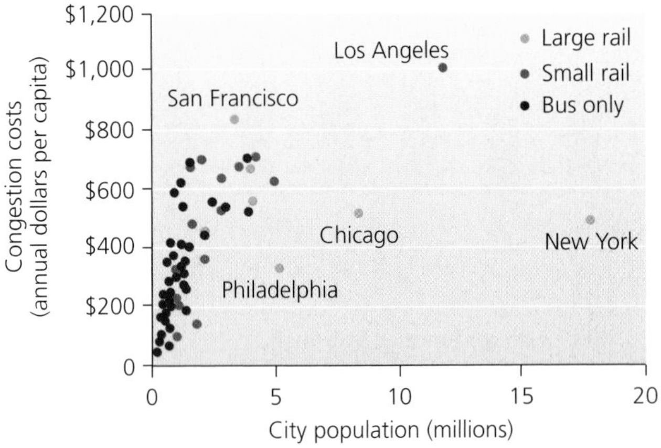

(b) Transport mode, congestion costs, and city size

Large-rail cities outperform bus-only cities in transport-oriented variables, but the benefits of rail (**a**) are not so clear-cut for small-rail cities. Per capita congestion costs increase with city size, except in large-rail cities (**b**). *Adapted from:* Litman, T. 2005, *Rail transit in America: A comprehensive evaluation of benefits.* Victoria, B.C.: Victoria Transport Policy Institute.

Fortunately, installing mass transit systems is not the only way to make urban transportation run more efficiently. Governments can also raise fuel taxes, tax inefficient modes of transport, reward carpoolers with carpool lanes, encourage bicycle use and bus ridership, and charge trucks for road damage. Overall, they can choose to minimize investment in infrastructure that encourages sprawl, and stimulate investment in renewed urban centers.

Parks and open space are key elements of livable cities

City dwellers often desire some sense of escape from the noise, commotion, and stress of urban life. Natural lands, public parks, and open space provide greenery, scenic beauty, freedom of movement, and places for recreation. These lands also keep ecological processes functioning by regulating climate, producing oxygen, filtering air and water pollutants, and providing habitat for wildlife. The as a major component of the transportation system; these include New York, Washington, D.C., Boston, San Francisco, Chicago, Philadelphia, and Baltimore, whose rail systems serve 12–48% of daily commuters.

Litman compared these three groups of cities for a number of variables. Several key results are summarized in part (a) in the accompanying figure. Compared with bus-only cities, large-rail cities had 36% fewer per capita traffic deaths each year. Residents of large-rail cities drove 21% fewer miles yearly than those of bus-only cities. Large-rail city residents also saved money on transportation, spending 14% less than bus-only city residents on transportation—equivalent to annual savings of $448. Transportation costs took up only 12% of their household budgets, compared to 14.9% of those of bus-only city residents.

The analysis also revealed benefits of large-rail systems for municipalities running the systems. Per passenger mile, large-rail cities paid only 42 cents in operating costs, versus 63 cents paid by bus-only cities. In addition, large-rail cities recovered far more of their costs than bus-only cities (38% cost recovery versus 24%).

Small-rail cities were intermediate in traffic deaths and vehicle mileage, as might be expected. However, they performed no better than bus-only cities in terms of operating costs per passenger mile or transit cost recovery. And interestingly, residents of small-rail cities fared slightly worse than residents of bus-only cities in annual transport expenditures and proportion of income spent on transportation. These results suggest that not all benefits of rail transit begin to accrue with small systems. Rather, for many factors, a rail system has to be large enough or accommodate enough riders to gain the economy of scale needed to provide benefits.

A similar pattern was found for annual per capita costs due to traffic congestion (part (b) in the figure). For bus-only and small-rail cities, congestion costs increased with city size. But for large-rail cities, congestion costs did not vary with size, and were lower than comparably sized cities.

Litman went on to quantify and assess benefits and costs of rail transit overall. He found that each year governments spend $12.5 billion on rail transit systems that they do not get back in revenues from fares. That means governments are subsidizing rail transit by $140 per resident of cities with rail systems.

Although these numbers are considerable, they are surpassed by the monetary benefits of rail systems. Each year, rail systems are estimated to save $19.4 billion in congestion costs, $22.6 billion in consumer transportation costs, $8.0 billion in roadway costs, $12.1 billion in parking costs, and $5.6 billion in accident costs, for a total of $67.7 billion in total annual savings. And that figure does not even account for indirect savings due to enhanced environmental quality.

A number of studies have critiqued rail transit and portrayed it as ineffective. The Litman study indicates that although not all rail systems are cost-effective, rail systems become more beneficial as they become larger and carry a greater proportion of a city's commuters.

animals and plants of urban parks and natural lands also serve to satisfy biophilia (▸ p. 328), our natural affinity for contact with other organisms.

Protecting natural lands and establishing public parks become more important as human societies become more urbanized, because many urban dwellers come to feel increasingly isolated and disconnected from nature. In the wake of urbanization and sprawl, people of every industrialized society in the world today have, to some degree, chosen to set aside land in public parks.

City parks were widely established at the turn of the last century

At the turn of the 20th century in urban America, civic improvement was garnering interest and support as politicians and citizens alike yearned for ways to make their crowded and dirty cities more livable. In the late 1800s, public parks began to be established in eastern U.S. cities, using aesthetic ideals borrowed from European parks, gardens, and royal hunting grounds. The

FIGURE 13.12 City parks were developed in many urban areas in the late 19th century to provide citizens aesthetic pleasure, recreation, and relief from the stresses of the city. Manhattan's Central Park, shown here, was one of the first.

lawns, shaded groves, curved pathways, and pastoral vistas we see today in many American city parks and cemeteries originated with these European ideals, as interpreted by the leading American landscape architect, Frederick Law Olmsted. Olmsted designed New York's Central Park in 1853 and a host of urban park systems afterwards (Figure 13.12). Olmsted and his followers sought to create landscapes that provided tranquility, greenery, and an escape from pollution and stress.

Two sometimes conflicting goals motivated the establishment and design of early city parks. On the one hand, the parks were meant to be "pleasure grounds" for the wealthy, who helped support their establishment financially and who would ride the parks' winding roadways in carriages. On the other hand, the parks were meant to alleviate congestion and allow some escape for the many poverty-stricken immigrants who lived in America's cities at the time. These park users were more interested in active recreation, such as ballgames, than in carriage rides. At times the aesthetic interests of the educated elite and the recreational interests of the laboring class came into conflict—a friction that survives today in debates over recreation in city and national parks.

East Coast cities, such as New York, developed parks early on, but cities further west were not far behind. In Chicago, civic boosters striving to overcome the city's reputation as a rough-and-ready frontier town of slaughterhouses and meatpacking plants were glad to have Daniel Burnham design parks and playgrounds and turn a lakeside swamp into stately Jackson Park. Burnham's *Plan of Chicago* also included a program for a greenbelt of forest preserves circling the city's outskirts (Figure 13.13). Thanks to these lands, Chicago-area residents today have places nearby to hike, fish, bird-watch, study nature, or just relax and escape urban noise and commotion. Further west, in San Francisco, William Hammond Hall transformed 2,500 ha (1,000 acres) of the peninsula's natural landscape of dunes into a verdant playground of lawns, trees, gardens, and sports fields. Golden Gate Park remains today one of the world's foremost city parks.

Portland's quest for urban parks began in 1900, when city leaders created a parks commission and then hired Frederick Law Olmsted's son, John Olmsted, to design a citywide park system. His 1904 plan recommended acquiring land to ring the city generously with parks, but no action was taken. Both Olmsted's plan and Edward Bennett's city plan recommended that the large forested ridge on the northwest side of the city be made a public park. The city instead planned to have it developed, but in the Depression, economic troubles, coupled with a few landslides, caused

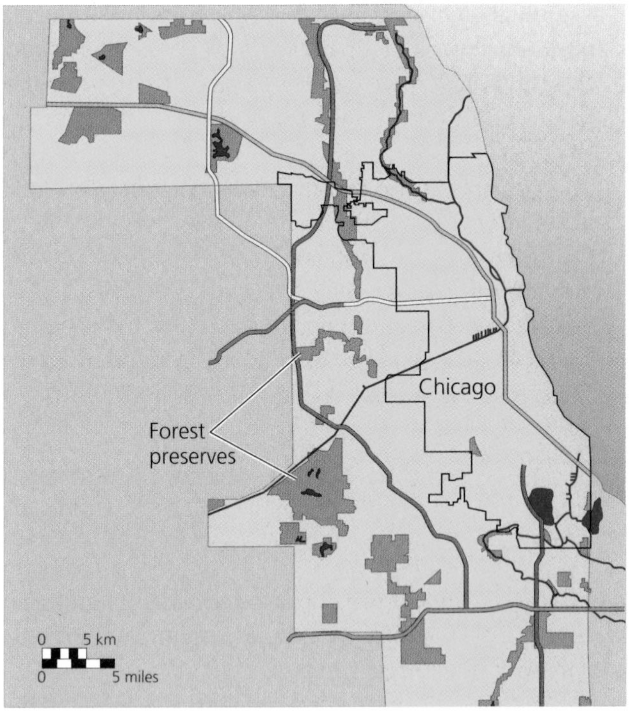

FIGURE 13.13 The forest preserves of Cook County, DuPage County, Lake County, and Will County, Illinois, wind through the suburbs surrounding the city of Chicago. This regional system features 40,000 ha (100,000 acres) of woodlands, fields, marshes, and prairie wildflowers, comprising the largest holding of locally owned public conservation land in the United States.

FIGURE 13.14 Urban community gardens like this one in Seattle provide city residents with a place to grow vegetables and also serve as greenspaces that beautify cities.

developers and investors to forfeit their holdings on the ridge to the city. Finally, citizen pressure resulted in the park's creation in 1948. At 11 km (7 mi) long, Forest Park is the largest city park in the United States.

Smaller public spaces are also important

Large city parks are a key component of a healthy urban environment, but even small spaces can make a big difference. Playgrounds are of immense value for providing places where children can be active outdoors and interact with their peers. Another type of urban space, community gardens, allows people to grow their own vegetables and flowers in a neighborhood setting (Figure 13.14). Portland, Seattle, Baltimore, Boston, and many other cities feature thriving community gardens.

Still another type of greenspace is the corridor or strip of land that connects parks or neighborhoods, often called a *greenway*. Greenways, often located along rivers, streams, or canals, may provide access to networks of walking trails. They can also protect water quality, boost property values, and serve as corridors for the movement of birds and wildlife. The Rails-to-Trails Conservancy has spearheaded the conversion of abandoned railroad rights-of-way into trails for walking, jogging, and biking. To date, nearly 24,000 km (15,000 mi) of 1,200 rail lines have been converted across North America.

Besides creating new types of urban spaces, many cities are working to enhance the "naturalness" of their parks, through ecological restoration (▸ p. 168), the practice of restoring native communities. In Portland parks, volunteer teams remove English ivy, an invasive plant that covers trees and smothers native plants on the forest floor. At some Chicago-area forest preserves, scientists and volunteers use prescribed burns (▸ p. 360) to restore patches of prairie native to the region. In San Francisco's Presidio, certain areas are being restored to the native dune communities that were wholly displaced by urban development.

Weighing the Issues:
Nature in the City

Do you feel it is important that urban residents have access to parks and natural areas within their cities and towns? Why or why not? What advantages strike you as most important? What types of parks or natural areas do you think are most helpful, and why?

Urban Sustainability

Urbanization and the ways that cities and suburbs function have immense consequences for the natural environment. Urban centers exert both positive and negative environmental impacts, and these impacts depend strongly on how we utilize resources, produce goods, transport materials, and deal with waste. Proponents of urban sustainability seek ways to increase the good that comes from living in urban centers, minimize the bad, and make cities sustainable.

Urban resource consumption brings a mix of environmental impacts

Most of us might guess that urban living involves greater consumption of resources than rural living and thus has a greater environmental impact. However, the picture is not so simple; instead, urbanization brings a complex mix of consequences.

Resource sinks Cities and towns are sinks for resources, having to import from widespread sources beyond their borders nearly everything they need to feed, clothe, and house their inhabitants. Urban and suburban areas are utterly dependent on large expanses of agricultural land to supply food and other crops and on undeveloped land to provide natural resources such as water, timber, metal ores, and mined fuels. Urban centers also need areas of natural land to provide ecosystem services, including purification of water and air, nutrient cycling, and waste treatment (▸ pp. 39–41). Major cities such as New York, Boston, San Francisco, and Los Angeles depend for their day-to-day survival on the water they pump in from faraway watersheds (Figure 13.15). As cities have grown, and as the material wealth of most societies has risen, the inexorable pull of resources from the countryside to the cities has become greater and greater. And as urban areas extend their reach, it becomes increasingly hard for urban residents isolated from natural lands to have a tangible sense of the environmental impacts of their choices.

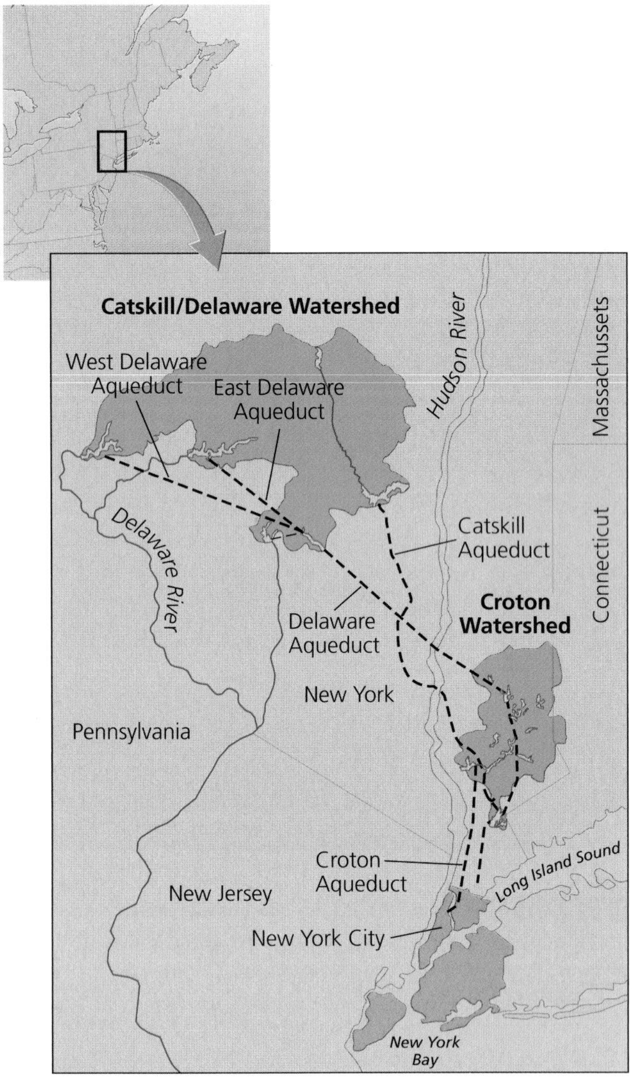

FIGURE 13.15 New York City pipes in its drinking water from upstate reservoirs in the Croton and Catskill/Delaware Watersheds. The city has gone to lengths to acquire, protect, and manage watershed land to minimize pollution of these water sources. When New York City was confronted in 1989 with an order by the Environmental Protection Agency to build a $6 billion filtration plant to protect its citizens against waterborne disease, the city opted instead to purchase and better protect watershed land, for a fraction of the cost.

The long-distance transportation of resources and goods requires a great deal of fossil fuel use, which has significant environmental impacts (Chapter 19). For this reason, the centralization of resource use that urbanization entails may seem a bad thing for the environment. However, imagine that all the world's 3 billion urban residents were instead spread evenly across the landscape. What would the transportation requirements be, then, to move all those resources and goods around to all those people? A world without cities would likely require *more* transportation to provide people the same level of access to resources and goods.

Efficiency Once resources have arrived at an urban center where people are densely concentrated, cities should be able to minimize per capita consumption by maximizing the efficiency of resource use and delivery of goods and services. For instance, providing electricity from a power plant for numerous urban houses close together is more efficient than providing electricity to far-flung homes in the countryside. The density of cities also facilitates the provision of many other social services that improve quality of life, including medical services, education, water and sewer systems, waste disposal, and public transportation.

More consumption Because cities draw resources from afar, their ecological footprints are much greater than their actual land areas. For instance, urban scholar Herbert Girardet calculated that the ecological footprint of London, England, extends 125 times larger than the city's actual area. By another estimate, cities take up only 2% of the world's land surface but consume over 75% of its resources. However, the ecological footprint concept is most meaningful when used on a per capita basis. So, in asking whether urbanization causes increased resource consumption, we must ask whether the average urban dweller has a larger footprint than the average rural dweller. The answer is yes, but urban and suburban residents also tend to be wealthier than rural residents, and wealth correlates best with resource consumption. Thus, although urban citizens tend to consume more than rural ones, the reason could be simply that they tend to be wealthier.

Urban centers preserve land

Although the ecological footprints of urban areas are large, because people are packed densely together in cities, more land outside cities is left open and undeveloped. This is, indeed, the idea behind urban growth boundaries. If cities did not exist, and if instead all 6.5 billion of us were evenly spread across the planet's land area, we would have much less room for agriculture, let alone wilderness, biodiversity, or privacy. There would be no large blocks of land left uninhabited by people or of habitat unfragmented for wildlife. The fact that half the human population is concentrated in discrete locations helps allow room for natural ecosystems to maintain themselves, continue functioning, and provide the ecosystem services on which all of us, urban and rural, depend.

Urban centers suffer and export pollution

Just as cities import resources, they export wastes. By exporting their waste, either passively through pollution or actively through trade, urban centers transfer the costs of their activities to other regions—and mask the costs from their own residents. Citizens of Toronto may not recognize that pollution from coal-fired power plants in their region exacerbates acid precipitation hundreds of miles to the east. Citizens of New York may not realize how much garbage their city produces if it is shipped to other states or nations for disposal.

However, not all waste and pollution leaves the city. Heavy metals, industrial compounds, and chemicals from manufactured products accumulate in soil and water. Airborne pollutants cause photochemical smog, industrial smog, and acid precipitation. Fossil fuel combustion releases carbon dioxide and other pollutants, leading to climate change. Urban residents also suffer noise pollution and light pollution. *Noise pollution* consists of undesired ambient sound. Excess noise degrades one's surroundings aesthetically, can induce stress, and at intense levels (such as with prolonged exposure to leafblowing, lawnmowing, and jackhammering) can harm hearing. *Light pollution* describes the way that city lights obscure the night sky, impairing the visibility of stars.

These various forms of pollution and the health threats it poses are not evenly shared among urban residents. Those who receive the brunt of the pollution are often those who are too poor to live in cleaner areas. Environmental justice concerns (▸pp. 36–38) center on the fact that a disproportionate number of people living near, downstream from, or downwind from factories, power plants, and other polluting facilities are people who are poor and, often, people of racial minorities.

Urban centers foster innovation

One of the greatest impacts of urbanization on environmental quality is also one of the most indirect and intangible. Cities promote a flourishing cultural life and, by mixing together diverse people and influences, spark innovation and creativity. The urban environment can promote education and scientific research, and cities have long been viewed as engines of technological and artistic inventiveness. This inventiveness can lead to solutions to societal problems, including ways to reduce the environmental impacts we have just discussed. For instance, research into renewable energy sources is helping us develop ways to replace fossil fuels. Technological advances have helped us reduce pollution. Wealthy and educated urban populations provide markets for low-impact goods, such as organic produce. Recycling programs help reduce the solid waste stream. Environmental education (Figure 13.16) is helping people choose their own ways to live cleaner, healthier, lower-impact lives. All these phenomena are outgrowths of the education, innovation, science, and technology that are part of urban culture.

Some seek sustainability for cities

Improving quality of life for people while also improving environmental quality—and making these gains permanent—is at the heart of the quest for sustainability. Unfortunately, many critics have noted that the model of a city that imports all its resources and exports all its wastes is not sustainable. Modern cities have a linear, one-way metabolism. Such linear models of production and consumption tend to destabilize environmental systems. Proponents of sustainability for cities urge us to reject the paradigm of an infinite quest for growth and, instead, stress the need to develop circular systems, akin to systems found in nature. Natural systems recycle materials and use renewable sources of energy. For these reasons, they are able to persist indefinitely.

Researchers in the field of **urban ecology** hold that cities can be viewed explicitly as ecosystems and that the fundamentals of ecosystem ecology and systems science (Chapter 7) apply to urban areas. Only by accepting that ecological principles apply to cities, they say, will we have hope of making them sustainable. Major urban ecology projects are ongoing in Baltimore and Phoenix. Researchers

FIGURE 13.16 Urban and suburban children take part in an environmental education program in a Chicago-area forest preserve. Although natural land is rare in urbanized areas, the education and innovation that urbanization often promotes can lead to solutions that reduce environmental impact.

funded by the National Science Foundation's Long Term Ecological Research (LTER) program are studying these two cities explicitly as ecosystems. The researchers are examining such topics as nutrient cycling of carbon and nitrogen, biodiversity patterns, and air and water quality. They are also exploring ways in which humans perceive their environment and react to environmental health threats.

To help cities maintain their standard of living while reducing their environmental impacts, urban sustainability advocates suggest that cities:

- Maximize efficient use of resources
- Recycle as much as possible
- Develop environmentally friendly technologies
- Account fully for external costs
- Offer tax incentives to encourage sustainable practices
- Use locally produced resources
- Use organic waste and wastewater to restore soil fertility
- Encourage urban agriculture

Increasing numbers of cities are adopting one or more of these strategies. For instance, urban agriculture is a growing pursuit in many urban areas, from Portland to Cuba (▸pp. 303–304) to Japan. Singapore produces all its meat and 25% of its vegetable needs within its city limits. In Berlin, Germany, 80,000 people grow food in community gardens, and 16,000 more are on waiting lists.

In general, experts advise that developed countries should invest in resource-efficient technologies to reduce their impacts and enhance their economies, whereas developing countries should invest in basic infrastructure to improve health and living conditions. Curitiba, Brazil shows the kind of success that can result when a city invests in well-planned infrastructure. Besides the highly effective bus transportation network described earlier, the city provides recycling, environmental education, job training for the poor, and free health care. Surveys show that its citizens are unusually happy and better off economically than people living in other Brazilian cities. Successes in places like Curitiba suggest that cities need not be unsustainable. Indeed, because they affect the environment in some positive ways and have the potential for efficient resource use, cities can and should be a key element in achieving environmental progress.

Conclusion

As half the human population has shifted from rural to urban lifestyles, the nature of our impact on the environment has changed. As urban and suburban dwellers, our impacts are less direct but often more far-reaching. Resources must be delivered to us over long distances, requiring the use of still more resources. Limiting the waste of those resources by making our urban areas more sustainable will be vital for the future. Fortunately, the innovative cultural environment that cities foster has helped us develop solutions to alleviate impact and promote sustainability.

Part of seeking urban sustainability lies in making urban areas better places to live. One key component of these efforts involves expanding transportation options to relieve congestion and make cities run more efficiently. Another lies in ensuring access to adequate park lands and greenspaces near and within our urban centers, to keep us from becoming wholly isolated from nature. Accomplishments in city and regional planning have made many American cities more livable than they once were, and we should be encouraged about such progress. Proponents of smart growth and the new urbanism believe they have solutions to the challenges posed by urban and suburban sprawl, although free-market theorists maintain that if people desire suburban lifestyles, governments should not stand in the way of sprawl. Continuing experimentation in cities from Portland, Oregon, to Curitiba, Brazil, will help us determine how best to ensure that urban growth improves our quality of life and does not degrade the quality of our environment.

REVIEWING OBJECTIVES

You should now be able to:

Describe the scale of urbanization

- The world's population is becoming predominantly urban. (p. 374)
- The shift from rural to urban living is driven largely by industrialization and is proceeding fastest now in the developing world. (pp. 374–375)
- Most of the world's metropolises are in developing nations, and nearly all future population growth will be in cities of the developing world. (pp. 375–376)
- The geography of urban areas is changing as cities decentralize and suburbs grow and expand. (pp. 376–378)

Assess urban and suburban sprawl

- Urban and suburban expansion has resulted in sprawl, that is, growth that covers large areas of land with relatively low density development. Both population growth and increased per capita land usage contribute to sprawl. (pp. 378–380)
- Sprawl has resulted from the home-buying choices of individuals who preferred suburbs to cities. (p. 380)
- Sprawl may lead to negative impacts involving transportation, pollution, health, land use, natural habitat, and economics. (pp. 380, 382–384)

Outline city and regional planning and land use strategies

- City and regional planning and zoning are key tools for improving the quality of urban life. (pp. 384–386)
- "Smart growth," urban growth boundaries, and the "new urbanism" attempt to recreate compact and vibrant urban spaces. (pp. 386–388)

Evaluate transportation options

- Mass transit systems can enhance the efficiency of urban areas, but bus and train systems of different sizes bring different benefits. (pp. 388–391)

Describe the roles of urban parks

- Urban park lands are vital for active recreation, soothing the stress of urban life, and keeping people in touch with natural areas. (pp. 390–393)

Analyze environmental impacts and advantages of urban centers

- Cities are resource sinks, and urban dwellers have high per capita resource consumption. However, cities also allow natural lands to be preserved. (pp. 394–395)
- Urban centers can maximize efficiency and help foster innovation that can lead to solutions for environmental problems. (p. 395)

Assess the pursuit of sustainable cities

- The linear mode of consumption and production is unsustainable, and more circular modes will be needed to create sustainable cities. (pp. 395–396)

TESTING YOUR COMPREHENSION

1. What factors lie behind the shift of population from rural areas to urban areas? What types of cities and countries are experiencing the fastest urban growth today, and why?
2. Why have so many city dwellers in the United States, Canada, and other nations moved into suburbs?
3. Give two definitions of *sprawl*. Describe five negative impacts that have been suggested to result from sprawl.

4. What are city planning and regional planning? Contrast planning with zoning. Give examples of some of the suggestions made by early planners, such as Daniel Burnham and Edward Bennett.
5. How are some people trying to prevent or slow sprawl? Describe some key elements of "smart growth." What effects, positive and negative, do urban growth boundaries tend to have?
6. Describe several apparent benefits of rail transit systems. What is a potential drawback?
7. How are city parks thought to make urban areas more livable? What types of smaller spaces in cities can serve some of the functions of parks?
8. Why do urban dwellers tend to consume more resources per capita than rural dwellers?
9. Describe the connection between urban ecology and sustainable cities.
10. Name two positive effects of urban centers on the natural environment.

SEEKING SOLUTIONS

1. Assess the reasons why urban populations are rising and rural populations are stable or falling. Do you think these trends will continue in the future, or might they change for some reason?
2. Evaluate the causes of the spread of suburbs and of the environmental, social, and economic impacts of sprawl. Overall, do you think the spread of urban and suburban development that many people label *sprawl* is predominantly a good thing or a bad thing? Do you think it is inevitable? Give reasons for your answers.
3. Would you personally want to live in a neighborhood developed in the style of the new urbanism? Would you like to live in a city or region with an urban growth boundary? Why or why not?
4. All things considered, do you feel that cities are a positive thing or a negative thing for environmental quality? How much do you feel we may be able to improve the sustainability of our urban areas?
5. Which is more ecologically sustainable: living in a high-rise apartment in a big city, or living on a 40-acre ranch abutting a national forest? Why?
6. You may soon be faced with a decision about where to live. You can live in the midst of a densely populated city; or, in a suburb where you have more space, but where development and sprawl may soon surround you for many miles; or, in a rural area with plenty of space but few cultural amenities. Where would you choose to live? Why?

INTERPRETING GRAPHS AND DATA

In the following graph, urban population density is used as an indicator of sprawl, and carbon emissions per capita provide some measure of the environmental impact of the transportation system or preferences for each of the cities represented.

1. Describe the relationship between urban density and carbon emissions, as shown in the graph.
2. Assuming that the standard of living is similar in these cities, to what might you attribute the relationship described in your answer to question 1?
3. If zoning ordinances successfully slowed urban sprawl and resulted in a doubling of urban population density in a city like Houston, Texas, what do you predict the change would be in carbon emissions per capita?

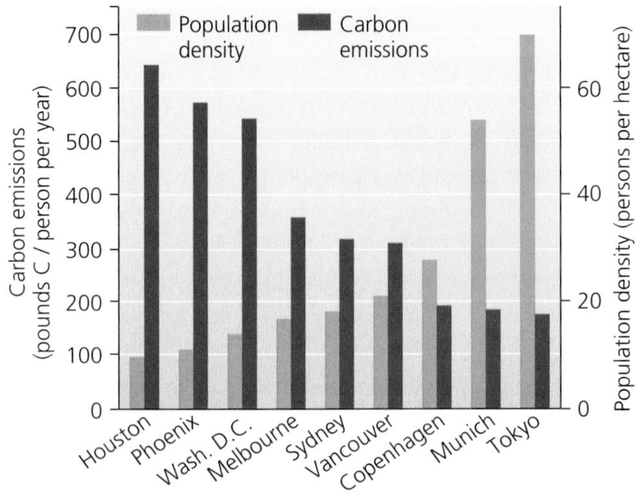

Population density versus carbon emissions from transportation in 1990. Data from Kenworthy, J., et al.,1999. *An international sourcebook of automobile dependence in cities,* Boulder, CO: University Press of Colorado, as cited by Sheehan, M. O. 2002. *What will it take to halt sprawl?* Washington D.C.: Worldwatch Institute.

CALCULATING ECOLOGICAL FOOTPRINTS

One way of altering your ecological footprint is to consider transportation alternatives. Each gallon of gasoline is converted to approximately 20 lb of carbon dioxide (CO_2) during combustion, and this CO_2 is then released into the atmosphere. Typical amounts of CO_2 released for each person per mile, through various forms of transportation, are listed in the table below, assuming typical fuel efficiencies.

For an average person who travels 12,000 miles per year, calculate the cumulative impact of choosing the alternatives to driving alone, and record your results in the table. Clearly, it is unlikely that any of us will walk or bicycle 12,000 miles per year or only travel in vanpools of eight people. In the last two columns, estimate how much of the 12,000 miles you think you might actually travel by each method, and then calculate the CO_2 emissions that you would be responsible for generating over the course of a year.

	CO_2 per person per mile	CO_2 per person per year	CO_2 emission reduction	Your estimated mileage per year	Your CO_2 emissions per year
Automobile (driver only)	0.825 lb	9,900 lb	0		
Automobile (2 persons)	0.413 lb				
Automobile (4 persons)	0.206 lb				
Vanpool (8 persons)	0.103 lb				
Bus	0.261 lb				
Walking	0.082 lb				
Bicycle	0.049 lb				
				Total = 12,000	

Take It Further

Go to www.aw-bc.com/withgott or the student CD-ROM where you'll find:

- Suggested answers to end-of-chapter questions
- Quizzes, animations, and flashcards to help you study
- *Research Navigator*™ database of credible and reliable sources to assist you with your research projects
- **GRAPHit!** tutorials to help you master how to interpret graphs
- **INVESTIGATEit!** current news articles that link the topics that you study to case studies from your region to around the world

14 Environmental Health and Toxicology

Alligator hatchling from Lake Apopka, Florida

Upon successfully completing this chapter, you will be able to:

- Identify the major types of environmental health hazards and explain the goals of environmental health
- Describe the types, abundance, distribution, and movement of synthetic and natural toxicants in the environment
- Discuss the study of hazards and their effects, including case histories, epidemiology, animal testing, and dose-response analysis
- Assess risk assessment and risk management
- Compare philosophical approaches to risk
- Describe policy and regulation in the United States and internationally

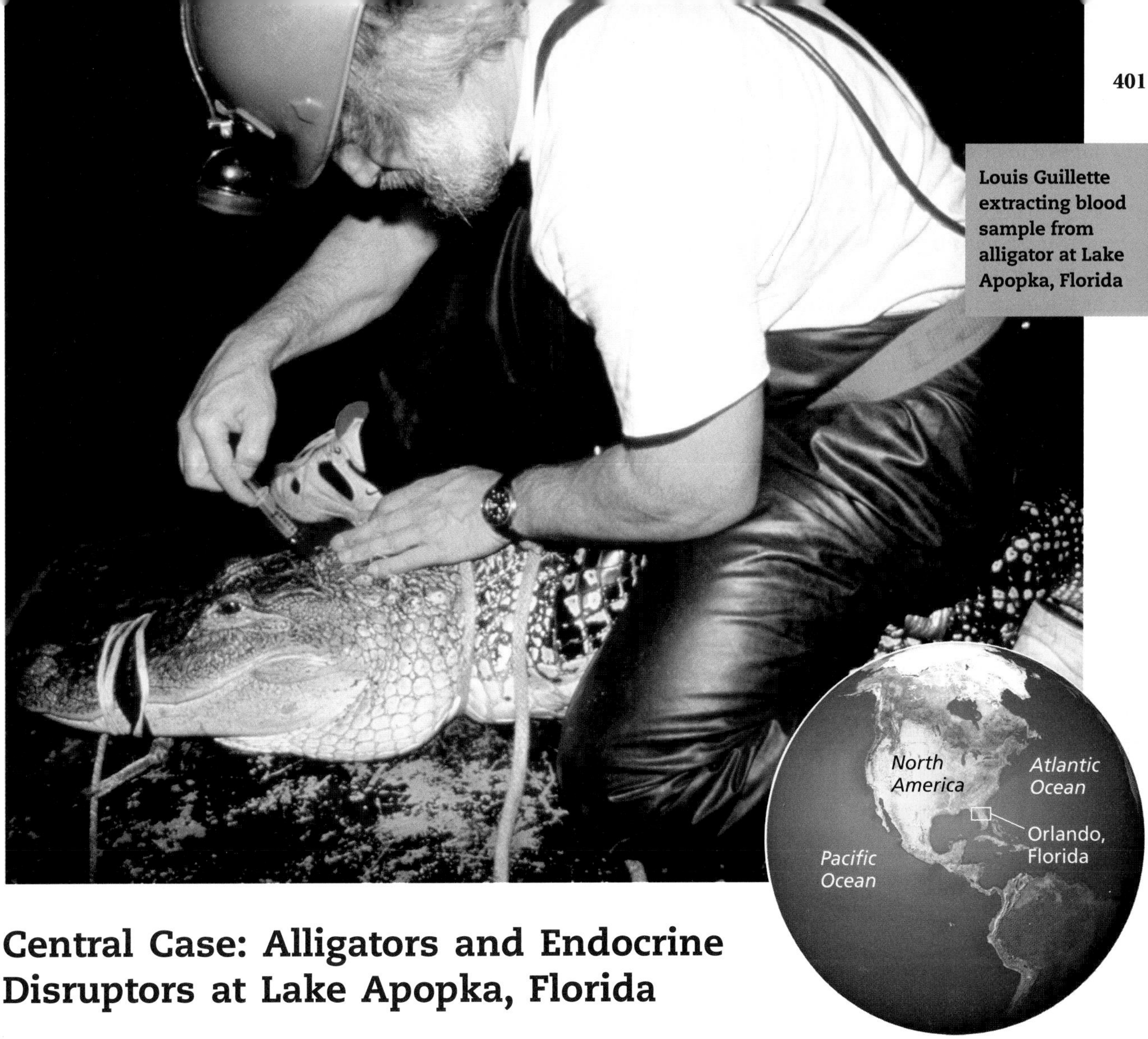

Louis Guillette extracting blood sample from alligator at Lake Apopka, Florida

Central Case: Alligators and Endocrine Disruptors at Lake Apopka, Florida

"Over the past 50 years we have all been unwitting participants in a vast, uncontrolled, worldwide chemistry experiment involving the oceans, air, soils, plants, animals, and human beings."

—United Nations Environment Programme, in a guide to the Stockholm Convention on Persistent Organic Pollutants

"Every man in this room is half the man his grandfather was."

—Reproductive biologist Louis Guillette to a U.S. Congressional committee in 1995, regarding declining sperm counts in men worldwide over the previous half-century

When biologist Louis Guillette began studying the reproductive biology of alligators in Florida lakes in 1985, he soon discovered that not all was well. Alligators from one lake in particular, Lake Apopka northwest of Orlando, showed a number of bizarre reproductive problems.

Females were having trouble producing viable eggs. Male hatchling alligators had severely depressed levels of the male sex hormone testosterone, and female hatchlings showed greatly elevated levels of the female sex hormone estrogen. The young animals had abnormal gonads, too. Testes of the males produced sperm at a premature age, and ovaries of the females contained multiple eggs per follicle instead of the expected one egg per follicle.

As Guillette and his co-workers tested and rejected hypotheses for the reproductive abnormalities one by one, they grew to suspect that environmental contaminants might be somehow responsible. Lake Apopka had suffered a major spill of the pesticides dicofol and DDT in 1980, and yearly surveys thereafter showed a precipitous decline in the number of juvenile alligators in the lake. In addition, the lake received high levels of chemical runoff from agriculture and was experiencing eutrophication (▸ pp. 188–189 and ▸ p. 453) from nutrient input from fertilizers.

The puzzle pieces fell together when a colleague of Guillette's told him that recent research was showing that some environmental contaminants mimic hormones and interfere with the functioning of animal endocrine (hormone) systems. At very low doses, these endocrine disruptors can affect animals, mimicking estrogen and feminizing males. Guillette realized that his alligator observations were similar to results from tests in which rodents were exposed to estrogen during embryonic development. He formulated the hypothesis that certain chemical contaminants in Lake Apopka were disrupting the endocrine systems of alligators during development in the egg.

To test his hypothesis, Guillette and his co-workers compared alligators from heavily polluted Lake Apopka with those from cleaner lakes nearby. They found that Lake Apopka alligators had abnormally low hatching rates in the years after the pesticide spill and, even as hatching rates recovered in the 1990s, continued to show aberrant hormone levels and bizarre gonad abnormalities. In addition, the penises of Lake Apopka male alligators were 25% smaller than those of male alligators from surrounding lakes.

Similar problems began cropping up in other lakes that experienced runoff of chemical pesticides. In the lab, researchers found that several contaminants detected in alligator eggs and young could bind to receptors for estrogen and exist in concentrations great enough to cause sex reversal of male embryos. One chemical in particular, atrazine—the most widely used herbicide in the United States—appeared to disrupt hormones by inducing production of aromatase, an enzyme that converts testosterone to estrogen. In 2003, Guillette reported preliminary findings that nitrate from fertilizer runoff may also act as an endocrine disruptor; when nitrate concentrations in lakes are above the standard for drinking water, juvenile male alligators have smaller penises and 50% lower testosterone levels.

Guillette's results have raised concern not only for alligator health but also for human health. Because alligators and humans share many of the same hormones, many scientists suspect that chemical contaminants could be affecting people, just as they have affected alligators.

Environmental Health

The study of impacts of human-made chemicals on wildlife and people is one aspect of the broad field of environmental health. The study and practice of **environmental health** assesses environmental factors that influence human health and quality of life. Those environmental factors include wholly natural aspects of the environment over which we have little or no control, as well as anthropogenic (human-caused) factors that spread through the environment and can affect people who do not cause them. Practitioners of environmental health seek to prevent adverse effects on human health and on the ecological systems that are essential to environmental quality and human well-being.

Environmental hazards can be physical, chemical, biological, or cultural

There are innumerable environmental health threats, or hazards, in the world around us. For ease of understanding, we can categorize them into four main types: physical, biological, chemical, and cultural. For each of these four types of hazards, there is some amount of risk that we cannot avoid—but there is also some amount of risk that we *can* avoid by taking precautions. Much of environmental health consists of taking steps to minimize the risks of encountering hazards and to minimize the impacts of hazards when we do encounter them.

Physical hazards Some *physical* processes that occur naturally in our environment can pose health hazards. These include discrete events such as earthquakes, volcanic eruptions, fires, floods, blizzards, landslides, hurricanes, and droughts. They also include ongoing natural phenomena, such as ultraviolet (UV) radiation from sunlight (Figure 14.1a). At excessive exposure, UV radiation damages DNA and has been tied to skin cancer, cataracts, and immune suppression in humans. We can do little to predict the timing of a natural disaster such as an earthquake, and nothing to prevent one. However, scientists can map geologic faults to determine areas at risk of earthquakes, engineers can design buildings in ways that help them resist damage, and citizens and governments can take steps to prepare for the aftermath of a severe quake.

Some common practices have increased our vulnerability to certain physical hazards. Deforesting slopes makes landslides more likely, for instance, and diking rivers can make flooding more likely in some areas while reducing flooding in others. We can reduce risk from such hazards by improving our forestry and flood control practices and by choosing not to build in areas prone to floods, landslides, fires, and coastal waves. For hazards such as exposure to UV light, we can reduce our exposure and risk by using clothing and sunscreen to shield our skin from intense sunlight.

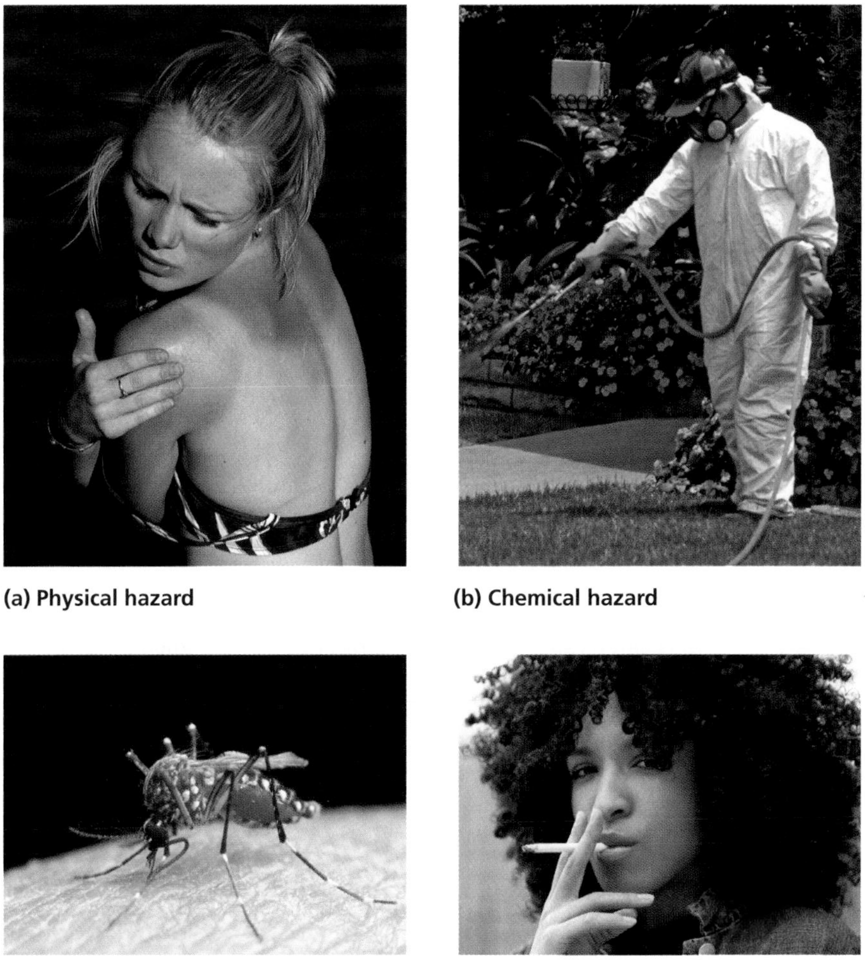

(a) Physical hazard

(b) Chemical hazard

(c) Biological hazard

(d) Cultural hazard

FIGURE 14.1 Environmental health hazards can be divided into four types. The sun's ultraviolet radiation is an example of a physical hazard (**a**). Excessive exposure increases the risk of skin cancer. Chemical hazards (**b**) include both artificial and natural chemicals. Much of our exposure comes from household chemical products, such as pesticides that some people apply to their lawns. Biological hazards (**c**) include other organisms that transmit disease. Some mosquitoes are vectors for certain pathogenic microbes, including those that cause malaria. Cultural or lifestyle hazards (**d**) include the decisions we make about how to behave, as well as the constraints forced upon us by socioeconomic factors. Smoking is a lifestyle choice that raises one's risk of lung cancer and other disease considerably.

Chemical hazards *Chemical* hazards include many of the synthetic chemicals that our society produces, such as disinfectants, pesticides (Figure 14.1b), and the compounds that contributed to reproductive problems for the alligators at Lake Apopka. Chemicals produced naturally by organisms also can be hazardous. Some hazardous chemicals are so widespread they are impossible to avoid, but with concerted effort, we may be able to reduce our exposure. Following our overview of environmental health, much of our chapter will focus on chemical health hazards and the ways people study and regulate them.

Biological hazards *Biological* hazards result from ecological interactions among organisms (Figure 14.1c). When we become sick from a virus, bacterial infection, or other pathogen, we are suffering parasitism by other species that are simply fulfilling their ecological roles. This is **infectious disease,** also called *communicable* or *transmissible disease.* Infectious diseases such as malaria, cholera, tuberculosis, and influenza (flu) all are considered environmental health hazards. As with physical and chemical hazards, it is impossible for us to avoid risk from biological agents completely, but we can take steps to reduce the likelihood of infection.

Cultural hazards Hazards that result from the place we live, our socioeconomic status, our occupation, or our behavioral choices can be thought of as *cultural* or *lifestyle* hazards. For instance, choosing to smoke cigarettes, or living or working with people who do, can greatly increase our risk of lung cancer (Figure 14.1d). Choosing to smoke is a personal behavioral decision, but exposure to secondhand smoke in the home or workplace may not be under one's control and may be influenced by socioeconomic constraints. Much the same might be said for drug use, diet and nutrition, crime, and mode of transportation. As advocates of environmental justice (▸ pp. 36–38) argue, such health factors as living in proximity to toxic waste sites or working unprotected with pesticides are often correlated with socioeconomic deprivation.

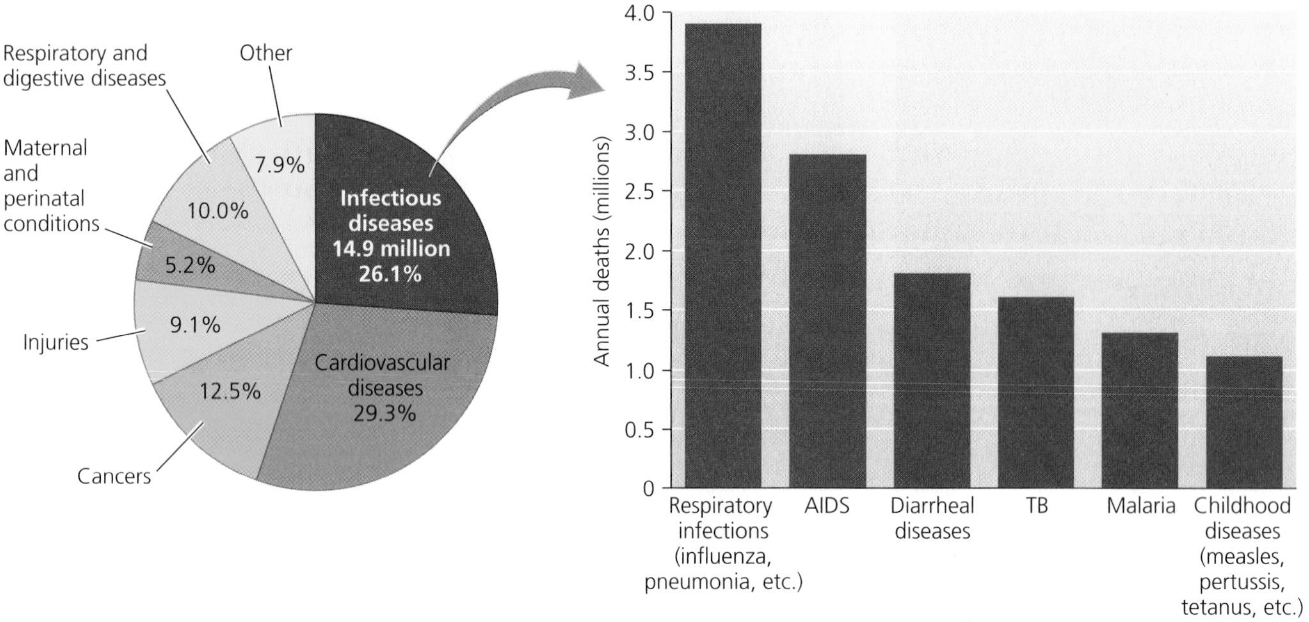

(a) Leading causes of death across the world, 2004 **(b) Leading causes of death by infectious disease, 2004**

FIGURE 14.2 Infectious diseases are the second-leading cause of death worldwide, accounting for over one-quarter of all deaths per year **(a).** Six types of diseases—respiratory infections, AIDS, diarrhea, tuberculosis (TB), malaria, and childhood diseases such as measles—account for 80% of all deaths from infectious disease **(b).** Data from World Health Organization, 2004.

Disease is a major focus of environmental health

Among the hazards people face, disease stands preeminent. Despite all our technological advances, we still find ourselves battling disease, which causes the vast majority of human deaths worldwide (Figure 14.2a). Many major killers such as cancer, heart disease, and respiratory disorders have genetic bases but are also influenced by environmental factors. For instance, whether a person develops asthma is influenced not only by the genes, but also by environmental conditions. Pollutants from fossil fuel combustion worsen asthma, and children raised on farms suffer less asthma than children raised in cities, studies have shown. Malnutrition (▸pp. 279–280) can foster a wide variety of illnesses, as can poverty and poor hygiene. Moreover, lifestyle choices can affect risks of acquiring some noninfectious diseases: Smoking can lead to lung cancer, and lack of exercise to heart disease, for example.

Infectious diseases account for 26% of deaths that occur worldwide each year—nearly 15 million people (Figure 14.2b). Some pathogenic microbes attack us directly, and sometimes infection occurs through a *vector*, an organism that transfers the pathogen to the host. Infectious diseases account for close to half of all deaths in developing countries, but for very few deaths in developed nations. This discrepancy is due to differences in hygiene conditions and access to medicine, which are tightly correlated with wealth. Public health efforts have lessened the impact of infectious disease in developed nations and even have eradicated some diseases. Nevertheless, other diseases, among them tuberculosis, acquired immunodeficiency syndrome (AIDS), and West Nile virus, are increasing (Figure 14.3).

To predict and prevent infectious disease, environmental health experts deal with the often complicated interrelationships among technology, land use, and ecology. One of the world's leading infectious diseases, malaria (which takes nearly 1.3 million lives each year), provides an example. The microscopic protist that causes malaria depends on mosquitoes as a vector. The protist can sexually reproduce only within a mosquito, and it is the mosquito that injects the protist into a human or other host. Thus the primary mode of malaria control has been to use insecticides such as DDT to kill mosquitoes. Large-scale eradication projects involving insecticide use and draining of wetlands have removed malaria from large areas of the temperate world where it used to occur, such as throughout the southern United States. However, types of land use that create pools of standing water in formerly well-drained areas can boost mosquito populations and allow malaria to reinvade an area.

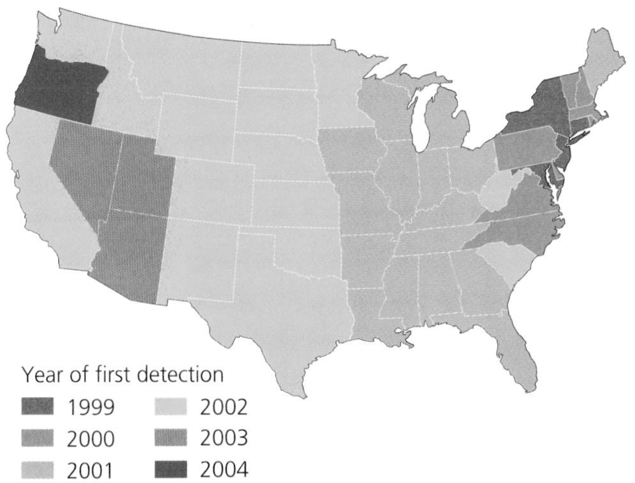

(a) Spread of West Nile virus

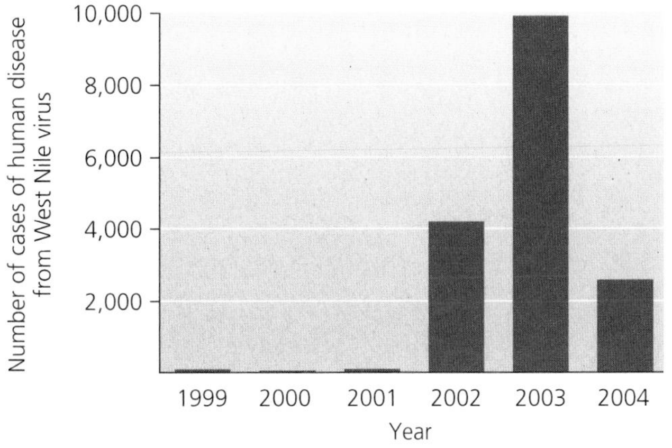

(b) Human cases of West Nile virus, 1999–2004

FIGURE 14.3 Modern public health efforts have greatly diminished the burden of disease in developed nations, but some infectious diseases are currently on the increase. One example is West Nile virus, an Old World virus that has spread rapidly through the Western Hemisphere since it was first detected in the New York City area in 1999. Some types of birds are most affected by this mosquito-transmitted virus, but over 15,000 human disease cases, including over 500 deaths, had been recorded in the United States through the end of 2004. In the map (**a**), U.S. states are color-coded to show the year that West Nile virus was first detected within their borders. The disease spread westward across the country in less than 5 years. The graph (**b**) shows the number of cases reported nationwide in each year. Data from Centers for Disease Control and Prevention, 2005.

Environmental health hazards exist indoors as well as outdoors

Outdoor hazards are generally more familiar to us, but we spend most of our lives indoors. Therefore, we must consider the spaces inside our homes and workplaces to be part of our environment and, as such, the sources of potential environmental hazards (Table 14.1). Radon is one indoor hazard (▸pp. 522–523). Radon is a highly toxic radioactive gas that is colorless and undetectable without specialized kits. Radon seeps up from the ground in areas with certain types of bedrock and can build up inside basements and homes with poor air circulation. The U.S. Environmental Protection Agency (EPA) estimates that slightly less than 1 person in 1,000 may contract lung cancer as a result of a lifetime of radon exposure at average levels for U.S. homes.

Lead poisoning represents another indoor health hazard. When ingested, lead, a heavy metal, can cause damage to the brain, liver, kidney, and stomach; learning problems and behavioral abnormalities; anemia; hearing loss; and even death. It has been suggested that the downfall of ancient Rome was caused in part by chronic lead poisoning. The elites of the power structure regularly drank wine sweetened with mixtures prepared in leaden vessels. Several historical personalities, such as the composer Beethoven, suffered maladies that some historians have attributed to lead poisoning. Today lead poisoning can result from drinking water that has passed through the lead pipes common in older houses. Even in newer pipes, lead solder was widely used into the 1980s and is still sold in

Table 14.1 Selected Environmental Hazards

Category	Hazards
Air	
	▶ Smoking and secondhand smoke
	▶ Chemicals from automotive exhaust
	▶ Chemicals from industrial pollution
	▶ Tropospheric ozone
	▶ Pesticide drift
	▶ Dust and particulate matter
Water	
	▶ Pesticide and herbicide runoff
	▶ Nitrates and fertilizer runoff
	▶ Mercury, arsenic, and other heavy metals in groundwater and surface water
Food	
	▶ Natural toxins
	▶ Pesticide and herbicide residues
Indoors	
	▶ Asbestos
	▶ Radon
	▶ Lead in paint and pipes
	▶ Toxicants in plastics and consumer products

stores. Lead is perhaps most dangerous, however, to children through its presence in paint. Until 1978, most paints contained lead, and interiors in most houses were painted with lead-based paint. Babies and young children often take an interest in peeling paint from walls and may ingest or inhale some of it. Today lead poisoning is thought to affect 3–4 million young children, fully one in six children under age 6.

Another indoor hazard is asbestos. There are several types of *asbestos,* each of which is a mineral that forms long, thin microscopic fibers. This structure allows asbestos to insulate heat, muffle sound, and resist fire. Because of these qualities, asbestos was used widely as insulation in buildings, as well as in many products. Unfortunately, its fibrous structure also makes asbestos dangerous when inhaled. When lodged in lung tissue, asbestos induces the body to produce acid to combat it. The acid scars the lung tissue but does little to dislodge or dissolve the asbestos. Consequently, within a few decades the scarred lungs may cease to function, a disorder called *asbestosis.* Asbestos can also cause types of lung cancer. Because of these risks, asbestos has been removed from many schools and offices (Figure 14.4). However, many people have argued that the dangers of exposure from asbestos removal exceed the dangers of leaving it in place.

Some indoor hazards likely have not yet come to light. One example of a recently recognized hazard is a group of chemicals known as polybrominated diphenyl ethers (PBDEs). These compounds provide fire-retardant properties and are used in a diverse array of consumer products, including computers, televisions, plastics, and furniture. They appear to be released during production and disposal of products and also to evaporate at very slow rates throughout the lifetime of products. These chemicals persist and accumulate in living tissue, and their abundance in the environment and in people in the United States is doubling every few years.

Like the chemicals affecting Louis Guillette's alligators, PBDEs appear to be endocrine disruptors; lab testing with animals shows them to affect thyroid hormones. Animal testing also shows limited evidence that PBDEs may affect brain and nervous system development and might possibly cause cancer. Concern about PBDEs rose after a study showed concentrations in the breast milk of Swedish mothers increasing exponentially from 1972 to 1997. The European Union decided in 2003 to ban PBDEs, and industries in Europe had already begun phasing them out. As a result, concentrations in breast milk of European mothers have fallen substantially. In the United States, however, there has so far been little movement to address the issue.

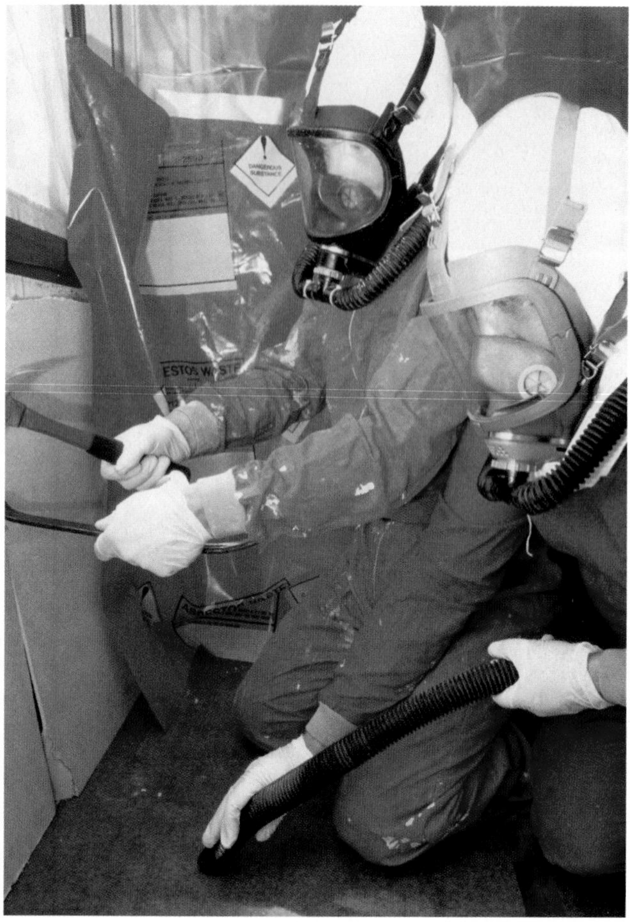

FIGURE 14.4 Asbestos was widely used in building insulation and other products. A cause of lung cancer and asbestosis, the substance has now been removed from many buildings in which it was used. Its removal poses risks as well, however, and removal workers must wear protective clothing and respirators.

Toxicology is the study of poisonous substances

Studying the health effects of chemical agents suspected to be harmful, such as PBDEs, is the focus of the field of **toxicology,** the science that examines the effects of poisonous substances on humans and other organisms. Toxicologists assess and compare substances to determine their *toxicity,* the degree of harm a chemical substance can inflict. The concept of toxicity among chemical hazards is analogous to that of *pathogenicity* or *virulence* of the biological hazards that spread infectious disease. Just as types of microbes differ in their ability to cause disease, chemical hazards differ in their capacity to endanger us. However, any chemical substance may exert negative effects if it is ingested in great enough quantities or if exposure is extensive enough. Conversely, a toxic agent, or **toxicant,** in a minute enough quantity may pose no health risk at all. These facts are often summarized in the catchphrase, "The dose makes the poison." In other words, a substance's toxicity depends not only on its chemical identity, but also on its quantity.

During the past century, our ability to produce new chemicals has expanded, concentrations of chemical contaminants in the environment have increased, and public concern for health and the environment have grown. These trends have driven the rise of **environmental toxicology,** which deals specifically with toxic substances that come from or are discharged into the environment. Environmental toxicology includes the study of health effects on humans, other animals, and ecosystems, and it represents one approach within the broader scope of environmental health.

Toxicologists generally focus on human health, using other organisms as models and test subjects. In environmental toxicology, animals are also studied out of concern for their welfare and because—like canaries in a coal mine—animals can serve as indicators of health threats that could soon affect humans. For example, the reproductive abnormalities of Lake Apopka alligators seem analogous to those seen in some Taiwanese boys. Studies showed that these boys were born to mothers who used cooking oil contaminated with toxicants called polychlorinated biphenyls (PCBs), which are by-products of chemicals used in transformers and other electrical equipment.

As we review the sometimes poisonous effects of human-made chemicals throughout this chapter, it is important to keep in mind that artificially produced chemicals have played a crucial role in giving us the standard of living we enjoy today. These chemicals have helped create the industrial agriculture that produces our food, the medical advances that protect our health and prolong our lives, and many of the modern materials and conveniences we use every day. It is important to remember these benefits as we examine some of the unfortunate side effects of these advances and as we search for better alternatives.

Toxic Agents in the Environment

The environment contains countless natural chemical substances that may pose health risks. These substances include oil oozing naturally from the ground; radon gas seeping up from bedrock; and toxic chemicals stored or manufactured in the tissues of living organisms—for example, *toxins* that plants use to ward off herbivores and

FIGURE 14.5 Synthetic chemicals, such as those in household products, are everywhere around us in our everyday lives. Some of these compounds may potentially pose environmental or human health risks.

toxins that insects use to defend themselves from predators. In addition, we are exposed to many synthetic (artificial, or human-made) chemicals.

Synthetic chemicals are ubiquitous in our environment

Synthetic chemicals are all around us in our daily lives (Figure 14.5). Thousands of synthetic chemicals have been manufactured (Table 14.2), and many have found their way into soil, air, and water. A 2002 study by the U.S. Geological Survey found that 80% of U.S. streams contain at least trace amounts of 82 wastewater contaminants, including antibiotics, detergents, drugs, steroids, plasticizers, disinfectants, solvents, perfumes, and other substances. The pesticides we use to kill insects and weeds (►pp. 282–283) on farms, lawns, and golf courses are some of the most widespread synthetic chemicals. As a result of all this exposure, every one of us carries traces of numerous industrial chemicals in our bodies.

Table 14.2 Estimated Numbers of Chemicals in Commercial Substances during the 1990s

Type of chemical	Estimated number
Chemicals in commerce	100,000
Industrial chemicals	72,000
New chemicals introduced per year	2,000
Pesticides (21,000 products)	600
Food additives	8,700
Cosmetic ingredients (40,000 products)	7,500
Human pharmaceuticals	3,300

Data from Harrison, P., and F. Pearce. 2000. *AAAS Atlas of Population and Environment*. Berkeley, CA: University of California Press.

This should not *necessarily* be cause for alarm. Not all synthetic chemicals pose health risks, and relatively few are known with certainty to be toxicants. However, of the roughly 100,000 synthetic chemicals on the market today, very few have been thoroughly tested for harmful effects. For the vast majority, we simply do not know what effects, if any, they may have.

Why are there so many synthetic chemicals around us? Let's consider pesticides and herbicides such as those that Louis Guillette found were disrupting sex hormones and feminizing male alligator embryos. Our many pesticides and herbicides were made possible by advances in chemistry during and following World War II. Mechanization of production and the growth of large corporations enabled immense quantities of synthetic chemicals to be produced. As material prosperity grew in Westernized nations in the decades following the war, people began using pesticides not only for agriculture, but also to improve the look of their lawns and golf courses and to fight termites, ants, and other insects inside their homes and offices. Pesticides were viewed as means toward a better quality of life.

It was not until the 1960s that people began to learn about the risks of exposure to pesticides. Many people began to speak up against indiscriminate pesticide use. Indeed, concern over pesticides was a catalyst that helped spur the environmental movement in the United States. The key event was the publication of Rachel Carson's 1962 book *Silent Spring* (▸ p. 67), which brought the pesticide dichloro-diphenyl-trichloroethane (DDT) to the attention of the public.

Silent Spring began the public debate over synthetic chemicals

Rachel Carson was a naturalist, author, and government scientist. In *Silent Spring*, she brought together a diverse collection of scientific studies, medical case histories, and other data that no one had previously synthesized and presented to the general public. Her message was that DDT in particular and artificial pesticides in general were hazardous to people's health, the health of wildlife, and the well-being of ecosystems. Carson wrote at a time when large amounts of pesticides virtually untested for health effects were indiscriminately sprayed over residential neighborhoods and public areas, on the assumption that the chemicals would do no harm to people (Figure 14.6). Most

FIGURE 14.6 Before the 1960s, the environmental and health effects of potent pesticides such as DDT were not widely studied or publicly known. Public areas such as parks, neighborhoods, and beaches were regularly sprayed for insect control without safeguards against excessive human exposure. Here children on a Long Island, New York, beach are fogged with DDT from a pesticide spray machine being tested in 1945.

consumers had no idea that the store-bought chemicals they used in their houses, gardens, and crops might be toxic.

Although challenged vigorously by spokespeople for the chemical industry, who attempted to discredit both the author's science and her personal reputation, Carson's book was a best-seller. Carson suffered from cancer as she finished *Silent Spring,* and she lived only briefly after its publication. However, the book helped generate significant social change in views and actions toward the environment. The use of DDT was banned in the United States in 1973 and has been banned in a number of other nations.

The United States still manufactures and exports DDT to countries that do use it, however. Many developing countries with tropical climates use DDT to control human disease vectors, such as mosquitoes that transmit malaria. In these countries, malaria represents a far greater health threat than do the toxic effects of the pesticide.

Weighing the Issues:
The Circle of Poison

It has been called the "circle of poison." Although the United States has banned the use of DDT, U.S. companies still manufacture and export the compound to many developing nations. Thus, it is possible that pesticide-laden food can be imported back into the United States. How do you feel about this? Is it unethical for one country to sell to others a substance that it has deemed toxic? Are there factors or circumstances that might change the view you take?

Toxicants come in several different types

Toxicants, whether they are natural or synthetic, can be classified into different types based on their particular effects on health. The best-known are **carcinogens,** which are chemicals or types of radiation that cause cancer. In cancer, malignant cells grow uncontrollably, creating tumors, damaging the body's functioning, and often leading to death. In our society today, the greatest number of cancer cases is thought to result from carcinogens contained in cigarette smoke. Carcinogens can be difficult to identify because there may be a long lag time between exposure to the agent and the detectable onset of cancer. Historically, much toxicological work focused on carcinogens. Now, however, we know that toxicants can produce many different types of effects, so scientists have many more endpoints, or health impacts, to look for.

FIGURE 14.7 The drug thalidomide turned out to be a potent teratogen. It was banned in the 1960s, but not before causing thousands of birth defects in babies born to mothers who took the product to relieve nausea during pregnancy. Butch Lumpkin was an exceptional "thalidomide baby" who learned to overcome his short arms and deformed fingers, becoming a professional tennis instructor in Georgia.

Mutagens are chemicals that cause mutations in the DNA of organisms (▸pp. 100–101). Although most mutations have little or no effect, some can lead to severe problems, including cancer and other disorders. If mutations occur in an individual's sperm or egg cells, then the individual's offspring suffer the effects.

Chemicals that cause harm to the unborn are called **teratogens.** Teratogens that affect the development of human embryos in the womb can cause birth defects. One example involves the drug thalidomide, developed in the 1950s as a sleeping pill and to prevent nausea during pregnancy. Tragically, the drug turned out to be a powerful teratogen, and its use caused birth defects in thousands of babies (Figure 14.7). Even a single dose during pregnancy could result in limb deformities and organ defects. Thalidomide was banned in the 1960s once scientists recognized its connection with birth defects. Ironically, today the drug shows promise in treating a wide range of diseases, including Alzheimer's disease, AIDS, and various types of cancer.

The human immune system protects our bodies from disease. Some toxicants weaken the immune system, reducing the body's ability to defend itself against bacteria, viruses, allergy-causing agents, and other attackers. Others, called **allergens,** overactivate the immune system, causing an immune response when one is not necessary. One hypothesis for the increase in asthma in recent years is that allergenic synthetic chemicals are more prevalent in our environment.

Still other chemical toxicants, **neurotoxins,** assault the nervous system. Neurotoxins include various heavy metals such as lead, mercury, and cadmium, as well as pesticides and some chemical weapons developed for use in war. A famous case of neurotoxin poisoning occurred in Japan, where a chemical factory dumped mercury waste into Minamata Bay between the 1930s and 1960s. Thousands of people in and around the town on the bay were poisoned by eating fish contaminated with the mercury. First the town's cats began convulsing and dying, and then people began to show odd symptoms, including slurred speech, loss of muscle control, sudden fits of laughter, and in some cases death. The company and the government eventually paid out about $5,000 in compensation to each affected resident.

Most recently, scientists have recognized the importance of **endocrine disruptors,** toxicants that interfere with the *endocrine system,* or hormone system. The endocrine system consists of chemical messengers *(hormones)* that travel though the body. Sent through the bloodstream at extremely low concentrations, these messenger molecules have many vital functions. They stimulate growth, development, and sexual maturity, and they regulate brain function, appetite, sexual drive, and many other aspects of our physiology and behavior. Hormone-disrupting toxicants can affect an animal's endocrine system by blocking the action of hormones or accelerating their breakdown. Many endocrine disruptors are so similar to some hormones in their molecular structure and chemistry that they "mimic" the hormone by interacting with receptor molecules just as the actual hormone would (Figure 14.8). One common type of endocrine disruption involves the feminization of male animals, as Louis Guillette found with alligators, and as other studies have indicated with fish, frogs, and other organisms. Feminization may be widespread because a number of chemicals appear to mimic the female sex hormone estrogen and bind to estrogen receptors.

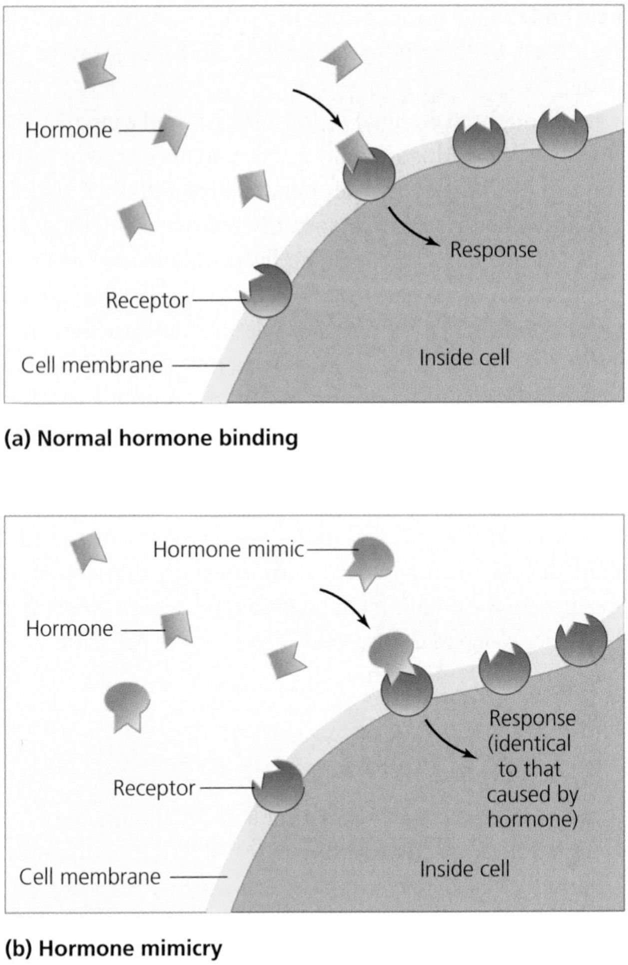

FIGURE 14.8 Many endocrine-disrupting substances act by mimicking the structure of hormone molecules. Like a key similar enough to fit into another key's lock, the hormone mimic binds to a cellular receptor for the hormone, causing the cell to react as though it had encountered the hormone.

Endocrine disruption may be widespread

Scientists first noted endocrine-disrupting effects as far back as the 1960s with the pesticide DDT. However, the idea that synthetic chemicals might be altering the hormones of animals was not widely appreciated until the 1996 publication of the book *Our Stolen Future,* by Theo Colburn, Dianne Dumanoski, and J. P. Myers. Colburn, a scientist with the World Wildlife Fund, had spent years bringing toxicologists, medical doctors, and wildlife biologists together to share their findings. She and her co-authors then presented the data and stories in *Our Stolen Future* (and on a Web page that provides updates on new research). Like *Silent Spring,* this book integrated scientific work from various fields and presented a unified picture that shocked many readers—and brought criticism from some scientists and from the chemical industry.

One line of research that has sparked debate is that of scientist Tyrone Hayes of the University of California at Berkeley, who studies frogs and has found in them gonadal abnormalities similar to those of Guillette's alligators. As

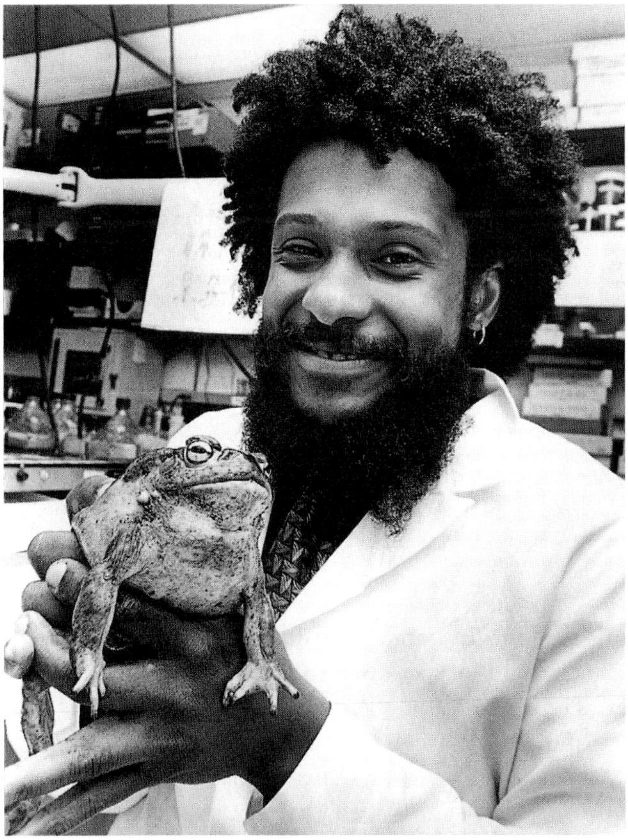

FIGURE 14.9 According to research by Tyrone Hayes of the University of California at Berkeley, frogs may suffer reproductive abnormalities from exposure to the best-selling herbicide atrazine. Industry-backed scientists have disputed the findings.

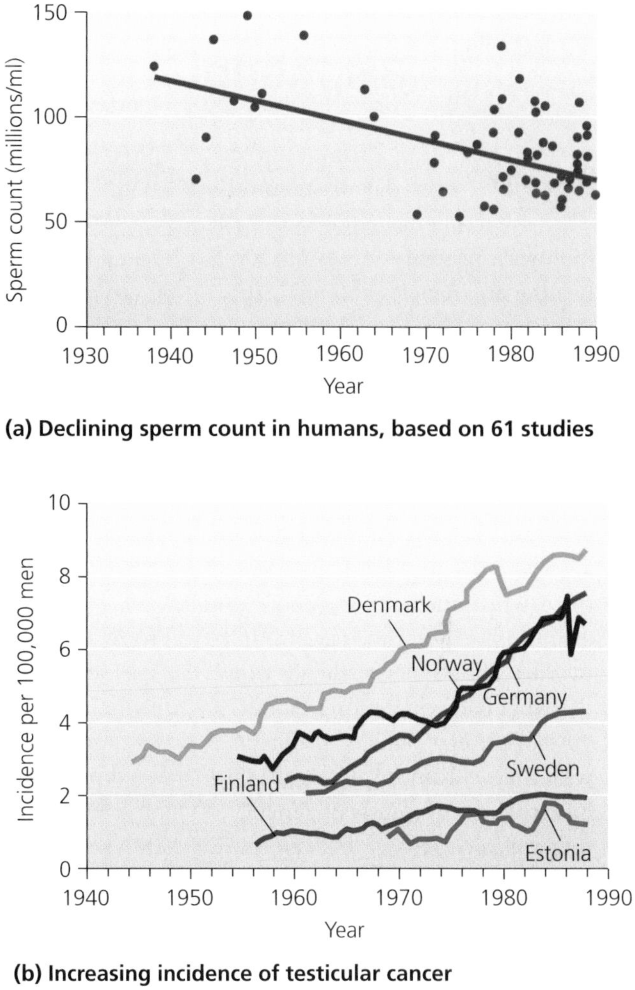

FIGURE 14.10 Research in 1992 synthesized the results of 61 studies that had reported sperm counts in men from various localities since 1938. The data were highly variable but showed a significant decrease in human sperm counts over time **(a)**. Many scientists have hypothesized that this decrease may result from exposure to endocrine-disrupting chemicals in the environment. Some have also hypothesized that endocrine disruptors could be behind an increased incidence of testicular cancer **(b)**. (a) Data from Carlsen, E., et al. 1992. Evidence for decreasing quality of semen during the last 50 years. *British Medical Journal* 305: 609–613, as adapted by Toppari, J., et al. 1996. Male reproductive health and environmental xenoestrogens. *Environmental Health Perspectives* 104(Suppl 4): 741–803. (b) Adami, H.O., et al. 1994. Testicular cancer in nine northern European countries. *International Journal of Cancer* 50: 33–38.

with the Lake Apopka alligators, Hayes has attributed the frog abnormalities to atrazine (Figure 14.9). In lab experiments, male frogs raised in water containing very low doses of the herbicide became feminized and hermaphroditic, developing both testes and ovaries. In field studies, leopard frogs across North America showed similar problems in areas of atrazine usage. Hayes found these effects at atrazine concentrations well below EPA guidelines for human health and at concentrations lower than commonly exist in U.S. drinking water. Thus, he suggests, people may be at risk.

To date, endocrine effects have been found most widely in nonhuman animals, but many scientists attribute the striking drop in sperm counts among men worldwide to endocrine disruptors (Figure 14.10a). In the most noted study, Danish researchers reported in 1992 that the number and motility of sperm in men's semen had declined by 50% since 1938. The research involved a review of 61 studies that included 15,000 men from 20 nations on six continents. Subsequent studies by other researchers—including some who set out to disprove the findings—have largely confirmed the results using other methods and other populations. Although there is tremendous geographic variation that remains unexplained, it seems that human sperm counts have declined notably in many parts of the world. Some researchers have also voiced concerns about rising rates of testicular cancer (Figure 14.10b), undescended testicles, and genital birth

The Science behind the Story

Bisphenol-A

Can a compound in everyday plastic products damage the most basic biological processes necessary for healthy pregnancies and births? A scientist in Ohio has found evidence that it does—all because one of her lab assistants reached for the wrong soap.

At a laboratory at Case Western Reserve University in August 1998, geneticist Patricia Hunt was making a routine check of her female lab mice, which included extracting and examining developing eggs from the ovaries. The results made her wonder what had gone wrong. About 40% of the eggs showed problems with their chromosomes, and 12% had irregular amounts of genetic material, a condition that can cause serious reproductive problems in mice and people alike.

Detective work revealed that a lab assistant had mistakenly washed the lab's plastic mouse cages and water bottles with an especially harsh soap, A-33, manufactured by Airkem Professional Products. The soap damaged the cages so badly that parts of them seemed to have melted. The cages were made from a type of plastic called polycarbonate, which contains the chemical bisphenol-A (BPA). BPA is used to manufacture thousands of plastic products, from baby bottles to food containers to auto parts. The plastics industry uses over 900 million kg (2 billion lb) of BPA every year, according to industry estimates.

The chemical, however, is an estrogen mimic and a long-suspected endocrine disruptor. Some studies link BPA to reproductive abnormalities in mice, such as low sperm counts and early sexual development. Other studies indicate that BPA can leach out of plastic products into water and food. One such study conducted by researchers for the magazine *Consumer Reports,* published in 1999, found that BPA seeped out of the plastic walls of heated baby bottles into infant formula. The plastics industry insisted that BPA was safe and fought back with its own studies. U.S. health

Dose and Meiosis Abnormalities in Mice Exposed to Bisphenol-A

BPA dose*	Chromosomal problems during cell division
Control: 0 mg/g	2/115 (1.7%)
20 ng/g (0.00002 mg)	10/172 (5.8%)
40 ng/g (0.00004 mg)	19/255 (7.5%)
100 ng/g (0.0001 mg)	5/46 (10.9%)

*Oral Reference Dose (or safe intake level) established by the U.S. Environmental Protection Agency is 0.05 mg/kg of body weight per day. The European Commission's Scientific Committee on Food's Tolerable Daily Intake is 0.01 mg/kg of body weight per day (recently revised down from U.S. EPA level of 0.05). Data from Hunt, P. A., et al. 2003. Bisphenol A exposure causes meiotic aneuploidy in the female mouse. *Current Biology* 13: 546–553.

defects in men. Other scientists have proposed that the rise in breast cancer rates (one in eight U.S. women today develops breast cancer) may also be due to hormone disruption, because an excess of estrogen appears to feed tumor development in older women.

Endocrine disruptors can affect more than just the reproductive system. Some impair the brain and nervous system. North American studies have shown neurological problems associated with PCB contamination. In one study in Michigan, mothers who ate Great Lakes fish contaminated with PCBs had babies with lower birth weights and smaller heads, compared to mothers who did not eat fish. These babies grew into children who showed weak and jerky reflexes and tested poorly in intelligence tests.

Endocrine disruption research has generated debate

Much of the research into hormone disruption has brought about strident debate. This is partly because a great deal of scientific uncertainty is inherent in any young and developing field. Another reason is that negative findings about chemicals pose an economic threat to the manufacturers of those chemicals. For instance, Tyrone Hayes's work has met with fierce criticism from scientists associated with atrazine's manufacturer, which stands to lose many millions of dollars if its top-selling herbicide were to be banned or restricted in the United States.

Indeed, our society has invested heavily in some chemicals that now are suspect. One example is bisphenol-A (see officials, caught in a scientific stalemate, declined to strengthen regulations on BPA.

Hunt, however, thought the chemical might be adversely affecting her mice. The mice in Hunt's lab were suddenly showing dramatic problems during meiosis, the division of chromosomes during the formation of eggs. The 40% of mouse eggs with meiotic problems represented a huge jump from the 1–2% rate of such defects seen before the cage-washing incident. The 12% with irregular amounts of genetic material, a dangerous cell division error called aneuploidy, would likely lead to nonviable pregnancies or birth defects. In humans, aneuploidy is the main cause of miscarriage and Down syndrome.

Deciding to recreate the BPA exposure experimentally, Hunt instructed researchers in her lab to intentionally wash polycarbonate cages and water bottles using varying levels of A-33. They then compared mice kept in damaged cages with plastic water bottles to mice kept in undamaged cages with glass water bottles. The developing eggs of mice exposed to BPA through the deliberately damaged plastic showed levels of meiotic problems similar to those in the original A-33 incident, whereas the eggs of mice in the control cages were normal. In an additional round of tests, three sets of female mice were given daily oral doses of BPA over 3, 5, and 7 days, and the same meiotic abnormalities were observed, although at lower levels. The mice given BPA for 7 days were most severely affected.

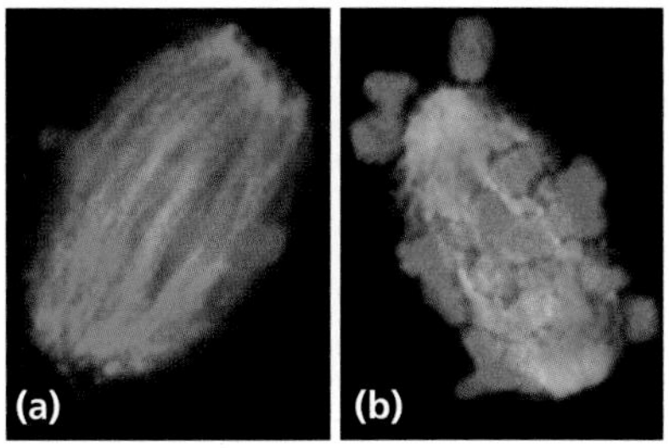

During normal cell division, chromosomes align properly (**a**). Exposure to bisphenol-A, however, causes abnormal cell division (**b**), where chromosomes scatter and are distributed improperly and unevenly between daughter cells.

Published in 2003 in the journal *Current Biology,* Hunt's findings quickly set off a new wave of concern over the safety of BPA. Hunt wrote, "We have observed meiotic defects in mice at exposure levels close to or even below those considered 'safe' for humans. Clearly, the possibility that BPA exposure increases the likelihood of genetically abnormal offspring is too serious to be dismissed without extensive further study."

Although Hunt's research focused on mice, and not humans, the findings were disturbing because sex cells of mice and humans divide and function in similar ways. Regulators in Europe have recently lowered the level of BPA intake they consider safe. In the United States, some scientists are now pushing for more research and renewed federal scrutiny into the safety of BPA.

"The Science behind the Story," above). An estrogen mimic, bisphenol-A occurs in a great variety of plastic products we use daily, from baby bottles to food cans to eating utensils to auto parts. It has even been used in a cavity-preventing coating for children's teeth. The chemical leaches out of many of these products into water and food, and recent evidence ties it to birth defects in lab mice. The plastics industry vehemently protests that the chemical is safe, however, and points to other research backing its contention.

Research results with bisphenol-A and mice, and those with atrazine and frogs, each showed effects at extremely low levels of the chemical. This is also the case with research on other known or purported endocrine disruptors. The likely reason is that the endocrine system is geared to respond to minute concentrations of substances (normally, hormones in the bloodstream). Because the endocrine system responds to minuscule amounts of chemicals, it may be especially vulnerable to effects from environmental contaminants that are dispersed and diluted through the environment and that reach our bodies in very low concentrations.

Toxicants may concentrate in surface water or groundwater

Toxicants are not evenly distributed in the environment, and they move about in specific ways (Figure 14.11). For instance, water, in the form of runoff, often carries toxicants from large areas of land and concentrates them in small volumes of surface water. The U.S. Geological Survey findings regarding surface waterways (▸p. 407)

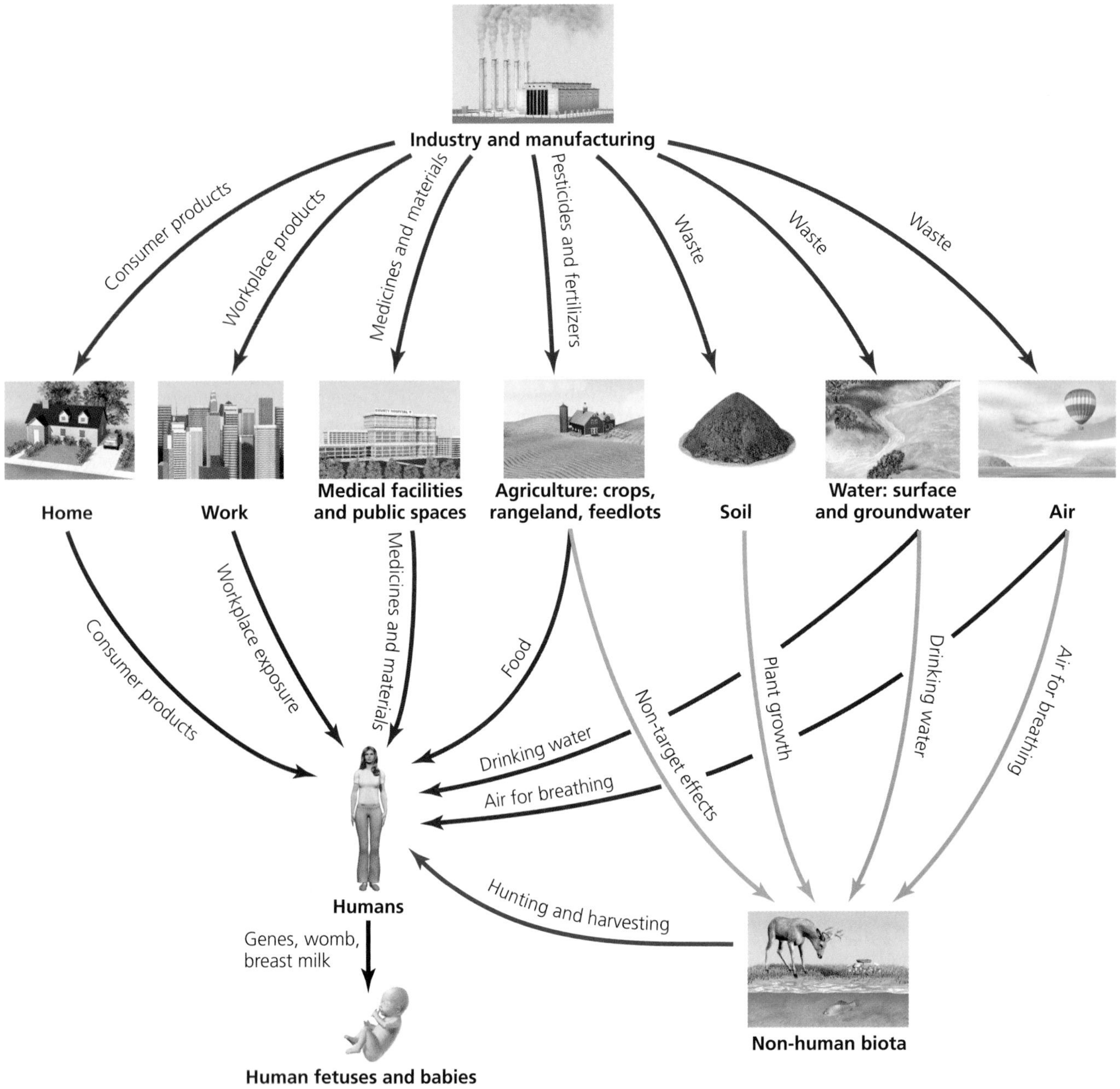

FIGURE 14.11 Synthetic chemicals take many routes in traveling through the environment. Although humans take in only a tiny proportion of these compounds, and although many compounds are harmless, humans—particularly babies—receive small amounts of toxicants from many sources.

demonstrated this concentrating effect. If chemicals can persist in soil, they can leach down into groundwater and contaminate drinking water supplies.

Many chemicals are soluble in water, and these are often the ones that are most accessible to organisms, entering organisms' tissues through drinking or absorption. For this reason, aquatic animals such as fish, frogs, and stream invertebrates are especially good indicators of pollution. When aquatic organisms become sick, we can take it as an early warning that something is amiss. We can then focus research to identify the problem, ideally before it harms humans or the ecological systems that support us. This is why findings that show impacts of low concentrations of pesticides on frogs, fish, and invertebrates have caused such a stir. Many scientists see such findings as a warning that humans could be next, because the pesticides that wash into streams and rivers also flow and seep into the water we drink, and drift through the air we breathe.

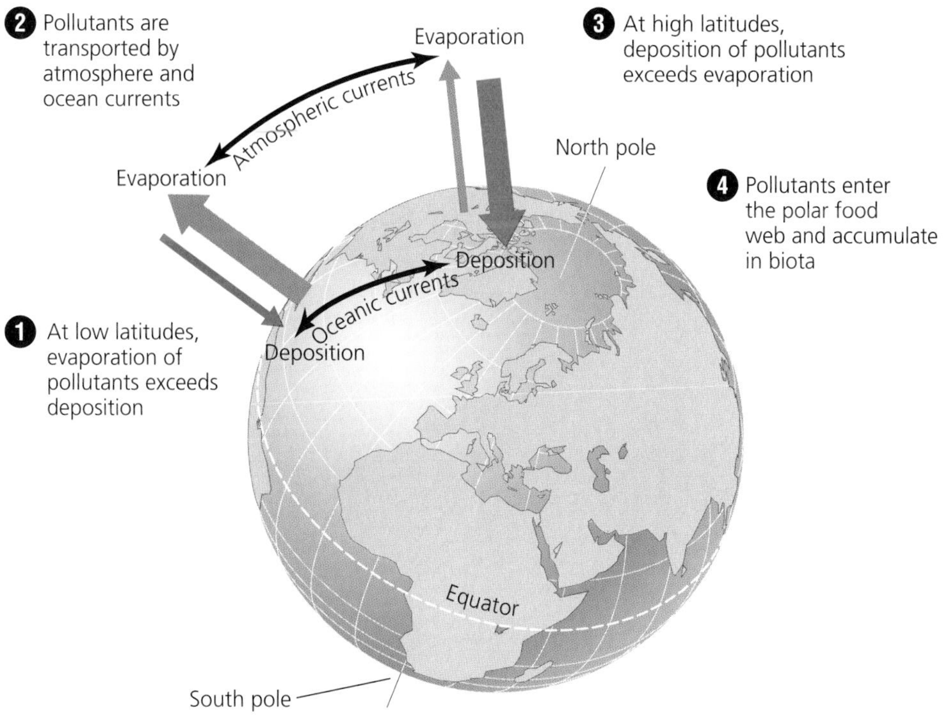

2 Pollutants are transported by atmosphere and ocean currents
Evaporation
Atmospheric currents
3 At high latitudes, deposition of pollutants exceeds evaporation
North pole
Evaporation
4 Pollutants enter the polar food web and accumulate in biota
Deposition
Oceanic currents
1 At low latitudes, evaporation of pollutants exceeds deposition
Deposition
Equator
South pole

FIGURE 14.12 In a process called *global distillation,* pollutants that evaporate and rise high into the atmosphere at lower latitudes, or are deposited in the ocean, are carried preferentially toward the poles by atmospheric currents of air and oceanic currents of water. For this reason, polar organisms take in more than their share of toxicants, despite the fact that relatively few synthetic chemicals are manufactured or used near the poles.

Airborne toxicants can travel widely

Because many chemical substances can be transported by air (Chapter 17), the toxicological effects of chemical use can occur far from the site of direct chemical use. For instance, airborne transport of pesticides is sometimes termed *pesticide drift.* The Central Valley of California is widely considered the most productive agricultural region in the world. But because it is a naturally arid area, food production depends on intensive use of irrigation, fertilizers, and pesticides. Roughly 143 million kg (315 million lb) of pesticide active ingredients are used in California each year, most of these in the Central Valley. The region's frequent winds often blow the airborne spray—and dust particles containing pesticide residue—for long distances. Families living in towns in the Central Valley have suffered health impacts, and activists for farmworkers maintain that hundreds of thousands of the state's residents are at risk. In the mountains of the Sierra Nevada, research has associated pesticide drift from the Central Valley with population declines in four species of frogs.

Synthetic chemical contaminants are ubiquitous worldwide. Despite being manufactured and applied mainly in the temperate and tropical zones, contaminants appear in substantial quantities in the tissues of Arctic polar bears, Antarctic penguins, and people living in Greenland. Scientists can travel to the most remote and seemingly pristine alpine lakes in British Columbia and find them contaminated with foreign toxicants, such as PCBs. The surprisingly high concentrations in polar regions result from patterns of global atmospheric circulation that move airborne chemicals systematically toward the poles (Figure 14.12).

Some toxicants persist for a long time

Once toxic agents arrive somewhere, they may degrade quickly and become harmless, or they may remain unaltered and persist for many months, years, or decades. The rate at which chemicals degrade depends on factors such as temperature, moisture, and sun exposure, and on how these factors interact with the chemistry of the toxicant. Toxicants that persist in the environment have the greatest potential to harm many organisms over long periods of time. A major reason people have been so concerned about toxic chemicals such as DDT and PCBs is that they have long persistence times. In contrast, the Bt toxin used in biocontrol and in genetically modified (GM) crops (▸ p. 284; Figure 10.13, ▸ p. 288) has a very short persistence time. The herbicide atrazine is variable in its persistence, depending on environmental conditions. There are large numbers of persistent synthetic chemicals because many are designed to be persistent. Those used in plastics, for instance, are used precisely because they resist breakdown.

Most toxicants eventually degrade into simpler compounds called *breakdown products.* Often these are less harmful than the original substance, but sometimes they are just as toxic as the original chemical, or more so. For instance, DDT breaks down into DDE, a highly persistent and toxic compound in its own right. Atrazine produces a large number of breakdown products whose effects have not been fully studied.

Toxicants may accumulate and move up the food chain

Toxicants that organisms absorb, breathe, or consume are subjected to the organism's metabolic processes. Some will be quickly excreted, and others will be degraded into harmless breakdown products. However, some will not be broken down but instead will remain in the body. Toxicants that are fat-soluble or oil-soluble (often organic compounds such as DDT and DDE) are absorbed and stored in fatty tissues. Others, such as methyl mercury, may be stored in muscle tissue. Such toxicants may build up in an animal in a process termed **bioaccumulation.**

Toxicants that bioaccumulate in the tissues of one organism may be transferred to other organisms as predators consume prey. When one organism consumes another, it takes in any stored toxicants and stores them itself, along with the toxicants it has received from eating many other prey. Thus with each step up the food chain, from producer to primary consumer to secondary consumer and so on, concentrations of toxicants can be greatly magnified. This process, called **biomagnification,** occurred most famously with DDT. Top predators, such as birds of prey, ended up with high concentrations of the pesticide because concentrations were magnified as DDT moved from water to algae to plankton to small fish to bigger fish and finally to fish-eating birds (Figure 14.13).

Biomagnification caused populations of many North American birds of prey to decline precipitously from the 1950s to the 1970s. The peregrine falcon was almost totally wiped out in the eastern United States, and the bald eagle, the U.S. national bird, was virtually eliminated from the lower 48 states. The brown pelican vanished from its Atlantic Coast range, remaining only in Florida, and the osprey and other hawks saw substantial population declines. Eventually scientists determined that DDT was causing these birds' eggshells to grow thinner, so that eggs were breaking while in the nest.

Although all these birds' populations have rebounded since the U.S. ban on DDT, such scenarios are by no means a thing of the past. The polar bears of Svalbard Island in Arctic Norway are at the top of the food chain and feed on seals that have biomagnified toxicants. Despite their remote Arctic location, Svalbard Island's polar bears show some of the highest levels of PCB contamination of any wild animals tested, as a result of biomagnification and the process of global distillation shown in Figure 14.12. The contaminants are likely responsible for the immune suppression, hormone disruption, and high cub mortality that the bears seem to be suffering. Cubs that survive receive PCBs in their mothers' milk, so that contamination persists and accumulates between generations.

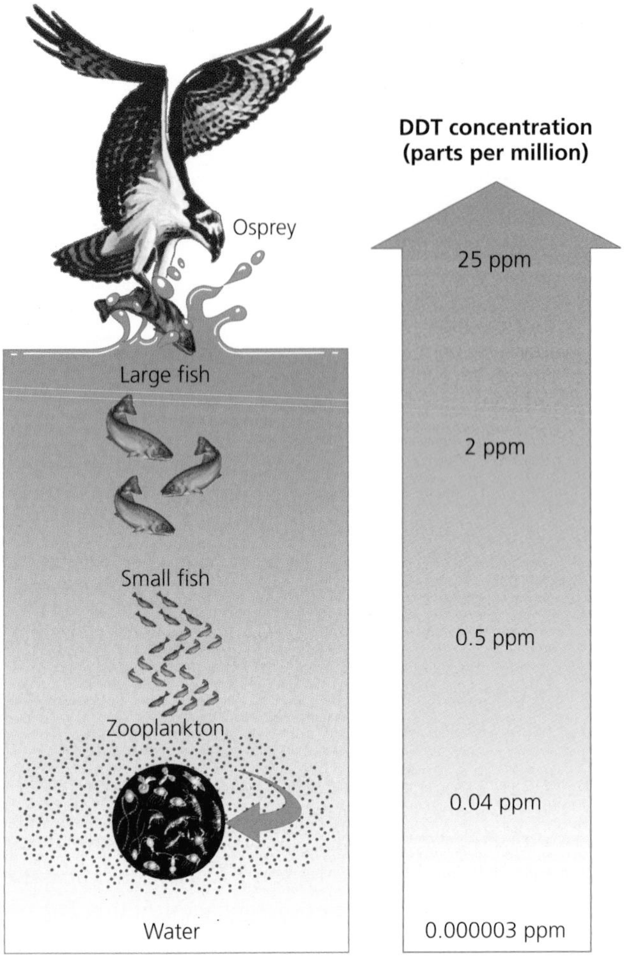

FIGURE 14.13 Fat-soluble compounds such as DDT bioaccumulate in the tissues of organisms. As animals at higher trophic levels eat organisms lower on the food chain, their load of toxicants passes up to each consumer. DDT moves from zooplankton through various types of fish, finally becoming highly concentrated in fish-eating birds such as ospreys.

Not all toxicants are synthetic

Although we have focused on synthetic chemicals thus far, it is vital to recognize that chemical toxicants also exist naturally in the environment around us and in the foods we eat. We have good reason as citizens and consumers to insist on being informed about risks synthetic chemicals may pose, but it is a mistake to assume that all artificial chemicals are unhealthy and that all natural chemicals are healthy. In fact, the plants and animals we eat contain many chemicals that can cause us harm. Recall that plants produce toxins to ward off animals that eat them. In domesticating crop plants, we have selected for strains with reduced toxin content, but we have not eliminated these dangers. Furthermore, when we consume animal meat, we take in toxins the animals have ingested from plants or animals they have eaten.

Scientists have actively debated just how much risk natural toxicants pose. Biochemist Bruce Ames of the University of California at Berkeley maintains that the amounts of synthetic chemicals in our food from pesticide residues are dwarfed by the quantities of natural toxicants. Ames also holds that the natural defenses against toxins that our bodies have evolved are largely effective against synthetic toxicants. Therefore, he reasons, synthetic chemicals are a relatively minor worry. Moreover, he contends, if fear of pesticide residues or increased cost to remove them causes people to eat fewer fruits and vegetables, then people will be at greater risk for disease, because a balanced diet that includes fruits and vegetables is important for health.

A respected senior scientist, Ames was a hero of environmentalists for his early work. The *Ames test,* a bacterial assay for mutagens, allows easy screening of suspected toxins. Today Ames is a target of criticism from many environmental advocates. His critics say that natural toxicants are usually more readily metabolized and excreted by the body than synthetic ones, that synthetic toxicants persist and accumulate in the environment, and that synthetic chemicals expose people (such as farmworkers and factory workers) to risks in ways other than the ingestion of food. What is clear is that more research is required in this area.

Weighing the Issues:
Natural and Synthetic Estrogen Mimics

Skeptics of the idea that synthetic chemicals act as endocrine disruptors point out that humans are exposed to phytoestrogens, natural estrogens from plants, which are ubiquitous in the environment. However, phytoestrogens are easily broken down by the body and are not stored in tissue. Do you think there is a distinction between synthetic and natural estrogen-mimicking chemicals? Should manufacturers of synthetic chemicals that mimic estrogen be held responsible for health impacts, given that natural chemicals can also mimic estrogen?

Studying Effects of Hazards

Determining health effects of particular environmental hazards is a challenging job, particularly because any given person or organism likely has a complex history of exposure to many hazards throughout life. Scientists rely on several different methods, including correlative surveys and manipulative experiments (▸pp. 15–17).

Wildlife studies use careful observations in the field and the lab

When scientists were zeroing in on the impacts of DDT, one key piece of evidence came from museum collections of wild birds' eggs from the decades before synthetic pesticides were manufactured. Eggs from museum collections had measurably thicker shells than the eggs scientists were studying in the field from present-day birds. Scientists have pieced together the puzzle of toxicant effects on alligators by taking measurements from animals in the wild, then doing controlled experiments in the lab to test hypotheses. With frogs and atrazine, scientists first measured toxicological effects in lab experiments, then sought to demonstrate correlations with herbicide usage in the wild.

Often the study of wildlife advances in the wake of some conspicuous mortality event. Off the California coast in 1998–2001, populations of sea otters fell noticeably, and many dead otters washed ashore. Field biologists documented the population decline, and specialists went to work in the lab performing autopsies to determine causes of death. The most common cause of death was found to be infection with the protist parasite *Toxoplasma,* which killed otters directly and also made them vulnerable to shark attack. *Toxoplasma* occurs in the feces of cats, so scientists hypothesized that sewage runoff containing waste from litter boxes was entering the ocean from urban areas and infecting the otters.

Human studies rely on case histories, epidemiology, and animal testing

In studies of human health, as in wildlife studies, we have gained much knowledge by studying sickened individuals directly. Medical professionals have long treated victims of poisonings, for instance, so the effects of common poisons are well known. Autopsies help us understand what constitutes a lethal dose. This process of observation and analysis of individual patients is known as a *case history* approach. Case histories have advanced our understanding of human illness, but they do not always help us infer

The Science behind the Story

Pesticides and Child Development in Mexico's Yaqui Valley

With spindly arms and big, round eyes, one set of pictures shows the sorts of stick figures drawn by young children everywhere. Next to them is another group of drawings, mostly disconnected squiggles and lines, resembling nothing. Both sets of pictures are intended to depict people. The main difference identified between the two groups of young artists: long-term pesticide exposure.

Children's drawings are not a typical tool of toxicology, but Elizabeth Guillette, an anthropologist married to Louis Guillette, wanted to try new methods. Guillette was interested in the effects of pesticides on children. She devised tests to measure childhood development based on techniques from anthropology and medicine. Searching for a study site, Guillette found the Yaqui Valley region of northwestern Mexico.

The Yaqui Valley is farming country, worked for generations by the indigenous group that gives the region its name. Synthetic pesticides arrived in the area in the 1940s. Some Yaqui embraced the agricultural innovations, spraying their farms in the valley to increase their yields. Yaqui farmers in the surrounding foothills, however, generally chose to bypass the chemicals and to continue following more traditional farming practices. Although differing in farming techniques, Yaqui in the valley and foothills continued to share the same culture, diet, education system, income levels, and family structure.

At the time of the study, in 1994, valley farmers planted crops twice a year, applying pesticides up to 45 times from planting to harvest. A previous study conducted in the valley in 1990, focusing on areas with the largest farms, had indicated high levels of multiple pesticides in the breast milk of mothers and in the

Drawings by children in the foothills

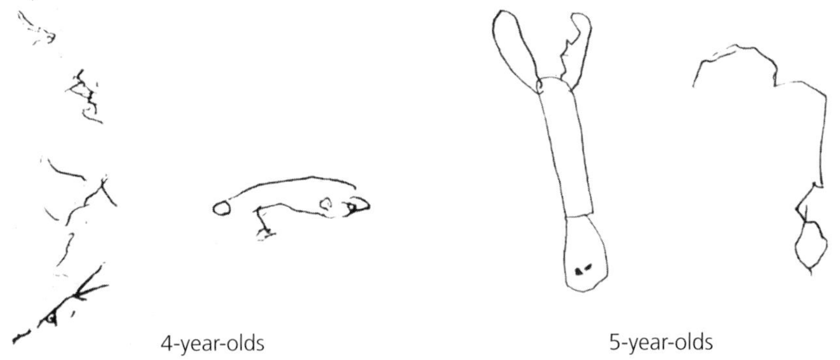

Drawings by children in the valley

Elizabeth Guillette's study in Mexico's Yaqui Valley offers a startling example of apparent neurological effects of pesticide poisoning. Young children from foothills areas where pesticides were not commonly used drew recognizable figures of people. Children the same age from valley areas where pesticides were used heavily in industrialized agriculture could draw only scribbles. Adapted from Guillette, E. A., et al. 1998. *Environmental Health Perspectives* 106: 347–353.

the effects of hazards people rarely come into contact with, of newly manufactured compounds, or of chemicals existing at low environmental concentrations and exerting minor long-term effects. Case histories also tell us little about probability and risk, such as how many extra deaths we might expect in a population due to a particular cause.

For such situations, which are common in environmental toxicology, epidemiological studies are necessary. **Epidemiological studies** involve large-scale comparisons among groups of people, usually contrasting a group known to have been exposed to some hazard and a group that has not. Epidemiologists track the fate of all people in the study, generally for a long period of time (often umbilical cord blood of newborn babies. In contrast, foothill families avoided chemical pesticides in their gardens and homes.

To understand how pesticide exposure affects childhood development, Guillette and fellow researchers studied 50 preschoolers aged four to five, of whom 33 were from the valley and 17 from the foothills. Each child underwent a half-hour exam, during which researchers showed a red balloon, promising to give the balloon later as a gift, and using the promise to evaluate long-term memory. Each child was then put through a series of physical and mental tests:

- Catching a ball from distances of up to 3 m (10 ft) away, to test overall coordination
- Jumping in place for as long as possible, to assess endurance
- Drawing a picture of a person, as a measure of perception
- Repeating a short string of numbers, to test short-term memory
- Dropping raisins into a bottle cap from a height of about 13 cm (5 in.), to gauge fine motor skills

The researchers also measured each child's height and weight but, because of lack of time and money, stopped short of taking blood or tissue samples to check for pesticides or other toxins. When all tests were completed, each child was asked what he or she had been promised and received a red balloon.

Although the two groups of children were not significantly different in height and weight, they differed markedly in other areas of development. Valley children were far behind the foothill children developmentally in coordination, physical endurance, long-term memory, and fine motor skills:

- From a distance of 3 m (10 ft), valley children had great difficulty catching the ball.
- Valley children could jump for an average of 52 seconds, compared to 88 seconds for foothill children.
- Most valley children missed the bottle cap when dropping their raisins, whereas foothill children dropped them into the caps far more often.
- Each group did fairly well repeating numbers, but valley children showed poor long-term memory. At the end of the test, all but one of the foothill children remembered that they had been promised a balloon, and 59% remembered it was red. However, of the valley children only 27% remembered the color of the balloon, only 55% remembered they'd be getting a balloon, and 18% were unable to remember anything about the balloon.

It was the children's drawings, however, that exhibited the most dramatic difference between valley and foothill children (see figure). The researchers determined each drawing could earn 5 points, with 1 point each for a recognizable feature: head, body, arms, legs, and facial features. The foothill children drew pictures that looked like people, averaging about 4.5 points per drawing. The valley children, in contrast, averaged 1.6 points per drawing; their scribbles resembled little that looked like a person. By the standards of developmental medicine, the four- and five-year-old valley children drew at the level of a two-year-old.

Some scientists greeted Guillette's study skeptically, pointing out that its sample size was too small to be meaningful. Others said that factors the researchers missed, such as different parenting styles or unknown health problems, could be to blame. Prominent toxicologists argued that without blood or tissue tests on the children, the study results couldn't be tied to agricultural chemicals. Regardless of these criticisms, Guillette maintains that her findings show that nontraditional study methods are a valid way to track the effects of environmental toxins and that pesticides present a complex long-term risk to human growth and health.

years or decades) and measure the rate at which deaths, cancers, or other health problems occur in each group. The epidemiologist then analyzes the data, looking for observable differences between the groups, and statistically tests hypotheses accounting for differences. When a group exposed to a hazard shows a significantly greater degree of harm, it suggests that the hazard may be responsible. This process is akin to a natural experiment (▸pp. 16–17), in which the experimenter takes advantage of the presence of groups of subjects made possible by some event that has already occurred. A slightly different type of natural experiment was conducted by anthropologist Elizabeth Guillette (see "The Science behind the Story," above).

The advantages of epidemiological studies are their realism and their ability to yield relatively accurate predictions about risk. The drawbacks include the need to wait a long time for results and the inability to address future effects of new products just coming to market. In addition, participants in epidemiological studies encounter many factors that affect their health besides the one under study. Epidemiological studies measure a statistical association between a toxicant and an effect, but they do not confirm that the toxicant *causes* the effect.

Manipulative experiments are needed to truly nail down causation. However, subjecting people to massive doses of toxicants in a lab experiment would clearly be unethical. So researchers have traditionally used other animals as subjects to test toxicity. Foremost among these animal models have been laboratory strains of rats, mice, and other mammals. Because of our shared evolutionary history with these mammals, their bodies function similarly to ours. Yet we are also different, of course. The extent to which results from animal lab tests apply to humans is always somewhat uncertain, and it can be expected to vary from one study to the next. Although some people feel the use of rats and mice for testing is unethical, animal testing enables scientific advances that would be impossible or far more difficult otherwise. However, new techniques (with human cell cultures, bacteria, or tissue from chicken eggs) are being devised that may one day replace some live-animal testing.

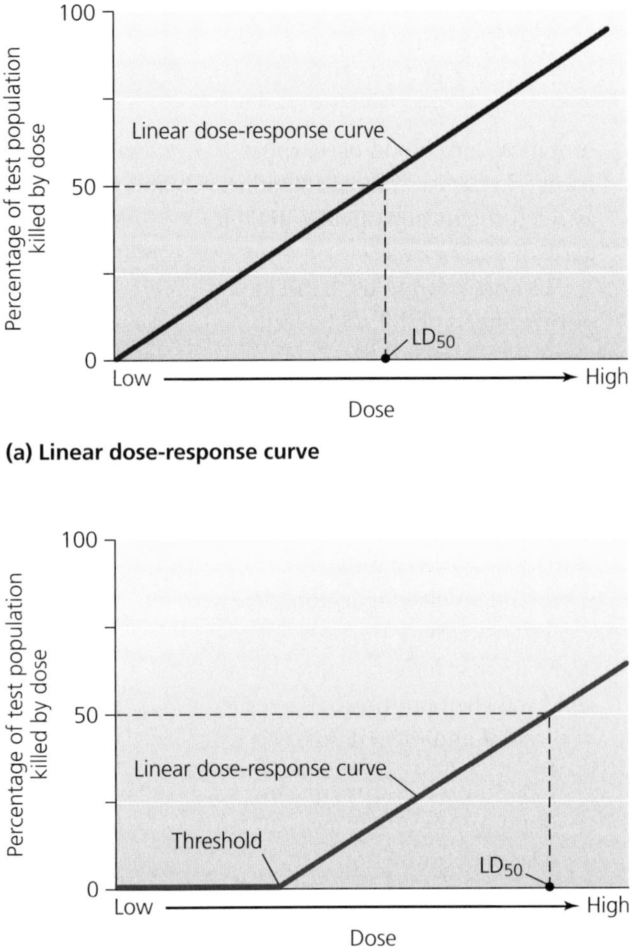

(a) Linear dose-response curve

(b) Dose-response curve with threshold

FIGURE 14.14 In a classic linear dose-response curve **(a)**, the percentage of animals killed or otherwise affected by the substance rises with the dose. The point at which 50% of the animals are killed is labeled the lethal-dose-50, or LD_{50}. For some toxic agents, a threshold dose **(b)** exists, below which doses have no measurable effect. Go to **GRAPHit!** at www.aw-bc.com/withgott or on the student CD-ROM.

Dose-response analysis is a mainstay of toxicology

The standard method of testing with lab animals in toxicology is dose-response analysis. Scientists quantify the toxicity of a given substance by measuring how much effect a toxicant produces at different doses or how many animals are affected by different doses of the toxic agent. The *dose* is the amount of toxicant the test animal receives, and the *response* is the type or magnitude of negative effects the animal exhibits as a result. The response is generally quantified by measuring the proportion of animals exhibiting negative effects. The data are plotted on a graph, with dose on the *x* axis and response on the *y* axis (Figure 14.14a). The resulting curve is called a **dose-response curve.**

Once they have plotted a dose-response curve, toxicologists can calculate a convenient shorthand gauge of a substance's toxicity: the amount of toxicant it takes to kill half the population of study animals used. This lethal dose for 50% of individuals is termed the $\mathbf{LD_{50}}$. A high LD_{50} indicates low toxicity, and a low LD_{50} indicates high toxicity. Of course, the experimenter may instead be interested in nonlethal health effects. A researcher often will want to document the level of toxicant at which 50% of the population of test animals is affected in whatever way is of interest in the study (for instance, what level of toxicant causes 50% of lab mice to lose their hair?). Such a level is called the effective-dose-50%, or $\mathbf{ED_{50}}$. Chemicals with an especially low LD_{50} or ED_{50} are assumed likely to be poisonous to humans, whereas those with an extremely high LD_{50} or ED_{50} are thought likely to be safe.

Sometimes responses occur only above a certain dose. Such a **threshold** dose (Figure 14.14b) might be expected if the body's organs can fully metabolize or excrete a toxicant at low doses but become overwhelmed at higher concentrations. It might also occur if cells can repair damage to their DNA only up to a certain point.

Scientists generally give lab animals much higher doses relative to body mass than humans would receive in the environment. This is so that the response is great enough to be measured, and so that differences between the effects of small and large doses are evident. Data from a range of doses help give shape to the dose-response curve. Once the data from animal tests are plotted, scientists generally extrapolate downward to estimate the effect of still-lower doses on a hypothetically large population of animals. This way, they can come up with an estimate of, say, what dose causes cancer in 1 mouse in 1 million. A second extrapolation is then required to estimate the effect on humans, with our greater body mass. Because these two extrapolations go beyond the actual data obtained, they introduce uncertainty into the interpretation of what doses are acceptable for humans. As a result, to be on the safe side, regulatory agencies set standards for maximum allowable levels of toxicants that are well below the minimum toxicity levels estimated from lab studies.

Knowing the shape of dose-response curves is crucial if one is planning to extrapolate from them to predict responses at doses below those that have been tested. And sometimes responses *decrease* with increased dose. Toxicologists are finding that some dose-response curves are U-shaped or J-shaped, or shaped like an inverted U. Such counterintuitive curves often appear to apply to endocrine disruptors. This likely occurs because the hormone system is geared to function with extremely low concentrations of hormones and is thus vulnerable to disruption by toxicants at extremely low concentrations. Inverted dose-response curves present a challenge for policymakers attempting to set safe environmental levels for toxicants. If many chemicals behave in these ways, we may have underestimated the dangers of these compounds, because many chemicals exist in very low concentrations over wide areas.

Individuals vary in their responses to hazards

Different individuals may respond quite differently to identical exposures to hazards. These differences can be genetically based, or can be due to a person's current condition. People in poorer health are often more sensitive to biological and chemical hazards. Sensitivity also can vary with sex, age, and weight. Because of their smaller size and rapidly developing organ systems, fetuses, infants, and young children tend to be much more sensitive to toxicants than are adults. The degree of this difference was not appreciated for many years. Regulatory agencies such as the EPA traditionally set standards for adults and extrapolated downward for infants and children. However, they have subsequently found that in many cases their linear extrapolations did not lower standards enough to protect babies adequately. Many critics today contend that despite improvements, regulatory agencies still do not account explicitly enough for risks to fetuses, infants, and children.

The type of exposure can affect the response

The risk posed by a hazard often varies according to whether a person experiences high exposure for short periods of time, known as **acute exposure,** or lower exposure over long periods of time, known as **chronic exposure.** Incidences of acute exposure are easier to recognize, because they often stem from discrete events, such as accidental ingestion, an oil spill, a chemical spill, or a nuclear accident. Lab tests and LD_{50} values generally reflect acute toxicity effects. However, chronic exposure is more common—and more difficult to detect and diagnose. Chronic exposure often affects organs gradually, as when smoking causes lung cancer, or when alcohol abuse induces liver or kidney damage. Pesticide residues on food or low levels of arsenic in drinking water (▸ pp. 456–457) also pose chronic risk. Because of the long time periods involved, relationships between cause and effect may not be readily apparent.

Mixes may be more than the sum of their parts

It is difficult enough to determine the impact of a single hazard on an organism, but the task becomes astronomically more difficult when multiple hazards interact. For instance, chemical substances, when mixed, may act in concert in ways that cannot be predicted from the effects of each in isolation. Mixed toxicants may sum each other's effects, cancel out each other's effects, or multiply each other's effects. Whole new types of impacts may arise when toxicants are mixed together. Such interactive impacts—those that are more than or different from the simple sum of their constituent effects—are called **synergistic effects.**

Examples of synergistic effects with toxic substances are numerous. With Florida's alligators, lab experiments have indicated that DDE can either help cause or inhibit sex reversal, depending on the presence of other chemicals. Mice exposed to a mixture of nitrate, atrazine, and aldicarb have been found to show immune, hormone, and nervous-system effects that were not evident from exposure to each of these chemicals alone. Wood frogs in the wild are increasingly suffering limb deformities, apparently the result of being parasitized by trematode flatworms. However, being near an agricultural field with pesticide runoff increases the rate of parasitic infection, because, as lab studies have shown, pesticides suppress the frog's immune response, making the amphibian more vulnerable to parasites.

Traditionally, environmental health has tackled effects of single hazards one at a time. In toxicology, the complex experimental designs required to test interactions, and the sheer number of chemical combinations, have meant that single-substance tests have received priority. This approach is changing, but scientists in environmental health and toxicology will never be able to test all possible combinations. There are simply too many hazards in the environment.

Risk Assessment and Risk Management

Policy decisions to ban chemicals or restrict their use are not made casually, but generally follow years of rigorous testing for toxicity. Likewise, strategies for combating disease and other environmental health threats are often based on extensive research. Policy and management decisions also reach beyond the scientific results on health to incorporate considerations about economics and ethics. And all too often, they are influenced by political pressure from powerful interests. The steps between the collection and interpretation of scientific data and the formulation of policy involve assessing and managing risk.

Risk is expressed in terms of probability

Exposure to an environmental health threat does not invariably produce some given effect. Rather, it causes some probability of harm, some statistical chance that damage will result. To understand the impact of an environmental health threat, a scientist must know more than just its identity and strength. He or she must also know the chance that an organism will encounter it, the frequency with which the organism may encounter it, the amount of substance or degree of threat to which the organism is exposed, and the organism's sensitivity to the threat. Such factors help determine the overall risk posed. Risk can be measured in terms of *probability*, a quantitative description of the likelihood of a certain outcome. The mathematical probability that some harmful outcome (for instance, injury, death, environmental damage, or economic loss) will result from a given action, event, or substance expresses the **risk** posed by that phenomenon.

Our perception of risk may not match reality

Every action we take and every decision we make involves some element of risk, some (generally small) probability that things will go wrong. Some actions and decisions are more risky than others, and we try our best in everyday life to behave in ways that minimize risk. Interestingly, our perceptions of risk do not always match statistical reality (Figure 14.15). People often worry unduly about negligibly small risks but happily engage in other activities that pose high risks. For instance, most people perceive flying in an airplane as a riskier activity than driving a car, but driving a car is statistically far more dangerous.

Psychologists agree that this difference between risk perception and reality stems from the fact that we feel more at risk when we are not controlling a situation and more safe when we are "at the wheel"—regardless of the actual risk involved. When we drive a car, we feel we are in control, even though statistics show we are at greater risk than as a passenger in an airplane. This psychology can account for people's great fear of nuclear power, toxic waste, and pesticide residues on foods—environmental hazards that are invisible or little understood, and whose presence in their lives is largely outside their personal control. In contrast, people are more ready to accept and ignore the risks of smoking cigarettes, overeating, and not exercising, all voluntary activities statistically shown to pose far greater risks to health.

Risk assessment analyzes risk quantitatively

The quantitative measurement of risk and the comparison of risks involved in different activities or substances together are termed **risk assessment.** Risk assessment is a way of identifying and outlining problems. In environmental health, it helps ascertain which substances and activities pose health threats to people or wildlife and which are largely safe. Assessing risk for a chemical substance

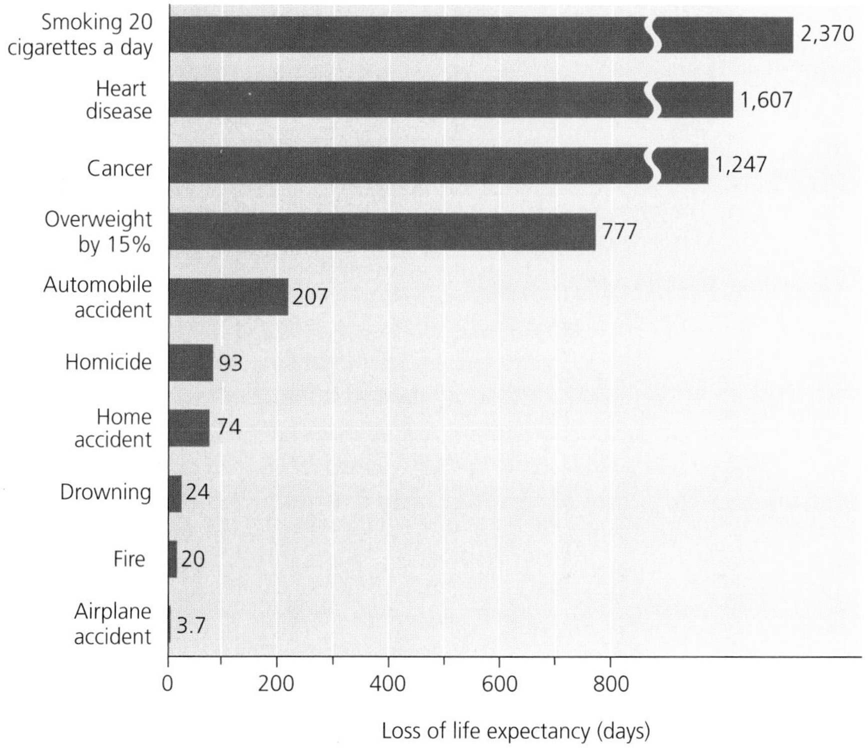

FIGURE 14.15 Our perceptions of risk do not always match the reality of risk. Listed here are several leading causes of death in the United States, along with a measure of the risk each poses. Risk is measured in days of lost life expectancy, or the number of days of life lost by people suffering the hazard, spread across the entire population—a measure commonly used by insurance companies. By this measure, one common source of anxiety, airplane accidents, poses 20 times less risk than home accidents, over 50 times less risk than auto accidents, and over 200 times less risk than being overweight. Data from Cohen, B. 1991. Catalog of risks extended and updated. *Health Physics* 61: 317–335.

involves several steps. The first steps involve the scientific study of toxicity outlined above—determining whether a given substance has toxic effects and, through dose-response analysis, measuring how effects on an organism vary with the degree of toxicant exposure. Subsequent steps involve assessing the individual's or population's likely extent of exposure to the substance, including the frequency of contact, the concentrations likely encountered, and the length of time the substance is expected to be encountered. Risk assessment studies are often performed by scientists associated with the industries that manufacture toxicants, which in many people's minds may undermine the objectivity of the process.

Weighing the issues: A Nationwide Health Tracking System?

When 16 children were diagnosed with leukemia in the small town of Fallon, Nevada, it raised alarm. The identification of such "cancer clusters" has prompted many U.S. scientists and policymakers to back the notion of creating a national environmental health tracking system. Some states already track environmental hazards and illnesses, but comparable statistics are not compiled everywhere. It would take an estimated $1 billion to integrate these disparate sets of information and to maintain records.

Do you think the potential benefits are worth it? How might this affect epidemiological studies? How might it affect risk assessments? Could it speed the identification of causes of poisonings or disease outbreaks? How might finding the causes of such events help to mitigate them or to prevent them in the future?

Risk management combines science and other social factors

Accurate risk assessment is a vital step toward effective **risk management,** which consists of decisions and strategies to minimize risk (Figure 14.16). In most developed nations, risk management is handled largely by federal agencies, such as the EPA, the Centers for Disease Control and Prevention (CDC), and the Food and Drug Administration (FDA) in the United States. In risk management, scientific assessments of risk are considered in light of economic, social, and political needs and values. The costs and benefits of addressing risk in various ways are assessed with regard to both scientific and nonscientific concerns. Decisions whether to reduce or eliminate risk are then made.

In environmental health and toxicology, comparing costs and benefits (▸ p. 42) is often difficult because the benefits are often economic and the costs often pertain to health. Moreover, economic benefits are generally known,

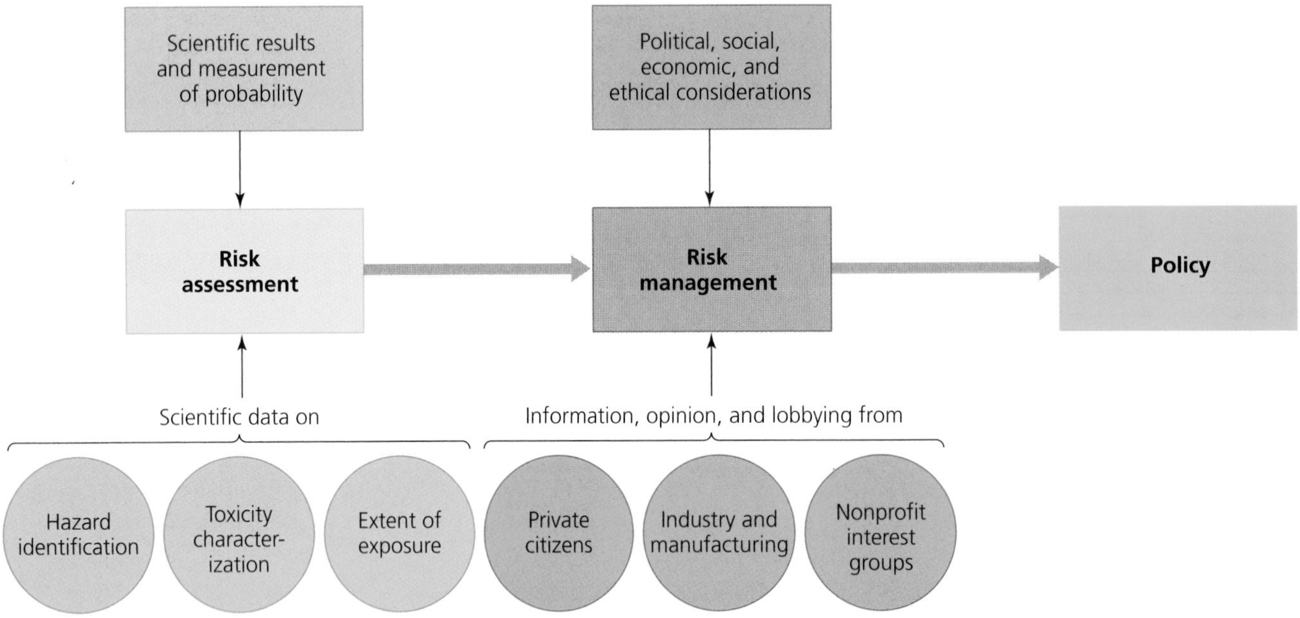

FIGURE 14.16 Risks must be assessed before policy steps can be taken to minimize them. Thus the first step in addressing the risk of an environmental hazard is risk assessment, a process of quantifying the risk of the hazard and comparing it to other risks. Once science identifies and measures risks, then risk management can proceed. In this process, economic, political, social, and ethical issues are considered in light of the scientific data from risk assessment. The consideration of all these types of information is designed to result in policy decisions that minimize the risk of the environmental hazard.

easily quantified, and of a discrete and stable amount, whereas health risks are hard-to-measure probabilities, often involving a very small percentage of people likely to suffer greatly and a large majority likely to experience little effect. When a government agency bans a pesticide, it generally means measurable economic loss for the manufacturer and the farmer, whereas the benefits accrue less predictably over the long term to some percentage of factory workers, farmers, and the general public. Because of the lack of equivalence in the way costs and benefits are measured, risk management frequently tends to stir up debate.

Philosophical and Policy Approaches

Because we do not know a substance's toxicity until we measure and test it, and because there are so many untested chemicals and combinations, science will never eliminate the many uncertainties that accompany risk assessment. In such a world of uncertainty, there are two basic philosophical approaches to categorizing substances as safe or dangerous (Figure 14.17).

Two approaches exist for determining safety

One approach is to assume that substances are harmless until shown to be harmful. We might nickname this the *innocent-until-proven-guilty approach.* Because thoroughly testing every existing substance (and combination of substances) for its effects is a hopelessly long, complicated, and expensive pursuit, the innocent-until-proven-guilty approach has the benefit of not slowing down technological innovation and economic advancement. However, it has the disadvantage of putting into wide use some substances that may later turn out to be dangerous.

The other approach is to assume that substances are harmful until they are shown to be harmless. This approach follows the precautionary principle first discussed in Chapter 10 in regard to genetically modified foods (►p. 290). This more cautious approach should enable us to identify troublesome toxicants before they are released into the environment, but it may also significantly impede the pace of technological and economic advance.

These two approaches are actually two ends of a continuum of possible approaches. The two endpoints differ

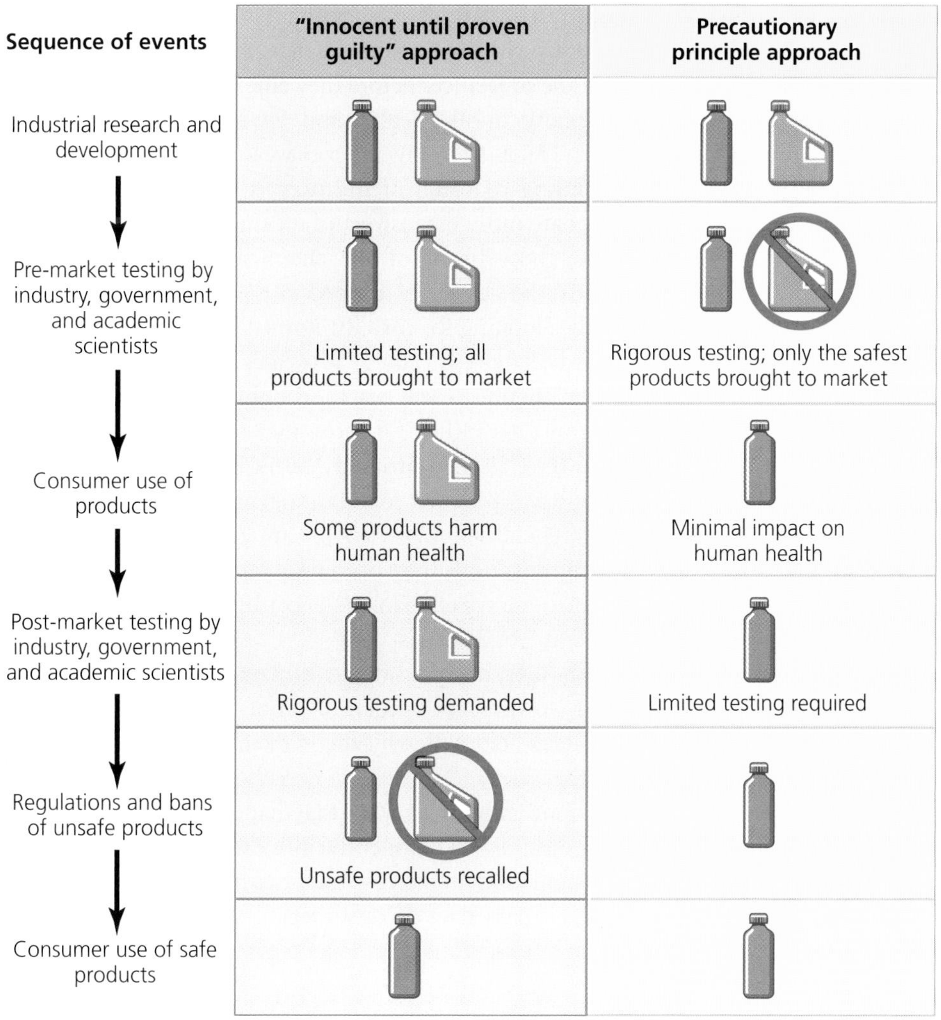

FIGURE 14.17 Testing a new chemical compound or product for toxicity rarely gives a black-or-white answer, and many tests must be run before a substance's properties are well understood. Given such uncertainty, two main approaches can be taken to introducing new substances to the market. In one approach, substances are innocent until proven guilty; they are brought to market relatively quickly after limited testing. Products reach consumers more quickly, but some fraction of them may cause harm to some fraction of people. In the other approach, the precautionary principle is adopted, and substances are brought to market cautiously, only after extensive testing. Products that reach the market should be safe, but many perfectly safe products will be delayed in reaching consumers.

mainly in where they lay the burden of proof—specifically, whether product manufacturers are required to prove safety or whether government, scientists, or citizens are required to prove danger.

Weighing the Issues: The Precautionary Principle

Given the substantial costs of testing chemicals for safety and the increasing concerns about their spread through the environment, should proof of safety be required by government prior to a chemical's release into the environment? Should the burden of proof fall to the company that stands to make a profit from a product's release? How do you think adopting the precautionary principle would affect the number of chemicals on the market?

Philosophical approaches are reflected in policy

One's philosophical approach has immediate and far-reaching impact on policy decisions, directly affecting what materials are allowed into our environment. Most nations follow a blend of the two approaches, but there is marked variation among countries. At the present time, European nations are to a great extent following the precautionary principle regarding the regulation of synthetic chemicals, whereas the United States is not. Although industry frequently complains that government regulation is cumbersome, environmental and consumer advocates criticize U.S. policies for largely following the innocent-until-proven-guilty approach. For instance, compounds involved in cosmetics require no FDA review or approval before being sold to the public.

In the United States, several federal agencies apportion responsibility for tracking and regulating synthetic chemicals. The FDA, under the Food, Drug, and Cosmetic Act of 1938 and its subsequent amendments, regulates foods and food additives, cosmetics, and drugs and medical devices. The EPA regulates pesticides under the Federal Insecticide, Fungicide, and Rodenticide Act of 1947 (FIFRA) and its amendments. The Occupational Safety and Health Administration (OSHA) regulates workplace hazards under a 1970 act. Several other agencies regulate other substances. Synthetic chemicals not covered by other laws are regulated by the EPA under the 1976 Toxic Substances Control Act (TSCA). Examining in more detail the policy process for pesticides and for toxic substances will provide an idea of how government and industry interact.

The EPA regulates pesticides and other substances

FIFRA was enacted as the post–World War II U.S. chemical industry was expanding, but before the environmental activism of the 1960s and 1970s that gave rise to so many environmental laws. As such, FIFRA was primarily intended not to protect public health or the environment, but to assure consumers that products actually worked as their manufacturers claimed. Subsequent amendments shifted the focus somewhat toward protecting health and safety and charged the EPA with "registering" each new pesticide that manufacturers propose to bring to market.

The registration process involves risk assessment and risk management. The EPA first asks the pesticide manufacturer to provide information, including the results of safety assessments the company has performed according to EPA guidelines. The EPA examines the company's research and all other relevant scientific research. It examines the product's ingredients and how the product will be used and evaluates whether the chemical poses risks to humans, other organisms, or water or air quality. The EPA then approves, denies, or sets limits on the chemical's sale and use. It also must approve the language used on the product's label.

Because the registration process takes economic considerations into account, critics say it allows hazardous chemicals to be approved if the economic benefits outweigh the hazards. Here the challenges of weighing intangible risks involving human health and environmental quality against the tangible and quantitative numbers of economics become apparent.

TSCA directed the EPA to monitor some 75,000 industrial chemicals manufactured in or imported into the United States, including PCBs and other compounds involved in plastics. The act gave the agency power to regulate these chemicals and ban them if they are found to pose excessive risk. TSCA was the first law to require screening of these substances before they entered the marketplace.

Many public health and environmental advocates view TSCA as being far too weak. They note that the screening required of industry is minimal and that to mandate more extensive and meaningful testing, the EPA must show proof of the chemical's toxicity. In other words, the agency is trapped in a Catch-22: To push for studies looking for toxicity, it must have proof of toxicity already. The result, these advocates say, is that most synthetic chemicals are not thoroughly tested before being put on the market. Of those that fall under TSCA, only 10% have been thoroughly tested for toxicity; only 2% have been screened for carcinogenicity, mutagenicity, or teratogenicity; fewer than 1% are government-regulated; and almost none have been tested for endocrine, nervous, or immune system damage, according to the U.S. National Academy of Sciences.

Industry's critics say chemical manufacturers should be made to bear the burden of proof for the safety of their products before they hit the market. Industry's supporters say that safety advocates will never be satisfied that industry has done enough. They say that mandating more toxicological research will hamper the introduction of products that consumers want, increase the price of products as research costs are passed on to consumers, and cause U.S. companies to move to nations where standards are more lax.

Toxicants are regulated internationally

In April 2003, European Union (EU) commissioners proposed legislation that would require chemical manufacturers to test and register 30,000 chemicals already in use and that would impose restrictions on 1,500 chemicals already considered hazardous. In announcing the policy, EU environment commissioner Margot Wallström said, "It is high time to place the responsibility where it belongs, with industry." Industry was not pleased and estimated that the law would cost it $7–8 billion over a decade. At the same time, some people on all sides agreed that the new policy would have the positive effect of spurring research into safer products and creating new markets for them. As of late 2005, the proposal was slowly making its way toward becoming law.

Action regarding chemical toxicants has also been taken in the form of international treaties. The Stockholm Convention on Persistent Organic Pollutants (POPs), introduced in 2001, appears on its way to ratification. POPs are toxic chemicals that persist in the environment, bioaccumulate in the food chain, and often can travel long

VIEWPOINTS

Chemical Product Testing: Industry or Government?

The testing of chemical products for safety can take the so-called innocent-until-proven-guilty or precautionary approaches. **Should manufacturers be held responsible for comprehensive testing of new chemical products before they are introduced to the public? What would be some of the advantages and disadvantages? How extensive a role should government play in the testing process?**

Testing Must Ensure Public Health

Like most things in life, the controversy over product testing arises because both approaches have valid advantages and disadvantages.

Allowing industry to follow the innocent-until-proven-guilty approach, with limited testing, reduces development costs for new chemical products and may lead to greater economic activity. If industry were required to comprehensively test chemical product safety before introduction to the public, chemical industry profits could fall and result in job loss and costly new product development. Consumer prices might rise to cover these costs.

However, comprehensive testing would lower the number of chemicals that adversely affect biological species.

Our definition of *innocent* is often too narrow. In the innocent-until-proven-guilty approach, *what the consumer actually buys is never tested,* because only ultra-pure active ingredients are tested. Surfactants and organic soaps ("other ingredients") are added to improve the active ingredients' lipid or water solubility, and these other ingredients are frequently very active biologically. Also, production contaminants are not tested and registered. Therefore, a so-called innocent product can cause cancer and reproductive defects.

The assumption of a linear dose response is also coming under increased scrutiny, especially at very low physiological doses, where hormonal, immune, and neurological processes respond. At much higher pharmacological doses, where toxicity testing is typically done, the responses of physiological systems to the same chemical can be very different.

Given the inherent inadequacies of the testing process and the uncertainty of the economic impacts, both government and industry should share the responsibility of testing to ensure public safety.

Warren Porter is a toxicologist and physiological ecologist at the University of Wisconsin–Madison. He evaluates the connections among climate, animal energetics, and behavior using statistical experimental design.

An Industry Perspective

Chemical manufacturers already take an active role in testing new chemicals. This process is part of current EPA regulations. Additionally, government often tests chemicals to elucidate either hazard or exposure. Generation of these data by government adds to the body of knowledge generated by industry.

Chemical risk depends on two factors: hazard (toxicity) and exposure. To evaluate a chemical, manufacturers typically start by conducting screening-level toxicological and environmental studies and proceed to more or higher-tier studies as warranted. There is no single comprehensive testing program that is appropriate for all industrial chemicals.

The Toxic Substances Control Act requires almost all new commercial substances to undergo Premanufacture Notification (PMN) review, and to describe this preliminary process as an innocent-until-proven-guilty approach is an oversimplification. When the EPA reviews a PMN, it considers the physical and chemical properties of the substance, structural similarity to other compounds of known toxicity, and potential for human exposure and environmental release. If there is no evidence of harm from preliminary testing, longer-term or more specialized testing may not be conducted. In some cases, the EPA may require additional testing to determine whether the chemical poses an unreasonable risk to human health or the environment. If the EPA finds that risk can be addressed by reducing exposure, it may enter into a binding agreement with the manufacturer to require exposure reduction activities, rather than additional laboratory testing.

Manufacturers often voluntarily conduct new studies to support the continued safe use of their chemicals. The 150 member companies of the American Chemistry Council represent about 90% of U.S. chemical production. These companies are committed to Responsible Care®, under which chemical manufacturers, as good stewards of their products, continue to test as new data and methodologies become available. There is a role for both government and industry in chemical testing, and it is important that the EPA and manufacturers work together in evaluating chemicals to improve health, safety, and the environment.

Marian K. Stanley is senior director for the American Chemistry Council, which she joined in 1990. She is responsible for managing chemical-specific issue groups in the Council's self-funded CHEMSTAR Department.

Explore this issue further by accessing **Viewpoints** at www.aw-bc.com/withgott.

Table 14.3 The "Dirty Dozen" Persistent Organic Pollutants (POPs) Targeted by the Stockholm Convention

Toxicant	Type	Description
Aldrin	Pesticide	Kills termites, grasshoppers, corn rootworm, and other soil insects
Chlordane	Pesticide	Kills termites and is a broad-spectrum insecticide on various crops
DDT	Pesticide	Widely used in the past to protect against malaria, typhus, and other insect-spread diseases; continues to be applied in several countries to control malaria
Dieldrin	Pesticide	Controls termites and textile pests; also used against insect-borne diseases and insects in agricultural soil
Dioxins	Unintentional by-product	Produced by incomplete combustion and in chemical manufacturing; released in some kinds of metal recycling, pulp and paper bleaching, automobile exhaust, tobacco smoke, and wood and coal smoke
Endrin	Pesticide	Kills insects on cotton and grains; also used against rodents
Furans	Unintentional by-product	Result from the same processes that release dioxins; also are found in commercial mixtures of PCBs
Heptachlor	Pesticide	Kills soil insects and termites, cotton insects, grasshoppers, and malaria-carrying mosquitoes
Hexachlorobenzene	Fungicide; unintentional by-product	Kills fungi that affect crops; released during chemical manufacture and from processes that give rise to dioxins and furans
Mirex	Pesticide	Combats ants and termites; also is a fire retardant in plastics, rubber, and electrical goods
PCBs	Industrial chemical	Used in industry as heat-exchange fluids, in electrical transformers and capacitors, and as additives in paint, carbonless copy paper, sealants, and plastics
Toxaphene	Pesticide	Kills insects on cotton, grains, fruits, nuts, and vegetables; kills ticks and mites on livestock

Data from United Nations Environment Programme (UNEP), 2001.

distances. The PCBs and other contaminants found in polar bears are a prime example. Because these contaminants so often cross international boundaries, an international treaty seemed the best way of dealing fairly with such transboundary pollution. The Stockholm Convention aims first to end the use and release of 12 of the POPs shown to be most dangerous, a group nicknamed the "dirty dozen" (Table 14.3). It sets guidelines for phasing out these chemicals and encourages transition to safer alternatives.

Conclusion

International agreements such as the Stockholm Convention represent a hopeful sign that governments will act to protect the world's people, wildlife, and ecosystems from toxic chemicals and other environmental hazards. At the same time, solutions can often come more easily when they do not arise from government regulation alone. To many minds, consumer choice, exercised through the market, may be the best way to influence industry's decision making. Consumers of products can make decisions that influence industry when they have full information from scientific research regarding the risks involved. Once scientific results are in, a society's philosophical approach to risk management will determine what policy decisions are made.

Whether the burden of proof is laid at the door of industry or of government, it is important to realize that we will never attain complete scientific knowledge of any risk. At some point we must choose whether or not to act on the information available. Synthetic chemicals have brought us innumerable modern conveniences, a larger food supply, and medical advances that save and extend human lives. Human society would be very different without them. Yet a safer and happier future, one that safeguards the well-being of both humans and the environment, depends on knowing the risks that some hazards pose and on having means in place to phase out harmful substances and replace them with safer ones.

REVIEWING OBJECTIVES

You should now be able to:

Identify the major types of environmental health hazards and explain the goals of environmental health

- Environmental health seeks to assess and mitigate environmental factors that adversely affect human health and ecological systems. (p. 402)
- Environmental health threats include physical, chemical, biological, and cultural hazards. (pp. 402–403)
- Disease is a major focus of environmental health. (pp. 404–405)
- Environmental hazards exist indoors as well as outdoors. (pp. 405–406)
- Toxicology is the study of poisonous substances. (pp. 406–407)

Describe the types, abundance, distribution, and movement of synthetic and natural toxicants in the environment

- Many thousands of potentially toxic substances exist all around us in varying degrees. (pp. 407–408)
- Toxicants may be of human or natural origin. They include carcinogens, mutagens, teratogens, allergens, neurotoxins, and endocrine disruptors. (pp. 409–413)
- Toxicants may enter and move through surface and groundwater reservoirs, or they may travel long distances through the atmosphere. (pp. 413–415)
- Some chemicals break down very slowly and thus persist in the environment. (pp. 415–416)
- Some organic poisons bioaccumulate and may move up the food chain, poisoning consumers on high trophic levels through the process of biomagnification. (p. 416)

Discuss the study of hazards and their effects, including case histories, epidemiology, animal testing, and dose-response analysis

- In case histories, researchers study health problems in individual people. (pp. 417–418)
- Epidemiology involves gathering data from large groups of people over long periods of time and comparing groups with and without exposure to the environmental health threat being assessed. (pp. 418–420)
- In dose-response analysis, scientists measure the response of test animals to various doses of the suspected toxicant. (pp. 420–421)
- Toxicity or strength of response may be influenced by the dose or amount of exposure, the nature of exposure (acute or chronic), individual variation among subjects, and synergistic interactions with other hazards. (pp. 421–422)

Assess risk assessment and risk management

- Risk assessment involves measuring risk quantitatively and comparing risks involved in different activities or substances. (pp. 422–423)
- Risk management integrates science with political, social, and economic concerns, in order to design strategies to minimize risk. (pp. 423–424)

Compare philosophical approaches to risk

- An innocent-until-proven-guilty approach assumes that a substance is not harmful unless it is shown to be so. (pp. 424–425)
- A precautionary approach entails assuming that a substance may be harmful unless proven otherwise. (pp. 424–425)

Describe policy and regulation in the United States and internationally

- The EPA, CDC, FDA, and OSHA are responsible for regulating environmental health threats under U.S policy. (pp. 425–426)
- European nations take a more precautionary approach than does the United States when it comes to testing chemical products. (pp. 426, 428)

TESTING YOUR COMPREHENSION

1. What four major types of health hazards does research in the field of environmental health encompass?
2. In what way is disease the greatest hazard that humans face? What kinds of interrelationships must environmental health experts study to learn about how diseases affect human health?
3. Where does most exposure to lead, asbestos, radon, and PBDEs occur? How has each of these exposure problems been addressed?
4. When did concern over the effects of pesticides start to grow in the United States? Describe the argument presented by Rachel Carson in *Silent Spring*. What policy

resulted from the book's publication? Is DDT still used?
5. List and describe the six types or general categories of toxicants described in this chapter.
6. How do toxicants travel through the environment, and where are they most likely to be found? What are the life spans of toxic agents? Describe the processes of bioaccumulation and biomagnification.
7. What are epidemiological studies, and how are they most often conducted?
8. Why are animals used in laboratory experiments in toxicology? Explain the dose-response curve. Why are high LD_{50} and ED_{50} levels considered safe and low LD_{50} and ED_{50} levels considered unsafe for humans?
9. What factors may affect an individual's response to a toxic substance? Why is chronic exposure to toxic agents often more difficult to measure and diagnose than acute exposure? What are synergistic effects, and why are they difficult to measure and diagnose?
10. How do scientists identify and assess risks from substances or activities that may pose health threats?

SEEKING SOLUTIONS

1. Describe some environmental hazards that you think you may be living with indoors. How do you think you may have been affected by indoor or outdoor environmental hazards in the past? What philosophical approach do you plan to take in dealing with these toxicants in your own life?
2. Why is it that research on endocrine disruption has spurred so much debate? What steps do you think could be taken to help establish more consensus among scientists, industry, regulators, policymakers, and the public?
3. Name some naturally occurring substances that can act as toxic agents. Explain the arguments of Bruce Ames and those of his critics regarding the prevalence and effects of natural toxicants.
4. Do you feel that laboratory-bred animals should be used in experiments in toxicology? Why or why not?
5. Discuss ways that we may cope with the uncertainty of risk assessment for synthetic chemicals in environmental health. Can you think of alternatives to taking one of the two philosophical approaches discussed in the chapter? Should these approaches apply to natural toxicants as well?
6. Describe what you have learned from this chapter regarding the policies of the United States and the European Union toward the study and management of the risks of synthetic chemicals. Which do you believe is better, the policies of the United States or the European Union, and why?

INTERPRETING GRAPHS AND DATA

To minimize their exposure to ultraviolet (UV) radiation and thus their risk of skin cancer, people have increased use of sunscreen lotions in recent decades. Recently, however, research has shown that chemicals in sunscreens may themselves pose some risk to human health. The compounds most commonly used as UV protectants are fat-soluble, environmentally persistent, and prone to bioaccumulation. Moreover, they exhibit estrogenic effects in laboratory rats (see Schlumpf et al., 2001). Although the benefits and risks of sunscreen use are not yet well understood, a hypothetical trade-off between the risk factors of UV exposure and sunscreen use illustrates the balancing act known as risk management.

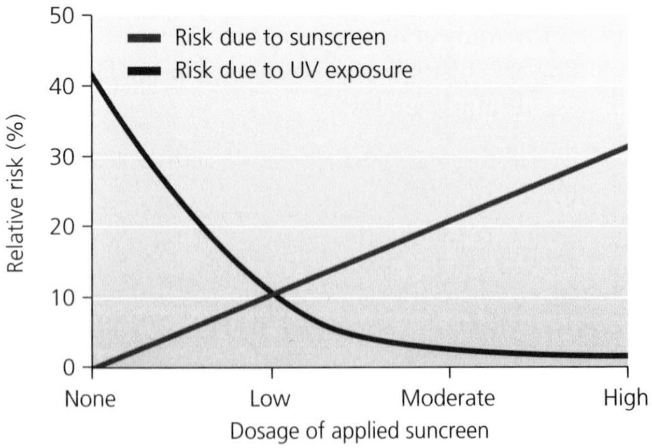

Hypothetical risk distributions for individuals using an estrogenic sunscreen to prevent skin cancer. Reference: Schlumpf, M., et al. 2001. *In vitro* and *in vivo* estrogenicity of UV screens. *Environmental Health Perspectives* 109: 239–244.

1. What dosage of applied sunscreen on the graph corresponds to the greatest risk due to UV exposure? What dosage corresponds to the greatest risk due to

chemicals in the sunscreen? Which of these two points on the graph is associated with the greater risk?

2. What dosage of applied sunscreen on the graph corresponds to the least risk due to UV exposure? What dosage corresponds to the least risk due to chemicals in the sunscreen? Which of these two points is associated with the greater risk?
3. The total risk to the individual is the sum of the two individual risks. What point on the graph corresponds to the greatest total risk? What sunscreen dosage corresponds to the least total risk? Based on the data shown here, how much sunscreen would you choose to apply the next time you go to the beach? Is there any other information you'd like to know before you change the way you use sunscreen? Can you think of any other cases that illustrate this sort of trade-off between dose-dependent risk factors?

CALCULATING ECOLOGICAL FOOTPRINTS

In 2001, the population of the United States was approximately 285 million, and the world's population totaled 6.16 billion. In that same year, pesticide use in the United States was approximately 1.20 billion pounds of active ingredient, and world pesticide use totaled 5.05 billion pounds of active ingredient. Pesticides include hundreds of chemicals used as insecticides, fungicides, herbicides, rodenticides, repellants, and disinfectants. They are used by farmers, governments, industries, and individuals. In the table, calculate your share of pesticide use as a U.S. citizen in 2001 and the amount used by (or on behalf of) the average citizen of the world.

	Annual pesticide use (pounds of active ingredient)
You	4.21
Your class	
Your state	
United States	
World (total)	
World (per capita)	

1. What is the ratio of your annual pesticide use to the world's per capita average? Refer back to the "Calculating Ecological Footprints" question for Chapter 1 (► p. 25), and find the ecological footprints of the average U.S. citizen and the average world citizen. Compare the ratio of pesticide usage with the ratio of the overall ecological footprints. How would you explain the difference?
2. Does the figure for per capita pesticide use for you as a U.S. citizen seem reasonable for you personally? Why or why not? Do you find this figure alarming or of little concern? What else would you like to know to assess the risk associated with this level of pesticide use?

Take It Further

Go to www.aw-bc.com/withgott or the student CD-ROM where you'll find:

- Suggested answers to end-of-chapter questions
- Quizzes, animations, and flashcards to help you study
- *Research Navigator*™ database of credible and reliable sources to assist you with your research projects
- GRAPHit! tutorials to help you master how to interpret graphs
- INVESTIGATEit! current news articles that link the topics that you study to case studies from your region to around the world

15 Freshwater Resources: Natural Systems, Human Impact, and Conservation

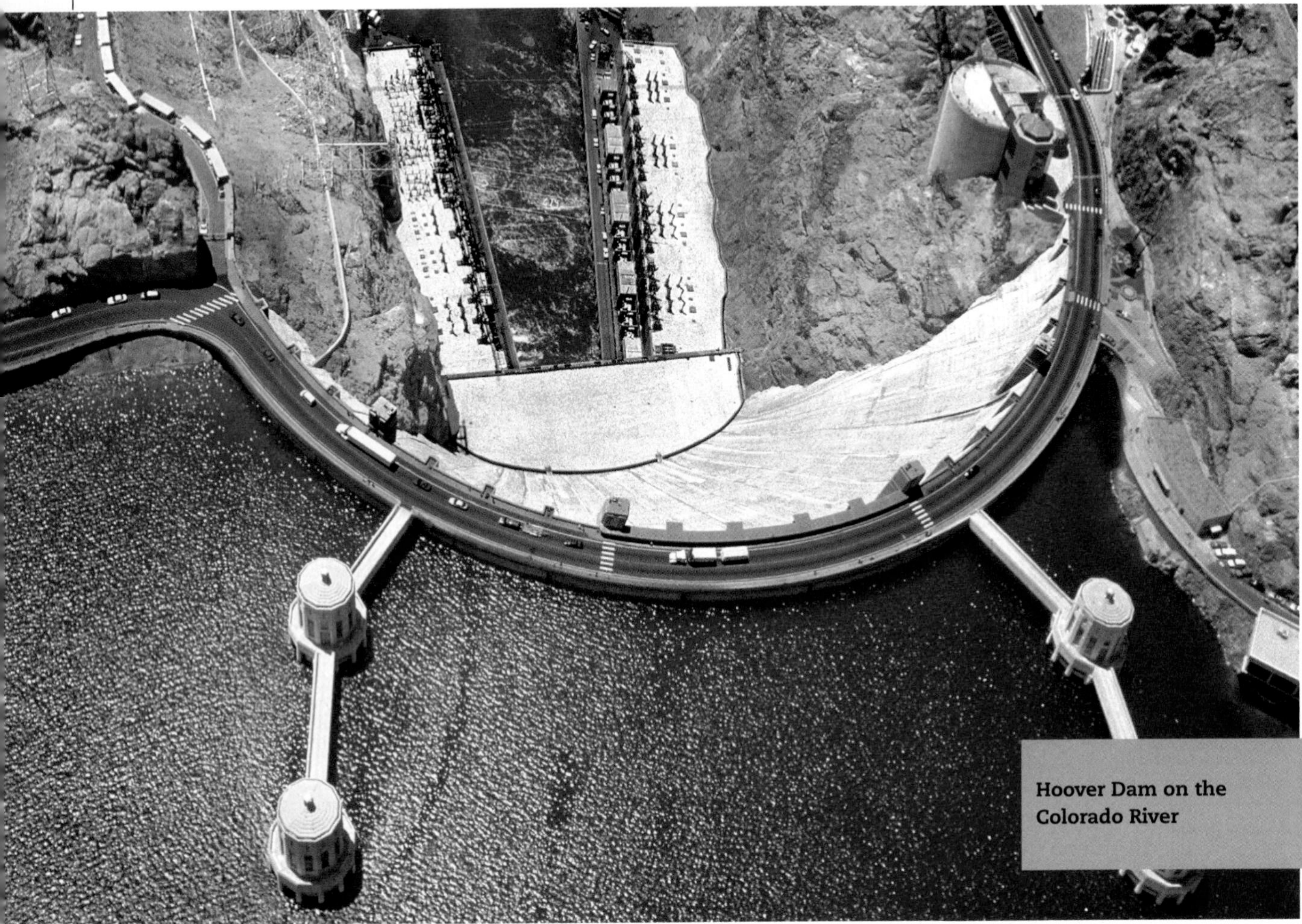

Hoover Dam on the Colorado River

Upon successfully completing this chapter, you will be able to:

- Explain the importance of water and the hydrologic cycle to ecosystems, human health, and economic pursuits
- Delineate freshwater distribution on Earth
- Describe major types of freshwater ecosystems
- Discuss how we use water and alter freshwater systems
- Assess problems of water supply and propose solutions to address freshwater depletion
- Assess problems of water quality and propose solutions to address water pollution
- Explain how wastewater is treated

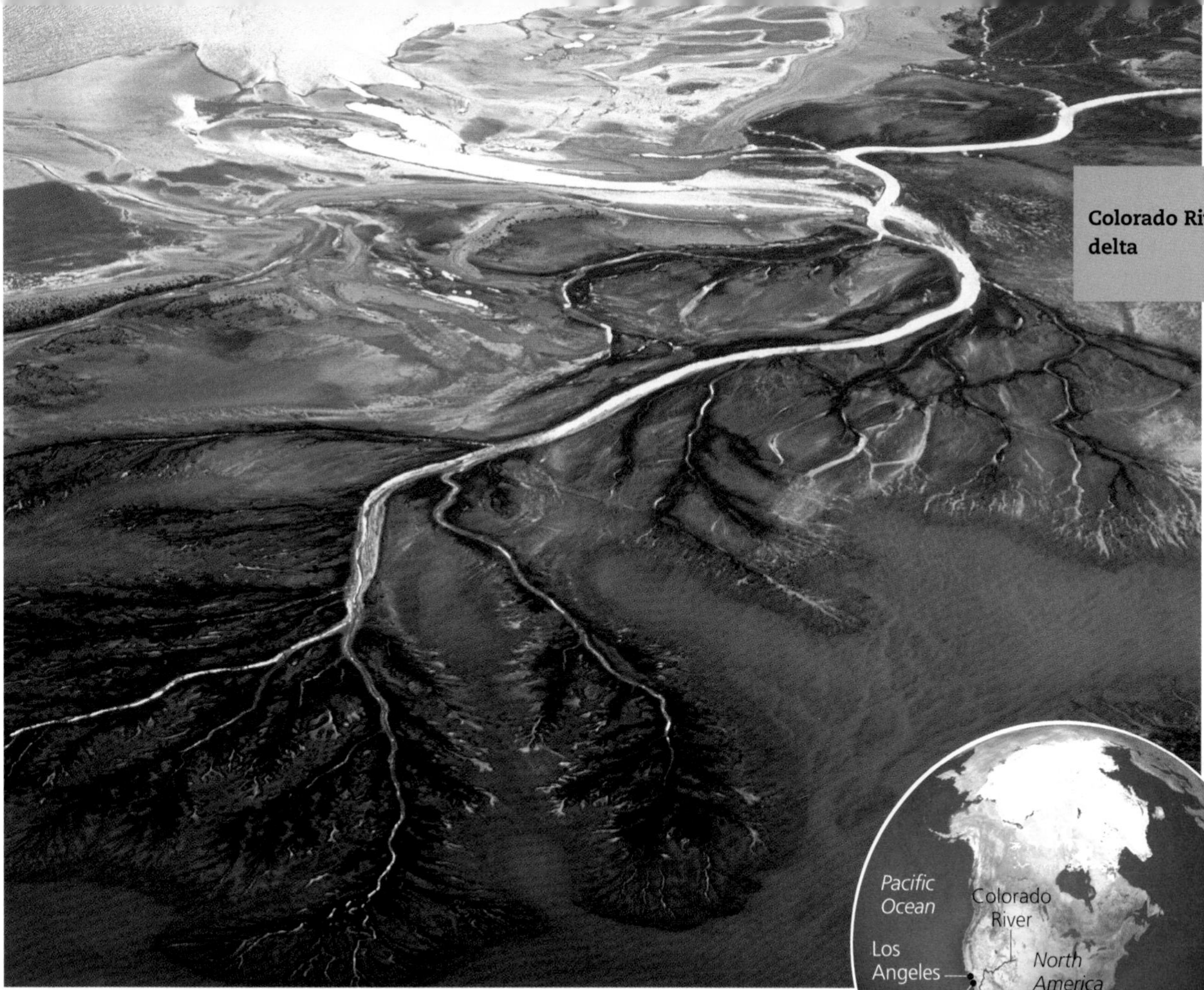

Colorado River delta

Central Case: Plumbing the Colorado River

"We've gone from being assured that we lived in this magical place where the rules of water didn't apply to [a] wake-up call about the fact that we do live in the California desert. People have lived in this false water utopia."

—BUFORD CRITES, CITY COUNCILOR, PALM DESERT CITY, CALIFORNIA

"Water promises to be to the 21st century what oil was to the 20th century: the precious commodity that determines the wealth of nations."

—FORTUNE MAGAZINE, MAY 2000

As the clock struck midnight on New Year's Eve, millions of Californians toasted the arrival of 2003 with champagne. But some people in the state that night had another liquid on their minds: water. Their fears were borne out the next day when the U.S. government followed through on its threat to cut off 15% of the water that California takes from the Colorado River.

Water is the lifeline for any civilization in an arid environment. Without generous supplies of freshwater delivered from elsewhere, southern California society as we know it could simply not exist. In ordering the New Year's Day cutoff, U.S. Interior Secretary Gale Norton was simply holding up her end of a deal that an irrigation district in California had scuttled. It may seem bizarre that a 3–2 vote of one county irrigation district could block enough water for 1.6 million households in Los Angeles and San Diego, but it was just the latest episode in the colorful history of California water politics and the battles among seven states jockeying for rights to what was once the West's wildest river.

The Colorado River begins in the high peaks of the Rocky Mountains, charges through the Grand Canyon, crosses the border into Mexico, and empties into the Gulf of California, draining 637,000 km^2 (246,000 mi^2) of southwestern North America. Its raging waters have chiseled through thousands of feet of bedrock, creating the Grand Canyon and leaving extraordinary scenery along its 2,330-km (1,450-mi) length. Today, however, only a small amount of water—often none at all—reaches the river's mouth. Instead, the waters of the Colorado River irrigate 7% of U.S. cropland, quench the thirst of over 20 million people, keep hundreds of golf courses green in the desert, and fill the swimming pools and fountains of Las Vegas casinos. The Colorado provides vital water to the rapidly growing metropolitan areas of the arid U.S. Southwest—Phoenix, Tucson, Las Vegas, San Diego, Los Angeles, and many others. The massive dams built across the river provide flood control and recreation, produce 12 billion kilowatt-hours of electricity from hydroelectric power each year, and provide irrigation that makes agriculture possible in this arid region.

For 80 years, the seven states along the Colorado have divided the river's water among themselves, guided by the Colorado River Compact they signed in 1922, which apportioned water to each state. California had long been permitted to exceed its allotment because Colorado, Wyoming, Utah, Nevada, New Mexico, and Arizona were not using all of their allotted portions. With the populations of these states booming, however, Interior Secretary Bruce Babbitt in 2000 pressured California to reduce its withdrawals by roughly 15% over 15 years.

California worked hard to get its agricultural districts, which controlled most water distribution in the state, to agree. At the last minute, however, the Imperial Irrigation District backed out of the agreement. The New Year's deadline passed, and 2003 saw the federal cutoff implemented. After 10 months of bickering and negotiation, the deal was patched up, and the cutoff was ended. Southern California's residents were able to continue living—at least for a little while longer—their mirage in the desert.

Freshwater Systems

"Water, water, everywhere, nor any drop to drink." The well-known line from Coleridge's poem *The Rime of the Ancient Mariner* describes the situation on our planet quite well. Water may seem abundant to us, but water that we can drink is actually quite rare and limited (Figure 15.1). Roughly 97.5% of Earth's water resides in the oceans and is too salty to drink or use to water crops. Only 2.5% is considered **freshwater,** water that is relatively pure, with few dissolved salts. Because most freshwater is tied up in glaciers, icecaps, and underground aquifers, just over 1 part in 10,000 of Earth's water is easily accessible for human use.

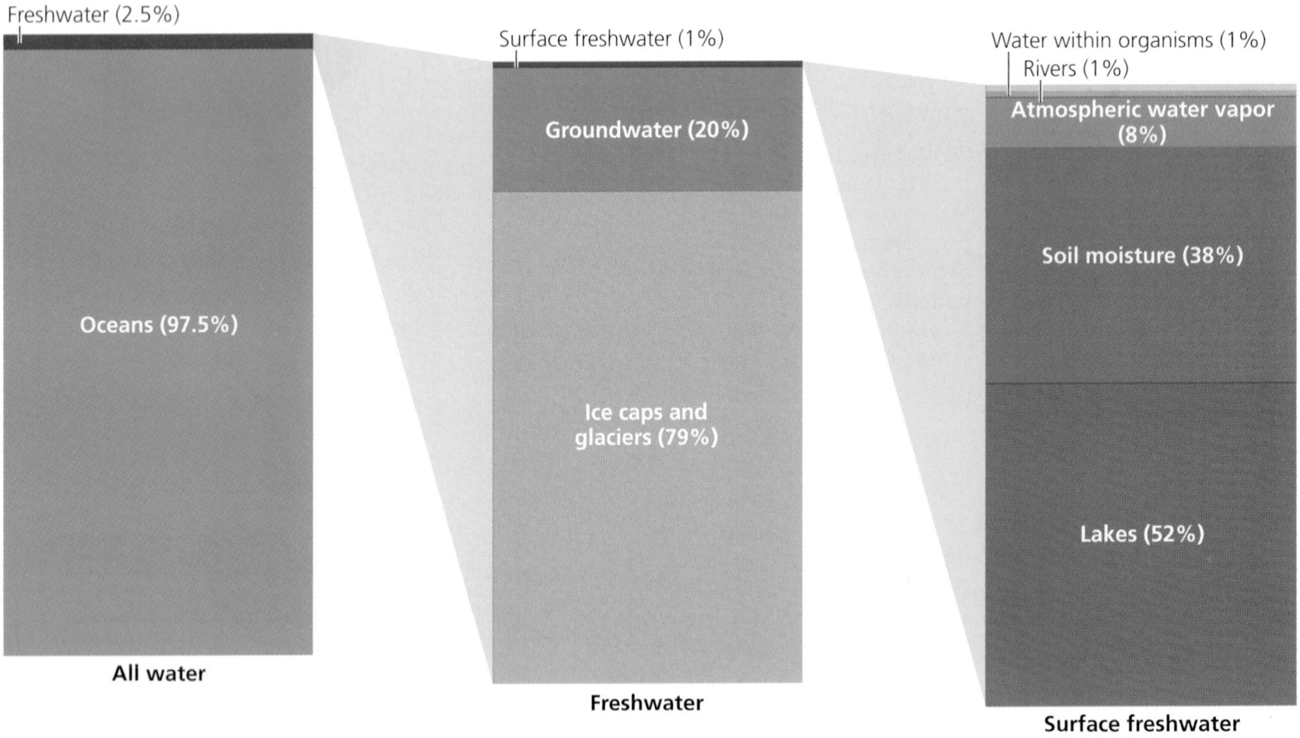

FIGURE 15.1 Only 2.5% of Earth's water is freshwater. Of that 2.5%, most is tied up in glaciers and ice caps. Of the 1% that is surface water, most is in lakes and soil moisture. Data from United Nations Environment Programme (UNEP) and World Resources Institute.

FIGURE 15.2 Rivers and streams flow downhill, shaping landscapes, as shown by an oxbow of this meandering river in Colorado.

Water is constantly moving among the reservoirs specified in Figure 15.1 via the *hydrologic cycle* (Figure 7.14, ▸ pp. 204–205). As water moves, it redistributes heat, erodes mountain ranges, builds river deltas, maintains organisms and ecosystems, shapes civilizations, and gives rise to political conflicts. Let's first examine the portions of the hydrologic cycle that are most conspicuous to us—surface water bodies—and take stock of the ecological systems they support.

Rivers and streams wind through landscapes

Water from rain, snowmelt, or springs runs downhill and converges in small channels, which join to form streams, creeks, or brooks. These watercourses join one another and eventually merge into rivers, whose water eventually reaches the ocean (or sometimes ends in a landlocked water body). As we learned in Chapter 3, a smaller river flowing into a larger one is a *tributary*, and the area of land drained by a river and all its tributaries is that river's *watershed*.

Rivers shape the landscape through which they run. The force of water rounding a river's bend gradually eats away at the outer shore, eroding soil from the bank. Meanwhile, sediment is deposited along the inside of the bend, where water currents are weaker. In this way, over time, river bends become exaggerated in shape (Figure 15.2). Eventually, a bend may become such an extreme loop (called an *oxbow*) that water erodes a shortcut from one end of the loop to the other, pursuing a direct course. The bend is cut off, and remains as an isolated, U-shaped water body called an *oxbow lake*.

Over thousands or millions of years, a river may shift from one course to another, back and forth over a large area, carving out a flat valley. Areas nearest a river's course that are flooded periodically are said to be within the river's **floodplain.** Frequent deposition of silt from flooding makes floodplain soils especially fertile. As a result, agriculture thrives in floodplains, and *riparian* (riverside) forests are productive and species-rich.

Diverse ecological communities exist in the water of rivers and streams. Algae and detritus support many types of invertebrates, from water beetles to crayfish. Insects as diverse as dragonflies, mayflies, and mosquitoes develop as larvae in streams and rivers before maturing into adults that take to the air. Fish consume aquatic insects, and birds such as kingfishers, herons, and ospreys dine on fish. Many amphibians spend their larval stages in streams, and some live their entire lives in streams. Salmon migrate from oceans up rivers and streams to spawn.

Wetlands include marshes, swamps, and bogs

Systems that combine elements of freshwater and dry land are enormously rich and productive. Often lumped under the term *wetlands,* such areas include different types of systems. Freshwater marshes (Figure 15.3) feature water shallow enough to allow plants to grow from the bottom and to rise above the water surface. Cattails and bulrushes

FIGURE 15.3 Shallow water bodies with ample vegetation are called wetlands, and include swamps, bogs, and marshes such as this one in Botswana, Africa.

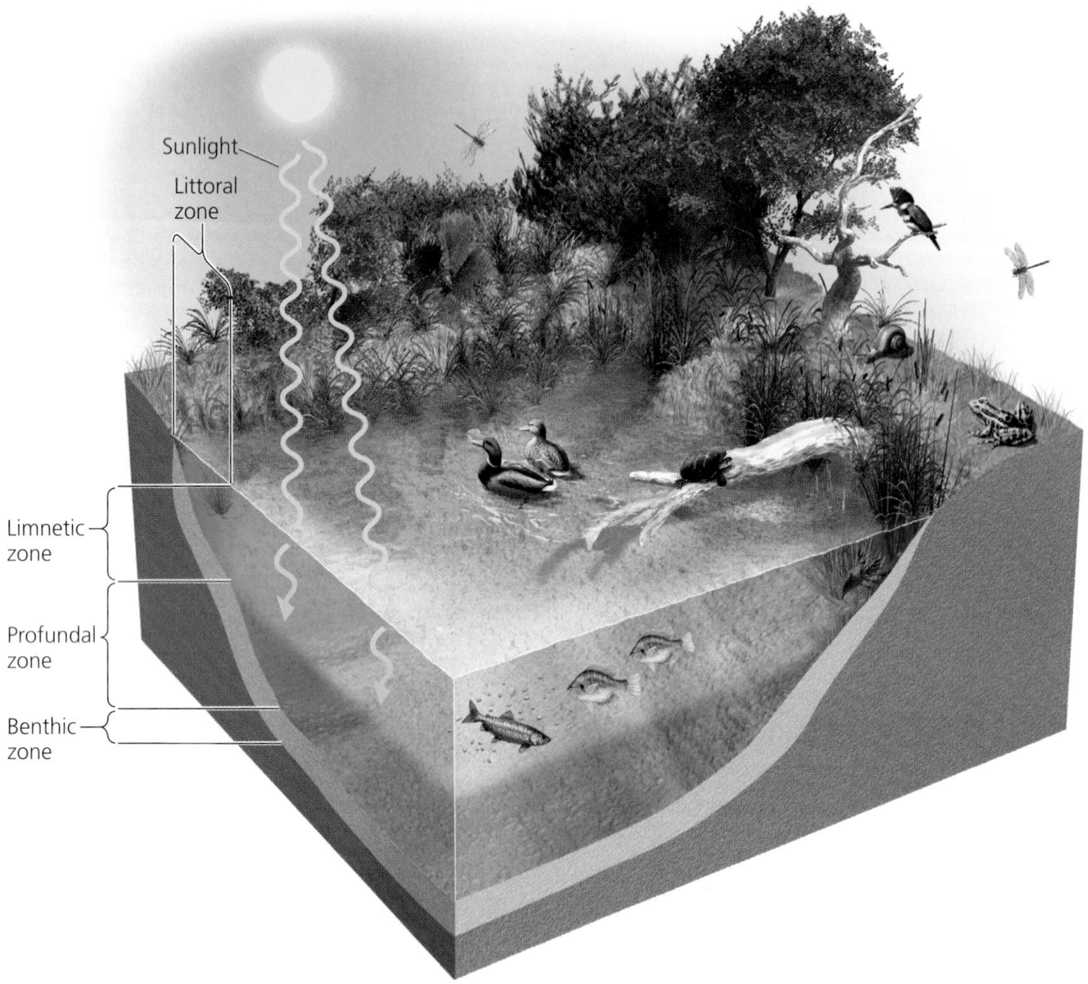

FIGURE 15.4 Lakes and ponds are open, still bodies of water consisting of different zones. Emergent plants grow around the shoreline in the littoral zone. The limnetic zone is the layer of open, sunlit water where photosynthesis takes place. Sunlight does not reach the deeper profundal zone. The benthic zone, which is the bottom of the water body, often is muddy, rich in detritus and nutrients, and low in oxygen.

are plants typical of North American marshes. Swamps also consist of shallow water rich in vegetation, but they occur in forested areas. The cypress swamps of the southeastern United States, where cypress trees grow in standing water, are an example. Swamps are also created when beavers build dams across streams with limbs from trees they have cut, flooding wooded areas upstream. Bogs are ponds thoroughly covered with thick floating mats of vegetation, and can represent a stage in aquatic succession.

All these types of wetlands are extremely valuable as habitat for wildlife. They also provide ecosystem services by slowing runoff, reducing flooding, recharging aquifers, and filtering pollutants. Wetlands have been extensively drained and filled by people, largely for agriculture (▸ pp. 362–363). It is estimated that southern Canada and the United States have lost well over half their wetlands since European colonization.

Lakes and ponds are ecologically diverse systems

Lakes and ponds are bodies of open standing water. Their depth varies greatly, and the physical conditions and types of life within them vary with depth and the distance from shore. As a result, scientists have described several zones typical of lakes and ponds (Figure 15.4).

The region ringing the edge of a water body is named the **littoral zone.** Here the water is shallow enough that aquatic plants grow from the mud and reach above the water's surface. The nutrients and productive plant growth of the littoral zone make it rich in invertebrates—such as insect larvae, snails, and crayfish—on which fish, birds, turtles, and amphibians feed. Many invertebrates live in the mud on the bottom, feeding on detritus or preying on one another. The bottom layer of a lake or pond is the **benthic zone.** The benthic zone extends along the bottom of the entire water body, from shore to the deepest point.

In the open portion of a lake or pond, away from shore, sunlight penetrates shallow waters of the **limnetic zone.** Because light enables photosynthesis and plant growth, the limnetic zone supports phytoplankton, which in turn support zooplankton, both of which are eaten by fish. Within the limnetic zone, sunlight intensity (and therefore water temperature) decreases with depth. The water's turbidity affects the depth of this zone; water that is clear allows sunlight to penetrate deeply, whereas turbid water does not. Below the limnetic zone is the **profundal zone,** the volume of open water that sunlight does not reach. This zone lacks plant life and thus is lower in dissolved oxygen and supports fewer animals. Aquatic animals rely on dissolved oxygen, and its concentration depends on the amount released by photosynthesis and the amount removed by animal and microbial respiration, among other factors.

Ponds and lakes change over time naturally, as streams and runoff bring them sediment and nutrients. *Oligotrophic* lakes and ponds, which have low-nutrient and high-oxygen conditions, may slowly give way to the high-nutrient, low-oxygen conditions of *eutrophic* water bodies. Eventually, water bodies may fill in completely by the process of aquatic succession (Figure 6.14, ▸ p. 166). As lakes or ponds change over time, species of fish, plants, and invertebrates adapted to oligotrophic conditions may give way to those that thrive under eutrophic conditions.

Some lakes are so large that they differ substantially in their characteristics from small lakes. These large lakes are sometimes known as inland seas. North America's Great Lakes are prime examples. Because they hold so much water, most of their biota is adapted to open water. Major fish species of the Great Lakes include lake sturgeon, lake whitefish, northern pike, alewife, bass, walleye, and perch. Lake Baikal in Asia is the world's deepest lake, at 1,637 m (5,370 ft) deep, and the Caspian Sea is the world's largest freshwater body, at 371,000 km^2 (143,000 mi^2).

Groundwater plays key roles in the hydrologic cycle

It is easy for us to understand surface water systems because we witness them all the time, but groundwater and its functions in the hydrologic cycle are more difficult to visualize (Figure 15.5). Any precipitation reaching Earth's land surface that does not evaporate, flow into waterways, or get taken up by organisms infiltrates the surface. Most percolates downward through the soil to become groundwater, which makes up one-fifth of Earth's freshwater supply and plays a key role in meeting human water needs.

Groundwater is contained within **aquifers,** porous, spongelike formations of rock, sand, or gravel that hold

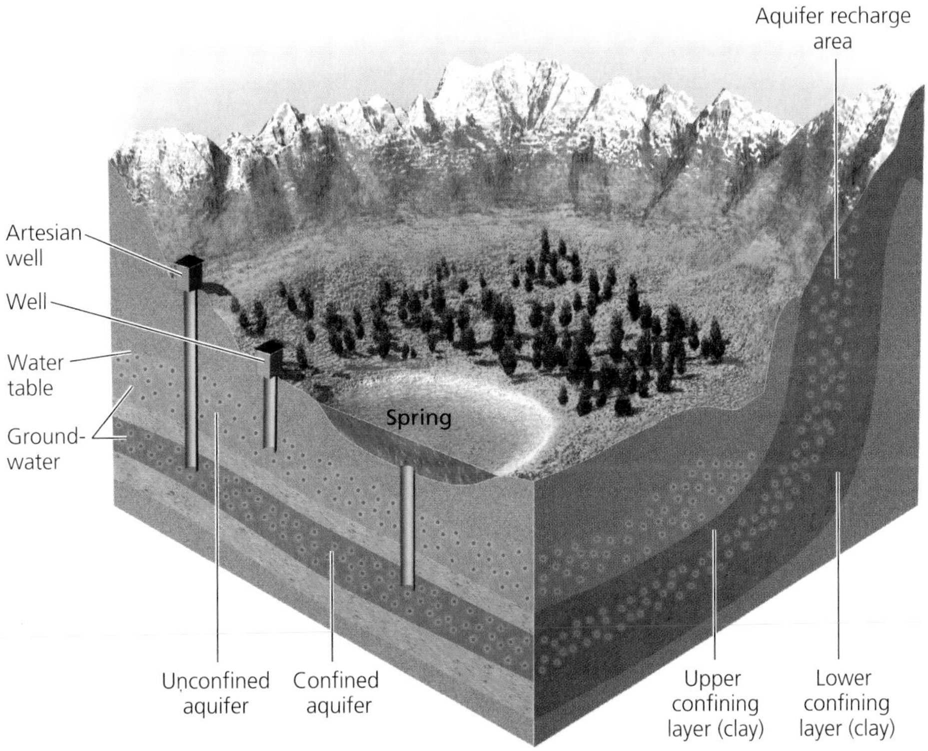

FIGURE 15.5 Groundwater may occur in unconfined aquifers above impermeable layers or in confined aquifers under pressure between impermeable layers. Water may rise naturally to the surface at springs and through the wells we dig. Artesian wells tap into confined aquifers to mine water under pressure.

water. An aquifer's upper layer, or zone of aeration, contains pore spaces partly filled with water. In the lower layer, or zone of saturation, the spaces are completely filled with water. The boundary between these two zones is the **water table.** Picture a sponge resting partly submerged in a tray of water; the lower part of the sponge is completely saturated, whereas the upper portion may be moist but contains plenty of air in its pores. Any area where water infiltrates Earth's surface and reaches an aquifer below is known as an *aquifer recharge zone.*

There are two broad categories of aquifers. A **confined aquifer,** or **artesian aquifer,** exists when a water-bearing porous layer of rock, sand, or gravel is trapped between upper and lower layers of less permeable substrate (often clay). In such a situation, the water is under great pressure. In contrast, an **unconfined aquifer** has no such upper layer to confine it. Thus its water is under considerably less pressure than that of a confined aquifer and it can be readily recharged by surface water.

Just as surface water becomes groundwater by infiltration and percolation, groundwater becomes surface water through springs (and human-drilled wells), sometimes keeping streams flowing when surface conditions are otherwise dry. Groundwater flows downhill and from areas of high pressure to areas of low pressure. A typical rate of groundwater flow might be about 1 m (3 ft) per day. Because of this slow movement, groundwater may remain in an aquifer for a long time. In fact, groundwater can be ancient. The average age of groundwater has been estimated at 1,400 years, and some is tens of thousands of years old. Nonetheless, volumes of groundwater are large enough that each day in the United States alone, aquifers release 1.9 trillion L (492 billion gal) of groundwater into bodies of surface water—nearly as much as the daily flow of the Mississippi River. The world's largest known aquifer is the Ogallala Aquifer, which underlies the Great Plains of the United States (Figure 15.6). It spans 453,000 km^2 (176,700 mi^2), is 370 m (1,200 ft) deep at its thickest point, and has a water-holding capacity of 3,700 km^3 (881 mi^3).

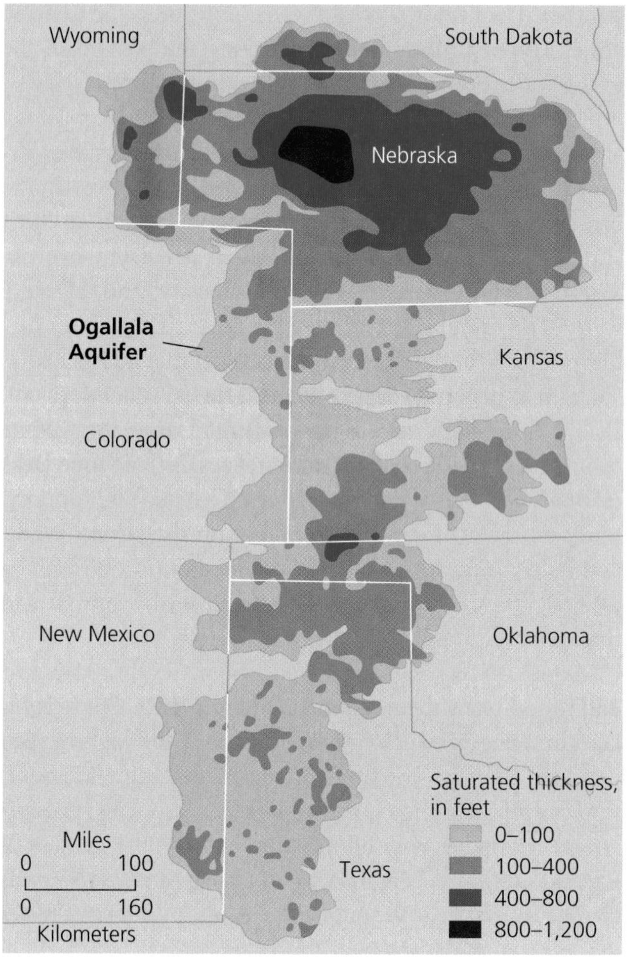

FIGURE 15.6 The Ogallala Aquifer is the world's largest aquifer, and it held 3,700 km^3 (881 mi^3) of water before pumping began. This aquifer underlies 453,000 km^2 (175,000 mi^3) of the Great Plains beneath eight U.S. states from South Dakota to Texas. Overpumping for irrigation is currently reducing the volume and extent of this aquifer.

Water is unequally distributed across Earth's surface

The Great Plains and its farmlands boast the massive Ogallala Aquifer as a source of freshwater, but many other areas are not so endowed. Different regions possess vastly different amounts of groundwater, surface water, and precipitation. Precipitation ranges from about 1,200 cm (470 in.) per year at Mount Waialeale on the Hawaiian island of Kauai to virtually zero in Chile's Atacama Desert.

People are not distributed across the globe in accordance with water availability. Many areas with high population density are water-poor (Figure 15.7), leading to inequalities in per capita water resources among and within nations. For example, Canada has 20 times more water for each of its citizens than does China. The Amazon River carries 15% of the world's runoff, but its watershed holds less than half a percent of the world's human population. Nations that hold back water in transboundary rivers exacerbate such natural inequities. Many densely populated nations, such as Pakistan, Iran, and Egypt, face serious water shortages. Asia possesses the most water of any continent but has the least water available per person, whereas Australia, with the least amount of water, boasts the most water available per person. Because of this mismatched distribution of water and

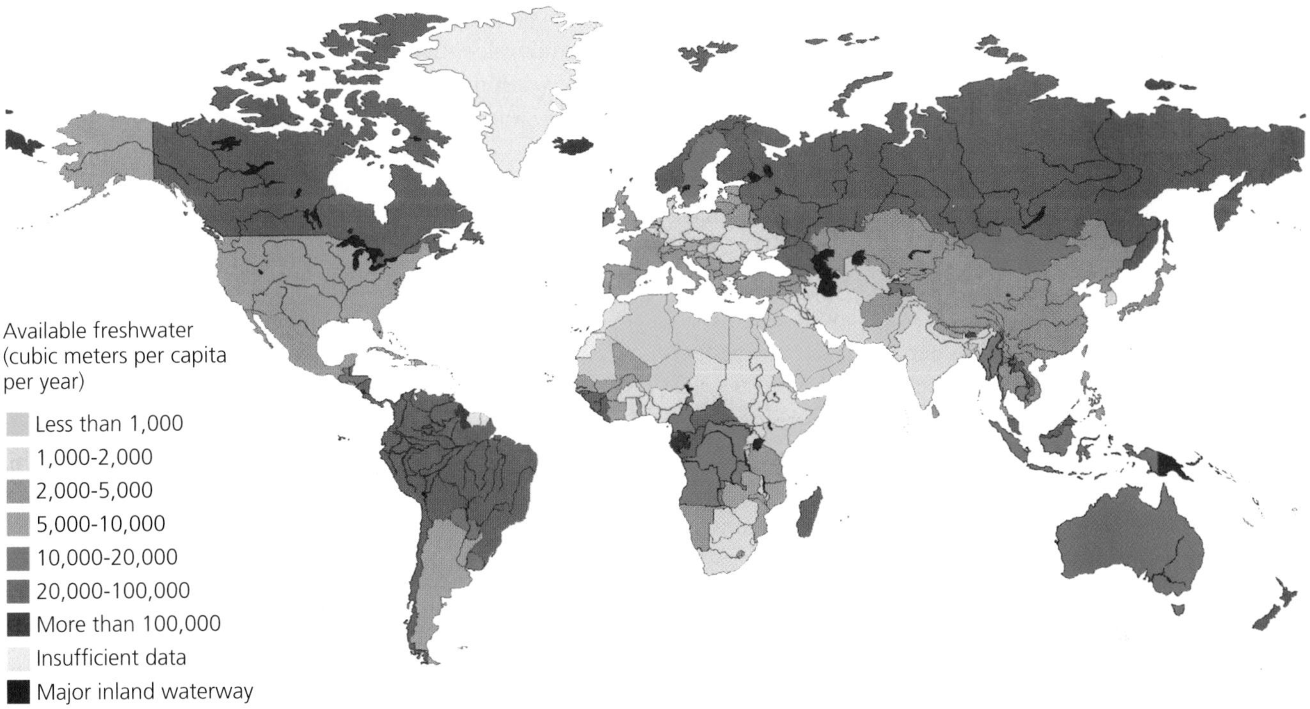

(a) Water-stressed nations: Available freshwater resources, 2000

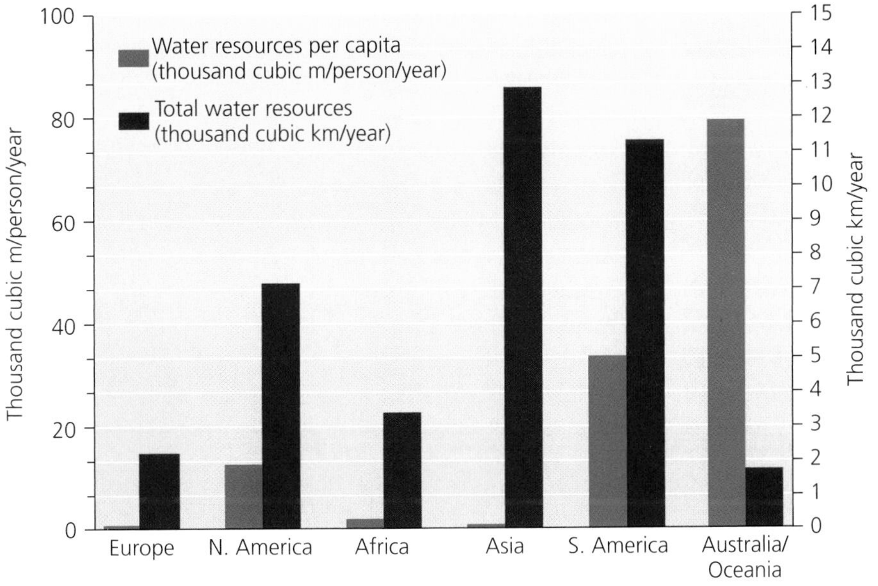

(b) Water-stressed nations: Total water resources and per capita water resources

FIGURE 15.7 Nations vary tremendously in the amount of freshwater per capita available to their citizens **(a)**. For example, Iceland, Papua New Guinea, Gabon, and Guyana have more than 100 times as much water as do many Middle Eastern and North African countries. There is almost as much variation among continents, which show great imbalances between total water resources and per capita water resources **(b)**. Heavily populated Asia has tremendous total water resources but extremely low amounts per capita, for example, whereas Australia and the oceanic island nations have little total water but high amounts per capita because of their low populations. Go to GRAPHit! at www.aw-bc.com/withgott or on the student CD-ROM. Data from (a) UNEP and World Resources Institute, as presented by Harrison, P., and F. Pearce. 2000. *AAAS atlas of population and the environment*. Berkeley, CA: University of California Press. (b) U.N. Sustainable Development Programme, 2002.

human population, one challenge for human societies has always been to transport freshwater from its source to where it is needed. In nearly every modern country, such transport is vital to equalizing access among people.

Freshwater is distributed unevenly in time as well as space. India's monsoon season brings concentrated storms in which half of a region's annual rain may fall in a few hours. Northwest China receives three-fifths of its annual precipitation during 3 months when crops do not need it. Uneven distribution of water across time is one reason people have erected dams to store water, so that it may be distributed when needed.

How We Use Water

In our attempts to harness freshwater sources for countless purposes and pursuits, we have achieved impressive engineering accomplishments. In so doing, we also have altered many environmental systems. It is estimated that 60% of the world's largest 227 rivers (and 77% of those in North America and Europe) have been strongly or moderately affected by artificial dams, canals, and diversions.

We are also using too much water. Data indicate that at present our freshwater consumption in much of the world is unsustainable, and we are depleting many sources of surface water and groundwater. At least 1.7 billion people live in regions of water scarcity (with less than 1,000 m^3 (35,000 ft^3) of water per person per year), and this number is expected to grow to at least 2.4 billion by 2025.

Water supplies our households, agriculture, and industry

Every one of us uses water at home for drinking, cooking, and cleaning. Farmers and ranchers use water to irrigate crops and water livestock. We use water in most manufacturing and industrial processes. The proportions of each of these three types of use—residential/municipal, agricultural, and industrial—vary dramatically among nations (Figure 15.8). Nations with arid climates tend to use more freshwater for agriculture, and heavily industrialized nations use a great deal for industry. Globally, we spend about 70% of our annual freshwater allotment on agriculture. Industry accounts for roughly 20%, and residential and municipal uses for only 10%.

Freshwater use can be consumptive or nonconsumptive. In **consumptive use,** water is removed from a particular aquifer or surface water body and is not returned. A large portion of agricultural irrigation and of many industrial and residential uses is consumptive. **Nonconsumptive use** of water does not remove, or only temporarily removes, water from an aquifer or surface water body. Using water to generate electricity at hydroelectric dams is an example of nonconsumptive use; water is taken in, passed through dam machinery to turn turbines, and released on the downstream side.

Inefficient irrigation wastes water

The green revolution (▸ pp. 250, 280–281) required significant increases in irrigation, and 60% more water is withdrawn for irrigation today than in 1960. During this period, the amount of land under irrigation has roughly doubled (Figure 15.9). Expansion of irrigated agriculture has kept pace with population growth; irrigated area per capita has remained stable for at least four decades at around 460 m^2 (4,900 ft^2).

Irrigation can more than double crop yields by allowing farmers to control the application of water when and where it is needed. The world's 274 million ha (677 million acres) of irrigated cropland make up only 18% of world farmland but yield fully 40% of world agricultural produce, including 60% of the global grain crop. Still, most irrigation remains highly inefficient. Only about 45% of the freshwater we use for irrigation actually is taken up by crops. Inefficient "flood and furrow" irrigation, in which fields are liberally flooded with water that may evaporate from standing pools, accounts for 90% of irrigation worldwide. Overirrigation leads to waterlogging and salinization, which affect one-fifth of farmland today and reduce world farming income by $11 billion (▸ pp. 264–265).

Many national governments have subsidized irrigation to promote agricultural self-sufficiency. Unfortunately,

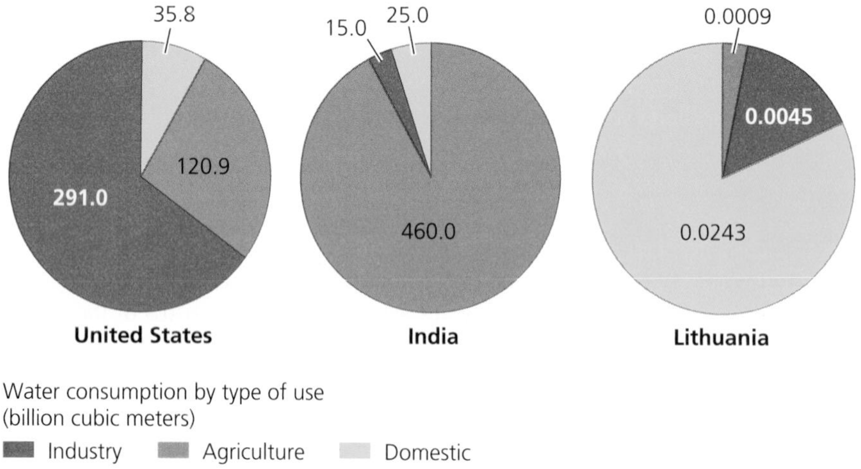

FIGURE 15.8 Nations apportion their freshwater consumption differently. Industry consumes most water used in the United States, agriculture uses the most in India, and most water in Lithuania goes toward domestic use. Data from World Bank, as presented by: Harrison, P., and F. Pearce. 2000. *AAAS atlas of population and the environment.* Berkeley, CA: University of California Press.

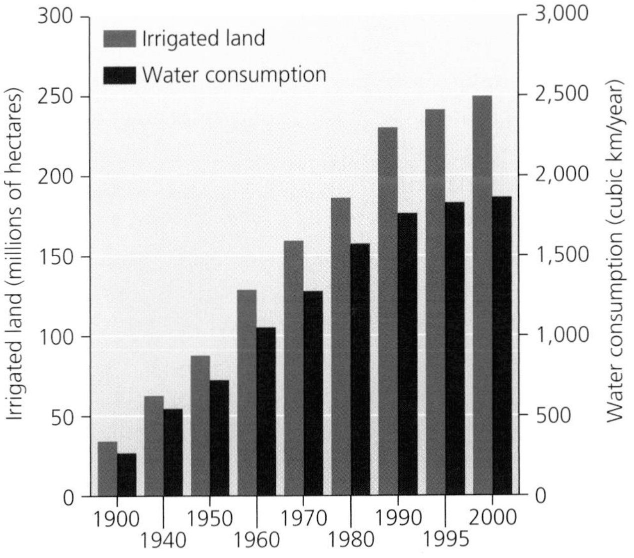

FIGURE 15.9 Throughout the 20th century, overall global water consumption rose in tandem with the area of land irrigated for agriculture. Data from United Nations Sustainable Development Commission on Freshwater, 2002.

inefficient irrigation methods in arid areas such as the Middle East are using up huge amounts of groundwater for little gain. Worldwide, roughly 15–35% of water withdrawals for irrigation are thought to be unsustainable. Figure 15.10 shows areas where agriculture is demanding more freshwater than can be sustainably supplied. In these areas, *water mining*—withdrawing water faster than it can be replenished—is taking place; aquifers are being depleted or surface water is being piped in from other regions.

We are depleting groundwater

Groundwater is more easily depleted than surface water because most aquifers recharge very slowly. One-third of Earth's human population—including 99% of the rural population of the United States—relies on groundwater for its needs. To obtain groundwater, many people in rural areas of developing countries walk long distances to haul water from local wells. In developed countries, most homes in rural areas have electrically powered wells.

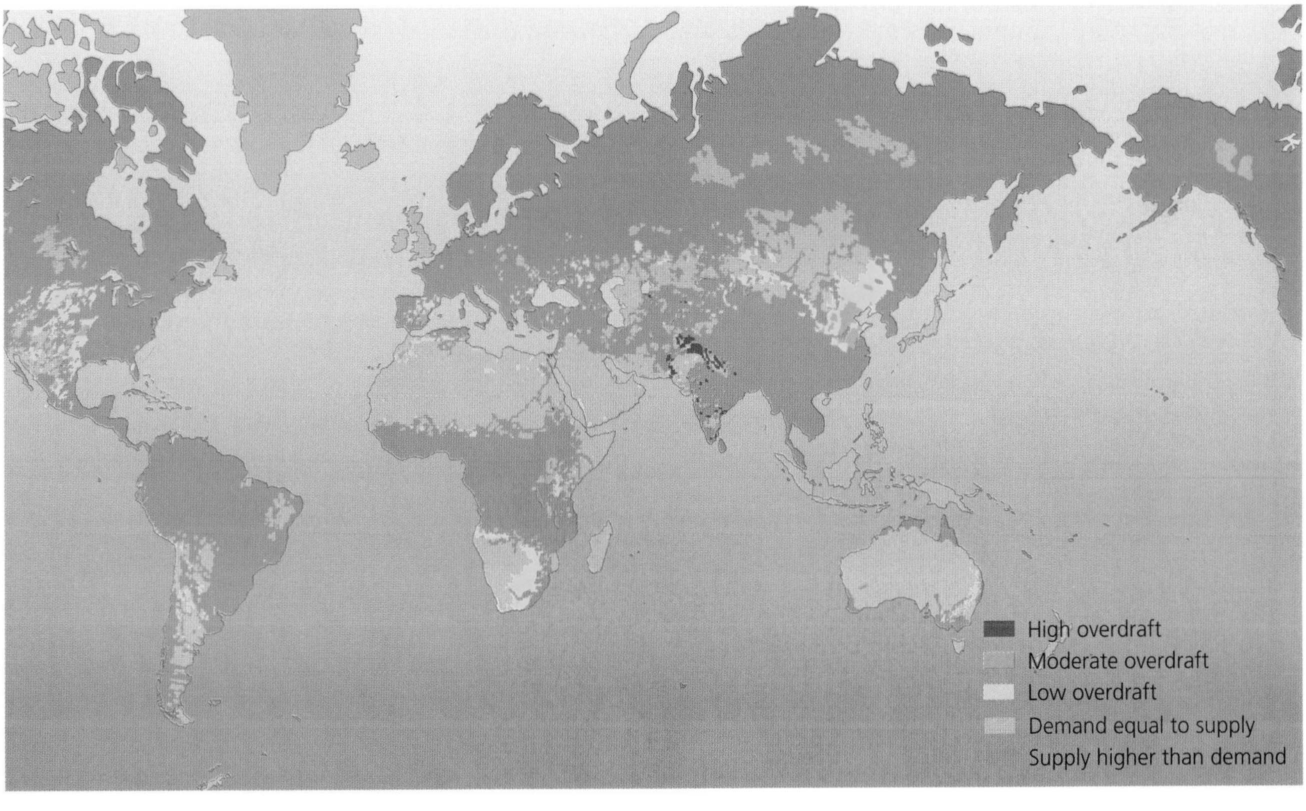

FIGURE 15.10 Globally, 15–35% of water withdrawals are estimated to be unsustainable. Mapped are regions where freshwater use for agriculture exceeds the available supply, requiring groundwater depletion or diversion of water from other regions. "High overdraft" equals more than 1 km^3/yr; "moderate overdraft" equals 0.1–1 km^3/yr; and "low overdraft" equals 0–0.1 km^3/yr. In areas of high overdraft, water tables are being drawn down 1.6 m/yr or more; in low overdraft areas, water tables are falling by less than 0.1 m/yr. Data from Millennium Ecosystem Assessment, 2005.

Homes and businesses in urban and suburban areas receive water through complex networks of pipes and pumps generally maintained by municipal governments.

If we compare an aquifer to a bank account, we are making more withdrawals than deposits, and the balance is shrinking. Globally, over the past 60 years we have been withdrawing groundwater in amounts that increase 2.5–3% annually, greater than the rate of population growth. Today we are extracting 160 km^3 (5.65 trillion ft^3) more water each year than is finding its way back into the ground. This degree of overpumping is equal to the quantity of water needed to produce 10% of the world grain supply. As aquifers are depleted, water tables drop. Groundwater becomes more difficult and expensive to extract, and eventually it may run out. In parts of Mexico, India, China, and other nations in Asia and the Middle East, water tables are falling 1–3 m (3–10 ft) per year. In the United States, by the late 1990s overpumping had drawn the Ogallala Aquifer down by 325 km^3 (11.5 trillion ft^3). This volume is equal to the yearly flow of 18 Colorado Rivers.

Overpumping of groundwater in coastal areas can cause salt water to intrude into aquifers, making water undrinkable. This has occurred widely in Middle Eastern nations and in localities as varied as Florida, Turkey, and Bangkok.

As aquifers lose water, their substrate can become weaker and less capable of supporting overlying strata, and the land surface above may subside. For this reason, cities from Venice to Bangkok to Beijing are slowly sinking. Mexico City's downtown has sunk over 10 m (33 ft) since the time of Spanish arrival; streets are buckled, old buildings lean at angles, and underground pipes break so often that 30% of the system's water is lost to leaks. Sometimes subsidence can occur suddenly in the form of **sinkholes,** areas where the ground gives way with little warning, occasionally swallowing people's homes (Figure 15.11). Once the ground subsides, soil becomes compacted, losing the porosity that enabled it to hold water. Recharging a depleted aquifer may thereafter become more difficult. It has been estimated that compacted aquifers under California's Central Valley have lost storage capacity equal to that of 40% of the state's artificial surface reservoirs.

Falling water tables can do vast ecological harm. Permanent wetlands exist where water tables are high enough to reach the surface, so when water tables drop, wetland ecosystems dry up. In Jordan, the Azraq Oasis covered 7,500 ha (18,500 acres) and enabled migratory birds and other animals to find water in the desert. The water table beneath this oasis dropped from 2.5 m (8.2 ft) to 7 m (23 ft) during the 1980s because of increased well use by the city of Amman. As a result, the oasis dried up altogether during the 1990s. Today international donors are collaborating with the Jordanian government to try to find alternate sources of water and restore this oasis.

We divert—and deplete—surface water to suit our needs

In areas near surface water, people have found it far easier to make use of rivers, streams, lakes, and ponds, diverting water from these sources to farm fields, homes, and cities.

FIGURE 15.11 When too much groundwater is withdrawn too quickly, the land above it may collapse in sinkholes, sometimes bringing buildings down with it.

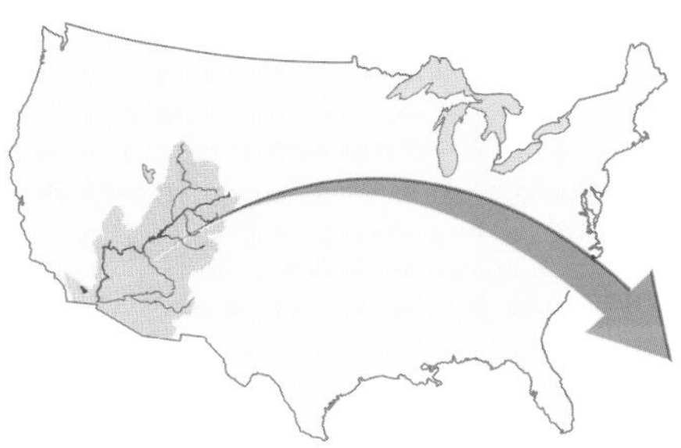

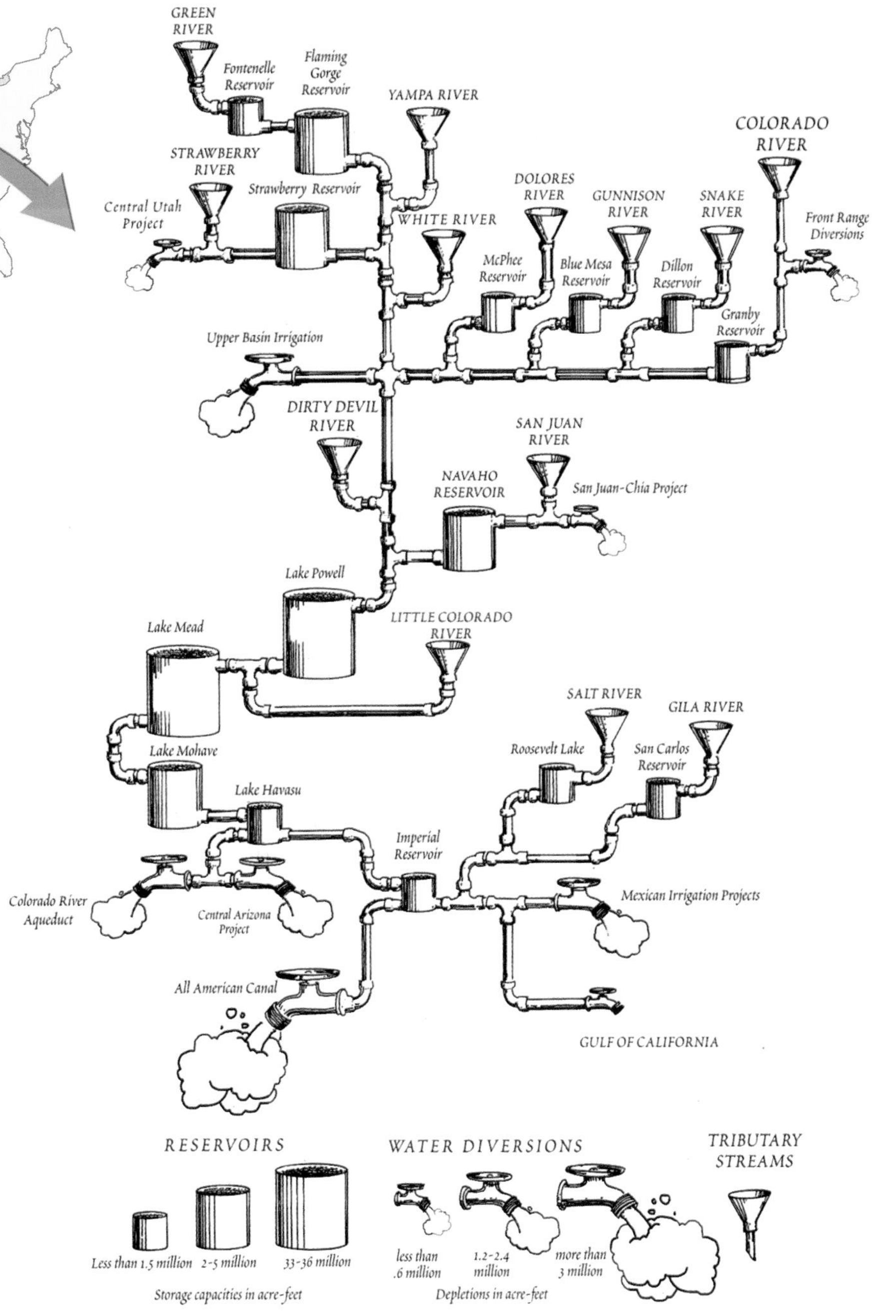

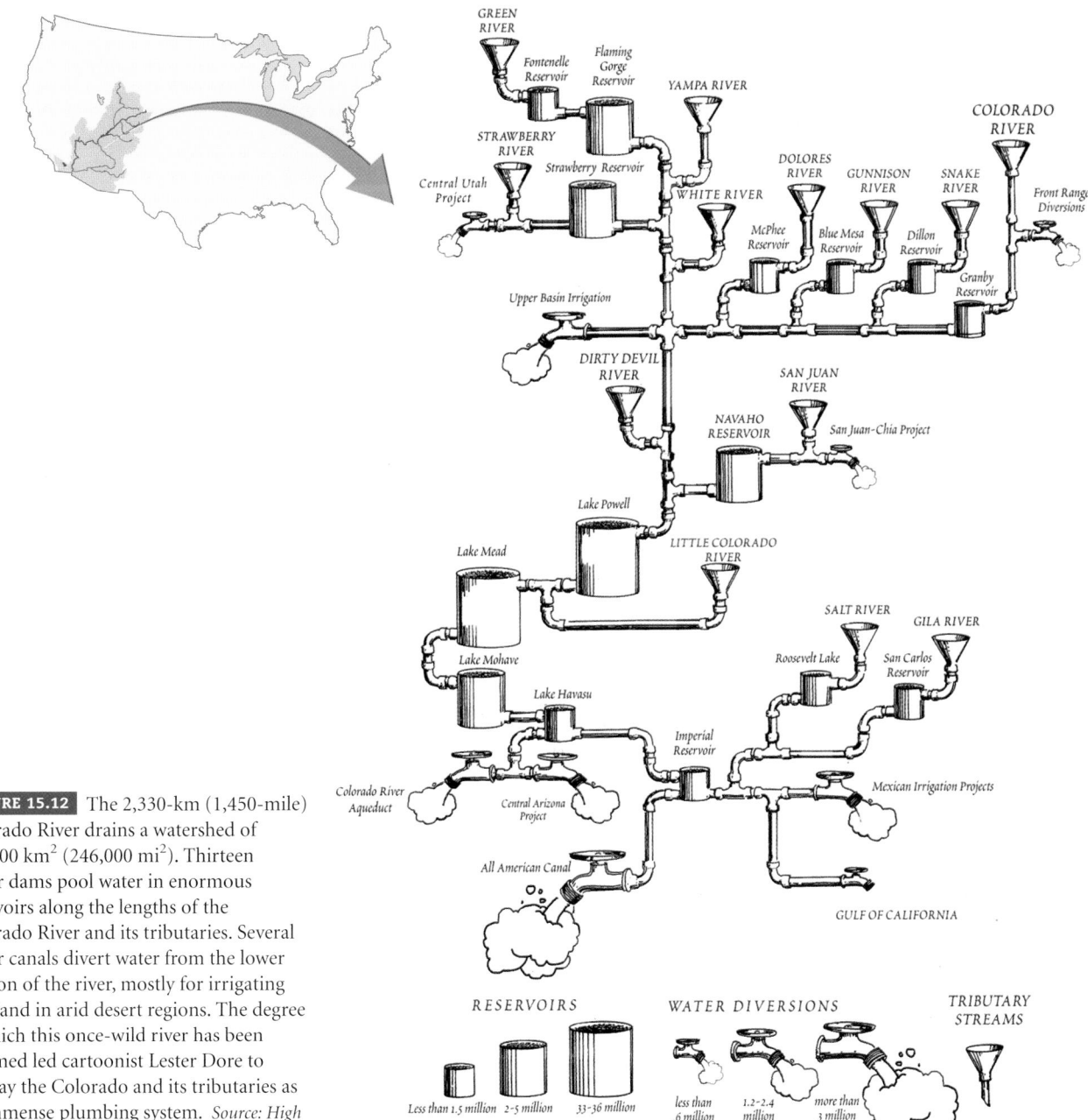

FIGURE 15.12 The 2,330-km (1,450-mile) Colorado River drains a watershed of 637,000 km^2 (246,000 mi^2). Thirteen major dams pool water in enormous reservoirs along the lengths of the Colorado River and its tributaries. Several major canals divert water from the lower portion of the river, mostly for irrigating farmland in arid desert regions. The degree to which this once-wild river has been dammed led cartoonist Lester Dore to portray the Colorado and its tributaries as an immense plumbing system. *Source: High Country News,* 10 November 1997.

The Colorado River's water is heavily diverted and utilized (Figure 15.12). Early on in its course, some Colorado River water is piped through a mountain tunnel and down the Rockies' eastern slope to supply the city of Denver. More is removed for Las Vegas and other cities and for farmland as the water proceeds downriver. When it reaches Parker Dam on the California-Arizona state line, large amounts are diverted into the Colorado River Aqueduct, which brings water to millions in the Los Angeles and San Diego areas via a long open-air canal. From Parker Dam, Arizona also draws water, transporting it in the large canals of the Central Arizona Project. Further south at Imperial Dam, water is diverted into the Coachella and All-American Canals, destined for agriculture, mostly in California's Imperial Valley. To make this desert bloom, Imperial Valley farmers soak the soil with subsidized water for which they pay one penny per 795 L (210 gal).

What water is left of the Colorado River after all the diversions comprises just a trickle making its way to the Gulf of California over the sediments of the once-rich delta (Figure 15.13). On some days, the water does not reach the Gulf at all.

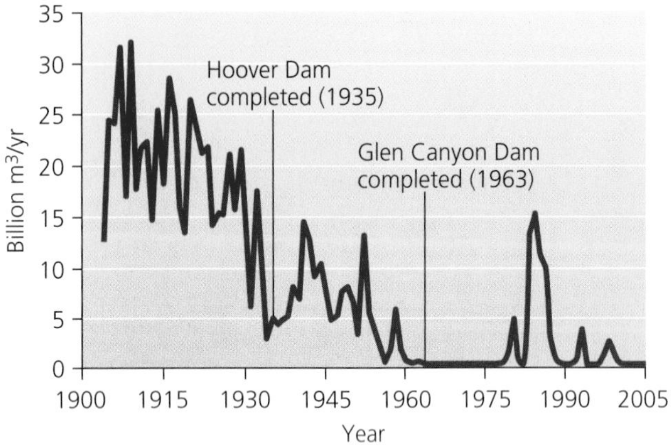

FIGURE 15.13 Flow at the mouth of the Colorado River has greatly decreased over the past century as a result of withdrawals, mostly for agriculture. The river now often runs dry at its mouth. Data from Postel, S., 2005. *Liquid assets: The critical need to safeguard freshwater ecosystems.* Worldwatch Paper 170. Washington, D.C.: Worldwatch Institute.

The Colorado's plight is not unique. Several hundred miles to the east, the Rio Grande also frequently runs dry, the victim of overextraction by both Mexican and U.S. farmers in times of drought. The story is even worse for China's Yellow River, so the Chinese government wants to build a massive aqueduct to supplement its flow with water from the Yangtze River. Even the river that has nurtured human civilization as long as any other, the Nile, now peters out before reaching its mouth.

Nowhere are the effects of surface water depletion so evident as at the Aral Sea. Once the fourth-largest lake on Earth, just larger than Lake Huron, it has lost four-fifths of its volume in 40 years and could soon disappear altogether (Figure 15.14). This dying inland sea, on the border of present-day Uzbekistan and Kazakhstan, is the victim of irrigation practices. The former Soviet Union instituted large-scale cotton farming in this region by flooding the dry land with water from the two rivers leading into the Aral Sea. For a few decades this action boosted Soviet cotton production, but it led the Aral Sea to shrink, while the irrigated soil became waterlogged and salinized. Today 60,000 fishing jobs are gone, winds blow pesticide-laden dust up from the dry lakebed, and what cotton grows on the blighted soil cannot bring the regional economy back. Scientists are struggling to save the Aral Sea, whose ecosystems have been seriously damaged.

Weighing the Issues: The Klamath Crisis

In 2001, angry farmers of Klamath County, Oregon, disobeyed a federal order to divert irrigation waters downstream to save two endangered species of fish. During that bone-dry year, there simply wasn't enough water to irrigate farmers' fields and also keep endangered salmon and suckerfish alive. The government had long ago begun allocating more water than was sustainable in the region. Although the homesteaders enjoyed federal incentives to settle and farm the region at the turn of the century, their children are now subject to a conflicting federal mandate—the Endangered Species Act. Can you think of solutions to the Klamath crisis? What would you do to reconcile these needs?

Will we see a future of water wars?

Freshwater depletion leads to shortages, and resource scarcity can lead to conflict. We have only to look to the Colorado River to see evidence of this. In 1933 the governor of Arizona, foreseeing that California's water diversion from the Colorado might endanger Arizona's future allotment, sent the state's National Guard to threaten the construction of Parker Dam. After a long standoff, the U.S. interior secretary halted the project to avoid hostilities while the issue was mediated in court. Arizona won the court case, but California got the dam authorized by Congress, and Arizona chose not to tackle the U.S. Army troops sent to protect the dam's construction.

Many predict that water's role in regional conflicts will increase as human population continues to grow in water-poor areas. A total of 261 major rivers, whose watersheds cover 45% of the world's land area, cross national borders, and transboundary disagreements are common. The World Water Commission's chairman, Ismail Serageldin, has remarked that "the wars of the twenty-first century will be fought over water."

On the positive side, many nations have cooperated with neighbors to resolve water disputes. India has struck cooperative agreements over management of transboundary rivers with Pakistan, Bangladesh, Bhutan, and Nepal. In Europe, international conventions have been signed by multiple nations along the Rhine and the Danube rivers. Such progress gives reason to hope that future water wars will be few and far between.

Dikes and levees are meant to control floods

We have applied a great deal of engineering brainpower and muscle to transport water from place to place, but we have also expended vast efforts to keep water in place. Flood prevention ranks high among reasons we control the movement of freshwater. People have always been attracted to riverbanks for their water supply and for the flat topography

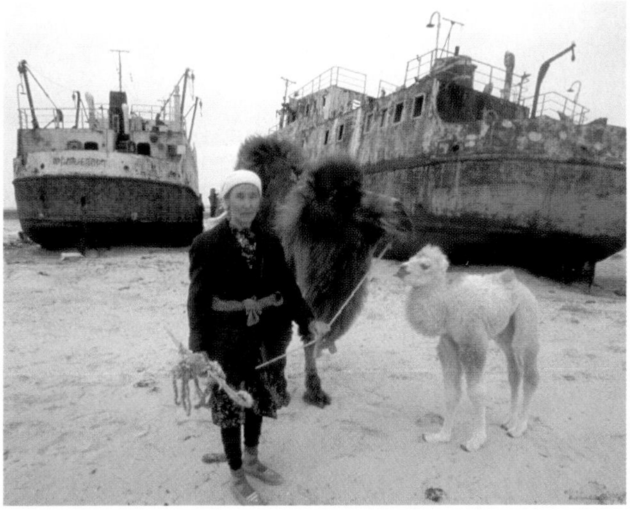

(a) Ships stranded by the Aral Sea's fast-receding waters

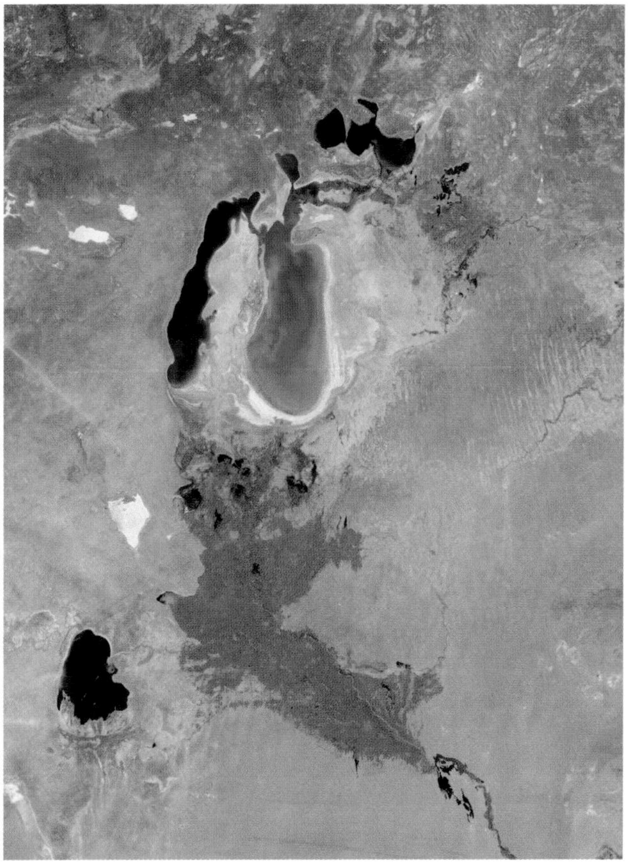

(b) Satellite view of Aral Sea, 2002

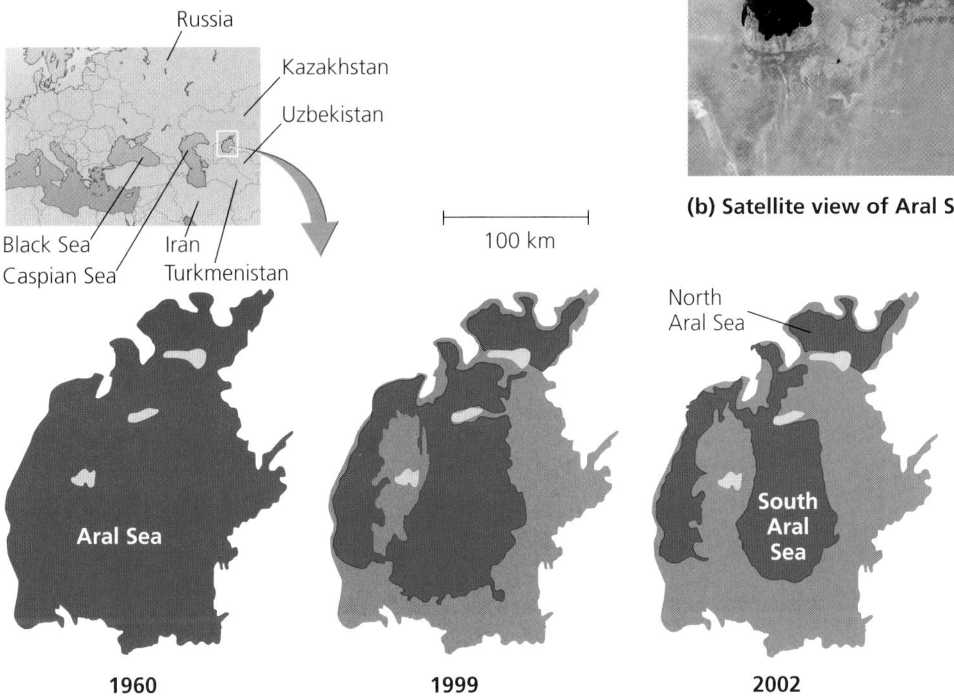

(c) The shrinking Aral Sea, then and now

FIGURE 15.14 Ships lie stranded in the sand (**a**) because the waters of Central Asia's Aral Sea have receded so far and so quickly (**b**). The Aral Sea was once the world's fourth-largest lake. However, it has been shrinking for the past four decades (**c**) and could disappear completely in the near future. The primary cause has been overwithdrawal of water to irrigate cotton crops.

and fertile soil of floodplains. Flooding is a normal, natural process due to snowmelt or heavy rain, and, as we have seen, floodwaters spread nutrient-rich sediments over large areas, benefiting both natural systems and human agriculture.

In the short term, however, floods can do tremendous damage to the farms, homes, and property of people who choose to live in floodplains. To protect against floods, individuals and governments have built *dikes* and *levees* (long raised mounds of earth) along the banks of rivers to hold rising water in main channels. Many dikes are small and locally built, but in the United States the Army Corps of Engineers has constructed thousands of miles of massive levees along the banks of major waterways (those that failed in New Orleans after Hurricane Katrina are examples). Although these structures prevent flooding at most times and places, they can sometimes exacerbate flooding

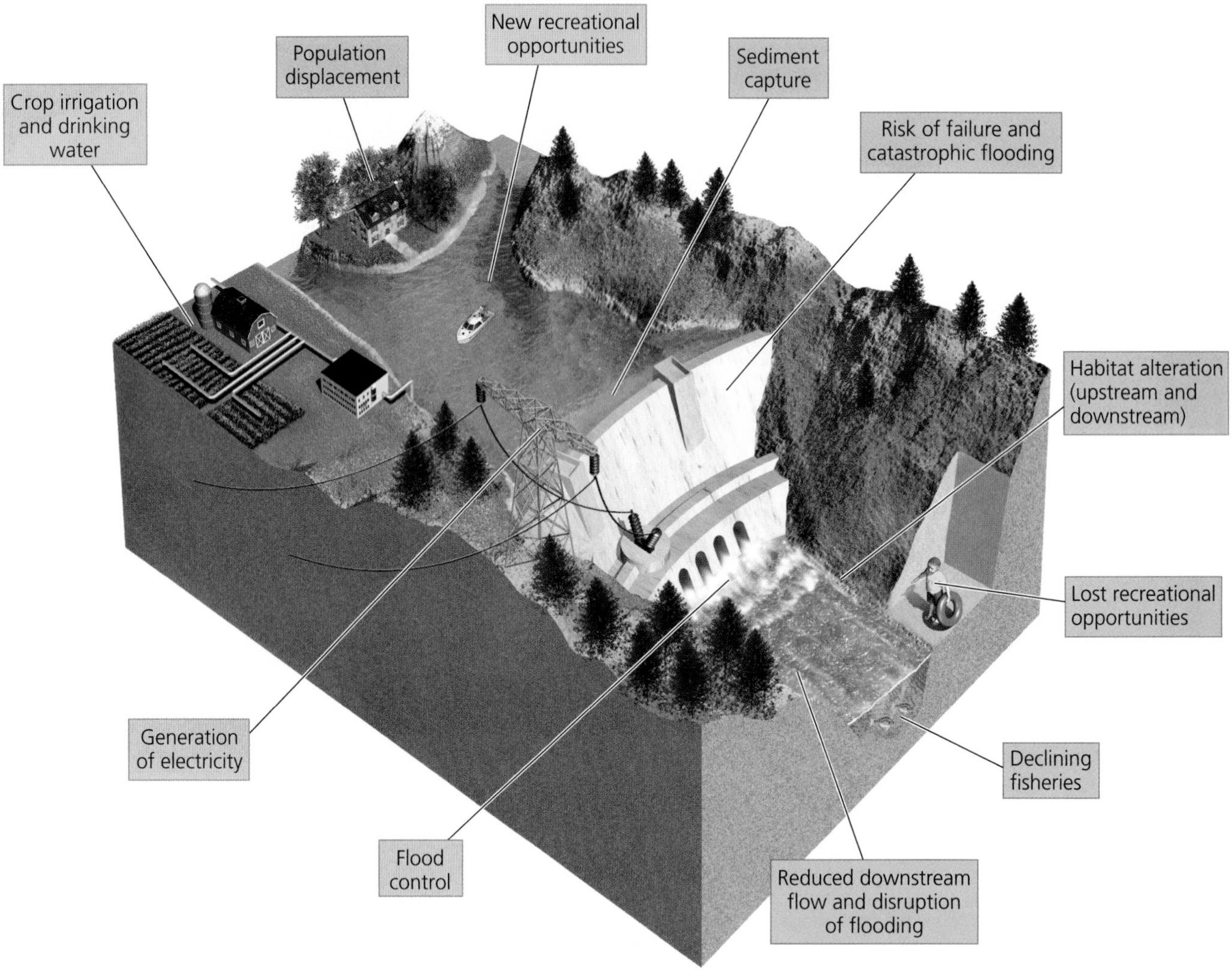

FIGURE 15.15 Damming rivers has many diverse consequences for people and the environment. The generation of clean and renewable electricity is one of several major benefits (green boxes) of hydroelectric dams. Habitat alteration is one of several negative impacts (red boxes).

because they force water to stay in channels and accumulate, building up enormous energy and leading to occasional catastrophic overflow events.

We have erected thousands of dams

A **dam** is any obstruction placed in a river or stream to block the flow of water so that water can be stored in a reservoir. We build dams to prevent floods, provide drinking water, facilitate irrigation, and generate electricity (Figure 15.15; Table 15.1). Power generation with hydroelectric dams is discussed in Chapter 20 (▸ pp. 612–615). Worldwide, more than 45,000 large dams (greater than 15 m, or 49 ft, high) have been erected across rivers in over 140 nations. Additionally, tens of thousands of smaller dams have been built. Virtually the only major rivers in the world that remain undammed and free-flowing run through the tundra and taiga of Canada, Alaska, and Russia, and in remote regions of Latin America and Africa.

The largest of our dams are some of the greatest engineering feats humans have produced. The two behemoths of the Colorado River, Hoover Dam and Glen Canyon Dam, stand 221 m (726 ft) and 216 m (710 ft) high, respectively, stretch for 379 m (1,244 ft) and 476 m (1,560 ft) across, and consist of 6.6 and 7.3 million tons of concrete and steel. Hoover Dam holds back 35.2 km^3 (8.4 mi^3) of water in a reservoir that stretches 177 km (110 mi) long

Table 15.1 Major Benefits and Costs of Dams

Benefits	Costs
▶ **Power generation.** Hydroelectric dams (▶ pp. 612–614) provide a great deal of inexpensive electricity to help run our economies and power our modern conveniences. Some nations gain nearly all their electricity from hydropower.	▶ **Habitat alteration.** Reservoirs flood riparian habitats and displace or kill riparian species. Dams modify rivers downstream, making them tranquil and low in dissolved oxygen, so that many fish adapted to fast-flowing rivers cannot survive. Shallow warm water downstream from a dam is periodically flushed with cold water from the reservoir, stressing or killing both cold- and warm-water species.
▶ **Emissions reduction.** Hydropower provides electricity without producing emissions that pollute the air. By replacing fossil fuel combustion as an electricity source, hydropower reduces air pollution, climate change, and their health and environmental consequences.	▶ **Decline in fisheries.** Fish that migrate up rivers to spawn encounter dams as a barrier. Although "fish ladders" have been built at many dams to allow passage, most fish do not make it. Population declines of salmon species have devastated fishing economies.
▶ **Crop irrigation.** Reservoir water can be withdrawn for irrigating crops. Irrigation is vital for agriculture in arid areas. By storing large amounts of water, reservoirs can release irrigation water when farmers most need it and can buffer regions against drought.	▶ **Population displacement.** Reservoirs generally flood fertile farmland, and have flooded many human settlements, even large modern cities. An estimated 40–80 million people globally have been displaced by dam projects over the past half century.
▶ **Drinking water.** Many reservoirs store water for municipal drinking water supplies. Such water is generally plentiful, reliable, and clean, provided that watershed lands draining into the reservoir are not developed or polluted.	▶ **Sediment capture.** As fast-flowing rivers give way to slow-flowing reservoirs, sediment falls through the water column and settles, trapped behind dams. Downstream floodplains and estuaries are no longer nourished, and reservoirs fill with silt.
▶ **Flood control.** Dams can prevent floods by storing seasonal surges, such as those following snowmelt or heavy rain. Water can then be released gradually to even out flows downstream. Flood prevention saves lives and prevents property damage.	▶ **Disruption of flooding.** In the long term, flooding is a valuable ecosystem service. Floods create productive farmland by depositing rich sediment. Without flooding, topsoil is lost and farmland deteriorates.
▶ **Shipping.** By replacing rocky river beds and rapids with deep placid pools, dams enable ships to transport goods over longer distances.	▶ **Risk of failure.** There is always a small risk that a dam could fail, causing massive property damage, ecological damage, and loss of life.
▶ **New recreational opportunities.** People can fish from boats and use personal watercraft on reservoirs in regions where such recreation was not possible before.	▶ **Lost recreational opportunities.** When a wild river is dammed, recreational opportunities such as tubing, whitewater rafting, and kayaking are lost.

and reaches 152 m (500 ft) deep. Glen Canyon Dam holds back 33.3 km^3 (7.9 mi^3) of water in a reservoir that stretches 300 km (186 mi) long and reaches 171 m (560 ft) deep. Together these reservoirs store four times more water than flows down the river in an entire year.

China's Three Gorges Dam is the world's largest

The complex mix of benefits and costs that dams produce is exemplified by the world's largest dam project. The Three Gorges Dam on China's Yangtze River, 186 m (610 ft) high and 2 km (1.3 mi) wide, was completed in 2003 (Figure 15.16a). When filled in 2009, the resulting reservoir should be 616 km (385 mi) long, as long as Lake Superior. The reservoir will hold over 38 trillion L (10 trillion gal) of water. It will generate hydroelectric power, enable boats and barges to travel farther upstream, and provide flood control. The power generation may be enough to replace dozens of large coal or nuclear plants.

One of the costs of the Three Gorges Dam, aside from its $25 billion construction price tag, is that its reservoir is flooding 22 cities and the homes of 1.13 million people, requiring the largest resettlement project in China's history (Figure 15.16b). The reservoir behind the dam is also inundating archaeological sites 10,000 years old and submerging productive farmlands and wildlife habitat. Moreover, the reservoir will slow the river's flow so that suspended sediment will settle and begin to fill the reservoir as soon as it is completed. Many scientists worry that the Yangtze's many pollutants will be trapped in the reservoir, making the water undrinkable. Indeed, high levels of bacteria were found in water as it began building up behind the dam. The Chinese government plans to sink $5 billion into building hundreds of sewage treatment and waste disposal facilities.

VIEWPOINTS

Dam Removal

Dams bring us many benefits, but also exert ecological and social impacts. **Have some dams outlived their usefulness, and if so, should we dismantle them?**

Dams for Today and Tomorrow

Our need for dams today is greater than ever because water is a finite resource. Over the past century, global water use has increased at twice the rate of population growth. Currently, world population is expected to grow by 50%, to 9 billion people in total, by 2050. Dams and reservoirs address the needs of a growing world by efficiently storing and regulating water for multiple uses. Our world relies on the benefits they bring—drinking water, flood control, power generation, irrigation, and recreation.

Ninety percent of the dams in the United States are small, local projects that lack controversy. Consider what dams do every day for millions of people. Along the Mississippi, 70% of America's grain exports are barged to the Gulf of Mexico. Dams support 55 million irrigated acres of crop and pasture land (mostly in the arid West). Dams and reservoirs carry water to millions of people via canals and aqueducts. And dams help communities avoid billions of dollars in flood damage.

With dams, however, come environmental concerns such as fish passage, changes to water quality, and altered habitats. The challenge for any community is to balance the economic, environmental, and social considerations in utilizing dams. Today's choices often reflect the values, needs, wealth, and options of different communities and countries. It should be no surprise that in a world where 1.7 billion people are without electricity, hydropower is being developed in 80 countries.

Our challenge is to make decisions that embrace what research, sound science, technological innovation, and engineering prowess offers. We can embrace these things while also staying true to our historic and evolving cultural, environmental, and economic values. We owe it to future generations to make thoughtful, responsible policy choices about dams that affect not only our way of life, but also theirs.

Thomas Flint is a fifth-generation farmer, actively farming in Grant County, Washington. He was elected to the Grant County Public Utility District board of commissioners in 2000, is founder and director of the public education effort known as AgFARMation, is a grassroots activist, and holds director's positions on the Black Sands Irrigation District and the Columbia Basin Development League.

The Case for Dam Removal

The dams currently in existence in the United States were built to provide a variety of services, including flood control, water supply, and hydropower, which runs mills and generates electricity. Although many dams continue to provide a useful service, large numbers are considered obsolete, providing no direct economic, safety, or social function. For example, many mill dams continue to stand across streams and rivers 100 to 200 years after the mill they powered went out of operation or was torn down. These dams should be considered for removal.

Regardless of size, all dams harm riparian environments. Dams block the free flow of water down a stream corridor and create a pool, or impoundment, behind them—an artificial lake in the middle of a stream community. Impounded waters often divide into layers by temperature and depth, with heated waters in the upper layer and oxygen-poor cooler water in the lower layer. The macroinvertebrates that fish depend on for food cannot survive under these lake conditions. Carp and non-native lake fish that can survive in hotter and oxygen-poor waters often displace trout and other cold-water stream species.

Dams block the movement of migratory fish and other aquatic species, preventing them from reaching upstream areas to feed, spawn, and successfully reproduce. Dams also block river sediments that would normally travel downstream and replenish beaches or gravel stream bottoms, where most macroinvertebrates live and where fish spawn.

Rivers are dynamic systems. They move within floodplains, exchanging nutrients, sediments, and interacting on many levels. When dams interrupt that exchange, river functions are impaired, and the fish and wildlife dependent on free-flowing river systems do not thrive as well. Once a dam has outlived its utility, it makes great sense to restore the river back to its original condition.

Sara Nicholas is associate director of dam programs for American Rivers and works out of their mid-Atlantic office in Harrisburg, Pennsylvania. She has a master of science degree in environmental science from the Yale School of Forestry.

Explore this issue further by accessing **Viewpoints** at www.aw-bc.com/withgott.

(a) The Three Gorges Dam in Yichang, China

(b) Displaced people in Sichuan Province, China

FIGURE 15.16 China's Three Gorges Dam, completed in 2003, is the world's largest dam (**a**). Well over a million people were displaced and whole cities were leveled for its construction, as shown here in Sichuan Province (**b**). The reservoir began filling in 2003 and will continue filling for several years.

Some dams are now being removed

People who feel that the costs of some dams have outweighed their benefits have been pushing for such dams to be dismantled. By removing dams and letting rivers flow free, these people say, we can restore riparian ecosystems, reestablish economically valuable fisheries, and revive river recreation such as fly-fishing and rafting. Increasingly, private dam owners and the Federal Energy Regulatory Commission (FERC), the U.S. government agency charged with renewing licenses for dams, have agreed. Roughly 500 dams have been removed in the United States, nearly 200 of them in the past decade. One reason is that many aging dams are in need of costly repairs or have outlived their economic usefulness.

The drive to remove dams first gathered steam in 1999 with the dismantling of the Edwards Dam on Maine's Kennebec River, which resulted from the first FERC determination that the environmental benefits of removing a dam outweighed the economic benefits of relicensing it. Within just a year after the 7.3-m (24-ft) high, 279-m (917-ft) long dam was removed, large numbers of 10 species of migratory fish, including salmon, sturgeon, shad, herring, alewife, and bass, ventured upstream and began using the 27-km (17-mi) stretch of river above the dam site. Some property owners along the former reservoir who had opposed the dam's removal had a change of heart once they saw the healthy and vibrant river that now ran past their property. More small dams—and perhaps some large ones—will come down in years ahead as over 500 FERC licenses come up for renewal in the next decade.

Solutions to Freshwater Depletion

Ensuring adequate quantities of freshwater is an endeavor as old as our species. Technological advances will not eliminate this challenge. Human population growth, expansion of irrigated agriculture, and industrial development doubled our global annual freshwater use between 1960 and 2000. We now use an amount equal to 10% of total global runoff. The hydrologic cycle makes freshwater a renewable resource, but if our usage exceeds what a lake, river, or aquifer can provide, we must either reduce our use, find another water source, or be prepared to run out of water.

Solutions can address supply or demand

To address freshwater depletion, we can aim either to increase supply or to reduce demand. Strategies for reducing demand include conservation and efficiency measures. Lowering demand is more difficult politically in the short term but may be necessary in the long term. In the developing world, international aid agencies are increasingly funding demand-based solutions over supply-based solutions, because demand-based solutions offer better economic returns and cause less ecological and social damage.

Strategies most often used to increase supply in a particular area have involved transporting water through pipes and aqueducts from areas where it is more plentiful or accessible. In some instances this has been done by mutual agreement, but in other instances water-poor regions have forcibly appropriated water from communities too weak to keep it for themselves. For instance, Los Angeles built itself up with water it appropriated from the Owens Valley, Mono Lake, and other rural and less-inhabited regions of California. Transporting water from place to place is not a just solution if it harms the region that loses the water. Another strategy is to develop new technologies to find or "make" more water.

Weighing the Issues: Reaching for Water

In 1941, the burgeoning metropolis of Los Angeles needed water and decided to divert streams feeding into Mono Lake, over 565 km (350 mi) away in northern California. As the lake level fell 14 m (45 ft) in 40 years, salt concentrations doubled and aquatic communities suffered. Other desert cities—such as Las Vegas, Phoenix, and Denver—are expected to double in population in coming decades. Where will they go for water to sustain their communities? What challenges might they face in trying to pipe in water from distant sources? How will people living in these source areas be affected? How else could these cities meet their future water needs?

Desalination "makes" more water

The best-known technological approach to generate freshwater is **desalination,** or *desalinization,* the removal of salt from seawater or other water of marginal quality. One method of desalination mimics the hydrologic cycle by hastening evaporation from allotments of ocean water with heat and then condensing the vapor—essentially distilling freshwater. Another method involves forcing water through membranes to filter out salts; the most common process of this type is called reverse osmosis.

Over 7,500 desalination facilities are operating worldwide, most in the arid Middle East (Figure 15.17) and some in small island nations that lack groundwater. The largest plant, in Saudi Arabia, produces 485 million L (128 million gal) of freshwater every day. However, desalination is currently quite expensive. In 1992 the world's largest reverse osmosis plant was completed along the Colorado River near Yuma, Arizona. It was intended to reduce the salinity of irrigation runoff reentering the river, but it proved too expensive to operate and closed after only 8 months.

FIGURE 15.17 Kuwaiti engineers walk along water intake pipes for a desalination plant on the Persian Gulf.

Agricultural demand can be reduced

Although supply-side solutions are worth pursuing, there is a need to reduce demand as well. Because most water use is for agriculture, it makes sense to look first to agriculture for ways to decrease demand. Farmers can improve efficiency by lining irrigation canals to prevent leaks, leveling fields to minimize runoff, and adopting efficient irrigation methods. Techniques to increase irrigation efficiency include low-pressure spray irrigation, which sprays water downward toward plants, and drip irrigation systems, which target individual plants and introduce water directly onto the soil (see Figure 9.17, ▸ p. 265). Both methods reduce water lost to evaporation and surface runoff. Low-pressure precision sprinklers in Texas have efficiencies of 80–95% and have resulted in water savings of 25–37%. Drip irrigation, which has efficiencies as high as 90%, could cut water use in half while raising yields by 20–90% and giving developing-world farmers $3 billion in extra annual income, it has been estimated. Such improvements in Jordan more than doubled crop yields from 1973 to 1986.

Choosing crops to match the land and climate in which they are being farmed can save huge amounts of water. Currently, crops that require a great deal of water, such as cotton, rice, and alfalfa, are often planted in arid areas with government-subsidized irrigation. As a result, the true cost of water is not part of the costs of growing the crop. Eliminating subsidies and growing crops in climates with adequate rainfall could greatly reduce water use in many parts of the world. Finally, selective breeding (▸ pp. 120–122; 248) and genetic modification (▸ pp. 287–295) can result in some crop varieties that require less water.

We can lessen residential, municipal, and industrial water use in many ways

We can reduce our household water use by installing low-flow faucets, showerheads, washing machines, and toilets. Automatic dishwashers, studies show, use less water than does washing dishes by hand. If our homes have lawns, it is best to water them at night, when water loss from evaporation is minimal. Better yet, replacing water-intensive lawns with native plants adapted to your region's natural precipitation patterns saves the most water. *Xeriscaping*, landscaping using plants adapted to arid conditions, has become a popular approach in much of the U.S. Southwest.

Industry and municipalities can take water-saving steps as well. Manufacturers have shifted to processes that use less water and in doing so have reduced their costs. Something as seemingly obvious as finding and patching leaks in pipes has saved some cities and companies large amounts of water—and money—once they have invested in the search. Boston and its suburbs reduced water demand by 31% between 1987 and 2004 by patching leaks, retrofitting homes with efficient plumbing, auditing industry, and promoting conservation to the public. This wildly successful program enabled Massachusetts to avoid an unpopular $500-million river diversion scheme.

Another water conservation practice is recycling municipal wastewater for irrigation and industrial uses, as some cities, among them St. Petersburg, Florida, are doing. In yet another approach, governments in Arizona and in England are capturing excess surface runoff during their rainy seasons and pumping it into aquifers.

Various economic approaches to water conservation are being debated

Many economists have suggested market-based strategies for achieving sustainability in water use, ending government subsidies of inefficient practices, and letting water become a commodity whose price reflects the true costs of its extraction. Others worry that making water a fully priced commodity would make it less available to the world's poor and increase the gap between rich and poor. Because industrial use of water can be 70 times more profitable than agricultural use, market forces alone could favor uses that would benefit wealthy and industrialized people, companies, and nations at the expense of the poor and less industrialized.

Similar concerns surround another potential solution, the privatization of water supplies. During the 1990s, some public water systems were partially or wholly privatized, and their construction, maintenance, management, or ownership transferred to private companies. This was done in hope of increasing the systems' efficiency, but many people worry that firms have little incentive to allow equitable access to water for rich and poor alike. Already in some developing countries, rural residents without access to public water supplies, who are forced to buy water from private vendors, end up paying on average 12 times more than those connected to public supplies.

Other experiences indicate that decentralization of control over water, from the national level to the local level, may help conserve water. In Mexico, the effectiveness of irrigation systems improved dramatically once they were transferred from public ownership to the control of 386 local water user associations.

Regardless of how demand is addressed, the ongoing shift from supply-side to demand-side solutions is beginning to pay dividends. In Europe, a new focus on demand (through government mandates and public education) has decreased public water consumption, and industries are becoming more efficient in their water use. The United States decreased its total water consumption by 10% from 1980 to 1995, thanks to conservation measures, even while its population grew 16%.

Freshwater Pollution and Its Control

The quantity and distribution of freshwater poses one set of environmental and social challenges. Safeguarding the *quality* of water involves another collection of environmental and human health dilemmas. To be safe for human consumption and other organisms, water must be relatively free of disease-causing organisms and toxic substances.

Although developed nations have made admirable advances in cleaning up water pollution over the past few decades, the World Commission on Water in 1999 concluded that over half the world's major rivers are "seriously depleted and polluted, degrading and poisoning the surrounding ecosystems, threatening the health and livelihood of people who depend on them." The largely invisible pollution of groundwater, meanwhile, has been termed a "covert crisis." Groundwater and surface water are polluted by numerous anthropogenic sources, including untreated sewage, agricultural runoff, and petroleum and chemical spills.

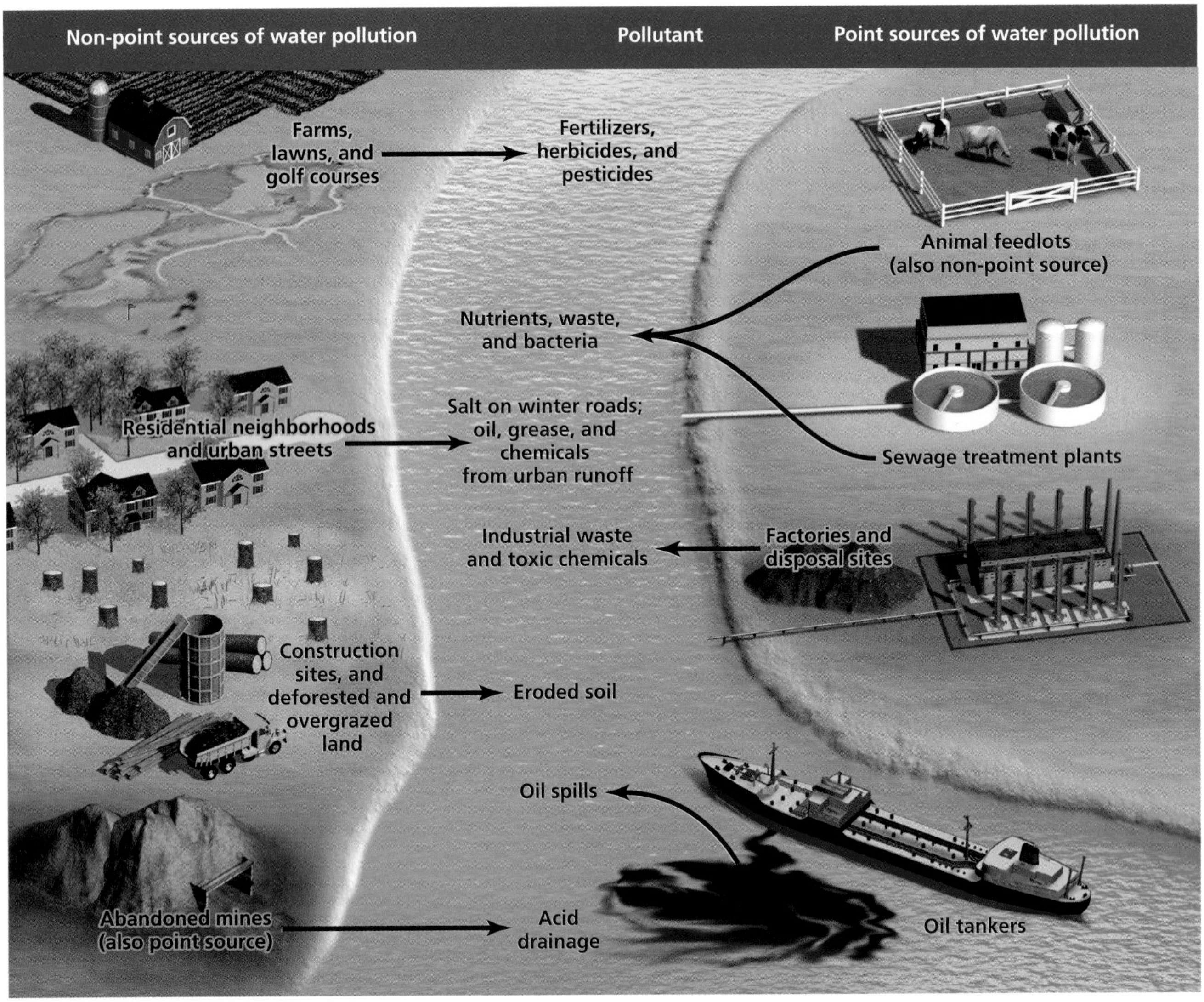

FIGURE 15.18 Point-source pollution comes from discrete facilities or locations, usually from single outflow pipes. Non-point-source pollution (such as runoff from streets, residential neighborhoods, lawns, and farms) originates from numerous sources spread over large areas.

Water pollution comes from point sources and from diffuse non-point sources

The term **pollution** describes any matter or energy released into the environment, whether from human activity or natural sources, that causes undesirable impacts on the health and well-being of humans or other organisms. Pollution can be physical, chemical, or biological, and can affect water, air, or soil.

Water pollution can be emitted from **point sources**—discrete locations, such as a factory or sewer pipe. Alternatively, it can consist of **non-point-source** pollution arising from multiple cumulative inputs over larger areas, such as farms, city streets, and residential neighborhoods (Figure 15.18). The U.S. Clean Water Act, by targeting industrial discharges, has addressed point-source pollution with some success, such that non-point-source pollution has a greater impact on water quality in the United States today. Many common activities give rise to non-point-source water pollution, such as applying fertilizers and pesticides to lawns, applying salt to roads in winter, and changing automobile oil.

Water pollution comes in many forms and can cause diverse impacts on aquatic ecosystems and human health. We can categorize pollution into several types, including nutrient pollution, biological pollution by disease-causing organisms, toxic chemical pollution, physical pollution by sediment, and thermal pollution.

Nutrient pollution We saw in Chapter 7 how nutrient pollution from fertilizers and other sources can lead to eutrophication and hypoxia in coastal marine areas, as exemplified by the Gulf of Mexico's dead zone (see Fig. 7.5, ▸ p. 189). Eutrophication proceeds in a similar fashion in freshwater systems, where phosphorus is usually the nutrient that spurs growth (▸ pp. 192–194). When excess phosphorus enters surface waters, it fertilizes algae and aquatic plants, boosting their growth rates and populations. Although such growth provides oxygen and food for other organisms, algae can cover the water's surface, depriving deeper-water plants of sunlight. As algae die off, they provide food for decomposing bacteria. Decomposition requires oxygen, so the increased bacterial activity drives down levels of dissolved oxygen. These levels can drop too low to support fish and shellfish, leading to dramatic changes in aquatic ecosystems.

Eutrophication (Figure 15.19) is a natural process, but excess nutrient input from runoff from farms, golf courses, lawns, and sewage can dramatically increase the rate at which it occurs. We can reduce nutrient pollution by treating wastewater, reducing fertilizer application, and planting vegetation to increase nutrient uptake.

(a) Oligotrophic water body

(b) Eutrophic water body

FIGURE 15.19 An oligotrophic water body **(a)** with clear water and low nutrient content may eventually become a eutrophic water body **(b)** with abundant vegetation and high nutrient content. Pollution of freshwater bodies by excess nutrients accelerates the process of eutrophication.

Pathogens and waterborne diseases Many disease-causing organisms (pathogenic viruses, protists, and bacteria) survive in surface water. Some enter inadequately treated drinking water supplies when these become contaminated with human or animal waste. Specialists monitoring water quality can tell when water has been contaminated by waste when they detect concentrations of fecal coliform bacteria, which live in the intestinal tracts of people and other vertebrates. These bacteria are usually not pathogenic themselves, but rather serve as indicators of fecal contamination, which may mean that the water holds other pathogens that can cause ailments such as giardiasis, typhoid, or hepatitis A.

Biological pollution by pathogens causes more human health problems than any other type of water pollution. A study of global water supply and sanitation issues by the World Health Organization (WHO) and the United Nations Children's Fund (UNICEF) in 2000 showed that despite advances in many parts of the world, major problems still existed. On the positive side, 4.9 billion people (82% of the population) had access to safe water as a result of some form of improvement in their water supply—an increase from 4.1 billion (79% of the population) in 1990. However, over 1.1 billion people were still without safe water supplies. In addition, 2.4 billion people had no sewer or sanitation facilities. Most of these people were Asians and Africans, and four-fifths of the people without sanitation lived in rural areas. These conditions contribute to widespread health impacts and 5 million deaths per year.

We have developed strategies for reducing the risks that water borne pathogens pose. Treating sewage constitutes one approach. Another is using chemical or other means to disinfect drinking water. Finally, hygienic measures can fight waterborne disease. Among these are public education to encourage personal hygiene and government enforcement of regulations to ensure the cleanliness of food production, processing, and distribution.

Toxic chemicals Our waterways have become polluted with toxic organic substances of our own making, including pesticides, petroleum products, and other synthetic chemicals. Many of these can poison animals and plants, alter aquatic ecosystems, and cause a wide array of human health problems, including cancer. In addition, toxic metals such as arsenic, lead, and mercury, and acids from acid precipitation and from acid drainage from mining sites (▸ p. 576), also cause negative impacts on human health and the environment. Health impacts of toxic chemicals are discussed in Chapter 14.

Legislating and enforcing more stringent regulations of industry can help reduce releases of these toxic inorganic chemicals. Better yet, we can modify our industrial processes to rely less on these substances.

Sediment Although floods build fertile farmland, sediment that rivers transport can also impair aquatic ecosystems. Mining, clear-cutting, land clearing for real estate development, and careless cultivation of farm fields all expose soil to wind and water erosion (▸ pp. 256–259). Although some water bodies, such as the Colorado River and China's Yellow River, are naturally sediment-rich, many others are not. When a clear-water river receives a heavy influx of eroded sediment, aquatic habitat can change dramatically, and fish adapted to clear-water environments may not be able to handle the change. We can reduce sediment pollution by more thoughtfully managing farms and forests and by avoiding large-scale disturbance of vegetation.

Heat and cold Because water's ability to hold dissolved oxygen decreases as temperature rises, some aquatic organisms may not survive temperature increases in surface waters. Many human activities can increase water temperatures. When water is withdrawn from a river or stream and used to cool industrial facilities, it transfers heat energy from the facility back into the surface water to which it is returned. People also raise surface water temperatures by removing streamside vegetation that shades water.

Too little heat can also cause problems. On the Colorado and many other dammed rivers, water at the bottoms of reservoirs is much colder than water at the surface. If dam operators release water from the depths of a reservoir, downstream water temperatures become much lower. In the Colorado River system, these low water temperatures have favored cold-loving invasive trout over an endangered native species of suckerfish.

Scientists use several indicators of water quality

Most forms of water pollution are not very visible to the human eye, so scientists and technicians measure certain physical, chemical, and biological properties of water to characterize its quality. Biological properties include the presence of fecal coliform bacteria and other disease-causing organisms, as discussed above. In addition, algae and aquatic invertebrates are commonly used as biological indicators of water quality.

Chemical properties include nutrient concentrations, pH (▸ p. 98), taste and odor, and hardness. Hard water contains high concentrations of calcium and magnesium ions, prevents soap from lathering, and leaves chalky deposits behind when heated or boiled. Another chemical characteristic is dissolved oxygen content. Dissolved oxygen is an indicator of aquatic ecosystem health because surface waters low in dissolved oxygen are less capable of supporting aquatic life.

Among physical characteristics, turbidity measures the density of suspended particles in a water sample. Fast-moving rivers that cut through arid or eroded landscapes, such as the Colorado and Yellow rivers, carry a great deal of sediment and are turbid and muddy-looking as a result. Water color can reveal particular substances present in a water sample. Some forest streams run the color of iced tea because of chemicals called tannins that occur naturally in decomposing leaf litter. Finally, temperature can be used to assess water quality. High temperatures can interfere with some biological processes, and warmer water holds less dissolved oxygen.

Groundwater pollution is a serious problem

Most efforts at pollution control have focused on surface water bodies. Yet increasingly, groundwater sources once assumed to be pristine have been contaminated by pollution from industrial and agricultural practices. Groundwater pollution is largely hidden from view and is extremely difficult to monitor; it can be out-of-sight, out-of-mind for decades until widespread contamination of drinking supplies is discovered.

Groundwater pollution can also be more difficult to manage than surface water pollution. Rivers flush their pollutants fairly quickly, but groundwater retains its contaminants until they decompose, which in the case of persistent pollutants can be many years or decades. The long-lived pesticide DDT, for instance, is still found widely in U.S. aquifers even though it was banned 35 years ago. Moreover, chemicals are broken down much more slowly in aquifers than in surface water or soils. Groundwater generally contains less dissolved oxygen, microbes, minerals, and organic matter, so decomposition is slower. For instance, concentrations of the herbicide alachlor are reduced by half after 20 days in soil but in groundwater this takes almost 4 years.

FIGURE 15.20 Leaky underground storage tanks are a major source of groundwater pollution. Under an EPA program, hundreds of thousands of these tanks are being unearthed and repaired.

There are many sources of groundwater pollution

Various chemicals that are toxic at high concentrations, including aluminum, fluoride, nitrates, and sulfates, occur naturally in groundwater. The poisoning of Bangladesh's wells by arsenic is one case of natural contamination (see "The Science behind the Story," ▸ pp. 456–457). However, groundwater pollution from human activity is widespread. Industrial, agricultural, and urban wastes—from heavy metals to petroleum products to industrial solvents to pesticides—can leach through soil and seep into aquifers. Pathogens and other pollutants can enter groundwater through improperly designed wells. Contamination even results from the intentional pumping of liquid hazardous waste below ground (▸ pp. 668–670).

Leakage from underground storage tanks is another contributor to groundwater pollution. These include septic tanks, tanks of industrial chemicals, and tanks of oil and gas. In the United States, the Environmental Protection Agency (EPA) has embarked on a nationwide cleanup program to unearth and repair leaky tanks before they do further damage to soil and groundwater quality (Figure 15.20). After more than a decade of work, the EPA in 2005 had confirmed leaks from 449,000 tanks, had initiated cleanups on 416,000 of them, and had completed cleanups of 324,000. Intercepting carcinogenic or otherwise toxic pollutants such as chlorinated solvents and gasoline before they reach aquifers is vital because once an aquifer is contaminated, it is extremely difficult to remediate.

Agriculture also contributes to groundwater pollution. Nitrate from fertilizers has leached into groundwater in Canada and in 49 U.S. states. Pesticides were detected in over half of the shallow aquifer sites tested in the United States in the mid-1990s, although generally below the standards set by the EPA for drinking water. Nitrate in drinking water has been linked to cancers, miscarriages, and "blue-baby" syndrome, in which infants suffocate. Agriculture can also contribute pathogens; in 2000, the groundwater supply of Walkerton, Ontario, became contaminated with the bacterium *Escherichia coli,* or *E. coli.* Two thousand people became ill, and seven died.

Manufacturing industries and military sites have been heavy polluters through the years. Although legislation has forced them to reduce discharges, groundwater can bear a toxic legacy long afterwards. During World War II, the U.S. Army operated the world's largest facility to produce trinitrotoluene (TNT) near St. Louis. Nitroaromatic by-products seeped into the drinking water for miles around. The site was high on the list to be cleaned up under the 1980 Superfund legislation (▸ pp. 670–671). At another Superfund site, the Hanford Nuclear Reservation in Washington State, radioactive waste has seeped into groundwater, some of it with a half-life of a quarter-million years.

Legislative and regulatory efforts have helped reduce pollution

As numerous as our freshwater pollution problems may seem, it is important to remember that many of them were worse a few decades ago. Citizen activism and government response during the 1960s and 1970s in the

The Science behind the Story

Arsenic in the Waters of Bangladesh

In the 1970s, UNICEF, with the help of environmental scientists at the British Geological Survey, launched a campaign to improve access to freshwater in Bangladesh. By digging thousands of small artesian wells, the designers of the program hoped to reduce Bangladeshis' dependence on disease-ridden surface waters. In the mid-1990s, however, scientists began to suspect that the wells dug to improve Bangladeshis' health were contaminated with arsenic, a poison that, if ingested frequently, can cause serious skin disorders and other illnesses, including cancer.

A medical doctor sounded the first alarm. In 1983, dermatologist K. C. Saha of the School of Tropical Medicine in Calcutta, India, saw the first of many patients from West Bengal, an area of India just west of Bangladesh, who showed signs of arsenic poisoning. Through a process of elimination, contaminated well water was identified as the likely cause of the poisoning. The hypothesis was confirmed by groundwater testing and by the work of epidemiologists, among them Dipankar Chakraborti of Calcutta's Jadavpur University.

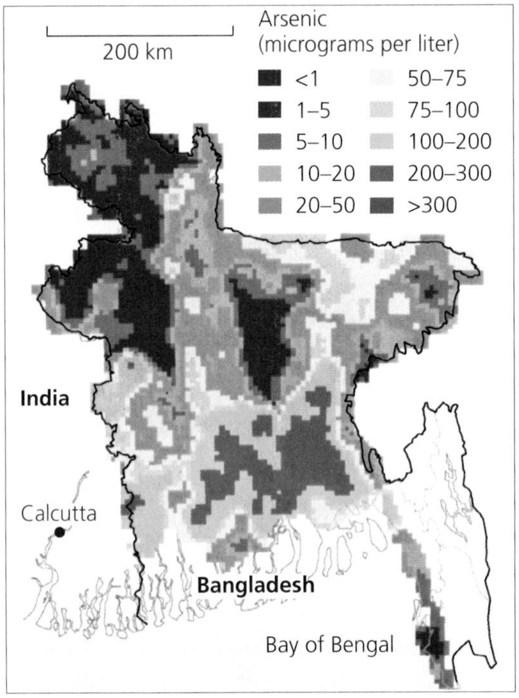

In a modern tragedy, thousands of wells dug for drinking water in Bangladesh at the urging of international aid workers turned out to be laced with arsenic. This map shows that the highest concentrations of arsenic—100–300 or more micrograms per liter of water—are found in the southern portion of the country. *Source:* Kinniburgh, D. G., and P. L. Smedley, eds. 2001. Arsenic contamination of groundwater in Bangladesh. Department of Public Health Engineering Bangladesh, British Geological Survey Report.

However, it was not until the late 1990s that large-scale testing of Bangladesh's wells began. By 2001, when the British Geological Survey and the government of Bangladesh published their final report, 3,524 wells had been tested. Of the shallow wells, those less than 150 m (490 ft) deep, 46% exceeded the World Health Organization's maximum recommended level of 10 µg/L (1.33×10^{-6} oz/gal) of arsenic. Extrapolating across all of Bangladesh, the scientists estimated that as many as 2.5 million wells serving 57 million people were contaminated. As the

United States resulted in legislation such as the Federal Water Pollution Control Act of 1972 (later amended and renamed the Clean Water Act in 1977). These acts made it illegal to discharge pollution from a point source without a permit, set standards for industrial wastewater, set standards for contaminant levels in surface waters, and funded construction of sewage treatment plants. Thanks to such legislation, point-source pollution in the United States has been reduced, and rivers and lakes are cleaner than they have been in decades.

Other developed nations have also reduced pollution. In Japan, Singapore, China, and South Korea, legislation, regulation, enforcement, and investment in wastewater treatment have brought striking water quality improvements. However, non-point-source pollution, eutrophication, and acid precipitation remain major challenges.

The Great Lakes of Canada and the United States represent a success story in fighting water pollution. In the 1970s these lakes, which hold 18% of the world's surface freshwater, were badly polluted with wastewater, fertilizers, and toxic chemicals. Algal blooms occurred along beaches, and Lake Erie was pronounced dead. Today, efforts of the Canadian and U.S. governments have paid figure shows, arsenic contamination is most prevalent in southern Bangladesh, but localized hot spots are found in northern regions of the country.

Scientists have not yet reached consensus on the chemical processes by which Bangladesh's shallow aquifers became contaminated. All agree that the arsenic is of natural origin; what remains unclear is how the low levels of arsenic naturally present in soils were dissolved in the aquifers in elemental and highly toxic form. One initial explanation, suggested by Chakraborti and his colleagues, placed most of the blame on agricultural irrigation. By drawing large amounts of water out of aquifers during Bangladesh's dry season, they argued, irrigation had permitted oxygen to enter the aquifers and prompted the release of arsenic from pyrite, a common mineral.

Other scientists contend that pyrite oxidation cannot explain most cases of arsenic contamination. In a 1998 paper in the journal *Nature,* Ross Nickson of the British Geological Survey and his colleagues suggested that arsenic was being released from iron oxides carried into Bangladesh by the Ganges River. They pointed to results of a hydrochemical survey that measured the chemical composition of aquifers throughout Bangladesh. Contrary to the predictions of the pyrite oxidation hypothesis, the survey found that arsenic concentrations tended to increase with aquifer depth and to be inversely correlated with concentrations of sulfur, a component of pyrite. Nickson and his colleagues concluded that highly reducing chemical conditions created by buried organic matter, such as peat, had probably leached arsenic from iron oxides over thousands of years.

Recently, Massachusetts Institute of Technology hydrologist Charles Harvey and colleagues suggested that irrigation may contribute to the arsenic problem after all, but not because of pyrite oxidation. In a 2002 paper in the journal *Science,* they described an experiment in which more than a dozen wells were dug near the capital city of Dhaka. Contrary to the pyrite oxidation hypothesis, and in agreement with Nickson and his colleagues, they found little evidence of a connection between sulfur or oxygen and arsenic.

However, they also found that they could increase arsenic concentrations by injecting organic matter, such as molasses, into their experimental wells. In the process of being metabolized by microbes, the molasses appeared to be freeing arsenic from iron oxides. A similar process might take place naturally, Harvey's team argued, when runoff from rice paddies, ponds, and rivers recharges aquifers that have been depleted by heavy pumping for irrigation. In support of this hypothesis, they found that much of the carbon in the shallow wells was of recent origin. Other scientists, however, have found arsenic in much older waters. This finding suggests that Bangladesh's arsenic problem may be caused by multiple hydrological and geological factors.

off. According to Environment Canada, releases of seven toxic chemicals have been reduced by 71%, municipal phosphorus has been decreased by 80%, and chlorinated pollutants from paper mills are down by 82%. Levels of PCBs and DDE are down by 78% and 91%. Bird populations are rebounding, and Lake Erie is now home to the world's largest walleye fishery. The Great Lakes' troubles are by no means over—sediment pollution is still heavy, PCBs and mercury still settle on the lakes from the air, and fish are not always safe to eat. However, the progress so far shows how conditions can improve when citizens push their governments to take action.

Drinking water is treated before it reaches your tap

Technological advances have also improved our ability to control pollution and mitigate its impacts. The treatment of drinking water and wastewater are widespread and successful mainstream practices in developed nations today. The U.S. EPA sets standards for over 80 drinking water contaminants, which local governments and private water suppliers are obligated to meet. Before being sent to your tap, water from a reservoir or aquifer is treated with chemicals to remove particulate matter; passed through filters of sand, gravel, and charcoal; and/or disinfected with small amounts of an agent such as chlorine.

It is better to prevent pollution than to mitigate it after it occurs

In many cases, solutions to pollution will need to involve prevention, not simply treatment and cleanup. One prominent expert has said that addressing groundwater pollution will require a complete overhaul of the way we live and dispose of waste, not just an "end-of-pipe" approach of piecemeal cleanups. Preventing pollution in the first place would seem to be the best strategy when one considers the other options for dealing with groundwater contamination:

- Filtering groundwater before distributing it can be expensive; facilities in the U.S. Midwest by one estimate spend $400 million annually just to remove the herbicide atrazine from water supplies.
- Pumping water out of an aquifer, treating it, then injecting it back in, repeatedly, takes an impracticably long time. Most Superfund sites with contaminated groundwater are using this method, but work has begun on just over 1% of all sites with heavily polluted groundwater, and the eventual cleanup bill has been estimated at $1 trillion.
- Restricting pollutants on lands above selected aquifers would alleviate contamination in those aquifers—but could simply shift pollution elsewhere.

Although a sea change in our attitudes, behavior, or technology may be necessary to clean up our water completely, there are many things ordinary people can do to help minimize freshwater pollution. One is to exercise the power of consumer choice in the marketplace by purchasing phosphorus-free detergents and other "environmentally friendly" products. Another is to become involved in protecting local waterways. Locally based "riverwatch" groups or watershed associations enlist volunteers to collect data and help state and federal agencies safeguard the health of rivers and other water bodies. Such programs are emerging in many countries throughout the world as citizens and policymakers increasingly demand clean water.

Wastewater and Its Treatment

Wastewater refers to water that has been used by people in some way. It includes water carrying sewage; water from showers, sinks, washing machines, and dishwashers; water used in manufacturing or cleaning processes by businesses and industries; and stormwater runoff.

Although natural systems can process moderate amounts of wastewater, the large and concentrated amounts generated by our densely populated areas can harm ecosystems and pose threats to human health. Thus, attempts are now widely made to treat wastewater before releasing it into the environment.

Municipal wastewater treatment involves several steps

In rural areas, **septic systems** are the most popular method of wastewater disposal. In a septic system, wastewater runs from the house to an underground septic tank, inside which solids and oils separate from water. The clarified water proceeds downhill to a drain field of perforated pipes laid horizontally in gravel-filled trenches underground. The wastewater they emit is decomposed there by microbes. Periodically, solid waste needs to be pumped from the septic tank and taken to a landfill.

In more densely populated areas, municipal sewer systems carry wastewater from homes and businesses to centralized treatment locations. There, pollutants in wastewater are removed by physical, chemical, and biological means (Figure 15.21).

Primary treatment, the physical removal of contaminants in settling tanks or clarifiers, generally removes about 60% of suspended solids from wastewater. Wastewater then proceeds to **secondary treatment,** in which water is aerated by being stirred up, and aerobic bacteria degrade organic pollutants. Roughly 90% of suspended solids are generally removed after secondary treatment.

FIGURE 15.21 Shown here is a generalized process from a modern, environmentally sensitive wastewater treatment facility. Wastewater initially passes through screens to remove large debris and into grit tanks to let grit settle (1). It then enters tanks called primary clarifiers (2), in which solids settle to the bottom and oils and greases float to the top for removal. Clarified water then proceeds to aeration basins (3) that oxygenate the water to encourage decomposition by aerobic bacteria. Water then passes into secondary clarifier tanks (4) for removal of further solids and oils. Next, the water may be purified by chemical treatment with chlorine, passage through carbon filters, and/or exposure to ultraviolet light (5). The treated water (called effluent) may then be piped into natural water bodies, used for urban irrigation, flowed through an artificial wetland, or used to recharge groundwater. In addition, most treatment facilities control odor in the early steps and use anaerobic bacteria to digest sludge removed from the wastewater. Sludge from digesters may be sent to farm fields as fertilizer, and gas from digestion may be used to generate electric power.

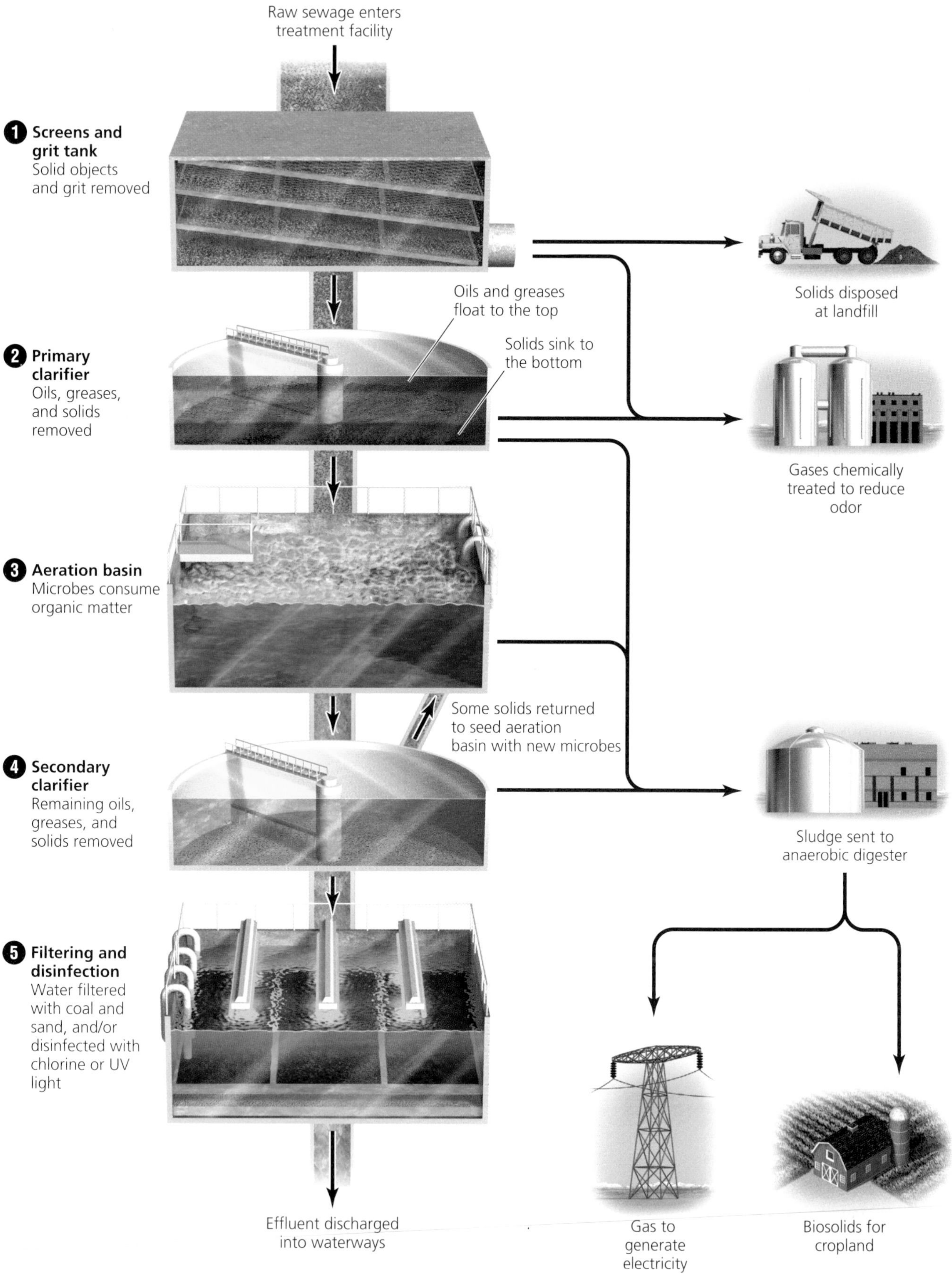
Raw sewage enters treatment facility
1 Screens and grit tank
Solid objects and grit removed
Solids disposed at landfill
Oils and greases float to the top
Solids sink to the bottom
2 Primary clarifier
Oils, greases, and solids removed
Gases chemically treated to reduce odor
3 Aeration basin
Microbes consume organic matter
Some solids returned to seed aeration basin with new microbes
4 Secondary clarifier
Remaining oils, greases, and solids removed
Sludge sent to anaerobic digester
5 Filtering and disinfection
Water filtered with coal and sand, and/or disinfected with chlorine or UV light
Effluent discharged into waterways
Gas to generate electricity
Biosolids for cropland

The Science behind the Story

Using Nature to Treat Our Wastewater

On a stretch of northern California's scenic Redwood Coast, a beautiful marsh hugs the waterfront lining the town of Arcata. Bulrushes, cattails, and wildflowers wave in the breeze. Locals come to jog or walk their dogs. Thousands of birds forage for food in the shallow waters. Few visitors would guess the whole thing is built on sewage.

About 450 km (280 mi) up the coast from San Francisco, the Arcata Marsh and Wildlife Sanctuary is a 121-ha (300-acre) site where science and nature merge to clean up pollution. The marsh is a human-engineered wetland created to filter wastewater. Aquatic plants and microbes perform secondary wastewater treatment, cleaning partially treated sewage water.

Arcata, a town of 16,000 people that sits on Humboldt Bay, built its treatment wetland after requirements of the Federal Water Pollution Control Act took effect in 1974. Arcata's first plan to meet the new requirements called for building a larger treatment plant that would cost more than $50 million and pipe wastewater far out into the ocean. Local residents opposed the plan, pushing for less expensive options more in step with the local environment. Two environmental scientists from nearby Humboldt State University, Robert Gearheart and George Allen, came forward, saying wetlands might provide the treatment Arcata needed.

Gearheart's and Allen's Marsh Pilot Project began in 1979 with 10 small constructed wetlands that together handled 10% of Arcata's wastewater. After 2 years, the pilot study revealed that wetlands could clean wastewater and also restore coastal wetland habitat. Arcata's civic leaders voted to use marshland to solve their sewage treatment problem. Wetland construction started in 1981, and the Arcata marsh system opened in 1986.

Treatment begins with heavy filtering at a conventional wastewater plant. To remove nitrogen compounds, pathogens, and suspended solids that initial sewage treatment fails to catch, Gearheart and Allen engineered a system of plants and microbes in the 20 ha (49 acres) of aeration ponds, where oxygen-breathing bacteria break down sewage, and in the six treatment and enhancement marshes, which receive the water for further filtering. The marshes are lined with a mix of aquatic plants, including duckweed, pennywort, and a native species, hardstem bulrush. The roots and stems of these plants form a dense network of underwater vegetation, hosting a wide variety of single-celled microbes and fungi. An especially productive zone lies in the plants' rhizome network, where tiny root hairs form a thick web that shelters algae, bacteria, and other microscopic sewage-treaters.

Some of the roots and microbes in this network trap suspended solids, holding them until the plants can use them as fertilizer. Other microbes break down nitrogen compounds such as nitrate through denitrification (▸ pp. 199–200). Still other microbes consume coliform bacteria, which may also die when exposed to sunlight in the marsh's

Finally, the clarified water is treated with chlorine, and sometimes ultraviolet light, to kill bacteria. Most often, the treated water, or effluent, is piped into rivers or the ocean following primary and secondary treatment. Sometimes, however, "reclaimed" water is used for lawns and golf courses, for irrigation, or for industrial purposes such as cooling water in power plants.

The solid material that is removed as water is purified throughout the treatment process is termed *sludge.* Sludge is sent to digesting vats, where microorganisms decompose much of the matter. The result, a wet solution of "biosolids," is then dried and either disposed of in a landfill, incinerated, or used as fertilizer on cropland. Each year about 6 million dry tons of sludge are generated in the United States.

Weighing the Issues: Sludge on the Farm

It is estimated that anywhere from 38% to over half of the sewage sludge, or biosolids, produced each year is used as fertilizer on agricultural lands. This practice makes productive use of the sludge, increases crop output, and conserves landfill space, but many people have voiced concern over accumulation of toxic metals, proliferation of dangerous pathogens, and odors. Do you feel this practice represents an efficient use of resources or an unnecessary risk? What further information would you want to know to inform your decision?

shallow waters. Scientists who monitor the marshes know the cleaning system is working when the roots and stems of marsh plants are slimy and slippery: evidence that cleansing microbes are growing and feeding and that the rhizome network is trapping suspended solids. After 2 months of treatment, the wastewater is released into Humboldt Bay.

The marshes are monitored carefully to avoid overwhelming the wetland's cleaning abilities. Too much nitrogen can lead to algal blooms. If initial sewage treatment does not catch contaminants such as lead, these toxins can accumulate and harm microbes or wildlife. Dissolved oxygen, temperature, and nitrogen content are measured regularly, and marsh plants are thinned every few years to keep them from choking out other forms of life in the wetland.

The system's low maintenance needs and simplicity have won awards and saved money. The cost of creating the marshes was approximately $7 million, far less than that of the conventional sewage plant that was first proposed. The marsh treatment system has also brought the Arcata waterfront back to life. More than 100,000 people visit the marsh each year, and more than 250 species of birds have been observed there. Other cities have used the Arcata Marsh and Wildlife Sanctuary as a model for their own wastewater treatment wetlands, and the marsh is a site for ongoing research on wetlands and solutions to water pollution problems.

The Arcata Marsh and Wildlife Sanctuary is the site of an artificially constructed wetland that helps treat this northern Californian city's wastewater.

Artificial wetlands can aid treatment

Natural wetlands already perform the ecosystem service of water purification, and wastewater treatment engineers are now manipulating wetlands and even constructing wetlands *de novo* to employ them as tools to cleanse wastewater. At the same time, these wetland areas serve as havens for wildlife and areas for human recreation.

A project in Arcata, California, was one of the first such attempts (see "The Science behind the Story," above). A more recent example is Sweetwater Wetlands in Tucson, Arizona. The artificial marshes constructed here in 1996 comprise a wetland oasis for birds and wildlife in this desert region while helping to recharge a depleted aquifer. The practice of treating wastewater with artificial wetlands is growing fast; today over 500 artificially constructed or restored wetlands in the United States are performing this service.

Conclusion

Citizen action, government legislation and regulation, new technologies, economic incentives, and public education are all enabling us to confront what will surely be one of the great environmental challenges of the new century:

ensuring adequate quantity and quality of freshwater for ourselves and for the planet's ecosystems.

Accessible freshwater comprises only a minuscule percentage of the hydrosphere, but we have grown used to taking it for granted. With our expanding population and increasing water usage, we are approaching conditions of widespread scarcity. Water depletion and water pollution are already taking a toll on the health, economies, and societies of the developing world, and they are beginning to do so in arid areas of the developed world. There is reason to hope that we may yet attain sustainability in our water usage, however. Potential solutions are numerous, and the issue is too important to ignore.

REVIEWING OBJECTIVES

You should now be able to:

Explain the importance of water and the hydrologic cycle to ecosystems, human health, and economic pursuits

- We depend utterly on drinkable water, and a properly functioning hydrologic cycle is vital to maintaining ecosystems and our civilization. (pp. 434–435)

Delineate freshwater distribution on Earth

- Of all the water on Earth, only about 1% is readily available for our use. (p. 434)
- Water availability varies in space and time, and regions vary greatly in the amounts they possess. (pp. 438–439)

Describe major types of freshwater ecosystems

- The main types of freshwater ecosystems include rivers and streams, wetlands, and lakes and ponds. (pp. 435–437)

Discuss how we use water and alter freshwater systems

- We use water for agriculture, industry, and residential use. The ratio of these uses varies among societies, but globally most is used for agriculture. (pp. 440–441)
- We pump water from aquifers and surface water bodies, sometimes at unsustainable rates. (pp. 441–442)
- We divert the flow of water with canals and irrigation ditches and attempt to control floods with dikes and levees. (pp. 442–446)
- Most of the world's rivers are dammed. Dams bring a diverse set of benefits and costs. Increasingly, people are calling for the removal of some dams. (pp. 446–449)

Assess problems of water supply and propose solutions to address freshwater depletion

- Water tables are dropping worldwide from unsustainable groundwater extraction. Surface water extraction has caused rivers to run dry and some water bodies to shrink. (pp. 441–445)
- Because of inequalities of water distribution amid shrinking supplies, we can expect heightened political tensions over water in the future. (p. 444)
- Solutions to expand supply, such as desalination, are worth pursuing, but not to the exclusion of finding ways to decrease demand. (pp. 449–450)
- Solutions to reduce demand include technology, approaches, and consumer products that increase efficiency in agriculture, industry, and the home. (pp. 450–451)

Assess problems of water quality and propose solutions to address water pollution

- Water pollution stems from point sources and nonpoint sources. (p. 452)
- Water pollutants include excessive nutrients, microbial pathogens, toxic chemicals, sediments, and thermal pollution. (pp. 453–454)
- Scientists who monitor water quality use biological, chemical, and physical indicators. (pp. 454–455)
- Groundwater pollution can be more persistent than surface water pollution. (pp. 455–456)
- Legislation and regulation have improved water quality in developed nations in recent decades. (pp. 456–457)
- Prevention of water pollution is better than mitigation. (p. 458)

Explain how wastewater is treated

- Septic systems are used to treat wastewater in rural areas. (p. 458)
- Wastewater is treated physically, biologically, and chemically in a series of steps at municipal wastewater treatment facilities. (pp. 458–460)
- Artificial wetlands can enhance wastewater treatment while restoring habitat for wildlife. (pp. 460–461)

TESTING YOUR COMPREHENSION

1. Define *groundwater*. What role does groundwater play in the hydrologic cycle?
2. Why are sources of freshwater unreliable for some people and plentiful for others?
3. Describe three benefits of damming rivers, and three costs. What particular environmental, health, and social concerns has China's Three Gorges Dam and its reservoir raised?
4. Why do the Colorado, Rio Grande, Nile, and Yellow rivers now slow to a trickle or run dry before reaching their deltas?
5. Why are water tables dropping around the world? What are some environmental costs of falling water tables?
6. Name three major types of water pollutants, and provide an example of each. List three properties of water that scientists use to determine water quality.
7. Why do many scientists consider groundwater pollution a greater problem than surface water pollution?
8. What are some anthropogenic (human) sources of groundwater pollution?
9. Describe how drinking water is treated. How does a septic system work?
10. Describe and explain the major steps in the process of wastewater treatment. How can artificial wetlands aid such treatment?

SEEKING SOLUTIONS

1. Discuss possible strategies for equalizing distribution of water throughout the world. Consider supply and transport issues. Have our methods of drawing, distributing, and storing water changed very much throughout history? How is the scale of our efforts affecting the availability of water supplies?
2. How can agricultural demand for water be decreased? Describe some ways that we can reduce household water use. How can industrial uses of water be reduced?
3. Have the provisions of the Clean Water Act been effective? Discuss some of the methods we can adopt, in addition to "end-of-pipe" solutions, to prevent contamination and ensure "water security."
4. How might desalination technology help "make" more water? Describe two methods of desalination. Where is this technology being used?
5. Your state's governor has put you in charge of water policy for the state. The aquifer beneath your state has been overpumped, and many wells have gone dry already. Agricultural production last year decreased for the first time in a generation, and farmers are clamoring for you to do something. Meanwhile, the state's largest city is growing so fast that more water is needed for the burgeoning urban population. What policies would you consider to restore your state's water supply? Would you try to take steps to increase supply, to decrease demand, or both? Explain why you would choose such policies.
6. Having solved the water depletion problem in your state, your next task is to deal with pollution of the groundwater that provides your state's drinking water supply. Recent studies have shown that one-third of the state's groundwater has levels of pollutants that violate EPA standards for human health. The federal government is threatening enforcement, and citizens are fearful for their safety. What steps would you consider taking to safeguard the quality of your state's groundwater supply, and why?

INTERPRETING GRAPHS AND DATA

Close to 75% of the freshwater used by people is used in agriculture, and about 1 of every 14 people live where water is scarce, according to a review by hydrologist J. S. Wallace. By the year 2050, scientists project that two-thirds of the world's population will live in water-scarce areas, including most of Africa, the Middle East, India, and China. How much water is required to feed over 6 billion people a basic dietary requirement of 2,700 calories per day? The answer depends on the efficiency with which we use water in agricultural production and on the type of diet we consume.

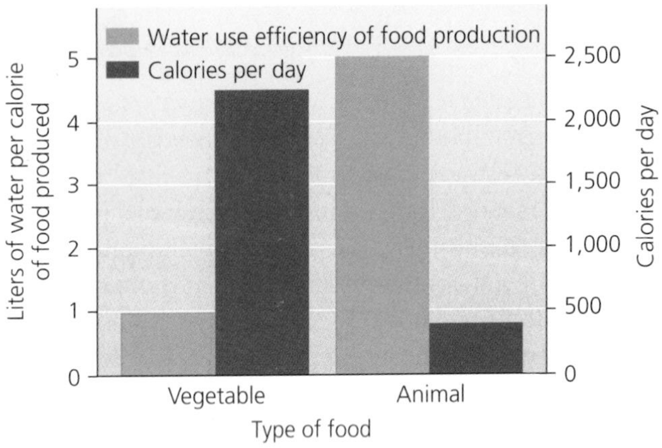

Amount of water needed to produce vegetable and animal food, and global average calories per day consumed of vegetable and animal food. Data from Wallace, J. S. 2000. Increasing agricultural water use efficiency to meet future food production. *Agriculture, Ecosystems and Environment* 82: 105–119.

1. How many liters of water are needed to produce 2,300 calories of vegetable food? How many liters of water are needed to produce 400 calories of animal food? How many liters of water are needed daily to provide this diet? Annually?
2. How many liters of water would be saved daily, compared to the diet in the graph, if the 2,700 calories were provided entirely by vegetables? Annually?
3. Reflect on the quote at the beginning of this chapter: "Water promises to be to the 21st century what oil was to the 20th century: the precious commodity that determines the wealth of nations." How do you think the demographic pressure on the water supply could affect world trade, particularly trade of agricultural products? Do you think it could affect prospects for peace and stability in and among nations? How so?

CALCULATING ECOLOGICAL FOOTPRINTS

In the United States, the EPA estimates that household water use averages 750 liters per person per day. One of the single greatest personal uses of water is for showering. Standard showerheads dispense 15 liters of water per minute, but so-called low-flow showerheads dispense only 9 liters per minute. Given an average daily shower time of 10 minutes, calculate the amounts of water used and saved over the course of a year with standard versus low-flow showerheads, and record your results in the table below.

	Annual water use with standard showerheads (liters)	Annual water use with low-flow showerheads (liters)	Annual water savings with low-flow showerheads (liters)
You	54,750	32,850	21,900
Your class			
Your state			
United States			

Data from U.S. EPA, 1995. *Cleaner water through conservation: Chapter 1—How we use water in the United States.* EPA 841-B-95-002.

1. What percentage of personal water consumption would you calculate is used for showering?
2. How much additional water would you be able to save by shortening your average shower time from 10 minutes to 8 minutes? To 5 minutes?
3. Can you think of any factors that are not being considered in this scenario of water savings? Explain.

Take It Further

Go to www.aw-bc.com/withgott or the student CD-ROM where you'll find:

- Suggested answers to end-of-chapter questions
- Quizzes, animations, and flashcards to help you study
- *Research Navigator*™ database of credible and reliable sources to assist you with your research projects
- GRAPHit! tutorials to help you master how to interpret graphs
- INVESTIGATEit! current news articles that link the topics that you study to case studies from your region to around the world

The Oceans: Natural Systems, Human Use, and Marine Conservation

16

Portion of the Florida Keys

Upon successfully completing this chapter you will be able to:

- Identify physical, geographical, chemical, and biological aspects of the marine environment
- Describe major types of marine ecosystems
- Outline historic and current human uses of marine resources
- Assess human impacts on marine environments
- Review the current state of ocean fisheries and reasons for their decline
- Evaluate marine protected areas and reserves as innovative solutions

Coral reef in the Florida Keys National Marine Sanctuary

Central Case: Seeding the Seas with Marine Reserves

"I saw the fisheries spiral down. There's less fish and they're smaller. It's not anything like it was 15 years ago."
—Don DeMaria, commercial fisherman, Florida Keys

"We have no water-quality problems. We have no problems with our reef. The sanctuary was shoved down our throats."
—Bettye Chaplin, Florida Keys realtor and co-founder of the Conch Coalition

Stretching southwest from the southern tip of Florida for 320 km (200 mi), the string of islands known as the Florida Keys hosts some of North America's richest marine and coastal ecosystems. These islands boast the world's third-largest barrier coral reef, as well as sea grass meadows, coastal mangrove forests, and estuaries. As more people have come to enjoy this remarkable natural environment, tourism in the Keys has grown to 3 million visitors each year.

As tourism was increasing, however, so was human impact. Overfishing, trash and sewage dumping, boat groundings, and careless anchoring were damaging the region's ecosystems while its waters received inputs of pesticides, oil, and heavy metals from roads, residential areas, and farms on the islands. Runoff rich in sediments, fertilizer, and nutrients from leaky septic systems were damaging sea grass beds and causing plankton blooms and eutrophication in Florida Bay, between the Keys and the mainland. All these impacts were harming the coral, living animals whose skeletons give structure to coral reefs, which are home to so much biodiversity. From 1966 to 2000, biologists documented declines in the diversity of living coral and the area covered by coral at two-thirds of the sites they monitored.

These impacts on water quality, sea grass, and coral reefs combined with overfishing to depress fish stocks. Since the 1970s, scientists and fishers alike had reported that the Keys' fish and lobster populations were declining and that the remaining fish and lobsters were smaller. Scientists monitoring the Keys' fish populations concluded that reef fish were severely overexploited by commercial and recreational fishing; a report held that

13 of 16 grouper populations, 7 of 13 snapper populations, and 2 of 5 grunt populations had been overfished.

To protect the area's natural and cultural resources, Congress in 1989 established the Florida Keys National Marine Sanctuary (Figure 16.1). This sanctuary incorporates several previously established protected areas and today safeguards 9,800 km^2 (3,800 mi^2) of marine habitat, including 530 km^2 (205 mi^2) of coral reef, and over 6,000 plant, fish, and invertebrate species. Oil exploration, mining, dumping, and large ships are banned. Fishing is allowed throughout most of the sanctuary, but 24 smaller areas recently zoned as reserves protect 65% of the sanctuary's shallow reef habitat and prohibit all harvesting of natural resources.

Although these 24 reserves amount to only 6% of the sanctuary's area, they have been a magnet for controversy. After the sanctuary was established, a group of Keys residents formed the Conch Coalition to protest the sanctuary and the proposal for no-fishing reserves within it. (A conch is a large marine snail that is a popular food item in the Keys.) The Conch Coalition, which eventually claimed 3,000 supporters, fought the proposal in public meetings and brought a lawsuit. Some Conch Coalition supporters hanged and burned in effigy the sanctuary superintendent and another sanctuary proponent during a protest. Today, however, as fish populations have begun to increase both inside and outside the reserves, most Keys residents have come to support the sanctuary and its no-fishing reserves.

Sanctuaries and reserves are types of *marine protected areas*, portions of ocean that are protected from some human activities. Whereas national parks and other types of protected areas (▸ pp. 364–369) have existed on land for over a century, the oceanic equivalent is quite new. "Most people think it's common sense on land to have areas where we don't hunt," James Bohnsack, a Florida-based National Marine Fisheries Service researcher, explained. "We're now trying to create natural water areas—to see the buffalo roam, so to speak. It's a major change of thinking that protects the ecosystem and biodiversity, but it also protects the fishery."

Oceanography

The oceans cover the vast majority of our planet's surface, and understanding them is crucial for understanding how our planet's systems work. The oceans influence global climate, teem with biodiversity, facilitate transportation

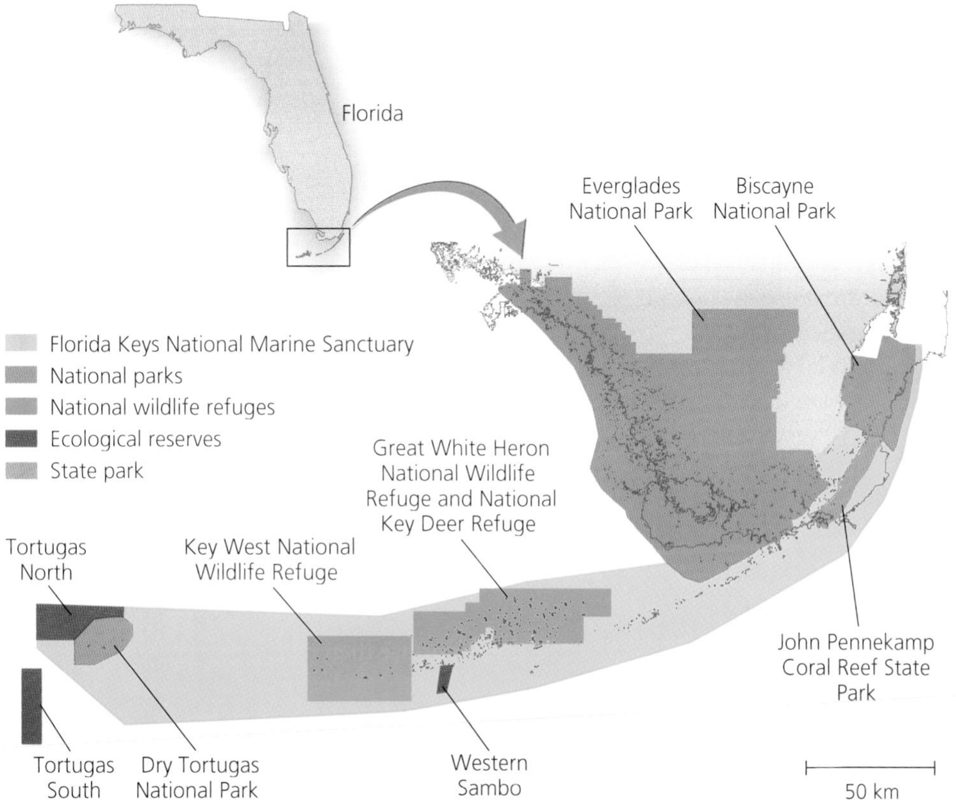

FIGURE 16.1 The Florida Keys National Marine Sanctuary includes over 9,800 km^2 (3,800 mi^2) of mangrove islands, coral reefs, and shallow waters, from the southeastern tip of Florida westward to the distant keys called the Tortugas. The sanctuary is adjacent to or encompasses three national parks, one state park, two national wildlife refuges, and three marine ecological reserves.

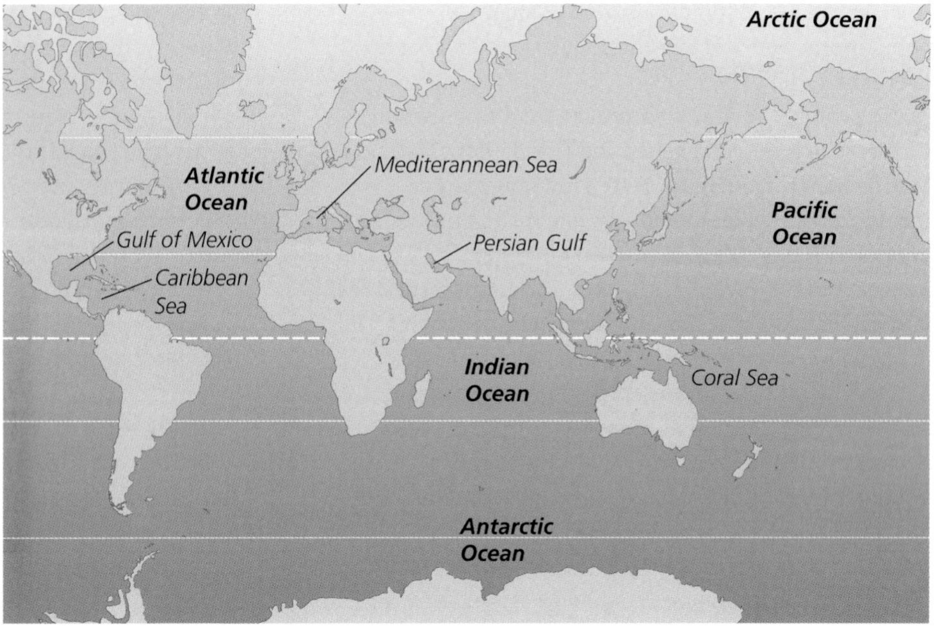

FIGURE 16.2 The world's oceans are connected in a single vast body of water but are given different names for convenience. The Pacific Ocean is the largest and, like the Atlantic and Indian Oceans, includes both tropical and temperate waters. The smaller Arctic and Antarctic Oceans include the waters in the north and south polar regions, respectively. Many smaller bodies of water are named as seas or gulfs; a selected few are shown here.

and commerce, and provide us many resources. Even landlocked areas far from salt water are affected by the oceans and the ways we interact with them. The oceans provide fish for people to eat in Iowa, supply oil to power cars in Ontario, and influence the weather in Nebraska. The study of the physics, chemistry, biology, and geography of the oceans is called **oceanography.**

Oceans cover most of Earth's surface

Although we generally speak of the world's oceans (Figure 16.2) in the plural, giving each major basin a name—Pacific, Atlantic, Indian, Arctic, and Antarctic—all these oceans are connected, comprising a single vast body of water. This one "world ocean" covers 71% of our planet's surface and contains 97.2% of its surface water. The oceans take up most of the hydrosphere, influence the atmosphere, interact with the lithosphere, and encompass a large portion of the biosphere, including at least 250,000 species. The world's oceans touch and are touched by virtually every environmental system and every human endeavor.

The oceans contain more than water

Ocean water contains approximately 96.5% H_2O by mass; most of the remainder consists of ions from dissolved salts (Figure 16.3). Ocean water is salty primarily because ocean basins are the final repositories for water that runs off the land. Rivers carry sediment and dissolved salts from the continents into the ocean, as do winds. Evaporation from the ocean surface then removes pure water, leaving a higher concentration of salts. If we were able to evaporate all the water from the oceans, the world's ocean basins would be covered with a layer of dried salt 63 m (207 ft) thick.

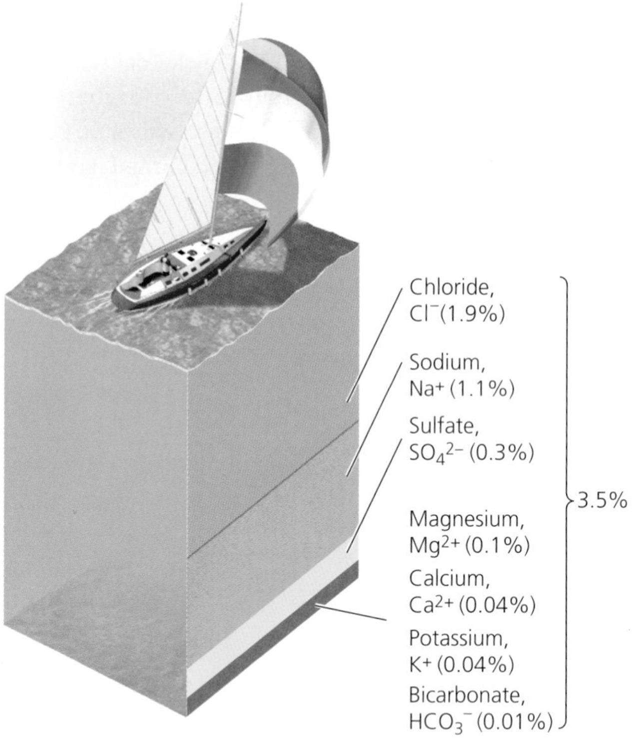

FIGURE 16.3 Ocean water consists of 3.5% salt, by mass. Most of this salt is NaCl in solution, so sodium and chloride ions are abundant. A number of other ions and trace elements are also present.

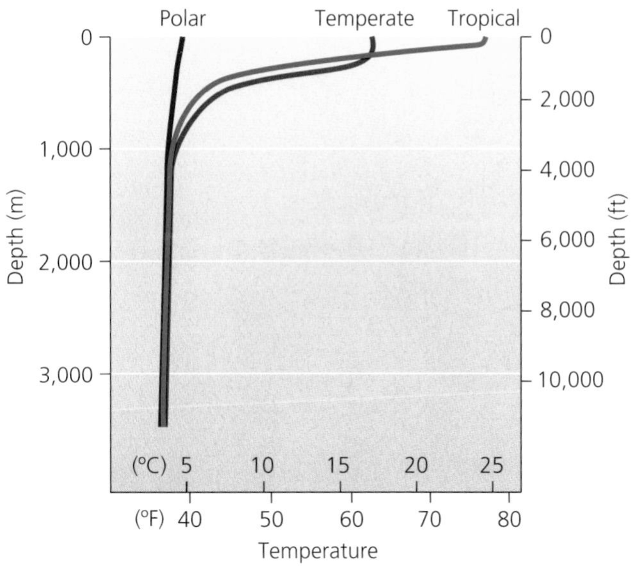

(a) Temperature profiles at polar, temperate, and tropical latitudes

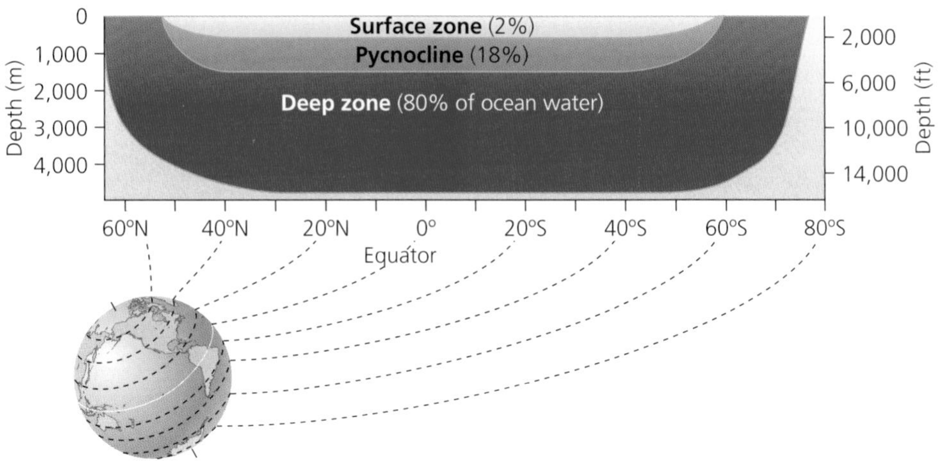

(b) Variation in the average depth of density zones with latitude

FIGURE 16.4 Ocean water varies in temperature and density with depth. Water temperatures (**a**) near the surface are warmer because of daily heating by the sun, and become rapidly colder with depth over the top 1,000 m (3,300 ft). This temperature differential is greatest in the tropics because of intense solar heating and is least in the polar regions. Deep water at all latitudes is equivalent in temperature. Density decreases unevenly with depth, giving rise to several distinct zones (**b**). Waters of the surface zone are well mixed and roughly equivalent in density, whereas density decreases rapidly with depth in the pycnocline. Waters of the deep zone resist mixing and are largely unaffected by sunlight, winds, and storms.

The salinity of ocean water generally ranges from 33 to 37 parts per thousand (ppt), varying from place to place because of differences in evaporation, precipitation, and freshwater runoff from land and glaciers. Salinity near the equator is low because this region has a great deal of precipitation, which is relatively salt-free. In contrast, at latitudes roughly 30–35 degrees north and south, evaporation exceeds precipitation, making the surface salinity of the oceans higher than average.

Besides the dissolved salts shown in Figure 16.3, nutrients such as nitrogen and phosphorus occur in seawater in trace amounts (well under 1 part per million) and play essential roles in nutrient cycling in marine ecosystems. Another aspect of ocean chemistry is dissolved gas content, particularly the dissolved oxygen on which many marine animals depend. Oxygen concentrations are highest in the upper layer of the ocean, reaching 13 ml/L of water. Roughly 36% of the gas dissolved in seawater is oxygen, which is produced by photosynthetic plants, bacteria, and phytoplankton, and by diffusion from the atmosphere.

Ocean water is vertically structured

Surface waters in tropical regions receive more solar radiation and therefore are warmer than surface waters in temperate or polar regions. In all regions, however, temperature declines with depth (Figure 16.4a). Water density by definition increases as salinity rises and as temperature falls. These relationships give rise to different layers of water; heavier (colder and saltier) water sinks, and lighter (warmer and less salty) water remains nearer the surface

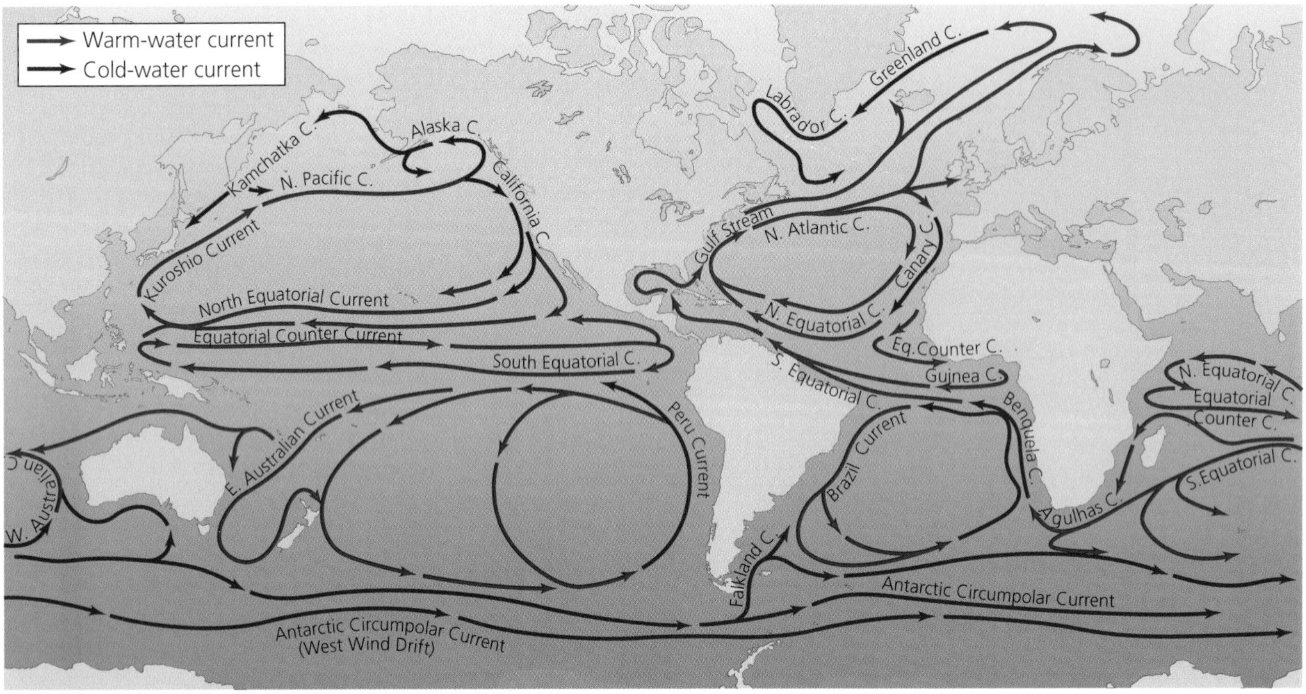

FIGURE 16.5 The upper waters of the oceans move in currents, which are long-lasting and predictable global patterns of water movement. Warm- and cold-water currents interact with the planet's climate system and have been used by people to navigate the oceans for centuries. *Source:* Garrison, T. S. 1999. *Oceanography,* 3rd ed. Belmont, CA: Wadsworth.

(Figure 16.4b). Waters of the surface zone are heated by sunlight each day and are stirred by wind such that they are of similar density throughout, down to a depth of approximately 150 m (490 ft). Below the zone of surface water lies the *pycnocline,* a region in which density increases rapidly with depth. The pycnocline contains about 18% of ocean water by volume, compared to the surface zone's 2%. The remaining 80% resides in the deep zone beneath the pycnocline. The dense water in this zone is sluggish and not affected by winds and storms, sunlight, and daily temperature fluctuations.

Despite the daily heating and cooling of surface waters, ocean temperatures are much more stable than temperatures on land. Midlatitude oceans experience yearly temperature variation of only around 10 °C (18 °F), and tropical and polar oceans are still more stable. The reason for this stability is that water has a very high *heat capacity,* a measure of the heat required to increase temperature by a given amount. It takes much more heat energy to increase the temperature of water than it does to increase the temperature of air by the same amount. High heat capacity enables the oceans to absorb a tremendous amount of heat from the atmosphere. In fact, the heat content of the entire atmosphere is equal to that of just the top 2.6 m (8.5 ft) of the oceans. By absorbing heat and later releasing it to the atmosphere, the oceans help shape Earth's climate (Chapter 18). Also influencing climate is the ocean's surface circulation, a system of currents that move in the pycnocline and the surface zone.

Ocean water flows horizontally in currents

Far from being a static pool of water, Earth's ocean is composed of vast riverlike flows (Figure 16.5) driven by density differences, heating and cooling, gravity, and wind. These surface **currents** move in the upper 400 m (1,300 ft) of water, horizontally and for great distances. These long-lasting patterns influence global climate and play key roles in the phenomena known as El Niño and La Niña (▸ pp. 534, 536–537). They also have been crucial in navigation and human history; currents helped carry Polynesians to Easter Island, Darwin to the Galapagos, and Europeans to the New World. Currents transport heat, nutrients, pollution, and the larvae of many marine species.

Some currents are very slow. Others, like the Gulf Stream, are rapid and powerful. From the Gulf of Mexico, the Gulf Stream moves eastward along the southern edge of the Florida Keys, then northward past Miami at a rate of 160 km per day (nearly 2 m/sec, or over 4.1 mi/hr). An

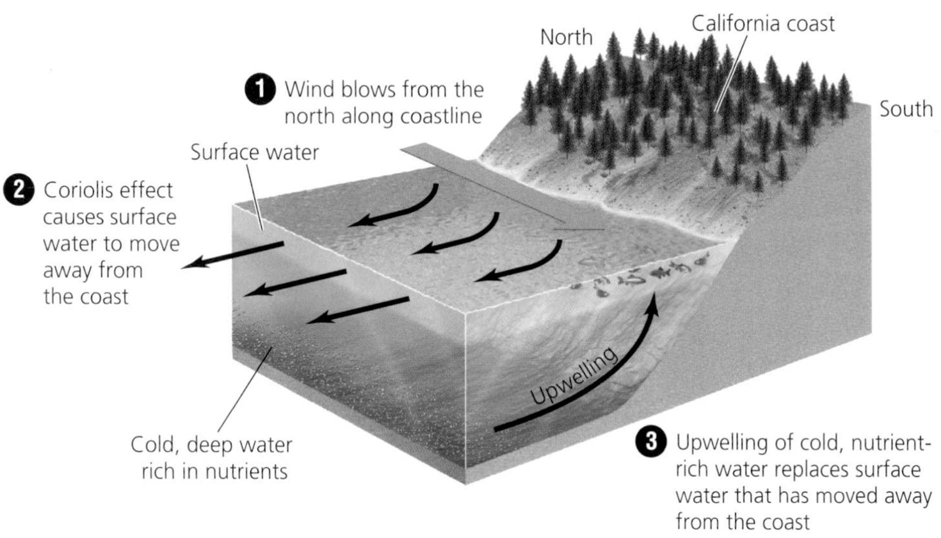
North
California coast
South
1 Wind blows from the north along coastline
Surface water
2 Coriolis effect causes surface water to move away from the coast
Upwelling
Cold, deep water rich in nutrients
3 Upwelling of cold, nutrient-rich water replaces surface water that has moved away from the coast

FIGURE 16.6 Upwelling is the movement of bottom waters upward. This type of vertical current often brings nutrients up to the surface, creating rich areas for marine life. For example, north winds blow along the California coastline, while the Coriolis effect draws wind and water away from the coast. Water is then drawn up from the bottom to replace the water that moves away from shore.

average width of 70 km (43 mi) across, the Gulf Stream continues across the North Atlantic, bringing warm water to Europe and moderating that continent's climate, which otherwise would be much colder.

Vertical movement of water affects marine ecosystems

Surface winds and heating also create vertical currents in seawater. **Upwelling,** the vertical flow of cold, deep water toward the surface, occurs in areas where horizontal currents diverge, or flow away from one another. Because upwelled water is rich in nutrients from the bottom, upwellings are often sites of high primary productivity (▸ pp. 192–193) and lucrative fisheries. Upwellings also occur where strong winds blow away from or parallel to coastlines (Figure 16.6). An example is the California coast, where north winds and the Coriolis effect (▸ pp. 504–505) move surface waters away from shore, raising nutrient-rich water from below and creating a biologically rich region. The cold water also chills the air along the coast, giving San Francisco its famous fog and cool summers.

In areas where surface currents converge, or come together, surface water sinks, a process called **downwelling.** Downwelling transports warm water rich in dissolved gases, providing an influx of oxygen for deep-water life. Vertical currents also occur in the deep zone, where differences in water density can lead to rising and falling convection currents, such as those seen in molten rock (▸ pp. 208–209) and in air (▸ pp. 501–502). The North Atlantic Deep Water (▸ p. 534) is an example of such circulation that has far-reaching effects on global climate.

Weighing the Issues: Why Understand Ocean Currents?

Mapping ocean currents is crucial to understanding where and how the larvae of many fish and marine invertebrates become distributed from place to place. Currents help determine where larvae will settle and mature into adults. Why do you think this information might be important for people to know? What else can be carried by currents? Describe another reason that an understanding of ocean currents can be helpful and important.

Seafloor topography can be rugged and complex

Although oceans are depicted on most maps and globes as smooth, blue swaths, parts of the ocean floor are just as complex as the terrestrial portion of the lithosphere. Underwater volcanoes shoot forth enough magma to build islands above sea level, such as the Hawaiian Islands. Steep canyons similar in scale to Arizona's Grand Canyon lie just offshore of some continents. The deepest spot in the oceans—the Mariana Trench, located in the South Pacific near Guam—is deeper than Mount Everest is high, by over 2.1 km (1.3 mi). Our planet's longest mountain range is under water—the Mid-Atlantic Ridge runs the length of the Atlantic Ocean (Figure 16.7).

We can gain an understanding of the major types of underwater geographic features by examining a stylized map that reflects *bathymetry* (the study of ocean depths)

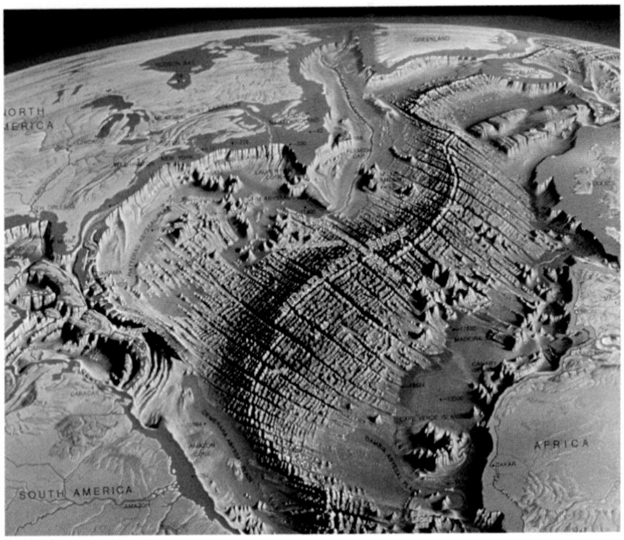

FIGURE 16.7 The seafloor can be every bit as rugged as continental topography. The spreading margin between tectonic plates at the Mid-Atlantic Ridge gives rise to a vast underwater volcanic mountain chain, cross-hatched by immense perpendicular breaks in the oceanic crust.

and topography (physical geography, or the shape and arrangement of landforms) (Figure 16.8). In bathymetric profile, gently sloping **continental shelves** underlie the shallow waters bordering the continents. Continental shelves vary in width from 100 m (330 ft) to 1,300 km (800 mi), averaging 70 km (43 mi) wide, with an average slope of 1.9 m/km (10 ft/mi). These shelves drop off with relative suddenness at the *shelf-slope break.* The *continental slope* angles somewhat steeply downward, connecting the continental shelf to the deep ocean basin below.

Most of the seafloor is flat, but volcanic peaks that rise above the ocean floor provide physical structure for marine animals and are frequently the site of productive fishing grounds. Some island chains, such as the Florida Keys, are formed by reef development and lie atop the continental shelf. Others, such as the Aleutian Islands curving across the North Pacific from Alaska toward Russia, are volcanic in origin, with peaks that rise above sea level. The Aleutians are also the site of a deep trench that, like the Mariana Trench, formed at a convergent tectonic plate boundary, where one slab of crust dives under another in the process of subduction (▸ p. 209).

Oceanic zones differ greatly, and some support more life than others. The uppermost 10 m (33 ft) of ocean water absorbs 80% of the solar energy that reaches its surface. For this reason, nearly all of the oceans' primary productivity occurs in the well-lit top layer, or *photic zone.* Generally, the warm, shallow waters of continental shelves are most biologically productive and support the greatest species diversity. Biological oceanographers, or marine biologists, tend to classify marine habitats and ecosystems into two types. Those occurring between the ocean's surface and floor are **pelagic,** whereas those that occur on the ocean floor are **benthic.** Each of these major areas contains several vertical zones (Figure 16.9).

FIGURE 16.8 A stylized bathymetric profile shows key geologic features of the submarine environment. Shallow regions of water exist around the edges of continents over the continental shelf, which drops off at the shelf-slope break. The relatively steep dropoff called the continental slope gives way to the more gradual continental rise, all of which are underlain by sediments from the continents. Vast areas of seafloor are flat abyssal plain. Seafloor spreading occurs at oceanic ridges, and oceanic crust is subducted in trenches. Volcanic activity along trenches often gives rise to island chains such as the Aleutian Islands. Features on the left side of this diagram are more characteristic of the Atlantic Ocean, and features on the right side of the diagram are more characteristic of the Pacific Ocean. Adapted from Thurman, H. V. 1990. *Essentials of oceanography,* 4th ed. New York: Macmillan.

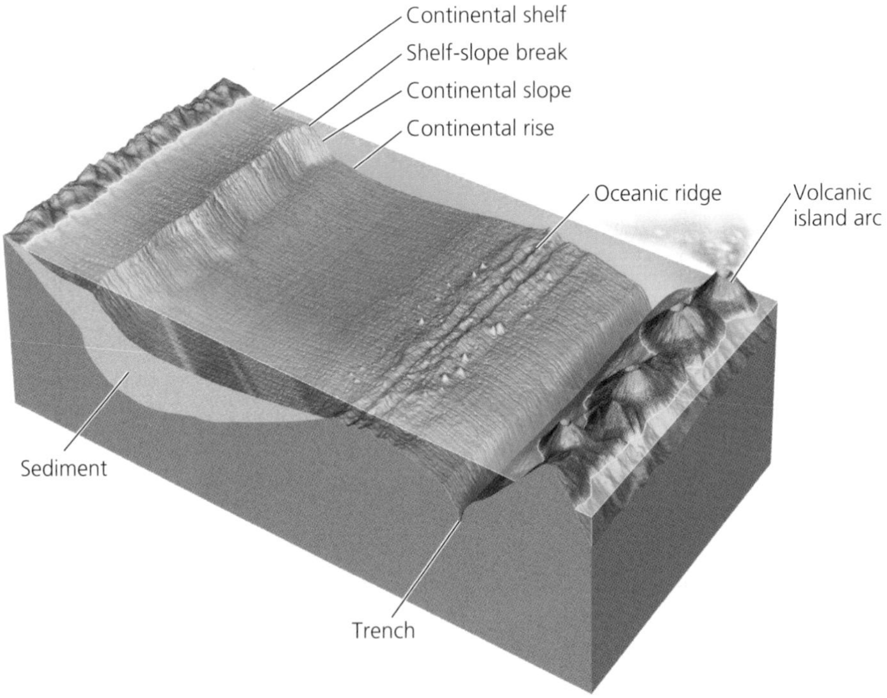

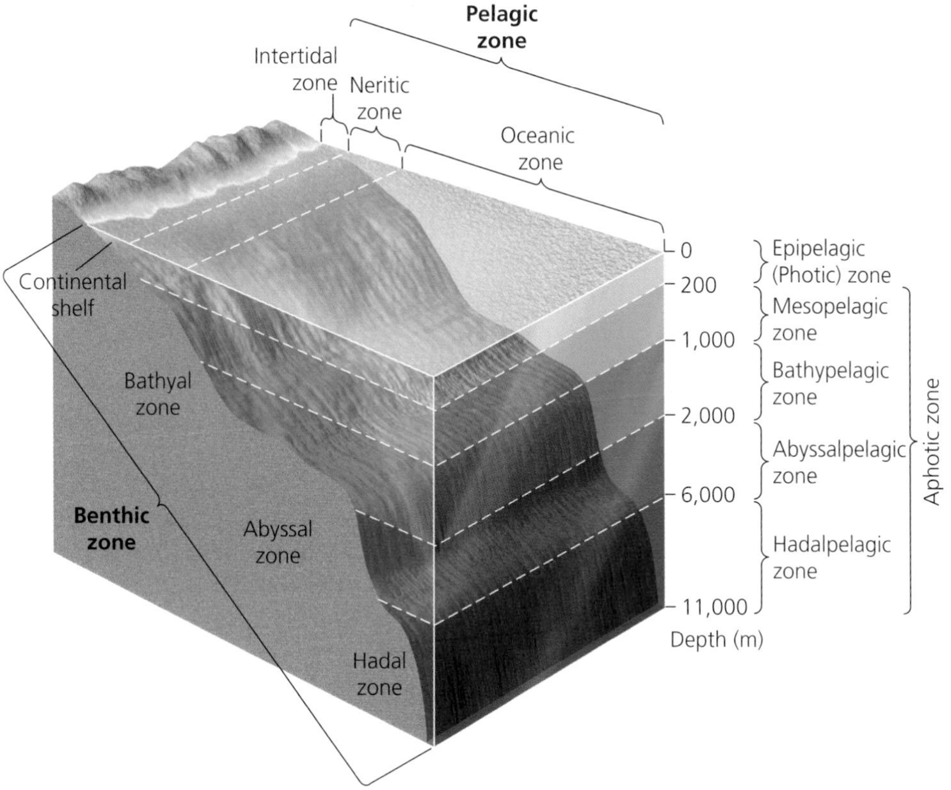

FIGURE 16.9 Oceanographers classify marine habitats into benthic (seafloor) and pelagic (open-water) categories and subdivide these major categories into zones based on depth. Pelagic waters extend from the epipelagic zone at and near the surface down to the hadalpelagic zone at depths below 6,000 m (19,700 ft). Near-surface waters that receive adequate light for photosynthesis are also often termed the *photic zone,* and waters above the continental shelves are also said to be in the *neritic zone.* Benthic zones range from the intertidal zone where the ocean meets the land, to the continental shelves, and down to the bathyal, abyssal, and hadal zones.

Marine Ecosystems

With their variation in topography, temperature, salinity, nutrients, and sunlight, marine environments feature a variety of ecosystems. Most marine ecosystems are powered by solar energy, with sunlight driving photosynthesis by phytoplankton in the photic zone. Yet even the darkest ocean depths host life.

Open-ocean ecosystems vary in their biological diversity

Biological diversity in pelagic areas of the open ocean is highly variable in its distribution. Plant productivity and animal life near the surface are concentrated in regions of nutrient-rich upwelling. Microscopic phytoplankton constitute the base of the marine food chain in the pelagic zone. These photosynthetic algae, protists, and cyanobacteria feed zooplankton, which in turn become food for fish, jellyfish, whales, and other free-swimming animals (Figure 16.10). Predators at higher trophic levels include larger fish, sea turtles, and sharks. In addition, many bird species feed at the surface of the open ocean, returning periodically to nesting sites on islands and coastlines.

In recent years biologists have been learning more about animals of the deep ocean, although tantalizing questions remain and many organisms are not yet discovered. In deep-water ecosystems, animals have adapted to

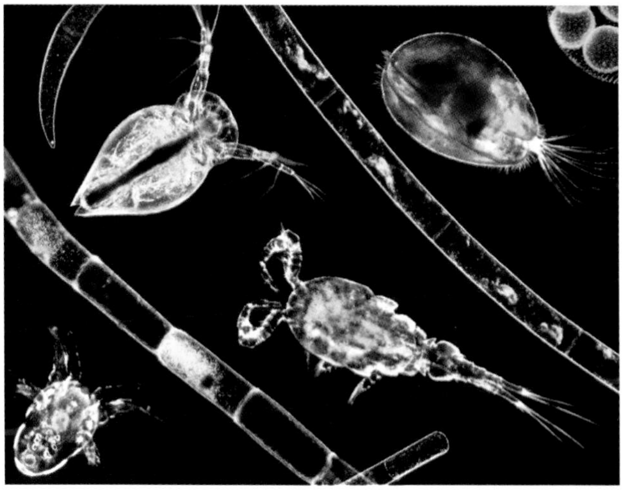

FIGURE 16.10 The uppermost reaches of ocean water contain billions upon billions of phytoplankton—tiny photosynthetic algae, protists, and bacteria that form the base of the marine food chain—as well as zooplankton, tiny animals and protists that dine on phytoplankton and comprise the next trophic level.

FIGURE 16.11 Life is scarce in the dark depths of the deep ocean, but the creatures that do live there often appear bizarre to us. The anglerfish lures prey toward its mouth with a bioluminescent (glowing) organ that protrudes from the front of its head.

FIGURE 16.12 Tall brown algae known as *kelp* grow from the floor of the continental shelf and provide structure with kelp forests, a key marine ecosystem. Numerous fish and other creatures eat kelp or find refuge among its fronds.

deal with extreme water pressures and to live in the dark without food from plants. Some of these often bizarre-looking creatures scavenge carcasses or detritus (organic particles) that fall from above. Others are predators, and still others attain food from symbiotic mutualistic (▸ pp. 155–156) bacteria. Some species carry bacteria that produce light chemically by bioluminescence (Figure 16.11).

Finally, as we explored in Chapter 4 (▸ pp. 106–107), some ecosystems form around hydrothermal vents, where heated water spurts from the seafloor, often carrying minerals that precipitate to form large rocky structures. Tubeworms, shrimp, and other creatures in these recently discovered systems use symbiotic bacteria to derive their energy ultimately from chemicals in the heated water rather than from sunlight. They manage to thrive within the amazingly narrow zones between scalding-hot and icy cold water.

Kelp forests harbor many organisms in temperate waters

Large brown algae, or **kelp** (often nicknamed *seaweed),* grow from the floor of continental shelves, reaching upward toward the sunlit surface. Some kelp reaches 60 m (200 ft) in height and can grow 45 cm (18 in.) in a single day. Dense stands of kelp form underwater forests on the continental shelves in many temperate waters (Figure 16.12). Kelp forests, with their complex structure, supply shelter and food for invertebrates and fish, which in turn provide food for higher-trophic-level predators, such as seals and sharks. Indeed, kelp forests were the setting for our discussion of keystone species in Chapter 6 (▸ pp. 159–165). Recall that sea otters control sea urchin populations, and when otters disappear, urchins overgraze the kelp, destroying the forests and creating "urchin barrens" in their place. Kelp forests also absorb wave energy and protect shorelines from erosion. People in Asian cultures eat some types of kelp, and kelp provides compounds known as alginates, which serve as thickeners in a wide range of consumer products, from cosmetics to paints to paper to soaps.

Coral reefs are treasure troves of biodiversity

In subtropical and tropical waters, kelp forests give way to coral reefs. A reef is an outcrop of rock, sand, or other material that rises near the surface of a relatively shallow body of salt water. A **coral reef** is a mass of calcium carbonate composed of the skeletons of tiny colonial marine organisms known as *corals.* A coral reef may occur as an extension of a shoreline, along a *barrier island* paralleling a shoreline, or as an *atoll,* a ring around a sunken island.

Corals are tiny invertebrate animals related to sea anemones and jellyfish. Some corals have hard external skeletons encompassing a soft hollow body, whereas others have flexible internal skeletons. Corals are sessile (stationary), attached to rock or existing reef and capturing

(a) Elkhorn coral

(b) Moray eel hiding behind seafan

(c) Partially bleached staghorn coral

FIGURE 16.13 Different species of reef-dwelling corals give rise to a variety of distinct structures on reefs in the Florida Keys and elsewhere. One example is elkhorn coral (**a**). Fish take refuge among corals, but reef crevices also hide predators, such as moray eels (**b**), that ambush them. Coral reefs face multiple environmental stresses from natural and human impacts, and many corals have died as a result of coral bleaching, in which corals lose their zooxanthellae. Such bleaching is evident in the whitened portions of this antler-shaped staghorn coral (**c**).

passing food with stinging tentacles. They also derive nourishment from symbiotic algae, known as *zooxanthellae,* that inhabit their bodies and produce food through photosynthesis. Most corals are colonial, and the colorful surface of a coral reef is made up of thousands or millions of densely packed individuals. As corals die, their skeletons remain part of the reef while new corals grow atop them, increasing the size of the reef.

The reefs of the Florida Keys National Marine Sanctuary host many types of coral (Figure 16.13a). Coral reefs absorb wave energy and protect the shore from damage. They also provide essential habitat for many of the Keys' 6,000 species of marine organisms. Globally, coral reefs host as much biodiversity as any other type of ecosystem. The likely reason is that coral reefs provide complex physical structure (and thus many habitats) in shallow nearshore waters, which are regions of high primary productivity. Besides the staggering diversity of anemones, sponges, hydroids, tubeworms, and other sessile invertebrates, innumerable molluscs, flatworms, starfish, and urchins patrol the reefs, and thousands of fish species find food and shelter in reef nooks and crannies. Larger predators, such as grouper and moray eels (Figure 16.13b), feed on the smaller fish.

Coral reefs are experiencing worldwide declines, however. Many reefs have undergone "coral bleaching," a process that occurs when zooxanthellae leave the coral, depriving it of nutrition (Figure 16.13c). Corals lacking zooxanthellae lose color and frequently die, leaving behind ghostly white patches in the reef. Coral bleaching is thought to result from increased sea surface temperatures associated with global climate change, from the influx of pollutants, from unknown natural causes, or from some combination of these factors. Nutrient pollution in coastal waters also promotes the growth of algae, which are blanketing reefs in the Florida Keys and many other regions. Coral reefs also sustain damage when divers stun fish with cyanide to capture them for food or for the pet trade, a common practice in waters of Indonesia and the Philippines. It is estimated that for each fish caught in this way, cyanide poisoning destroys one square meter of reef.

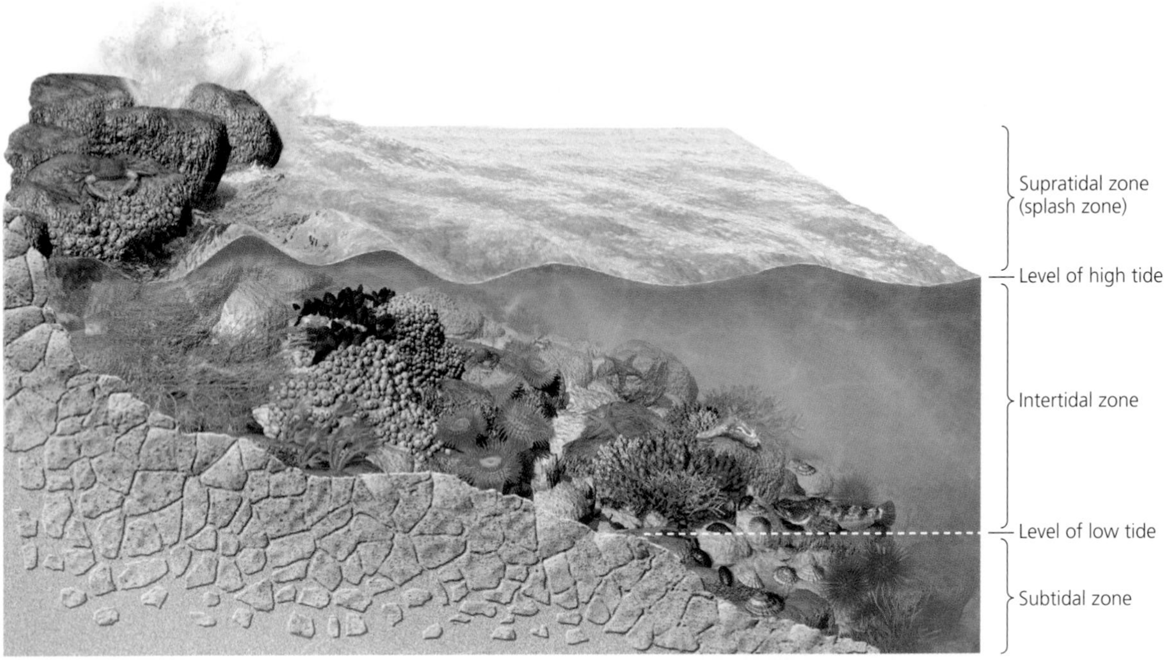

(a) Tidal zones

(b) Tidepools at low tide

FIGURE 16.14 The rocky intertidal zone is the region along a rocky shoreline between the lowest and highest reaches of the tides (**a**). This is an ecosystem rich in biodiversity, typically containing large invertebrates such as seastars (starfish), barnacles, crabs, sea anemones, corals, bryozoans, snails, limpets, chitons, mussels, nudibranchs (sea slugs), and sea urchins. Fish swim in tidal pools (**b**), and many types of algae cover the rocks. Areas higher on the shoreline are exposed to the air more frequently and for longer periods, so those organisms that can tolerate exposure best specialize in the upper intertidal zone. The lower intertidal zone is less frequently exposed and for shorter periods, so organisms less able to tolerate exposure thrive here.

A few coral species thrive in waters outside the tropics and build reefs on the ocean floor at depths of 200–500 m (650–1,650 ft). These little-known reefs, which occur in cold-water areas off the coasts of Norway, Spain, the British Isles, and elsewhere, are only now beginning to be studied by scientists. Already, however, many have been badly damaged by trawling, the fishing practice in which heavy nets are dragged over the seafloor (▸ p. 485). Norway and other countries are now beginning to protect some of these deep-water reefs.

Weighing the Issues: The Coral Crisis

Some scientists are exploring new technologies to enhance coral growth and the abundance of reef-inhabiting organisms, including chemical cues to attract larvae and special wavelengths of light to attract fish. Do you think such technology could offer adequate solutions to the global decline of coral reefs? Why or why not? What other solutions can you suggest to address the problems facing coral reefs? What challenges might each of these solutions encounter?

Intertidal zones undergo constant change

Where the ocean meets the land, **intertidal,** or **littoral,** ecosystems (Figure 16.14) lie along shorelines between the farthest reach of the highest tide and the lowest reach of the lowest tide. **Tides** are the periodic rising and falling of the ocean's height at a given location, caused by the gravitational pull of the moon and sun. High and low tides occur roughly 6 hours apart, although three overlapping tidal cycles make the timing and height of tides complex. Intertidal organisms spend part of each day submerged in water, part of the day dry and exposed to the air and sun, and part of the day being lashed and beaten by waves. Subject to tremendous extremes in temperature, moisture, sun exposure, and salinity, these creatures must also protect themselves from marine predators at high tide and terrestrial predators at low tide.

The intertidal environment is a tough place to make a living, but it is home to a remarkable diversity of organisms. Rocky shorelines can be full of life among the crevices, which provide shelter and pools of water (tide pools) during low tides. Sessile animals such as anemones, mussels, and barnacles live attached to rocks, filter-feeding on plankton in the water that washes over them. Urchins, sea slugs, chitons, and limpets eat intertidal algae or scrape food from the rocks. Starfish creep slowly along, preying on the filter-feeders and herbivores at high tide. Crabs clamber around the rocks, scavenging detritus. At low tide, birds, raccoons, and other land animals come by and dine on exposed animals.

The rocky intertidal zone is so diverse because environmental conditions, including temperature, salinity, and the presence or absence of water, change dramatically from the highest to the lowest reaches. This environmental variation gives rise to conspicuous bands formed by dominant organisms as they array themselves according to their habitat needs. Sandy intertidal areas, such as those of the Florida Keys, host less biodiversity, yet plenty of organisms burrow into the sand at low tide to await the return of high tide, when they emerge to feed.

Salt marshes cover large areas of coastline in temperate areas

Along many of the world's coastlines at temperate latitudes, **salt marshes** occur where the tides wash over gently sloping sandy or silty substrates. Rising and falling tides flow into and out of channels called *tidal creeks* and at highest tide spill over onto elevated marsh flats (Figure 16.15). Marsh flats grow thick with grasses, rushes, shrubs, and other herbaceous plants. Grasses such as those in the genera *Spartina* and *Distichlis* comprise the dominant vegetation in most salt marshes. Salt marshes boast very high primary productivity and provide critical habitat for shorebirds, waterfowl, and the adults and young of many commercially important fish and shellfish species. In many parts of the world, people have altered salt marshes

FIGURE 16.15 Salt marshes occur in temperate intertidal zones where the substrate is muddy, allowing salt-adapted grasses to grow. Tidal waters generally flow through marshes in channels called *tidal creeks,* amid flat areas called *benches,* sometimes partially submerging the grasses. In this salt marsh in Delaware, people have cut linear ditches to drain shallow areas to control mosquito populations.

to make way for coastal shipping, industrial facilities, farming, and other development.

Mangrove forests line coastlines in the tropics and subtropics

In tropical and subtropical latitudes, mangrove forests replace salt marshes along gently sloping sandy and silty coasts. **Mangroves** are trees with unique types of roots, some of which curve upward like snorkels to attain oxygen lacking in the mud, and some of which curve downward, serving as stilts to support the tree in changing water levels (Figure 16.16). Fish, shellfish, crabs, snakes, worms, and other organisms thrive among the root networks, and birds find habitat for feeding and nesting in the dense foliage of these coastal forests. Mangroves also provide materials that people use for food, medicine, tools, and construction. Mangroves just reach the southern edge of the United States, and the Florida Keys National Marine Sanctuary contains tens of thousands of acres forested by three species of mangrove, each adapted to specialize in a different tidal zone.

In south Florida and elsewhere, mangrove forests have been destroyed as people have converted coastal areas for residential, recreational, and commercial uses. It is estimated that people have eliminated half the world's mangrove forests and that mangrove forest area continues to decline by 2–8% per year. When mangroves are

FIGURE 16.16 Mangrove forests are important ecosystems along tropical and subtropical coastlines throughout the world. Mangrove trees, such as these at Lizard Island, Australia, show specialized adaptations for growing in salt water and provide habitat for many types of fish, birds, crabs, and other animals.

removed, coastal areas lose the ability to slow runoff, filter pollutants, and retain soil. As a result, offshore systems such as eelgrass beds and coral reefs are more readily degraded. Only about 1% of the world's remaining mangroves benefit from some sort of protection against development.

Freshwater meets salt water in estuaries

Many salt marshes and mangrove forests occur in or near **estuaries,** water bodies where rivers flow into the ocean, mixing freshwater with salt water. Biologically productive ecosystems, estuaries experience significant fluctuations in salinity as tidal currents and freshwater runoff vary daily and seasonally. For shorebirds and for many commercially important shellfish species, estuaries provide critical habitat. For anadromous fishes (those, like salmon, that spawn in freshwater and mature in salt water), estuaries provide a transitional zone where young fish make the passage from freshwater to salt water.

Estuaries around the world have been affected by urban and coastal development, water pollution, habitat alteration, and overfishing. The estuary of Florida Bay, where freshwater from the Everglades system mixes with salt water, has suffered pollution and a reduction in freshwater input due to irrigation and fertilizer use by sugarcane farmers, housing development, septic tank leakage, and other human impacts. Coastal ecosystems have borne the brunt of human impact because two-thirds of Earth's people choose to live within 160 km (100 mi) of the ocean.

Human Use and Impact

Our species has a long history of interacting with the oceans. We have long traveled across their waters, clustered our settlements along coastlines, and been fascinated by the beauty, power, and vastness of the seas. We have also left our mark upon them by exploiting oceans for their resources and polluting them with our waste.

Oceans provide transportation routes

The oceans have provided transportation routes for thousands of years and continue to provide affordable means of moving people and products over vast distances. Ocean shipping has accelerated the global reach of some cultures and has promoted interaction among long-isolated peoples. It has had substantial impacts on the environment as well. The thousands of ships plying the world's oceans today carry everything from cod to cargo containers to crude oil. Ships transport ballast water as well, which, when discharged at ports of destination, may transplant aquatic organisms picked up at ports of departure. Some of these species—such as the zebra mussel (Chapter 6)—establish and become invasive in their new homes.

We extract energy and minerals

We use the oceans as sources of commercially valuable energy. By the 1980s about 25% of our production of crude oil and natural gas came from exploitation of deposits

FIGURE 16.17 Crude oil and natural gas from beneath the seafloor are some of the economically valuable resources that we take from the oceans. Offshore petroleum drilling at platforms such as these off Santa Barbara, California, creates some of the many human impacts on the marine environment.

beneath the seafloor (Figure 16.17). According to recent estimates, offshore areas may contain as much as 2 trillion barrels of oil, roughly half the amount known to exist underground on land. The exploitation of oil and gas deposits in the Gulf of Mexico has triggered debate in recent years. The risk of an oil spill in the Florida Keys National Marine Sanctuary concerns many sanctuary supporters.

Ocean sediments also contain a novel potential source of fossil fuel energy. **Methane hydrate** is an ice-like solid consisting of molecules of methane (CH_4, the main component of natural gas) embedded in a crystal lattice of water molecules. Methane hydrates are stable at temperature and pressure conditions found in many sediments on the Arctic seafloor and the continental shelves. The U.S. Geological Survey estimates that the world's deposits of methane hydrates may hold twice as much carbon as all known deposits of oil, coal, and natural gas combined. Some people hope that methane hydrates can be developed as an energy source to power our civilization through the 21st century and beyond. However, a great deal of research remains before scientists and engineers can be sure of how to extract these energy sources safely. Destabilizing a methane hydrate deposit could lead to a catastrophic release of gas, which could cause a massive landslide and tsunami (tidal wave). This would also release huge amounts of methane, a potent greenhouse gas, into the atmosphere, exacerbating global climate change.

The oceans also hold potential for providing renewable energy sources that do not emit greenhouse gases. Engineers have developed ways of harnessing energy from waves, tides, and the heat of ocean water (▸ pp. 635–637). These promising energy sources are awaiting further research, development, and investment.

We extract minerals from the ocean floor, as well. By using large vacuum-cleaner-like hydraulic dredges, miners collect sand and gravel from beneath the sea. Also extracted are sulfur from salt deposits in the Gulf of Mexico and phosphorite from offshore areas near the California coast and elsewhere. Other valuable minerals found on or beneath the seafloor include calcium carbonate (used in making cement), silica (used as fire-resistant insulation and in manufacturing glass), and rich deposits of copper, zinc, silver, and gold ore. Many minerals are found concentrated in manganese nodules, small ball-shaped accretions that are scattered across parts of the ocean floor. It is estimated that over 1.5 trillion tons of manganese nodules exist in the Pacific Ocean alone and that their reserves of metal exceed all terrestrial reserves. The logistical difficulty of mining them, however, has kept their extraction uneconomical so far.

Marine pollution threatens resources

People have long made the oceans a sink for waste and pollution. Even into the mid-20th century, it was common for coastal cities in the United States to dump trash and untreated sewage along their shores. Fort Bragg, a bustling town on the northern California coast, boasts of its Glass Beach, an area where beachcombers can collect sea glass, the colorful surf-polished glass sometimes found on beaches after storms. Glass Beach is in fact the site of the former town dump, and besides well-polished glass, the perceptive visitor may also spot old batteries, rusting car frames, and all other manner of trash protruding from the bluffs above the beach.

Oil, plastic, industrial chemicals, and excess nutrients all eventually make their way from land into the oceans. Raw sewage and trash from cruise ships and abandoned fishing gear from fishing boats add to the input. The scope of trash in the sea can be gauged by the amount picked up each September by volunteers who trek beaches in the Ocean Conservancy's annual International Coastal Cleanup. In this nonprofit organization's 2004 cleanup, 300,000 people from 88 nations picked up 3.5 million kg (7.7 million lb) of trash from 17,700 km (11,000 mi) of shoreline.

Nets and plastic debris endanger marine life

Plastic bags and bottles, fishing nets, gloves, fishing line, buckets, floats, abandoned cargo, and much else that people transport on the sea or deposit into it can harm marine organisms. Because most plastic is not biodegradable, it can drift for decades before washing up on beaches. Marine mammals, seabirds, fish, and sea turtles may mistake floating plastic debris for food and can die as a result of ingesting material they cannot digest or expel. Fishing nets that are lost or intentionally discarded can continue snaring animals for decades.

Of 115 marine mammal species, 49 are known to have eaten or become entangled in marine debris, and 111 of 312 species of seabirds are known to ingest plastic. Sea turtles of five species in the Gulf of Mexico have died from consuming or contacting marine debris. Marine debris harms people, as well. A survey of fishers off the Oregon coast indicated that more than half had encountered equipment damage or other problems from plastic debris, and debris has caused over $50 million dollars in insurance payments.

Oil pollution comes not only from massive spills

Major oil spills, such as the *Exxon Valdez* spill in Prince William Sound, Alaska (Chapter 4), make headlines and cause serious environmental problems. Yet it is important to put such accidents into perspective. The majority of oil pollution in the oceans comes not from large spills in a few particular locations, but from the accumulation of innumerable widely spread small sources, including leakage from small boats and runoff from human activities on land. In addition, the amount of petroleum spilled into the oceans in recent years is equaled by the amount that seeps into the water from naturally occurring seafloor deposits (Figure 16.18a).

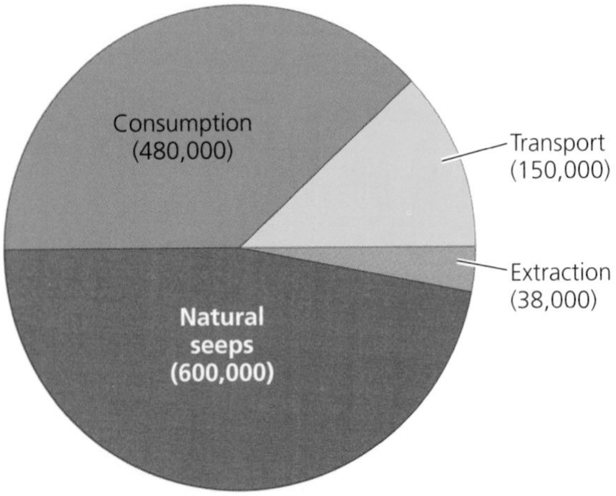

(a) Sources of petroleum input into oceans (metric tons)

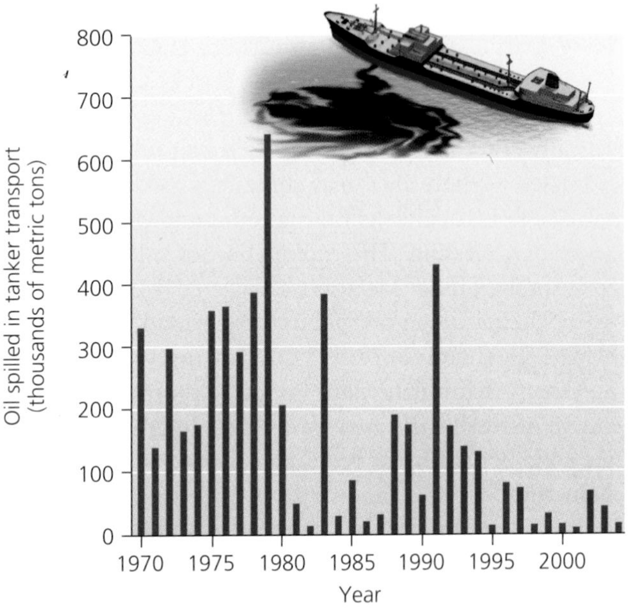

(b) Quantity of petroleum spilled from tankers, 1970–2004

FIGURE 16.18 Of the 1.3 million metric tons of petroleum spilled into the world's oceans each year, nearly half is from natural seeps **(a)**. Petroleum consumption by people accounts for 38% of total input, and this includes numerous diffuse non-point sources, especially runoff from rivers and coastal communities and leakage from two-stroke engines. Spills during petroleum transport account for 12%, and leakage during petroleum extraction accounts for 3%. Less oil is being spilled into ocean waters today in large tanker spills, thanks in part to regulations on the oil shipping industry and improved spill response techniques **(b)**. The figure shows cumulative quantities of oil spilled worldwide from nonmilitary spills over 7 metric tons. Data from: (a) National Research Council. 2003. *Oil in the sea III. Inputs, fates, and effects*. Washington, DC: National Academies Press. (b) International Tanker Owners Pollution Federation Ltd. 2005. *Oil tanker spill statistics: 2004*. London: ITOPF.

Nonetheless, minimizing the amount of oil we release into coastal waters is important, because petroleum pollution is detrimental to the marine environment and the human economies that draw sustenance from that environment. Petroleum can physically coat and kill marine organisms, and ingested chemical components in petroleum can poison marine life. In response to headline-grabbing oil spills, governments around the world have begun to implement more stringent safety standards for tankers, such as requiring industry to pay for tugboat escorts in sensitive and hazardous coastal waters, and to develop prevention and response plans for major oil spills.

The U.S. Oil Pollution Act of 1990 created a $1 billion prevention and cleanup fund and required that by 2015 all oil tankers in U.S. waters be equipped with double hulls as a precaution against puncture. The oil industry has resisted many such safeguards, and today the ship that oiled Prince William Sound is still plying the world's oceans, renamed the *Sea River Mediterranean,* and still featuring only a single hull. However, over the past three decades, the amount of oil spilled in U.S. and global waters has decreased (Figure 16.18b), in part because of an increased emphasis on spill prevention and response.

Excess nutrients cause algal blooms

Pollution from fertilizer runoff or other nutrient inputs can have dire effects on marine ecosystems, as we saw with the Gulf of Mexico's dead zone in Chapter 7. The release of excess nutrients into surface waters can spur unusually high growth rates and population densities of phytoplankton, causing eutrophication in freshwater and saltwater systems. Such problems have occurred in the Florida Keys, leading the U.S. Environmental Protection Agency (EPA) in 2001 to prohibit the discharge of sewage and other waste from boats into the waters of the Florida Keys National Marine Sanctuary.

Excessive nutrient concentrations sometimes give rise to population explosions among several species of marine algae that produce powerful toxins that attack the nervous systems of vertebrates. Blooms of these algae are known as **harmful algal blooms.** Some algal species produce reddish pigments that discolor surface waters, and blooms of these species are nicknamed **red tides** (Figure 16.19). Harmful algal blooms can cause illness and death among zooplankton, birds, fish, marine mammals, and humans as their toxins are passed up the food chain. They also cause economic loss for communities dependent on fishing or beach tourism. Reducing nutrient runoff into coastal waters can lessen the risk of these outbreaks, and health impacts can be minimized by monitoring to prevent human consumption of affected organisms.

As severe as the impacts of marine pollution can be, however, most marine scientists concur that the more worrisome dilemma is overharvesting. Unfortunately, the old cliché that "there are always more fish in the sea" appears not to be true; the oceans today have been overfished, and many fishing stocks have been largely depleted.

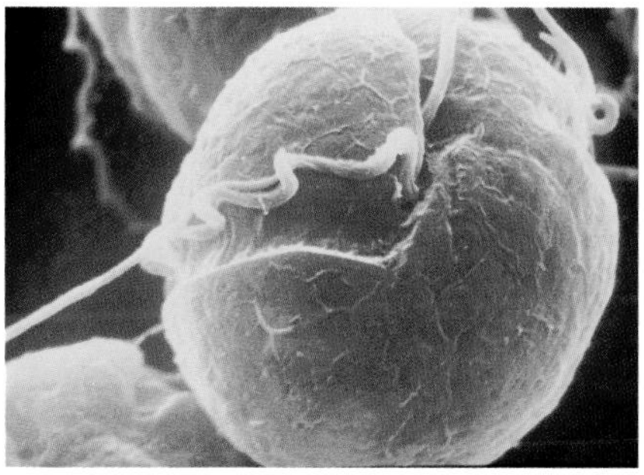

(a) Dinoflagellate (*Gymnodinium*)

FIGURE 16.19 In a harmful algal bloom, certain types of algae multiply to great densities in surface waters, producing toxins that can bioaccumulate and harm organisms. Red tides are a type of harmful algal bloom in which the algae, such as dinoflagellates of the genus *Gymnodinium* (**a**), produce pigment that turns the water red (**b**).

(b) Red tide, Gulf of Carpentaria, Australia

Emptying the Oceans

The oceans and their biological resources have provided for human needs for thousands of years, but today we are placing unprecedented pressure on marine resources. Half the world's marine fish populations are fully exploited, meaning that we cannot harvest them more intensively without depleting them, according to a 2004 U.N. Food and Agriculture Organization (FAO) report. An additional 25% of marine fish populations are overexploited and already being driven toward extinction, the FAO reported. Thus only one-quarter of the world's marine fish populations can yield more than they are already yielding without being driven into decline. Total global fisheries catch, after decades of increases, leveled off after about 1988 (Figure 16.20).

As our population grows, we will become even more dependent on the oceans' bounty. Existing fishing practices are not sustainable given present consumption rates, many scientists and fisheries managers have concluded. This makes it vital, they say, that we take immediate steps to modify our priorities and improve our use of science in fisheries management.

Overfishing is nothing new

People have harvested fish, shellfish, turtles, seals, and other animals from the oceans for millennia. Archaeological evidence from ancient coastal communities reveals shellfish-rich diets, and many sites around the world include vast middens, or piles, of discarded oyster and clam shells. Although much of this harvesting may have been sustainable, paleoecologists are learning that the depletion of marine animals by humans did not begin with today's industrialized fishing fleets. Rather, overfishing took a toll on a variety of marine species beginning centuries or millennia ago. It then accelerated during the colonial period of European expansion and intensified further in the 20th century.

A recent synthesis of historical evidence by marine biologist Jeremy Jackson and others revealed that ancient overharvesting likely affected ecosystems in astounding ways we only partially understand today. Several large animals, including the Caribbean monk seal, Steller's sea cow, and Atlantic gray whale, were hunted to extinction long enough ago that scientists never studied them or the ecological roles they played. Overharvesting of the vast oyster beds of Chesapeake Bay led to the collapse of its oyster fishery in the late 19th century. Eutrophication and hypoxia similar to that of the Gulf of Mexico (Chapter 7) have resulted, because there are no longer oysters to filter algae and bacteria from the water.

Florida Bay, according to Jackson and his colleagues, suffers today from the results of overhunting of green sea turtles in past centuries. The once-abundant turtles ate sea grass (often called turtle grass) and likely kept it cropped low, like a lawn. But with today's turtle population a tiny fraction of what it once was, sea grass grows thickly, dies, and rots, giving rise to disease like sea grass wasting disease, which ravaged Florida Bay in the 1980s.

A better-known case of historical overharvesting is the near-extinction of many species of whales. This resulted from commercial whaling that began centuries ago and was curtailed only in 1986. Although overfishing has a long history, the industrialized methods, new technology, and global reach of today's commercial fleets are making our impacts much more rapid and far-reaching than in past centuries.

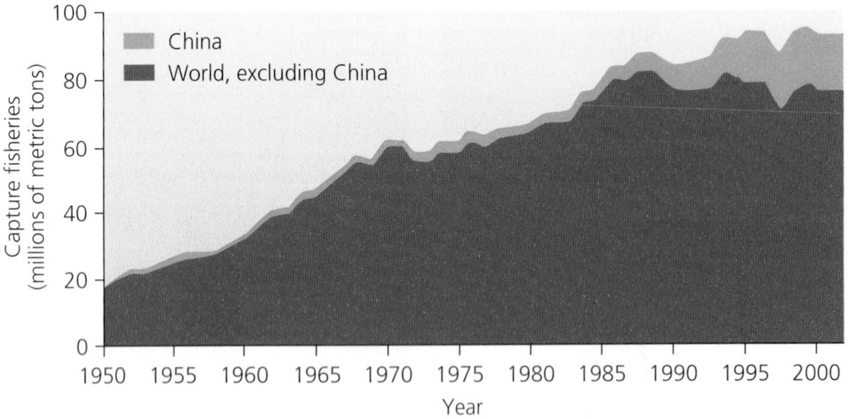

FIGURE 16.20 The total global fisheries catch has increased over the past half-century, but in recent years growth has stalled, and many fear that a global catch decline is imminent if conservation measures are not taken soon. The figure shows trends with and without China's data (see "The Science behind the Story," ▸ pp. 486–487). With China's data, global catch has leveled off since the mid-1990s. Without China's data, catch has decreased slightly since 1988. Go to GRAPHit! at www.aw-bc.com/withgott or on the student CD-ROM. Data from U.N. Food and Agricultural Organization (FAO). 2004. *World review of fisheries and aquaculture*. Rome: FAO.

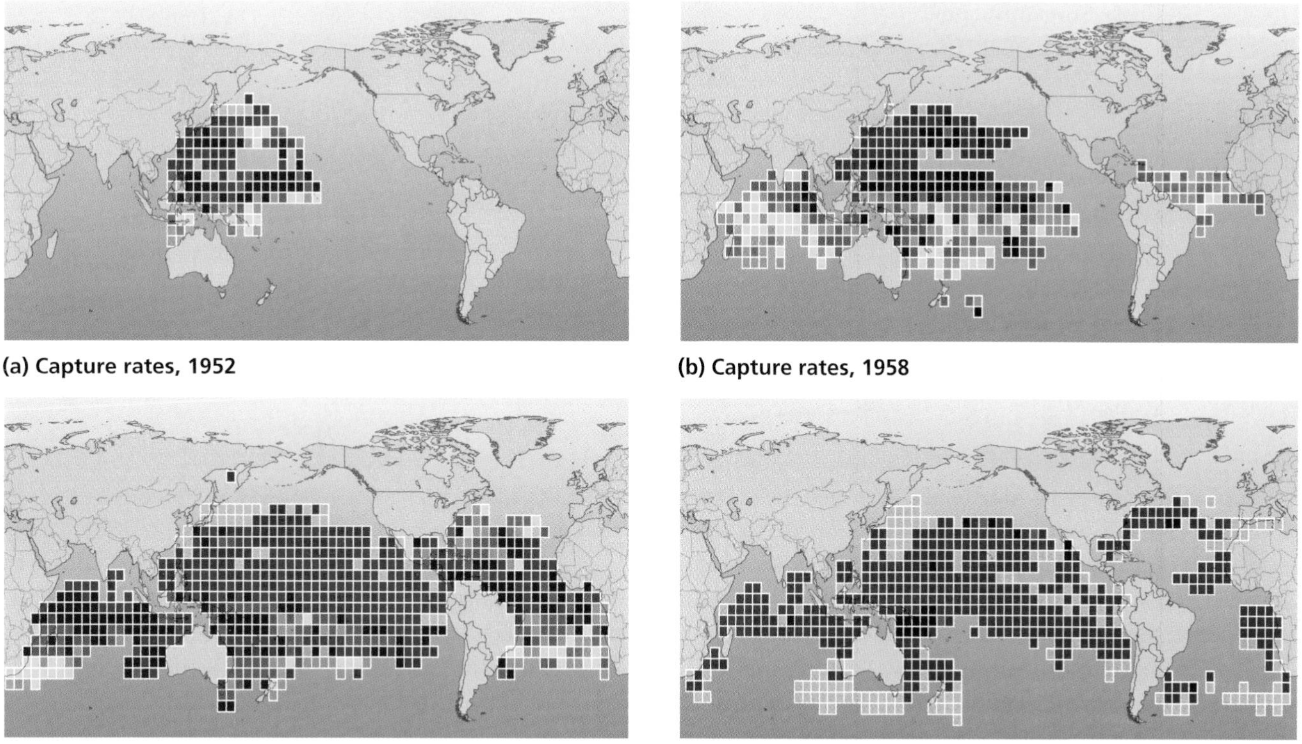

(a) Capture rates, 1952

(b) Capture rates, 1958

(c) Capture rates, 1964

(d) Capture rates, 1980

Number of fish caught per 100 hooks

<1 1-2 2-3 3-4 4-5 5-6 6-8 8-10 10+

FIGURE 16.21 As industrial fishing fleets reached each new region of the world's oceans, capture rates of large predatory fishes were initially high and then within a decade declined markedly. In the figure, reds and oranges signify high capture rates, and maroons and purples signify low capture rates. High capture rates in the southwestern Pacific in 1952 (**a**) gave way to low ones in later years. Excellent fishing success in the tropical Atlantic and Indian Oceans in 1958 (**b**) had turned mediocre by 1964 (**c**) and poor by 1980 (**d**). High capture rates in the north and south Atlantic in 1964 (**c**) gave way to low capture rates there in 1980 (**d**).

Modern fishing fleets deplete marine life rapidly

Today's industrialized fishing fleets can deplete marine populations quickly. In a 2003 study, fisheries biologists Ransom Myers and Boris Worm analyzed fisheries data from FAO archives, looking for changes in the catch rate of fish in various regions of ocean since they were first exploited by industrialized fishing. For one region after another, they found the same pattern: Catch rates dropped precipitously, with 90% of large-bodied fish and sharks eliminated within only a decade (Figure 16.21). Following that, populations stabilized at 10% of their former levels. This means, Myers and Worm concluded, that the oceans today contain only one-tenth of the large-bodied animals they once did. It also means that declines happened so fast in most regions of the world that scientists never knew the original abundance of these animals.

Many fisheries are collapsing today

The proportion of marine fish stocks that are overfished increased tenfold between 1950 and 1990. Many fisheries have collapsed, and others are in danger of collapsing. These collapses are ecologically devastating and also take a severe economic toll on human communities that depend on fishing.

A prime example of fishery collapse took place in the 1990s, affecting groundfish in the North Atlantic off the Canadian and U.S. coasts. The term *groundfish* refers to various species that live in benthic habitats, such as Atlantic cod, haddock, halibut, and flounder. These fish are major food sources that powered fishing economies in Newfoundland, Labrador, the Maritime Provinces, and the New England states for close to 400 years. Yet fishing pressure became so intense that most stocks collapsed in recent years, bringing fishing

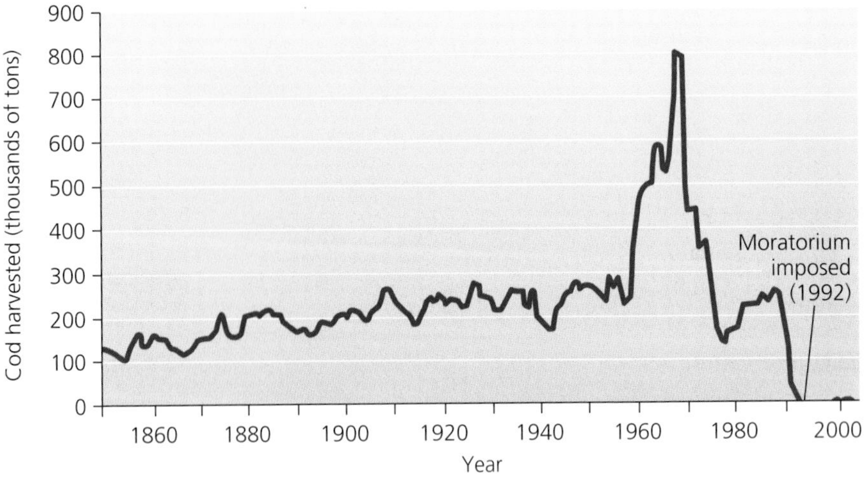

FIGURE 16.22 In the North Atlantic off the coast of Newfoundland, commercial catches of Atlantic cod increased with intensified fishing in the 1960s and 1970s. The fishery subsequently crashed, and moratoria imposed in 1992 and 2003 have not brought it back. Data from Millennium Ecosystem Assessment, 2005.

economies down with them (Figure 16.22). With Canada's cod stocks down by 99% and showing no sign of recovery, the Canadian government in 1992 ordered a complete ban on cod fishing in the Grand Banks region off Newfoundland and Labrador. The moratorium was partially lifted in 1998, but catches declined, so it was reimposed in 2003. To soften the economic blow, the government offered $50 million to affected fishers of the region.

The experience on the U.S. side of the border is showing that such bans can sometimes help restore depleted fisheries. When the groundfish fisheries of Georges Bank in the Gulf of Maine collapsed in the mid-1990s, three areas totaling 17,000 km^2 (6,600 mi^2) were closed to fishing. The closures worked; five years later, haddock, flounder, and yellowtail were recovering, and scallops rebounded strongly, attaining sizes 9–14 times as large as before the closures. Fishers began having better luck, especially just outside the closed regions. Unfortunately, cod have not recovered. Because adult cod eat fish that prey on young cod, some ecologists hypothesize that once the adult cod were depleted, intensified predation on young cod prevented the population from recovering.

Fisheries declines are masked by several factors

Although industrialized fishing has depleted fish stocks in region after region, the overall global catch has remained roughly stable for two decades (see Figure 16.20). You might wonder how this could be. The seeming stability of the total global catch can be explained by several factors that mask population declines. One is that fishing fleets have been traveling longer distances to reach less-fished portions of the ocean. They also have been fishing in deeper waters; average depth of catches was 150 m (495 ft) in 1970 and 250 m (820 ft) in 2000. Moreover, fishing fleets have been spending more time fishing and have been setting out more nets and lines—expending increasing effort just to catch the same number of fish.

Improved technology also helps explain high catches despite declining stocks. Today's Japanese, European, Canadian, and U.S. fleets can reach almost any spot on the globe with boats that can attain speeds of 80 kph (50 mph). They have access to an array of technologies that militaries have developed for spying and for chasing enemy submarines, including advanced sonar mapping equipment, satellite navigation, and thermal sensing systems. Some fleets rely on aerial spotters to find schools of commercially valuable fish, such as bluefin tuna.

Finally, another cause of misleading stability in global catch numbers is that not all data supplied to international monitoring agencies may be accurate (see "The Science behind the Story," ▸ pp. 486–487).

We are "fishing down the food chain"

Overall figures on total global catch tell only part of the story, because they do not include information on the species, age, and size of fish harvested. Careful analyses of fisheries data have revealed in case after case that as fishing increases, the size and age of fish caught decline. In addition, as particular species become too rare to fish profitably, fleets begin targeting other species that are in greater abundance. Generally this means shifting from large, desirable species to smaller, less desirable ones. Fleets have time and again depleted popular food fish such as cod and snapper and shifted their emphasis to species that were previously of lower value. Because this often entails catching species at lower trophic levels, this phenomenon has been termed "fishing down the food chain."

Some fishing practices kill nontarget animals and damage ecosystems

Some fishing practices catch more than just the species they target. **By-catch** refers to the capture of animals not meant to be caught, and it accounts for the deaths of many thousands of fish, sharks, marine mammals, and birds each year.

Boats that drag huge *driftnets* through the water (Figure 16.23a) capture substantial numbers of dolphins, seals, and sea turtles, as well as countless nontarget fish. Most of these end up drowning (mammals and turtles need to surface to breathe) or dying from air exposure on deck (fish breathe through gills in the water). Many nations have banned or restricted driftnetting because of excessive by-catch. The widespread death of dolphins in driftnets also motivated consumer efforts to label tuna as "dolphin-safe" if its capture uses methods designed to avoid dolphin by-catch. Such measures helped reduce dolphin deaths from an estimated 133,000 per year in 1986 to less than 2,000 per year since 1998.

Similar by-catch problems exist with *longline fishing* (Figure 16.23b), which involves dragging extremely long lines with baited hooks spaced along their lengths. Longline fishing kills turtles, sharks, and many albatrosses, magnificent seabirds with wingspans up to 3.6 m (12 ft). It is estimated that 300,000 seabirds of various species die each year from longline fishing.

Other fishing practices can directly damage entire ecosystems. *Bottom-trawling* (Figure 16.23c) involves dragging weighted nets over the floor of the continental shelf to catch such benthic organisms as scallops and groundfish. Trawling crushes many organisms in its path and leaves long swaths of damaged sea bottom. It is especially destructive to structurally complex areas, such as reefs, that provide shelter and habitat for many animals. Only in recent years has underwater photography begun to reveal the extent of structural and ecological disturbance done by trawling.

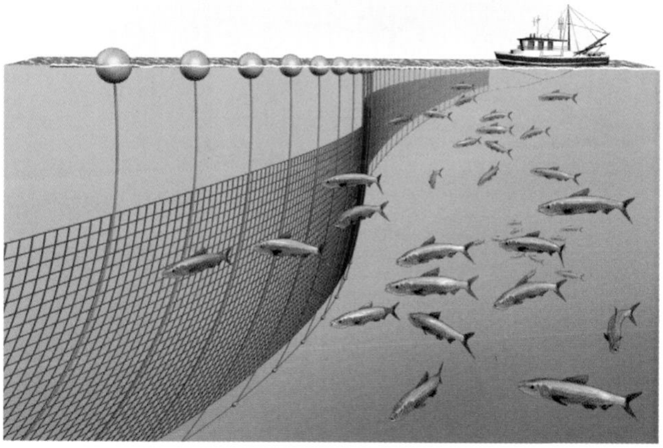

(a) Driftnetting

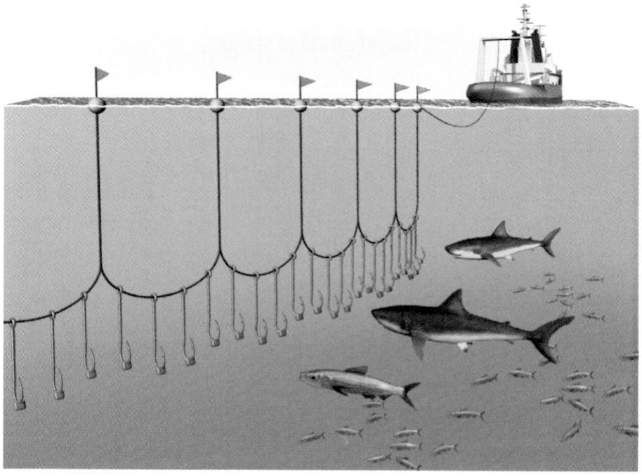

(b) Longlining

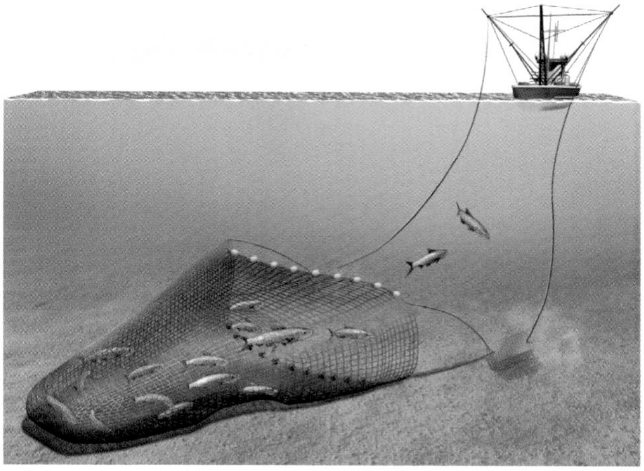

(c) Bottom-trawling

FIGURE 16.23 Commercial fishing fleets use several methods of capture. In driftnetting (**a**), huge nets are dragged through the open water to capture schools of fish. In longlining (**b**), lines with numerous baited hooks are pulled through the open water. In bottom-trawling (**c**), weighted nets are dragged along the floor of the continental shelf. All methods result in large amounts of by-catch, or capture of nontarget animals. Bottom-trawling can also result in severe structural damage to reefs and benthic habitats.

Consumer choice can influence fishing practices

To most of us, marine fishing practices may seem a distant phenomenon over which we have no control. Yet by exercising careful choice when we buy seafood, consumers can influence the ways fisheries function. Purchasing ecolabeled seafood products such as dolphin-safe tuna is one

The Science behind the Story

China's Fisheries Data

China is responsible for a larger share of the world's fisheries catch than any other nation, according to the United Nations Food and Agriculture Organization (FAO). Thus, China's catch data has a major impact on the FAO's attempts to assess the health of global fisheries. In 2001, two fisheries scientists published a paper suggesting that China had exaggerated its catch data by as much as 100% during the 1990s. The inflated catch data had led to complacency among the world's fisheries managers and policymakers, these scientists argued, because it led the FAO to overestimate the true amount of fish left in the ocean.

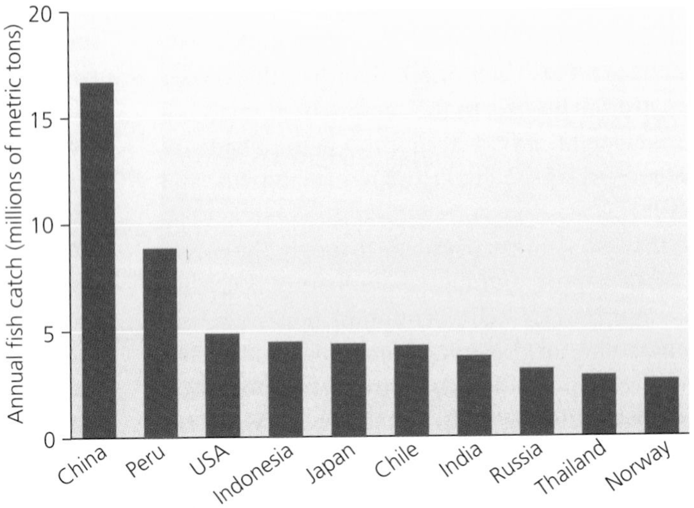

China's reported fish catches in recent years have dwarfed those of other nations, but a recent study asserted that Chinese fisheries managers had been falsely inflating their catches. Data from U.N. Food and Agriculture Organization, 2004.

The authors of the paper, University of British Columbia researchers Reg Watson and Daniel Pauly, had become suspicious of China's data for several reasons. First, China's coastal fisheries had been overexploited for decades, yet reported catches continued to increase. Second, China's "catch per unit effort" remained unchanged from 1980 to 1995, even though the abundance of fish had decreased. Finally, China's catch statistics suggested that its coastal waters were far more productive than ecologically similar areas fished by other nations.

In the November 29, 2001, issue of the journal *Nature,* Watson and Pauly reported the results of their study testing the hypothesis that China's apparent productivity was unrealistically high. Using a number of different databases, including FAO catch data collected since the 1950s, they built a statistical model to predict global fisheries catch. They divided the world's oceans into approximately 180,000 "cells," each spanning half a degree longitudinally and latitudinally. The cells were then characterized by factors that affect catch size, such as depth, primary productivity, fishing rights, and species distribution. The researchers then used these factors to predict total annual catch for each cell.

For most cells, the catch predicted by the model was similar to the catch reported by fleets that fished those waters. In large swaths of China's coastal waters, however, predicted catches were far smaller than

way to exercise choice, but in most cases consumers have no readily available information about how their seafood was caught. Thus, several nonprofit organizations have recently devised concise guides to help consumers make choices. These guides differentiate fish and shellfish that are overfished or whose capture is ecologically damaging from those that are harvested more sustainably. For instance, the Monterey Bay Aquarium provides a wealth of this information on its Web site.

Marine Conservation Biology

Because we bear responsibility and stand to lose a great deal if valuable ecological systems collapse, marine scientists have been working to develop solutions to the problems that threaten the oceans. Many have begun by taking a hard look at the strategies used traditionally in fisheries management.

reported catches. For instance, the model predicted that in 1999 China's total catch would be 5.5 million metric tons, whereas the figure reported by Chinese fisheries authorities was 10.1 million metric tons.

When Watson and Pauly proceeded to analyze global fisheries trends using the model's predictions rather than China's official numbers, they found that the total annual global catch had decreased by 360,000 metric tons since 1988, instead of increasing by 330,000 metric tons, as reported by the FAO. Their finding suggested that the global catch, rather than remaining stable through the 1990s, actually had begun to decline in the 1980s.

Fisheries managers and policymakers depend on the FAO for information about the state of the world's fisheries. By using China's apparently inflated numbers, Watson later wrote, the FAO had encouraged a global "mopping-up operation" in which the last of the world's fish were being decimated.

The Chinese government and the FAO both criticized the study. China's reported catch was "basically correct," said Yang Jian, director-general of the Chinese Bureau of Fisheries. Watson and Pauly's model, he suggested, failed to take into account unique aspects of China's fisheries, such as its large catch of crabs and jellyfish. Yang also noted that China had already begun to address the problems of overfishing and overreporting that did exist. In 1998, for instance, the Chinese government had promoted a "zero-growth policy" that capped China's total catch at 1998 levels.

For its part, the FAO noted that it was already treating China's statistics with caution. In its recent publications, global fisheries catches were being reported with and without China's data (see Figure 16.20). The FAO had also held several meetings with Chinese officials to discuss ways of reducing overreporting. Moreover, the FAO suggested, nothing in its reports had encouraged complacency. Although it had indeed reported that the global catch remained stable through the 1990s, it also noted that many individual fisheries had declined or collapsed and that others were being aggressively overfished.

Assigning blame for the state of the world's fisheries is not the most productive response to Watson and Pauly's findings, says fisheries scientist Andy Rosenberg of the University of New Hampshire. "This is a global problem, not a case of a few bad actors."

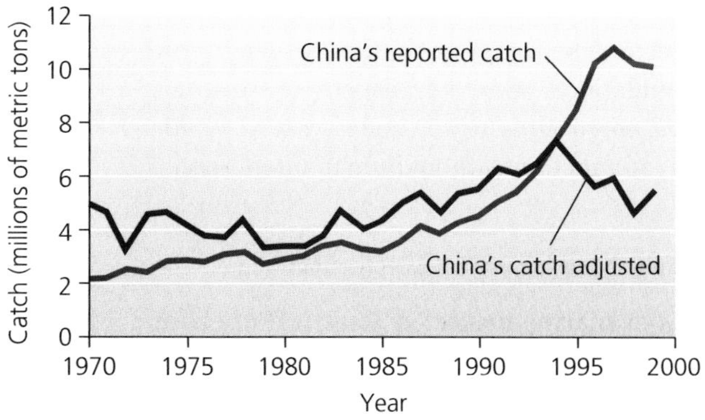

Fisheries biologists used models to estimate China's fish harvests, and found estimated amounts to be much lower than reported amounts. Data from Watson, R., and D. Pauly. 2001. Systematic distortions in world fisheries catch trends. *Nature* 414: 536–538.

Fisheries management has been based on maximum sustainable yield

For decades, fisheries management has sought to use scientific assessments to ensure sustainable harvests. Historically, fisheries managers have studied fish population biology and used that knowledge to regulate the timing of harvests, the techniques used to catch fish, and the scale of harvests. The goal was to allow for maximal harvests of particular populations while keeping fish available for the future—the concept of *maximum sustainable yield* (▸ pp. 345–346). If data indicated that current yields were unsustainable, managers might limit the number or total mass of that fish species that could be harvested or restrict the type of gear fishers could use.

Despite such efforts, a number of fish stocks have plummeted, and many marine scientists and managers now feel it is time to rethink fisheries management. One key change these reformers suggest is to shift the focus away from the individual fish species and toward viewing marine resources as elements of larger ecological systems. This means considering the effects of fishing practices on habitat quality, on interspecific interactions, and on other factors that may have indirect or long-term effects on populations. One key aspect of such an *ecosystem-based approach* is to set aside areas of ocean where systems can function without human interference.

We can protect areas in the ocean

Large numbers of **marine protected areas (MPAs)** have been established, most of them along the coastlines of developed countries. The United States now contains nearly 300 federally managed MPAs. However, despite the name, marine protected areas do not necessarily protect their natural resources, because nearly all MPAs allow fishing or other extractive activities. As a recent report from an environmental advocacy group put it, even national marine sanctuaries "are dredged, trawled, mowed for kelp, crisscrossed with oil pipelines and fiber-optic cables, and swept through with fishing nets."

Because of the lack of true refuges from fishing pressure, many scientists—and some fishers—now want to establish areas where no fishing is allowed. Such "no-take" areas have come to be called **marine reserves.** Designed to preserve entire ecosystems intact without human interference, marine reserves are also intended to improve fisheries. Scientists have argued that marine reserves can act as production factories for fish for surrounding areas, because fish larvae produced inside reserves will disperse outside and stock other parts of the ocean. By serving both purposes, proponents argue, marine reserves are a win-win proposition for environmentalists and fishers alike.

Weighing the Issues: Preservation on Land and at Sea

Almost 4% of U.S. land area is designated as wilderness, yet far less than 1% of coastal waters are protected in reserves. Why do you think it is taking so long for the preservation ethic to make the leap to the oceans?

Marine reserves have met forceful opposition

Many fishers don't like the idea of no-take reserves, however. Nearly every marine reserve proposed has met with pockets of intense opposition from people and businesses who use the area for fishing or recreation. Opposition comes from industrial fishing fleets that fish commercially as well as from individuals who fish recreationally. Both types of fishers are concerned that marine reserves will simply put more areas off-limits to fishing.

In the Florida Keys, property rights advocates who had opposed the sanctuary protested the 1998 establishment of reserve zones where fishing was to be prohibited. They rallied citizen opposition in local newspapers and filed a $27 million lawsuit against the federal government. In other parts of the world, such protests have become violent. Fishermen in the Galapagos Islands have rioted, looted, and destroyed the administration building at Galapagos National Park to protest fishing restrictions.

However, people in the Florida Keys have come to see that the reserves have not greatly threatened their access to areas. They have also witnessed improvements in the marine life around them. Surveys just 3 years after establishment of the no-take zones showed increases in the size and number of spiny lobsters and in the populations of three of four major reef fish. Keys fishers are catching more fish, and fishing revenues are up. As a result, many opponents of the reserve system have become supporters. Reserve workers estimated in 2000 that 70% of Keys residents supported the no-take reserves and that over 50% would support establishing more of them.

Reserves can work for both fish and fishers

In the past decade, data synthesized from marine reserves around the world have been indicating that reserves *do* work as win-win solutions that benefit ecosystems, fish populations, and fishing economies. In 2001, 161 prominent marine scientists signed a "consensus statement" summarizing the effects of marine reserves. Besides boosting fish biomass, total catch, and record-sized fish, the report stated, marine reserves yield several benefits. Within reserve boundaries, they

- Produce rapid and long-term increases in abundance, diversity, and productivity of marine organisms.
- Decrease mortality and habitat destruction.
- Lessen the likelihood of extirpation of species.

Outside reserve boundaries, marine reserves

- Can create a "spillover effect" when individuals of protected species spread outside reserves.
- Allow larvae of species protected within reserves to "seed the seas" outside reserves.

VIEWPOINTS

Marine Reserves

Can marine reserves or other forms of "no-fishing" zones help us solve problems facing the oceans today? Why or why not?

"No-Fishing" Zones Do Not Prevent Overfishing

If you close off part of the ocean from all forms of "take," there will obviously be more fish in the protected area. That is just common sense. However, a major problem with our oceans is severe overfishing. Fish are being killed faster than they can replenish themselves. There is a simple way to improve this situation—stop killing so many fish.

"No-fishing" zones don't accomplish this. What they do is to shift the fishing pressure to another area. Anything short of reducing the number of fish being killed, such as marine reserves, not only won't solve the problem, but also may give a false impression that something is being done, thus delaying action to actually solve the problem.

There are many traditional fishery management tools that can be used to prevent overfishing. These tools include bag limits, size limits, slot limits, closed seasons, protected species, and catch-and-release-only species. For commercial fishermen there are quotas, gear restrictions, trip limits, and limited entry, plus all the management tools listed above for recreational anglers.

A major problem for fishery managers has been politics. Commercial fishing interests have very effective lobbyists who have been instrumental in preventing or delaying needed management restrictions. If lobbying fails, they often challenge regulations through the court system. The result is management regulations that are not restrictive enough. Marine reserves will not solve this problem.

In cases where remote, pristine areas are meant to be preserved, perhaps marine reserves allowing only catch and release could be effective. Examples of this are the remote reefs in northern Hawaii and some snapper spawning areas far west of Key West, Florida. However, "no-fishing" zones by themselves will not rebuild fisheries.

Michael Leech joined the International Game Fish Association (IGFA) in 1983 and became president of IGFA in 1992. Under his leadership, the IGFA recently constructed their permanent world headquarters in Dania Beach, Florida. The IGFA Fishing Hall of Fame and Museum not only is a tourist attraction but also serves as the world center for most fishing-related information.

Marine Reserves Restore Ecosytems

Marine reserves are a powerful tool for protecting and restoring marine ecosystems. They are successful because they protect not only species but also habitats. In the past, there were innumerable naturally recurring marine reserves around the oceans—places that were too far from land, too deep, or too rocky to fish. Modern technology systematically eliminated those reserves, and today we protect far less than 1% of the oceans in established reserves.

Recent scientific studies have demonstrated that reserves provide a clear benefit to conservation of marine organisms: Biomass, density, individual size, and diversity all increase inside reserves. It is especially important to protect large marine organisms, which produce disproportionately more offspring than smaller ones. A 37-cm (14.6-in.) vermilion rockfish produces 150,000 young, whereas a 60-cm (23.6-in.) one produces 1.7 million young! Allowing individuals to get big and fat is very valuable.

Modeling results indicate that reserves substantially benefit fisheries, from both spillover (fish spilling from the reserve) and export (larvae produced inside the reserve and transported away by currents). However, not all species will recover immediately when a reserve is established. Species that grow slowly and reproduce late will need longer to colonize and recover.

Networks of marine reserves, connected by the movement of organisms, are likely to provide the best combination of conservation and fishery benefit. A network provides protection for a large total area and a long perimeter over which organisms can escape to reseed adjacent areas. Networks of reserves are not a panacea—they need to be coupled with good fishery management and pollution control—but marine reserves are one of the most promising tools available for solving the problems plaguing ocean environments today.

Jane Lubchenco is a professor of marine biology and zoology at Oregon State University. She works with policymakers, business leaders, private foundations, religious leaders, other scientists, governmental and nongovernmental organizations, and students to help figure out how to make a transition to sustainability. She is president of the International Council for Science and co-founded the Aldo Leopold Leadership Program, PISCO (Partnership for Interdisciplinary Studies of Coastal Oceans), and COMPASS (Communication Partnership for Science and the Sea).

Explore this issue further by accessing **Viewpoints** at www.aw-bc.com/withgott.

The Science behind the Story

Do Marine Reserves Work?

In November 2001, a team of fisheries scientists published a paper in the journal *Science*, providing some of the first clear evidence that marine reserves can benefit nearby fisheries. The team, led by York University researcher Callum Roberts, focused on reserves off the coasts of Florida and the Caribbean island of St. Lucia.

Following the establishment in 1995 of the Soufrière Marine Management Area (SMMA), a network of reserves intended to help restore St. Lucia's severely depleted coral reef fishery, Roberts and his colleague, Julie Hawkins, conducted annual visual surveys of fish abundance in the reserves and nearby areas. Within 3 years, they found that the biomass of five commercially important families of fish—surgeonfishes, parrot fishes, groupers, grunts, and snappers—had tripled inside the reserves and doubled outside them. Roberts and Hawkins also interviewed local fishers and found that those with large traps were catching 46% more fish per trip in 2000–2001 than they had in 1995–1996 and that fishers with small traps were catching 90% more. Roberts and his colleagues concluded that "in 5 years, reserves have led to improvement in the SMMA fishery, despite the 35% decrease in area of fishing grounds."

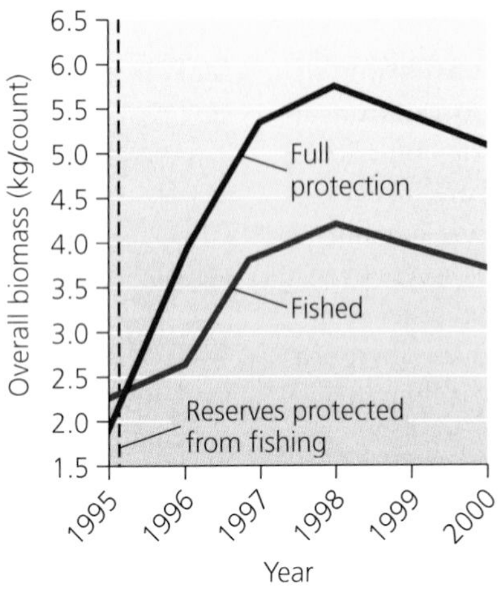

Established in 1995, the Soufrière Marine Management Area (SMMA), along the coast of St. Lucia, had a rapid impact. By 1998, fish biomass within the five reserves tripled and in adjacent, fished areas, it doubled. Data from Roberts, C., et al. 2001. Effects of marine reserves on adjacent fisheries. *Science* 294: 1920–1923.

Roberts and his coworkers also studied the oldest fully protected marine reserve in the United States, the Merritt Island National Wildlife Refuge (MINWR), established in 1962 as a buffer around what is today the Kennedy Space Center on Cape Canaveral, Florida. In a previous study, Darlene Johnson and James Bohnsack of the National Oceanic and Atmospheric Administration and Nicholas Funicelli of the United States Geological Service had found that the reserve contained more and larger fish than did nearby unprotected areas. This team also found that some of the reserve's fish appeared to be migrating to nearby fishing areas.

Bohnsack, Roberts, and their colleagues corroborated the evidence for migration by analyzing trophy records from the International

The consensus statement was backed up by research into reserves in the Caribbean and Florida (see "The Science behind the Story," above) and others worldwide. At Apo Island in the Philippines, biomass of large predators increased eightfold inside a marine reserve, and outside the reserve fishing improved. At two coral reef sites in Kenya, commercially fished and keystone species were up to 10 times more abundant in the protected area as in the fished area. At Leigh Marine Reserve in New Zealand, snapper increased 40-fold, and spiny lobsters were increasing by 5–11% yearly. Spillover from this reserve improved fishing and ecotourism, and—as in Florida— local residents who once opposed the reserve now support it.

The review of data from existing marine reserves as of 2001 revealed that just 1–2 years after their establishment, marine reserves

- Increased densities of organisms on average by 91%.
- Increased biomass of organisms on average by 192%.
- Increased average size of organisms by 31%.
- Increased species diversity by 23%.

From data like these and considerations of socioeconomic factors, the scientific consensus statement concluded that

Game Fish Association. They found that the proportion of Florida's record-sized fish caught near Merritt Island increased significantly after 1962. Nine years after the refuge was established, for instance, the number of spotted sea trout records from the Merritt Island area jumped dramatically. Bohnsack, Roberts, and their colleagues hypothesized that the reserve was providing a protected zone in which fish could grow to trophy size before migrating to nearby areas, where they were caught by recreational fishers.

Not everyone saw the St. Lucia and Merritt Island cases as proof that marine reserves could rescue depleted fisheries. In February 2002, several alternative interpretations were published as letters in *Science.* Mark Tupper, a fisheries scientist at the University of Guam, suggested that the St. Lucia results were relevant only to coral reef fisheries in developing nations, whereas Florida's boost in fish populations was due primarily to limits on recreational fishing. Karl Wickstrom, editor-in-chief of *Florida Sportsman* magazine, suggested that the increase in trophy fish near MINWR was caused by commercial fishing regulations and changes in how trophies were recorded and promoted. And Ray Hilborn, a fisheries scientist at the University of Washington, challenged the study's scientific methods. In the St. Lucia case, he pointed out, there had been no control condition.

In response, Roberts and his colleagues reaffirmed the validity of their results while acknowledging some limitations. They agreed with Tupper that marine reserves are not always effective and often need to be complemented by other management tools, such as size limits. "We agree that inadequately protected reserves are useless," they wrote, "but our study shows that well-enforced reserves can be extremely effective and can play a critical role in achieving sustainable fisheries."

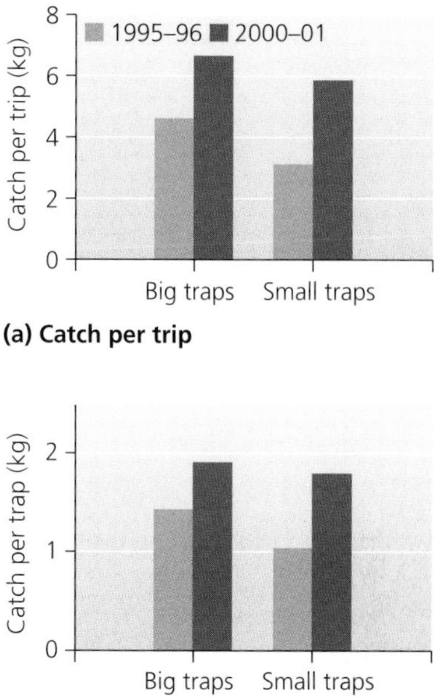

(a) Catch per trip

(b) Catch per trap

Roberts and his colleagues studied biomass of fish caught at the SMMA over two 5-month periods in 1995–1996 and 2000–2001. For fishers with big traps, catch increased by 46%, and for those with small traps, it increased by 90%. Per trap, catch increased by 36% for big traps and by 80% for small traps. Data from Roberts, C., et al. 2001. Effects of marine reserves on adjacent fisheries. *Science* 294: 1920–1923.

existing scientific information justifies the immediate application of fully protected marine reserves as a central management tool.

How should reserves be designed?

If marine reserves work in principle, the question becomes how best to design reserves and arrange them into networks. Scientists today are asking how large reserves need to be, how many there need to be, and where they need to be placed. Involving fishers directly in the planning process is crucial for coming up with answers to such questions. Of the 40 studies that have estimated how much area of the ocean should be protected in no-take reserves, estimates range from 10% to 65%, with most falling between 20% and 50%. Other studies are modeling how to optimize the size and spacing of individual reserves so that ecosystems are protected, fisheries are sustained, and people are not overly excluded from marine areas (Figure 16.24). If marine reserves are designed strategically to take advantage of ocean currents, many scientists say, then they may well seed the seas and help lead us toward solutions to one of our most pressing environmental problems.

Effects of Different-Sized Marine Reserves

	Consequences for conservation	Consequences for small fisheries	Consequences for commercial fisheries	Overall consequences
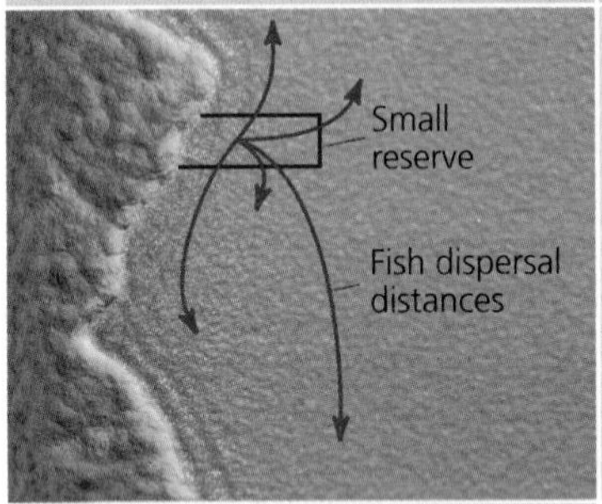 Small reserve; Fish dispersal distances	Reserve is not self-sustaining; most species are lost.	High periphery-to-area ratio makes many fish available to fishers outside reserve. This is unsustainable, however, because small reserves offer too little protection for fish populations.	Provides little or no boost to fish populations outside reserve, but does not appreciably decrease area of fishing grounds.	Too many fish wander out of reserve, yet fisheries experience little gain. Many populations are not sustained. Reserve fails to meet any stakeholder goals.
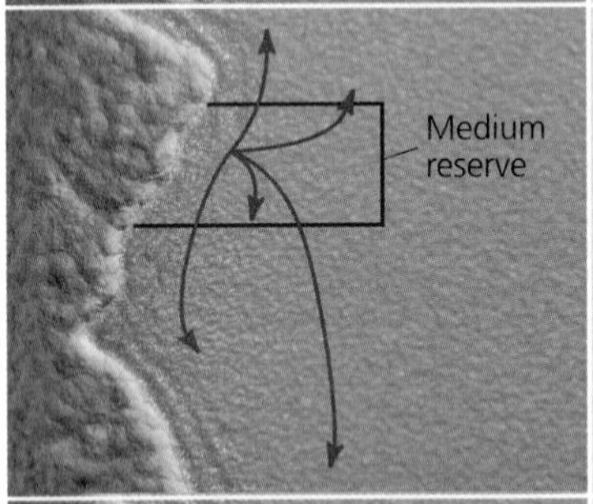 Medium reserve	Reserve is moderately self-sustaining; some species are lost.	Intermediate periphery-to-area ratio makes some fish available to fishers outside reserve, while offering a large enough area of refuge to protect other fish, thus sustaining many populations.	Provides a significant boost to fish populations outside reserve, with only a moderate decrease in area of fishing grounds.	Conservation is effective inside reserve, and fisheries gain fish while not losing too much fishing area. Good balance of benefits for all stakeholders.
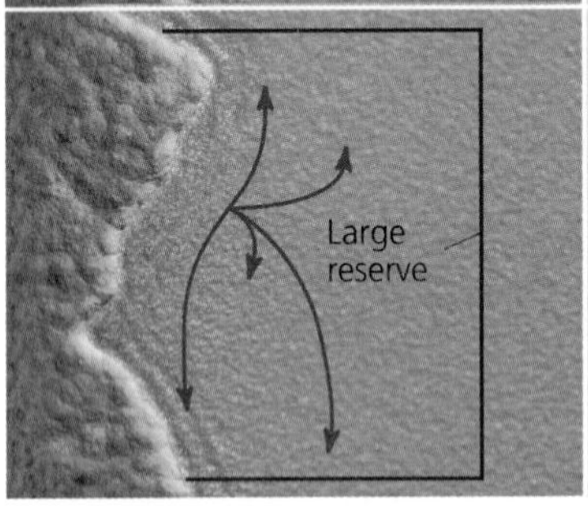 Large reserve	Reserve is completely self-sustaining; all species are retained.	Low periphery-to-area ratio makes relatively few fish available to fishers outside reserve; the "spillover" for fisheries is small in relation to the area protected in the reserve.	Provides relatively little boost to fish populations in regions around reserve, and severely decreases area of fishing grounds.	Preservation is effective inside reserve, but too few fish spill over into fishing grounds that have been much reduced in area. Outcome not acceptable to fisheries stakeholders.

FIGURE 16.24 Marine reserves of different sizes may have varying effects on ecological communities and fisheries. Young and adult fish and shellfish of different species can disperse different distances, as indicated by the red arrows in the figure. A small reserve may fail to protect animals because too many disperse out of the reserve. A large reserve may protect fish and shellfish very well but will provide relatively less "spillover" into areas where people can legally fish. Thus medium-sized reserves may offer the best hope of preserving species and ecological communities while also providing adequate fish to fishermen and human communities. Determining the actual size of such reserves requires a great deal of data on dispersal distances, water currents, community composition, animal behavior, and human economies. *Source:* Halpern, B. S., and R. R. Warner. 2003. Matching marine reserve design to reserve objectives. *Proceedings of the Royal Society of London B: Biological sciences* 270: 1871–1878.

Conclusion

Oceans cover most of our planet and contain diverse topography and ecosystems, some of which we are only now beginning to explore and understand. We are learning more about the oceans and coastal environments while we are intensifying our use of their resources. In so doing, we are coming to understand better how to use these resources without depleting them or causing undue ecological harm. In the Florida Keys and hundreds of other areas, scientists are demonstrating that setting aside protected areas of the ocean can serve to maintain natural systems and also to enhance fisheries. As historical studies reveal more information on how much biodiversity our oceans formerly contained and have now lost, we may increasingly look beyond simply making fisheries stable and instead consider restoring the ecological systems that used to flourish in our waters.

REVIEWING OBJECTIVES

You should now be able to:

Identify physical, geographical, chemical, and biological aspects of the marine environment

- Oceans cover 71% of Earth's surface and contain over 97% of its surface water. (p. 468)
- Ocean water contains 96.5% H_2O by mass and various dissolved salts. (p. 468–469)
- Colder, saltier water is denser and sinks. Water temperatures vary with latitude, and temperature variation is greater in surface layers. (pp. 469–470)
- Persistent currents move horizontally through the oceans, driven by density differences, sunlight, and wind. (pp. 470–471)
- Vertical water movement includes upwelling and downwelling, which affect the distribution of nutrients and life. (p. 471)
- Seafloor topography can be complex and rugged. (pp. 471–472)

Describe major types of marine ecosystems

- Major types of marine and coastal ecosystems include pelagic and deep-water open ocean systems, kelp forest, coral reefs, intertidal zones, salt marshes, mangrove forests, and estuaries. (pp. 473–478)
- Many of these systems are highly productive and rich in biodiversity. Many also suffer heavy impacts from human influence. (pp. 473–478)

Outline historic and current human uses of marine resources

- For millennia, people have drawn resources from the oceans and used ocean waters for transportation. (p. 478)
- Today we extract energy and minerals from the oceans, as well as using them for transportation. (pp. 478–479)

Assess human impacts on marine environments

- People pollute ocean waters with trash, untreated sewage, petroleum spills, plastic that harms marine life, and nutrient pollution that leads to harmful algal blooms. (pp. 479–481)
- Overharvesting is perhaps the major human impact on marine systems. (p. 482)

Review the current state of ocean fisheries and reasons for their decline

- Half the world's marine fish populations are fully exploited, 25% are already overexploited, and 25% can yield more without declining. (p. 482)
- People began depleting marine resources long ago, but impacts have intensified in recent decades. (p. 483)
- Global fish catches have stopped growing since the late 1980s, despite increased fishing effort and improved technologies. (p. 482, 484)
- Today's oceans hold only one-tenth the number of large animals that they did before the advent of industrialized commercial fishing. (pp. 483–484)
- As fishing intensity increases, the fish available become smaller. (pp. 484–485)
- Fishing practices such as driftnetting, longline fishing, and trawling capture nontarget organisms, called bycatch. (p. 485)
- Traditional fisheries management has not stopped declines, so many scientists feel that ecosystem-based management is needed. (pp. 487–488)

Evaluate marine protected areas and reserves as innovative solutions

- We have established fewer protected areas in the oceans than we have on land and most marine protected areas allow many extractive activities. (p. 488)
- No-take marine reserves can protect ecosystems while also boosting fish populations and making fisheries sustainable. (pp. 488–491)

TESTING YOUR COMPREHENSION

1. What proportion of Earth's surface do oceans cover? About how much salt does ocean water contain? How are water density, salinity, and temperature related in each layer of ocean water?
2. What factors drive the system of ocean currents? In what directions do ocean currents move, and how do such movements affect conditions for life in the oceans?

3. Where in the oceans are productive areas of biological activity likely to be found?
4. Describe three kinds of ecosystems found near coastal areas and the kinds of life they support.
5. Why are coral reefs biologically valuable? What are some possible ways in which they are being degraded by human impact? What is causing the disappearance of mangrove forests and salt marshes?
6. Describe three major forms of pollution in the oceans and the consequences of each.
7. Provide an example of how overfishing can lead to ecological damage and fishery collapse.
8. Explain the conclusion of the Myers and Worm study of 2003 (see Figure 16.21).
9. Name three industrial fishing practices that create bycatch and harm marine life, and explain how they do so.
10. How does a marine protected area differ from a marine reserve? Why do many fishers oppose marine reserves? Explain why many scientists say no-take reserves will be good for fishers.

SEEKING SOLUTIONS

1. What benefits do you derive from the oceans? How does your behavior affect the oceans? Give specific examples.
2. We have been able to reduce the amount of oil we spill into the oceans, but petroleum-based products such as plastic continue to litter our oceans and shorelines. Discuss some ways that we can reduce this threat to the marine environment.
3. What factors account for the trends in global fish capture over the past 20 years?
4. Consider what you know about biological productivity in the oceans, about the scientific data on marine reserves, and about the social and political issues surrounding the establishment of marine reserves. What ocean regions do you think it would be particularly appropriate to establish as marine reserves? Why?
5. Why does the 2001 scientific consensus statement argue for networks of reserves to be established? What role should science and scientists play in this kind of decision making? Discuss how you would engage a group of fishers, environmentalists, and scientists in a reserve-planning process.
6. You are mayor of a coastal town where some residents are employed as commercial fishers and others make a living serving ecotourists who come to snorkel and scuba-dive at the nearby coral reef. In recent years, several fish stocks have crashed, and ecotourism is dropping off as fish disappear from the increasingly degraded reef. Scientists are urging you to help establish a marine reserve around portions of the reef, but most commercial and recreational fishers are opposed to this idea. What steps would you take to restore your community's economy and environment?

INTERPRETING GRAPHS AND DATA

The accompanying graph presents trends in the status of ocean fisheries. Fully exploited fisheries (green line) are ones currently producing their maximum sustainable yield. Moderately exploited and underexploited fisheries (blue line) are ones that could be more heavily fished than they are and could produce larger, yet sustainable, yields. Overexploited, depleted, and recovering fisheries (red line) are ones that have been overfished.

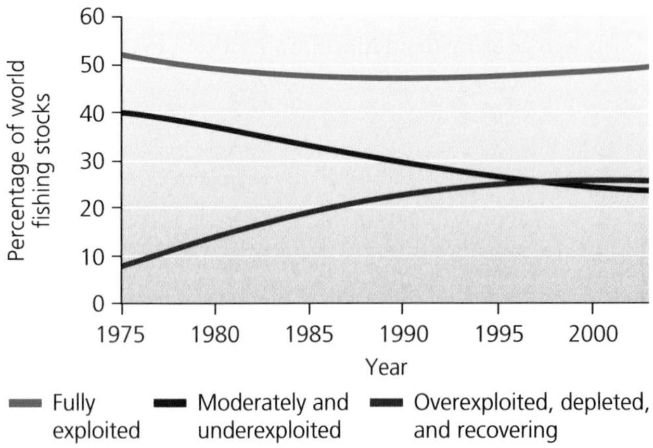

Global trends in the state of ocean fishing stocks from 1975 to 2003. Data from U.N. Food and Agriculture Organization (FAO), Fisheries Department. 2004. *The state of world fisheries and aquaculture: 2004.*

1. What is the sum of the percentages for any given year in the figure above? Explain why this is so.
2. Describe the trend from 1975 to 2003 in the percentage of fisheries categorized as overexploited, depleted, or recovering. Describe the trend from 1975 to 2003 in the percentage of fisheries categorized as moderately exploited or underexploited.
3. Based on the data in the graph, predict the likely trend in overall ocean fisheries production over the next 25 years. Explain your answer.

CALCULATING ECOLOGICAL FOOTPRINTS

The relationship between the ecological goods and services used by individuals and the amount of *land* area needed to provide those goods and services is relatively well developed. People also use goods and services from Earth's oceans, where the concept of *area* is less useful. It is clear, however, that our removal of fish from the oceans has an impact, or an ecological footprint, on remaining fish populations.

The table shows data on the mean annual per capita consumption from ocean fisheries for North America, China, and the world as a whole. Using the data provided, calculate the amount of fish each consumer group would consume at the annual per capita consumption rates for each of these three regions, and record your results in the table.

	Annual per capita consumption rate		
Consumer group	**North America 21.6 kg**	**China 27.7 kg**	**World 16.2 kg**
You			
Your class			
Your state			
United States	6.48×10^9 kg	8.31×10^9 kg	4.86×10^9 kg
World			

Data from U.N. Food and Agriculture Organization (FAO), Fisheries Department. 2004. *The state of world fisheries and aquaculture: 2004.* Data are for 2002, the most recent year for which comparative data are available.

1. Calculate the ratio of North America's per capita fish consumption rate to that of the world. Compare this ratio to the ratio of the per capita ecological footprints for the United States, Canada, and Mexico (see Figure 1.13, ▸ p. 19) versus the world average footprint of 2.2 ha/person/year. Can you account for similarities and differences between these ratios?
2. The population of China has grown at an annual rate of 1.1% since 1987, while over the same period fish consumption in China has grown at an annual rate of 8.9%. Speculate on the reasons behind China's rapidly increasing consumption of fish.
3. What ecological concerns do the combined trends of human population growth and increasing per capita fish consumption raise for you? What role might you play in contributing to these concerns or to their solutions?

Take It Further

Go to www.aw-bc.com/withgott or the student CD-ROM where you'll find:

- Suggested answers to end-of-chapter questions
- Quizzes, animations, and flashcards to help you study
- *Research Navigator*™ database of credible and reliable sources to assist you with your research projects
- GRAPHIt! tutorials to help you master how to interpret graphs
- INVESTIGATEit! current news articles that link the topics that you study to case studies from your region to around the world

17 Atmospheric Science and Air Pollution

View of Earth's atmosphere from space

Upon successfully completing this chapter, you will be able to:

- Describe the composition, structure, and function of Earth's atmosphere
- Outline the scope of outdoor air pollution and assess potential solutions
- Explain stratospheric ozone depletion and identify steps taken to address it
- Define acidic deposition and illustrate its consequences
- Characterize the scope of indoor air pollution and assess potential solutions

Police officer in London's "killer smog," 1952

Central Case: The 1952 "Killer Smog" of London

"You had this swirling, like somebody had set a load of car tires on fire."
—STAN CRIBB, EYEWITNESS TO THE DECEMBER 1952 "GREAT SMOKE" OF LONDON

"The modern field of environmental health owes much to the tragedy that befell Greater London some 50 years ago."
—DEVRA L. DAVIS, MICHELLE L. BELL, AND TONY FLETCHER, IN AN EDITORIAL FOR THE JOURNAL ENVIRONMENTAL HEALTH PERSPECTIVES

December 5, 1952, was a particularly cold Friday in London, and many residents stoked their coal stoves to keep the chill away. Few people took notice when thick, foul-smelling smog, a fog polluted with smoke and chemical fumes, first settled over the city. After all, London had been famous for its "pea souper" fogs for well over a century.

But the smog that December weekend was particularly thick; the city's air quality was ten times worse than usual for that year. Visibility was so poor that pedestrians could not see across the street. Transportation came to a standstill, and roads became clogged with abandoned cars. Schools closed, cattle asphyxiated, and an opera was halted because the audience could not see the stage.

A wind finally relieved Londoners of the miserable smog on Tuesday, December 9. By that time, however, authorities estimated that over 4,000 people had died, mostly from lung ailments such as bronchitis that were induced or aggravated by the pollution. A 2002 study by American researchers estimated that the actual death toll, including delayed cases that appeared over the next 2 months and were considered unrelated at the time, may have been as high as 12,000.

The "killer smog" of 1952 was remarkable, but hardly unique. Similar, although less severe, phenomena had occurred in London as early as 1813 and again in 1873, 1880, 1891, and 1948. Other such events have taken lives in Pennsylvania, New York, Mexico, and Malaysia. London's killer smog, together with other severe smog events, helped change the way the public viewed air pollution. Before the 1950s, most people considered urban smog a necessary burden. Today, we recognize the importance of clean air and view air pollution as an environmental challenge that can be overcome.

We have overcome much already; declines in air pollution represent some of the biggest successes of environmental policy. These declines have been due largely to limits on emissions brought about through legislation, such as the British Clean Air Acts of 1956 and 1968 and the U.S. Clean Air Acts of 1970 and 1990. Today, the air in many U.S. cities is cleaner, and the concentration of airborne particles around London averages one-tenth that of the 1950s.

However, much remains to be done. In 2002, a London governmental body estimated that vehicle emissions contribute to the premature deaths of 380 city residents each year. Furthermore, many cities in developing nations that have increasing populations, older technologies, and lax legislation and enforcement face conditions similar to those of 1950s London. For example, in 1995 air pollution in Delhi, India, was measured at 1.3 times London's average for the year 1952, and air pollution in Lanzhou, China, was measured at 2.7 times London's 1952 level. To understand what caused London's killer smog and how we can prevent recurrences of such events, we must examine both natural atmospheric conditions and human-made pollutants.

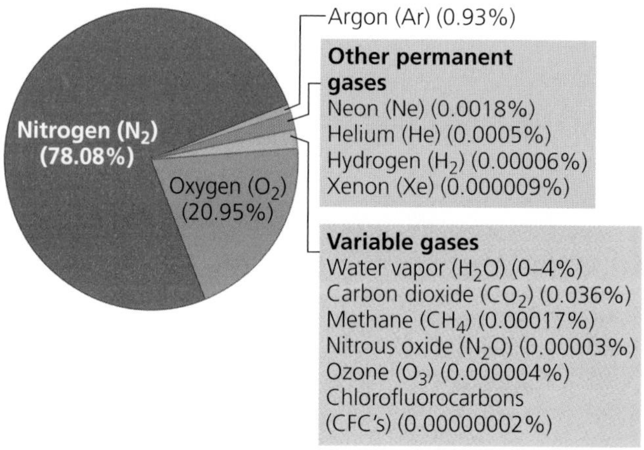

FIGURE 17.1 Earth's atmosphere consists mostly of nitrogen, secondarily of oxygen, and lastly of a mix of gases at dilute concentrations. Permanent gases are fixed in concentration. Variable gases vary in concentration as a result of either natural processes or human activities. Data from Ahrens, D.C. 1998. *Essentials of meteorology,* 2nd ed. New York: Wadsworth.

Atmospheric Science

Every breath we take reaffirms our connection to the **atmosphere,** the thin layer of gases that surrounds Earth. We live at the bottom of this layer, which provides us oxygen, absorbs hazardous solar radiation, burns up incoming meteors, transports and recycles water and nutrients, and moderates climate.

The atmosphere consists of roughly 78% nitrogen gas (N_2) and 21% oxygen gas (O_2). The remaining 1% is composed of argon gas (Ar) and minute concentrations of several other *permanent gases* that remain at stable concentrations. The atmosphere also contains a number of *variable gases* that vary in concentration from time to time or place to place, as a result of natural processes or human activities on Earth's surface (Figure 17.1).

Over Earth's long history, the atmosphere's chemical composition has changed. Oxygen gas began to build up in an atmosphere dominated by carbon dioxide (CO_2), nitrogen, carbon monoxide (CO), and hydrogen (H_2) about 2.7 billion years ago, with the emergence of autotrophic microbes that emitted oxygen as a by-product of photosynthesis (▸ p. 108). Just as these early organisms had a substantial impact on the Earth's atmosphere long ago, human activity is altering the quantities of some atmospheric gases today, such as carbon dioxide, methane (CH_4), and ozone (O_3). In this chapter and in Chapter 18, we will explore the atmospheric changes brought about by artificial pollutants, but we must first begin with an overview of Earth's atmosphere.

The atmosphere consists of several layers

The atmosphere that stretches so high above us and seems so vast is actually just a thin coating about 1/100th of Earth's diameter, like the fuzzy skin of a peach. This coating consists of four layers whose boundaries are not visible to the human eye, but which atmospheric scientists recognize by measuring differences in temperature, density, and composition (Figure 17.2).

The bottommost layer, the **troposphere,** blankets Earth's surface and provides organisms the air they need to live. The movement of air within the troposphere is also largely responsible for the planet's weather. Although it is thin (averaging 11 km [7 mi] high) relative to the atmosphere's other layers, the troposphere contains three-quarters of the atmosphere's mass, because air is denser near Earth's surface. On average, tropospheric air temperature declines by about 6 °C for each

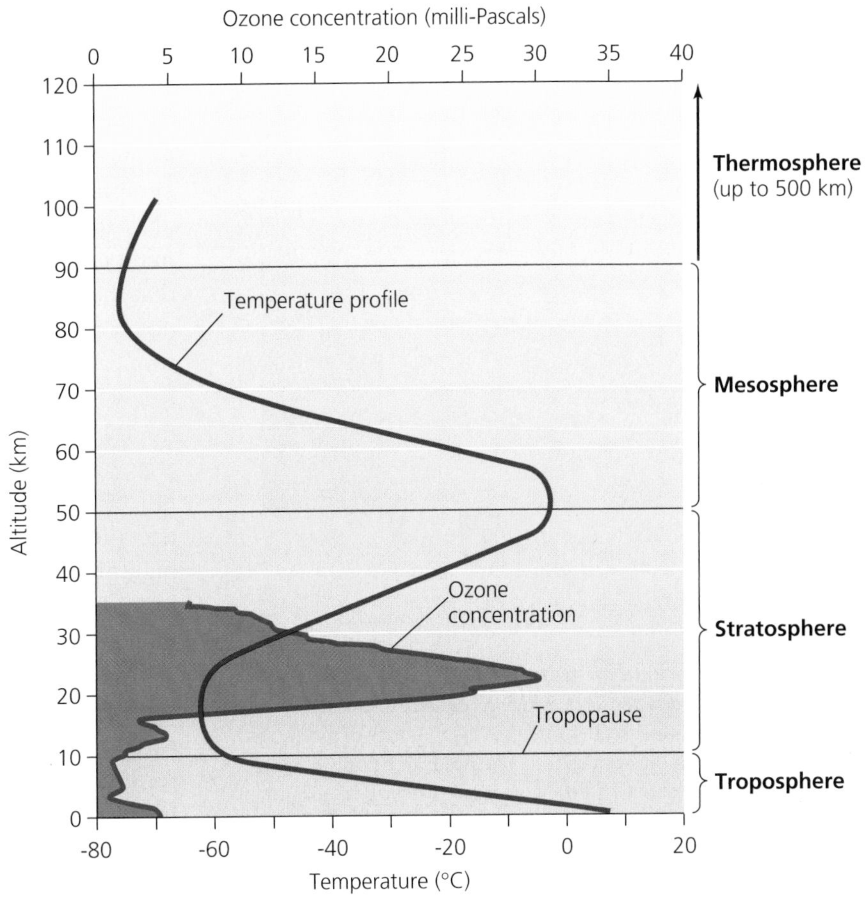

FIGURE 17.2 Some aspects of the atmosphere change with altitude across its four layers. Temperature drops with altitude in the troposphere, rises with altitude in the stratosphere, drops in the mesosphere, then rises again in the thermosphere. The tropopause separates the troposphere from the stratosphere. Ozone reaches a peak in a portion of the stratosphere, giving rise to the term *ozone layer.* Adapted from Jacobson, M. Z. 2002. *Atmospheric pollution: History, science, and regulation.* Cambridge: Cambridge University Press; Parson, E. A. 2003. *Protecting the ozone layer: Science and strategy.* Oxford: Oxford University Press.

kilometer in altitude (or 3.5 °F per 1,000 ft), dropping to roughly −52 °C (−62 °F) at its highest point. At the top of the troposphere, however, temperatures cease to decline with altitude, marking a boundary called the *tropopause.* The tropopause acts like a cap, limiting mixing between the troposphere and the atmospheric layer above it, the stratosphere.

The **stratosphere** extends 11–50 km (7–31 mi) above sea level. Similar in composition to the troposphere, the stratosphere is 1,000 times drier and less dense. Its gases experience little vertical mixing, so once substances (including pollutants) enter it, they tend to remain for a long time. The stratosphere attains a maximum temperature of −3 °C (27 °F) at its highest altitude and becomes cooler in its lower reaches. The reason is that ozone and oxygen absorb and scatter the sun's ultraviolet (UV) radiation (▸p. 106), so that much of the UV radiation penetrating the upper stratosphere fails to reach the lower stratosphere. Most of the atmosphere's minute amount of ozone is concentrated in a portion of the stratosphere roughly 17–30 km (10–19 mi) above sea level, a region that has come to be called Earth's **ozone layer.** The ozone layer greatly reduces the amount of UV radiation that reaches Earth's surface. Because UV light can damage living tissue and induce DNA mutations, the ozone layer's protective effects are vital for life on Earth.

Above the stratosphere lies the *mesosphere,* which extends 50–90 km (31–56 mi) above sea level. Air pressure is extremely low here, and temperatures decrease with altitude, reaching their lowest point (−90 °C, or −130 °F) at the top of the mesosphere. From here, the *thermosphere* extends upward to an altitude of 500 km (300 mi). In the thermosphere, molecules are so few and far between that they collide only rarely. As a result, heavier molecules (such as nitrogen and oxygen) sink, and light ones (such as hydrogen and helium) end up near the top of the thermosphere. This stands in contrast to the atmosphere's lower three layers, where frequent collisions among molecules keep the chemical composition (see Figure 17.1) relatively constant throughout.

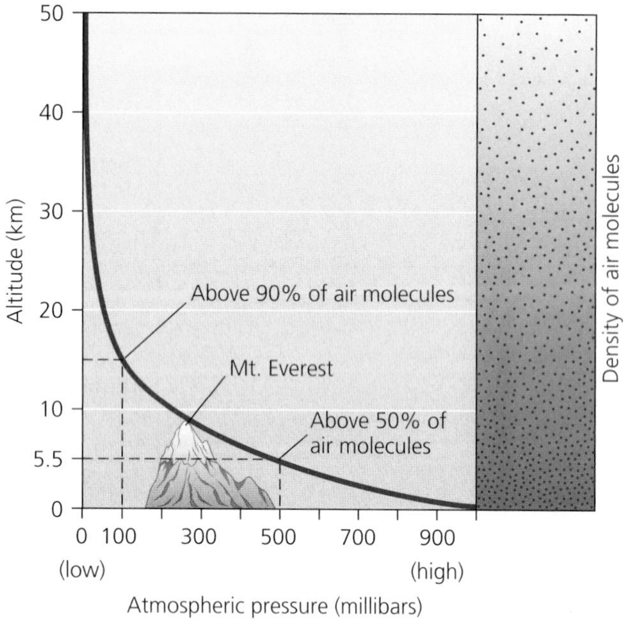

FIGURE 17.3 As one climbs higher through the atmosphere, gas molecules become less densely packed. As density decreases, so does atmospheric pressure. Because the vast majority of air molecules lie low in the atmosphere, one needs to be only 5.5 km (3.4 mi) high to be above half the planet's air molecules. Adapted from Ahrens, D. C. 1998. *Essentials of meteorology*, 2nd ed. New York: Wadsworth.

Atmospheric properties include temperature, pressure, and humidity

Although the lower atmosphere is stable in its composition, it is dynamic in its movement; air moves within it as a result of differences in the physical properties of air masses. Among these properties are pressure and density, relative humidity, and temperature. Gravity pulls gas molecules toward Earth's surface, causing air to be most dense near the surface and less so as altitude increases. **Atmospheric pressure,** which measures the force per unit area produced by a column of air, also decreases with altitude, because at higher altitudes fewer molecules are pulled down by gravity (Figure 17.3). At sea level, atmospheric pressure is 14.7 lb/in.^{2} or 1,013 millibar (mb). Mountain climbers trekking to Mount Everest, the world's highest mountain, can look up and view their destination from Kala Patthar, a nearby peak, at roughly 5.5 km (18,000 ft). At this altitude, pressure is 500 mb, and half the atmosphere's air molecules are above the climber and half are below. A climber who reaches Everest's peak (8.85 km [29,035 ft]), where the "thin air" is just over 300 mb, stands above two-thirds of the molecules in the atmosphere. When we fly on a commercial jet airline, at a typical cruising altitude of 11 km (36,000 ft), we are above roughly 80% of the atmosphere's molecules.

Another important property of air is **relative humidity,** the ratio of water vapor a given volume of air contains to the maximum amount it *could* contain at a given temperature. Relative humidity varies considerably from place to place and time to time. Average daily relative humidity in June in Phoenix, Arizona, is only 31% (meaning that the air contains just less than a third of the water vapor it possibly can at its temperature), whereas on the tropical island of Guam, relative humidity rarely drops below 88%. People are sensitive to changes in relative humidity because we perspire to cool our bodies. When relative humidity is high, the air is already holding nearly as much water vapor as it can, so sweat evaporates slowly and the body cannot cool itself efficiently. This is why high humidity makes it feel hotter than it really is. Low humidity speeds evaporation and makes it feel cooler than it really is.

The temperature of air also varies with location and time. At the global scale, temperature varies over Earth's surface because the sun's rays strike some areas more directly than others. At more local scales, temperature varies because of topography, plant cover, proximity of land to water, and many other factors.

Solar energy heats the atmosphere, helps create seasons, and causes air to circulate

Energy from the sun influences temperatures by heating air in the atmosphere. Solar energy also drives the atmosphere's air movement, helps create seasons, and influences both weather and climate. An enormous amount of solar energy continuously bombards the upper atmosphere—over 1,000 watts/m^2, many thousands of times greater than the total output of electricity generated by human society. Of that solar energy, about 70% is absorbed by the atmosphere and planetary surface, while the rest is reflected back into space (see Figure 18.1, ▸p. 531).

The spatial relationship between Earth and the sun determines the amount of solar radiation that strikes each point of Earth's surface. Sunlight is most intense when it shines directly overhead and meets the planet's surface at a perpendicular angle. At this angle, sunlight passes through a minimum of energy-absorbing atmosphere, and Earth's surface receives a maximum of solar energy per unit surface area. Conversely, solar energy that approaches Earth's surface at an oblique angle loses intensity as it traverses a longer distance through the atmosphere, and it is less concentrated when it reaches the surface.

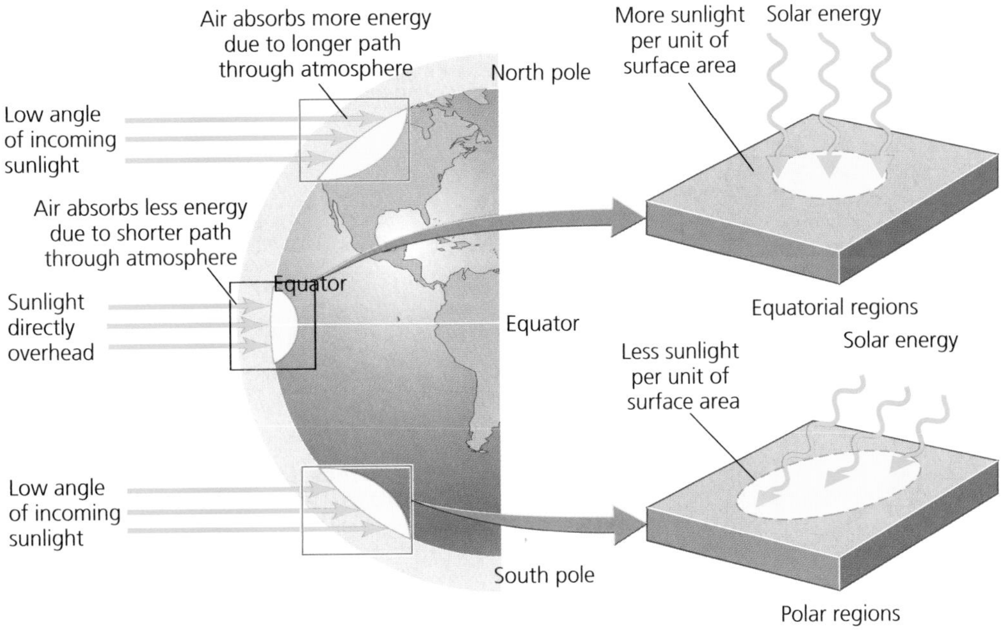

FIGURE 17.4 Because of Earth's curvature, polar regions receive on average less solar energy than equatorial regions. One reason is that sunlight gets spread over a larger area when striking the surface at an angle, a phenomenon that increases with distance from the equator. Another reason is that as sunlight approaches at a lower angle near the poles, it must traverse a longer distance through the atmosphere before reaching the surface, during which more energy is absorbed or reflected. These patterns represent year-round averages; the latitude at which radiation approaches the surface perpendicularly varies with the seasons (see Figure 17.5).

This is why, on average, solar radiation intensity is highest near the equator and weakest near the poles (Figure 17.4).

Because Earth is tilted on its axis (an imaginary line connecting the poles, running perpendicular to the equator) by about 23.5 degrees, the Northern and Southern Hemispheres each tilt toward the sun for half the year, resulting in the change in seasons (Figure 17.5). Regions near the equator are largely unaffected by this tilt; they experience about 12 hours each of sunlight and darkness every day throughout the year. Near the poles, however, the effect is strong, and seasonality is pronounced.

Land and surface water absorb solar energy, radiating some heat and causing some water to evaporate. Air near Earth's surface therefore tends to be warmer and moister than air at higher altitudes. These differences set into motion a process of **convective circulation** (Figure 17.6). Warm air, being less dense, rises and creates vertical currents. As air rises into regions of lower atmospheric pressure, it expands and cools. Once the air cools, it descends and becomes denser, replacing warm air that is rising. The air picks up heat and moisture near ground level and prepares to rise again, continuing the process. Similar convective circulation patterns occur within ocean waters (▸ p. 471), in columns of magma beneath Earth's surface (▸ p. 208), and even in a simmering pot of soup. Convective circulation influences both weather and climate.

The atmosphere drives weather and climate

In everyday speech, we often use the terms *weather* and **climate** interchangeably. However, these words have very distinct meanings. Both concepts involve the physical properties of the troposphere, such as temperature, pressure, humidity, cloudiness, and wind. However, **weather** specifies atmospheric conditions over short time periods, typically hours or days, and within relatively small geographic areas. **Climate,** in contrast, describes the pattern of atmospheric conditions found across large geographic regions over long periods of time, typically seasons, years, or millennia. Mark Twain once noted the distinction between climate and weather by saying, "Climate is what we expect, weather is what we get." For example, even very dry climates (such as that of the desert around Phoenix) occasionally have wet weather.

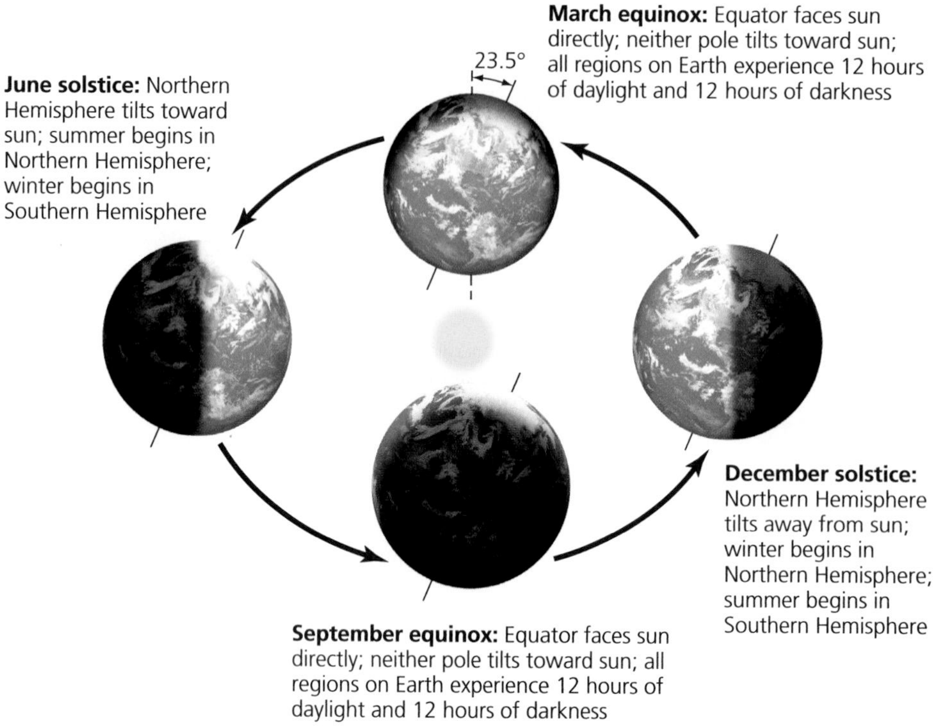
March equinox: Equator faces sun directly; neither pole tilts toward sun; all regions on Earth experience 12 hours of daylight and 12 hours of darkness
23.5°
June solstice: Northern Hemisphere tilts toward sun; summer begins in Northern Hemisphere; winter begins in Southern Hemisphere
December solstice: Northern Hemisphere tilts away from sun; winter begins in Northern Hemisphere; summer begins in Southern Hemisphere
September equinox: Equator faces sun directly; neither pole tilts toward sun; all regions on Earth experience 12 hours of daylight and 12 hours of darkness

FIGURE 17.5 The seasons occur because Earth is tilted on its axis by 23.5 degrees. As Earth revolves around the sun, the Northern Hemisphere tilts toward the sun for one half of the year, and the Southern Hemisphere tilts toward the sun for the other half of the year. In each hemisphere, summer occurs during the period in which the hemisphere receives the most solar energy because of its tilt toward the sun.

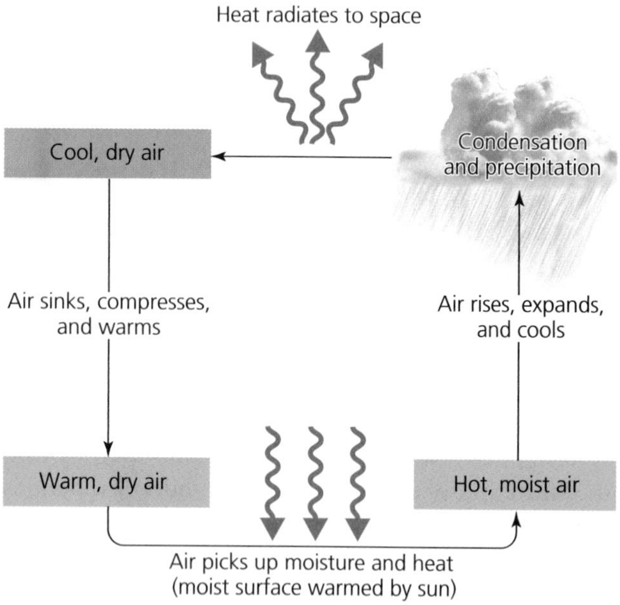
Heat radiates to space
Cool, dry air
Condensation and precipitation
Air sinks, compresses, and warms
Air rises, expands, and cools
Warm, dry air
Hot, moist air
Air picks up moisture and heat (moist surface warmed by sun)

FIGURE 17.6 Weather is driven in part by the convective circulation of air in the atmosphere. Air being heated near Earth's surface picks up moisture and rises. Once aloft, this air cools, and moisture condenses, forming clouds and precipitation. Cool, drying air begins to descend, compressing and warming in the process. Warm dry air near the surface begins the cycle anew.

Weather is produced by interacting air masses

Many changes in weather occur when air masses with different physical properties meet. The boundary between air masses that differ in temperature and moisture (and therefore density) is called a *front*. The boundary along which a mass of warmer, moister air replaces a mass of colder, drier air is termed a **warm front.** Some of the warm, moist air behind a warm front rises over the cold air mass and then cools and condenses to form clouds that may produce light rain. A **cold front** is the boundary along which a colder, drier air mass displaces a warmer, moister air mass. The colder air, being denser, tends to wedge beneath the warmer air. The warmer air rises, then cools and expands to form clouds that can produce thunderstorms. Once a cold front passes through, the sky usually clears, and the temperature and humidity drop (Figure 17.7).

Adjacent air masses may also differ in atmospheric pressure. An air mass with relatively high atmospheric pressure, or a **high-pressure system,** contains air that moves outward away from the center of high pressure as it descends. High-pressure systems typically bring fair weather. In a **low-pressure system,** air moves toward the low atmospheric pressure at the center of the system and spirals upward. The air expands and cools, and clouds and precipitation often result.

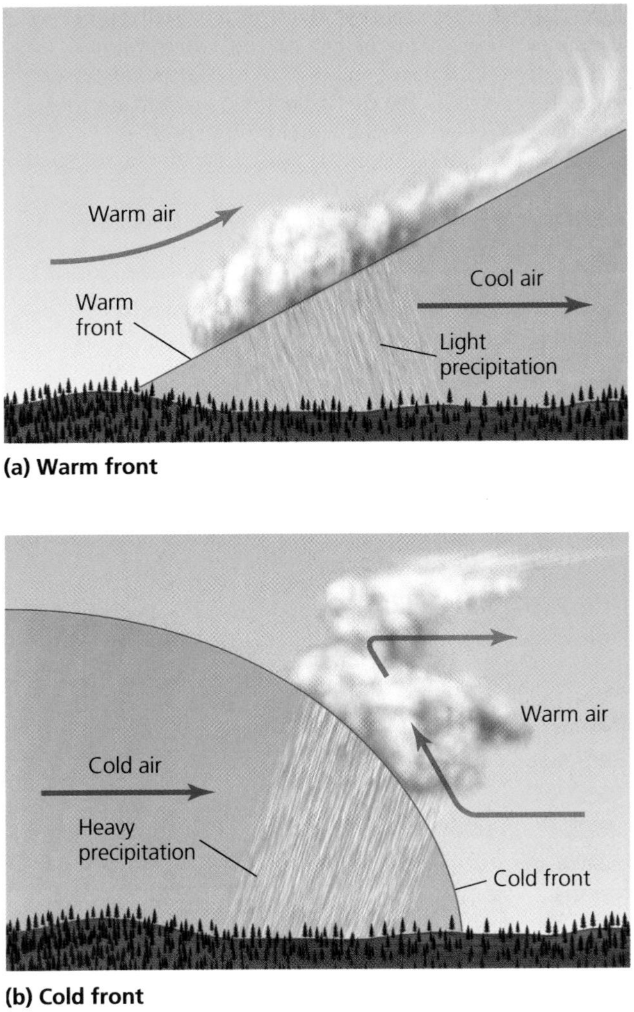

FIGURE 17.7 When a warm front approaches, warmer air rises over cooler air, causing light to moderate precipitation as moisture in the warmer air condenses (**a**). When a cold front approaches, colder air pushes beneath warmer air, and the warmer air rises, causing condensation and resulting in heavy precipitation (**b**).

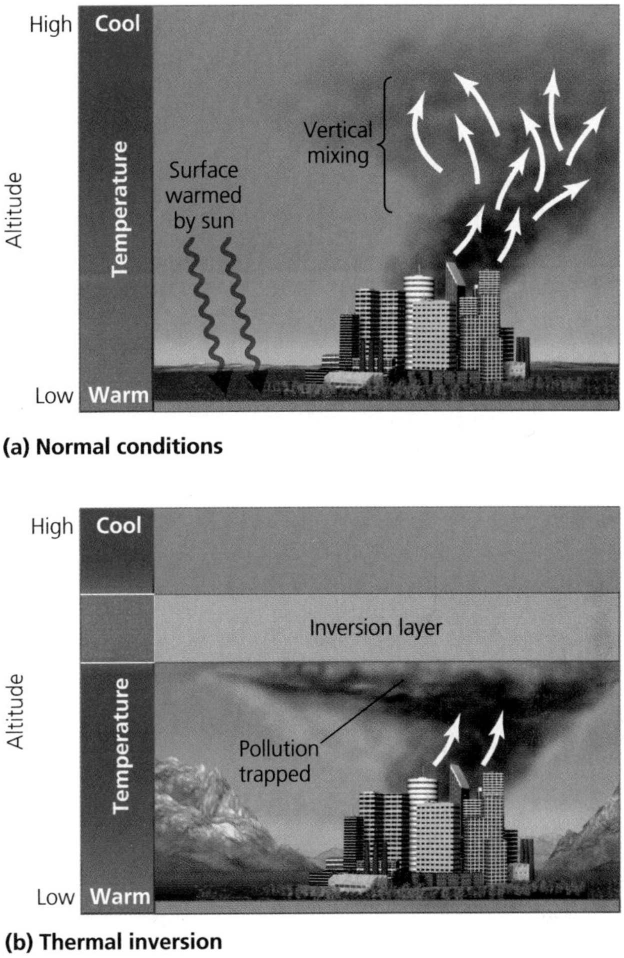

FIGURE 17.8 A thermal inversion is a natural atmospheric occurrence that can exacerbate air pollution locally. Under normal conditions (**a**), tropospheric temperature decreases with altitude, and air of different altitudes mixes somewhat freely, dispersing most pollutants upward and outward from their sources. During a thermal inversion (**b**), cool air remains near the ground underneath an "inversion layer" of warmer air. No mixing occurs, and pollutants are trapped within the cool layer near the surface.

Under most conditions, air in the troposphere decreases in temperature as altitude increases. Because warm air rises, this causes vertical mixing. Occasionally, however, a layer of relatively cool air occurs beneath a layer of warmer air. This departure from the normal temperature profile is known as a **temperature inversion,** or **thermal inversion** (Figure 17.8). The band of air in which temperature rises with altitude is called an **inversion layer** (because the normal direction of temperature change is inverted). The cooler air at the bottom of the inversion layer is denser than the warmer air at the top of the inversion layer, so it resists vertical mixing and remains stable. Thermal inversions can occur in different ways, sometimes involving cool air at ground level and sometimes producing an inversion layer higher above the ground. One common type of inversion occurs in mountain valleys where slopes block morning sunlight, keeping ground-level air within the valley shaded and cool.

Whereas vertical mixing normally allows ground-level air pollution to be diluted upward, thermal inversions trap pollutants near the ground. It was a thermal inversion that sparked London's killer smog. A high-pressure system settled over the city, acting like a cap on the air pollution and keeping it in place. Although London's 1952 smog event was the worst single such event recorded, inversions regularly cause smog buildup in many areas worldwide. This is a problem particularly in large metropolitan areas in hilly terrain, such as Los Angeles; Mexico City; Seoul, Korea; and Rio de Janeiro and São Paulo, Brazil.

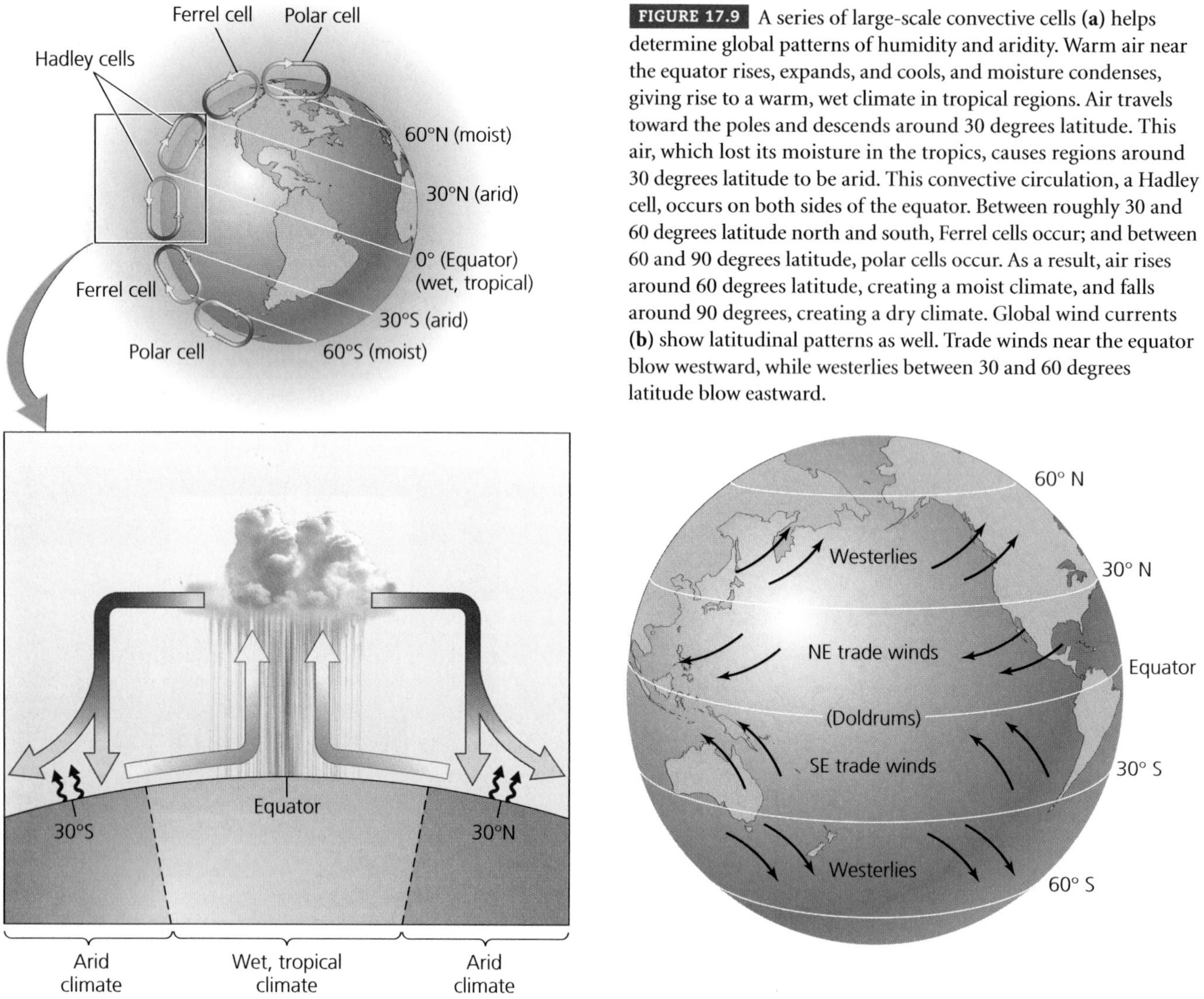

(a) Convection currents

(b) Global wind patterns

FIGURE 17.9 A series of large-scale convective cells (**a**) helps determine global patterns of humidity and aridity. Warm air near the equator rises, expands, and cools, and moisture condenses, giving rise to a warm, wet climate in tropical regions. Air travels toward the poles and descends around 30 degrees latitude. This air, which lost its moisture in the tropics, causes regions around 30 degrees latitude to be arid. This convective circulation, a Hadley cell, occurs on both sides of the equator. Between roughly 30 and 60 degrees latitude north and south, Ferrel cells occur; and between 60 and 90 degrees latitude, polar cells occur. As a result, air rises around 60 degrees latitude, creating a moist climate, and falls around 90 degrees, creating a dry climate. Global wind currents (**b**) show latitudinal patterns as well. Trade winds near the equator blow westward, while westerlies between 30 and 60 degrees latitude blow eastward.

Global climate patterns result from large-scale circulation systems

At larger geographic scales, convective air currents contribute to climatic patterns that are maintained over long periods of time (Figure 17.9a). Near the equator, solar radiation sets in motion a pair of convective cells known as **Hadley cells.** Here, where sunlight is most intense, surface air warms, rises, and expands. As it does so, it releases moisture, producing the heavy rainfall that gives rise to tropical rainforests near the equator. After releasing much of its moisture, this air diverges and moves in currents heading northward and southward. The air in these currents cools and descends back to Earth at about 30 degrees latitude north and south. Because the descending air has low relative humidity, the regions around 30 degrees latitude are quite arid, giving rise to deserts. Two pairs of similar but less intense convective cells, called **Ferrel cells** and **polar cells,** lift air and create precipitation around 60 degrees latitude north and south and cause air to descend at around 30 degrees latitude and in the polar regions.

These three pairs of cells account for the latitudinal distribution of moisture across Earth's surface: wet, tropical climates near the equator; arid climates near 30 degrees latitude; somewhat moist regions near 60 degrees latitude; and dry, arctic conditions near the poles. These patterns, combined with temperature variation, help explain why biomes tend to be arrayed in latitudinal bands (Figure 6.16, ▸p. 170).

The Hadley, Ferrel, and polar cells interact with Earth's rotation to produce the global wind patterns shown in Figure 17.9b. As Earth rotates on its axis, locations on the equator spin faster than locations near the poles. As a result, the north-south air currents of the convective cells appear to be deflected from a straight path, as some portions of the globe move beneath them more quickly than others. This apparent deflection is called the **Coriolis effect**, and it results in the curving global wind patterns evident in Figure 17.9b. Near the equator lies a region with few latitudinal winds known as the *doldrums.* Between the equator and 30 degrees latitude lie the *trade winds,* which blow from east to west. From 30 to 60 degrees latitude are the *westerlies,* which originate from the west and blow east.

People used these global circulation patterns for centuries to facilitate ocean travel by wind-powered sailing ships. Moreover, the atmosphere interacts with the oceans to affect weather, climate, and the distribution of biomes. For instance, winds and convective circulation in ocean water together maintain ocean currents (▸pp. 470–471), and trade winds weaken periodically, leading to El Niño conditions (▸pp. 534, 536–537). The atmosphere's interactions with other systems of the planet can be complex, but even a basic understanding of how the atmosphere functions can help us comprehend how our pollution of the atmo-sphere can affect ecological systems, economies, and human health.

Outdoor Air Pollution

Throughout human history, we have made the atmosphere a dumping ground for our airborne wastes. Whether from primitive wood fires or modern coal-burning power plants, people have generated significant quantities of **air pollutants,** gases and particulate material added to the atmosphere that can affect climate or harm people or other organisms. **Air pollution** refers to the release of air pollutants. In recent decades, government policy and improved technologies have helped us diminish *outdoor air pollution* (often called *ambient air pollution*) substantially in countries of the developed world. However, outdoor air pollution remains a problem, particularly in developing nations and in urban areas worldwide.

Natural sources can pollute

When we think of outdoor air pollution, we tend to envision smokestacks belching black smoke from industrial plants. However, natural processes produce a great deal of the world's air pollution. Some of these natural impacts can be exacerbated by human activity and land-use policies.

Winds sweeping over arid terrain can send huge amounts of dust aloft. In 2001, strong westerlies lifted soil from deserts in Mongolia and China. The dust blanketed Chinese towns, spread to Japan and Korea, traveled eastward across the Pacific Ocean to the United States, then crossed the Atlantic and left evidence atop the French Alps. Every year, hundreds of millions of tons of dust are blown westward by trade winds across the Atlantic Ocean from northern Africa to the Americas (Figure 17.10a). Fungal and bacterial spores carried along with the dust have been linked to die-offs in Caribbean coral reef systems. Although dust storms are natural, the immense scale of these events results from nonsustainable farming and grazing practices that strip vegetation from the soil and promote wind erosion. Continental-scale dust storms took place in the United States in the 1930s, when soil from the drought-stricken Dust Bowl states blew eastward to the Atlantic (▸p. 260).

Volcanic eruptions release large quantities of particulate matter, as well as sulfur dioxide and other gases, into the troposphere. Major eruptions may blow matter into the stratosphere, where it can remain for months or years. The 1980 eruption of Mount Saint Helens in Washington produced 1.1 billion m^3 (1.4 billion yd^3) of dust that circled the planet in 15 days (Figure 17.10b). The massive 1883 eruption on the Indonesian island of Krakatau blew enough dust into the atmosphere to cause a 1°C drop in global temperature (and produce gorgeous sunsets throughout the world).

The burning of vegetation also pollutes the atmosphere with soot and gases. Over 60 million ha (150 million acres) of forest and grassland burn in a typical year (Figure 17.10c). Fires occur naturally, but many today result from "slash-and-burn" forest clearing for farming and grazing in the tropics (▸p. 255). In 1997, a severe drought brought on by the 20th century's strongest El Niño event caused fires in Indonesia to rage out of control. Their smoke sickened 20 million Indonesians, caused cargo ships to collide, and brought about a plane crash in Sumatra. More than 170 million metric tons of carbon monoxide were released from these fires. These, along with tens of thousands of fires in drought-plagued Mexico, Central America, and Africa, released more carbon monoxide into the atmosphere during 1997–1998 than did the worldwide burning of fossil fuels.

Human activities create various types of outdoor air pollution

Human activity can exacerbate the severity of natural air pollution and can also introduce new sources of air pollution.

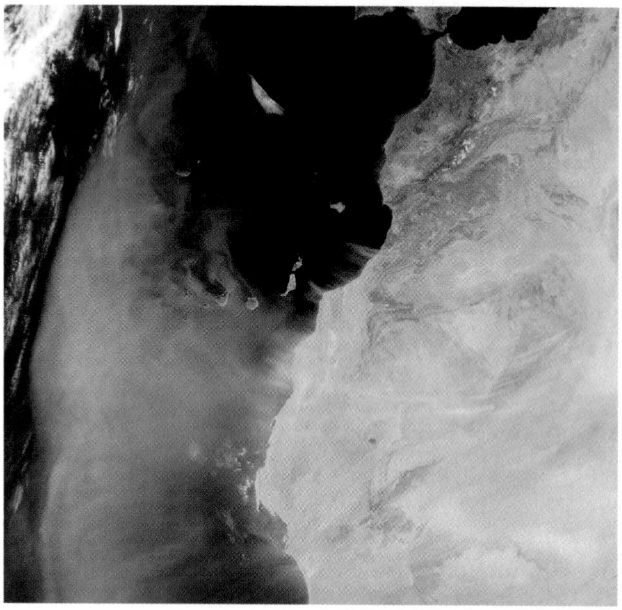

(a) Dust storm from Africa to the Americas

(b) Mount Saint Helens eruption, 1980

(c) Natural fire in California

FIGURE 17.10 Massive dust storms, such as this one blowing across the Atlantic Ocean from Africa to the Americas (**a**), are one type of natural air pollution. Volcanoes are another, as shown by Mount Saint Helens (**b**), which erupted in Washington State in 1980. A third cause is natural fires in forests and grasslands (**c**).

As with water pollution, air pollution can emanate from *point sources* or *non-point sources* (▸ p. 452). A point source describes a specific spot—such as a factory's smokestacks—where large quantities of pollutants are discharged. In contrast, *non-point sources* are more diffuse, often consisting of many small sources (such as millions of automobiles). In 1952 London, coal-burning power plants acted as point sources contributing to the killer smog, while millions of home fireplaces together comprised a potent non-point source. Pollution sources may be stationary or may emit pollutants while in motion, as do automobiles, aircraft, ships, locomotives, construction equipment, and lawn mowers.

Once pollutants are in the atmosphere at sufficient concentrations, they may do harm directly or they may induce chemical reactions that produce harmful compounds. **Primary pollutants,** such as soot and carbon monoxide, are pollutants emitted into the troposphere in a form that can be directly harmful or that can react to form harmful substances. **Secondary pollutants** are harmful substances produced when primary pollutants interact or react with constituents of the atmosphere.

Arguably the greatest human-induced air pollution problem today is our emission of greenhouse gases that contribute to global climate change. We will address this issue separately in Chapter 18.

Clean Air Act legislation has addressed pollution in the United States

To address air pollution in the United States, Congress has passed a series of laws, beginning with the Air Pollution Control Act of 1955. The Clean Air Act of 1963 funded research into pollution control and encouraged emissions standards for automobiles and stationary point sources, such as industrial plants. Subsequent amendments expanded the legislation's scope and established a nationwide air quality monitoring system.

In 1970, Congress thoroughly revised the law in what came to be known as the **Clean Air Act of 1970.** This legislation set stricter standards for air quality, imposed limits on emissions from new stationary and mobile sources, provided new funds for pollution-control research, and enabled citizens to sue parties violating the standards. Once some of these goals came to be viewed as too ambitious, amendments in 1977 altered some standards and extended some deadlines for compliance.

The **Clean Air Act of 1990** sought to strengthen regulations pertaining to air quality standards, auto emissions, toxic air pollution, acidic deposition, and stratospheric ozone depletion. It also introduced an emissions trading program (▸p. 73) for sulfur dioxide. Beginning in 1995, businesses and utilities were allocated permits for emitting this pollutant, and could then buy, sell, or trade these allowances with one another. Each year the overall amount of allowed pollution was decreased. This market-based incentive program has proven successful, and has spawned similar programs at state and regional levels and for other pollutants.

As a result of Clean Air Act legislation, the U.S. Environmental Protection Agency (EPA) sets nationwide standards for emissions of pollutants and concentrations of pollutants in ambient air throughout the nation. However, it is largely up to the states to monitor air quality and develop, implement, and enforce regulations within their boundaries. States submit implementation plans to the EPA for approval, and if a state's plans are not adequate, the EPA can take over enforcement in that state.

The EPA sets standards for "criteria pollutants"

The EPA and the states focus on six **criteria pollutants,** pollutants judged to pose especially great threats to human health—carbon monoxide (CO), sulfur dioxide (SO_2), nitrogen dioxide (NO_2), tropospheric ozone (O_3), particulate matter, and lead (Pb). For these, the EPA has established *national ambient air quality standards (NAAQS),* which are maximum allowable concentrations of these pollutants in ambient outdoor air. Through risk assessment procedures (▸pp. 422–423), these six pollutants were selected on the basis of criteria relating to human health.

Carbon monoxide Carbon monoxide is a colorless, odorless gas produced primarily by the incomplete combustion of fuels. In the United States in 2004, 87.2 million tons of CO were released, making it the most abundant air pollutant by mass. Vehicles account for about 62% of these emissions, but other sources include lawn and garden equipment (10%), forest fires (6%), open burning of industrial waste (3%), and residential wood burning (2%). Carbon monoxide poses risk to humans and other animals, even in small concentrations. It can bind irreversibly to hemoglobin in red blood cells, preventing the hemoglobin from binding with oxygen. U.S. emissions of CO have decreased in recent decades largely because of cleaner-burning motor vehicle engines.

Sulfur dioxide Like CO, sulfur dioxide is a colorless gas. Of the 15.2 million metric tons of SO_2 released in the United States in 2004, about 70% resulted from the combustion of coal for electricity generation and industry. During combustion, elemental sulfur (S) in coal reacts with oxygen gas (O_2) to form SO_2. Once in the atmosphere, SO_2 may react to form sulfur trioxide (SO_3) and sulfuric acid (H_2SO_4), which may then fall back to Earth in the form of acid precipitation.

Nitrogen dioxide Nitrogen dioxide is a highly reactive, foul-smelling reddish brown gas that contributes to smog and acid precipitation. Along with nitric oxide (NO), NO_2 belongs to a family of compounds called nitrogen oxides (NO_x). Nitrogen oxides result when atmospheric nitrogen and oxygen react at the high temperatures created by combustion engines. Of the 18.8 million tons of nitrogen oxides released in the United States in 2004, over half resulted from combustion in vehicle engines. Electrical utility and industrial combustion accounted for most of the rest.

Tropospheric ozone Although ozone in the stratosphere shields organisms from the dangers of UV radiation, O_3 from human activity forms and accumulates low in the troposphere and acts as a pollutant. In the troposphere, this colorless gas results from the interaction of sunlight, heat, nitrogen oxides, and volatile carbon-containing chemicals. Ozone is therefore categorized as a secondary pollutant. A major component of smog, O_3 can pose health risks as a result of its instability as a molecule; this triplet of oxygen atoms will readily release one of its threesome, leaving a molecule of oxygen gas and a free oxygen atom. The free oxygen atom may then participate in reactions that can injure living tissues and cause respiratory problems. Although concentrations fell by 11–18% (depending on how they were measured) in the United States from 1982 to 2001, tropospheric O_3 is the pollutant that most frequently exceeds the EPA standard.

Particulate matter Particulate matter is composed of solid or liquid particles small enough to be suspended in the atmosphere. Particulate matter includes primary pollutants such as dust and soot, as well as secondary pollutants such as sulfates and nitrates. Particulate matter can damage respiratory tissues when inhaled. Most particulate matter in the atmosphere is wind-blown dust (60%), but 2.5 million tons of particulate matter was released by human activities in the United States in 2004. Along with sulfur dioxide, it was largely the emission of particulate matter from industrial and residential coal-burning sources that produced London's 1952 killer smog and the deaths resulting from that episode.

Lead Lead is a heavy metal that enters the atmosphere as a particulate pollutant. The lead-containing compounds tetraethyl lead and tetramethyl lead, when added to gasoline, improve engine performance. However, the exhaust from leaded gasoline emits lead into the atmosphere, from which it can be inhaled or can be deposited on land and water. Lead can enter the food chain, accumulate within body tissues, and cause central nervous system malfunction, mental retardation among children, and a variety of other ailments. Once people recognized the dangers of lead, leaded gasoline was phased out in the United States and other industrialized nations, and U.S. lead emissions plummeted 93% from 1980 to 1990. Since then, lead emissions have remained steady and low. Today most lead emitted in the United States comes from industrial metal smelting. However, many developing nations continue to add lead to gasoline and experience significant lead pollution.

Agencies monitor pollutants that affect air quality

State and local agencies also monitor, calculate, and report to the EPA emissions of major pollutants that affect ambient concentrations of the six criteria pollutants. These include the four criteria pollutants that are primary pollutants (carbon monoxide, sulfur dioxide, particulate matter, and lead), as well as all nitrogen oxides (because NO reacts readily in the atmosphere to form NO_2, which is both a primary and secondary pollutant). Tropospheric ozone is a secondary pollutant only, so there are no emissions to monitor. Instead the EPA monitors emissions of volatile organic compounds, which can react to produce ozone and other secondary pollutants.

Volatile organic compounds (VOCs) are carbon-containing chemicals used in industrial processes such as dry-cleaning and manufacturing. One group of VOCs consists of hydrocarbons (► pp. 98–99) such as methane (CH_4, the primary component of natural gas), propane (C_3H_8, used as a portable fuel), butane (C_4H_{10}, found in cigarette lighters), and octane (C_8H_{18}, a component of gasoline). Human activity accounts for about half the VOC emissions in the United States, and the remainder comes from natural sources. For example, plants produce isoprene (C_5H_8) and terpene ($C_{10}H_{15}$), and animals produce methane. The largest sources of anthropogenic VOC emissions include industrial use of solvents (28%) and vehicle emissions (27%).

Air pollution has decreased markedly since 1970

Since the Clean Air Act of 1970, emissions of each of the six monitored pollutants have decreased, and total emissions of the six together have declined by 54% (Figure 17.11a).

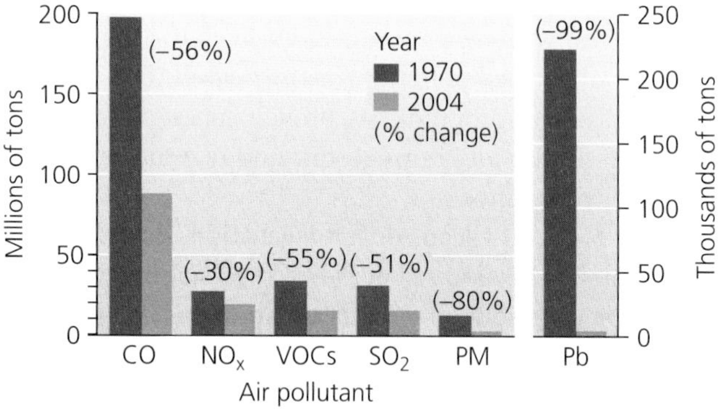

(a) Declines in six major pollutants

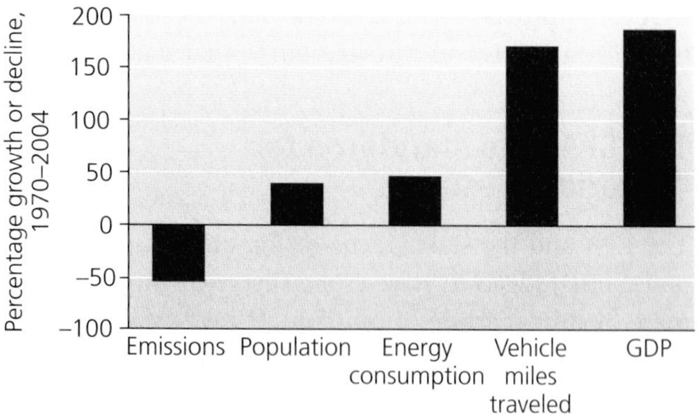

(b) Trends in major indicators

FIGURE 17.11 The EPA tracks emissions of several major pollutants into ambient air. Shown are emissions of four of the six "criteria pollutants," along with nitrogen oxides and volatile organic compounds. Each of these pollutants has shown substantial declines since 1970, and emissions from all six together have declined by 54% **(a)**. This decrease in emissions has occurred despite increases in U.S. population, energy consumption, vehicle miles traveled, and gross domestic product **(b)**. Go to GRAPHit! at www.aw-bc.com/withgott or on the student CD-ROM. Data from U.S. EPA, 2004.

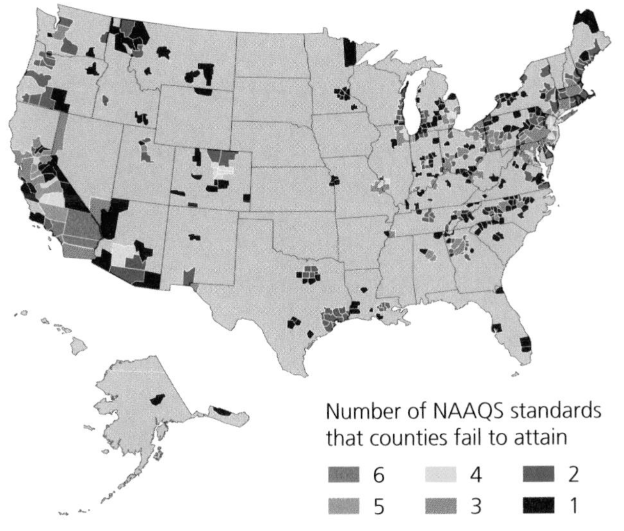

FIGURE 17.12 Nearly half of all Americans live in counties that periodically fail to meet the EPA's national ambient air quality standards (NAAQS) for at least one criteria pollutant. This map shows counties that failed to attain the standards for one (blue) through six (red) of the six criteria pollutants. Data are for April 2005, from U.S. EPA.

This has occurred despite substantial increases in the nation's population, energy consumption, miles traveled by vehicle, and gross domestic product (Figure 17.11b). Because of the success in reducing emissions, air quality in the United States has improved markedly. As just one indicator, EPA data show that the percentage of days U.S. citizens were exposed to unhealthy air dropped from 10% in 1988 to 3% in 2001.

The reduction of outdoor air pollutant levels since 1970 represents one of the greatest accomplishments in safeguarding human health and environmental quality in the United States. However, the nation still has plenty of room to improve. Outdoor air pollution remains a problem, and in recent years, EPA monitoring has found that nearly half of all Americans live in counties where at least one of the six criteria pollutants periodically reaches unhealthy levels (Figure 17.12).

Weighing the Issues: Your County's Air Quality

Locate where you live on the map in Figure 17.12. What is the status of your county's air quality, and how does your county compare to the rest of the nation? What factors do you think account for the quality of its air? Can you propose any solutions for reducing air pollution in your county?

Toxic substances are also major pollutants

Other chemicals known to cause serious health or environmental problems are classified as **toxic air pollutants.** These include substances known to cause cancer, reproductive defects, or neurological, developmental, immune system, or respiratory problems. Also included are substances that cause substantial ecological harm by affecting the health of nonhuman animals and plants. Some toxic air pollutants are produced naturally. For example, hydrogen sulfide gas (H_2S) gives the mud of swamps and bogs the odor of rotten eggs. However, most toxic air pollutants are produced by human activities, such as metal smelting, sewage treatment, and industrial processes. The 1990 Clean Air Act identifies 188 different toxic air pollutants, ranging from the heavy metal mercury (from coal-burning power plant emissions and other sources) to VOCs such as benzene (a component of gasoline) and methylene chloride (found in paint stripper). Among the 188 pollutants are 21 from mobile sources (such as diesel exhaust) and 33 "urban hazardous" pollutants judged to pose the greatest health risks in urban areas.

State and federal agencies do not monitor toxic air pollutants as extensively as they do the six criteria pollutants, but so far 300 monitoring sites are operating, and coverage is improving. The EPA estimates that because of Clean Air Act regulations, from 1990 to 1999 emissions of toxic air pollutants decreased by 30%.

Burning fossil fuels produces industrial smog

In response to the increasing incidence of fogs polluted by the smoke of Britain's industrial revolution, a British scientist coined the term *smog* long before the 1952 event in London. Today the term is used worldwide to describe unhealthy mixtures of air pollutants that often form over urban areas. The smog that enveloped London in 1952 was what we would today call **industrial smog,** or gray-air smog. When coal or oil is burned, some portion is completely combusted, forming CO_2; some is partially combusted, producing CO; and some remains unburned and is released as soot, or particles of carbon. Moreover, coal contains varying amounts of contaminants, including mercury and sulfur. Sulfur reacts with oxygen to form sulfur dioxide, which can undergo a series of reactions to form sulfuric acid and ammonium sulfate (Figure 17.13a). These chemicals and others produced by further reactions, along with soot, are the main components of industrial smog, and give the smog its characteristic gray color.

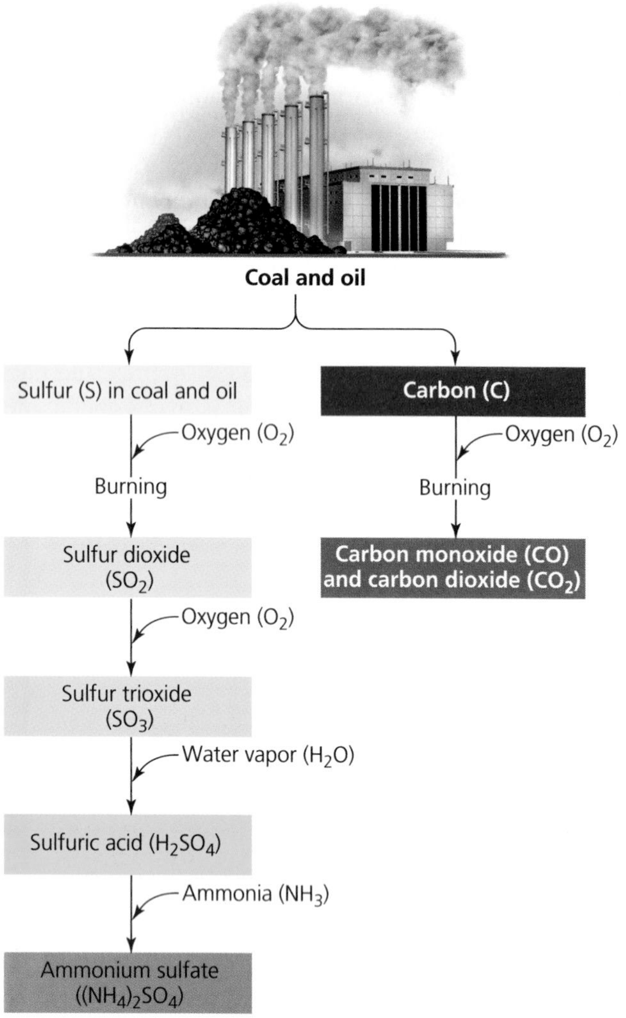

(a) Burning sulfur-rich oil or coal without adequate pollution control technologies

(b) Donora, Pennsylvania, at midday in the 1948 smog event

FIGURE 17.13 Emissions from the combustion of coal and oil in manufacturing plants and utilities without pollution control technologies can create industrial smog. Industrial smog consists primarily of sulfur dioxide and particulate matter, as well as carbon monoxide and carbon dioxide from the carbon component of fossil fuels. Sulfur contaminants in fossil fuels when combusted create the sulfur dioxide, which in the presence of other chemicals in the atmosphere can produce several other sulfur compounds **(a).** Under certain weather conditions, industrial smog can blanket whole towns or regions, as it did in Donora, Pennsylvania, shown here in the daytime during its deadly 1948 smog episode **(b).**

Industrial smog is far less common today in developed nations than it was 50–100 years ago. In the wake of the 1952 London episode and others, the governments of most developed nations began regulating industrial emissions to minimize the external costs (▸ pp. 43–44) they impose on citizens. However, in industrializing regions such as China, India, and Eastern Europe, heavy reliance on coal burning (both by industry and by citizens heating and cooking in their homes), combined with lax air pollution controls, produces industrial smog that poses significant health risks in many urban areas.

Although coal combustion supplies the chemical constituents for industrial smog, weather also plays a role, as it did in London in 1952. A similar event occurred 4 years earlier in Donora, Pennsylvania. A thermal inversion trapped smog containing particulate matter emissions from a steel and wire factory. Twenty-one people were killed, and over 6,000 people—nearly half the town—became ill (Figure 17.13b). In Donora's killer smog, air near the ground cooled during the night. Normally, morning sunlight warms the land and air, causing air to rise. However, because Donora is located in hilly terrain, too little sun reached the valley floor to warm and disperse the cold air. The resulting thermal inversion kept a pall of smog over the town long enough to impair visibility and cause serious health problems. Hilly topography such as Donora's is a factor in the air pollution of many other cities where surrounding mountains trap air and create inversions. This is true for the Los Angeles basin, which has long symbolized chronic smog problems in American popular culture. Modern-day Los Angeles, however, suffers from a different type of smog, one called photochemical smog.

Photochemical smog is produced by a complex series of atmospheric reactions

A photochemical process is one whose activation requires light. **Photochemical smog,** or brown-air smog, is formed through light-driven chemical reactions of primary pollutants and normal atmospheric compounds that produce a mix of over 100 different chemicals, tropospheric ozone

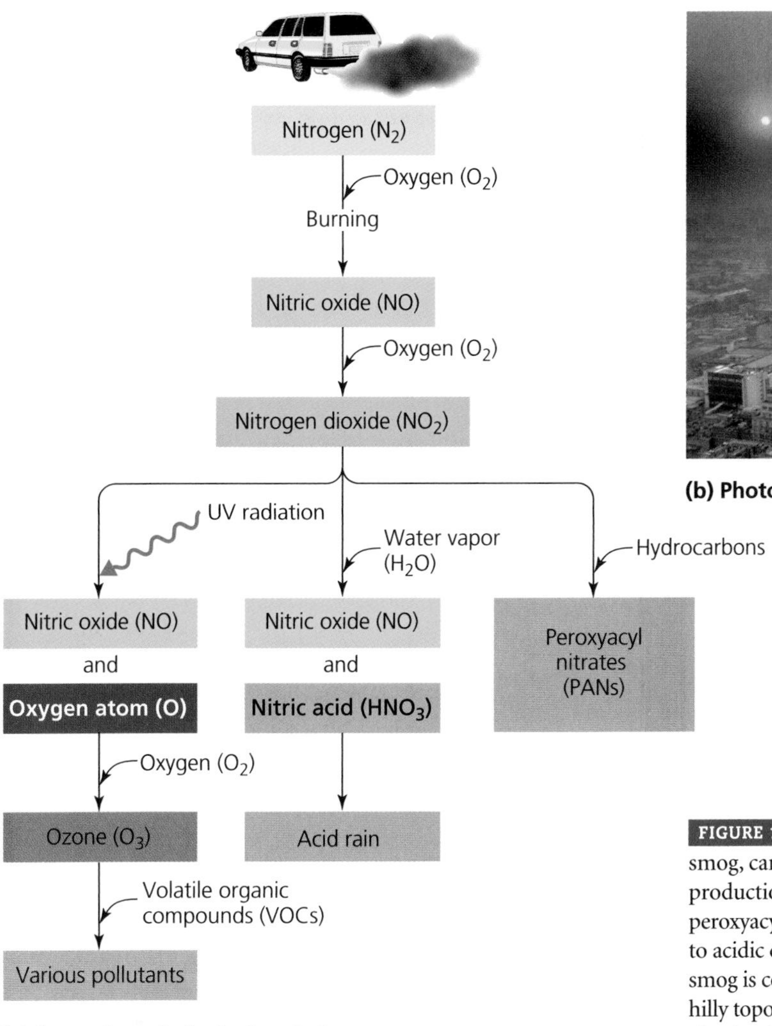

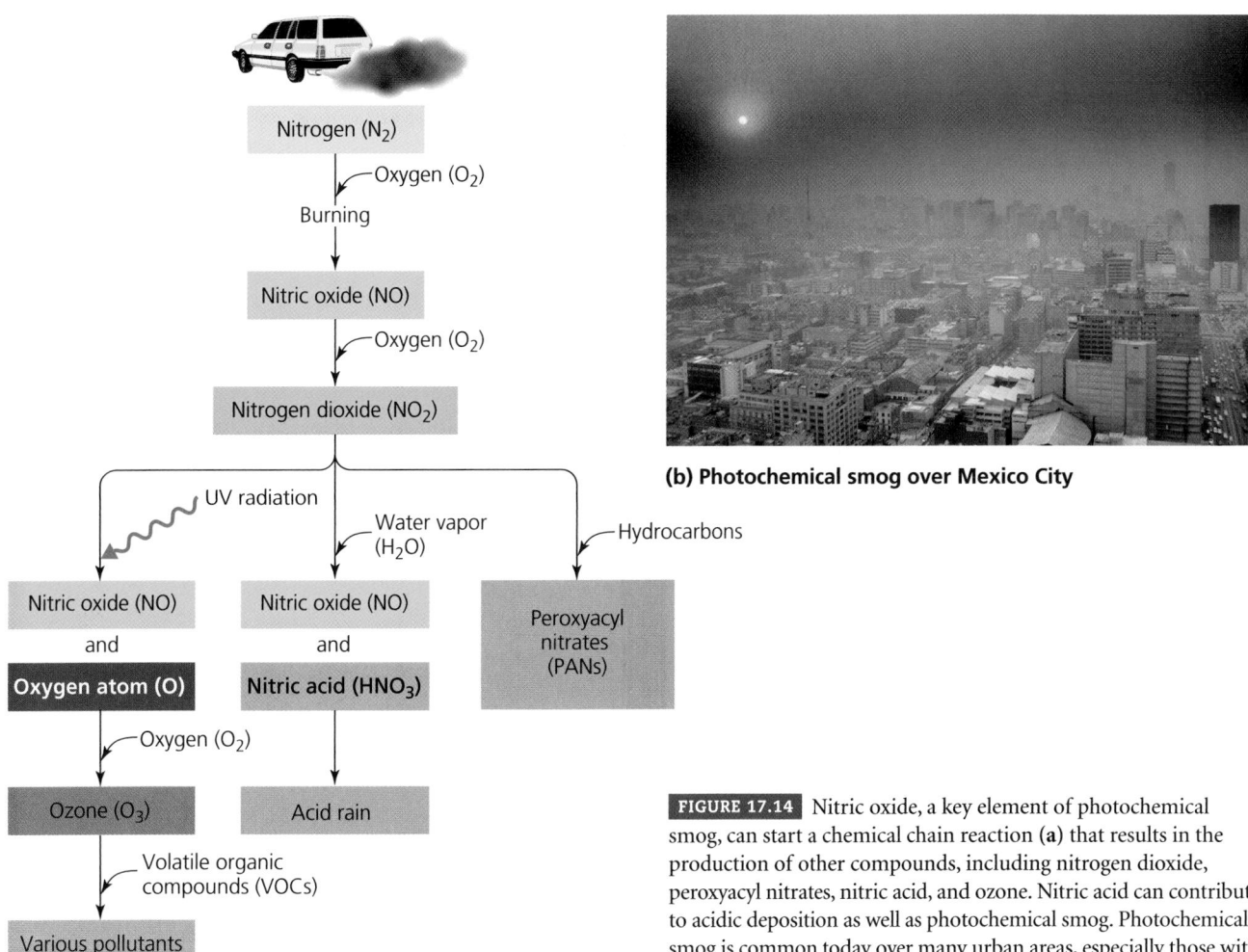

(a) Formation of photochemical smog

(b) Photochemical smog over Mexico City

FIGURE 17.14 Nitric oxide, a key element of photochemical smog, can start a chemical chain reaction (**a**) that results in the production of other compounds, including nitrogen dioxide, peroxyacyl nitrates, nitric acid, and ozone. Nitric acid can contribute to acidic deposition as well as photochemical smog. Photochemical smog is common today over many urban areas, especially those with hilly topography or frequent inversion layers. Mexico City (**b**) is one city that frequently experiences photochemical smog.

often being the most abundant among them (Figure 17.14a). High levels of NO_2 cause photochemical smog to form a brownish haze over cities (Figure 17.14b). Hot, sunny, windless days in urban areas provide perfect conditions for the formation of photochemical smog. Exhaust from morning traffic releases large amounts of NO and VOCs into a city's air. Sunlight then promotes the production of ozone and other constituents of photochemical smog. Levels of photochemical pollutants in urban areas typically peak in mid-afternoon and at sufficient levels can irritate people's eyes, noses, and throats. Air pollutants called *peroxyacyl nitrates,* created by the reaction of NO_2 with hydrocarbons, can induce further reactions that damage living tissues in animals and plants.

Photochemical smog afflicts many major cities, especially those with topography and weather conditions that promote it. In Athens, Greece, site of the 2004 Olympics, the problem had been bad enough that the city government provided incentives to replace aging automobiles. It also mandated that cars with odd-numbered license plates be driven only on odd-numbered days, and those with even-numbered plates only on even-numbered days. According to Greek officials, smog has been reduced by 30% since 1990 as a result.

Synthetic chemicals deplete stratospheric ozone

A pollutant at low altitudes, ozone is a highly beneficial gas at altitudes centering around 25 km (15 mi) in the lower stratosphere, where it is concentrated in the so-called *ozone layer.* Here, concentrations of ozone are only about 12 parts per million. However, ozone molecules are so effective at absorbing incoming ultraviolet radiation from the sun that this concentration is adequate to protect life on Earth's surface from the damaging effects of UV radiation.

In the 1960s, atmospheric scientists began wondering why their measurements of ozone were lower than theoretical models predicted. Researchers speculating that natural or artificial chemicals were depleting ozone finally

VIEWPOINTS

Air Pollution

Despite improvements in air quality, air pollution leading to photochemical smog remains a health and environmental concern in and around many major cities. **What is needed to reduce this type of pollution?**

A Policy Portfolio Approach to Fighting Smog

Federal and state regulations to control emissions from factories, power plants, and cars have reduced smog. Further progress will require more effective and affordable means to reduce smog "precursors," nitrogen oxides and volatile organic compounds, from old and new sources. Simply tightening the same regulations used in the past and applying them nationwide to the same sources is unlikely to significantly reduce smog for three reasons: control costs will be high and the marginal gains low for many "old" stationary sources already being regulated; emissions from many smaller sources, such as sport-utility vehicles, trucks, lawnmowers, and motorboats, are a growing concern; and the right mix of reductions in nitrogen oxide and volatile organic compound emissions depends on local circumstances.

There is no silver bullet. Regional air quality regulators will need to apply several different policy approaches, called a policy portfolio, to achieve further reductions. For large stationary sources, emissions trading is an economically attractive supplement to current technology-based regulations. Emissions trading works as follows: First, government regulators set a regional limit on nitrogen oxides. Then, regulators assign or auction to emissions sources only as many allowances (for example, one allowance equals one ton of nitrogen oxide emissions) as the overall limit (or "cap") permits. Lastly, emissions sources could take one of three actions: reduce their own emissions to stay within their allowances; sell or retire any excess allowances; or buy others' excess allowances to meet their own obligations. They would choose according to market prices, with the cheapest sources of reductions being made available first.

Small mobile sources that run on diesel fuel and gasoline are also a problem. For now, the federal government still needs to press manufacturers to build cleaner-burning vehicles and fuel makers to make cleaner fuels, as the EPA is doing. Meanwhile, regional regulators need to be given the flexibility and enforcement tools to tailor their efforts to highly variable regional climate and atmospheric conditions.

Debra Knopman is vice president and director of the Infrastructure, Safety, and Environment division of the RAND Corporation. This statement is based primarily on work she did at the Progressive Policy Institute, where she was director of the Center for Innovation and the Environment from 1995 to 2001.

To Reduce Smog, States Must Take Up the Slack

Nearly half the people in this country—136 million individuals—live in areas where smog levels exceed the Environmental Protection Agency's health standard. Unfortunately for the state governments that are responsible for those areas, a substantial amount of the pollution that forms smog in their cities and valleys originates from power plants in up-wind states.

In May 2005, EPA issued a regulation to address smog-forming pollution that crosses state lines. The rule will require states in the eastern half of the country to tighten limits on emissions of nitrogen oxides from their power plants.

The prescribed limits are too weak, however, to eliminate all of the interstate smog transport, much less clean up all of the smog. Without additional measures, 20 million people will live in areas exceeding EPA's smog standard in 2015—5 years after the latest smog cleanup deadline in the Clean Air Act.

To protect their citizens from asthma attacks, hospitalizations, and missed schooldays and workdays, state governments must pick up the slack. That will require new urban planning measures to reduce vehicle miles traveled, because tailpipe emissions contribute to smog. It will also require states to impose new emissions limits—more stringent than the federally prescribed ones—on their own power plants, oil refineries, and factories.

In an attempt to avoid those new limits, large polluters in 2004 and again in 2005 convinced congressional allies to attach to the federal energy bill a provision that would have extended the Clean Air Act's smog cleanup deadlines by a decade. Fortunately, the measure was stripped from the bill after state officials made clear they were not interested in ignoring the problem. That was a good start. Now there is a lot of work to do.

David McIntosh is a staff attorney at the Natural Resources Defense Council in Washington, D.C. He litigates and lobbies to ensure effective implementation of clean air laws and to counter efforts aimed at weakening those laws.

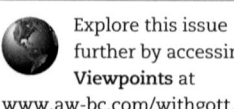

Explore this issue further by accessing **Viewpoints** at www.aw-bc.com/withgott.

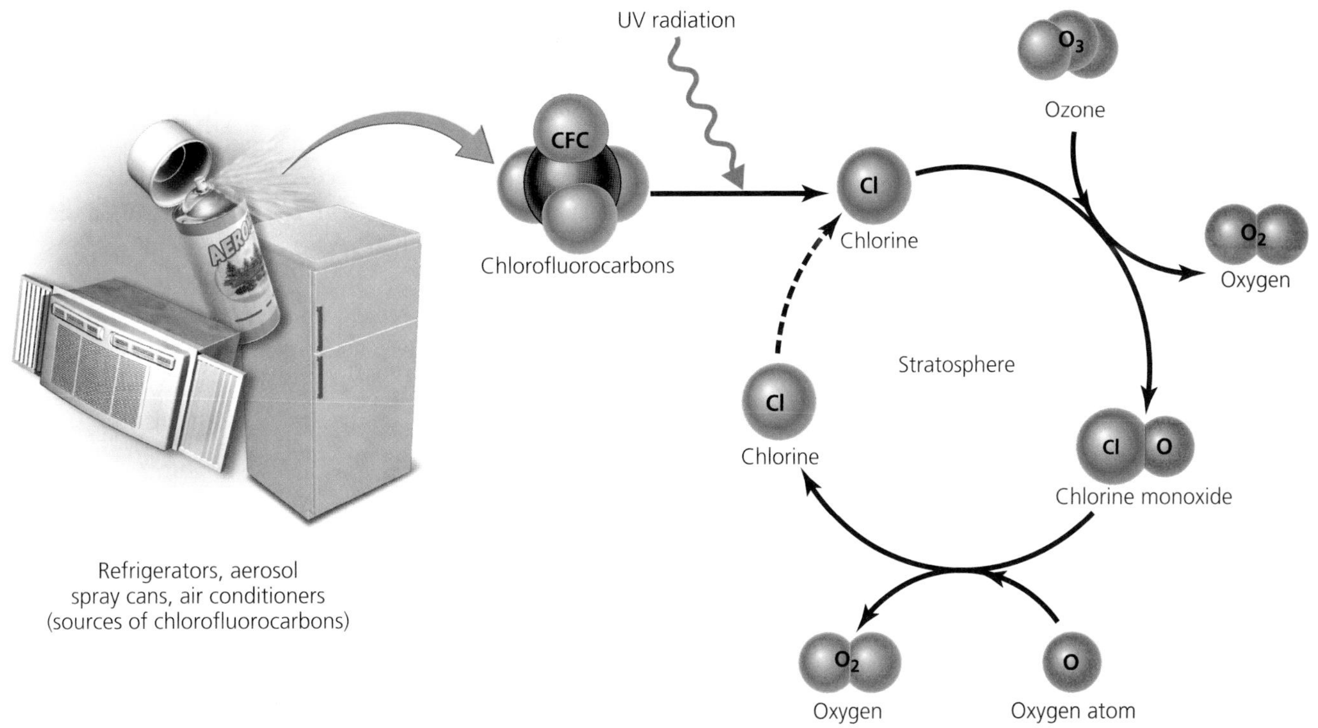

FIGURE 17.15 Chlorofluorocarbons lead to the destruction of ozone molecules in the presence of ultraviolet radiation. A chlorine atom released from a CFC molecule reacts with an ozone molecule, forming one molecule of oxygen gas and one chlorine monoxide (ClO) molecule. The oxygen atom in the ClO molecule will then bind with a stray oxygen atom to form oxygen gas, leaving the chlorine atom to begin the destructive cycle anew. The self-perpetuating nature of this process means that one CFC molecule can destroy a great many ozone molecules.

pinpointed a group of human-made compounds derived from simple hydrocarbons, such as ethane and methane, in which hydrogen atoms are replaced by chlorine, bromine, or fluorine. One class of such compounds, **chlorofluorocarbons (CFCs),** was being mass-produced by industry at a rate of a million metric tons per year in the early 1970s, and it was growing by 20% a year.

In 1974, atmospheric scientists Sherwood Rowland and Mario Molina showed that CFCs could deplete stratospheric ozone by releasing chlorine atoms that split ozone molecules, creating from each of them an O_2 molecule and a ClO molecule (Figure 17.15). Three years before Rowland and Molina's study, researcher J. E. McDonald had predicted that ozone loss, by allowing more UV radiation to reach the surface, would result in thousands more skin cancer cases each year. This caught the attention of policymakers, environmentalists, and industry alike (see "The Science behind the Story," ▸pp. 516–517).

Then in 1985, scientists from the British Antarctic Survey announced that stratospheric ozone levels over Antarctica had declined by 40–60% in the previous decade, leaving a thinned ozone concentration that was soon dubbed the *ozone hole* (Figure 17.16). Research over the next few years confirmed the link between CFCs and ozone loss in the Antarctic and indicated that depletion was also occurring in the Arctic, and perhaps globally. Already concerned about skin cancer, scientists were becoming anxious over the possible effects of increased UV radiation on ecosystems. Research was showing a multitude of effects, including harm to crops and to the productivity of ocean phytoplankton, the base of the marine food chain.

The Montreal Protocol addressed ozone depletion

In light of the science and the ongoing health and ecological concerns, international efforts to restrict CFC production finally bore fruit in 1987 with the **Montreal Protocol.** In this treaty, signatory nations (eventually numbering 180) agreed to cut CFC production in half. Five follow-up agreements strengthened the pact, deepening the cuts, advancing timetables for compliance, and addressing related ozone-depleting chemicals. Today the production and use

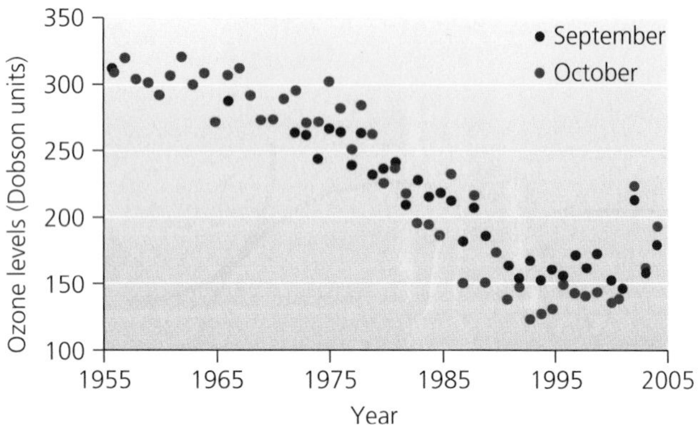

(a) Monthly mean ozone levels at Halley Bay, Antarctica

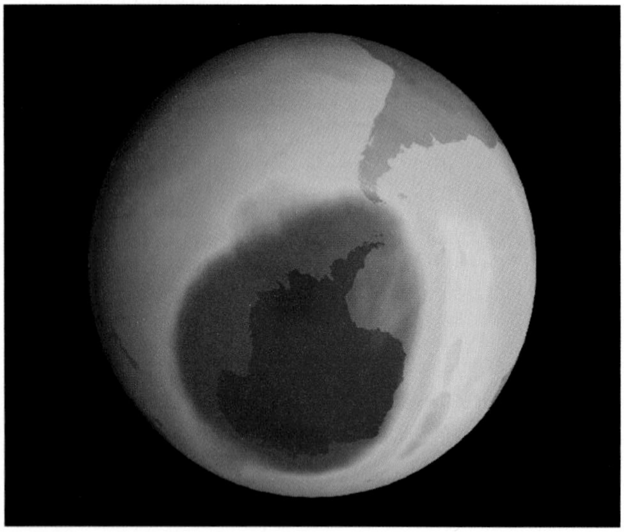

(b) The "ozone hole" over Antarctica, September 2000

FIGURE 17.16 The "ozone hole" consists of a region of thinned ozone density in the stratosphere over Antarctica and the southernmost ocean regions. It has reappeared seasonally each September in recent decades. Data from Halley Bay, Antarctica **(a)**, show a steady decrease in stratospheric ozone concentrations from the 1960s to 1990. Ozone-depleting CFCs began to be regulated under the Montreal Protocol in 1987, and ozone concentrations stopped declining. Colorized satellite imagery from September 6, 2000 **(b)**, shows the "ozone hole" (blue) at its maximal recorded extent to date. Data from British Antarctic Survey.

of ozone-depleting compounds has fallen by 95% since the late 1980s, and scientists can discern the beginnings of long-term recovery of the ozone layer (although much of the 5 billion kg of CFCs emitted into the troposphere has yet to diffuse up into the stratosphere, and CFCs are slow to dissipate or break down). Industry was able to shift to alternative, environmentally safer chemicals, which have largely turned out to be cheaper and more efficient. For these reasons, the Montreal Protocol and its follow-up amendments are widely considered the biggest success story so far in addressing any global environmental problem.

Environmental scientists have attributed this success primarily to two factors:

1. Policymakers engaged industry in helping to solve the problem, and government and industry worked together on developing technological fixes and replacement chemicals. This cooperation reduced the battles that typically erupt between environmentalists and industry.
2. Implementation of the Montreal Protocol after 1987 followed an adaptive management approach (▸p. 346), altering strategies midstream in response to new scientific data, technological advances, or economic figures.

Because of its success in addressing ozone depletion, the Montreal Protocol is widely seen as a model for international cooperation in addressing other pressing global problems, such as persistent organic pollutants (▸pp. 426, 428), climate change (▸pp. 550, 552), and biodiversity loss (▸pp. 335–336).

Weighing the Issues: International Cooperation to Solve Global Problems

The Montreal Protocol showed how international collaboration, together with technological advances, can drastically and rapidly address a pressing environmental problem. So far, however, global problems such as organic pollutants, climate change, and biodiversity loss have not seen the same degree of targeted action. Why do you think this is? Besides the effort to halt stratospheric ozone depletion, can you name other success stories in addressing major environmental problems? Are any on the horizon?

Acidic deposition represents another transboundary pollution problem

Just as the problem of stratospheric ozone depletion crosses political boundaries, so does another atmospheric pollution concern—acidic deposition. **Acidic deposition** refers to the deposition of acidic or acid-forming pollutants from the atmosphere onto Earth's surface. This can take place either by precipitation (commonly referred to as *acid rain,* but also including acid snow, sleet, and hail), by fog, by gases, or by the deposition of dry particles. Acidic deposition is one type of **atmospheric deposition,** which refers more broadly to the wet or dry deposition on land of a wide variety of pollutants, including mercury, nitrates, organochlorines, and others.

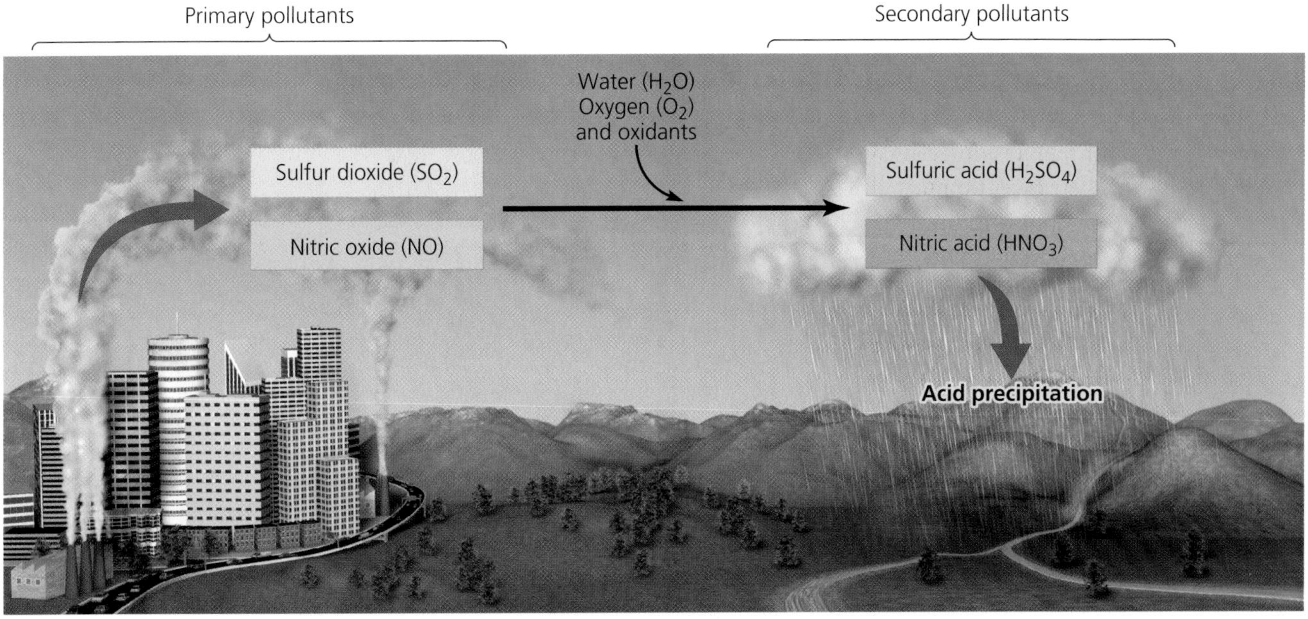

FIGURE 17.17 Acidic deposition can have consequences many miles downwind from its source. Emissions containing sulfur dioxide and nitric oxide from industries and utilities begin the process. Sulfur dioxide and nitric oxide can be transformed into sulfuric acid and nitric acid through chemical reactions in the atmosphere, and these acidic compounds descend to Earth's surface in rain, snow, fog, and dry deposition.

Acidic deposition originates primarily with sulfur dioxide and nitrogen oxides, pollutants produced largely through fossil fuel combustion by automobiles, electric utilities, and industrial facilities. Once emitted into the troposphere, these pollutants can react with water, oxygen, and oxidants to produce compounds of low pH (▸ p. 98), primarily sulfuric acid and nitric acid. Suspended in the troposphere, droplets of these acids may travel for up to days or weeks, sometimes covering hundreds or thousands of kilometers before falling in precipitation (Figure 17.17).

Acidic deposition can have wide-ranging, cumulative detrimental effects on ecosystems and on our built environment (Table 17.1). Acids can leach basic minerals such as calcium and magnesium from soil, changing soil chemistry and harming plants and soil organisms. Streams, rivers, and lakes may become significantly acidified from runoff. In fact, thousands of lakes in Canada, Scandinavia, the United States, and elsewhere now contain water acidic enough to kill fish. Fish can die when acidic conditions cause toxic aluminum to be more readily accessible to their tissues. Elevated aluminum in the soil hinders water and nutrient uptake by plants. In some regions of the United States, acid fog with a pH of 2.3 (equivalent to vinegar) can envelop forests for extended periods. Some forests in eastern North America have experienced widespread tree die-back from these conditions.

Besides harming trees, acid precipitation also may damage agricultural crops. Moreover, it can erode stone buildings, eat away at cars, and erase the writing from tombstones. Ancient cathedrals in Europe, monuments in

Table 17.1 Effects of Acidic Deposition on Ecosystems in the Northeastern United States

Acidic deposition in northeastern forests has . . .

- Accelerated leaching of base cations (ions that counteract acidic deposition) from soil
- Allowed sulfur and nitrogen to accumulate in soil
- Increased dissolved inorganic aluminum in soil, hindering plant uptake of water and nutrients
- Leached calcium from needles of red spruce, leading to tree mortality from wintertime freezing
- Increased mortality of sugar maples due to leaching of base cations from soil and leaves
- Acidified 41% of Adirondack, New York, lakes and 15% of New England lakes
- Lowered lakes' capacity to neutralize further acids
- Elevated aluminum levels in surface waters
- Reduced species diversity and abundance of aquatic life, and negatively affected entire food webs

Source: Adapted from Driscoll, C.T., et al. 2001. *Acid rain revisited.* Hubbard Brook Research Foundation.

The Science behind the Story

Identifying CFCs as the Main Cause of Ozone Depletion

For half a century after their invention in the 1920s, chlorofluorocarbons (CFCs) were thought to be useful, nontoxic, and environmentally friendly. In the early 1970s, however, scientists became concerned that CFCs could cause long-term damage to the ozone layer. By the late 1980s, evidence for such damage had become strong enough to justify a complete ban on CFC production. In their attempts to understand how CFCs influenced stratospheric ozone, scientists relied on a wide variety of data sources, including historical records, field observations, laboratory experiments, and computer models.

Stratospheric ozone and CFCs had each been the subject of much research before they were linked in the 1970s. Ozone was discovered in 1839, and its presence in the upper atmosphere was first proposed in the 1880s. In 1924, British scientist G. M. B. Dobson built what has become the standard instrument for measuring ozone from the ground. By the 1970s, the Dobson ozone spectrophotometer (see the figure) was being used by a global network of observation stations. However, scientists were unable to establish a clear picture of global trends in ozone concentrations because of natural variations in ozone levels and the difficulty of comparing data from different stations.

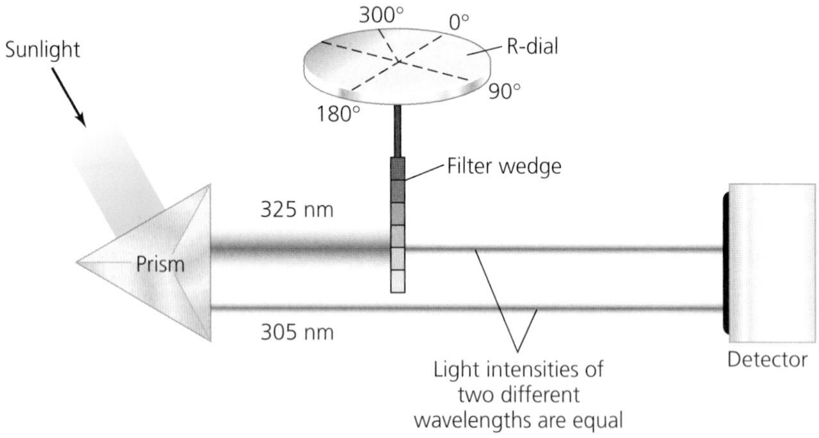

In the Dobson spectrophotometer, UV light passes through a prism that separates wavelengths of 325 nanometers (nm) and 305 nm, and then travels toward the detector. Because ozone absorbs wavelengths of 305 nm but not wavelengths of 325 nm, light that reaches the instrument after passing through the atmosphere contains more light of 325 nm wavelengths. The ratio between the intensities of the two wavelengths of light indicates the amount of ozone in the light's path between the sun and the spectrophotometer. To measure this ratio, the R-dial rotates, causing the filter wedge to block more and more 325-nm light, until the intensities of 325-nm and 305-nm light are equal. At that point, the reading on the R-dial is recorded, and a conversion is used to calculate the atmospheric ozone concentration. Figure adapted from University of Alaska, Fairbanks, http://ozone.gi.alaska.edu.

Research on CFCs also had a long history. First invented in 1928, CFCs were found to be useful as refrigerants, fire extinguishers, and propellants for aerosol spray cans. Starting in the 1960s, CFCs also found wide use as cleaners for electronics and as a part of the process of manufacturing rigid polystyrene foams. Research on the chemical properties of CFCs showed that they were almost completely inert; that is, they rarely reacted with other chemicals. Therefore, scientists surmised that, at trace levels, CFCs would be harmless to both people and the environment.

However, in June 1974, chemists F. Sherwood Rowland and Mario

Washington, D.C., temples in Asia, and stone statues throughout the world are experiencing billions of dollars of damage as their features gradually wear away (Figure 17.18).

Because the pollutants leading to acid deposition can travel long distances, their effects may be felt far from their sources—a situation that has led to political bickering among the leaders of states and nations. For instance, much of the pollution from power plants and factories in Pennsylvania, Ohio, and Illinois falls out in states to their east, including New York, Vermont, and New Hampshire, as well as in regions to the north, including Ontario, Quebec, and the maritime provinces of Canada. As Figure 17.19 shows, many regions of greatest acidification are downwind of major source areas of pollution.

Molina published a paper in the journal *Nature*, arguing that the inertness that made CFCs so ideal for industrial purposes could also have disastrous consequences for the ozone layer. More-reactive chemicals are broken down to their constituent atoms in the lower atmosphere. CFCs, in contrast, reach the stratosphere unchanged. Once CFCs reach the stratosphere, intense ultraviolet radiation from the sun breaks them into their constituent chlorine and carbon atoms. Each free chlorine atom, it was calculated, can catalyze the destruction of as many as 100,000 ozone molecules.

Rowland and Molina were the first to assemble a complete picture of the threat posed by CFCs, but they could not have reached their conclusions without the contributions of other scientists. British researcher James Lovelock had developed an instrument to measure extremely low concentrations of atmospheric gases, and American researchers Richard Stolarski and Ralph Cicerone had shown that chlorine atoms can catalyze the destruction of ozone.

Rowland and Molina's finding, which earned them the 1995 Nobel Prize in chemistry, helped spark discussion among scientists, policymakers, and industry leaders over limits on CFC production. As a result, the United States and several other nations banned the use of CFCs in aerosol spray cans in 1979. Other uses continued, however, and by the early 1980s global production of CFCs was increasing.

Then, in 1985, a new finding shocked scientists and spurred the international community to take further action. Scientists at a British research station in Antarctica had been recording ozone concentrations continuously since the 1950s. In May 1985, Joseph Farman and colleagues reported in *Nature* that Antarctic ozone concentrations had been declining dramatically since the 1970s. The decline exceeded even the worst-case predictions.

To determine what was causing the "ozone hole" over Antarctica, expeditions were mounted in 1986 and 1987 to measure trace amounts of atmospheric gases using ground stations and high-altitude balloons and aircraft. Together with other scientists, Dutch scientist Paul Crutzen, who would share the 1995 Nobel prize with Molina and Rowland, analyzed data collected on the expeditions and concluded that the ozone hole resulted from a combination of Antarctic weather conditions and human-made chemicals. In the frigid Antarctic winter, high-altitude clouds, or polar stratospheric clouds, were formed. In the spring, those clouds provided ideal conditions for CFC-derived chlorine and other chemicals to catalyze the destruction of massive amounts of ozone. The problem was exacerbated by the fact that prevailing air currents largely isolated Antarctica's atmosphere from the rest of Earth's atmosphere.

In the following years, scientists used data from ground stations and satellites to show that ozone levels were declining globally. In 1987, those findings helped convince the world's nations to agree on the Montreal Protocol, which aimed to cut CFC production in half by 1998. Within 2 years, however, further scientific evidence and computer modeling showed that more drastic measures would be needed if serious damage to the ozone layer were to be avoided. In 1990, the Montreal Protocol was strengthened to include a complete phaseout of CFCs by 2000. By 1998, the amount of chlorine in the atmosphere appeared to have leveled off, suggesting that the agreements had had the desired effect.

Acid deposition has not been reduced as much as scientists had hoped

Reducing acid precipitation involves reducing amounts of the pollutants that contribute to it. New technology has helped; "scrubbers" that filter pollutants in smokestacks have allowed factories to decrease emissions (▸pp. 655–656). As a result of declining emissions of SO_2, average sulfate precipitation in 1996–2000 was 10% lower than in 1990–1994 across the United States and 15% lower in the eastern states. However, because of increasing NO_x emissions, average nitrate precipitation increased nationally by 3% between these time periods.

FIGURE 17.18 Acidic deposition can harm vegetation, alter soil chemistry, affect soil- and forest-dwelling animals, and even eat away at statues and buildings.

A recent report by scientists at New Hampshire's Hubbard Brook research forest, where acidic deposition's effects were first demonstrated in the United States, disputed the notion that the problem of acid deposition is being solved (see "The Science behind the Story," ▸pp. 520–521). Instead, the report said, the effects are worse than first predicted, and the mandates of the 1990 Clean Air Act are not adequate to restore ecosystems in the northeastern United States. At Hubbard Brook, half the soil's calcium content has leached out, leaving the soil less able to neutralize future acid precipitation. This means that forests affected by acidification may take a long time to recover. An additional 80% reduction in sulfur emissions from electric utilities would be needed, the report estimates, to allow New Hampshire streams to recover in 20–25 years. The data on acid deposition show that although there have been many advances in the control of air pollution, more can clearly be done to alleviate outdoor pollution problems. The same can be said for indoor air pollution, a source of human health threats that is less familiar to most of us, but statistically more dangerous.

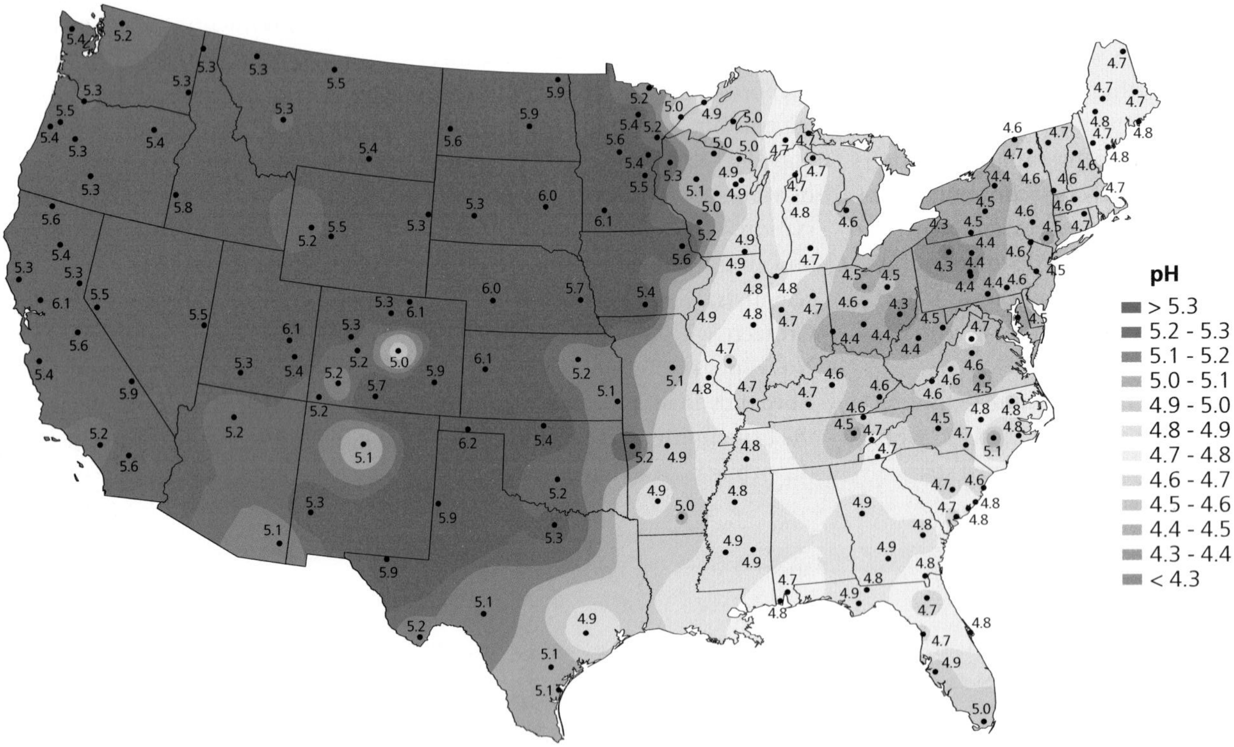

FIGURE 17.19 This map shows pH values for precipitation throughout the United States. Acid precipitation is most serious in parts of the Northeast and Midwest, generally downwind from (roughly east of) areas of heavy industrial development. Data from National Atmospheric Deposition Program. 2005. Hydrogen ion concentration as pH from measurements made at the Central Analytical Laboratory, 2003.

FIGURE 17.20 Indoor air pollution is an under-recognized health threat in both developed nations and developing nations. In the developing world, fires for cooking and heating are often built inside homes, as seen here in a South African kitchen **(a)**, exposing family members to particulate matter and carbon monoxide. In most regions of the developing world, indoor air pollution is estimated to cause upwards of 3% of all health risks. In this graph **(b)**, disability-adjusted life years (DALYs) indicate the burden of disease in total number of years of healthy life lost, including premature death and disability over a period of time, because of both indoor and outdoor air pollution. Indians suffer most severely from indoor air pollution, with approximately 650,000 years of life lost per million people, followed by sub-Saharan Africans, who suffer roughly 580,000 years loss of life per million people. *Source:* World Bank. 2002. Data from U.N. Development Programme, World Bank, Energy Sector Management Assistance Programme, 2002.

(a) South African family cooking indoors

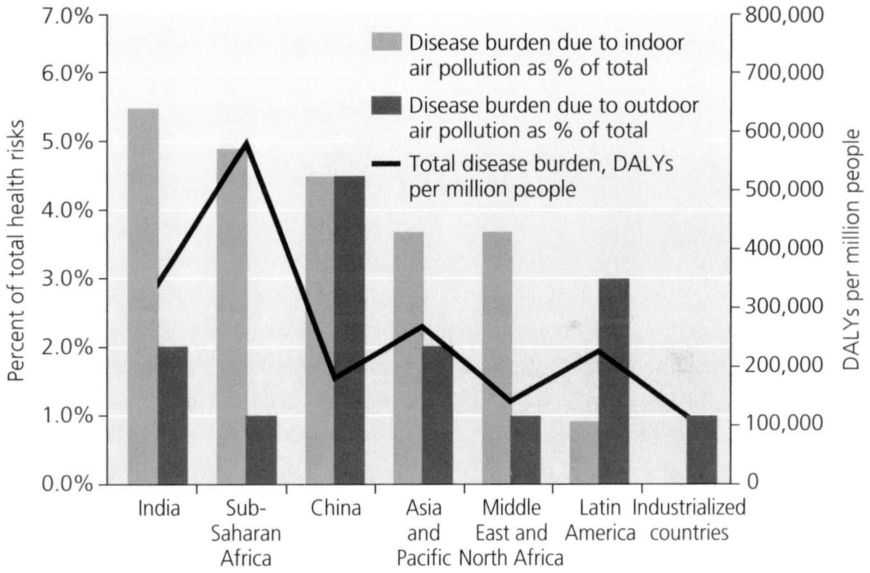

(b) Indoor and outdoor pollution health risk statistics

Indoor Air Pollution

Indoor air generally contains higher concentrations of pollutants than does outdoor air. As a result, the health effects from indoor air pollution in workplaces, schools, and homes outweigh those from outdoor air pollution (Figure 17.20). One estimate, from the U.N. Development Programme in 1998, attributed 2.2 million deaths worldwide to indoor air pollution and 500,000 deaths to outdoor air pollution. Indoor air pollution alone, then, takes roughly 6,000 lives each day.

If the impact of indoor air pollution seems surprising, consider that the average U.S. citizen spends at least 90% of his or her time indoors. Then consider that in the past half century a dizzying array of consumer products has been manufactured and sold, many of which we keep in our homes and offices and have come to play major roles in our daily lives. Many of these products are made of synthetic materials, and, as we saw in Chapter 14, novel synthetic substances are not comprehensively tested for health effects before being brought to market. Products such as insecticides and cleaning fluids can exude volatile chemicals into the air, as can solid materials such as plastics and chemically treated wood products.

In an ironic twist, some attempts to be environmentally prudent during the "energy crisis" of 1973–1974 worsened indoor air pollution in developed countries. To

The Science behind the Story

Acid Rain at Hubbard Brook Research Forest

The effects of acidic deposition are subtle, involving incremental changes in pH levels that take place over long periods of time and affect species with long life cycles, such as trees. For this reason, no single experiment can give us a complete picture of acidic deposition's effects. Nonetheless, one long-term study conducted in the Hubbard Brook Experimental Forest in New Hampshire's White Mountains has been critically important to our understanding of acidic deposition in the United States.

The effects of acidic deposition on trees can be seen in this forest on Mount Mitchell in western North Carolina.

Established by the U.S. Forest Service in 1955, Hubbard Brook was initially devoted to research on hydrology, the study of water flow through forests and streams. In 1963, in collaboration with scientists at Dartmouth University, Hubbard Brook researchers broadened their focus to include a long-term study of nutrient cycling in forest ecosystems. Since then, they have collected and analyzed weekly samples of precipitation. The measurements make up the longest-running North American record of acid precipitation and have helped shape U.S. policy on sulfur and nitrogen emissions.

At Hubbard Brook, only one form of acidic deposition—wet deposition—has been measured regularly. Wet deposition is the deposition of rain, snow, fog, or sea spray with low pH onto soil, plants, or water. Dry deposition, in contrast, is the deposition of airborne acidic particles. Throughout Hubbard Brook's 3,160 ha (7,800 acres), small plastic collecting funnels, 30 cm (1 ft) in diameter at their openings, channel precipitation into clean bottles, which researchers retrieve and replace each week. Hubbard Brook's laboratory measures acidity and conductivity, which indicates the amount of salts and other electrolytic contaminants dissolved in the water. Concentrations of sulfuric acid, nitrates, ammonia, and other compounds are measured elsewhere.

By the late 1960s, ecologists Gene Likens, F. Herbert Bormann, and others had found that precipitation at Hubbard Brook was several hundred times more acidic than natural rainwater. By the early 1970s, a number of other studies had corroborated their findings. Together, these studies indicated that precipitation from Pennsylvania to Maine had pH values averaging around 4, and that individual rainstorms showed values as low as 2.1—almost 10,000 times more acidic than ordinary rainwater.

reduce heat loss and improve energy efficiency, building managers sealed off most ventilation in existing buildings, and building designers constructed new buildings with limited ventilation and with windows that did not open. These steps may have saved energy, but they also worsened indoor air pollution by trapping stable, unmixed air—and its pollutants—inside.

Indoor air pollution in the developing world arises from fuelwood burning

Indoor air pollution has the greatest impact in the developing world. Millions of people in developing nations burn wood, charcoal, animal dung, or crop waste inside their homes for cooking and heating with little or no

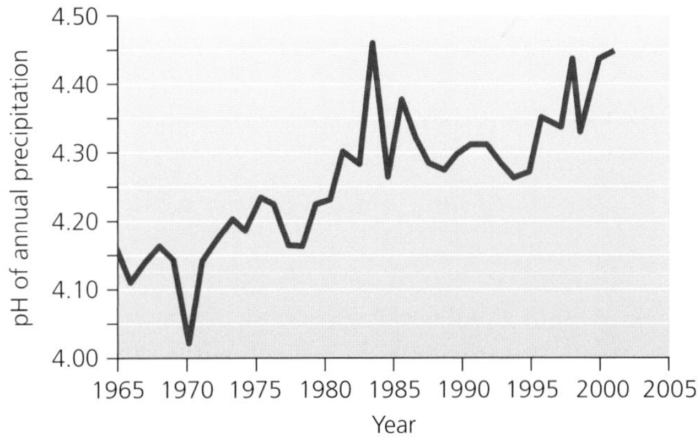

Over the past 40 years, precipitation at the Hubbard Brook Experimental Forest has become slightly less acidic. However, it is still far more acidic than is natural precipitation. Data from Likens, G. E. 2004. *Ecology* 85: 2355–2362.

In 1978, the National Atmospheric Deposition Program was launched to monitor precipitation and dry deposition across the United States. Initially consisting of 22 sites, including Hubbard Brook, the program now comprises more than 200, each of which gathers weekly data on acidic deposition and deposition of other substances. By the late 1980s, this program had produced a nationwide map of pH values. The most severe problems were found to be in the Northeast, where prevailing west-to-east winds were blowing emissions from fossil-fuel-burning power plants in the Midwest. Scientists hypothesized that when sulfur dioxide, nitrogen oxides, and other pollutants arrived in the Northeast, they were absorbed by water droplets in clouds, converted to acidic compounds such as sulfuric acid, and deposited on farms, forests, and cities in the form of rain or snow.

To some extent, the Clean Air Act of 1970 helped reduce acidic deposition in the Northeast. The accompanying figure shows the pH record for an area of Hubbard Brook known as Watershed 6. Between 1965 and 1995, average pH increased slightly, from about 4.15 to about 4.35. In 1990, as a consequence of the Hubbard Brook study and the nationwide research that followed, the Clean Air Act of 1970 was amended to further restrict emissions of sulfur dioxide and other acid-forming compounds. Nonetheless, acidic deposition continues to be a serious problem in the Northeast.

Some of the long-term consequences of acidic deposition are now becoming clear. In 1996, researchers reported that approximately 50% of the calcium and magnesium in Hubbard Brook's soils had been leached out. Meanwhile, acidic deposition had increased the concentration of aluminum in the soil, which can prevent tree roots from absorbing nutrients. The resulting nutrient deficiency slows forest growth and weakens trees, making them more vulnerable to drought and insects. It also reduces the ability of soil and water to neutralize acidity, making the ecosystem increasingly vulnerable to further inputs of acid.

In October 1999, researchers used a helicopter to distribute 50 tons of a calcium-containing mineral called wollastonite over one of Hubbard Brook's watersheds. Their objective was to raise the concentration of base cations to estimated historical levels. Over the next 50 years, scientists plan to evaluate the impact of calcium addition on the watershed's soil, water, and life. By providing a comparison to watersheds in which calcium remains depleted, the results should provide new insights into the consequences of acid rain and the possibilities for reversing its negative effects.

ventilation. In the process, they inhale dangerous amounts of soot and carbon monoxide. In the air of such homes, concentrations of particulate matter are commonly 20 times above U.S. EPA standards, the World Health Organization (WHO) has found. Poverty forces fully half the population and 90% of rural residents of developing countries to heat and cook with indoor fires. Some people will burn almost any available fuel, even discarded plastic, in indoor fires. In doing so, these people are in effect subjecting themselves to a daily dose of London's smog of 1952. Indoor air pollution from fuelwood burning, the WHO estimates, kills 1.6 million people each year, comprising over 5% of all deaths in some developing nations and 2.7% of the entire global disease burden.

Many people who tend indoor fires are not aware of the health risks. They do not have access to the statistics showing that chemicals and soot released by burning coal, plastic, and other materials indoors can increase risks of pneumonia, bronchitis, allergies, sinus infections, cataracts, asthma, emphysema, heart disease, cancer, and premature death. Many who are aware of the health risks are too poor to have viable alternatives.

Even in the developed world, recognizing indoor air pollution as a problem is still quite novel. Fortunately, scientists have identified the most deadly indoor threats. Particulate matter and chemicals from wood and charcoal smoke are the primary health risks in the developing world. In developed nations, the top risks are cigarette smoke and radon, a naturally occurring radioactive gas.

Tobacco smoke and radon are the most dangerous indoor pollutants in the developed world

The health effects of smoking cigarettes are well known in developed countries, but only recently have scientists quantified the risks of inhaling secondhand smoke. Secondhand smoke, or environmental tobacco smoke, is smoke inhaled by a nonsmoker who is nearby or shares an enclosed airspace with a smoker. Secondhand smoke has been found to cause many of the same problems as directly inhaled cigarette smoke, ranging from irritation of the eyes, nose, and throat, to exacerbation of asthma and other respiratory ailments, to lung cancer. This hardly seems surprising when one considers that environmental tobacco smoke consists of a brew of over 4,000 chemical compounds, many of which are known or suspected to be toxic or carcinogenic.

Women living with a spouse who smokes have a 24% greater chance of developing lung cancer from secondhand smoke, one study has indicated. Fortunately, the popularity of smoking has declined greatly in the United States and some other nations in recent years. The exposure of young children in the United States has decreased by almost half. A 1998 study found that 20% of young children had been exposed, as compared to 39% in 1986.

After cigarette smoke, radon gas is the second-leading cause of lung cancer in the United States, responsible for an estimated 20,000 deaths per year. Worldwide, the WHO estimates that radon may account for 15% of lung cancer cases. As we saw in Chapter 14 (►p. 405), radon is a radioactive gas resulting from the natural decay of uranium in soil, rock, or water, which seeps up from the ground and can infiltrate buildings. Radon is colorless and odorless, and it can be impossible to predict where it will occur without knowing details of an area's underlying geology (Figure 17.21). As a result, the only way to determine whether radon is entering a building is to measure radon with a test kit. Testing in 1991 led the EPA to estimate that 6% of U.S. homes exceeded the EPA's maximum recommended level for radon. Since the mid-1980s,

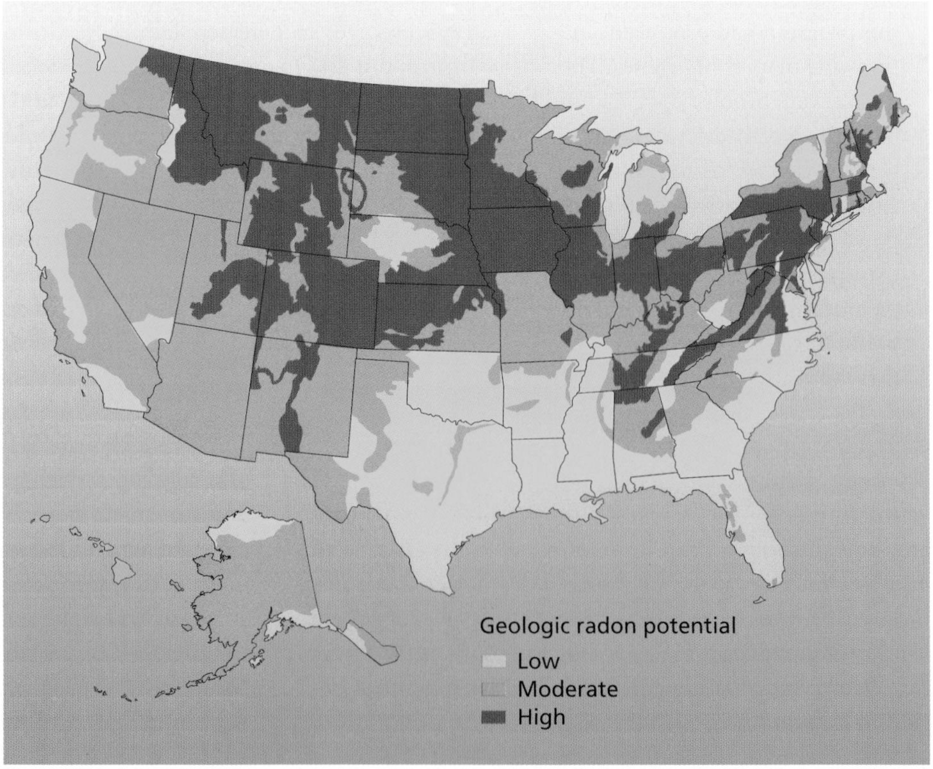

FIGURE 17.21 The risk from radon depends largely on underground geology. This map shows relative levels of risk from radon across the United States, but there is much fine-scale geographic variation from place to place not evident on the map. Testing your home for radon is the surest way to determine whether this colorless, odorless gas could be a problem in your home. Data from U.S. Geological Survey. 1993. Generalized geological radon potential of the United States, 1993.

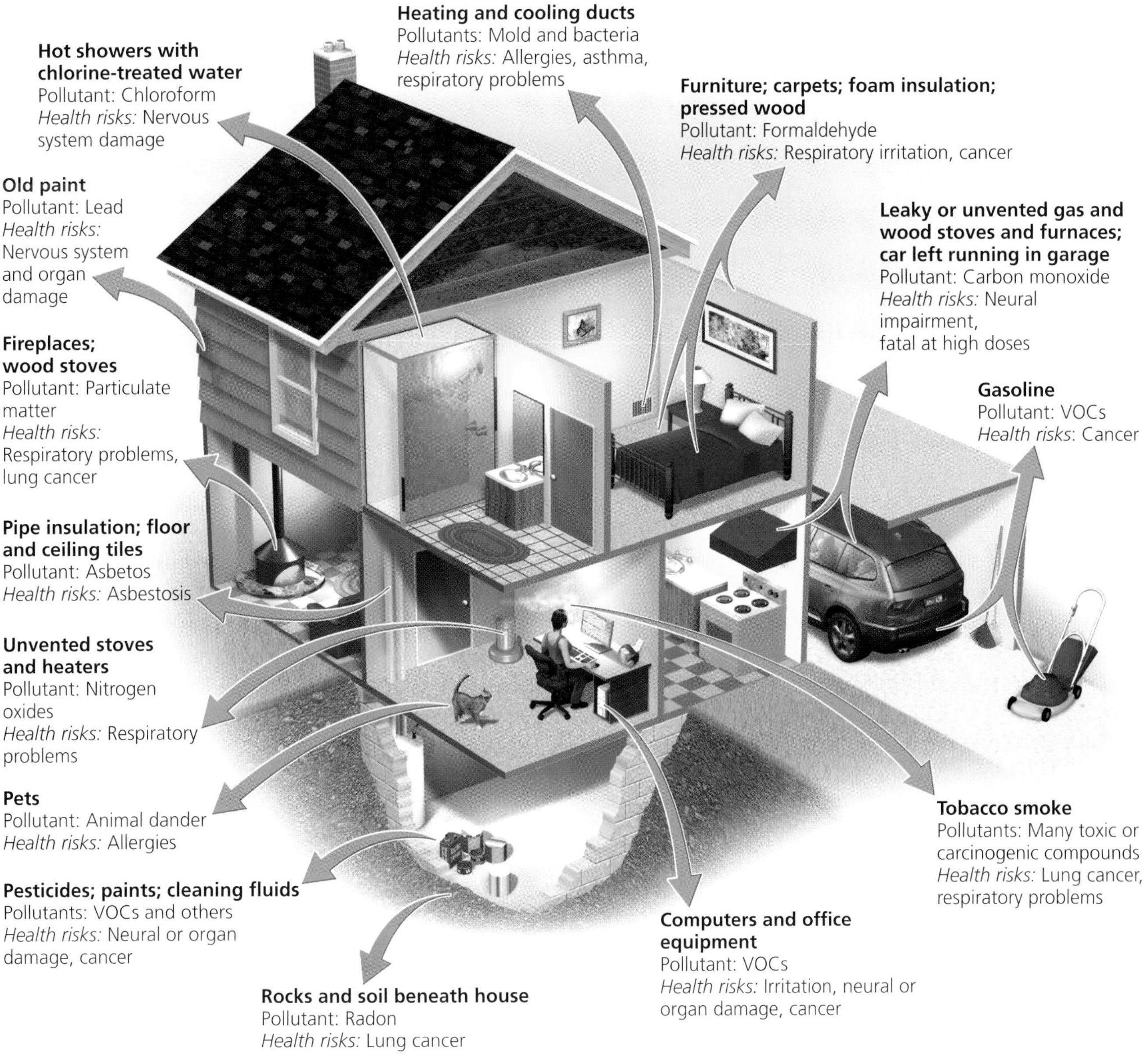

FIGURE 17.22 The typical U.S. home contains a variety of potential sources of indoor air pollution. Shown are some of the most common sources, the major pollutants they emit, and some of the health risks they pose.

18 million U.S. homes have been tested for radon, and 700,000 have undergone radon mitigation. New homes are now being built with radon-resistant features—over a million such homes since 1990.

Many volatile organic compounds pollute indoor air

In our daily lives at home, we are exposed to many indoor air pollutants (Figure 17.22). The most diverse indoor pollutants are VOCs. These airborne carbon-containing compounds are released by everything from plastics to oils to perfumes to paints to cleaning fluids to adhesives to pesticides. VOCs evaporate from furnishings, building materials, color film, carpets, laser printers, fax machines, and sheets of paper. Some products, such as chemically treated furniture, release large amounts of VOCs when new and progressively less as they age. Other items, such as photocopying machines, emit VOCs each time they are used.

Although we are surrounded by products that emit VOCs, they are released in very small amounts. EPA and European surveys have both measured overall levels of VOCs in buildings and have found them to be nearly always less than 0.1 part per million (ppm). This is, however, a substantially greater concentration than is generally found outdoors. When the EPA conducted a survey of U.S. office buildings between 1994 and 1998, it found 34 of the 48 VOCs it tested for, and found at least one type in 81% of buildings. All of these compounds existed at higher levels indoors than outdoors, suggesting that they originated from indoor sources.

The implications for human health of chronic exposure to VOCs are far from clear. Because they exist in such low concentrations and because individuals regularly are exposed to mixtures of many different types, it is extremely difficult to study the effects of any one pollutant. An exception is formaldehyde, which does have clear and known health impacts. This VOC, one of the most common synthetically produced chemicals, irritates mucous membranes, induces skin allergies, and causes other ailments. Formaldehyde is used in numerous products, but health complaints have mainly resulted from its leakage from pressed wood and insulation. The use of plywood has decreased in the last decade because of health concerns over formaldehyde.

VOCs also include pesticides, which we examined in Chapters 10 and 14. Three-quarters of U.S. homes use at least one pesticide indoors during an average year, but most are used outdoors. Thus it may seem surprising that the EPA found in a 1990 study that 90% of people's pesticide exposure came from indoor sources. Households that the agency tested had multiple pesticide volatiles in their air, at levels 10 times above levels measured outside. Some of the pesticides found had apparently been used years previously against termites, and then seeped into the houses through floors and walls. DDT, banned 15 years before the study, was found in five of eight homes, probably having been brought in on the soles of occupants' shoes from outdoors.

Living organisms can pollute indoor spaces

Tiny living organisms can be or produce indoor pollutants. In fact, they may be the most widespread source of indoor air pollution in the developed world. Dust mites and animal dander can exacerbate asthma in children. Some fungi, mold, and mildew (in particular, their airborne spores) can cause potentially severe health problems, including allergies, asthma, and other respiratory ailments. Some airborne bacteria can cause infectious disease. One example is the bacterium that causes Legionnaires' disease. Of the estimated 10,000–15,000 annual U.S. cases of Legionnaires' disease, 5–15% are fatal. Heating and cooling systems in buildings make ideal breeding grounds for microbes, providing moisture, dust, and foam insulation as substrates, as well as air currents to carry the organisms aloft.

Microbes that induce allergic responses are thought to be a major cause of building-related illness, a sickness produced by indoor pollution in which the specific cause may not be identifiable. When the cause of such an illness is a mystery, and when symptoms are general and nonspecific, the illness is often called *sick-building syndrome.* The U.S. Occupational Safety and Health Administration (OSHA) has estimated that 30–70 million Americans have suffered ailments due to the environment of the building in which they live. We can reduce the prevalence of sick building syndrome by using low-toxicity building materials and ensuring that buildings are adequately ventilated.

Weighing the Issues: How Safe Is Your Indoor Environment?

Think about the amount of time you spend indoors. Name the potential indoor air quality hazards in your home, work, or school environment. Are these spaces well-ventilated? What could you do to make the indoor spaces you use safer?

We can reduce indoor air pollution

Using low-toxicity material, monitoring air quality, keeping rooms clean, and providing adequate ventilation are the keys to alleviating indoor air pollution in most situations. In the developed world, we can try to limit our use of plastics and treated wood where possible and to limit our exposure to pesticides, cleaning fluids, and other known toxicants by keeping them in a garage or outdoor shed rather than in the house. The EPA recommends that we test our homes and offices for radon. Because carbon monoxide is so deadly and so hard to detect, many homes are equipped with detectors that sound an alarm if incomplete combustion produces dangerous levels of CO. In addition, keeping rooms and air ducts clean and free of mildew and other biological pollutants will reduce potential irritants and allergens. Finally, it is important to keep our indoor spaces as well ventilated as possible to minimize concentrations of the pollutants among which we live.

Remedies for fuelwood pollution in the developing world include drying wood before burning (which reduces the amount of smoke produced), cooking outside, shifting to less-polluting fuels (such as natural gas), and replacing inefficient fires with cleaner stoves that burn fuel more efficiently. For example, the Chinese government has invested in a program that has placed more fuel-efficient stoves in millions of homes in China. According to WHO studies, this is a relatively cost-efficient means of reducing the health impacts of indoor biomass combustion. Installing hoods, chimneys, or cooking windows can increase ventilation for little cost, alleviating the majority of indoor smoke pollution.

Conclusion

Indoor air pollution is a potentially serious health threat. However, by keeping informed of the latest scientific findings and taking appropriate actions, we as individuals can significantly minimize the risks to our families and ourselves. Outdoor air pollution has been addressed more effectively by government legislation and regulation. In fact, reductions in outdoor air pollution levels in the United States and other developed nations represent some of the greatest strides made in environmental protection to date. Much room for improvement remains, however, particularly in reducing acidic deposition and the photochemical smog resulting from urban congestion. Fortunately, developed nations no longer experience the type of pollution that Londoners suffered in their 1952 killer smog. Nevertheless, avoiding such high pollutant levels in the developing world will continue to pose a challenge as less-wealthy nations industrialize.

REVIEWING OBJECTIVES

You should now be able to:

Describe the composition, structure, and function of Earth's atmosphere

- The atmosphere consists of 78% nitrogen gas, 21% oxygen gas, and a variety of permanent and variable gases in minute concentrations. (p. 498)
- The atmosphere includes four principal layers: the troposphere, stratosphere, mesosphere, and thermosphere. Temperature and other characteristics vary across these layers. Ozone is concentrated in the stratosphere. (pp. 498–499)
- The sun's energy heats the atmosphere, drives air circulation, and helps determine weather, climate, and the seasons. (pp. 500–501)
- Weather is a short-term phenomenon, whereas climate is a long-term phenomenon. Fronts, pressure systems, and the interactions among air masses influence weather. (pp. 501–503)
- Global convective cells called Hadley, Ferrel, and polar cells create latitudinal climate zones. (pp. 504–505)

Outline the scope of outdoor air pollution and assess potential solutions

- Natural sources such as windblown dust, volcanoes, and fires account for much atmospheric pollution, but human activity can exacerbate some of these phenomena. (pp. 505–506)
- Human-emitted pollutants include primary and secondary pollutants from point and non-point sources. (p. 506)
- To safeguard public health, the U.S. EPA regulates six criteria pollutants (carbon monoxide, lead, nitrogen dioxide, tropospheric ozone, sulfur dioxide, and particulate matter), as well as volatile organic compounds and 188 toxic air pollutants. (pp. 506–509)
- Emissions in the United States have decreased substantially since the Clean Air Act of 1970, and ambient air quality is much improved. (pp. 508–509)
- Industrial smog like that which blanketed 1952 London is produced by fossil fuel combustion and is still a problem in urban and industrial areas of many developing nations. (pp. 509–510)
- Photochemical smog is created by chemical reactions of pollutants in the presence of sunlight. It impairs visibility and human health in urban areas. (pp. 510–511)

Explain stratospheric ozone depletion and identify steps taken to address it

- CFCs destroy stratospheric ozone, and thinning ozone concentrations pose dangers to life because they allow more ultraviolet radiation to reach Earth's surface. (pp. 511, 513–514, 516–517)
- The Montreal Protocol and its follow-up agreements have proven remarkably successful in reducing emissions of ozone-depleting compounds. (pp. 513–514, 517)

Define acidic deposition and illustrate its consequences

- Acidic deposition results when pollutants such as SO_2 and NO react in the atmosphere to produce strong acids that are deposited on Earth's surface. (pp. 514–515)
- Acidic deposition may occur a long distance from the source of pollution. (pp. 516, 518)
- Water bodies, soils, trees, and ecosystems all experience negative impacts from acidic deposition. (pp. 515–518, 520–521)

Characterize the scope of indoor air pollution and assess potential solutions

- Indoor air pollution causes far more deaths and health problems worldwide than outdoor air pollution. (p. 519)
- Indoor burning of fuelwood is the developing world's primary indoor air pollution risk. (pp. 519–522)
- Tobacco smoke and radon are the deadliest indoor pollutants in the developed world. (pp. 522–523)
- Volatile organic compounds and living organisms can pollute indoor air. (pp. 523–524)
- Using low-toxicity building materials, keeping spaces clean, monitoring air quality, and maximizing ventilation are some of the steps we can take to reduce indoor air pollution. (pp. 524–525)

TESTING YOUR COMPREHENSION

1. About how thick is Earth's atmosphere? Name one characteristic of each of the four atmospheric layers.
2. Where is the "ozone layer" located? How and why is stratospheric ozone beneficial for people, and tropospheric ozone harmful?
3. How does solar energy influence weather and climate? How do Hadley, Ferrel, and polar cells help to determine long-term climatic patterns and the location of biomes?
4. What factors led to the deadly smog in London in 1952? Describe a thermal inversion.
5. Name three natural sources of outdoor air pollution and three sources caused by human activity.
6. What is the difference between a primary and a secondary pollutant? Give an example of each.
7. What is smog? How is smog formation influenced by the weather? By topography? How does photochemical, or brown-air, smog differ from industrial, or gray-air, smog?
8. How do chlorofluorocarbons (CFCs) deplete stratospheric ozone? Why is this depletion considered a long-term international problem? What was done to address this problem?
9. Why are the effects of acidic deposition often felt in areas far from where the primary pollutants are produced?
10. Name five common sources of indoor pollution. For each, describe one way to reduce one's exposure to this source.

SEEKING SOLUTIONS

1. Consider London's "killer smog" of 1952 and modern urban pollution by photochemical smog. Describe several factors that make it particularly difficult to study causes of air pollution and to develop solutions.
2. How may human activity sometimes exacerbate natural forms of air pollution? Discuss two examples and potential solutions.
3. Describe how and why emissions of major pollutants have been reduced by over 50% in the United States since 1970, despite increases in population and economic activity.
4. International regulatory action has produced reductions in CFCs, but other transboundary pollution issues, including acidic deposition, have not yet been addressed as effectively. What types of actions do you feel are appropriate for pollutants that cross political boundaries?
5. Consider volatile chemicals, such as formaldehyde and VOCs, that may be emitted at very low levels over long periods of time from manufactured products. What do you think are the best ways to lessen the health impacts of such indoor pollutants?
6. You have just become the head of your county health department, and the EPA has informed you that your county has failed to meet the national ambient air quality standards for ozone, sulfur dioxide, and nitrogen dioxide. Your county is partly rural but is home to a city of 200,000 people and 10 sprawling suburbs. There are several large and aging coal-fired power plants, a number of factories with advanced pollution control technology, and no public transportation system. What steps would you urge the county government to take to meet the air quality standards? Explain how you would prioritize these steps.

INTERPRETING GRAPHS AND DATA

Since the Clean Air Act of 1970, total emissions of carbon monoxide, sulfur dioxide, nitrogen oxides, VOCs, lead, and particulate matter have dropped by over 50% while U.S. population, energy consumption, and economic productivity have all increased. As you learned from the "Interpreting Graphs and Data" feature in Chapter 3 (► pp. 86–87), lead emissions resulted mostly from the combustion of leaded gasoline and dropped precipitously once leaded gasoline was phased out. Consider the other pollutants listed above as you interpret the data in the graph.

1. Relative to 1970, what are the percentage changes by 2004 for each of the five variables graphed above? What are the percentage changes in the per capita values of each variable over this time period?
2. Fossil fuel combustion is the major source of most of the pollutants above. What is the percentage change in aggregate emissions of the six principal pollutants per unit of energy consumed in 2004, compared to 1970?
3. Do you think that additional reductions in emissions are likely to result primarily from changing technology (e.g., hybrid cars, advanced catalytic converters) or from changing behavior (e.g., driving fewer miles per person per year)? Use the data above to support your claim.

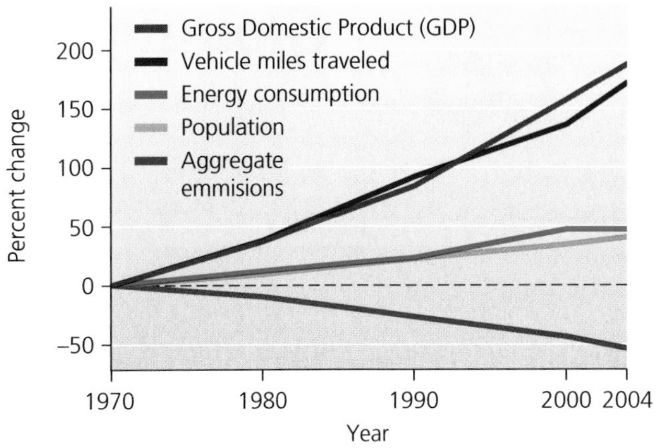

Trends from 1970 to 2004 in economic production, vehicle miles traveled, energy consumption, population, and aggregate emissions of the six principal air pollutants monitored by the EPA. Data from U.S. EPA. 2004. *Air emission trends: Continued progress through 2004.*

CALCULATING ECOLOGICAL FOOTPRINTS

According to EPA data, emissions of nitrogen oxides in the United States in 2000 were 24,899,000 tons, of which 13,251,000 tons were from the transportation sector. Of this amount, 8,150,000 tons came from on-road vehicles, with 5,859,000 tons of this total coming from light-duty cars and trucks. The U.S. Census Bureau estimated the nation's population to be 282,192,162 at mid-year in 2000 and projects that it will reach 300,000,000 by early 2007. Considering these data, calculate the missing values in the table below (1 ton = 2,000 lb).

	Total NO_x emissions (lb)	**NO_x emissions due to light-duty vehicles (lb)**
You		
Your class		
Your state		
United States		

Data from U.S. EPA. 2003. *National air quality and emissions trends report 2003,* Appendix A4.

1. By what percentage is the U.S. population projected to increase between 2000 and 2007? Do you think that NO_x emissions will increase, decrease, or remain the same over that period of time? Why? (You may want to refer to Figure 17.11.)
2. If you reduced by half the vehicle miles traveled for which you are responsible, how many pounds of NO_x emissions would you prevent? What percentage of your total NO_x emissions would that be?
3. How might you reduce your vehicle miles traveled by 50%? What other steps could you take to reduce the NO_x emissions for which you are responsible?

Take It Further

Go to www.aw-bc.com/withgott or the student CD-ROM where you'll find:

- Suggested answers to end-of-chapter questions
- Quizzes, animations, and flashcards to help you study
- *Research Navigator*™ database of credible and reliable sources to assist you with your research projects
- GRAPHit! tutorials to help you master how to interpret graphs
- INVESTIGATEit! current news articles that link the topics that you study to case studies from your region to around the world

18 Global Climate Change

Muratti Island of the Maldives

Upon successfully completing this chapter, you will be able to:

- Describe Earth's climate system and explain the variety of factors influencing global climate
- Characterize human influences on the atmosphere and global climate
- Delineate modern methods of climate research
- Summarize current consequences and potential future impacts of global climate change
- Evaluate the scientific, political, and economic debates concerning climate change
- Suggest potential responses to climate change

Fishermen of the Maldives

Central Case: Rising Temperatures and Seas May Take the Maldives Under

"The impact of global warming and climate change can effectively kill us off, make us refugees. . . ."
—Ismail Shafeeu, Minister of Environment, Maldives

"More people enjoy health and prosperity now than ever, and a warmer environment . . . should sustain life better than the current one does."
—Oil and Gas Journal, February 2001

A nation of low-lying coral islands, or atolls, in the Indian Ocean, the Maldives is known for its spectacular tropical setting, colorful coral reefs, and sun-drenched beaches. For visiting tourists it is a paradise, and for 320,000 Maldives residents it is home. But residents and tourists alike now fear that the Maldives could soon be submerged by the rising seas that are accompanying global climate change.

Nearly 80% of the Maldives' land area of 300 km^2 (116 mi^2) lies less than 1 m (39 in.) above sea level. In a nation of 1,190 islands whose highest point is just 2.4 m (8 ft) above sea level, rising seas could be a matter of life or death. The world's oceans rose 10–20 cm (4–8 in.) during the 20th century as warming temperatures expanded ocean water and as melting icecaps discharged water into the ocean. Current projections are that sea level will rise another 9–88 cm (3.5–35 in.) by the year 2100.

Higher seas are expected to flood large areas of land in the Maldives and to cause salt water to intrude into drinking water supplies. Moreover, if climate change produces larger and more powerful storms, these could worsen flooding and damage the coral reefs that are so crucial to the nation's tourism- and fishing-driven economy. Because of such concerns, the Maldives government has evacuated residents from several of the lowest-lying islands in recent years.

On December 26, 2004, the nation got a taste of what could be in store in the future. The massive *tsunami*, or tidal wave, that devastated coastal areas throughout the Indian Ocean hit the Maldives particularly hard. One hundred people were killed and 20,000 lost their homes, while schools, boats, tourist resorts, hospitals, and transportation and communication infrastructure were destroyed or badly damaged. On a per capita basis, the

Maldives suffered a greater economic shock from the tsunami than any other nation. The World Bank estimates that direct damage in the Maldives totaled $470 million, an astounding 62% of the nation's gross domestic product (GDP). Soil erosion, saltwater contamination of aquifers, and other environmental damage will result in still greater long-term economic losses.

The tsunami was caused *not* by climate change, but by an earthquake. Yet as sea level rises, the damage that such natural events—or ordinary storm waves—can inflict increases considerably. Maldives islanders are not alone in their predicament. Other island nations, from the Galapagos to Fiji to the Seychelles, are also fearing a future in which they may be constantly battling encroaching seawater. Mainland coastal areas of the world, such as the hurricane-battered coasts of Florida and Louisiana, will face similar issues. In one way or another, global climate change seems certain to affect each and every one of us for the remainder of our lifetimes.

Earth's Hospitable Climate

As we learned in Chapter 17, *weather* describes an area's short-term atmospheric conditions (over hours or days), including temperature, moisture content, wind, precipitation, barometric pressure, solar radiation, and other characteristics. *Climate* is an area's long-term pattern of atmospheric conditions. **Global climate change** describes changes in Earth's climate, involving aspects such as temperature, precipitation, and storm frequency and intensity. Although people often use the term *global warming* synonymously in casual conversation, *global warming* refers specifically to an increase in Earth's average surface temperature and thus is only one aspect of global climate change.

Our planet's climate has never been entirely stable and unchanging. However, the climatic changes taking place today are unfolding at an exceedingly rapid rate. Moreover, most scientists agree that human activities, notably fossil fuel combustion and deforestation, are largely responsible for the current modification of Earth's atmosphere and climate. Climatic changes will likely have adverse consequences for ecosystems and for millions of people, including residents of the Maldives, Florida, Louisiana, and other regions. For this reason, increasing numbers of scientists, policymakers, and ordinary citizens are seeking to take action to minimize and mitigate our impacts on the climate system.

The sun and the atmosphere keep Earth warm

Three factors exert more influence on Earth's climate than all others combined. The first is the sun. Without it, Earth would be dark and frozen. The second is the atmosphere. Without it, Earth would be as much as 33 °C (59 °F) colder on average, and temperature differences between night and day would be far greater. The third is the oceans, which shape climate by storing and transporting heat and moisture.

The sun is the source of most of the energy that Earth receives. Earth's atmosphere, clouds, land, ice, and water together reflect about 30% of incoming solar radiation back into space. The remaining 70% is absorbed by molecules in the atmosphere, clouds, or by land, water, or ice at Earth's surface (Figure 18.1).

"Greenhouse gases" warm the lower atmosphere

As Earth's surface absorbs solar radiation, the surface increases in temperature and emits radiation in the infrared portion of the spectrum (▸pp. 105–106). Some atmospheric gases absorb infrared radiation released from Earth's surface very effectively. These include water vapor, ozone, carbon dioxide (CO_2), nitrous oxide (N_2O), methane (CH_4), and halocarbons. Halocarbons are a diverse group of gases that include chlorofluorocarbons (CFCs; ▸pp. 513, 516–517) and hydrochlorofluorocarbons (HFCs). Such gases that absorb infrared radiation from Earth's surface are known as **greenhouse gases.** These gases subsequently re-emit infrared energy of slightly different wavelengths, warming the atmosphere (specifically the *troposphere;* ▸pp. 498–499) and the planet's surface. This warming of the troposphere and Earth's surface is known as the **greenhouse effect.** Despite its wide usage, the term *greenhouse effect* is actually a bit of a misnomer. The greenhouses we use for growing plants hold heat in place by preventing warm air from escaping. Atmospheric greenhouse gases, in contrast, do not trap air, but instead absorb, transform, and radiate heat.

The greenhouse effect is a natural phenomenon, and greenhouse gases (with the exception of the anthropogenic halocarbons) have been present in the atmosphere for billions of years. However, human activities have increased the concentrations of many greenhouse gases in the past 250–300 years, and we have thereby enhanced the greenhouse effect.

Not all greenhouse gases are equally effective in warming the troposphere and surface. *Global warming potential*

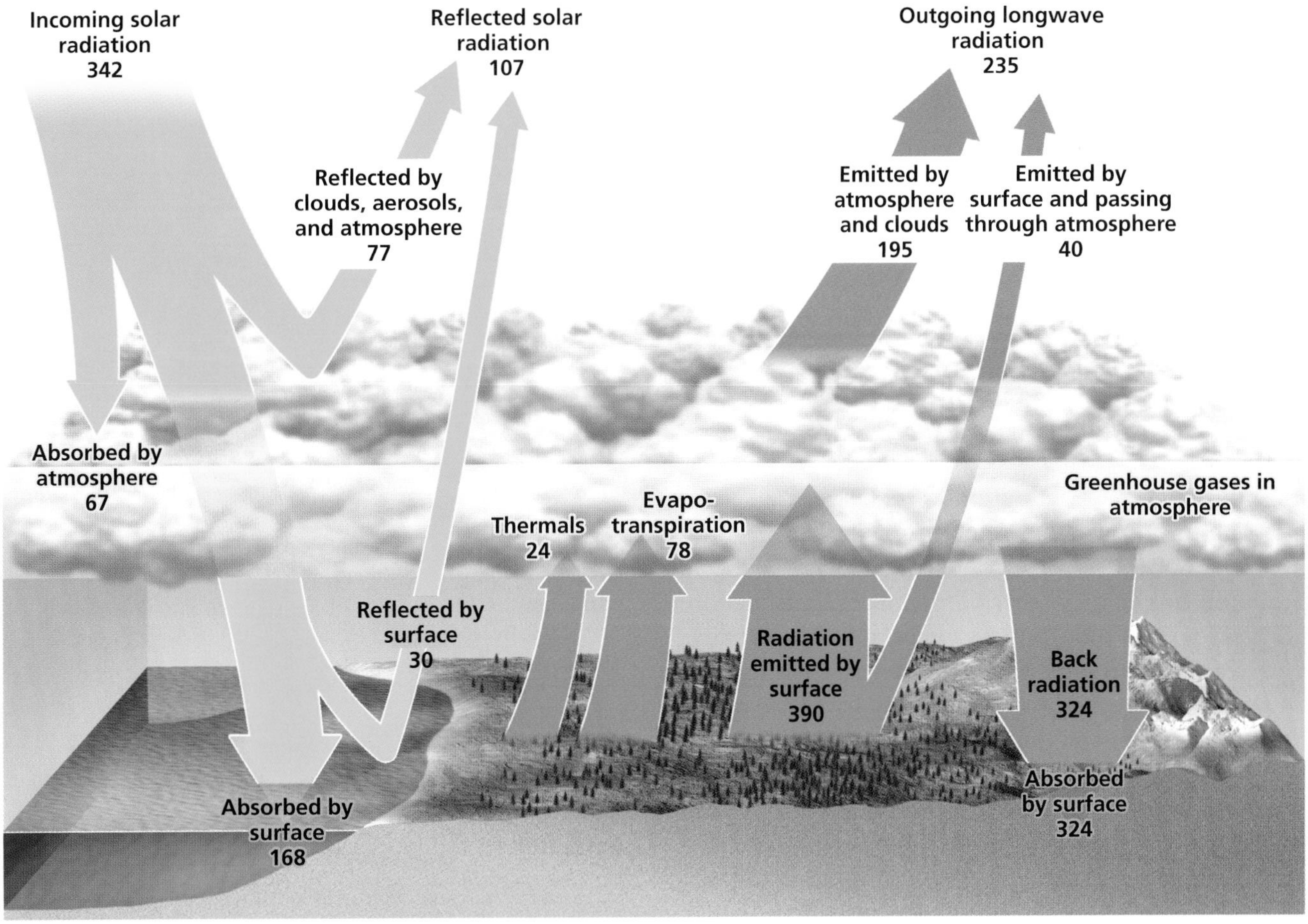

FIGURE 18.1 Earth's climate system is in rough equilibrium; our planet emits about the same amount of energy that it receives from the sun. As greenhouse gases accumulate in the atmosphere, however, they increase the amount of radiation that is emitted from the atmosphere back toward the surface. This illustration shows major pathways of energy flow in watts per square meter. Data from Kiehl, J. T., and K. E. Trenberth. 1997. Earth's annual global mean energy budget. *Bull. Amer. Meteorol. Soc.* 78: 197–208.

refers to the relative ability of one molecule of a given greenhouse gas to contribute to warming. Table 18.1 shows the global warming potential for several greenhouse gases. Values are expressed in relation to carbon dioxide, which is assigned a global warming potential of 1. Thus, a molecule of methane is 23 times as potent as a molecule of carbon dioxide, and a molecule of nitrous oxide is 296 times as potent as a CO_2 molecule.

Carbon dioxide is the primary greenhouse gas

Although carbon dioxide is not the most potent greenhouse gas on a per-molecule basis, its abundance in the atmosphere relative to gases such as methane and nitrous oxide means that it contributes more to the greenhouse effect. For this reason, changes in the atmospheric concentration of CO_2 are important. Human activities have contributed to a significant increase in the atmospheric concentration of carbon dioxide, from around 280 parts per million (ppm) as recently as the late 1700s to 316 ppm in 1959 to 378 ppm in 2004 (Figure 18.2). The atmospheric CO_2 concentration is now at its highest level in at least 400,000 years, and likely the highest in the last 20 million years. Moreover, it is increasing faster today than at any time in at least 20,000 years.

What has changed since the 1700s to cause the atmospheric concentration of this greenhouse gas to increase so rapidly? As you may recall from our discussion of the carbon cycle in Chapter 7 (▸pp. 195–197), and as we will see further in Chapter 19 (▸pp. 561–562), a great deal of carbon is stored for long periods in the upper layers of

Table 18.1 Global Warming Potentials of Four Greenhouse Gases

Greenhouse gas	Relative heat-trapping ability (in CO_2 equivalents)
Carbon dioxide	1
Methane	23
Nitrous oxide	296
Hydrochlorofluorocarbon HFC-23	12,000

Data from Intergovernmental Panel on Climate Change, 2001. *Climate change 2001: The scientific basis.*

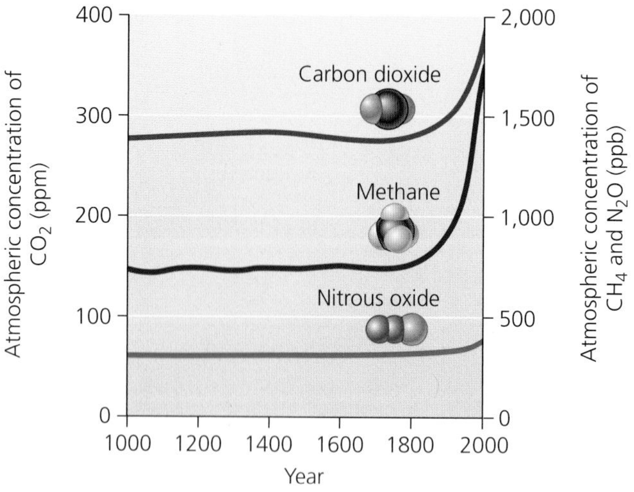

FIGURE 18.2 Global atmospheric concentrations of carbon dioxide (CO_2; top curve), methane (CH_4; middle curve), and nitrous oxide (N_2O; bottom curve) have increased dramatically since 1800. Data from Intergovernmental Panel on Climate Change. 2001. *Third assessment report.*

the lithosphere. The deposition, partial decay, and compression of organic matter (mostly plants) that grew in wetland or marine areas during the Carboniferous Period led to the formation of coal, oil, and natural gas in sediments from that time. In the absence of human activity, these carbon reservoirs would be practically permanent. However, over the past two centuries we have been burning increasing amounts of fossil fuels in our homes, factories, and automobiles. In so doing, we have transferred large amounts of carbon from one reservoir (the underground deposits that stored the carbon for millions of years) to another (the atmosphere). This flux of carbon from lithospheric reservoirs into the atmosphere is the main reason atmospheric carbon dioxide concentrations have increased so dramatically in such a short time.

At the same time, people have cleared and burned forests to make room for crops, pastures, villages, and cities. Forests serve as carbon sinks (▸p. 195), and their removal, especially in areas where they are slow to recover, can reduce the biosphere's ability to absorb carbon dioxide from the atmosphere. Therefore, deforestation has also contributed to increasing atmospheric carbon dioxide concentrations.

Other greenhouse gases add to warming

Carbon dioxide is not the only greenhouse gas increasing in the atmosphere. Methane is also on the rise. We release methane into the atmosphere by tapping into fossil fuel deposits, raising livestock that release methane as a metabolic waste product, disposing of organic matter in landfills, and growing certain types of crops, especially rice. Since 1750, atmospheric methane concentrations have increased by 151% (see Figure 18.2), and the current concentration is the highest in at least 400,000 years.

Nitrous oxide is another greenhouse gas whose atmospheric concentration has increased because of human activities. This gas, a by-product of feedlots, chemical manufacturing plants, auto emissions, and use of nitrogen fertilizers in modern agricultural practices, has risen by 17% since 1750 (see Figure 18.2).

Ozone concentrations in the troposphere have increased by 36% since 1750 as a result of processes described in Chapter 17 (▸pp. 507, 510–511). Halocarbon gases add greatly to the atmosphere's heat-absorbing ability on a per-molecule basis (see Table 18.1). Their overall contribution to global warming, however, has begun to slow because of the Montreal Protocol and subsequent controls (▸pp. 513–514).

Emissions of greenhouse gases from human activity in the United States consist mostly of carbon dioxide. Even after accounting for the greater global warming potential of other gases, carbon dioxide remains the major contributor to warming (Figure 18.3).

Water vapor is the most abundant greenhouse gas in the atmosphere. Its concentration varies, but if tropospheric temperatures continue to increase, the oceans and other water bodies should transfer increasingly more water vapor into the atmosphere. Such a positive feedback mechanism could amplify the greenhouse effect. Alternatively, increased water vapor concentrations could give rise to increased cloudiness, which might, in a negative feedback loop, slow global warming by reflecting more incoming solar radiation back into space. Depending on whether low- or high-elevation clouds resulted, they could either shade and cool Earth (negative feedback) or

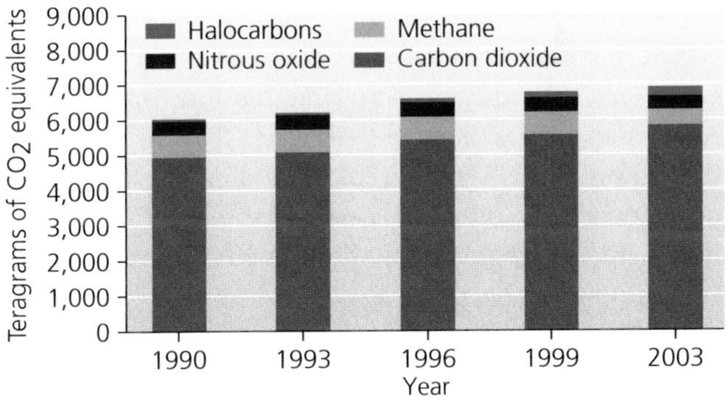

FIGURE 18.3 Emissions of six main greenhouse gases in the United States increased by 13.3% between 1990 and 2003. Even after accounting for the greater global warming potentials of methane, nitrous oxide, and three types of halocarbons, carbon dioxide accounts for nearly 85% of potential warming. Data from U.S. Environmental Protection Agency. 2005. *Inventory of U.S. greenhouse gas emissions and sinks: 1990–2003*. 430-R-05-003. Washington, D.C.: U.S. EPA.

else contribute to warming and accelerate evaporation and further cloud formation (positive feedback). Because of feedback loops (▸pp. 184–185), minor modifications of minor components of the atmosphere can potentially lead to major changes in climate.

Aerosols and other elements may exert a cooling effect on the lower atmosphere

Whereas greenhouse gases exert a warming effect, *aerosols*, microscopic droplets and particles, can have either a warming or cooling effect. Generally speaking, soot aerosols, also known as black carbon aerosols, can cause warming, but most tropospheric aerosols lead to climate cooling. Sulfate aerosols produced by fossil fuel combustion may slow global warming, at least in the short term. When sulfur dioxide enters the atmosphere, it undergoes a number of reactions, some of which lead to acid precipitation (▸p. 515). These reactions, along with volcanic eruptions, also contribute to the formation of a sulfur-rich aerosol haze in the upper atmosphere. Such a haze may reduce the amount of sunlight that reaches and heats Earth's surface. Major volcanic eruptions and the aerosols they release can also exert short-term cooling effects on Earth's climate on the scale of a few years.

The atmosphere is not the only factor that influences climate

Although the atmosphere shapes climate, the amount of energy released by the sun, as well as changes in Earth's rotation and orbit, also affect climate.

Milankovitch cycles During the 1920s, Serbian mathematician Milutin Milankovitch described three kinds of changes in Earth's rotation and orbit around the sun. These variations, now known as **Milankovitch cycles,** result in slight changes in the relative amount of solar radiation reaching Earth's surface at different latitudes over the long term (Figure 18.4). As these cycles proceed, they change the way solar radiation is distributed over Earth's surface. This, in turn, contributes to changes in atmospheric heating and circulation that have triggered climate variation, such as periodic glaciation episodes.

Oceanic circulation The oceans also shape climate. Ocean water exchanges tremendous amounts of heat with the atmosphere, and ocean currents move energy from one place to another. In equatorial regions, such as the area around the Maldives, the oceans receive more heat from the sun and atmosphere than they emit. Near the poles, the oceans emit more heat than they receive. Because cooler water is denser than warmer water, the cooling water at the poles tends to sink, and the warmer

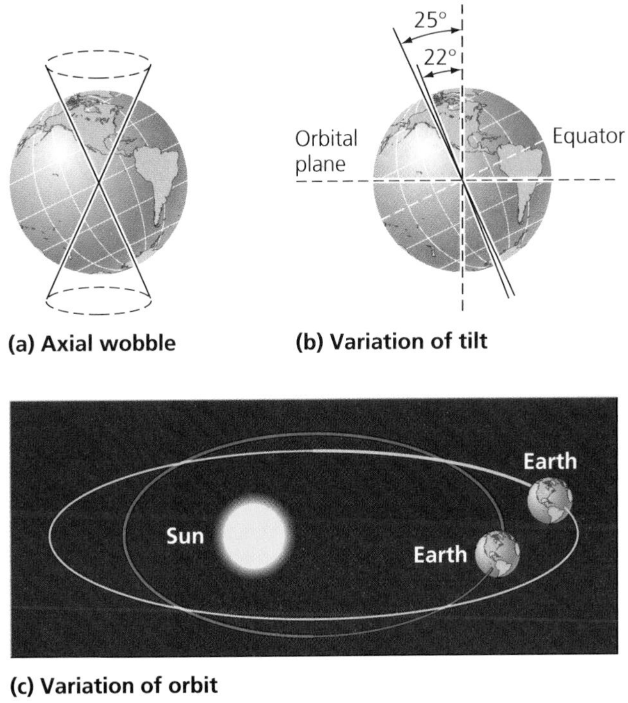

FIGURE 18.4 There are three types of Milankovitch cycles. The first is an axial wobble (**a**) that occurs on a 19,000- to 23,000-year cycle. The second is a 3-degree shift in the tilt of Earth's axis (**b**) that occurs on a 41,000-year cycle. The third is a variation in Earth's orbit from almost circular to more elliptical (**c**), which repeats itself every 100,000 years. These variations affect the intensity of solar radiation that reaches portions of Earth at different times, creating long-term changes in global climate.

surface water from the equator moves to take its place. This is one of the principles explaining global oceanic circulation (▸pp. 470–471).

The best-known interaction among ocean circulation, atmospheric circulation, and global climate involves the phenomena named El Niño and La Niña (see "The Science behind the Story," ▸pp. 536–537). **El Niño** conditions are triggered when equatorial winds weaken and allow warm water from the western Pacific to move eastward, preventing cold water from welling up in the eastern Pacific. In **La Niña** events, cold surface waters extend far westward in the equatorial Pacific. Both these events affect temperature and precipitation patterns around the world in complex ways. Scientists are exploring whether globally warming air and sea temperatures may be increasing the frequency and strength of El Niño events.

Another way in which climate and ocean currents interact involves the northern Atlantic Ocean. Here, warm surface water moves from the equator northward, carrying heat to higher latitudes, and keeping Europe much warmer than it would otherwise be. As the surface water of the North Atlantic releases heat energy and cools, it becomes denser and sinks. The deep part of this circulation pattern is called the *North Atlantic Deep Water (NADW).* Figure 18.5 portrays this circulation pattern as a sort of conveyor belt that moves water and heat from one place to another. Recently, scientists have realized that interrupting the NADW could trigger rapid climate change. If global warming causes large portions of the Greenland Ice Sheet to melt, freshwater runoff into the North Atlantic would increase. Surface waters would become less dense from such dilution and warming, because warm freshwater is less dense than cold salt water. This could potentially stop the NADW circulation, shutting down the northward flow of warm equatorial water. The entire North Atlantic region, including much of Europe, could cool rapidly as a result.

Methods of Studying Climate Change

To understand how climate is changing today, and to predict future changes, scientists must have a good idea of what climatic conditions were like thousands or millions of years ago. Environmental scientists have developed a number of methods to decipher clues from the past to learn about Earth's climate history.

Proxy indicators tell us about the past

Earth's ice caps and glaciers hold clues about past climate. Over the ages, these huge expanses of snow and ice have accumulated to great depths, preserving within them tiny bubbles of the ancient atmosphere (Figure 18.6). Scientists can examine these trapped air bubbles by drilling into the ice and extracting long columns, or cores. From

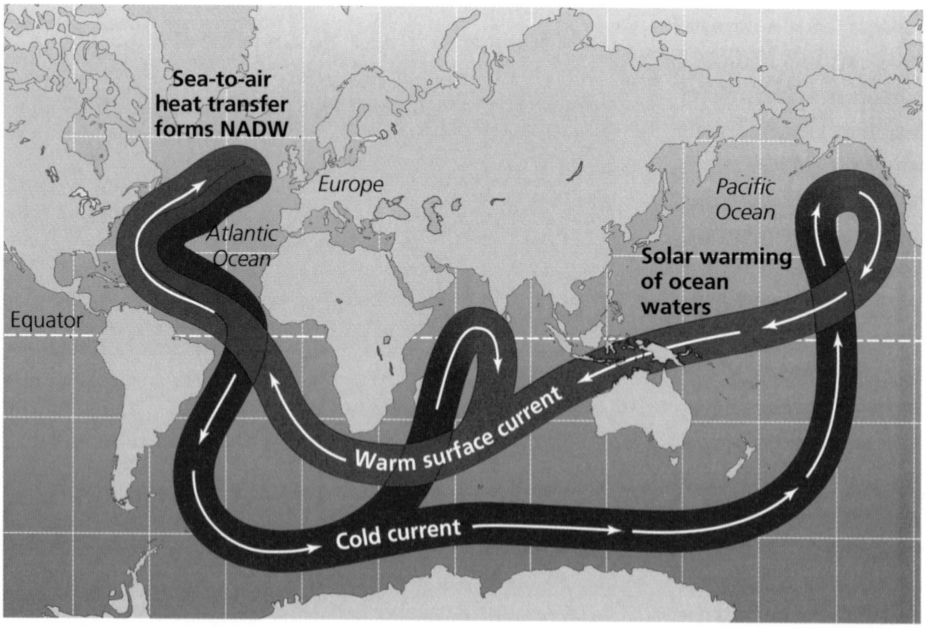

FIGURE 18.5 The atmosphere and the oceans interact. Warm surface currents carry heat from equatorial waters north toward Europe and Greenland, where they release heat into the atmosphere, then cool and sink into the deep ocean. This is how North Atlantic Deep Water (NADW) is formed. Melting of the Greenland Ice Sheet could potentially interrupt the flow of heat from equatorial regions and lead to rapid cooling of the North Atlantic and much of Europe.

(a) Ice core

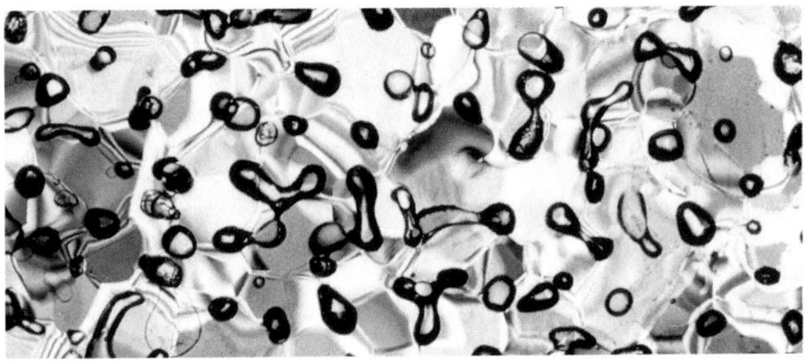

(b) Micrograph of ice core

FIGURE 18.6 In Greenland and Antarctica, scientists have drilled deep into ancient ice sheets and removed cores of ice like this one (**a**), held by Dr. Gerald Holdsworth of the University of Calgary, to extract information about past climates. Bubbles (black shapes) trapped in the ice (**b**) contain small samples of the ancient atmosphere.

these ice cores, scientists can determine atmospheric composition, greenhouse gas concentrations, temperature trends, snowfall, solar activity, and even (from trapped soot particles) frequency of forest fires.

Scientists also drill cores into beds of sediment beneath bodies of water. Sediments often preserve pollen grains and other remnants from plants that grew in the past, and as we saw with the study of Easter Island (▸ pp. 8–9), the analysis of these materials can reveal a great deal about the history of past vegetation. Because the types of plants that grow in an area depend on the area's climate, knowing what plants occurred in a given location can tell us much about the climate at that time (see "The Science behind the Story," ▸ p. 538). Sources of data such as pollen from sediment cores and air bubbles from ice cores are known as **proxy indicators**, types of indirect evidence that serve as proxies for direct measurement and that indicate the nature of past climate. Other types of proxy indicators include data culled from coral reefs and the tree rings of long-lived trees.

Direct atmospheric sampling tells us about the present

Studying present-day climate is more straightforward, because scientists can directly measure atmospheric conditions. The late Charles Keeling of the Scripps Institution of Oceanography in La Jolla, California, documented trends in atmospheric carbon dioxide concentrations starting in 1958 (Figure 18.7). Keeling collected four air samples from five towers every hour from his monitoring station at the Mauna Loa Observatory in Hawaii. Keeling's data show that atmospheric carbon dioxide concentrations have increased from 315 ppm to 378 ppm since 1958. Today Keeling's colleagues are continuing these measurements, building upon the single best long-term dataset we have of direct atmospheric sampling of any greenhouse gas.

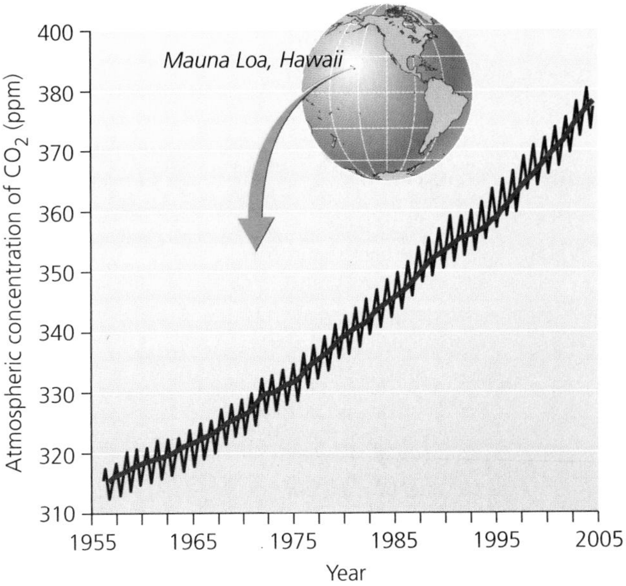

FIGURE 18.7 Atmospheric concentrations of carbon dioxide have risen steeply since 1958, when Charles Keeling began collecting these data at the Mauna Loa Observatory in Hawaii. The jaggedness apparent within the overall upward trend of the so-called "Keeling curve" represents seasonal variation caused by the Northern Hemisphere's vegetation, which absorbs more carbon dioxide during the northern summer, when it is more photosynthetically active. Go to GRAPHit! at www.aw-bc.com/withgott or on the student CD-ROM. Data from National Oceanic and Atmospheric Administration, Climate Prediction Center, 2005.

The Science behind the Story

Understanding El Niño and La Niña

Some scientists suspect that the warming of global air and ocean surface temperatures is increasing the frequency of the phenomenon known as El Niño. *El Niño* was originally the name that Spanish-speaking fishermen gave to an unusually warm surface current that sometimes arrived near the Pacific coast of South America around Christmastime. El Niño literally means "the little boy" or "the Christ child," and the current received this name because of its holiday arrival time. El Niño later became the term for the warming of the eastern Pacific that occurs every 2 to 7 years and depresses local fish and bird populations.

Under normal conditions, prevailing winds blow from east to west along the equator in the Pacific Ocean. These winds form part of a large-scale convective loop, or atmospheric circulation pattern; see part (a) of the first figure. These winds push surface waters westward and cause water to "pile up" in the western Pacific. As a result of these winds, water near Indonesia is, at times, about 50 cm (20 in.) higher and approximately 8 °C warmer than water along the coast of South America. The prevailing winds that move surface waters westward also make it possible for deep, cold, nutrient-rich water to form a nutrient-rich upwelling (▸ p. 471) along the coast of Peru.

When prevailing winds weaken, the warm water that has collected in the western Pacific flows "downhill" and eastward toward South America; see part (b) of the first figure. The influx of warm surface water and the

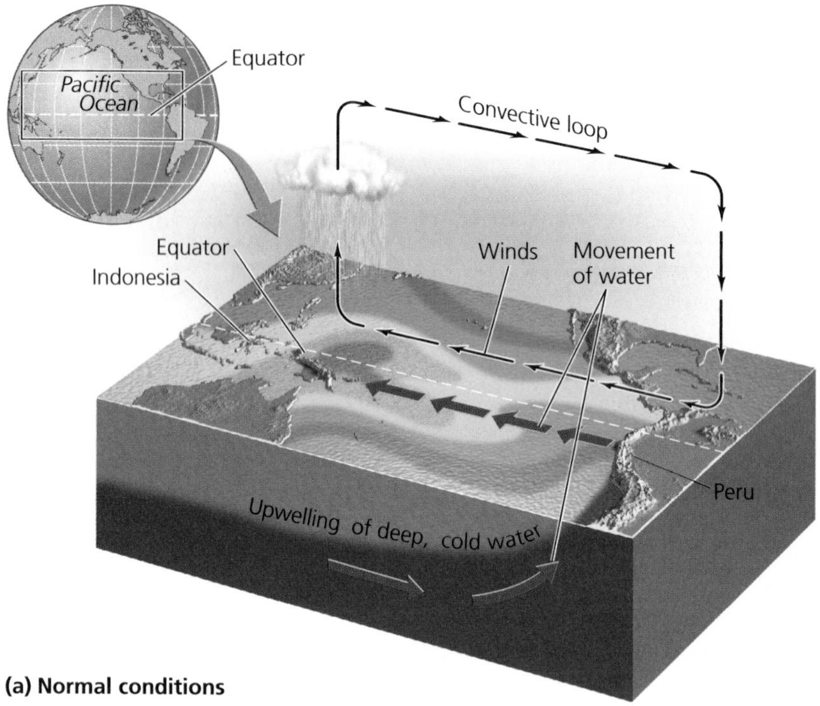

(a) Normal conditions

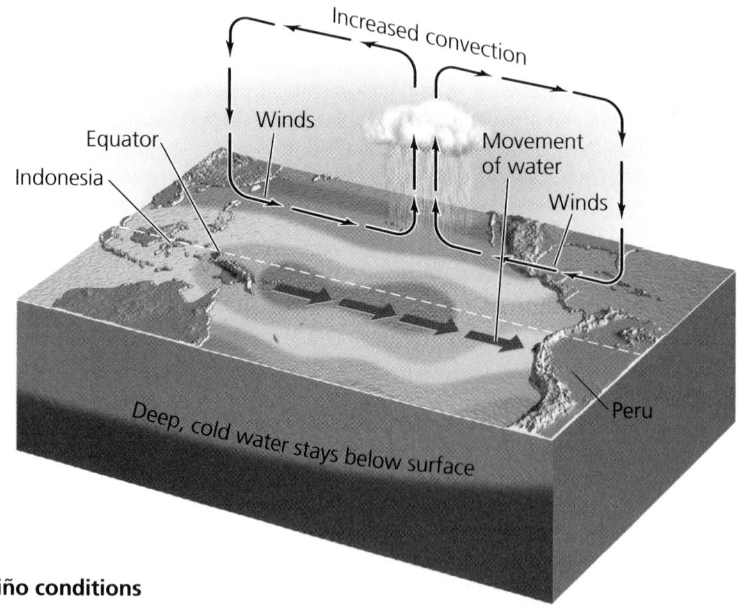

(b) El Niño conditions

In this view, Indonesia is on the left and Peru is on the right. During normal conditions (**a**), prevailing winds push warm surface waters toward the western Pacific. In this diagram, red and orange water is warmer than blue and green water. The red and orange waters in the west are also at a higher elevation than those in the east. During El Niño conditions (**b**), the prevailing winds weaken and no longer hold the warm surface waters in the western Pacific. As the warmer water "sloshes" back across the Pacific toward South America, precipitation patterns change.

absence of the normal prevailing winds suppress the upwelling along the coast of Peru, shutting down the delivery of nutrients that support the region's marine life and fisheries. This effect is felt along the entire Pacific Coast of the Americas, and it alters weather patterns around the world. Because warm surface water shapes weather by giving rise to heavy precipitation, its movement across the Pacific can create unusually intense rainstorms and floods in areas that are generally dry, and can cause drought and fire in regions that typically experience wet weather.

Scientists monitor the development of El Niño events with an array of wind- and temperature-sensing buoys, known as the Tropical Atmosphere Ocean Project, or TAO/TRITON, that are anchored across the equatorial Pacific; see part (a) of the second figure. These buoys measure temperature profiles during normal conditions (b) and El Niño events (c). Scientists use such data to determine the extent and severity of El Niño events. Awareness of El Niño conditions can enable governments and individuals to better prepare for extreme weather events and changes in ocean conditions.

La Niña, or "the little girl," is the opposite of El Niño and exerts impacts trending in the opposite direction. It is characterized by the presence of colder-than-normal surface water in the equatorial Pacific Ocean; see part (d) of the second figure. Scientists have made great strides recently in understanding El Niño and La Niña, but a great deal remains to be understood about their causes, their connections to climate change, and their effects on other environmental systems. They clearly show, however, that the atmosphere, oceans, and regional and global climate are tightly integrated systems.

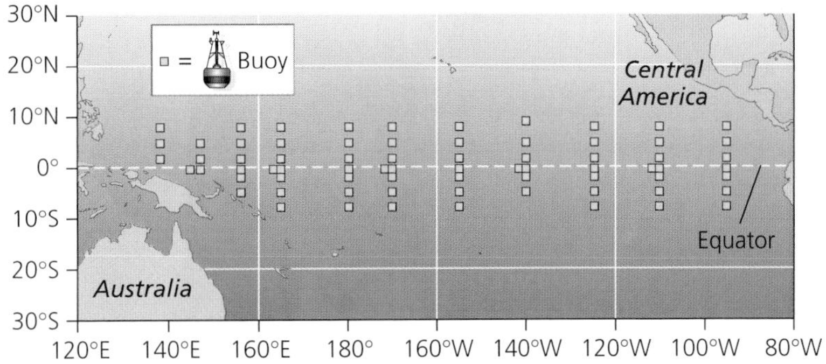

(a) Temperature-sensing buoys: TAO/TRITON array

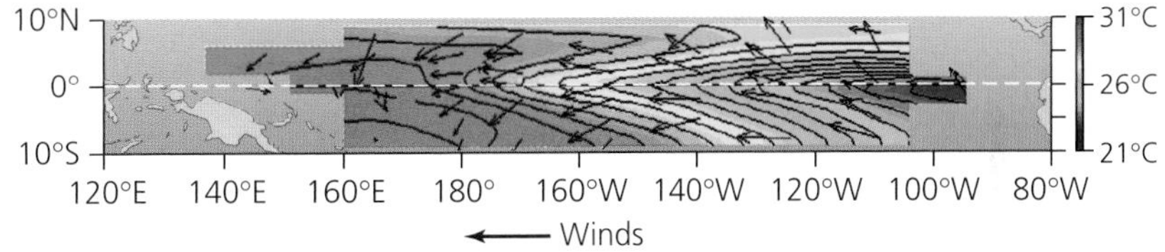

(b) Temperature during normal conditions (Dec. 1993 averages)

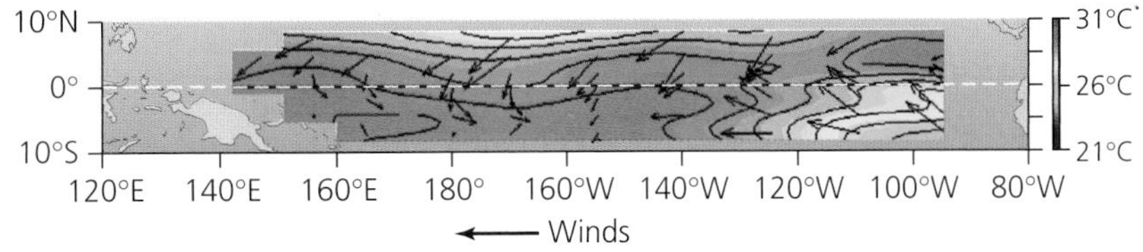

(c) Temperature during El Niño conditions (Dec. 1997 averages)

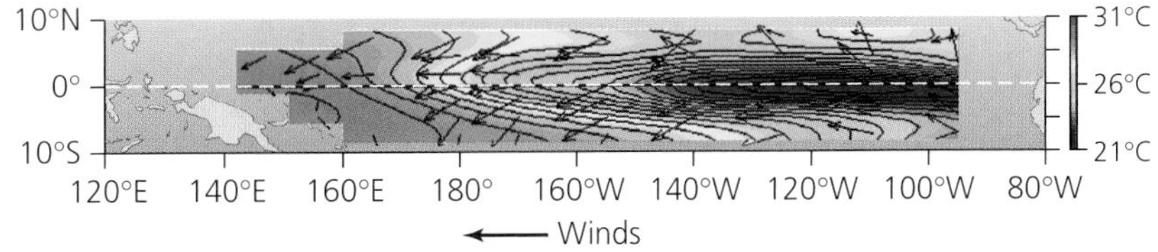

(d) Temperature during La Niña conditions (Dec. 1998 averages)

Scientists gather climate data in the Pacific Ocean by using anchored monitoring stations. These buoys (**a**) are distributed across the Pacific. During normal conditions (**b**), cold water reaches the surface of the eastern Pacific via an upwelling, represented by the blue tongue at the right of the graph. During an El Niño event (**c**), warmer water, shown in orange and red, is present in the eastern Pacific and the cold, blue tongue is absent. During a La Niña event (**d**), colder-than-normal water is present in the equatorial Pacific Ocean. Data from National Oceanic and Atmospheric Administration, Tropical Atmospheric Ocean Project, 2001.

The Science behind the Story

Scientists Use Pollen to Study Past Climate

Scientists can learn about climate by studying the types of vegetation that grew in particular places in the past. Cores taken from the sediments that collect at the bottoms of lakes and ponds contain pollen and other clues to the plants that lived in those locations at the time each layer was deposited. Tropical plant fossils and pollen tell scientists that warm, wet conditions were present in the past, whereas evidence of cold-loving plants reveal a cooler climate history.

In 1995, J. S. McLachlan of Harvard University and L. B. Brubaker of the University of Washington published a study on the past climate of Washington state's Olympic Peninsula. The researchers found that as climate and hydrological conditions changed, so did the area's vegetation. Their conclusions were based on the characteristics of pollen, fossils, and charcoal that had accumulated over time on the bottom of a small lake and in a cedar swamp, coupled with regional climate data from other sources. Such sediment deposits serve as a timeline. As scientists dig down through the layers of sediment, they peer farther back in time.

Pollen and larger plant parts preserved in sediments deposited on the bottoms of lakes and ponds over thousands of years can tell scientists about past plant communities and, by extension, about past climates. This sediment core sample was taken from Chesapeake Bay in Maryland.

The research team used a long metal cylinder to pull a 1,811-cm (59.4-ft) core of sediment from the middle of Crocker Lake and a 998-cm (32.7-ft) core of sediment from the middle of the cedar swamp. They wrapped the two cores in plastic and aluminum foil and stored them in large freezers at 4 °C (39.2 °F) to prevent decomposition. The team then began removing 1-mm thick slices every 10–30 cm (4–12 in.) along the length of the cores and analyzing the pollen and charcoal content of each slice. Pollen is easily transported by the wind and thereby can indicate the types of vegetation that occur over a fairly large area. Pollen, therefore, serves as a clue to regional climate.

The researchers identified between 300 and 1,000 pollen grains to genus and species within each slice. Larger plant fossils, such as those of cones, tree stems, and bark, were also classified. Because these items are not as easily transported as pollen, they are considered to indicate the vegetational and climatological history of smaller, more local, areas. Similarly, fine particles of charcoal, which, like pollen, are easily transported through the atmosphere, tell of the history of forest fires on a regional scale, whereas larger pieces suggest fire occurrence in the immediate area.

In combination with other available information, these clues enabled researchers to piece together a record of the general plant and climate history of the Olympic Peninsula.

Coupled general circulation models help us understand climate change

To understand how climate systems function, and to predict future climate change, scientists attempt to simulate climate processes with sophisticated computer programs. *Coupled general circulation models (CGCMs)* are computer programs that combine what is known about weather patterns, atmospheric circulation, atmosphere-ocean interactions, and feedback mechanisms to simulate climate processes (Figure 18.8). They couple, or combine, climate influences of the atmosphere and oceans in a single simulation. This requires manipulating vast amounts of data and complex mathematical equations—a task not possible until the advent of the supercomputer.

Over a dozen research labs around the world operate CGCMs. The models have been tested to see how closely they simulate known climate patterns when fed the appropriate historical data. In other words, if modelers enter past climate data, and the models produce accurate predictions of current climate, we have reason to believe the models may be realistic and accurate. Figure 18.9 shows temperature results from three such simulations.

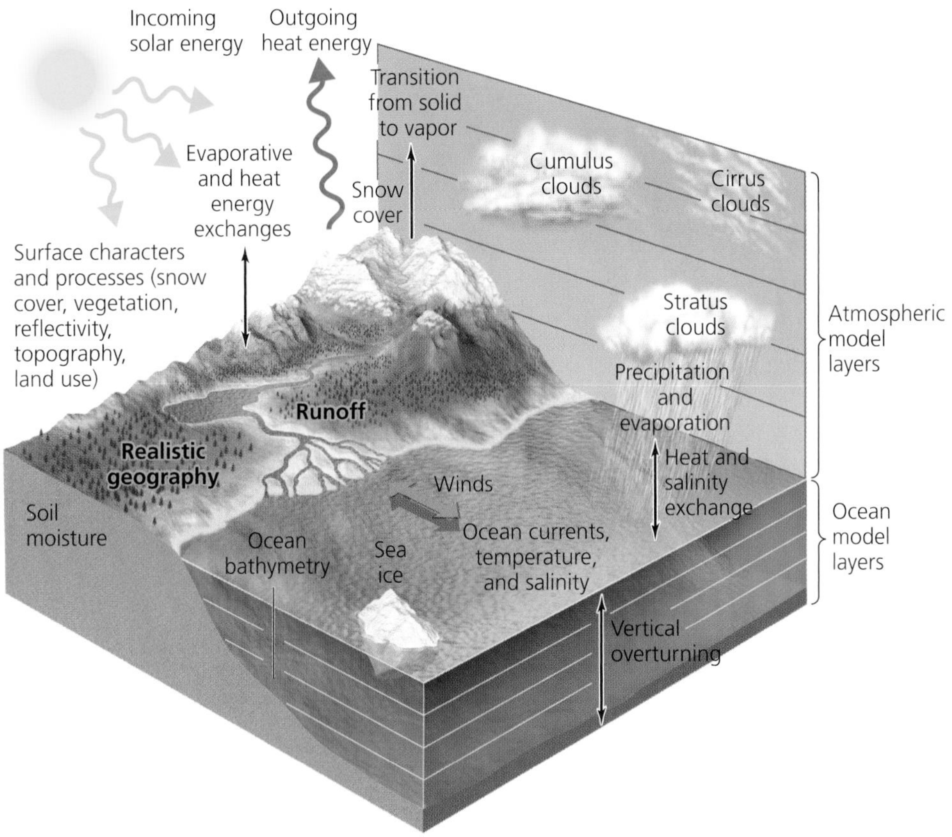

FIGURE 18.8 Modern climate models incorporate many factors, including processes involving the atmosphere, land, oceans, ice, and biosphere. Some of these factors are shown graphically here, but the actual models deal with them as mathematical equations in computer simulations.

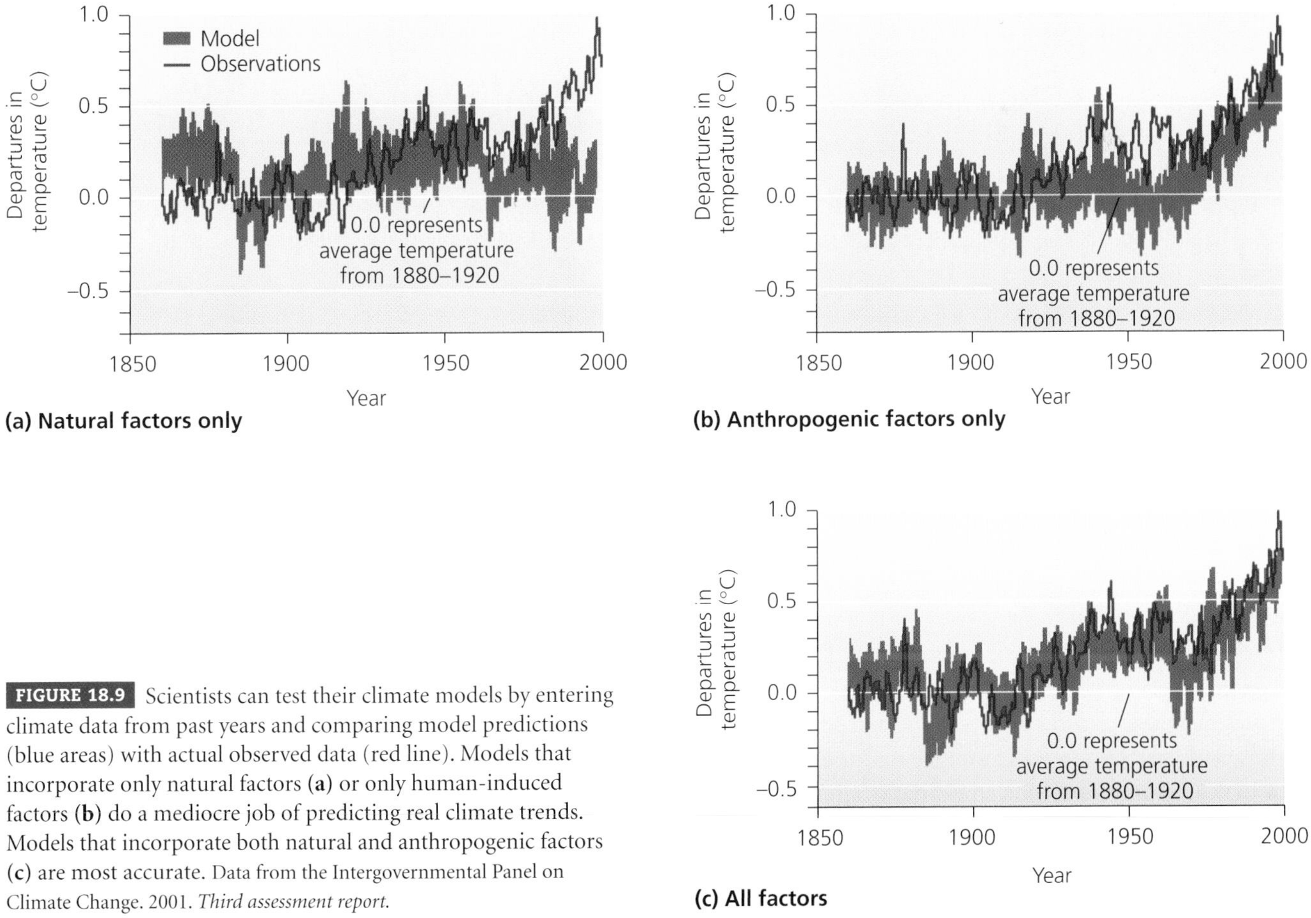

FIGURE 18.9 Scientists can test their climate models by entering climate data from past years and comparing model predictions (blue areas) with actual observed data (red line). Models that incorporate only natural factors (**a**) or only human-induced factors (**b**) do a mediocre job of predicting real climate trends. Models that incorporate both natural and anthropogenic factors (**c**) are most accurate. Data from the Intergovernmental Panel on Climate Change. 2001. *Third assessment report.*

The results in Figure 18.9a are based on natural climate-changing factors alone (such as volcanic activity and variation in solar energy). The results in Figure 18.9b are based on anthropogenic factors only (such as human greenhouse gas emissions and sulfate aerosol emissions). The results in Figure 18.9c are based on natural and anthropogenic factors combined, and this produces the closest match between predictions and actual climate. Results such as those in Figure 18.9c support the notion that both natural and human factors contribute to climate dynamics, and they also indicate that CGCMs can produce reliable predictions. As computing power increases and our ability to glean data from proxy indicators of past climate improves, CGCMs become increasingly reliable.

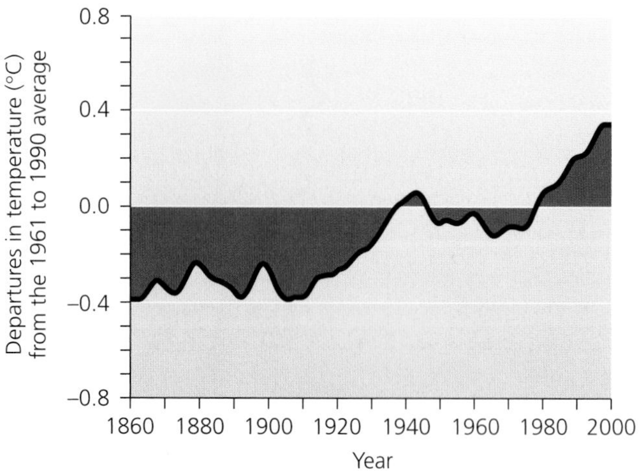

(a) Global temperature measured over the past 140 years

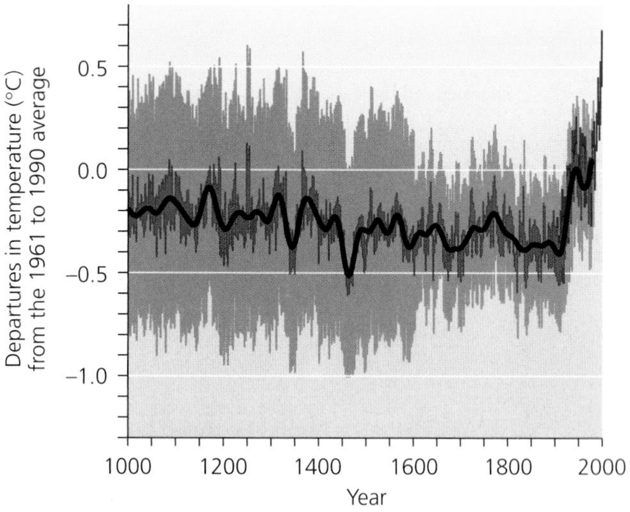

(b) Northern Hemisphere temperature over the past 1,000 years

FIGURE 18.10 Data from thermometers (**a**) show changes in Earth's average surface temperature since 1860. In (**b**), proxy indicators (blue line) and thermometer data (red line) together show average temperature changes in the Northern Hemisphere over the past 1,000 years. The gray-shaded zone represents the 95% confidence range. This record shows that 20th-century warming has eclipsed the magnitude of change during both the "Medieval Warm Period" (10th–14th centuries) and the "Little Ice Age" (15th–19th centuries). Go to GRAPHit! at www.aw-bc.com/withgott or on the student CD-ROM. Data from the Intergovernmental Panel on Climate Change, 2001. *Third assessment report.*

Climate Change Estimates and Predictions

The most thoroughly reviewed and widely accepted collection of scientific information concerning global climate change is a series of reports issued by the **Intergovernmental Panel on Climate Change (IPCC).** This international panel of atmospheric scientists, climate experts, and government officials was established in 1988 by the United Nations Environment Programme (UNEP) and the World Meteorological Organization. Its task is to assess information relevant to questions of human-induced climate change. In 2001 the IPCC released its *Third Assessment Report.* This summary of current global trends and probable future trends represents the consensus of atmospheric scientists around the world.

The IPCC report summarizes evidence of recent changes in global climate

The IPCC report's major conclusions concerning recent climate trends address surface temperature; snow and ice cover; rising sea level and warmer oceans; precipitation patterns and intensity; and effects on wildlife, ecosystems, and human societies. The IPCC report is authoritative but, like all science, deals in uncertainties. Its authors have therefore assigned statistical probabilities to each of its conclusions and predictions.

Besides the data on increases in atmospheric concentrations of greenhouse gases that we discussed earlier, the 2001 IPCC report presented a number of findings on how climate change has already influenced the weather, Earth's physical characteristics and processes, the habits of organisms, and our economies. The report concluded, for instance, that average surface temperatures had increased by 0.6 °C (1.0 °F) during the 20th century (Figure 18.10). It inferred that glaciers, snow cover, and ice caps were melting worldwide. It found that hundreds of species of plants and animals were being forced to shift their

Table 18.2 Major Findings of the IPCC Third Assessment Report, 2001

Weather indicators

- Earth's average surface temperature increased 0.6 °C (1.0 °F) during the 20th century (90–99% certainty)
- Cold and frost days decreased for nearly all land areas in the 20th century (90–99% certainty)
- Continental precipitation increased by 5–10% during the 20th century in the Northern Hemisphere (90–99% certainty), but it decreased in some regions
- The 1990s were the warmest decade of the past 1,000 years (66–90% certainty)
- The 20th-century Northern Hemisphere temperature increase was the greatest in 1,000 years (66–90% certainty)
- Droughts increased in frequency and severity (66–90% certainty)
- Nighttime temperatures increased twice as fast as daytime ones (66–90% certainty)
- Hot days and heat index increased (66–90% certainty)
- Heavy precipitation events increased at northern latitudes (66–90% certainty)

Physical indicators

- Average sea level increased 10–20 cm (4–8 in.) during the 20th century
- Rivers and lakes in the Northern Hemisphere were covered by ice 2 weeks less from the beginning to the end of the 20th century (90–99% certainty)
- Arctic sea ice thinned by 10–40% in recent decades, depending on the season (66–90% certainty)
- Mountaintop glaciers retreated widely during the century
- Global snow cover decreased by 10% since satellite observations began in the 1960s (90–99% certainty)
- Permafrost thawed in many regions
- El Niño events became more frequent, persistent, and intense in the past 40 years in the Northern Hemisphere
- Growing seasons lengthened 1–4 days per decade in the last 40 years in northern latitudes

Biological indicators

- Geographic ranges of many plants, insects, birds, and fish shifted toward the poles and upward in elevation
- Plants are flowering earlier, migrating birds are arriving earlier, animals are breeding earlier, and insects are emerging earlier in the Northern Hemisphere
- Coral reefs are experiencing bleaching more frequently

Economic indicators

- Global economic losses due to weather events rose 10-fold over the past 40 years, partly because of climate factors

Data from Intergovernmental Panel on Climate Change. 2001. *Third assessment report.*

geographic ranges and the timing of their life cycles. These findings and many more were based on the assessment of thousands of scientific studies conducted worldwide over many years, judged by the world's top experts in these fields. Some—but by no means all—of the report's major findings are shown in Table 18.2.

Sea-level rise and other changes interact in complex ways

Few of the processes and impacts examined in the IPCC report function in isolation, because Earth's environmental systems are connected and interact. For instance, warming temperatures are causing glaciers to shrink and disappear in many areas around the world. For the same reason, polar ice shelves that have been intact for millennia are breaking away and melting. As glaciers and ice shelves melt, an increased flow of water into the oceans is causing a rise in sea level. Sea level is also rising because ocean water is warming, and water expands in volume as its temperature increases. In fact, most sea-level rise is calculated to result from thermal expansion of seawater, rather than from runoff from melting glaciers and ice shelves.

Higher sea levels lead to beach erosion, coastal flooding, intrusion of saltwater into aquifers, and other impacts. In 1987, unusually high waves struck the Maldives and triggered a campaign to build a large seawall around Male, the nation's capital city. Known as "The

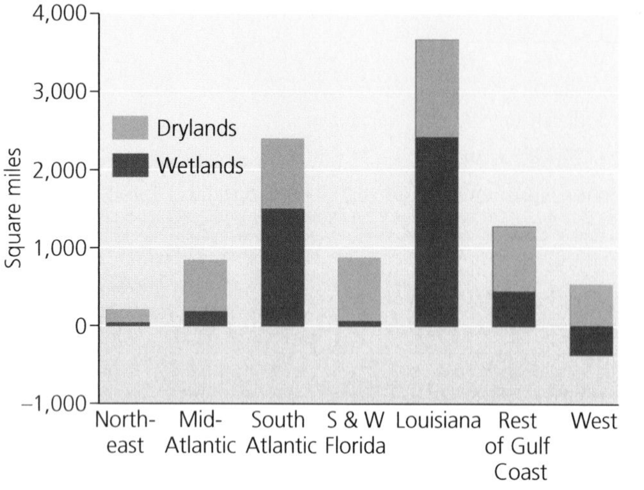

(a) U.S. coastal lands at risk from a 51-cm (20-in.) sea-level rise

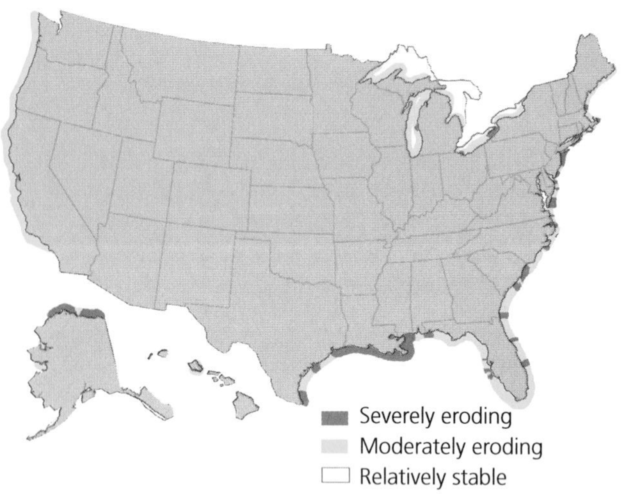

(b) Annual shoreline change

(c) Rescue from floodwaters of Hurricane Katrina

FIGURE 18.11 The United States will lose coastal areas as a result of a rise in sea level. A 51-cm (20-in.) sea level rise would inundate wetlands (red) and drylands (orange) on portions of all coasts (**a**). Coastal areas around the nation (**b**) are at risk of increased erosion due to sea-level rise. Rescues like this one (**c**) from the floodwaters of Hurricane Katrina's storm surge in coastal Mississippi in 2005 could become more frequent in a future world of higher sea levels. Data (a, b) from National Assessment Synthesis Team, U.S. Global Change Research Program. 2001. *Climate change impacts on the United States: The potential consequences of climate variability and change: Foundation.* Cambridge, U.K.: Cambridge University Press and USGCRP.

Great Wall of Male," the seawall is intended to break up incoming waves and dissipate their energy to prevent the destruction of buildings and roads during storm surges. A *storm surge* is a temporary and localized rise in sea level brought on by the high tides and winds associated with storms. The higher that sea level is to begin with, the further inland a destructive storm surge can reach. "With a mere 1-meter rise [in sea level]," Maldives' President Maumoon Abdul Gayoom warned in 2001, "a storm surge would be catastrophic, and possibly fatal to the nation."

The Maldives is not the only nation with such concerns. In fact, among island nations, the Maldives has fared better than many others. It saw sea level rise about 2.5 mm per year throughout the 1990s, but most Pacific islands are experiencing greater changes. Different regions experience different amounts of sea level change because land elevations may be rising or subsiding naturally, depending on local geological conditions.

The United States is not immune from coastal impacts—and 53% of U.S. residents live within the 17% of the nation's land that lies in coastal areas. Figure 18.11a shows the amounts of wetlands and dry lands in the United States that would be inundated by a 51-cm (20-in.) sea-level rise. Figure 18.11b shows coastal areas in the United States that are eroding because of sea-level rise and other factors. The vulnerability of parts of the nation to coastal flooding due to storm surges became tragically apparent in 2005 when Hurricane Katrina struck New Orleans and the Gulf Coast, followed shortly thereafter by Hurricane Rita. The flooding that devastated the region (Figure 18.11c) could have been even worse had sea level been higher.

The levees surrounding New Orleans that were breached by Katrina's floodwaters are now repaired, but large portions of the city will always remain below sea level. Areas that are now 2.1 m (7 ft) below sea level may be as much as 3.3 m (10 ft) below sea level by 2100, as the land subsides and sea level rises. Just outside the city, marshes of the Mississippi River delta are being lost rapidly, as dams upriver hold back the silt that used to maintain the delta, as land subsides due to the extraction of petroleum deposits, and as rising seas eat away at coastal vegetation. Approximately 2.5 million ha (1 million acres) of Louisiana wetlands have become open water since 1940. Continued loss of these wetlands will mean that New Orleans will have less protection against future storm surges. And coastal Louisiana is not the only U.S. region with these concerns. Houston, Texas, and Charleston, South Carolina are among the major cities most likely to be affected by increased sea level. In southern Florida, coastal wetlands and mangroves are being submerged, and salt water has intruded into aquifers, killing trees and threatening drinking water supplies.

Moreover, the record number of hurricanes and tropical storms in 2005—Katrina and 22 others—left many people wondering if global warming was to blame. Are warmer ocean temperatures spawning more hurricanes, or hurricanes that are more powerful or long-lasting? Scientists are not yet sure of the answers to these questions, but evidence from a number of recent studies indicates that warmer sea surface temperatures are likely increasing the destructive power of storms, and possibly their duration, but are apparently not increasing their frequency.

The IPCC and other groups project future impacts of climate change

Because the consequences of climate change could be substantial, the IPCC and other groups have attempted to predict future climate changes and their potential impacts. One such group is the U.S. Global Change Research Program (USGCRP), created by Congress in 1990 to coordinate federal climate research. In 2000–2001, the USGCRP issued a report highlighting the past and future effects of global climate change on the United States, where annual average temperatures increased by 0.6 °C (1.0 °F) during the 20th century.

The report used CGCMs to develop a series of predictions on impacts of climate change in the United States (Table 18.3). These models, such as the Hadley model designed by British researchers and the Canadian model designed by British Columbian researchers, allowed scientists to present unique graphical depictions summarizing predicted impacts of climate change on particular geographic areas (Figure 18.12).

Table 18.3 Some Predicted Impacts of Climate Change in the United States

- Average U.S. temperatures will increase 3–5 °C (5–9 °F) in the next 100 years
- Droughts and flooding will worsen, and snowpack will be reduced
- Drought and other factors could decrease crop yields, but longer growing seasons and enhanced CO_2 could increase yields
- Water shortages will worsen
- Greater temperature extremes will increase health problems and human mortality. Some tropical diseases will spread north into temperate latitudes
- Forest growth may increase in the short term, but in the long term, drought, pests, and fire may alter forest ecosystems
- Alpine ecosystems and barrier islands will begin disappearing
- Southeastern U.S. forests will break up into savanna/grassland/forest mosaics
- Northeastern U.S. forests will lose sugar maples
- Loss of coastal wetlands and real estate due to sea level rise will continue
- Melting permafrost will undermine Alaskan buildings and roads

Adapted from National Assessment Synthesis Team, U.S. Global Change Research Program. 2000. *Climate change impacts on the United States: The potential consequences of climate variability and change: Overview.* Cambridge, U.K.: Cambridge University Press and USGCRP.

Agriculture and forestry Drought and temperature extremes are among the threats that climate change may pose for farms and forests worldwide. Croplands presently near the limits imposed by heat stress and water availability could be pushed beyond their ability to produce food. If average temperatures increase by more than 3–4 °C, most tropical and subtropical areas will likely see decreased crop production, and midlatitude farmlands may also begin to see significant declines.

Conversely, warmer temperatures and longer growing seasons at higher latitudes could potentially increase agricultural productivity there. In fact, of 16 crops in the United States, the USGCRP's agricultural assessment team in 2002 predicted that yields of 13 would increase and only 1 would decrease because of climate change. The overall effect of a warmer climate on agricultural productivity is difficult to predict. Agricultural productivity might remain somewhat stable globally while increasing in some areas and decreasing in others. For this reason,

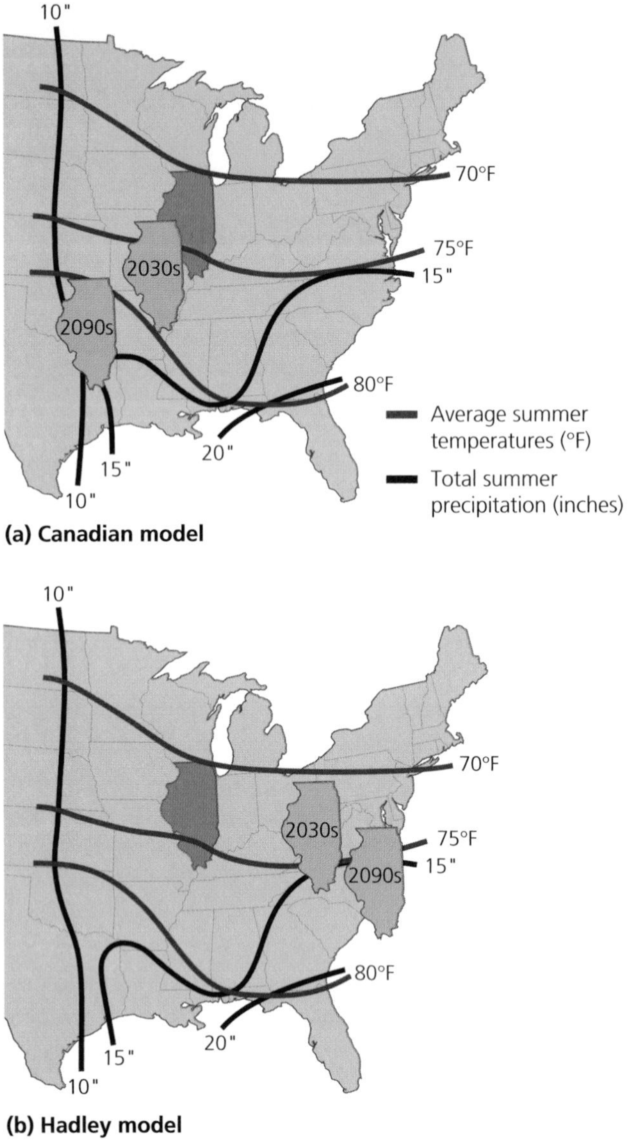

FIGURE 18.12 Using parameters of the Canadian and Hadley models, researchers were able to portray the predicted future climate of the state of Illinois. The Canadian model predicted Illinois to become warmer and drier, attaining in the 2030s a climate similar to Missouri's, and in the 2090s one similar to eastern Oklahoma's. The Hadley model predicted Illinois to become warmer and moister, attaining in the 2030s a climate similar to West Virginia's, and in the 2090s one similar to North Carolina's. Data from National Assessment Synthesis Team, U.S. Global Change Research Program. 2001. *Climate change impacts on the United States: The potential consequences of climate variability and change: Foundation.* Cambridge, U.K.: Cambridge University Press and USGCRP.

some people argue that global warming could benefit their nations, but others point out that it may increase inequities between developed and developing nations.

Research by scientists indicates that climate change could transform U.S. forests. In its 2000–2001 report, the USGCRP predicted that U.S. forests will become more productive, because additional carbon dioxide in the atmosphere will speed rates of photosynthesis. As productivity increases, however, the frequency and intensity of forest fires could increase by 10% or more. Forest communities should in general move northward and upward in elevation. Alpine and subalpine plant communities should become less common as the climate warms, because these mountaintop communities cannot move further upward in elevation. Although some forest types will likely decline, others, including oak-hickory and oak-pine forests, may expand in the eastern United States (Figure 18.13).

Effects on plant communities comprise an important component of climate change, because by drawing in CO_2 for photosynthesis, plants act as sinks for carbon. The widespread regrowth of forests in eastern North America (▸ pp. 350, 355) has offset an estimated 25% of U.S. carbon emissions during the past four decades. If climate change increases overall vegetative growth, this could partially mitigate carbon emissions, in a process of negative feedback. However, if climate change decreases overall growth (through drought or fire, for instance), then a positive feedback cycle could increase carbon flux to the atmosphere.

Weighing the Issues: Agriculture in a Warmer World

Some people maintain that a warmer climate would expand arable lands toward the poles and lead to greater agricultural production globally. In fact, the U.S. Corn Belt is already pushing into Canada. Locate the nations of Russia and Argentina on a world map, and hypothesize how such a poleward shift of agriculture might affect each of these nations. Now hypothesize how it might affect poorer nations near the equator that are already suffering from food shortages and agricultural problems. Thinking back to Chapters 9 and 10, name several factors that could potentially influence crop yields if climate continues changing.

Freshwater and marine systems In regions where climate change increases precipitation and stream flow, erosion and flooding could alter the structure and function of aquatic systems. Where agriculture and other human activities have modified the landscape, such flooding could increase pollution of freshwater ecosystems. In regions where precipitation decreases, lakes, ponds, wetlands, and streams would shrink, affecting the organisms that live in

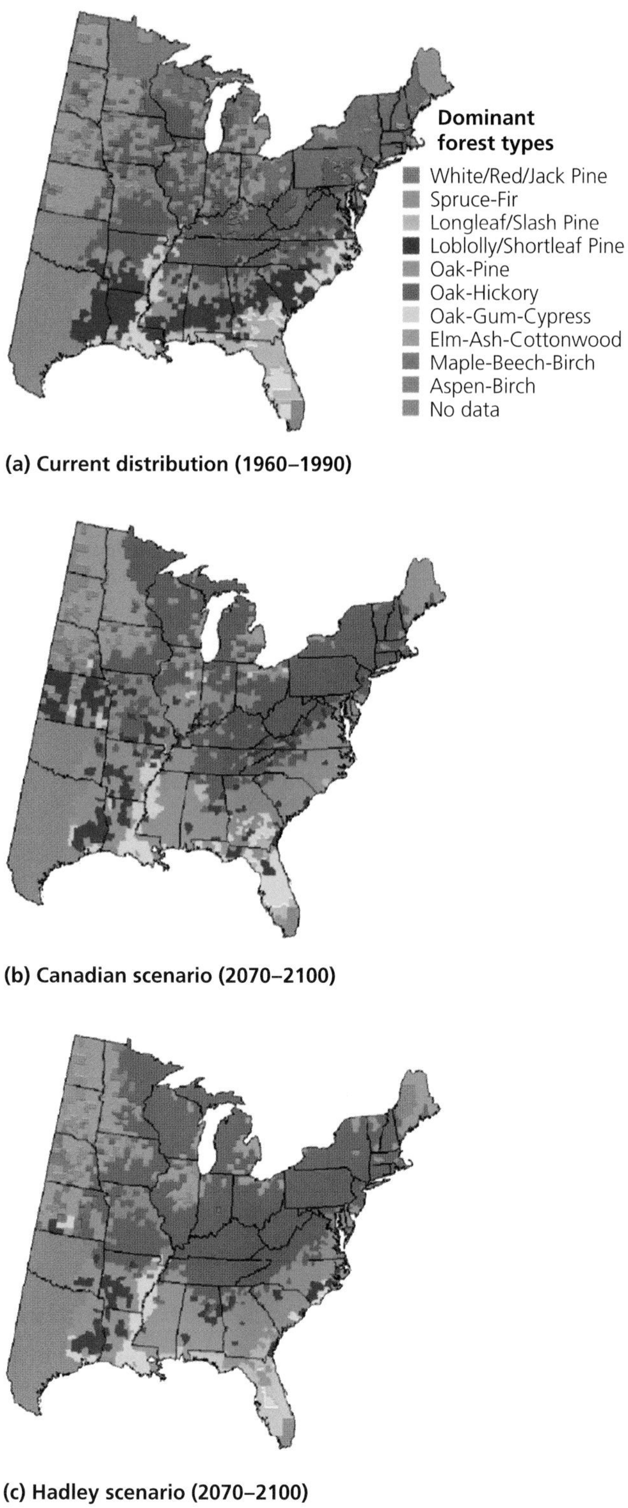

FIGURE 18.13 While some forest types will decline, others, including oak-hickory and oak-pine forests, may expand in the eastern United States. This figure shows predicted forest cover changes based on the Canadian and Hadley models. Data from National Assessment Synthesis Team, U.S. Global Change Research Program. 2001. *Climate change impacts on the United States: The potential consequences of climate variability and change: Foundation.* Cambridge, U.K.: Cambridge University Press and USGCRP.

those habitats, as well as human health and well-being. In 2001, about 1.7 billion people were living in areas with limited water supplies. By 2025, according to the IPCC, this number will likely increase to 5 billion out of a projected world population of 8.4 billion.

The Maldives is likely to suffer from water-related stresses, because its human population is expanding rapidly while rising seas threaten to bring salt water into the nation's wells, just as the 2004 tsunami did. The contamination of groundwater and soils by seawater is particularly threatening to island nations like the Maldives and coastal areas such as the Tampa, Florida, region, which depend on small lenses of freshwater that float atop saline groundwater.

Maldives residents also worry about damage to marine ecosystems, including coral reefs (▸ pp. 474–476). Coral reefs provide habitat for important food fish, reduce wave intensity and protect fragile coastlines from erosion, and provide popular snorkeling and scuba diving sites. Damage to reefs from storm surges would depress tourism and fishing.

Human health As a result of climate change, people could face increased exposure to an array of health problems:

- Heat stress resulting from high temperatures and humidity
- Respiratory ailments from air pollution, as hotter temperatures promote formation of photochemical smog (▸ pp. 510–511)
- Expansion of tropical diseases, such as malaria and dengue fever, into temperate regions
- Disease and sanitation problems when floods overcome sewage treatment systems
- Injuries and drowning if storms become more frequent or intense
- Hunger-related ailments as human population grows and demands on agricultural systems increase

Figure 18.14a shows USGCRP projections regarding the 21st-century July heat index across the United States. One model predicts that the heat index—a product of temperature and humidity—will be 14 °C (25 °F) hotter in much of the southeastern United States. Heat waves can lead to high mortality rates in American cities (Figure 18.14b). In Europe, heat stress killed 35,000 people in August 2003 during a record heat wave.

At the same time, other scientists, including some IPCC contributors, have argued that a warmer world will present fewer diseases and injuries that result from cold weather. The trade-off between an increase in warm-weather ailments and a decrease in cold-weather health problems remains one of the unknowns of global climate change.

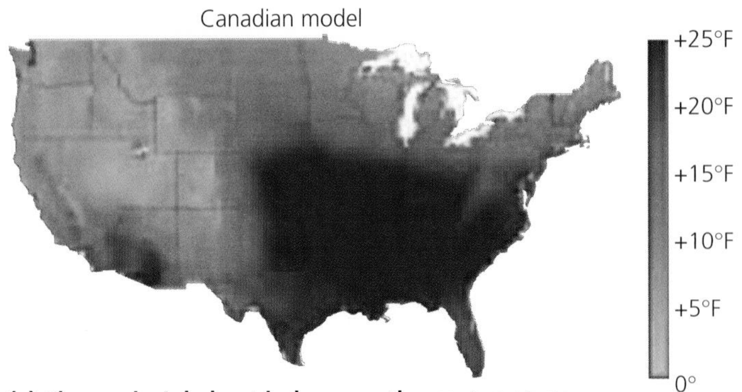

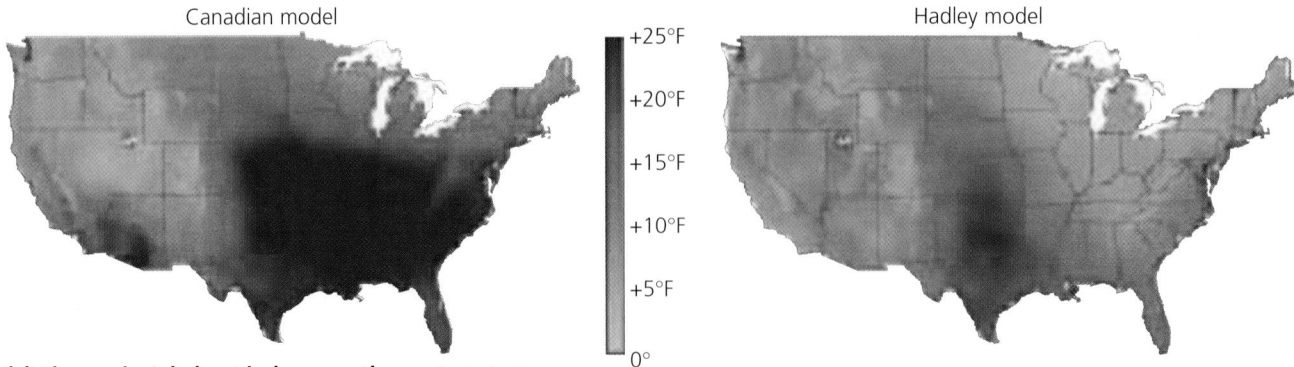

(a) Change in July heat index over the next century

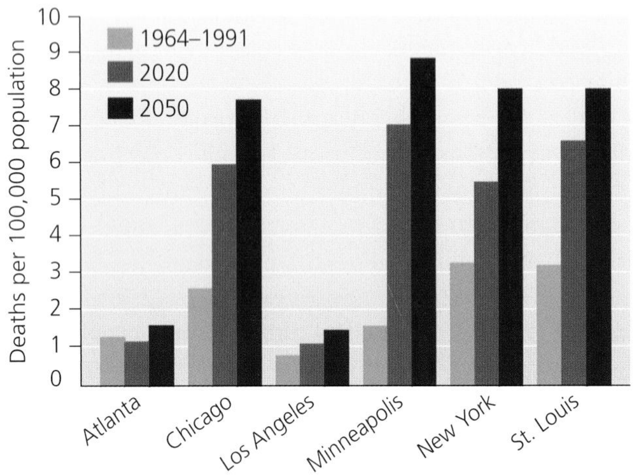

(b) Average summer mortality rates attributed to hot weather episodes

FIGURE 18.14 The Canadian model and the Hadley model provide two projections of the heat index (a product of temperature and humidity) for July across the United States for the upcoming century **(a)**. The Canadian model predicts that the July heat index will be 14 °C (25 °F) hotter than that of much of the present-day southeastern United States. Past and projected future mortality rates attributed to hot weather **(b)** are shown for several U.S. cities. Data from National Assessment Synthesis Team, U.S. Global Change Research Program. 2001. *Climate change impacts on the United States: The potential consequences of climate variability and change: Foundation.* Cambridge, U.K.: Cambridge University Press and USGCRP.

Debate over Climate Change

Virtually all environmental scientists agree that Earth's atmosphere and climate are changing. The great majority of them have concluded that human activity, particularly our emission of greenhouse gases, is the primary reason for this change. Despite this unusually strong scientific consensus, you have no doubt heard a great deal of debate over climate change. To understand why, it is most helpful to break down this debate into components that can be examined separately.

One component involves discussion within the scientific community about the details and mechanisms of climate change and the extent and nature of its likely effects on human welfare and environmental systems. This is part of the normal process of science as researchers gather evidence and test competing hypotheses, trying to get a full and complete picture of the truth. A second arena of debate involves people, primarily nonscientists, who contest the consensus findings and interpretations of the scientific community. Some of these so-called "greenhouse skeptics" have vested interests (▸ p. 30) in continuing the widespread use of fossil fuels, and some of them have significant sway over policymakers, particularly in the United States. A third arena of debate involves how our societies should respond to climate change. This is a wide-ranging discourse among scientists, economists, business leaders, policymakers, and others.

Scientists agree that climate change is occurring but disagree on many details

Although most scientists have concluded that our greenhouse gas emissions are altering the atmosphere and influencing climate, many details and aspects of climate change science remain uncertain, because climate systems are so complex. Climate scientists have not yet come to firm consensus on the roles played by clouds, water vapor, soot and sulfate aerosols, vegetative carbon sinks, and the oceans and NADW formation. Such uncertainties, coupled with the complexity of feedback mechanisms, make it difficult to predict the future.

Despite these uncertainties, the scientific community now feels that evidence for our role in influencing climate is strong enough that governments should take action to address greenhouse gas emissions. In June 2005, as the leaders of the "G8" industrialized nations met, the national academies of science from 11 nations (Brazil, Canada, China, France, Germany, India, Italy, Japan, Russia, the United Kingdom, and the United States) issued a joint statement urging these political leaders to take action. Such a broad consensus statement from the world's scientists was virtually unprecedented, on any issue. The statement read, in part:

> The scientific understanding of climate change is now sufficiently clear to justify nations taking prompt action. It is vital that all nations identify cost-effective steps that they can take now, to contribute to substantial and long-term reduction in net global greenhouse gas emissions. . . . A lack of full scientific certainty about some aspects of climate change is not a reason for delaying an immediate response that will, at a reasonable cost, prevent dangerous anthropogenic interference with the climate system.

Some challenge the scientific consensus

Despite such clear statements from the scientific community about the risks posed by climate change, many people, primarily nonscientists, have ignored or disputed mainstream scientific findings and interpretations. For instance, just days after the national academies' statement, the *Wall Street Journal* ran an editorial that asked, "What Warming?" and maintained that:

> The scientific case for [climate change] looks weaker all the time. . . . Since [1997], the case for linking fossil fuels to global warming has, if anything, become even more doubtful. The Earth currently does seem to be in a warming period, though how warm and for how long no one knows. In particular, no one knows whether this is unusual or merely something that happens periodically for natural reasons. Most global warming alarms are based on computer simulations that are largely speculative and depend on a multitude of debatable assumptions.

The editorial referenced several scientific studies that contested the data shown in Figure 18.10b, as well as other key findings in climate change science. It did not make clear the extent to which studies demonstrating climate change have earned broad scientific acceptance, whereas studies that deny climate change have not been widely accepted.

The media have played a large role in the public's understanding of climate change, and the U.S. media by tradition try to portray both sides of any issue. However, they often fail to make clear when spokespeople for one side of an issue have stronger evidence or have earned wider support than spokespeople for another side. Author Ross Gelbspan has maintained in two books that the American media have portrayed the climate change debate as much more even and two-sided than it actually is. By giving greenhouse skeptics equal time with mainstream scientists, he and many scientists have argued, the media have amplified the views of greenhouse skeptics out of proportion to their prevalence in the scientific community. Moreover, he documents that the relatively few greenhouse skeptics among scientists are often funded by industries that benefit from fossil fuel use, such as the petroleum and automobile industries.

Industry lobbying has also played a part in making policymakers hesitant to enact policy to reduce carbon emissions. Attempts to improve fuel efficiency for automobiles (▸ p. 582) have repeatedly failed in the U.S. Congress, and attempts to open the Arctic National Wildlife Refuge for oil drilling (▸ pp. 557–559) have repeatedly been made.

Weighing the Issues: Environmental Refugees

Citizens of the Maldives need only look southward to find an omen of their future. The Pacific island nation of Tuvalu, which has been losing 9 cm (3.5 in.) of elevation per decade to rising seas, may become the first casualty of global climate change. Appeals from its 11,000 citizens were heard by New Zealand, which began accepting these environmental refugees in 2003. Do you think this refugee flow casts doubt on the arguments of those who hold that climate change is uncertain? How might the perspective of a Tuvalu resident differ from that of a U.S. oil industry executive, and why? What steps might individual people of Tuvalu or the Maldives or coastal Florida take to protect their ways of life?

How should we respond to climate change?

Even if one accepts that climate change is real and poses significant ecological and economic threats, there is plenty of room for disagreement over appropriate responses. Political and economic debate over how we should respond stems from questions such as:

- Would the economic and political costs of reducing greenhouse gas emissions outweigh the costs of unabated emissions and resulting global climate change?
- Should industrialized nations, developing nations, or both take responsibility for reducing greenhouse gas emissions?
- Should steps to reduce emissions occur voluntarily or as a result of government regulation, political pressure, or economic sanctions?
- How should we allocate funds and human resources for reducing emissions and coping with climate change?

Strategies for Reducing Emissions

Since 1990, the generation of electricity, largely through coal combustion, has produced the largest portion (34%, as of 2000) of U.S. greenhouse gas emissions (Figure 18.15). Transportation ranks second at 27%, industry produces 19%, agriculture and residential sources produce 8% each, and commercial sources account for 5%. Tackling climate change will require reducing emissions from these sources, and strategies for doing so involve scientific, technical, political, and economic approaches.

FIGURE 18.15 Coal-fired electricity-generating power plants, such as this one in Maryland, are the largest contributors to U.S. greenhouse emissions.

Electricity generation is the largest source of U.S. greenhouse gases

From cooking and heating to the clothes we wear, much of what we own and do depends on electricity. Fossil fuel combustion generates over 70% of U.S. electricity. Coal alone accounts for 56%, along with most of the resultant greenhouse gas emissions. Reducing the volume of fossil fuels we burn to generate electricity would reduce greenhouse gas emissions, as would decreasing electricity consumption. There are two ways to reduce the amount of fossil fuels we use: (1) encouraging conservation and efficiency (▸ pp. 582–583, 585) and (2) switching to renewable energy sources (Chapters 20 and 21).

Conservation and efficiency Conservation and efficiency in energy use can arise from new technologies, such as high-efficiency lightbulbs and appliances, or from individual ethical choices to reduce electricity consumption. In one example of a technological solution, the U.S. Environmental Protection Agency (EPA) promotes energy conservation through its Energy Star Program. The Energy Star Program rates household appliances, lights, windows, fans, office equipment, heating and cooling systems, and appliances for their efficiency in using energy. Following are examples of the energy you can save by choosing Energy Star products:

- An Energy Star refrigerator can cut your CO_2 emissions by 100 kg (220 lb) annually.
- An Energy Star washing machine can cut your CO_2 emissions by 200 kg (440 lb) annually.
- Compact fluorescent lights can reduce the energy you use for lighting by 40%.
- Energy Star homes use highly efficient construction, duct work, insulation, heating and cooling systems, and windows to reduce energy use by as much as 30%.

Such technological solutions are popular, and they can be profitable for manufacturers while also saving consumers money. Alternatively, consumers can opt for lifestyle choices rather than technological fixes. For nearly all of human history, people managed without the electrical appliances that most of us take for granted today. It is entirely possible for each of us to simply choose to use fewer greenhouse-gas-producing appliances and technologies or to take practical steps to use electricity more efficiently.

Renewable sources of electricity Technologies that generate electricity without using fossil fuels represent another means of reducing greenhouse gas emissions. These include hydroelectric power (▸ pp. 446–447 and 612–615), geothermal energy (▸ pp. 633–635), photovoltaic cells (▸ pp. 625–626), and wind power (▸ pp. 627–633). We will examine renewable energy sources in more detail in Chapters 20 and 21.

Transportation is the second largest source of U.S. greenhouse gases

Can you imagine life without a car? Most Americans probably can't—a reason why transportation is the second largest source of U.S. greenhouse emissions. One-third of the average American city—including roads, parking lots, garages, and gas stations—is devoted to use by cars. The average American family makes 10 trips by car each day, and governments across the nation spend $200 million per day on road construction and repairs. Registered U.S. automobiles number over 220 million, and this figure is projected to surpass the human population of the country.

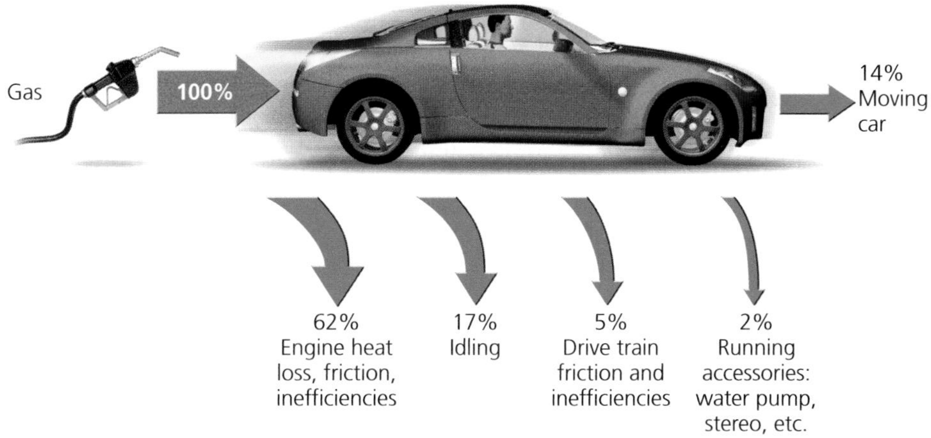

FIGURE 18.16 Conventional automobiles are extremely inefficient. Almost 85% of useful energy is lost, and only 14% actually moves the car down the road.

Unfortunately, the typical automobile is highly inefficient. Close to 85% of the fuel you pump into your gas tank does something other than move your car down the road. According to the U.S. Department of Energy, only about 13–14% of the fuel energy actually moves the vehicle and its occupants from point A to point B (Figure 18.16). Although more aerodynamic designs, increased engine efficiency, proper maintenance, and improved tire design help to reduce these losses, gasoline-fueled automobiles may always remain somewhat inefficient.

Automotive technology Advancing technology, however, is making possible a number of alternatives to the traditional combustion-engine automobile. These include hybrid vehicles that combine electric motors and gasoline-powered engines for greater efficiency (▸p. 583). They also include fully electric vehicles, alternative fuels such as biodiesel and compressed natural gas (▸pp. 609–610), and hydrogen fuel cells that use oxygen and hydrogen and produce only water as a waste product (▸pp. 637–642).

Driving less, and public transportation Despite these novel options, the high costs of automobile ownership and concerns regarding traffic and the environmental impacts of automobiles are leading many people to make lifestyle choices that reduce their reliance on cars. For example, many people are choosing to live nearer to their place of employment. Many others use buses, subway trains, and other modes of public transportation. Still others bike or walk to work or for their errands (Figure 18.17). Unfortunately, reliable and convenient public transit is not yet available in many U.S. communities. Making automobile-based cities and suburbs more friendly to pedestrian and bicycle traffic and improving people's access to public transportation stand as major challenges for progressive city and regional planners (▸pp. 388–391).

In a 2002 study, the American Public Transportation Association (APTA) concluded that increasing use of public transportation is the single most effective strategy for conserving energy and reducing environmental pollutants. Already, public transportation in the United States reduces fossil fuel consumption by 855 million gallons of gas (45 million barrels of oil) each year, the APTA has estimated. Yet according to the study, if U.S. residents increased their use of public transportation to the levels of

FIGURE 18.17 Sometimes the most effective solutions are the simplest. By choosing human-powered transportation methods, such as bicycles, we can greatly reduce our individual transportation-related greenhouse gas emissions. An increasing number of people are choosing to live closer to their workplaces and to enjoy the dual benefits of exercise and reduced emissions by walking or cycling to work or school.

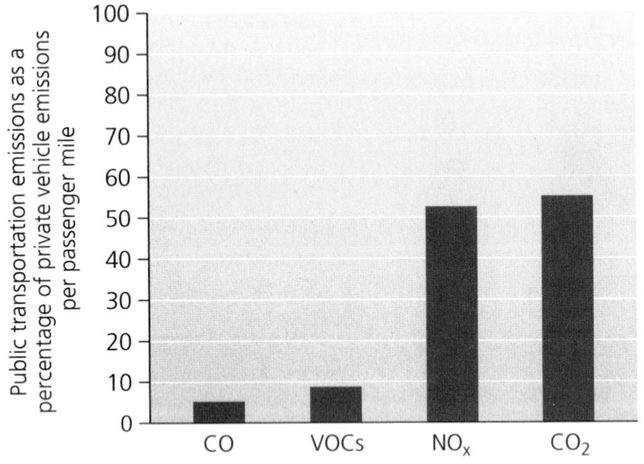

FIGURE 18.18 Relative to travel by private automobile, public transportation reduces almost all types of air pollution, including carbon monoxide (CO), volatile organic compounds (VOCs), nitrous oxides (NOx), and carbon dioxide (CO_2). Data from Shapiro, R. J., et al. 2002. *Conserving energy and preserving the environment: The role of public transportation.* American Public Transit Association.

Canadians (7% of daily travel needs) or Europeans (10% of daily travel needs), enormous amounts of energy and greenhouse emissions would be saved. These savings, the study indicated, could substantially cut air pollution, dependence on imported oil, and the nation's contribution to global climate change (Figure 18.18).

Some international treaties address climate change

In 1992, the United Nations convened the U.N. Conference on Environment and Development Earth Summit in Rio de Janeiro. Nations represented at the Earth Summit signed five documents, including the **U.N. Framework Convention on Climate Change (FCCC).** The FCCC outlined a plan for reducing greenhouse gas emissions to 1990 levels by the year 2000 through a voluntary, nation-by-nation approach.

By the late 1990s, it was apparent that a voluntary approach to slowing greenhouse gas emissions was not likely to succeed. Between 1990 and 2003, for example, U.S. greenhouse emissions (in CO_2 equivalents) increased by 13.3%.

However, some other nations have demonstrated that economic vitality does not require ever-increasing greenhouse gas emissions. For instance, Germany has the third most technologically advanced economy in the world and is a leading producer of iron, steel, coal, chemicals, automobiles, machine tools, electronics, textiles, and other goods—yet it managed between 1990 and 2003 to reduce its greenhouse gas emissions by 18.5%. In the same period, the United Kingdom cut its emissions by 13.0%.

After watching the seas rise and observing the refusal of most industrialized nations to cut their emissions, nations of the developing world—the Maldives among them—helped initiate an effort to create a binding international treaty that would *require* all signatory nations to reduce their greenhouse gas emissions. This effort led to the development of the Kyoto Protocol.

The United States has resisted the Kyoto Protocol

The **Kyoto Protocol** is an outgrowth of the FCCC. Drafted in 1997 in Kyoto, Japan, it mandates signatory nations, by the period 2008–2012, to reduce emissions of six greenhouse gases to levels equal to or lower than those of 1990 (Table 18.4). The treaty was to take effect once nations responsible for 55% of global greenhouse emissions ratified it, and in 2005, the Kyoto Protocol at last came into force after it was ratified by Russia, the 127th nation to ratify it.

The United States, the world's largest emitter of greenhouse gases, has continued to refuse to ratify the Kyoto Protocol. U.S. leaders have called the treaty unfair because it requires industrialized nations to reduce emissions but does not require the same of developing nations, even rapidly industrializing ones such as China and India.

Table 18.4 Emissions Reductions Required and Achieved

Nation	Required change,[1] 1990–2008/2012	Observed change, 1990–2003[2]
Russia	0.0%	−38.5%[3]
Germany	−21.0%	−18.5%
United Kingdom	−12.5%	−13.0%
France	0.0%	−1.9%
Italy	−6.5%	+8.3%
Japan	−6.0%	+12.8%
United States	−7.0%	+13.3%
Canada	−6.0%	+24.2%

[1]Percentage decrease in emissions (carbon-equivalents of six greenhouse gases) from 1990 to period 2008–2012, as mandated under Kyoto Protocol
[2]Actual percentage change in emissions (carbon-equivalents of 6 greenhouse gases) from 1990 to 2003. Negative values indicate decreases; positive values indicate increases.
[3]Data through 1999 (most recent available). Russia's substantial decrease was mainly due to economic contraction following the breakup of the Soviet Union.
Data from U.N. Framework Convention on Climate Change, National Greenhouse Gas Inventory Reports, 2005.

VIEWPOINTS

Global Climate Change

What is scientific research telling us about global climate change and its potential consequences for our society and the environment? How should humanity respond to climate change?

The Science Is Settled Enough to Justify Action

Many things are abundantly clear in climate change research. Human activities have undoubtedly led to the highest atmospheric levels of CO_2 and CH_4 seen in almost 1 million years, and both are powerful greenhouse gases that warm the surface of the planet. Temperatures have risen around 0.7 °C since around 1900, of which a large proportion is attributable to this enhanced greenhouse effect. The significant increases of heat in the oceans match climate model calculations of the increasing energy imbalance of the planet, and the almost global retreat of mountain glaciers is a graphic symptom of the increasing warmth of the atmosphere. Some further climate change is inevitable because the climate has yet to come into equilibrium with current greenhouse gas levels.

There are also many remaining uncertainties in climate science, particularly related to the role of aerosols and clouds. However, despite occasional claims to the contrary, the science is settled enough to justify action to reduce greenhouse gas emissions.

Due to the inertia of energy infrastructure in today's societies, CO_2 emissions will continue to rise for decades to come, and given the long atmospheric lifetime of CO_2, atmospheric concentrations will continue to increase for even longer. Model projections indicate that this could have significant and costly impacts on the environment and sea level if this continues unabated. Given the long time scales involved (in society and in climate itself), decisions made now will only start to have impacts many decades hence. Therefore, by the time serious effects are obvious, it may be too late to avoid the worst consequences.

Sensible policy approaches should combine investment in an increased resilience to climate changes, along with long-term efforts to reduce emissions. Cuts in other warming factors such as CH_4, black carbon, and tropospheric ozone would be positive steps for tackling both climate change and air pollution.

Gavin Schmidt is a climate modeller at the NASA Goddard Institute for Space Studies in New York. He is an associate editor for the *Journal of Climate* and was recently cited by *Scientific American* as one of the 50 Research Leaders of 2004.

Climate Changes Are Mostly Natural

Climate is never constant but varies, sometimes dramatically, on timescales ranging from years to eons. On the human scale, decades to centuries, the major cause seems to be cyclical variations of solar radiation. Since the end of the most recent ice age, some 10,000 years ago, there have been many such cycles. In recent history we have seen the Medieval Warming Period when England produced wines and Vikings colonized Greenland. Then followed the Little Ice Age, which disappeared only around 1850 A.D., about the time when thermometers first became available in much of the world. The global climate then warmed strongly until about 1940, followed by a cooling until 1975 that provoked great fears of a return to an ice age. All these changes are believed to be of natural origin, even though during this time there had been a steady increase in levels of atmospheric greenhouse gases from human activities.

Since 1979, weather satellites have reported a slight warming trend, which could be partially anthropogenic, as theoretical climate models suggest. Teasing out the small human contribution from the natural "noise" is difficult and the focus of ongoing scientific debate.

Nonetheless, we can draw certain conclusions. Anthropogenic global warming is not a significant problem, amounting to less than 1 °C by 2100. On the whole, it will be beneficial, with higher levels of carbon dioxide speeding growth of crops and forests. And realistically, there is little that can be done to stem the rise of emissions, especially from nations like China and India. The best policy is one of "no regrets"—energy conservation because it pays, and strengthening our ability to adapt by fighting poverty around the world.

S. Fred Singer is professor emeritus of environmental science at the University of Virginia, and president of the Science and Environmental Policy Project. He was the first director of the U.S. Weather Satellite Service, and served 5 years as vice-chair of the National Advisory Committee on Oceans and Atmospheres. With Dennis T. Avery, he is the co-author of *Unstoppable Global Warming—Every 1500 Years* (Rowman & Littlefield Publishers, 2006).

Explore this issue further by accessing **Viewpoints** at www.aw-bc.com/withgott.

Proponents of the Kyoto Protocol justify the differential requirements by pointing out that the industrialized world created the current problem and therefore should make the sacrifices necessary to solve it.

The United States' refusal to join international efforts to curb greenhouse emissions has generated resentment among its allies and has left it diplomatically isolated. In December 2005, as the world's nations met in Montreal to design a roadmap for how to build on Kyoto after 2012, the U.S. representative walked out of the meeting, and the Bush administration, according to the *New York Times,* was "repeatedly assailed by the leaders of other wealthy industrialized nations for refusing to negotiate."

Because resource use and per capita carbon dioxide emissions are far greater in the industrialized world, governments and industries in developed nations often feel they have more to lose, economically, from mandatory restrictions on emissions. Ironically, this fear neglects the equally likely probability that developed nations are the ones most likely to gain economically in such a situation. This is because they may be in the best position to invent and develop new technologies to power the world in a post-fossil-fuel era. Moreover, if an international "cap-and-trade" permit-trading system (▸ p. 73) in carbon dioxide emissions becomes successful, then U.S. industries that are left out may find themselves at a disadvantage relative to their international competitors.

Despite the U.S. government's refusal to ratify Kyoto, many state and local governments across the nation are expressing support for limits on greenhouse emissions. As of late 2005, nine states had announced voluntary commitments to emissions targets similar to those set forth by the Kyoto Protocol. These states comprised roughly one-quarter of the U.S. population and GDP. Several other states and several dozen municipalities were also poised to adopt policies to reduce their greenhouse gas emissions.

However, Kyoto Protocol critics and supporters alike acknowledge that even if every nation complied with the limits established in this treaty, greenhouse gas emissions would continue to increase—albeit more slowly than they would in the absence of the treaty.

Some feel climate change demands the precautionary principle

With regard to global climate change, as with many other environmental issues, we may never be entirely certain of the precise outcomes of our actions until after they have occurred, and perhaps not even then. With this uncertainty in mind, the drafters of the 1992 Rio Declaration included a passage invoking the precautionary principle (▸ p. 290):

> In order to protect the environment, the precautionary approach shall be widely applied by the States according to their capabilities. Where there are threats of serious or irreversible damage, lack of full scientific certainty shall not be used as reason for postponing cost-effective measures to prevent environmental degradation.

That is, advocates of the precautionary principle assert that if a threat is reasonably suspected, we should take precautionary action without waiting for full scientific certainty regarding cause and effect.

Weighing the Issues:
The Precautionary Principle

Critics of the precautionary approach say that it will impede economic growth and innovation. Advocates of the precautionary principle say the stakes are too high to gamble with climate. What do you think? Is the precautionary approach an appropriate guide for dealing with climate change? What role should economics play in the discussion?

Conclusion

We have seen that many factors, including human activities, can shape atmospheric composition and global climate. We have also seen that scientists and policymakers are beginning to understand anthropogenic climate change and its environmental, economic, and social consequences more fully. Although many policymakers and industrial leaders express no anxiety about climate change, many scientists and other policymakers are deeply concerned. As time passes, fewer experts are arguing that the changes will be minor. Sea-level rise and other consequences of global climate change will affect far-flung places such as the Maldives, but they will also influence populated mainland areas such as coastal Florida. By becoming familiar with climate science and with potential solutions to climate-related problems, you will become better equipped to interpret the many messages you will receive regarding global climate change in the coming years.

REVIEWING OBJECTIVES

You should now be able to:

Describe Earth's climate system and explain the variety of factors influencing global climate

- Climate is a homeostatic system that varies naturally with time. (p. 530)
- The sun provides most of Earth's energy and interacts with the atmosphere, land, and oceans to drive climate processes. (pp. 530–531)
- Earth absorbs about 70% of incoming solar radiation and reflects about 30% back into space. (pp. 530–531)
- "Greenhouse gases," such as carbon dioxide, methane, water vapor, nitrous oxide, ozone, and halocarbons, warm the atmosphere by absorbing infrared radiation and re-emitting infrared radiation of different wavelengths. (pp. 530–533)
- Milankovitch cycles influence climate in the long term. (p. 533)

Characterize human influences on the atmosphere and global climate

- By burning fossil fuels, deforesting landscapes, and manufacturing halocarbons, humans are increasing atmospheric concentrations of many greenhouse gases. Increased greenhouse gas emissions enhance the greenhouse effect. (pp. 530–532)

Delineate methods of modern climate research

- Geologic records, such as cores through ice or sediments, reveal information about past climatic conditions. (pp. 534–535, 538)
- Direct atmospheric sampling tells us about current composition of the atmosphere. (p. 535)
- Coupled general circulation models (CGCMs) serve to predict future changes in climate. (pp. 538–540)

Summarize current consequences and potential future impacts of global climate change

- The IPCC has comprehensively synthesized current climate research, and its periodic reports represent the consensus of the scientific community. (p. 540)
- Temperatures on Earth have warmed by an average of 0.6 °C (1.0 °F) over the past century. (pp. 540–541)
- Sea level has risen an average of 10–20 cm (4–8 in.) over the past century. (pp. 541–543)
- Other impacts include changes in precipitation, frequency of extreme weather events, and effects on plants and animals. (pp. 541–543)
- Various potential impacts of future climate change have been predicted, including physical, biological, ecological, and economic impacts. (pp. 543–546)

Evaluate the scientific, political, and economic debates concerning climate change

- Many details of climate change science remain uncertain because climate systems are so complex. Yet scientists broadly agree that climate change is occurring. (pp. 546–547)
- Despite the remaining uncertainties, the scientific community feels that evidence for humans' role in influencing climate is strong enough to justify governments taking action to reduce greenhouse emissions. (pp. 546–547)
- A few scientists and many nonscientists have resisted the scientific consensus, and these "greenhouse skeptics" have enjoyed a disproportionately large voice in the public debate. (p. 547)
- Policymakers in the United States and some other nations have resisted confronting climate change because of fears that reducing greenhouse emissions will be economically costly. (p. 547)

Suggest potential responses to climate change

- Conserving electricity, improving efficiency of energy use, and switching to renewable energy sources will help reduce fossil fuel consumption and greenhouse emissions. (p. 548)
- Encouraging new automotive technologies and public transportation systems should help reduce greenhouse emissions. (pp. 548–550)
- The Kyoto Protocol provides a first step for nations to begin addressing climate change. (pp. 550, 552)

TESTING YOUR COMPREHENSION

1. What happens to solar radiation after it reaches Earth? How do greenhouse gases warm the lower atmosphere?
2. Why is water vapor considered a greenhouse gas? How could an increase of water vapor create a positive or negative feedback effect?

3. How do scientists study the ancient atmosphere?
4. Has simulating climate change with computer programs been effective in helping us predict climate? How do these programs work?
5. How can rising sea levels, caused by global warming, create problems for people? How may rising sea levels affect marine ecosystems?
6. How might a warmer climate affect agriculture? How might a warmer climate affect forest distribution, according to recent research?
7. What are some likely negative impacts of warmer climate on human health? Could there be beneficial consequences?
8. Do all scientists agree that climate is indeed changing? On what counts do they currently disagree?
9. What are the largest two sources of greenhouse gas emissions in the United States, and why? In what ways can we try to reduce these emissions?
10. What roles have international treaties played in addressing climate change? Give two specific examples.

SEEKING SOLUTIONS

1. To determine to what extent current climate change is the result of human activity versus natural processes, which type(s) of scientific research do you think is (are) most helpful? Why?
2. As you have seen in many places in this book, people may draw dramatically different conclusions about the significance and implications of environmental change. Refer to the quotations presented at the beginning of this chapter's central case study. What questions might you ask of the editors of the *Oil and Gas Journal* and representatives of the Maldives' government to better understand their positions and their reasons for holding them?
3. Some people argue that it is appropriate, and even helpful, for former fossil fuel company executives to serve in high-level government positions where they can make decisions regarding energy policy, carbon dioxide emissions, and global climate change. Others say this presents a potential conflict of interest and should be discouraged. List some of the arguments justifying each side. What is your opinion?
4. Today, many people argue that we need "more proof," or "better science" before we commit to substantial changes in the way we live our lives. How much "science," or certainty, do you think we need before we have enough to guide our decisions regarding climate change? How much certainty do you need in your own life before you make a change? Should nations and elected officials follow a different standard? Do you believe that the precautionary principle is an appropriate standard in the case of global climate change? Why or why not?
5. Describe several ways that greenhouse gas emissions from transportation can be reduced. Which approach do you think is most realistic, which approach do you think is least realistic, and why?
6. Ismail Shafeeu, the Maldives' environment minister, believes that a solution to climate change will require political leadership at the international level on the part of the United States:

> We have to strive to point out the problems that are associated with climate change and the responsibilities we feel lie with countries such as the U.S. [because the developed countries burn most of the fossil fuel], particularly the U.S. as a global leader. We would like to see the U.S. take a more constructive approach to these problems than the one they are presently taking. . . . The general impression is that the government of the United States has this sense that it's surrounded by a lot of countries trying to destroy its economy, but I don't think any country is asking of the United States something that they have not offered to do themselves.

Do you agree with Shafeeu's statements? Why or why not? Why do you think the U.S. government has chosen not to ratify the Kyoto Protocol? Do you think it should ratify the treaty?

INTERPRETING GRAPHS AND DATA

We burn fossil fuels to generate electricity, to power vehicles for transportation, and as primary energy sources in the home, in businesses, and in industry. For each of these uses, trends in the emission of carbon dioxide from fossil fuel combustion in the United States are shown in the accompanying graph.

1. Calculate the approximate percentage changes in CO_2 emissions from transportation; electricity generation; and residential, commercial, and industrial primary energy use between 1980 and 2003.
2. Between 1980 and 2003, U.S. population increased by 28.4%, and the inflation-adjusted U.S. gross domestic product (GDP) doubled. What quantitative conclusions can you draw from these data about CO_2 emissions per capita? About CO_2 emissions per unit of total economic activity?
3. Imagine you are put in charge of designing a strategy to reduce U.S. emissions of CO_2 from fossil fuel combustion. Based on the data presented here, what approaches would you recommend, and how would you prioritize these? Explain your answers.

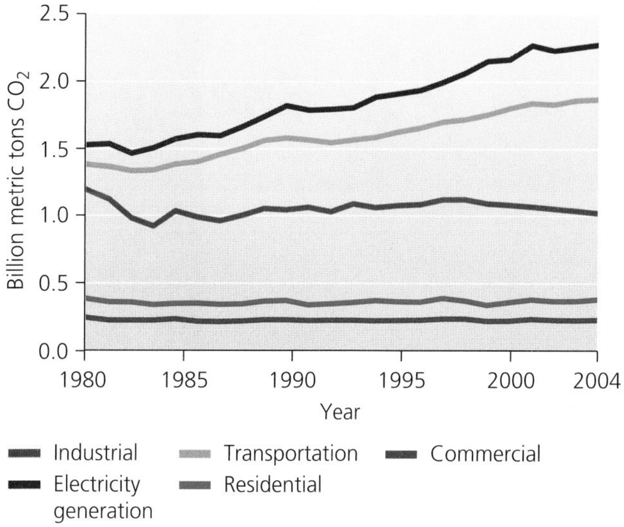

Emissions of CO_2 from fossil fuel combustion by end-use sector in the United States, 1980–2003. Data from U.S. Department of Energy, Energy Information Administration. 2004. *Annual Energy Review 2004,* Report No. DOE/EIA-0384.

CALCULATING ECOLOGICAL FOOTPRINTS

U.S. energy consumption from fossil fuels currently totals 306 gigajoules per person per year. Carbon emitted from fossil fuel combustion is "sequestered" when plants take up the emitted carbon dioxide through photosynthesis and store it as organic matter. Researchers have estimated that for each 100 gigajoules of fossil fuel burned, 1 hectare of ecologically productive land is required for carbon sequestration. Considering these data, calculate the component of the ecological footprint required to sequester carbon from fossil fuel emissions, and record your results in the table.

	Hectares of land to sequester carbon
You	3.06
Your class	
Your state	
United States	

Data from Wackernagel, M., and W. Rees. 1996. *Our ecological footprint: Reducing human impact on the earth.* Gabriola Island, British Columbia, Canada: New Society Publishers.

1. The land area of the United States is about 916 million hectares. What percentage of that land would need to be set aside to sequester all carbon from the fossil fuel consumption of 300 million Americans?
2. What is the environmental fate of carbon dioxide that is released from the combustion of fossil fuels and is not sequestered by plants? Why is this a concern?
3. Name four things you could do to lessen the modification of the global environment caused by your own personal energy consumption.

Take It Further

Go to www.aw-bc.com/withgott or the student CD-ROM where you'll find:

- Suggested answers to end-of-chapter questions
- Quizzes, animations, and flashcards to help you study
- *Research Navigator*™ database of credible and reliable sources to assist you with your research projects
- GRAPHit! tutorials to help you master how to interpret graphs
- INVESTIGATEit! current news articles that link the topics that you study to case studies from your region to around the world

19 Fossil Fuels: Energy and Impacts

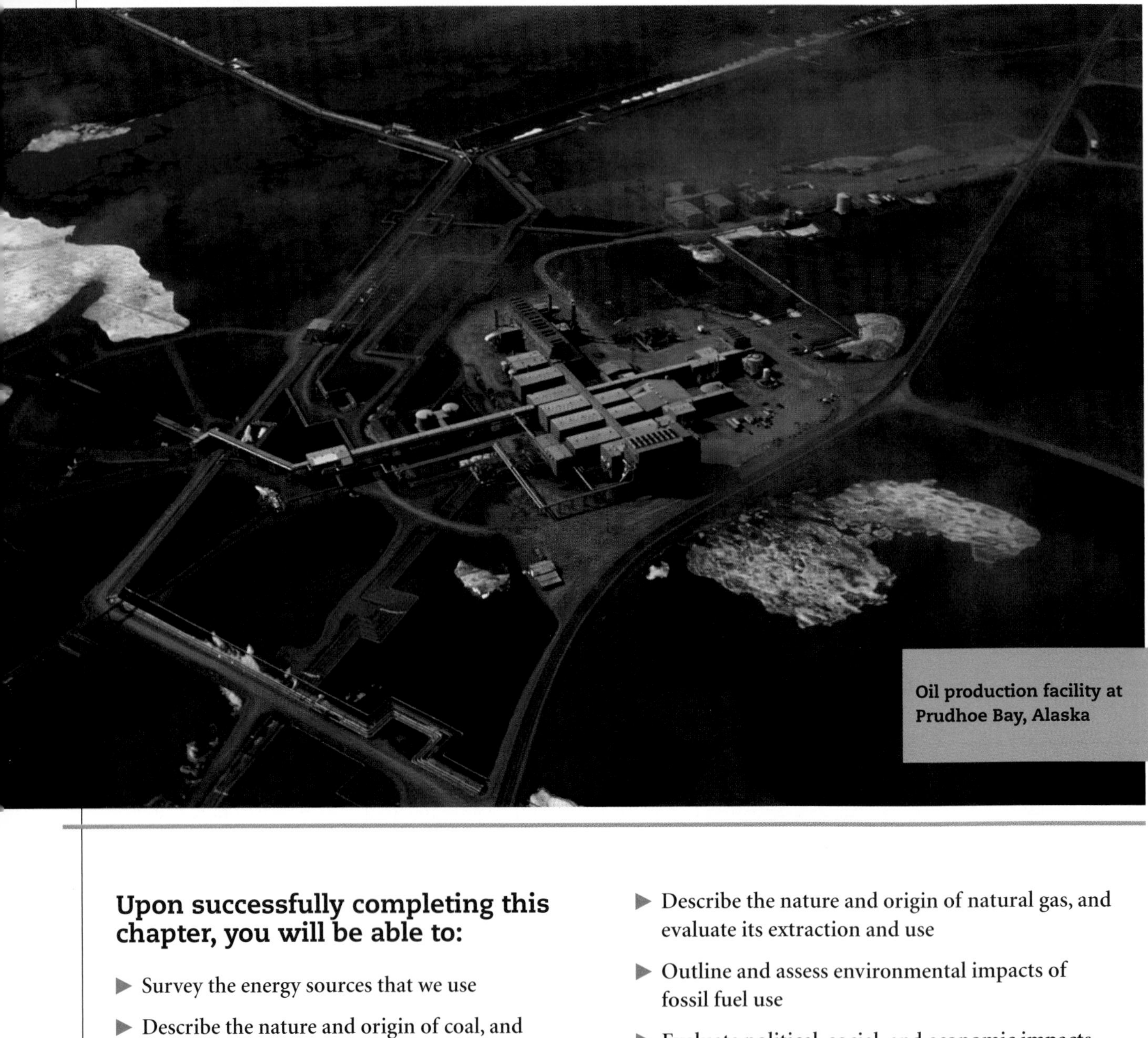

Oil production facility at Prudhoe Bay, Alaska

Upon successfully completing this chapter, you will be able to:

- Survey the energy sources that we use
- Describe the nature and origin of coal, and evaluate its extraction and use
- Describe the nature and origin of petroleum and evaluate its extraction, use, and future depletion
- Describe the nature and origin of natural gas, and evaluate its extraction and use
- Outline and assess environmental impacts of fossil fuel use
- Evaluate political, social, and economic impacts of fossil fuel use
- Specify strategies for conserving energy

Caribou crossing haul road at Prudhoe Bay

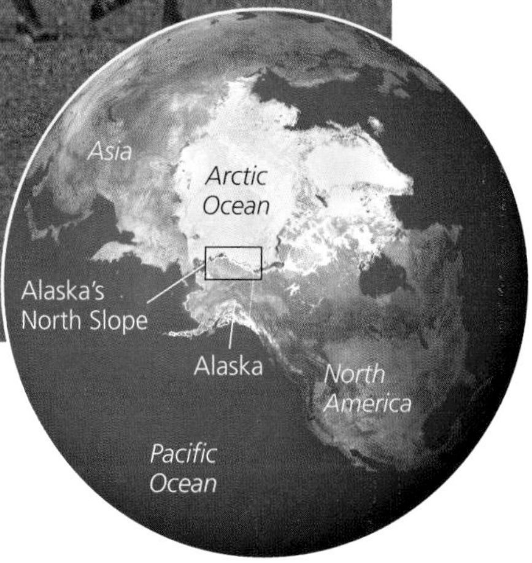

Central Case: Oil or Wilderness on Alaska's North Slope?

"The roar alone of road building, drilling, trucks, and generators would pollute the wild music of the Arctic and be as out of place there as it would be in the heart of Yellowstone or the Grand Canyon."
—Former President Jimmy Carter, 2000

"There is absolutely no indication that environmentally responsible exploration will harm the 129,000-member porcupine caribou herd."
—Alaska Senator Frank Murkowski, 2002

Above the Arctic Circle, at the top of the North American continent, the land drops steeply down from the jagged mountains of Alaska's spectacular Brooks Range and stretches north in a vast, flat expanse of tundra until it meets the icy waters of the Arctic Ocean. Few Americans have been to this remote region, yet it has come to symbolize a struggle between two values in our modern life.

For some U.S. citizens, Alaska's North Slope is one of the last great expanses of wilderness in their sprawling industrialized country—one of the last places humans have left untouched. For these millions of Americans, simply knowing that this wilderness still exists is of tremendous value. For millions of others, this land represents something else entirely—a source of petroleum, the natural resource that, more than any other, fuels our society and shapes our way of life. To these people, it seems wrong to leave such an important resource sitting unused in the ground. Those who advocate drilling for oil here accuse wilderness preservationists of neglecting the country's economic interests, whereas advocates for wilderness argue that drilling will sacrifice the nation's natural heritage for little gain.

Ever since oil was found seeping from the ground in this area a century ago, these two visions for Alaska's North Slope have competed. Now they exist side by side across three regions of this vast swath of land (Figure 19.1). The westernmost portion of the North Slope was set aside in 1923 by the U.S. government as an emergency reserve for petroleum. This parcel of land, the size of Indiana, is today called the National Petroleum

Reserve–Alaska and was intended to remain untapped for oil unless the nation faced an emergency. So far, most of this region's 9.5 million ha (23.5 million acres) remains undeveloped.

East of the National Petroleum Reserve are state lands that experienced widespread development and extraction after oil was discovered at Prudhoe Bay in 1968. Since drilling began in 1977, over 14 billion barrels (1 barrel = 159 L or 42 gal) of crude oil have been extracted from 19 oil fields spread over 160,000 ha (395,000 acres) of this region. The oil is transported across the state of Alaska by the 1,300-km (800-mi) trans-Alaska pipeline south to the port of Valdez, where it is loaded onto tankers.

East of the Prudhoe Bay region lies the Arctic National Wildlife Refuge (ANWR), an area the size of South Carolina consisting of federal lands set aside in 1960 and 1980 mainly to protect wildlife and preserve pristine ecosystems of tundra, mountains, and seacoast. This scenic region is home to 160 nesting bird species, numerous fish and marine mammals, grizzly bears, polar bears, Arctic foxes, timber wolves, musk oxen, and other animals. In most years, thousands of caribou arrive from the south to spend the summer, giving birth to and raising their calves. Because of the vast caribou herd and the other large mammals, ANWR has been called "the Serengeti of North America."

ANWR has been the focus of debate for decades. Advocates of oil drilling have tried to open its lands for development, and proponents of wilderness preservation have fought for its preservation. Scientists, oil industry experts, politicians, environmental groups, citizens, and Alaska residents have all been part of the debate. So have the two Native groups in the area, the Gwich'in and the Inupiat, who disagree over whether the refuge should be opened to oil development. The Gwich'in depend on

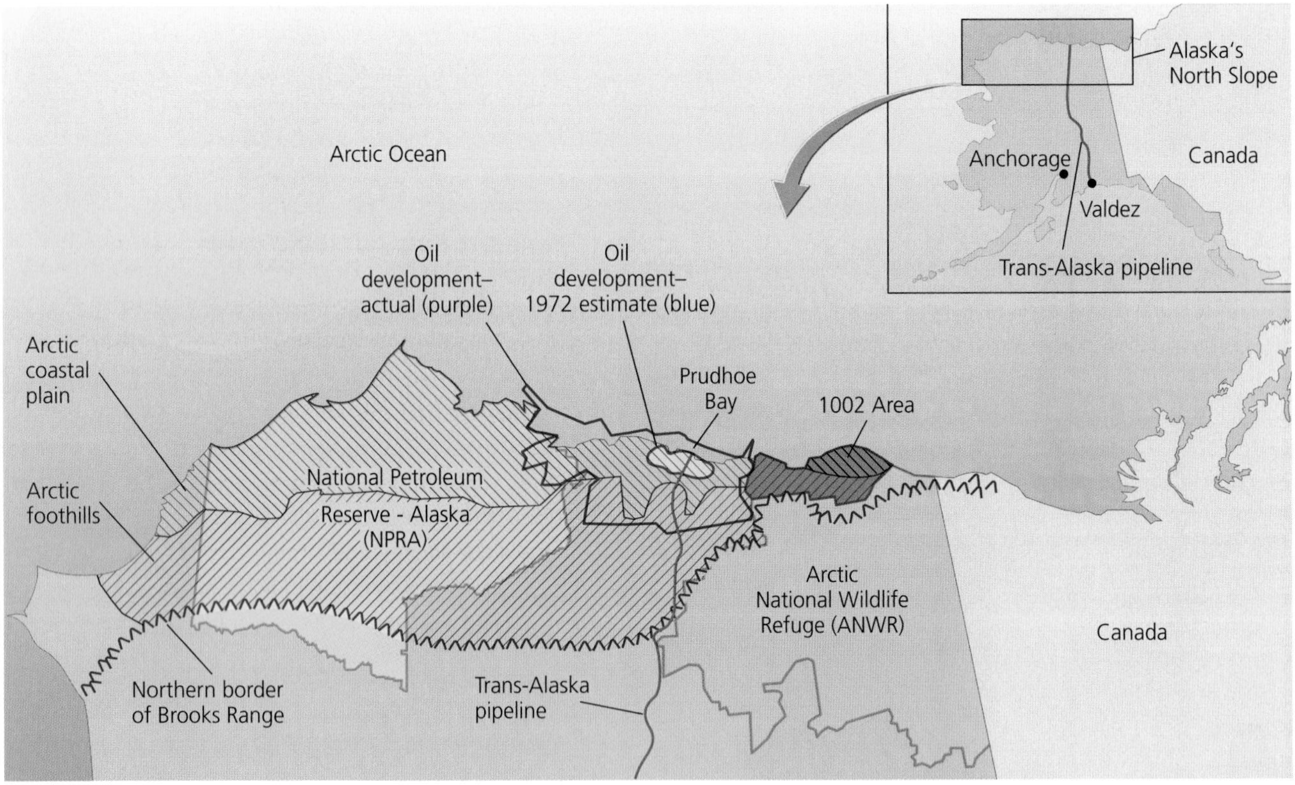

FIGURE 19.1 Alaska's North Slope is the site of both arctic wilderness and oil exploration. In the western portion of this region, the U.S. government established the National Petroleum Reserve–Alaska as an area in which to drill for oil if it is needed in an emergency. It is well explored but not widely developed. To the east of this area, the Prudhoe Bay region is the site of widespread oil extraction, and since 1977 it has produced over 14 billion barrels of oil. Oil development has expanded much farther than experts estimated it would in 1972. Farther east lies the Arctic National Wildlife Refuge, home to untrammeled Arctic wilderness and the focus of debate for years. Proponents of oil extraction and proponents of wilderness preservation have been battling over whether the 1002 Area of the coastal plain north of the Brooks Range should be opened to oil development.

hunting caribou and fear that oil industry activity will reduce caribou herds, whereas the Inupiat see oil extraction as one of the few opportunities for economic development in the area.

In a compromise in 1980, the U.S. Congress put most of the refuge off limits to oil but reserved for future decision-making a 600,000-ha (1.5-million-acre) area of coastal plain. This region, called the 1002 Area (after Section 1002 of the bill that established it), remains undeveloped for oil but can be opened for development by a vote of both houses of Congress. Its unsettled status has made it the center of the oil-versus-wilderness debate, and Congress has been caught between passionate feelings on both sides for a quarter of a century.

In 2005, the Republican-controlled Congress made several efforts to open the refuge to drilling. It added a provision approving drilling to a $450-billion military spending bill that included hurricane relief and other high-priority items. As this book went to press in late December, the House approved the bill but the Senate narrowly blocked it.

Behind the noisy policy debate over ANWR, scientists have attempted to inform the dialogue through research. Geologists have tried to ascertain how much oil lies underneath the refuge, and biologists have tried to predict the potential impacts of oil drilling on Arctic ecosystems. Moreover, scientists and nonscientists alike are debating the relevance of the oil beneath the refuge for the security and prosperity of the nation. We will examine these questions by revisiting Alaska's North Slope as we survey the fossil fuel energy we use to heat and light our homes, power our machinery, and provide the comforts, conveniences, and mobility to which technology has accustomed us.

Sources of Energy

The debate over drilling for oil in the Arctic National Wildlife Refuge is a thoroughly modern debate, pitting the culturally new concept of wilderness preservation against the desire to exploit a resource that has come to guide the world's economy only in the past 150 years. However, people have used—and fought over—energy in one way or another for all of our history.

We use a variety of energy sources

Earth receives energy from several sources, and people have developed many ways to harness the renewable and nonrenewable forms of energy available on our planet (Table 19.1). Most of Earth's energy comes from the sun. We can harness energy from the sun's radiation directly, but solar radiation also makes possible several other energy sources. Sunlight drives the growth of plants, from which we take wood as a fuel source. After their death, plants may impart their stored chemical energy to **fossil fuels** (such as oil, coal, and natural gas), which are highly combustible substances formed from the remains of organisms from past geological ages. Solar radiation also helps drive wind patterns and the hydrologic cycle, making possible other forms of energy, such as wind power and hydroelectric power.

A great deal of energy also emanates from Earth's core, enabling us to harness geothermal power. A much smaller amount of energy results from the gravitational pull of the moon and sun, and we are just beginning to harness the power from the ocean tides that these forces generate.

Table 19.1 Energy Sources We Use Today

Energy source	Description	Type of energy
Crude oil	Fossil fuel extracted from ground	Nonrenewable
Natural gas	Fossil fuel extracted from ground	Nonrenewable
Coal	Fossil fuel extracted from ground	Nonrenewable
Nuclear energy	Energy from atomic nuclei of uranium mined from ground and processed	Nonrenewable
Biomass energy	Chemical energy from photosynthesis stored in plant matter	Renewable
Hydropower	Energy from running water	Renewable
Solar energy	Energy from sunlight directly	Renewable
Wind energy	Energy from wind	Renewable
Geothermal energy	Earth's internal heat rising from core	Renewable
Tidal and wave energy	Energy from tidal forces and ocean waves	Renewable

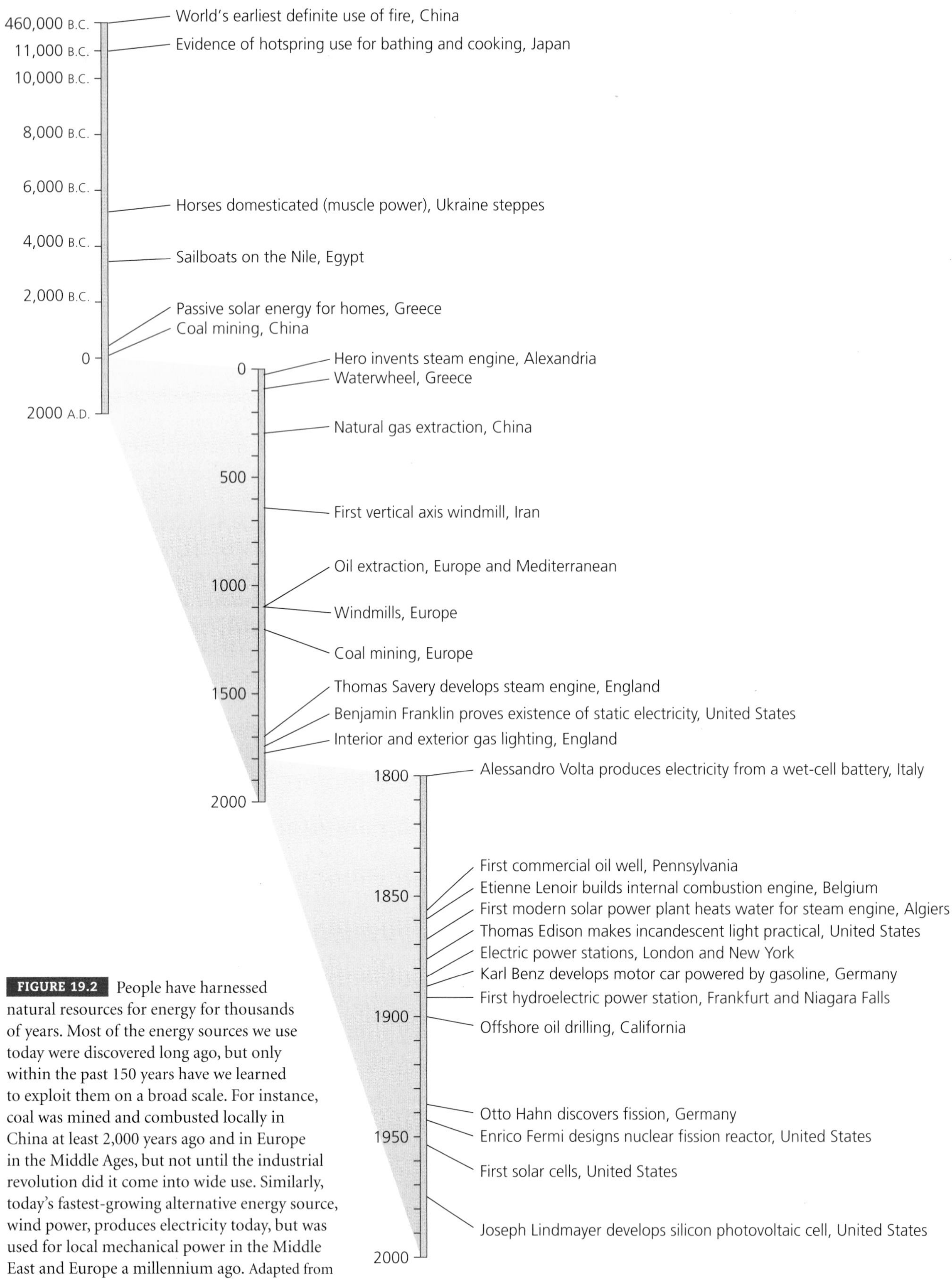

FIGURE 19.2 People have harnessed natural resources for energy for thousands of years. Most of the energy sources we use today were discovered long ago, but only within the past 150 years have we learned to exploit them on a broad scale. For instance, coal was mined and combusted locally in China at least 2,000 years ago and in Europe in the Middle Ages, but not until the industrial revolution did it come into wide use. Similarly, today's fastest-growing alternative energy source, wind power, produces electricity today, but was used for local mechanical power in the Middle East and Europe a millennium ago. Adapted from Energy timeline, Geothermal Education Office, 2000.

An immense amount of energy resides within the bonds among protons and neutrons in atoms, and this energy provides us with nuclear power.

Ever since our ancestors first discovered fire, people have extracted energy from natural resources to cook food and to light and heat dwellings (Figure 19.2). Wood products were our primary sources of energy for heating and cooking. We harnessed animals, wind, and water as energy sources for mechanical work in fields, mills, and granaries. The development of the steam engine in the 18th century allowed us to use concentrated, high-quality energy sources such as coal for a variety of mechanical purposes as economies industrialized. Meanwhile, deforestation in many areas of the world led to dwindling wood supplies, giving many societies incentive to shift to energy sources other than wood.

In the 20th century, fossil fuels became the dominant source of power in industrialized countries and then in developing nations. The high energy content of fossil fuels makes them efficient to burn, ship, and store. Besides providing for transportation, heating, and cooking, these fuels are used to generate electricity, a secondary form of energy that is easier to transfer over long distances and apply to a variety of uses.

As we first noted in Chapter 1 (▸ p. 4), energy sources such as sunlight, geothermal energy, and tidal energy are considered *renewable* because their supplies will not be depleted by our use of them. Other sources, such as timber, are renewable if we do not harvest them at too great a rate. In contrast, energy sources such as oil, coal, and natural gas are considered *nonrenewable,* because at our current rates of consumption we will use up Earth's accessible store of them in a matter of decades to centuries. Nuclear power as currently harnessed through fission of uranium (▸ p. 593) can be considered nonrenewable to the extent that uranium ore is in limited supply.

Although these nonrenewable fuels result from ongoing natural processes, the timescales on which they are created are so long that, once the fuels are depleted, they cannot be replaced in any time span useful to our civilization. It takes a thousand years for the biosphere to generate the amount of organic matter that must be buried to produce a single day's worth of fossil fuels for our society. To replenish the fossil fuels we have depleted so far would take many millions of years. For this reason, and because fossil fuels exert severe environmental impacts, renewable energy sources increasingly are being developed as alternatives to fossil fuels, as we will see in Chapters 20 and 21. Nonetheless, global consumption of the three main fossil fuels has risen steadily for years and is now at its highest level ever (Figure 19.3).

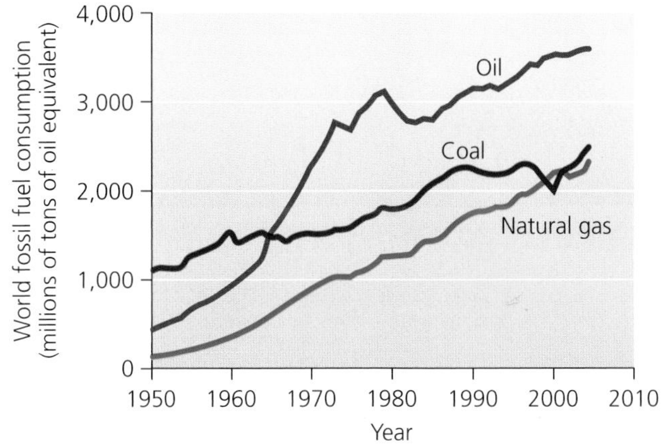

FIGURE 19.3 Global consumption of fossil fuels has risen greatly over the past half century. Natural gas is our fastest-growing fossil fuel today. Oil use rose steeply during the 1960s to overtake coal, and today it remains our leading energy source. Data from Worldwatch Institute. 2005. *Vital signs 2005.*

Fossil fuels are indeed fuels created from "fossils"

The fossil fuels we burn today in our vehicles, homes, industries, and electrical power plants were formed from the tissues of organisms that lived 100–500 million years ago. The energy these fuels contain came originally from the sun and was converted to chemical-bond energy as a result of plants' photosynthesis (▸ pp. 105–106). The chemical energy in these organisms' tissues was then concentrated as these tissues decomposed and their hydrocarbon compounds were altered and compressed (Figure 19.4).

Most organisms that die do not end up as part of a coal, gas, or oil deposit. A tree that falls and decays as a rotting log undergoes mostly **aerobic** decomposition; in the presence of air, bacteria and other organisms that use oxygen break down plant and animal remains into simpler carbon molecules that are recycled through the ecosystem. Fossil fuels are produced only when organic material is broken down in an **anaerobic** environment, one that has little or no oxygen. Such environments include the bottoms of deep lakes, swamps, and shallow seas. Over millions of years, organic matter that accumulated at the bottoms of such water bodies was converted into crude oil, natural gas, and coal. Which fuel formed in any given place depended on the chemical composition of the starting material, the temperatures and pressures to which the material was subjected, the presence or absence of anaerobic decomposers, and the passage of time.

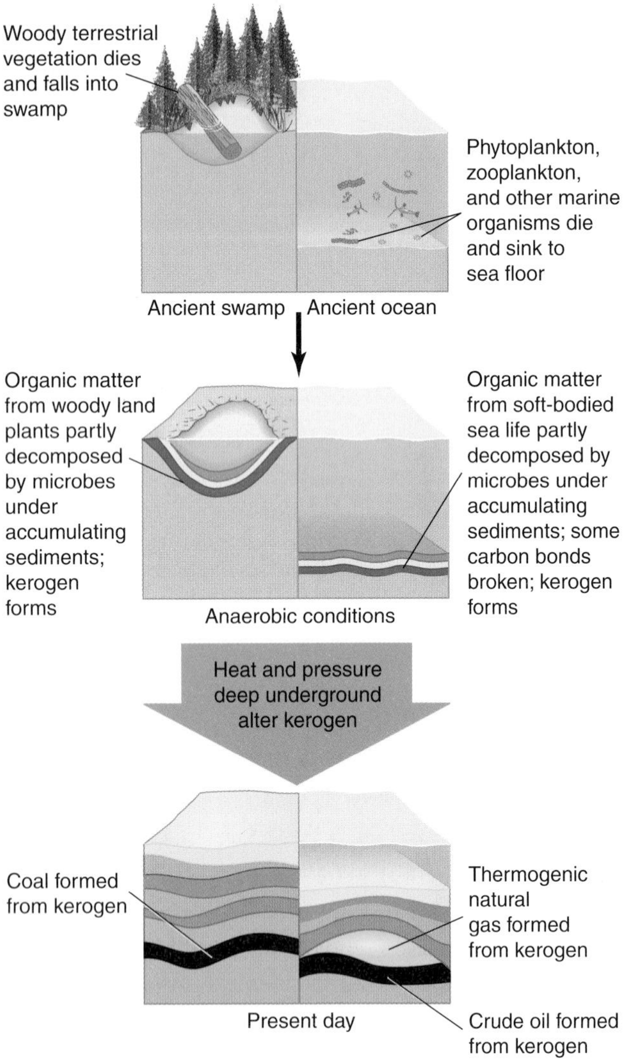

FIGURE 19.4 The fossil fuels we use for energy today consist of the remains of organic material from plants (and to a lesser extent, animals) that died millions of years ago. Their formation begins when organisms die and end up in oxygen-poor conditions, such as when trees fall into lakes and are buried by sediment, or when phytoplankton and zooplankton drift to the seafloor and are buried. Organic matter that undergoes slow anaerobic decomposition deep under sediments forms kerogen. Geothermal heating acts on kerogen to create crude oil and natural gas. Natural gas can also be produced nearer the surface by anaerobic bacterial decomposition of organic matter. Oil and gas come to reside in porous rock layers beneath dense, impervious layers. Coal is formed when plant matter is compacted so tightly that there is little decomposition.

Fossil fuel reserves are unevenly distributed

Fossil fuel deposits are localized and unevenly distributed over Earth's surface, so some regions have substantial reserves of fossil fuels whereas others have very few. How long each nation's fossil fuel reserves will last depends on how much the nation extracts, how much it consumes, and how much it imports from and exports to other nations. Nearly two-thirds of the world's proven reserves of crude oil lie in the Middle East. The Middle East is also rich in natural gas, but Russia contains more than twice as much natural gas as any other country. Russia is also rich in coal, as is China, but the United States possesses more coal than any other nation (Table 19.2).

Table 19.2 Nations with Largest Proven Reserves of Fossil Fuels

Oil (% world reserves)	Natural gas (% world reserves)	Coal (% world reserves)
Saudi Arabia, 22.1	Russia, 26.7	United States, 27.1
Iran, 11.1	Iran, 15.3	Russia, 17.3
Iraq, 9.7	Qatar, 14.4	China, 12.6
Kuwait, 8.3	Saudi Arabia, 3.8	India, 10.2
United Arab Emirates, 8.2	United Arab Emirates, 3.4	Australia, 8.6
Venezuela, 6.5	United States, 2.9	South Africa, 5.4
Russia, 6.1	Nigeria, 2.8	Ukraine, 3.8
Kazakhstan, 3.3	Algeria, 2.5	Kazakhstan, 3.4
Libya, 3.3	Venezuela, 2.4	Poland, 1.5
Nigeria, 3.0	Iraq, 1.8	Brazil, 1.1

Data from British Petroleum. 2005. *Statistical review of world energy 2005.*

Developed nations consume more energy than developing nations

Citizens of developed nations generally consume far more energy than do those of developing nations. Per person, the most-industrialized nations use up to 100 times more energy than do the least-industrialized nations (Figure 19.5). Moreover, developed and developing nations tend to apportion their energy use differently. Developing nations devote a greater proportion of energy to subsistence activities, such as food preparation, home heating, and food-growing, whereas industrialized countries use a greater proportion for transportation and industry (Figure 19.6). In addition, people in developing countries often rely on manual or animal energy sources instead of automated ones. For instance, rice farmers in Bali plant rice by hand, but industrial rice growers in California use airplanes. Because industrialized nations rely more on equipment and technology, they use more fossil fuels. In the United States, fossil fuels supply 89% of energy needs. Oil is the most heavily used fossil fuel, constituting 40% of U.S. energy use, compared with 25% for natural gas and 24% for coal.

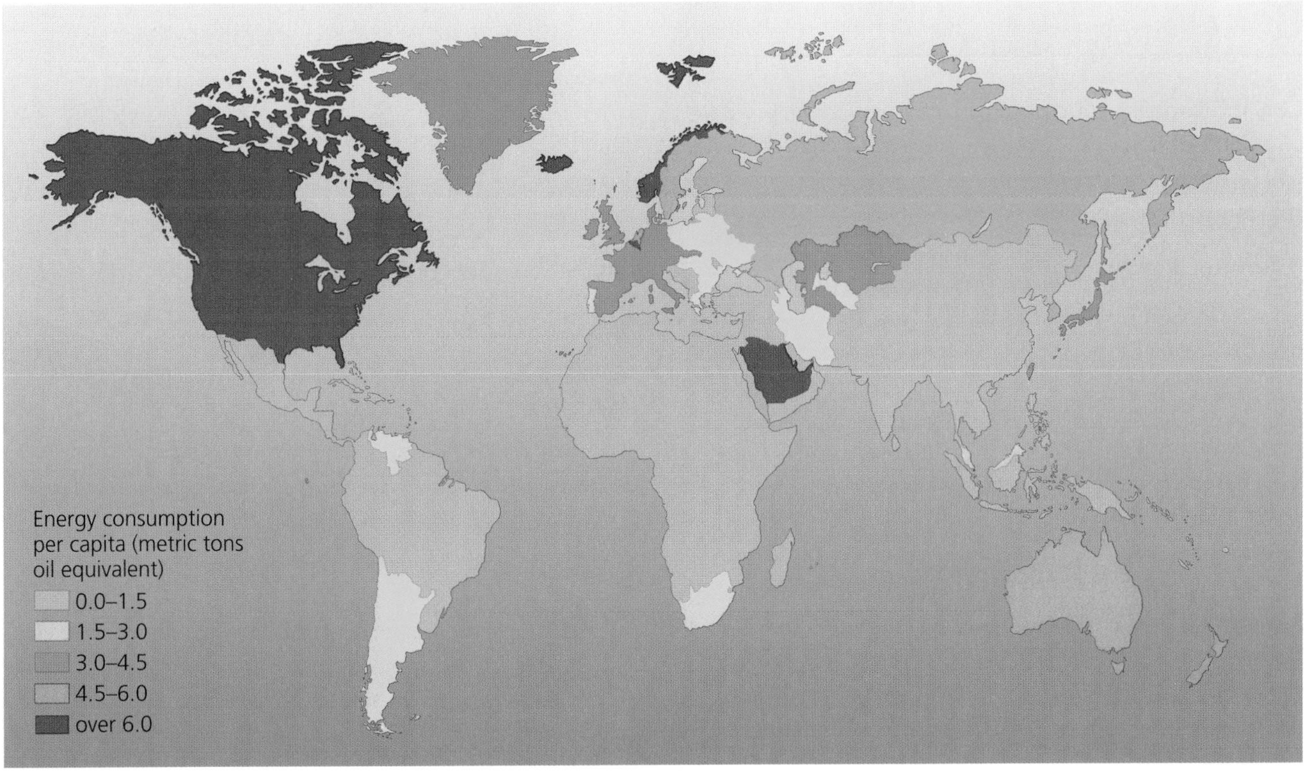

FIGURE 19.5 Regions vary greatly in their consumption of energy per person. People in developed nations consume the most. This map combines all types of energy, standardized to metric tons of "oil equivalent," that is, the amount of fuel needed to produce the energy gained from combusting one metric ton of crude oil. Data from British Petroleum. 2005. *Statistical review of world energy 2005.*

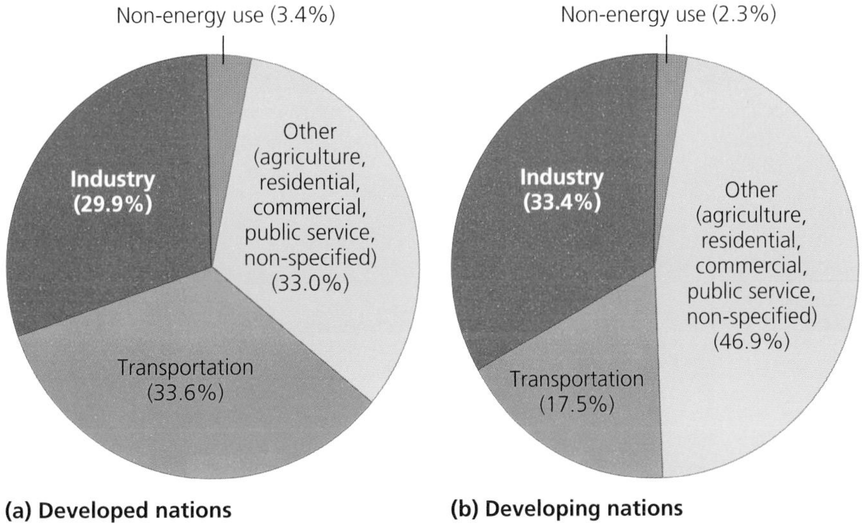

FIGURE 19.6 Developing nations and developed nations show somewhat different profiles of energy use. These data from 2002 show that developed nations of the Organisation for Economic Co-operation and Development (OECD) (**a**) devote nearly twice as much of their energy budgets to transportation as do non-OECD nations (**b**), whereas non-OECD nations devote relatively more energy to agricultural, residential, and other uses. The OECD includes 30 of the most industrialized nations from Europe and North America plus Japan, Korea, Turkey, Australia, and New Zealand. Non-OECD nations include some large, highly industrialized countries, such as Russia, China, Brazil, and Argentina, but consist mostly of smaller and developing nations. Data from International Energy Agency. 2004. *Key world energy statistics 2004.*

Coal

Coal is the world's most abundant fossil fuel. The proliferation 300–400 million years ago of swampy environments where organic material could be buried has resulted in substantial coal deposits throughout the world. **Coal** is organic matter (generally woody plant material) that was compressed under very high pressure to form dense, solid carbon structures (Figure 19.7). Coal typically results when little decomposition takes place because the material cannot be digested or appropriate decomposers are not present. One-quarter of the world's coal is located in the United States, and coal provides one-quarter of the world's commercial energy consumption.

Table 19.3 Top Producers and Consumers of Coal

Production (% world production)	Consumption (% world consumption)
China, 36.2	China, 34.4
United States, 20.8	United States, 20.3
Australia, 7.3	India, 7.4
India, 6.9	Japan, 4.3
South Africa, 5.0	Russia, 3.8
Russia, 4.7	South Africa, 3.4
Indonesia, 3.0	Germany, 3.1
Poland, 2.6	Poland, 2.1
Germany, 2.0	Australia, 2.0
Kazakhstan, 1.6	South Korea, 1.9

Data from British Petroleum. 2005. *Statistical review of world energy 2005.*

Coal use has a long history

People have used coal longer than any other fossil fuel. The Romans used coal for heating in the second and third centuries in Britain, as have people in parts of China for 2,000–3,000 years. In Arizona, today's Hopi Indians follow ancestral traditions by using coal to fire pottery, cook food, and heat their homes. Once commercial mining began in Europe in the 1700s, coal began to be used widely as a heating source. Coal found an expanded market after the invention of the steam engine, because it was used to boil water to produce steam. Coal-fired steam engines performed work previously done by people or horses, including manufacturing, harvesting, and powering trains and ships. The birth of the steel industry in 1875 increased demand still further because coal fueled the furnaces used to produce steel.

In the 1880s, people began to use coal to generate electricity. In coal-fired power plants, coal combustion converts water to steam, which turns a turbine to create electricity (see "The Science behind the Story," ▸ pp. 566–567). Today coal provides over half the electrical generating capacity of the United States. China and the United States are the primary producers and consumers of coal (Table 19.3).

Coal varies in its qualities

Coal varies from deposit to deposit in many ways, including in water content and the amount of potential energy it contains. Organic material that is broken down anaerobically but remains wet, near the surface, and not well compressed

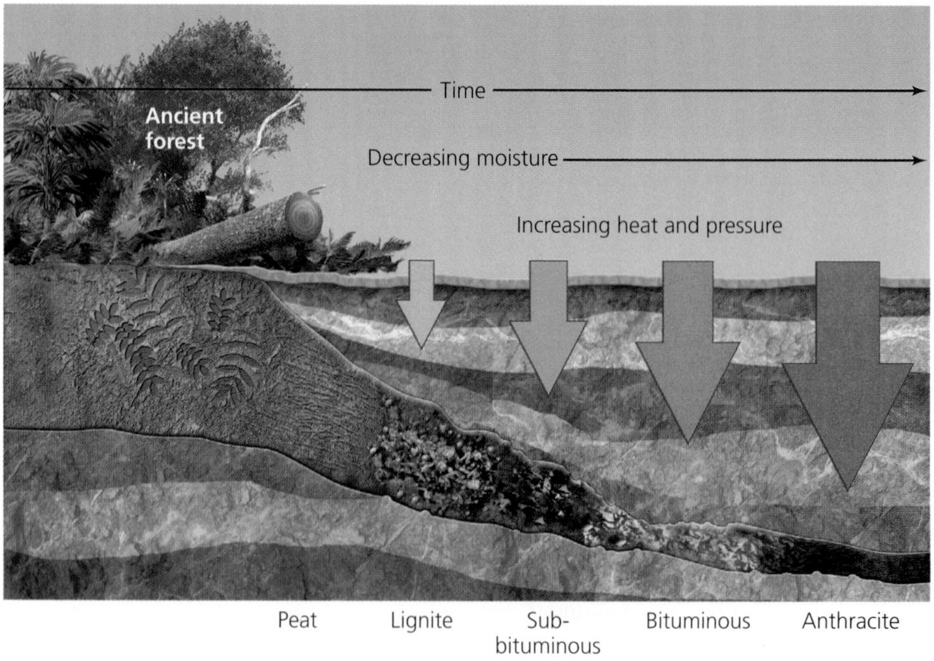

FIGURE 19.7 Coal forms as ancient plant matter is compacted underground. Scientists categorize coal into several types, depending on the amount of heat, pressure, and moisture involved in its formation. Anthracite coal is formed under greatest pressure, where temperatures are high and moisture content is low. Lignite coal is formed under conditions of much less pressure and heat, but more moisture. Peat is also part of this continuum, representing plant matter that is minimally compacted.

is called **peat.** A kind of precursor stage to coal, peat has been widely used as a fuel in Britain and other locations. As peat decomposes further, as it becomes buried more deeply under sediments, as pressure and heat increase, and as time passes, water is squeezed out of the material, and carbon compounds are packed more tightly together, forming coal. Scientists classify coal into four types: lignite, subbituminous, bituminous, and anthracite. Lignite is the least-compressed type of coal, and anthracite is the most-compressed type. The greater the compression, the greater is the energy content per unit volume (see Figure 19.7).

Most coal contains various impurities, including sulfur, mercury, arsenic, and other trace metals, and coal deposits vary in the amount of impurities they contain. For instance, sulfur content varies, depending in part on whether the coal was formed in freshwater or saltwater sediments. Coal in what is today the eastern United States tends to be high in sulfur because it was formed in marine sediments, where sulfur from seawater was present. When high-sulfur coal is burned, it produces sulfate air pollutants, which contribute to industrial smog and acidic deposition (▸pp. 509–510, 514–518). Combustion of coal high in mercury content emits mercury that can bioaccumulate in organisms' tissues, poisoning animals as it moves up food chains. Such pollution problems commonly occur downwind of coal-fired power plants. Scientists and engineers are seeking ways to cleanse coal of its impurities so that it can continue to be used as an energy source while minimizing impact on the environment (see "The Science behind the Story," ▸pp. 566–567).

Coal is mined from the surface and from below ground

We extract coal using two major methods (Figure 19.8). We reach underground deposits with **subsurface mining.** Shafts are dug deep into the ground, and networks of tunnels are dug or blasted out to follow coal seams. The coal is removed systematically and shipped to the surface. When coal deposits are at or near the surface, strip-mining methods are used. In **strip mining,** heavy machinery removes huge amounts of earth to expose and extract the coal. The pits are subsequently refilled with the soil that had been removed. Strip-mining operations can occur on immense scales; in some cases entire mountaintops are lopped off. This environmentally destructive process, called *mountaintop removal,* has become more common recently in the Appalachian Mountains. We will revisit some of coal's environmental impacts later in this chapter. Understanding these impacts is important because society's demand for this relatively abundant fossil fuel may soon rise as supplies of our most-used fossil fuel, oil, decline.

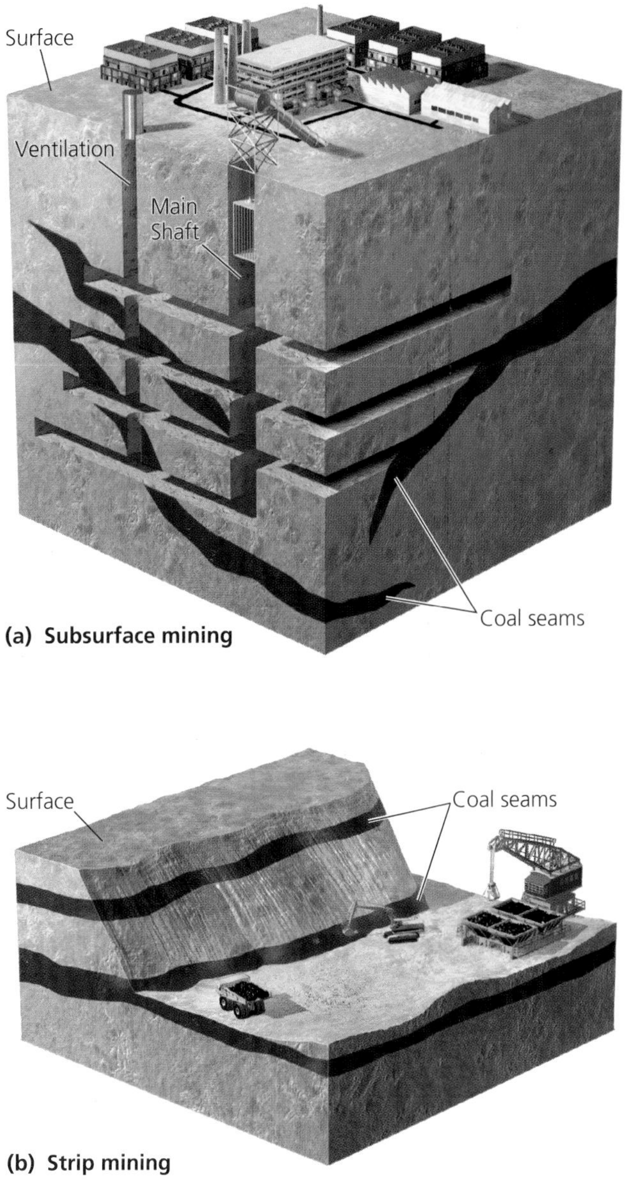

FIGURE 19.8 Coal is mined in two major ways. In subsurface mining **(a)**, miners work below ground in shafts and tunnels blasted through the rock; these passageways provide access to underground seams of coal. This type of mining poses dangers and long-term health risks to miners. In strip mining **(b)**, soil is removed from the surface, exposing coal seams from which coal is mined. This type of mining can cause substantial environmental impact.

Oil

Oil has been the world's most-used fuel since the 1960s, when it eclipsed coal. It now accounts for 37% of the world's commercial energy consumption. Its use worldwide over the past decade has risen roughly 16%.

The Science behind the Story

How Electricity Is Generated

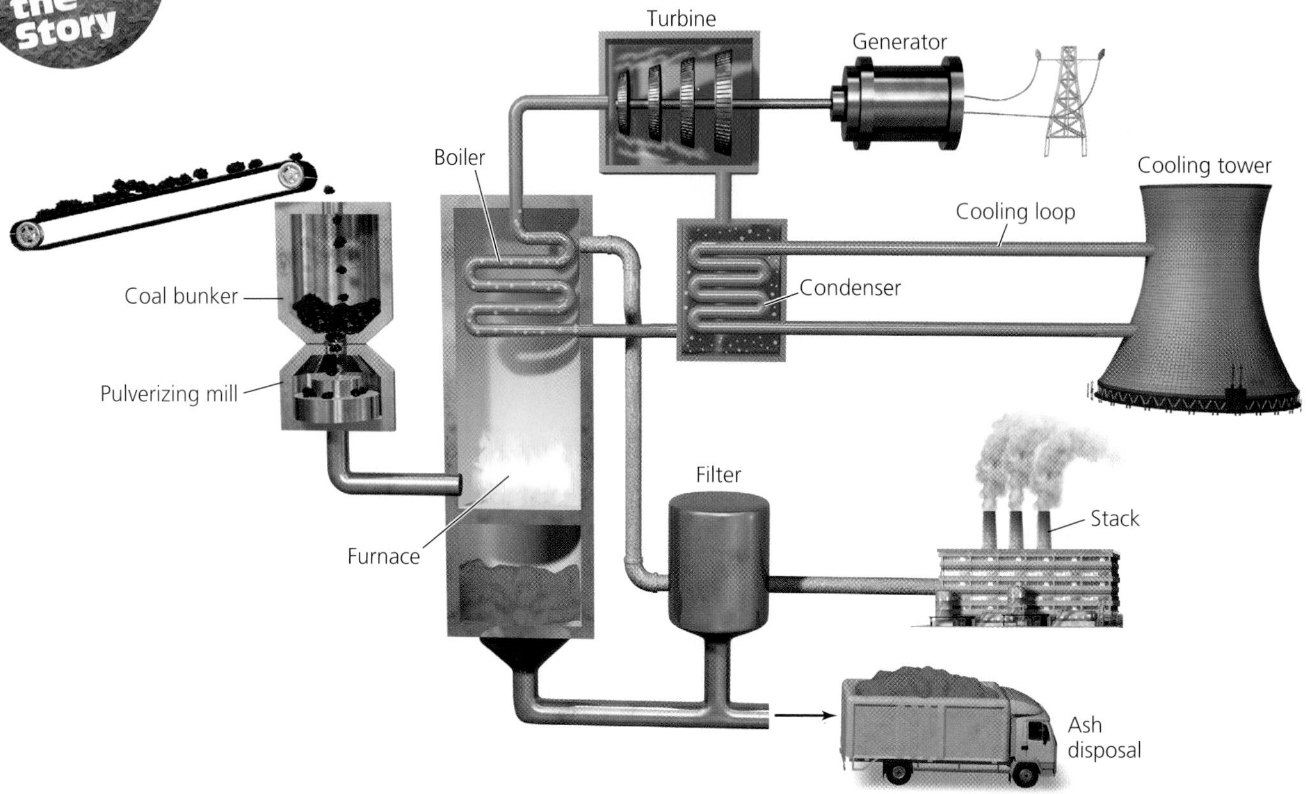

Coal is the primary fuel source used to generate electricity in the United States. Pieces of coal are pulverized and blown into a high-temperature furnace. Heat from the combustion boils water, and the resulting steam turns a turbine, generating electricity by passing magnets past copper coils. The steam is then cooled and condensed in a cooling loop and returned to the furnace. "Clean coal" technologies help filter out pollutants from the combustion process, and toxic ash residue is disposed of in hazardous waste disposal sites.

As the United States tries to balance its growing demand for electricity with rising concerns about environmental and health impacts of coal combustion, power plants continue to rely heavily on coal while scientists work to limit the pollution that use of this fuel creates.

Every new housing subdivision, cell phone, and DVD player means a greater draw on the power grid. To generate more electricity using domestic resources, coal has emerged as a central factor in recent U.S. energy policy.

Coal is used to generate electricity in a process that dates back more than a century (see the figure). Once mined, coal is hauled to power plants, where it is pulverized. The crushed coal is blown into a boiler furnace on a superheated stream of air and burned in a blaze of intense

Heat and pressure underground form petroleum

The sludgelike liquid we know as **crude oil,** or **petroleum,** tends to form within a window of temperature and pressure conditions often found 1.5–3 km (1–2 mi) below the surface. Crude oil is a mixture of hundreds of different types of hydrocarbon molecules characterized by carbon chains of different lengths (▸ pp. 98–99). A chain's length affects its chemical properties, which has consequences for human use, such as whether a given fuel burns cleanly in a car engine. Oil refineries sort the various heat—typical furnace temperatures often flare at 815 °C (1500 °F). Water circulating around the boiler absorbs the heat and is converted to high-pressure steam. This steam is injected into a **turbine,** a rotary device that converts the kinetic energy of a moving substance such as steam into mechanical energy. The turbine's key components are a drive shaft and a series of fanlike blades attached to the drive shaft. As steam from the boilers exerts pressure on the blades of the turbine, they spin, turning the drive shaft.

The drive shaft is connected to a generator, which also features two main components—a rotor that rotates and a stator that remains stationary. Generators have different designs, but they all make use of the same principle: Moving magnets adjacent to coils of copper wire cause electrons in the copper wires to move, generating alternating electric current. In some generators, the rotor consists of magnets and spins within a stator of coiled copper wire. As the turbine's drive shaft rotates, it causes the rotor to revolve, creating a magnetic field and causing electrons in the stator's copper wires to move, thereby creating an electrical current. This current flows into transmission lines that travel from the power plant out to the customers who use the plant's electricity.

Because coal is the country's largest domestic fossil fuel source but also a major source of pollution, U.S. policymakers have arrived at a dual approach: continuing to rely on coal while trying to make its use less toxic. The national energy legislation passed by Congress in 2005 provided $1.3 billion in tax credits to power plants that use state-of-the-art "clean coal" technologies, and set aside $2.5 billion over 10 years for further research into such technologies. These efforts largely focus on approaches to rid the generation process of toxic chemicals, either before or after the coal is burned. Technologies to do this include *scrubbers,* or materials based on minerals such as calcium or sodium that absorb and remove sulfur dioxide (SO_2) from smokestack emissions. Other approaches use chemical reactions to strip away nitrogen oxides (NO_X), breaking them down into elemental nitrogen and water. Multilayered filtering devices are used to capture tiny ash particles.

Research at the Niles Station power plant in Ohio found a way to remove 95% of SO_2 and 90% of NO_X from the plant's emissions. Researchers first installed smokestacks with efficient filter bags and captured significant amounts of ash and gases. The filters also proved effective at removing dangerous substances such as mercury and selenium. The captured ash and gases were reheated, and small amounts of ammonia were added to chemically convert NO_X into harmless nitrogen gas and water vapor. Further addition of oxygen to SO_2 in the smokestack gas converted the sulfur compounds into sulfuric acid, which the plant bottled and sold for use in fertilizer and manufacturing processes. The gas-scrubbing process worked well enough for the plant to begin using it permanently, and several power plants in Europe adopted the technology.

Some energy analysts and environmental advocates question a policy emphasis on clean coal, however. Coal, they maintain, is an inherently dirty way of generating power and should be replaced outright with cleaner energy sources. Others doubt that the Bush administration's push for clean coal technologies represents a true commitment to cleaner air, because the administration also eased Clean Air Act requirements that would have forced power companies to upgrade their pollution control technology. Without mandates, many power generators have been slow to adopt clean coal technology on their own. However, with the new funding from the 2005 energy legislation, research will press ahead.

hydrocarbons of crude oil, separating those intended for use in gasoline engines from those, such as tar and asphalt, used for other purposes.

The crude oil of Alaska's North Slope was formed when dead plant material (and small amounts of animal material) drifted down through coastal marine waters millions of years ago and was buried in sediments on the ocean floor. These organic remains were then transformed by time, heat, and pressure into the crude oil of today. The shelves of sedimentary rock that now lie beneath Alaska's coastal plain are believed to hold the largest remaining onshore petroleum deposits in North America.

Table 19.4 Top Producers and Consumers of Oil

Production (% world production)	Consumption (% world consumption)
Saudi Arabia, 13.1	United States, 24.9
Russia, 11.9	China, 8.6
United States, 8.5	Japan, 6.4
Iran, 5.2	Russia, 3.4
Mexico, 4.9	Germany, 3.3
China, 4.5	India, 3.2
Venezuela, 4.0	South Korea, 2.8
Canada, 3.8	Canada, 2.6
Norway, 3.9	France, 2.5
United Arab Emirates, 3.3	Italy, 2.4

Data from British Petroleum. 2005. *Statistical review of world energy 2005.*

The age of oil began in the mid-19th century

As long ago as 4,000 B.C., people used solid forms of oil (such as tar and asphalt) from deposits that were easily accessible at Earth's surface. The modern extraction and use of petroleum for energy began in the 1850s. In Pennsylvania, miners drilling for salt occasionally encountered oily rocks instead. At first, entrepreneurs bottled the crude oil from these deposits and sold it as a healing aid, unaware that crude oil is carcinogenic when applied to the skin and poisonous when ingested. Soon, however, a Dartmouth College scholar named George Bissell realized that this "rock oil" could be used to light lamps and lubricate machinery, and in 1854 Bissell started the Pennsylvania Rock Oil Company. By the time the firm developed its drilling technology and struck oil, Bissell was no longer at the company. Instead, Edwin Drake is credited with drilling the world's first oil well, in Titusville, Pennsylvania, in 1859. Over the next 40 years, Pennsylvania's oil fields produced half the world's oil supply and helped establish a fossil-fuel-based economy that would hold sway for decades to come.

Today our global society produces and consumes nearly 750 L (200 gal) of oil each year for every man, woman, and child. The United States consumes fully one-fourth of the world's oil. U.S. oil consumption has increased 16% in the past decade and shows little sign of abating. Table 19.4 shows the top oil-producing and oil-consuming nations.

Petroleum geologists infer the location and size of deposits

Because petroleum forms only under certain conditions, it occurs in isolated deposits. Once geothermal heating separates hydrocarbons from their source material and produces crude oil, this liquid migrates upward through rock pores, sometimes assisted by seismic faulting. It tends to collect in porous layers beneath dense, impermeable layers. Thus, oil deposits are not large black underground lakes, but instead consist of small droplets within holes in porous rock, like a hard sponge full of oil.

Geologists searching for oil (or other fossil fuels) drill rock cores and conduct ground, air, and seismic surveys (Figure 19.9) to map underground rock formations, understand geological history, and predict where fossil fuel deposits might lie. Using such techniques, geologists from the U.S. Geological Survey (USGS) in 1998 estimated, with 95% certainty, the total amount of oil underneath ANWR's 1002 Area to be between 11.6 and 31.5 billion barrels. The geologists' average estimate of 20.7 billion represents their best guess as to the number of barrels of oil the 1002 Area holds.

Some portion of oil will be impossible to extract using current technology and may have to wait for future advances in extraction equipment or methods. Thus, estimates are generally made of "technically recoverable"

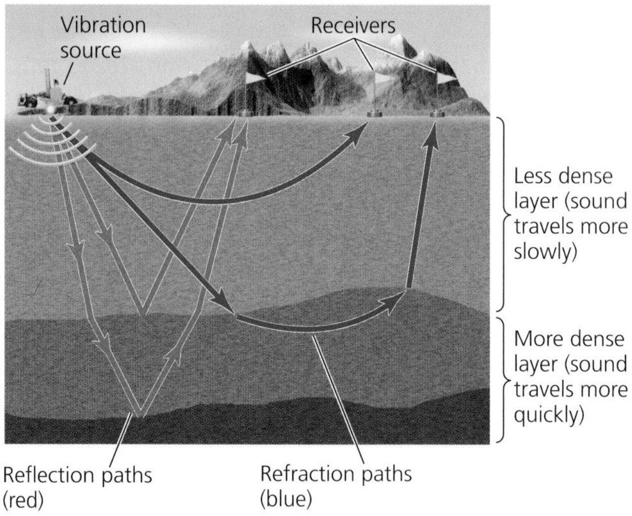

FIGURE 19.9 Petroleum geologists use seismic surveying to locate promising fossil fuel deposits. One method is to create powerful vibrations (by exploding dynamite, thumping the ground with a large weight, or using an electric vibrating machine) at the surface in one location and then measure how long it takes the seismic waves produced to reach receivers at other surface locations. Seismic waves travel more quickly through denser layers, and density differences in the substrate may cause waves to reflect off layers, refract, or bend. Scientists and engineers interpret the patterns of wave reception to infer the densities, thicknesses, and location of underlying geological layers—which in turn provide clues about the location and size of fuel deposits.

amounts of fuels. In its 1998 estimates, the USGS calculated technically recoverable amounts of oil under the 1002 Area to be between 4.3 and 11.8 billion barrels, with an average estimate of 7.7 billion barrels. Because some Native lands and state-owned offshore areas can be developed if the 1002 Area is developed, the USGS also surveyed these areas and estimated that 1.4–4.2 billion additional barrels likely lie beneath them.

However, oil companies will not be willing to extract these entire amounts. Some oil would be so difficult to extract that the expense of doing so would exceed the income the company would receive from the oil's sale. Thus, the amount a company chooses to drill for will be determined by the costs of extraction (and transportation), together with the current price of oil on the world market. Because the price of oil fluctuates, the portion of oil from a given deposit that is "economically recoverable" fluctuates as well. USGS scientists calculated that at a price of $30 per barrel, 3.0–10.4 billion barrels would be economically worthwhile to recover from the 1002 Area. The USGS did not present estimates for today's much higher prices, but as prices climb, economically recoverable amounts approach technically recoverable amounts.

Thus, technology sets a limit on the amount that *can* be extracted, whereas economics determines how much *will* be extracted. The amount of oil, or any other fossil fuel, in a deposit that is technologically and economically feasible to remove under current conditions is termed the **proven recoverable reserve** of that fuel.

We drill to extract oil

Once geologists have identified an oil deposit, an oil company will typically conduct *exploratory drilling.* Holes drilled during this phase are usually small in circumference and descend to great depths. If enough oil is encountered, extraction begins. Just as you would squeeze a sponge to remove its liquid, pressure is required to extract oil from porous rock. Oil is typically already under pressure—from above by rock or trapped gas, from below by groundwater, or internally from natural gas dissolved in the oil. All these forces are held in place by surrounding rock until drilling reaches the deposit, whereupon oil will often rise to the surface of its own accord.

Once pressure is relieved, however, oil becomes more difficult to extract and may need to be pumped out. Even after pumping, a great deal of oil remains stuck to rock surfaces. As much as two-thirds of a deposit may remain in the ground after **primary extraction,** the initial drilling and pumping of available oil (Figure 19.10a). Companies may then begin **secondary extraction,** in which solvents are used or underground rocks are flushed with water or steam to remove additional oil (Figure 19.10b). At Prudhoe Bay, seawater is piped in from the coast and pumped into wells to flush out remaining oil. Even after secondary extraction, quite a bit of oil can remain; we lack the technology to remove every last drop. Secondary extraction is more expensive than primary extraction, so many U.S. deposits did not undergo secondary extraction when they

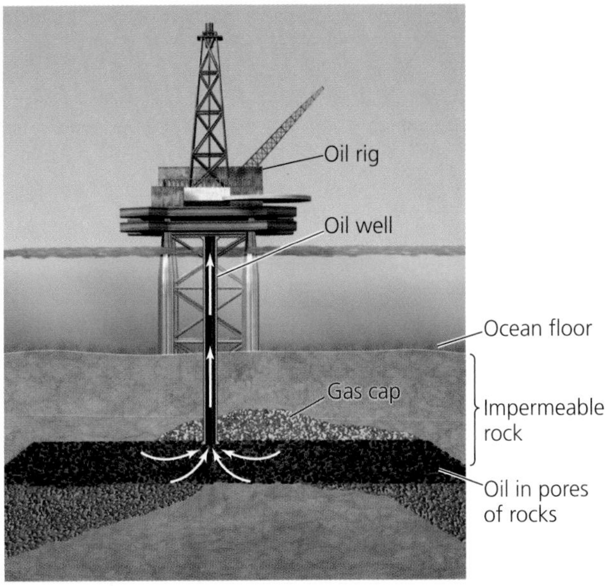

(a) Primary extraction of oil

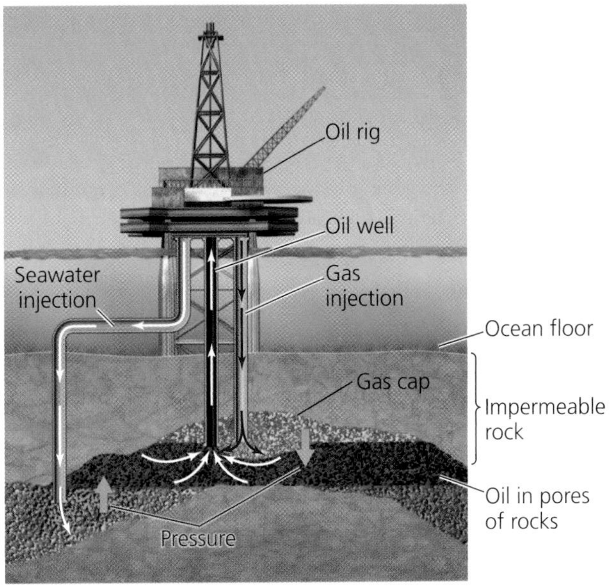

(b) Secondary extraction of oil

FIGURE 19.10 In primary extraction **(a)**, oil is drawn up through the well by keeping pressure at the top lower than pressure at the level of the oil deposit. Once the pressure in the deposit drops, however, material must be injected into the deposit to increase the pressure. Thus, secondary extraction **(b)** involves injecting seawater beneath the oil and/or gases just above the oil to force more oil up and out of the deposit.

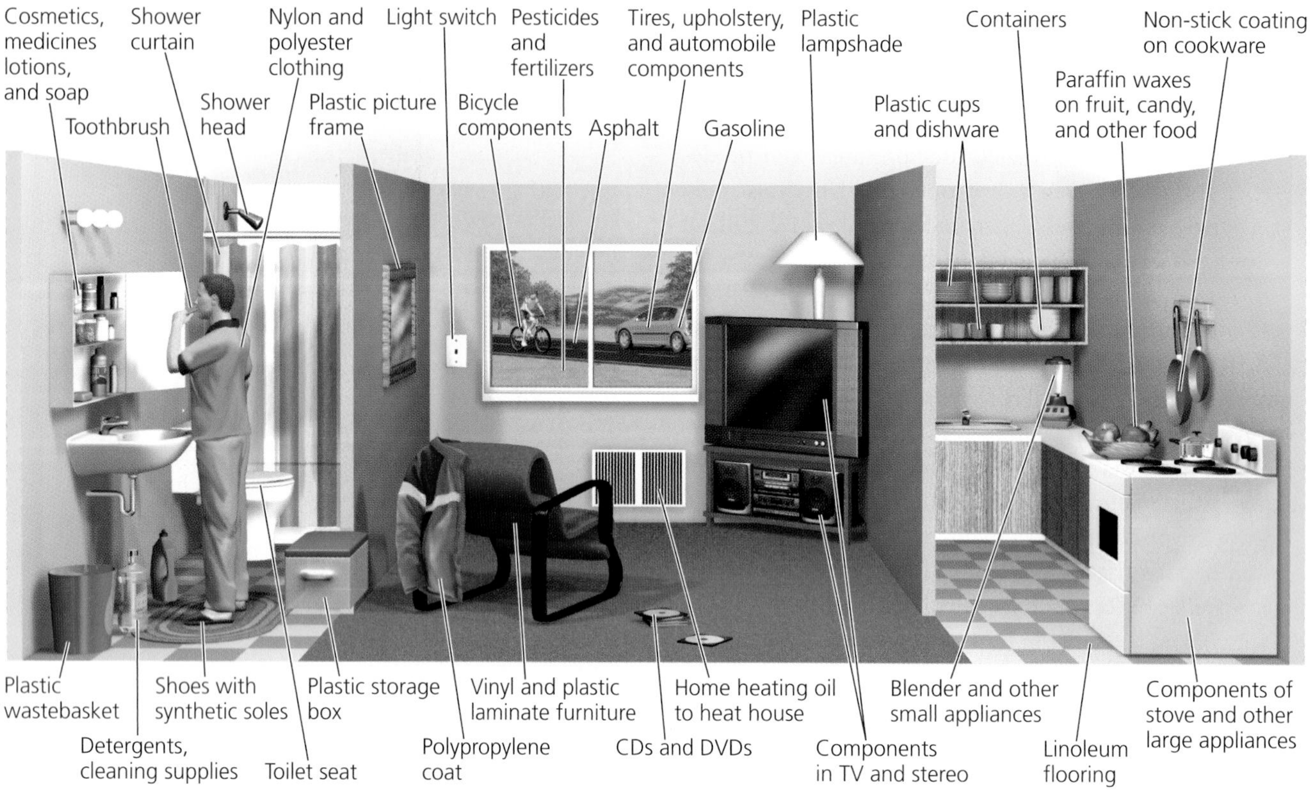

FIGURE 19.11 Petroleum products are everywhere in our daily lives. The gasoline and other fuels we use for transportation and heating are just a few of the many products we derive from petroleum. These products include many of the fabrics that we wear and most of the plastics that help make up countless items we use every day.

were first drilled because the price of oil was too low to make the procedure economical. When oil prices rose in the 1970s, many drilling sites were reopened for secondary extraction.

Offshore drilling produces much of our oil

Drilling for oil and natural gas takes place not just on land but also in the seafloor on the continental shelves. Offshore drilling has required development of technology that can withstand the forces of wind, waves, and ocean currents. Some drilling platforms are fixed standing platforms built with unusual strength. Others are resilient floating platforms anchored in place above the drilling site. Over 25% of the oil and gas extracted in the United States comes from offshore drilling sites, primarily in the Gulf of Mexico and off the southern California coast. This is why Hurricanes Katrina and Rita in 2005 caused such disruption to U.S. oil supplies. By damaging offshore oil platforms off the coasts of Louisiana and neighboring states and by damaging refineries onshore, the storms interrupted a substantial portion of the nation's oil supply, and prices rose accordingly.

Petroleum products have many uses

Once crude oil is extracted, it is put through refining processes (see "The Science behind the Story," ▸pp. 572–573). Because crude oil is a complex mix of hydrocarbons, we can create many types of petroleum products by separating its various components. Since the 1920s, refining techniques and chemical manufacturing have greatly expanded our uses of petroleum to include a wide array of products and applications, from lubricants to plastics to fabrics to pharmaceuticals. Today, petroleum-based products are all around us in our everyday lives (Figure 19.11).

Because petroleum products have become so central to our lives, many fossil fuel experts today are voicing extreme concern that oil production may soon decline as we continue to deplete the world's oil reserves.

We may have already depleted half our oil reserves

Many scientists and oil industry analysts calculate that we have already extracted half of the world's oil reserves. So far we have used up about 1 trillion barrels of oil, and most estimates hold that an additional 1 trillion barrels, or somewhat more, remain. To estimate how long this remaining oil will last, analysts calculate the **reserves-to-production ratio, or R/P ratio,** by dividing the amount of total remaining reserves by the annual rate of production (i.e., extraction and processing). At current levels of production (30 billion barrels globally per year), most analysts estimate that world oil supplies will last about 40 more years.

Many policymakers and industry spokespeople suggest that this means we will have little to worry about for 40 years. However, another group of scientists and analysts insists that we will face a crisis not when the last drop of oil is pumped, but when the rate of production first begins to decline. They point out that when production declines as demand continues to increase (because of rising global population and consumption), we will experience an oil shortage immediately. Because production tends to decline once reserves are depleted halfway, most of these experts calculate that this crisis will likely begin within the next several years.

To understand the basis of these concerns, we need to turn back the clock to 1956. In that year, Shell Oil geologist M. King Hubbert calculated that U.S. oil production would peak around 1970. His prediction was ridiculed at the time, but it proved to be accurate; U.S. production peaked in that very year and has continued to fall since then (Figure 19.12a). The peak in production came to be known as **Hubbert's peak.** In 1974, Hubbert analyzed data on technology, economics, and geology, to predict that global oil production would peak in 1995. It has kept growing past 1995, but many scientists using newer, better data today predict that at some point in the coming decade, production will begin to decline (Figure 19.12b). Discoveries of new oil fields peaked 30 years ago, and since then we have been extracting and consuming more oil than we have been discovering.

FIGURE 19.12 Because fossil fuels are nonrenewable resources, supplies at some point pass the midway point of their depletion, and annual production begins to decline. U.S. oil production peaked in 1970, just as geologist M. King Hubbert predicted decades previously; this highpoint is referred to as "Hubbert's peak" **(a).** Today many analysts believe global oil production is about to peak. Shown **(b)** is the latest projection, from a 2004 analysis by scientists at the Association for the Study of Peak Oil. Go to GRAPHIt! at www.aw-bc.com/withgott or on the student CD-ROM. Data from (a) Deffeyes, K. S. 2001. *Hubbert's peak: The impending world oil shortage.* (b) Colin J. Campbell and Association for the Study of Peak Oil, 2004.

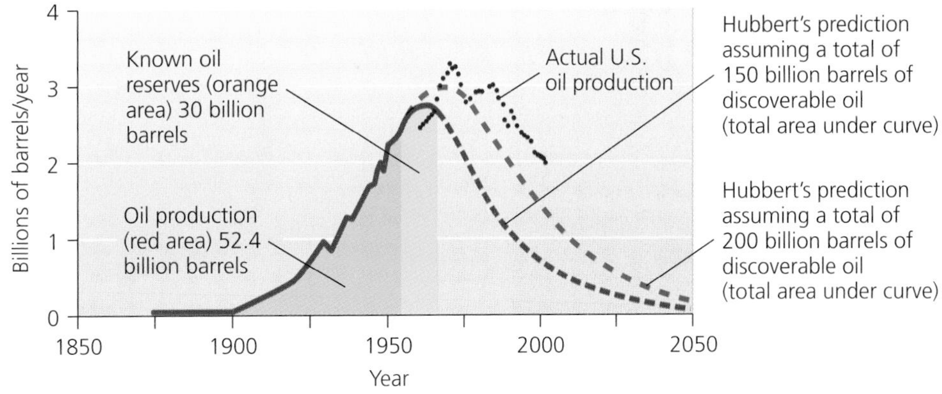

(a) Hubbert's prediction of peak in U.S. oil production, with actual data

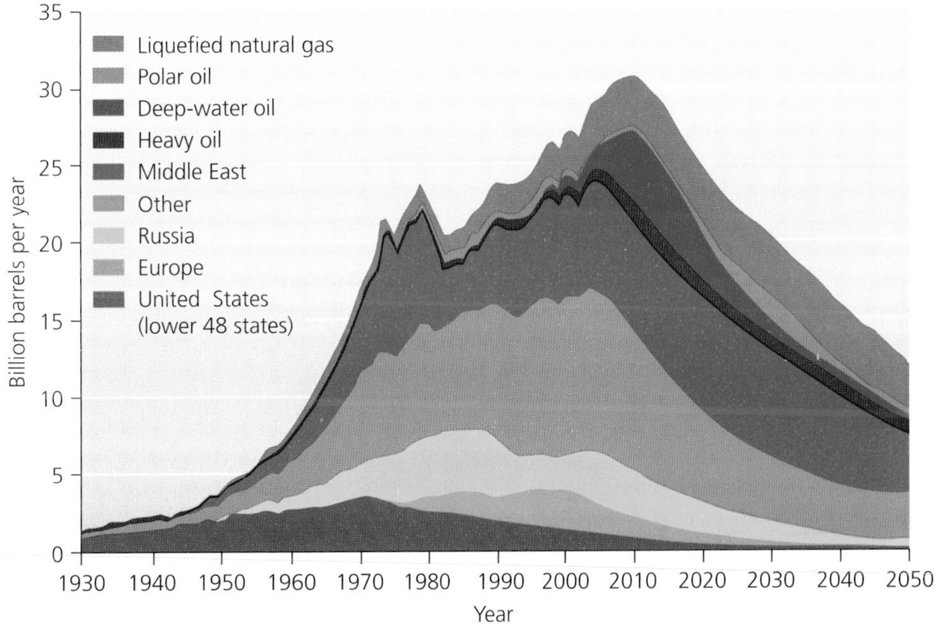

(b) Modern prediction of peak in global oil production

The Science behind the Story

How Crude Oil Is Refined

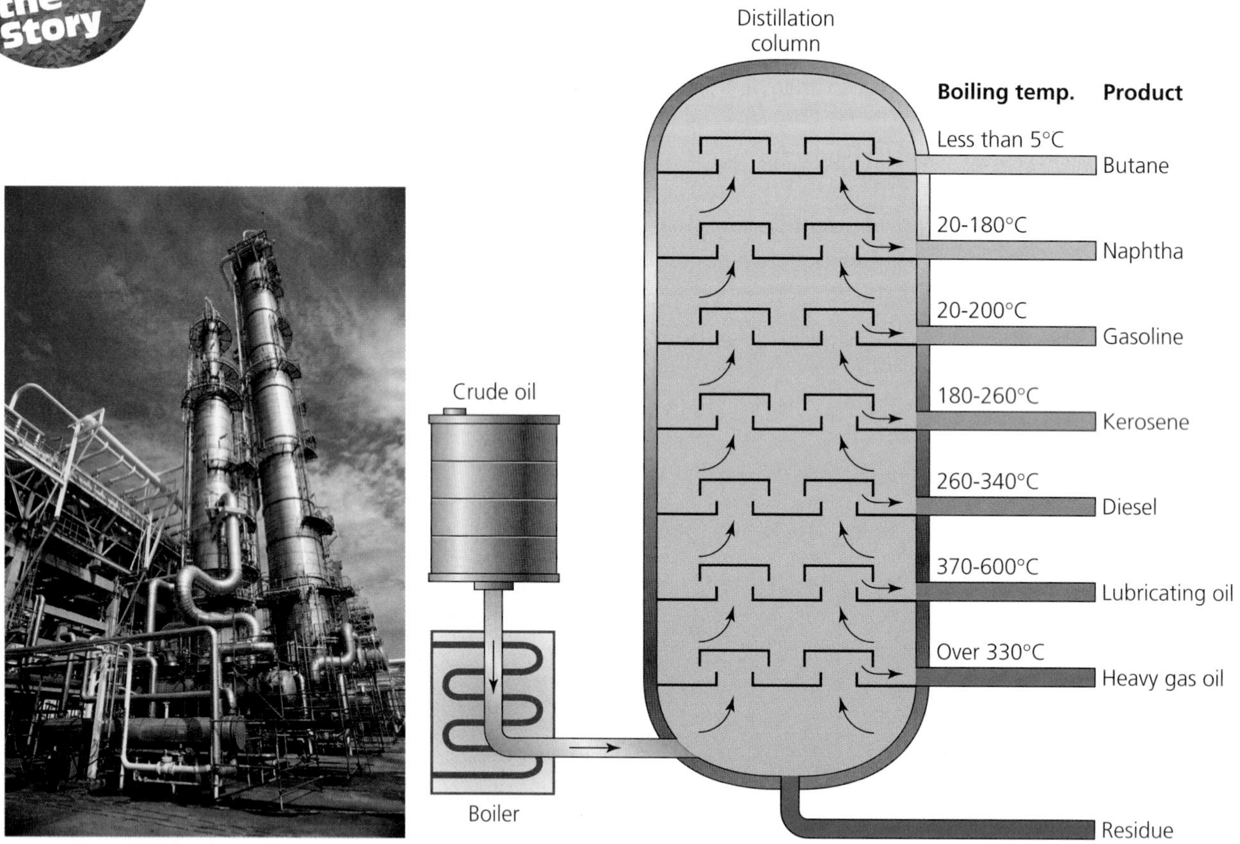

(a) Distillation columns **(b) Distillation process**

Crude oil is shipped to petroleum refineries **(a)**, where it is refined into a number of different types of fuel. Crude oil is boiled, causing its many hydrocarbon constituents to volatilize and proceed upward through a distillation column **(b)**. Constituents that boil only at the highest temperatures and condense readily once the temperature drops will condense at low levels in the column. Constituents that volatilize readily at lower temperatures will continue rising through the column and condense at higher levels, where temperatures are lower. In this way, heavy oils (generally consisting of long hydrocarbon molecules) are separated from lighter oils (generally those with short hydrocarbon molecules).

Crude oil is a complex mixture of thousands of kinds of hydrocarbon molecules. Through the process of **refining,** these hydrocarbon molecules are separated into classes of different sizes and chemically transformed to create specialized fuels for heating, cooking, and transportation and to create lubricating oils, asphalts, and the precursors of plastics and other petrochemical products. To maximize the production of marketable products while minimizing negative environmental impacts, petroleum engineers have developed a variety of refining techniques.

The first step in processing crude oil is *distillation,* or *fractionation.* This process is based on the fact that different components of crude oil boil at different temperatures. In refineries, the distillation process takes place in tall columns filled with perforated horizontal trays (see the first figure). The columns are cooler at the top than at the bottom. When heated crude oil is introduced into the column, lighter components rise as vapor to the upper trays, condensing into liquid as they cool, while heavier components sink to the lower trays. Light gases, such as butane, boil at less than 32 °C (90 °F), and heavier oils, such as industrial fuel oil, boil only at temperatures above 343 °C (650 °F).

Since the early 20th century, light gasoline, used in automobiles, has been in much higher demand than most other derivatives of crude oil. The demand for high-performance, clean-burning gasoline has also risen. To meet these demands, refiners have developed several techniques to convert heavy hydrocarbons into gasoline.

The general name for processes that convert heavy oil into lighter oil is *cracking.* One of the simplest methods is thermal cracking, in which long-chained molecules are broken into smaller chains by heating in the absence of oxygen. (The oil would ignite if oxygen were present.) Catalytic cracking, a related method, uses catalysts—substances that promote chemical reactions without being consumed by them—to control the cracking process. The result is an increase in the amount of a desired lighter product from a given amount of heavy oil. Today, the most widely used form of cracking is fluidized catalytic cracking, in which a finely powdered catalyst that behaves like a fluid is fed continuously into a reaction chamber with heavy oils. The products of cracking are then fed into a distillation column.

Refiners can also change the chemical composition of oil through a process called *catalytic reforming.* Catalytic reforming uses catalysts to promote chemical reactions that transform certain hydrocarbons that are slightly heavier than gasoline so that they can be blended with gasoline to obtain higher octane ratings. The octane rating reflects the amount of compression gasoline can undergo before it spontaneously ignites. An octane rating of 92 indicates that a gasoline blend is equivalent to a mixture of 92% octane, which spontaneously ignites only at very high compression levels, and 8% heptane, which ignites more easily. Other things being equal, high-compression engines are more powerful than low-compression engines, so gasoline that can withstand high levels of compression—that has a high octane rating—is preferred. Roughly 30–40% of U.S. gasoline is produced using catalytic reforming.

High-octane gasoline can also be produced by combining smaller hydrocarbons to synthesize larger molecules. In this process, called alkylation, molecules such as isobutylene and isobutene, which each have four carbon atoms, are combined in the presence of catalysts to form molecules with eight carbon atoms. The resulting high-octane alkylate can then be blended with lower-octane gasoline.

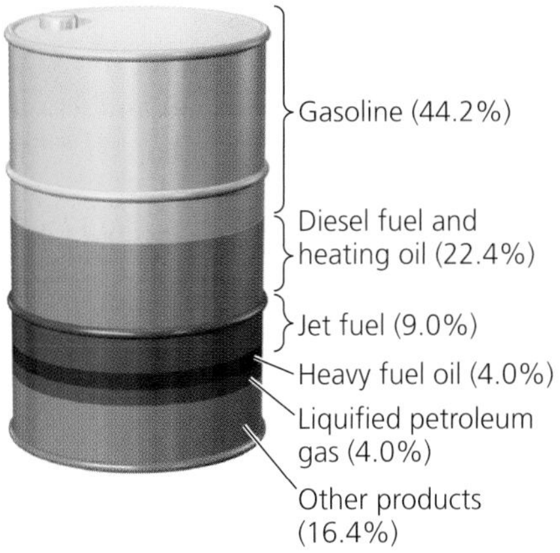

The refining process converts crude oil into a range of petroleum products. Shown are percentages of each major category of product typically produced from a barrel of crude oil. Adapted from U.S. Energy Information Administration.

Besides distilling crude oil and altering the chemical structure of some of its components, refineries also remove contaminants. Sulfur and nitrogen compounds, which can be harmful when released into the atmosphere, are the two most common contaminants in crude oil. Government regulations stemming from legislation such as the Clean Air Act (▸ pp. 506–507) have forced refiners to develop methods of removing such contaminants, particularly sulfur. Some methods successfully remove up to 98% of sulfur.

As a result of all these approaches, each barrel of crude oil is eventually converted into gasoline and a wide variety of other petroleum products (see the second figure).

Predicting an exact date for the coming decline in production is difficult. Nonetheless, it seems certain that Hubbert's peak for global oil will occur and that the divergence of supply and demand could have catastrophic economic consequences. To achieve a sustainable society, we will eventually need to switch to renewable energy sources (Chapters 20 and 21). Energy conservation (▸pp. 582–583, 585) can extend the time we have in which to make this transition. In the meantime, it seems most likely that we may come to rely more heavily on coal and on natural gas.

Weighing the Issues: The End of Oil

Physicist David Goodstein has calculated that the gap between rising demand and falling supply after world oil production begins to decline may amount to 5% per year. As a result, just 10 years after the production peak we will have only half the oil availability we had at the peak. He worries that we may not be able to modify our infrastructure and institutions fast enough to accommodate other energy sources before the economic impacts of an oil shortage undermine our ability to do so.

Do you think our society could adapt to a 50% decrease in oil availability over 10 years? How do you think we would most likely respond? How do you think we should respond?

Natural Gas

Natural gas is the fastest-growing fossil fuel in use today and provides for one-quarter of global commercial energy consumption. World supplies of natural gas are projected to last perhaps 60 years. This R/P ratio is slightly better than that of oil, but scientists expect that natural gas will likewise experience Hubbert's peak, such that production will decline after half the world's reserves are depleted.

Natural gas is formed in two ways

Natural gas consists primarily of methane, CH_4, and typically includes varying amounts of other volatile hydrocarbons. Natural gas can arise from either of two processes. *Biogenic* gas is created at shallow depths by the anaerobic decomposition of organic matter by bacteria. An example is the "swamp gas" you can sometimes smell when stepping into the muck of a swamp. In contrast, *thermogenic* gas results from compression and heat deep underground. As organic matter is buried more and more deeply under sediments, the pressure exerted by the overlying sediments grows, and temperatures increase. Carbon bonds in the organic matter begin breaking, and the organic matter turns to a substance called *kerogen,* which acts as a source material for both natural gas and crude oil. Further heat and pressure act on the kerogen to degrade complex organic molecules into simpler hydrocarbon molecules. At very deep levels—below about 3 km (1.9 mi)—the high temperatures and pressures tend to form natural gas.

Thermogenic gas may be formed directly, along with crude oil, or it may be formed from crude oil that is altered by heating. Biogenic gas is nearly pure methane, but thermogenic gas contains small amounts of a number of other hydrocarbon gases as well as methane. Most gas extracted commercially is thermogenic and found above deposits of crude oil or seams of coal, so its extraction often accompanies the extraction of those fossil fuels. In Alaska, where natural gas transport remains prohibitively expensive, gas captured during oil drilling is reinjected into the ground for potential future extraction.

One source of biogenic natural gas is the decay process in landfills, and many landfill operators are now capturing this gas to sell as fuel (▸ p. 657). This practice decreases energy waste, can be profitable for the operator, and helps reduce the atmospheric release of methane, a greenhouse gas that contributes to climate change (▸ p. 530).

Natural gas was long known but has only recently been widely used in homes

Throughout history, naturally occurring seeps of natural gas would occasionally be ignited by lightning and could be seen burning in parts of what is now Iraq, inspiring the Greek essayist Plutarch around A.D. 100 to describe their "eternal fires." Such natural seepages of gas from underground deposits were used to light some streets and buildings in the United States in the 1800s.

The first commercial extraction of natural gas took place in 1821, but during much of the 19th century its use was localized because technology did not exist to pipe gas safely over long distances. Natural gas was used to fuel streetlamps, but when electric lights replaced most gas lamps in the 1890s, gas companies began marketing gas for heating and cooking. The first long-distance gas pipeline was built in 1891 to carry gas from

Table 19.5 Top Producers and Consumers of Natural Gas

Production (% world production)	Consumption (% world consumption)
Russia, 21.9	United States, 24.0
United States, 20.2	Russia, 15.0
Canada, 6.8	United Kingdom, 3.6
United Kingdom, 3.6	Canada, 3.3
Iran, 3.2	Iran, 3.2
Algeria, 3.0	Germany, 3.2
Norway, 2.9	Italy, 2.7
Indonesia, 2.7	Japan, 2.7
Netherlands, 2.6	Ukraine, 2.6
Saudi Arabia, 2.4	Saudi Arabia, 2.4

Data from British Petroleum. 2005. *Statistical review of world energy 2005.*

deposits in Indiana to homes 195 km (120 mi) away in Chicago. After World War II, wartime improvements in welding and pipe building made gas transport safer and more economical, and during the 1950s and 1960s, thousands of miles of underground pipelines were laid throughout the United States. If laid end to end, our current network of natural gas pipelines would extend to the moon and back twice. Natural gas deposits are greatest in Russia and the United States, and these two nations lead the world in both gas production and gas consumption (Table 19.5).

Natural gas extraction becomes more challenging with time

To access some natural gas deposits, prospectors need only drill an opening, because pressure and low molecular weight drive the gas upward naturally. The first gas fields to be tapped were of this type. Most fields remaining today, however, require that gas be pumped to Earth's surface. In Texas, Kansas, California, and other areas of the United States, it is common to see a device called a horsehead pump (Figure 19.13). This pump moves a rod in and out of a shaft, creating pressure to pull both oil and natural gas to the surface. As with oil and coal, many of the most accessible natural gas reserves have already been exhausted. Thus, much extraction today makes use of sophisticated techniques to break into rock formations and pump gas to the surface. One such "fracturing technique" is to pump salt water under high pressure into the rocks to crack them. Sand or small glass beads are inserted to hold the cracks open once the water is withdrawn.

FIGURE 19.13 Horsehead pumps are used to extract natural gas as well as oil. They are a common feature of the landscape in areas such as west Texas. The pumping motion of the machinery draws gas and oil upward from below ground.

Other fossil fuels could be used in the future

Although natural gas, crude oil, and coal are the three fossil fuels that power our civilization today, other types of fossil fuels exist. *Oil sands* or *tar sands* are dense, hard, oily substances that can be mined from the ground. *Shale oil* is essentially kerogen, sedimentary rock filled with organic matter that was not buried deeply enough to form oil. *Methane hydrates* occur under the seafloor and were discussed in Chapter 16 (▸ p. 479). These sources are abundant, but technology for extracting usable fuel from them is largely undeveloped, and their extraction will likely remain extremely expensive. Many advocates of sustainability believe it would be a mistake to try to switch to these sources once our conventional fossil fuels become scarce. Such sources will require extensive mining and will emit at least as much carbon dioxide, methane, and other air pollutants as do coal, oil, and gas, so they will further the severe environmental impacts that fossil fuels are already causing.

Environmental Impacts of Fossil Fuel Use

Our society's love affair with fossil fuels and the many petrochemical products we have developed from them has boosted our material standard of living beyond what our ancestors could have dreamed, has eased constraints on travel, and has helped lengthen our life spans. It has had downsides as well, however, including harm to the environment and human health. Concern over these impacts is a prime reason many scientists, environmental advocates, businesspeople, and policymakers are increasingly looking toward renewable sources of energy that exert less impact on natural systems.

Fossil fuel emissions cause pollution and drive climate change

When we burn fossil fuels, we alter certain flux rates in Earth's carbon cycle (▸ pp. 195–197). We essentially take carbon that has been effectively retired into a long-term reservoir underground and release it into the air. This occurs as carbon from within the hydrocarbon molecules of fossil fuels unites with oxygen from the atmosphere during combustion, producing carbon dioxide. Carbon dioxide is a greenhouse gas, and CO_2 released from fossil fuel combustion has been inferred to warm our planet and drive changes in global climate (Chapter 18). Because global climate change may potentially have diverse, severe, and widespread ecological and socioeconomic impacts, carbon dioxide pollution is becoming recognized as the greatest environmental impact of fossil fuel use.

Fossil fuels release more than carbon dioxide when they burn. Methane is a potent greenhouse gas, and other air pollutants resulting from fossil fuel combustion can have serious consequences for human health and the environment. Deposition of mercury and other pollutants from coal-fired power plants is increasingly recognized as a substantial health risk. The burning of oil and coal in our power plants and vehicles releases sulfur dioxide and nitrogen oxides, which contribute to industrial and photochemical smog and to acidic deposition (▸ pp. 509–511, 514–518). Gasoline combustion in automobiles releases pollutants that irritate the nose, throat, and lungs. Some hydrocarbons, such as benzene and toluene, are carcinogenic to laboratory animals and likely also to people. In addition, gases such as hydrogen sulfide can evaporate from crude oil, irritate the eyes and throat, and cause asphyxiation. Crude oil also often contains trace amounts of known poisons, such as lead and arsenic. As a result, people working at drilling operations, refineries, and other jobs that entail frequent exposure to oil or its products can develop serious health problems, including cancer.

Fossil fuels pollute water as well as air. Atmospheric deposition of pollutants exerts many impacts on freshwater ecosystems (▸ p. 515). Moreover, oil from non-point sources, such as industries, homes, automobiles, gas stations, and businesses, runs off roadways and enters rivers and sewage treatment facilities to be eventually discharged into the ocean (▸ pp. 452, 480–481). Although most spilled oil results from these non-point sources, large catastrophic oil spills can have significant impacts on the marine environment. Crude oil's toxicity to most plants and animals frequently leads to high mortality among organisms exposed. This was the case with the *Exxon Valdez* spill in 1989 (Chapter 4), in which oil from Alaska's North Slope, piped to the port of Valdez through the trans-Alaska pipeline, exerted long-term damage to ecosystems and economies in Alaska's Prince William Sound.

Groundwater supplies can also be contaminated with oil, such as when leaks from oil operations penetrate deeply into soil. Of greater concern, as we saw in Chapter 15 (▸ p. 455), thousands of underground storage tanks containing petroleum products have leaked, posing dangers to the public by threatening drinking water supplies.

Coal mining affects the environment

The mining of coal can also have substantial impacts on natural systems and human well-being. Surface strip mining can destroy large swaths of habitat and cause extensive soil erosion. It also can cause chemical runoff into waterways through the process of **acid drainage.** This occurs when sulfide minerals in newly exposed rock surfaces react with oxygen and rainwater to produce sulfuric acid. As the sulfuric acid runs off, it leaches metals from the rocks, many of which are toxic to organisms in high concentrations. Acid drainage is a natural phenomenon, but its rate accelerates greatly when mining exposes many new rock surfaces at once.

Regulations in the United States require mining companies to restore strip-mined land following mining, but impacts are severe and long-lasting just the same. Most other nations exercise less oversight. Mountaintop removal (Figure 19.14) can have even greater impacts than conventional strip mining. When tons of rock and soil are removed from the top of a mountain, it is difficult to keep material from sliding downhill, where immense areas of habitat can be degraded or destroyed and creek beds can be polluted and clogged. Loosening of U.S. government restrictions in 2002 enabled mining companies to legally dump mountaintop rock and soil into valleys and rivers below, regardless of the consequences for ecosystems, wildlife, and local residents.

Whereas mountaintop removal threatens the welfare of nearby residents, subsurface mining raises the greatest health concerns for miners. Underground coal mining is one of our society's most dangerous occupations. Besides risking injury or death from collapsing shafts and tunnels and from dynamite blasts, miners constantly inhale coal dust in the enclosed spaces of mines, which can lead to respiratory diseases, including fatal black lung disease.

FIGURE 19.14 Strip mining in some areas is taking place on massive scales, such that entire mountain peaks are leveled, as at this site in West Virginia. Such "mountaintop removal" can cause enormous amounts of erosion into streams that flow from near the mine into surrounding valleys, affecting ecosystems over large areas, as well as the people who live there.

The costs of alleviating all these health and environmental impacts are high, and the public eventually pays them in an inefficient manner. The reason is that the costs are generally not internalized (▸ pp. 43–44) in the relatively cheap prices of fossil fuels.

Oil extraction can alter the environment

Besides the many impacts summarized above (and discussed in greater detail in other chapters), oil field development has also been shown to have environmental consequences. Drilling activities themselves have fairly minimal impact, but much more than drilling is involved in the development of an oil field (see the photo that opens this chapter, ▸ p. 556). Road networks must be constructed, and many sites may be explored in the course of prospecting. These activities can fragment habitats and can be noisy and disruptive enough to affect wildlife. The extensive infrastructure that must be erected to support a full-scale drilling operation typically includes housing for workers, access roads, transport pipelines, and waste piles for removed soil. Ponds may be constructed for collecting sludge, the toxic leftovers that remain after the useful components of oil have been removed.

Many onshore North American oil reserves are located in arctic or semi-arid areas. With their low rainfall and harsh conditions, plants grow slowly in tundra and semi-desert ecosystems. As a result, even minor changes can have long-lasting repercussions. For example, tundra vegetation at Prudhoe Bay still has not fully recovered from temporary roads last used 30 years ago during the exploratory phase of development.

Whether Prudhoe Bay's oil operations have had a negative impact on the region's caribou is widely debated. Surveys show that the region's summer population of caribou has increased in the 25 years since Prudhoe Bay was developed. Many supporters of drilling have used this trend to argue that oil development does no harm to caribou. Other studies, however, show that female caribou and their calves avoid all parts of the Prudhoe Bay oil complex, including the roads laid down to support it, sometimes detouring miles to do so. These studies also show that the reproductive rate of female caribou in the Prudhoe Bay region is lower than for those in undeveloped areas in Alaska. As a result, although the herd near Prudhoe Bay has increased over the past 25 years, it has not increased as much as have herds in some other parts of Alaska.

Because there is no way of knowing how the particular herd of the Prudhoe Bay region would have performed in the absence of development (that is, there is no control, as there would be for a manipulated experiment), it is difficult to draw conclusions about the actual impacts, if any, of oil development on caribou. In ANWR, the herd that visits the undeveloped 1002 Area rose from 100,000 animals in the 1970s to 178,000 in 1989, and has since declined to 123,000, all for unknown reasons.

FIGURE 19.15 Alaska's North Slope is home to a variety of large mammals, including grizzly bears, polar bears, wolves, arctic foxes, and large herds of caribou. Whether and how oil development may negatively affect these animals are highly controversial issues, and scientific studies are ongoing. The caribou herd near Prudhoe Bay has increased since oil extraction began there, but not by as much as have herds in other parts of Alaska. Grizzly bears such as the ones shown here can sometimes be found near, or even walking atop, the trans-Alaska pipeline.

Many scientists anticipate negative environmental impacts of drilling in ANWR

To predict the possible ecological effects of drilling in ANWR's 1002 Area, scientists have examined the effects of development on arctic vegetation, air quality, water quality, and wildlife, including caribou, grizzly bears, and a variety of bird species in Prudhoe Bay and other Alaska locales where the environment is similar to that of ANWR's coastal plain (Figure 19.15). Scientists have compared different areas, and they have contrasted single areas or populations before and after drilling, as described above with caribou. In addition, scientists have run small-scale manipulative experiments when possible, for example, to study the effects of ice roads and secondary extraction methods such as seawater flushing. In one way or another they have examined the effects of road building, oil pad construction, worker presence, oil spills, accidental fires, trash buildup, permafrost melting, offroad vehicle trails, and dust from roads.

Based on these studies, many scientists anticipate damage to vegetation and wildlife if drilling takes place in ANWR. Vegetation can be killed when saltwater pumped in for flushing deposits is spilled or when plants are buried under gravel pits or roads. Air and water quality can be degraded by fumes from equipment and drilling operations, burning of natural gas associated with oil extraction, sludge ponds, waste pits, and oil spills.

Other scientists, however, contend that drilling operations in ANWR would have little environmental impact. The roads would be built of ice that will melt in the summer, they point out, and most drilling activity would be confined to the winter, when caribou are elsewhere. Moreover, drilling proponents maintain, Prudhoe Bay is not an appropriate model for ANWR because Prudhoe Bay's development is larger than that projected for ANWR. Furthermore, they say, much of the technology used at Prudhoe Bay is now outdated, and ANWR would be developed with more environmentally sensitive technology and approaches.

Weighing the Issues: The Science of Oil's Arctic Impacts

Evidence from correlative and experimental studies indicates that oil development can harm some arctic plants and animals. However, some scientists say that ice roads, winter drilling, and new technologies will minimize or eliminate such impacts.

Given the evidence and arguments spelled out in the text, suggest several further scientific studies that you believe could be undertaken to help resolve the questions being debated. What evidence would it take to convince you that it is okay to drill, and what evidence would it take to convince you that it is not okay to drill? What would you decide if you had to decide based only on the evidence available now, as policymakers do?

Political, Social, and Economic Impacts of Fossil Fuel Use

The political, social, and economic consequences of fossil fuel use are numerous, varied, and far-reaching. Our discussion focuses on several negative consequences of fossil fuel use and dependence, but it is important to bear in mind that their use has enabled much of the world's population to achieve a higher material standard of living than ever before. It is also important to ask in each case whether switching to more renewable sources of energy would solve existing problems.

Oil supply and prices affect the economies of nations

The prospect of reaching Hubbert's peak for global oil is worrisome because oil is the substance that lubricates the world's economy. Nearly all our modern technologies and services depend somehow on oil. Hurricanes Katrina and Rita and the increased gasoline prices they caused recently served to remind us how much we rely on a steady and ever-increasing supply of petroleum. The hurricanes' economic impact should have come as no surprise, for we have experienced "oil shocks" before, particularly the "energy crisis" of 1973–1974 in the United States. By 1973, with domestic U.S. sources in decline, the nation was importing more and more oil, depending on a constant flow from abroad to keep cars on the road and industries running. Then in 1973, the predominantly Arab nations of the *Organization of Petroleum Exporting Countries (OPEC)* resolved to stop selling oil to the United States. OPEC wished to raise prices by restricting supply and opposed U.S. support of Israel in the Arab-Israeli Yom Kippur War. The embargo created panic in the West and caused oil prices to skyrocket (Figure 19.16), spurring inflation. Short-term oil shortages drove American consumers to wait in long lines at gas pumps.

In response to the embargo, the U.S. government enacted a series of policies designed to reduce reliance on foreign oil. The government called for developing additional domestic sources, such as those on Alaska's North Slope. However, it also called for resuming extraction at sites shut down after primary extraction had ceased being cost-effective, capping the price that domestic producers could charge for oil, importing oil from a greater diversity of nations, funding research into renewable energy sources, and enacting conservation measures we will discuss below. The government also established a stockpile of oil as a short-term buffer against future shortages. Stored deep underground in salt caverns in Louisiana, this is called the *Strategic Petroleum Reserve.* Currently the Reserve contains roughly 700 million barrels of oil; at present rates of U.S. consumption (20.7 million barrels/ day), this equals just over one month's supply.

Nations can become dependent on foreign oil

Putting all of one's eggs in one basket is always a risky strategy. The fact that so many nations' economies are utterly tied to fossil fuels means that those economies are tremendously vulnerable to supplies' becoming suddenly unavailable or extremely costly. Such reliance means that seller nations can potentially control the price of oil, forcing buyer nations to pay more and more as supplies dwindle. With the formation of OPEC in 1960, oil-producing nations tried to exercise even greater power over oil prices by regulating production (although OPEC so far has had only marginal success in this).

In the United States, concern over reliance on foreign oil sources has repeatedly driven the proposal to open

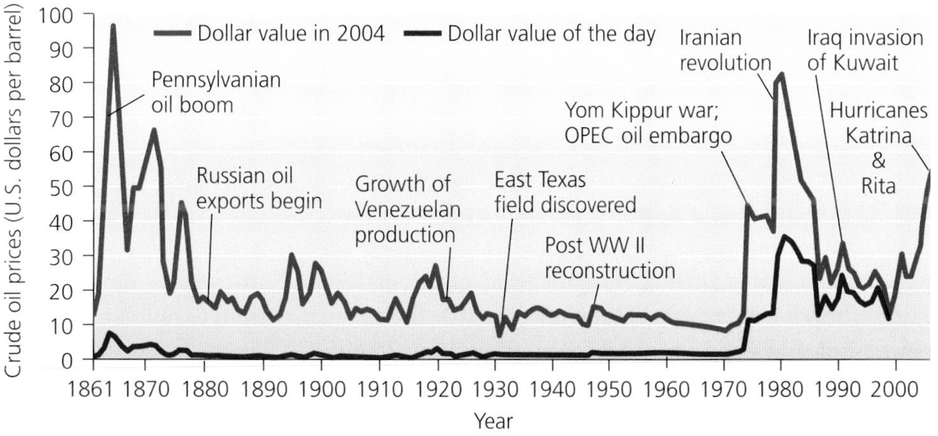

FIGURE 19.16 World oil prices have gyrated greatly over the decades, often because of political and economic events in oil-producing countries. The greatest price hikes in recent times have resulted from wars and unrest in the oil-rich Middle East. Data from U.S. Energy Information Administration.

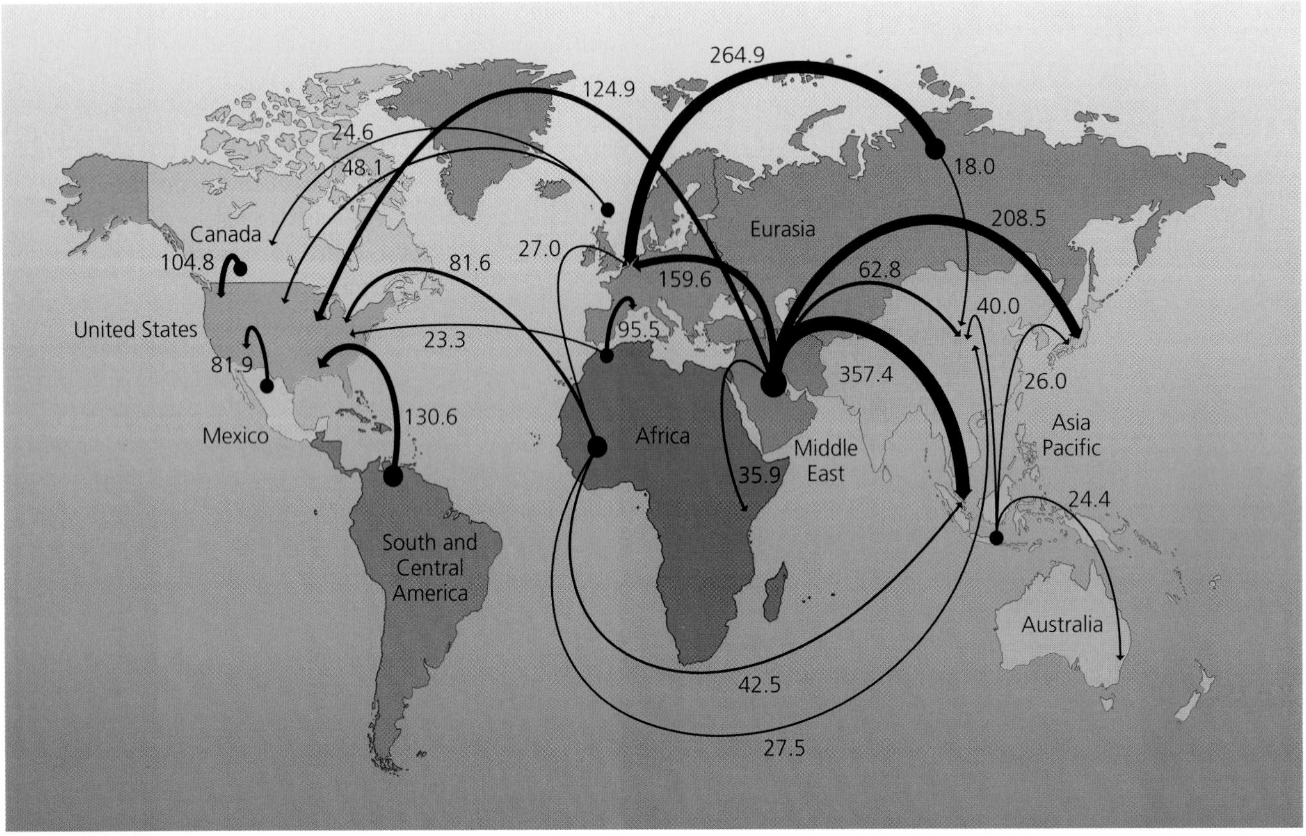

FIGURE 19.17 The global trade in oil is lopsided; relatively few nations account for most exports, and some nations are highly dependent on others for energy. The United States obtains most of its imported oil from Venezuela and Saudi Arabia, followed by Canada, Mexico, Nigeria, and the North Sea. Canada imports some North Sea oil while exporting more to the United States. Numbers in the figure represent millions of metric tons. Data from British Petroleum. 2005. *Statistical review of world energy 2005.*

ANWR to drilling, despite critics' charges that such drilling would likely do little to decrease the nation's dependence. The United States currently imports 65% of its crude oil. With the majority of world oil reserves located in the politically unstable Middle East, crises such as the 1973 embargo are a constant concern for U.S. policymakers. The United States has cultivated a close relationship with Saudi Arabia, the owner of 22% of world oil reserves, despite the fact that that country's political system allows for little of the democracy that U.S. leaders claim to cherish and promote. The world's third-largest holder of oil reserves, at 10%, is Iraq, which is why many people around the world believe the U.S.-led invasion of Iraq in 2003 was motivated primarily to secure access to oil.

To counter dependence on a few major supplier nations, the United States has diversified its sources of petroleum and receives much of it from non–Middle Eastern nations, including Venezuela, Canada, Mexico, and Nigeria. Major trade relations among nations and regions of the world are depicted in Figure 19.17.

As much as U.S. policymakers worry about dependence on foreign fossil fuels, several other nations are far more vulnerable in their energy dependence. Germany, France, South Korea, and Japan stand out as nations that consume far more energy than they produce and thus rely almost entirely on imports for their continued economic well-being. Figure 19.18 contrasts consumption and production for several selected nations.

Residents may or may not benefit from their fossil fuel reserves

The extraction of fossil fuels can be extremely lucrative; many of the world's wealthiest corporations deal in fossil fuel energy or related industries. These industries provide jobs to millions of employees and provide dividends to millions of investors. Development can potentially yield economic benefits for people who live in petroleum-bearing areas, as well. Since the construction of the trans-Alaska pipeline in the 1970s, the state of Alaska has received

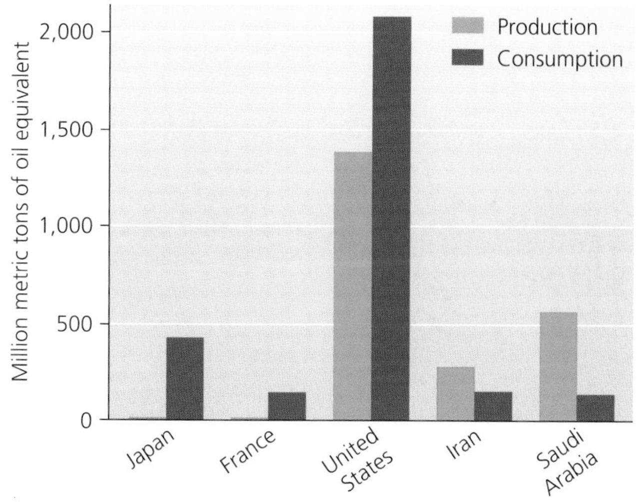

FIGURE 19.18 Japan, France, and the United States are among the nations that consume far more fossil fuel energy than they produce. Iran and Saudi Arabia, like many Middle Eastern nations, produce more fossil fuel energy than they consume and are able to export fuel to high-consumption countries. Data from British Petroleum. 2005. *Statistical review of world energy 2005.*

$60 billion in oil revenues. Alaska's state constitution requires that one-quarter of state oil revenues be placed in the Permanent Fund, which pays yearly dividends to all Alaska residents. Since 1982, each Alaska resident has received annual payouts ranging from $331 to $1,964.

Development of ANWR would add to this fund and create jobs. Some estimates anticipate creation of many thousands of jobs and billions of dollars of income. For this reason, most Alaska residents support oil drilling in ANWR. So do many Inupiat who live on the North Slope, because the income could pay for health care, police and fire protection, and other services that are currently scarce in this remote region.

Alaska's distribution of revenue among its citizenry is unusual, however. In most parts of the world where fossil fuels have been extracted, local residents have not seen great benefits, but instead have frequently suffered. When multinational corporations have extracted oil or gas in developing countries, paying those countries' governments for access, the money often has not trickled down to residents of the regions where the extraction takes place. Moreover, oil-rich developing countries such as Ecuador, Venezuela, and Nigeria tend to have few environmental regulations, and existing regulations may go unenforced if a government does not want to risk losing the large sums of money associated with oil development.

In Nigeria, oil was discovered in 1958 in the territory of the Ogoni, one of Nigeria's native peoples, and the Shell Oil Company moved in to develop oil fields. Although Shell extracted $30 billion of oil from Ogoni land over the years, the Ogoni still live in poverty, with no running water or electricity. The profits from oil extraction on Ogoni land have gone to Shell and to the military dictatorships of Nigeria. The development resulted in oil spills, noise, and constantly burning gas flares, all of which caused illness among people living nearby. From 1962 until his death in 1995, Ogoni activist and leader Ken Saro-Wiwa worked for fair compensation to the Ogoni for oil extraction and environmental degradation on their land. After years of persecution by the Nigerian government, Saro-Wiwa was arrested in 1994, given a trial widely regarded in the international human rights community as a sham, and put to death by military tribunal.

How will we convert to renewable energy?

Given that fossil fuel supplies are limited and that their use has health, environmental, political, and socioeconomic consequences, the nations of the world have several policy options for guiding future energy use. One option is to commit to using fossil fuels until they are no longer economically practical and to develop other energy sources only after supplies have dwindled. A second option is to fund development of alternative energy sources now and attempt to reduce our reliance on fossil fuels gradually. Third, we could try to end our fossil fuel use as soon as possible and hasten a switch to renewable alternatives.

Weighing the Issues: Will Capitalism Drive a Shift to Renewables?

Many experts say that governments need not take steps to hasten the readiness of renewable energy sources to replace fossil fuels as they decline, because forces of supply and demand in market capitalism will automatically take care of this shift. As fossil fuel supplies dwindle and prices rise, the argument goes, industries and consumers will switch to renewable sources as they become more economical. Do you agree with this outlook? Do you see any possible problems with it? Do you think governments should take steps to encourage the development of renewable energy sources? Why or why not?

In the meantime, it will benefit us to prolong the availability of fossil fuels as we make the transition to renewable sources. We can prolong our access to fossil fuels by instituting measures to conserve energy, primarily through lifestyle changes that reduce energy use and technological advances that improve efficiency.

Energy Conservation

Until our society reaches a point at which we are using solely renewable energy sources, we will face the gradual depletion of nonrenewable sources. In the face of dwindling fossil fuel resources, we will need to find ways to minimize the extent to which we expend energy. **Energy conservation** is the practice of reducing energy use to extend the lifetimes of our nonrenewable energy supplies, be less wasteful, and reduce our environmental impact.

Energy conservation has often been a function of economic need

In the United States, many people first saw the value of conserving energy following the OPEC embargo of 1973–1974. Policies enacted by the U.S. government in response to that event included conservation measures such as a mandated increase in the mile-per-gallon (mpg) fuel efficiency of automobiles and a reduction in the national speed limit to 55 miles per hour (at that time, the most efficient speed to drive a car).

Three decades later, many of the conservation initiatives developed after the 1973 oil crisis have been abandoned. Without high market prices and an immediate threat of shortages, people lack economic motivation to conserve. Government funding for research into alternative energy sources has decreased, speed limits have increased, and recent bills to raise the mandated average fuel efficiency of vehicles have failed in Congress. The average fuel efficiency of new vehicles has fallen from a high of 22.1 mpg in 1988 to 21.0 mpg in 2005 (Figure 19.19). This decrease is due to increased sales of light trucks (averaging 18.2 mpg), including sport-utility vehicles, relative to cars (averaging 24.7 mpg). Transportation accounts for two-thirds of U.S. oil use, and passenger vehicles consume over half this energy. Thus, the failure to improve vehicular fuel economy over the past 20 years, despite the existence of technology to do so, has added greatly to U.S. oil consumption.

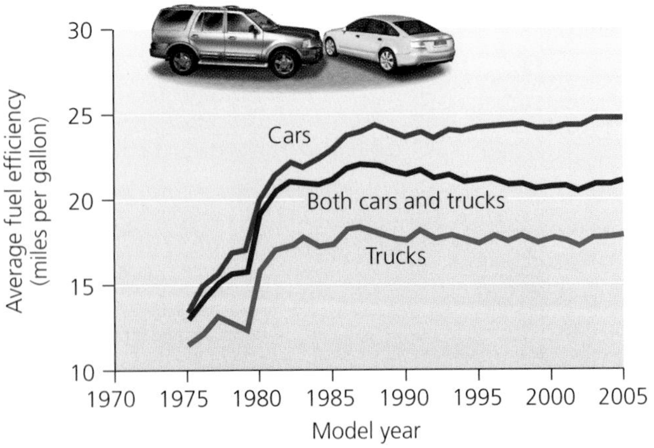

FIGURE 19.19 Fuel efficiency for automobiles in the United States rose dramatically in the late 1970s as a result of legislative mandates, but it has declined slightly since 1988. The decline is due to a lack of further legislation for improved fuel economy and to the increased popularity in recent years of sport-utility vehicles. Data from U.S. Environmental Protection Agency. 2005. *Light-duty automotive technology and fuel economy trends: 1975 through 2005.*

Weighing the Issues: More Miles, Less Gas

If you drive an automobile, what gas mileage does it get? How does it compare to the vehicle averages in Figure 19.19? If your vehicle's fuel efficiency were 10 mpg greater, and you drove the same amount, how many gallons of gasoline would you no longer need to purchase each year? How much money would you save? If all U.S. vehicles were mandated to increase fuel efficiency by 10 mpg, how much gasoline do you think the over 200 million Americans who drive could conserve? What other strategies can you think of to conserve fossil fuels, and how might they compare in effectiveness to a rise in fuel efficiency standards?

Many critics of oil drilling in the Arctic Refuge point out the vast amounts of oil wasted by our fuel-inefficient automobiles. They argue that a small amount of conservation would save the nation far more oil than it would ever obtain from ANWR. Indeed, the USGS's average estimate for recoverable oil in the 1002 Area, 7.7 billion barrels, represents one year's supply for the United States at current consumption rates. Spread over a period of extraction of many years, the proportion of U.S. oil needs that ANWR would fulfill appears strikingly small (Figure 19.20).

Personal choice and increased efficiency are two routes to conservation

Energy conservation can be accomplished in two primary ways. As individuals, we can make conscious choices to reduce our own energy consumption. Examples include driving less, turning off lights when rooms are not being used, turning down thermostats, and cutting back on the use of machines and appliances. For any given individual or business, reducing energy consumption can save money while also helping conserve resources.

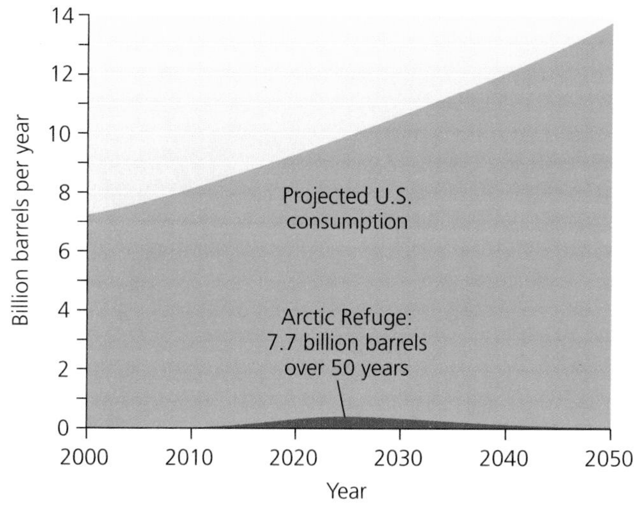

FIGURE 19.20 Opponents of oil drilling in the Arctic National Wildlife Refuge contend that the amount of oil estimated to be recoverable would make only a small contribution toward overall U.S. oil demand. In this graph, the best USGS estimate of oil from ANWR's 1002 Area is shown in red, in the context of total U.S. oil consumption, assuming that current consumption trends are extrapolated into the future and oil production takes place over many years. The actual ANWR contribution, if it comes to pass, would depend greatly on the amount of oil actually present under ANWR, the time it would take to extract it, and future trends in consumption. Adapted from Natural Resources Defense Council. 2002. *Oil and the Arctic National Wildlife Refuge*; and U.S. Geological Survey. 2001. *Arctic National Wildlife Refuge, 1002 Area, petroleum assessment, 1998, including economic analysis.*

We can also conserve energy as a society by making our energy-consuming devices and processes more efficient. In the case of automobiles, we already possess the technology to increase fuel efficiency far above the current average of 21 mpg. We could accomplish such improvement with more efficient gasoline engines, with alternative technology vehicles such as electric/gasoline hybrids (Figure 19.21), or with vehicles that use hydrogen fuel cells (▸ pp. 640, 642).

The efficiency of power plants can be improved through **cogeneration,** in which excess heat produced during the generation of electricity is captured and used to heat workplaces and homes and to produce other kinds of power. Cogeneration can almost double the efficiency of a power plant.

In homes and public buildings, a significant amount of heat is lost in winter and gained in summer because of inadequate insulation (Figure 19.22). Improvements in the design of homes and offices can reduce energy required to heat and cool them. Such design changes can involve the building's location, the color of its roof (light colors keep buildings cooler by reflecting the sun's rays), and its insulation.

Among consumer products, scores of appliances, from refrigerators to lightbulbs, have been reengineered through the years to increase energy efficiency. Even so, there remains room for improvement. Energy-efficient lighting, for example, can reduce energy use by 80%, and new federal standards for energy-efficient appliances have already reduced per-person home electricity use below what it was in the 1970s.

While manufacturers can improve the energy efficiency of appliances, consumers need to "vote with their wallets" by purchasing these energy-efficient appliances. Decisions by consumers to purchase energy-efficient products are crucial in keeping those products commercially available. The U.S. Environmental Protection Agency (EPA) estimates that if all U.S. households purchased energy-efficient appliances, the national annual energy expenditure would be reduced by

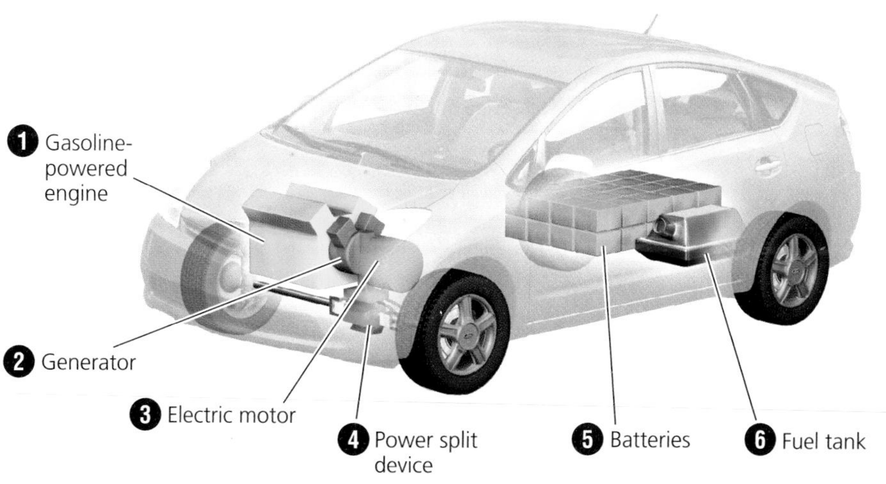

FIGURE 19.21 A hybrid car, such as the Toyota Prius diagrammed here, uses a small, clean, and efficient gasoline-powered engine (1) to produce power that the generator (2) can convert to electricity to drive the electric motor (3). The power split device (4) integrates the engine, generator, and motor, serving as a continuously variable transmission. The car automatically switches between all-electrical power, all-gas power, and a mix of the two, depending on the demands being placed on the engine. Typically, the motor provides power for low-speed city driving and adds extra power on hills. The motor and generator charge a pack of nickel-metal-hydride batteries (5), which can in turn supply power to the motor. Energy for the engine comes from gasoline carried in a typical fuel tank (6).

VIEWPOINTS

Drilling for Oil in ANWR

Should we drill for oil in the Arctic National Wildlife Refuge?

ANWR Oil Means Better Living, Not Eco-Apocalypse

Oil development has taken place in the Arctic for over 30 years. Originally there were fears for the environment, just as we are hearing today. The Inupiat feared harm to the caribou and to their lifestyles, while environmentalists claimed apocalypse, and that there were only 2 years of oil.

Well, "2 years" of oil turned into over 28; 3,000 caribou turned into over 32,000; and the Inupiat have turned into the number-one supporters of development. Technology has allowed a 140-acre drill pad to be reduced to 5 acres in size. Improvements in directional drilling now allow us to extract oil 8 miles from drill-point. The EPA, the North Slope Borough, and Alaska Department of Fish and Game monitor activities daily, so one can hardly claim an eco-apocalypse. Mis-information is rife in the ANWR debate.

Furthermore, the NIMBY argument doesn't work with ANWR; the people who live off the 1002 Area want this badly. To them, drilling in the 1002 Area will provide jobs, schools, clinics—a future. They have seen firsthand how their fears of 30 years ago were unfounded. They have seen their communities prosper.

The energy future of the United States should be multi-sourced. Wind turbines, solar panels, and hydrogen cells *all* need oil to build their working parts. None of these sources can produce plastic, bitumen, tires, paints, medicines, or a thousand other oil-based products we use daily. One can scream against oil, yet every one of us uses it every day in every aspect of our lives. We don't know any alternative. We should invest heavily in finding alternatives, but we should not think for a second that we can cut off our oil supply, domestic or foreign, and live as happily as we do now.

Adrian Herrera is a consultant for the lobby group Arctic Power, lobbying Congress to open the 1002 Area of ANWR to responsible oil development. Formerly, Herrera worked as an engineering assistant at Prudhoe Bay, and for an independent ecological research company tasked with monitoring the wildlife and environment in and around the oil fields. He is an Alaskan resident of over 30 years and has worked and lived in the Arctic extensively.

Drilling the Arctic Refuge Is Not a Solution to Our Energy Problems, It's a Distraction

The Arctic National Wildlife Refuge is one of the last unspoiled wild areas in the United States. Its 1.5-million-acre coastal plain is rich in biodiversity, home to nearly 200 species, including polar bears, musk oxen, caribou, and millions of migratory birds.

There is no way to drill in the refuge without permanently harming this unique ecosystem or destroying the culture of the native Gwich'in people, who have depended on caribou for thousands of years. The little oil beneath the refuge is scattered in more than 30 small deposits. To extract it, roads, pipelines, air strips, and other industrial infrastructure would be built across the entire area.

Drilling the Arctic Refuge would do nothing to lower gas prices or lessen our nation's dependence on imported oil. According to the U.S. Geological Survey, the refuge holds less economically recoverable oil than what Americans consume in a year, and it would take 8–10 years for that oil to reach the market. A recent U.S. Energy Department report found that oil from the Arctic Refuge would have little impact on the price of gasoline, lowering gas prices by less than a penny and a half per gallon—in 2025.

If we boosted the fuel economy performance of our cars and trucks just 1 mile per gallon annually over the next 15 years, we would save more than 10 times the oil that could be recovered from the refuge. We have the technology today to accomplish that goal.

The United States has 3 percent of the world's oil reserves but consumes 25 percent of all oil produced each year. We cannot drill our way to lower gas prices. By focusing on efficiency and alternative fuels, we can improve our energy security and preserve the Arctic Refuge for future generations.

Karen Wayland is the Natural Resources Defense Council's legislative director and an adjunct professor at Georgetown University. Dr. Wayland, who holds a dual Ph.D. in geology and resource development, was a legislative fellow for Sen. Harry Reid (D-Nev.) on nuclear waste, water, energy, and Native American issues before joining NRDC's staff.

Explore this issue further by accessing **Viewpoints** at www.aw-bc.com/withgott.

FIGURE 19.22 Many of our homes and offices could be made more energy-efficient. One way to determine how much heat a building is losing is to take a photograph that records energy in the infrared portion of the electromagnetic spectrum (▸ p. 106). In such a photograph, or *thermogram* (shown here), white, yellow, and red signify hot and warm temperatures at the surface of the house, whereas blue and green shades signify cold and cool temperatures. The white, yellow, and red colors indicate areas where heat is escaping.

$200 billion. On an individual consumer's level, studies show that the slightly higher cost of buying energy-efficient washing machines is rapidly offset by savings in water and electricity bills. On the national level, France, Great Britain, and many other developed countries have standards of living equal to that of the United States, but they use much less energy per capita. This disparity indicates that U.S. citizens could significantly reduce their energy consumption without decreasing their quality of life.

Some conservation measures were adopted in the energy bill endorsed by the Bush administration and passed by the U.S. Congress in 2005. Tax credits were offered to consumers who buy hybrid cars or improve energy efficiency in their homes, for example. However, these measures were a relatively minor component, and most of the bill's funding went toward subsidies for nuclear power, cleaner coal technology, ethanol, and wind power, as well as measures to increase production of oil and natural gas. The legislation did not address automotive fuel economy, and as the president signed the bill into law, gas prices rose to their highest inflation-adjusted levels in more than 20 years. Most analysts concluded that overall, the legislation did not notably change the direction of energy policy in the United States.

Both conservation and renewable energy are needed

It is often said that reducing our energy use is equivalent to finding a new oil reserve. Some estimates hold that effective energy conservation in the United States could save 6 million barrels of oil a day. Such a savings would likely far more than offset the energy produced by any oil under the Arctic National Wildlife Refuge while also reducing the negative impacts of fossil fuel extraction and use. Indeed, conserving energy is better than finding a new reserve, because it lessens impacts on the environment while extending our access to fossil fuels.

However, energy conservation does not add to our supply of available fuel. Thus, in the long term, conservation cannot be the sole solution to our energy dilemmas. Regardless of how much we conserve, we will still need energy, and it will need to come from somewhere. The only sustainable way of guaranteeing ourselves a reliable long-term supply of energy is to ensure sufficiently rapid development of renewable energy sources.

Conclusion

Over the past 200 years, fossil fuels have helped us build complex industrialized societies capable of exploring (and exploiting) all parts of the world and even venturing beyond our planet. Today, however, we are approaching a turning point in history: Our production of fossil fuels will begin to decline just as we become increasingly aware of the negative impacts of their use.

We can respond to this new challenge in creative ways, encouraging conservation and developing alternative energy sources. Or we can continue our current dependence on fossil fuels and wait until they near depletion before we try to develop new technologies and ways of life. The path we choose will have far-reaching consequences for human health and well-being, for Earth's climate, and for our environment.

The ongoing debate over the future of the Arctic National Wildlife Refuge is a microcosm of this debate over our energy future. Fortunately, there is not simply a trade-off between benefits of energy for us and harm to the environment, climate, and health. Instead, as evidence builds that renewable energy sources are becoming increasingly feasible and economical, it becomes easier to envision giving up our reliance on fossil fuels and charting a win-win future for humanity and the environment.

REVIEWING OBJECTIVES

You should now be able to:

Survey the energy sources that we use

- A variety of renewable and nonrenewable energy sources are available to us. (pp. 559–561)
- Since the industrial revolution, nonrenewable fossil fuels—including oil, natural gas, and coal—have become our primary sources of energy. (p. 561)
- Fossil fuels are formed very slowly as buried organic matter is chemically transformed by heat, pressure, and/or anaerobic decomposition. (pp. 561–562)

Describe the nature and origin of coal, and evaluate its extraction and use

- Coal is our most abundant fossil fuel. It results from organic matter that undergoes compression but little decomposition. (pp. 564–565)
- The first fossil fuel to be widely used for heating homes and powering industry, coal is used today principally to generate electricity. (pp. 564, 566–567)
- Coal comes in different types and varies in its composition. Combustion of coal that is high in contaminants emits toxic air pollution. (pp. 564–565)
- Coal is mined underground and strip-mined from the land surface. (p. 565)

Describe the nature and origin of petroleum and evaluate its extraction, use, and future depletion

- Crude oil is a thick liquid mixture of hydrocarbons that is formed underground under certain temperature and pressure conditions. (pp. 566–567)
- Scientists locate fossil fuel deposits by analyzing subterranean geology. Geologists estimate total reserves, as well as the technically and economically recoverable portions of those reserves. (pp. 568–569)
- Oil drilling often involves primary extraction followed by secondary extraction, in which gas or liquid is injected into the ground to help force up additional oil. (pp. 569–570)
- Petroleum-based products, from gasoline to clothing to plastics, are everywhere in our daily lives. (p. 570)
- We have nearly depleted half the world's oil. The remainder may last only 40 more years. Once production slows, the gap between rising demand and falling supply may pose immense economic and social challenges for our society. (pp. 570–571, 574)
- Components of crude oil are separated in refineries to produce a wide variety of fuel types. (pp. 572–573)

Describe the nature and origin of natural gas, and evaluate its extraction and use

- Natural gas consists mostly of methane and can be formed in two ways. (p. 574)
- Use of natural gas is growing rapidly. (pp. 574–575)
- Natural gas often occurs with oil deposits, is extracted in similar ways, and becomes depleted in similar ways. (pp. 574–575)

Outline and assess environmental impacts of fossil fuel use

- Airborne emissions of carbon dioxide from fossil fuel combustion drive global climate change. (p. 576)
- Emissions of fuel contaminants and volatile components of oil contribute to air pollution and pose human health risks. (p. 576)
- Oil is a major contributor to water pollution. (p. 576)
- Strip mining and mountaintop removal can devastate ecosystems locally or regionally, and acid drainage from coal mines pollutes waterways. (pp. 576–577)
- Development for oil and gas extraction exerts environmental impacts, although their severity is debated. (pp. 577–578)

Evaluate political, social, and economic impacts of fossil fuel use

- Today's societies are so reliant on fossil fuel energy that sudden restrictions in oil supplies can have major economic consequences. (pp. 579–580)
- Nations that consume far more fossil fuels than they produce are especially vulnerable to supply restrictions. (pp. 579–581)
- People living in areas of fossil fuel extraction do not always benefit from their extraction. (pp. 580–581)

Specify strategies for conserving energy

- Increases in automotive fuel efficiency could help us conserve immense amounts of oil. (p. 582)
- Energy conservation involves both personal choices and efficient technologies. These two forces interact through the market power of consumer choice. (pp. 582–585)
- Conservation helps lengthen our access to fossil fuels and reduce environmental impact, but to build a sustainable society we will also need to shift to renewable energy sources. (p. 585)

TESTING YOUR COMPREHENSION

1. Why are fossil fuels our most prevalent source of energy today? Why are they considered nonrenewable sources of energy?
2. How do developed and developing countries differ in their overall rates of energy consumption and in the ways they use energy?
3. How are fossil fuels formed? How do environmental conditions determine what type of fossil fuel is formed in a given location? Why are fossil fuels often concentrated in localized deposits?
4. How do scientists classify coal? What determines the potential energy contained in coal?
5. Describe how coal is used to generate electricity.
6. How have geologists estimated the total amount of oil beneath the Arctic National Wildlife Refuge (ANWR) 1002 Area? How is this amount different from the "technically recoverable" and "economically recoverable" amounts of oil?
7. How do we create petroleum products? Provide examples of several of these products.
8. Why is natural gas often extracted simultaneously with other fossil fuels? What constraints on its extraction does it share with oil?
9. What are some of the effects of fossil fuel emissions? Compare some of the contrasting views of scientists regarding the environmental impacts of drilling for oil in ANWR.
10. Describe two main approaches to energy conservation, and give a specific example of each.

SEEKING SOLUTIONS

1. Roughly how much oil is left in the world, and how much longer can we expect to use it? How effective were the conservation methods adopted by the United States in response to the "energy crisis" of 1973–1974? Do you think our use of oil and other fossil fuels will give rise to a future crisis? Why or why not? If so, what should we do to avoid such a situation in the future?
2. Compare the effects of coal and oil consumption on the environment. Which process do you think has ultimately been more detrimental to the environment, oil extraction or coal mining, and why? What steps could governments, industries, and individuals take to reduce environmental impacts?
3. Imagine we were living in the 1950s and the United States were facing an oil shortage. Do you think the nation would be debating whether to drill in ANWR? Why do you think so many people today are concerned about wildlife and about wilderness preservation?
4. If ANWR were located in a developing country, do you think the citizens and policymakers of that country would choose to drill there?
5. If the United States and other developed countries relinquished dependence on foreign oil and on fossil fuels in general, do you think that their economies would benefit or suffer? Might your answer be different for the short term and the long term? What factors come into play in trying to make such a judgment?
6. Contrast the experiences of the Ogoni people of Nigeria with those of the citizens of Alaska. How have they been similar and different? Do you think businesses or governments should take steps to ensure that local people benefit from oil drilling operations? How could they do so?

INTERPRETING GRAPHS AND DATA

The fossil fuels that we burn today were formed long ago from buried organic matter. However, only a small fraction of the original organic carbon remains in the coal, oil, or natural gas that is formed. Thus, it requires approximately 90 metric tons of ancient organic matter—so-called paleoproduction—to result in just 3.8 L (1 gal) of gasoline. The graph presents estimates of the amount of paleoproduction required to produce the fossil fuels humans have used each year over the past 250 years.

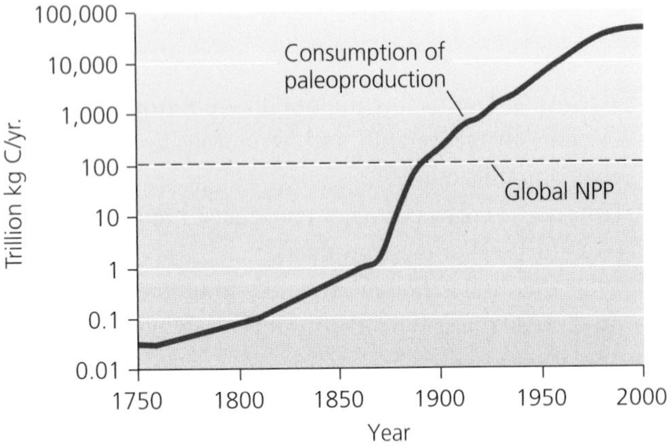

Annual human consumption of paleoproduction by fossil fuel combustion (red line), 1750–2000. The dashed line indicates current annual net primary production (NPP; ▸p. 192) for the entire planet. Data from Dukes, J. 2003. Burning buried sunshine: Human consumption of ancient solar energy. *Climatic Change* 61: 31–44.

1. Estimate in what year the consumption of paleoproduction represented by our combustion of fossil fuels surpassed Earth's current annual net primary production.
2. In 2000, approximately how many times greater than global net primary production was our consumption of paleoproduction?
3. If on average it takes 7,000 units of paleoproduction to produce 1 unit of fossil fuel, estimate the total carbon content of the fossil fuel consumed in 2000. How does this amount compare to global NPP?

CALCULATING ECOLOGICAL FOOTPRINTS

Wackernagel and Rees calculated the energy component of our ecological footprint by estimating the amount of ecologically productive land required to absorb the carbon released from fossil fuel combustion (see Chapter 18, "Calculating Ecological Footprints," ▸p. 555). For the average American, this translates into 2.9 of the 5.1 ha of our ecological footprint. Another way to think about our footprint, however, is to estimate how much land would be needed to grow biomass with an energy content equal to that of the fossil fuel we burn. Assume that you are an average American who burns 287 gigajoules of fossil fuels per year and that average terrestrial net primary productivity can be expressed as 160 megajoules/ha/year. Calculate how many hectares of land it would take to supply our fuel use by present-day photosynthetic production. A gigajoule is 10^9 joules; a megajoule is 10^6 joules.

	Hectares of land for fuel production
You	1,794
Your class	
Your state	
United States	

Data from Wackernagel, M., and W. Rees. 1996. *Our ecological footprint: Reducing human impact on the Earth.* Gabriola Island, British Columbia: New Society Publishers.

1. Compare the energy component of your ecological footprint calculated in this way with the 2.9 ha calculated using the method of Wackernagel and Rees. Explain why results from the two methods differ.
2. Earth's total land area is approximately 1.5×10^{10} ha. Compare this to the hectares of land for fuel production from the table.
3. How large a human population could Earth support at the level of consumption of the average American, if all of Earth's land were devoted to fuel production? Do you consider this realistic? Provide two reasons why or why not.

Take It Further

Go to www.aw-bc.com/withgott or the student CD-ROM where you'll find:

- Suggested answers to end-of-chapter questions
- Quizzes, animations, and flashcards to help you study
- *Research Navigator*™ database of credible and reliable sources to assist you with your research projects
- **GRAPHit!** tutorials to help you master how to interpret graphs
- **INVESTIGATEit!** current news articles that link the topics that you study to case studies from your region to around the world

Conventional Energy Alternatives

20

Cooling towers of nuclear power plant, Shropshire, U.K.

Upon successfully completing this chapter, you will be able to:

- Discuss the reasons for seeking alternatives to fossil fuels
- Summarize the contributions to world energy supplies of conventional alternatives to fossil fuels
- Describe nuclear energy and how it is harnessed
- Outline the societal debate over nuclear power
- Describe the major sources, scale, and impacts of biomass energy
- Describe the scale, methods, and impacts of hydroelectric power

Biomass power plant, Skellefteå, Sweden

Central Case: Sweden's Search for Alternative Energy

"Nowhere has the public debate over nuclear power plants been more severely contested than Sweden."
—Writer and analyst Michael Valenti

"If [Sweden] phases out nuclear power, then it will be virtually impossible for the country to keep its climate-change commitments."
—Yale University economist William Nordhaus

On the morning of April 28, 1986, workers at a nuclear power plant in Sweden detected suspiciously high radiation levels. Their concern turned to confusion when they determined that the radioactivity was coming not from their own plant, but through the atmosphere from the direction of the Soviet Union.

They had, in fact, discovered evidence of the disaster at Chernobyl, more than 1,200 km (750 mi) away in what is now the nation of Ukraine. Chernobyl's nuclear reactor had exploded two days earlier, but the Soviet government had not yet admitted it to the world.

As low levels of radioactive fallout rained down on the Swedish countryside in the days ahead, contaminating crops and cows' milk, many Swedes felt more certain than ever about the decision they had made collectively 6 years earlier. In a 1980 referendum, Sweden's electorate had voted to phase out their country's nuclear power program, shutting down all nuclear plants by the year 2010.

But trying to phase out nuclear power has proven difficult. Nuclear power today provides Sweden with one-third of its overall energy supply and nearly half its electricity. If nuclear plants are shut down, something will have to take their place. Aware of the environmental impacts of fossil fuels, Sweden's government and citizens do not favor expanding fossil fuel use. In fact, Sweden is one of the few nations that have managed to decrease use of fossil fuels since the 1970s—and it has done so largely by replacing them with nuclear power.

To fill the gap that would be left by a nuclear phaseout, Sweden's government has promoted research and development of renewable energy sources.

Hydroelectric power from running water was already supplying most of the other half of the nation's electricity, but it could not be expanded much more. The government hoped that energy from biomass sources and wind power could fill the gap.

Sweden has made itself an international leader in renewable energy alternatives, but because renewables have taken longer to develop than hoped, the government has repeatedly postponed the nuclear phaseout. Only one of the 12 reactors operating in 1980 has been shut down so far, and efforts to close a second one have generated sustained controversy.

Proponents of nuclear power say it would be fiscally and socially irresponsible to dismantle the nation's nuclear program without a ready replacement. And environmental advocates worry that if nuclear power is simply replaced by fossil fuel combustion, or if converting to biomass energy means cutting down more forests, the nuclear phaseout would be bad news for the environment. Moreover, Sweden has made international obligations to hold down its carbon emissions under the Kyoto Protocol (▸ pp. 550, 552), so its incentive to keep nuclear power is strong. Nuclear energy is free of atmospheric pollution and seems, to many, the most effective way to minimize carbon emissions in the short term.

In 2003, a poll showed 55% of the Swedish public in favor of maintaining or increasing nuclear power, and 41% in favor of abandoning it. But despite the mixed feelings over nuclear power, Swedes have little desire to return to an energy economy dominated by fossil fuels. A concurrent poll showed that 80% of Swedes supported boosting research on renewable energy sources—a higher percentage than in any other European country.

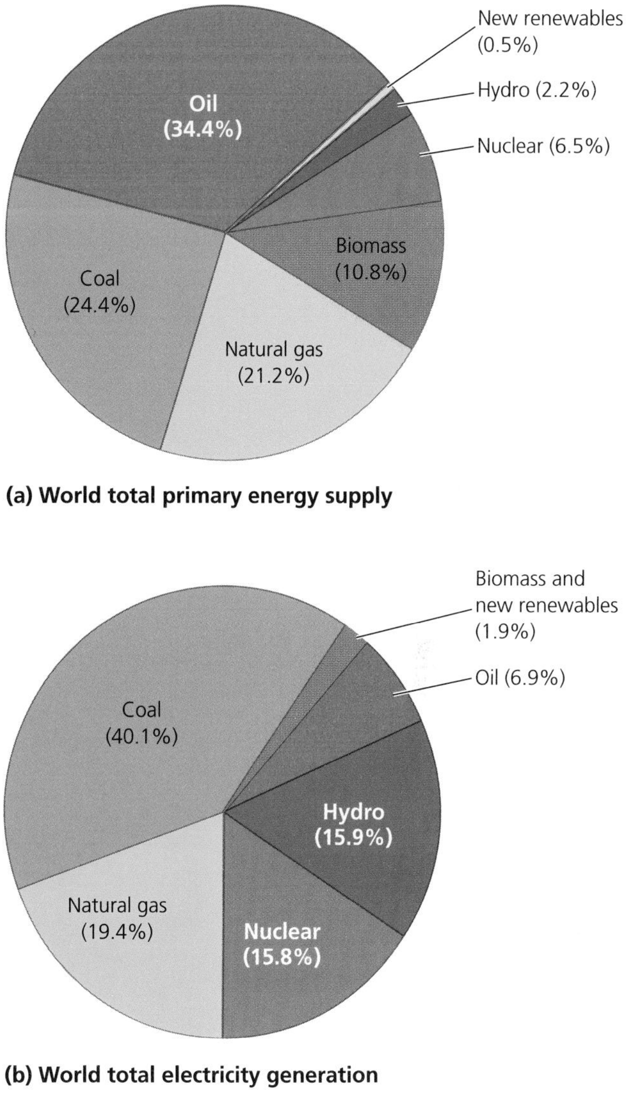

FIGURE 20.1 Fossil fuels account for 80% of the world's total supply of primary energy **(a)**. Nuclear and hydroelectric power contribute substantially to global electricity generation, although fossil fuels still power two-thirds of our electricity **(b)**. Data from International Energy Agency. 2005. *Key world energy statistics 2005*. Paris: IEA.

Alternatives to Fossil Fuels

Fossil fuels helped to drive the industrial revolution and increase our material prosperity. Today's economies are largely powered by fossil fuels; 80% of all primary energy comes from oil, coal, and natural gas (Figure 20.1a), and these three fuels also power two-thirds of the world's electricity generation (Figure 20.1b). However, these nonrenewable energy sources will not last forever. As we saw in Chapter 19, oil production is thought to be peaking, and easily extractable supplies of oil and natural gas may not last half a century more. Moreover, the use of coal, oil, and natural gas entails substantial environmental impacts, as described in Chapters 17, 18, and 19.

For these reasons, most scientists and energy experts, as well as many economists and policymakers, accept that the world's economies will need to shift from fossil fuels to energy sources that are less easily depleted and gentler on our environment. Developing alternatives to fossil fuels has the added benefit of helping to diversify an economy's mix of energy, thus lessening price volatility and dependence on foreign fuel imports.

People have developed a range of alternatives to fossil fuels. Most of these energy sources are renewable and cannot be depleted by use. Most have less impact on the environment than oil, coal, or natural gas. However, at this time most remain more costly than fossil fuels (at least in the short term), and many depend on technologies that are not yet fully developed.

Nuclear power, biomass energy, and hydropower are conventional alternatives

Three alternative energy sources are currently the most developed and most widely used: nuclear energy, hydroelectric power, and energy from biomass. Each of these well-established energy sources plays substantial roles in the energy and electricity budgets of nations today. We can therefore call nuclear energy, hydropower, and biomass energy "conventional" alternatives to fossil fuels.

In many respects, this trio of conventional energy alternatives makes for an eclectic collection. They are generally considered to exert less environmental impact than fossil fuels, but more environmental impact than the "new renewable" alternatives we will discuss in Chapter 21. Yet, as we will see, they each involve a unique and complex mix of benefits and drawbacks for the environment. Although nuclear energy is commonly termed a nonrenewable energy source and hydropower and biomass are generally described as renewable, the reality is more complicated. They are perhaps best viewed as intermediates along a continuum of renewability.

Conventional alternatives provide some of our energy and much of our electricity

Fuelwood and other biomass sources provide 10.8% of the world's primary energy, nuclear energy provides 6.5%, and hydropower provides 2.2%. The less established renewable energy sources account for only 0.5% (see Figure 20.1a). Although their global contributions to overall energy supply are still minor, alternatives to fossil fuels do contribute greatly to our generation of electricity. Nuclear energy and hydropower each account for nearly one-sixth of the world's electricity generation (see Figure 20.1b).

Energy consumption patterns in the United States (Figure 20.2a) are similar to those globally, except that the United States relies less on fuelwood and more on fossil fuels and nuclear power than most other countries. A graph showing changes in the consumption of fossil fuels and conventional alternatives in the United States over the past half century (Figure 20.2b) reveals two things. First, conventional alternatives play minor yet substantial roles in overall energy use. Second, use of conventional alternatives has been growing more slowly than consumption of fossil fuels.

Sweden and some other nations, however, have shown that it is possible to replace fossil fuels gradually with alternative sources. Since 1971, Sweden has decreased its fossil fuel use by 40%, and today nuclear power, biomass, and hydropower together provide Sweden with 60% of its total energy and virtually all of its electricity.

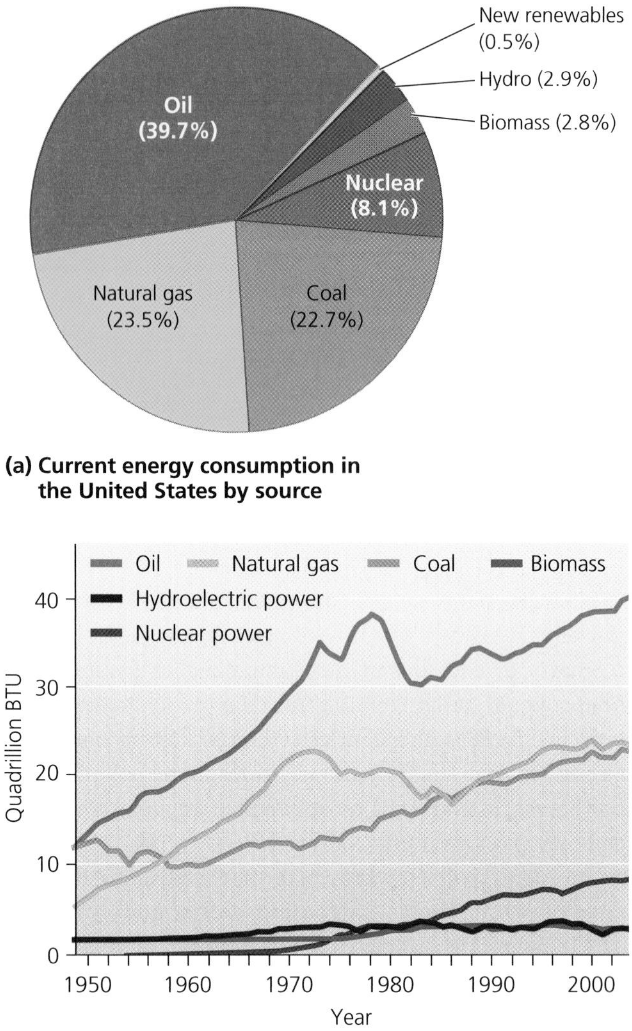

(a) Current energy consumption in the United States by source

(b) Changes in energy consumption in the United States over the past half-century

FIGURE 20.2 In the United States, fossil fuels account for over 85% of energy consumption, nuclear power for 8.1%, biomass for 2.8%, and hydropower for 2.9% **(a).** U.S. usage of the three fossil fuels has grown faster than biomass or hydropower over the past half century **(b).** Nuclear power grew considerably between 1970 and 2000. Data from Energy Information Administration. 2005. *Annual Energy Review 2004.* U.S. Department of Energy.

Nuclear Power

Nuclear power occupies an odd and conflicted position in our modern debate over energy. It is free of the air pollution produced by fossil fuel combustion, and so has long been put forth as an environmentally friendly alternative to fossil fuels. At the same time, nuclear power's great promise has been clouded by nuclear weaponry, the dilemma of radioactive waste disposal, and the long shadow of Chernobyl and other power plant accidents. As such, public safety concerns and the costs of addressing them have constrained the development and spread of nuclear power in the United States, Sweden, and many other nations.

First developed commercially in the 1950s, nuclear power has expanded 15-fold worldwide since 1970, experiencing most of its growth during the 1970s and 1980s. Of all nations, the United States generates the most electricity from nuclear power—nearly a third of the world's production— and is followed by France and Japan. However, only 20% of U.S. electricity comes from nuclear sources. A number of other nations rely more heavily on nuclear power (Table 20.1). For instance, France and Lithuania each receive roughly four-fifths of their electricity from nuclear sources.

Fission releases nuclear energy

Strictly defined, **nuclear energy** is the energy that holds together protons and neutrons within the nucleus of an atom. We harness this energy by converting it to thermal energy, which can then be used to generate electricity. Several processes can convert the energy within an atom's nucleus into thermal energy, releasing it and making it available for use. Each process involves transforming isotopes (▸ pp. 92–95) of one element into isotopes of other elements, by the addition or loss of neutrons.

Table 20.1 Top Consumers of Nuclear Power

Nation	Nuclear power consumed[1]	Number of plants[2]	Percentage of electricity generation from nuclear power plants[3]
United States	187.9	104	19
France	101.4	59	78
Japan	64.8	53	23
Germany	37.8	18	28
Russia	32.4	30	16
South Korea	29.6	19	37
Canada	20.5	16	13
Ukraine	19.7	13	45
United Kingdom	18.1	27	22
Sweden	17.3	11	50

Data from International Atomic Energy Agency, British Petroleum, and International Energy Agency.
[1]In million metric tons of oil equivalent, 2004 data.
[2]2003 data.
[3]2003 data.

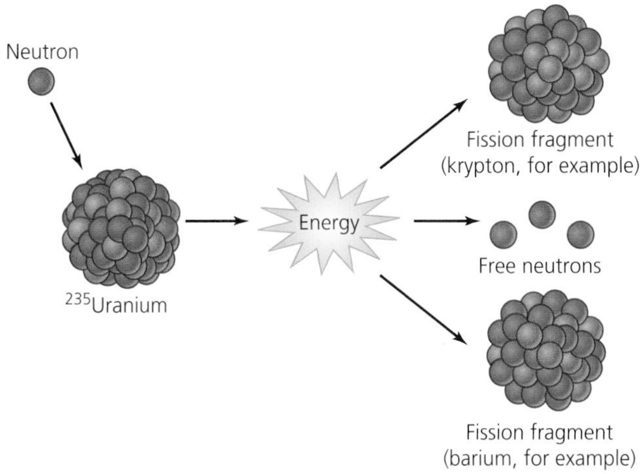

FIGURE 20.3 In nuclear fission, atoms of uranium-235 are bombarded with neutrons. Each collision splits uranium atoms into smaller atoms and releases two or three neutrons, along with energy and radiation. Because the neutrons can continue to split uranium atoms and set in motion a runaway chain reaction, engineers at nuclear plants must absorb excess neutrons with control rods to regulate the rate of the reaction.

The reaction that drives the release of nuclear energy in power plants is **nuclear fission,** the splitting apart of atomic nuclei (Figure 20.3). In fission, the nuclei of large, heavy atoms, such as uranium or plutonium, are bombarded with neutrons. Ordinarily neutrons move too quickly to split nuclei when they collide with them, but if neutrons are slowed down they can break apart nuclei. Each split nucleus emits multiple neutrons, together with substantial heat and radiation. These neutrons (two to three in the case of fissile isotopes of uranium-235) can in turn bombard other uranium-235 (^{235}U) atoms in the vicinity, resulting in a self-sustaining chain reaction.

If not controlled, this chain reaction becomes a runaway process of positive feedback that releases enormous amounts of energy—the same process that creates the explosive power of a nuclear bomb. Inside a nuclear power plant, however, fission is controlled so that only one of the two or three neutrons emitted with each fission event goes on to induce another fission event. In this way, the chain reaction maintains a constant output of energy at a controlled rate, rather than an explosively increasing output.

Nuclear energy comes from processed and enriched uranium

We generate electricity from nuclear power by controlling fission in **nuclear reactors,** facilities contained within nuclear power plants. But this is just one step in a longer process sometimes called the *nuclear fuel cycle.* This

FIGURE 20.4 Enriched uranium fuel is packaged into fuel rods, which are encased in metal and used to power fission inside the cores of nuclear reactors. In this photo, the fuel rods are visible arrayed in a circle within the blue water, at bottom.

process begins when the naturally occurring element uranium is mined from underground deposits, as we saw with the mines on Australian Aboriginal land in Chapter 2.

Uranium is an uncommon mineral, and uranium ore is in finite supply, which is why nuclear power is generally considered a nonrenewable energy source. Uranium is used for nuclear power because it is radioactive. Radioactive isotopes, or **radioisotopes,** emit subatomic particles and high-energy radiation as they decay into lighter radioisotopes, until they become stable isotopes. The isotope uranium-235 decays into a series of daughter isotopes, eventually forming lead-207. Each radioisotope decays at a rate determined by that isotope's **half-life,** the amount of time it takes for one-half of the atoms to give off radiation and decay. Different radioisotopes have very different half-lives, ranging from fractions of a second to billions of years. The half-life of ^{235}U is about 700 million years.

Over 99% of the uranium in nature occurs as the isotope uranium-238. Uranium-235 (with three fewer neutrons) makes up less than 1% of the total. Because ^{238}U does not emit enough neutrons to maintain a chain reaction when fissioned, we use ^{235}U for commercial nuclear power. So, mined uranium ore must be processed to enrich the concentration of ^{235}U to at least 3%. The enriched uranium is formed into small pellets of uranium dioxide (UO_2), which are incorporated into large metallic tubes called *fuel rods* (Figure 20.4) that are used in nuclear reactors. After several years in a reactor, enough uranium has decayed so that the fuel loses its ability to generate adequate energy, and it must be replaced with new fuel. In some countries, the spent fuel is reprocessed to recover what usable energy may be left. Most spent fuel, however, is disposed of as radioactive waste.

Fission in reactors generates electricity in nuclear power plants

For fission to begin in a nuclear reactor, the neutrons bombarding uranium are slowed down with a substance called a *moderator,* most often water or graphite. With a moderator allowing fission to proceed, it then becomes necessary to soak up the excess neutrons produced when uranium nuclei divide, so that on average only a single uranium atom from each nucleus goes on to split another nucleus. For this purpose, *control rods,* made of a metallic alloy that absorbs neutrons, are placed into the reactor among the water-bathed fuel rods. Engineers move these control rods into and out of the water to maintain the fission reaction at the desired rate.

All this takes place within the reactor core and is the first step in the electricity-generating process of a nuclear power plant (Figure 20.5). The reactor core is housed within a reactor vessel, and the vessel, steam generator, and associated plumbing are protected within a *containment building.* Containment buildings, with their meter-thick concrete and steel walls, are constructed to prevent leaks of radioactivity due to accidents or natural catastrophes such as earthquakes.

Breeder reactors make better use of fuel, but have raised safety concerns

Using ^{235}U as fuel for conventional fission is only one potential route to harnessing nuclear energy. A similar process, **breeder nuclear fission,** makes use of ^{238}U, which in conventional fission goes unused as a waste product. In breeder fission, the addition of a neutron to ^{238}U forms plutonium (^{239}Pu). When plutonium is bombarded by a neutron, it splits into fission products and releases more neutrons, which convert more of the remaining ^{238}U fuel into ^{239}Pu, continuing the process. Because 99% of all uranium is ^{238}U, breeder fission makes much better use of fuel, generates far more power, and produces far less waste.

However, breeder fission is considerably more dangerous than conventional nuclear fission, because highly reactive liquid sodium, rather than water, is used as a

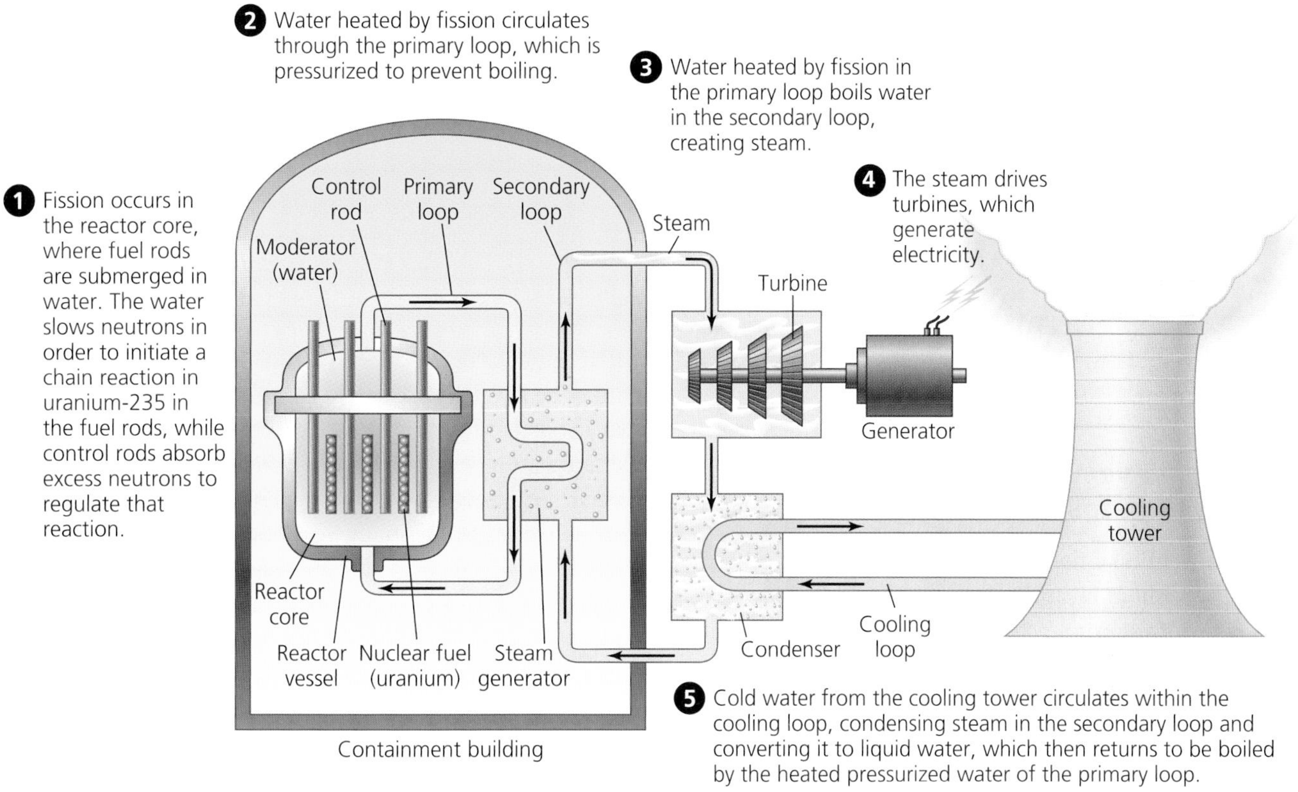

2 Water heated by fission circulates through the primary loop, which is pressurized to prevent boiling.
3 Water heated by fission in the primary loop boils water in the secondary loop, creating steam.
1 Fission occurs in the reactor core, where fuel rods are submerged in water. The water slows neutrons in order to initiate a chain reaction in uranium-235 in the fuel rods, while control rods absorb excess neutrons to regulate that reaction.
Control rod
Primary loop
Secondary loop
Moderator (water)
Steam
4 The steam drives turbines, which generate electricity.
Turbine
Generator
Cooling tower
Reactor core
Reactor vessel
Nuclear fuel (uranium)
Steam generator
Condenser
Cooling loop
Containment building
5 Cold water from the cooling tower circulates within the cooling loop, condensing steam in the secondary loop and converting it to liquid water, which then returns to be boiled by the heated pressurized water of the primary loop.

FIGURE 20.5 In a pressurized light water reactor, the most common type of nuclear reactor, uranium fuel rods are placed in water, which slows neutrons so that fission can occur (1). Control rods that can be moved into and out of the reactor core absorb excess neutrons to regulate the chain reaction. Water heated by fission circulates through the primary loop (2) and warms water in the secondary loop, which turns to steam (3). Steam drives turbines, which generate electricity (4). The steam is then cooled by water from the cooling tower and returns to the containment building (5), to be heated again by heat from the primary loop.

coolant, raising the risk of explosive accidents. Breeder reactors also are more expensive than conventional reactors. Finally, breeder fission can be used to supply plutonium to nuclear weapons programs. As a result, all but a handful of the world's breeder reactors have now been shut down.

Fusion remains a dream

For as long as scientists and engineers have been generating power from nuclear fission, they have been trying to figure out how they might use nuclear fusion instead. **Nuclear fusion**—the process responsible for the immense amount of energy that our sun generates and the force behind hydrogen or thermonuclear bombs—involves forcing together the small nuclei of lightweight elements under extremely high temperature and pressure. The hydrogen isotopes deuterium and tritium can be fused together to create helium, releasing a neutron and a tremendous amount of energy (Figure 20.6).

Overcoming the mutually repulsive forces of protons in a controlled manner is difficult, and fusion requires temperatures of many millions of degrees Celsius. Thus, researchers have not yet been able to develop this process for commercial power generation. Despite billions of dollars of funding and decades of research, fusion experiments in the lab still require scientists to input more energy than they produce from the process. Fusion's potentially huge payoffs,

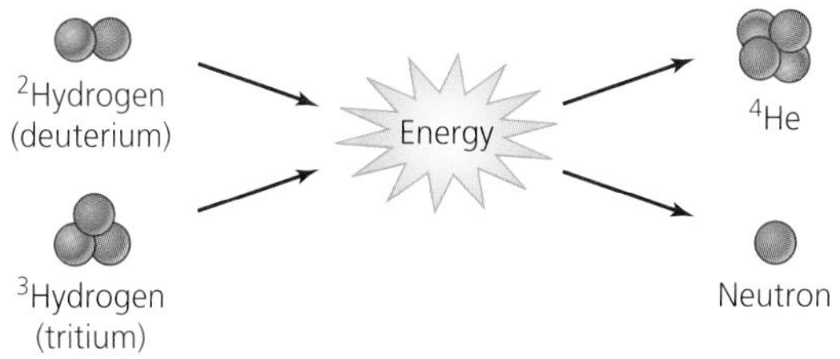

²Hydrogen (deuterium)
³Hydrogen (tritium)
Energy
⁴He
Neutron

FIGURE 20.6 In nuclear fusion, two small atoms, such as the hydrogen isotopes deuterium and tritium, are fused together, causing the loss of a neutron and the release of energy. So far, however, scientists have not been able to fuse atoms without supplying far more energy than the reaction produces, so this process is not used commercially.

though, make many scientists eager to keep trying. If one day we were to find a way to control fusion in a reactor, we could produce vast amounts of energy using water as a fuel. The process would create only low-level radioactive wastes, without pollutant emissions or the risk of dangerous accidents, sabotage, or weapons proliferation. A consortium of nations including Japan, several European countries, and the United States is currently seeking to build a prototype fusion reactor called the International Thermonuclear Experimental Reactor (ITER). Even if this multibillion-dollar effort succeeds, however, power from fusion seems likely to remain many years in the future.

Nuclear power delivers energy more cleanly than fossil fuels

Using conventional fission, nuclear power plants generate electricity without creating air pollution from stack emissions. In contrast, combusting coal, oil, or natural gas emits sulfur dioxide that contributes to acidic deposition, particulate matter that threatens human health, and carbon dioxide and other greenhouse gases that contribute to global climate change. Even considering all the steps involved in building plants and generating power, researchers from the International Atomic Energy Agency (IAEA) have calculated that nuclear power lowers emissions 4–150 times below fossil fuel combustion (see "The Science behind the Story," ▸ pp. 598–599). IAEA scientists estimate that at current global rates of usage, nuclear power helps us avoid emitting 600 million metric tons of carbon each year, equivalent to 8% of global greenhouse gas emissions.

Nuclear power has additional environmental advantages over fossil fuels, coal in particular. Because uranium generates far more power than coal by weight or volume, less of it needs to be mined, so uranium mining causes less damage to landscapes and generates less solid waste than coal mining. Additionally, in the course of normal operation, nuclear power plants are safer for workers than coal-fired plants.

However, nuclear power also has drawbacks. One is that the waste it produces is radioactive. Radioactive waste must be handled with great care and must be disposed of in a way that minimizes danger to present and future generations. The second main drawback is that if an accident occurs at a power plant, or if a plant is sabotaged, the consequences can potentially be catastrophic.

Given this mix of advantages and disadvantages (Table 20.2), most governments (although not necessarily most citizens) judged the good to outweigh the bad, and today the world has 439 operating nuclear plants.

Weighing the Issues: Choose your Risk

Given the choice of living next to a nuclear power plant or living next to a coal-fired power plant, which would you choose? What would concern you most about each option?

Table 20.2 Environmental Impacts of Coal-Fired and Nuclear Power

Type of Impact	Coal	Nuclear
Land and ecosystem disturbance from mining	*Extensive, on surface or underground*	Less extensive
Greenhouse gas emissions	*Considerable emissions*	None from plant operation; much less than coal over the entire life cycle
Other air pollutants	*Sulfur dioxide, nitrogen oxides, particulate matter, and other pollutants*	No pollutant emissions
Radioactive emissions	No appreciable emissions	*No appreciable emissions during normal operation; possibility of emissions during severe accident*
Occupational health among workers	*More known health problems and fatalities*	Fewer known health problems and fatalities
Health impacts on nearby residents	*Air pollution impairs health*	No appreciable known health impacts under normal operation
Effects of accident or sabotage	No widespread effects	*Potentially catastrophic widespread effects*
Solid waste	*More generated*	Less generated
Radioactive waste	None	*Radioactive waste generated*
Fuel supplies remaining	Should last several hundred more years	Uncertain; supplies could last longer or shorter than coal supplies

Note: More-severe impacts are in *italics*.

FIGURE 20.7 The Three Mile Island nuclear power plant near Harrisburg, Pennsylvania, was the site of the most serious nuclear power plant accident in U.S. history. The partial meltdown here in 1979 was a "near-miss"—radiation was released but was mostly contained, and no health impacts have been confirmed. The incident put the world on notice, however, that a major accident could potentially occur.

(a) The Chernobyl sarcophagus

(b) Technicians measuring radiation

FIGURE 20.8 The world's worst nuclear power plant accident unfolded in 1986 at Chernobyl, in present-day Ukraine (then part of the Soviet Union). As part of the extensive cleanup operation, the destroyed reactor was encased in a massive concrete sarcophagus (**a**) to contain further radiation leakage. Technicians scoured the landscape surrounding the plant (**b**), measuring radiation levels, removing soil, and scrubbing roads and buildings.

Nuclear power poses small risks of large accidents

Although scientists calculate that nuclear power poses fewer chronic health risks than does fossil fuel combustion, the possibility of catastrophic accidents has spawned a great deal of public anxiety over nuclear power. Two events were influential in shaping public opinion about nuclear energy.

The first took place at the **Three Mile Island** plant in Pennsylvania (Figure 20.7), where in 1979 the United States experienced its most serious nuclear power plant accident. Through a combination of mechanical failure and human error, coolant water drained from the reactor vessel, temperatures rose inside the reactor core, and metal surrounding the uranium fuel rods began to melt, releasing radiation. This process is termed a **meltdown,** and it proceeded through half of one reactor core at Three Mile Island. Area residents stood ready to be evacuated as the nation held its breath, but fortunately most radiation remained trapped inside the containment building.

The accident was brought under control within days, the damaged reactor was shut down, and multibillion-dollar cleanup efforts stretched on for years. Three Mile Island is best regarded as a near-miss; the emergency could have been far worse had the meltdown proceeded through the entire stock of uranium fuel, or had the containment building not contained the radiation. Although no significant health effects on residents have been proven in the years since, the event put safety concerns squarely on the map for U.S. citizens and policymakers.

Chernobyl saw the worst accident yet

In 1986 an explosion at the **Chernobyl** plant in Ukraine (part of the Soviet Union at the time) caused the most severe nuclear power plant accident the world has yet seen (Figure 20.8). Engineers had turned off safety systems to

The Science behind the Story

Assessing Emissions from Power Sources

Combusting coal, oil, or natural gas emits carbon dioxide and other greenhouse gases into the atmosphere, where they contribute to global climate change (Chapter 18). Reducing greenhouse emissions is one of the main reasons so many people want to replace fossil fuels with alternative energy sources.

But determining how different energy alternatives stack up in terms of emissions is a complex process. A number of studies have tried to quantify and compare emission rates of different energy types, but the varying methodologies used have made it hard to synthesize this information into a coherent picture.

Researchers from the International Atomic Energy Agency (IAEA) made an attempt at such a synthesis for the generation of electricity. Experts met at six meetings between 1994 and 1998 and reviewed the existing scientific literature, together with data from industry and government. Their goal was to come up with a range of estimates of greenhouse gas emissions for nuclear energy, each major fossil fuel type, and each major renewable energy source. IAEA scientists Joseph Spadaro, Lucille Langlois, and Bruce Hamilton then published the results in the *IAEA Bulletin* in 2000.

The researchers had to decide how much of the total life cycle of electric power production to include

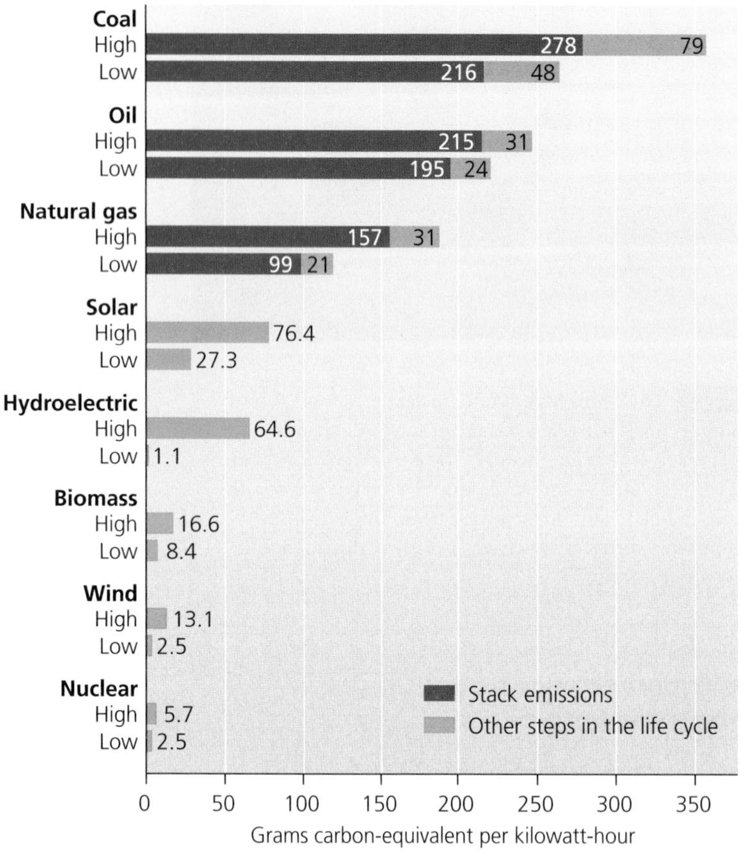

Coal, oil, and natural gas emit far more greenhouse gases than do renewable energy sources and nuclear energy. Red portions of bars represent stack emissions, and orange portions show emissions resulting from other steps in the life cycle. Data from Spadaro, J. V., et al. 2000. Greenhouse gas emissions of electricity generation chains: Assessing the difference. *IAEA Bulletin* 42(2).

conduct tests, and human error, combined with unsafe reactor design, resulted in explosions that destroyed the reactor and sent clouds of radioactive debris billowing into the atmosphere. For 10 days radiation escaped from the plant, while emergency crews risked their lives (some later died from radiation exposure) putting out fires. Most residents of the surrounding countryside remained at home for these 10 days, exposed to radiation, before the Soviet government belatedly began evacuating more than 100,000 people. In the months and years afterwards, workers erected a gigantic concrete sarcophagus around the demolished reactor, scrubbed buildings and roads, and removed irradiated materials, but the landscape for at least 30 km (19 mi) around the plant remains contaminated today.

The accident killed 31 people directly and sickened or caused cancer in many more. Exact numbers are uncertain because of inadequate data and the difficulty of determining long-term radiation effects (see "The Science behind the Story," ▸ pp. 604–605). It is widely thought that at least 2,000 people who were children at the time have in their estimates. Simply comparing the rotation of turbines at a wind farm to the operation of a coal-fired power plant might not be fair, because it would not reveal that greenhouse gases were emitted as a result of manufacturing the turbines, transporting them to the site, and erecting them there. Similarly, because uranium mining is part of the nuclear fuel cycle, and because we use oil-fueled machinery to mine uranium, perhaps emissions from this process should be included in the estimate for nuclear power.

The researchers decided to conduct a "cradle-to-grave" analysis and include all sources of emissions throughout the entire life cycle of each energy source. This included not just power generation, but also the mining of fuel, preparation and transport of fuel, manufacturing of equipment, construction of power plants, disposal of wastes, and decommissioning of plants. They did, however, separate stack emissions from all other sources of emissions in the chain of steps so that these data could be analyzed independently.

Different greenhouse gases were then standardized to a unit of "carbon equivalence" according to their global warming potential (▸ pp. 530–532). For instance, because methane is 21 times as powerful a greenhouse gas as carbon, each unit of methane emitted was counted as 21 units of carbon-equivalence. The researchers then calculated rates of emission per unit of power produced. They presented figures in grams of carbon-equivalent emitted per kilowatt-hour (gC_{eq}/kWh) of electric power produced.

The overall pattern they found was clear: Fossil fuels produce much higher emission rates than renewable energy sources and nuclear energy (see the figure). The highest emission rate for fossil fuels (357 gC_{eq}/kWh for coal) was 4.7 times higher than the highest emission rate for any renewable energy source (76.4 gC_{eq}/kWh for solar power). The lowest emission rate for any fossil fuel (106 gC_{eq}/kWh for natural gas) was nearly 100 times greater than the lowest rate for renewables (1.1 gC_{eq}/kWh for one form of hydropower). Overall, emissions decreased in the following order: coal, oil, natural gas, photovoltaic solar, hydroelectric, biomass, wind, and nuclear.

The majority of fossil fuel emissions were direct stack emissions from power generation, and the amounts due to other steps in the life cycle were roughly comparable to those from some renewable sources.

Within each category, emissions values varied considerably. This was due to many factors, including the type of technology used, geographic location and transport costs, carbon content of the fuel, and the efficiency with which fuel was converted to electricity.

However, technology was expected to improve in the future, creating greater fuel-to-electricity conversion efficiency and therefore lowering emissions rates. Thus, the researchers devised separate emissions estimates for newer technologies expected between 2005 and 2020. These estimates suggested that fossil fuels will improve but still will not approach the cleanliness of nuclear energy and most renewable sources.

Because the IAEA is charged with promoting nuclear energy, critics point out that the agency has clear motivation for conducting a study that shows nuclear power in such a favorable light. However, few experts would quibble with the overall trend in the study's data: Nuclear and renewable energy sources are demonstrably cleaner than fossil fuels.

contracted or will contract thyroid cancer from radioactive iodine. Estimates for the total number of cancer cases attributable to Chernobyl, past and future, vary widely, from a few thousand to hundreds of thousands.

Atmospheric currents carried radioactive fallout from Chernobyl across much of the Northern Hemisphere, particularly Ukraine, Belarus, and parts of Russia and Europe (Figure 20.9). Fallout was greatest where rainstorms brought radioisotopes down from the radioactive cloud. Parts of Sweden received high amounts of fallout. The accident reinforced the Swedish public's fears about nuclear power. A survey taken after the event asked, "Do you think it was good or bad for the country to invest in nuclear energy?" The proportion answering "bad" jumped from 25% before Chernobyl to 47% afterward.

The world has been fortunate not to have experienced another accident on the scale of Chernobyl in the two decades since. There have been smaller-scale incidents; for instance, a 1999 accident at a plant in Tokaimura, Japan, killed two workers and exposed over 400 others to leaked

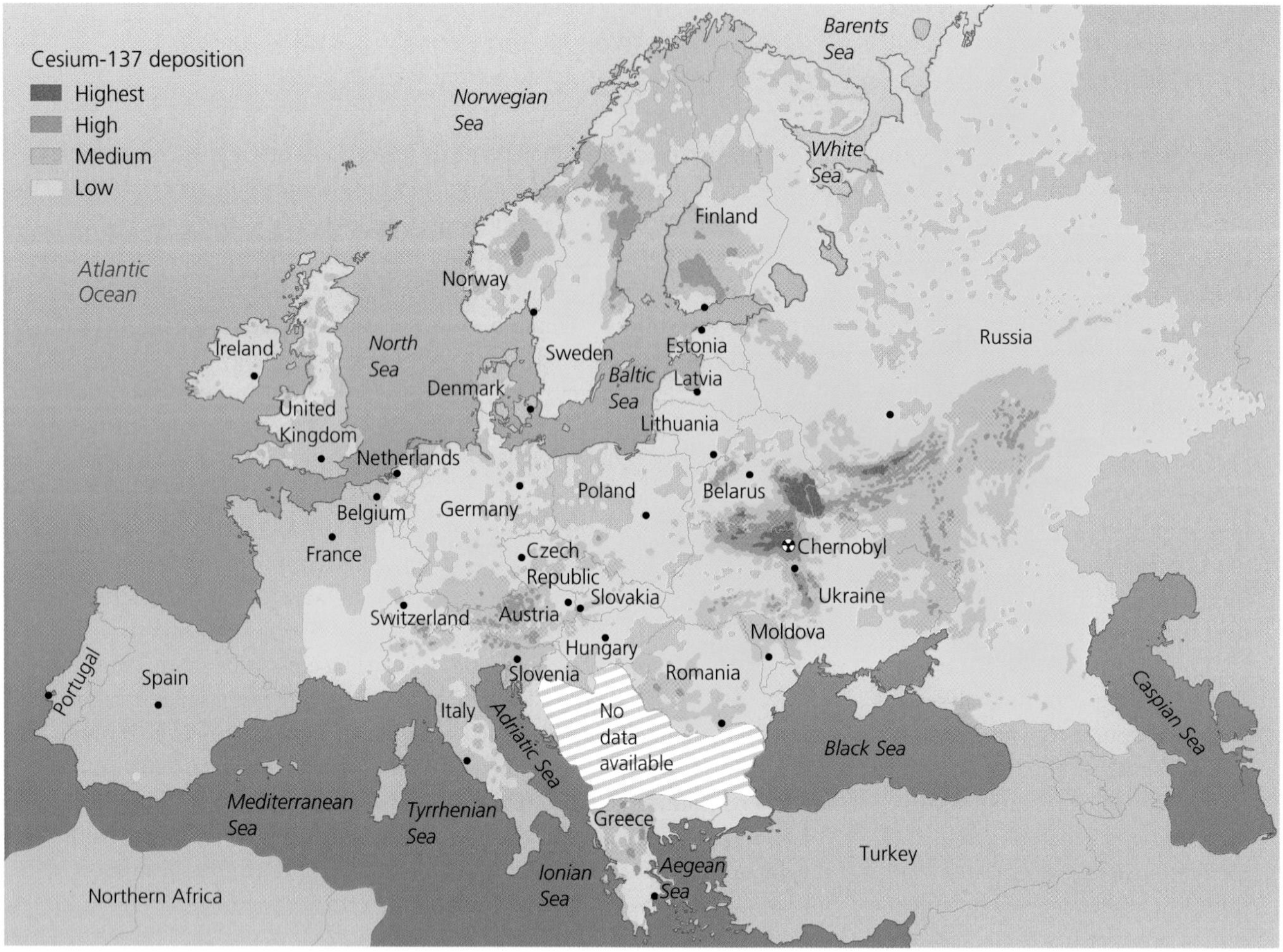

FIGURE 20.9 Radioactive fallout from the Chernobyl disaster was deposited across Europe in complex patterns resulting from atmospheric currents and rainstorms in the days following the accident. Darker colors in the map indicate higher levels of radioactivity. Although Chernobyl produced 100 times more fallout than the U.S. bombs dropped on Hiroshima and Nagasaki in World War II, it was distributed over a much wider area. Thus, levels of contamination in any given place outside of Ukraine, Belarus, and Russia were relatively low; the average European received less than the amount of radiation a person receives naturally in a year. Data from chernobyl.info, Swiss Agency for Development and Cooperation, Bern, 2005.

radiation. And as plants around the world age, they require more maintenance and are therefore less safe. New concerns have also surfaced. The September 11, 2001, terrorist attacks raised fears that similar airplane attacks could be carried out against nuclear plants. Moreover, radioactive material could be stolen from plants and used in terrorist attacks. This possibility is especially worrisome in Russia and other cash-strapped nations of the former Soviet Union, where hundreds of former nuclear sites have gone without adequate security for over a decade. In a cooperative international agreement, the U.S. government, through the "megatons to megawatts" program, has been buying up some of this material and diverting it to peaceful use in power generation.

Waste disposal remains a problem

Even if nuclear power generation could be made completely safe, we would still be left with the conundrum of what to do with spent fuel rods and other radioactive waste. Recall that fission utilizes ^{235}U as fuel, leaving the 97% of uranium that is ^{238}U as waste. This ^{238}U, as well as all irradiated material and equipment that is no longer being used, must be disposed of in a location where radiation will not escape and harm the public. Because the half-lives of uranium, plutonium, and many other radioisotopes are far longer than human lifetimes, this waste will continue emitting dangerous levels of radiation for thousands of years. Thus, radioactive waste must be placed in unusually stable and secure locations where radioactivity will not harm future generations.

Currently, nuclear waste from power generation is being held in temporary storage at nuclear power plants across the United States and the world. Spent fuel rods are sunken in pools of cooling water to minimize radiation leakage (Figure 20.10a). However, the U.S. Department of Energy (DOE) estimates that by 2010, three-fourths of U.S. plants will have no room left for this type of storage. Many plants are now expanding their storage capacity by storing waste in thick casks of steel, lead, and concrete (Figure 20.10b).

In total, U.S. power plants are storing over 49,000 metric tons of radioactive waste, enough to fill a football field to the depth of 3.3 m (10 ft). This waste is held at 125 sites spread across 39 states (Figure 20.11). A 2005 National Academy of Sciences report judged that most of these sites were vulnerable to terrorist attacks. The DOE estimates that over 161 million U.S. citizens live within 125 km (75 mi) of temporarily stored waste.

Because storing waste at numerous dispersed sites creates a large number of potential hazards, nuclear waste managers have long wanted to send all waste to a central repository that can be heavily guarded. In Sweden, that nation's nuclear industry has established a single repository for low-level waste near one power plant and is searching for a single disposal site deep within bedrock for spent fuel rods and other high-level waste. The industry hopes to decide on a site and get it approved by 2008.

In the United States, the multi-year search homed in on Yucca Mountain, a remote site in the desert of southern Nevada, 160 km (100 mi) from Las Vegas (Figure 20.12a, ▸ p. 603). Choice of this site followed extensive study by government scientists (Figure 20.12b), but Nevadans were not happy about the choice and fought against it. In 2002 the site was recommended by the president and approved by Congress, and the Department of Energy is now awaiting approval from the Nuclear Regulatory Commission. If given final approval, Yucca Mountain is expected to begin receiving waste from nuclear reactors, as well as high-level radioactive waste from military installations, in 2010. According to the design, waste would be stored in a network of tunnels 300 m (1,000 ft) underground, yet 300 m (1,000 ft) above the water table (Figure 20.12c) Scientists and policymakers chose the Yucca Mountain site because they determined that:

- It is unpopulated, lying 23 km (14 mi) from the nearest year-round residences.
- It has stable geology, with minimal risk of earthquakes that could damage the tunnels and release radioactivity.

(a) Wet storage

(b) Dry storage

FIGURE 20.10 Spent uranium fuel rods are currently stored at nuclear power plants and will likely remain at these scattered sites until a central repository for radioactive waste is developed. Spent fuel rods are most often kept in "wet storage" in pools of water (**a**), which keep them cool and reduce radiation release. Alternatively, the rods may be kept in "dry storage" in thick-walled casks layered with lead, concrete, and steel (**b**).

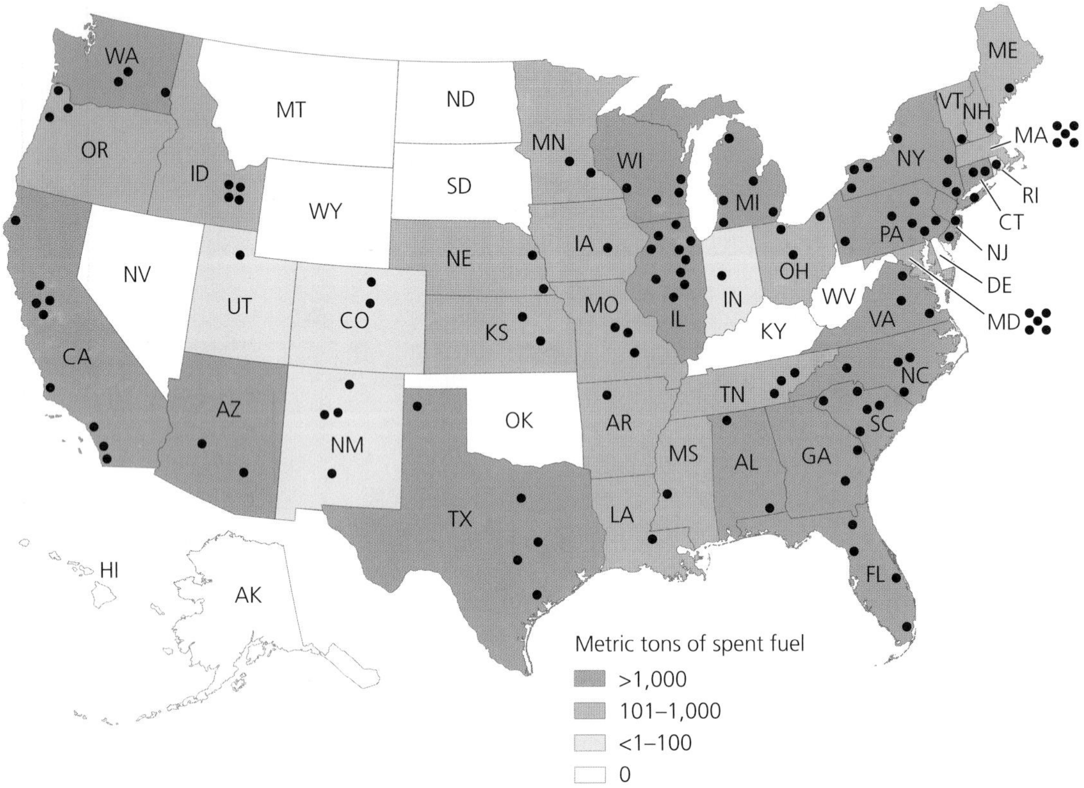

FIGURE 20.11 Nuclear waste from civilian reactors is currently stored at 125 sites in 39 states across the United States. In this map, dots indicate each storage site, and the four shades of color indicate the total amount of waste stored in each state. Data from: Office of Civilian Radioactive Waste Management, U.S. Department of Energy; and Nuclear Energy Institute, Washington, D.C.

- Its dry climate should minimize water infiltration through the soil, reducing chances of groundwater contamination.
- The water table is deep underground, making groundwater contamination less likely.
- The pool of groundwater does not connect with groundwater elsewhere, so that any contamination would be contained.
- The location, on federal land, can feasibly be protected from sabotage.

Some scientists, antinuclear activists, and concerned Nevadans have challenged these conclusions. For instance, they argue that earthquakes or volcanic activity could destabilize the site's geology. They also fear that fissures in the mountain's rock could allow rainwater to seep into the caverns.

A greater concern, in many people's minds, is that nuclear waste will need to be transported to Yucca Mountain from the 125 current storage areas and from current and future nuclear plants and military installations. Because this would involve many thousands of shipments by rail and truck across hundreds of public highways through almost every state of the union, many people worry that the risk of an accident or of sabotage is unacceptably high.

Weighing the Issues: How to Store Waste?

Which do you think is a better option—to transport nuclear waste cross-country to a single repository or to store it permanently at numerous power plants and military bases scattered across the nation? Would your opinion be affected if you lived near the repository site? Near a power plant? On a highway route along which waste was transported?

Multiple dilemmas have slowed nuclear power's growth

Dogged by concerns over waste disposal, safety, and expensive cost overruns, nuclear power's growth has slowed. Since the late 1980s, nuclear power has grown by 2.5% per year, about the same rate as electricity generation overall. Public anxiety in the wake of Chernobyl

(a) Yucca Mountain

(b) Scientific testing

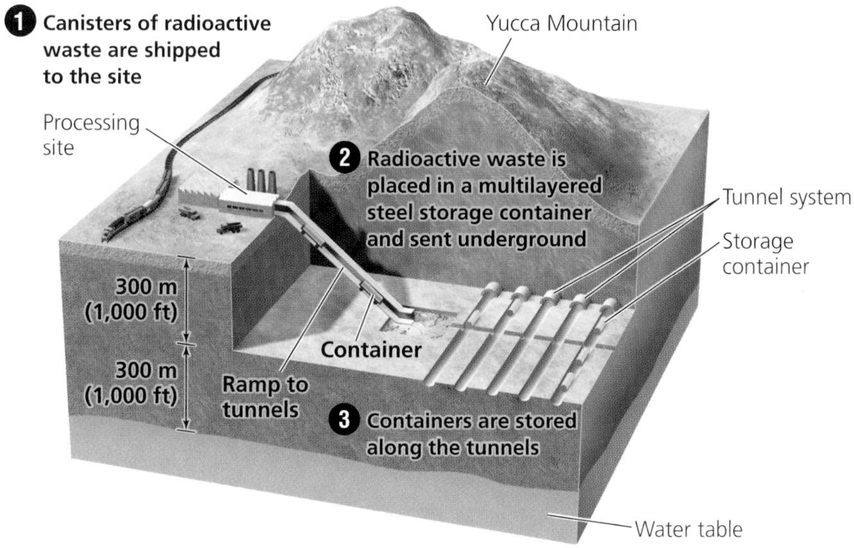

(c) Proposed design

FIGURE 20.12 Yucca Mountain (**a**), in a remote part of Nevada, awaits final approval as the central repository site for all the nuclear waste in the United States. Here (**b**), technicians are testing the effects of extreme heat from radioactive decay on the stability of rock. Waste would be buried in a network of tunnels deep underground yet still high above the water table (**c**).

made utilities less willing to invest in new plants. So did the enormous expense of building, maintaining, operating, and ensuring the safety of nuclear facilities. Almost every nuclear plant has turned out to be more expensive than expected. In addition, plants have aged more quickly than expected because of problems that were underestimated, such as corrosion in coolant pipes. The plants that have been shut down—well over 100 around the world to date—have served on average less than half their expected lifetimes. Moreover, shutting down, or decommissioning, a plant can sometimes be more expensive than the original construction. As a result of these economic issues, electricity from nuclear power today remains more expensive than electricity from coal and other sources. Governments are still subsidizing nuclear power to keep consumer costs down, but many private investors lost interest long ago.

Nonetheless, nuclear power remains one of the few currently viable alternatives to fossil fuels with which we can generate large amounts of electricity in short order. The International Energy Agency (IEA) predicts nuclear production will peak at 7% of world energy use and 17% of electricity production in 2010, then decline to 5% of world energy use and 9% of electricity production by 2030. The reason is that Asian nations are adding generating capacity, but three-quarters of Western Europe's capacity is scheduled to be retired by 2030. In Western Europe, not a single reactor is under construction today,

The Science behind the Story

Health Impacts of Chernobyl

In the wake of the nuclear power plant accident at Chernobyl in 1986, medical scientists from around the world rushed to study how the release of radiation might affect human health.

Determining long-term health impacts of a discrete event is difficult, so it is not surprising that the hundreds of researchers trying to pin down Chernobyl's impacts sometimes came up with very different conclusions. In an effort to reach some consensus, researchers at the Nuclear Energy Agency (NEA) of the Organization for Economic Co-operation and Development (OECD) reviewed studies through 2002 and issued a report summarizing what scientists had learned in the 16 years since the accident.

Doctors had documented the most severe effects among plant workers and firefighters who battled to contain the incident in its initial hours and days. Medical staff treated and recorded the progress of 237 patients who had been admitted to area hospitals diagnosed with acute radiation sickness (ARS). Radiation destroys cells in the body, and if the destruction outpaces the body's abilities to repair the damage, the person will soon die. Symptoms of ARS include vomiting, fever, diarrhea, thermal burns, mucous membrane damage, and weakening of the immune system by the depletion of white blood cells. In total, 28 (11.8%) of these people died from acute effects soon after the accident, and those who died had had the greatest estimated exposure to radiation.

IAEA scientists in 1990 studied residents of areas highly contaminated with radioactive cesium and

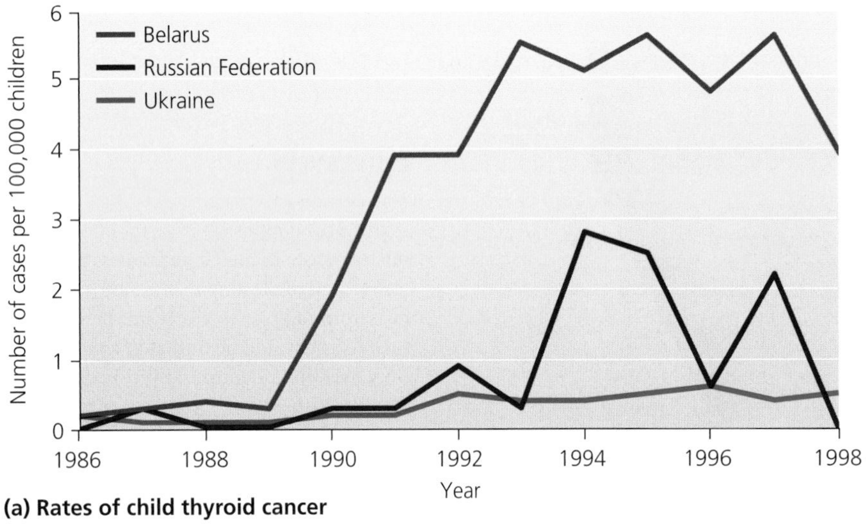

(a) Rates of child thyroid cancer

(b) Cancer patient with mother after surgery

The incidence of thyroid cancer (**a**) jumped in Belarus, Ukraine, and southwestern Russia starting 4 years after the Chernobyl accident released high levels of radioactive iodine isotopes. Many babies and young children (**b**) at the time of the accident developed thyroid cancer in later years. Most have undergone treatment and survived. Data (a) from Nuclear Energy Agency, OECD, 2002.

and Germany and Belgium, like Sweden, have declared an intention to phase out nuclear power altogether. However, France is committed to keeping its large share of nuclear-powered electricity. In Asia, Japan is so reliant on oil that it is eager to expand its options in the fastest and easiest way possible. China, India, and South Korea are expanding their nuclear programs to help power their rapidly growing economies. Indeed, Asia hosts 19 of the last 28 nuclear power plants to go into operation and 20 of the 31 plants now under construction.

compared their health with people of the same ages living in uncontaminated settlements nearby. Medical exams of 1,356 people showed no significant differences between the two groups or health abnormalities attributable to radiation exposure. However, the study was criticized for the quality of its data, for its small sample size, and for potential conflict of interest (the IAEA is charged with promoting the nuclear industry). In addition, the study was conducted only 4 years after the accident, before many cancers would be expected to appear.

Nonetheless, studies by the World Health Organization and others have come to similar conclusions. Overall, the NEA summary concluded, there is little evidence for long-term physical health effects resulting from Chernobyl (although psychological and social effects among displaced residents have been substantial). If cancer rates have risen among exposed populations, they have risen so little as to be statistically indistinguishable from normal variation in background levels of cancer.

The one exception is thyroid cancer, for which numerous studies have documented a real and perceptible increase among Chernobyl-area residents, particularly children (because children have large and active thyroid glands). The thyroid gland is where the human body concentrates iodine, and one of the most common radioactive isotopes released early in the disaster was iodine-131 (^{131}I).

Realizing that thyroid cancer induced by radioisotopes of iodine might be a problem, medical workers took measurements of iodine activity from the thyroid glands of hundreds of thousands of people—60,000 in Russia, 150,000 in Ukraine, and several hundred thousand in Belarus—in the months immediately following the accident. They also measured food contamination and had people fill out questionnaires on their food consumption. These data showed that drinking cows' milk was the main route of exposure to ^{131}I for most people, although fresh vegetables also contributed.

As doctors had feared, in the years following the accident rates of thyroid cancer began rising among children in regions of highest exposure (see the figure). The yearly number of thyroid cancer cases in the 1990s, particularly in Belarus, far exceeded numbers from years before Chernobyl. Multiple studies found linear dose-response relationships (▸ pp. 420–421) in data from Ukraine and Belarus. Fortunately, treatment of thyroid cancer has a high success rate, and as of 2002, only 3 of the 1,036 children cited in our figure had died of thyroid cancer. By comparing the Chernobyl-region data to background rates elsewhere, researchers calculated that Ukraine would eventually suffer 300 thyroid cancer cases more than normal and that the nearby region of Russia (with a population of 4.3 million) would suffer 349 extra cases, a 3–6% increase above the normal rate.

Critics pointed out that any targeted search tends to turn up more of whatever medical problem is being looked for. But the magnitude of the increase in childhood thyroid cancer was large enough that most experts judge it to be real. The rise in thyroid cancer, the NEA concluded, "should be attributed to the Chernobyl accident until proven otherwise."

Furthermore, thyroid cancer also appears to have risen in adults. Adult cases in Belarus in the 12 years before the accident totaled 1,392, but in the 12 years after Chernobyl totaled 5,449. In the most contaminated regions of Russia, thyroid cancer incidence rose to 11 per 100,000 women and 1.7 per 100,000 men, compared to normal rates of 4 and 1.1 for Russia as a whole. And although rates of childhood cancer may now be falling, rates for adults are still rising. As new cancer cases accumulate in the future, continued research will be needed to measure the full scope of health effects from Chernobyl.

In the United States, the nuclear industry stopped building plants following Three Mile Island, and public opposition scuttled many that were under construction. The $5.5-billion Shoreham Nuclear Power Plant on New York's Long Island was shut down just 2 months after its licensing because officials determined that evacuation would be impossible in this densely populated area should an accident ever occur. Of the 259 U.S. nuclear plants ordered since 1957, nearly half have been cancelled. At its peak in 1990, the United States had 112

VIEWPOINTS

Nuclear Power

Can nuclear power help reduce our reliance on fossil fuels? Should we revitalize and expand our nuclear power programs?

Nuclear Power: A Deadly and Needless Energy Source

Nuclear power is deadly and unnecessary. Disasters such as the 1979 Three Mile Island nuclear plant accident, the catastrophic Chernobyl plant explosion in 1986—and worse—are what will happen regularly if the United States and other nations move anew to build nuclear power plants.

The disastrous impacts of nuclear power are acknowledged in government documents. The U.S. Nuclear Regulatory Commission conducted a study in the 1980s, the *Calculation of Reactor Accident Consequences 2.* For the Indian Point 3 plant near New York, it calculated an accident causing 50,000 "peak fatalities," 141,000 "peak early injuries," 13,000 "peak cancer deaths," and $314 billion in property damage in 1980 dollars. The cost of a part of America left uninhabitable for millennia would be nearly $1 trillion today.

Nuclear power is so dangerous that there's a law called the Price-Anderson Act that limits a plant owner's liability for an accident, now $8 billion. If nuclear power is so safe, why the Price-Anderson Act?

The likelihood of an accident is far from "almost impossible," as atomic promoters once claimed. The NRC has conceded a 45% probability of a severe core melt accident every 20 years among the 100 U.S. atomic plants.

And it doesn't take an accident for a nuclear plant to spread radioactivity and contaminate and kill. There are "routine emissions" of radiation at every plant, as well as tons of lethal radioactive waste each plant produces annually, which must be isolated for thousands to millions of years. Moreover, in an age of terrorism, nuclear plants are sitting ducks.

We need not take the colossal risk of atomic power. Safe, clean, sustainable technologies are here today and can unhook us from fossil fuels: wind, solar, geothermal, and hydrogen power, among others. Let's have energy that won't kill us and our children—energy we can live with.

Karl Grossman is a professor at SUNY/College at Old Westbury and an investigative journalist who has authored books and hosted television programs about nuclear power.

Nuclear Power: Essential for Sustainable Global Development

We live in a world that has only *begun* to consume energy. During the next 50 years, as Earth's population expands from 6 billion toward 9 billion, humanity will consume more energy than the combined total used in all previous history.

With carbon emissions threatening human health and the stability of the biosphere, the security of our world requires a massive transformation to clean energy. This crisis is global. Today India and China are gaining rapidly on Europe and America in per capita energy consumption and climate-endangering emissions. Renewables such as solar, wind, and biomass can help. But only nuclear power offers clean energy on a massive scale.

Some "environmental" groups still spread misinformation about nuclear power. But here are the facts:

- *Safety.* Using 12,000 reactor-years of experience, nuclear power is the safest large-scale source of energy. The Chernobyl reactor used Soviet technology with no resemblance to today's technology.
- *Waste.* Nuclear power extracts enormous energy from tiny amounts of uranium. The small amounts of waste can be safely managed and placed in geological repositories with no long-term environmental harm.
- *Proliferation.* Civil nuclear power production does not foster nuclear weapons. Atomic bombs require sophisticated military programs, and worldwide IAEA safeguards prevent illicit diversion of nuclear material.
- *Cost.* Nuclear energy is already cost-competitive, and trends point to falling nuclear prices and rising fossil prices. A carbon tax would add to the nuclear advantage.
- *Usability.* Nuclear power is operating in countries representing two-thirds of total human population, and usage is expanding.

An informed public debate—focused on facts rather than myths—will demonstrate that nuclear energy is indispensable to sustainable global development.

John Ritch is director general of the World Nuclear Association. From 1994 to 2001, he was U.S. ambassador to the International Atomic Energy Agency and several other U.N. organizations. Previously, he served for 22 years as an advisor to the U.S. Senate Foreign Relations Committee.

Explore this issue further by accessing **Viewpoints** at www.aw-bc.com/withgott.

operable plants; today it has 104. However, the Bush administration advocates expanding U.S. nuclear capacity to decrease reliance on oil imports, and nuclear proponents point out that engineers are planning a new generation of reactors designed to be safer and less expensive.

Weighing the Issues:
More Nuclear Power?

Do you think the United States should expand its nuclear power program? Why or why not?

With slow growth predicted for nuclear power, and with fossil fuels in limited supply, where will our growing human population turn for additional energy? Increasingly, people are turning to renewable sources of energy: energy sources that cannot be depleted by our use. Although many renewable sources are still early in their stages of development, two of them—biomass and hydropower—are already well developed and widely used.

Biomass Energy

Biomass energy was the first energy source our species used, and it is still the leading energy source in much of the developing world. *Biomass* (▸ p. 158) consists of the organic material that makes up living organisms. People harness **biomass energy** from many types of plant and animal matter, including wood from trees, charcoal from burned wood, and combustible animal waste products such as cattle manure. Fossil fuels are not considered biomass energy sources because their organic matter has not been part of living organisms for millions of years and has undergone considerable chemical alteration since that time.

Fuelwood and other traditional biomass sources are widely used in the developing world

Over 1 billion people still use wood from trees as their principal power source. In developing nations, especially in rural areas, families gather fuelwood to burn in or near their homes for heating, cooking, and lighting (Figure 20.13). In these nations, fuelwood, charcoal, and manure account for fully 35% of energy use—in the poorest nations, up to 90%.

Fuelwood and other traditional biomass energy sources constitute nearly 80% of all renewable energy used worldwide. Considering what we have learned about the loss of forests (▸ pp. 349–354), however, it is fair to ask whether biomass should truly be considered a renewable resource. In reality, biomass is renewable only if it is not overharvested. At moderate rates of use, trees and other plants can replenish themselves over months to decades. However, when forests are cut too quickly, or when overharvesting leads to soil erosion and forests fail to grow back, then biomass is not effectively replenished.

FIGURE 20.13 Hundreds of millions of people in the developing world rely on fuelwood for heating and cooking. Wood cut from trees remains the major source of biomass energy used today. In theory biomass is renewable, but in practice it may not be if forests are overharvested.

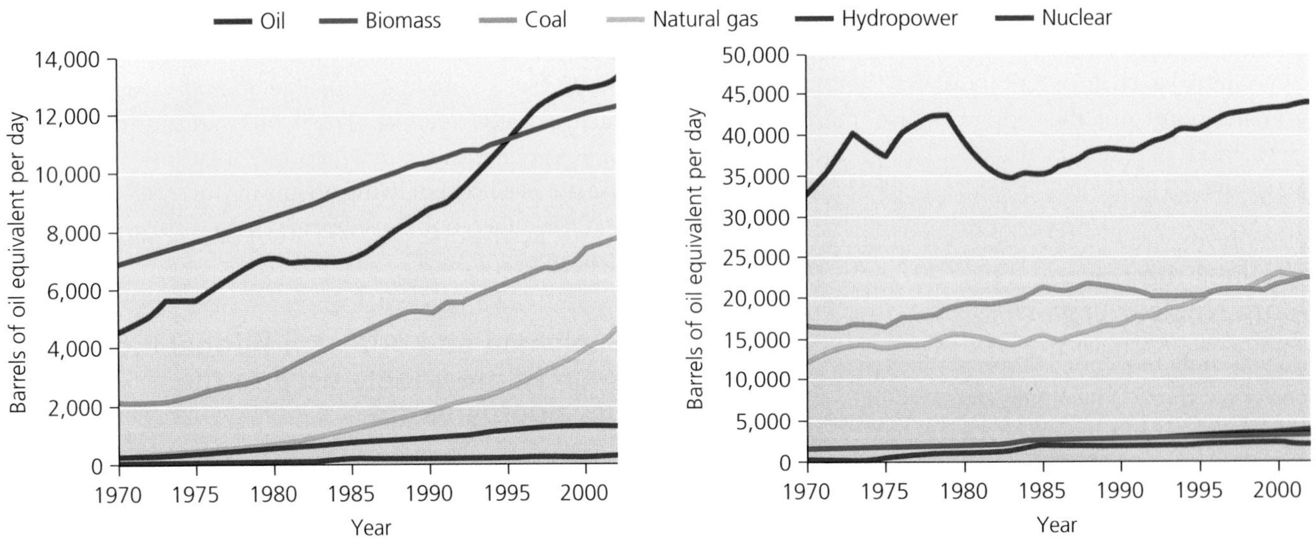

(a) Energy consumption in developing nations

(b) Energy consumption in developed nations in the OECD

FIGURE 20.14 Energy consumption patterns vary greatly between developing nations **(a)** and developed nations, here represented by nations of the Organisation for Economic Cooperation and Development (OECD) **(b).** Note the large role that biomass (primarily fuelwood) plays in supplying energy to developing countries. Data from Energy Information Administration, U.S. Department of Energy.

The potential for deforestation makes biomass energy less sustainable than other renewable sources, particularly as human population continues to increase.

As developing nations industrialize, fossil fuels are replacing traditional energy sources (Figure 20.14). As a result, biomass use is growing more slowly worldwide than overall energy use.

New biomass sources are being developed in industrialized countries

Besides the fuelwood, charcoal, and manure traditionally used, biomass energy sources in today's world include a number of sources for which innovative uses have recently been developed (Table 20.3). Some of these sources can be burned efficiently in power plants to produce **biopower**, generating electricity in the same way that coal is burned for power (▸ pp. 566–567). Other new biomass sources can be converted into fuels used primarily to power automobiles; these are termed **biofuels.** Because many of these novel biofuels and biopower strategies depend on technologies resulting from a good deal of research and development, they are being developed primarily in wealthier industrialized nations, such as Sweden and the United States.

Many of the new biomass resources are the waste products of preexisting industries or processes. For instance, the forest products industry generates large amounts of woody debris in logging operations and at sawmills, pulp mills, and paper mills (Figure 20.15). Sweden's efforts to encourage biomass energy have focused largely on using forestry residues. Because so much of the nation is forested and the timber industry is a major part of the national economy, plenty of forestry waste is available. Organic components of waste from municipal landfills also can provide biomass energy, as can animal waste from agricultural feedlots. Residue from agricultural crops (such as stalks, cobs, and husks from corn) could also soon become a major bioenergy source.

Some plants are now specifically grown for the purpose of producing biofuels. These bioenergy crops include various fast-growing grasses, such as bamboo, fescue, and switchgrass; grain and oil-producing crops,

Table 20.3 Major Sources of Biomass Energy

- ▸ Wood cut from trees (fuelwood)
- ▸ Charcoal
- ▸ Manure from domestic animals
- ▸ Crops grown specifically for biomass energy production
- ▸ Crop residues (such as corn stalks)
- ▸ Forestry residues (such as wood waste from logging)
- ▸ Processing wastes (such as solid or liquid waste from sawmills, pulp mills, and paper mills)
- ▸ Components of municipal solid waste

FIGURE 20.15 Forestry residues (here, from a Swedish logging operation) are a major source of biomass energy in some regions.

such as corn and soybeans; and fast-growing trees, such as poplar and cottonwood. For biofuels, willow trees are cultivated commercially in Sweden, sugarcane is grown in Brazil, and corn is raised in the United States.

Biofuels can power automobiles

Liquid fuels from biomass sources are helping to power millions of vehicles on today's roads, and some vehicles can run entirely on biofuels. The two primary types of such fuels developed so far are ethanol (for gasoline engines) and biodiesel (for diesel engines).

Ethanol is the alcohol in beer, wine, and liquor. It is produced as a biofuel by fermenting biomass, generally from carbohydrate-rich crops, such as corn, in a process similar to brewing beer. In fermentation, carbohydrates are converted to sugars and then to ethanol. Spurred by the 1990 Clean Air Act amendments and generous government subsidies, ethanol is widely added to gasoline in the United States to reduce automotive emissions. In 2004 in the United States, 14 billion L (3.7 billion gal) of ethanol were produced from corn. This amount has grown each year, as the number of U.S. ethanol plants approaches 100. A number of nations now have vehicles that can run on ethanol (Sweden has dozens of such public buses), and each of the U.S. "big three" automakers is now producing *flexible fuel vehicles* that run on E-85, a mix of 85% ethanol and 15% gasoline. In Brazil, half of all new cars are flexible-fuel cars, and ethanol from sugarcane accounts for 44% of all automotive fuel used.

Biodiesel is produced from vegetable oil, used cooking grease, or animal fat. The oil or fat is mixed with small amounts of ethanol or methanol (wood alcohol) in the presence of a chemical catalyst. In Europe, where most biodiesel is used, canola oil is often the oil of choice, whereas U.S. biodiesel producers use mostly soybean oil and recycled grease and oil from restaurants. Vehicles with diesel engines can run on 100% biodiesel. In fact, when Rudolf Diesel invented the diesel engine in 1895, he designed it to run on a variety of fuels, and he showcased his invention at the 1900 World's Fair using peanut oil. Since that time, petroleum-based fuel *(petrodiesel)* has been used because it has been cheaper. Because today's diesel engines have been designed to work with petrodiesel, some engine parts wear out more quickly with biodiesel, but one can use both and switch back and forth without making any modifications. Biodiesel can also be mixed with conventional petrodiesel; a 20% biodiesel mix (called B20) is common today.

Biodiesel cuts down on emissions compared with petrodiesel (Figure 20.16). Its fuel economy is almost as good, and it costs just 10–20% more. It is also nontoxic and biodegradable. Increasing numbers of environmentally conscious individuals in North America and Europe are fueling their cars with biodiesel from waste oils, and some buses and recycling trucks are now running on biodiesel. Governments are encouraging its use, too; Minnesota, for instance, mandates that all diesel sold must include a 2% biodiesel component, and already over 40 state and federal fleets are using biodiesel blends.

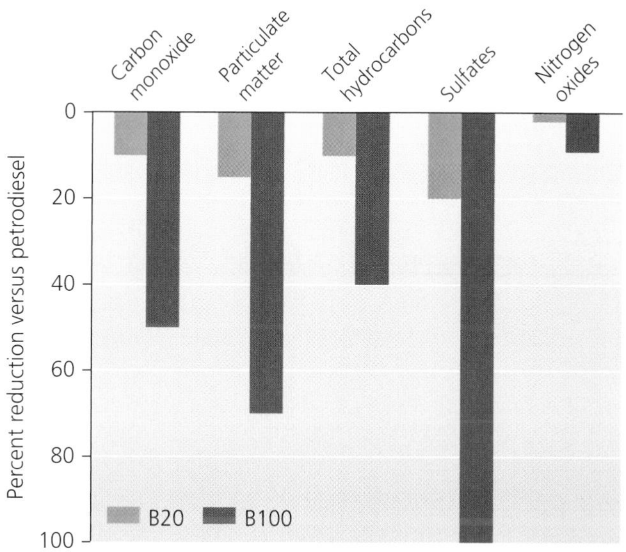

FIGURE 20.16 Burning biodiesel in a diesel engine emits fewer emissions than burning conventional petroleum-based diesel. Shown are the percentage reductions in several major automotive pollutants that one can attain by using B20 (a mix of 20% biodiesel and 80% petrodiesel) and B100 (pure biodiesel). Data from U.S. Environmental Protection Agency.

FIGURE 20.17 Each summer a group of alternative-fuel advocates goes on tour in the B.I.O. Bus, a bus fueled entirely on used vegetable oil from restaurants. Their "Bio Tours" sponsor events across North America that include music, dancing, and seminars on environmental sustainability and alternative fuels. Their motto: "Solar-powered sound, veggie-powered bus."

Some enthusiasts have taken biofuel use a step further. Eliminating the processing step that biodiesel requires, they use straight vegetable oil in their diesel engines. One notable effort is the "B.I.O. Bus," a bus fueled entirely on waste oil from restaurants that a group of students, environmentalists, and artists drives across North America (Figure 20.17). Each summer the group goes on tour with the bus, hosting festive events that combine music and dancing with seminars on environmental sustainability, and spreading the word about nonpetroleum fuels. In the summers of 2003, 2004, and 2005, the bus traveled 25,000 miles.

To run on straight vegetable oil, a diesel engine needs to be modified. Extra parts need to be added, so that there are tanks for both the oil and for petrodiesel, which is often needed to start the engine in cooler weather. Although these parts can be bought for as little as $800, it remains to be seen whether using straight vegetable oil might entail further costs, such as reduced longevity or greater engine maintenance.

Biopower generates electricity from biomass

We can harness biopower in various ways. The most frequently used strategy is simply to burn biomass in the presence of large amounts of air. This can be done on small scales with furnaces and stoves to produce heat for domestic needs, or on large industrial scales to produce district heating and to generate electricity. Power plants built to combust biomass operate in a similar way to those fired by fossil fuels; the combustion heats water, creating steam to turn turbines and generators, thereby generating electricity. Much of the biopower produced so far comes from power plants that generate both electricity and heating through cogeneration (▸ p. 583); these plants are often located where they can take advantage of waste material from the forest products industry.

Biomass is also increasingly being combined with coal in existing coal-fired power plants in a process called *co-firing*. Biomass is introduced with coal into a high-efficiency boiler that uses one of several technologies. Up to 15% of the coal can be substituted with biomass, with only minor equipment modification and no appreciable loss of efficiency. Co-firing can be a relatively easy and inexpensive way for fossil-fuel-based utilities to expand their use of renewable energy.

The decomposition of biomass by microbes also produces gas that can be used to generate electricity. The anaerobic bacterial breakdown of waste in landfills produces methane and other components, and this "landfill gas" is now being captured at many solid waste landfills and sold as fuel (▸ p. 657). Methane and other gases can also be produced in a more controlled way in anaerobic digestion facilities. This "biogas" can then be burned in a power plant's boiler to generate electricity.

The process of *gasification* can also provide biopower, as well as biofuels. In this process, biomass is vaporized at extremely high temperatures in the absence of oxygen, creating a gaseous mixture including hydrogen, carbon monoxide, and methane. This mixture can generate electricity when used in power plants to turn a gas turbine to propel a generator. Gas from gasification can also be treated in various ways to produce methanol, synthesize a type of diesel fuel, or isolate hydrogen for use in hydrogen fuel cells (▸ pp. 637–642). An alternative method of heating biomass in the absence of oxygen results in *pyrolysis*, which produces a mix of solids, gases, and liquids. This includes a liquid fuel called pyrolysis oil, which can be burned to generate electricity.

At small scales, farmers and ranchers can buy modular biopower systems that use livestock manure to generate electricity, and small household biodigesters are now providing portable and decentralized energy production. In many developing nations, particularly in Asia, international agencies have distributed hundreds of thousands of efficient biomass-fueled cooking stoves. At large scales, industries such as the forest products industry are using their waste to generate power, farmers are growing crops for biopower, and governments are encouraging biopower. In Sweden, one-sixth of the nation's energy supply now comes from biomass, and biomass provides more fuel for electricity generation than coal, oil, or natural gas. Pulp mill liquors are the main source, but solid wood waste, municipal solid waste, and biogas from digestion are all increasingly being used. In the United States, several dozen biomass-fueled power plants are now operating, and several dozen coal-fired plants are experimenting with co-firing.

Biomass energy brings environmental and economic benefits

Biomass energy has one overarching environmental benefit: It is essentially carbon-neutral, releasing no net carbon into the atmosphere. Although burning biomass emits plenty of carbon, the carbon released is simply the carbon that was pulled from the atmosphere by photosynthesis to create the biomass being burned. That is, because biomass is the product of recent photosynthesis, the carbon released in its combustion is balanced by the carbon taken up in the photosynthesis that created it. Therefore, when biomass sources take the place of fossil fuels, net carbon flux to the atmosphere is reduced. However, this holds only if biomass sources are not overharvested; deforestation will increase carbon flux to the atmosphere because less vegetation means less carbon uptake by plants for photosynthesis.

Biofuels can reduce greenhouse gas emissions that contribute to climate change in additional ways. Capturing landfill gas reduces emissions of methane, a potent greenhouse gas. And biofuels such as ethanol and biodiesel contain oxygen, which when added to gasoline or diesel helps those petroleum fuels to combust more completely, reducing pollution.

Shifting from fossil fuels to biomass energy also can have economic benefits. As a resource, biomass tends to be well spread geographically, so using it should help support rural economies and reduce many nations' dependence on imported fuels. Biomass also tends to be the least expensive type of fuel for burning in power plants, and improved energy efficiency brings lower prices for consumers.

By increasing energy efficiency and recycling waste products, use of biomass energy helps move our industrial systems toward greater sustainability. The U.S. forest products industry now obtains over half its energy by combusting the biomass waste it recycles, including woody waste and liquor from pulp mill processing.

Relative to fossil fuels, biomass also has benefits for human health. By replacing coal in co-firing and direct combustion, biomass reduces emissions of nitrogen oxides and particularly sulfur dioxide. The reason is that wood, unlike coal, contains no appreciable sulfur content. By replacing gasoline and petrodiesel and burning more cleanly, biofuels reduce emissions of various air pollutants.

Biomass energy also brings drawbacks

Biomass energy is not without negative environmental impacts, however. Burning fuelwood and other biomass in traditional ways for cooking and heating leads to health hazards from indoor air pollution (▸ pp. 519–522). In addition, harvesting fuelwood at an unsustainably rapid rate leads to deforestation, soil erosion, and desertification, damaging landscapes, diminishing biodiversity, and impoverishing human societies dependent on an area's resources. In arid regions that are heavily populated and support meager woodlands, fuelwood harvesting can have enormous impacts. Such is the case with many regions of Africa and Asia. In contrast, moister, well-forested areas with lower population densities, such as Sweden, stand less chance of deforestation.

Another drawback of biomass energy is that growing crops specifically to produce biofuels establishes monoculture agriculture, with all its impacts (▸ pp. 281–282), on precious land that might otherwise be used to grow food, be developed for other purposes, or be left as wildlife habitat. Fully 10% of the U.S. corn crop is used to make ethanol, and most U.S. biodiesel is produced from oil from soybeans grown specifically for this purpose (although researchers are studying how to obtain oil from algae, which could take up far less space). Growing

FIGURE 20.18 Ethanol is widely added to gasoline in the United States and some other nations, and 10% of the U.S. corn crop is used to produce ethanol. Ethanol makes for cleaner-burning fuel, but it requires a great deal of industrialized agriculture.

bioenergy crops also requires substantial inputs of energy. We currently operate farm equipment using fossil fuels, and farmers apply petroleum-based pesticides and fertilizers to increase yields. Thus, shifting from gasoline to ethanol for our transportation needs would not eliminate our reliance on fossil fuels. Moreover, growing corn for ethanol (Figure 20.18) yields only small amounts of ethanol per acre of crop grown; 1 bushel of corn creates only 9.4 L (2.5 gal) of ethanol. It is not efficient to grow high-input, high-energy crops merely to reduce them to a few liters of fuel; such a process is less efficient than directly burning biomass. Indeed, even using ethanol as a gasoline additive to reduce emissions has a downside; it lowers fuel economy very slightly, so that more gasoline needs to be used. For all these reasons, many critics have lambasted U.S. subsidies of ethanol, viewing them as effective in gaining politicians' votes in politically important farm states, but less effective as a path to sustainable energy use.

On the positive side, crops grown for energy typically receive lower inputs of pesticides and fertilizers than those grown for food. Researchers are currently refining techniques to produce ethanol from the cellulose that gives structure to plant material, and not just from starchy crops like corn. If these techniques can be made widely feasible, then ethanol could be produced primarily from low-quality waste and crop residues, rather than from high-quality crops.

Although biomass energy in industrialized nations currently revolves mostly around a few easily grown crops and the efficient use of waste products, the U.S. government sees a future of specialized crops that can serve as the basis for a wide variety of fuels and products. Meanwhile, use of traditional fuels in the developing world is expected to increase, and the IEA estimates that in the year 2030, 2.6 billion people will be using traditional fuels for heating and cooking in unsustainable ways. Like nuclear power and like hydropower, biomass energy use involves a complex mix of advantages and disadvantages for the environment and human society.

Weighing the Issues:
Ethanol

Do you think producing and using ethanol from corn or other crops is a good idea? Do the benefits outweigh the drawbacks? Can you suggest ways of using biofuels that would minimize environmental impacts?

Hydroelectric Power

Next to biomass, we draw more renewable energy from the motion of water than from any other resource. In **hydroelectric power,** or **hydropower,** the kinetic energy of moving water is used to turn turbines and generate electricity. We examined hydropower and its environmental impacts in our discussion of freshwater resources in Chapter 15 (▸ pp. 446–449), but we will now take a closer look at hydropower as an energy source.

Modern hydropower uses dams and "run-of-river" approaches

Just as people have long burned fuelwood, we have long harnessed the power of moving water. For instance, waterwheels spun by river water powered mills in past centuries. Today we harness water's kinetic energy in two major ways: with storage impoundments and by using "run-of-river" approaches.

Most of our hydroelectric power today comes from impounding water in reservoirs behind concrete dams that block the flow of river water, and then letting that water pass through the dam. Because immense amounts of water are stored behind dams, this is called the **storage** technique. If you have ever seen Hoover Dam on the Colorado River (▸ p. 432), or any other large dam, you have witnessed an impressive example of the storage

(a) Ice Harbor Dam, Washington

(b) Turbine generator inside MacNary Dam, Columbia River

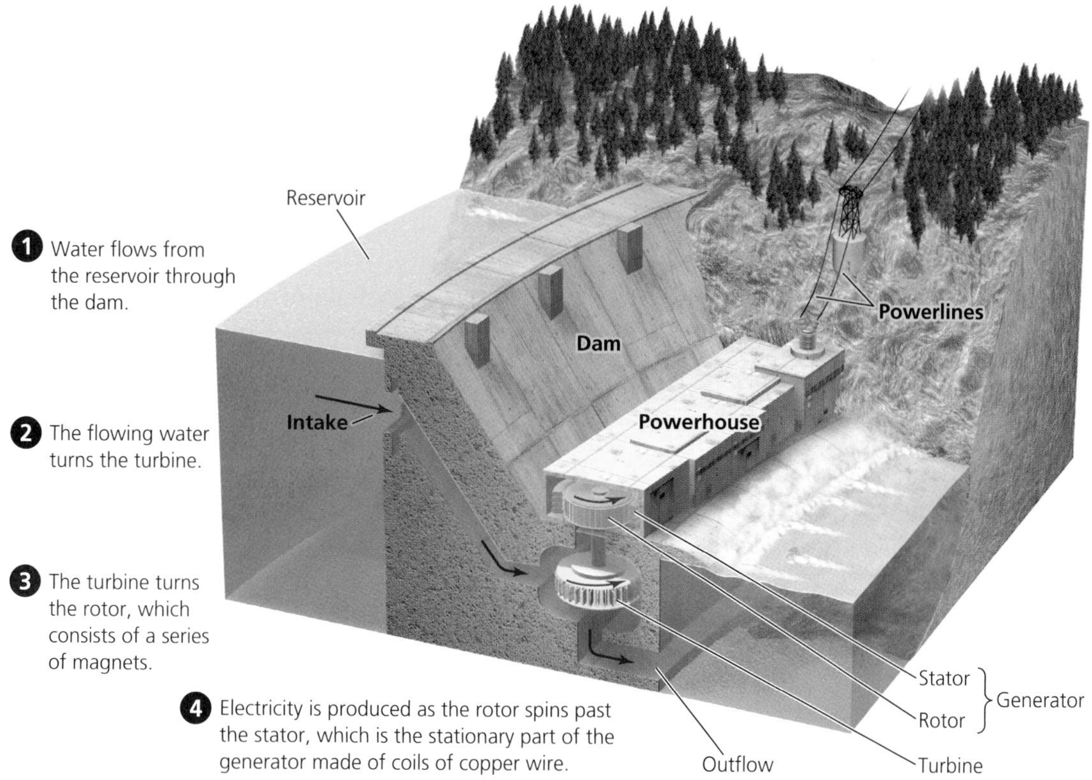
1 Water flows from the reservoir through the dam.
2 The flowing water turns the turbine.
3 The turbine turns the rotor, which consists of a series of magnets.
4 Electricity is produced as the rotor spins past the stator, which is the stationary part of the generator made of coils of copper wire.
Reservoir
Dam
Intake
Powerhouse
Powerlines
Stator
Rotor
Generator
Outflow
Turbine

(c) Hydroelectric power

FIGURE 20.19 Large dams, such as the Ice Harbor Dam on the Snake River in Washington (**a**), generate substantial amounts of hydroelectric power. Inside these dams, flowing water is used to turn turbines (**b**) and generate electricity. Water is funneled from the reservoir through a portion of the dam (**c**) to rotate turbines, which turn rotors containing magnets. The spinning rotors generate electricity as their magnets pass coils of copper wire. Electrical current is transmitted away through power lines, and the river's water flows out through the base of the dam.

approach to hydroelectric power. As reservoir water passes through a dam, it turns the blades of turbines, which cause a generator to generate electricity (Figure 20.19). Electricity generated in the powerhouse of a dam is transmitted to the electric grid by transmission lines, while the water is allowed to flow into the riverbed below the dam to continue downriver. The amount of power generated depends on the distance the water falls and the volume of water released. By storing water in reservoirs, dam operators can ensure a steady and predictable supply

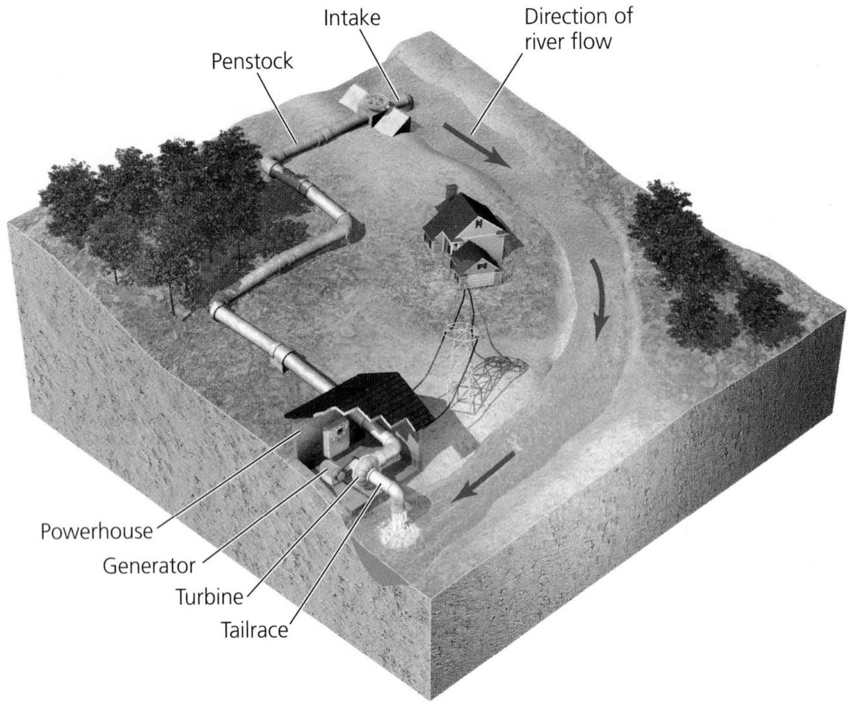

FIGURE 20.20 Run-of-river systems divert a portion of a river's water and can be designed in various ways. Some designs involve piping water downhill through a powerhouse and releasing it downriver, and some involve using water as it flows over shallow dams.

of electricity at all times, even during seasons of naturally low river flow.

An alternative approach is the **run-of-river** approach, in which electricity is generated without greatly disrupting the flow of river water. This approach sacrifices the reliability of water flow across seasons that the storage approach guarantees, but it minimizes many of the environmental impacts of large dams. The run-of-river approach can be followed using several methods, one of which is to divert a portion of a river's flow through a pipe or channel, passing it through a powerhouse and then returning it to the river (Figure 20.20). This can be done with or without a small reservoir that pools water temporarily, and the pipe or channel can be run along the surface or underground. Another method is to flow river water over a dam that is small enough not to impede fish passage, siphoning off water to turn turbines, and then returning the water to the river. Run-of-river systems are particularly useful in areas remote from established electrical grids and in regions without the economic resources to build and maintain large dams.

Hydroelectric power is widely used

Hydropower accounts for 2.2% of the world's primary energy supply and 16.2% of the world's electricity production. For nations with large amounts of river water and the economic resources to build dams, hydroelectric power has been a keystone of their development and wealth. Canada, Brazil, Norway, Austria, Switzerland, and many other nations today obtain large amounts of their energy from hydropower (Table 20.4). Sweden receives 14% of its total energy and nearly half its electricity from hydropower. In the wake of the nation's decision to phase out nuclear power, many people had hoped hydropower

Table 20.4 Top Consumers of Hydropower

Nation	Hydropower consumed[1]	Percentage of electricity generation from hydropower[2]
Canada	76.4	57.5
China	74.2	14.9
Brazil	72.4	83.8
United States	59.8	7.5
Russia	40.0	17.2
Norway	24.7	98.9
Japan	22.6	9.9
India	19.0	11.9
Venezuela	16.0	66.0
France	14.8	11.4

Data from British Petroleum and International Energy Agency.
[1]In million metric tons of oil equivalent, 2004 data.
[2]2003 data.

could play a still larger role and compensate for the electrical capacity that would be lost. However, Sweden has already dammed so many of its rivers that it cannot gain much additional hydropower by erecting more dams. Moreover, Swedish citizens have made clear that they want some rivers to remain undammed, preserved in their natural state.

In the United States, the great age of dam building for hydroelectric power (as well as for flood control and irrigation) began in the 1930s, when the federal government constructed dams as public projects, partly to employ people and help end the economic depression of the time. U.S. dam construction peaked in 1960, when 3,123 dams were completed in a single year.

Hydropower is clean and renewable

For producing electricity, hydropower has two clear advantages over fossil fuels. First, it is renewable; as long as precipitation falls from the sky and fills rivers and reservoirs, we can use water to turn turbines. When pressure from human population or consumption is intense enough, water supplies may be used faster than they are replenished, and hydropower may not be sustainable. However, in most cases hydropower continues today to provide the renewable energy that dams were built to provide.

The second advantage of hydropower over fossil fuels is its cleanliness. Because no carbon compounds are burned in the production of hydropower, no carbon dioxide or other pollutants are emitted into the atmosphere. Its cleanliness has won hydropower praise for protecting air quality, climate, and human health. Of course, fossil fuels are used in the construction and maintenance of dams. Moreover, recent evidence indicates that large reservoirs may release the greenhouse gas methane as a result of anaerobic decay in deep water. But overall, hydropower accounts for only a small fraction of the greenhouse gas emissions typical of fossil fuel combustion.

Hydropower has negative environmental impacts

Although it is renewable and produces little air pollution, hydropower does create other environmental impacts. Damming rivers destroys habitat for riverside wildlife as riparian areas above dam sites are submerged and those below dam sites often become starved of water. Because water discharge is carefully controlled to optimize electricity generation, the natural flooding cycles of rivers are disrupted. Suppressing flooding prevents river floodplains from receiving fresh nutrient-laden sediments. Instead, sediments become trapped behind dams, where they begin filling the reservoir. Dams also cause thermal pollution, because water downstream from dams may become unusually warm if water levels are kept unnaturally shallow. Moreover, periodic flushes of cold water may occur from the lower depths of the reservoir. Such thermal shocks, together with severe habitat alteration, have diminished or eliminated many native fish populations in dammed waterways throughout the world. In addition, dams generally block the passage of fish and other aquatic creatures, effectively fragmenting segments of the river and reducing biodiversity in each segment. All these ecological impacts generally translate into negative social and economic impacts on local communities. We discussed the environmental, economic, and social impacts of dams, and their advantages and disadvantages, more fully in Chapter 15 (▸ pp. 446–449).

Hydropower may not expand much more

Use of hydropower is growing slightly more quickly than overall energy use, and some gargantuan projects are being planned and carried out. China's recently completed Three Gorges Dam (▸ pp. 447, 449) is the world's largest. Its reservoir displaced over 1 million people, and the dam should soon be able to generate as much electricity as dozens of coal-fired or nuclear plants.

However, unlike other renewable energy sources, hydropower is not likely to expand very much more. One reason is that, as in Sweden, most of the world's large rivers that offer excellent opportunities for hydropower have already been dammed. Another reason is that people's awareness of the ecological impacts of dams has grown, and in some nations residents are fighting dam construction in their regions. Indeed, in Sweden, hydropower's contribution to the national energy budget has remained virtually unchanged for over 30 years. In the United States, 98% of rivers appropriate for dam construction already are dammed, many of the remaining 2% are protected under the Wild and Scenic Rivers Act, and many people now propose dismantling and removing some dams and restoring river habitats (▸ pp. 448–449).

Overall, hydropower will likely continue to increase in developing nations that have yet to dam their rivers, but in developed nations hydropower growth will likely slow or stop. The IEA forecasts that hydropower's share of electricity generation will decline between now and 2030, while the share of other renewable energy sources will triple, from 2% to 6%.

Conclusion

Given limited supplies of fossil fuels and their considerable environmental impacts, many nations have sought to diversify their energy portfolios with alternative energy sources. The three most developed and widely used alternatives so far are nuclear power, biomass energy, and hydropower.

Nuclear power showed promise at the outset to be a pollution-free and highly efficient form of energy. But high costs and public fears over safety in the wake of accidents at Chernobyl and Three Mile Island stalled its growth, and some nations are attempting to phase it out completely. Biomass energy sources include traditional ones such as fuelwood, as well as a variety of newer ones including several biofuels and various means of generating biopower. These sources can be carbon-neutral but are not all strictly renewable. Hydropower is a renewable, pollution-free alternative, but it is nearing its maximal use and can involve substantial ecological impacts. Although some nations, such as Sweden, already rely heavily on these three conventional alternatives, it appears that we will need further renewable sources of energy.

REVIEWING OBJECTIVES

You should now be able to:

Discuss the reasons for seeking energy alternatives to fossil fuels

- Fossil fuels are nonrenewable resources and we are gradually depleting them. (p. 591)
- Fossil fuel combustion causes air pollution that results in many environmental and health impacts and contributes to global climate change. (p. 591)

Summarize the contributions to world energy supplies of conventional alternatives to fossil fuels

- Biomass provides 10.8% of global primary energy use, nuclear power provides 6.5%, and hydropower provides 2.2%. (p. 591)
- Nuclear power generates 15.8% of the world's electricity, and hydropower generates 15.9%. (p. 591)
- "Conventional energy alternatives" are the alternatives to fossil fuels that are most widely used. They include biomass energy, nuclear energy, and hydroelectric power. In their renewability and environmental impact, these sources fall between fossil fuels and the less widely used "new renewable" sources. (p. 592)

Describe nuclear energy and how it is harnessed

- Nuclear power comes from converting the energy of subatomic bonds into thermal energy, using uranium isotopes. (pp. 592–593)
- Uranium is mined, enriched, processed into pellets and fuel rods, and used in nuclear reactors. (pp. 593–594)
- By controlling the reaction rate of nuclear fission, nuclear power plant engineers produce heat that powers electricity generation. (pp. 594–595)

Outline the societal debate over nuclear power

- Many advocates of "clean" energy support nuclear power because it lacks the pollutant emissions of fossil fuels. (pp. 596, 598–599)
- For many people, the risk of a major power plant accident, such as the one at Chernobyl, outweighs the benefits of clean energy. (pp. 597–600, 604–605)
- The disposal of nuclear waste remains a major dilemma. Temporary storage and single-repository plans each involve health, security, and environmental risks. (pp. 600–603)
- Economic factors and cost overruns have slowed the nuclear industry's growth. (pp. 602–607)

Describe the major sources, scale, and impacts of biomass energy

- The traditional source of biomass—wood from trees for cooking and for heating homes—remains the major one today, especially in developing nations. (pp. 607–608)
- Several processes are being developed to use biomass sources to generate electrical power. Many sources used are waste products from agriculture and forestry. (pp. 608, 610–611)
- Biofuels, which include ethanol and biodiesel, are now being used to power automobiles. Some crops are grown specifically for this purpose, and waste oils are also used. (pp. 608–610)
- Biomass energy theoretically adds no net carbon to the atmosphere and can use waste efficiently. But overharvesting of wood can lead to deforestation, and growing crops solely for fuel production is inefficient and ecologically damaging. (pp. 611–612)

Describe the scale, methods, and impacts of hydroelectric power

- Hydroelectric power is generated when water from a river runs through a powerhouse and turns turbines. (pp. 612–613)
- Dams and reservoirs and run-of-river systems are alternative approaches. (pp. 612–614)
- Hydropower produces no appreciable air pollution, but dams and reservoirs can greatly alter riverine ecology and local economies. (p. 615)

TESTING YOUR COMPREHENSION

1. How much of our global energy supply do nuclear power, biomass energy, and hydroelectric power contribute? How much of our global electricity do these three conventional energy alternatives generate?
2. Describe how nuclear fission works. How do nuclear plant engineers control fission and prevent a runaway chain reaction?
3. In terms of greenhouse gas emissions, how does nuclear power compare to coal, oil, and natural gas? How do hydropower and biomass energy compare?
4. In what ways did the incident at Three Mile Island differ from that at Chernobyl? What consequences resulted from each of these incidents?
5. What has been done so far about disposing of radioactive waste?
6. List five sources of biomass energy. What is the world's most-used source of biomass energy? How does biomass energy use differ between developed and developing nations?
7. Describe two biofuels, where each comes from, and how each is used.
8. Evaluate two potential benefits and two potential drawbacks of biomass energy.
9. Describe two major approaches to generating hydroelectric power.
10. Assess two benefits and two negative environmental impacts of hydroelectric power.

SEEKING SOLUTIONS

1. Given what you learned about fossil fuels in Chapter 19 and about some of the conventional alternatives discussed in this chapter, do you think it is important for us to minimize our use of fossil fuels and maximize our use of alternatives? If such a shift were to require massive public subsidies, how much should we subsidize this? What challenges or obstacles would we need to overcome to shift to such alternatives?
2. Nuclear power has by now been widely used for over three decades, and the world has experienced only one major accident (Chernobyl) responsible for any significant number of deaths. Would you call this a good safety record? Should we maintain, decrease, or increase our reliance on nuclear power? Why might safety at nuclear power plants be better in the future? Why might it be worse?
3. How serious a problem do you think the disposal of radioactive waste represents? How would you like to see this issue addressed?
4. There are many different sources of biomass and many ways of harnessing energy from biomass. Discuss one that seems particularly beneficial to you, and one with which you see problems. What biomass energy sources and strategies do you think our society should focus on investing in?
5. Discuss the advantages and impacts of hydropower. If there is a hydroelectric dam near where you live, consider what economic benefits and environmental impacts it has had in your area. Given this mix of effects, would you favor the construction of more dams and greater reliance on hydropower?
6. Imagine that you are the head of the national department of energy in a country that has just experienced a minor accident at one of its nuclear plants. A partial meltdown released radiation, but the radiation was fully contained inside the containment building, and there were no health impacts on area residents. However, citizens are terrified, and the media is playing up the dangers of nuclear power. Your country relies on its 10 nuclear plants for 25% of its energy and 50% of its electricity needs. It has no fossil fuel deposits and recently began a promising but still-young program to develop renewable energy options. What will you tell the public at your next press conference, and what policy steps will you recommend taking to assure a safe and reliable national energy supply?

INTERPRETING GRAPHS AND DATA

Recently, national security issues have received increased attention in energy policy discussions. Although nuclear reactors are potential targets for terrorists, they do lessen our dependence on petroleum imports. The United States produces reactor fuel but also relies on imports of uranium oxide from abroad, particularly from Canada, Australia, Russia, Kazakhstan, Uzbekistan, South Africa, and Namibia.

1. How much uranium oxide did the United States produce in 2004? How much did it export that year? How much did it import?
2. What was the net amount of uranium imported in 2004? How does this amount compare to domestic production for that same year? How do these amounts compare to the amounts in 1980?
3. What national security concerns do these data suggest to you? How might such concerns be alleviated?

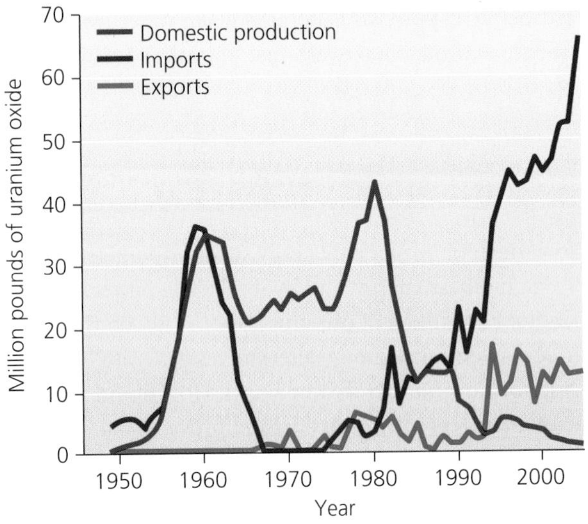

Production and trade of uranium oxide by the United States, 1949–2004. Data from Energy Information Administration. 2005. *Annual energy review 2004*. U.S. Department of Energy.

CALCULATING ECOLOGICAL FOOTPRINTS

Each of the conventional energy alternatives releases considerably less net carbon dioxide (CO_2) to the atmosphere than do any of the fossil fuels.

For each fuel source in the table, calculate net greenhouse gas emissions it produces while providing electricity to a typical U.S. household that uses 30 kilowatt-hours per day. Use data on emissions rates from each of the energy sources as provided in the figure from "The Science behind the Story" on ▸ p. 598. For each energy source, use the average of the high and low values given.

1. What is the ratio of greenhouse emissions from fossil fuels, on average, to greenhouse emissions from alternative energy sources, on average?
2. Why are there significant emissions from hydroelectric power, which is commonly touted as being a nonpolluting energy source?
3. Which energy source has the lowest emission rate? Would you advocate that the United States further develop that source? Why or why not?

Energy source	Greenhouse gas emission rate (g C_{eq} / kW-hr)	Time period		
		1 day	1 year	30 years
Coal	311	9,330	3,405,450	102,163,500
Oil				
Natural gas				
PV solar				
Hydroelectric				
Biomass				
Wind				
Nuclear				

Take It Further

Go to www.aw-bc.com/withgott or the student CD-ROM where you'll find:

- Suggested answers to end-of-chapter questions
- Quizzes, animations, and flashcards to help you study
- *Research Navigator*™ database of credible and reliable sources to assist you with your research projects
- GRAPHit! tutorials to help you master how to interpret graphs
- INVESTIGATEit! current news articles that link the topics that you study to case studies from your region to around the world

New Renewable Energy Alternatives

21

Iceland's Blue Lagoon and Svartsengi geothermal power plant

Upon successfully completing this chapter, you will be able to:

- Outline the major sources of renewable energy and assess their potential for growth
- Describe solar energy and the ways it is harnessed, and evaluate its advantages and disadvantages
- Describe wind energy and the ways it is harnessed, and evaluate its advantages and disadvantages
- Describe geothermal energy and the ways it is harnessed, and evaluate its advantages and disadvantages
- Describe ocean energy sources and the ways they can be harnessed, and evaluate their advantages and disadvantages
- Explain hydrogen fuel cells and assess future options for energy storage and transportation

Iceland's hydrogen buses arrive in Reykjavik

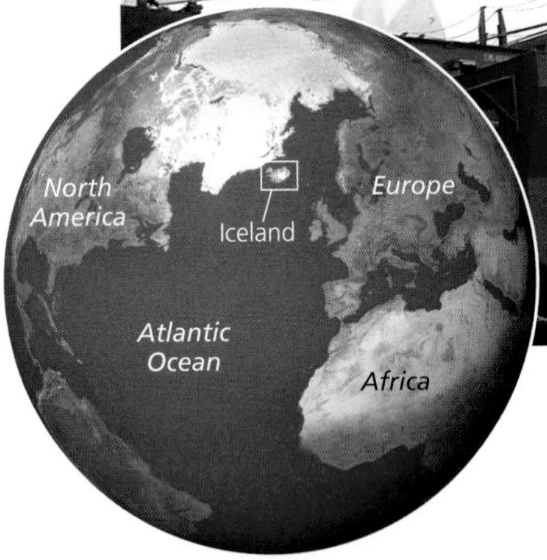

Central Case: Iceland Moves toward a Hydrogen Economy

"I believe that water will one day be employed as fuel, that hydrogen and oxygen which constitute it, used singly or together, will furnish an inexhaustible source of heat and light. . . . Water will be the coal of the future."

—Jules Verne, in *The Mysterious Island*, 1874

"Our long-term vision is of a hydrogen economy."

—Robert Purcell, Jr., executive director, General Motors, 2000

The Viking explorers who first set foot centuries ago on the remote island of Iceland were trailblazers. Today the citizens of the nation of Iceland are blazing a bold new path, one they believe the rest of the world may follow. Iceland aims to become the first nation to leave fossil fuels behind and convert to an economy based completely on renewable energy.

Iceland is essentially a hunk of lava the size of Kentucky that has risen out of the North Atlantic from the rift between tectonic plates known as the Mid-Atlantic Ridge. Most of Iceland's 290,000 people live in the capital city of Reykjavik, leaving much of the island an unpeopled and starkly beautiful landscape of volcanoes, hot springs, and glaciers.

The magma that gave birth to the island heats its groundwater in many places, giving Iceland some of the world's best sources of geothermal energy. Iceland has also dammed some of its many rivers for hydropower. Together these two renewable energy sources provide 73% of the country's energy supply and virtually all its electricity generation. Yet the nation, like most others, also depends on fossil fuels. Oil powers its automobiles, some of its factories, and its economically vital fishing fleet—and together these have given Iceland one of the highest per capita rates of greenhouse gas emission in the world. Because it possesses no fossil fuel reserves, all fossil fuels must be imported—a weak link in an otherwise robust economy that has given its citizens one of the highest per capita incomes in the world.

Enter Bragi Árnason, a University of Iceland professor popularly known as "Dr. Hydrogen." Árnason began arguing in the 1970s that Iceland could achieve independence from fossil fuel imports, boost its economy, and serve as a model to the world by converting from fossil fuels to a renewable energy economy based on hydrogen. He suggested zapping water with Iceland's cheap and renewable electricity in a chemical reaction called electrolysis, splitting the hydrogen from the oxygen and then using the hydrogen to power fuel cells that would produce and store electricity. The process is clean; nothing is combusted, and the only waste product is water.

In the late 1990s Árnason's countrymen began to listen. The nation's leaders decided to embark on a grand experiment to test the efficacy of switching to a "hydrogen economy." By setting an example for the rest of the world to follow, and by getting a head start at producing and exporting hydrogen fuel, these leaders hoped Iceland could become a "Kuwait of the North" and get rich by exporting hydrogen to an energy-hungry world, as Kuwait has done by exporting oil.

The leaders planning the shift to a hydrogen economy sketched a stepped transition in which fossil fuels would be phased out over 30–50 years. Conversion of the Reykjavik bus fleet to run on hydrogen fuel is the first step. After that, Iceland's 180,000 private cars would be powered by fuel cells, and then the fishing fleet would be converted to hydrogen. The final stage would be the export of hydrogen fuel to mainland Europe.

To make this happen, Icelanders in 1999 teamed up with corporate partners looking to develop technology for the future. Auto company Daimler-Chrysler is producing hydrogen buses, oil company Royal Dutch Shell is running hydrogen filling stations to fuel the buses, and Norsk Hydro is providing electrolysis technology.

In 2003, the world's first commercial hydrogen filling station opened in Reykjavik, and three hydrogen-fueled buses began operation. The public-private consortium, *Icelandic New Energy (INE)*, is monitoring the technology's effectiveness and the costs of developing infrastructure. Iceland's citizens are behind the effort; a recent poll showed 93% support among Icelanders.

Meanwhile, Daimler-Chrysler has introduced trios of hydrogen-fueled buses to nine other European cities. Hydrogen buses are also being developed by other companies and run in cities in Europe and throughout the world, from Tokyo to Chicago to Perth to Winnipeg. Hydrogen refueling stations are being demonstrated in Japan, Singapore, and the United States, and fuel-cell passenger automobiles are being tested in Japan and California. A global hydrogen economy could be closer than we suspect.

"New" Renewable Energy Sources

Iceland's bold drive toward a hydrogen economy is one facet of a global move toward renewable energy. Across the world, nations are taking different approaches to figure out how to move away from fossil fuels while ensuring a continued supply of energy for their economies.

In Chapter 20 we explored the two renewable energy sources that are most developed and widely used: biomass energy, the energy from combustion of wood and other plant matter, and hydropower, the energy from running water. These "conventional" alternatives to fossil fuels are renewable energy sources that can be depleted with overuse and that exert some undesirable environmental impacts.

In this chapter we explore a group of alternative energy sources that are often called "new renewables." These include energy from the sun, from wind, from Earth's geothermal heat, and from the movement of ocean water. These energy sources are not truly new. In fact, they are as old as our planet, and people have used them for millennia. They are commonly referred to as "new" because (1) they are not yet used on a wide scale in our modern industrial society, as are fossil fuels and the conventional renewable alternatives; (2) they are harnessed using technologies that are still in a rapid phase of development; and (3) it is widely believed that they will come to play a larger role in our energy use in the future.

The new renewables currently provide little of our power

The new renewable energy sources currently provide only 0.5% of our global primary energy supply. Fossil fuels provide 80% of the world's primary energy, nuclear energy provides 6.5%, and renewable energy sources account for 13.5%, nearly all of which is provided by biomass and hydropower (see Figure 20.1a, ▸ p. 591). The new renewables make a similarly small contribution to our global generation of electricity (see Figure 20.1b, ▸ p. 591). Less than 18% of our electricity comes from renewable energy, and of this amount, hydropower accounts for 90%.

Nations and regions vary in the renewable sources they use. In the United States, most energy supplied by renewable sources comes from hydropower and biomass, in nearly equal proportions. As of 2004, geothermal energy accounted for 5.6%, wind energy for 2.3%, and solar energy for 1.0% (Figure 21.1a). Of electricity generated in the United States from renewables, hydropower accounts

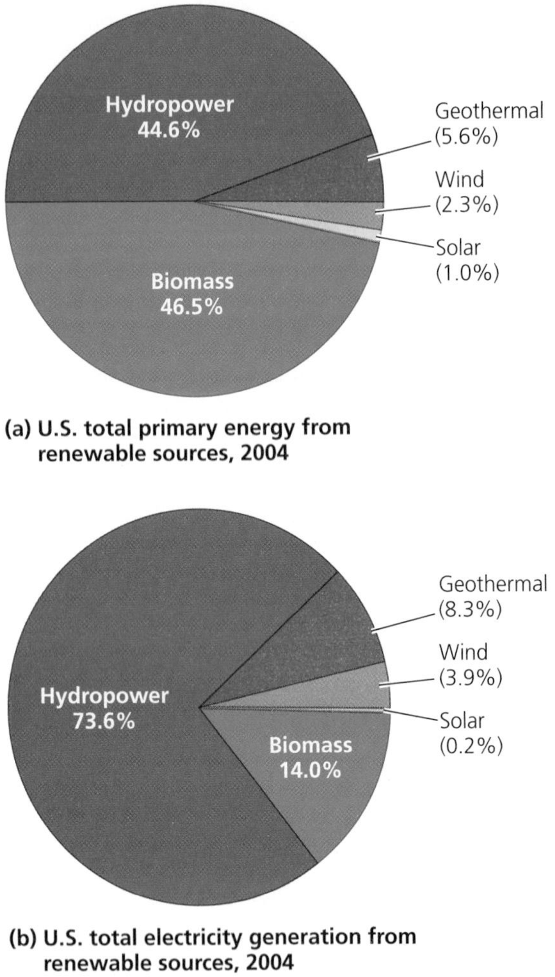

FIGURE 21.1 Only 6% of the total primary energy consumed in the United States each year comes from renewable sources. Of this amount, most derives from hydropower and biomass energy. Geothermal energy accounts for 5.6% of this amount, wind for 2.3%, and solar for 1.0% **(a)**. Similarly, only 9% of electricity generated in the United States comes from renewable energy sources, predominantly hydropower and biomass. Geothermal energy accounts for 8.3%, wind for 3.9%, and solar for only 0.2% **(b)**. Data from Energy Information Administration, U.S. Department of Energy. 2005. *Annual Energy Review, 2004.*

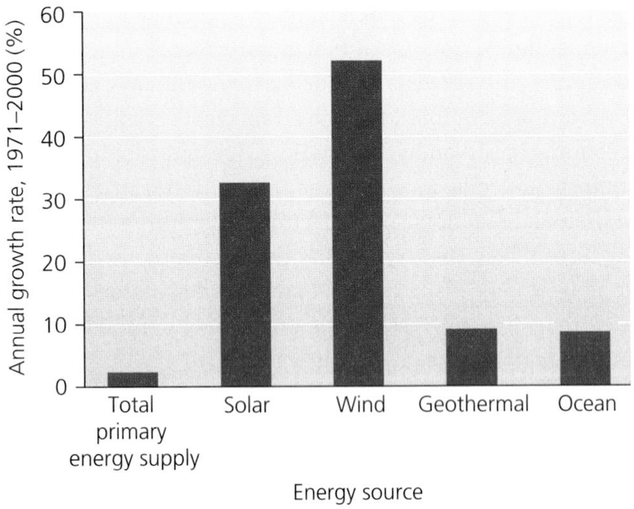

FIGURE 21.2 The "new renewable" energy sources are growing substantially faster than the total primary energy supply. Solar power has grown by 32% each year since 1971, and wind power has grown by 52% each year. Because these sources began from such low starting levels, however, their overall contribution to our energy supply is still small. Go to GRAPHIt! at www.aw-bc.com/withgott or on the student CD-ROM. Data from International Energy Agency Statistics, 2002.

for nearly 75%, and biomass 14%. As of 2004, geothermal power accounted for 8.3%, wind for 3.9%, and solar for just 0.2% (Figure 21.1b).

The new renewables are growing fast

Although they comprise only a minuscule proportion of our energy budget, the new renewable energy sources are growing at much faster rates than are conventional energy sources. Over the past three decades, solar, wind, geothermal, and ocean energy sources have grown far faster than has the overall primary energy supply (Figure 21.2). Among renewable sources, the leader in growth is wind power, which has expanded by about 50% *each year* over the past three decades. Because these sources started from such low levels of use, however, it will take them some time to catch up to conventional sources. For instance, the absolute amount of energy added by a 50% increase in wind power is still less than the amount added by just a 1% increase in oil, coal, or natural gas.

The rapid growth of the new renewables has been motivated by concerns over diminishing fossil fuel supplies and the environmental impacts of fossil fuel combustion (Chapter 19). As these concerns build, advances in technology are making it easier and less expensive to harness renewable energy sources. The new renewables promise several benefits over fossil fuels. As they replace fossil fuels, they help alleviate air pollution and the greenhouse gas emissions that drive global climate change (Chapter 18). Unlike fossil fuels, renewable sources are inexhaustible on time scales relevant to human societies. Developing renewables can also help diversify an economy's mix of energy, lowering price volatility and protecting against supply restrictions such as those caused by the 1973 oil embargo or by Hurricane Katrina in 2005 (▸ p. 579). New energy sources also can create new employment opportunities and sources of income and property tax for local communities, especially in rural areas passed over by other economic development. In many rural areas and developing countries, locally based renewable sources may prove cheaper to use for electricity than would extending the electricity grid infrastructure out from cities.

Rapid growth in renewable energy sectors seems likely to continue as population and consumption grow, global energy demand expands, fossil fuel supplies decline, and citizens demand cleaner environments. More governments, utilities, corporations, and consumers are now promoting and using renewable energy sources, and, as a result, the costs of renewables are falling.

The transition cannot be immediate, but it must be soon

If our civilization is to persist in the long term, it will need to shift to renewable energy sources. A key question is whether we will be able to shift soon enough and smoothly enough to avoid widespread war, social unrest, and further damage to the environment. The answer to this question will largely determine the quality of life for all of us in the coming decades.

We cannot switch completely to renewable energy sources overnight, because there are technological and economic barriers. Currently, most renewables lack adequate technological development and lack infrastructure to transfer power on the required scale. However, dramatic improvements in technology and infrastructure in recent years suggest that most of the remaining barriers are political. Renewable energy sources have received far less in subsidies, tax breaks, and other incentives from governments than have conventional sources. By one estimate, of the $150 billion in subsidies the U.S. government provided to nuclear, solar, and wind power in the past half century, the nuclear industry received 96%, the solar industry received 3%, and the wind industry less than 1%. For decades, research and development of renewable sources have suffered from the continuing availability of fossil fuels made inexpensive in part by government policy.

Many corporations in the fossil fuel and automobile industries understand that they will eventually need to switch to renewable sources. They also know that when the time comes, they will need to act fast to stay ahead of their competitors. However, in light of continuing short-term profits and unclear policy signals, companies have not been eager to invest in the transition. Under these circumstances, our best hope may be for a gradual transition from fossil fuels to renewable energy sources, one driven largely by economic supply and demand. However, if the transition proceeds too slowly—if we wait solely for economics to do its work, without government encouragement—diminishing fossil fuel supplies could outpace our ability to develop new sources, and we could find our economies greatly disrupted, and our environment greatly degraded. Thus, encouraging the speedy development of renewable energy alternatives holds promise for bringing us a vigorous and sustainable energy economy without the environmental impacts of fossil fuels.

Solar Energy

The sun provides energy for almost all biological activity on Earth (▸ pp. 105–106) by converting hydrogen to helium in nuclear fusion (▸ pp. 595–596). Each square meter of Earth's surface receives about 1 kilowatt of solar energy—17 times the energy of a lightbulb. As a result, an average-sized house whose roof is covered in panels that harness solar energy has enough roof area to generate all its power needs. The sun's raw energy is so strong that if only 0.1% of Earth's surface—roughly the combined area of New Mexico and South Dakota—were covered with solar panels, we would have enough solar energy to power all the world's electrical plants. The amount of energy Earth receives from the sun each day, if it could be collected in full for our use, would be enough to power human consumption for 27 years.

Clearly, the potential for using sunlight to meet our energy needs is tremendous. However, all this "free" energy from the sun cannot be harnessed just yet. We are still in the process of developing solar technologies and learning the most effective and cost-efficient ways to put the sun's energy to use.

The most commonly used way to harness solar energy is through **passive solar** energy collection. In this approach, buildings are designed and building materials are chosen to maximize direct absorption of sunlight in winter, even as they keep the interior cool in the heat of summer. This approach contrasts with **active solar** energy collection, which makes use of technological devices to focus, move, or store solar energy.

We tend to think of using solar power as a novel phenomenon, but people have chosen and designed their living sites with passive solar collection in mind for millennia. Moreover, solar energy was first harnessed with active solar technology in 1767, when Swiss scientist Horace de Saussure built a thermal solar collector to heat water and cook food. In 1891, U.S. inventor Clarence Kemp claimed the first commercial patent for a solar-powered water heater. Two California entrepreneurs bought the patent rights and outfitted one-third of the homes in Pasadena, California, with solar water heaters by 1897.

Passive solar heating is simple and effective

One passive solar design technique involves installing low, south-facing windows to maximize sunlight capture in the winter (in the Northern Hemisphere; north-facing windows are used in the Southern Hemisphere). Overhangs block light from above, shading these windows in the summer, when the sun is high in the sky and when cooling, not heating, is desired. Passive solar techniques also include the use of heat-absorbing construction materials. Often called *thermal mass,* these materials absorb heat, store it, and release it later. Thermal mass (of straw, brick, concrete, or other materials) most often makes up floors, roofs, and walls, but also can comprise portable blocks.

Thermal mass may be strategically located to capture sunlight in cold weather and radiate heat in the interior of the building. In warm weather, the mass should be located away from sunlight so that it absorbs warmed air in the interior to cool the building. Passive solar design can also involve planting vegetation in particular locations around a building. By heating buildings in cold weather and cooling them in warm weather, passive solar methods help conserve energy and reduce energy costs.

Active solar energy collection can heat air and water in buildings

One active method for harnessing solar energy involves using *solar panels* or *flat-plate solar collectors,* most often installed on rooftops. These panels generally consist of dark-colored, heat-absorbing metal plates mounted in flat boxes covered with glass panes. Water, air, or antifreeze solutions are run through tubes that pass through the collectors, transferring heat throughout a building. Heated water can be pumped to tanks to store the heat for later use and through pipes designed to release the heat into the building. Such systems have proven especially effective for heating water for residences.

Active solar energy is being used for heating, cooling, and water purification in Gaviotas, a remote town in the high plains of Colombia far from any electrical grid. Engineers have developed inexpensive solar panels that harvest enough solar energy, even under cloudy skies, to boil drinking water for a family of four. They also have designed, built, and installed a unique solar refrigerator in a rural hospital, along with a large-scale solar collector to boil and sterilize water. Their innovations show that solar power does not need to be expensive or confined to regions that are always sunny.

Concentrating solar rays magnifies the energy received

The strength of solar energy can be magnified by gathering sunlight from a wide area and focusing it on a single point. This is the principle behind *solar cookers,* simple portable ovens that use reflectors to focus sunlight onto food and cook it (see Figure 3.13, ▸ p. 71). Such cookers are becoming widespread and proving extremely useful in parts of the developing world.

The principle of concentrating the sun's rays has also been put to work by utilities in large-scale, high-tech approaches to generating electricity. In one approach, mirrors concentrate sunlight onto a receiver atop a tall "power-tower" (Figure 21.3). From the receiver, heat is

FIGURE 21.3 At the Solar Two facility operated by Southern California Edison in the desert of southern California, mirrors are spread across wide expanses of land to concentrate sunlight onto a receiver atop a "power-tower." Heat is then transported through fluid-filled pipes to a steam-driven generator that produces electricity.

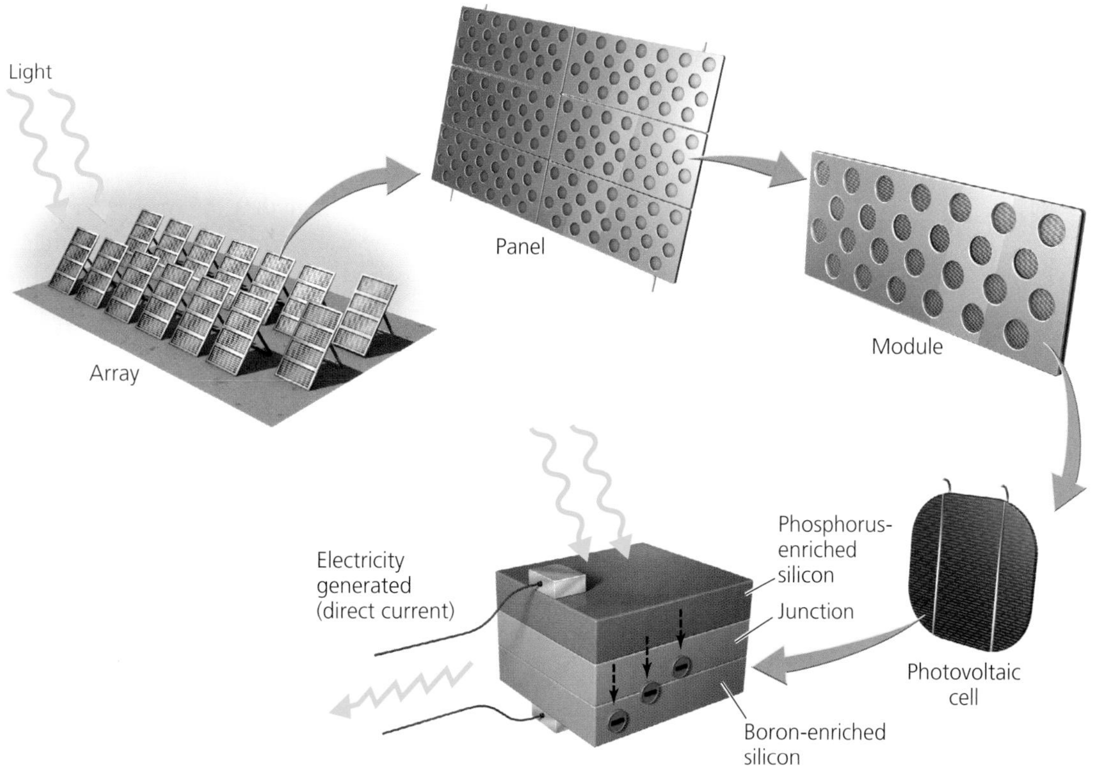

FIGURE 21.4 A photovoltaic (PV) cell converts sunlight to electrical energy. When sunlight hits a layer of silicon infused with phosphorus, electrons are released and travel toward the layer of silicon laced with boron. This movement of electrons induces an electric current, producing electricity. PV cells are grouped in modules, which can comprise panels, which can be erected in arrays.

transported by fluids (often molten salts) piped to a steam-driven generator to create electricity. These solar power plants can harness light from large mirrors spread across many hectares of land. The world's largest such plant so far—a collaboration among government, industry, and utility companies in the California desert—produces power for 10,000 households. Another approach is the use of solar-trough collection systems, which consist of mirrors that gather sunlight and focus it on oil in troughs. The superheated oil creates steam that drives turbines, as in conventional power plants.

Photovoltaic cells generate electricity directly from sunlight

A more direct approach to producing electricity from sunlight involves photovoltaic (PV) systems. **Photovoltaic (PV) cells** collect sunlight and convert it to electrical energy directly by making use of the *photovoltaic,* or *photoelectric effect,* first proposed in 1839 by French physicist Edmund Becquerel. This effect occurs when light strikes one of a pair of metal plates in a PV cell, causing the release of electrons, which are attracted by electrostatic forces to the opposing plate. The flow of electrons from one plate to the other creates an electrical current, which can be converted into alternating current (AC) and used for residential and commercial electrical power (Figure 21.4).

The plates of a PV cell are made primarily of silicon, enriched on one side with phosphorus and on the other with boron. Silicon is a semiconductor, so it conducts and controls the flow of electricity. Because of the chemical properties of boron and phosphorus, the phosphorus-enriched side has excess electrons, and the boron-enriched side has fewer electrons. When sunlight strikes the cell surface, it transfers energy and causes electrons to move. When wires connect the two sides, electricity is created as electrons flow from the phosphorus-enriched side to the boron-enriched side. Photovoltaic cells can be connected to batteries that store the accumulated charge until it is needed.

You may be familiar with small PV cells that power your watch or your calculator. Atop the roofs of homes and other buildings, multiple PV cells are arranged in modules. These modules can comprise panels, which can be gathered together in flat arrays. Increasingly, PV roofing tiles are being used instead of these arrays. PV roofing tiles look like normal roofing shingles but generate electricity by the photovoltaic effect. In some remote areas, such as Xcalak, Mexico, PV systems are being used in combination with wind turbines (▸ pp. 627–628) and a diesel generator to power entire villages.

Solar power is little used but fast growing

Although active solar technology dates from the 18th century, it was pushed to the sidelines as fossil fuels gained a stronger foothold in our energy economy. Even as solar technology was being refined for applications ranging from handheld calculators to spacecraft, it was not being developed for the roles that fossil fuels have filled. In recent U.S. history, funding for research and development of solar technology has been erratic. After the 1973 oil embargo, the U.S. Department of Energy funded the installation and testing of over 3,000 PV systems, providing a boost to companies in the solar industry. As oil prices declined, however, so did government support for solar power, and funding for solar has remained far below that for fossil fuels.

Largely because of the lack of investment, solar power currently contributes only a minuscule portion of our energy production. In 2004, solar accounted for only 0.06%—less than 6 parts in 10,000—of the U.S. primary energy supply, and only 0.02% of U.S. electricity generation. However, use of solar energy has grown by nearly one-third annually worldwide since 1971, a growth rate second only to that of wind power. Solar power is proving especially attractive in developing countries, many of which are rich in sun but poor in power infrastructure, and where hundreds of millions of people are still without electricity. Some multinational corporations that built themselves on fossil fuels are now investing in alternative energy as well. BP Solar, British Petroleum's solar energy wing, recently completed $30 million projects in the Philippines and Indonesia, and is working on a $48 million project to supply electricity to 400,000 people in 150 villages.

Sales of PV cells are growing fast—by 25% per year in the United States, for instance, and by 63% annually in Japan, which uses PV roofing tiles widely. Use of solar technology is widely expected to continue increasing as prices fall, technologies improve, and economic incentives are enacted. However, the very small amount of energy currently produced by solar power means that its market share will likely remain small for years or decades to come—unless governments, businesses, and consumers become more motivated by the benefits that solar energy can provide.

Solar power offers many benefits

The fact that the sun will continue burning for another 4–5 billion years makes it practically inexhaustible as an energy source for human civilization. Moreover, the amount of solar energy reaching Earth's surface should be enough to power our civilization once solar technology is adequately developed. Although these overarching benefits of solar energy are clear, the technologies themselves also provide benefits. PV cells and other solar technologies use no fuel, are quiet and safe, contain no moving parts, require little maintenance, and do not even require a turbine or generator to create electricity. An average unit can produce energy for 20–30 years.

Another advantage of solar systems is that they allow for local, decentralized control over power. Homes, businesses, and isolated communities can use solar power to produce their own electricity and may not need to be near a power plant or connected to the grid of a city.

In developing nations, solar cookers enable families to cook food without gathering fuelwood; as a result, they lessen people's daily workload and help reduce deforestation. In locations such as refugee camps, solar cookers are helping relieve social and environmental stress. The low cost of solar cookers—many can be built locally for $2–10 each—has made them available for purchase or donation in many impoverished areas.

In the developed world, most PV systems today are connected to the regional electric grid. As a result, owners of houses with PV systems can sell their excess solar energy to their local power utility through a process called *two-way metering.* The value of the power the consumer sells to the utility is subtracted from the consumer's monthly utility bill.

Finally, a major advantage of solar power over fossil fuels is that it does not pollute the air with greenhouse gas emissions and other air pollutants. The manufacture of photovoltaic cells *does* currently require fossil fuel use, but once it is up and running, a PV system produces no emissions. Consumers may be able to access online calculators to calculate the economic and environmental impacts of installing PV solar cells. For example, a calculator offered by BP Solar estimated that installing a 5-kilowatt PV system in a home in Dallas, Texas, can provide 51% of annual power needs, save $391 per year on energy bills, and prevent the emission of 9,023 lb of carbon dioxide from fossil fuel combustion. Even in overcast Seattle, Washington, a 5-kilowatt system can produce 32% of energy needs, save $336 per year, and prevent the emission of 6,419 lb of carbon dioxide.

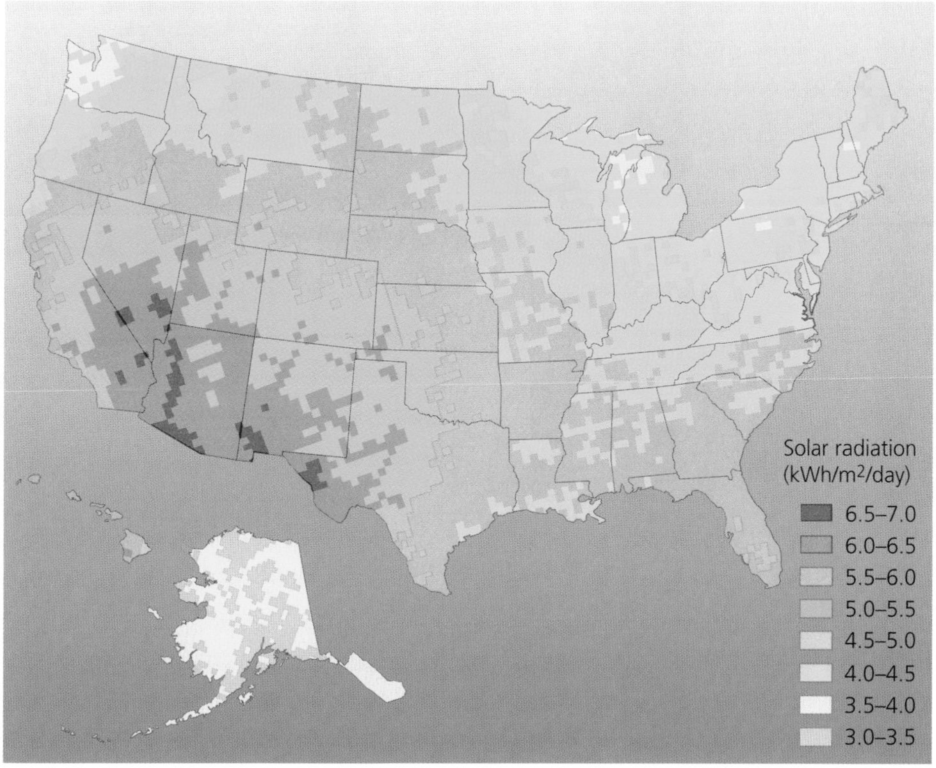

FIGURE 21.5 Because some locations receive more sunlight than others, harnessing solar energy is more profitable in some areas than in others. In the United States, many areas of Alaska and the Pacific Northwest receive only 3–4 kilowatt-hours per square meter per day, whereas most areas of the Southwest receive 6–7 kilowatt-hours per square meter per day. Data from National Renewable Energy Laboratory, U.S. Department of Energy, 2005.

Location and cost can be drawbacks for solar power

Solar power currently has two major disadvantages. One is that not all regions are sunny enough to provide adequate power, given current technology. Although Earth as a whole receives vast amounts of sunlight, not every location on Earth does (Figure 21.5). People in cities such as Seattle might find it difficult to harness enough sunlight most of the year to rely on solar power. Daily or seasonal variation in sunlight can also pose problems for stand-alone solar systems if storage capacity in batteries or fuel cells is not adequate or if backup power is not available from a municipal electricity grid.

The primary disadvantage of current solar technology— as with other renewable sources—is the up-front cost of investing in the equipment. The investment cost for solar is higher than that for fossil fuels, and indeed, solar power remains the most expensive way to produce electricity. Proponents of solar power argue that decades of government promotion of fossil fuels and nuclear power—which have received many financial breaks that solar power has not—have made solar power unable to compete. Because the external costs (▸ pp. 43–44) of nonrenewable energy have not been included in market prices, these energy sources have remained relatively cheap. Governments, businesses, and consumers thus have had little economic incentive to switch to solar and other renewables.

However, decreases in price and improvements in energy efficiency of solar technologies so far are encouraging, even in the absence of significant financial commitment from government and industry. At their advent in the 1950s, solar technologies had efficiencies of around 6%, while costing $600 per watt. Recent single-crystal silicon PV cells are showing 15% efficiency commercially and 24% efficiency in lab research, suggesting that future solar technologies may be more efficient than any energy technologies we have today. Solar systems have become much cheaper over the years and now can often pay for themselves in 10–15 years. After that time, they provide energy virtually for free as long as the equipment lasts. With future technological advances, some experts believe that the time to recoup investment could fall to 1–3 years.

Wind Energy

Wind energy can be thought of as an indirect form of solar energy, because it is the sun's differential heating of air masses on Earth that causes wind to blow. We can harness power from wind by using devices called **wind turbines,** mechanical assemblies that convert wind's kinetic energy, or energy of motion, into electrical energy.

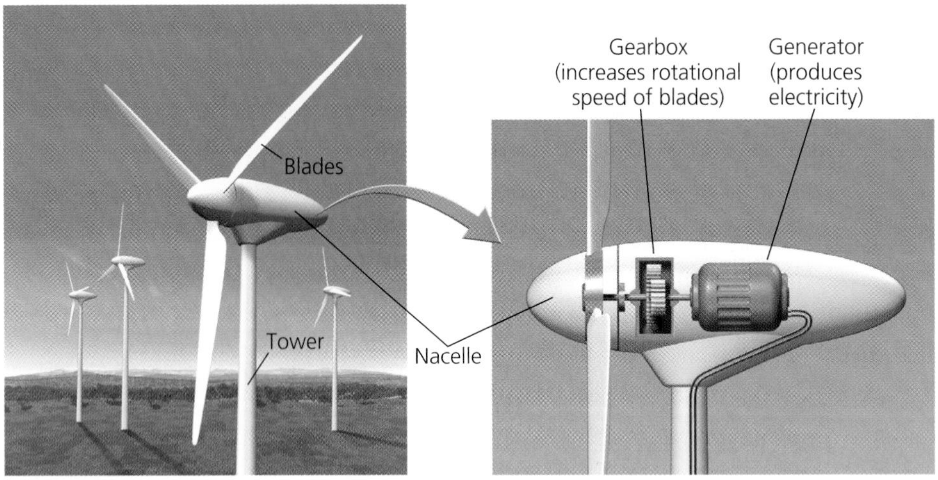

FIGURE 21.6 A wind turbine converts wind's energy of motion into electrical energy. Wind causes the blades of a wind turbine to spin, turning a shaft that extends into the nacelle that is perched atop the tower. Inside the nacelle, a gearbox converts the rotational speed of the blades, which can be up to 20 revolutions per minute (rpm) or more, into much higher rotational speeds (over 1,500 rpm). These high speeds provide adequate motion for a generator inside the nacelle to produce electricity.

Wind has long been used for energy

Today's wind turbines have their historical roots in Europe, where wooden windmills have been used for 800 years. The Netherlands in particular is known for its windmills, whose power has been used to pump water to drain wetlands and irrigate crops, and to grind grain into flour. In each application, wind causes a windmill's blades to turn, driving a shaft connected to several cogs that turn wheels, which either grind grain or pull buckets from a well. In the United States, countless ranches in arid areas of the West and Great Plains feature windmills that draw groundwater up to supply thirsty cattle.

The first wind turbine built to generate electricity was constructed in the late 1800s in Cleveland, Ohio, by inventor Charles Brush, who designed a turbine 17 m (50 ft) tall with 144 rotor blades made of cedar wood. But technology advanced slowly during the 20th century, and it was not until after the 1973 oil embargo that wind energy was funded by governments in North America and Europe. This moderate infusion of funding for research and development boosted technological progress, and the cost of wind power was cut in half in less than 10 years. Today wind power at favorable locations generates electricity for nearly as little cost per kilowatt-hour as do conventional sources, and modern wind turbines appear more like airplane propellers or sleek new helicopters than romantic old Dutch paintings.

Modern wind turbines convert kinetic energy to electrical energy

Wind blowing into a turbine turns the blades of the rotor, which rotate the machinery inside a compartment called a *nacelle,* which sits atop a tall tower (Figure 21.6). Inside the nacelle are a gearbox and a generator, as well as equipment to monitor and control the turbine's activity. Most of today's towers range from 40 to 100 m (131–328 ft) tall. Higher is generally better, to minimize turbulence (and potential damage) and to maximize wind speed. Most rotors consist of three blades and measure 42–80 m (138–262 ft) across. Turbines are designed to yaw, or rotate back and forth in response to changes in wind direction, ensuring that the motor faces into the wind at all times. Turbines can be erected singly, but they are most often erected in groups called wind parks, or *wind farms.* The world's largest wind farms contain several hundred or thousand turbines spread across the landscape.

Engineers have designed turbines to begin turning at specific wind speeds to harness wind energy as efficiently as possible. Some turbines create low levels of electricity by turning in light breezes. Others are programmed to rotate only in strong winds, operating less frequently but generating large amounts of electricity in short time periods. Slight differences in wind speed can yield substantial differences in power output, for two reasons. First, the energy content of a given amount of wind increases as the square of its velocity; thus if wind velocity doubles, energy quadruples. Second, an increase in wind speed causes more air molecules to pass through the wind turbine per unit time, making power output equal to wind velocity cubed. Thus a doubled wind velocity actually results in an eightfold increase in power output.

Wind power is the fastest-growing energy sector

Like solar energy, wind provides only a minuscule proportion of the world's power needs, but wind power is growing fast—nearly 30% per year globally between 2000 and 2004. Wind provided 3.9% of U.S. renewable electricity

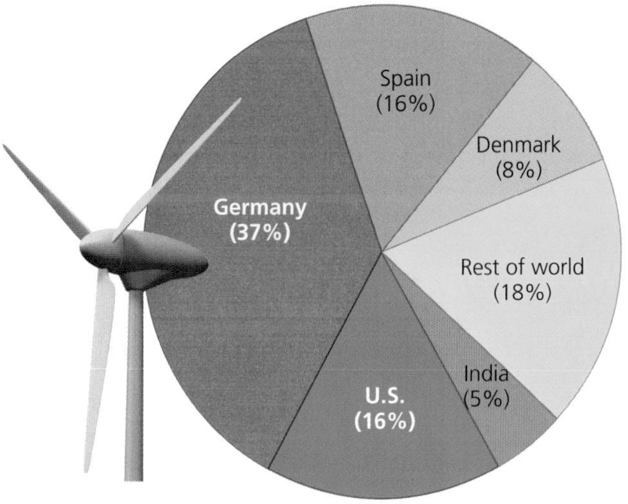

FIGURE 21.7 Most of the world's fast-growing wind power generating capacity is concentrated in a handful of countries. Tiny Denmark obtains the highest percentage of its energy needs from wind, but the larger nations of Germany, the United States, and Spain have so far developed more total wind capacity. Data from Global Wind Energy Council; and American Wind Energy Association. 2005. *Global wind energy market report*. AWEA.

generation in 2004—a small amount but nearly 20 times more than solar power. So far, wind energy production is geographically concentrated; only five nations account for 82% of the world's wind energy output (Figure 21.7). California and Texas account for two-thirds of the wind power generated within the United States. Denmark is a leader in wind power; there, a series of wind farms supplies over 20% of the nation's electricity needs. Experts agree that wind power's rapid growth will continue, because only a very small portion of this resource is currently being tapped. Meteorological evidence suggests that wind power could be expanded in the United States to meet the electrical needs of the entire country (see "The Science behind the Story," ▸ p. 632).

Offshore sites can be promising

Wind speeds on average are roughly 20% greater over water than over land. There is also less air turbulence over water surfaces than over land surfaces. For these reasons, offshore wind turbines are becoming popular Figure 21.8). Although costs to erect and maintain turbines in water are higher, the stronger, less turbulent winds produce more power and make offshore wind potentially more profitable. Currently, offshore wind farms are limited to shallow water, where towers are sunk into sediments singly or with a tripod configuration to stabilize them. However, in the future towers may be placed on floating pads anchored to the seafloor in deep water. At great distances from land, it may be best to store the generated electricity as hydrogen and then ship or pipe this to land (instead of building submarine cables to carry electricity to shore), but further research is needed on this option.

Denmark erected the first offshore wind farm in 1991. Over the next decade, nine more came into operation across northern Europe, where the North and Baltic Seas offer strong winds. The power output of these farms increased by 43% annually as larger turbines were erected. Several northern European nations are encouraging continued rapid

FIGURE 21.8 More and more wind farms are being developed offshore, because offshore winds tend to be stronger yet less turbulent. Denmark is a world leader in wind power, and much of it comes from offshore turbines. This Danish wind farm is one of several that provide over 20% of the nation's electricity.

growth in the near future. Wind advocates in Iceland are considering developing 240 offshore wind turbines in the nation's waters to meet future electricity demand for its hydrogen economy.

Wind power has many benefits

Like solar power, wind produces no emissions once the necessary equipment is manufactured and installed. As a replacement for fossil fuel combustion in the average U.S. utility generator, the U.S. Environmental Protection Agency has calculated that running a 1-megawatt wind turbine for one year prevents the release of more than 1,500 tons of carbon dioxide, 6.5 tons of sulfur dioxide, 3.2 tons of nitrogen oxides, and 60 lb of mercury. The amount of carbon pollution that all U.S. wind turbines together prevent from entering the atmosphere is greater than the cargo of a 50-car freight train, with each car holding 100 tons of solid carbon, each and every day.

Wind power appears considerably more energy-efficient than conventional power sources. One recent study, which compared the amount of energy that various types of technology produce to the amount they consume, found that wind turbines produce 23 times as much as they consume. For nuclear energy, the ratio was 16:1; for coal it was 11:1; and for natural gas it was 5:1. Wind farms also use less water than do conventional power plants.

Wind turbine technology can be used on many scales, from a single tower for local use to fields of thousands that supply large regions. Small-scale turbine development can help make local areas more self-sufficient, just as solar energy can. For instance, the Rosebud Sioux Tribe of Native Americans in 2003 set up a single turbine on their reservation in South Dakota. The turbine is producing electricity for 220 homes and brings the tribe an estimated $15,000 per year in revenue. Wind resources are rich in this region, and the tribe plans to develop a wind farm nearby in coming years.

Another societal benefit of wind power is that landowners can lease their land for wind development, which provides them extra revenue while also increasing property tax income for rural communities. A single large turbine can bring in $2,000–4,500 in annual royalties while occupying just a quarter-acre of land. Because each turbine takes up only a small area, most of the land can still be used for farming, ranching, or other uses.

Economically, wind energy involves up-front costs for the erection of turbines and the expansion of infrastructure to allow electricity distribution, but over the lifetime of a project it requires only maintenance costs. Unlike fossil fuel power plants, the turbines incur no ongoing fuel costs. Currently, startup costs of wind farms generally are higher than those of fossil-fuel-driven plants, but wind farms incur fewer expenses once they are up and running. Advancing technology is driving down the costs of wind farm construction; as large wind farms become more efficient, the cost of each unit of electricity produced is dropping.

Wind energy has some downsides

Unlike power sources that can be turned off and on at will, wind is an intermittent resource; we have no control over when wind will occur. This poses little problem, however, if wind is only one of several sources contributing to a utility's power generation. Moreover, several technologies are available to address problems posed by relying on intermittent wind resources. For example, batteries or hydrogen fuel can store energy generated by wind and release it later when needed.

Just as wind varies from time to time, it also varies from place to place. Some areas are simply windier than others. Global wind patterns combine with local topography—mountains, hills, water bodies, forests, cities—to create local wind patterns, and companies study these patterns closely before investing in a wind farm. Meteorological research has given us information with which to judge prime areas for locating wind farms. A map of average wind speeds across the United States (Figure 21.9a) shows that mountainous regions and areas of the Great Plains are best. Based on such information, the young wind power industry has located much of its generating capacity in states with high wind speeds (Figure 21.9b), and is seeking to expand in the Great Plains and mountain states. Provided that wind farms are strategically erected in optimal locations, an estimated 15% of U.S. energy demand could be met using only 43,000 km^2 (16,600 mi^2) of land (with less than 5% of this land area actually occupied by turbines, equipment, and access roads).

Good wind resources, however, are not always near population centers that need the energy. Thus, transmission networks would need to be greatly expanded. Moreover, when wind farms *are* proposed near population centers, local residents often oppose them. Turbines are generally located in exposed, conspicuous sites, and many people object to wind farms for aesthetic reasons, feeling that the structures clutter the landscape. Although polls show wide public approval of wind projects in regions where wind power has already been introduced, new wind projects often elicit the so-called *not-in-my-backyard (NIMBY)* syndrome. For instance, a proposal for North America's first offshore wind farm, in Nantucket Sound between Cape Cod and the islands of Nantucket and Martha's Vineyard, has faced stiff opposition from wealthy

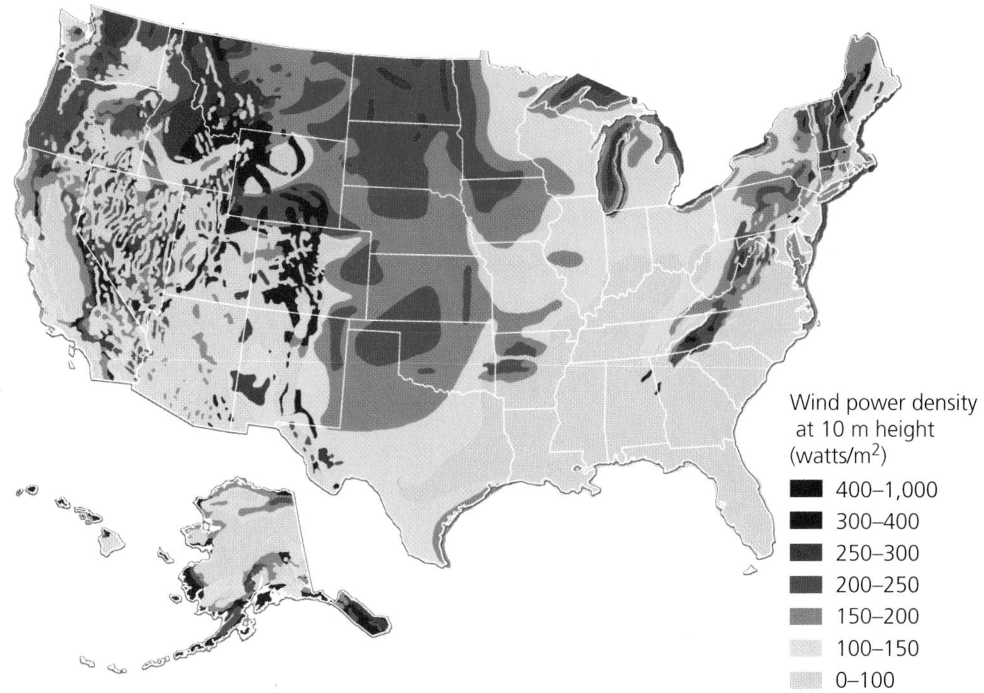

(a) Annual average wind power

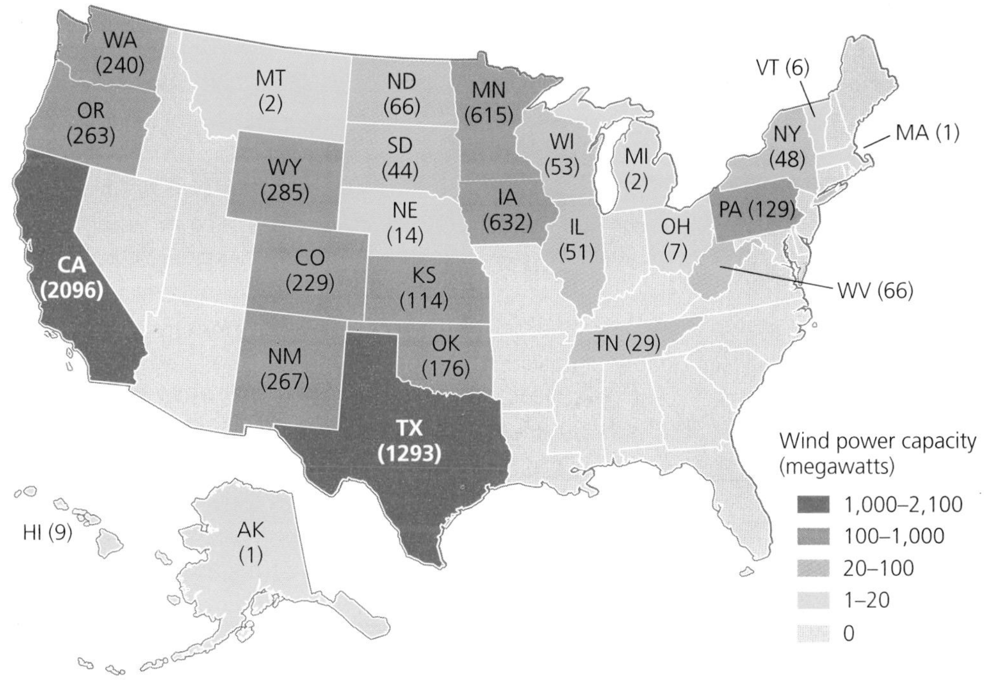

(b) Wind generating capacity, 2004

FIGURE 21.9 Wind's capacity to generate power varies according to wind speed. Meteorologists have measured wind speed to calculate the potential generating capacity from wind in different areas. The map in (**a**) shows average wind power in watts per square meter at a height of 10 m (33 ft) above ground across the United States. Such maps are used to help guide placement of wind farms. The development of U.S. wind power so far is summarized in (**b**), which shows the megawatts of generating capacity developed in each state through the end of 2004. *Sources:* (a) Elliott, D. L., et al. 1987. *Wind energy resource atlas of the United States.* Golden, CO: Solar Energy Research Institute; (b) National Renewable Energy Laboratory, U.S. Department of Energy, 2005.

The Science behind the Story

Idaho's Wind Prospectors

Where does wind translate into energy? In Idaho, where resource planners decided that the state's power future lies in generating electricity from wind. Now scores of Idahoans, from small farmers to Native American tribes, have joined in the search for gusts with energy potential.

By handing out wind-measuring devices to interested landowners, Idaho has turned its citizens into "wind prospectors" who pinpoint potential areas for wind farms. Idaho first launched the public wind prospecting program in 2001, after joining several other northwestern states in a regional research effort. People who join the program must collect data on wind speed and direction, share that data with the state, and agree to make it public.

A promising wind farm site requires some infrastructure, such as roads for erecting wind turbines and transmission lines for sending out power the turbines generate. But the single most important factor is the speed and frequency of the wind. Effective commercial wind farms have a steady flow of wind just above ground level, with regular gusts of at least 21 km/hr (13 mi/hr) at a height of about 50 m (164 ft).

Determining whether a site merits further study starts with analyzing existing data. In many parts of the developed world, decades worth of weather information have been compiled into computerized wind maps that indicate general wind conditions. In Idaho, energy planners provide prospectors with starter maps that divide the state into seven wind "classes" and reveal which general areas might have enough wind to make a wind farm worthwhile. Areas listed as "Class 3" or higher, with wind speeds of about 23 km (14.3 mi) per hour at 50 m (164 ft) above ground, offer the best possibilities.

To determine where to build wind farms, Idaho's wind prospectors use anemometers, which collect and relay wind data. Cup wheels rotate to indicate wind speed, and a vane turns on a vertical axis to reveal wind direction.

Such maps, however, may not provide enough detail about a specific location. A piece of property, for example, may sit in a Class 3 area but be sheltered by a small hill that blocks the wind. Knowing that kind of detail requires site-specific on-the-ground research.

To make such research possible, Idaho loans landowners in areas listed as Class 3 or higher devices called *anemometers* (see the figure), which measure wind speed and direction. The Idaho program uses a common cup anemometer, with an array of three or four hollow cups set to catch the wind and rotate around a vertical rod. The force of wind on the cups causes them to rotate at a speed proportional to the wind speed; the greater the wind, the faster the cups rotate. Wind direction is measured by a vane that turns on a vertical axis pointing directly into the wind. The cup wheel and wind vane are connected electrically to speed and direction dials, which relay wind data.

More than 80 landowners borrowed anemometers from the state in the Idaho program's first year, and sent data to the state every 60 days for review by energy planners and for subsequent posting online. The studies have generated new funding and wind farm plans in the state. In the fall of 2003, one farm near Idaho Falls won a \$500,000 federal grant to help build a 1.5-megawatt wind farm that could supply power for approximately 500 homes.

In eastern Idaho, five anemometers set up on Shoshone-Bannock tribal lands have revealed good prospects for a commercial wind farm on two Native American reservations. The research effort has shown that the lands are "world-class sites" for wind power, according to a state energy official. With average wind speeds in the 29 km/hr (18 mi/hr) range, further study of the tribal lands revealed possible sites for large-scale commercial wind farms, which could mean jobs and revenue for the reservations. Similar wind prospecting programs are now under way on other reservations, as well as in other states, including Utah, Oregon, Virginia, and Missouri.

area residents, even though many of these residents like to think of themselves as environmentalists.

Wind turbines are also known to pose a threat to birds and bats, which can be killed when they fly into the rotating blades. At California's Altamont Pass wind farm, which is located in a region with one of the densest populations of golden eagles in the country, turbines killed many eagles and other raptors during the 1990s. Studies since then at other sites have suggested that bird deaths may be a less severe problem than was initially feared. It has been estimated that roughly one to two birds are killed per turbine per year—far fewer than the millions already being killed annually by television, radio, and cell phone towers, tall buildings, automobiles, domestic cats, pesticides, and other human causes. Bat mortality may be higher, but more research is needed. The key for protecting birds and bats seems to be selecting sites that are not on flyways or in the midst of prime habitat for species that are likely to fly into the blades.

Weighing the Issues: Wind and NIMBY

If you could choose to get your electricity from a wind farm or a coal-fired power plant, which would you choose? How would you react if the electric utility proposed to build the wind farm that would generate your electricity atop a ridge running in back of your neighborhood, such that the turbines would be clearly visible from your living room window? Would you support or oppose the development? Why? If you would oppose it, where would you suggest the farm be located? Do you think anyone might oppose it in that location?

Geothermal Energy

Geothermal energy is one form of renewable energy that does not originate from the sun. Instead, it is generated from deep within Earth. The radioactive decay of elements amid the extremely high pressures deep in the interior of our planet generates heat that rises to the surface through magma (molten rock, ▸ p. 206) and through fissures and cracks. Where this energy heats groundwater, natural spurts of heated water and steam are sent up from below. Terrestrial geysers and submarine hydrothermal vents (▸ pp. 106–107) are the surface manifestations of these processes. Iceland is built from magma that extruded above the ocean's surface and cooled—magma from the Mid-Atlantic Ridge (▸ pp. 209, 471–472), the area of volcanic activity along the spreading boundary of two tectonic plates. Because of the geothermal heat in this region, volcanoes and geysers are numerous in Iceland. In fact, the word *geyser* originated from the Icelandic *Geysir,* the name for the island's largest geyser, which recently resumed its periodic eruptions after many years in dormancy.

Geothermal power plants use the energy of naturally heated water to generate power. Rising underground water and steam are harnessed to turn turbines and create electricity. Geothermal energy is renewable in principle (its use does not affect the amount of heat produced in Earth's interior), but the power plants we build to use this energy may not all be capable of operating indefinitely. If a geothermal plant uses heated water at a rate faster than the rate at which groundwater is recharged, the plant will eventually run out of water. This is occurring at The Geysers, in Napa Valley, California, where the first generator was built in 1960. In response, operators have begun injecting municipal wastewater into the ground to replenish the supply. More and more geothermal power plants throughout the world are now injecting water, after it is used, back into aquifers to help maintain pressure and thereby sustain the resource. A second reason geothermal energy may not always be renewable is that patterns of geothermal activity in Earth's crust shift naturally over time, so an area that produces hot groundwater now may not always do so.

Geothermal energy is harnessed for heating and electricity

Geothermal energy can be harnessed directly from geysers at the surface, but most often wells must be drilled down hundreds or thousands of meters toward heated groundwater. Generally, water at temperatures of 150–370 °C (300–700 °F) or more is brought to the surface and converted to steam by lowering the pressure in specialized compartments. The steam is then employed in turning turbines to generate electricity (Figure 21.10).

Hot groundwater can also be used directly for heating homes, offices, and greenhouses; for driving industrial processes; and for drying crops. Iceland heats most of its homes through direct heating with piped hot water. Iceland began putting geothermal energy to use in the 1940s, and today 30 municipal district heating systems and 200 small private rural networks supply heat to 86% of the nation's residences. Other locales are benefiting in similar ways; the Oregon Institute of Technology heats its buildings with geothermal energy for 12–14% of the cost it would take to heat them with

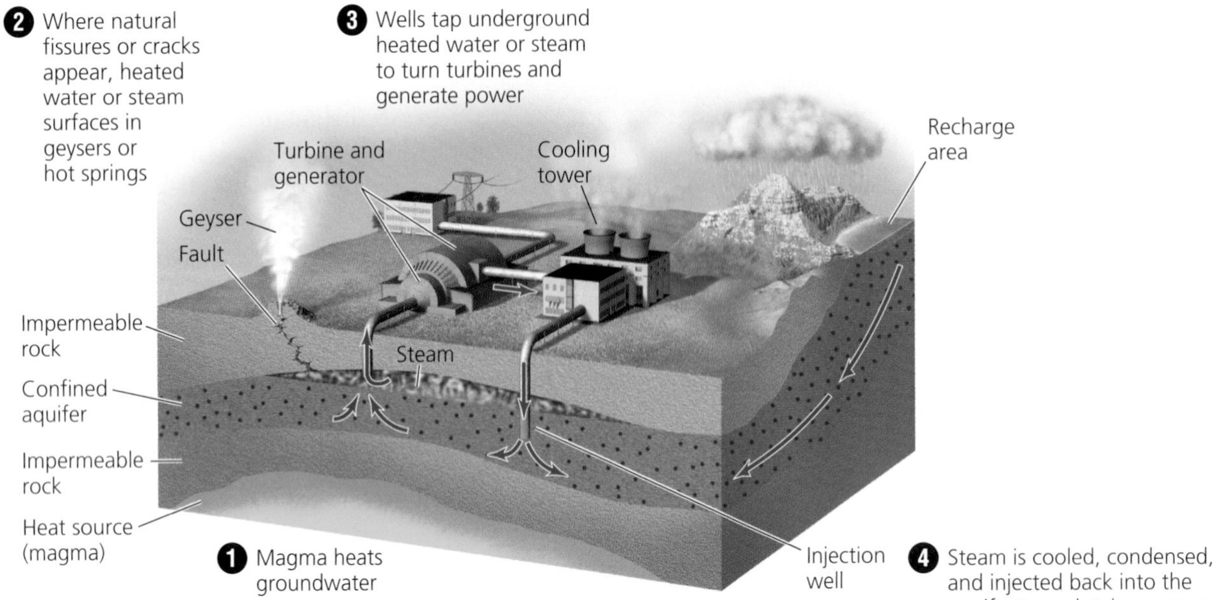

2 Where natural fissures or cracks appear, heated water or steam surfaces in geysers or hot springs
3 Wells tap underground heated water or steam to turn turbines and generate power
Turbine and generator
Cooling tower
Recharge area
Geyser
Fault
Impermeable rock
Steam
Confined aquifer
Impermeable rock
Heat source (magma)
1 Magma heats groundwater
Injection well
4 Steam is cooled, condensed, and injected back into the aquifer to maintain pressure

(a) Geothermal energy

(b) Nesjavellir geothermal power station, Iceland

FIGURE 21.10 With geothermal energy **(a)** magma heats groundwater deep in the earth (1), some of which is let off naturally through surface vents such as geysers (2). Geothermal facilities tap into heated water below ground and channel steam through turbines in buildings to generate electricity (3). After being used, the steam is often condensed and pumped back into the aquifer to maintain pressure (4). At Nesjavellir geothermal power station in Iceland **(b),** steam is piped from four wells to a condenser at the plant where cold water pumped from lakeshore wells 6 km (3.7 mi) away is heated. The water, heated to 83 °C (181 °F), is sent through an insulated 270-km (170-mi) pipeline to Reykjavik and environs, where residents use it for washing and space heating.

natural gas. Such direct use of naturally heated water is cheap and efficient, but it is feasible only in areas such as Iceland or parts of Oregon, where geothermal energy sources are available and near where the heat must be transported.

Thermal energy from water or solid earth can also be used to drive a heat pump to provide energy. Geothermal *ground source heat pumps* (GSHPs) use thermal energy from near-surface sources of earth and water rather than the deep geothermal heat for which utilities drill. Roughly half a million GSHPs are already used to heat U.S. residences. Compared to conventional electric heating and cooling systems, GSHPs heat spaces 50–70% more efficiently, cool them 20–40% more efficiently, can reduce electricity use by 25%–60%, and can reduce emissions by up to 72%. These pumps work because soil does not vary in temperature from season to season as much as air does. The pumps heat buildings in the winter by transferring heat from the ground into buildings; they cool buildings in the summer by transferring heat from buildings into the ground. Both types of heat transfer are accomplished by a single network of underground plastic pipes that circulate water. Because heat is simply moved from place to place rather than being produced using outside energy inputs, heat pumps can be highly energy-efficient.

Use of geothermal power is growing

Geothermal energy provides less than 0.5% of total primary energy used worldwide. It provides more power than solar and wind combined, but only a small fraction of the power from hydropower and biomass. Geothermal energy in the United States provides enough power to supply electricity to over 1.4 million homes. At the world's largest geothermal power plants, The Geysers in northern California, generating capacity has declined by more than 50% since 1989 as steam pressure has declined, but The Geysers still provide enough electricity to supply a million residents. Currently Japan, China, and the United States lead the world in use of geothermal power.

Geothermal power has benefits and limitations

Like other renewable sources, geothermal power greatly reduces emissions relative to fossil fuel combustion. Geothermal sources can release variable amounts of gases dissolved in their water, including carbon dioxide, methane, ammonia, and hydrogen sulfide. However, these gases are generally in very small quantities, and it has been estimated that geothermal facilities on average release only one-sixth of the carbon dioxide produced by plants fueled by natural gas. Geothermal facilities using the latest filtering technologies produce even fewer emissions. By one estimate, each megawatt of geothermal power prevents the emission of 7.8 million lb of carbon dioxide emissions and 1,900 lb of other pollutant emissions from gas-fired plants each year.

On the negative side of the ledger, geothermal sources, as we have seen, may not always be truly sustainable. In addition, the water of many hot springs is laced with salts and minerals that corrode equipment and pollute the air. These factors may shorten the lifetime of plants, increase maintenance costs, and add to pollution.

Moreover, use of geothermal energy is limited to areas where the energy can be tapped. Unless technology is developed to penetrate far more deeply into the ground, geothermal energy use will remain more localized than solar, wind, biomass, or hydropower. Places such as Iceland are rich in geothermal sources, but most of the world is not. In the United States, geysers exist in some areas, such as Yellowstone National Park, and hot groundwater and steam exist in various locations in the western part of the country. Nonetheless, many hydrothermal resources remain unexploited around the world, awaiting improved technology and governmental encouragement of their development.

Ocean Energy Sources

The oceans are home to several underexploited energy sources. Each involve continuous natural processes that could potentially provide sustainable energy for our needs. Of the three approaches developed so far, two involve motion and one involves temperature.

We can harness energy from tides and waves

Just as dams on rivers use flowing freshwater to generate hydroelectric power, some scientists, engineers, businesses, and governments are developing ways to use the motion of ocean water to generate electrical power. Two types of kinetic energy show the most promise so far: the energy of wave motion and the energy of tidal motion.

The rising and falling of ocean tides twice each day at coastal sites throughout the world can move large amounts of water past any given coastal point. Differences in height between low and high tides are especially great in long, narrow bays such as Alaska's Cook Inlet or the Bay of Fundy between New Brunswick and Nova Scotia. Such locations are best for harnessing tidal energy, which is accomplished by erecting dams across the outlets of tidal basins. The incoming tide flows through sluices past the dam, and as the outgoing tide passes through the dam, it turns turbines to generate electricity (Figure 21.11). Some designs allow for generating electricity from water moving in both directions. The world's largest tidal generating facility is the La Rance facility in France, which has operated for over 30 years. Smaller facilities now operate in China, Russia, and Canada. Tidal stations release few or no pollutant emissions, but they can have impacts on the ecology of estuaries and tidal basins.

Wave energy could be developed at a greater variety of sites than could tidal energy. The principle is to harness the motion of wind-driven waves at the ocean's surface and convert this mechanical energy into electricity. Many designs for machinery to harness wave energy have been invented, but few have been adequately tested. Some designs are for offshore facilities and involve floating devices that move up and down with the waves. Wave energy is greater at deep-ocean sites, but transmitting the electricity produced to shore would be expensive.

Other designs are for coastal onshore facilities. Some of these designs funnel waves from large areas into narrow channels and elevated reservoirs, from which water is then allowed to flow out, generating electricity as hydroelectric dams do. Other coastal designs use rising and falling

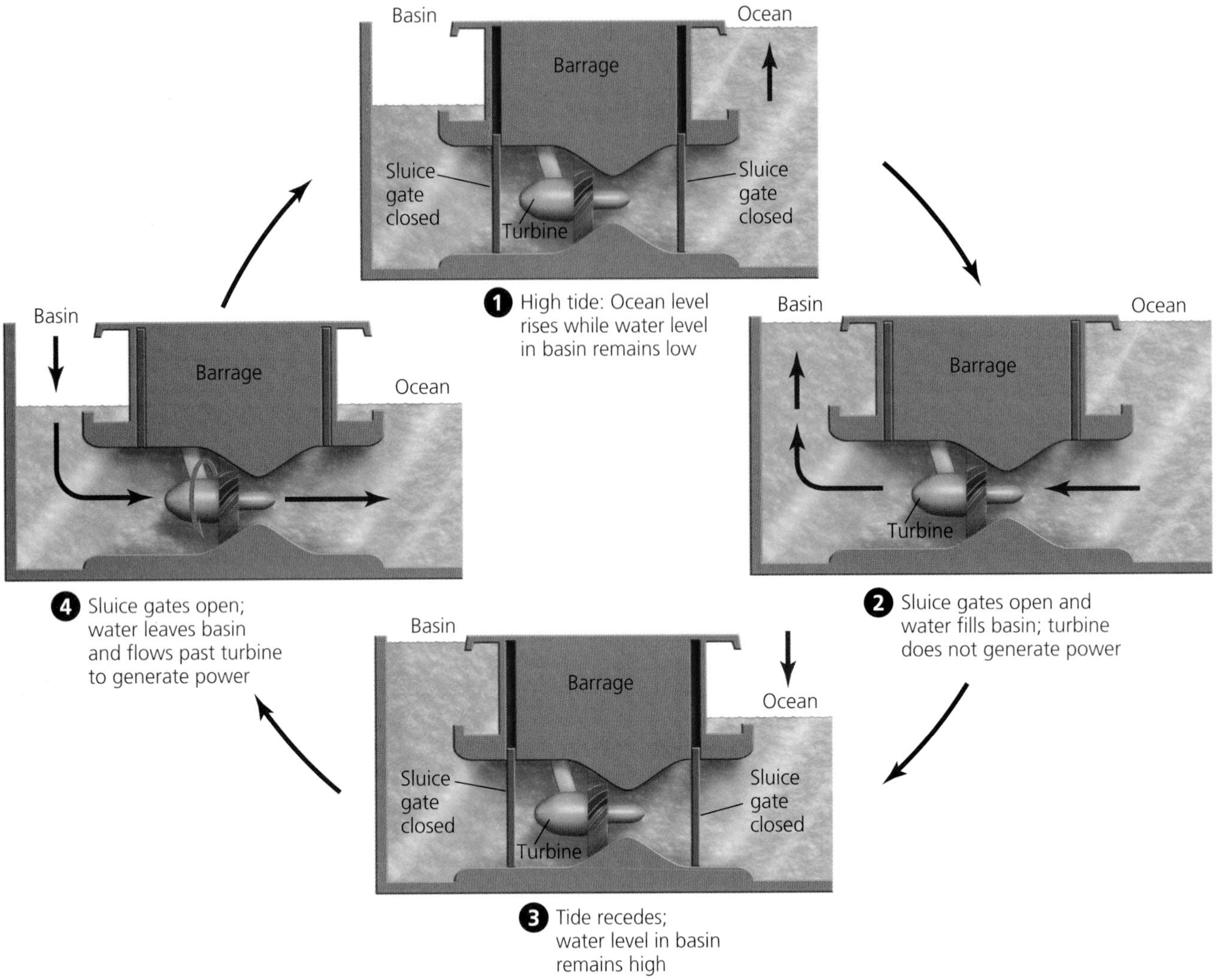

FIGURE 21.11 Energy can be extracted from the movement of the tides at coastal sites where tidal flux is great enough. One way of doing so is involves using bulb turbines in concert with the outgoing tide. At high tide, ocean water is let through the sluice gates, filling an interior basin. At low tide, the basin water is let out into the ocean, spinning turbines to generate electricity.

waves to push air into and out of chambers, turning turbines to generate electricity (Figure 21.12). No commercial wave energy facilities are operating yet, but some have been deployed as demonstration projects in several western European nations.

The ocean stores thermal energy

Besides the motion of tides and waves, other oceanic energy sources we have not yet effectively tapped include the motion of ocean currents, chemical gradients in salinity, and the immense thermal energy contained in the oceans. The concept of **ocean thermal energy conversion (OTEC)** has been most fully developed. Each day the tropical oceans absorb an amount of solar radiation equivalent to the heat content of 250 billion barrels of oil—enough to provide 20,000 times the electricity used daily in the United States. The ocean's sun-warmed surface is higher in temperature than its deep water, and OTEC approaches are based on this gradient in temperature.

In the *closed cycle* approach, warm surface water is piped into a facility to evaporate chemicals, such as ammonia, that boil at low temperatures. These evaporated gases spin turbines to generate electricity. Cold water piped in from ocean depths then condenses the gases so they can be reused. In the *open cycle* approach, the warm

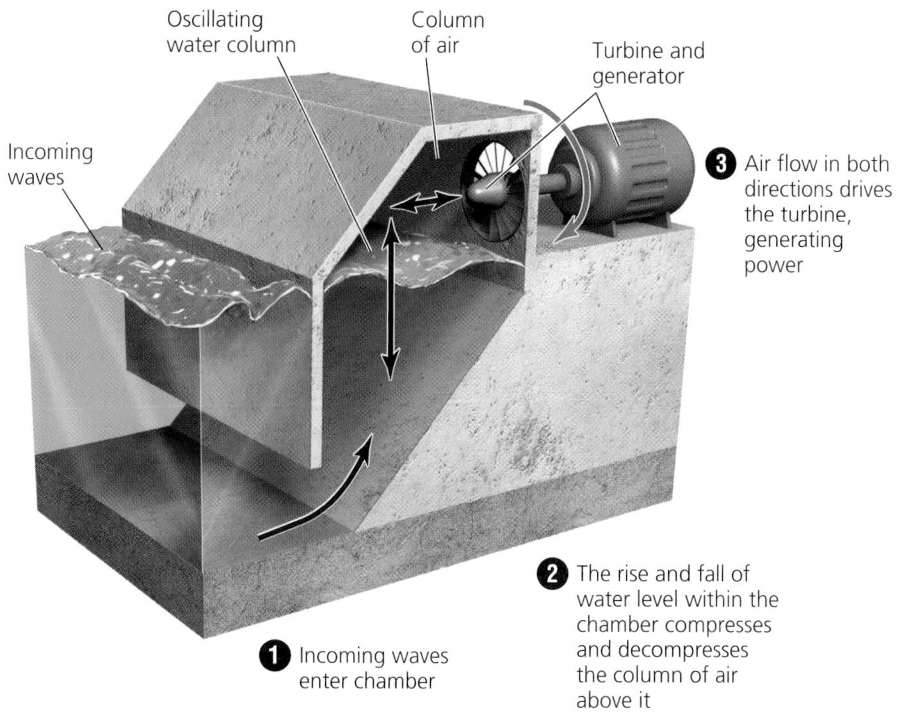
Oscillating water column
Column of air
Turbine and generator
Incoming waves
3 Air flow in both directions drives the turbine, generating power
2 The rise and fall of water level within the chamber compresses and decompresses the column of air above it
1 Incoming waves enter chamber

FIGURE 21.12 Coastal facilities can make use of energy from the motion of ocean waves. As waves are let into and out of a tightly sealed chamber, the air inside is compressed and decompressed, creating air flow that rotates turbines to generate electricity.

surface water is evaporated in a vacuum, and its steam turns the turbines and then is condensed by the cold water. Because the ocean water loses its salts as it evaporates, water can be recovered, condensed, and sold as desalinized freshwater for drinking or agriculture. Research on OTEC systems has been conducted in Hawaii and other locations, but costs remain high, and as of yet no facility is commercially operational.

Weighing the Issues: Your Island's Energy?

Imagine you have been elected the president of an island nation the size of Iceland, and your nation's congress is calling on you to propose a national energy policy. Unlike Iceland, your country is located in equatorial waters. Your geologists do not yet know whether there are fossil fuel deposits or geothermal resources under your land, but your country gets a lot of sunlight and a fair amount of wind, and broad, shallow shelf regions surround its coasts. Your island's population is moderately wealthy but is growing fast, and importing fossil fuels from mainland nations is becoming increasingly expensive.

What approaches would you propose in your energy policy? What specific steps would you urge your congress to fund immediately? What trade relationships would you seek to establish with other countries? What questions would you ask of your economic advisors? What questions would you fund your country's scientists to research?

Hydrogen

All the renewable energy sources we have discussed can be used to generate electricity more cleanly than can fossil fuels. As useful as electricity is to us, however, it cannot be stored easily in large quantities for use when and where it is needed. This is why vehicles rely on fossil fuels for power. The development of fuel cells and hydrogen fuel show promise to store energy conveniently and in considerable quantities and to produce electricity at least as cleanly and efficiently as renewable energy sources.

In the "hydrogen economy" that Iceland's leaders and many energy experts worldwide envision, hydrogen fuel, together with electricity, will serve as the basis for a clean, safe, and efficient energy system. This system will use as a fuel the universe's simplest and most abundant element. In this system, electricity generated from renewable sources that are intermittent, such as wind or solar energy, can be used to produce hydrogen. Fuel cells can then employ hydrogen to produce electrical energy as needed to power vehicles, computers, cell phones, home heating, and countless other applications.

Fuel cell technology has been used since the 1960s in NASA's space flight programs. Basing an energy system on hydrogen could alleviate dependence on foreign fuels and help fight climate change. For these reasons, governments are funding research into hydrogen and fuel cell technology, and automobile companies are investing in research and development to produce vehicles that run on hydrogen.

The Science behind the Story

Algae as a Hydrogen Fuel Source

As scientists search for new ways to generate energy, some are looking past wind farms and solar panels to an unlikely power source—pond scum. Algae are being studied as an innovative way to generate large amounts of hydrogen to move society toward a more sustainable energy future.

Hydrogen's benefits hinge on how hydrogen fuel is produced. Some methods release substantial amounts of carbon dioxide, and other, nonpolluting, processes can be costly. These drawbacks have kept scientists searching for new hydrogen sources.

At the University of California at Berkeley, plant biologist Anastasios Melis thought one possible hydrogen source might be a single-celled aquatic plant known to be a capable, if sporadic, hydrogen producer. The alga *Chlamydomonas reinhardtii* was known to emit small amounts of hydrogen for brief periods of time when deprived of light.

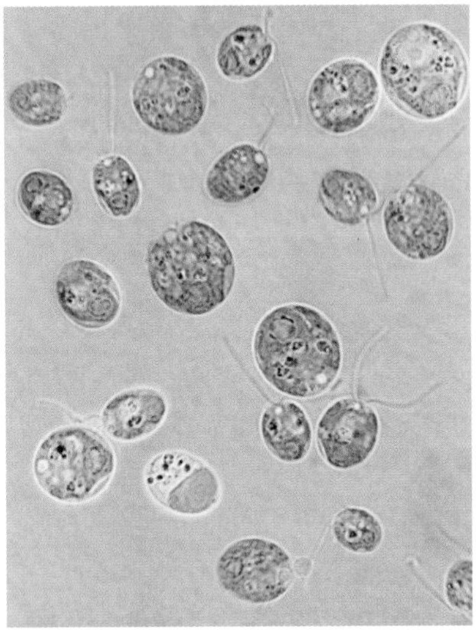

Could green algae such as this provide hydrogen for our energy needs?

Melis hypothesized that the alga might be encouraged to produce hydrogen in large amounts. He set up an experiment with energy experts at the National Renewable Energy Laboratory in Colorado, aiming to develop ways to tweak the alga's basic biological functions so that the plant produced greater quantities of hydrogen.

Green algae, like terrestrial green plants, photosynthesize, drawing in carbon dioxide and water, absorbing energy from light that converts those nutrients into food, and then expelling oxygen as a waste product. Additional nutrients from soil or water, and catalysts called *enzymes* within the plant, keep this process running smoothly. To conduct photosynthesis effectively, *Chlamydomonas reinhardtii* needs the element sulfur as a nutrient. The alga also contains an enzyme called *hydrogenase,* which can trigger the alga to stop producing oxygen as a metabolic by-product and start releasing hydrogen instead.

Hydrogen fuel may be produced from water or from other matter

Hydrogen gas (H_2) does not tend to exist freely on Earth; rather, hydrogen atoms bind to other molecules, becoming incorporated in everything from water to organic molecules. To obtain hydrogen gas for fuel, we must force these substances to release their hydrogen atoms, and this requires an input of energy. Several potential ways of producing hydrogen are being studied (see "The Science behind the Story," above). In **electrolysis,** the process being pursued by Iceland, electricity is input to split hydrogen atoms from the oxygen atoms of water molecules:

$$2H_2O \rightarrow 2H_2 + O_2$$

Electrolysis produces pure hydrogen, and it does so without emitting the carbon- or nitrogen-based pollutants of fossil fuel combustion. However, whether this strategy for producing hydrogen will cause pollution over its entire life cycle depends on the source of the electricity used for the electrolysis. If coal is burned to create the electricity, then the entire process will not reduce emissions compared with reliance on fossil fuels. If, however, the electricity is produced by some less-polluting renewable source, then hydrogen production by electrolysis would create much less pollution and greenhouse warming than reliance on fossil fuels. The "cleanliness" of a future hydrogen economy in Iceland or anywhere else would, therefore, depend largely on the source of electricity used in electrolysis.

The environmental impact of hydrogen production will also depend on the source material for the hydrogen. Besides water, hydrogen can be obtained from biomass and fossil fuels. Obtaining hydrogen from these sources generally requires less energy input, but results in emissions

Hydrogenase is normally active only after *Chlamydomonas reinhardtii* has been deprived of light. When the alga is deprived of light, the light-dependent reactions of photosynthesis ebb, little oxygen is produced, and hydrogenase is activated. When light returns and the alga begins producing oxygen again, hydrogenase is promptly deactivated, and its associated hydrogen release stops.

Melis's team wanted to activate hydrogenase so that more hydrogen would be produced. But simply keeping the algae in the dark would not escalate hydrogen production because the alga's metabolic functions slowed without light, resulting in small amounts of released hydrogen.

The researchers decided to try limiting the alga's oxygen output another way, by putting it on a sulfur-free, bright-light regimen. The lack of sulfur would hinder photosynthesis, limiting oxygen output enough to activate hydrogenase and trigger hydrogen production. The presence of light would keep the algae metabolically active and releasing large amounts of byproducts.

The researchers cultured large quantities of the algae in bottles in labs. Then they deprived the cultures of sulfur but kept the algae exposed to light for long periods of time—in some cases up to 150 hours. After the sustained light exposure, gas and liquids were extracted from the bottles and analyzed.

The analysis supported the team's hypothesis. Without sulfur or photosynthesis, the algae were not producing oxygen. This low-oxygen, or anaerobic, environment had induced hydrogenase, which spurred the algae to begin splitting water molecules and releasing gas. The plants had released amounts of hydrogen that were substantial relative to the size of the algal cultures. Hydrogen also dominated the alga's emissions—in gas collection analysis, approximately 87% of the gas was hydrogen, 1% was carbon dioxide, and the remaining 12% was nitrogen with traces of oxygen. The research teams published their findings in the journal *Plant Physiology* in 2000.

Many questions remain about algae-derived hydrogen, particularly how much fuel can be harvested continuously using this *photobiological* process. Nevertheless, the research results so far are helping to fuel the momentum of a future hydrogen economy.

Within 30 years, some federal energy experts predict that photobiological methods for generating hydrogen could be commonplace—meaning cars on future freeways might just be powered by pond scum.

of carbon-based pollutants. For instance, extracting hydrogen from the methane (CH_4) in natural gas entails producing one molecule of the greenhouse gas carbon dioxide for every four molecules of hydrogen gas:

$$CH_4 + 2H_2O \rightarrow 4H_2 + CO_2$$

Thus, whether a hydrogen-based energy system is environmentally cleaner than a fossil fuel system depends on how the hydrogen is extracted.

In addition, some new research suggests that leakage of hydrogen from the production, transport, and use of the gas at Earth's surface could potentially deplete stratospheric ozone and lengthen the atmospheric lifetime of the greenhouse gas methane. Research into these questions is ongoing, because scientists do not want society to switch from fossil fuels to hydrogen without first knowing the possible risks from hydrogen.

Weighing the Issues: Precaution over Hydrogen?

Some environmental scientists have recently warned that we do not yet know enough about the environmental consequences of replacing fossil fuels with hydrogen fuel. An increase in tropospheric hydrogen gas would deplete hydroxyl (OH) radicals, they hypothesize, possibly leading to stratospheric ozone depletion and global warming from increased concentrations of methane. Some scientists say such effects will be small; others say there could be further effects that are presently unknown. Do you think we should apply the precautionary principle to the development of hydrogen fuel and fuel cells? Or should we embark on pursuing a hydrogen economy before knowing all the scientific answers? What factors inform your view?

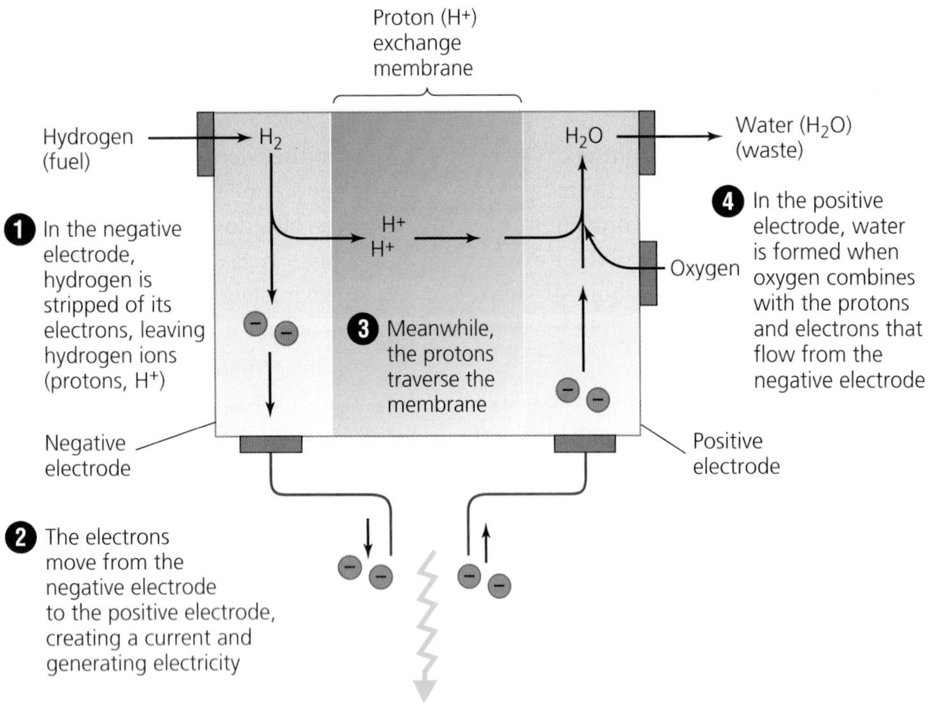
Proton (H^+) exchange membrane
Hydrogen (fuel)
H_2
H_2O
Water (H_2O) (waste)
1 In the negative electrode, hydrogen is stripped of its electrons, leaving hydrogen ions (protons, H^+)
H^+
H^+
Oxygen
4 In the positive electrode, water is formed when oxygen combines with the protons and electrons that flow from the negative electrode
3 Meanwhile, the protons traverse the membrane
Negative electrode
Positive electrode
2 The electrons move from the negative electrode to the positive electrode, creating a current and generating electricity

FIGURE 21.13 Hydrogen fuel drives electricity generation in a fuel cell, creating water as a waste product. Atoms of hydrogen are first stripped of their electrons (1). The electrons move from a negative electrode to a positive one, creating a current and generating electricity (2). Meanwhile, the hydrogen ions pass through a proton exchange membrane (3) and combine with oxygen to form water molecules (4).

Fuel cells produce electricity by joining hydrogen and oxygen

Once hydrogen gas has been isolated, it can be used as a fuel to produce electricity within fuel cells. The chemical reaction involved in a fuel cell is simply the reverse of that shown for electrolysis; an oxygen molecule and two hydrogen molecules each split so that their atoms can bind and form two water molecules:

$$2H_2 + O_2 \rightarrow 2H_2O$$

The way this occurs within one common type of fuel cell is shown in Figure 21.13. Hydrogen gas (usually compressed and stored in an attached fuel tank) is allowed into one side of the cell, whose middle consists of two electrodes that sandwich a membrane that only protons (hydrogen ions) can move across. One electrode, helped by a chemical catalyst, strips the hydrogen gas of its electrons, creating two hydrogen ions that begin moving across the membrane. Meanwhile, on the other side of the cell, oxygen molecules from the open air are split into their component atoms along the other electrode. These oxygen ions soon bind to pairs of hydrogen ions traveling across the membrane, forming molecules of water that are expelled as waste, along with heat. While this is occurring, the electrons from the hydrogen atoms have traveled to a device that completes an electric current between the two electrodes. The movement of the hydrogen's electrons from one electrode to the other creates the output of electricity.

Hydrogen and fuel cells have many benefits

As a fuel, hydrogen offers a number of benefits. We will never run out of hydrogen; it is the most abundant element in the universe. It can be clean and nontoxic to use, and—depending on the source of the hydrogen and the source of electricity for its extraction—it may produce few greenhouse gases and other pollutants. Pure water and heat may be the only waste products from a hydrogen fuel cell, along with negligible traces of other compounds. In terms of safety for transport and storage, hydrogen can catch fire, but if it is kept under pressure, it is probably no more dangerous than gasoline in tanks.

Hydrogen fuel cells are energy-efficient. Depending on the type of fuel cell, 35% to 70% of the energy released in the reaction can be used. If the system is designed to capture heat as well as electricity, then the energy efficiency of fuel cells can rise to 90%. These rates are comparable or superior to most nonrenewable alternatives.

Fuel cells are also silent and nonpolluting. Unlike batteries (which also produce electricity through chemical reactions), fuel cells will generate electricity whenever

VIEWPOINTS

Hydrogen and Renewable Energy

Is establishing a "hydrogen economy," as Iceland is trying to do, the best way to reduce the use of fossil fuels?

The Role of Renewable Energy for the Hydrogen Economy

Abundant, reliable, and affordable energy is an essential component of a healthy economy. Because hydrogen can be produced from a wide variety of domestically available resources and can be used in heat, power, and fuel applications, it is uniquely positioned to contribute to our growing energy demands, particularly for resource-constrained communities. However, if we are to realize the true benefits of a hydrogen economy, other renewables must play a substantial role in the efficient and affordable production of the hydrogen.

Several renewable options could make a substantial impact in the production of hydrogen: electrolysis powered by wind, photovoltaic, solar-thermal electric, hydropower, and geothermal energy; use of microorganisms and semiconductors to split water; and the thermal and biological conversion of biomass and wastes. Researchers around the globe are working on improving these renewable technologies. As a result, costs continue to drop. Technologies for renewable hydrogen production, coupled with advances in hydrogen production equipment (e.g., electrolyzers) can supply cost-competitive hydrogen and will ultimately play a substantial role in our energy supply.

In addition to the potential supply of affordable hydrogen, these technologies also offer a wide variety of opportunities for developing new centers of economic growth. Most investments in renewable energy are spent on materials and workmanship to build and maintain the facilities, rather than on costly energy imports. Therefore, funds are usually spent regionally and even locally, leading to new jobs and investments in local economies. Because of this synergistic relationship, the shift toward a hydrogen economy will naturally facilitate the advancement of renewable energy. By diversifying our energy supply, we will not only reduce our dependence on imported fuels, but also will benefit from cleaner technologies and investment in our communities.

Susan Hock directs the Electric and Hydrogen Technologies and Systems Center of the National Renewable Energy Laboratory. The center conducts research activities in four areas: distributed power systems integration, hydrogen technologies and systems, geographic information system analysis, and solar measurements and instrumentation.

Is Hydrogen the Answer?

We'll never use the last drop of oil, the last chunk of coal, the last cubic foot of natural gas, or the last pound of uranium. Eventually though, these fossil and nuclear fuels will become too expensive to extract, or politics will make one or more of them unavailable, leaving us to ask how we'll satisfy our voracious appetite in the future.

We should immediately apply all practical energy conservation strategies. Mother Nature is out there making more fossil fuels as we speak, but we don't have time to wait the few million years that will take. The short list of renewables: solar, wind, hydro, biomass, geothermal, waves, tides, and ocean thermal energy conversion. These are all relatively benign and abundant.

An alternative: hydrogen. It can either be burned or electrochemically used in fuel cells to provide useful energy. The by-product or "exhaust" is water. You start with water, get some energy, and end up with water, making it renewable. Another form of hydrogen energy is fusion, hydrogen atoms fusing to form helium plus a lot of energy, the way the sun does it. The catch? It takes about as much energy to extract hydrogen gas from water (by electrolysis) as you get back from your energy conversion device. Until it becomes cheaper (economically and in physical terms), fossil fuels will continue to rule the energy world. The breakthrough may involve using our renewable energy resources to separate hydrogen from other molecules.

Arguably, to reduce our dependence on fossil fuels, the priority list for this country should be:

1. Energy conservation
2. Wind
3. Passive solar
4. Biomass
5. Active solar
6. Hydrogen (chemical)
7. Hydroelectricity
8. Hydrogen (fusion)
9. Others (geothermal, tides, waves, ocean thermal)

Daryl Prigmore has studied energy and the environment since the late 1960s. After receiving bachelor and master of science degrees in mechanical engineering from Colorado State University, he spent 10 years in industry with a company developing solar, geothermal, and low-pollution automotive power systems. He has taught energy science classes for the past 23 years, 20 at the University of Colorado (Colorado Springs).

Explore this issue further by accessing **Viewpoints** at www.aw-bc.com/withgott.

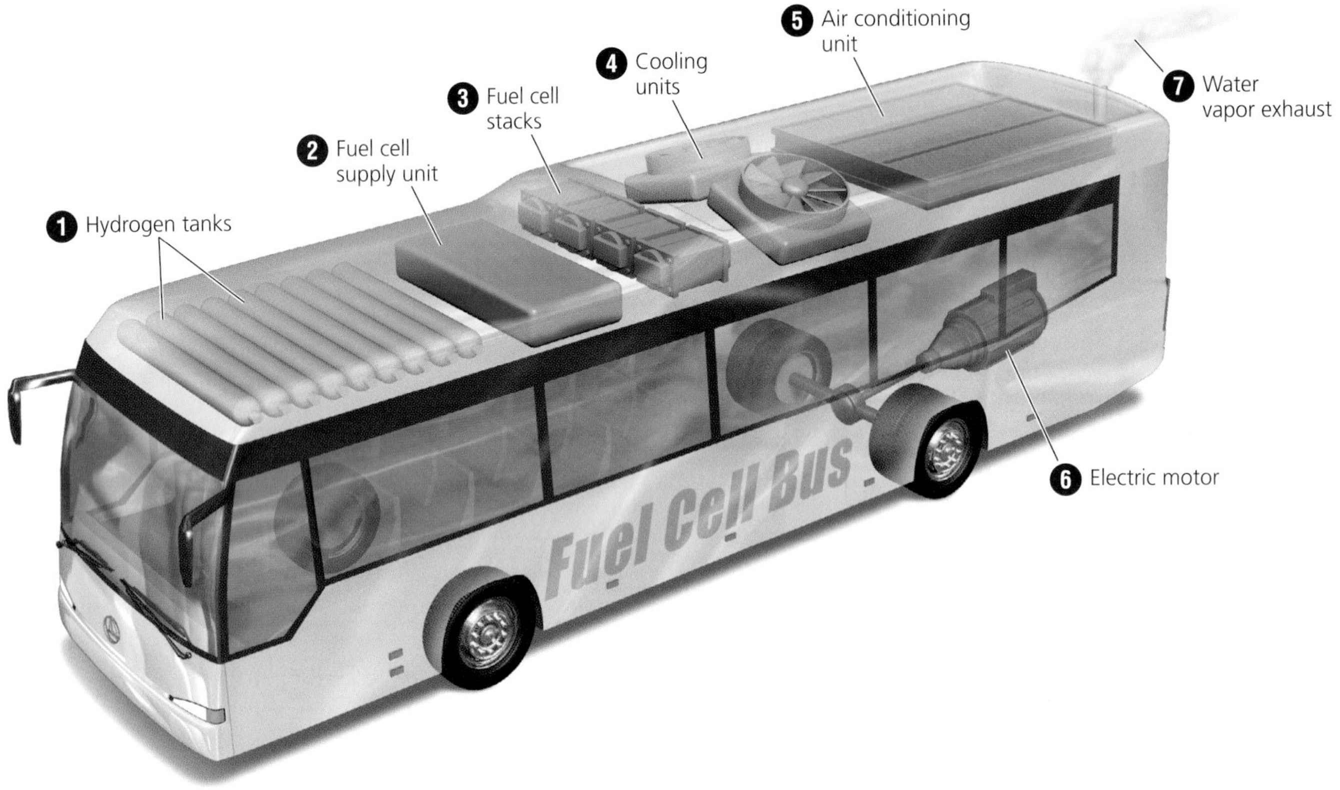

FIGURE 21.14 The hydrogen-fueled Citaro buses operating in Reykjavik and other European capitals are designed by Mercedes-Benz and Daimler-Chrysler. Hydrogen is stored in nine fuel tanks (1) The fuel cell supply unit (2) controls the flow of hydrogen, air, and cooling water into the fuel cell stacks (3). Cooling units (4) and the air conditioning unit (5) dissipate waste heat produced by the fuel cells. Electricity generated by the fuel cells is changed from direct current (DC) to alternating current (AC) by an inverter, and it is transmitted to the electric motor (6), which powers the operation of the bus. The vehicle's exhaust (7) consists simply of water vapor.

hydrogen fuel is supplied, without ever needing recharging. For all these reasons, hydrogen fuel cells are being used to power vehicles, including the buses now operating on the streets of Reykjavik and many other European, American, and Asian cities (Figure 21.14).

Conclusion

The coming decline of fossil fuel supplies and the increasing concern over air pollution and global climate change have convinced many people that we will need to shift to renewable energy sources that will not run out and will not pollute. Renewable sources with promise for sustaining our civilization far into the future without greatly degrading our environment include solar energy, wind energy, geothermal energy, and ocean energy sources. Moreover, by using electricity from renewable sources to produce hydrogen fuel, we may be able to use fuel cells to produce electricity when and where it is needed, helping convert our transportation sector to a nonpolluting, renewable basis.

Most renewable energy sources have been held back for a variety of reasons, including little funding for research and development, and artificially cheap market prices for nonrenewable resources that do not include external costs. Despite this, renewable technologies have progressed far enough to offer hope that we can shift from fossil fuels to renewable energy with a minimum of economic and social disruption. Whether we can also limit environmental impact will depend on how soon and how quickly we make the transition and to what extent we put efficiency and conservation measures into place.

REVIEWING OBJECTIVES

You should now be able to:

Outline the major sources of renewable energy and assess their potential for growth

- The "new renewable" energy sources include solar, wind, geothermal, and ocean energy sources. They are not truly "new," but rather are in a stage of rapid development. (pp. 621–622)
- The new renewables currently provide far less energy and electricity than the conventional renewables, hydropower and biomass energy—and only a small fraction of the energy and electricity we obtain from fossil fuels. (pp. 621–622)
- Use of new renewables is growing quickly, and this growth is expected to continue as people seek to move away from fossil fuels. (pp. 622–623)

Describe solar energy and the ways it is harnessed, and evaluate its advantages and disadvantages

- Energy from the sun's radiation can be harnessed using passive methods or by active methods involving powered technology. (pp. 623–624)
- Major solar technologies include solar panels, mirrors to concentrate solar rays, and photovoltaic cells. (pp. 624–626)
- Solar energy is perpetually renewable, and solar technology creates no emissions and allows for decentralized power. (p. 626)
- Solar radiation varies in intensity from place to place and time to time, and harnessing solar energy remains expensive. (p. 627)

Describe wind energy and the ways it is harnessed, and evaluate its advantages and disadvantages

- Energy from wind is harnessed using wind turbines mounted on towers. (pp. 627–628)
- Turbines are often erected in arrays at wind farms located on land or offshore. Wind farms are developed in locations with optimal wind conditions. (pp. 628–632)
- Wind energy is renewable, turbine operation creates no emissions, wind farms can generate economic benefits, and the cost of wind power is nearly competitive with that of electricity generated from fossil fuels. (p. 630)
- Wind is an intermittent resource and occurs at adequate strengths only in some locations. Turbines kill some birds and bats, and wind farms can face opposition from local residents. (pp. 630–633)

Describe geothermal energy and the ways it is harnessed, and evaluate its advantages and disadvantages

- Energy from radioactive decay in Earth's core rises toward the surface and heats groundwater. Energy from this heated water and steam is harnessed at the surface or by drilling at geothermal power plants. (pp. 633–634)
- Use of geothermal energy for direct heating of water, as well as for electricity generation, can be efficient, clean, and renewable. (pp. 633–635)
- Geothermal sources occur only in certain areas and may become exhausted if too much water is pumped out without being replenished. (p. 635)

Describe ocean energy sources and the ways they can be harnessed, and evaluate their advantages and disadvantages

- Major ocean energy sources include the motion of tides and waves and the thermal heat of ocean water. (pp. 635–637)
- Tidal and wave energy is perpetually renewable and holds much promise, but so far technologies have seen only limited development. (pp. 635–637)

Explain hydrogen fuel cells and assess future options for energy storage and transportation

- Hydrogen can serve as a fuel to store and transport energy, so that electricity generated by renewable sources can be made portable and used to power vehicles. (p. 637)
- Hydrogen can be produced through electrolysis, but it may also be produced by using fossil fuels—in which case its environmental benefits are greatly reduced. (pp. 638–639)
- There is some concern that releasing excess hydrogen could have negative impacts on the atmosphere. (p. 639)
- Fuel cells create electricity by controlling an interaction between hydrogen and oxygen, and they produce only water as a waste product. (pp. 640, 642)
- Hydrogen can be clean, safe, and efficient. Fuel cells are silent, are nonpolluting, and do not need recharging. (pp. 640, 642)

TESTING YOUR COMPREHENSION

1. About how much of our energy now comes from renewable sources? What is the most prevalent form of renewable energy we use? What form of renewable energy is most used to generate electricity?
2. What is causing renewable energy sectors to expand? What renewable source is experiencing the most rapid growth?
3. Describe how passive solar heating works. How does active solar heating work? Give examples of each.
4. Describe the photoelectric effect. Describe a photovoltaic (PV) cell, and explain one way these are used.
5. What are the environmental and economic advantages and disadvantages of solar power?
6. How do modern wind turbines generate electricity? How does wind speed affect the process?
7. What are the environmental and economic benefits of wind power? What are its disadvantages?
8. Define geothermal energy and explain how it is obtained and used. In what ways is it renewable, and in what way is it not renewable?
9. List and describe three approaches to obtaining energy from ocean water.
10. How is hydrogen fuel produced? Is this a clean process? What factors determine the amount of pollutants hydrogen production will emit?

SEEKING SOLUTIONS

1. Why might a hydrogen economy be closer than we think? Why might it instead not come to pass? Do you think water could be "the coal of the future"? Why or why not?
2. For each source of renewable energy discussed in this chapter, what factors are standing in the way of an expedient transition from fossil fuel use?
3. Explain how the use of new renewable energy sources can reduce fossil fuel emissions.
4. Do you think development and implementation of renewable energy resources to replace fossil fuels can be moved forward without great social, economic, and environmental disruption? What steps would need to be taken? Will market forces alone suffice to bring about this transition? Do you think such a shift will be good for the economy?
5. Iceland is giving itself many years to phase in its planned hydrogen economy. Do you think the United States could transition to a hydrogen economy more quickly, less quickly, or not at all? Why? What steps could the United States take to accelerate such a transition?
6. Imagine you are the CEO of a company that develops wind farms. Your staff is presenting you with three options, listed below, for sites for your next development. Describe at least one likely advantage and at least one likely disadvantage you would expect to encounter with each option. What further information would you like to know before deciding on which to pursue?
   - Option A. A remote rural site in North Dakota
   - Option B. A ridge-top site among the suburbs of Philadelphia
   - Option C. An offshore site off the Florida coast

INTERPRETING GRAPHS AND DATA

Of the new renewable energy alternatives discussed in this chapter, photovoltaic conversion of solar energy is the one that most areas of the United States could most easily adopt. The influx of solar radiation varies with time of day, time of year, and location, so all areas are not equally well suited. Today's photovoltaic technology is approximately 10% efficient at converting the energy of sunlight into electricity, but new technologies under development may increase that efficiency to as much as 40%.

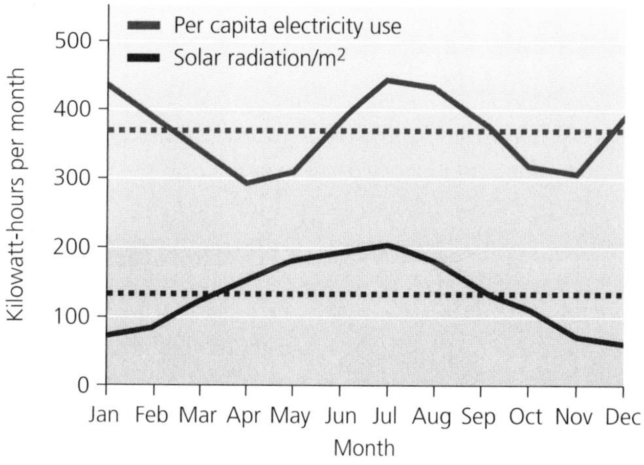

Average per capita residential use of electricity in the United States in 2004 (red line) and average influx of solar radiation per square meter for Topeka, Kansas (blue line). The dashed lines represent the yearly average values for each. Data from Renewable Resource Data Center, National Renewable Energy Laboratory, U.S. Department of Energy (DOE); and Energy Information Administration. 2005. *Annual energy review 2004.* DOE.

1. Given a 10% efficiency for photovoltaic conversion of solar energy, approximately how many square meters of photovoltaic cells would be needed to supply one person's residential electrical needs for a year, based on the yearly average values? How many square meters would be needed if efficiency were improved to 40%?
2. Given the same 10% conversion efficiency, approximately how many square meters of photovoltaic cells would be required to supply one person's residential electrical needs during the month of April? During July? How many square meters would be required to supply the average U.S. household of four people for each of those months?
3. Commercially available photovoltaic systems of this capacity cost approximately $20,000. The average cost of electricity in the United States is approximately 9¢ per kilowatt-hour. At these prices, how long would it take for the PV system to generate $20,000 worth of electricity? Calculate a combination of PV system cost and electricity cost at which the system would pay for itself in 10 years.

CALCULATING ECOLOGICAL FOOTPRINTS

Assume that average per capita residential consumption of electricity is 12 kilowatt-hours per day, that photovoltaic cells have an electrical output of 10% incident solar radiation, and that PV cells cost $800 per square meter. Now refer to Figure 21.5 on p. 627, and estimate the area and cost of the PV cells needed to provide all of the residential electricity used by each group in the table.

	Area of photovoltaic cells	Cost of photovoltaic cells
You	25	$20,000
Your class		
Your state		
United States		

1. What additional information do you need in order to increase the accuracy of your estimates for the areas in the table above?
2. Considering the distribution of solar radiation in the United States, where do you think it will be most feasible to greatly increase the percentage of electricity generated from photovoltaic solar cells?
3. The purchase price of a photovoltaic system is considerable. What other costs and benefits should you consider, in addition to the purchase price, when contemplating "going solar"?

Take It Further

Go to www.aw-bc.com/withgott or the student CD-ROM where you'll find:

- Suggested answers to end-of-chapter questions
- Quizzes, animations, and flashcards to help you study
- *Research Navigator*™ database of credible and reliable sources to assist you with your research projects
- GRAPHit! tutorials to help you master how to interpret graphs
- INVESTIGATEit! current news articles that link the topics that you study to case studies from your region to around the world

22 Waste Management

Containers en route to a recycling facility

Upon successfully completing this chapter, you will be able to:

- Summarize and compare the types of waste we generate
- List the major approaches to managing waste
- Delineate the scale of the waste dilemma
- Describe conventional waste disposal methods: landfills and incineration
- Evaluate approaches for reducing waste: source reduction, reuse, composting, and recycling
- Discuss industrial solid waste management and principles of industrial ecology
- Assess issues in managing hazardous waste

Fresh Kills Landfill, Staten Island, New York

Central Case: Transforming New York's Fresh Kills Landfill

"An extraterrestrial observer might conclude that conversion of raw materials to wastes is the real purpose of human economic activity."
—Gary Gardner and Payal Sampat, Worldwatch Institute

"Recycling is one of the best environmental success stories of the late 20th century."
—U.S. Environmental Protection Agency

The closure of a landfill is not the kind of event that normally draws politicians and the press, but the Fresh Kills Landfill was no ordinary dump. The largest landfill in the world, Fresh Kills was the primary repository of New York City's garbage for half a century. On March 22, 2001, New York City Mayor Rudolph Giuliani and New York Governor George Pataki were on hand to celebrate as a barge arrived on the western shore of New York City's Staten Island and dumped the final load of 650 tons of trash at Fresh Kills.

The landfill's closure was a welcome event for Staten Island's 450,000 residents, who had long viewed the landfill as a bad-smelling eyesore, health threat, and civic blemish. The 890-ha (2,200-acre) landfill featured six gigantic mounds of trash and soil. The highest, at 69 m (225 ft), was higher than the nearby Statue of Liberty.

New York City had grandiose plans for the site. It planned to transform the old landfill into a world-class public park—a verdant landscape of rolling hills and wetlands teeming with wildlife, and a mecca for recreation for New York's residents. The site certainly had potential. It was two-and-a-half times bigger than Manhattan's Central Park. It was the region's largest remaining complex of saltwater tidal marshes and freshwater creeks and wetlands, which still attracted birds and wildlife. And the mounds offered panoramic views of the Manhattan skyline and the rest of the region. The city sponsored an international competition to select a landscape architecture firm to design plans for the new park.

Meanwhile, with its only landfill closed, New York City began exporting its waste. The city began plans to develop an efficient network of stations to package and transfer the waste and ship it outward by barge and railroad. However, these plans soon fell apart amid neighborhood opposition, economic misjudgments, and accusations of political favoritism and mob influence. New York City instead found itself paying contractors exorbitant prices to haul its garbage away inefficiently, one truckload at a time. In the years following the Fresh Kills closure, trucks full of trash rumbled through neighborhood streets, carrying 12,000 tons of waste each day bound for 26 different landfills and incinerators in New York, New Jersey, Virginia, Pennsylvania, and Ohio. The city sanitation department's budget nearly doubled, and budget woes caused the city to scale back its recycling program. Some New Yorkers suggested reopening Fresh Kills.

The landfill *was* reopened, but not for a reason anyone could have foreseen. After the September 11, 2001, terrorist attacks, the 1.8 million tons of rubble from the collapsed World Trade Center towers, including unrecoverable human remains, was taken by barge to Fresh Kills, where it was sorted and buried. A monument will be erected at the site as part of the new park.

Today, plans for the park are forging ahead. Field Operations, the design firm that won the competition, completed a preliminary master plan in 2005, incorporating suggestions from the public. The plan involves everything from ecological restoration of the wetlands to construction of roads, ball fields, sculptures, and roller-blading rinks. People will be able to bicycle on trails paralleling tidal creeks of the region's largest estuary and reach stunning vistas atop the hills.

This undertaking—one of the largest public works projects in the world—will not be completed overnight. Designers, city officials, and Staten Island residents hope the first portions of the new park will open between 2008 and 2012. In the end, a longtime symbol of waste could be transformed into a world-class center for recreation and urban ecological restoration.

Approaches to Waste Management

As the world's human population rises, and as we produce and consume more material goods, we generate more waste. **Waste** refers to any unwanted material or substance that results from a human activity or process. For management purposes, waste is divided into several main categories. **Municipal solid waste** is nonliquid waste that comes from homes, institutions, and small businesses, whereas **industrial solid waste** includes waste from production of consumer goods, mining, agriculture, and petroleum extraction and refining. **Hazardous waste** refers to solid or liquid waste that is toxic, chemically reactive, flammable, or corrosive. It can include everything from paint and household cleaners to medical waste to industrial solvents. Another type of waste is *wastewater,* water we use in our households, businesses, industries, or public facilities and drain or flush down our pipes, as well as the polluted runoff from our streets and storm drains. We discussed wastewater in Chapter 15 (▸ pp. 458–461).

We have several aims in managing waste

Waste can degrade water quality, soil quality, and air quality, thereby degrading human health and the environment. Waste is also a measure of inefficiency, so reducing waste can potentially save industry, municipalities, and consumers both money and resources. Waste is also unpleasant aesthetically. For these and other reasons, waste management has become an important pursuit for cities, industries, and individuals.

There are three main components of **waste management:** (1) minimizing the amount of waste we generate, (2) recovering waste materials and finding ways to recycle them, and (3) disposing of waste safely and effectively. Minimizing waste at its source—called *source reduction*—is the preferred approach. There are several ways to reduce the amount of waste that enters the **waste stream,** the flow of waste as it moves from its sources toward disposal destinations (Figure 22.1). Manufacturers can use materials more efficiently. Consumers can buy fewer goods, buy goods with less packaging, and use those goods longer. Reusing goods you already own, purchasing used items, and donating your used items for others also help reduce the amount of material entering the waste stream.

Recovery (*recycling* and *composting*) is widely viewed as the next best strategy in waste management. Recycling involves sending used goods to facilities that extract and reprocess raw materials to manufacture new goods. Newspapers, white paper, cardboard, glass, metal cans, appliances, and some plastic containers have all become increasingly recyclable as new technologies have been developed and as markets for recycled materials have grown. Organic waste can be recovered through *composting,* or biological decomposition. Recycling is not a concept that humans invented; recall that all materials are recycled in ecosystems (▸ pp. 190–191). Recycling is a fundamental feature of the way natural systems function.

Regardless of how effectively we reduce our waste stream, there will likely always be some waste left to dispose of. Disposal methods include burying waste in landfills and burning waste in incinerators. In this chapter we first examine how these approaches are used to manage municipal solid waste, and then we address approaches for managing industrial solid waste and hazardous waste.

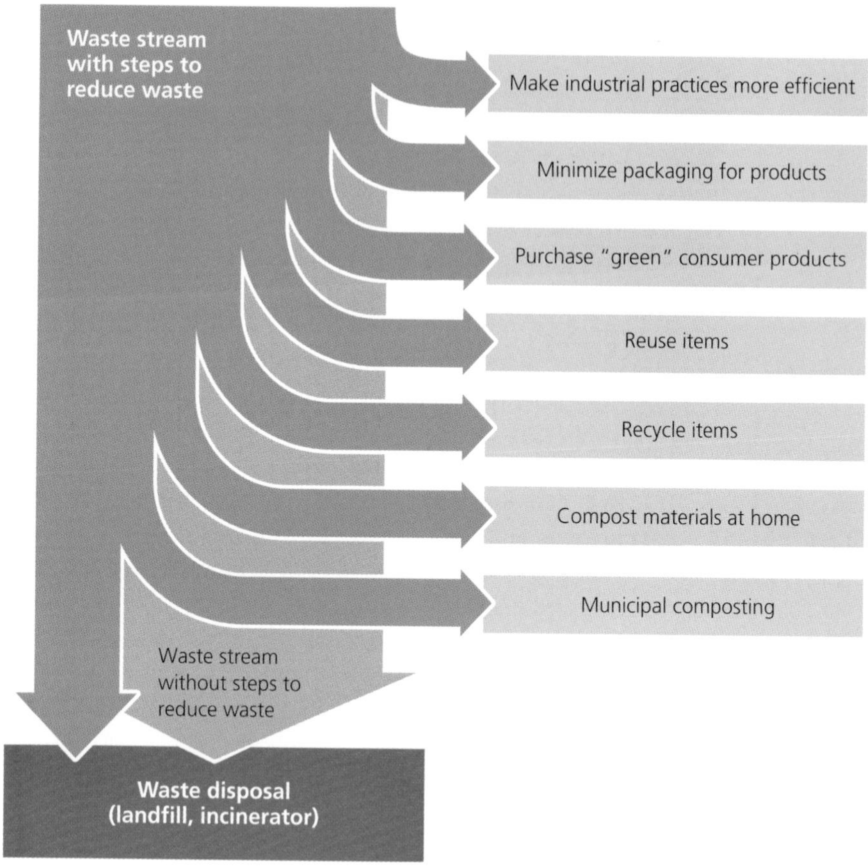

FIGURE 22.1 The most effective way to manage waste is to minimize the amount of material that enters the waste stream. To do this, manufacturers can increase efficiency and consumers can buy "green" products that have minimal packaging or are produced in ways that minimize waste. Individuals can compost food scraps and yard waste at home and can reuse items rather than buying new ones. When we are finished using products, many of us can recycle some materials and compost yard waste through municipal recycling and composting programs. For all remaining waste, waste managers attempt to find disposal methods that minimize impact to human health and environmental quality.

Municipal Solid Waste

Municipal solid waste is waste produced by consumers, public facilities, and small businesses. It is what we commonly refer to as "trash" or "garbage." Everything from paper to food scraps to roadside litter to old appliances and furniture is considered municipal solid waste.

Patterns in the municipal solid waste stream vary from place to place

In the United States, paper, yard debris, food scraps, and plastics are the principal components of municipal solid waste, together accounting for 70% of the waste stream (Figure 22.2). Even after recycling, paper is the largest component of U.S. municipal solid waste. Patterns differ in developing countries; there, food scraps are often the primary contributor to solid waste, and paper makes up a smaller proportion.

Most municipal solid waste comes from packaging and nondurable goods (products meant to be discarded after a short period of use). In addition, consumers throw away old durable goods and outdated equipment as they

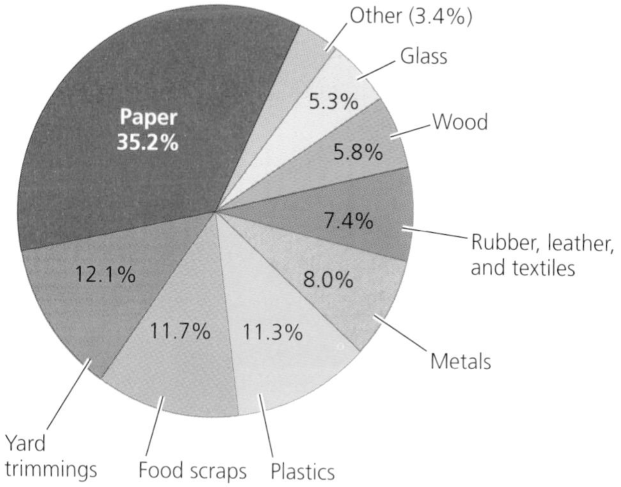

FIGURE 22.2 Paper products comprise the largest component of the municipal solid waste stream in the United States, followed by yard trimmings, food scraps, and plastics. In total, each U.S. citizen generates close to 1 ton of solid waste each year. Data for 2003, from U.S. Environmental Protection Agency, 2005. *Municipal solid waste generation, recycling, and disposal in the United States: Facts and figures for 2003.* EPA530-F-05-003. Washington, D.C.: EPA.

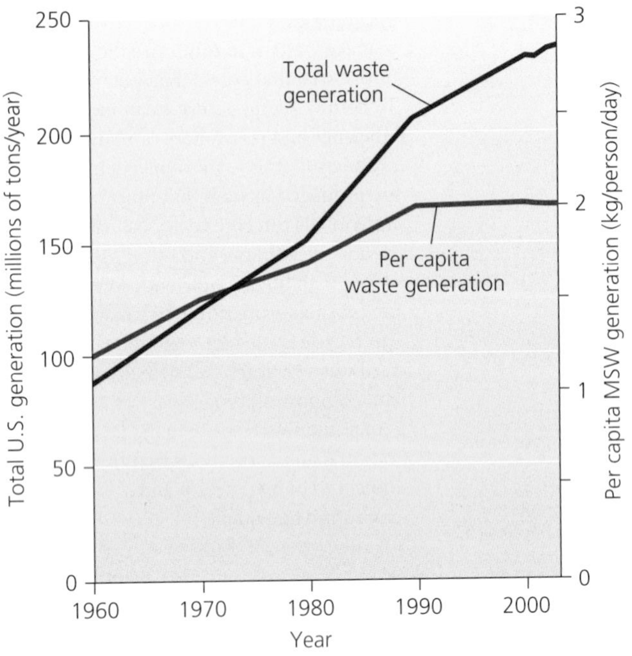

FIGURE 22.3 U.S. waste generation has increased by more than 2.7 times since 1960, and U.S. per-capita waste generation has risen by 66%. Per capita waste generation has leveled off in recent years largely because of recycling and source-reduction efforts. Data from U.S. Environmental Protection Agency, 2005. *Municipal solid waste generation, recycling, and disposal in the United States: Facts and figures for 2003.* EPA530-F-05-003. Washington, D.C.: EPA.

purchase new products. As we acquire more goods, we generate more waste. In 2003, U.S. citizens produced 236 million tons of municipal solid waste, almost 1 ton per person. This means that the average American generates about 2.0 kg (4.4 lb) of trash per day.

Following the United States in per capita solid waste generation are Canada, with 1.7 kg (3.75 lb) per day, and the Netherlands, with 1.4 kg (3.0 lb) per day. Of developed nations, Germany and Sweden produce the least waste per capita, generating just under 0.9 kg (2.0 lb) per day. Differences among nations result in part from differences in the cost of waste disposal; where disposal is expensive, people have incentive to waste less. The relative wastefulness of the U.S. lifestyle, with its excess packaging and reliance on nondurable goods, has caused critics to label the United States "the throwaway society."

People in developing nations, where consumption is lower, generate considerably less waste. One study (see Interpreting Graphs and Data, ▸ pp. 672–673) indicates that people of high-income nations waste more than twice as much as people of low-income nations. However, wealthier nations also tend to invest more in waste collection and disposal, so they are often better able to manage their waste proliferation and minimize impacts on human health and the environment.

Waste generation is rising in all nations

In the United States since 1960, waste generation has increased (Figure 22.3) by 2.7 times, and per capita waste generation has risen by 66%. Plastics, which came into wide consumer use only after 1970, have accounted for the greatest relative increase in the waste stream during the last several decades (Figure 22.4).

The rising consumption that has long characterized the United States and other wealthy nations is now proceeding rapidly in developing nations. To some extent,

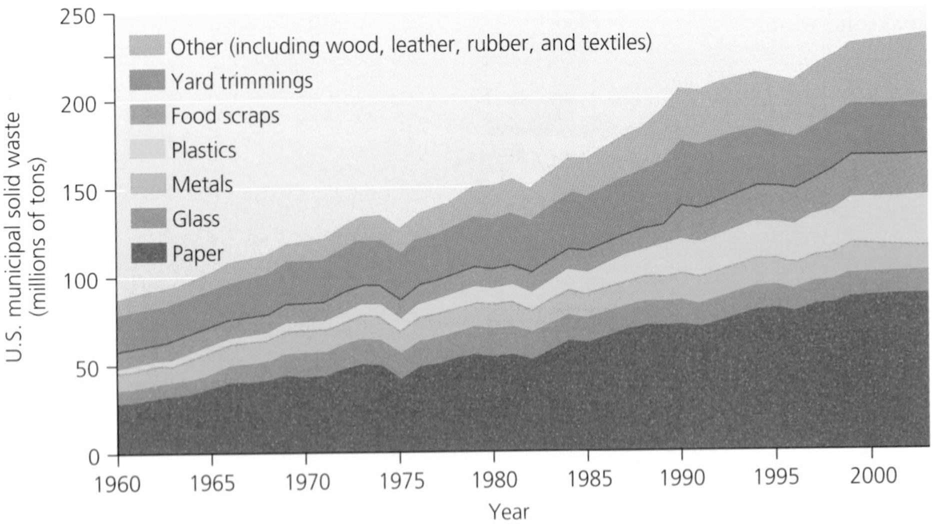

FIGURE 22.4 Amounts of all types of waste have grown in the United States over the past 4 decades, but plastic waste is the only type that has taken up a substantially greater share of the waste stream through time. Data from U.S. Environmental Protection Agency, 2005. *Municipal solid waste generation, recycling, and disposal in the United States: Facts and figures for 2003.* EPA530-F-05-003. Washington, D.C.: EPA.

FIGURE 22.5 Tens of thousands of people used to scavenge each day from the dump at Payatas, outside Manila in the Philippines, finding items for themselves and selling material to junk dealers for 100–200 pesos (U.S. $2–4) per day. That so many people could support themselves this way testifies to the immense amount of usable material needlessly discarded by wealthier portions of the population. The dump was closed in 2000 after an avalanche of trash killed hundreds of people.

this increase reflects rising material standards of living, but an increase in packaging is also to blame. Items made for temporary use, and poor-quality goods designed to be inexpensive, wear out and pile up quickly as trash, littering the landscapes of countries from Mexico to Kenya to Indonesia. Over the past three decades, per capita waste generation rates have more than doubled in Latin American nations and have increased more than fivefold in the Middle East. Like U.S. consumers in the "throwaway society," wealthy consumers in developing nations often discard items that can still be used. At many dumps and landfills in the developing world, in fact, poor people support themselves by selling items they scavenge (Figure 22.5).

In many developed nations, per capita generation rates have leveled off or decreased in recent years. For instance, note in Figure 22.3 that per capita waste production in the United States flattened out during the 1990s. This was due largely to the increased availability of recycling options. Recycling, composting, reduction, and reuse today are taking care of an increasingly larger portion of waste. We will examine these nondisposal approaches to waste management shortly, but let's first assess how we dispose of waste.

Open dumping of the past has given way to improved disposal methods

Historically, people dumped their garbage wherever it suited them. For example, until the mid-19th century, New York City's official method of garbage disposal was to dump it off piers into the East River. As population densities increased, municipalities took on the task of consolidating trash into open dumps at specified locations in order to keep other areas clean. To decrease the volume of trash, these dumps would be burned from time to time. Open dumping and burning still occur throughout much of the world.

As population and consumption rose in developed nations, waste increased and dumps grew larger. At the same time, expanding cities and suburbs forced more people into the vicinity of operating dumps and exposed them to the noxious smoke of dump burning. Reacting to opposition from residents living near dumps, and to the rising awareness of the health and environmental threats posed by unregulated open dumping and burning, many nations improved their methods of waste disposal. Most industrialized nations now bury their waste in lined and covered landfills, and burn their waste in incineration facilities.

In the 1980s in the United States, waste generation increased while incineration was restricted, and recycling was neither economically feasible nor widely popular. As a result, landfill space became limited, and there was much talk of a solid waste "crisis." New York and other East Coast urban areas felt this situation most acutely; Fresh Kills Landfill accepted its all-time annual peak of trash, 29,000 tons, in 1986–1987. Since the late 1980s, however, recovery of materials for recycling has expanded, decreasing the pressure on landfills (Figure 22.6). As of 2003, U.S. waste managers were landfilling 55.4% of municipal solid waste, incinerating 14.0%, and recovering 30.6% for composting and recycling.

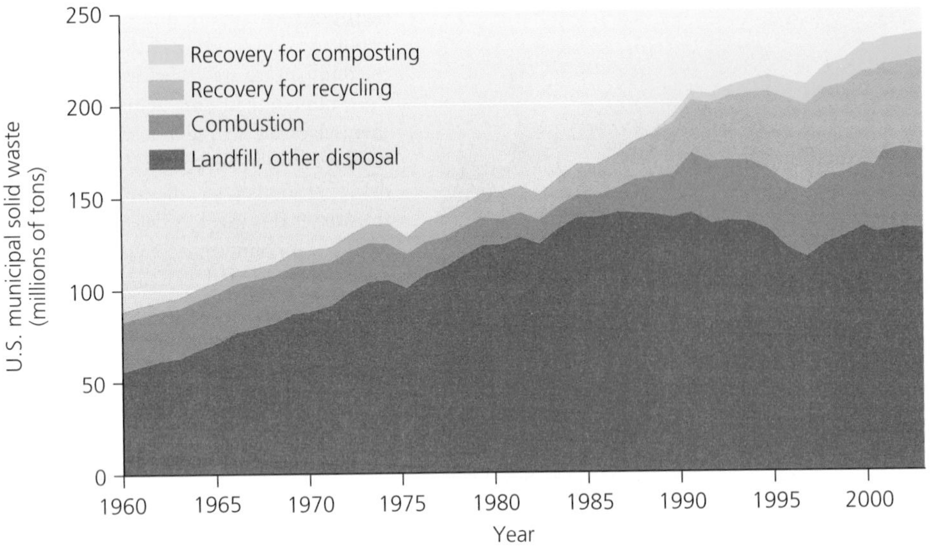

FIGURE 22.6 Since the 1980s, recycling and composting have grown in the United States, allowing a smaller proportion of waste to go to landfills. As of 2003, 55.4% of U.S. municipal solid waste was going to landfills and 14.0% to incinerators, while 30.6% was being recovered for composting and recycling. Go to **GRAPHit!** at www.aw-bc.com/withgott or on the student CD-ROM. Data from U.S. Environmental Protection Agency, 2005. *Municipal solid waste generation, recycling, and disposal in the United States: Facts and figures for 2003.* EPA530-F-05-003. Washington, D.C.: EPA.

Sanitary landfills are regulated by health and environmental guidelines

In modern **sanitary landfills,** waste is buried in the ground or piled up in large, carefully engineered mounds. In contrast to open dumps, sanitary landfills are designed to prevent waste from contaminating the environment and threatening public health (Figure 22.7). Although most municipal landfills in the United States are regulated locally or by the states, they must meet national standards set by the U.S. Environmental Protection Agency (EPA).

Guidelines set forth by the federal **Resource Conservation and Recovery Act (RCRA),** which was enacted in 1976 and amended in 1984, specify how waste should be added to a landfill. In a sanitary landfill, waste is partially decomposed by bacteria and compresses under its own weight to take up less space. Waste is layered along with

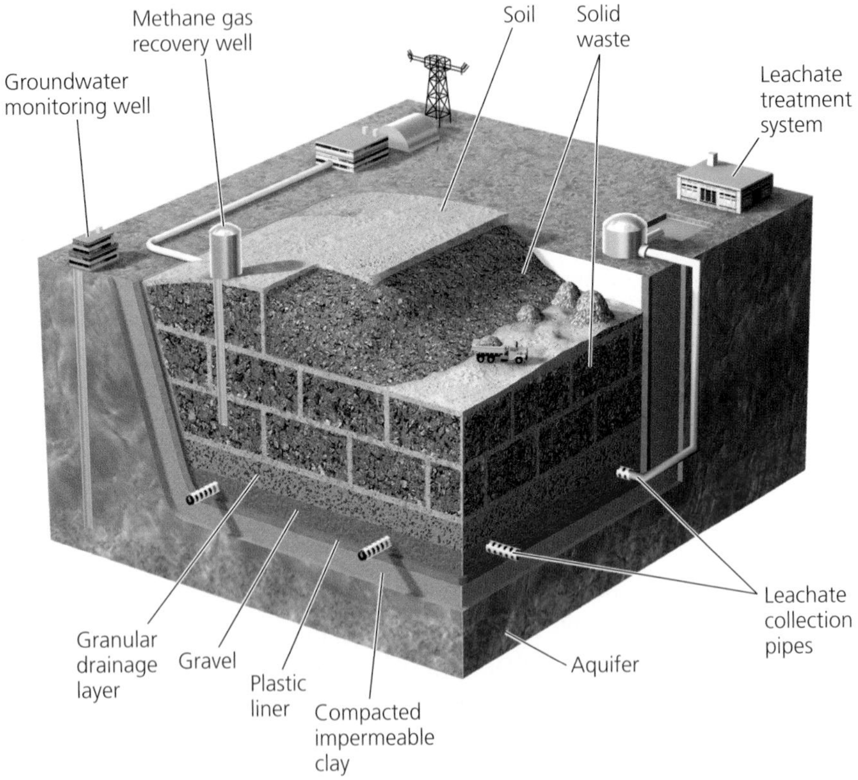

FIGURE 22.7 Sanitary landfills are engineered to prevent waste from contaminating soil and groundwater. Waste is laid in a large depression lined with plastic and impervious clay designed to prevent liquids from leaching out. Pipes of a leachate collection system draw out these liquids from the bottom of the landfill. Waste is layered along with soil until the depression is filled, and continues to be built up until the landfill is capped. Landfill gas produced by anaerobic bacteria may be recovered, and waste managers monitor groundwater for contamination.

soil, which speeds decomposition, reduces odor, and lessens infestation by pests. Limited infiltration of rainwater allows for biodegradation by aerobic and anaerobic bacteria. After a landfill is closed, it is capped with an engineered cover that must be maintained. This cap consists of a hydraulic barrier of plastic that prevents water from seeping down and gas from seeping up; a gravel layer above the hydraulic barrier, which drains water, lessening pressure on the hydraulic barrier; a soil barrier of at least 60 cm (24 in.), which stores water and protects the hydraulic layer from weather extremes; and a topsoil layer of at least 15 cm (6 in.), which encourages plant growth. Vegetation growing in the topsoil helps prevent erosion and returns some soil moisture to the atmosphere by transpiration (▸ p. 204).

Landfill engineers protect against environmental contamination in several ways. U.S. regulations require that landfills be located away from wetlands and earthquake-prone faults, and be at least 6 m (20 ft) above the water table. Regulations require that the bottom and sides of all sanitary landfills be lined with heavy-duty plastic and 60 to 120 cm (2 to 4 ft) of impermeable clay to help prevent contaminants from seeping into aquifers. Regulations also require that area groundwater be monitored regularly for contamination. Because landfills produce **leachate**—liquid that results when substances from the trash dissolve in water as rainwater percolates downward—sanitary landfills have systems to collect and treat it. Typically these systems consist of pipes running from the bottom of the landfill to collection ponds and treatment facilities. Landfill managers are required to maintain leachate collection systems for 30 years after a landfill has closed.

Although it was considered a model for advanced landfill technology at the time of its construction, the Fresh Kills Landfill predated most of the EPA guidelines. As a result, it caused some environmental contamination. However, engineers have retrofitted the landfill with clay liners and a sophisticated leachate collection system. Three of the six mounds have been capped with a "final cover," and the remaining mounds will soon be capped. Indeed, the city's investment of several hundred million dollars in an existing landfill to bring environmental protection measures up to modern standards was unprecedented. Because these safeguards need to be maintained and monitored for 30 years after closure, designs for a public park at Fresh Kills have had to work around these constraints.

Landfills can be transformed after closure

Today thousands of landfills lie abandoned. One reason is that waste managers have closed many smaller landfills and consolidated the trash stream into fewer, much larger, landfills. In 1988 the United States had nearly 8,000 landfills, but today it has fewer than 1,800.

FIGURE 22.8 Old landfills, once properly capped, can serve other purposes. A number of them, such as Cesar Chavez Park in Berkeley, California, shown here, have been developed into areas for human recreation.

A growing number of cities have been converting closed landfills into public parks (Figure 22.8). The Fresh Kills redevelopment endeavor will be the world's largest landfill conversion project, but it is hardly the first. Such efforts date back at least to 1938, when the site of an ash landfill at Flushing Meadows, in Queens, was redeveloped for the 1939 New York World's Fair. The site subsequently hosted the United Nations and the 1964–1965 World's Fair. Designated a park in 1967, today the site hosts Shea Stadium, the Queens Museum of Art, the New York Hall of Science, and the Queens Botanical Garden, as well as playgrounds, wetlands, and festival events.

Landfills have drawbacks

Despite improvements in liner technology and landfill siting, many experts believe that leachate will eventually escape from even well-lined landfills. Liners can be punctured, and leachate collection systems eventually cease to be maintained. Moreover, landfills are kept dry to reduce leachate, but the bacteria that break down material thrive in wet conditions. Dryness, therefore, slows waste decomposition. In fact, it is surprising how slowly some materials biodegrade when they are tightly compressed in a landfill. Innovative archaeological research has revealed that landfills often contain food that has not decomposed and 40-year-old newspapers that are still legible (see "The Science behind the Story," ▸ pp. 654–655).

The Science behind the Story

Digging Garbage: The Archaeology of Solid Waste

Garbage and *knowledge* are two words rarely put together. But when scientist William Rathje dons trash-flecked clothes and burrows into a city dump, he gleans valuable information about how modern Americans live.

By pulling tons of trash out of disposal sites over the course of decades, Rathje has turned dumpster-diving into a noteworthy field of scientific inquiry that he calls *garbology.* An archaeologist by training who has been called "the Indiana Jones of solid waste," Rathje has brought exacting archaeological techniques to the contents of trash cans.

"Garbologist" William Rathje has pioneered the study of our culture through the waste we generate.

As a professor at the University of Arizona in the early 1970s, Rathje wanted his students to learn a technique common among archaeologists—sorting through ancient trash mounds to understand the lives of past cultures. With few ancient civilizations or their trash close at hand, however, he arranged for his students to dig through their neighbors' garbage. In 1973, he gave that effort a name, "The Garbage Project," and began a methodical study of the contents of modern trash.

Rathje asked communities to divert some of their garbage trucks and bring trash to his study teams. With rakes and notebooks, the researchers sorted, weighed, itemized, and analyzed the refuse. They then visited the homes of the people who had generated the trash and asked residents about their shopping and consumption habits.

Then, in 1987, amid growing debates about how quickly U.S. landfills were filling up, Rathje decided to see what was taking up space in them. The Garbage Project headed to landfills with a truck-mounted bucket auger—a large drill commonly employed by

Another problem is finding suitable areas to locate landfills, because most communities do not want them in their midst. This not-in-my-backyard (NIMBY) reaction (▸ pp. 630, 633) is one reason why New York decided to export its waste and why residents of states receiving that waste are increasingly protesting. As a result of the NIMBY syndrome, landfills are rarely sited in neighborhoods that are home to wealthy and educated people with the political power to keep them out. Instead, they are disproportionately sited in poor and minority communities, as environmental justice advocates have frequently pointed out.

One famous case of long-distance waste transport illustrates the unwillingness of most communities to accept garbage. In Islip, New York, in 1987, the town's landfills were full, prompting town administrators to ship waste by barge to a methane production plant in North Carolina. Prior to the barge's arrival, however, it became known that the shipment was contaminated with 16 bags of medical waste, including syringes, hospital gowns, and diapers. Because of the medical waste, the methane plant rejected the entire load. The barge sat in a North Carolina harbor for 11 days before heading for Louisiana. Louisiana, however, would not permit the barge to dock. The barge geologists and construction crews to handle everything from excavating soil samples to creating new water wells. Rathje and his researchers dug into landfills around North America, boring as far as 30 m (100 ft) down, often drilling approximately 15 to 20 garbage "wells" at each site, with each well yielding up to 25 tons of trash.

Once excavated, landfill contents were sorted, weighed, and identified. Rathje's teams sometimes froze the trash before they worked with it to make the garbage easier to separate and to limit odor and flies. Smaller bits of trash were put through sieves and sometimes washed with water to make them easier to label. Rathje has excavated at least 21 dumps, including sites in Tucson, San Francisco, Chicago, Phoenix, New York City, and Philadelphia, uncovering a host of interesting facts in the process:

- *Not much rot.* Trash doesn't decay much in closed landfills, Rathje has found. In the low-oxygen conditions inside most closed dumps, trash turns into a sort of time capsule. Rathje's teams have found whole hot dogs in most digs, intact pastries that are decades old, and grass clippings that are often still green. Decades-old newspapers are legible and can be used to date layers of trash.
- *Paper rules.* Paper-based products make up more than 40% of most landfill content, and construction debris makes up about 20%. Newspapers are often a high-volume item, averaging about 14% of landfill space.
- *Plastic packaging no problem.* Rathje says plastic packaging is not the landfill problem many believe it to be. Plastic packaging makes up only about 4.5% of landfill content, and that figure has not increased substantially since the 1970s, Rathje reported in 1997. Fast-food packaging, polystyrene foam, and disposable diapers also aren't a major problem, making up only about 3% of landfill content. If all plastic packaging were to be replaced by containers made of glass, paper, steel, or similar materials, Rathje maintains, the packaging discarded by U.S. households would more than double.
- *Poison in small bottles.* Toxic waste comes in all sizes. If nail polish were sold in 55-gallon drums, its chemical composition would make it illegal to throw out in a regular dump. Nail polish, however, is tossed in small bottles. In 1991, after studying Tucson dumps, Rathje calculated that about 350,000 bottles were getting thrown away each year. Luckily, however, the potentially toxic ingredients in nail polish don't always spread far, he found. Paper, diapers, and other nontoxic garbage often absorb toxic materials in landfills and keep the poisons from leaching out.

Through garbology, Rathje has gleaned unique insights into how we can change our often-wasteful habits. Now a consulting professor at Stanford University, Rathje has emerged as a leading expert on how to reduce waste.

traveled toward Mexico, but the Mexican navy prevented it from entering that nation's waters. In the end, the barge traveled 9,700 km (6,000 mi) before eventually returning to New York, where, after several court battles, the waste was finally incinerated at a facility in Queens.

Incinerating trash reduces pressure on landfills

Just as sanitary landfills are an improvement over open dumping, incineration in specially constructed facilities can be an improvement over open-air burning of trash. **Incineration,** or combustion, is a controlled process in which mixed garbage is burned at very high temperatures (Figure 22.9). At incineration facilities, waste is generally sorted and metals removed. Metal-free waste is chopped into small pieces to aid combustion and then is burned in a furnace. Incinerating waste reduces its weight by up to 75% and its volume by up to 90%.

However, simply reducing the volume and weight does not rid trash of components that are toxic. The ash remaining after trash is incinerated therefore must be disposed of in hazardous waste landfills (▸ p. 667). Moreover, combustion can create new chemical compounds that can

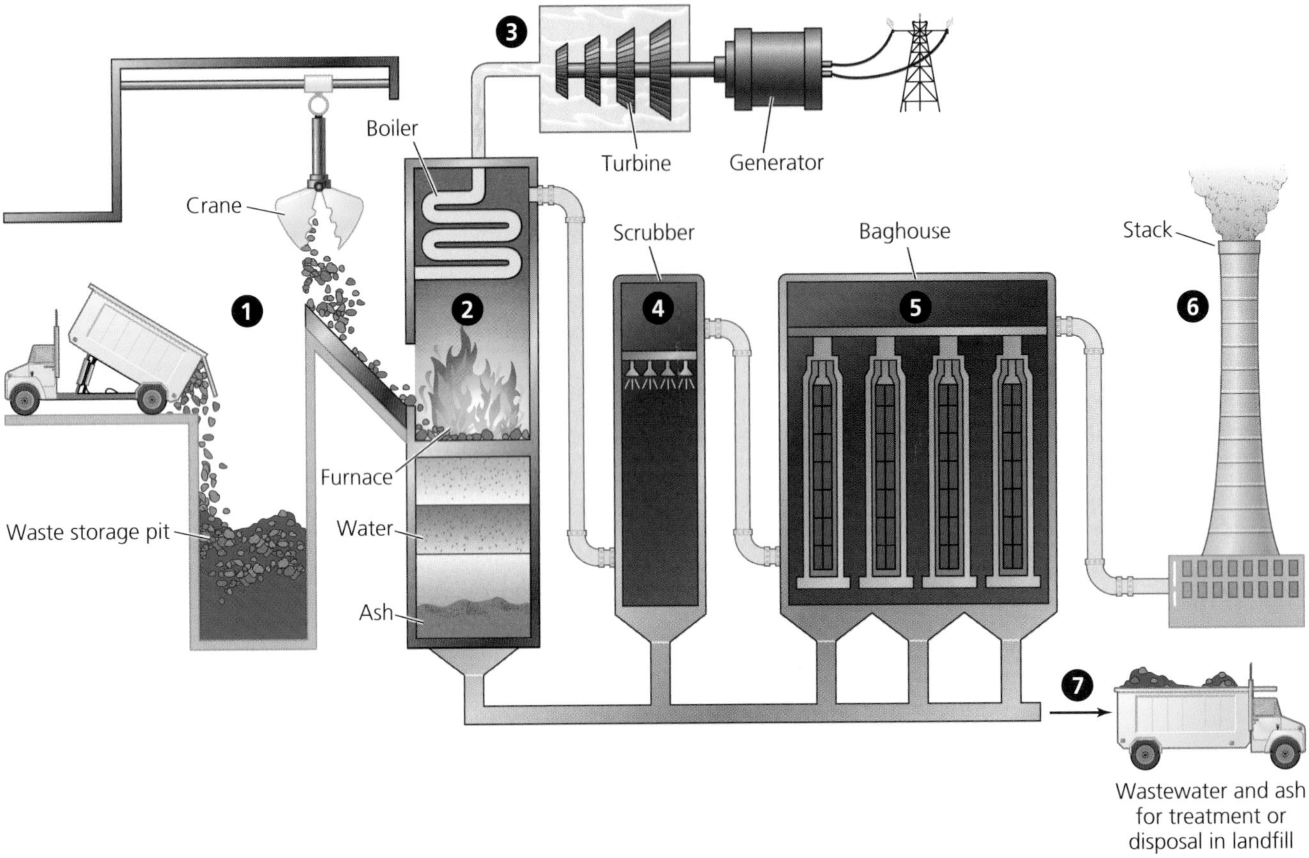

FIGURE 22.9 Incinerators reduce the volume of solid waste by burning it but may emit toxic compounds into the air. Many incinerators are waste-to-energy (WTE) facilities that use the heat of combustion to generate electricity. In a WTE facility, solid waste (1) is burned at extremely high temperatures (2), heating water, which turns to steam. The steam turns a turbine (3), which powers a generator to create electricity. In an incinerator outfitted with pollution control technology, toxic gases produced by combustion are mitigated chemically by a scrubber (4), and airborne particulate matter is filtered physically in a baghouse (5) before air is emitted from the stack (6). Ash remaining from the combustion process is disposed of (7) in a landfill.

be health hazards. When trash is burned, hazardous chemicals—including dioxins, heavy metals, and PCBs (Chapter 14)—can be released into the atmosphere. Such releases caused a backlash against incineration from citizens concerned about health hazards. Most developed nations now regulate incinerator emissions, and some have banned incineration outright.

As a result of real and perceived health threats from incinerator emissions, and of community opposition to these plants, several technologies have been developed to mitigate emissions. *Scrubbers* chemically treat the gases produced in combustion to remove hazardous components and neutralize acidic gases, such as sulfur dioxide and hydrochloric acid, turning them into water and salt. Scrubbers generally do this either by spraying liquids formulated to neutralize the gases or by passing the gases through dry lime. Particulate matter is physically removed from incinerator emissions in a system of huge filters known as a *baghouse*. These tiny particles, called fly ash, often contain some of the worst dioxin and heavy metal pollutants. In addition, burning garbage at especially high temperatures can destroy certain pollutants, such as PCBs. Even all these measures, however, do not fully eliminate toxic emissions.

Weighing the Issues: Environmental Justice?

Do you know where your trash goes? Where is your landfill or incinerator located? Who lives closest to the facility? Are the people in this neighborhood wealthy, poor, or middle-class? What race or ethnicity are they? Do you know whether the people of this neighborhood protested against the introduction of the landfill or incinerator?

Many incinerators burn waste to create energy

Incineration was initially practiced simply to reduce the volume of waste, but today it often serves to generate electricity as well. Most North American incinerators today are **waste-to-energy (WTE)** facilities that use the heat produced by waste combustion to boil water, creating steam that drives electricity generation or that fuels heating systems. When burned, waste generates approximately 35% of the energy generated by burning coal. Over 100 WTE facilities are operating across the United States (mostly in the northeast and south), with a total capacity to process nearly 100,000 tons of waste per day.

Revenues from power generation, however, are usually not enough to offset the considerable financial cost of building and running incineration facilities. Because it can take many years for a WTE facility to become profitable, many companies that build and operate these facilities require communities contracting with them to guarantee the facility a minimum amount of garbage. In a number of cases, such long-term commitments have interfered with communities' later efforts to reduce their waste through recycling and other waste-reduction strategies.

Landfills can produce gas for energy

Combustion in WTE plants is not the only way to gain energy from waste. Deep inside landfills, bacteria decompose waste in an oxygen-deficient environment. This anaerobic decomposition produces *landfill gas,* a mix of gases that consists of roughly half methane (▸ pp. 98–99, 532). Landfill gas can be collected, processed, and used in the same way as natural gas, one of our primary sources of fossil fuel energy (▸ pp. 574–575).

Today more than 330 operational projects collect landfill gas in the United States. Other countries take advantage of this resource as well. In Chile, four facilities in Valparaiso and Santiago supply 40% of the region's demand for natural gas. At Fresh Kills, landfill gas collection wells pull gas upward through a network of pipes by vacuum pressure. Landfill gas collected from Fresh Kills should soon provide enough energy for 25,000 homes. Where gas is not collected for commercial use, it is burned off in flares to reduce smells and greenhouse emissions.

Reducing waste is a better option than disposal

Reducing the amount of material entering the waste stream avoids costs of disposal and recycling, helps conserve resources, minimizes pollution, and can often save consumers and businesses money. Preventing waste generation in this way is known as **source reduction.**

Because much of our waste stream consists of materials used to package goods, manufacturers' choices about packaging have a major effect on the volume of the waste stream. Packaging serves worthwhile purposes—preserving freshness, preventing breakage, protecting against tampering, and providing information—but much packaging is extraneous. Consumers can exercise power by choosing minimally packaged goods, buying unwrapped fruit and vegetables, and buying food in bulk. Consumer preference can give manufacturers incentive to reduce packaging. In addition, manufacturers can use packaging that is more recyclable. They can also reduce the size or weight of goods and materials, as they already have with many items, such as aluminum cans, plastic soft drink bottles, and personal computers.

Increasing the longevity of goods also helps reduce waste. Consumers generally choose goods that last longer, all else being equal. To maximize sales, however, companies often produce short-lived goods that need to be replaced frequently. Thus, increasing the longevity of goods is largely up to the consumer. If demand is great enough, manufacturers will respond.

Weighing the Issues:
Reducing Packaging: Is It a Wrap?

Reducing packaging cuts down on the waste stream, but how, when, and how much should we reduce? Packaging can serve very worthwhile purposes, such as safeguarding consumer health and safety. Can you think of three products for which you would not want to see less packaging? Why? Can you name three products for which packaging could easily be reduced without ill effect to the consumer? Would you be any more or less likely to buy these products if they had less packaging?

Reuse is one main strategy for waste reduction

To reduce waste, you can save items to use again, or substitute disposable goods with durable ones. Habits as simple as bringing your own coffee cup to coffee shops or bringing sturdy reusable cloth bags to the grocery store can, over time, have substantial impact. You can also donate unwanted items and shop for used items yourself at yard sales and resale centers. Over 6,000 reuse centers

Table 22.1 Some Everyday Things You Can Do to Reduce and Reuse

- Donate used items to charity
- Reuse boxes, paper, plastic wrap, plastic containers, aluminum foil, bags, wrapping paper, fabric, packing material, etc.
- Rent or borrow items instead of buying them, when possible . . . and lend your items to friends
- Buy groceries in bulk
- Decline bags at stores when you don't need them
- Bring reusable cloth bags shopping
- Make double-sided photocopies
- Bring your own coffee cup to coffee shops
- Pay a bit extra for durable, long-lasting reusable goods rather than disposable ones
- Buy rechargeable batteries
- Select goods with less packaging
- Compost kitchen and yard wastes in a compost bin or worm bin (often available from your community or waste hauler)
- Buy clothing and other items at resale stores and garage sales
- Use cloth napkins and rags rather than paper napkins and towels
- Write to companies to tell them what you think about their packaging and products
- When solid waste policy is being debated, let your government representatives know your thoughts
- Support organizations that promote waste reduction

exist in the United States, including stores run by organizations that resell donated items, such as Goodwill Industries and the Salvation Army. Besides doing good for the environment, reusing items is often economically advantageous. Used items are quite often every bit as functional as new ones, and much cheaper. Studies from U.S. communities estimate that at least 2–5% of the waste stream consists of reusable items. Table 22.1 presents a sampling of actions that we all can take to reduce the waste we generate.

Composting recovers organic waste

Organic waste, such as yard trimmings and food scraps, can be reduced and recycled by composting. **Composting** is the conversion of organic waste into mulch or humus through natural biological processes of decomposition. The mulch or humus can then be used to enrich soil. Householders can place waste in compost piles, underground pits, or specially constructed containers. As wastes are added, heat from microbial action builds in the interior, and decomposition proceeds. Banana peels, coffee grounds, grass clippings, autumn leaves, and countless other organic items can be converted into rich, high-quality soil, given enough time, through the actions of earthworms, bacteria, soil mites, sow bugs, and other detritivores and decomposers. Home composting is a prime example of how we can live more sustainably by mimicking natural cycles and incorporating them into our daily lives.

Many municipalities are reducing waste through community composting programs—3,800 across the United States at last count. In these programs, food and yard waste are diverted from the waste stream to central composting facilities, where they decompose into mulch that community residents can use for gardens and landscaping. Nearly half of U.S. states now ban yard waste from the municipal waste stream, helping accelerate the drive toward composting. Approximately one-fifth of the U.S. waste stream is made up of materials that can be easily composted. Composting reduces landfill waste, enriches soil and helps it resist erosion, encourages soil biodiversity, makes for healthier plants and more pleasing gardens, and reduces the need for chemical fertilizers.

Recycling consists of three steps

Recycling, too, offers many benefits. **Recycling** consists of collecting materials that can be broken down and reprocessed to manufacture new items. Recycling diverted 55 million tons of materials away from incinerators and landfills in the United States in 2003.

The three basic steps in the recycling loop have given rise to the three elements of the commonly seen recycling symbol (Figure 22.10). The first step is collecting and

1
Collection and processing of recyclable materials by municipalities and businesses
2
Use of recyclables by industry to manufacture new products
3
Consumer purchase of products made from recycled materials

FIGURE 22.10 The familiar recycling symbol consists of three arrows to represent the three components of a sustainable recycling strategy: collection and processing of recyclable materials, use of the materials in making new products, and consumer purchase of these products.

processing used recyclable goods and materials. Communities may designate locations where residents can drop off recyclables or receive money for them. Many of these have now been replaced by the more convenient option of curbside recycling, in which trucks pick up recyclable items in front of houses, usually in conjunction with municipal trash pickup. Curbside recycling has grown rapidly, and its convenience has helped boost recycling rates. Nearly half of all Americans are now served by more than 9,000 curbside recycling programs across all 50 U.S. states.

Items collected are taken to **materials recovery facilities (MRFs),** where workers and machines sort items, using automated processes including magnetic pulleys, optical sensors, water currents, and air classifiers that separate items by weight and size. The facilities clean the materials, shred them, and prepare them for reprocessing. This is the second step in the recycling loop.

Once readied, these materials are used in manufacturing new goods. Newspapers and many other paper products use recycled paper, many glass and metal containers are now made from recycled materials, and some plastic containers are of recycled origin. Some large objects, such as benches and bridges in city parks, are now made from recycled plastics, and glass is sometimes mixed with asphalt (creating "glassphalt") for paving roads and paths. The pages in this textbook are made from recycled paper that is up to 20% post-consumer waste.

If the recycling loop is to function, consumers and businesses must complete the third step in the cycle by purchasing products made from recycled materials. Buying recycled goods provides economic incentive for industries to recycle materials, and for new recycling facilities to open or existing ones to expand. In this arena, individual consumers have power to encourage environmentally friendly options through the free market. Many businesses now advertise their use of recycled materials, a widespread instance of ecolabeling (▸ pp. 50–51). As markets for products made with recycled materials expand, prices continue to fall.

Recycling has grown rapidly and can expand further

The thousands of curbside recycling programs in place today have sprung up only in the last 20 years. Recycling in the United States has risen from 6.4% of the waste stream in 1960 to 23.5% in 2003 (and 30.6% if you include composting), according to EPA data (Figure 22.11). The EPA calls the growth of recycling "one of the best environmental success stories of the late 20th century."

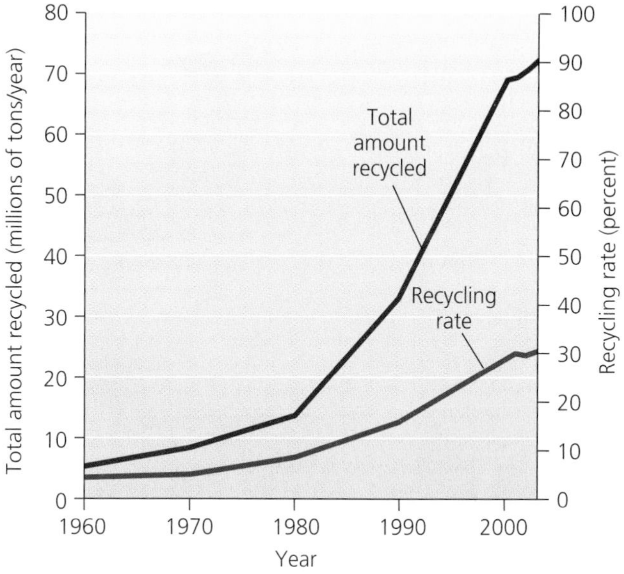

FIGURE 22.11 Recycling has risen sharply in the United States over the past 40 years. Today over 70 million tons of material is recycled, comprising more than 30% of the waste stream. Data from U.S. EPA, 2005.

Recycling rates vary greatly from one product or material type to another and from one location to another. Rates for different types of materials and products range from nearly zero to almost 100% (Table 22.2). Recycling rates among U.S. states also vary greatly, from less than 1% to nearly 50% (Figure 22.12).

Recycling's growth has been propelled in part by economic forces as established businesses see opportunities to save money and as entrepreneurs see opportunities to

Table 22.2 Recovery Rates for Various Materials in the United States

Material	Percentage that is recycled or composted
Auto batteries	93.0
Steel cans	60.0
Yard trimmings	56.3
Paper and paperboard	48.1
Aluminum cans	43.9
Tires	35.6
Plastic milk bottles	31.9
Plastic soft drink containers	25.2
Glass containers	22.0

Data are for 2003, the most recent year available, from U.S. Environmental Protection Agency, Dec. 2005.

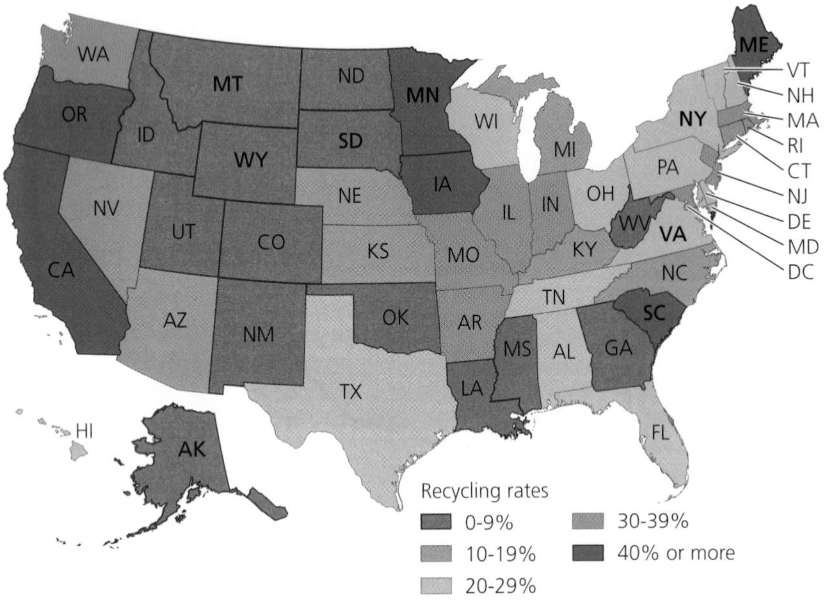

FIGURE 22.12 U.S. states vary greatly in the rates at which their citizens recycle. Data are for 2002 (with earlier data for Alabama, Alaska, and Montana), from Kaufman, S. M., et al., 2004. The state of garbage in America. *BioCycle* 45: 31–41.

start new businesses. However, recycling has also been driven by the desire of municipalities to reduce waste and by the satisfaction people take in recycling. These two forces have driven recycling's rise even though it has often not been financially profitable. In fact, many of the increasingly popular municipal recycling programs are run at an economic loss. The expense required to collect, sort, and process recycled goods is often more than recyclables are worth in the market. Furthermore, the more people recycle, the more glass, paper, and plastic is available to manufacturers for purchase, driving down prices.

Recycling advocates, however, point out that market prices do not take into account external costs (▸ pp. 43–44)—in particular, the environmental and health impacts of *not* recycling. For instance, it has been estimated that globally, recycling saves enough energy to power 6 million households per year. And recycling aluminum cans saves 95% of the energy required to make the same amount of aluminum from mined virgin bauxite, its source material.

As more manufacturers use recycled products, and as more technologies and methods are developed to use recycled materials in new ways, markets should continue to expand, and new business opportunities may arise. We are still at an early stage in the shift from an economy that moves linearly from raw materials to products to waste, to an economy that moves circularly, using waste products as raw materials for new manufacturing processes. The steps we have taken in recycling so far are central to this transition, which many analysts view as key to building a sustainable economy.

Weighing the Issues: Costs of Recycling and Not Recycling

Should recycling programs be subsidized by governments even if they are run at an economic loss? What external costs—costs not reflected in market prices—do you think would be involved in not recycling, say, aluminum cans? Do you feel these costs justify sponsoring recycling programs even when they are not financially self-supporting? Why or why not?

Financial incentives can help address waste

Waste managers have employed economic incentives as tools to reduce the waste stream. The "pay-as-you-throw" approach to garbage collection uses a financial incentive to influence consumer behavior. In these programs, municipalities charge residents for home trash pickup according to the amount of trash they put out. The less waste the household generates, the less the resident has to pay. Over 4,000 of these programs now exist in the United States. Besides reducing waste, they promote equity and fairness; people pay for disposal in accordance with how much they use the service.

"Bottle bills" represent another approach that hinges on financial incentive. Eleven U.S. states have these laws, which allow consumers to return bottles and cans to stores after use and receive a refund—generally 5 cents per bottle or can. The first bottle bills were passed in the 1970s to cut down on litter, but they have also served to decrease the

VIEWPOINTS

Recycling

Will we need to make recycling more economically profitable if it is to continue growing? What, if anything, should we do to encourage recycling? Or should we instead focus on other ways of managing waste?

How to Enhance Recycling?

Certainly recycling needs to be profitable if it is to survive and expand. The question is, how should it be made profitable?

Government subsidies are one common answer; many communities provide small subsidies to keep their recycling programs going. These subsidies will undoubtedly continue because people like municipal recycling programs. But subsidies are not likely to expand much beyond their current, modest level.

In some cases, new technologies are needed. Plastics recycling requires expensive sorting and separation of different plastics to create a high-quality product. With today's technology, recycling of unsorted, mixed plastics yields a low-value product with limited uses. New inventions that improve the sorting process or improve the quality of recycled mixed plastics could make plastics recycling much more profitable.

In other cases, we need new recycling programs and opportunities to keep up with changing lifestyles. Recycling rates are declining for beverage cans and bottles because so many beverages are consumed (and so many containers are discarded) away from home, at parks, beaches, and other public places. A system for recycling in public places could collect the growing quantities of beverage containers, newspapers, and other recyclable materials that are thrown out by people on the go.

Some products should be carefully recycled because it is hazardous to throw them out. Automobile batteries contain large amounts of lead, a toxic substance that should not be tossed in the trash. Some states have laws requiring a deposit on every battery that is sold—which makes it worthwhile to return a dead battery for a refund instead of discarding it. Similar approaches could and should be used with other potentially hazardous products.

Finally, it will never be possible to recycle everything. Along with continuing efforts to expand recycling, we must ensure that there are safe, clean opportunities for disposing of the remaining, nonrecyclable, wastes.

Frank Ackerman is an economist at Tufts University's Global Development and Environment Institute. He has advised the U.S. EPA and state and local agencies on waste management policies. His books include *Why Do We Recycle? Markets, Values, and Public Policies* (Island Press, 1997), and, with Lisa Heinzerling, *Priceless: On Knowing the Price of Everything and the Value of Nothing* (The New Press, 2004).

Recycling: A Mixed Bag

Recycling will continue as long as it is profitable. If it becomes more profitable, we will see more of it. Currently, 55% of all aluminum cans are recycled, a high rate compared to the 30% recycling rate that the Environmental Protection Agency says is average for solid waste. The reason: Aluminum companies can save money by using recycled materials because making cans from bauxite ore is expensive. Paper and cardboard are recycled to a great extent too, partly because cardboard can be made from many kinds of paper.

Plastics aren't recycled nearly as much (about 9%, according to the EPA). One reason is that different kinds of plastic resins can't be mixed. It is expensive for companies to separate the plastics before reprocessing them.

To increase recycling, governments would probably have to force people to recycle and require them to buy products made of recycled materials. To boost the city's recycling rate, Seattle has already made it illegal for residents to put recyclable materials in their regular trash.

This could make sense if recycling always saved resources. But does it?

Not always. The goal of recycling is to save resources, but mandatory recycling often wastes them. Additional trucks must go out into the community, using more energy and adding to air pollution. And reprocessing recyclables is just another form of manufacturing, which inevitably causes some pollution and waste.

Fortunately, there are additional ways to deal with trash. Modern landfills are scientifically engineered to keep garbage dry so that harmful leakage doesn't occur. And there is plenty of space. One widely cited estimate is that the United States could bury all its trash for the next century in a landfill 225 feet deep and 10 miles square.

We should recycle when it makes sense, but we shouldn't be afraid to use other means as well.

Jane S. Shaw is a Senior Fellow of the Property and Environment Research Center (PERC), a nonprofit institute in Bozeman, Montana, dedicated to improving environmental quality through property rights and markets. With Michael Sanera, she is coauthor of *Facts, Not Fear: Teaching Children about the Environment* (Regnery, 1999) and editor of the Greenhaven Press book series, *Critical Thinking about Environmental Issues*.

Explore this issue further by accessing **Viewpoints** at www.aw-bc.com/withgott.

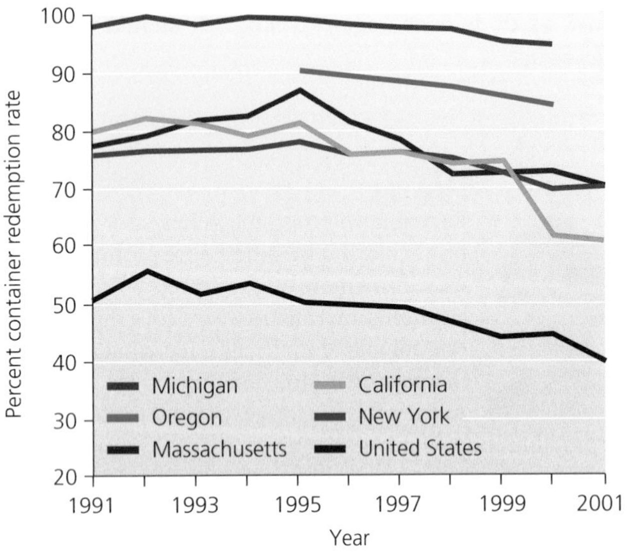

FIGURE 22.13 Data suggest that bottle bills increase recycling rates and that higher redemption amounts boost recycling rates further. The five states with bottle bills shown in this graph all have higher recycling rates than the United States as a whole. Michigan, the only state with a 10-cent deposit, has the highest recycling rate of all. The overall downward trend in rates in all states across the years is likely due to inflationary erosion of the value of deposits. Data from Gitlitz, J., and P. Franklin. 2004. *The 10¢ incentive to recycle.* Arlington, VA.: Container Recycling Institute.

waste stream. In states where they have been enacted, these laws have proved profoundly effective and resoundingly popular; they are recognized as among the most successful state legislation of recent decades (Figure 22.13). It is a testament to the lobbying influence of the beverage industries, which have traditionally opposed passage of bottle bills, that more states do not have such legislation.

States with bottle bills now face two challenges. One is to amend these laws to include new kinds of containers. In New York State, where 80 billion beverage containers have been redeemed since 1982, polls show 70% of the public favoring expanding their bottle bill to include more types of containers, and the legislature is currently considering the issue. The second challenge is to adjust refunds for inflation. In the three decades since Oregon passed the nation's first bottle bill, the value of a nickel has dropped such that today, the refund would need to be 22 cents to reflect the refund's original intended value. Proponents argue that increasing refund amounts will raise return rates, and available data support this view (see Figure 22.13).

One Canadian city showcases the shift from disposal to reduction and recycling

Edmonton, Alberta, has created one of the world's most advanced waste management programs. As recently as 1998, fully 85% of the city's waste was being landfilled, and space was running out. Today, just 35% goes to the new sanitary landfill, while 15% is recycled, and an impressive 50% is composted. Edmonton's citizens are proud of the program, and 81% of them participate in its curbside recycling program.

When Edmonton's residents put out their trash, city trucks take it to their new co-composting plant—at the size of eight football fields, the largest in North America (Figure 22.14a). The waste is dumped on the floor of the facility, and large items, such as furniture, are removed and landfilled. The bulk of the waste is mixed with dried sewage sludge for 1–2 days in five large rotating drums, each the length of six buses. The resulting mix travels on a conveyor to a screen that removes nonbiodegradable items. It is aerated for several weeks in the largest stainless steel building in North America (Figure 22.14b). The mix is then passed

(a) Composting facility, Edmonton, Alberta

(b) Aeration building, Edmonton composting facility

FIGURE 22.14 Edmonton, Alberta, boasts one of North America's most successful waste management programs. Edmonton's gigantic composting facility (**a**) is the size of eight football fields. Inside the aeration building (**b**), which is the size of 14 professional hockey rinks, mixtures of solid waste and sewage sludge are exposed to oxygen and composted for 14–21 days.

through a finer screen and finally is left outside for 4–6 months. The resulting compost—80,000 tons annually—is made available to area farmers and residents. The facility even filters the air it emits with a 1-m (3.3-ft) layer of compost, bark, and wood chips, which eliminates the release of unpleasant odors into the community.

Besides the co-composting facility and the sanitary landfill, Edmonton's program includes a state-of-the-art MRF that handles 30,000–40,000 tons of waste annually, a leachate treatment plant, a research center, public education programs, and a wetland and landfill revegetation program. In addition, 100 pipes collect enough landfill gas to power 4,000 homes, bringing thousands of dollars to the city and helping power the new waste management center. Five area businesses reprocess the city's recycled items. Newsprint and magazines are turned into new newsprint and cellulose insulation, while cardboard and paper are converted into building paper and shingles. Household metal is made into rebar and blades for tractors and graders, and recycled glass is used for reflective paint and signs.

Industrial Solid Waste

Solid waste generated by industry is another major contributor to the waste stream. Each year, U.S. industrial facilities generate about 7.6 billion tons of waste, according to the EPA, about 97% of which is wastewater. Thus, very roughly, 228 million or so tons of solid waste are generated by 60,000 facilities each year—an amount about equal to that of municipal solid waste. In the United States, industrial solid waste is defined as solid waste that is considered neither municipal solid waste nor hazardous waste under the Resource Conservation and Recovery Act.

U.S. waste managers differentiate between municipal and industrial solid waste largely because these categories are regulated differently. Whereas the federal government regulates municipal solid waste, state or local governments regulate industrial solid waste (although with federal guidance). Industrial waste includes more than just waste from factories. It includes waste from everything from manufactured consumer goods to mining activities to petroleum extraction to agricultural waste. Waste is generated at several points along the process from raw materials extraction to manufacturing to sale and distribution (Figure 22.15).

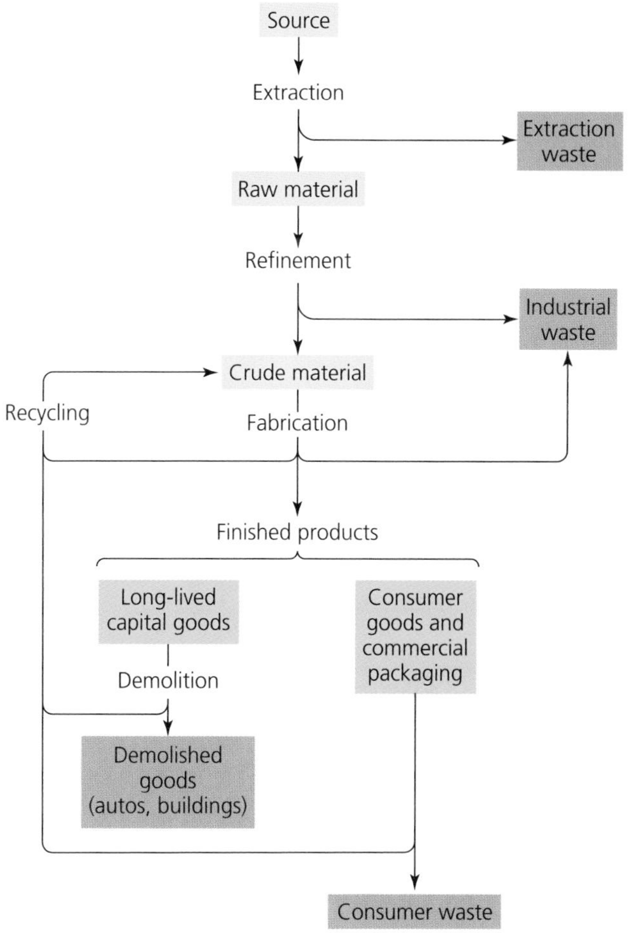

FIGURE 22.15 Industrial and municipal waste is generated at a number of stages throughout the life cycles of products. Waste is first generated when raw materials needed for production are extracted. Further industrial waste is produced as raw materials are processed and as products are manufactured. Waste results from the demolition or disposal of products once they are used by businesses and individuals. At each stage there are often opportunities for efficiency improvements, waste reduction, or recycling.

Regulation and economics both influence industrial waste generation

Most methods and strategies of waste disposal, reduction, and recycling by industry are similar to those for municipal solid waste. For instance, businesses that manage their own waste on site most often dispose of it in landfills, and companies must design and manage their landfills in ways that meet state, local, or tribal guidelines. Other businesses pay to have their waste disposed of at municipal disposal sites. Regulation varies greatly from state to state and area to area, but in most cases, state and local regulation of industrial solid waste is less strict than federal regulation of municipal solid waste. This is one reason that industry continues to suffer public criticism for its waste and pollution impacts. In many areas, industries are not required to have permits, install landfill liners or leachate collection systems, or monitor groundwater for contamination.

Government regulation is not the only factor that influences industrial waste generation. The amount of waste generated by a manufacturing process is one measure of its efficiency, because waste represents resources that are being lost. The less waste produced per unit or volume of product, the more efficient that process is, from a physical standpoint. However, physical efficiency is not always equivalent to economic efficiency. Often it is cheaper for industry to manufacture its products or perform its services quickly but messily. That is, it can be cheaper to generate waste than to avoid generating waste. In such cases, economic efficiency is maximized, but physical efficiency is not. The frequent mismatch between these two types of efficiency is a major reason why the output of industrial waste is so great.

Rising costs of waste disposal, however, enhance the financial incentive to decrease waste and increase physical efficiency. Once either government or the market makes the physically efficient use of raw materials also economically efficient, businesses have financial incentives to reduce their own waste.

Industrial ecology seeks to make industry more sustainable

In an effort to reduce waste, growing numbers of industries today are experimenting with industrial ecology. A holistic approach that integrates principles from engineering, chemistry, ecology, and economics, **industrial ecology** seeks to redesign industrial systems to reduce resource inputs and to minimize physical inefficiency while maximizing economic efficiency. Industrial ecologists would reshape industry so that nearly everything produced in a manufacturing process is used, either within that process or in a different one.

The larger idea behind industrial ecology is that industrial systems should function more like ecological systems, in which almost everything produced is used by some organism, with very little being wasted. Applied to human activity, this principle brings industry closer to the ideal of ecological economists, in which human economies attain sustainability by functioning in a circular fashion rather than a linear one (▸ pp. 45 and 689–690).

Industrial ecologists pursue their goals in several ways. For one, they examine the entire life cycle of a given product—from its origins in raw materials, through its manufacturing, to its use, and finally its disposal—and look for ways to make the process more ecologically efficient. This strategy is called **life-cycle analysis.**

Industrial ecologists also try to identify points at which waste products from one manufacturing process can be used as raw materials for a different process. For instance, used plastic beverage containers cannot be refilled because of the potential for contamination, but they can be shredded and reprocessed to make other plastic items, such as benches, tables, and decks. In addition, industrial ecologists examine industrial processes with an eye toward eliminating environmentally harmful products and materials. Finally, they study the flow of materials through industrial systems to look for ways to create products that are more durable, recyclable, or reusable. Goods that are currently thrown away when they become obsolete, such as computers, automobiles, and some appliances, could be designed to be more easily disassembled, and their component parts reused or recycled.

Attentive businesses are taking advantage of the insights of industrial ecology to reduce waste and save money while lessening their impact on the environment and human health. For example, American Airlines switched from hazardous to nonhazardous materials in its Chicago facility, decreasing its need to secure permits from the EPA. The company also used over 50,000 reusable plastic containers to ship goods, reducing packaging waste by 90%. Its Dallas–Fort Worth headquarters recycled enough aluminum cans and white paper in 5 years to save $205,000. Roughly 3,000 broken baggage containers were recycled into lawn furniture. A program to gather suggestions from employees resulted in over 700 ideas to reduce waste—and 15 of these ideas saved the company over $8 million in the first year of implementation.

Other efforts have gone further. The Swiss Zero Emissions Research and Initiatives (ZERI) Foundation sponsors dozens of innovative projects worldwide that attempt to create goods and services without generating waste. Although most are not fully closed-loop systems, they attempt to approach this ideal. In so doing, they cut down on waste while increasing output and income, and often they generate new jobs as well. One example involves breweries, currently being pursued in Canada, Sweden, Japan, and Namibia (Figure 22.16). Brewers in these projects take waste from the beer-brewing process and use it to fuel a series of other processes. The brewer as a result can make money from bread, mushrooms, pigs, gas, and fish, as well as beer, all while producing little waste.

Hazardous Waste

Solid waste from industrial and municipal sources is a problem largely because of the volumes in which it can accumulate. Hazardous waste is a problem because of its chemical nature. Public awareness of hazardous waste has increased greatly in recent decades, driven by highly publicized instances of toxic contamination at abandoned

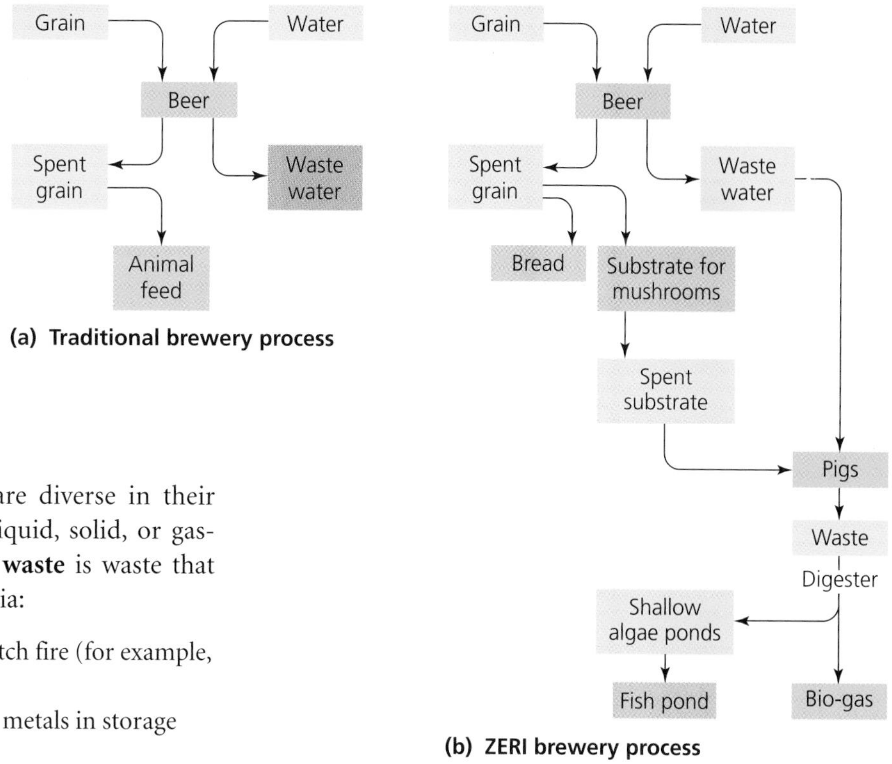

FIGURE 22.16 Traditional breweries (**a**) produce only beer while generating much waste, some of which goes toward animal feed. ZERI-sponsored breweries (**b**) use their waste grain to make bread and to farm mushrooms. Waste from the mushroom farming, along with brewery wastewater, goes to feed pigs. The pigs' waste is digested in containers that capture natural gas and collect nutrients used to nourish algae for growing fish in fish farms. The brewer derives income from bread, mushrooms, pigs, gas, and fish, as well as beer.

industrial sites. Hazardous wastes are diverse in their chemical composition and may be liquid, solid, or gaseous. By EPA definition, **hazardous waste** is waste that meets one of the following four criteria:

1. *Ignitability*. Substances that easily catch fire (for example, natural gas or alcohol).
2. *Corrosivity*. Substances that corrode metals in storage tanks or equipment.
3. *Reactivity*. Substances that are chemically unstable and readily react with other compounds, often explosively or by producing noxious fumes.
4. *Toxicity*. Substances that harm human health when they are inhaled, are ingested, or contact human skin.

Materials with these characteristics can harm human health and environmental quality. Flammable and explosive materials can cause ecological damage and atmospheric pollution. For instance, fires at large tire dumps in California's Central Valley have exacerbated air pollution and caused highway closures. Toxic wastes in lakes and rivers have caused fish die-offs and closed important domestic fisheries, such as those in Chesapeake Bay.

Hazardous wastes have diverse sources

Industry, mining, households, small businesses, agriculture, utilities, and building demolition are all sources of hazardous waste. Industry produces the largest amounts of hazardous waste, but in most developed nations industrial waste generation and disposal is highly regulated. This regulation has reduced the amount of hazardous waste entering the environment from industrial activities. As a result, households currently are the largest source of unregulated hazardous waste.

Household hazardous waste results from materials commonly used in and around the home that contain toxic, reactive, corrosive, or flammable ingredients. These include a wide range of items, such as paints, batteries, oils, solvents, cleaning agents, lubricants, and pesticides. U.S. citizens generate 1.6 million tons of household hazardous waste annually, and the average home contains close to 45 kg (100 lb) of it in sheds, basements, closets, and garages.

Although many hazardous substances become less hazardous over time as they degrade chemically, two classes of chemicals are particularly hazardous because their toxicity persists over time: organic compounds and heavy metals.

Organic compounds and heavy metals can be hazardous

In our day-to-day lives, we rely on the capacity of synthetic organic compounds and petroleum-derived compounds to resist bacterial, fungal, and insect activity. Items such as plastic containers, rubber tires, pesticides, solvents, and wood preservatives are useful to us precisely because they resist decomposition. We use these substances to protect our buildings from decay, kill pests that attack crops, and keep stored goods intact. However, the resistance of these compounds to decay is a double-edged sword, for it also makes them persistent pollutants. Many synthetic organic compounds are toxic because they can be readily absorbed through the skin of humans and other animals and can act as mutagens, carcinogens, teratogens, and endocrine disruptors (▸ pp. 409–410).

Heavy metals such as lead, chromium, mercury, arsenic, cadmium, tin, and copper are used widely in industry for wiring, electronics, metal plating, metal fabrication, pigments, and dyes. Heavy metals enter the environment when paints, electronic devices, batteries, and other materials are disposed of improperly. Lead from fishing weights and from hunters' lead shot has accumulated in many rivers, lakes, and forests. In older homes, lead from pipes contaminates drinking water, and lead paint remains a problem, especially for infants. Heavy metals are prone to bioaccumulate (▸ p. 416) when they are fat-soluble and break down slowly. In California's Coast Range, for instance, mercury washed downstream from abandoned mercury mines enters low-elevation lakes and rivers, is consumed by bacteria and invertebrates, and accumulates in increasingly larger quantities up the food chain, poisoning organisms at higher trophic levels and making fish unsafe to eat.

Computers, televisions, VCRs, cell phones, and other electronic devices represent major new sources of potential heavy metal contamination. These products have short lifetimes before people judge them obsolete, and most are discarded after only a few years. The amount of this electronic waste—sometimes called *e-waste*—is growing. In the United States alone, there are well over 300 million television sets, and the National Safety Council has estimated that 500 million computers will be retired between 1997 and 2007, creating several million tons of waste each year. Most e-waste is still disposed of in landfills as conventional solid waste, but recent research suggests that it should instead be treated as hazardous waste (see "The Science behind the Story," ▸ pp. 668–669).

Weighing the Issues:
Toxic Computers?

The cathode ray tubes in televisions and computer screens can hold up to 5 kg (8 lb) of heavy metals, such as lead and cadmium. These represent the second-largest source of lead in U.S. landfills today, behind car batteries. With a growing number of computer screens being purchased, the transition to high-definition television, and the rapid turnover of computers, what future waste problems might you expect? How do you think these products should be disposed of, and who should be responsible for their disposal?

Several steps precede the disposal of hazardous waste

For many years we discarded hazardous waste without special treatment. In many cases people did not know that certain substances were harmful to human health. In other cases their danger was known or suspected, but it was assumed that the substances would disappear or be sufficiently diluted in the environment. The resurfacing of toxic chemicals years after their burial at Love Canal in upstate New York demonstrated to the public that hazardous waste deserves special attention and treatment.

Today, several methods for disposing of hazardous waste have been developed. All are improvements on the old practice of open dumping, but none is completely satisfactory. A number of steps generally precede disposal. Since the 1980s, many communities have designated sites or special collection days to gather household hazardous waste, or facilities for the exchange and reuse of substances (Figure 22.17). Once consolidated in such sites, the waste is transported for treatment and ultimate disposal.

Under the Resource Conservation and Recovery Act, the EPA sets standards by which states are to manage hazardous waste. RCRA also requires large generators of hazardous waste to obtain permits, and mandates that hazardous materials be tracked "from cradle to grave." As hazardous waste is generated, transported, and disposed of, the producer, carrier, and disposal facility must each report to the EPA the type and amount of material generated; its location, origin, and destination; and the way it is being handled. This process is intended to prevent illegal dumping and to encourage the use of reputable waste carriers and disposal facilities.

FIGURE 22.17 Many communities designate collection sites or collection days for household hazardous waste. Here, workers handle waste from an Earth Day collection event near Los Angeles.

FIGURE 22.18 Unscrupulous individuals or businesses sometimes dump hazardous waste illegally to avoid disposal costs.

Because current U.S. law makes disposing of hazardous waste quite costly, irresponsible companies have sometimes been found guilty of illegally and anonymously dumping waste, creating health risks for residents and financial headaches for local governments forced to deal with the mess (Figure 22.18). However, high costs have also encouraged responsible businesses to heat hazardous waste to transform it into nonhazardous substances prior to disposal.

Many biologically hazardous materials can be broken down by incineration at high temperatures in cement kilns. Some hazardous materials can be treated by exposure to bacteria that break down harmful components and synthesize them into new compounds. Besides bacterial bioremediation, phytoremediation (▸ pp. 92–93) is also used. Various plants have now been bred or engineered to take up specific contaminants from soil, then break down organic contaminants into safer compounds or concentrate heavy metals in their tissues. The plants are eventually harvested and disposed of.

We have three main disposal methods for hazardous waste

Three primary means of hazardous waste disposal have been developed: landfills, surface impoundments, and injection wells. Design and construction standards for landfills that receive hazardous waste are stricter than those for ordinary sanitary landfills. Hazardous waste landfills must have several impervious liners and leachate removal systems, and must be located far from aquifers. Dumping of hazardous waste in ordinary landfills has long been a problem. In New York City, Fresh Kills largely managed to keep hazardous waste out, but most of the city's older landfills were declared to be hazardous sites because of past toxic waste disposal. Secure landfills for hazardous waste do nothing to lessen the hazards of the materials, but they do help keep these materials isolated from people, wildlife, and ecosystems.

A method for storing liquid hazardous waste, or waste in dissolved form, is in ponds or **surface impoundments.** To create a surface impoundment, a shallow depression is dug and lined with plastic and an impervious material, such as clay. Water containing dilute hazardous waste is placed in the pond and allowed to evaporate, leaving a residue of solid hazardous waste on the bottom (Figure 22.19). This process is repeated until the dry material is removed and transported elsewhere for permanent disposal. Impoundments are not ideal. The underlying layer can crack and leak waste. Some material may

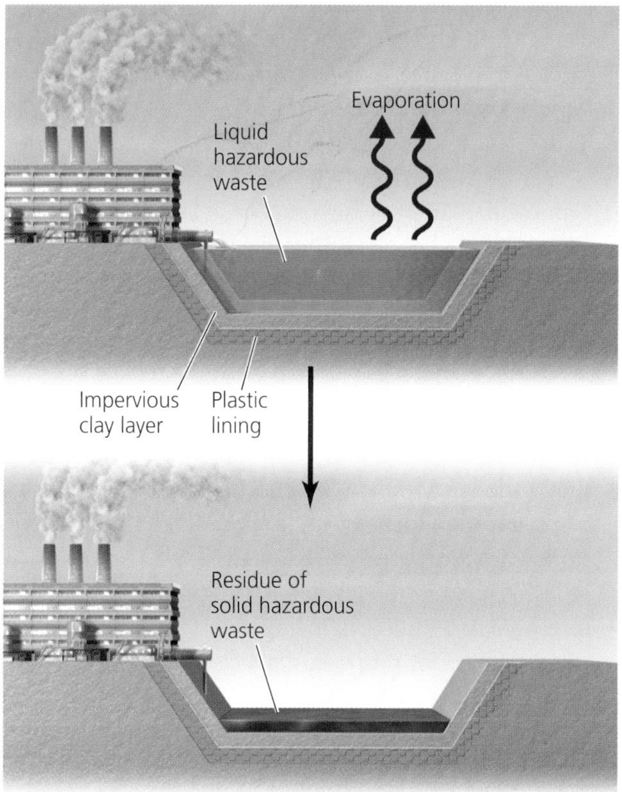

FIGURE 22.19 Surface impoundments are a strategy for temporarily disposing of liquid hazardous waste. The waste, mixed with water, is poured into a shallow depression lined with plastic and clay to prevent leakage. When the water evaporates, leaving a crust of the hazardous substance, new liquid is poured in and the process repeated. This method alone is not satisfactory, because waste can potentially leak, overflow, evaporate, or blow away.

The Science behind the Story

Testing the Toxicity of "E-Waste"

As we began to conduct more of our business, learning, and communication with computers and other electronic devices, many people predicted that our waste, particularly paper waste, would decrease. But instead, the proliferation of computers, printers, VCRs, fax machines, cell phones, and other gadgets has created a substantial new source of waste.

Most of this electronic waste, or "e-waste," is landfilled as conventional solid waste. However, most electronic appliances contain heavy metals that can cause environmental contamination and public health risks. For instance, over 6% of a typical computer is composed of lead.

At the University of Florida, Gainesville, Timothy Townsend's lab was funded by the EPA to determine whether e-waste is toxic enough to be classified as hazardous waste under the Resource Conservation and Recovery Act.

With students and colleagues, Townsend determined in 1999–2000 that cathode ray tubes (CRTs) from computer monitors and color televisions leach an average of 18.5 mg/L of lead, far above the regulatory threshold of 5 mg/L. Following this research, the EPA proposed classifying CRTs as hazardous waste, and several U.S. states banned these items from conventional landfills.

Discarded electronic waste can leach heavy metals and should be considered hazardous waste, researchers say.

Then in 2004, Townsend's lab group completed experiments on 12 other types of electronic devices. To measure their toxicity, Townsend's group used the EPA's standard test, the Toxicity Characteristic Leaching Procedure (TCLP). Designed to mimic the process by which chemicals leach out of solid waste in landfills, this test speeds up the process so that it can be measured in the lab. In the TCLP, waste is ground up into fine pieces and 100 g (3.5 oz) of it is put in a container with 2 L (0.53 gal) of an acidic leaching fluid. The container is rotated for 18 hours, after which the leachate is analyzed for its chemical content.

Researchers look for eight heavy metals—arsenic, barium, cadmium, chromium, lead, mercury, selenium, and silver—and determine for each whether their concentration in the leachate exceeds that allowed by EPA regulations. Of these eight elements, electronic devices contain notable amounts of four: cadmium, chromium, lead, and mercury.

To conduct the standard TCLP, Townsend's team ground up the central processing units (CPUs) of personal computers, creating a mix made up by weight of 15.8% circuit board, 7.5% plastic, 68.2% ferrous metal, 5.4% nonferrous metal, and 3.1% wire and cable. However, grinding up a computer into small bits is no easy task, and it is hard to obtain a sample that accurately represents all components and materials. So the researchers also designed a modified TCLP test in which they placed whole CPUs—with the parts disassembled but not ground up—in a rotating 55-gallon drum full of leaching liquid. Then they tested their 12 types of devices using a combination of the standard and modified TCLP methods.

The team's results are summarized in the figure. Lead was the only heavy metal found to exceed the EPA's regulatory threshold, but this threshold (5 mg/L) was exceeded in the majority of trials. Computer monitors leached the most lead (47.7 mg/L on average), as expected,

evaporate or be blown into surrounding areas. Heavy rainstorms may cause waste to overflow and contaminate nearby areas. For these reasons, surface impoundments are used only for temporary storage.

The third method is intended for long-term disposal. In **deep-well injection,** a well is drilled deep beneath the water table, reaching into porous rock. Once the well has been drilled, wastes are injected into it. The aim is that waste will accumulate in the porous rock and remain deep underground, isolated from groundwater and human contact (Figure 22.20). This idea seems attractive in principle, but in practice wells become because monitors include the cathode ray tubes already known to be a problem. However, laptops, color TVs, smoke detectors, cell phones, and computer mice also leached high levels of lead. Next came remote controls, VCRs, keyboards, and printers, all of which leached more lead on average than the EPA threshold, and did so in 50% or more of the trials. Whole CPUs and flat panel monitors were the only devices to leach less than 5 mg/L of lead on average, but even these exceeded the threshold more than one-quarter of the time.

The researchers found that items containing more ferrous metals (such as iron, copper, and zinc) tended to leach less lead. For instance, CPUs contain 68% ferrous metals (compared to only 7% in laptops), and laptops leached seven times as much lead as CPUs. Further experiments confirmed that ferrous metals were chemically reacting with lead and stopping it from leaching.

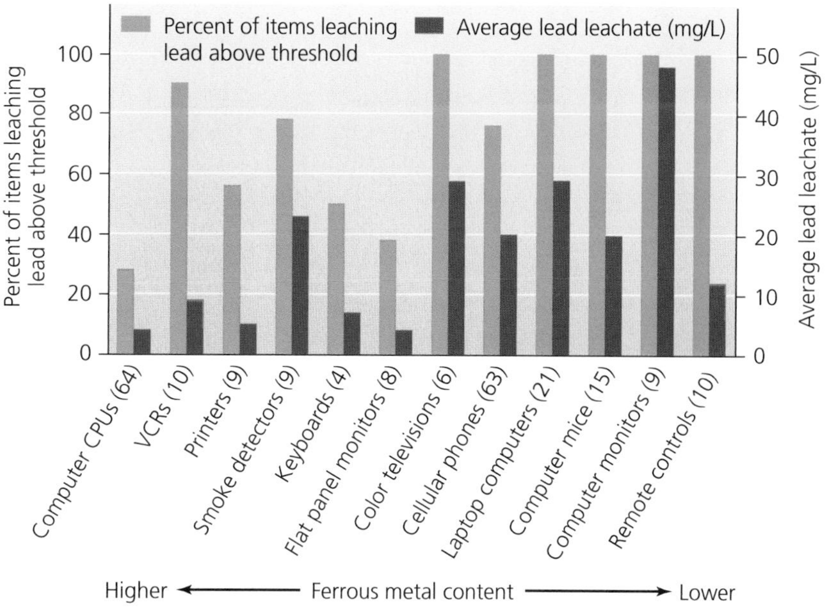

Some proportion of all 12 devices tested exceeded the EPA regulatory standard for lead leachate. Devices with higher ferrous metal content tended to leach less lead. Where both standard and modified TCLPs were used, results are averaged. Data from Townsend, T.G., et al., 2004. RCRA toxicity characterization of computer CPUs and other discarded electronic devices. July 15, 2004, report to the U.S. EPA.

Townsend says the work suggests that many electronic devices have the potential to be classified as hazardous waste because they frequently surpass the toxicity criterion for lead. However, EPA scientists must decide how to judge results from the modified TCLP methods, and must evaluate other research, before determining whether to alter regulatory standards.

Furthermore, lab tests may or may not accurately reflect what actually happens in landfills. So Townsend is also pursuing research to address this question. His team is filling columns measuring 24 cm (2 ft) wide by 4.9 m (16 ft) long with e-waste and municipal solid waste, burying them in a Florida landfill, and then testing the leachate that results.

As the EPA and more states move toward keeping e-waste out of conventional sanitary landfills, more computers and accessories are being recycled. The devices are taken apart, and parts are either reused or disposed of more safely. Although there are serious concerns about the health risks this may pose to workers doing the disassembly, recycling done responsibly seems likely to be the way of the future.

In many North American cities, businesses, nonprofit organizations, or municipal services now recycle used computers and related devices. So next time you upgrade to a new computer, TV, DVD player, VCR, or cell phone, check out what opportunities may exist in your area to recycle your old ones.

corroded and can leak wastes into soil, allowing them to enter aquifers. Currently the amount of waste disposed of in injection wells is declining, but roughly 34 billion L (9 billion gal) of hazardous waste continue to be placed in U.S. injection wells each year.

Radioactive waste is a special type of hazardous waste

Radioactive waste is particularly dangerous to human health and is persistent in the environment. The dilemma of disposal has dogged the nuclear energy industry and

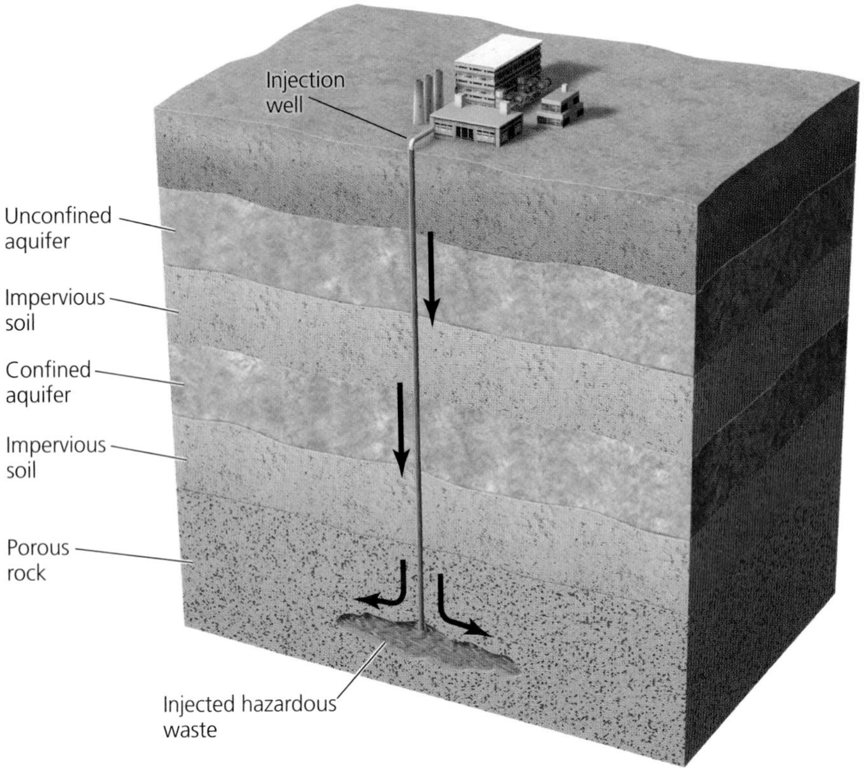

FIGURE 22.20 A seemingly more satisfactory way of disposing of liquid hazardous waste is to pump it deep underground, in deep-well injection. The well must be drilled below any aquifers, into porous rock separated by impervious clay. The technique is expensive, however, and leakage from the well shaft into groundwater may occur.

the U.S. military for decades. As we saw in our discussion of radioactive waste disposal in Chapter 20 (▸ pp. 600–603), Yucca Mountain in Nevada has recently been approved as the single-site repository for all U.S. nuclear waste.

Currently, a site in the Chihuahuan Desert in southeastern New Mexico serves as a permanent disposal site for radioactive waste. The Waste Isolation Pilot Plant (WIPP) is the world's first underground repository for transuranic waste from nuclear weapons development. The mined caverns holding the waste are located 655 m (2,150 ft) below ground in a huge salt formation thought to be geologically stable. Twenty years in the planning, WIPP became operational in 1999 and will receive thousands of shipments of waste from 23 other locations over the next three decades.

Contaminated sites are being cleaned up, slowly

Many thousands of former military and industrial sites remain contaminated with hazardous waste in the United States, Russia, and virtually every other nation on Earth. For most nations, dealing with these messes is simply too difficult, time-consuming, and expensive. In 1980 the U.S. Congress passed the Comprehensive Environmental Response Compensation and Liability Act (CERCLA). This legislation established a federal program to clean up U.S. sites polluted with hazardous waste from past activities. The EPA administers this cleanup program, called the **Superfund.** Under EPA auspices, experts identify sites polluted with hazardous chemicals, take action to protect groundwater near these sites, and clean up the pollution. The objective of the act is to charge responsible parties for cleanup of sites (according to the *polluter-pays principle*). For many polluted sites, however, the responsible parties cannot be found or held liable, and in such cases—roughly 1 out of 4 so far—Superfund activities are covered by taxpayers' funds. These funds come from the federal budget ($1.4 billion in 2005), and from a trust fund (which went bankrupt in 2004) established from a tax on chemical raw materials.

Once a Superfund site is identified, EPA scientists evaluate how close the site is to human habitation, whether wastes are currently confined or likely to spread, and whether the site threatens drinking water supplies. Sites that appear harmful are placed on EPA's National Priority List, ranked according to the level of risk to human health that they pose. Cleanup proceeds on a site-by-site basis as funds are available. Throughout the process, the EPA is required to hold public hearings to inform area residents of their findings and to receive feedback.

As of late 2005, 966 of the 1,547 Superfund sites on the National Priority List had been cleaned up. The average cleanup has cost $25 million and has taken 12–15 years. Many sites are contaminated with hazardous chemicals we have no effective way to deal with. In such cases, cleanups simply involve trying to isolate waste from human contact, either by building trenches and clay or concrete barriers around a site or by excavating contaminated material, placing it in industrial-strength containers, and shipping it to a hazardous waste disposal facility. For all these reasons, the current emphasis in the United States and elsewhere is on preventing hazardous waste contamination in the first place.

Conclusion

Our societies have made great strides in addressing our waste problems. Modern methods of waste management are far safer for people and gentler on the environment than past practices of open dumping and open burning. In many countries, recycling and composting efforts are making rapid strides. The United States has gone in a few decades from a country that virtually did not recycle to a nation in which 30% of all solid waste is diverted from disposal. The continuing growth of recycling, driven by market forces, government policy, and consumer behavior, shows potential to further alleviate our waste problems.

Despite these advances, our prodigious consumption habits have created more waste than ever before. Our waste management efforts are marked by a number of difficult dilemmas, including the cleanup of Superfund sites, safe disposal of hazardous and radioactive waste, and frequent local opposition to disposal sites. These dilemmas make clear that the best solution to our waste problem is to reduce our generation of waste. Finding ways to reduce, reuse, and efficiently recycle the materials and goods that we use stands as a key challenge for the new century.

REVIEWING OBJECTIVES

You should now be able to:

Summarize and compare the types of waste we generate

- Municipal and industrial solid waste, hazardous waste, and wastewater are major categories of waste. (p. 648)

List the major approaches to managing waste

- Source reduction, recovery, and disposal are the three main components of waste management. (pp. 648–649)

Delineate the scale of the waste dilemma

- Developed nations generate far more waste than developing nations, but waste everywhere is increasing as a result of growth in population, wealth, and consumption. (pp. 650–651)

Describe conventional waste disposal methods: landfills and incineration

- Sanitary landfills are an improvement over open dumping of the past, and extensive efforts guard against contamination of groundwater, air, and soil. Nonetheless, such contamination can occur. (pp. 651–655)
- Incinerators with pollution control technology are an improvement over open burning of the past, and these facilities remove the great majority of pollutants. Nonetheless, some pollutants escape, and highly toxic ash needs to be disposed of in landfills. (pp. 655–657)
- We can harness energy from landfill gas and electricity generation from incineration. (p. 657)

Evaluate approaches for reducing waste: source reduction, reuse, composting, and recycling

- Reducing waste before it is generated is the best waste management approach. Recycling materials comprises the next-best option. (p. 657)
- Industry can reduce waste and save money by increasing efficiency. (pp. 657–658)
- Consumers can take an array of simple steps to reduce their waste output. (pp. 657–658)
- Composting reduces waste while creating organic matter useful for gardening and agriculture. (p. 658)
- Recycling has grown in recent years and currently removes 23.5% of the U.S. waste stream. (pp. 658–663)

Discuss industrial solid waste management and principles of industrial ecology

- Although regulations differ, industrial waste management practices are similar to those for municipal solid waste. (pp. 663–664)
- Industrial ecology urges industrial systems to mimic ecological systems and provides ways for industry to increase its efficiency. (pp. 664–665)

Assess issues in managing hazardous waste

- Hazardous waste is regulated and monitored, yet illegal dumping remains a problem. (pp. 664–667)
- No fully satisfactory method of disposing of hazardous waste has yet been devised. (pp. 667–670)
- Cleanup of hazardous waste sites is a long and expensive process. (pp. 670–671)

TESTING YOUR COMPREHENSION

1. Describe five major methods of managing waste. Why do we practice waste management?
2. Why have some people labeled the United States "the throwaway society"? How much solid waste do Americans generate, and how does this amount compare to that of people from other countries?
3. Name several guidelines by which sanitary landfills are regulated. Describe at least five problems with landfills.
4. Describe the process of incineration or combustion. What happens to the resulting ash? What is one drawback of incineration?
5. What is composting, and how does it help reduce input to the waste stream?
6. What are the three elements of a sustainable process of recycling?
7. What are the goals of industrial ecology?
8. What four criteria are used to define hazardous waste? Why are heavy metals and synthetic organic compounds particularly hazardous?
9. What are the largest sources of hazardous waste? Describe three ways to dispose of hazardous waste.
10. What is the Superfund program? How does it work?

SEEKING SOLUTIONS

1. How much waste do you generate? Look into your waste bin at the end of the day and categorize and measure the waste there. List all other waste you may have generated in other places throughout the day. How much of this waste could you have avoided generating? How much could have been reused or recycled?
2. Some people have criticized current waste management practices as merely moving waste from one medium to another. How might this criticism apply to the methods now in practice? What are some potential solutions?
3. Of the various waste management approaches covered in this chapter, which ones are your community or campus pursuing, and which are they not pursuing? Would you suggest that your community or campus start pursuing any new approaches? If so, which ones, and why?
4. Can manufacturers and businesses benefit from source reduction if consumers were to buy fewer products as a result? How? Given what you know about industrial ecology, what do you think the future of sustainable manufacturing may look like?
5. Imagine you are the head of a major corporation that produces containers for soft drinks and a wide variety of other consumer products. Your company's shareholders are asking that you improve the company's image—while not cutting into profits—by taking steps to reduce waste. What steps would you consider taking?
6. Think of several industries or businesses in your community, as well as the ways these interact with facilities on your campus. Bearing in mind the principles of industrial ecology, can you think of any novel ways that these entities might mutually benefit from one another's services, products, or waste materials? Are there waste products from one business, industry, or campus facility that another might put to good use? Can you design an eco-industrial park that might work in your community or on your campus?

INTERPRETING GRAPHS AND DATA

Using 1990 data from 149 countries, David Beede, an economist at the U.S. Department of Commerce, and David Bloom, a professor of economics at Columbia University, examined global patterns in the generation and management of municipal solid waste (MSW). Beede and Bloom were particularly interested in the relationships among wealth, population size, and per capita generation of MSW. Their results are presented in the accompanying table.

Income category of nation	Total MSW generation		Population size		Percent-age of world total GDP	Pounds MSW per capita per day
	Tons/yr (billions)	% of world total	Millions of people	% of world total		
Low	0.658	46.3	3,091	58.5	18.7	1.17
Low-middle	0.160	11.2	629	11.9	9.9	
Upper-middle	0.212	14.9	748	14.2	16.5	
High	0.393	27.6	816	15.4	54.9	
All economies	1.422	100.0	5,284	100.0	100.0	

Data from Beede, D. N. and D. E. Bloom. 1995. The economics of municipal solid waste. *World Bank Research Observer* 10: 113–150.

1. Using the data for total MSW generation and for population size, calculate the pounds of MSW per capita per day for each category of nation.
2. Describe in general terms the relationship between wealth and per capita generation of MSW. Now create a graph that demonstrates this relationship, using the data from the table. Plot the values you calculated for per capita waste generation for each of the four wealth categories of nations. Can you offer at least one possible reason for the trend that you see?
3. Do you think it's possible for wealthy nations to reduce their per capita MSW generation to the rates of poorer nations? Why or why not?

CALCULATING ECOLOGICAL FOOTPRINTS

The 14th annual "State of Garbage in America" survey documents the prodigious ability of U.S. residents to generate municipal solid waste (MSW). According to the survey, on a per capita basis, South Dakotans generate the least MSW (3.72 lb/day), and Kansans the most (9.48 lb/day). The average for the entire country is 7.17 lb MSW per person per day. Compare this number to the data in the "Interpreting Graphs and Data" table. Now calculate the amount of MSW generated in 1 day and in 1 year by each of the groups indicated at each of the rates shown in the accompanying table.

Groups generating municipal solid waste	Per-capita MSW generation rates: U.S. average (7.17 lb/day)		South Dakota (3.72 lb/day)		"High-income" countries (2.64 lb/day)		World average (1.47 lb/day)	
	Day	Year	Day	Year	Day	Year	Day	Year
You	7.17	2,617.05						
Your class								
Hometown								
Your state								
United States								
World								

Data sources: Kaufman, S. M. et al. 2004.The state of garbage in America. *BioCycle* 45: 31–41.

1. Suppose your hometown of 50,000 people has just approved construction of a landfill nearby. Estimates are that it will accommodate 1 million tons of MSW. Assuming the landfill is serving only your hometown, for how many years will it accept waste before filling up? How much longer would a landfill of the same capacity serve a town of the same size in another industrialized ("high-income") country?
2. Why do you think U.S. residents generate so much more MSW than people in other "high-income" countries, when standards of living in those countries are comparable?

Take It Further

Go to www.aw-bc.com/withgott or the student CD-ROM where you'll find:

- Suggested answers to end-of-chapter questions
- Quizzes, animations, and flashcards to help you study
- *Research Navigator*™ database of credible and reliable sources to assist you with your research projects
- **GRAPHIt!** tutorials to help you master how to interpret graphs
- **INVESTIGATEIt!** current news articles that link the topics that you study to case studies from your region to around the world

23 Sustainable Solutions

Sri Lankan children planting tree seedlings on deforested hillsides around their village

Upon successfully completing this chapter, you will be able to:

- List and describe approaches being taken on college and university campuses to promote sustainability
- Explain the concept of sustainable development
- Discuss how protecting the environment can be compatible with promoting economic welfare
- Describe and assess key approaches to designing sustainable solutions
- Analyze the roles that consumption, population, and technology play in efforts to achieve sustainability
- Explain how time is limited but how human potential to solve problems is tremendous

Central Case: Ball State University Aims for Campus Sustainability

"We solemnly pledge to the peoples of the world and the generations that will surely inherit this Earth that we are determined to ensure that our collective hope for sustainable development is realized."
—Declaration signed by 193 nations at the World Summit, Johannesburg, South Africa, 2002

"We hope—surely we must believe—that our species will emerge from the environmental bottleneck in better condition than when we entered . . . taking as much of the rest of life with us as possible."
—Edward O. Wilson, *Consilience*, 1998

At Ball State University, "BSU" stands for more than the institution's name. It also stands for "Becoming a Sustainable University." For more than a decade, students, faculty, staff, and administrators at this 18,000-student campus in Muncie, Indiana, have made sustainability a goal. Today Ball State is a leader in the international movement for campus sustainability.

Proponents of campus sustainability view colleges and universities as microcosms of society at large, noting that institutions of learning consume resources, emit pollution, and exert other environmental impacts. They seek to change the ways these institutions operate and to reduce their ecological footprints.

Many colleges and universities have student-run environmental organizations, recycling programs, and courses or majors in environmental science or environmental studies. Ball State began taking steps beyond these back in 1994, when it initiated a program to promote environmental literacy among its faculty. Its "Green for Green" program pays faculty to take seminars on sustainability. These instructors are expected to integrate what they've learned into their teaching, promote changes in their departments, or conduct community outreach. To date, over 200 instructors from nearly all of the university's 47 departments have participated—over one-fifth of the BSU faculty.

Ball State accelerated its sustainability efforts after 1999, once the university's president signed the **Talloires Declaration,** a commitment to pursue sustainability. This document was composed in Talloires, France, in 1990 and has been signed by over 300 university presidents and chancellors from more than 40 nations.

To plan green initiatives at Ball State, faculty, staff, students, and community members formed the Council on the Environment (COTE). COTE wrote a guide for future goals and actions, meets monthly to plan initiatives, and works with BSU administrators to achieve its goals. A Center for Energy Research/Education/Service coordinates and implements COTE's initiatives.

One goal was to shift the university to recycled paper. In 2003, pressure from COTE led the university to replace its virgin stock paper with paper of 30% post-consumer recycled content. It also made 100% post-consumer recycled paper available, which a number of departments adopted. COTE estimates that, besides reducing landfill waste, switching to 30% recycled-content paper saves 3,000 trees each year and prevents the emission of 45,000 kg (100,000 lb) of CO_2 equivalents and 90 kg (200 lb) of nitrogen oxides.

Sustainability advocates are also greening Ball State's transportation system. Three gas-electric hybrid vehicles were added to the vehicle fleet in 2003. The next year, Ball State began running the entire bus fleet, as well as some other equipment, on B-20 biodiesel fuel. In 2005, Ball State added one hybrid bus to its shuttle fleet.

Other recent initiatives include composting, running Earth Day events, and urging the campus to eliminate mercury use and secondhand smoke exposure. COTE also sponsors annual awards programs honoring those at Ball State and in the community who contribute to sustainability efforts.

Most of these sustainability initiatives have cost money, and their proponents have had to argue their case to penny-conscious administrators time after time. But the university's administration has responded, often recognizing that many short-term costs are actually investments that will save money in the long term.

Beyond serving as a model for other schools, Ball State has launched the *Greening of the Campus* Conference Series. At each of these meetings, people involved in sustainability efforts from well over 100 colleges and universities have gathered to share their experiences. The university hosted the sixth of these biannual conferences in September 2005.

Sustainability on Campus

If we are to attain a sustainable civilization, we will need to make efforts at every level, from the individual to the household to the community to the nation to the world. Governments, corporations, and organizations must all encourage and pursue sustainable practices. Among the institutions that can contribute to sustainability efforts are colleges and universities.

We tend to think of colleges and universities as enlightened and progressive institutions that generate benefits for the communities that host them and for society as a whole. However, colleges and universities are also centers of lavish resource consumption. Institutions of higher education consist of extensive infrastructure, including classrooms, offices, research labs, and residential housing. They also have dining establishments, sports arenas, hospitals, vehicle fleets, and road networks. The 4,100 campuses in the United States interact with many thousands of businesses and spend over $200 billion each year on products and services. Thus, the ecological footprint of a typical college or university is substantial.

Reducing the size of this footprint is challenging. Colleges and universities tend to be bastions of tradition, where institutional habits are deeply ingrained and where bureaucratic inertia can often block the best of intentions for positive change. Nonetheless, faculty, staff, administrators, and especially students are progressing on a variety of fronts to make the operations of educational institutions more sustainable.

Why strive for campus sustainability?

You enrolled at your college or university to gain an education, not to transform the institution. Why, then, are increasing numbers of students promoting sustainable practices on their campuses? First, reducing the ecological footprint of a campus really can make a difference. The consumptive impact of educating, feeding, and housing hundreds or thousands of students is immense. Second, campus sustainability efforts can make more students aware of the need to address environmental problems, and students who act to promote campus sustainability can serve as models for others. Finally, the student who engages in sustainability efforts learns and grows as a result of the experience. The challenges, successes, and failures that you encounter can serve as valuable preparation for similar efforts in changing inertia-bound institutions in the broader society.

Support from faculty, staff, and administrators is crucial for success, but students are often the ones to initiate change. Students often feel freer than faculty or staff to express themselves. Students also arrive on campus with new ideas and perspectives, and they generally are less attached to traditional ways of doing things.

Campus efforts often begin with an audit

Campus sustainability efforts often begin with a quantitative assessment of the institution's operations. Such audits provide baseline information on what an institution is doing or how much it is consuming, and help set priorities and goals. For instance, Harford Community College in Maryland hired engineers and specialists to perform a complete audit of its energy use and pollutant emissions, which then served as the basis for setting goals. At the University of Vermont, graduate student Erika Swahn gathered data on heating, transportation, electricity, waste, food, and water use to calculate her school's ecological footprint—over 59,000 acres annually, or about 4.5 acres for each student, instructor, and staff member. Once changes are implemented, the institution can monitor progress by comparing new measurements to the audit's baseline data.

Audits can be extremely useful, but it's vital to target items that can lead directly to specific recommendations. If an audit produces lots of data but only generalized recommendations, such as "purchase energy-efficient appliances," then the work will not have been worth it. Such an audit should instead quantify the performance of individual appliances currently in operation so that decision makers can identify which ones to replace.

Recycling and waste reduction are the most common campus efforts

Campus sustainability efforts most frequently involve waste reduction, recycling, and composting. Depending on the type of material, 46–85% of schools have recycling services in place, according to the first comprehensive survey of campus sustainability efforts (Table 23.1). Waste management efforts are relatively easy to start and maintain because they offer many opportunities for small-scale improvements and because people generally enjoy recycling and reducing waste.

Students at Ohio University and Miami University in Ohio kicked off *RecycleMania,* a 10-week competition among schools to see which could recycle the most. The competition has expanded each year, and it involved 46 schools in 2005, when Miami University won the title for total amount recycled per capita, with 66.19 lb of material recycled per student. California State University at San Marcos won the title for recycling rate, recycling 43.65% of its waste.

Table 23.1 Frequency of Campus Sustainability Efforts

Activity	Percentage of schools performing activity*
Setting and reviewing goals	45
Staffing environmental programs with a coordinator	29
Water efficiency upgrades	72
Energy efficiency and conservation	63
Recycling: activity level and array of materials	65†
Sustainable landscaping	43
Managing transportation demand	24
Research and service opportunities for students	58
Professional development for faculty	50

*Based on voluntary responses from 891 schools.
†46–85%, depending on type of material.
Source: McIntosh, M., et al., 2001. *State of the campus environment: A national report card on environmental performance and sustainability in higher education.* National Wildlife Federation Campus Ecology; survey conducted by Princeton Survey Research Associates.

"Landfill on the lawn" events involve tipping dumpsters onto a central campus open space and sorting out recyclable items (Figure 23.1). When students at Ashland University in Ohio audited its waste, they found that 70%

FIGURE 23.1 In "landfill on the lawn" events, piles of waste are dumped onto a campus open space. Volunteers sort through the rubbish, separating out recyclables, as these students at the University of North Carolina at Greensboro are doing at a 2005 event. These events can offer a dramatic demonstration of just how many recyclable items are needlessly thrown away.

was recyclable and pressed their administration to support recycling programs. Several schools organize collections of items that might otherwise be thrown out. Clemson University in South Carolina in 2002 donated 11,500 lb of collected food, clothing, and other items to local charities.

In recent years, Ball State has also implemented a composting program for the school's bulky wood waste, such as surplus furniture and wood pallets. Rather than being thrown away, these items are shredded and made into mulch, which is used to nourish campus plantings.

Students find that administrators are more easily convinced to enact institutional changes that save money. At Ithaca College in New York, about 44% of the food waste generated annually on campus is composted. Disposal fees at the local landfill are $60/ton, so composting saves the college $11,500 each year in avoided fees. The compost is used on ornamental plantings on campus, and experiments showed that the plantings grew better with the compost mix than with chemical soil amendments.

Green building design is a key to sustainable campuses

Individual students can carry out waste reduction programs on small scales, but some campus administrators have invested in sustainable approaches on a much larger financial scale—in the construction of new buildings. Dozens of campuses now boast "green" buildings that simultaneously accomplish a number of advances toward sustainability. Typically, these buildings are constructed from certified sustainable building materials and incorporate design and technologies for energy efficiency and water efficiency, use of renewable energy, and reduction of indoor air pollution and emissions of outdoor pollutants.

As with any type of ecolabeling, there need to be agreed-upon standards for "green buildings." The leading set of standards for sustainable buildings are the *Leadership in Energy and Environmental Design (LEED)* standards, which developed and maintained by the nonprofit U.S. Green Building Council. These detailed standards guide the design and oversight of new construction or the renovation of existing structures on campuses and elsewhere.

One of the first and best-known green buildings on a college campus is the Adam Joseph Lewis Center for Environmental Studies at Oberlin College in Ohio (Figure 23.2). This building was constructed using materials that were recycled or reused, took little energy to produce, or were locally harvested, produced, or distributed. Some materials, such as carpeting, are leased and then returned to the company for recycling when they become worn out. The Lewis Center contains energy-efficient lighting, heating, and appliances. The building is also designed to maximize indoor air quality with a state-of-the-art ventilation system and paints, adhesives, and carpeting that emit few volatile organic carbons. The structure is powered largely by solar energy: PV panels on the roof provide active solar heating, and walls of south-facing glass and a tiled slate floor acting as a thermal mass provide passive solar heating (▸ p. 624). Over 150 sensors throughout the building monitor conditions such as temperature and air quality.

Wastewater generated in the Lewis Center is treated and recycled by a "living machine." This system consists of a series of tanks that mimic the stages of treatment at a wastewater treatment plant and finish with filtration in an

FIGURE 23.2 Oberlin College's Adam Joseph Lewis Center for Environmental Studies was one of the first green buildings on a U.S. campus, and it remains one of the most renowned.

FIGURE 23.3 In October 2005, 18 teams from colleges and universities across the United States and the world converged on the Mall in Washington, D.C., for the second Solar Decathlon. Each team erected an entire house, of the students' own design, fully powered by solar energy. The University of Colorado won the 2005 competition, edging out Cornell University and California Polytechnic State University. The next Solar Decathlon will be held in fall 2007.

artificial wetland. The treated wastewater is reused in toilets and to irrigate campus plantings. Oberlin's building is set in a landscape of orchards and gardens, with urban agriculture and lawns of grass specially bred to require less chemical care. Students have reforested areas around the building and have restored a wetland. These mimicked natural systems help slow and cleanse stormwater runoff.

The Lewis Center was expensive to construct, and Oberlin failed to incorporate sustainable building practices in later buildings. However, other institutions have adopted sustainable practices in multiple buildings and at less expense, and the movement for "green buildings" continues to expand. One newer example, opened in late 2005, is the Kirsch Center for Environmental Studies at DeAnza College in California. This LEED-certified building is energy-efficient, water-efficient, and climate-responsive; is built with recycled, renewable, and nontoxic materials; is solar-powered; and includes outdoor learning spaces and labs for pollution prevention, energy management, biodiversity, and geographic information systems (GIS; ▸ pp. 352–353).

Efficient energy and water use are vital

Building design is a primary way to reduce the amount—and cost—of energy and water used. In October 2005, teams of students from 18 universities competed in the second-ever *Solar Decathlon.* This remarkable event is sponsored by the U.S. Department of Energy's Office of Energy Efficiency and Renewable Energy, with several co-sponsors. Teams of students from each institution travel to the National Mall in Washington, D.C., bringing material for the solar-powered homes they have spent months designing. They erect the homes on the Mall, where they stand for 3 weeks (Figure 23.3). The homes are judged on 10 criteria, and prizes are awarded to winners in each category. One of the teams, from the University of Massachusetts at Dartmouth, donated its home to Habitat for Humanity for a Washington, D.C., family to live in. The team is also partnering with Habitat for Humanity to build several other homes with solar technology. The next Solar Decathlon will take place in fall 2007.

Solar energy plays a role on many campuses. The University of Vermont installed the largest solar panel installation in its state to generate part of the campus's power and to demonstrate the feasibility of solar energy. The 500 ft^2 array of panels provides enough electricity for nine desktop computers or 95 energy-efficient light bulbs for 10 hours each day. Other campuses are using wind power for their energy needs. MacAlester College in Minnesota even installed a wind turbine on its campus.

For those students with less inclination to build renewable energy technology themselves, any educational institution can invest in renewable energy by purchasing "green tags" that subsidize wind power and other renewable energy sources. College of the Atlantic in Maine offset 100% of its greenhouse gas emissions by buying green tags for renewable energy.

Students at Lewis and Clark College in Oregon took a dramatic step toward sustainable energy use in 2002, when they made their school the first in the nation to comply with the terms of the Kyoto Protocol (▸ pp. 550, 552).

After conducting a thorough energy audit of the college, student leaders found a variety of ways to reduce greenhouse gas emissions by the percentage that would be required of the United States under the Kyoto Protocol. Although U.S. leaders have repeatedly refused action to slow greenhouse gas emissions at the national level, citing economic expense, Lewis and Clark students found that becoming Kyoto-compliant on their campus cost only $10 per student per year.

To promote efficiency in water use, sustainability advocates at Manhattanville College in New York persuaded 217 people to reduce the length of their showers one day. On average, people cut their showers short by 30%, saving a total of 5,173 gallons of water. Water-saving technologies, such as living machines and "waterless urinals," also are being installed at a number of campuses. The University of British Columbia in Vancouver, Canada, has had great success in reducing energy and water use. The UBC student body increased by 15% between 1998 and 2003, yet in those 5 years the school reduced energy use by 6%, carbon dioxide emissions by 4%, and water use by 15% (enough water to fill the nearby Vancouver Aquarium 82 times)—all of which saved the university $2 million. Moreover, UBC is sinking $35 million into retrofitting its buildings with energy-efficient upgrades, which should save the school at least $2.5 million annually. In academic buildings, these upgrades should reduce energy use by 20%, water use by 45%, and carbon emissions by 15,000 metric tons each year.

Dining services and campus gardens let students eat sustainably

Campus food service operations can promote sustainable practices by buying organic produce, by purchasing food in bulk or food that involves less packaging, and by composting food scraps. In addition, buying locally grown or produced food supports local economies and cuts down on fuel consumption from long-distance transportation. Some college campuses even have gardens where students can help grow their own food (Figure 23.4).

Sterling College in Vermont provides a model for sustainable campus food service. Many foods are organic, grown by local farmers, or produced by Vermont-based companies; moreover, some foods are grown and breads are baked right on the Sterling campus. Food shipped in is purchased in bulk to reduce packaging. Dish soap is non-petroleum-based and biodegradable, and all kitchen scraps are composted, along with recycled unbleached paper products. By reducing unnecessary equipment use, the college saved money and slashed its electricity use by 23%.

FIGURE 23.4 A number of campuses now include gardens where students can grow organic vegetables that are used for meals in dining halls. Here, a student works in the garden at Middlebury College in Vermont.

Institutional purchasing can be influential

The kinds of purchasing decisions made in dining halls favoring local food, organic food, and biodegradable products are relevant across the entire spectrum of a campus's needs. Ball State's drive to convert to recycled paper is one of many examples in this area. When college and university purchasing departments preferentially buy certified sustainable wood, energy-efficient appliances, goods with less packaging, and other ecolabeled products, they send signals to manufacturers and increase the economic demand for such items.

At Chatham College in Pennsylvania, students chose to honor their school's best-known alumnus, Rachel Carson (▸ pp. 67, 408–409), by seeking to eliminate toxic chemicals used on campus. Administrators agreed to this effort, provided that alternative products to replace the toxic ones worked just as well and were not more expensive. Students brought in the CEO of a company that produced nontoxic cleaning products, who demonstrated to the janitorial staff that his company's products were superior. The university switched to the nontoxic products, which were also cheaper, and proceeded to save $10,000 per year. Chatham students also found a company offering paint without volatile organic compounds and negotiated with it for a free paint job and discounted prices on later purchases. Students also worked with grounds staff to eliminate herbicides and fertilizers used on campus lawns and to find alternative treatments.

Transportation alternatives are many

Many colleges and universities struggle with traffic and parking congestion, sprawl, commuting delays, and pollution from vehicle exhaust. Many are addressing these issues by establishing or expanding bus and shuttle systems; encouraging bicycling, walking, and carpooling; and introducing alternative fuels and vehicles to university fleets. Like Ball State, a number of campuses are using biodiesel fuels, including Hobart and William Smith Colleges in New York, the University of South Carolina, and the University of Vermont. Middlebury College in Vermont has gone farther than all others in this regard—thousands of miles, in fact. Middlebury students began Project Bio Bus, which crisscrosses North America each summer in a biodiesel bus spreading the gospel of this alternative fuel.

Other colleges and universities have also, like Ball State, added alternative vehicles to their fleets (Figure 23.5). Middlebury leases electric vehicles. The State University of New York College of Environmental Science and Forestry in Syracuse acquired six electric vehicles for on-campus use, a gas-electric hybrid car for the fleet, and a delivery van that runs on compressed natural gas. At the same time, it converted its buses to biodiesel.

The University of British Columbia is a leader in transportation efforts. For $20 a month, UBC's "U-Pass" program provides students with biking programs and facilities, expanded campus bus and shuttle services, unlimited use of some city transit systems, rides home in emergencies, merchant discounts, and priority parking spaces and ride-matching services for carpoolers. The program has boosted transit ridership to campus by 53% and has decreased single-person car use by 20%.

FIGURE 23.5 Most buses at the University of California at Davis run on compressed natural gas, and one runs on a novel mixture of natural gas and hydrogen. Several of these low-emission vehicles are shown here lined up in front of Freeborn Hall during a press conference celebrating their purchase.

FIGURE 23.6 Many schools have embarked on habitat restoration projects to beautify their campuses, provide wildlife habitat, restore native plants, and filter water runoff. Here, a student and a volunteer work to maintain a pond in the Cheeseman Environmental Study Area at DeAnza College in California, which showcases a sampling of plant communities native to the region.

Campuses are restoring native plants, habitats, and landscapes

No campus sustainability program would be complete without some effort to enhance the campus's natural environment. Along with recycling programs, programs for natural landscaping and habitat restoration make up the most widespread campus sustainability initiatives. Such programs remove invasive species, restore native plant species and communities, improve habitat for wildlife, enhance soil and water quality, reduce pesticide use, and create healthier, more attractive surroundings.

Native plant restoration has proven popular, from desert plants at Arizona State University in Tempe, to prairie plants at University of West Alabama, where a student crew has conducted controlled burns. Some schools, such as California's DeAnza College, feature demonstration gardens of native plant communities (Figure 23.6). Others, such as Loyola College in Maryland, focus on removing invasive species. Still others, such as Ohio State University and the New College of California in San Francisco, have rooftop gardens. At Clemson University, students work with churches and K-12 schools in the community and help them design landscaping plans. At Warren Wilson College in North Carolina, 40 student native plant enthusiasts and the landscaping supervisor used local plants to establish a stock of seeds, built a greenhouse for rearing plants, expanded their arboretum, and started propagating various species of grasses and wildflowers. The landscaping crew uses organic fertilizers and avoids using chemical pesticides. At Northland College in Wisconsin, sustainability advocates wanting to improve biodiversity and water quality have managed to re-landscape half of the campus. They have replaced invasive plants with native ones and have designed and planted a series of meadows that capture stormwater runoff and filter out pollution.

Weighing the Issues: Sustainability on Your Campus

Are any sustainability efforts being made on your campus? What efforts would you like to see on your campus? Do you foresee any obstacles to these efforts? How could these obstacles be overcome?

Organizations are available to assist campus efforts

Many campus sustainability initiatives are supported by organizations such as University Leaders for a Sustainable Future and the National Wildlife Federation's Campus Ecology program. These organizations act as information clearinghouses for campus sustainability efforts. Each year the NWF program recognizes the most successful campus sustainability initiatives, an award Ball State shared in 2003–2004 along with six other schools. With Ball State's Greening of the Campus conferences and the assistance of these organizations, it is easier than ever to start sustainability efforts on your own campus and obtain the support to carry them through to completion.

Sustainability and Sustainable Development

Efforts toward sustainability on college and university campuses parallel efforts in the world at large. As more people come to appreciate Earth's limited capacity to accommodate our rising population and consumption, they are voicing concern that we will need to modify our behaviors, institutions, and technologies if we wish to sustain our civilization and the natural environment on which it depends. In the quest for sustainability, the strategies pursued on campuses reflect those pursued in the wider society and also can serve as models.

When people speak of *sustainability*, what precisely do they mean to sustain? Generally they mean to sustain the natural environment, its biodiversity, and its ecological systems in a healthy and functional state—and also to sustain human civilization in a healthy and functional state. The short-term needs of human society and of environmental sustenance are often cast as being in opposition, but environmental scientists recognize that our civilization cannot exist without an intact and functional natural environment. The contributions of biodiversity (▸ pp. 324–328) and ecosystem goods and services (▸ pp. 39–41, 48–51) to human welfare are tremendous. Indeed, they are so fundamental (some would say infinitely valuable, thus literally priceless) that we have long taken them for granted.

Sustainable development involves environmental protection, economic welfare, and social equity

In recent years, people have increasingly drawn the connection between environmental quality and human quality of life. Moreover, we now recognize that very often it is society's poorer people who suffer the most from environmental degradation. This realization has led advocates of environmental protection, advocates of economic development, and advocates of social justice to work together toward common goals. This cooperative approach has given rise to the modern drive for sustainable development.

We first encountered sustainable development in our opening chapter (▸ p. 22), and offered the United Nations' definition: "Development that meets the needs of the present without compromising the ability of future generations to meet their own needs." This definition is taken from the U.N.-sponsored Brundtland Commission (named after its chair, Norwegian prime minister Gro Harlem Brundtland), which produced an influential 1987 report titled *Our Common Future.*

Prior to the Brundtland Report, most people aware of human impact on the environment might have thought "sustainable development" to be an oxymoron—a phrase that contradicts itself. *Development* involves making purposeful changes intended to improve the quality of human life, yet environmental advocates have long pointed out that development often so degrades the natural environment that it threatens the very improvements for human life that were intended. Today many people remain under the impression that protecting the environment is incompatible with serving people's needs. However,

FIGURE 23.7 The 2002 U.N.-sponsored World Summit on Sustainable Development in Johannesburg, South Africa, drew over 10,000 delegates from 200 nations to discuss and set sustainable development goals. In the photo, South African President Thabo Mbeki hugs a boy who performed in the welcoming ceremony. Delegates launched hundreds of initiatives and made millions of dollars of commitments at this conference, but most sustainability proponents said these efforts did not go nearly far enough.

efforts to develop sustainably, undertaken by individuals at all levels—from students on campus to international representatives at the United Nations (Figure 23.7)—are beginning to alter this perception and produce models of success. The question "Can we develop in a sustainable way?" may well be the single most important question in the world today. The myriad actions being taken at campuses such as Ball State's and by governments, businesses, industries, organizations, and individuals across the globe are giving people optimism that developing in sustainable ways is possible.

Environmental protection can enhance economic opportunity

How can protecting environmental quality be good for the economic bottom line? For individuals, businesses, and institutions, reducing resource consumption and waste often saves money—as many colleges and universities discover when they embark on sustainability initiatives. Sometimes savings accrue immediately, and other times an up-front investment results in long-term savings. For society as a whole, attention to environmental quality can also enhance economic opportunity for citizens by providing new types of employment. This contrasts with the common perception that environmental protection measures hurt the economy by costing people jobs. A prime example is the controversy over logging of old-growth forest in the Pacific Northwest (Chapter 12). Protection for the northern spotted owl (Figure 23.8) under the U.S. Endangered Species Act (▸ pp. 331–334) set limits on timber extraction in the northwestern United States. Proponents of logging rallied opposition to forest protection by claiming that the restrictions cost local loggers their jobs. Some job loss has indeed occurred, but loggers' jobs are far more at risk when timber companies cut trees at unsustainable rates and then leave a region, seeking mature forests elsewhere. As we saw in Chapter 12 (▸ pp. 350–351), this has happened in region after region throughout U.S. history. It is playing out in the Pacific Northwest now, as U.S. timber companies begin moving operations abroad.

The jobs-versus-environment debate frequently overlooks the fact that as some industries decline, others spring up to take their place. As jobs in logging, mining, and manufacturing have decreased in developed nations over the past few decades, jobs have greatly increased in many service occupations and in computer and other high-technology sectors. If we decrease our dependence

FIGURE 23.8 The northern spotted owl (*Strix occidentalis occidentalis*) has become a symbol of the "jobs-versus-environment" debate. This bird of the Pacific Northwest rainforest is considered endangered because of the logging of mature forests. Proponents of logging argue that laws protecting endangered species cause economic harm and job loss. Advocates of endangered species protection argue that unsustainable logging practices pose a larger risk of job loss.

on fossil fuels, experts predict, jobs and investment opportunities will open up in renewable energy sectors, such as wind power and fuel cell technology (Chapter 18).

Moreover, people desire to live in areas that have clean air and water, intact forests, and parks and open space. Environmental protection increases a region's attractiveness, drawing more residents and increasing property values and the tax revenues that help fund social services. As a result, those regions that act to protect their environments are generally the ones that retain and increase their wealth and quality of life.

Thus, environmental protection need not lead to economic stagnation, but instead is likely to enhance economic opportunity. A recent U.S. government review concluded that the economic benefits of environmental regulations greatly exceed their economic costs (see "The Science behind the Story," ▸ pp. 686–687). This connection is also suggested by the fact that both the U.S. and global economies have expanded rapidly in the past 30 years, the very period during which environmental protection measures have proliferated.

What accounts for the perceived economy-versus-environment divide?

If environmental protection and economic development are compatible and even mutually reinforcing, what, then, accounts for the conventional view that we cannot provide for people's needs and protect the environment simultaneously? One proximate explanation lies in the fact that economic development since the industrial revolution has so clearly diminished biodiversity, decreased habitat, and degraded ecological systems. A second proximate explanation lies in the fact that many people view command-and-control environmental policy (▸ p. 71) to pose excessive costs for industry and restrict rights of private citizens.

An ultimate explanation may lie in our species' long history. For the thousands of years that we lived as nomadic hunter-gatherers, population densities were low and consumption and environmental impact were limited. With natural resources in little danger of running out, humans were free to exploit them limitlessly and had little reason to adopt a conservation ethic. The establishment of sedentary agricultural societies beginning about 10,000 years ago, followed by the urbanization that continues today, has increased our impact while also causing us to overlook the connections between our economies and our environments. It is common to hear "humans and the environment" or "people and nature" being set in contrast, as though they were completely separate. Some philosophers venture to say that the perceived dichotomy between humans and nature is the root of all our environmental problems.

Humans are not separate from the environment

On a day-to-day basis, it is easy to feel disconnected from the natural environment, particularly in developed nations and large cities. We live inside houses; work in shuttered buildings; travel in cars, airplanes, and other enclosed vehicles; and generally know little about the plants and animals around us. Millions of urban citizens have never set foot in an undeveloped area. Just a few centuries or even decades ago, most of the world's people were able to name and describe the habits of the plants and animals that lived nearby. They knew exactly where their food, water, and clothing came from. Today it seems that water comes from the faucet, clothing from the mall, and food from the grocery store. It's little wonder we have lost track of the connections that tie us to the natural environment.

However, this psychological severance, this feeling of isolation, doesn't make our connections to the environment any less real. Consider a thoroughly un-"natural" (yet delicious!) invention of the human species: the banana split (Figure 23.9). Even in this triumph of human creation, seemingly concocted *de novo* at an ice cream shop, each and every element has ties to the resources of the natural environment and exerts environmental impacts. Once we learn to consider where the things we use and value each day actually come from, it becomes easier to see how humans are part of the environment. And once we reestablish this connection, it becomes readily apparent that our own interests are best served by preservation or responsible stewardship of the natural systems around us.

Because what is good for the environment can also be good for people, win-win solutions are very much within reach. We *can* have it both ways, if we learn from what science can teach us, think creatively, and act on our ideas. Ultimately, the only way for our species to "win"—to survive and thrive—is to conduct our activities in ways that sustain the processes and resources of our environment. If the environment "loses" because of our impacts, then so will we.

Weighing the Issues: Unavoidable Impacts?

Ecology teaches us that one species may exclude other species from resources. This has clearly been the case with humanity's recent spread across the planet.

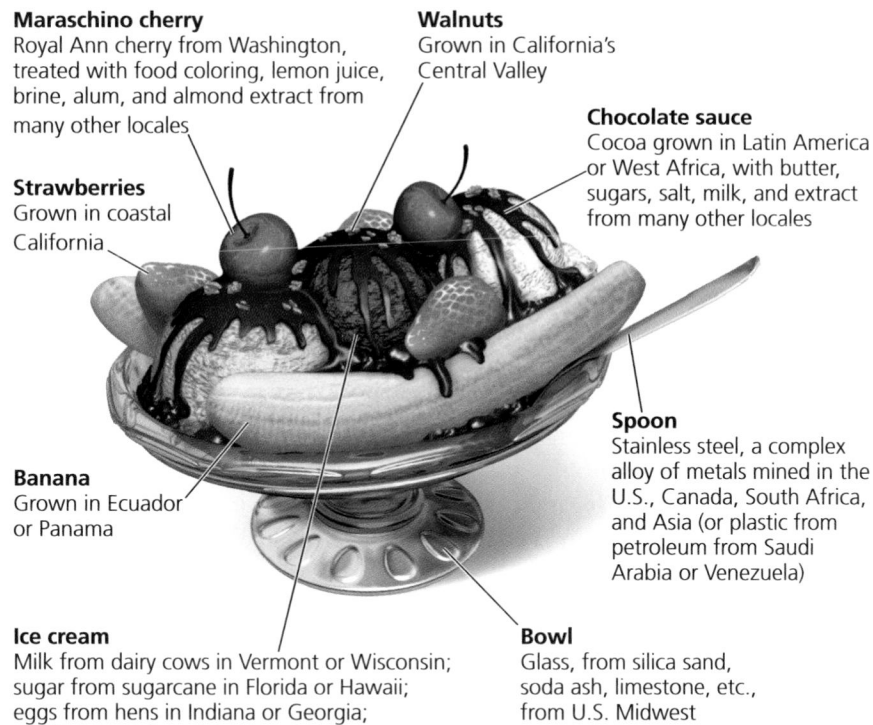

FIGURE 23.9 A banana split eaten at an ice cream shop in Tulsa, Denver, or Des Moines consists of ingredients from around the world, whose production has impacts on the environments of many faraway locations. Ice cream requires milk from dairy cows that graze pastures or are raised in feedlots on grain grown in industrial monocultures. Ice cream is sweetened with sugar from sugar beet farms or sugarcane plantations. The banana was shipped thousands of miles by oil-fueled transport from a tropical country, where it grew on a plantation that displaced rainforest and where it was liberally treated with fertilizers and fungicides. Fruits and nuts grown in California's Central Valley were irrigated generously with water piped in from the Sierras. The spoon originated with metal ores mined along with thousands of tons of soil and processed into stainless steel using energy from fossil fuels.

However, in the long scope of Earth's history, biodiversity has increased as organisms have evolved and adapted to environmental changes caused by other organisms. Does our society's development demand some amount of unavoidable harm to other species or to environmental processes? What kinds of impacts might be unavoidable, and what kinds might we be able to avoid?

Strategies for Sustainability

Sustainable solutions to environmental problems are numerous, and we have seen specific examples throughout this book. The challenges lie in being imaginative enough to think of solutions and then being shrewd and dogged enough to overcome whatever political or economic obstacles may lie in the path of their implementation. We will now summarize several broad strategies or approaches that can spawn sustainable solutions (Table 23.2).

We can refine our ideas about economic growth and quality of life

It is conventional among economists and the policymakers who heed their advice to speak of economic growth as an ultimate goal. Many politicians see nurturing an expanding economy as their prime responsibility. Yet economic growth is merely a tool with which we try to attain the real goal of maximizing human happiness. Thus if economic growth depends on an ever-increasing consumption of nonrenewable resources, then we may want to ask ourselves whether attaining happiness truly requires endlessly expanding the size of our economy (▸ pp. 44–47).

Another question we may ask is whether and how to incorporate external costs (▸ pp. 43–44) into the market

Table 23.2 Some Major Approaches to Sustainability

- Refine our ideas about economic growth and quality of life
- Reduce unnecessary consumption
- Limit population growth
- Encourage green technologies
- Mimic natural systems by promoting closed-loop industrial processes
- Think in the long term
- Enhance local self-sufficiency and be mindful of globalization
- Vote with our wallets
- Vote with our ballots
- Promote research and education

The Science behind the Story

Assessing Costs and Benefits of Environmental Regulations

Federal regulations that aim to protect environmental quality often result in clearer air or cleaner water. They also usually have financial implications for affected parties. Are such regulations worth the costs to industry, businesses, and consumers? In 2003, a federal study determined that the answer to that question is a definitive yes.

The study, from the White House Office of Management and Budget (OMB), weighed the costs of a range of recently created federal regulations against the economic benefits resulting from the regulations. The OMB examined 107 federal regulations enacted in the United States from 1992 to 2002, many of which dealt with environmental issues.

The study found that the reviewed regulations carried a sizable annual cost, estimated at \$36–42 billion. However, their estimated value in terms of public good totaled \$146–230 billion. Thus, the economic benefits of these regulations far exceeded their costs. Moreover, of all the regulations, those dealing with environmental protection were found to be especially cost-effective.

Most regulatory benefits were reflected in health and social gains resulting from clean air. Over the studied 10-year period, the benefits of clean-air regulations were found to be 6–10 times greater than the costs of complying with the regulations. The value of reductions in hospitalization and emergency room visits, premature deaths, and lost workdays resulting from improved air quality were estimated to be between \$118 billion and \$177 billion.

The authors of the OMB review relied on cost-benefit analyses previously made by experts in the EPA and other agencies, scrutinizing them for appropriateness and accuracy before using them. A look at one EPA regulation included in the OMB report—an air pollution regulation called the *Heavy Duty Engine/Diesel Fuel Rule*—shows how the EPA performed its own economic analysis and how the OMB double-checked the EPA's findings.

The diesel regulation aimed to address diesel engines that emit fine soot, sulfur compounds, and nitrogen oxides (NO_x). In the 1990s, the EPA regulation was intended to make diesel trucks and buses run more cleanly by lowering the fuel's sulfur content. However, achieving cleaner diesel fuel would mean substantial—and potentially costly—changes to the way that fuel is made and handled. To determine potential costs, EPA analysts examined how various parties involved with diesel fuel use—from fuel processors to engine manufacturers to vehicle operators—would need to change their processes or equipment to clean up the fuel. For example, a heavy-duty engine redesigned to run more efficiently on cleaner fuel would cost approximately \$4,600 more to operate, the analysts calculated. Total research and development costs for emission control were predicted to exceed \$600 million. The EPA asked for input from affected parties, scientists, and the public, finally determining that removing sulfur from diesel fuel would be neither cheap nor quick, with total annualized costs of about \$4.2 billion by 2030.

EPA analysts then considered the benefits of reduced diesel pollution. Using air quality data from more than 940 smog monitors around the country, the EPA estimated diesel-related air pollution levels with and without the proposed diesel regulation. The analysts examined public

prices of goods and services. Currently, goods and services are priced as though pollution and resource extraction involved no costs to society. If we can make our accounting practices reflect indirect negative consequences and provide a clearer view of the full costs and benefits of any given action or product, then the free market itself can become a force for improving environmental quality, our economy, and our quality of life. If we can incorporate external costs into our price structures, then market capitalism could potentially become the optimal tool for achieving prosperous and sustainable economies. Moreover, implementing green taxes (▸ pp. 72–73) and phasing out harmful subsidies could hasten the shift to sustainability, many economists maintain. The political obstacles to this are considerable, however. Such changes will require educated citizens to push for them and courageous policymakers to implement them.

Estimates of the Total Annual Benefits and Costs of Major Federal Rules, October 1, 1992 to September 30, 2002

Agency	Benefits (millions of 2001 dollars)	Costs (millions of 2001 dollars)
Agriculture	3,094–6,176	1,643–1,672
Education	655–813	361–610
Energy	4,700–4,768	2,472
Health & Human Services	9,129–11,710	3,165–3,334
Housing & Urban Development	551–625	348
Labor	1,804–4,185	1,056
Transportation	6,144–9,456	4,220–6,718
Environmental Protection Agency	120,753–193,163	23,359–26,604
Total	**146,812–230,896**	**36,625–42,813**

Source: U.S. Office of Management and Budget. 2003. *Informing regulatory decisions: 2003 report to Congress on the costs and benefits of federal regulations and unfunded mandates on state, local, and tribal entities.* Washington, D.C.: U.S. OMB.

health statistics, especially data related to respiratory illnesses such as asthma and chronic bronchitis. They then estimated how many cases of disease or premature death might be prevented if people breathed fewer diesel-related pollutants. They determined the economic effects of such health problems by determining the medical, workforce, and social costs. Each premature death was assigned an impact of $6 million, and each case of chronic bronchitis was pegged at $331,000. Many of the estimates were also regionalized to account for different economic conditions in various parts of the country.

In the end, the EPA analysis determined, the benefits of cleaner diesel fuel far exceeded the costs. If sulfur content of diesel were cut from 500 to 15 parts per million, then 2.6 million tons of smog-causing NO_x emissions would be eliminated each year. An estimated 8,300 premature deaths, 5,500 cases of chronic bronchitis, over 360,000 asthma attacks, and about 1.5 million lost workdays would be prevented each year. By 2030, the analysis determined, benefits would reach about $70 billion. Following this analysis, the Heavy Duty Engine/Diesel Fuel Rule took effect in early 2001. Requirements are being phased in gradually through 2010.

The OMB, in compiling its own review of federal rules, scrutinized the EPA study to determine whether the EPA estimates made economic sense. The OMB analysts concluded that most of the estimates did, indeed, make sense. The OMB analysts made some minor changes when calculating their own figures. For example, although the EPA said every 1-ton reduction in NO_x emissions represented a benefit of $10,200, the OMB used its own updated figures and determined NO_x reductions to be worth $5,500 per ton. Overall, however, the OMB confirmed the validity of the EPA estimates and used most of the EPA's economic data in the OMB report.

The OMB report, in totaling and highlighting the value of many recent environmental regulations, is now being used to evaluate and support efforts at environmental regulation around the country.

We can consume less

Economic growth is largely driven by consumption, the purchase of material goods and services (and thus the use of resources involved in their manufacture) by consumers (Figure 23.10). Our tendency to believe that more, bigger, and faster are always better is reinforced by advertisers seeking to sell more goods more quickly. Consumption has grown tremendously, with the wealthiest nations leading the way. The United States, with less than 5% of the world's population, consumes 30% of world energy resources and 40% of total global resources. U.S. houses are larger than ever, sports-utility vehicles are among the most popular automobiles, and many citizens have more material belongings than they know what to do with. We think nothing now of having home computers with high-speed Internet access, let alone the televisions, telephones, refrigerators, and dishwashers that were marvels just decades ago.

FIGURE 23.10 Citizens of the United States consume more than the people of any other nation. Unless we find ways to increase the sustainability of our manufacturing processes, this rate of consumption cannot be sustained in the long run.

Because many of Earth's natural resources are limited and nonrenewable, consumption cannot continue growing forever. Eventually, if we do not shift to sustainable practices of resource use, per capita consumption will drop for rich and poor alike as resources dwindle. Cornucopian critics often scoff at the notion that resources are limited, but we must remember that our perspective in time is limited and that our consumption is taking place within an extraordinarily brief slice of time in the long course of history (Figure 23.11). Our lavishly consumptive lifestyles are a brand-new phenomenon on Earth. We are enjoying the greatest material prosperity in all of human history, but if we do not find ways to make our wealth sustainable, the party may not last much longer.

Fortunately, material consumption is only one way to measure prosperity, and consumption alone does not reflect a person's quality of life. For many people in the United States and other developed nations, the accumulation of innumerable possessions has not brought contentment. Social critics have coined a word for the way material goods often fail to bring happiness to people affluent enough to afford them: **affluenza.** Economic growth is generally equated with "progress," but for many people, true progress consists of an increase in human happiness, not simply growth in material wealth. In the end we are, one would hope, more than simply the sum of what we buy.

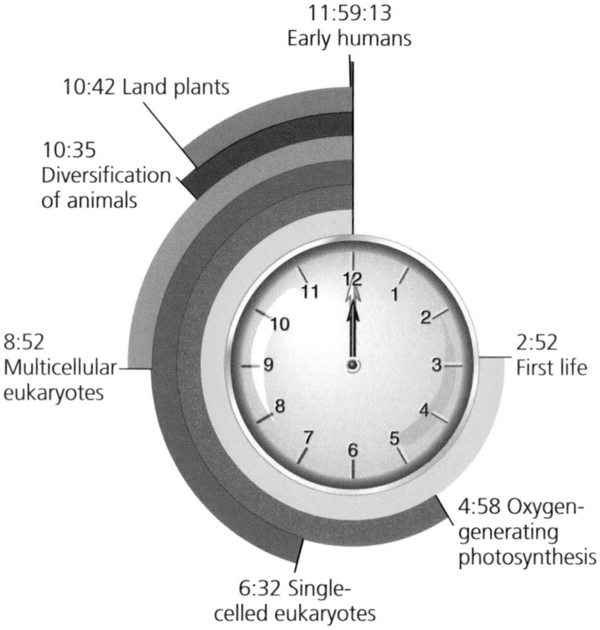

FIGURE 23.11 By viewing Earth's 4.5-billion-year history as a 12-hour clock, we can gain a better understanding of relative timescales across the immense span of geological time. *Homo sapiens,* as a species, has come into existence only during the final 1 or 2 seconds, around 11:59:59. The agricultural and industrial revolutions that have increased our environmental impacts have taken up only a minuscule fraction of a second.

We can reduce our consumption while enhancing our quality of life—squeezing more from less—in at least three ways. One way is to improve the technology of materials and the efficiency of manufacturing processes, so that industry produces goods using fewer natural resources. Another way is to develop a sustainable manufacturing system—one that is circular and based on recycling, in which the waste from a process becomes raw material for input into that process or other processes (▸ pp. 664–665). A third way is for each of us to modify our behavior, attitudes, and lifestyles to minimize consumption. At the outset, such choices may seem like sacrifices, but people who have slowed down the pace of their busy lives and freed themselves of an attachment to excess material possessions say it can feel tremendously liberating.

Population growth must eventually cease

Just as continued growth in consumption is not sustainable, neither is growth in the human population. We have seen (▸ pp. 136–137) that populations may grow exponentially for a time but eventually encounter limiting factors and level off at a carrying capacity. We have increased Earth's carrying capacity for our species with the help of technology, but our population growth cannot continue forever; sooner or later, the human population will stop growing. The question is how: through war, plagues, and famine, or through voluntary means as a result of wealth and education?

The demographic transition (▸ pp. 228–229) that many nations are undergoing provides reason to hope that population sizes will stabilize and begin to fall. This transition is already far along in many developed nations thanks to urbanization, wealth, education, and the empowerment of women. If the demographic transition occurs for today's developing nations, then there is hope that humanity may halt its population growth while creating more prosperous and equitable societies.

Technology can help us toward sustainability

It is largely technology—developed with the agricultural revolution, the industrial revolution, and advances in medicine and health—that has spurred our population increase. Technology has magnified our impacts on Earth's environmental systems. However, technology also can give us ways to reduce environmental impact. Technology is not an independent actor, but is a tool of human agency. The shortsighted use of technology may have gotten us into this mess, but wiser use of environmentally friendly, or "green," technologies can help get us out.

Recall the I=PAT equation (▸ pp. 220–221), which summarizes human environmental impact (I) as the interaction of population (P), consumption or affluence (A), and technology (T). Technology can exert either a positive or negative value in this equation. In recent years, technology has exacerbated environmental impact in developing countries as industrial technologies from the developed world have been exported to poorer nations eager to industrialize. In developed nations, meanwhile, green technologies have begun mitigating our environmental impact. Catalytic converters on cars have reduced emissions (Figure 23.12), as have scrubbers on industrial smokestacks (▸ p. 656). Recycling technology and advances in wastewater treatment are helping reduce our waste output. Solar, wind, and geothermal energy technologies are producing cleaner renewable energy. Countless technological advances such as these are one reason that people of the United States and western Europe today enjoy cleaner environments, although they consume far more, than people of eastern Europe and rapidly industrializing nations such as China.

Weighing the Issues: Renewable Energy and Sustainability

Environmental advocates who promote alternative energy sources such as solar, wind, and geothermal power often argue that, given adequate development, green technologies to exploit such renewable sources could provide us clean and limitless energy and thereby help bring a sustainable future. Do you agree? Or do you think abundant renewable energy might simply lead to further unsustainable development and cause further environmental impact? How might a shift to renewable energy sources influence the quest for sustainable development?

Industrial systems can mimic natural systems by recycling and become circular

As industries seek to develop green technologies and sustainable practices, they have available to them an excellent model: nature itself. Natural systems are sustainable; they've been around far longer than we have. As we saw in Chapter 7 and throughout this book, environmental systems tend to operate in cycles consisting of feedback loops and the circular flow of materials. In natural systems, output is recycled into input. In contrast, human manufacturing processes have run on a linear model, in which raw materials are input, go through processing, and create a product, while byproducts and waste are generated and disposed of.

Some forward-thinking industrialists are already making their industrial processes sustainable by transforming linear pathways into circular ones, in which waste is recycled and reused. For instance, several companies now produce carpets with materials that can be retrieved from the consumer when the carpet is worn out, and these materials are then recycled to create new carpeting (recall that

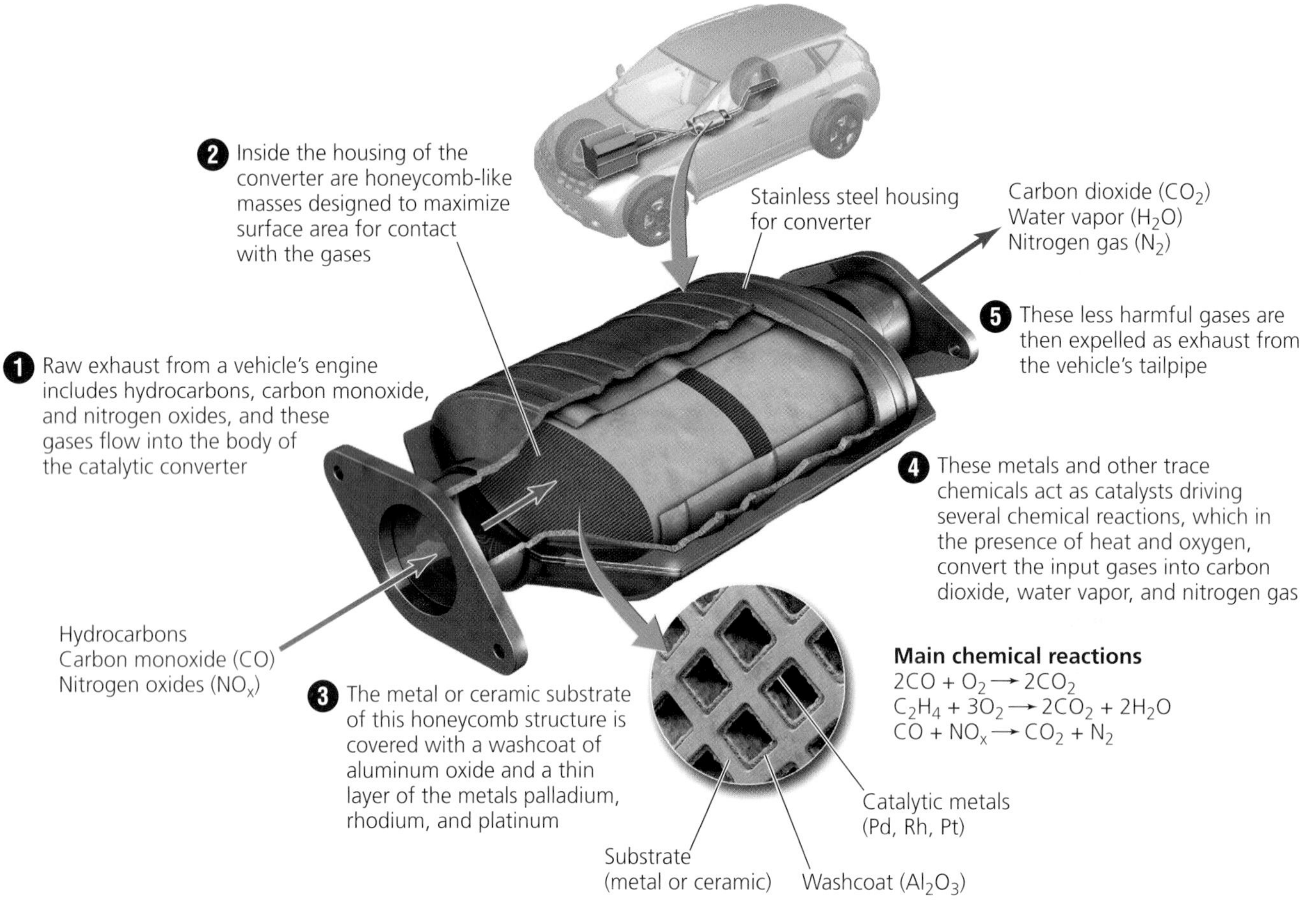

2 Inside the housing of the converter are honeycomb-like masses designed to maximize surface area for contact with the gases
Stainless steel housing for converter
Carbon dioxide (CO_2)
Water vapor (H_2O)
Nitrogen gas (N_2)
5 These less harmful gases are then expelled as exhaust from the vehicle's tailpipe
1 Raw exhaust from a vehicle's engine includes hydrocarbons, carbon monoxide, and nitrogen oxides, and these gases flow into the body of the catalytic converter
4 These metals and other trace chemicals act as catalysts driving several chemical reactions, which in the presence of heat and oxygen, convert the input gases into carbon dioxide, water vapor, and nitrogen gas
Hydrocarbons
Carbon monoxide (CO)
Nitrogen oxides (NO_x)
Main chemical reactions
$2CO + O_2 \rightarrow 2CO_2$
$C_2H_4 + 3O_2 \rightarrow 2CO_2 + 2H_2O$
$CO + NO_x \rightarrow CO_2 + N_2$
3 The metal or ceramic substrate of this honeycomb structure is covered with a washcoat of aluminum oxide and a thin layer of the metals palladium, rhodium, and platinum
Catalytic metals (Pd, Rh, Pt)
Substrate (metal or ceramic)
Washcoat (Al_2O_3)

FIGURE 23.12 The catalytic converter is a classic example of a green technology. This device filters air pollutants from vehicle exhaust and has helped bring about the vast improvement of air quality in the United States and other nations.

Oberlin College's green building took advantage of this). Some automobile manufacturers are planning cars that can be disassembled and recycled into new cars. Proponents of this industrial model see little reason why virtually all appliances and other products cannot be recycled, given the right technology. Their ultimate vision is to create industrial processes involving entirely closed loops, generating no waste.

We can base our decisions on long-term thinking

To be sustainable, a solution must work in the long term. Often the best long-term solution is not the best short-term solution, which explains why much of what human societies currently do is not sustainable. Policymakers in democracies often act for short-term good because they compete to produce immediate, positive results so that they will be reelected. This poses a major hurdle for addressing environmental dilemmas, because so many of these dilemmas are cumulative, worsen gradually, and can be resolved only over a long period. Often the costs for addressing environmental problems are short-term, whereas the benefits are long-term, giving a politician little incentive to tackle the problems. In such a situation, citizen pressure on policymakers becomes especially vital.

Businesses may act according either to long-term or short-term interests. A business committed to operating in a certain community for a long time has incentive to sustain environmental quality. However, a business merely attempting to make a profit and move on has little incentive to invest in environmental protection measures that involve short-term costs.

In 2002, the U.N. Environment Programme conducted analyses to predict the likely effects of a large-scale shift to sustainable development strategies over the 30 years until 2032. The U.N. scientists designed four scenarios that differed in the sustainability of their approaches and then compared projections for a variety of environmental and

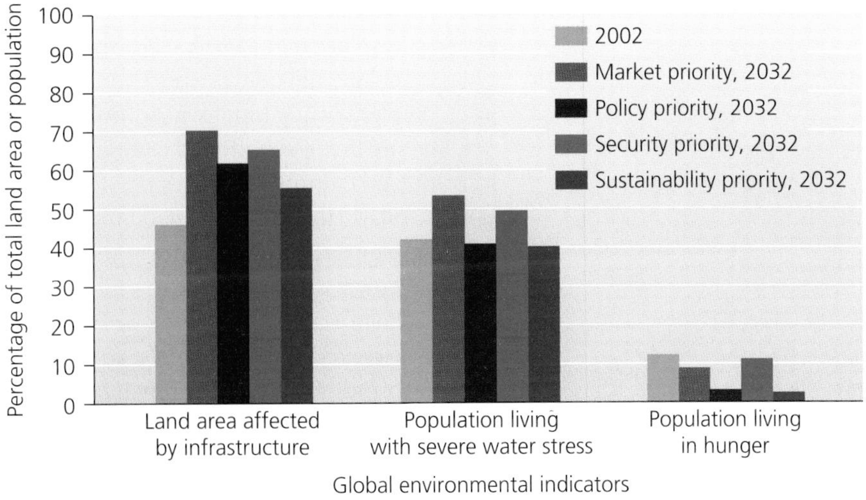

FIGURE 23.13 UNEP scientists projected the status of a number of environmental and socioeconomic indicators in 2032, across four different scenarios. The "markets first" scenario assumed that forces of market capitalism would have the greatest impact on global trends. The "policy first" scenario assumed that governments would take strong action to try to reach social and environmental goals. The "security first" scenario assumed a world of conflict in which security would be governments' highest priority. The "sustainability first" scenario assumed that a sustainable development paradigm would take hold, encouraging sustainable economic development, environmental protection, and social equity. The "sustainability first" approach was consistently predicted to result in the fewest negative impacts. Here it is shown to result in the least land affected by infrastructure, the fewest people living under water stress, and the fewest people living with hunger. Adapted from United Nations Environment Programme, 2002. *Global environment outlook 3 (GEO-3)*. Nairobi and London: UNEP and Earthscan Publications.

socioeconomic indicators under the four scenarios. For one indicator after another, the approach that prioritized sustainability was shown to result in the fewest negative impacts and the most positive impacts (Figure 23.13).

A similar review was conducted by scientists with the Millennium Ecosystem Assessment (▸ p. 20) and reported in 2005. These scientists also analyzed four approaches of their own design. Their analysis revealed that the two scenarios involving proactive protection and management of ecosystems best maintained ecosystem services (Figure 23.14a), whether the management was locally based or globally based. It also showed that an approach that prioritized security above all else was the only one that decreased human well-being (Figure 23.14b).

Many wish to promote local self-sufficiency and be mindful of globalization

As our societies become more globally interconnected, diverse impacts, both positive and negative, have resulted. Many proponents of sustainability believe that encouraging local self-sufficiency is an important element of building sustainable societies. When people are tied closely to the area in which they live, they tend to value the area and seek to sustain its environment and human communities. This line of argument has frequently been made regarding the growing and distribution of food, specifically in encouraging locally based organic or sustainable agriculture. It is estimated that the food Americans eat is transported an average of 2,400 km (1,500 mi) from the place it was grown to the place it is eaten.

Many people who focus their economic activity locally also critique globalization. However, as the ecological economist Herman Daly has pointed out, globalization means different things to different people. People who view it as a positive phenomenon generally accentuate the process by which people of the world's diverse cultures are increasingly communicating with one another and learning about one another's diversity. Books, airplanes, television, and the Internet have made people of every culture more aware of other cultures and thus more likely to respect and celebrate, rather than fear, differences among cultures.

In contrast, people who view globalization in a negative light generally accentuate the homogenization of the

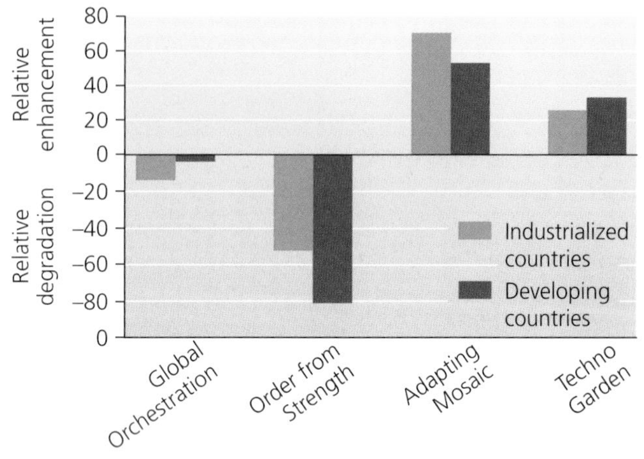

(a) Effect of scenarios on ecosystem services

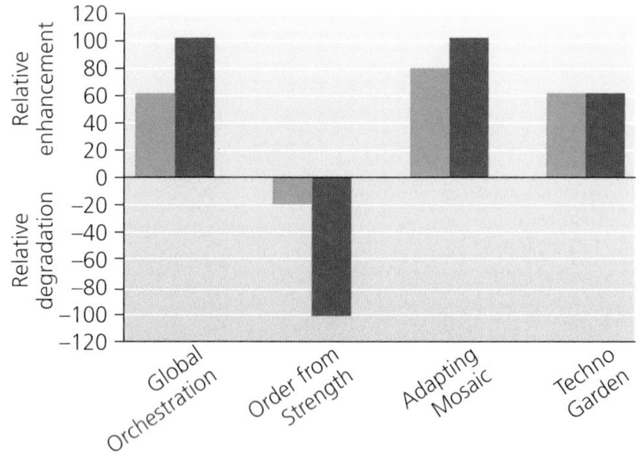

(b) Effects of scenarios on human well-being

FIGURE 23.14 Scientists with the Millennium Ecosystem Assessment analyzed impacts on ecosystem services and human welfare for each of four scenarios for the future. In the "global orchestration" scenario, a globally connected society focuses on trade and economic liberalization and takes a reactive approach to ecosystem problems, but also invests in education and infrastructure and strives to reduce poverty and inequality. In the "order from strength" scenario, nations focus on security in a fragmented world, trading regionally, investing little in public infrastructure, and taking a reactive approach to ecosystem problems. In the "adapting mosaic" scenario, regional watershed-scale ecosystems are the focus of political and economic activity, and local societies support strong institutions and proactively manage ecosystems. In the "technogarden" scenario, a globally connected world relies on green technology, taking a proactive and engineering approach to managing ecosystems and maximizing their delivery of ecosystem services. The assessment revealed that the adapting mosaic and technogarden approaches maintained ecosystem services best **(a)**, and that all but the "order from strength" approach promoted human well-being **(b)**. Adapted from Millennium Ecosystem Assessment, 2005.

world's cultures, in which a few cultures and worldviews displace many others. For instance, the world's many languages are going extinct more quickly than are species of plants and animals. Traditional ways of life in many areas are being abandoned as more people take up the material and cultural trappings of a few dominant cultures, particularly that of the United States. In recent years, more people have protested this homogenization and the growing power of large multinational corporations. In France, farm activist Jose Bove became a popular hero when he wrecked a McDonald's restaurant with his tractor to protest what many French farmers see as a threat to local French cuisine. In Seattle in 1999, thousands of protesters picketed the meeting of the World Trade Organization, which they viewed as a symbol of the homogenizing effects of Western market capitalism (Figure 23.15).

Daly and others argue that globalization entails a process in which multinational corporations attain greater and greater power over global trade while governments retain less and less. Most opponents of globalization consider businesses less likely than governments to encourage environmental protection, so they feel that globalization will hinder progress toward sustainability. Moreover, the U.S. model of market capitalism promotes a high-consumption lifestyle, which does indeed threaten efforts to advance toward sustainable solutions.

On the positive side, globalization may foster sustainability because Western democracy, as imperfect as it is, serves as a model for many people who live under repressive governments. Open societies allow for entrepreneurship and for the full flowering of creativity in academic research and in business. Millions of minds thinking and searching independently for solutions to problems are more likely to come up with sustainable solutions than the minds of a few who hold authoritarian power.

Weighing the Issues: Globalization

From your own experience, what advantages and disadvantages do you see in globalization? Have you personally benefited or been hurt by it in any way? In what ways might promoting local self-sufficiency be helpful for the pursuit of global sustainability, and in what ways might it not?

Many people enjoy eating at Vietnamese restaurants in the United States, yet many who protest globalization are quick to criticize the presence of McDonald's restaurants in Vietnam. Do you think this represents a double standard, or are there reasons the two are not comparable?

FIGURE 23.15 Thousands of protesters picketed the World Trade Organization's meeting in Seattle in 1999, criticizing the homogenizing effects of globalization, as well as relaxations in labor and environmental protections brought about by free trade.

Consumers vote with their wallets . . .

A capitalist free-market system driven by consumerism holds at least one great asset for sustainability: Consumers can exercise influence through what they choose to buy. When products produced sustainably are ecolabeled (▸ pp. 50–53), consumers can "vote with their wallets" by preferentially purchasing these products. Consumer choice has helped drive sales of everything from recycled paper to organic produce to "dolphin-safe" tuna. Individuals can multiply their own influence by promoting "green" purchasing habits at the institutions where they are employed or attend school. We have seen how purchasing power at colleges and universities has spurred sales of certified sustainable wood, organic food, energy-efficient appliances, and more. Motivated employees in businesses and government agencies can often promote change within those institutions by voicing their preferences in purchasing decisions.

. . . as well as with their ballots

Economic decisions are influential, but many of the changes needed to attain sustainable solutions require policymakers to usher them through. Policymakers respond to whoever exerts influence. Corporations and interest groups employ lobbyists to push politicians in one direction or another all the time. Citizens in a democratic republic have the same power, *if* they choose to exercise it. You can exercise this power at the ballot box, by attending public hearings, and by writing letters and making phone calls to policymakers. You may be surprised how little input policymakers receive from the public; sometimes a single letter or phone call can, in fact, make a difference.

Today's major environmental laws came about because citizens pressured their governmental representatives to do something about environmental problems. The raft of legislation enacted in the 1960s and 1970s in the United States and other nations might never have come about had ordinary citizens not stepped up and demanded action from policymakers. We owe it to our children and future generations to be engaged and to act responsibly now so that they have a better world in which to live. The words of anthropologist Margaret Mead are worth repeating: "Never doubt that a small group of thoughtful, committed people can change the world. Indeed, it's the only thing that ever has."

Promoting research and education is vital

None of these approaches will succeed fully if the public is not aware of their importance. An individual's decisions to reduce consumption, purchase ecolabeled products, or vote for candidates who support sustainable approaches will have limited effect unless a great many other people do the same. Individuals can influence large numbers of people by educating others with information and by serving as role models through their actions. The campus sustainability efforts at Ball State University and so many other colleges and universities accomplish both approaches. Moreover, the discipline of environmental science plays a key role in providing information people can use to make wise decisions about environmental problems. By promoting scientific research and by educating the public about environmental science, we can all assist in the pursuit of sustainable solutions.

VIEWPOINTS

Sustainability

How can we best pursue a goal of sustainability? What strategies and approaches are most needed and most effective?

GLOBALLY: Implementing Plan B

We are entering a new world. What we do not know is whether it will be a world of decline and collapse or a world of environmental restoration and economic progress.

Sustaining progress depends on restructuring the global economy, shifting from a fossil-fuel-based, automobile-centered, throwaway economy to one based on renewable energy sources, a diverse transportation system, and the comprehensive reuse and recycling of materials.

We can attain such a "Plan B" economy largely by restructuring taxes and subsidies to create a market that tells the ecological truth. Our modern economic prosperity is achieved in part by running up ecological deficits, costs that do not show up on the books, but that someone will eventually pay. Once we calculate the indirect costs of a product or service, we can incorporate them into market prices in the form of a tax, offsetting them with income tax reductions.

Sustaining progress also means eradicating poverty, stabilizing population, and restoring Earth's natural systems. Securing the public outlays needed to reach these goals depends on reordering fiscal priorities. A Plan B budget represents an additional annual expenditure of $161 billion—roughly one-third of the current U.S. military budget, or one-sixth of the global military budget.

Signs of the new economy are emerging all over the world—in the wind farms of Europe, the fast-growing U.S. fleet of gas-electric hybrid cars, the reforested hills of South Korea, the family planning program of Iran, the massive eradication of poverty in China, and the solar rooftops of Japan. What we need to do is doable.

The choice is ours—yours and mine. We can stay with business as usual and preside over an economy that continues to destroy its natural support systems until it destroys itself, or we can adopt Plan B and be the generation that changes direction, moving the world onto a path of sustained progress. The choice will be made by our generation, but it will affect life on Earth for all generations to come.

Lester R. Brown is founder and president of Earth Policy Institute, and founded Worldwatch Institute, serving as its president for 26 years. He helped pioneer the concept of environmentally sustainable development, a concept he uses in his design of an eco-economy. Brown has authored many books and received countless awards, including the 1987 United Nations Environment Prize and the 1994 Blue Planet Prize. In 1995, Marquis *Who's Who*, on the occasion of its 50th edition, selected Lester Brown as one of 50 Great Americans.

ON CAMPUS: A Positive Outlook for Sustainability

Colleges and universities, as centers of learning, have a responsibility to advance goals for global sustainability. Few experiences have as profound an impact on changing individuals and transforming the places where they live as education has. Courses, research, and actions focused on greening today's campuses will prepare and inspire students to go forward and enter a world in which the environment is inextricably linked to most local and global issues.

The challenges associated with becoming a more sustainable world can be somewhat daunting, but they are not insurmountable. As microcosms of society, institutions of higher education can demonstrate how to achieve goals for sustainability, which have an impact on and are transferable to other sectors of society. Models already exist on many campuses where carbon emissions are reduced, locally grown foods are served in dining halls, building design and construction is certified through LEED standards, and alternative energy is employed in the form of biofuels, wind turbines, and photovoltaics. Not only are colleges and universities recognizing that they have a responsibility, but many are leading the charge for the changes that are taking place.

However, administrators, employees, educators, and students need to understand more about the pathways leading to this change so that we can accelerate its pace. The template for a sustainable future will utilize an integrated, systemic approach, one that creates a shared institutional value, with a key component being the involvement of senior-level leadership. Actions must transcend traditional institutional boundaries and engage a diverse set of individuals, from the trustees to the grounds crew to the students. Collaborations will need to draw on varying types of expertise, resulting in shared outcomes and new networks. Achievements must be celebrated and setbacks should become fuel for reflection. The freedom to explore new ideas and innovation is essential, and these initiatives should leverage even greater change to bring about a more sustainable future.

Nan Jenks-Jay teaches and is director of environmental affairs at Middlebury College. She has created an integrated vision of campus sustainability by advancing Middlebury's exemplary environmental academic program, which received the EPA's Environmental Merit Award and the Vermont Governor's Award for Environmental Excellence. She has written extensively and speaks throughout the United States and abroad on topics related to the environment, sustainability, and higher education.

Explore this issue further by accessing **Viewpoints** at www.aw-bc.com/withgott.

Precious Time

The pace of our lives is getting faster. We do more, consume more, travel more, and often work more. Life's commotion can make it hard to give attention to problems we don't need to deal with on a daily basis. The world's sheer load of environmental dilemmas can feel overwhelming, and even the best intentioned among us may feel we have little time to devote to saving the planet.

However, time is getting short, and impacts on the natural systems we depend on are occurring quickly. Many human impacts continue to become more severe and widespread, including deforestation, overfishing, land clearing, wetland draining, and resource extraction. The window of opportunity for turning some of these trends around is getting short. Even if we can visualize sustainable solutions to our many problems, how can we possibly find the time to implement them before we have done irreparable damage to our environment and our own future?

We need to reach again for the moon

On May 25, 1961, U.S. President John F. Kennedy announced that within the decade the United States would be "landing a man on the moon and returning him safely to the Earth" (Figure 23.16). It was a bold and astonishing statement; the technology to achieve this unprecedented, almost unimaginable, feat did not yet exist. Kennedy's directive had powerful motivation behind it, however. The United States was caught up with the Soviet Union in the Cold War. In this competition for global hegemony, the two nations, held mutually at bay with nuclear weapons, each tried to prove their mettle by other means. The race for dominance in space became the centerpiece of the rivalry. Early victories in the "space race" went to the Soviet Union, which sent into orbit humanity's first satellite, followed it with others, and then sent the first human being into space. The United States, meanwhile, was crashing rockets.

The prospect of "losing" the space race prompted Kennedy's administration to set a national goal and an ambitious timeline to reach it. Congress followed through with funding, NASA performed the science and engineering, and in 1969 astronauts walked on the moon. The United States accomplished this milestone in human history by building public support for a goal and by giving its scientists and engineers the wherewithal to focus on developing technology and strategies to meet the goal.

FIGURE 23.16 U.S. President John F. Kennedy in 1961 called on Congress to fund a space program to send men to the moon and back again before 1970. Addressing our environmental problems and shifting our political, economic, and social institutions to a paradigm of sustainable development will require at least as much vision, resolve, and commitment. The fact that astronauts reached the moon in 1969 demonstrates the power of human ingenuity in meeting a challenge and provides some hope that we may yet be able to meet the bigger challenge of living sustainably on Earth.

The rapid and historic accomplishments of the United States and the Soviet Union during the space race show what societies can accomplish when they focus support for a chosen goal. Similar successes occurred when the United States confronted the Great Depression, threw itself into World War II, and conducted the Marshall Plan after the war to help rebuild western Europe.

Today humanity faces a challenge more important than any previous one—the challenge to achieve sustainability. Attaining sustainability is a larger and more complex process than traveling to the moon. However, it is one to which every single person on Earth can contribute; in which government, industry, and citizens can all cooperate; and toward which governments of all nations can work together. Unlike the space race, there is a real time limit in reaching sustainability. We cannot wait much longer to reduce our impact on the planet; if we delay, the damage may turn permanent. If America was able to reach the moon in a mere 8 years, then certainly humanity can begin down the road to sustainability with comparable speed. Human ingenuity is capable of it; it is just a question of rallying public resolve and engaging our governments, institutions, and entrepreneurs in the race.

We must pass through the environmental bottleneck

Human ingenuity and compassion give us reason to hope that we may achieve sustainability before doing too much damage to our planet and our prospects, but we must be realistic about the challenges that lie ahead. As we decrease the amount of natural capital we can draw on, we give ourselves and the rest of the world's creatures less room to maneuver. Until we implement sustainable solutions, we will be squeezing ourselves through a progressively tighter space, like being forced through the neck of a bottle. The key question for the future of our species and our planet thus becomes whether we can make it safely through this bottleneck. Biologist Edward O. Wilson (▸ p. 328) has written eloquently of this view:

> At best, an environmental bottleneck is coming in the twenty-first century. It will cause the unfolding of a new kind of history driven by environmental change. Or perhaps an unfolding on a global scale of more of the old kind of history, which saw the collapse of regional civilizations, going back to the earliest in history, in northern Mesopotamia, and subsequently Egypt, then the Mayan and many others. . . . People died in large numbers, often horribly. Sometimes they were able to emigrate and displace other people, making them die horribly instead. . . . Somehow humanity must find a way to squeeze through the bottleneck without destroying the environments on which the rest of life depends.

We must think of Earth as an island

We began this book with the vision of Earth as an island, and indeed that is what it is. Islands can be paradise, as Easter Island (▸ pp. 8–9) likely was when the Polynesians first reached it. But if people do not live sustainably on islands, they can turn paradises into desolate graveyards. When Europeans arrived at Easter Island, they witnessed the scene of a civilization that had depleted its resources, degraded its environment, and collapsed as a result. For the few people who remained of the once-mighty culture, life was difficult and unrewarding. They had lost even the knowledge of the history of their ancestors, who had cut trees unsustainably, kicking the base out from beneath their elaborate and prosperous civilization.

As Easter Island's trees disappeared, some individuals must have spoken out for conservation and for finding ways to live sustainably amid dwindling resources. And likely others ignored those calls and went on extracting more than the land could bear, assuming that somehow things would turn out all right. Indeed, whoever cut down the last tree from atop the most remote mountaintop could have looked out across the island and seen that it was the last tree. And yet that person cut it down.

It would be tragic folly to let such a fate occur to our planet as a whole. Yet if human impact on the environment continues growing, it is difficult to see how it can be avoided. By recognizing this, by deciding to shift our individual behavior and our cultural institutions in ways that encourage sustainable practices, and by employing sound science to help us achieve these ends, we may yet be able to live sustainably and happily on our wondrous island, Earth.

Conclusion

In every society facing the dilemma of dwindling resources and environmental degradation, there will be those who raise alarms and those who ignore them. Fortunately, in our global society today we have many thousands of scientists who study Earth's processes and resources. For this reason, we have access to an accumulated knowledge and an ever-developing understanding of our dynamic planet, what it offers us, and what impacts it can bear. The challenge for our global society today, our one-world island of humanity, is to support that science so that we can accurately judge false alarms from real problems and distinguish legitimate concerns from thoughtless denial. This science, this study of Earth and of ourselves, is what offers us hope for our future.

REVIEWING OBJECTIVES

You should now be able to:

List and describe approaches being taken on college and university campuses to promote sustainability

- Audits produce baseline data on how much a campus consumes and pollutes. (p. 677)
- The most common campus sustainability efforts involve recycling and waste reduction. (pp. 677–678)
- Sustainable building design is being pursued on a growing number of campuses. (pp. 678–679)
- There are numerous ways to reduce energy and water use, and these efforts often save money. (pp. 679–680)
- Providing local food is one way dining services can help in sustainability efforts. (p. 680)
- Colleges and universities can favor sustainable products in institutional purchasing. (p. 680)
- Institutions can use alternative fuels and vehicles in university fleets and encourage bicycling, walking, and public transportation. (p. 681)
- Habitat restoration is one of the most popular campus sustainability activities. (pp. 681–682)

Explain the concept of sustainable development

- Sustainable development entails environmental protection, economic development, and socioeconomic equity. (pp. 682–683)
- Proponents of sustainable development see no necessary tradeoff between economic development and environmental quality; rather, they feel that the two can enhance one another. (p. 683)

Discuss how protecting the environment can be compatible with promoting economic welfare

- Environmental protection and new green technologies and industries can create rich sources of new jobs. (pp. 683–684)
- Protecting environmental quality enhances a community's desirability and economy. (p. 684)

Describe and assess key approaches to designing sustainable solutions

- There are at least 10 general approaches that can inspire specific sustainable solutions. (pp. 685–693)

Analyze the roles that consumption, population, and technology play in efforts to achieve sustainability

- Growth in both population and per capita consumption will likely need to be halted if we are to create a sustainable society. (pp. 687–689)
- Technology has traditionally increased our environmental impact, but some new technologies can help reduce environmental impact. (pp. 689–690)

Explain how time is limited but how human potential to solve problems is tremendous

- Time for turning around our increasing environmental impacts is running short. (p. 695)
- The United States and other nations have met tremendous challenges before, so there is reason to hope that the world will be able to do what is needed to attain a sustainable society. (pp. 695–696)

TESTING YOUR COMPREHENSION

1. In what ways are campus sustainability efforts relevant to sustainability efforts in the broader society?
2. Name one way in which campus sustainability proponents have addressed each of the following areas: (1) recycling and waste reduction; (2) building using green technology; (3) energy and water efficiency; (4) dining services; (5) institutional purchasing; (6) transportation; (7) habitat restoration.
3. What do environmental scientists mean by *sustainable development*?
4. Why is it often thought that economic development and environmental protection are in direct opposition?
5. Describe three ways in which environmental protection can enhance economic well-being.
6. Why are many people now living at the highest level of material prosperity in history? Is this level of consumption sustainable? How can it feel good to consume less?
7. In what ways can technology help us achieve sustainability? How do natural processes provide good models of sustainability for manufacturing? Provide examples.
8. Why do many people feel that local self-sufficiency is important? What consequences of globalization may threaten sustainability? How can open democratic societies help to promote sustainability?
9. Explain Edward O. Wilson's metaphor of the "environmental bottleneck."
10. How can thinking of Earth as an island help prevent us from repeating the mistakes of other civilizations?

SEEKING SOLUTIONS

1. What efforts toward sustainable solutions have been made on your campus? What initiatives have succeeded, and why? What attempts have failed, and why? What lessons can you draw from these successes and failures?
2. What sustainability initiatives would you like to see attempted on your campus? If you were to take the lead in promoting such initiatives, how would you go about it? What obstacles would you expect to face, and how would you deal with them?
3. Choose one item or product that you enjoy, and consider how it came to be. What steps were involved in creating its components, and where did the raw materials come from? How was it manufactured? How was it delivered to you? Think of as many components of the item or product as you can, and determine how each of them was obtained or created.
4. Do you think that we can "have it both ways" by increasing our quality of life through development while also protecting the integrity of the environment? Can what is good for the environment also be good for humans? If so, how? Discuss examples from your course or from other chapters of this book that illustrate possible win-win solutions. Are there cases in which you think win-win solutions will not be possible?
5. Why do many observers find flaws in the conventional outlook that environmental protection necessarily is economically costly? Are you familiar with any cases in your community or at your college that bear on this issue? Describe such a case, and state what lessons you would draw from it.
6. Reflect on the experiences of prior human civilizations and how they came to an end. What is your prognosis for our current human civilization? Do you see a vast world of independent cultures all individually responsible for themselves, or do you consider human civilization to be one great entity? Is Earth an island, or is it a collection of many islands? If we accept that all humans depend on the same environmental systems for sustenance, what resources and strategies do we have to ensure that the actions of a few do not determine the outcome for all and that sustainable solutions are a common global goal?

INTERPRETING GRAPHS AND DATA

An undergraduate class at Penn State University conducted an ecological assessment of one of their biology laboratory buildings in response to the question: "How is this building like an ecosystem?" The result of their assessment was a 52-page report outlining ways to reduce the ecological footprint of the Mueller Laboratory Building in the areas of energy use, water use, communications/computing, furnishings/renovation, maintenance, and food. The students found ways to save an estimated $45,500 per year in the cost of energy alone, for a building that is occupied by 123 scientists and support staff. Their data on the current use and potential savings in the energy component of the ecological footprint are shown in the graph.

1. From the graph, estimate the amount of potential savings for each of the five areas identified in the Mueller Report. Approximately what percentage of the total electrical energy use of the building do these savings together represent?

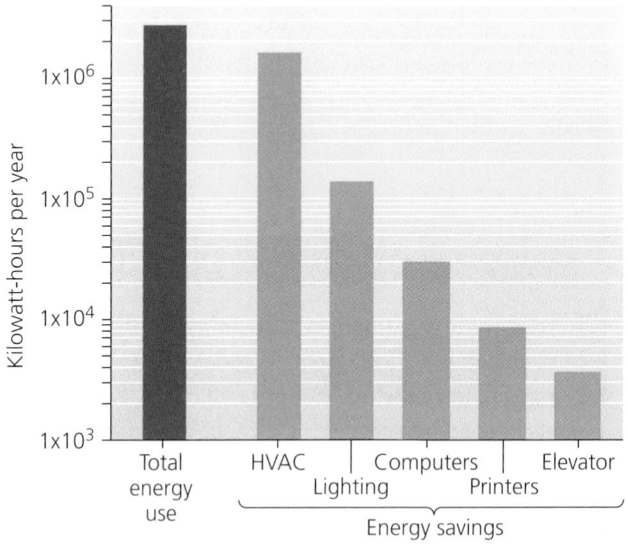

Total annual electrical energy use (red bar) and potential energy savings in five areas (orange bars) at Penn State University's Mueller Laboratory Building. The y axis is logarithmic, with each tick mark representing an increase equal to the value indicated just below it on the axis. *Data source:* Penn State Green Destiny Council, 2001. *The Mueller Report: Moving beyond sustainability indicators to sustainability action at Penn State.*

2. What was the approximate cost of electricity in cents per kilowatt-hour used to calculate the savings of $45,500? Do you think this cost may have changed since the report was issued in 2001? How and why?

3. How is the building where you take your course like an ecosystem? How is it not like an ecosystem? Is your building operating sustainably? What improvements do you think could be made to increase its sustainability?

CALCULATING ECOLOGICAL FOOTPRINTS

Of all the choices we make, our transportation choices have some of the greatest repercussions for environmental impact. According to the U.S. Department of Transportation, there are about 136 million passenger cars in the United States (excluding light trucks and SUVs), which travel an average of 12,200 miles per car per year at an average fuel economy of 22 mpg. These figures can be related to the size of our population (P) and the level of affluence (A) in the I = PAT equation (▸ pp. 220–221). The technology (T) term in this case can refer to the automotive technologies that determine fuel economy.

Federal CAFE (Corporate Average Fuel Economy) standards require a manufacturer's fleet of new passenger cars to average 27.5 mpg. With some hybrid gas/electric vehicles you may achieve 60 mpg. Using this information, calculate in the table the impact of choosing more fuel-efficient passenger car technologies. Estimate the number of cars owned by your classmates and by the residents of your hometown.

	Cars	Miles/yr	Gal/year @ 22 mpg	Gal/year @ 27.5 mpg	Gal/year @ 60 mpg	Gal saved @ 27.5 vs. 22 mpg	Gal saved @ 60 vs. 22 mpg
You	1	1.22×10^4	555	444	203	111	352
Your class							
Your hometown							
United States	1.36×10^8						

Data from Bureau of Transportation Statistics. 2005. *National Transportation Statistics 2004.* Washington D.C.: U.S. Department of Transportation.

1. You can reduce your personal environmental impact by choosing different technologies (T), as the table above demonstrates. If more people make these same reductions (P), the total reduction in environmental impact would be larger. What could you choose to do that would alter the value of the other independent variable in the I=PAT equation, namely A, which stands for your affluence, or the amount of resources you use?

2. How many gallons of gasoline would you estimate that you would save per year if you reduced the number of miles you drive per year by 20%? How could you reduce your annual mileage by 20% without affecting your "quality of life?"

3. Suppose the cost of gasoline is $2.50 per U.S. gallon. How much money would you save annually if your vehicle averaged 27.5 mpg instead of 22 mpg? If it averaged 60 mpg? In parts of western Europe, the price of gasoline is triple that in the United States. How much more would your annual driving costs be there at 22 mpg? Would such an additional cost influence your decision as to what type of vehicle to drive?

Take It Further

Go to www.aw-bc.com/withgott or the student CD-ROM where you'll find:

- Suggested answers to end-of-chapter questions
- Quizzes, animations, and flashcards to help you study
- *Research Navigator*™ database of credible and reliable sources to assist you with your research projects
- **GRAPHIt!** tutorials to help you master how to interpret graphs
- **INVESTIGATEIt!** current news articles that link the topics that you study to case studies from your region to around the world

Appendix A Some Basics on Graphs

Presenting data in ways that are clear and that help make trends and patterns visually apparent is a vital part of the scientific endeavor. Scientists' primary tool for presenting data and expressing patterns is the graph. Thus, the ability to interpret graphs is a skill that you will want to cultivate early in your study of the sciences. This appendix provides basic information on four of the most common types of graphs—line plots, bar charts, scatter plots, and pie charts—and the rationale for the use of each.

Line Plot

A line plot is drawn when a data set involves a sequence of some kind, such as a sequence through time or across distance (Figure A.1; see ▸ p. 140 and Figure 9.19, ▸ p. 266). Using a line plot allows us to see increasing or decreasing trends in the data. Line plots are appropriate when the variable measured by the *y* axis (the vertical axis) represents continuous numerical data, and when the variable measured by the *x* axis (the horizontal axis) represents either continuous numerical data or sequential categories, such as years.

One useful technique is to plot two data sets together on the same graph (Figure A.2; see Figure 6.5, ▸ p. 153). This allows us to compare trends in the two data sets to see whether and how they may be related.

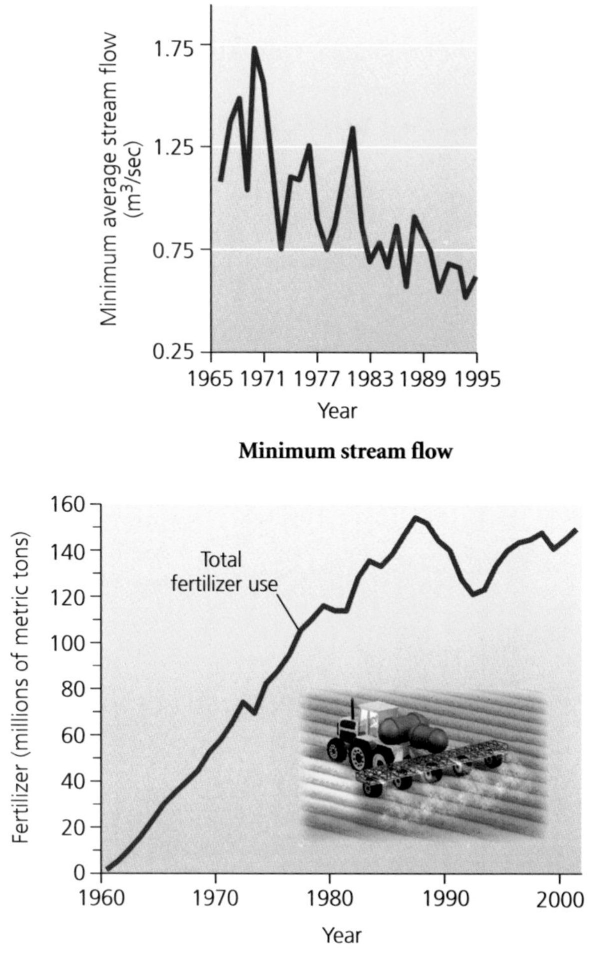

FIGURE A.1

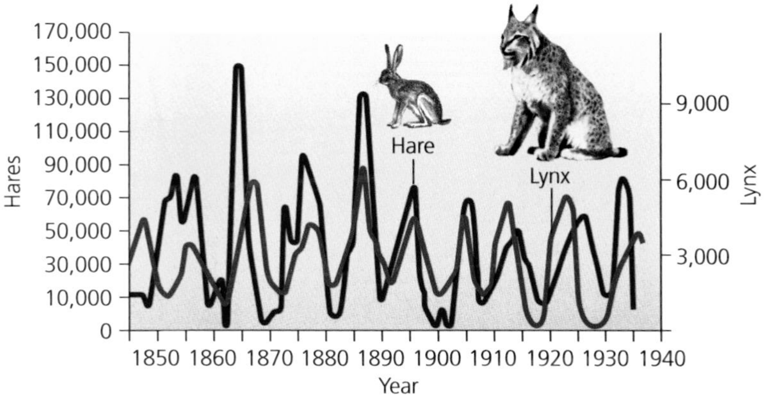

Population fluctuations in hare and lynx

FIGURE A.2

Bar Chart

A bar chart is most often used when one of the variables represents categories rather than numerical values (Figure A.3; see Figure 13.10a, ▸ p. 389). Bar charts allow us to visualize how a variable differs quantitatively among categories.

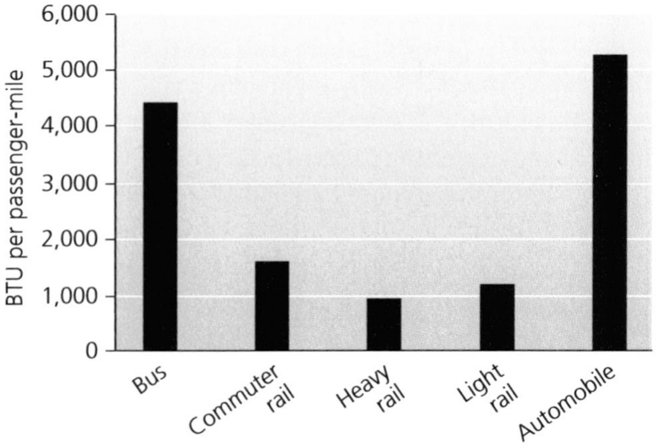

Energy consumption for different modes of transit

FIGURE A.3

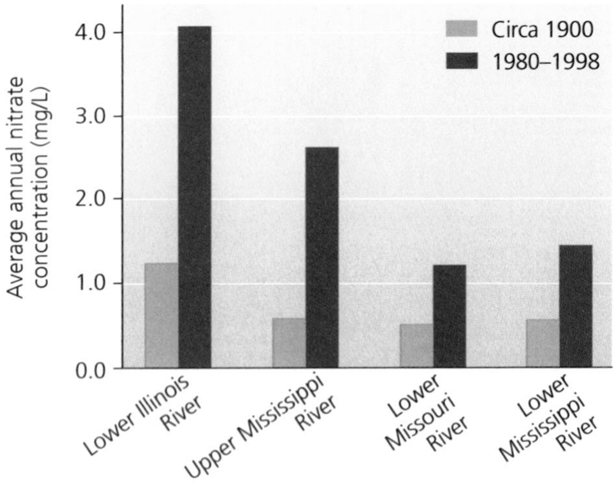

Nitrate concentrations in portions of the Mississippi River watershed

FIGURE A.4

It is often instructive to graph two or more variables together to reveal patterns and relationships (Figure A.4; see Figure 7.4a, ▸ p. 188). Most of the bar charts you will see in this book illustrate several types of information at once in this manner.

Bar charts are usually arrayed so that the bars extend vertically. Sometimes, however, a horizontal orientation may make for a clearer presentation. One special type of horizontally oriented bar chart is the age pyramid used by demographers (Figure A.5; see Figure 8.11, ▸ p. 226). Age categories are displayed on the y axis, with bars representing the population size of each age group varying in width instead of height.

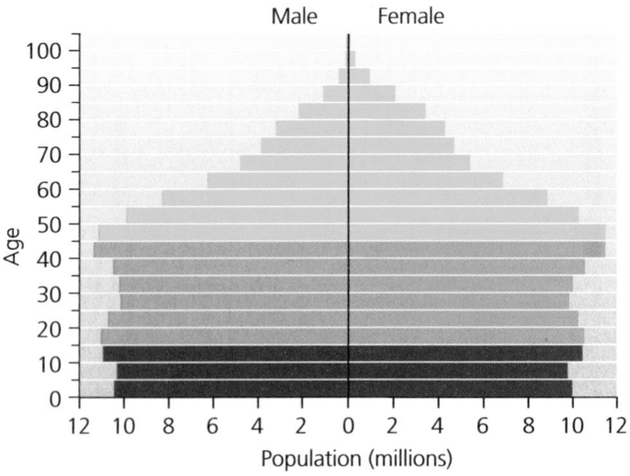

United States age structure, 2005

FIGURE A.5

Scatter Plot

A scatter plot is used most often when there is no sequential aspect to the data, and each data point is independent, having no particular connection to other data points (Figure A.6; see Figure 8.15, ▸ p. 232). Scatter plots allow you to visualize a broad positive or negative correlation between variables on the two axes.

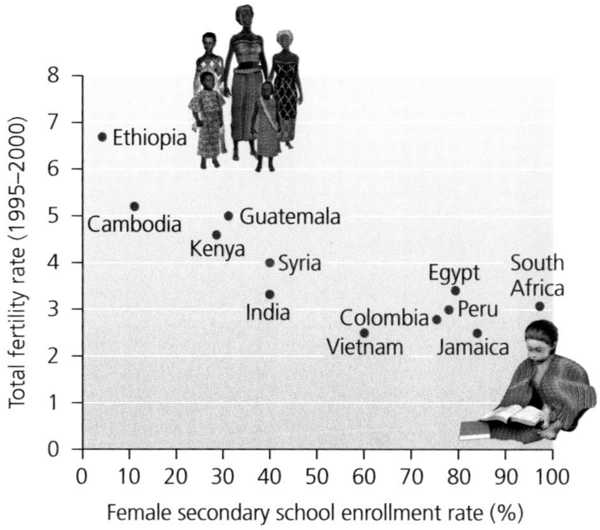

Fertility rate and female education

FIGURE A.6

Pie Chart

A pie chart is used when we wish to compare the proportions of some whole that are taken up by each of several categories (Figure A.7; see Figure 9.2, ▸ p. 247). A pie chart is appropriate when one variable is categorical and one is numerical. Each category is represented visually like a slice from a pie, with the size of the slice reflecting the percentage of the whole that is taken up by that category.

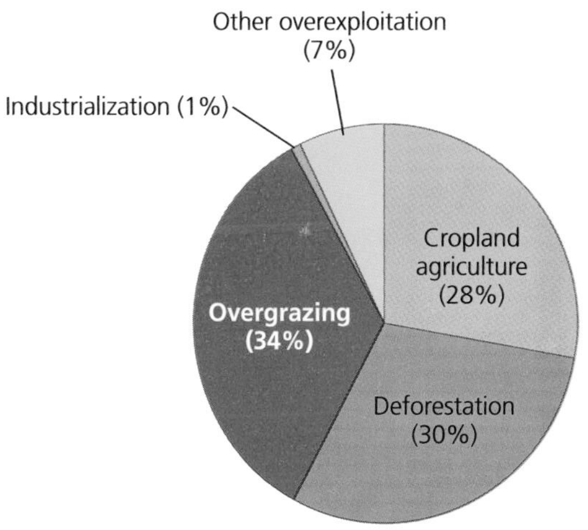

Causes of desertification

FIGURE A.7

But don't stop here. Take advantage of the GRAPHit! tutorials on the CD-ROM included with this text, or go to the Withgott/Brennan Companion Website at www.aw-bc.com/withgott. The GRAPHit! tutorials allow you to plot your own data, and help you further expand your comprehension of graphs.

Appendix B Metric System

Measurement	Unit and Abbreviation	Metric Equivalent	Metric to English Conversion Factor	English to Metric Conversion Factor
Length	1 kilometer (km)	= 1,000 (10^3) meters	1 km = 0.62 mile	1 mile = 1.61 km
	1 meter (m)	= 100 (10^2) centimeters	1 m = 1.09 yards	1 yard = 0.914 m
		= 1,000 millimeters	1 m = 3.28 feet	1 foot = 0.305 m
			1 m = 39.37 inches	
	1 centimeter (cm)	= 0.01 (10^{-2}) meter	1 cm = 0.394 inch	1 foot = 30.5 cm
				1 inch = 2.54 cm
	1 millimeter (mm)	= 0.001 (10^{-3}) meter	1 mm = 0.039 inch	
	1 micrometer (μm) [formerly micron (μ)]	= 0.000001 (10^{-6}) meter		
	1 nanometer (nm) [formerly millimicron (mμ)]	= 0.000000001 (10^{-9}) meter		
	1 angstrom (Å)	= 0.0000000001 (10^{-10}) meter		
Area	1 square meter (m^2)	= 10,000 square centimeters	1 m^2 = 1.1960 square yards	1 square yard = 0.8361 m^2
			1 m^2 = 10.764 square feet	1 square foot = 0.0929 m^2
	1 square centimeter (cm^2)	= 100 square millimeters	1 cm^2 = 0.155 square inch	1 square inch = 6.4516 cm^2
Mass	1 metric ton (t)	= 1,000 kilograms	1 t = 1.103 ton	1 ton = 0.907 t
	1 kilogram (kg)	= 1,000 grams	1 kg = 2.205 pounds	1 pound = 0.4536 kg
	1 gram (g)	= 1,000 milligrams	1 g = 0.0353 ounce	1 ounce = 28.35 g
			1 g = 15.432 grains	
	1 milligram (mg)	= 0.001 gram	1 mg = approx. 0.015 grain	
	1 microgram (μg)	= 0.000001 gram		
Volume (solids)	1 cubic meter (m^3)	= 1,000,000 cubic centimeters	1 m^3 = 1.3080 cubic yards	1 cubic yard = 0.7646 m^3
			1 m^3 = 35.315 cubic feet	1 cubic foot = 0.0283 m^3
	1 cubic centimeter (cm^3 or cc)	= 0.000001 cubic meter	1 cm^3 = 0.0610 cubic inch	1 cubic inch = 16.387 cm^3
		= 1 milliliter		
	1 cubic millimeter (mm^3)	= 0.000000001 cubic meter		
Volume (liquids and gases)	1 kiloliter (kl or kL)	= 1,000 liters	1 kL = 264.17 gallons	1 gallon = 3.785 L
	1 liter (l or L)	= 1,000 milliliters	1 L = 0.264 gallons	1 quart = 0.946 L
			1 L = 1.057 quarts	
	1 milliliter (ml or mL)	= 0.001 liter	1 ml = 0.034 fluid ounce	1 quart = 946 ml
		= 1 cubic centimeter	1 ml = approx. $\frac{1}{4}$ teaspoon	1 pint = 473 ml
			1 ml = approx. 15–16 drops (gtt.)	1 fluid ounce = 29.57 ml
				1 teaspoon = approx. 5 ml
	1 microliter (μl or μL)	= 0.000001 liter		
Time	1 second (s)	= $\frac{1}{60}$ minute		
	1 millisecond (ms)	= 0.001 second	1 second (s)	= $\frac{1}{60}$ minute
Temperature	Degrees Celsius (°C)	°F = $\frac{9}{5}$°C + 32	°C = $\frac{5}{9}$(°F − 32)	
Energy and Power	1 kilowatt-hour	= 34,113 BTUs = 860,421 calories		
	1 watt	= 3.413 BTU/hr		
		= 14.34 calorie/min		
	1 calorie	= the amount of heat necessary to raise the temperature of 1 gram (1cm^3) of water 1 degree Celsius		
	1 horsepower	= 7.457 × 102 watts		
	1 joule	= 9.481 × 10^{-4} BTU		
		= 0.239 cal		
		= 2.778 × 10^{-7} kilowatt-hour		
Pressure	1 pound per square inch (psi)	= 6894.757 pascal (Pa)		
		= 0.068045961 atmosphere (atm)		
		= 51.71493 millimeters of mercury (mm hg = Torr)		
		= 68.94757 millibars (mbar)		
		= 68.94757 (hectopascal hPa)		
		= 6.894757 kilopascal (kPa)		
		= 0.06894757 bar (bar)		
	1 atmosphere (atm)	= 101.325 kilopascal (kPa)		

Appendix C Periodic Table of the Elements

Representative (main group) elements: IA–IIA and IIIA–VIIIA. Transition metals: IIIB–IIB.

Period	IA	IIA	IIIB	IVB	VB	VIB	VIIB	VIIIB	VIIIB	VIIIB	IB	IIB	IIIA	IVA	VA	VIA	VIIA	VIIIA
1	1 H 1.0079 Hydrogen																	2 He 4.003 Helium
2	3 Li 6.941 Lithium	4 Be 9.012 Beryllium											5 B 10.811 Boron	6 C 12.011 Carbon	7 N 14.007 Nitrogen	8 O 15.999 Oxygen	9 F 18.998 Fluorine	10 Ne 20.180 Neon
3	11 Na 22.990 Sodium	12 Mg 24.305 Magnesium											13 Al 26.982 Aluminum	14 Si 28.086 Silicon	15 P 30.974 Phosphorus	16 S 32.066 Sulfur	17 Cl 35.453 Chlorine	18 Ar 39.948 Argon
4	19 K 39.098 Potassium	20 Ca 40.078 Calcium	21 Sc 44.956 Scandium	22 Ti 47.88 Titanium	23 V 50.942 Vanadium	24 Cr 51.996 Chromium	25 Mn 54.938 Manganese	26 Fe 55.845 Iron	27 Co 58.933 Cobalt	28 Ni 58.69 Nickel	29 Cu 63.546 Copper	30 Zn 65.39 Zinc	31 Ga 69.723 Gallium	32 Ge 72.61 Germanium	33 As 74.922 Arsenic	34 Se 78.96 Selenium	35 Br 79.904 Bromine	36 Kr 83.8 Krypton
5	37 Rb 85.468 Rubidium	38 Sr 87.62 Strontium	39 Y 88.906 Yttrium	40 Zr 91.224 Zirconium	41 Nb 92.906 Niobium	42 Mo 95.94 Molybdenum	43 Tc 98 Technetium	44 Ru 101.07 Ruthenium	45 Rh 102.906 Rhodium	46 Pd 106.42 Palladium	47 Ag 107.868 Silver	48 Cd 112.411 Cadmium	49 In 114.82 Indium	50 Sn 118.71 Tin	51 Sb 121.76 Antimony	52 Te 127.60 Tellurium	53 I 126.905 Iodine	54 Xe 131.29 Xenon
6	55 Cs 132.905 Cesium	56 Ba 137.327 Barium	57 La 138.906 Lanthanum	72 Hf 178.49 Hafnium	73 Ta 180.948 Tantalum	74 W 183.84 Tungsten	75 Re 186.207 Rhenium	76 Os 190.23 Osmium	77 Ir 192.22 Iridium	78 Pt 195.08 Platinum	79 Au 196.967 Gold	80 Hg 200.59 Mercury	81 Tl 204.383 Thallium	82 Pb 207.2 Lead	83 Bi 208.980 Bismuth	84 Po 209 Polonium	85 At 210 Astatine	86 Rn 222 Radon
7	87 Fr 223 Francium	88 Ra 226.025 Radium	89 Ac 227.028 Actinium	104 Rf 261 Unnilquadium	105 Db 262 Unnilpentium	106 Sg 263 Unnilhexium	107 Bh 262 Unnilseptium	108 Hs 265 Unniloctium	109 Mt 266 Unnilennium	110 Uun 269 Ununnilium	111 Uuu 272 Unununium	112 Uub 277 Ununbium		114		116		

Rare earth elements

Lanthanides	58 Ce 140.115 Cerium	59 Pr 140.908 Praseodymium	60 Nd 144.24 Neodymium	61 Pm 145 Promethium	62 Sm 150.36 Samarium	63 Eu 151.964 Europium	64 Gd 157.25 Gadolinium	65 Tb 158.925 Terbium	66 Dy 162.5 Dysprosium	67 Ho 164.93 Holmium	68 Er 167.26 Erbium	69 Tm 168.934 Thulium	70 Yb 173.04 Ytterbium	71 Lu 174.967 Lutetium
Actinides	90 Th 232.038 Thorium	91 Pa 231.036 Protactinium	92 U 238.029 Uranium	93 Np 237.048 Neptunium	94 Pu 244 Plutonium	95 Am 243 Americium	96 Cm 247 Curium	97 Bk 247 Berkelium	98 Cf 251 Californium	99 Es 252 Einsteinium	100 Fm 257 Fermium	101 Md 258 Mendelevium	102 No 259 Nobelium	103 Lr 262 Lawrencium

The periodic table arranges elements according to atomic number and atomic weight into horizontal rows called periods and vertical columns called groups.

Elements of each group in Class A have similar chemical and physical properties. This reflects the fact that members of a particular group have the same number of valence shell electrons, which is indicated by the group's number. For example, group IA elements have one valence shell electron, group IIA elements have two, and group VA elements have five. In contrast, as you progress across a period from left to right, properties of the elements change, varying from the very metallic properties of groups IA and IIA to the nonmetallic properties of group VIIA to the inert elements (noble gases) in group VIIIA. This reflects changes in the number of valence shell electrons.

Class B elements, or transition elements, are metals, and generally have one or two valence shell electrons. In these elements, some electrons occupy more distant electron shells before the deeper shells are filled.

In this periodic table, elements with symbols printed in black exist as solids under standard conditions (25 °C and 1 atmosphere of pressure), while elements in red exist as gases, and those in dark blue as liquids. Elements with symbols in green do not exist in nature and must be created by some type of nuclear reaction.

Appendix D Geologic Timescale

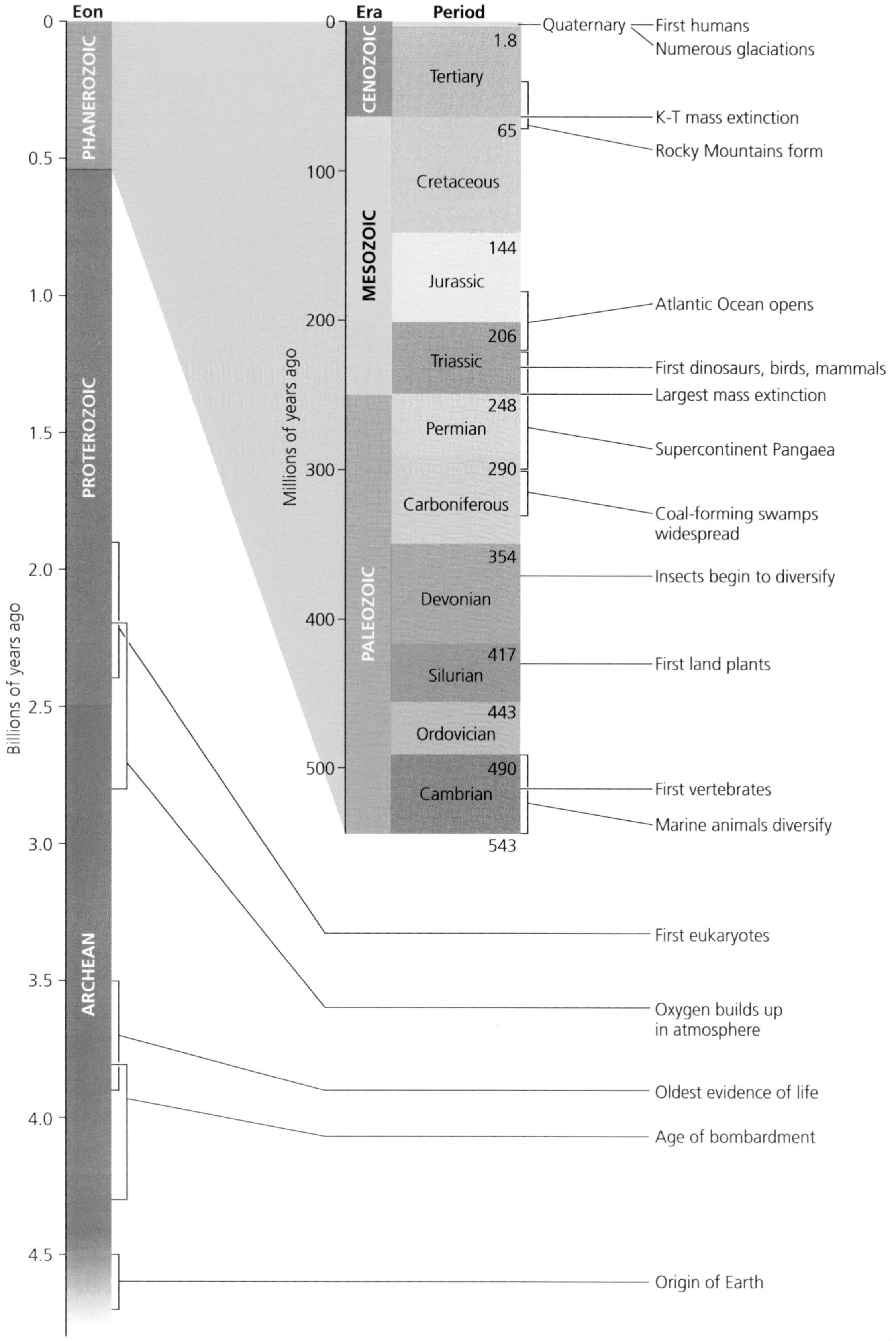

Glossary

abiotic factor Any nonliving component of the *environment.* Compare *biotic factor.*
acidic deposition The deposition of *acidic* or acid-forming pollutants from the *atmosphere* onto Earth's surface by *precipitation,* by fog, by gases, or by the deposition of dry particles.
acid drainage A process in which sulfide minerals in newly exposed rock surfaces react with oxygen and rainwater to produce sulfuric acid, which causes chemical *runoff* as it *leaches* metals from the rocks. Although acid drainage is a natural phenomenon, mining can greatly accelerate its rate by exposing many new rock surfaces at once.
acidic The property of a solution in which the concentration of hydrogen (H^+) *ions* is greater than the concentration of hydroxide (OH^-) ions. Compare *basic.*
active solar energy collection An approach in which technological devices are used to focus, move, or store solar energy. Compare *passive solar energy collection.*
acute exposure Exposure to a *toxicant* occurring in high amounts for short periods of time. Compare *chronic exposure.*
adaptive management The systematic testing of different management approaches to improve methods over time.
adaptive trait (adaptation) A trait that confers greater likelihood that an individual will reproduce.
affluenza Term coined by social critics to describe the failure of material goods to bring happiness to people who have the financial means to afford them.
aerobic Occurring in an *environment* where oxygen is present. For example, the decay of a rotting log proceeds by aerobic decomposition. Compare *anaerobic.*
age distribution The relative numbers of organisms of each age within a *population.* Age distributions can have a strong effect on rates of population growth or decline and are often expressed as a ratio of age classes, consisting of organisms (1) not yet mature enough to reproduce, (2) capable of reproduction, and (3) beyond their reproductive years.
age structure See *age distribution.*
agricultural revolution The shift around 10,000 years ago from a hunter-gatherer lifestyle to an agricultural way of life in which people began to grow their own crops and raise domestic animals. Compare *industrial revolution.*
agriculture The practice of cultivating *soil,* producing crops, and raising livestock for human use and consumption.
A horizon A layer of *soil* found in a typical *soil profile.* It forms the top layer or lies below the *O horizon* (if one exists). It consists of mostly inorganic mineral components such as *weathered* substrate, with some organic matter and humus from above mixed in. The A horizon is often referred to as *topsoil.* Compare *B horizon; C horizon; E horizon; R horizon.*
air pollutants Gases and particulate material added to the atmosphere that can affect *climate* or harm people or other organisms.
air pollution The act of polluting the air, or the condition of being polluted by *air pollutants.*
allergen A *toxicant* that overactivates the immune system, causing an immune response when one is not necessary.
amensalism A relationship between members of different *species* in which one organism is harmed and the other is unaffected. Compare *commensalism.*
anaerobic Occurring in an *environment* that has little or no oxygen. The conversion of organic matter to *fossil fuels (crude oil, coal, natural gas)* at the bottom of a deep lake, swamp, or shallow sea is an example of anaerobic decomposition. Compare *aerobic.*
anthropocentrism A human-centered view of our relationship with the *environment.*
aquaculture The raising of aquatic organisms for food in controlled *environments.*
aquifer An underground water reservoir.
artesian aquifer See *confined aquifer.*
artificial selection *Natural selection* conducted under human direction. Examples include the *selective breeding* of crop plants, pets, and livestock.
atmosphere The thin layer of gases surrounding planet Earth. Compare *biosphere; hydrosphere; lithosphere.*
atmospheric deposition The wet or dry deposition on land of a wide variety of pollutants, including mercury, nitrates, organochlorines, and others. *Acidic deposition* is one type of atmospheric deposition.
atmospheric pressure The weight per unit area produced by a column of air.
atom The smallest component of an *element* that maintains the chemical properties of that element.
autotroph (primary producer) An organism that can use the energy from sunlight to produce its own food. Includes green plants, algae, and cyanobacteria.

***Bacillus thuringiensis* (Bt)** A naturally occurring *soil* bacterium that produces a protein that kills many pests, including caterpillars and the larvae of some flies and beetles.
background rate of extinction The average rate of *extinction* that occurred before the appearance of humans. For example, the *fossil record* indicates that for both birds and mammals, one *species* in the world typically became extinct every 500–1,000 years. Compare *mass extinction event.*
basic The property of a solution in which the concentration of hydroxide (OH^-) *ions* is greater than the concentration of hydrogen (H^+) ions. Compare *acidic.*
bedrock The continuous mass of solid rock that makes up Earth's *crust.*
benthic Of, relating to, or living on the bottom of a water body. Compare *pelagic.*
benthic zone The bottom layer of water body. Compare *littoral zone; limnetic zone; profundal zone.*
B horizon The layer of *soil* that lies below the *E horizon* and above the *C horizon.* Minerals that leach out of the E horizon are carried down into the B horizon (or subsoil) and accumulate there. Sometimes called the "zone of accumulation" or "zone of deposition." Compare *A horizon; O horizon; R horizon.*
bioaccumulation The buildup of *toxicants* in the tissues of an animal.
biocentrism A philosophy that ascribes relative values to actions, entities, or properties on the basis of their effects on all living things or on the integrity of the *biotic* realm in general. The biocentrist evaluates an action in terms of its overall

impact on living things, including—but not exclusively focusing on—human beings.
biodiesel Diesel fuel produced by mixing vegetable oil, used cooking grease, or animal fat with small amounts of *ethanol* or methanol (wood alcohol) in the presence of a chemical catalyst.
biodiversity (biological diversity) The sum total of all organisms in an area, taking into account the diversity of *species,* their *genes,* their *populations,* and their *communities.*
biodiversity hotspot An area that supports an especially great diversity of *species,* particularly species that are *endemic* to the area.
biofuel Fuel produced from *biomass energy* sources and used primarily to power automobiles.
biogeochemical cycle See *nutrient cycle.*
biological control (biocontrol) The attempt to battle pests and weeds with organisms that prey on or parasitize them, rather than by using *pesticides.*
biological diversity See *biodiversity.*
biomagnification The magnification of the concentration of *toxicants* in an organism caused by its consumption of other organisms in which toxicants have *bioaccumulated.*
biomass energy *Energy* harnessed from plant and animal matter, including wood from trees, charcoal from burned wood, and combustible animal waste products, such as cattle manure. *Fossil fuels* are not considered biomass energy sources because their organic matter has not been part of living organisms for millions of years and has undergone considerable chemical alteration since that time.
biome A major regional complex of similar plant *communities;* a large *ecological* unit defined by its dominant plant type and vegetation structure.
biophilia A hypothetical phenomenon that E. O. Wilson defined as "the connections that human beings subconsciously seek with the rest of life."
biopower The burning of *biomass energy* sources to generate electricity.
bioremediation The attempt to clean up *pollution* by enhancing natural processes of biodegradation by living organisms.
biosphere The sum total of all the planet's living organisms and the *abiotic* portions of the *environment* with which they interact. Compare *atmosphere; hydrosphere; lithosphere.*
biosphere reserve A tract of land with exceptional *biodiversity* that couples preservation with *sustainable development* to benefit local people. Biosphere reserves are designated by UNESCO (the *United Nations* Educational, Scientific, and Cultural Organization) following application by local stakeholders.
biotechnology The material application of biological *science* to create products derived from organisms. The creation of *transgenic* organisms is one type of biotechnology.
biotic factor Any living component of the *environment.* Compare *abiotic factor.*
boreal forest A *biome* of northern coniferous forest that stretches in a broad band across much of Canada, Alaska, Russia, and Scandinavia. Also known as taiga, boreal forest consists of a limited number of *species* of evergreen trees, such as black spruce, that dominate large regions of forests interspersed with occasional bogs and lakes.
breeder nuclear fission A form of *nuclear fission* that uses uranium-238 instead of uranium-235 (which is used in the conventional method to harness *nuclear energy*). Breeder fission makes better use of fuel, generates more power, and produces less waste than does conventional nuclear fission, but it is considerably more dangerous.
Bureau of Land Management (BLM) Federal agency that owns and manages most U.S. rangelands. The BLM is the nation's single largest landowner; its 106 million ha (261 million acres) are spread across 12 western states.
by-catch That portion of a commercial fishing catch consisting of animals caught unintentionally. By-catch kills many thousands of fish, sharks, marine mammals, and birds each year.

capitalist market economy An *economy* in which buyers and sellers interact to determine which *goods* and *services* to produce, how much of them to produce, and how to distribute them. Compare *centrally planned economy.*
captive breeding The practice of capturing members of threatened and endangered *species* so that their young can be bred and raised in controlled *environments* and subsequently reintroduced into the wild.
carbohydrate An *organic compound* consisting of *atoms* of carbon, hydrogen, and oxygen.
carbon cycle A major *nutrient cycle* consisting of the routes that carbon *atoms* take through the nested networks of environmental *systems.*
carcinogen A chemical or type of radiation that causes cancer.
carrying capacity The maximum *population size* that a given *environment* can sustain.
cell The most basic organizational unit of organisms.
cellular respiration The process by which a *cell* uses the chemical reactivity of oxygen to split glucose into its constituent parts, water and carbon dioxide, and thereby release *chemical energy* that can be used to form chemical bonds or to perform other tasks within the cell. Compare *photosynthesis.*
centrally planned economy An *economy* in which a nation's government determines how to allocate resources in a top-down manner. Also called a "state socialist economy." Compare *capitalist market economy.*
chaparral A *biome* consisting mostly of densely thicketed evergreen shrubs occurring in limited small patches. Its "Mediterranean" *climate* of mild, wet winters and warm, dry summers is induced by oceanic influences. In addition to ringing the Mediterranean Sea, chaparral occurs along the coasts of California, Chile, and southern Australia.
chemical energy *Potential energy* held in the bonds between atoms.
chemosynthesis The process by which bacteria in hydrothermal vents use the *chemical energy* of hydrogen sulfide (H_2S) to transform inorganic carbon into *organic compounds.* Compare *photosynthesis.*
Chernobyl Site of a nuclear power plant in Ukraine (then part of the Soviet Union), where in 1986 an explosion caused the most severe *nuclear reactor* accident the world has yet seen. As with *Three Mile Island,* the term is often used to denote the accident itself.

chlorofluorocarbon (CFC) One of a group of human-made *organic compounds* derived from simple *hydrocarbons,* such as ethane and methane, in which hydrogen *atoms* are replaced by chlorine, bromine, or fluorine. CFCs deplete the protective *ozone layer* in the *stratosphere.*
C horizon The layer of *soil* that lies below the *B horizon* and above the *R horizon.* It contains rock particles that are larger and less *weathered* than the layers above. It consists of *parent material* that has been altered only slightly or not at all by the process of *soil* formation. Compare *A horizon; E horizon; O horizon.*
chronic exposure Exposure for long periods of time to a *toxicant* occurring in low amounts. Compare *acute exposure.*
city planning The professional pursuit that attempts to design cities in such a way as to maximize their efficiency, functionality, and beauty.
classical economics Founded by *Adam Smith,* the study of the behavior of buyers and sellers in a free-market *economy.* Holds that individuals acting in their own self-interest may benefit society, provided that their behavior is constrained by the rule of law and by private property rights and operates within competitive markets. See also *neoclassical economics.*
clay *Sediment* consisting of particles less than 0.002 mm in diameter. Compare *sand; silt.*
Clean Air Act of 1970 Revision of prior Congressional *legislation* to control *air pollution* that set stricter standards for air quality, imposed limits on emissions from new stationary and mobile sources, provided new funds for *pollution*-control research, and enabled citizens to sue parties violating the standards.
Clean Air Act of 1990 Congressional *legislation* that strengthened *regulations* pertaining to air quality standards, auto emissions, toxic *air pollution, acidic deposition,* and depletion of the *ozone layer,* while also introducing market-based incentives to reduce *pollution.*
clear-cutting The harvesting of timber by cutting all the trees in an area, leaving only stumps. Although it is the most cost-efficient method, clear-cutting is also the most damaging to the *environment.*
climate The pattern of atmospheric conditions found across large geographic regions over long periods of time. Compare *weather.*
climate diagram (climatograph) A visual representation of a region's average monthly temperature and *precipitation.*
closed system A *system* that is isolated an self-contained. Scientists may treat a system as closed to simplify some question they are investigating, but no natural system is truly closed. Compare *open system.*
coal A *fossil fuel* composed of organic matter that was compressed under very high pressure to form a dense, solid carbon structure.
cogeneration A practice in which the extra heat generated in the production of electricity is captured and put to use heating workplaces and homes, as well as producing other kinds of power.
cold front The boundary where a mass of cold air displaces a mass of warmer air. Compare *warm front.*
command and control An approach to protecting the *environment* that sets strict legal limits and threatens punishment for violations of those limits.
commensalism A relationship between members of different *species* in which one organism benefits and the other is unaffected. Compare *amensalism.*
community A group of *populations* of organisms that live in the same place at the same time.
community-based conservation The practice of engaging local people to protect land and wildlife in their own region.
community ecology The study of the interactions among *species,* from one-to-one interactions to complex interrelationships involving entire *communities.*
competition A relationship in which multiple organisms seek the same limited resource.
composting The conversion of organic *waste* into mulch or humus by encouraging, in a controlled manner, the natural biological processes of decomposition.
compound A *molecule* whose *atoms* are composed of two or more *elements.*
confined (artesian) aquifer A water-bearing, porous layer of rock, *sand,* or gravel that is trapped between an upper and lower layer of less permeable substrate, such as *clay.* The water in a confined aquifer is under pressure because it is trapped between two impermeable layers. Compare *unconfined aquifer.*
conservation biology A scientific discipline devoted to understanding the factors, forces, and processes that influence the loss, protection, and restoration of *biological diversity* within and among *ecosystems.*
conservation district One of many county-based entities created by the Soil Conservation Service (now the Natural Resources Conservation Service) to promote practices that conserve *soil.*
conservation ethic An *ethic* holding that humans should put *natural resources* to use but also have a responsibility to manage them wisely. Compare *preservation ethic.*
consumer See *heterotroph.*
consumptive use *Freshwater* use in which water is removed from a particular *aquifer* or surface water body and is not returned to it. *Irrigation* for *agriculture* is an example of consumptive use. Compare *nonconsumptive use.*
continental shelf The gently sloping underwater edge of a continent, varying in width from 100 m (330 ft) to 1,300 km (800 mi), with an average slope of 1.9 m/km (10 ft/mi).
contingent valuation A technique that uses surveys to determine how much people would be willing to pay to protect a resource or to restore it after damage has been done.
contour farming The practice of plowing furrows sideways across a hillside, perpendicular to its slope, to help prevent the formation of rills and gullies. The technique is so named because the furrows follow the natural contours of the land.
control The portion of an *experiment* in which a *variable* has been left unmanipulated, to serve as a point of comparison with the *treatment.*
controlled burn See *prescribed burn.*
controlled experiment An *experiment* in which the effects of all *variables* are held constant, except the one whose effect

is being tested by comparison of *treatment* and *control* conditions.
convective circulation A circular *current* (of air, water, magma, etc.) driven by temperature differences. In the atmosphere, warm air rises into regions of lower *atmospheric pressure,* where it expands and cools and then descends and becomes denser, replacing warm air that is rising. The air picks up heat and moisture near ground level and prepares to rise again, continuing the process.
conventional law International law that arises from conventions, or treaties, that nations agree to enter into. Compare *customary law.*
Convention on Biological Diversity An international treaty that aims to conserve *biodiversity,* use biodiversity in a *sustainable* manner, and ensure the fair distribution of biodiversity's benefits. Although many nations have agreed to the treaty (as of 2005, 188 nations had become parties to it), several others, including the United States, have not.
Convention on International Trade in Endangered Species of Wild Fauna and Flora (CITES) A 1973 treaty facilitated by the *United Nations* that protects endangered *species* by banning the international transport of their body parts.
convergent plate boundary Area where tectonic plates collide. Can result in *subduction* or mountain range formation.
coral reef A mass of calcium carbonate composed of the skeletons of tiny colonial marine organisms called corals.
core The innermost part of the Earth, made up mostly of iron, that lies beneath the *crust* and *mantle.*
Coriolis effect The apparent deflection of north-south air *currents* to a partly east-west direction, caused by the faster spin of regions near the equator than of regions near the poles as a result of Earth's rotation.
correlation A relationship among *variables.*
corridor A passageway of protected land established to allow animals to travel between islands of protected *habitat.*
cost-benefit analysis A method commonly used by *neoclassical economists,* in which estimated costs for a proposed action are totaled and then compared to the sum of benefits estimated to result from the action.
covalent bond A chemical bond in which the uncharged *atoms* in a *molecule* share *electrons.* For example, the uncharged atoms of carbon and oxygen in carbon dioxide form a covalent bond. Compare *ionic bond.*
criteria pollutants Six *air pollutants*—carbon monoxide, sulfur dioxide, nitrogen dioxide, tropospheric ozone, particulate matter, and lead—for which the *Environmental Protection Agency* has established maximum allowable concentrations in ambient outdoor air because of the threats they pose to human health.
cropland Land that humans use to raise plants for food and fiber.
crop rotation The practice of alternating the kind of crop grown in a particular field from one season or year to the next.
crude oil (petroleum) A *fossil fuel* produced by the conversion of *organic compounds* by heat and pressure. Crude oil is a mixture of hundreds of different types of *hydrocarbon* molecules characterized by carbon chains of different length.
crust The lightweight outer layer of the Earth, consisting of rock that floats atop the malleable *mantle,* which in turn surrounds a mostly iron *core.*
culture The overall ensemble of knowledge, beliefs, values, and learned ways of life shared by a group of people.
current The flow of a liquid or gas in a certain direction.
customary law International law that arises from long-standing practices, or customs, held in common by most *cultures.* Compare *conventional law.*

dam Any obstruction placed in a river or stream to block the flow of water so that water can be stored in a reservoir. Dams are built to prevent floods, provide drinking water, facilitate *irrigation,* and generate electricity.
Darwin, Charles (1809–1882) English naturalist who proposed the concept of *natural selection* as a mechanism for *evolution* and as a way to explain the great variety of living things. See also *Wallace, Alfred Russell.*
data Information, generally quantitative information.
deep ecology A philosophy established in the 1970s based on principles of self-realization (the awareness that humans are inseparable from nature) and *biocentric* equality (the precept that all living beings have equal value). Holds that because we are truly inseparable from our *environment,* we must protect all other living things as we would protect ourselves.
deep-well injection A *hazardous waste* disposal method in which a well is drilled deep beneath an area's *water table* into porous rock below an impervious *soil* layer. Wastes are then injected into the well, so that they will be absorbed into the porous rock and remain deep underground, isolated from *groundwater* and human contact. Compare *surface impoundment.*
deforestation The clearing and loss of forests.
demographic transition A theoretical model of economic and cultural change that explains the declining death rates and birth rates that occurred in Western nations as they became industrialized. The model holds that industrialization caused these rates to fall naturally by decreasing mortality and by lessening the need for large families. Parents would thereafter choose to invest in quality of life rather than quantity of children.
demography A *social science* that applies the principles of *population ecology* to the study of statistical change in human *populations.*
denitrifying bacteria Bacteria that convert the nitrates in *soil* or water to gaseous nitrogen and release it back into the *atmosphere.*
density-dependent factor A *limiting factor* whose effects on a *population* increase or decrease depending on the *population density.* Compare *density-independent factor.*
density-independent factor A *limiting factor* whose effects on a *population* are constant regardless of *population density.* Compare *density-dependent factor.*
deoxyribonucleic acid See *DNA.*
dependent variable The *variable* that is affected by manipulation of the *independent variable.*
desalination The removal of salt from seawater.
desert The driest *biome* on Earth, with annual *precipitation* of less than 25 cm. Because deserts have relatively little vegetation

to insulate them from temperature extremes, sunlight readily heats them in the daytime, but daytime heat is quickly lost at night, so temperatures vary widely from day to night and in different seasons.

desertification A loss of more than 10% of a land's productivity due to *erosion, soil* compaction, forest removal, *overgrazing,* drought, *salinization, climate* change, depletion of water sources, or other factors. Severe desertification can result in the actual expansion of desert areas or creation of new ones in areas that once supported fertile land.

divergent plate boundary Area where *magma* surging upward to the surface divides tectonic plates and pushes them apart, creating new *crust* as it cools and spreads. A prime example is the Mid-Atlantic ridge. Compare *transform plate boundary* and *convergent plate boundary.*

DNA (deoxyribonucleic acid) A double-stranded *nucleic acid* composed of four nucleotides, each of which contains a sugar (deoxyribose), a phosphate group, and a nitrogenous base. DNA carries the hereditary information for living organisms and is responsible for passing traits from parents to offspring. Compare *RNA.*

dose-response curve A curve that plots the response of test animals to different doses of a *toxicant.* The response is generally quantified by measuring the proportion of animals exhibiting negative effects.

downwelling In the ocean, the flow of warm surface water toward the ocean floor. Downwelling occurs where surface *currents* converge. Compare *upwelling.*

Dust Bowl An area that loses huge amounts of *topsoil* to wind *erosion* as a result of drought and/or human impact; first used to name the region in the North American Great Plains severely affected by drought and topsoil loss in the 1930s. The term is now also used to describe that historical event and others like it.

dynamic equilibrium The state reached when processes within a *system* are moving in opposing directions at equivalent rates so that their effects balance out.

ecocentrism A philosophy that considers actions in terms of their damage or benefit to the integrity of whole ecological *systems,* including both *biotic* and *abiotic* elements. For an ecocentrist, the well-being of an individual organism—human or otherwise—is less important than the long-term well-being of a larger integrated ecological system.

ecofeminism A philosophy holding that the patriarchal (male-dominated) structure of society is a root cause of both social and environmental problems. Ecofeminists hold that a *worldview* traditionally associated with women, which interprets the world in terms of interrelationships and cooperation, is more in tune with nature than a worldview traditionally associated with men, which interprets the world in terms of hierarchies and competition.

ecolabeling The practice of designating on a product's label how the product was grown, harvested, or manufactured, so that consumers buying it are aware of the processes involved and can differentiate between brands that use processes believed to be *environmentally* beneficial (or less harmful than others) and those that do not.

ecological economics A developing school of *economics* that applies the principles of *ecology* and *systems* thinking to the description and analysis of *economies.* Compare *environmental economics; neoclassical economics.*

ecological footprint The cumulative amount of land and water required to provide the raw materials a person or *population* consumes and to dispose of or *recycle* the *waste* that is produced.

ecological restoration Efforts to reverse the effects of human disruption of *ecological systems* and to restore *communities* to their "natural" state.

ecology The *science* that deals with the distribution and abundance of organisms, the interactions among them, and the interactions between organisms and their *abiotic environments.*

economics The study of how we decide to use scarce resources to satisfy the demand for *goods* and *services.*

economy A social *system* that converts resources into *goods* and *services.*

ecosystem All organisms and nonliving entities that occur and interact in a particular area at the same time.

ecosystem ecology The study of how the living and nonliving components of *ecosystems* interact.

ecosystem-based management The attempt to manage the harvesting of resources in ways that minimize impact on the *ecosystems* and ecological processes that provide the resources.

ecosystem service An essential service an *ecosystem* provides that supports life and makes *economic* activity possible. For example, ecosystems naturally purify air and water, cycle *nutrients,* provide for plants to be *pollinated* by animals, and serve as receptacles and *recycling* systems for the *waste* generated by our economic activity.

ecotone A transitional zone where *ecosystems* meet.

ecotourism Visitation of natural areas for tourism and recreation. Most often involves tourism by more-affluent people, which may generate *economic* benefits for less-affluent communities near natural areas and thus provide economic incentives for conservation of natural areas.

ED_{50} (effective dose–50%) The amount of a *toxicant* it takes to affect 50% of a *population* of test animals. Compare *threshold dose;* LD_{50}.

E horizon The layer of *soil* that lies below the *A horizon* and above the *B horizon.* The letter "E" stands for "eluviation," meaning "loss," and the E horizon is characterized by the loss of certain minerals through *leaching.* It is sometimes called the "zone of leaching." Compare *C horizon; O horizon; R horizon.*

El Niño The exceptionally strong warming of the eastern Pacific Ocean that occurs every 2 to 7 years and depresses local fish and bird *populations* by altering the marine *food web* in the area. Originally, the name that Spanish-speaking fishermen gave to an unusually warm surface *current* that sometimes arrived near the Pacific coast of South America around Christmas time. Compare *La Niña.*

electrolysis A process in which electrical current is passed through a *compound* to release *ions.* Electrolysis offers one way to produce hydrogen for use as fuel: Electrical current is passed through water, splitting the water *molecules* into hydrogen and oxygen *atoms.*

electron A negatively charged particle that surrounds the nucleus of an *atom.*
element A fundamental type of matter; a chemical substance with a given set of properties, which cannot be broken down into substances with other properties. Chemists currently recognize 92 elements that occur in nature, as well as more than 20 others that have been artificially created.
emergent property A characteristic that is not evident in a *system*'s components.
Emerson, Ralph Waldo (1803–1882) American author, poet, and philosopher who espoused transcendentalism, a philosophy that views nature as a direct manifestation of the divine, and who promoted a holistic view of nature among the public.
emigration The departure of individuals from a *population.*
Endangered Species Act (ESA) The primary *legislation,* enacted in 1973, for protecting *biodiversity* in the United States. It forbids the government and private citizens from taking actions (such as developing land) that would destroy endangered *species* or their *habitats,* and it prohibits trade in products made from endangered species.
endemic Native or restricted to a particular geographic region. An endemic species occurs in one area and nowhere else on Earth.
endocrine disruptor A *toxicant* that interferes with the endocrine (hormone) system.
energy conservation The practice of reducing *energy* use as a way of extending the lifetime of our *fossil fuel* supplies, of being less wasteful, and of reducing our impact on the *environment.*
energy An intangible phenomenon that can change the position, physical composition, or temperature of matter.
entropy The degree of disorder in a substance, *system,* or process. See *second law of thermodynamics.*
environment The sum total of our surroundings, including all of the living things and nonliving things with which we interact.
environmental economics A developing school of *economics* that modifies the principles of *neoclassical economics* to address environmental challenges. An environmental economist believes that we can attain *sustainability* within our current economic *systems.* Whereas ecological economists call for revolution, environmental economists call for reform. Compare *ecological economics; neoclassical economics.*
environmental ethics The application of *ethical standards* to environmental questions.
environmental health Environmental factors that influence human health and quality of life and the health of *ecological* systems essential to environmental quality and long-term human well-being.
environmental impact statement (EIS) A report of results from detailed studies that assess the potential effects on the *environment* that would likely result from development projects or other actions undertaken by the government.
environmental justice A movement based on a moral sense of fairness and equality that seeks to expand society's domain of ethical concern from men to women, from humans to nonhumans, from rich to poor, and from majority races and ethnic groups to minority ones.
environmental policy *Public policy* that pertains to human interactions with the *environment.* It generally aims to regulate resource use or reduce *pollution* to promote human welfare and/or protect natural systems.
Environmental Protection Agency (EPA) An administrative agency created by executive order in 1970. The EPA is charged with conducting and evaluating research, monitoring environmental quality, setting standards, enforcing those standards, assisting the states in meeting standards and goals, and educating the public.
environmental science The study of how the natural world works and how humans and the *environment* interact.
environmental studies An academic *environmental science* program that heavily incorporates the social sciences as well as the natural sciences.
environmental toxicology The study of *toxicants* that come from or are discharged into the *environment,* including the study of health effects on humans, other animals, and *ecosystems.*
environmentalism A social movement dedicated to protecting the natural world.
epidemiological study A study that involves large-scale comparisons among groups of people, usually contrasting a group known to have been exposed to some *toxicant* and a group that has not.
equilibrium theory of island biogeography A *theory* that was initially applied to oceanic islands to explain how *species* come to be distributed among them. Since its development, researchers have increasingly applied the theory to islands of *habitat* (patches of one type of habitat isolated within vast "seas" of others). Aspects of the theory include *immigration* and *extinction* rates, the effect of island size, and the effect of distance from the mainland.
erosion The removal of material from one place and its transport to another by the action of wind or water.
estuary An area where a river flows into the ocean, mixing *freshwater* with salt water.
ethanol The alcohol in beer, wine, and liquor, produced as a *biofuel* by fermenting biomass, generally from *carbohydrate*-rich crops such as corn.
ethical standards The criteria that help differentiate right from wrong.
ethics The study of good and bad, right and wrong. The term can also refer to a person's or group's set of moral principles or values.
eukaryote A multicellular organism. The *cells* of eukaryotic organisms consist of a membrane-enclosed nucleus that houses *DNA,* an outer membrane of lipids, and an inner fluid-filled chamber containing *organelles.* Compare *prokaryote.*
European Union (EU) Political and economic organization formed after World War II to promote Europe's economic and social progress. As of 2005, the EU consisted of 25 member nations.
eutrophication The process of *nutrient* enrichment, increased production of organic matter, and subsequent *ecosystem* degradation.
evaporation The conversion of a substance from a liquid to a gaseous form.
even-aged Condition of timber plantations—generally *monocultures* of a single *species*—in which all trees are of the same

age. Most *ecologists* view plantations of even-aged stands more as crop *agriculture* than as ecologically functional forests. Compare *uneven-aged.*

evolution Genetically based change in the appearance, functioning, and/or behavior of organisms across generations, often by the process of *natural selection.*

executive branch The branch of the U.S. government that is headed by the president and that includes administrative agencies. Among other powers, the president may approve (enact) or reject (veto) legislation and issue executive orders. Compare *judicial branch; legislative branch.*

experiment An activity designed to test the validity of a *hypothesis* by manipulating *variables.* See *manipulative experiment* and *natural experiment.*

exponential growth The increase of a *population* (or of anything) by a fixed percentage each year.

external cost A negative *externality;* a cost borne by someone not involved in an economic transaction. Examples include harm to citizens from water *pollution* or *air pollution* discharged by nearby factories.

externality A cost or benefit of a transaction that affects people other than the buyer or seller.

extinction The disappearance of an entire *species* from the face of the Earth. Compare *extirpation.*

extirpation The disappearance of a particular *population* from a given area, but not the entire *species* globally. Compare *extinction.*

feedback loop A circular process in which a *system*'s output serves as input to that same system. See *negative feedback loop; positive feedback loop.*

feedlot A huge barn or outdoor pen designed to deliver *energy*-rich food to animals living at extremely high densities. Also called a factory farm or concentrated animal feeding operation (CAFO).

Ferrel cell One of a pair of cells of *convective circulation* between 30° and 60° north and south latitude that influence global *climate* patterns. Compare *Hadley cell; polar cell.*

fertilizer A substance that promotes plant growth by supplying essential *nutrients* such as nitrogen or phosphorus.

first law of thermodynamics Physical law stating that *energy* can change from one form to another but cannot be created or lost. The total energy in the universe remains constant and is said to be conserved.

floodplain The region of land over which a river has historically wandered and periodically floods.

food security An adequate, reliable, and available food supply to all people at all times.

food web A visual representation of feeding interactions within an *ecological community* that shows an array of relationships between organisms at different *trophic levels.*

forestry The professional management of forests.

fossil The remains, impression, or trace of an animal or plant of past geological ages that has been preserved in rock or *sediments.*

fossil fuel A *nonrenewable natural resource,* such as *crude oil, natural gas,* or *coal,* produced by the decomposition and compression of organic matter from ancient life.

fossil record The cumulative body of *fossils* worldwide, which paleontologists study to infer the history of past life on Earth.

free rider A party that fails to invest in controlling *pollution* or carrying out other *environmentally* responsible activities and instead relies on the efforts of other parties to do so. For example, a factory that fails to control its emissions gets a "free ride" on the efforts of other factories that do make the sacrifices necessary to reduce emissions.

freshwater Water that is relatively pure, holding very few dissolved salts.

fundamental niche The full *niche* of a *species.* Compare *realized niche.*

gene A stretch of *DNA* that represents a unit of hereditary information.

gene bank See *seed bank.*

generalist A *species* that can survive in a wide array of *habitats* or use a wide array of resources. Compare *specialist.*

genetically modified (GM) organism An organism that has been *genetically engineered* using a technique called *recombinant DNA* technology.

genetic diversity A measurement of the differences in *DNA* composition among individuals within a given *species.*

genetic engineering Any process scientists use to manipulate an organism's genetic material in the lab by adding, deleting, or changing segments of its *DNA.*

Genuine Progress Indicator (GPI) An *economic* indicator introduced in 1995 that attempts to differentiate between desirable and undesirable economic activity. The GPI accounts for benefits such as volunteerism and for costs such as *environmental* degradation and social upheaval. Compare *Gross Domestic Product (GDP).*

geothermal energy Renewable *energy* that is generated deep within Earth. The radioactive decay of elements amid the extremely high pressures and temperatures at depth generate heat that rises to the surface in magma and through fissures and cracks. Where this energy heats *groundwater,* natural eruptions of heated water and steam are sent up from below.

global climate change Any change in aspects of Earth's *climate,* such as temperature, *precipitation,* and storm intensity. Generally refers today to the current warming trend in global temperatures and associated climatic changes.

good A material commodity manufactured for and bought by individuals and businesses.

greenhouse effect The warming of Earth's surface and *atmosphere* (especially the *troposphere*) caused by the *energy* emitted by *greenhouse gases.*

greenhouse gas A gas that absorbs infrared radiation released by Earth's surface and then warms the surface and *troposphere* by emitting *energy,* thus giving rise to the *greenhouse effect.* Greenhouse gases include carbon dioxide (CO_2), water vapor, ozone (O_3), nitrous oxide (N_2O), halocarbon gases, and methane (CH_4).

green revolution An intensification of the industrialization of *agriculture* in the developing world in the latter half of the 20th century that has dramatically increased crop yields produced per unit area of farmland. Practices include devoting large areas to *monocultures* of crops specially bred for high yields

and rapid growth; heavy use of *fertilizers, pesticides,* and *irrigation* water; and sowing and harvesting on the same piece of land more than once per year or per season.
green tax A levy on *environmentally* harmful activities and products aimed at providing a market-based incentive to correct for *market failure.* Compare *subsidy.*
Gross Domestic Product (GDP) The total monetary value of final *goods* and *services* produced in a country each year. The GDP sums all *economic* activity, whether good or bad, and does not account for benefits such as volunteerism or for *external costs* such as *environmental* degradation and social upheaval. Compare *Genuine Progress Indicator (GPI).*
gross primary production The *energy* that results when *autotrophs* convert solar energy (sunlight) to energy of chemical bonds in sugars through *photosynthesis.* Autotrophs use a portion of this production to power their own metabolism, which entails oxidizing *organic compounds* by *cellular respiration.* Compare *net primary production; secondary production.*
groundwater Water held in aquifers underground.
growth rate The net change in a *population*'s size, per 1,000 individuals. Calculated by adding the crude birth rate to the *immigration* rate and then subtracting the crude death rate and the *emigration* rate, each expressed as the number per 1,000 individuals per year.

Haber-Bosch process A process to synthesize ammonia on an industrial scale. Developed by German chemists Fritz Haber and Carl Bosch, the process has enabled humans to double the natural rate of *nitrogen fixation* on Earth and thereby increase *agricultural* productivity, but it has also dramatically altered the *nitrogen cycle.*
habitat The specific *environment* in which an organism lives, including both *biotic* and *abiotic factors.*
habitat use The process by which organisms select and use *habitats* from among the range of options they encounter.
Hadley cell One of a pair of cells of *convective circulation* between the equator and 30° north and south latitude that influence global *climate* patterns. Compare *Ferrel cell; polar cell.*
half-life The amount of time it takes for one-half the atoms of a *radioisotope* to emit radiation and decay. Different radioisotopes have different half-lives, ranging from fractions of a second to billions of years.
harmful algal bloom A *population* explosion of toxic algae caused by excessive *nutrient* concentrations.
hazardous waste *Waste* that is toxic, chemically reactive, flammable, or corrosive. Compare *industrial solid waste; municipal solid waste.*
herbivory The consumption of plants by animals.
heterotroph (consumer) An organism that consumes other organisms. Includes most animals, as well as fungi and microbes that decompose organic matter.
high-pressure system An air mass with elevated *atmospheric pressure,* containing air that descends, typically bringing fair *weather.* Compare *low-pressure system.*
homeostasis The tendency of a *system* to maintain constant or stable internal conditions.
horizon A distinct layer of *soil.* See *A horizon; B horizon; C horizon; E horizon; O horizon; R horizon.*
Hubbert's peak The peak in production of *crude oil* in the United States, which occurred in 1970 just as Shell Oil geologist M. King Hubbert had predicted in 1956.
hydrocarbon An *organic compound* consisting solely of hydrogen and carbon *atoms.*
hydroelectric power (hydropower) The generation of electricity using the *kinetic energy* of moving water.
hydrologic cycle The flow of water—in liquid, gaseous, and solid forms—through our *biotic* and *abiotic environment.*
hydropower See *hydroelectric power.*
hydrosphere All water—salt or fresh, liquid, ice, or vapor—in surface bodies, underground, and in the *atmosphere.* Compare *biosphere; lithosphere.*
hypothesis An educated guess that explains a phenomenon or answers a *scientific* question. Compare *theory.*
hypoxia The condition of extremely low dissolved oxygen concentrations in a body of water.

igneous rock One of the three main categories of rock. Formed from cooling *magma.* Granite and basalt are examples of igneous rock. Compare *metamorphic rock; sedimentary rock.*
immigration The arrival of individuals from outside a *population.*
incineration A controlled process of burning solid waste for disposal in which mixed garbage is combusted at very high temperatures. Compare *sanitary landfill.*
independent variable The *variable* that the scientist manipulates in a *manipulative experiment.*
industrial ecology A holistic approach to industry that integrates principles from engineering, chemistry, *ecology, economics,* and other disciplines and seeks to redesign industrial *systems* in order to reduce resource inputs and minimize inefficiency.
industrialized agriculture A form of *agriculture* that uses large-scale mechanization and *fossil fuel* combustion, enabling farmers to replace horses and oxen with faster and more powerful means of cultivating, harvesting, transporting, and processing crops. Other aspects include *irrigation* and the use of *inorganic fertilizers.* Use of chemical herbicides and *pesticides* reduces *competition* from weeds and *herbivory* by insects. Compare *traditional agriculture.*
industrial revolution The shift in the mid-1700s from rural life, animal-powered agriculture, and manufacturing by craftsmen to an urban society powered by *fossil fuels* such as *coal* and *crude oil.* Compare *agricultural revolution.*
industrial smog Gray-air smog caused by the incomplete combustion of *coal* or oil when burned. Compare *photochemical smog.*
industrial solid waste Nonliquid *waste* that is not especially hazardous and that comes from production of consumer goods, mining, *petroleum* extraction and *refining,* and *agriculture.* Compare *hazardous waste; municipal solid waste.*
industrial stage The third stage of the *demographic transition* model, characterized by falling birth rates that close the gap with falling death rates and reduce the rate of *population*

growth. Compare *pre-industrial stage; post-industrial stage; transitional stage.*
infectious disease A disease in which a pathogen attacks a host.
inorganic fertilizer A *fertilizer* that consists of mined or synthetically manufactured mineral supplements. Inorganic fertilizers are generally more susceptible than *organic fertilizers* to *leaching* and *runoff* and may be more likely to cause unintended off-site impacts.
integrated pest management (IPM) The use of multiple techniques in combination to achieve long-term suppression of pests, including *biocontrol,* use of *pesticides,* close monitoring of *populations, habitat* alteration, *crop rotation, transgenic* crops, alternative tillage methods, and mechanical pest removal.
intercropping Planting different types of crops in alternating bands or other spatially mixed arrangements.
interdisciplinary field A field that borrows techniques from several more traditional fields of study and brings together research results from these fields into a broad synthesis.
Intergovernmental Panel on Climate Change (IPCC) An international panel of *atmospheric* scientists, *climate* experts, and government officials established in 1988 by the *United Nations* Environment Programme and the World Meteorological Organization, whose mission is to assess information relevant to questions of human-induced *global climate change.* The IPCC's 2001 *Third Assessment Report* summarizes current and probable future global trends and represents the consensus of atmospheric scientists around the world.
intertidal Of, relating to, or living along shorelines between the highest reach of the highest *tide* and the lowest reach of the lowest tide.
invasive species A *species* that spreads widely and rapidly becomes dominant in a *community,* interfering with the community's normal functioning.
inversion layer In a *temperature inversion,* the band of air in which temperature rises with altitude (instead of falling with altitude, as temperature does normally).
ion An electrically charged *atom* or combination of atoms.
ionic bond A chemical bond in which oppositely charged *ions* are held together by electrical attraction. Compare *covalent bond.*
ionic compound (salt) An association of *ions* that are bonded electrically in an *ionic bond.*
IPAT model A formula that represents how humans' total impact (I) on the *environment* results from the interaction among three factors: *population* (P), affluence (A), and technology (T).
irrigation The artificial provision of water to support *agriculture.*
isotope One of several forms of an *element* having differing numbers of *neutrons* in the nucleus of its *atoms.* Chemically, isotopes of an element behave almost identically, but they have different physical properties because they differ in mass.

judicial branch The branch of the U.S. government, consisting of the Supreme Court and various lower courts, that is charged with interpreting the law. Compare *executive branch; legislative branch.*

kelp Large brown algae or seaweed that can form underwater "forests," providing habitat for marine organisms.
keystone species A *species* that has an especially far-reaching effect on a *community.*
kinetic energy *Energy* of motion. Compare *potential energy.*
K–selected Term denoting a *species* with low biotic potential whose members produce a small number of offspring and take a long time to gestate and raise each of their young, but invest heavily in promoting the survival and growth of these few offspring. *Populations* of K–selected species are generally regulated by *density-dependent factors.* Compare *r–selected.*
Kyoto Protocol An agreement drafted in 1997 that calls for reducing, by 2012, emissions of six *greenhouse gases* to levels lower than their levels in 1990. Although the United States has refused to ratify the protocol, it came into force in 2005 when Russia ratified it, the 127th nation to do so.

La Niña An exceptionally strong cooling of surface water in the equatorial Pacific Ocean that occurs every 2 to 7 years and has widespread climatic consequences. Compare *El Niño.*
land trust Local or regional organization that preserves lands valued by its members. In most cases, land trusts purchase land outright with the aim of preserving it in its natural condition. The Nature Conservancy may be considered the world's largest land trust.
landscape ecology An approach to the study of organisms and their *environments* at the landscape scale, focusing on geographical areas that include multiple *ecosystems.*
lava *Magma* that is released from the *lithosphere* and flows or spatters across Earth's surface.
LD_{50} (lethal dose–50%) The amount of a *toxicant* it takes to kill 50% of a *population* of test animals. Compare *ED_{50}; threshold dose.*
leachate Liquids that seep through liners of a *sanitary landfill* and leach into the *soil* underneath.
leaching The process by which solid materials such as minerals are dissolved in a liquid (usually water) and transported to another location.
legislation Statutory law.
legislative branch The branch of the U.S. government that passes laws; it consists of Congress, which includes the House of Representatives and the Senate. Compare *executive branch; judicial branch.*
Leopold, Aldo (1887–1949) American scientist, scholar, philosopher, and author. His book *The Land Ethic* argued that humans should view themselves and the land itself as members of the same *community* and that humans are obligated to treat the land *ethically.*
life-cycle analysis In *industrial ecology,* the examination of the entire life cycle of a given product—from its origins in raw materials, through its manufacturing, to its use, and finally its disposal—in an attempt to identify ways to make the process more *ecologically* efficient.
life expectancy The average number of years that individuals in particular age groups are likely to continue to live.
limiting factor A physical, chemical, or biological characteristic of the *environment* that restrains *population* growth.

limnetic zone In a water body, the layer of open water through which sunlight penetrates. Compare *littoral zone; benthic zone; profundal zone.*
lipid One of a chemically diverse group of *macromolecules* that are classified together because they do not dissolve in water. Lipids include fats, phospholipids, waxes, pigments, and steroids.
lithification The formation of rock through the processes of compaction, binding, and crystallization.
lithosphere The solid part of the Earth, including the rocks, *sediment,* and *soil* at the surface and extending down many miles underground. Compare *atmosphere; biosphere; hydrosphere.*
littoral See *intertidal.*
littoral zone The region ringing the edge of a water body. Compare *benthic zone; limnetic zone; profundal zone.*
loam *Soil* with a relatively even mixture of *clay-, silt-,* and *sand-*sized particles.
lobbying The expenditure of time or money in an attempt to influence an elected official.
logistic growth curve A plot that shows how the initial *exponential growth* of a *population* is slowed and finally brought to a standstill by *limiting factors.*
low-pressure system An air mass in which the air moves toward the low *atmospheric pressure* at the center of the system and spirals upward, typically bringing clouds and *precipitation.* Compare *high-pressure system.*

macromolecule A very large molecule, such as a *protein, nucleic acid, carbohydrate,* or *lipid.*
macronutrient A *nutrient* that organisms require in relatively large amounts. Compare *micronutrient.*
magma Molten, liquid rock.
malnutrition The condition of lacking *nutrients* the body needs, including a complete complement of vitamins and minerals.
Malthus, Thomas (1766–1834) British economist who maintained that increasing human *population* would eventually deplete the available food supply until starvation, war, or disease arose and reduced the population.
mangrove A tree with a unique type of roots that curve upward to obtain oxygen, which is lacking in the mud in which they grow, and that serve as stilts to support the tree in changing water levels. Mangrove forests grow on the coastlines of the tropics and subtropics.
manipulative experiment An *experiment* in which the researcher actively chooses and manipulates the *independent variable.* Compare *natural experiment.*
mantle The malleable layer of rock that lies beneath Earth's *crust* and surrounds a mostly iron *core.*
marine protected area (MPA) An area of the ocean set aside to protect marine life from fishing pressures. An MPA may be protected from some human activities but be open to others. Compare *marine reserve.*
marine reserve An area of the ocean designated as a "no-fishing" zone, allowing no extractive activities. Compare *marine protected area.*
marketable emissions permit A permit issued to polluters that allows them to emit a certain fraction of the total amount of *pollution* the government will allow an entire industry to produce. Polluters are then allowed to buy, sell, and trade these permits with other polluters. See also *permit-trading.*
market failure The failure of markets to take into account the *environment*'s positive effects on *economies* (for example, *ecosystem services*) or to reflect the negative effects of economic activity on the environment and thereby on people (*external costs*).
mass extinction event The extinction of a large proportion of the world's *species* in a very short time period due to some extreme and rapid change or catastrophic event. Earth has seen five mass extinction events in the past half-billion years.
materials recovery facility (MRF) A *recycling* facility where items are sorted, cleaned, shredded, and prepared for reprocessing into new items.
maximum sustainable yield The maximal harvest of a particular *renewable natural resource* that can be accomplished while still keeping the resource available for the future.
meltdown The accidental melting of the uranium fuel rods inside the core of a *nuclear reactor,* causing the release of radiation.
metamorphic rock One of the three main categories of rock. Formed by great heat and/or pressure that reshapes crystals within the rock and changes its appearance and physical properties. Common metamorphic rocks include marble and slate. Compare *igneous rock; sedimentary rock.*
methane hydrate An ice-like solid consisting of molecules of methane (CH_4) embedded in a crystal lattice of water molecules. Methane hydrates are being investigated as a potential new source of *energy* from *fossil fuels.*
micronutrient A *nutrient* that organisms require in relatively small amounts. Compare *macronutrient.*
Milankovitch cycle One of three types of variations in Earth's rotation and orbit around the sun that result in slight changes in the relative amount of solar radiation reaching Earth's surface at different latitudes. As the cycles proceed, they change the way solar radiation is distributed over Earth's surface and contribute to changes in *atmospheric* heating and circulation that have triggered the ice ages and other *climate* changes.
Mill, John Stuart (1806–1873) British philosopher who believed that as resources become harder to find and extract, *economic* growth will slow and eventually stabilize into a *steady-state economy.*
molecule A combination of two or more *atoms.*
monoculture The uniform planting of a single crop over a large area. Characterizes *industrialized agriculture.*
Montreal Protocol International treaty ratified in 1987 in which 180 signatory nations agreed to restrict production of *chlorofluorocarbons (CFCs)* in order to forestall stratospheric ozone depletion. Because of its effectiveness in decreasing global CFC emissions, the Montreal Protocol is considered the most successful effort to date in addressing a global *environmental* problem.
Muir, John (1838–1914) Scottish immigrant to the United States who eventually settled in California and made the

Yosemite Valley his wilderness home. Today, he is most strongly associated with the *preservation ethic.* He argued that nature deserved protection for its own inherent values (an *ecocentrist* argument) but also claimed that nature played a large role in human happiness and fulfillment (an *anthropocentrist* argument).

multiple use A principle that has nominally guided management policy for national forests over the past half century. The multiple use principle specifies that the forests be managed for recreation, wildlife habitat, mineral extraction, and various other uses.

municipal solid waste Nonliquid *waste* that is not especially hazardous and that comes from homes, institutions, and small businesses. Compare *hazardous waste; industrial solid waste.*

mutagen A *toxicant* that causes *mutations* in the *DNA* of organisms.

mutation An accidental change in *DNA* that may range in magnitude from the deletion, substitution, or addition of a single nucleotide to a change affecting entire sets of chromosomes. Mutations provide the raw material for evolutionary change.

mutualism A relationship in which all participating organisms benefit from their interaction. Compare *parasitism.*

National Environmental Policy Act (NEPA) A U.S. law enacted on January 1, 1970, that created an agency called the Council on Environmental Quality and required that an *environmental impact statement* be prepared for any major federal action.

national forest Public lands consisting of 191 million acres (more than 8% of the nation's land area) in many tracts spread across all but a few states.

National Forest Management Act *Legislation* passed by the U.S. Congress in 1976, mandating that plans for renewable resource management be drawn up for every national forest. These plans were to be explicitly based on the concepts of *multiple use* and *sustainable development* and be subject to broad public participation.

national park A scenic area set aside for recreation and enjoyment by the public. The national park system today numbers 388 sites totaling 78.8 million acres and includes national historic sites, national recreation areas, national wild and scenic rivers, and other types of areas.

national wildlife refuge An area set aside to serve as a haven for wildlife and also sometimes to encourage hunting, fishing, wildlife observation, photography, environmental education, and other public uses.

natural experiment An *experiment* in which the researcher cannot directly manipulate the *variables* and therefore must observe nature, comparing conditions in which variables differ, and interpret the results. Compare *manipulative experiment.*

natural gas A *fossil fuel* composed primarily of methane (CH_4), produced as a by-product when bacteria decompose organic material under *anaerobic* conditions.

natural rate of population change The rate of change in a *population's* size resulting from birth and death rates alone, excluding migration.

natural resource Any of the various substances and *energy* sources we need in order to survive.

natural science An academic discipline that studies the natural world. Compare *social science.*

natural selection The process by which traits that enhance survival and reproduction are passed on more frequently to future generations of organisms than those that do not, thus altering the *genetic* makeup of populations through time. Natural selection acts on genetic variation and is a primary driver of evolution.

negative feedback loop A *feedback loop* in which output of one type acts as input that moves the *system* in the opposite direction. The input and output essentially neutralize each other's effects, stabilizing the system. Compare *positive feedback loop.*

neoclassical economics A *theory* of *economics* that explains market prices in terms of consumer preferences for units of particular commodities. Buyers desire the lowest possible price, whereas sellers desire the highest possible price. This conflict between buyers and sellers results in a compromise price being reached and the "right" quantity of commodities being bought and sold. Compare *ecological economics; environmental economics.*

net primary production The *energy* or biomass that remains in an ecosystem after *autotrophs* have metabolized enough for their own maintenance through *cellular respiration.* Net primary production is the energy or biomass available for consumption by *heterotrophs.* Compare *gross primary production; secondary production.*

net primary productivity The rate at which *net primary production* is produced. See *productivity; gross primary production; net primary production; secondary production.*

neurotoxin A *toxicant* that assaults the nervous system. Neurotoxins include heavy metals, *pesticides,* and some chemical weapons developed for use in war.

neutron An electrically neutral (uncharged) particle in the nucleus of an *atom.*

new forestry A set of *ecosystem-based management* approaches for harvesting timber that explicitly mimic natural disturbances. For instance, "sloppy clear-cuts" that leave a variety of trees standing mimic the changes a forest might experience if hit by a severe windstorm.

new urbanism A school of thought among architects, planners, and developers that seeks to design neighborhoods in which homes, businesses, schools, and other amenities are within walking distance of one another. In a direct rebuttal to *sprawl,* proponents of new urbanism aim to create functional neighborhoods in which families can meet most of their needs close to home without the use of a car.

niche The functional role of a *species* in a *community.* See *fundamental niche; realized niche.*

nitrification The conversion by bacteria of ammonium ions (NH_4^+) first into nitrite ions (NO_2^-) and then into nitrate ions (NO_3^-).

nitrogen cycle A major *nutrient cycle* consisting of the routes that nitrogen *atoms* take through the nested networks of environmental *systems.*

nitrogen fixation The process by which inert nitrogen gas combines with hydrogen to form ammonium ions (NH_4^+),

which are chemically and biologically active and can be taken up by plants.
nitrogen-fixing Term describing bacteria that live in a *mutualistic* relationship with many types of plants and provide *nutrients* to the plants by converting nitrogen to a usable form.
nonconsumptive use *Freshwater* use in which the water from a particular *aquifer* or surface water body either is not removed or is removed only temporarily and then returned. The use of water to generate electricity in hydroelectric *dams* is an example. Compare *consumptive use.*
nonmarket value A value that is not usually included in the price of a *good* or *service.*
non-point source A diffuse source of *pollutants,* often consisting of many small sources. Compare *point source.*
nonrenewable natural resource A *natural resource* that is in limited supply and is formed much more slowly than we use it. Compare *renewable natural resource.*
Northwest Forest Plan A 1994 plan developed by the Clinton administration to allow logging of forests of western Washington, Oregon, and northwestern California with increased protection for *species* and *ecosystems.* The Northwest Forest Plan represented one of the first large-scale applications of *adaptive management.*
nuclear energy The *energy* that holds together *protons* and *neutrons* within the nucleus of an *atom.* Several processes, each of which involves transforming *isotopes* of one *element* into isotopes of other elements, can convert nuclear energy into thermal energy, which is then used to generate electricity. See also *nuclear fission; nuclear reactor.*
nuclear fission The conversion of the *energy* within an *atom*'s nucleus to usable thermal energy by splitting apart atomic nuclei. Compare *nuclear fusion.*
nuclear fusion The conversion of the *energy* within an *atom*'s nucleus to usable thermal energy by forcing together the small nuclei of lightweight *elements* under extremely high temperature and pressure. Developing a commercially viable method of nuclear fusion remains an elusive goal.
nuclear reactor A facility within a nuclear power plant that initiates and controls the process of *nuclear fission* in order to generate electricity.
nucleic acid A *macromolecule* that directs the production of *proteins.* Includes *DNA* and *RNA.*
nutrient An *element* or *compound* that organisms consume and require for survival.
nutrient cycle The comprehensive set of cyclical pathways by which a given *nutrient* moves through the *environment.*

oceanography The study of the physics, chemistry, biology, and geology of the oceans.
ocean thermal energy conversion (OTEC) A potential *energy* source that involves harnessing the solar radiation absorbed by tropical oceans in the tropics.
O horizon The top layer of *soil* in some *soil profiles,* made up of organic matter, such as decomposing branches, leaves, crop residue, and animal waste. Compare *A horizon; B horizon; C horizon; E horizon; R horizon.*
open system A *system* that exchanges *energy,* matter, and information with other systems. Compare *closed system.*
organelle A structure, such as a ribosome or mitochondrion, inside the *cell* that performs specific functions.
organic agriculture *Agriculture* that uses no synthetic *fertilizers* or *pesticides* but instead relies on biological approaches such as *composting* and *biocontrol.*
organic compound A *compound* made up of carbon *atoms* (and, generally, hydrogen atoms) joined by *covalent bonds* and sometimes including other *elements,* such as nitrogen, oxygen, sulfur, or phosphorus. The unusual ability of carbon to build elaborate molecules has resulted in millions of different organic compounds showing various degrees of complexity.
organic fertilizer A *fertilizer* made up of natural materials (largely the remains or wastes of organisms), including animal manure, crop residues, fresh vegetation, and compost. Compare *inorganic fertilizer.*
overgrazing The consumption by too many animals of plant cover, impeding plant regrowth and the replacement of biomass. Overgrazing can exacerbate damage to *soils,* natural *communities,* and the land's productivity for further grazing.
overnutrition A condition of excessive food intake in which people receive more than their daily caloric needs.
ozone layer A portion of the *stratosphere,* roughly 17–30 km (10–19 mi) above sea level, that contains most of the ozone in the *atmosphere.*

paradigm A dominant philosophical and theoretical framework within a scientific discipline.
parasitism A relationship in which one organism, the parasite, depends on another, the host, for nourishment or some other benefit while simultaneously doing the host harm. Compare *mutualism.*
parent material The base geological material in a particular location.
passive solar energy collection An approach in which buildings are designed and building materials are chosen to maximize their direct absorption of sunlight in winter, even as they keep the interior cool in the summer. Compare *active solar energy collection.*
peat A kind of precursor stage to *coal,* produced when organic material that is broken down by *anaerobic* decomposition remains wet, near the surface, and not well compressed.
peer review The process by which a manuscript submitted for publication in an academic journal is examined by other specialists in the field, who provide comments and criticism (generally anonymously), and judge whether the work merits publication in the journal.
pelagic Of, relating to, or living between the surface and floor of the ocean. Compare *benthic.*
permit-trading The practice of buying and selling government-issued *marketable emissions permits* to conduct environmentally harmful activities. Under such a system, the government determines an acceptable level of *pollution* and then issues permits to pollute. A company receives credit for amounts it does not emit and can then sell this credit to other companies.

pesticide An artificial chemical used to kill insects (insecticide), plants (herbicide), or fungi (fungicide).
petroleum See *crude oil.*
pH A measure of the concentration of hydrogen *ions* in a solution. The pH scale ranges from 0 to 14: A solution with a pH of 7 is neutral; solutions with a pH below 7 are *acidic,* and those with a pH higher than 7 are *basic.* Because the pH scale is logarithmic, each step on the scale represents a tenfold difference in hydrogen ion concentration.
phosphorus cycle A major *nutrient cycle* consisting of the routes that phosphorus *atoms* take through the nested networks of environmental *systems.*
photochemical smog Brown-air smog caused by light-driven reactions of *primary pollutants* with normal atmospheric *compounds* that produce a mix of over 100 different chemicals, ground-level ozone often being the most abundant among them. Compare *industrial smog.*
photosynthesis The process by which *autotrophs* produce their own food. Sunlight powers a series of chemical reactions that convert carbon dioxide and water into sugar (glucose), thus transforming low-quality *energy* from the sun into high-quality energy the organism can use. Compare *cellular respiration.*
photovoltaic (PV) cell A device designed to collect sunlight and convert it to electrical *energy* directly by making use of the *photoelectric effect.*
phylogenetic tree A treelike diagram that represents the history of divergence of *species* or other taxonomic groups of organisms.
Pinchot, Gifford (1865–1946) The first professionally trained American *forester,* Pinchot helped establish the U.S. Forest Service. Today, he is the person most closely associated with the *conservation ethic.*
pioneer species A *species* that arrives earliest, beginning the ecological process of *succession* in a terrestrial or aquatic *community.*
plate tectonics The process by which Earth's surface is shaped by the extremely slow movement of tectonic plates, or sections of *crust.* Earth's surface includes about 15 major tectonic plates. Their interaction gives rise to processes that build mountains, cause earthquakes, and otherwise influence the landscape.
point source A specific spot—such as a factory's smokestacks—where large quantities of *pollutants* are discharged. Compare *non-point source.*
polar cell One of a pair of cells of *convective circulation* between the poles and 60° north and south latitude that influence global *climate* patterns. Compare *Ferrel cell; Hadley cell.*
policy A rule or guideline that directs individual, organizational, or societal behavior.
pollination An interaction in which one organism (for example, bees) transfers pollen (male sex cells) from one flower to the ova (female cells) of another, fertilizing the female flower, which subsequently grows into a fruit.
pollution Any matter or *energy* released into the *environment* that causes undesirable impacts on the health and well-being of humans or other organisms. Pollution can be physical, chemical, or biological, and can affect water, air, or soil.
polymer A chemical *compound* or mixture of compounds consisting of long chains of repeated *molecules.* Some polymers play key roles in the building blocks of life.
population A group of organisms of the same *species* that live in the same area. Species are often composed of multiple populations.
population density The number of individuals within a *population* per unit area. Compare *population size.*
population dispersion See *population distribution.*
population distribution The spatial arrangement of organisms within a particular area.
population ecology Study of the quantitative dynamics of how individuals within a *species* interact with one another—in particular, why *populations* of some species decline while others increase.
population size The number of individual organisms present at a given time.
positive feedback loop A *feedback loop* in which output of one type acts as input that moves the *system* in the same direction. The input and output drive the system further toward one extreme or another. Compare *negative feedback loop.*
post-industrial stage The fourth and final stage of the *demographic transition* model, in which both birth and death rates have fallen to a low level and remain stable there, and *populations* may even decline slightly. Compare *industrial stage; pre-industrial stage; transition stage.*
potential energy *Energy* of position. Compare *kinetic energy.*
precautionary principle The idea that one should not undertake a new action until the ramifications of that action are well understood.
precipitation Water that condenses out of the *atmosphere* and falls to Earth in droplets or crystals.
predation The process in which one *species* (the predator) hunts, tracks, captures, and ultimately kills its prey.
prediction A specific statement, generally arising from a *hypothesis,* that can be tested directly and unequivocally.
pre-industrial stage The first stage of the *demographic transition* model, characterized by conditions that defined most of human history. In pre-industrial societies, both death rates and birth rates are high. Compare *industrial stage; post-industrial stage; transitional stage.*
prescribed (controlled) burns The practice of burning areas of forest or grassland under carefully controlled conditions to improve the health of *ecosystems,* return them to a more natural state, and help prevent uncontrolled catastrophic fires.
preservation ethic An ethic holding that we should protect the natural *environment* in a pristine, unaltered state. Compare *conservation ethic.*
primary extraction The initial drilling and pumping of available *crude oil.* Compare *secondary extraction.*
primary pollutant A hazardous substance, such as soot or carbon monoxide, that is emitted into the *troposphere* in a form that is directly harmful. Compare *secondary pollutant.*
primary succession A stereotypical series of changes as an *ecological community* develops over time, beginning with a lifeless substrate. In terrestrial *systems,* primary succession begins when a bare expanse of rock, *sand,* or *sediment* becomes newly

exposed to the atmosphere and *pioneer species* arrive. Compare *secondary succession.*
primary treatment A stage of *wastewater* treatment in which contaminants are physically removed. Wastewater flows into tanks in which sewage solids, grit, and particulate matter settle to the bottom. Greases and oils float to the surface and can be skimmed off. Compare *secondary treatment.*
producer See *autotroph.*
productivity The rate at which plants convert solar *energy* (sunlight) to biomass. *Ecosystems* whose plants convert solar energy to biomass rapidly are said to have high productivity. See *net primary productivity; gross primary production; net primary production.*
profundal zone In a water body, the volume of open water that sunlight does not reach. Compare *littoral zone; benthic zone; limnetic zone.*
prokaryote A typically unicellular organism. The *cells* of prokaryotic organisms lack *organelles* and a nucleus. All bacteria and archaeans are prokaryotes. Compare *eukaryote.*
protein A *macromolecule* made up of long chains of amino acids.
proton A positively charged particle in the nucleus of an *atom.*
proven recoverable reserve The amount of a given *fossil fuel* in a deposit that is technologically and economically feasible to remove under current conditions.
proxy indicator Indirect evidence, such as pollen from *sediment* cores and air bubbles from ice cores, of the *climate* of the past.
public policy *Policy* that is made by governments, including those at the local, state, federal, and international levels; it consists of *legislation, regulations,* orders, incentives, and practices intended to advance societal welfare. See also *environmental policy.*

radioisotopes Radioactive *isotopes* that emit subatomic particles and high-*energy* radiation as they "decay" into progressively lighter isotopes until becoming stable isotopes.
rangeland Land used for grazing livestock.
realized niche The portion of the *fundamental niche* that is fully realized (used) by a *species.*
recombinant DNA *DNA* that has been patched together from the DNA of multiple organisms in an attempt to produce desirable traits (such as rapid growth, disease and pest resistance, or higher nutritional content) in organisms lacking those traits.
recycling The collection of materials that can be broken down and reprocessed to manufacture new items.
Red List An updated list of *species* facing unusually high risks of *extinction.* The list is maintained by the World Conservation Union.
red tide A *harmful algal bloom* consisting of algae that produce reddish pigments that discolor surface waters.
refining Process of separating the *molecules* of the various *hydrocarbons* in *crude oil* into different-sized classes and transforming them into various fuels and other petrochemical products.
regional planning *City planning* done on broader geographic scales, generally involving multiple municipal governments.
regulation A specific rule issued by an administrative agency, based on the more broadly written statutory law passed by Congress and enacted by the president.
regulatory taking The deprivation of a property's owner, by means of a law or *regulation,* of most or all economic uses of that property.
relative humidity The ratio of the water vapor contained in a given volume of air to the maximum amount the air could contain, for a given temperature.
relativist An ethicist who maintains that *ethics* do and should vary with social context. Compare *universalist.*
renewable natural resource A *natural resource* that is virtually unlimited or that is replenished by the *environment* over relatively short periods of hours to weeks to years. Compare *nonrenewable natural resource.*
replacement fertility The *total fertility rate (TFR)* that maintains a stable *population* size.
reserves-to-production ratio (R/P ratio) The total remaining reserves of a *fossil fuel* divided by the annual rate of production (extraction and processing).
resilience The ability of an ecological *community* to change in response to disturbance but later return to its original state. Compare *resistance.*
resistance The ability of an ecological *community* to remain stable in the presence of a disturbance. Compare *resilience.*
resource management Strategic decision making about who should extract resources and in what ways, so that resources are used wisely and not wasted.
resource partitioning The process by which *species* adapt to *competition* by evolving to use slightly different resources, or to use their shared resources in different ways, thus minimizing interference with one another.
Resource Conservation and Recovery Act (RCRA) Congressional *legislation* (enacted in 1976 and amended in 1984) that specifies, among other things, how to manage *sanitary landfills* to protect against environmental contamination.
restoration ecology The study of the historical conditions of *ecological communities* as they existed before humans altered them.
revolving door The movement of powerful officials between the private sector and government agencies.
R horizon The bottommost layer of *soil* in a typical *soil profile.* Also called *bedrock.* Compare *A horizon; B horizon; C horizon; E horizon; O horizon.*
ribonucleic acid See *RNA.*
risk The mathematical probability that some harmful outcome (for instance, injury, death, *environmental* damage, or *economic* loss) will result from a given action, event, or substance.
risk assessment The quantitative measurement of *risk,* together with the comparison of risks involved in different activities or substances.
risk management The process of considering information from scientific *risk assessment* in light of economic, social, and political needs and values, in order to make decisions and design strategies to minimize *risk.*
RNA (ribonucleic acid) A usually single-stranded *nucleic acid* composed of four nucleotides, each of which contains a sugar (ribose), a phosphate group, and a nitrogenous base. RNA carries the hereditary information for living organisms and is responsible for passing traits from parents to offspring. Compare *DNA.*

rock cycle The very slow process in which rocks and the minerals that make them up are heated, melted, cooled, broken, and reassembled, forming *igneous*, *sedimentary*, and *metamorphic* rocks.
r–selected Term denoting a *species* with high biotic potential whose members produce a large number of offspring in a relatively short time but do not care for their young after birth. *Populations* of r–selected species are generally regulated by *density-independent factors.* Compare *K–selected.*
runoff The water from *precipitation* that flows into streams, rivers, lakes, and ponds, and (in many cases) eventually to the ocean.
run-of-river Any of several methods used to generate *hydroelectric power* without greatly disrupting the flow of river water. Run-of-river approaches eliminate much of the *environmental* impact of large *dams.* Compare *storage.*
Ruskin, John (1819–1900) British art critic, poet, and writer who criticized industrialized cities and their *pollution,* and who believed that people no longer appreciated the *environment*'s spiritual or aesthetic benefits.

salinization The buildup of salts in surface *soil* layers.
salt See *ionic compound.*
salt marsh Flat land that is intermittently flooded by the ocean where the *tide* reaches inland. Salt marshes occur along temperate coastlines and are thickly vegetated with grasses, rushes, shrubs, and other herbaceous plants.
salvage logging The removal of dead trees following a natural disturbance. Although it may be economically beneficial, salvage logging can be ecologically destructive, because the dead trees provide food and shelter for a variety of insects and wildlife and because removing timber from recently burned land can cause severe *erosion* and damage to *soil.*
sand *Sediment* consisting of particles 0.005–2.0 mm in diameter. Compare *clay; silt.*
sanitary landfill A site at which solid waste is buried in the ground or piled up in large mounds for disposal, designed to prevent the waste from contaminating the *environment.* Compare *incineration.*
savanna A *biome* characterized by grassland interspersed with clusters of acacias and other trees. Savanna is found across parts of Africa (where it was the ancestral home of our *species*), South America, Australia, India, and other dry tropical regions.
science A systematic process for learning about the world and testing our understanding of it.
scientific method A formalized method for testing ideas with observations that involves several assumptions and a more or less consistent series of interrelated steps.
secondary extraction The extraction of *crude oil* remaining after *primary extraction* by using solvents or by flushing underground rocks with water or steam. Compare *primary extraction.*
secondary pollutant A hazardous substance produced through the reaction of substances added to the *atmosphere* with chemicals normally found in the atmosphere. Compare *primary pollutant.*
secondary production The total biomass that *heterotrophs* generate by consuming *autotrophs.* Compare *gross primary production* and *net primary production.*
secondary succession A stereotypical series of changes as an *ecological community* develops over time, beginning when some event disrupts or dramatically alters an existing community. Compare *primary succession.*
secondary treatment A stage of *wastewater* treatment in which biological means are used to remove contaminants remaining after *primary treatment.* Wastewater is stirred up in the presence of *aerobic* bacteria, which degrade organic pollutants in the water. The wastewater then passes to another settling tank, where remaining solids drift to the bottom. Compare *primary treatment.*
second-growth Term describing trees that have sprouted and grown to partial maturity after virgin timber has been cut.
second law of thermodynamics Physical law stating that the nature of *energy* tends to change from a more-ordered state to a less-ordered state; that is, *entropy* increases.
sediment The eroded remains of rocks.
sedimentary rock One of the three main categories of rock. Formed when dissolved minerals seep through *sediment* layers and act as a kind of glue, crystallizing and binding sediment particles together. Sandstone and shale are examples of sedimentary rock. Compare *igneous rock; metamorphic rock.*
seed bank A storehouse for samples of the world's crop diversity.
septic system *A wastewater* disposal method, common in rural areas, consisting of an underground tank and series of drainpipes. Wastewater runs from the house to the tank, where solids precipitate out. The water proceeds downhill to a drain field of perforated pipes laid horizontally in gravel-filled trenches, where microbes decompose the remaining waste.
service Work done for others as a form of business.
sex ratio The proportion of males to females in a *population.*
shelterbelt A row of trees or other tall perennial plants that are planted along the edges of farm fields to break the wind and thereby minimize wind *erosion.*
silt *Sediment* consisting of particles 0.002–0.005 mm in diameter. Compare *clay; sand.*
sinkhole An area where the ground has given way with little warning as a result of subsidence caused by depletion of water from an *aquifer.*
SLOSS (Single Large or Several Small) dilemma The debate over whether it is better to make reserves large in size and few in number or many in number but small in size.
smart growth A *city planning* concept in which a community's growth is managed in ways that limit *sprawl* and maintain or improve residents' quality of life. It involves guiding the rate, placement, and style of development such that it serves the *environment,* the *economy,* and the community.
Smith, Adam (1723–1790) Scottish philosopher known today as the father of *classical economics.* He believed that when people are free to pursue their own economic self-interest in a competitive marketplace, the marketplace will behave as if guided by "an invisible hand" that ensures that their actions will benefit society as a whole.
social science An academic discipline that studies human interactions and institutions. Compare *natural science.*
soil profile The cross-section of a *soil* as a whole, from the surface to the *bedrock.*

soil A complex plant-supporting *system* consisting of disintegrated rock, organic matter, air, water, *nutrients,* and microorganisms.
source reduction The reduction of the amount of material that enters the *waste stream* to avoid the costs of disposal and *recycling,* help conserve resources, minimize *pollution,* and save consumers and businesses money.
specialist A *species* that can survive only in a narrow range of *habitats* that contain very specific resources. Compare *generalist.*
speciation The process by which new *species* are generated.
species A *population* or group of populations of a particular type of organism, whose members share certain characteristics and can breed freely with one another and produce fertile offspring. Different biologists may have different approaches to diagnosing species boundaries.
species diversity The number and variety of *species* in the world or in a particular region.
sprawl The unrestrained spread of urban or *suburban* development outward from a city center and across the landscape.
steady-state economy An *economy* that does not grow or shrink but remains stable.
storage Technique used to generate *hydroelectric power,* in which large amounts of water are impounded in a reservoir behind a concrete *dam* and then passed through the dam to turn *turbines* that generate electricity. Compare *run-of-river.*
stratosphere The layer of the *atmosphere* above the *troposphere* and below the mesosphere; it extends from 11 km (7 mi) to 50 km (31 mi) above sea level.
strip-mining The use of heavy machinery to remove huge amounts of earth to expose *coal* or minerals, which are mined out directly. Compare *subsurface mining.*
subduction The *plate tectonic* process by which denser ocean *crust* slides beneath lighter continental crust at a *convergent plate boundary.*
subsidy A government incentive (a giveaway of cash or publicly owned resources, or a tax break) intended to encourage a particular activity. Compare *green tax.*
subsistence economy A survival *economy,* one in which people meet most or all of their daily needs directly from nature and do not purchase or trade for most of life's necessities.
subsurface mining Method of mining underground *coal* deposits, in which shafts are dug deeply into the ground and networks of tunnels are dug or blasted out to follow coal seams. Compare *strip-mining.*
suburb A smaller community that rings a city.
succession A stereotypical series of changes in the composition and structure of an *ecological community* through time. See *primary succession; secondary succession.*
Superfund A program administered by the *Environmental Protection Agency* in which experts identify sites polluted with hazardous chemicals, protect *groundwater* near these sites, and clean up the *pollution.*
surface impoundment A *hazardous waste* disposal method in which a shallow depression is dug and lined with impervious material, such as *clay.* Water containing small amounts of hazardous waste is placed in the pond and allowed to evaporate, leaving a residue of solid hazardous waste on the bottom. Compare *deep-well injection.*
survivorship curve A graph that shows how the likelihood of death for members of a *population* varies with age.
sustainability A guiding principle of *environmental science* that requires us to live in such a way as to maintain Earth's systems and its *natural resources* for the foreseeable future.
sustainable agriculture *Agriculture* that does not deplete *soils* faster than they form.
sustainable development Development that satisfies our current needs without compromising the future availability of *natural resources* or our future quality of life.
sustainable forestry certification A form of *ecolabeling* that identifies timber products that have been produced using *sustainable* methods. Several organizations issue such certification.
symbiosis A *parasitic* or *mutualistic* relationship between different *species* of organisms that live in close physical proximity.
synergistic effect An interactive effect (as of *toxicants)* that is more than or different from the simple sum of their constituent effects.
system A network of relationships among a group of parts, elements, or components that interact with and influence one another through the exchange of *energy,* matter, and/or information.

Talloires Declaration A document composed in Talloires, France, in 1990 that commits university leaders to pursue *sustainability* on their campuses. It has been signed by over 300 university presidents and chancellors from more than 40 nations.
temperate deciduous forest A *biome* consisting of midlatitude forests characterized by broad-leafed trees that lose their leaves each fall and remain dormant during winter. These forests occur in areas where *precipitation* is spread relatively evenly throughout the year: much of Europe, eastern China, and eastern North America.
temperate grassland A *biome* whose vegetation is dominated by grasses and features more extreme temperature differences between winter and summer and less *precipitation* than *temperate deciduous forests.*
temperate rainforest A *biome* consisting of tall coniferous trees, cooler and less species-rich than *tropical rainforest* and milder and wetter than *temperate deciduous forest.*
temperature (thermal) inversion A departure from the normal temperature distribution in the *atmosphere,* in which a pocket of relatively cold air occurs near the ground, with warmer air above it. The cold air, denser than the air above it, traps *pollutants* near the ground and causes a buildup of smog.
teratogen A *toxicant* that causes harm to the unborn, resulting in birth defects.
terracing The cutting of level platforms, sometimes with raised edges, into steep hillsides to contain water from *irrigation* and *precipitation.* Terracing transforms slopes into series of steps like a staircase, enabling farmers to cultivate hilly land while minimizing their loss of *soil* to water *erosion.*

theory A widely accepted, well-tested explanation of one or more cause-and-effect relationships that has been extensively validated by a great amount of research. Compare *hypothesis.*
thermal inversion See *temperature inversion.*
Thoreau, Henry David (1817–1862) American transcendentalist author, poet, and philosopher. His book *Walden,* recording his observations and thoughts while he lived at Walden Pond away from the bustle of urban Massachusetts, remains a classic of American literature.
Three Mile Island Nuclear power plant in Pennsylvania that in 1979 experienced a partial *meltdown.* The term is often using to denote the accident itself, the most serious *nuclear reactor* malfunction that the United States has thus far experienced.
threshold dose The amount of a *toxicant* at which it begins to affect a *population* of test animals. Compare ED_{50}; LD_{50}.
tide The periodic rise and fall of the ocean's height at a given location, caused by the gravitational pull of the moon and sun.
topsoil That portion of the *soil* that is most nutritive for plants and is thus of the most direct importance to *ecosystems* and to *agriculture.* Also known as the *A horizon.*
total fertility rate (TFR) The average number of children born per female member of a *population* during her lifetime.
toxic air pollutant *Air pollutant* that is known to cause cancer, reproductive defects, or neurological, developmental, immune system, or respiratory problems in humans, and/or to cause substantial *ecological* harm by affecting the health of nonhuman animals and plants. The *Clean Air Act of 1990* identifies 188 toxic air pollutants, ranging from the heavy metal mercury to *volatile organic compounds* such as benzene and methylene chloride.
toxicant A substance that acts as a poison to humans or wildlife.
toxicology The scientific field that examines the effects of poisonous chemicals and other agents on humans and other organisms.
traditional agriculture Biologically powered *agriculture,* in which human and animal muscle power, along with hand tools and simple machines, perform the work of cultivating, harvesting, storing, and distributing crops. Compare *industrialized agriculture.*
transform plate boundary Area where two tectonic plates meet and slip and grind alongside one another. For example, the Pacific Plate and the North American Plate rub against each other along California's San Andreas Fault.
transgene A *gene* that has been extracted from the *DNA* of one organism and transferred into the DNA of an organism of another *species.*
transgenic Term describing an organism that contains *DNA* from another *species.*
transitional stage The second stage of the *demographic transition* model, which occurs during the transition from the *pre-industrial stage* to the *industrial stage.* It is characterized by declining death rates but continued high birth rates. See also *post-industrial stage.* Compare *industrial stage; post-industrial stage; pre-industrial stage.*
transpiration The release of water vapor by plants through their leaves.
treatment The portion of an *experiment* in which a *variable* has been manipulated in order to test its effect. Compare *control.*
trophic level Rank in the feeding hierarchy of a food chain. Organisms at higher trophic levels consume those at lower trophic levels.
tropical dry forest A *biome* that consists of deciduous trees and occurs at tropical and subtropical latitudes where wet and dry seasons each span about half the year. Widespread in India, Africa, South America, and northern Australia.
tropical rainforest A *biome* characterized by year-round rain and uniformly warm temperatures. Found in Central America, South America, southeast Asia, west Africa, and other tropical regions. Tropical rainforests have dark, damp interiors; lush vegetation; and highly diverse *biotic communities.*
troposphere The bottommost layer of the *atmosphere;* it extends to 11 km (7 mi) above sea level. See also *stratosphere.*
tundra A *biome* that is nearly as dry as *desert* but is located at very high latitudes along the northern edges of Russia, Canada, and Scandinavia. Extremely cold winters with little daylight and moderately cool summers with lengthy days characterize this landscape of lichens and low, scrubby vegetation.
turbine A rotary device that converts the *kinetic energy* of a moving substance, such as steam, into mechanical energy. Used widely in commercial power generation from various types of energy sources.

unconfined aquifer A water-bearing, porous layer of rock, *sand,* or gravel that lies atop a less-permeable substrate. The water in an unconfined aquifer is not under pressure because there is no impermeable upper layer to confine it. Compare *confined aquifer.*
undernutrition A condition of insufficient *nutrition* in which people receive less than 90% of their daily caloric needs.
uneven-aged Term describing stands of trees in timber plantations that are of different ages. Uneven-aged stands more closely approximate a natural forest than do *even-aged* stands.
United Nations (U.N.) Organization founded in 1945 to promote international peace and to cooperate in solving international economic, social, cultural, and humanitarian problems. Several agencies within it influence *environmental policy,* most notably the United Nations Environment Programme (UNEP), created in 1972.
United Nations Framework Convention on Climate Change (FCCC) International agreement to reduce *greenhouse gas* emissions to 1990 levels by the year 2000, signed by nations represented at the 1992 Earth Summit convened in Rio de Janeiro by the *United Nations.* The FCCC called for a voluntary, nation-by-nation approach, but by the late 1990s it had become apparent that it would not succeed. Its imminent failure sparked introduction of the *Kyoto Protocol.*
universalist An *ethicist* who maintains that there exist objective notions of right and wrong that hold across cultures and situations. Compare *relativist.*
upwelling In the ocean, the flow of cold, deep water toward the surface. Upwelling occurs in areas where surface *currents* diverge. Compare *downwelling.*

urban ecology A scientific field that views cities explicitly as *ecosystems.* Researchers in this field seek to apply the fundamentals of *ecosystem ecology* and *systems* science to urban areas.
urban growth boundary (UGB) In *city planning,* a geographic boundary intended to separate areas desired to be urban from areas desired to remain rural. Development for housing, commerce, and industry are encouraged within urban growth boundaries, but beyond them such development is severely restricted.
urbanization The shift from rural to city and *suburban* living.

variable In an *experiment,* a condition that can change. See *dependent variable* and *independent variable.*
volatile organic compound (VOC) One of a large group of potentially harmful organic chemicals used in industrial processes.

Wallace, Alfred Russell (1823–1913) English naturalist who proposed, independently of *Charles Darwin,* the concept of *natural selection* as a mechanism for *evolution* and as a way to explain the great variety of living things.
warm front The boundary where a mass of warm air displaces a mass of colder air. Compare *cold front.*
waste Any unwanted product that results from a human activity or process.
waste management Strategic decision making to minimize the amount of *waste* generated and to dispose of waste safely and effectively.
waste stream The flow of *waste* as it moves from its sources toward disposal destinations.
waste-to-energy (WTE) facility An incinerator that uses heat from its furnace to boil water to create steam that drives electricity generation or that fuels heating systems.
wastewater Any water that is used in households, businesses, industries, or public facilities and is drained or flushed down pipes, as well as the polluted *runoff* from streets and storm drains.
waterlogging The saturation of *soil* by water, in which the *water table* is raised to the point that water bathes plant roots. Waterlogging deprives roots of access to gases, essentially suffocating them and eventually damaging or killing the plants.
watershed The entire area of land from which water drains into a given river.
water table The upper limit of *groundwater* held in an *aquifer.*
weather The local physical properties of the *troposphere,* such as temperature, pressure, humidity, cloudiness, and wind, over relatively short time periods. Compare *climate.*
weathering The physical, chemical, and biological processes that break down rocks and minerals, turning large particles into smaller particles.
Whitman, Walt (1819–1892) American poet who espoused transcendentalism. See also *Emerson, Ralph Waldo* and *Thoreau, Henry David.*
wilderness area Federal land that is designated off-limits to development of any kind but is open to public recreation, such as hiking, nature study, and other activities that have minimal impact on the land.
wind turbine A mechanical assembly that converts the wind's *kinetic energy,* or energy of motion, into electrical energy.
wise-use movement A loose confederation of individuals and groups that coalesced in the 1980s and 1990s as a response to the increasing success of environmental advocacy. The movement favors extracting more resources from public lands, obtaining greater local control of lands, and obtaining greater motorized recreational access to public lands.
World Bank Institution founded in 1944 that serves as one of the globe's largest sources of funding for *economic* development, including such major projects as *dams, irrigation* infrastructure, and other undertakings.
World Trade Organization (WTO) Organization based in Geneva, Switzerland, that represents multinational corporations and promotes free trade by reducing obstacles to international commerce and enforcing fairness among nations in trading practices.
worldview A way of looking at the world that reflects a person's (or a group's) beliefs about the meaning, purpose, operation, and essence of the world.

zoning The practice of classifying areas for different types of development and land use.

Photo Credits

Part Opening Photos: Part One Bill Hatcher/National Geographic Collection **Part Two** AP Photo/Bill Haber

Chapter 1 Opening Photo NASA/Johnson Space Center **1.2b** Charles O'Rear/CORBIS **1.3a** Art Resource **1.3b** Bettman/CORBIS **1.4** John N. Smith **1.6** George Konig/Hulton Archive Photos **1.8** Joel W. Rogers/CORBIS **1.9a** George Bukenhofer **1.9b** Reuters New Media Inc./CORBIS **1.14** CORBIS **1.15** Reuters **1.16** Justin Sullivan/Getty Images **The Science behind the Story: Easter Island** Richard T. Nowitz/CORBIS

Chapter 2 Opening Photo Reuters New Media Inc./CORBIS **Case Study** The Gundjehmi Aboriginal Corporation **2.2a, b** The Gundjehmi Aboriginal Corporation **2.5** Library of Congress **2.6** CORBIS **2.7** CORBIS **2.8** Bettmann/CORBIS **2.10** Loomis Dean/Time Life Pictures/Getty Images **2.13** AFP/CORBIS **2.14** Larry Lee Photography/CORBIS **2.15a** Konrad Wothe/Minden Pictures **2.15b** Bill Hatcher/National Geographic Image Collection **2.15c** Bruce Forster/The Image Bank **2.15d** CORBIS **2.15e** Charles O'Rear/CORBIS **2.15f, g** Frans Lanting/Minden Pictures **2.16** Kristin Piljay

Chapter 3 Opening Photo CORBIS **Case Study** Denis Poroy/Associated Press **3.1** Lori Saldana **3.3** Annie Griffiths Belt/CORBIS **3.6** Rick Wilking/Reuters/Corbis **3.7a** Bettmann/CORBIS **3.7b** Museum of History & Industry/CORBIS **3.7c** University of Washington Libraries **3.8** Erich Hartmann/Magnum Photos, Inc. **3.9** Bettmann/CORBIS **3.10a** CWH, Associated Press **3.10b** Binod Joshi/Associated Press **3.11** Bettmann/CORBIS **3.13** Don Boroughs/The Image Works **3.14** Charles Mauzy/CORBIS **3.15** Martin Rugner/age fotostock **3.18** Lori Saldana **3.19** Jon Hrusa/AP Photo **3.20** Alex Wong/Getty Images **The Science behind the Story: Spotting Sewage via Satellite** NASA **The Science behind the Story: Assessing the Environmental Impact of Treating Transboundary Sewage** Metropolitan Wastewater Department

Chapter 4 Opening Photo AFP/CORBIS **Case Study** Exxon Corporation **4.1** AP Photo/Jack Smith **4.5** Digital Vision/Picture Quest **4.6a** Digital Vision/Picture Quest **4.13** Dorling Kindersley **4.16 left** Daniel Zheng/CORBIS **4.16 right** Anne-Marie Weber/Taxi **4.20** Jack Dykinga/Stone **4.21a** Ken MacDonald/SPL/Photo Researchers, Inc. **4.21b** Woods Hole Oceanographic Institute **4.22** Chip Clark **4.23** Dorling Kindersley **The Science behind the Story: Student Chemist** Mark Burrell

Chapter 5 Opening Photo Michael and Patricia Fogden/CORBIS **Case Study** Michael and Patricia Fogden/CORBIS **5.1a** Nicholas Athanas/Tropical Birding **5.1b** Michael Fogden/DRK Photo **5.1c** Michael &Patricia Fogden/CORBIS **5.1d** Michael Fogden/Photolibrary **5.3** Bishop Museum, Honolulu **5.7** The National History Museum, London, painting by J. Sibbick **5.8** Chip Clark **5.9** Lynda Richardson/CORBIS **5.11a** Wisconsin Historical Society **5.11b** G. I. Bernard/PhotoResearchers, Inc **5.12a** Keenan Ward/CORBIS **5.12b** Art Wolfe/The Image Bank **5.12c** PhotoDisc Green **5.18a** Mike Danzenbaker **5.18b** A. Witte/C. Mahaney/Getty Images **5.18c** John Waters/Nature Picture Library **5.19** Matthias Clamer/Stone **The Science behind the Story: Mass Extinction** Benjamin Cummings

Chapter 6 Opening Photo Robert Estall/Robert Harding World Imagery **Case Study** Wolfgang Polzer **6.1a** Peter Yates/Photo Researchers, Inc **6.1b** B. Runk/S. Schoenberger/Grant Heillman Photography **6.4** Michael & Patricia Fogden/CORBIS **6.6a** Peter Johnson/CORBIS **6.6b** Tom Brakefield/CORBIS **6.6c** Michael & Patricia Fogden/CORBIS **6.7** Tom Stack **6.8** Joke Stuurman–Huitema/Foto Nat/Minden Pictures **6.9** Michael & Patricia Fogden/CORBIS **6.18a** Pat O'Hara/CORBIS **6.19a** Philip Gould/CORBIS **6.20a** Charles Mauzy/CORBIS **6.21a** David Samuel Robbins/CORBIS **6.22a** O. Alamany & E. Vicens/CORBIS **6.23a** Wolfgang Kaehler/CORBIS **6.24a** Joe McDonald/CORBIS **6.25a** Darrell Gulin/CORBIS **6.26a** Liz Hymans/CORBIS **6.27a** Charles Mauzy/CORBIS

Chapter 7 Opening Photo Yann Arthus-Bertrand/CORBIS **Case Study** Owen Franken/CORBIS **7.7** D. W. Schindler **7.8** NASA **7.12 left** Milos Kalab/CMSP **7.12 right** Milos Kalab/CMSP **The Science behind the Story: Biosphere 2** James Marshall/CORBIS

Chapter 8 Opening Photo Robert Semenvik/CORBIS **Case Study** Louise Gibb/The Image Works **8.1** Reuters New Media Inc./CORBIS **8.10c** SETBOUN/CORBIS **8.12** David Turnley/CORBIS **8.19** Tiziana and Gianni Baldizzone/CORBIS **8.21a** Elie Bernager/Stone **8.21b** Ed Kashi/IPN/AURORA **8.23** Getty Images **The Science behind the Story: Causes of Fertility Decline in Bangladesh** Mark Edwards/Peter Arnold, Inc. **The Science behind the Story: AIDS Resistance Genes and the Black Death: An Unexpected Connection?** Jean-Loup Charmet/Photo Researchers, Inc

Chapter 9 Opening Photo Stephanie Maze/CORBIS **Case Study** Joanna B. Pinneo/AURORA **9.8** Dorling Kindersley **9.10.** Ron Giling/Peter Arnold, Inc **9.11a** Barry Runk/Stan/Grant Heilman Photography **9.11b** Grant Heilman/Grant Heilman Photography **9.11c, d** U.S. Department of Agriculture **9.13** Library of Congress **9.14** Ted Spiegel/CORBIS **9.15a** Sylvan Wittwer/Visuals Unlimited **9.15b** Kevin Horan/Stone **9.15c** Ron Giling/Peter Arnold **9.15d** Keren Su/Stone **9.15e** Yann Arthus-Bertrand/CORBIS **9.15f** U.S. Department of Agriculture **9.16** Michael Yamashita/IPN/Aurora **9.17a** Photo Disc Blue **9.17b** Carol Cohen/CORBIS **9.18** Richard Hamilton Smith/CORBIS **9.22** W. Perry Conway/CORBIS **9.23** Natalie Fobes/CORBIS **The Science behind the Story: Overgrazing and Fire Suppression** Malpai Borderlands Group

Chapter 10 Opening Photo Steve Satushek/Image Bank **Case Study** Macduff Everton/CORBIS **10.3** Alexandra Avakian/CORBIS **10.4** Art Rickerby/Time Life Pictures/Getty Images **10.6a** Jack Dykinga/Image Bank **10.6b** Barry Runk/Stan/Grant Heilman Photography **10.8a** Department of Natural Resources, Queensland, Australia **10.8b** Department of Natural Resources, Queensland, Australia **10.10** Photo Disc **10.11** Bob Rowan, Progressive Image/CORBIS **10.15** John Schmeiser **10.16** ALEXANDER JOE/AFP/Getty Images **10.17a** Native Seeds/SEARCH **10.17b** Hal Fritts, Native Seeds/SEARCH **10.19** Arthur C. Smith III/Grant Heilman Photography **10.23** Aqua Bounty Farms/AP **10.24** AP Wide World Photos **10.25** Martin Bourque **The Science behind the Story: Native Maize** Peg Skorpinski

Chapter 11 Opening Photo Maurice Hornocker **Case Study** Maurice Hornocker **11.12** Kathy Ferguson-Johnson/Photo Edit **11.15** Konrad Wothe/Photolibrary **11.19** Natalie Fobes/CORBIS **11.20** Tom & Pat Leeson/Photo Researchers, Inc. **The Science behind the Story: Amphibian Diversity** S. D. Biju **The Science behind the Story: Island Biogeography** Stock Connection, Inc./Alamy

Chapter 12 Opening Photo Daniel Dancer/Peter Arnold, Inc **Case Study** Garth Lenz **12.1** Dennis Hallinan/Alamy **12.5** CORBIS **12.9** Weyerhaeuser Company **12.10** Rob Badger/Getty Images **12.12** Jonathan Blair/CORBIS **12.13** Reuters/CORBIS **12.14** The Image Bank **12.15** Jim Brandenburg/Minden Pictures **12.16** Authenticated News/ Getty Images **12.17** North Carolina Museum of Art/CORBIS **12.18** David Hiser/Stone **12.20a** Gary Braasch/CORBIS **12.20b** James Zipp/Photo Researchers, Inc. **The Science behind the Story: Adaptive Management** Wolfgang Kaehler/CORBIS

Chapter 13 Opening Photo Rich Iwaski/age fotostock **Case Study** Phil Schermeister/CORBIS **13.4ab** U.S. EPA **13.5a** Lester Lefkowitz/CORBIS **13.5b** Bob Krist/CORBIS **13.5c** David R. Frazier Photolibrary, Inc. **13.5d** Aldo Torelli/Getty Images **13.6** Chicago Historical Society **13.9** Cooper Carry & Associates **13.11a** Steve Semler/Alamy **13.12** Bettmann/CORBIS **13.14** Bohemian Nomad Picture makers/CORBIS **13.16** Kim Karpeles **The Science behind the Story: Measuring the Impacts of Sprawl** Editorial Fotos/Alamy

Chapter 14 Opening Photo Howard K. Suzuki **Case Study** Howard K. Suzuki **14.1a** Ingram Publishing/Alamy **14.1b** Spencer Grant/Photo Edit **14.1c** Martin Dohrn/Photo Researchers, Inc. **14.1d** LWA-Dann Tardif/CORBIS **14.4** Photofusion Picture LibraryAlamy **14.5** Joel W. Rogers/ CORBIS **14.6** Bettmann/CORBIS **14.7** Bettmann/CORBIS **14.9** Peg Skorpinski **The Science behind the Story: Bisphenol-A** both From Hunt, PA, KE Koehler, M Susiarjo, CA Hodges, A Ilagan, RC Voight, S Thomas, BF Thomas and TJ Hassold. 2003. *Bisphenol A exposure causes meiotic aneuploidy in the female mouse.* Current Biology 13: 546–553 **The Science behind the Story: Pesticides and Child Development** Elizabeth A. Guillette

Chapter 15 Opening Photo Lester LefKowitz/CORBIS **Case Study** Galen Rowell/CORBIS **15.2** William Manning/CORBIS **15.3** Frans Lanting/Minden Pictures **15.11** Jim Tuten/Black Star **15.14a** Gilles Saussier/Liaison **15.14b** NASA Visible Earth **15.16b** Ian Berry/ Magnum Photos **15.17** AP Photo **15.19a** David Muench/CORBIS **15.19b** Andrew Brown, Ecoscene/CORBIS **15.20** David J. Cross/Peter Arnold, Inc. **The Science behind the Story: Wastewater** Gary Crabbe/ Enlightened Images

Chapter 16 Opening Photo Harvey Lloyd/Taxi **Case Study** Jeff Hunter/The Image Bank **16.7** Alcoa Technical Center **16.10** Laguna Design/Science Photo Library/Photo Researchers, Inc. **16.11** Bruce Robison/Minden Pictures **16.12** Ralph A. Clevenger/CORBIS **16.13a** Stephen Frink/CORBIS **16.13b** NOAA **16.13c** CORBIS **16.14b** Carol Cohen/CORBIS **16.15** Kevin Fleming/CORBIS **16.16** Mark A. Johnson/ CORBIS **16.17** Nik Wheeler/CORBIS **16.19a** David M. Phillips/Visuals Unlimited **16.19b** Bill Bachman/Photo Researchers, Inc.

Chapter 17 Opening Photo Bettmann/CORBIS **Case Study** Hulton Archive Photos **17.10a** NASA Earth Observatory **17.10b** Bettmann/ CORBIS **17.10c** Mark E. Gibson/CORBIS **17.13b** Pittsburgh Post-Gazette **17.14b** Allen Russell/Index Stock Imagery **17.16b** NASA/ Goddard Space Flight Center **17.18** Erik Schaffer/Ecoscene **17.20a** David Turnley/CORBIS **The Science behind the Story: Acid Rain** Will & Deni McIntyre/CORBIS

Chapter 18 Opening Photo Rossi, Guido Alberto/Image Bank **Case Study** Adam Woolfitt/CORBIS **18.6a** Ted Spiegel/CORBIS **18.6b** David Whillas, CSIRO Atmospheric Research **18.11c** Ben Sklar/AP Photo **18.15** Bettmann/CORBIS **18.17** Robert Brenner/Photo Edit **The Science behind the Story: Scientists Use Pollen** Lowell Georgia/ CORBIS

Chapter 19 Opening Photo Jim Wark/Alaska Stock **Case Study** Steve Armbrust/AlaskaStock.com **19.13** Daniel Esgro Photography/The Image Bank **19.14** Vivian Stockman **19.15** Steven Kazlowski/Alaska Stock **19.22** AGA Infrared/Photo Researchers, Inc. **The Science behind the Story: Crude Oil** CORBIS

Chapter 20 Opening Photo HAWKEYE/Alamy **Case Study** Skellefteå Kraft AB **20.4** argus/Peter Arnold, Inc. **20.7** John S. Zeedick/Getty Images **20.8a** Paul Fusco/Magnum Photos **20.8b** Reuters/CORBIS **20.10a** Roger Ressmeyer/CORBIS **20.10b** U.S. Department of Engergy/Photo Researchers, Inc. **20.12a** Nuclear Energy Institute **20.12b** U.S. Department of Energy **20.13** Chris Steele-Perkins/ Magnum Photos **20.15** Lars-Erik Larsson/SVEBIO **20.17** Brent Baker **20.18** Larry Lefever/ Grant Heilman Photography **20.19a** Earl Roberge/Photo Researchers, Inc. **20.19b** Michael Melford/The Image Bank **The Science behind the Story: Health Impacts of Chernobyl** Caroline Penn/CORBIS

Chapter 21 Opening Photo Roger Ressmeyer **Case Study** Islensk NyOrka **21.3** CORBIS **21.8** Jorgen Schytte/Peter Arnold, Inc. **The Science behind the Story: Idaho's Wind Prospectors** Pascal Goetgheluck/Science Photo Library **The Science behind the Story: Algae as a Hydrogen Fuel Source** Pascal Goetgheluck/Science Photo Library

Chapter 22 Opening Photo Mark Gibson/Index Stock **Case Study** City of New York Department of City Planning **22.5** Nigel Dickinson/ Stone **22.8** Lawrence Migdale/Pix **22.14ab** City of Edmonton **22.17** Joe Sohm/Alamy **22.18** Robert Brook/Photo Researchers, Inc. **The Science behind the Story: Digging Garbage: The Archaeology of Solid Waste** Louie Psihoyos/©psihoyos.com **The Science behind the Story: Testing the Toxicity of E-Waste** AP Photo/Keystone, Walter Bieri

Chapter 23 Opening Photo Mark Edwards/Peter Arnold, Inc. **Case Study** Ball State Photo Services **23.1** University of North Carolina at Greensboro **23.2** Oberlin College **23.3** Stefano Paltera/Solar Decathlon **23.4** Ari Joseph/Middlebury College **23.5** AP Photo/Steve Yeater **23.6** De Anza College, Environmental Studies Department **23.7** AP Photo/Dario Lopez-Mills **23.8** Galen Rowell/CORBIS **23.10** Costco Wholesale **23.15** AP/Wideworld Photos **23.16** Bettmann/ CORBIS

Selected Sources and References for Further Reading

Chapter 1

Bahn, Paul, and John Flenley. 1992. *Easter Island, Earth island.* Thames and Hudson, London.

Bowler, Peter J. 1993. *The Norton history of the environmental sciences.* W. W. Norton, New York.

Diamond, Jared. 2005. *Collapse: How societies choose to fail or succeed.* Viking, New York.

Ehrlich, Paul. 1968. *The population bomb.* 1997 reprint, Buccaneer Books, Cutchogue, New York.

Esty, Daniel C., et al., 2005. *2005 Environmental sustainability index: Benchmarking national environmental stewardship.* New Haven, Connecticut: Yale Center for Environmental Law and Policy.

Flenley, John, and Paul Bahn. 2003. *The enigmas of Easter Island.* Oxford University Press, New York.

Goudie, Andrew. 2000. *The human impact on the natural environment,* 5th ed. MIT Press, Cambridge, Massachusetts.

Hardin, Garrett. 1968. The tragedy of the commons. *Science* 162: 1243–1248.

Katzner, Donald W. 2001. *Unmeasured information and the methodology of social scientific inquiry.* Kluwer, Boston.

Kuhn, Thomas S. 1962. *The structure of scientific revolutions,* 2nd ed., 1970. University of Chicago Press, Chicago.

Lomborg, Bjorn. 2001. *The skeptical environmentalist: Measuring the real state of the world.* Cambridge University Press, Cambridge.

Malthus, Thomas R. *An essay on the principle of population.* 1983 ed. Penguin USA, New York.

Millennium Ecosystem Assessment. 2005. *Ecosystems and human well-being: General synthesis.* Millennium Ecosystem Assessment and World Resources Institute.

Musser, George. 2005. The climax of humanity. *Scientific American* 293(3): 44–47.

Ponting, Clive. 1991. *A green history of the world: The environment and the collapse of great civilizations.* Penguin Books, New York.

Popper, Karl R. 1959. *The logic of scientific discovery.* Hutchinson, London.

Porteous, Andrew. 2000. *Dictionary of environmental science and technology,* 3rd ed. John Wiley & Sons, Hoboken, New Jersey.

Redman, Charles R. 1999. *Human impact on ancient environments.* University of Arizona Press, Tucson.

Sagan, Carl. 1997. *The demon-haunted world: Science as a candle in the dark.* Ballantine Books, New York.

Schneiderman, Jill S., ed. 2003. *The Earth around us: Maintaining a livable planet.* Perseus Books, New York.

Siever, Raymond. 1968. Science: Observational, experimental, historical. *American Scientist* 56: 70–77.

Valiela, Ivan. 2001. *Doing science: Design, analysis, and communication of scientific research.* Oxford University Press, Oxford.

Van Tilburg, Jo Anne. 1994. *Easter Island: Archaeology, ecology, and culture.* Smithsonian Institution Press, Washington, D.C.

Venetoulis, Jason, et al., 2004. *Ecological footprint of nations 2004.* Redefining Progress, Oakland, California.

Wackernagel, Mathis, and William Rees. 1996. *Our ecological footprint: Reducing human impact on the earth.* New Society Publishers, Gabriola Island, British Columbia, Canada.

World Bank. 2005. *World development indicators 2005.* World Bank, Washington, D.C.

Worldwatch Institute. *State of the world 2005: Redefining global security.* Worldwatch Institute and W. W. Norton, Washington, D.C. and New York.

Worldwatch Institute. *Vital Signs 2005.* Worldwatch Institute and W. W. Norton, Washington, D.C. and New York.

Chapter 2

Balmford, Andrew, et al. 2002. Economic reasons for conserving wild nature. *Science* 297: 950–953.

Barbour, Ian G. 1992. *Ethics in an age of technology.* Harper Collins, San Francisco.

Brown, Lester. 2001. *Eco-economy: Building an economy for the Earth.* Earth Policy Institute and W. W. Norton, New York.

Carson, Richard T., Leanne Wilks, and David Imber. 1994. Valuing the preservation of Australia's Kakadu Conservation Zone. *Oxford Economic Papers* 46: 727–749.

Cole, Luke W., and Sheila R. Foster. 2001. *From the ground up: Environmental racism and the rise of the environmental justice movement.* New York University Press, New York.

Costanza, Robert, et al. 1997. The value of the world's ecosystem services and natural capital. *Nature* 387: 253–260.

Costanza, Robert, et al. 1997. *An introduction to ecological economics.* St. Lucie Press, Boca Raton, Florida.

Daily, Gretchen. 1997. *Nature's services: Societal dependence on natural ecosystems.* Island Press, Washington, D.C.

Daly, Herman E. 1996. *Beyond growth.* Beacon Press, Boston.

Daly, Herman E. 2005. Economics in a full world. *Scientific American* 293(3): 100–107.

Elliot, Robert, and Arran Gare, eds. 1983. *Environmental philosophy: A collection of readings.* Pennsylvania State University Press, University Park.

Field, Barry C., and Martha K. Field. 2001. *Environmental economics,* 3rd ed. McGraw-Hill, New York.

Fox, Stephen. 1985. *The American conservation movement: John Muir and his legacy.* University of Wisconsin Press, Madison.

Gardner, Gary, et al. 2004. The state of consumption today. Pp. 3–23 in *State of the world 2004.* Worldwatch Institute and W. W. Norton, Washington, D.C., and New York.

Gardner, Gary, and Erik Assadourian. 2004. Rethinking the good life. Pp. 164–179 in *State of the world 2004.* Worldwatch Institute and W. W. Norton, Washington, D.C., and New York.

Goodstein, Eban. 1999. *The tradeoff myth: Fact and fiction about jobs and the environment.* Island Press, Washington, D.C.

Goodstein, Eban. 2005. *Economics and the environment,* 4th ed. John Wiley & Sons, Hoboken, New Jersey.

Gundjehmi Aboriginal Corporation. Welcome to the Mirrar site. www.mirrar.net.

Hawken, Paul, Amory Lovins, and L. Hunter Lovins. 1999. *Natural capitalism.* Little, Brown, and Co., Boston.

Kolstad, Charles D. 2000. *Environmental economics.* Oxford University Press, Oxford.

Leopold, Aldo. 1949. *A Sand County almanac, and sketches here and there.* Oxford University Press, New York.

Millennium Ecosystem Assessment. 2005. *Ecosystems and human well-being: Opportunities and challenges for business and industry.* Millennium Ecosystem Assessment and World Resources Institute.

Nash, Roderick F. 1989. *The rights of nature.* University of Wisconsin Press, Madison.

Nash, Roderick F. 1990. *American environmentalism: Readings in conservation history,* 3rd ed. McGraw-Hill, New York.

O'Neill, John O., R. Kerry Turner, and Ian J. Bateman, eds. 2001. *Environmental ethics and philosophy.* Elgar, Cheltenham, U.K.

Pearson, Charles S. 2000. *Economics and the global environment.* Cambridge University Press, Cambridge.

Ricketts, Taylor, et al. 2004. Economic value of tropical forest to coffee production. *Proceedings of the National Academy of Sciences of the USA* 101: 12579–12582.

Sachs, Jeffrey. 2005. Can extreme poverty be eliminated? *Scientific American* 293(3): 56–65.

Singer, Peter, ed. 1993. *A companion to ethics.* Blackwell Publishers, Oxford.

Smith, Adam. 1776. *An inquiry into the nature and causes of the wealth of nations.* 1993 ed., Oxford University Press, Oxford.

Sterba, James P., ed. 1995. *Earth ethics: Environmental ethics, animal rights, and practical applications.* Prentice Hall, Upper Saddle River, New Jersey.

Stone, Christopher D. 1972. Should trees have standing? Towards legal rights for natural objects. *Southern California Law Review* 1972: 450–501.

Tietenberg, Tom. 2003. *Environmental economics and policy,* 4th ed. Addison Wesley, Boston.

Turner, R. Kerry, David Pearce, and Ian Bateman. 1993. *Environmental economics: An elementary introduction.* Johns Hopkins University Press, Baltimore.

Venetoulis, Jason, and Cliff Cobb. 2004. *The genuine progress indicator 1950–2002 (2004 update).* Redefining Progress, Oakland, California.

Wenz, Peter S. 2001. *Environmental ethics today.* Oxford University Press, Oxford.

White, Lynn. 1967. The historic roots of our ecologic crisis. *Science* 155: 1203–1207.

Chapter 3

Clark, Ray, and Larry Canter. 1997. *Environmental policy and NEPA: Past, present, and future.* St. Lucie Press, Boca Raton, Florida.

Dietz, Thomas, et al. 2003. The struggle to govern the global commons. *Science* 302: 1907–1912.

Fogleman, Valerie M. 1990. *Guide to the National Environmental Policy Act.* Quorum Books, New York.

Fox, Stephen. 1985. *The American conservation movement: John Muir and his legacy.* University of Wisconsin Press, Madison.

French, Hilary. 2000. Environmental treaties gain ground. Pp. 134–135 in *Vital Signs 2000.* Worldwatch Institute and W. W. Norton, Washington D.C., and New York.

Green Scissors, 2004. *Green Scissors 2004: Cutting wasteful and environmentally harmful spending.* Friends of the Earth, Taxpayers for Common Sense, and U.S. Public Interest Research Group.

Herzog, Lawrence A. 1990. *Where north meets south: Cities, space, and politics on the U.S.–Mexico border.* Center for Mexican-American Studies, University of Texas at Austin.

Houck, Oliver, 2003. Tales from a troubled marriage: Science and law in environmental policy. *Science* 302: 1926–1928.

Kraft, Michael E. 2003. *Environmental policy and politics,* 3rd ed. Longman, New York.

Kubasek, Nancy K., and Gary S. Silverman. 2004. *Environmental law,* 5th ed. Prentice Hall, Upper Saddle River, New Jersey.

Myers, Norman, and Jennifer Kent. 2001. *Perverse subsidies: How misused tax dollars harm the environment and the economy.* Island Press, Washington, D.C.

The National Environmental Policy Act of 1969, as amended (Pub. L. 91–190, 42 U.S.C. 4321–4347, January 1, 1970, as amended by Pub. L. 94–52, July 3, 1975, Pub. L. 94–83, August 9, 1975, and Pub. L. 97–258, § 4(b), Sept. 13, 1982). http://ceq.eh.doe.gov/nepa/regs/nepa/nepaeqia.htm.

Shafritz, Jay M. 1993. *The HarperCollins dictionary of American government and politics.* HarperCollins, New York.

Southwest Center for Environmental Research and Policy, and San Diego State University. Tijuana River Watershed Atlas Project. http://geography.sdsu.edu/Research/Projects/TWRP/tjatlas.html

Steel, Brent S., Richard L. Clinton, and Nicholas P. Lovrich. 2002. *Environmental politics and policy.* McGraw-Hill, New York.

Tietenberg, Tom. 2003. *Environmental economics and policy,* 4th ed. Addison Wesley, Boston.

Turner, R. Kerry, David Pearce, and Ian Bateman. 1993. *Environmental economics: An elementary introduction.* Johns Hopkins University Press, Baltimore.

United States Congress. House. H.R. 3378. 2000. The Tijuana River Valley Estuary and Beach Sewage Cleanup Act of 2000.

Vig, Norman J., and Michael E. Kraft, eds. 2002. *Environmental policy: New directions for the twenty-first century,* 5th ed. CQ Press, Congressional Quarterly, Inc., Washington, D.C.

Wilkinson, Charles F. 1992. *Crossing the next meridian: Land, water, and the future of the West.* Island Press, Washington, D.C.

Chapter 4

Alaska Department of Environmental Conservation. 1993. *The Exxon Valdez oil spill: Final report, State of Alaska response.* June, 1993.

Allen, K. C., and D. E. G. Briggs, eds. 1989. *Evolution and the fossil record.* John Wiley & Sons, Hoboken, New Jersey.

Atlas, Ronald M. 1995. Petroleum biodegradation and oil spill bioremediation. *Marine Pollution Bulletin* 31: 178–182.

Atlas, Ronald M., and Carl E. Cerniglia. 1995. Bioremediation of petroleum pollutants. *BioScience* 45: 332–338.

Berry, R. Stephen. 1991. *Understanding energy: Energy, entropy and thermodynamics for every man.* World Scientific Publishing Co.

Bragg, James R., et al. 1994. Effectiveness of bioremediation for the *Exxon Valdez* oil spill. *Nature* 368: 413–418.

Campbell, Neil A., and Jane B. Reece. 2005. *Biology,* 7th ed. Benjamin Cummings, San Francisco.

Fenchel, Tom. 2003. *Origin and early evolution of life.* Oxford University Press, Oxford.

Fortey, Richard. 1998. *Life: A natural history of the first four billion years of life on Earth.* Alfred Knopf, New York.

Gee, Henry. 1999. *In search of deep time: Beyond the fossil record to a new history of life.* Free Press, New York.

Hall, David O., and Krishna Rao. 1999. *Photosynthesis,* 6th ed. Cambridge University Press, Cambridge.

Lancaster, M., 2002. *Green chemistry.* Royal Society of Chemistry, London.

Manahan, Stanley E. 2004. *Environmental chemistry,* 8th ed. Lewis Publishers, CRC Press, Boca Raton, Florida.

McMurry, John E. 2003. *Organic chemistry,* 6th ed. Brooks/Cole, San Francisco.

National Response Team. *NRT fact sheet: Bioremediation in oil spill response.* U.S. EPA, www.epa.gov/oilspill/pdfs/biofact.pdf.

Nealson, Kenneth H. 2003. Harnessing microbial appetites for remediation. *Nature Biotechnology* 21: 243–244.

Ridley, Mark. 2003. *Evolution,* 3rd ed. Blackwell Science, Cambridge, Massachusetts.

United States Environmental Protection Agency. 2003. Oil program. www.epa.gov/oilspill.

Van Dover, Cindy Lee, 2000. *The ecology of deep-sea hydrothermal vents.* Princeton University Press, Princeton.

Van Ness, H.C. 1983. *Understanding thermodynamics.* Dover Publications, Mineola, New York.

Ward, Peter D., and Donald Brownlee. 2000. *Rare Earth: Why complex life is uncommon in the universe.* Copernicus, New York.

Wassenaar, Leonard I., and Keith A. Hobson. 1998. Natal origins of migratory monarch butterflies at wintering colonies in Mexico: New isotopic evidence. *Proceedings of the National Academy of the USA* 95: 15436–15439.

Chapter 5

Alvarez, Luis W., et al. 1980. Extraterrestrial cause for the Cretaceous-Tertiary extinction. *Science* 208: 1095–1108.

Barbour, Michael G., et al. 1998. *Terrestrial plant ecology,* 3rd ed. Benjamin/Cummings, Menlo Park, California.

Begon, Michael, Martin Mortimer, and David J. Thompson. 1996. *Population ecology: A unified study of animals and plants,* 3rd ed. Blackwell Scientific, Oxford.

Breckle, Siegmar-Walter. 1999. *Walter's vegetation of the Earth: The ecological systems of the geo-biosphere,* 4th ed. Springer-Verlag, Berlin, 1999.

Campbell, Neil A., and Jane B. Reece. 2005. *Biology,* 7th ed. Benjamin Cummings, San Francisco.

Clark K. L., et al. 1998. Cloud water and precipitation chemistry in a tropical montane forest, Monteverde, Costa Rica. *Atmospheric Environment* 32: 1595–1603.

Crump, L. Martha, et al. 1992. Apparent decline of the golden toad: Underground or extinct? *Copeia* 1992: 413–420.

Darwin, Charles. 1859. *The origin of species by means of natural selection.* John Murray, London.

Endler, John A. 1986. *Natural selection in the wild.* Monographs in Population Biology 21, Princeton University Press, Princeton.
Freeman, Scott, and Jon C. Herron. 2003. *Evolutionary analysis,* 3rd ed. Prentice Hall, Upper Saddle River, New Jersey.
Futuyma, Douglas J. 2005. *Evolution.* Sinauer Associates, Sunderland, Massachusetts.
Krebs, Charles J. 2001. *Ecology: The experimental analysis of distribution and abundance,* 5th ed. Benjamin Cummings, San Francisco.
Lawton, Robert O., et al. 2001. Climatic impact of tropical lowland deforestation on nearby montane cloud forests. *Science* 294: 584–587.
Molles, Manuel C., Jr. 2005. *Ecology: Concepts and applications,* 3rd ed. McGraw-Hill, Boston.
Nadkarni, Nalini M., and Nathaniel T. Wheelwright, eds. 2000. *Monteverde: Ecology and conservation of a tropical cloud forest.* Oxford University Press, New York.
Pounds, J. Alan. 2001. Climate and amphibian declines. *Nature* 410: 639.
Pounds, J. Alan, et al. 1997. Tests of null models for amphibian declines on a tropical mountain. *Conservation Biology* 11: 1307–1322.
Pounds, J. Alan, and Martha L. Crump. 1994. Amphibian declines and climate disturbance: The case of the golden toad and the harlequin frog. *Conservation Biology* 8: 72–85.
Pounds, J. Alan, Michael P. L. Fogden, and John H. Campbell. 1999. Biological response to climate change on a tropical mountain. *Nature* 398: 611–615.
Powell, James L. 1998. *Night comes to the Cretaceous: Dinosaur extinction and the transformation of modern geology.* W. H. Freeman, New York.
Raup, David M. 1991. *Extinction: Bad genes or bad luck?* W. W. Norton, New York.
Ricklefs, Robert E., and Gary L. Miller. 2000. *Ecology,* 4th ed. W. H. Freeman, New York.
Ricklefs, Robert E., and Dolph Schluter, eds. 1993. *Species diversity in ecological communities.* University of Chicago Press, Chicago.
Savage, Jay M. 1966. An extraordinary new toad (*Bufo*) from Costa Rica. *Revista de Biologia Tropical* 14: 153–167.
Savage, Jay M. 1998. The "brilliant toad" was telling us something. *Christian Science Monitor,* 14 September 1998: 19.
Smith, Thomas M., and Robert L. Smith. 2006. *Elements of ecology,* 6th ed. Benjamin Cummings, San Francisco.
Ward, Peter. 1994. *The end of evolution.* Bantam Books, New York.
Williams, George C. 1966. *Adaptation and natural selection.* Princeton University Press, Princeton.
Wilson, Edward O. 1992. *The diversity of life.* Harvard University Press, Cambridge, Massachusetts.
Whittaker, Robert H., and William A. Niering. 1965. Vegetation of the Santa Catalina Mountains, Arizona: A gradient analysis of the south slope. *Ecology* 46: 429–452.

Chapter 6

Breckle, Siegmar-Walter. 2002. *Walter's vegetation of the Earth: The ecological systems of the geo-biosphere,* 4th ed. Berlin: Springer-Verlag.
Bronstein, Judith L. 1994. Our current understanding of mutualism. *Quarterly Journal of Biology* 69: 31–51.
Chase, Jonathan M., et al., 2002. The interaction between predation and competition: A review and synthesis. *Ecology Letters* 5: 302.
Connell, Joseph H., and Ralph O. Slatyer, 1977. Mechanisms of succession in natural communities. *American Naturalist* 111: 1119–1144.
Drake, John M., and Jonathan M. Bossenbroek. 2004. The potential distribution of zebra mussels in the United States. *BioScience* 54: 931–941.
Estes, J.A., et al. 1998. Killer whale predation on sea otters linking oceanic and nearshore ecosystems. *Science* 282: 473–476.
Ewald, Paul W., 1987. Transmission modes and evolution of the parasitism-mutualism continuum. *Annals of the New York Academy of Sciences* 503: 295–306.
Gurevitch, Jessica, and Dianna K. Padilla. 2004. Are invasive species a major cause of extinctions? *Trends in Ecology and Evolution* 19: 470–474.
Krebs, Charles J. 2001. *Ecology: The experimental analysis of distribution and abundance,* 5th ed. Benjamin Cummings, San Francisco.
Menge, Bruce A., et al. 1994. The keystone species concept: Variation in interaction strength in a rocky intertidal habitat. *Ecological Monographs* 64: 249–286.
Molles, Manuel C. Jr. 2005. *Ecology: Concepts and applications.* 3rd ed. McGraw-Hill, Boston.
Morin, Peter J. 1999. *Community ecology.* Blackwell, London.
Power, Mary E., et al., 1996. Challenges in the quest for keystones. *BioScience* 46: 609–620.
Ricklefs, Robert E., and Gary L. Miller. 2000. *Ecology,* 4th ed. W. H. Freeman and Co., New York.
Shea, Katriona, and Peter Chesson, 2002. Community ecology theory as a framework for biological invasions. *Trends in Ecology and Evolutionary Biology* 17: 170–176.
Sih, Andrew, et al. 1985. Predation, competition, and prey communities: A review of field experiments. *Annual Review of Ecology and Systematics* 16: 269–311.
Smith, Robert L., and Thomas M. Smith. 2001. *Ecology and field biology,* 6th ed. Benjamin Cummings, San Francisco.
Springer, A.M., et al. 2003. Sequential megafaunal collapse in the North Pacific Ocean: An ongoing legacy of industrial whaling? *Proceedings of the National Academy of Sciences of the USA* 100: 12223–12228.
Strayer, David L., et al. 1999. Transformation of freshwater ecosystems by bivalves: A case study of zebra mussels in the Hudson River. *BioScience* 49: 19–27.
Strayer, David L., et al. 2004. Effects of an invasive bivalve (*Dreissena polymorpha*) on fish in the Hudson River estuary. *Canadian Journal of Fisheries and Aquatic Sciences* 61: 924–941.
Thompson, John N. 1999. The evolution of species interactions. *Science* 284: 2116–2118.
Weigel, Marlene, ed., 1999. *Encyclopedia of biomes.* UXL, Farmington Hills, Michigan.
Woodward, Susan L., 2003. *Biomes of Earth: Terrestrial, aquatic, and human-dominated.* Greenwood Publishing, Westport, Connecticut.

Chapter 7

Alling, Abigail, Mark Nelson, and Sally Silverstone. 1993. *Life under glass: The inside story of Biosphere 2.* Biosphere Press.
Capra, Fritjof. 1996. *The web of life: A new scientific understanding of living systems.* Anchor Books Doubleday, New York.
Carpenter, Edward J., and Douglas G. Capone, eds. 1983. *Nitrogen in the marine environment.* Academic Press, New York.
Committee on Environment and Natural Resources, 2000. *An integrated assessment: Hypoxia in the northern Gulf of Mexico.* CENR, National Science and Technology Council, Washington, D.C.
Ferber, Dan. 2004. Dead zone fix not a dead issue. *Science* 305: 1557.
Field, Christopher B., et al., 1998. Primary production of the biosphere: Integrating terrestrial and oceanic components. *Science* 281: 237–240.
Jacobson, Michael, et al. 2000. *Earth system science from biogeochemical cycles to global changes.* Academic Press.
Keller, Edward A. 2004. *Introduction to environmental geology,* 3rd ed. Prentice Hall, Upper Saddle River, New Jersey.
Larsen, Janet. 2004. Dead zones increasing in world's coastal waters. *Eco-economy update #41,* 16 June 2004. Earth Policy Institute, www.earth-policy.org/Updates/Update41.htm.
Mississippi River/Gulf of Mexico Watershed Nutrient Task Force. 2001. *Action plan for reducing, mitigating, and controlling hypoxia in the northern Gulf of Mexico.* Washington, D.C.
Mitsch, William J., et al. 2001. Reducing nitrogen loading to the Gulf of Mexico from the Mississippi River Basin: Strategies to counter a persistent ecological problem. *BioScience* 51: 373–388.
Montgomery, Carla. 2005. *Environmental geology,* 7th ed. McGraw-Hill, New York.

National Oceanic and Atmospheric Administration: National Ocean Service. 2000. Hypoxia in the Gulf of Mexico: Progress toward the completion of an integrated assessment. www.nos.noaa.gov/products/pubs_hypox.html.

National Science and Technology Council, Committee on Environment and Natural Resources. 2003. *An assessment of coastal hypoxia and eutrophication in U.S. waters*. National Science and Technology Council, Washington, D.C.

Rabalais, Nancy N., R. E. Turner, and D. Scavia. 2002. Beyond science into policy: Gulf of Mexico hypoxia and the Mississippi River. *BioScience* 52: 129–142.

Rabalais, Nancy N., R. E. Turner, and W. J. Wiseman, Jr. 2002. Hypoxia in the Gulf of Mexico, a.k.a. "The dead zone." *Annual Review of Ecology and Systematics* 33: 235–263.

Raloff, Janet. 2004. Dead waters: Massive oxygen-starved zones are developing along the world's coasts. *Science News* 165: 360–362. June 5, 2004.

Raloff, Janet. 2004. Limiting dead zones: How to curb river pollution and save the Gulf of Mexico. *Science News* 165: 378–380. June 12, 2004.

Ricklefs, Robert E., and Gary L. Miller. 2000. *Ecology,* 4th ed. W. H. Freeman and Co., New York.

Schlesinger, William H. 1997. *Biogeochemistry: An analysis of global change,* 2nd ed. Academic Press, London.

Skinner, Brian J., and Stephen C. Porter. 2003. *The dynamic earth: An introduction to physical geology,* 5th ed. John Wiley and Sons, Hoboken, New Jersey.

Smith, Robert L., and Thomas M. Smith. 2001. *Ecology and field biology,* 6th ed. Benjamin Cummings, San Francisco.

Stiling, Peter. 2002. *Ecology: Theories and applications,* 4th ed. Prentice Hall, Upper Saddle River, New Jersey.

Takahashi, Taro. 2004. The fate of industrial carbon dioxide. *Science* 305: 352–353.

Turner, R. Eugene, and Nancy N. Rabalais. 2003. Linking landscape and water quality in the Mississippi River Basin for 200 years. *BioScience* 53: 563–572.

Vitousek, Peter M., et al. 1997. Human alteration of the global nitrogen cycle: Sources and consequences. *Ecological Applications* 7: 737–750.

Whittaker, Robert H. 1975. *Communities and ecosystems,* 2nd ed. Macmillan, New York.

Chapter 8

Cohen, Joel E. 1995. *How many people can the Earth support?* W. W. Norton, New York.

Cohen, Joel E. 2003. Human population: The next half century. *Science* 302: 1172–1175.

Cohen, Joel E. 2005. Human population grows up. *Scientific American* 293(3): 48–55.

De Souza, Roger-Mark, et. al., 2003. Critical links: Population, health, and the environment. *Population Bulletin 58(3),* 48 pp. Population Reference Bureau, Washington, D.C.

Eberstadt, Nicholas. 2000. China's population prospects: Problems ahead. *Problems of Post-Communism* 47: 28.

Ehrlich, Paul R., and John P. Holdren. 1971. Impact of population growth: Complacency concerning this component of man's predicament is unjustified and counterproductive. *Science* 171: 1212–1217.

Ehrlich, Paul R., and Anne H. Ehrlich. 1990. The population explosion. Touchstone, New York.

Engelman, Robert, Brian Halweil, and Danielle Nierenberg. 2002. Rethinking population, improving lives. Pp. 127–148 in *State of the world 2002,* Worldwatch Institute and W. W. Norton, Washington D.C., and New York.

Greenhalgh, Susan. 2001. Fresh winds in Beijing: Chinese feminists speak out on the one-child policy and women's lives. *Signs: Journal of Women in Culture & Society* 26: 847–887.

Harrison, Paul, and Fred Pearce, eds. 2000. *AAAS atlas of population & environment.* University of California Press, Berkeley.

Hesketh, Therese, and Wei Xing Zhu, 1997. Health in China: The one child family policy: The good, the bad, and the ugly. *British Medical Journal* 314: 1685.

Holdren, John P. and Ehrlich, Paul R. 1974. Human population and the global environment. *American Scientist* 62: 282–292.

Kane, Penny. 1987. *The second billion: Population and family planning in China.* Penguin Books, Australia, Ringwood, Victoria.

Kane, Penny, and Ching Y. Choi. 1999. China's one child family policy. *British Medical Journal* 319: 992.

Mastny, Lisa. 2005. HIV/AIDS crisis worsening worldwide. Pp. 68–69 in *Vital signs 2005.* Worldwatch Institute and W. W. Norton, Washington, D.C. and New York.

Mastny, Lisa, and Richard P. Cincotta. 2005. Examining the connections between population and security. Pp. 22–41 in *State of the world 2005.* Worldwatch Institute and W. W. Norton, Washington, D.C. and New York.

McDonald, Mia, with Danielle Nierenberg. 2003. Linking population, women, and biodiversity. Pp. 38–61 in *State of the world 2003,* Worldwatch Institute and W. W. Norton, Washington D.C., and New York.

Meadows, Donella, Jørgen Randers, and Dennis Meadows. 2004. *Limits to growth: The 30-year update.* Chelsea Green Publishing Co., White River Junction, Vermont.

Notestein, Frank. 1953. Economic problems of population change. Pp. 13–31 in *Proceedings of the Eighth International Conference of Agricultural Economists.* Oxford University Press, London.

O'Brien, Stephen J., and Michael Dean. 1997. In search of AIDS-resistance genes. *Scientific American* 277: 44–51.

Population Reference Bureau. 2005. *2005 World Population Data Sheet.* Population Reference Bureau, Washington, D.C., and John Wiley & Sons, Hoboken, New Jersey.

Redefining Progress. Programs: Sustainability indicators. www.rprogress.org/newprograms/sustIndi/index.shtml.

Riley, Nancy E. 2004. *China's population: New trends and challenges.* Population Bulletin 59(2), 40 pp. Population Reference Bureau, Washington, D.C.

UNAIDS and World Health Organization. 2005. *AIDS epidemic update: December 2005.* UNAIDS and WHO, New York.

United Nations Economic and Social Commission for Asia and the Pacific. 2005. *2005 ESCAP population data sheet.* UNESCAP, New York.

United Nations Environment Programme. 2003. *Africa environment outlook: Past, present, and future perspectives.* UNEP, New York.

United Nations Population Division. 2004. *World population prospects: The 2004 revision.* UNPD, New York.

United Nations Population Fund. UNFPA, the 2005 World Summit and the millennium development goals. UNFPA. www.unfpa.org/icpd.

United Nations Population Fund. *State of world population 2005.* UNFPA, New York.

United States Census Bureau. www.census.gov. Wackernagel, Mathis, and William Rees. 1996. *Our ecological footprint: Reducing human impact on the earth.* New Society Publishers, Gabriola Island, British Columbia, Canada.

Chapter 9

Ashman, Mark R., and Geeta Puri. 2002. *Essential soil science: A clear and concise introduction to soil science.* Blackwell Publishing, Malden, Massachusetts.

Brown, Lester R. 2002. World's rangelands deteriorating under mounting pressure. *Eco-Economy Update #6,* 5 February 2002. Earth Policy Institute, www.earth-policy.org/Updates/Update6.htm.

Brown, Lester R. 2004. *Outgrowing the Earth: The food security challenge in an age of falling water tables and rising temperatures.* Earth Policy Institute, Washington, D.C.

Charman, P. E. V., and Brian W. Murphy. 2000. *Soils: Their properties and management,* 2nd ed. Oxford University Press, South Melbourne, Australia.

Curtin, Charles G. 2002. Integration of science and community-based conservation in the Mexico/U.S. borderlands. *Conservation Biology* 16: 880–886.

Diamond, Jared. 1999. *Guns, germs, and steel: The fates of human societies.* W. W. Norton, New York.
Diamond, Jared, and Peter Bellwood. 2003. Farmers and their languages: The first expansions. *Science* 300: 597–603.
Food and Agriculture Organization of the United Nations. 2001. Conservation agriculture: Case studies in Latin America and Africa. *FAO Soils Bulletin No. 78.* FAO, Rome.
Glanz, James. 1995. *Saving our soil: Solutions for sustaining Earth's vital resource.* Johnson Books, Boulder, Colorado.
Goudie, Andrew. 2000. *The human impact on the natural environment,* 5th ed. MIT Press, Cambridge, Massachusetts.
Fox, Stephen. 1985. *The American conservation movement: John Muir and his legacy.* University of Wisconsin Press, Madison.
Halweil, Brian. 2002. Farmland quality deteriorating. Pp. 102–103 in *Vital signs 2002.* Worldwatch Institute and W. W. Norton, Washington D.C., and New York.
Harrison, Paul, and Fred Pearce, eds. 2000. *AAAS atlas of population & environment.* University of California Press, Berkeley.
Jenny, Hans. 1941. *Factors of soil formation: A system of quantitative pedology.* McGraw-Hill, New York.
Larsen, Janet. 2003. Deserts advancing, civilization retreating. *Eco-Economy Update #23,* 27 March 2003. Earth Policy Institute, www.earth-policy.org/Updates/Update23.htm.
Millennium Ecosystem Assessment. 2005. *Ecosystems and human well-being: Desertification synthesis.* Millennium Ecosystem Assessment and World Resources Institute.
Morgan, R. P. C. 2005. *Soil erosion and conservation,* 3rd ed. Blackwell, London.
Natural Resources Conservation Service. 2001. *National resources inventory 2001: Soil erosion.* NRCS, USDA, Washington, D.C.
Natural Resources Conservation Service. Soils. NRCS, USDA. http://soils.usda.gov.
Pieri, Christian, et al. 2002. *No-till farming for sustainable rural development.* Agriculture & Rural Development Working Paper. International Bank for Reconstruction and Development, Washington, D.C.
Pierzynski, Gary M., et al. 2005. *Soils and environmental quality,* 3rd ed. CRC Press, Boca Raton, Florida.
Pretty, Jules, and Rachel Hine. 2001. *Reducing food poverty with sustainable agriculture: A summary of new evidence.* Center for Environment and Society, University of Essex. *Occasional Paper 2001–2.*
Richter, Daniel D. Jr., and Daniel Markewitz. 2001. *Understanding soil change: Soil sustainability over millennia, centuries, and decades.* Cambridge University Press, Cambridge.
Ritchie, Jerry C. 2000. Combining cesium-137 and topographic surveys for measuring soil erosion/deposition patterns in a rapidly accreting area. TEKTRAN, USDA Division of Agricultural Research, January 14, 2000.
Shaxson, T. F. 1999. The roots of sustainability, concepts and practice: Zero tillage in Brazil. *ABLH Newsletter ENABLE; World Association for Soil and Water Conservation (WASWC) Newsletter.*
Soil Science Society of America. 2001. Internet glossary of soil science terms. www.soils.org/sssagloss.
Stocking, M. A. 2003. Tropical soils and food security: The next 50 years. *Science* 302: 1356–1359.
Trimble, Stanley W., and Pierre Crosson. 2000. U.S. soil erosion rates—myth and reality. *Science* 289: 248–250.
Troeh, Frederick R., and Louis M. Thompson. 2004. *Soil and soil fertility,* 6th ed. Blackwell Publishing, London.
Troeh, Frederick R., J. Arthur Hobbs, and Roy L. Donahue 2004. *Soil and water conservation for productivity and environmental protection,* 4th ed. Prentice Hall, Upper Saddle River, New Jersey.
United Nations Convention to Combat Desertification, 2001. *Global alarm: Dust and sandstorms from the world's drylands.* UNCCD and others, Bangkok, Thailand.
United Nations Environment Programme. 2002. Land. Pp. 62–89 in *Global environment outlook 3 (GEO-3).* UNEP and Earthscan Publications, Nairobi and London.
Uri, Noel D. 2001. The environmental implications of soil erosion in the United States. *Environmental Monitoring and Assessment* 66: 293–312.
Wilkinson, Bruce H. 2005. Humans as geologic agents: A deep-time perspective. *Geology* 33: 161–164.

Chapter 10

Bazzaz, Fakhri A. 2001. Plant biology in the future. *Proceedings of the National Academy of the United States of America* 98: 5441–5445.
Brown, Lester R. 2004. *Outgrowing the Earth: The food security challenge in an age of falling water tables and rising temperatures.* Earth Policy Institute, Washington, D.C.
Buchmann, Stephen L., and Gary Paul Nabhan. 1996. *The forgotten pollinators.* Island Press/Shearwater Books, Washington, D.C./Covelo, California.
Commission for Environmental Cooperation. 2004. *Maize and biodiversity: The effects of transgenic maize in Mexico.* CEC Secretariat.
[Correspondence to *Nature,* various authors]. 2002. *Nature* 416: 600–602, and 417: 897–898.
Fedoroff, Nina, and Nancy Marie Brown, 2004. *Mendel in the kitchen: A scientist's view of genetically modified foods.* National Academies Press, Washington, D.C.
Food and Agriculture Organization of the United Nations. 2004. *The state of world fisheries and aquaculture, 2004.* FAO, Rome.
Gardner, Gary, and Brian Halweil. 2000. *Underfed and overfed: The global epidemic of malnutrition.* Worldwatch Paper #150. Worldwatch Institute, Washington, D.C.
Halweil, Brian. 2004. *Eat here: Reclaiming homegrown pleasures in a global supermarket.* Worldwatch Institute, Washington, D.C.
Halweil, Brian. 2005. Aquaculture pushes fish harvest higher. Pp. 26–27 in *Vital signs 2005.* Worldwatch Institute and W. W. Norton, Washington, D.C. and New York.
Halweil, Brian. 2005. Grain harvest and hunger both grow. Pp. 22–23 in *Vital signs 2005.* Worldwatch Institute and W. W. Norton, Washington, D.C., and New York.
Halweil, Brian, and Danielle Niereberg. 2004. Watching what we eat. Pp. 68–95 in *State of the world 2004.* Worldwatch Institute and W. W. Norton, Washington, D.C., and New York.
Harrison, Paul, and Fred Pearce, eds. 2000. *AAAS atlas of population & environment.* University of California Press, Berkeley.
International Food Information Council. 2004. Food biotechnology. IFIC, Washington, D.C. www.ific.org/food/biotechnology/index.cfm.
James, Clive. 2004. *Global status of GM crops, their contribution to sustainability, and future prospects.* International Service for the Acquisition of Agri-biotech Applications.
Kuiper, Harry A. 2000. Risks of the release of transgenic herbicide-resistant plants with respect to humans, animals, and the environment. *Crop Protection* 19: 773.
Liebig, Mark A., and John W. Doran. 1999. Impact of organic production practices on soil quality indicators. *Journal of Environmental Quality* 28: 1601–1609.
Losey, John E., Raynor, Linda S., and Carter, Maureen E. 1999. Transgenic pollen harms monarch larvae. *Nature* 399: 214.
Maeder, Paul, et al. 2002. Soil fertility and biodiversity in organic farming. *Science* 296: 1694–1697.
Mann, Charles C. 2002. Transgene data deemed unconvincing. *Science* 296: 236–237.
Manning, Richard. 2000. *Food's frontier: The next green revolution.* North Point Press, New York.
Miller, Henry I., and Gregory Conko. 2004. *The frankenfood myth: How protest and politics threaten the biotech revolution.* Praeger Publishers, Westport, Connecticut.
Nierenberg, Danielle. 2005. *Happier meals: Rethinking the global meat industry.* Worldwatch Paper #171. Worldwatch Institute, Washington, D.C.
Nierenberg, Danielle. 2005. Meat production and consumption rise. Pp. 24–25 in *Vital signs 2005.* Worldwatch Institute and W. W. Norton, Washington, D.C., and New York.

Nierenberg, Danielle, and Brian Halweil. 2005. Cultivating food security. Pp. 62–79 in *State of the world 2005.* Worldwatch Institute and W. W. Norton, Washington, D.C., and New York.

Nestle, Marion. 2002. *Food politics: How the food industry influences nutrition and health.* University of California–Berkeley Press, Berkeley.

Norris, Robert F., Edward P. Caswell-Chen, and Marcos Kogan. 2002. *Concepts in integrated pest management.* Prentice Hall, Upper Saddle River, New Jersey.

Paoletti, Maurizio G., and David Pimentel. 1996. Genetic engineering in agriculture and the environment: Assessing risks and benefits. *BioScience* 46: 665–673.

Pearce, Fred. 2002. The great Mexican maize scandal. *New Scientist* 174: 14 (15 June 2002).

Pedigo, Larry P., and Marlin E. Rice, 2006. *Entomology and pest management,* 5th ed. Prentice Hall, Upper Saddle River, New Jersey.

Pimentel, David. 1999. Population growth, environmental resources, and the global availability of food. *Social Research,* Spring 1999.

Pinstrup-Andersen, Per, and Ebbe Schioler, 2001. *Seeds of contention: World hunger and the global controversy over GM (genetically modified) crops.* International Food Policy Research Institute, Washington, D.C.

Polak, Paul. 2005. The big potential of small farms. *Scientific American* 293(3): 84–91.

Pringle, Peter. 2003. *Food, Inc.: Mendel to Monsanto—The promises and perils of the biotech harvest.* Simon and Schuster, New York.

Quist, David, and Ignacio H. Chapela. 2001. Transgenic DNA introgressed into traditional maize landraces in Oaxaca, Mexico. *Nature* 414: 541–543.

Ruse, Michael, and David Castle, eds. 2002. *Genetically modified foods: Debating technology.* Prometheus Books, Amherst, New York.

Schmeiser, Percy. Monsanto vs. Schmeiser. www. percyschmeiser.com/

Shiva, Vandana. 2000. *Stolen harvest: The hijacking of the global food supply.* South End Press, Cambridge, Massachusetts.

Smil, Vaclav. 2001. *Feeding the world: A challenge for the twenty-first century.* MIT Press, Cambridge, Massachusetts.

Stewart, C. Neal. 2004. *Genetically modified planet: Environmental impacts of genetically engineered plants.* Oxford University Press, Oxford.

Teitel, Martin, and Kimberly Wilson. 2001. *Genetically engineered food: Changing the nature of nature.* Park Street Press.

The Farm Scale Evaluations of spring-sown genetically modified crops. 2003. A themed issue from *Philosophical Transactions of the Royal Society of London B: Biological Sciences* 358(1439), 29 November 2003.

Tuxill, John. 1999. Appreciating the benefits of plant biodiversity. Pp. 96–114 in *State of the world 1999,* Worldwatch Institute and W. W. Norton, Washington D.C., and New York.

Westra, Lauren. 1998. Biotechnology and transgenics in agriculture and aquaculture: The perspective from ecosystem integrity. *Environmental Values* 7: 79.

Wolfenbarger, L. LaReesa 2000. The ecological risks and benefits of genetically engineered plants. *Science* 290: 2088.

Chapter 11

Balmford, Andrew, et al. 2002. Economic reasons for conserving wild nature. *Science* 297: 950–953.

Barnosky, Anthony D., et al. 2004. Assessing the causes of late Pleistocene extinctions on the continents. *Science* 306: 70–75.

Baskin, Yvonne. 1997. *The work of nature: How the diversity of life sustains us.* Island Press, Washington, D.C.

Bright, Chris. 1998. *Life out of bounds: Bioinvasion in a borderless world.* Worldwatch Institute and W. W. Norton, Washington D.C., and New York.

CITES Secretariat. "Convention on International Trade in Endangered Species of Wild Fauna and Flora. www. cites.org/

Convention on Biological Diversity. www.biodiv.org/

Daily, Gretchen C., ed. 1997. *Nature's services: Societal dependence on natural ecosystems.* Island Press, Washington, D.C.

Ehrenfeld, David W. 1970. *Biological conservation.* International Thomson Publishing, London.

Gaston, Kevin J., and John I. Spicer. 2004. *Biodiversity: An introduction,* 2nd ed. Blackwell, London.

Groom, Martha J., et al. 2005. *Principles of conservation biology,* 3rd ed. Sinauer Associates, Sunderland, Massachusetts.

Groombridge, Brian, and Martin D. Jenkins. 2002. *Global biodiversity: Earth's living resources in the 21st century.* UNEP, World Conservation Monitoring Centre, and Aventis Foundation; World Conservation Press, Cambridge, U.K.

Groombridge, Brian, and Martin D. Jenkins. 2002. *World atlas of biodiversity: Earth's living resources in the 21st century.* University of California Press, Berkeley.

Hanken, James. 1999. Why are there so many new amphibian species when amphibians are declining? *Trends in Ecology and Evolution* 14: 7–8.

Harris, Larry D. 1984. *The fragmented forest: Island biogeography theory and the preservation of biotic diversity.* University of Chicago Press, Chicago.

Harrison, Paul, and Fred Pearce, eds. 2000. *AAAS atlas of population & environment.* University of California Press, Berkeley.

Jenkins, Martin, 2003. Prospects for biodiversity. *Science* 302: 1175–1177.

Louv, Richard. 2005. *Last child in the woods: Saving our children from nature-deficit disorder.* Algonquin Books, Chapel Hill, North Carolina.

MacArthur, Robert H., and Edward O. Wilson. 1967. *The theory of island biogeography.* Princeton University Press, Princeton.

Mackay, Richard. 2002. *The Penguin atlas of endangered species: A worldwide guide to plants and animals.* Penguin, New York.

Maehr, David S., Reed F. Noss, and Jeffrey Larkin, eds. 2001. *Large mammal restoration: Ecological and sociological challenges in the 21st century.* Island Press, Washington, D.C.

Matthiessen, Peter. 2000. *Tigers in the snow.* North Point Press, New York.

Meegaskumbura, Madhava, et al. 2002. Sri Lanka: An amphibian hot spot. *Science* 298: 379.

Millennium Ecosystem Assessment. 2005. *Ecosystems and human well-being: Biodiversity synthesis.* Millennium Ecosystem Assessment and World Resources Institute.

Miquelle, Dale, Howard Quigley, and Maurice Hornocker. 1999. A habitat protection plan for Amur Tiger conservation: A proposal outlining habitat protection measures for the Amur Tiger. Hornocker Wildlife Institute.

Mooney, Harold A. and Richard J. Hobbs, eds. 2000. *Invasive species in a changing world.* Island Press, Washington, D.C.

Newmark, William D. 1987. A land-bridge perspective on mammal extinctions in western North American parks. *Nature* 325: 430.

Pimm, Stuart L., and Clinton Jenkins. 2005. Sustaining the variety of life. *Scientific American* 293(3): 66–73.

Primack, Richard B. 2004. *Essentials of conservation biology,* 3rd ed. Sinauer Associates, Sunderland, Massachusetts.

Quammen, David. 1996. *The song of the dodo: Island biogeography in an age of extinction.* Touchstone, New York.

Relyea, Rick, and Nathan Mills. 2001. Predator-induced stress makes the pesticide carbaryl more deadly to gray treefrog tadpoles. *Proceedings of the National Academy of Sciences, USA* 98: 2491–2496.

Rosenzweig, Michael L. 1995. *Species diversity in space and time.* Cambridge University Press, Cambridge.

Simberloff, Daniel S. 1969. Experimental zoogeography of islands: A model for insular colonization. *Ecology* 50: 296–314.

Simberloff, Daniel. 1998. Flagships, umbrellas, and keystones: Is single-species management passé in the landscape era? *Biological Conservation* 83: 247–257.

Simberloff, Daniel S., and Edward O. Wilson. 1969. Experimental zoogeography of islands: The colonization of empty islands. *Ecology* 50: 278–296.

Simberloff, Daniel S., and Edward O. Wilson. 1970. Experimental zoogeography of islands: A two-year record of colonization. *Ecology* 51: 934–937.

Soulé, Michael E. 1986. *Conservation biology: The science of scarcity and diversity.* Sinauer Associates, Sunderland, Massachusetts.

Takacs, David. 1996. *The idea of biodiversity: Philosophies of paradise.* Johns Hopkins University Press, Baltimore.
United Nations Environment Programme. 2002. Biodiversity. Pp. 120–149 in *Global environment outlook 3 (GEO-3).* UNEP and Earthscan Publications, Nairobi and London.
United Nations Environment Programme. 2003. Sustaining life on Earth: How the Convention on Biological Diversity promotes nature and human well-being. www.biodiv. org/doc/publications/guide.asp.
United States Fish and Wildlife Service. The endangered species act of 1973. Accessible online at http://endangered.fws.gov/esa.html.
Wilson, Edward O. 1984. *Biophilia.* Harvard University Press, Cambridge, Massachusetts.
Wilson, Edward O. 1992. *The diversity of life.* Harvard University Press, Cambridge, Massachusetts.
Wilson, Edward O. 1994. *Naturalist.* Island Press, Shearwater Books, Washington, D.C.
Wilson, Edward O. 2002. *The future of life.* Alfred A. Knopf, New York.
Wilson, Edward O., and Daniel S. Simberloff. 1969. Experimental zoogeography of islands: Defaunation and monitoring techniques. *Ecology* 50: 267–278.
World Conservation Union. 2005. IUCN Red List. www.iucnredlist.org/

Chapter 12

British Columbia Ministry of Forests. Introduction to Silvicultural Systems. www.for.gov.bc.ca/hfd/pubs/SSIntroworkbook/index.htm. British Columbia Ministry of Forests, Victoria, B.C.
Canadian Broadcasting Corporation. 1993. A little place called Clayoquot Sound. CBC broadcast, 13 April 1993. http://archives.cbc.ca/IDC-1-75-679-3918/Science_technology/clearcutting/clip6.
Clary, David. 1986. *Timber and the Forest Service.* University Press of Kansas, Lawrence.
Food and Agriculture Organization of the United Nations. 2005. *Global forest resources assessment.* FAO Forestry, Rome.
Foster, Bryan C., and Peggy Foster. 2002. *Wild logging: A guide to environmentally and economically sustainable forestry.* Mountain Press, Missoula, Montana.
Gardner, Gary. 2005. Forest loss continues. Pp. 92–93 in *Vital signs 2005.* Worldwatch Institute and W. W. Norton, Washington, D.C., and New York.
Harrison, Paul, and Fred Pearce, eds. 2000. *AAAS atlas of population & environment.* University of California Press, Berkeley.
Haynes, Richard W. and Gloria E. Perez, tech. eds. 2001. Northwest Forest Plan research synthesis. *Gen. Tech. Rep. PNW-GTR-498.* USDA Forest Service, Pacific Northwest Research Station, Portland, Oregon.
Jacobs, Lynn. 1991. *Waste of the West: Public lands ranching.* Lynn, Jacobs, Tucson, Arizona.
Landres, Peter, David R. Spildie, and Lloyd P. Queen. 2001. GIS applications to wilderness management: Potential uses and limitations. *Gen. Tech. Rep. RMRS-GTR-80.* USDA Forest Service, Rocky Mountain Research Station, Fort Collins, Colorado.
Myers, Norman, and Jennifer Kent. 2001. *Perverse subsidies: How misused tax dollars harm the environment and the economy.* Island Press, Washington, D.C.
National Forest Management Act of 1976. October 22, 1976 (P.O. 94–588, 90 Stat. 2949, as amended; 16 U.S.C.)
National Round Table on the Environment and the Economy, Environment Canada. Clayoquot Sound Biosphere Reserve. www.nrtee-trnee.ca/eng/programs/Current_Programs/Nature/Case-Studies/Clayoquot-Case-Study-Complete_e.htm.
Natural Resources Canada. 2005. *The state of Canada's forests, 2004-2005.* Natural Resources Canada, Ottawa.
Runte, Alfred. 1979. *National parks and the American experience.* University of Nebraska Press, Lincoln.
Sedjo, Robert A. 2000. *A vision for the US Forest Service.* Resources for the Future, Washington, D.C.
Singh, Ashbindu, et al. 2001. An assessment of the status of the world's remaining closed forests. United Nations Environmental Program, UNEP/DEWA/TR 01–2l, August 2001.
Smith, David M., et al. 1996. *The practice of silviculture: Applied forest ecology,* 9th ed. Wiley, New York.
Smith, W. Brad, et al. 2004. Forest resources of the United States, 2002. *Gen. Tech. Rep.* NC-241, North Central Research Station, USDA Forest Service, St. Paul, Minnesota.
Soulé, Michael E., and John Terborgh, eds. 1999. *Continental conservation.* Island Press, Washington, D.C.
Stegner, Wallace. 1954. *Beyond the hundredth meridian: John Wesley Powell and the second opening of the West.* Houghton Mifflin, Boston.
United Nations Environment Programme. 2002. Forests. Pp. 90–119 in *Global environment outlook 3 (GEO-3).* UNEP and Earthscan Publications, Nairobi and London.
USDA Forest Service. 2001. *U.S. forest facts and historical trends.* FS-696, March 2001.
U.S. National Park Service. 2002. *National Park Service statistical abstract 2002.* NPS Public Use Statistics Office, U.S. Department of the Interior, Denver, Colorado.

Chapter 13

Abbott, Carl. 2001. *Greater Portland: Urban life and landscape in the Pacific Northwest.* University of Pennsylvania Press.
Abbott, Carl. 2002. Planning a sustainable city. Pp. 207–235 in Squires, Gregory D., ed. *Urban sprawl: Causes, consequences, and policy responses.* Urban Institute Press, Washington, D.C.
Beck, Roy, et. al., 2003. *Outsmarting smart growth: Population growth, immigration, and the problem of sprawl.* Center for Immigration Studies, Washington, D.C.
Breuste, Jurgen, et al. 1998. *Urban ecology.* Springer-Verlag.
Cronon, William. 1991. *Nature's metropolis: Chicago and the great West.* W. W. Norton, New York.
Duany, Andres, et al. 2001. *Suburban nation: The rise of sprawl and the decline of the American dream.* North Point Press, New York.
Ewing, Reid, et al. 2002. *Measuring sprawl and its impact.* Smart Growth America.
Ewing, Reid, et al. 2003. Measuring sprawl and its transportation impacts. *Transportation Research Record* 1831: 175–183.
Girardet, Herbert. 2004. *Cities people planet: Livable cities for a sustainable world.* Academy Press.
Hall, Kenneth B. and Gerald A. Porterfield. 2001. *Community by design: New urbanism for suburbs and small communities.* McGraw-Hill, New York.
Horizon International. 2003. Efficient transportation for successful urban planning in Curitiba. www.solutions-site.org/artman/publish/printer_62.shtml.
Jacobs, Jane. 1992. *The death and life of great American cities.* Vintage.
Kalnay, Eugenia, and Ming Cai. 2003. Impact of urbanization and land-use change on climate. *Nature* 423: 528–531.
Kirdar, Uner, ed. 1997. *Cities fit for people.* United Nations, New York.
Litman, Todd. 2004. *Rail transit in America: A comprehensive evaluation of benefits.* Victoria Transport Policy Institute and American Public Transportation Association.
Logan, Michael F. 1995. *Fighting sprawl and city hall.* University of Arizona Press, Tucson.
Metro. www.metro-region.org.
New Urbanism. www.newurbanism.org.
Northwest Environment Watch. 2004. *The Portland exception: A comparison of sprawl, smart growth, and rural land loss in 15 U.S. cities.* Northwest Environment Watch, Seattle.
Portney, Kent. E. 2003. *Taking sustainable cities seriously: Economic development, the environment, and quality of life in American cities (American and comparative environmental policy).* MIT Press, Cambridge, Massachusetts.
Pugh, Cedric, ed. 1996. *Sustainability, the environment, and urbanization.* Earthscan Publications, London.

Sheehan, Molly O'Meara. 2001. *City limits: Putting the brakes on sprawl.* Worldwatch Paper #156. Worldwatch Institute, Washington, D.C.

Sheehan, Molly O'Meara. 2002. What will it take to halt sprawl? *World-Watch* (Jan/Feb 2002): 12–23.

Sprawl City. www.sprawlcity.org.

Stren, R., et al. 1992. *Sustainable cities: Urbanization and the environment in international perspective.* Westview Press, Boulder, Colorado, and San Francisco.

United Nations Environment Programme. 2002. Urban areas. Pp. 240–269 in *Global environment outlook 3 (GEO-3).* UNEP and Earthscan Publications, Nairobi and London.

United States Census Bureau. www.census.gov/

United States Environmental Protection Agency. Smart growth. www.epa.gov/smartgrowth.

Wiewel, Wim, and Joseph J. Persky., eds. 2002. *Suburban sprawl: Private decisions and public policy.* M. E. Sharpe, Armond, New York.

Chapter 14

Ames, Bruce N., Margie Profet, and Lois Swirsky Gold. 1990. Nature's chemicals and synthetic chemicals: Comparative toxicology. *Proceedings of the National Academy of the USA* 87: 7782–7786.

Bloom, Barry. 2005. Public health in transition. *Scientific American* 293(3): 92–99.

Cagen, S. Z., et al. 1999. Normal reproductive organ development in wistar rats exposed to bisphenol A in the drinking water. *Regulatory Toxicology and Pharmacology* 30: 130–139.

Carlsen, Elisabeth, et al. 1992. Evidence for decreasing quality of semen during past 50 years. *British Medical Journal* 305: 609–613.

Carson, Rachel. 1962. *Silent spring.* Houghton Mifflin, Boston.

Colburn, Theo, Dianne Dumanoski, and John P. Myers. 1996. *Our stolen future.* Penguin USA, New York.

Crain, D. Andrew, and Louis J. Guillette Jr. 1998. Reptiles as models of contaminant-induced endocrine disruption. *Animal Reproduction Science* 53: 77–86.

Guillette, Elizabeth A., et al. 1998. An anthropological approach to the evaluation of preschool children exposed to pesticides in Mexico. *Environmental Health Perspectives* 106: 347–353.

Guillette, Louis J. Jr., et al. 1999. Plasma steroid concentrations and male phallus size in juvenile alligators from seven Florida lakes. *General and Comparative Endocrinology* 116: 356–372.

Guillette, Louis J. Jr., et al. 2000. Alligators and endocrine disrupting contaminants: A current perspective. *American Zoologist* 40: 438–452.

Halweil, Brian. 1999. Sperm counts dropping. Pp. 148–149 in *Vital signs 1999.* Worldwatch Institute and W. W. Norton, Washington, D.C. and New York.

Hayes, Tyrone, et al. 2003. Atrazine-induced hermaphroditism at 0.1 PPB in American leopard frogs (*Rana pipiens*): Laboratory and field evidence. *Environmental Health Perspectives* 111: 568–575.

Hunt, Patricia A., et al. 2003. Bisphenol A exposure causes meiotic aneuploidy in the female mouse. *Current Biology* 13: 546–553.

Kolpin, Dana W., et al. 2002. Pharmaceuticals, hormones, and other organic wastewater contaminants in U.S. streams, 1999–2000: A national reconnaissance. *Environmental Science and Technology* 36: 1202–1211.

Landis, Wayne G., and Ming-Ho Yu. 2004. *Introduction to environmental toxicology,* 3rd ed. Lewis Press, Boca Raton, Florida.

Loewenberg, Samuel. 2003. E.U. starts a chemical reaction. *Science* 300: 405.

Millennium Ecosystem Assessment. 2005. *Ecosystems and human well-being: Health synthesis.* World Health Organization.

Manahan, Stanley E. 2000. *Environmental chemistry,* 7th ed. Lewis Publishers, CRC Press, Boca Raton, Florida.

McGinn, Anne Platt. 2000. *Why poison ourselves? A precautionary approach to synthetic chemicals* Worldwatch Paper #153. Worldwatch Institute, Washington, D.C.

McGinn, Anne Platt. 2002. Reducing our toxic burden. Pp. 75–100 in *State of the world 2002,* Worldwatch Institute and W. W. Norton, Washington, D.C. and New York.

McGinn, Anne Platt. 2003. Combating malaria. Pp. 62–84 in *State of the world 2003,* Worldwatch Institute and W. W. Norton, Washington, D.C. and New York.

Moeller, Dade. 2004. *Environmental health,* 3rd ed. Harvard University Press, 2004.

Nagel, S.C., et al. 1997. Relative binding affinity-serum modified access (RBA-SMA) assay predicts in vivo bioactivity of the xenoestrogens bisphenol A and octylphenol. *Environmental Health Perspectives* 105: 70–76.

National Center for Health Statistics, 2004. *Health, United States, 2004, with chartbook on trends in the health of Americans.* Hyattsville, Maryland.

National Center for Environmental Health; U.S. Centers for Disease Control and Prevention. 2005. *Third national report on human exposure to environmental chemicals.* NCEH Pub. No. 05-0570, Atlanta.

Pirages, Dennis. 2005. Containing infectious disease. Pp. 42–61 in *State of the world 2005.* Worldwatch Institute and W. W. Norton, Washington, D.C., and New York.

Renner, Rebecca. 2002. Conflict brewing over herbicide's link to frog deformities. *Science* 298: 938–939.

Rodricks, Joseph V. 1994. *Calculated risks: Understanding the toxicity of chemicals in our environment.* Cambridge University Press, Cambridge.

Salem, Harry, and Eugene Olajos. 1999. *Toxicology in risk assessment.* CRC Press, Boca Raton, Florida.

Spiteri, I. Daniel, Louis J. Guillette Jr., and D. Andrew Crain. 1999. The functional and structural observations of the neonatal reproductive system of alligators exposed *in ozo* to atrazine, 2,4-D, or estradiol. *Toxicology and Industrial Health* 15: 181–186.

Stancel, George, et al. 2001. "Report of the bisphenol A sub-panel." Chapter 1 in *National Toxicology Program's report of the endocrine disruptors low-dose peer review.* U.S. EPA and NIEHS, NIH.

Stockholm Convention on Persistant Organic Pollutants. www.pops.int/

United States Environmental Protection Agency. 2003. Pesticide registration program. www.epa.gov/pesticides/factsheets/registration.htm.

United States Environmental Protection Agency. 2003. Toxic Substances Control Act. www.epa.gov/region5/defs/html/tsca.htm.

United States Environmental Protection Agency. 2003. *EPA's draft report on the environment.* EPA 600-R-03-050. EPA, Washington, D.C.

Williams, Phillip L., Robert C. James, and Stephen M. Roberts, eds. 2000. *The principles of toxicology: Environmental and industrial applications,* 2nd ed. Wiley-Interscience, New York.

World Health Organization, 2004. *World health report 2004: Changing history.* WHO, Geneva, Switzerland.

Yu, Ming-Ho, 2004. *Environmental toxicology: Biological and health effects of pollutants,* 2nd ed. CRC Press, Boca Raton, Florida.

Chapter 15

American Rivers. 2002. *The ecology of dam removal: A summary of benefits and impacts.* American Rivers, Washington D.C., February 2002.

British Geographical Society and Bangladesh Department of Public Health Engineering. 2001. *Arsenic contamination of groundwater in Bangladesh.* Technical Report WC/00/19, Volume 1: Summary.

De Villiers, Marq, 2000. *Water: The fate of our most precious resource.* Mariner Books.

Gleick, Peter. H. 2003. Global freshwater resources: Soft-path solutions for the 21st century. *Science* 302: 1524–1527.

Gleick, Peter. H., et al. 2004. *The world's water 2004–2005: The biennial report on freshwater resources.* Island Press, Washington D.C.

Harrison, Paul, and Fred Pearce, eds. 2000. *AAAS atlas of population & environment.* University of California Press, Berkeley.

Harvey, Charles F., et al. 2002. Arsenic mobility and groundwater extraction in Bangladesh. *Science* 298: 1602–1606.

Institute of Governmental Studies, University of California, Berkeley. Imperial Valley-San Diego water transfer controversy. www.igs.berkeley.edu/library/htImperialWaterTransfer2003.html.

Jenkins, Matt. 2002. The royal squeeze. *High Country News* 35(1), January 20, 2003.

Jenkins, Matt. 2003. California's water binge skids to a halt. *High Country News* 34(17), September 16, 2002.
Marston, Ed. 2001. Quenching the big thirst. *High Country News* 33(10), May 21, 2001.
Millennium Ecosystem Assessment. 2005. *Ecosystems and human well-being: Wetlands and water synthesis*. Millennium Ecosystem Assessment and World Resources Institute.
Nickson, Ross, et al. 1998. Arsenic poisoning of Bangladesh groundwater. *Nature* 395: 338.
Postel, Sandra. 1999. *Pillar of sand: Can the irrigation miracle last?* W. W. Norton, New York.
Postel, Sandra. 2005. *Liquid assets: The critical need to safeguard freshwater ecosystems*. Worldwatch Paper #170. Worldwatch Institute, Washington, D.C.
Postel, Sandra, and Amy Vickers. 2004. Boosting water productivity. Pp. 46–67 in *State of the world 2004*. Worldwatch Institute and W. W. Norton, Washington, D.C. and New York.
Reisner, Marc. 1986. *Cadillac desert: The American West and its disappearing water*. Viking Penguin, New York.
Sampat, Payal. 2001. Uncovering groundwater pollution. Pp. 21–42 in *State of the world 2001*. Worldwatch Institute and W. W. Norton, Washington, D.C., and New York.
Sibley, George. 1997. A tale of two rivers: The desert empire and the mountain. *High Country News* 29(21), November 10, 1997.
Smith, Lingas, Rahman. 2000. Contamination of drinking water by arsenic in Bangladesh: A public health emergency. *Bulletin of the World Health Organization,* 78(9).
Stone, Richard. 1999. Coming to grips with the Aral Sea's grim legacy. *Science* 284: 30–33.
United Nations Environment Programme. 2002. Freshwater. Pp. 150–179 in *Global environment outlook 3 (GEO-3)*. UNEP and Earthscan Publications, Nairobi and London.
United Nations World Water Assessment Programme. 2003. *U.N. world water development report: Water for people, water for life.* Paris, New York, and Oxford, UNESCO and Berghahn Books.
United States Bureau of Reclamation, Lower Colorado Regional Office. www.usbr.gov/lc/region.
United States Environmental Protection Agency. 1998. *Wastewater primer*. EPA 833-K-98-001, Office of Wastewater Management, May 1998.
United States Environmental Protection Agency. 2003. *EPA's draft report on the environment*. EPA 600-R-03-050. EPA, Washington, D.C.
United States Environmental Protection Agency. 2003. *Water on tap: What you need to know*. EPA 816-K-03-007. Office of Water, EPA, Washington, D.C.
Wolf, Aaron T., et al. 2005. Managing water conflict and cooperation. Pp. 80–99 in *State of the world 2005*. Worldwatch Institute and W. W. Norton, Washington, D.C., and New York.
World Health Organization. 2000. *Global water supply and sanitation assessment 2000 report*. WHO, Geneva, Switzerland.
Youth, Howard. 2005. Wetlands drying up. Pp. 90–91 in *Vital signs 2005*. Worldwatch Institute and W. W. Norton, Washington, D.C., and New York.

Chapter 16

Ault, J. S., Bohnsack, J. A. and G.A. Meester. 1998. A retrospective (1979–1996) multi-species assessment of coral reef fish stocks in the Florida Keys. *Fishery Bulletin* 96: 395–414.
Baker, Beth. 1999. First aid for an ailing reef: Research in the Florida Keys National Marine Sanctuary. *BioScience* 49: 173–178.
Bellwood, David R., et al. 2004. Confronting the coral reef crisis. *Nature* 429: 827–833.
Causey, Billy D., Joanne Delaney, and Brian D. Keller. 2001. *The status of coral reefs of the Florida Keys*. Florida Keys National Marine Sanctuary.
[Correspondence to *Science,* various authors]. 2001. *Science* 295: 1233–1235.
Embassy of the People's Republic of China. 2001. Fishing statistics "basically correct," ministry says. Press release, 18 December 2001. http://saup.fisheries.ubc.ca/Media/Chinese_Embassy_18_Dec_2001.pdf.
Food and Agriculture Organization of the United Nations. 2002. Fishery statistics: Reliability and policy implications. FAO Fisheries Department, Rome. www.fao.org/DOCREP/FIELD/006/Y3354M/Y3354M00.HTM.
Food and Agriculture Organization of the United Nations. 2004. *The state of world fisheries and aquaculture, 2004*. FAO, Rome.
Garrison, Tom. 2005. *Oceanography: An invitation to marine science,* 5th ed. Brooks/Cole, San Francisco.
Gell, Fiona R., and Callum M. Roberts. 2003. Benefits beyond boundaries: The fishery effects of marine reserves. *Trends in Ecology and Evolution* 18: 448–455.
Halpern, Benjamin S., and Robert R. Warner. 2002. Marine reserves have rapid and lasting effects. *Ecology Letters* 5: 361–366.
Halpern, Benjamin S., and Robert R. Warner. 2003. Matching marine reserve design to reserve objectives. *Proceedings of the Royal Society of London B*: 270: 1871–1878.
Jackson, Jeremy B. C., et al. 2001. Historical overfishing and the recent collapse of coastal ecosystems. *Science* 293: 629–638.
Larsen, Janet. 2005. Wild fish catch hits limits—Oceanic decline offset by increased fish farming *Eco-economy indicators*. Earth Policy Institute, www.earth-policy.org/Indicators/Fish/2005.htm.
Mastny, Lisa. 2001. World's coral reefs dying off. Pp. 92–93 in *Vital signs 2001*. Worldwatch Institute and W. W. Norton, Washington D.C., and New York.
Myers, Ransom A., and Boris Worm. 2003. Rapid worldwide depletion of predatory fish communities. *Nature* 423: 280–283.
National Academy of Public Administration. 1999. *Protecting Our National Marine Sanctuaries*. Center for the Economy and the Environment, NAPA, Washington, D.C.
National Center for Ecological Analysis and Synthesis (NCEAS) and Communication Partnership for Science and the Sea (COMPASS), sponsors. 2001. *Scientific consensus statement on marine reserves and marine protected areas*. Available online at www.nceas.ucsb.edu/consensus.
National Oceanic and Atmospheric Administration (NOAA). Florida Keys National Marine Sanctuary. www. fknms.nos.noaa.gov.
National Research Council. 2003. *Oil in the sea III: Inputs, fates, and effects*. National Academies Press, Washington, D.C.
Norse, Elliott, and Larry B. Crowder, eds. 2005. *Marine conservation biology: The science of maintaining the sea's biodiversity.* Island Press, Washington, D.C.
Nybakken, James W., and Mark D. Bertness. 2004. *Marine biology: An ecological approach,* 6th ed. Benjamin Cummings, San Francisco.
Palumbi, Stephen. 2003. *Marine reserves: A tool for ecosystem management and conservation*. Pew Oceans Commission.
Pauly, Daniel, et al. 2002. Towards sustainability in world fisheries. *Nature* 418: 689–695.
Pauly, Daniel, et al. 2003. The future for fisheries. *Science* 302: 1359–1361.
Pew Oceans Commission. 2003. *America's living oceans: Charting a course for sea change*. A report to the nation. May 2003. Pew Oceans Commission, Arlington, Virginia.
Pinet, Paul R. 1999. *Invitation to oceanography,* 2nd ed. Jones & Bartlett, Boston.
Roberts, Callum M., et al. 2001. Effects of marine reserves on adjacent fisheries. *Science* 294: 1920–1923.
Sumich, James L., and John F. Morrissey, 2004. *Introduction to the biology of marine life,* 8th ed. Jones & Bartlett, Boston.
Thurman, Harold V., and Alan P. Trujillo, 2004. *Introductory oceanography,* 10th ed. Prentice Hall, Upper Saddle River, New Jersey.
United Nations Environment Programme. 2002. Coastal and marine areas. Pp. 180–209 in *Global environment outlook 3 (GEO-3)*. UNEP and Earthscan Publications, Nairobi and London.
United States Commission on Ocean Policy. 2004. *An ocean blueprint for the 21st century.* Final Report. Washington, D.C.

United States Department of Commerce. 1996. *Strategy for stewardship: Florida Keys National Marine Sanctuary final management plan/ environmental impact statement.* 3 vols. Dept. of Commerce, Washington, D.C.

United States Department of Commerce and United States Department of the Interior. Marine protected areas of the United States. www.mpa.gov.

Watson, Reginald, Lillian Pang, and Daniel Pauly. 2001. The marine fisheries of China: Development and reported catches. *Fisheries Centre Research Reports* 9(2). Fisheries Centre, University of British Columbia, Canada.

Watson, Reginald, and Daniel Pauly. 2001. Systematic distortions in world fisheries catch trends. *Nature* 414: 534–536.

Weber, Michael L. 2001. *From abundance to scarcity: A history of U.S. marine fisheries policy*. Island Press, Washington, D.C.

Chapter 17

Ahrens, C. Donald. 2003. *Meteorology today,* 7th ed. Brooks/Cole, San Francisco.

Akimoto, Hajime. 2003. Global air quality and pollution. *Science* 302: 1716–1719.

Bell, Michelle L., and Devra L. Davis. 2001. Reassessment of the lethal London fog of 1952: Novel indicators of acute and chronic consequences of acute exposure to air pollution. *Environmental Health Perspectives* 109(Suppl 3): 389–394.

Bernard, Susan M., et al. 2001. The potential impacts of climate variability and change on air pollution-related health effects in the United States. *Environmental Health Perspectives* 109(Suppl 2): 199–209.

Biscaye, Pierre E., et al. 2000. Eurasian air pollution reaches eastern North America. *Science* 290: 2258–2259.

Boubel, Richard W., et al., eds. 1994. *Fundamentals of air pollution,* 3rd ed. Academic Press, San Diego, California.

Bruce, Nigel, Rogelio Perez-Padilla, and Rachel Albalak. 2000. Indoor air pollution in developing countries: A major environmental and public health challenge. *Bulletin of the World Health Organization* 78: 1078–1092.

Cooper, C. David, and F. C. Alley. 2002. *Air pollution control,* 3rd ed. Waveland Press.

Davis, Devra. 2002. *When smoke ran like water: Tales of environmental deception and the battle against pollution.* Basic Books, New York.

Davis, Devra L., Michelle L. Bell, and Tony Fletcher. 2002. A look back at the London smog of 1952 and the half century since. *Environmental Health Perspectives* 110: A734.

Driscoll, Charles T., et al. 2001. *Acid rain revisited: Advances in scientific understanding since the passage of the 1970 and 1990 Clean Air Act Amendments.* Hubbard Brook Research Foundation. Science Links™ Publication. Vol. 1, no.1.

Ezzati Majid, and Daniel M. Kammen. 2001. Quantifying the effects of exposure to indoor air pollution from biomass combustion on acute respiratory infections in developing countries. *Environmental Health Perspectives* 109: 481–488.

Gardner, Gary. 2005. Air pollution still a problem. Pp. 94–95 in *Vital signs 2005.* Worldwatch Institute and W. W. Norton, Washington, D.C., and New York.

Godish, Thad. 2003. *Air quality,* 4th ed. CRC Press, Boca Raton, Florida.

Hunt, Andrew, et al. 2003. Toxicologic and epidemiologic clues from the characterization of the 1952 London smog fine particulate matter in archival autopsy lung tissues. *Environmental Health Perspectives* 111: 1209–1214.

Jacobson, Mark Z. 2002. *Atmospheric pollution: History, science, and regulation.* Cambridge University Press, New York.

Hoffman, Matthew J. 2005. *Ozone depletion and climate change: Constructing a global response.* SUNY Press, New York.

Kunzli, Nino, et al. 2000. Public-health impact of outdoor and traffic-related air pollution: A European assessment. *Lancet* 356: 795–801.

Lelieveld, Jos, et al. 2001. The Indian Ocean experiment: Widespread air pollution from South and Southeast Asia. *Science* 291: 1031–1036.

Likens, Gene E. 2004. Some perspectives on long-term biogeochemical research from the Hubbard Brook ecosystem study. *Ecology* 85: 2355–2362.

London, Stephanie J., and Isabelle Romieu. 2000. Health costs due to outdoor air pollution by traffic. *Lancet* 356: 782–783.

Molina, Mario J., and F. Sherwood Rowland. 1974. Stratospheric sink for chlorofluoromethanes: Chlorine atom catalyzed destruction of ozone. *Nature* 249: 810–812.

Pal Arya, S. 1998. *Air pollution: Meteorology and dispersion.* Oxford University Press, Oxford.

Parson, Edward A. 2003. *Protecting the ozone layer: Science and strategy.* Oxford University Press, Oxford.

United Nations Environment Programme. Montreal Protocol. http://hq.unep.org/ozone/Treaties_and_Ratification/2B_montreal_protocol.asp.

United Nations Environment Programme. 2002. Atmosphere. Pp. 210–239 in *Global environment outlook 3 (GEO-3).* UNEP and Earthscan Publications, Nairobi and London.

United States Environmental Protection Agency. 2003. *EPA's draft report on the environment.* EPA 600-R-03-050. Washington, D.C.

United States Environmental Protection Agency. 2003. *Latest findings on national air quality: 2002 status and trends.* EPA 454/K-03-001. Washington, D.C.

Wark, Kenneth, et al., 1997. *Air pollution: Its origin and control,* 3rd ed. Prentice Hall, Upper Saddle River, New Jersey.

World Health Organization. Indoor air pollution. WHO, Geneva, Switzerland. www.who.int/indoorair/en/index.html.

Chapter 18

Alley, Richard B. 2000. *The two-mile time machine: Ice cores, abrupt climate change, and our future.* Princeton University Press, Princeton, New Jersey.

Burroughs, William James. 2001. *Climate change: A multidisciplinary approach.* Cambridge University Press, Cambridge.

Drake, Frances. 2000. *Global warming: The science of climate change.* Oxford University Press, Oxford.

Dunn, Seth. 2001. Decarbonizing the energy economy. Pp. 83–102 in *State of the world 2001.* Worldwatch Institute and W. W. Norton, Washington, D.C., and New York.

Dunn, Seth, and Christopher Flavin. 2002. Moving the climate change agenda forward. Pp. 24–50 in *State of the world 2002.* Worldwatch Institute and W. W. Norton, Washington, D.C., and New York.

Gelbspan, Ross. 1997. *The heat is on: The climate crisis, the cover-up, the prescription.* Perseus Books, New York.

Gelbspan, Ross. 2004. *Boiling point: How politicians, big oil and coal, journalists, and activists are fueling the climate crisis—and what we can do to avert disaster.* Basic Books, New York.

Intergovernmental Panel on Climate Change. 2001. *IPCC third assessment report—Climate change 2001: The scientific basis.* World Meteorological Organization and United Nations Environment Programme.

Intergovernmental Panel on Climate Change. 2001. *IPCC third assessment report—Climate change 2001: Impacts, adaptations, and vulnerability.* World Meteorological Organization and United Nations Environment Programme.

Intergovernmental Panel on Climate Change. 2001. *IPCC third assessment report—Climate change 2001: Mitigation.* World Meteorological Organization and United Nations Environment Programme.

Intergovernmental Panel on Climate Change. 2001. *IPCC third assessment report—Climate change 2001: Synthesis report.* World Meteorological Organization and United Nations Environment Programme.

Intergovernmental Panel on Climate Change. 2001. *Technical Summary of the Working Group 1 report.*

Intergovernmental Panel on Climate Change. www.ipcc.ch/

Kareiva, Peter M., Joel G. Kingsolver, and Raymond B. Huey, eds. 1993. *Biotic interactions and global change.* Sinauer Associates, Sunderland, Massachusetts.

Karl, Thomas R., and Kevin E. Trenberth, 2003. Modern global climate change. *Science* 302: 1719–1723.
Lomborg, Bjorn. 2001. *The skeptical environmentalist: Measuring the real state of the world.* Cambridge University Press, Cambridge.
Mastny, Lisa. 2005. Global ice melting accelerating. Pp. 88–89 in *Vital signs 2005.* Worldwatch Institute and W. W. Norton, Washington, D.C., and New York.
Mayewski, Paul A., and Frank White. 2002. *The ice chronicles: The quest to understand global climate change.* University Press of New England, Hanover, New Hampshire.
McLachlan, Jason S., and Linda B. Brubaker. 1995. Local and regional vegetation change on the northeastern Olympic Peninsula during the Holocene. *Canadian Journal of Botany* 73: 1618–1627.
National Assessment Synthesis Team. 2000. *Climate change impacts on the United States: The potential consequences of climate variability and change.* U.S. Global Change Research Program. Cambridge University Press, Cambridge.
National Research Council, Board on Atmospheric Sciences and Climate, Commission on Geosciences, Environment, and Resources. 1998. *The atmospheric sciences: Entering the twenty-first century.* National Academies Press, Washington, D.C.
National Research Council, Committee on the Science of Climate Change, Division of Earth and Life Studies. 2001. *Climate change science: An analysis of some key questions.* National Academies Press, Washington, D.C.
Nordhaus, William D. 1998. Assessing the economics of climate change: An introduction. In Nordhaus, William D., ed., *Economic and policy issues in climate change.* Resources for the Future Press, Washington D.C.
Parmesan, Camille, and Gary Yohe. 2003. A globally coherent fingerprint of climate change impacts across natural systems. *Nature* 421: 37–42.
Real Climate. www.realclimate.org.
Root, Terry L., et al. 2003. Fingerprints of global warming on wild animals and plants. *Nature* 421: 57–60.
Sawin, Janet L. 2005. Climate change indicators on the rise. Pp. 40–41 in *Vital signs 2005.* Worldwatch Institute and W. W. Norton, Washington, D.C., and New York.
Schneider, Stephen H. and Terry L. Root, eds. 2002. *Wildlife responses to climate change: North American case studies.* Island Press, Washington, D.C.
Seinfeld, John H., and Spyros N. Pandis. 2006. *Atmospheric chemistry and physics,* 2nd ed. Wiley-Interscience, New York.
Shapiro, Robert J., Kevin A. Hassett, and Frank S. Arnold. 2002. *Conserving energy and preserving the environment: The role of public transportation.* American Public Transportation Association, July 2002.
Speth, James Gustave. 2004. *Red sky at morning: America and the crisis of the global environment.* Yale University Press, New Haven, Connecticut.
Stevens, William K. 1999. *The change in the weather: People, weather and the science of climate.* Delta Trade Paperbacks, New York.
Taylor, David. 2003. Small islands threatened by sea level rise. Pp. 84–85 in *Vital signs 2003.* Worldwatch Institute and W. W. Norton, Washington D.C. and New York.
Victor, David G. 2004. *Climate change: Debating America's policy options.* U.S. Council on Foreign Relations Press, Washington, D.C.
United Nations. United Nations Framework Convention on Climate Change. http://unfccc.int/2860.php.
United Nations. Kyoto Protocol. http://unfccc.int/resource/docs/convkp/kpeng.html.
United States Congress. House Committee on Science. 2001. Climate change: The state of the science. Hearing before the Committee on Science, House of Representatives, One Hundred Seventh Congress, first session, 14 March 2001.

Chapter 19

Association for the Study of Peak Oil and Gas. www.peakoil.net/
British Petroleum. 2005. *BP statistical review of world energy.* London, June 2005.
Campbell, Colin J. 1997. *The coming oil crisis.* Multi-Science Publishing Co., Essex, U.K.
Deffeyes, Kenneth S. 2001. *Hubbert's peak: The impending world oil shortage.* Princeton University Press, Princeton, New Jersey.
Deffeyes, Kenneth S. 2005. *Beyond oil: The view from Hubbert's peak.* Farrar, Straus, and Giroux, New York.
Douglas, D. C., P.E. Reynolds, and E. B. Rhode, eds. 2002. *Arctic Refuge coastal plain terrestrial wildlife research summaries. Biological science report.* USGS/BRD/BSR-2002-0001. United States Geological Survey, Washington, D.C.
Dunn, Seth. 2001. Decarbonizing the energy economy. Pp. 83–102 in *State of the world 2001,* Worldwatch Institute and W. W. Norton, Washington, D.C., and New York.
Energy Information Administration, U.S. Department of Energy. www.eia.doe.gov.
Energy Information Administration, U.S. Department of Energy. 1999. *Petroleum: An energy profile, 1999.* DOE/EIA-0545(99).
Energy Information Administration, U.S. Department of Energy. 2005. *Annual energy review 2004.* DOE/EIA-0384(2004). Washington, D.C.
Energy Information Administration, U.S. Department of Energy. 2005. *International energy annual 2003.* Washington, D.C.
Flavin, Christopher. 2005. Fossil fuel use surges. Pp. 30–31 in *Vital signs 2005.* Worldwatch Institute and W. W. Norton, Washington, D.C., and New York.
Freese, Barbara. 2003. *Coal: A human history.* Perseus Books, New York.
Goodstein, David. 2004. *Out of gas.* W. W. Norton, New York.
Holmes, Bob, and Nicola Jones. 2003. Brace yourself for the end of cheap oil. *New Scientist* (August 2, 2003): 9–11.
International Energy Agency. 2005. *Key world energy statistics 2005.* IEA Publications, Paris.
International Energy Agency. 2005. *World energy outlook 2005.* IEA Publications, Paris.
International Energy Agency. 2005. *Resources to reserves: Oil and gas technologies for the energy markets of the future.* IEA Publications, Paris.
Lovins, Amory B., et al. 2004. *Winning the oil endgame: Innovation for profits, jobs, and security.* Rocky Mountain Institute, Snowmass, Colorado.
Lovins, Amory B. 2005. More profit with less carbon. *Scientific American* 293(3): 74–83.
Nellemann, Christian, and Raymond D. Cameron. 1998. Cumulative impacts of an evolving oil-field complex on the distribution of calving caribou. *Canadian Journal of Zoology* 76: 1425–1430.
Pelley, Janet. 2001. Will drilling for oil disrupt the Arctic National Wildlife Refuge? *Environmental Science and Technology* 35: 240–247.
Powell, Stephen G. 1990. Arctic National Wildlife Refuge: How much oil can we expect? *Resources Policy* Sept. 1990: 225–240.
Prugh, Tom, et al. 2005. Changing the oil economy. Pp. 100–121 in *State of the world 2005.* Worldwatch Institute and W. W. Norton, Washington, D.C., and New York.
Ristinen, Robert A., and Jack J. Kraushaar, 1998. *Energy and the environment.* John Wiley and Sons, New York.
Russell, D. E. and P. McNeil. 2005. *Summer ecology of the Porcupine caribou herd.* Porcupine Caribou Management Board, Whitehorse, Yukon.
Sawin, Janet L. 2004. Making better energy choices. Pp. 24–45 in *State of the world 2004.* Worldwatch Institute and W. W. Norton, Washington, D.C. and New York.
Skinner, Brian J., and Stephen C. Porter. 2003. *The dynamic earth: An introduction to physical geology,* 5th ed. John Wiley and Sons, Hoboken, New Jersey.
United States Environmental Protection Agency. 2005. *Light-duty automotive technology and fuel economy trends: 1975 through 2005.* EPA420-R-05-001. EPA Office of Transportation and Air Quality, Washington, D.C.
United States Fish and Wildlife Service. 2001. Potential impacts of proposed oil and gas development on the Arctic Refuge's coastal plain: Historical overview and issues of concern. Web page of the Arctic National Wildlife Refuge, Fairbanks, Alaska. http://arctic.fws.gov/issues1.html.

United States Geological Survey. 2001. *The National Petroleum Reserve-Alaska (NPRA) data archive.* USGS Fact Sheet FS-024-01, March 2001.

United States Geological Survey. 2001. *Arctic National Wildlife Refuge, 1002 Area, petroleum assessment, 1998, including economic analysis.* USGS Fact Sheet FS-028-01, April 2001.

Walker, Donald A. 1997. Arctic Alaskan vegetation disturbance and recovery. Pp. 457–479 in *Disturbance and recovery in Arctic lands,* R.M.M. Crawford, ed. Kluwer Academic Publishers, Dordrecht, Netherlands.

Chapter 20

Aeck, Molly. 2005. Biofuel use growing rapidly. Pp. 38–39 in *Vital signs 2005.* Worldwatch Institute and W. W. Norton, Washington, D.C., and New York.

British Petroleum. 2005. *BP statistical review of world energy.* London, June 2005.

Chandler, David. 2003. America steels itself to take the nuclear plunge. *New Scientist* (August 9, 2003): 10–13.

Energy Information Administration, U.S. Department of Energy. www.eia.doe.gov.

Energy Information Administration. 2005. *Annual energy outlook 2005.* Washington, D.C.

Energy Information Administration, U.S. Department of Energy. 2005. *Annual energy review 2004.* DOE/EIA-0384(2004). Washington, D.C.

Energy Information Administration, U.S. Department of Energy. 2005. *International energy annual 2003.* Washington, D.C.

European Commission/International Atomic Energy Agency/World Health Organization. 1996. One decade after Chernobyl: Summing up the consequences of the accident. Summary of the conference results. Vienna, Austria, 8–12 April, 1996. EC/IAEA/WHO.

International Energy Agency. 2005. *Key world energy statistics 2005.* IEA Publications, Paris.

International Energy Agency. 2005. *World energy outlook 2005.* IEA Publications, Paris.

International Atomic Energy Agency. 2004. *Annual report 2003.* GC(48)/3.IAEA, Vienna, Austria.

International Atomic Energy Agency. *Nuclear power and sustainable development.*

IAEA Information Series 02-01574/FS Series 3/01/E/Rev.1. Vienna, Austria.

Klass, Donald L. 2004. Biomass for Renewable Energy and Fuels. In *The Encyclopedia of Energy,* Elsevier.

Lenssen, Nicholas. 2005. Nuclear power rises once more. Pp. 32–33 in *Vital signs 2005.* Worldwatch Institute and W. W. Norton, Washington, D.C., and New York.

Lovins, Amory B., et al. 2004. *Winning the oil endgame: Innovation for profits, jobs, and security.* Rocky Mountain Institute, Snowmass, Colorado.

Murray, Danielle. 2005. Ethanol's potential: Looking beyond corn. *Eco-economy Update #49,* 5 June 2005. Earth Policy Institute, http://www.earth-policy.org/Updates/2005/Update49.htm.

National Renewable Energy Lab, U.S. Department of Energy. www.nrel.gov.

Nuclear Energy Agency, OECD. 2002. *Chernobyl: Assessment of radiological and health impacts. (2002 Update of Chernobyl: Ten Years On).* OECD, Paris.

Nuclear Energy Agency. 2005. *NEA annual report 2004.* NEA, Organisation for Economic Co-operation and Development. OECD, Paris.

Office of Energy Efficiency and Renewable Energy, U.S. Department of Energy www.eere.energy.gov.

Organisation for Economic Co-operation and Development. 2000. *Business as usual and nuclear power.* OECD Publications, Paris.

REN21 Renewable Energy Policy Network. 2005. *Renewables 2005 global status report.* Worldwatch Institute, Washington, D.C.

Spadaro, Joseph V., Lucille Langlois, and Bruce Hamilton. 2000. Greenhouse gas emissions of electricity generation chains: Assessing the difference. *IAEA Bulletin* 42(2).

Swedish Bioenergy Association (SVEBIO). 2003. *Focus: Bioenergy.* Nos. 1–10. SVEBIO, Stockholm, 2003.

Swedish Energy Agency 2004. *Renewable electricity is the future's electricity.* Swedish Energy Agency, Eskilstuna, Sweden.

Swedish Energy Agency 2004. *Energy in Sweden: Facts and figures 2004.* Swedish Energy Agency, Eskilstuna, Sweden.

Swedish Energy Agency 2004. *The Swedish Energy Agency 2003.* Swedish Energy Agency, Eskilstuna, Sweden.

Swedish Energy Agency 2004. *Energy in Sweden 2003.* Swedish Energy Agency, Eskilstuna, Sweden.

U.N. Food and Agriculture Organization. *Biomass energy in ASEAN member countries.* FAO/ASEAN/EC. FAO Regional Wood Energy Development Programme in Asia, Bangkok, Thailand.

U.S. Environmental Protection Agency. Alternative fuels website. www.epa.gov/otaq/consumer/fuels/altfuels/altfuels.htm.

Chapter 21

American Wind Energy Association. 2005. *Global wind energy market report.* AWEA, Washington, D.C.

Ananthaswamy, Anil. 2003. Reality bites for the dream of a hydrogen economy. *New Scientist,* (November 15, 2003): 6–7.

Arnason, Bragi, and and Thorsteinn I. Sigfusson. 2000. Iceland—a future hydrogen economy. *International Journal of Hydrogen Energy* 25: 389–394.

Ásmundsson, Jón Knútur. 2002. Will fuel cells make Iceland the 'Kuwait of the North?' *World Press Review,* 15 February 2002.

Burkett, Elinor. 2003. A mighty wind. *New York Times magazine.* June 15, 2003.

Chow, Jeffrey, et al. 2003. Energy resources and global development. *Science* 302: 1528–1531.

DaimlerChrysler. 2003. *360 DEGREES/DaimlerChrysler Environmental Report 2003.* DaimlerChrysler AG, Stuttgart, Germany.

Dunn, Seth. 2000. The hydrogen experiment. *WorldWatch* 13: 14–25.

Energy Information Administration, U.S. Department of Energy. www.eia.doe.gov.

Energy Information Administration, U.S. Department of Energy. 2005. *Annual energy review 2004.* DOE/EIA-0384(2004). Washington, D.C.

Energy Information Administration, U.S. Department of Energy. 2005. *International energy annual 2003.* Washington, D.C.

Flavin, Christopher, and Seth Dunn. 1999. A new energy paradigm for the 21st century. *Journal of International Affairs* 53: 167–190.

Hirsch, Tim. 2001. Iceland launches energy revolution. *British Broadcasting Corporation News,* 24 December 2001.

Hydrogen & Fuel Cell Letter. 2003. World's first commercial hydrogen station opens in Iceland. *Hydrogen & Fuel Cell Letter* May 2003.

Idaho Wind Power Working Group for the Idaho Department of Water Resources Energy Division. 2002. *Idaho wind power development strategic plan.* Boise, Idaho.

Idaho Wind Power Working Group for the Idaho Department of Water Resources Energy Division. 2002. *Wind power potential in Idaho by county.* Boise, Idaho.

International Energy Agency. 2002. *Renewables in global energy supply: An IEA fact sheet.* IEA Publications, Paris.

International Energy Agency Renewable Energy Working Party. 2002. *Renewable energy . . . into the mainstream.* SITTARD, The Netherlands.

International Energy Agency. 2005. *Renewables information 2005.* IEA Publications, Paris.

Lovins, Amory B., et al. 2004. *Winning the oil endgame: Innovation for profits, jobs, and security.* Rocky Mountain Institute, Snowmass, Colorado.

Martinot, Eric, et al. 2002. Renewable energy markets in developing countries. *Annual Review of Energy and the Environment* 27: 309–48.

Martinot, Eric, Ryan Wiser, and Jan Hamrin. 2005. *Renewable energy markets and policies in the United States.* Center for Resource Solutions, San Francisco. www.martinot.info/Martinot_et_al_CRS.pdf.

Melis, Anastasios, et al. 2000. Sustained photobiological hydrogen gas production upon reversible inactivation of oxygen evolution in the green alga *Chlamydomonas reinhardtii. Plant Physiology* 122: 127–135.

National Renewable Energy Lab, U.S. Department of Energy. www.nrel.gov.

Office of Energy Efficiency and Renewable Energy, U.S. Department of Energy www.eere.energy.gov.

Office of Energy Efficiency and Renewable Energy, U.S. Department of Energy. 2005. *Wind power today: Federal wind program highlights.* DOE/GO-102005-2115. Washington, D.C.

Randerson, James. 2003. The clean green energy dream. *New Scientist* (August 16, 2003): 8–11.

Reeves, Ari, with Fredric Beck. 2003. *Wind energy for electric power: A REPP issue brief.* Renewable Energy Policy Project, Washington, D.C.

REN21 Renewable Energy Policy Network. 2005. *Renewables 2005 global status report.* Worldwatch Institute, Washington, D.C.

Ristinen, Robert A., and Jack J. Kraushaar, 1998. *Energy and the environment.* John Wiley and Sons, New York.

Rocky Mountain Institute webpage. Energy. RMI, Snowmass, Colorado. www.rmi.org/sitepages/pid17.php.

Sawin, Janet. 2004. *Mainstreaming renewable energy in the 21st century.* Worldwatch Paper 169. Worldwatch Institute, Washington, D.C.

Sawin, Janet L. 2005. Global wind growth continues. Pp. 34–35 in *Vital signs 2005.* Worldwatch Institute and W. W. Norton, Washington, D.C. and New York.

Sawin, Janet L. 2005. Solar energy markets booming. Pp. 36–37 in *Vital signs 2005.* Worldwatch Institute and W. W. Norton, Washington, D.C. and New York.

U.S. Department of Energy National Laboratory directors.1997. *Technology opportunities to reduce U.S. greenhouse gas emissions.* DOE, Washington, D.C.

Weisman, Alan. 1998. *Gaviotas: A village to reinvent the world.* Chelsea Green Publishing Co., White River Junction, Vermont.

World Alliance for Decentralized Energy. 2005. *World survey of decentralized energy 2005.* WADE, Edinburgh, Scotland.

Chapter 22

Allen, G. H. and R. A. Gearheart, eds. 1988. *Proceedings of a conference on wetlands for wastewater treatment and resource enhancement.* Humboldt State University, Arcata, California.

Ayres, Robert U., and Leslie W. Ayres. 1996. *Industrial ecology: Towards closing the materials cycle.* Edward Elgar Press, Cheltenham, U.K.

Beede, David N., and David E. Bloom. 1995. The economics of municipal solid waste. *World Bank Research Observer* 10: 113–150.

Diesendorf, Mark, and Clive Hamilton. 1997. *Human ecology, human economy.* Allen and Unwin, St. Leonards.

Edmonton, Alberta, City of. 2003. Waste management. www.edmonton.ca/portal/server.pt/gateway/PTARGS_0_2_104_0_0_35/http%3B/cmsserver/COEWeb/environment+waste+and+recycling/waste/

Energy Information Administration. Municipal solid waste. EIA, U.S. Department of Energy, Washington, D.C. www.eia.doe.gov/cneaf/solar.renewables/page/mswaste/msw.html.

Gitlitz, Jenny, and Pat Franklin. 2004. *The 10-cent incentive to recycle,* 3rd ed. Container Recycling Institute, Arlington, Virginia.

Graedel, Thomas E., and Braden R. Allenby, 2002. *Industrial ecology,* 2nd ed. Prentice Hall, Upper Saddle River, New Jersey.

Integrated Waste Services Association. WTE: About waste-to-energy. www.wte.org/waste.html. IWSA, Washington, D.C.

Kaufman, Scott, et al. 2004. The state of garbage in America. *Biocycle* 45: 31–41.

Lilienfeld, Robert, and William Rathje. 1998. *Use less stuff: Environmental solutions for who we really are.* Ballantine, New York.

Manahan, Stanley E. 1999. *Industrial ecology: Environmental chemistry and hazardous waste.* Lewis Publishers, CRC Press, Boca Raton, Florida.

McDonough, William, and Michael Braungart. 2002. *Cradle to cradle: Remaking the way we make things.* North Point Press, New York.

McGinn, Anne Platt. 2002. Toxic waste largely unseen. Pp. 112–113 in *Vital signs 2002.* Worldwatch Institute and W. W. Norton, Washington D.C., and New York.

New York City Department of Planning. Fresh Kills: Landfill to landscape. www.nyc.gov/html/dcp/html/fkl/ada/about/1_0.html/

New York City Department of Planning. Fresh Kills lifescape. www.nyc.gov/html/dcp/html/fkl/fkl_index.shtml.

New York City Department of Sanitation. 2000. Closing the Fresh Kills landfill. *The DOS Report,* Feb. 2000.

Rathje, William, and Colleen Murphy. 2001. *Rubbish! The archeology of garbage.* University of Arizona Press, March 2001.

Smith, Ronald S. 1998. *Profit centers in industrial ecology.* Quorum Books, Westport.

Socolow, Robert H., et al., eds. 1994. *Industrial ecology and global change.* Cambridge University Press, Cambridge.

United Nations Environment Programme. 2000. *International source book on environmentally sound technologies (ESTs) for municipal solid waste management (MSWM).* UNEP IETC, Osaka, Japan.

United States Environmental Protection Agency. 2005. *Municipal solid waste generation, recycling, and disposal in the United States: Facts and figures for 2003.* EPA530-F-05-003, EPA Office of Solid Waste and Emergency Response.

United States Environmental Protection Agency. Municipal solid waste. www.epa.gov/epaoswer/non-hw/muncpl.

Chapter 23

Bartlett, Peggy, and Geoffrey W. Chase, eds. 2004. *Sustainability on campus: Stories and strategies for change.* MIT Press, Cambridge, Massachusetts.

Brower, Michael, and Warren Leon. 1999. *The consumer's guide to effective environmental choices: Practical advice from the Union of Concerned Scientists.* Three Rivers Press, New York.

Brown, Lester. 2001. *Eco-economy: Building an economy for the Earth.* Earth Policy Institute and W. W. Norton, New York.

Brown, Lester. 2006. *Plan B 2.0: Rescuing a planet under stress and a civilization in trouble.* Earth Policy Institute and W. W. Norton, New York.

Creighton, Sarah Hammond. 1998. *Greening the ivory tower: Improving the environmental track record of universities, colleges, and other institutions.* MIT Press, Cambridge, Massachusetts.

Daly, Herman E. 1996. *Beyond growth.* Beacon Press, Boston.

Dasgupta, Partha, Simon Levin, and Jane Lubchenco. 2000. Economic pathways to ecological sustainability. *BioScience* 50: 339–345.

De Graaf, John, David Wann, and Thomas Naylor. 2002. *Affluenza: The all-consuming epidemic.* Berrett-Koehler Publishers, San Francisco.

Durning, Alan. 1992. *How much is enough? The consumer society and the future of the Earth.* Worldwatch Institute, Washington, D.C.

Erickson, Jon D., and John M. Gowdy. 2002. The strange economics of sustainability. *BioScience* 52: 212.

French, Hilary. 2004. Linking globalization, consumption, and governance. Pp. 144–163 in *State of the world 2004.* Worldwatch Institute and W. W. Norton, Washington, D.C., and New York.

Gardner, Gary. 2001. Accelerating the shift to sustainability. Pp. 189–206 in *State of the world 2001.* Worldwatch Institute and W. W. Norton, Washington, D.C., and New York.

Gardner, Gary, and Erik Assadourian. 2004. Rethinking the good life. Pp. 164–179 in *State of the world 2004.* Worldwatch Institute and W. W. Norton, Washington, D.C., and New York.

Gibbs, W. Wayt. 2005. How should we set priorities? *Scientific American* 293(3): 108–115.

Hawken, Paul. 1994. *The ecology of commerce: A declaration of sustainability.* HarperBusiness, New York.

Keniry, Julian. 1995. *Ecodemia: Campus environmental stewardship at the turn of the 21st century.* National Wildlife Federation, Washington, D.C.

Mastny, Lisa. 2002. Ecolabeling gains ground. Pp. 124–125 in *Vital signs 2002.* Worldwatch Institute and W. W. Norton, Washington D.C., and New York.

McMichael, A. J., et al. 2003. New visions for addressing sustainability. *Science* 302: 1919–1921.

Millennium Ecosystem Assessment. 2005. *Ecosystems and human well-being: General synthesis.* Millennium Ecosystem Assessment and World Resources Institute.

McIntosh, Mary, et al., 2001. *State of the campus environment: A national report card on environmental performance and sustainability in higher education.* National Wildlife Federation Campus Ecology.

Meadows, Donella, Jørgen Randers, and Dennis Meadows. 2004. *Limits to growth: The 30-year update.* Chelsea Green Publ. Co., White River Junction, Vermont.

National Research Council, Board on Sustainable Development. 1999. *Our common journey: A transition toward sustainability.* National Academies Press, Washington, D.C.

National Wildlife Federation. Campus Ecology. www.nwf.org/campusecology.

Office of Management and Budget, Executive Office of the President of the United States, Washington, D.C. 2003. *Informing regulatory decisions: 2003 report to Congress on the costs and benefits of federal and unfunded mandates on state, local, and tribal entities.* Washington, D.C., September 2003.

Sanderson, Eric W., et al. 2002. The human footprint and the last of the wild.*BioScience* 52: 891–904.

Schor, Juliet B., and Betsy Taylor, eds. 2002. *Sustainable planet: Solutions for the twenty-first century.* The Center for a New American Dream. Beacon Press, Boston.

Toor, Will, and Spenser W. Havlick. 2004. *Transportation and sustainable campus communities: Issues, examples, solutions.* Island Press, Washington, D.C.

United Nations. 2002. *Report of the World Summit on Sustainable Development, Johannesburg, South Africa, 26 August–4 September 2002.* United Nations, New York.

United Nations. 2002. *The road from Johannesburg: What was achieved and the way forward.* United Nations, New York.

United Nations Development Programme. 2002. *Human development report 2002.* Oxford University Press, Oxford.

United Nations Division for Sustainable Development. 1990. *Agenda 21.* Accessible online at www.un.org/esa/sustdev/documents/agenda21/index.htm.

United Nations Environment Programme. 2002. Outlook: 2002-2032. Pp. 319–400 in *Global environment outlook 3 (GEO-3).* UNEP and Earthscan Publications, Nairobi and London.

United States Environmental Protection Agency. 2000. *Regulatory impact analysis: Heavy-duty engine and vehicle standards and highway diesel fuel sulfur control requirements.* EPA420-R-00-026. EPA, Washington, D.C.

University Leaders for a Sustainable Future. www.ulsf.org.

Wackernagel, Mathis, Lillemor Lewan, and Carina Borgström-Hansson. 1999. Evaluating the use of natural capital with the ecological footprint. *Ambio* 28: 604.

Wilson, Edward O. 1998. *Consilience: The unity of knowledge.* Alfred A. Knopf, New York.

World Commission on Environment and Development. 1987. *Our common future.* Oxford University Press, Oxford.

Index

"Note to the reader: Italicized page numbers indicate highlighted text terms; tables are identified with a *t* following the page number and figures by *f*."

A horizon, *252*
AARP. *See* American Association of Retired Persons
abiotic factors, *3*, 189
  organisms and, 190–191
Aborigines, 27
acid fog, 515
acid rain, 515, 518. *See also* air pollution
  at Hubbard Brook Research Forest, 520–521
acidic, *98*
  definition of, 525
acidic deposition, 515*f*
  consequences of, 525–526
  definition of, *514*
  effects of, 515*t*
  harm caused by, 518*f*
  nitrogen oxides and, 515
  reduction of, 517–519
  sulfur dioxide and, 513
  as transboundary pollution, 514–517
acidity, 98
Ackerman, Frank, 661
active solar energy collection, *623–624*
  PV cells and, 625–626
acute exposure, to toxicants, *421*
Adam Joseph Lewis Center for Environmental Studies, green building design and, 678–679
adaptation, *119*. *See also* evolution; natural selection
  competition and, 152
adaptive management, *346*, 348
adaptive radiation, 121*f*
adaptive trait, *119*
adenine, 100
adenosine diphosphate (ADP), 197
adenosine triphosphate (ATP), 197
ADP. *See* adenosine diphosphate
aerobic decomposition, *561*
aerosols, cooling effect of, 533
aesthetic value, 48*t*, 49*f*, 50
affluence
  consumption from, 236
  human population and, 240
  TFR and, 230
affluenza, *688*
Africa. *See also specific countries*
  deforestation in, 350*f*
  HIV/AIDS and, 237–239
  population growth in, 236*f*
  TFR in, 228*t*, 232
age pyramid, 134, 134*f*
age structure, *134*, 224–226. *See also* demography
agricultural breeding, genetic engineering and, 288–290
agricultural land use, 361–364, 362*f*
  impacts of, 370
agricultural production
  challenges of, 19–20
  population growth and, 278–279
  problems with, 21
  reduced, 9
  revolution in, 19
agricultural revolution, *4*, 19, *250*. *See also* green revolution; pesticides
agriculture, 244–307. *See also* fertilizer; genetically modified (GM) crops; organism; pesticides; pollination; soil
  artificial selection and, 248
  beginning of, 248
  climate change and, 543–544
  contour farming in, *261–263*, *262f*
  crop rotation in, *261*, *262f*
  definition of, *246*
  drought and, 260*f*
  erosion and, 256–259
  feedlot, 296–298, 305
  fertilizers and, 266–267
  government agents and, 261*f*
  groundwater pollution by, 455
  historical developments in, 273
  industrialized, 248–250
  irrigation and, 264–265, 447*t*
  locally grown, 303
  monoculture and, *248–249*
  no-till, 245–246, *262f*
  organic, *300–304*, 301*t*, 304*f*
  origin of, 249*f*
  polyculture, 248–249
  population increase and, 248
  sustainable, 300–305
  swidden, 255*f*
  traditional, *248*
  water and, 440, 450–451
agroforestry, 263
AIDS. *See* HIV/AIDS
air
  temperature of, 500
  vertical mixing of, 503, 503*f*
air hazards, 405*t*
air pollutants
  decline in, 508*f*
  definition of, *505*
  EPA and, 527*f*
  toxic, *509*
air pollution, 11, 381, 496–527. *See also* Clean Air Act; greenhouse gases; pollution; smog
  biodiversity loss due to, 322–323
  decrease in, 508–509
  definition of, *505*
  indoor, 519–525
  major indicator trends of, 508*f*
  outdoor, 505–519
  policy for, 512
  sprawl and, 380
air quality
  monitoring of, 508
  in United States, 11
Alaska
  development of, 558
  drilling in, 557–559
  Exxon Valdez oil spill, 89–90, 112
  North Slope of, 557–559, 558*f*, 587*f*
algae, 105, 638*f*
  as hydrogen fuel source, 638–639
algal blooms, excess nutrients, 481
allelopathy, 156
allergens, *410*
alley cropping, 263
alligators, 410–411
  endocrine disruptors, 401–402, *410*
  hatching problems, 402
  reproductive biology of, 401–402
allopatric speciation, 122–124, 123*f*, 124*t*
alpine tundra, 176
altered communities, 168–170
alternative energy, 589–618
  new sources of, 619–645
altitude
  atmosphere and, 499*f*
  atmospheric pressure and, 500*f*
  patterns of, 177, 178*f*
Alvarez, Luis, 129
Alvarez, Walter, 128–129
ambient air pollution, 505–519
amensalism, *156*
American Association of Retired Persons (AARP), 80
American dream, sprawl and, 381
American Public Transportation Association (APTA), 549–550
amino acid, 100*f*
amphibian diversity, 322–323
  decline of, 322–323
amphibian extinction. *See* golden toad
anaerobic decomposition, *561*
Anderson, Jock R., 269
anemometers, 632, 632*f*
Angkor civilization, 7–10
anglerfish, 474*f*
animal(s). *See also* amphibian diversity; aquaculture; birds; communities; Endangered Species Act; evolution; feedlots agriculture
  known species of, 314*f*
  potential medicines from, 326*t*
animal products
  animal feed required for, 298*f*
  biodiversity loss due to, 323–324
  consumption of, 296–297
  hunting for, 324
  land required to raise, 298*f*
  water required to produce, 298*f*, 464*f*
animal rights, feedlot agriculture and, 297
animal testing, environmental hazards and, 417–420
Annan, Kofi, 218*f*
anthropocentrism, *32*, *32f*, 34, 53
anthropogenic health factors, 402
ANWR. *See* Arctic National Wildlife Refuge
APTA. *See* American Public Transportation Association
aquaculture, *299–300*
  approaches to, 305
  benefits of, 299
  disease spread by, 300
  negative impacts of, 299–300
  world production of, 299*f*
aquatic succession, 165–166
  primary, 166*f*
aquatic systems, biome-like patterns of, 177

aquifers, *205, 437–438*
sinkholes and, 442
unconfined, 437*f*
water loss consequences in, 442
Aral Sea, 444, 445*f*
Arctic National Wildlife Refuge (ANWR), 558–559, 577, 585
biodiversity and, 584
debates over, 558–559
development of, 581
drilling for oil in, 584
negative impacts of drilling in, 578
opponents to drilling in, 583*f*
arid desert, of Middle East, 7
arid land productivity, 259–260
Arkansas River, 150
Árnason, Bragi, 620–621
artesian aquifer, *438*
artesian wells, 437*f*
artificial selection, 12*f*, 121. *See also* evolution
agriculture and, 248
artificial wetlands, 204
wastewater treatment with, 460–461
asbestos, 406, 406*f*
Asia. *See also specific countries*
TFR in, 228*t*, 232
assumptions, of scientific method, 13
Atchafalaya River
Gulf of Mexico and, 188
runoff and, 202
atmosphere, *189*. *See also* biosphere; climate; weather
altitude and, 499*f*
carbon shifting to, 197
climate and, 501–502
composition of, 498, 525
definition of, 498
function of, 525
gaseous makeup of, 498*f*
insulation provided by, 530
layers of, 498–500
pressure and, 500*f*
properties of, 500
relative humidity of, *500*
structure of, 525
weather and, 501–502
atmospheric deposition, *514*
atmospheric pressure, *500*
atmospheric sampling, present conditions and, 535–537
atmospheric science, 498–505
atoll, 474
Maldives and, 529–530
atomic number, 91
atoms, *91–94*, 93*f*, 103*t*, 113, 130
bonding of, 94–96
compounds of, *95*
ATP. *See* adenosine triphosphate
atrazine, 411*f*
Australian national economy, uranium mining and, 27, 38
automotive technology, 549, 549*f*
biofuel and, 609–610
fuel efficiency and, 582, 582*f*
hybrid cars, 583*f*
autotrophs, *105*, 106*f*, 156, 192
B horizon, *253*
baby boom, 226*f*
Bacillus thuringiensis (Bt), *284*
persistence time of, 415
background extinction rate, 127, *316*, 339
bacteria
degradation by, 99*f*
denitrifying, *200*
environmental pollutants and, 92
hydrocarbons and, 89–90
nitrogen cycle and, 197–200
nitrogen-fixation and, *198–200*, 200*f*
bacteria levels, 76
baghouse, 656
Bajagua project, 74
Ball State University (BSU), 675–676
ballast water, 149
Bangladesh
arsenic in water of, 456–457
fertility decline in, 230
TFR in, 230–231, 230*f*
barrier island, 474
basic, *98*
bathymetric profile, 472*f*
bathymetry, definition of, 471–472
beaches, 65*f*
bacteria levels in, 76–77
bedrock, *251*
benefit
internal, 43
marginal, 42*f*
benthic ecosystems, *472*
zone of, 473*f*
benthic invertebrate animals, 158, 160
benthic zone, *437*
Bethlehem, uranium mining in, 29
bill, 81*f*
biocentric equality, 36
biocentrism, *32–33*, 32*f*
biocontrol, *284–285*, 284*f*
chemical methods and, 285
biodiesel, *609*
burning of, 609*f*
biodiversity, *20*, 50, 116–147, 121*f*, 308–341, *310*. *See also* conservation; evolution; extinction; species
ANWR and, 584
benefits of, 324–329, 339
causes for loss of, 319–324
challenges of, 19–20
concept of, 310*f*
conservation of, 141–143, 145, 327–328
coral reefs and, 474–476
Costa Rican, 122
determination of, 126
economic benefits of, 327
ecosystem function and, 325
ecosystem services and, 325
ecotourism and, 327
effects of, 340*f*
evolution and, 118–129, 145
extinction and, 126, 126*f*
food security and, 325–326
fragmentation and, 330
hotspots of, *336*, 337*f*
of intertidal zone, 477
leukemia treatments and, 326
levels encompassed by, 310–312
loss of, 318–319, 329, 339
measurements of, 313–314, 339
medicine and, 326–327, 326*t*, 327*t*
of open-ocean ecosystems, 473–474
organic agriculture and, 340*f*
parks and, 338, 368–369
protection of, 326*t*
reserves and, 368–369
scope of, 339
as source for medicines, 326–327
in tropical rainforests, 118*f*
uneven distribution of, 314–316
biofuel, *608*
automotives and, 609–610
biogenic gas, 574
landfills and, 574
biogeochemical cycles, 195–206
biogeography of islands, 329–331
biological conservation. *See* sustainability
biological control, *284–285*
biological diversity, *310*
evolution and, 122
of open-ocean ecosystems, 473–474
biological environmental hazards, 403, 403*f*
biological weathering, 251
biology. *See* conservation biology
biomass
electricity from, 610–611
human population and, 223–224
trophic levels and, 158
biomass energy, 158, 192, 211, 592, 607–612, 622*f*
benefits of, 611
derived from, *607*
drawbacks to, 611–612
forest residues as, 609*f*
new sources of, 608–609
scale of, 616
sources of, 608*t*, 616
biomes, 170–178, 170*f*, 179. *See also* communities; ecosystem; environmental systems
aquatic systems and, 177
communities and, 170–172
definition of, 170
biophilia, *327–328*
biopower, *608*
bioremediation, 112
bacteria and, *89–90*
oil spill and, *89–90*
biosolids, 460
biosphere, *189*
cost of services of, 50
Biosphere 2, 190–191
biosphere reserves, 367, 367*f*
biostimulation, 112
biotechnology, 276–307, *287*
transformations of, 290
biotic factors, *3*
biotic potential, 138–139
birds. *See also specific birds*
Easter Island and, 9
habitat reduction of, 319

birth, limits on, 5
birth rate, 135
    falling, *229*
bisphenol-A, 412–413
BLM. *See* Bureau of Land Management
blood pressure, sprawl and, 382
bonding, *95–96*
    hydrogen, 96
    ionic, *96*
Bonnicksen, Thomas M., 359
boreal forest, 176–177, 176*f*
Bosch, Carl, 200
Bottaccione Gorge, 128–129
bottle bills, 660–662, 662*f*
bottom-trawling, 485*f*
    by-catch of, 485
boundaries, of systems, 186–187
Brazil
    deforestation in, 352
    GM crops in, 293
    no-till agriculture in, 245–246
    WTO and, 83
breakdown products, 415–416
breast cancer rates, 411–412
breeder nuclear fission, 594–595
breweries, 665*f*
British Columbia, clear-cutting in, 343–344
Brown, Lester R., 694
Brundtland Report, 682–683
BSU. *See* Ball State University
Bt. *See Bacillus thuringiensis*
bubonic plague, 239
Bufo periglenes. See golden toad
Bureau of Land Management (BLM), *363*
bus system, in U. S., 388–389
buses, hydrogen-fueled, 642*f*
Bush administration, family planning and, 233–235
by-catch, *485*

C horizon, *253*
cadmium, 91
    removal of, 114
California, water scarcity in, 433–434, 450
California condor
    captive breeding of, 334
    preservation of, 334*f*
Cambrian period, 317
campaign contributions, *80*
campus gardens
    habitat restoration and, 681–682, 681*f*
    sustainable eating, 680
campus sustainability, 676–682, 697
    audit prior to implementation of, 677
    efforts of, 677*t*
    green building design, 678–679
    institutional purchases, 680
    Kyoto protocol and, 679–680
    landfill on lawn, 677*f*
    recycling and, 677–678
    student awareness and, 676–677
    transportation options, 681
    waste reduction and, 677–678
Canada
    age pyramid of 2005 of, 225*f*
    civil disobedience in, 343–344
    deforestation for growth of, 350–352
    logging in, 343–344, 349
cancer, *409*
    breast, 411–412
    lung, 37–38, 405
    testicular, 411–412, 411*f*
    thyroid, 604*f*
capitalist market economy, *38–39*
captive breeding, *334–335*
carbohydrates, *101,* 113
carbon
    chains of, 98–99
    sediment storage of, 195–197
carbon cycle, 195–197, 196*f*, 211
    fossil fuels and, 576
carbon dioxide, 532*t*. *See also* atmosphere; greenhouse gases
    atmospheric concentrations of, 535*f*
    global atmospheric concentrations of, 532*t*
    as primary greenhouse gas, 531–532
carbon dioxide emissions, 555*f*
    population density and, 398*f*
    sprawl and, 380
carbon monoxide, 507
carbon shifting, 197
carbon signature data, 95
carbon trapping, 213
    fossil fuels and, 195–197
    oceans and, 197
    sediments and, 195–197
carbon-based-life, 113
carcinogens, *409. See also* cancer
cardiovascular disease, 404*f*
Caribbean
    island biogeography and, 331*f*
    TFR in, 228*t*
Carlton, James T., 169
carnivores, tertiary consumers and, 157
carrying capacity, *136,* 145
    change in, 137–138
    of grazing land, 267–270
Carson, Rachel, 67*f*, 680
    DDT and, 408–409, 408*f*
    synthetic chemicals and, 408–409
case law, 63
Cassandras ideas, 21
    traditional economics and, 44–45
catalytic converter, 690*f*
categorical imperative, 30
CDC. *See* Centers for Disease Control
cell membranes, phosphorus cycle and, 197
cells, *102,* 103*t*, 130
    diagram of eukaryotic, 102*f*
    formation of, 109–110
    types of, 102–103
cellular respiration, chemical energy and, *107*
Centers for Disease Control (CDC), 423
Central America, 7–10
Central Park, 392
Centrally planned economy, *39*
CESPT. *See* Comisión Estatal de Servicios Publicos de Tijuana
CFCs. *See* chlorofluorocarbons
CGCMs. *See* coupled general circulation models
chaparral, 177, 177*f*
Chapela, Ignacio H., 294
chelating agents, 92
chemical elements, abundance of, 91*t*
chemical energy, 104
    celluar respiration and, *107*
chemical environmental hazards, 403, 403*f*
chemical pesticides, 282–283
chemical testing, 425*f*
    industry or government, 427
chemical weathering, 251
chemicals
    in commercial use, 408*t*
    synthetic, 407–409, 407*f*, 414*f*
    water solubility of, 414
chemistry. *See also* elements; energy
    energy, life and, 88–115
    environment and, 90–103, 112
chemoautotrophic hypothesis, 109, 113
chemosynthesis, 107–108, 113
Chernobyl nuclear reactor, 590–591
    accident at, 592–593, 597–600, 600*f*
    anxiety caused by, 602–603
    deaths and, 598
    health impacts of, 604–605
    sarcophagus of, 597
    thyroid cancer rates and, 604*f*
Chesapeake Bay, dead zone in, 195
Chicago
    forest preserve education program, 386*f*
    plan of, 384, 384*f*
    urban parks in, 392–393, 392*f*
children
    malnutrition in, 280*f*
    one-child policy and, 217–218, 240
    pesticide use and, 418–419
Chilean wine palm, 8
China. *See* People's Republic of China
chlorofluorocarbons (CFCs), *513*
    Montreal Protocol and, 513–514
    ozone depletion from, 516–517
    ozone reactions with, 513*f*
chlorophyll, 106
chloroplasts, 106
chronic exposure, to toxicants, *421*
CITES. *See* Convention on International Trade in Endangered Species of Wild Fauna and Flora
city parks, 391–393, 392*f*
city planning, 384–385, 397
    definition of, *384*
civil disobedience, 343–344
civilization
    fall of, 7
    past mistakes of, 7–10
    resources vital to, 344–345
clan warfare, 9
classical economics, *41,* 54
classification, 311*f*
clay, *254*
Clayoquot Sound
    clear-cutting, 343–344
    tree battle at, 343–344
Clean Air Act, 83, 85, 86, 498
    adequacy of, 518
    amendments to, 73, 521
    of 1990, 507
    of 1970, 507, 508–509

Clean Air Act—*continued*
- petroleum products and, 573
- smog cleanup deadline of, 512
- United States and, 506–507

Clean Water Act of 1977, 69, 70, 78, 79, 85, 149, 455–456
clear-cutting, 356*f*, 357*f*
- British Columbia and, 343–344
- of old growth forests, 343–344

Clements, Frederick, 166, 179
climate
- atmosphere and, 501–502
- definition of, *501*, 530
- factors of, 533–534

climate change, 20, 140. *See also* global climate change; greenhouse gases; sea level rise
- agriculture and, 543–544
- biodiversity loss due to, 324
- Canadian model of, 544*f*
- CGCMs and, 538–540
- debate over, 546–547
- desertification and, *259*, 260
- economic debates concerning, 553
- estimates of, 540–545
- forest changes and, 534–544, 545*f*
- forest loss and, 9
- freshwater systems and, 544–545
- future impacts of, 543–545
- Hadley model of, 544*f*
- human health and, 545–546
- international treaties on, 550
- Maldives and, 545
- marine systems and, 544–545
- methods of studying, 534–540
- potential responses to, 553
- precautionary principle and, 552
- predictions of, 540–545, 543*t*
- responses to, 547
- uncertainties of, 546–547

climate models, 539*f*
climate system, currents and, 470*f*
climatic changes, 530
climatographs, 171
climax community, 166
Cline, Timothy, 234
cloning, 334–335
closed cycle approach, 636–637
closed systems, 39, *187*
cloud forest, 117–118
- conservation of, 143–144
- golden toad and, 127

clumped distribution, 133*f*, *134*
coal, 564–565
- burning of, 510*f*
- consumers of, 564*t*
- definition of, *564*
- formation of, 564*f*
- history of use of, 564
- mining and, 565, 576–577
- origin of, 586
- peat and, 565
- producers of, 564*t*
- quality of, 564–565
- seams of, 565*f*
- smog from, 497–498
- subsidies, 72

Cobb, John, 47
coevolution, 154–155
cofiring, 610
cogeneration, *583*
cohesion, 96
cold front, *502*, 503*f*
cold pollution, 454
college campus sustainability. *See* campus sustainability
college ecological footprint reduction, 676
Colorado River
- diversion of, 443
- flow of, 444*f*
- location of, 433–434
- water scarcity and, 433–434
- watershed of, 443*f*

combustion, 655
Comisión Estatal de Servicios Publicos de Tijuana (CESPT), 81
command-and-control approach, *71*, 75, 85
commensalism, *156*
commons, tragedy of, 6
communities, *130*
- alteration of, 168–170
- biomes and, 170–172
- cohesiveness of, 166
- composition of, 139–141
- disturbance response of, 163
- ecology of, 148–181
- economic factors of, 142
- organism's roles within, 159–163
- restoration of, 169–170
- social factors of, 142

community
- climax, 166
- definition of, 156

community ecology, 156–170, 184
- scope of, *130*

community gardens, 393, 393*f*
community-based conservation, *336–337*
community-supported agriculture, 303
competition, *151–152*, 179
- adaptation and, 152

competitive exclusion, 151
compost, 266
composting, 648, 658, 671
- definition of, *658*
- growth of, 652*f*

compound, *95*
- ionic, *96*

concession, 352–353
conference presentations, 18
confined aquifer, *438*
conservation, 34–35
- of biodiversity, 141–143, 145
- biodiversity and, 327–328
- community-based, *336–337*
- economic strategies for, 337
- ecosystem services and, 338
- efforts of, 339
- of electricity, 548
- of energy, 582–585
- ethical aspects of, 327–328, 328*f*
- increased efficiency, 582–585
- international, 335–336
- keystone species and, 162
- marine, 465–495
- need for, 585
- personal choice and, 582–585
- of pollinators, 286
- treaties for, 335–336
- of water, 269, 451

conservation biology, 308–341, 329–337, 339
- definition of, *329*
- of marine ecosystems, 486–492

conservation districts, *261*
conservation ethic, *34*
Conservation Reserve Program, 272
constitutional amendments, environmental policy and, 65
consumers, *107*, 156–157
- ballot vote by, 693
- herbivores and, 157
- secondary, 157
- tertiary, 157
- wallet vote of, 693

consumption, environmental change and, 19
consumptive use, *440*
containment building, 594
contaminated site cleanup, 670–671
continental shelves, *472*
continental slope, 472
contingent valuation, *48*
contour farming, *261–263*, *262f*
contraceptive use, global per capita, 235*t*
control, *14*
control rods, 594
controlled burns, *360*
controlled experiment, *14*
convection currents, 504*f*
convective circulation, *501*
Convention on Biological Diversity, *335*
Convention on International Trade in Endangered Species of Wild Fauna and Flora (CITES), *335*
conventional law, *81–82*
convergent plate boundary, 209, 209*f*
coral reefs
- biodiversity and, 474–476
- definition of, *474*
- nourishment of, 475
- worldwide decline of, 475–476

corals, 474
- definition of, 474–475
- elkhorn, 475*f*
- seafan, 475*f*
- staghorn, 475*f*

core, *208*
Coriolis effect, 471*f*
- definition of, *505*

Cornucopian ideas, 21
correlation, *14–15*, 16
corridors, *369*
Costa Rica
- biodiversity in, 122
- bioprospecting in, 327
- climate change in, 140
- cloud forest of, 117–118
- conservation in, 142–143, 142*f*
- tropical rainforests and, 122

Costanza, Robert, 50–51, 51*f*

cost-benefit analysis, *42*
marginal curve of, 42*f*
of natural systems, 51
short-term, 43
costs
external, *43,* 44*f*
internal, 43
COTE. *See* Council on the Environment
Council on the Environment (COTE), 676
coupled general circulation models (CGCMs), 543
climate change and, 538–540
covalent bond, *95–96*
polar, 96
cover crops, 245
Cretaceous-Tertiary event (K-T event), 127–128
boundary of, 128
criteria pollutants, 507–508
definition of, *507*
crop diversity
food security from, 325–326
preservation of, 295–296, 305
crop rotation, *261, 262f*
crop-land matching, 450–451
crude birth rate, 135
crude death rate, 135
crude oil, 110–111
distillation columns and, 572*f*
formation of, *566–568*
oceans and, 479*f*
refinement of, 572–573
crust, *208*
Cryptosporidium, 60*t*
Cuba, organic agriculture in, 303–304, 304*f*
cultural environmental hazards, 403, 403*f*
cultural value, 48*t,* 49*f*
culture, 28–30
definition of, *28*
influence of, 53
perception of environment and, 28–29
worldview and, 29*f*
Curitiba, Brazil, efficient transport in, 388–389
current energy sources, 559*t*
currents
climate system and, 470*f*
definition of, *470*
mapping of, 471
speed of, 470–471
customary law, *81–82*
principles of, 82
Cuyahoga River, pollution in, 68, 68*f*
cyanobacteria, 105
biogeochemical cycles and, 195
oxygen production and, 110
cycles of populations, 153
cytosine, 100

Daly, Herman, 46–47
dams, 446–447, 446*f,* 613*f. See also* freshwater; surface water; water
benefits of, 447*t*
building of, 615
controversy around, 448
costs of, 447*t*
definition of, *446*
failure of, 447*t*
hydroelectric power and, 444–449, 612–615
impacts of, 446*f*
migratory fish and, 448
need for, 448
removal of, 448–449
riparian environments and, 448
tidal differences and, 635
Darwin, Charles, *119*
data, *15*
data analysis, replicates and, 15
DDT, 10*f,* 428*t*
Carson and, 67*f,* 408*f*
fat-solubility of, 416*f*
Lake Apopka, Florida, 401
malaria and, 10
mosquitoes and, 10
persistence time of, 415
dead zone, 183–184, 202–203, 210
known locations of, 194–195
map of, 202
seasonal timing of, 183
understanding of, 187–189
death
increases and, 5
leading causes of, 404*f*
death rate, 135
falling, *229*
December solstice, 502*f*
decomposers, 157–158
deep ecology, *36*
deep-well injection, 670*f*
of hazardous waste, 668–669
deforestation, 34, 272*f,* 331*f*
in Africa, 350*f*
definition of, *350*
in developing nations, 352–354
in Latin America, 350*f*
for national growth, 350–352
wood demand and, 349–350
demand, 42
demographic changes, 238–240
demographic fatigue, 239–240
demographic transition, 228–229, 229*f*
failure to advance through, 240
stages of, *228–229*
demography, 222–229
definition of, *223*
factors of, 223–226
fundamentals of, 241
Dene Nation, uranium mining and, 37
denitrification, 199–200
density, 132–133, 138
density-dependent factors, *138,* 151
density-independent factors, *138*
deoxyribonucleic acid (DNA), *100,* 110
mutations in, 123
phosphorus cycle and, 197
recombinant, *287*
structure of, 100*f*
dependent variable, *14*
deposition, 256
desalination, 450
Persian Gulf plant for, 450*f*
descriptive observational studies, 16–17
desert, 7, 174–175, 175*f*
grassland of, 178*f*
desertification, *259,* 260
detritivores, 157–158
Detroit, Michigan, 380
development, 22
of Alaska, 558
of ANWR, 581
approaches to, 379*f*
definition of, 682
sustainable, 22, 682–685
transit-oriented, 388–390
zoning and, *385–386*
dikes, 444–446
definition of, 445
dilemmas, 3
dining service sustainability, 680
dinoflagellate, 481*f*
directional selection, 119, 120*f*
directives, of EU, 83
disease, 404–405
disease pathogens, 154
aquaculture and, 300
dispersion, 133–134
disruptive selection, 119, 120*f*
distillation, 572–573
distribution
of age, *134*
of population, *133–134*
of water, 438–439
disturbance
community response to, 163
severe, 163–166
divergent plate boundaries, *208–209,* 209, 209*f*
diverging populations, 123
history of, 124
diversification of species, 124–126
DNA. *See* deoxyribonucleic acid
Dobson spectrophotometer, 516*f*
doldrums, 505
dolphin-safe tuna, 51
domesticated animals, 120–122, 122*f*
Donora, Pennsylvania, smog in, 510, 510*f*
dose-response analysis, 429
toxicology and, 420–421
dose-response curve, *420*
linear, 420*f,* 427
threshold in, 420*f*
double-canoes, Polynesia and, 9
downwelling, definition of, *471*
dreaming-track stories, 32
driftnetting, 485*f*
bycatch of, 485
drinking water, 447*t*
treatment of, 457–458
drought, agriculture and, 260*f*
due diligence, 82
dump scavenging, 651*f*
dust bowl, *260*
dust storm, 506*f*
Dutch elm disease, 167
dynamic equilibrium, *186*

E horizon, *252*
Earth
atmosphere of, 525, 530
axis of, 502*f*

Earth—*continued*
biomes of, 170–178
climate of, 530–534
climate system of, 531*f*, 553
core of, *208*
crust of, *208*, 208*f*
curvature of, 501*f*
early, 108, 113
Easter Island model of, 9
environmental systems of, 184–189
geography of, 207–209
history of, 688*f*
interacting systems of, 211
as island, 3–10, 696
mantle of, *208*
orbit of, 533*f*
origin of life and, 108–112, 108*f*
planet of life, 310–316
sun and, 530
to sun distance, 500–501
tilt variation of, 533*f*
warmth of, 530
water cycle of, 48
water supply on, 434*f*
working of, 206–210
Earth cyles
biogeochemical cycles, 195–206
carbon cycle, 195–197, 211, 576
hydrologic cycle, *204–205*, 205*f*, 437–438, 462
nitrogen cycle, *198*, 199*f*, 200–204, 211
phosphorus cycle, *197*, 198*f*, 211
rock cycle, 206–207, 207, 212
Earth Day, 68, 69*f*
Easter Island
birds on, 9
clan warfare on, 9
forest loss on, 9
forests of, 8
hauhau tree on, 9
lesson of, 8–9
as model for earth, 9
overpopulation and, 221
plant species on, 9
ecocentrism, 32*f*, *33*, 34
ecofeminism, 36
ecolabeling, *50–51*
ecological communities, 156–170
ecological economics, 40*f*, 54
ecological economists, *45*
ecological footprint, *6–7*, *7f*, *19f*
affluence and, 236
calculations of, 25
colleges and, 675
global, 236*f*
ecological organization, 130*t*, 145
levels of, 129–131
ecological restoration, 168–170, 179
ecological systems, zebra mussel impacts on, 150
ecological webs, 184
ecology, *16*
community, 148–181
deep, *36*
focuses of, 130
root of, 38
study of, 130–141
teachings of, 684–685
urban, *395–396*
economic activity
conventional view of, 40*f*
environmental protection and, 683–684
environmental view of, 40*f*
economic benefits, environmental costs and, 42
economic conservation strategies, 337
economic development, 20*t*
economic growth, 44, 54, 55
sustainability and, 685–687
sustainability of, 45
economic health, 54
economic need, energy conservation and, 582
economic progress, environmental quality and, 52
economics. *See also* natural resources; poverty
approaches to, 38–53
classical, 39–41, 54
definition of, *38*
environmental implications of, 38–53
environmental tradeoffs of, 38
ethics and, 28
government intervention in, 39
long-term effects and, 43
neoclassical, 41–44, 54
policy and, 60
root of, 38
sprawl and, 384
subsidies and, 71–72
types of, 40*f*
waste generation and, 663–664
economy
definition of, *38*
environment and, 39, 684, 697
types of, 38–39
ecosystem(s), 189–195, 211. *See also* biomes; communities
analysis of, 190–191
changes in, 20*t*
conceptualization of, 191
definition of, *130, 189*
degradation of, 62
economic value of, 50
erosion and, 256–259
forest, 347*f*
GM alterations of, 294
interactions in, 191–192
living and nonliving interactions in, 190–191
open-ocean, 473–474
types of, 193*f*
UV exposure to, 513
ecosystem degradation, eutrophication and, 188
ecosystem diversity, 310*f*, *312*, 339
ecosystem ecology, 182–214
scope of, *130*
ecosystem engineers, 162
ecosystem function, biodiversity and, 325
ecosystem productivity, nutrients and, 192–195
ecosystem services, *39*, 40*f*
biodiversity and, 325
conservation and, 338
cost of, 50
forests and, 348
Millennium Ecosystem Assessment and, 692*f*
monetary value of, 48–49
nonmarket values of, *48*, 48*t*
types of, 41*t*
value of, 51*f*
wetland, 436
ecosystem-based management, *346*
ecotones, *192*
ecotourism, *143*
biodiversity and, 327
cons of, 144
Florida Keys and, 466–467
human impacts and, 466–467
pros of, 144
ED_{50}. *See* effective-dose-50%
Edmonton, Alberta
composting facility in, 662*f*
reduction and recycling in, 662–663
educational value, 48*t*, 49*f*
effective-dose-50% (ED_{50}), *420*, 420*f*
Egypt, typical family in, 237*f*
Ehrlich, Paul, 5–6, 278
IPAT and, 220
EIS. *See* environmental impact statement
El Niño, *534*
conditions during, 536*f*
temperature during, 537*f*
understanding, 536–537
electricity
average per capita use in U.S. of, 645*f*
biomass and, 610–611
coal and, 566–567, 566*f*
cogeneration and, 583
conservation of, 548
generation of, 548, 566–567, 591*f*, 625–626
geothermal generation of, 633–634
nuclear fission creates, 594
renewable sources of, 548
electrolysis
equation for, *638*
hydrogen power and, 638
electronic waste, 668–669
electrons, 91*f*, 92, 92*f*
elements, *91–94*
emergent properties, *186*, 186*f*
Emerson, Ralph Waldo, *33*
emigration, *135*
emissions
decrease in air, 508*f*
of fossil fuels, 576
permits, 73*f*
reduction of, 447*t*, 550*t*
reduction strategies and, 548–552
routine radioactive, 606
Endangered Species Act (ESA), *331–334*
congress and, 333
endemic species, *127*, *336*
endocrine disruptors, *410*
research on, 412–413
widespread effects of, 410–412
energy. *See also* fossil fuels; hydroelectric power; nuclear power; photosynthesis
alternatives for, 589–645
biomass and, 158, 192, 211
challenges of, 19–20
chemical, *104*
climate system and, 531

conservation of, 582–585, 586
current U. S. consumption of, 592*f*
definition of, *103*
developed nations and, 562–563, 608*f*
developing nations and, 562–563, 608*f*
flow of, 113, 179, 191, 211
food webs and, *158–159*
fundamentals of, 103–108
geothermal, *107–108,* 633–635
from hydrogen, 637–642
incineration and, 657
from light, 105–106
from natural resources, 559–561, 560*f*
new alternatives for, 619–645, 622*f*
oceans and, 478–479
per capita consumption of, 563*f*
petroleum (*See also* petroleum)
quality of, 104–105
respiration and, 158
solar, 500–501, 623–627
sources currently used, 559*t,* 586
sources of, 559–563
tidal, 635–636
transfer of, 105*f,* 115, 158
types of, 103–104, 113
wind, 628–633
world supply of, 591*f*
energy consumption, for transit modes, 389*f*
enriched uranium, 593–594
Entamoeba histolytica, 60*t*
entropy, *104,* 105*f,* 113
environment
change in, 10
chemistry and, 90–103, 113
cleanup of, 52
definition of, *3,* 23
dependence on, 3
economic tradeoffs with, 38
economy and, 39, 684, 697
environmental ethics and, 30–31
ethics and, 28–30
factors that shape view of, 29–30
humans and, 3, 21, 684
justice and, *36,* 37*f*
neoclassical economics and, 42–44
perception of, 28–29
politics and, 30
population and, 220–222
quality of, 52
religion and, 30, 32
surroundings constitute, 3
sustainability and, 43
synthetic chemicals and, 407–408, 414*f*
toxic agents in, 407–417
environmental activists, 12*f*
environmental bottleneck, 696
environmental change, consumption and population and, 19
environmental costs, economic benefits and, 42
environmental economics, 26–56, 40*f*
fundamentals of, 54
environmental economists, *45*
environmental ethics, 26–56, 53
definition of, *30*
roots of, 33
specifics of, 30–38
environmental hazards, 402–403, 429
in air, 405*t*
animal testing for, 417–420
biological, 403, 403*f*
chemical, 403, 403*f*
cultural, 403, 403*f*
in food, 405*t*
indoor, 405–406, 405*t*
physical, 402, 403*f*
study of, 417–422
in water, 405*t*
wildlife studies of, 417
environmental health, 402–407, 429
anthropogenic factors of, 402
definition of, *402*
disease and, 404–405
goals of, 429
environmental impact
affluence and, 236
of fossil fuel use, 575–578, 596*t*
of hydropower, 615
of nuclear power, 596*t*
resource consumption and, 6
of urban resource consumption, 393–395
environmental impact statement (EIS), 68, 78
supplemental, 79
environmental justice, *36–38,* 403
movement of, 37*f,* 53
environmental organizations, of 19th century, 34*t*
environmental philosophers, 33–34
environmental policy, *59–60,* 85. *See also* policy
approaches to, 70–75, 85
constitutional amendments and, 65
creation process for, 75
decision making and, 57–87
EPA and, 68
implementation of, 61–62
innovation in, 62–63
international, 81–84
introduction to, 59–62
key laws of, 70*f*
need for, 62
organization and, 77–79
organizations shaping, 82–84
overview of, 62–70
pollution and, 67–68
problem solving and, 57–87
process of, 75–80, 85
public land management and, 66–67
social context of, 70
environmental problems, 10*f,* 60*f*
perception of, 10–11
solutions to, 20
environmental protection
economic opportunity, 683–684
support for, 33
trade barriers and, 84
Environmental Protection Agency (EPA), 36, 64*t,* 78
air pollutants monitored by, 527*f*
counties failing standards of, 509*f*
creation of, *68*
criteria pollutant standards set by, 507–508
fertilizer application and, 267
lung cancer, 405
PMN and, 427
environmental regulations, costs and benefits of, 686–687
environmental resistance, 137
environmental science
definition of, *3*
dilemmas of, 3
environmentalism and, 11–12
interactions of, 3
interdisciplinary solutions and, 11, 23
introduction to, 2–25
nature of, 10–12
past civilizations and, 7–10
practical applications of, 12
environmental scientists, 12*f*
environmental studies, *11*
environmental systems, 182–214. *See also* biomes
of Earth, 184–189
nature of, 211
perceptions of, 189
rock cycle and, 206–207, 207*f,* 212
environmental toxicology
definition of, *407*
human health and, 407
environmentalism
definition of, *11*
environmental science and, 11–12
environmentally harmful subsidies, 71–72
phasing out, 686
enzymes, 110
EPA. *See* U. S. Environmental Protection Agency
epidemiological studies, *418–420*
equilibrium theory of island biogeography, *329*
equitable resource use, 82
equity, policy and, 60–61
eradication, of invasive species, 169
erosion, 65*f,* 273
control practices for, 264
definition of, *251*
deposition, 256
global problem of, 259
increase in, 256
measuring, 256–257
mechanisms of, 257–259
no-till agriculture and, 246, *262f*
positive feedback loop and, 185
water and, *258f*
ESA. *See* Endangered Species Act
estrogen mimics, 417
estuaries, *478*
ethane, 99*f*
ethanol, *609,* 612*f*
ethical consideration, 31*f*
extension of, 31–33
of other species, 328
ethical standards, *30*
ethics
conservation, *34*
definition of, *30*
economics and, 28
environment and, 28
evolution of, 31*f*
policy and, 60
preservation, *34*
EU. *See* European Union

eukaryotes, *102,* 103
Europe, TFR in, 228*t*
European Environment Agency, EU and, 83
European Union (EU), 83
banning of PBDEs by, 406
directives of, 83
opposition to GM crops by, 293
eutrophic water bodies, 437, 453*f*
eutrophication, *188,* 210, 453
nitrogen and, 189*f*
phosphorus and, 192–193
evaporation, *204*
even-aged tree stands, *355,* 356*f*
evolution, 116–147. *See also* biodiversity; extinction; genetic diversity
biodiversity and, 118–129, 145
biological, *118*
natural selection and, *118*
predation and, 153
evolutionary arms race, 154–155
executive branch, *63,* 63*f,* 85
federal agencies of environmental policy in, 64*t*
executive orders, 63
existence value, 48*t,* 49*f*
experiment, *14*
exploitation, 152–155
exploratory drilling, for oil, 569
exponential growth, *136,* 136*f*
of human population, 185, 218–219
exposure, types of, 421
expressed preferences, 49
external cost, *43,* 44*f*
policy and, 61
externalities, *43,* 44*f*
policy and, 61
extinction, 21*f. See also* biodiversity; species
biodiversity and, 126, 126*f,* 316–324
of bird species, 318
current rates of, 317–318
definition of, *126*
justification of, 31
of mammal species, 318
as natural process, 316
reasons for, 319–324
recent wave of, 318*f*
of selected species, 325
speciation and, 126
of species, 145
vulnerability to, 126–127
extinction rates, island biogeography and, 330*f*
extirpation, *316*
extraterrestrial hypothesis, 109
extrusive igneous rock, 206
Exxon Valdez oil spill, 89–90, 112

facilitation, 156
families, 311
family planning, 233–235, 241
fertility reduction and, 232*f*
government control of, 234
international, 233–235
farmers' markets, 303*f*
fats, 101
FCCC. *See* U.N. Framework Convention on Climate Change
FDA. *See* Food and Drug Administration
federal land management agencies, 370
BLM, 363
Forest Service, 359
federal powers, 63*f*
Federal Water Pollution Control Acts of 1965 and 1972, 69, 70, 455–456
feedback loop, *184*
negative, *184,* 185*f*
feedlot agriculture, 305. *See also* grazing
animal rights and, 297
consumption increase and, 296–297
definition of, *297*
livestock and poultry, 296–298
female literacy, 232*f*
Ferrel cells, 504–505
definition of, *504*
fertility decline, 234
fertilizers, 266*f,* 267*f. See also* pesticides; water pollution
crop yield and, 266–267
definition of, *266*
nitrates and, 202–203
nitrogen from, 188
oleophilic, 112
phosphorus from, 188
slow-release, 112
synthetic, 266*f*
types of, 266
fire policy, 358–360, 360*f*
controlled burns, *360*
fire suppression, 270, 358–361, 360*f*
longterm, 360
first law of thermodynamics, *104*
fisheries
by-catch, *485*
capture rates, 483
collapse of, 483–484
current state of, 493
decline in, 447*t,* 493
ecosystem-based management approach to, 488
factors masking decline of, 484
management of, 487–488
management problems for, 489
oceans and, 482–486
technology and, 484
as vital resource for humans, 345
fishing, long haul, 183
fishing nets, 480
fishing practices, 485–486
consumer choice and, 486
fission, 600–602
nuclear energy and, *593*
flagship species, 335
flat-plate solar collectors, 624
Flint, Thomas, 448
flood control, 444–446, 447*t*
flooding, 444–446
disruption of, 447*t*
floodplain, definition, *435*
Florida bog frog, 127
Florida Everglades, restoration of, 168
Florida Keys
marine reserves in, 488, 492
overfishing in, 466–467
tourism in, 466–467
Florida Keys National Marine Sanctuary, 467, 467*f*
coral types in, 475
flux, 195
food. *See also* genetically modified crops; nutrients
future of, 276–307
genetic modification of, 287–295
hazards in, 405*t*
labeling of, 301
production of, 305, 307*f*
Food and Agriculture Organization, 272
Food and Drug Administration (FDA), 423
food chain, 158
fishing down, 484–485
toxicant accumulation in, 416
food production trends, 221*f*
global, 279*f*
food security, *278*
biodiversity and, 325–326
from crop diversity, 325–326
food webs, 159*f,* 179
biogeochemical cycles and, 195
energy flow and, *158–159*
of Great Lakes, 158
phosphorus and, 197
forest. *See also* rainforest
acidic deposition and, 515, 515*t*
of Easter Island, 8
ecologically valuable, 347–348
ecosystem of, 191–192
ecosystem services of, 348
even-aged, 355, 356*f*
fire policy and, 358–360
fragmentation of, 144
GIS surveys of, 352–353
as managed ecosystem, 358
management of, 346–361, 359, 369
mature, 347*f*
natural area of, 351*f*
organisms found within, 347–348
products of, 348–349
publicly-held, 358
as recreation area, 358
restoration of, 359
structural change in, 357
taxpayer-based subsidies for, 359
uneven-aged, 355
forest fires, 360*f*
forest loss
agriculture and, 255
climate change and, 9
forest preserves, 396*f*
Forest Service. *See* U. S. Forest Service
forestry, 342–371. *See also* deforestation; timber
climate change and, 543–544, 545*f*
definition of, 346–347
fundamentals of, 369
new, *358*
plantation, 355
resides of, 609*f*
soil impacts of, 270–271
sustainability of, *361*

fossil
definition of, *110*
fossil fuels and, 561–562
fossil fuels, *4,* 562*f,* 571*f*. *See also* coal; energy; hydrocarbons; natural gas; oil; petroleum
alternatives to, 591–592, 616
carbon trapping and, 195–197
developed nations and, 562–563, 563*f*
developing nations and, 562–563, 563*f*
economic impacts of, 579–581
emissions from, 576
energy of, 556–588, *559*
environmental impact of use of, 575–578, 596*t*
extraction of, 580–581
fossils and, 561–562
global consumption of, 561*f*
impacts of, 556–588, 586
nations with reserves of, 562
nuclear power and, 596–597
other sources of, 575
political impacts of, 579–581
PV cells and, 626
smog and, 509–510
social impacts of, 579–581
uneven distribution of reserves of, 562
fossil record, 110*f,* 111*f,* 113
definition of, *110*
life's history and, 110, 124
mass extinction events in, 316*f*
fractionation, 572–573
fragmentation, biodiversity and, 330
free rider, *61*
Freedom to Farm Act, 272
free-living parasites, 154
free-market, 41
Fresh Kills landfill, transformation of, 647–648
freshwater
accessible, 462
apportioning, 440*f*
available, 439*f*
conflict over, 444
definition of, *434*
demand for, 449–450
Gulf of Mexico and, 184
harnessing, 440
pollution of, 451–458
storage of, 434*f*
supply of, 449–450
unequal distribution of, 438–439, 462
as vital resource for humans, 344–345
freshwater depletion, solutions to, 449–451
freshwater ecosystems, 435–437, 462
freshwater resources, 432–464
freshwater systems, 434–439
climate change and, 544–545
fuelwood burning, 607*f*
biomass energy from, 607–608
developing nation's indoor air pollution, 520–522
fundamental niche, *151,* 151*f*
funding, 18
fungi, terrestrial plant species and, 155
fungicides, 282
fusion, 595–596
GAO. *See* U. S. General Accounting Office
garbology, 654–655
gasification, 610–611
gathering
agriculture and, 249*f*
cultures of, 32
Mirrar clan and, 28
GDP. *See* Gross Domestic Product
Geddes, Pete, 52
gender equality, 232–233
gene(s), *101,* 101*f*
gene banks, *295*
gene flow, 124
genera, 311
General Land Ordinances of 1785 and 1787, 66
General Mining Law of 1872, 72
generalists, *131*
genetic diversity, 310*f,* *312,* 339
within crop species, 325–326
genetic engineering, *287*
agricultural breeding and, 288–290
global experiment of, 294
science behind, 305
genetic modification, of food, 287–295
genetic trees, *124*
genetic variation, natural selection and, 119–120
genetically modified (GM) crops, 52, *287,* 289*t,* 294
controversies behind, 305
debates over, 291–295
early development of, 288*t*
global amounts of, 290
hurdles for, 292
impacts of, 290–291
morals and, 291
proponents of, 288–289
science behind, 305
testing of, 290–291
genetically modified (GM) organisms, *287*
genome, 101
Genuine Progress Indicator (GPI), 46, 46*f,* 47*f,* 55
Geographic Information Systems (GIS), 352–353
map layers using, 353*f*
UNEP and, 352–353
geographic isolation, 123
geography, of urban areas, 376–377
geological systems, 206–210
geology, phosphorus availability and, 197
geothermal energy, *107–108,* 622*f,* 633–635, 634*f,* 643
benefits of, 635
electricity from, 633–634
heating by, 633–634
limitations of, 635
United States and, 621–623
use of, 635
geyser, 633
Giardia lamblia, 60*t*
GIS. *See* Geographic Information Systems
Gleason, Henry, 166, 179
global banana split, 684, 685*f*
global climate change, 20, 482*f,* 528–555, 551, 553
agriculture and, 543–544
biodiversity loss and, 324
Canadian model of, 544*f*
CGCMs and, 538–540
coral bleaching and, 475
debate over, 546–547
definition of, *530*
estimates of, 540–545
extinction and, 319
forestry and, 543–544, 545*f*
freshwater systems and, 544–545
future impacts of, 543–545, 553
Hadley model of, 544*f*
human health and, 545–546
international treaties on, 550
Maldives and, 545
marine systems, 544–545
methods of studying, 534–540
precautionary principle and, 552
predictions of, 540–545, 543*t*
responses to, 547
uncertainties of, 546–547
global climate patterns, large-scale circulation systems, 504–505
global distillation, 415*f*
global environment, pressures on, 23
global fisheries catch, 482–483
global human population, 219*f*
annual growth rate of, 227*f*
global oil production, 571*f*
global pollution, 415*f*
global population, wealth and, 242
global population growth, 240
trajectories of, 224*f*
global temperature, 540*f*
global warming, 530
global warming potential, 530–531
of greenhouse gases, 532*t*
global water consumption, 441–442, 441*f*
global water shortage, 444, 445*f*
global wind patterns, 504*f*
globalization, 691–692
glucose, 101*f,* 106, 107
GM. *See* genetically modified crops
golden toad, 117–118, 143
climate change and, 141
extinction of, 127
habitat of, 130–131
shadow of, 144
good neighborliness, 82
goods, economic definition of, *38*
Goodstein, Eban, 52
government, policy and, 59–60, 71
government restriction, on land use, 385–385
GPI. *See* Genuine Progress Indicator
grain production, 279*f*
grants, 18
grass hedges, 256
grazing. *See also* overgrazing
soil degradation and, 267–270
unregulated, 6
Great Lakes, 149
food web of, 158
pollution reduction in, 456–457
green building design, 678
energy efficiency in, 679–680
water use in, 679–680
green for green, BSU and, 675
green manures, 264

green revolution, *250, 279–280*
benefits of, 280–282
goal of, 305
harm caused by, 280–282
irrigation and, 440–441
population and, 281–282
wheat and, 280*f*
green taxes, 50, *72–73,* 686
greenhouse effect
carbon dioxide contributes to, 531–532
definition of, *530*
greenhouse gases
definition of, *530*
electricity generation and, 548
emissions of, 533*f*
global climate change and, 324
lower atmosphere and, 530–531
nuclear power and, 596
other, 532–533
primary, 531–532
transportation and, 548–550
Greenpeace, 84, 84*f*
greenway, 393
Gross Domestic Product (GDP), 46, 46*f*, 47*f*, 55
ecosystem services cost and, 50
gross primary production, *192*
Grossman, Karl, 606
ground source heat pumps (GSHPs), 634
groundfish, 483–484
groundwater, *204–205,* 437*f*
average age of, 438
depletion of, 441–442
filtering of, 458
hydrologic cycle and, 437–438
overpumping of, 442
pollution of, 454–456
sinkholes and, 442*f*
groundwater pollution
by agriculture, 455
leaky underground storage tanks and, 455, 455*f*
sources of, 455
growth. *See also* human population; population growth
logistic, *136–137*
neoclassical economics and, 44
of new renewable sources of energy, 622–623
paradigm of, 44–45
population, 138*f*
sustainability of, 45
technology and, 45
growth rate, *135*
decline of, 240
limiting factors of, *136–137*
GSHPs. *See* ground source heat pumps
guanine, 100
Gulf of California, water supply to, 443
Gulf of Mexico
Atchafalaya River and, 188
dead zone in, 183–184, 187–189, 202–203, 210
eutrophication of, 188–189, 189*f*
freshwater inputs to, 184
nitrogen pollution in, 210
oxygen levels of, 188
phytoplankton blooms in, 194*f*
gully erosion, 257, *258f*
Haber, Fritz, 200
Haber-Bosch process, *200*
habitat, *130–131*
alteration of, 319, 447*t*
selection of, 131
use of, 131
habitat fragmentation, 331*f*, 339, 368*f*
corridors and, 369
SLOSS dilemma in, 369
habitat preservation, endangered species as, 335
Hadley cells, 504–505
definition of, *504*
Hadley model, 544*f*
half-life, *93, 594*
halocarbons, 530–531, 532–533
harbor seals, 165
Hardin, Garrett, 6, 61
Harmful Algal Bloom and Hypoxia Research and Control Act, 201–204
assessment report of, 204
harmful algal blooms, *481,* 481*f*
hauhau tree, 9
hazardous waste, 664–671, 667*f*. *See also* radioactive waste; waste
criteria of, *665*
deep-well injection of, 668–669, 670*f*
definition of, *648*
disposal of, 666–667
heavy metals as, 665–666
household, 666*f*
illegal dumping of, 667*f*
landfills and, 667
managing, 671
methods of disposal of, 667–669
organic compounds as, 665–666
radioactive waste and, 669–670
RCRA and, 666
sources of, 665
surface impoundments for, *667–668*
health effects
of PCB contamination, 412
sprawl and, 380–382
of uranium mining, 37
heat pollution, 454
Hepatitis A virus, 60*t*
herbicides, 282
herbivores
consumers and, 157
keystone predators and, 160
plants and, 155
herbivory, *155,* 179
heritability, 119
Herrera, Adrian, 584
heterotrophic hypothesis, 109, 113
heterotrophs, *107*
high-pressure system, *502*
historical sciences, 15
HIV/AIDS, 404*f*
cases of, 237*f*
casualties of, 238–239
dimensions of, 241
influence on population of, 237–238
patients with, 239*f*
potential biotreatments for, 326
resistance genes and, 238–239
severe demographic changes from, 238–240
Hock, Susan, 641
homeostasis, *186*
Homestead Act of 1862, 66, 66*f*
honeybees, 286*f*
horizon, *252*
hormonal mimicry, 410*f*
endocrine systems and, 401–402, 410
hormone binding, normal, 410*f*
host, 153–154
House of Representatives, 81*f*
household cleaners, 407*f*
household water, 440
housing developments
affordability of, 387
in suburbs, 378
Houston toad, 127
Hubbard Brook Research Forest, decrease in acidity of, 521*f*
Hubbert's peak, *571–574,* 571*f*
Hudson River, 149
Human Development Index, 47
human health
air pollution decrease and, 509
climate change and, 545–546
environmental toxicology and, 407
tobacco smoke and, 522–523
human life, 23
conditions of, 21
human population, 216–243. *See also* family planning
annual growth rate of, 219
biomass and, 223–224
density of, 224*f*
distribution of, 236*f*
food challenge for, 305
global, 219*f*
growth of, 4–6, 24*f*, 218–222, 240, 375
habitat fragmentation and, 368*f*
measuring, 226–227
pollution and, 221
range in, 223*f*
sex ratios in, 226
soil degradation and, 246–247
sprawl and, 380
trends in, 221*f*
humanity, footprint of, 7
humans
activities of, 21*f*
biodiversity loss and, 319–324
carbon and, 197
carrying capacity of, 222
diet of, 115
environment and, 3, 684
environmental ethics and, 30–31
environmental hazard studies concerning, 417–420
exponential growth of, 185
fisheries and, 345
freshwater and, 344–345
global climate, 553
habitat fragmentation and, 331*f*
hydrologic cycle and, 205–206
impacts of, 695–696
invasive species and, 179
long-term thinking by, 690–691
mass extinction events and, 317, 318*f*

microbes and, 155
minerals and, 345
nature and, 3, 222–229, 327–328
nitrogen cycle and, 200–204
nitrogen inputs by, 201*f*
oceans and, 478–481, 493
outdoor air pollution and, 505–506
petroleum products and, 570*f*
phosphorus cycle and, 197
potential of, 697
psychology of, 41–42
rangeland and, 345
reduced consumption by, 687–689
selection imposed by, 121–122, 122*f*
soil and, 344
sperm count decline in, 411–412, 411*f*
symbiosis in, 155
vital resources for, 344–345
well-being of, 20*t*, 692*f*
wildlife and, 345
humus, 251–252
hunter-gatherer cultures, 249*f*
ethics and, 32
hunting
agriculture and, 249*f*
for animal parts, 323–324
cultures of, 32
Mirrar clan and, 28
poverty and, 235
hurricane damage, 65*f*
Hurricane Katrina, 542*f*
energy alternatives and, 622–623
hybrid car, 583*f*
hydrocarbons, *98–99, 99f*
bacteria and, 89–90
liquidity of, 99
hydrochlorofluorocarbons (HFC-23), 532*t*
hydroelectric power, 591, 612–615
dams and, 444–449, 612–615
definition of, *612*
impacts of, 617
scale of, 617
use of, 614–615
hydrogen bond, 96, 96*f*
hydrogen fuel, 640*f*, 643
algae as, 638–639
buses running on, 642*f*
production of, 638–639
hydrogen ions, acidity and, 98
hydrogen power, 637–642
benefits of, 640–642
cleanliness of, 639
cost of, 641
economic benefits of, 641
electrolysis and, 638–639
Iceland and, 620–621
precautions of, 639
process of, 638–640
renewable energy sources and, 641
hydrologic cycle, *204–205, 205f*
groundwater and, 437–438, 462
human impacts on, 205–206
hydropower, 592, 622*f*
consumers of, 614*t*
definition of, *612*
expansion of, 615
negative environmental impacts of, 615
renewable energy and, 615
storage technique, *612–613*
hydrosphere, *189*
hydrothermal vents, 107, 108*f*
hypertension, sprawl and, 382
hypothesis
scientific method and, *14*
support for, 15
testing of, 15–17
hypoxia, *183*, 188, 202–203, 210
area of Gulf of Mexico zone of, 203*f*
frequency of, 203*f*
span of, 202

ice
floating of, 96–97, 97*f*
water and, 96–97
Iceland
geothermal energy in, 620, 634*f*
hydrogen power and, 620–621, 637
Icelandic New Energy (INE), 621
ICPD. *See* International Conference on Population and Development
Idaho's wind prospectors, 632
igneous rock, *206*, 207*f*
Illinois River, 149
illuviation horizon, 253
immigration, *135*
island biogeography and, 330*f*
immune system, 409–410
incineration, 655–656, 656*f*, 671
definition of, *655*
energy creation and, 657
scrubbers, 656
independent variable, *14*
Index of Sustainable Economic Welfare (ISEW), 47
India
GM crops in, 293
grain production in, 281*f*
water apportioning by, 440*f*
individual hazard response, 421
Indonesian government pest management, 285*f*
indoor air pollution, 519*f*
in developed world, 522–523
in developing world, 520–522
health risk statistics of, 519*f*
living organisms and, 524
potential sources of, 523*f*
poverty increases, 520–522
reduction of, 524–525
scope of, 526
VOCs and, 523–524
industrial ecology, *664*, 671
industrial fishing fleets, 483*f*
industrial output trends, 221*f*
industrial revolution, *4*
environmental philosophers and, 33–34
industrial smog, *509*
coal burning and, 510
industrial solid waste, *648*, 663–664, 663*f*, 671
industrial stage, *229*
industrial system, natural system and, 689–690
industrial water reduction, 451
industrialization, 5, *229*
urban center development and, 374
industrialized agriculture, *248–250*
industry, water and, 440
INE. *See* Icelandic New Energy
infant mortality, global, 235*t*
infectious disease, 403
deaths of, 404
leading causes of death, 404*f*
innocent-until-proven-guilty approach, 424, 427
innovation of urban centers, 395
inorganic compounds
chemosynthesis and, 107–108
soil and, 250*f*
inorganic fertilizers, *266*
insecticides, 282
insects, pollination and, 285–286, 286*f*
institutional purchasing, 680
Integrated Pest Management (IPM), *285*
popularity of, 285
intensive traditional agriculture, 248
interactions of species, 178
effects of, 151*t*
exploitative, 152–155
intercropping, *262f, 263*
interdisciplinary field, *11*, 23
interdisciplinary solutions, environmental science and, 11
interest groups, 71
Intergovernmental Panel on Climate Change (IPCC), *540*
future climatic impacts projected by, 543–545
major findings of, 541*t*
report of, 540–541
International Boundary and Water Commission (IBWC), 78
International Conference on Population and Development (ICPD), 234
International environmental policy, 81–84
International Organization for Standardization (ISO), 361
International Wastewater Treatment Plant (IWTP), 74, 78, 82*f*
international water apportionment, 440*f*
interpretations, of patterns in data, 16
interspecific competition, 151
intertidal zone, 473*f*, 476–477, 476*f*
biodiversity of, 477
at low tide, 476*f*
intraspecific competition, 151
intrusive igneous rock, 206
invasive species, *167–168*, 319–322
best known, 320*t*, 321*t*
control of, 169
eradication of, 169
explanation of, 319–322
humans and, 179
prevention of, 169
risk assessment and, 169
societal response to, 169
zebra mussel and, 167–168
inversion layer, *503*, 503*f*
invertebrate animals, 158
ionic bond, *96*
ionic compounds, *96*
ions, 93*f*, 94, 113
hydrogen, 98

IPAT model, 220–223, 689
affluence and, 236, 240
equation for, *220*
technology and, 240
IPCC. *See* Intergovernmental Panel on Climate Change
IPM. *See* Integrated Pest Management
iridium levels, 129
irrigation, *264–265*
drip, 450
green revolution and, 440–441
of land, 441*f*, 447*t*
low-pressure spray, 450
methods of, 265*f*
water reduction methods for, 450–451
water waste and, 440–441
ISEW. *See* Index of Sustainable Economic Welfare
island biogeography, 329–331, 330*f*, 339
Caribbean islands and, 331*f*
distance from mainland and, 330*f*
island size and, 330*f*
testing theory of, 332–333
ISO. *See* International Organization for Standardization
isolated populations, 123
isotopes, 92–94, 93*f*, 113
radiocarbon dating and, 94
isotopic ratios, plants and, 95*f*
isotopic signature, 94

Jabiluka uranium mine, 27–30
cancellation of, 28
contingent valuation and, 48
proposed site of, 29*f*
Jenks-Jay, Nan, 694
judicial branch, *63–64*, 63*f*, 85
June solstice, 502*f*
justice, environment and, *36*, 37*f*

Kakadu region, 27
contingent valuation and, 48
Kareiva, Peter, 338
kelp, 474*f*
definition of, 474
overgrazed, 162*f*
sea otters and, 162*f*, 164, 474
kerogen, 574
keystone, 162*f*
keystone species, 159–163, 162*f*, 179
conservation and, 162
removal of, 161–162
tiger as, 325
killer smog, deaths from, 497–498
killer whale, 164
kinetic energy, *103–104*, 104*f*, 105*f*, 113
Knopman, Debra, 512
Koehler, Matthew, 359
K-selected species, *139*
loss of, 323–324
traits of, 139*t*
K-T event. *See* Cretaceous-Tertiary event
Kyoto Protocol, 73
emissions reductions required by, 550*t*
FCCC and, *550*
United States resistance to, 550–552

La Niña, *534*
temperature during, 537*f*
understanding, 536–537
labeling food, 301
laissez-faire policies, 41
lake(s), 436*f*. *See also* hydrologic cycle; surface water
ecological system of, 436–437
zones in, 436–437
Lake Apopka, Florida, alligators and, 401–402, 410–411
Lake Erie, zebra mussel and, 149
land, irrigation of, 441*f*
land consumption per capita, 379
land ethic, 35–36, 272
The Land Ethic, 36
land management policy, 66–67
in U. S., 354*f*
wise-use, *366*
land policy, 269
land protection, 366–367
urban centers and, 395
land trusts, *367*
land use, 342–371
better management of, 363–364
government restriction of, 385–386
sprawl and, 382–383
land-crop matching, 450–451
landfill(s), 652–653, 671
biogenic natural gas and, 574
biomass energy from, 608
campus sustainability, lawns and, 677*f*
drawbacks of, 653–655
gas energy production in, 657
hazardous waste and, 667
leachate, *653*
NIMBY syndrome and, 654
transformation of, 653
U.S. usage of, 652*f*
landfill gas, 657
landscape ecology, *192*
multiple ecosystems in, 191–192
water flow and, 435*f*
language, limitations of, 3
large-scale circulation, 504–505
Latin America
deforestation in, 350*f*
TFR in, 228*t*
latitude, patterns of, 177
latitudinal gradient, 314–315, 315*f*
lava, *206*
law
conventional, *81–82*
customary law, *81–82*
IPAT and, 221
making of, 81*f*
layoffs, environmentally related, 52
LD_{50}. *See* lethal-dose-50%
leachate, from landfills, *653*
leaching, *252–253*
lead, 91, 508
lead emissions, 86*f*
lead poisoning, 405–406
Leadership in Energy and Environmental Design (LEED), 678, 694
leaf litter, 347–348

leapfrog development, 379*f*
Leech, Michael, 489
LEED. *See* Leadership in Energy and Environmental Design
legislation, *63*
legislative branch, *63*, 63*f*, 85
Leopold, Aldo, 35*f*, 71
holistic view of, 35
land ethic of, *35–36*, 272
precautionary principle and, 325
lethal-dose-50% (LD_{50}), *420*, 420*f*, 421
leukemia, biodiversity and, 326
levees, 444–446
definition of, 445
surrounding New Orleans, 543
lichens, 163
life expectancy, definition of, *228*
life origin, 108–112
carbon and, 113
hypotheses for, 108–109, 113
life-cycle analysis, *664*
life's history
fossil record and, 110
present-day organisms and, 111–112
light-dependent reactions, 106
light-independent reactions, 106
limestone, 129
limiting factors, *136–137*, 145
nitrogen and, 194
of primary productivity, 194
limnetic zone, *437*
lipids, *101–102*, 113
liquids, hydrocarbons and, 99
lithification, *206*
lithosphere, *189*
carbon shifting from, 197
littoral fish, 158, 161*f*
population size of, 161
littoral zone, *436–437*
livable cities
creation of, 384–393
parks and, 390–391
livestock
grazing practices for, 267–270, 268*f*
land for grazing, 363
Living Planet Index, 319*f*
loam, *254*
lobbying, *79–80*
local policy, 64–65
local self-sufficiency, 691–692
logging
in Canada, 349
commercial, 349
in early North America, 351*f*
environmentally sensitive practices for, 344
ESA and, 334
protection from, 359
in Russia, 349
salvage, *360*
logistic growth curve, 136–137, 145
London, England, killer smog in, 497–498
long haul fishing, 183
Long Term Ecological Research (LTER), 395–396
longlining, 485*f*
bycatch of, 485

Louv, Richard, 328
low-density single-use development, 379*f*
low-pressure system, *502*
LTER. *See* Long Term Ecological Research
Lubchenco, Jane, 489
lung cancer, EPA and, 405

macromolecules, *99–102,* 103*t,* 113. *See also* molecules
macronutrients, 192
Madagascar, age pyramid of 2005 of, 225*f*
magma, *206,* 207*f*
  geothermal energy and, 633–634
maladaptive trait, 119
malaria
  DDT and, 10
  deaths caused by, 404*f*
Maldives, 552
  climate change and, 545
  economic tsunami damage to, 530
  rising ocean level and, 529–530
  sea-level rise and, 541–543
  tsunami havoc on, 529–530
malnourishment, 279
malnutrition, 404
  in children, 280*f*
Malpai borderlands, 270–271
Malthus, Thomas, *4–5,* 5*f,* 119
  exponential growth and, 136
  growth and, 44–45
  population growth and, 220
mangrove forests, 477–478, 478*f*
  destruction of, 477–478
manipulative experiment, *15,* 16*f*
mantle, *208*
maquiladoras, 59, 76
  policy and, 61
March equinox, 502*f*
Mariana Trench, 471
marine conservation biology, 486–492
marine ecosystems, 473–478, 493. *See also* aquaculture; fisheries; oceans
  human impacts on, 493
  major types of, 493
  oil pollution and, 480–481
  pollution of, 479–480
  vertical water movement and, 471
marine habitats, 473*f*
marine life
  nets and, 480
  plastic debris, 480
  rapid depletion of, 483
marine protected areas (MPAs), 467, 488, 493
  marine reserves and, 488
  protests against, 467
marine reserves, 466–467, *488,* 493
  benefits of, 488–491
  data from, 488–491
  design of, 491–492
  ecosystem restoration through, 489
  networks of, 489
  opposition to, 488
  size effect of, 492*t*
  success of, 490–491
marine resource use, 493
marine systems, climate change and, 544–545
market failure, *49–50*
  methods to address, 50–53
  policy and, 61
market incentives, 73–75
marketable emissions permits, *73*
mass extinction events, 110, 127–128, *316–317,* 316*f,* 317*t*
  definition of, 127
  sixth of, 128–129
mass number, 92
mass transit establishment, 389–390
materials recovery facilities (MRFs), 659
mating, 123
matter, 212
  compounds of, 98–99
  hierarchy of, 103*t*
  recycling of, 191
maximum sustainable yield, 346*f,* 487
  resource management and, *345–346*
Maya, 7–10
McIntosh, David, 512
Meade, Margaret, 78
meat consumption, 296*f*
mechanical weathering, 251
medicine, biodiversity and, 326–327, 326*t,* 327*t*
meltdown, *597*
Mesopotamian marshes, restoration of, 168
mesosphere, 499
metamorphic rock, *207,* 207*f*
methane, 99*f,* 532*t*
  global atmospheric concentrations of, 532*t*
methane hydrates, 575
  definition of, *479*
methyl tert-butyl ether (MTBE), 86
Metro. *See* Metropolitan Service District
metropolitan areas, 375*t*
Metropolitan Service District (Metro), 374
microbes, 111*f*
  sick-building syndrome and, 524
micronutrients, 192
Middle Eastern countries, arid deserts of, 7
migratory fish, dams and, 448
Milankovitch cycles, *533*
Mill, John Stuart, 45–46
Millennium Ecosystem Assessment, *20,* 691
  conclusions of, 39
  ecosystem services and, 692*f*
  findings of, 20*t*
Mineral Lands Act of 1866, 66–67, 66*f*
minerals
  oceans and, 478–479
  as vital resource for humans, 345
mining, 66. *See also* uranium mining
  for coal, 565
  of water, 441
Mirrar clan
  consent of, 28
  culture of, 29
  external costs of, 43–44
  hunting and gathering of, 28
  uranium mine and, 27–30
Mississippi River, 149
  hypoxia and, 203*f*
  nitrate concentrations in, 188*f*
  nitrogen discharge from, 210
  nitrogen excess in, 184, 204
  nutrients and, 192–193
  phosphorus excess in, 184
  runoff and, 202
  as system, 187*f*
  tributaries of, 187
  watershed of, 187–189
Mississippi-Atchafalaya River basin, nitrogen use in, 188*f*
moderator, 594
modern climate models, 539*f*
modern climate research, methods, 553
modern market economies, nonmarket value and, 48*t*
molecules, 103*t,* 130
  components of, *94–96*
  compounds and, *95*
  formation of, *94–96*
  macromolecules and, *99–102*
monetary value assignment, 42
  to ecosystems goods, 48–49
monoculture, *248–249*
monosacchride, 101*f*
Monsanto Company, 291–292
Montane spruce-fir forest, 178*f*
Monteverde cloud forest, 117, 122, 130
  climate change and, 140, 140*f,* 141*f*
  conservation of, 143, 144
  golden toad and, 127
Montreal Protocol
  ozone depletion and, 513–514
  stipulations of, 513–514
moray eel, 475*f*
mortality, 135
  summer heat and, 546*f*
mosquitoes, DDT and, 10
Mount Saint Helens eruption, 506*f*
mountaintop removal, 565, 576
MPAs. *See* marine protected areas
MRFs. *See* materials recovery facilities
MTBE. *See* methyl tert-butyl ether
Muir, John, 34, 34*f*
multiple ecosystems, in landscape ecology, 191–192
multiple use policy, *358*
municipal solid waste, 649–663, 663*f*
  components of, 649*f*
  definition of, *648*
  improved disposal methods for, 651
  open dumping of, 651
  patterns in stream of, 649–650
municipal water reduction, 451
mutagens, *409*
mutations, *119*
  AIDS resistance and, 238–239
  DNA and, 123
  natural selection and, 145
mutualism, *155–156,* 156*f,* 179
Myers, Norman, 338

Nadkarni, Nalini M., 144
NADW. *See* North Atlantic Deep Water
Namib Desert, 175
Naphthalene, 99*f*
  bacterial degradation of, *99f*
natality, 135
National Environmental Policy Act (NEPA), *68,* 69*f,* 78, 85
national forest, *354*

National Forest Management Act, *358*
National Invasive Species Act of 1996, 168
National Park Service (NPS), 365
national parks, creation of, *364–365*
National Petroleum Reserve-Alaska, 557–558
National Rifle Association (NRA), 80
national wildlife refuge, 365
Native Americans, uranium mining and, 37
native clams, zebra mussels and, 150*f*
natural experiment, *16*
natural gas, 479*f*, 574–575. *See also* oil; petroleum
  consumers of, 575*t*
  extraction of, 575, 575*f*, 586
  formation of, *574*
  history of, 574–575
  producers of, 575*t*
natural pollution sources, 505
natural rate of population change, *228*
natural resources, 4*f*, 40*f*
  definition of, 23, 39
  energy from, 559–561, 560*f*
  humans and, 4
  nonrenewable, 4
  renewable, 4
  survival and, *4*
natural sciences, *11*
natural selection, *118*, 120*f*, 121*f*. *See also* artificial selection; competition; speciation
  crop pests and, 283*f*
  effect of, 120
  evidence of, 120–122
  genetic variation and, 119–120
  logic of, 119*t*
  pressures of, 120
  process of, 145
  types of, 119–120
natural systems
  industrial system and, 689–690
  preservation of, 51
nature
  animosity towards, 33
  humans and, 3, 222–229, 327–328
  male attitudes toward, 36
The Nature Conservancy, 84
Navajo
  lung cancer and, 37–38
  uranium mining and, 37
negative externalities. *See* externalities
negative feedback loop, *184*, 185*f*
neoclassical economics, *41–42*, 54
  assumptions of, 43–44
  environment and, 42–44
  growth and, 44
NEPA. *See* National Environmental Policy Act
neritic zone, 473*f*
net primary production, *192*
net primary productivity, *192*
  global map of, 193*f*
  for major ecosystem types, 193*f*
neurotoxins, *410*
neutrons, *91*, 91*f*, 92*f*
new alternative energy sources
  current use of, 621–622
  growth of, 622–623
new forestry, *358*
new urbanism, 387–388
  in Boca Raton, Florida, 388*f*
  definition of, 387–388
New York's Cental Park, 392
niche, 130–131, 151
  definition of, 131
  fundamental Vs realized, 151*f*
Nicholas, Sara, 448
NIMBY. *See* not-in-my-backyard syndrome
1990 Clean Air Act, 52
nitrate, 202
  peroxyacyl, 511
  water pollution and, 455
nitric acid, 200
nitric oxide, photochemical smog and, 511*f*
nitrification, *199–200*
nitrogen
  in air, 498*f*
  as bottleneck, 200
  dead zone and, 188
  eutrophication and, 189*f*
  excess of, 184
  fixation of, *198–200*
  human inputs of, 201*f*
  increased availability of, 201
  inorganic, 210
  pollution and, 201, 210
nitrogen cycle, 199*f*, 211
  bacteria and, 197–200
  human alteration of, 201
  humans and, 200–204
  plant growth and, *198*
nitrogen dioxide, 200, 507
  acidic deposition from, 515
nitrogen-fixing bacteria, *198–200*
nitrous oxide, 532–533, 532*t*
  global atmospheric concentrations of, 532*t*
Nixon, Richard, 69*f*
nonconsumptive use, *440*
nongovernmental organizations, 84
nonhuman entities, welfare of, 31
Nonindigenous Aquatic Nuisance Prevention and Control Act, 168
nonmarket values, *48*
non-point source pollution
  of air, 506
  of water, *452*, 452*f*
nonrenewable energy sources, 561
  fossil fuels and, 571*f*
nonrenewable natural resources, *4*
North America
  early logging practices in, 351*f*
  TFR in, 228*t*
North American, footprint of, 7
North Atlantic Deep Water (NADW), 534, 534*f*, 546
North Atlantic fisheries, 484*f*
North Slope of Alaska, 557–559, 558*f*, 578*f*
  crude oil in, 567
Northern Hemisphere temperature, 540*f*
northern spotted owl, 683*f*
Northwest Forest Plan, *348–349*
no-till agriculture, 245–246, *262f*
  benefits of, 263*t*
  critics of, 264
not-in-my-backyard (NIMBY) syndrome, 630–631
NPS. *See* National Park Service
NRA. *See* National Rifle Association
nuclear energy, 595*f*, 616
  fission and, *593*
nuclear fallout, soil and, 256–257
nuclear fission, *593*, 593*f*
  breeder, 594–595
nuclear fuel cycle, 593–594
nuclear fusion, 595–596, 595*f*
nuclear power, 592–607, 595*f*. *See also* Chernobyl nuclear reactor
  consumers of, 593*t*
  dangers of, 606
  dilemmas of, 602–607
  drawbacks to, 596
  environmental impact of, 596*t*
  facts about, 606
  fossil fuels and, 596–597
  greenhouse gases and, 596
  growth of, 602–607
  meltdown and, 597
  producers of, 593*t*
  social debate over, 616
  waste disposal and, 600–602
nuclear reactors, 593–594
nuclear waste, 601, 602*f*
nuclear weaponry, 592–593
nucleic acids, *100*, 113
nutrient cycles, *195*
nutrient pollution, 453
  coral reefs and, 475
nutrients
  circulation of, 195
  definition of, *192*
  ecosystem productivity and, 192–195
  macro, 192
  micro, 192
  Mississippi River and, 192–193

O horizon, *252*
Oak woodland, 178*f*
Oaxaca, Mexico, transgenic maize in, 277–278, 292–293
obesity, sprawl and, 380–382
observations, 16
ocean(s), 468*f*. *See also* beaches; coral reefs; fisheries; sea level rise
  carbon trapping and, 197
  coverage by, 468
  crude oil and, 479*f*
  currents and, 470*f*
  depth of, 469*f*
  energy and, 478–479
  energy sources in, 635–637, 643
  global fisheries catch from, 482
  human use and, 478–481, 493
  marine conservation and, 465–495
  minerals and, 478–479
  natural gas and, 479*f*
  as natural systems, 465–495
  nutrient excess in, 481
  overfishing in, 482–486
  Pacific, pollution in, 59*f*
  petroleum sources for input to, 480*f*

red tides and, *481*, 481*f*
salinity of, 469
sea-level rise in, 541–543
surface water of, 469–470
temperature of, 469*f*
thermal energy in, 636–637
as transportation routes, 478
water content of, 468–469, 468*f*
ocean floor, continental shelves and, *472*
ocean thermal energy conversion (OTEC), 636–637
ocean water
density of, 469–470
horizontal flow of, 470–471
upwelling of, 471*f*
vertical movement of, 471
vertical structure of, 469–470
zonation of, 469–470, 469*f*, 472
Oceania, TFR in, 228*t*
oceanic circulation, 533–534
oceanic zone, 473*f*
Oceanography, 467–473
offshore drilling of oil, 570
Ogallala Aquifer, 438*f*
Ohio River, 150
oikos, 38
oil, 101, 565–574. *See also* natural gas; petroleum
chemical makeup of, 90
consumers of, 581*f*, 586*t*
crude, 110–111
dependency on, 579–580
development of, 584
drilling for, 569–570, 584
end of, 574
exploratory drilling for, 569
extraction of, 577–578
global production of, 571*f*
global trade in, 580*f*
history of, 567
offshore drilling for, 570
prices of, 579, 579*f*
producers of, 568*t*, 581*f*
proven recoverable reserves for, 577–578
reserve depletion and, 570–574
uses of, 584
oil drilling, 577–578
debates concerning, 557–559
oil pollution, marine ecosystems and, 480–481
Oil Pollution Control Act, 90, 112
oil sands, 575
oil spills, 89–90
biodiversity loss due to, 322–323
wildlife population recovery and, 90
old-growth forests, clear-cutting of, 343–344
oleophilic fertilizers, 112
oligotrophic lakes, 437, 453*f*
oligotrophic ponds, 437
OMB. *See* White Hose Office of Management and Budget
omnivores, 114, 157
one-child policy benefits, 217–218
open cycle approach, 636–637
open systems, 39, *187*
open-ocean ecosystems, 473–474
open-water fish, 161, 161*f*
option value, 48*t*, 49*f*
orca, 164
diet of, 165
organ systems, 103, 103*t*
organelles, *102*, 103*t*
Organic Act of 1897, 359
organic agriculture, *300*, 302, 340*f*
biodiversity and, 340*f*
in Cuba, 303–304, 304*f*
government initiatives for, 301–302
increase in, 300–303
motivation for, 301
USDA criteria for, 301*t*
organic compounds, *98–99*, 113
carbon cycle and, 195–197
chemosynthesis and, 107–108
as hazardous waste, 665–666
soil and, 250*f*
volatility of, 523–524
organic fertilizers, *266*
organic matter, 253*f*
organisms
abiotic environments and, 190–191
cells of, 102–103
community role of, 159–163
diversity among, 121*f*
ethical obligation to other, 328
fate of newly introduced, 322
in forests, 347–348
GM, *287*
hierarchy of groupings of, 130*t*
hierarchy of matter within, 103*t*
indoor air pollution and, 524
known species of, 314*f*
latitudinal distribution of, 315*f*
macromolecules and, 102–103
niche breadth for, 131
present-day, 111–112, 113
speciation of, 122–123
species number scale of, 313*f*
organization, 130*t*, 145
levels of ecological, 129–131
organs, 103, 103*t*
origin of life, 108–112, 108*f*
hypotheses for, 108–109
OTEC. *See* ocean thermal energy conversion
O'Toole, Randal, 381
outdoor air pollution, 505–519
health risk statistics of, 519*f*
human activities and, 505–506
scope of, 525
overcrowding, 133
overfishing, 482–483
marine reserves and, 489
overgrazing, *268*, 270–271
effects of, 268*f*
overharvesting, biodiversity loss due to, 323–324
overnutrition, 279
overpopulation, 221
oxygen
in air, 498*f*
creation of, 106
water levels of, 188
ozone, 532–533
hole, 513, 514*f*
layer, *499*, 511–513
ozone depletion, 639
CFCs and, 516–517
Montreal Protocol and, 513–514

Pacific Ocean, pollution in, 59*f*
PACs. *See* political action committees
panspermia hypothesis, 109, 113
paper, global consumption of, 371
paper parks, 367
paradigm, *18*
of growth, 44–45
shift, 18
parasites, 153–154, 154*f*
disease pathogens, 154
sea lamprey and, 154
parasitism, *153*, 179
parasitoids, 154
parent material, *251*
weathering of, 251*f*
parks, 364–369
as biodiversity preservation, 338
in cities, 391–393, 392*f*, 397
design of, 368–369
international increase in, 367–368
livable cities and, 390–391
public land management and, 66–67
traditional reasoning for, 364
types of, 370
particulate matter, 508
passenger pigeon, 132
passive solar energy collection, *623*
past climate knowledge, pollen and, 538
pathogen pollution, 453
pathogenicity, 406
pathogens, 154
PBDEs. *See* polybrominated diphenyl ethers
PCB contamination, health effects of, 412
PCBs, 428*t*
persistence time of, 415
peat, 565
peer review, 17
politics and, 17
pelagic ecosystems, *472*
zone of, 473*f*
People's Republic of China
age pyramid of 2005 of, 225*f*
age structure of, 224–226
agricultural development in, 217, 222
agricultural domestication in, 248
economy of, 224–226
fisheries data from, 486–487
industrial development in, 217
one-child policy of, 217–218, 240
population growth and, 218
population growth trends in, 227*t*
population stability of, 227
projected 2030 age pyramid of, 225*f*
sex ratios in, 226
water supply to, 439
world's largest dam in, 447–449
permafrost, 176
permanent gases, 498*f*
permit-trading, *51–53*, 73
peroxyacyl nitrates, 511
persistent organic pollutants (POPs), 426–427, 428*t*

pest(s), 282–286
  management of, 285, 305
  natural selection and, 283*f*
pesticides, *282,* 282*t*. *See also* toxicants
  child development and, 418–419
  chronic exposure to, 421
  EPA and, 426
petroleum
  extraction of, 586
  formation of, *566–568*
  location and size of deposits, 568–569
  source of, 558–559
petroleum geologists, 568*f*
  size and location of deposit and, 568–569
petroleum products, 104–105
  humans and, 570*f*
  range of, 573*f*
  spillage of, 480*f*
  uses for, 570
pH scale, 98, 98*f*
phosphate, lakes and, 194, 194*f*
phospholipids, 101–102
phosphorus
  cycle of, *197,* 198*f,* 211
  dead zone and, 188
  excess of, 184
  food webs and, 197
  geology and, 197
  humans and, 197
  inorganic, 210
  reduction of, 203
photic zone, 472, 473*f*
photochemical smog, *510–511*
  formation of, 511*f*
  nitric oxide and, 511*f*
photoelectric effect, 625
photosynthesis, *105–106,* 113
  biogeochemical cycles and, 195
  products of, 106
photovoltaic (PV) cell, 625*f*
  active solar energy collection with, 625–626
  fossil fuels and, 626
  sales of, 626
  two-way metering and, 626
phylogenetic trees, *124,* 125*f*
physical environmental hazards, 402, 403*f*
physical weathering, 251
phytochelatins, 93
phytoplankton, 150, 188, 473*f*
  blooms of, 210
phytoplankton blooms, runoff and, 194*f*
phytoremediation, 92–93, 114
Pinchot, Gifford, *34,* 35*f*
Pine woodland, 178*f*
pioneer species, *163*
plantation forestry, 355
plants. *See also* algae; photosynthesis
  bioremediation and, 90–91
  cadmium and, 91
  growth of, 197–198
  herbivores and, 155
  isotopic ratios of, 95*f*
  lead and, 91
  phytoremediation, 92–93
  pollination of, 155
plastic debris, 480
plate tectonics, *207–209,* 208*f,* 212
  divergent plate boundaries and, *208–209*
point source pollution
  of air, 506
  of water, *452,* 452*f*
polar cells, 504–505
  definition of, *504*
polar covalent bond, 96
policy, *59,* 60*f*. *See also* environmental policy
  access to powerbrokers for, 79–80
  branches of government and, 63–64
  cause identification for, 76–77
  ecosystem degradation and, 61
  factors of, 70
  failure of, 71
  government and, 71
  implementation of, 61–62
  information to form, 60
  innovation in, 62–63
  making of, 60
  market failure and, 61
  organization and, 77–79
  organizations shaping, 82–84
  philosophical toxicology approaches and, 425–426, 429
  private participation and, 74
  process of, 75–80, 75*f*
  public health and, 61
  regulations and, 63
  scientific knowledge and, 13*f*
  shepherding solution to law and, 80
  societal context of, 85
  state and local, 64–65
  subsidies and, 71–72
  top-down approaches to, 71
  tragedy of commons and, 61
polio virus, 60*t*
political action committees (PACs), 80
political power, women and, 233f
politics
  environment and, 30
  peer review and, 17
pollen grains
  analysis of, 8
  past climate and, 538
pollination, 155
  by hand, 296*f*
  importance of, 305
  insects and, *285–286,* 286*f*
pollinators, 282–286
  conservation of, 286
  importance of, 305
polluter pays principle, 72, 670
pollution, 19–20. *See also* air pollution; water pollution
  acceptable levels of, 51–53
  air, 11
  biodiversity loss due to, 322–323
  challenges of, 19–20
  definition of, *452*
  environmental policy and, 67–68
  Federal Water Pollution Control Acts of 1965 and 1972, 69
  fossil fuels and, 576
  of freshwater, 451–458, 462
  global, 415*f*
  legislative efforts and, 455–457
  long-term effects and, 43
  of marine resources, 479–480
  natural sources of, 505
  nitrogen and, 201, 210
  non-point source, 452–454
  permit-trading, 51–53, 73
  point-source, 452–454
  population size and, 221
  prevention of, 458
  primary, *506*
  reduction of, 52, 455–457
  regulatory efforts and, 455–457
  of rivers, 62*f*
  secondary, *506*
  sprawl and, 380–381
  in Tijuana River, 59*f*
  tort law and, 69
  transboundary, 74
  trends in, 221*f*
  types of, 453–454
  in urban centers, 380, 395
polybrominated diphenyl ethers (PBDEs), 406
polyculture, 248–249
polymers, *99*
  proteins and, 100*f*
Polynesia, double-canoes and, 9
polysaccharide, 101*f*
ponds, 191–192, 436*f*
  zones in, 436–437
pool, 195
POPs. *See* persistent organic pollutants
population. *See also* human population
  age structure within, *134*
  birth rate of, 135
  carrying capacity, *136–138*
  changes in, 139–141
  characteristics of, 131–135, 145
  control of, 234
  crash of, 165
  cycles of, 153
  death rate of, 135
  definition of, *122*
  dispersion, *133–134*
  displacement of, 447*t*
  distribution, *133–134*
  dynamics of, 131–135
  green revolution and, 281–282
  HIV/AIDS influence on, 237–238
  isolated, 123, 124*t*
  limiting factors of, *136–137*
  natural rate of change in, *228*
  older, 226
  problem of, 220
  separation of, 123–124
  sex ratios within, *134*
  size of, *132,* 232
  society and, 229–240
  swings in size of, 139
  unregulated, 136
Population Bomb, The, 5
population decline, 9
population density, 132–133, 138, 223–224
  carbon emissions and, 398*f*
  dependent factors of, *138*

global, 235*t*
independent factors of, *138*
population distribution, *133–134, 223–224*
population ecology, 131–147
scope of, *120*
population growth, 5*f*, 138*f*, 145, 219*f*, 226–227, 240
agriculture and, 248
cessation of, 689
development problems and, 234
environmental change and, 19
global per capita, 235*t*
of humans, 4
hypothetical, 24*f*
Malthus and, 4–5
measuring, 226–227
poverty and, 235
reproductive freedom and, 222
tool making and, 222*f*
wealth gap and, 236–237
women's empowerment and, 230–233
Porter, Warren, 427
Portland, Oregon
comprehensive development plan of, 384–385
managing growth in, 373–374
population of, 376*f*
transit ridership in, 389*f*
UGB and, 386*f*
urban parks in, 392–393
positive feedback loop, *185*, 185*f*
erosion and, 185
post-industrial stage, *229*
potential energy, *103–104*, 104*f*, 105*f*, 113
poultry
debeaked, 297*f*
feedlot, 296–298
poverty, 241
exacerbation of, 20*t*
hunting of large mammals and, 235
indoor air pollution and, 520–522
population growth and, 235
sustainability and, 694
Powell, John Wesley, 363–364, 364*f*
power generation, 447*t*
dams and, 446–447
power source emissions, 598–599
power-tower, 624–625
PPV. *See* public-private partnership model
precautionary principle, 290
chemical testing and, 425
climate change and, 552
Leopold and, 325
precedents, *63–64*
precipitation, 171*f*, *204*
pH values of, 518*f*
predation, *153*, 179
avoidance of, 154*f*
evolutionary ramifications of, 153
predators, 153, 179
prey and, 153*f*, 184
predictions
of scientific method, *14*
testing of, 14
preferences
expressed, 49
revealed, 49
pre-industrial stage, *228–229*
Premanufacture Notification (PMN), 427
prescribed burns, *360*
present generation, obligations of, 30
preservation, 34–35
of California condor, 334*f*
of crop diversity, 295–296, 305
preservation ethic, *34*, 53
prey, 153
predators and, 153*f*, 184
Prigmore, Daryl, 641
primary consumers, 156–157, 158
primary extraction of oil, *569–570*, 569*f*
primary pollutants, *506*
primary producers, *105*
primary productivity, limiting factors of, 194
primary succession, *163*, 179
primary water treatment, *458*
primordial soup hypothesis, 109
Prince, Roger, 112
Prince William Sound, bioremediation in, 89–90
The principle of information and cooperation, 82
private participation, 74
problem
cause identification of, 76–77
environmental, 60*f*
with global population size, 220
government and, 60*f*
identification of, 22, 75–76
solutions for, 57–87
producers, 156, 158
biogeochemical cycles and, 195
productivity, *192*
profundal zone, *437*
prokaryotes, *102–103*
protected areas, 342–371
proteins, *100–101*, 113
polymers and, 100*f*
protons, *91*, 91*f*, 92*f*
proven recoverable reserve, 577–578
proxy indicators, past conditions and, 534–535
definition of, *535*
Prudhoe Bay
development of, 558
tundra vegetation at, 577
public health
efforts of, 405*f*
policy and, 62
public land management, 66–67
public policy, *59*
implementation of, 61–62
public-private partnership model (PPV), 74
PV. *See* photovoltaic cell
pycnocline, definition of, 470
pyrolysis, 611

qualitative data, 15
quality of life, 685–687
quantitative data, 15
Quaternary period, 317
questions, scientific method and, 13–14

R horizon, *253*
Rabalais, Nancy, 202–203
race to feed world, 278–282
radiation, 105, 113. *See also* nuclear energy
measurement of, 597*f*
routine emissions of, 606
radioactive fallout, 600*f*
radioactive spills, 28
radioactive waste, 601
hazardous waste and, 669–670
Superfund for, 670
radioactive waste disposal, 592–593
radioactive waste storage, 603*f*
radiocarbon dating, 94
radioisotopes, *93*, *594*
radon poisoning, 405
radon pollution, 522–523
risk of, 522*f*
rail transit, 390–391
benefits of, 390–391
benefits per capita of, 390*f*
cost of, 391
small-rail, large rail, 391
Rails-to Trails Conservancy, 393
rainforest
ESA and, 334
soil in, 255
random distribution, 133, 133*f*
rangeland, *246*
as vital resource for humans, 345
Ranger mine, 29*f*
Rathje, William, 654–655
RCRA. *See* Resource Conservation and Recovery Act
realized niche, *151*, 151*f*
reclaimed water, 460
recombinant DNA, *287*
creation of, 287*f*
recovery, 648
recycling, 21, 648, 658*f*, 659*f*, 661, 671
campus sustainability and, 677–678
costs of, 660
downside to, 661
growth of, 652*f*, 659–660
of matter, 191
natural, 40*f*
steps of, *658–659*
Red list, *318*
red tides, *481*
reduced agricultural production, 9
reduced tillage, 263–264
Rees, William, 6–7
refining oil, 572–573
reforestation, 264*f*
reformulated gas (RFG), 86
refugees, 227, 227*f*
regional planning, 384, *385*, 397
regulations, policy and, 63
regulatory taking, *65*
zoning and, 385–386
reintroduction, 334–335
relationships
causal, 14
feeding, 179
mutualistic, 155–156
among species, 124
relative humidity, of atmosphere, *500*
relativist, *30*
religion, environment and, 30, 32

renewable energy sources, 561
conversion to, 581
fossil fuels and, 622–623
hydrogen power and, 641
hydropower and, 615
need for, 585
new, 621–623
potential growth of, 643
Sweden and, 590–591
transition to, 623
renewable natural resources, *4,* 24
repeatability, 18
replacement fertility, *227–228*
replicate, *14*
data analysis and, 15
reproduction, 119
strategies for, 138–139
reproductive freedom, population growth and, 222
reproductive health care, 234
reserves, 364–369
design of, 368–369
international increase in, 367–368
marine, 488–489
of oil, 570–574
traditional reasoning for, 364
types of, 370
reserves-to-production ratio (R/P ratio), *571*
reservoir, 195
sink as, 195
source as, 195
residence time, 195
residential water reduction, 451
resilience, *163*
resistance, *163*
to HIV/AIDS, 238–239
pesticides and, 283
resource acquisition, 151
resource availability trends, 221*f*
Resource Conservation and Recovery Act (RCRA), 652–653
hazardous waste and, 666
resource consumption
environmental impacts and, 6
overuse and, 8
policy and, 60–61
of urban centers, 393–395
resource management, 344–346
adaptive, *346,* 348
approaches of, 369
definition of, *344*
ecosystem-based, 346
of forests, 346–361
maximum sustainable yield and, *345–346,* 346*f*
resource partitioning, *152*
resource protection, 52
resource replacement, 220
resource sinks, 394
resources
competition for, *151–152*
nonrenewable, 4
renewable, 4, 24
replacement of, 43
vital, 344–345
Resources for the Future, 45
respiration, 113
biogeochemical cycles and, 195
cellular, *107*
energy and, 158
restoration
of altered communities, 168–170
by college campuses, 681–682
ecological, 179
of Florida Everglades, 168
of forests, 359
of Mesopotamian marshes, 168
results, analysis and interpretation of, 15
reuse, 658*t*
revealed preferences, 49
revolution. *See also* green revolution
of agriculture, *4,* 19, *250*
industrial, 33–344
revolving door, *80*
RFG. *See* reformulated gas
ribonucleic acid (RNA), *100,* 110
phosphorus cycle and, 197
rights, conceptual expansion of, 31*f*
rill erosion, 257–258, *258f*
riparian environments, dams and, 448
risk, *422*
perceptions of, 422, 423*f*
philosophic approaches to, 424–425, 429
risk. *See also* hazardous waste; toxicology
risk assessment, 77, 424*f,* 429
definition of, *422–423*
invasive species and, 169
toxicology and, 422–424
risk management, 77, 429
science and social factors in, 423–424
toxicology and, 422–424
Ritch, John, 606
river(s)
communities within, 435
damming of, 446*f*
floodplain of, *435*
landscape formation and, 435
pollution of, 62*f*
running dry, 444
RNA. *See* ribonucleic acid
Roberts, Terry L., 210
rock cycle, 206–207, 207*f,* 212
rogue flows, 59
Roundup Ready, 289*t,* 291*f*
R/P ratio. *See* reserves-to-production ratio
r-selected species, *139*
traits of, 139*t*
runoff, *204,* 456–457
biodiversity loss due to, 322–323
of fertilizers, 188
minimizing, 450–451
phytoplankton blooms and, 194*f*
river pollution by, 202
sprawl and, 380
water pollution and, 451
run-of-river hydroelectric systems, *614,* 614*f*
Ruskin, John, *33*
Rwandan refugees, 227*f*

Sagan, Carl, 12
Sahara Desert, 175
salamanders, 192
Saldaña, Lori, 74, 78–79
salinization, *265*
prevention of, 265–266
Salmonella typhi, 60*t*
salt marshes, 477*f*
definition of, 477
grass species in, 477
salvage logging, *360*
San Diego
bacteria levels in, 76–77
public beaches in, 58
sewage flow tracking in, 77*f*
sewage pollution in, 58–59
sand, *254*
sanitary landfills, 652–653, 652*f*
definition of, *652*
drawbacks of, 653–655
RCRA and, 652–653
transformation of, 653
SAR. *See* synthetic aperture radar
SARA. *See* Species at Risk Act
Savage, Jay M., 117–118, 127
savanna, 174, 175*f*
scattered development, *379f*
Schindler, David, 193–194
Schmidt, Gavin, 551
science, *12. See also* environmental science; technology
applications of, 12
nature of, 12–18
policy and, 60
purpose of, 12
scientific inquiry, 12
scientific knowledge, application of, 13*f*
scientific method, 13–15, 14*f,* 23
assumptions of, 13
scientific process, 17–18, 17*f*
scientific value, 48*t,* 49*f*
scientists, weighing evidence and, 12
scrubbers, 689
incinerators and, 656
SCS. *See* Soil Conservation Service
sea floor, topography of, 471–472
sea lamprey, 154
sea otter, 162*f*
kelp and, 162*f,* 164, 474
keystone predator and, 162
population crash of, 165
urchin and, 164
sea-level rise, 541–543
U. S. coastal lands at risk during, 542*f*
seasons, Earth's tilted axis and, 502*f*
seaweed. *See* kelp
second law of thermodynamics, 104–105
secondary consumers, 157, 158
keystone predator and, 160
secondary extraction of oil, *569–570,* 569*f*
secondary pollutants, *506*
secondary production, 192
secondary succession, *163,* 163*f*
beginning of, 164–166
secondary water treatment, *458*
second-growth trees, *350*
sediment capture, 447*t*
sediment pollution, 454
sedimentary rock, *206,* 207*f*

sediments, *206*, 207*f*
carbon in, 195–197
seed banks, *295*
seeds from space hypothesis, 109
seed-tree management, 356, 357*f*
selection, 119–120, 120*f*
artificial, 121
evidence of, 120–122
process of, 145
self-replication, 109–110
Senate, 81*f*
Senate Bill 100, 373–374, 387
separation, mechanism of, 123–124
September equinox, 502*f*
septic systems, *458*
services, economic definition of, *38*
sewage pollution, 58–59
pathogens associated with, 61*t*
satellite spotting of, 76–77
sewage treatment facilities, 59, 459*f*
sex ratios, *134,* 226
cultural impacts on, 226
naturally occurring, 226
sexual reproduction, 119
shale oil, 575
Shaw, Jane S., 661
sheet erosion, 257, *258f*
shelf-slope break, 472
shelterbelts, *262f, 263*
shelterwood management, 356
Shigella spp., 60*t*
shipping, 447*t*
Siberian tiger, 309–310
conservation program for, 334
ethical obligation to, 328
overharvesting and, 323–324
sick-building syndrome, 524
Sierra Club, 34*f*, 80
silt, *254*
Simberloff, Daniel, 169
Singer, S. Fred, 551
single large or several small dilemma (SLOSS), 369
sink, 195, 394
sinkholes, 442, 442*f*
Sirota, Gary L., 74
SLOSS. *See* single large or several small dilemma
slow-release fertilizers, 112
sludge, 460
smallpox, 239
smart growth
definition of, *387*
principles of, 387*t*
of urban centers, 387
Smith, Adam, 39–41
smog
fossil fuel burning and, 509–510
industrial, *509*
photochemical, *510–511*
reduction of, 512
social sciences, *11*
society, population and, 229–240
soil, 246–250, 250*f*
characterization of, 253–255, 273
color of, 253–254, 253*f*
condition enhancement of, 246
contamination, 92
definition of, *246*
degradation of, 246–247, 247*f*, 255–272, 259*f*
erosion, 185, 246, 257–258
forestry and, 270–271
formation of, 251–252, 252*t*, 273
fundamentals of, 273
horizons of, *252*
mature, 253*f*
nuclear fallout and, 256–257
nutrient loss from, 274*f*
pH of, 254–255
porosity of, 254
productivity of, 269
profile of, *252–253*
protection of, 261–264
regional differences in, 255
structure of, 254
as system, 250–255
texture of, 254, 254*f*
topsoil, *252*
as vital resource for humans, 344
world conditions of, 247*f*
soil conservation, 21, 260–261, 269, 273
programs for, 271–272
Soil Conservation Act of 1935, 260
Soil Conservation Service (SCS), 260–261
solar cookers, 624–625
solar decathlon, 679, 679*f*
solar energy, 500–501, 622*f*, 623–627, 627*f*, 643
active collection of, *623*
concentrating rays magnifies reception of, 624–625
Earth's curvature and, 501*f*
passive collection of, *623–624*
United States, 621–623
solar panels, 624
solar power, 624*f*
benefits of, 626
drawbacks to, 627
growing use of, 626
solar radiation, 105, 113, 627*f*
solutions, 21
chemical, 96
Sonoran Desert, 175, 178*f*
source, 195
source reduction, 648, 657
Southeast Asia, 7–10
sparse street network, 379*f*
specialists, *131*
speciation, *122*
allopatric, 122–124
events of, 124–126
extinction and, 126
sympatric, 124
species. *See also* biodiversity; communities; extinction
coexistence, 151
definition of, *122*
distribution of, 313
economic factors of, 142
endemic, *127*
ethical obligation toward, 328
extinct tiger, 318
extinction rates of, 339
extinction vulnerability of, 126–127, 145
flagship, 335
as habitat umbrellas, 335
interactions of, 150–156, 151*f*, 178, 179
keystone, 159–163
knowledge of, 313
known, 314*f*
K-selected, *139*
relationships among, 124
reproductive strategies of, 138–139
selective extinction, 325
social factors of, 142
traits of, 139*t*
Species at Risk Act (SARA), 334
species diversity, 310*f*, *311,* 339
according to biome, 314
species richness, 311
latitudinal gradient of, 314–315
species-area curves, 330
Speck, Jeff, 381
sperm count decline, 411–412, 411*f*
splash erosion, 257, 258*f*
sprawl, 378–384, 397
amount of, 379–380
causes of, 379–380
criteria of, 382–383
definition of, *378*
economics and, 384
health effects of, 380–382
human population and, 380
impacts of, 380–384
land use and, 382–383
in Las Vegas, 378*f*
measuring, 382–383
negative connotations to, 380–384
pollution and, 380, 381
problems with, 381
quality of life and, 381
transportation and, 380
stabilizing selection, 119, 120*f*
stable isotopes, 93, 94
standard of living, 396
Stanley, Marian K., 427
starch, 101*f*
starfish, 161–162
keystone predator and, 162
state policy, 64–65
state socialist economy, 39
Staten Island, Fresh Kills landfill on, 647–648
steady-state economies, *45–47*
steroids, 102
storage technique for hydropower, *612–613*
stormwater runoff, 58
Strategic Petroleum Reserve, 579
stratosphere, *499,* 639
stratospheric ozone, depletion of, 511–513, 525
streams
communities within, 435
landscape formation and, 435
strip mining, 565, 565*f*, 576, 577*f*
student awareness, 676
studying health hazards, 417–422
subsidy, *71–72*
forest, 359
subsistence agriculture, 248

subsistence economy, *38*
subspecies, 311
  of tiger, 312*f*
subsurface mining, 565, 565*f*
suburban growth, 378–379
  sprawl and, 397
suburbs, *374*
  movement to, 377–378
  qualities of, 378
  space per person in, 378–379
  sprawl of, 381
succession, *163*
  aquatic, 165–166
  primary, 163
  process of, 179
  secondary, 163–166
  trajectory of, 166
sulfur dioxide, 507
  acidic deposition from, 515
  coal burning and, 510*f*
  cooling effect of, 533
summer mortality rates, 546*f*
sun
  Earth's orbit around, 533*f*
  energy of, 105–106, 623–627
  warmth of, 530
Superfund, *670*
supply, 42
supply-demand, 42
  classic curve of, 42*f*
surface freshwater, storage of, 434*f*
surface impoundments, for hazardous waste, *667–668*
surface water
  depletion of, 442–444
  diversion of, 442–444
survival, natural resources and, 4
survivorship curves, *135,* 135*f*
sustainability, 19–22, *22,* 54, 682–685
  attainment of, 47
  of college campuses, 676–682
  concept of, 23
  definition of, 682
  environment and, 43
  in forestry, *361*
  global plan B, 694
  goal of, 695–696
  major approaches to, 685*t*
  positive outlook for, 694
  poverty and, 694
  pragmatism of, 53
  solutions for, 674–699
  strategies for, 685–694
  technology and, 689
  urban, 393–396
  of urban centers, 395–397
sustainability education, 693
sustainability research promotion, 693
sustainable agriculture, 300–304
  definition of, *300*
  evaluation of, 305
  future role of, 304
sustainable development, 22, 682–685, 697
  concept of, 23
sustainable forestry certification, *361*
sustainable solutions, 674–699, 697
Sweden
  alternative energy and, 590–591
  fossil fuel usage decrease in, 592
  nuclear power phase out in, 590–591
swidden agriculture, 255*f*
Swiss Zero Emissions Research and Initiatives (ZERI), 664
Sylva, Douglas A., 234
symbiosis, *155*
sympatric speciation, 214
synergistic effects, 421–422
synthetic aperture radar (SAR), 76–77
synthetic chemicals, 407*f,* 408*t*
  Carson and, 408–409
  environment and, 407–408, 414*f*
  stratospheric ozone depletion by, 511–513
systems
  boundaries of, 186–187
  of classification, 311*f*
  closed, *187*
  defining properties of, 184–187
  definition of, *184*
  emergent properties of, *186,* 186*f*
  feedback loop of, *184*
  freshwater, 434–439
  geological, 206–210
  industrial, 689–690
  interactions between, 211
  natural, 689–690
  oceans, 465–495
  open, *187*
  output of, 184
  perception of, 189
  soil, 250–255

taiga, 176–177
takings clause, 65
Talloires Declaration, BSU and, 676
Tansley, Arthur, 190
tar sands, 575
tax breaks, 71
taxation, 72–73
taxonomists, 311, 311*f*
TCLP. *See* Toxicity Characteristic Leaching Procedure
technology
  of automotives, 549, 549*f*
  fisheries catch and, 484
  green, 690*f*
  growth and, 45
  hazardous waste from, 666, 668–669
  sustainability and, 689
  urban centers and, 396
  of wind turbines, 630
temperate deciduous forest, 172, 172*f*
temperate grassland, 172–173, 172*f,* 255
temperate latitude species, 315*f*
temperate rainforest, 173, 173*f*
temperature inversion, *503*
Templet, Paul, 210
Tennessee River, 150
teratogens, *409*
terracing, *262f, 263*
terrestrial plant species, fungi and, 155
tertiary consumers, 157, 158
  carnivores and, 157
  keystone predator and, 160
testicular cancer, 411–412, 411*f*
TFR. *See* total fertility rate
Thailand, family planning in, 233
thalidomide, 409*f*
theory, *18*
thermal inversion, *503,* 503*f*
thermal mass, 624
thermodynamics
  first law of, *104*
  second law of, 104–105
thermogenic gas, 574
thermogram, 585*f*
thermosphere, 499
Thoreau, Henry David, *33*
Three Gorges Dam, 447–449
  costs of, 447
Three Mile Island nuclear power plant, 597*f,* 605
threshold dose, *420–421*
thymine, 100
thyroid cancer rates, Chernobyl and, 604*f*
tidal creeks, 477
tidal differences, 635
  dams and, 635
tides
  definition of, *477*
  energy from, 635–636, 636*f*
tiger
  extinct species of, 318
  as keystone predator, 325
  sale of body parts of, 324*f*
  subspecies of, 312*f*
Tijuana
  sewage flow tracking in, 77*f*
  sewage problem in, 58–59
Tijuana River, 58, 59*f*
  pollution in, 59
Tijuana River Valley, 78
  as system, 187
Tijuana River Valley Estuary and Beach Sewage Cleanup Act, 77, 84
Tijuana Sana, 81
tillage, 269
timber
  cost-efficient, 356*f*
  extraction of, 354–355
  harvesting methods for, 355–357, 357*f,* 369
Timber Culture Act of 1873, 67
timber famine, national forest formation and, 354
timber selection management, 356, 357*f*
tissue, 103, 103*t*
toad. *See* golden toad
tobacco smoke pollution, 522–523
tool making, 222*f*
top predator. *See* keystone species
topography, of sea floor, 471–472
topsoil, *252*
tort law, 70
  pollution and, 69
total fertility rate (TFR), *227–228,* 241
  factors of, 228
  global per capita, 235*t*
  spread of AIDS and, 237
toxic agents. *See* toxicants

toxic air pollutants, *509*
toxic chemical pollution, 454
Toxic Substances Control Act, 427
toxicants, 406–407. *See also* pesticides
  abundance of, 429
  accumulation of, 416
  airborne, 415
  approaches to, 424–428
  distribution of, 429
  in environment, 407–417
  exposure methods and, 421
  food chain and, 416
  groundwater and, 413–414
  international regulation of, 426–428
  naturally occurring, 416–417
  persistence of, 415–416
  safety determination for, 424–425
  surface water and, 413–414
  synergistic effects of, 421–422
  types of, 409–410
Toxicity Characteristic Leaching Procedure (TCLP), 668–669, 669*f*
toxicology, 406–407
  children's drawings and, 418–419
  definition of, *406*
  dose-response analysis and, 420–421
  risk assessment and management in, 422–424
trade barriers, environmental protection and, 84
trade winds, 505
traditional agriculture, *248*
tragedy of commons, 6
  policy and, 61
  solutions to, 6
transboundary parks, 367
transboundary pollution, 74
  of acidic deposition, 514–517
  environmental impact of, 78–79
transboundary watershed, 58–59
transcendentalists, view of nature by, 33–34
transcription, 100, 101*f*
transform plate boundary, 209, 209*f*
transgenic maize, 277–278, 292–293
transgenic salmon, 300*f*
transitional stage, *229*
transit-oriented development, 388–390
translation, 100, 101*f*
transpiration, *204*
transportation
  automotive technology and, 549, 549*f*
  ease of, 388
  green alternatives for, 681
  greenhouse gases and, 548–550
  increased use of, 388–390
  oceans and, 478
  operating costs for modes of, 389*f*
  options for, 397
  public, 549
  railway, 390–391, 390*f*
  urban centers and, 380
treatment, *14*
tributary, definition of, 435
Troeh, Frederick R., 269
trophic levels, 157*f*, 179
  biomass and, 158
  definition of, *156*
  energy and, 156–158
  energy decrease and, 158
  toxicant accumulation in, 416*f*
tropical conservation parks, 338
tropical dry forest, 174, 174*f*
tropical latitude species, 315*f*
tropical rainforests, 173, 174*f*
  biodiversity in, 118*f*
  Costa Rica and, 122
troposphere, *498–499*, 530
tropospheric ozone, 507, 510–511
tsunami, Maldives and, 529–530
tuna, 51
tundra, 175–176, 176*f*
  Prudhoe Bay and, 577
turbine, 567, 613*f*
  geothermal, 633
  nacelle and, 628
  wind energy and, 628*f*
two-way metering, 626
type I survivorship, 135, 135*f*
type II survivorship, 135, 135*f*
type III survivorship, 135, 135*f*

U. S. Clean Water Act, 452
U. S. Environmental Policy. *See* environmental policy
U. S. Environmental Protection Agency (EPA), 36
U. S. Forest Service, 35, 354*f*
  forest management and, 358–360
  road-building subsidies and, 72
  timber harvesting and, 355
  U. S. tree data, 355*f*
U. S. General Accounting Office (GAO), 36–37
U. S.-Mexico border pollution, 74
UGB. *See* urban growth boundary
ultraviolet exposure, 402
  to ecosystems, 513
  ozone depletion and, 511–513
ultraviolet light, 516*f*
U.N. *See* United Nations
U.N. Framework Convention on Climate Change (FCCC), *550*
uncentered commercial strip development, 379*f*
unconfined aquifers, *437–438*, 437*f*
  definition of, *438*
underground storage tanks, groundwater pollution and, 455, 455*f*
undernourishment, 279
UNEP. *See* United Nations Environment Programme
uneven-aged tree stands, *355*
uniform distribution, 133, 133*f*
United Nations (U.N.), *82*
  secretary of, 218*f*
  World Summit, 683*f*
United Nations Environment Programme (UNEP), 82
  biodiversity and, 325
  conservation conventions and, 335–336
  desertification and, 259–260
  GIS and, 352–353
  Living Planet Index of, 318–319
  socioeconomic indicators and, 691*f*
United States
  air quality of, 11
  annual shoreline change in, 542*f*
  average annual wind power of, 631*f*
  average per capita electricity use in, 645*f*
  baby boom in, 226*f*
  bus systems in, 388–389
  Clean Air Act and, 506–507
  composting in, 652*f*
  current energy consumption in, 592*f*
  DDT manufacturing by, 409
  deforestation for growth of, 350–352
  deregulation of GM crops by, 294
  dust bowl and, *260*
  energy consumption patterns in, 592, 592*f*
  forest management in, 355*f*
  GDP of, 55–56
  GM land dedication of, 290*f*
  heat index future in, 546*f*
  Kyoto Protocol resistance from, 550–552
  landfill usage in, 652*f*
  material recovery rates in, 659*t*
  metropolitan area study in, 380
  pesticides used in, 282*t*
  recycling in, 652*f*, 659*f*
  recycling rates across, 660*f*
  resource consumption by, 687, 688*f*
  sunlight received throughout, 627*f*
  types of waste in, 650*f*
  typical family in, 237*f*
  uranium oxide production and trade by, 618*f*
  urban growth in, 376
  waste generation by, 650*f*
  water apportioning by, 440*f*
  wilderness areas in, 366*f*
United States Congress, ESA and, 333
universal solvent, water as, 97
universalist, *30*
unregulated areas, 6*f*
unregulated grazing, 6
upwelling, 471*f*
  definition of, *471*
uranium fuel rod storage, 601*f*
uranium mining, 27–30
  Australian national economy and, 27, 38
  in Bethlehem, 29
  health effects of, 37
  Jabiluka cancellation of, 28
  nuclear power and, 593–594
urban centers, 248
  consumption increase in, 394–395
  efficiency of, 374, 394
  environmental impacts of, 397
  geography of, 376–377
  innovation in, 395
  land preservation and, 395
  layout of, 377*f*
  least sprawling, 383*t*
  most sprawling, 383*t*
  pollution in, 380, 395
  population increase in, 374
  as resource sinks, 394
  rural areas interact with, 377*f*
  scale of, 375
  sprawl of, 397
  spread of, 378–379
  standard of living and, 396
  sustainability of, 395–396, 397

urban centers—*continued*
technology use in, 396
transportation in, 380
urban ecology, *395–396*
urban geography, factors influencing, 376–377
urban growth, 375–376
in the developing world, 376
environmental variables in, 376–377
reasons for, 375
urban growth boundary (UGB), *373,* 386*f*
housing prices within, 387
popularity of, 386–387
urban parks, role of, 397
urban planning, zoning and, *385–386*
urban populations, growth of, 375*f*
urban resource consumption, environmental impacts of, 393–395
urban sprawl, as fictitious problem, 381
urban sustainability, 393–396
urbanization, 5, 372–399
biodiversity loss due to, 319
definition of, *374*
scale of, 397
spatial patterns of, 377
urbanizing world, 374–378
urchin, sea otter and, 164
use value, 48*t*, 49*f*

values, 28
culture and, 29*f*
nonmarket, 48*t*
variables, *14*
variation, 119
genetic, 119–120
Vasil, Indra K., 294
vegetable oil fuel, 610, 610*f*
vegetation, 171*f*
vehicle emissions, 383
Venezuela, WTO and, 83
Venosa, Albert D., 112
vertebrates, known species of, 314*f*
vertical mixing, 503, 503*f*
vested interest, 30
Vibrio cholerae, 60*t*
virulence, 406
VOCs. *See* volatile organic compounds
volatile organic compounds (VOCs), *508*
indoor air pollution and, 523–524

Wackernagel, Mathis, 6–7, 25
Wallace, Alfred Russell, *119*
warm front, *502,* 503*f*
waste
definition of, *648*
reduction of, 649*f*
types of, 648, 671
waste acceptance, 40*f*
waste disposal, from nuclear power, 600–602
waste generation, 650*f*
economics and, 663–664
regulation of, 663–664
rise of, 650–651
waste management, 646–673
aims of, 648
approaches to, 648, 671
financial incentives and, 60–662
long-distance, 654–655
main components of, 648
waste reduction, 657, 658*t*
on college campuses, 677–678
reuse and, 657–658
waste stream, *648*
waste-to-energy (WTE), 657
wastewater
artificial wetlands and, 460–461, 461*f*
definition of, *458*
in green buildings, 678–679
municipal treatment of, 458–461
natural treatment of, 460–461
treatment of, 458–461, 459*f*, 462
water. *See also* freshwater; groundwater; hydrologic cycle; oceans; rivers; surface freshwater
characteristics of, 113, 211, 454
chemical solubility in, 414
chemical structure of, 96–97
cohesion of, 96
components in oceans, 468–469
conservation of, 269, 451
consumptive use of, *440*
cooperative usage agreements for, 444
depletion of, 462
desalinization of, 450
erosion and, *258f*
global withdrawals of, 441–442, 441*f*
Gulf of California supply of, 443
hazards in, 405*t*
hydrogen bonding of, 96–97
ice and, 96–97
mining of, 441
nonconsumptive use of, *440*
polar nature of, 97
quality protection of, 74, 451
reclaiming of, 460
reduction in use of, 450–451
storage of, 434*f*
total resources of, 439*f*
turbidity of, 454
unequal distribution of, 438–439
universal solvent and, 97*f*
uses of, 440–449, 462
water intake pipes, 150
water management, Mexico and U. S. and, 81
water pollution, 74, 452, 453–454, 462
biodiversity loss due to, 322–323
legislative efforts and, 455–457
nitrates and, 455
prevention of, 458
regulatory efforts and, 455–457
water quality indicators, 454
water table, *205,* 462
definition of, *437–438*
falling, 442
water vapor, 532–533
waterborne disease pollution, 453
waterlogging, *265*
watershed
definition of, 435
of Tijuana River, *58*
transboundary, 58–59
water-stressed nations, 439*f*
waves, energy from, 635–636, 637*f*
waxes, 102
Wayland, Karen, 584
wealth gap, 236–237, 241
global population and, 242
weather
air mass interactions and, 502–503
atmosphere and, 501–502
definition of, *501,* 530
driving force behind, 502*f*
weathering, *251*
of parent material, 251*f*
welfare, of nonhuman entities, 31
West Nile virus, 404, 405*f*
westerlies, 505
wetlands, 435*f*
artificial, 204, 460–461, 461*f*
draining of, 362, 363*f*
ecosystem services provided by, 436
ecosystems classified as, 435–436
loss of, 210
wheat
green revolution and, 280*f*
monoculture of, 281*f*
Wheelwright, Nathaniel, 144
White House Office of Management and Budget (OMB), 686–687
cost of federal rules, 687*t*
Whitman, Walt, *33*
Wilderness Act of 1964, 67
wilderness areas
establishment of, *365–366*
in U. S., 366*f*
wildfires, 270–271, 402, 506*f*
wildlife
environmental hazard studies of, 417
as vital resource for humans, 345
wildlife population revovery, oil spills and, 90
Wilson, Edward O., 20, 327–328, 328*f*
biophilia and, *327–328*
on environmental bottleneck, 675
island biogeography and, 332–333
wind energy, 622*f*, 627–633, 643
downsides to, 630–633
history of, 628
intermittent resource and, 630
turbines and, 628*f*
United States, 621–623
U.S. generational capacity of, 631*f*
wind farms, 628
wind power, 629*f*
anemometers and, 632, 632*f*
benefits of, 630
growth of, 628–629
U.S. average annual, 631*f*
wind prospectors, 632
wind turbine
animal casualties from, 633
energy transport distance and, 630
first electrical, 628
history of, 628
nacelle, 628
offshore placement of, 629–630, 629*f*
technology of, 630
wise-use movement, *366*

wolves
    keystone predator and, 12, 325
    reintroduction of, 335
    Yellowstone National Park and, 180
women, 241
women's empowerment
    political power and, 233*f*
    population growth rates and, 230–233
wood, demand for, 349–350
World Bank, *83*
    Greenpeace and, 84*f*
World Conservation Union, 317–318
    Red list of, 318
World Domestic Product, 25
world oil prices, 579*f*
World Trade Organization (WTO), 83–84
    Greenpeace and, 84*f*
    protests against, 692, 693*f*
worldview, 28–30
    culture and, 29*f*
    definition of, *28–29*
    factors in, 29–30
    influence of, 53
    perception of environment and, 28–29
    relativist, *30*
    universalist, *30*
WTE. *See* waste-to-energy
WTO. *See* World Trade Organization

xeriscaping, 451

Yellowstone
    environmental policy and, 67
    forest fire in, 360*f*
    geothermal energy and, 635
    reintroduction programs in, 335
    wolves and, 180
Yosemite National Park, 34*f*, 365*f*
Yosemite toad, 127
Yucca Mountain, 602–603

Zambia, GM crops and, 293–294
zebra mussels, 149–150, 320*t*
    impact of, 160
    invasive species and, 167–168
    native clams and, 150*f*
    problems with, 150
ZERI. *See* Swiss Zero Emissions Research and Initiatives
ZERI brewery process, 665*f*
zero-tillage agriculture, 245, *262f*
    benefits of, 263*t*
    erosion and, 246
    yield boosting and, 246
zinc removal, 114
zone of accumulation, 253
zone of deposition, 253
zones, types of development, 385–386
zoning
    regulatory taking and, 385–386
    urban guide determines, 385*f*
    urban planning and, *385–386*
zooplankton, 150, 188
zooxanthellae, 475